2019 IEEE 69th Electronic Components and Technology Conference (ECTC 2019)

Las Vegas, Nevada, USA
28-31 May 2019

Pages 1-784

IEEE Catalog Number: CFP19ECT-POD
ISBN: 978-1-7281-1500-9

**Copyright © 2019 by the Institute of Electrical and Electronics Engineers, Inc.
All Rights Reserved**

Copyright and Reprint Permissions: Abstracting is permitted with credit to the source. Libraries are permitted to photocopy beyond the limit of U.S. copyright law for private use of patrons those articles in this volume that carry a code at the bottom of the first page, provided the per-copy fee indicated in the code is paid through Copyright Clearance Center, 222 Rosewood Drive, Danvers, MA 01923.

For other copying, reprint or republication permission, write to IEEE Copyrights Manager, IEEE Service Center, 445 Hoes Lane, Piscataway, NJ 08854. All rights reserved.

*** This is a print representation of what appears in the IEEE Digital Library. Some format issues inherent in the e-media version may also appear in this print version.

IEEE Catalog Number: CFP19ECT-POD
ISBN (Print-On-Demand):   978-1-7281-1500-9
ISBN (Online):   978-1-7281-1499-6
ISSN: 0569-5503

Additional Copies of This Publication Are Available From:

Curran Associates, Inc
57 Morehouse Lane
Red Hook, NY 12571 USA
Phone: (845) 758-0400
Fax: (845) 758-2633
E-mail: curran@proceedings.com
Web: www.proceedings.com

Proceedings

The 2019 IEEE 69th Electronic Components and Technology Conference

Proceedings

IEEE 69th Electronic Components and Technology Conference

ECTC 2019

28 – 31 May 2019
Las Vegas, Nevada

Los Alamitos, California

Washington • Tokyo

2019 IEEE 69th Electronic Components and Technology Conference (ECTC)
ECTC 2019

Table of Contents

Foreword lvii
Executive Committee lix
Program Committee lx

Session 1: Wafer-Level Fan-Out Process Integration

3D-MiM (MUST-in-MUST) Technology for Advanced System Integration 1
    An-Jhih Su (Taiwan Semiconductor Manufacturing Company), Terry Ku
    (Taiwan Semiconductor Manufacturing Company), Chung-Hao Tsai (Taiwan
    Semiconductor Manufacturing Company), Kuo-Chung Yee (Taiwan
    Semiconductor Manufacturing Company), and Douglas Yu (Taiwan
    Semiconductor Manufacturing Company)

Construction of FO-MCM with C4 Bumps Built First Using Chip Last Assembly Technology 7
    Chih-Hsun Hsu (Siliconware Precision Industries Co., Ltd), Wen-Yang Li
    (Siliconware Precision Industries Co., Ltd), Chi-Jen Chen (Siliconware
    Precision Industries Co., Ltd), Yih-Jenn Jiang (Siliconware Precision
    Industries Co., Ltd), Jui-Feng Tai (Siliconware Precision Industries
    Co., Ltd), Chang-Fu Lin (Siliconware Precision Industries Co., Ltd),
    and C. Key Chung (Siliconware Precision Industries Co., Ltd)

Feasibility Study of Fan-Out Panel-Level Packaging for Heterogeneous Integrations 14

Cheng-Ta Ko (Unimicron Technology Corporation), Henry Yang (Unimicron Technology Corporation), John H. Lau (ASM Pacific Technology Ltd), Ming Li (ASM Pacific Technology Ltd), Curry Lin (Unimicron Technology Corporation), Chieh-Lin Chang (Dow Chemical Company), Jhih-Yuan Pan (Dow Chemical Company), Hsing-Hui Wu (Dow Chemical Company), Iris Xu (Jiangyin Changdian Advanced Packaging Co., Ltd.), Tony Chen (Jiangyin Changdian Advanced Packaging Co., Ltd.), Zhang Li (Jiangyin Changdian Advanced Packaging Co., Ltd.), Kim Hwee Tan (Jiangyin Changdian Advanced Packaging Co., Ltd.), Penny Lo (ASM Pacific Technology Ltd), R. So (ASM Pacific Technology Ltd), Y. H. Chen (Unimicron Technology Corporation), Nelson Fan (ASM Pacific Technology Ltd), Eric Kuah (ASM Pacific Technology Ltd), Marc Lin (Dow Chemical Company), Y. M. Cheung (ASM Pacific Technology Ltd), Eric Ng (ASM Pacific Technology Ltd), Cao Xi (Huawei Technologies Co. Ltd.), Rozalia Beica (Dow Chemical Company), Sze Pei Lim (Indium Corporation), N. C. Lee (Indium Corporation), Mian Tao (Hong Kong University of Science and Technology), Jeffery Lo (Hong Kong University of Science and Technology), and Ricky Lee (Hong Kong University of Science and Technology)

Ultra-Thin FO Package-on-Package for Mobile Application 21

Hsiang-Yao Hsiao (Institute of Microelectronics, A*STAR), Soon Wee Ho (Institute of Microelectronics, A*STAR), Simon Siak Boon Lim (Institute of Microelectronics, A*STAR), Leong Ching Wai (Institute of Microelectronics, A*STAR), Ser Choong Chong (Institute of Microelectronics, A*STAR), Pei Siang Sharon Lim (Institute of Microelectronics, A*STAR), Yong Han (Institute of Microelectronics, A*STAR), and Tai Chong Chai (Institute of Microelectronics, A*STAR)

Development of Wafer Level Process for the Fabrication of Advanced Capacitive Fingerprint Sensors Using Embedded Silicon Fan-Out (eSiFO(R)) Technology 28

Shuying Ma (Huantian Technology (Kunshan) Electronics Co., Ltd., China), Chengqian Wang (Huantian Technology (Kunshan) Electronics Co., Ltd., China), Fengxia Zheng (Huantian Technology (Kunshan) Electronics Co., Ltd., China), Daquan Yu (Huantian Technology (Kunshan) Electronics Co., Ltd., China), Hong Xie (Filipchip International, USA), Xiaobing Yang (Huantian Technology (Kunshan) Electronics Co., Ltd., China), Li Ma (Huantian Technology (Kunshan) Electronics Co., Ltd., China), Ping Li (Huantian Technology (Xi'an) Electronics Co., Ltd., China), Weidong Liu (Huantian Technology (Xi'an) Electronics Co., Ltd., China), Jambo Yu (Synaptics, USA), and Jason Goodelle (Synaptics, USA)

Three-Dimensional Integrated Circuit (3D-IC) Package Using Fan-Out Technology 35

Jun Kyu Lee (NEPES Corporation), Sang Yong Park (NEPES Corporation), Young Ho Kim (NEPES Corporation), Jae Cheon Lee (NEPES Corporation), Sung Hyuk Lee (NEPES Corporation), Chul Hyo Lee (NEPES Corporation), Yong Tae Kwon (NEPES Corporation), Chang Woo Lee (NEPES Corporation), Jong Heon Kim (NEPES Corporation), Nam Chul Kim (NEPES Corporation), and Yun Hyun Sung (NEPES Corporation)

Ultra High Density IO Fan-Out Design Optimization with Signal Integrity and Power Integrity 41
  *Keng Tuan Chang (Advanced Semiconductor Engineering, Inc.), Chih-Yi
  Huang (Advanced Semiconductor Engineering, Inc.), Hung-Chun Kuo
  (Advanced Semiconductor Engineering, Inc.), Ming-Fong Jhong (Advanced
  Semiconductor Engineering, Inc.), Tsun-Lung Hsieh (Advanced
  Semiconductor Engineering, Inc.), Mi-Chun Hung (Advanced Semiconductor
  Engineering, Inc.), and Chen-Chao Wang (Advanced Semiconductor
  Engineering, Inc.)*

Session 2: Next-Generation Wirebonding and Die Attach

SB²-WB: A New Process Solution for Advanced Wire-Bonding 47
  *Matthias Fettke (Pac Tech GmbH), Andrej Kolbasow (Pac Tech GmbH),
  Georg Friedrich (Pac Tech GmbH), Anna Palys (PacTech GmbH), Vinith
  Bejugam (Pac Tech GmbH), and Thorsten Teutsch (Pac Tech GmbH)*

Smart Wire Bond Solutions for SiP and Memory Packages 55
  *Basil Milton (Kulicke & Soffa, USA), Aashish Shah (Kulicke & Soffa,
  USA), Hui Xu (Kulicke & Soffa, USA), Odal Kwon (Kulicke & Soffa, USA),
  Gary Schulze (Kulicke & Soffa, USA), Ivy Qin (Kulicke & Soffa, USA),
  and Nelson Wong (Kulicke & Soffa, Singapore)*

Preparation and Application of Cu-Ag Composite Preforms for Power Electronic Packaging 63
  *Dongxiao Zhang (Wuhan University of Technology), Shengfa Liu (Wuhan
  University of Technology), Hui Xiang (Wuhan University of Technology),
  Li Liu (Wuhan University of Technology), Zhaoxia Zhou (Loughborough
  University), Stuart Robertson (Loughbrough University), Canyu Liu
  (Loughborough University), Zhiwen Chen (Wuhan University), and
  Changqing Liu (Loughborough University)*

Au-Rich/Sn-Bi Interconnection in Chip-on-Module Package 69
  *Jin Wang (Tsinghua University, China), Qian Wang (Tsinghua University,
  China), Xinnan Hou (GalaxyCore Inc., China), Ke Du (GalaxyCore Inc.,
  China), Lixin Zhao (GalaxyCore Inc., China), and Jian Cai (Tsinghua
  University, China)*

The Properties of Cu Sinter Paste for Pressure Sintering at Low Temperature 76
  *Jung-Lae Jo (Mitsui Engineered Materials Sector R&D Center, Mitsui
  Mining & Smelting Co., Ltd.), Kei Anai (Mitsui Engineered Materials
  Sector R&D Center, Mitsui Mining & Smelting Co., Ltd.), Sinichi
  Yamauchi (Mitsui Engineered Materials Sector R&D Center, Mitsui Mining
  & Smelting Co., Ltd.), and Takahiko Sakaue (Mitsui Engineered
  Materials Sector R&D Center, Mitsui Mining & Smelting Co., Ltd.)*

Low Temperature Sintering of Dendritic Cu Based Pastes for Power Semiconductor Device Interconnection 81

Gang Li (Shenzhen Institutes of Advanced Technology, Chinese Academy of Sciences, Shenzhen College of Advanced Technology, University of Chinese Academy of Sciences), Jilei Fana (Shenzhen Institutes of Advanced Technology, Chinese Academy of Sciences, Shenzhen College of Advanced Technology, University of Chinese Academy of Sciences), Siyuan Liao (Shenzhen Institutes of Advanced Technology, Chinese Academy of Sciences, Shenzhen College of Advanced Technology, University of Chinese Academy of Sciences), Pengli Zhua (Shenzhen Institutes of Advanced Technology, Chinese Academy of Sciences), Baotan Zhang (Shenzhen Institutes of Advanced Technology, Chinese Academy of Sciences), Tao Zhao (Shenzhen Institutes of Advanced Technology, Chinese Academy of Sciences), Rong Sun (Shenzhen Institutes of Advanced Technology, Chinese Academy of Sciences), and Ching-Ping Wong (Shenzhen Institutes of Advanced Technology, Chinese Academy of Sciences)

A New Development of Direct Bonding to Aluminum and Nickel Surfaces by Silver Sintering in air Atmosphere 87

Ly May Chew (Heraeus Deutschland GmbH & Co. KG, Germany), Tamira Stegmann (Hochschule Aschaffenburg, University of Applied Sciences, Germany), Erika Schwenk (Hochschule Aschaffenburg, University of Applied Sciences, Germany), Monique Dubis (Hochschule Aschaffenburg, University of Applied Sciences, Germany), and Wolfgang Schmitt (Heraeus Deutschland GmbH & Co. KG, Germany)

Session 3: RDL and Additive Manufacturing

Submicron-Scale Cu RDL Pattering Based on Semi-Additive Process for Heterogeneous Integration 94
Takamasa Takano (DNP Co., Ltd.), Hiroshi Kudo (DNP Co., Ltd.), Masaya Tanaka (DNP Co., Ltd.), and Miyuki Akazawa (DNP Co., Ltd.)

Sub-Micron RDL Patterning for Advanced Packaging 101
Ken-Ichiro Mori (Canon Inc.), Yoshio Goto (Canon Inc.), Yasuo Hasegawa (Canon Inc.), Seiya Miura (Canon Inc.), and Douglas Shelton (Canon U.S.A Inc.)

Optimization of Electrolytic Plating Processes for Challenging Fan-Out Panel Level Package Designs 106
Ralph Zoberbier (Atotech Deutschland GmbH, Germany), Britta Scheller (Atotech Deutschland GmbH), and Christian Ohde (Atotech Deutschland GmbH)

3D Printed Substrates for the Design of Compact RF Systems 113
Mohd Ifwat Mohd Ghazali (Universiti Sains Islam Malaysia , Michigan State University), Saikat Mondal (Michigan State University), Saranraj Karuppuswami (Michigan State University), and Premjeet Chahal (Michigan State University)

Fully Additively Manufactured Tunable Active Frequency Selective Surfaces with Integrated On-package Solar Cells for Smart Packaging Applications 119
Syed Abdullah Nauroze (Georgia Institute of Technology), Xuanke He (Georgia Institute of Technology), and Manos M. Tentzeris (Georgia Institute of Technology)

First Demonstration of a Low Cost/Customizable Chip Level 3D Printed Microjet Hotspot-Targeted
Cooler for High Power Applications 126
> Tiwei Wei (IMEC & KU Leuven, Belgium), Herman Oprins (IMEC, Belgium),
> Vladimir Cherman (IMEC, Belgium), Ingrid De Wolf (IMEC & KU Leuven,
> Belgium), Eric Beyne (IMEC, Belgium), and Martine Baelmans (KU Leuven,
> Belgium)

Rapid Production of Customized 3D Electronics Via Hybrid Additive Manufacturing Technology 135
> Ji Li (Key Laboratory of MEMS of the Ministry of Education, Southeast
> University), Yang Wang (Key Laboratory of MEMS of the Ministry of
> Education, Southeast University), Peiren Wang (Key Laboratory of MEMS
> of the Ministry of Education, Southeast University), Jiangling He (Key
> Laboratory of MEMS of the Ministry of Education, Southeast
> University), Handa Liu (Key Laboratory of MEMS of the Ministry of
> Education, Southeast University), and Gengzhao Xiang (Key Laboratory
> of MEMS of the Ministry of Education, Southeast University)

Session 4: Advancements in Automotive and Power Devices

Solid-Liquid InterDiffusion (SLID) Bonding, for Thermally Challenging Applications 141
> Knut E Aasmundtveit (University of South-Eastern Norway, Norway),
> Thi-Thuy Luu (Zimmer and Peacock, Norway), Hoang-Vu Nguyen (University
> of South-Eastern Norway, Norway), Andreas Larsson (University of
> South-Eastern Norway, Norway), and Torleif A Tollefsen (TEGma, Norway)

Fluxless Bonding Technique of Diamond to Copper Using Silver-Indium Multilayer Structure 150
> Roozbeh Sheikhi (University of California, Irvine), Yongjun Huo
> (University of California, Irvine), and Chin C. Lee (University of
> California, Irvine)

Formulation and Processing of Conductive Polysulfide Sealants for Automotive and Aerospace
Applications 157
> Bo Song (Georgia Institute of Technology), Fan Wu (Georgia Institute
> of Technology), Kyoung-sik Moon (Georgia Institute of Technology), and
> CP Wong (Georgia Institute of Technology)

Challenges and Approaches to Developing Automotive Grade 1/0 FCBGA Package Capability 163
> Rajen Dias (Amkor Technology, USA), Mike Kelly (Amkor Technology,
> USA), Devarajan Balaraman (Amkor Technology, USA), Hideaki Shoji
> (J-Devices Corporation, Japan), Tomio Shiraiwa (J-Devices, Japan),
> KwangSeok Oh (Amkor Technology, Korea), and JoonYoung Park (Amkor
> Technology, Korea)

Advanced Substrates for GaN-Based Power Devices 168
Anthony Cibié (Univ. Grenoble Alpes, CEA, LETI, France), Julie Widiez
(Univ. Grenoble Alpes, CEA, LETI, France), René Escoffier (Univ.
Grenoble Alpes, CEA, LETI, France), Denis Blachier (Univ. Grenoble
Alpes, CEA, LETI, France), Kremena Vladimirova (Univ. Grenoble Alpes,
CEA, LETI, France), Jean-Philippe Colonna (Univ. Grenoble Alpes, CEA,
LETI, France), Paul-Henri Haumesser (Univ. Grenoble Alpes, CEA, LETI,
France), Stéphane Bécu (Univ. Grenoble Alpes, CEA, LETI, France),
Perceval Coudrain (Univ. Grenoble Alpes, CEA, LETI, France), William
Vandendaele (Univ. Grenoble Alpes, CEA, LETI, France), Jérôme
Biscarrat (Univ. Grenoble Alpes, CEA, LETI, France), Charlotte Gillot
(Univ. Grenoble Alpes, CEA, LETI, France), Matthew Charles (Univ.
Grenoble Alpes, CEA, LETI, France), and Léa Di Cioccio (Univ. Grenoble
Alpes, CEA, LETI, France)

A New Reliable, Corrosion Resistant Gold-Palladium Coated Copper Wire Material 175
Sandy Klengel (Fraunhofer Institute for Microstructure of Materials
IMWS), Robert Klengel (Fraunhofer Institute for Microstructure of
Materials IMWS), Jan Schischka (Fraunhofer Institute for
Microstructure of Materials IMWS), Tino Stephan (Fraunhofer Institute
for Microstructure of Materials IMWS), Matthias Petzold (Fraunhofer
Institute for Microstructure of Materials IMWS), Motoki Eto (Nippon
Micrometal Corporation), Noritoshi Araki (Nippon Micrometal
Corporation), and Takashi Yamada (Nippon Micrometal Corporation)

Ultrasonic-Accelerated Intermetallic Joint Formation with Composite Solder for High-Temperature
Power Device Packaging 183
Hongjun Ji (Harbin Institute of Technology at Shenzhen), Mingyu Li
(Harbin Institute of Technology at Shenzhen), Weiwei Zhao (Harbin
Institute of Technology at Shenzhen), and Wenwu Zhang (Harbin
Institute of Technology at Shenzhen)

Session 5: Bonding Manufacturing Technologies

Comprehensive Study of Copper Nano-Paste for Cu-Cu Bonding 191
Ser Choong Chong (Institute of Microelectronics) and Pei Siang Lim
Sharon (Insttitute of Microelectronics)

Enhanced Performance of Laser-Assisted Bonding with Compression (LABC) Compared with Thermal
Compression Bonding (TCB) Technology 197
Kwang-Seong Choi (Electronics and Telecommunications Research
Institute), Yong-Sung Eom (Electronics and Telecommunications Research
Institute), Seok Hwan Moon (Electronics and Telecommunications
Research Institute), Jiho Joo (Electronics and Telecommunications
Research Institute), leeseul Jeong (Electronics and Telecommunications
Research Institute), Kwangjoo Lee (LG Chem), Jung Hak Kim (LG Chem),
Ju hyeon Kim (LG Chem), Gil-Sang Yoon (KITECH), Kwang-Hee Lee (Inha
University), Chul-Hee Lee (Inha University), Geun-Sik Ahn (Protec),
and Moo-Sup Shim (Protec)

A Study of 3D Packaging Interconnection Performance Affected by Thermal Diffusivity and Pressure
Transmission 204
  Jin-San Jung (Samsung Electronics Co., Ltd), Hyeong Gi Lee (Samsung
  Electronics Co., Ltd), Ji-Min Kim (Samsung Electronics Co., Ltd),
  Yong-Jin Park (Samsung Electronics Co., Ltd), Ji-In Yu (Samsung
  Electronics Co., Ltd), Yong Sung Park (Samsung Electronics Co., Ltd),
  Jun Su Lim (Samsung Electronics Co., Ltd), Hyun-Seok Choi (Samsung
  Electronics Co., Ltd), Sung-Il Cho (Samsung Electronics Co., Ltd),
  Dong wook Kim (Samsung Electronics Co., Ltd), and Sang-Ho An (Samsung
  Electronics Co., Ltd)

Vertical Laser Assisted Bonding for Advanced "3.5D" Chip Packaging 210
  Andrej Kolbasow (Pac Tech GmbH), Timo Kubsch (Pac Tech), Matthias
  Fettke (Pac Tech GmbH), Georg Friedrich (Pac Tech GmbH), and Thorsten
  Teutsch (PacTech)

Optimization of a BEOL Aluminum Deposition Process Enabling Wafer Level Al-Al Thermo-Compression
Bonding 218
  Sebastian Schulze (IHP - Innovations for High Performance
  Microelectronics), Matthias Wietstruck (IHP - Innovations for High
  Performance Microelectronics), Mirko Fraschke (IHP - Innovations for
  High Performance Microelectronics), Peter Kerepesi (EV Group, Inc.),
  Helmut Kurz (EV Group, Inc.), Bernhard Rebhan (EV Group, Inc.), and
  Mehmet Kaynak (IHP - Innovations for High Performance
  Microelectronics)

Self-Assembly Process for 3D Die-to-Wafer using Direct Bonding: A Step Forward Toward Process
Automatisation 225
  Jouve Amandine (CEA, LETI), Loïc Sanchez (CEA, LETI), Clément Castan
  (CEA, LETI), Maxence Laugier (CEA, LETI), Emmanuel Rolland (CEA,
  LETI), Brigitte Montmayeul (CEA, LETI), Rémi Franiatte (CEA, LETI),
  Frank Fournel (CEA, LETI), and Severine Cheramy (CEA, LETI)

A Single Bonding Process for Diverse Organic-Inorganic Integration in IoT Devices 235
  Tilo H. Yang (National Taiwan University), Yu-Shan Chiu (National
  Taiwan University), Hai-Yang Yu (National Taiwan University), Akitsu
  Shigetou (National Institute for Materials Science), and C. Robert Kao
  (National Taiwan University)

Session 6: Emerging Flexible Hybrid Electronics

Stretchable and Printable Medical Dry Electrode Arrays on Textile for Electrophysiological
Monitoring 243
  Yougen Hu (Shenzhen Institutes of Advanced Technology, Chinese Academy
  of Sciences), Hui Wang (Shenzhen Institutes of Advanced Technology,
  Chinese Academy of Sciences), Ommeaymen Sheikhnejad (AC2T Research
  GmbH), Yaoxu Xiong (Shenzhen Institutes of Advanced Technology,
  Chinese Academy of Sciences), Han Gu (Shenzhen Institutes of Advanced
  Technology, Chinese Academy of Sciences), Pengli Zhu (Shenzhen
  Institutes of Advanced Technology, Chinese Academy of Sciences),
  Guanglin Li (Shenzhen Institutes of Advanced Technology, Chinese
  Academy of Sciences), Rong Sun (Shenzhen Institutes of Advanced
  Technology, Chinese Academy of Sciences), and Ching-Ping Wong (Georgia
  Institute of Technology)

Screen-Printed Flexible Coplanar Waveguide Transmission Lines: Multi-physics Modeling and Measurement 249

> Nahid Aslani Amoli (Georgia Institute of Technology), Sridhar Sivapurapu (Georgia Institute of Technology), Rui Chen (Georgia Institute of Technology), Yi Zhou (Georgia Institute of Technology), Mohamed L. F. Bellaredj (Georgia Institute of Technology), Paul A. Kohl (Georgia Institute of Technology), Suresh K. Sitaraman (Georgia Institute of Technology), and Madhavan Swaminathan (Georgia Institute of Technology)

Inkjet-Printed Filtering Antenna on a Textile for Wearable Applications 258

> Hsuan-Ling Kao (Dept. of Electronic Engineering, Chang Gung University, Tao-Yuan, Taiwan), Chun-Hsiang Chuang (Dept. of Electronic Engineering, Chang Gung University, Tao-Yuan, Taiwan), and Cheng-Lin Cho (Dept. of Engineering and System Science, National Tsing Hua University, Hsin-Chu, Taiwan)

Mechanical and Electrical Characterization of FOWLP-Based Flexible Hybrid Electronics (FHE) for Biomedical Sensor Application 264

> Yuki Susumago (Tohoku University), Qian Zhengyang (Tohoku University), Achille Jacquemond (Tohoku University), Noriyuki Takahashi (Tohoku University), Hisashi Kino (Tohoku University), Tetsu Tanaka (Tohoku University), and Takafumi Fukushima (Tohoku University)

A Wearable Fingernail Deformation Sensing System and Three-Dimensional Finite Element Model of Fingertip 270

> Katsuyuki Sakuma (IBM Thomas J. Watson Research Center, U.S.A.), Bucknell Webb (IBM Thomas J. Watson Research Center, U.S.A.), Rajeev Narayanan (IBM Thomas J. Watson Research Center), Avner Abrami (IBM Thomas J. Watson Research Center), Jeff Rogers (IBM Thomas J. Watson Research Center), John Knickerbocker (IBM Thomas J. Watson Research Center), and Stephen J Heisig (IBM Thomas J. Watson Research Center)

Heterogeneous Integration of a Fan-Out Wafer-Level Packaging Based Foldable Display on Elastomeric Substrate 277

> Arsalan Alam (University of California, Los Angeles), Amir Hanna (University of California, Los Angeles), Randall Irwin (University of California, Los Angeles), Goutham Ezhilarasu (University of California, Los Angeles), Hyunpil Boo (University of California, Los Angeles), Yuan Hu (University of California, Los Angeles), Chee Wei Wong (University of California, Los Angeles), Timothy S Fisher (University of California, Los Angeles), and S. S. Iyer (University of California)

A Study on the Flexible Chip-on-Fabric (COF) Assembly using Anisotropic Conductive Films (ACFs) Materials 283

> Seung-Yoon Jung (KAIST) and Kyung-Wook Paik (KAIST)

Session 7: Advances in Flip Chip Packaging

7nm Chip-Package Interaction Study on a Fine Pitch Flip Chip Package with Laser Assisted Bonding and Mass Reflow Technology 289

> Ian Hsu (MediaTek), Chi-Yuan Chen (MediaTek), Stanley Lin (MediaTek), Ta-Jen Yu (MediaTek), NamJu Cho (JCET), and Ming-Che Hsieh (JCET)

Ultra Large Area SIPs and Integrated mmW Antenna Array Module for 5G mmWave Outdoor Applications ... 294
Pouya Talebbeydokhti (Intel Corporation), Sidharth Dalmia (Intel Corporation), Trang Thai (Intel Corporation), Raanan Sover (Intel Corporation), and Sharon Tal (Intel Corporation)

Hybrid Approach for Large Size FC-BGA to Enhance Thermal and Electrical Performance Including Power Delivery 300
Heeseok Lee (Samsung Electronics, Co. Ltd.), Yunhyeok Im (Samsung Electronics, Co. Ltd.), Junghwa Kim (Samsung Electronics, Co. Ltd.), Jisoo Hwang (Samsung Electronics, Co. Ltd.), James Jeong (Samsung Electronics, Co. Ltd.), Youngsang Cho (Samsung Electronics, Co. Ltd.), Heejung Choi (Samsung Electronics, Co. Ltd.), and Youngmin Shin (Samsung Electronics, Co. Ltd.)

Package-on-Package Micro-BGA Microstructure Interaction with Bond and Assembly Parameter 306
Pascale Gagnon (IBM Canada Limited), Clément Fortin (IBM Canada Limited), and Thomas Weiss (IBM Systems)

Low Cost Flip-Chip Stack for Partitioning Processing and Memory 314
Fabian Hopsch (Fraunhofer Institute for Integrated Circuits IIS, Division Engineering of Adaptive Systems EAS) and Andy Heinig (Fraunhofer Institute for Integrated Circuits IIS, Division Engineering of Adaptive Systems EAS)

Impact of Low Temperature Solder on Electronic Package Dynamic Warpage Behavior and Requirement 318
Wei Keat Loh (Intel Technology Sdn Bhd), Ron W. Kulterman (Flex Ltd), Haley Fu (iNEMI), and Chih Chung Hsu (CoreTech System (Moldex3D))

High Density Ultra-Thin Organic Substrates for Advanced Flip Chip Packages 325
Nokibul Islam (JCET), KH Tan (JCET), Seung Wook Yoon (JCET), and Tony Chen (JCET)

Session 8: Material and Process Trends in FOWLP and PLP

Laser Releasable Temporary Bonding Film with High Thermal Stability 330
Yong-suk Yang (3M), Kyo-sung Hwang (3M), and Robin Gorrell (3M)

Design and Demonstration of 1μm Low Resistance RDL Using Panel Scale Processes for High Performance Computing Applications 334
Bartlet DeProspo (3D Systems Packaging Research Center), Aya Momozawa (Tokyo Ohka Kogyo Co. LTD.), Atsushi Kubo (Tokyo Ohka Kogyo Co. LTD.), Chandrasekharan Nair (3D Systems Packaging Research Center), Varun Rajagoapal (3D Systems Packaging Research Center), Jenefa Kannan (Georgia Institute of Technology), Emanuel Surillo (3D Systems Packaging Research Center), Fuhan Liu (3D Systems Packaging Research Center), Mohananlingam Kathaperumal (3D Systems Packaging Research Center), and Rao Tummala (3D Systems Packaging Research Center)

Advances in Temporary Carrier Technology for High-Density Fan-Out Device Build-up 340
Arnita Podpod (IMEC), Alain Phommahaxay (IMEC), Pieter Bex (IMEC), John Slabbekoorn (IMEC), Julien Bertheau (IMEC), Abdellah Salahoueldhadj (IMEC), Erik Sleeckx (IMEC), Andy Miller (IMEC), Gerald Beyer (IMEC), Eric Beyne (IMEC), Alice Guerrero (Brewer Science Inc), Kim Yess (Brewer Science Inc), and Kim Arnold (Brewer Science Inc)

xiii

Development of Novel Low-Temperature Curable Positive-Tone Photosensitive Dielectric Materials with High Reliability 346

> Yutaro Koyama (Toray Industries, Inc.), Yu Shoji (Toray Industries, Inc.), Keika Hashimoto (Toray Industries, Inc.), Yuki Masuda (Toray Industries, Inc.), Hitoshi Araki (Toray Industries, Inc.), and Masao Tomikawa (Toray Industries, Inc.)

Highly Reliable Photosensitive Negative-Tone Polyimide with Low Cure Shrinkage 352

> Daisaku Matsukawa (Hitachi Chemical DuPont MicroSystems, Ltd, Japan), Hiroko Yotsuyanagi (Hitachi Chemical DuPont MicroSystems, Ltd, Japan), Shiori Sakakibara (Hitachi Chemical DuPont MicroSystems, Ltd, Japan), Noriyuki Yamazaki (Hitachi Chemical DuPont MicroSystems, Ltd, Japan), Tetsuya Enomoto (Hitachi Chemical DuPont MicroSystems, Ltd, Japan), and Takeharu Motobe (Hitachi Chemical DuPont MicroSystems, Ltd, Japan)

High Rate and Low Damage Etching Method as Pre Treatment of Seed Layer Sputtering for Fan out Panel Level Packaging 358

> Tetsushi Fujinaga (ULVAC, Inc.)

Investigation and Methods Using Various Release and Thermoplastic Bonding Materials to Reduce Die Shift and Wafer Warpage for eWLB Chip-First Processes 363

> Michelle Fowler (Brewer Science, Inc.), John P. Massey (Brewer Science, Inc.), Tanja Braun (Fraunhofer Institute IZM), Steve Voges (Fraunhofer Institute IZM), Robert Gernhardt (Fraunhofer Institute IZM), and Markus Wohrmann (Fraunhofer institute IZM)

Session 9: Wearables and Thin-Package Reliability and Chip Package Interaction

Effect of Charging Cycle Elevated Temperature Storage and Thermal Cycling on Thin Flexible Batteries in Wearable Applications 370

> Pradeep Lall (Auburn University), Amrit Abrol (Auburn University), Ben Leever (US AFRL), and Scott Miller (NextFlex)

Bladder Inflation Stretch Test Method for Reliability Characterization of Wearable Electronics 382

> Benjamin G Stewart (Georgia Institute of Technology) and Suresh K Sitaraman (Georgia Institute of Technology)

Study of BEOL Failure Mode in Flip Chip Packages at High Temperature Conditions 392

> Wei Wang (Qualcomm Technologies, Inc.), Yangyang Sun (Qualcomm Technologies, Inc.), Xuefeng Zhang (Qualcomm Technologies, Inc.), Lejun Wang (Qualcomm Technologies, Inc.), Lily Zhao (Qualcomm Technologies, Inc.), Mark Schwarz (Qualcomm Technologies, Inc.), Bill Stone (Qualcomm Technologies, Inc.), and Ahmer Syed (Qualcomm Technologies, Inc.)

A Novel Metal Scheme and Bump Array Design Configuration to Enhance Advanced Si Packages CPI Reliability Performance by Using Finite Element Modeling Technique 397

> Kuo-Chin Chang (Taiwan Semiconductor Manufacturing Company Ltd.), Mirng-Ji Lii (Taiwan Semiconductor Manufacturing Company Ltd.), Steven Hsu (Taiwan Semiconductor Manufacturing Company Ltd.), Hao-Chun Liu (Taiwan Semiconductor Manufacturing Company Ltd.), Yen-Kun Lai (Taiwan Semiconductor Manufacturing Company Ltd.), Sheng-Han Tsai (Taiwan Semiconductor Manufacturing Company Ltd.), and Chieh-Hao Hsu (Taiwan Semiconductor Manufacturing Company Ltd.)

Assessment of CMP Fill Pattern Effect on the Thermal Performance of Interconnects in Integrated Circuits BEOL 405

    Assaad Helou (Southern Methodist University), Peter Raad (Southern Methodist University), and Archana Venugopal (Texas Instruments, Inc.)

Three-Dimensional Simulation of the Thermo-Mechanical Interaction between the Micro-Bump Joints and Cu Protrusion in Cu-Filled TSVs of the High Bandwidth Memory (HBM) Structure 410

    Jie-Ying Zhou (South China University of Technology), Shui-Bao Liang (South China University of Technology), Cheng Wei (South China University of Technology), Wen-Kai Le (South China University of Technology), Chang-Bo Ke (South China University of Technology), Min-Bo Zhou (South China University of Technology), Xiao Ma (South China University of Technology), and Xin-Ping Zhang (South China University of Technology)

Study of Design Optimization Method for Ultra-Low Power Micro Gas Sensor 417

    Eiji Nakamura (IBM Research - Tokyo), Keiji Matsumoto (IBM Research - Tokyo), Andrea Fasoli (IBM Research - Almaden), Luisa Bozano (IBM Research - Almaden), and Hiroyuki Mori (IBM Research - Tokyo)

Session 10: Dicing and Encapsulation Technologies

A More Than Moore Enabling Wafer Dicing Technology 423

    Jeroen van Borkulo (ASM Pacific Technologies Inc), Rogier Evertsen (ASM Pacific Technologies Inc), and Richard van der Stam (ASM Pacific Technologies Inc.)

Plasma Dicing Integration Schemes for Scribe Lane Layout and the Impact on Die Strength 428

    David Parker (STMicroelectronics, France), Emmanuel Gourvest (STMicroelectronics, France), and Boris Bouillard (STMicroelectronics, France)

Advanced Dicing Technologies for Combination of Wafer to Wafer and Collective Die to Wafer Direct Bonding 437

    Fumihiro Inoue (IMEC, Belgium), Alain Phommahaxay (IMEC, Belgium), Arnita Podpod (IMEC, Belgium), Samuel Suhard (IMEC, Belgium), Hitoshi Hoshino (Disco Hi-Tech Europe, Germany), Berthold Moeller (Disco Hi-Tech Europe, Germany), Erik Sleeckx (IMEC, Belgium), Kenneth June Rebibis (IMEC, Belgium), Andy Miller (IMEC, Belgium), and Eric Beyne (IMEC, Belgium)

Active Control of NCF Fillet Shape for 3D CoW by Multi Beam Laser Bonder 446

    Keiko Ueno (Hitachi Chemical Co., Ltd.), Kazutaka Honda (Hitachi Chemical Co., Ltd.), Tsuyoshi Ogawa (Hitachi Chemical Co., Ltd.), and Toshihisa Nonaka (Hitachi Chemical Co., Ltd.)

Ultrafast Laser Scribe: An Improved Metal and ILD Ablation Process 453

    Julia Chiu (Intel Corporation, USA), Aaron Gore (Intel Corporation, USA), Tyler Osborn (Intel Corporation, USA), Daragh Finn (Electro Scientific Industries, USA), Zhibin Lin (Electro Scientific Industries, USA), David Lord (Electro Scientific Industries, USA), and Jon Mellen (Electro Scientific Industries, USA)

Reliability and Benchmark of 2.5D Non-Molding and Molding Technologies 461
>Yu-Hsiang Hsiao (Advanced Semiconductor Engineering, Group, Inc.,
>Kaohsiung, Taiwan), Che-Ming Hsu (Advanced Semiconductor Engineering,
>Group, Inc., Kaohsiung, Taiwan), Yi-Sheng Lin (Advanced Semiconductor
>Engineering, Group, Inc., Kaohsiung, Taiwan), and Chien-Lin Chang
>Chien (Advanced Semiconductor Engineering, Group, Inc., Kaohsiung,
>Taiwan)

Laser-Induced Trench Design, Optimisation and Validation for Restricting Capillary Underfill Spread
in Advanced Packaging Configurations 467
>Gul Zeb (Université de Sherbrooke), David Danovitch (Université de
>Sherbrooke), and Eric Turcotte (IBM Canada)

Session 11: Automotive and Harsh-Environment Reliability

Effect of Substrate Preheating Treatment on Thermal Reliability and Micro-Structure of Ag Paste
Sintering on Au Surface Finish 474
>Zheng Zhang (Institute of Scientific and Industrial Research, Osaka
>University), Chuantong Chen (Institute of Scientific and Industrial
>Research, Osaka University), Katsuaki Suganuma (Institute of
>Scientific and Industrial Research, Osaka University), and Seigo
>Kurosaka (C. Uyemura & Co., Ltd.)

Package Material Selection Criteria for High Temperature Automotive Applications 479
>Rene T.H. Rongen (NXP Semiconductors), A. Mavinkurve (NXP
>Semiconductors), G.M. O'Halloran (NXP Semiconductors), N. Owens (NXP
>Semiconductors), Y. Weber (NXP Semiconductors), P. Oberndorff (NXP
>Semiconductors), M-L Farrugia (NXP Semiconductors), E. van Olst (NXP
>Semiconductors), and M. van Soestbergen (NXP Semiconductors)

Solder Joint Reliability of Double-Side Mounted DDR Modules for Consumer and Automotive Applications... 486
>Dongji Xie (NVIDIA), Joe Hai (Nvidia Corp.), Zhongming Wu (Nvidia
>Corp.), and Manthos Economou (Nvidia Corp.)

Reliability Investigation of Extremely Large Ratio Fan-Out Wafer-Level Package with Low Ball Density
for Ultra-Short-Range Radar 493
>P.S. Huang (MediaTek Inc.), C.K. Yu (MediaTek Inc.), W.S. Chiang
>(MediaTek Inc.), M.Z. Lin (MediaTek Inc.), Y.H. Fang (MediaTek Inc.),
>M.J. Lin (MediaTek Inc.), N.W. Liu (MediaTek Inc.), Benson Lin
>(MediaTek Inc.), and Ian Hsu (MediaTek Inc.)

Fatigue Behaviour of Lead-Free Solder Joints Under Combined Thermal and Vibration Loads 498
>Meier Karsten (Technische Universität Dresden), Winkler Maria
>(Technische Universität Dresden), Leslie David (University of
>Maryland, Center for Advanced Life Cycle Engineering (CALCE)),
>Dasgupta Abhijit (University of Maryland, Center for Advanced Life
>Cycle Engineering (CALCE)), and Bock Karlheinz (Technische Universität
>Dresden)

Prognostication of Accrued Damage and Impending Failure Under Temperature-Vibration in Leadfree
Electronics 505
>Pradeep Lall (Auburn University), Tony Thomas (Auburn University),
>Jeff Suhling (Auburn University), and Ken Blecker (US Army ARDEC)

Electrochemical Impedance Spectroscopy (EIS) for Monitoring the Water Load on PCBAs Under Cycling
Condensing Conditions to Predict Electrochemical Migration Under DC Loads 515
  Simone Lauser (Robert Bosch GmbH), Theresia Richter (Robert Bosch
  GmbH), Verdingovas Vadimas (Denmarks Technical University), and Rajan
  Ambat (Denmarks Technical University)

Session 12: Advanced Photonic Devices and Packaging

Micro-Fabricated SERF Atomic Magnetometer for Weak Gradient Magnetic Field Detection 522
  Xiang Yue (Southeast University), Jintang Shang (Southeast
  University), and Chen Ye (Southeast University)

Novel Solder Pads for Self-Aligned Flip-Chip Assembly 528
  Yves Martin (IBM T.J.Watson Research Center, Yorktown Heights USA),
  Swetha Kamlapurkar (IBM T.J.Watson Research Center, Yorktown Heights
  USA), Nathan Marchack (IBM T.J.Watson Research Center, Yorktown Height
  USA), Jae-Woong Nah (IBM T.J.Watson Research Center, Yorktown Height
  USA), and Tymon Barwicz (IBM T.J.Watson Research Center, Yorktown
  Height USA)

Collective Curved CMOS Sensor Process: Application for High-Resolution Optical Design and Assembly
Challenges 535
  Bertrand Chambion (Univ. Grenoble Alpes, CEA, LETI), Christophe
  Gaschet (Univ. Grenoble Alpes, CEA, LETI), Marc Lombard (Univ.
  Grenoble Alpes, CEA, LETI), Maïlys Fernandez (Univ. Grenoble Alpes,
  CEA, LETI), Pierre Joly (Univ. Grenoble Alpes, CEA, LETI), Stéphane
  Caplet (Univ. Grenoble Alpes, CEA, LETI), Fabien Zuber (Univ. Grenoble
  Alpes, CEA, LETI), Aurélie Vandeneynde (Univ. Grenoble Alpes, CEA,
  LETI), Patrick Peray (Univ. Grenoble Alpes, CEA, LETI), Gilles
  Lasfargues (Univ. Grenoble Alpes, CEA), Marc Zussy (Univ. Grenoble
  Alpes, CEA, LETI), Jerôme Deschamps (Univ. Grenoble Alpes, CEA, LETI),
  Alexis Bedoin (Univ. Grenoble Alpes, CEA, LETI), and David Henry
  (Univ. Grenoble Alpes, CEA, LETI)

Integration and Characterization of InP Die on Silicon Interconnect Fabric 543
  Eric Sorensen (Center for Heterogeneous Integration and Performance
  Scaling (CHIPS) University of California, Los Angeles), Boris Vaisband
  (Center for Heterogeneous Integration and Performance Scaling (CHIPS)
  University of California, Los Angeles), SivaChandra Jangam (Center for
  Heterogeneous Integration and Performance Scaling (CHIPS) University
  of California, Los Angeles), Tim Shirley (Keysight Technologies), and
  Subramanian S. Iyer (Center for Heterogeneous Integration and
  Performance Scaling (CHIPS) University of California, Los Angeles)

Y-Branched Multimode/Single-Mode Polymer Optical Waveguides for Low-Loss WDM MUX Device:
Fabrication and Characterization 550
  Takaaki Ishigure (Keio University), Tomoki Nakayama (Keio University),
  Fukino Nakazaki (Keio University), and Hiroki Hama (Keio University)

Vertically Stacked and Directionally Coupled Cavity-Resonator-Integrated Grating Couplers for
Integrated-Optic Beam Steering 556
  Shogo Ura (Kyoto Institute of Technology, Japan), Junishi Inoue (Kyoto
  Institute of Technology, Japan), and Kenji Kintaka (AIST, Japan)

xvii

CiB(Chip in Board) Optical Engine Module Using Advanced Fan-Out Package Technology 563
   Sang Yong Park (NEPES Corporation), Ju Hyun Nam (NEPES Corporation),
   Ji Ni Shim (NEPES Corporation), Jun Kyu Lee (NEPES Corporation), Yong
   Tae Kwon (NEPES Corporation), Chang Woo Lee (NEPES Corporation), Jong
   Heon Kim (NEPES Corporation), and Nam Chul Kim (NEPES Corporation)

Session 13: Technologies Enabling 3D and Heterogeneous Integration

Active Interposer Technology for Chiplet-Based Advanced 3D System Architectures 569
   Perceval Coudrain (Univ. Grenoble Alpes, CEA, LETI, France), Jean
   Charbonnier (Univ. Grenoble Alpes, CEA, LETI, France), Arnaud Garnier
   (Univ. Grenoble Alpes, CEA, LETI, France), Pascal Vivet (Univ.
   Grenoble Alpes, CEA, LETI, France), Rémi Vélard (Univ. Grenoble Alpes,
   CEA, LETI, France), Andrea Vinci (Univ. Grenoble Alpes, CEA, LETI,
   France), Fabienne Ponthenier (Univ. Grenoble Alpes, CEA, LETI,
   France), Alexis Farcy (STMicroelectronics, France), Roselyne Segaud
   (Univ. Grenoble Alpes, CEA, LETI, France), Pascal Chausse (Univ.
   Grenoble Alpes, CEA, LETI, France), Lucile Arnaud (Univ. Grenoble
   Alpes, CEA, LETI, France), Didier Lattard (Univ. Grenoble Alpes, CEA,
   LETI, France), Eric Guthmuller (Univ. Grenoble Alpes, CEA, LETI,
   France), Giovanni Romano (Univ. Grenoble Alpes, CEA, LETI, France),
   Alain Gueugnot (Univ. Grenoble Alpes, CEA, LETI, France), Frédéric
   Berger (Univ. Grenoble Alpes, CEA, LETI, France), Jérôme Beltritti
   (STMicroelectronics), Therry Mourier (Univ. Grenoble Alpes, CEA, LETI,
   France), Mathilde Gottardi (Univ. Grenoble Alpes, CEA, LETI, France),
   Stéphane Minoret (Univ. Grenoble Alpes, CEA, LETI, France), Céline
   Ribière (Univ. Grenoble Alpes, CEA, LETI, France), Gilles Romero
   (Univ. Grenoble Alpes, CEA, LETI, France), Pierre-Emile Philip (Univ.
   Grenoble Alpes, CEA, LETI, France), Yorrick Exbrayat (Univ. Grenoble
   Alpes, CEA, LETI, France), Daniel Scevola (Univ. Grenoble Alpes, CEA,
   LETI, France), Didier Campos (STMicroelectronics), Maxime Argoud
   (Univ. Grenoble Alpes, CEA, LETI, France), Nacima Allouti (Univ.
   Grenoble Alpes, CEA, LETI, France), Raphaël Eleouet (Univ. Grenoble
   Alpes, CEA, LETI, France), César Fuguet Tortolero (Univ. Grenoble
   Alpes, CEA, LETI, France), Christophe Aumont (Univ. Grenoble Alpes,
   CEA, LETI, France), Denis Dutoit (Univ. Grenoble Alpes, CEA, LETI,
   France), Corinne Legalland (Univ. Grenoble Alpes, CEA, LETI, France),
   Jean Michailos (STMicroelectronics, France), Séverine Chéramy (Univ.
   Grenoble Alpes, CEA, LETI, France), and Gilles Simon (Univ. Grenoble
   Alpes, CEA, LETI, France)

Process Development of Power Delivery Through Wafer Vias for Silicon Interconnect Fabric 579
   Meng-Hsiang Liu (University of California, Los Angeles), Boris
   Vaisband (University of California, Los Angeles), Amir Hanna
   (University of California, Los Angeles), Yandong Luo (University of
   California, Los Angeles), Zhe Wan (University of California, Los
   Angeles), and Subramanian S. Iyer (University of California, Los
   Angeles)

Active Through-Silicon Interposer Based 2.5D IC Design, Fabrication, Assembly and Test 587
   Jayasanker Jayabalan (Institute of Microelectronics), Vivek
   Chidambaram Nachiappan (Institute of Microelectronics), Sharon Lim Pei
   Siang (Institute of Microelectronics), Wang Xiangyu (Institute of
   Microelectronics), Jong Ming Chinq (Institute of Microelectronics),
   and Surya Bhattacharya (Institute of Microelectronics)

xviii

System on Integrated Chips (SoIC(TM)) for 3D Heterogeneous Integration 594

Ming-Fa Chen (Taiwan Semiconductor Manufacturing Company (TSMC)),
Fang-Cheng Chen (Taiwan Semiconductor Manufacturing Company (TSMC)),
Wen-Chih Chiou (Taiwan Semiconductor Manufacturing Company (TSMC)),
and Doug C.H. Yu (Taiwan Semiconductor Manufacturing Company (TSMC))

Die-to-Wafer (D2W) Processing and Reliability for 3D Packaging of Advanced Node Logic 600

Luke England (GLOBALFOUNDRIES), Daniel Fisher (GLOBALFOUNDRIES), Katie
Rivera (GLOBALFOUNDRIES), Bill Guthrie (GLOBALFOUNDRIES), Ping-Jui Kuo
(Advanced Semiconductor Engineering (ASE)), Chang-Chi Lee (Advanced
Semiconductor Engineering), Che-Ming Hsu (Advanced Semiconductor
Engineering), Fan-Yu Min (Advanced Semiconductor Engineering),
Kuo-Chang Kang (Advanced Semiconductor Engineering), and Chen-Yuan
Weng (Advanced Semiconductor Engineering)

Enabling Ultra-Thin Die to Wafer Hybrid Bonding for Future Heterogeneous Integrated Systems 607

Alain Phommahaxay (IMEC), Samuel Suhard (IMEC), Pieter Bex (IMEC),
Serena Iacovo (IMEC), John Slabbekoorn (IMEC), Fumihiro Inoue (IMEC),
Lan Peng (IMEC), Koen Kennes (IMEC), Erik Sleeckx (IMEC), Gerald Beyer
(IMEC), and Eric Beyne (IMEC)

The Thermal Dissipation Characteristics of The Novel System-In-Package Technology (ICE-SiP) for
Mobile and 3D High-end Packages 614

Taejoo Hwang (Samsung Electronics Co., Ltd., Republic of Korea),
Dan(Kyung Suk) Oh (Samsung Electronics Co., Ltd., Republic of Korea),
Jaechoon Kim (Samsung Electronics Co., Ltd., Republic of Korea),
Euseok Song (Samsung Electronics Co., Ltd., Republic of Korea), Taehun
Kim (Samsung Electronics Co., Ltd., Republic of Korea), Kilsoo Kim
(Samsung Electronics Co., Ltd., Republic of Korea), Joungphil Lee
(Samsung Electronics Co., Ltd., Republic of Korea), and Taehwan Kim
(Samsung Electronics Co., Ltd., Republic of Korea)

Session 14: Fine-Pitch Solderless Bonding

Fine-Pitch (≤10 μm) Direct Cu-Cu Interconnects Using In-Situ Formic Acid Vapor Treatment 620

SivaChandra Jangam (University of California Los Angeles), Adeel Ahmed
Bajwa (Kulicke & Soffa Industries Inc), Umesh Mogera (University of
California Los Angeles), Pranav Ambhore (University of California Los
Angeles), Tom Colosimo (Kulicke & Soffa Industries Inc), Bob Chylak
(Kulicke & Soffa Industries Inc), and Subramanian Iyer (University of
California Los Angeles)

Low Temperature Cu Interconnect with Chip to Wafer Hybrid Bonding 628

Guilian Gao (Xperi Corporation, USA), Laura Mirkarimi (Xperi
Corporation, USA), Thomas Workman (Xperi Corporation, USA), Gill
Fountain (Xperi Corporation, USA), Jeremy Theil (Xperi Corporation,
USA), Gabe Guevara (Xperi Corporation, USA), Ping Liu (Xperi
Corporation, USA), Bongsub Lee (Xperi Corporation, USA), Pawel Mrozek
(Xperi Corporation, USA), Michael Huynh (Xperi Corporation, USA),
Catharina Rudolph (Fraunhofer Institute for Reliability and
Micro-Integration, IZM – ASSID, Germany), Thomas Werner (Fraunhofer
Institute for Reliability and Micro-Integration, IZM – ASSID,
Germany), and Anke Hanisch (Fraunhofer Institute for Reliability and
Micro-Integration, IZM – ASSID, Germany)

Cu Microstructure of High Density Cu Hybrid Bonding Interconnection 636
    *Seokho Kim (Samsung Electronics Co., Ltd., Korea), Pilkyu Kang
    (Samsung Electronics Co., Ltd., Korea), Taeyeong Kim (Samsung
    Electronics Co., Ltd., Korea), Kyuha Lee (Samsung Electronics Co.,
    Ltd., Korea), Joohee Jang (Samsung Electronics Co., Ltd., Korea),
    Kwangjin Moon (Samsung Electronics Co., Ltd., Korea), Hoonjoo Na
    (Samsung Electronics Co., Ltd., Korea), Sangjin Hyun (Samsung
    Electronics Co., Ltd., Korea), and Kihyun Hwang (Samsung Electronics
    Co., Ltd., Korea)*

Low-Resistance and high-Strength Copper Direct Bonding in no-Vacuum Ambient Using Highly
(111)-Oriented Nano-Twinned Copper 642
    *Jing Ye Juang (National Chiao Tung University, Taiwan), Kai Cheng Shie
    (National Chiao Tung University, Taiwan), Po-Ning Hsu (National Chiao
    Tung University, Taiwan), Yu Jin Li (National Chiao Tung University,
    Taiwan), K N Tu (University of California at Los Angeles), and Chih
    Chen (National Chiao Tung University, Taiwan)*

Sub-10μm Pitch Hybrid Direct Bond Interconnect Development for Die-to-Die Hybridization 648
    *John P. Mudrick (Sandia National Laboratories, USA), Jonatan A.
    Sierra-Suarez (Sandia National Laboratories, USA), Matthew B. Jordan
    (Sandia National Laboratories, USA), T. A. Friedmann (Sandia National
    Laboratories, USA), Robert Jarecki (Sandia National Laboratories,
    USA), and M. David Henry (Sandia National Laboratories, USA)*

Cu Pillar with Nanocopper Caps: The Next Interconnection Node Beyond Traditional Cu Pillar 655
    *Ramón A. Sosa (Georgia Institute of Technology - Packaging Research
    Center), Kashyap Mohan (Georgia Institute of Technology - Packaging
    Research Center), Luu Nguyen (Texas Instruments), Rao Tummala (Georgia
    Institute of Technology - Packaging Research Center), Antonia Antoniou
    (Georgia Institute of Technology), and Vanessa Smet (Georgia Institute
    of Technology - Packaging Research Center)*

Cu-Cu Bonding by Low-Temperature Sintering of Self-Healable Cu Nanoparticles 661
    *Junjie Li (Huazhong University of Science and Technology, China), Qi
    Liang (Huazhong University of Science and Technology, China), Chen
    Chen (Huazhong University of Science and Technology, China), Tielin
    Shi (Huazhong University of Science and Technology, China), Guanglan
    Liao (Huazhong University of Science and Technology, China), and
    Zirong Tang (Huazhong University of Science and Technology, China)*

Session 15: High-Bandwidth Packaging

Electrical Performance Limits of Fine Pitch Interconnects for Heterogeneous Integration 667
    *Ahmet C. Durgun (Assembly and Test Technology Development Intel
    Corporation), Zhiguo Qian (Assembly and Test Technology Development
    Intel Corporation), Kemal Aygun (Assembly and Test Technology
    Development Intel Corporation), Ravi Mahajan (Assembly and Test
    Technology Development Intel Corporation), Tim Tri Hoang (Programmable
    Solutions Group Intel Corporation), and Sergey Yuryevich Shumarayev
    (Programmable Solutions Group Intel Corporation)*

A High-Bandwidth Fine-Pitch 2.57Tbps/mm In-package Communication Link Achieving 48fJ/bit/mm Efficiency 674

Nicolas Pantano (IMEC, KULeuven), Geert Van der Plas (IMEC), Pieter Bex (IMEC), Philip Nolmans (IMEC), Dimitrios Velenis (IMEC), Marian Verhelst (KU Leuven), and Eric Beyne (IMEC)

A New SI-PI co-Simulation Approach for Efficient Consideration of Coupling Between PDN and SDN 682

Heesok Lee (Samsung Electronics, Co. Ltd.), Jisoo Hwang (Samsung Electronics, Co. Ltd.), Hoi-jin Lee (Samsung Electronics, Co. Ltd.), and Youngmin Shin (Samsung Electronics, Co. Ltd.)

Signal Integrity of Submicron InFO Heterogeneous Integration for High Performance Computing Applications 688

Chuei-Tang Wang (Taiwan Semiconductor Manufacturing Company Ltd.), Jeng-Shien Hsieh (Taiwan Semiconductor Manufacturing Company Ltd.), Victor C. Y. Chang (Taiwan Semiconductor Manufacturing Company Ltd.), Shih-Ya Huang (Taiwan Semiconductor Manufacturing Company Ltd.), T. Ko (Taiwan Semiconductor Manufacturing Company Ltd.), Han-Ping Pu (Taiwan Semiconductor Manufacturing Company Ltd.), and Douglas Yu (Taiwan Semiconductor Manufacturing Company Ltd.)

28GHz Through Glass Via (TGV) Based Band Pass Filter Using Through Fused Silica Via (TFV) Technology. 695

Renuka Bowrothu (University of Florida), Seahee Hwangbo (University of Florida), Todd Schumann (University of Florida), and Yong-Kyu Yoon (University of Florida)

Innovative Packaging Solutions of 3D Double Side Molding with System in Package for IoT and 5G Application 700

Mike Tsai (Siliconware Precision Industries Co., Ltd., Taiwan), Ryan Chiu (Siliconware Precision Industries Co., Ltd., Taiwan), Dick Huang (Siliconware Precision Industries Co., Ltd., Taiwan), Feng Kao (Siliconware Precision Industries Co., Ltd., Taiwan), Eric He (Siliconware Precision Industries Co., Ltd., Taiwan), J. Y. Chen (Siliconware Precision Industries Co., Ltd., Taiwan), Simon Chen (Siliconware Precision Industries Co., Ltd., Taiwan), Jensen Tsai (Siliconware Precision Industries Co., Ltd., Taiwan), and Yu-Po Wang (Siliconware Precision Industries Co., Ltd., Taiwan)

Enhancing Efficiency of Antenna-in-Package (AiP) by Through-Silicon-Interposer (TSI) with Embedded Air Cavity and Polyimide Dielectric Micro-Substrate 707

Yunna Sun (Shanghai Jiao Tong University), Yunting Sun (Shanghai Jiao Tong University), Jiangbo Luo (Shanghai Jiao Tong University), Huiying Wang (Shanghai Jiao Tong University), Zhuoqing Yang (Shanghai Jiao Tong University), Yan Wang (Shanghai Jiao Tong University), Guifu Ding (Shanghai Jiao Tong University), and Kwangwoo Han (Samsung Electronics Co.)

Session 16: Advanced Materials for High-Speed Electronics

Low-Loss Glass Substrates Formulated with a Variety of Dielectric Characteristics for Millimeter-Wave Applications 712

Kazutaka Hayashi (AGC Inc.), Nobutaka Kidera (AGC Inc.), and Yoichiro Sato (AGC Inc.)

Evaluation of Fine-Pitch Routing Capabilities of Advanced Dielectric Materials for High Speed Panel-RDL in 2.5D Interposer and Fan-Out Packages 718

Shreya Dwarakanath (Georgia Institute of Technology), P. Markondeya Raj (Florida International University), Amit Agarwal (Microchips, USA), Daichi Okamoto (TAIYO INK MFG. CO., LTD. Japan), Atsushi Kubo (Tokyo Ohka Kogyo Co., Ltd., Japan), Fuhan Liu (Georgia Institute of Technology), Mohan Kathaperumal (Georgia Institute of Technology), and Rao R. Tummala (Georgia Institute of Technology)

Attenuation of high Frequency Signals in Structured Metallization on Glass: Comparing Different Metallization Techniques with 24 GHz, 77 GHz and 100 GHz Structures 726

Letz Martin (SCHOTT AG, Germany), Jost Matthias (TU Darmstadt), Brandon T. Gore (Samtec, Colorado Springs), William J. Kozlovsky (Samtec, Colorado Springs), Romeo Premerlani (Varioprint AG), Alex Bruderer (Varioprint AG), Manuel Martina (Schweizer Electronic AG), Thomas Gottwald (Schweizer Electronic AG), Tetsuya Onishi (Grand Joint Technology Ltd.), Shigeo Onitake (KOTO Electric Co.), Siddharth Ravichandran (Packaging Research Center), Holger Maune (Institute for microwave engineering and photonics), and Mathias Mydlak (SCHOTT AG)

The Highly Effective EMI Shielding Materials for Electric and Magnetic Fields Over the Wide Range of Frequency in Near-Field Region 733

Yoon-Hyun Kim (Ntrium Inc.), Kisu Joo (Ntrium Inc.), Kyu Jae Lee (Ntrium Inc.), Jung Woo Hwang (Ntrium Inc.), Seung Jae Lee (Ntrium Inc.), Se Young Jeong (Ntrium Inc.), and Hyun Ho Park (Ntrium Inc.)

Low Loss NCF Material for High Frequency Device 740

Kazutaka Honda (Hitachi Chemical Co., Ltd.), Keiko Ueno (Hitachi Chemical Co., Ltd.), Tsuyoshi Ogawa (Hitachi Chemical Co., Ltd.), and Toshihisa Nonaka (Hitachi Chemical Co., Ltd.)

In-Situ Redox Nanowelding of Copper Nanowires with Surficial Oxide Layer as Solder for Flexible Transparent Electromagnetic Interference Shielding 746

Xianwen Liang (Chinese Academy of Sciences), Jianwen Zhou (Chinese Academy of Sciences), Gang Li (Chinese Academy of Sciences), Tao Zhao (Chinese Academy of Sciences), Pengli Zhu (Chinese Academy of Sciences), Rong Sun (Chinese Academy of Sciences), and Ching-ping Wong (Georgia Institute of Technology)

Compartmental EMI Shielding with Jet-Dispensed Material Technology 753

Xuan Hong (Henkel Corporation), Qizhuo Zhuo Zhuo (Henkel Corporation), Xinpei Cao (Henkel Corporation), Dan Maslyk (Henkel Corporation), Noah Ekstrom (Henkel Corporation), Juliet Sanchez (Henkel Corporation), Selene Hernandez (Henkel Corporation), and Jinu Choi (Henkel Corporation)

Session 17: Materials and Design for Reliability of Next-Generation Packages:

Highly (111)-Oriented Nanotwinned Cu for High Fatigue Resistance in Fan-Out Wafer-Level Packaging 758

Yu-Jin Li (National Chiao Tung University), Chih-Han Theng (National Chiao Tung University), I-Hsin Tseng (National Chiao Tung University), Chih Chen (National Chiao Tung University), Benson Lin (MediaTek Inc), and Chia-Cheng Chang (MediaTek Inc)

WLCSP Package and PCB Design for Board Level Reliability 763
  Jason Chiu (Taiwan Semiconductor Manufacturing Company, Ltd.), K.C.
  Chang (Taiwan Semiconductor Manufacturing Company, Ltd.), Steven Hsu
  (Taiwan Semiconductor Manufacturing Company, Ltd.), Pei-Haw Tsao
  (Taiwan Semiconductor Manufacturing Company, Ltd.), and M.J. Lii
  (Taiwan Semiconductor Manufacturing Company, Ltd.)

Assessing the Reliability of Highly Stretchable Interconnects for Flexible Hybrid Electronics 768
  Rajesh Sharma Sivasubramony (Binghamton University), Ashwin Varkey
  Zachariah (Binghamton University), Mohammed Alhendi (Binghamton
  University), Manu Yadav (Binghamton University), Peter Borgesen
  (Binghamton University), Mark D. Poliks (Binghamton University), Nancy
  C. Stoffel (GE Global Research), David M. Shaddock (GE Global
  Research), and Liang Yin (GE Global Research)

The How and why of Biased Humidity Tests with Copper Wire 777
  Amar Mavinkurve (NXP Semiconductors), R.T.H. Rongen (NXP
  Semiconductors), L. Goumans (NXP Semiconductors), M-L Farrugia (NXP
  Semiconductors), E. van Olst (NXP Semiconductors), Orla O'Halloran
  (NXP Semiconductors), and M. van Soestbergen (NXP Semiconductors)

Twist Testing for Flexible Electronics 785
  Justin H. Chow (Georgia Institute of Technology), Jeffrey Meth (DuPont
  Electronics and Imaging), and Suresh K. Sitaraman (Georgia Institute
  of Technology)

Mechanical Properties and Microstructural Fatigue Damage Evolution in Cyclically Loaded Lead-Free
Solder Joints 792
  Sinan Su (Auburn University), Mohd Aminul Hoque (Auburn University),
  Md Mahmudur Chowdhury (Auburn University), Sa'd Hamasha (Auburn
  University), Jeffrey C. Suhling (Auburn University), John L. Evans
  (Auburn University), and Pradeep Lall (Auburn University)

Reliability Studies of Silicon Interconnect Fabric 800
  Niloofar Shakoorzadeh (UCLA), Siva Chandra Jangam (UCLA), Kaysar Rahim
  (GlobalFoundries), Pranav Ambhore (UCLA), Han Chien (National Chiao
  Tung University), Amir Hanna (UCLA), and Subramanian S. Iyer (UCLA)

Session 18: Warpage and Material Performance

Improved Finite Element Modeling of Moisture Diffusion Considering Discontinuity at Material
Interfaces in Electronic Packages 806
  Lulu Ma (Lamar University), Rahul Joshi (AMD), Keith Keith Newman
  (AMD), and Xuejun Fan (Lamar University)

Study of Thermal Aging Behavior of Epoxy Molding Compound for Applications in Harsh Environments 811
  Adwait Inamdar (Robert Bosch GmbH), Alexandru Prisacaru (Robert Bosch
  GmbH), Martin Fleischman (Robert Bosch GmbH), Erick Franieck (Robert
  Bosch GmbH), Przemyslaw Gromala (Robert Bosch GmbH), Agnes Veres
  (Robert Bosch Kft), Csaba Nemeth (Robert Bosch Kft), Yu-Hsiang Yang
  (University of Maryland), and Bongtae Han (University of Maryland)

Warpage Variation Analysis and Model Prediction for Molded Packages 819
Yuling Niu (Qualcomm Technologies, Inc., USA), Wei Wang (Qualcomm Technologies, Inc., USA), Zhijie Wang (Qualcomm Technologies, Inc., USA), Karthik Dhandapani (Qualcomm Technologies, Inc., USA), Mark Schwarz (Qualcomm Technologies, Inc., USA), and Ahmer Syed (Qualcomm Technologies, Inc., USA)

Peridynamics for Predicting Thermal Expansion Coefficient of Graphene 825
Erdogan Madenci (University of Arizona), Atila Barut (University of Arizona), and Mehmet Dorduncu (University of Arizona)

Machine Learning Approach to Improve Accuracy of Warpage Simulations 834
Cheryl Selvanayagam (Singapore University of Technology and Design), Pham Luu Trung Duong (Singapore University of Technology and Design), Rathin Mandal (Advanced Micro Devices Inc.), and Nagarajan Raghavan (Singapore University of Technology and Design)

Study on Warpage of Fan-Out Panel Level Packaging (FO-PLP) Using Gen-3 Panel 842
Fa Xing Che (Institute of Microelectronics A*STAR), Kazunori Yamamoto (Institute of Microelectronics A*STAR), Vempati Srinivasa Rao (Institute of Microelectronics A*STAR), and Vasarla Nagendra Sekhar (Institute of Microelectronics A*STAR)

Mechanical Properties of Intermetallic Compounds at Elevated Temperature by Nanoindentation 850
Fan Yang (The Institute of Technological Sciences, Wuhan University, Wuhan, China), Sheng Liu (Wuhan University), Zhaoxia Zhou (Loughborough University), Zhiwen Chen (Wuhan University), Li Liu (Wuhan University of Technology), Canyu Liu (Loughborough University), and Changqing Liu (Loughborough University)

Session 19: MEMS, Sensors, and IoT

A MEMS Microphone in a FOWLP 855
Horst Theuss (Infineon Technologies AG), Christian Geissler (Infineon Technologies AG), Franz-Xaver Muehlbauer (Infineon Technologies AG), Claus von Waechter (Infineon Technologies AG), Thomas Kilger (Infineon Technologies AG), Juergen Wagner (Infineon Technologies AG), Thomas Fischer (Infineon Technologies AG), Ulf Bartl (Infineon Technologies AG), Stephan Helbig (Infineon Technologies AG), Alfred Sigl (Infineon Technologies AG), Dominic Maier (Infineon Technologies AG), Bernd Goller (Infineon Technologies AG), Matthias Vobl (Infineon Technologies AG), Matthias Herrmann (Infineon Technologies AG), Johannes Lodermeyer (Infineon Technologies AG), Ulrich Krumbein (Infineon Technologies AG), and Alfons Dehe (Hahn-Schickard)

Fan-Out Wafer Level Packaging - A Platform for Advanced Sensor Packaging 861

Tanja Braun (Fraunhofer IZM), Karl-Friedrich Becker (Fraunhofer Institute for Reliability and Microintegration), Ole Hoelck (Fraunhofer Institute for Reliability and Microintegration), Steve Voges (Fraunhofer Institute for Reliability and Microintegration), Ruben Kahle (Fraunhofer Institute for Reliability and Microintegration), Pascal Graap (Fraunhofer Institute for Reliability and Microintegration), Markus Wöhrmann (Fraunhofer Institute for Reliability and Microintegration), R. Aschenbrenner (Fraunhofer Institute for Reliability and Microintegration), Tanja Braun (Technical University Berlin), Marc Dreissigacker (Technical University Berlin), Martin Schneider-Ramelow (Technical University Berlin), and Klaus-Dieter Lang (Technical University Berlin)

3D-MID Evaluation and Validation for Space Applications 868

Etienne Hirt (Art of Technology AG), Klaus Ruzicka (Art of Technology AG), Benedikt Wigger (Hahn – Schickard, Mikromontage), Maximilian Barth (Hahn – Schickard), Rafat Saleh (Hahn – Schickard), Florian Janek (Hahn – Schickard), and Ernst Müller (University Stuttgart)

High-Temperature Pressure Sensor Package and Characterization of Thermal Stress in the Assembly up to 500 °C 878

Nilavazhagan Subbiah (University of Freiburg, Germany), Qingming Feng (University of Freiburg, IMTEK), Juergen Wilde (University of Freiburg, Germany), and Gudrun Bruckner (CTR AG, Austria)

Development of 3D WLCSP with Black Shielding for Optical Finger Print Sensor for the Application of Full Screen Smart Phone 884

Daquan Yu (Huantian Technology (Kunshan) Electronics Co., Ltd), Yichao Zou (Huantian Technology (Kunshan) Electronics Co., Ltd), Xirui Xu (Huantian Technology (Kunshan) Electronics Co., Ltd), Aihua Shi (Huantian Technology (Kunshan) Electronics Co., Ltd), Xiaobing Yang (Huantian Technology (Kunshan) Electronics Co., Ltd), and Zhiyi Xiao (Huantian Technology (Kunshan) Electronics Co., Ltd.)

Micro Fountain-Like Resonators 890

Jianfeng Zhang (Southeast University, China), Jintang Shang (Southeast University, China), Bin Luo (Southeast University, China), and Zhaoxi Su (Southeast University, China)

Novel Additively Manufactured Packaging Approaches for 5G/mm-Wave Wireless Modules 896

Tong-Hong Lin (Georgia Institute of Technology), Aline Eid (Georgia Institute of Technology), Jimmy Hester (Georgia Institute of Technology), Bijan Tehrani (Georgia Institute of Technology), Jo Bito (Texas Instrument), and Manos M. Tentzeris (Georgia Institute of Technology)

Session 20: Fanout and Heterogeneous Integration

Feasibility Study of Fan-Out Wafer-Level Packaging for Heterogeneous Integrations 903
*John Lau (ASMPT), Ming Li (ASMPT), Iris Xu (Jiangyin Changdian
Advanced Packaging Co., Ltd.), Tony Chen (Jiangyin Changdian Advanced
Packaging Co., Ltd.), Kim Hwee Tan (Jiangyin Changdian Advanced
Packaging Co., Ltd.), Zhang Li (Jiangyin Changdian Advanced Packaging
Co., Ltd.), Nelson Fan (ASMPT), Eric Kuah (ASMPT), Raymond So (ASMPT),
Penny Lo (ASMPT), Y. M. Cheung (ASMPT), Cao Xi (Huawei Technologies
Co. Ltd.), Rozalia Beica (Dow Chemical Company), Sze Pei Lim (Indium
Corporation), NC Lee (Indium Corporation), Cheng-Ta Ko (Unimicron
Technology Corporation), Henry Yang (Unimicron Technology
Corporation), YH Chen (Unimicron Technology Corporation), Mian Tao
(Hong Kong University of Science and Technology), Jeffery Lo (Hong
Kong University of Science and Technology), and Ricky Lee (Hong Kong
University of Science and Technology)*

Experiment of 22FDX(R) Chip Board Interaction (CBI) in Wafer Level Packaging Fan-Out (WLPFO) 910
*Jae Kyu Cho (GLOBALFOUNDRIES, USA), Jens Paul (GLOBALFOUNDRIES,
Germany), Simone Capecchi (GLOBALFOUNDRIES, Germany), Frank
Kuechenmeister (GLOBALFOUNDRIES, Germany), and Ta-Chien Cheng
(GLOBALFOUNDRIES, Singapore)*

FOWLP Design for Digital and RF Circuits 917
*Teck Guan Lim (Institute of Microelectronics, Singapore), David Soon
Wee Ho (Institute of Microelectronics, Singapore), Eva Wai Leong Ching
(Institute of Microelectronics, Singapore), Zihao Chen (Institute of
Microelectronics, Singapore), and Surya Bhattacharya (Institute of
Microelectronics, Singapore)*

Next Generation of 2-7 Micron Ultra-Small Microvias for 2.5D Panel Redistribution Layer by Using
Laser and Photolithography Technologies 924
*Fuhan Liu (Georgia Institute of Technology), Chandrasekharan Nair
(Georgia Institute of Technology), Gaurav Khurana (Georgia Institute
of Technology), Atom Watanabe (Georgia Institute of Technology),
Bartlet H. DeProspo (Georgia Institute of Technology), Atsushi Kubo
(Tokyo Ohka Kogyo Co. Ltd., Japan), Cheng Ping Lin (Panasonic
Corporation, Japan), Toshiyuki Makita (Panasonic Corporation, Japan),
Naoki Watanabe (Panasonic Industrial Devices Sales Company of America,
USA), and Rao R. Tummala (Georgia Institute of Technology)*

Multilayer RDL Interposer for Heterogeneous Device and Module Integration 931
*Yi-Hang Lin (Taiwan Semiconductor Manufacturing Company Ltd.), M.C.
Yew (Taiwan Semiconductor Manufacturing Company Ltd.), S.M. Chen
(Taiwan Semiconductor Manufacturing Company Ltd.), M.S. Liu (Taiwan
Semiconductor Manufacturing Company Ltd.), Pravin Kavle (Taiwan
Semiconductor Manufacturing Company Ltd.), T.M. Lai (Taiwan
Semiconductor Manufacturing Company Ltd.), C.T. Yu (Taiwan
Semiconductor Manufacturing Company Ltd.), F.C. Hsu (Taiwan
Semiconductor Manufacturing Company Ltd.), C.S. Chen (Taiwan
Semiconductor Manufacturing Company Ltd.), T.J. Fang (Taiwan
Semiconductor Manufacturing Company Ltd.), C.K. Hsu (Taiwan
Semiconductor Manufacturing Company Ltd.), K.C. Lee (Taiwan
Semiconductor Manufacturing Company Ltd.), C.H. Lin (Taiwan
Semiconductor Manufacturing Company Ltd.), P.Y. Lin (Taiwan
Semiconductor Manufacturing Company Ltd.), and Shin-Puu Jeng (Taiwan
Semiconductor Manufacturing Company Ltd.)*

Effects of Dielectric Curing Conditions on the Interfacial Adhesion of Cu RDL for Fan-Out Wafer
Level Packaging 937

Gahui Kim (Andong National University, Korea), Kirak Son (Andong
National University, Korea), Dogeun Kim (Korea Institute of Materials
Science, Korea), Seok-hyun Lee (SAMSUNG ELECTRONICS CO., LTD, Korea),
and Young-Bae Park (Andong National University, Korea)

Al-Al Direct Bonding with Sub-µm Alignment Accuracy for Millimeter Wave SiGe BiCMOS Wafer Level
Packaging and Heterogeneous Integration 942

Matthias Wietstruck (IHP - Leibniz Institut für innovative
Mikroelektronik), Sebastian Schulze (IHP - Leibniz Institut für
innovative Mikroelektronik), Bernhard Rebhan (EV Group E. Thallner
GmbH), Peter Kerepesi (EV Group E. Thallner GmbH), Helmut Kurz (EV
Group E. Thallner GmbH), Gerald Silberer (EV Group E. Thallner GmbH),
Josef Meiler (EV Group E. Thallner GmbH), Selin Tolunay Wipf (IHP -
Leibniz Institut für innovative Mikroelektronik), Christian Wipf (IHP
- Leibniz Institut für innovative Mikroelektronik), and Mehmet Kaynak
(IHP - Leibniz Institut für innovative Mikroelektronik, Sabanci
University)

Session 21: 5G, mm-Wave, and Antenna-in-Package

Vivaldi Antenna Array Fabricated Using a Hybrid Process 948
Vincens Gjokaj (Michigan State University), Cameron Crump (Michigan
State University), John Papapolymerou (Michigan State University),
John Albrecht (Michigan State University), and Premjeet Chahal
(Michigan State University)

Novel Multicore PCB and Substrate Solutions for Ultra Broadband Dual Polarized Antennas for 5G
Millimeter Wave Covering 28GHz & 39GHz Range 954
Trang Thai (Intel Corporation), Sidharth Dalmia (Intel Corporation),
Josef Hagn (Intel Corporation), Pouya Talebbeydokhti (Intel
Corporation), and Yossi Tsfati (Intel Corporation)

3D Glass Package-Integrated, High-Performance Power Dividing Networks for 5G Broadband Antennas 960
Muhammad Ali (Georgia Institute of Technology), Atom Watanabe (Georgia
Institute of Technology), Tong-Hong Lin (Georgia Institute of
Technology), Markondeya Raj Pulugurtha (Florida International
University), Manos M. Tentzeris (Georgia Institute of Technology), and
Rao R. Tummala (Georgia Institute of Technology)

Advanced Wafer Level PKG Solutions for 60GHz WiGig (802.11ad) Telecom Infrastructure 968
Dapeng Wu (Sivers IMA AB), Robin Dahlbäck (Sivers IMA AB), Erik
Öjefors (Sivers IMA AB), Mats Carlsson (Sivers IMA AB), Francis Chee
Peng Lim (STATS ChipPAC Pte. Ltd.), Yew Kheng Lim (STATS ChipPAC Pte.
Ltd.), Aung Kyaw Oo (STATS ChipPAC Pte. Ltd.), Won Kyung Choi (STATS
ChipPAC Pte. Ltd.), and Seung Wook Yoon (STATS ChipPAC Pte. Ltd.)

xxvii

Low-Loss Additively-Deposited Ultra-Short Copper-Paste Interconnections in 3D Antenna-Integrated Packages for 5G and IoT Applications 972

Atom O. Watanabe (Georgia Institute of Technology), Yiteng Wang
(Georgia Institute of Technology), Nobuo Ogura (Nagase & Co., LTD.,
Japan), P. Markondeya Raj (Florida International University), Vanessa
Smet (Georgia Institute of Technology), Manos M. Tentzeris (Georgia
Institute of Technology), and Rao R. Tummala (Georgia Institute of
Technology)

Advanced Thin-Profile Fan-Out with Beamforming Verification for 5G Wideband Antenna 977

Ricky Hsieh (Advanced Semiconductor Engineering (ASE), Inc., Taiwan),
Fu-Cheng Chu (Advanced Semiconductor Engineering (ASE), Inc., Taiwan),
Cheng-Yu Ho (Advanced Semiconductor Engineering (ASE), Inc., Taiwan),
and Chen-Chao Wang (Advanced Semiconductor Engineering (ASE), Inc.,
Taiwan)

Integrated Compact Planar Inverted-F Antenna (PIFA) with a Shorting Via Wall for Millimeter-Wave
Wireless Chip-to-Chip (C2C) Communications in 3D-SiP 983

Seahee Hwangbo (University of Florida), Renuka Bowrothu (University of
Florida), Hae-in Kim (University of Florida), and Yong-Kyu Yoon
(University of Florida)

Session 22: Advanced Substrates and Interconnect Technology

Temporary SiC-SiC Wafer Bonding Compatible with High Temperature Annealing 989
Fengwen Mu (The University of Tokyo), Tadatomo Suga (The University of
Tokyo), Miyuki Uomoto (Tohoku University), and Takehito Shimatsu
(Tohoku University)

Ultrathin Glass to Ultrathin Glass Bonding Using Laser Sealing Approach 995
Messaoud Bedjaoui (Univ. Grenoble Alpes, CEA-LETI, France), Johnny
Amiran (Univ. Grenoble Alpes, CEA-LETI, France), and Jean Brun (Univ.
Grenoble Alpes, CEA-LETI, France)

Development of Resins for Bumpless Interconnects and Wafer-On-Wafer (WOW) Integration 1002
Naoko Araki (Daicel Corporation), Shinji Maetani (Daicel Corporation),
Kim Young Suk (Disco Corporation), Shoichi Kodama (Disco Corporation),
and Takayuki Ohba (Tokyo Institute of Technology)

Development of Novel Photosensitive Dielectric Material for Reliable 2.1D Package 1009
Yune Kumazawa (Mitsubishi Gas Chemical Company, Inc.), Seiji Shika
(Mitsubishi Gas Chemical Company, Inc.), Shunsuke Katagiri (Mitsubishi
Gas Chemical Company, Inc.), Takuya Suzuki (Mitsubishi Gas Chemical
Company, Inc.), Tsuyoshi Kida (Mitsubishi Gas Chemical Company, Inc.),
and Shu Yoshida (Mitsubishi Gas Chemical Company, Inc.)

High Reliability Solder Resist with Strong Adhesion and High Resolution for High Density Packaging 1015
Sawako Shimada (TAIYO INK MFG. CO., LTD.), Kazuya Okada (TAIYO INK
MFG. CO., LTD.), Tomoya Kudo (TAIYO INK MFG. CO., LTD.), Chiho Ueta
(TAIYO INK MFG. CO., LTD.), and Yuya Suzuki (TAIYO INK MFG. CO., LTD.)

Method for Mitigating the Warpage of Ultra-Thin FC-CSPs by Controlling of EMC Properties 1022
Chika Arayama (Panasonic Corporation), Takahiro Akashi (Panasonic
Corporation), Yasunari Tomita (Panasonic Corporation), and Naoki
Kanagawa (Panasonic Corporation)

Innovative Socketable and Surface-Mountable BGA Interconnections 1028
Omkar Gupte (Georgia Institute of Technology - Packaging Research
Center), Kristie Teoh (Georgia Institute of Technology - Packaging
Research Center), Rao Tummala (Georgia Institute of Technology -
Packaging Research Center), Gregorio Murtagian (Intel Corporation),
and Vanessa Smet (Georgia Institute of Technology - Packaging Research
Center)

Session 23: High-Bandwidth 3D and Photonic Integration

A Highly Reliable 1.4µm Pitch Via-Last TSV Module for Wafer-to-Wafer Hybrid Bonded 3D-SOC Systems . 1035
Stefaan Van Huylenbroeck (IMEC), Joeri De Vos (IMEC), Zaid El-Mekki
(IMEC), Geraldine Jamieson (IMEC), Nina Tutunjyan (IMEC), Karthik Muga
(IMEC), Michele Stucchi (IMEC), Andy Miller (IMEC), Gerald Beyer
(IMEC), and Eric Beyne (IMEC)

Nanoscale Topography Characterization for Direct Bond Interconnect 1041
Bongsub Lee (Xperi Corporation), Pawel Mrozek (Xperi Corporation),
Gill Fountain (Xperi Corporation), John Posthill (Xperi Corporation),
Jeremy Theil (Xperi Corporation), Guilian Gao (Xperi Corporation),
Rajesh Katkar (Xperi Corporation), and Laura Mirkarimi (Xperi
Corporation, USA)

Fully-Filled, Highly-Reliable Fine-Pitch Interposers with TSV Aspect Ratio >10 for Future 3D-LSI/IC
Packaging 1047
Murugesan Murugesan (Tohoku University, Japan), Takafumi Fukushima
(Tohoku University, Japan), Kiyoharu Mori (T-Miro, Japan), Ai Nakamura
(T-Micro, Japan), Yisang Lee (T-Micro, Japan), Makoto Motoyoshi
(T-Micro, Japan), J.C Bea (Tohoku University), Shigeru Watariguchi
(Meltex, Japan), and Mitsumasa Koyanagi (Tohoku University, Japan)

3D Silicon Photonics Interposer for Tb/s Optical Interconnects in Data Centers with Double-Side
Assembled Active Components and Integrated Optical and Electrical Through Silicon Via on SOI 1052
Bogdan Sirbu (Fraunhofer Institute for Reliability and
Microintegration IZM Berlin), Yann Eichhammer (Fraunhofer Institute
for Reliability and Microintegration IZM Berlin), Hermann Oppermann
(Fraunhofer Institute for Reliability and Microintegration IZM
Berlin), Tolga Tekin (Fraunhofer Institute for Reliability and
Microintegration IZM Berlin), Jochen Kraft (ams AG, Austria), Victor
Sidorov (ams AG, Austria), Xin Yin (IMEC, Belgium), Johan Bauwelinck
(IMEC, Belgium), Christian Neumeyr (Vertilas GmbH, Germany), and
Francisco Soares (Fraunhofer Heinrich-Hertz-Institut HHI Berlin)

Flip-Chip III-V-to-Silicon Photonics Interfaces for Optical Sensor 1060

Yves Martin (IBM T. J. Watson Research Center, Yorktown Heights USA),
Jason S. Orcutt (IBM T. J. Watson Research Center, Yorktown Heights
USA), Chi Xiong (IBM T. J. Watson Research Center, Yorktown Heights
USA), Laurent Schares (IBM T. J. Watson Research Center, Yorktown
Heights USA), Tymon Barwicz (IBM T. J. Watson Research Center,
Yorktown Heights USA), Martin Glodde (IBM T. J. Watson Research
Center, Yorktown Heights USA), Swetha Kamlapurkar (IBM T. J. Watson
Research Center, Yorktown Heights USA), Eric J. Zhang (IBM T. J.
Watson Research Center, Yorktown Heights USA), William M.J. Green (IBM
T. J. Watson Research Center, Yorktown Heights USA), Victor
Dolores-Calzadilla (Fraunhofer Heinrich-Hertz Institute, Germany),
Ariane Sigmund (Fraunhofer Heinrich-Hertz Institute, Germany), and
Martin Moehrle (Fraunhofer Heinrich-Hertz Institute, Germany)

Extremely Low-Profile Single Mode Fiber Array Coupler Suitable for Silicon Photonics 1067

Mitsuharu Hirano (Sumitomo Electric Industries, Ltd., Japan), Akira
Furuya (Sumitomo Electric Industries, Ltd., Japan), Hideki Machida
(Sumitomo Electric Industries, Ltd., Japan), Koichi Koyama (Sumitomo
Electric Industries, Ltd., Japan), Yasunori Murakami (Sumitomo
Electric Industries, Ltd., Japan), and Kazunori Tanaka (Sumitomo
Electric Industries, Ltd., Japan)

Micro Lens Array Assembly for Optical Organic Substrate 1074

Patrick Jacques (IBM Bromont, Canada), Richard Langlois (IBM Bromont,
Canada), Élaine Cyr (IBM Bromont, Canada), Alexander Janta-Polczynski
(IBM Bromont, Canada), Paul Fortier (IBM Bromont, Canada), Koji Masuda
(IBM Tokyo Research, Japan), Masao Tokunari (IBM Tokyo Research), and
Hsiang-Han Hsu (IBM Tokyo Research)

Session 24: Advancements in Solder Joint Characterization and Reliability Evaluation

Effects of In and Zn Double Addition on Eutectic Sn-58Bi Alloy 1081

Shiqi Zhou (Osaka University, Japan), Yu-An Shen (Osaka University,
Japan), Tiffani Uresti (Texas A&M University at Qatar, Qatar), Vasanth
Shunmugasamy (Texas A&M University at Qatar, Qatar), Bilal Mansoor
(Texas A&M University at Qatar, Qatar), and Hiroshi Nishikawa (Osaka
University, Japan)

Microstructural Evolution in SAC+X Solders Subjected to Aging 1087

Jing Wu (Auburn University), Jeffrey C. Suhling (Auburn University),
and Pradeep Lall (Auburn University)

Microstructure Signature Evolution in Solder Joints, Solder Bumps, and Micro-Bumps Interconnection
in A Large 2.5D FCBGA Package During Thermo-Mechanical Cycling 1099

Arman Ahari (Portland State University), Andy Hsiao (Portland State
University), Greg Baty (Portland State University), Peng Su (Juniper
Networks, USA), and Tae-Kyu Lee (Portland State University)

Long-Term Reliability of Solder Joints in 3D ICs Under Near-Application Conditions 1106

Omar Ahmed (University of Central Florida), Golareh Jalilvand
(University of Central Florida), Hector Fernandez (University of
Central Florida), Peng Su (Juniper Networks), Tae-Kyu Lee (Portland
State University), and Tengfei Jiang (University of Central Florida)

Experimental Investigation of the Correlation between a Load-Based Metric and Solder Joint
Reliability of BGA Assemblies on System Level 1113

Fabian Schempp (Robert Bosch GmbH, University of Freiburg - IMTEK),
Marc Dressler (Robert Bosch GmbH), Daniel Kraetschmer (Robert Bosch
GmbH), Friederike Loerke (Robert Bosch GmbH), and Juergen Wilde
(University of Freiburg - IMTEK)

Fatigue Life Prediction Model Development for Decoupling Capacitors 1121

Krishna Tunga (IBM Corporation, USA), Joseph Ross (IBM Corporation,
USA), Kamal Sikka (IBM Corporation, USA), and Bakul Parikh (IBM
Corporation, USA)

A Study of Substrate Models and Its Effect On Package Warpage Prediction 1130

Van-Lai Pham (Binghamton University), Huayan Wang (Binghamton
University), Jiefeng Xu (Binghamton University), Jing Wang (Binghamton
University), Chrandeep Singh (Corning Inc.), and Seungbae Park
(Binghamton University)

Session 25: Wafer Level Packaging and Fan-In/Fan-Out Structures & Materials

3D Fan-Out Package Technology with Photosensitive Through Mold Interconnects 1140
Kentaro Mori (Toshiba Electronic Devices and Storage Corporation),
Soichi Yamashita (Toshiba Electronic Devices and Storage Corporation),
Takafumi Fukuda (Toshiba Development & Engineering Corporation),
Masahiro Sekiguchi (Toshiba Electronic Devices and Storage
Corporation), Hirokazu Ezawa (Toshiba Memory Corporation), and Shuzo
Akejima (Toshiba Electronic Devices and Storage Corporation)

Effects of the Materials Properties of Epoxy Molding Films (EMFs) on Fan-Out Packages (FOPs)
Characteristics 1146
Sangmyung Shin (KAIST), Hanmin Lee (KAIST), JunMo Kim (KAIST), Tae-Ik
Lee (KAIST), Taek-Soo Kim (KAIST), Youjin Kyung (LG Chem), Minsu Jeong
(LG Chem), Kwangjoo Lee (LG Chem), and Kyung-Wook Paik (KAIST)

Mechanism of Moldable Underfill (MUF) Process for RDL-1st Fan-Out Panel Level Packaging (FOPLP) ... 1152
Lin Bu (Institute of Microelectronics A*STAR), F. X. Che (Institute of
Microelectronics A*STAR), Vempati Srinivasa Rao (Institute of
Microelectronics A*STAR), and Xiaowu Zhang (Institute of
Microelectronics A*STAR)

Study of the Board Level Reliability Performance of a Large 0.3 mm Pitch Wafer Level Package 1159
Bernd Waidhas (Intel Deutschland GmbH), Jan Proschwitz (Intel
Deutschland GmbH), Christoph Pietryga (Intel Deutschland GmbH), Thomas
Wagner (Intel Deutschland GmbH), and Beth Keser (Intel Deutschland
GmbH)

Study of Board Level Reliability of eWLB (embedded Wafer Level BGA) for 0.35mm Ball Pitch 1165
Kang Hai Lee (STATS ChipPAC Pte. Ltd.), Yeow Kheng Lim (STATS ChipPAC
Pte. Ltd.), Seng Guan Chow (STATS ChipPAC Pte. Ltd.), Kang Chen (STATS
ChipPAC Pte. Ltd.), Won Kyung Choi (STATS ChipPAC Pte. Ltd.), Seung
Wook Yoon (STATS ChipPAC LTD PTE), NW Liu (Advanced Package
Technology, Mediatek Inc.), Yenyao Chi (Advanced Package Technology,
Mediatek Inc.), and Benson Lin (Advanced Package Technology, Mediatek
Inc.)

Board Level Reliability Study of Fan-Out Single Die Package with 350um Bump Pitch 1170
Chieh-Lung Lai (Siliconware Precision Industries Co., Ltd.), Gu-Yan
Lin (Siliconware Precision Industries Co., Ltd.), Tz-Yuan Chao
(Siliconware Precision Industries Co., Ltd.), Yih-Sin Chen
(Siliconware Precision Industries Co., Ltd.), and Feng-Lung Chien
(Siliconware Precision Industries Co., Ltd.)

The Analysis for Bump Resistance Improvement by Optimizing the Sputter Condition 1175
Ming-Sin Su (Taiwan Semiconductor Manufacturing Company Ltd.),
Chang-Ning Wang (Taiwan Semiconductor Manufacturing Company Ltd.),
Clair Tsai (Taiwan Semiconductor Manufacturing Company Ltd.), T. L.
Yang (Taiwan Semiconductor Manufacturing Company Ltd.), Rolance Yang
(Taiwan Semiconductor Manufacturing Company Ltd.), W. C. Wu (Taiwan
Semiconductor Manufacturing Company Ltd.), C. S. Liu (Taiwan
Semiconductor Manufacturing Company Ltd.), J. M. Chiu (Taiwan
Semiconductor Manufacturing Company Ltd.), Y. F. Chen (Taiwan
Semiconductor Manufacturing Company Ltd.), Ponder Pang (Taiwan
Semiconductor Manufacturing Company Ltd.), Harry Ku (Taiwan
Semiconductor Manufacturing Company Ltd.), Kirin Wang (Taiwan
Semiconductor Manufacturing Company Ltd.), C.H. Su (Taiwan
Semiconductor Manufacturing Company Ltd.), Steven Hsu (Taiwan
Semiconductor Manufacturing Company Ltd.), Calvin Lu (Taiwan
Semiconductor Manufacturing Company Ltd.), K. C. Liu (Taiwan
Semiconductor Manufacturing Company Ltd.), and Marvin Liao (Taiwan
Semiconductor Manufacturing Company Ltd.)

Session 26: High-Speed Signaling for High-Performance Computing and Memory

Hybrid Prepreg Conventional Build-Up Laminate for 112Gbit/s SerDes 1179
Kwang Won Choi (GLOBALFOUNDRIES US Inc.), Edmund Blackshear
(GLOBALFOUNDRIES US Inc.), Eric Tremble (GLOBALFOUNDRIES US Inc.),
David Stone (GLOBALFOUNDRIES US Inc.), Jean Audet (IBM Corporation,
Canada), and Keiichi Hirabayashi (Shinko Electric Industries Co.,
LTD., Japan)

PI/SI Analysis and Design Approach for HPC Platform Applications 1188
Sungwook Moon (Samsung Electronics Co. Ltd.), Chanmin Jo (Samsung
Electronics Co. Ltd.), and Seungki Nam (Samsung Electronics Co. Ltd.)

PoP LPDDR5 (6.4 Gbps) NTODT and 1-Tap DFE for Signal Integrity Enhancement 1194
Sunil Gupta (Qualcomm Technologies, Inc.)

OpenCAPI Memory Interface Signal Integrity Study for High-Speed DDR5 Differential DIMM Channel with Standard Loss FR-4 Material and SNIA SFF-TA-1002 Connector 1200
>
> Biao Cai (IBM), Jose Hejase (IBM), Kyle Giesen (IBM), Junyan Tang
> (IBM), Brian Connolly (IBM), KyuHyoun Kim (IBM), Daniel Dreps (IBM),
> Zhineng Fan (Amphenol ICC), Rocky Huang (Amphenol ICC), Luyun Yi
> (Amphenol ICC), Qiaoli Chen (Amphenol ICC), Yifan Huang (Amphenol
> ICC), and Stephen Smith (Amphenol ICC)

Effectiveness of Equalization and Performance Potential in DDR5 Channels with RDIMM(s) 1208
>
> Nanju Na (Xilinx) and Hing "Thomas" To (Xilinx)

Inductive Links for 3D Stacked Chip-to-Chip Communication 1215
>
> Xiao Sun (IMEC, Belgium), Nicolas Pantano (IMEC, Belgium), Kim
> Soon-Wook (IMEC, Belgium), Geert Van der Plas (IMEC, Belgium), and
> Eric Beyne (IMEC, Belgium)

System Co-design of a 600V GaN FET Power Stage with Integrated Driver in a QFN System-in-Package (QFN-SiP) 1221
>
> Jie Chen (Texas Instruments, Inc), Yong Xie (Texas Instruments, Inc),
> Django Trombley (Texas Instruments Incorporated), and Rajen Murugan
> (Texas Instruments Incorporated)

Session 27: Advanced Biosensors and Bioelectronics

Flexible Probe for Electrical Neural Signal Recording 1227
>
> Sajay Bhuvanendran Nair Gourikutty (Institute of Microelectronics,
> A*STAR, Singapore) and Ruiqi Lim (Institute of Microelectronics,
> A*STAR, Singapore)

Stretchable, Implantable Nanomembrane Biosensor for Wireless, Real-Time Monitoring of Hemodynamics .. 1233
>
> Robert Herbert (Georgia Institute of Technology) and Woon-Hong Yeo
> (Georgia Institute of Technology)

A Wearable Passive pH Sensor for Health Monitoring 1240
>
> Saikat Mondal (Michigan State University), Saranraj Karuppuswami
> (Michigan State University), Rachel Steinhorst (Michigan State
> University), and Premjeet Chahal (Michigan State University)

Novel Packaging Structure and Processes for Micro-TFB (Thin Film Battery) to Enable Miniaturized Healthcare Internet-of-Things (IoT) Devices 1246
>
> Bing Dang (IBM Research), Qianwen Chen (IBM Research), Leanna Pancoast
> (IBM Research), Yu Luo (IBM Research), Hongqing Zhang (IBM Systems),
> Jae-woong Nah (IBM Research), John Knickerbocker (IBM Research), Andy
> Shih (Front Edge Technologies Inc.), Po Wen Cheng (Front Edge
> Technologies Inc.), Kai Liu (Front Edge Technologies Inc.), Mengnian
> Niu (Front Edge Technologies Inc.), and Simon Nieh (Front Edge
> Technologies Inc.)

Screen Printed Temporary Tattoos for Skin-Mounted Electronics 1252
>
> Samuli Tuominen (Tampere University) and Matti Mantysalo (Tampere
> University)

Thermoset Polymers for Bioelectronic Interfaces - Engineering of Thermomechanical Properties 1258

Adriana Carolina Duran-Martinez (The University of Texas at Dallas),
Seyedmahmoud Hosseini (The University of Texas at Dallas), Daniel Del
Nero (The University of Texas at Dallas), Alexandra Joshi-Imre (The
University of Texas at Dallas), Walter E. Voit (The University of
Texas at Dallas), and Melanie Ecker (The University of Texas at
Dallas)

Direct Heterogeneous Bonding of SiC to Si, SiO2, and Glass for High-Performance Power Electronics
and Bio-MEMS 1266

Jikai Xu (Harbin Institute of Technology), Chenxi Wang (Harbin
Institute of Technology), Qiushi Kang (Harbin Institute of
Technology), Shicheng Zhou (Harbin Institute of Technology), and
Yanhong Tian (Harbin Institute of Technology)

Session 28: Embedded and Integrated Technologies

Development of Flexible Hybrid Electronics Using Reflow Assembly with Stretchable Film 1272

Weifeng Liu (Flex), William Uy (Flex), Alex Chan (Flex), Dongkai
Shangguan (Flex), Andy Behr (Panasonic), Takatoshi Abe (Panasonic),
and Fukao Tomohiro (Panasonic)

Highly Compact RF Transceiver Module Using High Resistive Silicon Interposer with Embedded Inductors
and Heterogeneous Dies Integration 1279

G. Pares (Univ. Grenoble Alpes, CEA), Michel Jean-Philippe (CEA),
Deschaseaux Edouard (CEA), Ferris Pierre (CEA), Serhan Ayssar (CEA),
and Giry Alexandre (CEA)

Process Induced Wafer Warpage Optimization for Multi-chip Integration on Wafer Level Molded Wafer 1287

Chen-Yu Huang (Siliconware Precision Industries Co., Ltd, Taiwan),
Daniel Ng (Siliconware Precision Industries Co., Ltd, Taiwan), Hung-Ho
Lee (Siliconware Precision Industries Co., Ltd, Taiwan), Vito Lin
(Siliconware Precision Industries Co., Ltd, Taiwan), Chang-Fu Lin
(Siliconware Precision Industries Co., Ltd, Taiwan), and C. Key Chung
(Siliconware Precision Industries Co., Ltd, Taiwan)

Improved Structure for Package Substrates with Embedded Thin-Film Capacitor 1294

Tomoyuki Akahoshi (Fujitsu Laboratories Ltd.), Daisuke Mizutani
(Fujitsu Laboratories Ltd.), Kei Fukui (Fujitsu Interconnect
Technologies Ltd.), Seigo Yamawaki (Fujitsu Interconnect Technologies
Ltd.), Hidehiko Fujisaki (Fujitsu Interconnect Technologies Ltd.),
Manabu Watanabe (Fujitsu Advanced Technologies Ltd.), and Masateru
Koide (Fujitsu Advanced Technologies Ltd.)

3D Packaging with Embedded High-Power-Density Passives for Integrated Voltage Regulators 1300

Teng Sun (Georgia Institute of Technology), Robert G. Spurney (Georgia
Institute of Technology), Atom Watanabe (Georgia Institute of
Technology), P. Raj Pulugurtha (Florida International University),
Himani Sharma (Georgia Institute of Technology), Rao Tummala (Georgia
Institute of Technology), and Furukawa Yoshihiro (Nitto Denko
Corporation)

A Novel Panel Level Double Side Embedded Package for Small Size Power Devices 1306
    Kunpeng Ding (Shenzhen Siptory Technologies Co., Ltd, China), Zhichao
    Wu (Institute of Microelectronics, Tsinghua University, China), Mian
    Huang (Shenzhen Siptory Technologies Co., Ltd, China), Bowei Zhang
    (Wuxi Sky Chip Interconnection Technology Co., Ltd, China), and Jian
    Cai (Institute of Microelectronics, Tsinghua University, China)

Chiplet Micro-Assembly Printer 1312
    Bradley B. Rupp (PARC), Anne Plochowietz (PARC), Lara S. Crawford
    (PARC), Matthew Shreve (PARC), Sourobh Raychaudhuri (PARC), Sergey
    Butylkov (PARC), Yunda Wang (PARC), Ping Mei (PARC), Qian Wang (PARC),
    Jamie Kalb (PARC), Yu Wang (PARC), Eugene M. Chow (PARC), and JengPing
    Lu (PARC)

Session 29: Electromigration and Innovative Reliability Test Methods

Effect of Intermetallic Compound Growth on Electromigration Failure Mechanism in Low-Profile Solder
Joints 1316
    Hossein Madanipour (University of Texas at Arlington), Yi-Ram Kim
    (University of Texas at Arlington), Choong-Un Kim (University of Texas
    at Arlington), Ninad Shahane (Texas Instruments, Inc.), Dibyajat
    Mishra (Texas Instruments, Inc.), and Luu Nguyen (Texas Instruments,
    Inc.)

Effect of Grain Orientation and Microstructure Evolution on Electromigration in Flip-Chip Solder
Joint 1324
    Xing Fu (Science and technology on reliability physics and application
    of electronic component laboratory, China), Bin Zhou (Science and
    technology on reliability physics and application of electronic
    component laboratory), Ruohe Yao (University of Technology), Yunfei En
    (Science and technology on reliability physics and application of
    electronic component laboratory), and Si Chen (Science and technology
    on reliability physics and application of electronic component
    laboratory)

High Electromigration Lifetimes of Nanotwinned Cu Redistribution Lines 1328
    I-Hsin Tseng (National Chiao Tung University), Yu-Jin Li (National
    Chiao Tung university), Benson Lin (MediaTek Inc), Chia-Cheng Chang
    (MediaTek Inc), and Chih Chen (National Chiao Tung University)

Non-destructive Failure Analysis of Various Chip to Package Interaction Anomalies in FCBGA Packages
Subjected to Temperature Cycle Reliability Testing 1333
    Vishnu V. B. Reddy (Georgia Institute of Technology), I. Charles Ume
    (Georgia Institute of Technology), Jaimal Williamson (Texas
    Instruments Inc., USA), and Luu Nguyen (Texas Instruments Inc., USA)

Assessment of Accelerometer Versus LASER for Board Level Vibration Measurements 1339
    Varun Thukral (NXP Semiconductors), M. Cahu (NXP Semiconductors),
    J.J.M. Zaal (NXP Semiconductors), J. Jalink (NXP Semiconductors), R.
    Roucou (NXP Semiconductors), and R.T.H. Rongen (NXP Semiconductors)

Effect of Process Parameters on the Long-Run Print Consistency and Material Properties of Additively Printed Electronics 1347

    Pradeep Lall (Auburn University), Nakul Kothari (Auburn University),
    Amrit Abrol (Auburn University), Jeff Suhling (Auburn University),
    Sudan Ahmed (Auburn University), Ben Leever (US AFRL), and Scott
    Miller (NextFlex)

A Viscoplastic-Based Fatigue Reliability Model for the Polyimide Dielectric Thin Film 1359

    Yu-Chen Chang (National Cheng Kung University), Tz-Cheng Chiu
    (National Cheng Kung University), Yu-Ting Yang (Advanced Semiconductor
    Engineering Group, Inc.), Yi-Hsiu Tseng (Advanced Semiconductor
    Engineering Group, Inc.), and Xi-Hong Chen (Advanced Semiconductor
    Engineering Group, Inc.)

Session 30: Assembly and Process Modeling

Explicit FE Failure Prediction of Interfaces and Interconnect in Potted Electronics Assemblies Subject to High-g Acceleration Loads 1366

    Pradeep Lall (Auburn University), Kalyan Dornala (Auburn University),
    Ryan Lowe (ARA Associates), and John Deep (US AFRL)

Numerical Simulation on the Formation Process of Metal Droplets by Pneumatic Diaphragm Drop-on Demand Technology 1377

    Kun Ma (Wuhan University), Sheng Liu (Wuhan University), Zhiwen Chen
    (Wuhan University), Li Liu (Wuhan University of Technology), Hao Zheng
    (China Ship Development and Design Center), and Yao Zhang (China Ship
    Development and Design Center)

On Curing-Induced Residual Stresses After Molding Processes: Mold Shrinkage, Chemical Shrinkage or Both? 1382

    Changsu Kim (University of Maryland), Sukrut Phansalkar (University of
    Maryland), Hyun-Seop Lee (University of Maryland), and Bongtae Han
    (University of Maryland)

Realistic Solder Joint Geometry Integration with Finite Element Analysis for Reliability Evaluation of Printed Circuit Board Assembly 1387

    Chun Sean Lau (Western Digital Corporation), Ning Ye (Western Digital
    Corporation), and Hem Takiar (Western Digital Corporation)

Multi-physics Modelling and Experimental Investigation – An Original Approach for Laser-Dicing/Grooving Process Optimization 1396

    Jeff Moussodji Moussodji (3IT-UdeS/C2MI/IBM), Oswaldo Chacon (IBM
    Canada Ltd), Francis Santerre (IBM Canada Ltd), and Dominique Drouin
    (3IT-Université de Sherbrooke)

Thermal Characteristics of Vertically-Integrated GaN/SiC-on-Si Assemblies: A Comparative Study 1405

    Kimmo Rasilainen (Chalmers University of Technology, Sweden), Per
    Ingelhag (Ericsson AB, Sweden), Peter Melin (Ericsson AB, Sweden),
    Torbjörn M. J. Nilsson (Saab AB, Sweden), Mattias Thorsell (Chalmers
    University of Technology, Sweden and Saab AB, Sweden), and Christian
    Fager (Chalmers University of Technology, Sweden)

Comprehensive Investigation on Warpage Management of FOPLP with Multi Embedded Ring Designs 1413
Chang-Chun Lee (National Tsing Hua University), Yan-Yu Liou (National Tsing Hua University), Pei-Chen Huang (National Tsing Hua University), Fussen Hsu (Unimicron Technology Corporation), Puru Bruce Lin (Unimicron Technology Corporation), Cheng-Ta Ko (Unimicron Technology Corporation), and Yu-Hua Chen (Unimicron Technology Corporation)

Session 31: Automotive and Power Packaging

Development of High Power and High Junction Temperature SiC Based Power Packages 1419
Gongyue Tang (Institute of Microelectronics, A*STAR), Leong Ching Wai (Institute of Microelectronics, A*STAR), Teck Guan Lim (Institute of Microelectronics, A*STAR), Yong Liang Ye (Institute of Microelectronics, A*STAR), Pal Singh Ravinder (Institute of Microelectronics, A*STAR), Lin Bu (Institute of Microelectronics, A*STAR), Boon Long Lau (Institute of Microelectronics, A*STAR), Tai Chong Chai (Institute of Microelectronics, A*STAR), Kazunori Yamamoto (Institute of Microelectronics, A*STAR), and Xiaowu Zhang (Institute of Microelectronics, A*STAR)

New Developments of Copper Plating Technology for Embedded Power Chip Packages Challenges 1426
Yung-Da Chiu (Advanced Semiconductor Engineering (ASE) Inc.), Shiu-Chih Wang (Advanced Semiconductor Engineering (ASE) Inc.), David Tarng (Advanced Semiconductor Engineering (ASE) Inc.), An-Tai Wu (Advanced Semiconductor Engineering (ASE) Inc.), Allenyl Chen (Advanced Semiconductor Engineering (ASE) Inc.), Louis Chen (Advanced Semiconductor Engineering (ASE) Inc.), and Chi-Tsung Chiu (Advanced Semiconductor Engineering (ASE) Inc.)

Innovative Flip Chip Package Solutions for Automotive Applications 1432
Tom Tang (Siliconware Precision Industries Co., Ltd. Taiwan), Bo-Siang Fang (Siliconware Precision Industries Co., Ltd. Taiwan), David Ho (Siliconware Precision Industries Co., Ltd. Taiwan), B.H. Ma (Siliconware Precision Industries Co., Ltd. Taiwan), Jensen Tsai (Siliconware Precision Industries Co., Ltd. Taiwan), and Yu-Po Wang (Siliconware Precision Industries Co., Ltd. Taiwan)

Reliability of Laminated Bond Structure Using (Cu, Ni)/Sn TLP Bonding with Al Interlayer for High Temperature Power Electronics Packaging 1437
Yanghe Liu (Toyota Research Institute of North America), Shailesh N. Joshi (Toyota Research Institute North America), and Ercan M. Dede (Toyota Research Institute North America)

Silver Sintering on Organic Substrates for the Embedding of Power Semiconductor Devices 1443
Alexander Schiffmacher (IMTEK University of Freiburg, Germany), Juergen Wilde (IMTEK University of Freiburg, Germany), Lorenz Litzenberger (IMTEK University of Freiburg, Germany), Till Huesgen (University of Applied Science Kempten, Germany), and Vladimir Polezhaev (University of Applied Science Kempten, Germany)

xxxvii

High Temperature Resistant Packaging Technology for SiC Power Module by Using Ni Micro-Plating
Bonding 1451

> Kohei Tatsumi (Waseda University), Isamu Morisako (Waseda University),
> Keiko Wada (Waseda University), Minoru Fukuomori (Waseda University),
> Tomonori Iizuka (Waseda University), Nobuaki Sato (Mitsui High-tec
> Inc.), Koji Shimizu (Mitsui High-tec Inc.), Kazutoshi Ueda (Mitsui
> High-tec Inc.), Masayuki Hikita (Kyushu Institute of Technology),
> Rikiya Kamimura (Kitakyushu Foundation for the Advancement of
> Industry, Science and Technology), Naoki Kawanabe (WALTS Co., LTD.),
> Kazuhiko Sugiura (DENSO Corporation), Kazuhiro Tsuruta (DENSO
> Corporation), and Keiji Toda (TOYOTA Motor Corporation)

Pb-Free, High Thermal and Electrical Performance Driven Die Attach Material Development for Power
Packages 1457

> Kim Byong Jin (AMKOR), Dong Su Ryu (AMKOR), HyeongIl Jeon (AMKOR),
> Muhammad Hadhari Hazellah (AMKOR), Weng Tuck Chim (AMKOR), and Jin
> Young Khim (AMKOR)

Session 32: Power and Panel Assembly

An RDL-First Fan-Out Panel-Level Package for Heterogeneous Integration Applications 1463

> Yu-Min Lin (Industrial Technology Research Institute (ITRI), Taiwan),
> Sheng-Tsai Wu (Industrial Technology Research Institute (ITRI),
> Taiwan), Chun-Min Wang (Unimicron Technology Corporation, Taiwan),
> Chia-Hsin Lee (Brewer Science, Taiwan), Shin-Yi Huang (Industrial
> Technology Research Institute (ITRI), Taiwan), Ang-Ying Lin
> (Industrial Technology Research Institute (ITRI), Taiwan), Tao-Chih
> Chang (Industrial Technology Research Institute (ITRI), Taiwan), Puru
> Bruce Lin (Unimicron Technology Corporation, Taiwan), Cheng-Ta Ko
> (Unimicron Technology Corporation, Taiwan), Yu-Hua Chen (Unimicron
> Technology Corporation, Taiwan), Jay Su (Brewer Science, Taiwan), Xiao
> Liu (Brewer Science, USA), Luke Prenger (Brewer Science, USA), and
> Kuan-Neng Chen (National Chiao Tung University, Taiwan)

High Yield Precision Transfer and Assembly of GaN μLEDs Using Laser Assisted Micro Transfer Printing.... 1470

> Goutham Ezhilarasu (University of California Los Angeles), Amir Hanna
> (University of California Los Angeles), Ajit Paranjpe (Veeco
> Instruments Inc., USA), and Subramanian Iyer (University of California
> Los Angeles)

High-Density Flexible Substrate Technology with Thin Chip Embedding and Partial Carrier Release
Option for IoT and Sensor Applications 1475

> Kai Zoschke (Fraunhofer IZM), Piotr Mackowiak (Fraunhofer Institute
> for Reliability and Microintegration), Ha-Duong Ngo (Fraunhofer
> Institute for Reliability and Microintegration), Christian Tschoban
> (Fraunhofer Institute for Reliability and Microintegration), Carola
> Fritsche (Fraunhofer Institute for Reliability and Microintegration),
> Kevin Kröhnert (Fraunhofer Institute for Reliability and
> Microintegration), Thorsten Fischer (Fraunhofer Institute for
> Reliability and Microintegration), Ivan Ndip (Fraunhofer Institute for
> Reliability and Microintegration), and Klaus-Dieter Lang (Technical
> University of Berlin)

Advance Embedded Packaging for Power Discrete Device 1485

Jia Ren Huo (Wuxi Sky Chip Interconnection Technology co., LTD), Song Guan Qiang (Wuxi Sky Chip Interconnection Technology co., LTD), Jing Jiang (Wuxi Sky Chip Interconnection Technology co., LTD), Wang Jun Tao (Wuxi Sky Chip Interconnection Technology co., LTD), and Ling Wen Kong (Wuxi Sky Chip Interconnection Technology co., LTD)

Large Panel Size Bonder with High Performance and High Accuracy 1492

Hubert Selhofer (Besi Austria GmbH), Andreas Mayr (Besi Austria GmbH), and Hugo Pristauz (Besi Austria GmbH)

Advances in high Speed Plating for Vertical Glass Panel Fine-Line Plating 1498

Christian Dunkel (Semsysco), Herbert Ötzlinger (Semsysco), Onishi Tetsuya (GJTech / Semsysco), and Raoul Schroeder (Semsysco)

Study of the Properties of AlN PMUT used as a Wireless Power Receiver 1503

Dan Gong (Xiamen University), Shenglin Ma (Xiamen University), Yihsiang Chiu (Peking University), Hungping Lee (J-Metrics Technology, Shenzhen), and Yufeng Jin (Peking University)

Session 33: Fan-Out, Flip Chip, and WLCSP

A Sequential Finite Volume Method / Finite Element Analysis of a Power Electronic Semiconductor Chip..... 1509

Mario Gschwandl (Polymer Competence Center Leoben GmbH, Austria), Peter Filipp Fuchs (Polymer Competence Center Leoben GmbH, Austria), Thomas Antretter (Montanuniversitaet Leoben, Austria), Martin Pfost (TU Dortmund University), Ivaylo Mitev (Polymer Competence Center Leoben GmbH, Austria), Tao Qi (Austria Technologie & Systemtechnik Aktiengesellschaft), Thomas Krivec (Austria Technologie & Systemtechnik Aktiengesellschaft), Angelika Schingale (CPT Group GmbH, Germany), and Michael Decker (CPT Group GmbH, Germany)

Failure Life Prediction of Wafer Level Packaging using DoS with AI Technology 1515

P. H. Chou (National Tsing Hua University), H. Y. Hsiao (National Tsing Hua University), and K.N. Chiang (National Tsing Hua University)

Thermal Cycling Simulation and Sensitivity Analysis of Wafer Level Chip Scale Package with Integration of Metal-Insulator-Metal Capacitors 1521

Yi Zhou (Georgia Institute of Technology), Liangbiao Chen (ON Semiconductor), Yong Liu (ON Semiconductor), and Suresh Sitaraman (Georgia Institute of Technology)

Effect of Time-Dependent Bulk Modulus on Reliability Assessment of Automotive Electronic Control Unit 1529

Hyun Seop Lee (University of Maryland), Bongtae Han (University of Maryland), and Przemyslaw Gromala (Robert Bosch GmbH, Germany)

Thermal and Mechanical Simulations for Fan-Out Wafer-Level Packaging Technology: Introduction of a "Solder Heatsink" 1535

Jean-Philippe Colonna (CEA-Leti, Université Grenoble Alpes, France), Loic Marnat (Université Grenoble Alpes), Mathilde Cartier (Université Grenoble Alpes), Gabriel Pares (Université Grenoble Alpes), and Dominique Noguet (Université Grenoble Alpes)

Wafer Level Warpage Modelling and Validation for FOWLP Considering Effects of Viscoelastic Material Properties Under Process Loadings 1543

zhaohui chen (Institute of Microelectronics, A*STAR (Agency for Science, Technology and Research)), Xiaowu Zhang (Institute of Microelectronics, A*STAR (Agency for Science, Technology and Research)), Sharon Pei Siang Lim (Institute of Microelectronics, A*STAR (Agency for Science, Technology and Research)), Simon Siak Boon Lim (Institute of Microelectronics, A*STAR (Agency for Science, Technology and Research)), Boon Long Lau (Institute of Microelectronics, A*STAR (Agency for Science, Technology and Research)), Yong Han (Institute of Microelectronics, A*STAR (Agency for Science, Technology and Research)), Ming Chinq Jong (Institute of Microelectronics, A*STAR (Agency for Science, Technology and Research)), Songlin Liu (A*STAR (Agency for Science, Technology and Research)), Xiaobai Wang (A*STAR (Agency for Science, Technology and Research)), and Yosephine Andriani (A*STAR (Agency for Science, Technology and Research))

Ultra-Thin Package Board Level Drop Impact Modeling and Validation 1550

Shu-Shen Yeh (Taiwan Semiconductor Manufacturing Company (TSMC)), P. Y. Lin (Taiwan Semiconductor Manufacturing Company (TSMC)), M. C. Yew (Taiwan Semiconductor Manufacturing Company (TSMC)), W. Y. Lin (Taiwan Semiconductor Manufacturing Company (TSMC)), K. C. Lee (Taiwan Semiconductor Manufacturing Company (TSMC)), C. C. Yang (Taiwan Semiconductor Manufacturing Company (TSMC)), J. H. Wang (Taiwan Semiconductor Manufacturing Company (TSMC)), P. C. Lai (Taiwan Semiconductor Manufacturing Company (TSMC)), C. K. Hsu (Taiwan Semiconductor Manufacturing Company (TSMC)), and Shin-Puu Jeng (Taiwan Semiconductor Manufacturing Company (TSMC))

Session 34: Emerging Materials and Processing

Flexible Graphene-Glass Fiber Composite Film with Ultrahigh Thermal Conductivity and Mechanical Strength as Highly Efficient Thermal Spreader Materials 1556

Xiaoliang Zeng (Center for Advanced Material Research Shenzhen Institutes of Advanced Technology, Chinese Academy of Sciences), Linlin Ren (Center for Advanced Material Research Shenzhen Institutes of Advanced Technology, Chinese Academy of Sciences), Rong Sun (Center for Advanced Material Research Shenzhen Institutes of Advanced Technology, Chinese Academy of Sciences), Jianbin Xu (Center for Advanced Material Research Shenzhen Institutes of Advanced Technology, Chinese Academy of Sciences), and Ching-Ping Wong (School of Materials Science and Engineering Georgia Institute of Technology Atlanta, USA)

Highly Thermal Conductive and Electrically Insulated Graphene Based Thermal Interface Material with Long-Term Reliability 1564

Nan Wang (SHT Smart High Tech AB), Ya Liu (Chalmers University of Technology), Shujing Chen (Shanghai University), Lilei Ye (SHT Smart High Tech AB), and Johan Liu (Chalmers University of Technology)

Further Enhancement of Thermal Conductivity through Optimal Uses of h-BN Fillers in Polymer-Based Thermal Interface Material for Power Electronics 1569

Han Jiang (Loughborough University), Han Zhou (Loughborough University), Stuart Robertson (Loughborough University), Zhaoxia Zhou (Loughborough University), Liguo Zhao (Loughborough University), and Changqing Liu (Loughborough University)

Wafer Level Integration of Thin Silicon Bare Dies Within Flexible Label 1575

Jean-Charles Souriau (Univ. Grenoble Alpes, CEA, LETI), Ahmad Itawi (Univ. Grenoble Alpes, CEA, LETI), and Laetitia Castagné (Univ. Grenoble Alpes, CEA, LETI)

Laser Sintering of Aerosol Jet Printed Conductive Interconnects on Paper Substrate 1581

Mohammed Alhendi (Binghamton University), Rajesh S. Sivasubramony (Binghamton University), Jack Lombardi (Binghamton University), Darshana L. Weerawarne (Binghamton University), Peter Borgesen (Binghamton University), Mark D. Poliks (Binghamton University), and Azar Alizadeh (General Electric Global Research)

In-Situ Investigation of Organic Additive Interactions in Copper Electroplating Solutions with Surface Enhanced Raman Spectroscopy (SERS) 1588

Nithin Nedumthakady (Georgia Institute of Technology), Bartlet DeProspo (Georgia Institute of Technology), Himani Sharma (Georgia Institute of Technology), Rahul Manepalli (Intel Corporation), Sashi Kandanur (Intel Corporation), Sajanlal Panikkanvalappil (Georgia Institute of Technology), Nasrin Hooshmand (Georgia Institute of Technology), and Rao Tummala (Georgia Institute of Technology)

C4 Compatible Ultra-Thick Cu On-chip Magnetic Inductor Architecture Integrated with Advanced Polymer/Cu Planarization Process 1595

C.H. Kuo (Taiwan Semiconductor Manufacturing Company, Ltd.), S.B. Yang (Taiwan Semiconductor Manufacturing Company, Ltd.), C.C. Kuo (Taiwan Semiconductor Manufacturing Company, Ltd.), Y.N. Chen (Taiwan Semiconductor Manufacturing Company, Ltd.), K.S. Yuan (Taiwan Semiconductor Manufacturing Company, Ltd.), G.C. Huang (Taiwan Semiconductor Manufacturing Company, Ltd.), C.N. Ke (Taiwan Semiconductor Manufacturing Company, Ltd.), Grace Chang (Taiwan Semiconductor Manufacturing Company, Ltd.), C.C. Hsu (Taiwan Semiconductor Manufacturing Company, Ltd.), H.L. Huang (Taiwan Semiconductor Manufacturing Company, Ltd.), Kirin Wang (Taiwan Semiconductor Manufacturing Company, Ltd.), Harry Ku (Taiwan Semiconductor Manufacturing Company, Ltd.), C.S. Chen (Taiwan Semiconductor Manufacturing Company, Ltd.), K.C. Liu (Taiwan Semiconductor Manufacturing Company, Ltd.), Alex Kalnitsky (Taiwan Semiconductor Manufacturing Company, Ltd.), and Marvin Liao (Taiwan Semiconductor Manufacturing Company, Ltd.)

Session 35: New Interconnects for Package Scaling

Development of 2.3D High Density Organic Package using Low Temperature Bonding Process with Sn-Bi Solder 1599

Shota Miki (SHINKO ELECTRIC INDUSTRIES CO., LTD.), Hiroshi Taneda (SHINKO ELECTRIC INDUSTRIES CO., LTD.), Naoki Kobayashi (SHINKO ELECTRIC INDUSTRIES CO., LTD.), Kiyoshi Oi (SHINKO ELECTRIC INDUSTRIES CO., LTD.), Koji Nagai (SHINKO ELECTRIC INDUSTRIES CO., LTD.), and Toshinori Koyama (SHINKO ELECTRIC INDUSTRIES CO., LTD.)

PowerTherm Attach Process for Power Delivery and Heat Extraction in the Silicon-Interconnect Fabric Using Thermocompression Bonding 1605

Pranav Ambhore (University of California, Los Angeles), Umesha Mogera (University of California, Los Angeles), Boris Vaisband (University of California, Los Angeles), Ujash Shah (University of California, Los Angeles), Timothy Fisher (University of California, Los Angeles), Mark Goorsky (University of California, Los Angeles), and Subramanian S. Iyer (University of California, Los Angeles)

Interconnect Scheme for Die-to-Die and Die-to-Wafer-Level Heterogeneous Integration for High-Performance Computing 1611

Rabindra Das (MIT Lincoln Laboratory), Vladimir Bolkhovsky (MIT Lincoln Laboratory), Christopher Galbraith (MIT Lincoln Laboratory), Daniel Oates (MIT Lincoln Laboratory), Jason Plant (MIT Lincoln Laboratory), Renée Lambert (MIT Lincoln Laboratory), Scott Zarr (MIT Lincoln Laboratory), Ravi Rastogi (MIT Lincoln Laboratory), Dmitri Shapiro (MIT Lincoln laboratory), Manuel Docanto (MIT Lincoln Laboratory), Terence Weir (MIT Lincoln laboratory), and Leonard Johnson (MIT Lincoln Laboratory)

Ultra Wide Micro Bumps Interconnection Matrix for High Energy Particle Detection: Process and Assembly 1622

Jean Charbonnier (CEA Leti), Myriam Assous (CEA-Leti), Thierry Mourier (CEA-Leti), Céline Ribière (CEA-Leti), Stéphane Minoret (CEA-Leti), Sophie Verrun (CEA-Leti), Pierre Tissier (CEA-Leti), Rémi Coquand (CEA-Leti), Mehmet Bicer (CEA-Leti), Fabienne Allain (CEA-Leti), Rémi Franiatte (CEA-Leti), and Gabriel Pares (CEA-Leti)

Growth Behavior and Orientation Evolution of Cu_6Sn_5 Grains in Micro Interconnect During Isothermal Reflow 1629

S. Chen (Dalian University of Technology), N. Zhao (Dalian University of Technology), Y.Y. Qiao (Dalian University of Technology), Y.P. Wang (Dalian University of Technology), H.T. Ma (Dalian University of Technology), and C.M.L. Wu (City University of Hong Kong)

Development of a no Reflow Cu Pillar Bump to Improve Chip/Package Interactions (CPI) Process and Reliability Performance 1635

Kuei Hsiao (Frank) Kuo (Siliconware Precision Industries Co., Ltd. (SPIL)), Jiunn Jie Wang (Siliconware Precision Industries Co., Ltd. (SPIL)), Yen Neng Wang (Siliconware Precision Industries Co., Ltd. (SPIL)), Feng Lung Chien (Siliconware Precision Industries Co., Ltd. (SPIL)), and Rick Lee (Siliconware Precision Industries Co., Ltd. (SPIL))

A Novel Interconnection Technology Using Ultra-Thin Under Barrier Metal for Multiple Chip-on-Chip Stacking Structure 1641

Takuya Nakamura (Sony Semiconductor Solutions), Kan Shimizu (Sony Semiconductor Solutions), Masataka Maehara (Sony Semiconductor Solutions), Toshihiko Hayashi (Sony Semiconductor Solutions), Kentaro Akiyama (Sony Semiconductor Solutions), Junichiro Fujimagari (Sony Semiconductor Solutions), Tomohiro Ohkubo (Sony Semiconductor Manufacturing), Atsushi Fujiwara (Sony Semiconductor Manufacturing), and Hayato Iwamoto (Sony Semiconductor Solutions)

Session 36: RF & Power Components and Modules

Multilayer Decoupling Capacitor using Stacked Layers of BST and LNO 1647

Todd Schumann (University of Florida), Sheng-Po Fang (University of Florida), Yong-Kyu Yoon (University of Florida), Jongmin Yook (Korea Electronics Technology Institute), and Dongsu Kim (Korea Electronics Technology Institute)

System Co-Design of a High Current (40A) Synchronous Step-Down Converter in an Innovative Multi-chip Module (MCM) LQFN-Type Packaging Technology 1653

Todd Harrison (Texas Instruments, Inc.), Jie Chen (Texas Instruments, Inc.), and Rajen Murugan (Texas Instruments, Inc.)

Integrating Solid State Protection with a RF-MEMS Switch for Achieving ESD Robustness 1660

Srivatsan Parthasarathy (Analog Devices, USA), Padraig Fitzgerald (Analog Devices, Ireland), Javier Salcedo (Analog Devices, USA), Ray Goggin (Analog Devices, Ireland), and Jean-Jacques Hajjar (Analog Devices, USA)

A Zero Height Small Size Low Cost RF Interconnect Substrate Technology For RF Front Ends For M.2 Modules And SiP 1666

Sidharth Dalmia (Intel Corporation), Kirthika Nahalingam (Intel Corporation), Swathi Vijayakumar (Intel Corporation), and Pouya Talebbeydokhti (Intel Corporation)

Open and Closed Loop Inductors for High-Efficiency System-on-Package Integrated Voltage Regulators 1672

Claudio Alvarez (Georgia Institute of Technology), Mohamed Bellaredj (Georgia Institute of Technology), and Madhavan Swaminathan (Georgia Institute of Technology)

RF Inductors Integrated in Organic Packaging 1680

Denis Mercier (CEA-Leti), Jean-Philippe Michel (CEA-Leti), Christine Raynaud (CEA-Leti), and Christophe Billard (CEA-Leti)

3D Printed Interposer Layer for High Density Packaging of IoT Devices 1687

Saikat Mondal (Michigan State University), Mohd. Ifwat Mohd. Ghazali (Universiti Sains Islam Malaysia), Kanishka Wijewardena (Michigan State University), Deepak Kumar (Michigan State University), and Premjeet Chahal (Michigan State University)

xliii

Session 37: Interactive Presentations 1

Comprehensive Solution for Micro Bump Coplanarity Control 1693
> Chun-Chen Liu (Taiwan Semiconductor Manufacturing Company), J.H. Chen
> (Taiwan Semiconductor Manufacturing Company), Y.N. Hsu (Taiwan
> Semiconductor Manufacturing Company), Rung-De Wang (Taiwan
> Semiconductor Manufacturing Company), Yu-Cheng Wang (Taiwan
> Semiconductor Manufacturing Company), Bin-En Ho (Taiwan Semiconductor
> Manufacturing Company), Y.H. Wu (Taiwan Semiconductor Manufacturing
> Company), Ponder Pan (Taiwan Semiconductor Manufacturing Company),
> Harry Ku (Taiwan Semiconductor Manufacturing Company), Kirin Wang
> (Taiwan Semiconductor Manufacturing Company), Calvin Lu (Taiwan
> Semiconductor Manufacturing Company), K.C. Liu (Taiwan Semiconductor
> Manufacturing Company), and Marvin Liao (Taiwan Semiconductor
> Manufacturing Company)

Structural Enhancement for a CMOS-MEMS Microphone Under Thermal Loading by Taguchi Method 1697
> Chun-Lin Lu (National Tsing Hua University) and Meng-Kao Yeh (National
> Tsing Hua University)

A Methodology to Correct in-Fixture Measurement of Impedance by a Machine Learning Model 1704
> Bo-Siang Fang (Siliconware Precision Industries Co., Ltd. (SPIL),
> Taiwan), Chia-Chu Lai (Siliconware Precision Industries Co., Ltd.
> (SPIL), Taiwan), Ying-Wei Lu (Siliconware Precision Industries Co.,
> Ltd. (SPIL), Taiwan), Kuan-Ta Chen (Siliconware Precision Industries
> Co., Ltd. (SPIL), Taiwan), Mike Tasi (Siliconware Precision Industries
> Co., Ltd. (SPIL), Taiwan), and Don-Son Jiang (Siliconware Precision
> Industries Co., Ltd. (SPIL), Taiwan)

Material and Structure Design Optimization for Panel-Level Fan-Out Packaging 1710
> Dao-Long Chen (Advanced Semiconductor Engineering, Inc.), Ian Hu
> (Advanced Semiconductor Engineering, Inc.), KarenYU Chen (Advanced
> Semiconductor Engineering, Inc.), Meng-Kai Shih (Advanced
> Semiconductor Engineering, Inc.), David Tarng (Advanced Semiconductor
> Engineering, Inc.), Dinos Huang (Advanced Semiconductor Engineering,
> Inc.), and JY On (Advanced Semiconductor Engineering, Inc.)

The Microstructure and Mechanical Property of the High Entropy Alloy as a low Temperature Solder 1716
> Li Pu (Beijing Institute of Technology), Quanfeng He (City University
> of Hong Kong), Yong Yang (City University of Hong Kong), Xiuchen Zhao
> (Beijing Institute of Technology), Zhuangzhuang Hou (Beijing Institute
> of Technology), K. N. Tu (University of California, USA), and Yingia
> Liu (Beijing Institute of Technology)

A Versatile Fan-Out Infrastructure Based on Die-Stencil Substrate Promoted by an Advanced
Multifunctional Temporary Bonding Material 1722
> Xiao Liu (Brewer Science, Inc.), Baron Huang (Brewer Science, Inc.),
> Hong Zhang (Brewer Science, Inc.), Lisa Kirchner (Brewer Science,
> Inc.), Arthur Southard (Brewer Science, Inc.), Rama Puligadda (Brewer
> Science, Inc.), and Tony Flaim (Brewer Science, Inc.)

Low Temperature and Pressureless Microfluidic Electroless Bonding Process for Vertical
Interconnections 1729
> Han-Tang Hung (National Taiwan University), Sean Yang (National Taiwan
> University), I-An Weng (National Taiwan University), Yan-Hao Chen
> (Unimicron Technology Corporation, Taiwan), and C. Robert Kao
> (National Taiwan University)

3D Integration of CMOS-Compatible Surface Electrode Ion Trap and Silicon Photonics for Scalable Quantum Computing 1735

Jing Tao (Nanyang Technological University), Yu Dian Lim (Nanyang Technological University), Hong Yu Li (Agency for Science, Technology and Research (A*STAR)), Nam Piau Chew (Nanyang Technological University), Anak Agung Alit Apriyana (Agency for Science, Technology and Research (A*STAR)), Lin Bu (Agency for Science, Technology and Research (A*STAR)), Peng Zhao (Nanyang Technological University), Luca Guidoni (Université Paris Diderot), and Chuan Seng Tan (Nanyang Technological University)

Integrated RTD Sensors for Maintaining Thermal Uniformity During TCB Process 1744

Salwa Ben Jemaa (Interdisciplinary Institute for Technological Innovation (3IT) Sherbrooke University), Julien Sylvestre (Interdisciplinary Institute for Technological Innovation (3IT) Sherbrooke University), and Pascale Gagnon (IBM Canada Bromont, QC, Canada)

Wireless Transfer of Power and Data Via a Single Resonant Inductive Link 1751

Shiang-Hwua Yu (National Sun Yat-sen University), Yi-Chen Hsieh (National Sun Yat-sen University), Chin-Wei Chan (National Sun Yat-sen University), I-Fang Lo (National Sun Yat-sen University), Heri Suryoatmojo (Institut Teknologi Sepuluh), and Lih-Tyng Hwang (National Sun Yat-sen University)

Adaptive Patterning of Optical and Electrical Fan-Out for Photonic Chip Packaging 1757

Ahmed Elmogi (Centre for Microsystems Technology, imec and Ghent University), Andres Desmet (Centre for Microsystems Technology, imec and Ghent University), Jeroen Missinne (Centre for Microsystems Technology, imec and Ghent University), Hannes Ramon (Ghent University-imec), Joris Lambrecht (Ghent University-imec), Peter De Heyn (imec), Marianna Pantouvaki (imec), Joris Van Campenhout (imec), Johan Bauwelinck (Ghent University-imec), and Geert Van Steenberge (imec and Ghent University)

Low Surface Reflectance Structure at Near Infrared Wavelength by Injection Molding 1764

Sho Yakabe (Sumitomo Electric Industries, Ltd.), Takuro Watanabe (Sumitomo Electric Industries, Ltd.), Takayuki Shimazu (Sumitomo Electric Industries, Ltd.), Ryohei Hokari (National Institute of Advanced Industrial Science and Technology), and Kazuma Kurihara (National Institute of Advanced Industrial Science and Technology)

A Novel Design of a Bandwidth Enhanced Dual-Band Impedance Matching Network with Coupled Line Wave Slowing 1770

Deepayan Banerjee (IIIT Delhi, India), Antra Saxena (IIIT Delhi, India), and Mohammad Hashmi (Nazarbayev University, Kazakhstan)

Effects of Electromigration on Microstructural Evolution and Mechanical Properties of Preferential Growth Intermetallic Compound Interconnects for 3D Packaging 1774

Mingliang L. Huang (Dalian University of Technology) and Lin Zou (Dalian University of Technology)

Telemetry for Implantable Biosensors 1782

Ryan B. Green (Virginia Commonwealth University) and Erdem Topsakal (Virginia Commonwealth University)

xlv

Ultra-Thin QFN-Like 3D Package with 3D Integrated Passive Devices 1789
Ayad Ghannam (3DiS Technologies S.A.S, France), Niek van Haare (Besi Netherlands, B.V., Netherlands), Julian Bravin (EV Group E.Thallner GmbH, Austria), Elisabeth Brandl (EV Group E.Thallner GmbH, Austria), Birgit Brandstätter (Besi Austria GmbH, Austria), Hannes Klingler (Besi Austria GmbH, Austria), Benedikt Auer (Besi Austria GmbH, Austria), Philippe Meunier (NXP Semiconductors, France), and Sebastiaan Kersjes (Besi Netherlands, B.V., Netherlands)

Low-Cost Non-TSV Based 3D Packaging Using Glass Panel Embedding (GPE) for Power-Efficient, High-Bandwidth Heterogeneous Integration 1796
Siddharth Ravichandran (Georgia Institute of Technology), Shuhei Yamada (Murata Manufacturing Co. Ltd, Kyoto, Japan), Fuhan Liu (Georgia Institute of Technology), Vanessa Smet (Georgia Institute of Technology), Mohanalingam Kathaperumal (Georgia Institute of Technology), and Rao Tummala (Georgia Institute of Technology)

Polylithic Integration of 2.5D and 3D Chiplets Using Interconnect Stitching 1803
Paul K. Jo (Georgia Institute of Technology), Ting Zheng (Georgia Institute of Technology), and Muhannad S. Bakir (Georgia Institute of Technology)

Characterization of the Current Mechanisms and Improved Leakage Current in Silver Doped Barium Strontium Titanate 1809
Todd Schumann (University of Florida), Kyoung-Tae Kim (University of Florida), Sheng-Po Fang (University of Florida), and Yong-Kyu Yoon (University of Florida)

High Temperature Aging Effects in SAC and SAC+X Lead Free Solders 1815
Mohammad S. Alam (Auburn University), KM Rafidh Hassan (Auburn University), Jeffrey C. Suhling (Auburn University), and Pradeep Lall (Auburn University)

Session 38: Interactive Presentations 2

Laundering Reliability of Electrically Conductive Fabrics for E-Textile Applications 1826
Jeffrey ChangBing Lee (iST-Integrated Service Technology Inc.), Weifeng Liu (FLEX Ltd.), ChangHo Lo (iST-Integrated Service Technology Inc.), and Cheng-Chih Chen (iST-Integrated Service Technology Inc.)

Preconditioning Technologies for Sputtered Seed Layers in FOPLP 1833
Johannes Weichart (Evatec AG), Jüergen Weichart (Evatec AG), Andreas Erhart (Evatec AG), and Kay Viehweger (Fraunhofer IZM ASSID)

Impact of Thermal Boundary Resistance on the Thermal Design of GaN-on-Diamond HEMTs 1842
Huaixin Guo (Nanjing Electronic Devices Institute), Yuechan Kong (Nanjing Electronic Devices Institute), and Tangsheng Chen (Nanjing Electronic Devices Institute)

Measuring the Electric Properties of Thin Film Shape Memory Polymers in Simulated Physiological Conditions 1848
Daniel Del Nero (The University of Texas at Dallas), Alexandra Joshi-Imre (The University of Texas at Dallas), and Walter Voit (The University of Texas at Dallas)

xlvi

Evaluation of WLP Dielectrics for High Voltage Applications 1853
  Markus Wöhrmann (Fraunhofer IZM), Michael Toepper (Fraunhofer IZM),
  Marcus Paeck (Fraunhofer IZM), and Klaus-Dieter Lang (Technical
  University Berlin)

Mitigating the Effects of Microvortices in high-Re Deterministic Lateral Displacement by Using
Symmetric Airfoil-Shaped Pillars .... N/A
  Brian Dincau (Washington State University Vancouver), Kawkab Ahasan
  (Washington State University Vancouver), and Jong-Hoon Kim (Washington
  State University Vancouver)

Plasma Dry Process Technology Development of Glass-Epoxy Film on the Silicon Substrate to Fabricate
RDL for Future GPU/AI Application 1865
  Takahide Murayama (ULVAC, Inc.), Muneyuki Sato (ULVAC, Inc.), Akiyoshi
  Suzuki (ULVAC, Inc.), Atsuhito Ihori (ULVAC, Inc.), Tetsushi Fujinaga
  (ULVAC, Inc.), and Yasuhiro Morikawa (ULVAC, Inc.)

Fully Solid-State Integrated Capacitors Based on Carbon Nanofibers and Dielectrics with Specific
Capacitances Higher Than 200 nF/mm2 1870
  Amin Saleem (Smoltek AB, Sweden), Rickard Andersson (Smoltek AB,
  Sweden), Maria Bylund (Smoltek AB, Sweden), Charlotte Goemare (Smoltek
  AB, Sweden), Guilhem Pacot (Smoltek AB, Sweden), Mohammed Kabir
  (Smoltek AB, Sweden), and Vincent Desmaris (Smoltek AB, Sweden)

Application of Fan-Out Panel Level Packaging Techniques for Flexible Hybrid Electronics Systems 1877
  Wei-Yuan Cheng (ITRI), Shau-Fei Cheng (ITRI), Chen-Tsai Yang (ITRI),
  Shau-Fei Cheng (ITRI), Wei-Han Chen (ITRI), Hsin-Cheng Lai (ITRI),
  Tai-Jui Wang (ITRI), and Yuh-Zheng Lee (ITRI)

Structuring of Laser Activated Polymers for Sensor Applications 1883
  Sebastian Bengsch (University Hanover), Marc Christopher Wurz
  (University Hanover), Kevin Cromwell (University Hanover), and
  Maximilian Aue (University Hanover)

A Deep Learning Approach for Volterra Kernel Extraction for Time Domain Simulation of Weakly
Nonlinear Circuits 1889
  Thong Nguyen (University of Illinois at Urbana Champaign), Xinying
  Wang (University of Illinois at Urbana Champaign), Xu Chen (University
  of Illinois at Urbana Champaign), and Jose Schutt-Aine (University of
  Illinois at Urbana Champaign)

224G Package Interconnect Design Study - Based on Artificial Neural Network Modeling Approach 1897
  Hui Liu (Intel Corporation), Qian Ding (Intel Corporation), and
  Penglin Liu (Intel Corporation)

Enhanced Reliability of a RF-SiP with Mold Encapsulation and EMI Shielding 1902
  Chan-Yuan Liu (Advanced Semiconductor Engineering, Inc., Taiwan),
  Jason Chien (Advanced Semiconductor Engineering, Inc., Taiwan),
  Yu-Chou Tseng (Advanced Semiconductor Engineering, Inc., Taiwan),
  Kuo-Hsien Liao (Advanced Semiconductor Engineering, Inc., Taiwan),
  Alex Chan (Advanced Semiconductor Engineering, Inc., Taiwan), Dao-Long
  Chen (Advanced Semiconductor Engineering, Inc., Taiwan), Meng-Kai Shih
  (Advanced Semiconductor Engineering, Inc., Taiwan), and Mark Gerber
  (Advanced Semiconductor Engineering, Inc., U.S.)

Study of the Effect and Mechanism of a Cap Layer in Controlling the Statistical Variation of Via Extrusion 1909

Golareh Jalilvand (University of Central Florida) and Tengfei Jiang
(University of Central Florida)

Three Dimensional Copper Foam-Filled Elastic Conductive Composites with Simultaneously Enhanced Mechanical, Electrical, Thermal and Electromagnetic Interference (EMI) Shielding Properties 1916

Tan Lu (Shenzhen Institutes of Advanced Technology, Chinese Academy of
Sciences), Han Gu (Shenzhen Institutes of Advanced Technology, Chinese
Academy of Sciences), Yougen Hu (Shenzhen Institutes of Advanced
Technology, Chinese Academy of Sciences), Tao Zhao (Shenzhen
Institutes of Advanced Technology, Chinese Academy of Sciences),
Pengli Zhu (Shenzhen Institutes of Advanced Technology, Chinese
Academy of Sciences), Rong Sun (Shenzhen Institutes of Advanced
Technology, Chinese Academy of Sciences), and Ching-Ping Wong (Georgia
Institute of Technology)

Vertical Interconnect Technology for Enlarging Capacity on Micro Solid Thin Film Rechargeable Battery 1921

Akihiro Horibe (IBM Research - Tokyo), Kuniaki Sueoka (IBM Research -
Tokyo), Takahiro Mori (IBM Research - Tokyo), Risa Miyazawa (IBM
Research - Tokyo), and Hiroyuki Mori (IBM Research - Tokyo)

Characterization of Fine Pitch Hybrid Bonding Pads using Electrical Misalignment Test Vehicle 1926

Imed Jani (CEA, LETI), Didier Lattard (CEA, LETI), Pascal Vivet (CEA,
LETI), Lucile Arnaud (CEA, LETI), Severine Cheramy (CEA, LETI), Edith
Beigné (CEA, LETI), Alexis Farcy (STMicroelectronics), Joris Jourdon
(STMicroelectronics), Yann Henrion (STMicroelectronics), Emilie
Deloffre (STMicroelectronics), and Halim Bilgen (STMicroelectronics)

Dynamic Characteristics Evaluation on NCF Under Challenging Conditions and Its Application 1933

Tomonori Nakamura (Shinkawa LTD, Japan), Hiromi Shibahara (Shinkawa
LTD, Japan), Osamu Watanabe (Shinkawa LTD, Japan), Tetsuya Utano
(Shinkawa LTD, Japan), Daisuke Tani (Shinkawa LTD, Japan), Sung
Chenhsiu (Shinkawa LTD, Japan), Toru Maeda (Shinkawa LTD, Japan), Doug
Day (Shinkawa LTD, Japan), Hidekazu Yagi (Dexerials Corporation,
Japan), Ryoji Kojima (Dexerials Corporation, Japan), Daichi Mori
(Dexerials Corporation, Japan), Tatsuo Nagamatsu (Dexerials
Corporation, Japan), and Junichi Kaneko (Dexerials Corporation, Japan)

Study of Electrical and Mechanical Characteristics of Inkjet-Printed Patch Antenna Under Uniaxial and Biaxial Bending 1939

Yi Zhou (Georgia Institute of Technology), Sridhar Sivapurapu (Georgia
Institute of Technology), Rui Chen (Georgia Institute of Technology),
Nahid Aslani Amoli (Georgia Institute of Technology), Mohamed
Bellaredj (Georgia Institute of Technology), Madhavan Swaminathan
(Georgia Institute of Technology), and Suresh K. Sitaraman (Georgia
Institute of Technology)

Effects of Oven and Laser Sintering Parameters on the Electrical Resistance of IJP Nano-Silver Traces on Mesoporous PET Before and During Fatigue Cycling 1946

G.S. Khinda (SUNY Binghamton), M.Z. Kokash (SUNY Binghamton), M.
Alhendi (SUNY Binghamton), M. Yadav (SUNY Binghamton), J.P. Lombardi
(SUNY Binghamton), D.L. Weerawarne (SUNY Binghamton), Mark D. Poliks
(SUNY Binghamton), P. Borgesen (SUNY Binghamton), and Nancy C. Stoffel
(General Electric Global Research Center)

Multilayer Glass Substrate with High Density Via Structure for All Inorganic Multi-chip Module 1952
  Toshiki Iwai (FUJITSU LABORATORIES LTD.), Taiji Sakai (FUJITSU
  LABORATORIES LTD.), Daisuke Mizutani (FUJITSU LABORATORIES LTD.),
  Seiki Sakuyama (FUJITSU LABORATORIES LTD.), Kenji Iida (FUJITSU
  INTERCONNECT TECHNOLOGIES LIMITED), Takayuki Inaba (FUJITSU
  INTERCONNECT TECHNOLOGIES LIMITED), Hidehiko Fujisaki (FUJITSU
  INTERCONNECT TECHNOLOGIES LIMITED), Akira Tamura (FUJITSU INTERCONNECT
  TECHNOLOGIES LIMITED), and Yoshinori Miyazawa (FUJITSU INTERCONNECT
  TECHNOLOGIES LIMITED)

The Poisson's Ratio of Lead Free Solder - The Often Forgotten But Important Material Property 1958
  KM Rafidh Hassan (Auburn University), Mohammad S. Alam (Auburn
  University), Jeffrey C. Suhling (Auburn University), and Pradeep Lall
  (Auburn University)

Additive Laser Metal Deposition Onto Silicon for Enhanced Microelectronics Cooling 1970
  Arad Azizi (Binghamton University (SUNY)), Matthias A. Daeumer
  (Binghamton University (SUNY)), Jacob C. Simmons (Binghamton
  University (SUNY)), Bahgat G. Sammakia (Binghamton University (SUNY)),
  Bruce T. Murray (Binghamton University (SUNY)), and Scott N. Schiffres
  (Binghamton University (SUNY))

Moisture Barrier, Mechanical, and Thermal Properties of PDMS-PIB Blends for Solar Photovoltaic (PV)
Module Encapsulant 1977
  Jinho Hah (Georgia Institute of Technology), Michael Sulkis (Georgia
  Institute of Technology), Chao Ren (Georgia Institute of Technology),
  Minsoo Kang (Georgia Institute of Technology), Kyoung-sik Moon
  (Georgia Institute of Technology), Samuel Graham (Georgia Institute of
  Technology), and C. P. Wong (Georgia Institute of Technology)

Session 39: Interactive Presentations 3

Modeling and Design of Power Distribution Network for a Heterogeneous Integrated Active Interposer
with Neuromorphic Computing Circuits 1983
  Min Miao (Beijing Information Science and Technology University),
  Tianfang Chen (Beijing Information Science and Technology University),
  Yang Yang (Peking University Shenzhen Graduate School), Jincan Zhang
  (Beijing Information Science and Technology University), Na Li
  (Beijing Information Science and Technology University), Kunkun Li
  (Beijing Information Science and Technology University), Liyuan Wang
  (Beijing Information Science and Technology University), Huan Liu
  (Peking University), Xiaole Cui (Peking University Shenzhen Graduate
  School), and Yufeng Jin (Peking University Shenzhen Graduate School)

PCB Microstrip Line Far-End Crosstalk Mitigation by Surface Mount Capacitors 1989
  Zhaoqing Chen (IBM Corporation)

New Cost-Effective Via-Last Approach by "One-Step TSV" After Wafer Stacking for 3D Memory
Applications 1996
  Masaya Kawano (Institute of Microelectronics, A*STAR), Xiangy-Yu Wang
  (Institute of Microelectronics, A*STAR), and Qin Ren (Institute of
  Microelectronics, A*STAR)

xlix

Microstructure and Property Changes in Cu/Sn-58Bi/Cu Solder Joints During Thermomigration 2003
  Yu-An Shen (Joining and Welding Research Institute (JWRI), Osaka
  University), Shiqi Zhou (Osaka University), Jiahui Li (City University
  of Hong Kong), K. N. Tu (UCLA), and Hiroshi Nishikawa (Joining and
  Welding Research Institute (JWRI), Osaka University)

Simulation and Experimental Validations of EM/TM/SM Physical Reliability for Interconnects Utilized
in Stretchable and Foldable Electronics 2009
  Chang-Chun Lee (National Tsing Hua University), Oscar Chuang (National
  Tsing Hua University), Chia-Ping Hsieh (National Taiwan University),
  Wei-Yuan Cheng (Industrial Technology Research Institute), and Steve
  Chiu (Industrial Technology Research Institute)

A Complex Integrated Circuit Structure Transformation, Modeling and Simulation Method 2016
  Daixing Wang (Peking University Shenzhen Graduate School), Wei Wang
  (Institute of Microelectronics Peking University), and Yufeng Jin
  (Peking University Shenzhen Graduate School)

A Study on the Oxygen Plasma Treatment on the Peel Adhesion Strength and Solder Wettability of
SnBi58 Based Anisotropic Conductive Films 2022
  Shuye Zhang (Harbin Institute of Technology), Mingliang Huang (Dalian
  University of Technology), Yang Wu (Dalian University of Technology),
  Ming Yang (Hisilicon Optoelectronics Co., Ltd), Tiesong Lin (Harbin
  Institute of Technology), Peng He (Harbin Institute of Technology),
  and Kyung-Wook Paik (Nano-Packaging and Interconnection Laboratory)

Numerical Analysis of the Influence of Polymeric Materials on a MEMS Package Performance Under
Humidity and Temperature Loads 2029
  Mahesh Yalagach (Polymer Competence Center Leoben GmbH, Leoben,
  Austria.), Peter Filipp Fuchs (Polymer Competence Center Leoben GmbH,
  Leoben, Austria.), Archim Wolfberger (Polymer Competence Center Leoben
  GmbH, Leoben, Austria.), Mario Gschwandl (Polymer Competence Center
  Leoben GmbH, Leoben, Austria.), Thomas Antretter (Montanuniversitaet
  Leoben, Institute of Mechanics, Leoben, Austria.), Michael Feuchter
  (Montanuniversitaet Leoben, Institute of Material Science and Testing
  of Polymers, Leoben, Austria.), Coen Tak (ams AG, Premstaetten,
  Austria), and Qi Tao (Austria Technologie & Systemtechnik
  Aktiengesellschaft, Leoben, Austria.)

Electromigration-Induced -Sn Grain Rotation in Lead-Free Flip Chip Solder Bumps 2036
  Mingliang L. Huang (Dalian University of Technology), Jiameng M. Kuang
  (Dalian University of Technology), and Hongyu Y. Sun (Dalian
  University of Technology)

Low-Cost MT-Ferrule-Compatible Optical Connector for Co-packaged Optics Using Single-Mode Polymer
Waveguide 2042
  Akihiro Noriki (National Institute of Advanced Industrial Science and
  Technology (AIST)), Takeru Amano (National Institute of Advanced
  Industrial Science and Technology (AIST)), Masatoshi Tsunoda (Kyocera
  Corporation), and Toshiaki Michihiro (Kyocera Corporation)

Characterization of Coated Silver Wire Bond Interface Using TEM 2048
  Murali Sarangapani (Heraeus Materials Singapore Pte. Ltd.,), Eric Tan
  Swee Seng (Heraeus Materials Singapore Pte. Ltd.,), and Jason Wong
  Chin Yeung (Heraeus Materials Singapore Pte. Ltd.,)

Research on Applied Reliability of BGA Solder Balls in Extreme Marine Environment 2054

Liyuan Liu (China Electronic Product Reliability and Environmental
Testing Research Institute), Tao Lu (China Electronic Product
Reliability and Environmental Testing Research Institute), Daojun Luo
(China Electronic Product Reliability and Environmental Testing
Research Institute), and Hui Xiao (China Electronic Product
Reliability and Environmental Testing Research Institute)

Influence of Single/Double Sweeping Mode and Sweeping Voltage Increment/Polarity on Measurement of
TSV Leakage Current 2061

Qinghua Zeng (Peking University), Jing Chen (Peking University), and
Yufeng Jin (Peking University)

Improving the Solder Wettability Via Atmospheric Plasma Technology 2067

Sagung Dewi Kencana (National Taiwan University of Science and
Technology), Yu-Lin Kuo (National Taiwan University of Science and
Technology), Yee-Wen Yen (National Taiwan University of Science and
Technology), Eckart Schellkes (Robert Bosch Taiwan Co., Ltd), and
Wallace Chuang (Robert Bosch Taiwan Co., Ltd)

Orthogonal Quilt Packaging 3D Integration for High-Energy Particle Detectors 2072

Jason Kulick (Indiana Integrated Circuits, LLC), Tian Lu (Indiana
Integrated Circuits, LLC), Edit Varga (Indiana Integrated Circuits,
LLC), Gary H. Bernstein (Indiana Integrated Circuits, LLC), Carlos
Ortega (Indiana Integrated Circuits, LLC), Christopher Kenney (SLAC
National Accelerator Laboratory), and Julie Segal (SLAC National
Accelerator Laboratory)

Carbonized Electrodes for Electrochemical Sensing 2073

Mohammad Aminul Haque (The University of Tennessee, Knoxville),
Nickolay V. Lavrik (Oak Ridge National Laboratory), Dale Hensley (Oak
Ridge National Laboratory), and Nicole McFarlane (The University of
Tennessee, Knoxville)

Moldability Challenges Associated with the Assembly of Thicker IC Packages for High Voltage and
Power Applications 2079

Sadia Naseem (Texas Instruments Inc.), Jack Chiang (Texas Instruments
Inc.), Megan Chang (Texas Instruments Inc.), Bob Lee (Texas
Instruments Inc.), and Jason Chien (Texas Instruments Inc.)

Highly Compact, Multiband Composite-Right/Left-Handed(CRLH) Transmission Line Based Stub for GPS
Applications 2085

Hae-In Kim (University of Florida), Seahee Hwangbo (University of
Florida), Renuka Bowrothu (University of Florida), and Yong-Kyu Yoon
(University of Flordia)

Session 40: Interactive Presentations 4

Die Thickness Optimization for Preventing Electro-Thermal Fails Induced by Solder Voids in Power
Devices 2091

Dario Vitello (STMicroelectronics), Andrea Albertinetti
(STMicroelectronics), and Marco Rovitto (STMicroelectronics)

3-T (8-T) Decoupling Capacitors for Improved PDN in LPDDR4/4X/5 System 2097

Sunil Gupta (Qualcomm Technologies, Inc.)

Improved Correlation Between Accelerated Board Level Reliability (BLR) Testing and Customer BLR Results Using a Hybrid Closed-Form/Finite Element Methodology 2103

Maxim Serebreni (DfR Solutions), Natalie Hernandez (DfR Solutions), Gil Sharon (DfR Solutions), Nathan Blattau (DfR Solutions), Craig Hillman (DfR Solutions), and Ken Symonds (Western Digital)

Fabrication and Reliability Demonstration of 3 μm Diameter Photo Vias at 15 μm Pitch in Thin Photosensitive Dielectric Dry Film for 2.5 D Glass Interposer Applications 2112

Daichi Okamoto (TAIYO INK MFG. CO., LTD.), Yoko Shibasaki (TAIYO INK MFG.CO.LTD), Daisuke Shibata (TAIYO INK MFG.CO.LTD), Tadahiko Hanada (TAIYO INK MFG.CO.LTD), Fuhan Liu (Georgia Institute of Technology), Mohanalingam Kathaperumal (Georgia Institute of Technology), and Rao R. Tummala (Georgia Institute of Technology)

Pre-Cure Modification of Electrically Conductive Adhesive for Low Temperature Interconnection 2117

Jinto George (University of Sherbrooke, Bromont, QC, Canada), David Danovitch (University of Sherbrooke, Bromont, QC, Canada), Alexandre Leblanc (IBM Canada Ltd, Bromont, QC, Canada), Eric Savage (IBM Canada Ltd, Bromont, QC, Canada), Michael Ayukawa (Redlen Technologies, Saanichton, BC, Canada), and Dexter Macaisa (Redlen Technologies, Saanichton, BC, Canada)

RDL-1st Fan-Out Panel Level Packaging (FOPLP) for Heterogeneous and Economical Packaging 2126

Nagendra Sekhar Vasarla (Institute of Microelectronics, A*STAR (Agency for Science, Technology and Research)), Vempati Srinivasa Rao (Institute of Microelectronics, A*STAR (Agency for Science, Technology and Research)), F. X. Che (Institute of Microelectronics, A*STAR (Agency for Science, Technology and Research)), Chong Ser Choong (Institute of Microelectronics, A*STAR (Agency for Science, Technology and Research)), and Kazunori Yamamoto (Institute of Microelectronics, A*STAR (Agency for Science, Technology and Research))

Epoxy Composites with Surface Modified Silicon Carbide Filler for High Temperature Molding Compounds. 2134

Fan Wu (Georgia Institute of Technology), Nicholas C Mitchell (Nicholas C), Bo Song (Georgia Institute of Technology), Kyoung-sik Moon (Georgia Institute of Technology), and CP Wong (Georgia Institute of Technology)

Ultra Low Resistivity and High Electrical Stability Silo-Ag ECAs Produced from Curing Chemistry Optimization for Flexible Electronics 2140

Xueqiao Wang (Georgia Institute of Technology), Bo Song (Georgia Institute of Technology), Kyoung-Sik Moon (Georgia Institute of Technology), and C. P. Wong (Georgia Institute of Technology)

Physics of Failure Based Simulation and Experimental Testing of Quad Flat No-Lead Package 2144

Jia-Shen Lan (National Sun Yat-sen University) and Mei-Ling Wu (National Sun Yat-sen University)

An Assessment of Electromigration in 2.5D Packaging 2150

Jiefeng Xu (The State University of New York at Binghamton), Scott McCann (The State University of New York at Binghamton), Huayan Wang (The State University of New York at Binghamton), Jing Wang (The State University of New York at Binghamton), VanLai Pham (The State University of New York at Binghamton), Stephen R. Cain (The State University of New York at Binghamton), Gamal Refai-Ahmed (The State University of New York at Binghamton), and S.B. Park (The State University of New York at Binghamton)

Diffusion Enhanced Drive Sub 100 °C Wafer Level Fine-Pitch Cu-Cu Thermocompression Bonding for 3D IC Integration 2156

Asisa Kumar Panigrahy (Gokaraju Rangaraju Institute of Engineering & Technology, Hyderabad, India), Satish Bonam (Indian Institute of Technology Hyderabad, India), Tamal Ghosh (Indian Institute of Technology Hyderabad, India), Siva Rama Krishna Vanjari (Indian Institute of Technology Hyderabad, India), and Shiv Govind Singh (Indian Institute of Technology Hyderabad)

Development of Sheet Type Molding Compounds for Panel Level Package 2162

Kenichi Ueno (Company), Kazuhiro Dohi (SANYU REC CO., LTD.), Yui Suzuki (SANYU REC CO., LTD.), and Masakazu Hirose (SANYU REC CO., LTD.)

Defect Detection for the TSV Transmission Channel Using Machine Learning Approach 2168

Huan Liu (Peking University), Runiu Fang (Peking University), Min Miao (Beijing Information Science and Technology University), Yang Yang (Shenzhen Graduate School, Peking University), and Yufeng Jin (Shenzhen Graduate School, Peking University)

Direct Printing of Heat Sinks, Cases and Power Connectors on Insulated Substrate Using Selective Laser Melting Techniques 2173

Rabih Khazaka (Safran SA), Donatien Martineau (Safran SA), Toni Youssef (Safran SA), Thanh Long Le (Safran SA), and Stephane Azzopardi (Safran SA)

Server CPU Package Design Using PoINT Architecture 2180

Arun Chandrasekhar (Intel), Vijaya Boddu (Intel), Erich Chuh (Intel), Krishna Bharath (Intel), Farzaneh Yahyaei-Moayyed (Intel), Srikrishnan Venkataraman (Intel), Sriram Srinivasan (Intel), Ram Viswanath (Intel), Huthasana Kalyanam (Intel), and Ritesh Jain (Intel)

Highly Reliable Die-Attach Silver Joint with Pressure-Less Sintering Process 2186

Sihai Chen (Indium Corporation, USA), William Shambach (Rochester Institute of Technology), Jordan Palmer (Rochester Institute of Technology), Christine Labarbera (Indium Corporation), Xuanyi Ding (Cornell University), and Ning-Cheng Lee (Indium Corporation)

3D Power Packaged Device Thermo-Mechanical Modeling and Stress Analysis After Reliability Trials 2194

Lucrezia Guarino (STMicroelectronics, Italy), Lucia Zullino (STMicroelectronics, Italy), Luca Cecchetto (STMicroelectronics, Italy), Fiorella Pozzobon (STMicroelectronics, Italy), and Antonio Andreini (STMicroelectronics, Italy)

Millimeter Wave Dual Polarization Design Using Frequency Selective Surface (FSS) for 5G Base-Station Applications 2200

Chi-Hau Yang (National Sun Yat-Sen University), Chung-Yi Hsu (National Sun Yat-Sen University), and Lih-Tyng Hwang (National Sun Yat-Sen University)

Direct Bonding of low Temperature Heterogeneous Dielectrics 2206

Serena Iacovo (imec), Ian Peng (imec), Alain Phommahaxay (imec), Fumihiro Inoue (imec), Patrick Verdonck (imec), Soon-Wook Kim (imec), Erik Sleeckx (imec), Andy Miller (imec), Gerald Beyer (imec), and Eric Beyne (imec)

Session 41: Student Interactive Presentations

Low Temperature Transient Liquid Phase (TLP) Bonding using Eutectic Sn-In Solder Anisotropic Condctive Films (ACFs) for Flexible Ultrasound Transducer 2213

Jae-Hyeong Park (KAIST), Jongcheol Park (NanoFab Center), and Kyung-Wook Paik (KAIST)

Room-Temperature Bonding with Pd Coated Cu Wire on Al Pads: Ball Bond Optimization with 2-Stage Methodology 2219

Nicholas Kam (University of Waterloo), Michael David Hook (University of Waterloo), Celal Con (KA Imaging), Karim S. Karim (University of Waterloo), and Michael Mayer (University of Waterloo)

On-Chip ESD Monitor 2225

Kannan Kalappurakal Thankappan (University of California, Los Angeles), Boris Vaisband (University of California, Los Angeles), and Subramanian S. Iyer (University of California, Los Angeles)

Preparation and Characterization of Electroplated Cu/Graphene Composite 2234

Xin Wang (Tsinghua University), Qian Wang (Tsinghua University), Jian Cai (Tsinghua University), Changming Song (Tsinghua University), Yang Hu (Tsinghua University), Yang Zhao (University of Science and Technology of China), and Yu Pei (University of Science and Technology of China)

Quantifying the Impact of RF Probing Variability on TRL Calibration for LTCC Substrates 2240

Ömer Faruk Yildiz (Hamburg University of Technology, Germany), David Dahl (Hamburg University of Technology, Germany), and Christian Schuster (Hamburg University of Technology, Germany)

Effects of NCF and UBM Materials on Electromigration Reliabilities of Sn-Ag Microbumps for Advanced 3D Packaging 2246

Kirak Son (Andong National University), Gahui Kim (Andong National University), Hyodong Ryu (Andong National University), Gyu-Tae Park (Amkor Technology Korea Inc.), Ho-Young Son (SK hynix Inc.), Nam-Seog Kim (SK hynix Inc.), Cheol-Woong Yang (Sungkyunkwan University), Young-Cheon Kim (Andong National University), Jeong Sam Han (Andong National University), and Young-Bae Park (Andong National University)

Ag Diffusion Control Through Sn on a Sequential Plating-Based Bumping Process 2252

Abderrahim EL Amrani (Université de Sherbrooke), Etienne Paradis (Université de Sherbrooke), David Danovitch (Université de Sherbrooke), and Dominique Drouin (Université de Sherbrooke)

liv

Mechanical Reliability Assessment of Cu_6Sn_5 Intermetallic Compound and Multilayer Structures in Cu/Sn Interconnects for 3D IC Applications 2258

    Jui-Yang Wu (National Taiwan University), C. Robert Kao (National Taiwan University), and Jenn-Ming Yang (University of California, Los Angeles)

A Study on the Anchoring Polymer Layer (APL) Anisotropic Conductive Films (ACFs) with Self-Exposed Conductive Particles Surface for Ultra-Fine Pitch Chip-on-Glass (COG) Applications 2266

    Dal-Jin Yoon (KAIST) and Kyung-Wook Paik (Korea Advanced Institute of Science and Technology)

Bending Properties of Fine Pitch Flexible CIF (Chip-in-Flex) Packages Using APL (Anchoring Polymer Later) ACFs (Anisotropic Conductive Films) 2272

    Ji-Hye Kim (KAIST, Korea), Dal-Jin Yoon (KAIST, Korea), and Kyung-Wook Paik (KAIST, Korea)

Effects of the Curing Properties and Viscosities of Non-Conductive Films (NCFs) on the Sn-Ag Solder Bump Joint Morphology and Reliability 2278

    HanMin Lee (KAIST, South Korea), SeYong Lee (KAIST, South Korea), SangMyung Shin (KAIST, South Korea), TaeJin Choi (Doosan Corporation Electro-Materials BG, South Korea), SooIn Park (Doosan Corporation Electro-Materials BG, South Korea), and Kyung-Wook Paik (KAIST, South Korea)

Experimental Investigations on Vertical Ultrasonic Assisted Low Temperature Sintering Process 2284

    Henning Seefisch (Leibniz Universität Hannover) and Jens Twiefel (Leibniz Universität Hannover)

Pressureless Transient Liquid Phase Sintering Bonding of Sn-58Bi with Ni Particles for High-Temperature Packaging Applications 2290

    Kyung Deuk Min (Sungkyunkwan University, Republic of Korea), Kwang-Ho Jung (Sungkyunkwan University, Republic of Korea), Choong-Jae Lee (Sungkyunkwan University, Republic of Korea), and Seung-Boo Jung (Sungkyunkwan University, Republic of Korea)

Epoxy/ Triazine Copolymer Resin System for High Temperature Encapsulant Applications 2296

    Jiaxiong Li (Georgia Institute of Technology), Chao Ren (Georgia Institute of Technology), Kyoung-sik Moon (Georgia Institute of Technology), and Ching-ping Wong (Georgia Institute of Technology)

Low Temperature Ag-Ag Direct Bonding Technology for Advanced Chip-Package Interconnection 2302

    Jiaqi Wu (University of California Irvine) and Chin C. Lee (University of California Irvine)

Reliability of Micro-Alloyed SnAgCu Based Solder Interconnections for Various Harsh Applications 2309

    Sinan Su (Auburn University), Francy John Akkara (Auburn University), Anto Raj (Auburn University), Cong Zhao (Auburn University), Seth Gordon (Auburn University), Sharath Sridhar (Auburn University), Sivasubramanian Thirugnanasambandam (Auburn University), Sa'd Hamasha (Auburn University), Jeffery Suhling (Auburn University), and John Evans (Auburn University)

Wideband Low-Profile Ka-Band Microstrip Antenna with Low Cross Polarization Using Asymmetry AMC Structure 2318

Mei Xue (Institute of Microelectronics of the Chinese Academy of Sciences), Weikang Wan (Institute of Microelectronics of the Chinese Academy of Sciences), Qidong Wang (Institute of Microelectronics of the Chinese Academy of Sciences), and Liqiang Cao (Institute of Microelectronics of the Chinese Academy of Sciences)

Automatic Transient Thermal Impedance Tester for Quality Inspection of Soldered and Sintered Power Electronic Devices on Panel and Tile Level 2324

Maximilian Schmid (Technische Hochschule Ingolstadt), Bhogaraju Sri Krishna (Technische Hochschule Ingolstadt), and Gordon Elger (Technische Hochschule Ingolstadt)

Time 0 Void Evolution and Effect on Electromigration 2331

Jiefeng Xu (The State University of New York at Binghamton), Scott McCann (Xilinx, Inc.), Huayan Wang (The State University of New York at Binghamton), VanLai Pham (The State University of New York at Binghamton), Stephen R. Cain (The State University of New York at Binghamton), Gamal Refai-Ahmed (Xilinx, Inc.), and S.B. Park (The State University of New York at Binghamton)

Quintuple Band Lambda/4 Stub by using Unbalanced Bridged CRLH Transmission Lines 2337

Renuka Bowrothu (University of Florida), Seahee Hwangbo (University of Florida), Haein Kim (University of Florida), and Yong-Kyu Yoon (University of Florida)

Product Level Design Optimization for 2.5D Package Pad Cratering Reliability During Drop Impact 2343

Huayan Wang (State University of New York at Binghamton), Jing Wang (State University of New York at Binghamton), Jiefeng Xu (State University of New York at Binghamton), Vanlai Pham (State University of New York at Binghamton), Ke Pan (State University of New York at Binghamton), Seungbae Park (State University of New York at Binghamton), Hohyung Lee (Xilinx Inc, USA), and Gamal Refai-Ahmed (Xilinx Inc, USA)

Microstructures of Pb-Free Solder Joints by Reflow and Thermo-Compression Bonding (TCB) Processes 2349

Youngja Kim (Samsung Electronics), Jinho Hah (Georgia Institute of Technology), Patxi Fernandez-Zelaia (Georgia Institute of Technology), Sangil Lee (Georgia Institute of Technology), Leroy Christie (ASM Pacific Technology), Paul Houston (Engent Inc.), Shreyes Melkote (Georgia Institute of Technology), Kyoung-Sik Moon (Georgia Institute of Technology), and Ching-Ping Wong (Georgia Institute of Technology)

Reduction of Ag Corrosion Rate During Decapsulation of Ag Wire Bond Packages 2359

Young-Ja Kim (Samsung Electronics, Korea), Jinho Hah (Georgia Institute of Technology), Kyoung-Sik (Jack) Moon (Georgia Institute of Technology), and C. P. Wong (Georgia Institute of Technology)

Author Index

Foreword

On behalf of the Program Committee and Executive Committee, it is our pleasure to welcome you to the 69th Electronic Components and Technology Conference (ECTC) which will be held at The Cosmopolitan of Las Vegas in Las Vegas, Nevada from May 28-31, 2019. This premier international conference is sponsored by the IEEE Electronic Packaging Society (EPS). The ECTC Program Committee has selected over 350 papers which will be presented in 36 oral sessions and five interactive presentation session including one interactive presentation session exclusively featuring papers by student authors. The oral sessions will feature selected papers on key topics such as fan-out packaging, wafer-level packaging, flip-chip packaging, 3D/TSV technologies, design for RF performance and signal/power integrity, thermal and mechanical modeling, optoelectronics packaging, materials and reliability. Interactive presentation sessions will showcase papers in a format that encourages more in-depth discussion and interaction with authors about their work.

Authors from over twenty countries are expected to present their work at the 69th ECTC, covering ongoing technology development within established disciplines or emerging topics of interest for our industry such as additive manufacturing, heterogeneous integration, flexible and wearable electronics.

ECTC will also feature six special sessions with invited industry experts covering several important and emerging topic areas. On Tuesday, May 29 at 10 a.m., W. Hong Yeo and Mikel Miller will chair a special session covering "Transient Electronics: A Green Revolution for Packaging". On the same day at 2 p.m., Rena Huang and Soon Jang will chair a session focused on Photonics on the Cutting-Edge of Technology Evolution. Tuesday evening will also include the ECTC Panel Session at 7:30 p.m. chaired by IEEE EPS President Avi Bar-Cohen and Karlheinz Bock, where young researchers will share their visions of future packaging technologies and participate in discussions with experts in the field.

This conference will also feature a Women's Panel and Reception jointly organized by ECTC and ITherm on Wednesday, May 29 at 6:30pm. This year, panelists from around the globe will share their perspectives on efforts to enhance the participation of women in engineering, and the panel will be chaired by Kristina Young-Fisher and Cristina Amon. On the same day at 7:30 p.m., Tanja Braun will chair the ECTC Plenary Session titled "Sensors and Packaging for Autonomous Driving". In this plenary session, experts will address the challenges and demands for sensors and packages for autonomous driving along the value chain. On Thursday, May 31 at 8 p.m., the IEEE EPS Seminar titled "Roadmap of IC Packaging Materials to Meet Next-Generation Smartphone Performance Requirements" will be moderated by Yasumitsu Orii and Sheigenori Aoki from the High-Density Substrates & Boards Technical Committee of the IEEE EPS Society.

Supplementing the technical program, ECTC also offers Professional Development Courses (PDCs) and the Technology Corner exhibits. Co-located with the IEEE iTHERM Conference this year, the 69th ECTC will offer eighteen PDCs, organized by the PDC Committee chaired by Kitty Pearsall and Jeffrey Suhling. The PDCs will take place on Tuesday, May 28 and are taught by distinguished experts in their respective fields. The Technology Corner will showcase the latest technologies and products offered by leading companies in the electronic components, materials, packaging and services fields. More than one hundred Technology Corner exhibits will be open Wednesday and Thursday starting at 9 a.m. ECTC also offers attendees numerous opportunities for networking and discussion with colleagues during coffee breaks, daily luncheons and nightly receptions.

Whether you are an engineer, a manager, a student or an executive, ECTC offers something unique for everyone in the microelectronics packaging and components industry. I invite you to make your plans now to join us for the 69th ECTC and be a part of all the exciting technical and professional opportunities. I also take this opportunity to thank our sponsors, exhibitors, authors, speakers, PDC instructors, session chairs, and program committee members, as well as all the volunteers who help make the 69th ECTC a success. We look forward to meeting you in Las Vegas, Nevada May 28 –31, 2019.

Nancy Stoffel
69th ECTC Program Chair
General Electric Research
stoffel@ge.com

Mark Poliks
69th ECTC General Chair
Binghamton University
mpoliks@binghamton.edu

2019 Executive Committee

General Chair
Mark Poliks
Binghamton University
mpoliks@binghamton.edu

Vice-General Chair
Christopher Bower
X-Celeprint Inc.
cbower@x-celeprint.com

Program Chair
Nancy Stoffel
GE Research
nstoffel1194@gmail.com

Assistant Program Chair
Rozalia Beica
DuPont
rozalia.beica@dupont.com

Web Administrator
Ibrahim Guven
Virginia Commonwealth University
iguven@vcu.edu

Jr. Past General Chair
Sam Karikalan
Broadcom Inc.
sam.karikalan@broadcom.com

Sr. Past General Chair
Henning Braunisch
Intel Corporation
braunisch@ieee.org

Sponsorship Chair
Wolfgang Sauter
GLOBALFOUNDRIES
wolfgang.sauter@globalfoundries.com

Finance Chair
Patrick Thompson
Texas Instruments, Inc.
patrick.thompson@ti.com

Publications Chair
Steve Bezuk
sbezuk@gmail.com

Publicity Chair
Eric Perfecto
eric.perfecto.us@ieee.org

Treasurer
Tom Reynolds
T3 Group LLC
t.reynolds@ieee.org

Exhibits Chair
Joe Gisler
Vector Associates
gisler.h.dr@ieee.org

Exhibits Co-Chair
Alan Huffman
Micross Advanced Interconnect Technology
alan.huffman@micross.com

Arrangements Chair
Lisa Renzi Ragar
Renzi & Company, Inc.
lrenzi@renziandco.com

EPS Representative
C. P. Wong
Georgia Institute of Technology
cp.wong@mse.gatech.edu

2019 Program Committee

Applied Reliability

Chair
Deepak Goyal
Intel Corporation
deepak.goyal@intel.com

Assistant Chair
Darvin R. Edwards
Edwards Enterprise Consulting, LLC
darvin.edwards1@gmail.com

Tim Chaudhry
Amkor Technology, Inc.

Tz-Cheng Chiu
National Cheng Kung University

Vikas Gupta
Texas Instruments, Inc.

Sandy Klengel
Fraunhofer Institute for Microstructure of Materials and Systems

Pilin Liu
Intel Corporation

Varughese Mathew
NXP Semiconductors

Toni Mattila
Aalto University

Keith Newman
AMD

Donna M. Noctor
Nokia

S. B. Park
Binghamton University

Lakshmi N. Ramanathan
Microsoft Corporation

René Rongen
NXP Semiconductors

Scott Savage
Medtronic Microelectronics Center

Jeffrey Suhling
Auburn University

Pei-Haw Tsao
Taiwan Semiconductor Manufacturing Company, Ltd.

Dongji Xie
NVIDIA Corporation

Assembly & Manufacturing Technology

Chair
Mark Gerber
Advanced Semiconductor Engineering Inc.
mark.gerber@aseus.com

Assistant Chair
Jin Yang
Intel Corporation
jin1.yang@ieee.org

Sai Ankireddi
Soraa, Inc

Christo Bojkov
Qorvo

Garry Cunningham
NGC

Habib Hichri
Suss Microtech Photonic Systems Inc.

Paul Houston
Engent

Li Jiang
Texas Instruments

Chunho Kim
Medtronic Corporation

Wei Koh
Pacrim Technology

Ming Li
ASM Pacific Technology

Debendra Mallik
Intel Corporation

Jae-Woong Nah
IBM Corporation

Valerie Oberson
IBM Canada Ltd

Chandradip Patel
Schlumberger Technology Corporation

Shichun Qu
Intersil, a Renesas Company

Paul Tiner
Texas Instruments

Andy Tseng
JSR Micro

Jan Vardaman
Techsearch International

Yu Wang
Sensata Technologies

Shaw Fong Wong
Intel Corporation

Wei Xu
Huawei

Tonglong Zhang
Nantong Fujitsu Microelectronics Ltd.

Emerging Technologies

Chair
Florian Herrault
HRL Laboratories, LLC
fgherrault@hrl.com

Assistant Chair
Benson Chan
Binghamton University
chanb@binghamton.edu

Isaac Robin Abothu
Siemens Healthineers

Meriem Akin
Robert Bosch GmBH

Vasudeva P. Atluri
Renavitas Technologies

Karlheinz Bock
Technische Universitat Dresden

Vaidyanathan Chelakara
Acacia Communications

Rabindra N. Das
MIT Lincoln Labs

Dongming He
Qualcomm Technologies, Inc.

TengFei Jiang
University of Central Florida

Jong-Hoon Kim
Washington State University Vancouver

Ahyeon Koh
Binghamton University

Ramakrishna Kotlanka
Analog Devices

Santosh Kudtarkar
Analog Devices

Kevin J. Lee
Qorvo Corporation

Zhuo Li
Fudan University

Chukwudi Okoro
Corning

Bharat Penmecha
Intel Corporation

C. S. Premachandran
GLOBALFOUNDRIES

Jintang Shang
Southeast University

Rohit Sharma
IIT Ropar

Nancy Stoffel
GE Research

Liu Yang
IBM

Jimin Yao
Intel Corporation

W. Hong Yeo
Georgia Institute of Technology

Hongqing Zhang
IBM Corporation

High-Speed, Wireless & Components

Chair
Wendem Beyene
Intel Corporation
wendem.beyene@intel.com

Assistant Chair
Lianjun Liu
NXP Semiconductor, Inc.
lianjun.liu@NXP.com

Amit P. Agrawal
Microsemi Corporation

Kemal Aygun
Intel Corporation

Eric Beyne
IMEC

Prem Chahal
Michigan State University

Zhaoqing Chen
IBM Corporation

Charles Nan-Cheng Chen
HiSilicon Technologies

Craig Gaw
NXP Semiconductor

Abhilash Goyal
Velodyne LIDAR, Inc.

Xiaoxiong (Kevin) Gu
IBM Corporation

Rockwell Hsu
Cisco Systems, Inc.

Lih-Tyng Hwang
National Sun Yat-Sen University

Bruce Kim
City University of New York

Timothy G. Lenihan
TechSearch International

Rajen M Murugan
Texas Instruments

Nanju Na
Xilinx

Dan Oh
Samsung

P. Markondeya Raj
Florida International University

Hideki Sasaki
Renesas Electronics Corporation

Li-Cheng Shen
Wistron NeWeb Corporation

Jaemin Shin
Qualcomm Corporation

Manos M. Tentzeris
Georgia Institute of Technology

Maciej Wojnowski
Infineon Technologies AG

Yong-Kyu Yoon
University of Florida

Interconnections

Chair
Wei-Chung Lo
ITRI
lo@itri.org.tw

Assistant Chair
Dingyou Zhang
Broadcom Inc.
dingyouzhang.brcm@gmail.com

Thibault Buisson
Yole Développement

Jian Cai
Tsinghua University

William Chen
Advanced Semiconductor Engineering, Inc.

David Danovitch
University of Sherbrooke

Rajen Dias
Amkor Technology, Inc.

Bernd Ebersberger
Infineon Technologies

Takafumi Fukushima
Tohoku University

Tom Gregorich
Zeiss Semiconductor Manufacturing Technology

Kangwook Lee
Amkor Technology Korea

Steward Lee

Li Li
Cisco Systems, Inc.

Changqing Liu
Loughborough University

Nathan Lower
Rockwell Collins, Inc.

James Lu
Rensselaer Polytechnic Institute

Voya Markovich
Microelectronic Advanced Hardware Consulting, LLC

Lou Nicholls
Amkor Technology, Inc.

Peter Ramm
Fraunhofer EMFT

Katsuyuki Sakuma
IBM Corporation

Lei Shan
IBM Corporation

Ho-Young Son
SK Hynix

Jean-Charles Souriau
CEA Leti

Chuan Seng Tan
Nanyang Technological University

Matthew Yao
GE Energy Management

Materials & Processing

Chair
Mikel Miller
EMD Performance Materials
mikel.miller@emdgroup.com

Assistant Chair
Tanja Braun
Fraunhofer IZM
tanja.braun@izm.fraunhofer.de

Yu-Hua Chen
Unimicron

Qianwen Chen
IBM Research

Bing Dang
IBM Research

Yung-Yu Hsu
Apple Inc.

Lewis Huang
Senju Electronic

C. Robert Kao
National Taiwan University

Chin C. Lee
University of California, Irvine

Alvin Lee
Brewer Science

Yi (Grace) Li
Intel Corporation

Ziyin Lin
Intel Corporation

Yan Liu
Medtronic Inc. USA

Daniel D. Lu
Henkel Corporation

Joon-Seok Oh
Samsung Electro-Mechanics

Praveen Pandojirao-S
Johnson & Johnson

Mark Poliks
Binghamton University

Dwayne Shirley
Inphi

Ivan Shubin
Oracle

Bo Song
HP Inc.

Yoichi Taira
Keio University

Lejun Wang
Qualcomm Technologies, Inc.

Frank Wei
Disco Japan

Kimberly Yess
Brewer Science

Myung Jin Yim
Apple

Hongbin Yu
Arizona State University

Packaging Technologies

Chair
Dean Malta
Micross Advanced Interconnect Technology
Dean.Malta@micross.com

Assistant Chair
Luke England
GLOBALFOUNDRIES
luke.england@globalfoundries.com

Daniel Baldwin
H.B. Fuller Company

Bora Baloglu
Amkor Technology

Jie Fu
Apple

Mike Gallagher
DuPont

Ning Ge
Consultant

Allyson Hartzell
Veryst Engineering

Kuldip Johal
Atotech

Beth Keser
Intel Corporation

Young-Gon Kim
Integrated Device Technology, Inc.

Andrew Kim
Intel Corporation

John Knickerbocker
IBM Corporation

Albert Lan
Applied Materials

John H. Lau
ASM Pacific Technology

Jaesik Lee
Nvidia

Markus Leitgeb
AT&S

Luu Nguyen
Texas Instruments Inc.

Deborah S. Patterson
Harbor Electronics, Inc.

Raj Pendse
Facebook FRL (Facebook Reality Labs)

Subhash L. Shinde
Notre Dame University

Joseph W. Soucy
Draper Laboratory

Peng Su
Juniper Networks

Kuo-Chung Yee
Taiwan Semiconductor Manufacturing Corporation, Inc.

Christophe Zincke
Advanced Semiconductor Engineering, Inc.

Photonics

Chair
Ping Zhou
LDX Optronics, Inc.
pzhou@ldxoptronics.com

Assistant Chair
Z. Rena Huang
Rensselaer Polytechnic Institute
zrhuang@ecse.rpi.edu

Mark Beranek
Naval Air Systems Command

Stephane Bernabe
CEA Leti

Fuad Doany
IBM Research

Gordon Elger
Technische Hochschule Ingolstadt

Takaaki Ishigure
Keio University

Ajey Jacob
GLOBALFOUNDRIES

Soon Jang
ficonTEC USA

Harry G. Kellzi
Teledyne Microelectronic Technologies

Richard Pitwon
Resolute Photonics Ltd

Alex Rosiewicz
A2E Partners

Henning Schroeder
Fraunhofer IZM

Andrew Shapiro
JPL

Masato Shishikura
Oclaro Japan

Masao Tokunari
IBM Corporation

Shogo Ura
Kyoto Institute of Technology

Stefan Weiss
II-VI Laser Enterprise GmbH

Feng Yu
Huawei Technologies Japan

Thomas Zahner
OSRAM Opto Semiconductors GmbH

Thermal/Mechanical Simulation & Characterization

Chair
Przemyslaw Gromala
Robert Bosch GmbH
Przemyslawjakub.gromala@de.bosch.com

Assistant Chair
Ning Ye
Western Digital
ning.ye@wdc.com

Christopher J. Bailey
University of Greenwich

Kuo-Ning Chiang
National Tsinghua University

Xuejun Fan
Lamar University

Nancy Iwamoto
Honeywell Performance Materials and Technologies

Pradeep Lall
Auburn University

Chang-Chun Lee
National Tsing hua University (NTHU)

Yong Liu
ON Semiconductor

Sheng Liu
Wuhan University

Erdogan Madenci
University of Arizona

Tony Mak
Wentworth Institute of Technology

Karsten Meier
Technische Universität Dresden

Erkan Oterkus
University of Strathclyde

Sandeep Sane
Intel Corporation

Suresh K. Sitaraman
Georgia Institute of Technology

Wei Wang
Qualcomm Technologies, Inc.

G. Q. (Kouchi) Zhang
Delft University of Technology (TUD)

Tieyu Zheng
Microsoft Corporation

Jiantao Zheng
Hisilicon

Interactive Presentations

Chair
Michael Mayer
University of Waterloo
mmayer@uwaterloo.ca

Assistant Chair
Pavel Roy Paladhi
IBM Corporation
Pavel.Roy.Paladhi@ibm.com

Swapan Bhattacharya
Engent Inc.

Rao Bonda
Amkor Technology

Mark Eblen
Kyocera International SC

Ibrahim Guven
Virginia Commonwealth University

Alan Huffman
Micross Advanced Interconnect Technology

Jeffrey Lee
iST-Integrated Service Technology Inc.

Nam Pham
IBM Corporation

Mark Poliks
Binghamton University

Patrick Thompson
Texas Instruments, Inc.

Kristina Young-Fisher
GLOBALFOUNDRIES

Professional Development Courses

Chair
Kitty Pearsall
Boss Precision, Inc.
kitty.pearsall@gmail.com

Assistant Chair
Jeffrey Suhling
Auburn University
jsuhling@auburn.edu

Vijay Khanna
IBM Corporation

Eddie Kobeda
Nypro, A Jabil Company

Lakshmi N. Ramanathan
Microsoft Corporation

2019 IEEE 69th Electronic Components and Technology Conference (ECTC)

3D-MiM (MUST-in-MUST) Technology for Advanced System Integration

An-Jhih Su, Terry Ku, Chung-Hao Tsai, Kuo-Chung Yee, and Douglas Yu
Research and Development, Taiwan Semiconductor Manufacturing Company
Ltd. 168, Park Ave. 2, Hsinchu Science Park, Hsinchu, Taiwan 30844, R.O.C.

Abstract—**An advanced 3D Multi-stack (MUST) system integration technology, 3D MUST-in-MUST (3D-MiM) fan out package, has been developed as next generation wafer-level fan-out package technology. 3D-MiM technology utilizes a more simplified architecture which eliminates BGAs between packages for system-level performance, power and form-factor (PPA) purpose. This technology also makes use of a modularized approach for both design and integration flow to improve design flexibility and integration efficiency. Known-good pre-stacked memory cube and/or logic-memory cubes are fabricated by leveraging the established integrated fan-out technology platform (InFO) in tools, materials, design rules, and processes to shorten development cycle time and achieve cost effectiveness. Two 3D-MiM fan-out examples are presented in this paper. The first 3D-MiM package integrates a SoC with 16 memory chips in a 15x15 mm² footprint with 0.5 mm package height (final BGA included) for mobile application. The other 3D-MiM package integrates 8 SoCs with 32 memory chips in a 43x28 mm² footprint to mimic a system integration of multiple logic cores and multiple memory chips for HPC applications.**

Keywords—3D MUST-in-MUST (3D-MiM); InFO; 5G/AI; WLSI; Data Bandwidth

I.   INTRODUCTION

Advanced system integration on logic and memory has been an important subject to packaging community for mobile and HPC applications in the 5G/AI era. Recent years, 3DIC integration technology has advanced from substrate level package to wafer level system integration (WLSI) [1-2]. WLSI leads the semiconductor industry into a new era of system scaling beyond Moore's Law. In mobile application, other than flip chip PoP (FC-PoP), the high density, high performance integrated fan-out (aka InFO-PoP) has been adopted in the leading mobile devices for its thin profile and best electrical/thermal performances [3-4]. In HPC and 5G/AI applications, the integration of high bandwidth memory (HBM) cubes with high power, high performance logic has been realized through a high density silicon interposer technology (aka CoWoS®) [5-6]. The HBM itself is fabricated by the TSV 3DIC stacking with micro-bump flip chip assembly [7].

In 5G and AI era, the massive data and data communication become bottlenecks in realizing the fast-growing applications of AI + IoT (AIoT), autonomous driving, and ubiquitous communication of data centers, servers, and edge devices. Fast data processing requires power

computing with low energy consumption. High speed data communication requires high data bandwidth with low communication latency. To achieve these requirements, an emerging system integration solution, 3D-MiM fan-out, for logic and memory is developed for its compact structure, high density interconnect, improved system performances, and competitive cost. Firstly, compact structure is crucial as it leads to a smaller footprint and thinner profile, with lower system energy consumption and improves system thermal and electrical performances. Secondly, high density interconnect allows the adoption of advanced node SoC and memory for high power computing, with high data bandwidth and low latency. Finally, competitive cost is achieved by leveraging well-established WLSI infrastructures for yield and capacity.

In this paper, the 3D-MiM package architecture and package design/integration are introduced first. Two 3D-MiM fan-out package demonstrations are presented with details. Finally, system performances of 3D-MiM fan-out and its benchmarks with other advanced packaging solution are addressed in full aspects.

II.   MEMORY MODULE INTEGRATION

Serving as an integration building block, the modularized memory unit integrates multiple vertically-stacked memory chips by InFO wafer level system integration, with memory I/O pads at one side of the chip. Embedded memory chips are shown in Fig. 1. The memory module is applicable to generic memory commodities, including LPDDR4/5, SRAM, Wide I/O in mobile, DDR in NB/PC and servers. Such memory module has great flexibility in horizontal footprint and extensibility in vertical stacking to meet design needs. Structure-wise, the memory module can be in 2-tiers, 3-tiers, or more tiers depending on the system needs. In each tier, there could be single or multiple memory chips. Each memory module can be tested and embedded as a known good module.

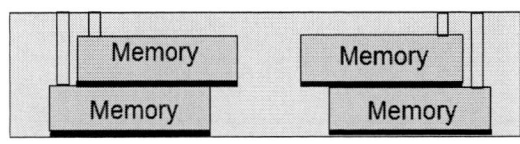

Figure 1. A MUST memory module (building blocks) used in 3D-MiM

III.   3D-MiM PACKAGE INTEGRATION

Two 3D-MiM packages are presented in this paper. The first 3D-MiM package was realized by InFO technology. First, integrate 16 memory chips in the first two fan-out tiers and then cascade a SoC in the third fan-out tier (Fig. 2). This 3D-MiM is designed as an alternative solution to the current FC-

978-1-7281-1500-9/19 $31.00 © 2019 IEEE

PoP and fan-out PoP, for use in mobile or computing devices requiring a thinner profile, with high memory capacity and memory bandwidth.

Similarly, the second, super large, 3D-MiM fan-out package was realized by integrating 32 memory chips in the first tier fan-out, and then integrating 8 SoCs in the second tier fan-out as shown in Fig. 3. This 3D-MiM fan-out is designed as a low cost alternative to the current 2.5DIC, for use in 5G/AI-driven HPC and server applications requiring high computing performance, memory bandwidth, and low power consumption and latency in applications such as machine learning, and AI training.

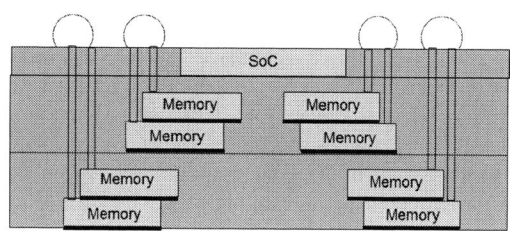

Figure 2. 3D-MiM fan-out package for edge computing

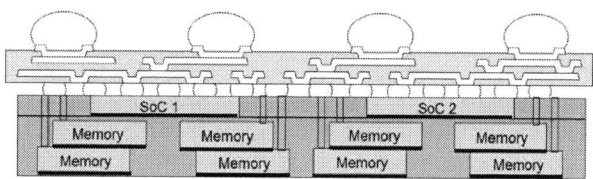

Figure 3. 3D-MiM fan-out package for 5G/AI-driven HPC and server applications

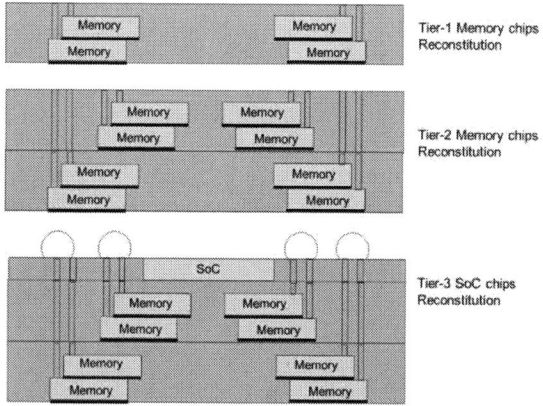

Figure 4. Integration flow of 3D-MiM fan-out package

Fig. 4 shows the brief integration steps of 3D-MiM fan-out. Each pre-fabricated and tested memory module acts as a known good die to be sequentially integrated to the system by WLSI process. Interconnects are extended from I/Os of memory module to I/Os of SoC chip through vertical via and horizontal redistribution layers. BGA or C4 solder bumps are then placed on top of redistribution layer to complete the wafer fabrication process. There are several advantages by

adopting the 3D-MiM fan-out technology. Firstly there are no wafer bumps and flip chip bonding during the 3D-MiM fan-out integration process to reduce the assembly complexity and avoid the chip-package-interaction (CPI) reliability challenges in flip chip assembly. Secondly much thinner package profile is achieved for improved form factor, thermal, and electrical performances. Thirdly the tools, materials, and capacity are shared for both mobile and HPC applications to streamline the design rules and shorten product development cycle time for high yield and cost effectiveness. Simplified structure and short I/O wiring length directly lead to improved power consumption, electrical, and thermal performances. This will be discussed in the latter sessions.

IV.   3D-MiM FAN-OUT DEMONSTRATION

A.   3D-MiM for mobile

Compact footprint and slim profile are required in mobile devices. Fig. 5a shows the cross-section image of the first 3D-MiM fan-out for mobile application. One SoC and 16 memory chips are all integrated in a single compact package with a footprint of 15x15 mm^2. For visual comparison, Fig. 5b shows a comparison of a 3D-MiM fan-out, and a fan-out PoP (FO-PoP).

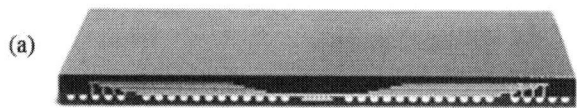

Figure 5. (a) 3D-MiM fan-out for mobile application, (b) Profile comparisons among U.S. Dime, 3D-MiM fan-out, and FO-PoP

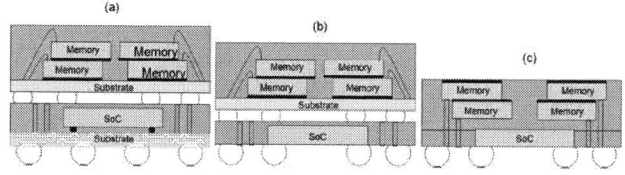

Figure 6. Package profile comparison for (a) FC-PoP, (b) InFO-PoP, and (c) 3D-MIM fan-out

This 3D-MiM fan-out represents an alternative solution to a typical mobile PoP, with thinner profile and shorter wiring between the SoC and the memories. In comparison to the wire-bond memory package, the fan-out stacked memory module, without BGA connectors and substrate, offers thinner profile and shorter wiring length to reduce the undesired electrical parasitics for power consumption, power/signal-integrity, and memory latency improvements. Comparing to FC-PoP and FO-PoP, the 3D-MiM fan-out eliminates memory

978-1-7281-1500-9/19 $31.00 © 2019 IEEE

package BGA connectors and substrate to achieve the much thinner profile as shown in Fig. 6. The typical package profile is around 1.2~1.4 mm for FC-PoP, 0.9~1.0 mm for FO-PoP. The 3D-MiM fan-out achieves 0.5 mm profile height with a clear outperformance over FC-PoP and FO-PoP.

Simple structure helps to achieve high integration yield for 3D-MiM fan-out, which was validated by daisy chain yield measurement. As shown in Fig. 7a, there are total four daisy chain routings for yield tracking. Chain-A connects the top tier memory I/O to SoC, Chain-B connects the 2nd tier memory I/O to SoC, Chain-C connects the 3rd tier memory I/O to SoC, and Chain-D connects the bottom tier memory I/O to SoC. Fig. 7b shows excellent full daisy chain yield from total 102 test samples illustrating the robustness of the 3D-MiM integration.

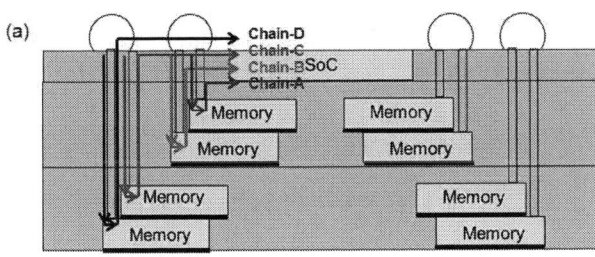

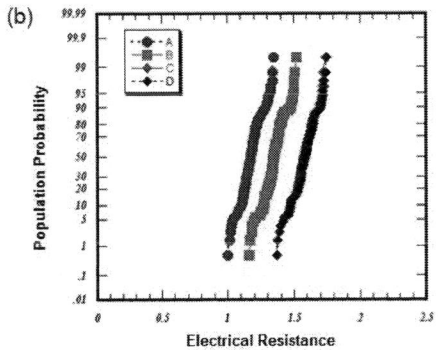

Figure 7. (a) 3D-MiM daisy chain routing for yield tracking, (b) Population probability vs. normalized electrical resistance for Chain-A to Chain-D

B. 3D-MiM for HPC

Integration of high power SoC(s) with advanced HBM cubes is essential in HPC application. High memory bandwidth and thermal design power (TDP) are two key performance requirements for advanced integration solutions. Here, we demonstrate a 3D-MiM fan-out as a new SoC-memory integration architecture to deliver memory data bandwidth comparable to SoC-HBMs integration using 3DIC TSV approach. The new 3D-MiM fan-out architecture realizes the integration of multiple logic cores with multi-tiers memory modules without using TSV. Fig. 8 shows the image of a 3D-MiM fan-out for HPC application, where 8 SoCs and 32 memory chips are integrated in a single compact package with footprint of 43x28 mm^2 and assembled on a substrate by conventional flip-chip process. In a simple architecture, the memory chips are pre-stacked by the mature InFO technology. Comparing to sophisticated 3DIC TSV stacking, the current memory module has no TSVs and micro-bumps/bonding process, yet achieves better system performances in electrical

and thermal aspects. Fig. 9 elucidates the process differences between 3DIC HBM and InFO memory cube fabrication. Comparing to HBM, it can be readily seen that memory module using InFO 3D technology has a thinner profile and less complicated integration steps. Other than that, precious memory silicon asset can be significantly reduced for cost saving since TSVs and keep out zone are not needed.

Figure 8. Image of 3D-MiM fan-out for HPC application

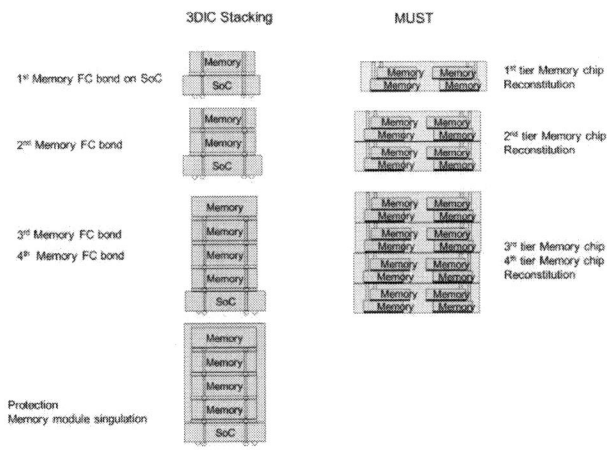

Figure 9. Comparison of integration flow between TSV HBM stacking and InFO memory module

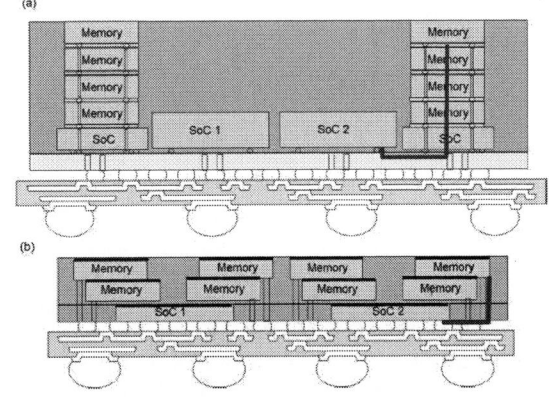

Figure 10. Pictorial schematics comparison between (a) 2.5D IC, and (b) 3D-MiM fan-out for multi-SoCs and multi-memory cubes integration (Note: blue line denotes the SoC-memory communication path)

978-1-7281-1500-9/19 $31.00 © 2019 IEEE

Fig. 10 depicts the differences between 2.5DIC and 3D-MiM fan-out in architecture, footprint, profile, and communication path of SoC-SoC, and SoC-memory. Unlike 3DIC memory cube, the InFO memory module allows the direct and parallel communication of each memory chip to the computing logic, which helps improving the memory data bandwidth between SoC and memory. In addition, in 3D-MiM fan-out, no SERDES controller IC is required, which reduce cost and design/integration complexity. It is observed that, with side-by-side layout, the communication wiring length between SoC and HBM cube of 2.5D IC is longer than that of 3D-MIM fan-out. Aside from that, the footprint and profile height of 2.5DIC are also greater than those of 3D-MiM fan-out. While many recent studies focused on near-memory computing (NMC) for improved memory data bandwidth, latency, and energy saving [8-9], 3D-MiM fan-out provides a low cost alternative for near-memory computing with new SoC-memory integration architecture. It is noted that to realize the 3D-MiM fan-out in HPC, the interconnects for SoC and memory chips need to be co-designed to achieve the desired goals.

V. SYSTEM PERFORMANCES BENCHMARK

Memory-intensive computing workloads such as real-time image/video recognition, real-time autonomous driving decision-making, edge AI training, big-data processing, and AI security control, require an advanced system integration to deliver high power computing and massive memory data pipeline in bandwidth. In this session, we present the important performance benchmarks between 3D-MiM fan-out and FC-PoP on memory data bandwidth, power integrity (PI), signal integrity (SI), and thermal characteristics.

A. Data Bandwidth

Fig. 11 shows the maximum memory data bandwidth that can be achieved by FC-PoP and 3D-MiM fan-out, respectively. Eight LPDDR5 chips are studied in the analysis. With eight LPDDR5 chips, the FC-PoP can achieve 51.2 GB/s, while 3D-MiM achieves 102.4GB/s, a 2X in memory data bandwidth due to high density interconnects. As more memory chips integrated, higher data bandwidth and larger data bandwidth differences between the FC-PoP and 3D-MiM fan-out are observed. For example, for sixteen LPDDR5 chips, FC-PoP remains 51.2GB/s, while 3D-MiM fan-out achieves 204.8 GB/s, a 4X in memory data bandwidth. This puts LPDDR5 memory module bandwidth near to HBM2 (250 GB/s) with a lower cost and lower power for mobile/edge computing. The reason lies in that, for FC-PoP, memory I/Os merge due to memory substrate layout design and BGA count limitation, which limit the communication channels to SoC. On the other hand, similar to wide I/O concept, the memory I/Os of 3D-MiM communicate with SoC independently, which enables a wider data bandwidth. It is clear that 3D-MiM fan-out unlocks the limitation of data bandwidth on FC-PoP by stacking more memory chips.

B. Electrical Performance

Improved power integrity, signal integrity, and latency are other critical performances to be considered in 5G/AI-driven mobile applications. Insertion loss, interconnect delay, and power integrity are essential electrical features to ensure the interconnect integrity in high speed, high frequency data transmission between SoC-memory chips. Fig. 12 compares the aforementioned performances between 3D-MiM fan-out and FC-PoP. On insertion loss, 3D-MiM fan-out offers near 3X lower in electrical loss than FC-PoP by shorter wiring length and less discontinuous interconnect from memory to SoC. On interconnect delay, 3D-MiM fan-out outperforms FC-PoP by 5X lower in latency as a result of shorter wiring length and lower wiring parasitics. On power integrity, 3D-MiM fan-out surpasses FC-PoP by 10X lower in PDN impedance due to much lower PDN inductance of wiring from memory to SoC.

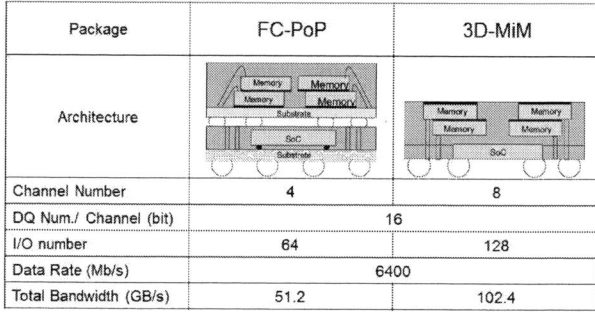

Package	FC-PoP	3D-MiM
Architecture		
Channel Number	4	8
DQ Num./ Channel (bit)	16	
I/O number	64	128
Data Rate (Mb/s)	6400	
Total Bandwidth (GB/s)	51.2	102.4

Figure 11. Maximum memory data bandwidth achieved by FC-PoP and 3D-MiM fan-out

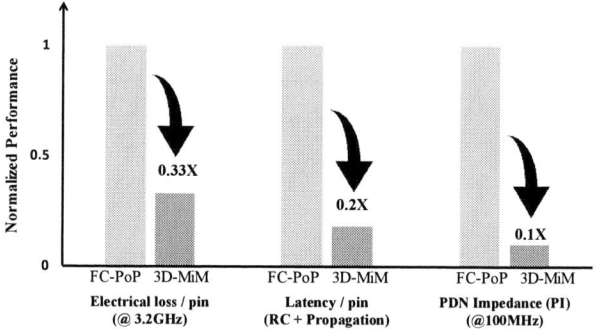

Figure 12. Comparisons of insertion loss, interconnect delay, and PDN impedance between 3D-MiM fan-out and FC-PoP

C. Thermal Performance

Thermal management becomes increasingly important in heterogeneous chips integration co-packaged in a small footprint. Thermal management becomes even challenging for SoC-memory in 3D chip integration due to a high thermal power density. Inadequate heat dissipation may cause chip over-heat and fail to function normally. Here, a computational fluid dynamics (CFD) model was used for two thermal analysis cases. One case is for mobile and the other is for HPC. On mobile case, the transient behavior comparison between 3D-MiM fan-out and FC-PoP was analyzed. In the

978-1-7281-1500-9/19 $31.00 © 2019 IEEE

analysis, underfill is applied between memory package and SoC package. The chipsets include SoC of 6W and memory chips of total 3.6W, co-packaged in a 15x15 mm^2 footprint. Fig. 13 shows the transient thermal response of SoC and memory for 3D-MiM fan-out and FC-PoP, respectively. To facilitate the understanding of transient behavior of SoC and memory as time elapsed, all temperature values have been normalized by the SoC temperature @ 10 sec. It is worthwhile knowing that the time scale in Fig. 11 is not linear. According simulation, the system reaches 97% level in temperature rise to the steady state for both packages after 600 seconds. From the simulation, the SoC of 3D-MiM fan-out has a much better thermal performance in transient characteristics than that of FC-PoP. A normalized temperature difference about 0.5 between 3D-MiM fan-out and FC-PoP was observed throughout 600 seconds. This thermal advantage allows 3D-MiM fan-out to adopt a SoC of higher power for power computing. Furthermore, thermal advantage improves SoC power consumption with a lower leakage current. It is important to observe that 3D-MiM helps to boost the SoC performance to higher frequency.

in Table I. The chipset includes 8-SoCs of total 320W and 32 memory chips of total 7.5W, co-packaged in a 3D-MiM fan-out with a 43x28 mm^2 footprint.

Table I. MODELING CONDITIONS FOR 1U SERVER SYSTEM

Modeling Conditions	
Rack size	434 mm x 495 mm x 44.45 mm
Fan area & flow rate	200 mm x 40 mm x 56 mm / 150CFM
Heat sink size	180 mm x 110 mm x 36 mm
Vapor chamber size	180 mm x 110 mm x 3 mm
SoC total power	320 W
Memory total power	7.5 W

As shown in Fig. 15 on temperature contours, the maximum temperatures are lower than 105C for SoC and 85C for memory, respectively. Simulation results provide the confidence of applying 3D-MiM fan-out in HPC application.

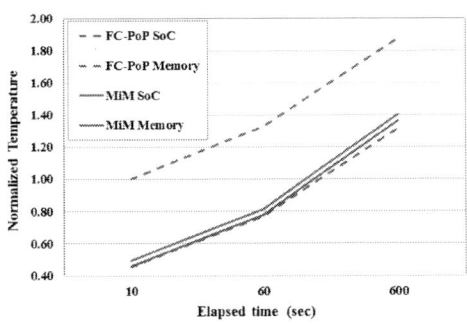

Figure 13. Transient responses of SoC and memory for 3D-MiM fan-out and FC-PoP, respectively, after 600 seconds.

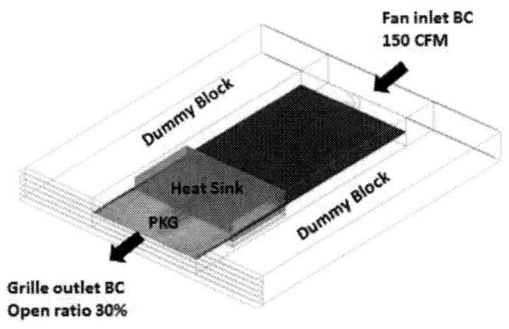

Figure 14. Schematic view of 1U server system for thermal modeling

On HPC application, the thermal requirement is very different from that of mobile application. Much high power of advanced node SoC imposes severe thermal challenging on SoC itself as well as on memory. To illustrate this, a CFD model was conducted to mimic the thermal management of 3D-MiM fan-out in a 1U server system as shown in Fig. 14. The key modeling conditions of 1U sever system is presented

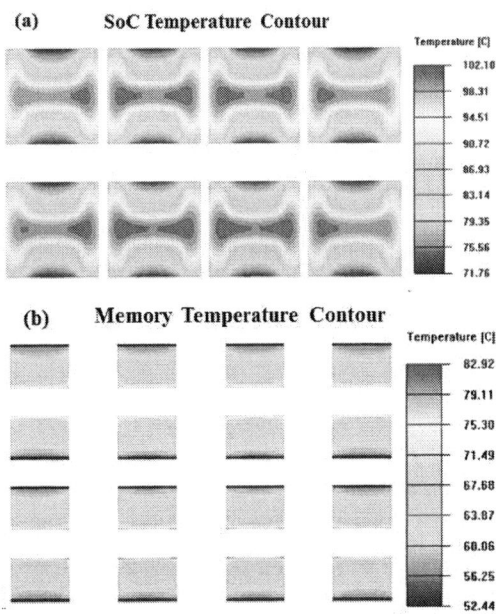

Figure 15. Temperature contours of (a) SoC, and (b) memory in 1U server system

VI. CONCLUSION

Near memory processing is attractive for high bandwidth, low latency, and power saving. We present new 3D-MiM, a next generation integrated fan-out technology, proposed to be an alternative heterogeneous integration solution to FC-PoP for mobile and 3DIC stacking for HPC applications, to realize in-package near memory computing. On mobile application, in comparison to the FC-PoP, the 3D-MiM fan-out offers thinner package profile (~0.5mm z-height), higher data bandwidth (2X~4X), with lower latency (0.2X) and thermal resistance to meet demands on future 5G/AI-driven edge computing. On HPC, in comparison to 3DIC HBM, the 3D-MiM fan-out offers a lower cost alternative with new memory to memory, and SoC to memory integration architecture with promising electrical and thermal performances. On manufacturing consideration, 3D-MiM fan-out technology leverages the well-established infrastructure of WLSI in capacity and materials, tools, processes, design rules for yield and competitive cost. 3D-MiM fan-out is an important part of the WLSI technology family in future 5G/AI-driven applications for years to come.

ACKNOWLEDGEMENT

Authors would like to acknowledge TSMC product engineering, quality and reliability, manufacturing, and R&D teams for their contributions to this paper

REFERENCES

1. Doug C.H. Yu, "Advanced System Integration Technology Trend", SEMICON Taiwan SiP Global Summit, Taipei, Taiwan, 2018/9/6
2. Doug C.H. Yu, "WLSI and Wafer Foundry Growth with Moore's Law and More-than-Moore, and Vice Versa", IWLPC Keynote speech, San Jose, CA, 2018/10/23
3. C. Tseng, C. Liu, C. Wu and D. Yu, "InFO (Wafer Level Integrated Fan-Out) Technology," in 2016 IEEE 66th Electronic Components and Technology Conference (ECTC), Las Vegas, NV, 2016.
4. H. Pu, H. J. Kuo, C. S. Liu and D. C. H. Yu, "A Novel Submicron Polymer Re-Distribution Layer Technology for Advanced InFO Packaging," in 2018 IEEE 68th Electronic Components and Technology Conference (ECTC), San Diego, CA, 2018.
5. W. S. Liao., et al. "A high-performance low-cost chip-on-Wafer package with sub-μm pitch Cu RDL," in 2014 Symposium on VLSI Technology (VLSI-Technology): Digest of Technical Papers, Honolulu, HI, 2014.
6. S. Y. Hou., et al. "Wafer-Level Integration of an Advanced Logic-Memory System Through the Second-Generation CoWoS Technology," IEEE Transactions on Electron Devices, vol. 64, no. 10, pp. 4071-4077, Oct 2017.
7. Hongshin Jun, "HBM (High Bandwidth Memory) for 2.5D", SEMICON Taiwan SiP Global Summit, Taipei, Taiwan, 2015/9/2-4
8. Gagandeep Singh et. al. "A Review of Near-Memory Computing Architectures: Opportunities and Challenges", Proceedings - 21st Euromicro Conference on Digital System Design, DSD 2018
9. Salessawi Ferede Yitbarek et. al. "Exploring Specialized Near-Memory Processing for Data Intensive Operations", 2016 Design, Automation & Test in Europe Conference & Exhibition

Construction of FO-MCM with C4 bumps built first using Chip last assembly technology

Chih-Hsun Hsu, Wen-Yang Li, Chi-Jen Chen, Yih-Jenn Jiang, Jui-Feng Tai, Chang-Fu Lin, C. Key Chung

Corporate R & D
Siliconware Precision Industries Co., Ltd
Taichung, 42881, Taiwan
AlfredXu@spil.com.tw

Abstract— Chip last assembly technology is complex and higher cost for fan-out wafer level package (FOWLP). But, this technology is fit well for very high density interconnection packages. This article presents chip last assembly technology using C4 bump-first for fan out multi-chip module (FO-MCM) package. The objective is to reduce cycle-time. A chip module with 28 x 30 mm was fabricated using 2 daisy-chain Si dies that bonded onto 2/2 μm line/space redistribution layers (RDLs). This module was then assembled on high density substrate with size of 70 x 70 mm. This FOMCM package is constructed using C4 first process.

C4 bumps were built on same side of the carrier after RDL was fabricated. The assemblies were protected and bonded on the carrier using temporary bond glue. The 1st carrier was then de-bonded. High I/O Si dies were attached onto the opposite side of the carrier followed by molding. The difference between C4 first and C4 last is the Si dies that were attached and molded with the carrier first then fabricated the C4 bumps.

C4 first process has the challenge is micro-pads pattern shift between Si dies. By increasing the RDL density, one could reduce the irregular of micro-pads pattern shift. Additionally, by reducing the thermal budget and using higher T_g of the temporary bond glue, the pattern shift was improved to less than 5 μm. Additionally, the wafer warpage of C4 first was found consistently warped at the same side, thus the process was easier to control as compared to C4 last.

The assembled FOMCM packages were then stressed for reliability tests. It passed 1000 hours of high temperature storage life test; MSL4 preconditioning with 1000 thermal cycles under B-conditions (-55~125 °C) and 192 hours unbiased high accelerated stress. Details of the results will be presented and discussed.

Keywords- FO-MCM; Chip last; Pattern shift; warpage

I. INTRODUCTION

In the modern age where information explodes every minute, high data rate transfer is ever demanded in data center market to satisfy people needs. This drives the data center for higher functionality. Larger die size and higher power consumption are observed. In order to handle huge amounts of data transfer yet lower energy consumption, one of the latest advanced electronic packages Fan-Out Wafer Level Package (FO-WLP) has been developed. FO-WLP is a substrate less package with thinner thickness, better thermal dissipation, and lower power consumption. This technology extends the terminal layout area outside of the die, which applicable for the high-density and high-performance devices, such as high performing computers, and networking systems [1].

The production flows on FO-WLP are typically characterized by two basic integration categories, one is die-first and the other is die-last (or called re-distribution layers, RDLs, first) [2]. In the die-first process, the known good die (KGD) is attached to a temporary carrier followed by mold embedded and de-bonding of the molded wafer from the carrier. Then, the RDLs are applied on the reconfigured molded wafer. In the die-first process, the molding often induces the wafer warpage and the die shift. This process, therefore, is not suitable for making an interconnection between the fine pad of the die and the fine line with fine pitch in the RDLs. In the die-last process, on the contrary, the RDLs are fabricated on a carrier using well-developed bonding and de-bonding technology [3,4,5] prior to the die bonding and the molding. In this scheme, a fine L/S pattern of RDLs can be fabricated on a flat surface and the dies with fine pitch bump can be interconnected to it. Die sizes and quantities may not be the limitation by using this scheme. Thus, die-last FO-WLP is more suitable for larger die size, multi-chip integration, and high-density RDL fabrication with finer L/S than die-first FO-WLP. This die-last FO-WLP is also more appropriate to apply in an OSAT industry with conventional flip-chip assembly flow for high-end servers and computers.

In the known die-last FO-WLP assembly technology, RDLs is processed first before the dies are individual attached and over-molded. Then, the molded wafer returns to the bumping process for C4 bumps finished (solder ball planting or copper pillar fabrication).This study named it as C4 last flow. This flow requires additional materials, process, equipment, and manufacturing floor space and incurs very high cost with long production time. This study, thus, evaluated the feasibility of the new process flow which was calling C4 first flow to reduce the production time and the cost based on the die-last FO-WLP assembly technology.

II. PLATFORM AND PROCESS FLOW

A. FO-MCM test vehicle information

In this study, the test vehicle information is listed in the table 1, including the configurations and dimensions of the FO-MCM package which can be categorized as FCBGA package type. As showed in table 1, the FO-MCM package with a dimension of 70x70mm was constructed by a FO chip of 30x28 mm using C4 bumps interconnection. This FO chip was built by two identical top dies, which were side-by-side bonded on fine-line micro-pad interconnecting with fine line RDL first fabricated on a carrier. The top die of 26x14mm is using a daisy-chain design (non-low dielectric constant). Besides, 4200 BGA balls mounting on the substrate bottom surface were using in the FO-MCM package. This FO-MCM package might be attached a metal heat sink to reduce package's coplanarity. Figure 1 shows the package top view and bottom surface view mounted 4200 BGA balls with 1 mm minimum pitch on the substrate bottom surface. The cross-section view of SEM (Scanning Electron Microscope) photos, was indicated in fig 2, show advanced RDL fabrication technology using the Line/Space of 2/2 um and the micro pad pitch of minimum 40 um in this package.

TABLE I. THE CONFIGURATION AND DIMENSION OF THE TOP DIES, BARE SUBSTRATE AND HEAT SINK USED IN FO-MCM TEST VEHICLE.

Top view	
	DC 01 / DC 02
X-section View	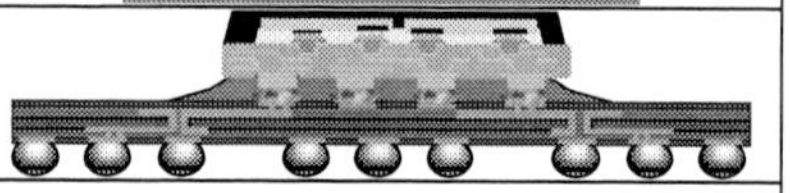
Package size	70 x 70 mm
BGA count	~4200
FO Size	30x28 mm
Die size	26x14 mm
3L PI/RDL thickness	~23um
Wafer Tech	Daisy Chain

Figure 2. Top view and bottom view of FO-MCM package

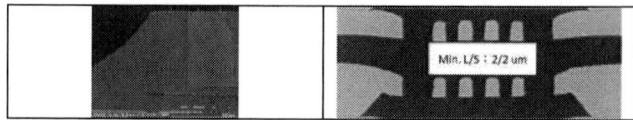

Figure 3. Cross-section view of FO-MCM package and fine-line RDL.

B. C4 first/C4 last flow based on Chip last technology

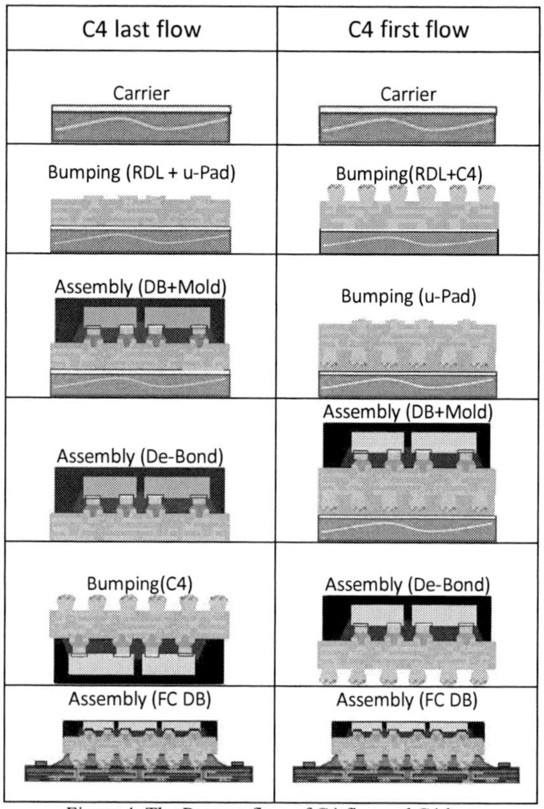

Figure 4. The Process flow of C4 first and C4 last

The two construction flows of FO-MCM package are shown in figure 4. As shown in figure 4, the 3 layers of RDLs with micro-pad finished were normally first built on the carrier in the bumping process. Then, the two dices were bonded on the micro-pad interconnecting with the fine line RDL on the carrier and following with the underfill dispensing and molding in the assembly process. The carrier was removed before the molding wafer was entry the bumping process again for the finish of C4 copper pillar bumps. Finally, the wafer was lapped and saw for the requirement size for the flip chip die bonded process. This manufacture procedure was called C4 last flow in this study.

The other process flow was called C4 first flow, as shown in figure 4. The different was the 3 layers of RDLs with C4 copper pillar bumps finished were built on the 1st carrier in the bumping process. To protect the C4 copper pillar bumps, the 2nd carrier with the temporary glue was bonded on the 1st carrier. The 1st carrier, subsequently, was removed and the 2nd carrier was moved on to construct the micro-pad to complete the needed structure in bumping process. After that, the two dies were bonded on the micro-pad and next with the underfill dispensing and molding on the 2nd carrier in the assembly process. At last, the 2nd carrier wafer was removed, and then lapped and saw for the required size for the flip chip die bonded process.

III. RESULTS AND DISCUSSION

Since the C4 copper pillar bumps were first finished accompanied with 3 layers RDLs on the 1st carrier in the C4 first flow. This study evaluated the temporary bonding material to protect the protrusion parts of the C4 copper pillar bumps and for the following building of u-Pads portion. According to the known technology, however, the glass transition (Tg) temperature of this bonding material was lower than the process thermal temperature. This low Tg of bonding material induce the pattern shift in high temperature process. Thus, this study investigated the 3 type of pattern shift in the process, including FO chip to FO chip shift, u-Pads shift, and C4 bumps shift.

A. FO chip to FO chip shift

TABLE II. PROPERTIES OF TEMPERARY BONDING MATERIALY, BUMPING CONDITION AND RESULT OF THIS STUDY.

	Material properties			bumping thermal reduction	Shift improved on Chip to Chip
	Tg (℃)	CTE (α1, <Tg) [ppm]	Young's modulus [GPa]		
Leg 1	90	90	0.15	N	--
Leg 2	140	90	0.53	N	69%
Leg 3	140	90	0.53	Y	95%
Note: Die to Die Shift improvement (%) = (Leg 1 - Leg 2) / Leg 1					

The properties of temporary bonding material were illustrated in table II. The pattern shift between one FO chip to another FO chip was over 70 um in the leg 1. This large shift seriously affects the alignment of stepper of bumping machine. When the Tg of temporary bonding material was increased from 90℃ to 140℃, as indicated in table II, the pattern shift between one FO chip to another FO chip was reduced 69%. The Tg temperature of the bonding material, as indicated in table II, is really lower than that of the u-Pad build stage in bumping process. It's known that the Tg of a material characterizes the range of temperatures over which this glass transition occurs. At low temperature below Tg, the bonding material occurs in a glassy state (it appears as a rigid and brittle solid). With increasing temperature above Tg, the bonding material is in the glass transition range and the molecules of the bonding material are showing more and more mobility and the polymer occurs in a soft-elastic, rubber-like state. That's, the low Tg of bonding material induces the pattern shift between one FO chip to another FO chip in the C4 first flow. The leg 3, as shown in table II, demonstrates the shift between one FO chip to another FO chip reduced up to 95% through the reduction on the thermal budget of the u-Pads stage in the bumping process. This result showed the feasibility condition was found in this study.

B. u-Pads shift

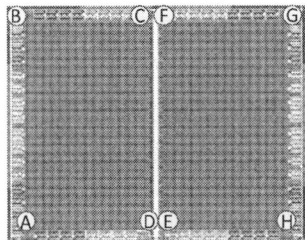

Figure 5. The scheme of 8 position inspected by X-ray inside FO chip

The pattern shift behavior of u-Pads was inspected by X-ray at 8 positions inside the FO chip after Die-Bonding. These 8 positions were paired corners of two DC silicon die, as indicated in figure 5. The results of u-Pads shift inside the FO chip was indicated in table III. Table III illustrated that the maximum and average shift of leg I were, respectively, 8.5 um and 5.2um with the RDL density designed between 40~48%. The maximum and average u-Pads shift were, respectively, increased to 15.8 um and 10.2 um when the RDL density was elevated to 49~57%. These results mean that high RDL density led to large u-Pads shift of the FO chip.

TABLE III. THE CONDITIONS AND u-PADS (BUMP/PAD) MISALIGNMENT OF THE FO CHIP

	RDL density	X-ray view	Bump Pad	u-Pads shift (um) Max	u-Pads shift (um) Ave
Leg I	40~48%			8.5	5.2
Leg II	49~57%			15.8	10.2

Figure 6 shows the results of u-Pads shift in leg I and leg II inside the FO Chip, the cross-section position of the FO chip, and the sampling FO chip position of the wafer. The wafer sampling positions included wafer edge and center. The wafer edge cross-section results, as indicated in figure 6.a, shows 2.0, 3.0, 2.6, 4.9 um shift, respectively in position A, B, G, and H in the Leg 1. The same u-Pads shift behavior was found in the wafer center by cross-section results, as was shown in figure 6.a. It's known that the CTE of silicon dice is extremely lower than that of the PI/RDL structure. This suggests that the u-Pads shift came from the expansion of PI/RDL structure while post reflow. It's also means that this PI/RDL structure which built on a carrier with temporary material need to further solve the u-Pads shifts of the FO chip in the C4 first flow.

978-1-7281-1500-9/19 $31.00 © 2019 IEEE

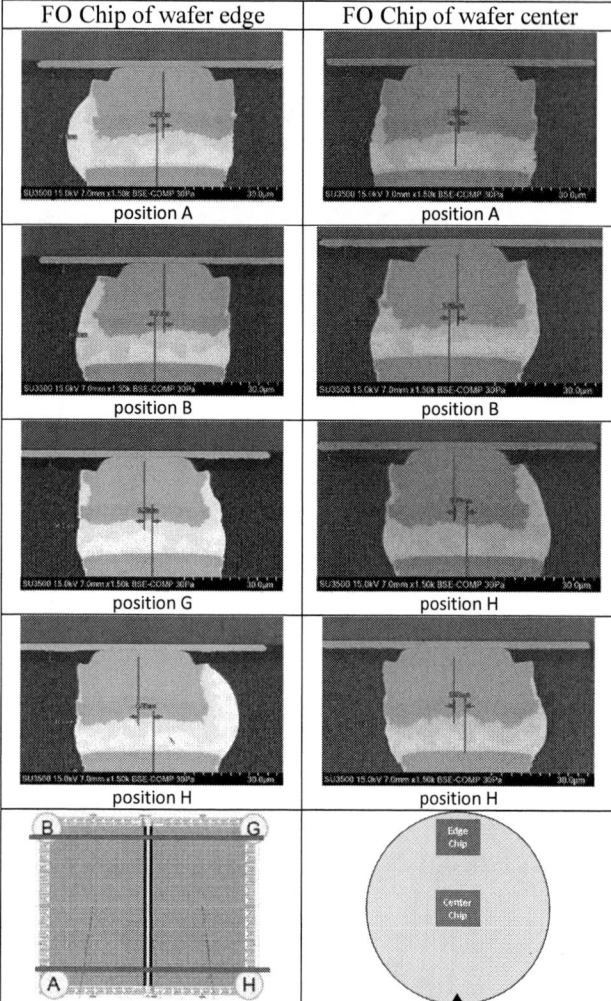

FO Chip of wafer edge	FO Chip of wafer center
position A	position A
position B	position B
position G	position H
position H	position H

Figure 6.a Cross-section view on leg 1 u-Pads shift of the FO chip

While the RDLs density was elevated to 49~57% in the leg 2, figure 6.b shows the wafer edge cross-section results of the u-Pads shift were 16.6, 16.2, 14.7, and 13.5 um, respectively, in position A, B, G, and H. The similar u-Pads shifts were also observed in the FO chip of the wafer center. This result confirmed the elevated RDL density increased the u-Pads shifts in this PI/RDL structure which built on a carrier with temporary bond material. Since the u-Pads shifts were found in both low and high RDLs density, the compensation method need to be used to fulfill the u-Pads shifts of the FO chip which fabricated on a carrier with temporary material

The cross-section on position C, D, E, and F of the FO chip, as shown in figure 5, and totally 5 FO chip in a wafer was collected to add more databases for statistical analysis. Figure 7.a shows the leg I u-Pads shift trend of the X-axis in five FO chip of the wafer. The X-axis shift of u-Pads was shown in this figure to demonstrate a random distribution inside the FO chip although the u-Pads shift was small of the

leg 1. The random distribution of u-Pads shift was found in five FO chip including both wafer edge and center. The same u-Pads shift behavior was also found in Y-axis of the five FO chips, as was shown in figure 7.b. That is, the compensation rule was not suitable for the leg I.

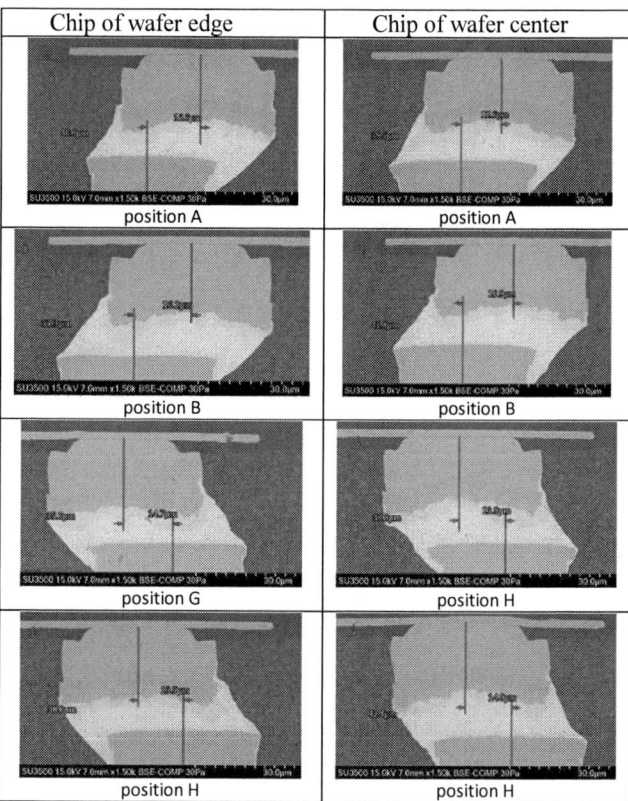

Chip of wafer edge	Chip of wafer center
position A	position A
position B	position B
position G	position H
position H	position H

Figure 6.b Cross-section view on leg 2 u-Pads shift of the FO chip

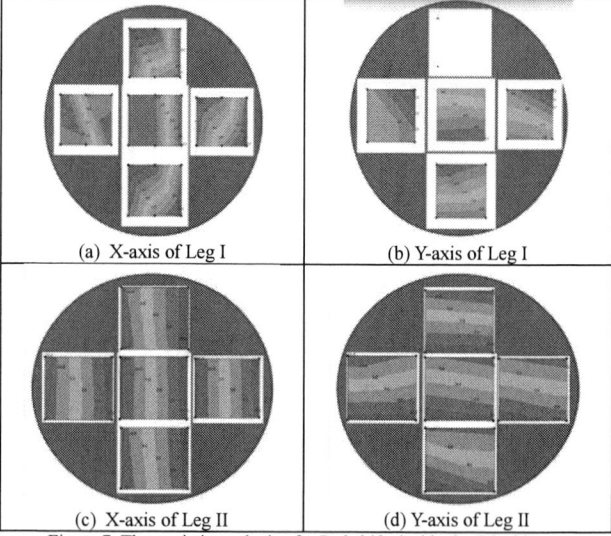

(a) X-axis of Leg I	(b) Y-axis of Leg I
(c) X-axis of Leg II	(d) Y-axis of Leg II

Figure 7. The statistic analysis of u-Pad shifts inside the FO chip

978-1-7281-1500-9/19 $31.00 © 2019 IEEE

Figure 7.c shows the Leg 2 u-Pads shift distribution of the X-axis in the five FO chip of the wafer. If the dividing line was the FO chip center, the symmetry of u-Pads shift was found between the right and left of the FO chip in this figure. The symmetry of u-Pads shift of the FO chip in Y-axis was also observed in figure 7.d. The high RDL density brought about the larger u-Pads shift of the FO chip, as indicated from comparison of Fig. 7.a with Fig. 7.c and Fig. 7.b with Fig. 7.d, however, this high RDL density created a regular u-Pads shift, indicating the compensation rule can be implemented in this FO chip by elevated RDLs density.

C. Shift of C4 copper pillars bumps

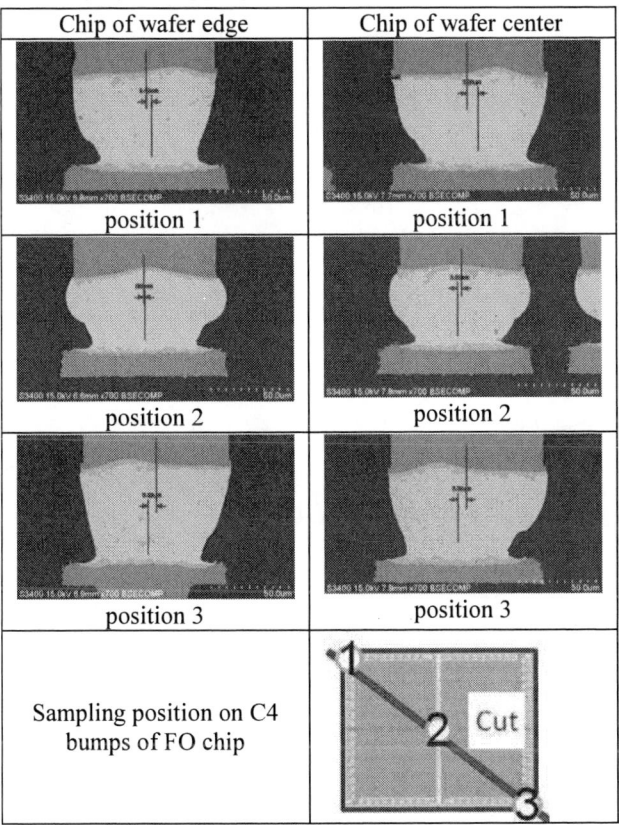

Chip of wafer edge	Chip of wafer center
position 1	position 1
position 2	position 2
position 3	position 3
Sampling position on C4 bumps of FO chip	Cut

Figure 8. Cross-section view of C4 bump area.

Figure 8 shows the cross-section position and results of C4 bumps area after the FO chip, which was built by the C4 first flow, was mounted on the substrate. The C4 bumps pattern shifts in the position 1, 2 and 3, as indicated in figure 8, were all in the criteria, including both wafer center and edge position. The C4 bumps shift seems not influence by the temporary bonding material process. This is because the C4 bumps were first construction on the 1st carrier. Besides, the same BOM was used in the C4 first flow to construct FO chip. The statistical analysis comparison on the C4 bumps shift between C4 first and C4 last flow was indicated in figure 9. This statistical result concluded that no difference on the C4 bumps shift between C4 first and C4 last flow.

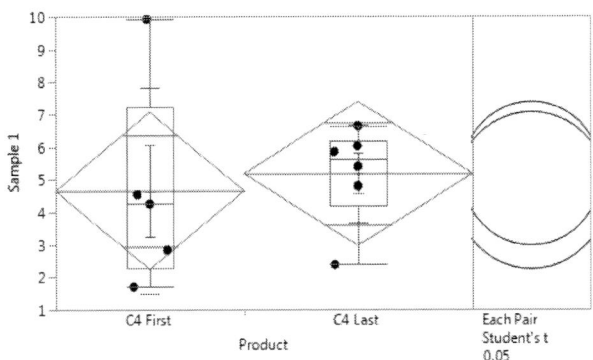

Figure 9. Comparsion on C4 bump shift between C4 first and C4 last flow.

D. Wafer form warpage trend

The FO-MCM platform of the C4 last flow has been qualified. This study, thus, used the same BOM material to collect the wafer wparpage of C4 first flow. The wafer warpage trend was showed in figure 10. Figure 10 shows the warpage of C4 last flow was elevated to 0.8mm at u-Pads finished stage in the bumping process. Then, the wafer warpage was dropped to -0.6mm after die-bonding. While post molding process, the wafer warpage was increased to 1.1 mm due to the shrinkage of the mold compound. This wafer warpage was dropped again to -1.8 mm while the carrier was removed (de-bond). This obviously drop of warpage was attributed to the wafer stress domination has changed from carrier to mold compound. Thus, considering the stress balance between carrier and mold compound is one task force in the known C4 last flow. The wafer warpage was decreased to -3.0mm after C4 plating in the following bumping process. This low wafer warpage was almost close equipment handling limit. It's known that the C4 stage was full of high temperature curing procedure in bumping process, including PI cure, wafer bake, other bumping thermal and etc. Thus, the selection of EMC character is the other considered task force for pass the high thermal exam of C4 last flow in bumping process [6,7].

In the C4 first flow, the warpage was obviously increased to 2.2mm at u-Pads finished stage in the bumping process, as was shown in figure 10. This wafer warpager behavior in C4 first flow was significant different as compared with that in the C4 last flow. This was resulted from the carrier replace and structure variance. The wafer warpage reduced to 0.8 mm after Die-Bonding. The down trend warpage is similar as compared with that in the C4 last flow. The wafer warpage then increased to 1.3mm while post molding and subsequently, elevated to 1.8 mm after the 2nd carrier was de-bond. After de-boned, the mold wafer was no needed to through any high thermal procedure in the C4 first flow. In other word, the wafer with the carrier passed through the mostly high thermal procedure including both bumping and assembly process. Since the carrier CTE was extra low, it plays an important rule and supports on wafer warpage domination in C4 first flow.

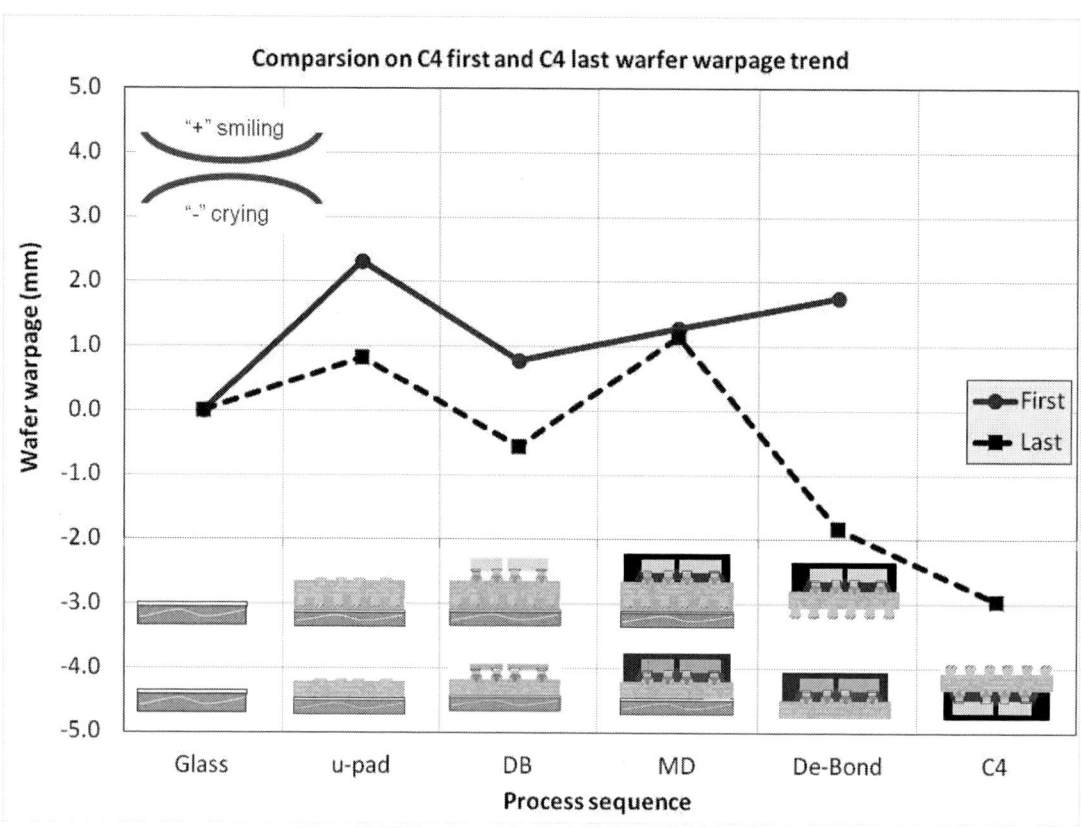

Figure 10. Wafer warpage trend of C4 first and C4 last flow.

E. Reliability Verification

TABLE IV. RELIABILITY VERIFICATION ON C4 FIRST BUILT FO-MCM PACKAGE

Reliability Test Items	Read point	Sample	Result
Moisture Sensitivity Level (MSL 4)	Precon	50/50	Pass
Temperature cycling test - 55C ~ +125C(TC-B)	1000 cycles	25/25	Pass
unbiased High Accelerated Stress test, 130C/85%RH, (uHAST)	192 hours	25/25	Pass
High Temperature Storage Life test, 150C (HTSL)	1000 hours	25/25	Pass

This study had presented the Fan-Out Multi-Chip Module (FO-MCM) structure of FCBGA package that constructed by C4 first flow as a case study for package level full reliability test items' proceeding. The reliability tests were followed JEDEC standard reliability [9] including moisture soaking level 4 (MSL4) pre-conditioning, and were subsequently used temperature cycling test (TCT) with -55℃ to 125℃/1000 cycle and unbiased high accelerated stress test (uHAST) with 130℃/192 hours. The high temperature storage life test (HTSL) with 150℃/1000 hours was also evaluated. The reliability passed result as shown in

Table IV and there was no abnormality related to assembly issues.

IV. CONCLUSION

C4 bumps first was successfully developed using FO-MCM package based on chip last assembly technology. The advantages of C4 fist process reduces thermal budget and cycle times. Both C4 bumps and μ-Pads were fabricated before wafer molding. The technology has better warpage control for both bumping and assembly processes. The μ-pad pattern shift between Si chips was solved by reducing the thermal budget and using higher T_g of the temporary bond glue. By comparing the C4 bumps shift between C4 first and C4 last process, there was statistically no difference. This C4 first that built on FO-MCM package was passed the reliability requirements. In conclusion, the evaluations of C4 flow fabricated FO-MCM package was verified in this study.

REFERENCES

[1] David Butler, Chris Jones, Steve Burgess, Tony Wilby and Paul Densley, "FAN-OUT WAFER PROCESSING IN THE HIGH DENSITY PACKAGING ERA", Proceedings of the International Wafer-Level Packaging Conference 2018.

[2] Hitoshi Onozeki, Hiromichi Aoki, Kohei Mizuno, Mitsuki Nakata, Naoya Suzuki, Tsuyoshi Ogawa, Toshihisa Nonaka., "STUDY OF FINE PITCH RDL FIRST FO-PLP/WLP", Proceedings of the International Wafer-Level Packaging Conference 2018.

[3] Hikaru MIZUNO, Hiroyuki ISHII, Hitoshi KATO, Takashi MORI, Hiroki ISHIKAWA, Yooichiroh MARUYAMA, Kenzo OHKITA,

978-1-7281-1500-9/19 $31.00 © 2019 IEEE

Koichi HASEGAWA., "UV laser releasable temporary bonding materials for FO-WLP"., 2018 International Conference on Electronics Packaging and iMAPS All Asia Conference (ICEP-IAAC).

[4] Thomas Uhrmann ; Matthias Pichler ; Julian Bravin ; Daniel Burgstaller ; Boris Povazay., "Laser Debonding Enabling Ultra-Thin Fan-Out WLP Devices"., 2018 7th Electronic System-Integration Technology Conference (ESTC).

[5] P. Andry ; R. Budd ; R. Polastre ; C. Tsang ; B. Dang ; J. Knickerbocker ; M. Glodde ., "Advanced wafer bonding and laser debonding"., 2014 IEEE 64th Electronic Components and Technology Conference (ECTC)

[6] Kouji Hamagucht, Hirokazu Noma, Hiroshi Takahashi, Naoya Suzuki, Toshihisa Nonaka., "Warpage study of FO-WLP build up by material properties and process"., 2016 6th Electronic System-Integration Technology Conference (ESTC)

[7] Yanhui Guo, Guohua Zhang, Jianfeng Wang., "Warpage Simulation and Optimization of Fan-Out Wafer level Package(FO-WLP) with TMV under Different Processes"., 2018 19th International Conference on Electronic Packaging Technology (ICEPT)

[8] Gary R. Trott; Aric Shorey., "Glass wafer mechanical properties: A comparison to silicon" 2011 6th International Microsystems, Packaging, Assembly and Circuits Technology Conference (IMPACT)

[9] JEDEC Solid State Technology Association, JESD22-A113D: Preconditioning of Nonhermetic Surface Mount Devices Prior to Reliability Testing.

Feasibility Study of Fan-Out Panel-Level Packaging for Heterogeneous Integrations

Cheng-Ta Ko[1], Henry Yang[1], John H Lau[2], Ming Li[2], Curry Lin[1], Chieh-Lin Chang[3], Jhih-Yuan Pan[3], Hsing-Hui Wu[3], Iris Xu[4], Tony Chen[4], Zhang Li[4], Kim Hwee Tan[4], Penny Lo[2], R. So[2], Y. H. Chen[1], Nelson Fan[2], Eric Kuah[2], Marc Lin[1], Y. M. Cheung[2], Eric Ng[2], Cao Xi[6], Rozalia Beica[3], Sze Pei Lim[5], N. C. Lee[5], Mian Tao[7], Jeffery Lo[7], and Ricky Lee[7]

[1]Unimicron Technology Corporation
[2]ASM Pacific Technology Ltd
[3]Dow Chemical Company
[4]Jiangyin Changdian Advanced Packaging Co., Ltd.
[5]Indium Corporation
[6]Huawei Technologies Co. Ltd.
[7]Hong Kong University of Science and Technology.
Ph: 886-920-109-826, Email: CT_Ko@unimicron.com

ABSTRACT

The design, materials, process, and fabrication of a heterogeneous integration of 4 chips by a FOPLP (fan-out panel-level packaging) with chip-first and dies face-down formation are investigated in this study. Emphasis is placed on the application of a new assembly process and materials for fabricating the RDLs (redistribution layers) of the FOPLP. The panel size is 508mm x 508mm. The epoxy molding compound (EMC) is a dry-film material and is molded by lamination method. The minimum metal line width and spacing is 10μm and they are fabricated by printed circuit board (PCB) method and equipment.

(1) Introduction

Semiconductor industry has identified five main growth engines (applications), namely (1) mobile, (2) high-performance computing (HPC), (3) automotive (especially for self-driving car), (4) internet of things (IoTs), and (5) big data (especially for cloud computing).

The system-technology drivers such as (1) 5G, (2) artificial intelligence (AI), and (3) machine learning (ML) are boosting the growths of these 5 semiconductor applications.

The packaging people are using various packaging methods such as wirebonding, flip chip, build-up substrate, package-on-package, wafer-level chip scale package, fan-out wafer/panel-level packaging, TSVs (through-silicon vias), 2.5D/3D IC integration, high bandwidth memory, system-in-package/heterogeneous integration, chiplets, and embedded multi-die interconnect bridge to house (package) the semiconductor devices for those 5 main applications. Recently, fan-out wafer/panel-level packaging [1-14] and heterogeneous integrations [15-19] obtain lots of tractions and are the focus of this study.

Figure 1 schematically shows a simple concept of heterogeneous integrations. It can be seen that heterogeneous integration uses packaging technology to integrate dissimilar chips, photonic devices, and/or components (side-by-side, stack, or both) with different materials and functions, and from different fabless design houses, foundries, wafer sizes, feature sizes and companies into a system or subsystem. How should these dissimilar chips talk to each other? The answer is: redistribution layers (RDLs) [20, 21]. How

should those RDLs be made? They can be made by fan-out wafer/panel level packaging method, which is the focus of this study.

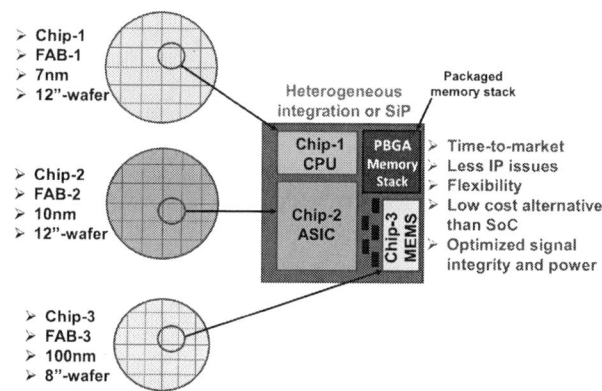

Figure 1 Schematic of a heterogeneous integration or SiP.

In general, for high-volume manufacturing (HVM), 70% of the RDLs of heterogeneous integrations should be on organic substrates and the metal line width and spacing are ≥10μm. (Most of these heterogeneous integrations are actually system-in-packages.) No more than 5% of the RDLs of heterogeneous integrations should be on organic substrates and the metal line width and spacing are <10μm.

In general, for HVM, 25% of the RDLs of heterogeneous integrations would be on other (non-organic) substrates such as the silicon substrates (either passive TSV-interposers or active TSV-interposers or both), silicon substrate (bridges), and ceramic substrate. The metal line width and spacing of the RDLs are usually very small and can go down to submicron values, which are out of the scope of this study.

In [1], we had demonstrated the chip-first and die face-down fan-out wafer-level packaging for heterogeneous integration. The metal line width and spacing of the RDLs were 20μm and 25μm. In this study, the metal line width and spacing of the RDLs are 10μm and 25μm. The dimensions of the package are 10mm x 10mm, which consists of one 5mm x 5mm chip and three 3mm x 3mm chips. The spacing

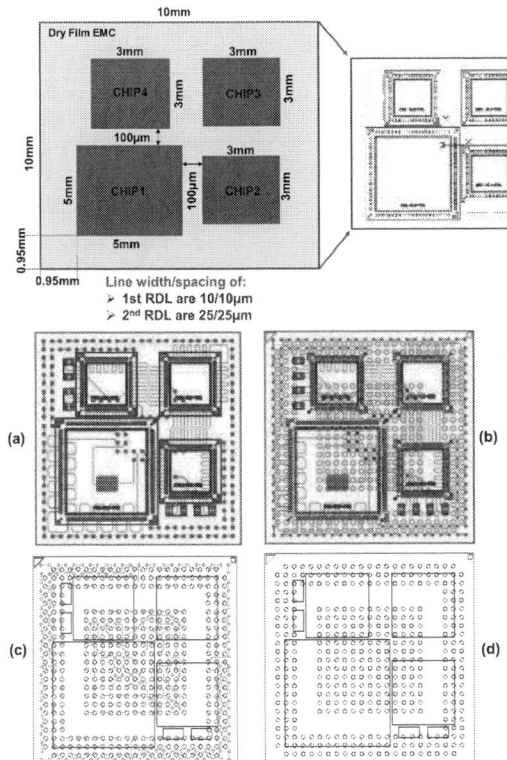

Figure 2 (Top) Test package (10mm x 10mm) layouts. (Bottom) (a) RDLs between test chips and RDL1. (b) RDLs between RDL1 and RDL2. (c) RDLs between RDL2 and PCB. (d) Footprint of the test package bottom.

between the large chip and the small chip is only 100μm. One practical application of this kind of package is for the application processor chipset, i.e., the large chip could be a processor and the small chips could be memories.

In order to have a very high-throughput and low-profile package and save the expensive EMC, a process called uni-substrate-integrated package (Uni-SIP) [1] is used to fabricate the RDLs. Unlike in [1], the panel dimensions in this study are 508mm x 508mm and a dry-film EMC is laminated on the reconstituted panel (instead of the liquid EMC with compression molding as in [1]).

The Uni-SIP process starts off by attaching the backside of the ECM-panel on both sides of a coreless panel substrate with an epoxy resin. (In this study, chips embedded in the dry-film EMC is called the ECM-panel.) The RDLs are formed on both surfaces of the ECM-panels. A new Ajinomoto build-up film (ABF) is used as the dielectric of the RDLs and is built up by semi additive process (SAP). De-smear is used to remove the smear due to laser drilling and rough the ABF surface. Electroless Cu is used to make the seed layer. Laser direct imaging (LDI) is used for opening the photoresist, and printed circuit board (PCB) Cu plating is used for making the metal of the RDLs. The solder mask is then applied on both sides of the ECM panel leaving pad openings for surface finishing. The ECM panels with build-up RDLs are finally separated from the coreless panel substrate mechanically.

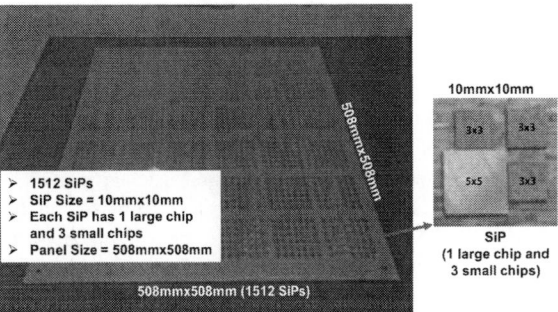

Figure 3 508mm x 508mm panel with 1512 SiPs.

It is followed by solder-ball mounting and dicing. This process adds a new dimension in high-throughput and thin fan-out packaging without using semiconductor equipment.

(2) Test Chip Layouts and Fabrications

The test chips, 5mm x 5mm x 150μm and 3mm x 3mm x 150μm are the same as those in [1].

(3) Test Heterogeneous Integration Package

The test package, Figure 2, looks the same as that in [1]. However, the major differences are: (1) a new dry-film EMC instead of a liquid EMC is used, (2) the EMC is fabricated by a lamination method instead of a compression method, (3) the metal line width and spacing is reduced from 20μm [1] to 10μm, (4) the panel size is increased from 340mm x 340mm to 508mm x 508mm, and (5) a new de-smear and electroless Cu plating process.

The spacing (gap) between the large chip and the small chip is 100μm. There are two RDLs of the test package. The RDLs between the chips and RDL1 is shown in Figure 2(a), between the RDL1 and RDL2 is shown in Figure 2(b), and between the RDL2 and the PCB is shown in Figure 2(c). Figure 2(d) shows the footprint of the test package. These packages are to be made from a 508mm x 508mm reconstituted panel as shown in Figure 3. There are 1,512 heterogeneous integration packages and each one with 4 chips.

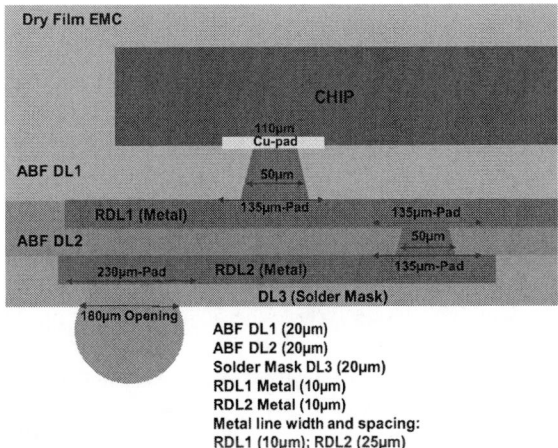

Figure 4 Schematic of the test package cross section.

Figure 4 schematically shows the cross-sectional view of the test package. It can be seen that there are two RDLs, and the thickness of the metal of RDL1 and RDL2 is 10μm. The metal line width and spacing of RDL1 are 10μm (which is different from 20μm of [1]), and those of RDL2 are 25μm. The dielectric layer thickness of DL1, DL2, and DL3 is 20μm. The via through the first dielectric layer (DL1), connecting the Cu contact pad of the test chips to the first RDL (RDL1) is 50μm in diameter. The pad diameter on the RDL1 is 135μm, which is connected to RDL2 through the via with a diameter of 50μm. Similarly, the pad diameter on the RDL2 is 135μm. Finally, 230μm solder-ball Cu pads are formed on RDL2. The opening of the solder mask (DL3) is 180μm. The solder ball size is 200μm, and the ball pitch is 0.4 mm.

(4) A New Uni-SIP Process

Figure 5 shows a new Uni-SIP process. The temporary carrier is made of an organic substrate with the coefficient of

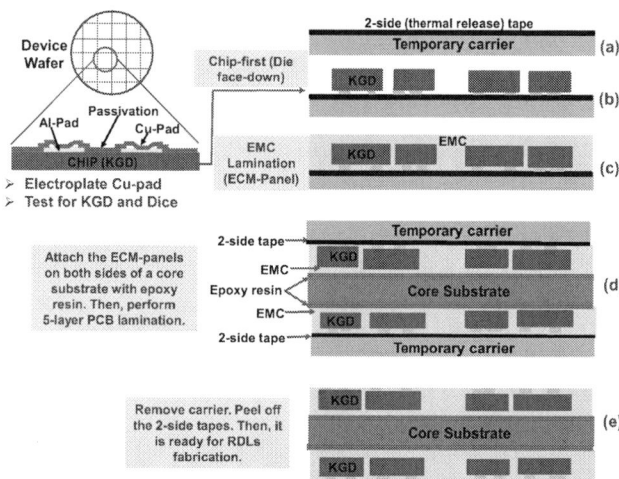

Figure 5 Key process steps of Uni-SIP.

thermal expansion (CTE) = $4 \times 10^{-6}/°C$. The thickness of the organic carrier is 1mm as shown in Figure 3. The advantages of 1mm-carrier are: (a) not easy to break when it is cleaned and reused, and (b) to increase the stiffness and resist to bending (warpage) of the reconstituted carrier during pick and place of the chips, the 2-stage vacuum lamination, and post mold cure (PMC) of the dry-film EMC. After a temporary carrier is chosen, then attach the 2-side thermal release tape (REVALPHA) provided by Nitto Denko on top of the carrier, Figure 5(a). It is followed by picking and placing the chips on top of the tape, Figure 5(b).

(a) Dry-Film Lamination of ECM-Panel

For large panel size such as the 508mm x 508mm, the liquid (previous) EMC is difficult to fill the full panel without flow marks and the EMC total thickness variation (TTV) between the corner and the center of the panel is large. The advantages of dry-film EMC are: (a) better flattening control, (b) more consistence TTV as full panel, (c) PCB process suitable, (d) higher throughput, and (e) less particle pollution. In this study, a dry-film EMC is used.

TABLE 1 Material Properties of the EMCs

EMC material	Previous	Current
	Liquid	Dry film
	EMC (R4507)	New EMC
Filler max cut (μm)	25	5
Filler contents (%)	85	80
Tg (DMA) (°C)	150	163
CTE 1 (ppm/K)	10	15
CTE 2 (ppm/K)	41	23
Young's modulus (GPa)	19	10

The material properties of the dry-film EMC are shown in Table 1. It can be seen that the filler content is 80% and the maximum size of the filler is 5μm. The Young's modulus of the EMC is 10GPa and the glass transition temperature (Tg) is 163°C. The dry-film EMC is laminated on top of the carrier and chips, Figure 5(c), by a 2-stage vacuum lamination machine as shown in Figure 6. The first stage is vacuum lamination (100°C and vacuum for 30s and press (0.68MPa)

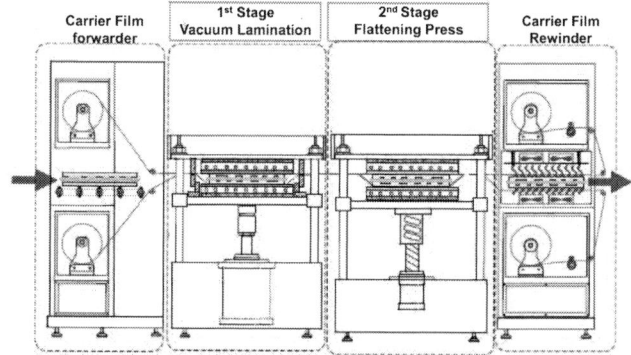

Figure 6 Nikko-material 2-stage vacuum lamination machine

for 30s), which is for the resin of the dry-film EMC to be confirmed. After vacuum lamination, it is followed by the second stage (flattening press: 100°C for 60s and press (0.54MPa)), which is for the surface of the dry-film EMC to be flattened. Then, cure the structure at 150°C for 1h + 180°C for 0.5h. The structure of Figure 8(c) is called ECM-panel.

(b) Lamination of Uni-SIP Structure

The next step is to attach an epoxy resin on both side of another organic core carrier (substrate). Then, attach the back-side of the ECM-panel to both sides of the core carrier as shown in Figure 5(d). It is followed by a 5-layer (two-side ECM-panels + core) PCB lamination with the same 2-stage vacuum lamination machine as shown in Figure 6 and the conditions are the same as the 2-layer ECM-panel. Then, remove the temporary carrier and peel of the tape, Figure 5(e). It is ready for making the RDLs on the pads on both sides of the chips.

978-1-7281-1500-9/19 $31.00 © 2019 IEEE

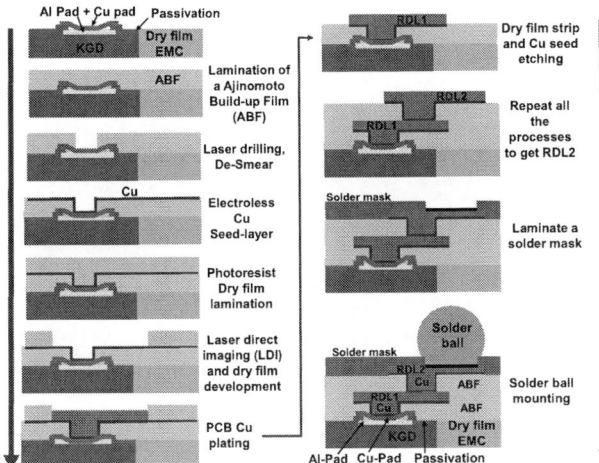

Figure 7 Key process steps in making the RDLs

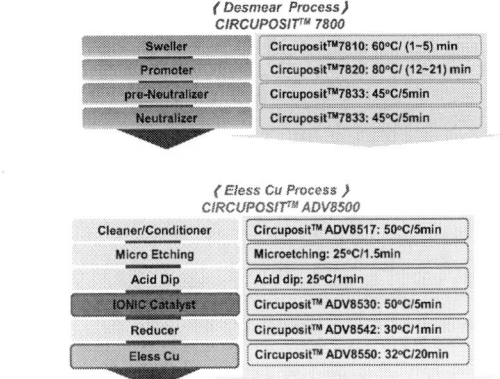

Figure 8 DOW's De-Smear process.

(c) Lamination of the New ABF

Figure 7 shows the process steps in fabricating the RDLs from the Cu pads on both sides of the Uni-SIP structure shown in Figure 5(e). First, laminate a new ABF (Table 2) on both sides of the chips. In order to control the panel warpage,

TABLE 2 Material Properties of ABF

Item	Old ABF	New ABF
Filler max cut (µm)	15	5
Filler contents (%)	82	80
CTE (30-150°C) (ppm/K)	7	15
Young's modulus (GPa)	7	10
Dielectric constant (Dk)	3.2	3.3

TABLE 3 DF and LDI combination comparison

Item	DF+LDI (Old)	DF+LDI (New)
DF thickness (µm)	25	12
Resolution (µm)	20	5 on glass

the selection of ABF and EMC materials is an important consideration. The CTE of ABF should be matched to the equivalent one of ECM-panel. Also, for fine metal line width and spacing ($\leq 10\mu m$) applications, the filler size should be small. In this study, the CTE ($15 \times 10^{-6}/°C$) of ABF ($5\mu m$) is significantly smaller. Furthermore, the previous DF (Dry Film) and LDI combination [1] only can get the limit resolution down to $20\mu m$. The new thinner DF ($12\mu m$) and more advanced LDI used in this study can get the limit resolution down to $5\mu m$ on glass.

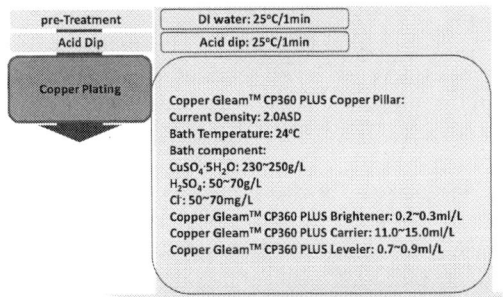

Figure 9 DOW's electroless Cu plating process.

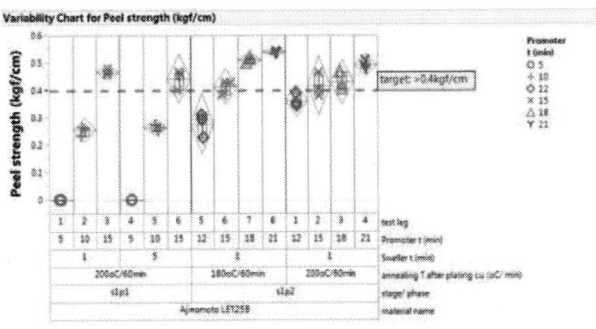

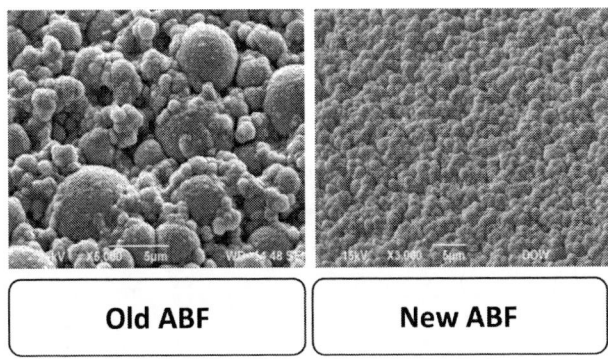

Figure 10 Surface roughness after de-smear process.

978-1-7281-1500-9/19 $31.00 © 2019 IEEE

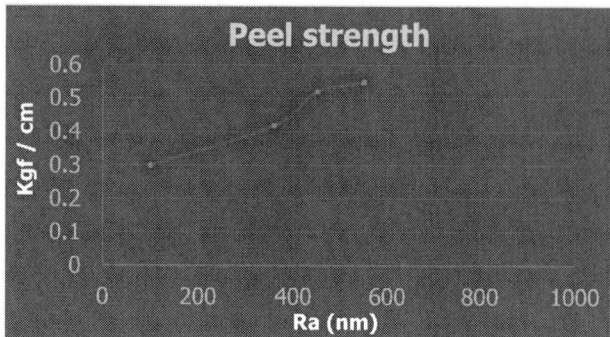

Figure 11 Design of experiments. Pull strength.

(d) De-Smear the ABF and Electroless-Cu Plating the Seed-Layer

After the lamination of the ABF on both sides of the chips, then laser drill the ABF and stop at the Cu pad (this is the reason for making the Cu pad on the device wafer) as shown in Figure 7. Before metallization, a post laser de-smear of the ABF dielectric is generally required for surface roughing to achieve sufficient adhesion of electroless-Cu to ABF. In this study, DOW's SAP solution – CIRCOPIST 7800 de-smear process (Figure 8) and CIRCUPOSIT ADV 8500 E-less Cu process (Figure 9) with some modified parameters are used for this study. As shown in Figure 10, the surface roughness of the new ABF is smoother than that of the old ABF, which will be beneficial to the fine metal line width/spacing formation on top of the ABF.

Peel strength is the key on the build-up metallization performance. It can be related to surface roughness (Ra) after de-smear process. There are three modified prominent factors: sweller dwell time, promoter dwell time and annealing time. After a detailed design of experiment performed by DOW, the peel strength results are shown in Figure 11. It shows that $180^{\circ}C$ annealing time with higher than 12-minute promoter dwell time leads to better performance – the peel strength is higher than $0.4kgf/cm^2$. Consequently, with the better performance on peel strength, the new condition is selected for subsequent development in this study. The average thickness of electroless-Cu film is $0.65\mu m$.

(e) LDI and PCB Cu-Plating

After electroless-Cu on both sides of the chips, laminate a new photoresist dry film with a thickness of $12\mu m$ on both sides of the Uni-SIP structure. It is followed by LDI photolithographic process and dry-film development. The LDI is with a positioning accuracy of +/-5μm. The wavelength of the UV laser is 365nm and the exposure energy is $200mJ/cm^2$. Cu plating is carried out using an electrolyte containing 240g/L of $CuSO_4$ and 60g/L of H_2SO_4. The Cu film is plated at a constant current density of 2.0ASD. After metallization, these fine traces are formed on the ABF dielectric according to the pre-defined photoresist pattern. Finally, the photoresist is stripped off and Cu seeding layer is etched to form the RDL1. The fabricated metal line width and spacing (on average) is $10\mu m$ according to the top-view measurements of the cross section shown in Figure 12. Repeat all the processes to get RDL2.

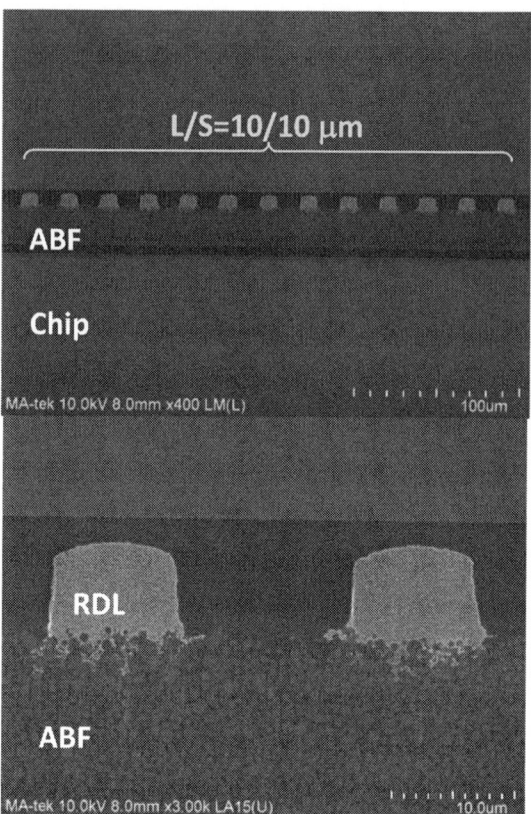

Figure 12 SEM images of the RDL1 (10μm metal line width and spacing).

Figure 13 shows the images of the heterogeneous integration package of 4 chips. It can be seen that the gap between the large chip (5mm x 5mm) and the small chip (3mm x 3mm) is 100μm and there is no void in the dry-film EMC.

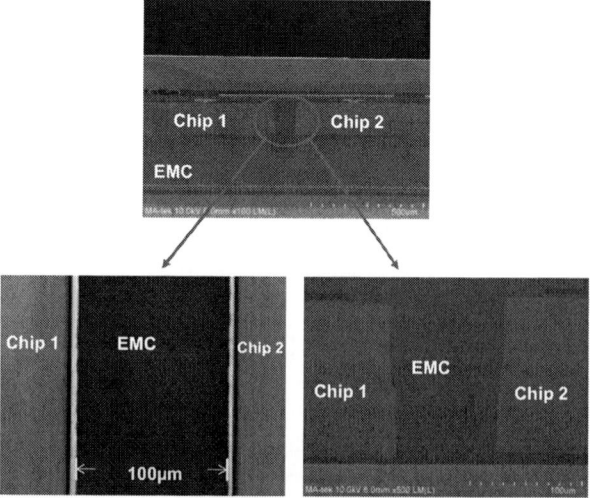

Figure 13 Images of the test package showing the 100μm gap between the large chip and small chip and there is no void in the dry-film EMC.

978-1-7281-1500-9/19 $31.00 © 2019 IEEE

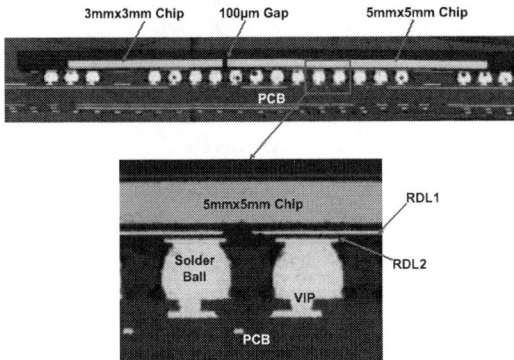

Figure 14 (Top) 508mm x 508mm panel with SiPs. (Bottom) Cross section view of the SiP assembly.

Figure 14 shows the SiPs fabricated by the FOPLP chip-first and die face-down on a 508mm x 508mm panel. A cross section of the SiP PCB assembly is shown in the bottom of Figure 14. It can be seen that the SiP has been properly fabricated.

Summary

The feasibility of design, materials, process, and fabrication of a thin heterogeneous integration of 4 chips by a FOPLP method has been demonstrated. Some important results and recommendations are summarized as follows.

➢ The 10μm RDL1 of the heterogeneous integration fan-out package has been successfully fabricated by an all PCB SAP and equipment.

➢ The lamination of a dry-film EMC has been successfully demonstrated on making the ECM-panel in a 2-stage vacuum lamination machine with the given important parameters.

➢ The lamination of the 5-layer (2 ECM-panel + core-substrate) Uni-SIP structure has been successfully fabricated. The debonding of the temporary carriers and removing of the two-side tape have been successfully demonstrated.

➢ A new ABF with a CTE ($15 \times 10^{-6}/^{\circ}$C) which is the same as the dry-film EMC and a very small filler size (5μm) has been selected.

➢ DOW's SAP solution – CIRCOPIST 7800 de-smear process and CIRCUPOSIT ADV 8500 E-less Cu process with some modified parameters have been successfully

demonstrated with smooth ABF surface and outstanding peel strength results.

➢ The present LDI + PCB Cu-plating on fabricating the 10μm metal line width and spacing of RDL1 have been successfully demonstrated.

➢ The PCB assembly of the heterogeneous integration package is almost done. The samples are going to be subjected to drop test and thermal cycling test and the results will be presented at the conference.

Acknowledgements

The authors would like to thank their upper managements from ASM, DOW, Huawei, Indium, JCAP, and Unimicron for their strong support of this consortium project. The constructive contributions from TJ Tseng, CM Lai, Casper Tsai, YM Chan, Leslie Chang, TW Lam, JW Dong, and Jiang Leon are greatly appreciated.

References

[1] Ko, CT, H. Yang, J. H. Lau, M. Li, M. Li, C. Lin, J. W. Lin, T. Chen, I. Xu, C. Chang, J. Pan, H. Wu, Q. Yong, N. Fan, E. Kuah, Z. Li, K. Tan, Y. Cheung, E. Ng, K. Wu, J. Hao, R. Beica, M. Lin, Y. Chen, Z. Cheng, S. Koh, R. Jiang, X. Cao, S. Lim, N. Lee, M. Tao, J. Lo, and R. Lee, "Chip-First Fan-Out Panel-Level Packaging for Heterogeneous Integration", *IEEE Transactions on CPMT*, September 2018, pp. 1561-1572.

[2] Lee, Y., W. Lai, I. Hu, M. Shih, C. Kao, D. Tarng, and C. Hung, "Fan-Out Chip on Substrate Device Interconnection Reliability", *IEEE/ECTC Proceedings*, May 2017, pp. 22-27.

[3] Lau, J. H., M. Li, Q. Li, I. Xu, T. Chen, Z. Li, K. Tan, X. Qing, C. Zhang, K. Wee, R. Beica, C. Ko, S. Lim, N. Fan, E. Kuah, K. Wu, Y. Cheung, E. Ng, X. Cao, J. Ran, H. Yang, Y. Chen, N. Lee, M. Tao, J. Lo, and R. Lee, "Design, materials, process, and fabrication of fan-out wafer-level packaging," *IEEE Transactions on CPMT*, Vol. 8, No. June 2018, pp. 991-1002.

[4] Lau, J. H., M. Li, D. Tian, N. Fan, E. Kuah, K. Wu, M. Li, J. Hao, Y. Cheung, Z. Li, K. Tan, R. Beica, T. Taylor, C.T. Lo, H. Yang, Y. Chen, S. Lim, N.C. Lee, J. Ran, X. Cao, S. Koh, and Q. Young, "Warpage and thermal characterization of fan-out wafer-level packaging," *IEEE Transactions on CPMT*, Vol. 7, No. 10, pp. 1729-1738, 2017.

[5] Lau, J. H., M. Li, N. Fan, E. Kuah, Z. Li, K. Tan, T. Chen, I. Xu, M. Li, Y. M. Cheung, Wu Kai, Ji Hao, R. Beica, T. Taylor, C. Ko, H. Yang, Y. Chen, S. Lim, N. Lee, J. Ran, K. Wee, Q. Yong, C. Xi, M. Tao, J. Lo, and R. Lee, "Fan-out wafer-level packaging (FOWLP) of large chip with multiple redistribution layers (RDLs)," *IMAPS Transactions, Journal of Microelectronics and Electronic Packaging*, Vol. 14, No. 4, pp. 123-131, 2017.

[6] Lau, J. H., M. Li, Y. Li, M. Li, I. Au, T. Chen, S. Chen, Q. Yong, J. Madhukumar, K. Wu, N. Fan, E. Kuah, Z. Li, K. Tan, W. Bao, S. Lim, R. Beica, C. Ko, and X. Cao, "Warpage measurements and characterizations of FOWLP with large chips and multiple RDLs," *IEEE*

978-1-7281-1500-9/19 $31.00 © 2019 IEEE

Transactions on CPMT, Vol. 8, No. 10, pp. 1729-1737, 2018.

[7] Lin, Y., S. Wu, W. Shen, S. Huang, T. Kuo, A. Lin, T. Chang, H. Chang, S. Lee, C. Lee, J. Su, X. Liu, Q. Wu, and K. Chen, "An RDL-First Fan-out Wafer Level Package for Heterogeneous Integration Applications", *IEEE/ECTC Proceedings*, May 2018, pp. 349-354.

[8] Lau, J. H., M. Li, M. Li, T. Chen, I. Xu, X. Qing, Z. Cheng, N. Fan, E. Kuah, Z. Li, K. Tan, Y. Cheung, E. Ng, P. Lo, K. Wu, J. Hao, S. Koh, R. Jiang, X. Cao, R. Beica, . Lim, N. Lee, C. Ko, H. Yang, Y. Chen, M. Tao, J. Lo, and R. Lee, "Fan-Out Wafer-Level Packaging for Heterogeneous Integration", *IEEE Transactions on CPMT*, 2018, September 2018, pp. 1544-1560.

[9] Lau, J. H., M. Li, Y. Lei, M. Li, I. Xu, T. Chen, Q. Yong, Z. Cheng, K. Wu, P. Lo, Z. Li, K. Tan, Y. Cheung, N. Fan, E. Kuah, X. Cao, J. Ran, R. Beica, S. Lim, NC Lee, C. Ko, H. Yang, Y. Chen, M. Tao, J. Lo, and R. Lee, "Reliability of Fan-Out Wafer-Level Heterogeneous Integration", *IMAPS Transactions, Journal of Microelectronics and Electronic Packaging*, Vol. 15, Issue: 4, October 2018, pp. 148-162.

[10] Ko, CT, H. Yang, J. H. Lau, M. Li, M. Li, C. Lin, J. W. Lin, C. Chang, J. Pan, H. Wu, Y. Chen, T. Chen, I. Xu, P. Lo, N. Fan, E. Kuah. Z. Li, K. Tan, C. Lin, R. Beica, M. Lin, X. Cao, S. Lim, NC Lee, M. Tao, J. Lo, and R. Lee, "Design, Materials, Process, and Fabrication of Fan-Out Panel-Level Heterogeneous Integration", *IMAPS Transactions, Journal of Microelectronics and Electronic Packaging*, Vol. 15, Issue: 4, October 2018, pp. 141-147.8-162.

[11] Lau, J. H., *Fan-Out Wafer-Level Packaging*, Springer, April 2018.

[12] Knickerbocker, J., R. Budd, B. Dang, Q. Chen, E. Colgan, L.W. Hung, S. Kumar, K. W. Lee, M. Lu, J.W. Nah, R. Narayanan, K. Sakuma, V. Siu, and B. Wen, "Heterogeneous Integration Technology Demonstrations for Future Healthcare, IoT, and AI Computing Solutions", *IEEE/ECTC Proceedings*, May 2018, pp. 1519-1522.

[13] Panigrahi, A., C. Kumar, S. Bonam, B. Paul, T. Ghosh N. Paul, S. Vanjari, and S. Singh, "Metal-Alloy Cu Surface Passivation Leads to High Quality Fine-Pitch Bump-Less Cu-Cu Bonding for 3D IC and Heterogeneous Integration Applications", *IEEE/ECTC Proceedings*, May 2018, pp. 1555-1560.

[14] Faucher-Courchesne, C., D. Danovitch, L. Brault, M. Paquet, and E. Turcotte, "Controlling Underfill Lateral Flow to Improve Component Density in Heterogeneously Integrated Packaging Systems", *IEEE/ECTC Proceedings*, May 2018, pp. 1206-1213.

[15] Hu, Y., C. Lin, Y. Hsieh, N. Chang, A. J. Gallegos, T. Souza, W. Chen, M. Sheu, C. Chang, C. Chen, K. Chen, "3D Heterogeneous Integration Structure Based on 40nm- and 0.18μm-Technology Nodes", *Proceedings of IEEE/ECTC*, May 2015, pp. 1646-1651.

[16] Bajwa, A., S. Jangam, S. Pal, N. Marathe, T. Bai, T. Fukushima, M. Goorsky, and S. S. Iyer, "Heterogeneous Integration at Fine Pitch (≤ 10μm) using Thermal Compression Bonding", *IEEE/ECTC Proceedings*, May 2017, pp. 1276-1284.

[17] Dittrich, M., A. Heinig, F. Hopsch, and R. Trieb, "Heterogeneous interposer based integration of chips with copper pillars and C4 balls to achieve high speed interfaces for ADC application", *Proceedings of IEEE/ECTC*, Mat 2017, pp. 643-648.

[18] Beal, A., and R. Dean, "Using SPICE to Model Nonlinearities Resulting from Heterogeneous Integration of Complex Systems", *IMAPS Proceedings*, October 2017 pp. 274-279.

[19] Lau J. H., *Heterogeneous Integrations*, Springer, April 2019.

[20] Lau, J. H., "Redistribution-Layers for Heterogeneous Integrations", *Chip Scale Review*, Vol. 23, January/February 2019, pp. 20-25.

[21] Lau, J. H., P. Tzeng, C. Lee, C. Zhan, M. Li, J. Cline, et al., "Redistribution Layers (RDLs) for 2.5D/3D IC Integration", *IMAPS Transactions, Journal of Microelectronic Packaging*, Vol. 11, No. 1, First Quarter 2014, pp. 16-24.

Ultra-thin FO Package-on-Package for Mobile Application

Hsiao Hsiang-Yao, Soon Wee Ho, Simon Siak Boon Lim, Wai Leong Ching, Chong Ser Choong, Sharon Lim Pei Siang, Han Yong and Chai Tai Chong

Institute of Microelectronics, A*STAR (Agency for Science, Technology and Research)
2 Fusionopolis Way, #08-02 Innovis Tower, Singapore 138634
hsiaohy@ime.a-star.edu.sg, +65-6770-5474

Abstract— Today, Package on Package is a major trend of three-dimensional fabrication for processors and high-performance memory applications in portable applications. Package-on-Package has the benefit of a mini packaging size with multi-functionality by stacking two different packs. However, an ordinary Printed circuit board substrate Package on Package has a weak point to meet the now low profile necessary of high-performance in the thin portable application. To overcome this weak point, the package has been introduced to the market by Fan Out Wafer Level, and this structure of the package allows I/O to be within the device surface and expand through the combination of form so that they can be accommodated more FOWLP. Ultra-thin Fan-out PoP was developed using RDL-first process flow. The developed Fan-out PoP has a package size of 15 x 15 mm² and thickness of 800 μm, and it consists of three embedded chips. The bottom package consists of a 10 x 10 mm² processor chip assembled to under bump metallization (UBM) of the bottom RDL layers. Vertical wire-bonds are integrated into the bottom package to act as vertical through mold interconnect (TMI) to the top RDL layers. The top package consist of two 7 x 11 mm² silicon chips assembled laterally on top of the bottom package and connected to the top RDL layer with low-loop wire-bonds. The top chips were encapsulated in epoxy mold compound to form an integrated PoP. RDL-first integration flow was used to fabricate the fan-out package whereby RDL, molding and chips assembly processes were performed on a carrier wafer to overcome warpage associated with conventional Mold-first process. The ultra-thin Fan-out PoP samples pass the reliability include the MST level 3, drop impact test and the Thermal Cycling. It also provides good thermal performance on packaging level and system level applied in mobile device.

Keywords-component; Fan-out; Cu RDL; TMI; molding

I. INTRODUCTION

FO-WLP technology is rapidly being adopted as advance packaging solution for semiconductor companies [1-4]. It offers a versatile and cost-effective approach for heterogeneous chips integration. FO-WLP technology also enables multi-chips packages to have a smaller form factor as compared to traditional microelectronic packages, and thus is readily implemented in mobile application where package profile thickness is a key consideration for the final product.

FO-WLP has been extensively implemented in mobile products, an example is the application processor module which is like TSMC InFO [5]. In TSMC InFO package, the processor chip is embedded in epoxy mold compound and redistribution layers are used to fan-out the chips I/Os to the board level solder ball interconnects. Cu pillars vertical interconnects are used to connect the top and bottom package. A molded memory package consisting of DRAM chips wire-bonded to a core-less organic substrate will be assembled on top of the InFO package with solder balls, hence forming the package on package (PoP). The thickness of application process module tends to be higher due to the height of the solder balls between the top mold memory module and bottom processor package.

An ultra-thin Fan-out PoP was developed in this work to further reduce the module thickness. In the ultra-thin package, the DRAM chips were stacked directly onto the bottom processor FO-WLP and connected to bottom processer chip through a series of ultra low-loop wire-bond, RDLs and vertical wire-bond interconnects. The entire top DRAM chip assemblies were encapsulated directly to the bottom FO-WLP forming an integrated ultra-thin Fan-out PoP. Due to the close proximity of top DRAM chips to bottom processor, thermal interaction between the top and bottom devices could lead to deterioration on overall performance. Therefore, 3D thermal simulation was also performed on ultra-thin Fan-out PoP in mobile phones.

II. TEST VEHICLE DESCRIPTION

Fig.1 and 2 show the test vehicle of ultra-thin Fan-out PoP. The package size is 15 x 15 mm² with one processor chip and two DRAM chips. The dimension of the processor chip with micro-bump is 10 x 10 mm² and it is assembled on the bottom side package with UBM of the bottom RDL layers. Vertical wire bonding interconnects are arranged along the top and bottom peripheral of the bottom package. The two DRAM chips with a size of 7 x 11 mm² are assembled on the top package and connected to the top RDL layer via low-loop wire-bonds. The entire assembly is encapsulated in epoxy mold compound to form an ultra-thin Fan-out PoP.

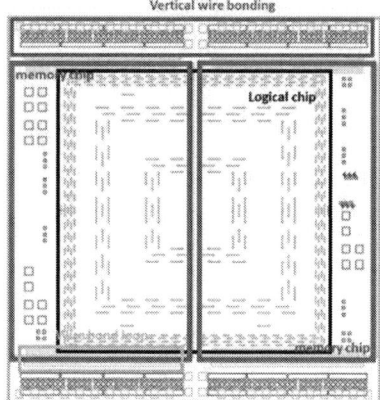

Figure 1. Plan view layout of the integrated ultra-thin Fan-out PoP showing locations of the processor chip, memory dies, TMI and wire-loops.

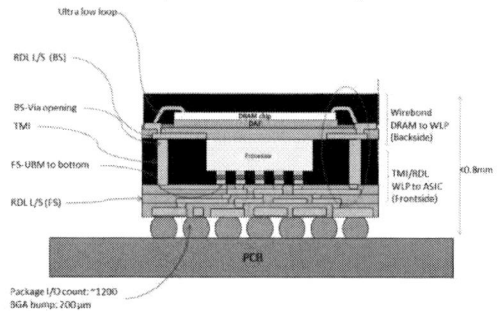

Figure 2. Cross-section illustration of the integrated ultra-thin Fan-out PoP

III. THERMAL SIMULATION RESULTS OF ULTRA-THIN FAN-OUT POP ON MOBILE PHONE

The thermal simulation model is constructed using ANSYS Multi-physics to simulate the thermal behavior when ultra-thin Fan-out PoP is applied in mobile phone. The 3D simulation modeling is shown in Fig. 3. The 3D simulation model includes a 5.5 inch display, top screen, backside cover, one heat spreader, one battery and a main motherboard with the ultra-thin Fan-out PoP. The 3D simulation mode is constructed, considering the outside natural convection environment. The ultra-thin Fan-out PoP is the main heat source in this model. Two main thermal dissipation paths: (1) top side through EMI shielding and insulation layer to heat spreader and screen; (2) bottom side through backside package to back cover. The device side frame and nearby battery can also aid in the conduction the heat to top and bottom surface. Fig. 4 shows the temperature distribution of motherboard with ultra-thin Fan-out PoP and processor and memory of ultra-thin Fan-out PoP with an applied power of 5W on the processor. The maximum temperature of processor and memory are 82.0 deg C and 81.5 deg C respectively. In order to maintain all chips below the temperature limit of 85 deg C, processor heating power of 5.3W can be dissipated, and total package heating power is 6.1W. For mobile phone thermal performance, the result is shown in Fig. 4(d). The maximum temperature of mobile device with ultra-thin Fan-out PoP is 45.7 deg C when 2W power is applied in the processor. Without violating the skin temperature limit, processor power is 2W and it can be applied in both packages. The total package power is 2.3W.

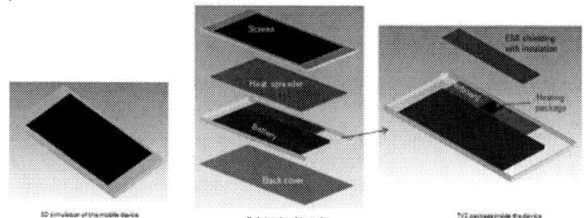

Figure 3. 3D thermal simulation model

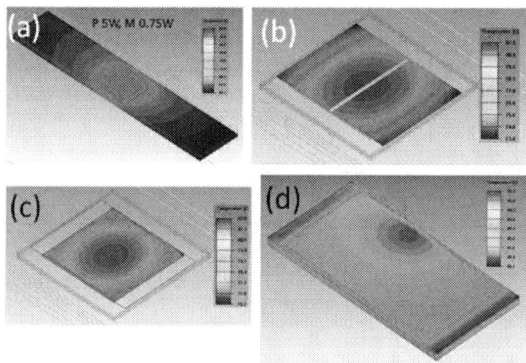

Figure 4. (a) temperature distribution on motherboard with ultra-thin Fan-out PoP, (b) temperature distribution on processor, (c) temperature distribution on DRAM chips, and (d) temperature distribution on mobile device level.

IV. PROCESS FLOW FOR INTEGRATED FO-POP

RDL-first integration flow was select as the fabrication method for the ultra-thin Fan-out PoP. The fabrication flow is shown in Fig. 5. 300 mm glass wafer was used as the carrier. A release layer was coated over the glass wafer. RDL and under bump metallization (UBM) were processed on the sacrificial layer using semi-additive process. Photo-dielectric was used as the interlayer passivation and vertical wire bond was fabricated over the RDL layers forming the TMIs. The processor chips with micro-bumps were assembled onto the carrier in a C2W arrangement. The top surface of the carrier is encapsulated with epoxy mold compound and the top surface of the TMI is revealed by back-grinding the excess over-mold. A UBM layer and wire bond pads were formed over the TMI for DRAM chips and low-loop wire bond. The DRAM chips assembly used die attach film. After the completion of the memory die assembly and low loop wire bonding, the second molding encapsulated the whole wafer. Finally, the carrier was de-bonded and the package singulation was performed on the ultra-thin Fan-out PoP wafer and solder balls interconnects were formed using solder ball drop process. The ultra-thin Fan-out PoP was then assembled to PCB for reliability testing.

978-1-7281-1500-9/19 $31.00 © 2019 IEEE

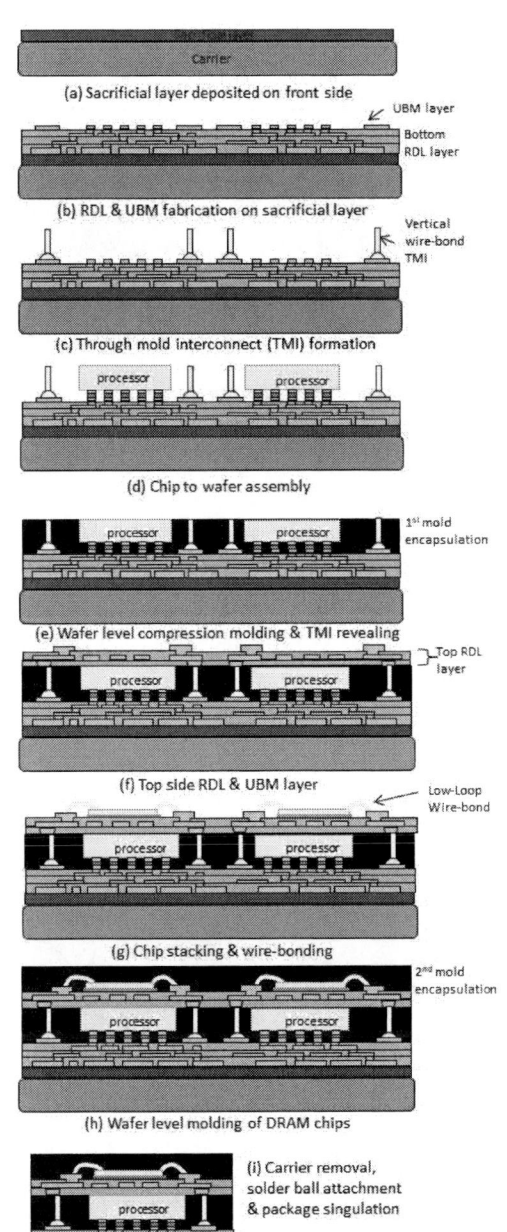

Figure 5. Integration flow of the ultra-thin Fan-out PoP

V. EXPERIMENTS AND DISCUSSION

A. *Multi-layer RDLs fabrication*

UBM and Cu RDLs layers were formed using a semi-additive process that includes seed layer deposition, photoresist mold patterning, electroplating, photoresist stripping and seed layer etching. Low temperature cured photo-dielectric material was used as the passivation layer. Photolithography steps which include spin coating, photo-mask exposure and developing were used to pattern contact vias on the photo-dielectric passivation. Cu RDL traces were formed using semi-additive process similar to that of the

UBM structure. The semi-additive process and photo-dielectric process were repeated to build up 2 layers of RDL. Fig. 6 shows the Cu/Ni/Cu UBM and 1st photo-dielectric passivation fabricated on top of the sacrificial layer. The UBM diameter is 200 µm and photo-via diameter for photo-dielectric is 80 µm. The sacrificial layer shows good thermal stability and chemical resistance during the RDL processing, without showing blisters or de-lamination. 2 layers of Cu redistribution lines are processed onto the UBM. The 1st Cu layer consists of 10 µm L/S traces and 2nd Cu layer consists of 5 µm L/S traces. Separating these two Cu metal layers is a 6 µm thick photo-dielectric layer with 5 um diameter photo-via openings.

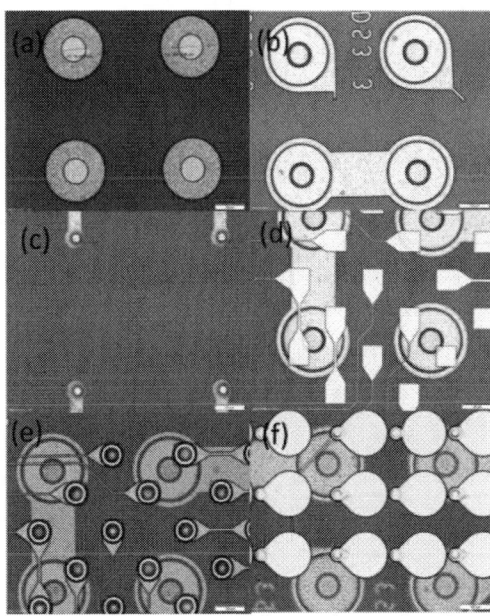

Figure 6. Optical images of the carrier wafer after RDL process: (a) images of BGA pads and 1st photo-dielectric, (b) 1st RDL routing, (c) 2nd passivation via opening, (d) 2nd RDL routing, (e) 3rd passivation via opening, and (f) UBM pad for vertical wire bond.

B. *TMI formation and Processor chip assemblely*

After Cu RDLs formation, the vertical wire bonding is employed to form TMIs. Vertical wire is being evaluated as interconnect for wafer level packaging. This is an alternate option besides copper pillar interconnect. IME fabricated team developed the through Mold interconnects process with vertical wire bonding and applied it for the FOWLP package [6]. The vertical loop is bonded to the metal pad on passivation layer (glass as a carrier) before molding. 2.0 mil diameter PdCu wires were bonded over the Cu/Ni/Au UBM pads to form vertical wire-bond TMIs. The average wire-bond height is about 270 µm and shear test results show good bond strength of 9.7 gf/mil². All shear test samples show wire-shear failure mode. The results of vertical wire bond are shown in Fig. 7. There are no pad peel and pad break during the bonding process.

Figure 7. The images of vertical wire bonding: (a) tilted image of vertical wire-bond, (b) SEM image of vertical wire bond, (c) wire ball on the Cu/Ni/Au UBM, and (d) tail section of vertical wire bond.

Processor chip with micro-bumps were assembled onto the glass wafer using thermo-compression reflow process. Fig. 8 shows the carrier after thermal compression assembly. There are good interconnects between micro-bump and Cu RDLs layer, and the CSAM result shows no void after processor chip is assembled.

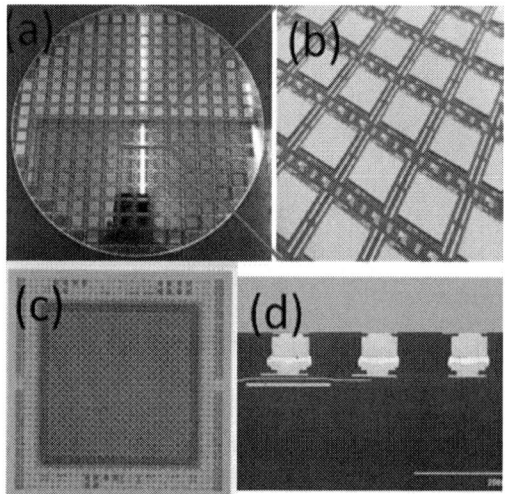

Figure 8. Images of the glass wafer after C2W assembly process: (a) image of whole glass wafer with processor chips, (b) magnified image of the RDL layers for chip assembly, (c) X-ray Image for POP package, and (d) Cross-section of the solder joints with Cu RDL at high magnification.

After the formation of the vertical wire bond, the wafer was encapsulated with epoxy mold compound. The mold tool chase used in this work has a minimum molding thickness of 400 μm. Post-Mold-Cure (PMC) was carried out in a hot oven in nitrogen environment. The top Cu surfaces of the vertical wire bond TMI were exposed by mechanical back grinding away 200 μm of excess over-mold. The Cu surface after mechanical exposure is shown in

Fig. 9(c) and (d). Grinding lines can be observed on the Cu surface but no Cu smearing that would lead to electrical shorts was observed.

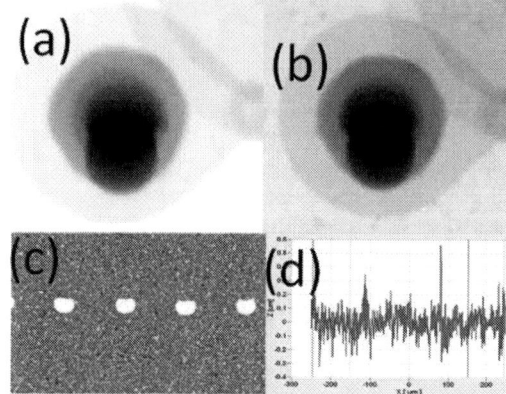

Figure 9. (a)-(b) X-ray result of the vertical wire-bond before and after molding process, (c) Cu TMI after mechanical backgrinding, and (d) Line scan data of vertical wire bond after backgrinding using contact profiler.

C. *Backside RDL and wire bond fingers were fabricated*

Top-side Cu/Ni/Au RDL was processed on the mold surface to form wire-bond fingers for top chips wire-bonding. The results are shown in Fig. 10.

Figure 10. Optical images of the glass wafer after backside RDL process: (a) image of the carrier wafer, (b) magnified image of the RDL layers and wire bond finger, (c) Cu/Ni/Au metallization on top of vertical wire-bond TMI, and (d) wire bond fingers for ultra-low loop wire-bond formation.

D. *Topside DRAM chips assembly*

DRAM chips were attached onto the top RDL layer with 20 μm die attach film (DAF) 20 μm thick die attach film (DAF were laminated onto the 50 μm thick DRAM wafer, following by singulation. DRAM chips were attached onto the top RDL layer using pick and place process. DAF shear test was also performed after MSL3 reliability test and passed reliability test showing no voids and delamination. The DRAM chips are bonded across the wafer using 3N bonding force at 100 deg C. The results are show in Fig. 11.

Figure 11. DAF process assembly of two memory chips to glass wafer.

After DAF process, wire-bond loops were formed by using reverse stitch bonding (RSSB) because the loop height of Forward loop stand-off stitch is higher than that of reverse stitch bonding. Ball-stud bumps were formed on the chip Al pads. Ball stud joints were formed on the wire-bond fingers on the mold side and looped to the ball-stud bumps of the Al chips to form a wedge joint. The wafer consists of two daisy chain chips on one package. There are 154 wires per chip and a total of 308 wires per package. The wire length is about 620μm. The Al chip bond pad pitch is 80um and the bond pad opening is 50μmx50μm. Ultra-low loop bonding completed with 0.8mil Au wire with reversed bond profile is shown in Fig. 12. Bond at lower temperature has better stability during bonding, 100 deg C was used in actual run. Wire sweep sampling inspection (after wire bonding process) was completed and the wire sweep is <6.51%. No pad peel is seen after shear and there is no pad lift or ball lift.

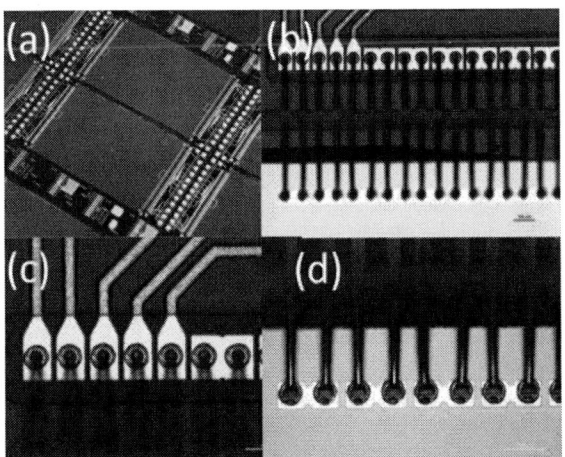

Figure 12. (a) Plan view image of two memory chips wire-bonded to Au fingers on top of the 1st mold, (b) magnified images of the reversed stitched loop, (c)

magnified image of the ball joints on Au fingers, and (d) wedge joints on ball studs on the Al pads.

Ball shear reading is 5.5gf/mil², although had increases the ball size (at bond finger) for better contact adhesion. The average wire-loop height was about 75μm. Wire-sweep measurements were conducted using X-ray inspection after 2nd wafer level molding. The maximum wire-sweep after molding is 6.5% of the wire-length. No wire shorting or breakage was observed after molding. The results are shown in Fig. 13.

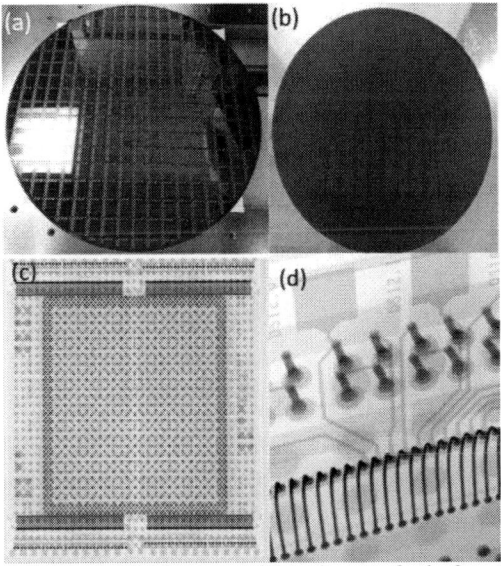

Figure 13. (a) Images of the glass wafer before 2nd molding process, (b) image of the carrier wafer after 2nd molding, (c) X-ray magnified image of PoP sample after 2nd molding, and (d) X-ray magnified image of vertical wire-bond TMI and Low-loop wire interconnects.

VI. RELIABILITY TEST RESULTS

Ultra-thin Fan-out PoP samples have been subjected to Moisture stress test (MST) level 3, thermal cycling and drop impact test. Reliability test parameters are provided in Table 1. SAM analysis and electrical measurement are used to check the failure after reliability testing. SAM analysis is used to check for void or delamination and 20% increase in daisy chain resistance is considered an electrical failure. Table 2 presents the reliability testing results of MST level 3, drop test and thermal cycling. All samples passed MST level 3 and thermal cycling testing in the component level. For drop impact testing, 22 samples passed the 30 cycle drop test. All sample remained physical intact without any sample detachment after 30 drops, shows in Fig. 15.

TABLE I. RELIABILITY TEST CONDITIONS

Test	Test Conditions
MST L3 (Component)	85 C/85% RH for 168 hrs +

Test	Test Conditions
	3X reflow, 260 deg C
Drop impact	JESD22-B111, G max :1500g, t=0.5 ms, Drop times: 30 drops
TC (Component level)	Thermal Cycling, -40 deg C to +125 deg C, 15mins. Dwell, 100, 500 and 1000 cycles

TABLE II.   RELIABILITY TEST RESULTS

Reliability Test	Test results (pass /total)
MST L3	22/22
Drop test (30 drops)	22/22
TC	22/22

Figure 14. CSAM images of component level reliability results (a) Time Zero Thru scan Image and (b) after MSL3 Thru scan Image.

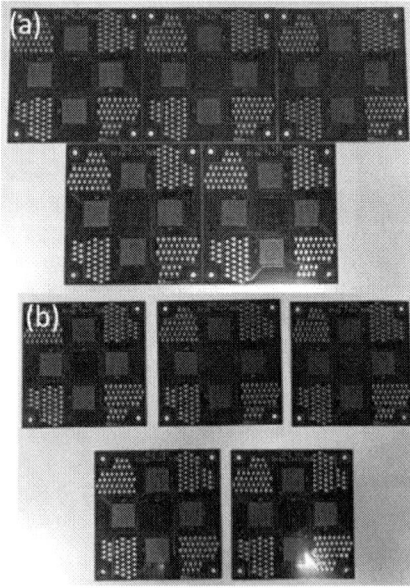

Figure 15. Drop Test Result: (a) image of the board level samples before drop test, and (b) image of the board level samples after drop test.

VII.   CONCLUSIONS

The ultra-thin Fan-out Package-on-Package for mobile application has made a few significant achievements. Some of the important results are summarized below:

1. Fan-out package-on-package (PoP) has been successfully developed using RDL-first integration flow.

2. The Cu redistribution lines are processed onto the UBM. Vertical wire-bonds are integrated into the bottom package to act as vertical through mold interconnect (TMI) to the top RDL layers.

3. The top package consist of two DRAM chips assembled laterally on top of the bottom package and connected to the top RDL layer with low-loop wire-bonds. The DRAM chips were encapsulated in epoxy mold compound to form an integrated PoP.

4. For the reliability testing, the ultra-thin Fan-out PoP samples passed MSL3, Temperature Cycle and drop test without electrical failure or delamination under CSAM inspection.

5. For the thermal simulation results, the ultra-thin Fan-out PoP in mobile device can provide good thermal performances when applied standard power for mobile devices.

6. The application to uses such as smartphone and wearable devices which has the limitation of the package height is expected.

ACKNOWLEDGMENT

This work has been carried out as part of the Multi-chip Fan-out Wafer level Packaging Development Line Consortium led by the Institute of Microelectronics (IME), a research institute of the Agency for Science, Technology and Research (A*STAR). The authors are grateful to members of Project as well as IME staffs who had contributed and made this work possible.

REFERENCES

[1] V. S. Rao, C. T. Chong, , S. W. Ho, M. Zhi, C.S. Choong, P. S. Lim, D. Ismael, and Y. Y. Liang, "Development of High Density Fan Out Wafer Level Package (HD FOWLP) with Multi-layer Fine Pitch RDL for Mobile Applications" 2016 IEEE 66th Electronic Components and Technology Conference (ECTC), pp. 1522-1529, 2016.

[2] J. H. Lau, M. Li, M. Li, T. Chen, I. Xu, X. Qing, Z. Cheng, N. Fan, E. Kuah, Z. Li, K. Tan, Y. Cheung, E. Ng, P. Lo, K. Wu, J. Hao, S. Koh, R. Jiang, X. Cao, R. Beica, S. Lim, N. Lee, C. Ko, H. Yang, Y. Chen, M. Tao, J. Lo, and R. Lee, "Fan-Out Wafer-Level Packaging for Heterogeneous Integration", IEEE Transactions on Components, Packaging and Manufacturing Technology, vol. 8, pp. 1544-1560, September 2018.

[3] X. Hua, H. Xu, Z. L. Zhang, D. Chen, K. H. Tan, C. M. Lai, J. H. Lau, M. Li, M. Li, E. Kuah, N. Fan, W. Kai and K. Cheung "Development of chip-first and die-up fan-out wafer-level packaging", 2017 IEEE 19th Electronics Packaging Technology Conference (EPTC), pp. S23-1–6, December 2017.

[4] M. Li, Q. Li, J. H. Lau, N. Fan, E. Kuah, K. Wu, K. cheung, Z. Li, K. H. Tan, I. Xu, D. Chen, R. Beica, C.T. Ko, H. Yang, S. P. Lim, J. Ran and C. Xi, "Characterizations of fan-out wafer-level packaging", 2017 International Microelectronics Assembly and Packaging Society, vol. 2017, pp. 557-562, October 2017.

[5] C. F. Tseng, C. S. Liu, C. H. Wu, and D. Yu, "InFO (wafer-level integrated fan-out) technology", 2016 IEEE 66th Electronic Components and Technology Conference (ECTC), pp. 1-6, June 2016.

[6] S. W. Ho, L.C. Wai, S.A. Sek, D. I. Cereno, B. L. Lau, H.Y. Hsiao, T. C. Chai, and V. S. Rao, "Through mold interconnects for fan-out wafer level package," 2016 IEEE 18th Electronics Packaging Technology Conference (EPTC), pp. 51-56, December 2016.

Development of Wafer Level Process for the Fabrication of Advanced Capacitive Fingerprint Sensors Using Embedded Silicon Fan-Out (eSiFO®) Technology

Shuying Ma[1], Chengqian Wang[1], Fengxia Zheng[1], Daquan Yu[1]*, Hong Xie[2], Xiaobing.Yang[1], Li Ma[1], Ping Li[3], Weidong Liu[3], Jambo Yu[4], Jason Goodelle[4]

[1]Huantian Technology (Kunshan) Electronics Co., Ltd., Kunshan, Jiangsu Province, P. R. China
[2]Flipchip International, USA
[3]Huantian Technology (Xi'an) Electronics Co., Ltd., Xi'an, Shanxi Province, P. R. China
[4]Synaptics, USA
Email: daquan.yu_ks@ht-tech.com; shuying.ma_ks@ht-tech.com

Abstract — **Typical wafer level chip scale package (WLCSP) is a standard fan-in structure. As the increase of I/O numbers, it can't provide enough area to redistribution layer (RDL). In recent years, an emerging technology of an embedded package with a fan-out area around the chip has been developed showing many advantages of higher I/O numbers, low cost, integration flexibilities and small form factor. Usually fan-out wafer level package (FOWLP) uses epoxy mold compound (EMC) materials, which faces lots of technical challenges such as warpage, CTE mismatch between each layer, expensive cost. In this paper, a completely new wafer level embedded silicon fan-out, named eSiFO® technology was reported to accomplish a capacitive finger print sensor packaging. In this innovative integrated device, an ASIC die with size of ~6mm² was thinned to 90um and then embedded into the silicon cavity to reconstruct a new wafer. In the whole processes, the warpage was less than 2 mm although more than 80% package area was covered by metal. For the wafer level process, ten masks were used and a yield of 98% was achieved. The device passed standard reliability test, including Pre-Con, TCT, uHAST and HTST.**

Keywords – WLP; Fan-out; eSiFO®; Fingerprint sensor; high reliability

I. INTRODUCTION

In 2013, Apple applied fingerprint sensor in iPhone 5s which triggered the wide adoption of finger print sensors in the mobile market. Since then, due to small size and high safety, fingerprint sensors become increasingly popular in smart phones, security cards and other ICs. Over the past several years, the market is almost from zero to a multi-billion dollars and the annual growth rate of fingerprint sensor shipments has reached more than 100%. Nowadays there are mainly three kinds of fingerprint sensors (FPS), i.e., capacitive, optical and ultrasonic, whereas capacitive FPS takes most of the market, especially in smartphones.

It seems that fingerprint has become the standard function of every smartphone and adds a lot of practical value like payment. However, the rapid increase in shipments has been accompanied by strong cost pressure. At present, the mainstream packaging technology for fingerprint sensors is

wire bond and molding process. However molding material can affect the performance of the sensor and increase the thickness of the package. In 2016, we developed via last TSV process for the packaging of FPS which has been successfully used in Huawei Mate 9. This technology enhanced the signal intensity and provided better user experience.

This paper presents innovative work on the fabrication of a disaggregated (i.e. sensor and control silicon separated) capacitive finger print sensor using Embedded Silicon Fan-Out (eSiFO®) technology, which has been successfully development and proved to be a high-performance, low-cost system integration technology [1-3]. Due to its high degree of difficulty and importance, we named this project as "Liaoning", the same as Chinese first aircraft carrier. In this new integrated device, a controller die with size of ~6mm² was thinned to 90um and then embedded into the cavity with 100um depth on the silicon carrier. Then two layer RDLs with 12/14µm minimum L/S were built on the surface as the capacitive part for signal sensing. In order to improve the sensing performance, shielding regions were strategically added and altogether 3362 micro bumps with 70 um pitch and 12µm height were fabricated on the first RDL to enhance signal emission. The sensor with a size of ~35mm², and total thickness of 0.28 mm was successfully developed. The warpage was less than 2 mm although more than 80% package area was covered by metal [5-8]. For the wafer level process, ten masks were used and a yield of 98% was achieved. The device passed standard reliability test, including Pre-Con, TCT, uHAST and HTST.

For the packaging of the new capacitive FPS, an LGA package with 13×13×0.65mm was designed and fabricated, where a two layer substrate was used and after wire bonding, over molding was applied. The reliability tests for the LGA package including Pre-Con, TCT, uHAST, HTST were carried out and all the packages pass the reliability tests. The present study shows that the new developed FPS with LGA package has great potential for wide applications.

II. LIAONING ESIFO® PRODUCT

Product Design

The layout of "Liaoning" is shown in Fig. 1. The target is to create an integrated chip, each of which consists of ASIC die and Pixel array interconnected by RDLs. Such integrated chips can be further processed to form WLCSP components or used just like standard silicon dies in conventional packages types. To interconnect the chips and the I/Os to the periphery of the reconstructed chip, two RDLs are designed and produced. In order to complete RDLs, Cu pillar and metal shielding layer, ten masks and six alignment marks were used. Minimum L/S is 12um/14um, which requires high accuracy lithography technology. In addition, more than 80% area was covered by metal layer. Many proposals were applied to reduce the internal stress, such as thin shielding layer thickness, stress release openings.

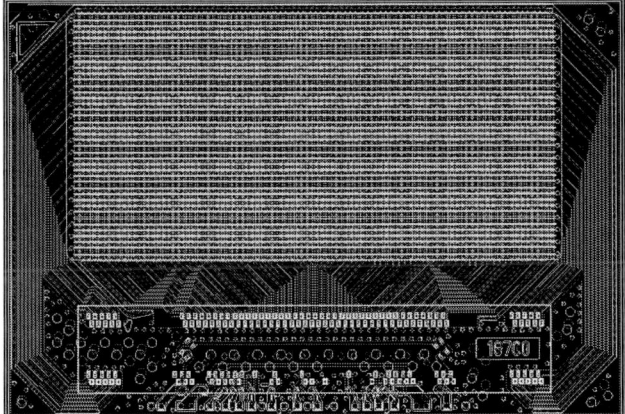

Figure. 1 Layout of Liaoning eSiFO® product.

Process Flow

The wafer-level process flow of "Liaoning" product is illustrated in Fig. 2. Cavities with accurate dimensions and depth were formed by dry etching [4]. Known-good-dies (KGDs) from different device wafers were thinned to target thickness and then were embedded into the cavities. The narrow trenches between the chips and side walls of the cavities are filled by a vacuum film lamination process. Meanwhile, the dry film can also act as passivation layer. The pads of the embedded die were exposed during lithography in the following step. At this point, a new wafer is reconstructed, with no significant differences compared to a standard processed silicon wafer. The process flow continues with standard WLP processes, such as RDL formation and bumping. The difficulties of this particular example were how to keep uniformity of the hundreds of copper bumps and how to release the metal stress due to large metal coverage area. In addition, as many as ten masks were used during the fabrication of "Liaoning" project. How to guarantee the alignment accuracy was also a big challenge. In the final design, pads were finalized with a Ni/Pd/Au finish to enable the use of the reconstructed chips in a LGA package with traditional wire bond interconnects. Some of the essential aspects in the process steps are discussed in the following text.

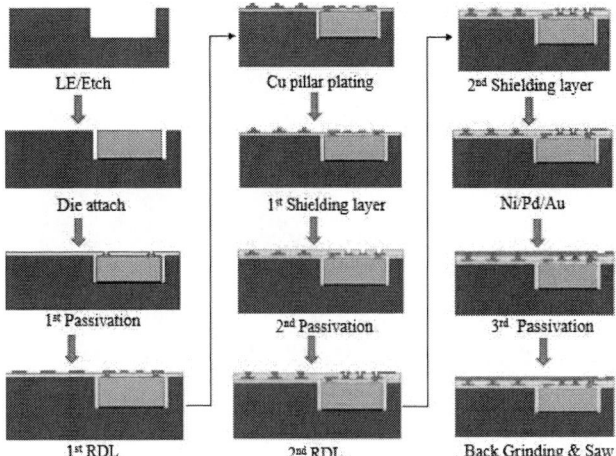

Figure. 2. Process flow of Liaoning eSiFO® product

1) Cavity Formation

In Liaoning project, an 8inch blank wafer was used as the carrier wafer. The cavities were formed by Bosch process. The spec of cavities is 100 with TTV of ±5μm. The length and width of cavities is ~20μm larger than die size on each side. Due to the embedded die size is ~6 mm², the etching of large silicon cavities with good TTV is challenging. Meanwhile, to avoid die tilt and cracks during die attach, smooth bottom surface without any grass and bumps was necessary. After process optimization, we collected 9 points in the whole wafer and the depth data was shown in TABLE 1. The average depth of the cavities is within the spec.

TABLE 1 DEPTH DATA OF CAVITY FORMATION

Wafer ID	E0	E1	E2	E3	E4	E5
Point 1	100.8	102.9	99	99.1	99.8	101.6
Point 2	100.6	101.1	99.2	99.6	100.8	99.5
Point 3	100.2	100.5	97.6	103	103	97.3
Point 4	103	102.1	100.6	102.1	101.8	100.1
Point 5	104.1	104.4	102.3	101.7	103.5	101.4
Point 6	104.6	104.3	102.6	99.6	102.4	102.5
Point 7	102.9	100.1	99	98.8	98.4	100.6
Point 8	101.4	100.3	99.8	98.5	99.2	99.8
Point 9	100.3	100.9	101	102.1	97.5	100
AVG.	102.0um	101.8um	100.1um	100.5um	100.7um	100.3um
SPEC	100±5um					

To achieve flat bottom with ~1μm roughness, we slowed down the etching rate. Fig. 3 shows the 3D microscope image of a cavity. No "grass" or "bumps" were found in the sidewall and bottom. The data of the footing was listed in TABLE 2. The footing variation in a single cavity is less than 5μm without counting the region of 15μm away from the sidewall.

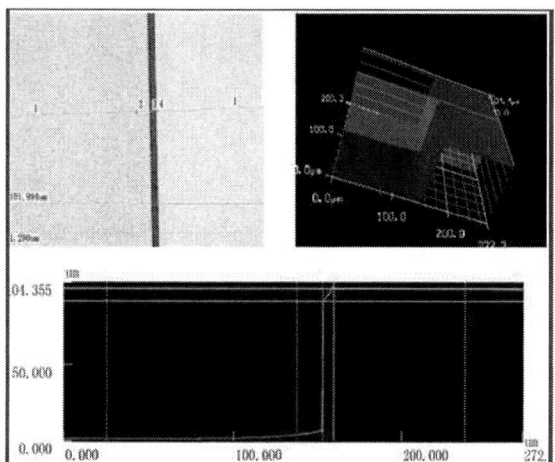

Figure. 3 3D microscope image of a cavity

TABLE 2 FOOTING DATA OF CAVITY FORMATION

Wafer ID	Point 1	Point2	Point 3	Point 4	Point 5	AVG.	SPEC
E3	2.06	4.44	3.90	2.07	4.33	3.36	≤5um
E4	2.52	2.71	1.69	3.86	1.69	2.49	≤5um

2) Die Attach

Due to the large die size of the embedded die, the nozzle with similar shape was custom-made to reduce the risk of rotation and shift. The thickness of the dies and the die-attach film (DAF) are accurately calculated and suitable to the depth of the cavities on the Si carrier wafers. We used a dedicated die attach tool from ASM, NUCLEUS, for C2W attachment with a high accuracy of ±4µm. Fig. 4 shows the image of a die after attachment. The gap is about 25-30µm for all dies on one wafer, with high uniformity and all in the design spec.

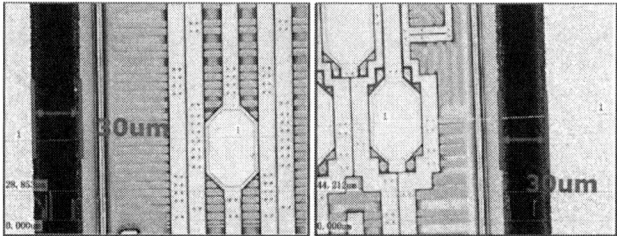

Figure. 4 Image after die attach process

TABLE 3 DIE SHIFT ANDROTATION OF DIE ATTACH

Wafer ID	Direction	Point 1	Point 2	Point 3	Point 4	Point 5
E0	X	0.5um	0.95um	1.25um	0.75um	1.25um
	Y	0.75um	1.5um	0.05um	1um	1.1um
E1	X	1.4um	0.7um	1.15um	2.4um	0.3um
	Y	0.6um	0um	1.5um	0um	0.4um
E2	X	2.25um	0.9um	0.85um	0.65um	1.15um
	Y	1.1um	2.5um	0.25um	1.5um	1um
E3	X	1.35um	0.8um	0.25um	2um	0.55um
	Y	0.25um	1.15um	0.35um	0um	1.25um

TABLE 3 shows the X & Y shift measurement results from 4pcs wafers. We collected 5 points in every wafer. According the data, we can see that the lateral misalignment and longitudinal displacement were within 3µm after process optimization.

3) First Passivation (PA1)

The formation of the first passivation layer is another essential step in the Liaoning process flow. The trench between the die and carrier wafer was filled with dry film by vacuum laminator. Meanwhile, a passivation layer on the reconstructed wafer surface was formed at the same time. The dry film material is very import because it has a big influence on the warpage and electrical performance. We tested a number of materials and selected one with well photo-patternable, low CTE and low dielectric constant. TABLE 4 shows the film thickness collected from 6 areas from the whole wafer after trench filling. The passivation layer thickness variation on the reconstructed wafer is less than 4µm.

TABLE 4 FILM THICKNESS AFTER TRENCH FILLING

Wafer ID	Point 1	Point 2	Point 3	Point 4	Point 5	Point 5
E0	19.8um	18.9um	19.9um	17.2um	18.2um	18.1um
E3	17.8um	20.4um	18.6um	18.5um	21.4um	16.7um

Due to the shape of the pad is a diamond, we designed an oval pad opening shape to achieve a largest pad opening size within the limited pad size. Optical microscopy images showing the wafer after the first passivation, pad opening, and cross-sectional scanning are shown in Fig. 5. No voiding or cracking was observed in the trench and the pad was well connected with RDL.

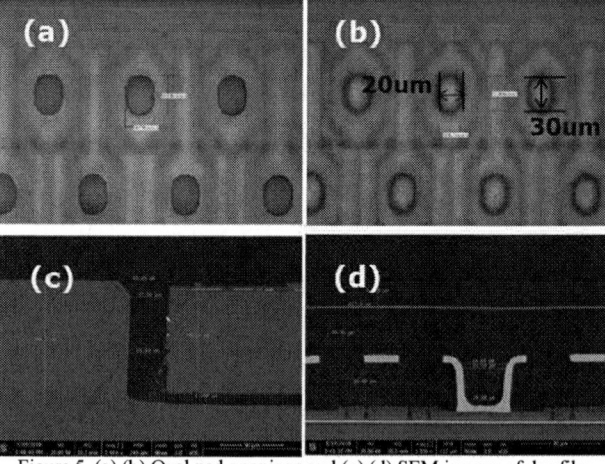

Figure 5. (a) (b) Oval pad openings and (c) (d) SEM images of dry film filling and pad openings

4) First RDL (RDL1)

After PA1 process, the following is RDL1 fabrication to realize electrical interconnection between ASIC chip and external package. RDL1 includes three main processes, lithography lead, Cu pillar and metal shielding processes. In lithography lead process, a seed layer with 0.1µm metal Ti layer and 0.3μm metal Cu layer was deposited by physical vapor deposition (PVD), then line patterns with lithography technology were completed, and finally the plating Cu process increased lead thickness to 3μm to ensure a low resistance circuit. The lead minimum L/S is 15μm/15μm, which should be controlled in ±3μm considering the ASIC signal accuracy requirements. From our measurement results shown in TABLE 5, all the data meets our design. The next step is fingerprint contact circuit fabrication - Cu pillar process. The Cu pillar process used the same seed layer with lithography lead process. It formed 82*41 (3362) oval shaped Cu pillars in a die whose x-section size is 40μm*30μm as presented in Fig. 6. The Cu pillar height uniformity, which should be in 12 ±2μm spec. The CU pillar uniformity plays an important role in FPS performance. From our statistical data shown in Fig. 5(d) measured by x-ray we can see that they all meet this requirement. Finally, metal shielding layer is finished to protect internal signal. We used the thin seed layer rather than plating Cu as metal shieling to reduce the internal stress. In addition, to avoid the internal stress from large area metal layer, a plurality of metal openings were formed. Fig. 7 shows more detailed size and morphology information of RDL1. It further confirms that the lead thickness, PA1 openings, metal layer layout, and Cu pillar size are all in the design specification.

TABLE 5 MEASUREMENT RESULTS OF LINE WIDTH AND SPACE

wafer	Data	Point1	Point 2	Point3	Point4	Point5	SPEC
E2	Line Width	14.7um	15.2um	14.8um	14.9um	15.1um	15±3um
	Line Space	15.3um	15.1um	15.3um	14.7um	14.8um	15±3um
E3	Line Width	14.5um	15.1um	15.3um	14.9um	15.2um	15±3um
	Line Space	16.3um	15.5um	14.3um	14.9um	14.5um	15±3um

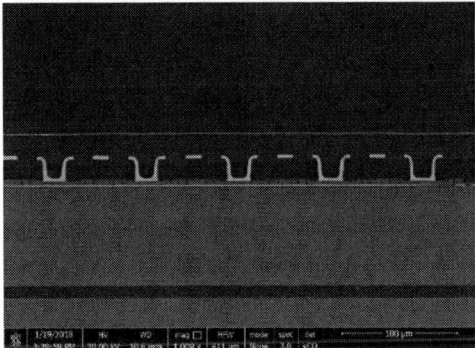

Figure 7. Cross section picture of RDL1

5) Second RDL (RDL2)

After RDL1 fabrication, a large number of 12μm high Cu pillars are formed. Therefore a a dry film lamination (PA2) process to form a plane to start RDL2 process is required. PA2 process was carried out same as PA1 process. Three kinds of pad opening should be developed including 30μm*20μm oval pad openings on Cu pillar, 30μm pad openings on lithography lead and metal shielding layer. Then using PVD, lithography, and plating Cu technologes achieved RDL2 metal layer. The minimum L/S of RDL2 metal layer is 12μm/14μm. As shown in Fig. 8, the L/S have high uniformity and are controlled in ±2μm, which is helpful to improve fingerprint recognition performance. Finally, metal shielding layer of RDL2 was completed to protect internal signal circuit.

Fig. 9 presents x-section pictures after RDL2 fabrication. Fig. 9(a) and (b) show x-section of two metal shielding layers. From the SEM pictures we can see the gap between these two layers is 23μm, and the metal shielding connection opening is 30μm, which are all in the design spec. Fig. 9(c) and (d) are Cu pillar connections. As shown in Fig. 9(d), the filling and coplanarity of PA2 is good. There is no bubble, void or Cu pillar exposure found. The opening of PA2 satisfies the electrical connection requirements. In addition, from the pictures we also can see that the Cu pillar height and diameter are 12μm and 30μm respectively. The Cu layers thickness of RDL1 and RDL2 are 3μm and 5μm respectively. They are all in accordance with our spec design. The overall profiles of RDL1 and RDL2 were presented in Fig. 10, which shows good uniformity and appearance. Further detailed information is demonstrated by SEM picture in Fig. 10(c). It shows good dry film filling, uniform pad openings, ordered Cu pillar array, and perfect metal shielding layers. Owing to successful developments of these processes, we obtained a good final functional test yield of 98%. The main yield loss distributes in the edge of wafer due to handling which can

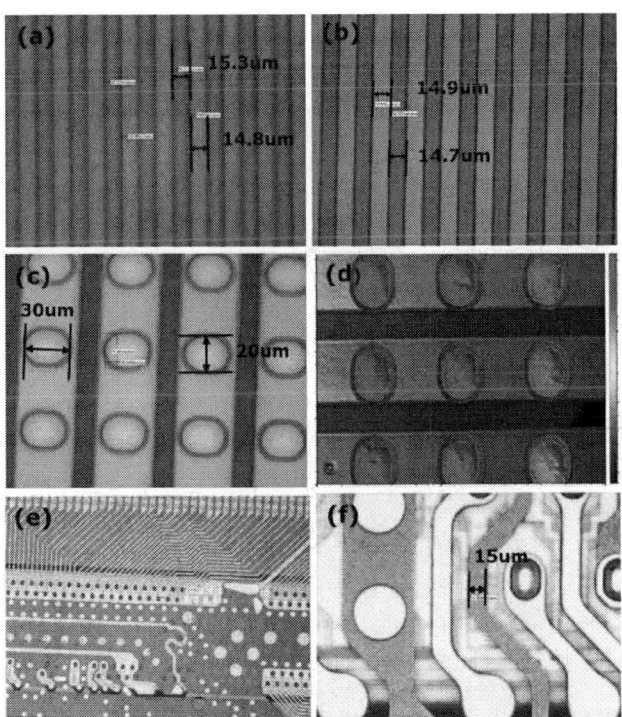

Figure 6. RDL1 fabrication process: (a) litho patterns and (b) plating Cu layer; (c) Cu pillar formation and (d) Cu pillar height data measured by x-ray; (e) metal shielding layer and (f) enlarged view of minimum lead width

easily result in particle contamination and damage to good die.

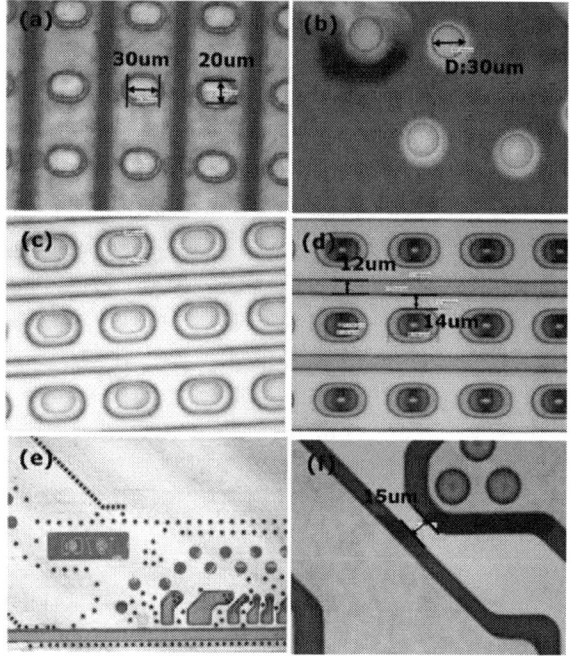

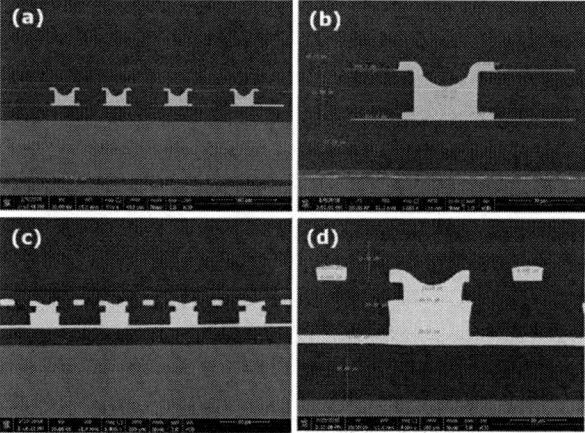

Figure 8. PA2 pad openings of (a) Cu pillar and (b) fist metal shielding layer; (c) (d) litho patterns and plating Cu of RDL2; (e) (f) pictures of second metal shielding layer

Figure 9. SEM pictures after RDL2, (a) overview of two metal shielding layers; (b) interconnection of metal shielding layers; (c) Cu pillar distribution; (d) Cu pillar connection with RDL2

After wafer level eSiFO® packaging, LGA packaging is required to improve reliability and compatibility with end customer system needs. Therefore, there should be an E-less Ni/Pd/Au plating process for the required wire bonding process. Considering the fine pitch of L/S, it is easy to cause lead shorts when E-less plating Ni because the Ni metal layer thickness should reach to 2μm. So we only do the E-less Ni/Pd/Au plating on bonding pads with a positive photoresist protection. After the E-less plating process this positive photoresist was stripped. Fig. 11(d) exhibits each metal layer thickness, including 2.2μm metal Ni layer, 161nm metal Pd

layer, and 388nm metal Au layer. In order to protect the RDL2 layer, a third passivation (PA3) was coated. PA3 used a spin coating method rather than vacuum lamination because there is no rough surface after RDL2. The pad opening of PA3 is a chamfered square whose length side is 60μm. Up to now, wafer level eSiFO® was finished and the overview image is presented in Fig. 12. To execute the following LGA packaging, the packaged wafer was diced to single dies.

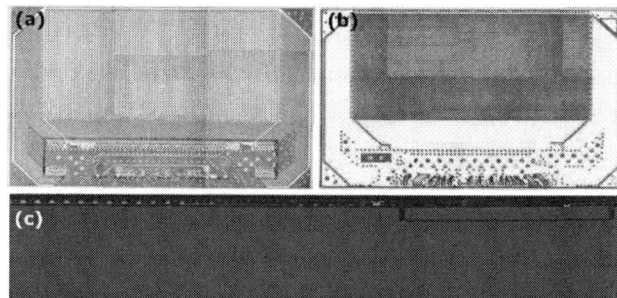

Figure 10. (a) RDL1 layout; (b) RDL2 layout; (c) x-section picture after RDL2 process

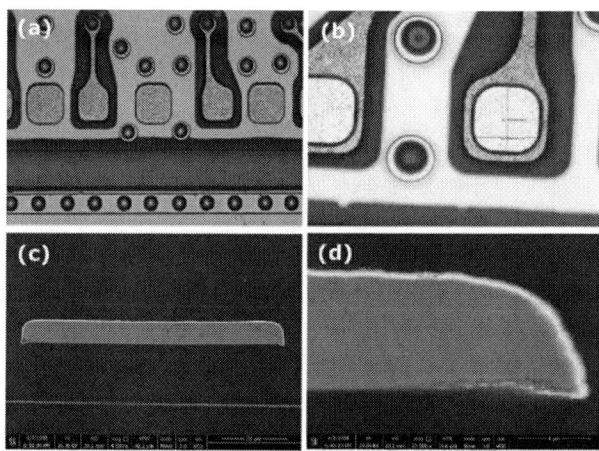

Figure 11. (a) Microscope picture after plating Ni/Pd/Au; (b) pad opening of PA3; (c) (d) cross section of bonding pad after plating Ni/Pd/Au

Figure 12. Overview after wafer level eSiFO® package

6)   LGA encapsulation

Finally, the chips are further packaged in an LGA format in order to be compatible with current end customer system needs. Before LGA encapsulation, the wafer was thinned to

978-1-7281-1500-9/19 $31.00 © 2019 IEEE

280μm from backside, and then dicing to single dies to do LGA encapsulation with two kinds of package sizes as shown in TABLE 6. Fig. 13 shows the microscope photos of single chip before LGA encapsulation and schematic diagram after LGA encapsulation. These two package sizes are 13mm*13mm*0.65mm with 50μm EMC clearance thickness and 13mm*13mm*0.69mm with 90μm EMC clearance thickness respectively as shown in Fig. 13(e) and (f).

TABLE 6. TWO LGA PACKAGE SIZE OF LGA

NO.	EMC Clearance (um)	DAF (um)	PCB (um)	Wire Loop Height(um)	Package size(mm)
1	50	20	300	40MAX	13*13*0.65
2	90	20	300	75MAX	13*13*0.69

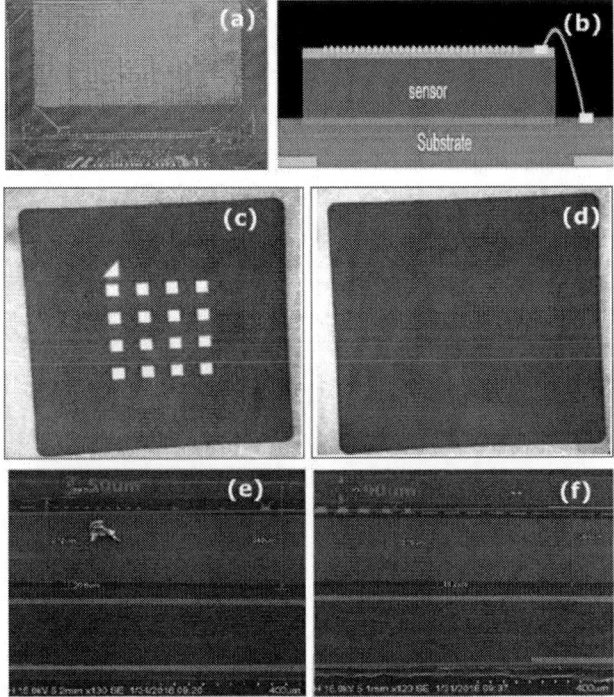

Figure 13. (a) Microscope photo of single die; (b) schematic diagram of LGA; (c) front side and (d) back side pictures after LGA package; (e) (f) cross-section images of final LGA package

LGA packages can result in large warpage which has great infuluence on FPS performance. So a mechanical simulation was used to analyze the warpage for the LGA package. Firstly, the 3D finite element model of LGA package was established and then analyzing the warpage of unit under the temperature load from 175 ℃ to 25 ℃. Fig. 14 shows the 3D model used for simulation and TABLE 7 showed the best simulation result for 50um and 90um EMC clearance. The simulation results demonstrate that 90μm thick EMC clearance has the lowest warpage [9].

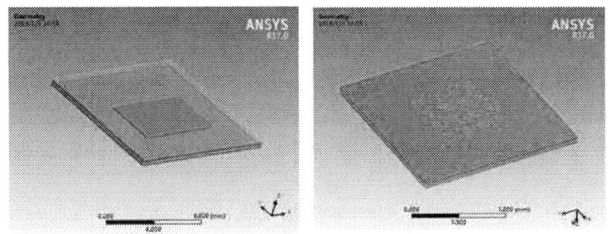

Figure 14. 3D finite element simulation model

TABLE 7. BEST SIMULATION RESULTS FOR DIFFERENT EMC CLEARANCE THICKNESSIF 50UM AND 90UM

No.	SB Thickness /mm	Core Type	Die Thickness /mm	EMC Thickness /mm	Unit Warpage /um	EMC clearance /um
1	0.3	HL-832N S-LC	0.28	G1250AH-F1-B/0.35	30.5(Cry)	50
2	0.3	HL-832N S-LC	0.28	G1250AH-F1-B/0.39	22.7(Cry)	90

7) Reliability Tests

Standard reliability tests were executed to qualify eSiFO® and LGA packages. 77samples were used for each reliability test. Pre-con was performed to simulate the effects of board assembly. Thermal cycling (TC-B) test (-55~125 ℃) reaching to 2000cyc, high temperature storage (HTS) test at 150 ℃ for 1000 hours, and un-bias highly accelerated stress test temperature storage (uHAST) test at 130 ℃ under 85% relative humidity (RH) and 33.3psia for 192 hours. No failures were found after all reliability tests. The reliability test conditions and results are summarized in TABLE 8.

TABLE 8 RELIABILITY TESTS AND RESULTS OF THE LGA PACKAGES

Test item	Condition	Read point	Samples	Result
Pre-con	MSL#3(30℃/60%RH)	192hrs	154	Pass
		500cyc		Pass
TC	TC-B(-55℃~125℃)	1000cyc	77	Pass
		2000cyc		Pass
uHAST	130℃/85%RH 33.3psia	96hrs	77	Pass
		192hrs		Pass
HTS	150℃	500hrs	77	Pass
		1000hrs		Pass

III. CONCLUSION

In this paper, a capacitive fingerprint sensor using eSiFO® technology is successfully developed. The maximum warpage was less than 2 mm although more than 80%

package area was covered by metal layer. For the wafer level process, ten masks were used. The final yield is more than 98% with high reliability performance. The results clearly demonstrate that eSiFO® is an effective as well as versatile technology platform, and has great potential for wide applications.

ACKNOWLEDGMENT

The authors would like to express their gratitude to the great support and collaboration from Synaptics Incorporated. The process support from equipment suppliers including ASM Pacific, and NAURA Technology Group Co., Ltd., Beijing is also highly appreciated. In addition, the authors would like to thank the support from various teams at different locations within Huatian Group.

REFERENCES

[1] Daquan Yu, et al., Embedded Si Fan Out: A Low Cost Wafer Level Packaging Technology Without Molding and De-bonding Processes, ECTC 2017.

[2] Shuying Ma, Jiao Wang, Fengxia Zhen, Zhiyi Xiao, Teng Wang, Daquan Yu, Embedded Silicon Fan-out(eSiFO): A Promising Wafer Level Packaging Technology for Multi-Chip and 3D System Integration., ECTC 2018.

[3] Cheng Chen, Teng Wang, Daquan Yu, Shuying Ma, Kai Zhu, Zhiyi Xiao, Lixi Wan, Reliability of Ultra-thin Embedded Silicon Fan-out(eSiFO) Package Directly Assembled on PCB for Mobile Applications., ECTC 2018.

[4] Z. Xiao, J. Fan, Y. Ren, Y. Li, X. Huang, D. Yu, W. Zhang, "Development of 3D Thin WLCSP Using Vertical Via Last TSV Technology with Various Temporary Bonding Materials and Low Temperature PECVD Process," Proc. 66th Electronic Components and Technology Conference (ECTC), IEEE Press, Jun. 2016, pp. 302-309.

[5] M. Brunnbauer, et. al. "Embedded wafer level ball grid array (eWLB)." 10th EPTC , 2008, pp 994-998.

[6] John H. Lau, Nelson Fan, Li Ming, " Design, material, process, and equipment of embedded fan-out wafer/panel-level packaging", Chip Scale Review, May/June, 2016, pp. 38-44

[7] Yaojian Lin, Chen Kang, Linda Chua, Won Kyung Choi and Seung Wook Yoon, "Advanced 3D eWLB-PoP(embedded Wafer Level Ball Grid Array-Package on Package) Technology", Proc. 66th Electronic Components and Technology Conference (ECTC), IEEE Press, Jun. 2016, pp. 1772-1777

[8] X. Zhang, JK Lin, S. W, S. Zhang, et. al., "Heterogeneous 2.5D integration on through silicon interposer". Applied Physics Reviews, 2015, 2(2):021308.

[9] C.C. Liu, et al., "High-performance integrated fan-out wafer level packaging (InFO-WLP): Technology and system integration", IEDM Dec.2013, pp.14.1.1-14.1.4.

Three-Dimensional Integrated Circuit (3D-IC) Package Using Fan-out Technology

Jun Kyu LEE, Sang Yong PARK, Young Ho KIM, Jae Cheon LEE, Sung Hyuk LEE, Chul Hyo LEE,
Yong Tae KWON, Chang Woo LEE, Jong Heon KIM, Nam Chul KIM, Yun Hyun SUNG

nepes Corporation

Gwhaksaneop-2ro, Ochang-eup, Cheongwon-gun, Cheongju-si, South Korea 28116

jklee0208@nepes.co.kr

Abstract— **Three-dimensional integrated circuit (3D-IC) and 2.5D IC with Si interposer are regarded as promising candidates to overcome the limitations of Moore's law due to their advantages of lower power consumption, smaller form factor, higher performance, and higher function density. To achieve 3D and 2.5D IC integrations, several key technologies are required, such as through-silicon via (TSV), wafer thinning and handling, as well as wafer/chip bonding. Among the 3D integrated technologies, TSV process technology is well-known for penetrating via hole inside the chip followed by metal filling. However, the mass production of dedicated chips for TSV purpose is not widely seen across the semiconductor market due to high investment cost and low productivity. Secondly, the conventional POP (Package-on-Package) has the potential risk for solder-joint defect caused by CTE mismatch between top package to bottom package, which may result in poor reliability.**

To solve these problems, we propose a form of stacked package solution based on Fan-Out Package Technology. This has the advantage of implementing a solder-less joint structure by executing die and via stack-up on panel repeatedly, which can simplify the process flow and improve productivity, reliability as well. In this paper, the development of 2-die stacked package using self-developed Artificial Intelligence (AI) chips was reported. The stacked package has a form factor of 6.75x6.75 mm, 0.78 mmt and ball I/O of 78ea, including two AI chips with size of 4.5x4.6 mm. The package was successfully constructed based on advanced fan-out platform technologies such as the stacking of known good dies under panel level, encapsulation and passivation by laminating thick dielectric material of 150 µm, the technology which is patterning fine pitch arrayed-via on die pad and deep via in fan-out zone at the same time, void-less via filling.

Keywords-Fanout package; 3D-IC; panel level processing; Deep-via stack-up; Artificial Intelligence

I. INTRODUCTION

With the miniaturization of electronic parts, the trend of package technology is gradually evolving toward satisfying all the effects of heat dissipation, high electrical characteristics, and high reliability while realizing multi-function and high integration with a small form factor. In the case of conventional 2D package technology, the wide area is required for the electrical and mechanical interconnection between chips, which is made of long metal wiring. It causes a higher parasitic capacitance and inductance, which

restrains the low power consumption and high signal bandwidth as well as noise margin, chip design flexibility, and package cost. On the other hand, the invention of three-dimensional (3D) integrated package technology enhances design flexibility, performance improvement, and cost effectiveness through high integration of chips having various functions in a small space [1-3].

As the solution for 3D packaging, most semiconductor packaging companies use package-on-package (POP) types to stack the individual packages. However, the interconnection between package to package is commonly carried out by soldering with solder balls, which causes a lot of assembly failures, low solder joint reliability due to CTE mismatching between top package and bottom package, as well as increase in overall package thickness (Fig. 2) [4].

Fan-out-based 3D-IC package is a solution for stacking heterogeneous chips through the panel level process without solder ball joint, so it can prevent assembly failures in advance and also expected to have excellent electrical characteristics, heat dissipation performance and thin package due to short interconnect length.

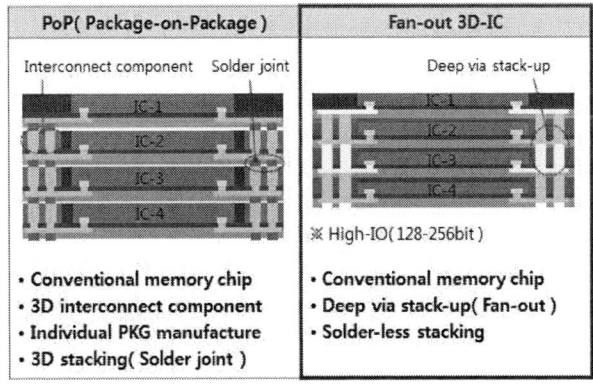

Figure 1. Conventional PoP (Left) and Fan-out 3D-IC (Right)

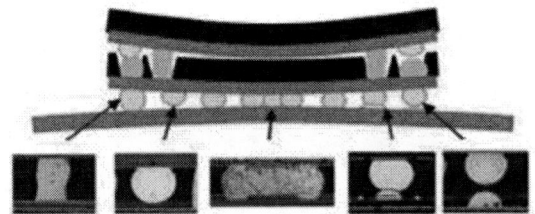

Figure 2. Solder Joint Failure Mechanism Due to Warpage Caused by Mismatch of PoP

The proposed 3D-IC package is using the Fan-Out platform technology to provide a three-dimensional connection at the panel level, which the 2nd chips are stacked on a conventional fan-out reconstituted panel with a built-in 1st chip and connected between the pad of the 2nd chips and redistributed layer on bottom panel through deep-via technology.

Figure 3 shows the process flow for the fan-out-based 3D-IC manufacturing. As a first step, the 1st chips are processed in the form of 300 mm panel through a typical fan-out process. Then, 2nd KGD chip attached with a DAF on the die backside are bonded to the panel in face-up direction, and thick photosensitive film is laminated on dies in place of traditional EMC (Epoxy Mold Compound) molding method. In the next step, the via patterns are formed simultaneously through the photolithography process on the 2nd die pad and the redistributed pad formed in the fan-out zone of the bottom panel, and the vias are connected by Cu RDL (Redistribution Layer). After the via hole is filled with EMC material, unnecessary EMC covering on Cu RDL is etched to expose the surface of Cu RDL. Then, the structure is completed through the PSV (Passivation) / UBM (Under Bump Metallurgy) / ball mount process. All of these process is done under 300 mm round platform, and finally the individual package is completed through process carrier removal and sawing process.

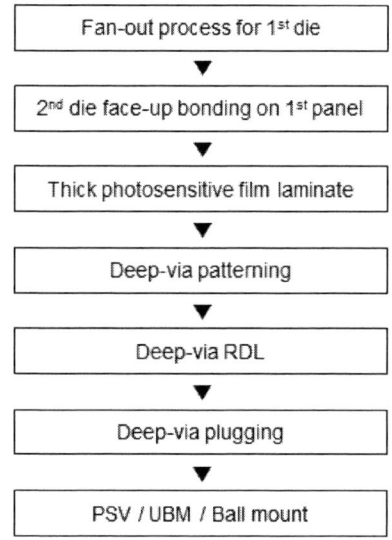

Figure 3. Process Flow of Fan-out 3D-IC Package

II. FANOUT-BASED 3D-IC PLATFORM TECHNOLOGY

A. Test Vehicle Design

The test vehicle was designed to the body thickness of 0.53 mm with a package size of 6.75x6.75 mm, and two artificial intelligence chips with the size of 4.5x4.6 mm, 576 neurons were stacked in one package.

Via pitch for 3D interconnection between top chips and bottom chips was 0.26 mm and the package terminal was composed of solder ball 78 ea with the pitch of 0.65 mm.

Table 1 shows the detailed specifications, and Figure 4 shows the actual RDL routing design for the 1st layer and 2nd layer and ball map.

In general, total profile of PoP structure using flip chip BGA is known to be about 1 mm, while the thickness of 0.53 mm of this structure is expected to be competitive in the aspect of thin profile, furthermore, it implies that overall profile for 4-die stacking application can be realized to be 1 mm or less.

Item	Specification
Device(Die)	NM500(Neuromorphic AI chip), 576 neurons
# of die stack	2-die stack
Die size	4.5x4.6 mm, 0.12 mmt
Package size	6.75x6.75 mm, 0.53 mmt (PKG body thickness)
Via pitch	0.26 mm
Ball pitch	0.65 mm, Ball I/O 78ea

Table 1. Package Specification

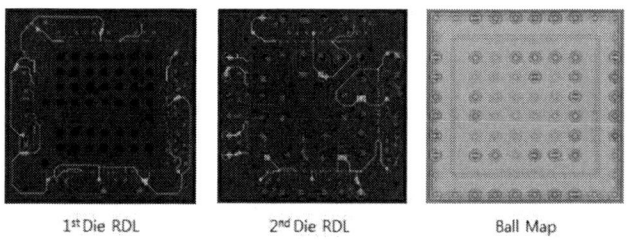

1st Die RDL   2nd Die RDL   Ball Map

Figure 4. RDL routing design and ball map

B. Manufacturing of 1st Die Reconstituted Fanout Panel

Base panel including the 1st die was fabricated through a typical fan-out process. The dies were mounted on a temporary carrier, and encapsulated by EMC in form of 300 mm round panel. After that, it was attached to the ceramic carrier, followed by build-up process in order to generate RCF (Reconfiguration) layer, RDL layer and PSV layer on the active side on dies. Figure 5 shows a 300 mm round basic panel with 1st die embedded and extended RDL from die pad. By using the base panel supported by the ceramic carrier, it is possible to control the panel warpage that could occur in the subsequent process such as 2nd die stacking and build-up process.

978-1-7281-1500-9/19 $31.00 © 2019 IEEE

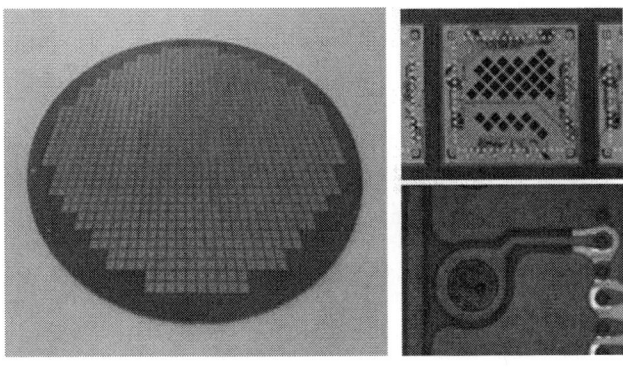

Figure 5. 1st die embedded fanout panel (Left) and RDL pattern of fanout zone (Right)

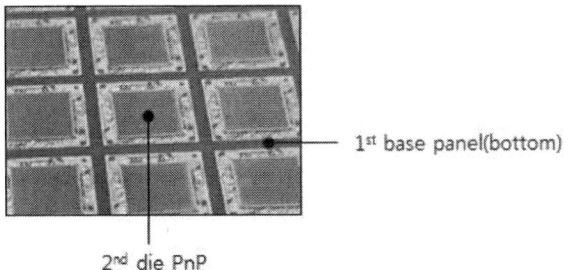

Figure 6. 2nd die bonding image on 1st die embedded fanout panel

C. Encapsulation using Thick Photosensitive Film

To encapsulate the structure bonded with the 2nd chips on the bottom panel, the laminating process was applied using thick and photosensitive insulating film. At this time, the surface of film after laminating should be planarized without voids between the chips to minimize impact during subsequent layer stacking and build-up process. In this work, two-step laminating method was selected to ensure void-less and flattening.

As shown in Figure 7, at the first step, the film was conformal laminated using the rubber based press in vacuum environment to remove the voids between the dies. Secondly, the panel was pressed additionally with SUS (Steel Use Stainless) stage for planarization. As a result, a flatness of less than ± 10 μm was secured in a 300 mm round panel as shown in Figure 8.

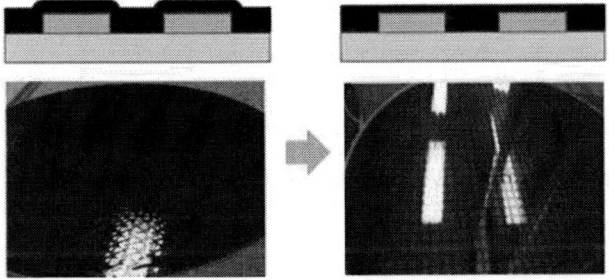

Figure 7. 2-step laminating : (A) Conformal Laminate, (B) Flattening

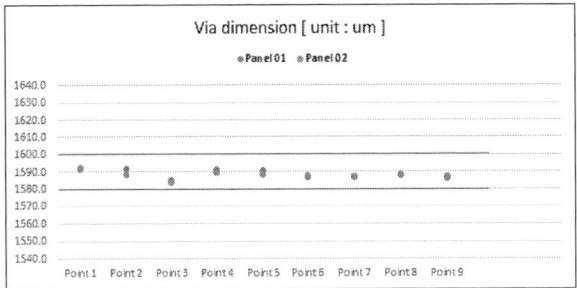

* Thickness include the process carrier and bottom panel.

Figure 8. Film Thickness Variation after 2-step Laminating

D. 3D Interconnection Technology: Deep-via Formation

After the thick photosensitive flim was laminated as described above, the respective vias for connecting the pad portion of the 2nd chip and the redistributed pad portion of the fan-out region of the bottom panel should be defined. The chip pad has a small via opening size of 20 ~ 30 μm and via depth of 30 μm in the material of Al, on the other hand, the via size and depth on Cu pad of the bottom panel were designed to be 150 μm and 160 μm respectively. To define the pattern of small via and deep-via simultaneously on different kinds of pads but under same package, epoxy-based photosensitive film was carefully selected which provides wide photo process window margin with excellent uniformity, resolution and side wall profiles.

Cross-section image illustrated in Figure 9 confirmed that heterogeneous-sized vias was well defined despite that different film thickness across package. The images seen in Figure 10 showed the top view images of small via and deep-via opening while Figure 11 revealed the actual measurement data of small via on the 2nd die and a deep via in the fan out zone on the bottom panel, indicating that both types of via openings were able to achieve and comply the target specification of 20 μm/30 μm (± 5 μm), 150 μm/160 μm (± 5 μm) respectively.

Figure 9. Cross-sectional View of Small Via and Deep-Via

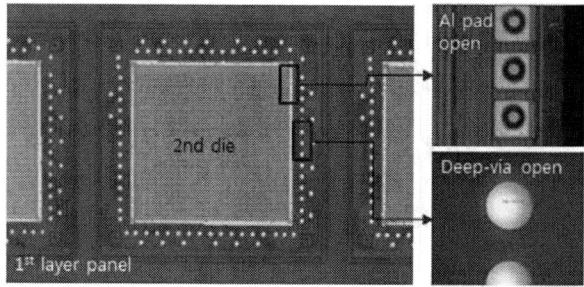

Figure 10. Via Opening Images of Small Via and Deep-Via

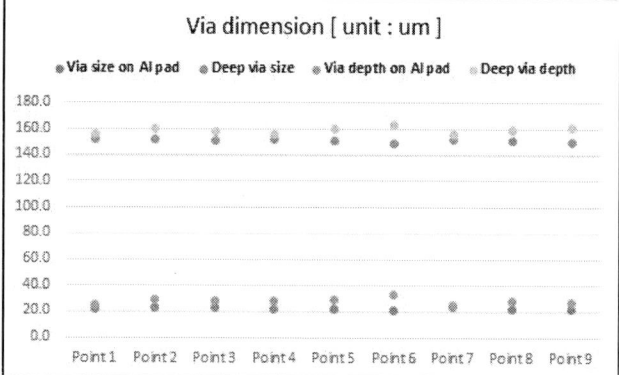

Figure 11. Structure and Specification for Small Via and DeepVia

E. 3D Interconnection Technology: Deep-via Plugging

In general, deep-via plugging process consists of the photo patterning of deep-via, Cu redistribution, filling the via hole with EMC, and exposure of RDL surface by mechanical etching as shown in Figure 12.

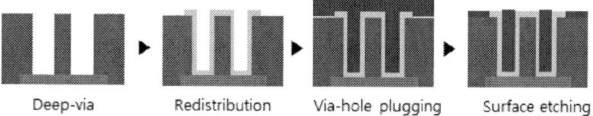

Figure 12. Process Flow of Deep-via Plugging

In this project, deep-via formation by fully plated with Cu was not considered because of the increased plating time may result reliability issues in the via structure due to excessive plating stress, and also may be disadvantageous to cover with the subsequent passivation layer due to increase of Cu overburden on the via top surface [5]. Therefore, Cu plating was made in the sidewall of via-hole, and the other was filled to EMC material with low CTE (Coefficient of Thermal Expansion) and low modulus properties to minimize the stress in the via.

To confirm about electrical characteristics of EMC-plugged via, a simulation test was conducted in comparing with Cu-filled via (Figure 13). As a result, the parasitic difference between two structures is not significant in terms of capacitance, inductance and resistance as shown in Table

2. Hence, it is predicted that the electrical characteristics do not deteriorate even if Cu plating is performed only on the via-hole wall.

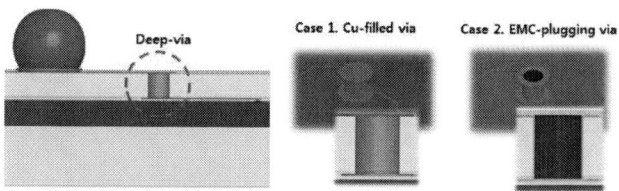

Figure 13. Electrical Simulation Modeling of Deep-Via Structure

Item	Target specification	Case 1 Full fill	Case 2 Deep via
Capacitance (pF)	0.15~0.7	0.33	0.35
Inductance (nH)	< 3	2.47	2.47
Resistance (Ohm)	45~85	83	83.3

Table 2. Electrical Performance Comparison between Cu-Filled Via and EMC-Plugging Via

Figure 14 shows Cu-plated image inside via-hole, the Cu step-coverage at via bottom side was confirmed to more than 50% of the via-top side. Also, EMC filling inside via hole and RDL exposure by the mechanical etching were performed successfully as shown in Figure 15.

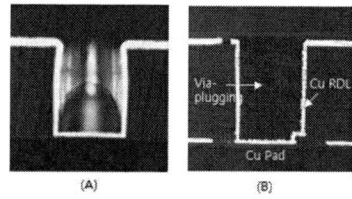

Figure 14. (A) Cu-Plated Along Deep-Via Side Wall, (B) Via-plugging

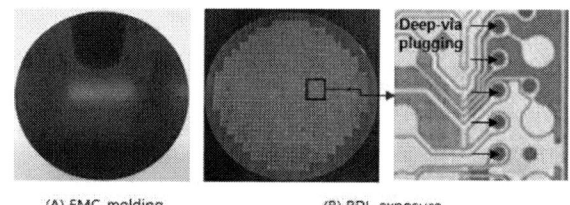

Figure 15. Deep-via plugging; (A) EMC molding (B) RDL exposure after mechanical etching

F. Prototypes

Dual-die-stacked 3D-IC package was completed through the passivation layer, UBM and ball mounting process on the via plugged panel as described above. Figure 16 shows the top, bottom and cross section view of the completed package. As shown in the cross-sectional image, it is confirmed that the connection of 1st die and 2nd die was successfully connected through the deep-via of the fan-out zone.

978-1-7281-1500-9/19 $31.00 © 2019 IEEE

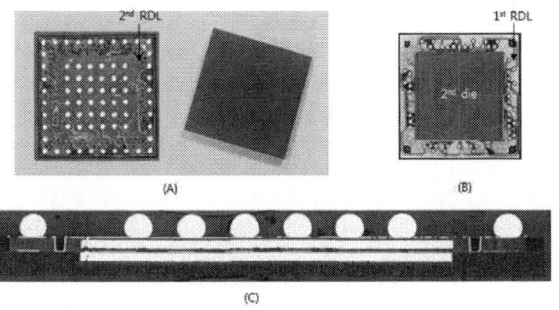

Figure 16. Prototype of Fan-Out 3D-IC ; (A) Top and Bottom View (B) Package Inside View (C) Cross-sectional View

III. PERFORMANCE

We used NM500 AI chip which is produced in-house. NM500 is a brain-based AI semiconductor chip based on neuromorphic consisting of neurons and nervous system.

Depending on the application, it is necessary to increase neurons to improve AI performance. However, a separate board is implemented and combined in order to increase the neuron by the conventional method. 3D-IC technology can be used to reduce the package form factor as shown in the Figure 17, while same performance can be achieved.

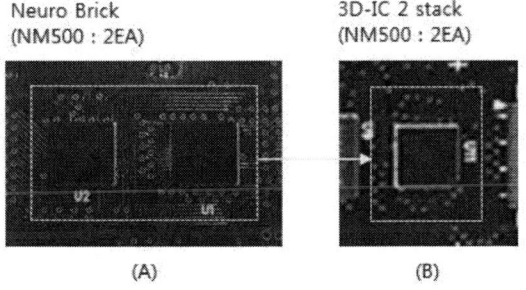

Figure 17. Comparison with Side by Side Array (A) and 3D-IC 2-Stack (B)

A. Functional Test

Figure 18 shows the test board to evaluate the 2-stacked NM500 prototype. The test environment consisted of socket board, NEUROSHIELD, Orange Board (Arduino) and PC. Test Board H/W is implemented using orange board (Arduino) embedded for operating and functional verification [6]. NEUROSHIELD converted the interface to SPI communication using bus I/O used in NM500. Test socket is a sub board for connecting the developed 3D-IC to the board.

Figure 18. 3D-IC Test Board Configuration

Figure 19 is the function test result for the prototype of 3D-IC. The learning by the input data was operated normally and the number of neurons was confirmed to increase to twice capacity of NM500 by designed.

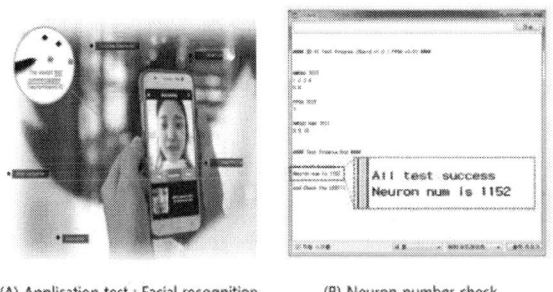

(A) Application test : Facial recognition   (B) Neuron number check

Figure 19. (A) Application ; Facial Recognition (B) Neurons Check

B. Reliability Result

The test vehicle with the daisy chain was fabricated for verifying the board level reliability, which has the ball pitch of 0.35 mm. Fig. 20 showed package after SMT process on test board. Board level TCT (Temperature cycling) test was carried out according to JESD22-A104E requirement.

From the Table 3 result, all the 45 units had passed 150 cycles without initial failure. Currently, the extended testing to secure the fatigue lifetime is in progress. The optimization of structure and design will be done according to the reliability results.

Figure 20. Package SMT on JEDEC Test Board

Test mode	Test condition	Sample size	Test result	Reference
Board Level Temperature Cycling Test	Temp. : -40℃~125℃ Ramp rate : 14℃/min Dwell time : 10min Duration : 150cycles	#1 15unit	PASS	JESD22-A104
		#2 15unit	PASS	
		#3 15unit	PASS	

Table 3. Board Level TC Result

IV. CONCLUSION

We have successfully developed a fan-out based 3D-IC packaging platform technology using a proprietary AI chip (NM500) based on the photosensitive thick film, deep-via formation & plugging element technology.

The extended fan-out technology has superior advantages compared to the conventional POP with solder joints interconnection in terms of package form factor, electrical and thermal performance since the solder joint for assembly purpose is replaced by the deep via patterning technology. In addition, it is expected to minimize the various process issues occurring during several dies stacking process and therefore bringing great advantage in productivity aspect.

This technology is expected to be one of the 3D package solutions that can be applied not only to the memory stack field, but also to the integration of composite ICs such as AP, memory, and sensors, etc.

As the next study, the development the 4-die stack of less than 150 μm via pitch will be continued to enhance the scalability of the application in the future.

ACKNOWLEDGMENT

The authors would like express their gratitude to the Technology Innovation Program (20000868, Development of AI 3D IC Fabrication Process Technology using a FO Package) funded By the Ministry of Trade, Industry & Energy (MOTIE, Korea) for their supporting.

REFERENCES

[1] John H. Lau., "TSV manufacturing yield and hidden costs for 3D-IC integration," 2010 Proceedings 60th electronic components and technology conference (ECTC), Las Vegas, NV, USA, pp. 1–4, June 2010.

[2] John H. Lau., "Evolution, challenge, and outlook of TSV, 3D IC integration and 3d silicon intergration" 2011 international Symposium on Advanced Packaging Materials (APM), Xiamen, China, pp. 1–4, Oct 2011.

[3] Cheng-Ta Ko, and Kuan-neng Chen "Wafer Level Bonding / Stacking technology for 3D integration," Microelectronics Reliability, Vol 50, pp. 481–488, April 2010

[4] Jong Heon Kim, Yong Tae Kwon, Young Ho Kwon and Yong Woon Yeo, "Fan out package: Performance and Scalability Perspective" ECTC 2018. vol. 68, pp. 1194–1199, 2018.

[5] Yun-Mook Park, "Development of Silicon module with Cu-filled TSV and Intergrated Passive Devices" IMAPS 2010 – 43rd International Symposium on Microeletronics, vol. 43, pp. 385-391.

[6] Tarig M. King, "Getting Started with AI for Testing," Midium.

2019 IEEE 69th Electronic Components and Technology Conference (ECTC)

Ultra High Density IO Fan-Out Design Optimization with Signal Integrity and Power Integrity

Keng Tuan Chang, Chih-Yi Huang, Hung-Chun Kuo, Ming-Fong Jhong, Tsun-Lung Hsieh, Mi-Chun Hung, Chen-Chao Wang

Integrated Design, Corporation Design Division, Corporate Research and Development, Advanced Semiconductor Engineering (ASE), Inc.,

No. 26, Chin 3rd Road Nantze Export Processing Kaohsiung 811, Taiwan

Email: gordon_chang@aseglobal.com

Abstract

With the development of internet and the rise of artificial intelligence industry, the high performance semiconductor integrated circuits have become a hot product in the semiconductor industry. The 2.5D IC package with ultra-high density I/O is the first structure applied on high performance computing (HPC) like GPU. Applied on GPU or HPC, there is an ASIC die and multiple HBM dice on silicon interposer. Between ASIC die and HBM die, there are lots of high speed signal lines between them and over hundreds of thousands of small vias. But the productivity of silicon interposer is always issue to realize the ultra-high density I/O products. To consider the productivity and performance, TSV-less structure like FOCoS (Fan-Out Chip on Substrate) is proposed by few years ago [1] [2] [3] [4] [5].There are Chip First FOCoS and Chip Last FOCoS for different process and application [6].

In this paper, a real case with an ASIC die and 2 HBM dice is designed in 2.5D IC and Chip Last FOCoS structures. In this real case, the interposer design and Fan-Out RDL is utilized SiP-id (System in Package intelligent design) design platform to accelerate the ultra-high density I/O routings. In addition, the electrical performance including signal integrity (SI) and power integrity (PI) are compared between 2.5D IC and Chip Last FOCoS. From the analysis results, the dynamic power noise between the two structures is showed in this paper and the electrical performance of HBM2 and 28Gbps SerDes I/Os are displayed as well.

I. Introduction

The big data calculations were originally used in space science or meteorological simulation and analysis which were performed by supercomputers. With the advancement of information industry, internet and communication technology, the rise of the Internet of Things, big data analysis, artificial intelligence, industrial 4.0, autonomous driving and even the development of robotic development applications, the demand for huge amounts of data computing has been everywhere.

Previous supercomputers were usually built with many computers and a large number of CPUs, having the advantages of high speed, high bandwidth and huge computing capability, but volume of supercomputer is very large.

As semiconductor design and manufacturing capabilities continue to improve, today's computers can use smaller volumes to perform as much or as much computing power as ever. The 2.5D IC, as shown in Figure 1, is an innovation in IC packaging to meet the demand for huge amounts of computing in a small size.

II. Ultra High Density Packages

The main feature of the 2.5D IC is to use a Silicon Interposer to connect the homogenous dice or heterogeneous dice of a fine pitch footprint to expand computing capabilities or data bandwidth of a single IC package. Now many products have been successfully developed and mass produced with 2.5D IC packages.

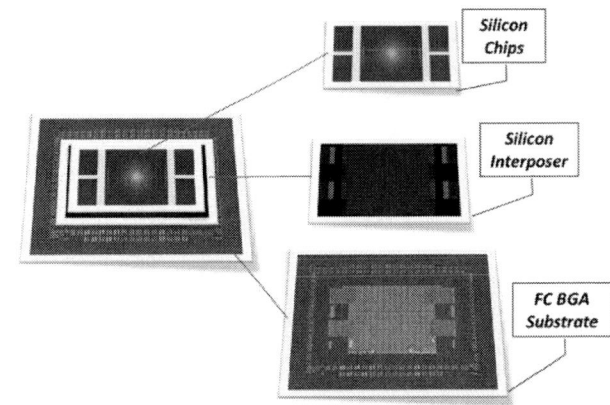

Fig. 1 2.5D IC Package Structure

Silicon interposer is manufactured using wafer fabrication process with Through-Silicon-Via (TSV), as shown in Figure 2. Therefore, the complexity of package manufacturing are always much higher than traditional packages. Because less than 1um RDL width and the diameter of via less can be designed on the silicon interposer [7] [8] [9], no other technology can completely replace the silicon interposer currently.

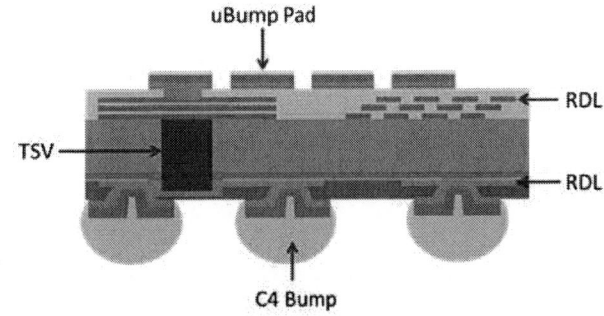

Fig. 2 TSV in 2.5D IC Interposer

The Bumping RDL process is another technology that can be used to design fine lines compared to conventional package

978-1-7281-1500-9/19 $31.00 © 2019 IEEE   41

substrates. Originally the bumping process is only applied to the wafer level package type, which is used as the routing layer for pins redistribution of the silicon IC, the width of the finest line is about 10um.

With the bumping technology is used in Fan-Out package type, and the development of Fan-Out PoP, Fan-Out SiP and integrated Fan-Out Chips on Substrate (FOCoS) packages, fine line requirements and technology development continues to improve, and the suitable applications for Fan-Out packages and the types of product designs are increasing.

For FOCoS which can be used to replace the homogenous integration IC package that can only be designed with 2.5D IC package, as shown in figure 3.

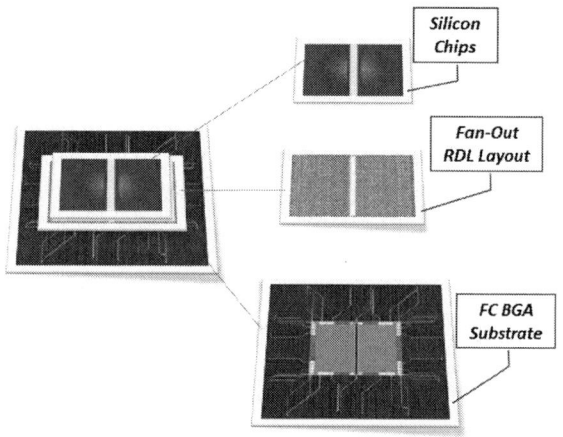

Fig. 3 Fan-Out Chip on Substrate

In this paper, FOCoS is used to design heterogeneous integration IC package which integrated ASIC and HBM (High Bandwidth Memory) dice. Chip Last FOCoS is suitable for ASIC integrated HBM chips IC package because Known-good RDL layer could be available. The spacing between I/O pins is getting smaller with the advancement of IC process technology and the increasing number of I/O pins for integrated multiple chips. The line width of the bumping technology RDL line is improved from the originally 10um to currently 2um. Table 1 is a comparison of the traditional FCBGA substrate, 2.5D IC interposer and the RDL layer design specification on FOCoS. According to the roadmap of the bumping technology, the line width of the RDL line will continue to shrink in the future.

Table 1. Design Rule Comparison

Design Rule	PKG Size	Cu Thickness	RDL L/S	Via Size	Hole Size
FC BGA Substrate	55 mm x 55 mm	15 um	15 um	50 um	150 um
Fan-Out RDL	30 mm x 30 mm	3 um	2 um	10 um	10 um
2.5D IC Interposer	30 mm x 30 mm	1 um	1 um	0.4 um	1 um

III. Design Tools and Flow

In addition to changes and improvements in ultra-high-density I/O package types and the corresponding process technology, the complete design flow and efficient design tools must be ready, as shown in figure 4. The traditional package design tools are difficult to design ultra high I/O package like Fan-Out RDL routings or 2.5D IC interposer routings including layout, design rule check (DRC) and layout versus netlist (LVS) tools. For example, each action of the layout tool becomes very slow even not work when the layout density becomes higher and higher; efficient design for manufacturing (DFM) check because the embedded DRC function in original package design tool doesn't support advanced design rule and new design drawing format; the complex signal interconnections cannot be compared with the original version of the customer's design document because the complete package design is much complexer than tranditional package design. Issue above are the challenges while the package design with ultra-high-density I/Os.

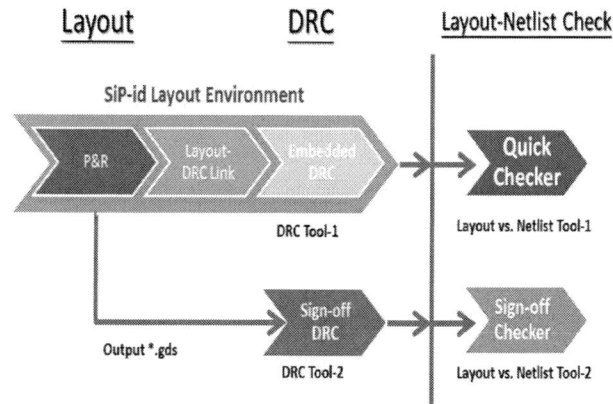

Fig. 4 SiP-id Package Design Environment

Addressing to problems above, the improvements for the design tool and the enhancement of the overall design process, as shown in figure 5, is to be planned and executed by the designer and EDA tool suppliers. Thus it results in a novel high-density I/O package design environment, as shown in figure 4. The novel high density I/O package design flow and tools include improved layout tool, new DRC tools and new LVS tools to make sure better design cycle time and design quality. Generally it takes about 3 months to finish a Fan-Out RDL layers design using the traditional package flow and tools for example. With new SiP-id design flow and tools, the design could be finished in 1 month design cycle time. It has about 3 times improvement.

978-1-7281-1500-9/19 $31.00 © 2019 IEEE

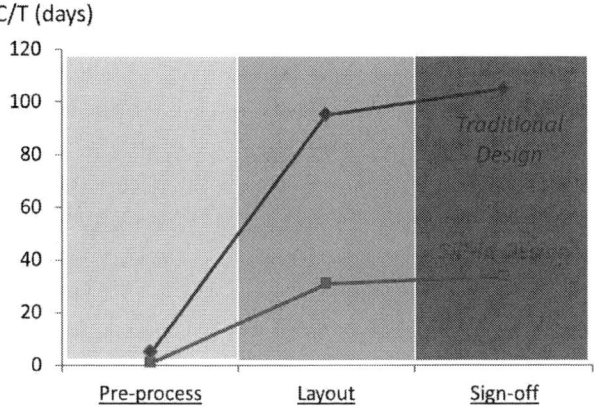

Fig. 5 Design Efficiency Improvement by SiP-id

IV. Design Requirement with HBM

HPC products incorporating High-Bandwidth-Memory (HBM) are common used for high-end GPUs, AI processors, high-end networking processors or state-of-the-art FPGA products. With 1 HBM, the bandwidth of the IC package could be higher than the PCB system with tens of traditional DRAM memory ICs, as well as shorter interconnections distance with better electrical performance between the processor and the memory. Figure 6 is for the JEDEC standard pinout. There are totally 1024 DQ signals and there are 48 DQ signals arranged repeatly in about 155um pitch. In other words, the 48 signals must to be faned-out in 155 um layout space. According to evaluation, it could utilize 2 RDL layers to Fan-Out all 48 signals, 24 signals in each RDL layer, if the RDL line width/space is 2um/2um with 16 um via size.

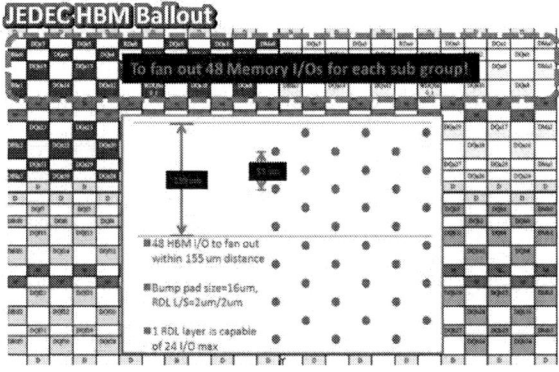

Fig. 6 Partial JEDEC HBM Footprint

Thus, the design strategy could be 2 RDL layers for die to die HBM I/O routings, 1 RDL layer as the ground layer between 2 HBM I/O routing RDL layers and 1 RDL layer for landing C4 pad, there are totally 4 RDL layers for the Fan-Out chip. Because RDL line width and space as well as via size are continuously to be reduced for Fan-Out technology, The design with processor integrated with HBM could be designed in FOCoS in more economical package, as shown in Figure 7.

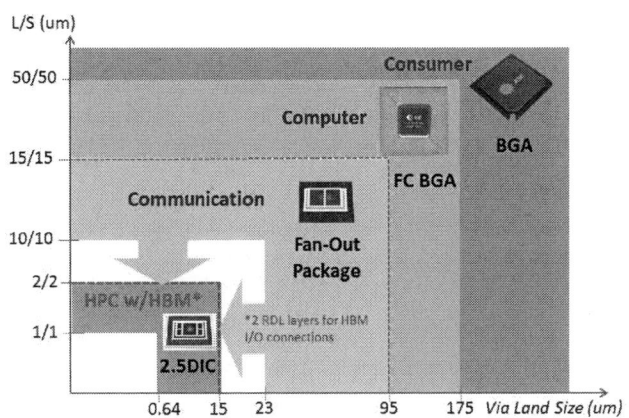

Fig. 7 Package Technology vs. Application

In addition, the process for FOCoS process can now achieve stack vias. The design with stacked vias can enable the package design engineers to have more design space and more design flexibility when designing high density packages. Figure 8 is for photographic images of the actual FOCoS product. In the pictures which shows both stacked vias and non-stacked vias structures.

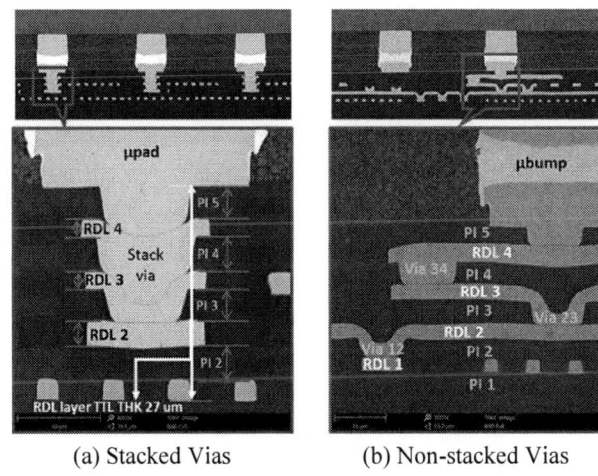

(a) Stacked Vias   (b) Non-stacked Vias

Fig. 8 SEM Images of FOCoS product

V. Electrical Characterization

From viewpoint of package structure and dimension, both 2.5D IC interposer and Fan-Out RDL layer can be designed to be much thinner than conventional BGA substrates. Since thin lines must be made with very thin copper thickness, the thickness is about 15 um for FC BGA substrate, while the thickness of the Fan-Out RDL copper is only 3~4 um, and the thickness of the 2.5D IC interposer is even smaller than 1um. So that the resistance value is very large for the line made by the thin line width plus the thin copper thickness. The resistance value of the FC BGA, Fan-Out RDL and 2.5D IC interposer per 1mm length is compared with table 2. For the resistance of Fan-Out RDL routing is about 70X of the

978-1-7281-1500-9/19 $31.00 © 2019 IEEE   43

resistance of FC BGA substrate, and the resistance is about 4X of the resistance of Fan-Out RDL. The resistance is very high then traditional FC BGA substrate. Therefore, when designing high speed I/O on Fan-Out RDL and 2.5D IC interposer, try to shorten the connection distance to avoid high resistance. The major routings of the I/O traces could be designed on the traditional FC BGA substrate to connect especially for power or ground connections. The design should have vertical direct connection between the micron bump to the C4 bump by stacked vias, so that the high resistance value will not cause product design failure.

Table 2. Resistance of 1 mm Trace Routing

Resistance	wire bond BGA	FC BGA	FOCoS RDL	2.5D IC Interposer
@ DC (mΩ/mm)	19.4	64.4	4314.7	18966
@ 1GHz (mΩ/mm)	106.1	233.1	4552.5	19080

Generally the test pattern is designed at the edge of wafer which is next to the actual FOCoS design in order to verify the electrical performance of this new interface of Fan-Out RDL. The test pattern has a single-end transmission line and a differential pair transmission line to confirm that the electrical simulation tool can achieve accurate results when building the model of HBM I/O and SerDes I/O. Figure 9 and figure 10 are for the comparison of analysis between the measurement and the simulation of single-end and differential pair transmission lines individually. According to results, no matter signle –end transmission line or differential transmission line, the simulation results are very close to the measurement results. Therefore, electrical simulation tools can be used to perform complex electrical performance analysis.

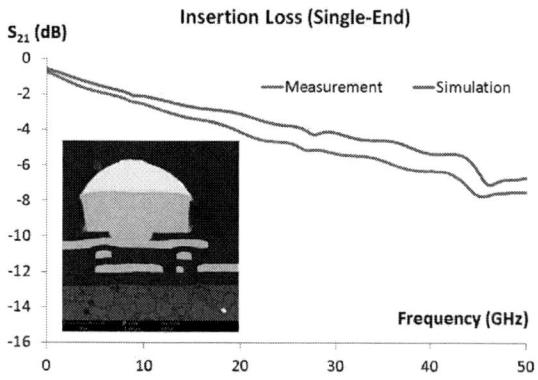

Fig. 9 Simulation vs. Measurement for Single-end TL in FOCoS

The design of a HPC product containing HBM, the most critical signals need to be carefully designed can be divided into HBM interconnection and SerDes signal of high-speed differential signal pair, as shown in Figure 11. Following sections will focus on signal integrity and power integrity analysis on these critical signals.

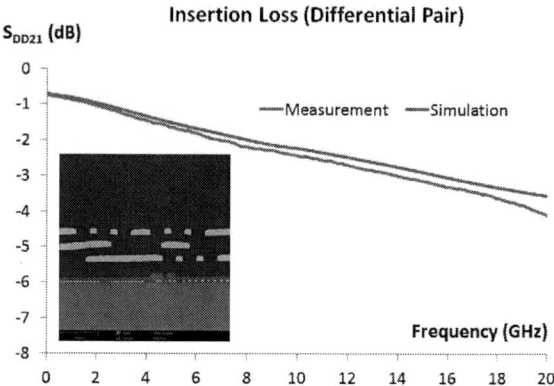

Fig. 10 Simulation vs. Measurement for Differential TL in FOCoS

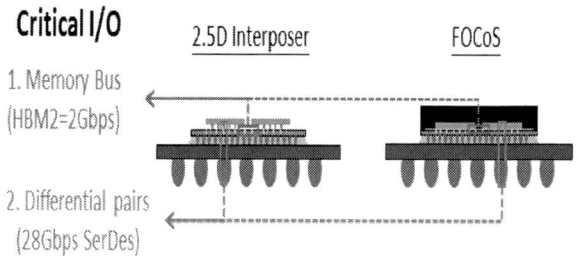

Fig. 11 Critical Signals in HPC Device

VI. Signal Integrity

Model extractions from electromagnetic solver and analysis on the eye diagram by system simulator to compare signal integrity performance of the HBM interconnection and high speed SerDes signal design [10]. The eye diagram analysis results for HBM signals is shown in figure 12. According to HBM eye diagram analysis results, it can be seen that the performance of the FOCoS design is the same as that of the 2.5D IC interposer for the analysis of HBM2 at 2Gbps speed. But at 3Gbps or 4Gbps speed, FOCoS performance is significantly better than 2.5D IC interposer because 2.5D IC interposer has large resistive transmission line loss than FOCoS package. Table 3 is the summary of eye height and eye widh for HBM eye diagram analysis.

Foe 28Gbps SerDes eye diagram simulation is shown as figure 13 and summary as table 4. Because the design from uBump of ASIC die side connected to C4 bump of substrae side is only stacked vias without resistive RDL routings, the eye diagram analysis results are good and almost equal for both 2.5D IC interposer and Chip Last FOCoS package design.

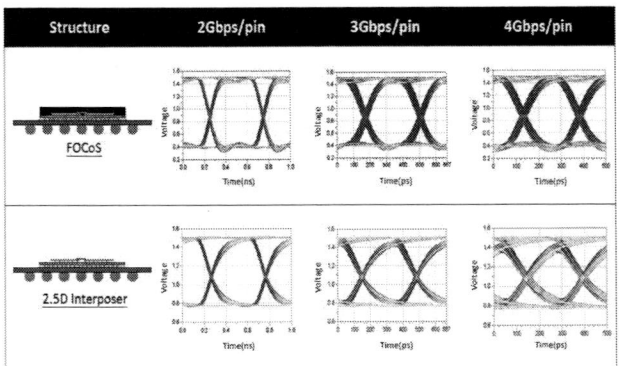

Fig. 12 Eye Diagram of HBM Interconnect

Table 3 Summary of HBM Eye Height and Eye Width

HBM Signal	Eye Height			Eye Width		
Package Design	2Gbps	3Gbps	4Gbps	2Gbps	3Gbps	4Gbps
2.5D IC Interposer	0.59 V	0.49 V	0.4 V	0.98 UI	0.94 UI	0.89 UI
FOCoS	0.97 V	0.99 V	0.95 V	0.98 UI	0.95 UI	0.94 UI

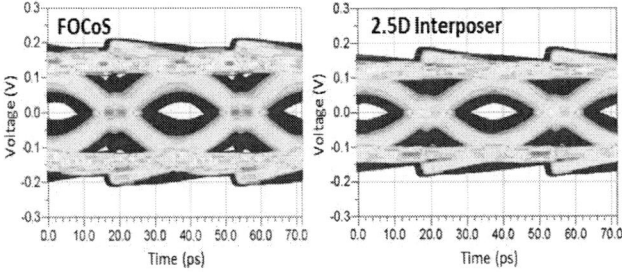

Fig. 13 Eye Diagram of 28Gbps SerDes Signals

Table 4 Summary of 28Gbps SerDes Eye Diagram

28Gbps SerDes	Eye Height	Eye Width
2.5D IC Interposer	0.072 V	0.47UI
FOCoS	0.082 V	0.48UI

VII. Power Integrity

Due to 1024 bits for HBM die, power noise is critical analysis item for HBM relative power system. Figure 14 shows the simulation results of HBM power/ground dynamic power noise including HBM core power VDDC and HBM I/O power VDDO network. Table 5 is the summary of VDDC and VDDO dynamic power

noise analysis results. Both the 2.5D IC interposer and FOCoS designs have similar power integrity designs because they both have tens of thousands of u-bump, via and C4 bumps the design and they are vertically and directly connected. In this dynamic power noise analsys, the decoupling capacitors are not included in the analysis yet.

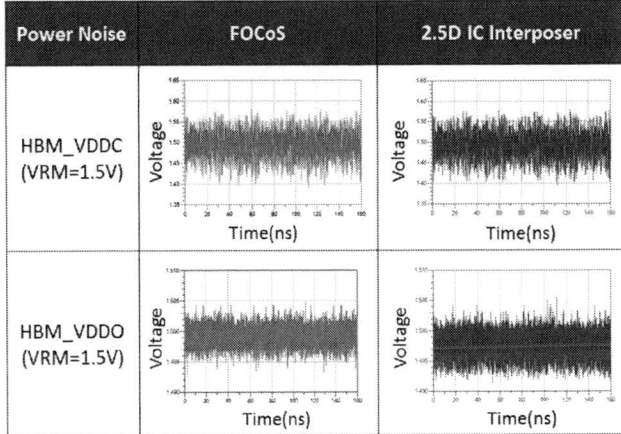

Fig. 14 Dynamic Power Noise Analysis

Table 5 Summary of Dynamic Power Noise

Power Net	Design	Noise_P2P (V)	Noise_P2P (%)
HBM_VDDC (VRM=1.5V)	2.5D IC Interposer	0.187	12.5
	FOCoS	0.189	12.6
HBM_VDDO (VRM=1.5V)	2.5D IC Interposer	0.014	0.9
	FOCoS	0.012	0.012

VIII. Conclusion

It could be expected that there will be more and more requirements for ultra high densisty package design including 2.5D IC interposer and chip last FOCoS design. Therefore the design tools for fast layout, design for manufacturing check including design rule check (DRC) and interconnections check for multi-chips are innovated and integrated. Through the novel design flow with innovated design tools, the design cycle time and quality are improved.

2.5D IC packaging is validated for ultra high density I/O device. FOCoS is an alternative package solution for ultra high density I/O device due to continously improvement of RDL trace width and space as well as the via size. In this paper, the package design for HPC application which includes ASIC die and HBM dice are implemented with SiP-id design platform and fabricated by chip last FOCoS process. From viewpoint of electrical analysis, chip last FOCoS has similar even better electrical performance than 2.5D IC interposer no matter signal integrity or power integrity analysis.

References

[1]. C. F. Tseng, C. S. Liu, C. H. Wu, and D. Yu, "InFO (Wafer Level Integrated Fan-Out) Technology," IEEE Electronic Components and Technology Conference, pp. 1 - 6, May 2016.

[2]. Y. C. Lee, W. H. Lai, I. Hu, M. K. Shih, C. L. Kao, D. Tarng, and C. P. Hung, "Fan-Out Chip on Substrate Device Interconnection Reliability Analysis," IEEE Electronic Components and Technology Conference, pp. 22 - 27, May 2017.

[3]. D. Hinter, M. Kolbehdri, M. Kelly, Y. R. Kim, W. C. Do, J. H. Bae, M. H. Chang, and A. R. Jo, "SLIMTM Advanced Fan-out Packaging for High Performance Multi-die Solutions," IEEE Electronic Components and Technology Conference, pp. 575 - 580, May 2017.

[4]. K. Chen, L. Chua, W. K. Choi, S. G. Chow, and S. W. Yoon, "28nm CPI (Chip/Package Integrations) in Large Size eWLB (Embedded Wafer Level BGA) Fan-Out Wafer Level Packages," IEEE Electronic Components and Technology Conference, pp. 581 - 586, May 2017.

[5]. C. Zwenger, G. Scott, B. Baloglu, M. Kelly, W. Do, W. Lee, and J. Yi, "Electrical and Thermal Simulation of SWIFTTM High-density Fan-out PoP Technology," IEEE Electronic Components and Technology Conference, pp. 1962 - 1967, May 2017.

[6]. S. Chen, S. Wang, J. Hunt, W. Chen, L. Liang, G. Kao, and A. Peng, "A Comparative of a Fan Out Packaged Product: Chip First and Chip Last," IEEE Electronic Components and Technology Conference, pp. 380 - 385, May 2016.

[7]. M. Ma, S. Chen, P. I. Wu, A. Huang, C. H. Lu, A. Chen, C. H. Liu, and S. L. Peng, "The development and the integration of the 5 μm to 1 μm half pitches wafer level Cu redistribution layers," IEEE Electronic Components and Technology Conference, pp. 1509 - 1514, May 2016..

[8]. V. S. Rao, C. T. Chong, D. Ho, D. M. Zhi, C. S. Choong, S. L PS, D. Ismael, and Y. Y. Liang, "Development of High Density Fan Out Wafer Level Package (HD FOWLP) with Multi-layer Fine Pitch RDL for Mobile Applications," IEEE Electronic Components and Technology Conference, pp. 1522 - 1529, May 2016.

[9]. Y. R. Kim, J. H. Bae, M. H. Chang, A. R. Jo, J. H. Kim, S. E. Park, D, Hinter, M. Kelly, and W. C. Do, "SLIMTM, High Density Wafer Level Fan-out Package Development with Submicron RDL," IEEE Electronic Components and Technology Conference, pp. 8 - 13, May 2017.

[10]. T. Wang and D. Yu, "Signal and Power Integrity Analysis on Integrated Fan-out (InFO_PoP) Technology for Next Generation Mobile Applications," IEEE Electronic Components and Technology Conference, pp. 1483 - 1488, May 2016.

SB²-WB: A new process solution for advanced wire-bonding

Matthias Fettke, Andrej Kolbasow, Georg Friedrich, Anna Palys, Vinith Bejugam and Thorsten Teutsch

PacTech GmbH

Am Schlangenhorst 7-9, 14641 Nauen, Germany

Phone: +49 (0)3321 4495-504, Fax: +49 (0)33214495-110, E-Mail: fettke@pactech.de

Abstract—**This paper describes a novel and innovative wire-bonding method which combines standard wire feeding application with the unique solder-jetting process; i.e. SB²-Jet. In contrast to conventional ultrasonic, thermosonic or thermocompression bonding, the laser-wire-bond connection, SB²-WB, is not welded but soldered. Neither pressure, ultrasonic vibration nor high temperatures are utilized. These technical advancements broaden the spectrum of wire-bonding applications. Besides the fundamental process explanation and the comparison to conventional wire-bonding methods, the results of initial reliability and stability tests on 50µm Au wire contacts bonded in wedge-wedge configuration are presented and discussed. For this comparison, the bonds were thermally, mechanically and electrically stressed. The impact on the bonds after thermal cycling (250TW) and vibration tests were microscopically inspected and metallurgically studied through cross-sectional polishing and FIB-SEM characterization. The mechanical load-capacity was quantified using a pull and shear test measurement system. The failure characteristics of the bonds during an ampacity test were analyzed by thermal imaging.**

Moreover, the fabrication of laser-wire-bond connections on a piezoceramic element as part of a PDC-ultrasonic sensor using 80µm insulated Cu wire and SAC_305 (760µm sphere diameter) solder alloy is described and qualified as an example of potential application. The mechanical strengths of the interconnections were measured using a shear tester, while the concomitant metallurgical properties were analyzed by X-ray and FIB-SEM. The opportunity to remove surrounding insulation material of a wire during the bonding process will be an additional topic of discussion.

Finally, a roadmap for this new technology and future prospects involving intended reliability and comparative stability studies are elucidated.

Keywords: **wire-bonding, wire-soldering, solder-jetting, heavy wire, Cu wire, dismantling, chip packaging**

I. INTRODUCTION

In the field of semiconductor packaging, wire-bonding is the most common bonding technology for creating electrical interconnects. More than three-quarters of all IC-elements in 2018 contained primary wire-bond interconnections, which indicates the dominant and growing interest in wire-bonding technologies [1].

However, the continuous growth of the global market for micro-electromechanical components and systems entails further miniaturization and augmented performance capabilities, requiring new technologies where the standard wire-bonding processes reach an impasse. Alternative process ideas are in great demand for the partial substitution of conventional wire-bond processes on the one hand and for broadening of application fields on the other hand [2].

The common goal is to facilitate the growth of new geometrical & physical bonding capabilities as well as to increase the available material selections and material combinations.

II. PROCESS EXPLANATION & DESCRIPTION

A. General process setting

The key components of the process setting are PacTech´s unique solder-jetting unit involving, a wire feeder and a 3D axis system. In Figure. 1, the principle configuration is illustrated.

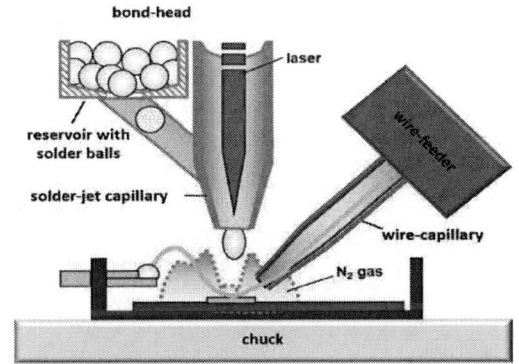

Figure 1: Process setting of laser-wire-bonding core unit

978-1-7281-1500-9/19 $31.00 © 2019 IEEE

The bond-head singulates solder sphere preforms whose sizes are in the range of 30μm-1200μm, and transfers each solder sphere into a ceramic capillary. Inside the capillary, a NIR-laser induces thermal energy and liquifies the solder sphere. The liquid droplet is then expelled out of the capillary by supporting N2-pressure [3].

The wire feeder system positions the bond wire in the target area of the solder-jetting capillary. During the feeding and motion of the system, the wire is guided through a wire capillary, which is located in close proximity to the solder-bond capillary. The wire can be pushed, pulled, unwound or fixated to enable precise positioning and diverse loop designs.

The bond-table moves in x and y directions and positions the substrates with respect to the target area of the solder-jetting unit. All bonding components are mounted on the z-axis unit, which moves to the required bond-height. Distance and angle adjustments of the feeder capillary are performed manually.

In Figure 2, the developed system arrangement and configuration of the work-area for a solder-wire-bond is shown.

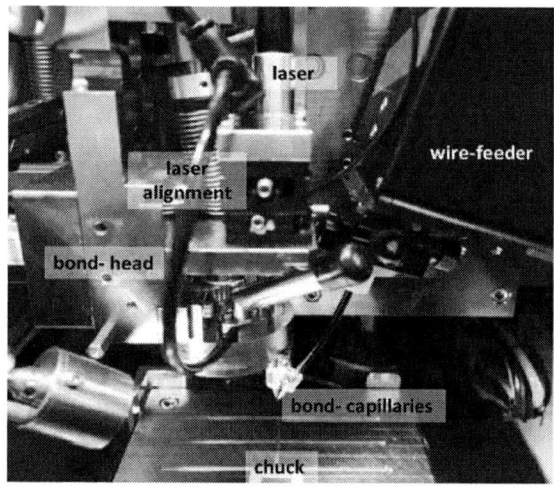

Figure 2: SB²-WB machine setup

B. General process flow

After moving the laser-wire-bonding unit to the target position of a contact pad, the wire is fed with a defined length. Depending on the adjusted position of the wire capillary, the wire is either pushed onto the pad or held in a floating condition above the pad. Parallelly, a solder droplet is prepared and applied. The liquid solder droplet bonds the wire onto the pad and creates a uniform wire-bump contact. The continuous supply of N_2 to this interface protects the involved materials against oxidation. Both process steps are performed almost synchronously. After contacting the first position, the axis-system moves to the subsequent bond location. During the movement, the wire unwinds and can be plastically manipulated to form different loop designs. Figure 3 presents multiple wire loop forms generated by the wire-bond system.

At the second position, apart from feeding, the same previous process steps are repeated and the wire is cut. To separate the wire, a cutting-blade is used. The in-built laser or a sharp edge of the capillary are alternative available options for the separation process.

Figure 3: Laser-wire-bond (200μm Au-coated Cu wire) results on a LTCC-connector device. Left image shows a "Gaussian shape" while right image shows an asymmetric rectangular shape

C. Relevant process parameters

The solder-wire-bond quality is primarily controlled by five adjustable key-parameters: a) laser-power, b) laser-pulse-width, c) nitrogen-pressure, d) bond-distance and e) wire position. Depending upon the interaction time, kinetic force, position and thermal-intensity, the bond-shape, surface-quality, the solder coverage-level on wire and pad, specific electrical resistance, tensile strength and the ensuing characteristics of IMC-layer are affected.

The variation of the parameters in combination with different proportions of wire diameter and solder volume allows diverse geometrical bond-contact formations. As shown in Figure 4, either the wire is floating and encapsulated by the solder bump (left) or the wire is positioned on the contact-pad and bonded in the form of a volcano-shape (right).

Figure 4: Examples of possible solder-wire-bond formations

D. Temporal & thermal qualification

The temporal and thermal analysis of the laser-wire-bonding process was conducted using a fast "Optris CTvideo" pyrometer. The measurement device contains of a pyroelectric-sensor type "3MH1-CF" with a time resolution of 2ms, spectral sensitivity of 1μm, temperature resolution of 0.1K and a measurement range of 150°C-1000°C. An emissivity (ε) of 0.25 (Sn) was selected [4]. A representative measurement result of a process cycle involving 760μm SAC_305 solder ball on a 80μm Cu wire connected to a Ag-plated ceramic substrate is illustrated in Figure 5.

In order to form a homogenous, stable and reliable material interconnection between the described solder, wire and substrate configuration, a minimum NIR laser-pulse energy of 1400mJ (140W/10ms) is required.

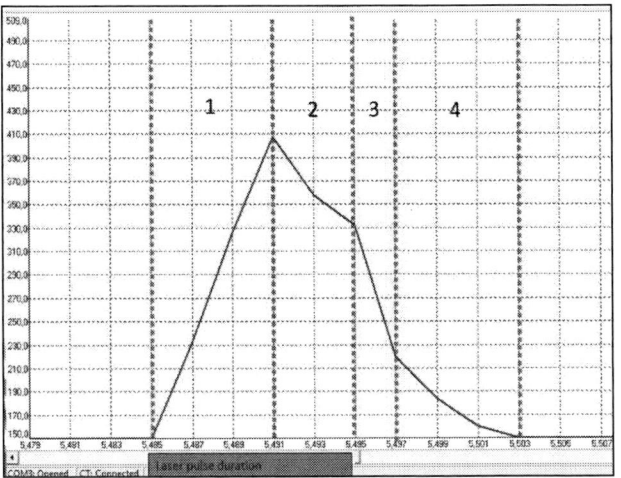

Figure 5: Temporal and thermal cycle of a SB²-WB bond (760µm, SAC_305 solder ball on 80µm Cu wire placed on an Ag-plated ceramic substrate)

<u>Section 1 (transformation)</u> of Figure 5 shows the thermal ramp-up of the capillary tip. Within 6ms the solder-jet capillary reaches its peak temperature of 410°C before the solder ball leaves the capillary.

<u>Section 2 (flight-phase)</u> represents the decreasing gradient (25K/ms) of the liquid solder ball temperature while dropping. During the flight-time, the droplet is continuously pumped with laser-energy for additional 4ms before laser switched off.

<u>Section 3 (recrystallization)</u> shows the cooling and interdiffusion phases of the solder ball after hitting on the substrate surface and solidifying.

<u>Section 4 (attenuation)</u> illustrates the typical polynomic thermal fading of the substrate-system.

E. A comparison with common wire-bond technologies

The presented curve of the thermal progression in Figure 5 shows extreme dynamic characteristics of the SB²-WB bonding process which is temporally similar to high-speed thermosonic bonding transducer times in the range of 8 to 12ms [5].

A simplified comparison between the key-characteristics of bonding processes for wires ≤ 50µm is shown in Table 1. The contents of this putative Table [2] are supplemented with new information including SB²-WB process data.

Parameter	Wire-Bond Technologies		
	TS	US	SB²-WB
ultrasonic power	yes	yes	no
bonding force	low (30-90cN)	low (25-45cN)	negligible (< 2µN)
temperature substrate	middle (100°C-220°C)	low (room temperature)	low (room temperature)
bonding time	short (30-100ms)	short (50-100ms)	ultra short (1-10ms)
preferred wire metal	Au, Ag, Cu, Pt, Pd	Al, Au, Cu	Au, Cu, Ag, Pd, Pt
preferred pad material	Al, Au, Cu	Al, Au, Cu	NiAu, Au, Cu
contamination	middle	middle	middle
speed	4-10 wire-bonds/s	2-3 wire-bond/s	3-4 wire-bonds/s*
min. pitch	35µm (15µm wire)	35µm (15µm wire)	40µm (15µm wire)*

* possible but not shown

Table 1: Comparison of bonding-parameter [2,6,7,8]

In contrast to the conventional bonding techniques, the bond substrate is not stressed by a mechanical load from the tool due to its contactless nature. Brittle materials such as LTCC, organic thin films <20µm used in MCM-D-type

modules, or resonating cantilever leads which are used in optoelectronic packages can be bonded with the new technique without any impediments [8,9,10].

The structure of the wire material dictates the concomitant mechanical and electrical properties. Manipulation or transformation of the initial bulk grain structure most often is detrimental to the performance characteristics of the wire. Conventional bonding processes changes the material structure locally either by forming the ball with EFO or by ultrasonic vibration during the deformation sequence on the interface. Diagnosis of this effect is possible by analysing the fracture-modes of mechanical stress tests where the wire breakage frequently occurs along the HAZ as "neck-break". An additional and frequent consequence of wire weakening during EFO ball generation is the forming of higher loop heights [11]. By soldering the wire using the method introduced herein, neither the diameter of the wire is altered nor bulk properties are affected by recrystallization processes. This method is advantageous especially for a continuous wire-bond connection of multiple landing platforms. The thermal load for the interface is extremely low and provides significant metallurgical advantages compared to other solder reflow processes due to the generation of only few microns thick acicular IMC-layer [12,13]. The mechanisms of various loop formations during cold deformation in the material are identical for common technologies as well as SB²-WB.

Aluminium still is a dominating material for wire-bond applications pertaining to mono- or multi-metallic and ribbon or thick wires-bonds. Stable and reliable welding results confirm the capability of the conventional wire-bonding processes to inter-diffuse the already existing ~5nm oxidized surface layers [14]. For packages requiring Al as the bond material, the wire-soldering process is not applicable.

Normal or folded ULL's as well as SSB loop formations enable minimum loop heights of <80µm. The bonding sequences takes longer and affects the productivity. The wire feeder position of a SB²-WB setting can be manipulated in a hemispherical workspace thus realizing wire-bond processes in the working range of 0°-90°. Consequently, wire-bond connections with a height corresponding to the used wire thickness is feasible [15].

Compared with a mono-metallic welded interconnection a soldered one is basically less stable and less reliable. Additionally, due to the melting temperatures of the used soft solder alloys (< 350°C), the wire-bond soldering process is not applicable for high temperature applications.

III. JOINT RELIABILITY

For the initial mechanical and electrical performance characterization of the developed wire-bond technique, TS and SB²-WB bonded wire samples were subjected to reliability tests and qualified. Common test material for both the processes was a 1mm² area and 140µm thick Si-chip with simple contact structure and 5µm ENEPIG pad plating [16,8]. The chip was mounted on a 4mm² area and 400µm thick "Kovar"-substrate with a 50µm Au plating. A 50µm Au wire (Heraeus BW AU HD2 WR) was bonded in a ball-wedge configuration to function as the connection between the 200µm octagonal pads on the chip and the "Kovar"-substrate surface. The TS bonds were formed on a "F&S-Bondtec series 58" system and qualified based on tests pertaining to mechanical, thermal and electrical stress (Figure 6). The metrology used in these tests included a pull and shear tester ("XYZTEC Condor Sigma" & "Dage BT 4000").

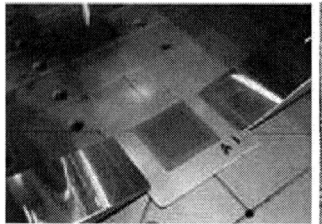 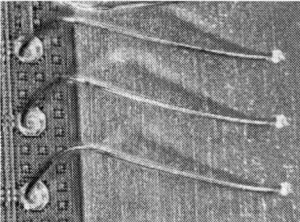

Figure 6: Test chip configuration with Au-wire-bonds on "F&S Bondtec system". Left image shows the bond situation. Right image shows the TS-bonds.

The SB²-WB-bonds were generated on a "PacTech SB²-WB" prototype system (Figure 7). For sufficient coverage, a 200μm SAC_305 solder sphere was used. Loop height and bond-length were adjusted to mimic TS-bond parameters and loop forms. The wire was soldered with respect to the pad-surface at a 30° angle.

 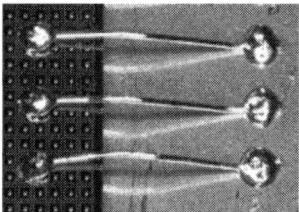

Figure 7: Test chip configuration with SB²-WB wire-bonds on "PacTech prototype system". Left image shows the bond situation. Right image shows the soldered wire-bonds.

The measured vertical strengths of the wire bond samples are presented in Figure 8.

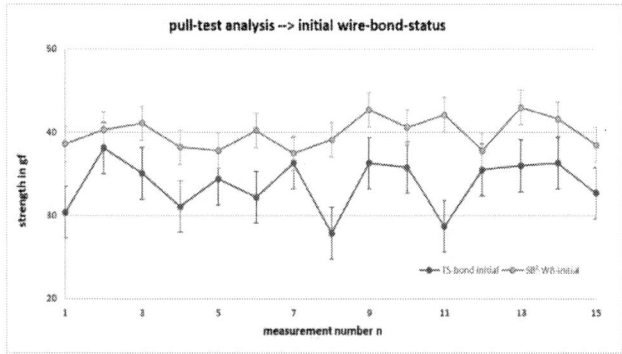

Figure 8: Pull test force measurement results of TS- and SB²-WB-bonds

The joint fractures were optically inspected and classified with a light microscope [17]. Figure 9 shows the resulting dominant fracture modes of the TS-bonds (left) and the SB²-WB-bonds (right) after pull test.

The average pull strengths for TS-bonds was found to be 33.8gf and 39.9gf for SB²-WB-bonds. While two different

fracture modes were identified for the TS-bonds, SB²-WB-bonds always showed a wire-break phenomenon around the hook-location. The failure-modes associated with TS-bonds in the initial tests were diagnosed as neck-break (73%) and wire-break (27%) respectively.

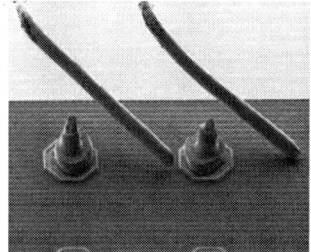 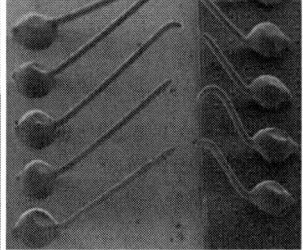

Figure 9: Dominant fracture modes after pull test. Left image shows neck-break fracture of TS-bonds. Right image shows fracture of SB²-WB-bonds.

A. Vibration test

In order to understand the mechanical performance and metallurgical stability of the soldered wire-bonds under stress, an initial vibration test was conducted which shed light on potential failure fractures. A "CTS RMS vibration tester" equipped with sinusoidal load feature was used to apply stress on the samples in x, y and z directions at a frequency range of 50-200Hz and an axis load of 5g for a time period of 8min.

The failure-mode distribution of the TS-bonds changed slightly to 50% neck-break and 50% wire-break and remained in 100% wire-break regime for the SB²-WB-bonds (Diagram 12). In both cases, the pull strength reduced (4.35% for TS- & 8.23% for SB²-WB-bonds), as illustrated in Diagram 13 that summarizes all measurement results.

In order to confirm that only the wires were weakening during the stress test and not the interface, additional shear tests on the SB²-WB-bonds were performed. The shear force measurement results before and after loading are presented in Figure 10.

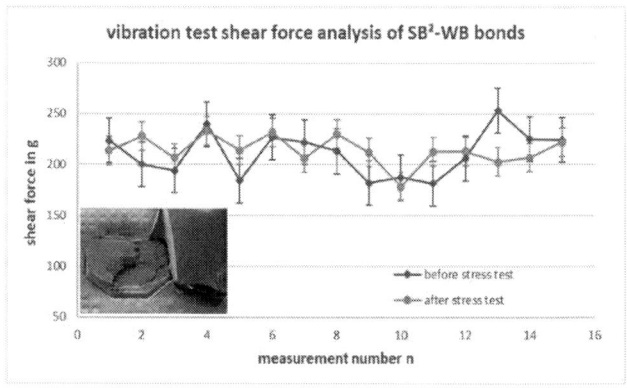

Figure 10: Shear force measurement results of SB²-WB-bonds before and after 3D vibration test

Moreover, the metallurgical properties at the interfaces of SB²-WB-bonds were analysed and inspected for fractures. The cross-sectional view of the interconnections before (left) and after tests (right) is shown in Figure 11.

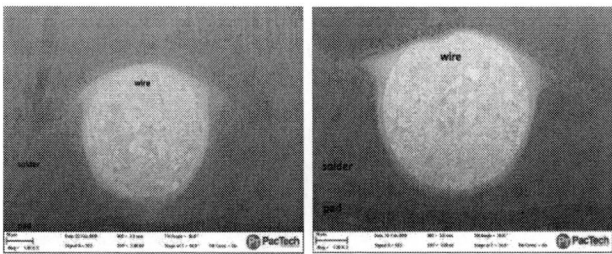

Figure 11: Cross-sectional views of SB²-WB-bonds before and after vibration test captured using SEM-FIB

For the examined range and configuration, the results show that both connections primarily lost the mechanical integrity in the bulk wire and not at the interface. Both bond-technologies withstood the massive mechanical load. Discrepancies in pull-tests results in the case of a wire breakage fracture mode have caused due to slight geometrical variances in the wire-bond build-up.

B. Climate test

The increased degradation of a soldered interface in comparison to a welded one is a widely recognized fact. To analyse and understand the degradation mechanism of the selected laser soldered wire-bond configuration in comparison to the TS-bonds, a temperature-cycle test in a 3-zone oven ("Vötsch VT 7012 S3") was performed. The temperatures ranged from -40°C in the cold-chamber to +125°C in the hot-chamber. The samples underwent 250 thermal cycles with a hold time of 30min in each chamber.

The failure-mode distribution of the TS-bonds indicated 60% neck-break, 37% wire-break and 3% ball-bond failure, as shown in figure 12. The inspection results of the SB²-WB bonds after pull test showed 90% wire-break and 10% wedge-metal-lift fracture-modes (Figure 12). The measurement results of the pull test are illustrated in Figure 13.

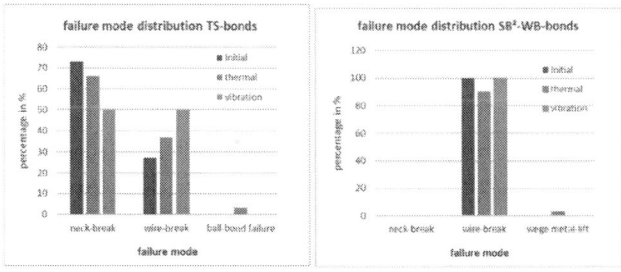

Figure 12: Distribution of fracture modes

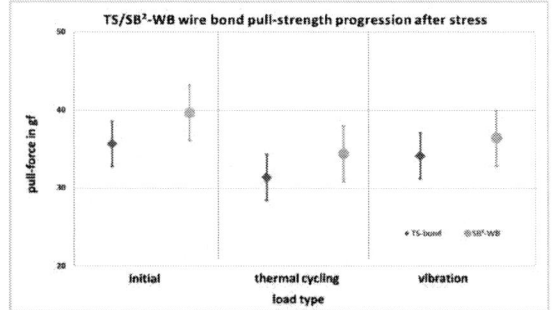

Figure 13: Pull strength measurement results

A reduction of the pull strength for TS-bonds of 12.7% and 13.2% for SB²-WB bonds was measured. Also in this case the fracture-modes of both connections primarily lose the mechanical integrity in the bulk wire and not at the interface.

Cross-sectional views of the laser soldered wire-bond interconnections after thermo-cycling tests are shown in Figure 14.

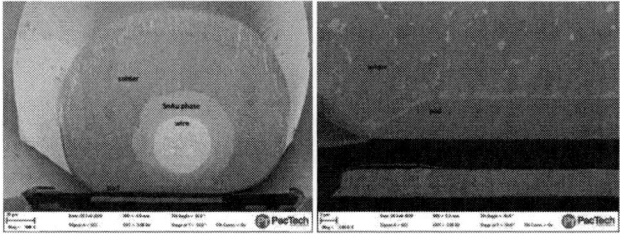

Figure 14: Cross-sectional views of SB²-WB-bonds after thermal cycling captured using SEM-FIB

After a period of 250 thermal cycles, an increase of the IMC-layer formation and an elongation of the SnAu diffusion zone were obvious, as expected. Neither cracks nor "Kirkendahl" voids were identified. The SnAu diffusion zone expanded by a factor of 4 and formed a stable and rotation symmetrical corona around the wire. The gradient of this micro structure shows a decreasing portion of Au towards the solder. The new SnAu material composition in this zone has a higher melting point than the basic solder material which might prevent further propagation of Au into the Sn.

During cross-sectional analysis of the TS-bond material structure no significant changes were identified after thermal cycling.

C. Ampacity test

Aside from evaluating mechanical and thermal performances, it is important to understand the electrical reliability especially for gaining insight into the process of laser-diode wire assembly on fragile GaAs devices where high load-currents are required. In contrast to TS-bonds, the solder-interface between wire and contact-pad presents an additional transition resistance. In the initial test, the wires were loaded with a ramp-up current until a break-down occurred in order to examine the dominant position of failure. To identify the weakest point, the tests were conducted and analysed with an infrared thermal imaging camera.

For this evaluation, two Au-coated "Kovar"-substrates were conjoined together with an epoxy insulating compound. A shift in the position generates a step in the package for the wire-bonds. On each of these test packages, four wires were bonded to connect both "Kovar"-elements for the voltage application. A black resist was applied on the wires to support thermal imaging. The test-setup is shown in Figure 15.

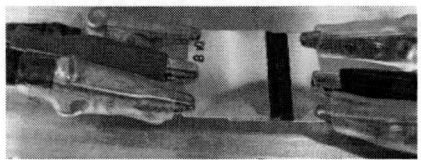

Figure 15: Test-setup for ampacity test sequence

978-1-7281-1500-9/19 $31.00 © 2019 IEEE

The current was increased step-wise and applied for 0.2s. The waiting time between the cycles was 7s and each current cycle was repeated once. As evident in Figure 16, wire breakage event occurs with the TS-bonds and SB²-WB-bonds at 15A and 20A respectively.

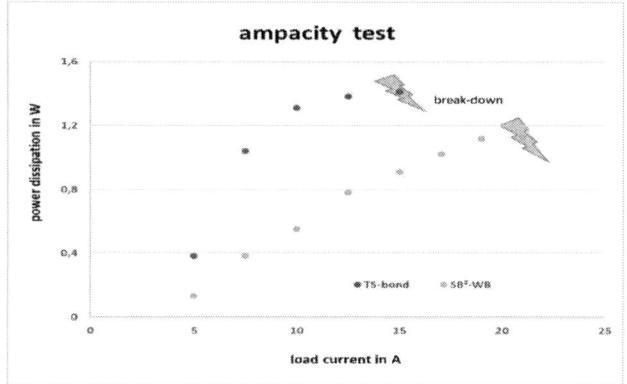

Figure 16: Current damage threshold measurement results

Representative thermal images during the test sequence for the TS-bonds (left) and for the soldered wire-bonds (right) are shown in Figure 17.

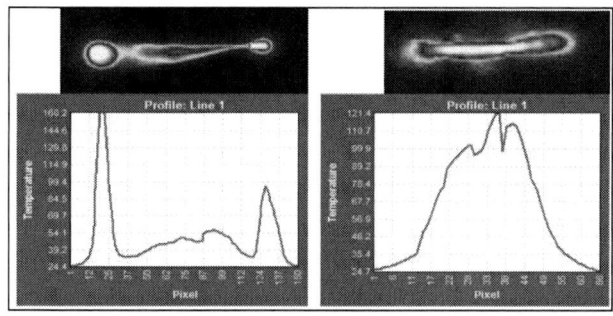

Figure 16. Thermal maps and corresponding temperature distribution of TS-bonds (left) and SB²-WB-bonds (right)

Each break-down occurred at the epicentre of the bent wire loop (Figure 18). The solder did not melt and lose contact at the wire interface as anticipated. Higher thermal mass of the selected solder volume, and consequently, larger available surface area for emitting thermal radiation prevented the incident of solder melting and component separation to occur. The thermal images show a more homogenous distribution of the heat in SB²-WB-bonds in contrast to the TS-bonds where distinct thermal peaks resemble the two different conjoined areas.

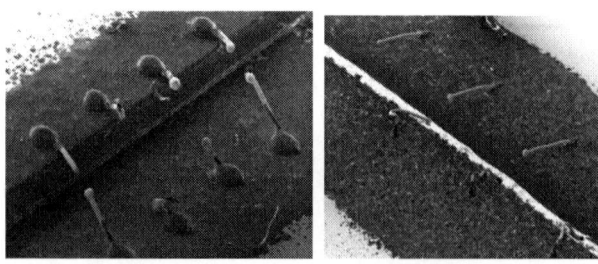

Figure 18: SEM pictures of wire-break fractures after electrical load. The left image shows SB²-WB-bonds and the right image shows TS-Bonds

Slight geometrical variances in the wire-bond build-up between TS and SB²-WB schemes is attributable to difference in specific resistances. This difference could be the underlaying root-cause for the differences in maximum load currents. Nevertheless, for the examined test range and configuration, the SB²-WB-bond configuration shows an enhanced electrical performance based on thermal distribution data.

IV. APPLICATION PROTOTYPE PDC-SENSOR

A. Process description

This application deals with the laser-wire-soldering of piezoelectric actuator, which is the most important part of a PDC ultrasonic sensor. The underlying principle is to connect the piezoelectric actuator located at the bottom of an alumina chassis with two contact pins of the electrical connector interface. This construct is schematically depicted in Figure 19.

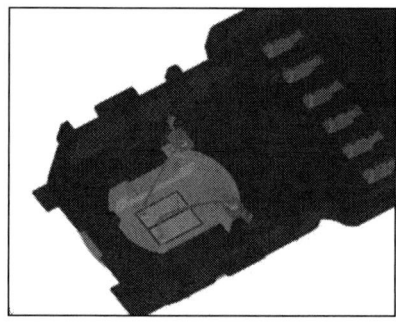

Figure 19: Model of bonded PDC-sensor package

The free work space available for bonding is 7.5mm x 10.68mm area with an orthogonal distance of 18mm between the pins and sensor surface. A wire length of 26mm with a diameter of 80µm is required to realize a reliable electrical contact with sufficient space for compensation of vibration. The wire material consists of Cu (2N) covered with a protective polyurethane resist coating. The contact-pad surface metallization of the PZT-piezo sensor element is electro-plated with a Ag-layer. The Cu contact pins are galvanically pre-tinned. In order to bond the Cu wire against the contact surfaces, SAC_305 solder with a spherical pre-form volume of 0.23mm³ is used.

This application is highly challenging and common wire-bonding technologies tend to fail due to either substrate breakage or insufficient bonding-strength [18].

B. Process realization

Particularly challenging steps involved the preparation of the bond-tool unit to contact the deep-seated sensor surface and removal of the polyurethane insulation coating around the wire during the bonding process. In order to prevent defects such as voids in the final bulk material caused by rapid evaporation of organic residues, the resist layer needs to be removed completely. This was realized using a precise pre-pulse with the built-in NIR laser-system. A sufficient thermal induced shrinkage was observed at a power density of 31J/m².

The first bond was formed on the sensor surface at an orthogonal wire direction with respect to the pin. The wire was then positioned on the pin. This was followed by bonding with the solder sphere and severing of the wire. The bonding was realized with a NIR laser-pulse energy of 1400mJ (140W/10ms). Figure 5 shows the thermal plot of the optimized process-window.

C. Analysis of process results

In order to evaluate the interconnection quality, the generated wire-bonds were analysed with X-ray and SEM-FIB. With X-ray, bulk solder was scanned for any parasitic air inclusions, cracks and inhomogeneities. Figure 20 presents the X-ray measurement results.

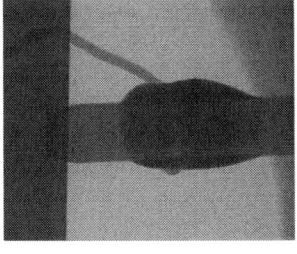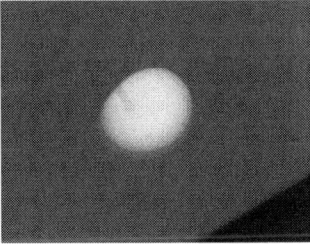

Figure 20: X-ray analysis of realized wire-bond interconnections. Left image shows bonded pin and right image shows the sensor surface bond

Both the bond-connections showed absence of any fractures or voids. The wire was stable and reliable surrounded by the 0.23mm³ solder volume. Metallurgical analysis of the solder wire interconnect using SEM-FIB is presented in Figure 21.

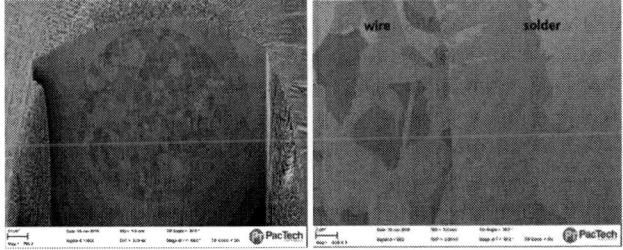

Figure 21.: SEM-FIB cross-section of Cu-SAC305 wire-bond interconnect

The cross-sectional view confirms a defect-free, high fidelity situation of the local interface. The IMC around the wire shows the preferred acicular metallurgic structure with a forming length of ca. 2-4µm.

An average resulting pull strength of 97.32gf in combination with solely wire-break fracture modes around the measurement hook-location confirms a mechanically stable wire joint structure.

V. SUMMARY & FUTURE PROSPECT

- For the examined qualification range, the results discussed herein show that solder-wire-bond connections are reliable and deliver similar performance compared to conventional wire-bond technologies.

- The SB²-WB wire-bond is a versatile configuration with regard to solder volume and alloy type and allows the determination of the fracture position based on design.

- A soldered interconnection in contrast to conventional wire bonding can cover the whole surface of a pad, thus providing benefits from mechanical and thermal standpoints, especially on rectangular pad designs.

- The initial test sequences outlined herein along with the results from mechanical, electrical and thermal characterisation reflects only few aspects of a more complex wire-bonding technology and require additional evaluation and studies.

- The wire-bond soldering process creates homogenous and stable halo formations through diffusion processes around the embedded wire. The SAC_305/Au system outlined in the current work needs additional examination to a) verify whether further propagation of the Au into the Sn occurs, b) to evaluate the point of saturation and c) to consider other material compositions like Au80Sn20.

- Slight variations in the wire-bond built-up explain the discrepancies in pull tests results in case of identical fracture modes.

- In order to reach the stability level of a welded interface, high-melting solder alloys (HTS) can enable the transition from soldering to brazing [19].

- The generation of a stable and reliable Cu wire solder connection for Ag plated LTCC substrates is possible using SB²-WB wire-bond technology.

Future work involves detailed analysis of the soldered wire-bond connections along with specification of optimum process capabilities. Applications pertaining to bonding of wire bundles, ribbons and optical fibres will be explored. Wires corresponding to a wide range of diameters (15µm-600µm) will be investigated. The roadmap of SB²-WB process technology is illustrated in Figure 22.

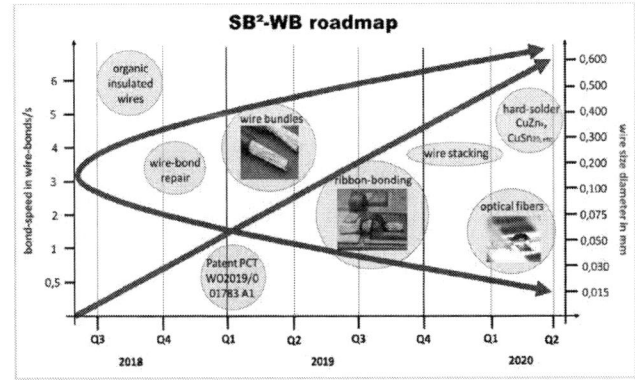

Figure 22.: SB²-WB process technology roadmap

VI. APPENDIXES

A. Abbreviations and Acronyms

SB² → Solder Ball Bumping

SB²-WB → Solder Ball Wire Bonding

TS → Thermosonic

PDC → Park Distance Control

PZT → Plumb Zirconate Titanate

LTCC → Low Temperature Co-fired Ceramic

MCM-D → Multi-Chip Module-Depositioned

EFO → Electronic Flame-Off

HAZ → Heat Affected Zone

ULL → Ultra Low Loop

SSB → Stand-Off-Stitch Bonds

HTS → High Temperature Solder

FHG-ENAS → Fraunhofer Institute for Electronic Nanos Systems

B. *List of references*

[1] TechSearch International, "Advanced Packaging Growth", 2017

[2] J.Phan, P.Fraud, "wire-bonding challenges in optoelectronic packagings", 1st SME Annual Manufacturing Technology Summit Deabron, 2004

[3] P.Kasulke, W.Schmidt, T.Oppert, „Solder Ball Bumper SB²-A flexible Manufacturing Tool for 3-dimensional Sensor and Microsystem Packages", 22th Inter. Electronics Manufacturing Technology Symposium, 1998

[4] User manual Optris CT

[5] Lee Levine, ASM International, "wire-bonding", EDFAAO, 2016

[6] ASIC Labor Heidelberg, "wire-bonding interconnections", workshop on Silicon Detectors, 2007

[7] Prof Dr. Salwani Mohd Daud, "IC Assembly, packaging and testing", DDE3253 Microelectronics

[8] Daniel Lu, C.P. Wong, "Materials for advanced packaging", Springer, 2009

[9] Lee R. Levine, "wire-bonding in optoelectronics", Advancing Microelectronic, 2002

[10] C.K.Charles Jr, K.J. Mach, R.L Edwards, "wire-bonding: Reinventing the process for MCM´s"

[11] U.Geißler, M.Schneider, "wire-bonding as dynamic process of hardening and softening", FHG-IZM

[12] Hiroshi Nishikawaa, Noriya Iwatab, "Formation and growth of intermetallic compound layers at the interface during laser soldering using Sn–Ag Cu solder on a Cu Pad", Journal of material processing technology, 2014

[13] W.Liu, C.Wang, M.Lie et al, "Comparison of AuSnx IMCs's Morphology", Distribution in Lead-free Solder Joints Fabricated by Laser and Hot Air Reflow Process", 6th Electronic Packaging Technology Conference, 2005

[14] H.Xu, C.Liu, V.Silberschmidt et al., "A micromechanism study of thermosonic gold wire-bonding on aluminium pad", Journal of Applied Physisc 108, 2010

[15] B.Chlyk, L.Levine, S.Babinetz et al., "Advanced Ultra-Low-Loop Wire Bonds", Semicon China, 2006

[16] Beng Teck Ng, Ganesh VP, Charles Lee, "Optimization of Gold Wire Bonding on Electroless Nickel Immersion Gold for high temperature Applications", IEEE, 2006

[17] Nordson Dage, "Finaler-Entwurf-Prüfdokument", 2016

[18] F.Seigneur, Y.Fournier, Th. Maeder et.al., "Laser soldering of piezoelectric actuator with minimal thermal impact", 2014

[19] S.Stein, J.Dippert, S.Roth et al., "Laser drop on demand micro joining for high temperature wire-bonding allications-system technology and mechanical joint performance"

Smart Wire Bond Solutions for SiP and Memory Packages

Basil Milton[1], Aashish Shah[1], Hui Xu[1], Odal Kwon[1], Gary Schulze[1], Ivy Qin[1] and Nelson Wong[2]

[1]Kulicke and Soffa Industries, 1005 Virginia Drive, Fort Washington, PA, USA
[2]Kulicke and Soffa Industries Pte Ltd, 23A Serangoon North Avenue 5, Singapore
bmilton@kns.com

Abstract— **Memory device market has grown 12% annually in the past five years. NAND and DRAM comprise majority of the package volume within this segment and each of them have an average annual growth rate of 7%. Ultra-thin dies stacked 32 times are making their way into mass production and very long overhang die structures are becoming increasingly prevalent in new package designs. Multi-Chip modules and SiP applications have also witnessed tremendous growth in recent years. Even with the emergence of several competing packaging technologies, wire bonding is by far the most widely used and cost-effective interconnect solution for SiP and memory device packaging. The dominance of wire bonding is maintained not only due its exceptional flexibility and low cost, but also due to new generations of smart processes that have reduced optimization time, improved reliability and ensured higher yield, all of which further contribute to the cost effectiveness. Modern NAND and DRAM packages have very stringent specifications and often have challenges in ball bond, looping and stitch bond formation. Extremely low loops lower than 35µm that have not been used in mass production before, are now desired to further reduce package height. For packages utilizing Stand-off Stitch bonding (SSB), height and flatness of SSB bump requires finer control. The overhanging die structures are prone to varying amounts of deflection during wire bonding, which in turn affects the quality of the bond and its intermetallic growth. Such dies have higher risk of die crack under the normal forces applied during wire bonding. These challenges increase wire bond process optimization time and the risk of part failure.**

This paper highlights key advancements in wire bond processes to overcome the aforementioned challenges in modern packages. New ultrasonic and bond force control mechanisms are introduced, and ball bond process improvements are demonstrated for both Au and Cu fine-pitch applications. Au wire remains predominant in memory applications and fine Cu wire is the preferred choice in high pin count SiP applications. Finite element analysis (FEA) of an overhang configuration is presented, compared against experimental results and used to develop and improve calibration routines for minimizing die crack and maximizing throughput. New looping processes that can better withstand die compliance and achieve ≤35 µm loop heights are also demonstrated. Bump formation is explored, and results integrated into smart response-based solutions for achieving a flat-topped bump with finer height control for 0.6 mil wire. All of the above developments significantly reduce overall optimization time. Furthermore, new on-bonder process monitoring capabilities are discussed and shown to improve quality controls, paving the way for industry 4.0 in wire bond process capabilities.

Keywords- Wire bonding, SiP, Memory, Industry 4.0

I. INTRODUCTION

The IC market was valued at $409.1 billion for 2018 by VLSI Research and this accounted for a 15.5% sequential change over 2017. The memory IC segment alone was valued at $159.2 billion [1]. High demand for mobile memory, wearable electronics, solid state drives and memory for automotive safety features continue to drive research and development of high performance memory packaging. Smartphones in particular, require thinner profiles every design cycle in order to make the final product slimmer, and also to create more room for the battery. As a result, there has been continuous development, and widespread adoption of package-on-package (PoP) and system-in-package (SiP) technologies in the semiconductor industry in recent years. High-end smartphones now require total package height under 0.80 mm [2]. According to TechSearch International, amongst all competing packaging technologies, 3D stacked IC with TSV provides the ultimate in package height reduction. However, TechSearch also concludes that persistent thermal issues and high packaging costs have pushed out high volume manufacturing (HVM) using this technology [2]. As proven time and again over the years, cost remains one of the important factors in determining the choice of interconnect for OSATs and IDMs alike. Wire bond technology, due to its low cost and increased flexibility continues to remain the preferred interconnect choice for most applications. New innovations in wire bond processes and its efficient integration in PoP/SiP technology has enabled package designers to meet the stringent footprints and ever-increasing performance requirements, all at a low cost. For instance, stacked die CSP for processor and wire-bonded memory packaging in the Apple Watch® has a total package height of 0.56 mm. NAND flash memory stack die CSP was restricted to 0.60 mm package height [2].

The use of new wire bond technology varies based on application. A typical PoP application is made up of logic in the bottom package and wire bonded memory on top. SiP modules on the other hand may include memory, controller, connectivity and MEMS components, all of which are most commonly wire-bonded. In flash memory, 16 stacks are the new norm and 32 stack nears HVM. To keep up with the lower profiles and high-performance demands of modern memory applications, higher I/O counts, more die stacks and the use of longer overhang structures is inevitable. These requirements generate new challenges for wire bond process

978-1-7281-1500-9/19 $31.00 © 2019 IEEE

engineers. For instance, higher I/O counts raise the bar of wire bond reliability in order to minimize defects and maintain high yields. Longer overhang structures can introduce resonance issues which are detrimental to ball bond formation and its reliability. Thinner packages necessitate development of new ultra-low wire loop profiles that can meet the stringent wire bond pull strength requirement. Furthermore, with increasing number of components in SiPs and the resultant increase in usage of SSB wire bonding, additional improvement in bump and stitch formation is desired to increase the overall strength of the weld. Lastly, there is also a higher level need to provide smarter wire bond solutions that can be monitored in real-time to detect defects and help optimize manufacturing processes both from a cost and quality standpoint. This paper presents the latest advancements in wire bond process development, related factory automation and addresses the challenges in each of the aforementioned areas.

II. LOOPING DEVELOPMENTS FOR SiP & MEMORY APPLICATIONS

A. Low Loop Capability

The pursuit of lower wire loop profiles is directly driven by shrinking form factors and the reduction in overall package height. The current loop height capability using Ag and Au fine bonding wire for non-compressed wire loops is 3× the wire diameter [3]. Compressed loops, on the other hand, can be further lowered up to 2× the wire diameter [3]. Non-compressed and compressed wire loop types are compared in Fig 1. Compressed loops take longer to optimize and may have weaker necks due to excessive deformation of the ball neck area. This results in lower pull strengths during destructive tests.

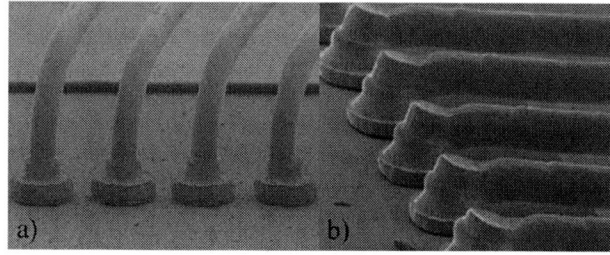

Figure 1.   SEM images of (a) Non-compressed loop; and (b) Compressed loop.

There is merit in developing lower non-compressed wire loops to achieve both a stronger neck and a lower loop height. Fig. 2 illustrates the latest development in ultra-low loop height capability on an extremely long overhang die. The test is done using 4N Au wire of 18 µm diameter with 2.8 mm long wires on overhang structures ranging from 0.6 mm to 1.1 mm. The looping motions are optimally designed to withstand large amounts of compliance of the overhang die. The maximum loop height and destructive neck pull results are tabulated in Table I. Average loop height of 31.95 µm is achieved with maximum recorded height under 2×

wire diameter (i.e. <36 µm). Minimum neck pull strength is >3 g.

Figure 2.   New improved non-compressed loop capability with loop height under 2× wire diameter on a 0.9 mm overhang configuration.

TABLE I.   PERFORMANCE DATA FOR NON-COMPRESSED LOW LOOP

	Loop Height [µm]	Neck Pull [grams]
Spec	< 36.00	> 3.00
Avg	31.95	3.49
Min	30.20	3.13
Max	33.70	3.78
St Dev	1.02	0.14
CpK		3.22

B. Automatic Loop Height Monitoring

Traditionally, process engineers have optimized loop shapes in various wire bond packages using the trial and error approach. This method is often time consuming and extremely dependent on the skills of the process engineer. There is also added uncertainty around the ability to make robust loop shapes. With the rapid proliferation of SiP and PoP technology and increasing complexity in package designs, there is a strong desire to shorten development cycles. As a result, there has been a push for smart wire loop solutions. Offline wire loop design and 3D clearance check tools have been introduced in recent years to aid process

978-1-7281-1500-9/19 $31.00 © 2019 IEEE

engineers to quickly design all the wire loops within a package [4]. Furthermore, response-based ProCu Loop™ solutions have demonstrated the capability of achieving the desired loop shape on wire bonders for a wide range of loop height and wire length [4, 5]. These advanced solutions shorten package development time and reduce the time to market. In addition, a new automatic loop height measurement feature has been developed to monitor the performance of these loops in real time to improve yield and quality.

Fig. 3 depicts a typical 3D flash memory stack. In this example, an 8 die stack is illustrated with wire loop chains connecting each die and eventually down to the substrate. Two such stacks can be attached on top of each other to create a 16 stack NAND chip. Similarly, 4 stacks may be placed on top of each other to form a 32 stack NAND chip and so on. Due to the presence of so many wire bond chains, tight loop height control, especially on the top-most die is essential to ensure good production yield. To address this need, a smart monitoring capability has been developed and integrated on the wire bonder to measure and monitor the maximum heights of individual loops on a real time basis during HVM. Depending on manufacturing specifications, this smart loop height monitoring can be enabled for select wires that are more prone to height variation or all wires, if comprehensive traceability is desired. Manufacturing process engineers can also program control limits beyond which the wire bonder can either issue a warning or terminate the bonder operation altogether so that the issue can be addressed.

The 8 stack memory configuration described above was simulated on a flat aluminum paddle of a QFP leadframe. Fig. 4(a) shows the top view of a portion of a bonded sample with all the wire bond chains. Fig. 4(b) shows the wire loop profiles. For this experiment, the top-most wire loop of each chain was optimized to achieve a target loop height of 100 µm. Five wire loops within the sample are configured for automatic loop height measurement, and the real time maximum loop height data is collected. This is followed by a controlled run of 42 samples. The results of the experiment are summarized in Fig 5. Each of the series represents the maximum loop height data for a given wire loop across the 42 samples. The flat lines on top and bottom of the plot indicate the upper/lower control limits of ±12.5 µm, respectively. The data suggests that the maximum loop height of all wires across the 42 samples are well controlled and are within the set control limits.

Figure 4. (a) Top view of a portion of wire bonded sample simulating a stack die memory application; (b) Loop profiles of the wire bond chains.

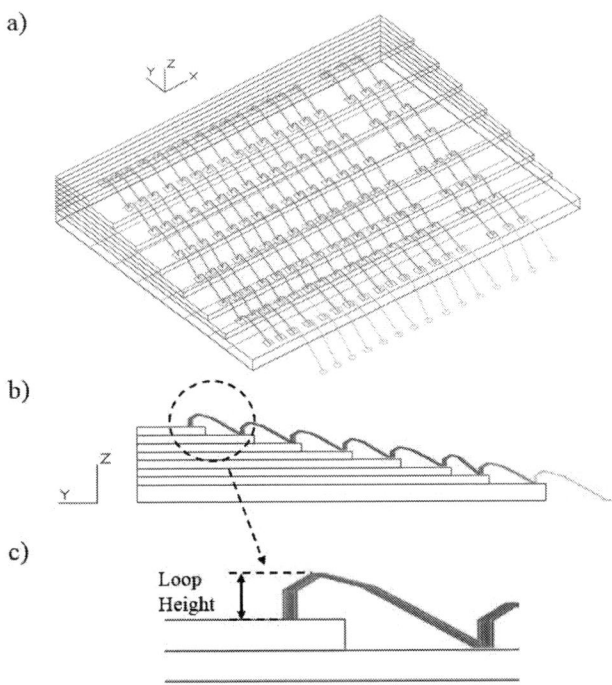

Figure 3. (a) Isometric view of a typical flash memory application with 8 wire bonded die stacks; (b) Side view of the wire loops; (c) Wire loop on the top die wherein loop height control is critical to total package height.

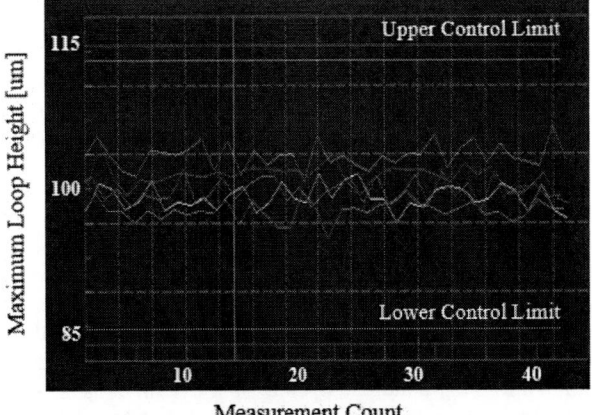

Figure 5. Loop height monitor data for 5 wire loops across 42 samples.

This smart loop height monitoring tool may also be used to measure the height at other critical areas along the wire loop. For instance, Fig. 6 illustrates a PoP application with logic die at the bottom packaged using flip chip technology and memory die on the top wire bonded on an overhang and inboard configuration. In this case, the clearance of the wire loop is of particular interest to prevent wire shorts. To ensure sufficient clearance during HVM, the loop height monitoring feature can be programmed at the die edge location as shown in Fig. 6 (c).

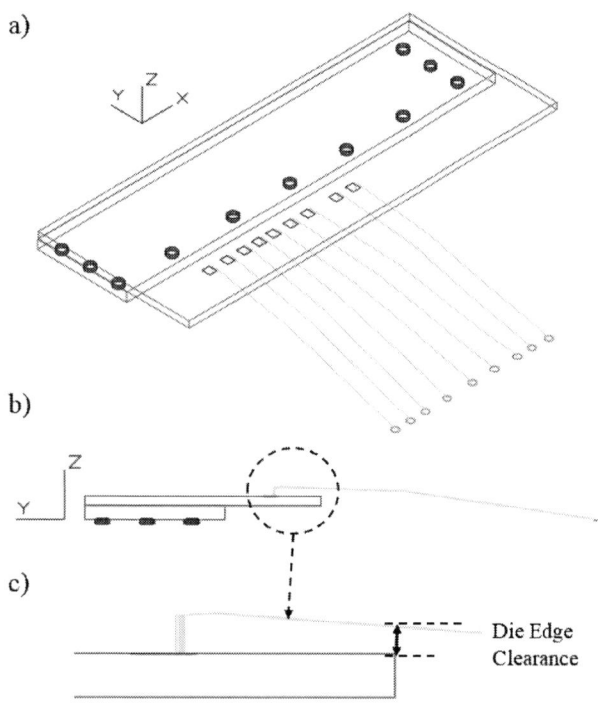

Figure 6. (a) Isometric view of a PoP application with flip chip die on bottom and overhang memory die on top; (b) Side view of very low and long wire loop profiles; (c) Illustration of die edge clearance.

III. BALL BOND IMPROVEMENTS FOR OVERHANG BONDING

PoP and memory devices commonly employ a vertical integration technique where multiple dies are stacked over one another. This method of stacking dies commonly creates a case where the top die overhangs the edge of the bottom die in a cantilever beam configuration, as shown by the schematic in Fig 7. Ball bonding on an overhang device is very challenging due to the resulting die deflection. An important consideration for overhang bonding is the concept of Max Deflection (D) and Max Force (F_N) for the overhang configuration. D is defined as the max limit of die deflection that can be safely accepted for the given overhang device. D in turn determines the maximum normal force F_N during wire bonding. For a conventional ball bonding process optimization, the bond pad opening is the main consideration to determine the process design such as bonded ball size, capillary design and wire diameter. For an overhang device,

the process design and specifications also need to take into account acceptable values of D and F_N.

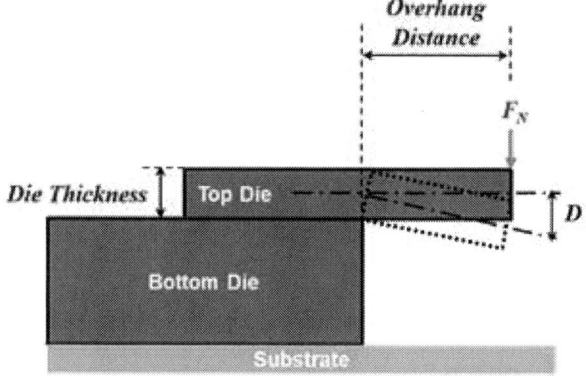

Figure 7. Schematic of an overhang die configuration

There are several challenges in overhang wire bonding for memory and SiP applications. They include loop interference, die crack and bond planarity. When wire loops are present directly below the overhang die to be bonded, D must be less than the gap between the top of loop and bottom of the overhang die to prevent interference and loop damage. Die crack depends on the maximum principal stress during bonding. Since die bending and shear stresses are proportional to applied force, F_N must be limited to prevent die crack. FEA models that represent the actual device as closely as possible can be useful to predict the failure stress and determine the F_N limit. Another limiting factor is bond planarity degradation, which may result in bonding issues such as improper bond formation (elliptical bonds, smashed bonds), poor bondability (low shear, non-uniform IMC), and inconsistent bump formation.

Another consideration for overhang bonding is to control the lift-off motion after bonding to prevent large die oscillation. A gentle lift-off from the bond surface ensures that the bonds and wire loops already made are not damaged. The oscillation profile of an overhang die is an intrinsic mechanical property of the overhang configuration. Optimizing the wire bonder lift-off motion can be very challenging and time consuming.

The latest RapidMEM® bonder includes a state-of-the-art smart process feature called *ProOverhang*. *ProOverhang* is an on-bonder adaptive machine learning and optimization tool that automatically learns the dynamic characteristics of each programmed bonding location on the overhang die. It uses these learned characteristics to automatically find the optimal settings of key parameters and recommends the maximum ball diameter that can be made on the given overhang configuration. The *ProOverhang* feature works with a suite of response based smart processes called *ProAu* and *ProAg* to deliver an easy to use framework for robust ball bonding process development on overhang devices.

Next, we will present the ball bonding results on the thin overhang device previously discussed and shown in Fig. 2. The overhang die thickness is 30 µm and distance is 0.9 mm. Ball bonding is performed on RapidMEM® bonder using 18

978-1-7281-1500-9/19 $31.00 © 2019 IEEE 58

μm diameter Au wire. The overhang configuration (thin die and long overhang distance) results in excessive die deflection and die oscillation. The die deflection causes the die planarity to be disturbed significantly resulting in improper bonded ball shape, and insufficient bonding (due to partial contact of Au ball in the Al pad). Proper control of die oscillation is very important in this case due to the stringent loop requirements as described in Section II (A). Moreover, the BGA substrate allows for a maximum heat block temperature of 150°C. This results in an actual bonding temperature of ~128°C on top of the overhang die, which is low for a typical Au-Al bonding process. The low temperature bonding combined with a difficult overhang die configuration makes the process optimization more challenging.

The bonding optimization is started by using the *ProOverhang* adaptive learning tool to learn the dynamic characteristics of each bonding location on the overhang die. The *ProOverhang* learning tool recommended a bonded ball diameter of 36 μm, which results in a max die deflection of 165 μm. The recommended ball diameter is then input to *ProAu*-2 process, which is then used to automatically calculate the default settings of all bonding parameters, e.g. FAB size, contact velocity, bond force, ultrasonic (US) current, time, etc. Fig. 8 shows the plot of bonded ball diameter against the wire # arranged along the edge of the die from one corner to the other. Inconsistent bonded ball diameter is observed, with two bonds (wire #7 and wire #27) showing excessive ball deformation compared to rest of the wires. Interestingly, the bonding location of these two wires is symmetric, with each of these two wires programmed at the same distance from each corner. Since the over-squashed bonds occur only at these locations, this issue can be attributed to structural resonance issue confined to those locations. To further understand the cause and type of resonance, an FE analysis is performed.

Finite element model to simulate vibrational resonance behavior of the overhang die is shown in Fig 9. Only the overhang die, and the die attach film are required to adequately capture the dynamic characteristics. Modal analysis is performed, which reports the mode "shape" (pattern of motion) and corresponding modal frequency. Any mode that is close to the bonding transducer's operating frequency will likely be excited during bonding. In this case, the mode illustrated in Fig. 10 (predicted frequency of 116 kHz vs transducer 120 kHz) is suspect. In Fig. 10, dark blue regions are points on the die which will experience minimum vertical vibration. Conversely, red regions experience maximum vertical vibration. Examining the location of over-squashed bonds in Fig. 8, it is found that they correlate very well to the minimum vibration areas predicted by the FE analysis. Bonding on stable (zero relative motion) locations is always desirable and is generally accomplished with more reasonable bond parameter settings compared to bonding on locations experiencing larger vertical vibration. Thus, one would expect for bonds made with similar parameters along the entire edge, the highest deformations would be seen at the blue, stable regions.

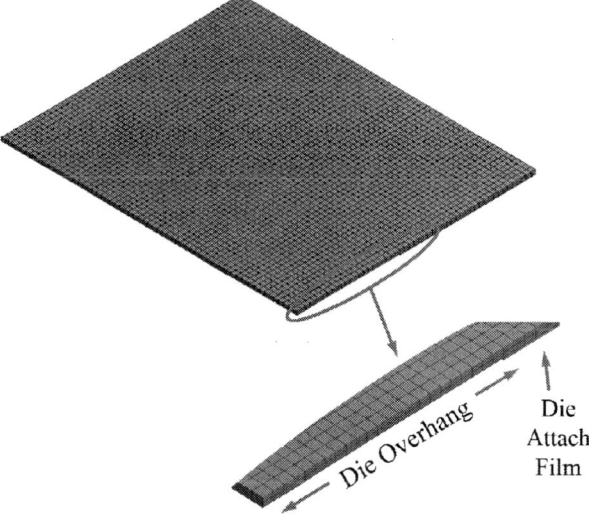

Figure 9.   Die resonance finite element model.

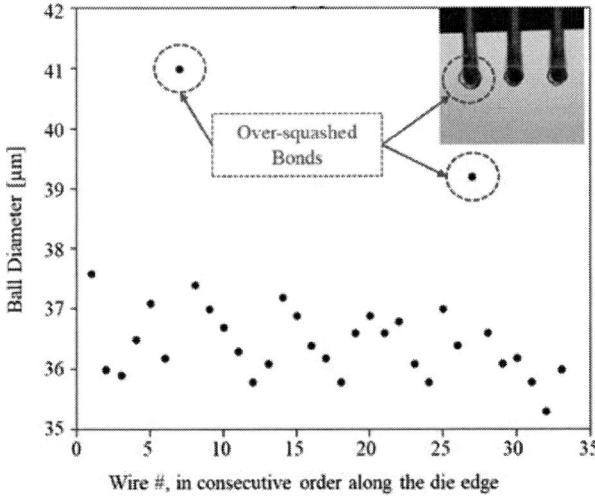

Figure 8.   Distribution of ball diameter vs wire #, arranged in consecutive order along the die edge from one corner to the other corner. Data shown is bonded using *ProAu*-2 process at 120 kHz US frequency.

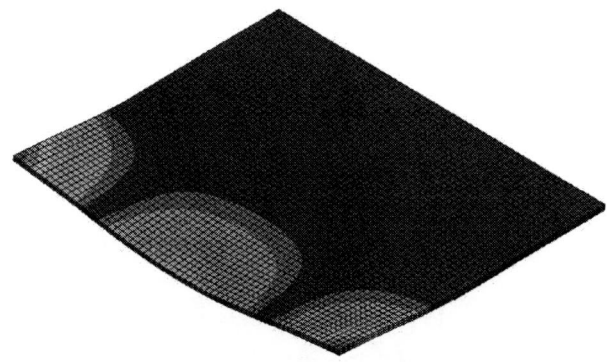

Figure 10. Predicted magnitude of vertical displacements.

By applying optimized bond force, the smash bond issue can be resolved as shown in Fig. 11. The corresponding ball shear data is shown in Fig. 12. In general, lower shear values are observed for bonds bonded at locations in the corner compared to the center of the overhang die edge. This behavior is expected with such an extreme case of overhang configuration, mainly attributed to the wide range of die deflection amounts at corner pads compared to the center pads.

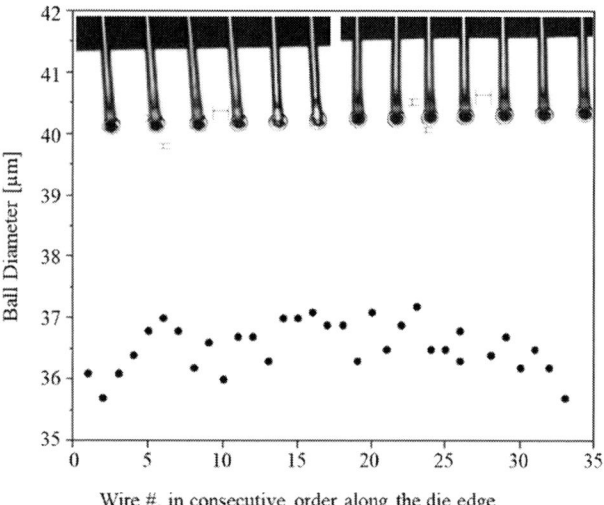

Figure 11. Distribution of ball diameter vs wire #, arranged in consecutive order along the die edge from one corner to the other corner with optimized forces.

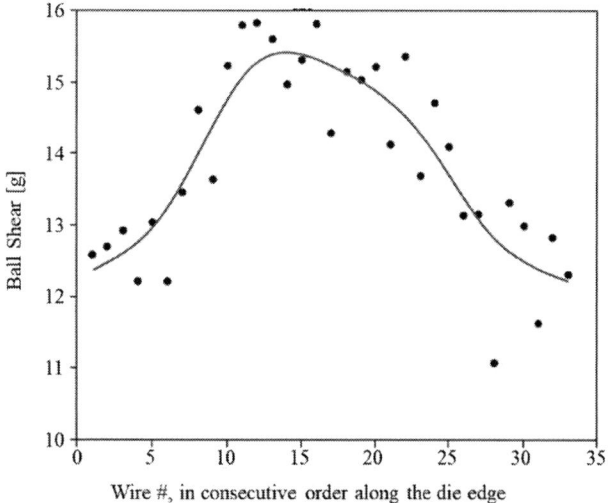

Figure 12. Distribution of ball shear vs wire #, arranged in consecutive order along the die edge from one corner to the other corner with optimized forces.

IV. SSB ADVANCEMENTS

Stand-off Stitch Bond (SSB) starts with placing a flat-topped bump ball, followed by a regular ball bond, and then stitch bond on top of the bump. SSB process has ball bonds at both ends of the wire and it provides die-to-die bonding capability essential to SiP and PoP technology. We discuss advancements in SSB bump and stitch formation in this section.

A. SSB Bump Optimization

The bump formation consists of bump bonding and bump shaping. The bump bonding optimization is similar to that of a forward ball bonding process. However, bump shaping process puts a lot of shear stress on the bump bond. If the bump bonding is not fully optimized, the bump may lift during the bump shaping process. To prevent bump lift, the optimal bonding parameters including ultrasonic current and bond time is typically higher than equivalent regular non-SSB ball bond process. Higher ultrasonic energy may cause other issues including higher Al splash, pad peeling and pad crack. This can make SSB bump bonding optimization on sensitive devices very challenging and it may result in a narrow process window. The use of response-based processes can simplify bump bonding process optimization for such challenging devices, reduce optimization time and improve the bonding performance.

The purpose of bump shaping is to cut the wire after bump bonding and to create a flat top surface for the subsequent stitch bond landing area. A clean, flat and consistent bump top finish is desired to ensure reliable stitch bond and loop performance. Wire sway is commonly encountered if bump shape is not optimized. Long tail and semi-long tail are the common and critical issues during bump shaping. Longer bump cutting distance is usually desired to eliminate the long tail issue. However, if the cutting distance is too large, the tail will break entirely, and the wire may fly out of the wire feed system of the bonder. Multiple servo motions and other process parameters may need to be optimized to generate a good bump cutting distance window. Another requirement is the fine control of bump height. Fig. 13 demonstrates the bump height control capability of latest bump shaping development. Low bump height is desired for ultra-low-loop applications. On the other extreme, high bump is used to ensure a good clearance between wire loop and die edge. Bump shaping is also required to generate a consistent wire tail for the free air ball (FAB) for the next bond. All of the above factors are taken into account to develop a response-based bump shaping process. The results from this new smart process for ultra-fine pitch applications are presented in section C.

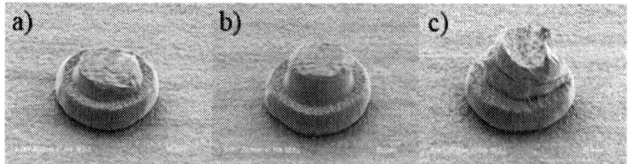

Figure 13. Bump height control: (a) low bump; (b) normal bump; and (c) high bump.

B. SSB Stitch Optimization

In a SSB process, the stitch bond is on the top of a shaped bump. There are many requirements for a good stitch bond on bump process. Besides the regular stitch bond specifications including no stitch peeling, no stitch lift, no SHTL, no long tail, good stitch remains and stitch strength, the SSB stitch bond is also required to minimize the growth of bump size and Al splash during stitch bonding. Therefore, SSB stitch bonding usually uses lower impact and bond forces, and lower ultrasonic energy as compared to conventional stitch bonds. Too low bond energy may cause stitch lift, low stitch strength and other issues. The SSB stitch bonding optimization can be very challenging. Results of stitch bond optimization of a fine pitch application is presented in the next session.

C. Fine Pitch SSB Capability Study

Fine pitch SSB capability study on thick Al pad for use in SiP applications was performed using 0.6 mil (15 μm) diameter AuPd coated Cu wire. A device with 2.8 μm thick Al pad and bond pad opening of 40 μm was used. The first process step was bump bonding, and there is 10% US energy process window with ProCu6 response-based smart process. The new bump shaping process has a large cutting distance window, as shown in Table II. ~30 μm diameter bump with ~36 μm maximum Al splash is achieved prior to stitch bond. With optimization of stitch bond process, ~32 μm diameter bump with ~39 μm maximum Al splash was achieved, as shown in Fig. 14 which demonstrated good feasibility for this process.

TABLE II.   0.6 MIL Cu BUMP BONDING ULTRASONIC WINDOW TEST RESULTS

BSA	60	65	70	75	80	85	90	95
Nsop	Lift	OK	OK	OK	OK	OK	OK	OK
Splash Y Avg	33.0	34.0	35.0	35.5	37.1	37.7	38.4	39.4
Splash Y Max	34.5	35.0	36.1	37.5	38.7	39.3	40.2	42.5
Ball Dia	29.7	29.8	29.8	29.9	29.7	29.7	30.0	29.9
Ball Ht	8.6	8.8	8.9	9.0	9.0	8.8	8.5	8.4

TABLE III.   0.6 MIL Cu BUMP CUTTING DISTANCE WINDOW TEST RESULTS

Cut Dist.	0.55	0.63	0.71	0.79	0.87	0.95	1.03	1.11
Bump lift	OK	OK	OK	OK	OK	OK	OK	OK
Shtl	-	OK	OK	OK	OK	OK	OK	OK
Long Tail	Few	OK	OK	OK	OK	OK	OK	OK

Figure 14. SEM images of 0.6mil Cu wire SSB: ~30 μm diameter bump with ~36 μm Al splash prior to stitch bond, and ~32 μm diameter bump with ~39 μm Al splash after stitch bond.

Au wire ultra-fine pitch SSB capability for memory applications was also studied using 0.6 mil (15 μm) diameter wire and described in detail in [6]. Unlike Cu wire, the Au bump does not have much Al splash. The finest SSB bump with 0.6 mil Au wire that can withstand the bump formation is ~29 μm, as shown in Fig. 15. The bump size increases to ~31 μm after the 2nd bond bonded on top of the bump.

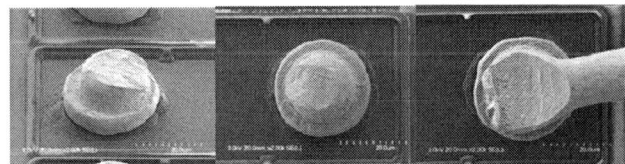

Figure 15. SEM photographs of 0.6 mil Au wire SSB: ~29 μm diameter bump bond prior to stitch bond, and ~31 μm dia bump after stitch bond. [6]

V. CONCLUSIONS

Looping advancements outlined in this paper have reduced the maximum loop height of non-compressed wire loops to under 2× the wire diameter enabling package designers to pursue even lower total package heights. The new loop height monitoring feature integrated into the latest generation of K&S wire bonders provides a smart capability to monitor looping performance in real-time during HVM. Furthermore, new ball bond improvements allow process engineers to overcome challenges in overhang configurations and quickly optimize the ball bond process for the most challenging PoP applications. The new capabilities in bump formation and SSB complements the increasing use of SiP technology. The response-based processes developed in the last few years along with the new on-bonder process monitoring capabilities are a significant step forward in achieving factory automation with regards to wire bonding. All of these advancements will most certainly improve quality controls, maximize yield and deliver exceptional value towards industry 4.0.

ACKNOWLEDGMENT

The authors would like to thank Romeo Olida and Nestor Mendoza from K&S for their assistance in collecting looping and ball bonding data respectively for the overhang configuration described in this paper. JH Yang and Jeffrey Timlin from K&S provided valuable inputs for SSB processes. We would also like to thank Hung Ly from K&S for supporting SEM imagery.

References

[1] "The chip insiders graphics file," VLSI Research, CA, November 27, 2018.

[2] E. Jan Vardaman, "Memory packaging challenges for the new era," TechSearch International, 2017.

[3] I. Qin, T. Rockey, B.Milton, G. Schulze, R. Olida, A. Chang et al., "Optimizing Ag wire bonding for memory devices," 2017 IEEE 19th Electronics Packaging Technology Conference (EPTC), Singapore, 2017, pp. 1-6. doi: 10.1109/EPTC.2017.8277432

[4] B. Milton, O. Kwon, C Huynh, I. Qin and B. Chylak, "Wire bonding looping solutions for high density system-in-package," 2017 IMAPS International Symposium on Microelectronics, Raleigh, NC, USA, October 2017. doi: 10.4071/isom-2017-WP41_151

[5] I. Qin, B. Milton, G. Schulze, C. Huynh, B. Chylak and N. Wong, "Wire Bonding Looping Solutions for Advanced High Pin Count Devices," 2016 IEEE 66th Electronic Components and Technology Conference (ECTC), Las Vegas, NV, 2016, pp. 614-621. doi: 10.1109/ECTC.2016.222

[6] H. Xu, A. Shah, B. Milton, and I. Qin "Wire Bonding Advances for Multi-Chip and System in Package Devices." 2018 IMAPS International Symposium on Microelectronics, Pasadena, CA, USA, October 2018, pp 000583-000588, 10.4071/2380-4505-2018.1.000583

Preparation and Application of Cu-Ag Composite Preforms for Power Electronic Packaging

Dongxiao Zhang[1], Shengfa Liu[1], Hui Xiang[1], Li Liu[1*]

[1] School of Materials Science and Engineering,
Wuhan University of Technology
Wuhan, Hubei, China
l.liu@whut.edu.cn

Zhaoxia Zhou[3], Stuart Roberson[2,3], Canyu Liu[2], Changqing Liu[2*]

[2] Wolfson School of Mechanical, Electrical and Manufacturing Engineering,
Loughborough University,
Loughborough, LE11 3TU, UK
[3] Loughborough Materials Characterisation Centre,
Loughborough, LE11 3TU, UK
c.liu@lboro.ac.uk

Zhiwen Chen[4*]

[4] The Institute of Technological Science,
Wuhan University,
Wuhan, Hubei, China
zhiwen.chen@whu.edu.cn

Abstract—Nano-silver and nano-copper can be effectively sintered at lower temperatures, to form the joints with high melting point meeting the needs of the interconnections of high temperature electronics. However, It has been concerned that the high cost and electromigration associated with the use of nano-silver, as well as the oxidation with nano-copper can somewhat hinder the further development and uses of these materials. In this work, we have developed micro-scale Cu-Ag core-shell composite particles by electroless plating, combining the excellent properties of copper and silver, which can potentially solve the identified problems. The mechanism of pretreatment and electroless plating in the process of preparation of these particles was explained. The composite preform was then prepared instantaneously using the particles by the electromagnetic pressing, which can significantly improve the quality of the joints formed in comparison with the direct powder sintering. Finally, the bonded structure using the preform was successfully prepared through low temperature sintering at 250 °C under a pressure of 3 MPa. Analyses of the sintered joints showed that the attachment layer had a uniform microstructure with the potential for high reliability under high temperature applications.

Keywords- High-temperature electronics package; Cu-Ag composite particle; Electromagnetic pressing; Microstructure.

I. INTRODUCTION

Insulated Gate Bipolar Transistor (IGBT) modules have been developed rapidly and widely used as main switching devices for high power electronics since the 1980s [1]. In IGBT modules, die attachments with solder alloys play a vital role in determining the functional performance and reliability of power devices. Especially, with the increasing power input, die attachments are required to withstand elevated temperature up to 250°C, which exceeds the melting temperature of conventional lead-free solders [2,3]. In the past decade, many high temperature Pb free solders were proposed including Au-based, Zn-based and Bi-based solders [4]. However, the high processing temperatures required for these solders

potentially causes the damage of the attached devices and hinders their application. Therefore, it is appreciable to fabricate robust die attachments at a relative low processing temperature in order to minimize the heat effect on the devices during the soldering process. There has been extensive research conducted, primarily based on two approaches: i) transient liquid phase bonding (TLPB) at low temperature to form intermetallic compounds (IMCs) e.g. Cu-Sn, Ag-Sn systems, to serve at higher temperature [5, 6]; ii) sintering of nanoparticles (NPs) (e.g. Ag or Cu). However, the IMCs formed through TLPB are relatively brittle, tend to have voids which can significantly deteriorate the mechanical integrity of the joints. Sintered nano-silver exhibits good electrical and mechanical properties, but is prone to migrate under thermal and electrical conditions, leading to void agglomeration. While sintering with nano-copper, the oxidation of the Cu particles can become a barrier which is difficult to remove during sintering. The copper oxide layer limits the integrity of the bond.

To eliminate the drawbacks of sintering with Ag or Cu NPs alone, sintering with combined Cu-Ag core-shell composite particles can be beneficial. This approach has the potential to mitigate the problems associated with the poor oxidation resistance of copper and the electrical migration of silver [7,8]. Furthermore Cu-Ag systems do not exhibit brittle IMCs. However, the porosity of Cu-Ag joints tends to be high and cannot be fully densified even with high pressure and long sintering times.

In this work, Cu-Ag composite particles were prepared through electroless plating, which is a simple and scalable process suitable for industrial applications. Then, the composite powders were prepared into a homogeneous preform by electromagnetic pressing; this process provides a new solution for preparing composite preform efficiently and simply. The particle morphology and composition was studied with scanning electron microscopy (SEM). The interfacial reaction between Cu and Ag were also examined with the aid of SEM. Focused Ion Beam (FIB) and Transmission Electron Microscopy (TEM).

978-1-7281-1500-9/19 $31.00 © 2019 IEEE

II. EXPERIMENTAL PROCEDURES

A. The preparation of Cu-Ag Composite particles

In this work, the spherical-like copper particles in a size range of 1-10 µm were selected and coated to produce Cu-Ag core-shell composite particles by electroless plating. Firstly, these copper particles were cleansed in 10% dilute sulfuric acid for 5 min to remove oxides. Next, these particles were added to 30 g/L SnCl$_2$ hydrochloric acid solution for sensitization. These particles were added to 0.5 g/L PdCl$_2$ hydrochloric acid solution for activation. Finally, the copper particles were transferred to the chemical solution immersed in a water bath ($\sim$60 $\pm$2 ℃) for Ag coating up to 20 min. The formulation of the plating chemistry and the plating conditions are shown in Table 1.

TABLE. I The formulation of the plating chemistry and the plating conditions

Chemical reagents	Concentration (g/L)
PVP	1~1.5
AgNO₃	7~12
EDTA disodium salt	20~40
Reducing agent	1~5
Plating Temperature	55~65℃
Magnetic stirring rate	700 rpm

B. The preparation of Cu-Ag composite preform

After electroless Ag plating, Cu-Ag particles were then directly used to make the preforms by electromagnetic pressing with required dimension in milliseconds. In this work, we performed an electromagnetic pressing experiment using 0.36 g of Cu-Ag composite particles each time to prepare preforms of 100 $\pm$ 20 µm. In order to obtain the preform with good quality, the constant capacitance of 440 µf maintained , to perform electromagnetic suppression experiments at voltages of 2800 V, 3200 V 3600 V, 4000 V and 4400 V. The schematic diagram of the electromagnetic pressing is shown in Fig. 1. The energy provided by electromagnetic suppression can be calculated according to the Equation:

$$W = \frac{1}{2}CU^2 \qquad (1)$$

Where W is the energy, C is the capacitance, U is the voltage.

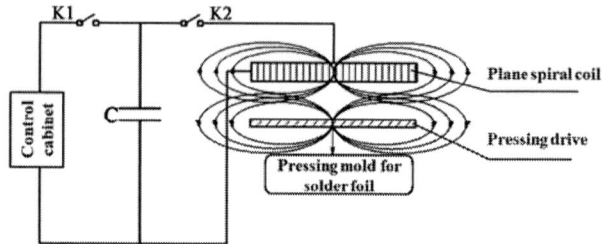

Fig.1 Schematic diagram of electromagnetic particles pressing.

C. Sintering process

The schematic diagram of the Cu/Ag-Cu preform/Cu sandwich structure is shown in Fig.2. Copper sheets (3 mm thickness, 99.9 % purity) were used as substrates, while the Cu-Ag preforms were placed between the copper substrates and bonded. The size of Cu sheets was 3 mm × 3 mm × 3 mm. These copper sheets were used as surrogates for chips to fabricate the joints. Prior to sintering process the Cu sheets and the preform were soaked in diluted hydrochloric acid solution and then cleaned with distilled water and ethanol. The whole sintering process was carried out at 250 ℃ with a pressure of 3 MPa, successful joining was completed in 60 minutes. The entire process was performed in a vacuum furnace (10 Pa).

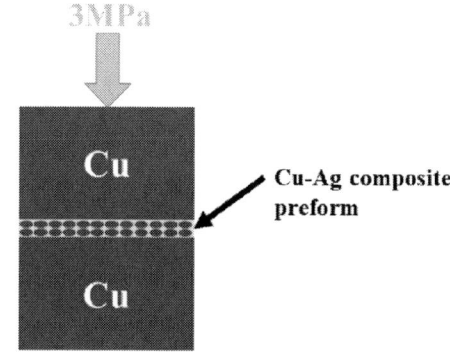

Fig.2 Schematic diagram of the Cu/Cu-Ag preform/Cu sandwich structure

D. Characterizations

The surface morphologies of Cu-Ag composite particles and the Cu-Ag composite preform were observed by scanning electron microscopy (SEM, Zeiss ULTRA plus), the compositions of the surfaces were analyzed by energy dispersive spectroscopy (EDS) incorporated with the SEM. In order to observe the internal microstructure of the Ag-Cu preform and sandwich structures, samples were cold-mounted, ground and final polished with 1 µm colloidal alumina particle. The interfacial microstructure and the composition at these interfaces were observed by SEM and EDS, respectively. However, due to the low thickness (< 1 µm) of the interdiffusion layer between the pre-form and copper, the resolution available to the SEM technique is not sufficient to clarify the reaction mechanisms between Cu and electroless Ag coatings. In particular the probe size of the high accelerating voltage beam required for EDS could not be used to accurately resolve the interfacial reactions. Therefore, a dual-beam focus ion beam (FIB, FEI Nova 600 Nanolab Dual Beam) was used to fabricate cross section lamella samples for transmission electron microscopy (TEM), (FEI Tecnai F20 G2 S scanning TEM (STEM) (FEI, USA)) analysis at a nano-scale. Using the TEM in-depth material analysis can be conducted at the Cu/electroless Ag interfaces. The porosity of the composite preform was quantitatively analysed by Image-J image analysis software.

III. RESULTS AND DISSCUSSIONS

A. Formation of Cu-Ag core-shell composite particles

The Ag electroless plating formation is summarized in Fig. 3. It is also worth mentioning that this synthetic process was performed in an aqueous solution, which avoids the use of expensive organic solvents. Generally, there is oxide or organic oil on the surface of Cu particles. Therefore, a series of pre-treatments of the Cu surface is necessary prior to the electroless Ag plating to for the Cu-Ag composite microspheres.

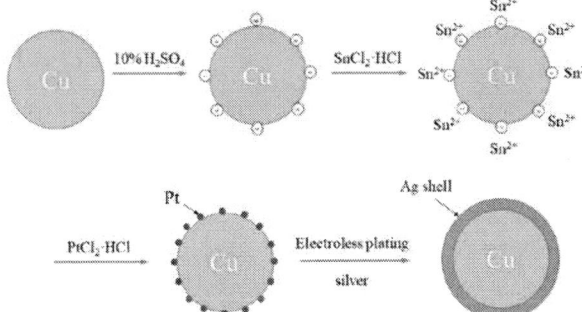

Fig.3 Schematic illustration for the fabrication of Cu-Ag composite particles.

The pretreatments can also add additional features such as aspirates on the surface of Cu particles to increase the roughness, hence the contact area and hydrophilic characteristics of the Cu surfaces. The adhesion between the Cu surface and the coating is therefore enhanced [10]. The rougher surface might result from the partial dissolution of Cu on the exterior of the surface by sulfuric acid. Consequently, the adsorption of Sn^{2+} onto roughened Cu is facilitated because of the larger surface area. Moreover, the roughened Cu surface hinders the removal of the final metal layer and results in high interfacial adhesion, known as the anchor effect [11]. Next, the Cu particles were sensitized through the adsorption of Sn^{2+} onto the surfaces of the microspheres via electrostatic interactions followed by the colloidal Pt with a negative charge adsorbed onto the surface of the Cu particles [12]. These metallic Pt seeds provide the catalytic activity

required for the subsequent deposition of silver. Such a two-step activation process can avoid the agglomeration of resulting products during electroless silver plating owing to removal of excess catalyst ions or atoms before immersion into the plating bath.

Finally, Cu-Ag composite microspheres were obtained by electroless silver plating. This process differs from the normal chemical reduction of Ag^+ in an aqueous solution, in which Ag^+ ions are present everywhere [13]. Because of the self-catalyzing effect of silver, the silver-plating bath is unstable. The Ag^+ can be easily reduced to Ag and cause agglomeration because of their high redox potential. Thus, EDTA disodium salt, an inexpensive and green complexing agent was used in the electroless plating bath to ensure uniform nucleation of silver onto the Cu particles in this work. The EDTA disodium salt can form a stable coordination compound with Ag^+, which reduces the concentration of free Ag^+ in the solution and can reduce the rate of Ag+ reduction. When the formation and growth rate of the silver nucleus decreases, the silver nucleus can be deposited on the surface of the copper powder when it is not grown, which can improve the deposition coating and form a stable and uniform coating. When the copper powder is completely coated with silver, the substitution reaction is stopped. At this time, the remaining Ag^+ in the solution are further reacted with the reducing agent to form silver coated on the surface of the copper powder to increase the thickness of the silver layer.

Fig.4 shows a surface morphology of Cu-Ag composite particles as a typical sample in this work. It can be seen from Figure 5 that the silver-plated copper powder was relatively integrated, the coating is relatively uniform, the surface was smooth and there was no separate granular silver. Only a very small number of points were uneven.

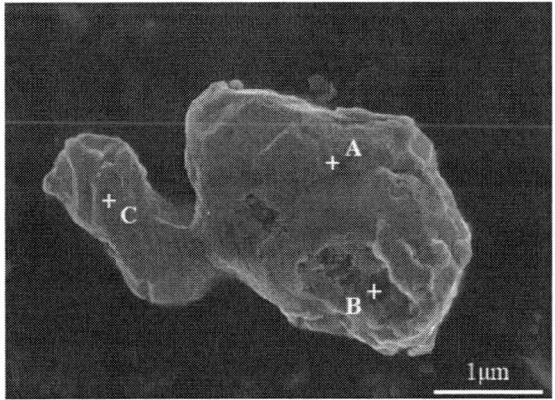

Fig.4 Surface micrographs of Cu-Ag composite particles

Meanwhile, the chemical composition of the Cu-Ag composite surface in Fig. 4 were analyzed by EDS. The corresponding composition in these location are listed in TABLE II. It is clear that the surface of the copper particles was coated relatively completely with silver, but the thickness of the coating was partially different. At the thicker point of the coating, the silver content is above 35 wt.%; and at the thinner point of the coating, the silver content is only 12 wt.%. On the one hand, this may be due

978-1-7281-1500-9/19 $31.00 © 2019 IEEE

to the subsequent small amount of oxidation after copper surface treatment. On the other hand, because during the reduction process, the silver particles grow on the surface of the plating layer to form a sheet-like structure. However, the adhesion between the silver plating layers was poor and peeling may occur during the subsequent experiments.

TABLE. II SEM quantitative of the three points marked in Fig.4

	Cu (wt.%)	Ag (wt.%)
A	67.6	32.4
B	87.8	12.2
C	63.5	36.5

B. Preparation of Cu-Ag composite preform

In this work, the Cu-Ag composite preform with a thickness of about 100 μm was instantaneously prepared by electromagnetic pressing in the range of 2800 V to 4400 V. Compared with the conventional static pressing method, the electromagnetic preparation process is simpler and more efficient. This demonstrated that electromagnetic pressing can be a new and novel method with great potential in preparing preform with powder.

In order to improve the compactness of the composite preform for improving the subsequent bonding strength, a denser composite preform can be obtained at a high voltage. But when the voltage is too high, the internal stress of the bonding piece increases, and defects such as microcracks may occur, even macroscopic fracture. Through our experiments, a Cu-Ag composite integrity preform with high density can be prepared at a voltage of 4400 V. Fig 5 shows the cross-sectional morphology and surface morphology of the composite prepared at 4400 V. In the cross-sectional morphology, we can clearly see that inside the preform, most of the particles were connected without voids, and only a small number of particles were surrounded by voids. As measured by Imagine J, the estimated porosity inside the preformed is 0.78 %, which greatly decrease compared to the direct powder sintered joint [14]. In the surface morphology, we find that the surface of the composite preform was relatively integrated, and no obvious microcracks exist, which proves that a good combination was formed between the particles.

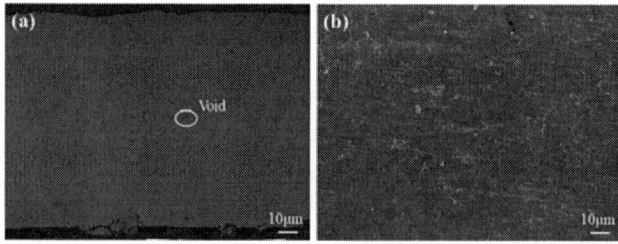

Fig.5 (a) Cross-section and (b) surface SEM images of the Cu-Ag preform

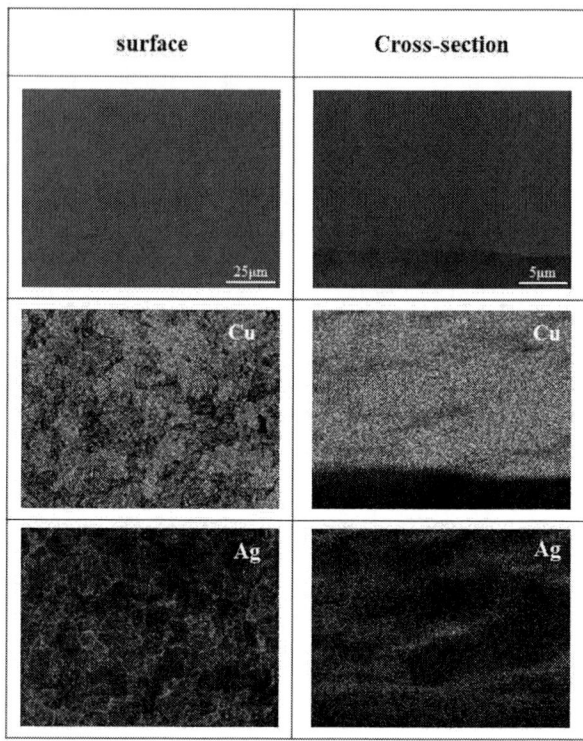

Fig. 6 Elemental distribution of Cu and Ag on the Cu-Ag preform surface and cross-section

For Cu-Ag composite preforms with the voltage of 4400 V, the elemental distribution of the cross-sectioned and surface were investigated through EDS as shown in Fig.7. From the EDS results of surface and cross-section, it can be seen that the elements are evenly distributed, and the connection between the particles was very tight. This indicates that the silver on the surface between the particles has a more obvious diffusion phenomenon, which further forms an effective connection between the particles. While Ag and Cu may have diffused into each other to some extent, the boundaries are still very clear. This may be due to the low solid solubility between copper and silver at room temperatures, so it is difficult to cause significant diffusion between Ag and Cu only by large instantaneous energy.

C. Interfacial microstructure of the sintering Cu-Ag composite/Cu joint

Fig.7 shows a cross-section micrograph of Cu/Cu-Ag composite preform/Cu sandwich structure as a typical Ag-Cu joint in this work. It can be observed that, after sintering, the entire composite preform layer had no obvious defects in the SEM, and the entire structure became denser, the elements would be further diffused. but the boundary between the particles and the particles marked by the silver layer can still be found, indicating that no complete alloying occurs between the silver and the copper in the composite preform. For the Cu-Ag composite preform/Cu interface, an effective connection was formed between the composite preform and the copper substrate, and no obvious defects were observed at the connection interface. To verify the integrity of this the joint, the enlarge view of religion b at the Cu/Ag-Cu interface is shown in Fig. 3 b). Generally, the joint was crack-free and compact, showing good adhesion characteristics. However, no significant elemental diffusion layer was identified at the interface under the SEM.

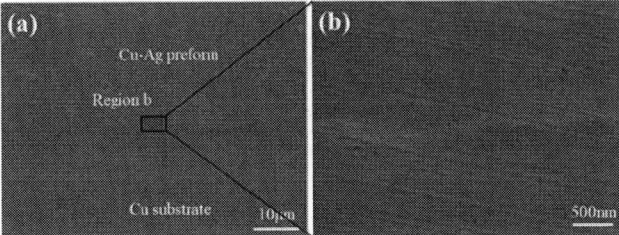

Fig. 7 The SEM images of the cross-sectioned Cu/Cu-Ag preform joint: (a) Overview of the joint, b) enlarge interface view of Region b.

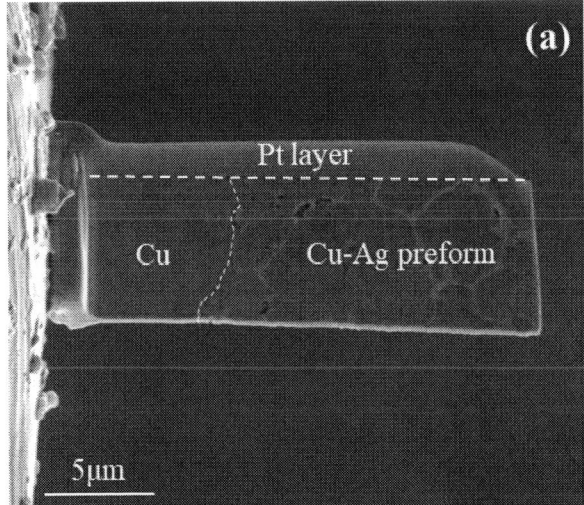

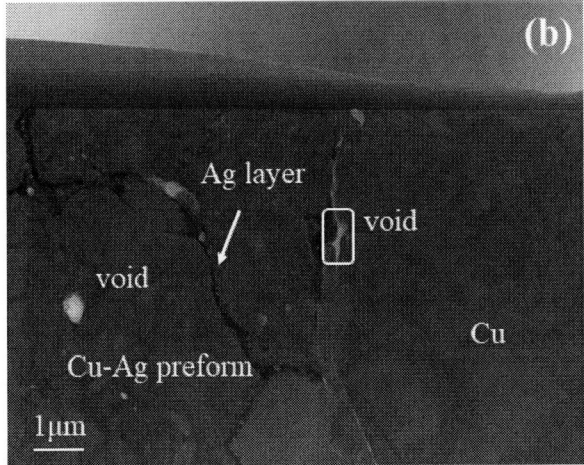

Fig. 8 FIB specimen at the Cu-Ag composite preform/Cu interface from: a) FIB image; b)TEM bright-field image.

Due to the large probe size of SEM equipment, the accuracy in quantitative composition of the diffusion layers from EDS analysis is limited, especially when the diffusion thickness is below 1 μm in Cu-Ag joints. Y. Morisida *et al.* reported that solid solution of Ag and Cu could be observed near the bonding interface between the Ag and the Cu nanoparticles [15]. To prepare a TEM sample for precise composition results, FIB machining was used to fabricate a thin film (thickness < 100 nm) at the Cu-Ag composite preform/Cu substrate interface (marked in Fig. 7(a)) for TEM analysis. The microstructure of FIB sample is illustrated in Fig. 8.

As shown in Fig. 8 (a), this TEM sample consists of Cu-Ag composite preform and Cu substrate. The top platinum (Pt) layer was for protection of the area of interests during FIB milling. The direction of the FIB milling was from top to bottom side. The thickness of the TEM sample was approximately 100 nm, enabling precise and accurate results of TEM analysis including high resolution transmission electron microscopy (HRTEM) and EDS analysis at a nano-scale.

From the TEM bright-field image in Fig. 8 (b), of the prefabricated bonded layer, a denser bond was formed between the composite particles. And smaller crystal grains were distributed in the copper particles. We believe that this may be because during the sintering process, the internal stress released by electromagnetic pressing caused the recrystallization of copper grains together with the temperature, which is beneficial to diffusion. According to the corresponding EDS results, the Cu content of the Ag coating layer was 15.3 %wt. in spite of the 250 °C bonding temperature. The high activity of Cu-Ag preform may have led to the remarkable nanostructural solid solution, which seems rather useful for the bonding application and also for the creation of various new materials as well. However, some small voids (< 0.5 μm) were also formed near the Cu/Ag interface, as shown in Fig. 8 (b), which was highly likely attributed to the peeling of the Ag coating and the surface breakage of the preform during the sintering process due to excessive energy.

IV. CONCLUSIONS

Based on the results obtained in this study, some preliminary findings can be summarised as follows:

- Electroless Cu-Ag core-shell composite particles was successfully developed and utilised in high temperature interconnects. The formation of the electroless Ag plating on Cu particles was elaborated in detail. EDS compositional analysis illustrated that silver content in the composite particles was 12-35 wt.%.

- An integrated composite preform with compact microstructure was successfully fabricated by electromagnetic pressing using voltage of 4400 V under a capacitance of 440 μf. The composite preform prepared by electromagnetic pressing have a dense structure with a porosity of approximately 0.78 %.

- The Cu-Ag composite preform was used to successfully form an effective joint structure with Cu substrate of micro-sized defect free at temperature of 250 °C. However, when observed at a submicron scale, although the copper crystal grains in the preform layer were refined, some small voids (<0.5 μm) were found at the interfaces. A nano-level transition layer was also detectable at the copper-silver interfaces, which indicates that a certain degree of solid solution has been formed between Cu and Ag.

ACKNOWLEDGMENT

This research was supported by Hubei Provincial Natural Science Foundation of China (Grant No. 2018CFB212) and the open funding via State Key Laboratory of Materials Processing and Die & Mould Technology (Grant No. P2018-018). The appreciation is also due to the Fundamental Research Funds for the Central Universities (Grant No. 173101001) and Student Innovation and Entrepreneurship Training Program (Grant No. 201810497227).

The authors also acknowledge the research grants on "quasi-ambient bonding (QAB)" (Ref: EP/R032203/1) and "heterogeneous integration (HI)" (Ref: EP/R004501/1) funded by EPSRC of the United Kingdom.

REFERENCES

[1] Khazaka R, Mendizabal L, Henry D, Hanna R. "Survey of high-temperature reliability of power electronics packaging components", IEEE Transactions on Power Electronics, vol. 30, 2015, pp. 2456-2464.

[2] S. Fu, Y. Mei, X. Li, P. Ning, G.Q. Lu, "Parametric study on pressureless sintering of nanosilver paste to bond large-area (>100 mm²) power chips at low temperatures for electronic packaging", Journal of Electronic Materials, vol. 44, 2015, pp. 3973-3984.

[3] Brian J. Grummel, Zheng John Shen, Habib A. Mustain, and Allen R. Hefner, "Thermo-Mechanical Characterization of Au-In Transient Liquid Phase Bonding Die-Attach", IEEE Transactions on Components, Packing and Manufacturing Technology, vol. 3, No. 5, 2013, pp. 716-723.

[4] Vemal R. Manikam and Kuan Y. Cheong, "Die Attach Materials for High Temperature Applications: A Review", IEEE Transactions on Components, Packing and Manufacturing Technology, vol. 1, No. 4, 2011, pp. 457-478.

[5] JF. Li, PA. Agyakwa, CM. Johnson. "Interfacial reaction in Cu/Sn/Cu system during the transient liquid phase soldering process", Acta Materialia, vol. 59, 2011, pp. 1198–1211.

[6] JF. Li, PA. Agyakwa, CM. Johnson. "Kinetics of Ag3Sn growth in Ag–Sn–Ag system during transient liquid phase soldering process", Acta Materialia, vol. 58, 2010, pp. 3429–3443.

[7] JQ Wang and S Shin. "Sintering of multiple Cu–Ag core–shell nanoparticles and properties of nanoparticle sintered structures", RSC Advances, vol. 7, 2017, pp. 21607–21617.

[8] X Yu, JJ Li, TL Shi, CL Cheng, JH Fan, SY Cheng, TX Li, GL Liao, ZR Tang. "A green approach of synthesizing of Cu-Ag core-shell nanoparticles and their sintering behavior for printed electronics", Journal of Alloys and Compounds, vol. 724, 2017, pp. 365-372.

[9] CT Chen, S Noh, H Zhang, CY Choe, JT Jiu,S Nagao, K Suganuma. "Bonding technology based on solid porous Ag for large area chips", Scripta Materialia, vol. 146, 2018, pp. 123-127

[10] M. Sanles-Sobrido, M. Banobre-López, V. Salgueirino, M.A. Correa-Duarte, B.Rodríguez-González, J. Rivas, L.M. Liz-Marzán, "Tailoring the magnetic properties of nickel nanoshells through controlled chemical growth", Journal of Materials Chemistry, vol. 20, No, 35, 2010, pp. 7360-7365.

[11] L. Li, B. Liu, "Study of Ni-catalyst for electroless Ni–P deposition on glass fiber", Materials Chemistry and Physics, vol. 128, 2011, pp. 303-310.

[12] Y. Ma, Q. Zhang, "Preparation and characterization of monodispersed PS/Ag composite microspheres through modified electroless plating", Applied Surface Science, vol.258, 2012, pp. 7774– 7780

[13] X. Yin, L. Hong, B.H. Chen, "Role of a Pb²⁺ Stabilizer in the Electroless Nickel Plating System: A Theoretical Exploration", Journal of Physical Chemistry B, vol.108, 2004, 10919-0929.

[14] S, Fu, Y. Mei, G.Q. Lu, X. Li, G. Chen, X. Chen, "Pressureless sintering of nanosilver paste at low temperature to join large area (>100mm²) power chips for electronic packaging", Materials Letters, vol. 128, 2014, pp. 42-45.

[15] Y. Morisada, T. Nagaoka, M. Fukusumi, Y. Kashiwagi, M. Yamamoto and M. Nakamoto, " A Low-Temperature Bonding Process Using Mixed Cu–Ag Nanoparticles" Journal of Electronic Materials, vol. 39, No. 8, 2010, pp. 1283-1288.

Au-Rich/Sn-Bi interconnection in Chip-on-Module package

Jin Wang[1], Qian Wang[1], Xinnan Hou[2], Ke Du[2], Lixin Zhao[2], Jian Cai[1, *]

[1]Department of Microelectronics and Nanoelectronics, Tsinghua University, Beijing 100084, China
[2]GalaxyCore Inc., Shanghai 201203, China
[*]Email: jamescai@tsinghua.edu.cn

Abstract—**A new package structure for CMOS image sensor (CIS) is proposed as COM (Chip on Module), based on a novel soldering wire bonding technique whose second bond point is achieved by soldering with the help of eutectic Sn-57Bi solder paste, forming the Au-rich/Sn-Bi interconnection which is the focus of this paper. The reliability of the interconnection has been confirmed by functional tests, pull tests and interfacial observation after high temperature and humidity storage (THS) reliability tests at 85°C/85%RH for 120h, 240h, 480h and 1000h. All COM packaged samples have passed the camera functional tests after aging. The tensile strengths of the gold wires, which have finished soldering wire bonding, always maintain above 9 grams during aging, which is higher than the standard strength in conventional wire bonding (8 grams). Furthermore, as for the interfacial structures in Au/Sn-Bi couple, AuSn and $AuSn_2$ are found at the interface, whose thickness can be described as a parabolic relationship with the annealing time, suggesting a bulk diffusion controlling mechanism. Besides, the effects of 0.3%~0.35% Ag addition, in Sn-57Bi solder, on the reliability of the interconnection have also been investigated. Owing to the barrier effects of Ag and Ag_3Sn on the diffusion of Au atoms into Sn-contained solder, a more reliable joint has been obtained during aging, with a higher tensile strength (above 19 grams), thinner IMC thickness (20%~30% thinner) and unobvious Kirkendall voids at the interface.**

Keywords-Chip on Module package; soldering wire bonding; Au-rich/Sn-Bi interconnection; tensile strengths; IMC

I. INTRODUCTION

CMOS Image Sensors (CIS) have been widely used in consumer products such as digital cameras, mobile phones, computers, scanners, security systems and so on [1]. Packages for CIS need more considerations beyond those encountered for conventional ICs, including an optical window, precision mechanical alignment of the die and protection of imaging area from contamination during manufacture [2]. Traditional packages for CIS mainly include Chip Scale Package (CSP) and Chip on Board (COB) [3, 4]. CSP, as shown in Fig. 1(a) with pad on chip rerouted to form BGA type, is mostly used in low-end products, such as 2M and 5M pixels cameras. The top glass, which is used to protect the chip from contamination, is directly attached on the chip and leads to poor optical performance by blocking a part of light and also affecting the reflection of the light. The solder balls are assembled on the substrate, resulting in two-thousandths of long-term failure for HVM

due to the mismatch of thermal expansion coefficients between solder balls and substrate. For high-end products as 8M or higher pixel cameras, COB in Fig. 1(b) is often applied, which use conventional die-attach and wire bonding to connect the chip to the substrate. Compared with CSP, the optical performance is improved without the glass attached directly on chip and the reliability is also improved by converting solder ball to gold wire connection. But the particle problem still needs carefully handling during manufacture which may bring higher costs. In order to further improve efficiency and reduce costs, Chip on Module (COM) package, based on COB, has been proposed, as shown in Fig. 1(c). Au wires, in wire bonding process, are firstly bonded onto the chip through thermo-sonic as the first bond point, but cut off just before the second bond point be carried out. The impending wires on the chip are then coated with solder paste and connected to the PCB by soldering as the second bond point. Compared with COB, COM could avoid particle contamination, improve optical performance with less tilt, decrease the module size and lower the costs with simplified process.

The second bond point in COM wire bonding process is achieved by soldering. Eutectic Sn-Bi solder paste has been selected because of its low melting point (138°C) in this proposal and Au/Sn-Bi solder joint is obtained after reflow. However, the combination of Au and Sn is known to form one of the fast diffusion systems at solid-state temperature [5] and it is necessary to ensure the reliability of the Au-rich/Sn-Bi interconnection at the second bond point to identify the feasibility and as well the reliability of COM package. There have been many studies on Au/Sn interdiffusion at solid-state temperature before. Nakahara et al. [6] investigated the interfacial reactions of the Au/Sn system at room temperature and found that the ζ-solid solution, AuSn, $AuSn_2$ and $AuSn_4$ phases were observed at the interface. It was also found that the reactions between Au and Sn at solid-state temperature were diffusion-controlled, which can be mathematically described as a power function of the annealing time, and the diffusion rate was very rapid even at room temperature [6-8]. Some studies on the effect of Bi on Au/Sn-Bi interdiffusion during aging were also published. Yee-Wen Yen et al [9] studied the interfacial reactions of Sn–xBi alloys reacting with the Au substrate and found that the growth-rate constant of IMCs increased with less Bi content in the Sn–Bi alloys because the increase in the Bi content could make the large (Bi) phase segregate in the solder/Au interface and decrease IMC growth. In this work,

978-1-7281-1500-9/19 $31.00 © 2019 IEEE

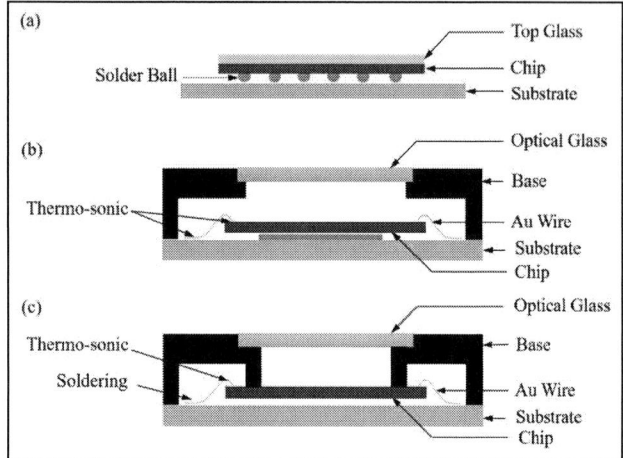

Figure 1. Schematic of packages for CIS: (a) CSP; (b) COB; (c) COM.

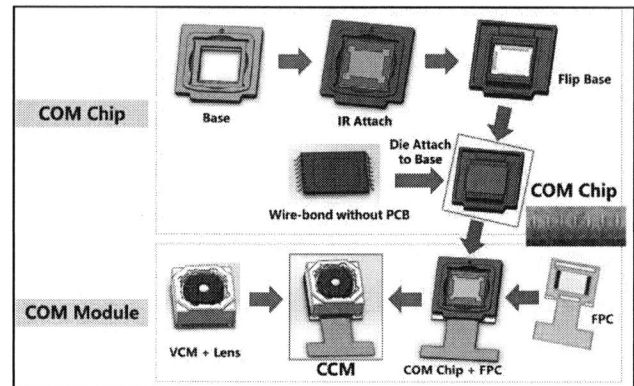

Figure 2. Manufacturing process of COM package

the reliability of Au-rich/Sn-Bi interconnection was investigated by conducting the high temperature and humidity storage (THS) tests at 85°C/85%RH for COM packaged samples and executing measurements such as functional tests, pull tests and microstructure observation of interfacial IMC.

Moreover, many studies have found that a small amount addition of silver in Sn-Bi eutectic solder can help to improve the ductility of the solder by a factor more than three which was attributed to the substantial refinement of the solidification microstructure [10, 11]. However, previous studies were mainly focused on the mechanical properties reinforcement of Sn-Bi solder with Ag added but researches about the effect of Ag on the diffusion of Au/Sn-Bi couple and the growth-rate of interfacial IMC are rarely found. In this paper, 0.3%~0.35% Ag was added into the eutectic Sn-57Bi solder, following with reflowing, aging and measurements as above. Focused on the tensile strengths and the interfacial IMCs growth during aging, the reliability of the Au-rich/Sn-Bi with and without Ag was compared.

In this paper, the Au-rich/Sn-57Bi interconnection as the second bond joint in COM package has been investigated. The tensile strengths of the interconnection, the composition and growth mechanism of the interfacial IMCs, and the effect of silver on the diffusion have been discussed, to ensure the reliability of this novel soldering wire bonding in COM using eutectic Sn-57Bi solder with or without Ag addition.

II. EXPERIMENTS

A. Manufacturing Process of COM Package

The manufacturing process of COM package for CIS is shown in Fig. 2. Firstly, a base for package has been made and IR Glass is attached to the front of the base. Then a novel soldering wire bonding technique is performed on the chip without PCB, which is implemented by cutting off the gold wire at the position of the second bond point, and hanging up the wire as the lead for further connection with PCB. The first bond point on chip is completed by thermo-sonic which is the same as the conventional wire bonding in COB. Then, attach the front of the chip to the base to protect the image sensor from contamination. By now, COM chip has been finished and the above steps need to be carried out in the clean room. Following, dip the end of the hanging gold wire into the solder paste, as shown in the inset in Fig. 2 with the solder coated at the end of the wires. Connect the gold wire to PCB by soldering with the help of the solder at the end. Finally, assemble the Voice Coil Motor (VCM) and lens to obtain the final CMOS Camera Module (CCM) product. The most different part between the wire bonding process in COB and COM is the connection of the second bond point, which is achieved by thermo-sonic in COB whereas by soldering in COM.

As mentioned, eutectic Sn-57Bi has been chosen to realize the connection between gold wire and PCB. The reason for selection of this solder is that it has a lower eutectic temperature as 138°C which can reduce the manufacturing temperature. Two kinds of Sn-Bi solder pastes, solder A and solder B, are used in this work, whose compositions are different slightly. Solder A is pure eutectic Sn-57Bi, whereas in solder B, there is a small amount of 0.3%~0.35% Ag addition. Au-rich/solder interconnection is finally implemented by mass reflow at 180°C for 30s. The interface of the solder joint during aging (at solid-state temperature) has been investigated to identify the feasibility and as well the reliability of COM package. Moreover, by comparing the interconnection of two solders, the role of silver addition in the solder has been obtained for further reliability improvement.

B. High Temperature and Humidity Storage Reliability Tests (THS)

In order to access the reliability behavior of COM package, samples after package are subjected to JEDEC high temperature and humidity storage (THS) reliability tests at 85°C/85%RH for 120h, 240h, 480h and 1000h respectively. The specific process is as follows. Firstly, measure the function of each packaged sensor sample before the reliability test. After the samples get by, place them into the test chamber. Adjust the temperature of the test chamber to 85°C at a rate of 1°C/min, and the humidity is kept at 85%RH. Keep 120h, 240h, 480h and 1000h respectively in the condition of 85°C/85%RH without power on. Finally, after leaving the samples at room temperature for 2 hours, various measurements for the samples under different conditions: without the reliability test, reliability test for120h, 240h, 480h and 1000h, are performed. Comparing diverse performances of the packaged device, the reliability of this new package structure, COM, is evaluated.

C. Measurements

To sum up, there are two groups of variables to be compared and analyzed in this paper. One is the reliability test time kept at 85°C/85%RH, including 0h, 120h, 240h, 480h and 1000h. The other one is the different composition of Solder A and B (Solder A is eutectic Sn-Bi, whereas Solder B has 0.30%-0.35% silver additionally). Different samples under different conditions are named from Group 1 to Group 10, as shown in Table I.

After reliability tests, various measurements have been conducted on the packaged samples. Firstly, camera functional tests for the packaged image sensor have been carried out to verify that the integrated circuit and interconnection from chip to package are operating correctly during reliability tests, therefore, to predict its lifetime and estimate whether it meets the life requirements (usually is less than 3 years) of mobile product such as the smartphones and tablets. Secondly, since the most different part between COB and COM is the second bond point in wire bonding process, the reliability of Au-rich/Sn-Bi interconnection is our focus to attention. Pull tests for the gold wires, after the novel soldering wire bonding, are performed in all groups of samples and the tensile strengths with aging time are obtained. We also observe the fracture positions along the gold wire by optical microscope and the proportion of the failures at the second bond point is also analyzed. Finally, Scanning Electron Microscopy (SEM) and Energy Dispersive Spectrum (EDS) are then performed to observe the interfacial structure at the second bond of the samples under different conditions from Group 1 to 10 to further investigate the reliability. The composition and thickness of the interfacial IMC are observed and measured and the growth rate of each IMC is calculated mathematically which can be fitted as a power function of the annealing time. A theoretical analyzation of the IMC growth mechanism is then carried out by comparing various groups of samples, thus providing a theoretical basis for a more reliable connection in the future.

TABLE I. GROUPS OF SAMPLES UNDER DIFFERENT CONDITIONS

Groups of Samples under Different Conditions		Different Conditions	
		Solder A	*Solder B*
As-reflowed		Group 1	Group 6
High Temperature and Humidity Storage Reliability Tests	120h	Group 2	Group 7
	240h	Group 3	Group 8
	480h	Group 4	Group 9
	1000h	Group 5	Group 10

III. RESULTS AND DISCUSSION

A. Functional Tests

200 samples were randomly picked that have completed the COM package in each group and carried out the camera functional tests. It is found that the samples with solder A and solder B have all passed the functional tests and there is no failure even after 1000h-aging. According to the reliability test specifications by consumers, the life expectancy of COM package can be estimated to be no less than 5 years, which is certainly enough for the mobile products such as the smartphones and tablets whose life requirements are usually less than 3 years.

B. Pull Tests

5 samples from each group were selected and each sample has 32 gold wires that have finished the soldering wire bonding and completed the connection between chip and PCB. Bond tester is used to pull the bonded gold wire and obtain the tensile strengths. Fig. 3 shows the average tensile forces of the gold wire in the samples from Group 1 to 10. For Solder A which has barely eutectic Sn-Bi, the tensile forces maintain above 22 grams and basically unchanged after aging for 120h and 240h. While after aging for 480h, it has a small drop to 20 grams and the standard deviation is increasing with a larger data dispersion. However, after 1000h-reliability test, the tensile force decreases dramatically to 9.7 grams with the standard deviation increasing to 5.1 grams. On the other hand, for Solder B which has Ag addition in the solder, the average tensile forces are no less than 21 grams until the samples is aged for 1000h when it slightly decreased to 19 grams, meanwhile there is no significant fluctuations in the standard deviation. By comparison, the difference of tensile strengths between Solder A and B mainly starts at 480h with A decreasing slightly while B basically maintaining the same,

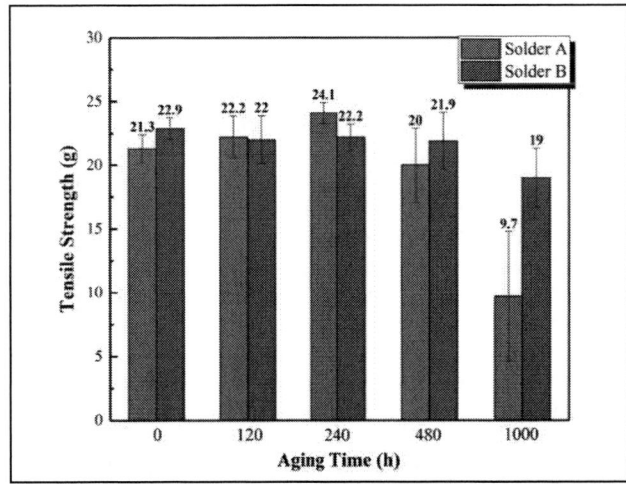

Figure 3. Average tensile forces of the gold wire in the samples from Group 1 to 10.

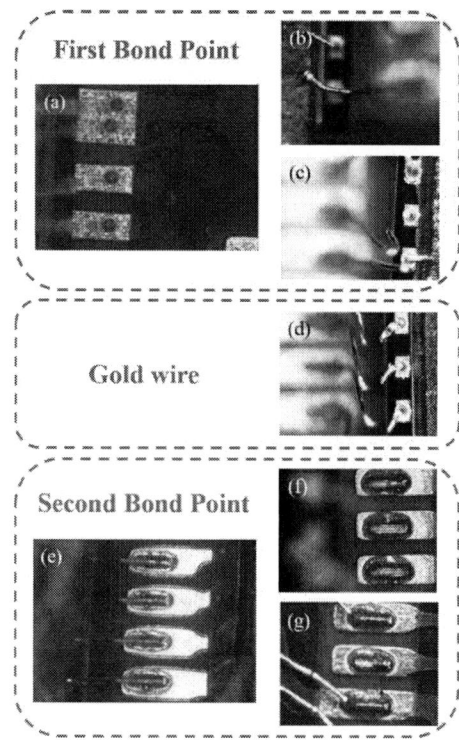

Figure 4. The fracture positions after pull tests: (a-c) at the first bond point; (d) inside the gold wire; (e-g) at the second bond point. (a) and (e) shows the perfect interconnection at the first and the second bond point before aging, respectively.

decreasing sharply to a half but B only falling slightly. Besides, it is worth mentioning that the standard tensile force for 30μm gold wire, in conventional wire bonding by thermo-sonic in COB, is only 8 grams, which is less than half the strength in our soldering wire bonding, illustrating the superiority of the novel soldering wire bonding in COM.

In order to find the reasons for the differences between Solder A and B, the fracture positions along the gold wire after pull tests are observed by optical microscope and also analyzed statistically. The failure positions can be classified into three types: at the first bond point (by thermo-sonic), the gold wire and at the second bond point (by soldering), as shown in Fig. 4. Fig. 4(a) and (e) respectively show the perfect interconnection at the first and the second bond point before aging. After reliability tests and pull tests, the failures at the first bond point may have been caused by a disconnection between the gold wire and the pad in Fig. 4(b), or a fracture in the neck of the gold wire in Fig. 4(c). Fig. 4(d) shows the break inside the gold wire after pull tests. As for the second bond point, the fractures mainly occur between the gold wire and the solder paste as shown in Fig. 4(f) and (g). Further, we investigate the proportion of the failures at the second bond point, as shown in Fig. 5, which can explain the differences between Solder A and B reasonably. As for Solder A, the fractures mainly occur inside the gold wire or at the first bond point after aging for 120h and 240h, with a proportion of about 98%. The situation is also satisfied with Solder B at the same time, corresponding with the tensile

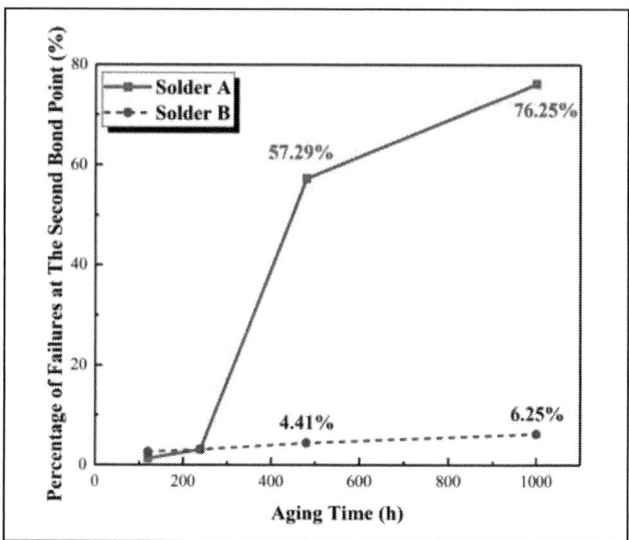

Figure 5. The percentage of failures at the second bond point after pull tests with Solder A and Solder B.

strengths being roughly the same using Solder A and B (above 21 grams) in Fig. 3. However, it changed after 480h- and 1000h-reliability tests. The proportions of the second bond point failures have raised to 57.29% and 76.25% using Solder A, respectively, consistent with the strengths falling to 20 grams and 9.7 grams in Fig. 3. While in Solder B, there is only a small increase in the proportion from 3.13% to 6.25%, which exactly explains why the strengths remain basically the same (above 19 grams) even after 1000h-aging. In conclusion, the sharp decrease in strength in Solder A after 1000h-aging is mainly due to the reduced strength at the second bond point after aging. While in Solder B, after aging for 1000h, the strength of the second bond joint is still higher than that of the first bond point and the gold wire, leading to basically nondecreasing tensile strengths. It is evidential that the second bond joint is more reliable using Solder B with Ag addition than Solder A.

C. Interfacial Structure

To explain why Solder A is less robust than Solder B at the second joint point, the interfacial structures between the gold wire and eutectic Sn-57Bi solder from Group 1 to 10 was investigated by Scanning Electron Microscopy (SEM) and Energy Dispersive Spectrum (EDS). The composition,

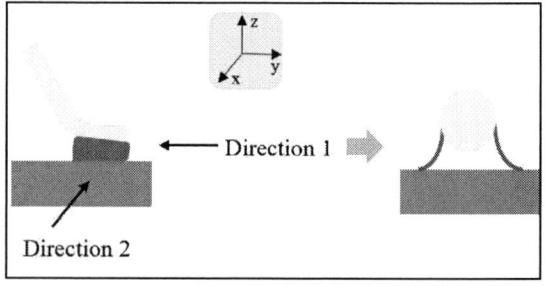

Figure 6. Schematic of the connection between gold wire and pads on PCB from direction 1 and 2.

morphology and thickness of each intermetallic compound (IMC) at the interface are obtained and measured. Two directions were observed in each sample, as shown in Fig. 6, for more reliable results.

Fig. 7 shows the interfacial structures of samples from Group 1 to 10 with Solder A in Fig. 7(a-e) and Solder B in Fig. 7(f-j) respectively. Through the longitudinal comparison of the pictures, we can obtain the growth rule of interfacial IMC with the aging time. And the differences of the interfacial structures between Solder A and B under the same aging time can also be analyzed by horizontal contrast of the pictures.

Firstly, we focused on the IMC growth with the aging time in Solder A in Fig. 7(a-e). The composition of IMC at the interface is always AuSn and $AuSn_2$, and $(Au,Bi)Sn_4$ can be found both at the interface (Fig. 7(c, d)) and in the solder (Fig. 7(a, e)). In Fig. 7(a), the thickness of AuSn and $AuSn_2$ are 1.58μm and 1.07μm respectively as-reflowed. During aging, AuSn and $AuSn_2$ are both growing by the rapid diffusion of Au atoms into the Sn-Bi solder and reacting with Sn atoms to form IMCs. After aging for 240h in Fig. 7(c), AuSn and $AuSn_2$ has been grown to 3.27μm and 3.88μm respectively at the same time, suggesting that AuSn and $AuSn_2$ are growing proportionally during aging. Meanwhile, Kirkendall voids are found in abundance at the interface between Au and AuSn because of the diffusion of Au. When aging for 480h, Sn in the solder has been largely consumed and the interfacial IMC is about to expend to the bottom pad on PCB in Fig. 7(d), with AuSn and $AuSn_2$ growing to 5.42μm and 5.29μm respectively. Furthermore, when the reliability test runs to 1000 hours, in Fig. 7(e), AuSn and $AuSn_2$ are extending to 5.29μm and 6.81μm separately and have connected to the bottom pad over a large area. The solder has also changed to a pure Bi phase, reflecting the growth of IMC has almost finished. Besides, a seam of Kirkendall voids are also observed at the interface, which is bad for the reliability of the interconnection.

The growth of IMC in solder B with Ag addition can be obtained in Fig. 7(f-j). Except for AuSn, $AuSn_2$ and $(Au,Bi)Sn_4$, spherical Ag and Ag_3Sn are also found at the interface (Fig. 7(g, i, j)) or dispersed in the solder (Fig. 7(h)). Before aging, AuSn and $AuSn_2$ are measured to be 0.54μm and 2.14μm respectively in Fig. 7(f). However, during aging, it is interesting to find that $AuSn_2$ is growing much faster than AuSn. For example, in Fig. 7(h) after 240h-aging, AuSn has only grown to 0.97μm, whereas $AuSn_2$ has extended to 4.89μm, which is different from the situation in Solder A above. After aging for 480h in Fig. 7(i), with 1.39μm AuSn and 6.24μm $AuSn_2$ at the interface, there is still a large amount of Sn in the solder and the reactions is not complete yet. Furthermore, after 1000h-aging in Fig. 7(j), the interfacial AuSn and $AuSn_2$ has grown rapidly to 1.60μm and 6.56μm correspondingly, expending to the bottom pad on PCB partly. Moreover, the composition of the solder has changed from the original lamellar eutectic Sn-Bi phase to a pure Bi phase with only little Sn remained, proving that Sn phase in the solder has almost finished the reactions with Au and Ag. The remaining Bi phase cannot react but act as a

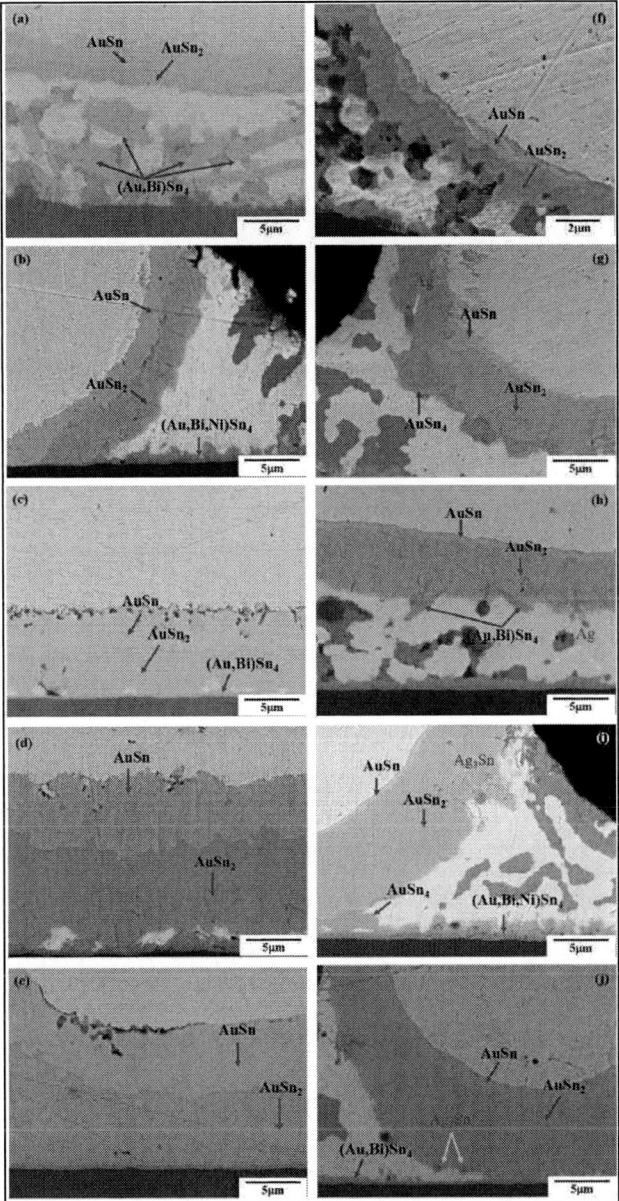

Figure 7. The interfacial structures of samples aging for (a) 0h, (b) 120h, (c) 240h, (d) 480h and (e) 1000h with Solder A and (f) 0h, (g) 120h, (h) 240h, (i) 480h and (j) 1000h with Solder B.

barrier to prevent the diffusion reactions. Surprisingly, almost no Kirkendall voids are found at the interface in Fig. 7(f-j), even after 1000h-aging.

To further evaluate the growth mechanism of Au-rich/Sn-Bi couple with Solder A and B, the average thickness of each IMC at the interface, including AuSn and $AuSn_2$, has been measured and analyzed. From the cross-sectional graphs in Fig. 7, the average thickness l_i of layer i was evaluated at each aging time by Nano Measure software. For convenience sake, the AuSn and $AuSn_2$ layers are denoted with subscripts of $i = 1$ and 2 respectively. The total thickness l of the intermetallic layer is obtained from the equation $l = l_1 + l_2$.

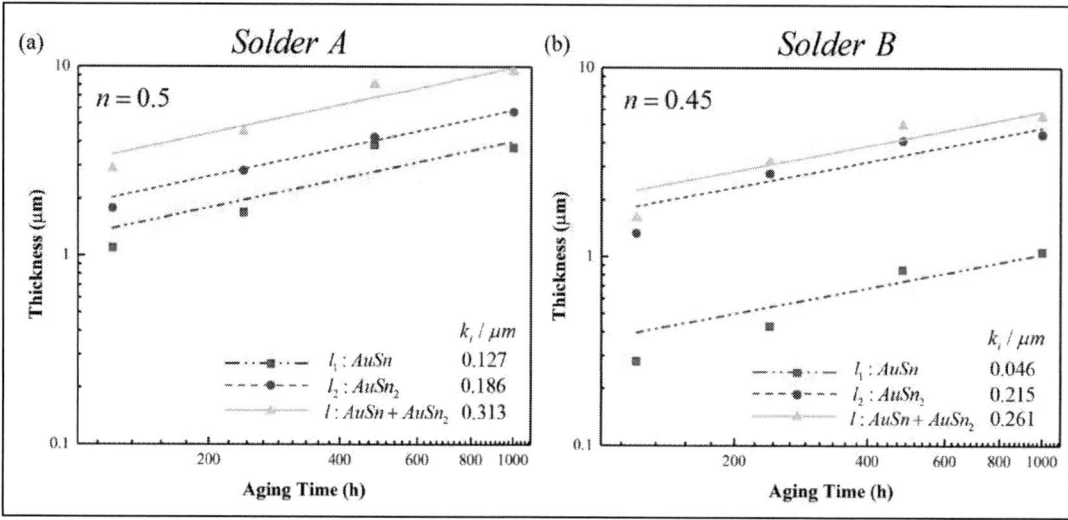

Figure 8. The thickness l_i of layer i vs. the annealing time t for the Au-rich/Sn-Bi diffusion couple using (a) Solder A and (b) Solder B. Straight lines indicate the calculations from (1).

The results for l, l_1 and l_2 are plotted in Fig. 8. Here, the ordinate indicates the logarithm of the thickness l_i, and the abscissa shows the logarithm of the aging time t. Fig. 8(a) indicates the result with Solder A, whereas Fig. 8(b) shows that with Solder B. As can be seen, the thicknesses l, l_1 and l_2 monotonically increase with increasing aging time t. Furthermore, the plotted points for each thickness are located well on a straight line. Therefore, l_i is mathematically described as a function of t by the equation

$$l_i - l_{i0} = k_i \times (t/t_0)^n. \qquad (1)$$

Here, t_0 is unit time, 1h. It is adopted to make the ratio t/t_0 dimensionless. The proportionality coefficient k_i has the same dimension as the thickness l_i, but the exponent n is dimensionless. l_{i0} represents the initial thickness of layer i before aging. The values of n and k_i were simultaneously determined from the plotted points in Fig. 8 by a least squares method. The determination gives $k = 0.313\mu m$, $k_1 = 0.124\mu m$ and $k_2 = 0.189\mu m$ with $n = 0.5$ in Solder A, and $k = 0.261\mu m$, $k_1 = 0.046\mu m$ and $k_2 = 0.215\mu m$ with $n = 0.45$ in Solder B. Using these values, the thickness l_i was calculated as a function of the annealing time t from (1). The results are shown as parallel straight lines in Fig. 8. As can be seen, most of the experimental points for each thickness l_i lie well on the corresponding straight line.

From Fig. 8, it is obvious that the IMC in Solder A is growing faster than that in Solder B by comparison. Through the previous references [6-8], we know that the value of n fitted from the experiments data can be used as the pointer to the controlling mechanism for IMC growth. When the IMC thickness and aging time are consistent with the parabolic relationship, indicating the obtained value for $n = 0.5$, the growth mechanism is bulk diffusion controlling. When n takes a value between 1/4 and 1/2, it can be inferred that the grain boundary diffusion contributes to the reactive diffusion and the grain growth occurs at certain rates which cause the volume fraction of grain boundary monotonically decreasing with time. In this paper, we obtain the values of n as 0.5 and 0.45 with Solder A and Solder B, respectively. Naturally, we can speculate that it is the bulk diffusion controlling mechanism in Solder A. But for Solder B, whether the minor difference of the values for n is caused by the difference in diffusion controlling mechanism or just by the data error, will be further confirmed by observing the grain size at different aging times in Solder B. Moreover, we also find that the growth rate difference between AuSn and $AuSn_2$ is small in Solder A in Fig. 8(a), whereas in Solder B in Fig. 8(b), $AuSn_2$ is growing much faster than AuSn. With the same n value of AuSn and $AuSn_2$ layers, it is further confirmed that the same rate-controlling mechanism works in the AuSn and $AuSn_2$ layers.

In conclusion, we can find the differences of the interfacial structures between Solder A and Solder B from the analyzation above. First of all, as for the composition of the IMCs, there are Ag and Ag_3Sn observed at the interface or in the solder when using Solder B which has Ag addition. Secondly, the total interfacial IMC is growing much faster in Solder A than in Solder B, which can be explained as follows. Dispersed Ag and Ag_3Sn which is produced by the reaction between Ag and Sn in solder B act as grain refiners in the matrix of the material, enhancing the barrier effect of Bi atoms and slowing the diffusion of Au atoms into the solder. Besides, Ag and Ag_3Sn can also act as the barriers by themselves to cause a deceleration of Au diffusion. The slower the diffusion speed of Au is, the slower the interface IMC grows. So, a thinner interfacial IMC (20%~30% thinner) can be obtained in Solder B with the help of Ag addition in the solder. The third difference focus on the growth of AuSn in these two solders. In solder A, AuSn can catch up with $AuSn_2$, whereas in Solder B the layer of AuSn is basically no growing. It can be speculated that Ag blocked the growth of AuSn to a greater extent than that of $AuSn_2$. Because of the slower diffusion speed of Au in Solder B, the number of Au atoms diffusing to the reaction interfaces, between AuSn and

$AuSn_2$, or between $AuSn_2$ and the solder, is greatly reduced. Majority of the Au atoms would react with the Sn in the solder first to form $AuSn_4$ with the lowest activation energy [6], which would convert to $AuSn_2$ followingly. Only a few Au atoms left would react with $AuSn_2$ to form $AuSn$, leading to less $AuSn$ growing. More theoretical reasons can be explained further by calculating the reaction activation energy of $AuSn$ and $AuSn_2$. The final difference is the Kirkendall voids at the interface of Au and $AuSn$, which are found in abundance in Solder A after 240h-, 480h- and 1000h-aging, whereas are rarely observed in Solder B even after 1000h-aging. This is also an evidence of a slower diffusion of Au atoms in Solder B. Besides, the differences we discussed above theoretically explain why the second bond joint, as the Au-rich/Sn-Bi interconnection, is more reliable in Solder B with Ag addition than in Solder A.

IV. CONCLUSIONS

Chip on Module package for CIS has been proposed based on a novel soldering wire bonding whose second bond point is achieved by soldering, using eutectic Sn-Bi solder paste with or without Ag addition and forming the Au-rich/Sn-Bi interconnection. The COM packaged samples have all passed the functional tests during THS reliability tests for 120h, 240h, 480h and 1000h, confirming the reliability of COM.

The tensile strengths of the gold wire in the novel soldering wire bonding have also been measured after aging. Using Solder A without Ag addition, due to the decreased strength at the second bond point, the tensile strengths are sharply reduced from 20 grams to 9.7 grams after 1000h-aging. While in Solder B with Ag addition, the strengths of the gold wire always maintain above 19 grams even after 1000h-aging, confirming that the second bond joint is more reliable using Solder B than A.

Interfacial structures of the second bond joint as Au-rich/Sn-Bi couple have been observed during aging. $AuSn$ and $AuSn_2$ are found at the interface and $(Au,Bi)Sn_4$ at the interface or in the solder. The thickness of interfacial $AuSn$ and $AuSn_2$ has also been analyzed mathematically which can be described as a function of the annealing time by the equation $l_i - l_{i0} = k_i \times (t/t_0)^n$ with the exponent n is equal to 1/2, illustrating the bulk diffusion controlling mechanism. Moreover, with less amount of IMC and Kirkendall voids in Solder B, the great effects of silver on diffusion reactions have also been analyzed. Dispersed Ag and Ag_3Sn at the interface or in the solder act as grain refiners in the matrix of the material, enhancing the barrier effect of Bi atoms, and slowing the diffusion of Au atoms into the solder. Besides, Ag and Ag_3Sn themselves also act as barriers to cause a deceleration of Au diffusion. Therefore, thinner interfacial IMC (20%~30% thinner) and unobvious voids can be obtained at the interface, leading to a higher tensile strength after reliability tests and a more reliable interconnection for COM package.

ACKNOWLEDGMENT

The authors would like to thank Ms. Rong Wang from the State Key Laboratory of Tribology in Tsinghua University for her help on SEM and EDS tests.

REFERENCES

[1] M Bigas, E Cabruja, J Forest, and J Salvi, "Review of CMOS image sensors," Microelectronics journal, vol. 37, May 2006, pp. 433-451, doi:10.1016/j.mejo.2005.07.002.

[2] K Sengupta, R Sundahl, S Kawashima, et al. "Packaging requirements and solutions for CMOS imaging sensors," Electronics Manufacturing Technology Symposium, 1998. Twenty-Third IEEE/CPMT, IEEE Press, Oct. 1998, pp. 194-198, doi:10.1109/IEMT.1998.731075.

[3] C Y Chen, Y C Chao, D S Liu, and Z W Zhuang, "Design of a novel chip on glass package solution for CMOS image sensor device" Microelectronics Reliability, vol. 46, Aug. 2006, pp. 1326-1334, doi:10.1016/j.microrel.2005.12.003.

[4] J Wang, X N Hou, J M Li, et al. "Reliability Evaluation of Interconnection in COM Package" 2018 19th International Conference on Electronic Packaging Technology (ICEPT), IEEE Press, Aug. 2018, pp. 1528-1532, doi:10.1109/ICEPT.2018.8480759.

[5] B F Dyson, "Diffusion of gold and silver in tin single crystals," Journal of Applied Physics, vol 37, Jan. 1966, pp. 2375-2377, doi:10.1063/1.1708821.

[6] S Nakahara, R J McCoy, L Buene, and J M Vandenberg, "Room temperature interdiffusion studies of Au/Sn thin film couples," Thin Solid Films, vol. 84, Oct. 1981, pp. 185-196, doi:10.1016/0040-6090(81)90468-5.

[7] H Xu, V Vuorinen, H Dong, and M Paulasto-Kröckel, "Solid-state reaction of electroplated thin film Au/Sn couple at low temperatures," Journal of Alloys and Compounds, vol. 619, Jan. 2015, pp. 325~331, doi:10.1016/j.jallcom.2014.08.245.

[8] T Yamada, K Miura, M Kajihara, N Kurokawa, K Sakamoto, "Kinetics of reactive diffusion between Au and Sn during annealing at solid-state temperatures," Materials Science and Engineering: A, vol. 390, Jan. 2005, pp. 118~126, doi:10.1016/j.msea.2004.08.053.

[9] Y W Yen, W K Liou, C M Chen, et al. "Interfacial reactions in the Sn–xBi/Au couples," Materials Chemistry and Physics, vol. 128, July 2011, pp. 233-237, doi:10.1016/j.matchemphys.2011.03.004.

[10] M McCormack, H S Chen, G W Kammlott, and S Jin, "Significantly improved mechanical properties of Bi-Sn solder alloys by Ag-doping," Journal of Electronic Materials, vol. 26, Aug. 1997, pp. 954-958, doi:10.1007/s11664-997-0281-7.

[11] S Sakuyama, T Akamatsu, K Uenishi, and T Sato, "Effects of a third element on microstructure and mechanical properties of eutectic Sn–Bi solder," Transactions of The Japan Institute of Electronics Packaging, vol. 2, 2009, pp. 98-103, doi:10.5104/jiepeng.2.98.

The properties of Cu sinter paste for pressure sintering at low temperature

Jung-Lae Jo, Kei Anai, Sinichi Yamauchi and Takahiko Sakaue
Mitsui Engineered Materials Sector R&D Center
Mitsui Mining & Smelting Co., Ltd.
1333-2, Haraichi, Ageo-shi, Saitama, Japan
e-mail: j_jo@mitsui-kinzoku.com

Abstract—This study introduces a newly developed pressure type Cu sinter paste as a joining material substituting for Ag and our investigation into its joining properties and reliability. The Cu sinter paste we developed as a filler by using low temperature sinterable copper particles was able to be sintered at 280℃ from 5 to 20 minutes. Pressure was below 10 MPa under N$_2$ atmosphere. The result of bonding strength showed high shear strength over 50 MPa between the dummy die and Ag/Cu substrate. The cohesive shear modes were shown between the Ag plated layer on the dummy die and Cu paste indicating good enough strength for initial joint reliability. The image of Scanning Acoustic Tomograph also indicated good joining layer with no voids. Ag metallized SiC-SBD and Cu bared TO-247 leadframes were bonded with copper pastes. The sound acoustic analysis was also compared with before and after joint area of Cu paste. Reliability tests showed good performance with these packages. The details will be discussed in this study.

Keywords-Copper sinter, Power device, Reliability test, Transient heat measurement, Sound acoustic analysis, Thermal cycling test, High temperature storage test, Pressure cooker test, Ag to Cu joint, TO-247, SiC

I. INTRODUCTION

Power control systems are in the limelight of eco-friendly vehicles such as electron vehicles (EV). The SiC is noted for its many advantages such as higher band gap and superior heat resistance compared to conventional Si [1-2]. Power packages with SiC face many challenges due to downsizing and high efficiency for their higher power density. To increase the operation temperature and downsizing of power semiconductors, the reliability of joined layer should be emphasized more than conventional bonding materials such as solders, especially in automotive applications [3-4].

The lead-free solders are good materials for electro devices joining for their good wettability [5], high electrical performance and good reliability in Si packages [6-7]. The Si packages have been optimized with reflow process with good melting point of solder material. All the substrate materials such as printed circuit boards are also optimized with solder materials. High reliability is required for bonding materials used for power devices as their operation temperature increases due to the trend for downsizing, high efficiency and high power. However, it is difficult to ensure high reliability with solder, the current mainstream material. Sintering materials, therefore, are recommended as alternatives for their high reliability as bonding materials.

Much research has been reported about silver sinters as bonding materials with their high reliability [8-9]. Silver has many advantages including low electrical resistance and high thermal conductivity. However, they come with some concerns such as anti-migration property [10], high cost and handling property.

Copper materials are inexpensive and provide better anti-migration property than silver sinter materials. However, it is difficult to sinter at low temperature because copper nanoparticles are prone to oxidize more than silver nanoparticles.

In our study, copper paste with submicron which we have developed with our unique technique can be sintered under 280 ℃ in nitrogen atmosphere under pressure. The sinter achieved high shear strength without oxidation. In addition, reliability results of the copper paste as bonding material will be shown in this paper.

These preliminary reliability results note that copper pastes have a superior potential as new bonding materials and alternatives to solder alloys in power electronics fields.

II. MATERIALS AND PROCEEDURE

A. The copper paste

The new copper paste for pressure sintering was developed by Mitsui Mining & Smelting Co., Ltd. The paste includes submicron Cu particles as low temperature sinterable fillers for the paste. The Cu particles synthesize our unique technique and can be sintered at low temperature. In this study, the copper paste can be sintered at a low temperature below 280 ℃ with pressure. Fig. 1 shows the appearance of copper paste after fabrication.

Figure 1.   The appearance of copper paste.

B. Test vehicles

Fig. 2 shows the silicon carbide Schottky diode die (SiC-SBD) in (a) and TO-247 Cu leadframe in (b). The dimensions of SiC-SBD were 4.9 mm by 4.9 mm with 380 μm thickness. The topmost surface of anode was Al and cathode was Ag. The bare Cu TO-247 was prepared as substrate with 2mm thickness.

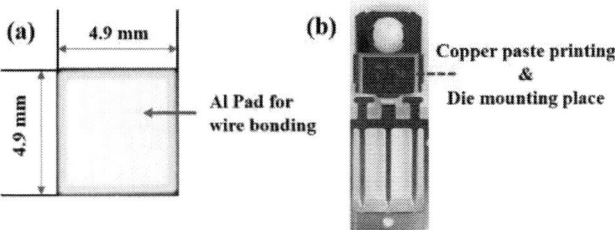

Figure 2.   (a) Die and (b) TO-247

C. Experimental procedure

Fig. 3 shows the flows of schematic diagram for pressure sintering. TO-247 leadframe was cleaned by 5% volume sulfuric acid before sintering. SiC-SBD was cleaned by ethanol on Ag metalized side only.

Copper paste was mixed by planetary centrifugal mixer before printing. After mixing, the copper paste was printed on a TO-247 leadframe by screen printing. The thickness of printing mask was 100 μm and the material was SUS.

The copper paste printed on TO-247 leadframe was dried for 20 minutes in a constant oven for previous treatment before pressure sintering. After drying, the SiC-SBDs were mounted on dried copper pastes manually. After that, a relaxation sheet was placed on the mounted SiC die to prevent the SiC die from breakage. TO-247 has a low level thickness difference between the mounted area and the electrode. Spacer blocks were used for cancelation of thickness when pressure sintering. Finally, the pressure sintering was achieved with a pressure sinter machine.

The sintering temperature was at 280℃ for 5 minutes under nitrogen and the 9 MPa pressure was applied. The detailed process flow of pressure sintering is shown in Fig. 3 The sintering profile is shown in Fig. 4.

The initial joint strength was evaluated by shear test before wire bonding. After sintering, wire bonding was carried out for electrical path. Wired packages were then molded with molding compound. The fabricated final package is shown in Fig. 5.

All the packages were analyzed with Sound Acoustic Tomograph (SAT) in order to evaluate the joint layer. 75 MHz probe was used in this study.

The electrical and thermal properties will be investigated in our future work. In this paper, mainly the shear strength and SAT results will be discussed.

D. Reliability test

To evaluate in a similar environment for real packages, all the samples were fabricated in close proximity to the actual package, including wire bonding and mold compound.

1) High temperature storage (HTS)

The condition of HTS was 1,000 hours at 175℃.

The main purpose of HTS is to evaluate the long-term reliability of the package during high temperature aging.

2) Pressure cooker test (PCT)

PCT was carried out under 2 barometric pressures, 121℃ and 100% relative humidity for 96 hours. PCT is a method to accelerate moisture absorption in packages, and the corrosion of the metal layer could be affected on delamination between molding compound and package by moisture absorption of molding compound and increasing electrical resistance by corrosion in general.

3) Thermal cycling test (TCT)

TCT condition was extremely severe with ΔT=205℃ for targeting automotive. The condition of TCT was - 55℃ to 150 ℃ during 1000 cycles. The first cycle ranged from room temperature to minus temperatures. The one cycle was 40 minutes and dwell time was 10 minutes at each top temperature. A large mismatch of coefficient of thermal expansion (CTE) would be expected among mold compound, SiC and sintered copper paste.

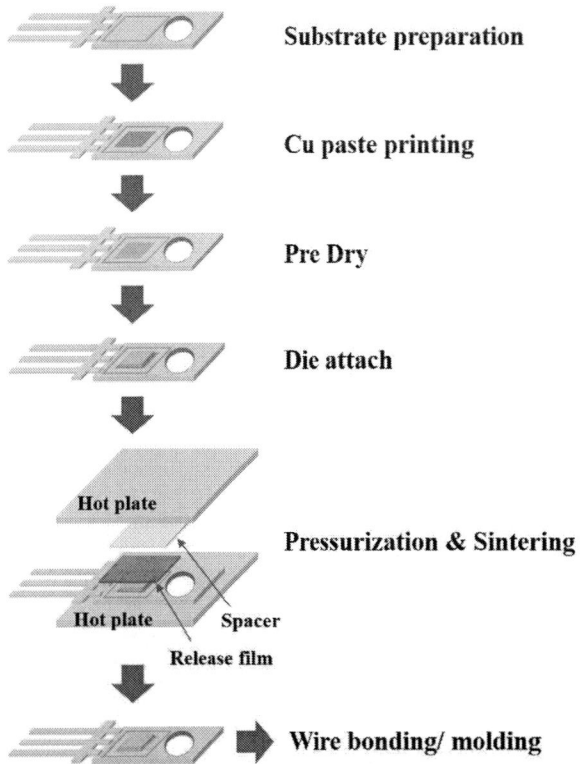

Figure 3.   The schematic diagram of pressure copper sintering.

978-1-7281-1500-9/19 $31.00 © 2019 IEEE

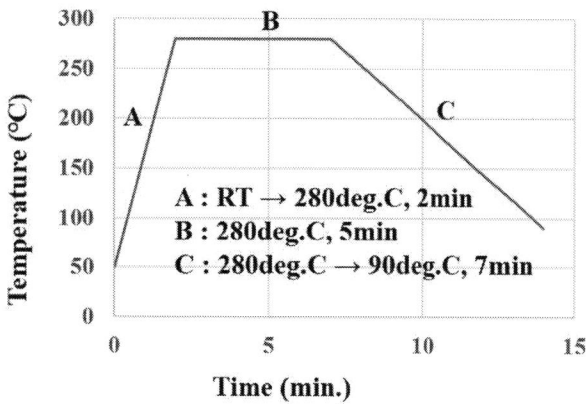

A : RT → 280deg.C, 2min
B : 280deg.C, 5min
C : 280deg.C → 90deg.C, 7min

Figure 4. Pressure copper paste sintering profile.

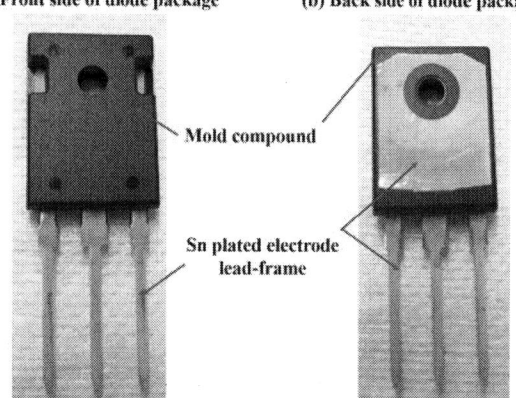

(a) Front side of diode package   (b) Back side of diode package

Mold compound

Sn plated electrode lead-frame

Figure 5. The appearance of the fabricated package.

III. RESULTS & DISCCUSION

A. SAT and shear test result

To verify the initial state of the joint, SAT analysis and shear test were carried out after pressure sintering. Fig. 6 shows the results of SAT after sintering. SAT results indicate that uniform joint layer could be achieved without void between SiC and TO-247 leadframe. The high shear strength value is well coincident with SAT results as shown in Table 1.

We also evaluated shear test under high temperature at 260℃. The shear strength value under high temperature was 55.6 MPa as shown in Table 1. The shear strength was slightly reduced by deterioration under high temperature. It is caused due to softening of joint layers at high temperature. Although the shear strength decreased slightly, it is remarkable that high joint strength was maintained between SiC and TO-247 leadframe. For reference, shear strength of lead-free solders is below 45 MPa at room temperature. Moreover, at 100 ℃, the shear strength of lead-free solders deceases down to 25 MPa. [11-12].

TABLE I. THE SHEAR STRENGTH OF AMBIENT AND 260℃

Condition	Shear strength (MPa)
Ambient temperature	>80
260°C	56

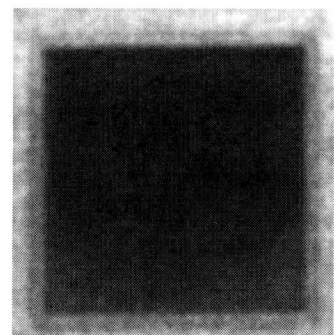

Figure 6. The SAT image after copper sintering.

B. Cross-sectional analysis in initial sample

The cross-sectional image after sintering is shown in Fig. 7 (a). The copper pastes sintered and joined well with Ag metallized layer of SiC side. Further analysis was conducted on the joint area with blue rectangle in Fig. 7 (a) by EDS. As shown in Fig. 7 (b), the EDS data indicate the mutual diffusion occurred between Ag layer and Cu paste during pressure sintering. The high shear strength might be described as this mutual diffusion.

C. Reliability test.

All packages were analyzed by SAT before and after reliability tests to compare the joint layer of copper paste.

1) HTS results

Fig. 8 clearly indicates that there was no change in the joint layer before and after HTS by SAT. Meanwhile the delamination of molding compounds occurred before HTS.

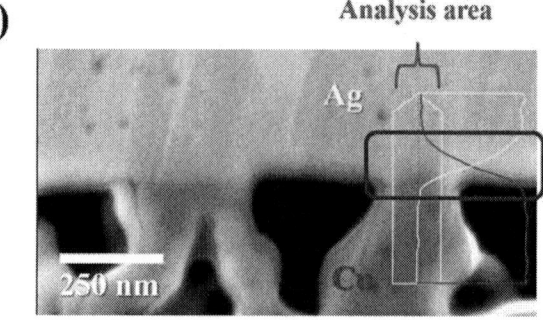

(a)   Analysis area

Ag

250 nm   Cu

978-1-7281-1500-9/19 $31.00 © 2019 IEEE

(b)

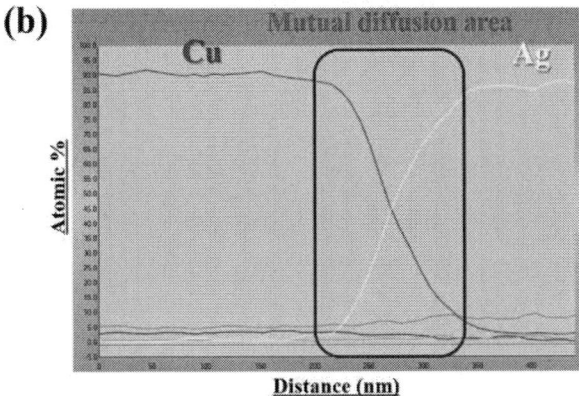

Figure 7.   (a) The cross sectioned image between Ag metallization and sintered copper, and (b) the EDS analysis of designated area shown in blue square (a).

In case of solder alloy based on tin, as is well known, the growth of IMC is accompanied by high temperature aging. Even more IMC is formed as the aging time increases. IMC decreases solder shear strength by forming a brittle joint interface [13]. The condition of copper sintered layer after 1000 hour of HTS will be described in subsequent articles.

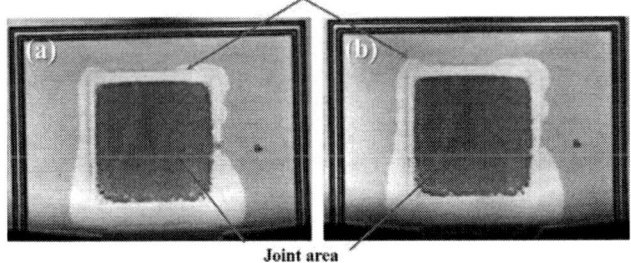

Figure 8.   SAT result of (a) Before HTS and (b) after 1000 hours HTS.

2) PCT

The results of PCT show the same results as HTS by SAT and thermal transient result. In this study, there was no applied bias during the PCT test. Of course, PCT has been performed without inducing voltage. We have just confirmed that there was not any delamination on the joint area throughout 96 hours of PCT, meanwhile, the molding compound deteriorated by PCT as shown in Fig. 9.

3) TCT results

Fig. 10 shows SAT results of before TCT (a) and after 1,000 cycle TCT (b). The delamination of molding compound is found as shown in Fig. 10. In spite of delamination, there is no significant change in SAT results before and after TCT. Note that joint deterioration could not be shown after 1,000 cycles of TCT.

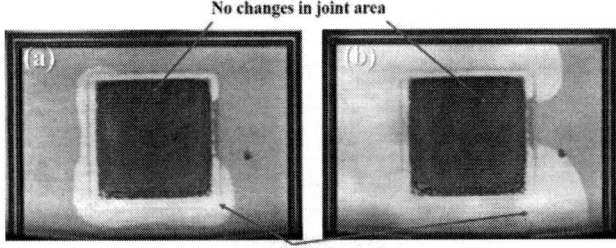

Figure 9.   PCT results of (a) before TCT and (b) after 96 hours.

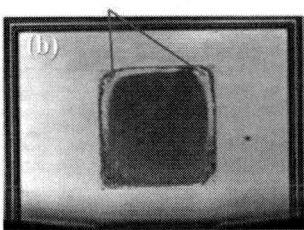

Figure 10. SAT result of (a) Before TCT and (b) after 1000 cycles TCT.

The reliability test results strongly imply copper sintering is a bonding material suitable for power devices and long-term reliability greater than conventional bonding materials such as lead-free solders. Although manufacturing in the field of electronics has been optimized including surface mount technology, a lot of work remains with pressure copper sintering for optimization for HVM, cost reduction and experience with processes for manufacturing. Nonetheless, this study might be the first step for developing possibility for power devices especially in automobiles.

IV.   SUMMARY

The low temperature pressure copper sintering was investigated with SAT, shear test and cross-sectional analysis. Also, the reliability tests were performed with the joint between SiC and TO-247. The conditions of reliability tests were targeted for power devices.

The results of HTS and PCT show excellent joint reliability by SAT analysis. No significant changes of joint area were observed in TCT results, either. This research can be achieved with a new copper paste. The suggested copper paste can be sintered for just 5 minutes below 280 ℃ with 9 MPa pressure and under nitrogen atmosphere. Over 80 MPa shear strength was obtained with the above sintering condition. Furthermore, high temperature the shear test result was 56 MPa at 260℃.

These results show the possibility of achieving reliability in the power device field. Without doubt, future work remains such as the mechanism of copper sintering on other surface finishes. We have already planned to sinter with the variety surface finishes used in power devices.

The thermal and electrical properties also will be evaluated before and after reliability tests in the near future.

REFERENCES

[1] M. Bhatnagar and B.J. Baliga, "Comparison of 6H-SiC, 3C-SiC, and Si for power devices," IEEE Transactions on Electron Devices, vol. 40, pp. 645-655, March 1993.

[2] H. Matsunami, "Current SiC technology for power electronic devices beyond Si," Microelectronic Engineering, vol. 83, pp. 2-4, January 2006

[3] C. Göbl and J. Faltenbacher, "Low temperature sinter technology die attachment for power electronic applications," Proc. 2010 6th International Conference on Integrated Power Electronics Systems (CIPS2010), IEEE Press, March 2011, p10.1

[4] Luis Alberto Navarro Melchor, ".Evaluation of die attach materials for high temperature power electronics appictions and analysis of the Ag particles sintering solution," Doctoral thesis, Universitat Autònoma de Barcelona, 2015

[5] Gabriel Takyi, Peter Kojo Bernasko, "Investigation of Wettability of Lead Free-Solder on Bare Copper and Organic Solder Preservatives Surface Finishes," International Journal of Materials Science and Applications, vol. 4, Issue 3, pp. 165-172, April 2015

[6] Weiping Liu, Ning-Cheng Lee, Adriana Porras, Ming Ding, Anthony Gallagher, Austin Huang, Scott Chen and Jeffrey ChangBing Lee, "Achieving high reliability low cost lead-free SAC solder joints via Mn or Ce doping," 2009 59th Electronic Components and Technology Conference, IEEE press, June 2009, doi: 10.1109/ECTC.2009.5074134

[7] Jung-Lae Jo, Soon-Bum Kim, Tae-Eun Kim, Yeo-Hoon Yoon, Ho Jeong Moon, Hyung-Gil Back, Tae-Je Cho and Sa-yoon Kang, "Mechanisms of improving thermal cycling reliability with pad finish & solder alloying effect on solder grains," Proc. KCS 2015, [The 22nd Korean Conference on Semiconductors (KCS 2015)], pp. 5-9

[8] Fang Yu, Jinzi Cui, Zhangming Zhou, Kun Fang, R. Wayne Johnson and Michael C. Hamilton, "Reliability of Ag Sintering for Power Semiconductor Die Attach in High-Temperature Applications," IEEE Transactions on Power Electronics, vol. 32, Issue 9 , September 2017

[9] Fen Chen, Sihai Chen, Guangyu Fan, Xue Yan, Chris LaBarbera, Lee Kresge and Ning-Cheng Lee, "Pressureless Sintering of Nano-Ag Paste with Low Porosity for High Power Die Attach," PCIM Asia 2015; International Exhibition and Conference for Power Electronics, Intelligent Motion, Renewable Energy and Energy Management, VDE press, July 2015

[10] Simeon J. Krumbein, "metallic electromigration phenomena," 33rd Meeting of the IEEE Holm Conference on Electrical Contacts, September 1987

[11] R. Mahmudi, A. Maraghi1 and A. R. Geranmayeh, "High-temperature shear strength and hardness of cast lead-free solders," Metallic Materials, vol. 55, pp. 211–216, June 2017

[12] A.R. Geranmayeh, R. Mahmudi and M. Kangooie, "High-temperature shear strength of lead-free Sn–Sb–Ag/Al2O3 composite solder," Materials Science and Engineering: A, vol. 528, Issue 12, pp. 3967-3972, May 2011

[13] Liu Yang, Sun Fenglian and Yang Miaosen, "Shear strength and brittle failure of low-Ag SAC-Bi-Ni solder joints during ball shear test," 2013 14th International Conference on Electronic Packaging Technology, March 2014, doi: 10.1109/ICEPT.2013.6756574

Low temperature sintering of dendritic Cu based pastes for power semiconductor device interconnection

Gang Li[a,b 1], Jilei Fan[a,b 1], Siyuan Liao[a,b], Pengli Zhu[a,*], Baotan Zhang[a], Tao Zhao[a], Rong Sun[a], Ching-Ping Wong[a]

[a]Shenzhen Institute of Advanced Electronic Materials, Shenzhen Institutes of Advanced Technology, Chinese Academy of Sciences.
[b] Shenzhen College of Advanced Technology, University of Chinese Academy of Sciences, Shenzhen, China.
Shenzhen, P.R.China.
pl.zhu@siat.ac.cn

Abstract—As novel high temperature interconnect materials, copper nanoparticle based conductive pastes have received more and more attention due to their good electrical conductivity, thermal conductivity and reliability after sintering. However, the copper nanoparticles are easily oxidized in ambient condition, which not only causes a decrease in electrical conductivity of the sintered Cu, but also increases the sintering temperature, thereby deteriorating the performance of the device. In our work, we demonstrated a facile and scalable approach for synthesizing of hierarchical dendritic Cu with micro/nano structures with 1-3 μm in stem diameter and 6-15 μm in length based on a galvanic displacement reaction between the Al foil and Cu precursor ($CuSO_4$). Then we evaluate the applications of dendritic Cu based paste as a low temperature interconnect material. Cross-sectional results showed that the initial sintering temperature of dendritic Cu can be reduced to 80°C under the aid of pressure. Moreover, the dendritic Cu paste showed a better low temperature sinterability compared with Cu nanoparticle-based paste and they had fused to create a low porosity layer under a sintering temperature of 150°C. The shear strength of bonded joints sintered at 350°C, 10MPa was 20MPa, and there was no obvious degradation even they were subjected to thermal shock test for 1000 cycles. The excellent mechanical properties provide a solution for device interconnection of third-generation semiconductors.

Keywords-dendritic Cu paste; shear strength; high temperature interconnect

I. INTRODUCTION

The development of the third-generation semiconductors represented by GaN/SiC puts higher demands on interconnect materials [1]. The applications of conventional die attach materials such as Sn based solder [2] and electrically conductive adhesive[3] are greatly limited since they cannot withstand the high operational temperature. Nanosized metallic particles based conductive pastes have received more attention due to the characteristics of low-temperature sintering and high-temperature service. Among the metallic nanoparticles studied as interconnection material, current researches mainly focus on the nano Ag based pastes due to their high electrical and thermal conductivity [4, 5]. However, they also have some drawbacks such as serious electromigration phenomenon and high cost. Due to the much lower cost and comparable electrical conductivity of Cu compared with Ag, as well as high reliability with no intermetallic reactions, Cu-based pastes have become ideal alternatives to silver pastes[6, 7].

However, the easy oxidation of Cu nanoparticles not only increases the sintering temperature, but also reduces the electrical conductivity after sintering. Meanwhile, organic stabilizers are usually required to avoid the spontaneous oxidation, but these passivation layers can only be removed completely at elevated temperature, which will have adverse effects on sintering inevitably. Therefore, improving the low-temperature sinterability of Cu nanoparticles is a prerequisite for their applications as die-attaching materials in power devices.

The sinterability of metallic particles not only depends on the size and size distribution, but also is closely related to the morphology structure. For example, Cheng Yang et al demonstrated that the unique three-dimensional fractal geometrical configuration of fractal Ag micro-dendrites render them excellent low-temperature sintering characteristic[8, 9]. In our work, we have proposed a simple and large-scale synthetic method of dendritic Cu filler with excellent low-temperature sinterability. The dendritic Cu can be formulated to pastes suitable for printing by optimizing the solvent, dispersant combination. The rheological properties of Cu pastes before sintering and the mechanical shear strength of boned Cu-Cu joints after sintering as well as reliability of sintered joints after aging tests were investigated in detail. The die shear test results showed that a bonding strength of about 20MPa was achieved at a sintering temperature of 350°C and an applied pressure of 10 MPa under vacuum atmosphere. These performances showed no obvious degradation even after extreme thermal cycles from -40°C-125°C, repeated for 1000 cycles. These results suggested that the dendritic Cu based pastes with excellent mechanical properties are promising die-attachment materials for the high temperature power devices packaging.

II. EXPERIMENTAL

A. Materials

Absolute ethanol (AR) ,Sodium chloride (AR, 99.5%), anhydrous copper sulfate (AR, 99.0%) ,2-(2-ethoxyethoxy) ethyl acetate .Aluminum foil (A1423) with a thickness of 0.018mm. Deionized water prepared by laboratory.

B. Preparation of dendritic Cu and their paste

Before the synthesis, Al foil with a size of 4x4 cm^2 was ultrasonically cleaned in ethanol to remove the surface impurities. A typical preparation of Cu dendritic structure is as follows [10]: 2.0 g anhydrous copper sulfate (CuSO₄) was added to 100 ml of deionized water, and then 1.46 g sodium chloride (NaCl) was added under magnetic stirring. The previously cleaned Al foil was immersed into the mixture solution, and reacted at room temperature for 30 min. Finally, the formed product floated on the water, and they were collected by decanting the upper pellucid solution, washed with absolute ethanol several times and dried under vacuum for 6 hours for further measurements and paste preparation.

The dendritic Cu based paste was formulated as follows: the solvent 2-(2-ethoxyethoxy) ethyl acetate and dendritic Cu powder was mixed together using a Speedmixer (FlackTec.Inc) at a speed of 2000 rpm for 10min to achieve uniform dispersion of Cu in the solvent, and then the mixture was subjected to degassed to remove the trapped air from the mixture. The amount of the dendritic Cu filler was maintained at 60wt%. To prepare samples for shear strength tests, the Cu paste was printed on Cu a larger substrate with size of 5mm*5mm and a smaller Cu with size of 3mm*3mm was coated onto the larger Cu substrate and sintered under vacuum at various conditions (temperature, pressure and time) using a home-made hot-press sintering machine.

C. Characterization

The phase composition of the dendritic Cu was determined by XRD (Rigaku Model D/MAX-2500V/PC) at a scan rate of 10 min⁻¹ from 10° to 80° with a Cu Kα radiation (λ =1.5418 Å). The size and morphologies of the dendritic Cu were observed by Scanning Electron Microscope (SEM, FEI Nova Nano SEM450) and TEM (FEI F20). SEM was also used to investigate the top and crossed-sectioned microstructures of the sintered joints. Rheometer (Anton Paar MCR 302) was used to measure the rheological behavior of the dendritic Cu paste under a shear rate from 1 s⁻¹ to 100 s⁻¹. TGA measurement was carried out with a Thermal Analysis (TA) Instruments SDT-Q600 analyzer under nitrogen atmosphere from 25°C to 800°C at a heating rate of 10 °C/min to study the thermal behavior of Cu paste. The sintering behavior of the dendritic Cu was investigated by differential scanning calorimetry (DSC, TA Instrument, Q20) at a heating rate of 10 °C/min from 25°C to 350 °C under a nitrogen atmosphere. The shear strength of the sintered joints was measured by DAGE 4000.

D. Characterization of dendritic Cu filler.

Figure 1a is an optical photograph of the preparation of dendritic Cu based on the galvanic replacement reactions between aluminum foil and copper sulfate. Cu²⁺ ions can easily be reduced to metallic Cu due to the large difference of the redox potentials between the Al³⁺/Al (1.66 eV) and Cu²⁺/Cu (0.34 eV). As the reaction progresses, the blue color of the copper sulfate gradually fades until it was colorless, which indicates that the Cu²⁺ has been completed reduced to Cu. At this time, a large amount of red-colored product floats on the surface of the aqueous. This is mainly due to the fact that surfactants or polymers are not used during the preparation process, and the hydrophobic property of the product makes it hardly dispersed well in water. Figure 1b shows the XRD spectrum of the dried product, three sharp diffraction peaks located at 43°, 51° and 74° are corresponding to the {111}, {200}, {220} planes of face-centered cubic (fcc) Cu respectively. In addition, there are no peaks of Cu₂O and CuO observed in the XRD pattern, indicating that the obtained product is pure Cu. As shown from the SEM in the Figure 1c, the as-prepared Cu exhibits well-defined dendritic micro/nano structure. The central trunks and the primary branches have length ranging from 6-15 μm and 1-3 μm. TEM in the Figure 1d provides more detailed structure characteristics, the branches are parallel to each other, and each branch is composed of a large number of nanoparticles. Generally, oriented attachment mechanism is proposed to account for the formation of the unique Cu dendritic structure, the intensity of {111} plane is much stronger than those of {200}, {220} planes, indicating that the growth direction of the dendritic Cu is along ⟨111⟩.

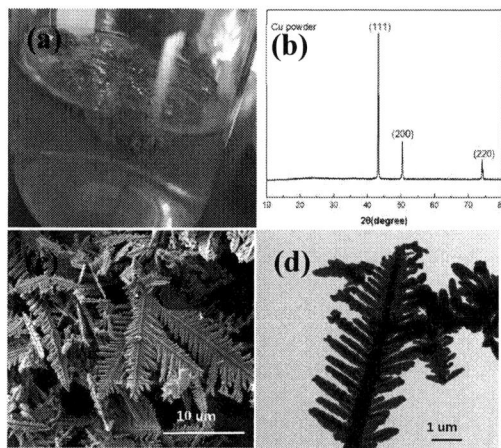

Figure 1: (a) optical image of preparation of dendritic Cu based on galvanic displacement reaction. (b) XRD (c) SEM and (d) TEM images of as-prepared dendritic Cu.

E. Physical properties charateration of dendritic Cu based paste

The rheological property of copper paste is an important factor affecting its printability and processability. There is no obvious agglomeration observed in the paste, which indicates that the copper dendrites are dispersed well in the solvent. It can be seen from Fig. 2 that the as-prepared copper paste exhibits an obvious thixotropic shear thinning behavior characterized by a decreased viscosity with an increased shear rate, which is mainly due to orientation arrangement of dendritic Cu filler under external shear forces. When the shear rate is 100/s, the viscosity is as low as 6.5 Pa·S. In addition, the copper paste has a high thixotropic index of 2.7 ($\eta_{1.5/s}/\eta_{15/s}$), which can prevent from spreading effectively after printing. The copper paste can be screen printed on the copper substrate and shows a good wettability on the copper substrate.

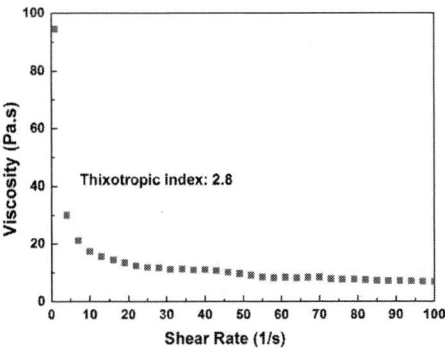

Figure 2. Rheological property of dendritic Cu based paste under shear rate from 1/s to 100/s.

Figure 3 shows the TGA curves of copper paste under nitrogen atmosphere. It can be seen that the curve declines sharply between 100-200 °C, due to the volatilization of the solvent diethylene glycol monomethyl ether acetate (boiling point 217 °C). The loading of functional phase dendritic Cu 60%, which is consistent with the initial feeding amount. Too high Cu loading results in a much higher viscosity, however, too low Cu loading leads to a decrease in the densification degree, which both deterioate the final mechanical performance.

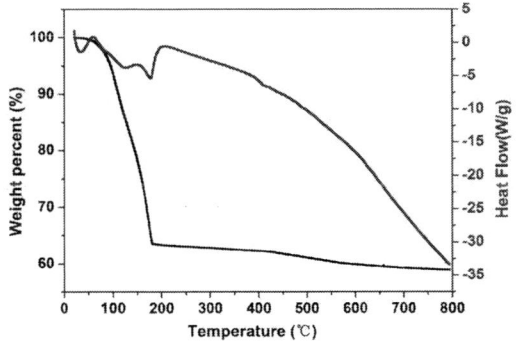

Figure 3. TGA curve dendritic Cu based paste under N_2 atmosphere.

Figure 4 is the DSC curve of the copper dendritic under nitrogen atmosphere. It can be seen that there are two obvious endothermic peaks between 40°C-350°C. The first peak located at about 80°C corresponds to the solvent volatilization of residual ethanol on the dendritic Cu surface. The second endothermic peak located at 300°C is likely due to the spontaneous sintering of dendritic copper. This result shows that the dendritic Cu has a high sintering temperature without the assitant of pressure, which can also be seen in the subsequent SEM results. The poor sinterability is related to the low diffusion rate of surface atoms as well as the prescence of of oxide on the surface, although it is difficult to be detected sometimes.

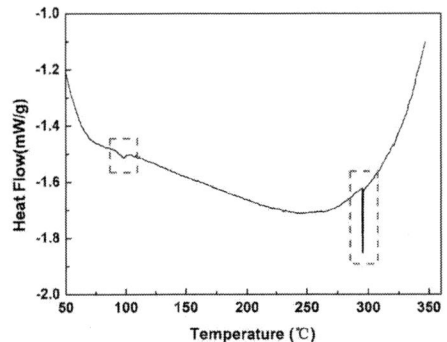

Figure 4. DSC curve of dendritic Cu based paste under N_2 atmosphere.

F. Adhesion strength of sintered Cu paste

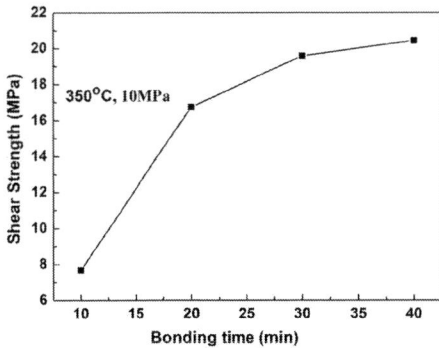

Figure 5. Shear strength of dendritic Cu based paste under different sintering time.

The bongding strength of the sintered metal paste is related to sintering process parameters such as sintering time, pressure, and temperature, etc[11]. We first examined the effect of sintering time on the bonding strength of sintered Cu paste under a temperature of 350°C and a pressure of 10MPa. As shown in the Figure 5, the shear strength increased with the sintering time. When the sintering time was 10 min, the shear strength was only 8 MPa. The shear strength increased 2.5 times (20 MPa) when the sintering time prolonged to 30 min. However, the shear strength did not change significantly and reached a saturation value when the sintering time further increased to 40min, which indicated that the sintering process has been completed. Therefore, the optimum sintering time was 30min.

Next, we investigated the effect of sintering temperature on the shear strength of sintered Cu paste. It can be seen from Figure 6 that under the pressure of 10 MPa, the shear strength increased with the increase of sintering temperature. The shear strengths at 150 °C, 200 °C, 250°C, 300 °C and 350 °C were 1.8MPa, 3.6MPa, 6.5MPa, 9.7MPa, 19.6MPa, respectively. The shear strength was significantly improved from 300°C to 350°C, which indicated that the sintering degree was greatly improved within this temperature range. The same trend was observed under the pressure of 25 MPa.

By contrast, higher pressure lead to higher shear strength, and this tendency was more pronounced at high temperatures. Moreover, the bonding strength was very low in the case of pressureless sintering, even at a high temperature of 350°C.

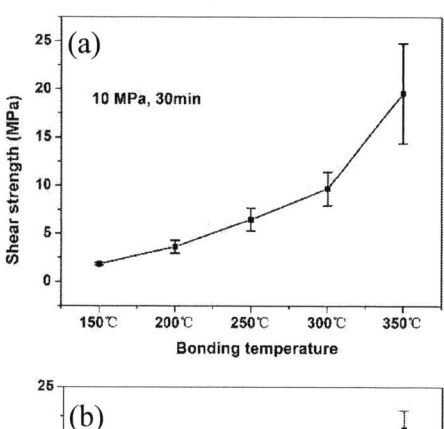

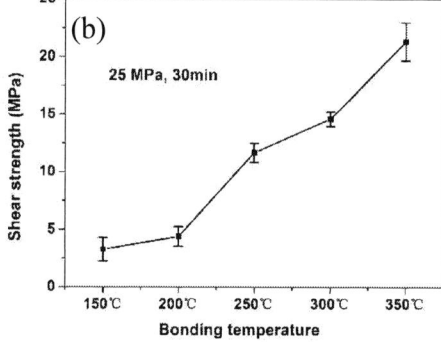

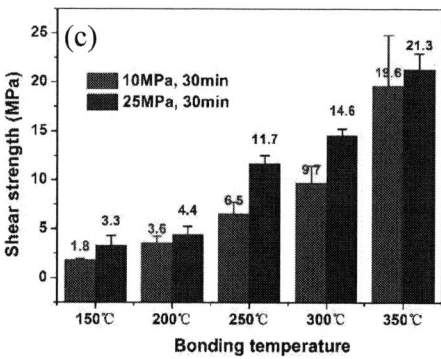

Figure 6. Shear strength of dendritic Cu based paste under different sintering pressure (a) 10MPa, (b) 20MPa for 30min and (c) their results comparation.

We also investigated the effect of preheat treatment on the bonding strength. Figure 7 showed that the shear strength was significantly decreased after the preheating the paste at 160°C for 5min. The shear strength was only about 2.0 MPa under sintering at 350 °C for 30 min and 10MPa, which was one tenth of that of the sample that was not subjected to preheated. Usually, in order to achieve to obtain a high shear strength, the solvent must be removed completed during the whole sintering process. However, our results showed that

the excessive evaporation of the solvent before the sintering is not favorable for the denstification of Cu paste, possibaly due to the deterioration of wettability of paste on the substrate. Therefore, the sintering profile should be carefully controled to optimize the sintering behavior and to achieve excellent bonding strength.

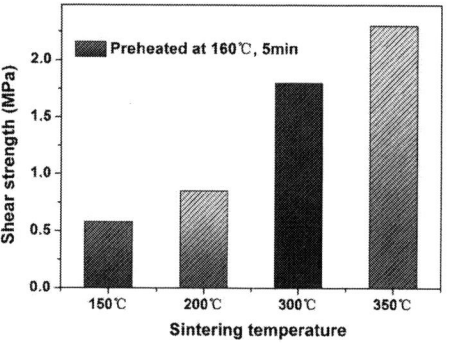

Figure 7. Effect of preheat on the bonding strength of dendritic Cu based paste under different sintering temperature.

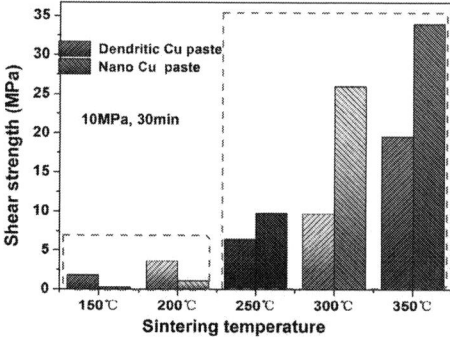

Figure 8. Comparation of shear strength of dendritic Cu and nano Cu based paste sintered under 10MPa for 30min.

Figure 8 showed the comparation of bonding strength of dendritic Cu and nano-copper based paste. It can be seen that the shear strength of dendritic copper paste sintered at 150°C and 200°C was 1.7 MPa and 5 MPa respectively, which was much higher than that of nano-copper (0.3 MPa and 1.1 MPa), indicating that the dendritic Cu has good low-temperature sinterability. However, at high sintering temperatures, the shear strength of dendritic Cu paste is lower than that of the nanoparticles paste. This result can be interpreted as follows: usually, a layer of potecting agent is present on the nanoparticles surface, which is necessary to control the size during the synthsis process and prevent the oxidation during the stroage stage. However, it is unfavorable to its sintering, due to the prevention of the atommic diffusion. Therefore, an innitial sintering temperature exists, which related closely to the decomposition temperature of the capping agent [12]. In contrast, there was no orginic capping agent used during the prepartion of dendritic Cu, therefore, its sintering behvavior can proceed at lower temperature. However, once the

978-1-7281-1500-9/19 $31.00 © 2019 IEEE   84

sintering temperature exceed the critical temperature, the sintering rate of nanoparticle was much higher than that of dendritic Cu due to their much larger surface energy.

G. Microstructure evolution of sintred Cu

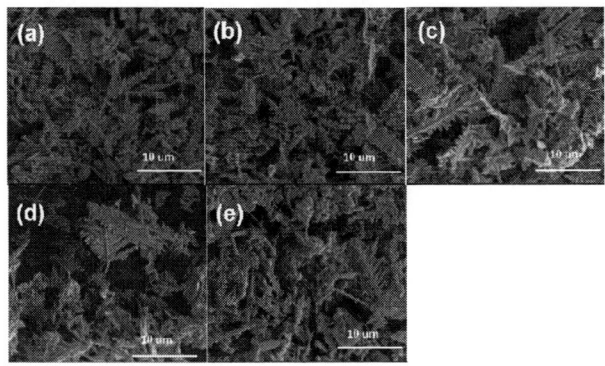

Figure 9. The microstructure evolution of dendritic Cu heated up to (a)150°C, (b)200°C, (c)250°C, (d)300°C and (c) 350°C respectively under ambient pressure.

In order to understand the sintering properties of dendritic Cu more clearly, it is necessary to study the microscopic morphology envotion of dendritic Cu with the increase of sintering temperature. Figure 9 shows the microstructure changes with sintering temperature under ambient conditions. It can be seen that the dendritic structure does not change significantly until the sintering temperature is increased to 300°C. At this temperature, the dendritic structure was destoryed with varying degree, indicating that the sintering began to occur, which was consistent with the above DSC result. When the temperature increased to 350°C, there was no dendritic structure observed in the SEM, indicating the sintering proceed greatly. Moreover, there was a high porosity, which indicate that the dendritic struture exhibited poor sinterability without pressure. In contrast, at a pressure of 10 MPa, its dendritic structure disappeared even at a low temperature of 150°C. Besides, the porosity decreased and densification degree increased as the sintering temperature increased, which well explained the increase in shear strength with the increase of temperature. In order to confirm whether the destory of dendritic struture was caused by the high pressure, the morphology of was dendritic Cu paste was sintered at room temperature under the same pressure, however, the dendritic structure was clearly observed in the SEM, indicating the high pressure is not enough to triggrer the deformation and sintering of dendritic Cu. With the temperature increased to 80°C, the dendritic structure was not well-defined, indicating the sintering has occurred. This results showed that the innitial sintering temperaure of dendritic can be decreased significantly with the assistant of pressure.

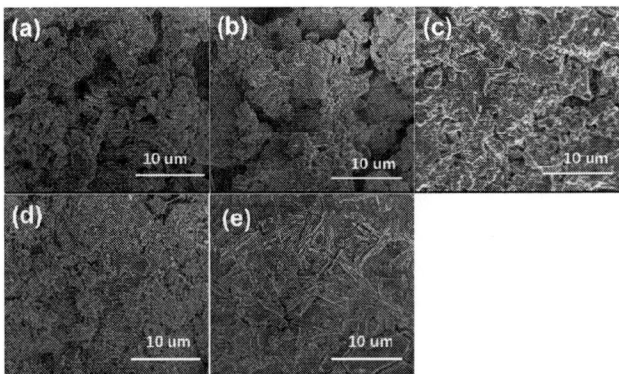

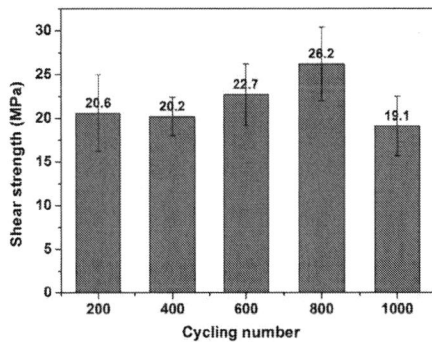

Figure 10. Microsture evolution of dendritic Cu with the increase of sintering temperature under a pressure of 10MPa.

H. Reliability of sintered dendritic Cu bonded joints

Figure 11. Shear strength of Cu paste sintered at 350°C, 10MPa, 30min with various thermal shock cycling.

Figure 11 shows the shear strength of Cu paste sintered at 350°C, 10MPa, 30min versus temperature cycling number. It can be seen that the average shear strength had no obvious change throughout the entire cycling process, which was still about 20MPa after 1000 cycles, indicating the high reliability of sintered Cu joint.

III. CONCLUSIONS

Dendritic Cu structure has been prepared and their application as low temperature interconnect materials was investigated. Proper optimizations of process parameters (sintering pressure, time and temperature) can yield high shear strength of the bonded joint. Compared with traditional nanoparticle-based paste, the dendritic Cu paste showed good low temperature sinterability. However, the shear strength of the dendritic Cu sintered bonded joints at high temperature was inferior to that of that of nanoparticle. The sintered Cu joints exhibit excellent reliability without obvious degradation in shear strength even after 1000 thermal shock cycling.

ACKNOWLEDGMENT

The authors would like to acknowledge the National Natural Science Foundation of China (61704182), SIAT

CAS-CUHK Joint Laboratory of Materials and Devices for High Density Electronic Packaging.

Author contributions: Gang Li and Jilei Fan contributed equally.

REFERENCES

[1] V. R. Manikam and C. Kuan Yew, "Die Attach Materials for High Temperature Applications: A Review," *IEEE Transactions on Components, Packaging and Manufacturing Technology,* vol. 1, no. 4, pp. 457-478, 2011.

[2] H. S. Chin, K. Y. Cheong, and A. B. Ismail, "A Review on Die Attach Materials for SiC-Based High-Temperature Power Devices," *Metallurgical and Materials Transactions B,* vol. 41, no. 4, pp. 824-832, 2010.

[3] M. Arifin, N. Wivanius, N. F. Prebianto, "Epoxy Adhesive as Die Attach Material in Semiconductor Packaging: A Review,"in *2018 International Conference on Applied Engineering(ICAE)* ,2018.pp.1-5 :IEEE.

[4] K. S. Siow, "Are Sintered Silver Joints Ready for Use as Interconnect Material in Microelectronic Packaging?," *Journal of Electronic Materials,* vol. 43, no. 4, pp. 947-961, 2014.

[5] S. A. Paknejad and S. H. Mannan, "Review of silver nanoparticle based die attach materials for high power/temperature applications," *Microelectronics Reliability,* vol. 70, pp. 1-11, 2017.

[6] B. H. Lee, M. Z. Ng, A. A. Zinn, and C. L. Gan, "Evaluation of Copper Nanoparticles for Low Temperature Bonded Interconnections," (in English), *Proceedings of the 22nd International Symposium on the Physical and Failure Analysis of Integrated Circuits (Ipfa 2015),* pp. 102-106, 2015.

[7] A. A. Zinn, R. M. Stoltenberg, J. Chang, Y. L. Tseng, S. M. Clark, and D. A. Cullen, "A Novel NanoCopper-Based Advanced Packaging Material," (in English), *Proceedings of the 2016 Ieee 18th Electronics Packaging Technology Conference (Eptc),* pp. 844-849, 2016.

[8] C. Yang *et al.*, "Fractal dendrite-based electrically conductive composites for laser-scribed flexible circuits," *Nat Commun,* vol. 6, p. 8150, Sep 3 2015.

[9] R. Yang *et al.*, "Low-Temperature Fusible Silver Micro/Nanodendrites-Based Electrically Conductive Composites for Next-Generation Printed Fuse Links," (in English), *Acs Nano,* vol. 11, no. 8, pp. 7710-7718, Aug 2017.

[10] R. Bakthavatsalam *et al.*, "Solution chemistry-based nano-structuring of copper dendrites for efficient use in catalysis and superhydrophobic surfaces," *RSC Advances,* vol. 6, no. 10, pp. 8416-8430, 2016.

[11] K. S. Siow and Y. T. Lin, "Identifying the Development State of Sintered Silver (Ag) as a Bonding Material in the Microelectronic Packaging Via a Patent Landscape Study," *Journal of Electronic Packaging,* vol. 138, no. 2, 2016.

[12] Y. Jianfeng, Z. Guisheng, H. Anming, and Y. N. Zhou, "Preparation of PVP coated Cu NPs and the application for low-temperature bonding," *Journal of Materials Chemistry,* vol. 21, no. 40, 2011.

A new development of direct bonding to aluminum and nickel surfaces by silver sintering in air atmosphere

Ly May Chew[1], Tamira Stegmann[2], Erika Schwenk[2], Monique Dubis[2], Wolfgang Schmitt[1]

[1] Heraeus Deutschland GmbH & Co. KG
Hanau, Germany

[2] Hochschule Aschaffenburg, University of Applied Sciences
Aschaffenburg, Germany

Abstract—Considering the promising properties of silver such as high thermal and electrical conductivity as well as high melting points, low temperature silver sinter technology has attracted growing attention in recent years especially for the applications required high power and high operating temperature. Current silver sinter technology required plating of precious metal finishing on the substrates prior to sintering process in order to form a strong sinter joint. Direct bonding on non-precious metal surfaces by silver sintering is therefore of great interest, since the precious metal finishing on substrate is no longer necessary, which will lead to the reduction of manufacturing cost. This paper explores the development of a safe-to-use micro-silver sinter paste for direct bonding on aluminum and nickel surfaces by pressure sintering. In this study, Ag metallized Si dies were attached on nickel-plated direct copper bonding substrates and high purity aluminum plates by silver sintering process with a pressure of 10 MPa at 250 °C for 3 min in air atmosphere. The cross-sectional SEM images of sintered samples indicate that a dense sintered joint was formed on the surface of Ni and Al. After die shear test, SEM-EDX was conducted on the fracture surface of Ni and Al substrates and the results confirmed that silver sintered layer was formed on the surface of Ni and Al. The EDX analysis results further illustrate an interdiffusion of Ag/Ni and Ag/Al appeared at the interface located between substrates and sintered layer. Silver sintered joint with high bonding strength was formed on Ni and Al surfaces and the average die shear strength remained above 30 N/mm² after 500 h storage at 250 °C. Cohesive break in the sintered layer was found for both Ni and Al samples before and after high temperature storage where sintered layer can be found on both the die backside and the substrate surface indicating that good adhesion on Al and Ni surfaces was achieved with the newly developed silver sinter paste.

Keywords-silver sintering; non-precious metal surfaces; interdiffusion; aluminum oxide

I. INTRODUCTION

Low temperature silver sinter material possesses several advantages such as lead-free, high melting temperature as well as high efficiency in thermal and electrical conduction. Accordingly, silver sinter material was generally considered to be a highly reliable interconnect material especially for the applications required continuous operations under extreme high temperatures [1-3]. Lead-free solder materials and conventional electronics adhesives containing conductive

components are not suitable to be applied as interconnect material under such a harsh working environment due to the low melting points.

Die attach application by silver sinter technology has been intensively studied in recent years [4-7] and it has been reported that silver sinter joint exhibits higher reliability compared with lead-free solder materials. As a result, it is strongly believed that the reliability of entire power module can be increased significantly if silver sinter material is used as interconnect material for both die-to-substrate and substrate-to-baseplate attachment. The commonly used baseplate materials are Al, Cu or AlSiC, whereas, Ni and Cu ceramic substrates with or without additional metallization are usually used in high temperature power electronics module. It is well-known that direct bonding on the surface of non-precious metal such as Al and Ni using the existing interconnect materials is relatively difficult, since the native oxide layer formed on the surface can prevent the formation of strong bonding [1,8]. This is one of the reasons substrate-to-baseplate attachment by silver sintering has only drawn a little interest. Therefore, a technique which is able to directly bond on non-precious metal surfaces would be of great interest. Recently, several researches [8-9] have been published reporting other possible techniques used for direct bond on aluminum. Fu et al. [8] carried out a study revealing the possibility of forming Ag-Al joint with high bonding strength by solid-state bonding processes, in which the native aluminum oxide layer on the surface of Al substrates does not necessarily need to be removed prior to the sintering process.

Our previous study [10] has clearly demonstrated that by using Heraeus commercial silver sinter paste ASP 338-28, sintered joint with stable and reliable bonding strength is able to be formed on Ag, Au and Cu surfaces. A high bonding strength of the sintered joint (die shear strength > 35 N/mm²) was achieved. No delamination was observed after 2000 cycles of variation of the cycling temperature between -40 °C and 150 °C. However, according to our recent research [11], ASP 338-28 sinter paste seems not to be able to deliver the same performance on Al and Ni surfaces as on Ag, Au and Cu surfaces. In present work, we reported a newly developed silver sinter paste which can directly bond on Al and Ni surfaces. The die shear strengths of the sintered samples before and after high temperature storage were measured to characterize the bonding strength of the formed sintered joint. Various microscopic and spectroscopic methods were utilized to study the microstructure of silver sintered joint and the chemical composition on the fracture surface after die shear test.

II. EXPERIMENTAL

A. Materials

LTS 342-54 is our newly developed sinter paste, which consists of micro-scale silver ranged from 1 to 10 µm and it was used as interconnect material in this study. 2 x 2 mm and 4 x 4 mm silicon dies with Ag backside metallization were utilized for die attach. Substrates used in this study are (1) direct copper bonding (DCB) substrates with the material combinations of 0.3 mm Cu / 0.38 mm Al_2O_3 / 0.3 mm Cu plated with 3-7 µm Ni and (2) 5 mm thick Al plate with Al purity of 99.5 %.

B. Sample preparation

Sintered samples were prepared via the process flow illustrated in Fig. 1. The process is briefly described as following:

(1) stencil was used to print silver sinter paste (LTS 342-54) on Ni DCB and Al plate.

(2) the solvents in LTS 342-54 was evaporated in an oven at 120 °C for 10 min.

(3) the silicon die was attached to the substrate by applying a force of 400 g for 3 seconds. During die attachment, the substrate was heated constantly at 130 °C.

(4) sinter press, which was heated constantly at 250 °C, was used to sinter the samples with a pressure of 10 MPa in air atmosphere for 3 min.

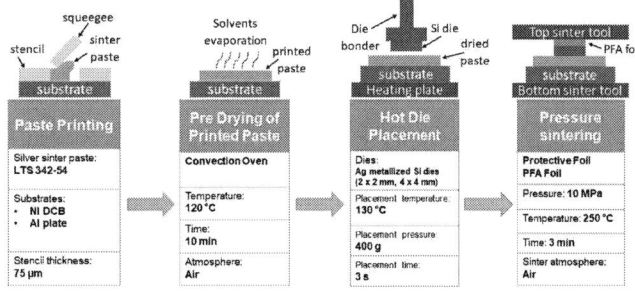

Figure 1. Process flow chart.

2 x 2 mm Si dies were attached on Ni DCB, while 4 x 4 mm Si die were mounted to Al plate, as shown in Fig. 2(a) and 2(b), respectively.

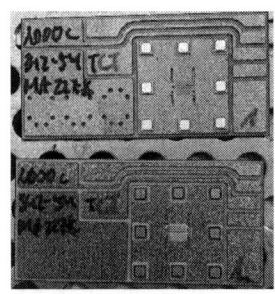 

| (a) | (b) |

Figure 2. (a) 2 x 2 mm Si die attached on Ni DCB, (b) 4 x 4 mm Si die attached on Al plate.

C. Characterization

In this study, the sintered samples were additionally stored at 250 °C for 250 h and 500 h for the investigation on the reliability of the silver sintered joint.

Die shear strength of the samples before and after high temperature storage (HTS) was measured employing a Nordson Dage 4000 plus to evaluate the bonding strength of silver sintered joints. As a standard method, the die shear measurement is conventionally applied to determine the shear strength of bonding materials by measuring the force applied to a semiconductor die attached to a substrate

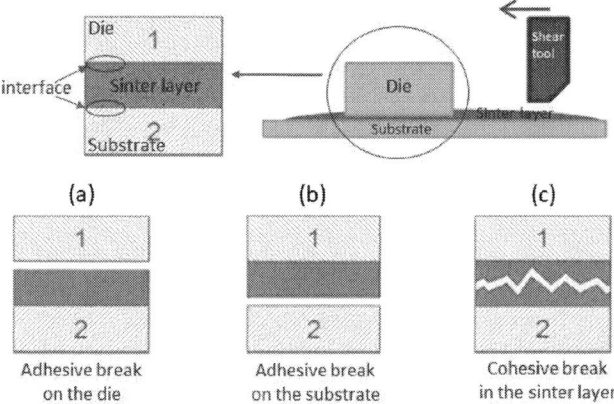

Figure 3. Schematic diagram of die shear test and possible die shear failure modes.

Fig. 3 shows the principle of die shear measurement and the possible die shear failure modes. In the case of low bonding strength of sintered joint, the sintered layer is weakly bonded on either semiconductor die or on substrate, which can lead to the adhesive break as shown in Fig. 3(a) and (b). On the other hand, in the case of forming strong bonds between die and sintered layer as well as between substrate and sintered layer, the cohesive break mode as shown in Fig. 3(c) can be observed after die shear measurement.

Cross section of the sintered samples was carried out before and after die shear test. Scanning electron microscopy (SEM) equipped with energy dispersive X-ray spectroscopy (EDX) was applied to investigate the cross-sectional microstructure and to determine the chemical composition of the interface between silver sintered joint and substrate.

978-1-7281-1500-9/19 $31.00 © 2019 IEEE

III. RESULTS AND DISCUSSION

A. SEM images of cross section combined with EDX analysis of sintered samples

Fig. 4(a) shows the cross-sectional SEM image of sintered Ni DCB sample. A silver sintered layer with low porosity was formed between Ag metallized Si die and Ni DCB. In the high magnification SEM image, it can be clearly seen that silver sintered layer is strongly bonded onto the Ni DCB. EDX line scan was performed to further evaluate the chemical composition at the interface between silver sintered joint and Ni DCB. According to the EDX result, a thin Ni layer is found to be present on the top of Cu layer. Phosphorus is known to be the residual from Ni plating process is also found in the thin Ni layer. An interdiffusion of Ag and Ni occurring at the interface between silver sintered joint and Ni DCB can also be clearly seen in the EDX line scan result.

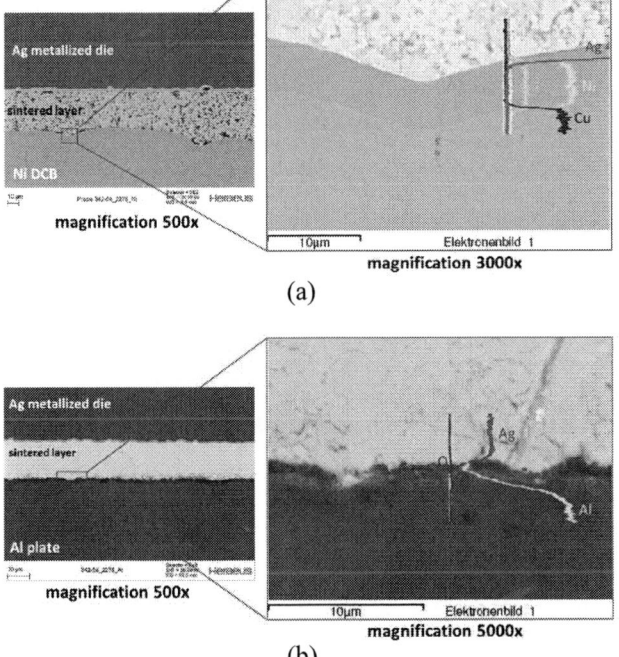

(a)

The cross-sectional SEM image of sintered Al plate sample is displayed in Fig. 4(b). A very dense silver sintered layer was formed between Ag metallized die and Al plate. A thin dark interface between sintered layer and Al plate can be clearly seen in the high magnification SEM image. It is likely that the thin dark interface is the native aluminum oxide layer generated on the surface. The EDX result indicates that an interdiffusion of Ag and Al occurred at the interface between silver sintered joint and Al plate, which is similar to the sintered Ni DCB sample. Furthermore, it can be seen that Ag diffused towards Al plate and simultaneously Al diffused into sintered layer through the native oxide layer demonstrating

(b)

Figure 4. Cross-sectional SEM images of sintered (a) Ni DCB, (b) Al plate samples.

that silver sintered joint was created on Al plate through the aluminum oxide layer.

B. Die shear test of the sintered samples before and after high temperature storage at 250 °C

Die shear test was conducted on the sintered Ni DCB and Al plate samples before and after 250 h and 500 h storage at 250 °C. Fig. 5 displays the die shear strength of samples before and after HTS. 24 Si dies were sheared to create an individual box plot. It can be seen in Fig. 5 that an average die shear strength of 19 N/mm² was obtained for Ni DCB samples before HTS. The average die shear strength increased to 36 N/mm² after 250 h storage. This phenomenon can be attributed to the continuous proceeding of the sintering process during HTS. It is believed that the sintering process is not yet completed under the mild sintering process condition we used in this work. Furthermore, it is worth noting that the die shear strength after 250 h and 500 h storage is rather similar which is in agreement with our previous study [10]. It is very likely that the sintering process is completed after a certain time of storage at 250 °C. As a result, there is no further increase in the die shear strength.

The average die shear strength for Al plate samples (57 N/mm²) before HTS is higher than for Ni DCB samples (19 N/mm²). This observation can be explained by the density of sintered layer formed on Al plate is higher than that on Ni DCB as can be seen in Fig. 4. A slight decreased in the die shear strength was observed for the samples after 250 h storage. However, there is no significant difference in the die shear strength before and after HTS. The average die shear strength remained above 45 N/mm² indicating that sintered joint with high bonding strength was formed on Al surface even after 500 h storage at 250 °C.

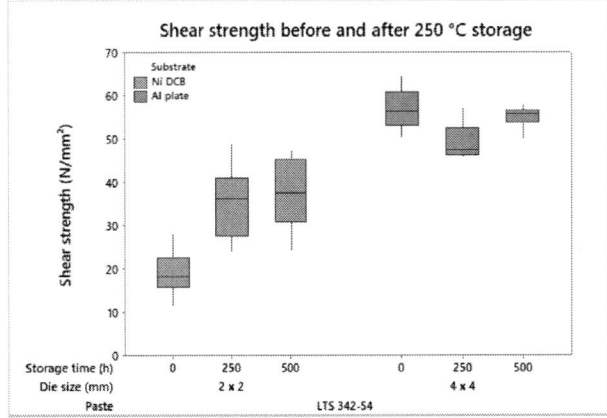

Figure 5. Comparison of die shear strength before and after storage at 250 °C for Ni DCB samples (green box plot) and Al plate samples (blue box plot).

978-1-7281-1500-9/19 $31.00 © 2019 IEEE

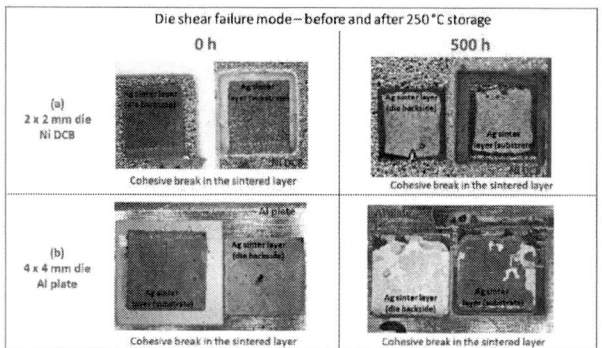

Figure 6. Die shear failure mode before and after 500 h storage at 250 °C of (a) Ni DCB samples and (b) Al plate samples.

Fig. 6 shows the die shear failure mode before and after 500 h storage at 250 °C. Similar failure mode was found for Ni DCB and Al plate samples before and after HTS where cohesive break in the silver sintered layer was observed as can be seen in Fig. 6. This observation demonstrates that sintered joint was generated on die backside as well as on Ni and Al surfaces.

C. SEM images of cross section combined with EDX analysis of fracture surface of sintered sample after die shear test

SEM-EDX was conducted on the fracture surface of substrate after the die was removed to confirm that the silver sintered joint was formed on Ni and Al surfaces. The SEM images of cross section as well as EDX analysis of fracture surface of Ni DCB and Al plate are presented in Fig. 7(a) and (b), respectively. Part of the silver sintered layer remained on the surface of Ni DCB after the Si die was removed which can be seen in Fig. 7(a). Corresponding with the die shear failure mode (Fig. 6), it is shown that silver sintered layer was formed on the Ni DCB surface. Furthermore, the presence of both Ni and Ag in the thin interface (approximately 3 μm) between silver sintered layer and Ni DCB, which is shown in the inset in Fig. 7(a), indicates an interdiffusion of Ag and Ni between sintered layer and Ni DCB.

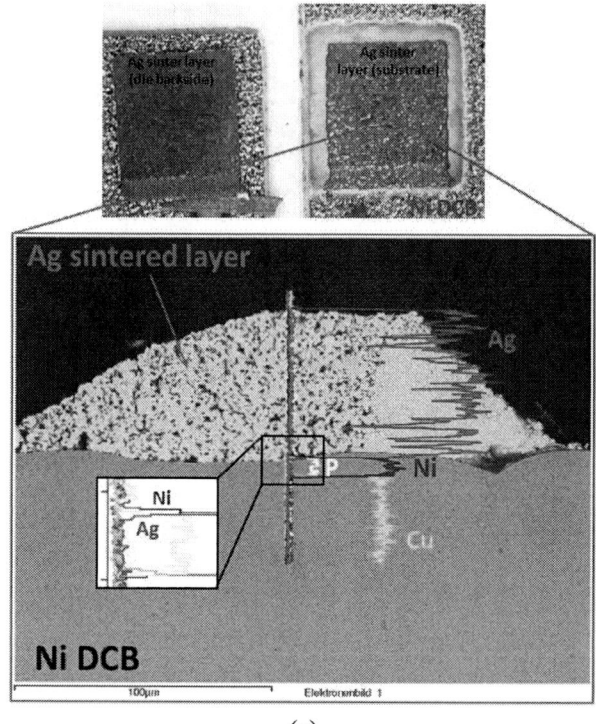

(a)

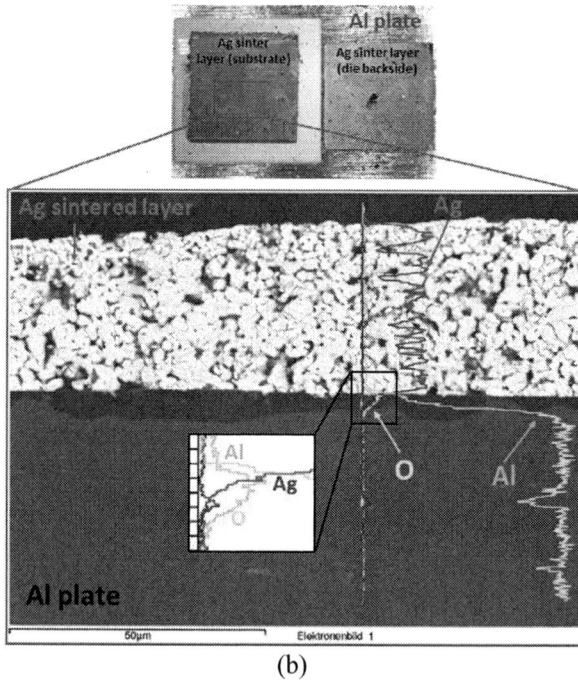

(b)

Figure 7. Cross-sectional SEM images and EDX analysis of fracture surface of as-sintered (a) Ni DCB, (b) Al plate samples after die shear test.

As shown in Fig. 7(b), similar observation for Ni DCB sample was found, where part of the silver sintered layer remained on the surface of Al plate after the removal of the Si die indicating that silver sintered joint was formed on the surface of Al plate. It is clearly shown in the inset in Fig. 7(b) that Ag, Al and O are found to be in the thin dark layer located between silver sintered joint and Al plate. Presumably, the thin dark layer is mainly composed of aluminum oxide. Fig. 7(b) indicates that the interdiffusion of Ag and Al occurred in the thin dark layer.

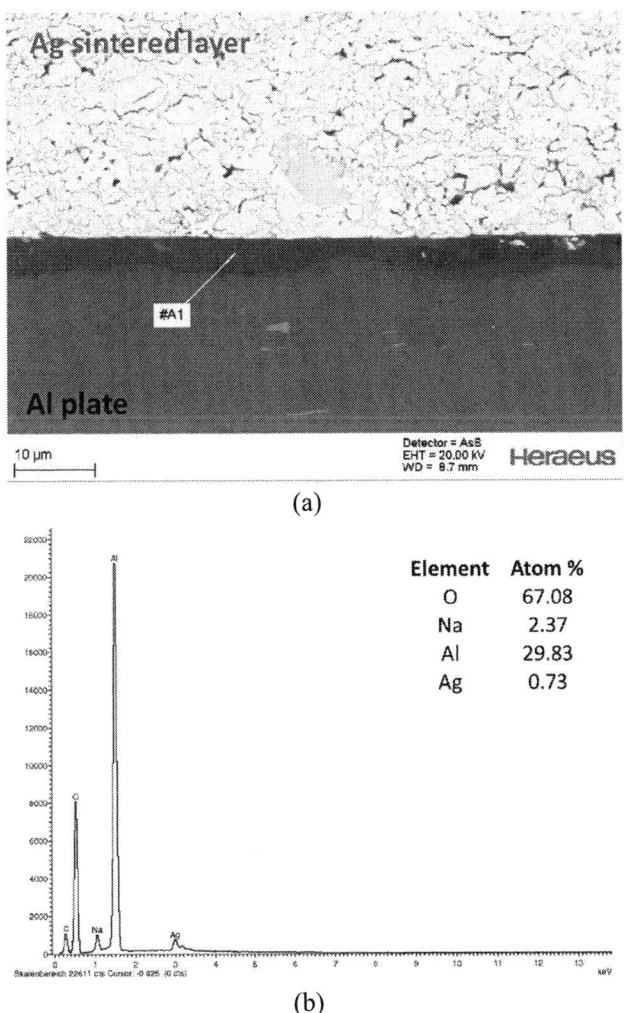

(b)

Figure 8. (a) cross-sectional SEM image of Al plate sintered sample and (b) EDX analysis of area marked as #A1 in (a).

Fig. 8(a) shows the cross-sectional SEM image of Al plate sintered sample. The area marked as #A1 in the thin dark layer was analyzed by EDX to confirm the proposed hypothesis. The EDX analysis result shown in Fig. 8(b) illustrates that the area marked as #A1 contains mainly Al and O indicating that the thin dark layer is aluminum oxide with a tiny amount of diffused Ag.

D. SEM images and EDX analysis of fracture surface of 500 h storage samples after die shear test

As mentioned above, cohesive break in the silver sintered layer was observed for Ni DCB and Al plate samples after 500 h storage at 250 °C where silver sintered layer is formed on both the die backside and substrate surface as can be seen in Fig. 6. It is worth noting that distinct white and grey areas are observed on the die backside and the substrate surface. It is suggested that the white area contains mainly silver, whereas, the grey area consists of silver oxide. After die shear test of Ni DCB sample which is stored at 250 °C for 500 h, the fracture surface of Ni DCB was analyzed by EDX.

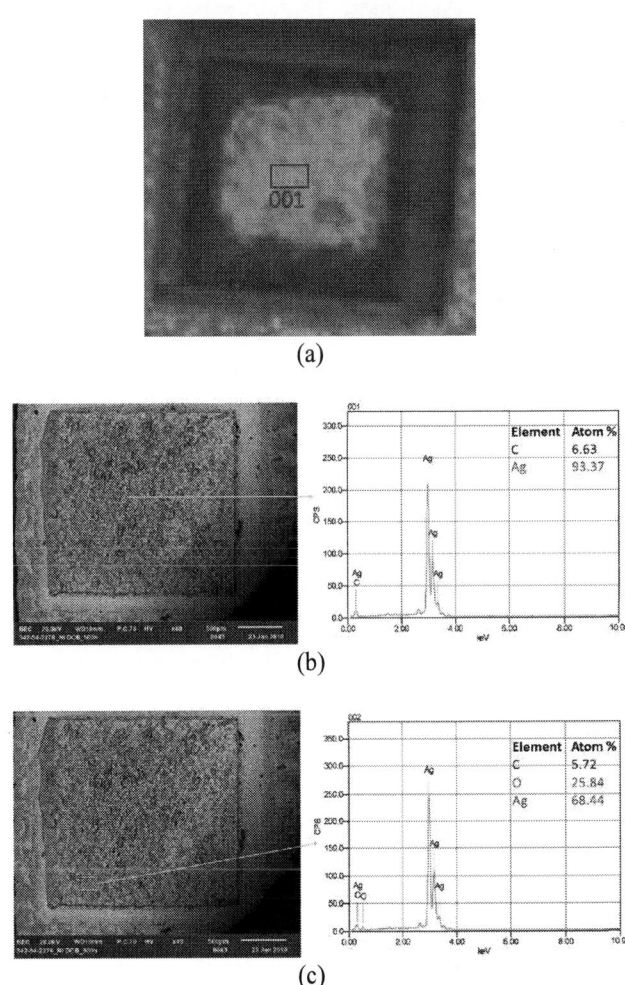

Figure 9. Ni DCB sample after 500 h storage at 250 °C (a) image of fracture surface of Ni DCB after die was sheared from Ni DCB, (b) EDX analysis of area marked as 001, (c) EDX analysis of area marked as 002.

Fig. 9(a) shows the image of fracture surface of Ni DCB. EDX analysis was performed on the white area marked as 001 and on the grey area marked as 002 in Fig. 9(a). The EDX analysis result presented in Fig. 9(b) shows that the white area contains 93.37 at. % Ag. It can be concluded that silver is the main element in the white area. In contrast, the EDX analysis result indicates that the grey area composed of 68.44 at. % Ag and 25.84 at. % O as displayed in Fig. 9(c) indicating that the grey area contains mostly silver oxide.

In addition to Ni DCB sample, EDX analysis was conducted on the fracture surface of die backside after the die was sheared from Al plate sample which is stored at 250 °C for 500 h. Similarly, the white area marked as 001 and the grey area marked as 002 on the fracture surface of die backside as shown in Fig. 10(a) are analyzed by EDX. In addition to elements such as C, O, Si and Ag, approximately 2 at % Al is detected on the fracture surface of die backside. It is likely that Al diffused towards die backside through the silver sintered layer during HTS. Fig. 10(b) shows the EDX analysis result of white area where 80.44 at. % Ag and 7.64 at. % O are detected demonstrating that silver is the main phase in the white area. The EDX analysis result shown in Fig. 10(c) indicates that the grey area contains mainly silver oxide where 55.38 at. % Ag and 33.63 at. % O are detected.

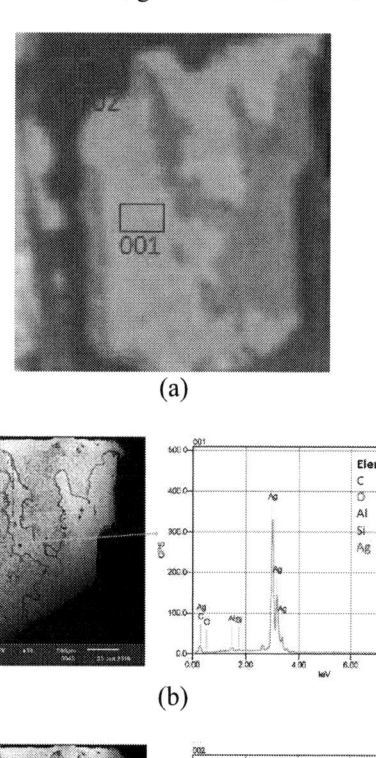

(a)

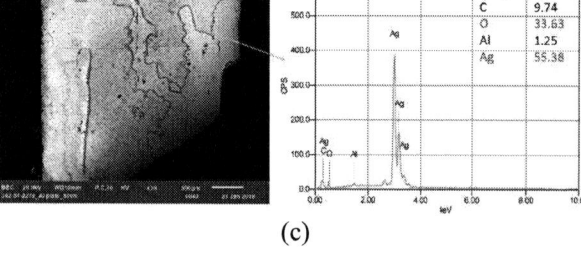

(b)

(c)

Figure 10. Al plate sample after 500 h storage at 250 °C (a) image of fracture surface of die backside after die was sheared from Al plate, (b) EDX analysis of area marked as 001, (c) EDX analysis of area marked as 002.

IV. SUMMARY

The present work reported a recently developed silver sinter paste, which enable the direct bonding on Al and Ni surface by sintering process with pressure higher than 10 MPa in air atmosphere without pretreatment of Ni and Al surfaces. A dense sintered layer was obtained for both Ni DCB and Al plate samples as can be seen in the cross-sectional SEM images. The EDX analysis results illustrate a diffusion between Ni and Ag as well as between Al and Ag appeared at the interface between substrate surface and silver sintered layer. Additionally, it can be clearly seen from the EDX analysis results that silver sintered joint was formed on Al plate through the aluminum oxide layer. The cross-sectional SEM images of fracture surface after the removal of die further prove that sintered joint was generated on Al and Ni surfaces. Cohesive break in the sintered layer was observed for both Al plate and Ni DCB samples before and after HTS indicating that good adhesion on Ni and Al surfaces was achieved. Interestingly, distinct white and grey areas on surface of substrate and the die backside are observed from the die shear failure mode of Ni DCB and Al plate samples which were stored at 250 °C for 500 h. The EDX analysis results proposed that silver is the main phase in the white area and the grey area consists of mainly silver oxide. An average die shear strength > 30 N/mm² was observed for all samples after 500 h storage demonstrating that sintered joint with high bonding strength was achieved on Al and Ni surfaces using the newly developed silver sinter paste.

ACKNOWLEDGMENT

The authors gratefully acknowledge the financial support from the German Federal Ministry of Education and Research (BMBF) for the project "Korrosionsfeste Sinterverbindungstechnologie für korrosionsgefährdete Anwendungen (KorSikA)".

REFERENCES

[1] K. S. Siow, Y. T. Lin, "Identifying the development state of sintered silver (Ag) as a bonding material in the microelectronic packaging via a patent landscape study", Journal of Electronic Packaging, Transactions of the ASME, Vol. 138, June 2016, 020804.

[2] K. S. Siow, M. Eugenie, "Patent landscape and market segments of sintered silver as die attach materials in microelectronic packaging", 37th International Electronics Manufacturing Technology Conference, Sept. 2016.

[3] L. C. Wai, W. W. Seit, E. O. J. Rong, M. Z. Ding, V. S. Rao, D. R. MinWoo, "Study on silver sintered die attach material with different metal surfaces for high temperature and high pressure (300 °C/30kpsi) applications", IEEE 15th Electronics Packaging Technology Conference, Dec. 2013, pp. 335-340.

[4] S. Kraft, A. Schletz, M. Maerz, "Reliability of silver sintering on DBC and DBA substrates for power electronic applications", 7th International Conference on Integrated Power Electronics Systems, March 2012, pp. 1-6.

[5] T. Krebs, S. Duch, W. Schmitt, S. Kötter, P. Prenosil, S. Thomas, "A breakthrough in power electronics reliability – new die attach and wire

bonding materials", IEEE 63rd Electronics Components and Technology Conference, May 2013, pp. 1746-1752.

[6] W. Schmitt, L. M. Chew, "Silver sinter paste for SiC bonding with improved mechanical properties", IEEE 67th Electronic Components and Technology Conference, May 2017, pp. 1560-1565.

[7] G. Lu, M. Wang, Y. Mei, X. Li, "Advanced die-attach by metal-powder sintering: the science and practice", 10th International Conference on Integrated Power Electronics Systems, March 2018, pp. 594-602.

[8] S. Fu, C. C. Lee, "Direct silver to aluminum solid-state bonding processes", Materials Science & Engineering A, Vol. 722, March 2018, pp. 160-166.

[9] T. Morita, Y. Yasuda, E. Ide, A. Hirose, "Direct bonding to aluminum with silver-oxide microparticles", Materials Transaction, Vol. 50, No. 1, 2009, pp. 226-228.

[10] L. M. Chew, W. Schmitt, C. Schwarzer, J. Nachreiner, "Micro-silver sinter paste developed for pressure sintering on bare Cu surfaces under air or inert atmosphere", IEEE 68th Electronic Components and Technology Conference, May 2018, pp. 323-330.

[11] L. M. Chew, W. Schmitt, M. Dubis, "High bonding strength of silver sintered joints on non-precious metal surfaces by pressure sintering under air atmosphere using micro-silver sinter paste", 20th Electronics Packaging Technology Conference, Dec. 2018.

Submicron-scale Cu RDL Pattering Based on Semi-additive Process for Heterogeneous Integration

Takamasa Takano, Hiroshi Kudo*, Masaya Tanaka, Miyuki Akazawa,
Yumi Okazaki, Haruo Iida, Kouji Sakamoto, Daisuke Kitayama, Shouhei Yamada, and Satoru Kuramochi
DNP Co., Ltd.
250-1, Wakashiba, Kashiwa-shi, Chiba-ken 277-0871, Japan
*e-mail: kudou-h10@mail.dnp.co.jp

Abstract— Use of dry plasma etching rather than wet etching of a Cu-seed layer in a semi-additive process enabled more precise dimension controllability in the patterning of submicron-scale Cu traces due to no shift in the width of the traces. This controllability is comparable to that of competitive fabrication technologies such as that for damascene-based Cu redistribution layers. The dry etching enabled the patterning of Cu traces with an aspect ratio as high as 4.2 (L/S=0.7/0.7 μm, 3.0 μm in height) without any failures such as electrical shorts between traces. Simulation showed that an increase in the aspect ratio effectively reduced signal transmission loss due to a reduction in conductor loss. The dry etching provided very smooth surfaces on the Cu-trace side-wall (roughness as low as 0.05 μm). This further reduced the signal transmission loss compared to that of wet-etched Cu traces. Submicron-scale patterning of Cu traces using dry etching enables flexible design of redistribution layer lines in terms of signal integrity, in addition to increasing the number of signal I/Os cost effectively.

Keywords- redistribution layer; 2.5D interposer; wafer-level packaging; semi-additive process; Cu damascene; transmission loss; signal integrity

I.   INTRODUCTION

Chip-to-chip lateral interconnections for heterogeneous integrations (e.g., FOWLP, 2.5D interposers) are increasingly critical components in high-performance computing, mobile, and IoT sensing devices, which require high-density, high-speed signal lines and highly reliable electrical isolation, in addition to low-power consumption [1–3].

A key factor in achieving such interconnections is the redistribution layer (RDL) process used. The semi-additive process (SAP) is the most cost-effective of the potential RDL processes in terms of high-volume production. However, a serious concern regarding SAP is that using wet-etching chemistry to remove the Cu-seed layer causes an unwanted trace-width shift, resulting in increased electrical resistance variability [4]. This increase becomes noticeable and degrade signal integrity when Cu traces are reduced to submicron scale.

To prevent shifts in the trace width, we examined dry etching of the Cu-seed/Ti barrier layers in the patterning of submicron-scale Cu traces in comparison with the use of wet etching. The difference between their etching mechanisms is illustrated in Fig. 1. With wet etching, the reactive chemical species etch the electroplated Cu (Ep-Cu) isotropically instead of etching the Cu-seed layer. This etching shift causes variability in the electrical resistance of the Cu traces.

With dry etching using a plasma chemical, the Cu-seed layer is selectively and mostly anisotropically etched due to the use of voltage-accelerated ion species. This minimizes the etching shift, resulting in low variability in the electrical resistance.

Section II describes the patterning of Cu traces using dry etching. Section III characterizes submicron-scale Cu traces, focusing on electrical resistance and isolation. Section IV explains the depth of focus (DOF) in the patterning of submicron-scale Cu traces. Section V describes the simulation of signal integrity including signal transmission loss. Section VI concludes the paper with a summary of the key points.

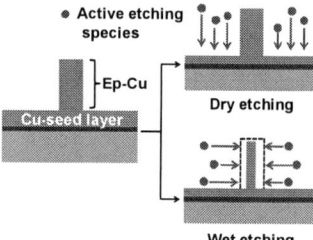

Figure 1. Mechanisms of dry and wet etching of Cu-seed layer in SAP.

II.   PATTERNING OF CU TRACES

SEM images of Cu traces (L/S=1/1 μm) patterned using dry etching on the ground level layer (Cu grid pattern) of an RDL structure are shown in Fig. 2. The close-up image clearly shows that there was no etching residue between the traces, meaning that active etching species reached the Cu-seed/Ti barrier layer at the bottom although the aspect ratio was as high as 3.0.

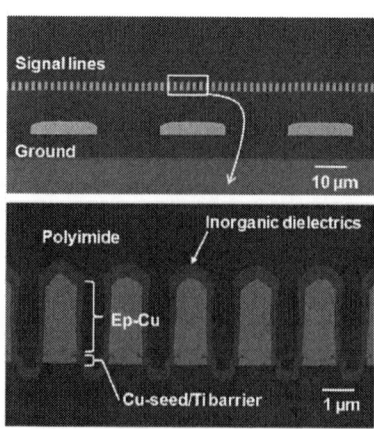

Figure 2. SEM images of Cu traces patterned using dry etching.

978-1-7281-1500-9/19 $31.00 © 2019 IEEE

The close-up image also shows that the Cu traces were covered with inorganic dielectrics, which prevent the Cu from diffusing into the polyimide and thus mechanically strengthen the RDL structure against thermal stressing. This enhanced copper redistribution (EnCoRe) layer structure was previously reported [5–7].

More precise dimension control of submicron-scale Cu traces is required to achieve better signal integrity. In SAP, wet etching of the Cu-seed layer is a critical step in determining the dimensions of the Cu traces, in addition to the photolithography process. The SEM top-view image in Fig. 3 (a) of Cu traces (L/S=1/1 μm) patterned using wet etching shows that the isotropic wet etching caused a large shift in the trace width. In contrast, there was no shift in the width of Cu traces patterned using dry etching, as shown by the image in Fig. 3 (b)). This was due to the anisotropic etching.

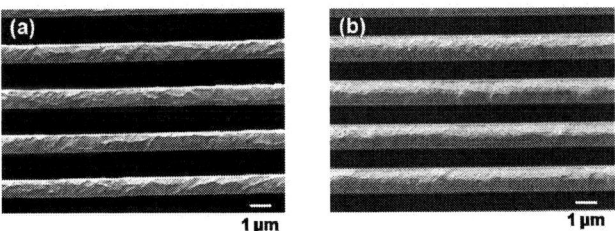

Figure 3. SEM top-view images of Cu traces (L/S=1/1 μm) patterned using wet etching (a) and dry etching (b).

We compared the expected trace width with the width measured from the SEM images. As shown in Fig. 4, the wet etching shifted the trace width about 0.1 μm on each side, which greatly increased the variability in electrical resistance. For the dry etching, the measured trace width was in good agreement with the expected one, showing that there was no shift in the trace width.

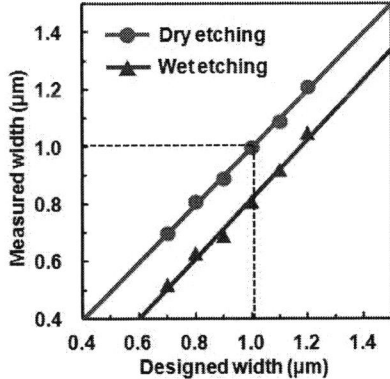

Figure 4. Trace-width shift caused by etching of Cu-seed layer.

SEM images observed at an angle of 30° for Cu traces (L/S=1/1 μm) patterned using wet and dry etching chemistries are shown in Figs. 5 (a) and (b), respectively. The wet etching roughened the side-wall of the Cu traces as much as 0.3 μm due to the differences in orientation of the Cu grains on the side-wall. Such roughness degrades signal

integrity at high frequencies, as discussed later. The dry etching resulted in a very smooth surface on the trace side-wall since the active etching ion species did not attack the Cu grains on the side-walls.

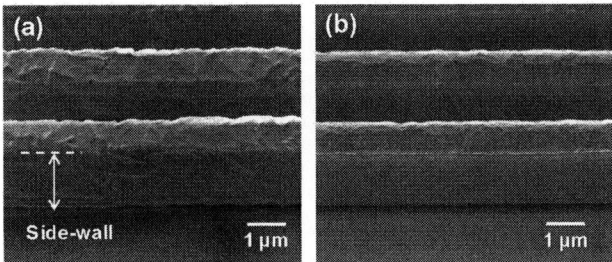

Figure 5. SEM images of Cu traces (L/S=1/1 μm) patterned using wet (a) and dry (b) etching observed at 30°.

III. ELECTRICAL CHARACTERIZATION OF SUBMICRON-SCALE CU TRACES

The variabilities in the electrical resistance of Cu traces patterned using wet and dry etching are plotted in Figs. 6 (a) and (b), respectively. The wet etching resulted in large variability and caused tens of percent of open failures due to peeling of the Cu traces with L/S=0.8/0.8 μm. In contrast, the dry plasma etching resulted in much less variability even though the trace width was as narrow as 0.7 μm. There were no open failures.

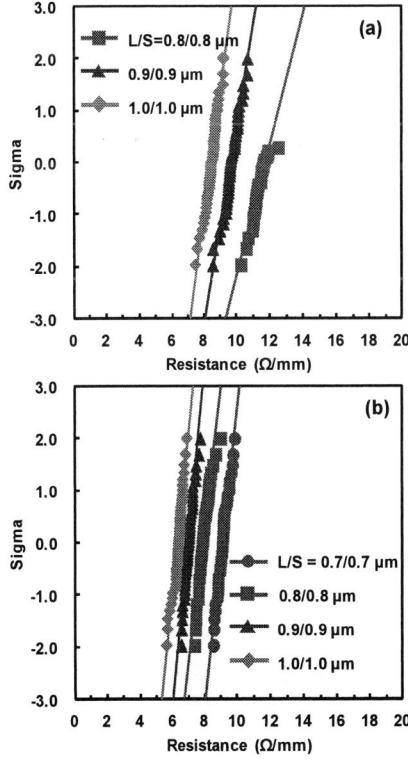

Figure 6. Variability in electrical resistance of Cu traces patterned using wet (a) and dry (b) etching.

The relative standard deviation of the electrical resistance as a function of the trace width is plotted in Fig. 7. For the wet etching, the variability in the resistance increased with a decrease in the trace width. The maximum relative standard deviation of the wafer-level electrical resistance was as high as 13%. For the dry plasma etching, the variability was 5% over the 0.7 to 1.0 μm width range.

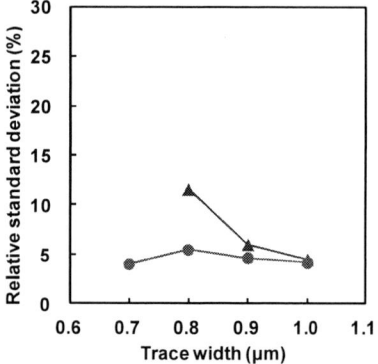

Figure 7. Relative standard deviation of electrical resistance of Cu traces.

The measured and expected electrical resistances of the Cu traces as a function of the trace width are plotted in Fig. 8. With the wet etching, the difference in electrical resistance between measured and expected increased with a decrease in the trace width. This was due to the shift in the trace width shown in Fig. 4. With the dry etching, the measured resistance closely matched the expected one. This was due to the widths of the Cu traces not shifting.

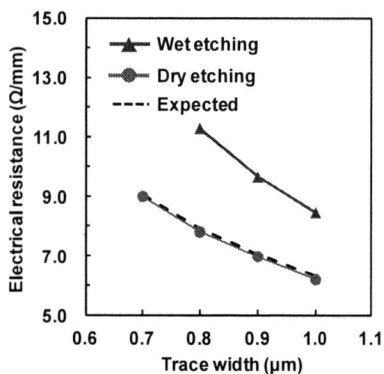

Figure 8. Measured and expected electrical resistances of Cu traces.

Reliable electrical isolation in the patterning of submicron-scale Cu traces is required because incomplete electrical isolation caused by etching residue induces ion migration. The increase in the electric field with the downsizing of Cu traces accelerates this ion migration, seriously degrading isolation reliability.

The variability in the dielectric resistance of the submicron-scale Cu traces patterned using wet and dry etching is plotted in Fig. 9. The variability was greater with the wet etching, indicating the presence of electrical shorts

between the traces. Additional wet etching would reduce the number of shorts but would also increase the trace width shift, resulting in increased variability in the electrical resistance. Wet etching is thus a less-than-optimal process for patterning submicron-scale Cu traces. With the dry plasma etching, the dielectric resistance remained as high as 1.0E8 Ω. There were no electrical shorts and no failures due to etching residue even though the trace width was as narrow as 0.7 μm.

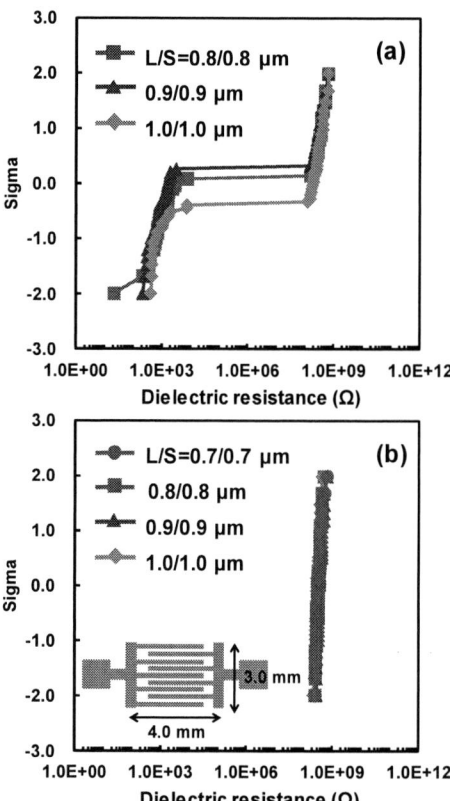

Figure 9. Variability in dielectric resistance of Cu-comb monitor patterned using wet (a) and dry (b) etching.

IV. DEPTH OF FOCUS IN PATTERNING SUBMICRON-SCALE CU TRACES

The depth of focus (DOF) is a major factor in determining the resolution capability of a photo lithography tool. The DOF is defined as the range between the upper and lower limits of the resolution capability at the center of the best focus. Submicron-scale patterning of Cu traces requires the use of a stepper lithography tool with high-resolution capability, but the resolution required greatly reduces the DOF [8]. This means that local and global roughness on the buildup RDL layers degrades the resolution capability.

The electrical resistance of a 1-μm-wide Cu trace patterned using dry etching is plotted against the DOF in Fig. 10. The electrical resistance varied about 10% over a DOF range of ±1.5 µm. Beyond ±1.5 µm, open failures occurred due to DOF mismatch.

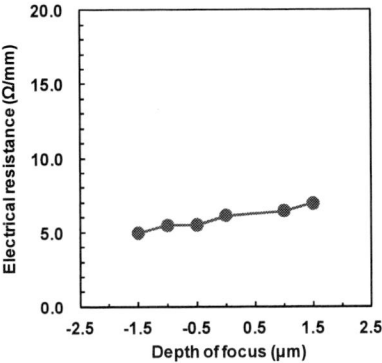

Figure 10. Electrical resistance of 1-µm-wide Cu trace patterned using dry etching against depth of focus.

The relationship between the DOF and the designed trace width is plotted in Fig. 11. The DOF decreased with the width. When the width was 0.7 µm, the DOF was 2.0 µm. This means that local and global roughness must be controlled to be within ±1.0 µm in the submicron-scale patterning of Cu traces.

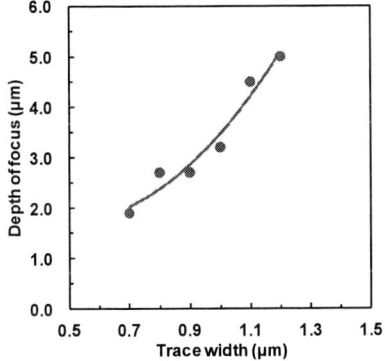

Figure 11. Depth of focus against designed trace width.

An effective way, for example, to reduce local roughness is to use low-shrinkage polyimide [1]. An effective way to reduce both local and global roughness is to optimize the design and dimensions of Cu-grid/mesh RDLs, which are commonly used for the ground and power supply. An optimized Cu grid pattern is shown in Fig. 12. Dielectric areas (25 µm^2) are arrayed using a 200-µm pitch in the X and Y directions. Residual gas in the dielectric layer is released through the dielectric areas, preventing the Cu grid from bulging during thermal RDL fabrication processing. This optimized Cu grid pattern enables high signal integrity in addition to reducing both local and global roughness.

The local and global roughnesses of optimized Cu-grid RDLs consisting of power-supply and ground-level layers over a range of 20 mm are plotted in Fig. 13 (b). This range is nearly equal to the exposure area of the stepper lithography tool. The measured local roughness, which was generated in the dielectric square area (25 µm^2), was as

small as ±1.0 µm (Fig. 13(a)). The measured global roughness over the 20 mm range was as small as ±1.0 µm. Both roughnesses are within the range of the required DOF and thus enable patterning of submicron-scale Cu traces.

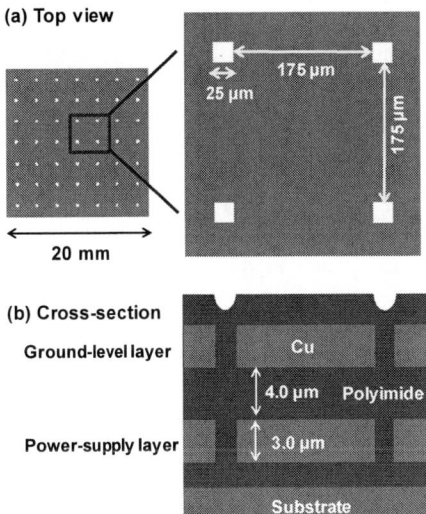

Figure 12. Top view (a) and cross-sectional (b) schematics of Cu-grid RDLs consisting of power-supply and ground-level layers.

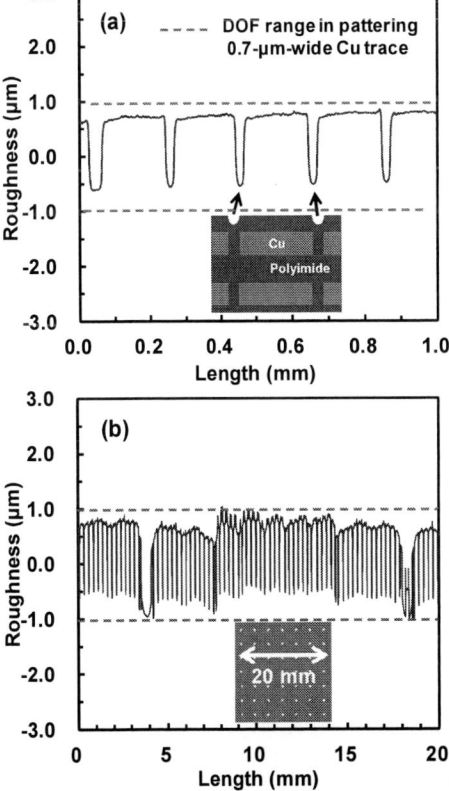

Figure 13. Local (a) and global (b) roughnesses of RDLs consisting of power-supply and ground-level layers.

V. SIGNAL INTEGRITY

Reducing signal transmission loss is a critical factor in terms of signal integrity since downsizing Cu traces increases conductor loss. The simulated signal transmission loss for a 0.7-μm-wide Cu trace is plotted in Fig. 14 for trace heights of 1.0, 2.0, and 3.0. The patterning of Cu traces 3.0 μm in height was described in this report while a height of 1.0 μm is typical for damascene-based RDLs. The trace length is assumed to be 5.0 mm, which is the typical distance between the HBM and GPU chips in a 2.5D interposer.

The signal transmission loss decreases greatly with an increase in the height of a Cu trace due to the reduction in conductor loss. At a signal frequency of 40 GHz, corresponding to the fifth-order harmonics of 6th-generation HBM (16 Gb/s) operation, the transmission losses for the trace heights of 1.0 μm and 3.0 μm were −5.5 and −3.3 dB, respectively. This shows that the signal transmission capability of a Cu trace with a height of 3.0 μm is 1.7 times that of one with a height of 1.0 μm. This is a great advantage compared to damascene-based Cu RDLs.

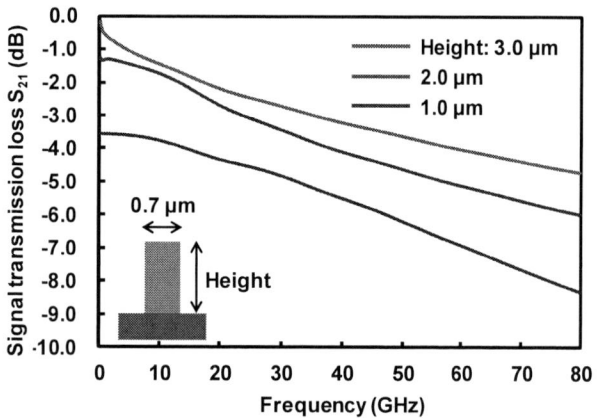

Figure 14. Effect of trace height on signal transmission loss for 0.7-μm-wide Cu trace.

As we mentioned in Fig. 5, the Cu traces patterned using dry etching had a smooth side-wall surface while those patterned using wet etching had a rough one. The surface roughness for submicron-scale Cu traces increases the signal transmission loss substantially [9]. We simulated the signal transmission loss for both dry- and wet-etched Cu traces (0.7 μm in width, 3.0 μm in height, 5.0 mm in length) and also simulated the signal transmission loss for a no-roughness Cu trace as a control condition. The surface roughnesses of the dry- and wet-etched traces were 0.05 and 0.3 μm, respectively, as measured from close-up SEM images. As shown in Fig. 15, the signal transmission loss of the dry-etched Cu trace was comparable to that of the no-roughness trace, meaning there was no increase in the transmission loss. At a signal frequency of 40 GHz, the signal transmission loss of the dry-etched Cu trace was −3.1 dB, which is 21% lower than that of the wet-etched one.

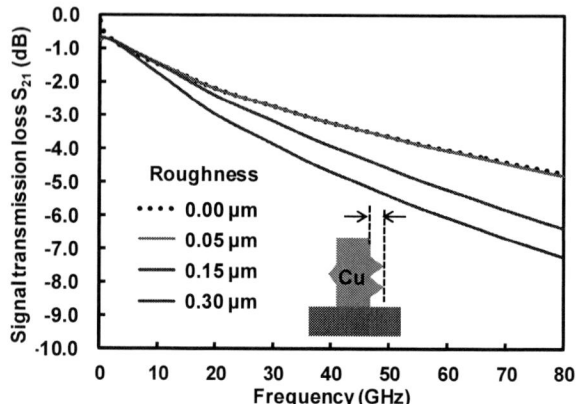

Figure 15. Effect of side-wall roughness on signal transmission loss for 0.7-μm-wide Cu trace.

Cross-talk will become a serious problem when the Cu trace pitch is reduced to less than 2.0 μm. In addition, an increase in the aspect ratio results in higher cross-talk. Optimizing the configuration of the signal lines has been shown to effectively reduce cross-talk. Sandwiching aggressor signal lines between ground lines appears to be the most effective configuration. [10, 11]

We evaluated the effectiveness of this approach for the two configurations shown in Fig. 16 using simulation. In configuration A, there are two victim signal lines and one aggressor signal line, none of which are sandwiched between ground lines. In configuration B, there are two levels of line layers. In the upper level, two victim signal lines sandwich a ground line, and in the lower level, two ground lines sandwich an aggressor signal line. For both configurations, the impedance was adjusted to 50 Ω, and the trace length was assumed to be 5.0 mm.

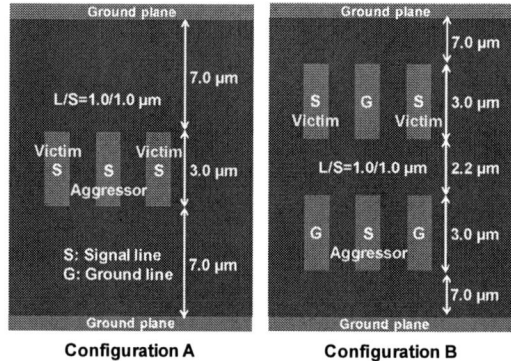

Figure 16. Two potential signal-line configurations for reducing cross-talk.

The simulated near-end and far-end cross-talks between the victim and aggressor signal lines are plotted in Figs. 17 (a) and (b). These results clearly show that sandwiching aggressor signal lines between ground lines effectively reduces both near-end and far-end cross-talk. This reduction is due to the shielding effect of the ground lines. The simulated near-end cross-talk with configuration

B was about −40 dB up to 80 GHz, just 2.9% that with configuration A.

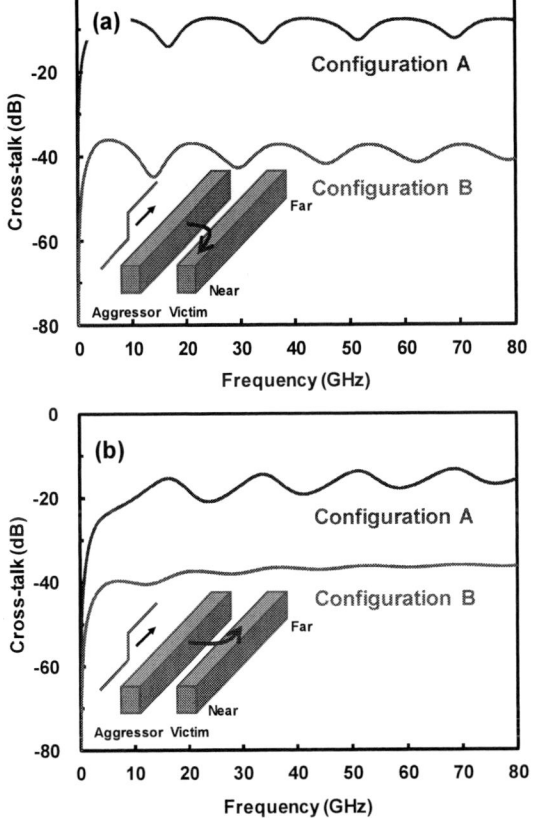

Figure 17. Simulated near-end (a) and far-end (b) cross-talk for configurations A and B.

VI. CONCLUSION

Using dry etching rather than wet etching to etch a Cu-seed layer in a semi-additive process was shown to enable more precise dimension controllability in the patterning of submicron-scale Cu traces with a high aspect ratio. The electrical resistance of the patterned Cu traces was in good agreement with the expected one, and the wafer-level variability was as small as 5% because there was no shift in the trace width. The patterned Cu traces were completely electrically isolated. When the trace width was 0.7 μm, the depth of focus of the photolithography tool was 2.0 μm. Optimizing the design of Cu-grid RDLs consisting of ground and power supply layers resulted in local and global roughnesses within the range of the depth of focus, enabling submicron-scale patterning.

Using dry etching increased the aspect ratio of the Cu traces up to 4.2 (L/S=0.7/0.7 μm, 3.0 μm in height) without any failures such as open failures and electrical shorts. Simulation showed that an increase in the aspect ratio greatly reduced signal transmission loss due to a reduction in conductor loss. The dry-etched Cu traces had very a smooth side-wall surface (roughness of 0.05 μm). This

further reduced the signal transmission loss 21% compared to that of the wet-etched Cu traces. The simulation also showed that sandwiching an aggressor signal line between ground lines effectively reduced cross-talk. Submicron-scale patterning of Cu traces using dry etching provides a flexible way to design redistribution layer lines in terms of signal integrity, in addition to increasing the number of signal I/Os cost effectively.

REFERENCES

[1] Han-Ping Pu, H. J. Kuo, C. S. Liu, and Douglas C. H. YuB, "A Novel Submicron Polymer Re-Distribution Layer Technology for Advanced InFO Packaging", IEEE 68th Electronic Components and Technology Conference (ECTC), pp. 45-51, 2018.

[2] Jinyoung Kim, Ikjun Choi, JunHyeong Park, Jae-Ean Lee, TaeSung Jeong, Jungsoo Byun, YoungGwan Ko, Kangheon Hur, Dae-Woo Kim, and Kyung Suk Oh, "Fan-Out Panel Level Package with Fine Pitch Pattern", IEEE 68th Electronic Components and Technology Conference (ECTC), pp. 52-57, 2018.

[3] Vadim Heyfitch, Shen Dong, Nanju Na, Hong Shi, Jaspreet Gandhi, Jane Xi, and Susan Wu, "High Bandwidth Memory Interface on Organic Substrate: Challenges to Electrical Design", IEEE 68th Electronic Components and Technology Conference (ECTC), pp. 1289-1294, 2018.

[4] Hiroshi Kudo, Yuuki Aritsuka, Tanaka Masaya, Ryohei Kasai, Jyunichi Suyama, Mitsuhiro Takeda, Yumi Okazaki, Haruo Iida, Daisuke Kitayama, Kouji Sakamoto, Takamasa Takano, Miyuki Akazawa, Hiroaki Sato, Shouhei Yamada, and Satoru Kuramochi, "Introduction of Sub-2-micron Cu traces to EnCoRe enhanced copper redistribution layers for heterogeneous chip integration", International Conference on Electronics Packaging and iMAPS All Asia Conference (ICEP-IAAC), pp. 399-404, 2018..

[5] Hiroshi Kudo, Takamasa Takano, Masaya Tanaka, Ryohhei Kasai, Jyunichi Suyama, Miyuki Akazawa, Mitsuhiro Takeda, Hiroshi Mawatari, Toshio Sasao, Yumi Okazaki, Naoki Oota, Susumu Tashiro, Haruo Iida, Kouji Sakamoto, Hiroyuki Sato, Daisuke Kitayama, Shouhei Yamada, and Satoru Kuramochi, "A Characterized Redistribution Layer Architecture for Advanced Packaging Technologies," IEEE 66th Electronic Components and Technology Conference (ECTC), pp. 2063-2067, 2016.

[6] Hiroshi Kudo, Ryohei Kasai, Jyunichi Suyama, Mitsuhiro Takeda, Yumi Okazaki, Haruo Iida, Daisuke Kitayama, Toshio Sasao, Kouji Sakamoto, Hiroaki Sato, Shouhei Yamada, and Satoru Kuramochi, "Demonstration of High Electrical Reliability of Sub-2 Micron Cu Traces Covered with Inorganic Dielectrics for Advanced Packaging Technologies," IEEE 67th Electronic Components and Technology Conference (ECTC), pp. 1849-1854, 2017.

[7] Hiroshi Kudo, Ryohei Kasai, Jyunichi Suyama, Mitsuhiro Takeda, Yumi Okazaki, Haruo Iida, Daisuke Kitayama, Kouji Sakamoto, Hiroaki Sato, Shouhei Yamada, Miyuki Akazawa, and Satoru Kuramochi, "Demonstration of high electromigration resistance of enhanced sub-2 micron-scale Cu redistribution layer for advanced fine-pitch packaging," IEEE CPMT Symposium Japan (ICSJ), pp. 5-8, 2017.

[8] Fuhan Liu, Hirokazu Ito, Rui Zhang, Bartlet H DeProspo, Fabian Benthaus, Hisanori Akimaru, Kouichi Hasegawa, Venky Sundaram, and Rao R Tummala, "Low Cost Panel-Based 1-2 Micron RDL Technologies with Lower Resistance than Si BEOL for Large Packages," IEEE 68th Electronic Components and Technology Conference (ECTC), pp. 613-618, 2018.

[9] C. H. Yu, L. J. Yen, C. Y. Hsieh, J. S. Hsieh, Victor C. Y. Chang, C. H. Hsieh, C. S. Liu, C. T. Wang, KC Yee, and Doug C. H. Yu, "High Performance, High Density RDL for Advanced Packaging," IEEE 68th Electronic Components and Technology Conference (ECTC), pp. 587-593, 2018.

[10] Masaya Kawano, Chum-Mei Wang, Hong-Yu Li, Mian-Zhi Ding, Sharon Pei-Siang Lim, Teck-Guan Lim, Zi-Hao Chen, and Fa-Xing Che, "High Density TSV-Free Interposer (TFI) Packaging with Submicron Cu Damascene RDLs for Integration of CPU/GPU and HBM," IEEE 68th Electronic Components and Technology Conference (ECTC), pp. 1885-892, 2018.

[11] Po-Hao Chang, Chia-Yuan Hsieh, Chun-Wei Chang, Chih-Lun Chuang, and Chen-Feng Chiang, "Signal and Power Integrity Analysis of InFO Interconnect for Networking Application," IEEE 68th Electronic Components and Technology Conference (ECTC), pp. 1720- 725, 2018.

Sub-micron RDL patterning for Advanced Packaging

Ken-Ichiro MORI[1], Douglas SHELTON[2], Yoshio GOTO[1], Yasuo HASEGAWA[1], Seiya MIURA[1]

[1]Semiconductor Production Equipment PLM Center 1, Canon Inc., Utsunomiya, Japan
[2]Industrial Products Division, Canon U.S.A., Inc., San Jose, USA
mori.ken-ichiro@mail.canon

Abstract— **More than Moore strategies have been a hot topic for more than a decade, and Redistributed line (RDL) layers in next generation interposers will require sub-micron patterning. A key lithography challenge for fine-RDL in interposer applications is providing a sufficient Depth of Focus (DoF) to accurately resolve sub-micron features. To meet this demand, Canon's FPA-5520iV steppers can now provide new projection optics offering Numerical Aperture (NA) 0.24 imaging and 52 x34 mm exposure field. FPA-5520iV steppers with NA 0.24 provide excellent 0.8 μm resolution performance throughout all imaging fields.**

Keywords- Lithography, fine-RDL, inteconnection, Fan-out Wafer Level Packaging, FOWLP, Fan-out Panel Level Packaging, FOPLP, Silicon Interposer, Organic Interposer

I. INTRODUCTION

More Than Moore strategies have been a hot topic for more than a decade, starting with 3D integration using Through Silicon Vias (TSVs) and evolving into today's promising technology of Heterogeneous Integration using interposers and fine-RDL. Because advanced GPUs and FPGAs used in autonomous driving require wideband interconnection with memories, RDL layers in next generation interposers will require sub-micron patterning.

A key lithography challenge for fine-RDL in interposer applications is providing a sufficient DoF to accurately resolve sub-micron features. Front-End-of-Line (FEOL) lithography tools feature large NA optical systems that do not provide enough DoF to resolve sub-micron patterns over large interposer topography. In addition to DoF issues, FEOL lithography tools have no solutions for warped wafer handling which is a major challenge in packaging processes. On the other hand, traditional Back-End-of-Line (BEOL) lithography tools struggle to resolve very fine patterning due to their extremely low NA and leveling systems that can't reliably position wafers in tight DoF range during exposure. From this background, the demand for new advanced fine patterning BEOL stepper is growing.

Canon has a unique position in the lithography tool market. In addition to our many years' experience in the FEOL lithography tool business, we have enjoyed strong growth in the BEOL stepper market since we shipped the first BEOL exposure system (stepper), FPA-5510iV, in 2011. For seven years Canon has developed many solutions to improve productivity and yields, culminating in the 2016 release of the FPA-5520iV stepper. (Figure 1)

Because of this background, Canon has a responsibility to contribute to fine RDL interposer technology by developing a lithography tool that is optimized for the processes. Canon developed new projection optics offering NA 0.24 imaging and 52 x34 mm exposure field and Canon started shipping FPA-5520iV steppers with this new lens in the end of 2018.

In this paper, Canon reports a study of photography challenges to realization of sub-micron RDL for chip-to-chip wide bandwidth interconnections. In addition, Canon presents updates of optional systems available for the FPA-5520iV stepper that help enable advanced packaging processes.

Figure 1.   Figure 1. Canon FPA-5520iV i-line stepper

II. CHALLENGES OF HIGH-RESOLUTION FOR ADVANCED PACKAGING

Current advanced packaging processes mainly use design rules calling for 5 μm L/S RDL patterns that connect a chip's pads and ball bumps. However, 5 μm L/S RDL are not cost-effective for multi-chip interconnections because these processes require multiple interconnect layers for high-bandwidth, chip-to-chip connections.

RDL line width reduction is a key challenge to expand the advanced packaging market to multi-chip interconnections by FOWLP or interposers, including interconnections between SoC and DRAM, split die connection of FPGA, and interconnections between image sensors and SoC. Next generation advanced packaging requires 1.0 μm RDL and its future applications are targeting 0.8 μm RDL.

A. New projection optics for fine RDL

To meet these requirements, Canon has developed a new projection optical system featuring a high NA and wide-field that is best suited for sub-micron RDL. These new projection optics are a new option for FPA-5520iV steppers, offering NA 0.24 imaging and a 52 x 34 mm exposure field. A summary of FPA-5520iV stepper standard and optional specifications are presented in Table 1 including the optional 0.8 μm resolution option that requires the optional 0.24 NA.

FPA-5520iV steppers with NA 0.24 optics provide excellent 0.8 μm resolution performance throughout all imaging fields thanks to Canon's wave-front aberration based projection optics manufacturing methods, wave-front engineering, and on-axis optical tilt focus sensor.

To show the importance of NA with respect to imaging results, Canon conducted an examination comparing the associated DoF with NA values varying from 0.18 to 0.57 when imaging 0.8 μm features. Figure 2 shows that large NA values deliver the lowest ultimate resolution although for a 0.8 μm process the usable DoF is less than 5 μm for the highest NA examples. To provide a larger process window and maximize DoF, the Canon simulations show that NA 0.24 exposure conditions can deliver < 7 μm of DoF for 0.8 μm imaging which is suitable for fine RDL processes.

TABLE I.   FPA-5520iV I-LINE STEPPER SPECIFICATIONS

Wafer size	Φ300 mm
Resolution	≤ 1.5 μm / ≤ 0.8 μm (option)
NA	0.15 − 0.18 / 0.15 − 0.24 (option)
Reduction ratio	2:1
Exposure field	52 x 34 mm
Exposure wavelength	365 nm (i-line)
Single machine overlay accuracy	Front ≤ 0.15μm Backside ≤ 0.5μm (option)

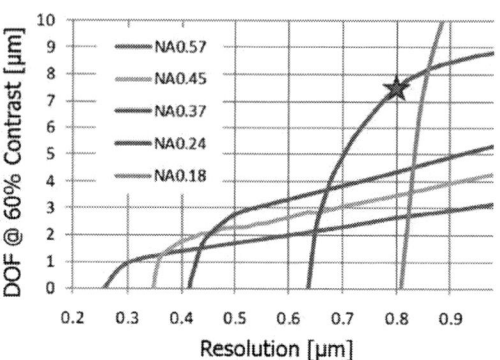

Figure 2.   NA 0.24 is an optimum condition for 0.8 μm imaging

B. Wave-front engineering in optics manufacturing

To achieve sub-micron resolution, precision technology is required to design and fabricate low aberration lenses and the simulation results in Figure 3 shows the relationship between resist image profiles and coma aberration.

The resist profile strongly depends on the lens aberration, especially in 0.8 μm patterning, indicating the importance of strictly managing the lens aberration for sub-micron resolution. Canon employs manufacturing technology developed over many years of FEOL experience in wave-front engineering and manufacturing to produce a stable, supply of low-aberration lenses.

Figure 4 explains a schematic diagram of Canon's Phase Measurement interferometer (PMi) lens aberration measurement system. The PMi system can quickly measure lens aberration by obtaining the interference image of the ideal light wave and the light wave passed through the lens. The PMi enables high precision lens aberration analysis and high quality lens assembly to realize and realizes stable low-aberration lens manufacturing.

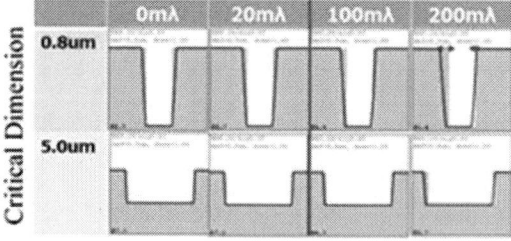

Figure 3.   Simulation result of impact of Coma aberration

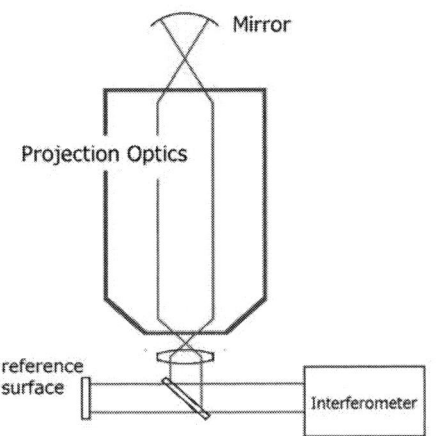

Figure 4. Schematic diagram of Phase Measurement interferometer

C. Die-by-die tilt and focus compensation

In FOWLP, an RDL layer is formed on wafers that are warped largely due to the molding process. Wafer warpage reduces focus margin and reduction of focus margin due to warpage is a big issue in submicron resolution since DOF is very small. FPA-5520iV steppers are equipped with die by die tilt & focus compensation systems as a solution for this issue.

Figure 5 is a schematic diagram of the tilt focus system. Measurement light is projected obliquely incident from the side of the lens to each shot and the reflected light from the wafer is detected by sensor. Focus is calculated from the position information of the reflected light.

The real-time focus system provides very accurate focus compensation as the focus position within each shot can be measured immediately before exposure. This system also does not need to measure wafer topography in advance, so high productivity is achieved. This system benefits from the Canon dioptric lens design that allows a large gap between the projection lens and wafer (> 20 mm), which is difficult to realize with mirror optical systems that are common in BEOL exposure applications.

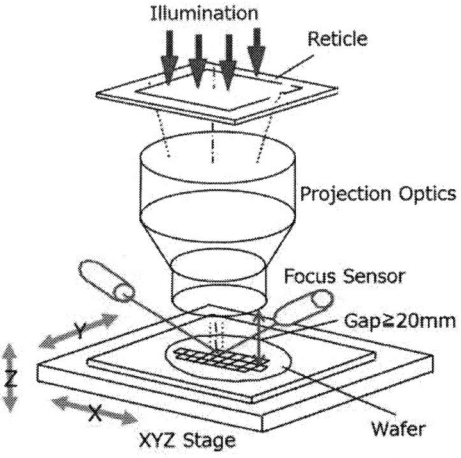

Figure 5. Schematic diagram of the tilt focus system

D. Photo resist patterning study

Canon has collaborated with resist companies and materials makers to study 0.8 µm resolution performance and Figure 6, Figure 7 and Figure 8 demonstrate the high-resolution imaging performance of the FPA-5520iV steppers.

Figure 6 displays a plot of printed feature sizes (Critical Dimension, CD) vs. Focus position for 0.8 µm L/S patterns. Using 1.095 µm thick PFi-38 A7 i-line photoresist, a DoF evaluation was conducted examining 15 images heights across a 52 x 34 mm field. This result showed that NA 0.24 was suited for 0.8 µm L/S imaging, with an NA 0.24 condition yielding a DoF of about 8 µm for 0.8 µm L/S patterns that should provide sufficient focus margin for fine RDL processes.

Figure 7 displays resist profile cross sections for 0.8 µm L/S patterns with NA0.24 & NA0.18. Using 1.48 µm thick TDMR®-AR1100 LB (TOKYO OHKA KOGYO CO.LTD.) i-line photoresist, NA0.24 condition provides a larger profile DoF than NA0.18 condition for 0.8 µm L/S patterns. This resist profile cross-section result showed a DoF of about 8 µm for 0.8 µm L/S pattern, and NA0.24 is more preferable than NA 0.18 for 0.8 µm L/S patterns.

Figure 8 displays resist profile cross-section for 0.8 µm L/S patterns on a Cu sputter wafer with NA0.24. Using 2.5 µm thick JSR i-line photoresist, NA0.24 condition provides good resist profiles on Cu sputter wafers.

Another advantage of the new projection optic system is the large 52 x 34 wide imaging field. The large-field and high-resolution imaging benefits of the FPA-5520iV can help speed the development of large size FOWLP and interposers manufacturing and performance improvements of FPGA and GPU devices.

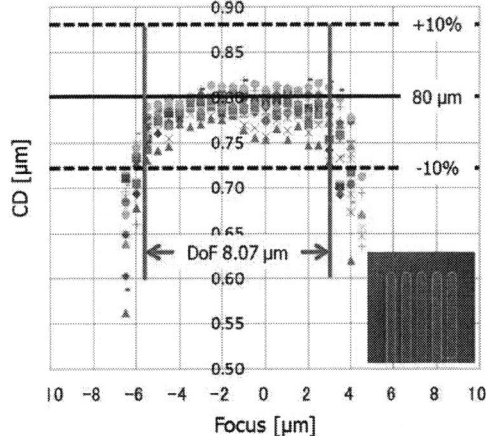

Figure 6. DoF evaluation for 0.8µm Line & Space pattern

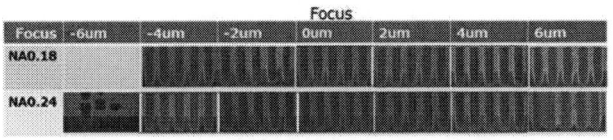

Figure 7. NA 0.24 & NA0.18 resolution comparison for 0.8 µm Line/Space imaging

978-1-7281-1500-9/19 $31.00 © 2019 IEEE

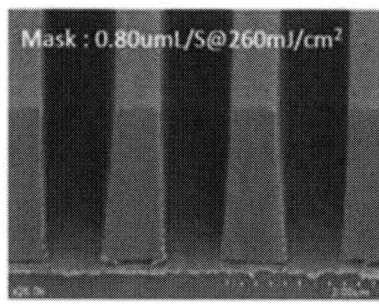

Figure 8. NA 0.24 resolution testing profile for 0.8 μm Line/Space imaging on Cu sputter wafers (2.5 μm thick resist)

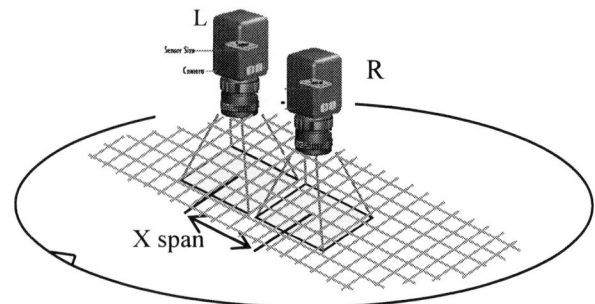

Figure 9. Schematic diagram of Grid PA

III. UPDATE OF OPTIONAL SYSTEMS FOR FPA-5520iV STEPPERS SUPPORTING FOWLP

A. Warped wafer handling system

Warped wafer handling which is a key challenge for BEOL exposure tools and the FPA-5520iV is equipped with a new wafer handling system that offers a significant improvement over first generation FPA-5510iV steppers that were capable of handling up to 500 μm of wafer warpage. The latest FPA-5520iV steppers are designed to handle more than 5 mm of wafer warpage in order to process FOWLP reconstituted wafers that typically exhibit more warpage than silicon wafers.

B. Die rotation detection and compensation system

In the chip first process, die shift and die rotation occur in the molding process. In addition, since the wafer notch as a reference in alignment is generally formed after molding, the positional accuracy of notch is poor. In a standard wafer handling sequence, steppers will execute a mechanical pre-alignment step which references a wafer notch. However, Fan-out wafers with poor notch accuracy are difficult to align to which can lower productivity.

To solve this problem, FPA-5520iV steppers adopt a Grid Pre-alignment (Grid PA) function that aligns not only to a wafer notch, but also to the wafer chip image reference. An illustration of the Grid PA alignment system is provided in Figure 9.

During the Grid PA sequence, a wafer image is acquired with cameras arranged symmetrically along the X axis with respect to the wafer. Left (L) and right (R) measurements of grid Y position are performed and the differences between the L/R Y values are calculated relative to the distance between the cameras, Grid PA estimates and compensates the globally for chip rotation relative to the notch reference. With this function it is possible to expose a fan-out wafer where die-rotation has occurred without lowering productivity.

C. Through-Silicon Alignment

The FPA-5520iV steppers can also be equipped with an optional Through-Silicon Alignment (TSA) System that can uses infrared light to view through full-thickness silicon wafers to measure the position of alignment marks on the backside of the wafer. The TSA system provides the FPA-5520iV with a selection of infrared wavelength bands and center wavelengths that may be optimized to increase alignment accuracy. Principles of the TSA system are illustrated in Figure 10.

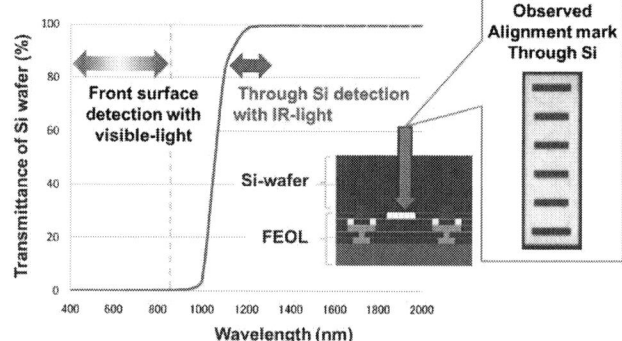

Figure 10. Principles of Though Si Alignment

D. Panel handling (Under study)

While maintaining 0.8 µm pattern fidelity across a large exposure area is a key requirement of fine RDL processes, Fan-Out adopters seek to transition from wafer based to panel based processes to improve productivity and overall costs [1]. To support high-resolution FOPLP processes, Canon has studied a panel handling system for FPA-5520iV steppers.

Although FPA-5520iV steppers cannot directly process Gen 4.5 panels that can be up to 730 x 920 mm in size, we developed a substrate handling system for R&D purpose to handle and process panels sizes up to 365 x 306.7 mm or approximately 1/6 of a Gen 4.5 panel. FPA-5520iV steppers can provide 0.8 µm resolution on 365 x 306.7 mm rectangular substrates as illustrated in Figure 11, providing unique advantages supporting fine RDL & FOPLP processes. Canon is willing to do collaborative works with panel market players. Details of the allowable substrate range for the FPA-5520iV panel handling system are included in Table 2.

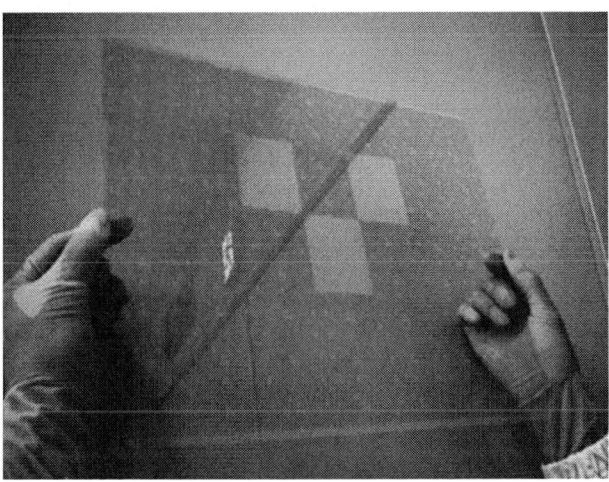

Figure 11. Example of panel exposure

TABLE II.   SUBSTRATE SPECIFICATION FOR PANEL HANDLING SYSTEM

Item	Specification
Material	Glass
Maximum panel size	X 306.7 mm / Y 365.0 mm
Thickness	0.5 mm
Warpage	≤ 1.0 mm

IV.   CONCLUSION

To meet the demands of sub-micron RDL and advanced packaging applications, Canon developed the FPA-5520iV stepper that can be equipped to deliver 0.8 µm resolution on 300 mm wafers. In addition, Canon developed many original options for the packaging process. These could be realized by applying Canon's advanced FEOL and precision optical technology.

Canon keeps meeting the challenging requirements for advanced packaging with Canon's extensive technology of exposure systems including the FPA-5520iV stepper. Canon remains committed to continued innovation enabling current and future FOWLP and interposer processes.

REFERENCES

[1]   C. Chu, R. Rice, "FOPLP Standardization Survey Results", SEMI Standards, San Jose, CA, Nov. 2017

[2]   Shin-Ichiro Hirai, Nobuyuki SAITO, Yoshio GOTO, Hiromi SUDA, Ken-Ichiro MORI, Seiya MIURA "A study of vertical lithography for high-density 3D structures" Proc. SPIE 8326, Optical Microlithography XXV, 83261E (13 March 2012)

[3]   Masaki MIZUTANI, Shin-Ichiro HIRAI, Ichiro KOIZUMI, Ken-Ichiro MORI, Seiya MIURA "A STUDY OF VERTICAL LITHOGRAPHY FOR HIGH-DENSITY 3D STRUCTURES" Proc. SPIE 8683, Optical Microlithography XXVI, 86831F (12 April 2013)

[4]   Yoshio GOTO, Kosuke URUSHIHARA, Bunsuke TAKESHITA, Ken-Ichiro MORI "A study of Sub-micron Fan-out Wafer Level Packaging solutions", THA2-103, iMAPS 2018 (11 Occtober 2018), unpublished.

Optimization of electrolytic plating processes for challenging fan-out panel level package designs

Ralph Zoberbier[1], Britta Scheller[1], Dr. Christian Ohde[2]

[1]Atotech Deutschland GmbH, Feucht, Germany
Equipment Electronics Sales and Marketing
ralph.zoberbier@atotech.com

[2]Atotech Deutschland GmbH, Berlin, Germany
Global Product Team Panel/Pattern Plating
christian.ohde@atotech.com

Abstract — The ever increasing demand for higher performance, lower cost and thinner end user devices like smartphones require intense developments and innovation in all areas of the electronic component design including the substrate and device packaging. While on one hand design rules increase by complexity, manufacturing technologies are challenged to support both technical requirements and cost targets.

Fan-out wafer level packaging is still considered a key enabler to support the technology roadmaps. This packaging platform had been introduced to the industry years ago and became a widely adopted packaging technology with various platforms now available in the market by key players of the industry. With the basic concept of creating an artificial substrate the traditional limitation of the available maximum wafer size becomes obsolete and the introduction of larger scale substrate in panel format seems straight forward, mainly driven by cost reduction purposes and system in package SiP applications which would benefit from the larger area. Batch processing of larger substrates promise to reduce manufacturing costs significantly by concept. One can think that panel based processing in areas like PVD, lithography and electroplating is a standard manufacturing technology, established in display and PCB applications and could be easily adopted. However the transfer of the traditional manufacturing methods require a high level of modifications in materials, processes and equipment solutions to become compatible with latest wafer level packaging applications and requirements.

Electrolytic metal deposition is a key process step in the manufacturing of vertical and horizontal interconnections used in today's PCBs and IC substrates on one hand and advanced packaging applications on the other hand. Historically both application areas were clearly defined and separated by different requirements in feature sizes and substrate formats, supported by tailored equipment and process solutions. PCBs and IC substrates is typically based on organic large scale substrates with rather large features while advanced packaging technology is wafer based with the capability to incorporate fine features down to a few microns.

Both areas currently merge and define a new application segment. This segment combines the request of small feature sizes with the manufacturability on large scale substrates. In this respect, the specific challenges in the area of electrolytic metal deposition is the creation of feature sizes down to 2μm L/S with heterogeneous feature density on large substrates up to 600mm at excellent metal thickness uniformity and high plating speed.

The paper discusses latest studies and conclusions in critical performance areas of the plating process such as electrolyte fluid dynamics, impact of anode design and pulse reverse rectification. An overview and comparison of typical plating systems will be presented. Finally, latest test results of optimized process conditions will be discussed in detail with different feature sizes providing data of within die and within substrate uniformity. All tests are done either on panel or coupon level.

The latest developments and findings of the discussed panel based plating technology will support the industry to transfer fan-out process technology from wafer to panel resulting in lower manufacturing costs.

Keywords- FanOut, IC substrates, plating, pattern plating, panel based packaging, electro-chemical deposition, ECD

I. INTRODUCTION

Fan-out packaging was introduced to the market many years ago as a key enabler in the packaging industry to provide important technology advantages such as smaller footprint, thinner packages, lower thermal resistance, no need for a laminate substrate, less restrictions in bumps size and shorter connections. With its introduction, fan-out technology was based on newly reconstituted wafers and was mainly used for communication devices. Over the years the industry introduced the technology to a growing number of applications like baseband processors, advanced application processors, power management, radar modules and audio codec. With these adoptions newly developed features like through mold via and multiple RDL layers were required to ready the technology for the constantly increased complexity.

Fan-out packaging follows generally two manufacturing approaches, such as chip-first or chip-last which defines the sequence of the manufacturing the RDL traces either before or after chip assembly [1]. In both options the technology is based on a reconstituted or carrier substrate. As the technology is not limited to the actual wafer size any kind of size or shape of the substrate is theoretically possible. The means of using a larger substrate size than a 300mm wafer promises to provide significant cost reductions due to economies of scale and is driving the industry to move from wafer level to panel level packing. The industry targets approx. 30% of cost savings when transferring from wafer scale to a panel scale of a size of 600x600mm.

Currently first panel packaging manufacturing lines are set up, under qualification or already in pilot production with panel sizes like 600x600, 510x515, 470x370mm. Multiple industry consortia have been kicked-off in Germany, Singapore, Taiwan, China and Japan to develop a cost effective packaging process and to address the challenges that are critical to achieve a reasonable yield level on panel processing.

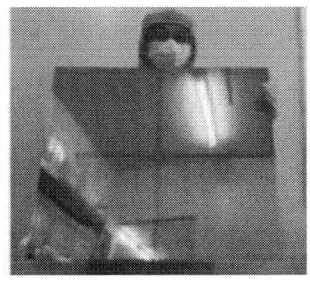

Figure 1.   Picture of 600x600mm panel, plated on Atotech MultiPlate P600 system, Atotech Tech Center Taiwan

Electrolytic Cu deposition is a key process step in the manufacturing of RDL traces and vertical interconnections like bumps. New developments and innovative process optimizations is required to fulfil the technical specifications and cost targets of the panel level packing process. Key elements and latest research results will be presented and discussed in the following sections.

II.   TECHNOLOGY CHALLENGES FOR ELECTROLYTIC METAL DEPOSITION

The main purpose of an electro-plating process is the deposition of metals such as Cu, Ni or SnAg to build horizontal or vertical interconnections in a device or package. During the process, the metals are deposited either evenly across the full surface or into a mold of photo resist or dielectric materials which is built on a metal seed layer and patterned by precedent photo lithography processes. Typically the substrate acts as cathode, electrically connected at the edge of a substrate in a sealed dry area.

Today's fan-out packaging features contain a wide variation of size and shape. A combination of Cu RDL traces with typical L/S of 8-10µm are combined with large pad structures that can easily be in a range of 250-350µm, with target Cu thicknesses of 10-20µm. While package density increased over time, the requirement for 2µm L/S becomes more and more important to avoid multiple RDL layers

which would result in a significant cost increase. The key challenge for the plating process is to provide a very uniform Cu deposition across the full substrate featuring such structure variations.

In addition micro vias are used for vertical interconnections from one RDL layer to the next. While conformal plating was used in the past, micro-via filling processes are used today which requires a larger amount of metal deposition in the via compared to the horizontal interconnections like RDLs at relatively low metal deposition on the surface. Different feature sizes and shape have a significant impact on the overall plating uniformity and process fine tuning and optimization is needed to achieve best possible results. Of course, a large substrate size is an additional challenge as anode technology and fluid delivery are becoming more complex in panel scales up to 650mm compared to a round 200 or 300mm wafer. Typical features of a 2 layer RDL stack are illustrated in figure 2.

Figure 2.   Schematic illustration of Cu plated features in a dual RDL package design, including µVias, RDL traces and Cu pads

The transition to panel scale packaging is mainly driven by the cost reduction potential to process a larger area at a time. This becomes even more important when package sizes will increase or when multi-chip modules are being built like in SiP applications. The applicable plating speed is the key process parameter that will define the overall throughput and cost of ownership of the process. It is clearly desired to design chemistry and equipment solutions that result in highest possible current densities which are still capable to meet the technical requirements and to reduce process time as best as possible. Plating speed is considered an important factor especially when very thick Cu deposition is needed like for tall Cu pillar applications in PoP applications.

Figure 3.   PTI FO- Package example with tall Cu pillar, RDL and µvia structures [3]

Finally the complexity and challenge of the plating process for PLP is the development of chemistry, equipment and process solutions that deliver a wafer like plating performance on large scale substrates, with a big variation of feature sizes and shapes at highest possible plating speeds.

978-1-7281-1500-9/19 $31.00 © 2019 IEEE

III. Process Optimization for a panel Based Cu plating process

The optimization of a plating process requires a key set of technology features that are important to achieve good uniformity levels at high plating speeds. Tailored electrolyte systems with equipment supported features result in excellent plating performance and speed, even on very large substrates as targeted for PLP applications. The selection of the plating technology defines advantages or limitations in the process, costs and need to be considered wisely. Latest research and investigations show the importance of pulse reverse rectification and anode segmentation as key enabler and will be discussed in detail.

A. Overview and comparison of typical plating methods

In the area of copper plating there are 3 techniques known in the market which are Cu plating with soluble anodes, with insoluble anodes and with insoluble anodes coupled with a Fe redox system. The main components of the electrolyte, which is needed as the basis of copper plating is quite similar for all three copper plating techniques. The main components are sulfuric acid H_2SO_4, copper sulfate $CuSO_4$, Chloride Cl^- and the organic additives. $FeSO_4$ is additionally added into the system in case insoluble anodes coupled with a Fe redox system is applied, which gives many advantages that are explained in a later section.

1) Cu plating with soluble anodes

Copper plating in mass production started with the use of soluble anodes. In theory this approach is quite straight forward as all the copper, which is plated at the cathode, is dissolved at the same time and amount at the anode:

Anode: $Cu^0 \rightarrow Cu^{2+} + 2\ e^-$
Cathode: $Cu^{2+} + 2\ e^- \rightarrow Cu^0$

The technology is mainly used in lower end applications due to several reasons. The anodes need to be replaced constantly as Cu dissolves during the plating process over time. In addition surface conditions change during the dissolving process which builds a $Cu^{2+}/Cu^+/$organic additives coating that is not stable during idle times and risks to create particles and which impact additive concentrations.

2) Cu plating with inert anodes and CuO

Inert anodes are also called MMO – mixed metal oxide or DSA – dimensional stable anodes. They do not change their dimensions during plating (DSA) as there is no dissolution of the anode itself (inert anode). This supports low maintenance of the anodes itself which is important to meet long term reliability of the process. As the anodic oxidation reaction is still required during plating, but copper cannot be dissolved at the anode a different reaction takes place. Water is oxidized, which is indicated below:

Anode: $2\ H_2O \rightarrow 4\ H^+ + 2\ e^- + O_2$
Cathode: $Cu^{2+} + 2\ e^- \rightarrow Cu^0$

The composition of the electrolyte is constantly changing as Cu is deposited at the cathode and protons are created at the anode. A Cu replenishment is needed that increases the Cu concentration and consumes the newly created protons. Cu replenishment can be either performed by copper oxide or copper carbonate.

Cu-Replenishment: $CuO + 2\ H^+ \rightarrow H_2O + Cu^{2+}$

In addition the oxygen evolution at the anodes requires a relatively high potential compared to the copper dissolution (1.0 V vs. Ag/AgCl). Furthermore oxygen radicals are formed prior the oxygen evolution. Both facts require a separation between anode and cathode. Costly membranes are used for this separation.

3) Inert anodes with Fe redox system

The iron redox system combines the advantages of the previously mentioned plating methods but addresses the mentioned limitations. In order to keep the anode potential low during the plating process, an alternative reaction to oxygen evolution has to take place. For this reason ferrous ions are added to the electrolyte to completely draw the electrons needed for the anode reaction from the oxidation of Fe^{2+} to Fe^{3+}. Due to the normal potential of that reaction (+ 0.55 V vs. Ag/AgCl) the anode potentials remain very low resulting in low consumption of organics.

Anode: $2\ Fe^{2+} \rightarrow 2\ Fe^{3+} + 2\ e^-$
Cathode: $Cu^{2+} + 2\ e^- \rightarrow Cu^0$
Cu-Replenishment: $2\ Fe^{3+} + Cu^0 \rightarrow Cu^{2+} + 2\ Fe^{2+}$

With the technology illustrated in figure 4, anode sludge as well as oxygen formation is prevented thus no segmentation or membrane technology is needed. This fact enables the use of a high flow spray systems into the anode design which is a key element to optimize the fluid dynamics and to achieve optimized copper plating distribution.

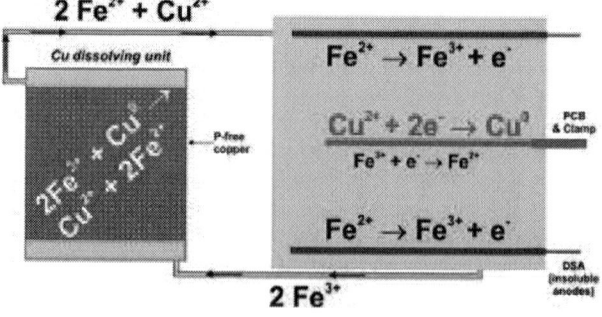

Figure 4.   Illustration of a Fe-Redox plating system

During the chemistry development of the iron containing electrolyte it has been seen that a 3 additive system is needed to fulfill the challenging requirements of latest PLP requirements for fine line applications, while a 2 additive system was developed to provide highest current densities up to 20ASD/cm² for thick Cu plating applications like tall Cu pillar. The Fe redox system is a well-established plating technology that also provides the possibility of pulse reverse rectification which has been identified as important parameter for process optimization. In conclusion, the Fe redox systems was identified to be a promising choice to develop a reliable and high performance PLP plating process.

B. Reverse pulse rectification

Typical designs of fan-out packages include a variation of different feature sizes like RDL traces and pad structures as shown in figure 5.

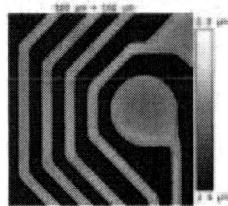 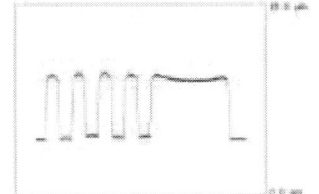

Figure 5. Cross section of plated pads with and without reverse pulse rectification

A very uniform Cu thickness distribution is very challenging for the plating process and requires specific technologies like reverse pulse rectification. This method allows a free programmable setting of pulse parameters which is not possible in DC plating. During a set of experiments it was found that this feature can be used to align Cu thickness deposition especially when different feature sizes (in that case of a 15 μm line & a 150 μm pad) are incorporated in the layout as illustrated in Figure 6. For better visibility the pad height in the following diagram is normalized to 100 % and the line height is shown in relation to that.

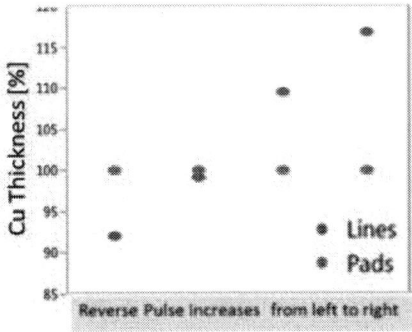

Figure 6. Cross section of plated pads with and without reverse pulse rectification.

The diagram shows that with increased reverse pulse rectification Cu thickness of lines increase while the Cu thickness of the plated pads remain.

In addition it was found that reverse pulse rectification can be applied to shape lines or pads. Typically very straight sidewalls and a flat top is desired. Especially when high current densities shall be applied to reduce process times, reverse pulse rectification can be used to adjust the top side of the features. A general impact on feature shape optimization is illustrated in figure 7.

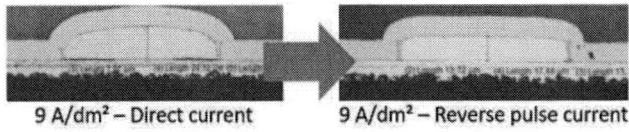

Figure 7. Cross section of plated pads with and without reverse pulse rectification

C. Anode segmentation

When moving from round shaped wafer formats to large rectangular substrates that easily can exceed a size of 600x600mm, anode design and technology becomes even more important to achieve the required distribution uniformity. In general, when applying voltage to anode and cathode, the voltage difference will be the largest at anode or cathode edges. This will result in higher currents in these areas and higher Cu deposition as illustrated in figure 8.

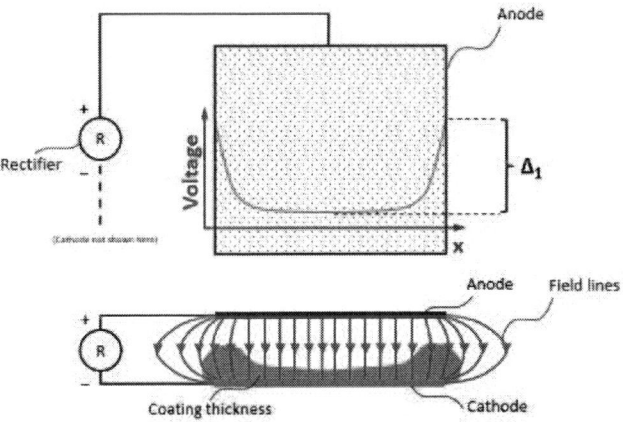

Figure 8. Illustration of voltage difference and field lines for square substrates in panel plating applications

Segmented anode solutions are applied to resolve this effect. Individual rectifier control of each of the multiple segments offers an important degree of freedom and parameter to optimize the overall plating process.

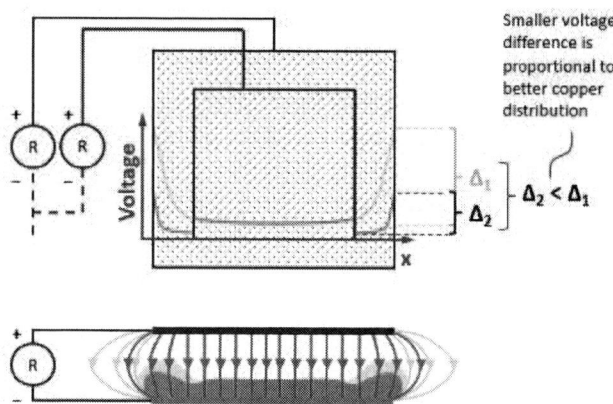

Figure 9. Illustration of voltage difference and field lines for square substrates with segmented anodes

D. Fluid delivery system

As Fe redox systems have no need for anode/cathode separation, which typically would result in a membrane or diaphragm, a direct through-anode fluid pressure system can be used. The system is based on hundreds of jets that are incorporated into the anode system itself and has been optimized in terms of nozzle diameter, nozzle shape and locations across the area. The system provides a very uniform fluid flow that is important to achieve very good uniformity distribution across a large area (figure 10). A physical cathode agitation with unlimited movement paths is an additional process parameter to adjust the fluid flow pattern and to optimize the overall distribution performance.

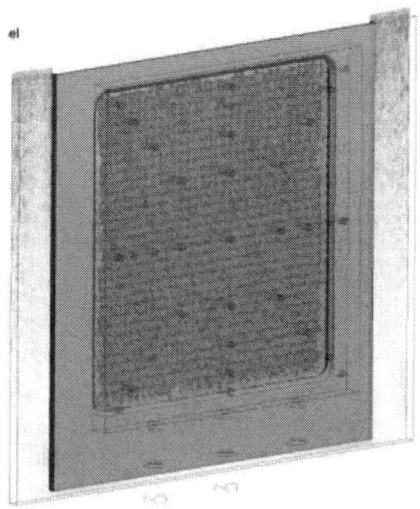

Figure 10. Illustration of fluid velocity at panel level based on a tailored fluid pressure system

IV. PROCESS OPTIMIZATION AND RESULTS

A prerequisite for achieving the challenging industry plating targets for a panel level packaging design with within panel uniformity/distribution (WIPD) below 10% is an optimization on blanket panels. This performance can be considered as best case scenario as fan out designs will further impact the overall distribution results and need to be compensated with additional fine tuning of the process with the mentioned process parameters or equipment features. The process flow for the optimization of panel plating distribution is shown in figure 11. Panel sizes of 370x470mm and 600x600mm was used for process development and final optimization, following current industry requests.

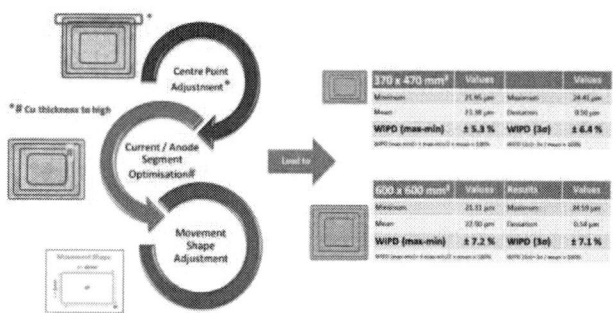

Figure 11. WIPD optimization procedure in panel plating

First the centre points of the substrate to the actual plating window of the anodes need to be adjusted. The second step is the current adjustments of the segmented anodes that are needed to fine tune the overall plating distribution. On top of that, the substrate agitation profile is important to optimize the plating uniformity.

Following this procedure it was possible to reduce the WIPD for formats up to 600 × 600 mm² close to ±7 % and further optimization is ongoing. These measurements were done by automated 4 point probe resistivity measurements and started 5 mm form the sealing with a grid of 1 cm (e.g. over 3,000 measurement points for 600 × 600 mm² panels).

After the optimization in panel plating mode is completed the With-In-Unit-Distribution needs to be improved, which means to equalize the plating height of different features in the same unit. Normally the WIUD can hardly be changed, the main knobs for this optimization are the chemical parameters and the current density. As mentioned before, Reverse Pulse Plating, not possible in DC plating systems, has been identified as an additional, very important process parameter to adjust feature shapes and Cu thicknesses especially in designs with various feature sizes and shapes. The final distribution result of a fan out design is highly design dependent and requires iterative process optimization.

Process optimization was performed on a 370x470mm panel for a RDL first type fan-out process with feature L/S of 8/4µm and 8/100µm at Cu target thickness of 7µm. 25 units fully spread over the entire substrate were evaluated and resulted in a WIPD of <5% after following the optimization process.

Latest plating tests on 600x600mm panels in a typical fan-out pattern with Cu target thickness of 8μm resulted in 9.5% uniformity distribution. Details are shown in figure 12.

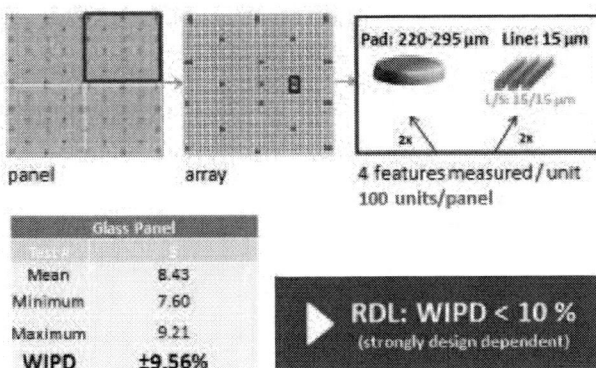

Glass Panel	
Mean	8.43
Minimum	7.60
Maximum	9.21
WIPD	**±9,56%**

> RDL: WIPD < 10 %
> (strongly design dependent)

Figure 12. Process results of a typical fan-out package design heterogenous feature sizes with 8.5μm target Cu thickness on 600x600mm panels

The process optimization has been done for the last 12 months on various panel formats (250 × 250 mm²; 370 × 470 mm²; 510 × 415 mm²; 510 × 515 mm² & 600 × 600 mm²) and could achieve WIPD results of <10% in any case.

In addition to RDL process optimization, tall pillar plating is under development and research. The industry target for this application is to plate tall pillars with the same speed as it was demonstrated in wafer level applications. Especially for these high thicknesses plating speed is the major factor that drives throughput of the equipment and therefore total cost of ownership.

The applicable plating speed depends strongly on pillar dimension and aspect ratio. For aspect ratios of 1:1 industry targets to apply larger than 20 ASD/cm², which corresponds to a plating rate of 4.5μm/min. Process time to create pillar sizes of 200μm would drop to less than 45 min.

First results indicate that this target is realistic. The results shown below were achieved in coupon tests, where each coupon had 9 units, which were individually investigated. The total amount of 30 tall pillars was measured by LEXT, which is an optical measurement tool to determine the pillar height and shape.

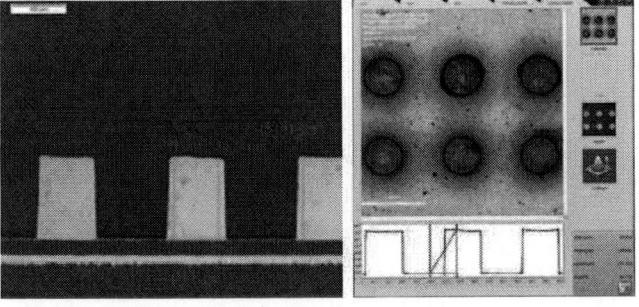

Figure 13. Plating results of 150 μm Cu pillars, cross sections left and LEXT measurement right

– Pillar height	150 μm	
– Pillar diameter	120 μm	
– Dry film height	168 μm	
– **Effective CD**	**>20 A/dm²**	

> Pillar plated rectangular
> Pillar height achieved: **155 μm ± 5 μm**

○ = minimum measurement areas with 6 pillars each

→ Minimum measurement of 30 pillars per coupon

Figure 14. Initial test results of Cu pillar plating optimization on coupon level

Initial coupon tests resulted in a WICD of +/-3%. Further investigations and process developments take place on panel level. First tests show feasibility of WIPD below 10%, detailed analysis and results will be presented in a separate paper.

V. CONCLUSION

Electro-plating has to be considered as a critical process step to establish panel level packaging as the next step in the evolution for fan-out packaging technology. On the one hand, ECD is a process that can support the cost reduction targets through economies of scale when moving from wafer to large panel scale processing. However it needs special attention in the selection of plating technology, its chemistry and requires specific process optimization procedures and features. Different plating technologies were reviewed in detailed and critical process parameters such as revere pulse rectification were evaluated and presented.

Finally latest plating process results of typical fan-out designs on 370x470mm and 600x600mm were demonstrated. The combination of a three additive Fe redox electrolyte with segmented anode technology and direct fluid control achieves a distribution uniformity close to 7% in panel and below 10% in pattern plating on large scale substrates. First results of high speed Cu pillar plating on coupon level was presented. Additional panel scale tests for Cu pillar applications are ongoing.

VI. ACKNOWLEDGEMENT

The authors want to thank our business partners for providing FO-PLP substrates that allowed in depth studies and process tests. The authors also thank Holger Schulz, Ray Weinhold, Bernd Schmitt, Thomas Reike, Torsten Küssner, Bert Lin, Oden Hsu, Bobby Chen, Bruce Lin, additional Atotech colleagues all around the globe and Sandra Niemann, who should recover soon, for supporting application tests and research work that was included in the paper.

REFERENCES

[1] Jinyoung Kim et. al., Samsung Electro-Mechanics Co., Ltd., "Fan-out Panel Level Package with Fine Pitch Pattern," in 2018 IEEE 68th Electronic Components and Technology Conference

Article in a journal:

[2] C. Melvin et. al., Atotech Deutschland GmbH, "Fan-out packaging: A key enabler for optimal performance in mobile devices" Chip Scale Review, January 2017, 40-43.

[3] Powertech Inc., David Fang, "Panel Level Fan-Out Moves Into HVM", SEMI presentation 2018

3D Printed Substrates for the Design of Compact RF Systems.

Mohd Ifwat Mohd Ghazalil[1,2], Saikat Mondal[2], Saranraj Karuppuswami[2], and Premjeet Chahal[2]

[1]*Faculty of Engineering and Built Environment, University Sains Islam Malaysia (USIM), Nilai, Negeri Sembilan, Malaysia*
[2]*Department of Electrical and Computer Engineering, Michigan State University, East Lansing, MI, USA*
chahal@msu.edu

Abstract—**In this paper, Additive Manufacturing (AM) using 3D printing has been shown as a potential candidate for realizing customized compact solutions for RF packaging applications. Cost effective 3D printing based packaging solutions with customized substrates and air gaps allow easier integration of multiple RF components with lower substrate losses. Using a damascene-like conductor patterning process and a LEGO-like assembly process, an amplifier coupled to an air-substrate based patch antenna is demonstrated in a single integrated package.The antenna overlays the amplifier circuit leading to a compact design. The proposed customization of substrates and 3D printing strategies can be extended to multiple-system level stacking for SOP/SIP packaging customized for applications such as 5G.**

Keywords-**3D printing, 5G, Compact systems, Electronics packaging.**

I. INTRODUCTION

A growing demand for hand-held devices with many functionalities has led to the development of compact systems through novel packaging solutions. A common approach for high density packaging involves stacking multiple layers of electronic components such as integrated chips in case of System-in-Packages (SiP) or active and passive components in case of System-on-Packages (SoP) [1]. These techniques provide electrical connectivity between multiple layers through conductive vias that reduces the overall form factor of the package and at the same time, improves the system's performance through the use of shorter interconnects which have lower loss due to reduced parasitics [2]. SiP and SoP techniques for RF systems have complex circuitry with multiple components integrated together on a single substrate and with multiple substrates stacked inside a single enclosure. In order to realize low loss RF SoP/SiP structure, a number of techniques are available in literature and are popularly used such as utilizing low loss dielectric substrates, fabricating on air-cores or integrating embedded air cavities along with the RF circuitry [3]. The existing SiP and SoP packaging techniques are robust but also face many challenges in terms of customization of substrate heights for multiple pre-packaged components, complexity in realizing good electrical connectivity using vias, and poor structural stability in realizing an air based substrate. To overcome these limitations, additive manufacturing by 3D printing can be adapted as an alternative fabrication

technique for RF electronics packaging. 3D printing allows seamless integration of multiple RF components without any shape or size restrictions. 3D printing allows rapid customization of complex structures due to the inherent freedomto print in all directions. With recent advances in 3D printing technology, multi-material printing can be employed to realize cost-effective RF packaging solutions [4].

3D printing is preferred to traditional micromachining in realizing multi-layer structures due to the advantages such as low cost, light-weight, generates less waste, and the ability to revise a design quickly and manufacture on demand [5]. Recently, 3D printing has been employed for fabricating a plethora of planar and non-planar RF and microwave components such as antennas, wave guides, resonators, and microstrip circuits [6]–[8]. In [9], a damascene-like metal patterning process was introduced for patterning conductive traces onto the 3D printed non-planar structures. In [7], a LEGO-like assembly process was introduced for realizing air substrate based low loss RF components. 3D printing has been employed to embed different active and passive components using a solder-free technique in [10]. Ability to customize the design along the z-axis allows fabrication of non-planar components for miniaturization of the structure. Moreover, utilizing 3D printing minimizes the misalignment error between different stacked layers due to the availability of high resolution 3D printers. Typically, sensitive RF components are shielded using metallic enclosures to prevent signal interference. By employing 3D printing, customized enclosures can be fabricated with air substrate based components printed directly on the surface of the enclosures for further miniaturization of the packaged system. Figure 1 shows an example of the envisioned fully 3D printed customizable packaging enclosure with multiple components such as an air substrate based antenna and many active and passive devices for 5G applications.

In this paper, a compact 3D printed module consisting of an amplifier coupled to an air substrate based patch antenna and associated passive components is demonstrated. The antenna partially overlays the amplifier circuit leading to the realization of a compact design. The process demonstrated in this paper, in which components are integrated in the vertical direction, can be adapted for miniaturization and self-enclosure of RF modules similar to existing SiP or SoP

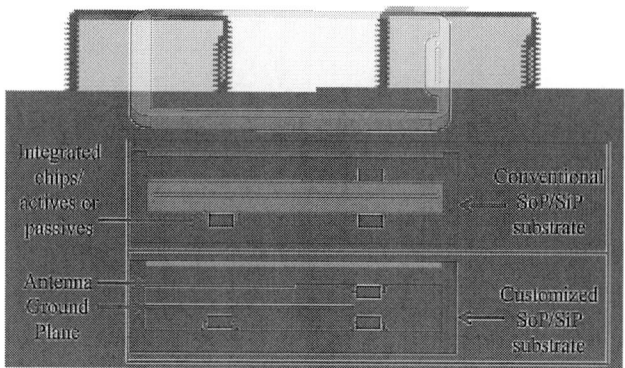

Figure 1. An example of the envisoned 3D printed electronic packages with customizable substrate for 5G applications.

packaging solutions.

II. DESIGN AND FABRICATION

The performance of the individual sub-modules are analyzed first followed by the integrated module. All the components are printed using a commerically available 3D-printer (Object Connex350) that uses a photopolymer resin VeroWhitePlus® with dielectric constnt ($\epsilon_r \approx 2.8$) and loss tangent (tan$\delta \approx 0.04$) [7]. The 3D printed structures are metallized first with a blanket sputtering of a thin seed layer of Titanium (Ti - 500 nm) and Copper (Cu - 5000 nm) followed by electroplating of Cu to achieve an additional Cu thickness of 5-6 μm. The metal layer is patterned using a damascene-like process as explained in [9]. The areas on the structure where metal is to be retained are printed 200 μm lower than the rest and the unwanted metal on the elevated areas is removed using a mechanical polishing process. Figure 2 shows the flow of a damascene-like process.

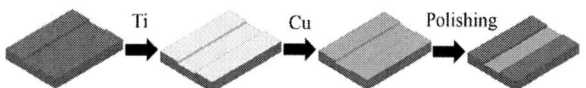

Figure 2. Patterning conductive traces on 3D printed substrate using a damascene-like process.

A. Air-Substrate based Patch Antenna Sub-Module

A simple patch antenna operating at 5.4 GHz is chosen for integrating with an amplifier to demonstrate the RF module. The antenna uses air as a substrate and is connected to the amplifier circuitry using a conductive post. The post is connected to the top and bottom layers using conductive silver paste. 3D printing allows fabricating using air as a substrate by printing the antenna in two separate parts (top and bottom layers) with support pillars on one side and corresponding holes on the other side. The two pieces are snapped together using a LEGO-like assembly process.

Figure 3 shows an example schematic of the LEGO-like assembly process.

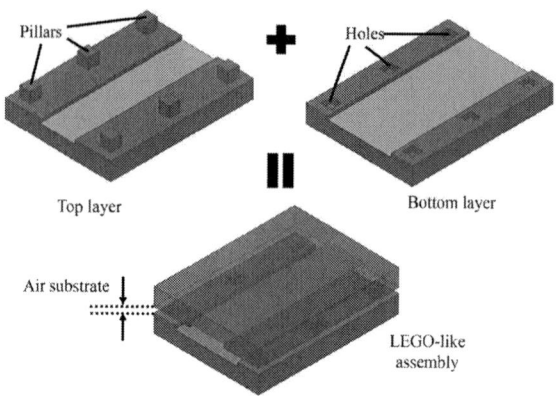

Figure 3. An example of a LEGO-like assembly process for air substrate based circuits.

Figure 4A shows the schematic of the air substrate based patch antenna along with the dimensions. Figures 4B and C shows the image of the structure before and after metallization and Figure 4D shows the image after assembly .

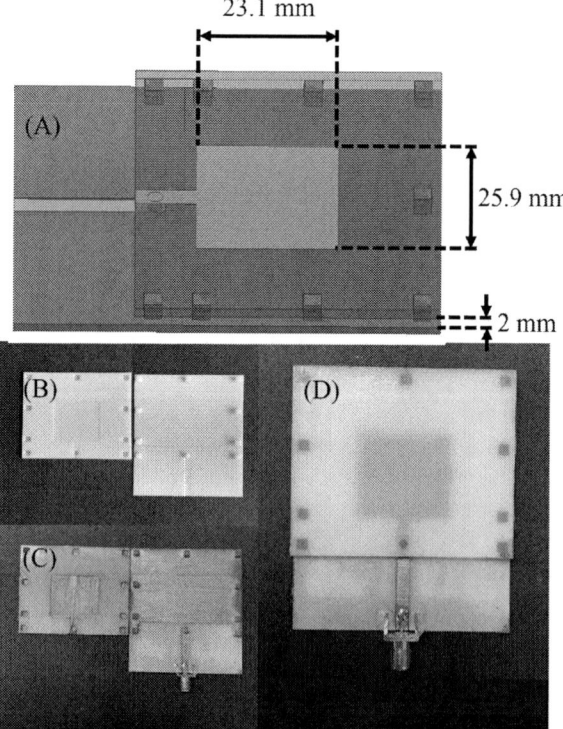

Figure 4. 3D printed air substrate based patch antenna (A) Schematic with dimensions, (B) and (C) structure before and after metallization, respectively, and (D) image after LEGO-like assembly.

978-1-7281-1500-9/19 $31.00 © 2019 IEEE

Figure 5 shows the simulated and measured reflection co-efficients and the radiation patterns.

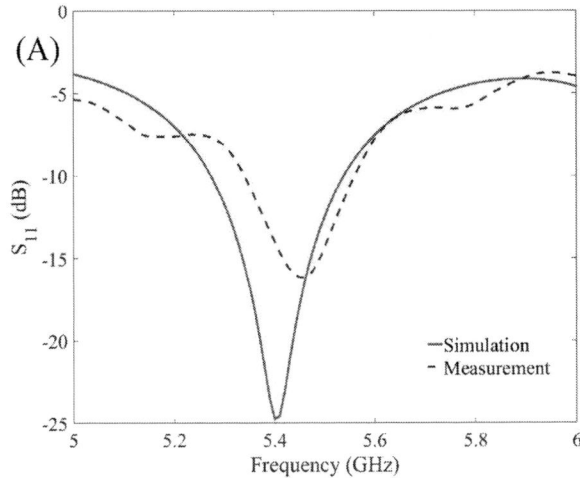

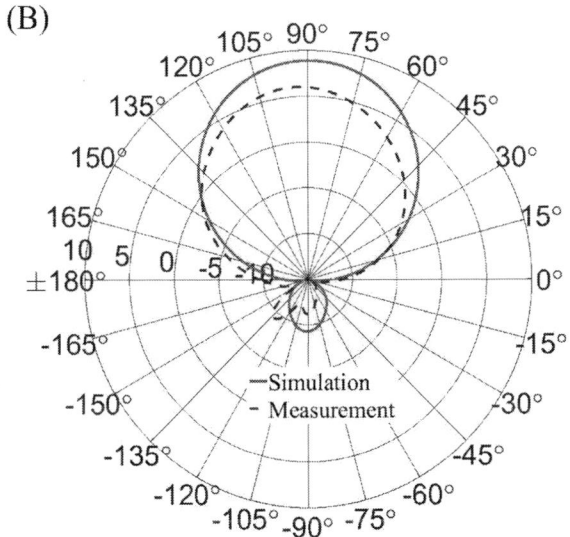

Figure 5. Simulated and measured (A) reflection coefficients and (B) radiation patterns of the air substrate based patch antenna.

The simulated and measured results matches closely and the slight mismatch is due to the loss associated with slight reduction in air gap due to warping of the substrate during the sputtering process as well as the poor conductivity of the silver paste. The fabricated patch operates at 5.4 GHz with a gain of 6.04 dBi.

B. Amplifier Sub-Module

The amplifier is designed next using a commercially available surface mount wide band amplifier chip from Mini Circuits(GVA-84+). Figure 6A shows the basic amplifier circuit design consisting of a bias circuit with an inductor at the input and DC blocking capacitors at both the input and

the output of the amplifier. Figure 6B shows the schematic of the amplifier circuit in a 3D printed configuration.

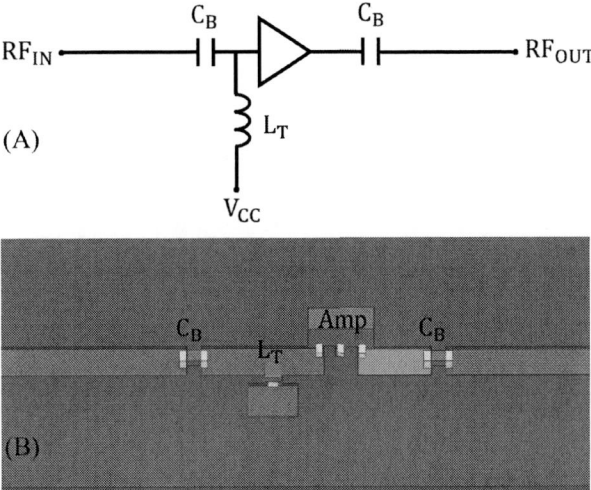

Figure 6. Schematic of the (A) amplifier circuit with lumped components, and (B) 3D printed configuration.

In order to validate the design, two different amplifier circuits are simulated using Agilent's Advanced Design System (ADS). For the first circuit, the amplifier is terminated using a matched 50 Ω, and for the second circuit, the matched termination is replaced by the S-parameters of the designed patch antenna. FEM solver High Frequency Structural Simulator (HFSS®) is used to design and simulate the S-parameters of the patch antenna. Figure 7 shows the simulated reflection coefficients at the input port of the amplifier circuit for both the terminations.

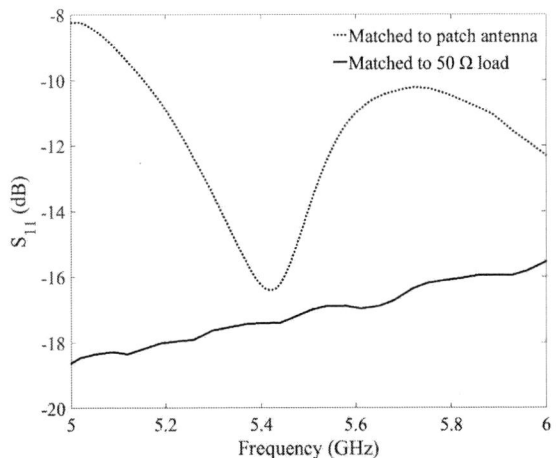

Figure 7. Simulated reflection coefficients of the amplifier

It can be seen from the Figure 7 that the chosen amplifier shows a good performance over 5-6 GHz band when terminated by a matched load and when terminated by the patch, the amplifier's band of operation is restricted to that of the antenna's operational bandwidth. Figure 8 shows the image of the 3D printed amplifier.

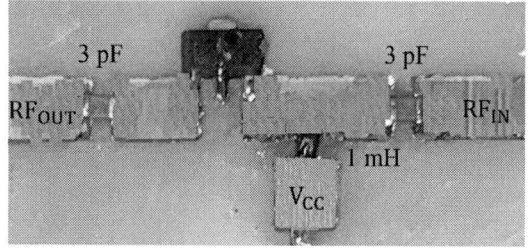

Figure 8.   Image of the fabricated 3D printed amplifier

C. Integrated RF Module

The amplifier and the air substrate based patch antenna are combined together forming a single RF module. Figure 9A shows the schematic of the module, Figure 9B and C shows the fabricated structure before and after metallization. Figure 9D shows the assembled module. Here, the power received by the patch antenna is amplified and measured on a spectrum analyzer

III. RESULTS

Two sets of experiments are performed to evaluate the performance of the RF module. The experiments are performed in the anechoic chamber where the transmitter antenna is placed at a distance of 118 inches from the receiver module. For both the experiments, a commercial wide band Vivaldi antenna is used as a transmitter connected to an RF source. The device under test is connected to a DC source for biasing, and a spectrum analyzer for measuring the data. Figure 10 shows the measurement setup.

The first set of experiments is performed to analyze the effect of integrating an amplifier with the air substrate based patch antenna as a function of frequency with a fixed input power of 5.5 dBm. For the measurements, first, an air substrate based patch antenna without the amplifier is connected and the received power is measured. Second, the integrated module, which includes the antenna and the amplifier is connected and received power is measured as a function of frequency with a fixed bias voltage of 4.6 V. Figure 11 shows the measured results for the first set of experiments.

As expected, the amplifier improves the signal strength in comparison to the direct received signal from the patch. The integrated module provides a good gain at the resonance frequency of the patch antenna (5.4 GHz). The second set of experiments is performed to analyze the effect of bias

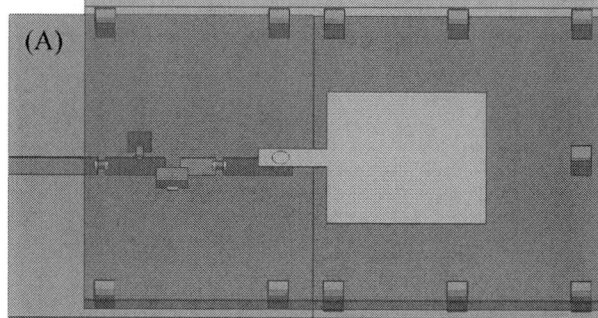
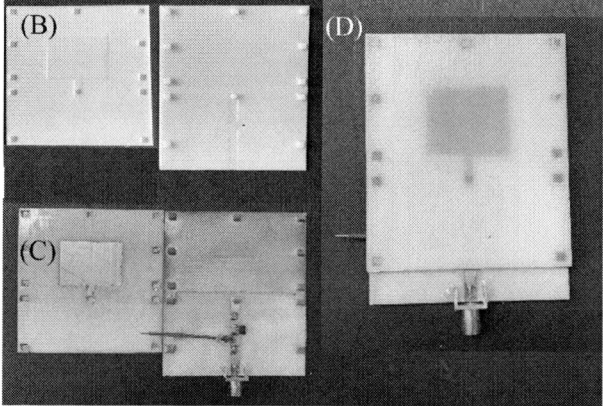

Figure 9.   3D printed integrated RF module: schematic (A), structure before and after metallization (B) and (C), respectively, and , (D) integrated module after the assembly process.

Figure 10.   Measurement setup.

voltage on the received signal. The input frequency and signal power are fixed at 5.4 GHz and 5.5 dBm, respectively, and the received signal is measured as a function of bias voltage. Figure 12 shows the measured gain using a spectrum analyzer as a function of bias voltage.

The amplifier performs the best beyond 4 V as seen from Figure 12. A combination of the two experiments helps in quantifying the performance of the integrated RF module

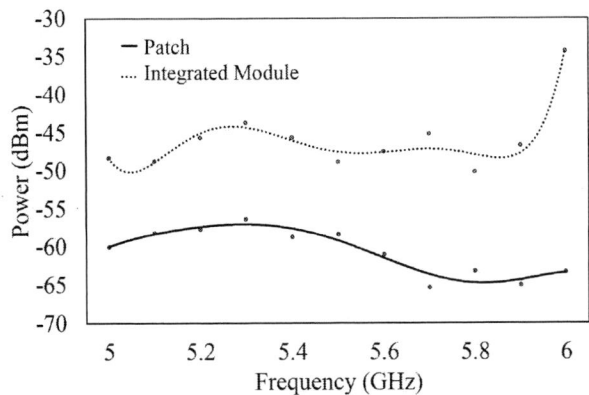

Figure 11. Measured received power as a function of input frequency.

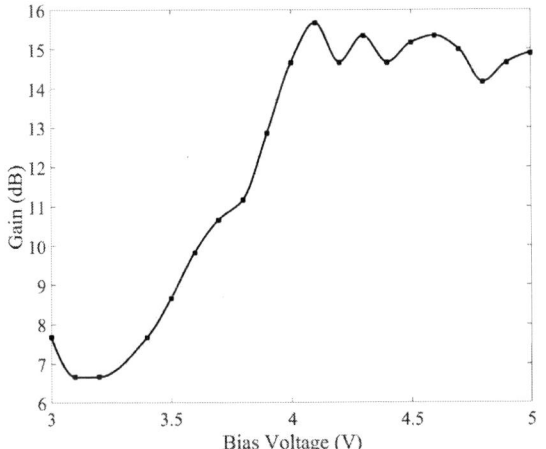

Figure 12. Measured gain as a function of bias voltage.

showcasing the ability of 3D printing to fabricate a compact RF module for 5G and other applications.

IV. DISCUSSION

In this paper, a 3D printed RF system for 5G communication application is presented in which multiple components can be incorporated onto a custom printed plastic substrate. The vertical stacking of components allows reduction in the overall form factor of the system leading to a simpler SiP or SoP solution for realizing compact system designs. Although 3D printing is beneficial in realizing multi-layer structures, there are still some challenges associated using these techniques such as surface roughness, warpage due to heating, parasitics associated with soldering the SMT components, and the high dielectric loss associated with the photopolymer VeroWhitePlus®. For reducing surface roughness, mechanical polishing can be incorporated as an intermediate step before metallization. Warpage during

sputtering process can be reduced by using a high temperature print polymer. Another critical factor that effects the performance of the RF circuits is the parasitics associated due to soldering on the surface mount components onto the 3D printed plastic. This can be reduced by introducing solder-free embedding techniques where different active and passive components can be incorporated as part of the substrate with corresponding vias for maintaining electrical connection. The effect of the dielectric loss associated with the 3D printed substrate can be reduced by utilizing LEGO-like assembly technique and by using air as a substrate over the whole structure. For a practical purpose, realizing fully 3D printed SoP/SiP solutions requires a combination of multiple printing techniques capable of printing metals, dielectrics, and nano-particles as well as advancement in print resolution, developing low loss print polymers, and automated assembly techniques.

V. CONCLUSION

This paper shows the potential of 3D printing in realizing customized electronics packaging. Multilayer SoP/SiP solutions with customized substrate heights and air gaps using 3D printing allows realization of compact high performance RF systems. RF module with a receiving antenna and an amplifier sub-module is fabricated using multiple 3D printing strategies such as damascene-like metal pattering process, LEGO-like assembly, and air-based substrates and demonstrated for 5G communication applications. The microchip along with the passive components are surface mounted onto the 3D printed amplifier module by direct soldering. The receiver antenna is designed to operate at 5.4 GHz and the performance of the integrated module is analyzed over a 5G frequency band. The proposed technique can be extended to multilayer system stacking for designing compact RF systems.

VI. ACKNOWLEDGMENT

The authors would like to thank Mr. Brian Wright of the MSU ECE shop for sputtering and electroplating the 3D printed samples. The authors are also grateful to other members of the Electromagnetic Research Group (EMRG at MSU) for helpful discussions..

REFERENCES

[1] S. K. Lim, "Physical design for 3D system on package," *IEEE Design & Test of Computers*, vol. 22, no. 6, pp. 532–539, 2005.

[2] P. Wu, F. Liu, J. Li, C. Chen, F. Hou, L. Cao, and L. Wan, "Design and implementation of a rigid-flex RF front-end system-in-package," *Microsystem Technologies*, vol. 23, no. 10, pp. 4579–4589, 2017.

[3] F. Liu, V. Sundaram, H. Chan, G. Krishnan, J. Shang, J. Dobrick, J. Neill, D. Baars, S. Kennedy, and R. Tummala, "Ultra-high density, thin core and low loss organic system-on-package (SOP) substrate technology for mobile applications," in *2009 59th Electronic Components and Technology Conference*, pp. 612–617, IEEE, 2009.

[4] N. Arnal, T. Ketterl, Y. Vega, J. Stratton, C. Perkowski, P. Deffenbaugh, K. Church, and T. Weller, "3D multi-layer additive manufacturing of a 2.45 GHz RF front end," in *2015 IEEE MTT-s International Microwave Symposium*, pp. 1–4, IEEE, 2015.

[5] D. Espalin, D. W. Muse, E. MacDonald, and R. B. Wicker, "3D printing multifunctionality: structures with electronics," *The International Journal of Advanced Manufacturing Technology*, vol. 72, no. 5-8, pp. 963–978, 2014.

[6] E. Macdonald, R. Salas, D. Espalin, M. Perez, E. Aguilera, D. Muse, and R. B. Wicker, "3D printing for the rapid prototyping of structural electronics," *IEEE access*, vol. 2, pp. 234–242, 2014.

[7] M. I. M. Ghazali, S. Karuppuswami, A. Kaur, and P. Chahal, "3-D printed air substrates for the design and fabrication of RF components," *IEEE Transactions on Components, Packaging and Manufacturing Technology*, vol. 7, no. 6, pp. 982–989, 2017.

[8] M. I. M. Ghazali, J. A. Byford, S. Karuppuswami, A. Kaur, J. Lennon, and P. Chahal, "3D printed out-of-plane antennas for use on high density boards," in *2017 IEEE 67th Electronic Components and Technology Conference (ECTC)*, pp. 1835–1842, IEEE, 2017.

[9] J. A. Byford, M. I. M. Ghazali, S. Karuppuswami, B. L. Wright, and P. Chahal, "Demonstration of RF and microwave passive circuits through 3-D printing and selective metalization," *IEEE Transactions on Components, Packaging and Manufacturing Technology*, vol. 7, no. 3, pp. 463–471, 2017.

[10] M. I. M. Ghazali, S. Karuppuswami, S. Mondal, and P. Chahal, "Embedded active elements in 3D printed structures for the design of RF circuits," in *2018 IEEE 68th Electronic Components and Technology Conference (ECTC)*, pp. 1062–1067, IEEE, 2018.

Fully Additively Manufactured Tunable Active Frequency Selective Surfaces with Integrated On-package Solar Cells for Smart Packaging Applications

Syed A. Nauroze, Xuanke He, Manos M. Tentzeris
School of Electrical & Computer Engineering
Georgia Institute f Technology
Atlanta, GA, USA
Email: nauroze@gatech.edu

Abstract—A first-of-its-kind fully inkjet-printed electronically tunable active flexible frequency selective surface (FSS) using varactors with integrated on-package solar cells is presented in this paper. Each varactor is biased by a dedicated on-package inkjet-printed solar cell using a low-temperature fabrication process. The solar cell changes its output voltage with variation in light intensities that eventually leads to a change in capacitance of the varactor and overall frequency response of the FSS. The proposed design eliminates the use of labor intensive biasing network, bulky power supply and micro-controllers to tune the FSS frequency response. These structures presents an autonomous, ultra low cost on-package RF shielding mechanism for next-generation of system-on-chip packaging applications that can be tuned on-demand at different frequency bands by simple variation of incident light intensities. Thus making them useful for a wide range of terrestrial, outer-space and EMI shielding applications.

Keywords-Additive manufacturing; solar cells; Frequency selective surfaces; inkjet printing; varactors; tunable FSS; tunable RF structures;

I. Introduction

The recent trend towards integrated mm-wave multi-chip modules and system-on-chip systems require efficient shielding to reduce electromagnetic interference (EMI) from external sources as well as nearby modules. Traditionally, this is achieved by either using fine conductive wire mesh or sheet to cover the modules [1], [2]. However, such shields are usually very bulky, occupy substantial amount of useful chip area and requires substantial amount of via holes to avoid any floating ground. Moreover, these techniques do not allow selective transmission of specific frequency bands that is typically required for systems with an on-chip antenna modules. This can be easily realized by using on-package inkjet-printed spatial filtering structures such as frequency selective surfaces (FSSs).

FSSs typically consist of periodic arrangements of resonant elements that are printed/etched on a thin substrate. These structures exhibit band-pass or band-stop filtering properties depending on the size, type and shape of the resonant elements as well as their inter-element distances [3]. Recently, they have

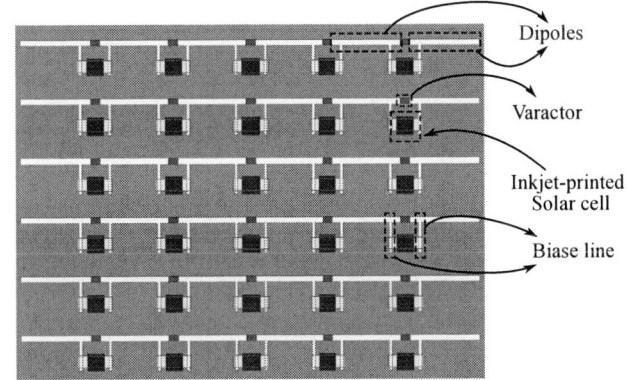

(a)

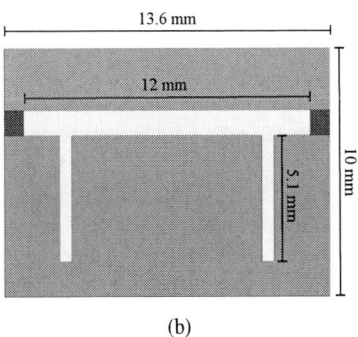

(b)

Figure 1. (a) FSS with integrated inkjet-printed solar cell and varactor and (b) its unit cell

found many applications such as antenna radar cross section reduction, smart skins, metamaterials, absorbers and radome design to name a few. However, conventional FSSs are narrow band and can only realize a fixed f requency r esponse which may vary due to environmental changes, installation or fabrication errors. Moreover, their inability to change their frequency response limit their use in most real-life applications. That is why a significant amount of research and development has been done in the past two decades to realize tunable FSSs that can change their frequency behavior in response to external stimulus. Some of the most common tuning mechanisms for FSS structures include changing the electrical properties or geometrical shape of the substrate [4], [5], [6] and integrating varactors [7], diodes or MEMS switches [8] with the resonant element of the FSS. The latter approach has the advantage of realizing rapid frequency tunability. However, one of the key disadvantages of this approach is the requirement of a biasing network with a dedicated source which significantly complicates the overall design as the size of the FSS becomes large.

This paper introduces a fully inkjet-printed electronically tunable flexible FSS using varactors that are biased by a dedicated inkjet-printed solar cell source. The solar cells bias the varactors at different voltage levels as the light intensity is varied, resulting in different values of capacitance across each varactor. Thus eliminating use of bulky power supplies and micro-controllers that are typically used in conventional electronically tunable FSS structures. These structures can be easily integrated with smart encapsulations to realize tunable RF shielding that can vary its frequency response on-demand according to transmission requirements. The use of additive manufacturing technologies such as inkjet-printing also facilitates rapid and mass production of tunable FSS-based shielding for mm-wave applications that can change their electromagnetic behavior on-demand across different frequency bands by simply varying incident light intensity. These autonomous, ultra-low cost on-package tunable RF-shielding structures are especially suited for rugged or unpredictable EM radiation environments where passive shielding would become insufficient to block a wide band of unwanted EM radiations or frequency bands. The bias network for typical electronically tunable FSS structures presents a key challenge for their use in autonomous moving platforms and loose electrical connection to the power supply with movement. However, the proposed design overcome this problem by directly inkjet-printing the bias network and solar cells across each resonant element. Its tunable nature also facilitates mitigation of any installation or fabrication errors typically associated with traditional electronically tunable FSS structures.

The inkjet-printed solar cells presented in this paper can be realized using low temperature process on lightweight flexible substrates allowing for conformal adhesion to curved or irregular surfaces, allowing for easy solderless process and reduced breakages of solder joints during flexing. These factors grant massive reductions in complexity and a vast increase in functionality that can be utilized in weight sensitive vehicles that move in and out of sunlight regularly like spacecrafts or aircrafts.

The rest of the paper is organized as follows. First the design and modeling of FSSs with integrated solar cell will be discussed in section II. Then a detailed fabrication of inkjet-printed solar cells and FSSs will be highlighted in section III. Next, the simulation and measurement results of the proposed

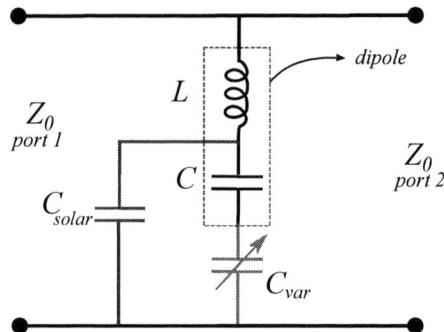

Figure 2. Equivalent circuit model of proposed FSS with integrated solar cells

structures will be discussed in detail in section IV and finally we will conclude with some final remarks.

II. FREQUENCY SELECTIVE SURFACES WITH INTEGRATED SOLAR CELLS DESIGN AND MODELING

An ideal frequency selective surface is an infinitely large 2D or 3D periodic array of resonant elements whose electromagnetic behavior is primarily determined by the size, shape and inter-element distance between the resonant elements. The resonant elements can be broadly categorized either as open (e.g. dipole, Jerusalem cross), closed (loops), plate/solid interior (rectangular/circular patch) or combination of the three [3]. In general, FSSs with open-shaped or closed-shaped conductive resonant elements resonates when their arm length and perimeter is comparable to $\lambda/2$ respectively, where λ is the wavelength of the operating frequency [3]. Typically, single-layer FSSs with such resonant elements exhibit first-order band-stop frequency response where the bandwidth is determined by the shape and inter-element distances between resonant elements. On the other hand, FSS structures with their complementary slot-type resonant elements present a first-order band-pass filtering response. Higher order filtering response can be realized by using broadband resonant elements or using multi-layer configuration for the FSS.

It is important to note here that while the ideal FSS assumes

Figure 3. Layer stackup for the single solar cell, showing the layers of material that is all fully inkjet printed. The inverted solar cell has a top electrode that is PEDOT:PSS, which is a transparent conductor allowing light to pass through.

978-1-7281-1500-9/19 $31.00 © 2019 IEEE

an infinite structure, it is impossible to realize it in reality. The finite size of the FSS structure introduces unequal current distribution across each resonant element and introduces edge effects. This however, can be mitigated by using FSS structures that are large enough to encloses the main beam of the antenna. Moreover, the electromagnetic response of the overall FSS structure can be accurately calculated by using Floquet theorem on its unit cell with master/slave boundary conditions. This ensures electric field at the master boundary to be the same as the slave boundary with a phase delay that is determined by the size of the unit cell.In this paper, a single-layer dipole-based FSS structure is realized where each neighbouring resonant element is connected by a varactor biased by an inkjet-printed solar cell as shown in Fig. 1a. The unit cell consists of a 12mm long dipole elements that are connected to printed solar cells with a 5 mm line on both sides of the dipole, resulting in a π-shaped resonant element as shown in Fig. 1b.

The overall frequency response of dipole-based FSS can be approximated by an equivalent series LC network connected across two ports with intrinsic impedance ($Z_o = 377\Omega$) as shown in Fig. 2. The values of the inductance (L) and capacitance (C) are determined by the length of the dipole and inter-element distance respectively. While the varactor can be represented by a series variable capacitor whose value depends on its biasing voltage provided by the solar cell. Note that substrate effect on the overall performance of the FSS structure can be ignored in this case as the substrate thickness is very small compared to the operating wavelength **[1]**. In order to mathematically calculate the values of L and C, we first evaluate impedance of the freestanding FSS (Z_{FSS}) and set it equal to zero:

$$Z_{FSS} = jL\omega + \frac{1}{jC\omega} = 0$$
$$\omega_r = \frac{1}{\sqrt{LC}} \tag{1}$$

where, ω_r is the central design (resonant) frequency. Next, we use the full-wave simulator such as Ansoft HFSS to determine the null in Z_{FSS} (ω_r) and set $C = 1/(L\omega_r^2)$. This would ensure that the equivalent-circuit model has the same resonant frequency as the full-wave simulation and only differs in bandwidth. The bandwidth is then optimized by iteratively varying the value of L such that the overall equivalent-circuit response has the minimum Euclidean distance from the full-wave simulation. For this purpose, two frequency points on the lower and higher edges of the -10dB bandwidth were sufficient for good agreement of equivalent-circuit model and simulation results.

III. Fabrication of FSS with Integrated Solar Cells

The use of inkjet printed solar cells on the same substrate as the FSS enables on package level integration of tunable structures in one compact manufacturing process. All parts of the structure, with the exception of the Kapton substrate was

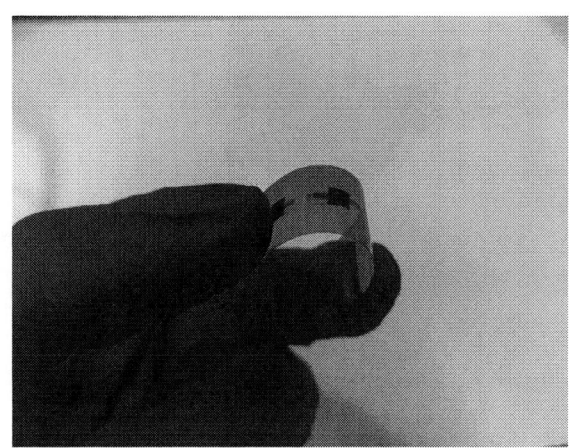

Figure 4. Fabricated inkjet-printed solar cells

inkjet printed on the Fujifilm Dimatix 2800 series inkjet printer using Dimatix 16 nozzle 10pL printheads. Commercial silver nanoparticle (SNP) and poly(3,4-ethylenedioxythiophene) polystyrene sulfonate (PEDOT:PSS) inks were used as the electrode layer material. With the inverted solar cell, the transparent PEDOT:PSS are on top and the SNP on the bottom allowing light to penetrate through to the active layers beneath to create electron-hole pairs. Both ready made inks were inkjet printable.

The active layer consists a mix of electron donor and acceptor materials. These are Poly(3-hexylthiophene-2,5-diyl) (P3HT) and Phenyl-C61-butyric acid methyl ester (PCBM) as these materials promise good power conversion efficiency and inkjet printability [9]. The materials are soluble in (1,2)orthodichlorobenzene (oDCB), however due to the low boiling point of the solvent, the materials quickly clogs up the nozzles of the inkjet printer. Additional high temperature boiling point solvent, Mesitylene was use to dilute the solution to allow it to be able to allow the ink to maintain the liquid phase during printing and storage. The active layer ink was also filtered using a 1 μm porous filter to get rid of any possible precipitates that formed and printed well at 2kHz jetting frequency.

Polyethylenimine (PEI) was prepared due to the fact that it can improve the performance of solar cells by modifying the work function between electrodes [10]. Due to the resistive effects of polyethylenimine, only a thin layer can be deposited before the performance of the solar cell degrades. Thus, branched PEI with a molecular weight of 70,000 was diluted to 0.4 % by weight in DI water, and further diluted to 0.2 % in glucose to modify the viscosity for the solution to be inkjet printable.

Fabrication steps were similar to the procedure documented in [11]. The bottom electrode of the solar cell was printed using 3 layers of SNP ink and sintered at 180^oC. Then, 1 layer of PEI is printed on the SNP at a stage temperature of 60^oC and was dried in a convection oven at 90^oC for 1 hour. Following a 1 min UV ozone treatment, the 4 layers of the active material was printed with a 5 min wait in between layers

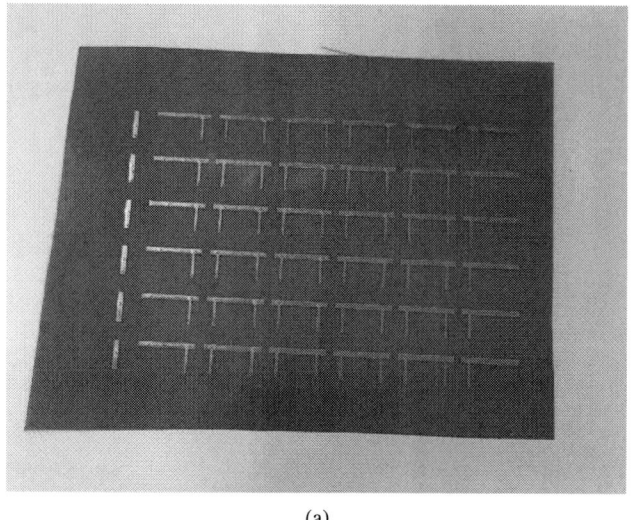

(a)

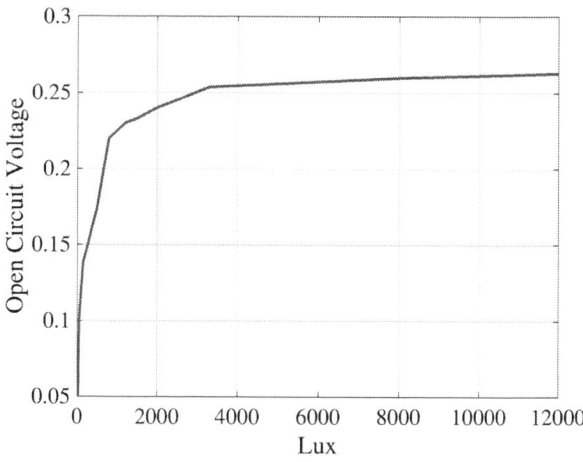

Figure 6. Single solar cell response as a function of illuminance.

at a stage temperature of 45^oC. The entire sample was then sintered again at 120^oC. Another 3 min of UV ozone treatment were done to enhance the wetting of the PEDOT:PSS on the active layer material and 4 layers of PEDOT:PSS was used as the top electrode. The final stackup is shown in Fig.3. The there is a 3 mm x 3 mm overlap between the electrodes of the SNP and the PEDOT:PSS which is the equivalent area of the solar cell. Preliminary testing of the solar cell demonstrates around 0.26V open circuit voltage at 12,000 lux and 0.15V open circuit voltage at room light levels (350 lux) using a halogen lamp as the exciting light source. The printed solar cells demonstrates good flexibility on Kapton substrates as shown in Fig 4. The open circuit voltage is displayed in Fig. 6. These solar cells were integrated into the FSS as shown in Fig. 5, where the electrodes of the solar cell were built into the FSS structure. Varactors were added for further tunability of the FSS and were connected using silver paste in Fig. 5(c).

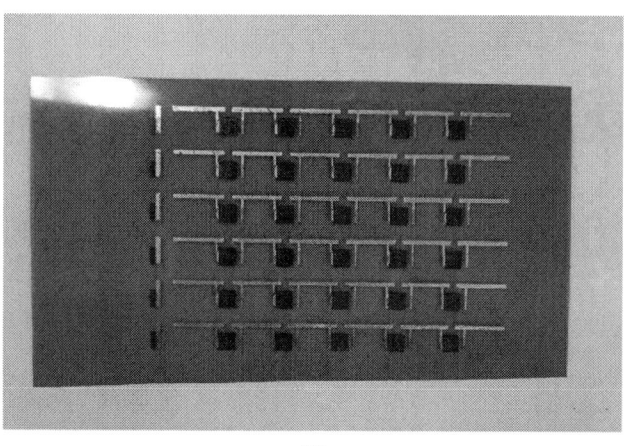

(b)

(c)

Figure 5. Fabricated FSS (a) with only resonant elements and feed lines (b) with integrated inkjet-printed solar cells (c) complete FSS structure with inkjet-printed solar cells and varactors

IV. SIMULATION AND MEASUREMENT RESULTS

In order to exploit the periodicity of the FSS structure, the unit cell shown in Fig. 1b was designed and simulated in Ansoft HFSS with Floquet ports exciation and master/slave boundary conditions as shown in Fig.7a. Thereby significantly reducing the overall simulation time and computational resources. It is important to note here that due to the presence of biasing lines connected across the dipole, it transforms into a π-like structure. The resultant structure would exhibit resonance in both TE and TM modes. Hence in the TE mode, the dominant resonant structure would be the dipole while on the TM mode the biasing lines would be resonant. The biasing lines would act like small dipoles (of length 5.1mm). However, since the resonant frequency of biasing lines is very far from the main dipole, it would not affect the filtering response in the desired frequency band.

The simulation results were verified using a bistatic measurement setup that consists of two broadband horn antennas in line-of-sight of each other and the FSS placed in middle of

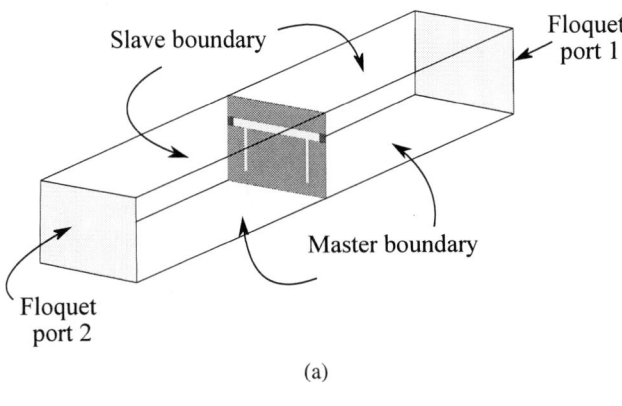

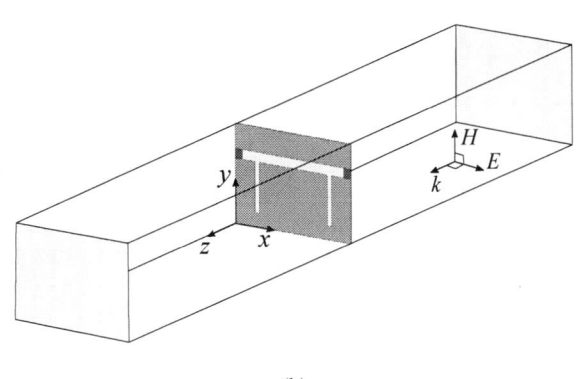

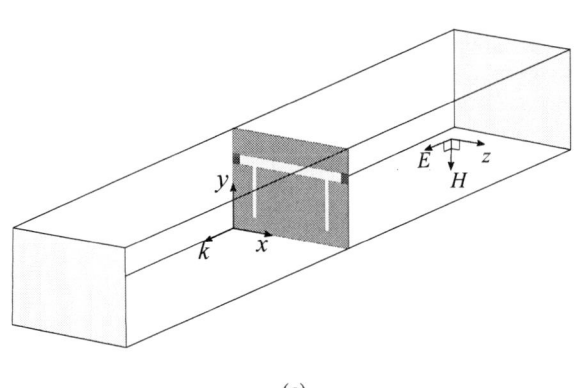

Figure 7. Simulation setup of FSS with integrated varactors. (a) Boundary conditions with Floquet port excitation (b) TE-mode (c) TM-mode excitation.

them as shown in Fig. 8. The horn antennas were connected to Anristu MS46522B-040 vector network analyzer (VNA) using coaxial cables. The distance between the antennas was kept large enough so that the FSS structure was in their far-field and was excited by a plane wave. The path between the FSS and the two antennas was de-embedded using type-D network extraction on the VNA.

The simulated and measured results for the proposed FSS (with and without inkjet-printed solar cells) in TE-mode and TM-mode are shown in Fig. 9 & 10 respectively. It can be

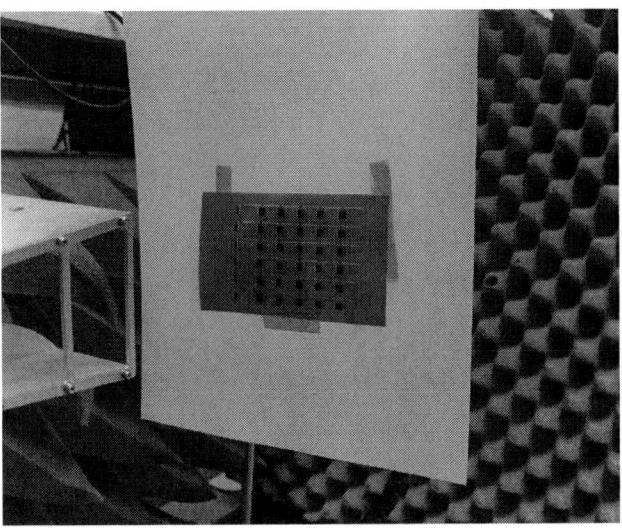

Figure 8. Measurement setup of FSS

easily seen that in the TE-mode, the FSS without inkjet-printed solar cells resonates at around 7GHz & 14GHz which represents fundamental and higher-mode (second harmonic) excitation respectively. The 5 dB difference between the simulated and measured results can be attributed towards the finite size of the FSS structure which results in higher losses due to edge effects [12]. This can be easily mitigated by using horn antennas with narrow beamwidth or larger FSS size. It can be seen that the frequency response slightly shifts to the right for the FSS with inkjet-prited solar cells. This is because the solar cells can be modelled as an additional capacitor in parallel to the capacitor in the LC network shown in Fig. 2 resulting in decrease in resonant frequency.

In order to tune the frequency response of the proposed FSS structures, the capacitance of the SMV2019 varactors must be varied from 2.3pF to 0.25 pF by changing biasing voltage from

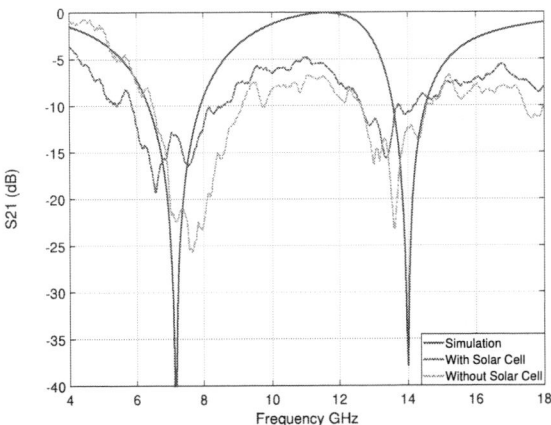

Figure 9. Simulated and measured frequency response of single-layer FSS with and without integrated solar cells (TE-mode)

978-1-7281-1500-9/19 $31.00 © 2019 IEEE

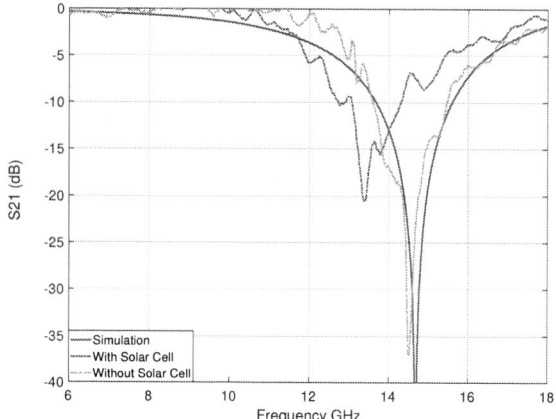

Figure 10. Simulated and measured frequency response of single-layer FSS with and without integrated solar cells (TM-mode)

0V to 20V respectively. In this work, the biasing is realized by the inkjet-printed solar cells that can produce a maximum of 0.26V at 12,500 lux. However, this voltage is not large enough to significantly vary the frequency response of the proposed FSS structure. However, this problem can be resolved by using larger solar cells, combining multiple solar cells in series or more sensitive varactors. The simulated results for proposed FSS structure (for TE-mode) for different values of the varactor capacitance is shown in Fig. 11. The increase in bandwidth is attributed towards additional capacitance across the LC network. Moreover, note that the frequency response of the overall structure will remain unchanged for TM-mode due to absence of any tuning element.

V. CONCLUSION

The paper presents a novel methodology to realize fully inkjet-printed electronically tunable frequency selective sur-

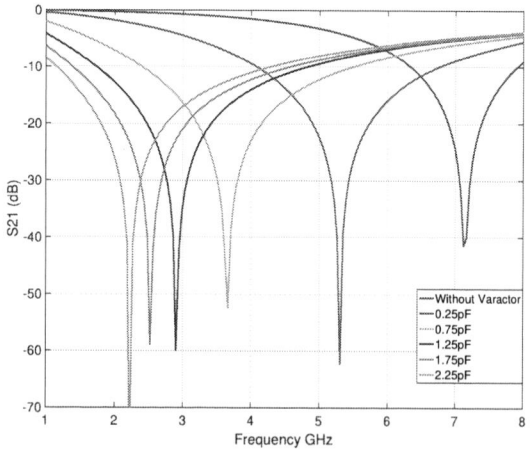

Figure 11. Simulated frequency response of single-layer FSS with integrated solar cells and varactors (TE-mode)

faces with an on-package solar cell and varactors. The proposed design addresses some of the major issues with the traditional electronically tunable FSS structures that includes complicated biasing network, dedicated (bulky) power supply, use of microcontrollers (to realize tunability) and manual placement of lumped components across each individual elements making them expensive, laborious and prone to breakage for large curved surfaces which limits their use in practical applications. By inkjet-printing biasing network and dedicated solar cell for each varactor, the design can be rapidly scaled-to-large-numbers and facilitates easy tuning mechanism by simply changing the light intensities across the solar cell. These ultra low-cost, autonomous, fully inkjet-printed FSS structure features fast on-the-fly tunable RF-shielding in a moving platform or unpredictable EM radiation environment without a dedicated bulky power supply or inefficient bias network that typically breaks or tangles due to movement. A detailed analysis and fabrication process of the structure along with an equivalent circuit modeling of the structure is also presented. The simulation results match very closely to the measured results. The proposed structure truly represents the uniqueness and substantial advantage of additive manufacturing technologies as such structures would almost be impossible to realize using traditional manufacturing technologies.The design can be further improved in the future by using larger solar cells to realize higher biasing voltage for the varactor that can lead to many interesting application such as tunable beamsteering reflect arrays that can change their frequency response on-demand by simply varying light intensities across different regions of the FSS structure.

ACKNOWLEDGMENT

The authors would like to thank the National Science Foundation (NSF), Semiconductor Research Corporation (SRC), and the Defense Threat Reduction Agency (DTRA) for their support during this project.

REFERENCES

[1] E. Unal, A. Gokcen, and Y. Kutlu, "Effective electromagnetic shielding," *IEEE Microwave magazine*, vol. 7, no. 4, pp. 48–54, 2006.

[2] I. S. Syed, Y. Ranga, L. Matekovits, K. P. Esselle, and S. G. Hay, "A single-layer frequency-selective surface for ultrawideband electromagnetic shielding," *IEEE Transactions on Electromagnetic Compatibility*, vol. 56, no. 6, pp. 1404–1411, 2014.

[3] B. A. Munk, *Frequency selective surfaces: theory and design.* John Wiley & Sons, 2005.

[4] T. Chang, R. J. Langley, and E. A. Parker, "Frequency selective surfaces on biased ferrite substrates," *Electronics Letters*, vol. 30, no. 15, pp. 1193–1194, 1994.

[5] K. Fuchi, J. Tang, B. Crowgey, A. R. Diaz, E. J. Rothwell, and R. O. Ouedraogo, "Origami tunable frequency selective surfaces," *IEEE antennas and wireless propagation letters*, vol. 11, pp. 473–475, 2012.

[6] S. A. Nauroze, L. S. Novelino, M. M. Tentzeris, and G. H. Paulino, "Continuous-range tunable multilayer frequency-selective surfaces using origami and inkjet printing," *Proceedings of the National Academy of Sciences*, vol. 115, no. 52, pp. 13 210–13 215, 2018.

[7] T. K. Chang, R. J. Langley, and E. A. Parker, "Active frequency-selective surfaces," *IEE Proceedings-Microwaves, Antennas and Propagation*, vol. 143, no. 1, pp. 62–66, 1996.

978-1-7281-1500-9/19 $31.00 © 2019 IEEE

[8] B. Schoenlinner, L. C. Kempel, and G. M. Rebeiz, "Switchable rf mems ka-band frequency-selective surface," in *2004 IEEE MTT-S International Microwave Symposium Digest (IEEE Cat. No. 04CH37535)*, vol. 2. IEEE, 2004, pp. 1241–1244.

[9] C. Hoth, S. Choulis, P. Schilinsky, and C. Brabec, "High photovoltaic performance of inkjet printed polymer:fullerene blends," *Advanced Materials*, vol. 19, no. 22, pp. 3973–3978. [Online]. Available: https://onlinelibrary.wiley.com/doi/abs/10.1002/adma.200700911

[10] L. Yan, Y. Song, Y. Zhou, B. Song, and Y. Li, "Effect of pei cathode interlayer on work function and interface resistance of ito electrode in the inverted polymer solar cells," *Organic Electronics*, vol. 17, pp. 94 – 101, 2015. [Online]. Available: http://www.sciencedirect.com/science/article/pii/S1566119914005424

[11] J. Bito, "Applications of additive manufacturing for ambient rf energy harvesting and wireless power transfer systems," Ph.D. dissertation, Georgia Institute of Technology, 2017.

[12] B. A. Munk, *Finite antenna arrays and FSS*. John Wiley & Sons, 2003.

First Demonstration of a Low Cost/Customizable Chip Level 3D Printed Microjet Hotspot-Targeted Cooler for High Power Applications

T.-W. Wei[1,2], H. Oprins[1], V. Cherman[1], I. De Wolf[1,3], E. Beyne[1], M. Baelmans[2]

[1]imec, Leuven, Belgium, [2]Dept. Mech. Eng. KU Leuven, Leuven, Belgium,
[3]Dept. Materials Eng. KU Leuven, Leuven, Belgium
email: tiwei.wei@imec.be

Abstract—This work presents the modeling, design, demonstrator fabrication and experimental characterization of a customized chip level direct liquid impingement jet cooler for hotspot cooling, fabricated with high-resolution stereolithography technology. The study, using a dedicated thermal test vehicle, demonstrates that 3D printing enables the design and low-cost fabrication of high-performance impingement coolers matching hotspot power dissipation patterns. The hotspot targeted cooler can improve the thermal resistance by 36% compared with full nozzle array cooling for the same flow rate.

I. INTRODUCTION

With the increasing trend of the heat flux as well as the scaling down of the transistor size, thermal management becomes more and more challenging due to the performance and reliability degradation with elevating chip temperature. For practical microprocessor or power electronic devices, the heat flux on the chip is mostly non-uniform showing various hotspot patterns with different localized heat flux values. Therefore, controlling the maximum temperature of the hotspot always determines the thermal design in the thermal management of electronic devices and packages [1]. On the other hand, the control of the temperature uniformity across the whole chip is also important since a larger temperature gradient can increase the thermal stress and reduce the electronic reliability and circuit imbalances in devices [2].

Single phase liquid cooling has been regarded as an effective and practical solution for high power electronics and high-performance systems. In general, the research studies regarding the liquid cooling of the chip temperature are mostly focused on the maximum temperature reduction and chip temperature uniformity. However, the hotspot management on chip level is very challenging due to the complexity of the cooler fabrication. In literature [2,3], an extensive overview of the hotspot target cooling techniques is discussed. Most of the cooling solutions are based on Si processing, including hotspot-targeted embedded liquid cooling by adapting the channel density with the hotspot [2], micro-gaps with variable pin fin clustering [4], thermoelectric cooling (TEC) [5], alternating current electrothermal flow (ACET) cooling [3] and electrowetting droplet on hotspot [6]. However, the drawbacks of these techniques are low energy conversion efficiency, low heat flux pumping capacities, high cost and introduction of additional thermal resistances in the heat flow path.

Polymer based bare die liquid micro-jet impingement cooling is an efficient cooling technique for high power electronics, especially for hotspot management. The

advantage of microjet cooling is that it can directly target the hotspot by placing jet nozzles on top of the hotspot. Previously [7], we have first introduced a chip level 3D-shaped polymer cooler with sub-mm nozzle diameters fabricated using mechanical micromachining. A cross section of this cooling concept is shown in Fig. 1. The performance benchmarking study shows that low cost polymer based 4×4 microjet coolers can achieve similar thermal performance at lower required pumping power than much more expensive silicon and ceramic based microjet coolers [7].

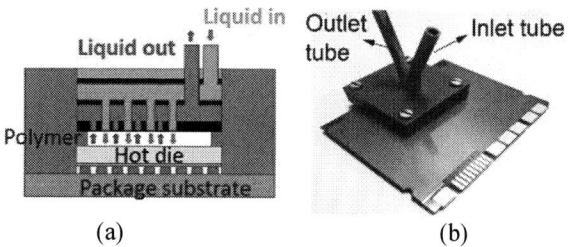

(a)   (b)

Figure 1. Cross section of the bare die impingement jet cooling concept (a) and (b) micromachined cooler for the 8x8 mm² chip in the 14x14 mm² BGA package [7].

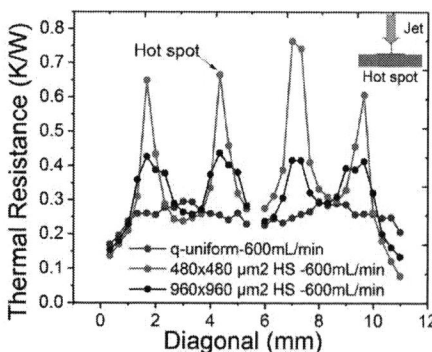

Figure 2. Measurement results with aligned nozzles and hotspot locations for different hotspot area [8].

Hotspot measurements for this 4×4 micromachined cooler, assembled on the thermal chip show that the temperature peak is significantly higher for hotspot power dissipation, even though impingement cooling achieves a very good cooling performance [8], as shown in Fig.2, Therefore, a customized hotspot targeted cooling design is needed for high performance chips with non-uniform power distribution. The main idea of hotspot targeted cooling is to focus the cooling solution, at the location where it is needed. In terms of impingement cooling, this means to concentrate the impinging of the liquid coolant and the high power areas in the chip. This cooling concept is illustrated in Fig.3. For the hotspot targeted impingement cooler, the location of the inlet nozzles that eject the coolant onto the chip should be aligned with the location of the hotspot.

As shown in the concept, isolated inlet nozzles have to be built to target at the hotspot. However, the use of micromachining techniques to fabricate such a cooler increases the manufacturing complexity since the nozzles targeted at the hotspot have to be drilled one after one and also the micro-milling process for the inlet and outlet plenum that connects all inlets and outlets becomes challenging. And also, the cooler has to be fabricated with different parts by micromachining. The assembly of different parts increases the risks of the water leakage. Therefore, we introduced the use of additive manufacturing to fabricate polymer impingement coolers to increase the design options for polymer coolers for more complex geometries [9]. Additive manufacturing enables to use low cost materials for the cooler fabrication, to print the whole geometry in one piece and to customize the design to match nozzle array to the chip power map. The details of the internal fluidic channels and inlet/outlet nozzles for the 3D printed 4x4 jet array cooler are shown in Fig.4.

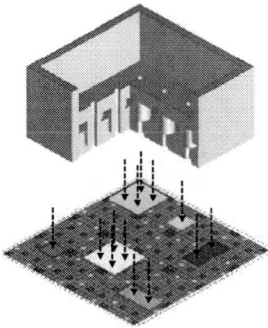

Figure 3. Concept of hotspot targeted liquid impingement cooling.

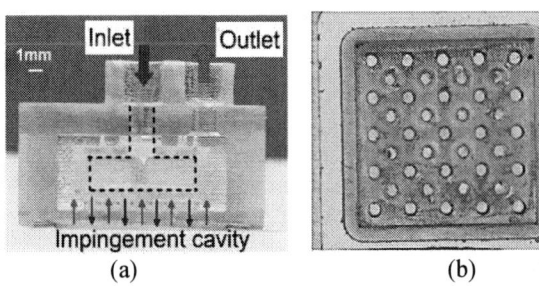

(a)   (b)

Figure 4. Fabrication of the 3D printed cooler [9]. Detail of the channels in cross-section of the cooler (a); bottom view of the nozzle plate with inlet/outlet nozzles (b).

In this paper, we present the design, fabrication and experimental characterization of a highly efficient customizable and low cost hotspot targeted cooling solution for high power electronics fabricated using additive manufacturing. In order to evaluate the thermal performance, the printed cooler is assembled to a 8×8 mm^2 thermal test chip [10] with an array of 32×32 temperature sensors and programmable heat dissipation patterns. The possible power dissipation patterns range from customized hotspot patterns to quasi-uniform power dissipation on the chip area.

The first section of the paper (Section 2) presents the manufacturing tolerance analysis with the 3D printed cooler, including the impact of the nozzle diameter variation, nozzle angle deviation, and the nozzle-to-chip distance deviations. Next in in Section 3, the design and fabrication of the hotspot targeted cooler demonstrators with two different test cases are introduced in detail. Then in section 4, the thermal performance between the uniform nozzle array cooling and hotspot targeted cooling are characterized and compared based on the advanced thermal test chips. Finally, the CFD full cooler level model is validated with experimental measurements. Based on the validated CFD model, the velocity and pressure information inside the hotspot targeted cooler are extracted. Moreover, the thermal-hydraulic trade-off analysis is summarized based on the system considerations for constant flow rate, pressure drop or pumping power.

II. MANUFATURING TOLERANCE ANALYSIS

A. Manufacturbaility analysis of SLA

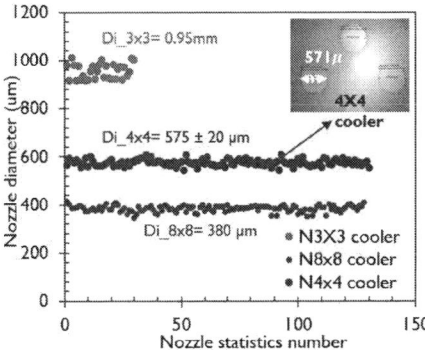

Figure 5. Nozzle diameter variation measurements with 3×3, 4×4 and 8×8 array cooler.

First, the fabrication tolerance and the deviation of the printed geometry from the nominal design are evaluated for the used high-resolution Stereolithography (SLA) technology. Microscopy is used to measure the nozzle diameter. Fig. 5 shows the measurement results of the printed nozzles diameters for different types of coolers (3×3, 4×4 and 8×8 nozzle arrays) that have been designed and fabricated for the 8 × 8 mm^2 test chip. The deviation between the measured printed nozzle (575 μm ± 10 μm) and the nominal design value of 600 μm is only 5% The measurements of the printed nozzle diameters for different cooler nozzle arrays, show that 3D printing is capable of producing the coolers with a small nominal diameter of 300 μm for the 8×8 array on the 8×8 mm^2 chip area, with high reproducibility. The unit cooling cell is 1×1 mm^2 with a nozzle diameter ratio of 0.3. This nozzle diameter ratio is defined as the nozzle diameter divide by unit cell length.

978-1-7281-1500-9/19 $31.00 © 2019 IEEE

B. Impact of nozzle diameter deviation

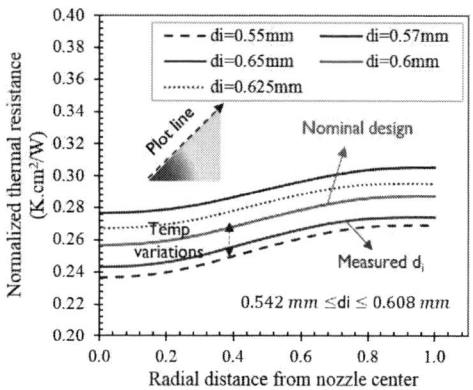

Figure 6. Impact of the nozzle diameter deviations on the temperature distributions for 2×2 mm² cooling unit cell area with nominal design nozzle diameter of 600 μm. (flow rate = 600 ml/min, Q = 50 W).

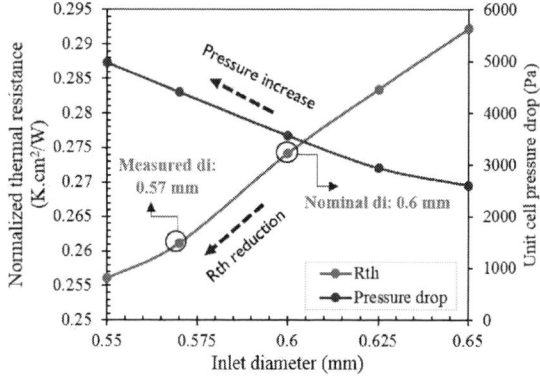

Figure 7. Impact of the inlet/outlet nozzle diameter on the averaged chip temperature and pressure drop for 2×2 mm² cooling unit cell area with nominal design nozzle diameter of 600 μm. (flow rate =600 ml/min, Q = 50 W).

In order to understand the impact of the 3D printing fabrication tolerance on the cooler thermal/hydraulic performance, the impact for a nozzle geometry of a 4×4 cooler with cooling unit cell area of 2×2 mm², and a nominal nozzle diameter of 600 μm, is investigated numerically. A unit cell computation fluid dynamics (CFD) modeling approach is used to assess the impact of the geometry deviation on the temperature and pressure drop based on the 4×4 array cooler. The used CFD software package is ANSYS Fluent. The deviation between the measured printed nozzle (575 μm ± 20 μm) and the nominal design value of 600 μm is only 5% for 4×4 array cooler. Fig.6 shows that the normalized thermal resistance will drop down for a decrease of the nozzle diameter at a constant flow rate. The reason is that the inlet nozzle velocity will increase due to the reduction of the nozzle diameter for the fixed flow rate. For the impingement jet cooling, the chip temperature is dominated by the stagnation point where the inlet jet nozzles are targeted. The stagnation

temperature in the temperature profile shows about 7.7% variation for the nozzle diameter ranging from 0.55 mm to 0.6 mm for the 4×4 cooler. The reduction of the nozzle diameter can reduce the chip temperature, however, at the expense of a higher pressure drop. The thermal and hydraulic comparison between the nominal design and actual measured values are illustrated in Fig.7. The modeling study shows that the nozzle diameter deviation of 5% at flow rate of 600 ml/min results only in a 4.7% reduction for the averaged chip temperature and 23% higher for the pressure drop.

C. Impact of nozzle angle deviation

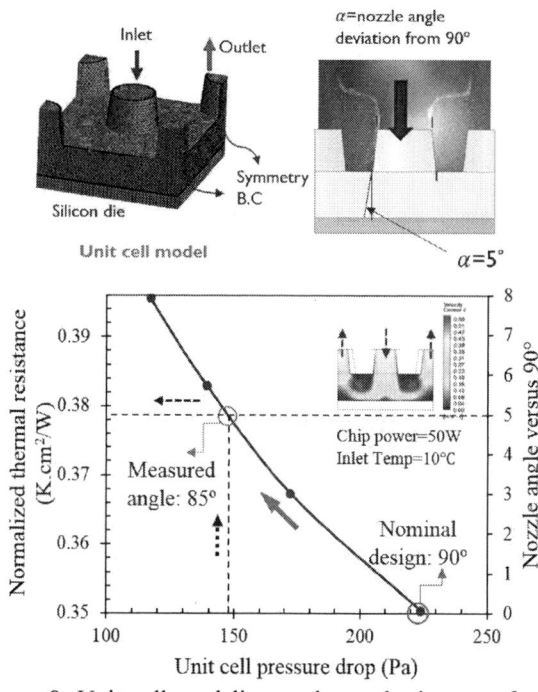

Figure 8. Unit cell modeling study on the impact of nozzle angle on the thermal and hydraulic performance for 2×2 mm² cooling unit cell area with nominal design nozzle diameter of 600 μm. (flow rate =300ml/min, Q=50W)

The cross-section pictures of the printed cooler show that the nozzle shapes are slightly tapered instead of straight. The tapered nozzle can reduce the cooling performance due to the less concentrated flow targeted at the stagnation point, resulting in a higher local chip temperature. At the other hand, the tapered inlet nozzle and outlet nozzle shape can help to reduce the pressure drop. As shown in Fig.8, the modeling study shows that a nozzle diameter deviation of 5° (85° instead of 90°) only results in a 8% difference for the averaged chip temperature but caused a 34.2% reduction for the local pressure drop on the unit cell level.

D. Impact of nozzle-to-chip distance deviation

In order to define the deviation of the nozzle-to-chip distance H, the groove depth, the thickness of the O-ring and the fabrication tolerance of the cavity height should be considered. For the cavity height and groove, the actual depth is about 0.65 mm compared to the nominal design value of 0.6 mm. The groove at the cooler bottom is designed for the O-ring assembly. The thickness of the O-ring is 1 mm, which will be placed on the organic substrate. The chip thickness is 0.2 mm. The thickness of the micro-bump used to connect the thermal test chip and organic substrate is 0.02 mm. Taking account into the O-ring thickness without compression, the distance between the nozzles and chip cooling surface is 0.78 mm. Therefore, the nozzle-to-chip backside distance variation is expected to between 0.6 mm to 0.8 mm.

The impact of the nozzle-to-chip distance above the chip cooling surface is shown in Fig. 9. The modeling study shows that the impact on the thermal resistance is negligible beyond 0.6 mm while the impact on the cooler pressure drop will result in a difference of 1.1 % between the range of 0.6 mm and 0.8 mm. Therefore, the nozzle-to-chip distance variations shows less impact on the chip averaged temperature when the nozzle-to-chip distance ratio is above H/L > 0.25.

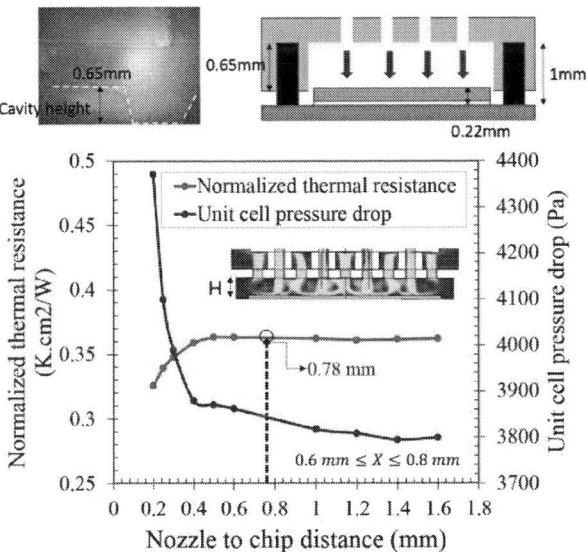

Figure 9. Impact of the nozzle-to-chip distance variations for the for 2×2 mm^2 cooling unit cell area with nominal design nozzle diameter of 600 μm. (flow rate =300ml/min, Q=50W).

III. 3D PRINTED HOTSPOT COOLER DEMONSTRATOR

A. Thermal test chip with HS

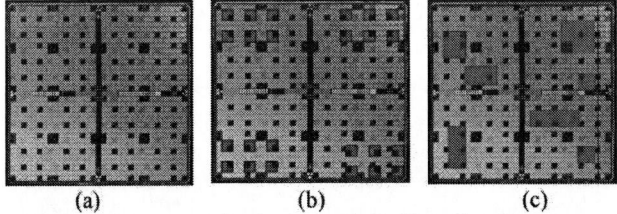

Figure 10. Test cases for the hotspot cooling: a) reference case with quasi-uniform cooling; b) test case 1 with regular pattern; c) test case 2 with various hotspot size;

In order to investigate the hotspot targeted cooling, two hotspot case studies have been defined: test case 1: regular hotspot pattern and test case 2: various hotspot sizes. These power dissipation maps are generated with the programmable test chip and are all shown in Fig.10. For the thermal test chip, the chip area is 8×8 mm^2 while the chip heated area is 75% of the chip surface. The thermal test chip is divided into a 32 × 32 array of 240 × 240 μm^2 square cells with additional peripheral circuits with I/O and control cells in the central cross of the chip [9]. There are 32×32 array of 'thermal pixel' cells with a diode as temperature sensors. The sensor temperature sensitivity is calibrated as -1.55 mV/ºC. Moreover, there are 832 cells indicated as 'heater cells' within the 32 × 32 array. The heater cells are programmable since each cell is individually controlled by a local transistor. The single heater cell is equipped with two 200 × 100 μm2 metal meander heaters in the back-end of line (BEOL). The maximal measured power for the single heater cell is 47.6mW at 1V, resulting in a maximal heat flux per cell about 82.6 W/cm^2.

For test case 1 with the regular hotspot pattern, there are 72 heater cells turned on. The total heater area is 4.15 mm^2, with a total measured chip power of 4.1 W. For test case 2 with various hotspot size pattern, the total number of the activated heater cells is 127, with a corresponding heater area of 7.32 mm^2 and a total measured heat power 5.5 W. These two different hotspot patterns are designed to mimic hotspot scenarios in power electronic devices.

B. HS cooler design

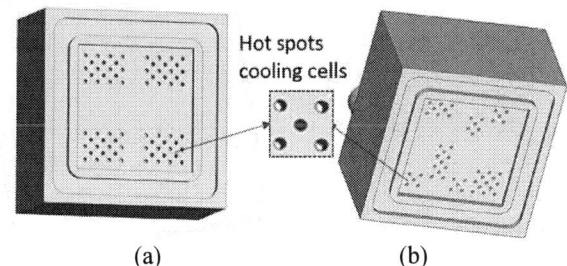

Figure 11. Designs of the hotspot cooler with 300 μm nozzles diameter: regular pattern (a) and various hotspot sizes (b) with 1×1 mm^2 cooling unit cell area.

Based on the manufacturing capabilities of the 3D printing technology, two different hotspot coolers based on the geometry of a cooling cell area of 1×1 mm^2 have been designed to match with the two hotspot test cases shown in Fig.10 (b) and (c). For the 8×8 mm^2 chips, this means that the nozzle array is based on the 8×8 array of cooling cells. In the hotspot targeted cooler designs, the inlet nozzles are kept at the locations of the hotspot, and removed at the locations where not power generation is present. Fig.11 shows the design details of the dedicated hotspot cooler for the two test cases with a nominal nozzle diameter of 300 μm. The design software used in this study is VariCAD. Moreover, the internal fluidic channels are illustrated in Fig.12, showing the flow directions inside the cooler. The nozzles within the cooling unit cells are designed specifically and aligned with the hotspot while no designed cooling cells are in the "cold region".

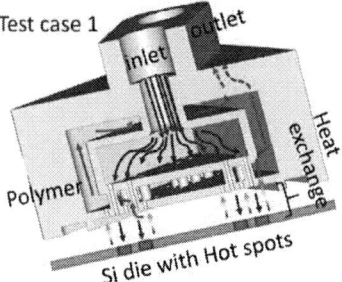

Figure 12. Interior view of 3D hotspot targeted cooler for the regular hotspot pattern.

C. Fabrication of 3D printed cooler

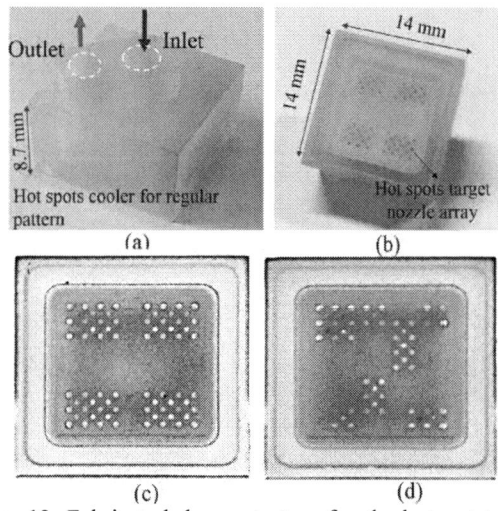

Figure 13. Fabricated demonstrators for the hotspot targeted cooler: a) top view and (b) bottom view of the printed cooler; c) and d) with bottom view of fabricated nozzle patterns.

The objective is to compare uniform liquid impingement cooling over the whole chip surface with dedicated cooling only at locations where it is needed. Fig. 13 shows the printed coolers for the two hotspot test cases. The total cooler size is 14 mm×14 mm×8.7 mm matched with the package substrate size. The bottom view of the two hotspot targeted coolers reveals the location of the corresponding to the hotspot patterns of the test cases. The groove for the placement of O-ring can be also seen in the figure.

IV. COOLER THERMAL CHARACTERIZATION

The dedicaed hotspot cooler is assembled on the advanced thermal test chip with programmable heater cells and 32×32 array of temperature sensors, illustrated in Fig.14(b). The O-ring is placed inside the groove to seal the cooler shown in Fig.14(a). The hotspot cooler with test board is finally assembled into the liquid flow loop with dedicated pressure sensor, heat exchanger and flow rate controller. The inlet

temperature is set as 10 °C. The chip temperature sensors allow absolute temperature measurements with an accuracy of ± 2-3 °C.

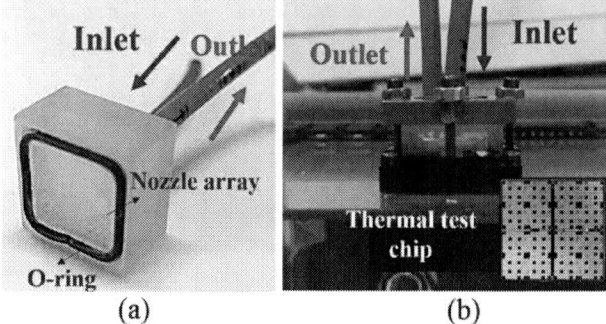

Figure 14. Cooler assembly: (a) hotspot targered cooler for regular pattern with O-ring placement; (b) assembly of the cooler on the thermal test chip and test PCB board.

For the characterization of the hotspot targeted coolers, the measured temperature maps for both coolers have been compared to the temperature maps obtained with the full array coolers at the same flow rate for the corresponding hotspot power dissipation maps, as shown in Fig.15. Fig. 16 shows the temperature profiles for the array cooler and the hotspot targeted cooler for diagonal/vertical line scan across the measured temperature maps of Fig. 15. These temperature measurements show a peak temperature reduction of 16% shown in Fig.16(a) and 24% shown in Fig.16(b) at a flow rate of 900 mL/min/cm^2 compared to the full array cooler for the targeted hotspot coolers of test case 1 and test case 2 respectively.

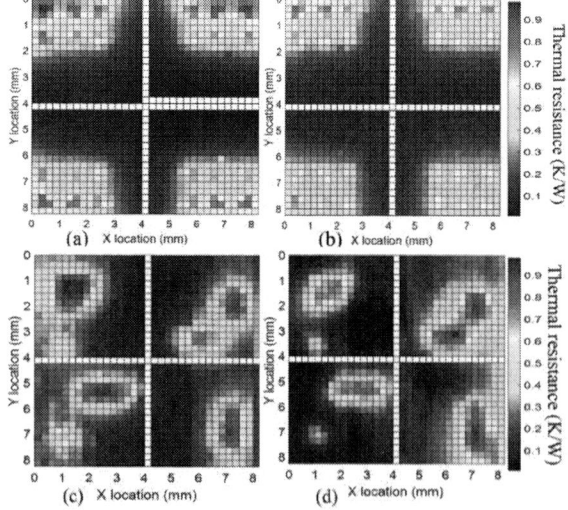

Figure 15 Chip temperature distribution measurements for (a) nozzle array uniform cooling and (b) HS targeted cooling for regular pattern at the flow rate of 600ml/min; Measurements of (c) nozzle array uniform cooling and (d) HS target cooling for test case 2 at same flow rate of 1000 ml/min.

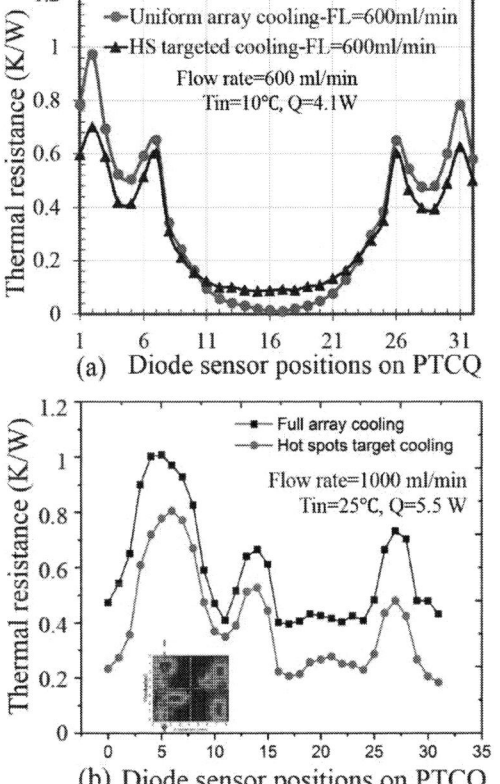

(a) Diode sensor positions on PTCQ

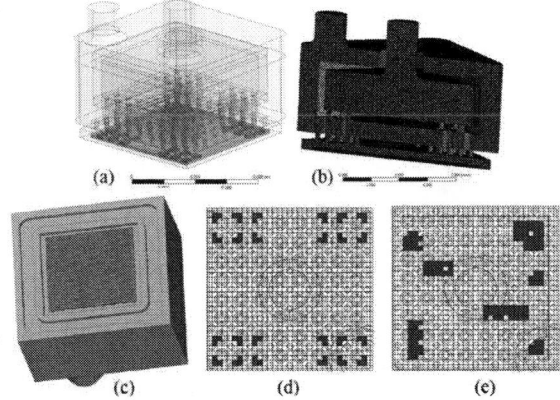

(b) Diode sensor positions on PTCQ

Figure 16. Experimental measurements of HS cooling with (a) test case 1 at the flow rate of 600 ml/min and (b) test case 2 at the flow rate of 1000 ml/min.

V. SYSTEM LEVEL MODELING ANALYSIS

A. Full cooler level model

Figure 17 Full cooler level CFD model: (a) transparent view of the cooler geometry; (b) cross section of the mesh of test case 1; (c) indication of the thermal test die on the full cooler; (d) detailed heater cells of test case 1 and test case 2.

In order to extract more detailed temperature, velocity and pressure drop information inside the dedicated cooler, a full cooler level CFD model is built including the details of the hotspot heater cells, as shown in Fig.17. The element size for the fluid domain is set as 0.12 mm while the meshing size is 0.04 mm for the solid domain in order to include sufficient detail of the heater cells. The boundary layer is defined as the impingement jet region, which is the interface between the liquid and chip surface. The first layer thickness of the boundary layer is set as 1e-3 mm with 10 layers above fluid/solid interface, with layer growth rate 1.2. The total number of elements is 1.6M. In order to capture the temperature map of the hotspot cooling, a high level of details in the thermal model of the heaters is needed for the high heat removal rates obtained in high cooling performance impingement cooler [8].

B. CFD Model validation

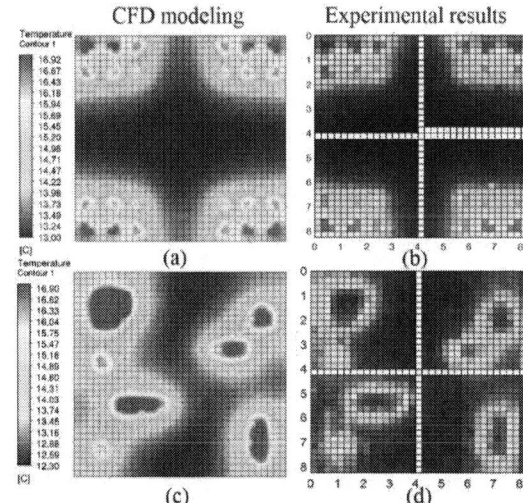

Figure 18. Temperature map distribution of CFD modleing versus experimental measurement: test case 1 (a) with CFD modeling and b) experimental results (flow rate =600 ml/min, Q=4.1W); test case 2 with (c) CFD modeling and (d) experimental results (flow rate =600 ml/min, Q=5.5 W).

As illustrated in Fig.18, the temperature distribution comparisons between the full cooler level CFD modeling and experimental results for the regular pattern are illustrated. For the test case 1, the comparisons between the experiments and CFD modeling results show that the averaged chip temperature is less than 2.5% difference compared with the experimental results with uniform nozzle array cooling, while the difference is 8.5% for hotspot cooling. For the averaged chip temperature in the test case 2, the comparison shows about 18.3% difference comparing with the experimental data for uniform nozzle array cooling. The asymmetrical temperature map shown in Fig.19(a) and (b) is due to the placement of the outlet tube connector, which is located at the one side of the cooler.

The comparison of hotspot cooling with various hotspot size in Fig.19(c) also indicates that the full CFD cooler level model can capture the major trends of the experimental results. In general, the modeling curve for HS array cooling shows good agreement with the experimental curve. Based on the acceptable errors of the full CFD cooler model comparing with experimental data, the CFD model with hotspot cooler is successfully validated. This means that we can use the validated CFD model for future design improvements of the printed cooler and to assess the trade-off between the thermal performance improvement and the pressure drop penalty in the hotspot targeted cooler.

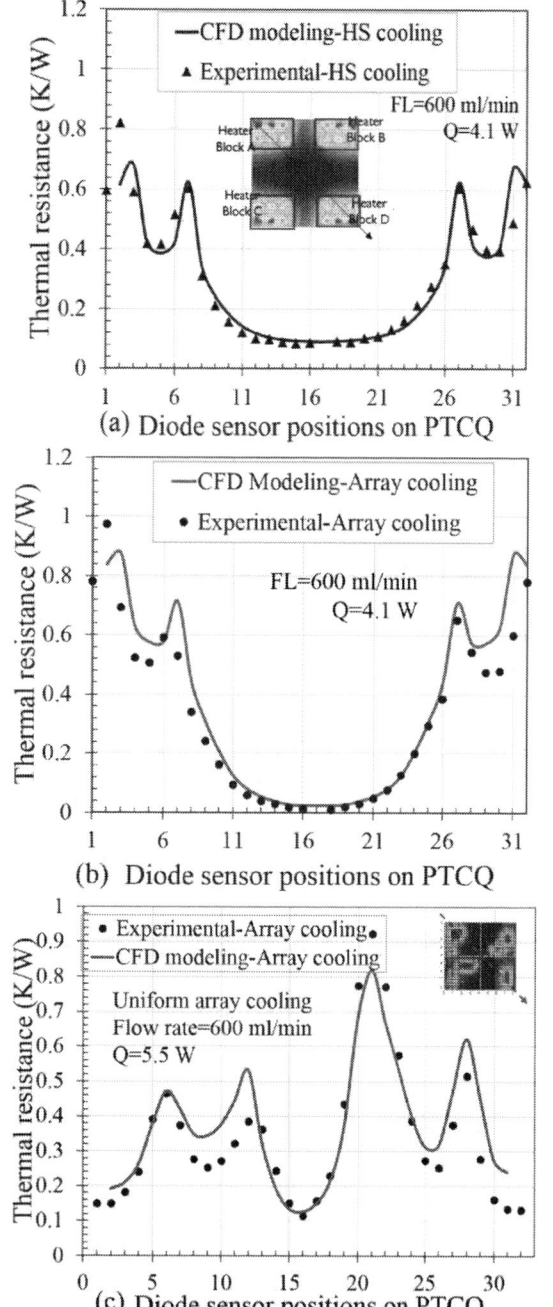

Figure 19. Experimental CFD model validation for cooler of test case 1 with (a) hotspot targeted cooling and (b) nozzle array uniform cooling; (c) experimental validation for test case 2 with nozzle array uniform cooling.

C. Hydraulic performance analysis

From the above experiments and modeling studies, we can see that the hotspot target cooling can achieve lower temperature by concentrating the coolant, but higher pressure is required to push the same flow rate through smaller number of nozzles. Therefore, more information about the nozzle velocity and pressure are needed. In the following section, the validated CFD model is first used to extract the hydraulic behavior inside the coolers. The flow streamlines inside the hotspot cooler for test case 1 and test case 2 are both shown in Fig.20. The flow streamline inside the distributor shows more flow recirculation for the hotspot targeted cooling since the flow is concentrated into the reduced number of inlet nozzles.

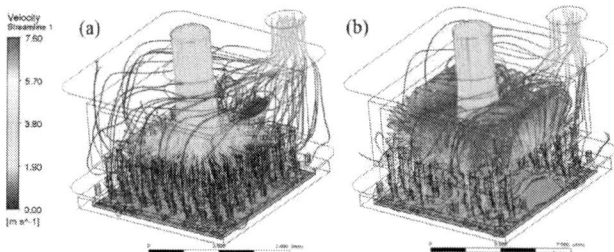

Figure 20. Flow streamlines inside the cooler for (a) uniform nozzle array cooling and (b) hotspot targeted cooling. (flow rate =600ml/min)

Moreover, the validated CFD model is also used to extract the internal velocity and pressure information for different nozzle arrays as shown in Fig.21. The pressure drop between inlet and outlet of the cooler for the nozzle array uniform cooling with regular pattern is 0.16 bar while the pressure drop is 0.57 bar for hotspot targeted cooler. The comparison indicates that the pressure drop increases by a factor of 3.56X for the same flow rate. Moreover, the flow distribution for every inlet nozzle is shown in Fig.21 (c). The velocity is plotted across the nozzle plate region by covering the inlet/outlet velocity. The flow distribution for full array cooler shows about 57% maximum deviation in the center while the maximum deviation is 22% in other regions compared with the theoretical velocity of 2.2 m/s. The large deviation in the center is due to the inlet impinging flow coming from the top. For the hotspot cooler, the velocity for the nozzles are with maximum difference of 13.5% comparing with theoretical velocity value of 5.9 m/s.

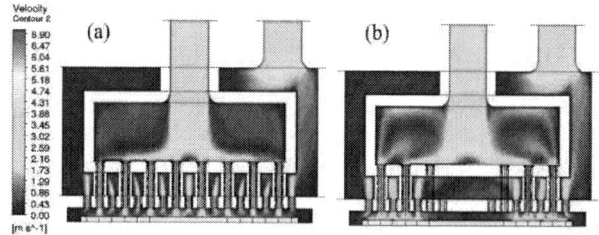

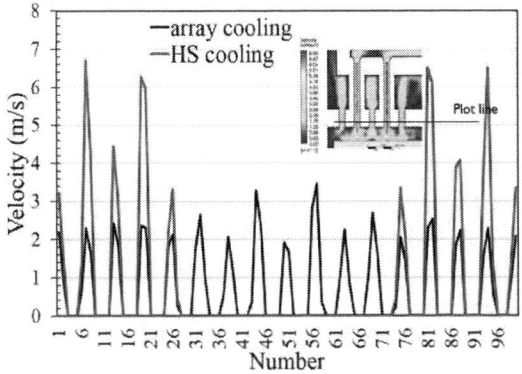

Figure 21. Velocity information extractions for regular HS pattern: cross section of the velocity distribution for (a) nozzle array cooling and (b) hotspot targeted cooling; (c) inlet velocity profile comparison.

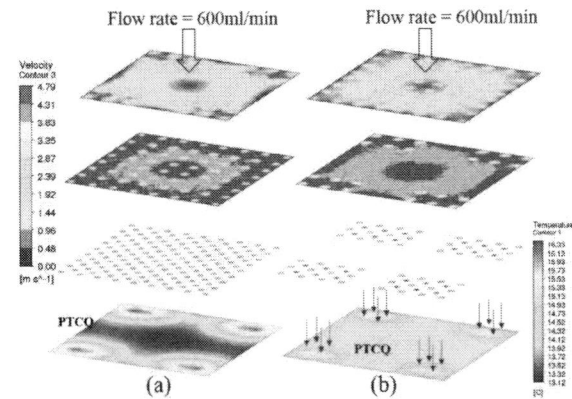

Figure 22 Flow and thermal interactions with (a) nozzle array cooling and (b) hotspot targeted cooling.

As shown in the flow and thermal interaction analysis in Fig.22, we can see that, hotspot targeted cooler with locally high flow rate can locally reduce the HS temperature with the constant total flow rate.

D. Thermal – hydraulic trade-off analysis

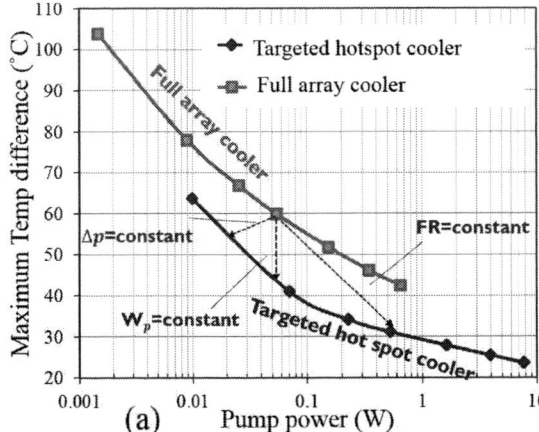

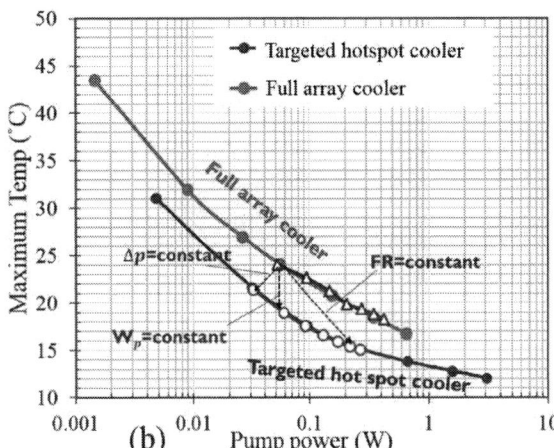

Figure 23. Thermal/hydraulic performance trade-off comparison between the hotspot cooler and the full array cooler for (a) test case 1 and (b) test case 2.

In the last section, the comparison between the hotspot targeted cooler (test case 1) and the full array cooler is shown in Fig.23(a). The performance of the two coolers is shown as curves in terms of the hotspot peak temperature ΔT_{max} and the required pumping power for a range of low rates. A lower position of the curve, closer to the origin, corresponds to a better performance. The performance of the coolers is now compared for different constraints [11]:

1. Same pressure drop over the cooler
2. Same flow rate
3. Same pumping power

For the same pressure drop of 10 kPa, the ΔT_{max} can drop by a factor of 1.08 for the hotspot targeted cooler compared to the full array cooler, but it would use 2.5x less flow rate and pumping power. For the same flow rate at 100 ml/min, the ΔT_{max} can reduce by a factor of 2, but it requires 10x larger pressure drop; For the same pumping power at 0.05W, the ΔT_{max} drops by 35% compared to the full array cooler.

The comparison for test case 2 is shown in Fig.23(b). For the same pressure drop at 10 kPa, the ΔT_{max} can drop by a factor of 1.1, but with less flow rate; For the same flow rate at 100ml/min, the ΔT_{max} can reduce by 1.54x, but with larger pressure drop; For the same pump power at 0.05W, the ΔT_{max} drop by 1.36x. In general, for all considerations, the hotspot cooler curve shown in blue is below the full array cooler curve (in red), indicating that the hotspot targeted cooling is more energy efficient compared with the full array cooler for specific hotspot patterns.

CONCLUSIONS

This work demonstrates the hotspot targeted cooler for high power devices with localized high heat flux regions, for the first time using 3D printing, which enables the fabrication of a matching design of the cooler nozzle array to the power map, and at the same time offers a huge reduction in the cooler size, matching the footprint of the power electronics packaging. Two types of hotspot cooler with a small nominal diameter of 300 μm for a cooling unit cell area of 1×1 mm^2 are demonstrated with high reproducibility. The fabrication tolerance has been assessed and its impact on the thermal and hydraulic performance has been evaluated by using unit cell CFD models. The modeling study shows that there is a benefit with respect to significant pressure drop reduction due to the slightly tapered fabricated nozzle wall while this results in a slight reduction of thermal performance. The thermal impact of the nozzle-to-chip distance variation is negligible for cavity heights with $H/L > 0.25$ due to the cooling saturation.

The hotspot cooling experimental measurements show a peak temperature reduction of 16% and 24% at a flow rate of 900 mL/min/cm^2 compared to the full array cooler for the targeted hotspot coolers of test case 1 and test case 2 respectively. Moreover, the experimentally validated full cooler level CFD models for both test cases are used to investigate the thermo-hydraulic behavior of the cooler for different flow constraint conditions. The trade-off charts between the maximum temperature difference and required pumping power prove that the hotspot targeted cooler exhibits a more energy efficiency cooling performance compared with the uniform nozzle array cooler for the hotspot test cases for all considered comparison constraints.

ACKNOWLEDGMENTS

This work was performed as part of the imec Industrial Affiliation Program on 3D System Integration and has been strongly supported by the imec partners and the imec Reliability, Electrical testing, Modeling and 3D technology teams.

REFERENCES

[1] Avram Bar-Cohen, "On-chip thermal management and Hotspot remediation, Nano-Bio- Electronic", Photonic and MEMS Packaging. 2009, pp 349-429.

[2] C S Sharma, "Energy efficient hotspot-targeted embedded liquid cooling of electronics", Applied Energy, Volume 138 2015, pp. 414-422.

[3] Golak Kunti, "Alternating Current Electrothermal Flow for Energy Efficient Thermal Management of Microprocessor Hotspot", Applied Physics, 2018

[4] DanishAnsari, Kwang-YongKim, "Hotspot thermal management using a microchannel-pinfin hybrid heat sink", International Journal of Thermal Sciences, Volume 134, December 2018, pp.27-39.

[5] G. J. Snyder, M. Soto, et al., "Hotspot cooling using embedded thermoelectric coolers," 22nd Annual IEEE Semiconductor Thermal Measurement And Management Symposium, Dallas, TX, 2006, pp. 135-143.

[6] G Bindiganavale, et al., "Study of hotspot cooling using electrowetting on dielectric digital microfluidic system", 2014 IEEE 27th International Conference on Micro Electro Mechanical Systems (MEMS), San Francisco, CA, 2014, pp. 1039-1042.

[7] T.-W. Wei, H. Oprins, et al., "High efficiency direct liquid jet impingement cooling of high power devices using a 3D-shaped polymer cooler," 2017 IEEE International Electron Devices Meeting (IEDM), San Francisco, CA, 2017, pp. 32.5.1-32.5.4.

[8] Tiwei Wei, Herman Oprins, et al., "Experimental characterization and model validation of liquid jet impingement cooling using a high spatial resolution and programmable thermal test chip", Applied thermal engineering, 2019, *submitted.*

[9] T.-W. Wei, H. Oprins, et al., "3D Printed Liquid Jet Impingement Cooler: Demonstration, Opportunities and Challenges," 2018 IEEE 68th Electronic Components and Technology Conference (ECTC), San Diego, CA, 2018, pp. 2389-2396.

[10] H. Oprins, et al., "Characterization and Benchmarking of the Low Intertier Thermal Resistance of Three-Dimensional Hybrid Cu/Dielectric Wafer-to-Wafer Bonding". ASME. J. Electron. Packag. 2017;139(1):011008-011008-9.

[11] Tiwei Wei, Herman Oprins, et al., "Nozzle array scaling analysis of the thermal performance of liquid jet impingement coolers for high performance electronic applications", The International Heat Transfer Conferences (IHTC), 2018.

Rapid Production of Customized 3D Electronics via Hybrid Additive Manufacturing Technology

Ji Li*, Yang Wang, Peiren Wang, Jiangling He, Handa Liu, Gengzhao Xiang

Key Laboratory of MEMS of the Ministry of Education, Southeast University
Nanjing, China
Email: j.li5@seu.edu.cn

Abstract — **Here a novel hybrid additive manufacturing technology is proposed. This method integrates multimaterial fused deposition modelling 3D printing and selective electroless plating. Fused deposition modeling is used to print three-dimensional substrate composed by platable and non-platable plastics in accordance with CAD model. With surface treatment steps, only platable part can absorb electroless plating catalysts and thereby metal film can be selectively deposited. This provides a simple solution to realize freeform 3D circuitry patterning. Electronic components are then mounted on corresponding positions. A full-functional customized 3D electronics is created. The adhesion of electroless metal film can achieve the highest grade (5B) of tape test, and the minimal resistivity obtained is 5 $\mu\Omega \cdot$ cm with copper film. A 3D LED blinking circuitry was fabricated to testify the feasibility and application potential of the technology.**

Keywords-3D electronics, hybrid additive manufacturing, 3D printing , fused deposition modelling, electroless plating

I. INTRODUCTION

Nowadays, many applications, in particular medical equipment, automotive industry, consumer goods, aerospace and aviation, require high-value, customized, end-use electronics with complex 3D structures that can be manufactured rapidly and cost-effectively. This creates challenges to traditional planar manufacturing technologies (e.g. PCB technology). Due to the design and fabrication flexibility, hybrid additive manufacturing (AM) technology has been regarded as a potential solution. Compare with common AM processes that can only fabricate structural parts with single type material, hybrid AM technologies combine traditional AM processes with other digital manufacturing technologies such as CNC machining, direct writing (DW), laser-based processing. This enables the rapid development and direct production of 3D full-functional electronic products with complex geometrical structures [1].

Earlier hybrid AM works focused on combining direct writing (DW) with various AM processes to build 3D structural electronics. DW is a versatile and multi-scale group of processes that are able to deposit various materials onto different types of substrates based on a digital pattern design [2]. Most commonly used DW processes are dispensing process, inkjet printing, aerosol jet printing. Ryan Wicker's group built the first hybrid AM apparatus by installing material extrusion unit to a stereolithography (SL) 3D printer. A series of 3D electronics such as RF antenna [3], Hall-effect sensor [4], and electronic game die [5] have been created with this apparatus. A similar fused deposition modelling (FDM) platform was also developed to for fabricating 3D electronic products with higher mechanical strength [4]. Our previous work also combined digital light projection (DLP) stereolithography process with material extrusion process for fabricating 3D electronics [6]. Johander et al adopted Inkjet printing to plot electrical conductive interconnections on the exterior of a 3D substrate made by shape deposition manufacturing (SDM) [7]. Aerosol jet was applied by Optomec Inc. and Stratasys Ltd. to integrate antennas, sensors and circuits on/within the wings of a FDM printed UAV [8]. This could realize ~10 μm wide conductors [9]. Chang et al. used Aerosol jet to embed strain gauges into 3D parts made by PolyJet process [10]. DW processes generally have advantages in plotting conductive features planar surface. However, it is difficult for them to realized three-dimensional freeform circuitry. Furthermore, the conductive inks widely used in DW processes are normally a mixture of polymer binders and conductive particles. Generally, their mechanical strength and electrical conductivity are much lower than bulky metallic materials, which largely limited the performance of 3D electronics.

This paper proposes a new hybrid AM solution that combines multimaterial fused deposition modeling 3D printing and selective electroless plating. Metallic film can be selectively plated on dual-material 3D substrate. This provides great versatility for the design and rapid fabrication of customized electronic products with complex geometry structures and multiple material combinations and high end-use functionality. A 3D LED blinking circuitry was fabricated to testify the feasibility and application potential of the technology.

II. EXPERIMENTAL METHOD

A. Process Chain

The fabrication process of proposed technology is demonstrated as Fig. 1.

---

* Corresponding author.

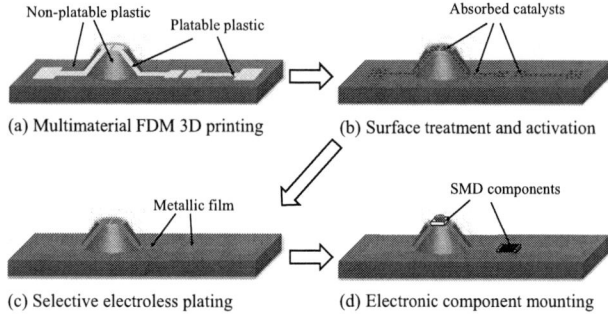

Figure 1. Fabrication process of proposed technology.

First, the 3D structure is fabricated using a FDM multimaterial 3D printer. The circuit base and substrate are made of platable and non-platable plastics, respectively. After that, the 3D structure is ultrasonically cleaned in isopropanol (C_3H_8O), and then etched in chrome-sulfuric solution (400g/L chromium trioxide (CrO_3), 400 g/L sulfuric acid (H_2SO_4)) at 60°C. The test samples are etched from 3 to 10 min. The strong etching solution partially modifies the surface of FDM plastic making the surface of platable plastic hydrophilic and rough. This enhances the absorption chance of electroless plating catalysts. After cleaning with deionised water (DI) water, the structure is sensitized and activated in a Pd/Sn catalyst colloids (Noviganth Activator PL, purchased from Atotech GmbH, Germany) for 2 min at room temperature, and a DI water rinsing is performed. Following that, the structure is dipped in ADHEMAX® Accelerator 1 solution (Atotech GmbH, Germany) for 15 mins at 45°C. This acceleration step reduces Pd^{2+} ions to Pd particles with Sn^{2+} ions, and dissolves excess tin colloids to release Pd catalysts. Furthermore, this step is capable of removing the Pd particles on the non-platable part. Following that, the structure is DI water cleaned and then immersed in electroless copper bath. The reaction was performed at 65 °C and pH13. The plating duration was varied from 5 minutes to 2 hours. Finally, electronic components are mounted on their corresponding place fulfilling the functionality of the electronic products.

B. Characterization Methods

Optical microscopy measurements were performed using a Nikon SMZ745T microscope. Carl Zeiss Ultra Plus FE-SEM system was employed for scanning electron microscopy (SEM) investigation. An X-Max 20 energy dispersive X-Ray spectroscopy (EDS) from Oxford Instrument integrated with Ultra Plus FE-SEM was used for chemical composition analysis. Four-wire sensing measurement was conducted via a UT620B digital micro ohmmeter (UNI-T Inc., China) in 4-wire measurement mode.

III. RESULTS AND DISCUSSIONS

A. Surface Modification of FDM Plastics

In preliminary experiments, we tested four conventional FDM plastics including polylactic acid (PLA), Acrylonitrile-butadiene-styrene (ABS), polycarbonate (PC), and polyethylene terephthalate glycol-modified (PETG). It was found that ABS, PC, and PLA could be plated, while there was no metal film deposited on PETG surface. However, the metal layer on PC and PLA shew poor adhesion and could be easily erased even with finger. Therefore, ABS and PETG were adopted as plateable and non-plateable plastics, respectively, in the following experiments.

Fig. 2 shows the surface morphology modification of ABS and PETG before and after etching step. Dense pores were generated on ABS surface etched for only 3 mins (Fig. 2b). Expanding etching time, the density and dimension of these micro-pores increased (Fig. 2c). The entire surface of ABS was nearly eaten away for 10 min etching creating a coarse appearance (Fig. 2d). In contrast, PETG demonstrated limited roughening effect even after 10 min processing (Fig. 2h). There are few pores found on the surface.

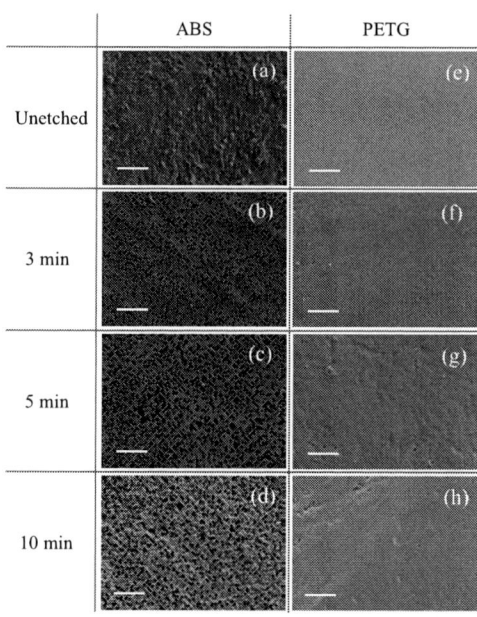

Figure 2. SEM photos of ABS and PETG surfaces before and after etching.

The different molecular structure of ABS and PETG plastics probably caused their different etching effects. According to Fig. 3, ABS plastic consists of two polymer phase: a polybutadiene phase and a styrene–acrylonitrile phase. The chrome-sulfuric etchant can remove both phases. However, because of C=C bond oxidization the polybutadiene phase is much easier to be dissolved. Thus, micro-pores could be found on ABS surface. For PETG plastic, there is no easily broken C=C bond so it is more stable in chrome-sulfuric solution. Moreover, PETG has mono phase chemical structure, thus the etching reaction is occurred homogeneous on the surface achieving less roughening effects (Fig. 2f, g, and h).

The micro-pores in ABS surface provide interlock points for plated metallic film. This can improve the adhesion between plated metallic film and ABS surface. To fully exploit this effect, etching duration need to be precisely controlled to create a modified surface with well-preserved styrene–acrylonitrile matrix and totally dissolved polybutadiene phase.

978-1-7281-1500-9/19 $31.00 © 2019 IEEE

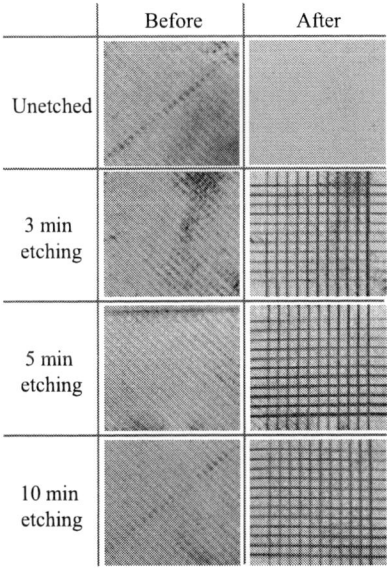

Figure 3. Chemical structure unit of ABS and PETG.

B. Adhesion of Plated Metallic Film

Tape test was performed to identify the effects of etching duration on the adhesion of plated metallic film. The test is conducted according to ASTM D3359-09 standard. A matrix with 100 squares was cut into the metallic film. After that, Intertape 51596 tape, a pressure-sensitive tape, was attached on the matrix and then peeled off. Inspecting the ratio of the peeled metallic film, the adhesion strength can be identified as the following criteria: 0B (65% or more of the matrix is peeled), 1B (the ratio of removed matrix is between 35% to 65%), 2B (the ratio of removed matrix is between 15% to 35%), 3B (the ratio of removed matrix is between 5% to 15%), 4B (the maximum ratio of removed matrix is 5%), and 5B (the whole matrix is intact).

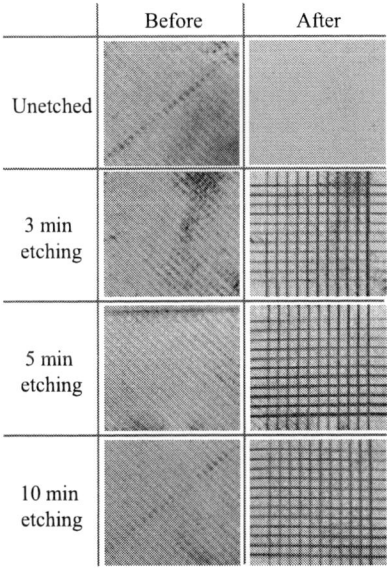

Figure 4. Tape test result of electroless copper film on ABS surface.

According to Fig. 4, for unetched ABS substrate, the whole copper film was peeled off, and thus its adhesion grade was 0B. With 3 min etching, less than 5% copper film was detached implying a 4B adhesion rate. When etching time was longer than 5 min, there was nearly no copper teared off, so the highest adhesion score 5B was obtained. Thus, for ABS plastic, 5 min etching at 60 °C is sufficient to generate enough rooting points for the metallic film providing high-strength adhesion (Fig. 2c).

According to tape test and SEM imaging results, a combination of PETG and ABS plastics was found to be ideal for realizing metallization selectivity.

C. Fabrication accuracy of multimaterial FDM 3D Printing

In order to improve the integration of electronic products the linewidth of conductor has to be miniaturized. Therefore, minimal linewidth that can be realized by this hybrid AM technology need to be identified. The design of test samples is shown in Fig. 5. There are 3 ABS tracks that were embedded inside a PETG base. Five different linewidths: 1 mm, 0.8 mm, 0.6 mm, 0.5 mm, and 0.4 mm, were tested in this paper. Due to the 0.4 mm-diameter nozzles used in our FDM printer, the minimal linewidth was designed to be 0.4 mm. Any further reduction of linewidth could cause errors when slicing CAD models to create digital control code for driving FDM printer. The printed samples were plated as described in Section II and the finished samples were shown in Fig. 5d.

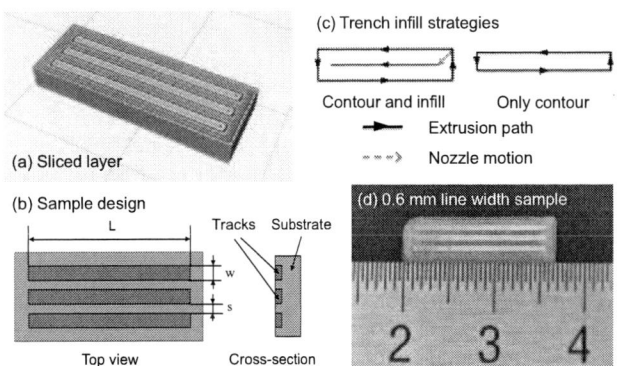

Figure 5. Schematic drawing and fabrication technique of the samples.

Fig. 5 demonstrates the difference between measured and design linewidths of the printed conductors. In general, increasing design value could dramatically reduce fabrication error. For the samples with the linewidth of 0.4 mm and 0.5 mm, the actual values were approximately one fifth larger than the design values. Greater standard deviations were also observed on those samples compared with wider counterparts. For 0.6 mm-, 0.8mm-, and 1mm-wide samples, the fabrication error was smaller than 0.04 mm. Their standard deviation was decreased to ±0.015 mm. The greater fabrication errors found on narrower conductors was majorly caused by the printing mode and the nozzle orifice. For a specific printing layer, a PETG outline was first plotted by the main extruder, and then the trenches was filled with melted ABS by the secondary extruder (Fig. 5a, c). Therefore, when the linewidth was close to the nozzle diameter, melted ABS was over-filled into the PETG trenches. At the same time, the moving nozzle squashed the filled ABS and thus resulted in the overflow of ABS to PETG part. Therefore, 0.6 mm linewidth was adopted to be the linewidth used in electrical conductivity characterization in order to maintain the fabrication precision. Moreover, the minimal distance between neighboring conductors was limited to 0.5 mm to avoid any short connecting.

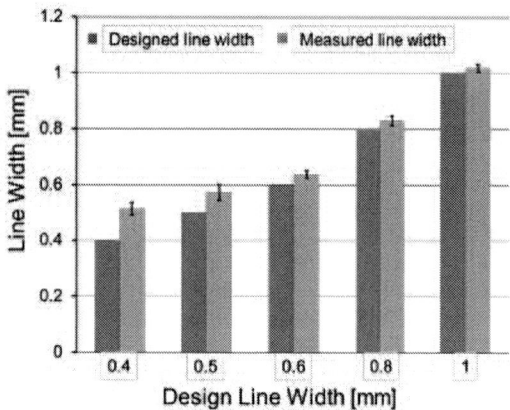

Figure 6. Comparison between measured and design linewidths.

D. Chemical Composition of Electroless Copper Film

To fabricate high performance 3D electronic systems, the electrical conductivity of circuit interconnection must be improved. Thus, electroless copper plating was employed here to obtain highly conductive copper layer on ABS surface. Theoretically, the resistivity of bulk copper is as low as 1.68 $\mu\Omega\cdot$cm. Moreover, the heat deflection temperature of PETG and ABS plastics are about 65 °C and 80 °C, which limits the maximum reaction temperature of electroless copper bath. Thus, alkaline electroless copper plating with reaction temperature of 65 °C was adopted here. The composition of alkaline electroless copper bath is shown in the table below.

TABLE I.   COMPOSITION OF ALKALINE ELECTROLESS COPPER BATH

Chemicals	Concentration
Copper sulfate pentahydrat ($CuSO_4\cdot5H_2O$)	0.04 M
Glyoxylic acid ($C_2H_2O_3$)	0.2 M
Ethylenediaminetetraacetic acid disodium salt (EDTA·2Na)	0.1 M
2,2'-Dipyridyl ($C_{10}H_8N_2$)	10 mg/L
Potassium ferrocyanide ($K_4Fe(CN)_6\cdot3H_2O$)	10 mg/L
Polyethylene glycol ($HO(CH_2CH_2O)nH$)	100 mg/L

The electroless copper bath is composed of copper sulfate pentahydrat ($CuSO_4\cdot5H_2O$) as the copper ion source, glyoxylic acid ($C_2H_2O_3$) as the reducing agent, ethylenediaminetetraacetic acid disodium salt (EDTA·2Na) as the complexing agent, polyethylene glycol ($HO(CH_2CH_2O)nH$) as the surface activator, 2,2'-Dipyridyl ($C_{10}H_8N_2$) and potassium ferrocyanide ($K_4Fe(CN)_6\cdot3H_2O$) as the stabilizing agents. The over-all reaction of the electroless copper plating is:

$$Cu^{2+} + 2CHOCOOH + 4OH^- \rightarrow Cu^0 + 2HC_2O_4^- + 2H_2O + H_2 \uparrow \tag{1}$$

The chemical composition of electroless copper film was investigated via EDS inspection. The test sample was plated at 65°C and pH13 for 1 h. A typical EDS spectra is shown in

Fig. 7. The metallic film consisted of copper (0.94 keV/Lα, 8.04 keV/Kα; and 8.92 keV/Kβ) and its content was about 92%. Other parts were carbon (0.28 keV/ Kα) and oxygen (0.52 keV/ Kα), which had a content of about 7% and 1%, respectively. The signal of carbon and oxygen probably came from ABS substrate due to the electron beam penetration through the copper film during EDS test.

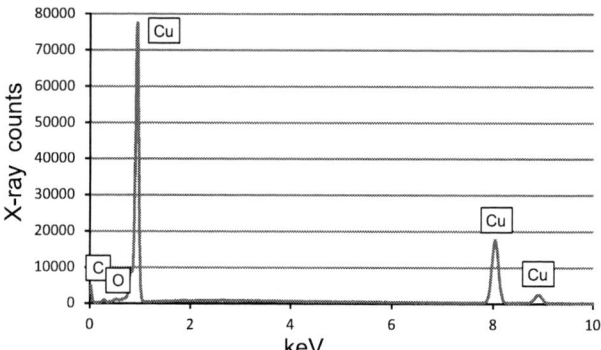

Figure 7. EDS inspection result of electroless copper layer.

E. Electrical Resistivity of Copper layer

Thus, the effects of electroless plating time on the electrical conductivity were characterized via investigating test samples plated for 5, 10, 30, 60, and 120 minutes, respectively.

Four-wire sensing measurement was employed to identify the electrical resistivity of electroless copper layer obtained at different plating times. Four-wire sensing measurement is based on Ohm's law. A constant current is driven through the resistor, and the voltage drop at both sides is measured to obtain the resistance as:

$$R = \frac{U}{I} \tag{2}$$

The resistivity of a resistor ρ (Ω cm) can be acquired as Pouillet's law:

$$\rho = \frac{RA}{L} = \frac{Rwd}{L} \tag{3}$$

where l (cm), d (cm), w (cm), R (Ω), and A (cm²) are the length, thickness, width, resistance, and cross-section area of the resistor, respectively. SEM and optical microscopy were used to obtain the value of l, d, and w.

The front surface and the cross-section of copper conductor were SEM investigated as shown in Fig. 8. The copper film was dense and continuous, and there is no skip-plate island on the surface of copper film (Fig 8b). Micro-pores created by etching step were observed on cross-section image (Fig. 8a) which provided interlock points between electroless copper layer and the ABS substrate. This enhanced the adhesion strength of copper layer.

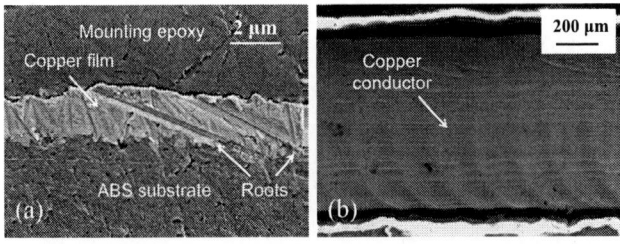

Figure 8. SEM images of front surface and cross-section of electroless copper layer.

The temperature and pH value of electroless copper bath were set to be 65 °C and 13, respectively, which were conventional settings used in plating industry. As shown in Fig. 9, the copper film thickness increased power-exponentially with the increase of reaction time. This indicated that the plating rate decreased gradually with the consumption of electroless plating reagents.

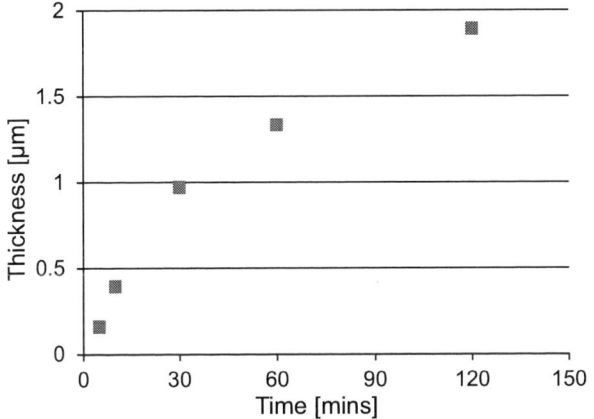

Figure 9. The thickness coper film changing with electroless plating time.

According to Fig. 10, the resistivity had a dramatic reduction in the beginning of plating, and then turn to be constant of about 5 $\mu\Omega\cdot$cm after 60 mins reaction. The main reason of this phenomenon was that the copper layer at this stage was not thick enough to cover the coarse ABS surface becoming a whole conductor. This resulted in poor conductivity of the copper layer. With continuous copper deposition, ABS surface was fully covered and thereby the resistivity turned to be constant. In accordance with the tendency of resistivity, the resistance dropped remarkably in the first 30 mins and then declined linearly with the thickness increase of copper layer. This means the resistance of the copper film can be customized via varying electroless plating time.

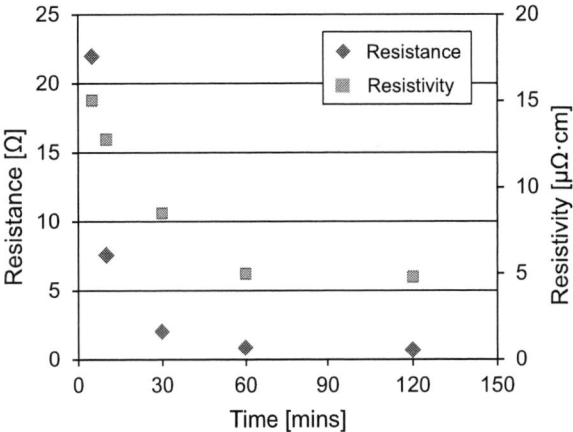

Figure 10. The electrical performance of copper conductors changing with electroless plating time.

IV. DEMONSTRATORS

In order to prove the feasibility of the hybrid AM technology mentioned above, a three-dimensional LED twinkling circuitry as shown in Fig. 10 was built. The circuitry includes 3 SMD resistors, 1 SMD capacitor, 1 SMD LED, and a 555 timer.

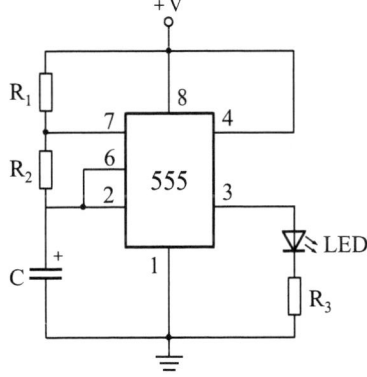

Figure 10. Working principle of the LED blinking circuitry.

Three representative structures including plane, pyramid and trench were arranged in a three-dimensional base. ABS paths that used as circuit base were placed on the surfaces of these three structures (Fig. 11b). The models of ABS circuit base and PETG substrate were designed individually with CAD software. After that, both models were combined and sliced with Cura software (Fig. 11). The 3D substrate was 3D printed using multimaterial FDM printer and then electrolessly copper plated for 1 hour. The copper circuitry was selectively deposited on the exteriors of the plane, the pyramid and the trench implementing three-dimensional interconnection. All SMD components were mounted on the 3D substrate by silver conductive adhesive. Eventually, the LED on the top of pyramid twinkled in a frequency of approximate 1 Hz with 9V voltage power (Fig. 12).

978-1-7281-1500-9/19 $31.00 © 2019 IEEE

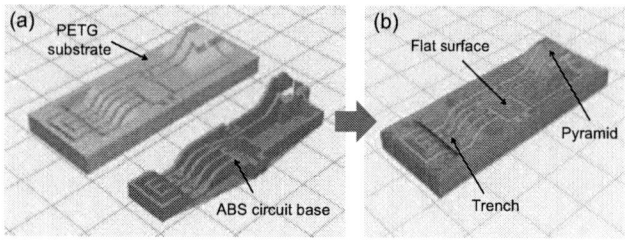

Figure 11. (a) 3D designs of 3D dual-material structure; (b) merged and sliced models.

The 3D LED twinkling circuitry proved that the hybrid AM technology could provide great design and fabrication flexibility, which is capable of realizing high-value customized electronic products with complex three-dimensional geometries.

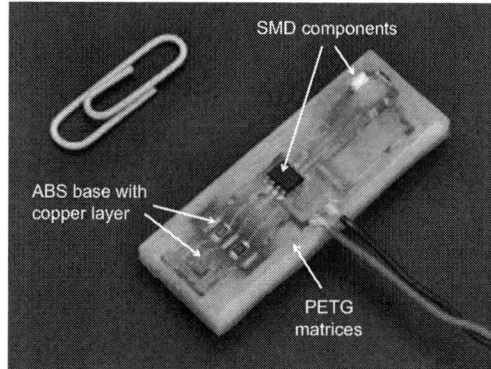

Figure 12. 3D LED blinking circuitry.

V. CONCLUSIONS

This work proposed a new hybrid AM method integrating fused deposition modelling 3D printing and electroless copper plating. It facilitates the rapid design and fabrication of bespoke full-functional 3D electronic products.

The combination of PETG and ABS plastics has facilitated desirable selective metallization and high-strength adhesion of plated copper layer. The finest linewidth of conductor is around 0.5 mm via using 0.4 mm-orifice FDM nozzle. To achieve high conductive 3D electrical interconnection, electroless copper plating was performed and the effect of plating time on the resistivity of copper film was investigated. It was found that the deposition rate reduced power-exponentially with the increase of plating time and the resistivity turn to be constant after 1h reaction. Therefore, the resistance of copper conductor could be customized based on practical requirements. With SEM imaging and four-wire sensing measurement, the resistivity of copper conductor was obtained to be approximate 5 $\mu\Omega \cdot$ cm.

A 3D LED twinkling circuitry was fabricated to demonstrate the application potential of the proposed hybrid AM method. Its capability of rapidly customizing 3D electronics has been fully testified.

ACKNOWLEDGEMENTS

This work was supported by the National Natural Science Foundation of China (Grant No. 61504024), the Fundamental Research Funds for the Central Universities (Grant No. 2242017K40060 and 2242019k1G002), the Innovative and entrepreneurial talent plan of Jiangsu province, China (Grant No. 1106000206), and the science and technique program foundation for elite returned overseas Chinese scholars, Nanjing, China.

REFERENCES

[1] E. MacDonald and R. Wicker, "Multiprocess 3D printing for increasing component functionality," Science (80-.)., vol. 353, no. 6307, 2016.

[2] A. Piqué, Direct-Write Technologies for Rapid Prototyping: Applications to Sensors, Electronics, and Passivation Coatings. 2002.

[3] A. J. Lopes, E. MacDonald, and R. B. Wicker, "Integrating stereolithography and direct print technologies for 3D structural electronics fabrication," Rapid Prototyp. J., vol. 18, no. 2, pp. 129–143, 2012.

[4] D. Espalin, D. W. Muse, E. MacDonald, and R. B. Wicker, "3D Printing multifunctionality: Structures with electronics," Int. J. Adv. Manuf. Technol., vol. 72, no. 5–8, pp. 963–978, Mar. 2014.

[5] E. MacDonald, R. Salas, D. Espalin, M. Perez, E. Aguilera, D. Muse, and R. B. Wicker, "3D printing for the rapid prototyping of structural electronics," IEEE Access, vol. 2, pp. 234–242, Dec. 2014.

[6] J. Li, T. Wasley, T. T. Nguyen, V. D. Ta, J. D. Shephard, J. Stringer, P. Smith, E. Esenturk, C. Connaughton, and R. Kay, "Hybrid additive manufacturing of 3D electronic systems," J. Micromechanics Microengineering, vol. 26, no. 10, p. 105005, 2016.

[7] P. Johander, S. Haasl, K. Persson, and U. Harrysson, "Layer Manufacturing as a Generic Tool for Microsystem Integration," in Proceedings of the Third International Conference on Multi-Material Micro Manufacture, 2007, pp. 3–5.

[8] J. A. Paulsen, M. Renn, K. Christenson, and R. Plourde, "Printing conformal electronics on 3D structures with aerosol jet technology," in Proceedings of the 2012 Future of Instrumentation International Workshop (FIIW), 2012, pp. 47–50.

[9] K. B. Perez and C. B. Williams, "Design Considerations for Hybridizing Additive Manufacturing and Direct Write Technologies," in Proceedings of ASME 2014 International Design Engineering Technical Conferences & Computers and Information in Engineering Conference, 2014, pp. 1–12.

[10] Y.-H. Y. Chang, K. Wang, C. Wu, Y. Chen, C. Zhang, and B. Wang, "A facile method for integrating direct-write devices into three-dimensional printed parts," Smart Mater. Struct., vol. 24, no. 6, p. 065008, 2015.

Solid-Liquid InterDiffusion (SLID) Bonding, for Thermally Challenging Applications

Knut E. Aasmundtveit[1], Thi-Thuy Luu[2], Hoang-Vu Nguyen[1], Andreas Larsson[1,3] and Torleif A. Tollefsen[4]

[1]Department of Microsystems, University of South-Eastern Norway, 3184 Borre, Norway
[2]Zimmer and Peacock AS, 3183 Horten, Norway
[3]Techni AS, 3184 Borre, Norway
[3]TEGma AS, 3015 Drammen, Norway
Knut.Aasmundtveit@usn.no

Abstract—Solid-Liquid InterDiffusion (SLID) bonding is particularly suited for high-temperature applications, since SLID bonds can tolerate higher temperatures than the bonding temperature. SLID uses a layered binary metal structure, which reacts to high-temperature stable intermetallics at normal solder temperatures. Hence, high-temperature stability is achievable for a process at moderate bonding temperatures. Alternatively, low-temperature SLID bonding (bonding down to ~100 °C) allows bonding of temperature-sensitive components and materials, without restricting the application temperature range.

Cu–Sn is the most mature SLID system. We show optimized Cu–Sn SLID bonding for vacuum encapsulation of MEMS devices. Au–Sn SLID has superior oxidation resistance, and we demonstrate that Au–Sn SLID has excellent reliability when bonding thermally mismatched chips and substrates. We demonstrate Ni–Sn SLID bonding, as well as the low-temperature alternatives Au–In and Au–In–Bi.

For Cu–Sn, Au–Sn and Au–In SLID, we show experimental evidence for the high-temperature stability predicted from phase diagrams.

Keywords-Microelectronics/ microsystem assembly, harsh environments, transient liquid phase.

I. INTRODUCTION

Solid-Liquid InterDiffusion (SLID) bonding has the distinct property that the bondline remelts at higher temperatures than the bonding temperature. It is therefore very suitable for high-temperature applications, also at temperatures where regular solder bonds would melt [1-4]. It is a versatile technique that can be used for interconnects, die attach or sealing, and flux-free bonding is achievable [1]. SLID bonding will survive high-temperature processes such as getter activation, hence it is also well suited for vacuum sealing. SLID bonding can eliminate the need for a hierarchy of solder alloys with different melting temperatures for bonding of stacks or for bonding at different levels of integration, hence simplifying manufacturing processes. Using appropriate metal systems, low-temperature SLID bonding can be used for temperature-sensitive materials such as polymers or poled piezoelectrics, where the resulting bond can still be reliable at somewhat elevated temperatures [5].

SLID bonds have a well-defined metallurgy and typical bondline thickness in the micrometer range. It is also very well suited for fine-pitch bonding [2, 6, 7]. SLID bonding is compatible with wafer-level, as well as chip-level, bonding.

SLID bonding technology typically uses a thin-layered metal system, with one high-temperature stable metal (such as Cu, Ni or Au) and one low-temperature melting metal (such as Sn or In). When bonding at normal soldering temperatures, all the low-temperature melting metal is consumed in reactions, isothermally solidifying to intermetallic compounds (IMCs) with high melting temperatures. The desired bondline (obtained by designing a dedicated thickness ratio of the two metal layers) is usually one that is in a thermal equilibrium state, implying that no further reactions will occur during the lifetime of the device. This enhances the reliability of the bond, particularly for high-temperature applications, since there will be no further growth of IMC layers (this being a typical failure mechanism for solder joints). The technique is also known as Transient Liquid Phase (TLP) bonding [8-10].

Cu–Sn is the most mature SLID system, using metals and processing temperatures comparable to that of conventional soldering [2]. Significant research has also been done in Au–Sn SLID [11, 12], which is a system excellently suited for harsh environments, but any combination of a high-temperature melting metal and a low-temperature melting metal with high-temperature stable IMCs are candidates for SLID bonding. The process temperature will then be determined by the low-temperature melting metal, and the high-temperature survivability by the resulting IMC and the high-temperature metal.

This paper presents the extensive work on various SLID systems in our group [13]. We have optimized flux-free Cu–Sn SLID bonding for wafer-level hermetic encapsulation of MEMS devices, being ready for industrial implementation [14-17]. We have verified the high reliability of Au–Sn SLID even for thermally mismatched bonding pairs, and we have identified the resulting phases in Au–Sn SLID bonds. We have also used Au–Sn SLID bonding as proof-of-concept for ultrasound transducer manufacturing, where the property of thin-layered metallurgy is more important than the high-temperature stability. Furthermore we have explored newer SLID systems: Ni–Sn, that has potential for higher thermal stability than Cu–Sn and Au–Sn; Au–In SLID (allowing lower bonding temperatures); Au–In–Bi SLID (allowing sub-100 °C bonding temperatures). We have verified experimentally the high-temperature stability of three of the SLID systems: Cu–Sn, Au–Sn and Au–In, quantifying the bond strength at temperatures well above the melting temperatures of Sn and In.

II. EXPERIMENTAL

A. Wafers, substrates and chips for SLID bonding

Our most commonly used test vehicles are silicon dies or wafers. Si test vehicles were the relevant choice for our work on Cu–Sn SLID for wafer-level vacuum encapsulation of MEMS devices, and they are versatile test structures for demonstration of bonding, with a range of manufacturing processes readily adapted for Si substrates. Note that whereas SLID provides very strong bonds, IMCs tend to be brittle. Hence, care must be taken for SLID bonds to tolerate thermomechanical stress. In the case of Si-to-Si bonding, the thermal match of the bonding partners minimizes this thermomechanical stress. Our Au–Sn SLID bonding was performed on thermally mismatched bonding pairs, to represent die attach of power devices to realistic substrates: SiC dies to Cu clad Si_3N_4 substrates, in order to investigate the ability of Au–Sn SLID to withstand thermomechanical stress. For Au–Sn SLID as well as Au–In–Bi SLID, relevant substrates for ultrasound transducer manufacturing were bonded: Piezoelectric PZT (Lead Zirconate Titanate) to a Resonant Backing Layer (acoustically hard material), allowing acoustic testing of the assembly.

B. Metallization

Cu, Sn, Au and Ni layers for SLID bonding were deposited by electroplating, on top of sputtered adhesion/ seed layers, with electroplated layer thicknesses ranging from 1 µm to 10 µm. Electroplating is compatible with large-scale manufacturing, performed at wafer-level. It allows patterning for interconnects or seal rings in an additive process. For Au–In SLID bonding, Au layers were sputtered (0.2 µm and 0.8 µm) and In layers evaporated (1.3 µm). For Au–Sn and Au–In–Bi SLID, eutectic preforms were used as the low-temperature melting metal: $Au_{80}Sn_{20}$ (melting point 278 °C) and $In_{66}Bi_{34}$ (melting point 73 °C), the ratios here given as wt%. Preforms are versatile for fundamental characterization of the SLID bond, and they represent available manufacturing processes for die attach and for lamination (as in ultrasound transducer fabrication). However, preforms are not compatible with patterning for interconnects or wafer-level seal rings. For such applications, the preferred metallization is electroplating, sputtering or evaporation. We have also demonstrated Au–In–Bi SLID bonding, using evaporated eutectic In–Bi. For the different SLID systems, the layer thicknesses are designed in order to achieve the desired phases after bonding. The different metallization schemes for the different SLID process are summarized in Table I, in chapter VIII.

C. Bonding

Cu–Sn and Au–In SLID bonding is performed at wafer-level, whereas the other SLID systems are bonded at chip level, in both cases with an applied, moderate bonding force to ensure good wetting of the bonding partners while the liquid phase is present. Bonding profiles (two examples are shown in Figure 2 below) includes heating to a bonding temperature above the melting point of the low-melting metal, where the temperature is kept constant for a holding time long enough to convert the metal system to the desired phase composition (normally being a thermal equilibrium state), before the temperature is lowered to room temperature. The holding time ranges from 6 minutes for Au–Sn SLID bonded at 300-350 °C, to hours for the low-temperature (~100 °C) bonded Au–In–Bi. Note that the moderate bonding force is needed only in the initial part of the bonding, and long bonding times may be executed as annealing in a batch process. Usually, a short holding time is introduced in the bonding profile, at a temperature before the melting temperature of the low-melting metal is reached. This ensures initial formation of a uniform layer of IMC by solid-state diffusion, which is a key factor for obtaining uniform growth of IMC layers during isothermal solidification, crucial for minimizing void formation in the bondline [18]. All our SLID bonding is flux-free, eliminating the need for post-bond cleaning and allowing SLID bonds to be used for vacuum encapsulation without risking residual flux jeopardizing vacuum levels.

D. Characterization

SLID bond strength is characterized by shear testing, both conventional room temperature testing as well as shear testing while the bonded sample is heated to defined temperatures. The bondline morphology and microstructure is characterized by cross-section microscopy (optical and SEM) and EDX (Energy-Dispersive X-ray spectroscopy), as well as XRD (X-Ray Diffraction) and EBSD (Electron BackScatter Diffraction). The SLID systems used for ultrasound transducer prototyping (Au–Sn and Au–In–Bi) have been characterized by electrical impedance spectroscopy, for evaluating the acoustic integrity of the assembly.

III. CU–SN

A. Cu–Sn SLID process

Cu–Sn is by far the SLID system that has got most attention in research and industry, and the fundamentals of the SLID system is well understood. Cu and Sn are low-cost materials well known in industry, hence Cu–Sn is the SLID system most easily adapted for mass production. The desired bond is a layered structure of Cu / Cu_3Sn / Cu. According to the Cu–Sn phase diagram (Figure 1), such a bond will remain solid up to ~700 °C, and the bond is in thermal equilibrium for temperatures below ~350 °C [19].

The initial electroplated Cu and Sn layers are designed with a thickness ratio

$$\frac{t_{Cu}}{t_{Sn}} > \frac{3M_{Cu}\rho_{Sn}}{M_{Sn}\rho_{Cu}} = 1.3 \,, \qquad (1)$$

in order to ensure complete conversion to Cu_3Sn. Typically, a significantly larger thickness ratio is selected, in order to have well-defined layers of remaining Cu on both sides of the central Cu_3Sn layer. This will give a margin to ensure that all Sn is consumed to form Cu_3Sn, and a continuous Cu layer will avoid unwanted reactions between residual Sn and the adhesion layers. Cu layers will also give ductility to the bond, compared with the hard and brittle Cu_3Sn.

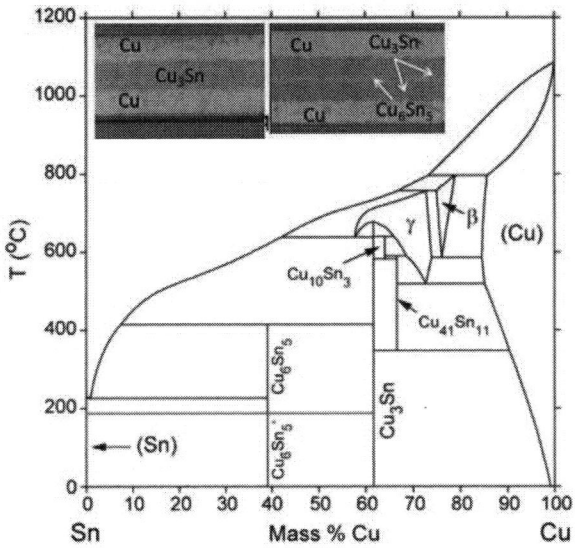

Figure 1: Cu–Sn phase diagram. The insets show cross-section of SLID bonds: Complete conversion to Cu_3Sn (left), and incomplete conversion with remaining Cu_6Sn_5 layer (right).

In our Cu–Sn SLID bonds, we typically use 5 μm Cu thickness, and 1-2 μm Sn thickness. The same layer structure is used for both bonding partners, which is ideal for a wafer-bonding process: An optimized bonding profile (heating rate and bonding force vs time) allows breaking of the thin tin oxide layer to ensure good wetting during bonding, which results in strong bonds with a minimum of voids. For chip-level bonding, Sn would be plated only on the chip side, to avoid Cu–Sn interdiffusion on the heated substrate side prior to bonding.

Bonding is performed at temperatures above the melting point of Sn (232 °C), typically at 250-300 °C. The first IMC to solidify is Cu_6Sn_5, which further reacts with Cu to form Cu_3Sn in the final bond.

B. Cu–Sn SLID wafer-level process optimization

In order to optimize the wafer-level bonding process for industrial implementation, we developed a simulation model to predict the thickness of the different layers (Cu_3Sn, Cu_6Sn_5 and remaining Sn) as function of time during the actual bonding profile [20]. We obtained the input data for the temperature-dependent kinetics of diffusion and reactions from Cu–Sn bi-layer experiments, through annealing at different times and temperatures, cross-section microscopy and EDX spectroscopy. Figure 2 shows the simulation results for two different bond profiles: One with 10 minutes holding time, where a remaining, sub-micron layer of Cu_6Sn_5 is predicted, and another with 30 minutes holding time, were full conversion to Cu_3Sn is predicted. The cross-section micrographs show that these predictions correspond well to

the experimental results. Note that the simulation model takes into account diffusion and reactions in all parts of the bond profiles: the heat-up, the holding time at the bonding temperature, as well as the cool-down.

Two outputs of this simulation model are particularly important for industrial process optimization: The times t_1 and t_2, where all pure Sn, respectively all Cu_6Sn_5, has been consumed. Thus, t_1 signifies the time when the entire bondline is solid at the bonding temperature, and the assembly can be removed from the bonder for the remaining part of the bond profile to be carried out as a batch anealling process. t_2 signifies the time when the desired Cu / Cu_3Sn / Cu structure has been obtained, and the bond is in thermal equilibrium. This is therefore the minimum total time (adding the time in the bonder and the annealing time) needed to obtain the final bond, for a chosen layer thickness and bonding temperature/ bonding profile.

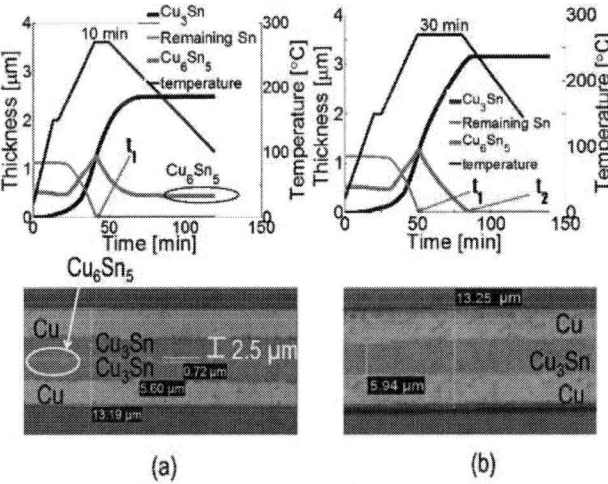

Figure 2: Different bond profiles for Cu–Sn SLID wafer-level bonding, and corresponding cross-section optical micrographs [20].

IV. AU–SN

A. Au–Sn SLID process

Au–Sn is a particularly interesting SLID system for applications where the excellent oxidation resistance or the ductility of Au justifies the higher cost. The Au–Sn system is more complex and exhibits more IMCs than its Cu–Sn counterpart [21] (see Figure 3). The desired bond is a layered structure where one of the Au–rich phases (β or ζ) is sandwiched between two Au layers. Such a bond will be solid for temperatures up to 522-532 °C, and a bond with β as the IMC will be in thermal equilibrium. Our work has shown that the IMC formed in a Au–Sn SLID bond is the ζ phase, and the conversion to the thermal equilibrium state will only take place after long-term annealing [22].

The Au–Sn system has a eutectic point at 29 at% Sn (20 wt% Sn), with eutectic melting point 278 °C. A SLID process with initial layer structure Au / Au–Sn eutectic / Au will have a direct reaction to the the SLID bond structure outlined above. If we use pure Sn as the low-melting metal,

978-1-7281-1500-9/19 $31.00 © 2019 IEEE   143

there will be intermediate reactions to several IMCs, one of which has a fairly high melting point of 419 °C. It is therefore advantageous to use the eutectic composition as the low-melting metal in the initial layer structure. The corresponding bonding temperature must be above 278 °C.

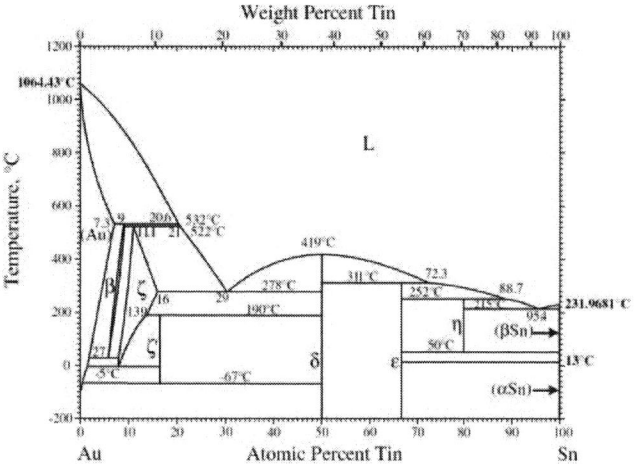

Figure 3: Au–Sn phase diagram, reprinted by permission from Springer Nature [21].

We have used eutectic Au–Sn preforms as the initial material in our Au–Sn SLID bonding, placed between electroplated Au layers on each of the bonding partners. This is a versatile bonding technique, relevant for die attach and for lamination of materials. The layer thickness used depends on the eutectic preform available: For a 7.5 µm thick preform, we have used 10 µm thick Au on both sides, in order to ensure remaining Au on both sides of the IMC. We typically use bonding temperature in the range 300-350 °C.

B. Au–Sn SLID die attach of power device

We bonded SiC dies to Cu clad Si_3N_4 substrates, as demonstrator for die attach of power devices to realistic substrates [22]. The chips and substrates are thermally mismatched: SiC and Si_3N_4 have similar CTE of 4.2 ppm/K while the thick (150 µm) Cu layers have a CTE of 17 ppm/K. It is crucial to test the reliability during thermal cycling for such an assembly, since the thermomechanical stress will be significant. The layers of ductile Au can absorb thermomechanical stress, whereas the brittle Au–Sn IMCs are susceptible of fracture if the stress is too high. The bond shear strength was tested for virgin samples, and for samples after reliability testing: high-temperature storage (HTS) at 250 °C up to 6 months, as well as thermal cycling (1000 cycles, 0 to 200 °C).

Figure 4 shows the measured bond shear strength. The as-bonded samples are extremely strong, with a shear strength of the order of 140 MPa. The reliability testing of combined high temperature storage and thermal cycling reduces the bond strength, but it is still at a very high level, around 70 MPa. Hence, we conclude that Au–Sn SLID is well suited for handling thermomechanical stress when bonding thermally mismatched dies and substrates.

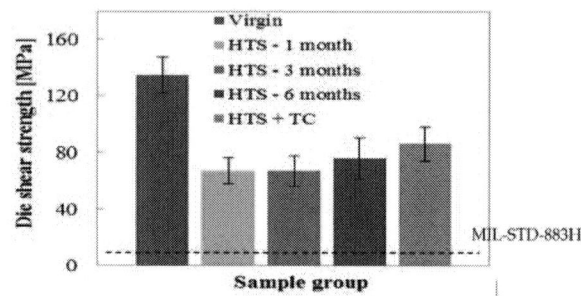

Figure 4: Die shear strength for virgin Au–Sn SLID samples, for high temperature stored (HTS) samples (250 °C for 1, 3 and 6 months), and samples exposed for 3 months HTS + 1000 thermal cycles (TC) between 0 and 200 °C. Reprinted by permission from Springer Nature [22].

C. Au–Sn SLID, lamination for transducers

To demonstrate the ability to use Au–Sn SLID for lamination of materials, we bonded substrates of the piezoelectric material PZT (Lead Zirconate Titantate) to a RBL (Resonant Backing Layer) substrate, as a demonstrator for an ultrasound transducer [23]. This particular bonding requires thin, uniform bondlines of well-defined metallurgy, hence the suitability of SLID. Figure 5 shows a cross-section of a successfully bonded assembly, demonstrating that Au–Sn SLID is suitable also for bonding rough substrates (roughness > 1 µm) over a large area (5 x 5 mm²). The bondline shows certain, but limited, voiding. The amount of voiding can be controlled by careful optimization of the bonding parameters. We have shown by electrical impedance spectroscopy that the voids obtained do not jeopardize the acoustic properties of the assembly [23].

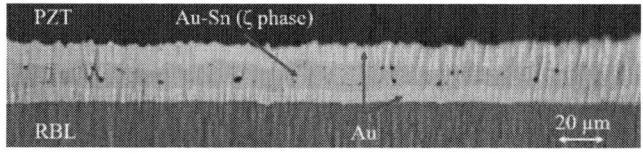

Figure 5: Cross-section optical micrograph of Au–Sn SLID bonded/ laminated acoustic materials, demonstrating the suitability of Au–Sn SLID for large-area lamination for transducers, also for substrates with rough surfaces.

V. Ni–Sn

The Ni–Sn SLID system is much less explored than its Cu–Sn and Au–Sn counterparts. The Ni–Sn phase diagram shows three IMCs: Ni_3Sn_4, Ni_3Sn_2, and Ni_3Sn, which are solid to very high temperatures (798 °C, 1280 °C and 1189 °C, respectively) [24]. According to the phase diagram [25], a thermal equilibrium SLID bond would consist of a layered Ni / Ni_3Sn / Ni structure, that would be solid for temperatures up to 1139 °C (where there is a eutectic point between Ni and Ni_3Sn), and be in thermal equilibrium for temperatures up to 948 °C (where Ni_3Sn undergoes a solid-state phase transition). Hence, Ni–Sn SLID has potential for applications for extreme temperatures, well above the temperatures where Cu–Sn SLID would be applicable. Ni and Sn are low-cost materials well known to industry. However, Ni–Sn shows a much slower kinetics than Cu–Sn [26], bringing challenging in optimization of manufacturing processes.

We have demonstrated Ni–Sn SLID bonds, and verified the slow kinetics reported by previous workers [27]. Although the thermal equilibrium data in the phase diagrams suggest similar bonding temperatures to be used for Ni–Sn as for Cu–Sn (250-300 °C), this will call for extensive bonding times. Figure 6 shows a cross-section micrograph of a Ni–Sn SLID bond performed at 360 °C for 20 minutes. The Ni_3Sn layers have only grown to thickness of 0.8 μm, the dominating IMC in the bondline being Ni_3Sn_2.

The bondline shows significant voiding, and the level of voiding varies greatly for different positions along the bondline. This can partly be explained by the high volume contraction when Ni and Sn reacts to IMCs (15-17% for reaction to Ni_3Sn), compared to the somewhat lower values for the Cu–Sn and Au–Sn systems (~10% and ~4%, respectively). Note, however, that optimization of the bonding parameters for Ni–Sn SLID bonding is challenging due to the slow kinetics. We expect such an optimization to result in lower voiding levels, as has been our experience in our more extensive research in Cu–Sn and Au–Sn SLID bonding.

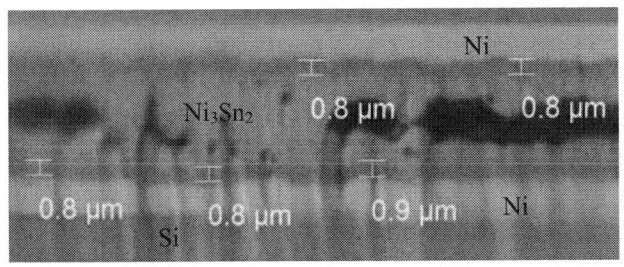

Figure 6: Cross-section optical micrograph of Ni–Sn SLID bonded at 360 °C for 20 minutes. The dark gray phase with thickness measurements, is the Ni_3Sn phase. The black regions are voids [27]. © 2016 IEEE

We obtain high strength of our Ni–Sn SLID bonds, in the order of 40 MPa. Also the more severely voided joints show high strength. When normalizing to the actual bonded area, we have achieved effective shear strengths as high as 230 MPa.

VI. Au–In

In has a low melting point of 156 °C, whereas the numerous Au–In IMCs all have melting temperatures of 450 °C or higher [28]. Au–In SLID bonding is therefore an excellent candidate for low-temperature bonding that still has a good thermal stability. Low-temperature bonding is important for temperature-sensitive materials, as well as for minimizing thermomechanical stress induced in the manufacturing process. This can also be achieved by traditional low-temperature solders, but only allowing a temperature range during the lifetime of the device well below the melting point of the solder. Au–In SLID will not have this limitation.

Au and In show high interdiffusion already at room temperature, and a thin-layer Au–In structure can react to IMCs within the timescale of hours [29, 30]. In a realistic process where metallization takes place days or weeks before bonding, a layered Au / In structure must be designed with significant In surplus to ensure a fresh In surface to be bonded. This implies thin Au layers, and the resulting SLID bond will not be the typical SLID bond that is a thermal equilibrium layered Au / IMC / Au structure. Since a low-temperature SLID process is not expected to be used for applications going to extreme temperatures, it is acceptable to deviate from the thermal equilibrium requirement that usually defines SLID.

We demonstrate successful wafer-level Au–In bonding, bonded at 180 °C [31]. Figure 7 shows a cross-section micrograph of a resulting bond, with two Au–In IMCs. The shear strength of the bonds is good, in the order of 30 MPa. Testing the die shear strength for higher temperatures, up to 300 °C (see Figure 9), we prove that the bonds are indeed strong well above the melting temperature of In. The strength at 300 °C is even higher than at room temperature, explained by a solid-state phase transformation between different IMCs [31].

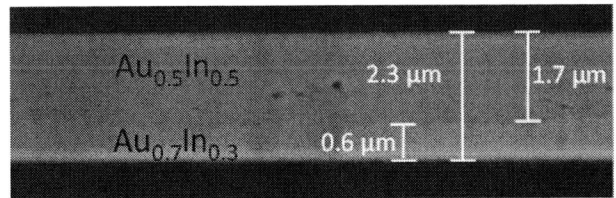

Figure 7: Cross-section SEM micrograph of Au–In bonded sample. Two Au–In intermetallic compounds (IMCs) are present in the final bond-line. Reprinted by permission from Springer Nature [31].

VII. Au–In–Bi

Several materials and applications will need even lower bond temperatures than what can be achieved by Au–In SLID bonding. Examples are polymers, poled piezoelectric and ferromagnetic materials. By using binary alloys as the low-temperature metal in SLID bonds, the bonding temperature may be pushed to even lower values.

In–Bi has a melting point of 73 °C for the eutectic point (79 at% In), and a variety of In–Bi IMCs melting in the range 89-110 °C [32]. The pure metals melt at 156 °C (In) and 271 °C (Bi). The Au–Bi phase diagram shows that these two metals are not miscible, neither do they react to IMCs that are stable at room temperatures [19]. The lowest melting point in the Au–Bi system is at 241 °C. Hence, for a low-temperature SLID bonding using Au as the high-temperature metal and In–Bi eutectic as the low-temperature metal, the expected reactions to IMCs are the Au–In IMCs described in the previous chapter. No Au–Bi reactions are expected.

We demonstrate Au–In–Bi SLID bonding, using a eutectic In–Bi preform sandwiched between electroplated Au layers, with bonding temperatures as low as 90 °C [33]. However, the resulting bond is a multi-phase bond, including $BiIn_2$ that has a melting point of 89 °C. This implies that a higher remelting temperature cannot be expected. To avoid the presence of low-melting In–Bi IMCs, a bonding temperature higher than 110 °C should be selected. Figure 8 shows a cross-section micrograph of a Au–In–Bi SLID bond performed at 115 °C, showing a Au / Au–In IMCs / Au layered structure. Inspection along the length of the bondline reveals inclusions of elemental Bi, but no In–Bi or Au–Bi IMCs. This finding is as expected from analysis of the phase diagrams, and predicts a thermal stability up to 271 °C, where the Bi inclusions would melt. Considering that these inclusions do not form a continuous path through the bondline, we would actually expect the bond to remain solid for temperatures up to the melting point of Au–In IMCs (~450 °C).

A particular case where low-temperature bonding is required, is for lamination of ultrasound transducers where the piezoelectric material is poled prior to assembly. Polarization will be lost if the material is exposed to temperatures above the Curie temperature, typically around 150 °C for PZT. We have demonstrated Au–In–Bi bonding of PZT to RBL [5]. Electrical impedance spectroscopy verifies that the polarization of PZT is intact after bonding, and that the voiding level is acceptable for the acoustic properties of the assembly.

Bonding at low temperatures with relatively thick preforms results in extensive bonding times, in the order of hours. We have demonstrated that shorter bond times can be achieved by thermal evaporation of thin films of eutectic In–Bi on gold layers for SLID bonding [5]. The integrity of these bonds has also been verified by electrical impedance spectroscopy.

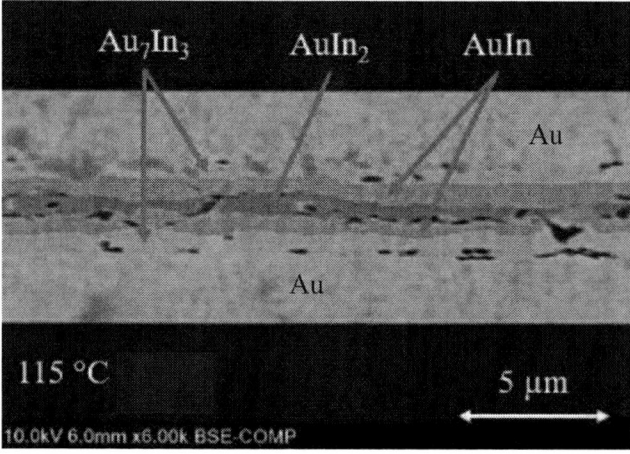

Figure 8: Cross-section SEM micrograph of Au–In–Bi SLID bonded Si dies, bonded at 115 °C. The Au layers are bonded by Au–In IMC, and Bi appears as inclusions of elemental Bi at several positions along the bondline (not in the particular part of the bondline shown here).

VIII. COMPARING THE SLID SYTEMS

A. Properties of different SLID systems, as bonded by USN

Table I summarizes the different SLID systems discussed in this paper, justifying the different research directions we have undertaken for the different systems, and comparing the main results. All our SLID systems show bond strength comparable with or higher than what is typically obtained with Sn-based solder (30-40 MPa [34, 35]). The highest bond strength is obtained for Au–Sn SLID. This is partially explained by the oxidation resistance of Au and eutectic Au–Sn, ensuring a bond process with excellent wetting.

B. Shear strength vs temperature

We have tested experimentally the high-temperature stability of SLID for three systems: Cu–Sn, Au–Sn and Au–In. The bond shear strength is measured while heating the assembly to a specified temperature, up to 300 °C [36]. Figure 9 shows this shear strength vs temperature, demonstrating that all three bonds remain solid to temperatures well above the melting points of Sn and In. For Au–Sn SLID, we have shown that the bond is solid at 400°C, but without quantifying the bond strength [37].

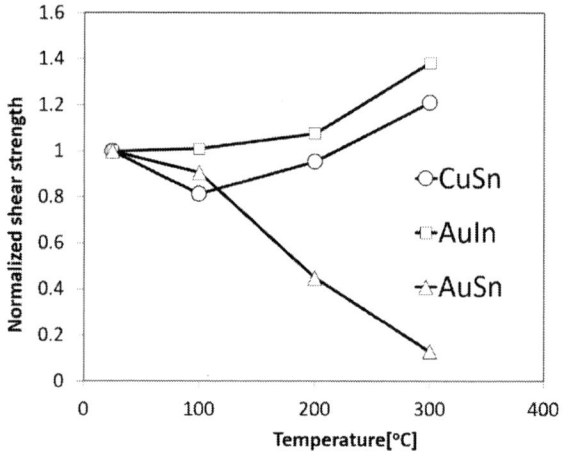

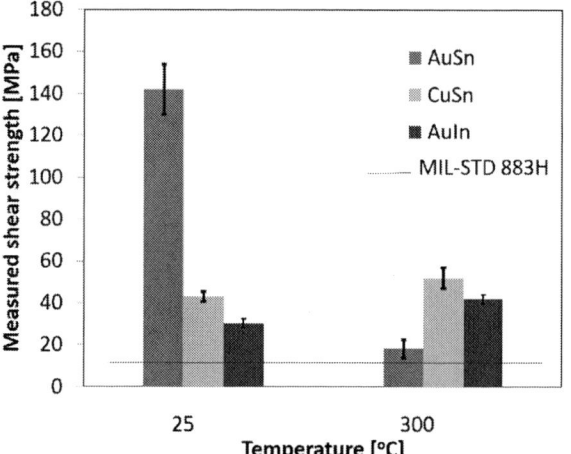

Figure 9: Normalized shear strength vs. die shear temperature and actual shear strength at RT and 300°C [36]. © 2014 IEEE

The different strength vs temperature behaviour of the three SLID systems can be explained from the phase diagrams. A Cu/ Cu₃Sn system is stable up to 350 °C, with no phase transitions, consistent with the moderate variations of Cu–Sn SLID strength with temperature. Au–In SLID shows a remarkable increase in strength between 200 °C and 300 °C. At the lower temperatures, the Au–In bond fracture typically occurs in the original bond interface, indicating that wetting of the bonding partners was the limiting factor for bond strength. Our Au–In bonds consist of AuIn and Au_7In_3 layers, which undergoes a solid-state phase transformation to the Ψ phase above 224 °C [31]. For the layer thickness design of our Au–In SLID bonds, this involves Au–In interdiffusion across the original bond interface, strengthening the bond. The strong bonds at 300 °C show cohesive fractures, no longer following the original bond interface. For Au–Sn SLID, the strength decreases significantly at the higher temperatures, but the strength at 300 °C (20 MPa) is still much higher than what is required by Mil-Std 883H, and comparable with room-temperature strength of Sn-based solders. The decrease in strength is explained by the composition-dependent melting point of the non-stoichometric ζ phase (see Figure 3). We expect that an annealed Au–Sn SLID bond consisting of a Au / β / Au structure, would show a more constant strength vs temperature.

A related technique where we also show stability at higher temperatures than the process temperature, is off-eutectic Au–Ge bonding [38], where we measure the bond remelt temperature to 460 °C, to be compared with the eutectic melting temperature of 361 °C. In this process the high-temperature stability is not based on IMCs, but rather on a microstructure ensuring that molten eutecticum is present in isolated pockets that do not form a continuous path throughout the bondline.

TABLE I.   SUMMARY OF SLID SYSTEMS

	Cu–Sn	Au–Sn	Ni–Sn	Au–In	Au–In–Bi
USN contributions	Flux-free, wafer-level bonding for hermetic sealing Simulation model for IMC kinetics Industrial implementation Strength vs temperature	Composition determined Excellent reliability (HTS & TC, also with CTE mismatch) Strength vs temperature	Bond demonstrated and characterized	Wafer-level bonding demonstrated and characterized Strength vs temperature	Bond demonstrated and characterized (mechanically and acoustically)
Process temperature	> 232 °C	> 278 °C	> 232 °C	> 156 °C	> 73 °C (>110 °C for high-temperature stability)
Obtained bond strength	70 MPa	140 MPa	>40 MPa (nominally) > 200 MPa in the actually bonded areas	30 MPa (breaking at bond interface) 40 MPa at 300 °C	50 MPa
Metallization used in USN research	Electroplated Cu and Sn	Electroplated Au, Au–Sn eutectic preform	Electroplated Ni and Sn	Sputtered Au, evaporated In	Electroplated Au, In–Bi eutectic preform Demonstration: evaporated In–Bi eutecticum
Application demonstrator	Wafer-level vacuum encapsulation of MEMS	Die attach of power devices Lamination of transducers	Bond demonstration	Wafer-level bond demonstration	Lamination of transducers
Special attributes	Low-cost, mature process	Chemical inertness Absorbs thermomechanical stress	Extreme high-temperature stability	Low-temperature bonding	Extreme low-temperature bonding

IX. Conclusion/Outlook

SLID bonding is a versatile technique that gives high-temperature stability of bonded joints. It can be used for high-temperature / harsh environments applications, for products that must tolerate high-temperature manufacturing processes, for low-temperature bonding that do not impose severe temperature restrictions on the final product, as well as to eliminate a hierarchy of solder alloys with different melting temperatures. It can also be used for fine-pitch interconnects, although that has not been the research direction of our group.

We have shown Cu–Sn SLID bonding for wafer-level vacuum encapsulation of MEMS devices, Au–Sn SLID bonding for harsh environment applications, and demonstrated Ni–Sn, Au–In and Au–In–Bi SLID bonding, experimentally verifying the high-temperature stability of Cu–Sn, Au–Sn and Au–In SLID.

Our future SLID work will further optimize bonding processes for the metal systems described in this paper, we will explore novel metal systems for SLID, and we will apply our SLID bonding expertise for new, relevant applications.

Acknowledgments

The Research Council of Norway is acknowledged for the support to the Norwegian Micro- and Nano-Fabrication Facility, NorFab, project number 245963/F50. The work presented here was funded by the Research Council of Norway through project numbers 235302/O70, 193108/S60, 244915, 208929 and 38068. The authors thank Zekija Ramic and Anh-Tuan Thai Nguyen, both USN, for laboratory assistance.

References

[1] N. Hoivik and K. E. Aasmundtveit, "Wafer-Level Solid-Liquid Interdiffusion Bonding," in *Handbook of wafer bonding*, P. Ramm, J. J.-Q. Lu, and M. M. V. Taklo, Eds., ed. Hoboken, NJ, USA: John Wiley & Sons, 2011, pp. 181-214.

[2] H. Huebner, S. Penka, B. Barchmann, M. Eigner, W. Gruber, M. Nobis, *et al.*, "Microcontacts with sub-30 μm pitch for 3D chip-on-chip integration," *Microelectronic Engineering*, vol. 83, pp. 2155-2162, Nov-Dec 2006.

[3] L. Bernstein, "Semiconductor joining by solid-liquid-interdiffusion (SLID) process.," *Journal of the Electrochemical Society*, vol. 113, pp. 1282-88, 1966.

[4] L. Bernstein and H. Bartholomew, "Applications of Solid-Liquid Interdiffusion (SLID) Bonding in Integrated-Circuit Fabricatoin," *Transactions of the Metallurgical Society of AIME*, vol. 236, pp. 405-411, 1966.

[5] K. E. Aasmundtveit, T. Eggen, T. Manh, and H. V. Nguyen, "In-Bi low-temperature SLID bonding for piezoelectric materials," *Soldering & Surface Mount Technology*, vol. 30, pp. 100-105, 2018.

[6] J. M. Lannon, C. Gregory, M. Lueck, J. D. Reed, C. A. Huffman, and D. Temple, "High Density Metal-Metal Interconnect Bonding for 3-D Integration," *IEEE Transactions on Components, Packaging and Manufacturing Technology*, vol. 2, pp. 71-78, 2012.

[7] Y. X. Liu, Y. C. Chu, and K. N. Tu, "Scaling effect of interfacial reaction on intermetallic compound formation in Sn/Cu pillar down to 1 μm diameter," *Acta Materialia*, vol. 117, pp. 146-152, Sep 2016.

[8] D. S. Duvall, W. A. Owczarski, and D. F. Paulonis, "TLP Bonding: A New Method for Joining Heat Resisant Alloys," p. 203, 1974.

[9] G. O. Cook and C. D. Sorensen, "Overview of transient liquid phase and partial transient liquid phase bonding," *Journal of Materials Science*, vol. 46, pp. 5305-5323, Aug 2011.

[10] W. D. Macdonald and T. W. Eagar, "TRANSIENT LIQUID-PHASE BONDING," *Annual Review of Materials Science*, vol. 22, pp. 23-46, 1992.

[11] V. Vuorinen, A. Rautiainen, H. Heikkinen, and M. Paulasto-Krockel, "Optimization of contact metallizations for reliable wafer level Au-Sn bonds," *Microelectronics Reliability*, vol. 64, pp. 676-680, Sep 2016.

[12] T. A. Tollefsen, A. Larsson, O. M. Lovvik, and K. Aasmundtveit, "Au-Sn SLID Bonding-Properties and Possibilities," *Metallurgical and Materials Transactions B-Process Metallurgy and Materials Processing Science*, vol. 43, pp. 397-405, Apr 2012.

[13] K. E. Aasmundtveit, T.-T. Luu, H.-V. Nguyen, A. Larsson, and T. A. Tollefsen, "Intermetallic Bonding for High-Temperature Microelectronics and Microsystems: Solid-Liquid Interdiffusion Bonding," in *Intermetallic Compounds - Formation and Applications*, M. Aliofkhazraei, Ed., ed Open Access: Intech Open, 2018.

[14] N. Hoivik, W. Kaiying, K. Aasmundtveit, G. Salomonsen, A. Lapadatu, G. Kittilsland, *et al.*, "Fluxless wafer-level Cu-Sn bonding for micro- and nanosystems packaging," in *Electronic System-Integration Technology Conference (ESTC), 2010 3rd*, 2010, pp. 1-5.

[15] F. Forsberg, A. Lapadatu, G. Kittilsland, S. Martinsen, N. Roxhed, A. C. Fischer, *et al.*, "CMOS-Integrated Si/SiGe Quantum-Well Infrared Microbolometer Focal Plane Arrays Manufactured With Very Large-Scale Heterogeneous 3-D Integration," *Selected Topics in Quantum Electronics, IEEE Journal of*, vol. 21, pp. 1-11, 2015.

[16] F. Forsberg, N. Roxhed, A. C. Fischer, B. Samel, P. Ericsson, N. Hoivik, *et al.*, "Very large scale heterogeneous integration (VLSHI) and wafer-level vacuum packaging for infrared bolometer focal plane arrays," *Infrared Physics & Technology*, vol. 60, pp. 251-259, 2013.

[17] A. Lapadatu, T. Simonsen, G. Kittilsland, B. Stark, N. Hoivik, V. Dalsrud, *et al.*, "Cu-Sn Wafer Level Bonding for Vacuum Encapsulation of Microbolometers Focal Plane Arrays," in *ECS Transactions*, 2010, pp. 73-82.

[18] H. Etschmaier, H. Torwesten, H. Eder, and P. Hadley, "Suppression of Interdiffusion in Copper/Tin Thin Films," *Journal of Materials Engineering and Performance*, vol. 21, pp. 1724-1727, 2012/08/01 2012.

[19] T. B. Massalski(ed.), *Binary Alloy Phase Diagrams*: ASM International, 1992.

[20] T. T. Luu, A. Duan, K. E. Aasmundtveit, and N. Hoivik, "Optimized Cu-Sn wafer-level bonding using intermetallic phase characterization," *Journal of Electronic Materials*, vol. 42, pp. 3582-3592, 2013.

[21] H. Okamoto, "Au-Sn (Gold-Tin)," *Journal of Phase Equilibria and Diffusion*, vol. 28, pp. 490-490, Oct 2007.

[22] T. A. Tollefsen, A. Larsson, M. M. V. Taklo, A. Neels, X. Maeder, K. Hoydalsvik, *et al.*, "Au-Sn SLID Bonding: A Reliable HT Interconnect and Die Attach Technology," *Metallurgical and Materials Transactions B-Process Metallurgy and Materials Processing Science*, vol. 44, pp. 406-413, Apr 2013.

[23] H. V. Nguyen, M. Tung, T. Eggen, and K. E. Aasmundtveit, "Au-Sn Solid-Liquid Interdiffusion (SLID) bonding for mating surfaces with high roughness," in *2016 6th Electronic System-Integration Technology Conference (ESTC)*, 2016, pp. 1-6.

[24] J. M. Liu, C. P. Guo, C. R. Li, and Z. M. Du, "Thermodynamic re-assessment of the Ni-Sn system," *International Journal of Materials Research*, vol. 104, pp. 51-59, Jan 2013.

[25] H. Okamoto, "Ni-Sn (nickel-tin)," *Journal of Phase Equilibria and Diffusion*, vol. 29, pp. 297-298, Jun 2008.

[26] S. Bader, W. Gust, and H. Hieber, "RAPID FORMATION OF INTERMETALLIC COMPOUNDS BY INTERDIFFUSION IN THE CU-SN AND NI-SN SYSTEMS," *Acta Metallurgica Et Materialia*, vol. 43, pp. 329-337, Jan 1995.

[27] A. Larsson, T. A. Tollefsen, and K. E. Aasmundtveit, "Ni-Sn solid liquid interdiffusion (SLID) bonding - Process, bond characteristics and strength," in *2016 6th Electronic System-Integration Technology Conference (ESTC)*, 2016, pp. 1-6.

[28] H. Okamoto, "Au-In (gold-indium)," *Journal of Phase Equilibria and Diffusion*, vol. 25, pp. 197-198, April 01 2004.

[29] W. Zhang and W. Ruythooren, "Study of the Au/In Reaction for Transient Liquid-Phase Bonding and 3D Chip Stacking," *Journal of Electronic Materials*, vol. 37, pp. 1095-1101, 2008/08/01 2008.

978-1-7281-1500-9/19 $31.00 © 2019 IEEE

[30] Y. M. Liu and T. H. Chuang, "Interfacial reactions between liquid indium and Au-deposited substrates," *Journal of Electronic Materials,* vol. 29, pp. 405-410, 2000/04/01 2000.

[31] T.-T. Luu, N. Hoivik, K. Wang, K. E. Aasmundtveit, and A.-S. B. Vardoy, "Characterization of Wafer-Level Au-In-Bonded Samples at Elevated Temperatures," *Metallurgical and Materials Transactions a-Physical Metallurgy and Materials Science,* vol. 46A, pp. 2637-2645, Jun 2015.

[32] H. Okamoto, "Bi-In (Bismuth-Indium)," in *Binary Alloy Phase Diagrams.* vol. 1, T. B. Massalski, Ed., ed: ASM International, 1990, pp. 748-751.

[33] K. E. Aasmundtveit, T. A. V. Nguyen, and H. V. Nguyen, "In-Bi low-temperature SLID bonding," in *2016 6th Electronic System-Integration Technology Conference (ESTC)*, 2016, pp. 1-5.

[34] S. K. Kang, "Development of Lead (Pb)-free Interconnection Materials for Microelectronics," *Metals and Materials International,* vol. 5, no. 6, pp. 545-549, 1999.

[35] T. Siewert, S. Liu, D. R. Smith, and J. C. Madeni. Properties of Lead-Free Solders [Online]. Available: http://www.boulder.nist.gov/div853/lead_free/solders.html

[36] K. E. Aasmundtveit, T. T. Luu, A. S. B. Vardøy, T. A. Tollefsen, K. Wang, and N. Hoivik, "High-temperature shear strength of solid-liquid interdiffusion (SLID) bonding: Cu-Sn, Au-Sn and Au-In," in *Electronics System-Integration Technology Conference (ESTC), 2014,* 2014, pp. 1-6.

[37] K. E. Aasmundtveit, T. T. Luu, H.-V. Nguyen, R. Johannessen, N. Hoivik, and K. Wang, "Au-Sn fluxless SLID bonding: Effect of bonding temperature for stability at high temperature, above 400 degC," in *Electronic System-Integration Technology Conference (ESTC), 2010 3rd*, 2010, pp. 1-6.

[38] A. Larsson and C. B. Thoresen, "Off-Eutectic Au–Ge Die-Attach — High-Temperature Stability," *Trans. Components, Packaging and Maufacturing Technologies,* vol. Submitted, 2019.

Fluxless bonding technique of diamond to copper using silver-indium multilayer structure

Roozbeh Sheikhi, Yongjun Huo and Chin C. Lee

Electrical Engineering and Computer Science
Materials and Manufacturing Technology
University of California, Irvine, CA 92697-2660, USA
Email: rsheikhi@uci.edu

Abstract— In this study we report on successful bonding of chemical vapor deposition (CVD) grown diamond to Cu using a multi-layer Ag-In structure. To manage the large coefficient of thermal expansion (CTE) mismatch between copper and diamond, Ag-rich Ag-In solution is chosen as the final phase in joint. In our previous investigations, we have shown that Ag-In solid solution exhibit superior mechanical properties, such as low yield strength, high tensile strength, and large elongation. Here, we show that by using a fluxless process at vacuum, mechanically robust joints can be formed at 180 °C between copper and diamond. Numerous samples that were bonded with proposed structure show acceptable shear strength and by performing a post bond annealing at 250 °C for 192 hours, we were able to achieve a joint almost fully composed of Ag solid solution with In, with significantly increased shear strength. The deposited multi-layer structure is examined using scanning electron microscopy (SEM) coupled with focused ion beam (FIB) prior to bonding. Following the bonding, samples are sheared and fracture surfaces are examined using energy dispersive X-ray spectroscopy (EDX). Our studies show that Cr/diamond interface, which is the metallization scheme on diamond is a weak interface in the bond design and as the joint becomes stronger by conversion of Ag-In intermetallic compounds into (Ag), more delamination occurs in the Cr/diamond interface. Additionally, it is reported that annealing the Cr/diamond interface can effectively improve its adhesion.

Keywords-CVD diamond; copper; fluxless bonding; Ag-In solution; diamond metallization

I. INTRODUCTION

Effective thermal management is crucial for efficient and reliable performance of optoelectronic and electronic devices. CVD diamond is among the best materials for efficient conduction and removal of the highly concentrated heat in the device. Diamond possesses a very unique combination of thermal and electrical properties, it has a high thermal conductivity of approximately 2000 W/ m·K and a low dielectric constant. These characteristics along with its relatively low heat capacity, make diamond and diamond-based materials, the optimal choice for heat-spreading purposes by integrating them as an intermediate layer between a heat source and a heat sink in electronic devices [1].

Moreover, recently, the cost of diamond grown by chemical vapor deposition has been on a decline, making them more affordable [2]. Diamond exhibits a very low coefficient of thermal expansion (1×10^{-6} / °C). When it comes to bonding diamond to common metallic heat sinks such as copper with a much higher CTE value (16.5×10^{-6} / °C), selection of bonding material and process becomes more demanding. These bonding techniques should result in a rigid yet ductile joint capable of managing the remarkable CTE mismatch between bonding components [3]. Numerous thermal interface materials (TIM) and lead free solders are being used for purposes of making interconnection and heat dissipation in electronics packaging, however thermal resistance of lead free solders are roughly one order of magnitude lower than commercially available TIMs, making them more advantageous [4,5]. Among them, indium and Au-Sn solders have been used extensively for the specific case of diamond and Cu bonding. Indium is a ductile solder, which sounds promising for managing high CTE mismatch values, however long term mechanical reliability issues are associated with the use of In. High temperature creep and its high diffusion into metals such as Cu are among these issues [6,7]. On the other hand, Au-Sn is a more rigid solder with less long term reliability concerns, however growth of intermetallic compounds in these solders can adversely affect their functionality [8].

Our research group has implemented Ag-In joints for bonding a wide array of materials during the past few decades [9,10]. These joints can be achieved at low temperatures, while being capable of withstanding high operating temperatures. Post-bond annealing of these joints results in formation of Ag rich Ag-In solution, which exhibits high tensile strength and low yield strength while having a solidus temperature beyond 800 °C [11,12]. These compelling attributes of Ag-In fluxless bonding can address the limitations of conventional bonding techniques.

In this paper we report on the technique developed for bonding diamond to Cu, using a multi-layer Ag-In structure along with detailed study of bond microstructure based on the fracture surface examination of sheared samples. Based on these findings, weak interfaces within the joint are identified and appropriate processing steps are introduced, in order to enhance the mechanical strength of the joint.

II. EXPERIMENTAL PROCEDURE

Figure 1 shows the designed bonding structure. The structure is designed so that the final joint composition would be (Ag) with 18 at% In. Diamond used in this study was cut with laser into 3 mm × 3 mm dies. 10 mm × 10 mm Cu blocks are used as the substrate. Diamond's surface is cleaned with isopropanol and then a seed layer of Cr 30 nm/Au 50 nm/ Ag 1μm is deposited on its surface using E-beam evaporation in a single vacuum cycle. Cr is a strong carbide forming element, thus it can provide good adhesion between diamond and metallic films and the Au layer prevents Cr oxidation. The final Ag layer acts as a seed layer for the subsequent electroplating process. Since diamond is a material with very poor electrical conductivity, such a conductive layer is needed for the electroplating process. After E-beam evaporation, diamond coated with the seed layer is annealed at 300 °C for 1 hour, this annealing step improves the adhesion between Cr and diamond, through formation of Cr carbide compounds. Subsequently a 4 μm Ag layer is electroplated on top of seed layer and the diamond is ready for bonding process.

The Cu substrate is fine polished using 0.5 μm suspended diamond slurries, fine polishing of Cu substrate will result in decreased roughness of the final electroplated layer and thus facilitates the bonding process by providing more contact area. Subsequent to polishing, a 15 μm Ag and 8 μm indium is electroplated on the Cu. The backside of Cu substrate is coated with lacquer to prevent silver and indium deposition during electroplating process. The Ag plating solution is a cyanide-free, mildly alkaline at pH 10.5. The In plating solution is a sulfamate indium bath at pH of 1–3.5. The bath temperature is room temperature for both plating solutions. The current density is 13 mA/cm² for Ag and 22 mA/cm² for indium plating. indium is prone to oxidation in room temperature, thus in previous studies on Ag-In joints, a thin cap layer has been used to suppress the In oxidation, however in this study we eliminate the cap layer, our results show that successful bonding can be carried out even without the cap layer. It has been reported that indium oxides grow on the surface while storage in room temperature. Indium oxide layer thickness can reach a maximum of 80-100 Angstrom and afterwards it will act as a passive layer preventing further oxidation. We believe that pressure during bonding will rupture the thin indium oxide layer, Moreover since bonding is carried out in vacuum, further oxidation is very limited.

In this fluxless bonding process, the diamond chip is placed on top of Cu substrate and held by a fixture with 200 psi (1.37 MPa) static pressure to ensure intimate contact. The assembly is mounted on a graphite platform in a vacuum furnace that is pumped to 65 mtorr to suppress In oxidation during bonding [13]. The platform is heated and the sample temperature is monitored by a miniature thermal couple. The bonding temperature is set at 180 °C with a dwell time of 15 min. The heater is then turned off and the assembly cools naturally in 65 mtorr vacuum. It takes about 90 minutes to cool to room temperature. No flux is used. Bonded samples were then annealed at 250 °C for 192 hours (in air) and cooled down naturally in air.

Due to hardness of diamond, preparation of metallographic cross-sections using conventional methods is not possible, therefore bonded samples were sheared and the fracture surface was thoroughly studied using SEM/EDX. Also SEM coupled with focused ion beam (FIB) is used to evaluate the thickness of films deposited on diamond prior to bonding.

III. RESULTS AND DISCUSSION

A. Deposition of layered structure

1) Deposition of layered structure on diamond

As stated, initially a thin seed layer is deposited onto the diamond in a single cycle of E-beam evaporation, and then a thick layer of Ag is electroplated on top of the seed layer. Figures 2(a) and (b) show these layers respectively.

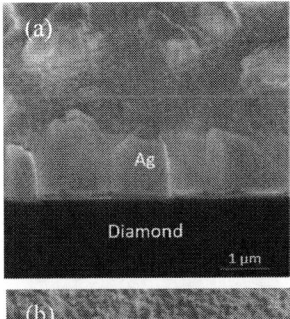

Figure 2. SEM image of FIB trenched layers on diamond (a) seed layer on diamond (b) electroplated Ag on seed layer

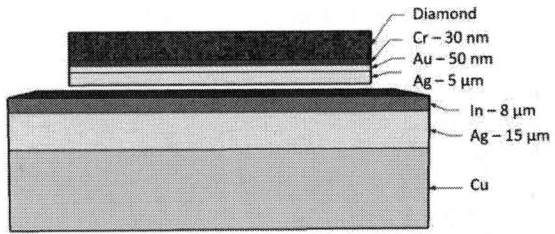

Figure 1. Multi-layer structure used for bonding, not to scale

The thickness of layers achieved is very close to the design values.

In a layered bonding structure such as the one used in this study, adhesion between the metallization layer and bonding component, plays a key role. If the adhesion at these interfaces such as Cr/Diamond is weak, then functionality of the bond will be degraded no matter how strong the joint (Ag-In) is. Therefore in order to improve the adhesion at Cr/diamond interface, an annealing step at 300 °C was performed. In order to examine the effectiveness of this step, diamond coated with the seed layer was annealed at 300 °C for 1 hour and subsequently a thick Ag layer was electroplated on top of it. The same process except for the annealing step was done for another sample. It was observed that the cohesive stress associated with deposition of a thick electroplated layer overcomes the adhesion strength between Cr and diamond and thus peeling occurs at the Cr/diamond interface. Therefore an annealing step is necessary in order to have a reliable metallization layer.

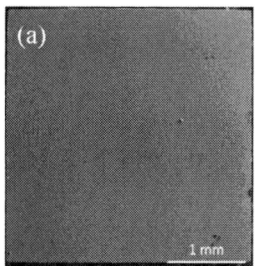 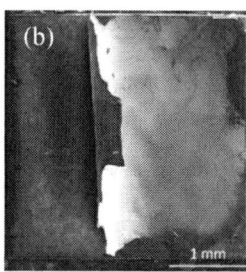

Figure 3. Top view SEM image of thick electroplated silver layer on diamond (a) diamond and seed layer annealed at 300 °C for 1 hour prior to electroplating (b) diamond and seed layer not annealed prior to electroplating silver

2) Deposition of layered structure on Cu

Initially, as received Cu blocks were used for the electroplating process. Cross sectional study of the surface revealed that after electroplating the Ag layer, the top surface is rather rough and may limit the contact area during the bonding. Thus Cu substrate was fine polished using 0.5 µm suspended diamond slurries and then electroplated. This polishing step decreases the final roughness since the electroplated layer adapts the roughness of substrate. The calculated RMS (root mean square average profile height) for electroplated Ag layer on polished and not polished Cu substrate is 0.8 µm and 3.4 µm respectively. A comparison between two cases is depicted in figure 4.

B. Bonding

According to Ag-In phase diagram and our previous studies on this system, as indium is electroplated on Ag, even at room temperature it starts to diffuse into the Ag grain boundaries and form $AgIn_2$. As temperature rises during bonding, more $AgIn_2$ forms with the consumption of In. While reaching 156 °C, the remaining indium will melt, wet the contact area and react with the neighboring Ag. As 166 °C is reached, $AgIn_2$ decomposes into molten (In) and Ag_2In, providing additional molten phase for further reaction. This will continue until all the molten phase is consumed at the bonding temperature. Since Ag_2In does not melt until 650 °C, the joint solidifies at bonding temperature of 180 °C [10]. Thus this bonding process can be categorized as a transient liquid phase bonding process.

Figures 5(a) and (b) show the overall fracture surface of sample bonded at 180 °C for 15 minutes and subsequently sheared. Two very distinct regions can be detected on the Cu side, see figure 5(a). EDX analysis showed that the area labeled as A contains Cr, Au, Ag and In. Knowing that Cr does not diffuse into the surrounding layers; it can be concluded that at these regions, Cr/Diamond delamination has occurred. Figure 5(c) shows the higher magnification view of region B in figure 5(a). EDX shows $63 \geq Ag$ at% ≥ 68 and $37 \geq In$ at% ≥ 32. This

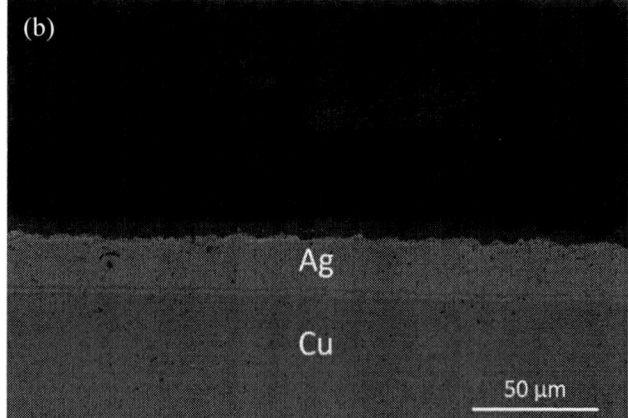

Figure 4. Cross sectional SEM image of Ag electroplated on Cu (a) as received Cu used for electroplating (b) Cu fine polished prior to electroplating

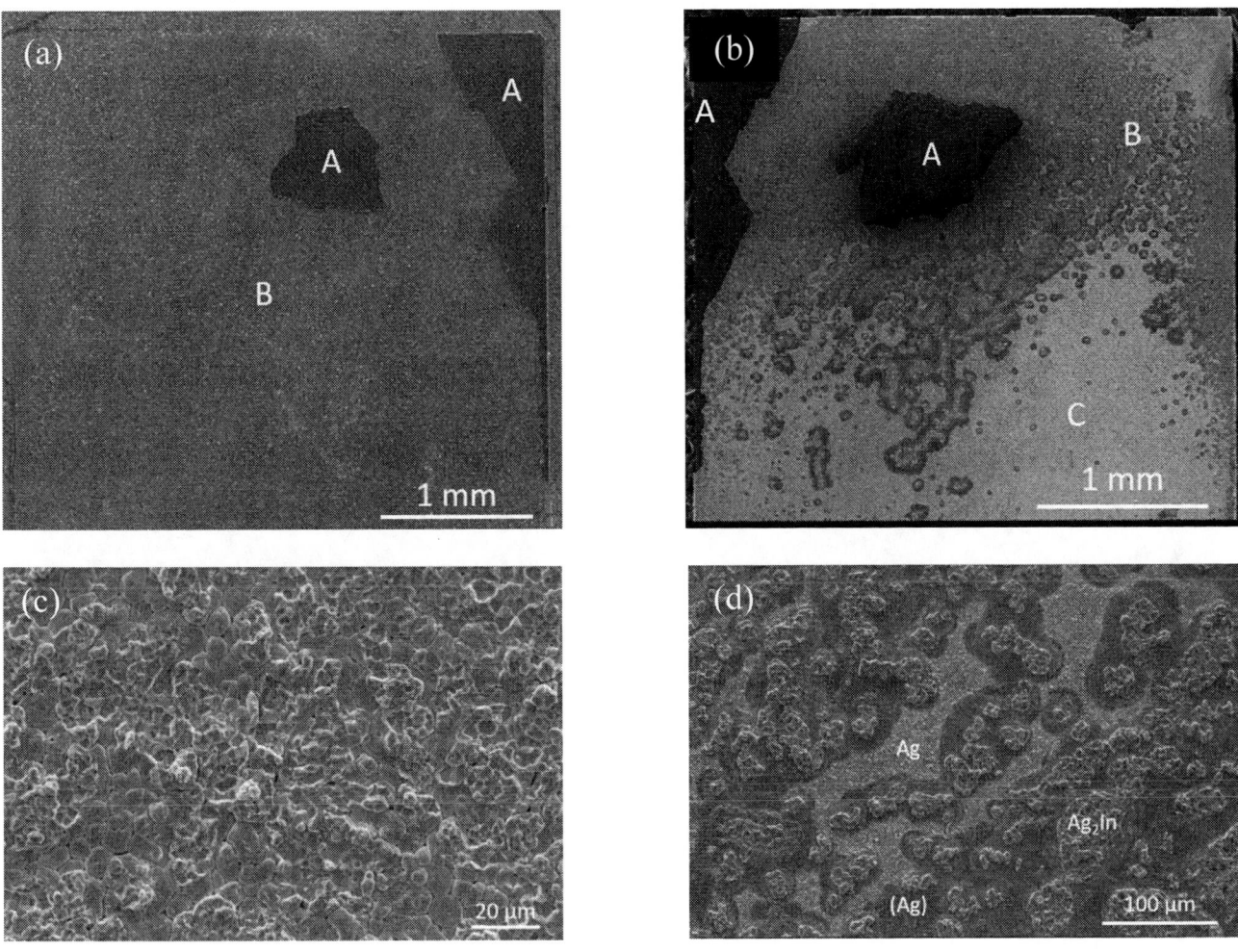

Figure 5. SEM image of fracture surface after shear test, (a) overall view of Cu substrate (b) overall view of diamond surface (c) magnified image of region B in figure 5(a), (d) magnified image of region B in figure 5(c).

ratio is very close to that of Ag_2In. Since EDX has a penetration depth of a few microns, some deviation of the exact stoichiometric ratio is acceptable. Figure 5(b) shows the diamond surface after shear test. Region A only showed Carbon, which confirms the initial finding that Cr/diamond delamination has occurred at that region. There are 2 more distinct regions that can be identified on the diamond fracture surface. Region A was studied by SEM and EDX, it contains a mixture of Ag_2In, (Ag) and Ag. Figure 5(d) shows magnified view of this area. As for region C, no indium was detected. This indicates that the bonding/reaction has not fully occurred. Lack of sufficient molten phase during the bonding should be the main reason.

C. Post-bond annealing

Fracture surface of sheared samples without annealing showed internal fracture of Ag_2In along with interphase delamination between Ag_2In and (Ag) + Ag. Thus further annealing steps are required to achieve the final (Ag) phase that has superior mechanical properties compared to that of Ag_2In. By having (Ag) as the single phase within joint, crack propagation within Ag_2In would be eliminated, also there will be no interphase boundaries that can facilitate crack propagation upon shear.

According to figure 5, it can be concluded that when reaching the bonding temperature, amount of molten phase is limited. This will result in areas within the joint that do not react. In order to address this issue, prior to electroplating indium on to the Cu/Ag coupons, the Cu/Ag coupons were annealed at elevated temperature for 3 hours, this step increase the Ag grain size and slows the diffusion of In into Ag grain boundaries. Therefore potentially increasing the amount of molten (In) needed for complete wetting and subsequent reaction. After the bonding process, samples are annealed at 250 °C for 192 hours in air to achieve the desired (Ag) as the final phase.

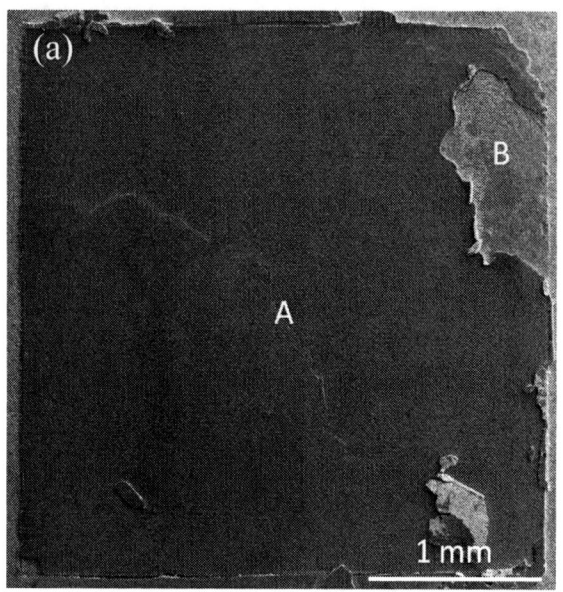 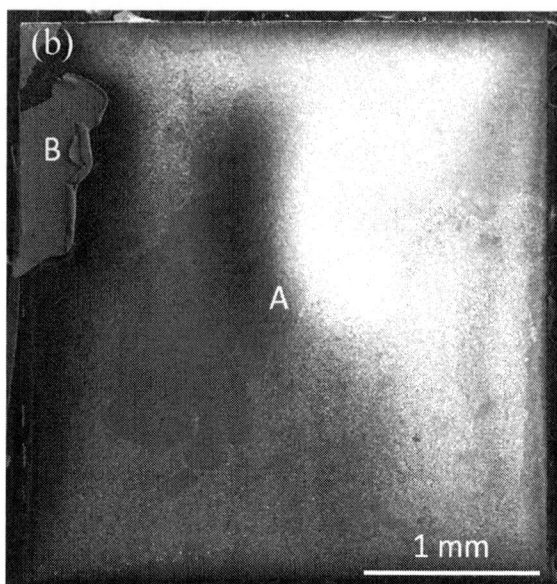

Figure 6. SEM image of fracture surface after shear test for sample bonded and subsequently annealed at 250 °C for 192 hours. (a) Cu substrate (b) diamond

Table 1. Detected elements using EDX in different areas in figure 6 (a) and (b), all values are in at%

		Ag	In	Cr	Au	Cu	C
Cu	A	82.1	4.9	8.9	4.1	-	-
	B	79	19.4	-	-	1.6	-
Diamond	A	-	-	-	-	-	100
	B	70.3	29.7	-	-	-	-

Annealed samples showed very high shear strength. Figure 6 shows the fracture surfaces of annealed sample.

It can be observed that major portion of the joint remains intact on the Cu substrate, while the delamination is almost fully occurring at the Cr/Diamond interface. Table 1 shows the detected elements using EDX for region A and B on both the diamond and Cu surfaces. A mixture of Cr, Au, Ag and indium is detected in region A on the Cu side, indicating the occurrence of delamination at Cr/diamond interface. This is well aligned with our initial notion that Cr/diamond interface is the weak link within the joint structure. As for region B in figure 6, the delamination is happening within the joint, on the Cu side, the indium is in the range for (Ag) and very close to the initial design values of In concentration (the design value is 18 at%), see table 1. Moreover small amount of Cu is detected in this region, indicating that delamination is occurring close to Cu substrate, this is very important, since none of samples prior to annealing showed such behavior. As for the corresponding region (B) on the diamond side, a mixture of Ag_3In and (Ag) seems to be

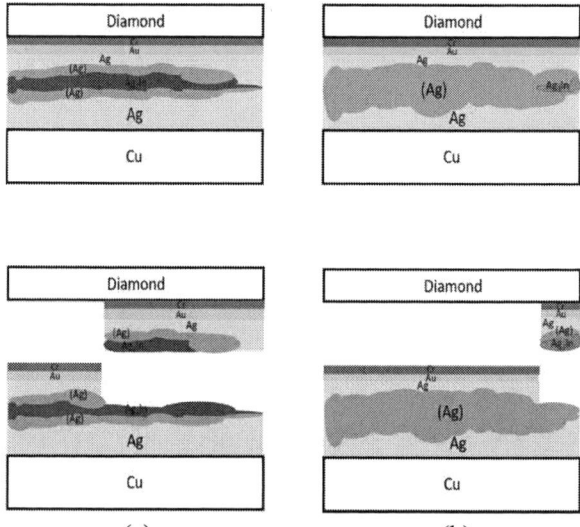

(a)   (b)

Figure 7. Schematic representation of proposed joint microstructure and fracture behavior upon shear for (a) as bonded sample (b) Annealed after bonding

the existing phase. The exact ratio for Ag_3In is not seen in table 1, because EDX has a penetration depth of a few microns and Ag_3In is only on the top surface while beneath it is (Ag) and Ag. Thus what we see in terms of concentration using EDS is the average for some depth within the joint.

In conclusion, regarding the annealed sample, it can be said that delamination at Cr/diamond interface is dominant and the whole joint is either (Ag) or a multi-layer structure of (Ag) and Ag. Furthermore, a small portion of delamination occurs within the joint along the remaining Ag_3In and (Ag) interface.

Figure 7 provides the proposed microstructure for each case based on the findings of fracture surface studies. For the as bonded sample Ag_2In is formed at areas near the initial interface. Ag_2In is surrounded by (Ag). In some areas of the sample no reaction occurs and thus pure Ag remains on the diamond side. When the as bonded sample is sheared, all the weak interfaces take part in the delamination, i.e., Cr/diamond, within Ag_2In and Ag_2In / (Ag) interface. All these interfaces are boundaries between different phases which can inherently be weak points in terms of adhesion. As for fracture within Ag_2In phase, since Ag_2In is an intermetallic compound, brittle fracture within the phase is rather expected.[14]

By annealing, Ag_2In is converted into (Ag) and small amount of Ag_3In. It is expected that if the annealing time was longer, no Ag_3In would remain in the joint. When sheared, no delamination occurs within the (Ag) phase, this can be explained by the fact that Ag solid solution with In, exhibits high tensile strength and low yield strength. Thus there are no weak interfaces within the joint and crack needs to propagate along the Cr/diamond interface. This is true for majority of the joint. However the small amount of Ag_3In remaining in the joint is weakly adhered to (Ag) (weaker than Cr/diamond) so that fracture through this boundary is also observed.

IV. CONCLUSION

The CVD grown diamond was successfully bonded to the Cu substrate using a multi-layer Ag-In structure in a fluxless process. The bonding temperature is 180C and bonding takes place in vacuum. Study of fracture surface of sheared joints revealed that a mixture of Ag_2In, (Ag) and Ag is present in the joint and fracture occurs within intermetallic phase (Ag_2In) and also along its boundary with other existing phases. It was also noted that Cr as a metallization layer adheres relatively weak to the diamond. An annealing process at 300 °C was studied and proved to be effective for improving the Cr/diamond adhesion. However, Cr/diamond delamination still takes place in the shear test of bonded samples. In an effort to eliminate the Ag_2In phase and achieve the (Ag) which has superior mechanical properties, annealing of bonded samples at 250 °C for 192 hours was carried out. The annealed samples show increased shear strengths. Based on the fracture surfaces, it is concluded that almost all of the joint is converted into (Ag). Fracture surface of these samples also revealed that delamination occurs almost entirely at the Cr/diamond interface, which is an indication of how strong the joint is.

V. REFERENCES

[1] J. Asmussen and D. K. Reinhard, *Diamond films handbook*. Marcel Dekker, 2002.

[2] A. L. Moore and L. Shi, "Emerging challenges and materials for thermal management of electronics," *Materials (Basel).*, vol. 17, no. 4, 2014.

[3] K. Yoshida and H. Morigami, "Thermal properties of diamond/copper composite material," *Microelectron. Reliab.*, vol. 44, no. 2, pp. 303–308, Feb. 2004.

[4] R. Prasher, "Thermal Interface Materials: Historical Perspective, Status, and Future Directions," *Proc. IEEE*, vol. 94, no. 8, pp. 1571–1586, Aug. 2006.

[5] J. Cho, R. Sheikhi, S. Mallampati, L. Yin, and D. Shaddock, "Bismuth-Based Transient Liquid Phase (TLP) Bonding as High-Temperature Lead-Free Solder Alternatives," in *2017 IEEE 67th Electronic Components and Technology Conference (ECTC)*, 2017, pp. 1553–1559.

[6] R. Darveaux and I. Turlik, "Shear deformation of indium solder joints," *IEEE Trans. Components, Hybrids, Manuf. Technol.*, vol. 13, no. 4, pp. 929–939, 1990.

[7] M. E. Kassner, C. S. Campbell, and R. Ermagan, "Large-Strain Softening of Aluminum in Shear at Elevated Temperature: Influence of Dislocation Climb," *Metall. Mater. Trans. A*, vol. 48, no. 9, pp. 3971–3974, Sep. 2017.

[8] C. C. Lee, C. Y. Wang, and G. S. Matijasevic, "A new bonding technology using gold and tin multilayer composite structures," *IEEE Trans. Components, Hybrids, Manuf. Technol.*, vol. 14, no. 2, pp. 407–412, Jun. 1991.

[9] R. W. Chuang and C. C. Lee, "Silver-indium joints produced at low temperature for high temperature devices," *IEEE Trans. Components Packag. Technol.*, vol. 25, no. 3, pp. 453–458, Sep. 2002.

[10] Y.-Y. Wu and C. C. Lee, "The Strength of High-Temperature Ag–In Joints Produced Between Copper by Fluxless Low-Temperature Processes," *J. Electron. Packag.*, vol. 136, no. 1, p. 011006, Jan. 2014.

[11] Y.-Y. Wu, D. Nwoke, F. D. Barlow, and C. C. Lee, "Thermal Cycling Reliability Study of Ag–In Joints Between Si Chips and Cu Substrates Made by Fluxless Processes," *undefined*, 2014.

[12] Y. Huo and C. C. Lee, "The growth and stress vs. strain characterization of the silver solid solution phase with indium," 2016.

[13] C. C. Lee, D. T. Wang, and W. S. Choi, "Design and construction of a compact vacuum furnace for scientific research," *Rev. Sci. Instrum.*, vol. 77, no. 12, p. 125104, Dec. 2006.

978-1-7281-1500-9/19 $31.00 © 2019 IEEE

[14] R. Sheikhi and J. Cho, "Growth kinetics of bismuth nickel intermetallics," *J. Mater. Sci. Mater. Electron.*, vol. 29, pp. 19034–19042, 1234.

Formulation and Processing of Conductive Polysulfide Sealants for Automotive and Aerospace Applications

Bo Song, Fan Wu, Kyoung-sik Moon, C.P. Wong
School of Materials Science and Engineering
Georgia Institute of Technology
Atlanta, GA, 30332
cp.wong@mse.gatech.edu

Abstract—Polysulfide rubbers are versatile elastomers that are widely used in electronic and automotive industries. Specifically, polysulfides are used as sealants for aircraft fuel tanks liners due to their superior properties. To dissipate electrostatic charges that are generated from the motion of highly flammable fuels, polysulfide sealants need to be made electrically conductive. In this report, a series of conductive polysulfide sealants were made through roll milling processes. The effect of fillers on conductivity was systematically investigated by varied sizes, shapes, and surface coatings. Filler-polymer interfaces were studied and modified to present better electrical properties. Conductive polysulfide can achieve a high conductivity over 10^5 S/m using high aspect ratio silver nanowires at low filler loading with specific treatment.

Keywords-polysulfide sealants; electrically conductive adhesives; aircraft fuel tank; sealing and bonding; nanomaterials; harsh environment

I. INTRODUCTION

Polysulfides represent a class of polymers containing chains of sulfur atoms. The first synthetic polysulfide rubbers were introduced by Thiokol Corp in 1927, and liquid form of polysulfides were produced in 1940s that offers increased versatility for applications. Polysulfides show excellent chemical resistance, high physical strength, low temperature serviceability, good weatherability, good elasticity to relieve stress, and strong adherence to many substrates, such as glass, woods, stainless steel, aluminum, titanium, and composites [1, 2]. These attributes make them exceptionally suitable as high-performance sealants in automotive, electronic, construction and chemical industries. For example, polysulfides have been applied in modern curtain walls, building exterior joints, glazing of windshield and rear automotive lights. Specifically, polysulfide sealants containing ~37% of sulfur are used to seal aircraft fuel tanks due to its superior chemical resistance to kerosene-type jet fuels, hydraulic oils, de-icing fluids, and corrosion products. Other advantages of PS sealants include their resistance to structural loading, fuel inertia, pressure changes, and thermal cycling (-55 to 55°C). One major limitation of polysulfide is the relatively high density since extra sealant mass leads to additional fuel consumption. The quality of sealed joints in integral fuel tanks is critical to ensure safe operation while addressing cost and environmental implications [3]. Generally, aircraft integral fuel tanks are located inside wing structures in accordance with design principles. The major sealing joints include joint edge sealing, interfay sealing, overcoat, and wet assembly. A sealant weight of ~100 kg is needed per wing to seal total fuel volume up to 150,000 liters. Typical sealant applications in an integral fuel tank is illustrated in Fig.1 [4].

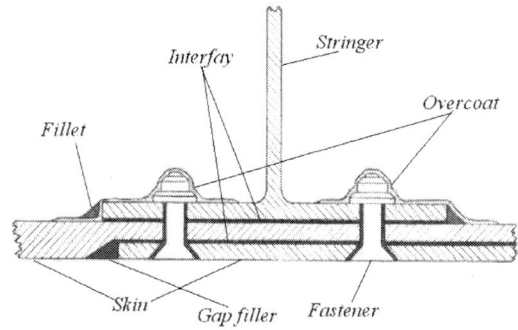

Figure 1. Typical applications for integral aircraft fuel tank sealing.

The sealants currently used in aerospace fuel tanks are mainly two-part metal oxide cured liquid polysulfides. Liquid polysulfides are synthesized by polycondensation of bis-(2-chloroethyl)-formal and sodium polysulfide (NaS_x) [5]. Small amount of trichloropropane is added to create crosslinking or branching sites that regulates modulus and elasticity. The polymer then goes through chain length reduction process to yield liquid polysulfide with lower molecular weight. The polymerization reaction and chain length reduction are shown in Fig. 2. Polysulfides are cured by converting mercaptan groups (-SH) to disulfide bonds (-S-S-) bonds, which link the short chains of liquid polymer to long chains of rubbery polymer. Oxidizing agents are commonly used as curing agents, such as lead dioxide, manganese dioxide (activate grade), calcium peroxide, ammonium dichromate, cumene hydroperoxide, and p-quinone dioxime [6]. The two-part fuel tank sealants are mostly manganese dioxide cured, in which mercaptan groups are oxidized by coordination and the manganite ions serve as a base to deprotonate mercaptan terminals [1].

Generally, viscosity and working time are two key properties that define tank sealants. There are three classes of tank sealants based on viscosity. Class B sealant have paste consistency to allow for filling into small crevices of fixture. Class B can be applied using an extrusion gun or a spatula. Class A sealant has lower viscosity and can be applied by brush. Class C sealant has intermediate viscosity, which can be applied by either brushing or rolling on faying surfaces. Working or application time defines the workable time of the material before it gets sufficiently cured. Typical application time are ½ h, 2 h, 4 h, and 6 h [7]. In addition, tack-free time and cure time are important parameters for material handling: tack free time is determined as the point when the surface no long feels sticky, and cure time is the length for sealant to reach a certain hardness [8]. All these parameters are based on standard condition with a temperature of 25°C and 50% relative humidity. Increased temperature and humidity levels tend to accelerate the curing process and shorten these times. Before applying sealants, it is essential to thoroughly clean the surface and remove possible contaminants. Adhesion promotors or primers may be needed for durability of adhesion.

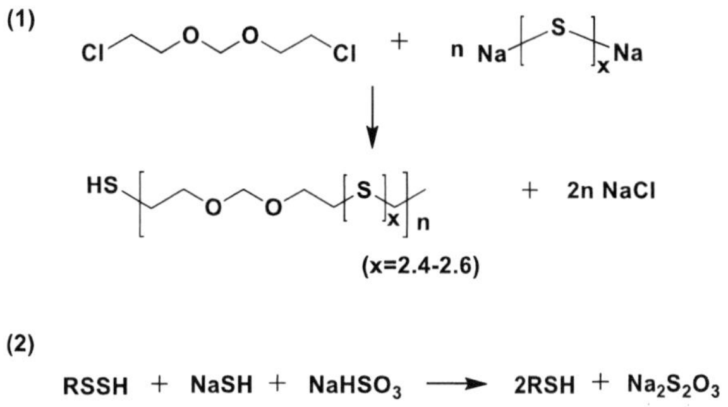

Figure 2. (1) Polymerization and (2) chain length reduction reaction of polysulfides.

The movement of flammable fuels in automotive and aircraft fuel tanks can lead to the generation of electrostatic charges. If not eliminated, these charges can cause the risk of sparking or explosion. To prevent these charge buildup, fuel lines need to be made electrically conductive. One effective way is to incorporate conductive fillers into polymer resins to make electrically conductive adhesives. A variety of carbon nanomaterials, such as carbon black, carbon nanotubes, fibers, and graphene have been explored as conductive fillers [9-11]. These fillers can provide moderate conductivity and serve as reinforcing agent to improve overall strength, modulus, hardness, and abrasion resistance [12]. In the case of harsh environment conditions, higher conductivity of polysulfide sealants is required. For example, ultra-low resistive paths across some electrical bonding joints is important to carry large current or dissipate excessive energy. Metallic and alloy fillers have high instinct conductivity, but their interactions with polymer system need to be studied and optimized for electrical purpose. In this report, various conductive polysulfide sealants were formulated through filler screening and their structural, thermal, and electrical properties were investigated.

II. EXPERIMENTAL

A. Materials

Two-part polysulfide sealants were acquired from PPG Industries. Majority of silver fillers and other metallic fillers were provided by AMES Goldsmith Corp. Carbon fillers were obtained from Asbury Carbons Inc. and Orion Engineered Carbons.

B. Processing

Commercial fillers were used as received. Silver nanowires were made by a polyol process in ethylene glycol solvent. Silver dendrites were made by a one-step solution method using silver nitrate as the precursor. Both products were washed with DI water and ethanol repeatedly. For surface treatment, silver fillers were stirred in different surfactant solutions and purified by centrifuge. Conductive polysulfide sealants were prepared by blending two-part polymer and fillers through a three-roll mill. The mixed pastes were stencil printed on substrates and cured at room temperature.

C. Characterization

Surface morphology and polymer/filler distribution of samples were characterized by scanning electron microscopy (SEM) with an energy dispersive X-ray spectroscopy (EDX) source. Thermal stability was studied by thermal gravimetric analysis (TGA) in N_2. Hardness was determined using a shore A Durometer. Tensile test was conducted on an Instron Mirotester. Conductivity of the sealants were measured on designed test coupons using a four-point probe.

III. RESULTS AND DISCUSSION

The conductivity of polymer composites with fillers can be realized if the loading of fillers exceeds the percolation threshold, in which fillers are connected to form a conductive network. The percolation threshold has a large variation depending on filler types. Various fillers were utilized to modify the conductivity of polysulfide (PS) sealants, including carbon (C), copper (Cu), nickel (Ni), titanium (Ti), silver (Ag), and their hybrids. For carbon fillers, carbon black (CB), multi-walled carbon nanotubes (MWCNT), graphite (G) were selected, and the conductivity of corresponding sealants is shown in Fig. 3a. The conductivity of PS/CB was 10 S/m at 5 vol.% loading, and maximum conductivity of ~30 S/m was achieved when the loading increased. The high-structure CB particles are more efficient to boost conductivity compared to MWCNTs and graphite. Considering the thermal stability, CB is thermally stable, while PS has an onset degradation temperature around 220°C. The addition of CB showed negligible change of thermal profiles (Fig. 3b). From the cross-section SEM image (Fig. 3c), the conductive CB were distributed in PS resin to form conductive networks.

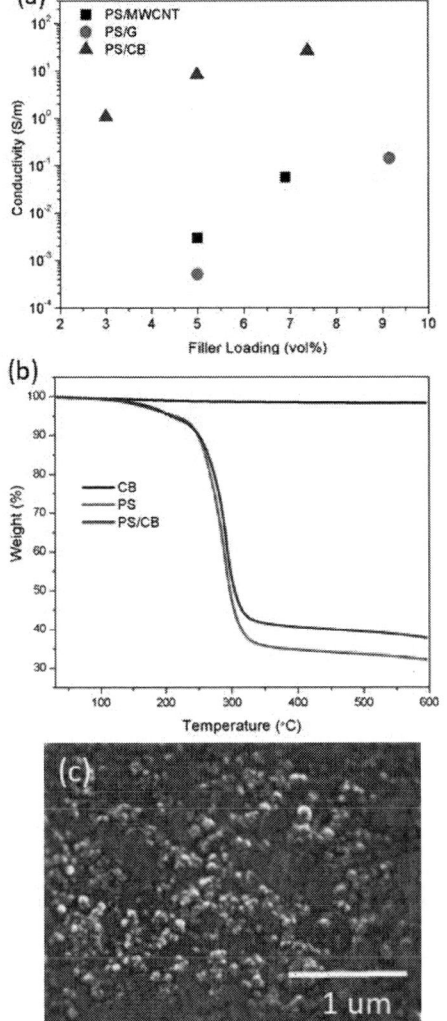

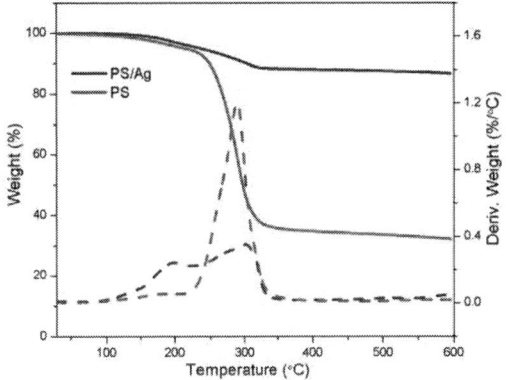

Figure 4. TGA and differential TG curves of PS and PS/Ag sealants.

sulfide coordination bonds and the decomposition of short-chain sulfide adducts. To improve the conductivity of PS/Ag, post-cure thermal treatment was performed. We observed that PS/Ag showed a rapid conductivity increase up to 10^5~10^6 S/m by annealing within a short period of time. The PS filled with larger-size flakes typically showed a higher conductivity than that of Ag particles.

In addition to Ag filler, Ni (Ni particles and Ag/Ni) and Cu fillers (Cu flakes and Ag/Cu) and were also employed due to their high intrinsic conductivity. Similar to PS/Ag, PS/Ni showed large resistance after RT cure. After thermal treatment, PS/Ni presented a high conductivity over 10^3 S/m, and a maximum conductivity of 4×10^4 S/m was achieved for PS/Ag/Ni (Fig. 5a). It can be seen in Fig. 5b that PS/Ni had lower onset degradation temperature compared to PS but only one peak was observed in derivative TGA plot (possibly due to weaker Ni-sulfide interaction). The commercial Cu flakes had oxide layers on the surface, which tremendously brought down conductivity. Thermal treatment would induce further oxidation, which make Cu inadequate fillers for PS.

Figure 3. (a) Conductivity of PS/MWCNT, PS/G, and PS/CB at different filler loadings; (b) TGA plots of PS, CB, and PS/CB; (c) cross-section SEM image of PS/CB sealant.

Among the metallic fillers, silver has been widely used to provide high conductivity for electrical interconnection. However, all PS/Ag sealants exhibited high resistance (MΩ) after cure, regardless of filler size and surface coatings. The reason can be explained that silver fillers have strong interactions (chelation) with sulfide groups, which passivate the silver surface and block the electron conduction pathway.

Previous study revealed that Ag could bond (crosslink) to sulfide groups, restricting the chain mobility of PS and results in an increase in glass transition temperature (T_g) [13]. The shore A hardness was determined to be 50 and 68 for PS and PS/Ag respectively, indicating a complete cure for both sealants. Unlike PS/CB, a two-step thermal degradation profile was observed from deriv. TGA plots (Fig.4). The second peak matched with the decomposition of polymer chain, while the first peak might correspond to the breakage of Ag-

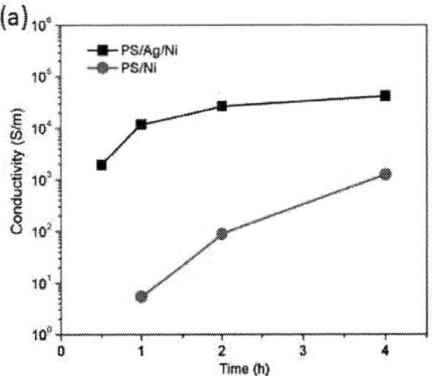

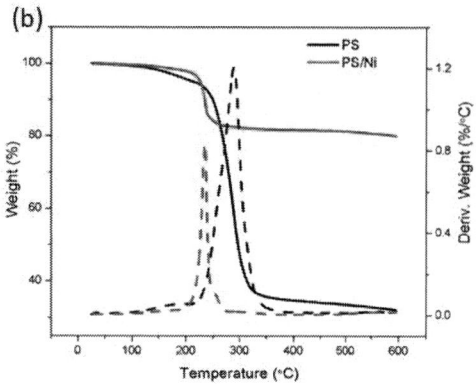

Figure 5. (a) Conductivity of PS/Ni and PS/Ag/Ni versus time under thermal treatment; (b) TGA and differential TG curves of PS and PS/Ni sealants.

To select metallic fillers with minimized interactions with PS, we have referred to the HSAB theory for "hard" and "soft" Lewis acids and bases [14]. According to the theory, "hard" applies to species with small size, high charge states, low polarizability, high electronegativity (bases); while "soft" applies to species with big size, low charge states, and strong polarizability. "Soft-soft" and "hard-hard" react faster and form stronger bonds. As shown in Fig. 6, polysulfide (S rich) serves as a soft base, Ag serves as a soft acid; they are highly reactive towards each other to form strong coordination bonds, blocking the electron transfer between fillers. Therefore, it is necessary to consider fillers that are intermediate or hard acids. Al and Ti was proposed as hard acid fillers. However, Al flakes are easily oxidized to form a thin Al_2O_3 layer on the outer surface. Surface oxidation is also an issue for Ti. The conductivity of PS/Ti barely increased to 1.5 S/m after thermal treatment. Based on these results, while carbon fillers can deliver moderate conductivity to PS after RT cure, Ag fillers are still considered to still be the best candidate that are capable to deliver high conductivity.

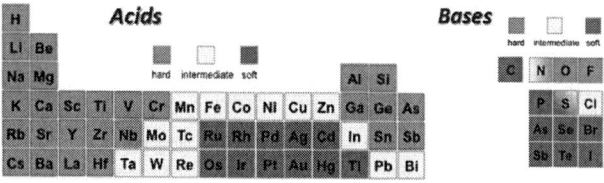

Figure 6. Trends for hard-soft acids and bases.

The effect of Ag fillers was further investigated from three aspects, namely, shapes, surface coatings, and loading levels. In addition of silver flakes and particles, 2D nanowires and 3D dendrites were used to lower down percolation threshold [15]. High conductivity at low volume fraction of fillers is critical to ensure mechanical properties of sealants. Generally, commercial Ag have lubricant layers on surface to improve filler dispersion in polymer matrix [16]. In this work, a thin lubricant layer can serve to protect Ag fillers from being coordinated

with sulfide groups in the sealants. Therefore, synthesized AgNWs and Ag dendrites were also surface treated with lubricants to enable filler dispersion and reduce Ag-S interactions.

AgNWs with different surface coatings were incorporated into the PS and the corresponding electrical conductivity was measured. Despite the variation in surface coatings, Ag-S chelation was strong enough to replace the original surfactants and diminish the conductivity. For AgNWs treated with fatty acid A, a conductivity of 7×10^5 S/m was achieved at 10.9 vol% after thermal treatment (Fig. 7a). The effectiveness of surfactants for conductivity improvement was compared: longer-chain fatty acids can enable better dispersion of fillers in PS and form a bulkier passivation layer to decrease Ag-S interactions; while multi-functional acids had greater reactivity towards AgNWs to form a stronger interfacial bonding. Due to the high performance of surfactant A, Ag dendrite fillers were modified by acid A before mixing with PS. PS/Ag dendrite at low filler loading could reach a conductivity of 1.5×10^5 S/m after thermal treatment (Fig. 7b). It was observed that conductivity was slightly decreased upon further annealing, which indicated that high aspect ratio nanostructures might be susceptible to surface oxidation.

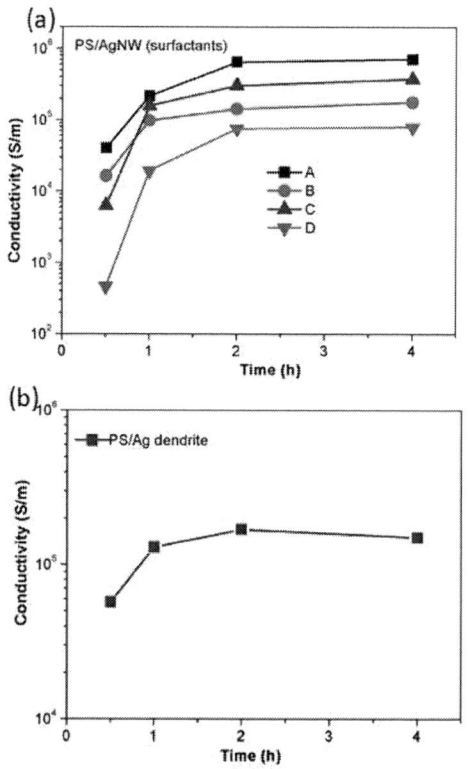

Figure 7. Conductivity change of (a) PS/AgNWs treated with different surfactants and (b) Ag dendrite as a function of time.

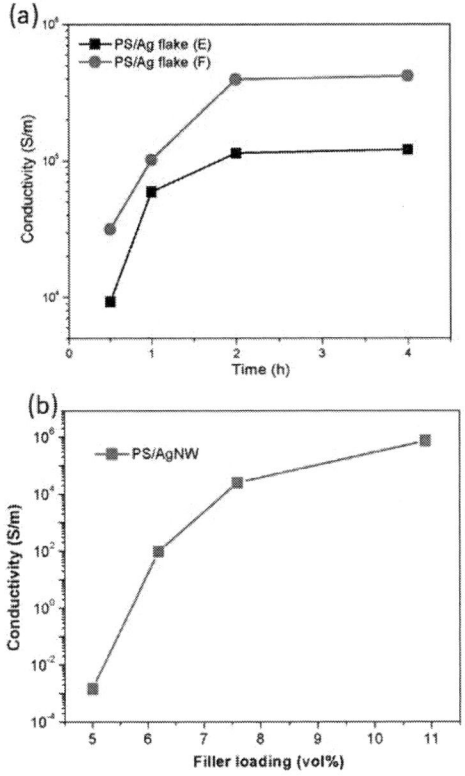

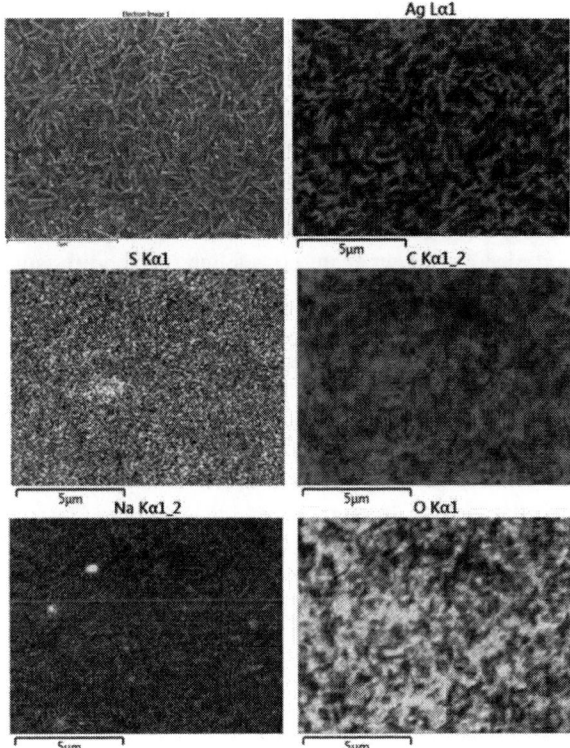

Figure 9. Elemental mapping of PS/AgNW.

Figure 8. (a) Conductivity change of PS/Ag flakes with varied surface coatings; (b) conductivity of PS/AgNW at different filler loadings.

For Ag flakes at low filler loadings, the effect of surface coatings became more prominent. As shown in Fig. 8a, PS filled with 10.9 vol% silver flakes (surfactant E) had a conductivity of $1.2×10^5$ S/m after thermal treatment, while fillers with another surfactant (F) can further improve conductivity up to $4.2×10^5$ S/m. The conductivity values for PS/AgNWs at different loading levels were displayed in Fig. 8b. Conductivity increased from $1.4×10^{-3}$ S/m at 5 vol.% filler loading to $7.1×10^5$ S/m at 10.9 vol.%. The percolation threshold for PS/AgNWs was at 6-8 vol.%, lower than those for Ag flakes and Ag dendrites.

The distribution of AgNWs and other essential elements in the sealants was studied by EDX. Fig.9 shows the elemental mapping of the PS/AgNW composites. It can be seen that AgNWs were evenly distributed within PS, and some other elements included in the sealant formulation (e.g. Al, Mg, Si, Ca) can also be identified [8]. The flexibility of the PS-based films was investigated by tensile tests. The stress-strain curves (Fig. 10) showed that PS sealant was highly stretchable that can withstand tensile strain over 400%. The addition of 5 vol% fillers still maintained the good elasticity of PS. Particularly, PS/CB exhibited tremendously increased tensile strength and Young's modulus.

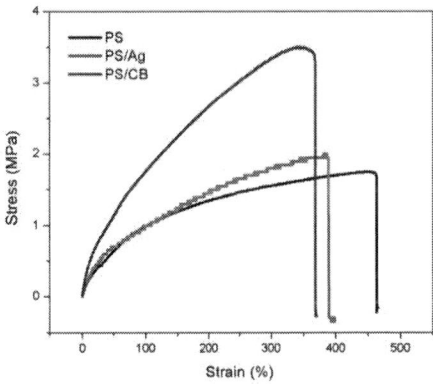

Figure 10. Stress-strain curves of PS, PS/Ag, and PS/CB sealant films.

IV. CONCLUSION

Different types of carbon and metallic fillers were utilized to modify commercial polysulfide sealants to improve conductivity. Polysulfide filled with 5 vol% carbon black can achieve conductivity of 30 S/m after RT cure. Several metallic fillers, including Cu, Ni, Ti, Al, Ag and their hybrids were employed. It was discovered that that metallic fillers could form coordination bonds with the sulfide groups in polysulfide, which block the electron conduction pathway. Post cure thermal treatment was essential to dramatically enhance the conductivity. Particularly, silver fillers with different shapes, surface coatings, and loading levels were discussed. Results showed that polysulfide filled with high aspect ratio AgNWs can achieve maximum conductivity of 7×10^5 S/m, and the percolation threshold can be as low as 6-8 vol%. To further improve conductivity with minimal thermal treatment, an in-depth understanding of the filler/polymer interactions will be necessary. Theoretical models can be used to estimate the thermodynamic and kinetic properties that drive interfacial reactions, and optimized coatings can be applied to modify surface chemistry.

ACKNOWLEDGMENT

This research work was supported by The Boeing Company.

REFERENCES

[1] J. R. Panek, Polysulfide sealants and adhesives, Handbook of Adhesives, Springer, 1990, pp. 307-315.

[2] O.Syao, G. Malysheva, "Properties and application of rubber-based sealants", Polymer Science Series D, 7, 2014, pp. 222-227.

[3] L.H. Lee, "Adhesives, sealants, and coatings for space and harsh environments", Adhesives, Sealants, and Coatings for Space and Harsh Environments, Springer, 1988, pp. 5-29.

[4] P. Trotter, "An Introduction to Tank Sealant", EAA Chapter 130.

[5] H. Lucke, Aliphatic polysulfides, Publisher Huthing & Wepf, Verlag Basel, 1994.

[6] G. Lowe, "The cure chemistry of polysulfides", International journal of adhesion and adhesives, vol. 17, 1997, pp. 345-348.

[7] A. Hutchinson, "Requirements of sealed joint systems for demanding applications-an engineering perspective", FEICA Conference, Brussels, Sept. 14, 2017.

[8] A. Usmani, Chemistry and technology of polysulfide sealants, Polymer-Plastics Technology and Engineering, 19, 1982, 165-199.

[9] E.T. Bannink Jr, "Conductive bonded/bolted joint seals for composite aircraft", U.S. Patents 4556591, 1985.

[10] B. Song, Z. Wu, Y. Zhu, K.S. Moon, C. Wong, "Three-dimensional graphene-based composite for flexible electronic applications", Electronic Components and Technology Conference (ECTC), 2015 IEEE 65th, 2015, pp. 1803-1807.

[11] D.J. Kovach, K.L. Stromsland, D.L. Heidlebaugh, J.A. Ward, A.M. Brown, D.K. Dabelstein, "Edge seals for composite structure fuel tanks", U.S. Patents 8900496, 2014.

[12] E.H. Park, Reinforced elastomeric seal, U.S. Patents 7658387, 2010.

[13] B. Song, J. Li, F. Wu, S. Patel, J. Hah, X. Wang, K. S. Moon, C. P. Wong, "Processing and characterization of silver-filled conductive polysulfide sealants for aerospace applications", Soft matter, vol. 14, 2018, pp. 9036-9043.

[14] R.G. Pearson, "Hard and soft acids and bases", Journal of the American Chemical Society, 85 (1963) 3533-3539.

[15] B. Song, F. Wu, K. S. Moon, R. Bahr, M. Tentzeris, C. Wong, "Stretchable, Printable and Electrically Conductive Composites for Wearable RF Antennas", Electronic Components and Technology Conference (ECTC), 2018 IEEE 68th, 2018, pp. 9-14.

[16] B. Song, K. S. Moon, C. Wong, "Stretchable and Electrically Conductive Composites Fabricated from Polyurethane and Silver Nano/Microstructures", Electronic Components and Technology Conference (ECTC), 2017 IEEE 67th, 2017, pp. 2181-2186.

Challenges and Approaches to Developing Automotive Grade 1/0 FCBGA Package Capability

Rajen Dias, Mike Kelly, Devarajan Balaraman
Advanced Package and Technology Integration
Amkor Technology
Tempe, USA
rajen.dias@amkor.com

Hideaki Shoji, Tomio Shiraiwa
Package Development
J-Devices Corporation
Hakodate, Japan
hideaki.shoji@j-devices.co.jp

KwangSeok Oh, JoonYoung Park
K5 Materials and Process Development
Amkor Technology
Incheon, Korea
kwangseok.oh@amkor.co.kr

Abstract— Automotive Grade 1 and 0 package requirements, defined by Automotive Electronics Council (AEC) Document AEC-100, require more severe temperature cycling and high temperature storage conditions to meet harsh automotive field requirements, such as a maximum 150°C device operating temperature, 15-year reliability and zero-defect quality level. Moreover, increased integration of device functionality to meet the new automotive requirements for in-vehicle networking, autonomous driving, infotainment and sensor integration are driving increases in die and package sizes. This paper provides an update on flip chip ball grid array (FCBGA) package development as quality and reliability requirements increase for larger and larger package form factors and approaches that should be taken to meet Grade 1/0 requirements. Package quality and wear-out failure modes and mechanisms experienced during extended reliability testing in Automotive Grade 2 and 3 package qualifications have identified thermomechanical stress and material degradation at high temperatures as key factors for focus in Grade 1/0 development. To achieve higher grade levels, key package substrate materials such as core, solder resist and build-up layers need to be evaluated as well as assembly materials such as underfills materials may need improvement.

Mechanical simulation data of key material properties such as coefficient of thermal expansion (CTE), modulus of elasticity (E1) and glass transition temperature (Tg) of the substrate and assembly materials are used to provide guidance for the selection of substrate and assembly materials used in the design of experiments to meet Auto Grade 1 and 0 reliability requirements.

Taguchi mechanical simulations results show that use of low CTE materials for the substrate core and build up material was beneficial in preventing SR cracking, UF cracking and bump cracking. Reliability stress results on design of experiments based on inputs from simulation resulted in developing a substrate and assembly material set that meets AEC100 solder resist (SR) Grade 1 and 0 package requirements on a 45-mm x 45-mm FCBGA.

Keywords-FCBGA, Auto-grade, AEC, FEM, Taguchi

I. INTRODUCTION

Increased digital processing content in automobiles, ranging from infotainment applications to Advanced Driver Assist System (ADAS), is necessitating a thorough evaluation of the reliability performance of flip chip ball grid array (FCBGA) semiconductor packages. Depending on the function, the component level reliability requirements are classified by the Automotive Electronics Council (AEC) as Grade 3, Grade 2, Grade 1 or Grade 0. The requirements for these grade levels are summarized in TABLE I. The requirements for Temperature Humidity Bias (THB), biased and unbiased Highly Accelerated Stress Test (HAST) remain the same across all automotive grades – 96 hours at 130°C /85% humidity or 264 hours at 110°C /85% humidity. All reliability testing is followed by electrical testing at room and hot temperature.

TABLE I.   AUTOMOTIVE GRADE-LEVEL REQUIREMENTS [1]

Stress	Condition	Duration			
		Grade 0	Grade 1	Grade 2	Grade 3
Temperature-Humidity-Bias (THB)	85°C/85%RH	1000 h	1000 h	1000 h	1000 h
Biased Highly Accelerated Stress Test (HAST)	110°C/85%RH	264 h	264 h	264 h	264 h
Unbiased Highly Accelerated Stress Test (uHAST)	110°C/85%RH	264 h	264 h	264 h	264 h
	130°C /85%RH	96 h	96 h	96 h	96 h
Temperature Cycling (TC)	C: -65°C-150 °C		500x		
	H: -55°C-150°C	2000x	1000x		
	B: -55 °C-125°C			1000x	500x

	175⁰C	1000x	500x		
High Temperature Storage Life (HTSL)	150⁰C	2000x	1000x	1000x	1000x
	125⁰C			1000x	1000x
Power Temp Cycle (PTC) (Required for Parts rated at >1W)	-40⁰C-150⁰C	1000x			
	-40⁰C-125⁰C		1000x		
	-40⁰C-105⁰C			1000x	1000x

It becomes apparent that the stringent temperature cycling and high temperature storage requirements for Auto G1/0 drive the need for careful selection of both substrate and assembly materials. Larger temperature ranges of -65⁰C to 150⁰C exacerbate many of the failure mechanisms associated with temperature cycling, such as fatigue cracking of solder bumps as well as cracking of polymeric materials such as underfills and solder resist (SR) films. Similarly, the high temperature storage test conditions (HTSL: 150⁰C and 175⁰C) result in permanent changes in the material properties that can result in performance degradation of packages. The long cycle times for reliability data collection also means failures in qualification could result in substantial product qualification delays. Hence, judicious use of thermomechanical modeling is needed through-out the development and qualification process.

The common material degradation mechanisms encountered in FCBGA packages at different reliability stresses are summarized in TABLE II.

TABLE II.   MATERIAL DEGRADATION MECHANISMS

	TCx	HTS
Solder	Bump cracking	Solder consumption, IMC growth
Underfill	Cracking	Oxidation, cracking
TIM[1]	X	Delamination
Substrate	PTH cracking, micro-via cracking, SR cracking (leading to trace cracking)	

1 thermal interface material

Polymeric materials undergo permanent changes when subjected to high temperatures for extended periods of time. Depending on the ambience, this may include material oxidation as well as mechanical property changes resulting in embrittlement. The presence of humidity can also lead to loss of adhesion at the die passivation and substrate solder mask interfaces. Lin et al have studied the evolution of tensile properties and creep behavior of underfills under isothermal aging at various temperatures and developed a 4-parameter empirical model to predict the property changes [2]. The monotonic increase in modulus and tensile strength of the underfills with isothermal aging at temperatures above and below the Tg of the underfill was reported along with an increase in elongation to break suggesting an overall toughening of the underfill material.

Likewise, solder joints in the FCBGA packages undergo thermal aging resulting in growth of relatively brittle intermetallic layers, adversely impacting the mechanical behavior of solder joints [4].

In addition, there is a strong interaction between design rules and materials. Often, the best materials available in the market cannot meet the reliability requirements of Auto-G1/0

without restrictive design rules for substrates and assembly. The design rules, while enabling the demonstration of higher reliability, may adversely impact the electrical performance and/or result in higher cost. As a result, there is a continuous effort to re-enable restrictive design rules over time through materials and process improvements.

Lastly, due to the mission critical functions of the automotive components, there is emphasis on defect elimination and containment as well. This is best achieved by stringent quality control measures which include failure mode effects analyses (FMEAs), materials and process control plans across the entire packaging supply chain. Thus, there are three components to achieving Automotive Grade 1/0 certification – materials selection, substrate and assembly design rules and robust manufacturing systems (see Figure 1.). The paper focuses on materials selection via thermomechanical simulations and assembly experiments and reliability testing.

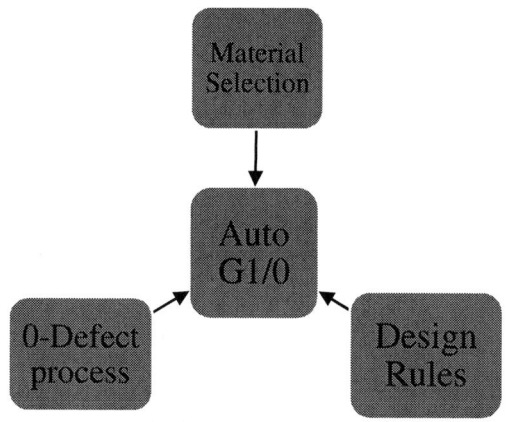

Figure 1.   Components of Automotive Grade1/0.

II.   MECHANICAL SIMULATION STUDIES

The impact of material properties of the package substrate and assembly materials on solder resist cracking, underfill cracking and chip interconnect cracking was studied using the Taguchi method of analysis. The main material properties investigated were coefficient of thermal expansions (CTE1, CTE2), elastic modulus (E1) and glass transition temperature (Tg) of the substrate core material, build up material and solder resist material as well as the underfill material in assembly. Three levels (low, medium, high) of each material property were simulated using commercially available materials. For example, three core materials with CTEs of 4, 7.6 and 15 ppm/C were compared. TABLE III. shows the material properties investigated.

TABLE III.   PROPERTIES OF SUBSTRATE AND ASSEMBLY MATERIALS

Factor Level	Core α1 (ppm/°C)	Core E1 (GPa)	BU α1 (ppm/°C)	BU E1 (GPa)	SR α1 (ppm/°C)	SR α2 (ppm/°C)	SR E1 (GPa)	SR Tg (°C)	UF α1 (ppm/°C)	UF α2 (ppm/°C)	UF E1 (GPa)	UF Tg (°C)
1	4	20	10	4	9	50	4	100	20	80	3.8	100
2	7.6	24	23.8	8.5	20	96	8.4	125	35	100	8	125
3	15	40	40	25	60	150	17	150	52	135	11	150

The Taguchi simulation method was used to determine the impact of the material properties for each of the three failure mechanisms: solder resist cracking, underfill (UF) cracking

978-1-7281-1500-9/19 $31.00 © 2019 IEEE

and chip interconnect bump cracking. Figure 2. shows the impact of varying material properties on the propensity for solder resist cracking at the die corner solder bumps. The Taguchi output is a signal to noise ratio (S/N) with 0 being equivalent to no stress. Higher numbers (absolute value) indicate higher stress. Figure 2. shows that SR and UF material have a large influence on SR cracking. Low modulus of the SR and low CTE of the UF results in improving resistance to SR stresses.

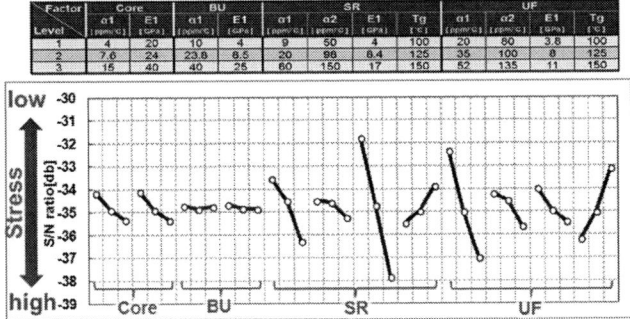

Figure 2. Impact of material properties on SR cracking.

Figure 3. shows the impact of varying material properties on the propensity for UF cracking at the die corners. The core material and UF material have a large influence on UF cracking. Low CTE core materials and underfills with low CTE, low modulus and high Tg lower the propensity for UF cracking.

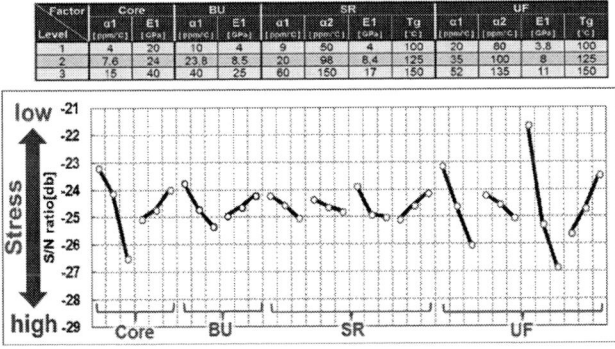

Figure 3. Impact of material properties on UF cracking.

Figure 4. shows the impact of varying material properties on the propensity for chip interconnect bump cracking. The substrate core and buildup materials and UF material have a large influence on bump cracking. Low CTE core material, low modulus buildup material and underfills with high modulus can be used to lower bump cracking risks.

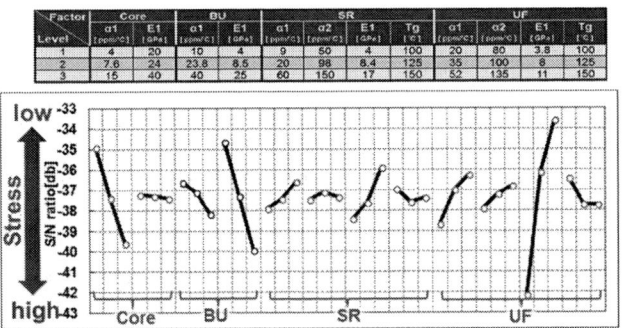

Figure 4. Impact of material properties on chip-interconnect bump cracking.

From the results shown in Figure 2, 3 and 4, it becomes apparent that optimizing substrate and assembly material properties to reduce risk of one mechanism may adversely increase risk of another failure mechanism, resulting in necessary tradeoffs. Figure 5. shows a composite for all three fail mechanisms. In general, material properties need to be optimized for the dominant failure mechanism observed for a specific package geometry while at the same time also modified to prevent another failure mechanism from becoming dominant. Based on material property trends shown in Figure 4, general recommendations would be to use a low CTE core material with low modulus buildup material and with a solder resist that has low CTE and modulus. For underfills, while a low CTE and high Tg are beneficial, modulus has a strong influence. A low modulus UF is recommended for UF and SR crack prevention but significantly increases bump cracking risk for which a high modulus UF is desired.

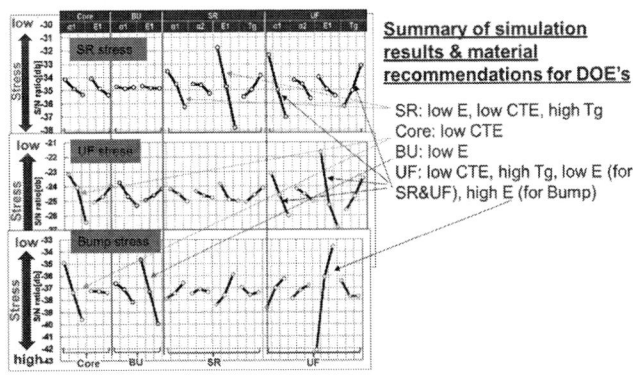

Figure 5. Comparison of impact of material properties on all three mechanisms (SR cracking, UF cracking and bump cracking).

Results and recommendations from the Taguchi simulations were used as a guide in selection of current and new materials for design of experiments (DOEs) to meet AEC 100 Grade 0 and 1 reliability requirements.

III. SUBSTRATE AND ASSEMBLY MATERIALS DOE'S

A daisy chain test vehicle (TV) was used for assessing different commercially available substrate and assembly materials. The TV die was 19.2 mm x 19.2 mm with copper phosphate (CuP) bumps at 165-um pitch. The TV die was flip chip attached to a 45-mm x 45-mm, 2-2-2 organic substrate with a 0.8-mm core and a 1-mm thick copper heat spreader attached to the package substrate. SAC 305 solder was used on both CuP bumps and BGAs. A schematic of the package crross-section is shown in Figure 6.

Figure 6. Cross-section schematic of the 45-mm FCBGA package used in this study.

The DOE consisted of evaluating substrates with two core materials (core 1, core 2), two buildup (BU) materials (BU1, BU2) and two solder resist materials (SR1, SR2). Three substrate types were fabricated using a combination of core, buildup and SR materials as shown in TABLE IV. The material properties of the substrate materials are shown in TABLE V.

TABLE IV. SUBSTRATE MATERIALS USED FOR THE THREE TYPES OF SUBSTRATES FABRICATED.

Substrate Type	Core	Build up	Solder Resist
Substrate type 1	Core 2	BU 2	SR 1
Substrate type 2	Core 1	BU 1	SR 1
Substrate type 3	Core 1	BU 1	SR 2

TABLE V. MATERIAL PROPERTIES OF SUBSTRATE MATERIALS USED IN THE DOES

Substrate Material	CTE 1 (ppm/C)	CTE2 (ppm/C)	Modulus (E1) (GPa)	Tg (C)
Core 1	6	3	23	260
Core 2	9	4	23	260
BU 1	23	78	8	154
BU 2	46	120	7	156
SR 1	38	115	5	130
SR 2	33	90	4	175

FCBGA assembly was performed using four different underfill materials (UF1, UF2, UF3 and UF4). The other assembly materials such as thermal interface material (TIM) and copper lid attach material were kept the same for all builds on the three types of substrates. The physical properties of the four UF materials evaluated are shown in TABLE VI. .

TABLE VI. PHYSICAL PROPERTIES OF THE FOUR UNDERFILLS EVALUATED

Underfill Material	CTE 1 (ppm/C)	CTE 2 (ppm/C)	Modulus (E1) (GPa)	Tg (C)
UF1	28	105	10	90
UF2	28	89	10	130
UF3	22	89	9	154
UF4	29	96	8	156

DOE samples assembled were subjected to reliability testing after Level 3 preconditioning was performed. Reliability stress tests done were a) Unbiased HAST (uHAST) (110°C/85% RH) for 264hrs, b) temperature cycle condition H (-55°C to 150°C) for 1000 and 2000 cycles and c) High temperature storage (175°C) for 500 and 1000 hours. Open/short (OS) tests were done on some legs. Approximately 10 samples/leg were removed after preconditioning and after each of the reliability readouts and visually inspected for fillet cracking after lid removal. Scanning acoustic tomography (SAT) analysis was done to identify any underfill delamination and units were cross-sectioned and planar polished to inspect for solder resist cracking, bump cracking, and underfill cracking at each readout.

IV. DOE RESULTS

One leg with substrate type 1 (low CTE core, low CTE buildup) and with UF1 passed O/S testing after all reliability testing to AEC-100 grade 0 (264 hrs uHAST, 2000 cycles TCH, 1000-hrs HTS at 175°C). Optical, SAT, X-section and planar polishing did not reveal any evidence of damage such as solder resist cracking, fillet cracking or bump cracking after 264 hrs UHAST, 2000 cycles TCH and 500 hrs HTS at 175°C. Representative images of the inspections are shown in Figure 7. However, small UF fillet surface cracks were seen on units after 1000 hrs HTS at 175°C. Cross-section analysis showed that the fillet surface cracks were contained within a thin discolored region of the underfill fillet. The discolored region is believed to be a thin skin of oxidation-damaged underfill epoxy. Some of the underfills did not exhibit this surface fillet cracks.

 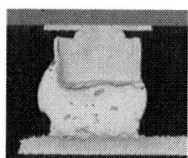 

Optical image of die corner showing no UF fillet cracks after 2000 cycles of TCH.

Optical image of die corner CuP bump showing no SR,UF or bump cracks after 2000 cycles of TCH.

Optical image of die corner after planar grind to SR surface showing no SR cracks between bumps after 2000 cycles of TCH.

Figure 7. Optical images of units after 2000 cycles TCH showing no evidence of SR, UF or bump cracking.

V. DISCUSSION AND CONCLUSIONS

Both Taguchi mechanical simulations results and DOE reliability results show that use of low CTE materials for the substrate core and build up material was beneficial in preventing SR cracking, UF cracking and bump cracking. For underfill properties, mechanical simulation results indicate that a compromise needs to be made between the use of a high modulus material for bump crack resistance and a low modulus material needed for solder resist and underfill cracking resistance. The DOE results showed that underfill (UF1) with an intermediate modulus performed the best and met Grade 1 and 0 reliability requirements. The anticipated benefits of using a high Tg UF material could not be realized in the DOE results.

A new phenomenon of underfill fillet surface cracks seen only after 1000 hrs of HTS at 175°C for some underfills is believed to be related to underfill epoxy and/or hardener degradation by high temperature oxidation. While these surface cracks do not result in electrical failures, further investigation is necessary to understand the kinetics of the oxidation and to partner with underfill material suppliers to formulate underfills with more resistance to oxidation. Alternate underfills did not exhibit this surface oxidation.

In conclusion, Taguchi mechanical simulation results were predictive in determining the impact of substrate and assembly material properties on the propensity for solder resist cracking, underfill cracking and bump cracking. Reliability stress results on design of experiments based on inputs from simulation resulted in developing a substrate and assembly material set that meet AEC100 Grade 1 and 0 package requirements on a 45-mm x 45-mm FCBGA package.

© 2019, Amkor Technology, Inc. All rights reserved.

REFERENCES

[1] AEC - Q100 Rev - H: Failure Mechanism Based Stress Test Qualification For Integrated Circuits (base document).

[2] C. Lin, J. C. Suhling, P. Lall, , "Isothermal aging induced evolution of the material behavior of underfill encapsulants," 59th Electronic Components and Technology Conference, 2009.

[3] Z. Lin, V. Subramanian, P. Malatkar, N. Ananthakrishnan, "Understanding Underfill Degradation in Reliability Testing Conditions for ADAS Package Development," 68th Electronic Components and Technology Conference, 2018.

[4] V.L. Nguyen, C-S. Chung, H-K. Kim, "Mechanical Behavior of Sn-3.0Ag-0.5Cu/Cu Solder Joints After Isothermal Aging," Journal of Electronic Materials, vol.45, No.1, 2016, pp 125-135.

Advanced substrates for GaN-based power devices

Anthony Cibié, Julie Widiez, René Escoffier, Denis Blachier, Kremena Vladimirova, Jean-Philippe Colonna, Paul-Henri Haumesser, Stéphane Bécu, Perceval Coudrain, William Vandendaele, Jérome Biscarrat, Charlotte Gillot, Matthew Charles, Léa Di Cioccio

Univ. Grenoble Alpes, CEA, LETI
38000 Grenoble, France
julie.widiez@cea.fr

Abstract—**In this paper, we present an approach to enhance thermal dissipation of GaN on silicon high electron mobility transistors (HEMTs) and Schottky barrier diodes (SBDs). The initial silicon substrate was removed and replaced by a copper substrate using a transfer process based on polymer bonding to a temporary wafer. Electrical and thermal characterizations were performed to study the impact of the transfer. The thermal measurements showed a lower temperature for the HEMTs transferred onto a copper substrate in the case of a 1 ms pulse at 0.34 W/mm.**

Keywords—*AlGaN/GaN, power devices, substrate transfer, HEMTs, SBDs, bonding, copper, thermal management.*

I. INTRODUCTION

GaN-based devices have gained considerable interest for power electronics and high-frequency applications due to their improved semiconductor properties compared to silicon, in particular their wide band-gap value, high breakdown electric field and high saturation electron velocity. To allow rapid commercialization and to benefit from the available silicon manufacturing facilities, the current trend is to fabricate the GaN-based devices on 200 mm silicon substrates. Silicon is an excellent choice for large scale integration but when used as a receiver substrate for GaN devices it also affects their electrical and thermal performances [1]. Therefore, insulating substrates with better thermal properties than silicon are desirable in order to improve the thermal management and to allow an enlarged operating range and better electrical performance of the GaN devices. Several post processing substrate transfer solutions have been proposed [2]. GaN devices have been transferred from their native substrate onto substrates with higher thermal conductivity, such as copper [3, 4] or diamond [5, 6], in order to enhance thermal dissipation. Other research studies are focusing on improving the electrical performance of the GaN devices by removing the Si substrate and transferring the AlGaN/GaN HEMTs to a glass wafer [7] or by developing a transfer method allowing to fabricate N-face GaN/AlGaN devices [8].

In this paper, we present our work allowing to transfer GaN-based power devices initially fabricated on a silicon carrier on another receiver substrate (copper substrate or electrically-isolated substrate). The devices were fabricated on 200 mm GaN on silicon substrate which was first temporarily bonded to a glass wafer to assure the handling of the GaN layer and to allow the selective etching of the silicon substrate. Then, the selected receiver substrate is deposited or bonded directly at the bottom of the GaN-based devices. The described process sequence is fast, cheap and compatible with traditional (8 or even 12 inches) process technologies. The obtained practical results demonstrated that the developed GaN device transfer process preserves the electrical performance of the devices while taking advantage of the better thermal properties of substrates with higher thermal conductivity than silicon.

II. DEVICES FABRICATION

III-N heterostructure devices were fabricated on 200 mm GaN on silicon wafers grown using metal organic vapor phase epitaxy (MOVPE). First, a nucleation layer of aluminum nitride (AlN) was grown directly on the 1 mm thick p-type (111) Si substrate. Then, a thick AlGaN-based buffer layer was grown before a thin unintentionally doped GaN layer. Next, a 1 nm AlN spacer and an $Al_{0.25}Ga_{0.75}N$ barrier were grown to form the two-dimensional electron gas (2DEG) at the AlGaN/GaN interface, before a final 10 nm in-situ SiN passivation was grown.

Normally-off AlGaN/GaN HEMTs and SBDs were fabricated through a CMOS compatible integration flow. Both devices used a partial recess of the AlGaN barrier. Concerning the HEMTs process, the recess depth reached 35 nm to cut the 2DEG channel. 30 nm thick aluminum oxide (Al_2O_3) and titanium nitride (TiN) were used as the gate oxide and the gate metal respectively. The recess strategy is common in such architectures and offers the benefit of reducing the turn-on voltage of the devices allowing more current to be delivered in the on-state. The nominal width of the SBD devices was 52 mm. 10 A 650 V AlGaN/GaN HEMTs with gate width (W_G) of 100 mm, gate length (L_G) of 1 μm, gate-source spacing (L_{GS}) of 2 μm, and gate–drain spacing (L_{GD}) of 15 μm were used for this study.

III. SUBSTRATE TRANSFER

The AlGaN/GaN power devices were transferred from the Si substrate onto a copper substrate using a post-processing double-flip technology (Fig. 1). The transfer process was performed with AlGaN/GaN power devices fabricated on a (111) Si substrate (Fig. 1-a). First, the top surface of the wafer was bonded to a glass substrate with a temporary polymer bonding (Fig. 1-b). The silicon wafer was then thinned down using a combination of grinding (down to 40 μm Si thickness) and a dry plasma etching step selective to the AlN nucleation layer (Fig. 1-c). In order to conduct a large number of studies with devices fabricated on the same wafer, the wafers were diced into 2.2 x 2.2 cm² chips using the transparency of the glass substrate. After the dry etching step, 50 nm thick Ti, 60 nm thick TiN and finally 200 nm thick Cu layers were deposited on the AlN layer by physical vapor deposition (Fig. 1-d). This last conductive layer was used as a seed for the 80 μm-thick copper growth made by electrochemical deposition (Fig. 1-e). Finally, the glass substrate was removed using a polymer removal to regain access to the devices (Fig. 1-f).

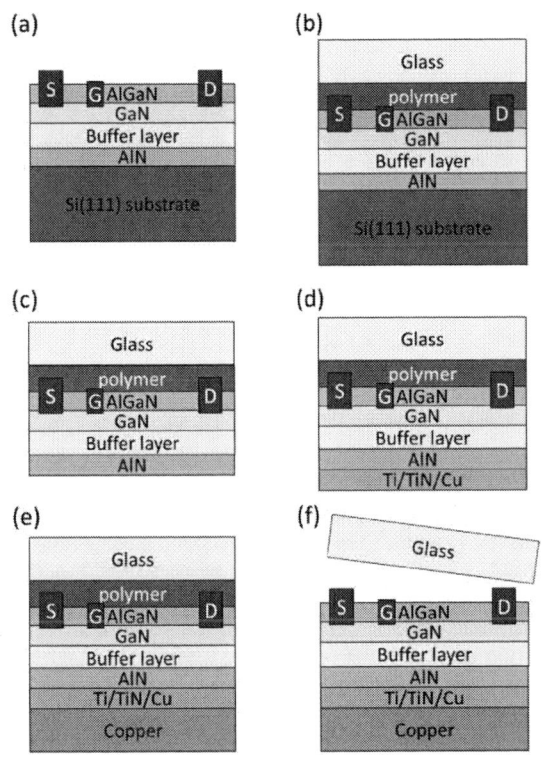

Figure 1. Schematics of the substrate transfer process developped in this work.

This substrate transfer process combines the easy use of standard silicon processing with freedom in the choice of the final substrate. It should also be noted that this process is wafer-level compatible by adding an "anti-sticky" layer in between the polymer glue and the temporary substrate to easily remove the latter.

In order to have a better understanding of the impact of the substrate transfer on the electrical properties of AlGaN/GaN devices, we performed three studies: the transfer of power devices directly onto copper as described above (Fig. 2-a), onto copper without complete removal of all the silicon (Fig. 2-b), and onto an electrically insulating ceramic glue (Fig. 2-c). To process the second study case, the dry etching was not completed and 10 μm of Si was left before the metal deposition. For the third study case, a ceramic glue called "CeramaBond™865" was used to bond the devices to another Si substrate. This glue is an aluminum nitride-based ceramic adhesive with good electrical insulating properties (dielectric strength ≈ 7400 kV/m) but with relatively poor thermal properties (thermal conductivity ≈ 70 W/mK) compared to silicon (≈ 150 W/mK) or copper (≈ 400 W/mK).

Fig. 3-a shows a FIB-SEM cross-section of a HEMT transferred onto copper. No damage is observed either at the component level or III-N stack level which validates our process from a structural point of view. Fig. 3-b is a zoom of the interface between the AlN layer and the deposited Ti/TiN/Cu layers. It shows a good interface quality, although some small defects are visible at the TiN/Cu interface. This well-known voiding phenomenon is a stress-driven vacancy diffusion mechanism [9].

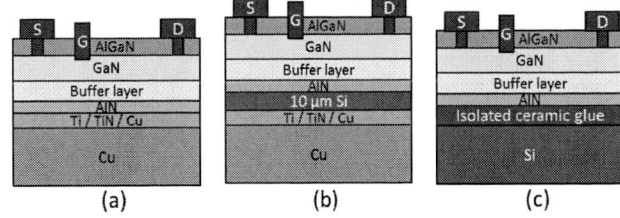

Figure 2. Schematics of the different substrate transfers performed for this study. From left to right: onto copper, onto copper with 10 μm of Si remaining, onto an electrically insulating ceramic glue.

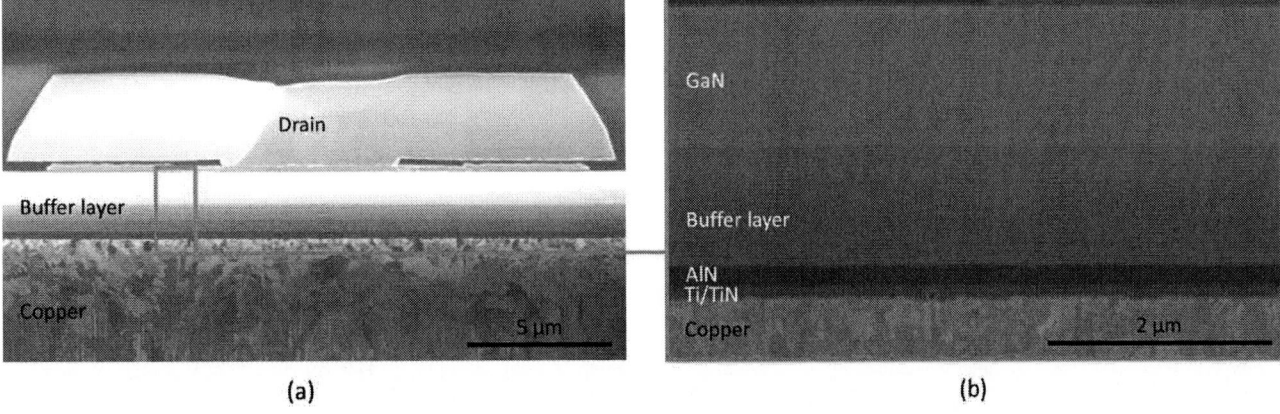

Figure 3. (a) FIB-SEM cross-section of a HEMT transferred onto copper. (b) FIB-SEM cross-section zoom on the III-N stack and the metallic layers.

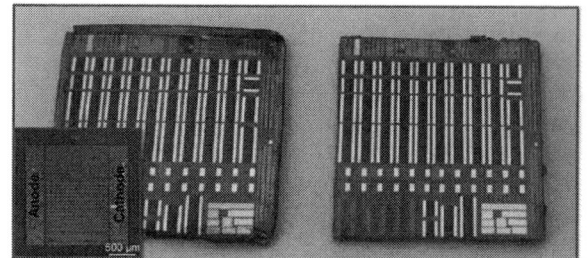

Figure 4. Photo of the Schottky diodes transferred onto copper (left) and onto the electrically insulating substrate (right). Inset: top view microscope image of one SBD transferred onto copper.

The resulting chips including the SBDs devices transferred onto the copper substrate and onto the ceramic glue are shown in Fig. 4.

IV. ELECTRICAL RESULTS

A. Direct biased characterization

Pulsed $I_{DS}(V_{DS})$ and $I_{DS}(V_{GS})$ characteristics were performed with 200 µs pulse width. The substrate was grounded for all the measurements presented in this paper. Fig. 5 compares the $I_{DS}(V_{DS})$ curves of AlGaN/GaN HEMT before (on silicon, blue dots) and after the transfer (on copper, red triangles) for different gate voltages (V_{GS} = 2, 4 and 6 V). It should be noted that the same HEMT device is used prior and after the substrate transfer. We see that the drain current increases after the transfer onto copper (Fig. 5). Fig. 6 compares the $I_{DS}(V_{GS})$ characteristics of AlGaN/GaN HEMT before (on silicon, blue dots) and after the transfer (on copper, red triangles) at V_{DS} = 0.5 V. The threshold voltage (V_{th}), defined as the voltage at I_{DS} = 10^{-5} A/mm, shifts negatively by 0.6 V (from 1.9 V on Si to 1.3 V on Cu). If the V_{th} difference is freed by shifting the post-transfer curve by ΔV = 0.43 V, we still observe the drain current increase after the transfer onto Cu (dashed curve in Fig. 6).

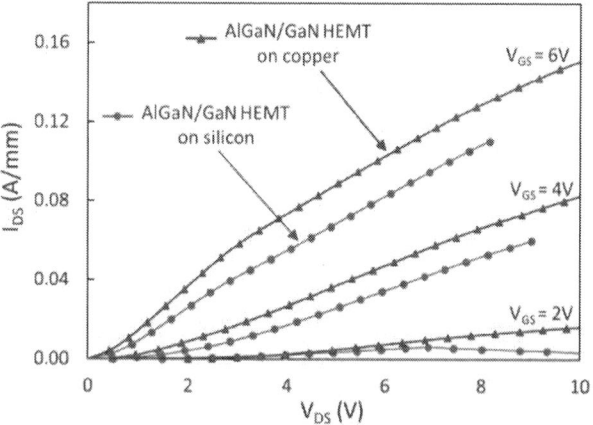

Figure 5. Pulsed $I_{DS}(V_{DS})$ characteristics of the same AlGaN/GaN HEMT on Si (blue dots) and after the transfer onto Cu substrate (red triangles).

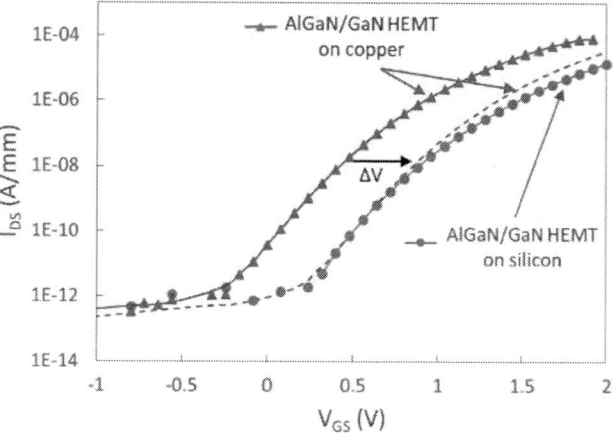

Figure 6. Pulsed $I_{DS}(V_{GS})$ characteristics of the same AlGaN/GaN HEMT before transfer (on silicon) and after the transfer (on copper substrate). V_{DS} = 0.5 V. The dashed curve is the $I_{DS}(V_{GS})$ curve of the HEMT on copper shifted by 0.43 V.

Sheet resistance values (R_{sh}) were also extracted on the same Van Der Pauw pattern before and after the transfer: we measure a decrease of R_{sh} from 676 Ω/sq on Si substrate to 480 Ω/sq on Cu substrate. Similar R_{sh} reduction observations were reported in previous studies [4, 7, 8]. This R_{sh} reduction indicates that the 2DEG density is increased during our substrate transfer process.

A possible explanation of the negative V_{th} shift and of the drain current increase may be a modification of the stress/strain inside the III-N epitaxial layers during the substrate transfer process. This stress/strain modification would cause an increase of the electron density in the channel and thus a reduction of the R_{sh} value. This R_{sh} reduction would explain the increase of the drain current and the reduction of the threshold voltage value (less voltage is needed to "open" the channel).

To better understand the origin of these phenomena, the initial Si substrate of a processed HEMTs wafer was thinned down by grinding to a Si thickness of 400 µm then diced into samples and thinned down again to different final Si thicknesses (250 µm and 100 µm) by plasma etching steps (Fig. 7). We observe similar R_{sh} and threshold voltage (ΔV_{th} = -0.67 V) reductions between the reference devices made on 1 mm thick Si substrate and the thinned devices on 100 µm thick Si substrate (Fig. 7 and Fig. 8). We can conclude that the R_{sh} and V_{th} reductions are related to the substrate thinning during our substrate process and could be due to a modification of the mechanical strain inside the active layers. Raman spectroscopy measurements are in progress to confirm this hypothesis. A previous study corroborates this assumption by showing a strong correlation between the V_{th} shift and the GaN strain/stress [10].

978-1-7281-1500-9/19 $31.00 © 2019 IEEE

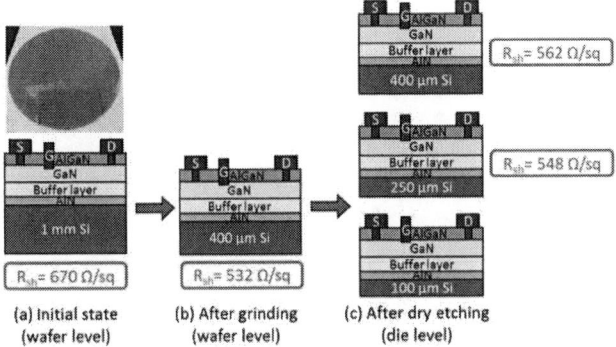

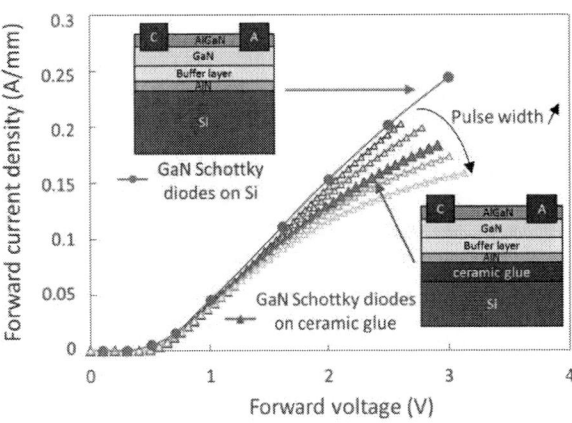

Figure 7. Schematics of the Si substrate thinned-down experiment to intentionnally modify the mecanical stress/strain in the III-N layers. Measured R_{sh} values are specified.

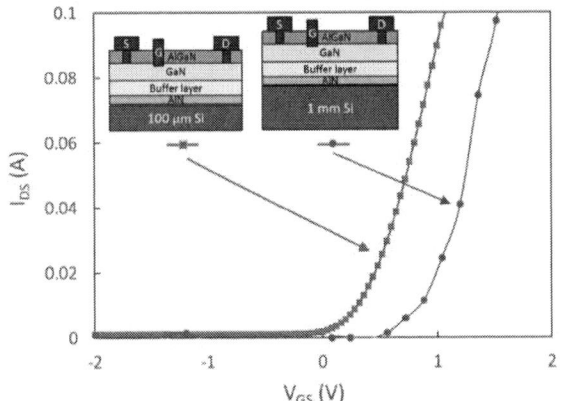

Figure 8. Comparison of pulsed $I_{DS}(V_{GS})$ characteristics of an AlGaN/GaN HEMT as a function of the Si substrate thickness: 1 mm thick Si (blue dots) *vs* 100 μm thick Si (red squares). V_{DS} = 0.5 V.

Forward I(V) characteristics of the same 52 mm SBD before and after transfer onto the ceramic glue (Fig. 2-c) were also compared at a pulse width of 200 μs (Fig. 9, solid symbols). The current decreases after the substrate transfer. This current degradation is related to the poor thermal conductivity of the ceramic glue (60-80 W/mK). The glue, deposited directly on the AlN layer, acts as a thermal barrier. The heating of the component is more important inducing a current degradation by self-heating effect. The pulse width was varied from 50 μs to 500 μs (Fig. 9, open symbols). The longer the pulse time, the more the current degrades. Indeed, the pulse time increase leads to a more significant heat generation by Joule effect. On the other hand, for a small pulse width (50 μs), the I(V) characteristic of the diode transferred onto the isolating glue becomes similar to the one on silicon. This confirms that the current reduction is due to a higher heating after transfer onto the insulating substrate.

B. Off-current characterization

Off-state characterizations were also performed on the HEMTs and the SBDs devices and compared before and after transfer onto the different substrates. Fig. 10 shows the I_{DS} and I_{GS} leakage currents of the same AlGaN/GaN HEMT before and after the transfer onto copper for V_{DS} ranging from 0 V up to 450 V and for V_{GS} = -2 V.

Figure 9. Forward I-V characteristics of the same SBD before transfer (on Si) and after tranfer on a ceramic glue. The "standard" characterizations are performed with a pulse width of 200 μs (solid symbols). The anode and the substrate are grounded. To study the self-heating effect, the pulse width was varied from 50 μs to 500 μs.

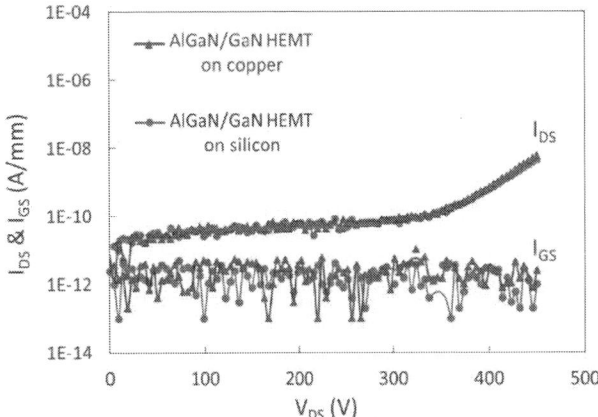

Figure 10. I_{DS} and I_{GS} leakage currents of the same AlGaN/GaN HEMT before and after the transfer from silicon to copper substrate. The measurement was performed up to V_{DS} = 450V. V_{GS} = -2 V. No change was observed after the transfer.

No change of the drain current was observed after the transfer, showing the good isolation quality of the buffer layer. However, in order not to damage the HEMT component before transfer, the drain voltage was limited at a maximum value of 450 V. As the SBDs are free from the possible breakdown of the gate insulator, we performed leakage current measurements at higher voltages (up to a cathode voltage V_C of 700 V) on these devices (Fig. 11). For V_C higher than 370 V, we observe a higher leakage current on copper compared to that for the device on silicon. This may be due to the fact that in the copper substrate configuration, the depletion layer of the silicon substrate is suppressed. Indeed, several recent studies have shown that the AlN/Si interface and the Si substrate play more and more important roles in the vertical leakage as the quality of the epi-stack is improved. Lomgobardi *et al.* [11] showed that the electrons generation that forms the leakage is mostly due to impact ionization at the surface of the Si substrate. Yacoub *et al.* [12] confirmed the existence of the inversion layer at the AlN/Si interface and revealed a wide depletion

978-1-7281-1500-9/19 $31.00 © 2019 IEEE

region in the Si substrate (p-doped with $N_A = 10^{15}$ cm^{-3}), Li *et al.* [13] showed that the carrier concentration of the Si substrates strongly impacts the vertical leakage characteristics. To see the impact of the AlN/substrate interface, we performed off-state current characterizations on the SBDs transferred onto copper but with 10 µm of Si left in between the AlN layer and the Cu layer (Fig. 2-b). The leakage characteristic shows exactly the same evolution as in the reference configuration *i.e.* with SBDs on the (111) Si substrate (Fig. 12). This experiment indicates that the presence of a remaining Si layer in between the III-N layers and the copper layer is essential to avoid breakdown voltage degradation when transferring power devices onto copper substrate. However, we are not currently able to know if the 10 µm-thick Si layer plays a direct role (with the addition of a depletion region) or if the AlN/Cu interface is degraded by our etching process adding traps in the structure. To answer this question, complementary experiments should be done: the bonding of a Si substrate after the removal of the initial Si substrate will directly indicate whether our transfer process degrades or not the AlN/Si interface.

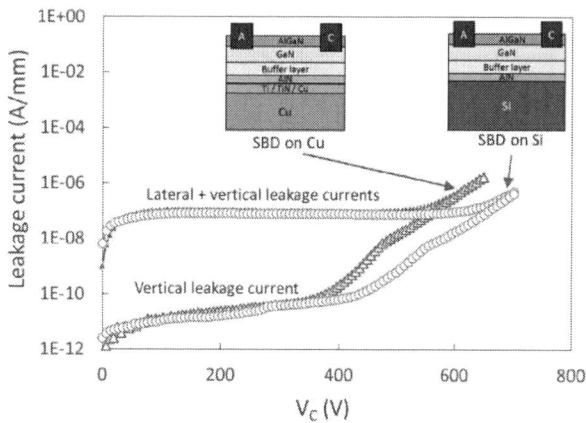

Figure 11. Leakage currents of the same GaN SBD before and after transfer onto copper. The measurement was performed up to $V_C = 700$ V For $V_C > 370$ V, the vertical leakage current on copper increases faster than on silicon substrate.

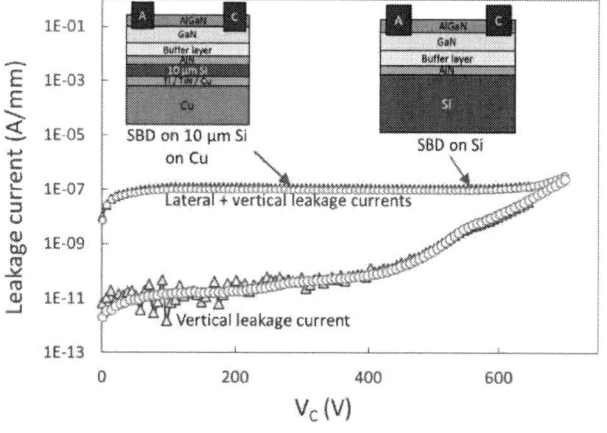

Figure 12. Comparison of leakage currents of the same GaN SBD before transfer (on Si) and after transfer on Cu with 10 µm of the initial Si substrate remaining. No change was observed after the transfer.

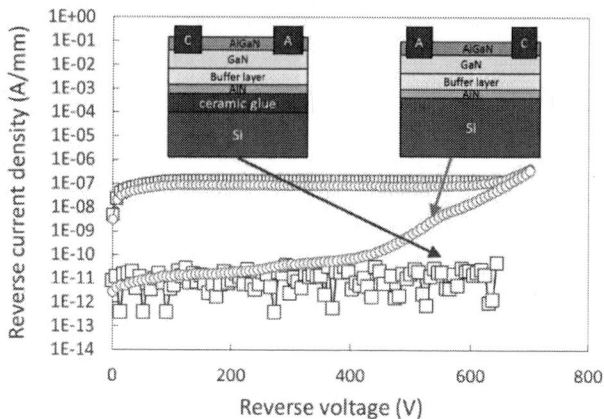

Figure 13. Reverse I-V characteristics of the same SBD on Si and on an isolating glue. The anode and the substrate were grounded.

To complete the study, reverse I-V characteristics of the same 52 mm SBD before transfer (*i.e.* on silicon) and after transfer onto the ceramic glue were compared (Fig. 13). The insulating glue acts as a barrier to vertical current and thus improves the breakdown voltage.

V. THERMAL RESULTS

A. *Thermal measurement protocol*

Thermal measurements were performed on AlGaN/GaN HEMTs before and after the substrate transfer, i.e. respectively onto silicon and onto copper substrates. We used an InSb infrared camera placed on top of the devices to measure their surface temperature. The acquisition time of the camera is limited to 10 ms but the temperature evolution within the device is much faster. To overcome this problem, we used the trigger delay method. It allows acquisitions with much smaller steps, ranging from 1 µs to 10 µs, while synchronizing the gate voltage with the software that controls the camera. The principle of the measurement is shown in Fig. 14.

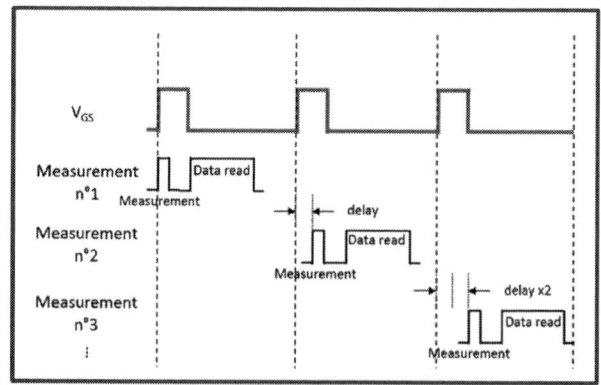

Figure 14. Principle of the trigger delay method used for the thermal measurements.

This approach consists in applying successive and identical voltage pulses on the gate (blue line in Fig. 14) in order to switch the transistor cyclically. A temperature measurement is performed at each cycle with a delay compared to the measurement of the previous cycle. All measured data are then gathered together to obtain the evolution of the temperature during a single pulse. It is then possible to achieve thermal characterizations with a measurement step as low as 1 μs. A thermal calibration of the camera was realized from 25°C to 45°C using a thermal chuck. Thus, the power applied on the devices was controlled to keep the maximum temperature below 45°C in order to stay in the calibration range. Besides, the measurement setup is restricted to measurements with a frequency below 2 kHz.

B. Samples preparation

Due to the non-uniformity of the 80 μm-thick copper layer deposited by electrochemical deposition (a higher thickness of Cu was deposited all around the sample edges compared to the center), the thermal evacuation through the chuck was degraded. To solve this problem, all the samples were glued to a Cu plate using a thermal paste for the thermal experiments (schematics of Fig. 17).

C. Thermal simulations

Finite element simulations were performed in order to assure a comparison with the practical measurements and to help us understand the observed phenomena. The table in Fig. 15 indicates the physical properties of the materials used for the simulations. The evolution of the temperature in the case of a 20 kHz switching frequency for a GaN device on silicon and on copper substrate is shown in Fig. 16. A better thermal dissipation of the HEMTs components transferred onto the copper substrate is demonstrated.

Materials	Thermal conductivity (W.m^{-1}.K^{-1})	Density (g.cm^{-3})	Thermal capacity (J.g^{-1}.K^{-1})
SiO2	1.4	2.2	0.73
GaN	160	6.15	0.5
AlGaN	30	4	0.6
Si	130	2.33	0.7
Cu	400	9	0.38
Thermal paste	29	5	0.5

Figure 15. Physical properties of the materials used in the simulations.

D. Experimental results

IR thermal images were compared between a 100 mm GaN on Si HEMT and a 100 mm GaN on Cu HEMT (Fig. 17). The same channel temperature average was measured for both devices but with a power 4 times higher in the case of the GaN on Cu HEMT device. A 0.34 W/mm and 1 ms pulse was applied on the devices with a period of 150 ms. The dissipated power has been determined in order to limit the heating of the devices in the calibration range of the experimental setup from 25 °C to 45 °C. A maximum temperature of 41 °C was measured for the device on silicon substrate compared to 35 °C for the device on copper substrate (Fig. 18). This result indicates a better thermal dissipation for the device on copper substrate compared to the one on silicon substrate. From a technological point of view, the experimental results confirm the good properties of

the deposited copper and validate the success of the developed transfer process. Although those first results are promising, further investigations are needed to investigate the operation of the transferred devices at higher frequencies.

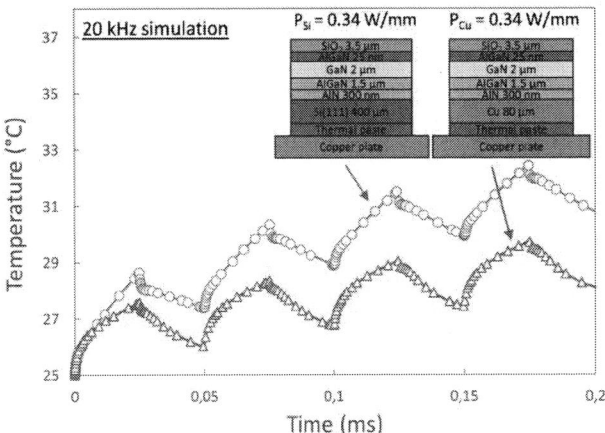

Figure 16. Simulation of the evolution of the temperature for a device on silicon (blue dots) and on copper substrate (red triangles) for a frequency of 20 kHz with a power of 0.34 W/mm.

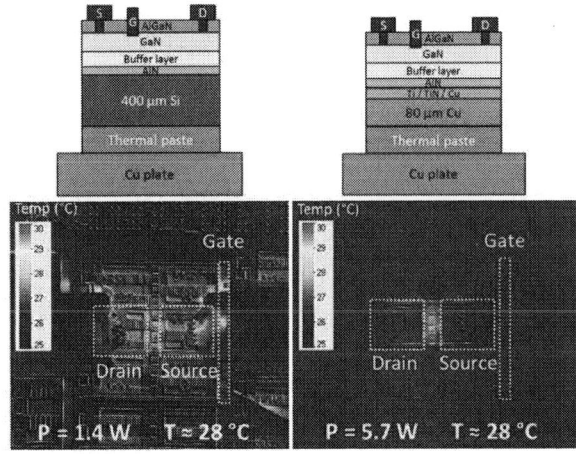

Figure 17. Comparison of IR thermography images for a 100 mm GaN on Si HEMT (left) to that of a 100 mm GaN on Cu HEMT (right).

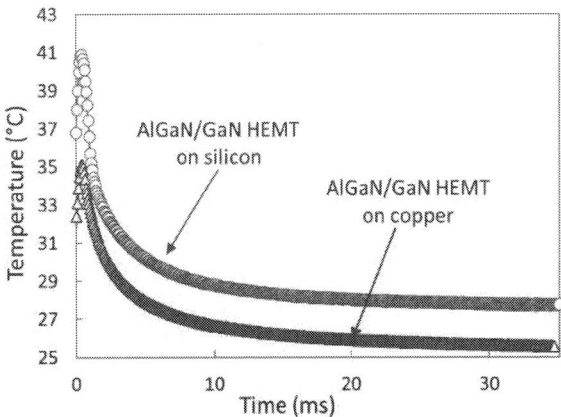

Figure 18. Evolution of AlGaN/GaN HEMT surfaces temperature on silicon (blue dots) and on copper (red triangles) substrates after a 1 ms pulse of 0.34 W/mm on both devices.

VI. CONCLUSION

We successfully transferred GaN on silicon power devices (both HEMTs and SBDs) from 200 mm (111) Si substrate onto different substrates: directly onto copper, onto copper without complete removal of all the silicon, and onto an electrically insulating substrate. The transfer onto a copper substrate induces a drain current increase, a R_{sh} decrease and a negative V_{th} shift compared to the non-transferred devices. This may be due to a modification of the stress/strain value inside the III-N stack. The off-state characterizations indicate that the presence of a remaining Si layer in between the III-N layers and the copper layer is essential to avoid breakdown voltage degradation when transferring the GaN-based power devices onto copper substrate. A better thermal dissipation was shown for AlGaN/GaN HEMTs transferred onto copper for a 0.34 W/m dissipation and 1 ms pulse. This work paves the way of a better understanding of the impact of the substrate on the electrical and thermal properties of the GaN-based power devices.

REFERENCES

[1] A. Nigam, T. N. Bhat, S. Rajamani, S. Bin Dolmanan, S. Tripathy, and M. Kumar. "Effect of self-heating on electrical characteristics of AlGaN/ GaN HEMT on Si (111) substrate, " AIP Advances, vol. 7, no. 8, August 2017.

[2] R. Dekker, P.G.M. Baltus, H.G.R. Maas, "Substrate transfer for RF technologies", IEEE Transactions on Electron Devices, vol. 50, no. 3, March 2003.

[3] M. Hiroki, K. Kumakura, Y. Kobayashi, T. Akasaka, T. Makimoto, and H. Yamamoto. "Suppression of self-heating effect in AlGaN/GaN high electron mobility transistors by substrate transfer technology using h-BN" Applied Physics Letters, vol. 105, no. 19, November 2014.

[4] M. Hiroki, K. Kumakura, H. Yamamoto, "Enhancement of performance of AlGaN/GaN high-electron-mobility transistors by transfer from sapphire to a copper plate", Japanese Journal of Applied Physics 55, 05FH07, 2016.

[5] T. Liu, Y. Kong, L. Wu, H. Guo, J. Zhou, C. Kong, and T. Chen. "3-inch GaN-on-Diamond HEMTs With Device-First Transfer Technology". IEEE Electron Device Letters, vol. 38, no. 10, pp. 1417–1420, October 2017.

[6] K.K.Chu et al., "High-Performance GaN-on-Diamond HEMTs Fabricated by Low-Temperature Device Transfer Process", 2015 IEEE Compound Semiconductor Integrated Circuit Symposium (CSICS), Oct. 2015. doi: 10.1109/CSICS.2015.7314511.

[7] B. Lu and T. Palacios. "High Breakdown (> 1500 V) AlGaN/GaN HEMTs by Substrate-Transfer Technology". IEEE Electron Device Letters, vol. 31, no. 9, pp. 951–953, September 2010.

[8] J. W. Chung, E. L. Piner, and T. Palacios, "N-face GaN/AlGaN HEMTs fabricated through layer transfer technology," IEEE Electron Device Letters, vol. 30, no. 2, pp. 113–116, February 2009.

[9] P. Gondcharton, F. Baudin, L. Benaissa, and B. Imbert, "Mechanisms overview of Thermocompression Process for Copper Metal Bonding," in MRS Proceedings, vol. 1559, mrss13-1559-aa08-07, 2013. doi:10.1557/opl.2013.718.

[10] A. F. Wilson, A. Wakejima, and T. Egawa, "Influence of GaN Stress on Threshold Voltage Shift in AlGaN/GaN High-Electron-Mobility Transistors on Si under Off-State Electrical Bias", Applied Physics Express 6, 086504, 2013.

[11] G. Longobardi et al., "On the vertical leakage of GaN-on-Si lateral transistors and the effect of emission and trap-to-trap-tunneling through the AlN/Si barrier", Proceedings of the 29th International Symposium on Power Semiconductor Devices & ICs, Sapporo, 2017.

[12] H. Yacoub et al., "The effect of the inversion channel at the AlN/Si interface on the vertical breakdown characteristics of GaN-based devices," Semicond. Sci. Technol., vol. 29, no. 11, p. 115012, 2014. doi:10.1088/0268-1242/29/11/115012.

[13] X. Li, M. Van Hove, M. Zhao, B. Bakeroot, S. You, G. Groeseneken, and S. Decoutere. "Investigation on Carrier Transport Through AlN Nucleation Layer From Differently Doped Si(111) Substrates". IEEE Transactions on Electron Devices, vol. 65, no. 5, pp. 1721–1727, May 2018.

A new reliable, corrosion resistant gold-palladium coated copper wire material

Sandy Klengel, Robert Klengel, Jan Schischka,
Tino Stephan, Matthias Petzold
Assessment of electronic system integration
Fraunhofer Institute for Microstructure of Materials
IMWS
Halle, Germany
sandy.klengel@imws.fraunhofer.de

Motoki Eto, Noritoshi Araki, Takashi Yamada
Nippon Micrometal Corporation
158-1 Sayamagahara
Iruma-City, Saitama 358-0032, Japan
meto@nmc-net.co.jp

Abstract— **In this paper, we present studies for the new gold-palladium coated copper wire "EX1R" with focus to application in the automotive industry. This wire material was developed to prevent halide induced interface corrosion and also sulfur induced pitting corrosion. The copper base material was alloyed with special elements to systematically adjust the intermetallic compound formed at the interface to the pad metallization and to define the microstructure of the copper base material to prevent sulfur induced corrosion. We show results of high resolution analyses (Scanning Electron Microscopy, Transmission Electron Microscopy, nanospot-EDS-mapping, electron-beam diffraction) for the characterization of the intermetallic compounds formed between bond wire and die metallization of different wire types in comparison to the new APC wire "EX1R". Also the running corrosion mechanisms at the interconnection area will be investigated in detail. The results show that the newly developed gold-palladium coated copper wire "EX1R" shows a much higher reliability, robustness and stability especially for harsh environments with regard to halide induced corrosion.**

Keywords- APC wire, wire bonding, corrosion resistance, reliable packaging, automotive electronics packaging, high temperature electronics, materials for harsh environments

I. INTRODUCTION

Wire bonding is still the most common interconnect technology for semiconductors. During the last years many efforts have been continuing to increase the integration density and to minimize the packaging size. Copper wire was used to substitute gold wire materials. In the past years, bare copper, palladium coated (PCC) and gold-palladium coated (APC) copper wires were developed especially to meet the requirements in automotive electronics industry for harsh environments. Copper wire is readily oxidized in air and forming gas is necessary during wire bond processing. Also the resistance of bare copper wire bonded systems against high temperature and humidity stress is a concern and has widely been studied. Several publications show that commonly available copper wires are good for application

under high temperatures but they are weak regarding halide induced corrosion in the contact interface between copper and aluminum [1],[2],[3]. Benefits of PCC and APC wires regarding low cost, long-term storage, bonding characteristics and corrosion resistance brought them to the majority of the market share. The APC wires show an improved behavior against halide induced corrosion but for them high temperature induced pitting corrosion induced by sulfur containing environments and mold compounds is limiting for reliability [4], [5]. In order to apply copper wire to automotive devices, the wire material has to meet the demands for long term reliability specific to automobiles including stable operation under harsh environment with high temperature and humidity. To meet these demands, a new type of APC wire has been developed and introduced to the market [5],[6].

In this paper, we present studies for the new APC wire "EX1R" with focus to the automotive industry. This wire material was developed to prevent halide induced interface corrosion and also sulfur induced pitting corrosion. The base material was alloyed with special elements to systematically adjust the intermetallic compound (IMC) formed at the interface to the pad metallization and to define the microstructure of the copper base material to prevent sulfur induced corrosion. We show high resolution analyses results (Scanning Electron Microscopy SEM, Transmission Electron Microscopy TEM, nanospot-EDS-mapping for element identification, electron-beam diffraction for determination of intermetallic compounds) for the characterization of the intermetallic compounds formed between bond wire and die metallization of different wire types. Also the running corrosion mechanisms at the interconnection area and at the wire itself will be investigated in detail. The results show that the newly developed APC wire "EX1R" shows a much higher reliability especially for harsh environments with regard to halide induced corrosion.

978-1-7281-1500-9/19 $31.00 © 2019 IEEE

II. FAILURE MODE " HALIDE INDUCED CORROSION OF INTERMETALLIC COMPOUNDS"

In copper- aluminum system different intermetallic compounds will be formed after wire bonding and developed during thermal ageing. In general copper rich and aluminum rich IMCs are known. Figure 1 shows the binary phase diagram of the Al/Cu system, indicating possible IMCs in the temperature range below T=300 °C [7]. Several publication reported additional intermetallic compounds. Table 1 summarizes these results.

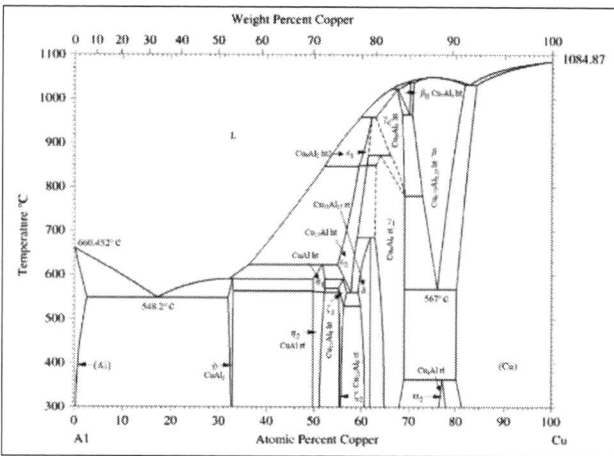

Figure 1: Binary phase diagram of system Cu-Al [7]

TABLE 1: INTERMETALLIC COMPOUNDS FOR THE SYSTEM CU-AL

Intermetallic compound	Composition	reference
Cu_4Al	Al: 20 at% Cu: 80 at%	[8]
Cu_9Al_4	Al: 30,8 at% Cu: 69,2 at%	[7]
Cu_3Al_2	Al: 40 at% Cu: 60 at%	[7]
Cu_4Al_3	Al: 47,7 at% Cu: 52,3 at%	[7]
CuAl	Al: 50 at% Cu: 50 at%	[7]
Cu_2Al_3	Al: 60 at% Cu: 40 at%	[11], [10]
$CuAl_2$	Al: 66,6 at% Cu: 33,3 at%	[9]

Figure 2 shows a scheme of the location of the IMCs for bare copper wire bonded contacts on aluminum pad metallization. As reported many times before, the copper-aluminum intermetallic compounds can be attacked by halides leading to a limited reliability of the complete wire bonded system. This is most likely caused by galvanic corrosion as suggested afterwards.

$$Cu_mAl_n + 3nCl^- = nAlCl_3 + mCu + 3ne^- \quad (1)$$

$$2AlCl_3 + 3H_2O = Al_2O_3 + 6HCl \quad (2)$$

During this corrosion reaction the copper rich IMC Cu_9Al_4 or Cu_3Al_2 is decomposed to amorphous aluminum oxide and dispersed Cu particles.

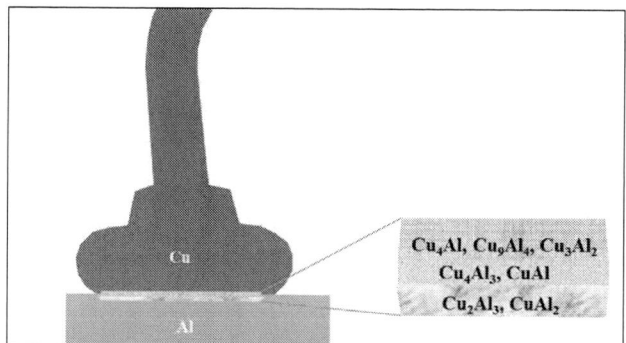

Figure 2: Location of the intermetallics for bare copper wire bonded contacts on aluminum pad metallization

Lim et al. investigated the electrical impedance behavior and the corrosion currents of different Cu-Al-IMCs in chlorine solutions . They showed that corrosion currents of the IMC shift to higher values with increasing aluminum content indicating that the aluminum rich IMC are less resistant against chloride attack [12].

Theoretical calculation done by H. Abe et al. [13] was used to predict possible IMC creation and reactivity between chlorine ions and single intermetallic compounds. By use of chemical model simulation technique, generation of copper rich and copper poor IMC was indicated, while copper rich IMC is most likely corroded by chlorine ions during HAST testing.

Boettcher et al. [3] explained that effect by the self-passivation of aluminum. They concluded that the copper rich IMC have a weak self-passivation oxide, compared to aluminum rich IMC. Thus, the copper rich IMC should be more prone for corrosive attack by chlorine.

The findings in [3] and [13] correlate with effects seen in molded packages during accelerated stress testing.

The simulation of Abe et al. [13] showed also that the addition of palladium in copper-aluminum system would improve the corrosion resistance. With palladium coated copper the generation of single copper rich IMCs was inhibited and the IMC growth slowed down in general. The simulation also indicated that a palladium layer could act as barrier layer for chlorine ion penetration and work as reaction partner with HCl forming $PdCl_2$.

Similar results were also published by C. Lee et al. [14] showing that interfacial crack line propagation and IMC growth is slowed down by the use of Pd-coated Cu wire compared to bare Cu wire. They assumed that Pd agglomerates are formed at the interface between copper wire and copper rich IMC. I. Qin et al. [15] found that Pd substitutes Cu in copper rich Cu_9Al_4 intermetallic compound leading to a higher corrosion resistance. Figure 3 shows the proposed IMC system for PCC wires on aluminum pad metallization.

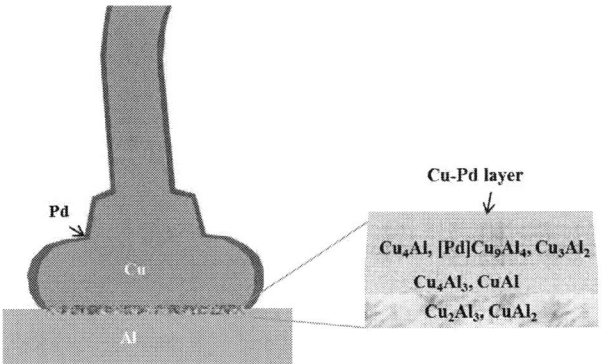

Figure 3: Location of the intermetallics of PCC wire bonded contacts on aluminum pad metallization

III. EXPERIMENTAL

A. Samples and reliability testing

Within our studies, three types of test specimen were used for the different research objectives: a bare copper wire and a conventional APC wire serving as reference samples in comparison to the optimized APC wire "EX1R". All samples had a wire diameter of d=20µm. For reliability testing the wires were bonded to a silicon die with aluminum metallization (ball bond contact) and a nickel-palladium-gold plated lead frame (stitch bond contact) and finally encapsulated by a chlorine containing (c=21 ppm) mold compound. Afterwards the package was tested for t=192hrs by HAST at T=130°C and H=85% rH with U=4V bias.

B. Investigation methods

For microstructural analysis the test specimen after HAST testing were prepared in cross section using a JEOL cross section polisher (JEOL GmbH, Freising, Germany). The cross section polisher uses an argon beam to mill cross sections with a high surface quality and to obtain cleanly polished surfaces of the interface between the wire bond materials and substrate metallization, see Figure 4.

Following the preparation, high resolution microstructural analyses were performed by scanning electron microscopy (SEM) and energy-dispersive X-ray Spectroscopy (EDS) using a Zeiss Supra-55 VP (Carl Zeiss Microscopy GmbH, Oberkochen, Germany) with an EDAX-Trident System (AMETEK EDAX GmbH, Wiesbaden, Germany). For high resolution analyses of the running corrosion mechanism and intermetallic compounds, electron-transparent lamellas were prepared out of the interface between wire material and substrate metallization, see Figure 5. Then, Transmission Electron Microscopy (TEM) was

performed working with an FEI Tecnai G2 F20 (FEI Company, Eindhoven, The Netherlands) operated at U=200 keV for EDS line scans and with an FEI Titan[3] 80-300 (FEI Company, Eindhoven, The Netherlands) operated at U=300 keV for EDS mappings. For identification of the intermetallic compounds nano-beam-diffraction analyses were done at the FEI Tecnai and FEI Titan[3] additionally.

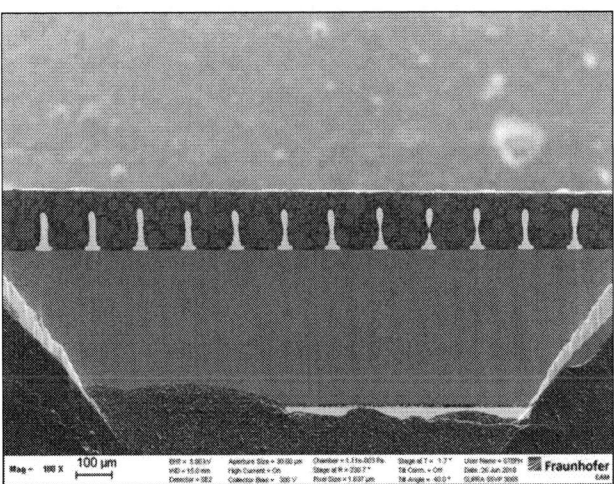

Figure 4: Overview of wire bonded contacts prepared in cross section by high rate ion polishing

Figure 5: Workflow for TEM preparation of area with intermetallic compounds in wire bonded contacts

IV. RESULTS

SEM analyses were performed at the interface between wire bond contact and silicon die metallization after high rate ion polishing. Bare copper wire shows a nearly complete corroded interface and conventional APC wire also shows a fairly corroded interface. The optimized APC wire "EX1R" shows a starting corrosion at the outer areas of the ball bond contact. The corrosion stops after a few micrometer. Comparable TEM analyses of the interface area show Cu_2Al_3 IMC at the pad metallization side for bare copper and also for the optimized "EX1R" wire. These IMCs were not attacked by the corrosion process. In the corroded area of the copper wire Cu_9Al_4 residues are detectable. The optimized "EX1R" shows a palladium modified Cu_9Al_4 intermetallic compound at these area that prevents the corrosive attack. Additionally alloying element is enriched on top of the copper rich IMC working as a diffusion barrier and ion catcher.

A. Scanning Electron Microscopy after HAST

Figure 6 shows the results of SEM analyzes comparing bare copper wire, conventional APC wire and optimized APC wire "EX1R". The bare copper wire has a completely corroded interface between the wire bond material and silicon die metallization, and the conventional APC wire also has a fairly corroded interface. The IMC that is next to the aluminum metallization seems not attacked. The corrosion area is only 50-100 nm thick but results in a complete delamination of the ball bond contact. The interface of the optimized APC wire "EX1R" and the pad metallization shows only a starting corrosion at the outer areas. The corrosion stops after a few µm and the main interconnection area is free from any corrosion processes.

Compared with the reactions taking place at the interface of the bare copper wire and the conventional APC wire a significantly higher reliability of "EX1R" wire is recognizable.

In addition, nano-scale voids are observed in the vicinity of bonded ball surface for all three wire types. These voids are caused by the flame-off process for FAB formation and are considered not affecting the corrosion process; previous experiment indicates that the positions of those voids do not change before and after HAST testing.

B. Transmission Electron Microscopy:Bare copper wire

Figure 7 shows a high resolution TEM image of the interface between bare copper wire material and aluminum pad metallization with a corroded zone formed during HAST testing. A typical matrix of corrosion products containing aluminum, copper and oxygen are detectable. Aluminum and oxygen form an amorphous matrix filled with fine dispersed copper particles. Partly residues of the attacked IMC are visible that are not completely corroded yet. It can be clearly seen that the aluminum near intermetallic is not attacked by the corrosive progress.

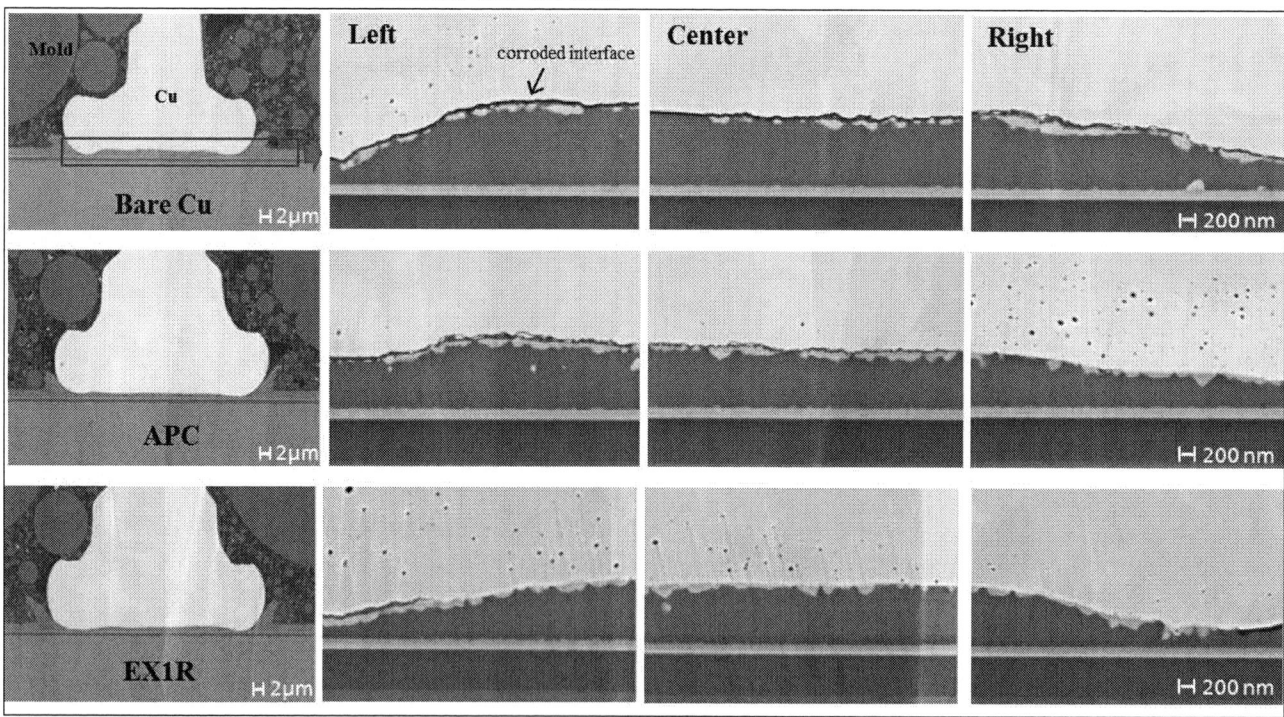

Figure 6: SEM results of the interface between bare Cu, conventional APC, EX1R wire and aluminum pad metallization after HAST. Bare copper shows a completely corroded interface while for the conventional APC wire faily corroded interface was found. The optimized APC wire "EX1R" shows only starting corrosion processes at the outer areas of the contact that stop after a few micrometer.

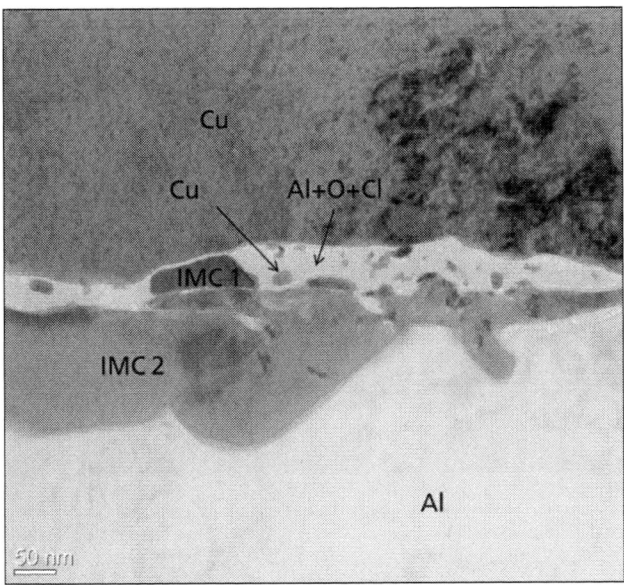

Figure 7: High resolution TEM image of interconnection area between bare copper wire and aluminum pad metallization. Two different IMC are detectable. The copper near IMC is decomposed to an amorphous matrix containing aluminum, oxygen and chlorine and only residues of this IMC are visible.

High resolution EDS mappings of the corroded area are shown in Figure 8. Chlorine and fluorine are detectable as corrosion inducing halides. No sulfur was detectable in the corroded area after HAST testing.

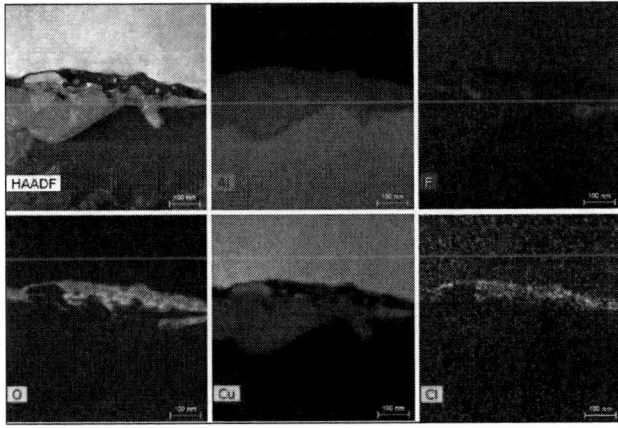

Figure 8: Nano-spot EDS mapping of the interface between bare copper wire and aluminum pad metallization showing the presence of aluminum, oxygen, fluorine and chlorine at the corroded area with precipitated copper particles.

Furthermore, EDS quantification in combination with the results of electron diffraction analyses were performed to analyze the IMCs at the interconnection area of the bare copper wire. The intermetallic compound that seems to be not attacked by the corrosive progress could be identified as Cu_2Al_3, see Figure 9.

The results of high resolution EDS and diffraction analyses for IMC 2, that is attacked by the corrosive progress, are shown in Figure 10. This IMC could be identified as Cu_9Al_4. Compared to Cu_2Al_3 IMC, Cu_9Al_4 is copper enriched and thus in agreement with previous studies, prone for corrosive attack.

Element	[wt%]	[at%]
Al	39.09	60.13
Cu	60.96	39.87

Figure 9: TEM diffraction analyses and EDS quantification results identified the aluminum near IMC 2 as Cu_2Al_3 (data from literature: trigonal lattice structure, Al ~60at%, Cu ~40at% [11])

Element	[wt%]	[at%]
Al	20.08	37.17
Cu	79.92	62.82

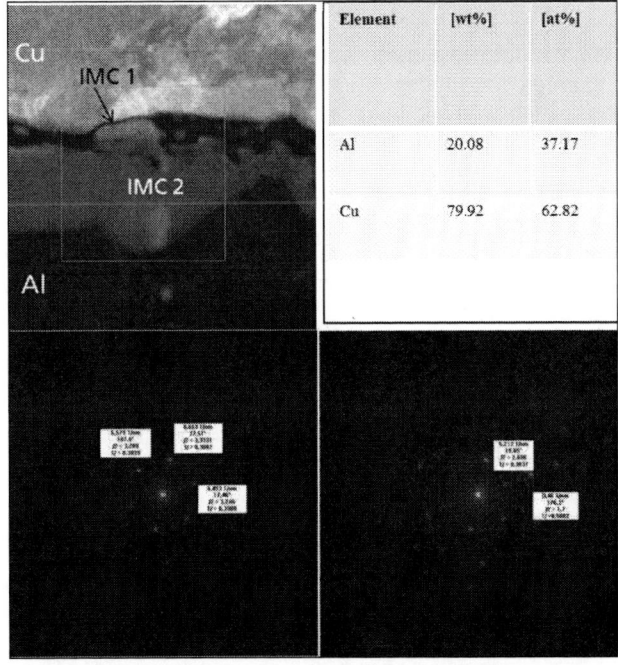

Figure 10: TEM diffraction analyses and EDS quantification results identified the copper near IMC 1 as Cu_9Al_4 (data from literature: cubic lattice structure, Al ~30at%, Cu ~70at% [7])

C. Transmission Electron Microscopy: EX1R wire

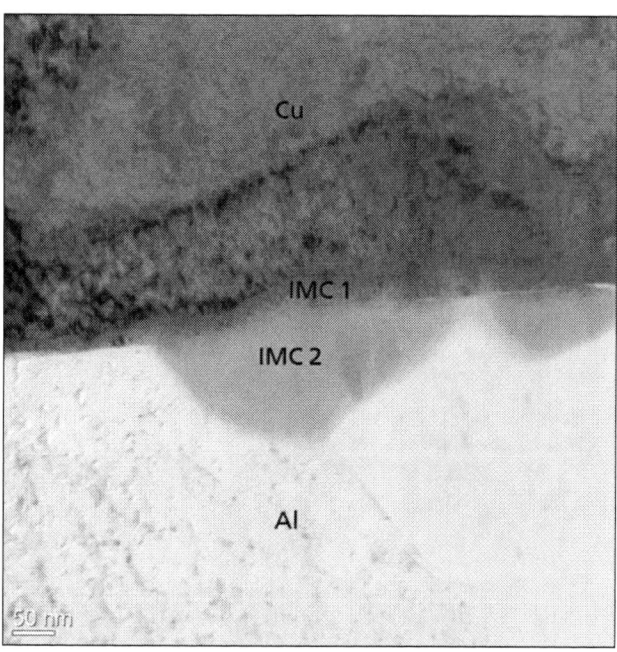

Figure 11: High resolution TEM image of interconnection area between EX1R wire and aluminum pad metallization. Two different intermetallic compounds are visible. No corrosion processes are detectable.

Figure 11 shows a high resolution image by TEM of the interface between copper wire material and aluminum pad metallization for the optimized APC wire "EX1R" after HAST testing. The interconnection area shows two IMCs. One is located near to copper and one is located near to aluminum. Although there is a clear evidence of the corrosion inducing halides chlorine and fluorine, see Figure 12, the copper near intermetallic of "EX1R" is not attacked. Presence of palladium and alloy element is also clearly seen at the Cu side interface. Quantification of the EDS results and electron diffraction analyses were done to identify the IMCs at the interconnection area of the optimized APC wire "EX1R". IMC 2, near aluminum, could be identified as Cu_2Al_3, see Figure 13.

The results of EDS quantification and electron diffraction analyses for IMC 1, that is more located to the copper side, are shown in Figure 14. This IMC may be considered as palladium modified Cu_9Al_4. Detailed nano-spot EDS analyses revealed also that the alloying element is situated on top of IMC 1 forming an additional inhibition layer.

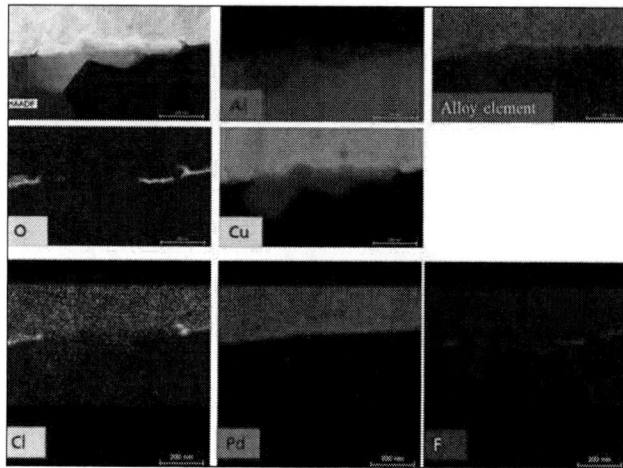

Figure 12: Nano-spot EDS mapping of the interface between EX1R wire and aluminum pad metallization showing the presence of oxygen, fluorine and chlorine but no indication for corrosive attack of the intermetallic compounds

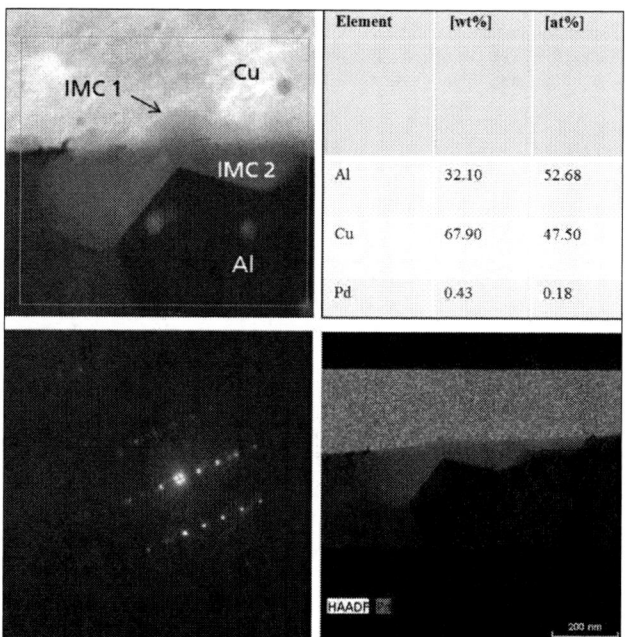

Element	[wt%]	[at%]
Al	32.10	52.68
Cu	67.90	47.50
Pd	0.43	0.18

Figure 13: TEM diffraction analyses and EDS quantification results identified the aluminum near IMC 2 as Cu_2Al_3 (data from literature: trigonal lattice structure, Al ~60at%, Cu ~40at% [11]). There is no palladium detectable in this IMC.

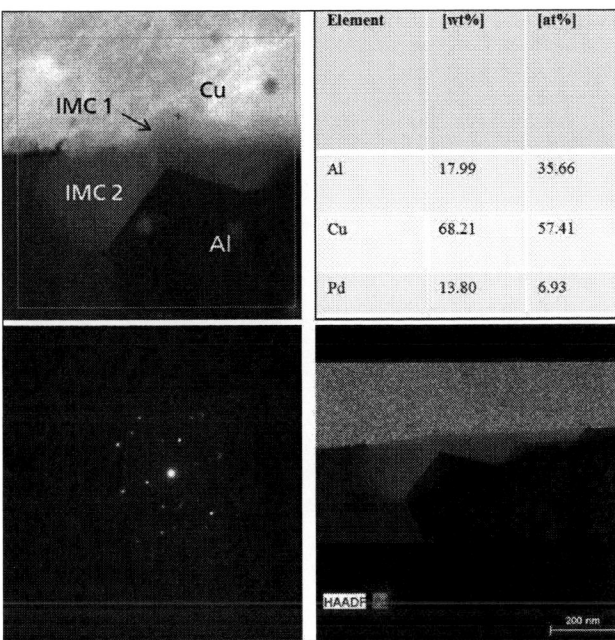

Element	[wt%]	[at%]
Al	17.99	35.66
Cu	68.21	57.41
Pd	13.80	6.93

Figure 14: TEM diffraction analyses and EDS quantification results identified the copper near IMC 1 as modified Cu_9Al_4 (data from literature for Cu_9Al_4: cubic lattice structure, Al ~30at%, Cu ~70at% , copper partly replaced by palladium [15])

V. SUMMARY AND CONCLUSION

We presented a study of the new APC bond wire "EX1R" with potential to be applied for robust automotive applications. We show results for its interconnection stability in HAST conditions in comparison to a bare copper and conventional APC wire. The wire material of "EX1R" was developed to prevent halide induced interface corrosion and also sulfur induced pitting corrosion. The base material was alloyed with special elements to systematically adjust the intermetallic formed at the interface to the pad metallization and to define the microstructure of the copper base material to prevent sulfur induced corrosion.

SEM analyses were done after HAST testing at the interface between wire bond contact and silicon die metallization after high rate ion polishing for all samples. Bare copper wire shows a nearly complete corroded interface and also at the interconnection of the conventional APC wire a fairly corrosive attack is detectable. The optimized APC wire "EX1R" shows a starting corrosion at the outer areas of the ball bond contact. But therefore, the corrosion stops after a few micrometer. Comparable TEM analyses of the interface area shows Cu_2Al_3 IMC at the die metallization side for bare copper and optimized APC wire "EX1R", that was not attacked by the corrosion for both samples. In the corroded area near the bare copper wire interface, Cu_9Al_4 residues are detectable. The optimized "EX1R" shows a palladium modified Cu_9Al_4 intermetallic compound at these area that slows down the corrosive attack. Comparable results were published by März et al. for Au-Al IMC systems

[17]. For APC wire "EX1R" the alloying element is enriched on top of the copper rich intermetallic compound additionally and works as a diffusion barrier and ion catcher. Both effects lead to a high corrosion resistance of the intermetallic compound layer between wire material and pad metallization.

As halides and sulfur are common impurities in microelectronic packaging (e.g. chlorine in low ppm range in mold compounds, fluorine in low ppm range on aluminum pad metallization and sulfur in adhesion promoters in mold compounds) it is very important to use a modified wire material that will not react with these substances. The results show that the optimized APC wire "EX1R" and its IMCs on aluminum metallization are nearly inert to halide induced corrosion processes in microelectronic packaging, indicating a much more higher reliability compared to common APC or bare copper wire types.

Further studies will be extended to high temperature testing at T=175°C where sulfur-induced corrosion attack becomes prominent.

VI. REFERENCES

[1] Hang, C. J., et al. "Growth behavior of Cu/Al intermetallic compounds and cracks in copper ball bonds during isothermal aging". *Microelectronics reliability,* 2008, 48. Jg., Nr. 3, S. 416-424.

[2] Liu, Hai, et al. "Reliability of copper wire bonding in humidity environment." *2011 IEEE 13th Electronics Packaging Technology Conference.* IEEE, 2011.

[3] Boettcher, Tim, et al. "On the intermetallic corrosion of Cu-Al wire bonds." *2010 12th Electronics Packaging Technology Conference.* IEEE, 2010.

[4] Han, Mingchuan, et al. "Copper wire bond pad/IMC interfacial layer crack study during HTSL (high temperature storage life) test." *2016 IEEE 18th Electronics Packaging Technology Conference (EPTC).* IEEE, 2016.

[5] M. Eto, T. Haibara, R. Oishi, T. Yamada, T. Uno, T. Oyamada "Thermal bond reliability of high reliability new palladium-coated copper wire," *Electronic components and Technology Conference (ECTC)*, 2017, pp.1297-1302.

[6] M. Eto, T. Haibara, R. Oishi, T. Yamada, T. Uno, T. Oyamada "Newly developed high reliability palladium coated copper wire for automotive application," *European Microelectronics Packaging Conference (EMPC)*, 2017.

[7] Kah, Paul, et al. "Factors influencing Al-Cu weld properties by intermetallic compound formation." *International Journal of Mechanical and Materials Engineering* 10.1 (2015): 10.

[8] Leach, JS Llewelyn. " On the structure of a phase formed in copper-aluminum alloys at low temperatures." *J INST MET* 92 (1964): 93-94.

[9] A. Meetsma, J. L. de Boer, and S. van Smaalen, "Refinement of the crystal structure of tetragonal Al2Cu," *Journal of Solid State Chemistry*, vol. 83, pp.370-372, 1989.

[10] Pelzer, Rainer, et al. "Growth behavior and physical response of Al-Cu intermetallic compounds." *2014 IEEE 16th Electronics Packaging Technology Conference (EPTC).* IEEE, 2014.

[11] P. Ramachandrarao and M. Laridjani, "A metastable phase Al3Cu2," *Journal of Materials Science,* vol. 9, pp. 434-437, 1974.

[12] Lim, Adeline BY, et al. "Evaluation of the corrosion performance of Cu–Al intermetallic compounds and the effect of Pd addition." *Microelectronics Reliability* 56 (2016): 155-161.

[13] Abe, Hidenori, et al. "Cu wire and Pd-Cu wire package reliability and molding compounds." *2012 IEEE 62nd Electronic Components and Technology Conference*. IEEE, 2012.

[14] Lee, Chu-Chung Stephen, et al. "Copper versus palladium coated copper wire process and reliability differences." *2014 IEEE 64th Electronic Components and Technology Conference (ECTC)*. IEEE, 2014.

[15] Qin, Ivy, et al. "Molded reliability study for different Cu wire bonding configurations." *2013 IEEE 63rd Electronic Components and Technology Conference*. IEEE, 2013.

[16] U, Hui. Thermosonic ball bonding: a study of bonding mechanism and interfacial evolution. 2010. Doktorarbeit.

[17] März, B., et al. "Investigation of the palladium distribution in the intermetallic phase region of Au-Al wire bond interconnects." *2012 4th Electronic System-Integration Technology Conference*. IEEE, 2012.

Ultrasonic-Accelerated Intermetallic Joint Formation with Composite Solder for High-Temperature Power Device Packaging

Wenwu Zhang[a,b], Hongjun Ji[a,b,*], Mingyu Li[a,b], Weiwei Zhao[a,b]

a. State Key Laboratory of Advanced Welding and Joining, School of Materials Science and Engineering, Harbin Institute of Technology at Shenzhen, Shenzhen 518055, Guangdong, China

b. Flexible Printing Electronic Technology Center, Harbin Institute of Technology at Shenzhen, Shenzhen 518055, Guangdong, China

Address: HIT Campus, Shenzhen University Town, Xili, Nanshan, Shenzhen 518055, P.R. China

Corresponding author, Tel.: +86 755 26033281 Fax: +86 755 26033504; e-mail: jhj7005@hit.edu.cn,

Abstract-A novel and effective method of ultrasonic-assisted reflow is an effective solution to time-consuming problem of traditional reflow. In this study, three kinds of hybrid solder paste composed by different size of Ni particles were synthesized to obtain the high-melting-point joint for high temperature operation under ultrasonic-assisted reflow. The microstructure evolution, shear strength, and fracture morphology in joints before and after aging were systematically investigated. A high-melting-point and high-strength joint of sole IMCs with a few residual Ni was obtained at 250 °C in a short time (only 10 s) under lower ultrasonic power of 200 W, and the shear strength was up to 45.24 MPa in response to ordered arrangement that small Ni particles filled in the gap of large Ni particles. In addition, the microstructure, shear strength, and fracture morphology of joints formed by three hybrid solder pastes after aging treatment of 72 h, 170 h, and 360 h at 200 °C are distinct dramatically. Especially, the joint composed by mixing Ni particles remained higher shear strength of 49.75 MPa after 360 h aging, which ascribed to the dense joint without defects.

Keywords-Composite solder paste, Ultrasonic-assisted reflow, Microstructure, Shear strength, Fracture

I. INTRODUCTION

With the development of electronic industry, IC chip is following the trend of miniaturization, high integration and high power will in the future. At the same time, the demand for IC chips, with they applied for the field of aerospace, nuclear industry, automotive electronics, artificial intelligence, has been put forward [1-3]. Thus, the higher performance and reliability requirements will be dominated in the harsh and sophisticated environment, such as high temperature, high pressure, corrosive atmosphere, large current and so on. Traditional devices of Si-substrate for first generation semiconductor are confronted with many challenges in these fields, but the emergence of third generation semiconductor which are representative of SiC and GaN makes it possible for operating normally in high temperature and harsh environment [4-6]. Achieving effective mechanical support, physical protection and electrical interconnection for IC chip device systems is becoming extreme urgent so that the development of reliable electronic packaging

technology is playing a key role in the third semiconductor devices [7]. Especially, the interconnect interlayer of packaging materials has proposed higher requirements for excellent performance. Nowadays, the primary solders of IC chip packaging are Sn-based lead-free solders, the working temperature of which were below 200 °C. Therefore, it is very important to exploit a novel solder or package technology applying for high temperature service.

Regarding electronic package at high temperature, there are several common methods applied in the high temperature field, mainly including Sn-based solders with high temperature [8], nanoparticles sintering [9], direct solid-phase bonding [10], and transient liquid phase (TLP) [11]. Traditional high-temperature lead-based solders using in the field of consumer electronics, such as containing hazardous element solders of Sn95Pb and Sn90Pb, were prohibited by the Restriction of Hazardous Substances (RoHS) promulgated by European Union in 2003 [12]. During the subsequent period, the Sn80Au high-temperature lead-free solder was developed subsequently, which worked as an alternative material for the lead-based solder with the advantages of good wettability, high thermal conductivity, and high strength [13,14], but the demand for cost-effective has limited its large particle application on account of the high content of Au. It is well-known that nanoparticles with high surface energy has lower sintering melting-point, and their sintering temperature is much lower than that of the bulk material. Therefore, the nanoparticle sintering has the competitive advantage of die bonding at low temperature to work at high temperature. At present, sintering of Ag nanoparticles with low melting-point and high conductivity and ductility has been systematically studied by famous researchers from all over the world [15-17]. However, their wide industry application was an insurmountable barrier for the require of cost-effective. Thus, Cu nanoparticles of relative cost-economic with the optimistic conductivity was developed rapidly for chip bonding [18]. However, Cu nanoparticles undergoes rapid oxidation in air forming a thin amorphous CuO and Cu2O layer with a thickness of 2-6 nm that results in the loss of electrical conductivity and makes it inapplicable for conductive materials [19].

Direct solid-phase bonding technology with the advantage of cost-effective, operation-simple, and high reliability is focused by many researchers, its principle is that it is broken by the oxide film on the surface of the connecting metal under the external pressure, and then metal atoms produce the diffuse with each other under high temperature with small holes reducing and grain boundary migration occurring. Finally, the bonding layer extends to the direction of volume with the micro holes disappearing, and forming a reliable macroscopic solder joint [20, 21]. However, this method was limited by the demand for high surface cleanliness, vacuum degree of metals, and even extreme pressure. Transient liquid phase bonding technology has attracted extensive attentions with low processing temperature, effective cost and simple operation. The principle is that a liquid metal reacts with another solid metal by atomic diffusion in a certain temperature during, and then intermetallic compound (IMC) in joint will be generated in a plenty of period. This method has a promising application in the packaging process of high-power electronic devices. However, TLP bonding technology is a comparative time-consuming course at low temperature, which makes its completely reduce production efficiency in industrialization process of modern electronic products of high integration and power [22-24].

Recently, a novel and effective bonding method has emerged from the TLP technologies under the background of ultrasonic effects, ultrasonic-assisted reflow takes full advantage of TLP and ultrasound wave to improve effectively production efficiency. This method has been proven to achieve successfully die bonding at low temperature in a short time, which attributed to cavitation and acoustic streaming with the benefits of reducing holes, promoting wetting, and refining grain [25-27]. In this paper, the die and the Ni substrate were bonded together by the Sn/Ni hybrid solders composed of different size of Sn/Ni particles under the ultrasonic-assisted reflow. The microstructure, shear strength, and fracture of joints or aged joints formed by the hybrid solders with and without ultrasound was systematically investigated.

II. EXPERIMENTAL PROCEDURES

According to our previous study, excessive Ni particles would result in a decrease in the wettability of the solder, so the mass ratio of Ni and Sn particles in the solder was set as 1:2, and the hybrid solder paste was prepared by mixing Ni particles, Sn particles, and flux. The function of the flux is to remove oxide, reduce surface tension, and promote the wetting spread of solder. Based on previous reports [28], the content of flux in solder paste is generally suitable for 10% - 15% of the total mass. In this experiment, the CR-32 flux containing 10% of total mass was purchased by the TIANJIU company. The solder paste was designed for the forms of Sn/Ni composite solder, which were composed of Sn particles (40μm), large Ni particles (35μm) and small Ni particles (5μm) obtained by

gas atomization. The structure of Die/Sn+Ni composite solder paste/Ni substrate is like a sandwich structure, the die with dimension of 5×5 mm2 was deposited with 8 μm thick Ni, and the size of Ni substrates is 10×10×1 mm3. The principle of ultrasonic-assisted reflow was shown in Fig. 1 (a). The bonding temperature was set at 250 °C and the heating and cooling rate was about 24 °C/min and 48 °C/min, respectively. The traditional reflow time was set for 10 min, the ultrasonic-assisted reflow time was just 10 s and other parameters are as follows: the ultrasonic vibration was in the horizontal direction, and ultrasonic frequency, pressure and power were fixed as 35 kHz, 0.15 MPa and 200 W, respectively. The temperature curve of ultrasonic-assisted reflow is shown in Fig. 1 (b).

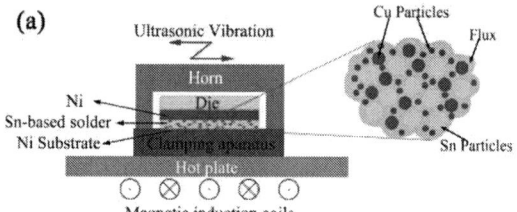

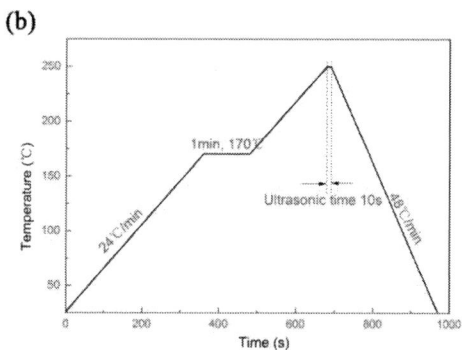

Figure 1. (a) Schematic of the bonding sandwich structure of die/hybrid solder paste/substrate, (b) the hear temperature curve of ultrasonic-assisted reflow.

The ultrasound wave was applied by HX-3008SC produced by Shenzhen Xinke company. The obtained samples by ultrasonic-assisted reflow mounted in epoxy were manually ground from 180 to 4000 grade silicon carbide papers, and then polished by 1.0μm diamond and 0.05μm Al2O3 polishing agents. The joint microstructures were observed by Scanning Electron Microscope (SEM, Hitachi S-4700) equipped with an Energy Dispersive X-ray Spectrometer (EDS) detector. To observe the interior microstructures, they were deeply etched by 10% hydrochloric acid ethanol solutions for 15 s. Their shear strength was measured by a shear tester with a shear speed of 200μm/min and 5 samples for each condition were tested. In order to clarify fracture mode and determine the phases in the joint, the sheared samples were further analyzed by SEM, EDS and X-ray diffraction (XRD, Rigaku D/max-2500PC, Cu Kα) immediately.

III. RESULTS AND DISCUSSTION

In our previous studies found that the joint of low young's modulus composed of a small amount of Ni3Sn4 with residual Ni can be obtained using Sn/Ni hybrid particles solder pastes at 250 ℃ under ultrasonic power of 500W [28], ultrahigh power ultrasound will be detrimental for the chip in the practical application process. However, the reaction rate of Sn/Ni is affected directly by ultrasonic power, which will result in the existence of residual Sn so that high working temperature of chip will be obviously reduced to a certain extent. Therefore, the aim to my paper added the specific area to accelerate reaction rate by mixing different size of Ni particles under low ultrasonic power (200 W), and the joint formed of a sole IMC with a bit of Ni will be obtained working at high temperature electronic devices.

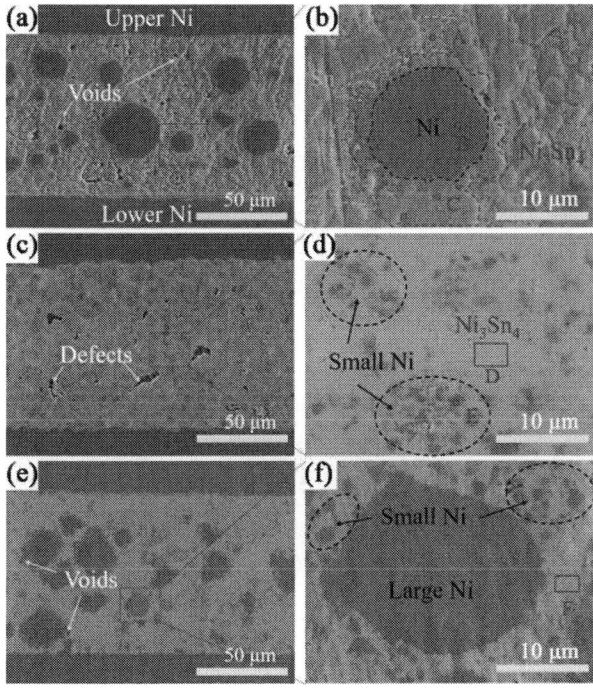

Figure 2. SEM images of the microstructure of cross-sectional joints soldered by ultrasound at 200 W for 10 s with different size of particles and the corresponding local enlarged SEM images, (a) and (b) large Ni particle, (c) and (d) small Ni particle, (e) and (f) mixed Ni particle

Fig.2 shows SEM images of microstructure of the cross-sectional joints formed by three kinds of composite solder paste at 250℃ for 10 s under ultrasonic power of 200 W and the corresponding particle enlarged images. Fig. 2 (a) and (b) are microstructure of joint composed by large Ni particles, it can be seen that there were three phases existing in the joints from the EDS analysis in Table 1 of A, B, and C, these phases of which are the most possibly Sn, Ni and Ni3Sn4 from the ratio of atomic composition, respectively. At this moment, it is clear that there was a large amount of residual Sn without disappearance in joints. In addition, a few of defects distributed in joints due to there were many shrinkage voids during the process of solidification. Notably, a great deal of fragmented Ni3Sn4 IMCs was randomly scattered in joints, that is probably ascribed to ultrasonic effects. It is

well-known that the crystal nucleation and growth of Ni3Sn4 IMCs occurred at the surface of Ni particles, these growing grains would be broken and distributed at random in joints under the ultrasonic effects. Fig. 2(c) and (d) describes microstructure images of joint formed by small Ni particles. According to EDS analysis of D and E at Table 1, it is found that a plenty of residual small and random-sharp Ni existed in joints, which attributed to adequate reaction rate of small Ni particles with large specific surface area. However, there were a few of residual Sn in joints in that a large number of IMCs wrapped Ni, and hindering the atomic diffusion between Sn and Ni, that resulted in completely consuming Sn. Comparing to Fig. 2 (a) and (b), the number of Ni3Sn4 IMCs were much more than that of joints formed by small Ni particles, but there were several big voids in joints for the reason that a great a plenty of IMCs and residual small Ni declined the wettability. The dense joint without defects was obtained from Fig.2 (e) and (f), it can be seen that the almost-perfect joint of sole Ni3Sn4 IMCs and residual Ni without residual Sn can solve the problem of high temperature operation. Contrasting with Fig. 2 (a)-(f), this joint with moderate reaction rate and wettability has no defects since the fact that the enormous number of small Ni particles wrapping with a thick layer of Ni3Sn4 IMCs exactly right filled in the gap of with large Sn/Ni particles so that a dense joint without defects was obtained.

Table 1. EDS analysis for the fracture SEM images under ultrasonic-assisted reflow

Region	Sn	Ni	Phase
A	97.03	2.97	Sn
B	4.88	95.12	Ni
C	59.93	40.07	Ni3Sn4
D	58.75	41.25	Ni3Sn4
E	1.44	98.56	Ni
F	59.02	41.98	Ni3Sn4

With regard to the long-term high temperature operation of chip, the demand for aging resistance of joints is a significant performance index. Thus, the aging resistance of joints formed by three kinds of hybrid solder pastes was tested systematically under ultrasonic-assisted reflow at 200 ℃ after aging 72 h, 170h, and 360 h. Fig 3 exhibits SEM images of cross-sectional under ultrasonic-assisted reflow at different aging period. It can be observed from Fig. 3 (a)-(c), the thickness of Ni3Sn4 IMCs layer increased by degree with adding of aging time owing to atomic diffusion between Sn and Ni in a long time. Additionally, the residual Sn and Ni has gradually declined, and the number of IMCs increased gradually. Fig. 3 (d)-(f) shows SEM images of cross-sectional joint of small Ni particles. During the period of aging, the residual Sn and Ni has gradually disappeared, the number of voids had steady growth from 72 h to 170 h and followed the slow increase in the subsequent time of 190 h. That is probably ascribed to the difference of mutual diffusion rate between Sn and Ni. It is well-known that the rate that Sn atom diffuses into Ni atom with high chemistry potential is much more than that of Ni diffusing into Sn, which was encountered in exhausting rapidly residual Sn to leave a large number of voids during the formation of Ni3Sn4 IMCs layer. For SEM images of cross-sectional joints in Fig. 3 (g)-(i), indicating that there was no difference for the microstructure before and after aging. And this is further evidence that residual Sn was depleted completely and Ni3Sn4 IMCs and residual Ni were steady phases without no transformation at 200 ℃. This joint composed

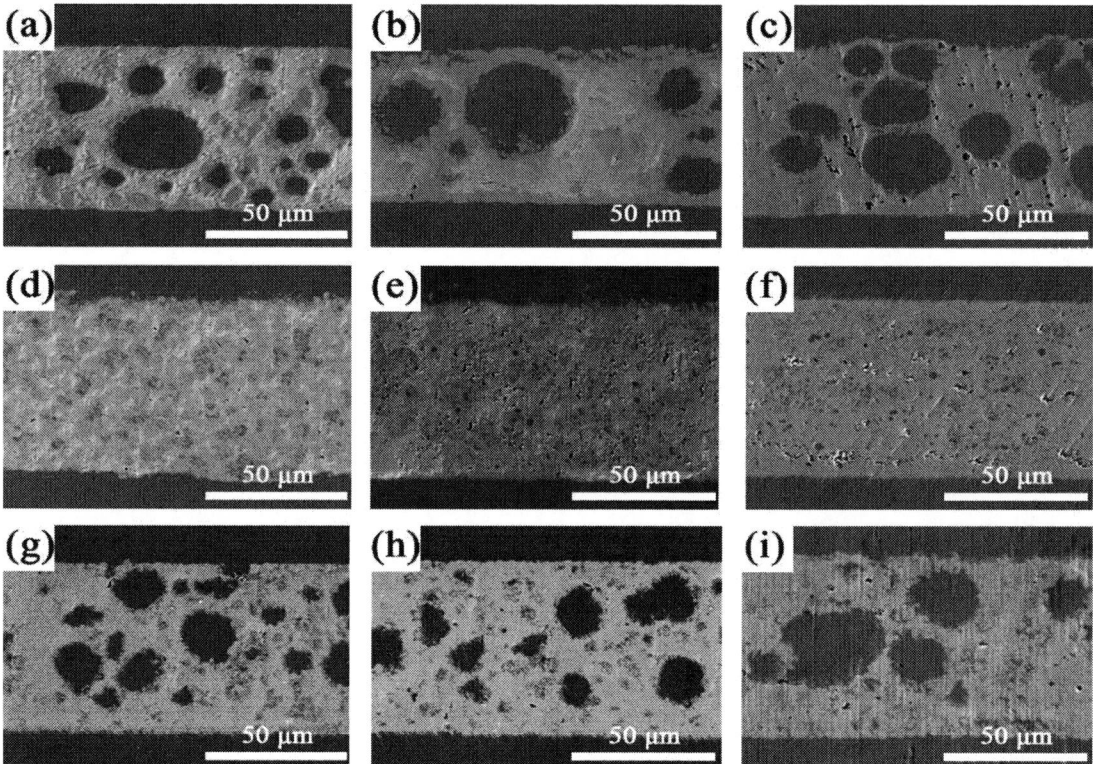

Figure 3. SEM images of the microstructure under ultrasonic-assisted reflow at 200 W for 10 s after aging 72 h, 170 h, and 360 h, (a), (b) and (c) for large Ni particle, (d), (e) and (f) for small Ni particle, (g), (h) and (i) for mixed Ni particle

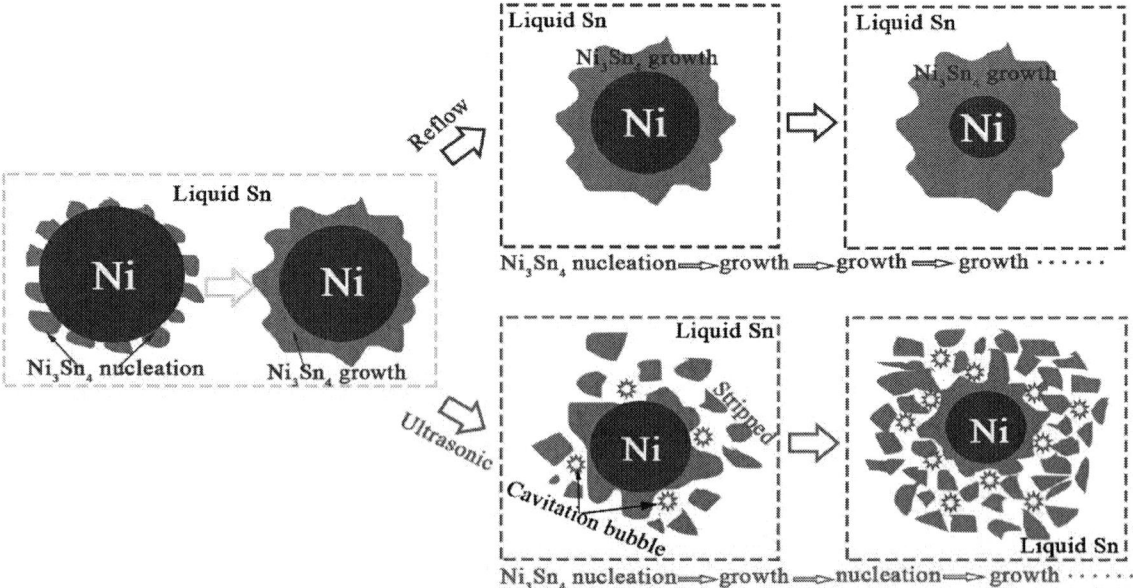

Figure 4. Schematic diagram of Ni3Sn4 IMC growth during the process of ultrasonic-assisted reflow and traditional reflow.

978-1-7281-1500-9/19 $31.00 © 2019 IEEE

of sole Ni3Sn4 IMC and a little residual Ni have the promising practical application in high-temperature electronic device. That is because the melting point of Ni and Ni3Sn4 are 1453°C and 794.5°C, respectively, according to the Sn/Ni binary phase diagram. Therefore, the microstructure of joints composed by mixing Ni particles was more in line with the requirements than other two kinds of hybrid solder paste. interconnect in normal environment.

Fig. 4 gives a breakdown of the growth of Ni3Sn4 IMCs during the process of reflow, aging, and ultrasonic-assisted reflow. It is obvious that the reaction rate depends on atomic diffusion distance between Sn and Ni during the reflow and aging, but the formed layer of Ni3Sn4 IMCs is an obstacle for the atom diffusion, which will be time-consuming to obtain a high-melting-point joint applying the high temperature field of chip. With regard to ultrasonic-assisted reflow, the ultrahigh reaction rate led to produce a joint of sole IMCs with a few residual Ni due to the acoustic cavitation and streaming effects accelerated the atom diffusion and dissolution. Additionally, the direct-contact mechanism of solid/liquid interface was subjected to high-intensity ultrasonic, resulting in the fracture of Ni3Sn4 IMC that surrounds Ni particles with acoustic cavitation bubbles collapsing. These Ni3Sn4 IMCs rapidly tripped from Ni surface due to the ultrasonic streaming effect, which result in exposing highly reactive surface of Ni to accelerate the reaction of Ni and liquid Sn. The mechanism of growth-tripping-recycling of Ni3Sn4 IMCs alternately and simultaneously occurred until the reaction of Sn/Ni was completely exhausted. That is why the reaction of Sn/Ni by ultrasonic-assisted reflow was rapidly completed to form high-melting-point joint in a short time.

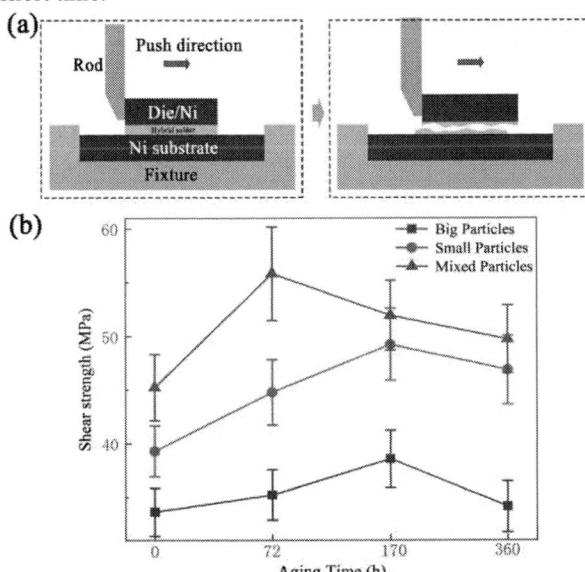

Fig 5. (a) schematic diagram of testing shear strength, (b) The trend of shear strength that accompanies the increase of aging time with different hybrid solder pastes under ultrasonic-assisted reflow

Fig. 5 (a) shows the schematic diagram of testing shear strength. During the test of shear strength, the

sample should be fixed in the fixture and then the Angle of the tested sample should be adjusted to the position paralleled to the push rod in the thrust test. With the increase of the load, the value of shear strength is the maximum strength that joint can withstand until the fracture. Fig. 5 (b) exhibits shear strength of joints with the increasing of aging time, it is clear that the change trends of shear strength are all added firstly and then declined gradually. Therefore, the following analysis is the reason for the change trends of shear strength in view of the morphology of fracture.

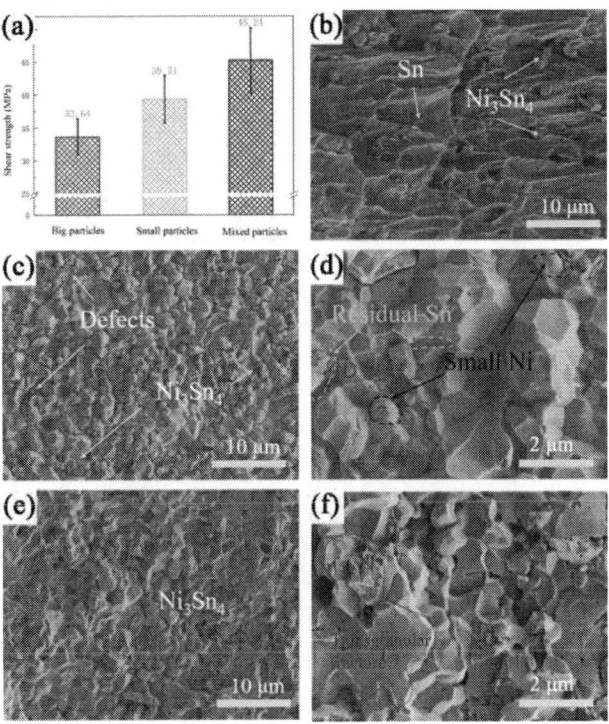

Fig 6. (a) The value of shear strength accompanies a change of Ni particles, and the corresponding SEM images after testing, (b) large particles, (c) small particles, (d) enlarged image for (c), (e) mixed particles, (f) enlarged image for (e)

Table 2. EDS analysis for the fracture SEM images under ultrasonic-assisted reflow

Region	Sn	Ni	Phase
A	98.32	1.68	Sn
B	59.68	40.32	Ni₃Sn₄
C	59.59	40.41	Ni₃Sn₄
D	76.36	23.64	Sn, Ni₃Sn₄
E	19.46	80.54	Ni, Ni₃Sn₄
F	48.83	51.17	Ni₃Sn₄
G	46.35	53.65	Ni₃Sn₄

Fig 6 shows shear strength values and SEM morphology images of fracture in joints formed by three kinds of hybrid solder paste at 250 ℃ for 10 s under ultrasonic power of 200 W. It is apparent that the shear strength value of joint composed by mixing Ni particles was the maximum at 45.23 MPa in Fig. 6 (a), most of values by traditional reflow is much lower than that of it from our previous works [6, 23, 29]. According to the foregoing microstructure and Fig. 6 (b), the shear strength of joint formed by large Ni particles is the lowest for the primary reason that an awful lot of residual Sn happened plastic deformation. However, the plastic slip track of Sn accompanied by a great deal of shape-grain Ni_3Sn_4 IMCs on the basis of EDS analysis of B in Table 2. Thus, it is deduced that this mode of fracture is plastic fracture based on Sn and there are many Ni_3Sn_4 IMCs distributing randomly in joint as a function of dispersion strengthening under the acoustic cavitation and streaming effects. Fig. 6 (c) and (d) show SEM morphology and particle enlarged

image of fracture in joint by small Ni particles. Obviously, it is different for fracture morphology of former that the brittle fracture of Ni_3Sn_4 IMCs was the dominate mechanism, but a quantitative number of residual Sn and defects remained in joint from EDS analysis of C, D, and E in Table 2. This result is in conformity with the structure in Fig. 2 (c) and (d), so the lower shear strength primarily caused by certain residual Sn and defects. With regard to Fig. 6 (e) of the microstructure and EDS analysis, it is only observed that the sole Ni_3Sn_4 IMCs with a little residual Ni consisted of the high-melting-point joint. Notably, the fracture modes were dominated by intergranular fracture and supplemented by transgranular fracture, which results in a dense and high-strength joint with no defects.

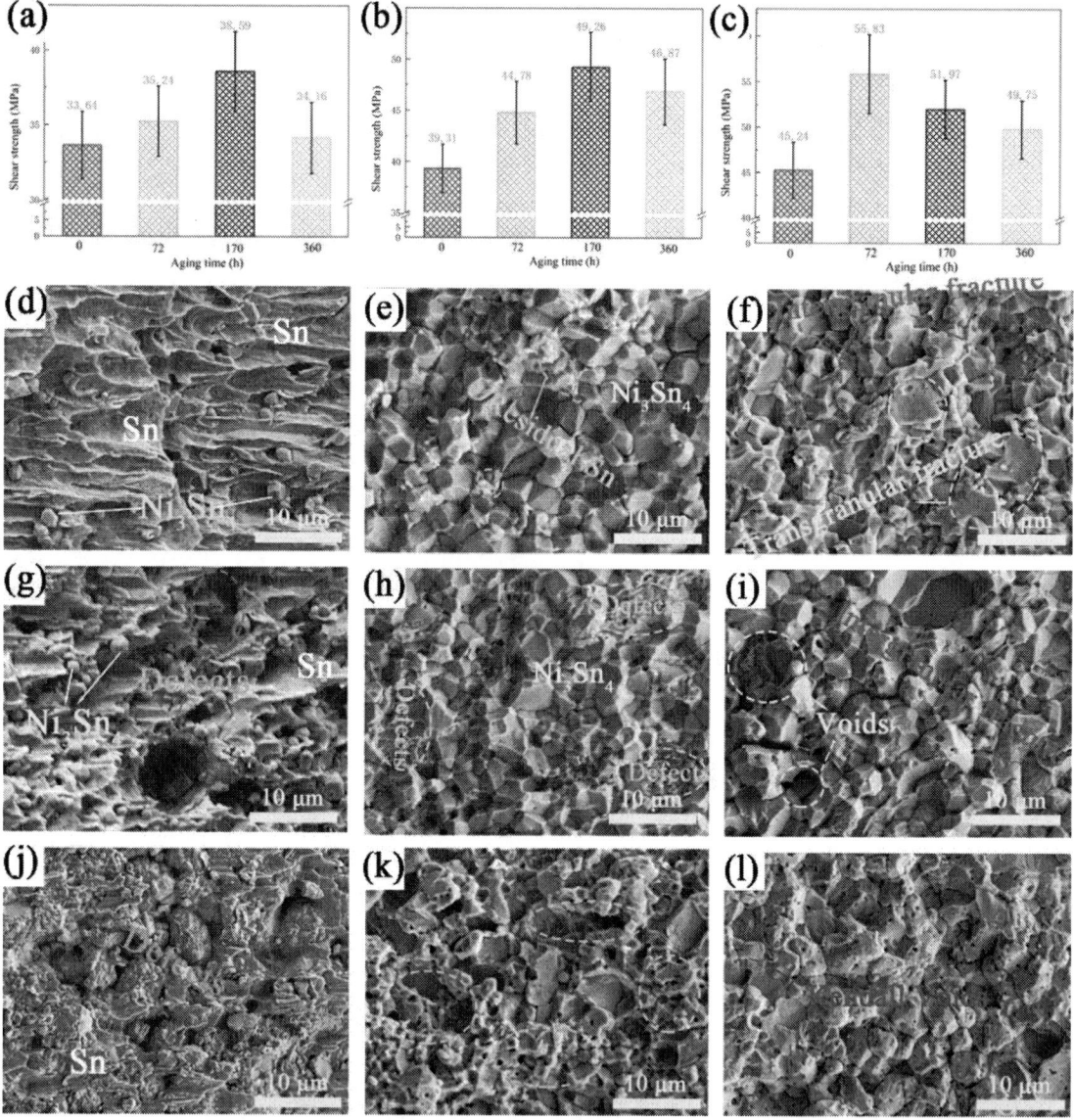

Fig 7. The value of shear strength accompanies the time of aging 0, 72 h, 170h, and 360h at (a) large particles, (b) small particles, (c) mixed particles; and the corresponding SEM images of fracture with different hybrid solder pastes after testing, large particles of aging 72 h, 170 h, and 360 h for (d), (g), and (j), small particles of aging 72 h, 170 h, and 360 h for (e), (h), and (k), mixed particles of aging 72 h, 170 h, and 360 h for (f), (i), and (l)

Fig.7 shows shear strength values and SEM morphology images of fracture in joints after aging of 72 h, 170h, and 360 h under ultrasonic-assisted reflow. It can be seen from Fig. 7 (a) that the shear strength increased gradually during the adding of aging time, and the value reached the maximum 38.59 MPa at 170 h, but fell to 34.16 MPa at 360 h. Combined with the mode of fracture and corresponding microstructure, the number of Ni3Sn4 IMCs increased with the increase of aging time owing to atomic diffusion between Sn and Ni in a long time, which obviously contributed to more short and thin plastic fractures of Sn in Fig. 7 (d), (g), and (j). But shear strength at 360 h fell slightly, that is likely ascribed to many defects against forming dense joints. Fig.7 (b) exhibits that shear strength value of joints composed by small Ni particles after aging of 72 h, 170h, and 360 h, there is the similar trend to that of large Ni particles. According to Fig.3 (d)-(f), the residual Sn and Ni has gradually disappeared, the number of voids had steady growth and followed the slow increase with the rise of aging period. Corresponding to fracture in Fig. 7 (e), (h), and (k), it is clear that the fracture modes were dominated by intergranular fracture, but it is observed that there was a short amount of residual Sn to reduce shear strength from Fig.7 (k), Meanwhile, the shear strength at 360 h was 3.59 MPa lower than that of joint at 170 h on account of more defects. Finally, with regard for Fig.7 (c), it is different that the shear strength went down rapidly from 72 h to 360 h, and the value of shear strength at 72 h was up to the maximum 55.83 MPa. The increase of shear strength from 0 to 72 h originated from release of internal stress produced by ultrasonic-assisted reflow, and the following dramatic decline in 288 h put down to ascribe to a large number of micro-voids which was called 'Kendall Void' formed by nonequilibrium diffusion between Sn and Ni atom from Fig.7 (l).

IV. CONCLUSIONS

In this study, three kinds of hybrid solder paste composed by different size of Ni particles were synthesized to obtain the anticipant high-melting-point joint for high temperature operation under ultrasonic-assisted reflow. The microstructure evolution, shear strength, and fracture morphology in joints with the adding of aging time were systematically analyzed. A high-melting-point and high-strength joint of sole IMCs with a few residual Ni was obtained at 250 ℃ in a short time (only 10 s) under lower ultrasonic power of 200 W, and the shear strength was up to 45.24 MPa in response to ordered arrangement that small Ni particles filled in the gap of large Ni particles. In addition, the microstructure and shear strength of joints formed by three hybrid solder pastes after aging treatment of 72 h, 170 h, and 360 h at 200 ℃ are distinct dramatically. Especially, the joint composed by mixing Ni particles remained higher shear strength of 49.75 MPa after 360 h aging, and the fracture modes were dominated by intergranular fracture and

supplemented by transgranular fracture, which results in a dense and high-strength joint with no defects.

ACKNOWLEDGMENT

The study was supported by the National Natural Science Foundation of China (NSFC 51775140). Part of the work was also supported by the Guangdong Province Natural Science Foundation (2017A030313302), the Shenzhen Science and Technology Plan Projects No. JCYJ20170307150122514.

REFERENCES

[1] K. S. Tan, N. M. Noordin, K. Y. Cheong. An overview of die-attach material for high temperature applications, AIP Conf. Proc. 1865(2017) 050011-1–050011-5.

[2] T. Funaki, J. C. Balda, J. Junghans, et al. Power conversion with SiC devices at extremely high ambient temperatures, Power Electro. IEEE Trans. 22(2007) 1321-1329.

[3] V. Chidambaram, H. B. Yeung, G. Shan. Development of metallic hermetic sealing for mems packaging for harsh environment applications, J. Electron. Mater. 41(2012) 2256-2266.

[4] P.G. Neudeck, R.S. Okojie, L.Y. Chen, High-temperature electronics-a role for wide bandgap semiconductors? J. Pro. IEEE. 90(2002) 1065-1076.

[5] E. Dalton, G. Ren, J. Punch, et al. Accelerated temperature cycling induced strain and failure behavior for BGA assemblies of third generation high Ag content Pb-free solder alloys, Mater. Des. 154(2018) 184-191.

[6] H. Ji, Y. Qiao, M. Li, Rapid formation of intermetallic joints through ultrasonic-assisted die bonding with Sn-0.7Cu solder for high temperature packaging application. Scr. Mater. 110(2016) 19-23.

[7] H. S. Chin, K. Y. Cheong, A. B. Ismail. A review on die attach materials for SiC-based high-temperature power devices, Metal. Mater. Trans. B, 41(2010) 824-832.

[8] Z. L. Li, H. J. Dong, X. G. Song, et al. Rapid formation of Ni3Sn4 joints for die attachment of SiC-based high temperature power devices using ultrasound-induced transient liquid phase bonding process, Ultrason. Sonochem. 36(2017) 420.

[9] Li J, Johnson C M, Buttay C, et al. Bonding strength of multiple SiC die attachment prepared by sintering of Ag nanoparticles, J. Mater. Pro. Technol. 215(2015) 299-308.

[10] N. Hirokazu, K. Takumi, K. Hiroyuki, et al. Compensation of Surface Roughness Using an Au Intermediate Layer in a Cu Direct Bonding Process, J. Electron. Mater. 47(2018) 5403-5409.

[11] J. F. Li, P. A. Agyakwa, C. M. Johnson. Interfacial reaction in Cu/Sn/Cu system during the transient liquid phase soldering process, Acta Mater. 59(2011) 1198-1211.

[12] K. Zeng, K. N. Tu. Six cases of reliability study of Pb-free solder joints in electronic packaging technology, Mater. Sci. Eng. R, 38(2002) 55-105.

[13] Y. T. Lai, C. Y. Liu. Study of wetting reaction between eutectic AuSn and Au foil, J. Electron. Mater. 35(2006) 353-359.

[14] Y. Oda, N. Fukumuro, S. Yae. Intermetallic compound growth between electroless nickel/electroless palladium/immersion gold surface finish and Sn-3.5Ag or Sn-3.0Ag-0.5Cu solder, J. Electron. Mater. 47(2018) 1-5.

[15] S. Magdassi, M. Grouchko, O. Berezin, et al. Triggering the sintering of silver nanoparticles at room temperature, ACS Nano. 4(2010) 1943-1948.

[16] A. Hu, J. Y. Guo, H. Alarifi, et al. Low temperature sintering of Ag nanoparticles for flexible electronics packaging, Appl. Phys. Lett. 97(2010) 153117-1-3

[17] J. Yan, G. Zou, A. P. Wu, et al. Pressureless bonding process using Ag nanoparticle paste for flexible electronics packaging, Scr. Mater. 66(2012) 582-585.

[18] J. D. Liu, H. T. Chen, M. Y. Li, et al. Highly conductive Cu-Cu joint formation by low-temperature sintering of formic acid-treated cu nanoparticles, ACS Appl. Mater. Interfaces. 8(2016) 33289-33298.

[19] Y. Gao, W. Li, C. Chen, et al. Novel copper particle paste with self-reduction and self-protection characteristics for die attachment of power semiconductor under a nitrogen atmosphere, Mater. Des. 160(2018) 1265-1272.

[20] J. Liu, H. Chen, M. Li, et al. Highly conductive Cu-Cu joint formation by low-temperature sintering of formic acid-treated cu nanoparticles, ACS Appl. Mater. Interfaces. 8(2016) 33289-33298.

[21] Y. Ma, A. Roshanghias, A. Binder. A comparative study on direct Cu–Cu bonding methodologies for copper pillar bumped flip-chips, J. Mater. Sci. Mater. Electron. 29(2018): 9347-9353.

[22] J. F. Li, P. A. Agyakwa, C. M. Johnson. Interfacial reaction in Cu/Sn/Cu system during the transient liquid phase soldering process, Acta Mater. 59(2011) 1198-1211.

[23] H. Shao, A. Wu, Y. Bao, et al, Rapid Ag/Sn/Ag transient liquid phase bonding for high-temperature power devices packaging by the assistance of ultrasound, Ultrason. Sonochem. 37(2017) 561-570.

[24] B. S. Lee, S. K. Hyun, J. W. Yoon. Cu–Sn and Ni–Sn transient liquid phase bonding for die-attach technology applications in high-temperature power electronics packaging, J. Mater. Sci. Mater. Electron. 28(2017) 7827-7833.

[25] Z. Li, M. Li, Y. Xiao, et al. Ultrarapid formation of homogeneous Cu6Sn5 and Cu3Sn intermetallic compound joints at room temperature using ultrasonic waves. Ultrason. Sonochem. 21(2014) 924-930.

[26] A. T. Tan, A. W. Tan, F. Yusof. Effect of ultrasonic vibration time on the Cu/Sn-Ag-Cu/Cu joint soldered by low-power-high-frequency ultrasonic-assisted reflow soldering, Ultrason. Sonochem. 34(2017) 616-625.

[27] D. Zhao, K. Zhang, J. Cui, et al. Effect of ultrasonic vibration on the interfacial IMC three-dimensional morphology and mechanical properties of Sn-2.5Ag-0.7Cu-0.1RE-0.05Ni/ Cu halogen free solder joints, J. Mater. Sci. Mater. Electron. 29(2018) 18828-18839.

[28] H. Ji, M. Li, S. Ma, et al. Ni3Sn4-composed die bonded interface rapidly formed by ultrasonic-assisted soldering of Sn/Ni solder paste for high-temperature power device packaging, Mater. Des. 108(2016) 590-596.

[29] H. Pan, J. Huang, H. Ji, M. Li. Enhancing the solid/liquid interfacial metallurgical reaction of Sn+Cu composite solder by ultrasonic-assisted chip attachment, J. Alloys. Compd. 784 (2019) 603-610.

Comprehensive Study of Copper Nano-paste for Cu-Cu Bonding

Ser Choong Chong, Sharon Lim Pei Siang

Institute of Microelectronics,

A*STAR (Agency for Science, Technology and Research),

2 Fusionopolis Way, #08-02 Innovis Tower, Singapore 138634

Tel: 67705724 Fax: 67745747

Email: chongsc@ime.a-star.edu.sg

Abstract—Cu-Cu bond is a prefer solution for shrinking interconnect pitch as it offered lower interconnect resistance with no intermetallic compound risk. Major challenge in Cu-Cu bond is the formation of copper oxide and also the long thermal process. Several approaches had been proposed in the industry to overcome the challenges are diffusion bonding, ultra-sonic bonding, Cu nano-paste, use of Formic acid and argon plasma for copper surface treatment and bit grinding process to enable good Cu-Cu bond.

In this paper, we explore the use of Cu nano-paste as the bonding medium between Cu bump and Cu pad. Cu nano-paste is an attractive solution for Cu-Cu bond as it deploy similar approach as conventional solder bump flip chip process. The flux used in conventional solder bump flip chip process is replaced by Cu nano-paste and the conventional reflow oven is replaced by Formic Acid Chamber. We had demonstrated good formic acid process in dealing with Cu nano-paste in terms of low resistance value and prestige copper surface.

Keywords- Cu nano-paste, Cu micro-bump. Formic acid.

1. INTRODUCTION

Cu-Cu bonding approach is getting more popular as the demand for a higher performance device that in turns needs to pack several thousand I/O into a small real estate area on the die [1, 2]. The other popular approach for higher performance device is to adopt 2.5D or 3D packaging scheme [3]. Solder interconnect is another bonding technique that suffered from issues like solder merging, solder squeezing or solder void when the bump diameter is getting smaller and smaller. As such, Cu-Cu bonding is getting more focus as a promising interconnect approach for ultra-fine pitch devices.

Cu-Cu bonding can be achieved either through the use of thermal energy, mechanical energy, ultra-sonic energy and paste material [4, 5, 6, 7, 8, 9, 10 & 11]. The use of thermal, mechanical or ultrasonic energy requires the Cu bump to be of certain quality such as good bump co-planity, Cu surface free of oxide, and smooth surface roughness. These requirements complicate the bonding process and induce additional processes such as bit grinding of the bump to ensure good bump co-planity, surface treatment to remove oxide layer and to ensure smooth surface, Chemical-Mechanical Polishing to ensure smooth copper surface. Use of paste material will reduce the impact of bump height co-planity, oxidation, and surface roughness in the formation of Cu-Cu bond.

This paper presents the work done on Cu-Cu bonding with the use of a nano-paste material. The paste material is a low temperature sintering nano-copper based paste. This material need to be subjected to sintering process during the bonding process in order to prevent or remove the formation of oxide layer on the nano-copper particle. Extensive study of the oxide removal process and the copper fusing process are conducted to understand the impact of the sintering process parameters such as chamber pressure between 30sccm and 950sccm, temperature between 100°C and 250°C; and the duration of the process between 10sec and 350sec.

In this paper, we have successfully demonstrated good formic acid process of the nano-copper paste for Cu-Cu bonding.

2. COPPER NANO-PASTE INTRODUCTION

The copper nano-paste consist of nano-size copper filler. The small surface area of the nano-size copper filler enable it to have much lower reactive temperature as compare to bulk copper. The copper filler is susceptible to oxidation at elevated temperature. The paste need to go through a sintering process whereby the present of Formic acid will remove the oxide layer surrounding the copper nano-particles before it could fused with adjacent copper particle to form the Cu-Cu bond. Figure 2.1 showed the nano-paste before and after sintering process.

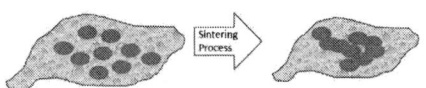

Figure 2.1: Nano-particles fused together after going through Sintering process.

The application of copper nano-paste in forming the joint between the copper bump and copper pad is similar to application of flux in conventional solder bump flip chip process. The copper nano-paste replaced the flux in the dipping process and the sintering process replaced the reflow process as shown in Figure 2.2. The chip with the copper bump is dipped into the dipping station with nano-paste that had pre-defined thickness. The chip with the nano-paste coated copper bump is aligned and bond to the substrate. The substrate with the chip is send through the sintering process to form the Cu-Cu bond.

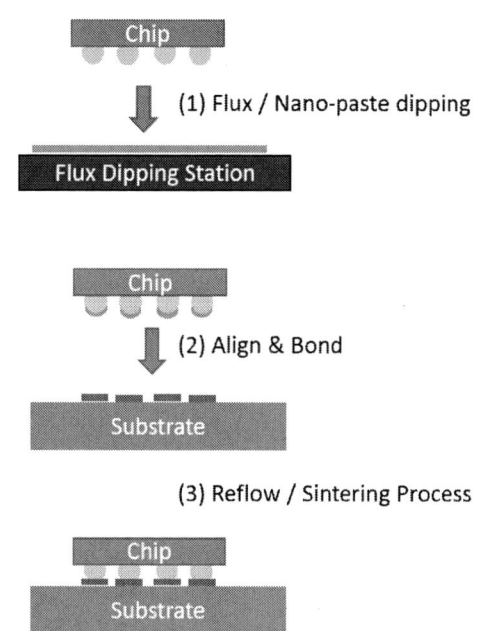

Figure 2.2: Process flow using either flux or nano-paste

3. SINTERING PROCESS INTRODUCTION

Sintering of copper nano-paste involved the introduction of Formic Acid, Nitrogen, Vacuum and heat. Formic Acid is use to remove the oxide layer surrounding the copper particles before it melts and fused with adjacent copper particle.

The sample with copper nano-paste is placed inside a chamber. The chamber is first vacuum to evacuate the air inside it before nitrogen is input into it. The Formic Acid is then introduced into the chamber with the sample heated up to preset temperature. The copper nano-paste will reacts with Formic Acid to remove oxide layer and elevated temperature to fuse together and formed Cu-Cu bond. Figure 3.1 illustrate the Sintering process for the sample.

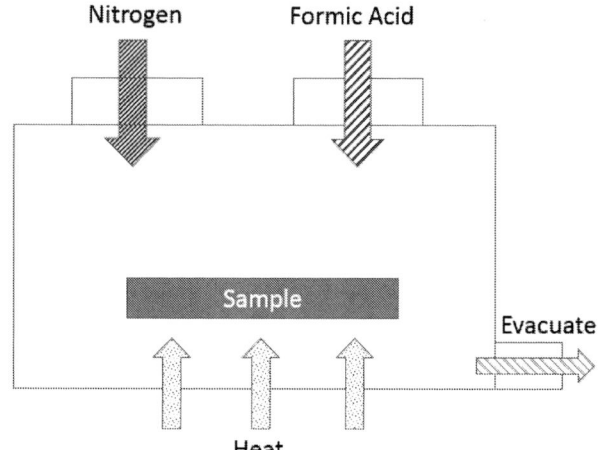

Figure 3.1: Sintering Process

4. EXPERIMENT APPROACHES

Sintering process of copper nano-paste involved two main events that will determine the success of the process. The two main events are the removal of copper oxide and the fusing of the copper particles to form metallic bond.

Therefore, two experiments are conducted to understand these two events that is happening to copper nano-paste. First experiment is to understand the effectiveness of sintering process in removal of copper oxide. The 2nd experiment is to understand the effectiveness of sintering process for the copper particles fusing inside the copper nano-paste to achieve Cu-Cu bond with good electrical performance.

The first experiment is carried out with the preparation of the test–couple. It is prepared by depositing a thin layer of copper on silicon wafer. The copper-coated wafer is then diced into smaller piece of 10mm x10mm to form the test-couple for the 1st experiment as shown in Figure 4.1. The copper coated die is first subjected to thermal process to allow a thick copper oxide to grow on the copper surface.

Figure 4.1: Silicon die coated with Copper

978-1-7281-1500-9/19 $31.00 © 2019 IEEE

Study of the sintering process in terms of processed temperature, formic acid chamber pressure and the treatment duration with respect to surface roughness, oxide thickness and contact angle is conducted. Surface roughness is measured using Atomic Force Microscopic (AFM), oxide thickness is determined by Auger Electron Spectroscopy (AES) and contact angle by contact angle measurement tool. Figure 4.2 illustrate the flow of first experiment.

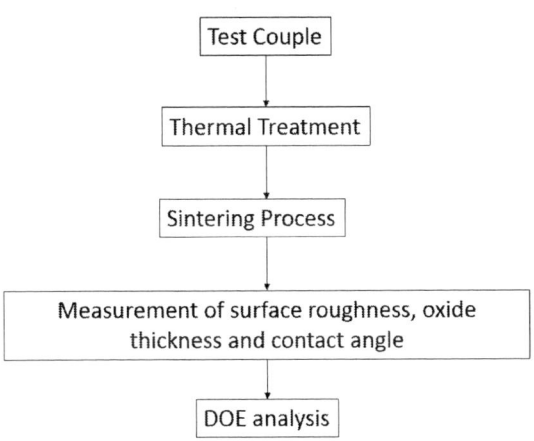

Figure 4.2: 1st experiment flow process

The second experiment for the copper nano-paste sintering process is carried out by depositing the nano paste on a silicon carrier. The silicon carrier with the copper nano-paste is then subjected to formic acid with different process temperature, formic acid chamber pressure and the process duration. The copper nano-paste is inspected visually before and after the formic acid process and also measured the resistance across the nano-paste after the sintering process. Figure 4.3 illustrate the flow process of 2nd experiment.

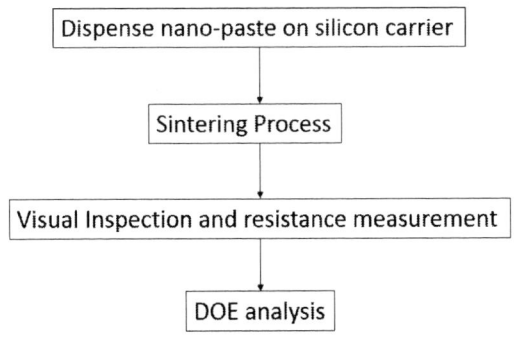

Figure 4.3: 2nd experiment flow process

JMP DOE software is used to analysis the various response with respect to processed temperature, formic acid chamber pressure and the process duration for both experiments.

5. 1ST EXPERIMENT RESULTS

The test couple with coated copper is subjected to 125°C for 15min and 30min to determine the best condition to prepare the test couple for the experiment. Auger Electron Microscopy is used to determine the thickness of the oxide layer at time zero, 15min at 125°C and 30min at 125°C. Figure 5.1 showed the typical Auger plot for the oxide thickness measurement.

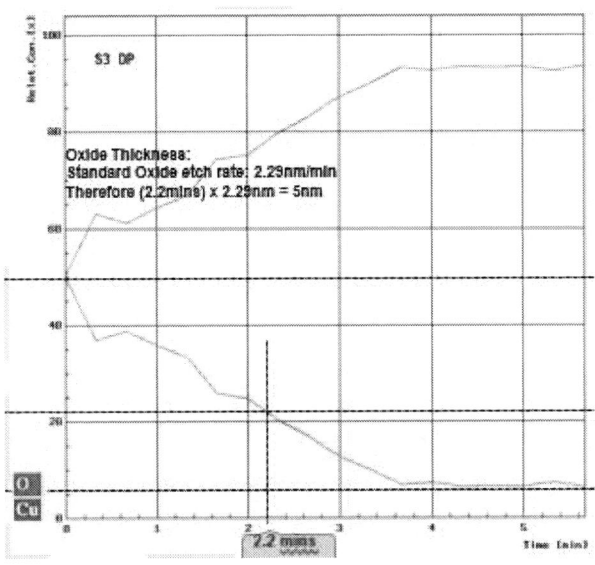

Figure 5.1: Typical Auger Plot for oxide thickness measurement.

The oxide thickness measurement results are tabulated in Table 5.1. It indicated that the oxide thickness is 4nm at time zero and increased to 40nm after 30min at 125°C.

Table 5.1: Oxide Thickness at time zero and after thermal treatment

No	Duration at 125 Celsius	Oxide Thickness
1	0 min	4nm
2	15min	10nm
3	30min	40nm

The duration of 30min is selected to prepare the test couple for the subsequent Formic Acid DOE run. Different process temperature from 100°C to 200°C, formic acid chamber pressure from 450 mbar to 950 mbar and process duration from 10 to 50 sec were tested out as tabulated in Table 5.2.

Table 5.2: DOE Run for 1st experiment

DOE	Process Temperature, T (Celsius)	Formic Acid Chamber Pressure, P (mbar)	Process duration, t (sec)
1	100	450	30
2	100	700	10
3	100	950	50
4	150	450	10
5	150	700	30
6	150	700	50
7	150	950	30
8	200	450	50
9	200	700	30
10	200	950	10
11	200	950	50

Table 5.3: DOE Results for 1st experiment

DOE	Surface Roughness (nm)	Oxide Thickness (nm)	Contact angle (degree)
1	15.4	44.8	99.2
2	14.8	43.3	98.3
3	14.6	46.4	91.2
4	15.6	40.8	74.2
5	14.5	43.9	72.8
6	14.3	45.1	69.8
7	14.5	38.3	65.0
8	14.2	2.75	60.5
9	13.9	2.98	60.8
10	14.2	2.75	60.7
11	13.9	2.63	60.7

The oxide thickness, surface roughness and contact angle are taken as the response for the experiment. Figure 5.2 showed the typical surface roughness plot and Figure 5.3 showed the typical contact angle measurement.

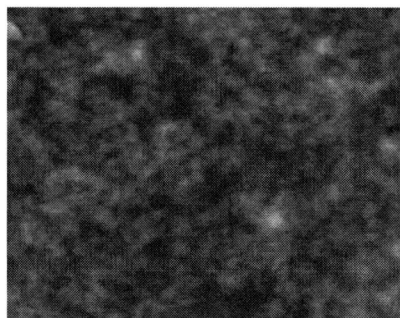

Figure 5.2: Typical AFM plot for the surface roughness

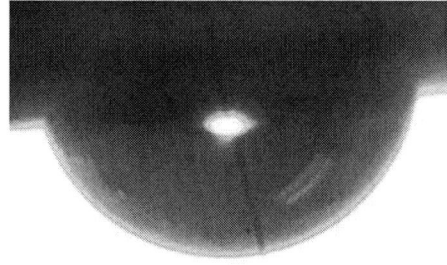

Figure 5.3: Typical contact angle measurement

Table 5.3 tabulated the responses for the DOE run. It showed that the surface roughness is almost 14nm, oxide thickness can be reduce down to 3nm and the contact angle down to 60 degree for certain process parameters combination.

Further analysis of the results indicated that the most significant factor influence the processing of copper nano-paste is the processing temperature as shown in Figure 5.4.

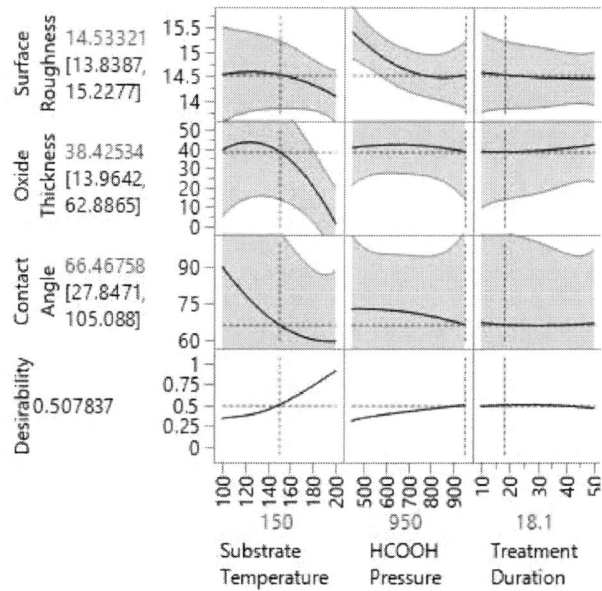

Figure 5.4: DOE analysis of 1st experiment

6. 2ND EXPERIMENT RESULTS

The silicon carrier with deposited copper nanopaste was subjected to Formic Acid process. Different process temperature from 100°C to 175°C, Formic acid chamber pressure from 30mbar to 400mbar and duration from 50 to 350 sec were tested out as tabulated in Table 6.1. The visual inspection and resistance measurement on the copper nano-paste are taken as response for this DOE run. Bright color of the copper nano-paste is considered grade 1 whereas dull color is considered as grade 1 as shown in Figure 6.1.

Grade 1   **Grade 4**

Figure 6.1: Color code for visual inspection

Table 6.1: DOE Run parameters for 2nd experiment

DOE	Process Temperature, T (Celsius)	Formic Acid Chamber Pressure, P (mbar)	Process duration, t (sec)
1	175	400	200
2	250	400	350
3	250	215	50
4	250	30	200
5	100	400	350
6	100	215	200
7	100	400	50
8	100	30	350
9	175	178	350
10	136	50	50
11	175	215	200

Analysis of the results for 2nd experiment indicated that the most significant factor influencing the processing of copper nano-paste is still the processing temperature as shown in Figure 6.2.

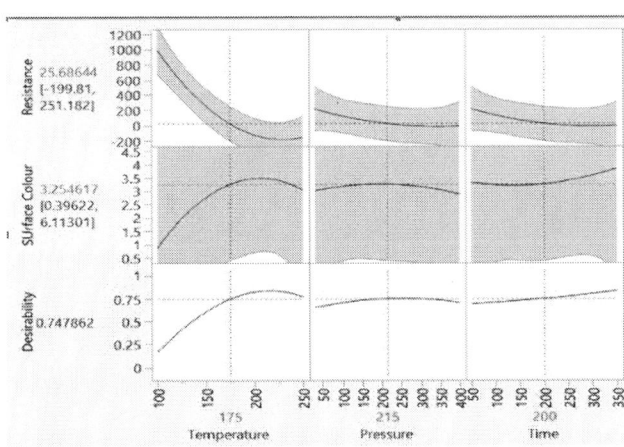

Figure 6.2: DOE analysis of 2nd experiment

Additional test was done with the best parameters but with Nitrogen gas instead of Formic Acid. Figure 6.3 showed the dull color of the copper nanopaste after sintering process with Nitrogen. The dull colour copper nanopaste registered open circuit.

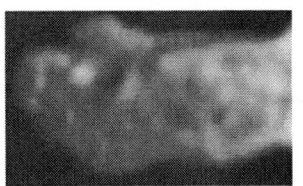

Figure 6.3: Dull colour of copper nano-paste after Nitrogen sintering process

7. CONCLUSION

We have successfully demonstrated good formic acid process of the nano-copper paste for Cu-Cu bonding. Some of the important observations are as followed

(a) Higher processing temperature of at least 200°C is necessary to reduce the oxide layer down to less than 10nm.
(b) Duration of the treatment process is not significant toward oxide removal.
(c) Formic Acid Chamber pressure is not significant toward copper line resistance but higher pressure is prefer in order to have shiny copper surface.

8. REFERENCES

1. S. Yang, H. T. Hung, Y. B. Chen, C. R. Kao, "Low-Temperature, Pressureless Cu-to-Cu Bonding By Electroless Ni Plating", International Microsystems Packaging Assembly and Circuits Technology Conference 2016, pp. 111-114.

2. Luca Del Carro, Jonas Zürcher, Sebastian Gerke, Thomas Brunschwiler, "Morphology of low-temperature all-copper interconnects formed by dip transfer, ECTC 2017, pp 961-967

3. J.U. Knickerbocker, P.S. Andry, B. Dang, R.R. Horton, C S. Patel, R.J. Polastre, K. Sakuma, E.S. Sprogis, C.K. Tsang, B.C. Webb, and S.L. Wright, "3D Silicon Integration", ECTC 2008, pp 538-543

4. Taiji Sakai, Nobuhiro Imaizumi, and Toyoo Miyajima, "Low temperature Cu-Cu direct bonding for 3D-IC by using fine crystal layer", IEEE CPMT Symposium 2012, pp 1-4

5. Yoshiyuki Arai, Masatsugu Nimura, and Hajime Tomokage, "Cu-Cu Direct Bonding Technology Using Ultrasonic Vibration for Flip-chip Interconnection", ICEP-IAAC 2015, pp468-472

6. Junjie Li, Tielin Shi, Xing Yu, Chaoliang Cheng, Jinhu Fan, Guanglan Liao,Zirong Tang, "Low-temperature and low-pressure Cu-Cu bonding by pure Cu nanosolder paste for wafer-level packaging", ECTC 2017, pp 976-981

7. Lan Peng, Lin Zhang, Ji Fan, Hong Yu Li, Dau Fatt Lim, and Chuan Seng Tan, "Ultrafine Pitch (6 μm) of Recessed and Bonded Cu–Cu Interconnects by Three-Dimensional Wafer Stacking", IEEE Electron Device Letters, Vol 33, No. 12, December 2012, pp 1747-1749

8. Ser Choong Chong, Ling Xie, Sunil Wickramanayaka, Vasarla Nagendra Sekhar, Daniel Ismael Cereno, "Ultra-fine pitch Cu-Cu bonding of 6μm bump pitch for 2.5D application, EPTC 2016, December 2016, pp 102-106

9. Ser Choong Chong, Jie Li Aw, Daniel Ismael Cereno, Li Yan Siow, Chee Guan Koh, David Witarsa, Srinivas Vempati, Tai Chong Chai, "Fine pitch solder-less bonding using ultrasonic technique", EPTC 2012, pp 420-425

10. Ling Xie; Sunil Wickramanayaka, Ser Choong Chong, Vasarla Sekhar, Daniel Ismael, Yong Liang Ye, "6μm Pitch High Density Cu-Cu Bonding for 3D IC Stacking", ECTC 2016, pp 2126-2133

11. Ling Xie; Sunil Wickramanayaka, Ser Choong Chong, Vasarla Sekhar, Daniel Ismael, "High Throughput Thermal Compression Bonding of 20μm Pitch Cu Pillar with Gas Pressure Bonder for 3D IC Stacking", ECTC 2016, pp 108-114

Enhanced Performance of Laser-Assisted Bonding with Compression (LABC) Compared with Thermal Compression Bonding (TCB) Technology

Kwang-Seong Choi, Yong-Sung Eom,
Seok Hwan Moon, Jiho Joo, leeseul Jeong
ICT Materials and Components Laboratory
Electronics and Telecommunications Research Institute
218 Gajeong-ro, Yuseong-gu, 34129, Korea
kschoi@etri.re.kr

Kwangjoo Lee, Jung Hak Kim, Ju hyeon Kim
LG Chem R&D Campus Daejeon
188, Munji-ro, Yuseong-gu, Daejeon, 34122, Korea
kwangjoolee@lgchem.com

Gil-Sang Yoon
Molds & Dies R&D Group, KITECH
156 Gaetbeol-ro, Yeonsu-gu, 21999, Korea
seviaygs@kitech.re.kr

Kwang-Hee Lee, Chul-Hee Lee
Department of Mechanical Engineering
Inha University
100 Inha-ro, Michuhol-gu, Incheon, Korea
chulhee@inha.ac.kr

Geun-Sik Ahn, Moo-Sup Shim
Protec
Dongan-gu, Anyang-si, Gyeonggi-do, Korea
msshim@protec21.co.kr

Abstract— **A LABC (Laser-Assisted Bonding with Compression) bonder and NCF (Non-Conductive Film) were developed to increase the productivity of the bonding process for the advanced microelectronic packaging technology. The design features of a LABC make its UPH above 1,000. The NCF was applied to both of LAB and TCB (Thermal Compression Bonding Technology). The 780μm-thick daisy chain top and bottom chips with the minimum pitch of 30μm and bump number of about 27,000 were prepared and tested to verify the LABC and NCF technology. The effects of the laser power on the joints quality after the LABC bonding process were investigated and compared with the joints formed by the TCB technology. Finally, the SAT (Scanning Acoustic Tomography) images of the test vehicles before and after the TCO (Pressurized oven) were observed to check the voids in the NCF after the LABC bonding process.**

Keywords-laser-assisted bonding with compression (LABC), non-conductive film (NCF), bonding performance, throughput, thermal compression bonding (TCB)

I. INTRODUCTION

The bonding performance for the advanced microelectronic packaging technologies such as fan-out packaging, 2.5D interposer and 3D-stacked through-silicon via (TSV) technology has been considered one of the key technical parameters increasing the penetration of these technologies in the commercial areas. Rather than the conventional flip chip bonding, the thermal compression bonding (TCB) technology combined non-conductive paste (NCP) or non-conductive film (NCF) materials was adopted as one of technical solutions because it can reduce greatly the faults caused by the fine pitch of the interconnections of these technologies [1-4]. Its low throughput, however, has led to other technical innovations such as thermo-compression bonding (TCB) with multi-table and multi-bond head system, but its productive is still around UPH 2,000, which is much lower than that of the convention mass reflow process [5,6].

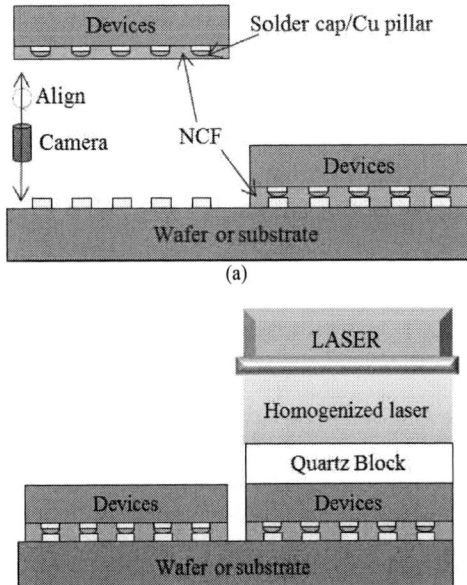

Fig. 1. Process Flow of Laser-Assisted Bonding with Compression (LABC):
(a) pre-bonding and (b) post-bonding process

To overcome such a drawback of TCB technology, laser assisted bonding (LAB) was proposed [6-13]. It features fast ramp-up and –down of the packages or devices to be

978-1-7281-1500-9/19 $31.00 © 2019 IEEE

interconnected on a substrate and vertical directional temperature distributions so that its productivity and reliability can be comparable with that of TCB technology. However, LAB usually required flux material, cleaning and underfill process like the conventional flip chip bonding process. It could be difficult to apply LAB technology for the interconnection of the devices or packages with large warpage or fine-pitch interconnection.

In this paper, laser-assisted bonding with compression (LABC) technology with NCF was proposed to accomplish the productivity and process reliability at the same time. To increase the throughput, the bonding procedure was divided into two-step, that is, a pre-bonding and post-bonding; pre-bonding for the attachment of the devices on a substrate and post-bonding for the compression and laser irradiation process, as shown in Fig. 1. The cycle time of the pre-bonding and post-bonding were designed 0.5 and about 3 seconds, respectively, with the result that a UPH over 1,000 could be easily obtained using a single table and two bonding heads. A quartz block as a header was used to deliver a pressure to the devices because of its extremely low absorption of the laser during the bonding process.

Newly developed NCF for LABC was designed not only to be applicable to the conventional TCB technology but also to have stability on a hot stage during pre-bonding and to show solder wetting and fast curing with no void and optimal fillet during post-bonding despite the highly fast ramp-up of the bonding temperature in the LABC technology.

To evaluate the bonding performance a daisy chain test vehicle with the minimal bump pitch of 30μm was prepared. Its dimensions were 8.5x13mm². The number of the bumps per test vehicle was about 27,000. The align accuracy and microstructures using the LABC and the TCB were compared. Finally, the each process impacts on the microstructures of each bonded test vehicle were investigated.

II. DESIGN OF NCF, DAISY CHAIN TEST VEHICLE, AND LABC EQUPMENT

A. Design of NCF

Newly developed NCF was designed for both the conventional TCB technology and the LABC technology. The NCF sheet was consisted of cover, NCF, and base film. For both TCB and LABC technology, thermodynamic profile and viscosity of NCF was designed as shown in Fig. 2-3. Thermodynamic profile of NCF was measured by DSC (Differential Scanning Calorimetry, TA instruments). The viscosity of NCF was measured by rheometer (ARES-G2, TA instruments).

The film was laminated on the bumped wafer by vacuum laminator after cover film peel-off. Proper vacuum lamination temperature, pressure and time were applied to prevent air trap around the bumps and dicing lines. After the lamination processes, the base film was peeled-off without delamination. The base film peel-off process was shown in Fig. 4(a) and NCF laminated 12-inch bumped wafer after

base film peel-off was shown in Fig. 4(b). Voids or delamination were not present on the wafer.

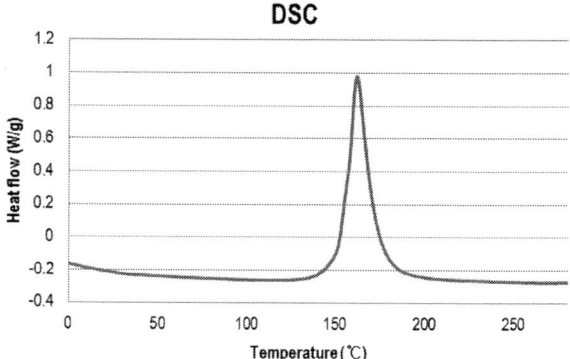

Fig. 2. DSC curve of NCF with temperature

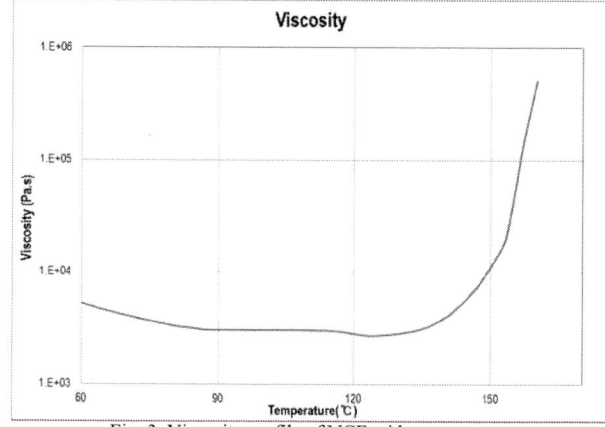

Fig. 3. Viscosity profile of NCF with temperature

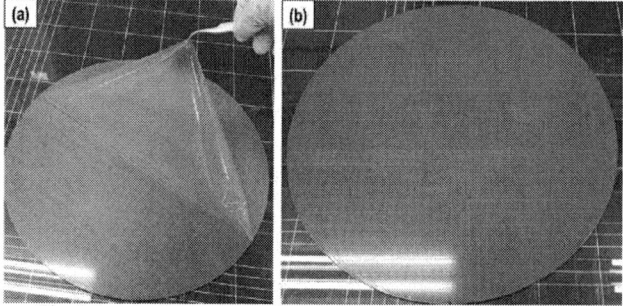

Fig. 4. Lamination process of NCF (a) during the base-film peel-off (b) after the base-film peel off

NCF laminated bumped wafer was diced by blade. After dicing, each chip was diced. Fig. 5 showed a cross-section view of NCF laminated bumped chip where there was no chipping in the side view.

The alignment mark on the wafer could be recognized through NCF. The NCF laminated chips were clear enough to be recognized both in the bonding device and optical microscope as shown in Fig. 6.

B. Daisy Chain Test Vehicle

Daisy chain wafers were designed to verify the performance of a NCF film and LABC bonder. Table 1 shows the specifications of the daisy chain chips. Fig. 7 shows the SEM image of an array Cu pillar/solder cap on a top chip. The bump pitch of each bump was designed from about 70μm to 30μm. The SEM image of the array of UBMs (Under Bump Metallization) on the bottom chip was shown in Fig. 8. The diameters of the bump and UBM were designed 18μm, and 20μm, respectively.

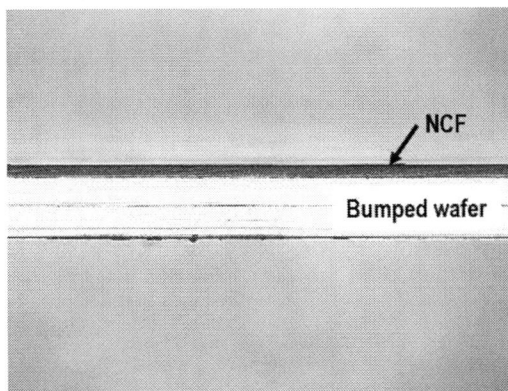

Fig. 5. Side view of NCF on bumped wafer after dicing

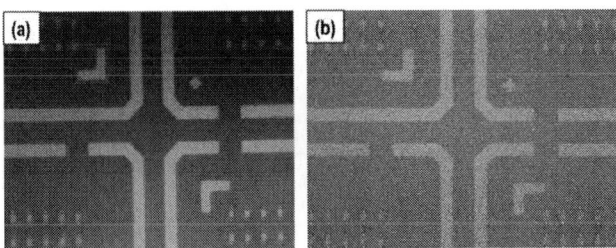

Fig. 6. Recognition of NCF (a) in the bonding device and (b) in the optical microscope

Table 1. Specifications of daisy chain test vehicles

Type	Dimensions	Surface finish of pads	# of pads
Top	11x7x0.78mm³	Cu pillar/Ni/SnAg	27,000
Bottom	13x8.5x0.78mm³	Ni/Au	

Fig. 7. SEM image of array of Cu pillar/solder cap on top chip

C. LABC Equpment

In order to implement the LABC technology and high throughput described above, we designed and manufactured the equipment consisting of Pre Bonder and Post Bonder. Fig. 9 is a design of LABC equipment designed to verify LABC technology and its high throughput.

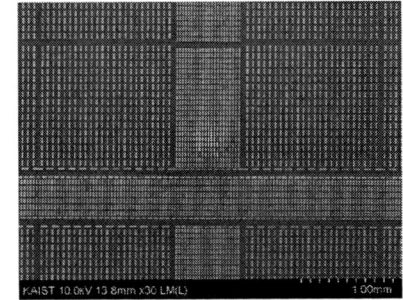

Fig. 8. SEM image of array of UBM on bottom chip

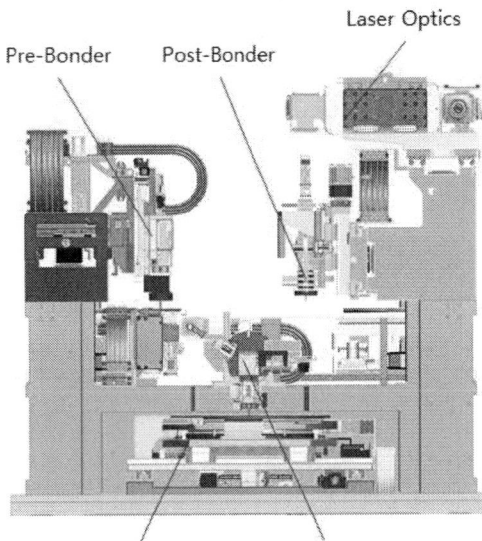

Fig. 9. Design of LABC Equipment

Pre Bonder is composed of two heads in one gantry. It plays a role of bonding a NCF-laminated die to a substrate wafer in a short time at low temperature and pressure. It is advantageous to improve the UPH by constructing 3 or more multiple heads in 1 gantry. However, it consists of 2 heads considering the complexity of equipment, manufacturing cost, mounting accuracy, and productivity balance with post-bonder. Fig. 10 shows the pre-bonder heads in detail and Fig. 11 shows the manufactured bonder system with 2 pre-bonder heads.

Glass Chip was used to verify the performance and long-term stability of the pre-bonder. The test was performed on the 300mm (x) * 200mm (y) bonding area. Fig. 12 shows the measured bonding accuracy results in the XY region. The achieved accuracy was ± 1.61μm at 3σ in X direction, and ± 1.32μm at 3σ in Y direction, respectively.

Post Bonder is composed of a laser source, optics parts for the laser irradiation, and parts for applying high bonding pressure. It bonds a pre-bonded die on a substrate wafer using laser and pressure in a short time. When the dies are bonded, it is pressurized by using three piezo actuators.

When the bonding force is applied, three force sensor is used to precisely control the bonding pressure while piezo actuators measures the actual pressure in real time to control the fine adjustment of displacement of the die to be bonded. Laser is converted into the uniform quadrature parallel light by laser optics, and it is irradiated at the center of the pressure module composed of the piezo actuators and force sensors. The irradiated laser passes through a transparent quartz tool at the bottom of the pressure module and reaches the die to be bonded. Fig. 13 and Fig. 14 show the designed post-bonder system and the manufactured post-bonder head, respectively.

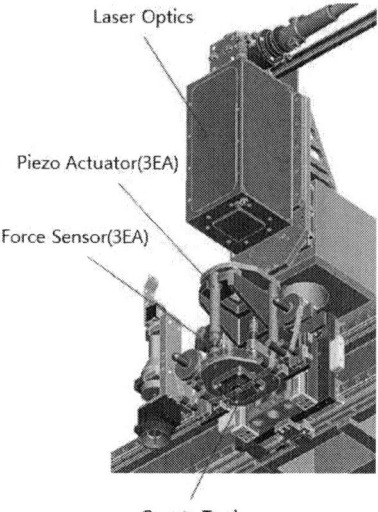

Fig. 13. Design of post-bonder head

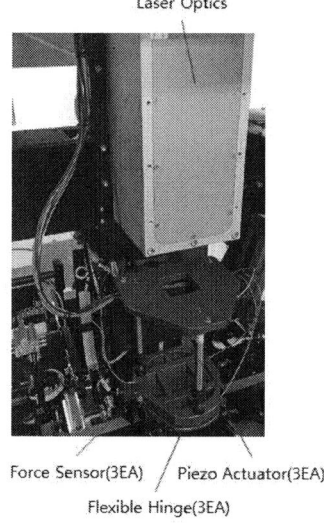

Fig. 14. Photograph of post-bonder head

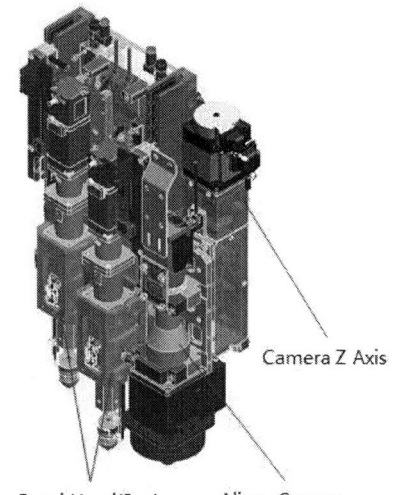

Fig. 10. Design of gantry table with 2 pre-bonder heads

Fig. 11. Photograph of gantry table with 2 pre-bonder heads

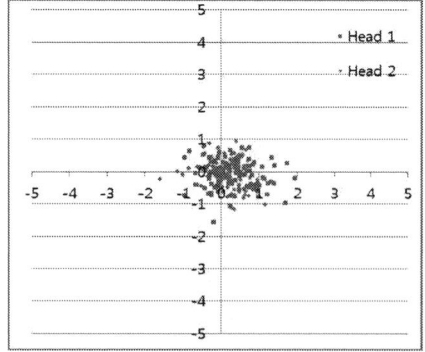

Fig. 12. Measured bonding accuracy of pre-bonder with 2 bonding heads using glass chip

III. PERFORMANCE COMPARISONS BETWEEN LABC AND TCB

A. Development of Bonding Profile of LABC

To establish the bonding profile for the bonding process, a thermocouple between a top and a bottom chip was prepared as shown in Fig. 15. The test vehicle was placed on the post-bonder and the laser was irradiated on it with the variations of the power. The laser exposure time was 5 sec. The measured temperature increased with the laser exposure time, and decreased rapidly just after the stop of the laser irradiation. The measured peak temperatures according to the laser power of 80W, 100W, 120W, 140W, 160W, and 180W were 128℃, 167℃, 190℃, 220℃, 242℃, 260℃, and 280℃, respectively.

978-1-7281-1500-9/19 $31.00 © 2019 IEEE   200

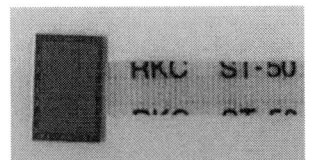

Fig. 14. Thermocouple between top and bottom chips

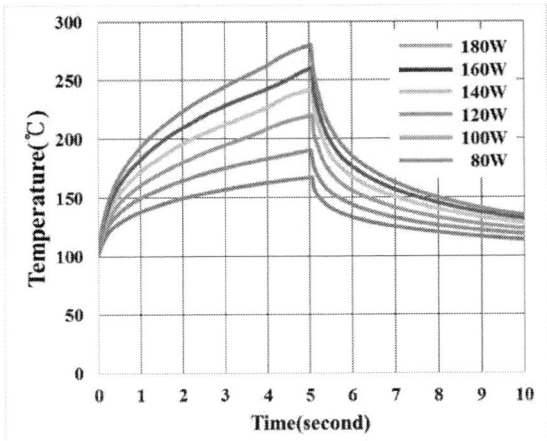

Fig. 15. Measured temperature variations with laser power.

B. Performance Comparisions between LABC and TCB

We performed the bonding process using the LABC bonder and compared its bonding results with that base on the TCB technology. The temperature of the stages in the pre-bonder and post-bonder was maintained at 90℃. The laser power was varied from 80W to 180W to investigate the effects of the laser power on the quality of the solder joints after the bonding process. The bonding force was 70N during the post-bonding process.

Fig 16 shows the bonding profile used in a TCB bonder for the comparison. Its cycle time was about 15sec. Its cycle time was long compared with the LABC because it took time to ramp up and cool down the bonding head of the TCB bonder. The NCF used in the TCB bonding process was the same one used in the LABC process.

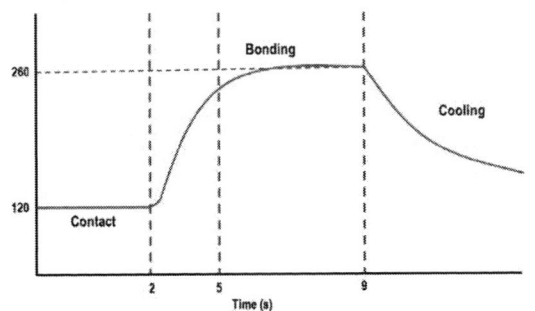

Fig. 16. Bonding profile of TCB bonding process

Fig. 17 shows the SEM images of the joints observed after the LABC and TCB bonding process. The solder joints formed by the LABC with the laser power of 80W, 100W, 120W, 160W, and 180W were shown in Fig. 17 (a), (b), (c),

(d), and (e), respectively. The joints formed by the TCB were shown in Fig. (f). It should be noted that the UBMs on the bottom chip were fabricated Ni/Au, solder was wetted on the UBMs under the low laser power. There was NCF between the torn solders in the joints formed under the laser power of 120W. In case of 180W irradiation, the solder joints show the robustness of the solder joints. The bonding results by the TCB were good, though the alignment was not good, but it could be easily improved. There were no voids in the NCF regardless of the bonding method. The NCF film was proved to be used in the LABC as well as in the TCB bonding process.

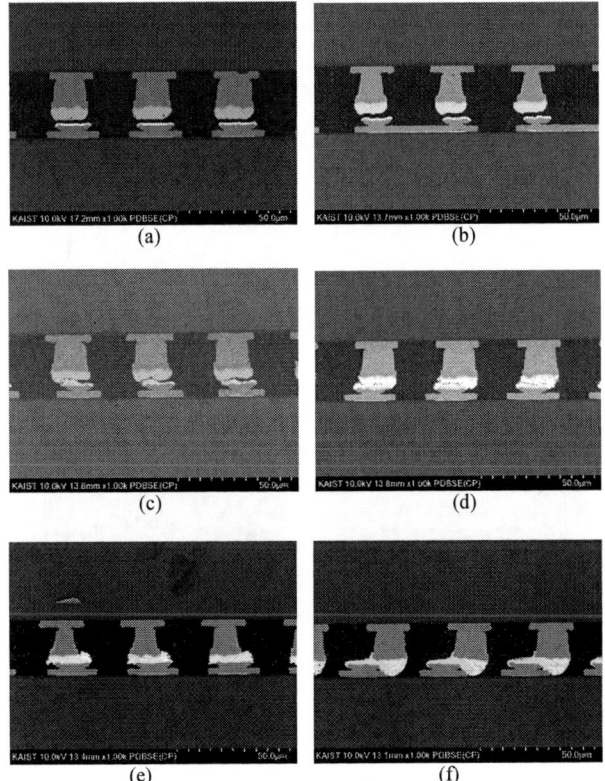

Fig. 17. SEM images (BSE) of joints with laser power of (a) 80W, (b) 100W, (c) 120W, (d) 160W, (e) 180W,and (f) joints formed by TCB.

These behaviors observed in the LABC bonding process can be explained as follows. The bonding energy used in the LAB is the absorbed the laser by the silicon [7]. The absorbed energy is conducted to the joints area below the silicon through the lattice vibrations, that is, thermal energy. That means the bonding mechanism can depend on the design of the thermal path of interconnects to be bonded.

Fig. 18 shows the schematic diagram explaining the bonding mechanism during LABC, especially for the case of the bonding under the laser irradiation with low power. When laser is irradiated on the post-bonding step, the heat is conducted to the Cu pillar/solder cap because of its higher thermal conductivity compared with the NCF. The temperature of the NCF near the Cu pillar/solder cap increases so that the viscosity of the NCF is decreased, leading the solder cap and UBM can be contacted. Because

978-1-7281-1500-9/19 $31.00 © 2019 IEEE

of the elevated temperature of the NCF, the oxide layer on the solder cap is reduced. The solder can be wetted on the UBMs on the bottom chip. The temperature of the whole NCF is increased because of the conduction of the heat from all materials surrounding the NCF, but the temperature is not so high. The viscosity of the NCF is getting low and NCF is expanding because of its coefficient of thermal expansion (CTE). The solder is apart because of the stress caused by the expansion of the NCF. On the other hand, when the laser power is high, the temperature of the whole NCF is high enough to reduce the viscosity much so that the stress caused by the expansion of the NCF can be negligible. This mechanism leads to us another way to obtain robust interconnection during the LABC: Increase the pressure during the LABC so that the NCF cannot be expanded in the step of (f) in Fig. 18. This method has an advantage of low laser power.

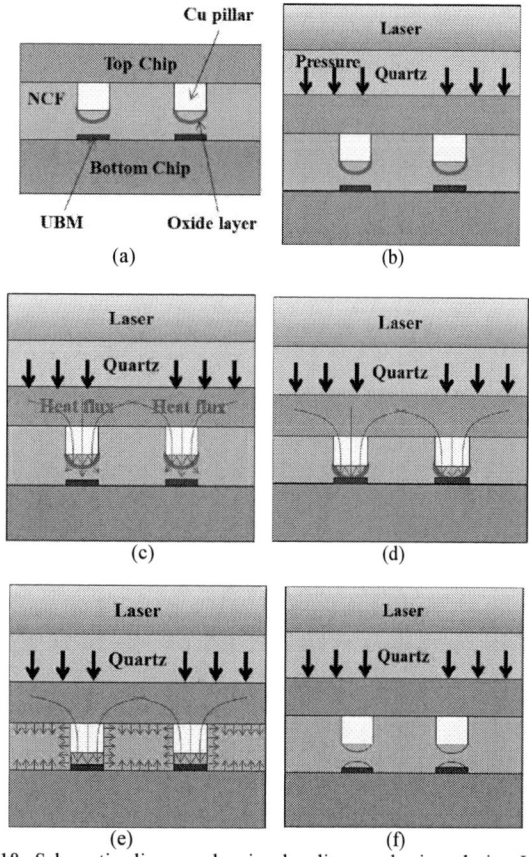

Fig. 18. Schematic diagram showing bonding mechanism during LABC bonding with laser irradiation of low power; (a) test vehicle on pre-bonding step (b) laser irradiation on post-bonding step (c) heat conduction to NCF adjacent to solder cap, (d) contact of solder cap and UBM (e) reduction of oxide layer on solder cap and heat conduction to NCF film and (f) thermal expansion and separated solder

Fig. 19 shows the morphologies of the IMC (intermetallic compound) formed during LABC and TCB. The IMC formed during TCB bonding was thicker than that formed during LABC bonding process.

Since the bonding time is too short, the curing of the NCF was not completed during the LABC bonding process. We applied the PCO (pressurized oven cure) to the bonded test vehicle to achieve the full curing of the NCF. The conditions of the PCO was 1hour duration at 125℃ and 2hours duration at 180℃. The SAT (Scanning Acoustic Tomography) images of the test vehicle bonded under the laser power of 120W, 140W, 160W, and 180W were observed before and after the PCO as shown in Table 2. There were some voids in the NCF of the test vehicle with 120W, 140W, and 160W. These voids were removed during the PCO so that no voids were found in the test vehicles. No voids are found in case of the laser power of 180W after the LABC bonding process.

In case of TCB technology, the duration time at the peak temperature is longer than the LABC bonding process, there is no need of the PCO process. Table 3 shows the SAT images of the test vehicles after the TCB bonding process. No voids in the NCF were observed.

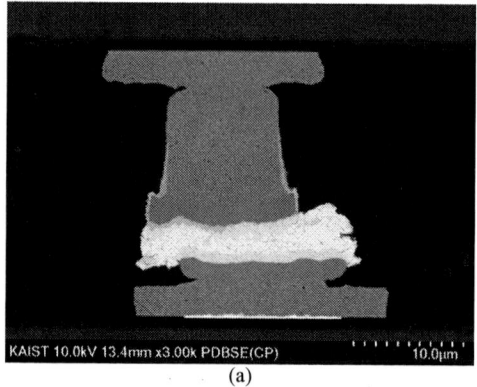

(a)

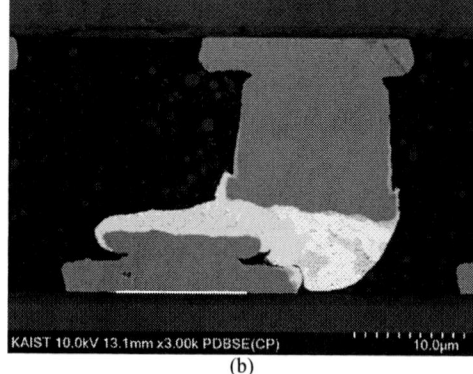

(b)

Fig. 19. Comparison of thickness of IMC formed during (a) LABC and (b) TCB bonding process

In this study, the thickness of the top chip was above 700μm. Usually the chip thickness is around 300μm. Thinner chip means less volume to be heated so that rapid increase of the temperature of the interconnection to be bonded can be achievable. We will apply the LABC to the thinner chip, and expect the irradiation time can be reduced less than 3 seconds.

Table 2. SAT images of test vehicles after LABC bonding process before and after PCO

Before PCO			
120W	140W	160W	180W

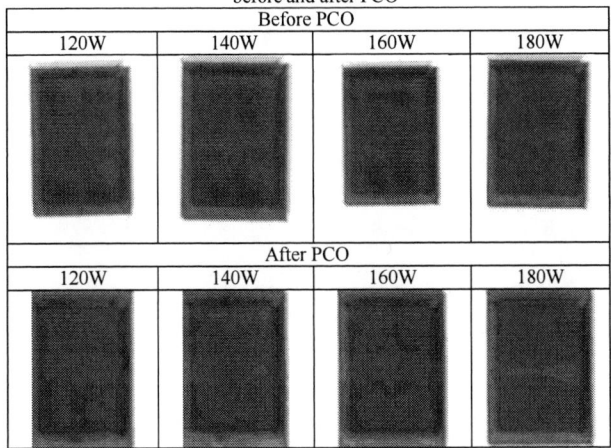

After PCO			
120W	140W	160W	180W

Table 3. SAT images of test vehicles after TCB bonding process

IV. CONCLUSIONS

A LABC bonder and NCF film were developed to increase the productivity of the bonding process. The LABC consists of pre-bonder and post-bonder, and its UPH can reach above 1,000. The NCF was found out it can be applicable to LABC and TCB. The 780μm-thick daisy chain top and bottom chips with the minimum pitch of 30μm and bump number of about 27,000 were successfully bonded using the LABC and NCF film. The irradiation time was 5 second. It can be reduced less than 3 second for the thinner chip. A mechanism was proposed explaining the bonding process during the LABC bonding, featuring the importance of the thermal design of the interconnection to be bonded. With this result of that, the bonding with low laser power can be possible, which will be examined in the future.

ACKNOWLEDGMENT

This work was supported by the Korea Institute of Energy Technology Evaluation and Planning(KETEP) and the Ministry of Trade, Industry & Energy(MOTIE) of the Republic of Korea (No. 10082367, 20183010014310 and 20000352). The authors would like to thank Ji Young Lee for her support in the sample preparation.

REFERENCES

[1] Kazutaka Honda, Akira Nagai, Makoto Satou, Shinsuke Hagiwara, Satoru Tuchida, Hidenori Abe, "NCF for Pre-Applied Process in Higher Density Electronic Package Including 3D-Package," Proc. IEEE Electronic Components and Technology Conf. (ECTC 12), 2012, pp. 385-392.

[2] Toshihisa Nonaka, Yuta Kobayashi, Noboru Asahi, Shoichi Niizeki and Koichi Fujimaru, "High Throughput Thermal Compression NCF Bonding," Proc. IEEE Electronic Components and Technology Conf. (ECTC 14), 2014, pp.913-918.

[3] Kazuyuki Matsumura, Masao Tomikawa, Yohei Sakabe and Yoichi Shiba, "New Non Conductive Film for high productivity process," IEEE CPMT Symposium Japan (ICSJ 15), 2015, pp.19-20.

[4] SeokGeun Ahn, HwanKyu Kim, Dong Wook Kim, David Hiner, KeunSoo Kim, TaeKyeong Hwang, MinJae Lee, DaeByoung Kang, and JuHoon Yoon, "Wafer Level Multi-Chip Gang Bonding Using TCNCF," Proc. IEEE Electronic Components and Technology Conf. (ECTC 16), 2016, pp. 122-127.

[5] Kohei Seyama, Shoji Wada, Yuji Eguchi, Tomonori Nakamura, Doug Day, Shigetoshi Sugawa, "Design and Application of Innovative Multi-Table and Bond Head Drive System on Thermal Compression Bonder with UPH over 2000," Proc. IEEE Electronic Components and Technology Conf. (ECTC 18), 2018, pp. 392-400.

[6] DongSu Ryu, "Advanced Interconnect with Laser Assisted Bonding," Proc. Semicon Taiwan, 2015.

[7] Yanggyoo Jung, Dongsu Ryu, Minho Gim, Choonghoe Kim, Yunseok Song, Jinyoung Kim, Juhoon Yoon, Choonheung Lee, "Development of Next Generation Flip Chip Interconnection Technology using Homogenized Laser-Assisted Bonding," Proc. IEEE Electronic Components and Technology Conf. (ECTC 16), 2016, pp. 392-400.

[8] Tomonori Nakamura, Farhan Shafiq, Tetsuya Otani, Osamu Watanabe, Toru Maeda, Yoshihito Hagiwara, Keiji Honjo, Tamotsu Owada, Daichi Mori, Outa Egashira, Tatsuo Nagamatsu, "Improvement of C2W collective bonding reliability and UPH through innovations in machine, materials and methods," Proc. IEEE Electronic Components and Technology Conf. (ECTC 17), 2017, pp. 108-115.

[9] Kwang-Seong Choi, Wagno Alves Braganca Junior, Keon-Soo Jang, Hyun-Cheol Bae, and Yong-Sung Eom, "Development of Stacking Process for 3D TSV (Through Silicon Via) Structure using Laser," Proc. International Symposium on Microelectronics (IMAPS 17), 2017, pp. 67-71.

[10] Luca Del Carro, Thomas Brunschwiler, Martin Kossatz, Lucas Schnackenberg, Matthias Fettke, Ian Clark, "Laser sintering of dip-based all-copper interconnects," Proc. IEEE Electronic Components and Technology Conf. (ECTC 18), 2018, pp. 279-286.

[11] Luke A. Wentlent, Mohammed Genanu, Thaer Alghoul, "Effects of Laser Selective Reflow on Solder Joint Microstructure and Reliability, " Proc. IEEE Electronic Components and Technology Conf. (ECTC 18), 2018, pp. 425-433.

[12] Kwang-Seong Choi, Wagno Alves Braganca Junior, leeseul Jeong, Keon-Soo Jang, Seok Hwan Moon, Hyun-Cheol Bae, Yong-Sung Eom, Min Kyo Cho, and Seung Il Chang, "Interconnection Process using Laser and Hybrid Underfill for LED Array Module on PET Substrate," Proc. IEEE Electronic Components and Technology Conf. (ECTC 18), 2018, pp. 1561-1567.

[13] Wagno Alves Braganca Junior, Yong-Sung Eom, Keon-Soo Jang, Seok Hwan Moon, Hyun-Cheol Bae, and Kwang-Seong Choi, "Collective Laser-assisted Bonding Process for 3D TSV Integration with NCP, ETRI J. Accepted, 2019.

A Study of 3D Packaging Interconnection Performance affected by Thermal Diffusivity and Pressure Transmission

Jin-San Jung, Hyeong Gi Lee, Ji-Min Kim, Yong-Jin Park, Ji-In Yu, Yong Sung Park, Jun Su Lim, Hyun-Seok Choi, Sung-Il Cho, Dong wook Kim, Sang-Ho An

Package Engineering Team
Samsung Electronics Co., Ltd
Gyeonggi-do, Republic of Korea
Jinsans.jung@samsung.com

Abstract—**3D packaging technology has been considered as one of the best candidates to improve the system performance by implementing high I/O density as well as providing shortest signal channel path with given package form factor. However, it is difficult to uniformly control the bonding thickness and the precisely align the bumps other than thermo compression (TC) bonding to enable 3D packaging. Moreover, high chip cost and possibly low productivity of TC bonding are main business reasons to prevent this attractive technology from prevailing the mass production environment. To address these well-known technical issues of TC bonding, non-conductive film are proposed for especially high vertical stack with small bump pitch and also minimum chip to chip distance required packages such as high bandwidth memory. In this article, we investigated key process parameters to understand how to optimize bonding process to ensure excellent joint quality for highly dense 3D packages products.**

Keywords-Thermo-compresssion bonding; Heat Flow; Multi Post Bonding; Multi Chip Stack Bonding; NCF properies

I. INTRODUCTION

The cutting-edge industries such as an automotive car and artificial intelligence system need to utilize and process a tremendous data in a short time. To provide their service to customers, those industries have consistently required high performing ICs, becoming more severe. To meet their requirement, in recent, the semiconductor companies have developed a structure called as Through Silicon Via (TSV) penetrating silicon chip and producing an electrical connection with the shortest signal path. It enables to achieve high I/O density and high speed data transfer, and several leading companies are competing to develop high performing 3D IC package and dominate the market in advance.

Thermo compression (TC) bonding has been paid attention as a promising technique to produce electrical connections between TSVs in 3D IC package, by adjusting factors such as bonding temperature, duration and pressure level [1-3]. In order to bond a chip, a TC bonding consists of (1) a chip alignment, (2) heating and (3) cooling stages [4]. First, a chip bonder is carefully aligned to the position of a target chip, and starts heating up to form interconnections. Figures 1 and 2 show the shape of a bump before and after TC bonding process, respectively, with a schematic

explanation for TC bonding method. Various conditions and techniques have been attempted to obtain enough margins for risks such as bump short and non-wet [4 -7], and it results in successful bump interconnections under the condition of mass production. However, one of the most critical issues is that the conventional TC bonding is quite time-consuming process as every chip should be bonded layer by layer, which brings about the lack of production capacity.

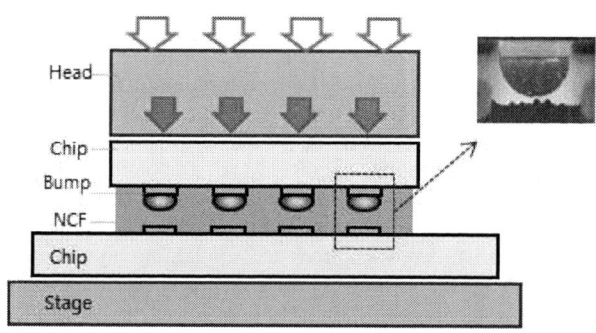

Figure 1. Schematics of the conventoinal TC bonding and bump shape before bonding process

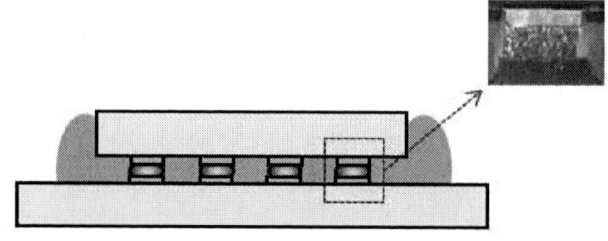

Figure 2. Bump shape after bonding process

This study proposes a multi post bonding (MPB) technique which can dramatically increase the capacity of the conventional TC bonding. Major feature of the proposed MPB is that it consists of multi-pre-bonding steps only with a single post-bonding step. This feature considerably reduces the bonding time, making it possible to increase the production capacity. Even though the proposed MPB has clear advantages in capacity increase, most companies have not adopted it for mass production because of a critical drawback, the temperature differences between top/bottom and center/peripheral regions, which can cause quality issues in interconnection. This study aims to (1) establish physical mechanisms of the proposed MPB, (2) improve joint gap difference between top and bottom chips by controlling non-conductive film (NCF) property, bonding temperature and force, and (3) experimentally validate the proposed mechanism through a real-functioning high-bandwidth-memory (HBM) chip.

II. HEAT TRANSFER SCHEME IN TC BONDING

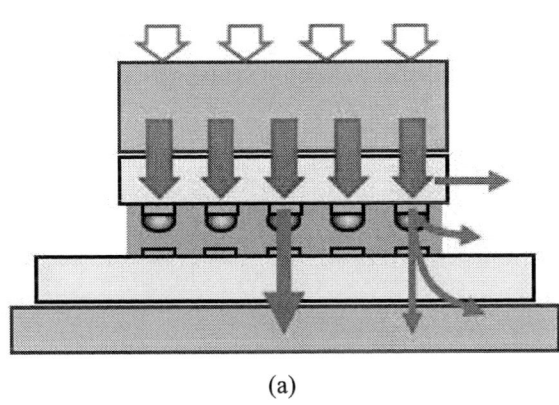

(a)

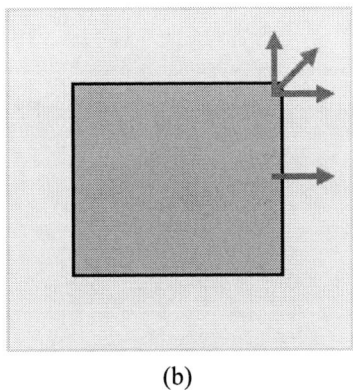

(b)

Figure 3. Heat dissipation paths at different locations in TC bonding: (a) side and (b) top views.

In TC bonding, the provided heat energy from a bonding tool at a top transfers through a chip to a heating stage under a constant temperature. During the heat flow, there are several dissipation paths such as an adjacent chip at the bottom, the heating stage and the surrounding air by convection inside equipment. As depicted in Figures 3 and 4, depending on the location in a chip, the influence of the mentioned heat dissipation factors varies. At a central part, for instance, heat flow is restricted to the downward where the previously-bonded chip is located at. On the other hand, at the corner part, heat flow occurs not only to the downward but also to the side edges where the heat can be readily dissipated to the heating stage, and in addition, the convection from the surrounding air rolls as a dissipation path.

For this reason, temperature variations are produced between the central and corner parts, affecting the bump formation [8]. By adopting a finite element method using commercial software ABAQUS 6.10 Standards, the temperature difference was analytically investigated and Figure 4 shows the results through a temperature contour map at different locations in a chip.

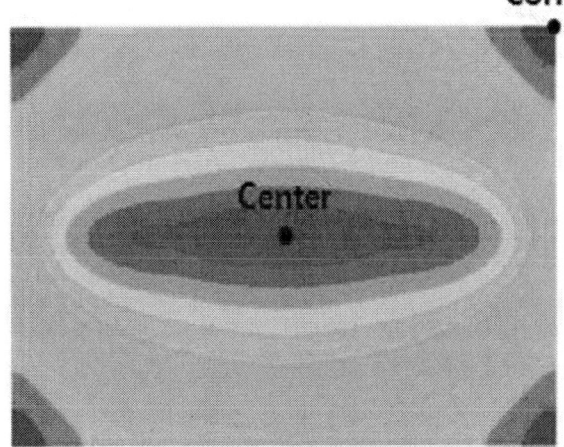

Figure 4. Temperature contour map in a chip acquired using a finite element simulation

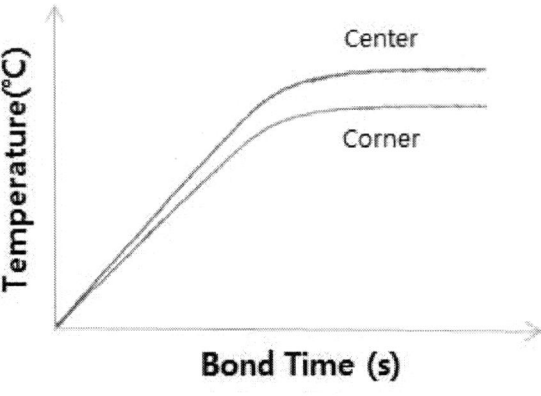

Figure 5. Temperature profile measured at the center of a chip

According to the measured temperature at the central and corner part of a chip, the difference in both the rates of temperature rise and the maximum temperature was observed as shown in the Figure 5. Even though it causes the difference in the joint gap height at the central and corner part, the produced deviation of the joint gap is ignorable without serious quality issues.

III. FUNDAMENTALS OF MULTI POST BONDING (MPB)

TC bonding that applies the heat and pressure to chip for high quality interconnection is the important process to form the fine pitch and high density solder bump joints in 3D packaging technologies. [9] Non-conductive film (NCF) was used as underfill material to fill the fine pitch gap without voids. Therefore, heat and pressure of TC bonding using NCF that was related to NCF flow time should be optimized to achieve the good electrical interconnection. In TC bonding, NCF is used to effectively disperse the stress applied to the chip and prevent the adhesive overflow. NCF can be pre-laminated to chips by applying a little heat and pressure. [10, 11] Current TC bonding requires a lot of time because bonding chips one by one. So TC bonding has been a bottle-neck process in the entire package assembly. It is efficient to divide the bonding process into aligning (pre-bonding) and TC step (post-bonding) to improve the TC bonding productivity. [12, 13] In addition, by applying Multi Post Bonding (MPB) method which performs TC bonding at a time after pre-bonding several chips, the productivity of TC bonding can be maximized. Figure 6 shows the MPB process flow. As shown in Figure 6 (a), NCF-laminated chips were pre-bonded on the wafer. Another chip was stacked on the bonded chip in the same way as shown in Figure 6 (b), (c). In Figure 6 (d), the bumps were precisely interconnected by TC bonding, applying heat and pressure using MPB method. NCF properties and TC bonding conditions should be controlled more carefully than conventional TC bonding to bond the multiple chips at once.

When the several chips were stacked vertically, the temperature difference during the MPB occurs because the main heat source existed on the chips. The influence of the temperature difference is greater using a chip with a high thermal conductivity, such as a Si wafer. [8, 14] Figure 7 shows that there is a temperature difference for each position in the stacked chips. As the distance from the heat source increases, the temperature decreases. Furthermore, the temperature of the chip corner was lower than the center of the chip due to the heat loss to the surroundings.

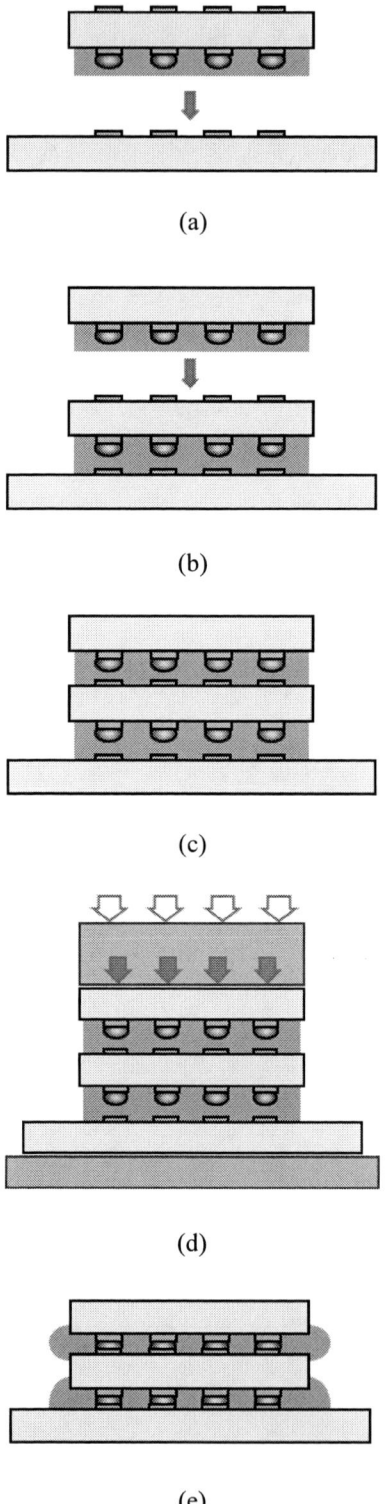

(a)

(b)

(c)

(d)

(e)

Figure 6. Multi post bonding process flow (a) First chip pre bonding, (b) Second chip pre bonding, (c) Structure after second chip pre bonding, (d) Multi Post Bonding (MPB), (e) Structure after MPB

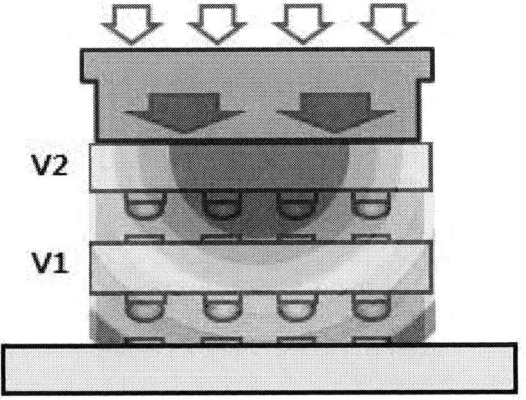

Figure 7. Schematic explanations for temperature difference varying with the location of the stacked chip

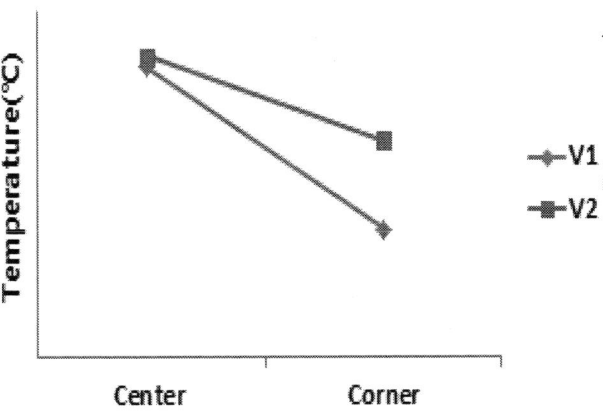

Figure 8. Temperature measured at the center and corner of upper and bottom chip (V1 is bottom chip temperature, V2 is upper chip temperature)

The actual temperature that was measured using a thermocouple inside of the stacked chips was shown in Figure 8. It was confirmed that the temperature difference occurred according to the distance from the heat source and the position in the chip.

The temperature difference between upper chip and bottom chip during multi chip stack bonding makes the joint height gap more intense than conventional single chip bonding. However, current mass production conditions are not a direct problem because the Joint gap height deviation caused by the temperature difference is not large.

IV. EXPERIMENT OF IMPROVEMENT SOLUTION

A. NCF property

NCF viscosity and modulus decreased as the temperature increased by molecular vibration of epoxy chains in NCF. The modulus of NCF greatly affected the pressure dispersion applied to the chip during TC bonding. The viscosity of NCF has a great influence on the difference of joint height [15]. The temperature variation that occurs during TC bonding induces the deviation of the viscosity and causes the joint height difference. Therefore, by controlling the viscosity level of NCF itself, it is possible to control the difference of joint height. Figure 9 shows the difference in joint height between upper chip and bottom chip according to NCF viscosity level. Adjusting the NCF viscosity reduced the difference of joint height up to 50%.

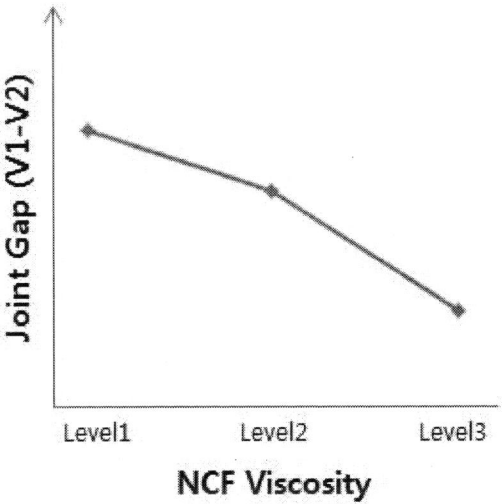

Figure 9. Difference of corner bump joint height between upper and bottom chip according to NCF viscosity

B. Bonding Temperature

In the MPB process, the temperature applied to the bottom chip is lowered than the upper chip and the solder is melted relatively later, which causes a difference in joint height between the upper and bottom chip. Rapid melting of the solder on the bottom chip will reduce the difference in joint height. Increasing the bonding temperature can increase the time that solder of the bottom chip is melted. Reducing the time difference between the upper and bottom chip solder melting can decrease the difference in joint height. Figure 10 shows the difference in joint height between upper and bottom chip according to the bonding temperature.

978-1-7281-1500-9/19 $31.00 © 2019 IEEE   207

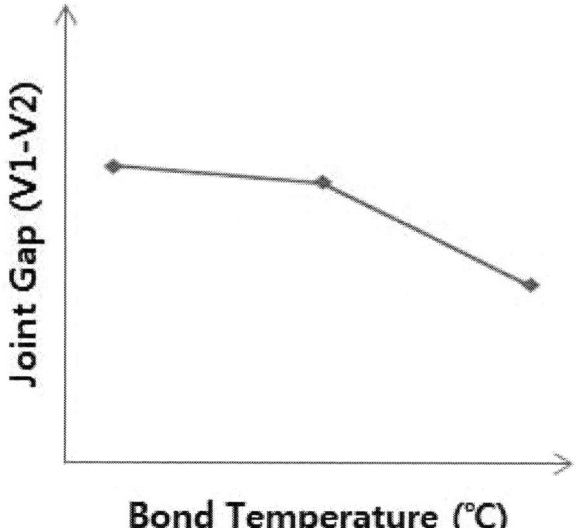

Figure 10. Difference of corner bump joint height between upper chip and lower chip according to bonding temperature

V. CONCLUSION

A Multi Post Bonding (MPB) technique has been taken attention as one of the most promising techniques to increase the thermo-compression (TC) bonding productivity of high-bandwidth-memory (HBM). In spite of clear advantages of the multi post bonding technique, most companies have not applied it for mass production due to critical risks such as poor solder wetting by the joint gap difference between adjacent chips.

This paper discussed the fundamentals of a conventional TC bonding, and defined the joint gap difference issue caused by temperature difference in a chip. It was observed that the difference becomes more severe between the upper and lower chips in MPB, due to the difference in temperature applied to upper and lower chip. To solve this problem, the NCF viscosity was adjusted to distribute the pressure to bumps evenly, and the bonding temperature was raised to achieve a simultaneous melting of the solders at the upper and lower chips. It is expected that the MPB technique can replace a time-consuming conventional TC bonding in the near future.

REFERENCES

[1] P. Gagnon, C. Bergeron, R. Langlois, S. Barbeau, S. Whitehead, C. Tyberg, R. Robertazzi, K. Sakuma, M. Wordeman, and M. Scheuermann, "Thermo-compression bonding and mass reflow assembly processes of 3D logic die stacks", Proc. 2017 IEEE 67th Electronics Components and Technology Conference (ECTC), IEEE Press, 2017, pp. 116-122, doi: DOI 10.1109/ECTC.2017.53

[2] S.G. Ahn, H.K. Kim, D.W. Kim, D. Hiner, K.S. Kim, and T.K. Hwang, "Wafer Level Multi-Chip Gang Bonding Using TCNCF", Proc. 2016 IEEE 66th Electronics Components and Technology Conference (ECTC), IEEE Press, 2016, pp. 122-127, doi: DOI 10.1109/ECTC.2016.171

[3] P. Bex, T. Wang, M. Lofrano, V. Cherman, G. Capuz, E. Sleeckx, and E. Beyne, "Thermal Compression Bonding: Understanding Heat Transfer by In-situ Measurements and Modeling", Proc. 2017 IEEE 67th Electronics Components and Technology Conference (ECTC), IEEE Press, 2017, pp. 392-398, doi: DOI 10.1109/ECTC.2017.49

[4] H.G. Lee, Y.W. Choi, J.W. Shin and K.W. Paik, "Effects of Thermocompression Bonding Parameters on Cu Pillar/Sn-Ag Microbump Solder Joint Morphology using Nonconductive Films", IEEE Trans. Compon., Packag., Manuf. Technol., vol. 7, 2017, pp. 450-455, doi: 10.1109/TCPMT.2016.2641040

[5] S.T. Lu, J.Y. Juang, H.C. Cheng, Y.M. Tsai, T.H. Chen, and W.H. Chen, "Effects of bonding parameters on the reliability of finepitch Cu/Ni/SnAg micro-bump chip-to-chip interconnection for threedimensional chip stacking", IEEE Trans. Device Mater. Rel., vol. 12, Jun. 2012, pp. 296–305, doi: 10.1109/TDMR.2012.2187449.

[6] C.J. Zhan, C.C. Chuang, J.Y. Juang, S.T. Lu and T.C. Chang, "Assembly and Reliability Characterization of 3D Chip Stacking with 30 m Pitch Lead-Free Solder Micro Bump Interconnection", Proc. 2010 Proceedings 60th Electronic Components and Technology Conference (ECTC), IEEE Press, June 2010, pp. 1043-1049, doi: 10.1109/ECTC.2010.5490829

[7] A.B.Y. Lim, A. Rezvani, R. D. Bacay, T. Colosimo, O. Yauw, H. Clauberg and B. Chylak, "High Throughput Thermo-compression Bonding with Pre-applied Underfill for 3D Memory Applications," Proc. 2016 IEEE 18th Electronics Packaging Technology Conference (EPTC), IEEE Press, Dec. 2016, pp. 427-434, doi: 10.1109/EPTC.2016.7861515.

[8] N. Aasahi and M. Nimura, "Heat Transfer Analysis in the Thermal Compression Bonding for CoW Process," Proc. International Conference on Electronics Packaging (ICEP 2016), IEEE Press, April 2016, pp. 640-643, doi: 10.1109/ICEP.2016.7486908

[9] Y. M. Lin, C. J. Zhan, K. S. Kao, C.W. Fan, S. C. Chung, Y. W. Huang, S. Y. Huang, J. Y. Chang, T. F. Yang, J. H. Lau and T. H. Chen, "Low temperature bonding using non-conductive adhesive for 3D chip stacking with 30μm-pitch micro solder bump interconnections", Proc. 2012 IEEE 62nd Electronic Components and Technology Conference (ECTC), July 2012, pp. 1656-1661, doi: 10.1109/ECTC.2012.6249060.

[10] T. Nagamatsu, K. Honjo, K. Ebisawa, T. Ishimatsu, T. Saito, D. Mori, D. Motomura and H. Yagi, "Use of Non-conductive Film (NCF) with Nano-Sized Filler Particles for Solder Interconnect: Research and Development on NCF Material and Process Characterization" Proc. 2016 Proceedings 66th Electronic Components and Technology Conference (ECTC), IEEE Press, August 2016, pp. 923-928, doi: 10.1109/ECTC.2016.224.

[11] N. Asahi, Y. Miyamoto, M. Nimura, Y. Mizutani and Y, Arai, "High productivity thermal compression bonding for 3D-IC" proc. 2015 International 3D Systems Integration Conference (3DIC), IEEE Press, November 2015, pp. 129-133, doi: 10.1109/3DIC.2015.7334577.

[12] T. Nonaga, Y. Kobayashi, N. Asahi, S. Niizeki, K. Fujimaru and Y. Arai, "High throughput thermal compression NCF bonding", proc. 2014 IEEE 64th Electronic Components and Technology Conference (ECTC), IEEE Press, Semptember 2014, pp. 913-918, doi: 10.1109/ECTC.2014.6897396.

[13] H. Clauberg, A. Marte, Y. Yang, J. Eder, T. Colosimo, D. Buergi, A. Rezvani and B. Chylak, "High Productivity Thermocompression Flip Chip Bonding", Proc. 2015 IEEE 65th Electronic Components and Technology Conference (ECTC), July 2015, pp. 22-29, doi: 10.1109/ECTC.2015.7159566.

[14] R.Daily, G. Capuz, T. Wang, P. Bex, H. Struyf, E. Sleeckx, C. Demeurisse, A. Attard, W. Eberharter, and H. Klingler, "3D IC Process Development for Enabling Chip-on-Chip and Chip on Wafer

978-1-7281-1500-9/19 $31.00 © 2019 IEEE

Multi-Stacking at assembly", Proceedings of International Conference on Electronics Packaging (ICEP), April 2015, pp. 56-60. doi: 10.1109/ICEP-IAAC.2015.7111000

[15] H.M. Lee, S.Y. Lee, J.H. Park, C.K. Chung, K.W. Jang, I. Kim, S.W. Choi and K.W. Paik, "A Study on the Curing properties and Viscosities of Non-Conductive Films (NCFs) for Sn-Ag Solder Bump Flip Chip Assembly", Proc. 2018 IEEE 68th Electronics Packaging Technology Conference (EPTC), IEEE Press, June 2018, pp. 2464-2469, doi: 10.1109/ECTC.2018.00371.

Vertical laser assisted bonding for advanced "3.5D" chip packaging

Andrej Kolbasow, Timo Kubsch, Matthias Fettke, Georg Friedrich and Thorsten Teutsch

PacTech GmbH

Am Schlangenhorst 7-9, 14641 Nauen, Germany

Phone: +49 (0)3321 4495-504, Fax: +49 (0)33214495-110, E-Mail: fettke@pactech.de

Abstract — **In this work the processes of laser assisted bonding (LAB) is compared to thermal compression bonding (TCB). Their respective advantages and disadvantages regarding the assembly of flip chip stacks are compared. It is found, that the LAB allows for faster processing, negligible compression force and creates less internal stress in the chip stack. The concept of "3.5D" stacking is introduced. This new concept allows for the vertical bonding of chips/semiconductors to the sides of a chip stack. The vertically bonded parts can be used to contact the individual layers, which eliminates the necessity for through silicon vias (TSVs).**

Keywords - 3D-packaging, Laser assisted bonding (LAB), Thermal compression bonding (TCB), Silicon interposer, System on Package (SOP), Laser beam modulation, Inter metallic phase (IMC-layer), vertical Flip Chip bonding

I. INTRODUCTION

Growing performance, further miniaturization and increasing system density are major technical drivers for the global semiconductor market to improve and develop new chip-designs and packaging-concepts. 3D-IC and 2.5D TSV (through silicon via) packaging technologies are common concepts of tackling this challenge by manufacturing multi-layer packages such as HBM (Hight Bandwidth Memory) and HMC (Hybrid Memory Cube). Caused by the continuous demand of increased performance, density of interconnects such as TSVs, micro bumps and Cu pillars between and inside of each layer increases significantly. Parasitic effects such as capacitance, inductance, resistance and EMI (electromagnetic interference) compatibility bring daily challenges for developing new packaging concepts, designs and manufacturing technologies [1]. Beside design and concept challenges, economic and reliable manufacturing technologies play an important role by achieving new technological standards. In this study we will introduce new, economic and reliable concepts, to bond vertical semiconductors against a chip package at all four sides to build a new "3.5D" chip package (see Figure 1).

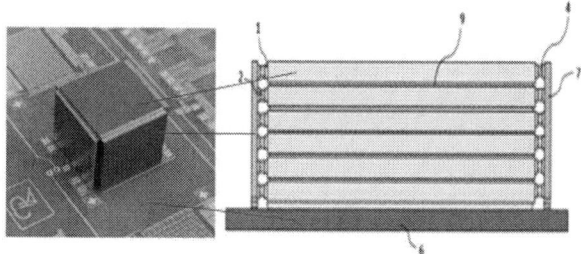

Figure 1: Stack with vertically bonded semiconductor devices.

The major goal is to redistribute TSV structures to the die edges and realize the layer interconnects via vertically bonded chips.

The presented assembly technology enables the possibility to use interconnects at the die edges or top / bottom combination and gives future designers the possibility to reduce or eliminate interconnect density from die main area and move it to the package edges. Die layouts can be simplified and parasitic effects of interconnects minimized. In this study, metrology comparison of generated interconnects by TCB and LAB process will be shown. Shear-tests, cross-section, X-ray, EDX and thermal aging analyses will provide reliability data for further discussions. Finally, the concept of "3.5D" stacking will be outlined and first assembly results presented.

II. MATERIALS AND TEST COMPONENTS

For the performance characterization of the two different bond technologies discussed in this work the test materials presented in Table 1 were selected for qualification tests.

	Description	Picture	Specification
Sample A	Si- interposer		Size: 14mm x 14mm x 110µm UBM: ENIG Type: interposer for SOP applications
Sample B	PCB		Size: 20mm x 20mm x 1mm UBM: ENIG Type: Multilayer PCB
Sample C	Si Pac Tech test die		Size:14mm x 14mm x 110µm UBM: ENIG Type: Pac Tech Silicon Test Die
Sample D	Si Pac Tech test die		Size:14mm x 14mm x 110µm UBM: ENIG Type: Pac Tech multi project PCB board
Material A	MEMS		Size: 0.3mm X 0.3mm X 190µm UBM: ENIG Type: MEMS
Material B	SAC 305 Solder balls 250µm - Solder preform		Size: 350µm +/- 5µm Melting point: 493.15K Alloy: Sn96.5% Ag3% Cu0.5%

Table 1: Substrate and material overview.

III. PROCESS DESCRIPTION OF CHIP BONDING TECHNOLOGIES

A 3D-axis system, a bond chuck and a bond tool are a bonding system's central basic components. The common goal is the realization of a reliable interconnection between two substrates. The main difference between laser assisted bonding and thermal compression bonding is the mechanism of inducing the required energy into the devices for sufficient soldering.

Before the bonding sequence is started, the daughter substrate is picked up by the bonding unit via vacuum, optically measured and finally aligned to the mother substrate located on the bond chuck.

A. General process flow of PacTech's LAB system (laser assisted bonding)

A modulated near infrared laser beam is used to heat up the daughter substrate to be placed on the mother substrate. At its bottom, the daughter substrate features a set of pre-soldered interconnects, which are wetted with flux (see Figure 2).

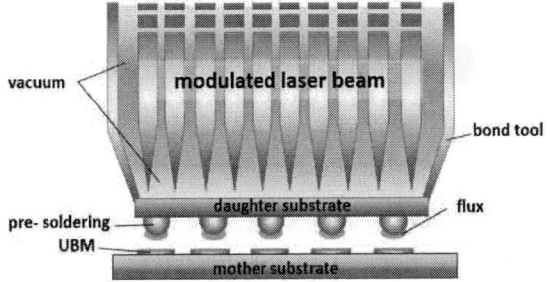

Figure 2: Basic process setting of PacTech`s LAB bonding process.

In this particular case, the beam modulation is performed in two steps. Step one is the transformation of the initial gaussian profile into a top-hat profile. In step two, the beam is resized into the desired shape by means of an adjustable, rectangular aperture (see Table 2).

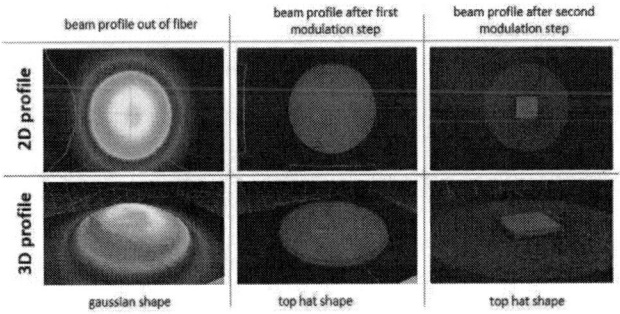

Table 2: Basic illustration of laser beam modulation.

Size, shape and energetic homogeneity of the laser beam are depending on the optical configuration of the used lens system. The modulated laser beam passes through the aperture of the ceramic bonding tool. During the bonding process, a force and IR sensor monitors and records the process conditions in order to keep the process parameters in the defined process range. A positive thermal gradient over time of up to 1773.0K/s within a ±0.3K range can be achieved based on the component size and material. Before the component to be placed touches the mother substrate, the solder is liquefied by the energy thermally induced by the laser beam. Finally, the component is placed on the mother substrate and bonded. No significant force is needed, since the solder is liquefied before touchdown [2, 3].

B. General process flow setting of a TCB system (thermal compression bonding)

An electrical coil is used to heat a ceramic or metallic bonding tool. The principle configuration of a TCB bonding setup is illustrated in Figure 3.

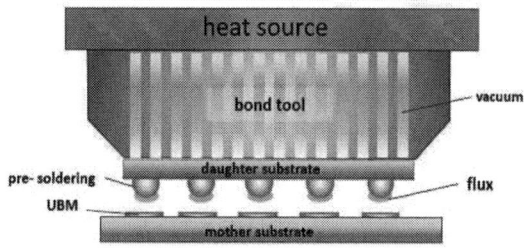

Figure 3: Process setting of TCB bonding unit.

The heat distributes throughout the tool, which then heats the component to be placed. The quality of the heat distribution depends on the tool design, size and material. During the approach, the component is kept below the melting temperature of the solder. A force sensor is used to detect the touchdown at substrate surface before increasing the heat to liquefy the solder that is already in contact with the mother substrate. Heat ramping speeds of about 473K/s are possible. Due to thermal losses at the heated bond tool, an offset calibration is continuously necessary before and while bonding [8,9,10] .

IV. COMPARISON BETWEEN THERMO COMPRESSION BONDING AND LASER ASSISTED BONDING

A. *Investigation of possible warpage effect by bonding Si-interposers on PCB substrate*

In order to analyze and compare possible warpage effects between the TCB and the LAB process, three sets of samples have been assembled for each process (see Table 3). Afterwards samples have been measured using a "Keyence VR 3000" 3D-Profilometer. The results were confirmed using a Keyence LK-G3000 laser sensor. Table 3 describes the test configuration that has been used for sample making. As shown, the main bonding parameters are significantly different between both processes.

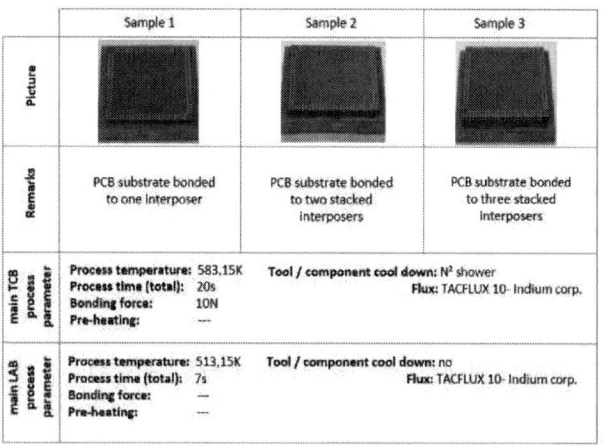

Table 3: Test description and specification for substrate warpage evaluation.

The most drastic differences are found in the bonding force, the peak temperature and the process duration. All samples have been pre-soldered by PacTech's SB² (Solder Ball Bumping) solder jet process [4]. Spherical solder preforms with a diameter of 350µm of an alloy of SAC305 (Sn 96.5%, Ag 3.0%, Cu 0.5%) have been used to generate proper solder

depots on the interposer pads. On the 190µm octagonal pads a mean solder bump height of 300µm was measured after solder jetting. Table 4 shows solder height deviation after pre-soldering process by SB² on interposer.

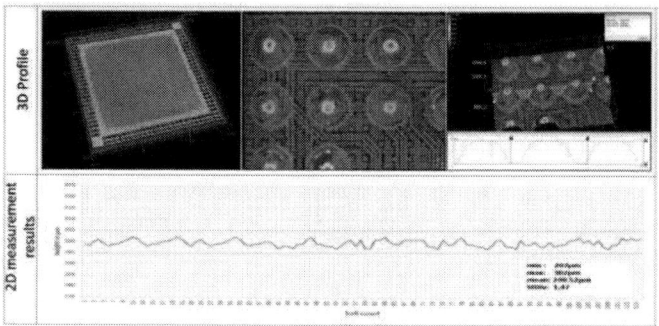

Table 4: Solder height measurement results by 3D profiling and 2D measurement.

After the pre-soldering process, three samples for each of the bonding technologies have been prepared. Figure 4 and 5 show the observed thermal energy profile during the bonding process.

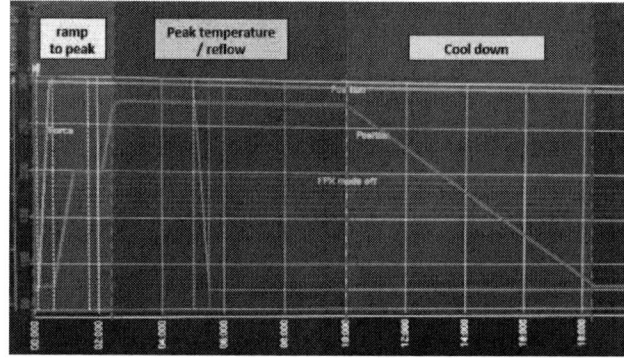

Figure 4: Thermal energy profile during the TCB process recorded by temperature sensor at the tool heating unit.

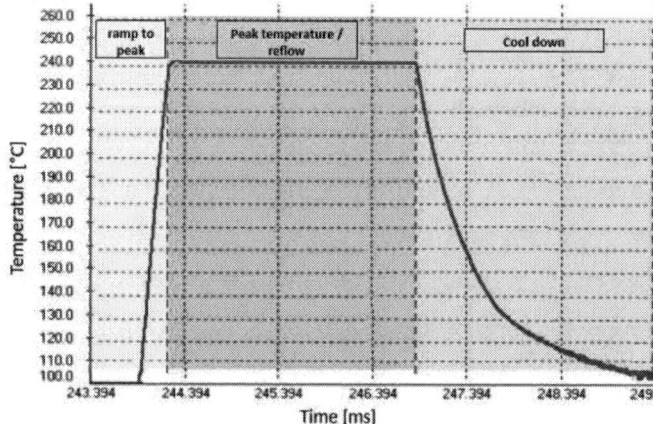

Figure 5: Thermal energy profile during PacTech's LAB process recorded by IR-sensor on device surface.

As shown in Figure 4 and 5, there are major differences in time, ramping speed and cooling characteristic. This is mainly caused by the fashion in which the interposer is heated. During the TCB bonding, the system needs to heat up the mass of the bonding tool in order to heat the interposer. In the LAB process, the heat is directly created at the interposer surface via interaction with the laser radiation. As described in Table 3, in contrast to the TCB process, the LAB system requires no N_2 for rapid cooling since there is no significant mass to cool down.

It is also to be noted, that the thermal profile displayed in Figure 5 has been measured at the top of the bonding tool. The usual static temperature offset between the top and the bottom of the tool has been taken into account ($\approx 67K$ in the process at hand). The temporary drop in temperature as the warm tool touches the cold sample, can however not be accounted for as the heat transmission through the tool is too slow. In contrast, during the LAB process, the temperature is measured optically, directly at the sample.

A common problem in chip stacking is internal stress and strain in the stack. The stress is created during the bonding process. It is caused by thermal expansion and shrinkage of the component before and after mechanically fixing it to the underlying stack. A less than ideal thermal profile of the process may also negatively influence the amount of internal stress.

The stress can, in combination with mechanical influences such as vibration, lead to the breaking of solder bonds and/or the components in the stack. This can occur at any point in the device's lifetime and needs to be omitted. To investigate the internal stress of a component, its warpage is analyzed.

In order to quantify the warpage, height maps of the top interposer's surface have been taken using the above-mentioned 3D profilometer. The results have then been re-gauged, to let the four corners for each interposer be at height 0. Then the point of maximum elevation was determined, and its height difference to the corners measured.

The Tables 5 and 6 give an overview of the measured warpage in the assembled samples ordered by their number of layers and their bonding method.

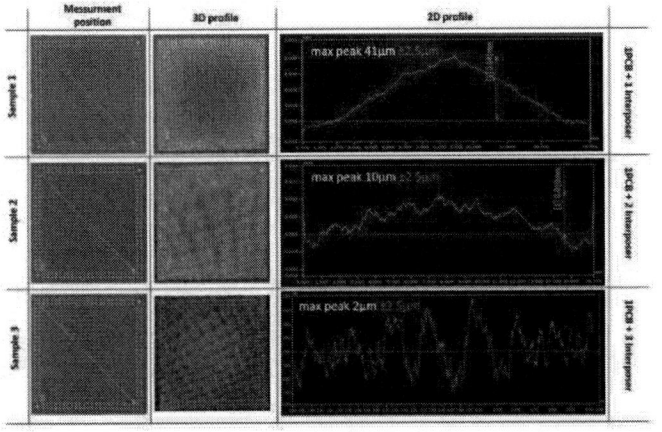

Table 5: Warpage measurement of TCB bonded samples.

For the TCB process a clear surface warpage is identifiable, which shrinks with a growing number of layers in the stack. In the first layer bonded to the substrate, a significant warpage can be seen (see Table 5). While the edges of the chip are fixed to the underlying substrate, its center bulges up by about 40μm

in a rotationally symmetric shape. This clearly points at internal stress in the sample and interconnects. In the second layer, the difference is less pronounced and in the third, no significant warpage has been found. As more and more layers are added to the stack, their influences may cancel each other out mechanically, which would lead to the reduction of warpage in the stack. Another explanation may be, that repeated thermal cycling due to the adding of further layers relaxes the components and therefore the stack as a whole. This form of thermal annealing would not only reduce the measurable warpage but also the internal stress causing it.

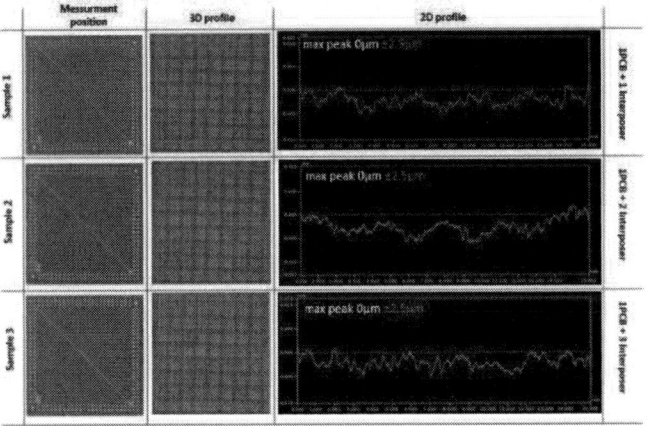

Table 6: Warpage measurement of LAB bonded samples.

In the samples, created using the LAB process, no warpage larger than the measurement uncertainty of ±2.5μm could be observed. To confirm this observation, another set of single layer samples was produced using the LAB and TCB process under identical conditions. This second set was measured as well and shows the same warpage behavior.

B. Comparison of IMC (intermetallic compound) layer characteristic between TCB and LAB process

In the following an analysis about the formation and aging of the IMC layers, which form during the TCB and LAB process, is discussed.

IMC formation is an essential requirement for a stable and reliable electrical and mechanical interconnection [5]. However, as IMCs age, they grow and become more brittle, which can cause a variety of problems such as cracks, delamination and reduced conductivity [6].

The resilience of the IMCs created via the different bonding processes against thermal aging is investigated. Therefore, in addition to creating the samples, some of them were also exposed to a temperature cycle ranging from -40°C to 125°C over a duration of 35min. To simulate the aging of the part during its lifetime, each sample underwent the temperature cycle 200 times. This is to provoke the formation of weak spots such as micro cracks or bump lift.

The respective results are discussed below. Table 7 displays an overview over the cross sections of bumps generated via TCB and LAB with a special focus on the IMC layer.

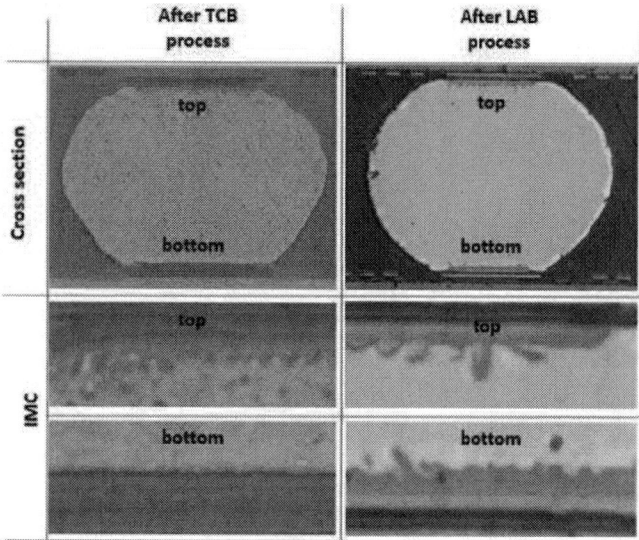

Table 7: Cross-sections of IMC layers of TCB generated bonds.

As can be seen in Figure 6 and 7, two different IMC layers have formed during the TCB process. This is because, in the case of the upper interface, the pre-soldering has been realized using a laser implicit process (PacTech's SB² process). The lower interface was created by the respective bonding process.

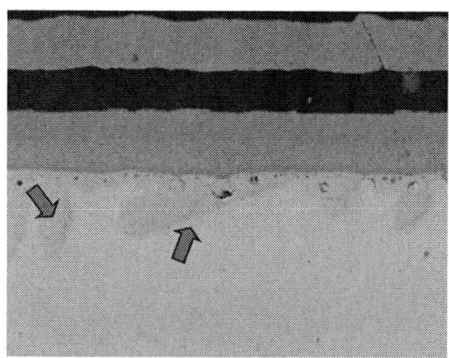

Figure 6: IMC layer of bonds generated by SB² and TCB process (top).

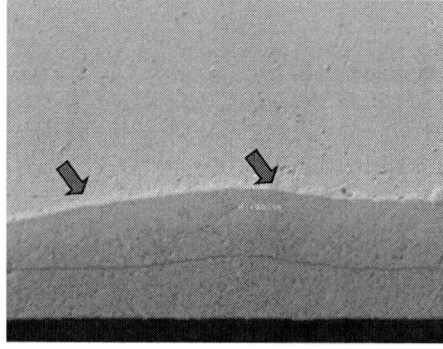

Figure 7: IMC layer of 0.9µm generated by TCB bond process before thermo cycling. (bottom).

In comparison, the IMC produced by only the TCB process are thinner than that with only the SB² process or the SB² and TCB processes combined. The TCB IMCs form a smooth layer, while the ones created by SB² and TCB feature a more acicular structure.

The Figure 7 shows an SEM image of the only TCB IMC after creation and Figure 8 after 200 thermo cycles.

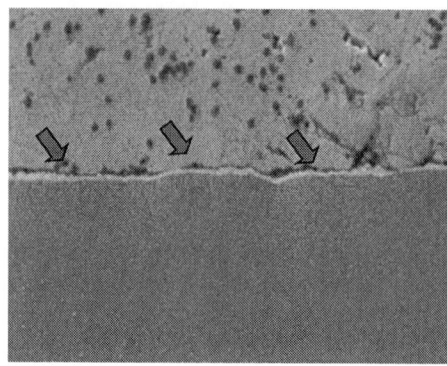

Figure 8: IMC layer of bond generated by TCB process after thermo cycling with micro cracks.

Interestingly, the IMC layers created by TCB have grown significantly more during the thermo cycles than those using the laser implicit process for pre-soldering. While the layer, generated by the only TCB process has grown during the cycling as shown at Figure 9 the IMC layer thickness doubles, the SB² and TCB layer barely changed as shown in Figure 10.

Figure 9: IMC layer of 1.9µm generated by TCB bond process after thermo.

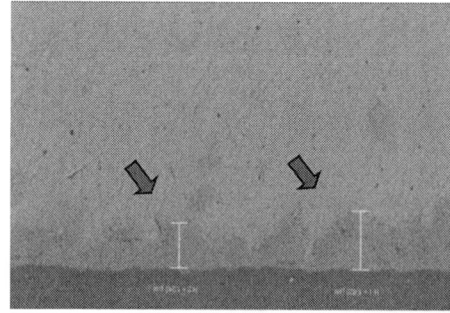

Figure 10: IMC layer of SB² + TCB bond process generated after thermo cycling.

As can be seen in Figures 11 and 12, the IMC's behavior is similar for samples featuring the LAB process with and without the laser implicit pre-soldering before thermal cycling.

Figure 11: IMC layer of 1.9μm generated by LAB bond process before thermo cycling.

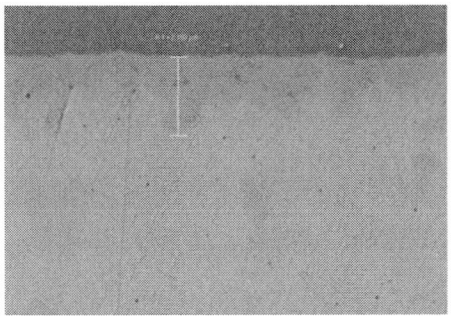

Figure 12: IMC layer of 2.2μm generated by SB² + LAB bond process generated before thermo cycling.

The following thermo cycle test had no negative impact either as shown in Figure 13. No signs of micro cracks or other defects have been found as in the TCB processed ones shown in Figure 8. However, despite these promising results, more reliability tests need to be performed to further investigate the IMC layer created in the LAB process. Based on the data at hand, it can already be assumed, that the LAB process generates a sufficient and resilient IMC layer.

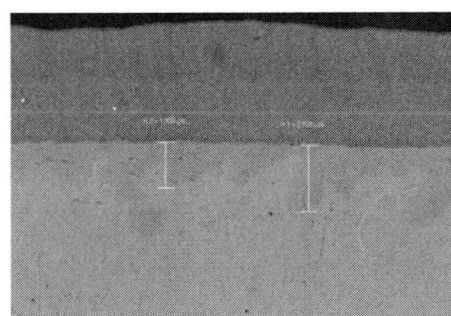

Figure 13: IMC layer of SB² + LAB bond process generated after thermo cycling.

V. CONCEPT OF "3.5D" STACKING BY USING PACTECH'S LAB PROCESS

A challenge, besides the prevention of heat spread and parasitic effects inside a complex 3D package with up to 32 layers is the production of reliable TSV interconnects. Most of the layer interconnections of 3D chip stacks are realized by TSV technologies or wire bonding processes. Both ways are cost intensive and include up to 320 process steps (masking, etching, sputtering, etc.). The risk of quality rejects increases with the number of layers and the depth of TSV structures. The realization of a vertical bonding technology for placing active or passive semiconductor elements like interposer devices has the potential to overcome these limitations. "3.5D" stacking technology allows for the reduction or complete elimination of the described challenges as the layers in the stack are contacted using the vertically bonded components and TSVs become obsolete. This does not only create more space on each individual layer, it also allows for more a dense stacking of the layers. All 4 sides of a chip-stack can be used for this vertical placement and bonding.

The prerequisite for the vertical connection of a functional group to a chip stack or chip package is the presence of lateral contact surfaces. These are to be considered and produced in the design and manufacturing of the microchip. Ideally, all of the otherwise on-surface contacts of a microchip may be routed to the side surface. This is extremely advantageous in further reducing the assembly height, since the contact surfaces, built-up of pillars or solder bumps, can be omitted.

The thickness of the stacked chips as well as the minimal pitch of the vertically bonded component are then the limiting factors for the clearance between the layers in the stack.

If this approach is projected onto a wafer production chain, TSV structures are etched along the chip edges in a first step. These are metallized together with the contact conductor tracks, thus creating the contact pad in the form of a TSV.

The chip stack is then fixed to the carrier substrate while the required heat is provided via laser from the bottom of the substrate. The illustrated LAB process, optimized for soldering the stack as a whole to the substrate, is particularly suitable for this step.

The wafers are bonded to each other after thinning and mated using a thin film to form stacks with an arbitrary number of layers. Subsequently, the wafer stack is sawed along the TSV structures and the chip stacks are separated. These stacks can now be connected to a vertical functional module (see Figure 14).

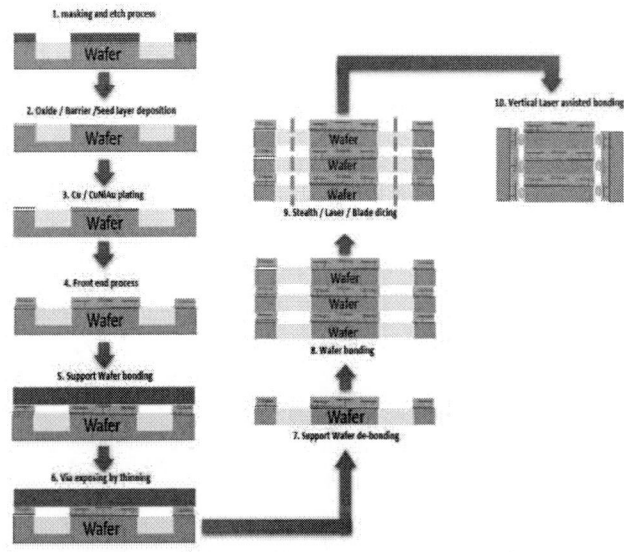

Figure 14: Possible process chain for vertical laser assisted bonding [7].

978-1-7281-1500-9/19 $31.00 © 2019 IEEE

Like in the sections II and III, where the principles of the LAB and the TCB process have been explained respectively, the concept of the vertical chip bonding will be outlined in this section using a process example. Figure 15 shows a schematic representation of the process.

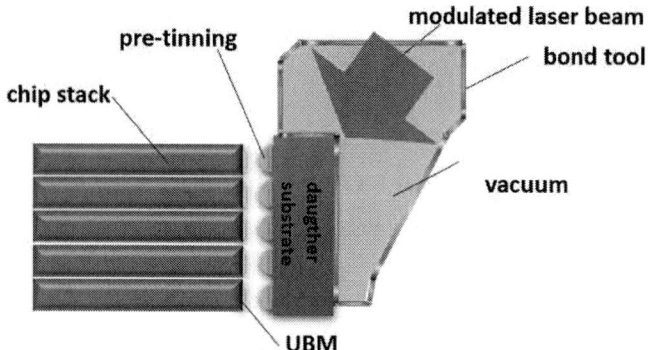

Figure 15: Schematic representation of the vertical chip bonding process.

For the initial assembly tests of vertically bonded components at the sides of a "3.5D" chip stack, the LAB process is used. This is mainly because it has a smaller thermal impact on the assembly as a whole. The bonding system performing the task is PacTech's "Laplace-Bonder". The optic imaging system, as well as the tooling, have been redesigned and adapted to fulfill the requirements of a 45° orientation. A special transfer station allows for the handover of the horizontally stored chip to the 45° tilted tool. A vacuum is created inside the tool to fix the chip to its bottom. After determining the reference positions, the axis system moves the tool to the height of the carrier substrate with a remaining distance of a few micron between the chip and the placed stack. Prior to the bonding process, the chip had been prepared with solder depots of 80μm size via solder jetting. In analogy to the LAB process described in section III the chip is vertically bonded to the side of the chip stack. During the bonding, the laser hits the chip's surface at an incident angle of 45°. The result of the initial bonding tests is displayed in Figure 16 and 17.

Figure 16: Chip stack with vertically bonded components at each side with dimensions of 12mm x 12mm x 12mm

Figure 17: "3.5D" MEMS stack with dimensions of 0.9mm x 0.9mm x 0.9mm.

VI. SUMMARY & OUTLOOK

In this work, the TCB and LAB processes have been compared to each other regarding their properties and functionalities. It was found, that they significantly differ in important process parameters such as the thermal heat gradient over time and the overall process duration. In all aspects investigated, the performance, flexibility and speed of the LAB process either matched or surpassed that of the TCB process. Regarding the formation of the IMC layers and their resilience against aging as well as the avoidance of internal stress in a chip stack, the LAB process produced significantly better results than the TCB process. Nevertheless, the TCB process may require further process parameter optimization to insure proper IMC formation.

The concept of "3.5D" stacking was introduced, outlined and explained. It allows for vertical chip bonding, a technique, with which a microelectronic component can be vertically bonded to the side of an existing chip stack. All four sides of a chip stack can be contacted to generate a 3.5D package. This vertical chip bonding can not be achieved through the traditional TCB bonding process. It can, however, be performed using the LAB process described in this work.

In the future, "3.5D" stacking will make it possible to contact the individual layers in a chip stack via vertically bonded components and greatly reduce, if not eliminate the need for TSVs. This allows for taller stacks with more functionality, as conventional stacks need to reserve chip space in the lower layers for TSVs to contact the upper layers. Further, the persisting problem of heat dissipation is addressed, as the current bearing and therefore heat producing contacts are moved to the edges of the stack, where they can be cooled more easily. Finally, as chip designers embrace the possibilities of this new tool of manufacturing, completely new designs will become possible.

VII. REFERENCES

[1] Santosh Kumar. *3D-IC and 2.5D TSV Interconnect for Advanced Packaging: 2016 Business Update* (Sep. 2016) Yole Development.

[2] Thorsten Teutsch et al. *LAPLACE-A New Assembly Method using Laser Heating for Ultra Fine Pitch Devices* (Jan. 2003) researchgate.net.

[3] Thomas Oppert. *Flip Chip Processes – Electroless UBM, Wafer Level Solder Sphere Transfer, Laser Solder Jetting & Laser Chip Bonding* (Nov. 2018) IMAPS UK Die Attach Workshop.

[4] Thomas Oppert. *Implementing laser heating for next generation packaging mass productions and beyond* (Sep. 2017) 1Executive Forum on Laser Technologies.

[5] Peter Kojo Bernasko. *Study of Intermetallic Compound Layer Formation, Growth and Evaluation of Shear Strength of Lead-Free Solder Joints* (2012) University of Greenwich. pp.31-32.

[6] Beáta Šimeková et al. *Growth of the IMC at the interface of SnAgCuBi (Bi = 0,5; 1,0) solder joints with Cu substrate* (2012) Tehnicki Vjesnik. nr.19, pp.107-110.

[7] Daniel Lu et al. *Materials for Advanced Packaging* (2009) Springer Nature.

[8] V. Jadhav et al. *Flip chip assembly challenges using high density, thin core carriers* (2005) Proc. 55th Electronic Components and Technology Conference. pp.314,319.

[9] Jie Li Aw et al. *Development of bonding process for high density fine pitch micro bump interconnections with wafer level underfill for 3D applications* (2013) Proc 63th Electronics Packaging Technology Conference. pp.543-548.

[10] T. Colosimo et al. *High Productivity Thermal-Compression Flip Chip Bonding* (Oct. 2014) In Proc. International Microelectronics Assembly and Packaging Society. pp.100-106

Optimization of a BEOL Aluminum Deposition Process Enabling Wafer Level Al-Al Thermo-Compression Bonding

S. Schulze[1], M. Wietstruck[1], M. Fraschke[1], Peter Kerepesi[3], Helmut Kurz[3], Bernhard Rebhan[3], M. Kaynak[1,2]

[1]IHP, Im Technologiepark 25, 15236 Frankfurt (Oder), Germany
Email: schulze@ihp-microelectronics.com
[2]Sabanci University, Electronics Engineering, 34956 Tuzla, Istanbul
[3]EV Group, DI E. Thallner Straße 1, St. Florian/Inn 4782, Austria

Abstract—The main challenges for Al-Al wafer bonding are the fast oxidation and the high roughness of the Al surface. This paper describes an optimized Al sputter-deposition process reducing the surface roughness to values below 2 nm. Based on this, a wafer level Al-Al thermo-compression bonding process is presented, where a surface treatment and the subsequent bonding are both performed in a high vacuum cluster. The patterned wafers were bonded with temperatures between 300 and 500 °C for 1 h using a bonding force of 60 kN. Scanning acoustic microscopy and transmission electron microscopy studies revealed a reliable bonding, an accurate alignment and a high uniformity for 200 mm wafers. The electrical characterization of contact chains with Al bonding pad sizes of 20×20 µm² showed resistances lower than 50 mΩ per contact, which might indicate areas of oxide-free bonding.

Keywords- Aluminum; Al surface roughness, Al-Al thermocompression bonding; wafer level bonding

I. INTRODUCTION

Metal-to-metal thermo-compression bonding is known as one of the key technology for achieving wafer level bonding in different applications such as micro-electro-mechanical systems (MEMS) packaging and 3-D heterogeneous integration. In comparison to more common known techniques such as chip-to chip and chip-to-wafer bonding, wafer-to-wafer bonding has the advantages of high throughput; thus lower cost and accurate alignment [1, 2]. Furthermore, the metal-to-metal thermo-compression bonding provides not only the mechanical attachment between wafers but also enables a good electrical and thermal conductivity [3]. Despite the fact that metals like gold and copper are mainly used for wafer level bonding, aluminum is also a promising candidate because of its easy compatibility to complementary metal-oxide-semiconductor (CMOS) processes. The main challenges for Al-Al bonding are the fast oxidation and the roughness of the Al surface. Additionally, low temperature (<500 °C) Al-Al wafer level bonding, which is crucial for post-CMOS processing, is still desired by the industry.

In the present paper, the surface roughness of the sputter-deposited aluminum bonding pads is reduced to values below 2 nm by optimizing the aluminum deposition parameters. On the basis of that, a wafer level Al-Al thermo-compression bonding process at temperatures between 300 and 500 °C with a very low contact resistance, accurate alignment and high uniformity for 200 mm wafers is demonstrated.

II. OPTIMIZATION OF THE AL SURFACE ROUHGNESS

A key parameter affecting the quality of the Al-Al bonding is the surface roughness of the bonding pads [4, 5, 6, 7]. A surface roughness of 1 nm or less is expected to achieve a high bond strength after Al-Al bonding. However, with the standard Al deposition techniques, the surface roughness values are far above this requirement, typically between 5 and 8 nm. The aim of this study is to reduce the surface roughness of the sputter-deposited aluminum pads by optimizing the aluminum deposition parameters. The Al surface morphology was investigated on blanked 200 mm wafers using an atomic force microscope (AFM Park NX20) in the true non-contact mode™, measuring 5×5 µm² areas at various wafer positions. For each sample the mean roughness value (R_a) was computed, which is the arithmetical average of the deviations from the mean line of the roughness profile. The aluminum films of 99.5 % Al and 0.5 % Cu were deposited in a DC magnetron sputter chamber mounted onto a fully automated Applied Materials™ Endura multi-chamber single wafer tool. The Al physical vapor deposition (PVD) process was carried out with argon sputter gas, a pressure of 2.2 mTorr and a DC power of 10 kW. The deposition temperature was varied between 100 and 400 °C.

At first, the influence of different underlayers on the Al surface roughness was investigated (Fig. 1). As a starting point, a bonding layer with a thickness of 1.9 µm aluminum was deposited at 10 kW and 200 °C directly on top of the standard Ti/TiN underlayer. A high mean surface roughness of 5.1 nm was measured, which is obviously not sufficient for uniform and reliable Al-Al thermo-compression wafer level bonding. Depositing the

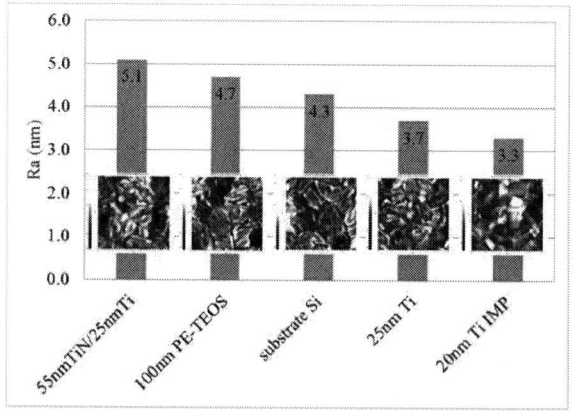

Figure 1. Al surface roughness R_a from the wafer center for different underlayers.

978-1-7281-1500-9/19 $31.00 © 2019 IEEE

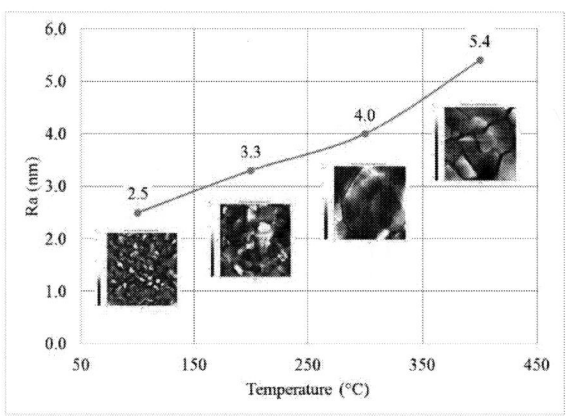

Figure 2. Al surface roughness R_a from the wafer center for different Al deposition temperatures.

1.9 µm aluminum on 100 nm plasma-enhanced tetraethyl orthosilicate (PE-TEOS) oxide lowered the mean surface roughness to 4.7 nm. The deposition of aluminum directly on top of the silicon substrate further reduced the Al surface roughness to 4.3 nm. The lowest R_a values were obtained for titanium underlayers. A 25 nm thin Ti PVD layer resulted in a mean surface roughness of 3.7 nm. The Al surface roughness was further improved to 3.3 nm by using 20 nm Ti IMP™ (ionized metal plasma) instead of the standard Ti, which is consistent with [8]. As can be seen from the AFM images in Fig. 1, also the Al grain sizes appear slightly smaller for the Ti IMP™ underlayer. Smaller grain sizes lead to a higher amount of grain boundary diffusion paths, which might result in a better bond ability, especially at reduced bonding temperatures [7, 9].

Secondly, 1.9 µm aluminum films were sputter-deposited on the 20 nm Ti IMP™ underlayer and the Al deposition temperature was varied between 100 and 400 °C. All other parameters were kept constant. Fig. 2 shows the mean roughness R_a with the corresponding AFM images for the deposition temperatures 100, 200, 300 and 400 °C. It is visible that the deposition temperature has a significant influence on the Al surface conditions. A higher deposition temperature results in larger grains and a higher surface roughness. At lower deposition temperatures the grains are getting smaller and the surface roughness is decreasing, which is expected to promote a higher bond quality. The lowest mean surface roughness of

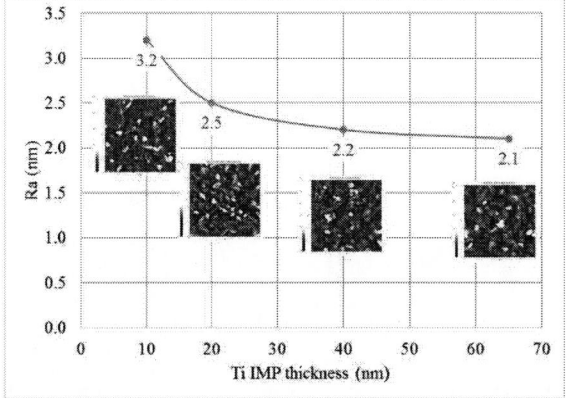

Figure 3. Al surface roughness R_a from the wafer center for different Ti IMP™ thicknesses.

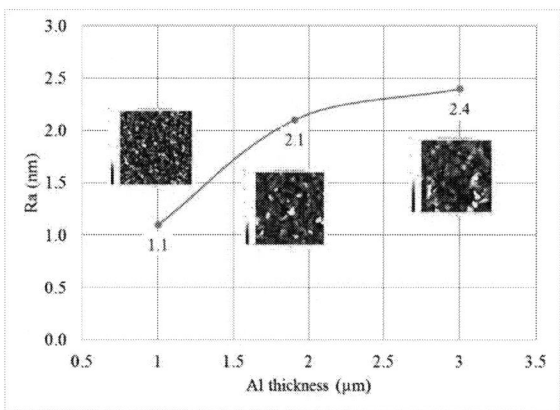

Figure 4. Al surface roughness R_a from the wafer center for different Al thicknesses.

2.5 nm and the smallest grains are achieved with a deposition temperature of 100 °C.

Again, special attention was paid to the relationship between the Ti IMP™ underlayer and the Al surface properties. Here, the underlayer thickness was varied between 10 nm and 65 nm before the 1.9 µm thick Al layers were sputter-deposited on top with the optimized deposition temperature of 100 °C. As depicted in Fig. 3, a thin Ti layer of only 10 nm results in a slightly higher mean surface roughness R_a of 3.2 nm. Increasing the Ti thickness to 20 nm reduces R_a to 2.5 nm. A further increase of the Ti thickness still shows a slight improvement of the Al surface properties. With a Ti layer of 40 nm the mean surface roughness R_a is decreased to 2.2 nm. The lowest surface roughness of 2.1 nm is obtained with a Ti IMP™ underlayer of 65 nm. These results are in good agreement with [10] and [11], where a critical thickness of the Ti underlayer was required to optimize the surface structure of the subsequently deposited Al film. This critical thickness was necessary to develop a strong (0002) texture in the Ti underlayer, which than led to a high (111) texture in the aluminum film. A stronger oriented (111) Al layer correlated with a smoother surface. As evident from the AFM images in Fig. 3 there is no perceptible difference between the grain sizes for different underlayer thicknesses.

Finally, the Al thickness itself was varied in the range of 1 to 3 µm. On the 65 nm Ti IMP™ underlayer aluminum films with a thickness of 1 µm, 1.9 µm and 3 µm were sputter-deposited at 100 °C. As illustrated by Fig. 4, a thicker Al layer of 3 µm results in a higher mean surface roughness R_a of 2.4 nm, whereas the reduction of the Al thickness down to 1 µm considerably lowers the surface roughness to 1.1 nm. From the AFM images in Fig. 4 it can be seen that also the grain sizes are smaller for the 1 µm aluminum film. According to [12], an aluminum bonding layer thickness of 1 µm is very well suited for reliable Al-Al thermo-compression bonding.

In summary, we significantly reduced the Al surface roughness from initially 5.1 nm down to <2 nm. The lowest mean surface roughness R_a of 1.1 nm and the smallest grain sizes were obtained for the sputter deposition of a 1.0 µm Al film on 65 nm Ti IMP™ using a substrate temperature of 100 °C. These optimized Al bonding layers provide the best prerequisites for successful Al-Al thermo-compression bonding.

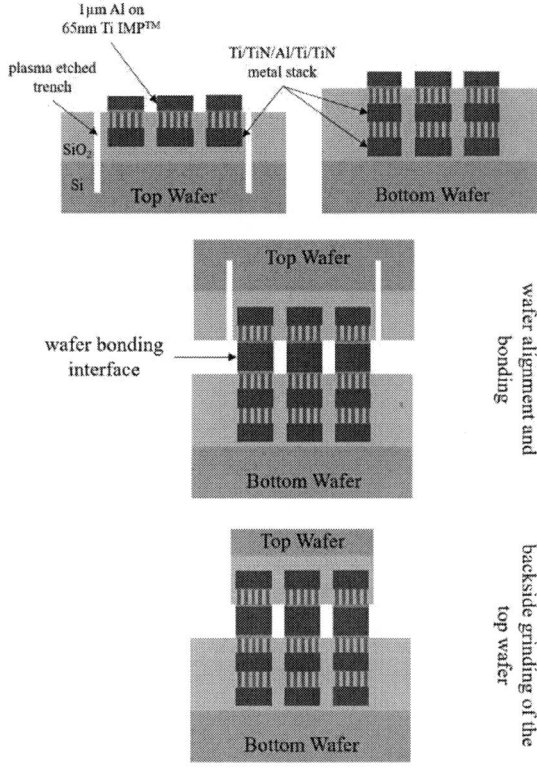

Figure 5. Fabrication process for Al to Al bonded wafers.

III. WAFER LEVEL AL-AL THERMO-COMPRESSION BONDING

A. Wafer fabrication

In the present work, Al-Al thermo-compression bonding of two patterned wafers is demonstrated. Fig. 5 illustrates the fabrication process for Al-Al bonded wafers. The bottom silicon wafer consists of two standard Ti/TiN/Al/Ti/TiN redistribution layers, which are isolated

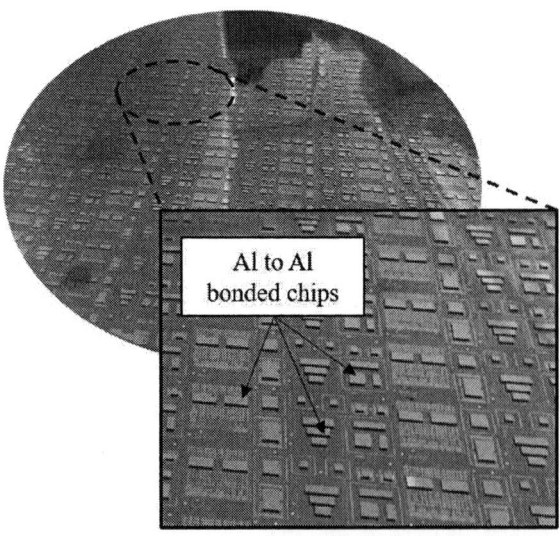

Figure 6. Photograph image of a 200 mm wafer with the bonded chips on top using the Al-Al wafer bonding technique.

by SiO$_2$ and vertically connected to the upper Al bonding pads with tungsten vias. For the bonding pads the optimized deposition parameters were used. The 1 µm aluminum film was sputter-deposited on top of 65 nm Ti IMP™ at 100 °C. Subsequently, the bonding layer was patterned by dry etching to obtain bonding pads with sizes ranging from 20x20 up to 80x80 µm^2. The top silicon wafer consists of only one metal layer and the Al bonding pads, which are mirrored to the bottom wafer. The plasma dicing before grinding approach was applied to the top wafer to separate the individual dice. Here, deep vertical trenches were etched up to a depth of approximately 200 µm into the silicon. The actual bonding was carried out in an EVG® ComBond® 200 mm automated high vacuum wafer bonding system, which is described in detail in the next section. After bonding the backside of the top wafer was ground until the non-bonded chips could be removed. Here, rough-grinding was done with a #320 mesh[13], followed by fine grinding with a #2000 mesh diamond wheel. A photograph image of the final bottom wafer with the Al-Al bonded chips on top is shown in Fig. 6.

B. Surface pretreated Al-Al Wafer Bonding

The wafer bonds were produced in an EVG® ComBond® 200 mm automated high vacuum wafer bonding system. The ComBond® system uses a high vacuum handling cluster with several modules. At the beginning, the wafers were handled to the ComBond® activation module. Here, a gentle dry etch process is applied, which removes the oxide on the Al bonding pads and only negligibly increases the surface roughness [14]. Without exposure to air, the wafers were cooled and subsequently handled to the vacuum align module. Using

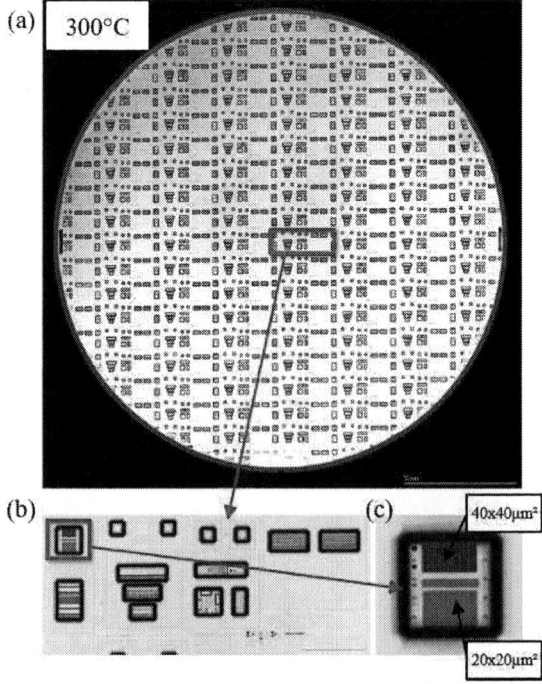

Figure 7. C-SAM image of a 200 mm wafer bonded at 300 °C for 1 h with 60 kN. (a) Overview image (b) detailed scan of the Al-Al bonded chips (c) detailed scan of a bonded chip with 40x40 µm^2 and 20x20 µm^2 Al bonding pads.

infrared alignment the bottom wafer and the flipped top wafer were adjusted to each other and transferred into the bond module. Here, the wafers were bonded for 1 h with a bonding temperature of 300, 400 or 500 °C. The bonding force was set to 60 kN for all samples. Afterwards, the samples were cooled and the post-bond alignment accuracy was evaluated. Here, an alignment accuracy as low as 0.5 µm was measured.

C. Characterization of the Al-Al Wafer Bonding

The bonded wafer pairs were characterized by C-mode scanning acoustic microscopy (C-SAM), scanning electron microscopy (SEM) and transmission electron microscopy (TEM). Additionally, the electrical performance of the Al-Al wafer bonding was analyzed using daisy chain (DC) test structures.

Directly after bonding and before the backside grinding of the top wafer, C-SAM was used to inspect the bonding quality and bonding uniformity of the 200 mm wafers. This non-destructive method creates images by generating a pulse of ultrasound. The beam of ultrasound is focused onto the bonded interface using water as coupling medium. A two-dimensional gray-level image is created by scanning a transducer over the sample and recording the intensity of the reflected echo [15]. The dark appearing portions in the image indicate a high bond quality with little defects. In contrast, the bright regions represent non-bonded areas and voids.

Fig. 7 shows a C-SAM image of a 200 mm Al-Al bonded wafer at 300 °C. Beside the overview image (Fig. 7a), detailed scans of the bonded chips are presented (Fig. 7b, c). Fig. 7c illustrates the C-SAM image of a bonded chip with 40x40 µm² and 20x20 µm² Al bonding pads. The dark appearing structures in Fig. 7a suggest a uniform bonding across the entire wafer. Also the detailed

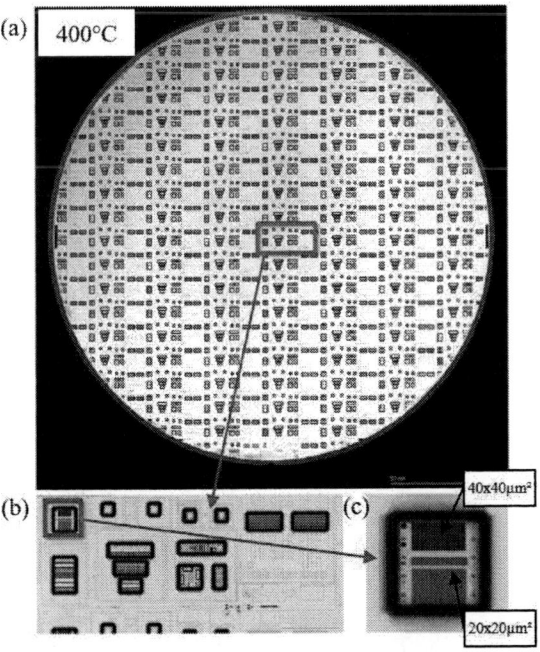

Figure 8. C-SAM image of a 200 mm wafer bonded at 400 °C for 1 h with 60 kN. (a) Overview image (b) detailed scan of the Al-Al bonded chips (c) detailed scan of a bonded chip with 40x40 µm² and 20x20 µm² Al bonding pads.

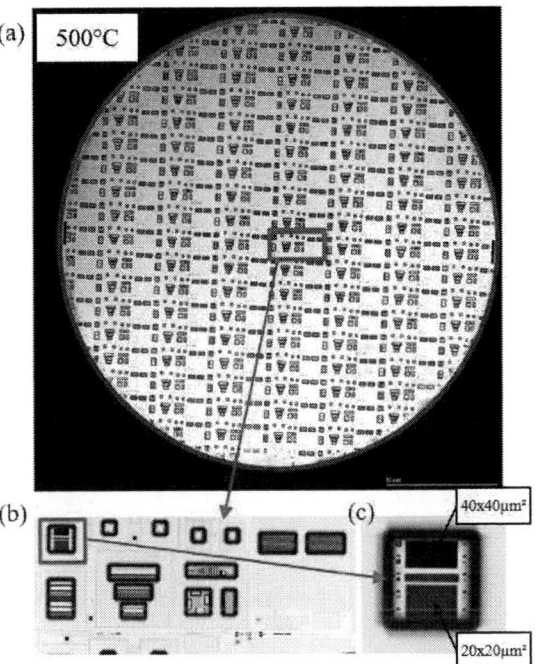

Figure 9. C-SAM image of a 200 mm wafer bonded at 500 °C for 1 h with 60 kN. (a) Overview image (b) detailed scan of the Al-Al bonded chips (c) detailed scan of a bonded chip with 40x40 µm² and 20x20 µm² Al bonding pads.

scan in Fig. 7c predicts a high bonding quality with just little defects. Fig. 8 and 9 present the corresponding C-SAM images for the bonding temperatures of 400 and 500 °C, respectively. Also the bonds produced at 400 and 500 °C show a good bonding at any position on the wafer. Comparing the different temperatures, there is almost no difference in the C-SAM images. Only the detailed scan of the 500 °C bond in Fig. 9c appears slightly darker, which might be interpreted as a higher bond quality.

To characterize the bonding interface FIB/SEM cross sections were obtained at various wafer positions. The cross sections revealed no major differences between the bonding temperatures. Thus, only an exemplary image for a bonding temperature of 400 °C is presented in Fig. 10, which was captured at the center of the wafer. It shows the

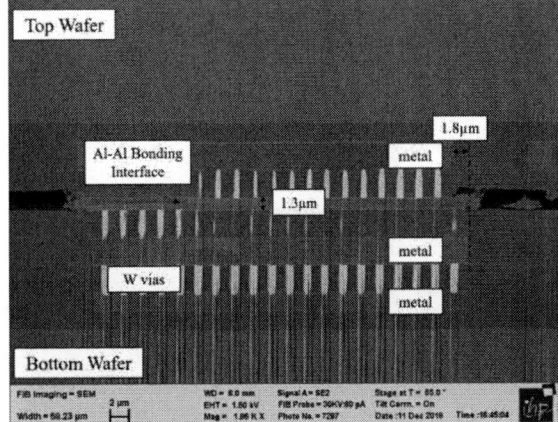

Figure 10. FIB/SEM cross section with 40x40 µm² Al pads bonded at 400 °C for 1 h with 60 kN.

978-1-7281-1500-9/19 $31.00 © 2019 IEEE

bonding interface of 40x40 µm² Al pads and the Ti/TiN/Al/Ti/TiN redistribution layers of the top and bottom wafer, which are vertically connected with tungsten vias. Not all tungsten vias are clearly visible, because they are partly covered by SiO$_2$, which was not completely milled by FIB. The cross section shows that the top and bottom wafer are properly aligned to each other. A lateral shift of only 1.8 µm was measured. This is slightly higher than the lowest measured value of 0.5 µm, which was obtained at another wafer. The Al-Al bonding interface looks very smooth and no major defects are observable. This indicates strong Al-Al bonding. Furthermore, it can be seen that the Al bonding layers are highly compressed from the initially two deposited and 1 µm thick layers down to 1.3 µm. The excessive aluminum was squeezed out of the bonding interface and is visible at the edge of the bonding pads. The highly

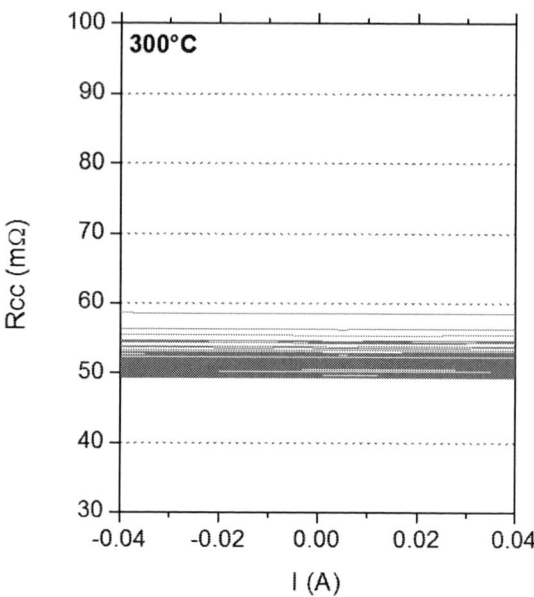

Figure 12. Chain resistances per 20x20 µm² contact for the different chips of an Al-Al bonded wafer at 300 °C for 1 h with 60 kN.

compressed aluminum implies that a bonding force lower than the applied 60 kN would apparently be enough to successfully bond the aluminum layers, which should be explored in future work.

To investigate the bonding interface in more detail TEM studies were performed with a FEI Tecnai Osiris system operated at 200 kV in STEM mode. Again, only an exemplary image for a bonding temperature of 400 °C is presented. Fig. 11 shows a TEM cross section of the Al-Al bonding interface obtained in the bright field (BF) mode, where the image is formed from the direct electron beam. The Al-Al bonding interface is visible by a thin dark line between both bonding pads. Nevertheless, no major defects are appearing at the interface. Fig. 11b displays a

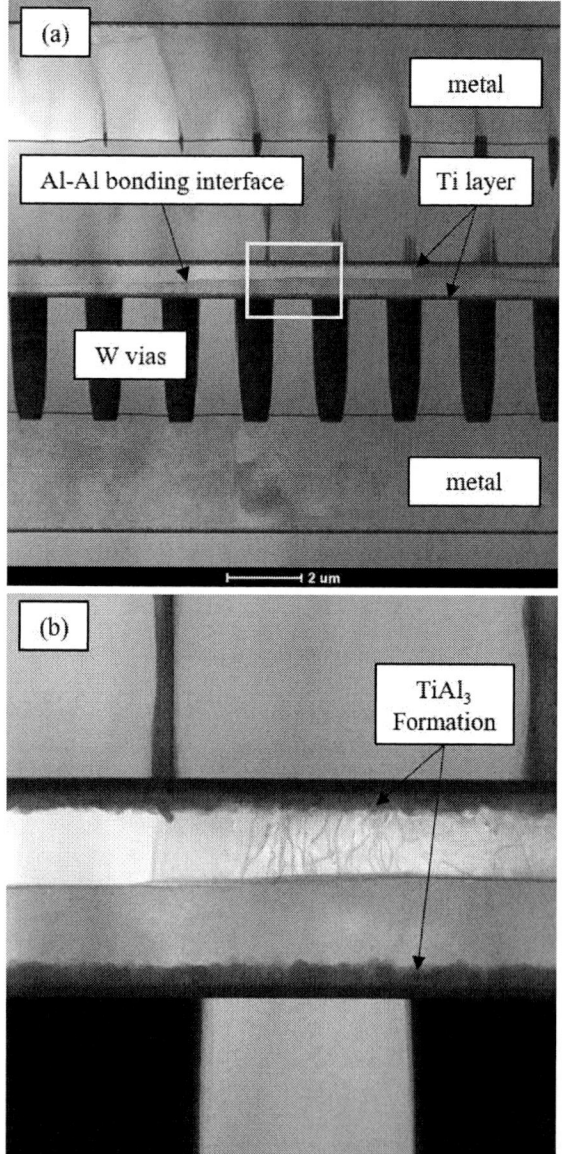

Figure 11. (a) TEM cross section of the Al-Al Interface bonded at 400 °C for 1 h with 60 kN. (b) Close-up of the Al-Al bonding interface.

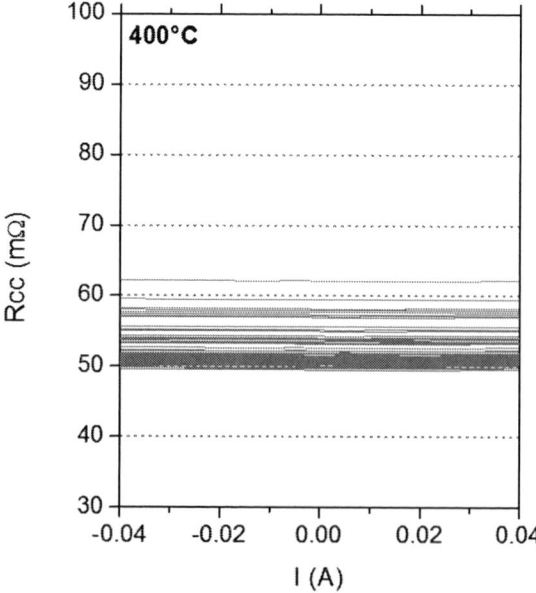

Figure 13. Chain resistances per 20x20 µm² contact for the different chips of an Al-Al bonded wafer at 400 °C for 1 h with 60 kN.

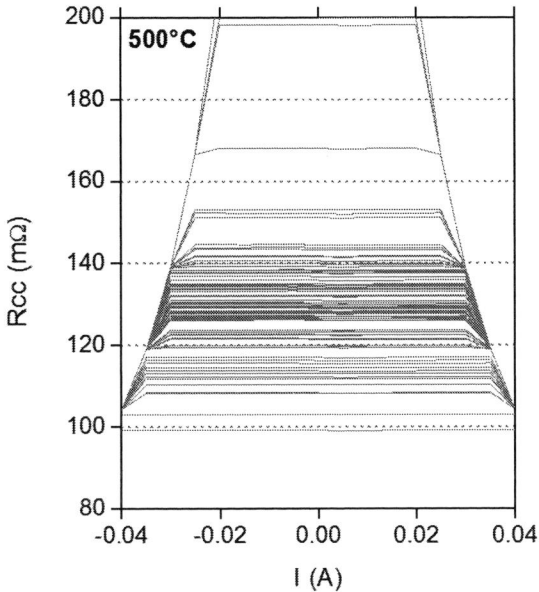

Figure 14. Chain resistances per 20x20 µm² contact for the different chips of an Al-Al bonded wafer at 500 °C for 1 h with 60 kN.

close-up of the marked region in Fig. 11a. Also the magnified picture shows a well-bonded layer with no larger voids. Additionally, the TEM images display the Ti IMP™ layers below the aluminum pads. Originating from the Ti underlayers, a blurry dark region is extending into the aluminum. This might suggest the formation of TiAl₃ at the Ti/Al interface [16], which occurs at higher bonding temperatures (>400 °C) and should be studied in further investigations. The TEM pictures of the sample bonded at 500 °C (not shown) indicated an even stronger TiAl₃ reaction.

The electrical performance of the Al-Al bonded wafers was analyzed using daisy chain (DC) test structures with 480 elements. The size of each Al contact pad was 20x20 µm². The measurements were carried out on 91 different chips across the wafer. Fig. 12 displays the chain resistances per contact for an Al-Al bonded wafer at 300 °C. Each line presents one of measured 91 chips. The resistances are varying between 49 and 59 mΩ with a yield of 87 %. A very low mean value of 51.9 mΩ was calculated. Fig. 13 shows the corresponding resistances per contact for a bonding temperature of 400 °C. A mean value of 52.3 mΩ and yield of 86 % were measured, which are very similar to those of the samples bonded at 300 °C. Also the variation of the resistances between 49 and 62 mΩ is comparable. In contrast, the wafers bonded at 500 °C show a larger scattering and much higher resistances (Fig. 14) with values above 100 mΩ per contact. This could be due to a stronger TiAl₃ reaction at the Ti/Al interface during the 500 °C bonding process, which results in higher resistances since the bulk resistivity of Al is much lower than that of TiAl₃. It is evident from the results that a bonding temperature of 300 °C is sufficient to achieve a reliable Al-Al wafer bonding with low contact resistances. Future work should therefore include a further reduction of the bonding temperature.

IV. CONCLUSION

Low temperature Al-Al wafer level bonding requires oxide-free and smooth aluminum surfaces. In this study we optimized the Al surface roughness to values below 2 nm. The best surface conditions with a surface roughness R_a of 1.1 nm were obtained for a 1 µm thick Al film sputter-deposited at 100 °C on a 65 nm Ti IMP™ underlayer. On that basis, we presented a surface pre-treated Al-Al wafer bonding process. Here, a gentle dry etch was applied to remove the native oxide on the Al bonding pads. Without vacuum break the subsequent wafer bonding was carried out at temperatures between 300 and 500 °C with a pressure of 60 kN. It was shown that a temperature of 300 °C is sufficient to achieve strong Al-Al wafer bonding. The electrical characterization of contact chains with Al bonding pad sizes of 20×20 µm² revealed resistances lower than 50 mΩ per contact, which might imply areas of oxid-free bonding. In summary, a wafer level Al-Al thermo-compression bonding process with a very low contact resistance, accurate alignment and high uniformity for 200 mm wafer has been demonstrated. Our results present an excellent starting point for future investigations, which will include a further reduction of the bonding temperature and pressure.

REFERENCES

[1] Tan CS, Gutmann RJ, Reif LR, editors. Wafer level 3-D ICs process technology. Springer Science & Business Media; 2009 Jun 29.

[2] Panigrahi AK, Ghosh T, Vanjari SR, Singh SG. Dual Damascene Compatible, Copper Rich Alloy Based Surface Passivation Mechanism for Achieving Cu-Cu Bonding at 150 Degree C for 3D IC Integration. InElectronic Components and Technology Conference (ECTC), 2017 IEEE 67th 2017 May 30 (pp. 982-988). IEEE.

[3] Yu J, Wang Y, Moore RL, Lu JQ, Gutmann RJ. Low-Temperature Titanium-Based Wafer Bonding Ti/Si, Ti/SiO2, and Ti/Ti. Journal of the Electrochemical Society. 2007 Jan 1;154(1):H20-5.

[4] Rebhan, B., Wimplinger, M. and Hingerl, K., 2014. Impact factors on low temperature Cu-Cu wafer bonding. ECS Transactions, 64(5), pp.369-377.

[5] Dragoi V, Mittendorfer G, Burggraf J, Wimplinger M. Metal thermocompression wafer bonding for 3D integration and MEMS applications. ECS Transactions. 2010 Oct 1;33(4):27-35.

[6] Panigrahi AK, Bonam S, Ghosh T, Vanjari SR, Singh SG. Low temperature, low pressure CMOS compatible Cu-Cu thermo-compression bonding with Ti passivation for 3D IC integration. InElectronic Components and Technology Conference (ECTC), 2015 IEEE 65th 2015 May 26 (pp. 2205-2210). IEEE.

[7] Rebhan, B. and Hingerl, K., 2015. Physical mechanisms of copper-copper wafer bonding. Journal of Applied Physics, 118(13), p.135301.

[8] Spinler, S., Schmidbauer, S. and Klotzsche, J., 2000. Surface roughness reduction for AlCu-based metallization: implications from an integrative point of view. Microelectronic engineering, 50(1-4), pp.311-319.

[9] Rebhan B, Hinterreiter A, Malik N, Schjølberg-Henriksen K, Dragoi V, Hingerl K. Low-Temperature Aluminum-Aluminum Wafer Bonding. ECS Transactions. 2016 Aug 24;75(9):15-24

[10] Voutsas AT, Hibino Y, Pethe R, Demaray E. Structure engineering for hillock-free pure aluminum sputter deposition for gate and source line fabrication in active-matrix liquid crystal displays. Journal of Vacuum Science & Technology A: Vacuum, Surfaces, and Films. 1998 Jul;16(4):2668-77.

[11] Kamijo A, Mitsuzuka T. A highly oriented Al [111] texture developed on ultrathin metal underlayers. Journal of applied physics. 1995 Apr 15;77(8):3799-804.

978-1-7281-1500-9/19 $31.00 © 2019 IEEE

[12] Froemel J, Baum M, Wiemer M, Roscher F, Haubold M, Jia C, Gessner T. Investigations of thermocompression bonding with thin metal layers. InSolid-State Sensors, Actuators and Microsystems Conference (TRANSDUCERS), 2011 16th International 2011 Jun 5 (pp. 990-993). IEEE.

[13] Liu JH, Pei ZJ, Fisher GR. Grinding wheels for manufacturing of silicon wafers: a literature review. International Journal of Machine Tools and Manufacture. 2007 Jan 1;47(1):1-3.

[14] Hinterreiter AP, Rebhan B, Flötgen C, Dragoi V, Hingerl K. Surface pretreated low-temperature aluminum–aluminum wafer bonding. Microsystem Technologies. 2018 Jan 1;24(1):773-7.

[15] Gordon GA, Canumalla S, Tittmann BR. Ultrasonic C-scan imaging for material characterization. Ultrasonics. 1993 Sep 1;31(5):373-80.

[16] Bresolin C, Pirotta S. TiAl3 formation kinetic in sputtered Ti/AlCu0. 5% thin films. Microelectronic engineering. 2002 Oct 1;64(1-4):125-30.

978-1-7281-1500-9/19 $31.00 © 2019 IEEE   224

Self-Assembly process for 3D Die-to-Wafer using direct bonding: A step forward toward process automatisation

Amandine Jouve, Loïc Sanchez, Clément Castan, Maxence Laugier, Emmanuel Rolland, Brigitte Montmayeul, Rémi Franiatte , Frank Fournel, Severine Cheramy

Univ. Grenoble Alpes, CEA, LETI
38000 Grenoble, France
Amandine.jouve@cea.fr

Abstract— Die-To-Wafer hybrid bonding process is foreseen by major microelectronic industrials as essential for the success of future memory, HPC or photonic devices. However, compared to Wafer-To-Wafer bonding, the process flow is much more complex and the die assembly throughput is significantly reduced, which do not facilitate process industrialization. CEA-LETI is been working since several years on development of self-assembly process in order to overcome the throughput issue. The use of water drop is very promising to self-align several thousand of dies per hour. This paper presents the latest developments to improve self-assembly robustness and move a step forward toward manufacturing. Firstly, a new material benchmark enabled to identify a foundry compatible hydrophobic material. Secondly, we demonstrated that the die and substrate realization can be fully achieved on automatized equipments. Finally, the process setup on standard pick and place equipment significantly improved the self-assembly process robustness: 98% of the dies were perfectly bonded, and 63% presented an alignment lower than 1µm with a low 5µm die edge topography, which is compatible with hybrid bonding design.

Keywords- 3D integration; Die-To-Wafer stacking; self-assembly; direct bonding

I. INTRODUCTION

The continuous search for smallest interconnections for the High density 3D circuits has led to the development of the direct hybrid bonding, now in production for imager applications. Wafer-to-Wafer direct hybrid bonding has now reached a certain level of maturity with functional demonstrator achieved with only 1,44 µm interconnection pitch [1; 2]. However the Die-to-Wafer (D2W) direct hybrid bonding is today at its early stage due to the complexity of the process flow for die preparation (figure 1) and the limited assembly throughput due to high accuracy requirement (figure 2). Indeed, when dealing with die stacking for hybrid bonding, the post dicing surface preparation become very challenging due to fact that die handling strategies compatibles with plasma or wet treatment tools have to be found.

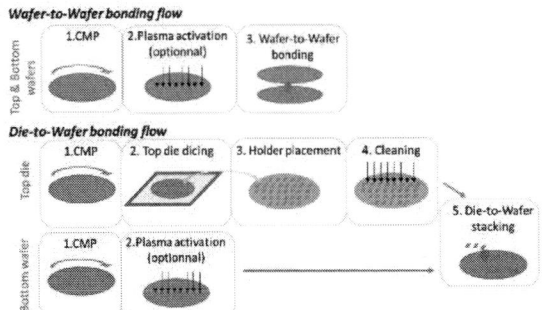

Figure 1. Simplified process flows for Wafer-to-Wafer and Die-to-Wafer direct bonding.

Today tape handling or wafer-type holders seem to be the most mature solutions [3,4]. The other challenge faced today with D2W direct bonding is throughput. Indeed, when dealing with sub-5µm interconnections, the placement accuracy should be +/- 1µm. However, at such high accuracy the tool throughput is lower than 1000 Die-per-Hour (DpH) (figure 2).

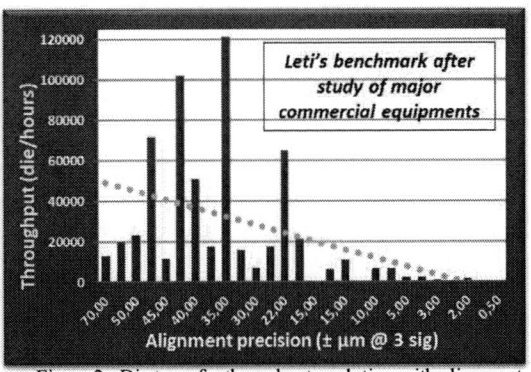

Figure 2. Die to wafer throughput evolution with alignment precision [ref]

If solutions to increase throughput are today under evaluation by tool supplier with double head use for example, breakthrough technical processes have to be developed if industrial throughputs over 5000 DpH need to be reached with one single equipment.

In that frame, CEA-LETI worked since 2010 on an alternative self-assembly process compatible with direct hybrid bonding. Self-Assembly process is based on the use of water droplet to align top on bottom die. The driving force for alignment is the liquid capillary tension (mostly water), which will act as a spring effect until the minimization of the system energy, that is to say the minimum liquid/air surface. This process would considerably reduces die pre-alignment criterion because the die could be placed on its corresponding substrate with a much lower accuracy (>100µm). Throughput is significantly improved according to graph figure 2.

If the self-assembly process could solve the throughput issues faced by standard D2W, it rises additional challenges. Indeed, the water confinement to the bonded surface is key for a proper self-assembly with alignment accuracy control. This water confinement is ensured by a topography and/or chemical contrast between active die surface and it edges.

Only few publications can be found on self-assembly process for direct D2W bonding [5]. The work driven by S.Mermoz et al. enabled to identify a first hydrophobic polymer and developed « Mixed » structures presenting both hydrophobic die edges and hydrophilic die surface which was good starting point to obtain sub-micronic self-alignment and relative defect free bonding interface after bonding [6]. They finally demonstrated the feasibility of electrical interconnections with such process. However, they also mentioned that the hydrophobic polymer material used at that time was not compatible with all pre-bonding treatments required for direct copper/oxide hybrid bonding. Water confinement was not optimal[7]. Secondly, the reproducibility was limited by the manual process flow for substrate preparation as well as the use of a homemade equipment for die placement [8].

The objective of this paper is to improve self-assembly process robustness. In order to do so, we have searched for a foundry-compatible hydrophobic compound more resistant to surface pre-bonding treatments, transfer the substrate preparation process steps that were previously done manually on fully automated tools. We demonstrated the compatibility of self-assembly stacking process with automated equipment.

After a review of the whole self-assembly flow and its related challenges, the first part of the article is dedicated to the hydrophobic material benchmark.

In the second part of the article, we demonstrated that 300mm and 200mm wafers for self-assembly process could be fabricated using only foundry compatible process steps.

Finally, we assessed the self-assembly process performances when using more automated stacking tool (D2W or collective bonding) to fully evaluate the industrial potential of this technic.

II. SELF-ASSEMBLY PROCESS FLOW

A. Self-Alignment theory

Self-assembly process is based on the use of small water drops to align and bond dies to a wafer substrate and has already been described in previous publications [6-8]. Chip alignment results from surface tension minimization of the water drop. After self-alignment step, hybridation is achieved thanks to direct bonding after the drop evaporation as shown on figure 3.

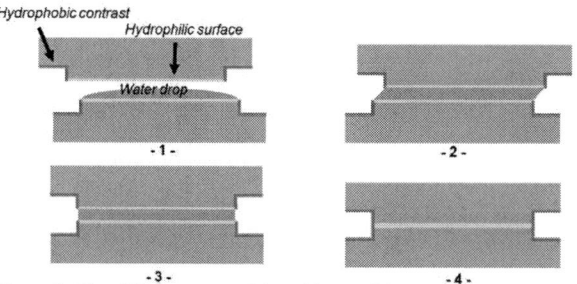

Figure 3. The different steps of the self-assembly process. -1- Inaccurate pre-alignment with Pick&Place tool (around 200 µm), -2-Drop spreading and chip is realigned with submicron accuracy (around 400 nm); -3-Drop evaporation, -4-Direct bonding.

A good fluid containment is required in order to get high accuracy in chip-to-wafer self-assembly process. The aim of this containment is to avoid fluid overflowing which will induces misalignment. Creation of fluid's restricted areas is obtained by physical and/or chemical contrast. The physical contrast is based on canthotaxis effect.

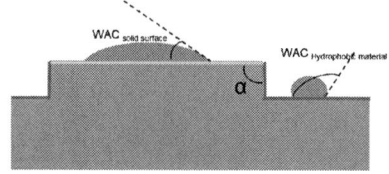

Figure 4. Angle definitions for water confinement condition equation.

By combining both canthotaxis and chemical effects, we can obtain a better water confinement. The liquid will not overflow from the top solid surface as long as condition described by eq. (1) is true:

$$WCA_{\text{solid surface}} < \alpha_{\text{topo}} + WCA_{\text{hydrophobic surface}} \qquad (1)$$

With WCA standing for Water Contact Angle (figure 4). Therefore, a step surrounding the active die surface to be bonded is always patterned, and a hydrophobic material is selectively deposited on the slopes to ensure a robust self-assembly process. This strategy implies challenges in order to fit industrial manufacturing processes. These challenges are described in the paragraph below.

B. Self-Assembly industrial process flow challenges

To evaluate the technical challenges to be solved in order to make the self-assembly process compatible with manufacturing, the complete integration flow has to be reviewed. Figure 5 presents a complete self-assembly process flow, without the stacking step details:

The first requirement to achieve the self-assembly process in high Volume manufacturing foundries, is to ensure that the hydrophobic material, the processes to patterned it and the cleaning recipes should be cleanroom compatible.

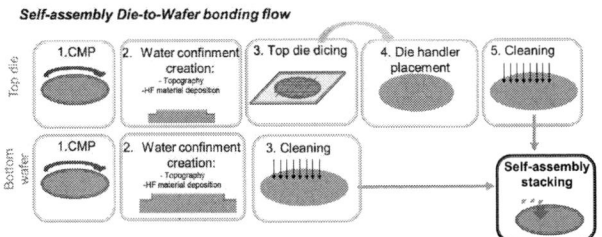

Figure 5. Self-assembly process flow.

In order to ensure a defect-free direct bonding the top die surface needs to be highly hydrophilic [6]. Therefore the hydrophobic compound has to be easily patternable to only remain around the dies. However, one of the most challenging technical requirement for self-assembly is the hydrophobic material withstand the pre-bonding cleaning processes. The surface preparation process consists in removing the organic compound and remaining particles. Additional treatment is used to remove metal oxide and make the silicon oxide surface perfectly hydrophilic (WAC≤5°). After all these treatments, the hydrophobic material surface preserves an hydrophobic behavior (that is to say, WAC≥90°). The best cleaning processes identified today are: (1) UV/O3 treatment to remove organic compounds, (2) a specific plasma treatment can be used to remove the metal oxide for hybrid bonding surfaces (as H_2/He for instance), (3) a megasonic cleaning using a MegPie from Prosys in a EVG301 tool from EvGroup to remove particles and to obtain a high hydrophilicity on oxide surface, and. These cleaning processes are compatible with the bottom wafer as well as with the die placed in a specific silicon holder.

It is also important to notice from the scheme in figure 6 that the topography used for canthotaxis effect creation remains limited. Indeed, as the BEOL levels might be used for chip-to-chip communication, the self-assembly plot topography should ideally be built in the additional levels specifically created for direct hybrid bonding. Depending on final product contact resistance targeted, we foresee the plot topography specification to be between 5 to 1 μm.

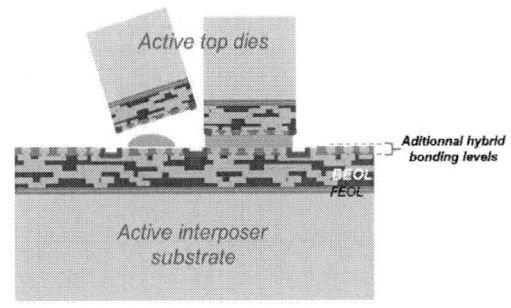

Figure 6. Schematic view of a Self-assembly process integration on a functional test vehicle.

III. EXPERIMENTAL

This part presents the work achieved at CEA-LETI in order to improve the self-assembly robustness including new material benchmark, optimization of lift-off process and evaluation of automated stacking equipment.

A. Test vehicle presentation & characterizations

In order to work on process robustness improvement, we used only plain oxide surface dies. The test chips were prepared from 200 or 300mm base wafers. Each die presents a pad dimension of 8x8mm. For the bottom substrate, the pads were spaced by 8mm. On each die, two verniers were etched on the die for post-bonding alignment accuracy measurements. After silicon etching a 6μm thick PECVD oxide was deposited.

The topography required for self-assembly around the die was then fabricated thanks to lithography/etching processes. This structure enable to cover a very wide range of topographies. In this study, pad topographies of 2μm, 5μm and 15μm were evaluated.

Post-bonding accuracy measurements were performed with the verniers patterns thanks to an infrared microscope (Figure 6).

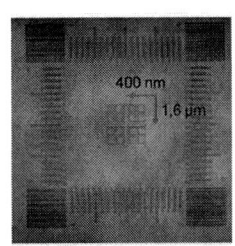

Figure 6. Image of the Vernier patterns used to evaluate post-bonding misalignment.

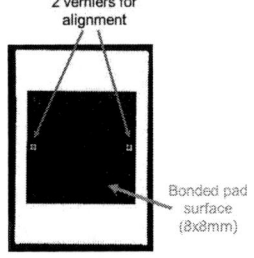

Figure 7. Post bonding acoustic microscopic image of the die. In black is the bonded surface of the pad.

978-1-7281-1500-9/19 $31.00 © 2019 IEEE

In addition to alignment, we evaluated the bonding quality thanks to interfacial acoustic microscopy analysis. An example is shown in figure 7. This technique enables to check bonding interface defects appearing as "white" patterns.

B. Hydrophobic materials patterning solutions

Several methods exist to selectively pattern the hydrophobic material. Among them, the lift-off & the anisotropic dry etching was identified as the most compatible with foundry manufacturing environment (figure 8).

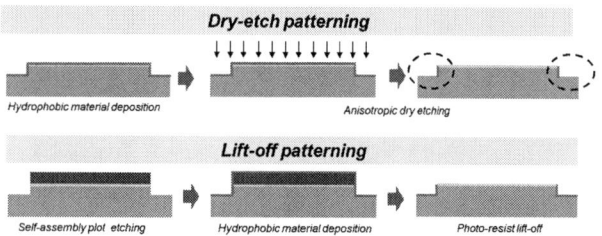

Figure 8. Scheme of the 2 hydrophobic material patterning technics (rem).

Evaluation of both technique limitations was done using the hydrophobic polymer described in previous works.

We observed very low self-alignment yields with the anisotropic dry etch samples. Indeed, in that case, self-alignment was observed only for a pad height over 14µm, and very limited surface preparation time of the UV/O$_3$ treatment, which limits our organic removal performance. In the contrary, encouraging self-alignment and bonding yield was obtained with lift-off process: 17/20 dies were self-aligned with success and 13/20 dies presented defect-free bonding interface using only 5µm pad height. The hydrophobic polymer shape at pad surface after surface preparation can be responsible for this important process performance variation. Indeed in the case of anisotropic dry etching the hydrophobic polymer is partially etched (100-150nm depth) at pad top surface which eliminates the canthotaxis effect and reduces water confinement performances (figure 9).

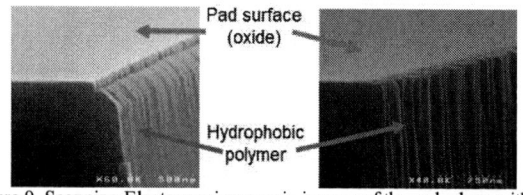

Figure 9. Scanning Electron microscopic images of the pad edges with the residual hydrophobic polymer (left) Dry etch patterning flow (Right) Lift-off patterning.

Moreover, the UV/O3 treatment rapidly etch the hydrophobic polymer compound. This explains why shorter UV/O3 treatments were required with dry etched dies to limit hydrophobic polymer degradation and to conserve self-alignment. However, the SAM images in figure 10 tend to show that a shorter surface preparation treatment is insufficient to perfectly clean the die surface and interfacial defects remain.

Figure 10. Post bonding acoustic microscopic image of the die with hydrophobic polymer (left) Dry etch patterning flow (Right) Lift-off patterning.

Therefore, considering these performances, the process optimization work exposed in this paper only integrate substrates built with lift-off approach.

C. New hydrophobic material benchmark

1) Hydrophobic polymer limitations

The choice of hydrophobic material is essential for self-assembly success and deeper characterizations demonstrated polymeric material limitations and poor compatibility with an industrial flow.

Indeed, post-dicing cleaning process for particles used so far water only, which was insufficient for robust particles removal. Therefore, we tested ammonia based particle cleaning recipes on hydrophobic polymer; however, the material delaminated with this chemistry (figure 11).

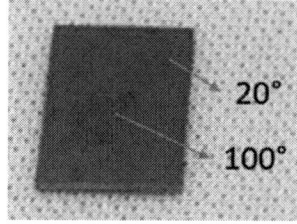

Figure 11. Hydrophobic polymer limited resistance to ammonia cleaning: material delamination and loss of hydrophobicity.

Furthermore, the limited resistance of the polymer to UV/O3 implied that a 200nm thick layer of hydrophobic polymer was deposited to conserve a proper water confinement after UV/O3 treatment. However, this relatively thick layer limits the lift-off process performances, and an automated wafer-level lift-off process was very difficult to develop. Indeed, in previous publications by S.Mermoz et al., the resist lift-off was always done manually in acetone trays, which generated many defects and resist redeposition issues. Furthermore, acetone cannot be used for production; therefore before initiating a new hydrophobic material

978-1-7281-1500-9/19 $31.00 © 2019 IEEE

benchmark, hydrophobic material compatibility with automated lift-off systems was checked.

Six different lift-off processes have been evaluated on automated SSEC wet process equipment with hydrophobic polymer: Two resists thicknesses (840nm and 8μm), two cleaning chemistries (DMSO and CLK820, both w/wo pressure jet), as well as well as two hydrophobic polymer thicknesses (200 and 50nm thicknesses): All failed. Indeed, it can be seen on figure 12 that either photo-resist remains at pad surface due to limited solvent infiltration through the thick hydrophobic polymer; either the hydrophobic polymer delaminated when using harsher solvent pressure jet.

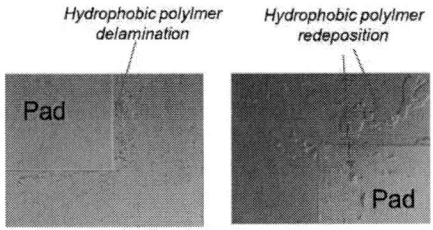

Figure 12. Microscopic observation of automated wafer-level lift-off attempts with hydrophobic polymer.

As a conclusion, this polymer was relevant for the early feasibility demonstrations, but was not suitable for proper robust and industrial self-assembly flow.

Therefore we searched for a foundry-compatible hydrophobic compound more resistant to surface pre-bonding treatments and suitable with standard lift-off systems.

2) Hydrophobic material selection

A total of five spin-coatable products were evaluated. The evaluation has consisted in measuring the product remaining hydrophobicity after the optimum surface treatments: UV/O3 and MegPie cleaning with ammonia.

The variation of the hydrophobicity with UV/O3 exposure duration for the best 2 products, named A and B is shown in the figure 13. It has to be noted that, whatever the UV/O3 duration, the UV treatment was always followed by a wet cleaning (MegPie with ammonia).

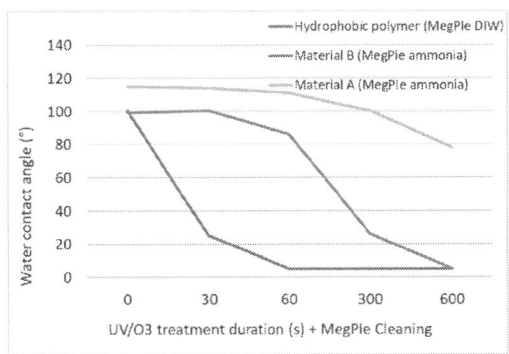

Figure 13. Hydrophobic material benchmark. Evolution of material hydrophobicity with UV/O3 pre-bonding treatment duration.

As the longest exposure time is the best for organic compound removal, the two products named A and B are more efficient than the previous hydrophobic polymer. Indeed, the treatment A conserve a hydrophobicity of almost 80° after 10min exposure (with additional wet cleaning). This value was not modified if the material was firstly exposed to an automated lift-off with DMSO solvent.

The thickness of these new hydrophobic layers is very thin, in the range of few nanometers, as shown in the case of treatment B in the figure 14:

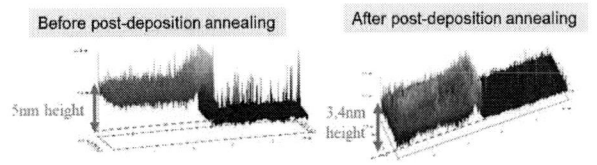

Figure 14. Hydrophobic material thickness characterization (material B).

These low thickness values will ease the lift-off process, as shown in the next paragraph.

D. Lift-off process improvement

1) Automated lift-off process development

We achieved automated lift-off in SSEC wet processing tool using DMSO solvent. The A or B material are transparent, thus the layer integrity is impossible to check with microscopy. Therefore, the presence of hydrophobic product has been checked thanks to water deposition tests and hydrophobicity value measurements.

It is important to note that to ensure the complete removal of the polymer post lift-off two parameters were adjusted: the solvent dispense pressure as well as the post-material deposition annealing temperature. Finally, we were able to obtain post lift-off resist free surfaces as shown in figure 15:

 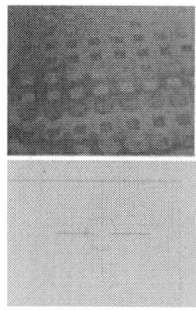

Figure 15. Hydrophilic material lift-off process optimization (Left) Photo-resist residues when hydrophilic material has been annealed (right) Residue-free lift-off obtained when hydrophilic material is not annealed.

If the optical inspection confirms the absence of residues, the water contact angle measurements confirmed excellent hydrophobicity contrast between the top and the bottom of the pad post lift-off process (see figure 16):

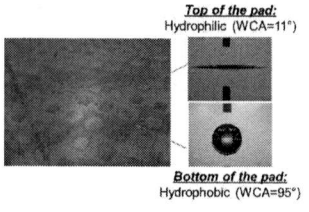 

Figure 16. Characterization of the A material residual hydrophobicity (Left) Water Contact Angle measurement (Right) Wafer-Level water dispense confirm the excellent water confinement on pad top surface only.

A fine AFM surface characterization of the top plot was carried out in order to check if the oxide surface roughness was not increased by lift-off process. We measured a sub-0.5nm roughness (figure 17), which confirmed the remaining oxide roughness compatibility with direct bonding.

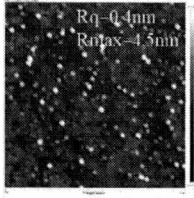

Figure 17. Post-lift off Oxide pad surface roughness characterizationby AFM.

2) Top die lift-off specificities

The previous paragraph exposed our improvement of the lift-off process achieved at wafer-level. However, could we use the same process for the top dies?

Indeed, for top die preparation an additional challenge is to minimize the silicon particle contamination induced by the dicing. As the dicing can be achieved at various process steps we have evaluated the lift-off feasibility of 3 different flows described in the figure 18.

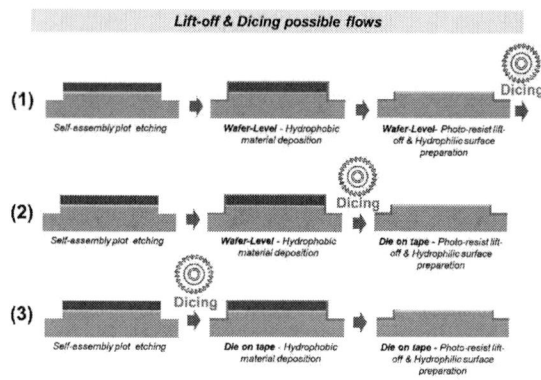

Figure 18. Scheme of the 3 possible process flow for top die preparation.

The sawing generates many particles that can stick to the hydrophobic treatment surface and causes subsequent bonding defects if not removed. Indeed, we experimentally observed that, if the dicing occurred after hydrophobic material deposition (like in the cases (1) and (2) of figure 18), many particles remained on the hydrophobic surface. In these cases, particle contamination was so high, that the Megpie cleaning even if with ammonia was non-efficient, as shown on the figure 19 below.

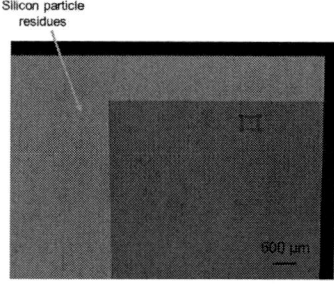

Figure 19. Silicon particles observation on a hydrophobic material when dicing process occurs after lift-off.

Therefore, we decided to modify our flow in order to ensure a lower particle contamination before surface preparation and stacking.

This new flow (number (3) in figure 18) consisted in achieving the dicing before the hydrophobic polymer deposition. If this solution prevent the remaining dicing particles on the hydrophobic surface, it means that the hydrophobic material deposition now occurs on singulated dies. This increases drastically the flow complexity by adding hydrophobic material deposition and lift-off at die-level.

However, we demonstrated that such processes were doable at die-level by preliminary placing the dies on a polymer tape and metallic frame (see figure 20). After dicing, the dies were placed on a tape thanks to a die sorter equipment. The selected tape was chemically compatible with the hydrophobic material solvent. Industrial spin-coating modules compatible with tape and frames already

978-1-7281-1500-9/19 $31.00 © 2019 IEEE   230

exist (FX300 module from Brewer Science). As we demonstrated earlier that the annealing of the thermal treatment A and B was not necessary and even could be eliminated to improve lift-off, there was no need to have a thermally stable tape.

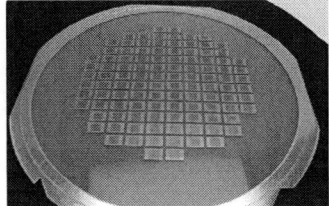

Figure 20. Self-assembly dies placed on a tape for further hydrophobic material deposition and lift-off process steps.

Lift-off on tape is currently under development at LETI using the FX300 system, however very encouraging preliminary results were already been obtained with excellent post lift-off water confinement (see figure 21).

Figure 21. Hydrophobic contrast characterization of the dies after lift-off process on tape. High water confinement demonstrated on the die pad.

E. Toward stacking process automatisation

In order to reach an industrial robustness, a self-assembly stacking system should meet the following requirements:
- Water dispense: control / no evaporation – uniformity across the wafer
- Direct bonding compatibility: Limited particle contamination
- Die placement reproducibility: Height, rotation and planarity control.
- High-speed water evaporation system in order not to reduce throughput

It has to be noted that self-assembly process tools could be today either single D2W tools (very similar with pick and place or flip-chip tools) or collective bonding system with a global positioning of all the top die simultaneously. Today both solutions have pros and cons: single D2W systems would help water volume control and heterogenous integrations, but on the other side, collective bonding would be a tremendous gain for throughput.

1) Automated Die-To-Wafer system evaluation

For early demonstrations a homemade specific equipment composed of infra-red high speed camera (300 frames per second, suction tool and X-Y-Z translation support) was designed [6-8]. However, if this system was perfect to explore process feasibility the reproducibility was limited due to many manual steps: Water volume measurement and placement, top die placement (vacuum centering as well as z positionning). Therefore, the die topography had to be over 50µm to ensure water confinement and reproducibility.

In order to increase the number of die stacked, and to investigate with precision self-assembly capabilities for sub-20µm topography structures, we decided to move forward toward automatization by implementing our process on a fully automated pick and place equipment: the BESI APM2200+ (figure 22).
This tool enabled us to apply a reproducible water volume thanks to an automated syringe dispense. A 2µL of water was deposited on each die. The top die was automatically placed before deposition with the following reproducible parameters: 200µm height above bottom substrate which ensure a water drop contact before releasing the die. In order to simulate high-speed D2W placement, intentional misalignment was performed: the rotation of the top die was 10° and the x and y translation error were of 100µm.

Figure 22. Automated pick & place process development: Water and die placement picture.

However, this tool was not designed originally for direct bonding, as cold have been the latest generation of SETFC1 system for example [2]. This specific system implements microenvironment to limit particle contamination and specific low contact piking head which enables to take the top die directly from the preparation holder. Therefore, when using the BESI tool the dies were firstly loaded upside-down in specific cassettes to avoid any contact of the picking head on the pad surface (see figure 23). This could increase the contamination.

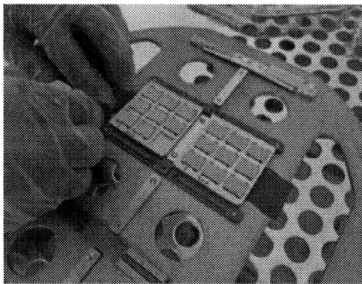

Figure 23. Picture of the self-assembly die cassettes. Dies are placed on their frontside to enable a backside picking by BESI system.

The self-assembly capacity of 3 hydrophobic layer types were evaluated : The polymer, the hydrophobic treatment A and B. The surface preparations for product A and B was similar: UV/O3 and NH4OH Megpie cleaning, whereas for the old polymer a water Megpie cleaning was applied after Ozone treatment.

We tested several pad topography heights for each material ranging from 2 to 15µm. For each configuration, at least 40 dies of the same type were bonded in a raw to ensure reproducibility. An example of 200mm bottom wafer populated with top dies is shown in figure 24.

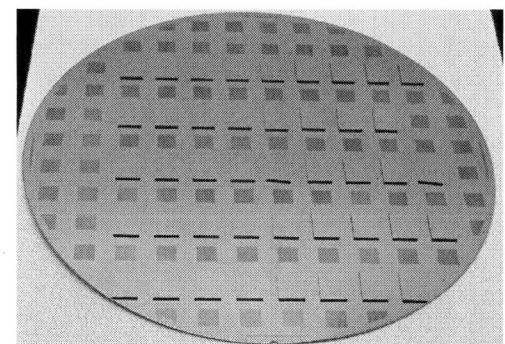

Fig.24. 200mm wafer populated with more than 40 dies placed in BESI pick & place tool by self-assembly process.

After stacking and interfacial water evaporation, the bonded wafers were annealed for 2h at 200°C before acoustic microscopy interfacial inspection and further alignment measurements.

The table 1 presents the self-assembly results for the various configurations tested. The following parameters were considered: Bonding quality (defect visible or not at bonded interface with acoustic microscopy), self-alignment occurring or not, and if misalignment lower than 1µm was measured.

Materials	Die edge topograpy	Number of die stacked	Bonded dies	In situ optical self-alignment checking	Alignment accuracy < 1µm
Polymer	5 µm	36	97%	92%	11%
	15 µm	36	94%	66%	20%
B material	2 µm	62	89%	30%	10%
	5 µm	34	88%	70%	53%
	15 µm	81	94%	56%	44%
A material	5 µm	43	98%	70%	63%
	15 µm	44	91%	88%	78%

TABLE I. SELF- SELF-ASSEMBLY D2W RESULTS ON BESI TOOL

It can be seen that for all the configurations, the number of bonded die was always higher than 88%, confirming that our surface preparation recipes were excellent. An example of acoustic microscopy inspection for material B is shown figure 25. It can be seen that only one die presented a significant defect, probably due to an entrapped particle.

However, it can be seen that the self-alignment performances strongly depend on the die configurations. Indeed, the first parameter influencing the self-alignment performances is the hydrophobic material type. Indeed, it can be seen in the last two columns of the table I above that self-alignment performances <1µm is significantly better with the material A. With this material almost 80% of the dies were self-aligned with success for 15µm pad topography, and 63% with only 5µm pad topography. As discussed in section II.B., this topography can be compatible with a functional demonstrator realization.

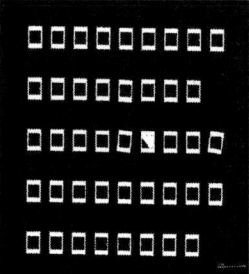

Figure 25. Die to wafer bonding interface evaluation by SAM. No bonding defects are visibles (Hydrophilic treatment A, 15µm pad topography).

These results confirmed the industrial potential of the self-assembly process applied to direct bonding, because even with a non-dedicated automated equipment very encouraging results were obtained in term of bonding quality and self-alignment yield. This result confirmed the relevancy of our newly selected material A as well as wafer and top die preparation integration flow described in the section III.

The next step would be to test lower pad topography for material A type. However, it is important to underline that in order to go further in the self-assembly alignment performance qualification using lower pad topographies, the

978-1-7281-1500-9/19 $31.00 © 2019 IEEE

tool has to be optimized and dedicated to this process. The following points need to be optimized: New loader system compatible with die on tape, Microenvironment for water evaporation control, double head for throughput increase, post stacking water evaporation rate acceleration with specific environment. To achieve such progresses, a close collaboration with a tool manufacturer is mandatory.

2) Collective bonding system exploration

If a $\approx$100-200µm die pre-alignment is required for D2W self-assembly, this process could probably reach a throughput of several thousand off dies/hour if we refer to curve in figure 2 on industrial plateform. However, a major breakthrough for throughput canbe reached with self-assembly adapted to collective bonding.

Indeed the most efficient self-assembly process flow is to use the die holder (or tape) required for surface reactivation prior to bonding as die feeder for collective bonding.

In that case, the water is deposited at wafer-level on the substrate base wafer. Water uniformity deposition control becomes an important challenge, as well as substrate planarity. Moreover, in the case of collective bonding the die should not be loaded onto a tape but into specific cavity holders to prevent any adhesion issues.

We started at LETI to investigate the potential of such collective bonding technique by using the previous manual system [6-8]. Our protocol enables to stack simultaneously 16 dies (4x4). The pad topography selected here was 15µm and the dies and substrate were prepared with the flow exposed in previous section IIB using material A. In our experiment, the dies were loaded into cavity holder wafers (see figure 26). The top substrate (square) was handled by a vacuum pipette over the dies. The water was manually dispensed with a micropipette (2µL).

The early results are very encouraging considering the number of manual steps in our stacking process: All dies presented a perfect bonding interface, alignment accuracy lower than 1µm was obtained on 12/16 dies (75% yield).

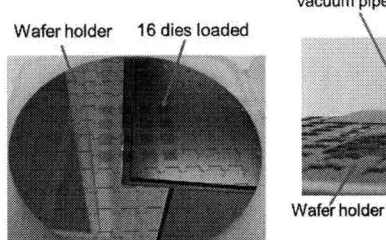

Figure 26. Collective self-assembly direct bonding feasibility demonstration. (Left) The Die holder loaded with the 16 self-assembly dies (Right) The top substrate put in contact with water and bottom dies thanks to the vacuum pipette.

Considering the high potential of collective bonding, LETI is currently developing a semi-automated collective bonding system in order to achieve complete wafer- level bonding. Discussions are ongoing with a major tool manufacturer in order to develop automated modules and complete tool during the next 2 years.

IV. CONCLUSION

The results presented in this paper demonstrated the industrial potential of self-assembly process with significant technical improvements of previous work. Indeed, the water confinement control and surface preparation was drastically improved thanks to a new spin-coatable hydrophobic material selection. This improvement is also due to the development of fully automated lift-off process for bottom substrate preparation. The top die lift-off process is currently under development, with the use of support tape and early results are encouraging. Finally the process setup on standard BESI APM2200+ pick and place equipment significantly improved the self-assembly process robustness: 98% of the dies were perfectly bonded, and 63% presented an alignment lower than 1µm with a low 5µm die edge topography which is compatible with hybrid bonding design. A collective bonding bench is also being developed to prospect on the throughput potential of this process.

CEA-LETI main objective is today to pursue the self-assembly process development, especially with collective bonding systems, in close collaboration with equipment supplier in order to better determined the overall process limitations (topography height, throughput and bonding yield).

REFERENCES

[1] J. Jourdon, S. Lhostis, S. Moreau, J. Chossat, M. Arnoux, C. Sart, Y. Henrion, P. Lamontagne, L. Arnaud, N. Bresson, V. Balan, C. Euvrard, Y. Exbrayat, D. Scevola, E. Deloffre, S. Mermoz, A. Martin, "Hybrid bonding for 3D stacked image sensors: impact of pitch shrinkage on interconnect robustness", 2018 IEE International Electron Device Meeting, pp. 157-160.

[2] P.Metzer, N.Raynaud, A.Jouve, N.Bresson, L.Sanchez, F.Fournel, S.Cheramy , "New flip-chip bonder dedicated to direct bonding for production environment", Proc. Of 2018 7th Electronic System-Integration Technology Conference (ESTC).

[3] G.Gao, L.Mirakarimi, T.Workman, G.Guevara, J.Theil, C.Uzoh, G.Fountain, B. Lee, P.Mrozek, M.Huynh, R.Katkar, "Development of Low Temperature direct bonding interconnect technology for Die-to-Wafer and die-to-die applications-Stacking, yield improvement, reliability assessment", Proc. Of the 2018 Internationnal Wafer-Level Packaging Conference (IWLPC).

[4] A. Jouve et al., « 1µm Pitch Direct hybrid bonding with <300nm Wafer-to-Wafer overlay accuracy", Proc. of S3S conference, 2017

[5] T. Fukushima et al., "Evaluation of Alignment Accuracy on Chip-to-Wafer Self-Assembly and Mechanism on the Direct Chip Bonding at Room Temperature", 2010 IEEE International 3D Systems Integration Conference (3DIC), 2010.

[6] L.Sanchez et al. « Chip to Wafer Bonding Technologies for High Density 3D Integration », Proc. Of IEEE Electronic Components and Technology Conference, 2012.

[7] S.Mermoz et al. , « Impact of Containment and Deposition Method on sub-Micron Chip-to-Wafer Self-Assembly Yield », Proc. Of

IEEE International 3D Systems Integration Conference (3DIC), 2011.

[8] S. Mermoz et al, "High density chip-to-wafer integration using self-assembly on the performances of directly interconnected structures made by direct copper/oxide bonding", IEEE 15th Electronics Packaging Technology Conference, 11-13 December 2013.

2019 IEEE 69th Electronic Components and Technology Conference (ECTC)

A Single Bonding Process for Diverse Organic-Inorganic Integration in IoT Devices

T. H. Yang[1,2*], Y. S. Chiu[1,2], H. Y. Yu[1], A. Shigetou[2], and C. R. Kao[1]

[1]Department of Materials Science and Engineering, National Taiwan University, Taipei 10617, TAIWAN
[2]National Institute for Materials Science (NIMS), Tsukuba, Ibaraki 305-0044, JAPAN
*email: f03527057@ntu.edu.tw

Abstract—**Material hybridization between organic and inorganic materials is crucially important for the development of IoT devices. Especially for wearable and flexible IoT electronics, which are commonly integrated by transfer printing process, organic-inorganic bonding is indispensable for the integration of diverse electronic components. Existing hybrid bonding technologies like laser-assisted bonding or friction stirring welding achieve organic-inorganic bonding using high temperatures as high as melting point of polymers; however, these causes severe material deterioration. Thus, hybrid bonding must be achieved at low temperatures. Here we report a novel hybrid bonding method at the solid-state level and under the atmospheric pressure. Inorganic materials like tin were bonded to polyimide via the ethanol-assisted vacuum ultraviolet (E-VUV) irradiation process, where specimen surface were exposed to a vacuum-ultraviolet (VUV)-irradiated ethanol vapor atmosphere before bonding. VUV-induced re-assembly of ethanol vapor molecules was used to develop hydroxyl-carrying alkyl chains through coordinatively-bonded carboxylate onto tin, whereas numerous hydroxyl-carrying alkyls were created on polyimide. Triggering dehydration via these hydroxyls by merely heating to 150 °C for a few minutes produced robust organic-inorganic reticulated complexes at the tin/polyimide interface. Interface observation via transmission electron microscopy (TEM) shows that the bonded tin/polyimide was extremely compact without readily visible voids. A great number of nano-grains of organic-inorganic complexes were observed in the polyimide side but located ca. 35 nm away from the initial interface, indicating that tin interdiffusion into the polyimide side occurred during hybrid bonding and thus enhanced bondability. The hybrid interface is believed robust due to the strong organic-inorganic nano-grains. Finally, the E-VUV process was experimentally proven to possess broad applicability to diverse inorganic materials, such as aluminum, iron, titanium, and silicon. Adhesion mechanism of E-VUV process was proposed in this study. The E-VUV bonding strategy is expected to be utilized in micro-assembly of flexible and wearable/implantable IoT electronics.**

Keywords—*human-machine interaction, wearable devices, heterogeneous integration, organic-inorganic bonding, direct bonding*

I. INTRODUCTION

In the recent years, there are increasing attentions on wearable IoT devices that are used as human-machine interactive (HMI) systems in terms of contact lens, ultrasound sensors, photonics, etc [1-7]. They are basically required being integrated directly with human bodies, that is, attached on human skin or tissues, for accurate signal resolution. Although existing commercially-developed wearable electronics are able to provide reliable and sufficient functionalities for medical diagnosis, but achieving portable, comfortable, precise, and continuous monitoring of physiological status still remains challenging due to their cumbersome device design like bulky structures and rigid substrate materials. Although integrated circuits (ICs) have been miniaturized and their performances have been greatly improved, the concept of component integration still remains conventional using encapsulation in rigid packages, or using hard components that cannot afford deformation, which results in poor integration with human bodies due to huge mismatch in mechanical stiffness between skins and device substrates. **Fig. 1** compares young's modulus of tissues, skin, organic, inorganic materials commonly used in electronic devices [1, 8]. There are huge differences in young's modulus $10^5 \sim 10^6$ (Pa) between skin and inorganic materials like silicon and copper; therefore, integration of silicon-based devices on skin can barely be achieved. To reduce mechanical mismatch, a promising solution is to integrate electronic components on an organic substrate that has mechanical stiffness extremely close to skins, which provides more comfortable attachment and higher signal resolution due to conformal adhesion. For this purpose, transfer-printing technology [5, 9, 10], that allows pre-fabricated components to be brought to another substrate via temporarily substrates, is employed to achieve successive stacking of electronic functional components into multi-layered architectures on ultra-soft substrates. Bonding is subsequently conducted after components are brought to destination organic substrates. Although various fabrication procedures and handling temporarily materials in transfer printing process are near industrially-matured; however, existing bonding methods for inorganic and organic materials are still far from ideal. Due to totally different natures between inorganic and organic materials, using intermediate organic adhesives [9, 10] between them is somehow inevitable. However, they tend to be degraded due to severe operation

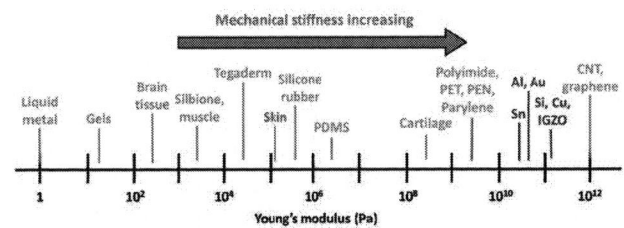

Figure 1. Comparison of young's modulus between skin, tissues, organic, and inorganic materials commonly used in microelectronics, obtained from [1].

978-1-7281-1500-9/19 $31.00 © 2019 IEEE   235

environments, resulting in great reduction in device performance. To solve this, an alternative solution is to eliminate intermediate adhesion layers via direct bonding [8, 11, 12]. To date, organic-inorganic direct bonding has been realized by using laser-assisted bonding that employs high temperatures near the melting points to enhance plasticity and interdiffusion across heterogeneous interfaces [13]. However, such high temperatures may be accompanied by considerable thermo-mechanical damages. Therefore, bonding temperature must be reduced to low temperatures, for example, around the glass-transition temperatures (T_g) of polymers [14, 15].

In a recent study, we have realized aluminum-polyimide bonding at the solid state level by utilizing a novel low-temperature hybrid bonding technique known as ethanol-assisted vacuum ultraviolet (E-VUV) irradiation process [12]. The E-VUV-bonded aluminum/polyimide interface exhibited robust and anti-hydrolysis characteristics. In this study, we further examined the feasibility of E-VUV process to more hybrid material combinations. Considering inorganic materials like titanium, tin, iron, and silicon are commonly used in electronic packaging and structural materials, the E-VUV feasibility to these materials were thoroughly investigated. Adhesion mechanism of the E-VUV process was also discussed in this study.

II. EXPERIMENTAL PROCEDURE

A. Materials & Specimen Preparation

In this study, five kinds of hybrid bonding, tin-, aluminum-, titanium-, iron-, and silicon-polyimide were conducted. A polyimide (PI) film with 25 μm thickness (PMDA-ODA, DuPont Kapton) was prepared to dimensions of 15×15 mm². A polycrystalline tin (Sn) stick (Sn99.3Cu, Sasaki Solder Industry Inc.) were prepared into a plate with dimensions of 10×10×2 mm³. A polycrystalline aluminum film with a thickness of ca. 140 nm was pre-deposited onto a silicon die using an electron-beam evaporation system (UEP-3000-2C, Ulvac Inc.) and prepared to dimensions of 10×10 mm². High purity iron (Fe) and titanium (Ti) plates with thickness of 1.5 mm were prepared to dimensions of 10×10 mm². A silicon (Si) die with thickness 0.5 mm was cut from a high purity Si wafer, and prepared to dimensions of 10×10 mm². Prior to use, all the inorganic specimens were ultrasonically cleaned for 3 min each in acetone, ethanol, and deionized water. All the polyimide films were ultrasonically cleaned for 3 min each in ethanol and deionized water.

B. Ethanol-Assisted Vacuum Ultraviolet (E-VUV) Irradiation Treatment and Hybrid bonding

After cleaning, all sample were introduced into the hybrid bonding apparatus. The equipment has been depicted in our pioneering study [11, 12], so its details are omitted here. The hybrid bonding apparatus mainly comprises three chambers with different functions, including VUV irradiation, XPS analysis, and hybrid bonding. The VUV chamber, where specimens were treated by ethanol-assisted vacuum ultraviolet (E-VUV) process, was pumped until the background vacuum reached 10^{-4} Pa to limit the unfavorable influence of residual gas molecules on the chemical binding

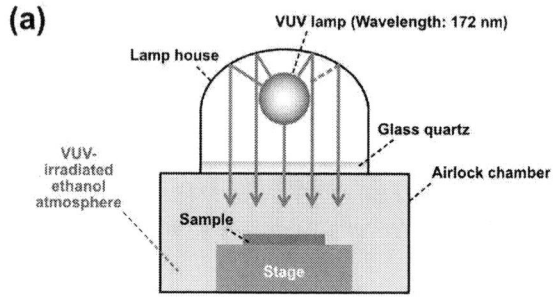

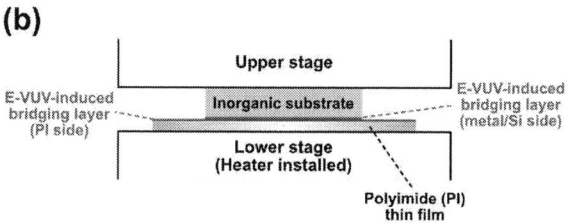

Figure 2. (a) Illustration showing the E-VUV surface treatment in the VUV chamber. (b) Illustration showing hybrid bonding in the bonding chambers.

condition. An ethanol-containing (99.5 vol%, Wako Chemicals, Ltd.) nitrogen atmosphere was then introduced into the VUV chamber using an ultrasonic atomizer until the gas pressure reached 9×10^4 Pa (ca. 0.9 atm). **Fig. 2(a)** illustrates how the E-VUV process was conducted in the VUV chamber. During E-VUV modification, chamber temperature was kept at 24°C. A VUV source (UER20H-172VA, Ushio Inc.) with a wavelength of 172 nm was employed to provide VUV irradiation and its incident power per area was set ca. 5 mW/cm². The distance from the VUV light source to the specimen surface was set ca. 70 mm.

In our recent works, the main factor to growth behavior of the E-VUV-induced layer has been verified to be the amount of exposure to ethanol, Γ (kg·s/m³), which represents the product of the ethanol vapor density (kg/m³) and the VUV irradiation time (s). The evolution of the bridge layer was closely correlated with the amount of exposure to ethanol. In the present study, we used the same definition to estimate the Γ value (for which details have been discussed in our previous study [12]).

$$\Gamma = 0.14RH\% \cdot t \qquad (1)$$

Recording the percentage relative humidity (RH%) and the VUV irradiation time (s) enabled us to approximate the ethanol exposure during the E-VUV treatment.

After the E-VUV irradiation treatment reached pre-determined exposure (for which the value was determined based on the results of surface analysis), the E-VUV-treated specimens were transferred into the hybrid bonding chamber. **Fig. 2(b)** illustrates how hybrid bonding was conducted. A bonding pressure ca. 3~4 MPa was given to the hybrid couple for sufficient surface contact, followed by isothermal bonding at 150 °C for 10 min.

C. Surface Analysis

After the E-VUV treatment, the samples were subjected to surface analysis to identify growth behavior of the E-VUV-modified layers. An x-ray photoelectron spectroscopy (XPS) and an attenuated total reflection Fourier transmission infrared spectroscopy (ATR-FTIR) were conducted. All XPS investigation was conducted in the XPS chamber installed with the hybrid bonding apparatus. Measurements were taken using a nondestructive angle-resolved method, with the Mg Kα (1256 eV) source operating at 15 kV and 27 mA. The takeoff angles (θ) of the angle-resolved analyses were determined to be 15, 30, and 45°, where θ = 90° corresponds to the normal to the surface, to investigate the vertical distribution of the E-VUV-induced surface species. After XPS spectrum acquisition, we conducted a series of XPS spectrum analyses using PHI Multipak software. As for ATR-FTIR measurements, we conducted a Thermo Scientific Nicolet 4700 spectrometer in total external reflectance mode. An infrared beam was transmitted through KBr windows onto the sample, and reflected at an incident angle of 80° to a mercury-cadmium-tellurium (MCT) detector cooled with liquid nitrogen. The spectrometer was kept purged with nitrogen gas during the measurement. Spectra were collected at a resolution of 2 cm^{-1} and with the co-addition of 128 single scans.

D. Interface observation

For observation of the interfacial microstructures and chemical configurations, a transmission electron microscopy (TEM) and an electron energy loss spectroscopy (EELS) were conducted. we prepared TEM samples by using a focus ion beam (FIB) system (JIB-4000, JEOL) with a Ga liquid-crystal ion source. TEM and EELS investigations were conducted using a FEI Tecnai G2 F30 installed with parallel EELS (PEELS) operated at 300 kV. The energy resolution was measured to have a full width at half maximum (FWHM) of 1.3 eV on the zero-loss peak. To record the energy-loss near-edge structure (ELNES), we employed a dispersion of 0.1 eV/channel, an entrance aperture diameter of 2 mm, and a cumulative time of 10.24 s per acquisition.

III. RESULTS AND DISCUSSION

A. Influence of E-VUV on Inorganic Surfaces

Provided tin is a metal material widely used in electronic packaging [16, 17], we first analyzed the effects of E-VUV on polycrystalline Sn plates. We confirmed the difference in the chemical bonding status after VUV irradiation at various exposures by using XPS and ATR-FTIR. **Fig. 3** to **Fig. 5** show the evolution of the XPS and ATR-FTIR spectra (see also peak assignments in **Table 1** and **2**). At low exposure, organic contaminants were almost removed, as shown in spectrum Γ_1 in **Fig. 3**, which was contributed mainly by the reaction of H and O radicals with the organic contaminants to turn them into gaseous CO_2 and H_2O [18, 19]. After the native organic adsorbates almost disappeared, the solder surfaces gradually underwent hydration via the formation of hydroxyl groups on the Sn sites (spectrum Γ_1 in **Fig. 4**). With increased exposure to E-VUV, we found that the VUV-induced molecules with

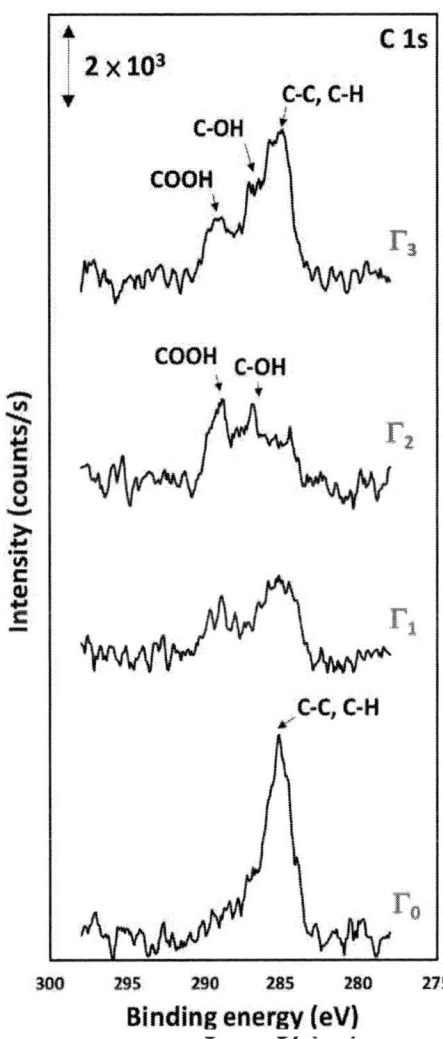

Figure 3. Evolution of XPS C1s spectra for the Sn99.3Cu solder surface subjected to the E-VUV process. The exposure (Γ) of ethanol with a numerical subscript from 0 to 3 represents 0, 159.6, 1230.6, and 2175.6 (kg·s/m^3), respectively.

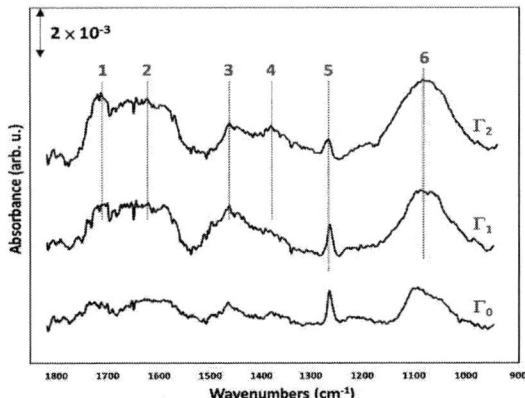

Figure 4. ATR-FTIR spectra for the Sn99.3Cu solder surface subjected to the E-VUV process. Γ_0 to Γ_2 indicate 0, 180.6, and 2776.2 (kg·s/m^3), respectively.

carboxylic groups were deprotonated through these hydroxyl groups on the solder surface to form coordinatively-bonded carboxylate species (see spectrum Γ_2 in **Fig. 3**; see also spectrum Γ_1 and Γ_2 in **Fig. 4**). We further verified the coordination type of these carboxylate ligands by measuring the peak separation between symmetric and asymmetric stretching vibrations of carboxylate in the ATR-FTIR spectra, as shown in **Table 1**. Despite the peak of symmetric stretching vibrations of carboxylate overlapped with -CH$_2$- groups slightly, the peak separation (Δ) still could be measured to be ca. 162 cm^{-1}, indicating that the carboxylates were coordinated to two Sn cations in the surface tin oxide (SnO$_2$) in the conformation of a bridging bidentate ligand [20-22]. It is also found that considerable hydroxyls were formed at the chain terminals of those carboxylates (spectrum Γ_2 and Γ_3 in **Fig. 3**). As E-VUV exposure further increased, acyclic alkyl and ester groups were then produced (spectrum Γ_3 in **Fig. 3**, spectrum

Γ_2 in **Fig. 4**, and **Fig. 5**; see also peak assignments in **Table 2**). This phenomenon was probably attributed to the hydroxyl groups on the carboxylates that underwent esterification through the gaseous molecules carrying carboxyls [12], which resulted in alkyl grafting. Another possible reaction path for the increase in the C-C signal was that VUV irradiation produced many unsaturated carbon atoms on the surface carboxylates by cleaving C-H bonds, so that the unsaturated carbon atoms attracted unsaturated gaseous alkyl groups through unpaired electrons and were then transformed into saturated C-C bonds.

On the basis of XPS and ATR-FTIR results, the evolution of surface functional groups induced by the E-VUV process could be obtained, as illustrated in **Fig. 6**. The E-VUV process was experimentally proven to be effective for the generation of acyclic species carrying hydroxyl or alkyl end groups onto the Sn99.3Cu surface through bridging carboxylate. According to our recent finding [12], a hybrid interface that carried numerous carboxylate groups firmly bonded on inorganic side exhibited an anti-hydrolysis characteristic. Thus, it is expected that the E-VUV process was applicable to various inorganic materials for the creation of water-tolerant inorganic/organic interfaces through anti-hydrolytic carboxylates.

B. Interfacial Microstructures and Chemical Configurations

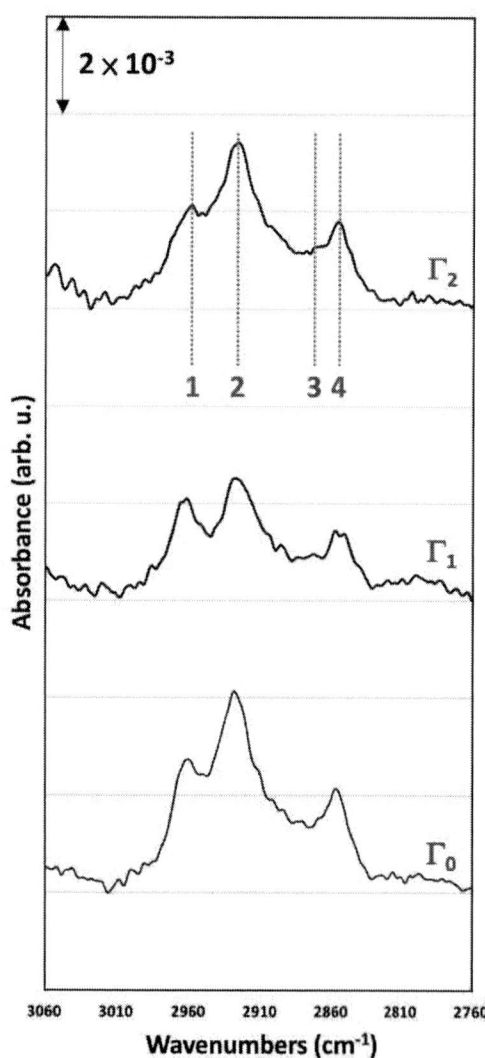

Figure 5. ATR-FTIR spectra for the Sn99.3Cu solder surface subjected to the E-VUV process. Γ_0 to Γ_2 indicate 0, 180.6, and 2776.2 (kg·s/m^3), respectively.

TABLE I.   PEAK POSITIONS AND CORRESPONDING ASSIGNMENTS OF THE NUMBERS MARKED IN FIG. 4.

No.	Wavelength (cm^{-1})	Assignment
1	1712	$\nu(C = O)$ of COOH
2	1626	$\nu_{as}(COO^-)$
3	1464	$\nu_s(COO^-)$ $\delta(CH_2)$
4	1383	$\delta(CH_3)$
5	1267	$\delta(Sn - OH)$ (terminal)
6	1084	(O–C–C) bending modes of –CO$_2$R ester

TABLE II.   PEAK POSITIONS AND CORRESPONDING ASSIGNMENTS OF THE NUMBERS MARKED IN FIG. 5.

No.	Wavelength (cm^{-1})	Assignment
1	2958	$\nu_{as}(CH_3)$
2	2626	$\nu_{as}(CH_2)$
3	2873	$\nu_s(CH_3)$
4	2856	$\nu_s(CH_2)$

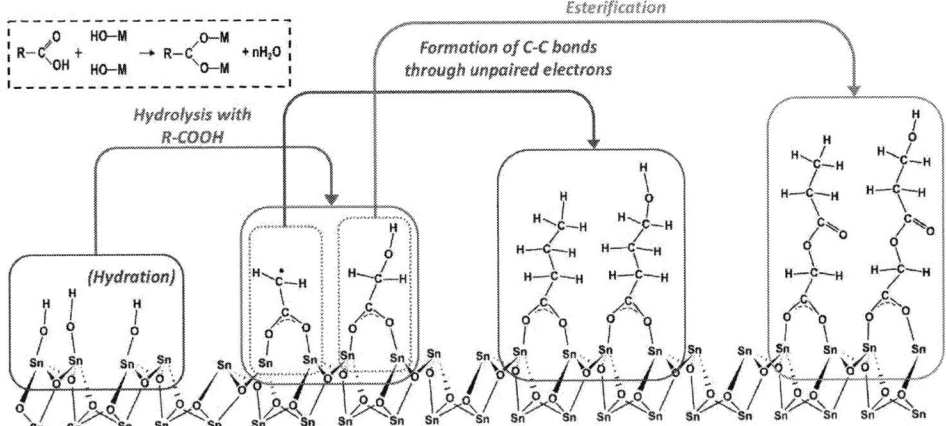

Figure 6. Illustration showing how the bridging layer forms on the Sn99.3Cu surface with the E-VUV process.

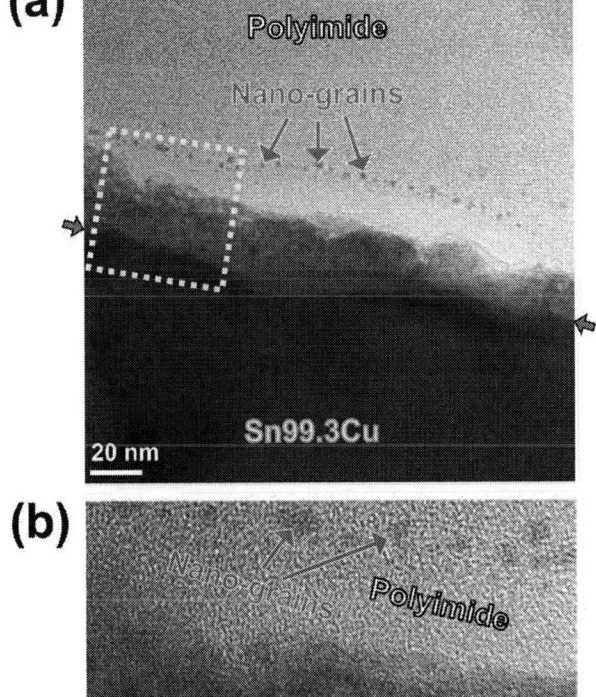

Figure 7. TEM investigation of bonded interface achieved by E-VUV process. (a) polyimide/Sn99.3Cu interface. (b) Zoom-in image of yellow dash box in (a).

Since we have previously discussed influences of the VUV process on PI surfaces [11, 12, 23], so the experimental results are omitted here. The main influence induced by the E-VUV process was that a great number of hydroxyl groups were produced on the polyimide surface if it was treated with optimized exposures [12]. Considering that hydroxyl group is a critical species to trigger dehydration condensation between two surfaces; thus, in this work we conducted the E-VUV treatment at a close ethanol exposure on PI surfaces based on the pioneering experimental results [12].

After the E-VUV treatment, Sn99.3Cu-PI hybrid bonding was conducted at 150 °C for 10 min, followed by cooling in the air. **Fig. 7** shows high-resolution TEM images of Sn99.3Cu-PI interface. It is observed that the Sn99.3Cu-PI interface looked compact without voids, indicating the E-VUV method was effective for the Sn99.3Cu-PI bonding via the triggering of dehydration reaction through hydroxyls between the Sn99.3Cu/PI interface at 150 °C. A continuous diffusion layer was extended from the Sn99.3Cu side to the PI side, and its thickness was measured to be 20~25 nm. Compared with the results about Al-PI bonding demonstrated in our previous report [12], the Sn99.3Cu/PI interfacial region was relatively thicker, which indicates that the interfacial Sn might have better mobility than the interfacial Al in the E-VUV-achieved heterogeneous interface. In addition, a great number of grain-like speckles were found to be arranged in a row 35~40 nm away from the initial surface, as shown in **Fig. 7(a)**, which implies a composition-gradient layer through successive interdiffusion from Sn99.3Cu to PI via the ultrathin bridge layer was formed. **Fig. 7(b)** shows the enlarged image of the yellow-dash box in **Fig. 7(a)**. Evidently, numerous nano-grains with different crystal orientations were observed in the continuous diffusion region.

In order to identify chemical configurations of the Sn99.3Cu/PI interface, a series of EELS analyses were conducted. **Fig. 8** demonstrates the EEL spectra obtained in the Sn99.3Cu matrix and near the bonded interface. Although the substantial overlap of the O-K edges with the $M_{4,5}$ delayed edges of tin, it has been reported [24, 25] that the dominant fine structure observed above 530 eV was mainly attributed to oxygen states, whereas the delayed $M_{4,5}$ edges of tin made the

978-1-7281-1500-9/19 $31.00 © 2019 IEEE   239

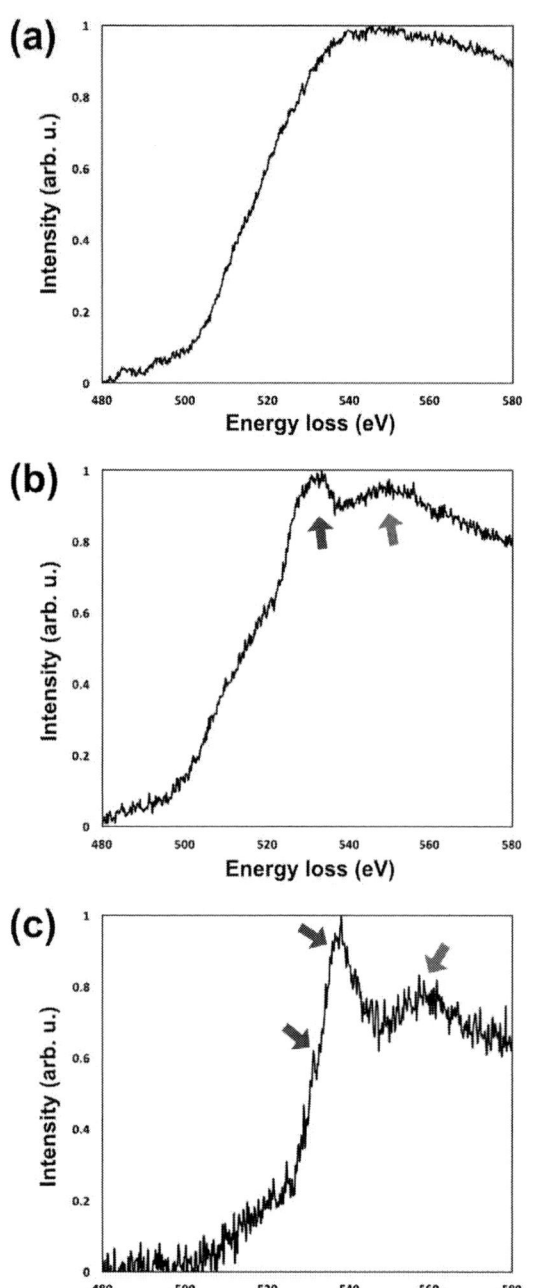

Figure 8. EELS Sn $M_{4,5}$- and O-edges from (a) the Sn99.3Cu matrix, (b) the continuous diffusion region, and (c) the grain-like speckles region ca. 35 nm far from the initial interface.

main contribution to the intensity. The oxygen edges thus were able to reflect more clearly the different structural environments in tin compounds than the tin edges. Compared with the pure Sn spectrum presented in **Fig. 8(a)**, the O-K edge characteristics could be observed in the diffusion region and the grain-like speckle region, indicating that the interface Sn were almost oxidized and linked with the surrounding PI molecules via those oxygen bridge, as shown in **Fig. 8(b)** and

(c). The characteristic peaks around 530~540 eV (pointed by blue arrays), which were supposed to be the most important fingerprint of tin oxide, were not exactly corresponded to a certain tin oxide crystal. Instead, the O-K edges exhibited a combination of tetrahedral and triangular oxygen, which might imply the existence of more than one chemical (and/or crystallographic) site for oxygen. Therefore, it is considered that the interfacial Sn and surrounding molecules of PI formed organic-inorganic complexes though Sn-O-C bonds. The broad peak within 540~570 eV (pointed by red arrays) was considered as the subsidiary evidence since it was mainly due to an increase in the variance of neighboring Sn-O bond lengths. The formation of Sn-O-C complex bonds resulted in the different Sn-O bond lengths, and the broad peak was thus formed. On the basis of our recent experimental results [12], it is expected that the formation of the organic-inorganic reticulated nano-grains via such the metal-oxygen-carbon complex bonds would enhance the interfacial adhesion strength due to robust covalent/ionic bonding.

C. Applicability of E-VUV to Diverse Hybrid Combinations

Other organic/inorganic combinations of Al/PI, Fe-PI, Ti-PI, and Si-PI (not shown here) were successfully bonded by means of the E-VUV method, as demonstrated in **Fig. 9**. As seen in these TEM images, the E-VUV process was demonstrated to possess broad applicability to various inorganic materials widely utilized in lightweight structural materials and microelectronics. The bonded interfaces were compact without visible voids. Moreover, substantial interdiffusion was considered to occur within the bonded interfacial region since many lattice fringes were observed at the PI side, as shown in **Fig. 9(a)** and **(b)**. Some of them were extended from the inorganic matrix into the PI side, on which numerous randomly arranged grains were produced, indicating that the crystal orientations newly formed during hybrid bonding first followed those of the inorganic matrix, and recrystallization was then occurred to produce considerable nano-grains without preferred orientation. According to our recent research results [12], such kind of interfacial microstructures would possess excellent ability to resist crack propagation due to crack deflection by those robust organic-inorganic nano-grains.

D. Bonding Mechanism and Merits of E-VUV Process

The E-VUV process enabled robust organic-inorganic direct bonding within a few minutes mainly via the triggering of dehydration condensation between organic and inorganic surfaces. As the ethanol exposure was low, native organic contaminants on inorganic surfaces were almost removed so that the coordinatively-bonded carboxylates could be formed and firmly adhered on inorganic oxide surface. Numerous surface hydroxyls were then formed on these coordinatively-bonded carboxylates. On the other hand, a great number of hydroxyl-carrying alkyl groups were grafted on the polyimide surface during the E-VUV treatment. The triggering of dehydration reaction between the surface hydroxyls produced strong covalent/ionic bonds between the two target surfaces during hybrid bonding, and thus enhanced bondability significantly.

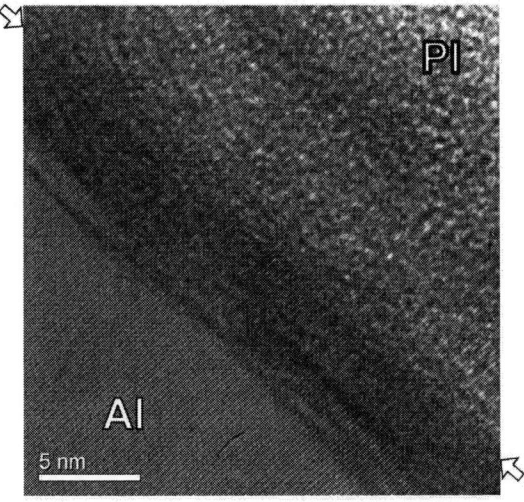

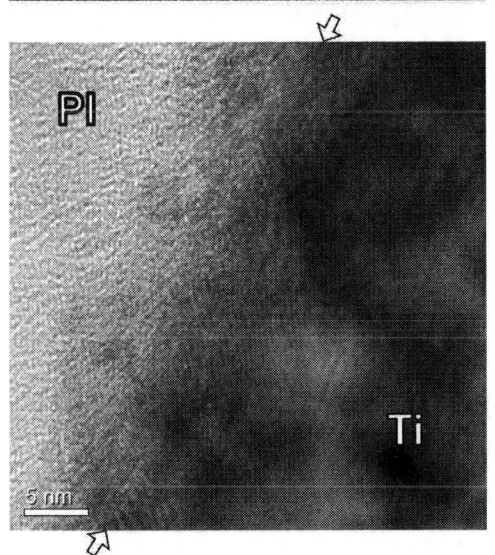

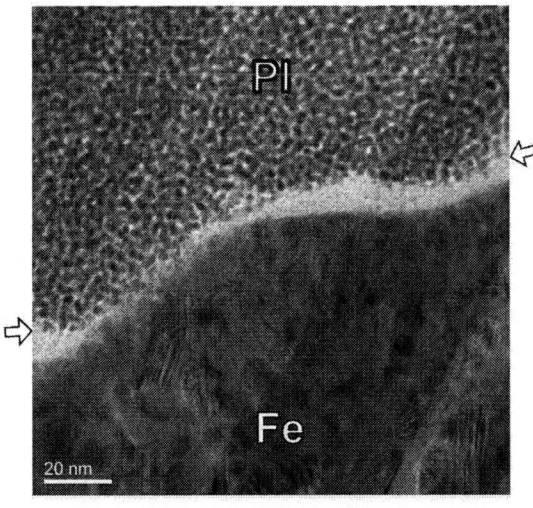

Figure 9. TEM investigation of bonded (a) polyimide/aluminum, (b) polyimide/titanium, and (c) polyimide/iron interfaces achieved by E-VUV process.

The E-VUV bonding process had the following features. First, the proposed process was highly time-efficient because the whole bonding process was completed within a few minutes. The as-bonded interfacial strength was proven to be strong [12], so post-annealing was not necessary. Second, since bonding process was capable of being conducted at relatively low temperatures compared with the glass-transition temperature (T_g) of polymers, unfavorable thermo-mechanical damages to electronic components could be effectively suppressed. Third, the proposed process was conducted under non-vacuum conditions and in a non-toxic environment; therefore, it has potential to be utilized in existing industrial production lines, such as roll-to-roll processes. Production complications and costs are also expected to be significantly reduced. Finally, the proposed process meets the growing needs of heterogeneous integration structures, where homogeneous and heterogeneous materials appear on the same plane, and bonding are required being achieved in a single bonding process [26, 27]. The E-VUV process is expected to be combined with existing assembly process, such as transfer printing process, to be utilized in flexible electronic packaging.

IV. CONCLUSION

In this study, several kinds of organic-inorganic hybrid bonding like Sn99.3Cu-, Al-, Si-, Fe-, and Ti-polyimide were achieved at 150 °C and under the ambient atmosphere using the ethanol-assisted vacuum ultraviolet (E-VUV) irradiation process. The XPS and ATR-FTIR results reveal that the E-VUV process produced highly hydrophilic modification layer with anti-hydrolysis characteristic on the inorganic surfaces by creating hydroxyl-carrying alkyl chains through coordinatively-bonded bridging carboxylate groups. Chemically strong adhesion between these coordinatively-bonded carboxylates and the inorganic surfaces was considered to ensure high interfacial strength of hybrid interface. The thickness of interfacial diffusion layer was measured to be smaller than 20 nm in all the bonding combinations. The E-VUV-bonded heterogeneous interfaces were demonstrated extremely compact without readily visible voids, and considerable nano-grains with random crystal orientations were found to be formed in the interfacial region during hybrid bonding. With the help of EELS analysis, these nano-grains were identified to be organic-inorganic complexes mainly via oxygen bridges. Combined with our recent study [12], such the interfacial microstructures are expected to possess high fracture toughness due to crack deflection by these robust organic-inorganic complex nano-grains. Since the E-VUV process featured low process temperature, non-vacuum process atmosphere, high time-efficiency, and broad applicability to diverse metal, semiconductor, and polymer materials, it has potential to be utilized in future flexible electronic assembly for IoT applications.

ACKNOWLEDGMENT

This study was supported by the Ministry of Education, Culture, Sports, Science and Technology of Japan (MEXT), Japan Society for the Promotion of Science (JSPS) through a grant MEXT/JSPS Grant-in-Aid for Scientific Research (A) (No. 17H01275), and the National Institute for Materials Science (NIMS)-National Taiwan University (NTU) International Joint Graduate Program (IJGP). T. H. Y. thanks Ms. Chia-Ying Chien in the Instrumentation Center of NTU for the skilled TEM operation. The authors appreciate specimen preparation from the Namiki Foundry (NIMS) and the Nanotechnology Platform (NIMS).

REFERENCES

[1] Y. Liu, M. Pharr, and G. A. Salvatore, "Lab-on-skin: a review of flexible and stretchable electronics for wearable health monitoring," ACS nano, 2017, vol. 11, pp. 9614-9635.

[2] A. Vásquez Quintero, R. Verplancke, H. De Smet, and J. Vanfleteren, "Stretchable electronic platform for soft and smart contact lens applications," Advanced Materials Technologies, 2017, vol. 2, 1700073.

[3] W. Gao et al., "Fully integrated wearable sensor arrays for multiplexed in situ perspiration analysis," Nature, 2016, vol. 529, pp. 509-514.

[4] J. Park et al., "Soft, smart contact lenses with integrations of wireless circuits, glucose sensors, and displays," Science Advances, 2018, vol. 4, eaap9841.

[5] S. Xu et al., "Soft microfluidic assemblies of sensors, circuits, and radios for the skin," Science, 2014, vol. 344, pp. 70-74.

[6] T. Yokota et al., "Ultraflexible organic photonic skin," Science Advances, 2016, vol. 2, e1501856.

[7] S. Luo and T. Liu, "Graphite nanoplatelet enabled embeddable fiber sensor for in situ curing monitoring and structural health monitoring of polymeric composites," ACS Applied Materials & Interfaces, 2014, vol. 6, pp. 9314-9320.

[8] T.H. Yang, C.Y. Yang, A. Shigetou, and C.R. Kao, "A single bonding process for homogeneous and heterogeneous bonding in flexible electronics," Proceeding of 2019 IEEE International Conference on Electronics Packaging (ICEP), Niigata, Japan, Apr. 17-19, 2019; IEEE, 2019.

[9] D.H. Kim, R. Ghaffari, N. Lu, J.A. Rogers, "Flexible and stretchable electronics for biointegrated devices," Annual Review of Biomedical Engineering, 2012, vol. 14, pp. 113-128.

[10] A.M. Hussain and M.M. Hussain, "CMOS-technology-enabled flexible and stretchable electronics for internet of everything applications," Advanced Materials, 2016, vol. 28, pp. 4219-4249.

[11] H.W. Yang, C.R. Kao, and A. Shigetou, "Fast atom beam- and vacuum-ultraviolet-activated sites for low-temperature hybrid integration," Langmuir, 2017, vol. 33, pp. 8413-8419.

[12] T.H. Yang, C.R. Kao and A. Shigetou, "Inorganic-organic solid-state hybridization with high strength and anti-hydrolysis interface," Scientific Reports, 2019, vol. 9, 504.

[13] M. Wahba, Y. Kawahito, and S. Katayama, "Laser direct joining of AZ91D thixomolded Mg alloy and amorphous polyethylene terephthalate," Journal of Materials Processing Technology, 2011, vol. 211, pp. 1166-1174.

[14] W.D. Callister Jr., Materials Science and Engineering, An Introduction, 7th ed., New York: Wiley, 2007.

[15] D.W. van Krevelen and K. te Nijenhuis, Properties of Polymers, 4th ed., Amsterdam: Elsevier, 2009, pp. 148.

[16] H. Y. Chuang et al., Critical concerns in soldering reactions arising from space confinement in 3-D IC packages," IEEE Transactions on Device and Materials Reliability, 2012, vol. 12, pp. 233-240.

[17] T.H. Yang, H.Y. Yu, Y.W. Wang, and C.R. Kao, "Effects of aspect ratio on microstructural evolution of Ni/Sn/Ni microjoints," Journal of Electronic Materials, 2019. vol. 48, 2019, pp. 9-16.

[18] W. Kern, Handbook of semiconductor wafer cleaning technology, New Jersey: Noyes, 1993.

[19] A. Shigetou, J. Mizuno, and S. Shoji, "Vacuum ultraviolet (VUV) and vapor-combined surface modification for hybrid bonding SiC, GaN, and Si substrates at low temperature and atmospheric pressure," Proc. 2015 IEEE 65th Electronic Components and Technology Conference (ECTC), San Diego, CA, U.S.A., May 26-29, 2015; IEEE, 2015, doi: 10.1109/ECTC.2015.7159796.

[20] G.B. Deacon and R.J. Phillips, "Relationships between the carbon-oxygen stretching frequencies of carboxylato complexes and the type of carboxylate coordination," Coordination Chemistry Reviews, 1980, vol. 33, pp. 227-250.

[21] A.A. Davydov and C.H. Rochester, Infrared spectroscopy of adsorbed species on the surface of transition metal oxides, Chichester: Wiley, 1990.

[22] J.N. Brönsted, "Acid and basic catalysis," Chemical Reviews, 1928, vol. 5, pp. 231-338.

[23] T.H. Yang, C.R. Kao, and A. Shigetou, "Organic/inorganic interfacial microstructures achieved by fast atom beam bombardment and vacuum ultraviolet irradiation," Proceeding of 2018 IEEE International Conference on Electronics Packaging (ICEP), Mie, Japan, Apr. 17-21, 2018; IEEE, 2018, doi: 10.23919/ICEP.2018.8374342.

[24] M.S. Moreno, R.F. Egerton, and P.A. Midgley, "Differentiation of tin oxides using electron energy-loss spectroscopy," Physical Review B, 2004, vol. 69, 233304.

[25] M.S. Moreno, R.F. Egerton, J.J. Rehr, and P.A. Midgley, "Electronic structure of tin oxides by electron energy loss spectroscopy and real-space multiple scattering calculations," Physical Review B, 2005, vol. 71, 035103.

[26] M. Ohyama et al., "Evaluation of hybrid bonding technology of single-micron pitch with planar structure for 3D interconnection," Microelectronics Reliability, 2006, vol. 59, pp. 134-139.

[27] A. Shigetou and T. Suga, "Vapor-assisted surface activation method for homo-and heterogeneous bonding of Cu, SiO_2, and polyimide at 150 °C and atmospheric pressure," Journal of electronic materials, 2012, vol. 41, pp. 2274-2280.

978-1-7281-1500-9/19 $31.00 © 2019 IEEE

Stretchable and Printable Medical Dry Electrode Arrays on Textile for Electrophysiological Monitoring

Yougen Hu[1*], Hui Wang[1,2], Ommeaymen Sheikhnejad[3], Yaoxu Xiong[1,2], Han Gu[1,4], Pengli Zhu[1], Guanglin Li[1], Rong Sun[1*], Ching-Ping Wong[5]

[1] Shenzhen Institutes of Advanced Technology, Chinese Academy of Sciences, Shenzhen 518055, China
[2] Shenzhen College of Advanced Technology, University of Chinese Academy of Sciences, Shenzhen 518055, China
[3] AC2T research GmbH, Viktor-Kaplan-Straße 2, 2700 Wiener Neustadt, Austria
[4] Nano Science and Technology Institute, University of Science and Technology of China, Suzhou 215123, China
[5] School of Mechanical Engineering, Georgia Institute of Technology, 771 Ferst Drive, Atlanta, Georgia 30332, USA
e-mail: yg.hu@siat.ac.cn; rong.sun@siat.ac.cn

Abstract—Medical electrodes are medical devices that facilitate the transfer energy of ionic currents in the body into electrical currents that can be amplified, studied, and used for diagnosis of medical condition of the patients. Multi-channel and wearable electrode arrays are even more challenging for the medical dry electrodes to simply and accurately achieve electrophysiological signals such as electromyography (EMG). There are a few challenges in materials preparation and device fabrication have to be solved: i) highly conductive and highly conformal electrode materials in order to reduce the contact impedance with human skin; ii) facile patterning and integration procedures of large area multi-channel electrodes with high throughput and low costs; iii) good adhesion between electrodes and substrate materials to avoid delamination after multiple use. Herein, we demonstrate a simple and low-cost fabrication process of stretchable and printable medical dry electrode arrays based on highly conductive composites to simultaneously resolve all the above challenges. The proposed electrode arrays (2 rows × 8 columns = 16 channels) are screen printed on thermoplastic polyurethanes (TPU) film substrate and bonded with textile by hot melt adhesive. The electrodes are composed of flexible conductive composites of TPU/Ag formed by mixing silver micro-flakes with TPU/DMF solution and followed by heat curing at 80 ℃ for 2 h. The TPU/Ag conductive composite is intrinsic stretchable with high electrical conductivity up to 2.76×10^6 S m^{-1} at 85 wt% silver loading. The as-prepared multi-channel electrode arrays are easily attached onto human body such as arm, leg and forehead, etc, and they demonstrate excellent performance in accurately detecting forearm EMG signals for various motions.

Keywords-flexible dry electrode; electromyography (EMG); screen printing; thermoplastic polyurethane; flexible conductive polymer composites

I. INTRODUCTION

The electrophysiological signals (i.e., electrocardiogram (ECG), electroencephalogram (EEG), electromyography (EMG), electrooculography (EOG), *etc.*) from human body are of great value for early examination, diagnosis, and rehabilitation monitoring of a great deal of brain-, heart-, and muscle-related diseases [1-5]. Medical electrodes are a key element of wearable devices for future health care and the internet of humans, which will facilitate the transfer energy

of ionic currents in the body into electrical currents that can be amplified, studied, and used for diagnosis of medical condition of the patients [6, 7]. On the basis of technology, the medical electrodes are divided into wet electrodes, needle electrodes and dry electrodes. Compared with wet electrodes and needle electrodes, dry electrodes offer higher patient comfort and provide higher signal quality. Stretchable dry electrodes have drawn more and more attention in recent years due to they can conformally contact with the skin surface without the assistance of conductive gel and can adapt to human movement, which will improve the signal acquisition stability and signal-to-noise ratio [8-10]. Therefore, there is an urgent need to develop and facilitate innovative ways of fabricating high-performance stretchable medical dry electrodes.

Multi-channel and wearable electrode arrays are even more challenging for the medical dry electrodes to simply and accurately achieve electrophysiological signals. There are a few challenges in materials preparation and device fabrication have to be solved: i) highly conductive and highly conformal electrode materials in order to reduce the contact impedance with human skin; ii) facile patterning and integration procedures of large area multi-channel electrodes with high throughput and low costs; iii) good adhesion between electrodes and substrate materials to avoid delamination after multiple use [11-14].

Herein, we demonstrate a simple and low-cost fabrication process of stretchable and printable medical dry electrode arrays based on highly conductive composites to simultaneously resolve all the above challenges. The proposed electrode arrays are screen printed on thermoplastic polyurethanes (TPU) film substrate and bonded with textile by hot melt adhesive. The electrodes are composed of flexible conductive composites of TPU/Ag formed by mixing silver micro-flakes with TPU/DMF solution and followed by heat curing at 80 ℃ for 2 h. The TPU/Ag conductive composite is intrinsic stretchable with high electrical conductivity up to 2.76×10^6 S m^{-1} at 85 wt% silver loading. The as-prepared multi-channel electrode arrays are easily attached onto human body such as arm, leg and forehead, etc, and they demonstrate excellent performance in accurately detecting forearm EMG signals for various motions. As a result, the developed flexible dry electrodes have promising practical applications in healthcare, soft electronics, sports physiology and rehabilitation.

II.   EXPERIMENTAL SECTION

A. Fabrication of the flexible medical dry electrode arrays

Firstly, a thin TPU film was fabricated by spin coating TPU/N, N-dimethylformamide (DMF) mixture solution on a silicon (Si) wafer under 1000 rpm for 12 s followed by solidifying at 80 ℃ for 1 h to completely volatilize DMF solvent. To aid in visualization, 3 wt% of carbon black (CB) was added into the above TPU/DMF solution. Then the conductive composite paste of TPU/Ag was formed by mixing silver micro-flakes (2-5 µm) with TPU/DMF (25 wt%) solution. The TPU/Ag paste was subsequently programmable and scalable deposited on to the black TPU film by a facile screen printing method followed by heat curing at 80 ℃ for 2 h to achieve the 2 rows × 8 columns array patterns (16 channels). Finally, the printed TPU/Ag electrode on TPU film was peeled off from Si wafer, and boned its back with textile (gauze roller bandage) by hot melt adhesive under 150 ℃ for 30 min. The schematic diagram of the printed electrode arrays on textile is shown in Figure 1.

The electrode arrays are designed with 16 units (2 rows× 8 columns) otherwise known as channels to adequately acquire spatial information. The diameter of the single printed electrode unit is 4 mm, and the center distances between two neighboring electrode units are 8 mm in row direction and 15 mm in column direction, respectively, to reduce crosstalk between neighboring electrode units. In addition, the width and the spacing of printed conductive lines are 0.5 mm and 0.5 mm, respectively, as shown in Figure 2.

B. Characterization of the flexible medical dry electrodes

The morphologies and microstructures of the printed TPU/Ag electrodes were characterized by a field emission scanning electron microscope (FE-SEM, FEI Nova Nano SEM 450). The mechanical performance of the samples was measured by an electronic universal testing machine. The electrical resistance of the electrode was measured by a digital multimeter (Agilent 34401A) and the corresponding electrical conductivity is calculated by the following equation:

$$\sigma = \frac{1}{R} \times \frac{l}{t \times w} \qquad (1)$$

where σ is the electrical conductivity, R is the electrical resistance, l is the length, w is the width and t is the thickness of the measured samples, respectively.

C. Acquisition of EMG signals

Before attached the EMG electrode to the body, the skin was wiped with alcohol to remove stratum corneum and completely air dried before testing, and a little electrode gel was dropped to the circular electrode surface. A commercial EMG acquisition system was used to record the EMG signals

data from human body. The acquisition system and the printed electrode arrays were connected by a standard flexible flat cable with 16 pins (the center distance between two adjacent pins is 1.0 mm). Each electrode array with 16 electrode units otherwise known as channels was applied during recording EMG signals. At the same time, a reference electrode was attached over the wrist of the non-dominant arm. LabVIEW and Matlab were used to convert the analog signals to digital data for recording and analysis. All experiments in this study involving human subjects were conducted following the guidelines provided by Shenzhen Institutes of Advanced Technology, Chinese Academy of Sciences.

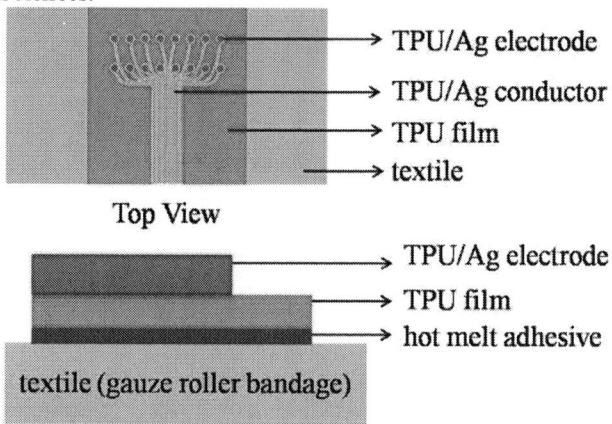

Figure 1. Schematic diagrams of the printed medical dry electrode.

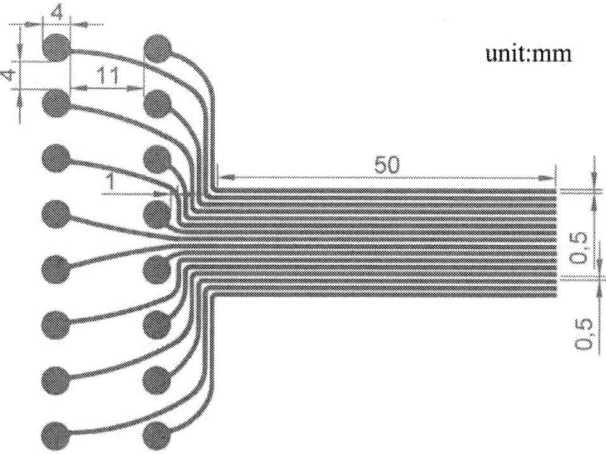

Figure 2. Schematic illustration of the printed electrode patterns (2 rows × 8 columns array).

III.   THE PERFORMANCE OF THE FLEXIBLE INTERLOCKED PRESSURE SENSORS

Figure 3 shows the SEM images of the printed TPU/Ag composites. It can be seen from Figure 3a that the width of the conductive TPU/Ag lines is ~520 µm and the spacing between neighboring lines is ~480 µm, respectively, indicating a slightly lateral spreading of the printed lines

978-1-7281-1500-9/19 $31.00 © 2019 IEEE

compared with the design values of 0.5 mm and 0.5 mm for line width and spacing, respectively. Figure 3b shows the partial magnifying SEM image of the edge of a printed TPU/Ag line. There is an obvious border between the TPU film (containing a little of carbon black) and the TPU/Ag line. And the TPU/Ag composites tightly combined with TPU layer. Figure 3c and 3d the enlarged SEM images of the printed TPU/Ag conductive line with different magnifications. It can be seen that the silver micro-flakes were uniformly and compactly embedded in the TPU polymer matrix, providing adequate conductive paths and high electrical conductivity for the printed TPU/Ag electrodes and lines. In order to quantitatively evaluate the electrical performance of the printed TPU/Ag electrodes, the mixed paste of TPU solution and Ag flakes was applied to a polytetrafluoroethylene (PTFE) mould with desired length, width and depth and smoothed by a blade, followed by heating curing at 80 ℃ for 2 h to obtain the TPU/Ag strips. According to the geometric dimension (30 × 8 × 0.04 mm of length × width × thickness) and electrical resistance (0.0534 Ω) of the TPU/Ag samples, the electrical conductive of the TPU/Ag with 85 wt% of Ag flakes is calculated by equation (1) to be a high value of about 2.76×10^6 S m^{-1}.

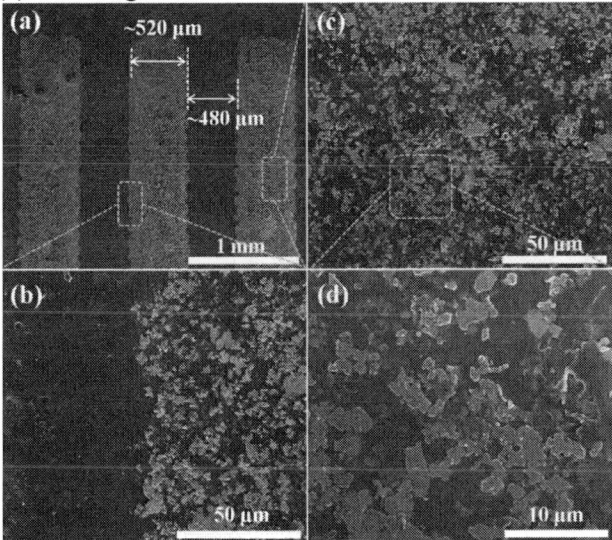

Figure 3. SEM images of the printed TPU/Ag conductive composites.

The cross sectional SEM images were also observed and shown in Figure 4. It can be clearly seen from Figure 4a that there are three layers of the printed PTU/Ag patterns including hot melt adhesive layer, TPU substrate layer and TPU/Ag conductive layer with thickness of about 35, 60 and 30 μm, respectively. In addition, the centre thickness of the printed TPU/Ag composites is larger than that of two sides, exhibiting a trapezoid-like shape due to the slight lateral spreading of the printed paste before heat curing. The magnified SEM image of the cross-section in Figure 4b shows the microstructures of the interface region between the TPU/Ag layer and the TPU layer. Notably, there is no delamination in the interface region (white dotted line region in Figure 4b), revealing strong interface bonding between the printed TPU/Ag and the TPU substrate. It can be attributed

to their same chemical composition of TPU and micro-dissolution in the interface region by DMF solvent in the TPU/Ag paste before DMF was completely removed during heat curing. This strong bonding avoids delamination of the TPU/Ag electrode/conductive lines and TPU substrate, exhibiting excellent stretchability for the whole printed EMG electrodes.

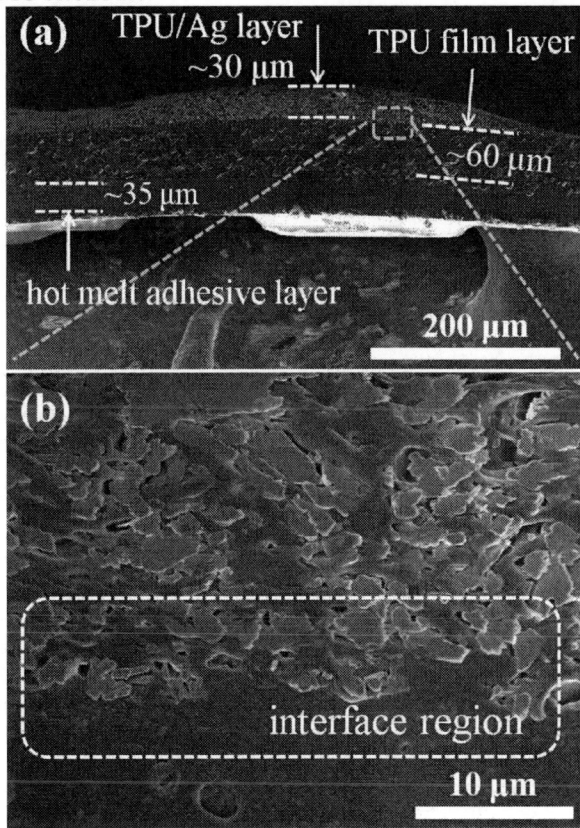

Figure 4. Cross-sectional SEM images of the printed TPU/Ag conductive composites on TPU film.

Figure 5 shows the stress-strain curve of the TPU/Ag strip with 85 wt% silver loading. It can be clearly observed that there is a large breaking strain of ~246%, indicating excellent intrinsic stretchability of the TPU/Ag electrodes and excellent mechanical flexibility. And a stress yield phenomenon was occurred at a low strain of 3.5%, that is the stress of the TPU/Ag decreases beyond strain of 3.5%. With futher increasing of the strain, the stress graduately increases to its maximum value of 12.8 MPa until break at maximum strain of ~246%. Insert in Figure 5 is a magnified curve upon 1% strain of the stress-strain curve. The elastic modulus of the TPU/Ag can be calculated to be 733.9 MPa by a linear fitting with excellent fit coefficient (R^2=0.97555). The large stretchability and elastic modulus demonstrate excellent flexibility and durability of the TPU/Ag electrodes in practical applications.

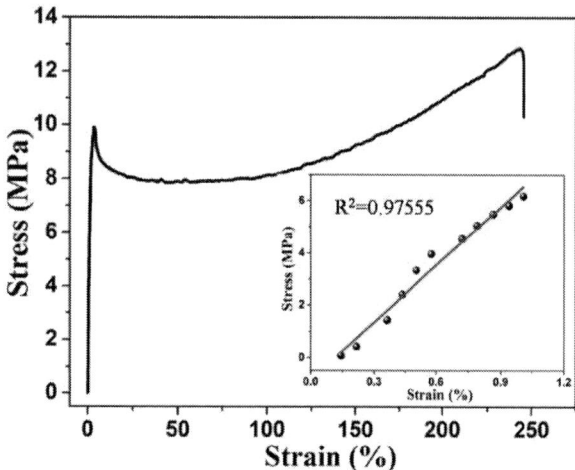

Figure 5. Stress-strain curve of the TPU/Ag strip.

Figure 6 shows the photos of the as-prepared printed TPU/Ag electrode array with 16 units and location of the electrode for EMG signals recording. The EMG electrode was placed over the circumference of the left brachioradialis muscle of the subject's dominant forearm by itself of the textile of gauze roller bandage. It should be emphasized that it is very convenient to attach the electrode array onto forearm and connect it to the data acquisition unit by a commercial flexible flat cable. Before attaching the electrode, only a small amount of gel was applied to the surface of the circular electrode arrays to fill the air gap between electrodes

and skin. Moreover, there was no relative motion between the electrode and skin due to tightly bond of the gauze roller bandage, even when the man was walking or waving hand.

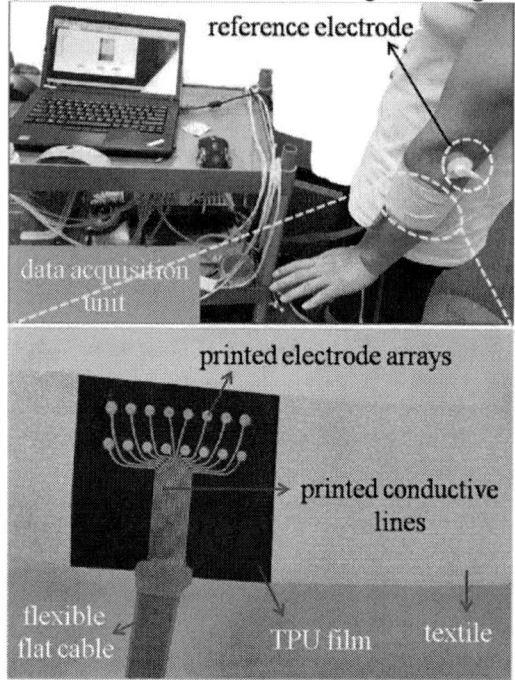

Figure 6. Photos of the printed TPU/Ag electrode arrays and location of the electrode for EMG signals recording.

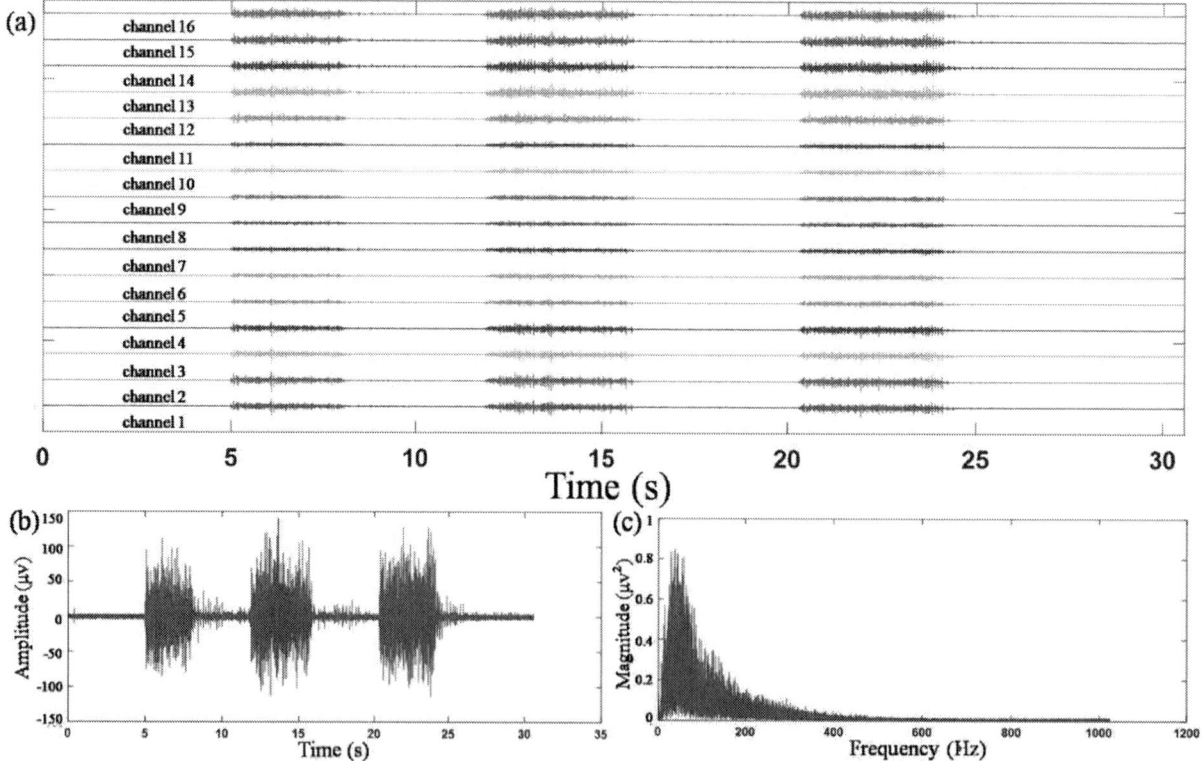

Figure 7. Measurement performance of the printed EMG electrode array.

EMG signals were recorded in real time to evaluate the practical properties of the printed electrode array in monitoring and distinguishing various motions. An able-bodied subject (AB) (males; age 25 years; right-hand dominated) participated in the study. Four typical hand classes, namely hand close (HC), hand open (HO), wrist extension (WE), and wrist flexion (WF) were involved in the study, and also one inactive movement of rest (RE) was tested as control group. Prior to EMG data collection, subject was well informed about the experimental details and given several minutes to get familiar with the procedure. During the experiment, the subjects were promoted to perform the four motion classes demonstrated on a computer screen in front of them. The experiment was composed of 4 sessions. In each session, one of motion classes was displayed in a sequential manner and repeated three times. Each motion was hold for 2 s and then a rest of 3 s was scheduled between two sequential motions. Thus each session would produce 6 s EMG recordings for each movement. To minimize muscle and mental fatigue, the subject was instructed to elicit muscle contractions with a moderate force determined by themselves. A rest of about 2 minutes was scheduled between the sessions. The EMG signal was recorded at a sampling rate of 2048 Hz and filtered with a band-pass filter with a frequency of 10 Hz to 500 Hz.

Take the motion task hand open as an example, the waveform and spectrogram of the EMG signals were analyzed. The detected EMG waveforms of the electrode array with 16 channels were shown in Figure 7a. It can be clearly seen that all the 16 electrode channels successfully recorded the EMG signals with a high signal-to-noise ratio, indicating all these electrode units contacted well with the skin surface even during hand moving. In addition, there is highly similar of the EMG signals for the same electrode channel during three times of HO motion, revealing excellent consistency of the EMG signals. The enlarged view for EMG waveform of the channel 14 (a random choice as representative of the 16 channels) was shown in Figure 7b, and its spectrogram was shown in Figure 7c. The accuracy of signals obtained by the printed electrode array was confirmed by amplitude-frequency spectrum from these two enlarged views.

IV. CONCLUSIONS

In this study, a simple and low-cost fabrication process of stretchable and printable medical dry electrode arrays based on highly conductive composites of TPU/Ag are presented. The TPU/Ag electrode arrays and corresponding conductive lines are easily screen printed on a thin TPU film and bonded to a textile bandage by heat melt adhesive. The TPU/Ag electrode is intrinsic stretchable with a large elongation of 246% at break and high initial electrical conductivity of 2.76×10^6 S m^{-1} at 85 wt% silver loading. Strong interface bonding force between the printed TPU/Ag and the TPU substrate improves the mechanical flexiblity and durability of the printed electrode. The as-prepared multi-channel electrode arrays on textile are easily attached onto human body such as forearm and they demonstrate excellent performance in accurately detecting EMG signals

for various motions of hand open, hand close, wrist extension and wrist flexion, etc. The accurate and adequate spatial EMG signals information will great contribute to diagnosis medical condition of the patients or human-machine interface control.

ACKNOWLEDGMENT

The authors thank the financial support from the National Natural Science Foundation of China (Grant No. 61701488), Shenzhen Basic Research Plan (Grant No. JCYJ20170818162548196), the National key R&D project from minister of science and technology of China (Grant No. 2016YFA0202702), and Guangdong Provincial Key Laboratory (Grant No. 2014B030301014).

REFERENCES

[1] M. A. Lopez-Gordo, D. Sanchez-Morillo, and F. P. Valle, "Dry EEG Electrodes,". Sensors, vol. 14, Jul. 2014, pp. 12847-12870, doi: 10.3390/s140712847.

[2] H. Jin, Y. S. Abu-Raya, and H. Haick, "Advanced Materials for Health Monitoring with Skin-Based Wearable Devices,". Advanced Healthcare Materials, vol. 6, Mar. 2017, pp. 1700024, doi: 10.1002/adhm.201700024.

[3] S. Park, S. W. Heo, W. Lee, D. Inoue, Z. Jiang, K. Yu, H. Jinno, D. Hashizume, M. Sekino, T. Yokota, K. Fukuda, K. Tajima, and T. Someya, "Self-Powered Ultra-Flexible Electronics via Nano-Grating-Patterned Organic Photovoltaics,". Nature, vol. 561, Sept. 2018, pp. 516-521, doi: 10.1038/s41586-018-0536-x.

[4] S. S. Yao, A. Myers, A. Malhotra, F. Lin, A. Bozkurt, J. F. Muth, and Y. Zhu, "A Wearable Hydration Sensor with Conformal Nanowire Electrodes," Advanced Healthcare Materials, Jan. 2017, vol. 6, pp. 1601159, doi: 10.1002/adhm.201601159.

[5] S. Kim, L. K. Jang, M. Jang, S. Lee, J. G. Hardy, and J. Y. Lee, "Electrically Conductive Polydopamine-Polypyrrole as High Performance Biomaterials for Cell Stimulation in Vitro and Electrical Signal Recording in Vivo," ACS Applied Materials & Interfaces, Sept. 2018, vol. 10, pp. 33032-33042, doi: 10.1021/acsami.8b11546.

[6] J. Kim, and W. S. Kim, "A Paired Stretchable Printed Sensor System for Ambulatory Blood Pressure Monitoring," Sensors and Actuators A: Physial, Feb. 2016, vol. 238, pp. 329-336, doi:10.1016/j.sna.2015.12.030.

[7] D. P. Qi, Z. Y. Liu, Mei. Yu, Y. X. Tang, J. H. Lv, Y. C. Li, J. Wei, B. Liedberg, Z. Yu, and X. D. Chen, "Highly Stretchable Gold Nanobelts with Sinusoidal Structures for Recording Electrocorticograms," Advanced Materials, Apr. 2015, vol. 27, pp. 3145-3151, doi: 10.1002/adma.201405807.

[8] T. Q. Trung, and N. E. Lee, "Flexible and Stretchable Physical Sensor Integrated Platforms for Wearable Human-Activity Monitoring and Personal Healthcare," Advanced Materials, Feb. 2016, vol. 28, pp. 4338-4372, doi: 10.1002/adma.201504244.

[9] F. Stauffer, M. Thielen, C. Sauter, S. Chardonnens, S. Bachmann, K. Tybrandt, C. Peters, C. Hierold, and J. Voros, "Skin Conformal Polymer Electrodes for Clinical ECG and EEG Recordings,". Advanced Healthcare Materials, Jan. 2018, vol. 7, pp. 1700994, doi: 10.1002/adhm.201700994.

[10] E. Bihar, T. Roberts, E. Ismailova, M. Saadaoui, M. Isik, A. Sanchez-Sanchez, D. Mecerreyes, T. Hervé, J. B. D. Graaf, and G. G. Malliaras, "Fully Printed Electrodes on Stretchable Textiles for Long-Term Electrophysiology," Advanced Materials Technology, Feb. 2017, vol. 2, pp. 1600251, doi: 10.1002/admt.201600251.

[11] P. Zheng, H. Y. Zhuo, Y. X. Zou, W. Guo, H. Wu, and Z. Li, "Highly-Conductive Stretchable Electrically Conductive Composites by Halogenation Treatment and Its Application in Stretchable Electronics," IEEE 68th Electric Components and Technology

978-1-7281-1500-9/19 $31.00 © 2019 IEEE

Conference (ECTC), IEEE Press, Aug. 2018, pp. 1744-1750, doi: 10.1109/ECTC.2018.00262.

[12] K.-I. Jang, H. N. Jung, J. W. Lee, S. Xu, Y. H. Liu, Y. Ma, J.-W. Jeong, Y. M. Song, J. Kim, B. H. Kim, A. Banks, J. W. Kwak, Y. Yang, D. Shi, Z. Wei, X. Feng, U. Paik, Y. G. Huang, R. Ghaffari, and J. A. Rogers, "Ferromagnetic, Folded Electrode Composite as a Soft Interface to the Skin for Long-Term Electrophysiological Recording," Advanced Functional Materials, sept. 2016, vol. 26, pp. 7281-7290, doi: 10.1002/adfm.201603146.

[13] T. Kim, J. Park, J. Sohn, D. Cho, and S. Jeon, "Bioinspired, Highly Stretchable, and Conductive Dry Adhesives Based on 1D-2D Hybrid Carbon Nanocomposites for All-in-One ECG Electrodes," ACS Nano, Mar. 2016, vol. 10, pp. 4770-4778, doi: 10.1021/acsnano.6b01355.

[14] D. Son, J. Kang, O. Vardoulis, Y. Kim, N. Matsuhisa, J. Y. Oh, J. W. F. To, J. Mun, T. Katsumata, Y. Liu, A. F. McGuire, M. Krason, F. Molina-Lopez, J. Ham, U. Kraft, Y. Lee, Y. Yun, J. B.-H. Tok, and Z. Bao, "An Integrated Self-Healable Electronic Skin System Fabricated via Dynamic Reconstruction of a Nanostructured Conducting Network," Nature Nanotechnology, Aug. 2018, vol. 13, pp. 1057-1065, doi: 10.1038/s41565-018-0244-6.

Screen-Printed Flexible Coplanar Waveguide Transmission Lines: Multi-Physics Modeling and Measurement

Nahid Aslani Amoli[1], Sridhar Sivapurapu[1], Rui Chen[2], Yi Zhou[2], Mohamed L. F. Bellaredj[1], Paul A. Kohl[1], Suresh K. Sitaraman[2], and Madhavan Swaminathan[1]

School of Electrical and Computer Engineering, Georgia Institute of Technology, Atlanta, GA 30332-0250, USA
[1]Center for Co-Design of Chip, Package, and System (C3PS)
G. W. Woodruff School of Mechanical Engineering, Georgia Institute of Technology, Atlanta, GA 30332-0405, USA
[2]Computer-Aided Simulation of Packaging Reliability (CASPaR) Lab
aslani@gatech.edu

Abstract— **Flexible hybrid electronics (FHE) is a promising technology enabling many applications in biomedical, communication, energy harvesting and internet of things (IoT) areas. To realize FHE applications, the components and devices used in the mentioned technologies need to be electrically characterized under various flexible conditions such as stretching, bending, twisting, and folding. Also, the strain analysis from the mechanical point of view needs to be conducted to justify the reliable applications of FHE under different flexible scenarios. In this paper, the design and electrical characterization of coplanar waveguides (CPWs) in flexible substrates such as Kapton polyimide and polyethylene terephthalate (PET) under uniaxial bending are studied and discussed. The fabricated lines were measured using a vector network analyzer (VNA) up to 8 GHz under both flat and bending conditions. Finite-element models (FEM) of CPW lines were created in ANSYS HFSS to capture the effect of bending on the CPW frequency response. In addition, the variations in the trace width and separations along the CPW lines were modeled accurately to capture the variations in the fabrication process and their effect on the CPW S-parameters in the flat condition. The finite element analysis of strain variation during bending was also performed and the relationship between strain variation and CPW performance was investigated. The bending of the CPW lines was carried out using two parallel plates that had a gap distance varying from 40 mm to 140 mm. The S-parameters were monitored in-situ while the substrate was under bending. The experimental results were compared against simulated results under both flat and bent configurations. Based on the conducted studies, correlation was achieved for the flat and bending scenarios between measurement and simulation results. Also, it was observed that the CPW line has better matching and lower losses compared with the flat case and tensile bending cases.**

Keywords- compressive bending; flexible substrates; screen-printed CPW; strain analysis; tensile bending

I. INTRODUCTION

Flexible hybrid electronics (FHE) including both conventional silicon electronics and printed electronics helps realize many applications in wearable technologies used in biomedical, sensing, energy harvesting and communication areas [1]. For the wearable and implantable electronics used

in the healthcare monitoring, the progress in the materials science made it possible to design mechanically compliant electronics [2]. As an example, a microwave CPW transmission line was used as a 40-GHz radio frequency (RF) biosensor in [3] to characterize the dielectric parameters of cancer cells. The various technologies for printing sensors and electronics over flexible substrates including screen-printing and inkjet printing were reviewed and discussed in [4]. Both printing technologies have been widely used to print the microwave devices and components such as antennas, inductors, microstrip and CPW transmission lines on the flexible substrates, for example Kapton and PET.

The RF characterization of CPW transmission lines printed on both rigid and flexible substrates have been reported in several literatures. In [5], the ink-jet printed CPW lines using conductive nano-silver ink on a quartz glass were electrically characterized over the frequency range of 30 kHz to 6 GHz. The derived attenuation from the measured results was reported as 0.15 dB/mm for 10 cm long CPW lines. The authors in [6] have measured and simulated the ink-jet printed CPW lines with silver conductor ink on both flexible Kapton substrates and an Alumina substrate. The line losses of 1.5 dB/mm and 0.8 dB/mm at 60 GHz were reported for the same CPW lines printed on these two substrates, respectively. Also, in [7], a 1mm long CPW line, ink-jet printed on the Kapton substrate, was investigated for high frequency applications up to 67 GHz in both flat and tensile bending conditions. The line losses of 0.3 dB/mm at 67 GHz when sintering at 300°C was measured in the flat case. The bending of CPW line was performed on a curvature with radii of 15 mm and 25 mm. A minimal variation was observed in the measured attenuation constant of CPW line before and during bending condition. The microwave characterization of ink-jet printed CPW lines on flexible PET substrates over the frequency range of 10 MHz to 20 GHz was explored in [8] and using a space-mapping technique for linking the measured results with simulations, they could extract the conductivity of the ink and dielectric constant of the spacer used during measurements [9]. They also applied their proposed approach on the ink-jet printed CPW lines on Kapton substrates and showed that their methodology can detect small changes in material properties caused by changes in fabrication parameters [10]. The printed electronics over the flexible substrates can be subjected to

978-1-7281-1500-9/19 $31.00 © 2019 IEEE

different shape variations such as bending, twisting, stretching and folding, depending on the considered application [1, 2]. Therefore, both electrical and mechanical characterization of FHE under different flexible conditions needs to be carried out to ensure a desirable and reliable performance.

There are some methods for bending test of FHE, such as adaptive curvature flexure, mandrel bending test, etc. The main advantage of adaptive curvature flexure over mandrel bending test is that the flexible substrate can be subjected to different radii of curvature through moving parallel plates using just one test setup. The distance between two parallel plates, called panel separation (PS), can be adjusted for having different radii of curvature [11]. Using this bending test, an aerosol jet printed (AJP) microstrip transmission line was characterized both electrically and mechanically through finite element modeling in the tensile and compressive bending tests and the measured results were compared against simulation. It was shown that there is little change in DC resistance of the transmission line during bending and the change in either return loss or insertion loss at different panel separations (40 mm – 100 mm) compared to the flat condition was not noticeable [12].

Compared to previous work [5-10], this paper presents three main contributions on screen-printed CPW lines over flexible substrates such as Kapton and PET. First, the effect of inherent variability involved in the fabrication process on the CPW frequency response was investigated through line profile measurements and multi-section modeling of the CPW line in ANSYS HFSS [13]. Second, as opposed to [7], the CPW line with a length of 166.37 mm was measured under different bending (tensile and compressive) scenarios using adaptive curvature bending for a wide range of panel separations (40 mm – 140 mm) and the results were then correlated with measurements using ANSYS HFSS. Third, the strain variations during bending were studied using a 2D finite-element model of CPW line in ANSYS [14] and its impact on the electrical properties of the CPW line was explored. To the best of our knowledge, the above-mentioned contributions have not been reported in the open literature.

This paper is organized as follows. In section II, the dimensions of CPW lines screen-printed on Kapton and PET substrates are presented. The electrical characterization of CPW lines including measurement and simulation together with the impact of process variations on CPW frequency response (flat case) are described in section III. Section IV presents the measurement and simulation results of CPW lines in both tensile and compressive bending cases along with the analysis of strain variations during bending conditions. Finally, the concluding remarks are given in section V.

II. DIMENSIONS OF CPW LINES

The CPW lines were designed and screen-printed at Georgia Tech and DuPont, respectively. Figure 1 shows the fabricated CPW lines (L1 and L2) along with TRL (Through-Reflect-Line) structures on a 5 mil Kapton and PET substrates. The DuPont 5028 silver conductor [15] was used to print both the signal (trace) and ground planes. All the results presented in this paper are related to the longest CPW line, L2, with a length of 166.37 mm. The fabricated TRL structures (L3, L4,

and Open/Short), used in the de-embedding process to remove the effect of SMA connectors from the measurement results, did not work as expected. Therefore, the de-embedding of measurement results presented later was not implemented.

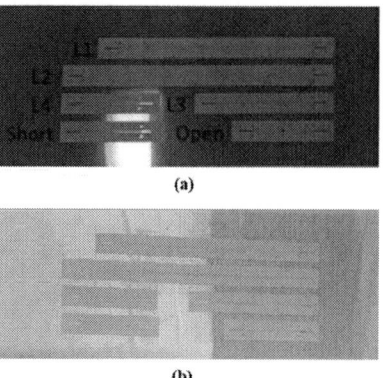

(a)

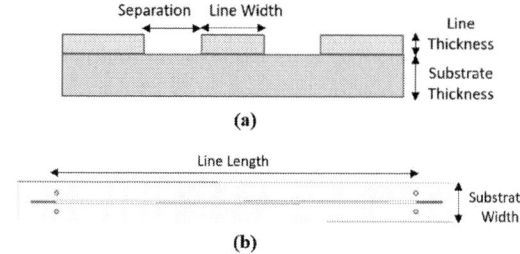

(b)

Figure 1.   The screen-printed CPW lines on (a) Kapton and (b) PET substrates.

A cross-sectional view and a top view of the CPW line is depicted in Fig. 2. As shown, the CPW line consists of a center conductor (trace) with two finite-dimension ground planes on each side of a slot (separation or gap).

Figure 2.   The cross-section (a) and the top view (b) of the CPW lines.

The dimensions of CPW lines are listed in Table I. The line width and separations were measured using an optical microscope while cross-section dimensions such as the thickness and surface roughness were measured using a Veeco Dektak 150 surface profilemeter. Because of variations in the line width and separations along the CPW lines after fabrication, the average values for these parameters are presented in Table I. To determine the conductivity of fabricated CPW lines, the DC resistance of the lines was measured using a digital multimeter (DMM). Then, the conductivity was calculated using $\sigma = L/(R.A)$ where σ is the line conductivity, L is the line length, R is the measured DC resistance, and A is the cross-sectional area of the CPW line. It is worth noting that the cross-sectional area of CPW lines is assumed as a rectangular area which equals the product of the line width and the line thickness as listed in Table I. This assumption is used in the transmission line modeling (for simplification) as mentioned in [5-10, 12].

However, such cross-sectional area does not typically occur in the fabricated CPW lines and in practice, it is a trapezoidal cross-section as shown in Fig. 3.

TABLE I. DIMENSIONS OF CPW LINES

CPW Parameters	Value	
	Substrate: Kapton	Substrate: PET
Substrate length (mm)	196.418	196.418
Substrate width (mm)	14.82	14.82
Line length (mm)	166.37	166.37
Line width (μm)	762.07	764
Line thickness (μm)	10	10
Separation (μm)	106.68	90.3
Surface roughness (μm)	1	1
Line conductivity (S/m)	$2.974*10^6$	$2.9*10^6$

The calculated conductivities presented in Table I are estimations of the actual conductivities of CPW lines which have a trapezoidal cross-section. These conductivities are used for CPW line simulations in ANSYS HFSS as a constant value for the entire frequency range because the average values of the line width and thickness were used in the calculations.

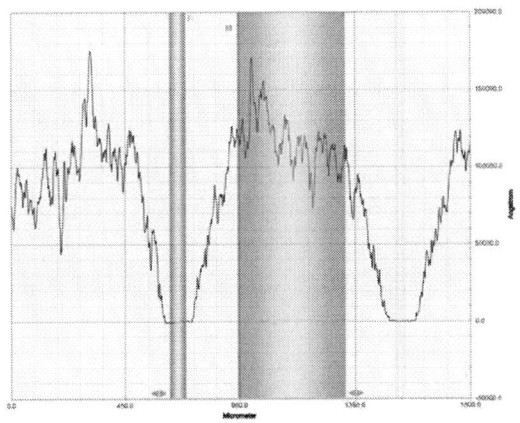

Figure 3. Measured trace profile of the CPW line with Kapton substrate.

The relative permittivity and the loss tangent of the Kapton substrate are 3.5 and 0.0026 at 1 kHz [16], respectively, while these values for the PET substrate characterized using the ring resonator method in the frequency range of 0 to 6 GHz are 2.9 [17] and 0.0314 [18], respectively.

III. MEASUREMENT AND MODELING OF CPW LINES (FLAT CASE)

A. Measurement Results

To measure the S-parameters of CPW lines, an Agilent E8363B VNA was used and calibrated using an HP 85052D Calibration Kit where SOLT (Short-Open-Load-Through) calibration was used to calibrate until the edge of the SMA connectors. The VNA cables were connected to each side of CPW lines using a VLF40-002 SMA which is a vertical

launch connector. The CPW lines with mounted SMAs are depicted in Fig. 4.

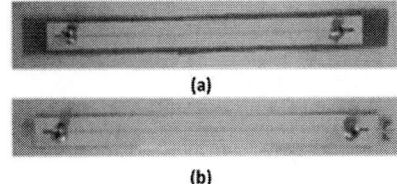

Figure 4. CPW lines with mounted SMAs screen-printed on (a) Kapton and (b) PET substrates.

Figure 5 shows the measured S-parameters of CPW lines in the flat condition while floating in the air in the frequency range of 10 MHz to 8 GHz. As seen, the CPW line printed on the Kapton substrate is less lossy than the CPW line printed on the PET substrate. This is because of the higher loss tangent of PET substrate compared to that of Kapton substrate which has a great impact at higher frequencies. Also, the line conductivity printed on PET substrate is a little bit smaller than that printed on the Kapton substrate. The measured losses for the CPW lines printed on the PET and Kapton substrates are 0.09057 dB/mm and 0.0749 dB/mm at 8 GHz, respectively.

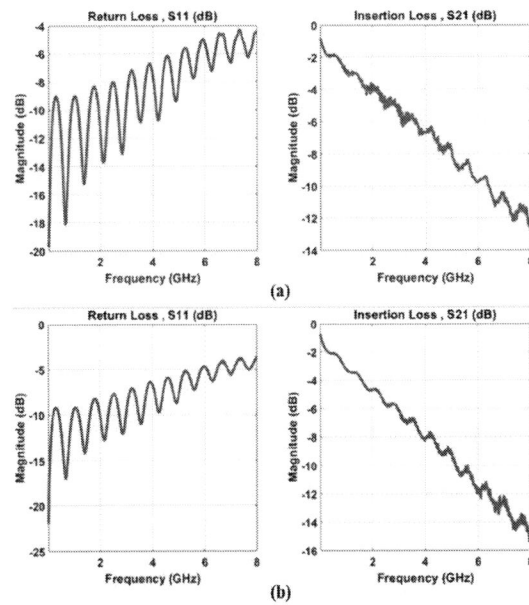

Figure 5. Measured return loss (S_{11}) and insertion loss (S_{21}) of CPW lines screen-printed on (a) Kapton and (b) PET substrates in the flat case.

The skin depth of the printed silver ink on the Kapton and PET substrates at 8 GHz were calculated as 3.263 μm and 3.304 μm, respectively, which are more than the measured surface roughness listed in Table I. Therefore, the surface roughness doesn't have a significant impact on the attenuation of CPW lines. The measured return loss for both lines is above -10 dB approximately in the whole frequency range which shows a high impedance mismatching for these lines.

978-1-7281-1500-9/19 $31.00 © 2019 IEEE

B. Simulation and Measurement Correlation

The finite element models of CPW lines were created in ANSYS HFSS [13] using the parameters listed in Table I and the material properties of both substrates mentioned earlier. The simulation results were compared against the measured ones for both CPW lines as shown in Fig. 6. A good agreement between the simulated and measured results is observed for both CPW lines. The discrepancies seen in the comparison of insertion loss results may be due to the conductivity value used in the simulation models which is an estimate for the actual conductivity value of the lines, as explained in section II. The conductivity values listed in Table I were calculated with the assumption of the rectangular cross-section of the line which is a simplification for the trapezoidal cross-section as shown in Fig. 3. In addition, the process variations contribute to these discrepancies which are discussed in the following sections.

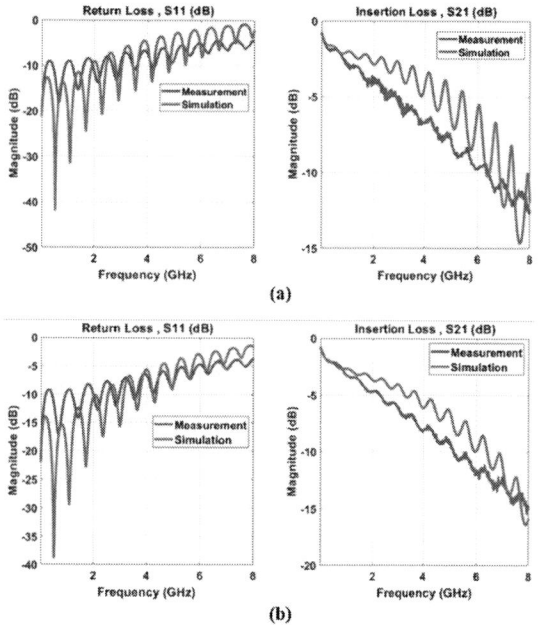

Figure 6. Correlation between the measurement and simulation results for CPW lines printed on (a) Kapton and (b) PET substrates in the flat case.

C. Study of Fabrication Process Variations

The process variations in the fabricated CPW lines appear in the form of variations in the line width and separations along the lines. To demonstrate the effect of process variations on the CPW S-parameters, we measured the profile of the CPW line printed on the PET substrate at 20 equal-length sections as shown in Fig. 7, in which sections 1 and 20 represent the leftmost side and the rightmost side of the CPW line. The maximum change in the line widths among 20 sections is 107.1 μm which is 14% of the average line width listed in Table I. As shown in Fig. 2 (a), there are two separations along the CPW lines while we listed one single value for both separations in Table I. It means that we assumed a symmetric structure for the CPW line which is not valid in practice. From the 20 measured profiles, the maximum change

in the left and the right separations are 106.66 μm and 108 μm, respectively, which are 118.12% and 119.6% of the average separation value listed in Table I, respectively. As we traverse along the line from section 1 to section 20, the trend of measured data points for both separations is increasing while that for the line width is decreasing.

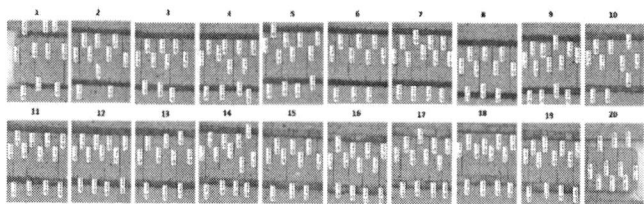

Figure 7. Measured profiles of CPW line printed on PET substrate at 20 equal-length sections.

The finite element model of the CPW line with 20 measured sections was created in ANSYS HFSS and the obtained results for the CPW line printed on the PET substrate were compared against the results shown in Fig. 6 (b). As seen, a noticeable improvement in the correlation of the insertion loss is observed compared to the simulation result without considering the 20-section profile which is expected. As the number of sections of the CPW profile measurements increases, we get closer to the actual geometry of the CPW line. As a result, a better correlation is achieved between the simulation and measurement results. Because the characteristic impedance of the CPW lines mainly depends on the line width and separation, the variations in the line width and separation along the line cause impedance variation or mismatch which results in more reflections along the line as shown in Fig. 8.

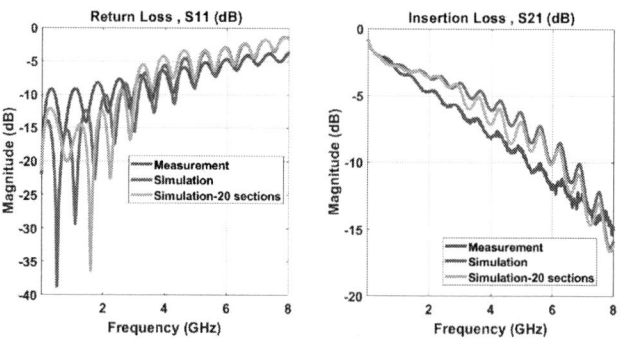

Figure 8. Comparison of simulation results from 20-section CPW line printed on PET substrate with the results of Fig. 6 (b).

IV. MEASUREMENT AND MODELING OF CPW LINES FOR BENDING CASES

In this section, the bending measurements performed on the CPW line printed on PET substrate are only presented and discussed as the same trend for bending measurement results was observed for the CPW line printed on Kapton substrate. After performing SOLT calibration on the VNA, the bending measurements of the CPW line were carried out using the adaptive curvature bending test setup shown in Fig. 9. The

CPW line is placed between two plates where the bottom plate is fixed while the top plate is moved up and down to have different panel separations. There are two flat sections on either side of the CPW line which are where the VNA cables are connected to the SMAs via a cutout in the plate. The laboratory setup for both tensile and compressive bending measurements of the CPW line is shown in Fig. 10.

The plates made of high-density polyethylene (HDPE) material have a relative permittivity of 2.3 and loss tangent of 0.0005 [13]. Under tensile bending, the CPW line is on the convex side (outside) surface of the substrate, while under compressive bending, the CPW line is on the concave side (inside) surface of the substrate.

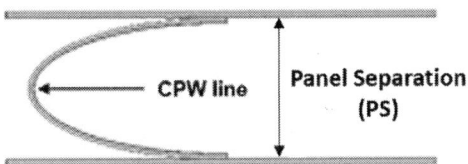

Figure 9.   Illustration of the adaptive curvature bending test setup.

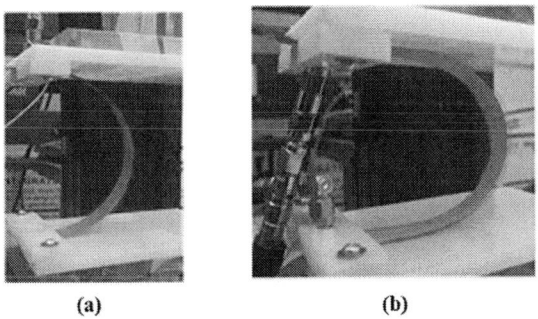

Figure 10. The laboratory setup for the CPW line (a) tensile and (b) compressive bending measurements.

Both bending measurements were carried out at the panel separations of 40 mm to 140 mm. These measurements were compared against the flat measurement result of the CPW line to observe the effect of bending on the CPW frequency response. To have an accurate comparison, two flat measurements of the CPW line, as shown in Fig. 11, were also performed in which small parts from both ends of the CPW line were attached to the HDPE panels and the remaining part of the line is floating in the air.

These two flat measurement results were compared with those shown in Fig. 5 (b) in which the CPW line was freely floating in the air without those panels. It shows that there is a very negligible difference between these three cases such that the results of Fig. 5 (b) can be considered as the reference case to compare with bending measurement results.

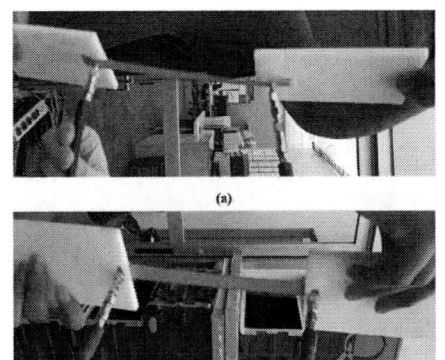

Figure 11. Two flat measurements of the CPW line using the HDPE panels.

A. Measurement Results

1) Tensile Bending: The comparison of tensile bending measurement results of the CPW line with the flat case is shown in Fig. 12. As seen, the frequency response of the CPW line changes while undergoing tensile bending. Although, the return loss is not better than 10 dB (matching criterion) for the flat case and for all bending cases over the whole frequency range, it is observed that the return loss of all bending cases is better than that of the flat case, indicating a better match.

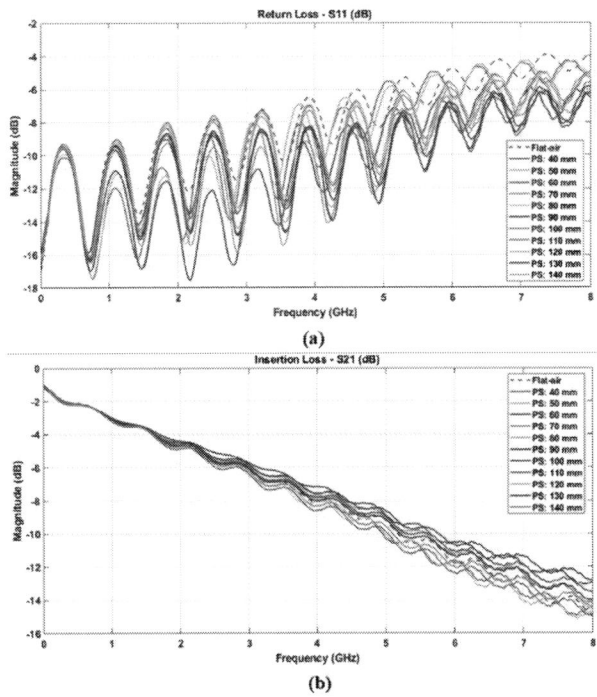

Figure 12. Comparison of flat and tensile bending measurement results for the CPW line, (a) Return loss, (b) Insertion loss.

To have a better view for the change in insertion loss during bending, the magnitudes of insertion loss for all bending cases were normalized to that of the flat case (-14.93

dB) at 8 GHz as shown in Fig. 13. As seen, the maximum and the minimum change in the insertion loss occur at the panel separation of 110 mm and 130 mm, respectively. In other words, the maximum and the minimum deviation in the insertion loss compared to the flat case at 8 GHz are 2.05 dB and 0.09 dB, respectively. Also, the range of variation in the insertion loss among all bending cases is 2.14 dB at 8 GHz. From Fig. 12, it seems that no clear relationship can be found for the change in results for different panel separations.

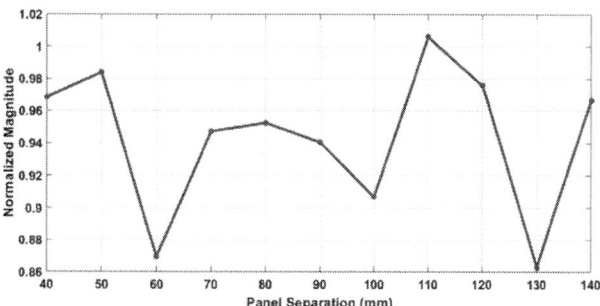

Figure 13. Normalized magnitude of insertion loss for all tensile bending cases with respect to the flat case at 8 GHz.

2) Compressive Bending: Fig. 14 compares the compressive bending measurement results of the CPW line with the flat case. As seen, there is a small difference between the results of different panel separations. Also, the measured return loss of all compressive bending cases compared with the flat case is better than 10 dB over the entire frequency range.

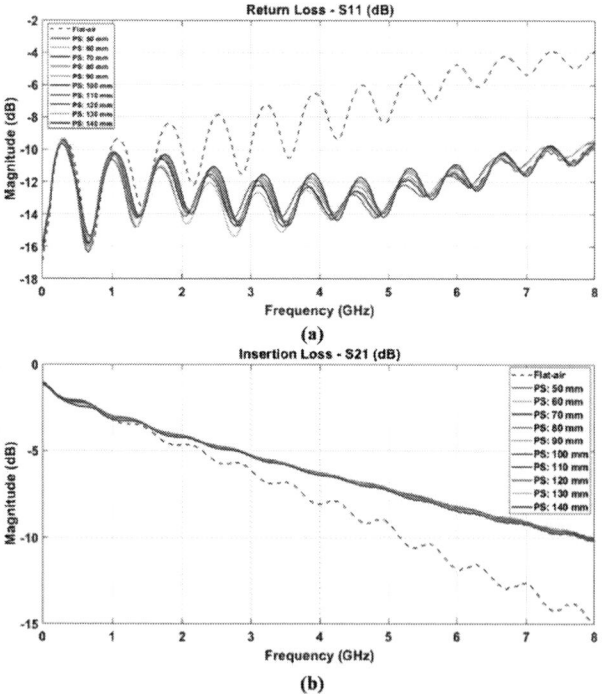

Figure 14. Comparison of flat and compressive bending measurement results for the CPW line, (a) Return loss, (b) Insertion loss.

However, the maximum deviation in the insertion loss between the flat case and all compressive bending cases is around 4.93 dB at 8 GHz which shows 33% improvement in the insertion loss. The reason behind the better insertion loss in all compressive cases is not clear yet and needs to be investigated further. Comparing Fig. 12 and Fig. 14 shows that the CPW line under compressive bending exhibits better matching and lower insertion loss compared with the flat case and all tensile bending cases as well.

B. Study of Strain Variations for the Bending Cases

To better understand the effect of bending on the electrical properties of the CPW line, a 2D mechanical finite element analysis of strain variations during tensile bending was carried out in ANSYS 18.1 [14], as shown in Fig. 15.

From these simulations, it was found that the change in the average relative resistance during bending is very small such that the line conductivity under flat condition can be used for the tensile and compressive bending simulations in HFSS as well. This finding was also observed in [12, 19] for the microstrip transmission line under bending scenarios.

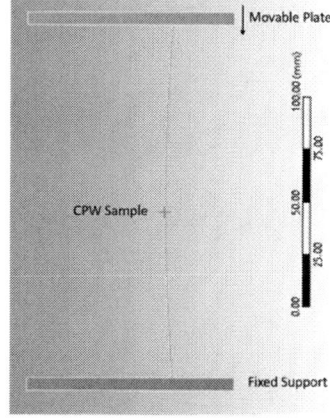

(a)

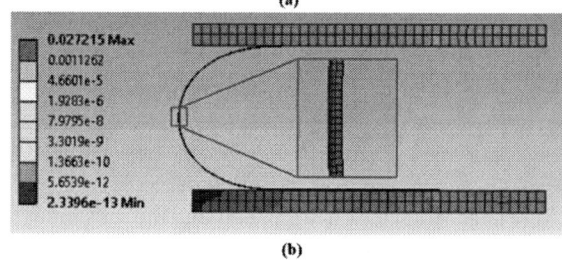

(b)

Figure 15. The 2D model for mechanical finite-element analysis of CPW line. (a) the simulation setup. (b) maximum principal strain distribution at the panel separation of 40 mm while CPW line under tensile bending.

The relative resistance (R/R_0) is calculated using (1) as follows [19]:

$$\frac{R}{R_0} = \int_0^L \frac{(1+\epsilon(s))^2 \rho(\epsilon(s))}{L} ds. \qquad (1)$$

Where L is the length of the line, $\rho(\epsilon)$ is the relationship between resistivity and tensile strain, and $\epsilon(s)$ is the tensile strain at position s along the CPW line. The function $\rho(\varepsilon)$ is the same as that used in [19] for the screen-printed microstrip line which is $\rho = \rho_0(1+20.5\ \varepsilon)$.

The relationship between resistivity and compressive strain is not determined yet and needs to be investigated further in the future work. However, as the strain under compressive bending is less than that under tensile bending, the change in relative resistance would be smaller than that under tensile bending [12].

Table II presents the finite element analysis results of the average strain, the average relative resistance and the maximum strain obtained at the panel separations of 40 mm to 140 mm for tensile bending of the CPW line. As we traverse along the CPW line, the maximum strain occurs at the middle point of the line for panel separations of 40 mm-100 mm, while for those of 110 mm-140 mm, the maximum strain occurs at the clamps (SMA locations at both ends). The maximum increase in the relative resistance is 2.43% which occurs at the panel separation of 40 mm. It corresponds to a relative conductivity of 0.9763 which shows a relative conductivity decrease of 2.37%. Therefore, the conductivity of the CPW line (L2) for tensile bending at this panel separation reduces to 2.83e6 S/m as compared to 2.9e6 S/m under flat conditions. In the following subsection, the tensile and compressive bending simulation results are presented and compared with the corresponding measurement cases.

TABLE II.   ESTIMATED CHANGE IN AVERAGE RELATIVE RESISTANCE FOR CPW LINE UNDER TENSILE BENDING

Panel Separation (mm)	Average Strain	Average Relative Resistance	Maximum Strain
40	0.001184196	1.024276026	0.0037
50	0.001172991	1.024046307	0.003
60	0.0011623	1.023827153	0.0025
70	0.001151256	1.023600754	0.0021
80	0.001145209	1.023476784	0.0018
90	0.001144059	1.023453217	0.0016
100	0.001141791	1.023406712	0.0013
110	0.001139223	1.023354081	0.0014
120	0.001136177	1.02329162	0.0019
130	0.001132146	1.023208994	0.0027
140	0.001125645	1.02307572	0.0039

C. Simulation and Measurement Correlation

The adaptive curvature bending setup for both tensile and compressive bending simulations was modeled in ANSYS HFSS accurately to capture the conditions of the measured CPW line in the laboratory. Figure 16 shows the HFSS models for both tensile and compressive bending simulations of the CPW line. The HFSS simulations were carried out in the

frequency range of 10 MHz-8 GHz with 401 points for interpolating frequency sweep at the mentioned range of the panel separation. The correlation of simulation results against the measurement ones is only discussed for three of panel separations which are 60 mm, 90 mm and 120 mm for both bending cases. The same discussions can be applied to the results for other panel separations as well.

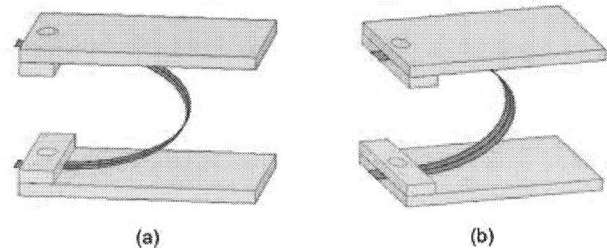

Figure 16. HFSS models for (a) tensile and (b) compressive bending simulations of the CPW line.

1) Tensile Bending: The comparison of simulation and measurement results at panel separations of 60 mm, 90 mm and 120 mm is presented in Fig. 17. Good agreement between the simulation and measurement results of the insertion loss is observed for all panel separations shown in Fig. 17. Also, for the return loss, good correlation is achieved between the simulation and measurement results by 3 GHz where some small deviation between the results start appearing and continues up to 8 GHz.

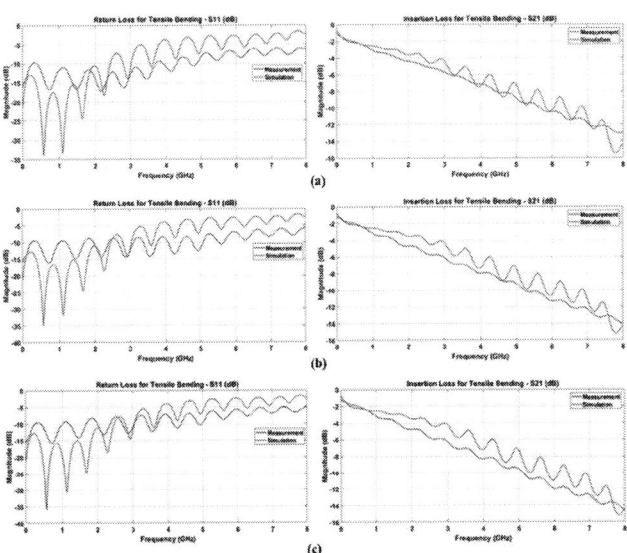

Figure 17. Correlation between the measurement and simulation results for tensile bending of the CPW line at panel separations of (a) 60 mm, (b) 90 mm, and (c) 120 mm.

2) Compressive Bending: The comparison of simulation and measurement results at panel separations of 60 mm, 90 mm and 120 mm is shown in Fig. 18. Once again, good agreement between simulation and measurement results for insertion loss is observed for all panel separations up to 6 GHz. Beyond this frequency, the simulation results start deviating from measurements such that the deviation in results reaches around 5 dB at 8 GHz. One reason for this deviation could be that HDPE panels at both ends of the CPW line for compressive bending behave as another layer of dielectric attached to the PET substrate. This added layer appears to have a larger impact at higher frequencies in the simulation results. For return loss, a good correlation between simulation and measurement results was achieved for all panel separations up to 3 GHz. But, the simulation results start diverging from measurements after 3 GHz such that the resulting discrepancy reaches around 7 dB at 8 GHz. The process variations along the CPW line discussed in section III can be another reason for the result discrepancies at higher frequencies.

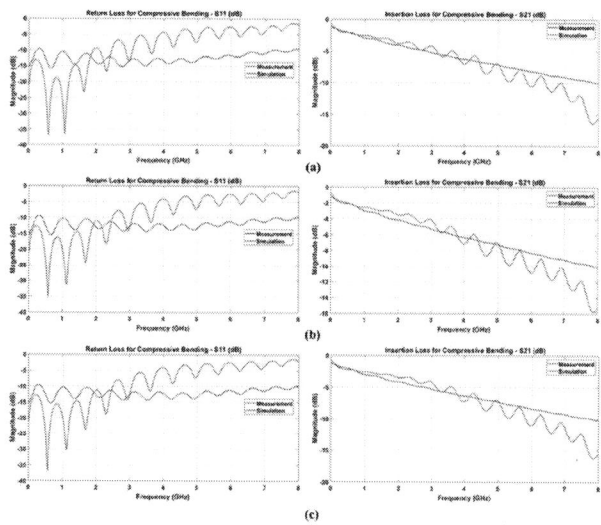

Figure 18. Correlation between the measurement and simulation results for compressive bending of the CPW line at panel separations of (a) 60 mm, (b) 90 mm, and (c) 120 mm.

V.   CONCLUSION

The CPW lines screen-printed on two flexible substrates, Kapton polyimide and PET were electrically characterized and the measurement results were compared against simulation results from the FEM models created in ANSYS HFSS. The insertion loss of CPW lines printed on PET and Kapton substrates was measured as 0.09057 dB/mm and 0.0749 dB/mm at 8 GHz, respectively. However, the measured insertion loss for the AJP microstrip line [12] and screen-printed microstrip line [19], both printed on Kapton polyimide substrate, was reported as 0.125 dB/mm and 0.04 dB/mm at 8 GHz, respectively. The simulation results agreed reasonably well with measurements in the flat case. The effect of process variations on the CPW frequency response was investigated through multi-section modeling of the CPW line in HFSS. Simulation results verified the impact of process variations which exhibits itself in the form of variations in the line width and separation. By increasing the number of sections, improvement in the correlation of insertion loss was expected. Also, both tensile and compressive bending measurements and simulations of the CPW line using the adaptive curvature bending were performed at different panel separations ranging from 40 mm to 140 mm and the analysis of the strain variations during bending cases showed that there is a little change in the average relative resistivity of the CPW line. The discrepancies observed in the simulation to measurement correlation for all bending cases showed that there is still room for improvement which needs to be investigated further in future work. Compressive bending measurements of the CPW line demonstrated that the CPW line has better matching and lower losses compared with the flat case and tensile bending cases.

ACKNOWLEDGMENT

This material is based, in part, on research sponsored by Air Force Research Laboratory under agreement number FA8650-15-2-5401, as conducted through the flexible hybrid electronics manufacturing innovation institute, NextFlex. The U.S. Government is authorized to reproduce and distribute reprints for Governmental purposes notwithstanding any copyright notation thereon. The views and conclusions contained herein are those of the authors and should not be interpreted as necessarily representing the official policies or endorsements, either expressed or implied, of Air Force Research Laboratory or the U.S. Government.

REFERENCES

[1]   G. Tong, Z. Jia, and J. Chang, "Flexible hybrid electronics: review and challenges," IEEE International Symposium on Circuits and Systems (ISCAS), May. 2018, pp. 1-5.

[2]   R. Herbert, J. Kim, Y. S. Kim, H. M. Lee, and W. Yeo, "Soft material-enabled, flexible hybrid electronics for medicine, healthcare, and human-machine interfaces," Materials Journal, vol. 11, pp. 1-33, Feb. 2018.

[3]   Y. F. Chen, H. W. Wu, Y. H. Hong, and H. Y. Lee, "40 GHz RF biosensor based on microwave coplanar waveguide transmission line for cancer cell (HepG2) dielectric characterization," Biosensors and Bioelectronics, vol. 61, pp. 417-421, 2014.

[4]   S. Khan, L. Lorenzelli, and R. S. Dahiya, "Technologies for printing sensors and electronics over large flexible substrates: a review," IEEE Sensors Journal, vol. 15, no. 6, pp. 3164-3185, June 2015.

[5]   B. Shao et al., "High frequency characterization of inkjet printed coplanar waveguides," 12th IEEE Workshop on Signal Propagation on Interconnects (SPI), 2008, pp. 1-4.

[6]   S. Swaisaenyakorn, P. R. Young, and M. Shkunov, "Characterization of ink-jet printed CPW on Kapton substrates at 60 GHz," Loughborough Antennas and Propagation Conference (LAPC), 2014, UK, pp. 676-678.

[7]   M. M. Belhaj et al., "Inkjet printed flexible transmission lines for high frequency applications up to 67 GHz," 2014 9th European Microwave Integrated Circuit Conference, Dec. 2014, pp. 584-587.

[8]   A. Sahu et al., "Microwave characterization of ink-jet printed CPW on PET substrates," 86th ARFTG Microwave Measurement Conference, Dec. 2015, Atlanta, GA, USA, pp. 1-4.

[9]   A. Sahu, V. Devabhaktuni, A. Lewandowski, T. M. Wallis, and P. H. Aaen, "CAD-assisted microwave characterization of ink-jet printed

CPW on PET substrates," in Proc. 88th ARFTG Microwave Measurement Conference, Austin, TX, USA, Dec. 2016, pp. 1-4.

[10] A. Sahu et al., "Robust microwave characterization of inkjet-printed coplanar waveguides on flexible substrates," IEEE Trans. On Instrumentation and Measurement, vol. 66, no. 12, Dec. 2017, pp. 3271-3279.

[11] R. Chen, J. Chow, C. Taylor, J. Meth, and S. K. Sitaraman, "Adaptive curvature flexure test to assess flexible electronic systems," Proc. 68th IEEE Electronic Components and Technology Conference (ECTC), Aug. 2018, pp. 236-242.

[12] S. Sivapurapu et al., "Multi-physics modeling & characterization of aerosol jet printed transmission lines," 2018 IEEE MTT-S International Conference on Numerical Electromagnetic and Multiphysics Modeling and Optimization, Reykjavik, Aug. 2018.

[13] ANSYS, "Ansys HFSS ver. 2018.2." [Online]. Available: http://www.ansys.com.

[14] ANSYS, "Ansys 18.1." [Online]. Available: http://www.ansys.com.

[15] Dupont, "5028 Silver Conductor," 5028 datasheet, Aug. 2014.

[16] DuPont, "DuPont Kapton HN," DuPont Kapton HN datasheet, Aug. 2018.

[17] DuPont, "DuPont Teijin Films," DuPont Teijin Films datasheet, 2006.

[18] D. Betancourt, and J. Castan, "Printed antenna on flexible low-cost PET substrate for UHF applications," *Progress in Electromagnetics Research C*, vol. 38, pp. 129-140, Mar. 2013.

[19] S. Sivapurapu et al., "Multi-physics modeling & characterization of components on flexible substrates," unpublished.

Inkjet-Printed Filtering Antenna on a Textile for Wearable Applications

Hsuan-Ling Kao*, Chun-Hsiang Chuang
Dept. of Electronic Engineering
Chang Gung University
Department of Dermatology
Chang Gung Memorial Hospital
Tao-Yuan, Taiwan
*e-mail: snoopy@mail.cgu.edu.tw

Cheng-Lin Cho
Dept. of Engineering and System Science
National Tsing Hua University
Hsin-Chu, Taiwan
Dept. of Electronic Engineering
Chang Gung University
Tao-Yuan, Taiwan

Abstract—**This paper has presented an inkjet-printed filtering antenna on a textile substrate for wearable electronics. In order to improve the porous of textiles, screen-printed interface layer (Fabink-UV-IF1) was used to obtain the uniformity of surface on an textile. The optimal wettability of interface layer for subsequent inkjet printing was obtained by using post-baking. Inkjet-printed conditions of silver film on a textile with interface layer were examined by surface morphologies and electrical properties. Based on optimal conditions, inkjet-printed filtering antenna was design and implemented using silver ink. Two open stubs were used to produces two transmission zeros near the passband. The frequency band was 2.5 – 4.6 GHz with the maximal gain of - 0.42 – -3.1 dBi for voltage standing wave ratio was less than 2. Two rejection frequencies at 1.6 and 5.7 GHz were found because of two open stubs. The characteristics of two bending directions for the filtering antenna were also studied. The size of the filtering antenna was 37.5 mm × 23 mm × 0.49 mm. The results demonstrated the integrity of inkjet-printed technology which combination of wearable and printed electronic applications.**

Keywords- Screen printing, inkjet printing, textiles, filtering antenna.

I. INTRODUCTION

Smart clothing by wireless sensors technologies has been proposed to monitor the health for elderly or homecare patient to provide independent living. Fabric-based electronics are required to meet low-profile, low-cost, light-weight, and flexible. Inkjet-printing is a direct writing technology without any lithographic fabrication. Additionally, it also provides more freedom on various substrates, practically in flexible substrates [1]-[2]. Inkjet printing provides high resolution than embroidering [3] and weaving [4] on textiles for fabric-based electronic applications. One of the challenges for inkjet printing on textiles is rough and porous surface. Therefore, coated interface layer on fibrous substrates before printing is proposed [5]-[6]. In this work, the polyurethane-based ultraviolet curable paste was used as an interface layer to fill the porous in textiles and improve the surface roughness. Then, inkjet-printed filtering antenna was presented on a textile substrate for wearable electronics. Two open stubs were used to produces two transmission zeros near the passband. The characteristics of two bending directions for

the filtering antenna were also studied. The results demonstrated the integrity of inkjet-printed technology which combination of wearable and printed electronic applications.

II. FABRICATION PROCESS

Fig. 1 illustrates the steps of inkjet-printed silver film on a textile. Polyurethane-based ultraviolet curable paste (Fabink UV-IF1, Smart Fabric Inks, Ltd.) was used as an interface layer by screen-printing technology [7]. Silver-nanoparticle-ink (Advance Nano Products) [8] was used by inkjet-printing technology. The steps of inkjet-printed silver film on a textile can be summarized as follows. A 0.5 mL Fabink was dripped by dropper inside the desired mesh on the textile. The Fabink was pushed by a squeegee to obtain the interface layer. In order to flatness Fabink, a pressure was applied by two glasses. Then, a 30-minute UV exposure was applied on Fabink. Repeated screen-printing as follows dripping, pushing, and UV-exposed were used to fill and cover the fabric grid. The uniform surface was obtained after two screen printing process. Before printing, a 150°C 60-min post-baking was treated the surface of Fabink to obtain optimal contact angle for inkjet printing. Then, multiple passes of silver film were printed onto the Fabink at 60°C using 10 pL drop volumes. Finally, the sintering at 150 °C for 1 h in an oven was preformed to improve the conductivity of silver film. The thickness of the Fabink after two screen-printing was 490 μm with permittivity (ε_r) and loss tangent (tan δ) of 3.74 and 0.15, respectively.

III. RESULTS AND DISCUSSION

A. Inkjet-printed silver film on a textile

Wetting is an important parameter for inkjet-printing technology. Surface treatment affects the wetting should be examined. Fig. 2 shows the water contact angle (WCA) as a function of post-baking temperature for the Fabink. The WCA was increased with the temperature increased. A suitable condition of WCA was 79° for inkjet-printing technology [9] and the post-baking temperature of Fabink was no larger than 150 °C. of the silver film was decreased from 44.3 to 46.6 mΩ after 5,000 cycles. It proves the bending behavior of silver film on a Therefore, 150 °C and 60 min was used to treat the surface of Fabink. Fig. 3 shows the surface morphologies of silver film on a textile with

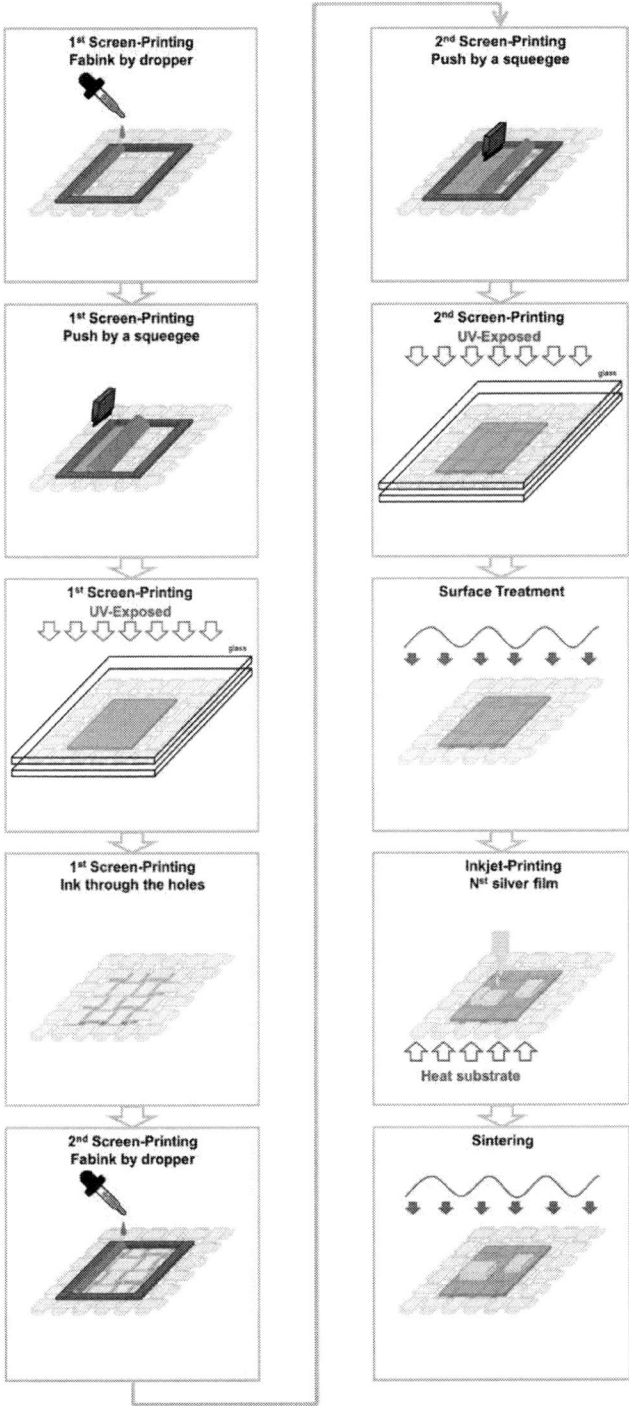

Figure 1. Process flow of inkjet printing silver film on a textile.

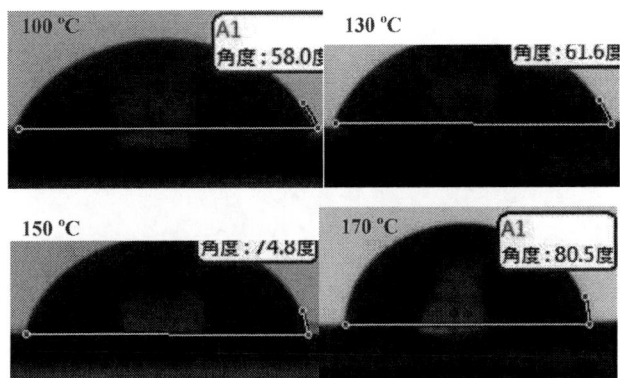

Figure 2. Water contact angle on Fabink at various hard-bake temperatures.

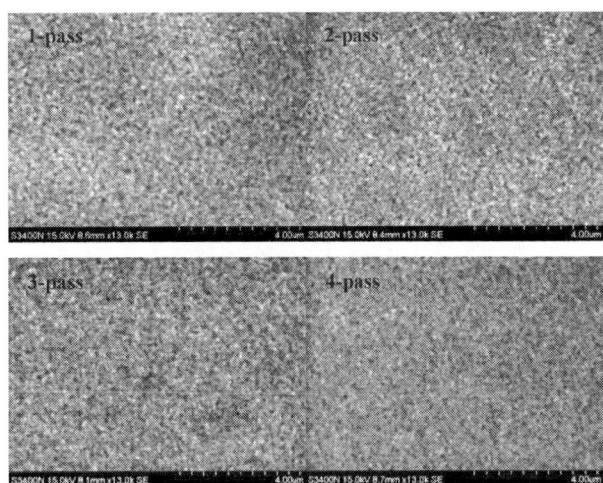

Figure 3. Surface morphologies of inkjet printed silver film with 1 to 4 passes printing.

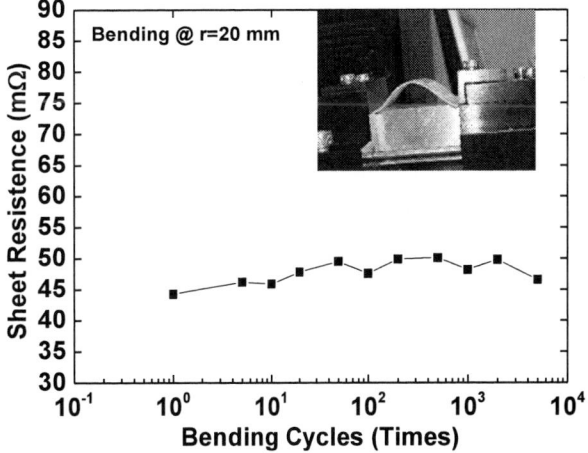

Figure 4. Sheet resistance of silver film for 5,000 dynamic bending on bending vehicle with radius of 20 mm.

1 to 5 passes. Little variation was observed in surface morphologies from 1 to 3 passes and saturated after 3-pass printing. Multi-pass printing provides appropriate conductivity and thickness for silver film for RF performance. Therefore, 3-pass is used in this work with conductivity and thickness of 4.15×10^6 S/m and 5.35 μm, respectively. The bending cycles from 1 to 5,000 for silver film on an textile was tested by applying tensile strain with a curvature radius of 20 mm. The sheet resistance of the silver film at various bending cycles is shown in Fig. 4. The sheet resistance textile is acceptable which is suitable for wearable

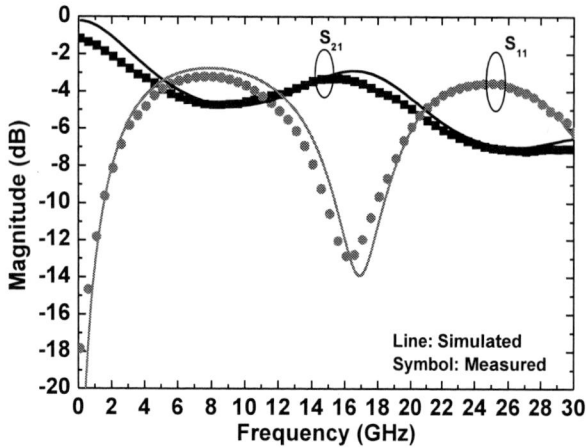

Figure 5. S-parameters of the 5,000 μm silver line on the textile.

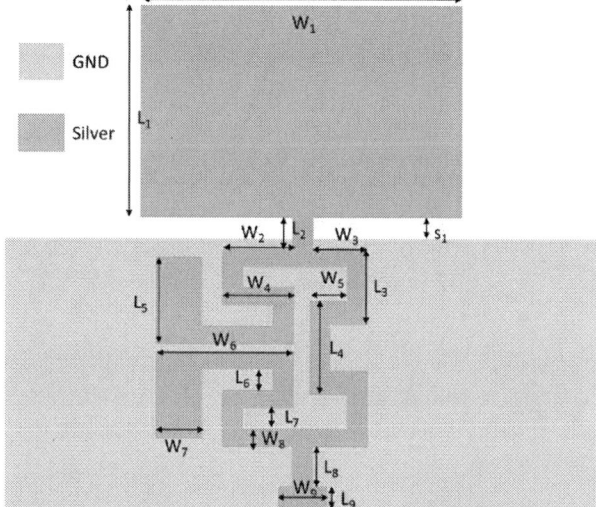

Figure 6. Layout of filtering antenna.

applications. Fig. 5 shows the insertion (S_{21}) and return losses (S_{11}) of the silver line with length of 5,000 μm and width of 100 μm on a textile. The insertion losses were 3.5 dB at the resonant frequencies of 16.2 GHz, respectively. Simulated and measured results were consistent which also confirmed the ε_r and tan δ of 3.73 and 0.15, respectively, for the textile.

B. Inkjet-printed filtering antenna on a textile

The filtering antenna was implemented on T/C fabric by using inkjet printing technology. Fig. 6 shows the structure of filtering antenna which consists of patch antenna and a meandered feeding line with two open stubs. Ansoft HFSS was used to perform the electromagnetic (EM) simulation. Fig. 7 shows the surface current distribution of the filtering antenna based at 1.7, 3.6, and 5.8 GHz. The strong current distribution on the radiating element was found at 3.6 GHz. At 1.7 and 5.8 GHz, the weak current were observed on the radiating element. The rejection radiation was suppressed by two open stubs to achieve filtering antenna. After fine

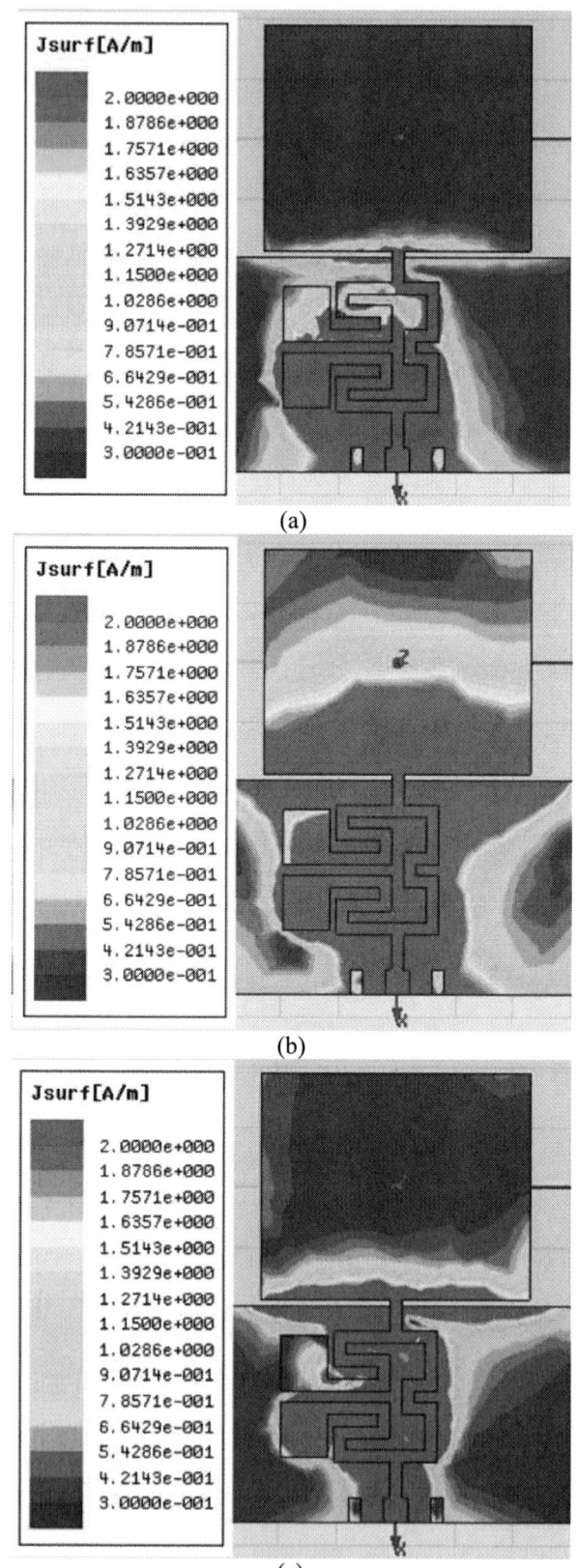

Figure 7. The surface current distribution of the filtering antenna at the frequencies of (a) 1.6 GHz, (b) 3.6 GHz, and (c) 5.7 GHz.

978-1-7281-1500-9/19 $31.00 © 2019 IEEE

Figure 8. Photographs of inkjet-printed filtering antenna on a textile.

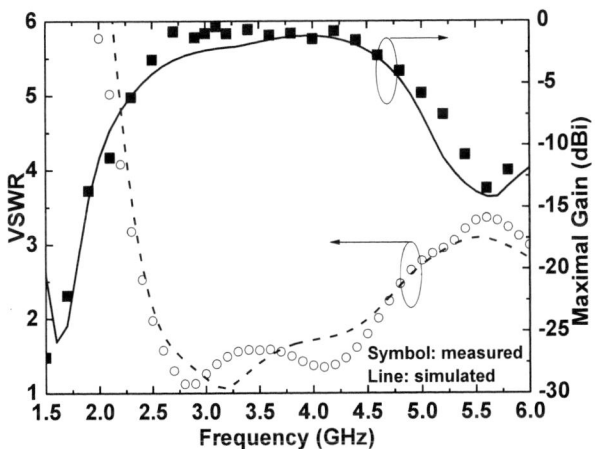

Figure 9. Simulated and measured input VSWR and maximal gain of filtering antenna.

tuning, the optimal dimension values of filtering antenna were as follows: $L_1 = 19$, $L_2 = 2.5$, $L_3 = 5$, $L_4 = 3$, $L_5 = 4.6$, $L_6 = 1$, $L_7 = 1$, $L_8 = 3$, $L_9 = 2$, $W_1 = 23$, $W_2 = 4.8$, $W_3 = 3$, $W_4 = 4.8$, $W_5 = 2$, $W_6 = 9.3$, $W_7 = 4$, $W_8 = 1$, $W_9 = 2$, and $s_1 = 0.5$ mm. Fig. 8 shows photographs of the filtering antenna. The volume of the antenna with a director was 37.5 mm × 23 mm × 0.49 mm. The radiation patterns of fabricated filtering antenna were measured inside a far-field anechoic chamber Fig. 9 shows the simulated and measured voltage standing wave ratio (VSWR) and maximal gain at frequency from 1.5 to 6 GHz. The frequency band was 2.5 − 4.6 GHz for a VSWR was less than 2, representing a fraction bandwidth of 59% at the central frequency of 3.55 GHz. The maximal gain was -0.42 − -3.1 dBi in the range of frequency band. Two transmission zeros were found at 1.6 and 5.7 GHz. Figs. 10 and 11 show the simulated and measured in yz-plane and zx-plane radiation patterns at 2.9, 3.6, and 4.2 GHz. The radiation patterns in yz-plane are near omnidirectional like a conventional monopole antenna and the radiated power is nearly constant over the entire frequency range. The zx-plane is perpendicular to yz-plane. The radiation patterns in zx-plane are forming a dipole shape and almost symmetric. The overall agreement between the simulated results and measured data were observed. The discrepancy between the simulated and measured cross-polarization patterns caused by back-radiation from filer, SMA feed connector, and coaxial cable.

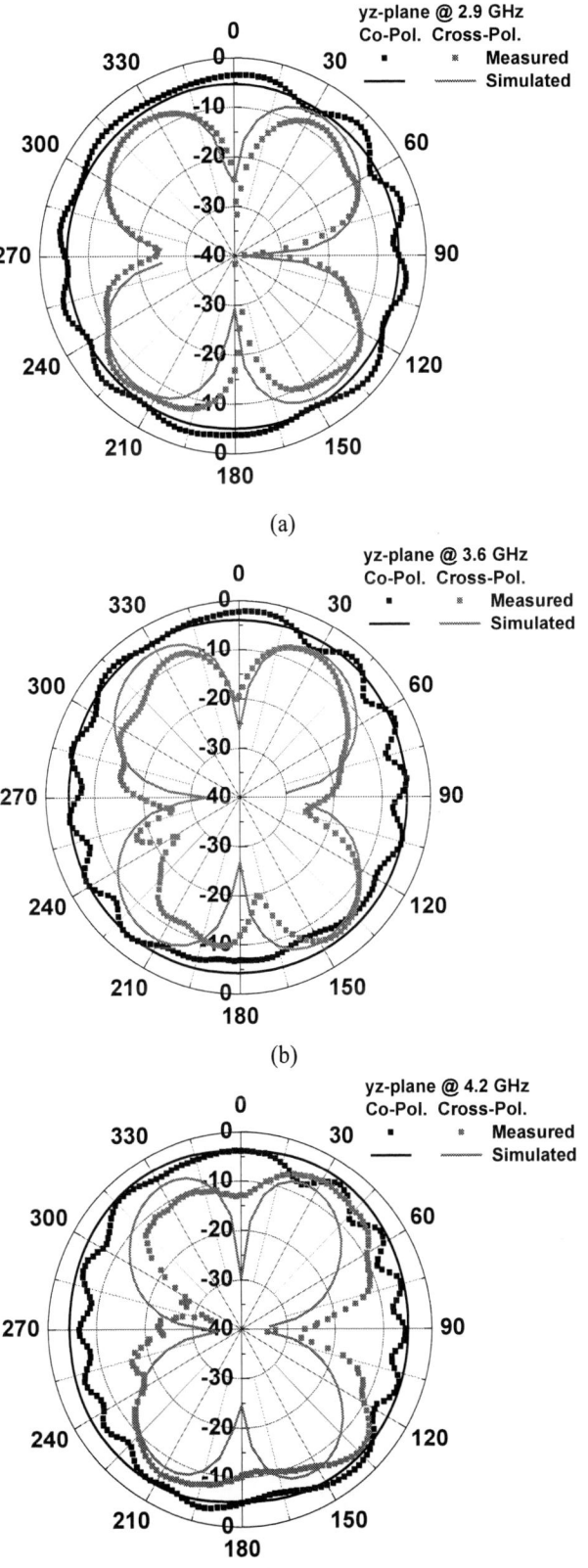

(a)

(b)

(c)

Figure 10. Simulated and measured radiation patterns of the proposed antenna in yz-plane at (a) 2.9 GHz, (b) 3.6 GHz, and (c) 4.2 GHz.

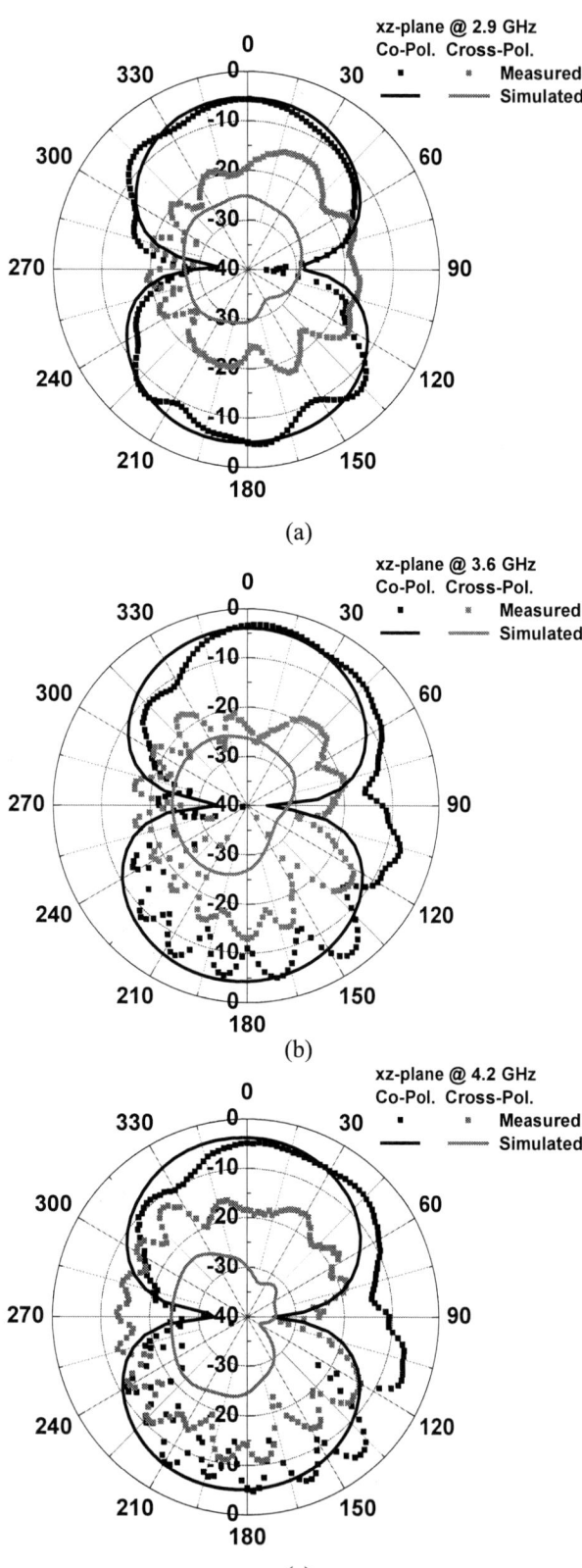

(a)

(b)

(c)

Figure 11. Simulated and measured radiation patterns of the proposed antenna in xz-plane at (a) 2.9 GHz, (b) 3.6 GHz, and (c) 4.2 GHz.

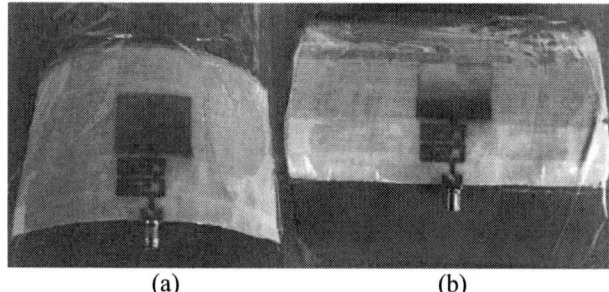

(a)   (b)

Figure 12. Bending directions (a) transverse and (b) longitudinal of inkjet-printed filtering antenna.

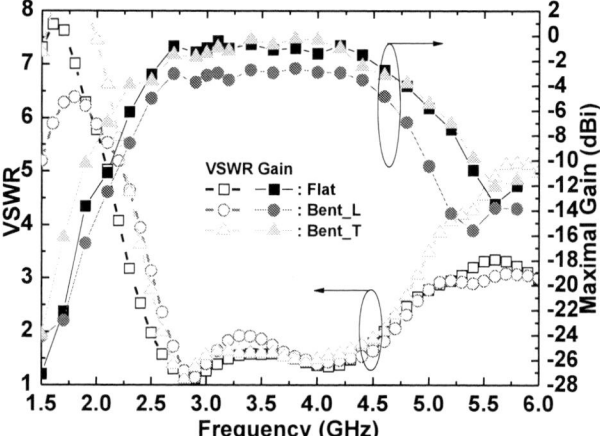

Figure 13. Measured input VSWR and maximal gain of the antenna before and after bending.

C. Bending effect of Inkjet-printed filtering antenna on a textile

Fig. 12 shows photographs of two bending directions of the filtering antenna. The longitudinal bending (Bent_L) was oriented parallel to, whereas the transverse bending (Bent_T) was oriented perpendicular to, the direction of current path (from filter to patch). The bending was applied by a curvature radius of r=50 mm. Fig. 13 illustrates the input VSWR and gain of the filtering antenna before and after bending. The VSWR was a slight change for flat and transverse bending because the filter was nearly flat while bending. The maximal gain, bandwidth, and two transmission zeros were almost similar for transverse bending. The transverse bending only slightly affected the resonant frequency. However, a 0.2 GHz frequency shift in low frequency and a higher VSWR in the middle of frequency band were observed for longitudinal bending. The maximal gain was decreased from -0.42 dBi to -2.61 dBi and the bandwidth was decreased 0.1-0.2 GHz under longitudinal bending. Two transmission zeros were shift to 1.7 and 5.4 GHz for longitudinal bending, respectively. This is because the resonance path (the direction of the current) was more affected by longitudinal bending than transverse bending [10]. Fig. 13 compares the measured co-polarization radiation patterns at 2.9, 3.6, and 4.2 GHz in yz-plane. The lower gain was found under transverse bending, resulting in poor resonance. Near omnidirectional radiation patterns were

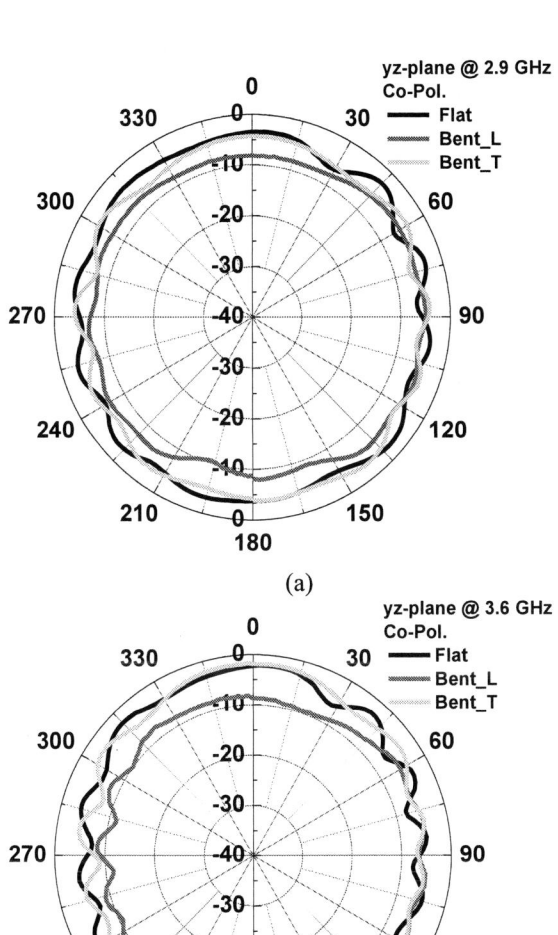

(a)

(b)

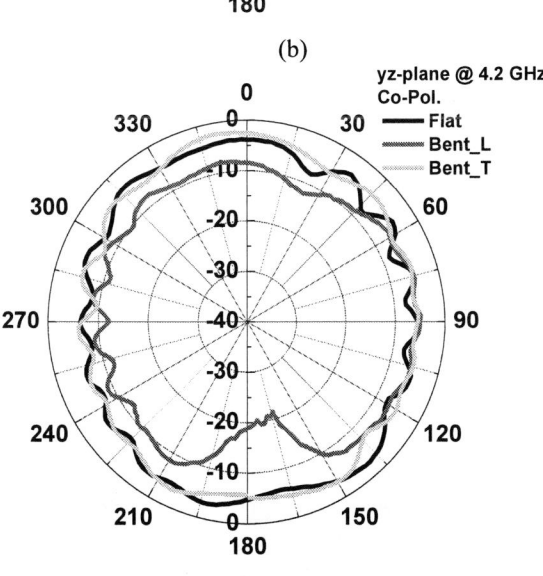

(c)

Figure 14. Measured co-polarization radiation patterns of the proposed antenna in yz-plane at (a) 2.9 GHz, (b) 3.6 GHz, and (c) 4.2 GHz.

observed before and after bending. The stable radiation patterns of flexible antenna, even at bending conditions, provide easy and efficient communicate between transmitters and receivers in all directions. It indicates that the bending effect of the wearable antenna should be carefully considered.

IV. CONCLUSION

This paper presented a filtering antenna on a textiles by inkjet printing technology. Two open stubs were used in feeding line to produce two transmission zeros. The frequency band of filtering antenna was 2.5 – 4.6 GHz with the maximal gain of -0.42 – -3.1 dBi. The maximal gain under longitudinal bending was -2.61 dBi should be carefully considered. Near omnidirectional radiation patterns were observed before and after bending. Inkjet-printed silver film on a textile is suitable for wearable electronics.

ACKNOWLEDGMENT

The authors thank the Centre for Reliability Sciences and Technologies at Chang Gung University for their help. This work was partially supported by the Ministry of Science and Technology, Taiwan (no. 106-2221-E-182-062 & 107-2221-E-182-042) and the Chang Gung Memorial Hospital, Taiwan (BMRP957 & CMRPD2H0311).

REFERENCES

[1] R.C. Roberts, and N.C. Tien, "Multilayer passive RF microfabrication using jet-printed au nanoparticle ink and aerosol-deposited dielectric," *Intl. Conf. on Solid-State Sensors, Actuators and Microsystems*, 2013, pp.178-181.

[2] E. Halonen, T. Viiru, K. Östman, A. L. Cabezas, and M. Mäntysalo, , "Oven Sintering Process Optimization for Inkjet-Printed Ag Nanoparticle Ink," *IEEE Trans. Compon., Packag. Manufact. Technol.*, vol. 3, no. 2, pp. 350-356, Feb. 2013.

[3] S. Zhang, R. Seager, A. Chauraya, W. Whittow, Y. Vardaxoglou, "Textile Manufacturing Techniques in RF Devices," *Loughborough Antennas and Propagation Conf.* , 2014, pp. 182-186.

[4] S. Wu, P. Liu, Y. Zhang, H. Zhang, X. Qin, "Flexible and conductive nanofiber-structured single yarn sensor forsmart wearable devices," *Sensors and Actuators B: Chemical*, vol. 252, 2017, pp. 697-705.

[5] H. Saghlatoon, L. Sydänheimo, L. Ukkonen, M. Tentzeris, "Optimization of Inkjet Printing of Patch Antennas on Low-Cost Fibrous Substrates," *IEEE Antennas and wireless propagation letters*, vol. 13, 2014, pp. 915-918.

[6] R. Goncalves, S. Rima, R. Magueta, P. Pinho, A. Collado, A.Georgiadis, J. Hester, N. Borges Carvalho and M. M. Tentzeris, "RFID-Based Wireless Passive Sensors Utilizing Cork Materials," *IEEE Sensors Journal*, vol. 15, no. 12, 2015, pp. 7242-7251.

[7] "Fabink-UV-IF1004," *Smart Fabric Inks Ltd*, 2014.

[8] "ULTRALAM 3850," *Rogers Corporation*, Rogers, CT, 2012.

[9] H.-L. Kao, C.-L. Cho, L.-C. Chang, C.-S. Yeh, B.-W. Wang, H.-C. Chiu, "Inkjet Printing RF Bandpass Filters on Liquid Crystal Polymer Substrates," *Thin Solid Films*, vol. 544, 2013, pp. 64–68.

[10] L. Song, Y. Rahmat-Samii, "A Systematic Investigation of Rectangular Patch Antenna Bending Effects for Wearable Applications," *IEEE Trans. on Antennas and propagation*, vol. 66, , no. 5, 2018, pp. 2219-2228.

Mechanical and Electrical Characterization of FOWLP-Based Flexible Hybrid Electronics (FHE) for Biomedical Sensor Application

Yuki Susumago[1], Qian Zhengyang[1], Achille Jacquemond[1,2], Noriyuki Takahashi[1], Hisashi Kino[3], Tetsu Tanaka[1,4], and Takafumi Fukushima[1]

1 Department of Mechanical Systems Engineering, Graduate School of Engineering, Tohoku University
2 Génie Mécanique, INSA (Institut National des Sciences Appliquées) Lyon
3 FRIS (The Frontier Research Institute for Interdisciplinary Sciences), Tohoku University
4 Graduate School of Biomedical Engineering, Tohoku Univesity
6-6-12 Aza-Aoba, Aramaki, Aoba-ku, Sendai 980-8579, Japan Sendai, Miyagi, 980-8579, Japan
Email: link@lbc.mech.tohoku.ac.jp

Abstract— **A new trans-nail FHE (flexible hybrid electronics) system with photoplethysmographic (PPG) sensors is proposed and characterized from both aspects of mechanical and electrical properties in this study. The unique FHE structure is consisting of an elastomer as a flexible substrate in which Si LSI dielets having photodiode and LED driver circuits etc. are embedded based on a FOWLP concept. Stress buffer layers (SBLs) as a key material are inserted between interdielet wirings and the substrate to mitigate mechanical stress and enhance wire reliability. The impact of the Young's moduli of the SBLs on the repeated bendability of the FHE systems is described. In addition, we evaluate the electrical properties of the LSI dielets between before and after bending for comparison.**

Keywords- Flexible hybrid electronics (FHE); Trans-nail PPG sensor; Biocompatible; FOWLP; Bendable interconnect; Metallization on PDMS

I. INTRODUCTION

Typical FHE process requires printable wiring technologies and ultrathin dies (20 μm or less in thickness) formed by backside grinding to make them bendable to follow curved profiles [1]-[4]. Since the ultrathin dies can be technologically fabricated down to 7 μm or less [5] [6], several pioneering works utilize the ultrathin dies for 3D integration with TSV (Through-Silicon Via). However, one concern is characteristic degradation induced by mechanical stress. Although it depends on the applications, we have been raising that problem with thinned DRAM chips since 2013 [7]. Since Si is a very sensitive material to mechanical stress, Si is widely used as stress sensors. Another issue is that the mechanical strength of the ultrathin dies is determined by thinning processes (chemical mechanical polishing, plasma etching, dry polishing, kai-dry polishing, poly grinding, ultra-poly grinding, #2000, etc.) [8]. Cost effective mechanical grinding with dry polishing and standard simple saw dicing are not enough to relief the mechanical stress and remove micro-cracks at the chip edges. A spalling technique [9] and transfer technologies

from SOI wafer [10]-[12] are other interesting approaches to fabricate ultrathin dies without the mechanical thinning processes.

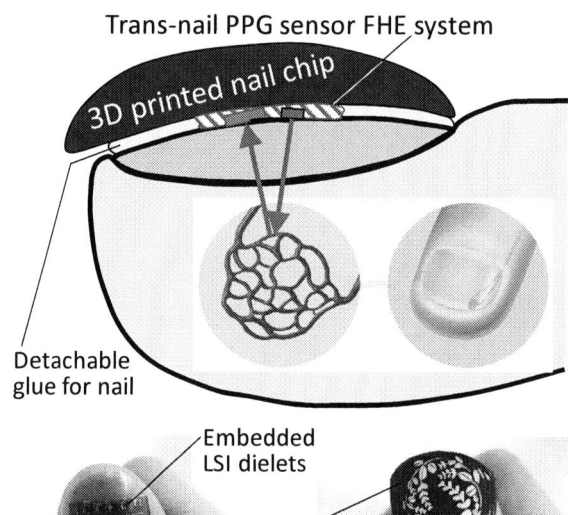

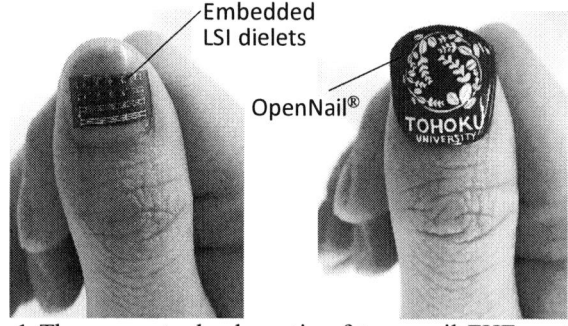

Fig. 1 The conceptual schematic of trans-nail FHE system with PPG sensors.

However, ultrathin die assembly on flexible substrates is still challenging due to the limited processes without high-stress thermal compression bonding. Small dielets ranging in thickness from 50 to 200 μm are employed in our FHE works [13] [14]. The dielets are embedded in flexible substrate by using advanced FOWLP and fine-pitch wirings are also formed in wafer-level processing. The biggest advantage of our FHE is that no bonding processes are required. We believe SBL acts as an enabler to mitigate

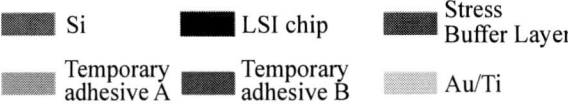

1. Temporary adhesive formation and LSI chips alignment

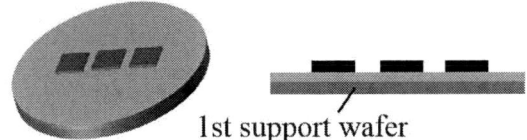

1st support wafer

2. PDMS compression molding

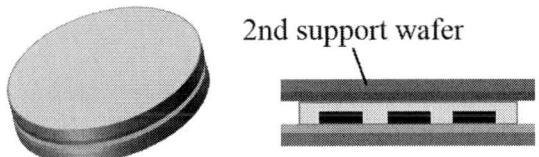

2nd support wafer

3. 1st support wafer debonding

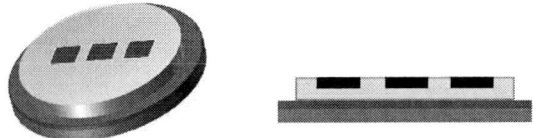

4. Stress Buffer Layer formation

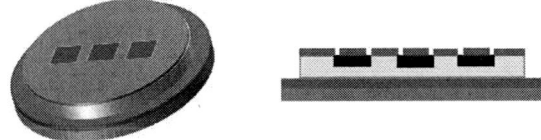

5. Au wiring formation

6. 2nd support wafer debonding

Fig. 2 A process flow of FHE fabrication in this study

mechanical stresses when bending. In this work, we design the SBLs, and evaluate the impact of their mechanical and electrical properties on the bendability of the FHE systems with a real LSI dielets. We have previously proposed trans-nail photoplethysmographic (PPG) sensor system [15] and verified that the PPG recording is successfully monitored in both the reflection and transmission modes with an external LED chip and a thick LSI die having photodiodes and an LED driver designed on it. Nails are the only part not sweating on human body. Thus, the nails are the best part on which flexible devices are sitting. Here, FHE fabrication embedding the LSI dielet is demonstrated based on FOWLP processes with a biocompatible PDMS substrate for achieving the trans-nail FHE system integration with PPG sensors. The electrical properties of the embedded LSI dielets in our FHE are also characterized.

II. EXPERIMENTAL

Fig. 1 shows the conceptual schematic of the trans-nail FHE system with PPG sensors. 3D-printed nail chips are fabricated by OpenNail® technologies from Toshiba Digital Solutions Corporation. The tailor-made nail chips are designed to fit their own nail curves with various curvature radius with a 3D scanner. Fig. 2 shows the process flow of our FHE fabrication. First, embedding of 100-μm-thick Si dielets (1mm x 1mm) or a 200-μm-thick LSI dielet (2.5 mm x 2.5 mm) having photodiode and micro-LED driver circuits in a biomedical grade PDMS were implemented with two carrier Si wafers. These dielets were gently placed on the 1st carrier wafer in a face-down configuration. Then, the PDMS monomers were poured on the dielet-on-wafer structure, followed by vacuum defoaming and the subsequent wafer-level compression molding with the 2nd carrier. After debonding of the 1st carrier, the dielets are embedded on the 2nd carrier and planarized without any mechanical processes. Prior to the following metallization processes, a SBL or two SBLs are coated on the PDMS/dielets. By using standard photolithography with metal sputtering/wet etching, 200-nm-thick Au wirings were formed on the SBL(s) at the wafer-level. Finally, the FHE systems were debonded from the 2nd carrier. Fig. 3 shows the cross-sectional schematic of our new FHE.

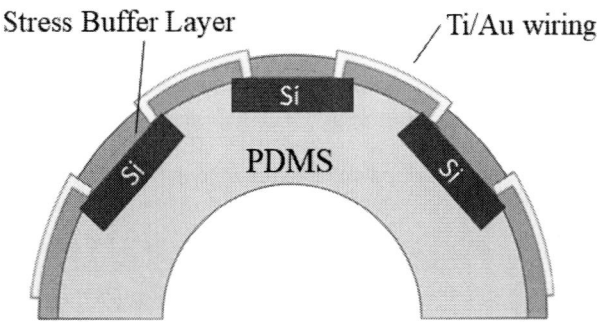

Fig. 3 Cross-sectional structure of our FHE with embedded dielets

978-1-7281-1500-9/19 $31.00 © 2019 IEEE

III. RESULTS AND DISCUSSION

A. Trans-nail FHE system fabrication with PPG sensors

SBL formed underneath the wirings can compensate the Young's modulus mismatch between the metal and PDMS. In this work, the following three SBLs were employed: soft polyurethane (PU), hard PU, and Parylene C. These properties are listed in Table 1. The lowest Young's modulus is 0.5 MPa for PDMS, whereas the wire Au has high Young's modulus of 80 GPa. Since the PDMS is an elastic material, the Au wires are supposed to be deformed when high tensile or compressive stresses are applied to the PDMS. The small elongation at yield of Au leads to plastic deformation by which the wires are not restored to their original shape. The failure would be solved by SBL that can absorb the large stress resulted from PDMS deformation.

As seen from Fig.4 (a)-(b), Au wirings even with a width of 100 μm and a length of 10 mm are disconnected after 20 cycle bending with a bending radius of 20 mm when the single soft PU layer is used as a SBL. In contrast, Au wirings formed on the hard PU layer as a SBL exhibit high bendability. After 30 cycle bending, the 10-μm-width Au wirings are still connected. On the other hand, Au wirings with a 1-μm-thick Parylene C underneath the soft PU show high bendability, compared to that with the single soft SBL. Finally, the integration of 4 by 4 Si dielets in PDMS are demonstrated and they are interconnected each other. The bending results with the double layer SBL consisting of a soft PU and Parylene C indicates that one hard layer is essential for our FHE at least when we use extremely soft elastomer-based substrates such as PDMS.

The double layer SBL plays another important role in mechanical reliability. Fig. 5 shows the enlarged schematics of the edge of embedded dielets in PDMS. Parylene C as the first layer SBL is conformably deposited on the embedded dielets after compression molding with PDMS curing and the subsequent transferring to the 2nd carrier. The stand-off height can be lowered by decreasing the curing temperature [14]. However, Au wirings have to be climbing over a step, the height is 1 μm or more, formed with the sidewall of the dielets perpendicular to the PDMS even when PDMS is cured at room temperature. On the other hand, the second layer spin-on SBL can planarize the steep steps to give gradual slopes. Such a structure mitigates stress concentration and prevents Au wirings being broken when bending. Consequently, the combination of soft PU and Parylene C exhibits good bending endurance. It can be said that the highest reliability will be given by spin-on hard SBL(s). On the other hand, the Parylene C deposition process should not be skipped because the room-temperature deposition helps fine pattern formation without wrinkles and micro-cracks on PDMS with metals and photoresists. The Parylene C compensates CTE, elongation, and Young's modulus mismatches between the PDMS and the other materials on it.

TABLE I. THERMOMECHANIAL PROPERTIES FOR SBL CANDIDATES AND THE OTHER MATERIALS USED IN OUR FHE SYSTEM

Material	CTE (ppm/K)	T_g (°C)	Young's modulus (GPa)
PDMS	300	-200	0.0005
Soft PU	100	50	0.9
Hard PU	70	60	2.3
Parylene C	35	90	2.76
Ti	8.4	-	106
Au	14.2	-	79
Si	3	-	190

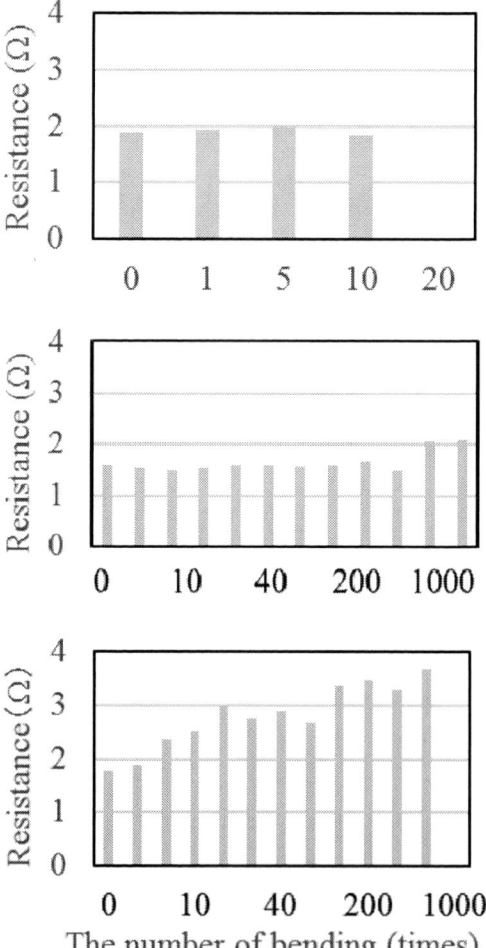

Fig. 4 Bending property comparison between three SBLs: soft PU (a), hard PU (b), soft PU with thin Parylene-C (c).

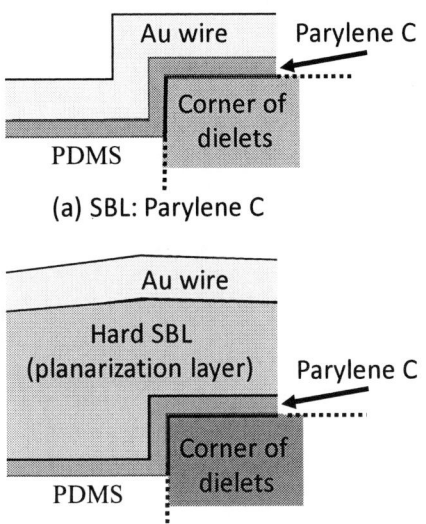

(a) SBL: Parylene C

(b) SBL: Hard SBL on Parylene C

Fig. 5 Cross-sectional schematic of wire shape formed at the edge of an embedded dielet in PDMS.

Fig. 6 shows the picture of an embedded LSI dielet in PDMS. The FHE system embedding the LSI dielet with PPG sensors can be mounted on curved human nail and used for trans-nail pulse-wave monitoring system. As seen from Figs. 7 and 8, current–voltage characteristics of photodiodes with various incident lights are evaluated. The photodiode function is confirmed with a 660 nm incident light and the light sensitivities before and after mounting are 0.0219 and 0.0237 A/W, respectively. There is no significant change before and after bending. As seen from Fig. 9, no degradation is also observed on the output characteristics of the CMOS before and after LSI dielet embedding and fan-out RDL formation.

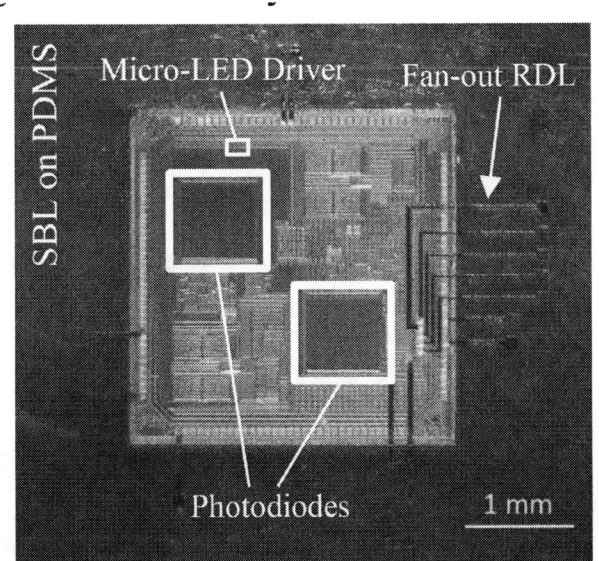

Fig. 6 A photo of an embedded dielet in PDMS for trans-nail FHE system with PPG sensors.

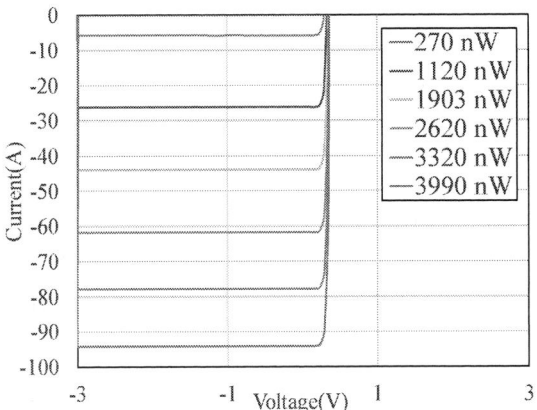

Fig. 7 I-V characteristics of photodiode on an embedded die in FHE.

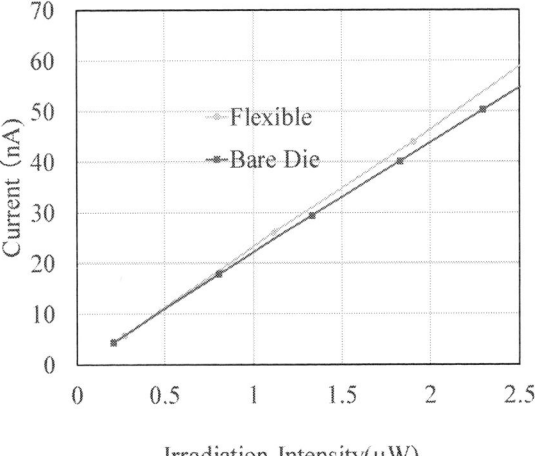

Fig. 8 Comparison of photocurrent as a function of incident light intensity of photodiodes between bare die and embedded die in FHE.

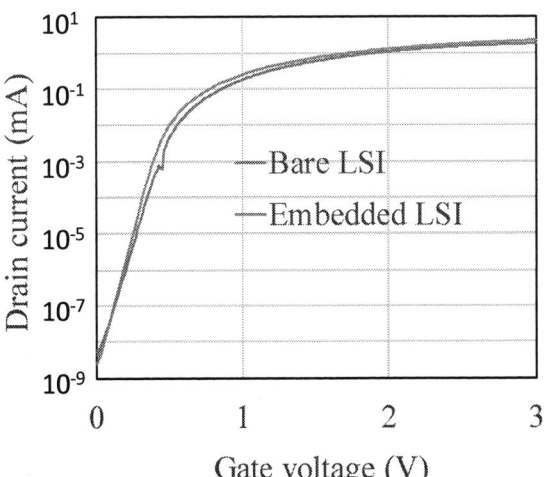

Fig. 9 Comparison of I_D-V_G characteristics of CMOS between bare die and embedded die in FHE.

B. Stress simulation

Stress simulation was performed with ANSYS Workbench 19.0. Through the simulation, the hard PU and soft PU were compared in terms of equivalent von Mises stress in the sample. In the simulation, the Poisson's ratio was 0.22 for Si, 0.42 for Au, and 0.49 for the PDMS, hard PU and soft PU. The results of the simulation are shown in Fig. 9 The brighter color indicates higher stress. The biggest stress is expressed in red that is appeared close to the dielet edges. The maximum von Mises stress was 137.18 and 118.51 MPa for hard PU and soft PU SBLs, respectively. Compared to the hard SBL, the soft SBL has larger stress on wiring fabricated on PDMS, which is in good agreement with the experiment results.

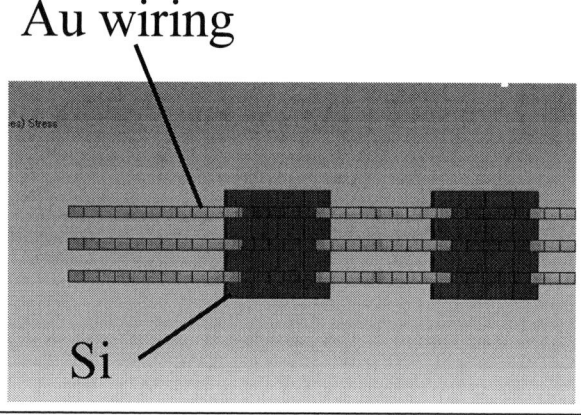

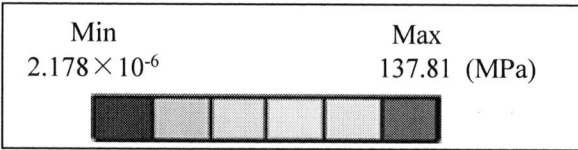

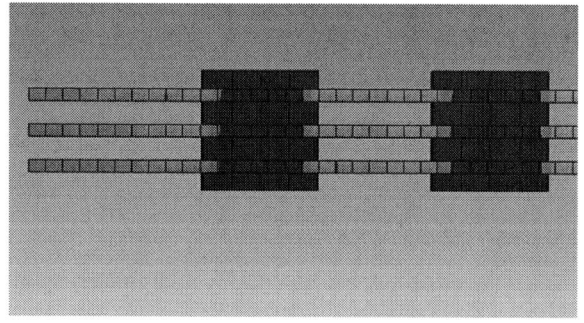

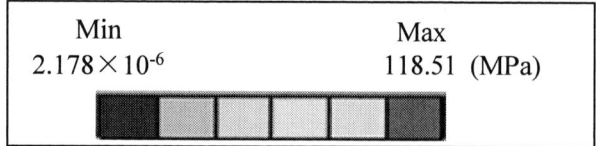

Fig. 9 Stress distribution in the 100-µm-width Au wires with a soft PU SBL (top) and a hard PU SBL (bottom).

IV. CONCLUSIONS

A trans-nail FHE system with PPG sensor embedding a real LSI dielet in PDMS was successfully fabricated using a hard SBL. 10-µm-width Au was formed on the SBL/PDMS and 40-µm-width Au RDLs exhibit high bendability with a bending cycle of beyond 1,000. This heterogeneous integration scheme enables high-performance and scalable FHE to create highly-integrated wearable systems.

Acknowledgement

This work was performed in the Micro/Nano-machining research and education Center (MNC) and Jun-ichi Nishizawa Research Center at Tohoku University. This work was supported by JSPS KAKENHI (Grants-in-Aid for Scientific Research) Grant Number 18K18841 of Challenging Research (Pioneering). We would like to acknowledge Toshiba Digital Solutions Corporation for 3D-printed nail chip fabrication by OpenNail® technologies. We wish also to thank Nippon Polytech Corp. R&D Department for their matrial supports.

REFERENCES

[1] Darrell E. Leber, Brian N. Meek, Seth D. Leija, Dale G. Wilson, Richard L. Chaney, Douglas and R. Hackler, "Electromechanical Reliability Testing of Flexible Hybrid Electronics Incorporating FleX Silicon-On-Polymer Ics ", *Proc. 2016 IEEE Workshop on Microelectronics and Electron Devices (WMED)*, (2016), pp. 1 – 4.

[2] Robert Herbert, Jong-Hoon Kim, Yun Soung Kim, HyeMoon Lee, and Woon-Hong Yeo, "Soft Material-Enabled, Flexible Hybrid Electronics for Medicine, Healthcare, and Human-Machine Interfaces", *Materials*, vol. 11, no. 187 (2018) 33 pages, doi:10.3390/ma11020187.

[3] Nagarajan Palavesam, Sonia Marin, Dieter Hemmetzberger, Christof Landesberger, Karlheinz Bock, and Christoph Kutter, "Roll-to-roll processing of film substrates for hybrid integrated flexible electronics", *Flex. Print. Electron.*, vol. 3 (2018) 014002, doi.org/10.1088/2058-8585/aaaa04.

[4] Jeroen van den Branda, Margreet de Koka, Marc Koetsea, Maarten Cauweb, Rik Verplanckeb, Frederick Bossuytb, Michael Jablonskib, and Jan Vanfleteren, "Flexible and stretchable electronics for wearable health devices", *Solid-State Electronics*, vol. 113 (2015) pp. 116–120, doi.org/10.1016/j.sse.2015.05.024.

[5] Takafumi Fukushima, Yusuke Yamada, Hirokazu Kikuchi, and Mitsumasa Koyanagi, "New Three-Dimensional Integration Technology Using Self-Assembly Technique", *IEEE International Electron Devices Meeting (IEDM) Tech. Dig.*, (2005), pp.348-351, doi.10.1109/IEDM.2005.1609347

[6] Y. S. Kim, S. Kodama, Y. Mizushima, T. Nakamura, N. Maeda, K. Fujimoto, A. Kawai, K. Arai, and T. Ohba, "A Robust Wafer Thinning down to 2.6-µm for Bumpless Interconnects and DRAM WOW Applications", *IEEE International Electron Devices Meeting (IEDM) Tech. Dig.*, (2015), pp.189-192, doi.10.1109/IEDM.2015.7409653

[7] K. W. Lee, S. Tanikawa, M. Murugesan, H. Naganuma, H. Shimamoto, T. Fukushima, T. Tanaka, and M. Koyanagi, "Degradation of Memory Retention Characteristics in DRAM Chip by Si Thinning for 3-D Integration", *IEEE Electron Device Lett.*, vol. 34, no. 8 (2013) pp. 1038-1040, doi. 10.1109/LED.2013.2265336. .

[8] Murugesan Mariappan, Takafumi Fukushima, Jichoel C. Bea, Kang-Wook Lee, and Mitsumasa Koyanagi, "Mechanical Characteristics of Thin Die/Wafers in Three-Dimensional Large-Scale Integrated Systems", *IEEE TRANSACTIONS ON SEMICONDUCTOR*

MANUFACTURING, vol. 27, no. 3 (2014), pp. 341-346, doi. 10.1109/TSM.2014.2316917

[9] D. Shahrjerdi, S. W. Bedell, A. Khakifirooz, K. Fogel, P. Lauro, K. Cheng, J. A. Ott, M. Gaynes, and D. K. Sadana, "Advanced Flexible CMOS Integrated Circuits on Plastic Enabled by Controlled Spalling Technology", *IEEE International Electron Devices Meeting (IEDM) Tech. Dig.*, (2012), pp.92-95.

[10] Pengfei Sun, Benjamin Mimoun, Edoardo Charbon, and Ryoichi Ishihara "A Flexible Ultra-Thin-Body SOI Single-Photon Avalanche Diode", *IEEE International Electron Devices Meeting (IEDM) Tech. Dig.*, (2013), pp.284-287.

[11] Zhang Cang-Hai, Yang Yi, Wang Yu-Feng, Zhou Chang-Jian, Shu Yi, Tian He, and Ren Tian-Ling, "A Novel Fabrication Method for Flexible SOI Substrate Based on Trench Refilling with Polydimethylsiloxane", *Chinese Phys. Lett.* vol. 30 (2013) 086201, doi.10.1088/0256-307X/30/8/086201.

[12] Kohei Sakaike, Muneki Akazawa, Akitoshi Nakagawa, and Seiichiro Higashi, "Meniscus-force-mediated layer transfer technique using single-crystalline silicon films with midair cavity: Application to fabrication of CMOS transistors on plastic substrates", *Jap. J. Appl. Phys.*, vol. 54 (2015) 04DA08, doi. 10.7567/JJAP.54.04DA08.

[13] Tak Fukushima, Arsalan Alam, Saptadeep Pal, Zhe Wan, Siva Jangam, Goutham Ezhilarasu, Adeel Bajwa, and Subramanian Iyer, "FlexTrate®" - Scaled Heterogeneous Integration on Flexible Biocompatible Substrates Using FOWLP", *Proc. the 67th Electronic Components and Technology Conference*, (2017) pp. 649-654, doi. 10.1109/ECTC.2017.226.

[14] Takafumi Fukushima, Arsalan Alam, Amir Hanna, Siva Chandra Jangam, Adeel Ahmad Bajwa, and Subramanian S. Iyer, "Flexible Hybrid Electronics Technology Using Die-First FOWLP for High-Performance and Scalable Heterogeneous System Integration", *IEEE TRANSACTIONS ON COMPONENTS, PACKAGING AND MANUFACTURING TECHNOLOGY*, vol. 8, no. 10 (2018) pp. 1738-1746, doi. 10.1109/TCPMT.2018.2871603.1.

[15] Zhengyang Qian, Yoshiki Takezawa, Kenji Shimokawa, Hisashi Kino, Takafumi Fukushima, Koji Kiyoyama, and Tetsu Tanaka, "Development of integrated photoplethysmographic recording circuit for trans-nail pulse-wave monitoring system", *Jap. J. Appl. Phys.*, vol. 57 (2018) 04FM11, doi.10.7567/JJAP.57.04FM11.

A Wearable Fingernail Deformation Sensing System and Three-Dimensional Finite Element Model of Fingertip

Katsuyuki Sakuma, Bucknell Webb, Rajeev Narayanan, Avner Abrami, Jeff Rogers, John Knickerbocker, and Stephen J. Heisig

IBM Thomas J. Watson Research Center
Yorktown Heights, U.S.A.
e-mail: ksakuma@us.ibm.com

Abstract—This paper describes the sensor, electronics, software, modeling, and characterization of a fingernail-mounted RF-connected wearable strain sensor system that measures nail deformation from finger movement. Applications to health monitoring and human computer interfaces in homes, hospitals, and workplaces are discussed. The mechanical deformation of a fingertip pressed or drawn against a plate is demonstrated using a three-dimensional finite-element linear-elastic model to predict the signal level, optimum sensor locations and the type and location of deformation expected for different finger motions. The 3D finite-element linear elastic model is derived from X-ray images of a human finger but generalized and parameterized to allow new models to be created by scaling internal and external parameters such as skin thickness and nail and finger shape to predict sensor system performance for a more general human population. Our analysis finds that a single sensor mounted in the center of the nail will respond to typical grip pressures on the fingertip with readily detectible strain amplitudes but that a multi-sensor array will be sensitive to more general haptic phenomena such as the direction and magnitude of frictional loads and loading of the distal phalangeal joint. It is shown that depending on finger use and loading the nail exhibits shifts in direction, location and sign of strain over the fingernail surface. Measurement data from a simple multi-sensor array is shown to be useful in distinguishing between load conditions, however additional sensors are required for full determination.

Keywords- Wearable, Flexible, Wireless, Sensor, Spalling, Finite element model, Fingernail, Biological tissues, Machine learning, CNN, LSTM, Neural network

I. Introduction

According to Stanford Encyclopedia of Philosophy, touch is recognized as the first sense that humans develop [1]. Touch allows us to perceive and understand the world around us by communicating distinct sensations to the brain through specialized neurons in the skin. Fingertips are a key human sense organ providing information about applied pressure, temperature, vibration, pain, and facilitating human communication. Several academic and industry researchers have developed sensor-based technologies to examine the finger and hand movements. For example, in [2-4], the authors show a glove type wearable sensor that can be used to monitor the finger movement. However, use of gloves inhibits direct contact between fingers and objects touched leading to loss of precision tactile sensing from fingertips. Sensors attached directly to the skin are becoming popular to monitor the health of muscles and nerves [5-8]. With skin-based sensors there are

growing concerns of allergic reaction and infection, especially among an elderly population with fragile skin. In this paper, we report on a new fingernail based wearable wireless sensor and processing methods for monitoring fingernail deformation to quantify and characterize hand activities [9]. Fingers transmit the majority of grip force when using a power grip for grasping heavy objects or a precision grip for holding small, delicate objects. Hand movements such as grasping an object and precision fingertip movements often result in slight deformation of the fingernail. Our wearable fingernail sensor system can detect micron level distortion of the fingernail and transmit the data to a wearable/portable computing device (smartwatch, smartphone, etc). Machine learning models are able to classify delicate hand activities by analyzing the continuous data stream [10]. It was also revealed that bending of the fingernail can be correlated with grip force [9]. Unlike skin-based sensors, this sensor doesn't need to touch the skin and so avoids allergic and infection related complications.

Figure 1. A photograph of wearable wireless fingernail deformation sensor.

II. Anatomy of Human Finger and Fingernail

Humans are members of the taxonomic order Primates and possessing flat fingernails (at least one) is one of the characteristic features distinguishing primates from other mammals. When fingers are used for touching and grasping, forces are applied through the finger pads and the fingernails are deformed. It is useful to understand the basic anatomy of the hand and finger to understand the structures behind fingertip dynamics. Fig. 2 shows X-ray images of the right-hand fingers of one of the authors. Each finger has three phalanges (DP: Distal phalanx, MP: Middle phalanx, and PP: Proximal phalanx) and these bones are connected and stabilized by tendons, muscles, and ligaments across the joints.

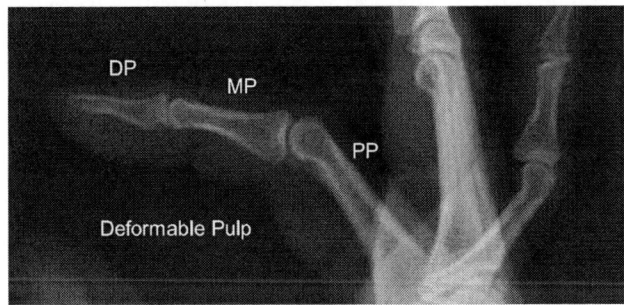

Figure 2.   X-ray image of right-hand fingers from the side.

Fig. 3 shows a schematic drawing of fingertip and scanning electron micrograph (SEM) images of one of the author's fingernails. The fingernail unit is composed of the nail plate, proximal nail fold, lateral nail fold, nail bed, hyponychium, and nail matrix [11, 12]. The nail matrix is the region where new nail plate cells are generated. The nail plate is made of up of three layers of keratin, a fibrous structural protein. The interaction between distal phalanx, deformable fingertip pulp, and nail plate causes complicated fingernail deformation during finger movements. From the top surface planar view (Fig. 3(b)), overlapping keratin layers are observed. The three layers of the nail plate have a total thickness of about 350-450 μm. The growth of human fingernail is approximately 3 mm in a month on average, but it slows with aging and vascular disease [13]. Therefore, the strain sensor moves quite slowly when attached to the nail plate and extended use would not require frequent replacement. It remains to be seen what the effect on machine learning models is as the sensor moves distally with growth.

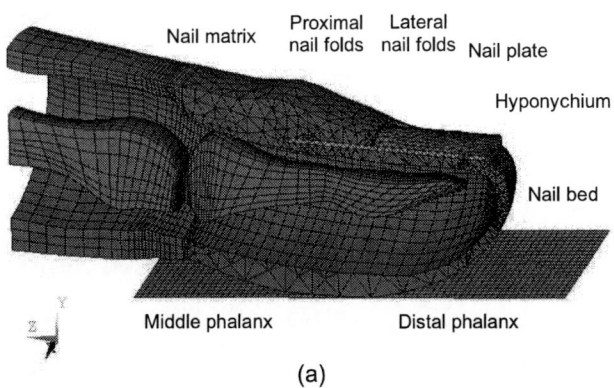

(a)

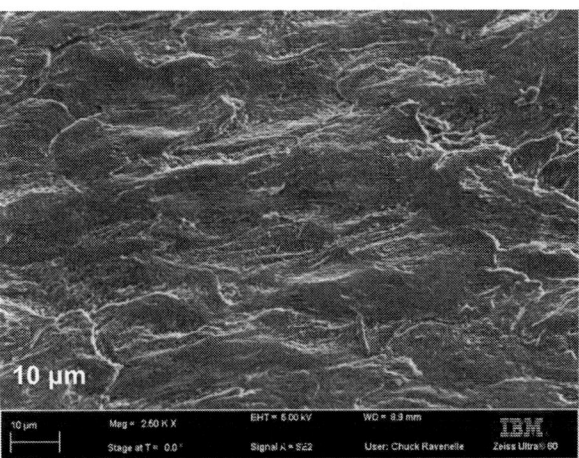

Figure 3.   (a) Schematic drawing of fingertip and (b) scanning electron micrograph (SEM) images of a nail plate: top surface of a human nail plate showing the overlapping layers of keratin.

III.   MODELING

A parameterized 3D finite element model of the last joint and fingertip has been constructed to relate pressures on the fingertip during activity to strains and deformations of the fingernail. The parameterized form reduces the description of a finger, thumb or toe to a modest table of numbers allowing an existing model structure to be modified in a systematic way to understand variations likely in populations. This paper will compare results from finger used in this study with other cases and with Digital Image Correlation (DIC) images of the finger and nail shape changes for different motions of the finger pressed against a flat high friction surface.

Fingertips are complex structures which were simplified to enable this study. Firstly, the structure was reduced to four uniform isotropic materials: bone, nail, skin/epidermis, and interior flesh (Table 1) [14-16]. Previous modeling studies adopted similar approximations because the exact mechanics of the finger tissue is not known. Tendons that control joint motion were replaced by two simple cases – a rigid joint with no motion or a motion constrained to a slight bend by the deformation of nearby elastic material as when the fingertip is allowed to flex backward under upward pressure.

TABLE I.   MECHANICAL PROPERTIES OF MATERIALS IN THE MODEL [14-16]

Material	Young's modulus (MPa)	Poisson's ratio
Human fingernail	3000	0.3
Distal phalanx	8000	0.3
Soft tissue	150	0.45
Epidermis	250	0.45

A last approximation, which enabled model parameterization, was to describe the finger by a series of sections transverse to the finger axis (z direction) where each material boundary is described by an ellipse with a dimension and center in the y direction (perpendicular to the nail) and a width in the finger transverse direction (x, mostly parallel to nail.) The finger structure is then defined by lofting over the sections. The nail itself is described as an extruded section with transverse and axial radii of curvature, width and thickness and y location and tilt. This construction allows a model to be built based on each finger's dimensions and then modified to look at different nail thicknesses or curvatures or the effects of thicker skin or other finger or bone shapes. For instance, fingernails may have either a positive or negative radius of axial curvature while each finger or thumb has a different shape.

Fig. 4 shows model section images of index fingers from three different people. Experimental data for the person in case A will be presented below. In each case measurements from top, side, and end views of the fingers are used to define the parameters defining the exterior finger shape and nail curvatures and dimensions. Next, the interior skin boundary is defined using typical skin thickness values. The thickness distribution and average thickness becomes a parameter for model comparison studies. After placement of the nail into the model based on exterior and thickness measurements and estimates of the nail base location, an expected bone shape and location is entered by scaling bone images from finger X-rays (Fig. 2). The bone shape would be more ideally determined by X-ray images of the actual fingers modeled but these were not available. Model variants show this uncertainty in bone shape and location mean the bending response of the nail will be smaller if the nail is located closer to the bone or if the bone is wider than modeled, and larger if not. However, the uncertainty in bone position is substantially constrained by plausible anatomy.

Measurements by Digital Image Correlation (DIC) of the motion and shape changes in the finger of person A while pressed against a flat surface have been reported previously [9]. In the DIC technique, separate binocular images of a speckle-painted finger are compared and correlated to determine the 3D location of each point on the finger surface as the shape changes with finger motion. These measurements have been used to validate two elements of the model results: first, the total motion of the finger and nail and second, results where rotation and translation of the finger and nail is subtracted leaving only changes in the nail shape relative to the average nail plane. This second case is important since the strain gauges tested to date only detected nail shape changes and not rotations and translations.

Fig. 5 compares model and measured total motion results for a motion where the finger is pressed down (about 1 mm-2 mm) against a nearly level plate (10 degree angle) and moved for two cases: (A) to the left by 2 mm, rotating the skin of the finger to move the nail to the right relative to the bone while pulling the left side of the nail downward. Good agreement is seen for this case. (B) forward by 1 mm. Smaller motions are seen in both model and experiment for the forward motion, but the model agreement is less accurate. Part of the discrepancy may be because precise isolated forward and backward motions were difficult to execute consistently. In addition, models show that if the finger joint is allowed to bend backward, the nail near the nail base will be elevated making the result more like the measured shape.

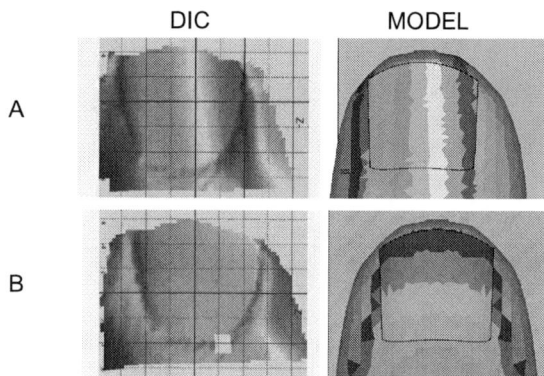

Figure 5. Comparison of DIC and Model results for finger motions: (A) down and then left by 2 mm, scale 100 μm/contour (B) down and forward by 1 mm, scale 40 μm/contour.

Fig. 6 compares model and measured changes where the tilt and motion of the nail is removed from the DIC and model data to leave changes in the actual nail shape. Shape changes bend the nail which create the strains measured by the nail-mounted strain sensors. The DIC shape change results agree in magnitude and shape with the model results for this finger, supporting the assumptions made in the model.

Comparison of results for finger models for different people illustrates variations in finger sensor performance that may be seen over human populations. Person A differs from the other cases in having a more turned-up finger, and a short and nearly flat nail in the axial direction. The finger of person

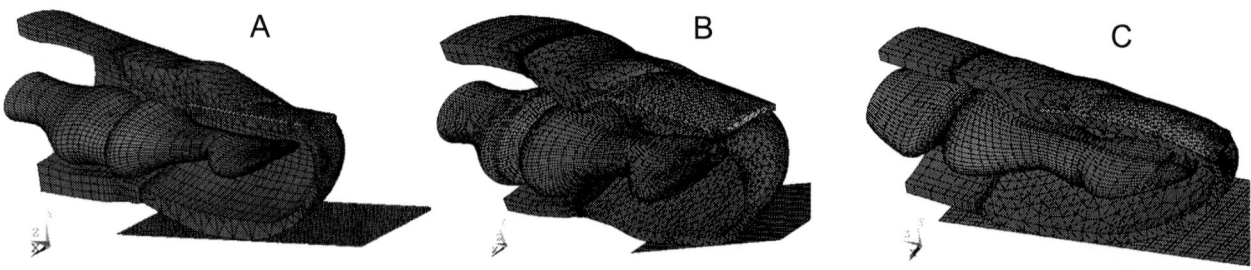

Figure 4. Three-dimensional meshed finite element models of human index fingers. Partial sections shown with interior flesh material omitted for

B is broader and straighter but the nail shape, while longer, is otherwise like A. Person B has also been modeled with a more extensive nail base. Person C has a tapering finger and a significant axial nail radius of curvature (65 mm versus 200 mm.) Fig. 7 shows axial and transverse strain maps for the case where the finger is pressed down 2 mm onto the flat surface with the pressure contact angle of 10 degrees as shown in Fig. 4. Between cases, the transverse strain shows the least variation suggesting that the transverse component may be the most useful when studying grip and finger interactions. The axial strain maps are markedly different between finger models. Additional modeling and comparison with actual data is required to verify how much of this difference is intrinsic to population variations.

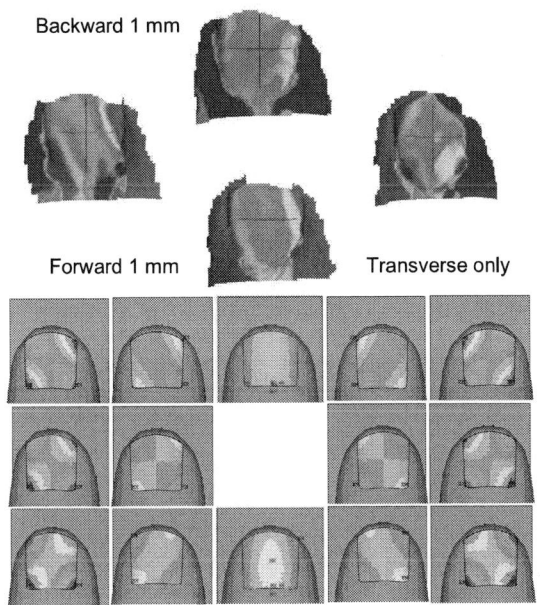

Figure 6. Top: DIC images for forward and backward motions of 1 mm and transverse motions of 2 mm with tilt and translation removed, scale 10 μm/contour. Bottom: Equivalent model results, scale 6 μm/contour.

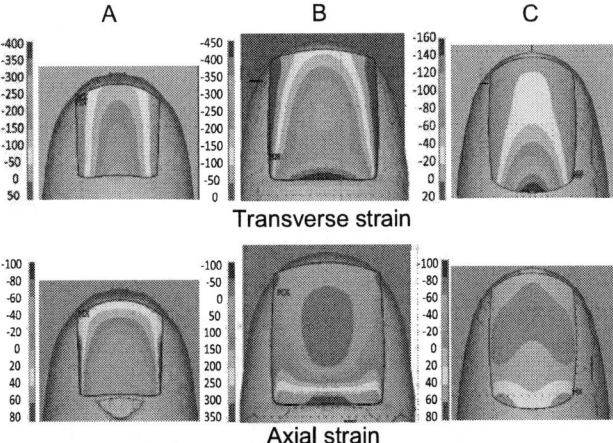

Figure 7. Model nail strain components: transverse (top) and axial (bottom) for index fingers on persons A, B, & C when the finger is pressed flat 2 mm into a surface. Scale: ppm.

As the contact angle shifts from normal downward pressure as in the previous figures to more concentrated at the tip of the finger, the peak transverse stresses also move toward the fingertip, as shown in Fig. 8, comparing the transverse strain model results for the model of finger B over the range from 10 degrees up to 60 degrees. Overall strain magnitudes are inconsistent because the finger pressures were not kept constant. A transverse sensor placed on axis between the midpoint and outer third of the nail will experience compressive strains over the entire range of angles.

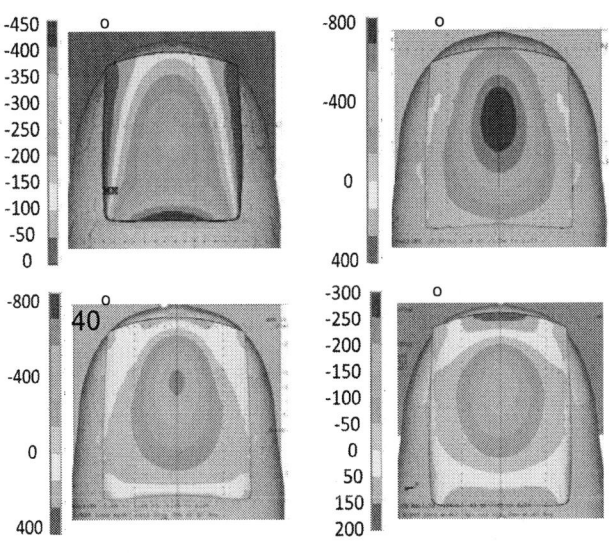

Figure 8. Transverse strain for finger B as the pressure angle of contact is varied between 10 and 60 degrees. Scale: ppm.

These models illustrate that there are substantial changes in strain distributions over the nail surface depending on finger and nail shape and direction of pressure and motions of the finger relative to the pressure surface. For example, pressures and motions off axis produce off center strains. For a single transverse sensor example, a good location is on axis between the midpoint and outer third of the nail but it is clear a full understanding of position and type of grip would be helpful to more fully map the strain distribution over the nail.

IV. FLEXIBLE THIN SILICON STRAIN GAUGE SENSOR

The fabrication sequence for highly sensitive flexible strain gauge sensor using doped silicon is shown in Fig. 9. Compared to several methods to produce a thin film of semiconductor material, our method has advantages such as fewer processing steps and the entire spalling process can be done at room temperature. The process flow has been revealed previously [17] and merely summarized here. A cross-sectional view illustrating a doped silicon substrate is shown in Fig. 10. Doping can be selective or non-selective such that the entire substrate is doped. Different CMOS device components can be fabricated on the same substrate before starting the spalling process. After patterning a first photoresist layer (Fig. 9(a)), a metal ohmic contact material is

978-1-7281-1500-9/19 $31.00 © 2019 IEEE

deposited on the substrate (Fig. 9(b)). The remaining first photoresist is removed utilizing the resist stripping process for lifting off the metal ohmic contact material from the photoresist layer (Fig. 9(c)). A seed layer (Ti) is sputtered and a thick stressor layer (Ni with a stress of 550MPa) is then electroplated after the second photoresist layer is patterned (Fig. 9(d, e)). A first flexible handle tape is applied to the top of the stressor layer to effect spalling (Fig. 9(f)). Then a thin layer of doped silicon can be spalled from the bulk silicon substrate in the regions where the second photoresist layer is not unmasked (Fig. 9(g) and Fig. 10(a)). Lifting handle tape causes a crack to form in the substrate at the abrupt edge of Ni layer. Second flexible handle tape is applied to the backside of the thin layer of spalled silicon (Fig. 9(h)) and the first handle tape, a second photoresist layer, stressor layer, and seed layer are removed (Fig. 9(i)). Fig. 10(b) shows a thin layer of spalled doped silicon after a handle tape, the second photoresist layer, stressor layer, and seed layer are removed. The thickness of the spalled silicon is approximately 40 μm. The change in electrical resistance of the spalled silicon when force is applied is measured by bending the silicon using a three-point bend test tool. The results indicate that the controlled spalling method can be used to fabricate a flexible thin silicon strain gauge sensor.

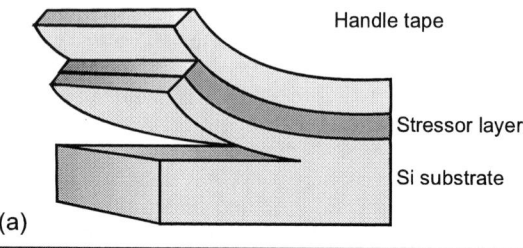

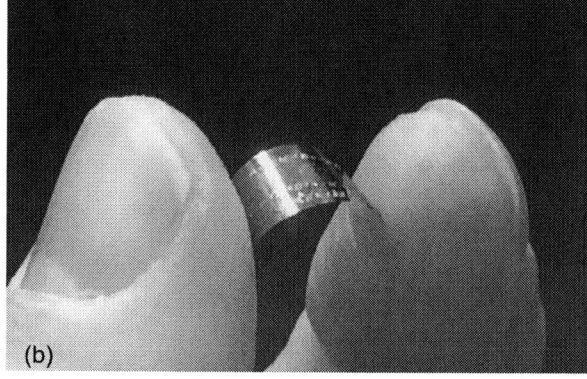

Figure 10. (a) Schematic illustration of controlled spalling for fabricating flexible thin silicon sensor and (b) a 40 μm-thin silicon strain gauge sensor formed by controlled spalling process.

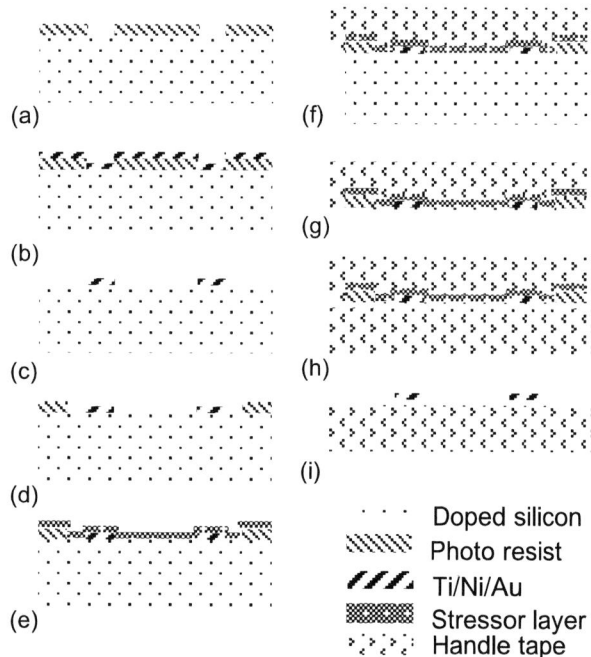

Figure 9. Fabrication sequence for flexible thin silicon strain sensor by using controlled spalling method.

V. WEARABLE WIRELESS FINGERNAIL DEFORMATION SENSOR

Fig. 11 shows a block diagram illustrating the simplified architecture for a wearable fingernail deformation sensor and the structure of the total system. The system is composed of a strain sensor, an accelerometer, a micro-controller, a radio frequency (RF) control module, a portable device, and cloud-based machines running the analytics code [9].

When a human grabs an object with their hand, the fingertips are pressed directly against the object, creating a deformation of the fingernail. Using the piezo-resistive property, a strain gauge built on a flexible substrate and attached to the fingernail using nail glue captures the micro-strains (deformation) of the fingernail. The prototype device uses printed circuit boards. Traditionally strain gauges are built using metal alloys such as copper-nickel alloys, nickel-chrome alloys, or using doped (p- or n-type) semiconductor materials as described in the previous section in this paper. Our overall system also includes a three-axis accelerometer to monitor finger movement. The micro-controller is configured to receive amplified raw data from the strain sensor which are combined with the accelerometer signals. The RF module establishes a wireless communication link (BLE: Bluetooth Low Energy) and transmits the sensor streams to a receiver. The receiver can be any portable device including a smartphone, tablet, laptop, or wearable computing device such as a smartwatch, etc. This receiver in our experiments is a smartwatch that runs some analytics and plots data for testing and immediate user feedback. This device also sends raw data and analytics output to a cloud based platform via

WiFi or LTE. The data is then saved in a database. This archived data is used for trend analysis, training of new models and longitudinal study.

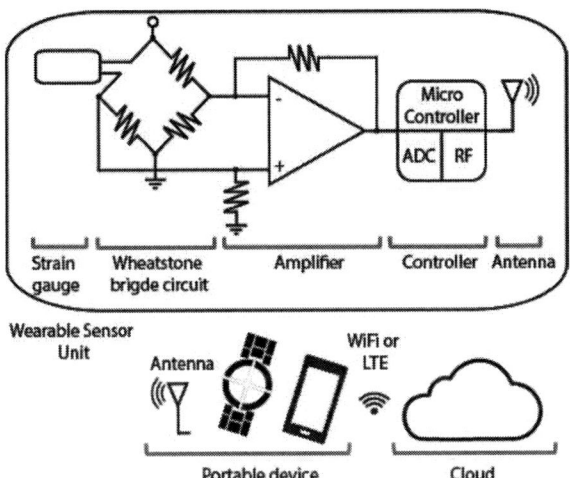

Figure 11. Wearable wireless fingernail deformation sensor system.

VI. HAND ACTIVITY DETECTION

A metal foil strain gauge sensor was chosen for use in the hand activity detection experiment. The sensor is flexible enough so that it doesn't completely damp nail flexion. It is robust enough to survive the nail environment, and sensitive enough to monitor fingernail deformation. As an optimization it could be replaced by tiny flexible silicon strain gauge sensors described in section V to allow the fingernail to move more freely. The wearable wireless device is designed to be worn on a finger and not to interfere with most normal hand activities (Fig. 1).

For the activity classification task, we used a combination CNN (Convolutional Neural Networks) / LSTM (Long-short term memory) neural network architecture [9]. The stack consisted of one CNN layer, a max pool layer, an LSTM, two dense layers and finally a soft-max layer. Our model used the Adam optimization algorithm to minimize the cross-entropy loss function. To generalize without overfitting, regularization was done by means of a 30% dropout layer. The input to the neural net model is a two second window that slides by 0.25

seconds consisting of three accelerometer tracks and three strain tracks. All data tracks were resampled to 50 Hz and pass band filtered to allow data between 0.25 Hz and 15 Hz. This removed the low frequency offsets and high-frequency noise. Segments of the continuous data were labeled by task. Finally, the model was trained on data from one six-minute training session. Fig. 12 shows hand activity classification. Task probability corresponding to four different activities: lifting a can, turning a key, screwing and unscrewing a nut, and rest are shown in Fig. 12(a). The task prediction confusion matrix for hand activities is shown in Fig. 12(b). It is clearly seen in the figure that the four different hand activities were distinguished with high probability during the validation session.

VII. SUMMARY AND CONCLUSION

This work demonstrates a novel wearable wireless sensor system that attaches to a fingernail and the algorithms developed to quantify forces on the fingers and classify hand activities.

- A 3D finite element model of finger was constructed to study fingernail deformation. Our modeling analysis indicates the axial strain maps are markedly different between finger models, although a simple nail-center mounted single sensor would respond to typical grip pressures on the fingertip with readily detectible strain amplitudes.
- A flexible 40 µm-thick silicon strain gauge sensor using a controlled spalling method is also fabricated as a potential sensor option.
- The data provided with sensors are transferred to a portable device/computer and are analyzed using machine learning. Neural network models identify human hand movements based on the pattern of signals and distinguish different hand activities.
- A wearable wireless fingernail deformation sensor has a potential to be applied to a broad range of application in the fields such as personalized precision healthcare monitoring and a new type of human-computer user interface.

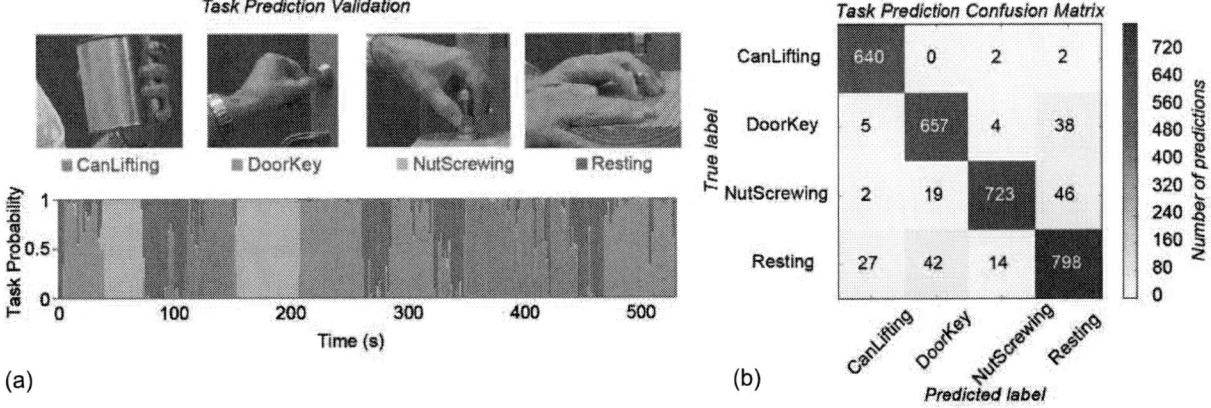

Figure 12. Hand activities (three hand-grips and resting) detection. (a) Task probability corresponding to four different hand activities: lifting a can, turning a key, screwing and unscrewing a nut, and rest, and (b) Task prediction confusion matrix of predicted hand activities for the test set.

ACKNOWLEDGMENT

The authors would like to sincerely thank Joseph W. Ligman, G. Blumrosen, J. Dicarlo, S. Lukashov, M. Agno, V. Caggiano, and M. Stellitano for their contribution in developing hardware, software, and system, and also thank V. Khanna, S. Li, and D. Yannitty for their help with DIC measurements. We would also like to thank C. Ravenelle and H. Zhang for construction analysis in fingernail. We also would like to acknowledge management support and encouragement from P. Andry and A. Royyuru of IBM during this research.

REFERENCES

[1] Fulkerson, Matthew, "Touch", *The Stanford Encyclopedia of Philosophy* (Spring 2016 Edition), Edward N. Zalta (ed.), URL = <https://plato.stanford.edu/archives/spr2016/entries/touch/>.

[2] M. Nishiyama and K. Watanabe, "Wearable Sensing Glove with Embeded Hetero-Core Fiber-Optic Nerves for Unconstrained Hand Motion Capture," *IEEE Trans. Instrum. Meas.* **58** (12) 3995-4000 (2009).

[3] Y. Park, J. Lee, and J. Bae, "Development of a Wearable Sensing Glove for Measuring the Motion of Fingers Using Linear Potentiometers and Flexible Wires," *IEEE Trans. Ind. Inform.* 11(1), 198-206 (2015).

[4] F. Lorussi, N. Carbonaro, D. De Rossi, R. Paradiso, P. Veltink, and A. Tognetti, "Wearable Textile Platform for Assessing Stroke Patient Treatment in Daily Life Conditions," Front. *Bioeng. Biotechnol.* **4** (28), 2016.

[5] S.P. Lee, G. Ha, D.E. Wright, Y. Ma, E. Sen-Gupta, et al., "Highly flexible, wearable, and disposable cardiac biosensors for remote and ambulatory monitoring," *NPJ Digit Med*, 2018; 1: 2

[6] K. F. Lei, Y. Z. Hsieh, Y. Y. Chiu, and M. H. Wu, "The Structure Design of Piezoelectric Poly (vinylidene Fluoride) (PVDF) Polymer-Based Sensor Patch for the Respiration Monitoring under Dynamic Walking Conditions," *Sensors* **15**, 18801–18812 (2015).

[7] A. Moin, A. Zhou, A. Rahimi, S. Benatti, A. menon, et al., "An EMG gesture recognition system with flexible high-density sensors and brain-inspired high-dimensional classifier", *Proc. IEEE Int. Symp. Circuits Syst. (ISCAS)*, pp. 1-5, May 2018.,

[8] Y.-S. Kim, J. Lu, B. Shih, A. Gharibans, Z. Zou, "Scalable Manufacturing of Solderable and Stretchable Physiologic Sensing Systems," *Advanced Materials*, **29** (39), 1701312 (2017).

[9] K. Sakuma, A. Abrami, G. Blumrosen, S. Lukashov, R. Narayanan, J. W. Ligman, V. Caggiano, and S. J. Heisig, "Wearable Nail Deformation Sensing for Behavioral and Biomechanical Monitoring and Human-Computer Interaction," *Sci. Rep.* **8**, 18031 (2018).

[10] K. Sakuma, G. Blumrosen, J. J. Rice, J. Rogers, and J. Knickerbocker, "Turning the Finger into a Writing Tool," *To appear in the proceeding of the International Conference of the IEEE Engineering in Medicine and Biology Society (EMBC)*, 2019.

[11] E. Haneke, "Surgical anatomy of the nail apparatus," *Dermatol Clin.* **24**(3) 291-6, (2006).

[12] P. Rich, "Nail Surgery". In: Dermatology, 2nd ed, Bolognia JL, Jorizzo JL, Rapini RP (Eds), Mosby, New York 2006.

[13] P. Fleckman, "Structure and function of the nail unit," In: Scher RK, Daniel CR III, eds. Nails: Diagnosis, Therapy, Surgery. Oxford, UK: Elsevier Saunders; 2005:14.

[14] L. Farren, S. Shayler, and A.R. Ennos, "The fracture properties and mechanical design of human fingernails," *J. Exp. Biol.* 207, 735 (2004).

[15] G. J. Gerling, I. I. Rivest, D. R. Lesniak, J. R. Scanlon and L. Wan, "Validating a Population Model of Tactile Mechanotransduction of Slowly Adapting Type I Afferents at Levels of Skin Mechanics, Single-unit Response and Psychophysics," *IEEE Transactions on Haptics*, 7(2), pp.216-228, 2014.

[16] Subrata Pal, "Mechanical Properties of Biological Materials". In: Design of Artificial Human Joints & Organs, Springer Science, New York, pp. 23-40, 2014.

[17] K. Sakuma, H. Hu, S. W. Bedell, B. Webb, S. Wright, K. Lazko, M. Agno, and J. Knickerbocker, "Flexible Piezoresistive Sensors Fabricated by Spallling Technique," *IEEE International Flexible Electronics Technology Conference (IFETC)*, 2018.

Heterogeneous Integration of a Fan-Out Wafer-Level Packaging Based Foldable Display on Elastomeric Substrate

A. Alam, A. Hanna, R. Irwin, G. Ezhilarasu, H. Boo, Y. Hu, C. W. Wong, T. S. Fisher, and S. S. Iyer

Center for Heterogeneous Integration and Performance Scaling (CHIPS)
Samueli School of Engineering
University of California, Los Angeles
Los Angeles, CA 90095, USA
email: arsalanalam89@g.ucla.edu

Abstract— We describe a Fan-Out Wafer-Level Packaging (FOWLP) integration process that is used to build an extremely flexible heterogeneous integration platform called "FlexTrate™". We integrated a daisy chain connected 10×20 array of 1 mm² Si dies over a 35 mm × 18 mm area using vertically corrugated Cu interconnects of 40 µm pitch and ~5 µm thickness. The system is reliable even upon bending to 1 mm bending radius for over 1000 bending cycles. We demonstrate a 37 mm × 52 mm foldable display with 1 mm² InGaN LEDs using this technology. Cyclic mechanical bending (1 mm bending radius), optical, and thermal reliability of integrated display are investigated.

Keywords- FOWLP; foldable display; heterogeneous integration; vertically corrugated, flexible and fine pitch interconnects

I. Introduction

Flexible Hybrid Electronics (FHE) [1] and Fan-Out-Wafer-Level Packaging (FOWLP) [2] are key technologies for enabling heterogeneous integration [3]. The FHE approach to make flexible systems relies on integration of ultra-thin bendable inorganic dies on a flexible organic substrate typically using printed interconnects to connect the inorganic dies. However, in this approach an additional bonding process is required to attach the inorganic dies on the organic substrates [4]. The printed interconnects have greater interconnect pitches and higher sheet resistance than conventional lithographically defined metal interconnects [5]. Moreover, extreme bending of large thinned inorganic dies induces strain on the devices of these dies. The mechanical induced strain, if not equalized, can lead to performance degradation of high-performance CMOS dies [6].

Our approach to flexibility was inspired by adapting FOWLP to FHE, where we have used a highly flexible elastomeric molding compound, Polydimethylsiloxane (PDMS), to allow our platform to bend down to extreme bending radii (≤ 1 mm). We call this platform FlexTrate™ [7, 8]. In our approach, the dies are embedded in PDMS and do not require any additional high temperature bonding process. While most FHE applications rely on the use of one large Application Specific Integrated Circuit (ASIC) die, as shown in Fig. 1 (a), our approach is based on a modular "dielet" approach. In the dielet approach, instead of using one large ASIC die, one can use multiple high yielding smaller size dies connected by lithographically defined metal interconnects at fine pitches (comparable to fat wire level) to

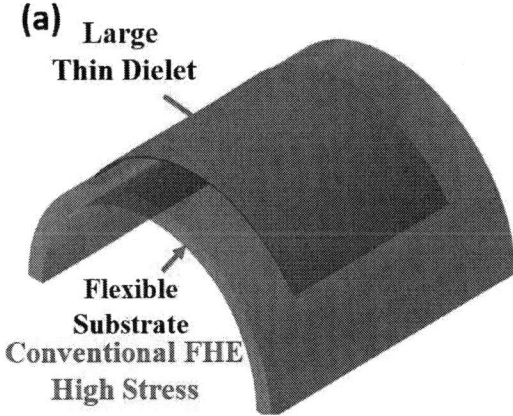

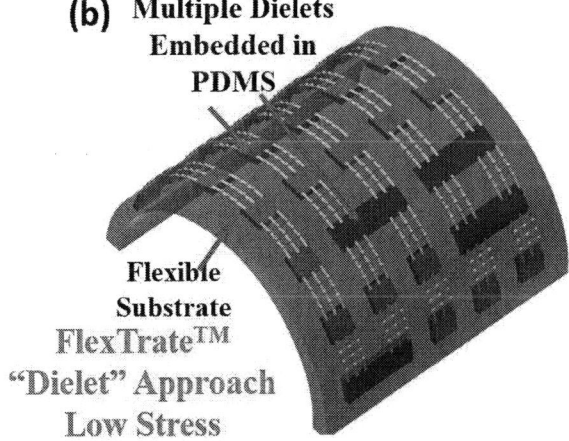

Fig. 1. (a) Schematic of a single large thin die bonded on flexible substrate in conventional FHE approach, where the die undergoes high stress upon bending to small bending radius, (b) schematic of "dielet" approach in FlexTrate™, showing multiple dielets connected at fine interconnect pitch.

to allow for both higher mechanical flexibility and "on-chip" like communication as shown schematically in Fig. 1 (b). We compared the tensile bending induced strain via Finite Element Analysis (FEA), for two thin dies of different sizes (10 *vs.* 1 mm²) that are both bent down to 2 mm bending radius, the large (10 mm²) thin die experiences up to 0.6 % strain, while the small sized (1 mm²) die only sees 0.005 % strain, which is two orders of magnitude lower. The flexibility of FlexTrate™ can be optimized both as a function of inter-dielet spacing, dielet size, and dielet thickness, as demonstrated in [9]. The modular approach in system design also offers enhanced isolation to mitigate failures in wiring and can allow for redundancy in system design. Moreover, in conventional FOWLP there are process challenges of die co-planarity, die tilt, and die shift [10]. We overcome these challenges using the favorable thermo-mechanical properties of our elastomeric molding compound (PDMS), demonstrating die co-planarity and die tilt of < 1 μm and die-shift of < 6 μm across the 100 mm wafer, which is the lowest die-shift value in literature [10, 11].

In this paper, we demonstrate a large area (37 mm × 52 mm) foldable 7-segment display integrating III-V InGaN LED dielets that are connected using 40 μm pitch vertically corrugated Cu interconnects. A corrugated interconnect is a non-planar interconnect topography that mitigates metal buckling upon release of the elastomeric molding compound from the handling substrate after processing [11]. We performed cyclic mechanical bending tests on a test vehicle of 200 Si dielets connected in a daisy chain, where the Si dielet size is identical to the LED dielets (1 mm²). The reliability of integrated corrugated Cu interconnects was tested for 10, 5, 4, 3, 2, and 1 mm bending radius for 1000 cycles each, showing less than 7 % change in average resistance from initial resistance after bending at 1 mm radius. The current (mA) vs. voltage (V), output power (mW) vs. input current (mA), and intensity (a.u.) vs. wavelength (nm) of the integrated LEDs were tested before and after bending at 1 mm radius for 1000 cycles and show no change post bending. Moreover, thermal dissipation experimental data and FEA have shown that the non-planar corrugated Cu interconnects can effectively cool down to ambient temperature 12.3 % faster compared to planar Cu interconnects occupying identical die area due to higher surface-to-volume ratio.

II. CYCLIC MECHANICAL BENDING RELIABILITY OF DAISY CHAIN CONNECTED 200 DIELETS

We previously published our FEA study to evaluate the minimum optimal dielet thickness and placement pitch for 1 mm × 1 mm × 0.2 mm dielets on FlexTrate™ [9]. A 15 mm × 4 mm × 0.500 mm of PDMS with three embedded Si dielets, assuming quarter-symmetry, was bent to 5 mm radius. The strain was measured from the edge of the outer dielet to the edge of the center dielet along the line of symmetry on the top surface of PDMS. The peak strain in Si dielets as a function of Si dielet thickness for different inter-dielet spacing showed an order of magnitude higher strain in 50 μm thick dielets than ≥ 100 μm thick dielets. This is why we chose 200 μm thick dielets for the integrated Si dielets. We had also examined the results for the strain profile in PDMS between two neighboring dielets at different placement pitches. We observed that the minimum strain value in PDMS, at the midpoint between two Si dielets, more than doubled for inter-dielet spacing of < 800 μm [9]. Therefore, here we choose 1.8 mm as the minimum placement pitch for integrating the 10 × 20 array of 1 mm² dielets.

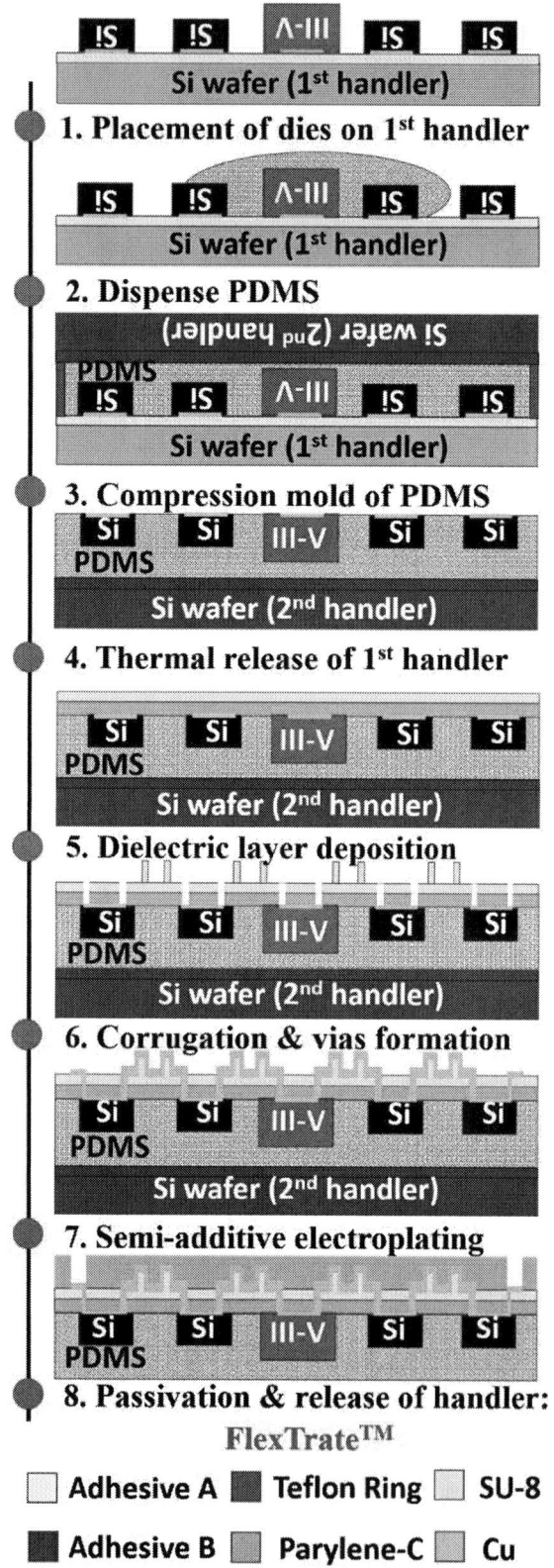

Fig. 2. Process flow for fabrication of FlexTrate™.

$\Delta R_{avg} < 7\%$ and $\sigma_{max} < 5.5\%$ of R_{avg}, with no breaking of interconnects. This shows the reliability of FlexTrate™ upon bending to small bending radii.

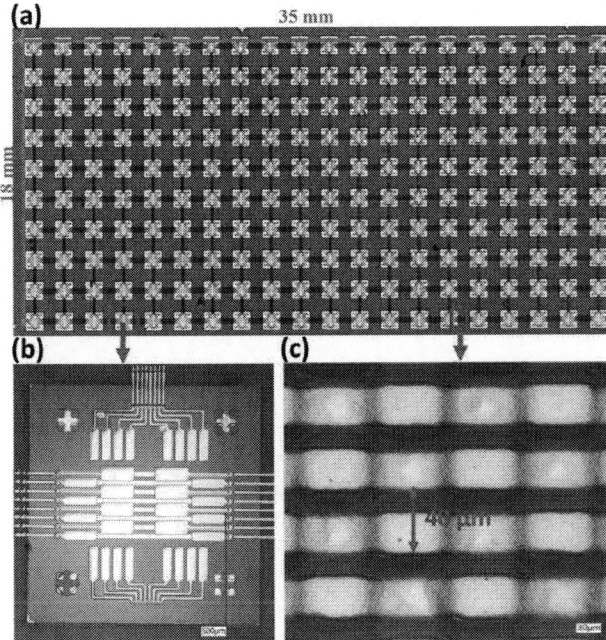

Fig. 3. (a) Image of 10×20 array for daisy chain connected dielets on FlexTrate™. Zoomed in image of (b) a dielet with pads, and (c) corrugated interconnects with 40 μm pitch.

The process flow for fabrication of FlexTrate™ is shown in Fig. 2, which is a die-first FOWLP process [7, 11]. This is a different approach from the capillary self-assembly process demonstrated in [12, 13]. We were able to achieve ~1 μm of placement accuracy while placing the dielets on the 1st handler using a pick and place tool. We successfully integrated a 10 × 20 array of 1 mm² dielets at 1.8 mm die pitch with vertically corrugated Cu interconnects (~5 μm thickness) at 40 μm interconnect pitch, connected in daisy chain, as shown in Fig. 3 (a). The 200 dielets were integrated over a 35 mm × 18 mm area. Zoomed-in images of a Si dielet with pads and corrugated interconnects at 40 μm pitch are shown in Figs. 3 (b, c), respectively. The alignment marks near the corners of the dielets, (Fig. 3 (b)) were helpful in fine alignment during the different stages of the fabrication process. The 10 × 20 array of dielets has total of 1520 interconnects and 3200 vias. We measured the resistance using a four-point probe across the 20 dielets for each of the 10 rows and plotted the average resistance (R_{avg}), as shown in Fig. 4 (a). The R_{avg} linearly increases with the number of dielets as expected, with maximum standard deviation (σ_{max}) = 0.23 Ω. The linear increase in resistance signifies that we were able to achieve good alignment for connecting the 40 μm pitch metal lines on the dies with 40 μm pitch corrugated interconnects on FlexTrate™ over 35 mm ×18 mm area. We then measure end-to-end resistances across 20 dielets for a total of 80 connections Before Release (BR) and After Release (AR) of FlexTrate™, as well as after cyclic bending at 10, 5, 4, 3, 2, and 1 mm bending radius for 1000 cycles, each. We plotted the R_{avg} with standard deviation (σ), as shown in Fig. 4 (b). We observe that the

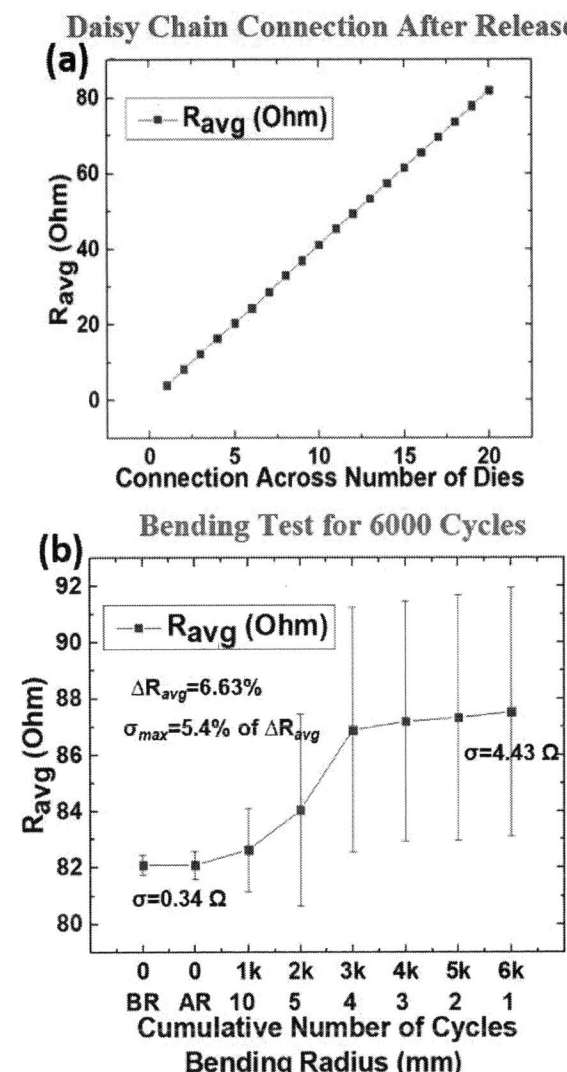

Fig. 4. Mechanical reliability plot for 10×20 dielets connected at 40 μm interconnect pitch on FlexTrate™ in a daisy chain demonstrating (a) R_{avg} (Ohm) vs connections across number of dielets, where the average resistance (σ_{max} = 0.23 Ω) linearly increases with increase in measurement across the number of dielets and (b) R_{avg} (Ohm) for different bending conditions including Before Release (BR), After Release (AR) of FlexTrate™ and down to 1 mm bending radius where the sample has undergone total of 6000 bending cycles, and the $\Delta R_{avg} < 7\%$ with 100 % interconnect yield post bending.

III. HETEROGENEOUS INTEGRATION FOR FOLDABLE DISPLAY

We integrated 42 commercially available, 335 μm thick, 1 mm², InGaN blue/green LEDs (DA 1000, Cree) on FlexTrate™ over to demonstrate a 7-segment foldable display over 37 mm × 52 mm area, as shown in Fig. 5 (a). The LEDs in each segment are placed using 3 mm placement pitch to allow for higher flexibility. Moreover, we used 200 μm thick, 1 mm² Si dielets with alignment marks to help us

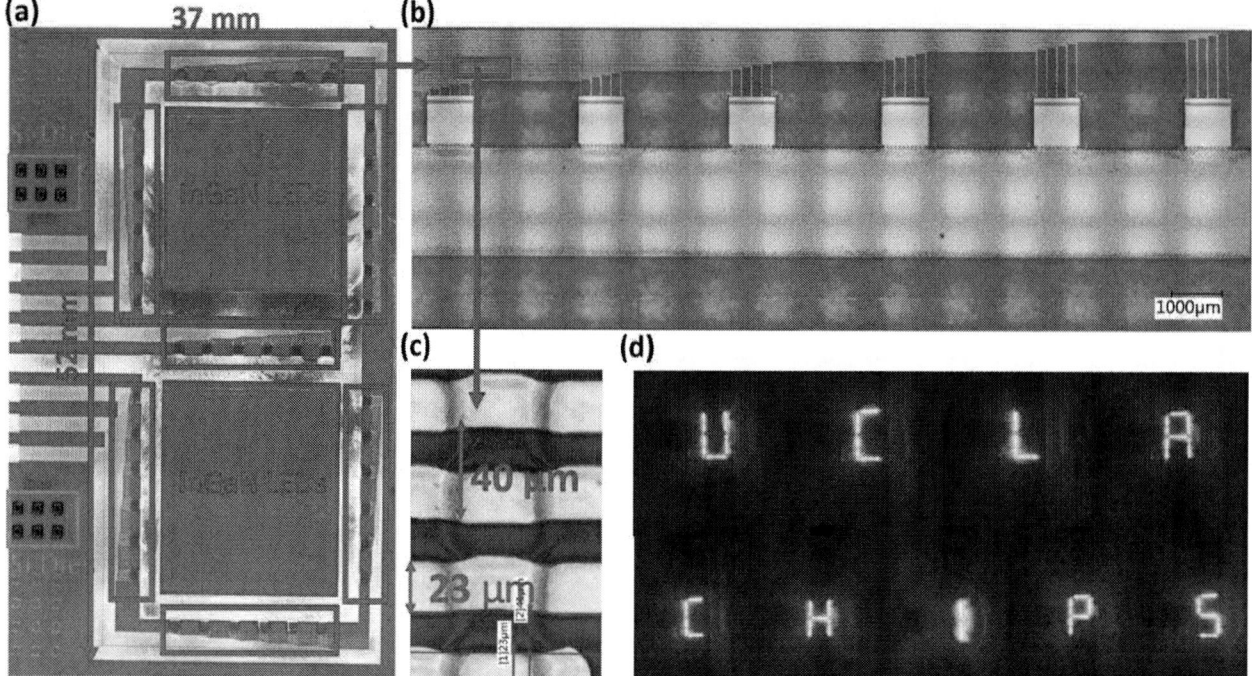

Fig. 5. (a) Foldable display with 42 LEDs on FlexTrate™ in the form of 7-segment display along with embedded Si dielets for accurate alignment purpose demonstrating a truly heterogenous integration, (b) zoomed in image of a segment of the 7-segment display consisting of six LEDs, (c) zoomed in image of 40 μm pitch corrugated interconnects, and (d) programming of 7-segment display to display "CHIPS UCLA", as an example, on the foldable display. Image of foldable display (e) during folding and (f) after folding, where the green LEDs remain illuminated throughout the folding process.

for alignment during the fabrication process, as shown in Fig. 5 (a). Each segment of the 7-segment display has 6 LEDs connected in parallel using corrugated Cu interconnects at 40 μm pitch, as shown in Fig. 5 (b, c) respectively. We soldered a 2.54 mm pitch connector to the pads and externally programmed the 7-segment display using a microcontroller. The LEDs were powered using a power supply under current compliance of 200 mA. A display of "UCLA CHIPS" is shown in Fig. 5 (d). Furthermore, we demonstrate complete folding of the display integrated with green LEDs, as shown in Fig. 6.

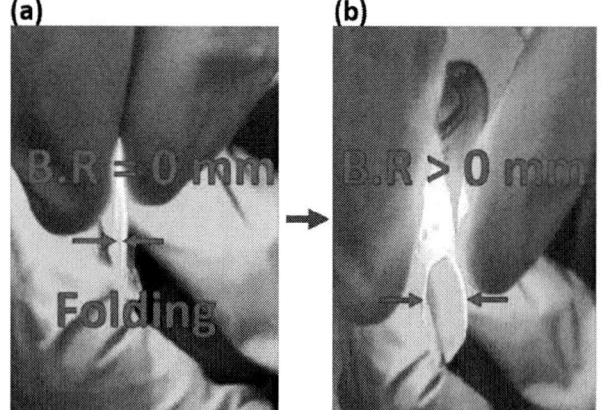

Fig. 6. Image of foldable display (a) during folding, as the bending radius (B.R) = 0 mm and (b) after folding when B. R > 0 mm. The green LEDs remain illuminated throughout the folding process.

A. Cyclic Mechanical Bending Reliability

We measured the (a) current (mA) *vs.* voltage (V), (b) output power (mW) *vs.* input current (mA), and (c) intensity (a.u.) *vs.* wavelength (nm) at 100 mA for the blue LEDs integrated on FlexTrate™ before and after bending at 1 mm bending radius for 1000 bending cycles, as shown in Fig. 7 (a, b, and c), respectively. The output power of the LEDs was measured using lenses to collimate the light to a photodiode power sensor (S120VC). The spectrum from the LEDs were measured using a spectrometer (Ocean Optics USB4000) which covers the range of 200 - 1100 nm. The measurement was done through free space and the photodiode, which has a detector wavelength range of 200–1100 nm, was set to measure the specific wavelength acquired from the spectrometer. The current *vs.* voltage characterization demonstrates that the resistance of the interconnects connecting the LEDs do not see any significant change upon bending. The output power *vs.* input current and intensity (a.u.) *vs.* wavelength characterizations demonstrate that the PDMS has similar transparency before and after bending. Moreover, post bending, the LEDs have similar functionality, Cu interconnects, pads, and dielets do not delaminate, demonstrating that the package is very reliable even under extreme cyclic bending. Such a system can be scaled up to manufacture high density, full-colored, highly flexible displays in the future.

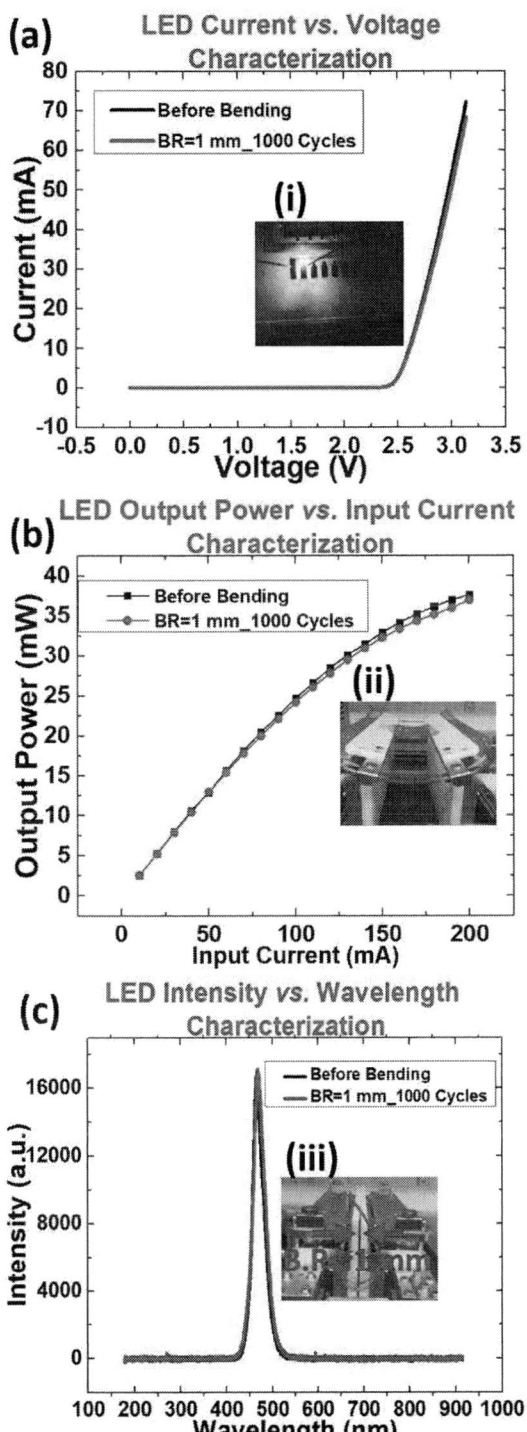

Fig. 7. Plot of characterization of the blue LEDs on the fabricated foldable display before and after bending to 1 mm bending radius for 1000 bending cycles, for (a) current (mA) *vs.* voltage (V), (b) output power (mW) *vs.* input current (mA), and (c) intensity (a.u.) *vs.* wavelength (nm) at 100 mA, overall demonstrating negligible change in impedance of fine pitch corrugated interconnects, similar transparency of foldable substrate and reliability of the blue LEDs post bending. The inset of Fig. 7. (a) shows the blue LED illuminated during measurement. The insets of Figs. 7 (b, c) show the bending mechanism when the plates are open, and fully bent to 1 mm bending radius, respectively, to demonstrate 1 mm bending.

B. Thermal Dissipation Analysis

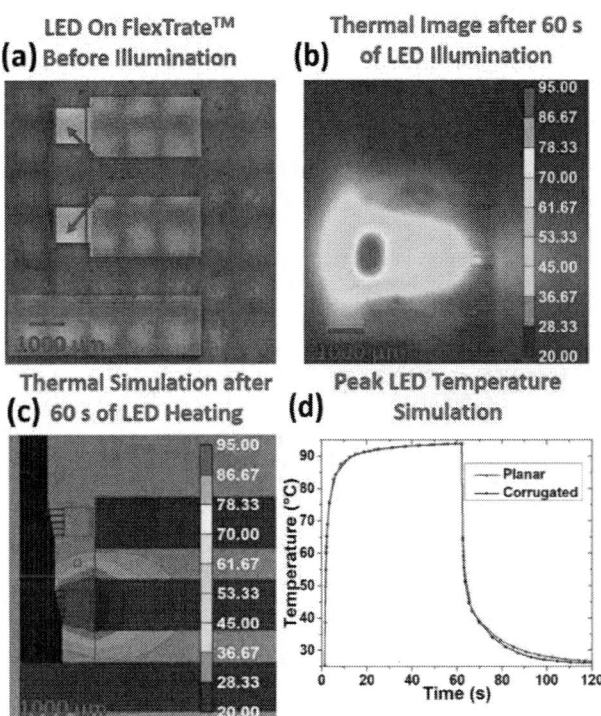

Fig. 8. (a) Optical image of the sample under test. Thermal image after 60 s of illumination at 100 mA during (b) experiment and (c) simulations, which demonstrate the heat spreading along corrugated interconnects, and (d) simulations evaluating peak temperature of LED over time for planar *vs.* corrugated interconnects, where the temperature reaches ambient in 114 s with planar interconnects *vs.* 100 s with corrugated interconnects.

We analyzed the dissipation of heat through the foldable display on FlexTrate™. The image of the sample with LEDs before illumination is shown in Fig. 8 (a). One LED was illuminated by passing 100 mA current for 60 s to reach steady state temperature. The current supply was then removed, and the sample was allowed to cool for 60 s. For FEA, a model based on the experimental setup was used. A convection coefficient of 20 W/m² was applied to the outward-exposed faces of the model. A temperature boundary was applied to the center die using the peak temperature measured in the first 60 s of the experiment. The temperature boundary was then removed, and the thermal dissipation was studied over the next 60 s. Figs. 8 (b, c) show the surface thermal profile after 60 s of illumination of the LED for experiment and simulation, respectively. The thermal image was taken using a FLIR A655sc high resolution infrared camera. We observe that the surface temperature distribution from the simulation results are similar to the experimental results. The simulation was performed again with planar interconnects replacing corrugated interconnects. Fig. 8 (d) shows the simulated peak temperature in the LED comparing the heat dissipation through planar *vs.* corrugated interconnects. Based on the experimental and simulated data, the Cu pads and interconnects act as heat spreaders. Heat dissipation occurs primarily through the Cu. In addition, the use of corrugated

Cu interconnects increases the surface-to-volume ratio allowing for 12.3 % faster cooling of the LED to ambient temperature compared to planar interconnects.

IV. CONCLUSION

We successfully integrated a 10×20 array of 1 mm^2 dielets and demonstrated foldable display by heterogeneous integration of 1 mm^2 Si and InGaN LED dielets on our platform. Both the systems had fine pitch (40 μm) corrugated interconnects and were successfully bent down to 1 mm bending radius for over 1000 bending cycles. FEA has shown that 800 μm was the minimum optimal dielet-dielet spacing for 200 μm thick, 1 mm^2 dielets on FlexTrate^{TM}. Moreover, the thermal analysis, showed that fine pitch corrugated interconnects also act as efficient heat dissipation channels. Through FlexTrate^{TM}, we can target wearable and implantable applications that require high-performance flexible systems, such as multi-channel surface electromyography (sEMG) system, optogenetics for neural implants, and so on.

ACKNOWLEDGMENT

The authors acknowledge the contributions of A. Powell from Cree Inc. for providing the LEDs, K&S for the pick and place tool, and UCLA CNSI and Nanolab cleanroom staffs. This work was supported in part by the Semiconductor Research Corporation (SRC), AFRL/NBMC, DARPA, UCLA CHIPS Consortium and the UC system.

REFERENCES

[1] R. H. Reuss, G. B. Raupp, and B. E. Gnade, "Special issue on advanced flexible electronics for sensing applications," Proceedings of the IEEE, vol. 103, pp. 491-496, 2015.

[2] M. Brunnbauer et al., "An embedded device technology based on a molded reconfigured wafer," 56th Electronic Components and Technology Conference (ECTC), 2006, pp.547-551.

[3] S. S. Iyer, "Heterogeneous Integration for Performance and Scaling", IEEE Trans. Compon., Packag., Manuf. Technol., vol. 6, pp. 973-983, Jul. 2016.

[4] H. Andersson, J. Sidén, V. Skerved, X. Li and L. Gyllner, "Soldering Surface Mount Components Onto Inkjet Printed Conductors on Paper Substrate Using Industrial Processes," in IEEE Transactions on Components, Packaging and Manufacturing Technology, vol. 6, no. 3, pp. 478-485, March 2016, doi: 10.1109/TCPMT.2016.2522474

[5] J. Perelaer et al., "Printed electronics: the challenges involved in printing devices, interconnects, and contacts based on inorganic materials," J. Mater. Chem., vol. 20, 2010, pp. 8446-8453.

[6] N. Wacker et al., "Stress analysis of ultra-thin silicon chip-on-foil electronic assembly under bending," Semicond. Sci. Technol., vol. 29, no. 9, p. 095007, Aug. 2014, doi: 10.1088/0268-1242/29/9/095007

[7] T. Fukushima et al., ""FlexTrate^TM" — Scaled Heterogeneous Integration on Flexible Biocompatible Substrates Using FOWLP," 2017 IEEE 67th Electronic Components and Technology Conference (ECTC), Orlando, FL, 2017, pp. 649-654, doi: 10.1109/ECTC.2017.226

[8] T. Fukushima, A. Alam, A. Hanna, S. C. Jangam, A. A. Bajwa and S. S. Iyer, "Flexible Hybrid Electronics Technology Using Die-First FOWLP for High-Performance and Scalable Heterogeneous System Integration," in IEEE Transactions on Components, Packaging and Manufacturing Technology, vol. 8, no. 10, pp. 1738-1746, Oct. 2018, doi: 10.1109/TCPMT.2018.2871603

[9] G. Ezhilarasu, A. Hanna, R. Irwin, A. Alam and S. S. Iyer, "A Flexible, Heterogeneously Integrated Wireless Powered System for Bio-Implantable Applications using Fan-Out Wafer-Level Packaging," 2018 IEEE International Electron Devices Meeting (IEDM), San Francisco, CA, 2018, pp. 29.7.1-29.7.4., doi: 10.1109/IEDM.2018.8614705

[10] G. Sharma, A. Kumar, V. S. Rao, S. W. Ho and V. Kripesh, "Solutions Strategies for Die Shift Problem in Wafer Level Compression Molding," in IEEE Transactions on Components, Packaging and Manufacturing Technology, vol. 1, no. 4, pp. 502-509, April 2011, doi: 10.1109/TCPMT.2010.2100431.

[11] A. Hanna et al., "Extremely Flexible (1mm Bending Radius) Biocompatible Heterogeneous Fan-Out Wafer-Level Platform with the Lowest Reported Die-Shift (<6 μm) and Reliable Flexible Cu-Based Interconnects," 2018 IEEE 68th Electronic Components and Technology Conference (ECTC), San Diego, CA, 2018, pp. 1505-1511, doi: 10.1109/ECTC.2018.00229

[12] T. Fukushima et al., "Process Integration for FlexTrateTM," 2018 International Flexible Electronics Technology Conference (IFETC), Ottawa, ON, 2018, pp. 1-2, doi: 10.1109/IFETC.2018.8584029

[13] T. Fukushima et al., "Self-Assembly Technologies for FlexTrate™," 2018 IEEE 68th Electronic Components and Technology Conference (ECTC), San Diego, CA, 2018, pp. 1836-1841, doi: 10.1109/ECTC.2018.00275

A Study on the Flexible Chip-on-Fabric (COF) Assembly using Anisotropic Conductive Films (ACFs) Materials

Seung-Yoon Jung, and Kyung-Wook Paik*
Nano Packaging and Interconnect Lab. (NPIL)
Department of Materials Science and Engineering
Korea Advanced Institute of Science and Technology (KAIST)
Daejeon, South Korea
*e-mail: kwpaik@kaist.ac.kr

Abstract— In this study, flexible Chip-on-Fabric (COF) assemblies using anisotropic conductive films (ACFs) and cover layer structure were demonstrated. Fabric substrates were fabricated by Cu pattern lamination method with additional electroless nickel immersion gold (ENIG) metal finish before laminating onto the fabrics. Thermo-compression (T/C) bonding method was used to bond the 50μm-thick Si chip on the fabric substrates. After T/C bonding, stable ACFs joint interconnection was formed between chips and substrates. Polymer cover layer structure was applied on the flip-chip surface of the COF packages. After polymer cover layer structure was applied, minimum bending radius before chip crack drastically decreased down to 10 mm radius. In addition, a dynamic bending test was performed to evaluate the dynamic bending reliability of the COF assemblies with polymer cover layer structure, and cross-section SEM analysis and digital image correlation (DIC) method were used to analyze the bending test results.

Keywords-Cu pattern-Laminated fabric substrates; Chip-on-Fabric; ACFs interconnection; Cover layer structure; Bending property; Dynamic bending test;

I. INTRODUCTION

Smart e-textiles fabrics are considered as one of the substrates for future wearable devices. To fabricate these wearable devices, electronic components should be mechanically attached and electrically interconnected to the fabrics with patterned electrical circuits such as PCBs (Printed Circuit Boards). As a result, Si chip interconnection to fabric substrates are needed. In this regard, Chip-on-Fabric (COF) interconnection using ACFs is one of the promising interconnection methods for fabric-based wearable devices.

For electrical interconnection, fine-pitch and flexible electrical circuits should be formed on fabrics. In the previous studies, a novel Cu pattern lamination method for fabric substrates has been introduced [1]. Cu foil was firstly attached to the B-stage adhesive films, and patterning process was followed to form electrical circuits on the adhesive films. And then, these metal circuits were laminated onto the fabrics. As a result, fabric substrates with fine-pitch and highly flexible Cu patterns have been successfully demonstrated.

ACFs are well-known film-type adhesive interconnection materials consisting of polymer resin and conductive particles. Compared with other interconnection methods such as soldering and connectors, ACFs can provide stable mechanical and excellent electrical interconnection using low temperature bonding process and flexibility. In addition, recent studies have demonstrated that ACFs interconnection can be a promising solution for flexible chip packages, because ACFs can endure both static and dynamic bending stresses [2-4]. Furthermore, by applying cover layer structure on the flip chip assembled surface, higher flexibility can be obtained [2-3]. Using ACFs materials and the cover layer structure, highly flexible Chip-on-Fabric packages can be realized.

In this study, 500μm-pitch Chip-on-Fabric (COF) packages with a polymer layer structure were demonstrated using Cu pattern-laminated fabric substrates and ACFs interconnection materials. Fabric substrates were prepared by the Cu pattern lamination method using two types of B-stage adhesive films. Before laminating on the fabric, electroless nickel immersion gold (ENIG) metal finish was performed on the Cu surface to prevent Cu oxidation. And 50 um-thick Si chip was bonded to the Cu pattern of fabric substrates using ACFs and a thermo-compression (T/C) bonding method. And then, the polymer cover layer structure consisting of polyimide and adhesive films was applied to the COF packages to protect the Si chip under bending deformation. First, the effects of the fabric substrates types on the ACFs joint properties such as electrical daisy chain resistance, contact resistance and peel adhesion strength will be investigated. In addition, 4-point bending test was performed to investigate the effects of the fabric substrates types and cover layer structure on the static bending properties of COF packages. Finally, a dynamic bending test was performed to evaluate the dynamic bending properties of COF packages with the polymer cover layer structure using two types of fabric substrates.

II. EXPERIMENTAL METHODS

A. Materials and Test vehicles

50μm-thick Si chips were designed in a peripheral array format with a size of 10 mm x 10 mm. There were Cu/Ni/Au bumps with a size of 300 μm x 300 μm x 12 μm on the I/O pads having 500 μm pitch.

To fabricate the fabric substrates, 12 μm-thick Cu foil, 40μm-thick B-stage adhesive films and commercially

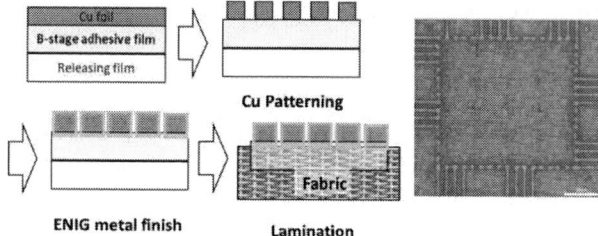

Figure 1. Schematic diagrams of fabric substrates fabrication process and the actual fabric substrate with metal patterns

available polyester/rayon woven fabrics were used. In this study, two types of B-stage adhesive films were prepared by adjusting the amount of the styrene-isoprene-styrene block copolymer (SIS) elastomer. Table 1 summarized the materials properties of B-stage adhesive films.

Table 1. Materials properties of B-stage adhesive films

	Film A (Fabric sub. A)	Film B (Fabric sub. B)
Film modulus	10.9 MPa	183.6 MPa
Flexural modulus of adhesive/fabric laminate	37.63 MPa	80.59 MPa
Adhesion strength of Cu-Fabric laminate	582.7 gf/cm	815 gf/cm

ACFs were consisted of epoxy-based polymer resin and 20 μm diameter Au/Ni coated polymer balls as conductive particles. ACFs thickness was 30μm, and the conductive particle contents were 20 wt %.

B. Fabric substrates fabrication

Fabric substrates were prepared using Cu pattern lamination method as previously described in Fig. 1.[1] After Cu circuit was formed on the B-stage adhesive film, ENIG metal finish was performed on the Cu electrodes to prevent Cu oxidation, and then ENIG/Cu circuits on the adhesive films were laminated onto the fabrics by a vacuum lamination method. In the fabric substrates, there were one daisy-chain resistance test pattern and 4-point Kelvin structures to measure the continuity of all patterns and single ACFs joint contact resistances.

C. Fabrication of Chip-on-Fabric packages with a polymer cover layer structure

To fabricate flip-chip COF packages, ACFs were laminated onto the thin Si chip and bonded onto fabric substrates using a thermo-compression (T/C) bonding method at 210 °C for 10 seconds at 1MPa.

After COF bonding, a polymer cover layer was applied using 135μm thick polyimide (PI) cover films and cover adhesive films with various thicknesses. Cover films were laminated onto the surface of the Si chip using a roll

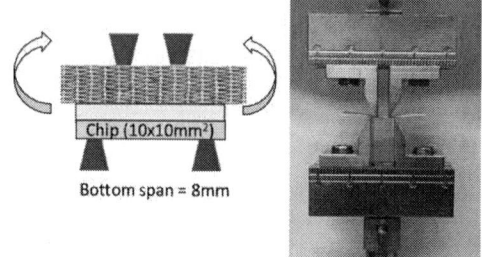

Figure 2. 4-point static bending test of COF packages

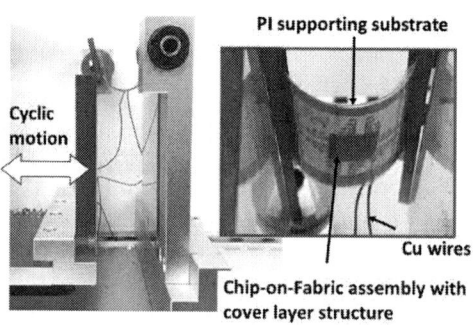

Figure 3. Dynamic bending test setup of COF packages

laminator, and then vacuum lamination was performed at 110 °C for 5 minutes at 0.14 MPa nitrogen gas pressure.

D. ACFs joint properties characterization

For electrical property characterization, daisy chain resistances and contact resistances were measured. And 90-degree peel adhesion test was also conducted to measure the adhesion strength of COF packages by pulling fabric substrates at 10 mm/min. speed.

ACFs joint morphology of COF packages was observed by optical microscope (OM) and Scanning Electron Microscope (SEM).

E. COF Bending property evaluation

Static bending property of various COF packages was evaluated by 4-point bending test as shown in Fig. 2. The minimum bending radius was measured before chip fracture.

And a dynamic bending test was conducted to evaluate the ACFs joint reliability under cyclic bending deformation as shown in Fig. 3. The bending radius was fixed at 12 mm, and the test speed was 60 cycles/min. Two types of the fabric substrates were used in COF packages having the same polymer cover layer structure. During bending tests, in-situ daisy chain resistances were measured at the bent state per every 5 cycles.

Cross-sectional SEM images were observed for failure analysis. Digital image correlation (DIC) method was used

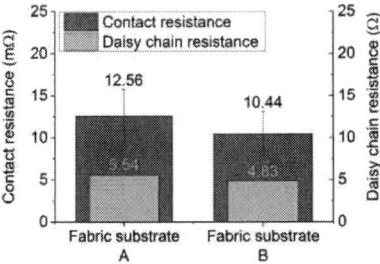

Figure 4. Daisy chain and contact resistances of COF assembly using various fabric substrates

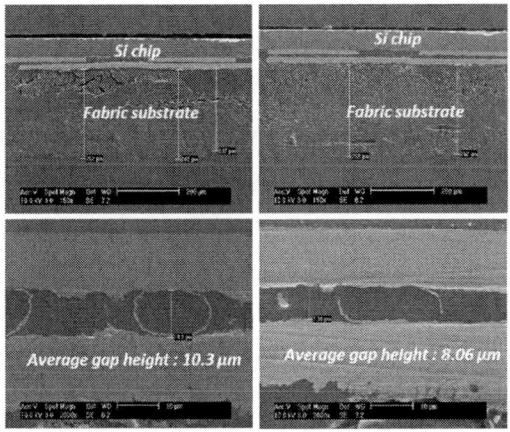

Fabric substrate A   Fabric substrate B

Figure 5. ACFs joint morphology of COF assembly using two types of fabric substrates

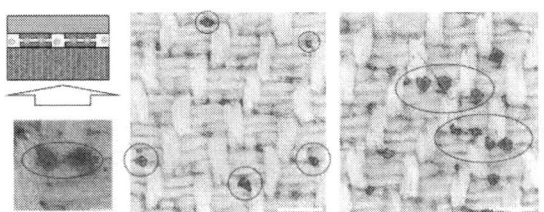

Fabric substrates A   Fabric substrates B
ACFs resin area ratio = 1.24%   ACFs resin area ratio = 3.45 %

Figure 6. Optical microscope images of the fabrics at the backside of the COF assembly

to measure the bending strain of the COF packages [4]. Encapsulated-COF packages were grinded without any external mold, and then silica speckle was sprayed onto the grinded surface. SEM images of the COF joints before and after 12 mm rod bending were taken. And the bending strains were calculated by measuring the speckle displacement on the SEM images.

Figure 7. SEM image of the top surface of two fabric substrates

III. RESULTS AND DISCUSSION

A. ACFs joint morphology and electrical properties of COF joints

Fig. 4 shows the contact and daisy chain resistances of the COF packages using two types of fabric substrates. Fabric substrates A showed higher resistances than the fabric substrates B. This resistance differences were due to the gap height difference, as shown in cross-section SEM images of Fig. 5.

In addition, ACFs resin was observed at the backside of the fabric substrates as shown in Fig. 6. This suggested that the ACFs resin permeated into the pores of fabric substrates during the T/C bonding process.

When the fabric substrates were fabricated, the resin of the B-stage adhesive films flowed into the fabric materials [1]. Therefore the surface of the resulting fabric substrates had porous structure as shown in Fig. 7. Since the fabric substrates B had more porous surface than substrate A, the ACFs resin permeation may be easier. Since the gap height of the ACFs joint was mainly determined by the ACFs resin flow behavior, the gap height difference can be explained by the degree of ACFs resin permeation into the fabrics.

B. Adhesion property of COF joints

Fig. 8 shows the peel strengths of the COF assemblies using two types of fabric substrates. COF A showed lower peel strength than COF B.

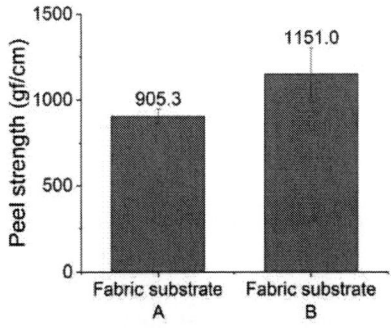

Figure 8. Peel strengths of the COF assemblies using various fabric substrates

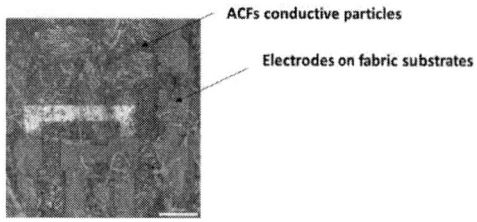

Figure 9. Top-view image of the Si chip surface after the peel test

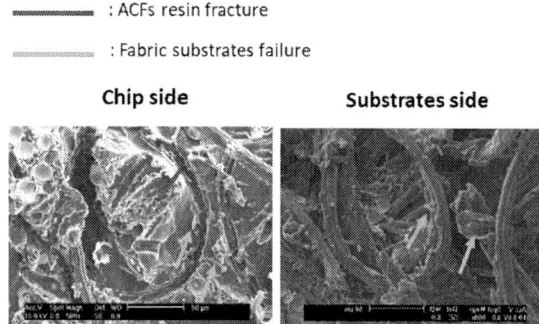

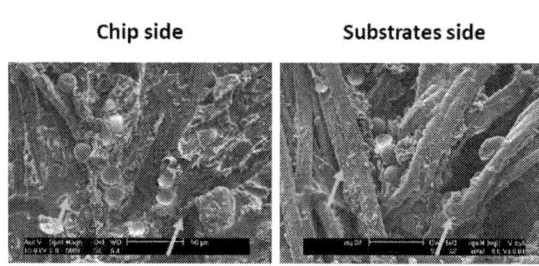

Figure 10. Top-view SEM images of the fracture surface of COF assemblies after the peel test

Top-view optical microscope and SEM images showed that the fabric substrates failure instead of ACFs joint failure occurred after the peel test as shown in Fig. 9 and 10. Fractured ACFs resin was observed between fibers, which also supported the ACFs resin permeation into the substrates.

Since the fracture occurred on the interfaces between Cu electrode and fabric materials in the fabric substrates, the main reason for the peel strength difference was from the adhesion strength of the fabric substrates which depended on the B-stage adhesive films used as described in Table 1.

C. Effects of polymer cover layer thickness on the static bending property of the COF packages

Fig. 11 shows the minimum bending radius of the COF packages as a function of total cover layer thicknesses. Without cover layer structure, COF packages showed over 20 mm minimum bending radius. However as the cover layer thickness increased, minimum bending radius

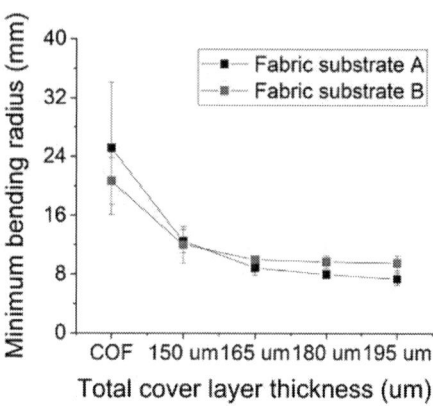

Figure 11. The minimum bending radius of the COF assemblies using various fabric substrates and polymer cover layer thicknesses

gradually decreased down to 7.4 mm for COF A and 9.5 mm for COF B.

According to the previous studies on the Chip-in-Flex (CIF) packages, cover layer structure on the Si chip reduced the bending stresses on the chip by shifting neutral axis position towards the chip center [2]. As a result, thick cover layer structure led to better bending property of COF packages. In addition, the encapsulated-COF A showed better bending properties, which might be presumably due to the lower modulus of the fabric substrate.

D. Dynamic bending fatigue test results

Fig. 12 shows the daisy chain resistance changes with increased bending cycles. For a convex bending, both encapsulated-COF packages showed stable daisy chain resistances up to 100,000 cycles. On the other hand, for a concave bending, encapsulated-COF B showed about 80% increase of the joint resistance, while encapsulated-COF A showed no resistance increase.

Fig. 13 shows the cross-section SEM images of ACFs joint before and after the dynamic bending test for both convex and concave bending conditions. For the convex bending case, no ACFs joint damage was observed. However, for the concave bending condition, ACFs resin delamination was observed for encapsulated-COF B. Because cyclic tensile stress was applied to the ACFs joint, the concave bending test caused ACFs joint damage especially for COF B.

To analyze the concave bending test results, bending strain on the ACFs joint was measured using SEM-DIC method as shown in Fig. 14. Encapsulated-COF B showed higher bending strain than encapsulated-COF A, which was consistent with the 4-point static bending test results. As a result, better bending reliability was obtained when the fabric substrates A were used.

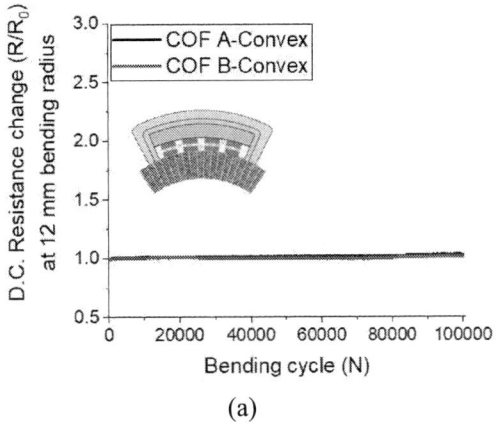

(a)

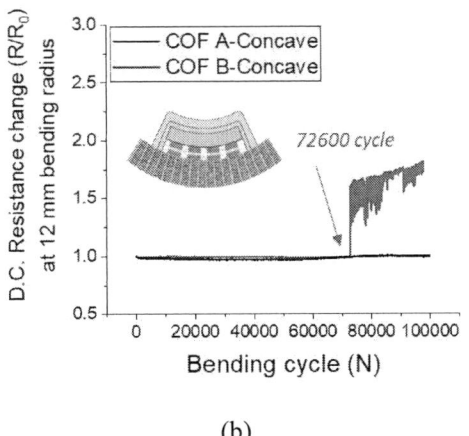

(b)

Figure 12. Dynamic bending test results of encapsulated-COF A & B packages at (a) convex and (b) concave bending conditions

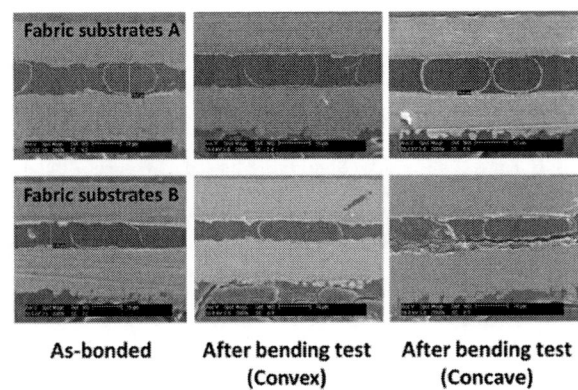

Figure 13. SEM images of ACFs joint after 100,000 cycle dynamic bending test

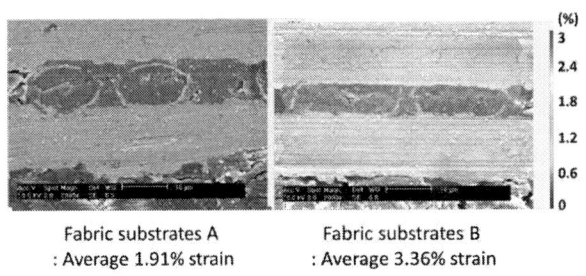

Figure 14. Bending strain mapping results of the two encapsulated-COF packages by the DIC analysis

IV. CONCLUSION

In this study, a novel Chip-on-Fabric assembly was demonstrated using Cu pattern-laminated fabric substrates and ACFs materials. After T/C bonding, ACFs joint was well formed between chip and fabric substrates. During the fabric substrates fabrication, the resin of the B-stage adhesive films permeated into the fabrics, and some part of the fabric substrates had porous structure. Therefore, the amount of ACFs resin permeated into the fabric substrates determined the final ACFs gap height and resulting electrical resistances. In addition, it was found that the adhesion strength of COF packages depended on the adhesion properties of the fabric substrates.

Flexible COF packages were achieved by applying the polymer cover layer structure on top of flip-chip surface. After 195 μm-thick cover layer structure was applied, encapsulated-COF packages A and B showed the minimum bending radius of 7.4 mm and 9.5 mm respectively. In addition, encapsulated-COF packages A showed stable joint resistance after 100,000 cycles of dynamic bending at 12 mm bending radius for both convex and concave bending conditions.

As a result, highly flexible Chip-on-Fabric assemblies were achieved using ACFs materials, fabric substrates A, and 195μm cover layer structure. This encapsulated-COF packages can be a promising solution for fabric-based wearable devices.

ACKNOWLEDGMENT

This work was supported by Wearable Platform Materials Technology Center (WMC) funded by the National Research Foundation of Korea(NRF) Grant of the Korean Government(MSIP) (No. 2016R1A5A1009926)

978-1-7281-1500-9/19 $31.00 © 2019 IEEE

REFERENCES

[1] S.-Y. Jung, and K. W. Paik, "A Study on the Fabrication of Electrical Circuits on Fabrics using Cu pattern Laminated B-stage adhesive Films for Electronic Textile Applications", Proc. 67th Elec. Compon. Technol. Conf. (ECTC), 2017.

[2] J. H. Kim, T. I. Lee, J.W. Shin, T. S. Kim and K. W. Paik, "Bending Properties of Anisotropic Conductive Films Assembled Chip-in-Flex Packages for Wearable Electronics Applications", in IEEE Trans. Compon. Packag. Manufac. Technol., vol.6, no.2, p208-215, 2016.

[3] K.L. Suk, and K.W. Paik, "Embedded chip-in-flex (CIF) packages using wafer level package (WLP) with pre-applied anisotropic conductive films (ACFs)", in Proc. IEEE Elec. Compon. Technol. Conf. (ECTC), San Diego, CA, May 26-29, 2009, pp. 1741-1748

[4] J. H. Kim, T. I. Lee, T. S. Kim and K. W. Paik, "The Effect of Anisotropic Conductive Films Adhesion on the Bending Reliability of Chip-in-Flex Packages for Wearable Electronics Applications," in IEEE Trans. Compon. Packag. Manufac. Technol., vol. 7, no. 10, p. 1583-1591, 2017.

7nm Chip-Package Interaction Study on a Fine Pitch Flip Chip Package With Laser Assist Bonding and Mass Reflow Technology

Ian Hsu, Chi-Yuan Chen, Stanley
Lin, Ta-Jen Yu
Package Technology Division
MediaTek, Inc.
Hsinchu, Taiwan
Ian.hsu@mediatek.com

NamJu Cho
Research & Development Division
JCET STATS ChipPAC Pte. Ltd.
Incheon, South Korea

*Ming-Che Hsieh
Field Applications Engineering
JCET STATS ChipPAC Pte. Ltd.
Singapore
mc.hsieh@statschippac.com

Abstract— **Due to the rapid growth in new technological features in mobile applications, new packaging solutions smaller form factor package designs, lower power consumption and other efficiency enhancements are required for the 7nm node silicon devices. Flip chip technology such as fcCSP (flip chip Chip Scale Package) has been widely adopted as the primary (or preferred) solution for mobile devices to satisfy these challenging requirements. The flip chip CSP package offers a cost-effective solution through the combination of Sn/Ag bumped copper (Cu) pillars, the use embedded trace substrate (ETS) technology along with mass reflow chip attach and molded underfill (MUF) processes.. While mass reflow chip attach process provides a cost-effective solution for flip chip assembly, there is nonetheless a high risk of bump to trace shorting especially as the need increases for finer bump pitch designs, with reduced copper line width and line spacing (LW/LS) for the escaped traces. To reduce this risk, we are exploring the use of laser assisted bonding (LAB) methodology to study the 7nm chip-package interaction (CPI) of a fcCSP with a 60μm bump pitch and escaped trace designs in this paper. For the purpose of measuring the extremely low-k (ELK) performance in a 14x14mm fine pitch fcCSP with 7nm node silicon live die, the thunder test, two-times mass reflow followed by a quick temperature cycling (QTC), and the hammer test, a multi-reflow process with a peak temperature of 260°C have been utilized. The results show that although both chip attach methodologies can pass the normal requirements of the thunder and hammer tests, the utilization of LAB technology can further enhance the strength of ELK, resulting in better yield performance. From these results, we believe that LAB not only can guarantee assembly yield but also ensure less ELK damage risk in the evaluated 7nm node silicon fcCSP. Futhermore we have shown that LAB technology is suitable for the 7nm node silicon devices along with the bump pitch reduction using finer LW/LS substrate with escaped traces design.**

Keywords-7nm silicon node, chip-package interaction, laser assisted bonding, mass reflow, embedded trace substrate, quick temperature cycling test, hammer test

I. INTRODUCTION

Advances in semiconductor devices for emerging technologies and the complexities in sustaining Moore's Law are pushing the industry to 7nm silicon node technology. In addition to reducing the die size, there is also a need to enhance the power efficiency to support portable and mobile applications. Because of these increasing demands for improved performance, wider bandwidths, and reduced power usage as well as increased functionality in mobile devices, the capabilities of finer bump pitch and finer LW/LS substrate manufacturing are needed. The flip chip CSP has been viewed as a viable option when higher input/output (I/O) counts are needed and when moving from traditional bond-on-capture (BOC) pad with cored substrates to bond-on-lead (BOL) with coreless substrates, especially due to the lower cost requirement and therefore has seen wide adoption for in mobile devices [1-3]. With the implementation of ETS technology in substrate manufacturing, not only can one be assured to meet the low cost requirements, but one is also able to implement finer bump pitch and LW/LS demands [4, 5]. Moreover, in order to deliver a high performance packaging and cost-effective solution, the design of flip chip interconnect with Cu pillar bumps using BOL design, mass reflow chip attach and MUF process need to be utilized. Even though mass reflow chip attach with Cu pillar bumps technology is a cost-effective process for flip chip package assembly, there is a large risk of a bump to trace shorts, especially when the designs require finer bump pitch of less than 60μm and finer LW/LS utilizing escaped traces. In this paper, LAB chip attach methodology is introduced to improve the joint soldering and reduce the bump to trace shorting risk in a flip chip assembly with fine bump pitch and ETS with escaped traces design. LAB technology uses localized heating on the silicon die which can reduce the thermal stress during the chip attach bonding process. Additionally, because only the silicon die is being heated, better substrate warpage control using vacuum hold down was observed during the solder joint process as compared to the mass reflow chip attach process. Moreover, LAB can increase throughput by more than 2-times UPH (units per hour) compared to current thermal compression bonding (TCB) technology. This results in a a high productivity rate that is nearly equivalent to that of the mass reflow manufacturing processes [6]. Hence, it is worthwhile to study if LAB chip attach methodology can also reduce the risk of ELK layer damage and thereby improve the assembly yield and become the optimum chip attach process for fine pitch flip chip CSP for advanced silicon node technology.

For the purpose of realizing the ELK performance in a 14x14mm flip chip CSP with 7nm advanced silicon node technology, the flip chip attach processes of LAB is evaluated against mass reflow for joint soldering in a MIP (Mediatek

Innovation Package) design which incorporates Cu pillar BOL with ETS to achieve low cost, finer bump pitch and finer LW/LS to meet the technical requirements [7]). The schematics of LAB chip attach processes are shown in Figure 1. The Cu pillar bump technology with 60μm bump pitch and a 2-layer ETS with finer LW/LS and escaped traces was adopted to achieve the high I/O solution. After performing LAB and mass reflow flip chip attach processes, the thunder test, with two-times mass reflow process and QTC (temperature range of -40°C to 60°C) and the hammer test, with multi-reflow process (peak temperature at 260°C), was studied on 7nm ELK performance in a flip chip CSP package with this specific fine bump pitch and substrate technology. By comparing the ELK reliability performance using LAB and mass reflow, we were also able to validate the extension of QTC and hammer test. Based on these results, it is clear that the LAB chip attach methodology generates better reliability performance than conventional mass reflow process. Through review of these results, it shows that not only can LAB and mass reflow technologies achieve bump pitch reduction with a finer LW/LS substrate and escaped traces, but also that silicon node reduction is possible with resulting device performance improvements.

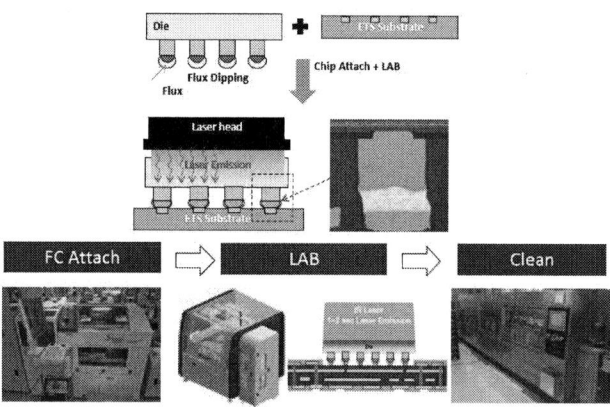

Figure 1. Illustrations of LAB chip attach methodologies.

II. CHIP PACKAGE INTERACTION (CPI) STUDY IN FCCSP WITH LAB AND MR PROCESS

As higher performance and increased functionality are needed in portable and mobile applications, there continues to be a drive to reduce bump pitch in flip chip CSP. Although the mass reflow chip attach process has been demonstrated to be a cost effective solution in flip chip CSP, there is a substantial risk of a bump to trace shorting during the mass reflow flip chip assembly process due to the space limitation of smaller bump pitch, especially when the escaped traces design is adopted. As the LAB process uses laser emission to heat up the die and Cu pillar bumps through the laser head prior to the chip attach process, it not only can increase the UPH as compared to TCB methodologies that require electrical heating of the bond head, but can also yield a high productivity rate that is nearly equivalent to that of the mass reflow manufacturing processes. In addition, as the substrate is not being directly heated lower thermal stress is achieved

during bonding by localizing the heating to silicon die reducing the risk of damage to the ELK die. Moreover, better substrate warpage control is obtained using vacuum hold down during solder joining process. Hence, the LAB technology can be viewed as a viable flip chip package chip attach methodology and cost-effective solution for advanced silicon node devices with fine bump pitch as well as for large die size where high aspect ratios are used.

In order to realize the ELK performance in 7nm silicon node technology by using LAB and mass reflow chip attach methodologies, a 14x14mm flip chip CSP test vehicle with die size of ~110mm^2 with the following features: fine Cu pillar bump pitch of 60μm, under bump metallization (UBM) with a size of ~1800μm^2, copper bump height of 55μm, a 2-layer ETS with total a substrate thickness of 100μm, CuOSP surface treatment on the bottom ball pads and lead-free ball of 0.4mm ball pitch. After die attach of the 7nm node silicon flip chip CSP, the overall maximum package thickness was set to be less than 0.8mm at which point the sample was subjected to the thunder test to detect if there is any white bump phenomenon with associated with ELK delamination failure. In Table I, the thunder test results are reported and show that both LAB and mass reflow chip attach process can pass QTC up to 60x without any failures detected. Furthermore, figures 2 and 3, illustrates C-Mode Scanning Acoustic Microscopy (C-SAM) inspection and Scanning Electron Microscope (SEM) cross-sectional results that clearly show the absence of white bump phenomenon, i.e. ELK damage and also the lack of bump to trace shorting. Hence, the methodologies of LAB and mass reflow chip attach processes can be cost effective processes that enable reliable ELK performance in this 7nm flip chip CSP.

TABLE I.   QTC TEST RESULT FOR 60μM BUMP PITCH EVALUATION WITH LAB AND MR REFLOW PROCESS

Assembly Plan for Thunder test

RUN #	Reflow type.	Rel.	S/S	Substrate	Bump Composition
1	MR	Thunder Test	20ea x 3 lots	2L/0.1mmT/UHD	Cu/Ni/SnAg
2	LAB		20ea x 3 lots		

Thunder Test Result (MR 2x + QTC)

	Thunder Test			
Reflow Type	T0	Thunder Test QTC30x	Thunder Test QTC60x	QTC60x REPORT
MR	Passed	Passed	Passed	Passed
LAB	Passed	Passed	Passed	Passed

-. Conduct each test w/o UF process.
-. Min. requirement is 30x and target thunder test is pass 60x

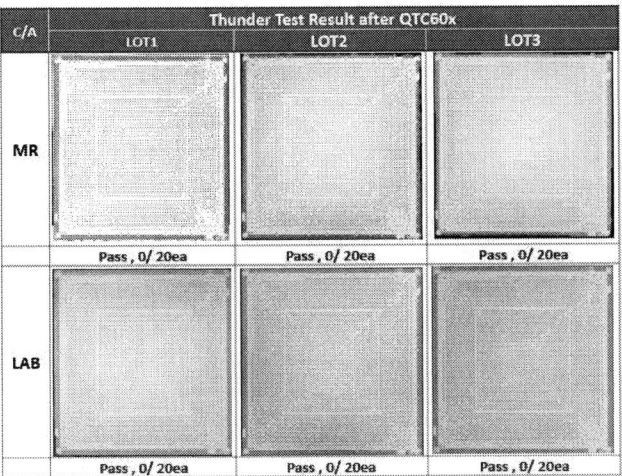

C/A	Thunder Test Result after QTC60x		
	LOT1	LOT2	LOT3
MR			
	Pass , 0/ 20ea	Pass , 0/ 20ea	Pass , 0/ 20ea
LAB			
	Pass , 0/ 20ea	Pass , 0/ 20ea	Pass , 0/ 20ea

Figure 2. QTC60x results through C-SAM inspection

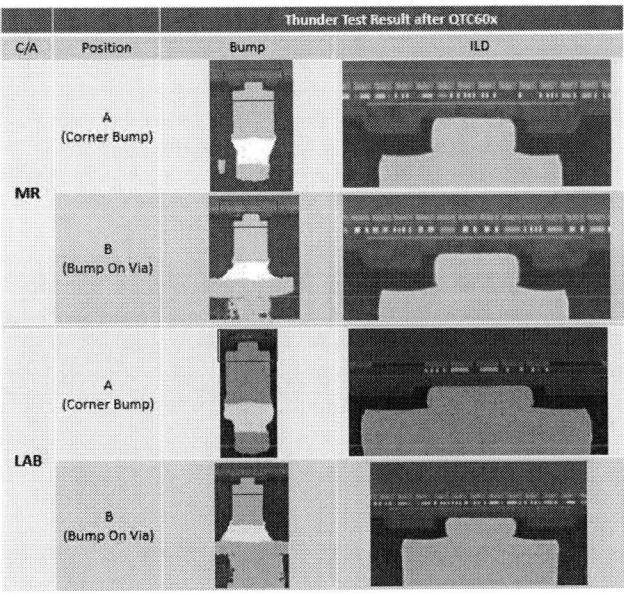

Figure 3. SEM images after QTC60x with MR and LAB process.

In addition to the thunder test, the hammer test was also evaluated to validate the robust chip attach processes 7nm flip chip CSP for ELK based on similar performance inspection. The corresponding results are listed in Table II, which shows both LAB and mass reflow chip attach process can both pass the hammer test up to 5-times reflow process without any failure observed. Figure 4 illustrates the corresponding SEM cross-sectional images and we did not find any abnormality and failure in ELK layer.

TABLE II.   HAMMER TEST RESULT FOR 60µM BUMP PITCH
EVALUATION WITH LAB AND MR REFLOW PROCESS

Assembly Plan for Hammer test

RUN #	Reflow type.	Rel.	S/S	Substrate	Bump Composition
3	MR	Hammer Test	20ea x 3 lots	2L/0.1mmT/UHD	Cu/Ni/SnAg
4	LAB		20ea x 3 lots		

Hammer Test (Multi-Reflow Test) Result:

Reflow Type	Multi-Reflow Test				
	T0	Hammer Test (3X)	3X C-SAM	Hammer Test (5X)	5X C-SAM
MR	Passed	Passed	Passed	Passed	Passed
LAB	Passed	Passed	Passed	Passed	Passed

-. Need to confirm Thunder test result is passed, then do multi-reflow test.
-. Conduct each test w/o UF process.
-. Min. requirement of multi-reflow test is 5X.

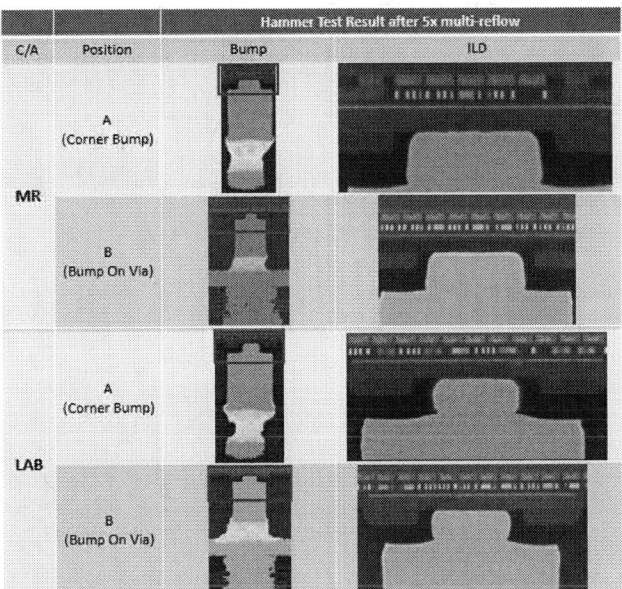

Figure 4. SEM images after 5x multi-reflow with MR and LAB process.

III.   CPI STUDY RESULT WITH EXTENDED THUNDER AND HAMMER TEST

For the sake of further validating the ELK performance between LAB and MR chip attach methodologies, extended thunder and hammer test were performed until an abnormality or failure was observed in the 7nm flip chip CSP devices. The corresponding test results showed that the LAB chip attach process can pass the QTC test up to 100x without any failure detected, while the mass reflow chip attach process failed at QTC100x, see Table III. Moreover, the hammer test results, also shown in Table III, illustrates that the examined 7nm fcCSP can pass 13-times multi-reflow processes when bonded using the LAB chip attach process but fails for the dies attached using mass reflow process. The corresponding SEM cross-sectional images for extended thunder and hammer tests are shown in figures 5 and 6, respectively, and indicate that no ELK damage is found with the LAB chip attach process, but that ELK failure is observed when mass reflow chip attach process is used . Based on these results, we can clearly state that the LAB methodology utilized can reduce the ELK damage risk as compared to the mass reflow methodology. It is believed that the LAB chip attach processes can not only

978-1-7281-1500-9/19 $31.00 © 2019 IEEE   291

guarantee higher assembly yields but also ensure less ELK damage risk in a 7nm flip chip CSP with 60μm bump pitch and finer substrate LW/LS design in the future.

TABLE III.   EXTENDED THUNDER AND HAMMER TEST RESULT WITH LAB AND MR REFLOW PROCESS

Thunder Test Extension Result (MR 2x + QTC)

	Thunder Test			
Reflow Type	Thunder Test QTC60x	Thunder Test QTC80x	Thunder Test QTC100x	QTC100x REPORT
MR	Passed	Passed	Failed	Failed
LAB	Passed	Passed	Passed	Passed

Hammer Test (Multi-Reflow Test) Extension Result:

	Multi-Reflow Test				
Reflow Type	Hammer Test (5X)	Hammer Test (9X)	Hammer Test (11X)	Hammer Test (13X)	Hammer Test 13x REPORT
MR	Passed	Passed	Passed	Failed	Failed
LAB	Passed	Passed	Passed	Passed	Passed

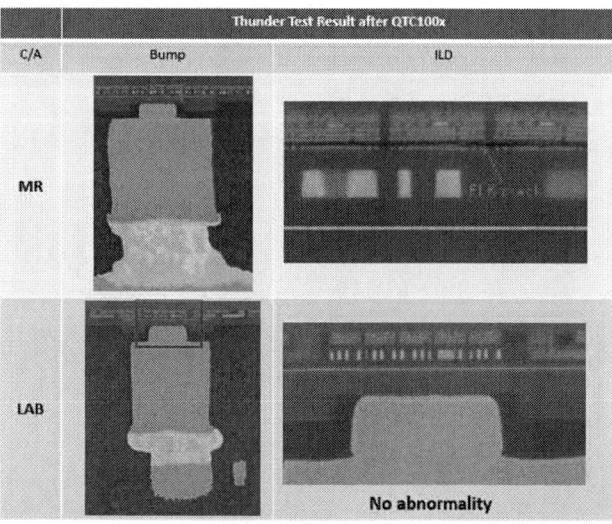

Figure 5. SEM images after QTC100x with MR and LAB process.

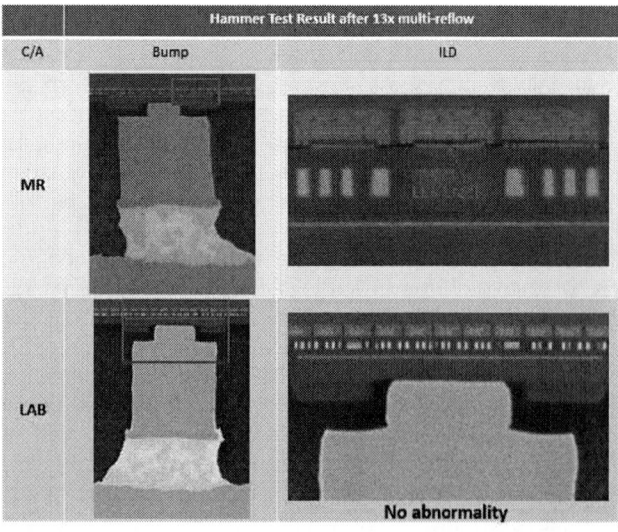

Figure 6. SEM images after 13x multi-reflow with MR and LAB process.

IV.   LONG-TERM RELIABILITY TEST

In order to study the package assembly performance for this 7nm flip chip CSP with 60μm bump pitch and fine LW/LS MIP design, a long-term reliability test was performed. Table IV lists the scheduled sample sizes after open-short (O/S) testing in this reliability qualification. Figure 7. illustrates the Through Scanning Acoustic Microscopy (T-SAM) results of a 7nm flip chip CSP with mass reflow chip attach process after pre-condition of moisture sensitivity level 3 (MSL3), unbiased highly accelerated stress test (μHAST) of 192 hours with MSL3, thermal cycling test condition B (TCB, -55°C~125°C) of 1000 cycles with MSL3 and high temperature storage test (HTST) at 150°C of 1000 hours with MSL3; with no white bump phenomenon, i.e. ELK damage, observed. Figure 8 shows the SEM cross-sectional images after the final reliability test read point (μHAST 92 hours, TCB 1000x and HTST 1000 hours). From figures. 7 and 8, it can be concluded that the 7nm flip chip CSP with 60μm bump pitch and mass reflow chip attach process passes all reliability test items without any defect observed through T-SAM and SEM inspection. Hence, it is believed that mass reflow flip chip attach methodology examined in this paper can guarantee the illustrated 7nm flip chip CSP (with 60μm bump pitch and two escaped traces in a 2-layer ETS) with good assembly quality yield and with less ELK damage risk.

Long-term reliability of a 7nm flip chip CSP with LAB chip attach process is currently in progress and the evaluation results will be shared in the future. Based on the above findings, the authors believe with the use of LAB methodology, this 7nm flip chip CSP with 60μm bump pitch and fine LW/LS MIP design can pass all reliability qualifications without any failure.

TABLE IV.   SAMPLE SIZES IN LONG-TERM RELIABILITY TEST

O/S Test Result

Leg#	EOL	After MSL3	After HTST 168 hrs	After HTST 500 hrs	After HTST 1000 hrs
Leg#1	0/44	0/44	0/44	0/44	0/44

Leg#	EOL	After MSL3	After TCB 200X	After TCB 500X	After TCB 1000X
Leg#2	0/45	0/45	0/45	0/45	0/45

Leg#	EOL	After MSL3	After HAST 48 hrs	After HAST 96 hrs
Leg#3	0/44	0/44	0/44	0/44

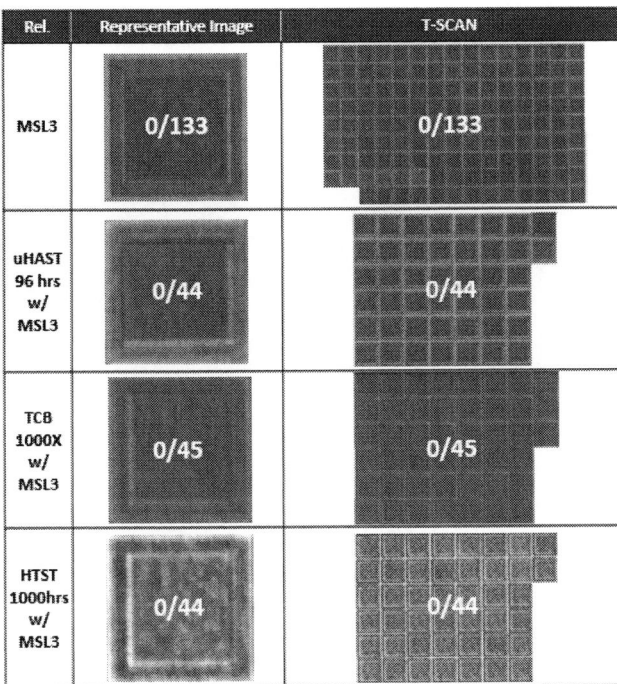

Figure 7. T-SAM result for long-term package reliability in 7nm fcCSP with MR chip attach process.

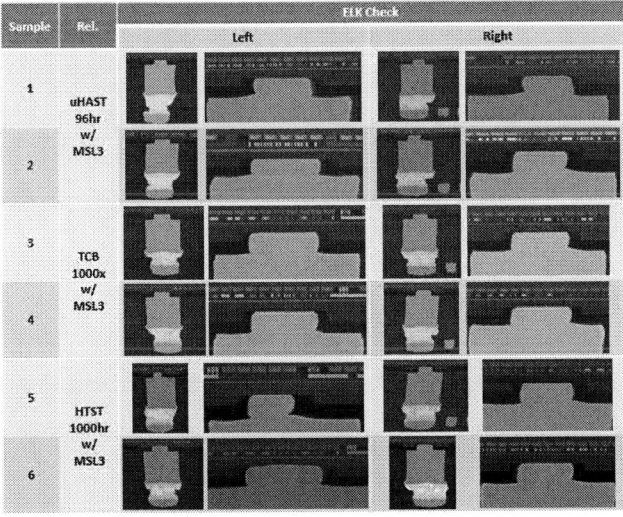

Figure 8. SEM cross-sectional result for long-term package reliability in 7nm fcCSP with MR chip attach process.

V. CONCLUSIONS

In this 7nm CPI study of a 14x14mm flip chip CSP with 60μm Cu pillar bump pitch, a 2-layer ETS and fine LW/LS in a MIP design, we have compared LAB and mass reflow chip attach methodologies.. The purpose of this work is to validate a reliable and cost-effective LAB process and compare it to a mass reflow chip attach process in a 7nm flip chip CSP. To detect white bump phenomena, i.e. ELK damage, we performed two critical tests: the thunder test, 2-times MR and a temperature cycling range of -40°C to 60°C, as well as the hammer test, multi-reflow processes and peak temperature of 260°C, . It was found that both LAB and mass reflow chip attach processes can reduce the ELK damage risk based on the results of this work. To further validate LAB as a robust chip attach methodology with improved package assembly yield and reduced ELK damage, we ran extended thunder and hammer test utilizing mass reflow chip attach process as the control. The results indicate that LAB technology achieves enhance package assembly yield and improved ELK performance over the mass reflow chip attach process. Moreover, the long-term reliability tests with mass reflow chip attach process have verified the package assembly yield and performance while the reliability tests by using LAB methodology are currently in progress. These results show that the robust flip chip CSP assembly processes, LAB and MR technology, discussed in this paper, can achieve package performance improvements and enable cost-effective packaging solutions. Furthermore, once the advanced silicon node, fine bump pitch, fine LW/LS as well as large die size with high aspect ratio in a flip chip CSP is considered, it is believed the LAB technology could be the optimum flip chip package chip attach methodology to implement reliable assembly yield without any ELK damage issues.

REFERENCES

[1] R. D. Pendse, K. M. Kim, K. O. Kim, O. S. Kim and K. Lee, "Bond-on-Lead: A novel flip chip interconnection technology for fine effective pitch and high I/O density," Electronic Components and Technology Conference (ECTC), pp. 16-23, 2006.

[2] S. Movva, S. Bezuk, O. Bchir, M. Shah, M. Joshi, R. Pendse, et,al., "CuBOL (Cu-Column on BOL) technology: A low cost flip chip solution scalable to high I/O density, fine bump pitch and advanced Si-nodes," Electronic Components and Technology Conference (ECTC), pp. 601-607, 2011.

[3] M. C. Hsieh, "Advanced Flip Chip Package on Package Technology for Mobile Applications," International Conference on Electronic Packaging Technology (ICEPT), 2016.

[4] H. Eslampour, Y. Kim, S. Park, T. Lee, "Low Cost Cu Column fcPoP Technology," Electronic Components and Technology Conference (ECTC) 2012.

[5] E. Ouyang, et al., "Improvement of ELK reliability in flip chip packages using Bond-on-Lead (BOL) interconnect structure," International Microelectronics and Packaging Society Proceedings (IMAPS), 2010.

[6] Chi-Yuan Chen, Ian Hsu, Stanley, Doosam Park, M. C. Hsieh, "Laser Assisted Bonding Technology Enabling Fine Bump Pitch in Flip Chip Package Assembly," Electronic System-IntegrationTechnologies Conference (ESTC), 2018.

[7] Chi-Yuan Chen, Ian Hsu, Stanley, KeonTaek Kang, M. C. Hsieh, "Various Chip Attach Evaluations in a Fine Bump Pitch and Substrate Flip Chip Package," Electronic Components and Technology Conference (ECTC) 2018.

Ultra Large Area SIPs and Integrated mmW Antenna Array Module for 5G mmWave Outdoor Applications

Pouya Talebbeydokhti, Sidharth Dalmia, Trang Thai
Intel Component and Devices Group (iCDG)
Intel Corporation
Santa Clara, CA USA
Pouya.talebbeydokhti@intel.com
Sidharth.dalmia@intel.com
Trang.thai@intel.com

Sharon Tal, Raanan Sover
Intel Component and Devices Group (iCDG)
Intel Corporation
Israel
Raanan.sover@intel.com
Sharon.tal@intel.com

Abstract— over the past three decades, the massive growth of mobile data traffic through internet and mobile networks has constantly increased. This growth requires target peak rates of 10Gb/s for both indoor and outdoor applications and the only way to achieve such high data rate is by operating at the Millimeter-wave frequencies. Millimeter-wave technology brings a new standard of wireless communication in various areas including mobile devices, automotive, IoT (Internet of Things), medical, military and many others. The evolution of 5G wireless communications requires the development of low cost, lightweight, and low profile SIPs and antennas in a package that are capable of maintaining high performance over a wide bandwidth. This paper presents an all-in-one module solution consists of large dual-band area SiP (System-in-Package) and dual-polarized patch array antenna for 5G mmW outdoor applications. The authors show the development of large arrays up to 4 x 8 and 8 x 8 to cover the low and high bands for 5G mmwave.

Keywords-component; 5G; Ultra Large SIP; mmW Antenna Array; Outdoor application

I. INTRODUCTION

In the fifth generation mobile communications system (5G), the high frequency bands over 6 GHz (6-100 GHz) will be key factor for the reason of providing very wide frequency bandwidths. 5G networks aim to achieve ubiquitous communication between anybody and anything, anywhere and at any-time. However, mmWave frequencies cannot travel through buildings and other obstacles. Moreover due to high absorption rate of moisture and plants they suffers from high path losses. This introduces small cell networks instead of the current large high power cell towers, small cells uses thousands of low power mini base stations to transmit the signal around obstacles. Figure 1 shows the small cell structure. There are many challenges in designing a 5G mmW wireless system, such as 1) Ultra Broad-bandwidth 2) Dual-polarization phased array antenna design, 3) high gain, 4) Small cell Networks 5) Beamforming and beamtracking, 6) Low Dk/Df material, 7) Robust multilayer packing and fabrication.

This paper presents an all-in-one module solution consists of large dual-band area SiP (System-in-Package) and dual-polarized patch array antenna for 5G mmWave outdoor applications. Each SiP will have multiple RFIC and the Radio Front End Module (RFEM) components, and will be placed on top of each antenna board. The SIP will be assembled on each antenna board using a BGA and controls the operating of each patch antenna. The height of dielectric material and volume is increased in the antenna board stack-up to achieve wide bandwidth. We have also introduced many different techniques to help with overall warpage and manufacturability of the novel module of dimension 50mm x 32.5mm and larger systems.

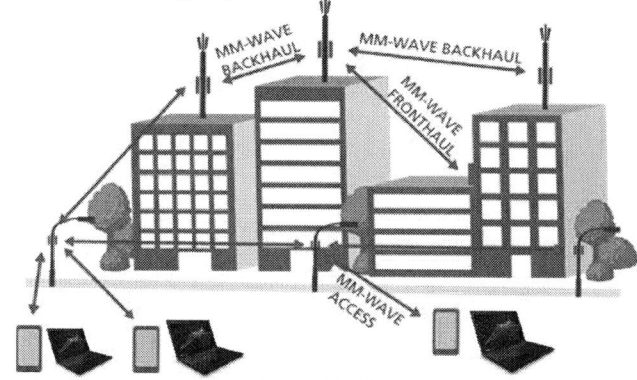

Figure 1. Small Cell Network

II. INTRODUCTION TO MMWAVE WIGIG PRODUCTS

Today's mmWave 5G packages have similar architecture requirements as the past Wi-Gig packages since they are both operating in mmwave frequency and there is a need for antenna in a package. Intel used Wi-Gig and the 802.11ad unlicensed bands to deliver wireless docking for PCs and small form factor devices. The frequency range of such products is between 57-64GHz. The requirement of having antenna in a package in previous generations presented different challenges in material stack-ups and thickness. The authors used complex 10 layers and 12 layer stack-ups with asymmetric metal densities and laminate thicknesses to achieve the function of routing dense Si in substrates and concurrently having low metal density based antenna structures. The authors could achieve the performance for the desired frequency range and the relatively small sized modules in this manner. However, the technology of using 10-14 layer stack-ups with uneven unbalanced metal layers combined with thin and ultra-thick layers is not feasible at the lower frequencies needing coverage from 25-45GHz. As

978-1-7281-1500-9/19 $31.00 © 2019 IEEE

a result the authors present a novel technique for creating ultra large SIPs with mmwave RFIC that are assembled and mounted onto much larger antenna boards to provide a platform for large 4x8 and 8x8 antenna arrays.

III. ULTRA LARGE SIP CONFIGURATION

Figure 2 shows the main invention and focus of this paper where the authors have portioned a very large 5G mmWave antenna array into separate components. The partition is made using very large SIPs that included multiple die that generated the RF signals at mmwave frequencies.

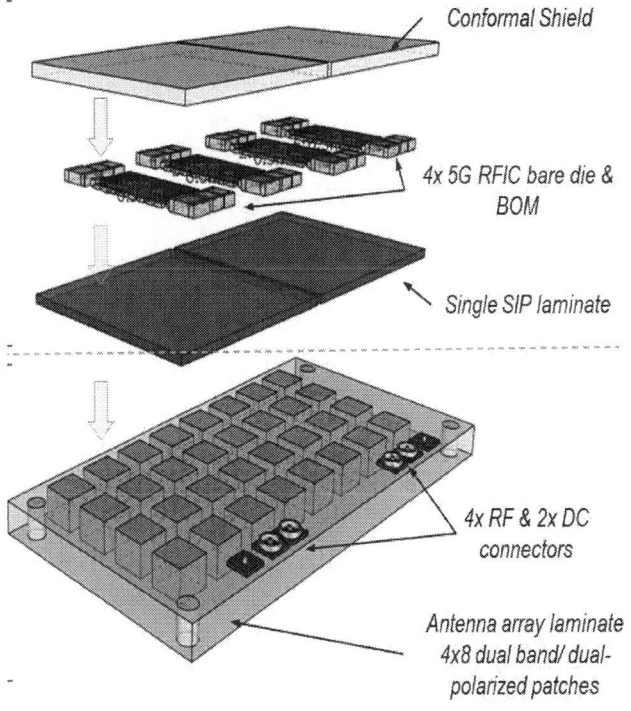

Figure 2.   SIPS and their components mounted onto antenna boards

The signals form the RFIC fed the BGA balls of the SIP via the means of very balanced and dense 6L-10L substrate. Figure 3 shows example of an 8 layer stack-up of the ultra large SIP which will be used for assembly of two flip-chip RFIC and BGA (Ball Grid Array) balls for connection to the antenna PCB. This SIP can include many other components such as DC-DC converters, decoupling capacitors, inductors, and filters. The multi-layer substrate is composed of a core and prepreg material (build-up) layers. The core material is to provide structural robustness in the multilayer substrate. The number of prepreg layers depends on the architecture of the antenna and RFIC routing.

Layer Name	Stackup Thickness	Via/Pad
FSR	20 SM	
L1	15 Cu	
L1-L2	35 - 45 PPG	60/120
L2	15 Cu	
L2-L3	35 - 45 PPG	60/120
L3	15 Cu	
L3-L4	35 - 45 PPG	60/120
L4	15 Cu	
LTH	55 - 65 Core	65/120
L5	15 Cu	
L5-L6	35 - 45 PPG	60/120
L6	15 Cu	
L6-L7	35 - 45 PPG	60/120
L7	15 Cu	
L7-L8	35 - 45 PPG	60/120
L8	15Cu	
BSR	20 SM	

Figure 3.   An example of 8L stackup for SIP

Figure 4 shows the top view of the complex assembly onto the antenna board.

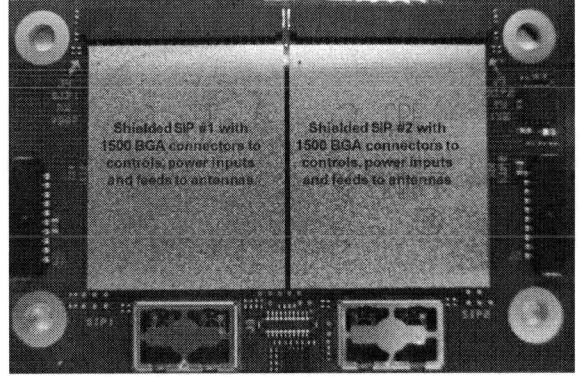

Figure 4.   Appearance of 2 large SIPs, BOM components on large Antenna Board

For manufacturing stability, the number of metal layer above and below the core material should be symmetric. This is why for the SIP stack-up we chose to have a symmetric 8 layer stack-up that only uses a 55-65μm core and 35-45μm PPG. In order to have good insertion loss and return loss at mmWave frequency all the RF transmission lines were designed as a stripline. This requires a solid GND on top and bottom of all RF lines to achieve good performance. Since at least four metal layers were required above the core material for power feeding lines and other signal lines, another four layers should be placed below the core material. Due to the skin depth of mmWave frequencies, a hyper very-low profile (HVLP) Cu foil with thickness of 15μm was used to achieve good performance with low loss. mmWave frequencies are extremely sensitive to any discontinuity and will impact the loss and performance especially with an 8L substrate, to achieve such low loss we used 60/120µm drill/pad for prepreg layers and 65/120µm drill/pad for core layers. To achieve the required z-height and form factor we used transfer molding with conformal shielding instead of the conventional metal shield.

The SIPs had a BGA footprint with over 1500 I/Os each and provided feeding to the antenna by means of very well designed BGA feed points showing <20dB return loss upto 50GHz. The BGA feeds also include several Amps of current supplied, control lines to steer the beams using on die phase shifters. These BGA connections need reliable connection. The planarity of the BGA at time "zero" and during reflow needs to be <=100µm to ensure that the 0.4mm BGA pitch with 200-250µm ball diameter works well. Figure 5 shows the 3000 connections over 2 SIPS with excellent solder coverage.

Figure 5.   XRAY of the entire mmwave module board

IV.   ANTENNA CONFIGURATION

Although in Wi-Gig and 5G mmWave applications, an antenna module and IC's are usually packaged together to reduce any potential loss and co-design process, in this work since the antenna design had to be dual polarized it required us to have a multi-core multi-prepreg PCB that has different stack-up and architecture than the RFIC substrate.

As previously mentioned at mmWave the free space path loss increases and therefore, to enable a feasible link budget for cellular networks, beamforming based on antenna arrays that provide high gain are implemented as a solution. Specifically, large arrays are required for the base station applications. Additionally, base stations often need to support broadband operation with dual polarization in multi-frequency ranges (27.5-29.5Hz and 37-40GHz). To meet these requirements of large high gain antenna array, 1) dual polarization, broadband dual frequencies, 2) consistent radiation pattern across the operation frequencies for ease of beam scanning control, and 3) stacked patch antenna which

offers a unique advantage are introduced. However, stacked patch antenna that provides dual resonant frequencies with large bandwidth in each operation frequency band requires thick substrate and multiple layers. For high volume production in standard substrate technology with acceptable yield, such solution targeting the consumer market has not yet been attempted. In this section, we present the design and fabrication of a unique multicore PCB/Substrate large area cost effective dual linear polarization broadband dual frequencies (28GHz and 39GHz) stacked patch antenna 4x8 array for mobile and base-station applications

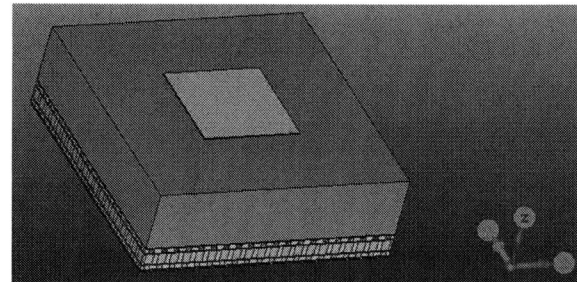

Figure 6.   Topology of the dual band dual polarization antenna concept

In order to accommodate all operation frequency bands for an all-in-one solution as well as all the performance requirements of the antennas, we have utilized the stacked patch antenna concept to achieve dual frequency and orthogonal dual linear polarization. There are three patches to form two wideband resonant frequency bands. A capture of the antenna unit is displayed in Figure 6. This concept allows the antenna to achieve very high cross polarization discrimination ratio (>25dB), which is critical in the MIMO operation of the dual polarization network.

All components of the 4x8 antenna array are built on 10 layers of low cost BT laminates that results in a board dimension of 30mm x 50mm x 1.5mm using standard PCB/substrate technology. The board was constructed using a multicore approach to reduce warpage and misalignment between layers, to improve yield, and consequently reduce cost and lead time. The stack-up consists of three core layers of BT laminates. The core layers are combined by the prepreg layers in between using a combination of different prepreg thicknesses. Here, stitching through vias were implemented as a technique to address the warpage and misalignment for such a big and thick board. Figure 7 shows the assembly process flow of the SIP onto the antenna board. This process flow only shows the assembly of the ultra large SIP onto the antenna board using BGA connection.

V. FABRICATION AND MEASURMENTS RESULTS

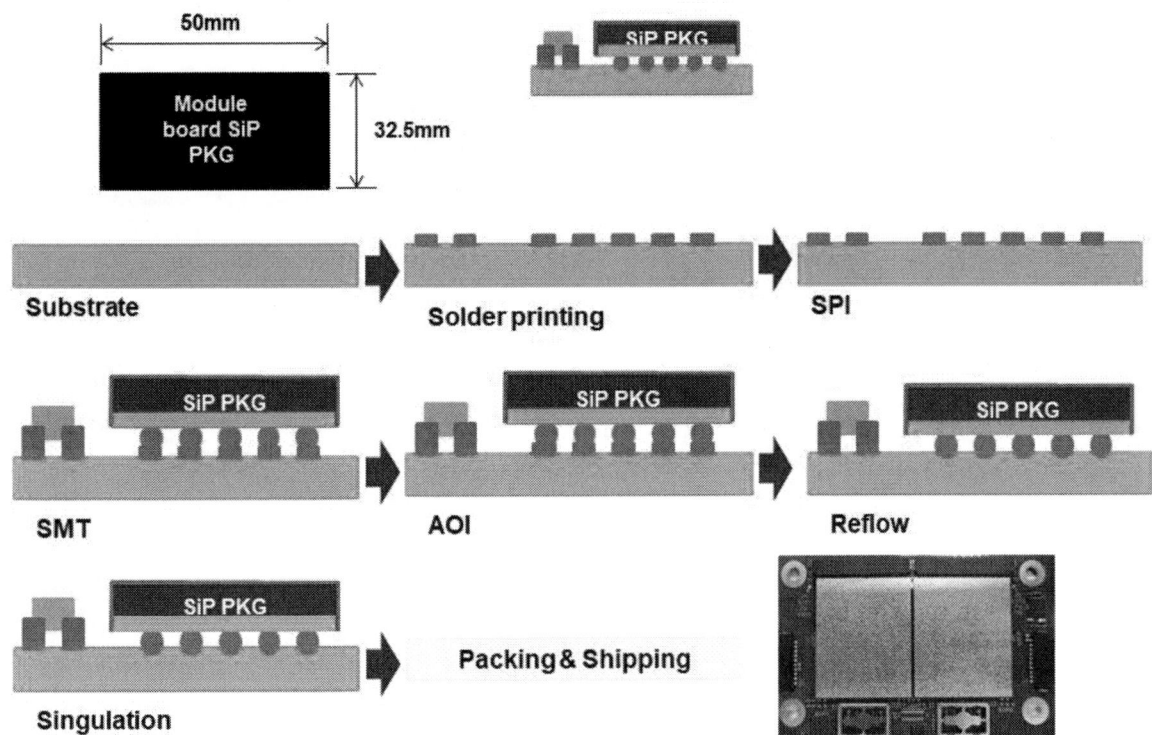

Figure 7. Shows the assembly process flow of large SIP onto Antenna board

Manufacturing and fabrication of the 10 layer antenna board was a challenging task as such a stack-up with multiple cores and multiple prepregs never been done in the industry. We faced many issues during the manufacturing process due to the complicated and complex architecture of the antenna board. To accommodate both high density circuit layers and low density high volume antenna structure, we developed a stackup that is composed of three portions with different core layers as previously mentioned. First of all, the top portion consists of four metal layers that supports the ultra-high-density circuit with hundreds of micro vias. Secondly due to the broadband patch antennas, we introduced a single thick core (without metal) that provides stiffness and mechanical symmetry for the stack up. This thick core layer joins the top portion and the bottom portion of the laminate stack ups with minimal risk of warpage. The final portion contains all of the 32 antenna patches and is the bottom four metal layers of the board. There are no vias in these layers, only metal patches that are parasitically coupling to each other. This portion employs two cores to enable the flexibility in antenna design as well as to maintain stiffness for the required volume. Note that each portion maintains its own layer thickness symmetry and the balance of its copper distribution in different layers. With help of our suppliers and sample builds ahead of the actual production build, we were able to achieve a substrate yield of 80% or better for both the SIP substrate and the antenna board.

As the SIP is 20x20mm and the antenna board is 50x35mm in size, this required working very closely with the suppliers to control the warpage and achieve a coplanarity of 50µm or better. Mechanical simulations were performed on antenna board and SIP due to large size and unbalanced stack-up. Results showed that both packages are within JEDEC warpage/coplanarity standard specification.

The performance of a single element antenna with 4 ports shows that both the low band (LB) and high band (HB) ports have excellent return loss (>10dB) and are in the required range. The isolation between V and H polarization of each band is better than 40dB. The antenna gain for a single element is about 5dB across the operation frequencies. The through vias placed in the East and West direction of the board are also arranged into a hard surface structure (frequency selective high impedance surface) to improve the radiation beam width to the side.

Figure 9 shows the module board with bare copper exposed for antenna port measurements where element 15 was investigated to validate the performance and fabrication of the antennas in this complex module board. Since element 15 is placed in the middle of the array, making it excellent validation for the unit cell design.

To fully understand the manufactured dimensions, we performed x-ray analysis on cross-section for random parts and   from   random

L2 measurement location   ## L3 measurement location

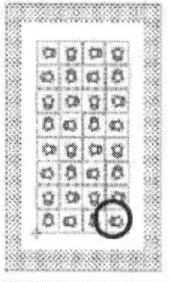 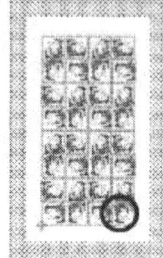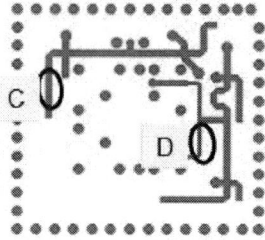

Layer		L2		L3	
Location		A	B	C	D
Target		109um	50um	109um	50um
Raw data	#1	106.69	51.67	108.04	47.67
	#2	105.98	46.97	106.04	45.61
	#3	106.69	48.32	106.69	46.97
	#4	108.75	49.67	106.69	47.67
	#5	106.04	52.38	105.98	45.61
Avg.		**106.83**	**49.802**	**106.688**	**46.706**
Std.		1.13	2.26	0.83	1.04
Max.		108.75	52.38	108.04	47.67
Min.		105.98	46.97	105.98	45.61

Figure 8.   X-section showing the line width for the critical RF transmission lines in layer 2 and 3.

Figure 9.   The module board with bare copper exposed for antenna ports measurements

lots and all results showed that our manufactured dimensions aligned with our (the target) design. The layer to layer alignment was pass with no significant shift in alignment between layers. Also the 200µm drill holes around the antenna board were shown to have no significant alignment issues shown in Figure 10.

The reflection coefficients of the two HB ports (for two polarizations) of two different samples are shown in Figure 11 with overlay of simulation results of the same antenna element. Similarly, the measured reflection level of the two

LB ports as shown in Figure 12. Note that in these measurements and simulations of a single element, all other RF ports on the board are left open. The results in Figures 11 and 12 show an excellent agreement between measurement and simulation with both exhibiting the expected resonant peak, additionally there is virtually no variation between two random samples.

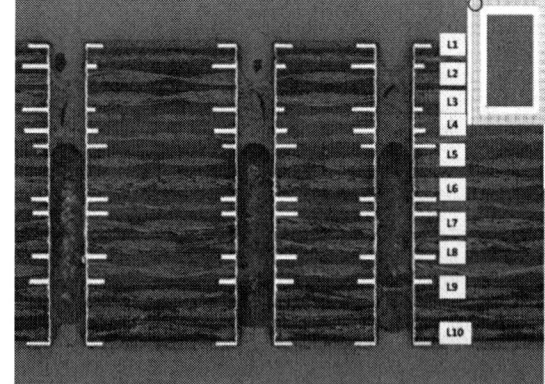

Figure 10. Cross section of mmwave module showing alignment of 10 layer stack-up

978-1-7281-1500-9/19 $31.00 © 2019 IEEE   298

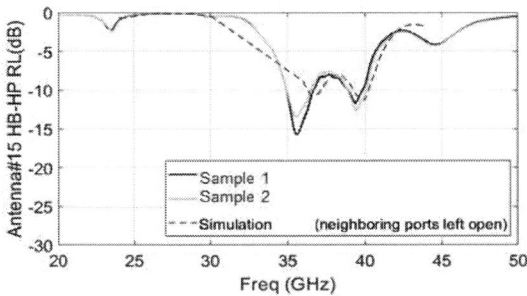

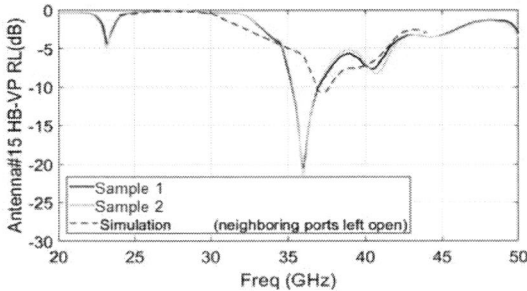

Figure 11. a) Top graph shows the reflection coefficients of two samples at the HB ports for horizontal polarizations; b) Bottom graph shows the reflection coefficients of two samples at the HB ports

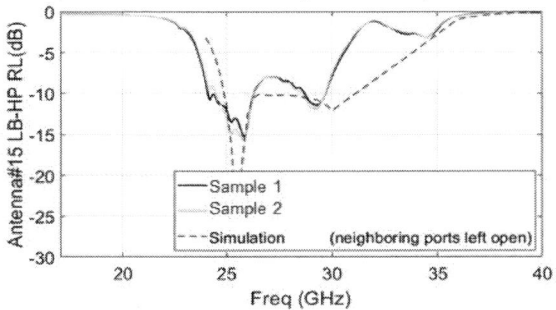

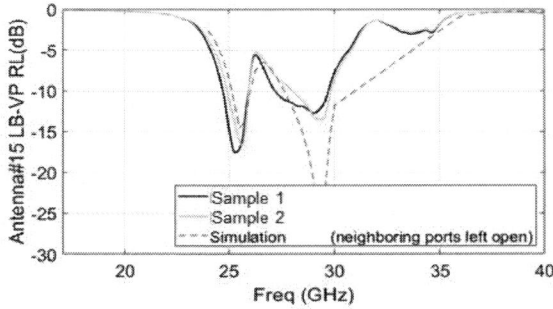

Figure 12. Reflection coefficients of two samples at the LB port for horizontal polarization; b) vertical polarizations in the bottom graph.

Finally, we also evaluated the line gap, especially between the traces and the vias in the ultrahigh density layers shown in Figure 8. Such a gap has a significant impact on the performance in mmWave circuit. The results showed that the difference between the target and the actual gap is 3.3µm or less. For the largest, first of its kind, mmWave module, we have demonstrated exceptional yield for all parameters from alignment to trace/feature size to trace gap.

VI. CONCLUSION

We present the design and fabrication of a unique multicore PCB/Substrate large area cost effective dual linear polarization broadband dual frequencies (28GHz and 39GHz) stacked patch antenna 4x8 array for mobile and base-station applications. The results are a high performance RFEM for a large array (4x8 elements) that can support all 5G millimeter wave communication standards with dual polarization and high cross polarization discrimination ratio (>25dB) to enable excellent MIMO operation.

ACKNOWLEDGMENT

The authors gratefully acknowledges the support of our internal partners (ATTD/CPTD/Intel Lab) as well as the collaboration of our OSAT partners and also special thanks to Rich, Quynh and Josef.

REFERENCES

[1] https://www.intel.com/content/www/us/en/products/wireless/ad-products/wireless-gigabit-11100vr.html

[2] X. Gu *et al.*, "A multilayer organic package with 64 dual-polarized antennas for 28GHz 5G communication," *IEEE International Microwave Symposium (IMS)*, pp. 1899-1901, Honolulu, HI, 2017

[3] D. Liu, J. A. G. Akkermans, H.-C. Chen, and B. Floyd, "Packages with integrated 60-GHz aperture-coupled patch antennas," in *IEEE Transactions on Antennas and Propagation*, vol. 59, no. 10, pp. 3607-3616, Oct. 2011.

[4] M. Fakharzadeh, "A compact 4 by 1 patch array antenna-in-package for 60 GHz applications," in Proceedings of *IEEE International Symposium on Antennas and Propagation*, pp. 1-2, Jul. 2012.

[5] T. Y. Lin, T. Chiu, and D. C. Chang, "Design of dual-band millimeter-wave antenna-in-package using flip-chip assembly," IEEE Transactions on Components, Packaging and Manufacturing Technology, vol. 4, pp.385-391, 2014.

[6] L. Li, X. Chen, Y. Zhang, L. Han, and W. Zhang, "Modeling and design of microstrip patch Antenna-in-Package for integrating the RFIC in the inner cavity," in *IEEE Antennas and Wireless Propagation Letters*, vol. 13, pp. 559-562, 20

978-1-7281-1500-9/19 $31.00 © 2019 IEEE

Hybrid Approach for Large Size FC-BGA to Enhance Thermal and Electrical Performance including Power Delivery

Heesok Lee, Yunhyeok Im, Junghwa Kim, Jisoo Hwang, James Jeong, Youngsang Cho, Heejung Choi, and Youngmin Shin

System LSI Division,
Samsung Electronics, Co. Ltd.
1-1 Samsungjeonja-ro, Hwaseong-si, Gyeonggki-do, Korea
hees.lee@samsung.com

Abstract—**A new approach to flip chip package with hybrid packaging technologies is presented, which will be pretty much promising package solution as a large size FC-BGA (flip chip ball grid array) for SOC (system on a chip). The proposed new approach can be utilized to enhance power integrity and heat spreading and to reduce the FC-BGA package height with keeping and not degrading the mechanical property of conventional FC-BGA. The presented new package based on hybrid approach is implemented by the combination of thick-core based substrate and thin-core (or coreless) substrate. While most of area in package is mechanically supported by 800um thick core, SOC die is mounted on thin substrate by proper flip chip bonding. The interconnection between thick-core based substrate and thin-core (or coreless) substrate is achieved by wire bonding by conventional packaging process or RDL (redistribution layer) by fan-out package technology. Comparative study on the proposed new structure with conventional FCBGA is presented in terms of power integrity, signal integrity, and thermal resistance. Through several case studies, it is demonstrated that the presented hybrid approach is an adequate package solution for cost-effective thin FCBGA enhancing power integrity and thermal performance.**

Keywords-Flip Chip, BGA, hybrid, fan-out, RDL, POP

I. INTRODUCTION

Although silicon die size of IC (integrated circuit) has been decreased with the advance of silicon processing technology, the size of package for several applications including D-TV (digital TV) system and server has not been able to be decreased. Since the package size for those applications should be determined by ball pitch given by system-level PCB (printed circuit board) and ball count given by product specification, the package size has been kept to be large around 40mm x 40mm, for which large size FC-BGA (flip chip ball grid array) package with ball pitch larger than 0.8mm has been utilized. Due to large package size, there is a strong need to use thick core-based organic substrate guaranteeing more rigidity to control the warpage of large FC-BGA. [1-3]

FC-BGA package larger than 25mm x 25mm is generally constructed by the substrate with the over 800um thick core, because the warpage caused by solder ball reflow during surface mounting process in system-level PCB assembly should be carefully determined for the large size FC-BGA. However, over 800um thick core has been a critical barrier limiting signal and power delivery for corresponding silicon die. As presented by [4-6], coreless substrate has been proposed to FC-BGA to improve signal and power integrity. Generally, the advantages of coreless substrate include shorter transmission path length for better insertion loss, less cross talk, lower cost and flexible substrate routings. For shorter transmission path length given by the vertical interconnect in thinner coreless substrate, signal delivery characteristics of via is negligible to the major routings of signals in certain copper layer. Therefore, the insertion loss improvement is expected and easily achieved with coreless substrate or thin core based organic substrate. Additionally, the power delivery can be remarkably improved by using coreless or thin core substrate instead of thick core substrate.

Although it is better to use thin core substrate for improving signal and power integrity of FC-BGA, there has been still challenges in BLR and mechanical propertie of FC-BGA to utilize thin core substrate instead of thick core substrate. For example, to keep mechanical behavior of FC-BGA, capacitor embedded in thick core has been introduced as given by [8]. To enhace power integrity performance of core circuits, the authors of [8] presented component-type capacitor embedded in core of organic substrate. The number of the embedded decoupling capacitors was optimized to maximize the PDN (power delivery network) performance and cost effectiveness. PDN design with thick core substrate is one of the critical design challenges with FC-BGA package as addressed by [8]. However, embedded capacitor with thick core whose thickness is larger than 800um is not popular.

Although thinner core or coreless substrate for FC-BGA has been reported in several literatures including [4-6] to improve signal integrity and power integrity, coreless substrate has not been popular. Because large size FC-BGA fabricated by thin core substrate or coreless substrate frequently suffers from warpage control issue, system-level assembly yield problem, and solder joint failures due to residual stresses.

In this paper, authors present a new approach for FC (flip chip) package with hybrid packaging technologies which can improve large size FC-BGA package for SOC (system on a chip) in terms of power delivery and thermal resistance. In the following sections, it will be addressed that the new approach is promising package solution to enhance power integrity and heat spreading and to reduce the FC-BGA package height without degrading the mechanical characteristics of conventional FC-BGA. Comparative study on the new approach with conventional FC-BGA will be presented in terms of power integrity, signal integrity, and thermal resistance. Through several case studies with various FC-BGA specifications, it will be demonstrated that the presented hybrid approach would be a promising package solution for cost-effective thin FC-BGA while having enhanced power integrity and thermal performance with acceptable signal integrity design. In the following second section, the proposed package structures will be introduced with introducing the assembly process flow.

II. A Hybrid Approach for FC-BGA

The presented hybrid approach will utilize the combination of multiple substrates to fabricate one FC-BGA package, for which thick-core substrate and thin-core substrate or coreless substrate are employed simultaneously and connected mechanically, electrically, and thermally. While most of area in the FC-BGA package suggested by Fig.1 and Fig.3 is mechanically supported by a rigid substrate whose core is thicker than 800um, SOC die is mounted on thin substrate by proper FC (flip chip) bonding. For ease of presentation, while thick core substrate will be denoted by sub_0, the coreless substrate or thin core substrate will be called by sub_1 in this paper. It is expected that sub_0 gives mechanical strength to support large size of FC-BGA and sub_1 should be proper for power delivery improvement and making FC-BGA thinner.

The interconnection between sub_0 and sub_1 is implemented by bonding wires in conventional assembly process or RDL (redistribution layer) in fan-out packaging technology. While the hybrid structure for FC-BGA based on bonding wires interconnecting sub_0 and sub_1 will be called by hybrid_1 (H1), the structure for FC-BGA using RDL (re-distribution layer) given by fan-out packaging process will be called by hybrid_2 (H2). In the following sub-sections, the package structure and the assembly process flows will be introduced for H1 and H2, respectively.

A. Hybrid_1 (H1) ;FC-BGA with wire bonding technology

As shown in Fig.1, hybrid_1 (H1) includes sub_0 and sub_1. While SOC die is mounted on sub_1 by proper FC bonding assembly process, H1 has bonding wires to connect sub_0 and sub_1 as presented previously. Since the SOC should be mounted on sub_1, the fine pitch interconnection to support fine pitch flip chip bump of SOC is needed only for sub_1. Surely, the major factor determining the design rule of sub_0 is the number of IOs of SOC. Sometimes, SOC with fixed number of IO is considered to be shrunk by using advanced silicon fabrication technology. However, the conventional FC-BGA needs fine-pitch design rule for interconnect in its thick-core substrate, which sometimes results in serious cost adder in packaging. In the other hand, the proposed FC-BGA with hybrid approach can keep the cost of thick core substrate called by sub_0 just by changing

the design rule of thin core substrate called by sub_1 finer. It is expected that H1 can be a proper FC-BGA package to support die size shrink achieved by advanced silicon fabrication node with reasonable cost adder in packaging.

The entire assembly process of H1 given by Fig.1 is presented in Fig.2. The first step is the preparation of sub_0, as shown in Fig. 2. After a perforation or hole is fabricated on sub_0 by using mechanical punching or laser drilling process, sub_0 with hole is placed on the rigid carrier. The carrier is temporary supporting layer, which will be removed before solder ball attach process at the end of assembly process. There is temporary adhesive layer between sub_0 and the carrier. In the hole, sub_1 is placed and temporarily fixed on the adhesive layer over rigid carrier. Sub_1 should be properly fixed on the temporary carrier for the interconnection process including flip chip bonding and wire bonding. After completing wire bonding process for connecting sub_0 and sub_1, the package should be encapsulated by epoxy molding compound to protect bonding wires and give mechanical support connecting sub_0 and sub1. If necessary, metal plate can be placed on the mold over die as a heat slug, which is given by Fig. 1 (c) Based on the structure given in Fig.1 and Fig.2, we can expect that the thickness of silicon die can be increased with the fixed FC-BGA height. The thermal benefit given by increasing silicon die thickness will be discussed in the following section. In addition, multiple dies can be integrated in one FC-BGA package with flip chip bonding and wire bonding as shown in Fig 1(d). As also introduced by [9], silicon interposer can be also employed to improve integration density with FC-BGA given in Fig.1 (d)

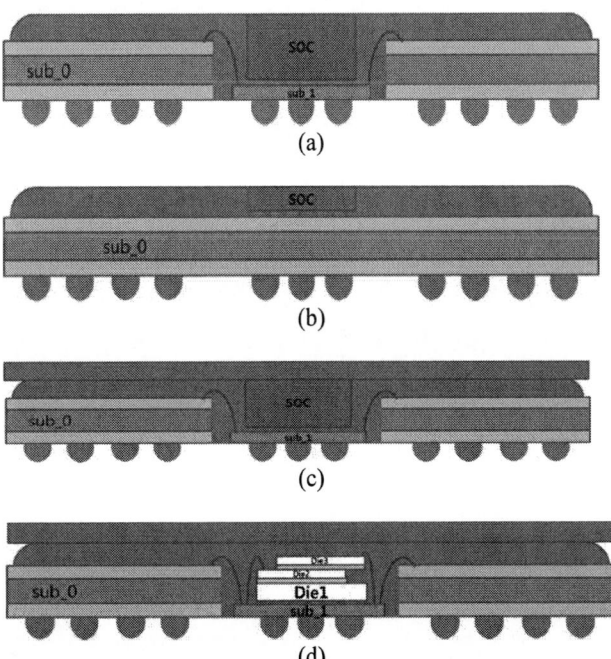

Figure 1. (a) The cross-section of FC-BGA based on hybrid structure with wire bonding, which is represented by hybrid_1 (H1) (b) conventional FC-BGA (c) the proposed hybrid structure called by H1 with heat-slug (d) SIP including multiple dies based on H1.

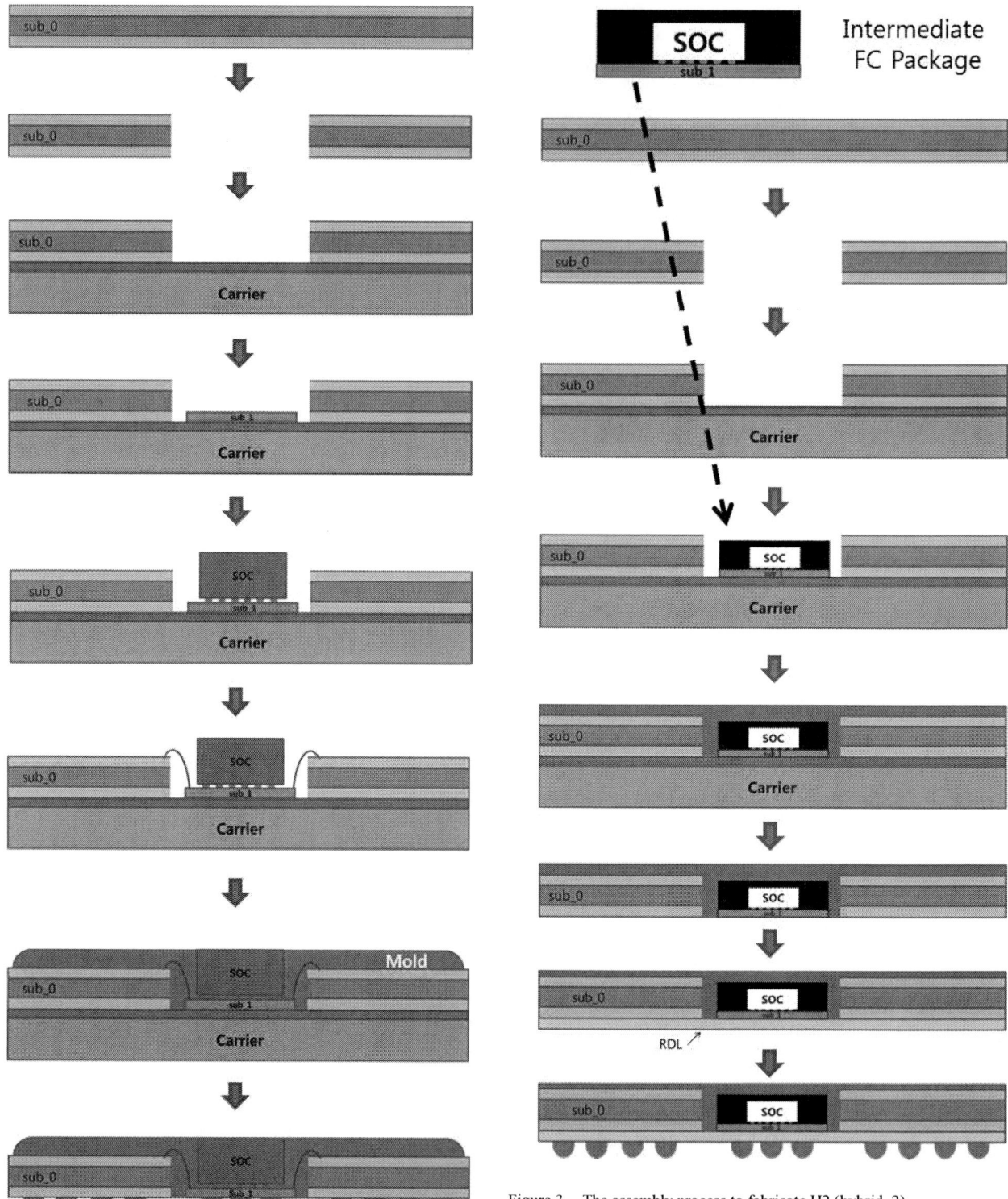

Figure 2. The assembly process to fabricate H1 (hybrid_1) as proposed.

Figure 3. The assembly process to fabricate H2 (hybrid_2).

978-1-7281-1500-9/19 $31.00 © 2019 IEEE

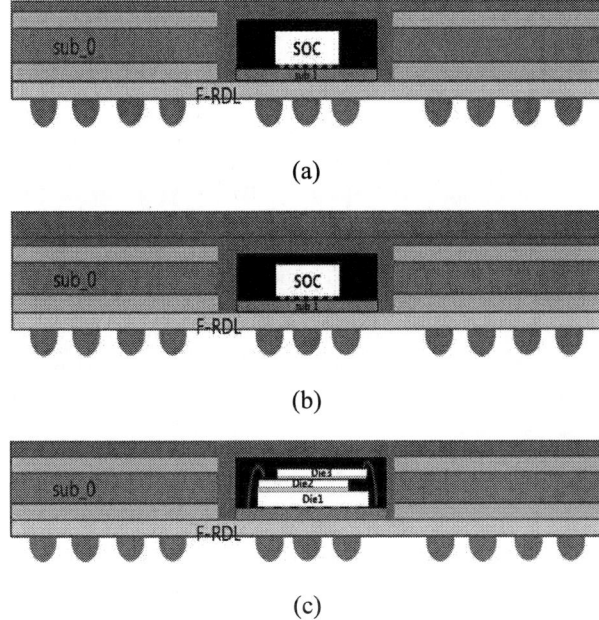

(a)

(b)

(c)

Figure 4. The cross-section of FC-BGA based on hybrid structure with RDL given by fan-out packaging technology. While (a) has no heat slug, (b) has heat slug attached on (a). As shown in (c), multiple dies can be integrated in the intermediate package even with wire bonding.

B. Hybrid_2 (H2) ; FC-BGA based on RDL of fan-out technology)

In this sub-section, hybrid_2 (H2) will be introduced. As shown in Fig. 3 and 4, H2-based FC-BGA includes thick-core substrate called by sub_0 and thin-core (or coreless) substrate called by sub_1 as presented for H1. While SOC die is mounted on sub_1 by proper FC bonding assembly process after placing sub_1 on the temporary carrier, SOC die should be bonded on sub_1 with separated FC assembly process. Since the SOC should be mounted on sub_1, the fine pitch interconnection to support fine pitch flip chip bump of SOC is needed only for sub_1. Surely, the major factor determining the design rule of sub_0 is the number of IOs of SOC.

In terms of interconnection between sub_0 and sub1, while bonding wires are used for H1, RDL is fabricated on sub_0 and sub1 to connect two or several substrate, where sub1 is an element of sub-package (or intermediate package) as shown in Fig. 3 and Fig. 4. Generally, RDL achieved by fan-out technology provides more IO rather than wire bonding. Fan-out packaging for implementing RDL can be FO-PLP (fan-out panel level package) introduced by [10]. It is expected that H2 can a proper package to support both IO count increase and die size shrink achieved by advanced silicon fabrication node with reasonable cost adder.

The entire assembly process of H2 shown in Fig.4 is presented by Fig.3. The first half of assembly process is almost same to that of H1. After a perforation is prepared on sub_0, sub_0 is placed on the rigid carrier. Intermediate package including SOC and sub_1 shown in Fig.3 should be fabricated before being placed on the temporary rigid carrier.

Encapsulating intermediate package and sub_0 on the temporary front-side carrier is needed to mechanically connect sub_0 and intermediate package before RDL process. After removing the front-side carrier, RDL layer is fabricated to interconnect sub_0 and sub_1. If necessary, additional multiple RDLs can be added to support the increased number of IOs. After completing RDL process, solder balls are attached. If necessary, metallic plate can be placed over mold for improving heat spreading or mechanical support, as presented by Fig. 3(b). As presented with H1 structure with Fig. 1 (d), multiple dies can be integrated as shown in Fig. 4 (c). While large FO-BGA introduced by [7] uses RDL to interconnect multiple silicon dies, RDL is used for interconnect sub_0 and the intermediate package in H2-based FC-BGA. Although the intermediate package in H2-based FC-BGA might be proper package types supporting single die or multiple dies, those study will not be included at this time.

III. THERMAL AND ELECTRICAL ANALYSIS

A. Thermal Perspective

Thermal resistance is presented for comparison of conventional FC-BGA and H1-based FC-BGA. As shown in Fig.5, the heat spreading capability of H1-based FC-BGA gives lower thermal resistance than the conventional FC-BGA. JEDEC still air thermal resistance, Rja on 4layer of system-level PCB was used by Ansys IcePAK. Heat source size is 3mm x 3mm to represent hot spot on SOC die in FC-BGA.

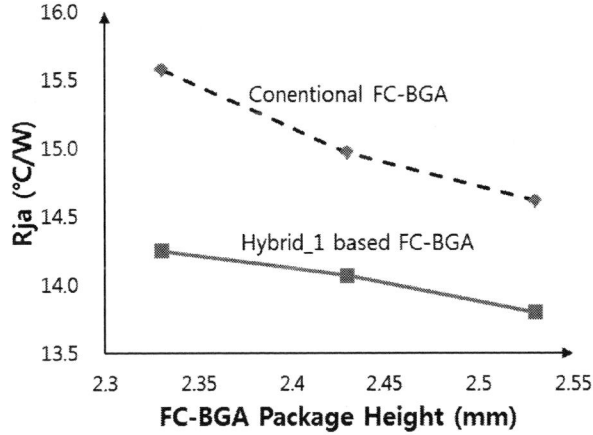

Figure 5. Thermal resistance of conventional FC-BGA and H1-based FC-BGA is compared.

We can explain lower thermal resistance of H1-based FC-BGA with thicker die and thinner substrate. First, for an identical package height, H1-based FC-BGA can have thicker die than conventional FC-BGA as shown in Table 1. Thick SOC die can spread heat to the horizontal direction of the die easily because silicon is good heat spreader itself. This reduces spreading thermal resistance when heat is concentrated in small region of SOC die. Second, H1-based FC-BGA can have thinner substrate than conventional FC-BGA. Thin substrate is able to reduce vertical thermal

978-1-7281-1500-9/19 $31.00 © 2019 IEEE

303

resistance because dielectric layer of organic substrate blocking heat conduction can be minimized. Thermal resistance can be reduced by 8.5% with H1-based FC-BGA having 2.33mm of package total height.

		Conventional FC-BGA	H1-based FC-BGA	
			Outer Area	Die-area
Heat Slug (Cu)		400um	400um	
TIM		50um	50um	
SOC die		100um		800um
Bump		80um		50um
Substrate	Sub_0	1.1mm (800um core)	1.1mm	
	Sub_1			400um
Solder ball		500um	500um	
PKG total height		2.33mm	2.33mm	
Rja		15.6°C/W	14.2°C/W (8.5% lower)	

Table 1. Vertical Structire used in the thermal analysis given by Fig.5.

B. Power Delivery Perspective

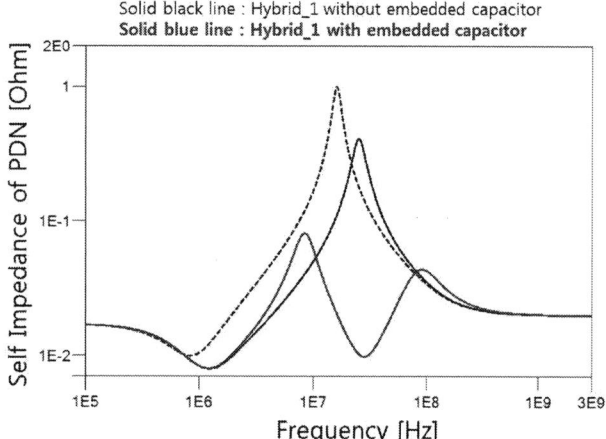

Figure 6. Self impedance of PDN (power delivery network) of conventional FC-BGA and H1-based FC-BGA is compared. With FC-BGA based on H1, PDN of two cases are presented. One has embedded capacitor in sub_1, whose self impedance is presented by solid blue line. The other has no embedded capacitor in sub_1, whose self impedance is presented by solid black line.

Fig.6 presents the power delivery characteristics of three cases including one conventional FC-BGA and two H1-based FC-BGA packages. As given, H1-based FC-BGA employing the thin substrate presented by sub_1 for PDN design of core circuits gives lower impedance of PDN than the conventional FC-BGA using the thick substrate presented by sub_0. In addition, sub_1 can easily include embedded capacitor to reduce the self-impedance of PDN, which has been popular for SOC package used in mobile phone. As presented in Fig.6, addition of embedded capacitor in thin substrate presented by sub_1 can be promising solution for enhancing the performance of PDN.

IV. APPLICATION TO POP

In this section, several examples of POP (package on package) with FC-BGA based on H1 or H2 will be presented. As shown in Fig. 7, H1-based FC-BGA is given for the base package of POP. Multiple DRAM package can be mounted on the top side of sub_0 in H1-based FC-BGA. While the cross-section of POP with H1-based FC-BGA is presented by Fig.7, the top views for two case studies are given in Fig. 8. In two case studies given by Fig.8, two different LPDDR4 DRAM packages providing 2 channels and 4 channels, respectively, are used.

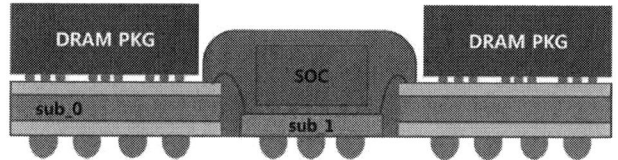

Figure 7. The cross-section of H1-based FC-BGA is presented. The presented POP (package on package) with FC-BGA has multiple DRAM packages on the top side of sub_0.

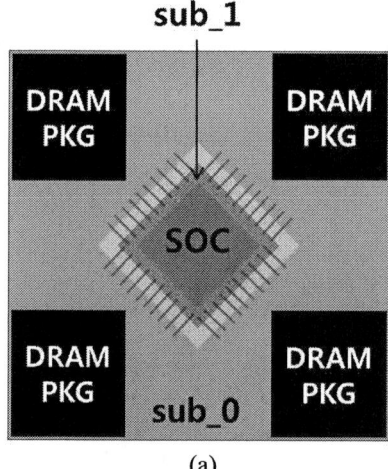

(a)

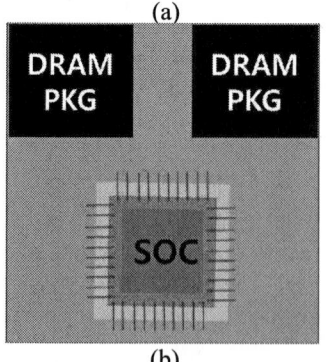

(b)

Figure 8. Top views of H1-based FC-BGA for (a) 40mm x 40mm POP with 4 DRAM packages and (b) 35mm x 35mm POP with 2 DRAM packages are given. The size of FC-BGA, SOC die size, and DRAM package size will be given by table.2

978-1-7281-1500-9/19 $31.00 © 2019 IEEE

	POP in Fig.8 (a)	POP in Fig.8 (b)
SOC die Size	9mm x 9mm	9mm x 9mm
DRAM PKG size	11.5mm x 13mm	12.4mm x 12.4mm
Number of DRAM PKG	4	2
Number of channel in SOC for LPDDR4	8	8
Number of DQs	128	128
POP Size	40mm x 40mm	35mm x 35mm

Table 2. Information of POP presented in Fig.8

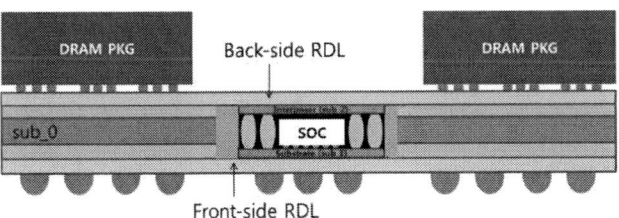

Figure 9. The cross-section of H2-based FC-BGA is presented. The presented POP (package on package) with FC-BGA has multiple DRAM packages on the top side of sub_0.

SOC PKG type for POP		H1-based FC-BGA given by Fig.7	H2-based FC-BGA given by Fig.9
	IO clock frequency		
Margin in Eye-open size	2.75GHz	+9% point	+11% point
	3.20GHz	+2% point	+5%point

Table 3. Comparison of signal integrity analysis result with two FC-BGA package types. Margin in eye-open size represents the difference betwwen minimum eye open size minus and required specification in SI simulation.

In addition to H1-based FC-BGA for POP, Fig.9 shows H2-based FC-BGA, in which I-POP (interposer-POP) introduced by [10] can be used for intermediate package. For implementing POP with H2-based FC-BGA, back-side RDL can be added as shown in Fig.9. Through back-side RDL, signal lines for LPDDR can be implemented to achieve proper SI performance.

The SI (signal integrity)-perspective comparison is presented in the following table 3 for H1-based FC-BGA and H2-based FC-BGA presented by Fig.7 and Fig.9, respectively. In this analysis, LPDDR4 DRAM package is used. Although the maximum IO frequency provided by LPDDR4 is lower than 2.7 GHz, we use 2.75GHz and 3.2GHz for the extended application beyond LPDDR4 in the presented simulation study. As expected, H2-based FC-BGA utilizing RDL instead of bonding wire used in H1-based FC-

BGA has better SI with the increase of eye open size. Since the difference might be acceptable with some applications, H1-based FC-BGA can be used for cost-effective POP. In the application requiring more margin and stability, H2-based FC-BGA would be proper with reasonable cost adder.

V. CONCLUSION

A new approach to large size FC-BGA with hybrid packaging technologies has been presented, which is expected to be an attractive package solution for SOC (system on a chip) in several consumer applications including D-TV. The proposed new approach is proper to enhance power integrity and thermal spreading and to reduce the FC-BGA package height with keeping and not degrading the mechanical property of conventional FC-BGA.

REFERENCES

[1] F. Tung, C. Chen, M.Lu, J. Tsai, and Y.-P. Wang, "Challenges of Large Body FCBGA on Board Level Assembly and Reliability," Proc. IEEE 68th Electronic Components and Technology Conference (ECTC), 2018, pp. 1962 – 1967

[2] F. Tung, M. Lu, A. Lan, and S. Pan, "Assembly Challenges for 75×75mm Large Body FCBGA with Emerging High Thermal Interface Material (TIM)", Proc. IEEE 67th Electronic Components and Technology Conference (ECTC), 2017, pp. 130-135

[3] P. H. Tsao, Steven Hsu, Y.L. Kuo, J.H. Chen, A. Chang, H. P. Pu, L.H. Chu, and M.J. Lii, "Cu Bump Flip Chip Package Reliability on 28nm Technology", Proc. IEEE 66th Electronic Components and Technology Conference (ECTC), 2016, pp. 1148-1153

[4] C.-Y. Huang, C.-C. Wang, T.-L. Hsieh, and C.-Y. Tsai, "Design and electrical performance analysis on coreless flip chip BGA substrate," Proc. IEEE Electrical Design of Advanced Packaging and Systems Symposium (EDAPS), 2017, pp.1 - 3

[5] G. W. Kim, J. H. Yu, C. W. Park, S. J. Hong, J. Y. Kim, G. Rinne, and C. Lee, "Evaluation and verification of enhanced electrical performance of advanced coreless flip-chip BGA package with warpage measurement data," Proc. IEEE 62nd Electronic Components and Technology Conference (ECTC), 2012, pp.897 – 903

[6] G.W. Kim, S. J. Lee, J. H. Yu, G. I. Jung, J. Y. Kim, N. Karim, H. Y. Yoo, and C. Lee, "Advanced coreless flip-chip BGA package with high dielectric constant thin film embedded decoupling capacitor," Proc. IEEE 61st Electronic Components and Technology Conference (ECTC), 2011, pp. 595 - 600

[7] C. K. Yu, W.S. Chiang, P.S. Huang, M.Z. Lin, Y.H. Fang, M.J. Lin, Cooper Peng, Benson Lin, and Michael Huang, "Reliability Study of Large Fan-Out BGA Solution on FinFET Process," Proc. IEEE 68th Electronic Components and Technology Conference (ECTC), 2018, pp.1623 - 1627

[8] G.W. Kim, M. S. Min, M. L. Yang, A. Gundurao, E. You, H. Gill, S. Cha, Y. Kim, S. You, S. Lee, and W. Ryu, "Package embedded decoupling capacitor impact on core power delivery network for ARM SoC application," Proc. IEEE 64th Electronic Components and Technology Conference (ECTC), 2014, pp. 354 – 359

[9] H. Lee, Y.-S. Choi, E. Song, K. Choi, T. Cho, and S. Kang, "Power Delivery Network Design for 3D SIP Integrated over Silicon Interposer Platform," Proc. IEEE 57th Electronic Components and Technology Conference (ECTC), 2007, pp. 1193 – 1198.

[10] T. Hwang, D. Oh, E. Song, K. Kim, J. Kim, and S. Lee, "Study of Advanced Fan-Out Packages for Mobile Applications" Proc. IEEE 68th Electronic Components and Technology Conference (ECTC), 2018, pp. 343 - 348

978-1-7281-1500-9/19 $31.00 © 2019 IEEE

Package-on-Package micro-BGA microstructure interaction with bond and assembly parameters

Pascale Gagnon, Clément Fortin
IBM Canada Limited
Bromont, Québec, Canada
e-mail: pgagnon@ca.ibm.com

Thomas Weiss
IBM Systems
Hopewell Junction, NY 12533, US
e-mail: tomweiss@us.ibm.com

Abstract— **Package on Package (PoP) technology can play a vital role in advanced packaging for various reasons. The use of a thin interposer allows for tighter wiring ground rules and can reduce die-to-die latency when placing either multiple die on a single interposer or placing multiple interposers on a larger carrier laminate. This assembly technology can also be of benefit for heterogeneous integration applications.**

The main challenge for bond and assembly process development with organic to silicon bonding, particularly with thin organic interposers, is to overcome solder join defects (non-wets, bridges) created by highly warped components. To address this issue, two different assembly process flows were developed, namely process A (chip first) and process B (laminate first) with PoP type test vehicle hardware. The assembly was composed of a 23X29mm die with 185μm pitch SAC solder connected to an organic interposer of 37.5X37.5mm dimension and 0.7mm thickness. The bottom side of the interposer, with either CuOSP or NiPdAu pads, was soldered to a 68.5X68.5mm carrier laminate of 1.8mm thickness and Cu pads with SAC 305 μBGA solder balls of 0.4mm pitch.

To ensure bond and assembly integrity, over 30 cross-sections were performed on the modules from these two process sequences. μBGA's were of first interest as warpage issues forced a thorough optimisation of joining parameters. Microstructure appears with undeniable variations between process sequences and interposer pad surface finishes. Cross-polarized microscopy observation and etching of cross-sections, with over 2500 μBGA characterized, showed that process flow A microstructure was mostly constituted by very large and few grains of beta-Sn phase compared to mainly small interlaced grains for process flow B. IMC's with "chip first" (process flow A) assembly were composed of rare AgSn crystals and several CuSn filaments, which were also observed on "laminate first" (process flow B), but only with NiPdAu interposer pads. Also, for process sequence B with NiPdAu pads, thick plates of AgSn were observed, as opposed to very thin sheets with CuOSP pads.

Interim cross-section in the bond and assembly process flows allowed the formulation and verification of a hypothesis on the mechanism that leads to a "laminate first" atypical microstructure. Cooling rates during reflow and interposer pad surface finish will change all the kinetics of intermetallic nucleation and produce eccentric growth. A thorough mechanism understanding will allow a proper process

parameter intervention to modulate a desired microstructure from a reliability stand point.

Keywords; Package-on-package; PoP; interposer; micro-BGA; microstructure; Cu_6Sn_5; Ag_3Sn

I. INTRODUCTION

Advanced packaging has taken a major space in process development these last few years. As Moore's Law gets closer and closer to approaching its limits, industry leaders are looking more and more to advanced packaging as a way to continuously improve performance. The desire to minimize the distance between components within a package as well as the desire to shrink the die size, thus enhancing yield, and putting more of them on the same package has led to several new packaging techniques. Package-on-package (PoP) is one such technique that is being studied in this paper. The assembly of a PoP implies many challenges in terms of component warpage harmonizing. Coefficient of thermal expansion (CTE) mismatch between organic carrier laminate, silicon die and epoxy underfill creates a mechanical puzzle to be resolved with joining integrity and residual stresses in mind. μBGA connections between a thin interposer, on which a die is joined with C4s, and a carrier laminate will be subjected to high stresses during their operational life. Multiple reflows during the assembly steps will affect the microstructure of those μBGAs [12] and potentially alter their reliability performance. Characterization and comprehension of μBGA microstructure interaction with process flow development of PoP is therefore of first importance and will be the object of this study, performed in a real production environment.

II. EXPERIMENTAL

Two process flows are studied in this experiment. The first one namely "process A" is a "die first" process, represented in Fig. 1. Warpage related problems during the process development of the first process flow led to the development of a second process flow namely "process B" also described in Fig. 1. μBGA microstructure from process B was noticed to be largely different than what was obtained with process A during first routine verification.

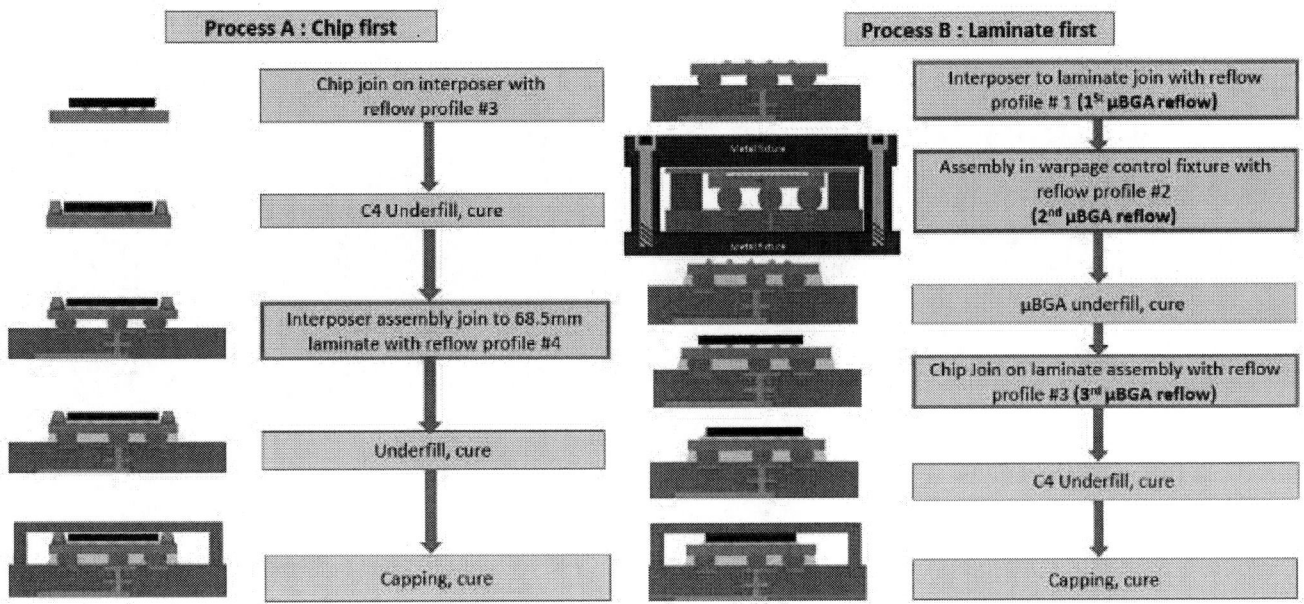

Figure 1. PoP process flow A (chip first) and process flow B (laminate first).

This first observation prompted a vast comparison study of β-Sn grains and intermetallics (IMC). After assessing the deltas between both processes at the end of assembly, interim microstructure was studied to allow a better understanding of the microstructure between process B's multiple reflows.

Both PoP studied processes are described in Fig. 1 to better understand what assembly steps μBGA are exposed to. Process A starts by joining the chip C4's onto the organic interposer in a furnace reflow. A stiffener is attached to reduce the warpage of the sub-assembly for the subsequent joining on the carrier laminate with μBGA connections during a second reflow in a furnace, followed by the underfilling of the μBGA's.

Process B begins by directly joining the bare interposer onto the carrier laminate in a furnace reflow. Due to the variable incoming interposer warpage, this first reflow would not suffice to obtain 100% good joining. A second reflow in a custom-made fixture for warpage control was instituted to heal partial wets and opens. μBGA's were then underfilled before a third reflow occurred on the package with the objective of joining the die C4's on a nicely flat laminate-interposer sub-assembly.

8400 μBGA SAC305 solder balls between the 0.7mm thin organic interposer (37.5X37.5mm) and the 1.8mm thick carrier laminate (68.5X68.5mm) are joined with two different interposer pad surface finishes: NiPdAu and CuOSP. NiPdAu is used for both processes A and B and CuOSP is used only on process B parts studied in this experiment. The pitch of these μBGAs is 0.4mm while their diameter is 0.2mm. The die of 23X29X0.785 mm dimensions is joined with SAC solder C4s with 185μm pitch.

Process A anticipated warpage mismatch between the interposer and the carrier laminate after the chip underfilling came out much higher than expected. An average value of 200μm on the 37.5X37.5mm bottom surface resulted in a

major open connection problem for μBGAs. Not only is the warpage at room temperature higher than desired, the thermal warpage of the interposer/die subassembly during subsequent joining to the carrier laminate is an issue as well. Fig. 2 shows the digital image correlation (DIC) curve of two sub-assembly samples constituted of a die underfilled on the interposer with a perimeter stiffener. The highly convex (-) room temperature warpage translates into a concave shape (+) warpage at reflow temperature, preventing the peripheral μBGAs from joining the carrier laminate, thus creating opens all around the edges.

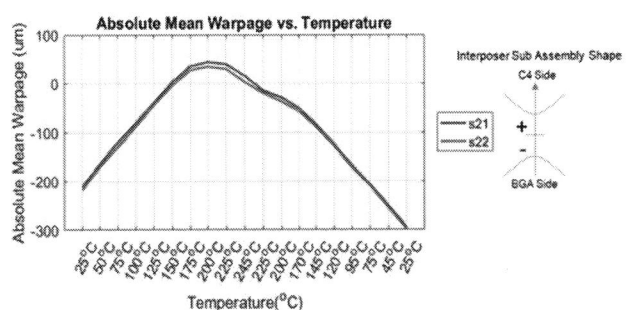

Figure 2. Digital image correlation curve of two sub-assembly interposer samples.

The assembly flow B was developed to resolve the open issue. With this new method, μBGAs see three reflows to complete the PoP. Fig. 3 shows and compares the reflow profiles that take place during both process A and B. Process A PoP uses reflow profiles #3 and #4, but μBGAs will only see profile #4. Process B uses, in order, profile #1, #2 and #3. All cooling rates are well under 1°C/s, but profile #2 has the slowest with 0.24°C/s due the high thermal mass of the warpage control fixture.

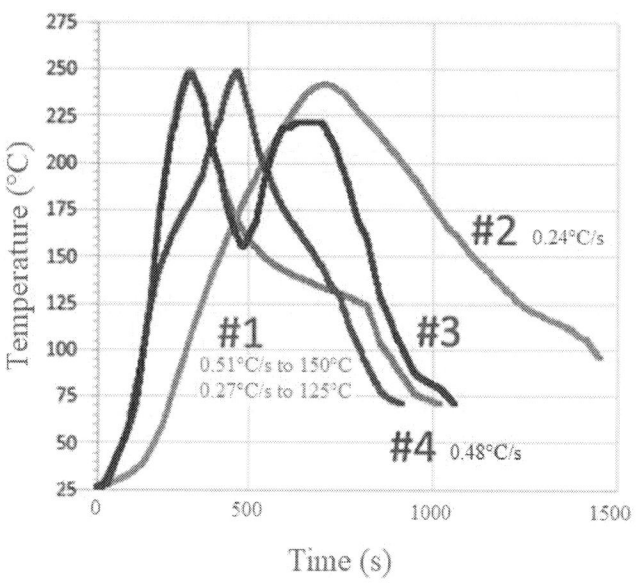

Figure 3. Reflow profile curves for process A and process B joining, performed in nitrogen furnace.

III. RESULTS

A comparison of the microstructure between process A and B on 16 modules was performed along with 38 cross-sections. The observations of more than 2700 µBGA have been analyzed to assess joining integrity and understand the potential reliability impact of a process flow modification.

Fig. 4 shows β-Sn grains of process A µBGA that went through one reflow profile with a cooling rate of 0.48°C/min (Fig. 3, profile #4). All the µBGA observed are made up with very few and large grains, sometimes of the "beach ball" type. NiPdAu interposer pads were joined to SAC305 solder balls on the carrier laminate for process A.

Figure 4. 200X cross-polarized light µBGA's typical β-Sn grains of process A on NiPdAu interposer pads after one reflow profile (Fig.3, #4).

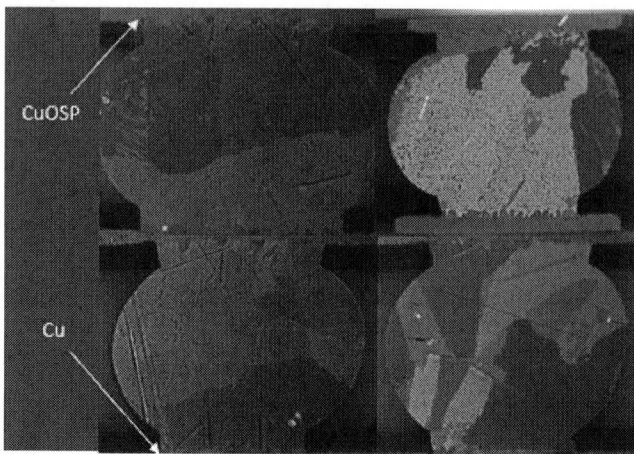

Figure 5. 200X cross-polarized light µBGA's β-Sn grains of process B on CuOSP interposer pads after 3 reflows (Fig. 3, #1, #2, #3).

Cross-sectioned µBGA's of process B parts that were assembled with CuOSP surface finish on the interposer bottom side are presented in Fig. 5. β-Sn shows a much different grain aspect for these three times reflowed parts. Beach ball crystals are still present but with a higher count and multiple orientations. At least 50% of µBGA observed include three to six grains while the remainder shows only one or two.

With process B and NiPdAu interposer pads displayed in Fig. 6, more than half of µBGA consist in a mix of very small interlaced grains and bigger crystals with at least four different orientations. We do still find a small proportion of µBGA showing only few grains, but we see a clear change in microstructure with Ni compared to Cu. A high volume of solder (compared to a C4) and a high Ag content tend to go in the opposite direction of promoting interlaced grains [3]. Also, the presence of interlaced crystals indicates a lower solidification temperature [3]. Therefore, the presence of interlaced grains in 200µm diameter µBGA with 3%Ag

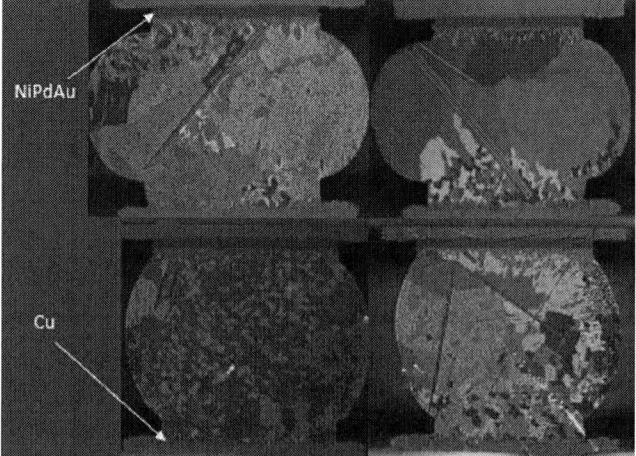

suggests that the undercool phenomena was very severe.

Figure 6. 200X cross-polarized light µBGA's β-Sn grains of process B on NiPdAu interposer pads after 3 reflows (Fig.3, #1, #2, #3).

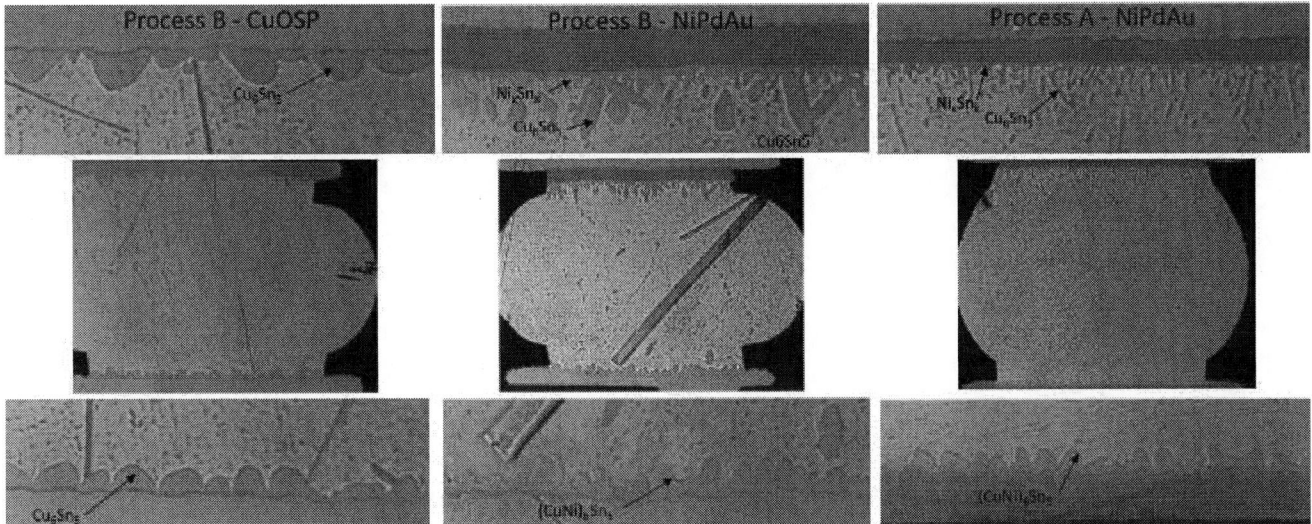

Figure 7.   200X white light µBGA's closer look at solder interfaces.

Fig.7 offers a closer look at carrier laminate and interposer interfaces.   Very smooth and large scallops of Cu_6Sn_5 intermetallic are observed on both Cu sides for process B CuOSP parts after three reflows.   Thin Ag_3Sn are also emerging from the Cu_6Sn_5 surface.   With a Ni top joining surface, process B – NiPdAu, $(CuNi)_6Sn_5$ is forming from the Cu bottom side with smaller scallop shaped intermetallics [12].   At the Ni interface, after having gone through three reflows, Ni_xSn_x IMC is forming, from which Cu_6Sn_5 with a large needle appearance is growing.   For µBGA formed with a one reflow process (process A), the same $(CuNi)_6Sn_5$ small scallops are created at the Cu interface as with the three reflow process, but very fine needles occupy the Ni top surface.

Fig. 8 goes deeper demonstrating Ni_xSn_x IMC at Ni interface on etched µBGAs, to be covered with Cu_6Sn_5 IMC that grow with subsequent reflows.

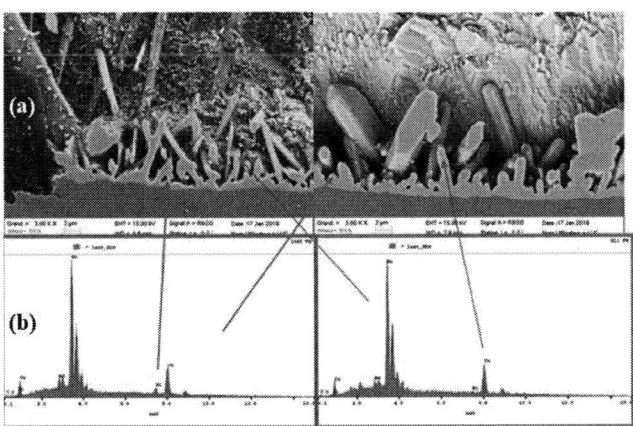

Figure 8.   Process B NiPdAu etched µBGA, (a) 3000X scanning electro microscopy (SEM) and (b) Energy dispersive X-ray spectrometry (EDS) after one reflow (left) and three reflows (right).

In order to further observe the microstructure differences between assembly processes and interposer pad metallurgy, cross-sectioned µBGA have been etched in a HNO_3 + HCl

solution in methanol as shown in Fig. 9.   Etching revealed intermetallic differences between one and three reflow processes as well as between CuOSP and NiPdAu interposer pads. Row a) shows that process A µBGA's contain small and scarce Ag-Sn. Row b) µBGA's from process B with NiPdAu hold huge and very thick Ag-Sn and several Cu-Sn filaments. With µBGA formed from process B with CuOSP on row c), the etching operation exposed large but thin Ag-Sn sheets. Surprisingly, these last µBGA contained no filament shape Cu-Sn.

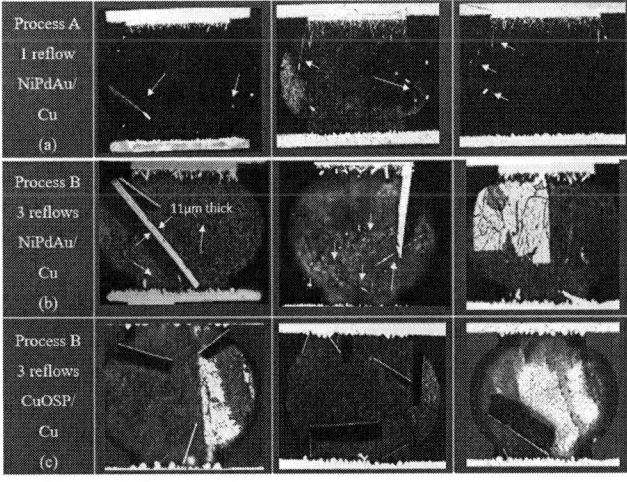

Figure 9.   SEM images of etched µBGA 2-5min. in 5% HNO_3 + 2% HCl solution in methanol : (a) process A, one reflow, NiPdAu/NiPdAu, (b) process B, three reflows, NiPdAu/NiPdAu, (c) process B, three reflows, CuOSP/NiPdAu.

To verify the observation summary done on etched parts, a meticulous survey was performed whose results are compiled in Table 1. 1388 µBGA were examined to inventory Ag_3Sn plates and sheets and Cu_6Sn_5 wires. Fig. 10 shows an example of sheet compilation on a 200X optical image. Cross-

978-1-7281-1500-9/19 $31.00 © 2019 IEEE   309

sections were performed not only after complete assembly, but also after the second reflow for process B, to verify the microstructure obtained after the new slower cooling reflow profile developed for the warpage control fixture. The minimum length for a plate or sheet to be considered was 30µm. Wires were much less prevalent, so they were accounted for under a more general criterion.

Statistically, from post polishing cross-sections, we can conclude that Ag_3Sn plates or sheets are more abundant under same conditions (process B, three reflows) with CuOSP than with NiPdAu pad. This same conclusion applies for more reflows with the same surface finish. For instance, Table 1 shows that an average of 1.6 to 2.5 sheets or plates per µBGA are present after two reflows with CuOSP pads, while a µBGA that went through three reflows contains 3.6 to 4.7 Ag_3Sn IMC. Rare Ag_3Sn are registered with NiPdAu, although statistically more after three reflows with process B than after one reflow with process A.

Cu_6Sn_5 were also observed in some µBGA, although scarcer. Unexpectedly, rod shaped Cu_6Sn_5 are observed on the CuOSP parts, but mostly only after the second reflow. After the third reflow, only traces of them are left. Moreover, they are always more prevalent in the center of the part (N-S cross-section area).

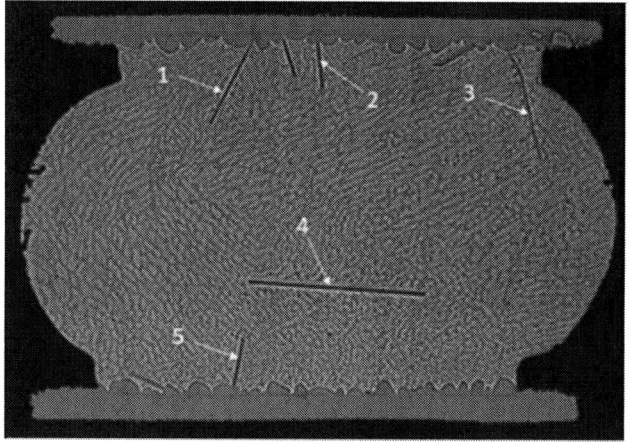

Figure 10. Optical 200X image of Process B with CuOSP µBGA showing Ag_3Sn sheet quantity.

Figure 11. Summary table of µBGA microstructure features for all study's findings.

Process	Interposer pad surface finish	Reflow number	β-Sn grains	IMC/Etch	Cu_6Sn_5 at interposer pad	Ag_3Sn	Cu_6Sn_5	Diagram
A	NiPdAu	1				Very few thin sheets	Wires	
B	CuOSP	2				Thin sheets	Big rods	
B	CuOSP	3				Lots of thin sheets	Few small rods	
B	NiPdAu	3				Few large plates	Wires	

TABLE I.   COMPILED OBSERVATIONS OF Ag_3Sn IMC ON CROSS-SECTIONED µBGA POST POLISHING WITH WHITE LIGHT 200X OPTICAL INSPECTION

a. No rod

b. No rod and lots of wires

c. Less than one rod per ten µBGA

d. One rod every µBGA

Subsequent part assembly to replicate process B with CuOSP pads was performed using a faster cooling rate 3rd reflow (0.48°C/s vs 0.24°C/s). The same observations compilation as in Table 1 resulted in a slightly lower amount of Ag_3Sn sheets than for the slower cooling rate with 3.5 sheet/µBGA, on 94 µBGA inspected. Fig. 11 shows a summary of this study's findings.

IV. DISCUSSION

Intermetallic growth of Ag_3Sn plates in a solder joint can affect reliability as crack propagation along interfaces between Ag_3Sn plate and β-Sn phase are reported during accelerated thermal cycling [2]. Process A, process B – NiPdAu and process B – CuOSP gave very different results in terms of Ag_3Sn IMC size and distribution. Process A has a more promising microstructure in terms of reliability but yielded as low as 0% thru bond & assembly depending on laminate lot and individual interposer warpage. The following questions arise: with the same solder composition, volume and interfaces (for process A vs process B-NiPdAu), why is the microstructure so different and why, even after the same thermal excursions, did process B parts with different surface finish not produce the same result? More specifically, why are so many Ag_3Sn sheets observed?

Firstly, process A uses the faster cooling rate profile (Fig. 4, #4) and the delay between the solidus and the undercool solidification is relatively short compared to the other process B profiles. Therefore, a literature review [6, 8] allows one to conclude that there is not enough time for nucleated Ag_3Sn to grow into large sheet or plate or for Cu_6Sn_5 needle to develop into rods.

Also, during the 2nd reflow which occurs in a high thermal mass fixture for warpage control, the solder cools at a very slow rate, 0.24°C/s (Fig. 4, profile #2). As liquid solder cools down, there is a nucleation multiplication of Ag_3Sn at Cu_6Sn_5 scallop surfaces, up to a point where the Ag_3Sn nuclei cover all the Cu_6Sn_5 intermetallic layer over the CuOSP pad [8, 10, 11]. From this Ag_3Sn nuclei coating, Ag_3Sn sheets are initiated and free to grow until β-Sn solidification [1]. Ag_3Sn nuclei barrier has also a shielding effect that obstructs the organization of Cu_6Sn_5 based rods and wires [11] in this high Cu content solder, which also increases undercool [7]. The large amount of thin Ag_3Sn sheets is revealed by Fig. 12 with an etched µBGA from process B after 3 reflows.

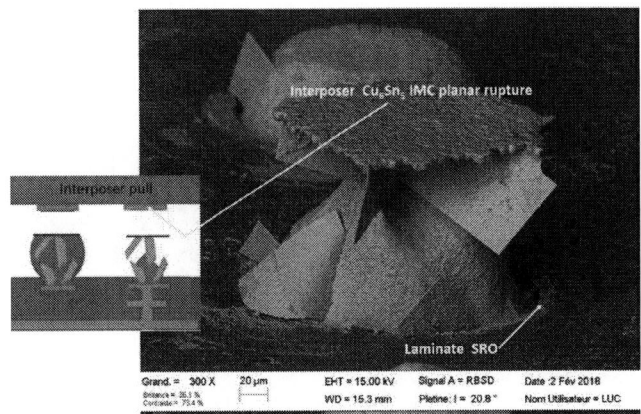

Figure 12. SEM images of pulled interposer and etched µBGA 2-5min. in 5% HNO3 + 2% HCl solution in methanol for process B CuOSP part.

Fig. 13 shows also an etched µBGA that enables the 3D representation of a typical process B on CuOSP µBGA. It confirms that thin Ag_3Sn sheets are present in a large amount. An interesting observation that β-Sn dendrites boundaries are also found directly on the sheets, confirms their post solidification.

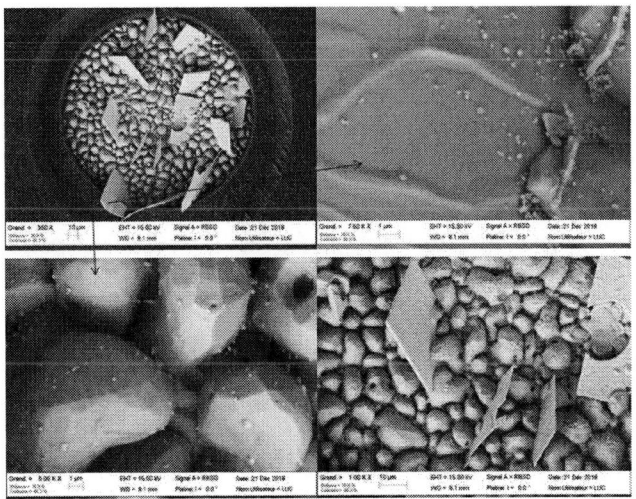

Figure 13. SEM images of pulled interposer and etched µBGA 2-5min. in 5% HNO3 + 2% HCl solution in methanol for process B CuOSP part.

Furthermore, assemblies with process flow B using NiPdAu organic interposer pads generate µBGA thick Ag_3Sn plates, up to 11µm thickness, and Cu_6Sn_5 wires. Based on several publications [8, 9, 10, 11, 12], the presence of Ni and Au change all the kinetics of intermetallic nucleation by the introduction of Ni atoms in the Cu_6Sn_5 IMC from the scallop layers to form $(CuNi)_6Sn_5$. This Ni biased compound affects the Ag_3Sn nuclei initiation and thus limits Ag_3Sn sheet formation [12]. Consequently, Ag atoms are constrained to solidify on very few Ag_3Sn crystals, which will grow substantially thicker and longer in a very rich Ag content alloy. Moreover, β-Sn interlaced grains in Fig. 7 and reference [3] suggest that the solder will not yet solidify during the valley around 150°C of the 3rd reflow. Therefore, it allows eons of time for the Ag3Sn plates to grow.

Fig. 14 shows SEM images for NiPdAu µBGA and Fig. 15 shows EDS analysis of thick Ag3Sn plates, which are found in much smaller quantities than in CuOSP µBGA. β-Sn grain boundary is observed on the plate, leaving SnCuAu small precipitates along them. SnCuAu are also analyzed by EDS on Fig. 15. On every side of this boundary, different β-Sn grain orientation is visible as shown on the lower right image of Fig. 14.

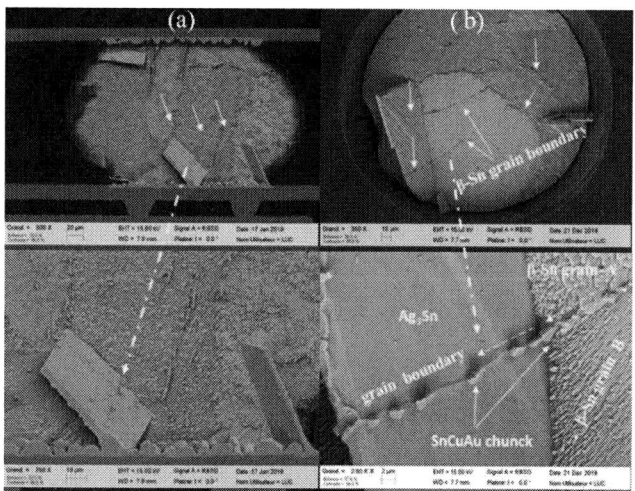

Figure 14. SEM images of (a) cross-section and (b) Z-section of etched µBGA 2-5min. in 5% HNO₃ + 2% HCl solution in methanol for process B NiPdAu part.

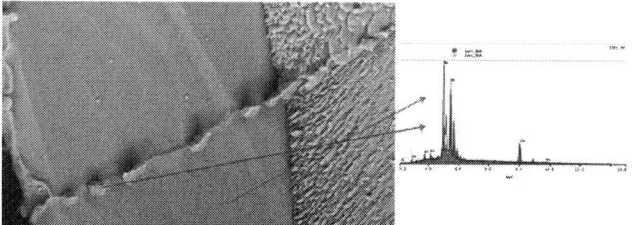

Figure 15. EDS analysis of Ag3Sn plate and SnCuAu precipitates.

To shed more light on the µBGA microstructure after three reflows, interim cross-sections have been performed. Fig. 16 is showing process B CuOSP µBGAs after a 2nd reflow. It is the first and only time that Cu₆Sn₅ in such big rods shape are observed in this experiment. The confirmation of the mechanism that explains why big rods are found after two reflows but not after three remains to be proven, but the following hypothesis can be laid out. After the first reflow, Cu consumption happens at both pad interfaces, producing already a copper sursaturation and a lot of Cu₆Sn₅ readily available when the 2nd reflow occurs. Profile #2 on Fig. 3 gives a very long and unusual time above 217°C were even more Cu will be consumed from the pad interfaces. Then will begin an ultra-slow cooling of the β-Sn at 0.27°C/s with whom Cu₆Sn₅ rods will have all the time and Cu needed to grow in a very high undercool situation. During the 3rd reflow with profile #3 (Fig. 3), Cu₆Sn₅ rods are dissolved while the Cu pads continues to be consumed, introducing again more Cu in the solder. Therefore, a completely new dynamic takes place during the solidification process preventing from creating new big Cu₆Sn₅ rods. Ag3Sn nuclei over scalloped Cu₆Sn₅ could impede the start of such big rods. Another alternate explanation would be that the 3rd reflow happens in a higher-pressure solder, being trapped into the underfill at this point. The pressure could affect the solidification process, but also could limit the temperature delta between the top and the bottom of the µBGA. Therefore, it could push even lower the solidification temperature.

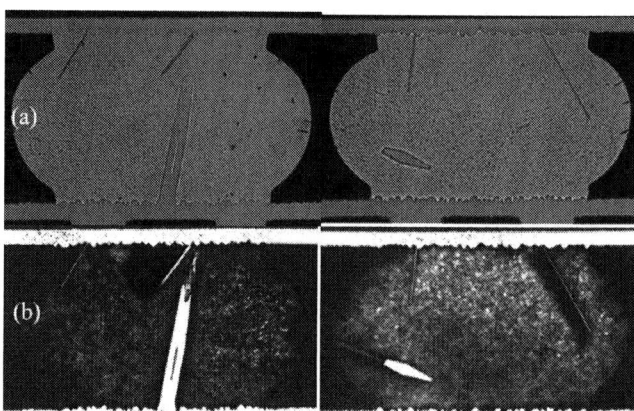

Figure 16. Process B CuOSP 100X optical image after (a) polishing and (b) etching 2-5min. in 5% HNO₃ + 2% HCl solution in methanol

To improve the reliability success potential of process B, Ag3Sn plates need to be controlled to a much lower level. With the experiments performed, the characterizations and observations of a large population of µBGA as well as with the review of literature, the following process modifications need to be applied : a) reduction of the Ag content in the µBGA, b) augmentation of the cooling rate of the 2nd reflow by reducing the thermal mass of the warpage control fixture as well as the cooling rate of the 3rd reflow and c) reduction of the time above solidification temperature in the 3rd reflow by lowering the valley temperature under the solidification temperature.

V. CONCLUSION

This study has compared the μBGA microstructure of a PoP chip first assembly (process A) with a laminate first assembly (process B), the latter being also compared between NiPdAu and CuOSP interposer pad interface. The following can be concluded:

a) Multiple reflows, high Ag content, slow cooling rate and high undercool situation all contributes to the formation of large and abundant Ag_3Sn plates or sheets intermetallics that can have negative reliability impact [2].

b) With CuOSP pads, there is paradoxically nearly no Cu_6Sn_5 wires in the μBGAs compared to NiPdAu. A shielding effect is created by Ag_3Sn nuclei during a first slow cool down, preventing the wires to grow, but allowing a large Ag_3Sn sheet amount to form.

c) With NiPdAu pads, the introduction of Ni creates a new IMC compound $(CuNi)_6Sn_5$ over the Cu_6Sn_5 scalopped IMC at pad interfaces, affecting the formation of Ag_3Sn nuclei [12], thus limiting the number of Ag_3Sn sheets and allowing the few of them to grow into very thick plates.

d) The presence of very large Cu_6Sn_5 rods after two reflows but not after three could be caused by the new solidification conditions created after three reflows such as a higher Cu content, a higher undercool situation, more Ag_3Sn nuclei creating a shielding effect and/or a pressurized solder caused by the underfilled environnement. More study need to be pursued to narrow down and confirm an hypothesis on the Cu_6Sn_5 rod shaped disparition after three reflows.

ACKNOWLEDGMENT

The authors want to thank the following persons for their participation and support in this study: Geneviève Beaulieu, Marcel Charest, Josée Fontaine, Nathalie Matte, Julien Paré and Luc Patry.

REFERENCES

[1] S. K. Kang, P. A. Lauro, D.-Y. Shih, D. W. Henderson, and K. J. Puttlitz, "Microstructure and mechanical properties of lead-free solders and solder joints used in microelectronic applications," IBM J. Res. & Dev. Vol. 49 No.4/5 July/September 2005.

[2] S. K. Kang, W. K. Choi, D.-Y. Shih, D. W. Henderson, T. Gosselin, A. Sarkhel, C. Goldsmith and K. J. Puttlitz, "Formation of Ag_3Sn plates in Sn-Ag-Cu alloys and optimization of their alloy composition," proc. IEEE, ECTC 2003.

[3] G. Park, B. Arfaei, M. Benedict, E. Cotts, M. Lu, and E. Perfecto, "The dependence of the Sn grain structure of Pb-free solder joints on composition and geometry," 62nd ECTC conference, IEEE, 2012.

[4] S.-K. Seo, S. K. Kang, M. G. Cho, D.-Y. Shih, and H. M. Lee, "The microstructure and crystal orientation of Sn-Ag ans Sn-Cu solders affected by their interfacial reactions with Cu and Ni(P)," IBM research report RC24757 (W0903-021), Electrical Engineering, March 6, 2009

[5] H.-T. ee, Y.-F. Chen, T.-F. Hong, and K.-T. Shih, "Effect of cooling rate on Ag_3S formation in Sn-Ag based lead-free solder," 11th EPTC conference, IEEE, 2009.

[6] C. P. Lin, K. Wan, T. Sun, C.-M. Chen, and R. Lee, "Effects of reflow cooling rate on the growth of Ag_3Sn platelets and deformation of solder balls," 66th ECTC conference, IEEE, 2016.

[7] X.-P. Li, J.-M. Xia, H.-B. Qin, X.-Q. He, and X.-P. Zhang, "Study of critical factors influencing the solidification of undercooling behavior of Sn-3.0Ag-0.5Cu (SAC) lead-free solder and SAC/Cu joints", International conference on electronic packaging & high density packaging, IEEE, 2012.

[8] L. Qu, H. Ma, H. Zhao, N. Zhao, A. Kunwar, and M. Huang, "The nucleation of Ag_3Sn and the growth orientation relationships with Cu_6Sn_5", 14th International conference on electronic packaging technology, IEEE, 2013.

[9] G.-S. Xu, J.-B. Zeng, M.-B. Zhou, and X.-P. Zhang, "Undercooling and solidification behavior of Sn-Ag-Cu solder balls and Sn-Ag-Cu/UBM joints", 14th International conference on electronic packaging technology, IEEE, 2013.

[10] M.L. Huang, F. Yang, N. Zhao, and Y.C. Yang, "Synchrotron radiation real-time in situ study on dissolution and precipitation of Ag_3Sn plates in sub-50μm Sn-Ag-Cu solder bumps", Journal of alloys and compounds 602, 281-284, Science Direct, 2014.

[11] B. Guoa, A. Kunwar, N. Zhao, J. Chen, Y. Wang, and H. Ma., "Effect of Ag_3Sn nanoparticles and temperature on Cu_6Sn_5 IMC growth in Sn_xAg/Cu solder joints," Materials research bulletin 99, 239-248, 2018.

[12] P. Liu, P. Yao, and J. Liu, "Effects of multiple reflows on interfacial reaction and shear strength of SnAgCu and SnPb solder joints with different PCB surface finishes," Journal of alloys and compounds 470, 188-194, Science Direct, 2009.

[13] N. Murad, S. R. Aisha, and M. Ishak, "Effects of cooling rates on microstructure, wettability and strength of Sn3.8Ag0.7Cu Solder Alloy", Procedia engineering 184, 266-273, 2017.

[14] H.R. Ma, S. Li, M. J. Yao, Y.P Wang, J. Chen, N. Zhao, and H.T. Ma, "The effect of cooling rate on growth kinetics of interfacial IMCs during multiple reflows," 18th International conference on electronic packaging technology, IEEE, 2017.

[15] Y. Zhong, M. Huang, J. Deng, H. Ma, W. Dong, and N. Zhao, "Study on precipitation and dissolution of interfacial Cu_6Sn_5 during thermomigration," China semiconductor technology international conference, IEEE, 2016.

[16] X. Liu, M. Huang, Y. Zhao, C. M. L. Wu and L. Wang, "The adsorption of Ag3Sn nano-particles on Cu-Sn intermetallic compounds of Sn-3Ag-0.5Cu/Cu during soldering," Journal of alloys and compounds 492, 433-438, 2013.

978-1-7281-1500-9/19 $31.00 © 2019 IEEE

Low Cost Flip-chip Stack for
Partitioning Processing and Memory

Fabian Hopsch
Fraunhofer Institute for Integrated Circuits IIS,
Division Engineering of Adaptive Systems EAS
Dresden, Germany
fabian.hopsch@eas.iis.fraunhofer.de

Andy Heinig
Fraunhofer Institute for Integrated Circuits IIS,
Division Engineering of Adaptive Systems EAS
Dresden, Germany
andy.heinig@eas.iis.fraunhofer.de

Abstract—**This paper presents a novel chip-to-chip packaging approach. Most nowadays systems comprise a least a processing unit and some memory for storing data. Rising bandwidth requirements, bring up new types of memory. This is achieved by increasing the clocking and/or using wider interfaces. But raised frequency leads to more effort on the communication channel in terms of signal integrity. The packaging approach presented here, results in a very short communication channel between memory and processor and with achieving high-performance while keeping the communication effort low. The requirements for the IO-cells are reduced, because of this short channel, leading to very small interfaces. All together the solution is a low-cost flip-chip stacking in terms of design and packaging costs.**

Keywords- Wafer-level-package fan-in, copper pillars, chip-to-chip packaging, flip-chip integration

I. INTRODUCTION

Most electronic systems comprise a processing unit and some memory as basis components. Devices with a high compute power also demand a lot of embedded memory. In many cases this embedded memory is integrated within the same IC. This is maybe not the best solution for each case, since processing unit and embedded memory have different requirements. So a technology could be optimized for performance or memory but not both. Also the demand for metal stack is distinct. Processing units need a lot of metal layers for proper routing and memory typically comes out with a lot less, because of the regular arrangement. Because of these differences it can be suitable to divide processing and memory during production and merge both during assembly.

To cope with the rising demand of memory bandwidth novel memory interfaces are developed. Besides the classical double-data-rate (DDRx) memories, this are the Hybrid-Memory-Cube (HMC) [1] and the High-Bandwidth-Memory (HBM) [2]. From the stack-up perspective both approaches look very similar. Both are using an interface logic die and stacked memory on top. But HMC uses high-speed SerDes communication with few channels, HBM uses wide channels with low speed. So both solution are in the same range of total bandwidth, but the effort for HMC for establishing a proper communication channel is much higher than for HBM.

Both memory solutions seem to be still under development or even out of business like for the HMC. But die stacking as

a packaging option is already used in some commercial solution like the AMD Fury [3]. But die stacking is not the best solution looking at dedicated application field, where low profile systems are required [4]. And we also see the integration requirements for 5G applications to include antennas into the package [5].

In this paper a novel flip-chip partitioning of processing and memory is presented and an example low-cost chip-scale package will be described, that is currently in production. At the current state processors can handle more data than can be provided by memory. This means a the memory connection is a bottleneck for processing data. This bottleneck was reduced by introducing more and more cache memory in different levels.

The system comprises a set of processing units. Each unit has it dedicated memory. Furthermore the system includes the organization structures for interaction between the processing units and to the outer world. In classical approaches the memory would be integrated within the same IC next to the processing unit. With the system presented here the memory is integrated into a separate IC. By flip-chip stacking of the processor and the memory very short connections arises. That lead to low latency and low access times, which is a very good performance for cache memory. Since the connections are very short, the drivers can be kept small.

In the example system presented in this paper the processing IC is in a 22nm technology. It has a size of 10x10 mm with a 10 layer metal stack. As this is a production sample the memory has been produced on the same reticle, but could be manufactured with reduced back-end-of-line of 5-6 layers or even in another technology. The memory IC has a size 2.3x 4.9 mm and is equipped with copper pillars. The processing IC has a wafer-level-package fan-in bumping interface with balls with 250um diameter and 400um pitch. The ball grid array has two gaps. In each gap area a memory IC is flip-chip mounted respectively. The system comprises one processing unit and two memory units in one chip-scale-package and can be mounted directly on a substrate like a printed circuit board. The processing and memory IC are designed in such a way, that each processing unit has directly its embedded memory lying underneath.

Such a system build-up has some advantages compared to classical approaches. Compared to putting the memory next to the processing unit the spatial distance with the stack-up

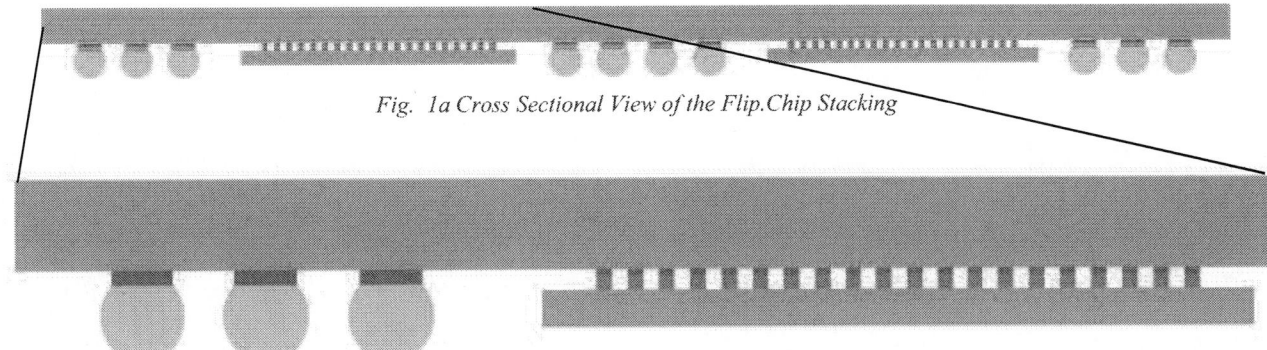

Fig. 1a Cross Sectional View of the Flip.Chip Stacking

Fig. 1b Detail of Cross Sectional View

becomes even smaller, leading to potentially lower latencies. Since memory devices are often very dense from a technology perspective, they are also a main yield delimiter. Using the stacked approach enable the possibility to sort out defective memory. Also the footprint of this system in chip-scale package is reduced. These advantages comes with the drawback of rising assembly effort and a more complex system build-up. To cope with these requirements an Assembly Design Kit (ADK) is used [6], which enables the co-design of different IC and package technologies within one design environment.

The paper is organized as follows. Paragraph II presents the general system build-up as a novel technology option. The following paragraph discusses the used components as processing unit and memory, but the packaging solution is not limited to these. Furthermore the paper closes with a conclusion and outlook.

II. SYSTEM BUILD-UP DESCRIPTION

The system consists of three single ICs. The main IC is 10 by 10 millimeter large and includes processing power. The other components are two identical memory ICs. The principal sketch of the system is depicted in Fig. 1, but not at scale. Fig 1a shows the complete system, while 1b part of the system in more detail.

Top side of the system is the processing IC. For this demonstrator it is manufactured in a 22nm FD-SOI technology. This IC has a standard metal stack with wafer-level package fan-in (WLP-FI) redistribution layer (RDL) to enable balling of the IC. The second component is a memory IC. For this demonstrator the memory IC is manufactured in the same technology, but in general the usage of another technology option or even another technology is possible.

The memory IC also includes a standard technology metal stack and uses copper pillars as balling option. The integration scheme is as follows. Both IC types are manufactured in the corresponding technology and metal stack. The memory ICs are transferred to the balling process to equipped them with copper pillars. In this case copper pillars with a diameter of 50 um and a stand-off height before assembly of around 58 um are used. After the copper pillar process the wafer with the memory ICs is thinned down to around 100um. So that the total thickness of the memory IC with copper pillars is in the region of 150um. Final step is than separation by sawing the wafer into individual dies. Than the memory ICs are ready for assembly.

After manufacturing the processor IC using a standard metal stack, the WLP-FI processing is done. For that on top of the last aluminum layer two RDL layers are processed to enable balling of the IC. For the demonstrator the RDL structures have dimensions to fit for 250um solder balls. In general also other balling sizes are possible. The 250um are a good compromise to enable proper integration of the system for example on a standard printed circuit board technology, but in the same time has enough stand-off height for the memory integration. This also involves thinning of the memory IC to a certain level, that still allows proper and easy handling.

After the RDL processing the ICs are equipped with solder balls using a wafer level balling process. After balling the memory ICs are flip-chip assembled at the corresponding positions at the processor IC. Final step is than separation of the single systems by sawing the wafer. By flipping the system is ready for integration for example on a printed circuit board.

The Fig. 1b shows a detail of the system. On the left hand side the solder balls can be seen. These have a diameter of 250um and are arranged in a regular grid with 400um pitch. The system uses two memory ICs to be able to establish solder balls also in the middle of the system. This enables proper electrical interfaces for the processor and also provides suitable stability across the system.

The system concept is based on standard technology options with minor adoption of the WLP-FI flow to allow at the same time interfaces for the solder balls and also assembly spots for the copper pillars.

III. PARTS OF THE SYSTEM

A. Processing Part

For the demonstrator of this paper a processing unit was implemented in a 22 nm fully-depleted silicon-on-insulator (FD-SOI) technology. It is used as a demonstrator, but the flip-chip technology is not limited to this technology.

In Fig. 2 the bottom view of the processing unit shows the WLP-FI. It is an integrated circuit with a chip size of 10x10 mm². The spare areas are for assembly of the memory dies.

978-1-7281-1500-9/19 $31.00 © 2019 IEEE   315

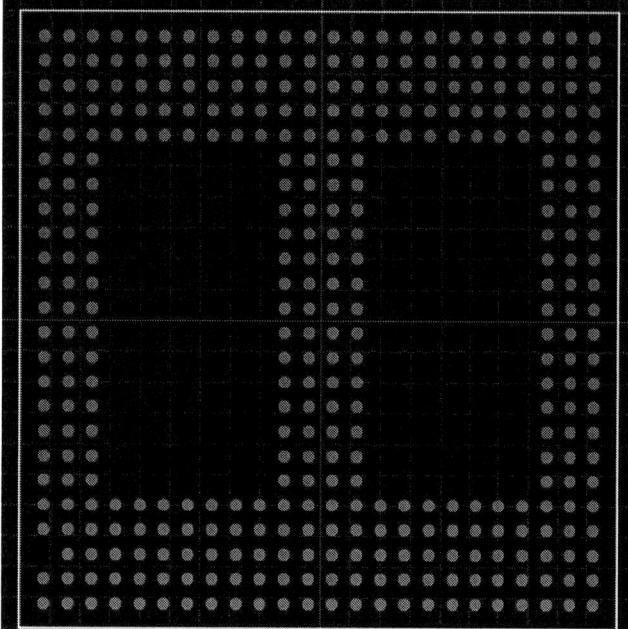

Fig. 2 Bottom view of the processing unit showing the WLP-FI interface

The WLP-FI technology is using two redistribution layers, that are processed on the last metal of the IC technology. These RDL layers are used to establish interfaces for 250um diameter solder balls, that are attached to the system using a wafer-level balling technology. The solder balls are arranged in a regular grid with 400 um pitch. This leads to a total number of 379 solder balls, while one is left out from the regular grid to provide a proper orientation marker for following integration steps.

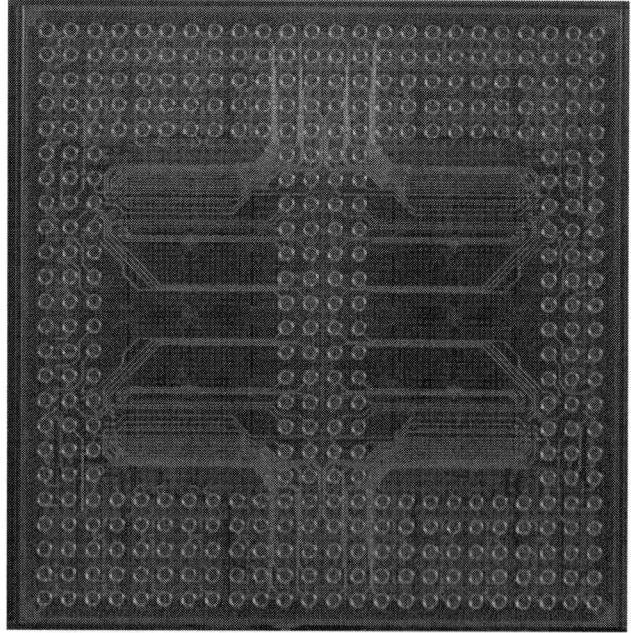

Fig. 3 Production view of RDL routing for WLP-FI

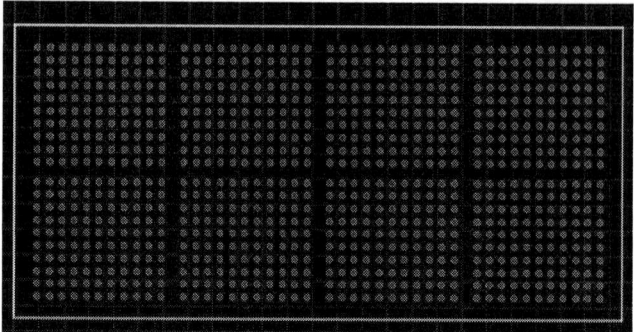

Fig. 4 Memory IC with Copper Pillar Interface

In the two spare areas without any balls the interfaces for the memory devices are placed. This involves minor adoption of the WLP-FI technology to enable on the one side the integration of solder balls, but also provides proper interfaces for copper pillars.

A picture after production of the RDL routing is depicted in Fig. 3. It is used to connect all processing units to an interface for communication with the outer world after assembly on a printed circuit board (PCB) for example. The processing unit itself comprises a set of identical processing unit each with a dedicated interface to the memory. From the outside these processing units communicate across the 250um diameter solder balls, while the communication to the memory is done using an interface, where the copper pillars of the memory dies are assembled.

B. Memory Part

The memory IC is about 2.3 by 5 millimeter large and involves a regular arranged grid of memory cells. In Fig. 4 the memory IC with its copper pillar interface is depicted. As can be seen the memory IC is organized as 4x2 independent memory channels. Each of these channels is connected to a dedicated processing unit at the processor IC.

The memory IC is equipped with copper pillars. These have a diameter of 50um and a stand-off height before assembly of around 58 um. The copper pillars are arranged in a regular grid for each memory channel with 10x11 copper pillars with a pitch of 100um in both directions.

In Fig. 5 an assembly of these copper pillars on RDL can be seen. As measured for this situation the final thickness of the copper pillar and the RDL pad after the assembly process is around 70 um. This value together with the thickness of the silicon of the memory IC concludes to a total thickness in the area of 150 um.

Fig. 5 Copper Pillar Assembly on RDL

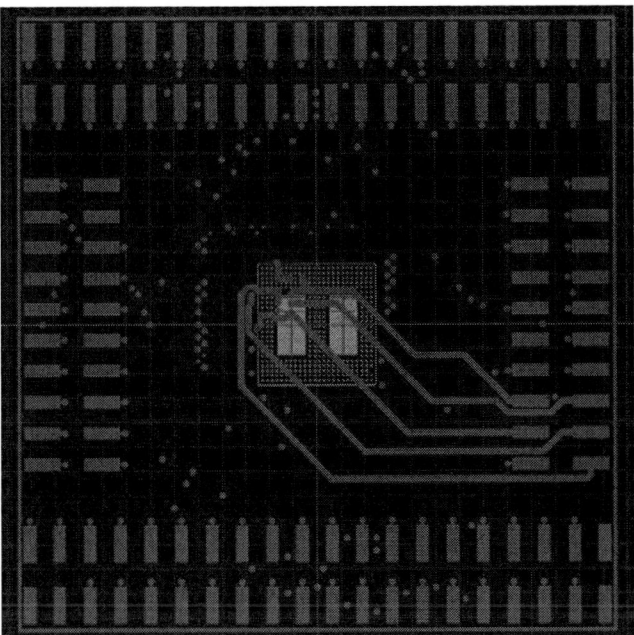

Fig. 6 Example PCB for the integration of the low-cost system

For the demonstrator the memory IC is manufactured in the same technology as the processor IC. But it is also possible to use different technologies, but the physical interface must be comparable to the used copper pillars. Also the usage of micro-bumps or also Cu-to-Cu bonding could be an option.

C. Printed Ciruit Board

To validate the proper assembly of the low-cost flip-chip stack of processor and memory a printed circuit board was designed. This board is used for assembly test and also for first commissioning of the system. In Fig. 6 a sketch of the PCB is shown. The PCB has a size of 50x50 millimeter and in the center the footprint for the system is placed.

Using flip-chip stacking the system is assembled using the 250 um diameter solder balls. The signals from the solder balls are routed to connectors on the PCB in each of the four cardinal directions.

IV. CONCLUSION AND OUTLOOK

This paper presented a novel flip-chip stack for partitioning processing power and memory in a low cost manner. The connection between processor and memory are very short and also small physical interfaces are used. This results in low latency and low access time. Also the drivers can be kept very small. This results in reduction in area and therefore cost reduction or the integration of more functionality on the same area.

As the system is currently in production, characterization will be the next step and a gain-effort analysis will be done to identify which applications are most suitable for such a build-up. Furthermore the adoptions of the design and manufacturing steps beyond the provided standard technology options are a matter of future work in the sense of commercialization and technology offering for wide adoption by customers.

The low-cost packaging approach is not limited to a solution presented here in this paper, but can be widely-adopted in various application domains. So the next step will also involve demonstrators beyond the pure processor and memory system, by integrating analogue or radio-frequency components of different technologies.

ACKNOWLEDGMENT

This work has been funded within the project USeP under label 100317397 by the European Union and the Free State of Saxony as part of the European Regional Development Fund (ERFD). Furthermore, we would like to thank the colleagues from Fraunhofer IZM Berlin, Fraunhofer IZM-ASSID Dresden and from Globalfoundries Dresden for their involvement in various steps for realizing this system.

REFERENCES

[1] J. T. Pawlowski, "Hybrid memory cube (HMC)," 2011 IEEE Hot Chips 23 Symposium (HCS), Stanford, CA, 2011, pp. 1-24. doi: 10.1109/HOTCHIPS.2011.7477494

[2] DITTRICH, Michael; HEINIG, Andy; HOPSCH, Fabian. Electrical Characterization of a High Speed HBM Interface for a Low Cost Interposer. In: 2018 IEEE 68th Electronic Components and Technology Conference (ECTC). IEEE, 2018. S. 2068-2073.

[3] B. Black, "Die-stacking is happening: AMD Fury X GPU" in 3D ASIP, Dec 2015

[4] F. Hopsch, A. Heinig, and M. Boettcher, "Very-Thin System-in-Package Technology for Structural Analysis," Smart System Integration. Barcelona, April 2019

[5] HOPSCH, Fabian; HEINIG, Andy. Antenna Integration Technologies for 5G Car-Application. In: 2018 International Wafer Level Packaging Conference (IWLPC). IEEE, 2018. S. 1-5.

[6] A. Heinig and R. Fischbach, "Enabling automatic system design optimization through Assembly Design Kits," 2015 International 3D Systems Integration Conference (3DIC), Sendai, 2015, pp. TS8.31.1-TS8.31.5. doi: 10.1109/3DIC.2015.7334602

Impact of Low Temperature Solder on Electronic Package Dynamic Warpage Behavior and Requirement

Wei Keat Loh
Intel Technology Sdn Bhd
Malaysia
wei.keat.loh@intel.com

Ron W. Kulterman
Flex Ltd, Austin Tx.
USA
ron.kulterman@flex.com

Chih Chung Hsu
CoreTech System (Moldex3D)
Taiwan
jimhsu@moldex3d.com

Haley Fu
iNEMI
Shanghai
haley.fu@inemi.org

Abstract—Low temperature solder (LTS) adoption has increased over the years with the primary focus on reducing energy consumption. The fundamentals of dynamic warpage are well established in SMT industry, but there is minimal literature on the application of low temperature solders, specifically for SAC BGA with low temperature solder paste. iNEMI has conducted a wide range of dynamic warpage characterization of different electronic package types, namely package on package, fine pitch BGA, large FCBGA package with and without lids, and a wide selection of PBGA packages. The database generated has been reviewed to understand the impact of low temperature solder on dynamic warpage considerations which avoids the typical maximum peak and valley warpage present at higher reflow temperatures. Additional consideration should be given to the rate and magnitude of contour change during LTS solidification phase that can induce a solder hot tear defect. It was found that mostly the PBGA and FCBGA package require greater attention and optimization for LTS adoption.

Keywords-component; warpage, low tempeature solder, FCBGA, PBGA, POP, FBGA

I. INTRODUCTION

Electronic packaging has been evolving rapidly during the last decades to enable new devices and new market segments. , Many of this made possible by component suppliers developing electronic packages that meet stringent requirements. Electronic package warpage needs to be managed to ensure the final product can be easily assembled on the board by original device manufacturers (ODMs). The drive to greener and lower energy consuming SMT processes has motivated the industry to assess the impact of low temperature solder on electronic devices.

Since the adoption of lead free Tin-Silver-Copper (SAC) solder more than a decade ago, managing electronic package dynamic warpage has been a major focus area. Process conditions include higher reflow temperatures and lower ductility of SAC compared to leaded solder systems in SMT. Some electronic package constructs which demonstrate a higher sensitivity to temperature change such as Plastic Ball Grid Array (PBGA) packages can suffer yield detracting SMT defects like solder bridging and non-contact opens. Advances in the areas of alloyed solder material has led to the introduction of low temperature solders (LTS) which melt and solidify at a range of temperature from 130°C to 200°C, creating a new opportunity to lower the reflow temperature which inherently reduce the energy demand and may favor the dynamic interaction between package warpage and solder phase formation.

An iNEMI Project focusing on Bismuth-Tin (BiSn) based LTS Process and Reliability (LTSPR) was formed in 2015 to address the challenges of various LTS pastes covering a multiple alloys and chemical formulations manufactured by suppliers. The projects had three phases [1]. The LTS process considered was SAC BGA attached on the package with LTS paste screen printed on the board for SMT. Phase I entailed the selection of applicable pastes and a determination of the SMT processing ability of twelve formulation of LTS pastes in three metallurgy categories were selected for characterization. These metallurgies include (i) three types BiSn based eutectic pastes with 0%, 0.4%, and 1.0%wt Ag, (ii) five non-eutectic ductile BiSn pastes and (iii) four kinds of Joint Reinforced Pastes (JRP) resin containing BiSn eutectic solder compositions pastes. The JRP provide polymeric reinforcement of cured resin at the base of the formed joints; whereas the chemical formulation of the first two categories were very important for paste wetting and joint forming performance [2]. Subsequently, these thirteen pastes, which include the SAC as reference, been applied into the SMT processing parameters defined to focus on the characteristics of the mixed SAC–BiSn solder joints formed. High density BGAs, QFN and Sockets with SAC BGA were assembled to PCBs using BiSn solder pastes at peak reflow temperatures significantly below the liquidus temperatures of SAC solder balls. The challenges of LTS adoption and SMT parameters optimization opportunity were discussed and proposed in [3] based on the experimental results. One of the key defect highlighted in adopting BiSn based LTS paste is the hot tearing defect of mixed SAC-BiSn alloy solders which was observed in FCBGA joints located at the die shadow area when using four of the twelve LTS BiSn pastes. Hot tearing is thought to be caused by an interaction of Bi stratification at the package interface coupled with dynamic warpage of the FCBGA package as it cools from peak reflow temperatures. It can also occurred without Bi stratification if the solidification temperature overlaid with the package substrate package warpage shape inversion temperature [3]. Hot tearing defect is a critical concern and solutions may exist to enable LTS in SMT process as an attempt to mitigate yield and reliability associate to high dynamic warpage behavior. Extension to this assembly study of LTS, Phase III evaluated the predefined mechanical shock test of Package on Package (PoP) with SAC BGA coupled with BiSn based and JRP based paste joints. The study shows that BiSn and JRP based were weaker than homogeneous SAC BGA on the mechanical shock test condition highlighted in [4]. With extensive evaluation of LTS metallurgy system, there is a need to better understand and quantify package warpage behavior, This paper leverages many types of electronic package's dynamic warpage characteristics and the results from the LTSPR project to understand the implication of LTS to the electronic package dynamic warpage behavior and requirements.

II. PLETHORA OF DYNAMIC WARPAGE DATABASE

TABLE I.   PACKAGE TECHNOLOGY CONSIDERED

Package Type	Design	Schematic drawing of package construction
PoP	Overmold Through Mold Via (OM TMV)	
	Expose Die Mold TMV (EDM TMV)	
	Bare Die PoP	
	Interposer PoP	
	Pre-stack PoP package	
	MCeP®	
	PoP Memories	
SiP	Overmold Multiple Chip Package (MCP)	
FBGA	Overmold single die package	
FCBGA	with single or multi dies	
FCBGA with Lid	Organic and ceramic substrate	
PBGA	Ranges	

Since iNEMI has conducted a wide range of dynamic warpage characterization of different electronic package types as shown in TABLE I. , namely the Package on Package (PoP), System in Package (SIP), fine pitch BGA (FBGA), large FCBGA package with and without lids and a wide selection of PBGA packages, the data collected [5]-[7] is further analyzed to establish the impact of LTS on dynamic

warpage requirement. The dynamic warpage data was collected as discussed in [8]. The high temperature warpage requirement for SAC system has been well understood and some been tabulated in design guides, JEDEC with the failure mode associated to SAC system typically limited to solder bridging, non-contact open and head on pillow (HoP). The requirement of dynamic warpage proposed for low temperature solder is not only governed by the maximum peak and valley warpage values at peak reflow temperature but the rate and magnitude of change during solidification phase of low temperature solder also needs to be evaluated, as highlighted in [3]. This shows that SMT defect for LTS system can relate in a different way to dynamic warpage of the package.

III. IMPACT OF SAC SOLDER AND LOW TEMPERATURE SOLDER ON DYNAMIC WARPAGE

A. Reflow Profile and Critical Temperature Points

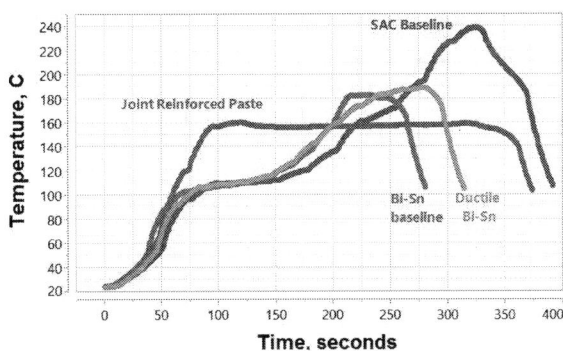

Figure 1.   Typical reflow soldering profiles for SAC baseline, BiSn baseline, Ductile BiSn and Joint reinforced (resin) pastes.[2]

Most BGA packages come with SAC based solder ball attached and the implementation of LTS is mainly focusing on the use of BiSn based solder paste for PCB assembly. The typical reflow profiles depicted in Figure 1. overlays the SAC baseline reflow profile with the other three LTS paste types employed during the iNEMI LTSPR studies. The SAC baseline, eutectic BiSn baseline, ductile BiSn, and JRP were studied to provide fundamental understanding of the solder joints formations. Typical SAC based SMT reflow targets three key thermal ramp ranges of temperature, namely (i) Initial ramp, (ii) Soak approaching reflow, (iii) Rapid Ramp to melting and peak temperature, and (iv) Final Cool ramp down. Initial ramp is intended to bring the PCBA as a whole up to sufficient temperature to activate the flux chemistry so as to clean and prepare the solder pad surfaces for wetting. The slower soak time is timed to allow sufficient duration for the flux to react but not so long as to fully consume it and expose the solder pad surfaces to oxidation before wetting of liquidus solder. The soak also brings the entire PCBA thermal mass to near reflow temperature without significant temperature deltas across. The rapid ramp typically starts within 10°C to 20°C of solder melting temperature, 217°C for SAC305 and continues to peak reflow temperature which is

978-1-7281-1500-9/19 $31.00 © 2019 IEEE

typically around 245°C to 260°C. The Time above Liquidus (TAL) is critical to ensure all solder volumes reach liquidus and all chemistry volatiles and voids can escape from the molten solder. Too short of a TAL can create HoP and non-contact open defects along with excessive voiding. Too long can damage the PCB and/or components and promote excessive IMC growth which in turn can increase the brittleness of the formed joints. The rapid cool down is targeted at improving grain structures, increasing throughput, and minimizing IMC generation at the joint interfaces of PCB and package pads. The LTS, namely as BiSn and Ductile BiSn, is mimicking the SAC baseline reflow profile but with lower peak temperature of approximately 182°C while the LTS solidification temperature ranges from 150°C to 125°C. The JRP reflow profile is uniquely different from the rest as the long soak requires to condition the resin and molten solder differently. However, the JRP paste used in [2] was not promising. Hence, JRP is not a key focus in this discussion.

B. Dynamic Warpage Interaction with SMT Defects

Electronic package dynamic warpage behavior has been the key assessment for this study. The typical warpage behavior that changes with temperature is shown in Figure 2. In this graph, there are three general dynamic warpage characteristics for Package A, B and C. Package A dynamic warpage is best resembling a typical PBGA package while the Package B and C resemble a typical FCBGA package. Package A started off as a concave shape (-ve) and transitions to convex as the temperature increases. Package B and C started off convex shape (+ve) and transition to concave shape. The difference between Package B and C is mainly the

magnitude of warpage at higher temperature where Package C has higher peak temperature warpage.

Based on the earlier discussion on the SAC and LTS reflow profile, the dynamic warpage requirements or metrics may be different from LTS system. Firstly is the reduction of peak reflow warpage from 260°C to ~182°C; secondly the reduction of warpage at a point of initiation of solder solidification of LTS at ~150°C as compared to eutectic solidification of SAC solder at ~220°C; third is the warpage change during the LTS solidification phase between 150°C to 125°C where the solder may be deformed as it cools and lastly is the warpage change slope and shape inversion from convex to concave or vice versa. These temperature points were considered because of the availability of existing data of dynamic warpage for various packages. Each of these metrics is mainly for quantification of impact of LTS on dynamic warpage requirement and perhaps explain the benefits and considerations of adopting LTS paste based on the dynamic warpage response of the package . Apart from dynamic warpage behavior, the package rigidity at LTS solidification phase temperature is expected to be higher compared to SAC solidification temperature due to the increased structural stiffness of the package at lower temperature which exert greater tensile and shear stresses on the solder joints. This could potentially lead to hot tearing defects mentioned earlier. Although the PCB board warpage is also part of the equation and also needs to be evaluated, it is out of the scope of this paper. The following sections demonstrate the impact of LTS on warpage metrics mentioned which is an extension of the prior dynamic warpage collected in earlier phases of iNEMI project listed here [5]-[7].

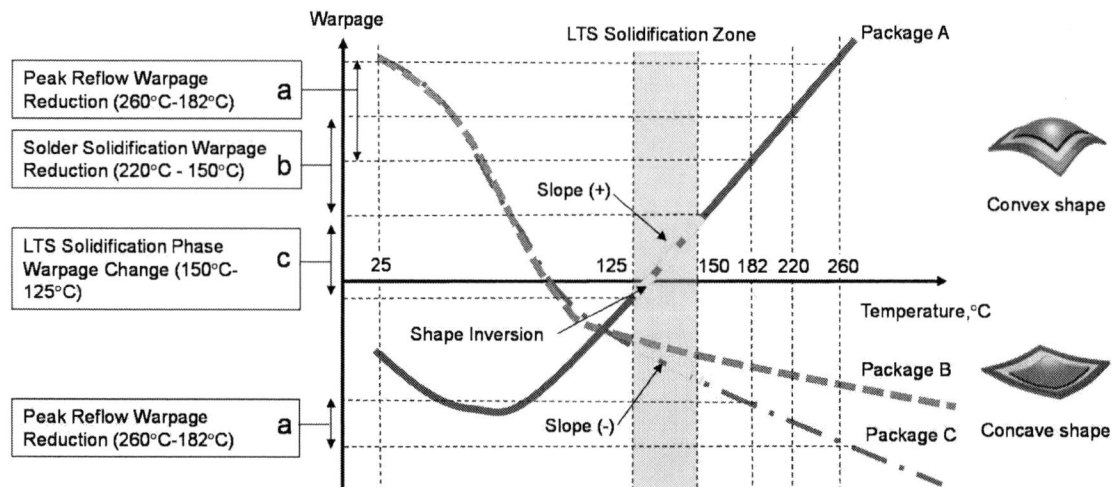

Figure 2. Typical Dynamic Warpage Behavior and LTS key characteristic

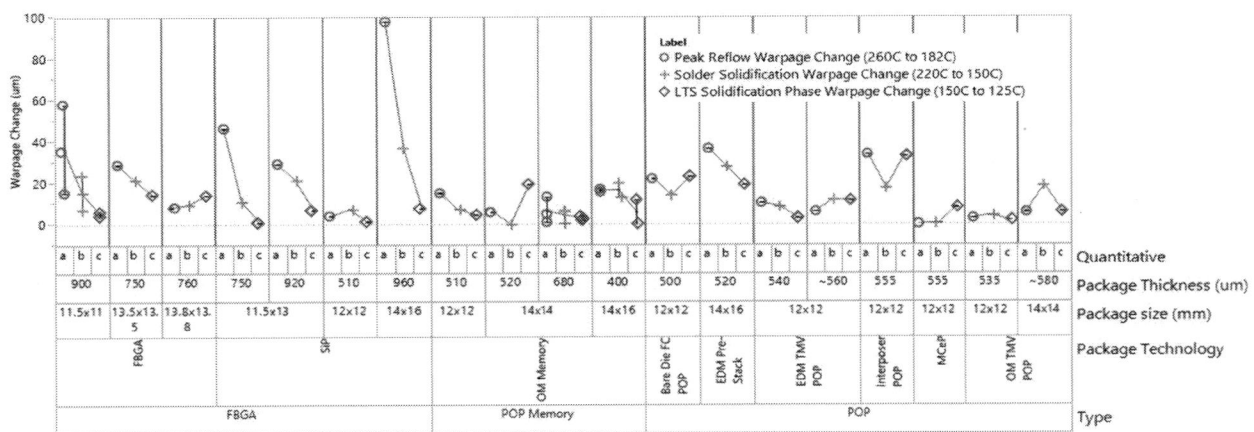

Figure 3. Package on Package and FBGA families warpage magnitude change

IV. RESULTS AND DISCUSSION

By leveraging the past collected dynamic warpage data across various package technologies, each of the warpage changes as a result of reduced peak reflow temperature, solidification temperature and solidification phase of LTS were processed. This processed data provides an overview of the changes in warpage from a SAC solder system to a LTS system. In the following sections, each of the package technology families is presented by comparing the magnitude changes. There are other factors like design, materials used and package construction which can result in very different dynamic warpage characteristics and it's behavior using LTS. This is not included in this study Package on Package and FBGA families

The dynamic warpage of various types of Package on Package (PoP), PoP memories and FBGA are categorized together due to it small package size of less than 20x20mm. The raw dynamic warpage data can be referred to [5][6]. The original data was collected at different temperature readout and some was provided by participating companies which was processed with best fit interpolation for specific temperature readout required to compute the warpage changes discussed. 0shows the magnitude based on the three quantification methods. Each data point here represents an average value for a given type of package technology with similar package attributes. For FBGA package, the peak reflow warpage change (O marker) ranges from ~100um to ~5um; while for solidification warpage change (+ marker) ranges from ~25um to ~5um and lastly the LTS solidification warpage change (◊ marker) ranges up to 20um. This suggest that a LTS system can have a impact to the warpage change and the adoption of LTS has minimal impact to the SMT yield quality because of minimal overall change of warpage magnitudes.

As for PoP memory, the warpage quantification of change for these is minimal with less than a 20um change. Since PoP memory was designed or optimized to pair up with a bottom package, the warpage of the PoP memory can have a bi-modal distribution of shapes (concave and convex) hence the average magnitude change is kept to a minimum. The adoption of LTS on PoP memory attach is still not widely practiced and hence

the assessment here may not be relevant. For PoP package, the peak and solidification warpage magnitude change range within 40um. There are a few type of PoP packages namely the MCeP and OM-TMV that demonstrated less than 10um warpage changes suggesting that the LTS reflow has little effect on package dynamic warpage behavior. This can be explained due to the unique balancing of package design and material selection used to keep the dynamic warpage minimal for better adoption to PoP memories as well. For Interposer PoP packages, the solidification phase warpage change is relatively higher compared to the rest which may suggest the increased potential for hot tearing during the solidification phase of LTS.

A. FCBGA families

FCBGA package technology consists of two distinct categories namely the single die version represented as FCBGA and multi-chip package which is represented as FCBGA MCP as shown in Figure 4. For FCBGA, the package size ranges from 20x20mm up to 37.5x27.5mm while the package thickness ranges from 800um up to 3.7mm. The peak and solidification warpage change for package size of less than 29x29mm is less than 30um. For package size greater than 29x29mm, the warpage change is up to 60um. As for the LTS solidification phase warpage change, the FCBGA package shows up to 60um change as well which is similar to the change of peak reflow and solidification warpage change. This may indicate the higher risk of hot tearing which requires further validation. In general, the thinner and larger package exhibited higher warpage magnitude change.

For FCBGA MCP, the peak reflow warpage change is higher which ranges from ~90um to ~135um. This is higher than the single die version mentioned earlier because the larger and thinner package geometry. The data shows about 60um warpage change at solidification temperature. The LTS solidification phase warpage change is about 40 to 60um for the 40x24mm package size. This warpage change may post some challenges to manage the deflection of the substrate while stretching the solder joint to induce the hot tearing defect, especially when fine BGA pitch is adopted.

B. FCBGA with Lids families

The FCBGA package with Lids database is shown in Figure 5. which consists of organic and ceramic substrates. The change peak reflow warpage for these type of packages seems minimal despite the large package size ranging up to 55x55mm coupled with package thickness from 1.13mm to 4mm. The peak reflow warpage change of up to 50um and the warpage during the solder solidification is below 25um. The change of warpage during the LTS solidification phase also kept below 20um. Unlike packages without Lids, the warpage changes

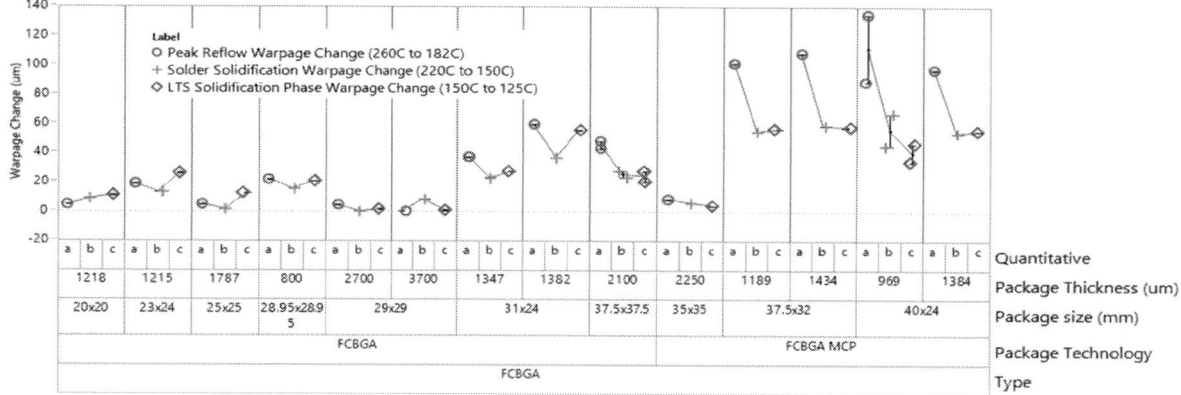

Figure 4.   FCBGA families warpage magnitude change.

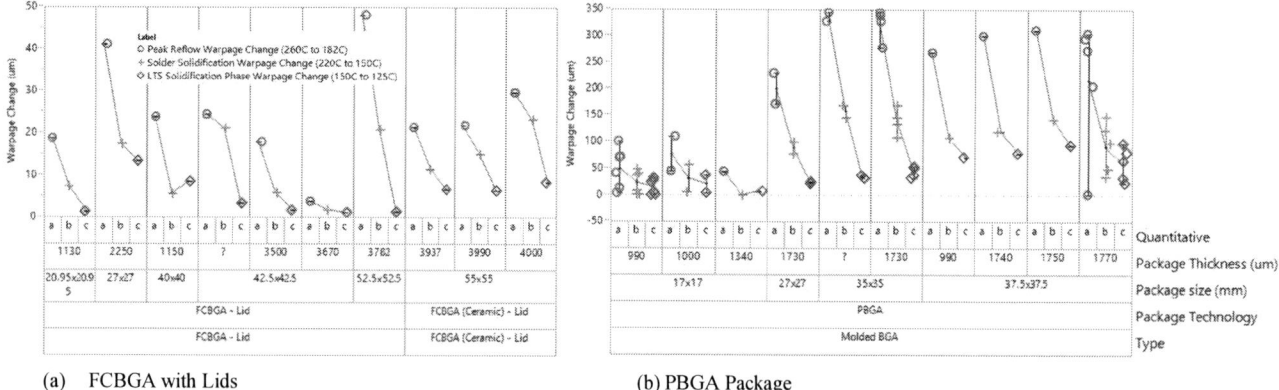

(a)   FCBGA with Lids   (b) PBGA Package

Figure 5.   FCBGA-Lids and PBGA Warpage Magnitude Change.

from SAC system to LTS system is minimal. With such large packages, the ball pitch and ball size used are typically larger and hence the magnitude change can be considered a fraction of solder ball height. This may suggest the LTS adoption in FCBGA with Lids may not have gained significant impact to warpage change nor will it result in higher SMT yield lost. The adoption of LTS for such package seem seamless but further validation may required.

C. Plastic Molded Ball Grid Array (PBGA)

PBGA package constructs consist of significant volume of mold as an encapsulation over the wire-bonded dies. The dynamic warpage of PBGA is unlike those typical of the FCBGA package where the warpage starts off either in convex or concave shape at room temperature and turns to convex shape at reflow temperature [6]. Figure 5. shows the

quantification used to understand the considerations in adopting LTS. Firstly, the peak reflow warpage changes up to 350um for the larger package size of 35x35mm and above. For smaller PBGA packages of 17x17mm, the peak reflow warpage can be changed up to 120um. The solder solidification warpage change ranges up to 160um for package sizes greater than 27x27mm. For 17x17mm package size, the solder solidification warpage change is within 50um. The LTS solidification phase warpage change can range up to 100um especially for those larger package sizes. The change of peak reflow and solidification warpage is highest among the package technology considered here. This seems to support the challenges of large PBGA package technology in adopting to higher reflow temperature during the transition from PbSn to SAC reflow[9]. On the other hand, the warpage change during the LTS solidification phase may post some challenge in managing the potential hot tearing phenomena which requires further validation and actual SMT assessment.

D. Overall Warpage Change due to LTS reflow profile

Due to the diverse type of electronic packaging technology considered here, comparing the magnitude change can be a challenge as each package has its unique dynamic warpage characteristic. Hence the percentage of warpage change in adopting to a LTS solder system with reference to the SAC peak warpage reflow at 260°C is considered here. Figure 6. shows the percentage of warpage change for these families of package technology. The markers denote the trending of the warpage change where not all types of packages experience a reduction of warpage as a result of adopting LTS. For PoP and FBGA packages, the warpage change ranges from negligible impact to a factor of ~1000%. This extreme change is due to the warpage at LTS is a few magnitude order higher than the warpage at 260°C[7]. This, however, reflects the change of warpage percentage and not the absolute warpage magnitude. Majority of the package warpage change is as low as 10% to about 100%. For FCBGA and FCBGA MCP packages, the majority of the packages considered show about 20% to 100% warpage reduction and a few show warpage increases as well. This reduction seems reasonable as the results of 30% to 50% warpage reduction was reported in [2]. As for FCBGA with lids, the percentage of warpage reduction is about 20% to 50%. The even lesser change for large packages is mainly due to the constraint from the lids coupled with thicker constructions. Lastly for PBGA packages, the majority of them shows a warpage reduction from 40% to 250%. This higher warpage reduction seen in PBGA is attributed to the higher sensitivity of the package to temperature as a result of the higher volume of mold encapsulations used.

E. LTS Solidification Phase Warpage Slope and Shape Inversion

In earlier sections, the warpage reduction and change were presented to provide an overview of the impact of LTS. Two other parameters that can be considered are the LTS solidification phase warpage slope and the shape inversion during the cooling down from 150°C to 125°C. Figure 7. shows the magnitude of the slope for each of the package technology presented earlier. The slope can range from ~-2um/°C to ~+4um/°C. The shape inversion describes the change of shape from convex to concave or vice versa. The magnitude shows the aggressiveness of warpage change with temperature during the solidification phase of LTS. At this range of temperature which is below or around the glass transition temperature of the substrate or mold, the stiffness of the package is higher which result in greater reaction force being exerted on the LTS joints. The result of this stresses can promote hot tearing defects. Based on the graph, the PBGA package with larger package size followed by FCBGA MCP packages exhibits increased risk of experiencing solder hot tearing due to higher slope and in some cases coupled with shape inversion. The + and - signage used on the slope axis will determine the region of hot tearing. For positive slope, the warpage change is towards concave. Hence the hot tearing region may appear at corner joints of the package. On the other hand, the hot tearing may appear in region below the package center or underneath the die shadow region because warpage shape change is toward convexity. The shape inversion can elevate the risk level of hot tearing which was presented in [2] where FCBGA package have seen such hot tearing underneath the die shadow at which cooling of the solder joints are relatively slower compared to peripheral joints. Also the shadow of the silicon die or dies also defines a line of sudden change in stiffness. Region beyond the die shadow is expected to allow for freer relative movement which may reduce the stresses on the joints. Hence, the adoption of LTS with range of solidification temperature requires further assessment to understand the mechanism and region of risk of hot tearing.

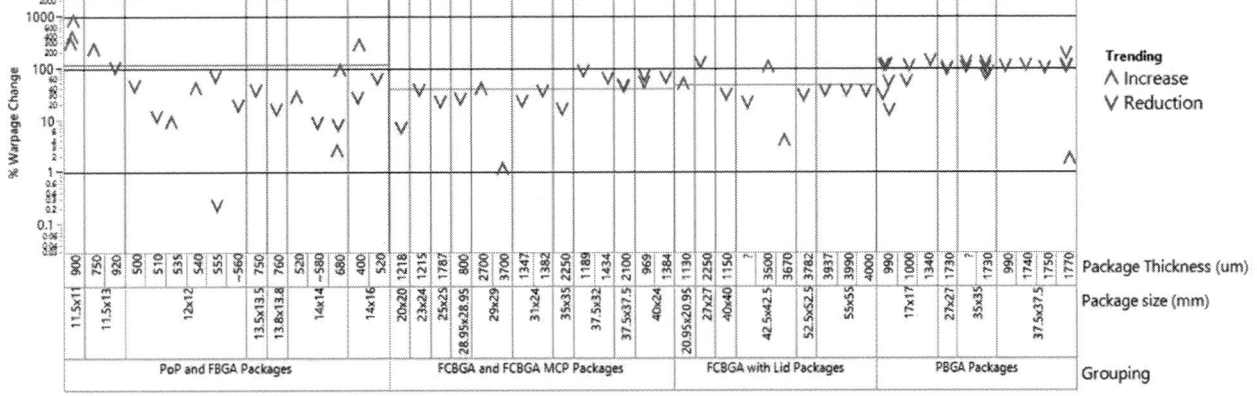

Figure 6.   Overall peak reflow waragpe reduction percentage

978-1-7281-1500-9/19 $31.00 © 2019 IEEE

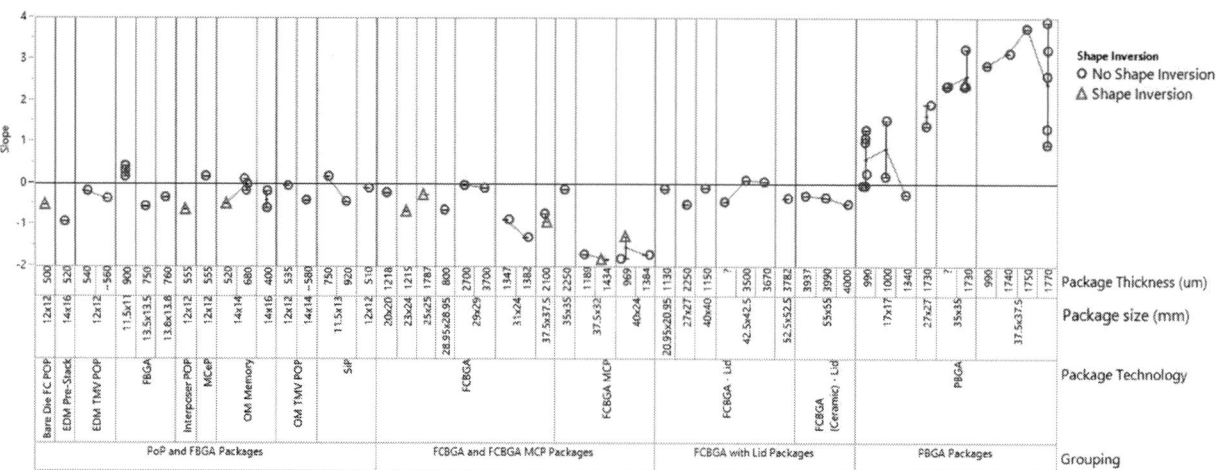

Figure 7.   LTS Solidification Phase Warpage Slope at 150°C to 125°C and Shape Inversion

V.   SUMMARY

Based on increasing adoption of LTS solder systems in products built, the consequences of dynamic warpage of electronic packages can be different from eutectic solder such as SAC systems. In most cases, the peak reflow warpage and solder solidification warpage reduces in magnitude, depending of package type and design construction. The LTS solidification phase happens in a range of temperature which concurrently coupled with the change of warpage magnitude and shape that elevate the risk of hot tearing. Such phenomena requires additional assessment in the industry. As part of providing an overview of the assessment, the warpage change during LTS solidification phase and the slope provide some indication of the risk of getting the hot tearing defect at either at package center or corner or edges. Based on the database gathered, the challenges to using LTS for PBGA packages need to be carefully considered. The percentage of warpage change for each of the package families provide an estimation of the impact of LTS and not all packages experience a reduction of warpage magnitude as a result of adopting LTS. The slope of warpage change and package shape inversion during LTS solidification phase can play a role in the potential of inducing hot tearing defects. The SMT risk assessment for these packages requires an effort from the SMT industry to characterize and understand the underlying risk of hot tearing and other SMT defects that encompass the effect of multiple reflow process and percentage of LTS paste to SAC solder volume. With initial work by iNEMI LTSPR project, there could be a solution in place to manage these defects through other means of optimizing SMT parameters as mentioned in [2].

ACKNOWLEDGMENT

This project is a joint effort of iNEMI members: Flextronics, Intel, Moldex3D and the iNEMI LTS project outcomes. We appreciate the supports from industry and companies who provided the samples.

REFERENCES

[1] Scott Mokler, Raiyo Aspandiar. *et.al.* –"The Application of BI-based Solders for Low Temperature Reflow to Reduce Cost while Iimproving SMT Yields in Client Computing Systems" Proceedings of SMTA International, Sep. 25 - 29, 2016, Rosemont, IL, USA

[2] H. Fu, R. Aspandiar *et.al.,* "iNEMI Project on Process Development of BiSn-Based Low Temperature Solder Pastes", Proceedings of the 2017 SMTA International Conference, Rosemont, IL, September 2017, 207-220.

[3] H. Fu, R. Aspandiar *et.al.* "iNEMI Project on Process Development of BiSn-Based Low Temperature Solder Pastes – Part II: Characterization of Mixed Alloy BGA Solder Joints", Proceedings of the 2018 Pan Pacific Microelectronics Symposium, Hawaii, February, 2018.

[4] Haley Fu, Jagadeesh Radhakrishnan, *et.al.* "iNEMI Project on Process Development of BiSn-Based LOW Temperature Solder Pastes Part III: Mechanical Shock Tests on POP BGA Assemblies" ICEP 2018

[5] Wei Keat Loh, Ron Kulterman, Tim Purdie, Haley Fu and Masahiro Tsuriya, "Package-on-Package (PoP) Warpage Characteristic and Requirement", 2015 17th Electronics Packaging Technology Conference

[6] Wei Keat Loh, Ron Kulterman, Tim Purdie, Haley Fu and Masahiro Tsuriya, "Recent Trend of Package Warpage Characteristic" , Conference: 2015 International Conference on Electronic Packaging and iMAPS All Asia Conference (ICEP-IAAC), April 2015

[7] Wei Keat Loh, Ron Kulterman, Haley Fu, Masahiro Tsuriya, "Recent trends of package warpage and measurement metrologies"

Conference: 2016 International Conference on Electronics Packaging (ICEP), April 2016

[8] Wei Keat Loh, Ron Kulterman, Haley Fu, "Comparison of advanced package warpage measurement metrologies", IEEE 37th International Electronics Manufacturing Technology (IEMT) September 2016

[9] Impact of molding parameters on PBGA warpage characteristic. - By: Loh Wei Keat; Quah Chin Aik; Lee Chee Kan; Lee Chek Loon. 2010 34th IEEE/CPMT International Electronic Manufacturing Technology Symposium (IEMT) , 2010, p1-4

High Density Ultra-Thin Organic Substrates for Advanced Flip Chip Packages

Nokibul Islam
Field Application Engineering,
JCET Group
Tempe, AZ, USA
e-mail: nokibul.islam@statschippac.com

SW Yoon
R&D, Singapore
JCET Group
Singapore

KH Tan
R&D, JCAP Bump House
JCET Group
China

Tony Chen
R&D, JCAP Bump House
JCET Group
China

Abstract—Advanced semiconductor packaging requirements for higher and faster performance in a thinner and smaller form factor continue to grow for mobile, network and consumer devices. While the increase in device input/output (I/O) density is driven by the famous "Moore's Law", the packaging industry is experiencing opposing trends for more complex packaging solutions while the expected cost targets are moving in a downward direction. Fine line/space is one of the key requirements for high I/O count packages where die-to-die or die-to-memory integration is needed. Si based inorganic interposers have typically been used in this area for the last few years. However, Si interposers are expensive and supply chain issues can be a serious bottleneck. Although there are few organic interposer solutions in the market, none of these solutions addressed clearly the entire needs of flip chip packages.

Demand for high-speed flip chip packages creates an opportunity for highly integrated, multi-chip modules (MCM's) and 2.5D/3D silicon (Si) interposer packages. These packages are emerging very slowly due to the higher costs often associated with infrastructure and supply chain challenges that can happen before a technology is mature. Achieving both increased margins in the power delivery and increased functionality in next generation high-speed applications requires extremely efficient, low loss package designs with an ultra-thin core or coreless substrate with fine line and space. As the substrate gets thinner, it becomes very flexible and one of the biggest assembly challenges for ultra-thin coreless substrates is to keep the substrate flat during the assembly process while still maintaining yield targets. Other issues with thin substrates are related to post assembly such as handling, long-term package reliability and functionality in the application field. The work presented in this paper describes key factors for mitigating several assemblies related issues in the manufacturing line, including package warpage/coplanarity, and selecting the optimum processes and materials for the ultra-thin coreless substrate, called uFOS (Ultra Format Organic Substrate), for flip chip packages with high assembly yields and lower cost.

uFOS potential application spaces including die-to-die and package-to-package using 2.5D or 2.1D (embedded high-density film or eHDF) methods will be explored using an ultra-thin coreless substrate as an interposer, as opposed to the traditional Si interposer. Various pros and cons along with relative cost data will be discussed.

Several design of experiments (DOE) for ultra-thin substrates are being carried out to achieve the objective of the work. Multiple test vehicles have been designed using a flip chip package with an ultra-thin coreless buildup substrate utilizing various assembly materials and processes. The detailed process and some reliability data will be published. More work will be carried out to expand the scope of the technology for multi-chip module (MCM) die and 2.5D integration. Some initial eHDF data will be published as well.

Keywords-component; flip chip, ultra- thin substrate, 2.1D, 2.5D, assembly, warpage

I. Introduction

As chip technology gets more and more advanced along with the aim toward product miniaturization, the need to reduce the chip package form factor while increasing chip performance has become critical to enabling more advanced chip technology and product miniaturization. Traditional flip chip packages are made with multi-layer build up or coreless substrates. However, conventional cored substrates have lots of through holes in the core, creating obstacles to high speed and high frequency applications. The curvature and unevenness of the core laminate have become the limiting factors in achieving the fabrication of high-density substrates. Moreover, some applications require very thin substrates with robust routing density. Cost is the other parameter that cannot be neglected at the end of the day. Considering all the challenges, a new enabling technology must address this issue. Another potential need is for die-to-die or die-to-package high-density interconnection where typically nonorganic based interposers (Si interposer) are used. A unique feature of the Si interposer is its capability for very fine (sub-micron level) routing density. However, this solution is expensive and has a limited supply chain. The ultra-thin substrate is one such enabler for more advanced chip technology and product miniaturization which potentially can be used for thin packages with homogeneous as well as heterogonous integration (system-in-package or SiP) and for die-to-die or die-to-package (memory) integration. According to some literature [1-3], the high-density thin film substrate can extend SiP technology to a new version of a thinner and higher performance package. The challenges facing ultra-thin substrate manufacturing along with suggestions for overcoming these challenges will be presented here.

II. Ultra-Thin Substrate (uFOS)

The salient feature of an ultra-thin substrate is its thickness, which can be an order of magnitude thinner than normal flip chip laminate or build up substrates. Figure 1 shows a typical ultra-fine uFOS substrate currently in development. With such a thin film based substrate, the chip package height can be reduced significantly, which enables the adoption of high-speed flip chip for thin application areas. uFOS has the capability to make a much finer routing density that potentially can be used to route ASIC-to-memory, or die-to-die integration within the same package. Currently expensive Si interposers are being used in the market to integrate die-to-memory of MCM dies.

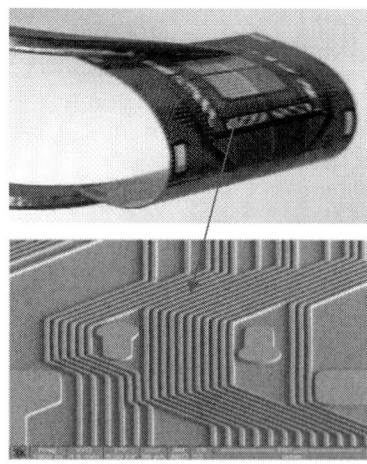

Figure 1: Typical RDL 1st ultra-fine line space (3/3) uFOS

There are various potential application areas for ultra-thin uFOS described in Figure 2.

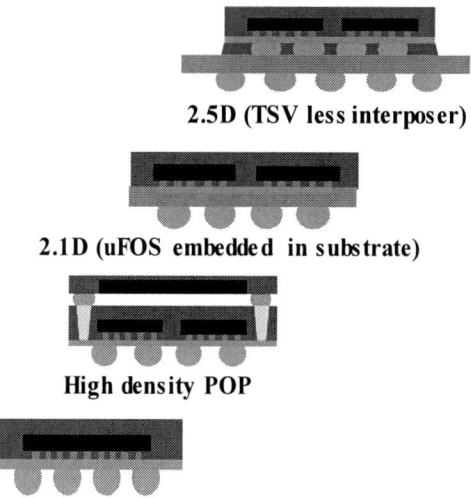

Figure 2: Potential application types for uFOS

III. Pros and Cons

Pros:
1. Much thinner form factor with high-density interconnection.
2. High performance applications
3. Wide range (low pin count to very high pin count package) of application spaces
4. Only known-good-die will be used due to RDL 1st and die last process
5. Much higher yield due to die last process
6. Cheaper solution for die-to-die and die-to-memory integration
7. Could be used as 2.1D or 2.5D with TSV less process
8. Shorter cycle time due to good control of uFOS substrate

Cons:
1. Handling issue due to thinness of the substrate (currently only developed on 200mm wafers)
2. Die attach assembly challenge due to high warpage
3. Currently line/space limited to 2/2 (recently demonstrated) for ASIC to memory integration
4. Reticle size limitation

Fine line space is one of the key attributes of uFOS. Multiple programs and DOEs are initiated to achieve a very robust process for 2/2 or smaller line/space. Figure 3 shows current line/space comparison chart of uFOS over other standard substrate and PCB. Figure 3 also shows a uFOS with 2/2 line space embedded on a 3 layer standard coreless substrate. uFOS pros and cons over other substrate types are also tabulated in Table 1 below. It is clear from the table that uFOS sweet spots are not for sub-micron routing density. Rather it is for $\geq$ 2/2 density, where most of the applications are currently in the industry.

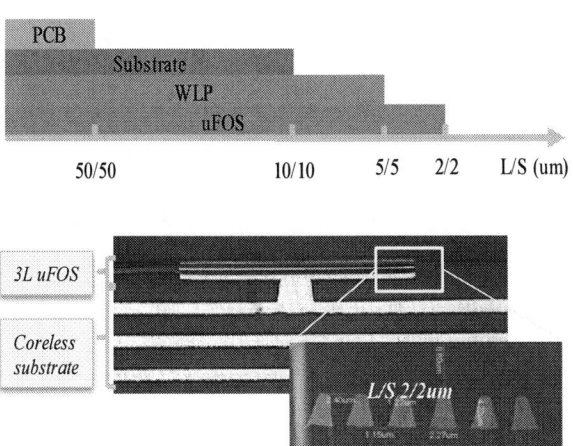

Figure 3: Routing density comparison chart

Item	uFOS	Build up substrate	Fan out RDL	2.5D Si Interposer
Metal-layer count	≤8	≤20	≤4	5~6
Min L/S(um)	2/2	15/15	5/5	0.5/0.5
Total Thickness (um)	20~80 (metal layer 2~8)	≥100	20~50 (metal layer 2~5)	100~150
CTE of Dielectric	3	≥18	20~60	5~6
Cost	Middle	Low	Middle	High

Table 1: Routing density comparison for various substrate types

IV. uFOS Assembly Process

uFOS were manufactured at several development phases as described below.

Feasibility build: One of the critical phases in manufacturing was the feasibility build with various design and process parameters. Several key assembly process parameters were reviewed in detail. A comprehensive DOE with assembly process and material BOM parameters was conducted to figure out how smoothly fine line space uFOS can be manufactured. As of now, a glass carrier with a 200mm diameter is used for RDL process. Simplified process steps are shown in figure 4 below. Some of the typical challenges encountered in the processes are metal debris, uneven PI thickness, via crack, excessive warpage, etc. Initially DOE started with 10/10-line space to check process repeatability and quality. 3/3 and 2/2-line space

capability have already been demonstrated. Figure 5 shows a SEM image of 2/2 line space multilayer uFOS.

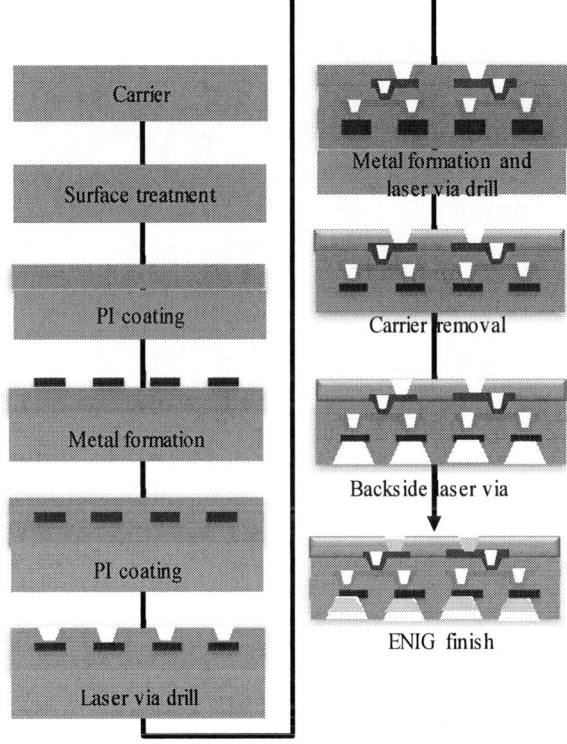

Figure 4: Typical uFOS process flow

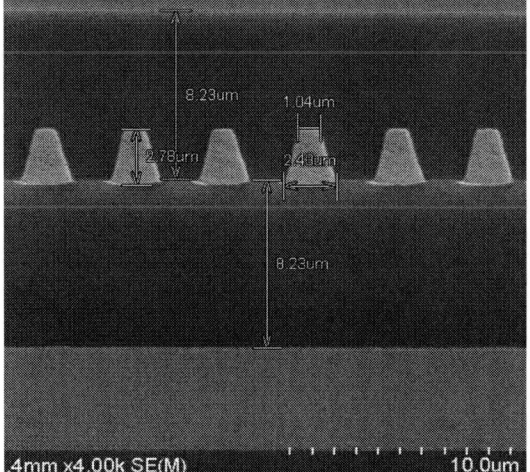

Figure 5: Typical SEM cross-section picture of 2/2 line space uFOS

A TV (test vehicle) was designed with 17x17mm six layers uFOS substrate at 15/15 line and space with 7x5mm flip chip die. Total uFOS thickness was about 55um. The package schematic is shown in Figure 6. Multiple DOE have been conducted to fine tune the flip chip assembly process to make sure package warpage is within the limit.

978-1-7281-1500-9/19 $31.00 © 2019 IEEE

After successful package assembly, component level data was collected as well to verify if any package level anomalies were observed prior to package level full qualification.

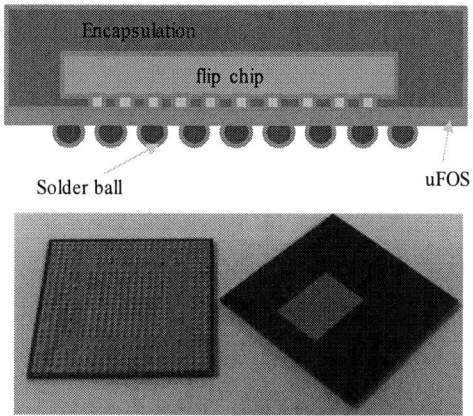

Figure 6: Assembled flip chip molded package with uFOS

V. 2.1D/2.5D

One of the key initiatives is to take advantage of uFOS high density routing and potentially use it for 2.1D (eHDF) or 2.5D applications. In the 2.1D program, fine line space uFOS is embedded in a multi-layer ABF type substrate so die-to-die or die-to-package level interconnection bridge can be achieved instead of using an expensive Si interposer type of package. It is also referred to as a "Hybrid substrate".

The micro-bridge uFOS substrate will be embedded only for the area where fine interconnection is required. Potential applications are FPGA, die-to-die or die-to-memory for graphics, networking, or any other die partitioning applications. At 2/2-line space with multilayer RDL, uFOS can be used for high bandwidth memory (HBM) integration with an ASIC. Other key advantages are high frequency performance, better thermal, etc. A representative picture of 2.1D uFOS is shown in Figure 7 where uFOS micro bridge is embedded in standard multi-layer ABF substrate.

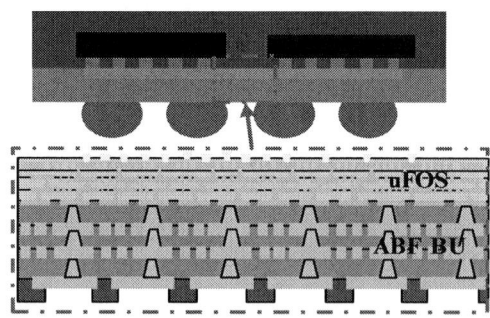

Figure 7: Representative cross-section picture of eHDF with uFOS in ABF substrate

A 21X21mm eHDF TV was selected to demonstrate fine line space (3/3) uFOS process capability. It was with a six layer uFOS with a total thickness of 38um. Comprehensive mechanical reliability data were collected from the TV. Based on the limited sample data, no big anomaly or abnormality were observed (Table 2). A full package level qual with JEDEC standard sample is under consideration.

Reliability items	Criteria	Sample Size	Result
Multi-reflow Ball shear (0x, 1x, 5x, 10x)	Spec> 3.875(kg/mm2) (BGA open:0.35mm)	20	Pass
Pre-condition: MSL3 30°C, 60%, 192HR, + Reflow*3	JEDEC-J-STD-020D.1	60	Pass
HAST (unbiased) 130°C, 85%RH, 96HR	JESD22-A118	60	Pass
PKG. TCT -40°C/125°C 1000cycles Test condition: G Soak mode: 2	JESD22-A104C	50	Pass
PCB(FR4) Board Level TCT -40°C/125°C, 1000cycles Test condition: G Soak mode: 4	JESD22-A104C	50	Pass
PCB(HDI) Board Level TCT -25°C/90°C, 1000cycles, 15min/cycle		20	Pass
Board Level Drop Test (*) (0.5ms, 1500G, 30 drops)	JEDEC22-B111	30	Pass
4-point Cyclic bending test (*) (2mm displacement, 1Hz, 200K cycles)	JEDEC22-B113	20	Pass

Table 2: Mechanical reliability data for an eHDF uFOS

Traditional expensive Si interposers are widely used in 2.5D packages. The limited supply chain is another bottleneck in Si interposer type package. Typically, foundries maintain Si interposer business for 2.5D packages. As mentioned before, a minimum 2/2 line space with 3 layers RDL is needed. Currently uFOS demonstrated the capability up to 2/2 line space which can replace the expensive Si interposer. Moreover, uFOS low Dk, Df properties will be an added benefit for high-speed applications. A comparison of 2.5D structures between a traditional Si interposer package and uFOS is shown in Figure 8. Lots of initiative have been taken to qualify 2.5D uFOS package. Detailed assembly and reliability data are expected to be published by year-end.

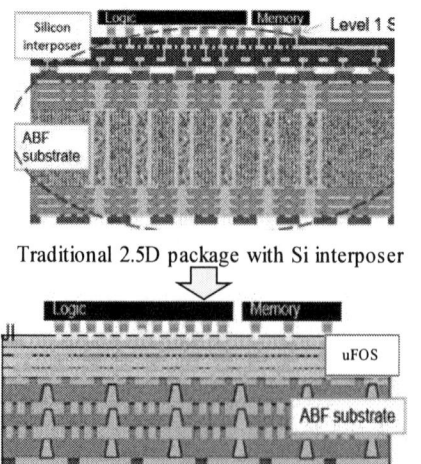

Traditional 2.5D package with Si interposer

2.5D package with uFOS embedded in ABF substrate

Figure 8: 2.5D with Si interposer vs uFOS

VI. Challenges with Hybrid Substrates

One of the main challenge with hybrid coreless substrates is to maintain flatness of the thin substrate during manufacturing process. Excessive substrate warpage can easily lead to non-wetting of the package during the die attach process. Other warpage related failures such as via or trace crack can be observed as well. In order to mitigate such issues, an embedded stiffener (e-STF) in the last layer of coreless substrate is introduced during the substrate manufacturing process. Metallic pieces are impregnated with prepreg to enhance the substrate's overall stiffness. A picture of a typical hybrid coreless substrate is shown in Figure 9. Significant warpage improvement was observed when comparing with and without stiffener conditions.

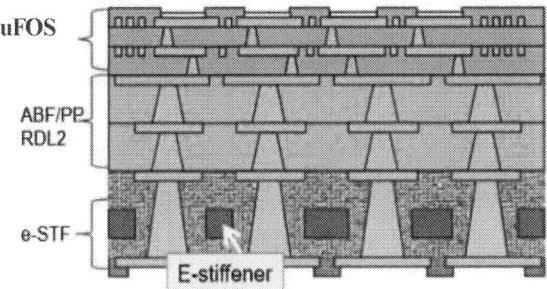

Figure 9: Thin hybrid substrate with e-STF

VII. Future work

Initial process and reliability data proved that uFOS is one of the promising high-density ultra-thin organic substrates that can be potentially used for a wide variety of applications from consumer mobile and handheld, to high-density high performance memory, networking and data center applications. Future work primarily focus on complex fine pitch interconnect where a cost benefit is the primary motivation.

A significant amount of work still needs to be conducted to make sure uFOS is much more robust and economical for various applications. Adoption of uFOS is expected to be very rapid once comprehensive package and board level qualification data are available.

Acknowledgement

The authors would like to thank the JCAP team in China and ABF, the substrate supplier for JCAP, for their continued guidance in the study. Our gratitude to Nanya PCB supplier in Taiwan, Dr. Qiqun Hu at Splus. The authors want to express gratitude to the individuals at our partner companies that helped design and support the advanced technology.

References

[1] Fred Lee et al, "Ultra-Thin Substrate Assembly Challenges for Advanced Flip Chip Package" IWLPC, Oct 18-20th, 2016. San Jose, CA.

[2] Chih Kuang Yang et al, "Systems on Film Type Substrates: Process, Structure, and Real Modules" 48th Annual International Symposium on Microelectronics, *IMAPS*, Oct 27-29th, 2015. Orlando, FL.

[3] Zihan Wu et al, "Modeling, Design and Fabrication of Ultra-thin and Low CTE Organic Interposers at 40μm I/O Pitch", *Electronic Components and Technology Conference, 2015. ECTC 2015. 65th,* San Diego, CA, pp. 301-307, May 26th-29th, 2015.

[4] Katsumi Kikuchi et al, "High Performance fcBGA Based on Ultra-Thin Packaging Substrate", NEC Journal of Adv Technology, summer, 2005, pp. 222-228, NEC, Japan.

Laser Releasable Temporary Bonding Film with High Thermal Stability

Yong-suk Yang, Kyo-sung Hwang, Robin Gorrell *

3M Korea, 7, Samsung 1-ro 5-gil, Hwaseong-si, Gyeonggi-do, Korea.
*3M Corporation, 6801 River Place Blvd., Austin, TX 78726-9000, United States.

Abstract— A temporary bonding film using a unique, laser releasable pressure sensitive adhesive construction has been formulated to advance wafer level fan-out process technology development. The film format and excellent thermal and dimensional stability of the temporary bonding film are particularly well suited to address the developing needs of panel level fan-out processes. The development of wafer level packaging process technology provides many benefits, including increased density and performance, greater design flexibility, simplification of the supply chain and process, and improved yield and overall cost of the semiconductor package. Specifically, in the case of fan-out, technology advances allowing wafer thinning to as little as 30 microns, backside patterning, and handling of the wafer or panel during copper RDL processing have become possible.

The thinning and handling processes on the molded wafer or panel are accompanied by large thermal, physical and chemical stresses. These external stimuli, which are difficult to overcome with non-rigid semiconductor materials such as poly-silicone or organic encapsulants, can significantly reduce the yield of these processes. The most effective approach to protecting the semiconductor package substrate from these process stresses is the use of a support carrier. The combination of a rigid carrier and a semiconductor device reduces or eliminates physical stresses in the process and can control the warping and expansion or contraction of the substrate during thermal exposures. Various carrier materials can be used, including thick silicon wafers, glass, ceramics, metals and others, with the choice of carrier dependent on a multitude of factors such as suitability for the carrier release process to be used, carrier CTE, reusability, and cost.

3M's laser releasable temporary bonding film supports bonding between semiconductor device and carrier with excellent adhesion. The benefits of process simplicity and the high level of material heat tolerance achieved with the 3M technology are well aligned with the current direction of packaging technology, with the film format allowing for processing of either wafers or large panels. 3M's laser releasable temporary bonding film employs a glass carrier. Laser scanning through the glass carrier enables separation of the carrier without physical stress, allowing thin materials to be easily removed from the carrier without physical damage. The heat resistance of the 3M temporary bonding film of up to 250C for an extended period allows precise passivation layer curing and sputtering processes. With the 3M temporary bonding film, process designers will be able to implement package production processes with a wide process margin and superior quality.

Key Words: Fan-out WLP, Temporary bonding de-bonding film, RDL-first, RDL-last, Laser releasable film

I. WAFER LEVEL PACKAGING AND FAN-OUT WLP

Wafer Level Packaging Technologies are being widely adopted in the semiconductor packaging industry based on a variety of benefits, such as enabling smaller package form factors, improved thermal and electrical performance, lower power consumption and lower manufacturing cost. Wafer Level Packaging technologies have led the market beyond existing packaging technologies such as FBGA and QFN in terms of performance and cost [1]-[2]. The fan-out process has also brought semiconductor micro-patterning technology into the realm of packaging. Redistribution layers (RDL), implemented with micro-patterning technology and the wafer reconstitution / molding process can provide smaller chips and more I / O, greatly contributing to maximizing the performance of the package and reducing the production cost. In the fan-out process, PCB interposers are no longer needed. The fan-out process can be divided into RDL-first and RDL-last when considering the RDL build-up process. In the case of an RDL-first process, where the redistribution layers are formed first, and then combined with device chips, the RDL quality does not affect chip loss. This is due the ability to inspect or test the RDL and not populate non-yielding sites. However, the RDL-first process itself is more complicated, requiring a high level of adhesive surface uniformity to achieve good RDL quality [1]-[3].

On the other hand, RDL-last is a process of directly building RDL on a redistribution wafer prepared by picking and placing singulated chips on a temporary carrier that is overmolded with specially formulated, low CTE epoxy molding compound. Cleaning of the adhesive residue after de-bonding is easy and the process itself is simpler than the RDL-first process. However, as shown in Figure 1 below, serious warpage of the mold wafer can occur during curing of the passivation layers, and is one of the biggest difficulties of the current RDL-last process [3].

Figure 1, Delamination between a carrier and a device due to severe device warpage during high temperature processing. If the gripping force of the temporary bonding film is insufficient, severe device warpage occurs during the high temperature process.

978-1-7281-1500-9/19 $31.00 © 2019 IEEE

II. 3M Laser Releasable Temporary Bonding Film

The target format for fan-out technology is transitioning from wafer to panel, particularly at OSATs that may not have existing wafer processing infrastructure, and where the cost advantage of large panel formats up to 600 x 600 mm is significant when compared to 200 or 300 mm diameter wafers. However, in general, spin-coatable materials that have heretofore been the mainstay of temporary bonding / de-bonding are difficult to apply in panel applications. On the other hand, an adhesive in film format is free from these limitations. Unlike liquid spin-coatable materials, where the bonding material and release material must be applied to the substrate and carrier through a separate process, film-based materials can be easily applied in a simple roll lamination process.

3M has developed a family of glass-carrier based, laser detachable temporary bonding / de-bonding solutions for the fan-out process (RDL-first process and RDL-last process).

Laser releasable temporary bonding film (a multi-layered product described in the next section) provides exceptional panel / wafer grip performance for the RDL-last process. Strong bending that occurs in thick mold wafers can be effectively controlled even at high temperatures over 200C. In addition, passivation curing processes requiring up to 230C for as long as 4 hours can be carried out. Support for high temperature processing for up to 10 hours is possible if ramp up and ramp down are included.

Laser releasable temporary bonding film (a single layered product) is a laser releasable adhesive film that can be de-bonded with low stress. The laser releasable film supports the RDL-first process with excellent thermal stability and surface roughness uniformity. Pre-aging of the film at 150C for 30 minutes can increase the adhesion and physicochemical stiffness of the film to the carrier. The maximum thermal stability of the laser releasable layer after thermal aging is up to 300C. The RDL layer is directly built-up on the adhesive through the metal deposition and patterning processes and the passivation curing process. After the RDL process is completed, the die is placed onto the RDL that was built. The laser releasable temporary bonding film (single layered product) can effectively support the RDL build-up process and die bonding and molding processes without delamination [5].

III. RDL-Last (Chip First) Process

3M Laser releasable temporary bonding film (multi-layered product) is a 3 in 1 structure as shown in Figure 2 below. Among the three layers, the bonding layer forms the bond to the substrate, and is composed of a high molecular weight thermoplastic elastomer, which shows excellent adhesion to substrates, and with exceptional thermal stability. This allows the temporary bonding film to support multiple passivation curing processes without delamination based on its excellent high temperature wafer grip force. After process completion, the adhesive layer can be fully removed without leaving residue as shown in Figure 3. The middle layer is a core film for providing structural support,

and selected to minimize CTE and warpage of the bonded substrate and carrier. The bottom layer is a pressure sensitive adhesive which is bonded to the glass carrier and also provides the laser release function.

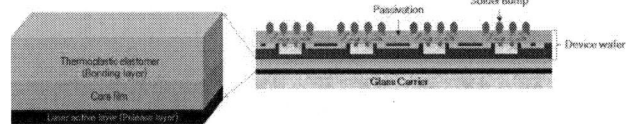

Figure 2, 3M Laser releasable temporary bonding de-bonding film (multi-layered product) for supporting RDL-last process

Figure 3, Passivation curing simulation test (230C X 2hrs X 2 Times) for actual bonding wafer with 3M Product, A) Bonded device with carrier by using 3M Product, B) Maintaining grip force of 3M Product for glass carrier during passivation layer curing process, C) Mold wafer after carrier de-bonding & adhesive peeling; Device wafer without warpage & adhesive residue.

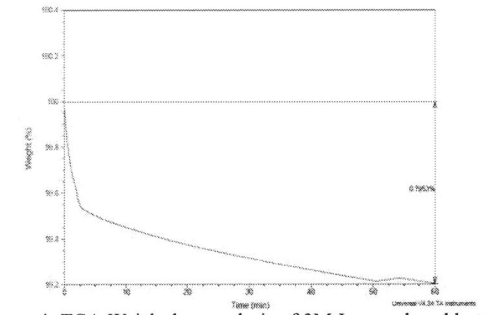

Figure 4, TGA Weight loss analysis of 3M Laser releasable temporary bonding film (multi-layered product): 0.793% @ 230C for 1hr, N2 Gas

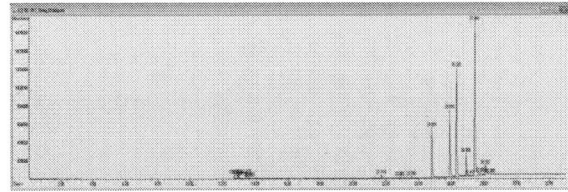

Figure 5, Residue Analysis on mold wafer by GC-Mass

As shown in Figure 3, after a passivation curing simulation test at 230C for 4 hours, the laser releasable temporary bonding film (multi-layered product) showed excellent wafer bonding performance with strong wafer grip force. Thermal defects were not found on the release layer surface. In addition, the temporary bonding film has demonstrated weight loss values of less than 1% in TGA weight loss analysis after simulated passivation curing as shown in Figure 4.

To check whether adhesive residue remained on the surface of the mold wafer after thermal processing, the mold wafer was cut and put into a capsule containing solvent and then heated at a high temperature for 4 hours. The solvent was analyzed precisely by gas chromatography – mass spectrometry (GC-MS) for the elements constituting the adhesive. As a result of the evaluation, the corresponding elements were not found as shown in Figure 5. We could confirm that the temporary bonding film did not leave adhesive residue after the passivation layer curing process.

IV. RDL-FIRST PROCESS

The RDL-first process can enable better RDL quality and lower production cost compared to the RDL-last process. As shown in Figure 6, the RDL-first process builds up the RDL directly on the release adhesive surface, so the adhesive strength and surface roughness can directly affect the quality of the RDL [4].

Figure 6, 3M Laser releasable temporary bonding film (Single layered product) for supporting RDL-First process

As shown in Figure 7 below, the laser releasable temporary bonding film (single layered product) has a laser active ingredient dispersed to a nano-scale level and a high molecular weight elastomer binder, providing exceptional thermal stability and excellent laser de-bonding performance. Numerous polar functional groups in the binder resin also play a major role in enhancing interfacial adhesion between the carrier and the film. Extra functional groups that are not involved in carrier attachment can be crosslinked to each other through a pre-aging process under high-temperature to provide higher thermal stability and chemical resistance [5].

The excellent surface uniformity of the laser releasable temporary bonding film (single layered product) allows high quality RDL build-up. Generally, to support the RDL process without any problems, an Ra value of less than 0.5µm is required.

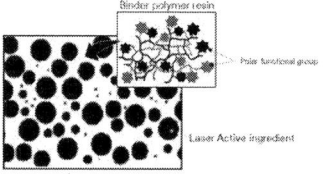

Figure 7, Design concept of 3M Laser releasable temporary bonding film (single layered product)

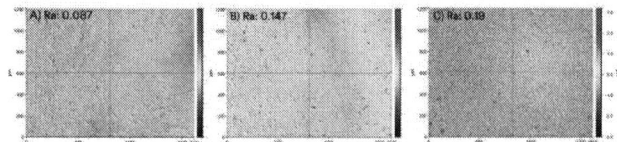

Figure 8, Surface roughness analysis of 3M Laser releasable temporary bonding film (single layered product) by Bruker confocal, A) Ra value after glass lamination, B) Ra value after metal deposition (Cu), C) Ra value after thermal aging (@ 210C for 2hrs)

Condition	Reference**	3M TBDB Film
RT		
Aging for 1hr at 300C*		

Table 1, Metal deposition process simulation test of 3M Laser releasable temporary bonding film (single layered product) @ 300C for 1hr in Air, *Using the flash for clear picture, **Reference sample: Initial version laser releasable film

As shown in Figure 8, the initial surface roughness value of the laser releasable temporary bonding film (single layered product) after lamination onto the carrier is very good at 0.087µm. Even after metal deposition and thermal aging, the Ra value does not exceed 0.2µm, which is also very good. the process for metal deposition is usually carried out under high temperature and vacuum conditions above 200C. The high temperature conditions under vacuum can cause much larger damage to the organic material than in the atmosphere. We conducted a thermal aging test for the laser releasable temporary bonding film (single layered product) at 300C to simulate this [6].

As shown in the lower right image of Table 1, no delamination or voiding was observed at the interface between carrier and film, despite the high temperature and atmospheric conditions of 300C. If the adhesive's adhesion or thermal stability were insufficient, voiding or delamination may occur such as shown in the reference sample. Generally, a 300C temperature has been used for simulating the metal deposition process, suggesting that no problems are anticipated in a typical metal deposition process with high vacuum condition of 220C [6].

V. DE-BONDING PROCESS WITH LASER SCANNING

The laser releasable temporary bonding films allow robust carrier de-bonding processes for lasers of various wavelengths. Currently, the temporary bonding films can be successfully de-bonded using 308, 355, and 1,064 nm wavelength lasers. De-bonding is possible without damaging the device with low stress as shown in Figure 9.

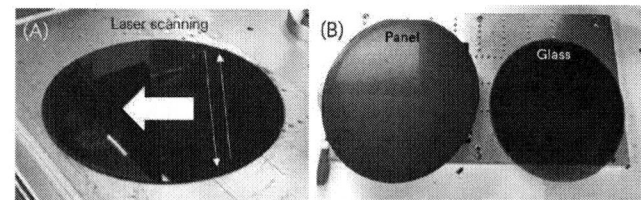

Figure 9, Laser de-bonding processing, (A) Laser scanning, (B) Demounted device wafer from a carrier

In addition, 3M laser releasable temporary bonding films can completely block any leaked laser energy during the debonding process. The laser releasable temporary bonding film exhibits superior laser blocking performance compared to spin-coatable materials. We evaluated the laser energy blocking performance of the temporary bonding film for the most intrusive IR Laser (1,064 nm wavelength) under conditions equivalent to the actual de-bonding process. For reference, a range of process conditions can be acceptable for the de-bonding process, even with laser equipment of the same wavelength. For example, if the laser beam width is wide, lower energy can be required. In general, a narrow beam is often used in a laser marking process and has a higher energy per unit area than a laser with wider beam width. In this evaluation, we used a beam of narrow width rather than the normally recommended de-bonding conditions to evaluate the laser blocking performance of 3M products under the most severe conditions. Of course, the condition evaluated (0.4mm X 20W) is also a condition that results in acceptable de-bonding.

As shown in Figure 10 and Table 2, a 20W laser dosage under the above evaluation conditions was used to carry out the de-bonding process. Unlike the spin coated reference material, the temporary bonding film provided complete shielding from the laser energy as shown by the laser energy detector. Superior barrier performance of the temporary bonding film to laser energy provides a much wider process margin in the design of the fan-out process, and in selection of materials.

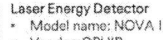
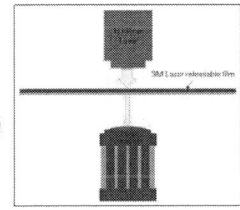

Figure 10, Laser energy blocking performance test, Measure transmittance of laser energy by using Laser energy detector (NOVA II)

Sample	Reference			3M Laser releasable film		
Scanning Energy (W)	10	15	20	10	15	20
Detected Energy (W)	3.7	2.7	2.1	0	0	0

Table. 2, Laser blocking performance evaluation result of 3M Laser releasable temporary bonding film

CONCLUSION

The fan-out process has brought semiconductor micro-patterning technology into the realm of packaging. Redistribution Layers (RDL), implemented with micro-patterning technology and the wafer reconstitution / molding process can provide smaller chips and more I / O, greatly contributing to maximizing the performance of the package

and reducing the production cost. Differences in the RDL process, which is the core of the fan-out process, result in differing process simplification, quality, manufacturing yield and cost. No matter the RDL process used, the support of a suitable temporary bonding / de-bonding material is required.

The RDL-first process, which can reduce chip loss and improve RDL quality, is achieved through the use of a single release layer with uniform and low surface roughness, as well as excellent adhesion [4].

In the case of an RDL-last process, with a simpler process enabled by building-up the RDL directly on the reconstituted wafer, it is necessary to employ a multi-layered type material which can prevent the warpage of the device.

The 3M laser releasable temporary bonding film enables the de-bonding process with low stress, and without device damage. The multi-layered film provides outstanding wafer grip force and the thermal stability required for an RDL-last process.

ACKNOWLEDGMENT

The authors would like to acknowledge YJ Park of 3M Korea Analysis lab for analytical supports.

REFERENCES

[1] Michelle Fowler, Christopher Apanius, and Kimberly Yess in Brewer Science, Inc., "High-temperature survivability and the processes it enables", Chip Scale Review, vol. 22, Number 6, pp. 14–17, Nov. 2018. *(references)*

[2] Ruben Fuentes, August Miller, and Jonathan Micksch in Amkor Technology, Inc., "Package assembly design kits bring value to semiconductor designs", Chip Scale Review, vol. 22, Number 6, pp. 39–42, Nov. 2018.

[3] John H. Lau et al., "Design, Materials, Process, Fabrication, and Reliability of Fan-Out Wafer-Level Packaging", IEEE TRANSACTIONS ON COMPONENTS, PACKAGING AND MANUFACTURING TECHNOLOGY.

[4] Vempati Srinivasa Rao et al., "Development of High Density Fan Out Wafer Level Package (HD FOWLP) with Multi-layer Fine Pitch RDL for Mobile Applications", 2016 IEEE 66th Electronic Components and Technology Conference, 978-1-5090-1204-6/16 $31.00 © 2016 IEEE, DOI 10.1109/ECTC.2016.203, pp. 1522-1529.

[5] Hong Zhang et al. in Brewer Science, Inc., "Novel Temporary Adhesive Materials for RDL-First Fan-Out Wafer-Level Packaging", 2018 IEEE 68th Electronic Components and Technology Conference, 2377-5726/18/$31.00 ©2018 IEEE, DOI 10.1109/ECTC.2018.00289, pp. 1925-1930.

[6] Chia-Hsin Lee et al. in Brewer Science, Inc., "Optimization of laser release process for throughput enhancement of fan-out waferlevel packaging", 2018 IEEE 68th Electronic Components and Technology Conference, 2377-5726/18/$31.00 ©2018 IEEE, DOI 10.1109/ECTC.2018.00273, pp. 1818-1823.

Design And Demonstration of 1µm Low Resistance RDL Using Panel Scale Processes for High Performance Computing Applications

Bartlet DeProspo, Aya Momozawa*, Atsushi Kubo*, Chandrasekharan Nair, Varun Rajagoapal, Jenefa Kannan, Emanuel Surillo, Fuhan Liu, Mohananlingam Kathaperumal, Rao Tummala
3D Systems Packaging Research Center, Georgia Institute of Technology, Atlanta, GA, USA
*Tokyo Ohka Kogyo Co. LTD.
Email: bdeprospo@gatech.edu

Abstract— This paper presents for the first time the latest challenges and solutions to enable electronics packaging redistribution layer (RDL) scaling to 1µm and beyond. The focus on RDL scaling for this paper is on how to scale semi-additive processing (SAP) for next generation high performance computing applications such as 2.5D Interposers. This paper combines novel next generation photoresist materials developed by Tokyo Ohka Kogyo Co., LTD. (TOK) and process innovations to the traditional SAP process. Traditionally, challenges in scaling SAP are related to seed layer etching and photoresist materials selection. This paper address both of these challenges by exploring various seed layer metals and the potential impact they can have on the SAP process flow for enabling SAP scalability as well as novel photoresist development with TOK. Scaling dry films to have similar performance to a matching liquid photoresist is demonstrated in this paper. The 3D Systems Packaging Research Center remains one of the leaders in package RDL scaling and this paper discusses at length recent advancements that have been made to enable silicon like RDL scaling on glass panels.

Keywords- RDL, High Performance Computing, Interposers

I. INTRODUCTION

Driven by emerging markets that are focused around the Internet of Things (IOT), 5G, artificial intelligence, cloud computing and autonomous driving high performance computing is experiencing unprecedented growth and demand. Increasing the logic to memory bandwidth as well as reducing power consumption at low cost are the key performance metrics for technology advancements in this area. To enable increases in logic to memory bandwidth critical dimension (CD) of the RDL

traces must be reduced to meet these demands. For reducing power consumption high aspect ratio traces are required to offset power losses due to capacitive impacts. Traditional approaches to reducing CD are to leverage back-end-of-line (BEOL) technologies to reach on chip like dimensions [1]. However, utilizing this technology results in a low aspect ratio that is prone to high RC losses in the package. To reduce power consumption BEOL processing tries to leverage low dielectric constant materials to reduce capacitive losses and high numbers of traces to distribute line resistance. However, these types of dielectrics at packaging feature CD's can be expensive due to time of deposition and loss in reliability due to mechanical instability. Traditionally packaging has struggled to scale low-cost high performance processes as shown in figure 1. The goal of this paper is to demonstrate that the materials and processes are in place to support next generation chipsets that have CD's of 7nm and below. In this figure there is a clear RDL gap that exists between what package foundries are traditionally capable of and what wafer foundries are traditionally capable of.

II. SAP SCALING

Traditionally in electronics packaging the process by

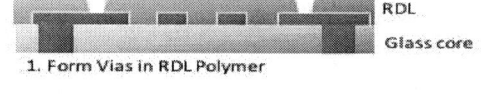

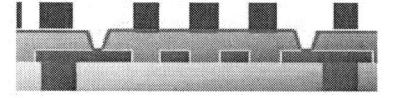

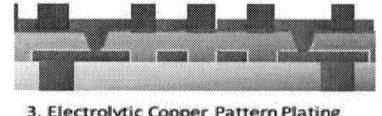

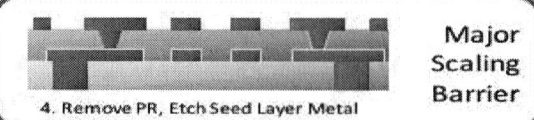

Figure 2. SAP Process flow with critical scaling barrier highlighted in blue.

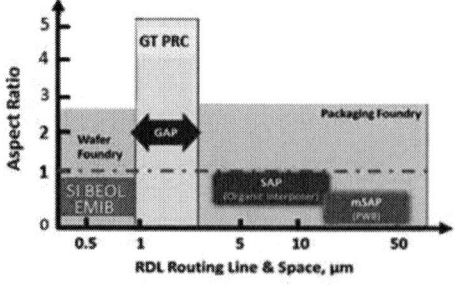

Figure 1. Gap between Si BEOL and Packaging RDL

which RDL are fabricated is through Semi-Additive Process (SAP). This process flow is shown in figure 2 with the limiting process step highlighted. As the demands for high performance computing applications continue to increase the requirement of RDL critical dimension (CD) continues to be pushed. Seed layer etching and photoresist processes remain the critical barriers to scaling the SAP process for next generation electronic packaging architectures. As the CD of RDL is pushed to 1μm and below seed layer etch remains an increasingly complex problem. Traditional seed layers are approximately 100 – 200 nm for sputtered seed and up to 500 nm for electroless seed layers traditionally deposited onto polymers such as ABF and other polymers used in electronics packaging [2]. This remains a problem because during the seed layer etch removal process to ensure that the entire seed layer is removed one must target a thickness greater than that of the seed layer that was deposited. Traditionally it hasn't been a problem as the CD is large enough that the thickness of the seed layer is <10% of the final CD of the line. However, as the linewidth approaches 1μm (and <1 μm) RDL continues the seed layer thickness exceeds 10% and can be as high as 20% of the final CD target. At 1μm, the seed layer etch process often results in the lines to flop over due to undercut from the seed layer etch. It has also been demonstrated that the etch chemistries remove the seed layer faster than they remove the plated copper likely due to the grain structure differences between the plated copper and the sputtered as deposited copper. The second most critical scaling parameter for electronics packaging is the photoresist itself. The only technology capable of fabricating line and spaces of 1μm and below is back-end-of-line (BEOL) copper damascene processing. This process from a materials standpoint uses ultra-thin chemically amplified liquid resists on the wafer scale to etch fine line and spaces into the dielectric material, usually SiO_2. This process does not have a seed layer etch problem nor does it have a resist scaling problem due to the nature of the process flow. To scale SAP to match the CD capability of BEOL processing photoresists that are required need to be thicker and still maintain performance. In conjunction with this material requirement the lithographic tooling must also be able to meet a minimum resolution of 1μm or below [3]. However, scaling numerical aperture (NA) causes a loss in depth of focus (DOF) which makes maintaining CD throughout the thickness of these resists difficult for electronics packaging fabrication. Ultimately, without key innovations in technology for SAP processing the tradeoffs must be weighed from photoresist requirements, to seed layer etching technologies and the lithographic tooling requirements. The remaining sections of this paper discuss and explore various approaches that can be explored to compensate for critical transitions required to scale SAP to 1μm[4].

III. SAP PROCESS FLOW

This section describes in detail the process flow that was utilized for all the experiments that will be discussed throughout the remainder of the paper unless otherwise specified.

(1) Dielectric Film Lamination

The glass is precleaned with acetone, IPA, methanol and O2 plasma. A vapor silane is applied to the surface prior to lamination. The 10 μm T61 is laminated on a glass panel using a vacuum hot press laminator. The temperature of lamination is 130 °C and is cured at 180 °C.

(2) Metallization

A conductive metal seed of 50 nm thick titanium and 150 nm copper is deposited using sputtering. Additionally, samples that have other seed metals such as chromium are deposited in thicknesses of 200 nm.

(3) TOK Photoresist Processing

TOK photoresists are then spin coated or laminated onto the top of the conductive seed metal. All the films were optimized to be deposited with a thickness of 7 μm. TMMR®P-W1000T which is a non-chemically amplified liquid resist was spin coated and then pre-baked. The other films used in this study were a PC series developed by TOK. This PC series is a chemically amplified photoresist that is available as a liquid and dry film. The PC-0471W-F8 (PC dry film) as received has a thickness of 8μm and after a pre-bake step it shrinks to 7μm. The PC-0471W (PC Liquid) photoresist is spin coated to achieve a thickness of 7μm and is also pre-baked. All these films are ready for exposure after the completion of their pre-bake steps.

(4) Exposure and Post-Bake:

After the photoresist has been pre-baked they are then exposed in various photolithographic steppers. Each stepper is used to deliver a range of exposure doses across the surface of the panel. These exposure doses are varied based on the recommendations from TOK. This provides a wide range of exposure doses across each of the panels to minimize the possibility for panel to panel variations. Following the exposure, the photoresist is subjected to a post-bake step.

(5) Development

The development process used was puddle development where the wafers are coated in 2.38% TMAH and then spun two times for 30s each.

Any changes to the process flow or additional data that is presented in this paper will be discussed in detail in the later sections.

IV. RESULTS AND DISCUSSION

In the following section, discussion will be dedicated to, A) the photoresist material selection and discussion, and B) the overall process with a focus on seed layer.

The first aspect of this paper is photoresist material selection and discussion. For the purposes of this paper a small photoresist matrix was selected. Based on scaling requirements and understanding of fundamental transition points in electronics packaging three photoresists were selected. The first is a positive non-chemically amplified liquid photoresist by TOK called TMMR®P-W1000T. The next is PC-0471W which is a new photoresist material developed by TOK which is a positive chemically amplified photoresist that is available in liquid and dry film formats. This matrix allows us to examine when the limit for non-chemically amplified photoresists is reached as well as dry film photoresists. For the dry-film a positive tone chemically amplified photoresist was selected as opposed to the more traditional negative dry-film due to inherent swelling challenges that exist for negative tone dry-films. These photoresists typically, in the semiconductor industry, are shown to have a poor

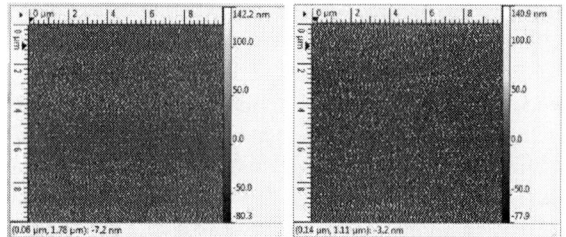

Figure 3. Surface AFM scans of a) PC Dry film and b) PC Liquid resist after pre-bake conditions

resolution. To make sure that each of the photoresists could be compared accurately, and more specifically the liquid and dry-film photoresists, certain process optimizations were performed. The liquid photoresist was spun on using traditional spin coating processes. The dry film was vacuum laminated onto the surface of the conductive seed metal using 0.3 MPa of force at a slightly elevated temperature. It was expected that due to the nature of processing the photoresist into a dry-film that the surface roughness should be greater compared to the

liquid version of the resist. To examine this potential difference in surface topography various methods were investigated, however AFM surface roughness measurements performed using a Veeco Dimension 3100 Atomic Force Microscope (AFM) are shown in Figure 3. The AFM was run in both tapping and contact mode however, tapping mode was chosen based off initial results that suggested contact mode was smearing the photoresist. However, after pre-bake it was observed that the surface roughness for the dry film was reported with an arithmetical mean deviation of the assessed profile (R_a) of 35.7nm. While for the liquid film the measured R_a value was found to be 34.2nm. As a result, we found that while there was a difference between the two films of 1.5nm overall the difference was not significant enough to produce performance differences for the two versions of the film. Additionally, the root mean squared (R_q) surface roughness values were also recorded. For PC dry film the R_q value was recorded to be 42.1nm and for PC liquid it was recorded to be 40.4nm for a difference of 1.66nm between the two versions of the film. This difference was deemed to be insufficient in dictating any type of performance difference between the two films. This proves that from a processing standpoint that liquid photoresists are capable of being converted into dry film formats that are traditionally used in electronics packaging specifically for large panels. Ultimately, for large panel processing dry film photoresists show known inherent challenges, from a physical aspect, in terms of scaling panel based RDL.

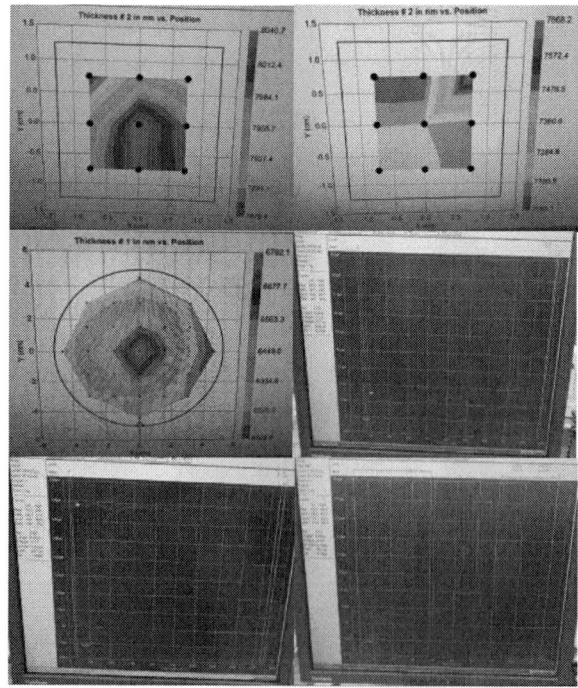

Figure 4. Ellipsometry scans of a) PC liquid b) PC Dry and c) TMMR®P-W1000T as well as profilometry scans to validate thickness of d) PC Liquid e) PC Dry and f) TMMR®P-W1000T

The second piece of the photoresist that had to be optimized was the thickness of the resist for the PC liquid as well as the TMMR®P-W1000T resist. The PC dry film was only available in 8μm as received and needed to be measured after pre-bake to see the final thickness. A target height of 7μm was selected for each of the photoresists, the PC dry film, PC liquid and TMMR®P-W1000T. The reason behind this selection was because the goal of scaling RDL using SAP on large panels is to have similar density but higher performance than silicon BEOL RDL. A height of 7μm gives plenty of tolerance for electroplating of copper of up to 6μm in thickness. This means that panel based RDL is much higher aspect ratio, while maintaining high density, ultimately lowering the resistance of the RDL traces by a factor of 4x when compared to the 1:1, width to height, RDL traces that have been demonstrated even in the most advanced silicon packages. To validate the thickness of each of the films both ellipsometry and profilometry were used. Ellipsometry measurements were conducted using a Woollam M2000 Ellipsometer and for profilometry a Tencor P15 Profilometer as shown in figure 4. The ellipsometry scans for the liquid films were repeated multiple times and the representative 7μm profilometry scans are shown. PC Liquid series ellipsometry and profilometry scans figure 4a and 4d allowed for optimization in the spin speed to be carried out to ensure that the final thickness of the PC liquid after pre-bake was 7μm. The PC dry film representative scans are show in figure 4b and 4e. PC dry film after pre-bake shows a final thickness on profilometer scans of 7μm this is consistent with the ellipsometry scans except for some areas where there were high spots likely due to processing variability at Georgia Tech and not due to the film quality itself. Finally, TMMR®P-W1000T ellipsometry and profilometry scans are shown in figures 4c and 4f, and

verify that this film was also 7μm in thickness. This optimization was done to make sure that when comparing the three-photoresist matrix there is no performance differences induced from differences in surface roughness or thickness.

The three photoresists were patterned using a FPA-5510iv Stepper tool by Cannon with a NA of 0.18 at i-line wavelength. The three photoresists are show in figure 5 at optimum dose. The optimum dose was selected by examining top down CD-SEM images for the photoresists at doses varying to 20% plus or minus outside of the recommended dose from TOK. After varying the dose for each of the photoresists and examining them from top down CD-SEM the optimum dose was then cross sectioned. A FEI Nova Nanolab 200 FIB/SEM was used to perform the cross sectioning of the photoresist. A Pd/Au seed was deposited at low power and over a long time to ensure that no damage was induced to the photoresist. After which a 1nA cut is performed to clear out the bulk of the photoresist. After this cut a 0.7nA cleaning cut is performed. Finally, a 0.3nA cleaning cut is performed in order to remove any residue and to cross section the photoresist with little to no melting from the ion beam. As can be seen in figure 5 both the chemically amplified liquid and dry photoresists performed better than that of the non-chemically amplified photoresist. The CD change from bottom to top of the non-chemically amplified resist at optimum dose, which is substantially larger than the optimum dose for the chemically amplified films, was 800nm. While for the chemically amplified films the CD change from bottom to top was between 200 and 400nm. For scaling RDL processing on panels to 1μm and beyond due to CD broadening a chemically amplified photoresist is required. The photoresists reported in this paper provide much better CD control and a reliable 1μm CD resolution of across large panels.

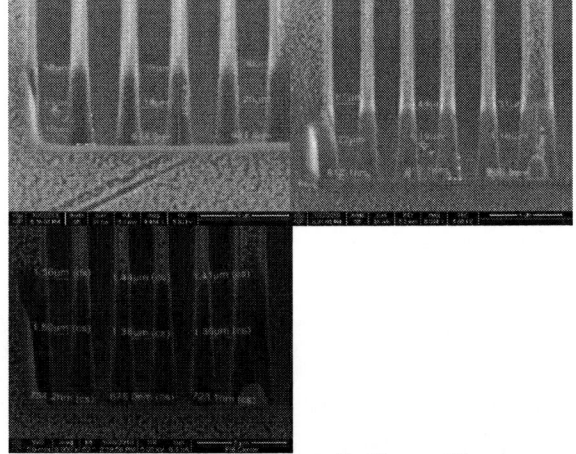

Figure 5. Focused Ion Beam Scanning Electron Microscopy (FIB/SEM) Cross sectional images of a)PC dry film b) PC liquid resist and c) TMMR®P-W1000T

	370nm	385nm	405nm
Cr	33.92%	23.22%	25.77%
Ti/Cu	47.17%	44.91%	46.49%

Table 1. Reflectance values of the seed layer as a function of wavelength

Previously discussed in this paper is the challenge of seed layer and its impact on scaling of SAP as a process to 1μm and beyond. From a tooling standpoint it is feasible to imagine that companies who make seed layer etch tools are working on technological developments that would allow for better control over the etching of the seed. However, the problem remains that unless the seed in the bulk can be selectively etched without impacting the seed layer under the traces than the physical challenges will always be present. Seed layers of up to 200nm are required in order to have low enough resistance for electrolytic plating of the copper traces. If the seed layer physically can not be shrunk to ensure electrolytic plating can take place than other solutions need to be explored. One of the aspects that is studied in this paper but that is not readily studied in literature is the impact of seed layer reflectance on photoresist performance. The changing of the seed layer to an entirely different material can provide ways to preferentially etch the seed metal in the bulk layers that is no longer desired after plating has taken place. For the purpose of this study chromium was selected as an alternative seed metal to the traditional titanium, copper seed layer that is used in electronics packaging today. To measure the reflectance values of the conductive seed layers at the litho tool exposure wavelength, a NanoSpec 3000 Reflectometer was used. The i-line wavelength that was used for the exposure of these photoresists is 365nm however, due to the limit of the reflectometer 370nm wavelength light was chosen to measure the reflectance of these seed layers reliably. As the wavelength is increased to h-line the seed layer reflectance delta between chromium and a titanium copper seed layer has a difference ranging between 13 – 21% as shown in table 1. A 150 nm thick Copper layer on top of a 50 nm Titanium seed layer is layer is compared to a 200nm chromium only seed layer. Surface AFM scans were performed to ensure that the surface of the seed layers was comparable. The AFM roughness scans for the

conductive seed layers are shown in figure 6. The measured R_a and R_q value for the titanium and copper seed layer was measured to be 3.4 and 4.3nm respectively. For chromium the R_a value was measured to be 0.8nm and the R_q value was 1.0nm. While these values are different, since both measured values are below 5nm it is reasonable that the roughness of these seed layers will not have any significant impact on photoresist performance.

After the conductive seed layer reflectance values were collected and the surface scanned with an AFM the photoresist was processed. The photoresists were spin coated or laminated in the same fashion that they were on the titanium copper seed. The photoresists were then exposed on the same Canon 0.18 i-line stepper tool and developed using the same process. The corresponding cross-sectional FIB images are shown in figure 7. When looking first at the PC dry film images and CD measurements for the two different seed layers it was observed that the minimum CD for the dry film was slightly larger on the chromium seed layer. However, an inverse effect was observed for the PC liquid where the CD profile was narrower on the chromium seed as compared to the titanium copper seed. For the TMMR®P-W1000T photoresist the mid and top height CD's were similar but the bottom CD for TMMR®P-W1000T was wider on chromium than it was on the titanium copper seed layer. Regardless of which seed layer was utilized the minimum CD that was able to be reached was 1μm. Thus, a pathway to differential chemical etching is presented for the first time in this paper. This technology would theoretically allow for scaling of SAP as a process to 1μm and beyond for electronics packaging. This is assuming that a chemical etch is available for the different seed layer that is used in the SAP scheme moving forward.

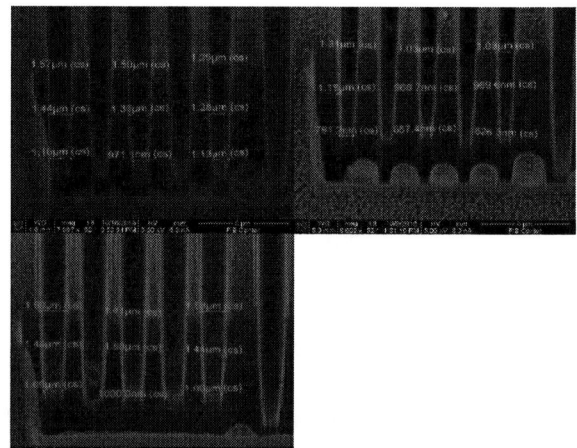

Figure 7. FIB/SEM Cross sectional images of photoresist spin coated on a 200nm chromium seed at optimum dose for a) PC dry film, b) PC-liquid resist and c) P-W1000T

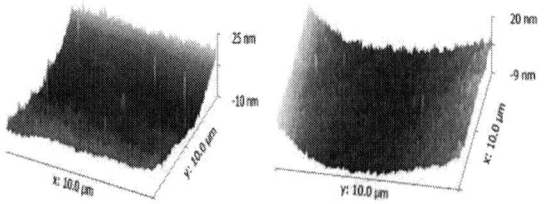

Figure 6. AFM roughness scans on blank a) Ti/Cu and b) Cr

V. Summary

In summary, there is a demand for maintaining traditional electronics packaging fabrication process in the supply chain for next generation RDL. Historically the only process capable of fabricating RDL that have a CD of 1µm are below is silicon BEOL processing. This process scheme utilizes ultra-thin liquid positive tone chemically amplified photoresists that are processed and transferred to a SiO_2 like dielectric. However, the aspect ratio is usually limited to one and due to the inherent nature of silicon as a core material the loss of these traces can be very high. To run signals across large body size packages on a substrate like this will require areas of the package devoted to repeaters to ensure that the signal arrives at the receiver without degradation in the signal quality. This paper demonstrates the feasibility and manufacturability of high density, low loss RDL traces with a focus on two critical process. The first being the photoresist as a material. Materials developed by TOK demonstrated that 1µm CD are possible in electronics packaging fabrication process schemes. These photoresists are available in liquid and dry film formats which is extremely beneficial to the packaging industry which has historically utilized dry-film exclusively. However, these dry films have been demonstrated to have performance similar to that of liquid photoresists allowing for scalability of SAP as a process using either liquid or dry film. The second aspect of this paper was to examine a potential processing scheme change to overcome the seed layer scalability challenges that exist in electronics packaging. This paper presented for the first time that by using a different conductive seed metal than copper the photoresist could still be placed onto the conductive seed layer with no degradation in photoresist performance in achieving a CD of 1 µm.

VI. Acknowledgements

The authors would like to acknowledge the continuous help and support of TOK for their help with photoresist development and processing, Intel for their input and guidance on this work, as well as the students, research scientists and Professors at the 3D Systems Packaging Research Center at Georgia Institute of Technology. This work could not have been possible without the help of the GT-PRC consortium members.

References

[1] Y. Kim, J. Bae, M.Chang, A. Jo, H. Kim, S. Park, D. Hiner, M. Kelly, and W. Do, "Slim™, High Density Wafer Level Fan-out Package Developmetn with Submicron RDL," 9-13, 2017 IEEE 67th Electronic Components and Technology Conference

[2] R. Furuya, H. Lu, F. Liu, H. Deng, T. Ando, V. Sundaram, R, Tummala, Demonstration of 2µm RDL wiring using dry film photoresists and 5µm RDL via by projection lithography for low-cost 2.5D panel-based glass and organic interposers, 1488-1493, 2015 IEEE 65th Electronic Components and Technology Conference

[3] F. Liu, H. Ito, R. Zhang, B. DeProspo, F. Benthaus, H. Akimaru, K. Hasegawa, V. Sundaram, R. Tummala, "Low Cost Panel-based 1-2 Micron RDL Technologies with Lower Resistance than Si BEOL for Large Packages" 2018 IEEE 68th Electronic Components and Technology Conference.

[4] B. DeProspo, F. Liu, C. Nair, A. Kubo, F. Wei, Y. Chen, V. Sundaram, R. Tummala. "FirstDemonstration of Silicon-Like >250 I/O Per mm Per Layer Multilayer RDL on Glass Panel Interposers by Embedded Photo-Trench and Fly Cut Planarization" In *Electronic Components and Technology Conference (ECTC), 2018*

Advances in Temporary Carrier Technology for High-Density Fan-Out Device Build-up

Arnita Podpod, Alain Phommahaxay, Pieter Bex,
John Slabbekoorn, Julien Bertheau, Abdellah
Salahouelhadj, Erik Sleeckx, Andy Miller, Gerald
Beyer and Eric Beyne[1]

imec
Leuven, Belgium
podpod@imec.be

Alice Guerrero, Kim Yess, Kim Arnold

Brewer Science, Inc.
Rolla, MO, USA
alice.guerrero@brewerscience.com

Abstract— **As the need for higher degrees of function integration on chips continues to rise, chip-to-chip connection density exponentially increases. The continuous push for denser interconnects has brought conventional FO-WLP to its limit. A novel FO-WLP concept has been proposed to enable 20-μm pitch interconnect chip-to-chip. To achieve this density and to further scale it down, a critical element is ultra-precise die-to-die positioning in the micron range. Advances in temporary bonding materials and carrier systems are required to enable such applications.**

Keywords: Wafer bonding, Wafer thinning, Temporary bonding material, adhesive, Die attach, Fan-out, Carrier debonding, Fan-Out Wafer-Level Package, Heterogenous Integration, Flip-Chip, Warpage, Die shift

I. INTRODUCTION

Temporary wafer bonding on a carrier was first introduced and developed with the introduction of 3D stacked IC technology and the emergence of through-silicon vias, which requires processing of ultrathin substrates. Various temporary bonding solutions have been developed to satisfy the need for thin wafer handling first-generation materials, including thermoplastic, laser-ablated, or chemically dissolvable adhesives, are readily available on the market and enable early 3D demonstrator processing. The solutions today enable low-stress carrier separation from device substrates at the end of the process while properly maintaining a thin substrate on the carrier throughout multiple and various backside process steps.

More recently, with the rise of fan-out wafer-level packaging, the temporary wafer bonding technology has been further derived and developed to accommodate the new constraints caused by processing reconstructed over-molded substrates. One of the major challenges when dealing with such substrate is stress holding. Indeed, one of the main differences compared to silicon is the large mechanical property mismatch between the reconstructed device and carrier. This main constraint is pushing the technology further. New temporary adhesive solutions must now cope with high and varying stress levels throughout the various process steps, while trying to maintain the wafer geometry (warp, bow) within SEMI standards.

II. HIGH DENSITY INTERCONNECT FAN-OUT CONCEPT

Imec is developing a novel 300-mm fan-out wafer-level packaging (FO-WLP) concept to enable 20-μm pitch chip-to-chip connection density within a package. This concept is based on flip-chip on FO-WLP and has been first described in [1] and [2]. The target final structure is depicted in Fig. 1 and is built using the assembly flow proposed in Fig. 2.

The high-density connectivity is realized by a silicon bridge consisting of a standard back end of line (BEOL) interconnect layer combined with fine pitch microbumps.

In order to obtain proper connectivity at such scale, the relative in-plane and out-of-plane tolerances between dies at step 2 and 3 in Fig. 2. are in the micron range.

This is not achievable with usual chip-first approaches where dies are simply placed and indexed by a die bonder. The precision required in this case is an order of magnitude higher. A comprehensive list of challenges is discussed in [2].

To reach the x, y, and z die position precision, we are proposing to use semiconductor-grade carrier substrates combined with aligning the dies to a temporary carrier with fiducials (Fig. 2.3).

This contribution will therefore focus on the requirements for such temporary carrier systems.

Figure 1.   High density interconnect FO-WLP concept

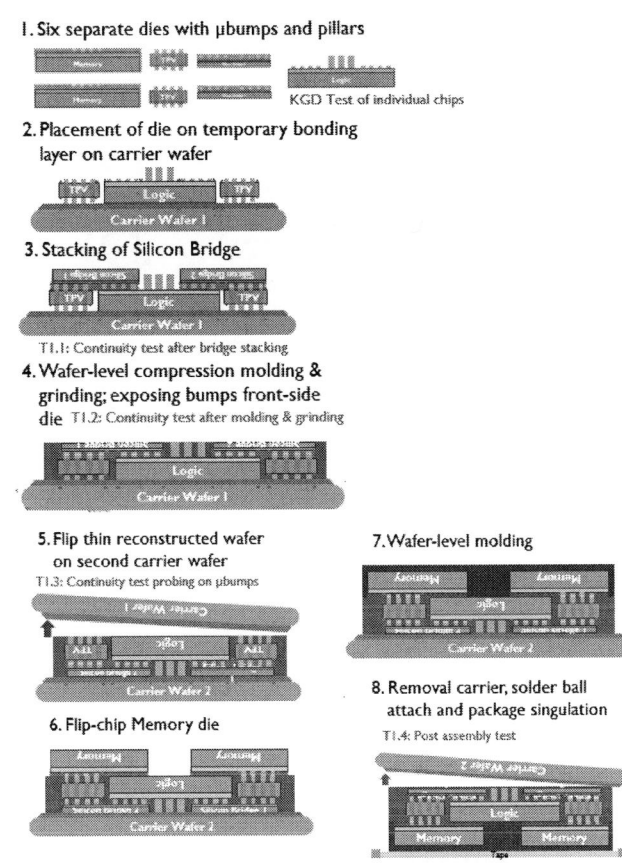

Figure 2.   Schematic of the overall assembly flow

III.   CARRIER SYSTEM REQUIREMENTS

In order to complete the whole assembly process flow described in Fig. 2, two temporary carrier substrates are used. The requirements derived from the assembly flow are very specific.

The first carrier system is required to assemble chips with high inter-die alignment precision within a few microns prior to over-molding from Fig. 2.1) to 2.5). The first adhesive material must therefore be:

- Transparent to enable pattern recognition for alignment
- Able to tack dies at low temperature for high precision die to carrier placement and survive a thermocompression bonding condition
- Capable of maintaining dies in place during the wafer over-molding step to minimize die shift
- Contributing to wafer bow and warp reduction during the entire process [1]

As chips have been originally placed face down onto the first adhesive material, re-accessing the original device frontside for testing requires a mold substrate flip. Upon completion of the wafer reconstruction, a second carrier is therefore paired to enable the removal of the first one (Fig. 2 step 5). A major requirement for this secondary carrier system, therefore, is to enable high selectivity during the first carrier removal.

IV.   FAN-OUT STRUCTURE BUILD UP VALIDATION

The evaluation process flow depicted in Fig. 3 has been put in place in order to identify and benchmark various carrier system candidates and assess their suitability for imec advanced FO-WLP.

This type of experiment enables the determination of:

- Die shift
- Wafer bow and warp evolution
- Potential material interaction
- The compatibility of both carrier systems for selective mold substrate debond

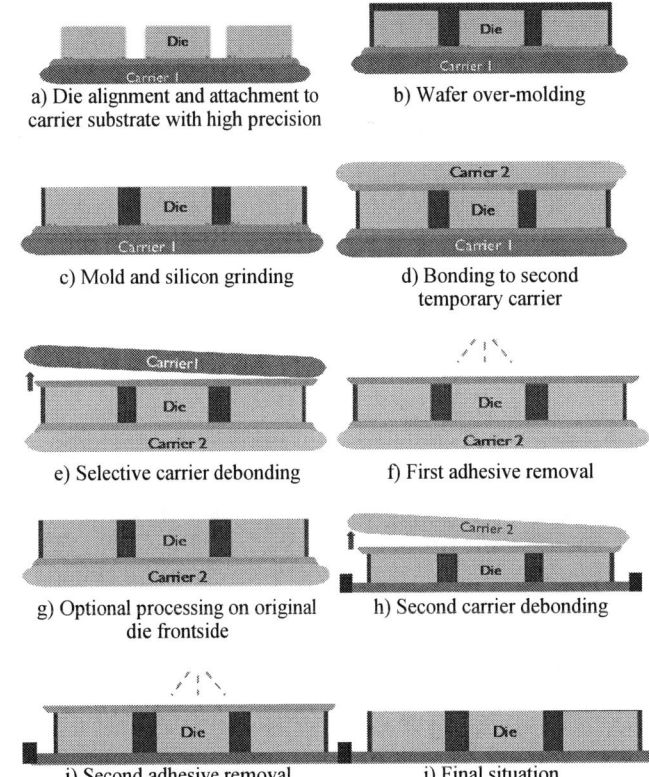

a) Die alignment and attachment to carrier substrate with high precision

b) Wafer over-molding

c) Mold and silicon grinding

d) Bonding to second temporary carrier

e) Selective carrier debonding

f) First adhesive removal

g) Optional processing on original die frontside

h) Second carrier debonding

i) Second adhesive removal

j) Final situation

Figure 3.   Schematic overview of carrier system evaluation method

In a first phase, several carrier systems have been evaluated for precise die attachment. We have previously reported the possibility to bond dies on the BrewerBOND® 305 bonding material even at room temperature [3]. This enables more precise die-to-carrier alignment as we eliminate thermal expansion issues.

We have characterized the adhesion properties of various bonding materials at various attachment temperatures. Shear force required to detach the 5-mm x 5-mm die typically ranges above 20 N, even for low-temperature bonding conditions (Fig. 4).

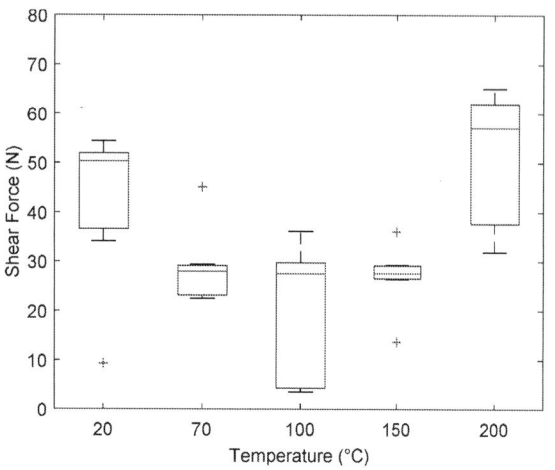

Figure 4. Die strength on a BrewerBOND® 305 bonding material layer for various die attach temperature [3]

We have therefore evaluated the BrewerBOND® 305 bonding material and BrewerBOND® 530 mechanical release material as a first carrier system on regular silicon substrate.

As the results can be influenced by the mold material and wafer over-molding technique, we have explored both granular and liquid form of materials, respectively described as M1 and M2 in the subsequent part.

Alignment markers needed for the die bonder and the overlay metrology have been first prepared by a standard Cu damascene process.

Infrared overlay metrology has been performed after die population (Fig. 5) to register the location of the dies prior to over-molding. At this step, the dies are placed within 2 µm precision (Fig. 6 a) and b)).

The substrates are then over-molded with two different materials (Fig. 3 b)). The excess mold material is then removed by grinding (Fig. 7), enabling IR metrology and the assessment of the combined adhesive and mold impact onto the die shift Fig. 6 c) and d).

An average shift after mold of <170 µm has been measured for the granular mold material M1, whilst <20 µm has been achieved with the liquid mold material M2.

Additionally, wafer bow and warp evolution has been monitored and data plotted in Fig. 8. For both mold material M1 and M2, the combination with this carrier system leads to extremely low bow, less than 200 µm, after grinding the mold substrate to 200 µm. Such values are well below the ones reported in literature [3-6]. Such low numbers enable processing the wafers in regular silicon equipment.

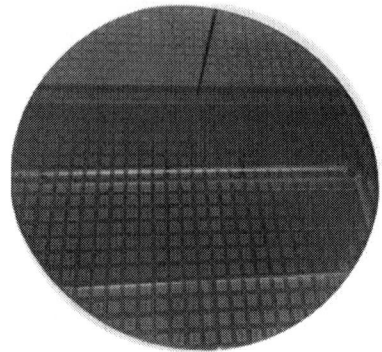

Figure 5. Photograph picture of patterned dies after population to a 300-mm silicon carrier with alignment fiducials

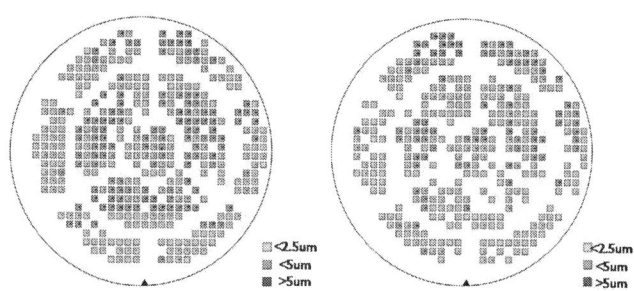

a) Die to carrier overlay before molding with material M1

b) Die to carrier overlay before molding with material M2

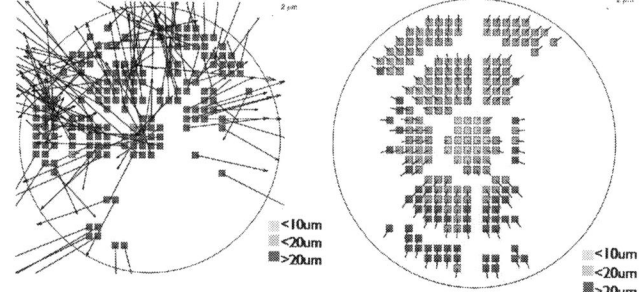

c) Die to carrier overlay after molding with material M1

d) Die to carrier overlay after molding with material M2

Figure 6. Evolution of die to carrier wafer overlay during overmolding with two different type of mold materials

Figure 7. Photograph picture of the reconstructed substrate after mold and silicon backgrind

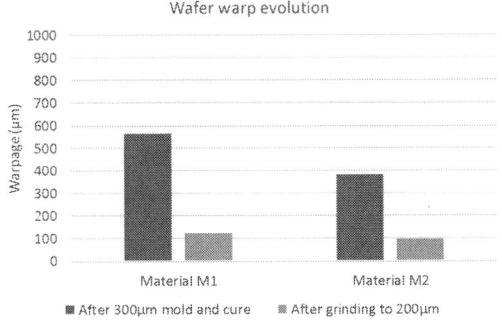

Figure 8. Wafer warpage evolution for both mold materials

V. CARRIER TRANSFER OF RECONSTRUCTED SUBSTRATES

The various steps taken to achieve a selective carrier debond are depicted in Fig. 3 d) to f). First, a second carrier is bonded onto the reconstructed wafer stack. The initial carrier substrate is then mechanically debonded without affecting the second bond line. The adhesive is then removed by cleaning with solvent.

We have previously reported the feasibility of selective carrier debonding with thinned mold substrates [7] between two mechanical debondable carrier systems. The materials used to perform the experiments are summarized in Table I.

TABLE I. CARRIER COMBINATION PREVIOUSLY REPORTED [7]

Element	Carrier 1	Carrier 2
Carrier substrate	Silicon	Silicon
Release material	BrewerBOND® 530 material mechanical release layer	BrewerBOND® 530 material mechanical release layer
Adhesive material	30 µm BrewerBOND® 305 bonding layer	Experimental bonding material A
Relative adhesion	+	+++

TABLE II. CARRIER DEBONDING SELECTIVITY FOR VARIOUS DEBOND METHODS [2]

Carrier 1 / Carrier 2	Thermal-mechanical debonding	Chemical-assisted mechanical debonding	Laser-assisted debonding	Direct mechanical debonding
Thermal-mechanical debonding	Low	Low	High	High
Chemical-assisted mechanical debonding	Low	Low	High	High
Laser-assisted debonding	Low	Low	High	High This contribution
Direct mechanical debonding	Low	Low	High	Medium [7]

As force is inherently applied during mechanical debond, removing the first carrier substrate can affect the second one if a mechanical release layer is also present. Hence, enough adhesion strength and release force difference are required to achieve high selectivity. This can become problematic in the case where higher stress mold materials are processed.

Several alternative debonding methods can be selected to achieve such selectivity. The possible combinations are summarized in Table II.

In our case, the choice of first carrier system is fixed by the requirement of having alignment marks on the carrier substrate indexed according to a lithographic pattern. We will therefore move forward with a mechanical-debondable silicon substrate.

However, the requirements for the second carrier substrate are not that stringent. Hence all alternative options are possible: thermal mechanical, chemical or laser-assisted debonding. For throughput and productivity reasons we have chosen a laser-assisted debonding approach for the second carrier system.

Two alternative carrier systems have been evaluated with the reconstructed substrates. The materials used have been summarized in Table III.

TABLE III. ALTERNATIVE SECONDARY CARRIER OPTIONS

Carrier 2	System A	System B
Substrate type	Corning SGW3 (Eagle XG) Glass	Corning SGW3 (Eagle XG) Glass
Thickness	700 µm	700 µm
Release material	BrewerBOND® 701 material laser release layer	Experimental multifunctional polymeric material B
Adhesive material	Experimental bonding material A	
Bonding Temperature	~200°C	~120°C
Laser debond wavelength	308 nm	308 nm

978-1-7281-1500-9/19 $31.00 © 2019 IEEE

In the case of System A, we have been combining the adhesive previously reported in [7] with a glass substrate and the BrewerBOND® 701 laser release material, whilst for System B we are using a multifunctional material providing the adhesion and laser release function.

In both cases, the targeted laser wavelength is 308 nm. Since the bonding the material requires temperature, we have chosen the Corning Eagle XG glass (SGW3) with a coefficient of thermal expansion (CTE) of 3 ppm/°C, matching the first silicon carrier.

For the two materials, wafer bonding has been carried out using the Suss Microtec XBS300 temporary bonding platform whilst the mechanical carrier removal has been performed using the Suss Microtec DB12T debonder.

Debonding and cleaning the adhesive (Fig. 3 e) to g)) did not affect the second carrier system as shown in Fig. 9.

We have previously reported that mold wafer warpage evolves with the thermal history of the wafer [1]. The thermal budget necessary to rebond the overmolded substrate should therefore be kept as low as possible. Hence the development of low-temperature bonding material such as the one used in carrier system B can provide further bow reduction compared to conventional systems as indicated in Fig. 10, comparing both results. The slight increase in wafer warpage compared to the silicon case is also expected due to the lower Youngs modulus and thickness of glass used compared to silicon. For both cases, numbers below 500 μm are still low compared to usual fan-out wafer-level package (FO-WLP) cases.

Finally, debonding the glass carrier substrate and cleaning the novel adhesive materials have been confirmed. For both carrier systems, the wafers have followed the steps h) to j) shown in Fig 3 using a Suss Microtec XBC300-Gen2 laser debonder cluster. The release of both BrewerBOND® 701 material and multifunctional materials occurred at an energy of around 200 mJ/cm² for a 308 nm excimer laser combined with the type of glass used. Traces of the adhesive materials were then removed by solvent. The final result shown in Fig. 11 does not indicate any damage or residues on the 200-μm. thin reconstituted substrate.

VI. CONCLUSION

Achieving ultra-fine pitch chip-to-chip interconnection densities in a fan-out wafer-level package requires new material innovation. The combination of advanced temporary bonding and release technology with advanced molding materials is leading to die shift and wafer bow levels that have not been achieved up to now. Processing of such reconstructed wafer in standard 300-mm equipment is now in reach and would enable further pitch scaling opportunities.

Figure 9. Photograph picture of the reconstructed substrate selective carrier debonding (dies facing now up)

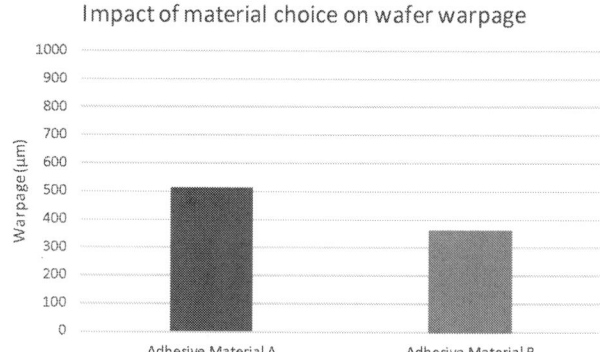

Figure 10. Warpage comparison after selective carrier debond

Figure 11. Photograph picture of the 200μm thick reconstructed substrate after laser debonding of glass carrier substrate (System B) and adhesive stripping

ACKNOWLEDGMENT

The authors would like to thank Sebastian Tussing, Walter Spiess, Frank Lauterbach, Mike Soules, Hans Mathee and Stefan Lutter from SÜSS MicroTec, Koen Kennes, Jakob Visker, and Tom Cochet from imec, and Tadashi Kubota from Towa for their support during the experiments.

This work was conducted within the frame of the imec 3D System Integration Industrial Affiliation Program and within a Joint Development Project between imec and Brewer Science, Inc., between imec and SÜSS MicroTec Lithography GmbH, and between imec and TOWA.

REFERENCES

[1] A. Podpod, J. Slabbekoorn, A. Phommahaxay, F. Duval, A. Salahouelhadj, M. Gonzalez, K. Rebibis, A. Miller, G. Beyer, E. Beyne, "A Novel Fan-Out Concept for Ultra-High Chip-To-Chip Interconnect Density with 20-µm Pitch", 2018 IEEE 68th Electronic Components and Technology Conference (ECTC), San Diego, CA, 2018, pp. 370-378.

[2] A. Podpod, D. Velenis, A. Phommahaxay, P. Bex, F. Fodor, EJ. Marinissen, K. Rebibis, A. Miller, G. Beyer, E. Beyne, "High Density and High Bandwidth Chip-To-Chip Connections with 20µm Pitch Flip-Chip on Fan-Out Wafer Level Package", 2018 International Wafer Level Packaging Conference (IWLPC), San Jose, CA, 2018, pp. 1-5.

[3] F.Che, D. Ho, M. Ding, X. Zhang, "Modeling and Design Solutions to Overcome Warpage Challenge for Fan-Out Wafer Level Packaging (FO-WLP) Technology", 17th Electronics Packaging Technology Conference, pp. 2-8, 2015.

[4] Y. Han, M. Ding, B. Lin, C. Choong, "Comprehensive Investigation of Die Shift in Compression Molding Process for 12 Inch Fan-Out Wafer Level Packaging", 66th Electronic Components and Technology Conference, pp. 1605-1610, 2016

[5] J. Mazuir, V. Olmeta, M. Yin, G. Pares, A. Planchais, K. Inal, M. Saadaoui, "Evaluation and optimization of die shift in embedded wafer-level packaging by enhancing the adhesion strength of silicon chips to carrier wafer", 13th Electronics Packaging Technology Conference, pp. 747-751, 2011

[6] J. Lau, M. Li, D. Tian et.al, "Warpage and Thermal Characterization of Fan-Out Wafer-Level Packaging", IEEE 67th Electronics Components and Technology Conference, pp. 902-908, 2017

[7] A. Phommahaxay, A. Podpod, J. Slabbekoorn, E. Sleeckx, G. Beyer, E. Beyne, A. Guerrero, D. Bai, K. Arnold, "Advances in Temporary Bonding and Release Technology for Ultrathin Substrate Processing and High-Density Fan-Out Device Build-up", 2018 IEEE 68th Electronic Components and Technology Conference (ECTC), San Diego, CA, 2018, pp. 985-992.

Development of Novel Low-temperature Curable Positive-Tone Photosensitive Dielectric Materials with High Reliability

Yutaro Koyama[*], Yu Shoji, Keika Hashimoto, Yuki Masuda, Hitoshi Araki, and Masao Tomikawa

Electronic & Imaging Materials Research Laboratories
Toray Industries, Inc.
Shiga, Japan
Yutaro_Koyama@nts.toray.co.jp

Abstract—Novel low-temperature curable positive-tone photosensitive polyimide (posi-PSPI) with high reliability has been developed as dielectric layers for copper redistribution layers (RDLs) in Fan-Out wafer/panel level packages (FOWLP, FOPLP). The posi-PSPI shows high tolerance to thermal cycle test, high temperature storage test and Cu migration test. In order to achieve these properties, we investigated both segments of flexible and rigid molecular skeletons within the base polymer backbone. Through a modification of suitable flexible segment contributed to Cu migration resistance with its assumed characteristics to have better flow coverage of Cu patterns. In addition to segmental modification, we also came to realize that a balance between flexible and rigid segment was an important factor for the stabilization of elongation under freezing temperature and thermal cycle test. Furthermore, we have also investigated an additive within the material such as anti-oxidant. This additive suppressed the voids from generating between Cu and Polyimide, which are the initial cause of delamination. This phenomenon of void formation was due to rapid speed of Cu oxide diffusion during a high temperature storage test.

The posi-PSPI offers fine pattern with good sensitivity by photolithographic system. It can also be processed by laser direct imager (LDI) instead of i-line stepper or aligner, and the patterned material made by photolithography can be reworked by organic solvents. In addition, this posi-PSPI showed high adhesion to various substrates, such as Si, Cu, Mold resin, and PI itself. These features certify that this material is suitable for applications of FOWLP/FOPLP.

Keywords-polyimides; redistribution layers; photo-sensitive; low temperature curable; high reliability; fine resolution; high adhesion.

I. INTRODUCTION

Fan-Out Package (FO-package) is utilized for one of advanced packages due to its low profile, good electric properties, and good heat dissipation property [1]. The FO-packages are applied for application processor (AP), power management IC (PMIC), and high frequency device for smart phones [2].

Re-distribution layers (RDL) of FO-package are composed of photo definable insulators and Cu patterns. From the point of reliability of the insulator, photo definable and low temperature curable polyimide is required for the insulator.

Requirement of the insulating polyimide is getting higher according to package trend in the direction of high reliability, fine resolution and low cost. In this work, we developed a low temperature curable posi-PSPI having high thermal reliability and fine resolution. We observed the polyimide structure affected reliability of the insulator in thermal cycle and biased HAST tests. We also verified the aptitude of our material to lower the cost processing FOPLP through RDL fabrication of 500mm X 600mm glass panel consisted of posi-PSPI and Cu plating [3].

II. EXPERIMENTAL SECTION

1. Polyimide Preparation

Polyimide resin consisted of polycondensation of tetracarboxylic dianhydrides and diamines. N-methyl-2-pyrridone (NMP, Mitsubishi Chem.) was placed in a 4 neck flask with a mechanical stirrer, thermometer and nitrogen inlet. Then fixed amount of diamines were added to the flask and the flask was heated up to 60 °C under nitrogen flow. The fixed amount of tetracarboxylic dianhydrides were added to the diamines solution with NMP. The mixture was stirred for 1 hour at 60 °C, then heated to 180 °C. Polycondensation reaction was carried out at 180 °C for 4 hrs.

After cooling the polyimide solution to room temperature, the solution was poured into a pot of water to precipitate the polyimide. After collecting the filtrated precipitation of polyimide, precipitated polyimide was washed by water three times. Washed polyimide was then dried at 50°C for 72hours in a convection oven. Molecular weight of polyimides were measured by GPC (510, Waters) at 40 °C. NMP with 1M LiCl and 0.5wt% H3PO4 was used as an eluent. The molecular weight of polyimides were distributed to 20K to 40K at Mw. This molecular weight range is suitable from the point of photo lithographic performance.

2. Photosensitive polyimide solution preparation

Photosensitive polyimide solution was obtained by the following procedures. First, 10g of polyimide was measured then dissolved in a Gamma-butyrolactone (GBL, Mitsubishi Chem,) at 30wt% concentration. Next, 10wt% of diazonaphthoquinone compound (TKF-280) and 4wt% of cross-linker (MW-100) were added to the solution. Lastly, the obtained solution was filtered by PTFE filter prior to use.

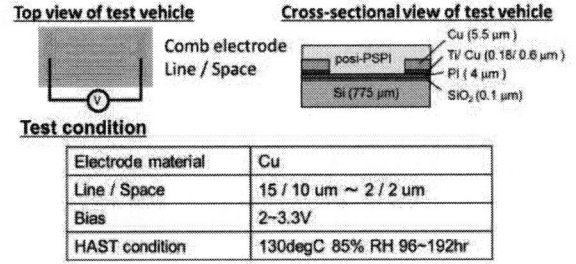

Figure 1. Chemical structure of TKF-280

Figure 2. Chemical structure of MW-100

3. Pattern formation

The obtained photosensitive polyimide solution was coated on an 8-inch Si wafer by a spin-coater (ACT-8 equipped with a hot plate, Tokyo Electoron), then soft-baked at 120 °C for 3 minutes on a hot plate. The film was exposed by an i-line stepper (Nikon, NSR-2005i9C) and laser direct imager from 200 mJ cm^{-2} to 800 mJ cm^{-2}. After lithographic exposure, the exposed film was developed by 2.38% tetramethylammonium hydroxide aqueous solution (TMAHaq.) at 23 °C. Finally, the polyimide patterned wafer was formed after cured in a clean oven (CLH-21CD (V)-S, KOYO THERMOSYSTEMS Co., Ltd) at a condition of 200 °C for 1hour under N$_2$ atmosphere.

4. Cu migration Measurement

The photosensitive polyimide solution was coated onto TEG with a comb-type electrode (Philtech) by spin coating method then cured at 200 °C for 1hour under nitrogen flow.

Electric resistance change (Biased HAST test) were monitored by SIR-13 (Kusumoto Chem.) under 130 °C 85% HAST condition at 2~3.3V electric field strength.

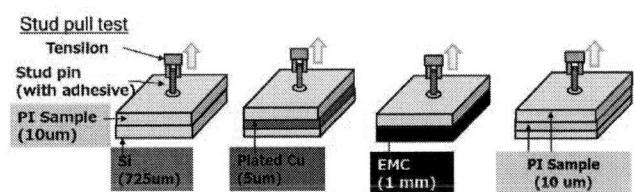

Figure 3. Test method and condition of Cu migration (top view, cross sectional view, conditions)

5. Film elongation Measurement

The photosensitive polyimide solution was coated on a Si wafer and cured at 200 °C for 1hour under nitrogen atmosphere. Then obtained polyimide film was removed from the wafer using 47% HF aqueous solution. Elongation of the film was measured by tensilon RTM-100 (Orientec) at room temperature or at -55 °C.

6. Cu and resin interface observation

The photosensitive polyimide solution was coated and cured on Cu on the TEG as previous explained. The cured TEG was stored in thermal shock chamber (espec) alternating between -55 °C to 125 °C for 500~1000cycles (Thermal Cycle test) or stored in 150 °C convection oven under air flow for 500~1000hrs (High Temperature Storage test).

Interface between Cu and polyimide was polished by cross section polisher IB-09010CP (JEOL) and observed by SEM (S-4800 Hitachi High Tech).

7. Evaluation method of adhesion.

Stud pull test was carried out to the patterned PSPI on various substrates. The photosensitive polyimide solution was coated onto Si, plated Cu, Mold resin (EMC), and polyimide itself respectively, then cured. A stud pin (90116 PHOTO TECHNICA) was placed onto the respective substrates with same adhesive. After reliability test such as thermal cycle test and high temperature storage test, each stud pin was pulled out by using tensilon RTM-100.

Figure 4. Stud pull test with various substrates

III. RESULTS AND DISCUSSTION

1. Development of Highly Reliable Photosensitive Polyimide

In our previous study, we developed polyimides which have high elongation with soft segment [4, 5]. These material have high elongation at room temperature, however, FO package in the next generation requires higher reliability. For example, high tolerances to thermal cycle test, high temperature storage test and biased HAST test are required. We investigated both segments of flexible and rigid molecular skeleton in the base polymer backbone.

978-1-7281-1500-9/19 $31.00 © 2019 IEEE

1-1. Stress and Elongation property of flexible and rigid polyimides during thermal cycle test

We investigated what properties of dielectric materials are effective to increase durability to thermal cycle test in which dielectric materials are exposed alternatingly between -55~125 °C for 500~1000cycles. Figure 5 shows wafer residual stress associated with cooling atmosphere. Figure 6 shows stress-strain curve of conventional polyimide film measured at -55°C and 125 °C.

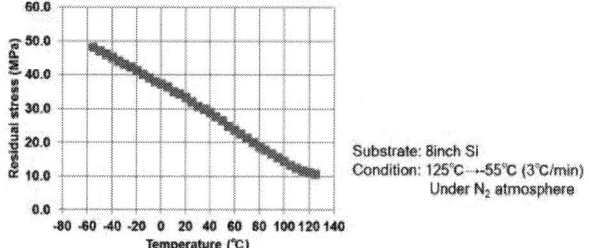

Figure 5. wafer residual stress of conventional PI

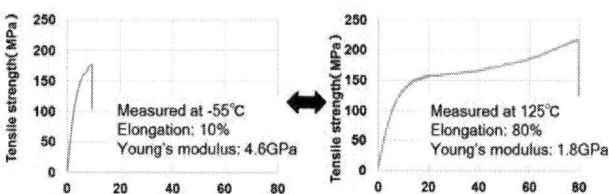

Figure 6. Stress - Strain Curves of conventional PI

Residual stress on a 8inch Si wafer severely increased from 125°C toward -55°C and at the freezing temperature, mechanical property of polyimide decreased. This result can also happen to the real packages and be considered as causes of defects such as crack of polyimide.

Thus dielectric layers are required high elongation not only at room temperature but also at freezing temperature for reliability during thermal cycle test.

We modified the flexible and rigid segment and optimized introduction amount of both segments (figure 7).

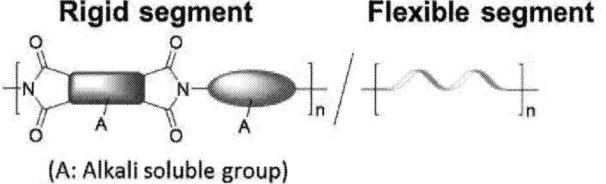

Figure 7. Design of polyimide with rigid and flexible segments

Obtained new polyimide showed higher elongation at freezing temperature and good performance with our TEG during thermal cycle test.

Observed cross section of the TEG is shown in figure 8. It is considered that entanglement of flexible segment and molecular interaction between polyimide structures of the rigid segment are both effective to elongation at freezing temperature and balance of these segments are important for high reliability.

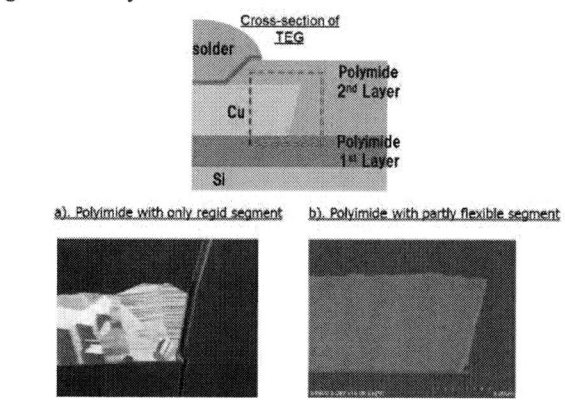

Figure 8. Cross section view of Cu/Polyimide interface after Thermal Cycle Test (1000cycles)

1-2. Biased HAST of flexible and rigid polyimides

We reported that even soft polyimide has high durability to bias HAST test [5]. As shown in figure 9, rigid polyimide has crack during HAST and Cu wiring shorted out, whereas soft polyimide insulate Cu for 96hrs and no change of resistance was detected

Generally, soft polyimide tends to go through a molecular relaxation during HAST, which activates the molecules for it to have low dielectric properties. However, since our particular soft polyimide has a relatively high Tg, where molecular relaxation does not occur. Thus soft polyimide with sufficiently high Tg has better reliability than rigid polyimide.

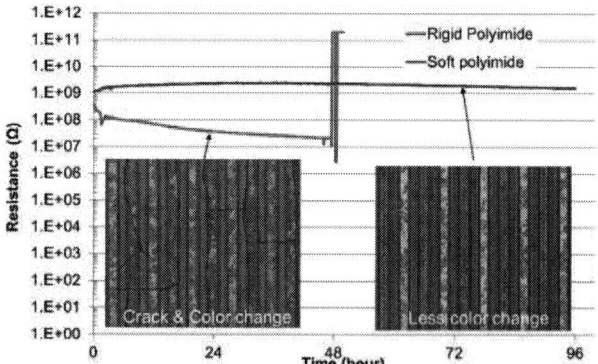

Figure 9. Bias HAST result with soft polyimide and rigid polyimide

We also tested Cu migration resistance of our new polyimide which flexible and rigid segments are modified as mentioned above and the result is shown in figure 10.

Our new polyimide showed excellent insulating property with line & space = 2 & 2um Cu comb electrode. This result is due to high Tg (270°C), good mechanical property, and high coatability of the new material.

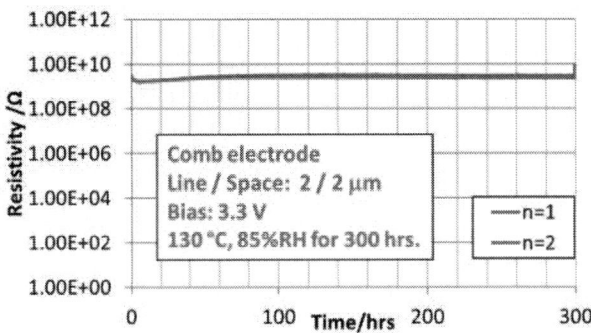

Figure 10. Bias HAST result with developed posi-PSPI

1-3. Countermeasure of delamination between Cu and Polyimide interface during high temperature storage

Packages are placed under the conditions of 150°C, air, for 500～1000hrs in high temperature storage test and main concerns are corrosion of Cu and delamination between Cu and insulation material.

Our previous report indicated that Cu in RDLs is oxidized mainly by oxygen which come from the air through polyimide overcoat during high temperature storage test [5] and we found that an additive having the functional group with copper interface was effective to suppress generation of voids [4].

We proceed further improvement with other additives and composition of polyimide solution. The expected mechanism of void formation and delamination between Cu and polyimide is shown in figure 11. We found it was important to suppress diffusion of Cu ion toward polyimide overcoat not only decreasing oxygen supply.

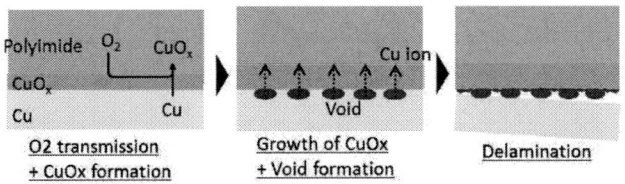

Figure 11. Mechanism of delamination during high temperature storage

Cross sectional view of our improved posi-PSPI composition is shown in figure 12. Thickness of CuOx was suppressed and the size of voids decreased. There is no delamination at the side of Cu. This result is due to reduction of oxygen supply and suppression of Cu ion diffusion.

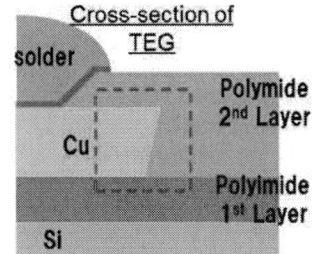

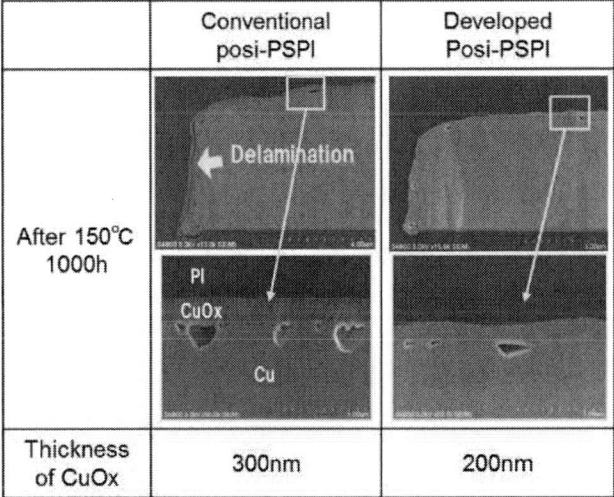

Figure 12. Cross-sectional view of developed posi-PSPI and Cu interface after high temperature storage test

1-4. Properties of new positive tone - photosensitive polyimide

We designed the reliable posi-PSPI with knowledges about polymer backbone and composition gained from the investigations as above. Newly developed meterial showed excelent reliability after thermal cycle test, biased HAST test, and high temperature storage test as shown in figure 8, 10, and 12 respectively.

Futhermore, this material offered fine pattern that was specific to positive-tone photosensitivity and we tested pattern formation with laser direct imager (LDI) instead of i-line stepper and aligner. Fine patterns are formed with optimized scanning condition as shown in figure 13 and 14. In addition, this material can be removed by various organic solvent such as acetone and 1-Methoxy-2-propanol (PGME) before curing. This easy rework ability will be an advantage in process cost reduction.

978-1-7281-1500-9/19 $31.00 © 2019 IEEE   349

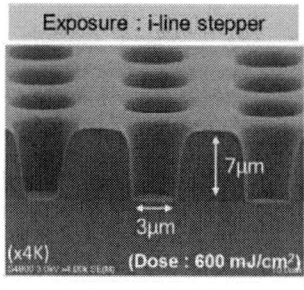 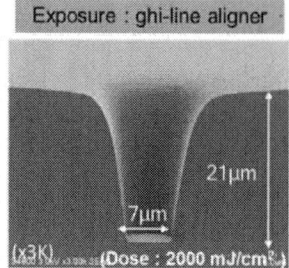

Prebaking :	120 °C/3min
Exposure :	i-line stepper / ghi-line aligner
Development :	45 sec x 2 Puddles (2.38 TMAH aq.)
Cure :	50°C→110°C/30min→220°C/60min

Figure 13. Patterning process examples of developed posi-PSPI with stepper or aligner

Exposure : Laser Direct Imager

Prebaking :	120 °C/3min
Exposure :	1.4W (700mJ)
Scan speed :	200mm/sec
Development :	45 sec x 2 Puddles (TMAH aq.)
Cure:	200 °C/60min

Figure 14. Patterning process example of developed posi-PSPI with LDI

Adhesion strength of the posi-PSPI to various metal or mold resin measured by stud pull method are shown in Figure 15. Adhesion requirements for dielectric material is to have a high adhesion to Si, Cu, polyimide itself, and EMC when utilized within FO package. Our posi-PSPI showed higher adhesion strength to each substrates than conventional polyimide which consist of rigid polyimide segment only. These results shows that additives within our material initiate adhesion to substrates and our polyimide structure itself, and soft segments within our polyimide backbone assist the imide structures to be aligned with the surfaces of the various substrates.

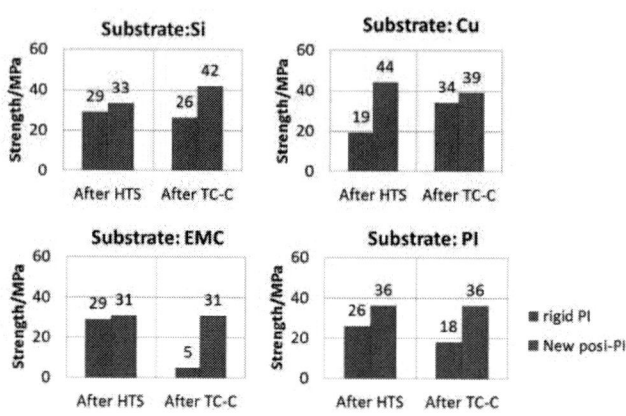

Figure 15. Adhesion strength on various substrates measured by stud pull method

2. Fabrication of Redistribution Layer on glass panel for Panel level Fan Out package

2-1. Slit coating

When dealing with glass panel process, thickness uniformity after slit coating is one of the key points for high yield. Polyimide vernish needs lower viscosity at the time of coating in order for it to be slit coated onto a panel. During the coating process, slit coater's vacuum plays an important role by evaporating the solvent away from the vernish, leaving more ratio of solid contents left on the surface of the panel. We adjusted the composition of photosensitive polyimide solution suitable to fit the slit coating process. The result of composition adjusted slit coating uniformity showed good uniformity less than 1% [4].

2-2. Cu plating

We tried all-plating process in which Cu plating was formed on plated Ni/Cu seed layer instead of sputtering. Fabrication of redistribution layer can be processed without sputtering apparatus. Pre-treatment of polyimide surface with O_2 ashing was effective to enhance adhesion between polyimide and plated Cu.

Utilizing these processes above, we obtained 40um line & space on 500mm X 600mm glass panel, and 2um line & space on 8inch glass wafer (Figure 17).

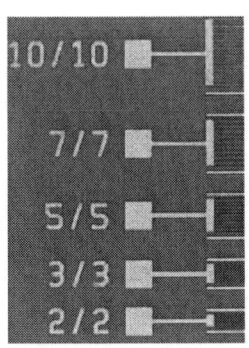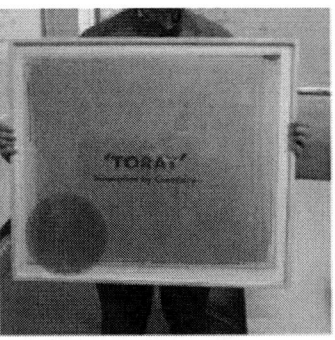

Figure 17. Redistribution pattern on a glass panel and 8inch glass wafer

IV. SUMMARY

We investigated the effect of polyimide structure having flexible and rigid segment on resistance to biased HAST and thermal cycle test. Polyimide having partial flexible segment showed better reliability than polyimide having only rigid segment. This is due to the properties of polyimide such as Cu adhesion, flow ability, and stable elongation property at freezing temperature. We also improved the composition of polyimide solution with additives to suppress void formation during high temperature storage. We successfully developed reliable positive-tone photosensitive polyimide which showed excellent reliability, adhesion to various substrates, and fine resolution.

Creating RDL by using posi-PSPI and plating Cu was successfully completed with surface treatment of O_2 ashing. We tried all-plating process which didn't need sputtering apparatus. RDL patterns with 40um line and space on 500mm X 600mm glass substrate; and 2um line and space on 8inch glass wafer were successfully obtained.

ACKNOWLEDGEMENT

Authors are great thankfulness to Okuno Chemical and "System Integration Platform Organization Standards and Research Center of Three Dimensional Semiconductors at Fukuoka University"

REFERENCES

[1] S.W. Yoon, J.A. Caparas, Y. Lin, and P.C. Marimuthu, Advanced Low Profile PoP Solution with Embedded Wafer Level PoP (eWLB-PoP) Technology, Electronic Components and Technology Conference 2012,62nd, pp.1250-1254.

[2] M.Brunbauer, E.Fergut, G.Beer, T.Meyeer, H.Hedler, J. Belonio, E. Nomura, K. Kikuchi, K. Kobayashi, An Embedded Device Technology Based on a Molded Reconfigured Wafer, Electronic Components and Technology Conference, 2006. Proceedings. 56th,pp.547-551.

[3] .http://www.yole.fr/iso_album/illus_panellevelpackaging_playersposi tioning_yole_nov_1.jpg

[4] Y.Shoji, H. Araki, Y Koyama, Y.Masuda, K, Hashimoto, K.Isobe, R. Okuda, M.Tomikawa, Higher Reliability for Low-Temperature Curable Positive-Tone Photosensitive Dielectric Materials, 2017 IEEE CPMT Symposium Japan (ICSJ), DOI: 10.1109/ICSJ.2017.8240099

[5] Hitoshi Araki, Yu Shoji, Yuki Masuda, Keika Hashimoto, Kazuyuki Matsumura, Yutaro Koyama, Masao Tomikawa. "Fabrication of Redistribution Structure Using Highly Reliable Photosensitive Polyimide for Fan Out Panel Level Packages", 2018 International Wafer Level Packaging Conference (IWLPC), 2018 DOI: 10.23919/IWLPC.2018.8573286

Highly Reliable Photosensitive Negative-tone Polyimide with Low Cure Shrinkage

Daisaku Matsukawa, Hiroko Yotsuyanagi, Shiori Sakakibara, Noriyuki Yamazaki, Tetsuya Enomoto, Takeharu Motobe

Technology Development Center
Hitachi Chemical DuPont MicroSystems, Ltd
13-1, Higashi-cho 4-chome, Hitachi-shi, Ibaraki, Japan
daisaku.matsukawa@hdms.co.jp

Abstract—A novel polyimide (PI) with improved shrinkage and lithographic performance was developed by re-designing key components of the formulation. It should be noted that this PI, due to lower shrinkage during cure, can achieve a flatter surface as compared to a conventional PI even in a multiple layer structure. In addition, lithographic performance of this PI is almost comparable to that of positive tone materials where a 3 μm via opening at 5 μm cured thickness is achievable. The side wall slope angle of the via opening can also be controlled by optimizing process conditions.

Keywords-component; polyimide; photosensitive; shrinkage; resolution

I. Introduction

Photosensitive negative-tone polyimides (PIs) have been widely used as dielectrics for re-distribution layers in wafer level packages (WLP) and as protection layers in semiconductor ICs as they can simplify the manufacturing process and ensure high reliability due to their good mechanical properties, high thermal stability, and high electrical properties [1]-[3]. Furthermore, fan out wafer level packaging (FOWLP) and fan out panel level packaging (FOPLP) are recently being applied in power devices, RF devices as well as application processers from the viewpoint of a smaller form factor including package size and profile, higher electrical performance and lowering of the manufacturing cost [4]-[6]. FOWLP and FOPLP do not use an interposer for packaging but instead utilize molding compound technology where low temperature curable photosensitive PIs with high reliability are required to prevent thermal degradation of the molding compound as well as reducing warpage of the molded wafers or panels [7]-[10]. In addition, multiple RDL structures and narrower L/S designs are being adapted to rearrange the I/O pad design in the limited package size indicating that both lower shrinkage after cure and higher lithographic performance are important requirements for photosensitive PIs [11].

Hitachi Chemical DuPont MicroSystems (HDMS) has been developing and supplying high temperature curable (> 300 °C) photosensitive PIs since 1998 [12] and, more recently, has also developed low temperature curable photosensitive PIs for FOWLP/FOPLP applications. In order to further meet current performance requirements that includes low shrinkage, high planarity after cure and high reliability at curing temperatures ranging from 200 to 250 °C, HDMS has been focusing on improving the performance of low temperature curable PIs [13].

In this paper, a novel low temperature curable photosensitive negative-tone solvent developable PI will be introduced for FOWLP and FOPLP applications that has improved shrinkage and lithographic performance obtained by re-designing key components of the formulation. It should be noted that the novel PI can produce flatter surfaces even in multiple layer structures as compared to conventional PIs due to its lower shrinkage during cure (20 % cure shrinkage). In addition, lithographic performance of the PI is almost comparable to that of positive tone materials where a 3 μm via opening with a 5 μm cured thickness is achievable.

II. Experimental

A. Preparation of photosensitive PI varnish

Solvent, PI precursor, photo-initiator, crosslinker, and others were mixed together and the resulting photosensitive negative-tone PI varnish filtered before use.

B. Evaluation of lithographic performance

Photosensitive PI varnishes were coated on 6-inch Si, Cu or Al wafers and baked at a given temperature such as 105 °C/ 2 min + 115 °C/ 2 min (Tokyo Electron Act 8). The coated films (8 μm thickness) were exposed through a mask using an i-line stepper (Canon FPA-3000iW) and developed with cyclopentanone. The wafers were then cured from 175 °C to 250 °C for 2 h (Koyo μ-TF) and the lithographic performance of the photosensitive PI evaluated using an optical microscope for top view and FIB for cross section (Hitachi-hightech SMI-500).

C. Evaluation of shrinkage during process

Shrinkage during processing was evaluated by measuring film retention as well as by inspecting cross sections of the test vehicle. The film retention during processing was calculated by using equation (1) below where TH(A) and TH(B) refer to the thickness after the A and B processes, respectively.

$$\text{Shrinkage (A/B)(\%)} = (TH(B) - TH(A))/TH(B) \times 100 \quad (1)$$

Flatness after cure was evaluated by preparing cross sections of a PI layer on a patterned substrate after cure. In this experiment, a patterned PI layer with a 10 µm thickness after cure was first processed on a Si substrate. A second PI layer was then coated, exposed, and cured on the test vehicle and cross sections of the cured test vehicle inspected using a SEM (Hitachi-hightech TM-3030).

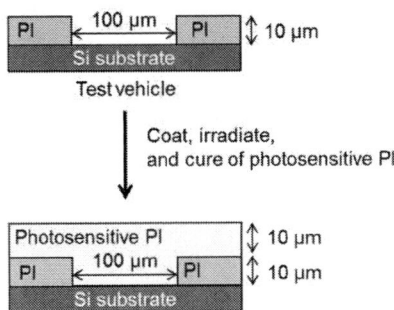

Figure 1. Evaluation method of flatness after cure.

D. Evaluation of cured film properties

Thermal and mechanical properties of the cured PI film such as Tg, Td5, elongation, and modulus were measured from free-standing cured PI films prepared by peeling off the PI films from Si wafers with a diluted HF treatment.

Chemical resistance was evaluated by dipping the patterned film after cure on a Si wafer into Dynastrip 7700 (resist stripper) at 70 °C for 90 min or into Propylene Glycol Monomethyl Ether (PGME)/ Propylene Glycol Monomethyl Ether Acetate (PGMEA) at 25 °C for 1 h. After rinsing with de-ionized water and drying, the appearance of the treated film was inspected using an optical microscope.

Adhesion after the pressure cooker test (PCT) was evaluated using the stud pull test. Adhesion performance after high temperature storage (HTS) at 150 °C was also carried out on the test vehicle (Fig. 2) and cross sections prepared after HTS were inspected by SEM.

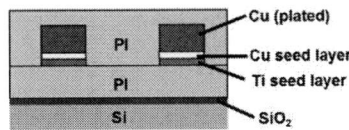

Figure 2. Cross section of test vehicle for HTS

Insulation performance of the PI was evaluated using a test vehicle with a comb-type Cu electrode (Fig. 3). For this test, the PI was coated on the test vehicle and cured at 200 °C for 2 h. The electrical resistance change during biased highly accelerated stress testing (bHAST) was monitored for up to 300 h (130 °C, 85 %RH, 3.3 V DC voltage between anode and cathode). After bHAST testing, cross sections were prepared and SEM-EDX analysis conducted to determine if any Cu diffusion occurred from the electrode to the PI.

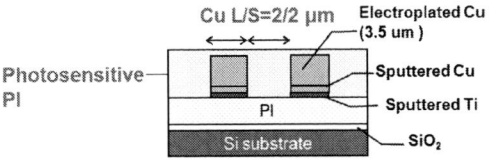

Figure 3. Top view and cross section of test vehicle for insulation performance.

III. RESULTS AND DISCUSSIONS

A. Design concept of photosensitive PI with low shrinkage

Generally, negative-tone PIs have superior cured film properties as compared to positive-tone materials and is the reason why negative-tone PIs are widely used in semiconductor applications. However negative-tone PIs have the disadvantage of higher shrinkage during the cure process and which can cause severe topographical and warpage issues especially in FOPLP applications.

There are two possible root causes for high shrinkage with negative-tone PIs. The first root cause is the orientation of polyimide during cure due to high interaction between benzene rings while the second is the degradation and volatilization of crosslinked moiety during the cure process. In the second case, it is important to select an appropriate cross linker for low shrinkage and, after much investigation, a photosensitive PI with low cure shrinkage was developed.

B. Shrinkage performance

The shrinkage of the new PI from pre-bake to cure was evaluated. The results are shown in Fig. 4 and Table 1 and where, due to the effect of low shrinkage, the new PI showed a lower film thickness change as compared to the conventional PI.

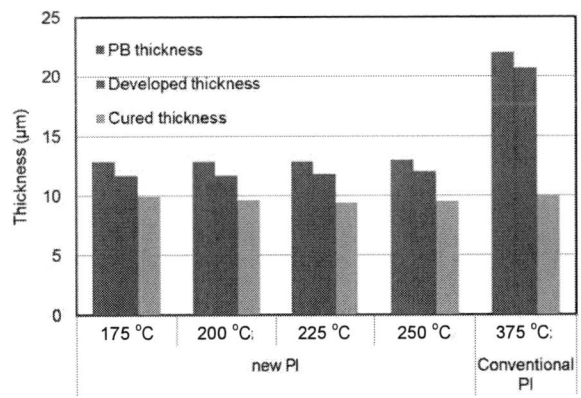

Figure 4. Thickness change during each process. (Target cured thickness: 10 µm)

TABLE I. SHRINKAGE OF PI IN EACH PROCESS

Material	New PI				Conventional PI
Cure temp. (°C)	175	200	225	250	375
Shrinkage (Dev/PB)(%)	10	11	10	9	6
Shrinkage (Cure/Dev)(%)	15	18	20	21	52
Shrinkage (Cure/PB)(%)	24	27	28	28	55

In order to confirm the lower shrinkage performance, a multi-coat experiment as outlined in Section II.C was conducted. Cross sections of the new and conventional PI after multi-coating is shown in Fig. 5 and the results indicated that the concave gap of the new and conventional PI were 2.9 um and 6.7 um, respectively. The new PI showed better flatness than the conventional PI due to lower shrinkage and which is important in the fabrication of advanced packaging structures.

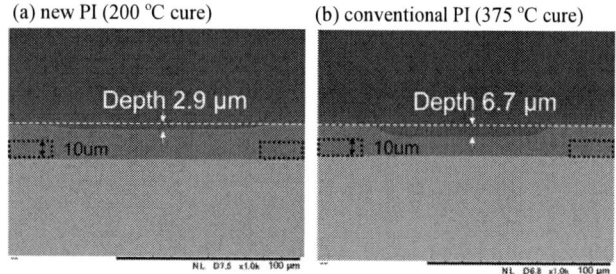

(a) new PI (200 °C cure)   (b) conventional PI (375 °C cure)

Figure 5. Cross section of (a) new PI and (b) conventional PI after multi cure

C. Lithographic performance

The lithographic performance of the new PI was evaluated and where film retention and resolution versus irradiation dose are shown in Fig. 6. The results show that, although the new PI had lower film retention than the conventional PI, the resolution of the new PI was higher than that of the conventional PI. Furthermore, the new PI showed a higher resolution over an exposure dose range from 300 to 1000 mJ/cm^2 indicating a wider exposure dose margin.

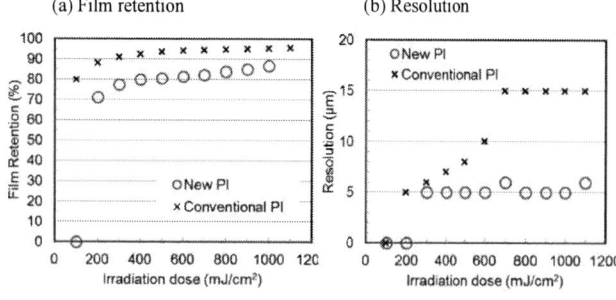

(a) Film retention   (b) Resolution

Figure 6. (a) Film retention and (b) resolution of photosensitive PIs on Si wafer (target cured thickness: 5 µm)

Lithographic performance of the new PI on Si, Cu, and Al wafers was also evaluated. The results are shown in Fig. 7 where the new PI showed high resolution even when plated Cu or Al wafers were used. In addition, the new PI also showed a smooth pattern profile after cure which is useful for metal sputtering processes.

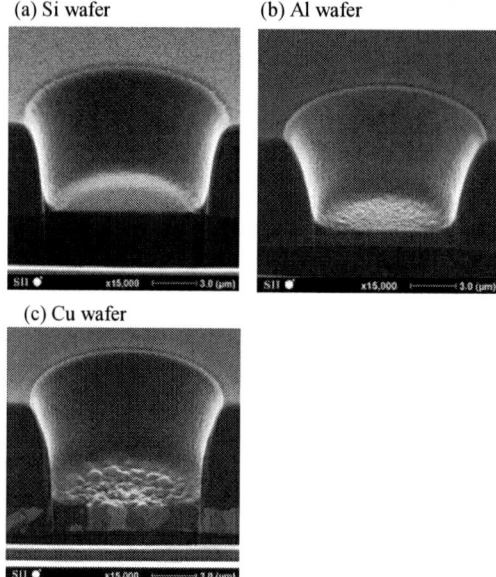

(a) Si wafer   (b) Al wafer

(c) Cu wafer

Figure 7. Cross section (10 µm via) of new PI on (a)Si, (b) Al and (c) Cu wafer

The impact of via size differences on the new PI was also evaluated. Cross sections of vias are shown in Fig. 8 and where it was confirmed that the new PI showed smooth pattern profile regardless of via size.

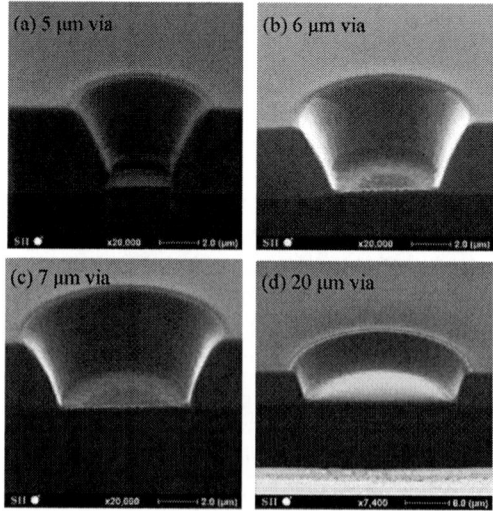

(a) 5 µm via   (b) 6 µm via

(c) 7 µm via   (d) 20 µm via

Figure 8. Cross section of new PI on Si wafer after 200 °C cure

Generally negative tone PIs, when compared to positive tone PBOs, have an advantage in cured film properties but a

978-1-7281-1500-9/19 $31.00 © 2019 IEEE

disadvantage in lithographic performance. To improve the lithographic performance of PI materials, new photo packages were evaluated and where an improvement in lithographic performance was finally achieved. A comparison of the lithographic performance between the new PI and a conventional alkaline positive tone polybenzoxazole (PBO) is shown in Fig. 9 and where the new PI showed slightly higher resolution than conventional PBO due to the new photo package. The lithographic performance of the new PI was comparable to that of positive tone PBO where a 3 µm via opening at 5 µm cured thickness with a smooth pattern profile was obtained.

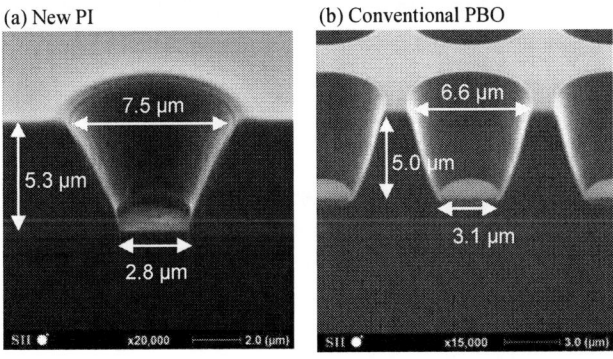

Figure 9. Cross section of (a) new PI and (b) conventional PBO on Si wafer after 200 °C cure

D. Cured film properties

Cured film properties of the new PI and conventional PI were evaluated. The thermal properties of new PI are shown in Fig. 10 and were found to be reasonably stable over a cure temperature range of 175 to 250 °C.

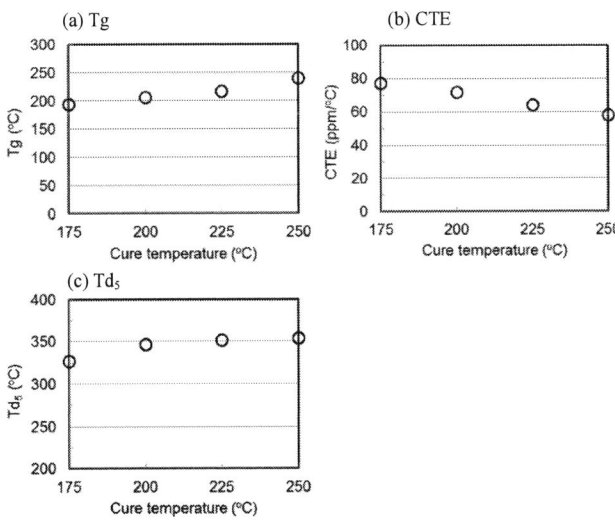

Figure 10. (a) Tg, (b) CTE, and (c) Td5 of new PI after cure

Mechanical properties of the new PI were also evaluated (Fig. 11) and where the new PI also showed a stable performance over a wide cure temperature range from 175 to 250 °C.

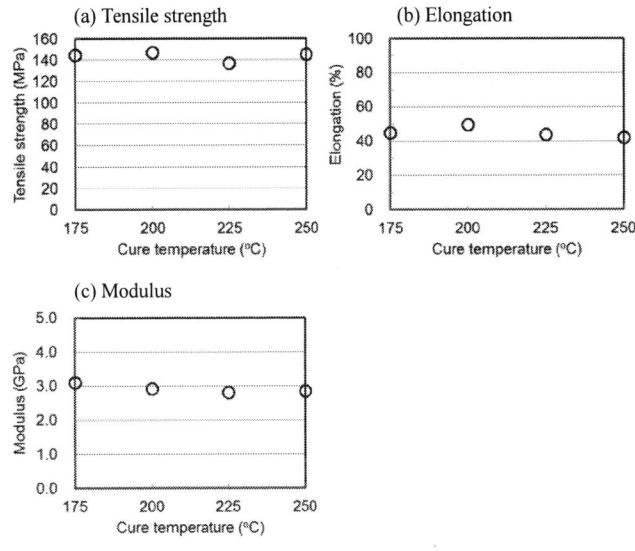

Figure 11. (a) Tensile strength, (b) elongation, and (c) modulus of new PI after cure

Table 2 summarizes the mechanical, thermal and electrical properties of the new and conventional PI and where the new PI showed a stable electrical performance (dielectric constant and dissipation factor) over a wide cure temperature range.

TABLE II. TYPICAL CURED FILM PROPERTIES

Item	New PI				Conventional PI
Cure temp. (°C)	175	200	225	250	375
Tensile strength (MPa)	145	147	137	145	200
Elongation (%)	45	50	45	42	45
Modulus (GPa)	3.1	2.9	2.8	2.9	3.5
Tg (°C)	194	206	217	239	320
CTE (ppm/°C)	78	72	64	58	35
Td_5(°C)	327	347	352	354	480
Water absorption (%)	1.9	1.6	1.6	1.9	1.1
Breakdown voltage (kV/mm)	324	368	374	311	250
Dielectric constant at 10 GHz	3.3	3.2	3.2	3.3	3.3
Dissipation factor at 10 GHz	0.019	0.021	0.021	0.022	0.009

978-1-7281-1500-9/19 $31.00 © 2019 IEEE

E. Reliability performance

The new PI was evaluated for adhesion to Si, SiN, SiO$_2$, and Cu wafers both before/after PCT (Fig 12). The results indicate that the adhesion strength of the new PI was not changed after PCT and showed a similar behavior to the conventional PI. The impact of curing temperature on Cu adhesion (Fig. 13) of the new PI was also evaluated and where the new PI showed high adhesion to Cu regardless of curing temperature.

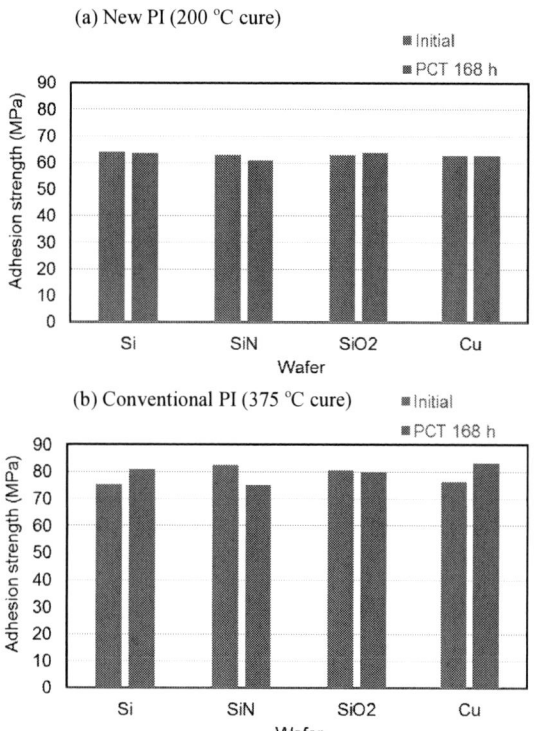

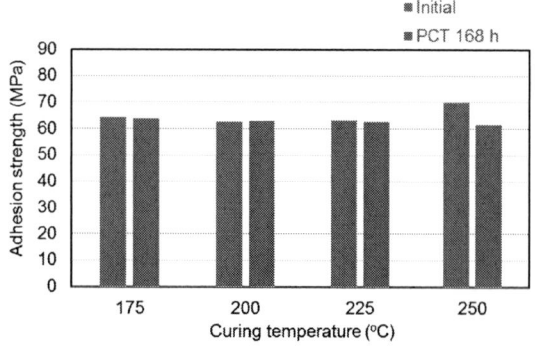

Figure 12. Adhesion of (a) new PI and (b) conventional PI on Si, SiN, SiO$_2$, and Cu after cure

Figure 13. Impact of curing temperature on Cu adhesion of new PI

The HTS stability of the new PI and conventional PI was also evaluated. Cross sections of the test vehicle after HTS are shown in Fig. 14 and where small voids were confirmed in the Cu electrode with both the new PI and conventional PI.

However, no delamination was observed after HTS indicating that the HTS stability of the new PI is the same as that of the conventional PI even though the curing temperature of the new PI is noticeably lower.

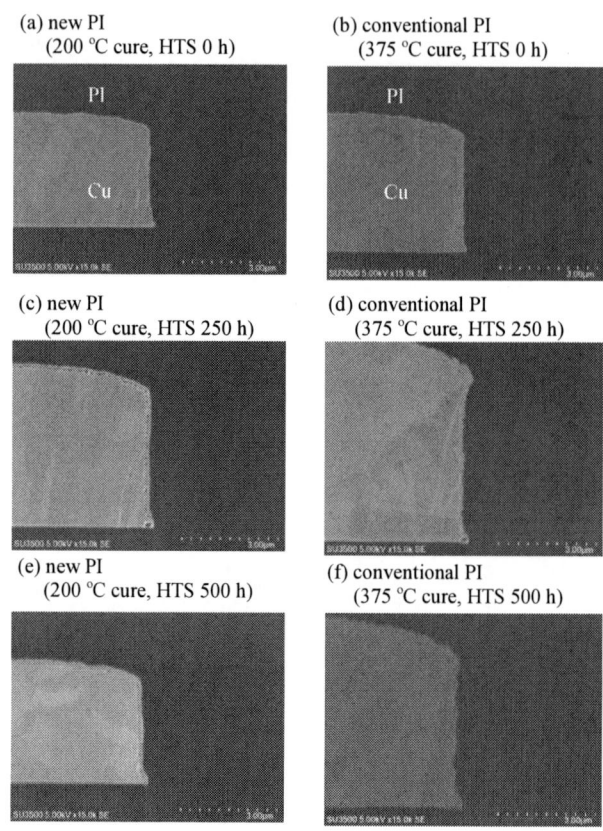

Figure 14. Cross section of test vehicle after HTS (temp.: 150 $^\circ$C)

The insulation performance of the new PI was evaluated by monitoring resistance change during bHAST using 2 μm L/S pattern followed by EDX analysis after bHAST. The results are shown in Fig. 15 and Fig. 16 and where no short circuiting with the new PI was observed after 300 h. SEM-EDX also revealed that there was no Cu diffusion after bHAST even though a test vehicle with 2 μm L/S was used indicating the high reliability performance of the new PI.

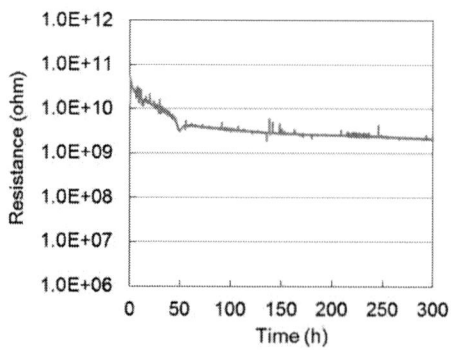

Figure 15. bHAST results of new PI cured at 200 $^\circ$C. Test conditiions: 130 $^\circ$C/85 %RH/3.3 V DC.

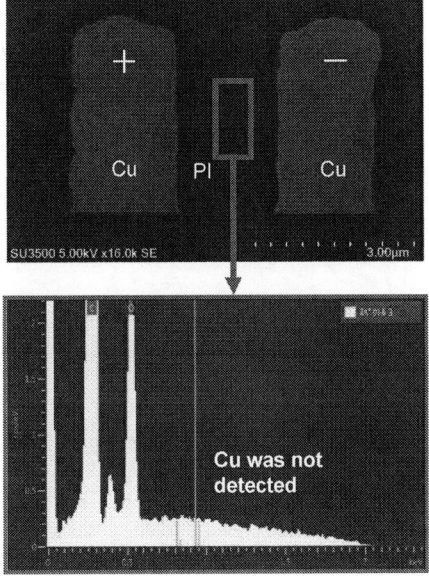

Figure 16. SEM-EDX results of new PI after bHAST for 300 h.
Test conditiions: 130 °C/85 %RH/3.3 V DC.

IV. CONCLUSION

A novel low temperature curable, negative-tone solvent developable PI was developed where, by re-designing key components of the formulation, the new PI showed low shrinkage and high resolution. Due to low shrinkage, the new PI had a flatter surface even in a multiple layer structure as compared to the conventional PI. Lithographic performance of this PI was almost comparable to that of positive tone materials and where a 3 µm via opening with a 5 µm cured thickness was achievable. In addition, the new PI showed a wide cure temperature window from 175 °C to 250 °C with respect to cured film properties and adhesion.

REFERENCES

[1] J. Kusunoki and T. Hirano, "Low temperature curable photosensitive dielectric materials for WLP applications" J. Photopolym. Sci. Technol. No.2, (2005), pp.321-325.

[2] T. Minegishi, et al., "200°C Curable Heat-resistant Photodefinable Material for Next Generation Semiconductor Packages," Hitachi Chemical Technical Report, No.52 (2009), pp.13-16.

[3] J. Hunt, 'Polymer Innovations for Advanced Packaging Applications,' he Symposium on Polymers for Microelectronics, Wilmington, DE, May. 2014.

[4] C. F. Tseng, C. S. Liu, C. H. Wu, and D. Yu, "Info (wafer level integrated fan-out) technology," 66th Electronic Components and Technology Conference, 2016, pp.1-6.

[5] H Pu, H. J. Kuo, C. S. Liu and D. C. H. Yu, 'A Novel Submicron Polymer Re-distribution Layer Technology for Advanced InFO Packaging,' 68th Electronic Components and Technology Conference, 2018, pp.45-51.

[6] J. Kim, I. Choi, J. Park, J. Lee, T. Jeong, J. Byun, Y. Ko, K. Hur, D Kim and K Oh, 'Fan-out Panel Level Package with Fine Pitch Pattern,' 68th Electronic Components and Technology Conference, 2018, pp.52-57.

[7] D. Matsukawa, A. Yoshizawa, T. Enomoto, K. Mizuno, N. Matsuie, and M. Ohe, 'Improvement in Dissolution Contrast of Positive-tone Photo-definable Poly(benzoxazole) Materials,' IEEE CPMT Japan, Kyoto, Nov. 2015.

[8] T. Sasaki, "Low temperature curable polymide for advanced package," J. Photopolym. Sci. Technol., vol.29, No.3, 2016, pp.379-382.

[9] D. Matsukawa, T. Nakamura, T. Enomoto, N. Yamazaki, M. Ohe, T. Motobe, M. Nishimura, 'Low Temperature Curable PI/PBO for Advanced Packaging,' iMAPS Device Packaging, Scottsdale, AZ, Mar. 2017.

[10] D. Matsukawa, H. Yotsuyanagi, S. Sakakibara, N. Yamazaki, T. Enomoto, T. Motobe, 'Novel Low Temperature Curable Photosensitive Negative-tone Polyimide with Higher Resolution and Reliability' iMAPS, Pasadena, CA, Oct. 2018.

[11] D. Nawrocki, A. Cooper, T. Koizumi, S. Inagaki, N. Kawamoto, N. Honda, K. Markt, M. Bernal, 'Novel Low Temperature Curable Photo-Patternable Polyamide with High Planarity for Wafer Level Fan-Out Packaging (WLFO)' iMAPS, Pasadena, CA, Oct. 2018.

[12] T. Motobe, M. Ohe, N. Yamazaki, T. Enomoto, 'Next Generation Photosensitive Dielectric Materials for Advanced Packaging Applications' J. Photopolym. Sci. Technol., vol. 31, No. 4, 2018, pp. 451-456.

[13] T. Enomoto, S. Abe, D. Matsukawa, T. Nakamura, N. Yamazaki, N. Saito, M. Ohe and T. Motobe, 'Recent Progress in Low Temperature Curable Photosensitive Dielectrics,' International Conference on Electronics Packaging, Yamagata, Japan, 2017.

High rate and low damage etching method as pre treatment of seed layer sputtering for fan out panel level packaging

Tetsushi Fujinaga
Institute for semiconductor and electronics technologies
ULVAC, Inc.
Chigasaki-city, Japan
e-mail : tetsushi_fujinaga@ulvac.com

Abstract— **This paper reports advanced pre treatment method before seed layer sputtering for Fan Out Panel Level Packaging (FOPLP). To realize high performance semiconductor devices, not only miniaturization of semiconductor chip but also minimizing packaging wiring length is also important. Fan out technology can take more I/O numbers than Fan In technology, so it is one of solution for short distance wiring, low power consumption and high density packaging. This technology originally started with wafer level process, but now its technology is going to spread to larger substrate like over 600mm square[1][2]. Enlarging substrate size is good way to suppress cost of ownership of manufacturing semiconductor devices. FOPLP is a kind of collaboration with front end technology which has fine pitch line and space and back end technology of packaging to realize high density and low cost semiconductor devices. In this technology, seed layer formation for re-distribution layer (RDL) is important to product fine pitch line and space wiring. Dielectric layer between top and bottom wiring is mainly polyimide called photosensitive imageable dielectric (PID) which can make pattern without photoresist. And to form good seed layer on polyimide with sputtering, pre treatment of polyimide is critical. We modified pre treatment for seed layer with sputtering in terms of productivity, adhesion and contact resistance.**

Keywords-component; FOPLP; Sputtering; Pre treatment; etching

I. Introduction

Fan out wafer level packaging (FOWLP) is one of best solution to realize high density wiring and short packaging height. In this technology, RDL formed in larger area than integrated circuit (IC) chip with fine line and space wiring pitch. This process has roughly separated 2 process. One is chip-first and the other is chip-last. In chip first process, IC chip encapsulation by epoxy mold compound (EMC) is first and then RDL formed on molded chip. Chip-last is opposite process. First, RDL formed on dielectric layer on support substrate. After RDL formation, IC chip is mounted and encapsulated. In each process RDL formed on dielectric layer like PID. And RDL formation process follows below steps to form high density wiring. First step is seed layer formation. Second step is photoresist coating, exposure, development and cure. Third step is Cu electroplating. Next is photoresist removal. Last step is seed layer etching. With

this flow, thick electroplated Cu patterning by wet etching is not necessary. Over a few micrometer Cu wet etching enlarges wiring space. But this process flow needs only a few hundred nanometer seed layer etching. It doesn't degrades wiring space and shape.

In case of using sputtering method to form seed layer, one of key points is pre treatment before sputtering. Seed layer is deposited on organic dielectric material like polyimide (PI). To get enough adhesion on PI, PI surface treatment is necessary. And PI has some Via to connect top and bottom wiring. There's metal pad at the bottom of Via At pre treatment step, both PI surface treatment to get enough adhesion and metal pad surface cleaning to reduce contact resistance between metal pad and seed layer has to be done simultaneously.

In sputtering system, several types of pre treatment system e.g. capacitive coupled plasma (CCP), dual frequency capacitive coupled plasma (Dual CCP), and inducted coupled plasma (ICP) are built in. In case of CCP etching, radio frequency (RF) power is applied to substrate, and ions are compelled to substrate by self bias. In Dual CCP etching system, RF power is applied to not only substrate but also opposite plate to control ion density and bias power. ICP generates high density plasma by RF induced coil and also compels ions by RF biased stage. For FOPLP, applying ICP etching is difficult in terms of good uniformity in over 600mm square area. And Dual CCP requires 2 set of RF power supply and electrode. CCP has advantage in terms of system cost and complexity. On the other hand, demerit of CCP is low etching rate comparison to Dual CCP and ICP. If Applied RF power increases to get higher etching rate, the risk of arching between electrode and substrate and inside panel wiring also increase. So in this paper, modification of CCP for FOPLP and examination each characteristics of seed layer will be reported.

II. Expeiments

In this study, Vertical transfer type sputtering system is used for seed layer formation showed in Figure 1. This system has degas chamber, pre treatment chamber, Ti sputtering chamber and Cu sputtering chamber. Substrates are held vertically to avoid particle contamination and they're transferred on carrier. Infrared (IR) lamp heats substrate in degas chamber under vacuum condition. In pre

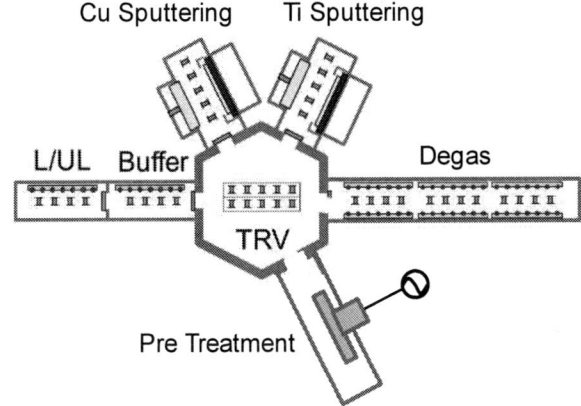

Figure 1. Top view of panel level sputtering system

treatment chamber, CCP discharge generates with RF biased stage. Generally CCP anode is planer grounded plate, but we examined anode plate configuration to get high etching ratio. Ti and Cu sputtering are done with DC magnetron sputtering with planer target.

Anode examination variation are shown in Figure 2. Just planer grounded plate is reference. Second one is shower plate type anode like CVD. Third one is also shower plate type, but shower hole diameter is different between first and second plate. And we also verified chamber shield potential. One is ground potential and the other is floating. To evaluate pre treatment etching ratio of each configuration, thermal SiO2 deposited silicon wafer is used. Ellipsometer is used to measure thermal SiO2 thickness before and after etching. RF frequency is 13.56MHz and electrode size is 600mm square. RF power was adjusted up to 1.5kW. Process pressure is controlled by inserted Ar gas flow between 0.1Pa to 2.2Pa.

Adhesion strength is tested with crosscut ASTM D3559 and pulling test JIS Z1522 before and after highly accelerated temperature and humid stress test (HAST 135 degree centigrade and 85% RH). Seed layer thickness is Ti 100nm and Cu 200nm. Pre treatment amount is normalized by etching rate of thermal SiO2. Etching amount is controlled from 0nm (no etching) to 40nm. Pulling test is done after 25um electroplated Cu formed on sputtering seed layer. Pulling angle was 90 degree and pulling width was 1cm.

And to check pre treatment cleaning ability against the surface at the bottom of Via, test pattern is prepared. Contact resistance is measured with 4-wire resistance measurement at single Via. Chain pattern which Via and wiring are connected in series is also evaluated. Test pattern and size are shown in Figure 3. Test pattern and formation flow is showed in Figure 4. Base substrate is 4inch diameter silicon wafer with 1um thermal SiO2. At first, photoresist is coated and patterned to form bottom wiring with lift-off process. Next AlCu sputtering is done as bottom wiring and lifted off. Then PID material which model is Toray xxx is coated. Via through PID to bottom wiring is formed with mask exposure, development and cure. After PID curing, descum with oxygen plasma is performed to clean the scum at the bottom of Via. Second photoresist is formed on PID. Sputtering of

seed layer and top wiring with degas and pre treatment is done on photoresist patterned PID. Seed layer is Ti 100nm and Cu 200nm stuck and top wiring 1um sputter Cu deposited on seed layer directly. Photoresist is lifted off and contact resistance is measured by Loresta which uses 4-wire resistance measurement [3].

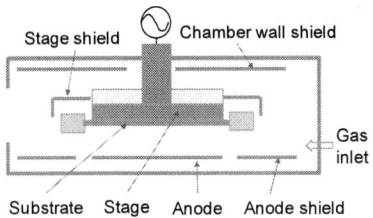

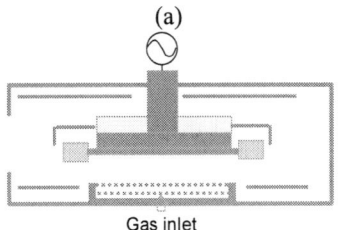

(a)

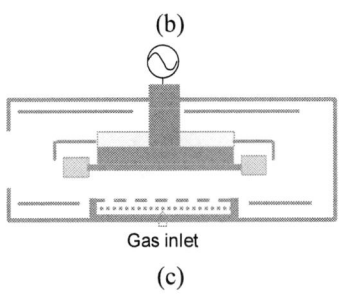

(b)

(c)

Figure 2. CCP etching chamber configuration
(a) conventional planer anode
(b) shower plate type anode
(c) shower plate type anode with different hole diameter

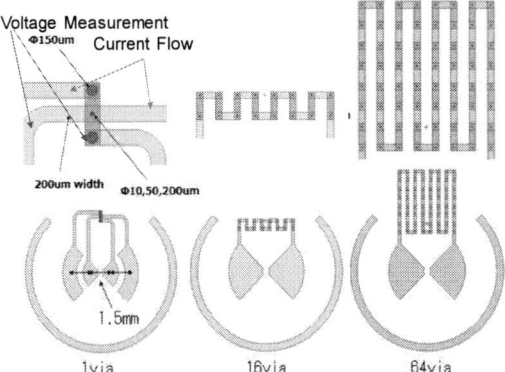

Figure 3. Test pattern diagram for contact resistance measurement

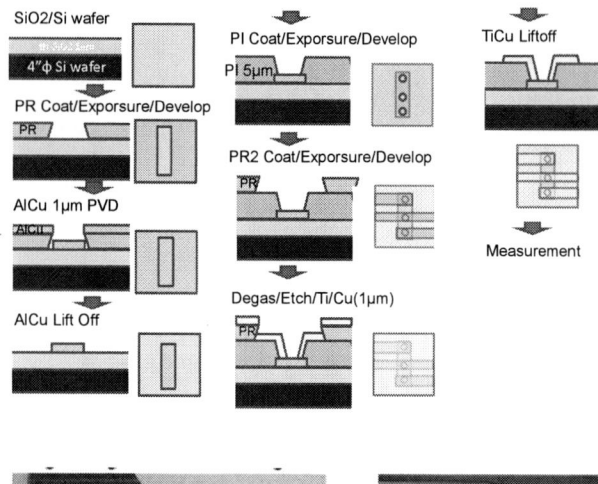

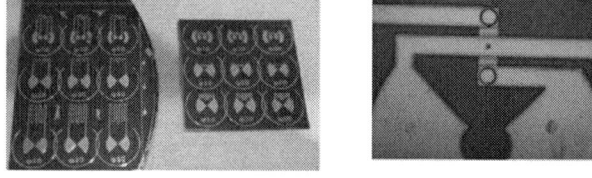

Figure 4. Contact resistance measurement pattern preparation flow and pattern image

III. RESULTS AND DISCUSSION

A. Modification of CCP ethcing system

In Table I, etching rate results in each configuration are shown. In case of vertical transfer sputtering system, carrier is necessary to hold panel. But carrier is generally made of metal which RF power runs even if panel is electrically isolated. So conventional CCP etching with carrier degrades etching rate because of RF power loss from substrate to carrier. To prevent this loss and concentrates RF power in front of only substrate, configuration (C) in Figure 2 is effective. This configuration is named hollow anode. Its anode structure is 2 stage and each plate has small holes like shower plate. First stage diffuses Ar gas uniformly with small holes. Second stage generates high density plasma with large holes with certain hollow space and high gas pressure. This anode acts like hollow cathode.

And further etching rate improvement, chamber shield potential was also optimized. By changing shield around stage and chamber wall from ground to floating, etching rate improved to 7.1nm/min with hollow anode. Etching rate of conventional CCP with carrier is about 3nm/sec, but just applying hollow anode improves etching rate to 5nm/min. Furthermore, shield potential control improves etching rate to 7nm/min with only one electrode CCP.

Table I. Etching rate result in each anode configuration

	Configuration		
	(A) Planer anode	(B) Shower anode	(C) Hollow anode
E/R	3.1nm/min	3.0nm/min	5.1nm/min

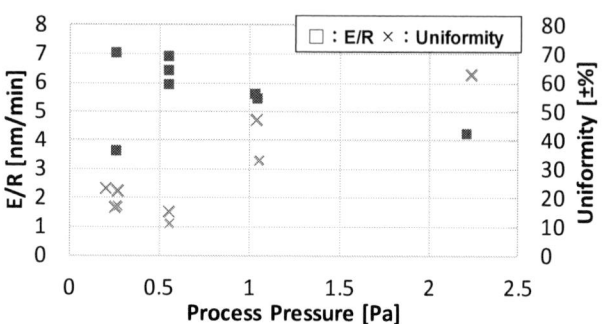

Figure 5. Etching rate and uniformity dependence on process pressure

Uniformity is modified by process pressure. Etching rate and uniformity dependence on process pressure is shown in Figure 5. There's optimized pressure around 0.5Pa in terms of etching rate and etching uniformity.

Optimized configuration with hollow anode and process pressure, 7.1nm/min in thermal SiO2 etching rate and 10% uniformity achieved in 600mm square panel..

B. Adhesion on PI

Adhesion strength is evaluated with crosscut and pulling test. Figure 6 shows crosscut result with CCP etching with hollow anode etching time split. Even if without pre treatment etching, crosscut passed after 168hour HAST. With this result, it's difficult to judge pre treatment is necessary or not and optimized or not. So comparison between with and without per treatment is done with pulling test.

Adhesion strength by pulling test dependence on HAST time is shown in Figure 7. Pulling strength evaluation on PI, without pre treatment shows lower adhesion strength. With 20nm etching adhesion strength slightly decreased, but it shows higher adhesion strength than without pre treatment. And peeled interface analysis after 168hour HAST is also evaluated. Depth profile in each side which PI film and peeled seed layer side are shown in Figure 8. There's carbon peak from seed layer Ti side, so peeled interface is inside PI, not PI/seed Ti interface. This result indicates seed layer adhesion on PI still has high peel strength after 168 hour HAST.

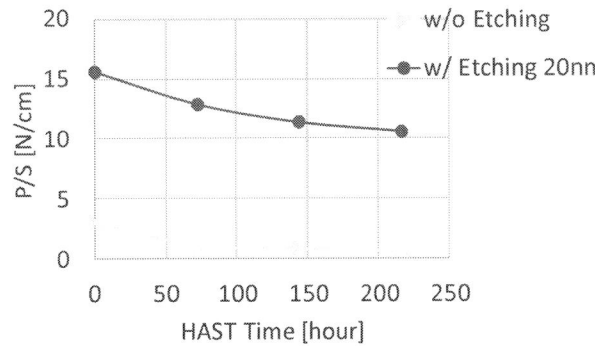

Figure 6. Crosscut test result on PI

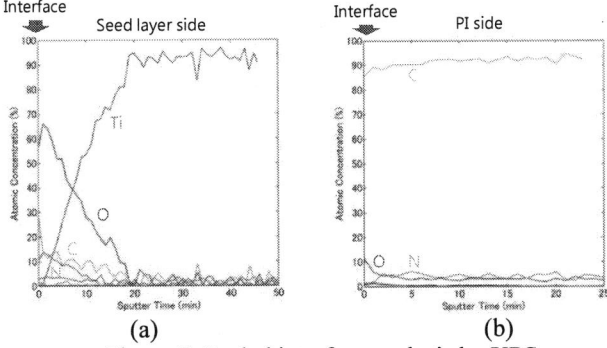

Figure 7. Adhesion strength along HAST with each etching amount

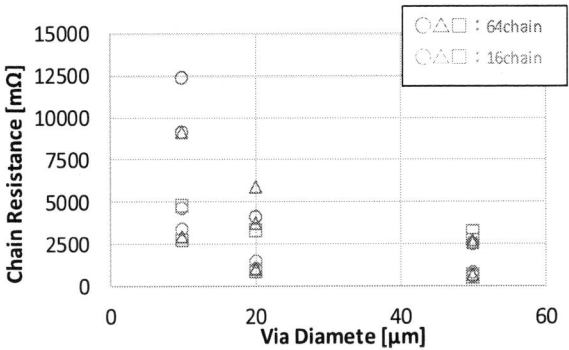

Figure 9. Contact resistance measurement result in chain pattern in each Via diameter

C. Contact resistance results

Contact resistance results in single Via with 4-wire resistance measurement are shown in Table II. Process condition is 30min 150 degree centigrade degas in vacuum, CCP etching with hollow anode same etching condition as thermal SiO2 20nm etching, Ti 100nm sputtering and Cu 1000nm sputtering. As described before, usually seed layer Cu thickness is around 200nm. But in this test, seed layer works as also instead of electroplated Cu, so Cu layer thickness is 1000nm. With 20um diameter Via, contact resistance was 0.21m ohm, so surface at the bottom of Via is enough cleaned with CCP and hollow anode etching.

And in Figure 9, contact resistance in chain pattern are shown. Along Via diameter increasing, contact resistance decreases. Also contact resistance depends on chain number. Long chain shows higher contact resistance. Calculating wiring resistivity with AlCu 3.2u ohm cm resistivity and Cu 2.0u ohm cm resistivity is 832mohm in 16chain and 3328m ohm in 64chain. Chain resistance with 50um Via shows almost same value as wiring resistivity. It also indicates contact resistance in each Via is enough low value with 20nm etching condition. .

Figure 8. Peeled interface analysis by XPS
(a) Seed layer (Ti) side (b) PI side

Table II. Contact resistance measurement result in each Via diameter with 4-wire resistance measurement

	Via Diameter		
	10um	*20um*	*50um*
Contact resistance	0.45mΩ	0.21mΩ	0.08mΩ

IV. CONCLUSION

In this study, advanced pre treatment method is developed by modifying anode structure and shield potential control. Hollow anode structure improved RF loss and concentrated CCP plasma in front of pane. Etching rate achieved 7nm/min and 10% uniformity with single CCP 600mm square electrode and hollow anode. Adhesion strength on PI material shows enough strength after seed layer sputtering with this advanced pre treatment. And contact resistance between bottom metal pad and seed layer is evaluated. With this pre treatment, of CCP and hollow anode combination, contact resistance showed enough low values.

REFERENCES

[1] T. Braun, et. al., "Material and process trends for moving from FOWLP to FOPLP", Proceding of IEEE 17th Electronics Packaging and Technology Conference (EPTC), December 2015

[2] Karl-Friedrich Becker, Tanja Braun, S. Raatz, M. Minkus, V. Bader, J. Bauer, R. Aschenbrenner, R. Kahle, L. Georgi, S. Voges, M. Wöhrmann, and K.-D. Lang (2016) On the Way from Fan-out Wafer to Fan-out Panel Level Packaging. International Symposium on Microelectronics: October 2016, Vol. 2016, No. S2, pp. S1-S23.

[3] Da-Quan Yu, Tai Chong Chai, Meei Ling Thew, Yue Ying Ong, Vempati Srinivasa Rao, Leong Ching Wai, John H. Lau, "Electromigration study of 50 μm pitch micro solder bumps using four-point Kelvin structure", Proceeding of 59th Electronic Components and Technology Conferencr (ECTC), May 2019

Investigation and Methods Using Various Release and Thermoplastic Bonding Materials to Reduce Die Shift and Wafer Warpage for eWLB Chip-First Processes

Michelle R. Fowler, John P. Massey
Wafer-Level Packaging Material Division
Brewer Science, Inc.
Rolla, MO, USA
mfowler@brewerscience.com

Tanja Braun, Steve Voges, Robert Gernhardt,
Markus Wohrmann
Fraunhofer Institute IZM
Berlin, Germany
Tanja.Braun@izm.fraunhofer.de

Abstract— Today's fan-out wafer-level packaging (FOWLP) processes use organic substrates composed of epoxy mold compound (EMC) created using a thermal compression process. EMC wafers are a cost-effective way to achieve lower profile packages without using an inorganic substrate to produce chip packages that are thinner and faster without the need for interposers or through-silicon-vias (TSVs). One approach using embedded die technology (eWLB) for FOWLP is a chip-first (mold-first) die assembly in a face-down configuration on an intermediate carrier wafer. The ideal chip attachment scheme should minimize lateral movement of the die during over-mold (die shift) and also minimize vertical deformation of the bonding material which results in die protrusion (stand-off). An ideal die attach material should provide adequate adhesion to the EMC wafers without inducing excessive substrate warp, while permitting a suitable debonding process, including complete residue removal. The attachment scheme must also survive any thermal, mechanical, and chemical processes that are performed prior to carrier release. The bonding materials must also have sufficient adhesion to the EMC material to overcome such stress without bond failure. Preventing lateral die shift and deformation due to the coefficient of thermal expansion (CTE) mismatch between the carrier substrate and EMC material can be accomplished using an appropriate bonding material.

Using a chip-first die attach process, this investigation will address die shift and deformation using various release and thermoplastic bonding materials. Combinations using different EMC products and carriers with various coefficient of thermal expansion (CTE) are also included. Successful pairs will then undergo carrier release using either mechanical release or laser ablation release technology.

Keywords: die-shift; embedded die; die stand-off; chip-first; epoxy mold compound; wafer bonding

I. INTRODUCTION

Various advanced packaging technologies such as packaging with through-silicon vias (TSV), 3D system-in-package (SiP), package-on-package (PoP), wafer-level chip-scale packaging (WLCSP) and fan-out wafer-level packaging (FOWLP) continue to evolve. These technologies offer miniaturization with increased component density, lower cost of ownership, improved performance, flexibility and faster time-to-market. Embedding known good die (KGD) in polymer encapsulants using FOWLP is one technology that enables multiple chips to be integrated into a single package, adding more redistribution layers (RDLs) and facilitating heterogeneous system integration [1,2]. Embedded wafer-level ball grid array (eWLB), developed by Infineon [3], the InFO package by TSMC [4], and Freescale's Redistributed Chip Package (RCP) [5] are various approaches utilizing FOWLP technology.

Two process flows are currently in use for fan-out wafer-level packaging, RDL-first (chip-last) and mold-first (chip-first), which offers two process configurations: chip face-down (eWLB) or chip face-up (InFO). Both approaches begin with die assembly on an intermediate carrier using a temporary adhesive layer. After assembly, mold-first face-down processing continues with over-molding and debonding of the molded wafer from the temporary carrier. The redistribution layer is typically based on thin film technology and applied on the reconfigured molded wafer.

In the chip-first face-up approach, dies are Cu bumped and placed face-up on the temporary carrier. After over-molding, grinding of the epoxy mold compound (EMC) allows access to the now-exposed Cu bumps. The redistribution layer is applied and the wafer is released from the temporary carrier and diced for package singulation.

For the "RDL-first" approach, the redistribution layer is applied to an intermediate carrier and the bumped dies are assembled by chip-to-wafer bonding on the RDL. The assembly is underfilled then over-molded and the molded wafer, including the RDL, has to be released from the carrier.

There continue to be several challenges facing FOWLP technology: wafer warpage, die-shift and die stand-off (protrusion) after EMC processing. In this study, a chip-first, face-down approach will be used to evaluate various factors and how they impact this process.

II. EXPERIMENT

The following sections describe in detail the experiments conducted for this work. This is a continuation of work previously published during the IMAPS October 2018 conference [6].

A. Process Flow

This work is focused on a "Chip-first" face-down process. Hence, reconfigured wafer preparation followed this manufacturing sequence. BrewerBOND® and WaferBOND® temporary adhesive and release materials are spin coated onto 200-mm glass wafers. Chips are then placed on the bonding material. Each wafer is over-molded with EMC and thermally cured. Molded wafers are then debonded either mechanically or by laser ablation and cleaned to remove residual adhesive material. For evaluation, wafers are finally characterized for die shift, warpage, and die stand-off measurement (Figure 1).

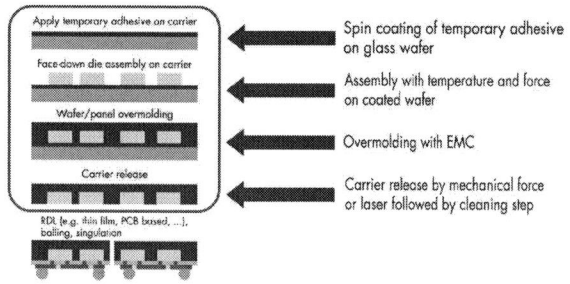

Figure 1: BrewerBOND® release and bonding material as the temporary die attach materials in a "mold-first" face-down process.

B. Test Vehicle

Material evaluations were made using 200-mm glass wafers. Three dies were placed per package: one large die (9x9x0.2 mm³) and two smaller dies (3x2x0.2 mm³). Partial assembly was done on the wafers guaranteeing enough information for process and material evaluation. Package and wafer assembly are depicted in Figure 2. Mold thickness was set to 300 µm, resulting in a 100 µm over-mold layer on the chip backside.

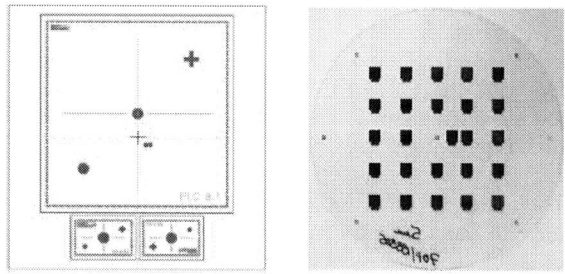

Figure 2: Test vehicle; left: package configuration, right: wafer layout after die attachment.

C. Design of Experiment

Previously published work reported that an optimum bonding material film thickness of 10 µm gave the least amount of die shift and die stand-off after mold. Additionally, the combination of release material with the bonding material had no influence on wafer warpage. Optimized assembly parameters determined from previous work include:

1. Best die attach result achieved using 100°C assembly head temperature
2. Assembly force of 3.5 kg
3. Carrier substrate continuously heated at 75°C
4. Bonding material film thickness at 10 µm

For this work, a 4-level full factorial designed experiment was conducted using two different bonding materials in combination with two release materials (Table 1). The influence of coefficient of thermal expansion (CTE) on substrate warpage and die shift is also explored using glass carriers with differing CTE. Both a liquid and granular epoxy mold compound are used for the manufacture of the reconfigured wafers. All test samples are also evaluated for debond performance using either a mechanical release or laser ablation technology. In addition to the bonding and release materials a standard thermal release tape was used as a reference using the same assembly parameters.

TABLE 1. Experiment Design

Factors	Levels	Response
Release Material	BrewerBOND® 510 BrewerBOND® 701	Die Shift, Stand-off, Warpage, Performance
Bonding Material	BrewerBOND® 305 WaferBOND® HT-10.10	
Glass Carrier CTE	Low 31.7×10^{-7}/°C High 75.8×10^{-7}/°C	
EMC Type	EMC 1 granular EMC 2 liquid	
Control: Thermal release tape		

III. MATERIALS

In the following sections materials used for the investigations are described in detail.

A. Glass Carrier Wafers

All glass carrier wafers used were 200-mm notched rounds. Eagle XG glass with a CTE of 31.7 ppm/°C (low CTE) and a substrate thickness of 0.7 mm were obtained from Stemmerich, Inc. Gorilla glass carriers were obtained from Corning Glass with a CTE of 75.8 ppm/°C (high CTE) and a substrate thickness of 0.7 mm. All carriers were used as received without additional cleaning or a dehydration bake.

B. Release material

Glass carrier wafers were spin coated with release material (as defined by the experiment) using a manual, static dispense:

1. BrewerBOND® 510 release material was spin applied at 1250 rpm using a 3000 rpm/s acceleration for a time of 35 s. The wafers were hotplate baked using a 0.2 mm proximity bake at 220°C for 1 min. Target film thickness after bake was 50 Å.
2. BrewerBOND® 701 release material was spin applied at 2500 rpm using a 5000 rpm/s acceleration for a time of 60 s. Using a bake plate at 300°C, the wafers were contact baked for 5 min. Target film thickness after cure was 150 nm.

C. Bonding Material

Bonding material was spin coated onto glass carrier wafers that were previously coated with release material. Using a manual, static dispense process, materials were coated as follows:

1. BrewerBOND® 305-30 bonding material was spin applied at 2200 rpm using a 1000 rpm/s acceleration for a spin time of 45 s. Using a hotplate, the wafers were contact baked at 60°C for 3 min followed by a bake at 160°C for 3 min with a final bake at 220°C for 3 min.
2. Wafer BOND® HT-10.10 bonding material was spin applied using a cast speed of 2200 rpm with an acceleration of 3000 rpm/s for 30 s. Using a contact bake process, the wafers were baked on a hotplate at 120°C for 3 min followed by a bake at 180°C for 4 min.

As a reference, a standard release tape was used with a release temperature of 170°C.

D. Epoxy Mold Compound

For this work, two epoxy mold compound materials were evaluated. Processing followed product data sheet recommendations. A summary of product properties is shown in Table 2.

TABLE 2. Properties for Epoxy Mold Compound

Properties	EMC 1	EMC 2
Type	Granular	Liquid
Filler content	90%	88%
Filler top cut	55 μm	55 μm
Gel time	40 s	N/A
CTE1	7 ppm	8 ppm
CTE2	28 ppm	34 ppm
T_g	175°C	165°C
Flexural modulus @ 25°C	30 GPa	22 GPa
In-mold cure	125°C, 10 min	125°C, 10 min
Post-mold cure	150°C, 60 min	150°C, 60 min

IV. EQUIPMENT

The following sections describe the various equipment used for this work.

A. Spin coat and bake equipment

Glass wafers were manually coated with both release and bonding materials using a Cee® X-Pro 2 work station equipped with a coat bowl and bake plates with programmable lift pins.

B. Assembly and encapsulation

Chips were assembled using a Datacon 2200 evo die bonder on carriers prepared with the release and bonding materials and as reference with thermo-release tape. Compression molding was done with a 120-ton press from TOWA Y-120 with a mold cap thickness of 300 μm followed

by debonding after post-mold cure. Figure 3 shows the principle of compression molding machines with mold cavity in the upper tooling and the reconfigured wafer in the lower tooling.

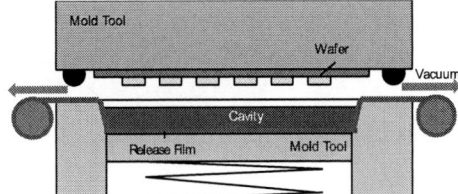

Figure 3. Principle sketch of cavity down compression molding with liquid molding compound.

Compression molding was done at a constant temperature, typically in the range of 120°C – 130°C and under pressure and vacuum to achieve a homogeneous encapsulation without voiding or air entrapment. A compression molding process diagram illustrating temperature, mold pressure and cavity vacuum over time is depicted in Figure 4.

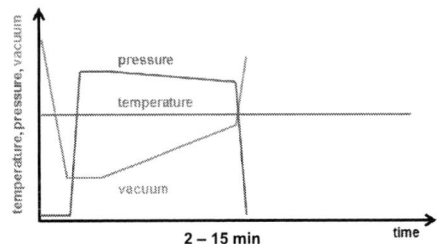

Figure 4. Compression molding process diagram.

C. Die Shear

A Nordson DAGE 4000 die shear bond tester was used to evaluate the adhesion of the die after assembly on the bonding materials.

D. Debonding

A mechanical release (used with BrewerBOND® 510 material) or laser release (used with BrewerBOND® 701 material) debond process was used to remove the glass carrier from the EMC wafer. Both an EVG® mechanical debonder and an EVG® Solid State Laser system were used to debond the glass carriers.

E. Cleaning

After debonding, residual bonding material was removed from the EMC surface. BrewerBOND® 305 material was removed using a tape peel process. A solvent rinse process using WaferBOND® remover material was used to remove residual WaferBOND® HT-10.10 material (Figure 5).

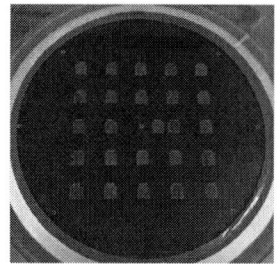

Figure 5. Embedded EMC after carrier debond and cleaning. Left: after tape peel removal of BrewerBOND® 305 material. Right: after cleaning with solvent to remove WaferBOND® HT-10.10 material.

F. Analytics

Die positions were measured with a Mahr OMS 600 optical measuring system after carrier release. Warpage measurements were done using a cyberSCAN VANTAGE, a laser-based, non-contact inspection system and die stand-off was measured using a HOMMEL Tester T8000.

V. RESULTS

Using the process flow, carrier wafers are manufactured and dies assembled onto the temporary bonding material. Adhesion of the dies to the bonding material was evaluated and the carriers were over-molded with EMC. Carrier wafers were then debonded from the EMC using either a mechanical release or laser ablation process. Residual bonding material was then removed from the reconstituted wafer using a tape peel or solvent rinse process. Measurements and analysis of die-shift and wafer warpage was then conducted (Table 3).

A. Die Adhesion

To test the adhesion of the dies to the bonding material, a die shear test was performed using a Nordson DAGE 400 die shear bond tester (Figure 6).

TABLE 3. Material Combinations

Carrier	Release Material	Bonding Material	EMC
High CTE	BrewerBOND® 510	BrewerBOND® 305	1
High CTE	BrewerBOND® 510	BrewerBOND® 305	2
High CTE	BrewerBOND® 701	BrewerBOND® 305	1
High CTE	BrewerBOND® 701	BrewerBOND® 305	2
High CTE	BrewerBOND® 510	WaferBOND® HT-10.10	1
High CTE	BrewerBOND® 510	WaferBOND® HT-10.10	2
High CTE	BrewerBOND® 701	WaferBOND® HT-10.10	1
High CTE	BrewerBOND® 701	WaferBOND® HT-10.10	2
Low CTE	BrewerBOND® 510	BrewerBOND® 305	1
Low CTE	BrewerBOND® 510	BrewerBOND® 305	2
Low CTE	BrewerBOND® 701	BrewerBOND® 305	1
Low CTE	BrewerBOND® 701	BrewerBOND® 305	2
Low CTE	BrewerBOND® 510	WaferBOND® HT-10.10	1
Low CTE	BrewerBOND® 510	WaferBOND® HT-10.10	2
Low CTE	BrewerBOND® 701	WaferBOND® HT-10.10	1
Low CTE	BrewerBOND® 701	WaferBOND® HT-10.10	2
High CTE			1
High CTE	Thermal Release Tape		2
Low CTE			2

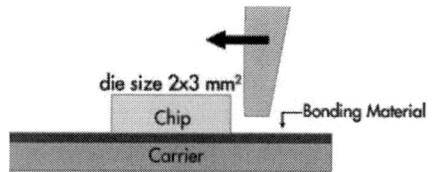

Figure 6. Die Shear Testing

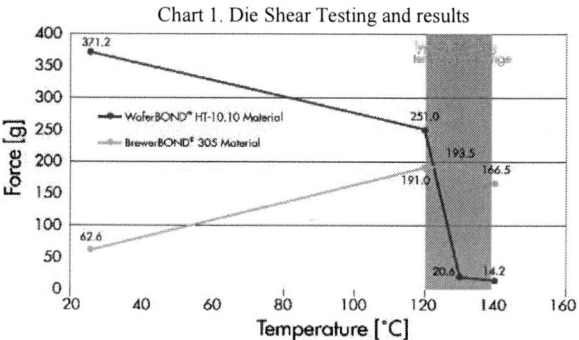

Chart 1. Die Shear Testing and results

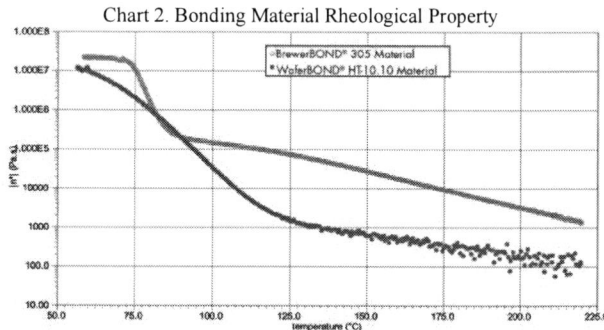

Chart 2. Bonding Material Rheological Property

During the compression mold process, as temperature and pressure increase, the bonding material softens, which may improve the wetting of the dies to the bonding material improving die adhesion as in the case of the WaferBOND® HT-10.10 material. However, under these same conditions, a decrease in film viscosity also results in a loss of film integrity allowing for die movement and loss. The BrewerBOND® 305 material demonstrated low die adhesion at temperatures <80°C, as the film at these temperatures has not yet softened. Die adhesion improved and stabilized through the mold process temperature range.

B. Optical Inspection

It was discovered that WaferBOND® HT-10.10 material had a strong interaction with liquid EMC 2. Every combination using this blend was adversely affected (Figure 7). After compression molding, all test samples were optically inspected for mold residues found on the top of the dies known as flash residues (Figure 8 and 9). Flash residues occur when the die has insufficient adhesion to the bonding material allowing the EMC to wick below the die and wet the die surface. No to minimal edge flashing was observed for combinations using BrewerBOND® 305 material (Figure 10 and 11).

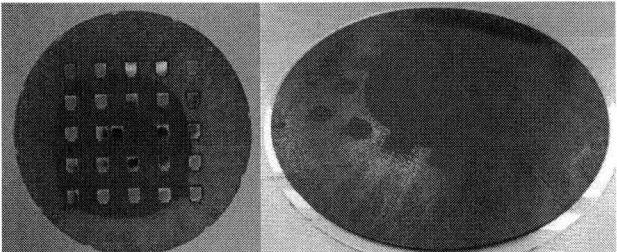

Figure 7. Visual interaction between WaferBOND® HT-10.10 bonding material and liquid EMC 2

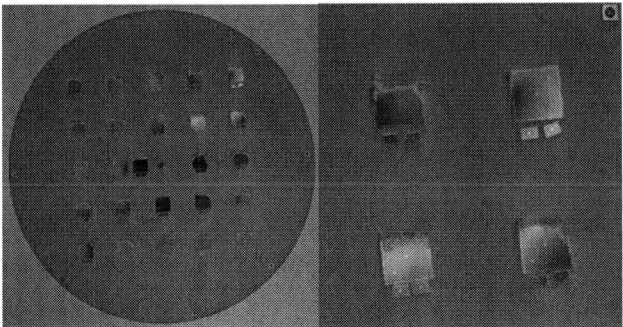

Figure 8 and 9. Visual inspection WaferBOND® HT-10.10 bonding material showing flash at die edges and die shift when used with granular EMC 1.

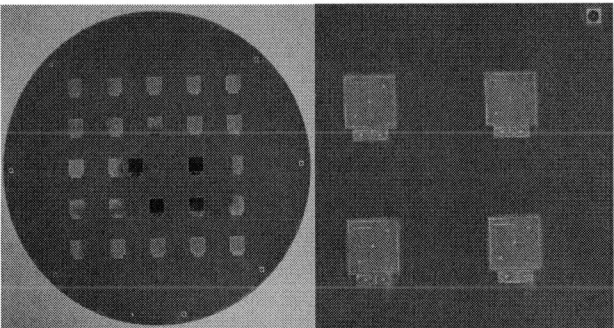

Figure 10 and 11. Visual inspection BrewerBOND® 305 bonding material showing no flash or die shift after molding with granular EMC 1.

C. Die Shift

Die shift is a major challenge for FOWLP and describes movement of the chip after placement during compression molding, debonding and cooling of the reconfigured wafer. Dies are placed onto the carrier using a certain temperature. The carrier is then heated to the mold temperature and the entire assembly expands with the CTE of the carrier material. During this process, the epoxy molding compound (EMC) is cured and the dies are fixed in the EMC. As it thermally crosslinks, the EMC material will shrink, leading to a change in volume of the molded wafer.

Due to the different CTEs of the materials involved and in combination with the temperature profile of the different process steps, dies will shift from their originally assembled position. In addition to the described effect, sliding of the dies during molding can occur depending upon the die size and adhesion of the dies to the release tape and the flow behavior of the EMC and the related forces on the dies during compression molding.

Linear die shift can be compensated for with an assembly adaptation. Dies are assembled at the "wrong" position and shift to the "correct" position during molding. However, the general placement tolerance of the assembly equipment and a random shift during molding cannot be compensated for using this method. After molding, the final position of the die is now defined and impacts the RDL yield and how the die will align with the die pads.

In Chart 3, die shift behavior for all material combinations is summarized. Because the WaferBOND® HT-10.10 bonding material had a strong interaction with EMC 2, that data set was eliminated from the final analysis. WaferBOND® HT-10.10 bonding material demonstrated very low adhesion at molding temperature, leading to mold flash on dies, flying dies, and high die shift. The least amount of die shift was found for combinations using EMC 1 and BrewerBOND® 305 bonding material. Glass carrier and release material had little effect on die shift (Chart 4). After performing high-temperature processing, the reconfigured wafer was cooled and die shift was largely dominated by the CTE of the EMC.

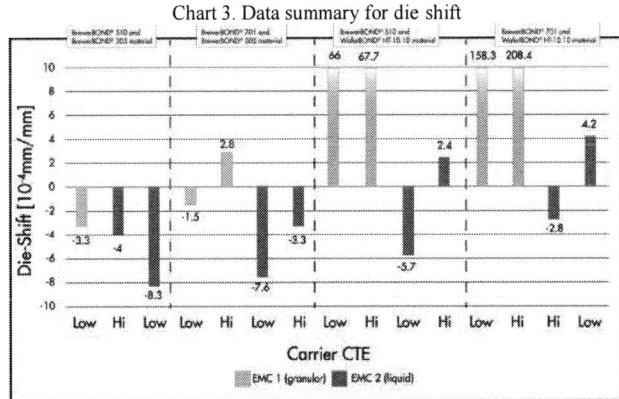

Chart 3. Data summary for die shift

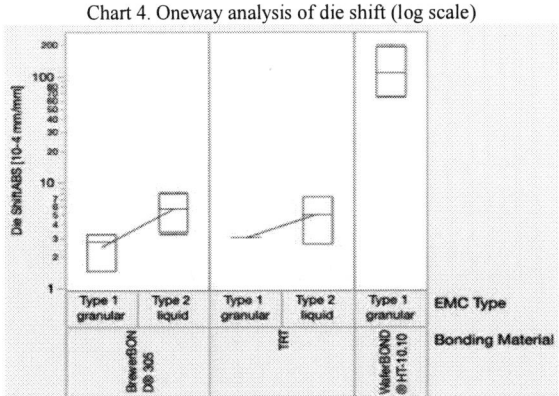

Chart 4. Oneway analysis of die shift (log scale)

Die Stand-off

Die stand-off or die protrusion is caused from shrinkage of the epoxy mold compound during the thermal cure and compression molding process. Embedded die whose surface extends above or below the EMC surface limits multichip integration and RDL scaling.

Chart 5 summarizes the data obtained during this study for die stand-off. Using JMP statistical software, a Oneway analysis of die stand-off was done for bonding material (Chart 6).

BrewerBOND® 305 material showed the least amount of die stand-off for all combinations. Measurements <2 µm were obtained for both EMC 1 (granular) and EMC 2 (liquid). Combinations using EMC 2 and WaferBOND® HT-10.10 material also show low die stand-off, however this data has been removed from the final analysis due to intermixing of the materials. Higher die stand-off was recorded for combinations using WaferBOND® HT-10.10 material with EMC 1 and for the thermal release tape.

The effect of carrier type in combination with EMC type is illustrated in Chart 7. Less variation is found for EMC 2 regardless of carrier type. Carrier type is not a significant factor impacting die stand-off.

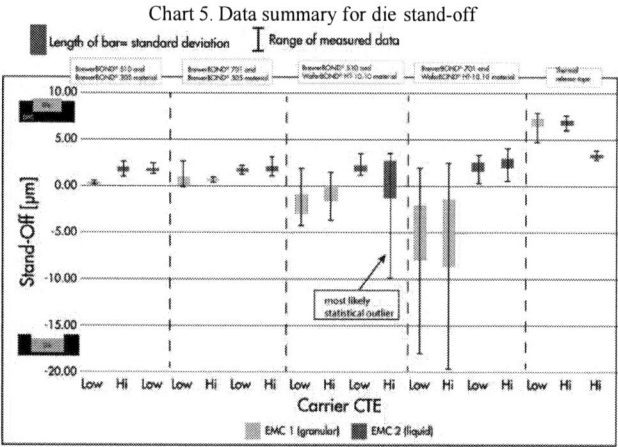

Chart 5. Data summary for die stand-off

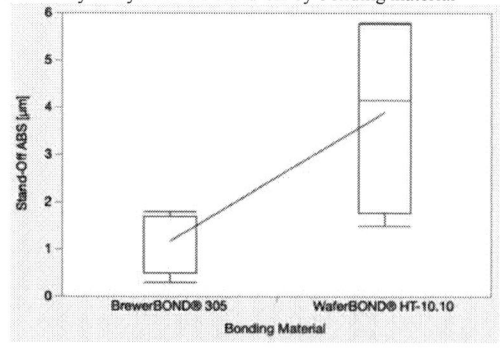

Chart 6. Oneway analysis of die stand-off by bonding material

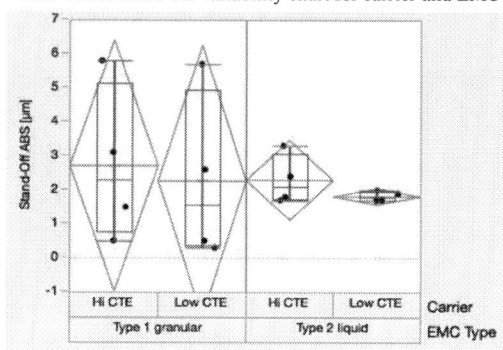

Chart 7. Die stand-off variability chart for carrier and EMC

D. Warpage

There is a large variety of epoxy mold compounds that are currently available for use in fan-out processes. The properties of Youngs's modulus, CTE, and glass transition temperature (T_g) have a significant effect on EMC warpage [7]. Wafer warpage can be reduced by lowering the Young's modulus, CTE, and increasing the T_g of the molding compound.

For this study, Chart 8 summarizes the data generated for wafer warpage. Compared to the thermal release tape, both bonding materials show lower warpage when used in combination with EMC 1 (granular). A Oneway analysis of bonding material and EMC type in Chart 9 shows the least amount of warpage for WaferBOND® HT-10.10 material combinations. This would be consistent with low-T_g materials having the least amount of film stress that can contribute to stack warpage.

Chart 10 illustrates the impact of carrier type on stack warpage. Wafer warpage is less for combinations using EMC 1. The data indicates that EMC type is the significant factor influencing wafer warpage and that all other factors are less significant.

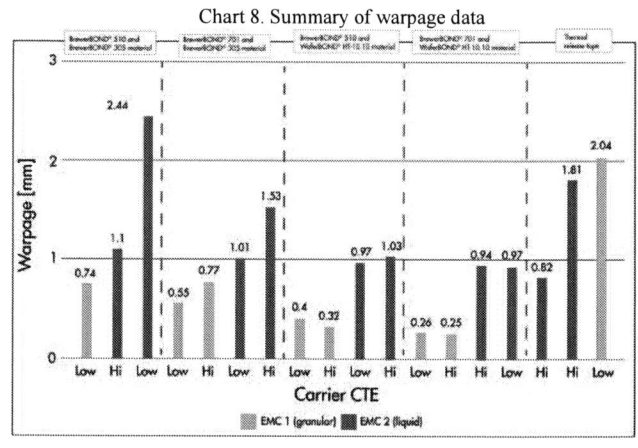

Chart 8. Summary of warpage data

Chart 9. Oneway analysis of warpage

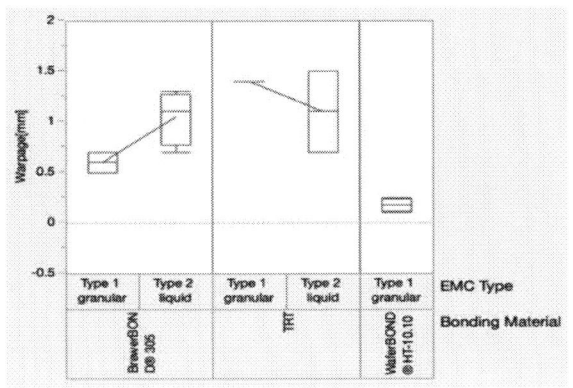

Chart 10. Oneway analysis of warpage for carrier type

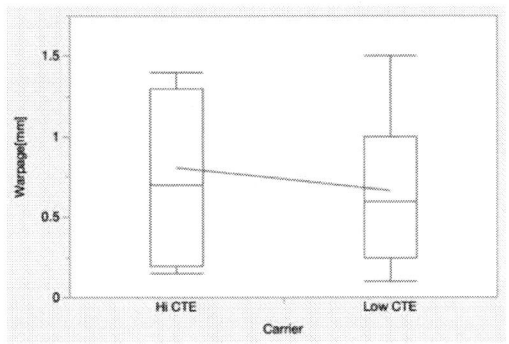

VI. Summary

In this study we evaluated various factors used in a die attach process for chip-first FOWLP. Epoxy mold compound, release and bonding material and carrier CTE effects on die shift, die stand-off, and warpage after debonding were evaluated.

WaferBOND® HT-10.10 bonding material had the highest amount of die shift when used with EMC 1. Due to intermixing between the WaferBOND® HT-10.10 material and EMC 2, this data set was removed from the final analysis. The least amount of die shift was found for EMC 2 when used with all factors. As the T_g for both epoxy mold compounds was not significantly different, additional material properties such as Young's modulus should be considered as a factor in future studies.

Die stand-off and protrusion of the die from the EMC was also evaluated. BrewerBOND® 305 material demonstrated the least amount of stand-off for all combinations. The combination of EMC 1 with WaferBOND® HT-10.10 material showed the highest amount of variability, most likely due to the low T_g of the bonding material. When used with EMC 2, WaferBOND® HT-10.10 material showed improved performance. However, most of this data was excluded due to the intermixing of EMC 2 with the

WaferBOND® HT-10.10 material. Carrier type did not have a significant effect on die stand-off.

It is well known that warpage after mold is primarily a function of the properties of the molding compound. Overall, warpage was reduced for all combinations using EMC 1. The carrier CTE or bonding material had little effect on warpage.

For all responses, the release material in combination with the bonding material had no significant effect. All bonded pairs were successfully debonded and cleaned prior to measurement.

Within this study, the replacement of thermal release tape by temporary adhesives in combination with room temperature laser or mechanical debonding were successfully demonstrated. With careful material and carrier selection, low mold wafer warpage, die stand-off, and die shifting can be achieved. The proposed technology using temporary release and bonding materials could be a suitable process alternative to thermal release tape for FOWLP chip-first, face-down approach.

ACKNOWLEDGEMENT

The authors gratefully acknowledge process and tooling support for this work from Fraunhofer Institute (IZM Berlin, Germany), EV Group (St. Florian Am Inn, Austria), and SUSS MicroTec (Sternenfels, Germany).

REFERENCES

[1] T. Thomas, F. Yuwen Lin, K.-F. Becker, T. Braun, E. Jung, R. Aschenbrenner, H. Reichl; State-of-the-art of 3D SiP technology; Proceedings of IMAPS Poland 2009.

[2] F.X.Che, D. Ho, M. Z. Ding, X. Zhang; "Modeling and design solutions to overcome warpage challenge for fan-out wafer level packaging (FO-WLP) technology," 17th ECTC conference 2015

[3] T. Meyer, G. Ofner, S. Bradl, M. Brunnbauer, R. Hagen; "Embedded wafer level ball grid array (eWLB)," Proceedings of EPTC 2008, Singapore.

[4] C. F. Tseng, C. S. Liu, C. H. Wu, D. Yu; "InFO (Wafer level integrated fan-out) technology," Proceedings of ECTC 2016, Las Vegas, USA.

[5] B. Keser, C. Amrine, t. Duong, O. Fay, S. Hayes, G. Leal, W. Lytle, D. Mitchell, R. Wenzel; "The redistributed chip package; A breakthrough for advanced packaging," Proceedings of ECTC 2007, Reno, Nevada, USA.

[6] M. Fowler, J. Massey, M. Koch, K. Edwards, T. Braun, S. Voges, R. Gernhardt, M. Wohrmann; "Advances in temporary bonding and debonding technologies for use with wafer-level system-in-package (WLSiP) and fan-out wafer-level packaging (FOWLP) processes," IMAPS Conference October 2018, Pasadena, CA, USA

[7] K. Kwon, Y. Lee, J. Kim, J. Y. Chung, Y.Park, D. Lee, S. K. Kim, "Compression molding encapsulants for wafer-level embedded active devices", 2017 IEEE 67th ECTC Conference

Effect of Charging Cycle Elevated Temperature Storage and Thermal Cycling on Thin Flexible Batteries in Wearable Applications

Pradeep Lall[1], Amrit Abrol[1], Ben Leever[2], Scott Miller[3]

[1]Auburn University, NSF-CAVE3 Electronics Research Center
Department of Mechanical Engineering, Auburn, AL 36849
[2]US Air Force Research Labs, Wright-Patterson AFB, OH
[3]NextFlex Manufacturing Institute, San Jose, CA
Tele: +1(334)844-3424; E-mail: lall@auburn.edu

Abstract— The increase in demand for long use time for high-power electronics has made the need for estimating the state of a charge, and capacity for power sources extremely important. Development of a prognostics health management framework for flexible electronics and flexible components is still in its nascent stages due to unique need of flexible electronics. Little to no information exists on the effects of the depth-of-discharge, charging profile, environmental use conditions and other varying test parameters such as varying load (across test conditions) in the the degradation in capacity and remaining useful life for flexible lithium-ion power sources. Flexible electronic systems need a thin, robust power system with the capability to maintain state of charge while sustaining dynamic stresses resulting from human motion and exposed to human body temperature. Further, flexible electronics in wearable applications may be exposed to environment extremes and thermal cycling resulting from changes in environmental temperature and use-conditions. In this research study, flexible lithium-ion batteries have been studied under a variety of conditions including charge-discharge cycles, variations in the depth-of-discharge, changes in the charge current, and changes in the system load. In addition, effects of simultaneous thermal stresses and repeated cyclic events have been studied on output parameters such as efficiency, capacity and charge-discharge time. Regression based modeling technique has been used to estimate the battery capacity deterioration as a function of number of cycles, operating temperature, and depth-of-discharge.

Keywords-flexible batteries, flexible hybrid electronics, lithium-ion, state-of-charge, capacity degradation, constant current, constant voltage, battery efficiency, stresses of daily motion, Power Sources, Memory Effects, Switch-Mode Charging, Shallow Charge, Deep Charge.

I. INTRODUCTION

Lithium-ion batteries are popular choice for power sources in portable electronics applications owing to their high energy density and high volumetric density. Lithium-ion power sources have found applications in electric vehicles, laptops, smartphones IoT, and wearable electronics. Electronics in wearable products may be often not charged to full capacity and may not be discharged completely prior to re-charge cycle. Further, the shallow charge cycles may outnumber the deep-charge cycles in operation of the portable product. Furthermore, the combined effects of distinct bending load(s) and operating temperatures can significantly attenuate the life of flexible Li-Ion batteries in foldable wearable electronics. The power sources may be operate under sustained exposure to human body temperature superimposed on ambient temperature extremes, mechanical loads under stresses of daily motion, and various depths of charge-discharge cycles. There are significant technical challenges in meeting the power-needs of wearable electronics including maximizing capacity in thinner form-factors, flexibility and robustness under the operating and usage environment.

A number of prior studies have focused on use parameters in lithium-ion batteries with a goal of improving capacity retention techniques by altering electrolyte salts or developing different material based cathode/anode but mostly for the rigid batteries. Three major chemical processes which degrade Li-ion rigid batteries are - electrolyte decomposition, transition metal dissolution and lithium plating. Sasaki, at.al. [2013] has studied the presence of memory effect, in $LiFePO_4$ based Li-Ion batteries when exposed to partial charge-discharge and how it affects the state of charge. Operating temperature of Li-Ion batteries is also a concern as it can change the performance of the power source and thermal runaway reaction can easily cause the battery to explode, resulting in catastrophic failure when operating at high temperatures [Feng 2016]. Techniques such as XRD and SEM have also been used before to study the effect of aging process of electrode's binder and electrolyte interface [Bodenes 2013]. Jin [2016] presented a protocol, which reduces electrolyte interface resistances and shows an improvement in capacity retention.

There is scarcity of literature on the use of lithium-ion batteries in flexible electronics. It is unclear if the degradation mechanisms in rigid lithium-ion batteries affect the flexible power sources. Further the sensitivity of the capacity degradation to the use parameters and the environmental conditions is unknown. Recently [Lall 2018, Lall 2018] investigated the effects of deep-discharge and bending loads of up to 15° on a 45mAh single cell flexible battery. Low charge-discharge rate, and small bending loads was found to help retain the life of the flexible Li-ion power sources. Capacity reduction was much less at higher operating temperature even when high charge-discharge rate and full-depth cycling were used. Another study conducted by [Lall 2017] demonstrated the effects of large bending loads of up to 150° on flexible Li-ion batteries and put forth that harsh bending loads can drastically affect the energy storage parameters such as capacity and efficiency of the power source thereby resulting in complete cracking of the electrodes and the outer pouch. Application of a flexible and robust electrode to increase the Li storage capacity was demonstrated by [Deng 2017]. The authors showcased that by constructing a 3D macro-porous structure of molybdenum sulfide on carbon cloth, the volume expansion issues during the charging-discharging process can be reduced. Enhancement of the li-ion battery performance has been explored through flexible hetero-structures for electrodes [Zhang 2017], development of halide ion based Co_3O_4 nano-sheets [Yao 2018] and prolusion of 3D graphene foam based nano-architecture to improve the performance of flexible Li-ion batteries [Mo 2017].

There is a lack of method to assess the acceleration factor and correlate the accelerated test conditions with use conditions. In addition, no established capacity degradation models exist which can capture the true behavior of flexible power sources in use applications. In the present day modern applications such as portable electronics, shallow discharge cycles easily outnumber the deep discharge cycles in operation and while simultaneously being subjected to daily stresses of motion; thus the importance of investigating the effect of depth-of-discharge with different operating parameters. This research work is focused on establishing the survivability and behavior of flexible Li-ion batteries under accelerated life conditions replicative of real world applications.

In this research study, the combined effects of deep-charge, shallow-charge, distinct bending load(s), operating temperatures, elevated temperature storage, and thermal cycling have been characterized for Li-Ion batteries. Thin flexible battery cells were cycled through multiple charge-discharge cycles under simultaneous bending loads plus thermal stresses. Output parameters such as efficiency, power, capacity and charge-discharge time have been analyzed for battery state assessment. In addition, flexible lithium-ion batteries have been analyzed for effects of long-term calendar aging and varying thermal loads by subjecting the test samples to: (1) unpowered aging at 50°C ranging from 10 days-120 days (2) unpowered exposure to thermal cyclic environment, 10°C-50°C, ranging from 50 loops to 550 loops.

This test framework puts to use a battery state assessment analyzer (BSAA) [Lall 2017, Lall 2018, Lall 2018] developed at CAVE3 electronics research center, Auburn University. The BSAA has the ability to store the battery parameters such as voltage, current and resistance online and later computes the capacity, efficiency and power offline.

BSAA has four major components including – (1) programmable source measurement unit (2) electronic load (3) multi-channel data-acquistion and (4) LabVIEW user interface for equipment control and interface. The test samples (Li-ion power sources) have been tested and evaluated at 1C and 0.5C rates. Discharge during the lifecycle tests is characterized by the depth of discharge (x) from a full charge state whereas during the charging phase the batteries are charge to a full charge always. Therefore the charge phase of the test is always conducted from (100-x)-percent charge state to 100-percent charge state, where x is the depth-of-discharge. Depth-of-discharge for shallow cycling is smaller than the depth-of-discharge for a deep cycle which could be close to 100-percent. In order to replicate real-world shallow cycling, a number of different discharge cut off voltages have been used within the test matrix. The accelerated tests run continuously for 180 charge-discharge cycles, where in each cycle voltage, current and resistance values are stored every 2 seconds, which provide significant amount of data points for offline evaluation of parameters such as capacity, efficiency and charge-discharge time and power. A regression based capacity degradation and a cycle to failure model has been developed. The effects of shallow cycling at different C-rates have been characterized.

II. TEST VEHICLE

In this study, state-of-art li-ion pouch form-factor flexible power sources have been used. The nominal rated capacity of these sources is 65mAh. The dimensions of the Li-ion test samples are 60mm x 35mm x 0.5mm as shown in Figure 1.

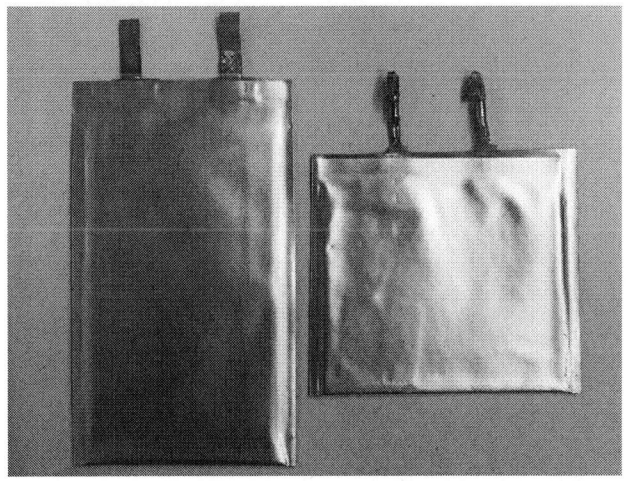

Figure 1:- Test vehicle used for this investigation (left)

The Li-ion power source is comprised of a positive electrode cathode, a negative electrode anode and an electrolyte as a conductor. The cathode is composed of a metal oxide and the anode is made of a carbon based material mostly graphite. The test samples used in this investigation have a graphite based anode; the cathode is made of $LiMnNiCoO_2$ while the electrolyte is a combination of Li based salts in an organic solvent, Ether. During the charge-cycle, the negative electrode (anode) undergoes oxidation i.e. loses electrons and creates positive ions accompanied with a reduction reaction at the cathode. During the discharge phase and a similar event occurs but in reversed direction. A layer by layer structure of the battery test sample is shown in Figure 2. Li-Ion samples used in this investigation consist of only one microcell. Each microcell is further comprised of two one-sided anode (graphite) layers, one two-sided cathode (LiCoO₂) layers, two separators, one positive current collector and one negative current collector.

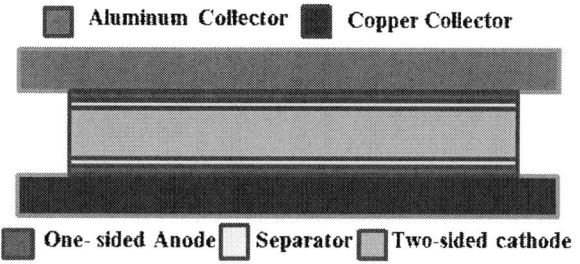

Figure 2: - Layer by layer configuration of the test sample

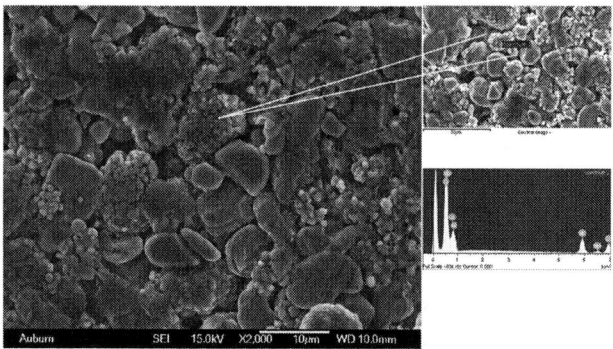

Figure 3: - x2000 zoomed in cathode SEM image and EDX analysis for elemental composition

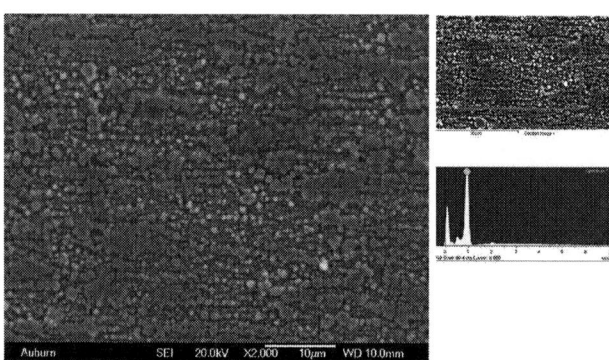

Figure 4: - SEM image of Copper current collector

High resolution scanning electron microscopy images were obtained to visualize and study the internal structure of the Li-ion test samples. The LiMnO₂ forms a α-NaFeO₂

structure more commonly known as the distorted rock salt structure. Energy dispersive X-ray analysis was done on the cathode to characterize the different elements which make up the electrode. Since Lithium is a relatively light metal and X-ray florescent yield probability is low therefore it remains undetected. Figure 3 shows an SEM micrograph of the cathode inside the li-ion power source. Figure 3 shows the elemental composition of the LiMnNiCoO₂ cathode. The separator layer resides between the anode and the cathode. Figure 4 shows the SEM micrograph of the copper current collector inside the li-ion power source. The specification use range for the li-ion batteries during the charging process is 0°C to 45°C and during the discharging process is -20°C to 60°C. In most rechargeable power sources especially the Li-ion ones, both the electrodes have the capability to intercalate Li^+ ions. During the charging period LiMnNiCoO₂ oxidizes i.e. loses electrons and in turn Li^+ ions are released into the electrolyte. Mixture of solvents and Lithium salts, the electrolyte, is already rich in Li^+ ions; therefore the arriving Li^+ ions from the cathode replace the existing Li^+ ions which in turn travel to the negative electrode. The graphite based anode absorbs the arriving Li^+ ions where are they combine with the electrons lost by the positive electrode. The direction of the Li^+ ions is reversed when in the discharging phase which makes the electrochemical reaction reversible in nature.

III. EXPERIMENTAL TEST SETUP AND RELIABILTIY TEST MATRIX

The depth-of-discharge or shallow discharging experiment was conducted using the BSAA. The BSAA has the ability to store the battery parameters such as voltage, current and resistance online and later compute the capacity, efficiency and power offline.

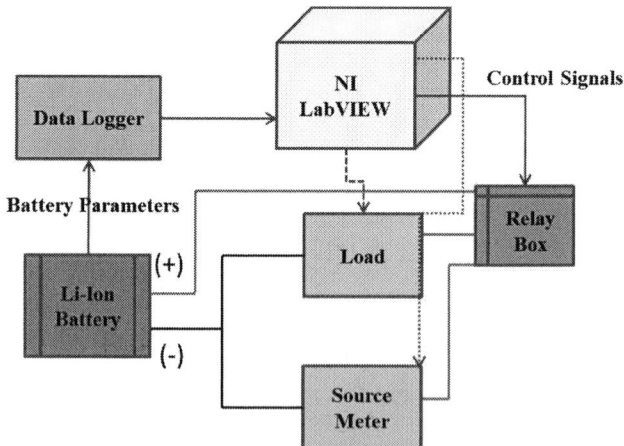

Figure 5: - Schematic of BSAA [Lall 2018]

BSAA has four major components a programmable source meter and DC electronic load, a multi-channel data acquisition system, LabVIEW user interface. A LabVIEW test protocol was developed which can control the number of cycles, cut off voltages, end of charge current and end of charge voltage. The user can also control the presence/absence of bending load during discharging via the module, which was demonstrated in [Lall 2018]. Figure 5 shows the schematic of test setup used for depth-of-discharge testing. The developed module is responsible for transmitting both the discharging and the charging profiles to the DC load and source meter respectively. The test setup connections and other required components have been previously explained in detail in previous studies [Lall 2018].

TABLE 1: - BATTERY CONTROL PARAMETERS

Control Parameter	Value
Charging Cutoff Voltage	4.2V
Charging Cutoff Current	4mA
Discharging Cutoff Voltage	A function of DoD
Charging Current Rate	1C,0.5C
Discharging Current rate	1C,0.5C
Charging Current	60mA,30mA
Discharging Current	60mA,30mA

TABLE 2: - TEST MATRIX FOR 1C

Test Temp.	Delta discharge	Rate	Cutoff
25°C	30%	1C	3.75V
25°C	45%	1C	3.5V
25°C	60%	1C	3.3V
25°C	80%	1C	3V
25°C	100%	1C	2.75V

TABLE 3: - TEST MATRIX FOR 0.5C

Test Temp.	Delta discharge	Rate	Cutoff
25°C	30%	0.5C	3.75V
25°C	45%	0.5C	3.5V
25°C	60%	0.5C	3.3V
25°C	80%	0.5C	3V
25°C	100%	0.5C	2.75V

4.2V 4V 3.9V 3.75V 3.65V 3.5V 3.3V 3.1V 3V 2.9V 2.7V
Full Cycle————————————————➤Shallow Cycle

Table 1 lists the battery control parameters which need to be maintained during the entirety of the test. The term called "delta discharge", has been introduced which is used in this paper to describe the depth-of-discharge being conducted on the flexible battery. A total of five delta discharge variations have been studied. For the charging phase all the samples have been charged to a 100-percent. While the discharging depths vary from test to test, the Li-ion samples are always charged to a value of 4.2V, which is the charging cutoff voltage. Five-replicates of the Li-ion power source have been tested at each condition of the different discharging depths. Table 2 and Table 3 show the reliability test matrices developed for the investigation of Li-ion power sources under varying depth of discharges with full charging. The effect of

different C rates with different depth-of-discharge have been investigated.

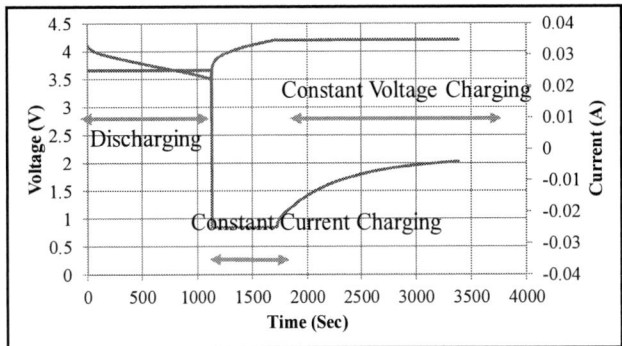

Figure 6:- Charging/discharging voltages for 30-percent Cycle.

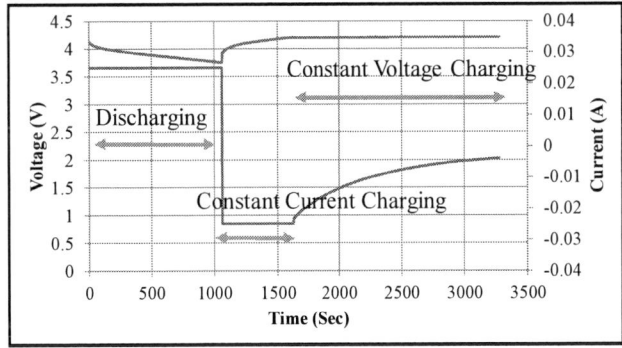

Figure 7:- Charging/discharging voltages for 45-percent Cycle.

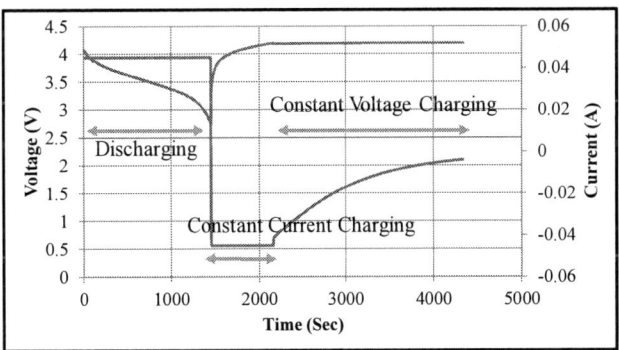

Figure 8:- Charging/discharging voltages for 100-percent Cycle.

During the accelerated life tests power source parameters such as capacity, efficiency charge-discharge time, state-of-charge and power have been evaluated and analyzed for each of the sampled readings. Table 2 and Table 3 show the test configurations for which each of the C-rates including 0.5C and 1C have been used. Each charging cycle in the repeated charge-discharge cyclic tests has three different stages. In the first stage, battery is discharged from a fully-charged state to the rates depth-of-discharge. The second stage initiates once the rated depth-of-discharge is achieved. In the second stage, the battery is charged using constant current charging till the battery voltage reaches a value of 4.2V. The third stage initiates when the battery voltage of 4.2V is achieved. In the third stage, constant voltage charging is continued till the charge current drops to 4mA. Figure 6, Figure 7, Figure 8 show the discharge and charge cycles for 30-percent, 45-percent and 100-percent cycles respectively.

IV. EVOLUTION OF THE DISCHARGE-CHARGE CURVES

Evolution of the discharge-charge curves under different depths-of-discharge are shown in Figure 9 to Figure 16. Li-ion battery output parameters namely capacity, efficiency, charge-discharge time, state-of-charge and power were evaluated by recording the in-situ characteristic VI curves. Figure 9 to Figure 12 show the voltage discharge-charge curves vs time for various depths-of-discharge from 30-percent to 80-percent. Figure 13 to Figure 16 show the current curves vs time for various depths-of-discharge from 30-percent to 80-percent.

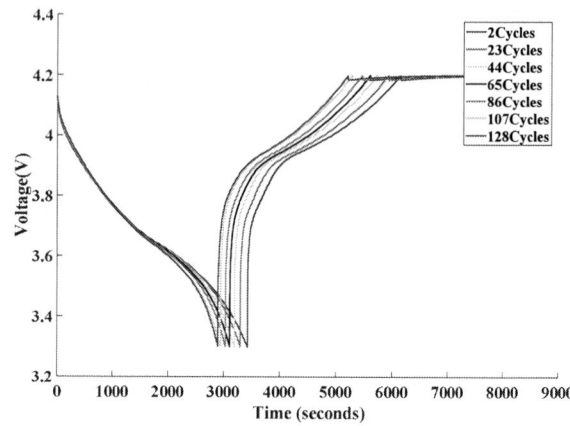

Figure 11: - Voltage characteristic curve during discharging and charging for a depth-of-discharge of 60-percent from full charge

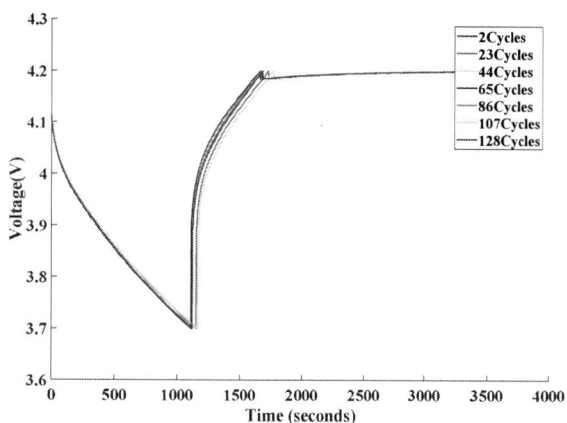

Figure 9: - Voltage characteristic curve during discharging and charging for a depth-of-discharge of 30-percent from full charge

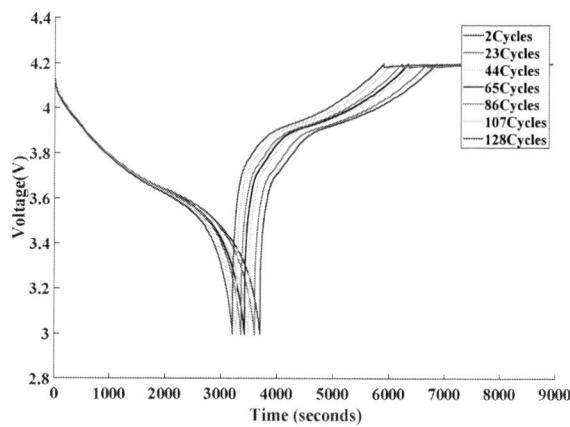

Figure 12: - Voltage characteristic curve during discharging and charging for a depth-of-discharge of 80-percent from full charge

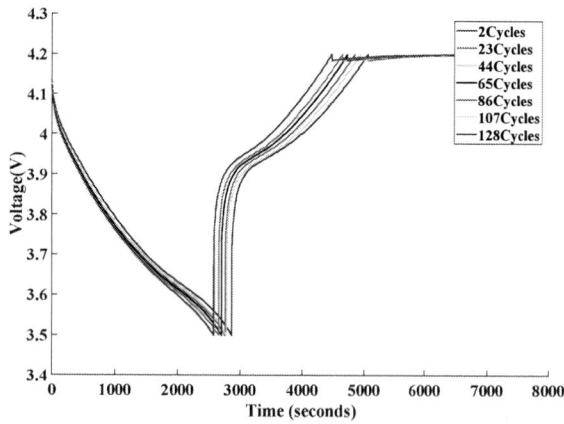

Figure 10: - Voltage characteristic curve during discharging and charging for a depth-of-discharge of 45-percent from full charge

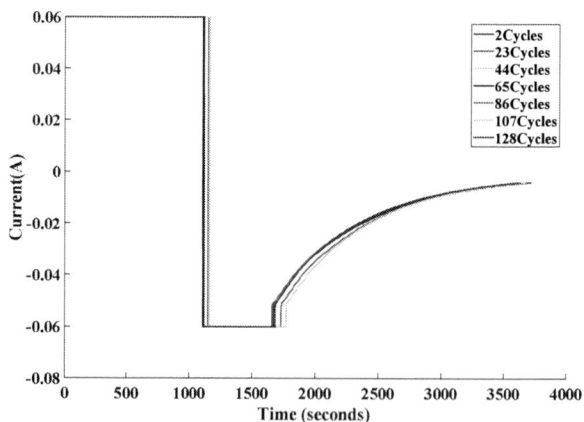

Figure 13: -Current characteristic curve during discharging and charging for a depth-of-discharge of 30-percent from full charge

978-1-7281-1500-9/19 $31.00 © 2019 IEEE

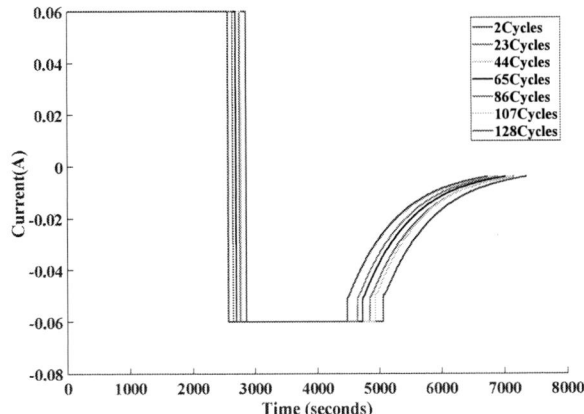

Figure 14: - Current characteristic curve during discharging and charging for a depth-of-discharge of 45-percent from full charge

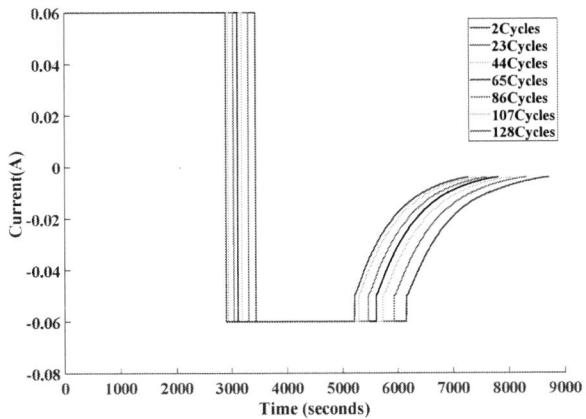

Figure 15: - Current characteristic curve during discharging and charging for a depth-of-discharge of 60-percent from full charge

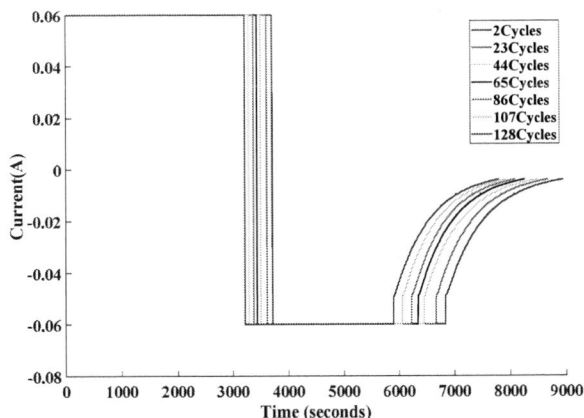

Figure 16: - Current characteristic curve during discharging and charging for a depth-of-discharge of 80-percent from full charge

The leftward shift in both the voltage curves and the current curves indicates the capacity degradation with the increase in discharge-charge cycles at the test conditions. A comparison

of the voltage and current curves indicates that leftward shift in both the voltage curves and the current curves is increases with the depth-of-discharge. Shallow depth-of-discharge have lower degradation in capacity than full depth-of-discharge.

V. EFFECT OF DEPTH-OF-DISCHARGE, CHARGE CURRENT AND OPERATING TEMPERATURE

Variation in depth-of-discharge occurs in operation when the battery is subjected to a discharge cycle to a partially discharged state instead of a deeply depleted state. The battery has been discharged from a fully-charged state to a partially state of charge. The level of discharge has been quantified using the depth-of-discharge. Data is presented on depth-of-discharge of 30, 45, 60, 80 and 100 percent. The 100-percent discharge corresponds to the deeply depleted battery, while the lower values of discharge correspond to a shallow discharge.

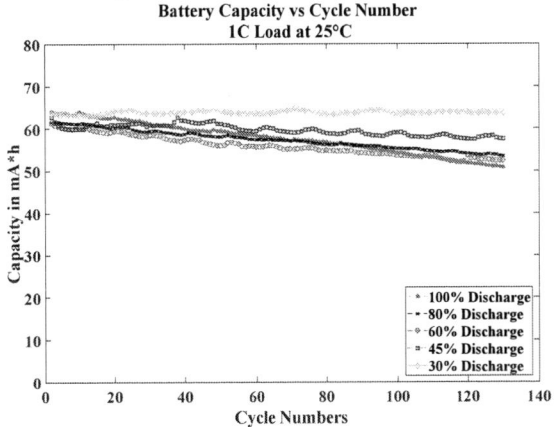

Figure 17: -Capacity vs the number of cycles as a function of the Depth-of-Discharge at 1C Load

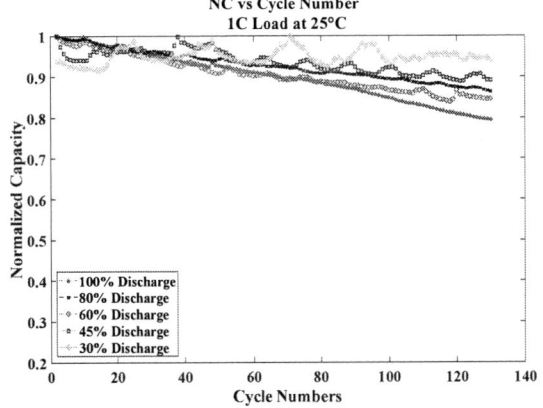

Figure 18: -Normalized capacity vs the number of cycles as a function of the Depth-of-Discharge at a 1C load

Data on the charge state, battery capacity, efficiency, charge-discharge current, charge-discharge voltage has been gathered to understand the effect of use parameters and environmental use conditions. In-situ data for voltage,

978-1-7281-1500-9/19 $31.00 © 2019 IEEE

current and resistance was recorded every 2 seconds until the end of the accelerated life cycle test which is 130 charge-discharge cycles. Battery discharge capacity has units of mAh and is calculated from the product of discharge current, measurement interval and the state-of-charge. Data was acquired at two different C-rates to capture the effect of the charging current on the rate of discharge. The values of charge current include 1C and 0.5C where the charge-discharge current was set to 60mA and 30mA respectively.

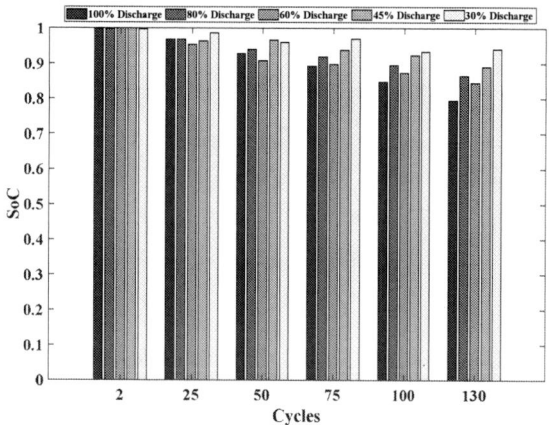

Figure 19: -State-of-Charge vs the number of cycles as a function of the Depth-of-Discharge

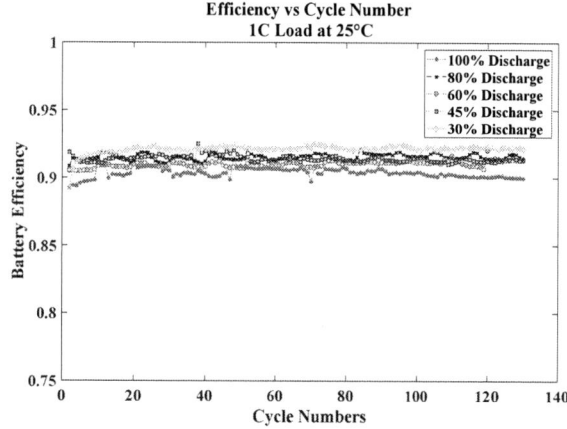

Figure 20: - Battery efficiency vs the number of cycles as a function of the Depth-of-Discharge at 1C load

Figure 17 shows the battery discharge capacity vs the number of cycles as a function of the depth-of-discharge. It can be seen that the battery degrades faster with the increase in the depth of discharge with the increase in the number of charge-discharge cycles. The reduction in the rate of capacity degradation with the decrease in the depth of discharge is expected because the shallow-cycle helps preserve the electrodes by reducing the expansion and contraction of the electrodes resulting from release-and-acceptance of Li-ions. Further, a higher depth-of-discharge results in deterioration of the electrode-electrolyte interface with the increase in charge-discharge cycles. Figure 18 shows the normalized

capacity curves as a function of the depth-of-discharge. A 100-percent discharge curve shows more deterioration in capacity as opposed to the 30-percent discharge curve.

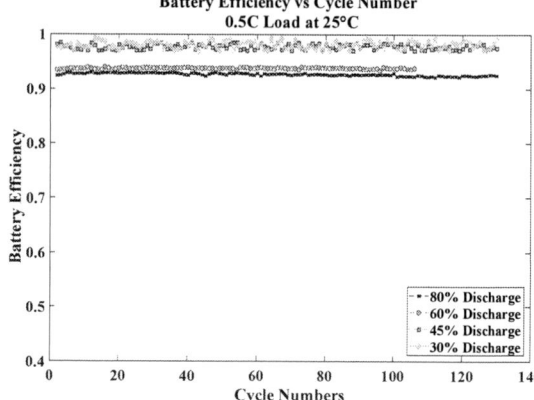

Figure 21: Battery efficiency vs the number of cycles as a function of the Depth-of-Discharge at 0.5C load

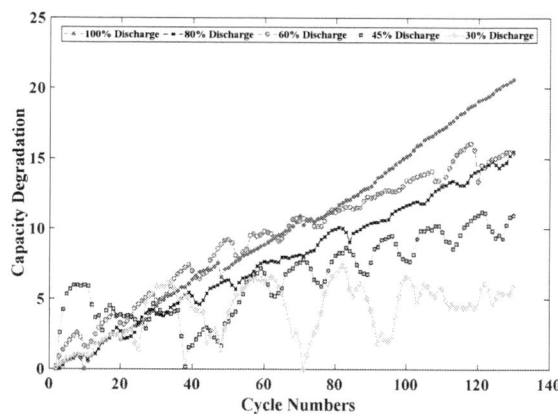

Figure 22: Capacity Degradation in percent as a function of the Depth-of-Discharge at 1C load

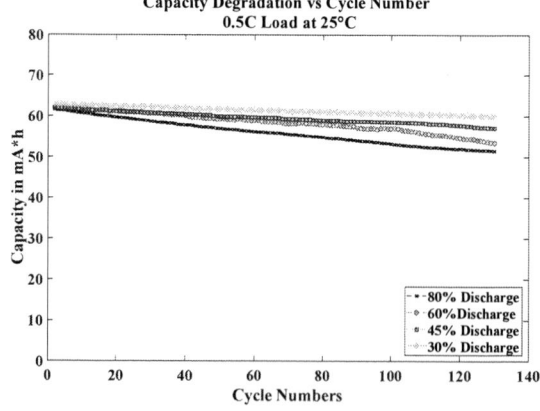

Figure 23: Capacity vs the number of cycles as a function of the Depth-of-Discharge at 0.5C Load

978-1-7281-1500-9/19 $31.00 © 2019 IEEE

State-of-charge (SoC) is an important parameter studied for rechargeable batteries. SoC is defined as the ratio of the useable charge to the maximum charge that can be stored within the power source. It is well known that Li-ion transport between the electrodes accompanies the charging and discharging of the power source. The anode has maximum concentration of Li-ions and cathode has the minimum concentration at the highest possible level of the SoC. Upon the completion of the circuit an internal redox reaction occurs enabling the anode to lose some electrons by creating cations, while the cathode gains a few electrons. Following the same trend the state-of-charge is at its lowest when the battery is discharged and the SoC is the lowest value.

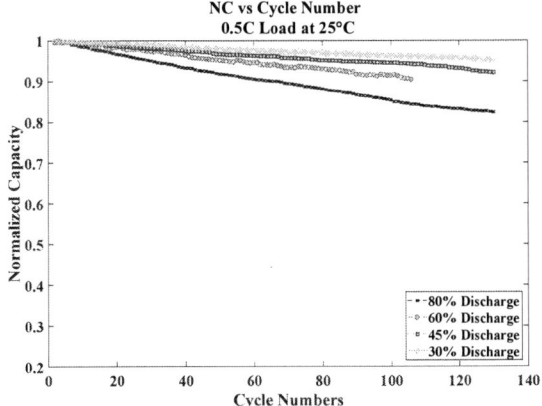

Figure 24: Normalized capacity vs the number of cycles as a function of the Depth-of-Discharge at 0.5C load

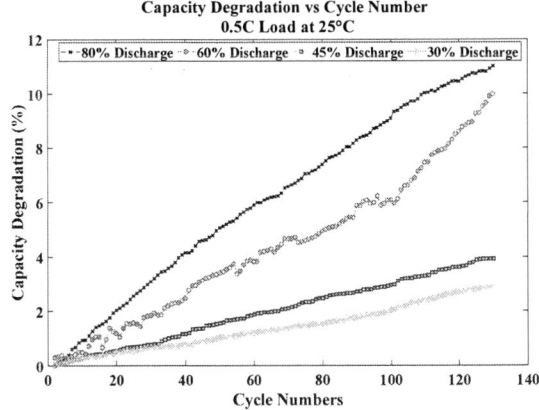

Figure 25: Capacity Degradation as a function of the Depth-of-Discharge at 0.5C load

For an ideal case, the battery would start with a full or a 100-percent state-of-charge when the battery is fully charged and after a complete discharge-and-charge cycle will end with 100-percent SoC when fully charged at the end of the cycle. Figure 19 shows that all test samples start with the same SOC but as the accelerated life cycle test progresses batteries which were subjected to only 30-percent discharge are able to retain the higher SoC at the end of the test in comparison with batteries which were subjected to higher depths-of-discharge. Since the depth-of-discharge is measured from a fully charged state of the battery during the discharge phase of the discharge-charge cycle, thus the state-of-charge is measured in terms of the depleted charge at the end of the discharge cycle. In each of the sub-groups at any number of discharge-charge cycles, the deterioration in the SoC is the highest for the higher depths-of-discharge. The trend holds true for all the sub-groups of discharge-charge cycles indicated by the upward sloping trend of the bar-graphs (Figure 19).

The battery efficiency has been quantified for each of the discharge-charge cycles using the ratio of the output energy dissipated across the electronic load to the input energy used to charge the battery during the charge cycle (Figure 20). The online measurements recorded for voltage and current at each measurement interval are multiplied and then integrated for each cycle. The aforementioned process is carried out for all cycles till the end of the accelerated life cycle test. The battery efficiency of the discharge-charge cycles does not show an upward or downward sloping trend for any of the depth-of-discharge conditions. However, the lower depth-of-discharge show a higher efficiency indicated by the highest efficiency values for the 30-percent depth-of-discharge. The highest depth-of-discharge of 100-percent showed the lowest efficiency during the discharge-charge cycles. Figure 21 shows the battery efficiency versus discharge-charge cycles for a charge current of 0.5C at an operating temperature of 25□. The trends of higher efficiency for shallow depth-of-discharge observed at the charging current of 1C also hold true for the charging rate of 0.5C. The efficiency is the highest for the depth-of-discharge of 30-percent and lowest for the depth-of-discharge of 100-percent.

Figure 22 shows the capacity degradation (measured in percent) for different depth-of-discharge test conditions for a charge current of 1C at 25□. The capacity degradation is the highest for the highest depth-of-discharge for the discharge-charge cycles indicated by the highest slope for the 100-percent depth-of-discharge and the lowest slope for the 30-percent depth-of-discharge. The capacity degradation ranges between 5-percent and 20-percent for depths-of-discharge between 30 to 100-percent. Figure 23 and Figure 24 show the capacity and the normalized capacity for a lower charge current of 0.5C at a charge temperature of 25□. Figure 25 shows the capacity degradation for various depth-of-discharge for charging current of 0.5C at operating temperature of 25□. The trends of capacity degradation are consistent with the trends observed for the higher charging current of 1C, where the higher depth-of-discharge exhibit the highest capacity degradation.

VI. EFFECT OF THERMAL AGING PRIOR TO DEPLOYMENT

In this section, the effect of sustained exposure of the Li-ion batteries to high-temperature storage prior to deployment in end application has been studied. The batteries have been subjected to a sustained temperature of 50□ for period ranging from 10-days to 115-days to simulate environmental storage in non-climate controlled environments. The capacity degradation and efficiency evolution with the progression of the discharge-charge cycles has been studied. All the battery test samples were evaluated at a C-rate of 1C where the discharge-charge current was set to 60mA. Online data was recorded every 2 seconds till the end of the accelerated life cycle test at 130 cycles.

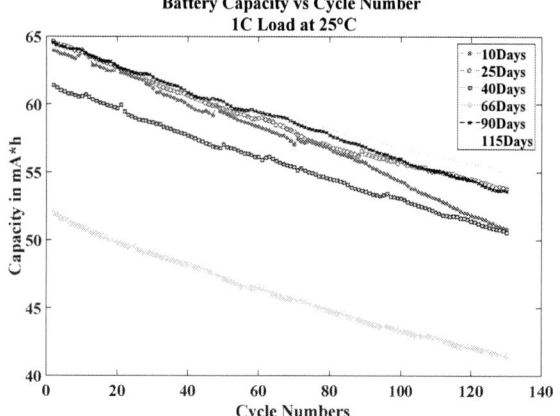

Figure 26: - Capacity vs the number of cycles of repeated lifecycle test after long term calendar aging

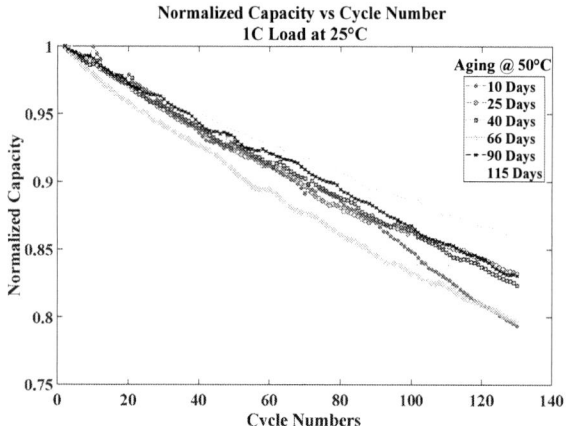

Figure 27: -Normalized capacity vs the number of cycles of repeated lifecycle test after long term calendar aging

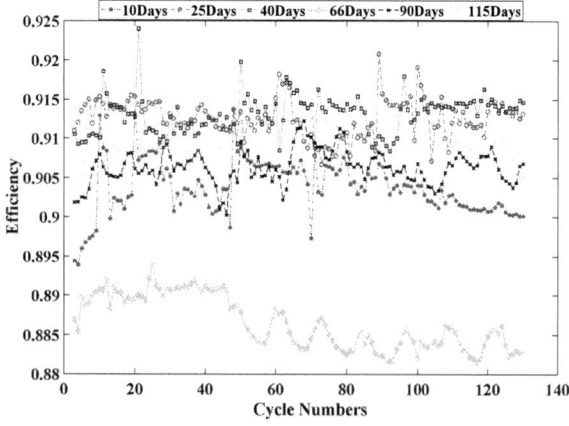

Figure 28: - Battery efficiency after long term calendar aging

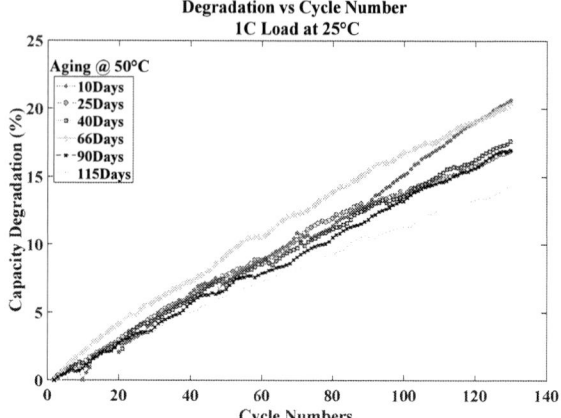

Figure 29: -Capacity degradation with respect to number of discharge-charge cycles after long term calendar aging

Figure 26 shows the battery discharge capacity vs the number of cycles as a function of aging days. It can be seen that from 10-days calendar aging till 66-days of calendar aging the initial battery capacity decreases but the 90 days and 115 days aged battery population does not fall in this trend. Thus, there is a reversal in the reduction in initial capacity with the higher durations of thermal aging. In order to see the effect of thermal aging on the damage progression in terms of battery capacity, the normalized capacity has been computed. The differences in the rate of damage progression can be seen in the differences in the slopes of the curves. Figure 27 shows the effect of thermal aging on the normalized capacity of the battery with the increase in the number of discharge-charge cycles at a charging current of 1C and operational temperature of 25□. The 66-days aged population which showed the highest degradation in the initial capacity also exhibited the highest rate of damage progression. The reversal in the initial battery capacity seen in the thermal-aged samples can also be seen in the rate of capacity degradation with the discharge-charge cycles – as the 110 day population shows a lower rate of damage progression.

The battery efficiency has been quantified for each of the discharge-charge cycles using the ratio of the output energy dissipated across the electronic load to the input energy used to charge the battery during the charge cycle. The online measurements recorded for voltage and current at each measurement interval are multiplied and then integrated for each cycle. The computation of efficiency is carried out for all cycles till the end of the accelerated life cycle test. Figure 28 shows the computed power source efficiencies for different aging days. The initial efficiency shows a reversal with the increase in the number of days of thermal aging. The degradation in efficiency increases till 66-days of thermal aging. However, the degradation in efficiency shows a reversal with the 90 days and 115 days populations showing a higher efficiency. Figure 29 show the capacity degradation for the different thermal aging conditions. The capacity degradation shows a reversal with the higher levels of thermal aging. The capacity degradation rate depicted by the slope of the curves increases monotonically till 66-days of thermal aging. However, higher levels of thermal aging results in the reduction on the capacity degradation rate depicted by the lower slope of the curves for 90-days and 115-days of thermal aging.

VII. EFFECT OF THERMAL CYCLING

Batteries may be subjected to thermal cycling as part of operational life as a result of environmental temperature variations, sustained exposure to human body temperature, and power cycling of the device. In this part of the study, the batteries have been subjected to thermal cycles between 10-50□. The thermal cycles in this section are called "loops" in order to distinguish them from the cycles imposed during the discharge-charge cycles. A total of 50-loops to 550-loops have been imposed on the batteries prior to the discharge-charge cycling. The degradation in the capacity with the discharge-charge cycles is shown in Figure 30. The capacity shows monotonic increase in the rate of capacity degradation with the discharge-charge cycles indicated by the slope of the curves.

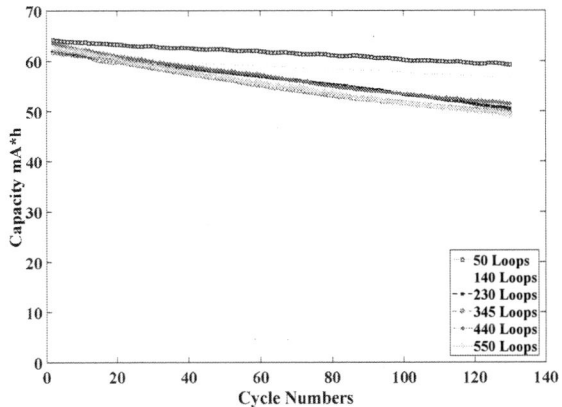

Figure 30: - Capacity vs the number of cycles of repeated lifecycle test after long term thermal cyclic storage

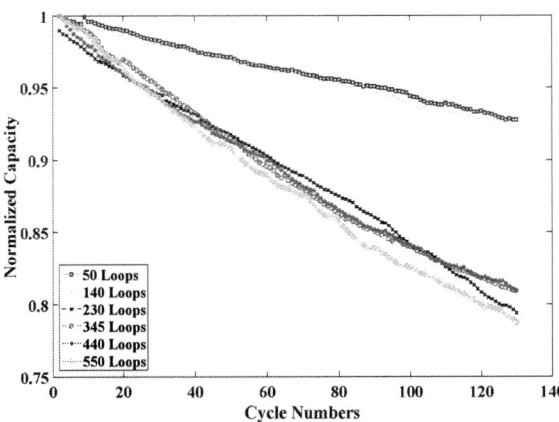

Figure 31: -Normalized capacity vs the number of cycles of repeated lifecycle test after long term thermal cyclic storage

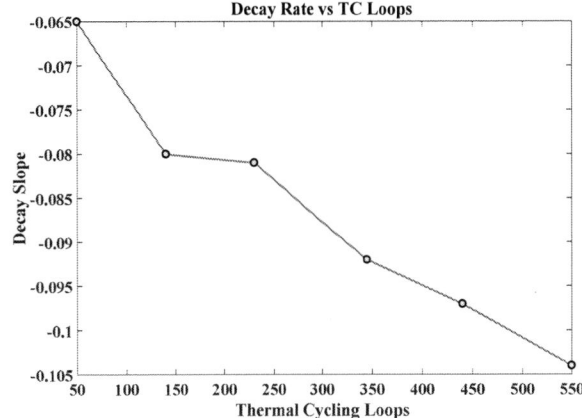

Figure 32: - Decay rate of capacity curves as a function of thermal cycling loops

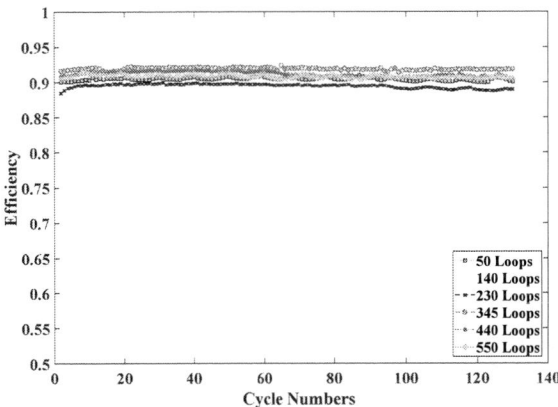

Figure 33: -Battery efficiency after long term thermal cyclic storage

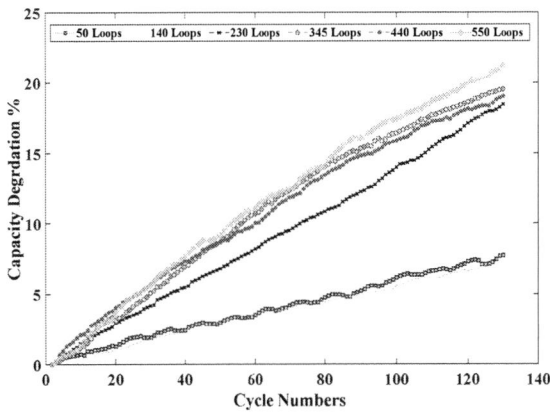

Figure 34: Capacity degradation vs the number of cycles of repeated lifecycle test after long term thermal cyclic storage

Figure 31 shows the effect of thermal cycles on the normalized capacity degradation. The trends in rate of capacity degradation can be seen more clearly in the normalized capacity curves. The rate of capacity degradation appears to be very non-linear with the slopes of the battery populations changing dramatically from 50-loops, 140-loops to populations that were exposed to higher levels of thermal aging for 230-loops, 345-loops, 440-loops, and 550-loops.

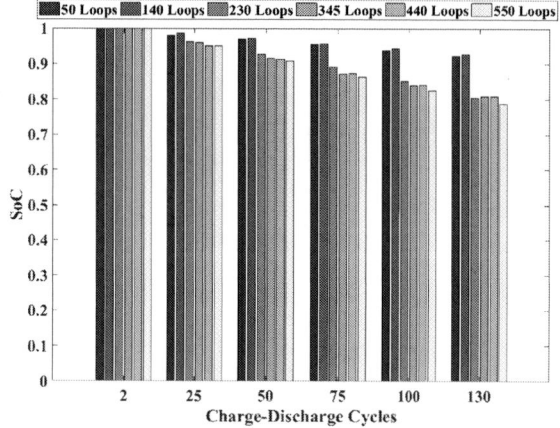

Figure 35: -State-of-Charge evolution as a function of thermal cycling loops

The increase in the rate of capacity degradation is monotonic with the slopes for the normalized capacity in battery populations exposed to 550-loops being the highest. The slopes of normalized capacity curves are also shown as decay rates in Figure 32. The discharge capacity decay slope after exposure to 550 loops is 1.6 times larger than the decay slope after 50 loops. Figure 33 shows the efficiency of the batteries after exposure to different levels of thermal cycling. The battery efficiency does not show evolution with the exposure to discharge-charge cycles irrespective of the number of thermal cycles imposed during the pre-conditioning phase of the test. The capacity degradation is shown for discharge-charge cycles for a number of different thermal cycling conditions is shown in Figure 34. The rate of capacity degradation increases with the increase in the number of thermal cycle loops – indicated by the increase in the slope of the curves. The increase in the rate of degradation is monotonic from the 50-loops condition to the 550-loops condition. The highest degradation after 130 discharge-charge cycles was in the neighborhood of 22-percent, while the lowest degradation was in the neighborhood of 7-percent. Figure 35 shows the comparison of the degradation after thermal aging for various levels of discharge-charge cycles. The downward sloping direction of the bar-graph groups indicates that the degradation increases in a monotonic manner with increase in duration of thermal cycling, irrespective of the number of discharge-charge cycles.

VIII. ACCELERATION FACTOR MODEL

In this section, the effect of the operational parameters on the capacity degradation of the flexible battery have been used to predict the effect of number of discharge-charge cycles, depth-of-discharge, charging current, and operational temperature. A power-law model has been developed using multiple linear regression. An Arrhenius dependence on operational temperature has been assumed.

$$S* = \left(e^{\frac{A_1}{T\_kelvin}} \right).\left(ncycles\_num^{A_2} \right)$$
$$.\left(depth\_discharge\_perc^{A_3} \right)$$
$$.\left(charge\_current\_mA^{A_4} \right)$$
$$.\left(bend\_dv^{A_5} \right)$$

(1)

Where S* is the defined as the capacity degradation in percent; T_kelvin is the operating temperature in kelvin; ncycles_num is the number of discharge-charge cycles imposed on the battery; depth_discharge_per is the depth of discharge from the fully charged condition of the battery expressed in percent; charge_current_mA is the charge and discharge current in the discharge-charge cycles expressed in mA; bend_dv is a dummy variable indicating the presence or absence of bending deformation imposed on the battery (0 denotes no deformation; 1 denotes deformation). In addition, A_1 to A_5 are constants which indicate the influence of each of the input variables. The equation has been simplified into the following form for the purpose of analysis,

$$\ln(S*) = \frac{A_1}{T\_kelvin} + A_2 \ln(ncycles\_num)$$
$$+ A_3 \ln(depth\_discharge\_perc)$$
$$+ A_4 \ln(charge\_current\_mA)$$
$$+ A_5 \ln(bend\_dv)$$

(2)

Table 4 shows the results of multiple linear regression. The regression model has low VIF values for all the constants which reflect to the fact that there are no multicollinearity issues within the battery dataset. In addition, all the coefficients have low p-values less than 0.05 indicating that the coefficient values are statistically significant from zero at 95-percent confidence interval. The model explains about 74.96 percent of the variability in the data-set.

TABLE 4:-REGRESSION MODEL VALUES

Term	Coef.	SE Coef.	T-Value	P-Value	VIF
Constant	-2.617	0.213	-12.31	0	
(1/T_kelvin)	25.17	1.51	16.66	0	1.25
current_mA	0.5426	0.0110	49.27	0	1.28
bend_dv	-0.5735	0.0140	-41.06	0	1.22
Depth_dischrge_perc	0.6986	0.0437	15.98	0	1.37
ncycles_num	0.4577	0.0131	35.04	0	1.01

TABLE 5:-MODEL SUMMARY

S	R-sq	R-sq(adj)	R-sq(pred)
0.3620	80.71%	80.64%	80.50%

The resulting regression equation from the model has been shown in Equation (3).

$$S^* = A \left(e^{\frac{25.17}{T\_kelvin}} \right) . \left(ncycles\_num \right)^{0.46} \tag{3}$$

$$. \left(depth\_discharge\_perc \right)^{0.67}$$

$$. \left(charge\_current\_mA \right)^{0.54}$$

$$. \left(bend\_dv \right)^{-0.57}$$

Where A is a pre-coefficient which lumps the unquantified parameters. Positive values of the exponents indicate a higher value of capacity degradation. Thus, higher values of charge and discharge current correspond to higher capacity degradation. The presence of low amounts of bending reduce the percentage of capacity degradation. A lower depth-of-discharge corresponds to a lower value of capacity degradation. Higher number of charge and discharge cycles corresponds to higher capacity degradation. Higher ambient temperatures correspond to lower capacity degradation.

IX. SUMMARY AND CONCLUSIONS

In this paper, the effects of shallow and deep cycling, thermal aging pre-conditioning, and thermal cycling preconditioning has been investigated on the battery capacity, normalized battery capacity, efficiency, and the capacity degradation in discharge-charge cycling. Experimental observations indicate that the battery capacity decreases with increase in the number of discharge-charge cycles. The capacity degradation is higher at higher charge-discharge currents in the charging cycle. In addition, exposure to higher duration of thermal cycling pre-conditioning results in higher degradation in the capacity, and normalized capacity and the higher rate of degradation versus discharge-charge cycles. Higher durations of thermal cycling preconditioning exhibit the higher decay rate in the discharge-charge cycles. Further, higher durations of thermal aging pre-conditioning result in a non-monotonic change in the initial capacity, and the rate of change of capacity versus discharge-charge cycles. Data on thermal aging pre-conditioned populations indicates that the capacity and the rate of degradation of capacity first increases and then reduces with the increase in thermal aging exposure duration. The results from the experiments have been used to develop a power-law model which can be used for assessment of acceleration factors between the accelerated test conditions and the use conditions. The model explains nearly 74-percent of the variation in the data-set.

ACKNOWLEDGMENTS

The project was sponsored by the NextFlex Manufacturing Institute under PC 2.5 Project titled – Mechanical Test Methods for Flexible Hybrid Electronics Materials and Devices. This material is based, in part, on research sponsored by Air Force Research Laboratory under agreement number FA8650-15-2-5401, as conducted through the flexible hybrid electronics manufacturing innovation institute, NextFlex. The U.S. Government is authorized to reproduce and distribute reprints for Governmental purposes notwithstanding any copyright notation thereon. The views and conclusions contained herein are those of the authors and should not be interpreted as necessarily representing the official policies or endorsements, either expressed or implied, of Air Force Research Laboratory or the U.S. Government.

REFERENCES

[1] Lall, P.; Abrol, A.; Leever, B.; Marsh, J., "Flexible Power-Source Survivability Assurance under Bending Loads and Operating Temperatures Representative of Stresses of Daily Motion," 2018 17th IEEE Intersociety Conference on Thermal and Thermomechanical Phenomena in Electronic Systems (ITherm), San Diego, CA, 2018

[2] [2] Lall, P.; Abrol, A.; Leever, B.; Marsh, J., "Effect of Shallow Cycling on Flexible Power-Source Survivability under Bending Loads and Operating Temperatures Representative of Stresses of Daily Motion," 2018 IEEE 67th Electronic Components and Technology Conference (ECTC), San Diego, CA, 2018

[3] Lall, P.; Zhang, H., "Test Protocol for Assessment of Flexible Power Sources in Foldable Wearable Electronics under Stresses of Daily Motion during Operation," 2017 IEEE 67th Electronic Components and Technology Conference (ECTC), Orlando, FL, 2017, pp. 804-814

[4] [4] Lall, P.; Zhang, H., "Prognostication of remaining useful-life for flexible batteries in foldable wearable electronics," 2016 IEEE International Conference on Prognostics and Health Management (ICPHM), Ottawa, ON, 2016, pp. 1-10.

[5] B. Scrosati, "Plastic lithium ion batteries," IECEC-97 Proceedings of the Thirty-Second Intersociety Energy Conversion Engineering Conference (Cat. No.97CH6203), Honolulu, HI, 1997, pp. 5-8 vol.1

[6] B. Scrosati and J.Garche, "Lithium batteries: Status, prospects and future." Journal of Power Sources 195.9 (2010): 2419-2430

[7] Leng, F., Tan, C. M., & Pecht, M., "Effect of Temperature on the Aging rate of Li-Ion Battery Operating above Room Temperature". 2015, Scientific Reports, 5, 12967.

[8] B. Scrosati, F. Croce, and S. Panero. "Progress in lithium polymer battery R&D." Journal of Power Sources 100.1 (2001): 93-100

[9] Wang, J., Jin, D., Zhou, R., Li, X., Liu, X., Shen, C., Xie, K., Li, B., Kang, F. & Wei, B. "Highly Flexible Graphene/Mn3O4 Nanocomposite Membrane as Advanced Anodes for Li-Ion Batteries" ACS NANO 2016, 10, 6227-6234

[10] T. Sasaki, Y. Ukyo, and P. Novak. " Memory effect in lithium-ion battery." Nature Materials 12, 569-575 (2013)

[11] P. Balbuena and Y.Wang." Lithium-ion batteries." Solid Electrolyte Phase, London

[12] V. Sangwan, R. Kumar and A. K. Rathore, "State-of-charge estimation for li-ion battery using extended Kalman filter (EKF) and central difference Kalman filter (CDKF)," 2017 IEEE Industry Applications Society Annual Meeting, Cincinnati, OH, USA, 2017, pp. 1-6

[13] S. Jin., J. Li., C. Daniel., D. Mohanty., S. Nagpure. & D.Wood "The state of understanding of the lithium-ion-battery graphite solid electrolyte and its relationship to formation cycling" Carbon, vol. 105,pp. 52-76, Aug. 2016

Bladder Inflation Stretch Test Method for Reliability Characterization of Wearable Electronics

Benjamin G Stewart and Suresh K Sitaraman

Computer-Aided Simulation of Packaging Reliability Laboratory
The George W. Woodruff School of Mechanical Engineering
Flexible Wearable Electronics Advanced Research
suresh.sitaraman@me.gatech.edu

Abstract— **The recent development of electronic materials that can maintain electrical performance while undergoing large applied strains have demonstrated potential for use in a new breed of electronic systems. The rapid development of these electronic systems that are flexible, stretchable, and/or wearable necessitates the concurrent development of robust mechanical and electrical test methods to improve their design and reliability. In this paper, one such mechanical test method is discussed in which a stretchable electronic test coupon is mounted onto an inflatable bladder of known geometry to induce multiaxial strains, while in-situ 4-point resistance measurement is employed to assess the device's performance and electromechanical integrity. The material combination of a stretchable screen-printed silver ink cured onto a thermoplastic polyurethane (TPU) substrate is studied given the proclivity for the use of TPU in wearable devices. A dome-shaped bladder configuration is employed in this work to study the performance of printed conductors under biaxial stretching. Various monotonic and cyclic loading regimes are employed to characterize the fatigue behavior and maximum use conditions of the samples. Volume of water displaced into the bladder during inflation is measured and correlated to the induced multiaxial strains on the mounted devices using 3D digital image correlation. Relationships between resistance and applied multiaxial strains are presented. Experimental results are compared with literature, and plausible extensions of the test method including direct printing on the bladder material are discussed.**

Keywords- flexible electronics, wearable electronics, stretching, biaxial, fatigue, resistance, strain, TPU, and 3D DIC

I. INTRODUCTION

The advent of stretchable electronic materials has enabled a new breed of systems and devices, which can stretch and conform to dynamic three-dimensional contoured surfaces. These flexible and stretchable electronic systems have applications in many industries ranging from health care to the military to e-textiles and so on. Novel and innovative device designs which seek to create systems with various electrical, physical, and chemical sensors take advantage of these new extremely compliant materials to seamlessly integrate with the human body, allowing for continuous monitoring of vital signs, brain activity, muscular movement, and other human activities and functions [1-17].

A specific area of emphasis for these stretchable wearable devices is biocompatibility, focusing on the development and use of electronic materials which have thicknesses, elastic moduli, stiffnesses, and areal masses to match with human skin. The human epidermis can experience repeated strains of over 20% and has a modulus in the range of hundreds of kPa to tens of MPa [6, 7, 14, 17-20]. Effective epidermal electronic devices utilize substrates, adhesives, and conductors with similar moduli which can withstand cyclic strains in these ranges. Commonly used substrates for these stretchable wearable devices include many thin film polymers, including thermoplastic polyurethane (TPU), polydimethylsiloxane (PDMS), silicone elastomer, and polyvinyl chloride (PVC), among others [2, 8, 10, 14]. Effective stretchable conductors must maintain excellent conductivity while undergoing repeated, large deformations. This functionality is achieved through the use of additive printing of electrically conductive inks [21]. Conductive composite inks are of increasing interest, as they can achieve simultaneous stretchability and high electrical performance through a distribution of discrete conductive metallic filler particles within a highly elastic polymer matrix. When strained, these composite inks rearrange the distribution of particles to provide conductive pathways via electron tunneling [22] (at low filler volume fractions) or through a percolation mechanism [20, 23, 24] (at higher filler volume fractions).

Some of the most common electronics printing methods include gravure, offset, flexographic, inkjet, aerosol jet, and screen printing, which deposit the conductive materials onto a planar surface or substrate [26]. Recent advances in printing technology have also enabled omnidirectional and multi-axis printing to allow for fabrication of electronic components onto 3D curvilinear surfaces [11, 12].

Depending on the intended application, stretchable electronics may be deformed repeatedly over hundreds to thousands of cycles. Therefore, one of the most essential design considerations for these systems is the characterization of the reliability of these devices under fatigue loading regimes in addition to monotonic loading conditions. Although many twisting, bending, stretching, and folding tests are being developed and established at Georgia Tech and elsewhere [27-38] to understand the electromechanical behavior and performance of stretchable and wearable devices, there are additional considerations that must be addressed. For example, most bending tests apply bending about a single axis. Similarly, existing tensile tests are limited to uniform uniaxial or planar biaxial stretching conditions. Human testing is an assessment option for wearable devices in use-case loading scenarios, but these kinds of examinations often require extended and constraining approval processes from governing bodies of research, making it an impractical option for rapid design and prototype scenarios. Additionally, many of these established test methods have not been specifically optimized for in-situ electrical and mechanical characterization of stretchable electronic systems. The limited scope of existing test methods illustrates a need for a laboratory test method in which complex, multi-axial, use-case loading conditions associated with stretchable and wearable devices can be emulated without the need for human testing.

II. EXPERIMENTAL METHODS

To address this need, a mechanical test method is proposed in which an electronic sample consisting of a stretchable substrate with a printed stretchable conductor is adhered to the surface of an inflatable bladder of known geometry. Upon inflation of the bladder, multiaxial strains are induced in the sample, and in-situ resistance of the stretchable conductor is concurrently measured. Similar tests have been employed in literature [39-41], but none has formalized a general test methodology. This test method will hereafter be referred to as the Bladder Inflation Stretch (BIS) test method for mechanical characterization of stretchable and wearable electronics, as introduced previously by Stewart et al *[42]*.

A. General Test Set-Up

The general BIS test set-up consists of four primary components: a bladder, a pump, a data acquisition and control system, and a stretchable electronic sample. The first step of the test procedure is to characterize the deformation behavior of the bladder to understand the strain distribution that will be applied to the device. This step can also be performed in reverse, wherein the bladder geometry can be designed in such a way as to produce a specific strain distribution. Next, the stretchable electronic sample is mounted to the bladder using a stretchable adhesive. Once the adhesive is given sufficient time to cure, it will enable the transfer of strains from the surface of the bladder to the device. A digital multimeter, pressure regulator, digital pressure transducer, and various strain-sensing technologies can then be configured to track the electromechanical performance of the device. The connection schematic for the BIS test is shown in Figure 1.

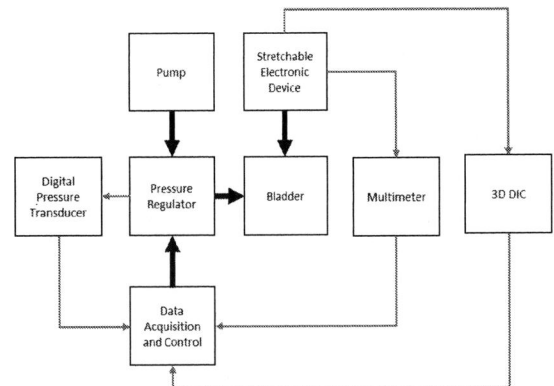

Figure 1. Connection schematic for BIS test

For the initial iteration of the BIS test, we selected a reversible flow transfer pump with a maximum flowrate of 2.5 liters per minute for water (Traceable Products Variable-Flow Chemical Transfer Pump). A digital multimeter was selected for four-wire resistance measurement and to enable easy interface with our data acquisition system (Keithley 2401 SourceMeter). A high accuracy in-line pressure transducer was chosen and integrated with our test set up to track the pressure inside the bladder (Omega PX409). To maintain versatility in testing capability, several distinct testing configurations can be employed, and this paper discusses a Dome Test Configuration.

B. Dome Test Configuration

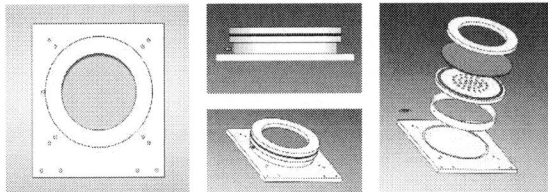

Figure 2. Model of dome configuration of BIS test.

This test configuration involves the inflation of a bladder into a dome-like geometry akin to a blister test for characterizing adhesion strength between thin films. In this approach, the bladder material is cut into a planar sheet and placed over the top of a cavity, through which pressure is applied. This causes the bladder to inflate upwards and to experience large magnitudes of surface strain, which can be transferred directly to the adhered, stretchable electronic sample. To ensure that an adequate pressure can be reached without leaking from the edges of the bladder sheet, a ring-like clamp and set of O-rings are utilized along the edges on either side of the bladder material. The clamp and the O-rings effectively create a seal and force the pressure to act on the center of the bladder material (See Figure 2).

In our fabricated proof-of-concept design, depicted in Figure 3, hinges were installed to alter the position of the plane from which the blister deforms, allowing for much easier in-situ imaging. This configuration offers a great degree of control due to the highly constrained nature of the system. The samples in this configuration can essentially be mounted and tested on any material which can be cut, and which can hold the moderate inflation pressures, and thus, this set-up offers a great deal of versatility in testing capability.

Figure 3. Proof-of-concept design for dome configuration of BIS test.

C. Materials and Geometry Selection

1) Bladder

Careful consideration must be taken when selecting the appropriate bladder material. The optimal selection will be sufficiently compliant and deformable to allow for inflation at moderate pressures, while at the same time being stiff enough to dominate the deformation behavior and adequately transfer strains to the device. For our tests, we chose neoprene rubber, a hyperelastic material with an approximate secant elastic modulus and thickness combination (between 1 and 3 MPa [43] and 0.6 to 1.2 mm, respectively) which we determined would be a good fit for testing many stretchable wearable electronic materials. Although all rubbers have some complex viscoelastic and hysteretic material behaviors, neoprene experiences relatively simple time-dependent behaviors that can be accurately accounted for. Neoprene rubbers are traditionally fabricated using dip molding or spreading techniques, allowing for a multitude of options for geometries. Circular sections were cut from neoprene sheets for the bladder material. The circular sections, when inflated, will form a dome shape and will produce relatively equibiaxial surface strains on the flexible electronic sample. These bladder geometries will provide an easy comparison to traditional planar biaxial tests and other test methods for validation of results. Obvious extensions can be made to vary the thickness across a bladder geometry to produce a non-uniform strain distribution.

Prior to testing and making strain and resistance measurements, the bladder geometry was cyclically inflated several times to allow for hysteretic stabilization.

2) Substrate

For the substrate of our stretchable electronic sample device, we chose to use a spin-coated TPU film given its proclivity for use in wearable stretchable devices. We chose a standard thickness of two mil (50 microns), with an approximate secant elastic modulus of 50 MPa. TPU is also hyperelastic and has time-dependent and strain-rate-dependent behaviors, but the associated material models are well understood and well documented [44].

3) Adhesive

An appropriate adhesive for mounting the sample onto the bladder was selected in consulting literature as well as adhesive vendors. As discussed earlier, the adhesive must be strong as well as stretchable enough to transmit the strains from the surface of the bladder to the device without failure or delamination. Our selected adhesive is two mil (50 micron) in thickness and can cure to 100% of its adhesion strength in 24 hours at room temperature.

4) Conductor

For our conductor, we selected a screen-printed stretchable silver flake polymer matrix composite ink, cured at 160 °C for 5 minutes. As highlighted earlier, these conductive composite materials offer many advantages in stretchability and electrical performance over alternative materials and maintain conductivity while deformed via interparticle connectivity and percolation mechanisms. The thickness of our printed traces range between 5 and 15 microns, and the silver flakes utilize a bimodal flake size distribution, with flakes ranging between 1 and 4 microns across and a flake volume fraction of approximately 70%.

5) Trace Design

The geometry of our trace, as depicted in Figure 4, was designed with the intent to approximate features and geometries frequently seen in wearable strain sensors and other devices i.e. serpentine shaped interconnections [2, 4, 5, 9, 14, 15, 17]. We chose a 1 mm trace width in our area of interest and included four-pad connections to allow for easy in-situ four-wire resistance measurement. Only the area of interest (indicated inside dashed lines in Figure 4) is adhered to the bladder, allowing the four-wire connections to hang free.

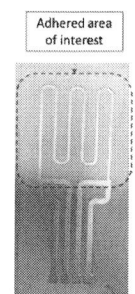

Figure 4. Trace design and adhered area of interest.

D. Strain Determination

The most essential component of this test method is determination of reliable estimates of the strain distribution experienced by both the bladder and the stretchable electronic sample during inflation. One way to achieve this is to use three-dimensional digital image correlation (3D DIC).

The first step in performing 3D DIC is to coat the material of interest (in this case, the entire bladder and test sample) with a light coating of paint in a stochastic speckle pattern. The aperture, focus, and lighting for one or more high-definition cameras are appropriately selected to take a series of high-fidelity, high-contrast photographs of the area of interest throughout deformation. A computer algorithm then looks at each picture pixel by pixel depicting the stochastic speckle pattern and identifies areas of interests or facets. The algorithm then seeks to identify the same distinct facets in each photograph in the sequence and continues until it can identify the same facets in every picture, regardless of the amount of deformation that has taken place. Finally, the algorithm uses the change in distance between facets throughout deformation to estimate a fairly accurate strain distribution across the surface of the material.

In two dimensional or planar deformations, only one camera is needed to fully capture the strain distribution. However, to accurately characterize strain across a dynamic, three-dimensional, curvilinear surface such as our inflating bladder, at least two cameras, called a stereo camera system, will be needed. Each camera is oriented at an angle to the area of interest, and by repeating the same algorithm, only this time with the added complexity of finding matching facets between the left and right camera angles, an accurate depiction of the three-dimensional strains can be determined. Figure 5 illustrates the 3D DIC set-up.

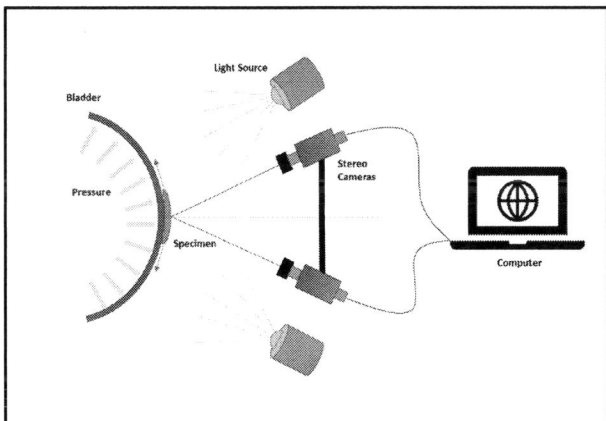

Figure 5. Diagram of 3D DIC test set-up.

Through the use of 3D DIC, we can determine the strain distribution that will occur on the surface of the bladder when inflated. Though 3D DIC is extremely helpful, it requires very expensive equipment, and is not a good fit for the intended application of the BIS stretch test method. Therefore, the intention of the use of 3D DIC in this context is to establish a relationship between the volume of fluid displaced into a given bladder of known geometry with surface strain so that expensive 3D DIC characterization is not required. Once the deformation behavior of a known bladder geometry has been completed, the applied strain distribution is known, and can be reliably reproduce without DIC verification.

Similar strain determination approaches have been used to track surface strains across the surfaces of silicone elastomer [45], and latex membranes [46] inflated as in the blister test configuration.

Figure 6 shows the 3D DIC system (ARAMIS AN1704 3D Measurement System) that was used in the current work. Figure 7 shows the DIC results for the bladder without and with the sample adhered. DIC results were continuously measured for all levels of inflation within each test to fully characterize the bladder deformation behavior.

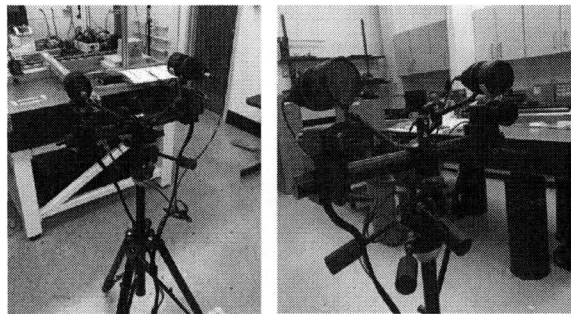

Figure 6. Photographs of ARAMIS AN1704 3D Measurement System used for 3D DIC

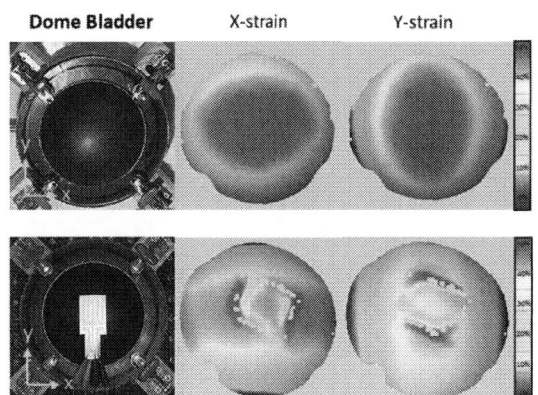

Figure 7. 3D DIC strain distribution results for the maximum deformation case of the dome configuration without (top row) and with (bottom row) the adhered sample.

The final step for determination of strains for the BIS test is to develop a meaningful relationship between the volume of fluid displaced into the bladder during inflation and the measured biaxial strain data collected using 3D DIC. To do this, we first divided our trace into segmented elements to better capture the effect that biaxial strain conditions would have on each portion or our device, as depicted in Figure 8. From our DIC data we have experimentally measured strains in the horizontal and vertical directions for all 19 elements on this trace.

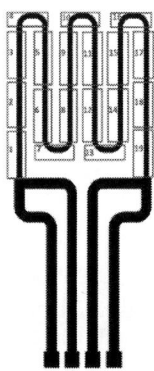

Figure 8. Trace design with enumerated elemental segmentation.

We then performed a series of uniaxial stretch tests both longitudinally (stretching in the same direction as the path of the printed trace) and laterally (stretching direction is perpendicular to the path of the trace) to determine which components of strain would contribute most to the change in resistance. The summary of our findings for this directional electromechanical performance characterization appear in Figure 9.

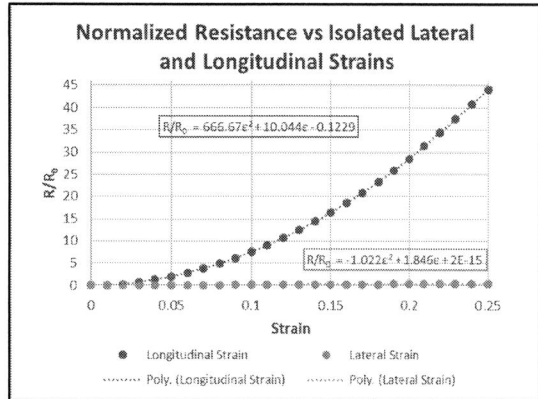

Figure 9. Normalized resistance vs isolated longitudinal and lateral strains to demonstrate larger impact of longitudinal strains on resistance over lateral strains.

From these findings, it is clear that isolated lateral stretching has almost no contribution to the change in resistance of this ink as compared to the longitudinal stretching. When longitudinal and lateral stretching act together, as in our biaxial inflation, there is a possibility that their combined effect is greater than any one mode of loading, but because the effect of isolated lateral stretching is multiple orders of magnitude less significant than longitudinal stretching, we determined its effect to be negligible. Based on this determination, we ascertained that we can effectively express the strains in each element of our segmented trace design by the longitudinal component of strain alone (which can by strain in the x or y direction depending on the element's orientation).

We then determined two separate expressions of strain which serve to aggregate and express the effect that a nonuniform biaxial strain distribution would have on the length of the active area of our trace:

EFFECTIVE LONGITUDINAL STRAIN

$$\varepsilon^* = \frac{\sum_{i=1}^{n} l_i (\varepsilon_l)_i}{\sum_{i=1}^{n} l_i}$$

This expression of strain is an effective strain, and it represents a weighted average of the longitudinal strains for each element along the trace. In this equation $(\varepsilon_l)_i$ is the elemental longitudinal strain, and l_i is the elemental length. This strain determination would be of interest to a designer that is focused more on the effective resistance change across the entire length of the trace. This could be helpful to an engineer interested in power transmission from one component to another on a stretchable device.

MAXIMUM LONGITUDINAL STRAIN

$$\varepsilon_{max} = MAX_{i \to n}[(\varepsilon_l)_i]$$

This expression of strain is a max strain, and it represents the maximum value of longitudinal strain for any of the elements along the trace. In this equation $(\varepsilon_l)_i$ is once again the elemental longitudinal strain. This strain determination would be of interest to a designer that is focused more on the location of electromechanical failure of a stretchable device during extreme use case loading conditions.

Using the above expressions, effective and maximum longitudinal strains can be calculated for every stage of inflation and volume displacement, and then fit a polynomial curve to express the relationship between the strains and the volume displaced as a simple equation. Figures 10 and 11 show the determination of these relationships for each of the bladders used for the BIS test.

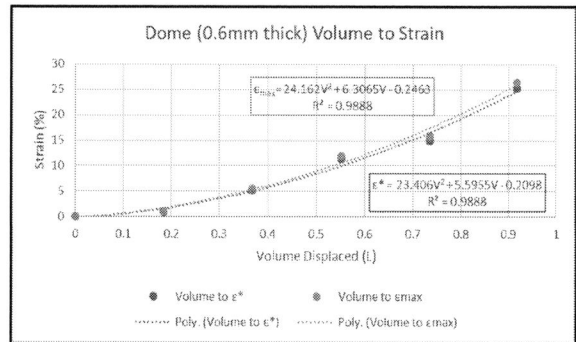

Figure 10. Relationship between the volume displaced into the 0.6 mm thick Dome bladder and the effective and the maximum strain of the trace.

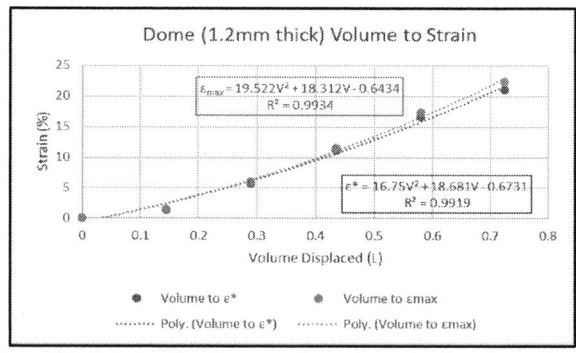

Figure 11. Relationship between the volume displaced into the 1.2 mm thick Dome bladder and the effective and the maximum strain of the trace.

Relationships have been developed between displaced volume of inflation fluid and the surface strains. BIS test results for the electromechanical performance of stretchable devices under various loading scenarios can now be displayed.

III. SUMMARY OF RESULTS

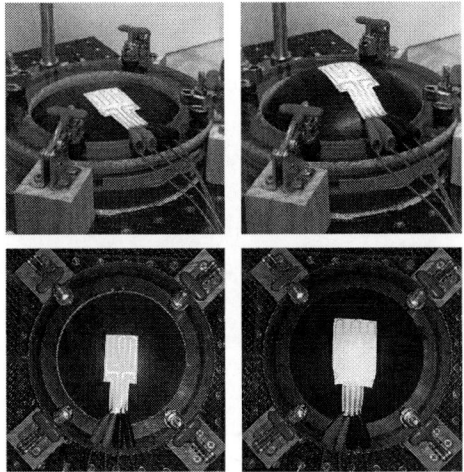

Figure 12. Images of dome bladder inflated and uninflated with an adhered sample and concurrent in-situ resistance measurement.

A. Monotonic Inflation Tests

The first set of tests involved monotonic inflation to electrical failure of the 0.6 mm thick Dome bladder. Resistance is continuously measured and plotted along with effective strain versus time, as well as vs effective strain (See Figures 13 and 14).

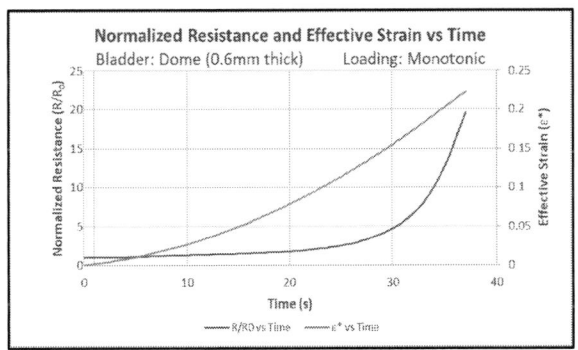

Figure 13. Monotonic inflation of the 0.6 mm thick Dome bladder, normalized resistance and effective strain plotted versus time.

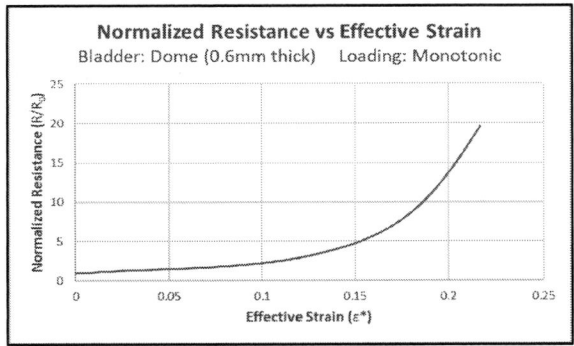

Figure 14. Normalized resistance versus effective strain for monotonic inflation of the 0.6 mm thick Dome bladder.

Both uniaxial stretch test (Figure 9) and biaxial stretch test (Figure 14) show similar variation of normalized resistance with effective strain.

B. Cyclic Inflation Tests

The next set of tests involved cyclic inflation to 30% strain for the 0.6 mm thick Dome bladder and to 7% for the 1.2 mm thick Dome bladder for approximately 20 cycles. The maximum strains were determined by the maximum allowable strains before delamination of the sample from the bladder during the monotonic test iterations. Resistance was continuously measured and is plotted along with effective and max strain versus time for each (See Figures 15-18).

a) Dome (0.6 mm thick)

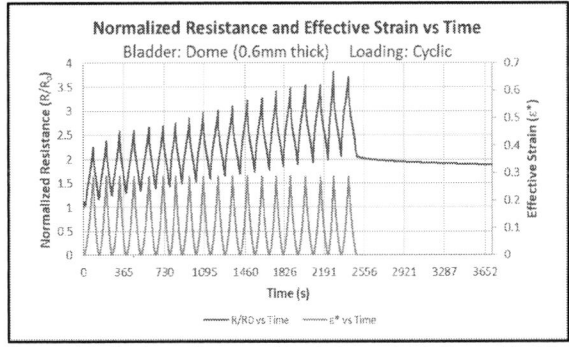

Figure 15. Cyclic inflation of the 0.6 mm thick Dome bladder, normalized resistance and effective strain plotted versus time.

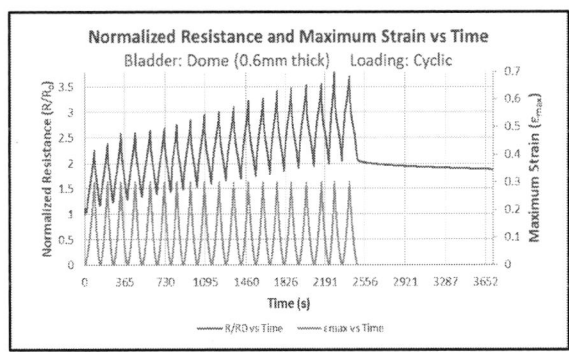

Figure 16. Cyclic inflation of the 0.6 mm thick Dome bladder, normalized resistance and maximum strain plotted versus time.

The normalized resistance is plotted versus both effective and maximum strain for cyclic inflation testing of the 0.6 mm Dome bladder, displayed in Figures 15 and 16. The resistance and strain data seem to track each other well, and there appears to be no lag between the two measurements. The resistance is shown to increase with each cycle until reaching a maximum level of 2 times the initial resistance after 20 cycles. This behavior potentially indicates damage evolution in the conductive ink throughout the cyclic loading regime. Through the 20 cycles tested, there was no significant difference between the effective strain and the maximum strain variation. The resistance measurements continued for 2 hours (only the first 20 minutes are displayed) after the final deflation, and the normalized resistance value shows a gradual decrease to a steady state resistance value of about 1.5 times the initial resistance. The elevated resistance after this period of time indicates the potential presence of irreversible damage in the ink. However, this damage needs to be determined or validated through imaging or other failure analysis techniques.

b) Dome (1.2 mm thick)

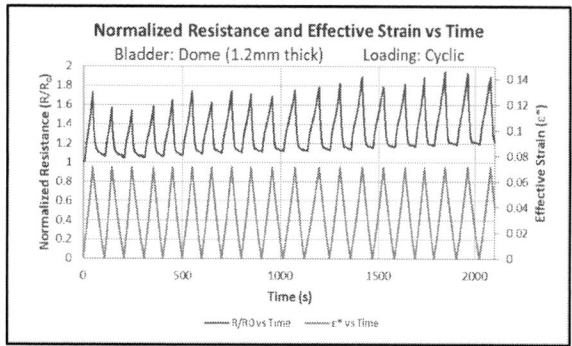

Figure 17. Cyclic inflation of the 1.2 mm thick Dome bladder, normalized resistance and effective strain plotted versus time.

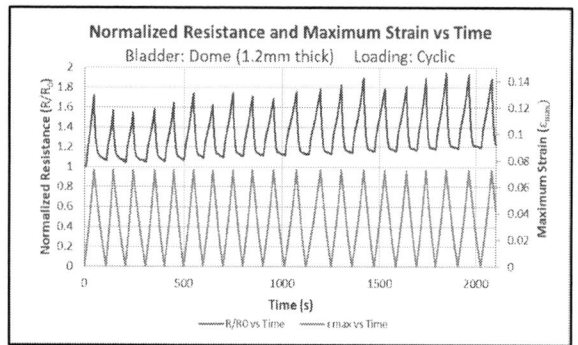

Figure 18. Cyclic inflation of the 1.2 mm thick Dome bladder, normalized resistance and maximum strain plotted versus time.

The normalized resistance is plotted versus both effective and maximum strain for cyclic inflation testing of the 1.2 mm Dome bladder, displayed in Figures 17 and 18. Once again, the resistance continues to increase over cyclic inflation. For this bladder configuration, deflation and continued resistance monitoring were not studied.

C. Inflate-and-Hold Tests

The final set of tests involved inflation to 6% strain for both bladders, then holding at that constant strain and maintaining a constant pressure for 20 minutes. Resistance was continuously measured and is plotted along with effective and max strain versus time for each (See Figures 19-22).

a) Dome (0.6 mm thick)

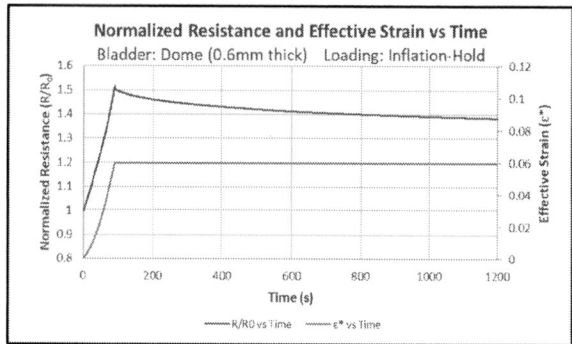

Figure 19. Inflate-and-hold test of the 0.6 mm thick Dome bladder, normalized resistance and effective strain plotted versus time.

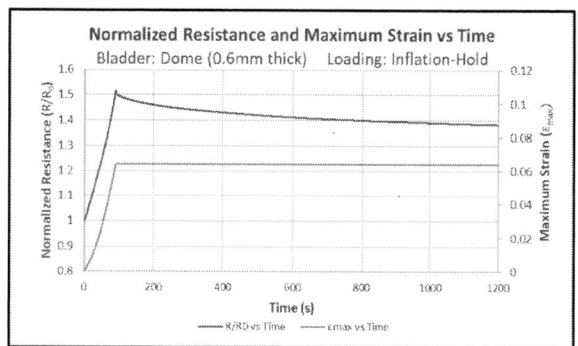

Figure 20. Inflate-and-hold test of the 0.6 mm thick Dome bladder, normalized resistance and maximum strain plotted versus time.

b) Dome (1.2 mm thick)

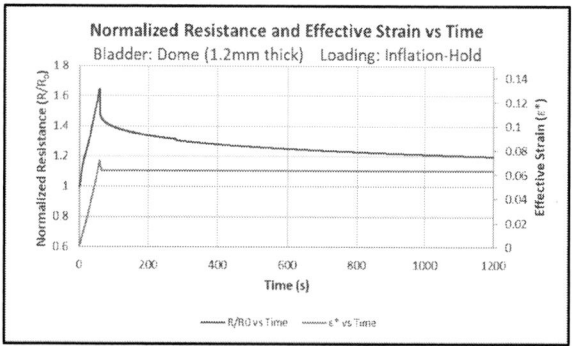

Figure 21. Inflate-and-hold test of the 1.2 mm thick Dome bladder, normalized resistance and effective strain plotted versus time.

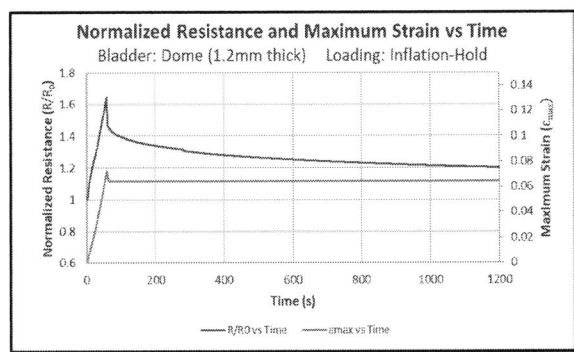

Figure 22. Inflate-and-hold test of the 1.2 mm thick Dome bladder, normalized resistance and maximum strain plotted versus time.

The normalized resistance is plotted versus both effective and maximum strain for inflate-and-hold testing, displayed in Figures 19-22. The normalized resistance value of ~1.6 occurs at a strain value of about 6% as indicated by Figures 9 and 14. A 6% strain is chosen as a preliminary loading scenario as this is a typical strain magnitude for many wearable applications. Future tests will employ inflation and hold to larger strains.

It can be seen from each of the plots that after loading to the selected strain value and associated pressure, the resistance experiences an equivalent increase. As the strain and pressure are held constant, the resistance begins to gradually decrease over time. This behavior can potentially be attributed to relaxation of the ink and TPU, causing partial recovery of conductivity. However, additional testing and characterization need to be performed to determine the cause for this recovery.

II. DISCUSSION

A. Comparison to Literature

As is generally seen in literature, the resistance change generally tracks the strain variation throughout each test. Similar stretching tests have been performed on similar materials sets, and they represent an interesting point of comparison for the results presented in this work. In one study [47], several stretchable silver flake polymer matrix composite inks were uniaxially tested on TPU to cyclic strains of 20%. More than 90% failed within the first 10 cycles, while the best of the inks tested was tested to 500 cycles while experiencing an increase in initial resistance of around 30 times. A similar newly developed silver flake based conductive composite ink was tested biaxially on a PDMS substrate, and it experienced an increase to just 10 times the initial resistance [48]. It is clear from these examples that the performance of the devices is heavily dependent on the substrate and ink material combination. Our monotonic results suggest similar performance as the inks discussed in the first study [47], with effective monotonic biaxial strains of 15-20% showing resistance increases of about 30 times the initial resistance.

Other sources sought to quantify the performance of these composite inks and their dependence on the substrate onto which it they are printed and tested. In one study [24], a silver flake based composite ink was printed onto a polyvinyl chloride (PVC) film substrate as well as on two different stretchable fabrics. These samples were then strained uniaxially to ~40%, and resistance increases were recorded. The sample printed onto the PVC experienced a resistance increase of 20 times the starting resistance, while fabric 1 and fabric 2 samples experienced increases of 5 and 200 times, respectively.

A potential cause of failure for our samples and similar material combinations is a phenomenon known as strain localization [49, 50]. When a freestanding polymer/metal composite ink film is subjected to a strain, the material will experience highly localized straining, causing conductive failure at relatively low global strains. However, when the composite conductive ink is printed onto a polymer substrate of sufficient stiffness, this strain-localization mechanism is ameliorated, and the global strain is distributed evenly across the conductor. On the other hand, if the polymer substrate (such as thin TPU) is not stiff enough, the conductor will behave as if it is a freestanding film and the strain will localize. The rate of failure, failure mode, and location are therefore functions of the elastic moduli and thicknesses of the chosen conductors and substrates. This is an important design consideration for stretchable and wearable electronic systems. In the current test, the presence of a bladder does alter the stiffness of the flexible electronic substrate and the conductive ink, and thus, could modify the strain-localization behavior. However, during use, the flexible electronic systems are usually adhered to host surfaces which can alter the strain-localization behavior in the conductor. Printing directly on an inflatable TPU and other stretchable substrates will be considered in the future. Such direct printing will remove the strain-localization modification due to the presence of additional support elements such as the bladder and the adhesive.

B. Additional Considerations

There are other material behaviors and physical mechanisms that should be considered when attempting to test and characterize these stretchable electronic systems. Materials that exhibit the requisite compliance to be compatible with wearable systems (i.e. TPU, neoprene rubber, etc.) often have complex material behavior, including hyperelasticity, and time dependent behaviors, such as viscoelasticity and hysteresis [44]. These material regimes often diverge significantly from linear elasticity and should be appropriately accounted for in testing regimes.

Other failure modes should be considered for the BIS test aside from failure in the conductor, including slippage and/or delamination via cohesive or adhesive failure of the adhesive layer bonding the sample and the bladder. We did not witness any delamination failure in our test given the full 24 hours to

978-1-7281-1500-9/19 $31.00 © 2019 IEEE

cure the adhesive, and restricting test regimes to less than 50% strain. Additionally, the adhesive proved difficult to remove from the extremely compliant samples used in this test, and removal of the sample from the bladder often damaged the sample significantly. For this reason, the BIS test should most often be considered a destructive test.

III. FUTURE WORK AND CONCLUSIONS

A. Future Work

The discussed BIS test has several possible future extensions:

First of all, given the versatile design of the dome configuration, many sample designs and material types are compatible for inflation testing. Samples of different material combinations such as different composite ink compositions and/or the uses of other stretchable substrates such as PDMS, silicone elastomers, etc. will be considered for future tests. The conductive ink can be directly printed on an inflatable substrate, and thus eliminating the need for the rubber and the adhesive. The inflation fluid and bladder material can also be modified to change the dielectric constant of the system for radio frequency applications for wearable technology. Additionally, the bladder materials and geometries can be modified and designed to prescribe specific strain distributions when inflated, emulating potential use case loading conditions [51]. Anthropomorphic bladder geometries are also possible.

Our current ongoing work includes: 1) establishment of standardized sample designs and allowable materials for the BIS test, 2) the development of robust finite-element models of the bladder inflation with an electronic sample, 3) direct comparison of results from planar biaxial testing to BIS test results for validation, 4) design of bladder to approximate anthropomorphic strains, paired with 3D DIC to compare strains between BIS test bladder and human movement, and 5) testing and analysis of a novel stretchable wearable device using the BIS test method.

B. Conclusion

In this work, a BIS test for mechanical and electrical characterization of stretchable and wearable electronics was proposed. The test consists of the controlled inflation of a rubber bladder of known geometry to induce biaxial strains in a device adhered to the bladder surface. Monotonic, cyclic, and inflate-and-hold tests were performed on a stretchable electronic test sample composed of TPU and silver flake based conductive composite ink to demonstrate the capability of the test. Normalized resistance and strain plots were presented for each test. In summary, the BIS test is a simple, versatile, inexpensive laboratory test method capable of

applying use-case loading conditions and serves as an alternative test method to study stretchable and wearable components and devices.

ACKNOWLEDGMENT

This paper is based, in part, on research sponsored by Air Force Research Laboratory under agreement number FA8650-15-2-5401, as conducted through the flexible hybrid electronics manufacturing innovation institute, NextFlex. The U.S. Government is authorized to reproduce and distribute reprints for Governmental purposes notwithstanding any copyright notation, thereon. The views and conclusions contained herein are those of the authors and should not be interpreted as necessarily representing the official policies or endorsements, either expressed or implied, of Air Force Research Laboratory or the U.S. Government. Also, the authors would like to thank Dr. Rudy Ghosh and NovaCentrix for fabricating the test coupons as well as Jason Dalton and the 3M Company for providing technical expertise and material samples for the selection of adhesives.

REFERENCES

[1] Adly, N., et al., Printed microelectrode arrays on soft materials: from PDMS to hydrogels. npj Flex. Electron, 2018. 2: p. 1-9.

[2] Amjadi, M., et al., Stretchable, skin‐mountable, and wearable strain sensors and their potential applications: a review. Advanced Functional Materials, 2016. 26(11): p. 1678-1698.

[3] Bandodkar, A.J., et al., All‐Printed Stretchable Electrochemical Devices. Advanced Materials, 2015. 27(19): p. 3060-3065.

[4] Bossuyt, F., T. Vervust, and J. Vanfleteren, Stretchable electronics technology for large area applications: fabrication and mechanical characterization. IEEE Transactions on components, packaging and manufacturing technology, 2013. 3(2): p. 229-235.

[5] Brosteaux, D., et al., Design and fabrication of elastic interconnections for stretchable electronic circuits. IEEE Electron Device Letters, 2007. 28(7): p. 552-554.

[6] Chen, Y., et al., Advances in materials for recent low-profile implantable bioelectronics. Materials, 2018. 11(4): p. 522.

[7] Cheng, H., Inorganic dissolvable electronics: Materials and devices for biomedicine and environment. Journal of Materials Research, 2016. 31(17): p. 2549-2570.

[8] Choi, S., et al., Recent advances in flexible and stretchable bio‐electronic devices integrated with nanomaterials. Advanced Materials, 2016. 28(22): p. 4203-4218.

[9] Fan, J.A., et al., Fractal design concepts for stretchable electronics. Nature communications, 2014. 5: p. 3266.

[10] Guo, S.Z., et al., 3D printed stretchable tactile sensors. Advanced Materials, 2017. 29(27): p. 1701218.

[11] Harris, K., A. Elias, and H.-J. Chung, Flexible electronics under strain: a review of mechanical characterization and durability enhancement strategies. Journal of materials science, 2016. 51(6): p. 2771-2805.

[12] Huang, Y., et al., Assembly and Application of 3D Conformal Electronics on Curvilinear Surface. Materials Horizons, 2019.

[13] Kim, D.-H., et al., Materials for stretchable electronics in bioinspired and biointegrated devices. MRS bulletin, 2012. 37(3): p. 226-235.

[14] Kim, D.-H., et al., Epidermal electronics. science, 2011. 333(6044): p. 838-843.

[15] Rogers, J.A., T. Someya, and Y. Huang, Materials and mechanics for stretchable electronics. science, 2010. 327(5973): p. 1603-1607.

[16] Trung, T.Q. and N.E. Lee, Flexible and stretchable physical sensor integrated platforms for wearable human - activity monitoringand personal healthcare. Advanced materials, 2016. 28(22): p. 4338-4372.

[17] van den Brand, J., et al., Flexible and stretchable electronics for wearable health devices. Solid-State Electronics, 2015. 113: p. 116-120.

[18] Amjadi, M., Y.J. Yoon, and I. Park, Ultra-stretchable and skin-mountable strain sensors using carbon nanotubes–Ecoflex nanocomposites. Nanotechnology, 2015. 26(37): p. 375501.

[19] Dąbrowska, A., et al., Materials used to simulate physical properties of human skin. Skin Research and Technology, 2016. 22(1): p. 3-14.

[20] Merilampi, S., T. Laine-Ma, and P. Ruuskanen, The characterization of electrically conductive silver ink patterns on flexible substrates. Microelectronics reliability, 2009. 49(7): p. 782-790.

[21] Yang, C., C.P. Wong, and M.M. Yuen, Printed electrically conductive composites: conductive filler designs and surface engineering. Journal of Materials Chemistry C, 2013. 1(26): p. 4052-4069.

[22] Sevkat, E., et al., A statistical model of electrical resistance of carbon fiber reinforced composites under tensile loading. Composites Science and Technology, 2008. 68(10-11): p. 2214-2219.

[23] Burda, I., et al., Low-cost scalable printing of carbon nanotube electrodes on elastomeric substrates: Towards the industrial production of EAP transducers. Sensors and Actuators A: Physical, 2018. 279: p. 712-724.

[24] Merilampi, S., et al., Analysis of electrically conductive silver ink on stretchable substrates under tensile load. Microelectronics Reliability, 2010. 50(12): p. 2001-2011.

[25] Cohen, M.H., J. Jortner, and I. Webman. The electronic properties of inhomogeneous materials; metal - nonmetal transitions. in AIP Conference Proceedings. 1978. AIP.

[26] Khan, S., L. Lorenzelli, and R.S. Dahiya, Technologies for printing sensors and electronics over large flexible substrates: a review. IEEE Sensors Journal, 2015. 15(6): p. 3164-3185.

[27] Chen, R., et al. Adaptive Curvature Flexure Test to Assess Flexible Electronic Systems. in 2018 IEEE 68th Electronic Components and Technology Conference (ECTC). 2018. IEEE.

[28] Elwi, T.A., et al., Effects of twisting and bending on the performance of a miniaturized truncated sinusoidal printed circuit antenna for wearable biomedical telemetry devices. AEU-International Journal of Electronics and Communications, 2011. 65(3): p. 217-225.

[29] Gunda, M., P. Kumar, and M. Katiyar, Review of mechanical characterization techniques for thin films used in flexible electronics. Critical Reviews in Solid State and Materials Sciences, 2017. 42(2): p. 129-152.

[30] International, A., ASTM D882-12, Standard Test Method for Tensile Properties of Thin Plastic Sheeting. 2012: ASTM International.

[31] International, A., ASTM E143-13, Standard Test Method for Shear Modulus at Room Temperature. 2014.

[32] International, A., ASTM F1842-15, Standard Test Method for Determining Ink or Coating Adhesion on Flexible Substrates for a Membrane Switch or Printed Electronic Device. 2015.

[33] International, A., ASTM E3147-15, Standard Test Method for Evaluating the Reliability of Surface Mount Device (SMD) Joints on a Flexible Circuit by a Rolling Mandrel Bend. 2017.

[34] IPC, IPC-2292: Design Standard for Printed Electronics on Flexible Substrates. 2018.

[35] Kim, B.-J., J.-H. Lee, and Y.-C. Joo, Effect of cyclic outer and inner bending on the fatigue behavior of a multi-layer metal film on a polymer substrate. Japanese Journal of Applied Physics, 2016. 55(6S3): p. 06JF01.

[36] Lambricht, N., T. Pardoen, and S. Yunus, Giant stretchability of thin gold films on rough elastomeric substrates. Acta materialia, 2013. 61(2): p. 540-547.

[37] Li, H.U. and T.N. Jackson, Flexibility testing strategies and apparatus for flexible electronics. IEEE Transactions on Electron Devices, 2016. 63(5): p. 1934-1939.

[38] Park, S.I., et al., Theoretical and experimental studies of bending of inorganic electronic materials on plastic substrates. Advanced Functional Materials, 2008. 18(18): p. 2673-2684.

[39] Bentil, S., K. Ramesh, and T. Nguyen, A dynamic inflation test for soft materials. Experimental Mechanics, 2016. 56(5): p. 759-769.

[40] Klein, S.A., et al., Mechanical testing for stretchable electronics. Journal of Electronic Packaging, 2017. 139(2): p. 020905.

[41] Su, Y., et al., Mechanics of stretchable electronics on balloon catheter under extreme deformation. International Journal of Solids and Structures, 2014. 51(7-8): p. 1555-1561.

[42] Stewart, B.G., I. Bower, and S.K. Sitaraman, Bladder Inflation Method for Mechanical Testing of Stretchable Electronics and Wearable Devices, in IPC APEX EXPO 2019. 2019, IPC: San Diego, CA.

[43] Celina, M., et al., Correlation of chemical and mechanical property changes during oxidative degradation of neoprene. Polymer degradation and Stability, 2000. 68(2): p. 171-184.

[44] Qi, H.J. and M.C. Boyce, Stress–strain behavior of thermoplastic polyurethanes. Mechanics of Materials, 2005. 37(8): p. 817-839.

[45] Machado, G., D. Favier, and G. Chagnon, Membrane curvatures and stress-strain full fields of axisymmetric bulge tests from 3D-DIC measurements. Theory and validation on virtual and experimental results. Experimental mechanics, 2012. 52(7): p. 865-880.

[46] Murienne, B.J. and T.D. Nguyen, A comparison of 2D and 3D digital image correlation for a membrane under inflation. Optics and lasers in engineering, 2016. 77: p. 92-99.

[47] Mohammed, A. and M. Pecht, A stretchable and screen-printable conductive ink for stretchable electronics. Applied Physics Letters, 2016. 109(18): p. 184101.

[48] Matsuhisa, N., et al., Printable elastic conductors with a high conductivity for electronic textile applications. Nature communications, 2015. 6: p. 7461.

[49] Li, T. and Z. Suo, Deformability of thin metal films on elastomer substrates. International Journal of Solids and Structures, 2006. 43(7-8): p. 2351-2363.

[50] Lu, N., et al., Metal films on polymer substrates stretched beyond 50%. Applied Physics Letters, 2007. 91(22): p. 221909.

[51] Skouras, M., et al. Computational design of rubber balloons. in Computer Graphics Forum. 2012. Wiley Online Library.

978-1-7281-1500-9/19 $31.00 © 2019 IEEE

Study of BEOL Failure Mode in Flip Chip Packages at High Temperature Conditions

Wei Wang, Yangyang Sun, Xuefeng Zhang, Lejun Wang, Lily Zhao, Mark Schwarz, Bill Stone, Ahmer Syed

Qualcomm Technologies, Inc.
5775 Morehouse Drive
San Diego, CA 92121
E-mail: weiwng@qti.qualcomm.com

Abstract-As copper pillar bumps have been widely used in flip chip packages to meet the performance demand for denser IO bump and finer pitch, chip-package interaction (CPI) has been critical to achieving high yield rate and package reliability. With the increasing requirement for advanced technology nodes in high-performance devices, low-K (LK), extreme low-K (ELK) and ultralow-K (ULK) dielectric materials have been introduced in Back-End-Of-Line (BEOL) interconnects of silicon process. However, this poses a significant challenge for copper pillar bump interconnects in flip-chip packages since copper pillar is much stiffer than solder bump. During the cooling stage of chip attach reflow, a high stress could be induced in the fragile dielectric materials due to the thermal mismatch between the chip and substrate without any underfill protection right after chip attach. Intensive work has been studied to mitigate the CPI stress caused by the reflow process, such as package design optimization, cooling rate control, and assembly jig implementation, etc. Besides the BEOL dielectric delamination during chip attach, it was observed that the chip level LK/ELK failure could also occur even for assembled packages which already passed final test. It could happen during reliability tests or SMT process when packages go through a high temperature condition. This work presents a study of this failure mode and related mechanical mechanism.

Flip chip using copper pillar bumps were tested in MSL 3 conditions to detect the failure mode. A multi-scale finite element model was developed to evaluate the CPI stress under different package conditions to understand the failure mechanism. First, a global model was built to include the bump array with individual bumps; then a detailed local bump model at fine scale with layered BEOL and copper pillar structure was built at the possible failed bump location to simulate stress on the dielectric layers. The simulation results have a good correlation with the test results to interpret the failure mode. The major factors impacting this failure mode include underfill material properties, polyimide opening (PIO), die thickness, UBM size, etc. are also studied. The simulation demonstrates the impact of package design associated with this failure mode.

Keywords-Flip Chip; Copper Pillar Bump; Dielectric; ELK; Delamination; Simulation; Stress

I. INTRODUCTION

The increasing demand for higher I/O counts and finer bump pitch to meet high performance requirement of electronic devices drives copper pillar bump interconnection in flip chip package technology. However, utilizing copper pillar bumps faces new challenges compared to solder bumps since it increases the thermal-mechanical stress in the dielectric material in BEOL. During the chip attach process, the temperature changes from about 260 °C to 25 °C, the large temperature change could generate a high stress in the BEOL structure due to CTE mismatch between chip and substrate. Passivation crack and BEOL delamination could occur during chip attach [1]. Package design and process control need to be implemented to reduce the stress level to mitigate the ELK delamination risk [2] [3].

Capillary underfills (CUF) and molded underfills (MUF) have been applied to protect the bumps during flip chip package assembly [4]. For assembled packages, CPI reliability tests are utilized to test the package reliability. Previous studies have been done to understand the impact of CUF or MUF materials on ELK stress [5] [6]. Dielectric delamination could occur when the high tensile stress exceeds the dielectric material strength. Recently, it was observed that even though packages have passed CPI hammer tests, dielectric materials could also experience high tensile stress even for assembled packages during multiple reflow tests or SMT process. This failure mode shows different failure mechanism as demonstrated in chip attach. This study aims to understand the failure mode of assembled flip chip packages at high temperature.

To understand the BEOL failure occurred during CPI reliably tests and/or SMT process, design of experiments (DOE) was conducted for copper pillar bump flip chip packages to expose the BEOL delamination using CPI reliability tests. It was found that this failure model was mainly driven by underfill expansion at high temperature conditions. Three-dimensional finite element models were developed to analyze the dielectric stress under different test conditions. Simulation modeling confirms that underfill material of high CTE has a significant impact on dielectric material stress. The simulation result shows good correlation with the test results. Finally, the impact of major factors on the dielectric stress in this failure mode is further studied by simulation models.

II. FAILURE MODE AT HIGH TEMPERATURE

Two DOEs using pillar flip chip packages with two technology nodes and BEOL structures are tested under MSL3 conditions to study the BEOL failure mode at high temperature conditions. Oblong shape copper pillar bumps were used in the DOEs. The long axis diameter and short axis diameter are 70 μm and 40 μm, respectively. The test chips were first mounted on embedded trace substrates (ETS). Then CUF and MUF were used in the assembly to protect the bumps. Figure 1 shows the schematic of CUF and MUF packages.

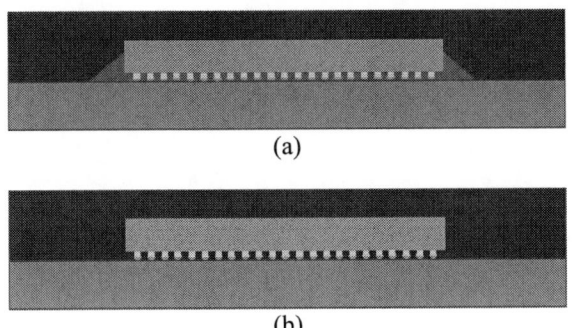

(a)

(b)

Figure 1. Copper pillar flip chip package using CUF (a) and MUF (b)

In the CPI reliability tests, MSL3 was first used to evaluate the BEOL strength of assembled packages. BEOL failure such as ELK delamination was inspected by C-SAM (C-mode SAM) and further confirmed by FIB cross-section check. The legs which passed MSL3 test were further tested under TCB condition and all passed TCB 1000 cycles.

During the MSL3 tests, the packages experience a high temperature over 200 °C. Underfill material expands at high temperatures due to thermal expansion. For copper pillar flip chip packages, underfill expansion could generate a high peel stress at the BEOL above the copper bumps, resulting in dielectric material delamination on top of the bumps.

Table 1: DOE 1 MSL 3 Test Results

Legs	Die thickness (μm)	Underfill	Test Result
Leg 1	100	CUF A	Fail
Leg 2	100	CUF B	Pass
Leg 3	260	CUF B	Fail

In DOE 1, two CUF materials and two die thicknesses were designed. The test results are summarized in Table 1. Leg 1 using CUF A and 100 μm thick die failed at MSL3 while Leg 2 using CUF B and 100 μm thick die passed the MSL 3 tests and TCB 1000 cycles. However, Leg 3 failed at MSL 3 for CUF B and 260 μm thick die. C-SAM inspection detected abnormal bumps after MSL3 as shown in Fig. 2(a). It shows that the failed bumps occurred at the core bump area which indicates the failure is not driven by DNP impact. Cross-section FA analysis further confirmed that the ELK delamination occurred at the top ELK layer as shown in Fig. 2(b).

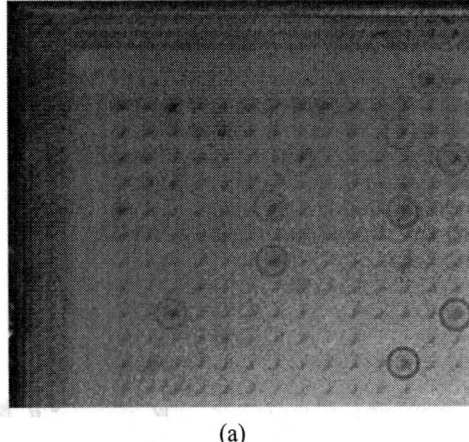

(a)

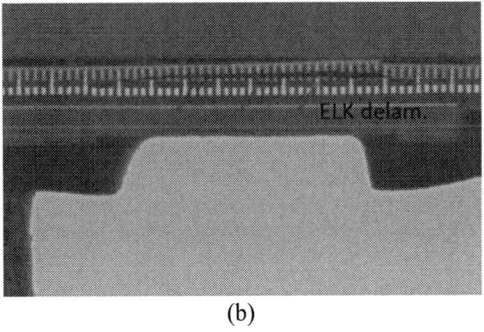

(b)

Figure 2. (a) C-SAM inspection and (b) Cross-section FA image of failed bumps

Three-dimensional finite element models have been developed for the copper pillar flip chip packages and compared the ELK stress in Fig. 3. It showed the highest ELK stress in the failed packages for CUF A and 100 μm thick die. ELK stress is reduced for Leg 2 with CUF B and 100 μm thick die which passed the MSL3 test. While ELK stress is increased again for Leg 3 with CUF B and 260 μm thick die. Therefore, it demonstrated that the ELK delamination in this failure mode is mainly driven by underfill expansion. A thicker die can further increase the ELK delamination risk.

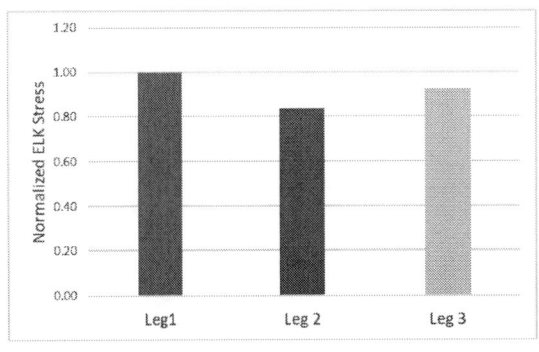

Figure 3. ELK stress at high temperature for different underfill and die thickness

Figure 4 plots the ELK stress contour at 200 °C. It shows the high stress is concentrated within the PIO region and gradually diminishes out of the PIO since polyimide (PI) provides a stress buffer between BEOL and copper pillar bumps.

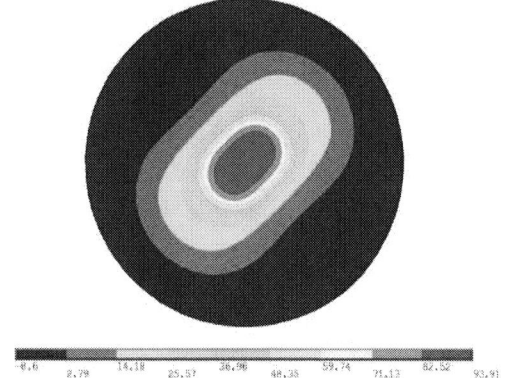

Figure 4. ELK stress contour at 200 °C

In DOE 2, the packages have the same die thickness while CUF and MUF materials are both used. DOE 2 test results are summarized in Table 2. For high CTE CUF A and CUF B, ELK delamination was detected while for low CTE MUF, it passed the tests.

Table 2: DOE 2 MSL 3 Test Results

Legs	Die thickness (µm)	Underfill	Test Result
Leg 1	100	CUF A	150/160
Leg 2	100	CUF B	30/240
Leg 3	100	MUF	0

Figure 5 shows FIB image for a failed bump in the test. It clearly shows the ELK delamination in the BEOL.

Figure 5. ELK delamination in FIB

Simulation results showed the ELK stress in Leg 1 and Leg 2 are both higher than Leg 3 in Fig. 6. The stress comparison is correlated with the test failure data.

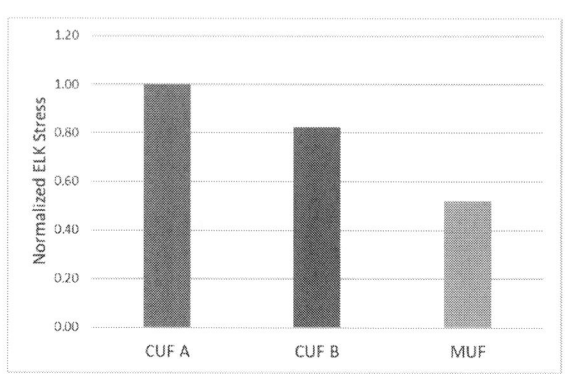

Figure 6. ELK stress at high temperature for different underfill materials

From the DOE, it shows that this failure mode is mainly driven by underfill expansion and the ELK stress is affected by package design factors, e.g., die thickness. In the next section, simulation results are presented to study the impact of major factors on the dielectric failure.

III. IMPACT OF FACTORS ON FAILURE MODE AT HIGH TEMPERATURE

A. Impact of Underfill Mateiral on ELK Stress

ELK stress is highly affected by underfill materials when packages experience high temperature conditions. Thermal expansion of underfill material induces a tensile stress at the dielectric materials. Modeling results of two underfill types are plotted, with one material having a higher CTE than the other. The ELK stress is reduced by 17% for CUF B compared to CUF A.

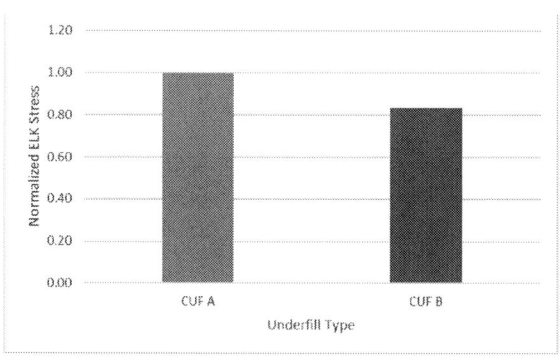

Figure 6. Effect of underfill material on ELK stress

B. Impact of Die Thickess

ELK stress is affected by the die thickness. Thicker die can further increase the ELK stress. Figure 7 shows the ELK stress is increased by 10% for 260 µm die compared to 100 µm die.

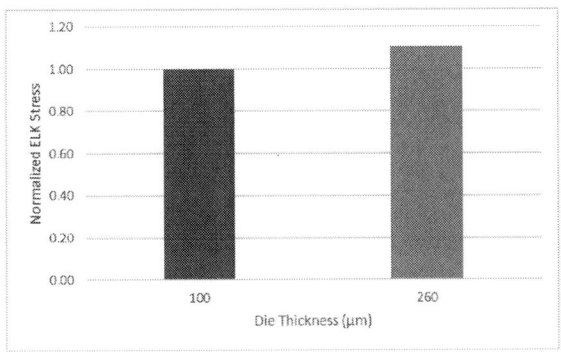

Figure 7. Effect of die thickness on ELK stress

C. Impact of PIO on ELK Stress

During reflow process, underfill expansion generates tensile stress in dielectric material. This tensile stress is concentrated within the PIO area and gradually diminishes out of the PIO since PI provides a stress buffer between BEOL and copper pillar. At high temperature, tensile stress is induced at the PIO region due to thermal expansion of underfill in the out-of-plane direction since the copper bumps are in contact with the metal pad within PIO region. Thus, a larger PIO can reduce ELK stress due to larger contact area. This impact of PIO on ELK stress is opposite to its impact during die attach where a smaller PIO helps to reduce ELK stress [1]. Modeling results shows the impact of PIO on ELK stress. It shows a 10×20 µm oblong shape PIO can increase ELK stress by 25% compared to 20×30 µm oblong shape PIO.

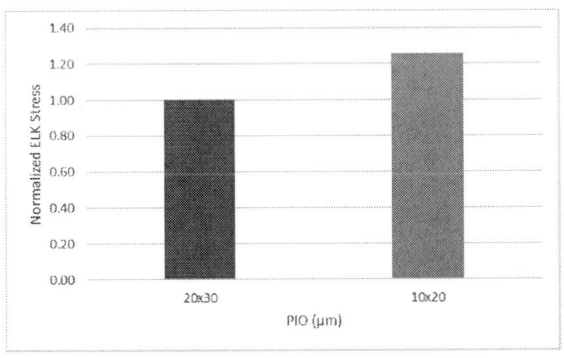

Figure 8. Effect of PIO on ELK stress

D. Impact of IMC Thickness

During solder reflow process, IMC grows between the copper pillar bump and copper trace or pad. Previous studies have shown that IMC grows with the number of reflow cycles [2]. Since IMC is stiffer than solder materials, thicker IMC could add more stress into the BEOL which reduces the ELK delamination margin. Figure 9 shows the ELK stress is increased by about 12% for 7 µm thick IMC and about 20%

for 12 µm thick IMC compared to 2 µm IMC even for low CTE MUF.

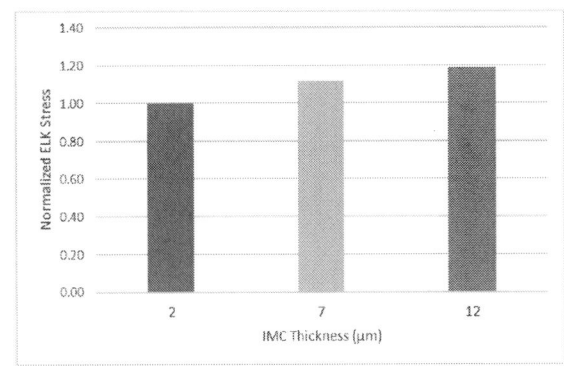

Figure 9. Effect of IMC Thickness on ELK stress

E. Impact of UBM Size on Stress

UBM size also impacts the ELK stress. Large UBM can reduce stress. Figure 10 shows 50×80 µm oblong shape UBM can reduce the ELK stress by 8% compared to 40×70 µm oblong shape UBM.

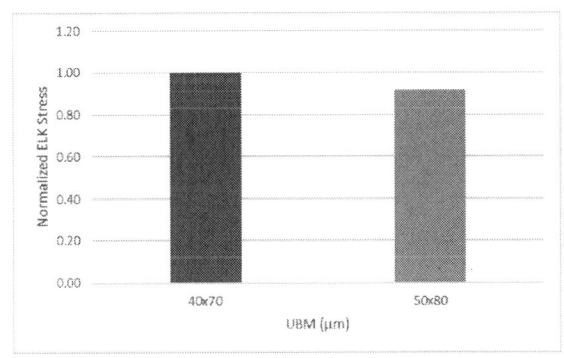

Figure 10. Effect of UBM on ELK stress

IV. CONCLUSIONS

As LK/ELK dielectric materials have been introduced in advanced technology nodes, copper pillar flip chip packages face big challenges to mitigate dielectric delamination risk. In the past, most of efforts were focused on chip attach stage to prevent dielectric delamination and increase CPI margin. In this work, it shows dielectric delamination could also occur at high temperature even for fully assembled packages. The DOE test data shows that underfill expansion is the major failure mechanism for this failure mode. This failure mode was further analyzed and validated through simulation models. Simulation modeling demonstrates that high CTE underfills generate high ELK stress, resulting in dielectric delamination in assembled packages. Furthermore, simulation shows that package design and assembly process both impact the ELK stress for this failure mode. Simulation results are presented to understand the major factors affecting the ELK stress associated with this failure mode.

978-1-7281-1500-9/19 $31.00 © 2019 IEEE

ACKNOWLEDGMENT

The authors would like to thank the packaging team of Qualcomm Technologies, Inc., foundries, OSAT partners and all other collaborators for their help on this work.

REFERENCES

[1] W. Wang, D. Zhang, Y. Sun, D. Rae, L. Zhao, J. Zheng, M. Schwarz, M. Shah, and A. Syed, "Study of Polyimide in Chip Package Interaction for Flip-Chip Cu-Pillar Packages," Proc. IEEE Electronic Components and Technology Conference, 2018, pp. 1039-1043.

[2] A. Bao, L. Zhao, Y. Sun, M. han, G. Yeap, S. Bezuk, P. Holmes, C. Alcira, X. Zhang and K. Lee, "Challenges and opportunities of chip package interaction with fine pitch Cu pillar for 28nm," Proc. IEEE Electronic Components and Technology Conference, 2014, pp. 47-49.

[3] N.Chang, A. Lan, M.Liao, E. Chen, "ELK Delaminate Improvement Methodology on Cu Pillar interconnect BOP Structure," Proc. IEEE Electronic Components and Technology Conference, 2014, pp. 81-84.

[4] M. Joshi, R. Pendse, V. Pandey, T.K. Lee, I.S. Yun, Y.C. Kim, and H.R. Lee, "Molded Underfill (MUF) Technology For Flip Chip Packages In Mobiel Applications," Proc. IEEE Electronic Components and Technology Conference, 2010, pp. 1250-1257.

[5] P. C. Kuo, C. H. Wang, K. K. Ho, K. M. Chen, C. Y. Wu, and C. L. Yang, "14 nm chip package interaction development with Cu pillar bump flip chip package," Proc. IEEE Electronic Components and Technology Conference, 2015, pp. 30-34.

[6] C.Y. Wu, C. H.Wang, K. K. Ho, K. M. Chen, P. C.Kuo, C. L. Yang, "Chip Package Interaction Development of Flip Chip CSP Package with CuPillar Bump on Lead for Advanced Node Chip," 11th International Microsystems, Packaging, Assembly and Circuits Technology Conference (IMPACT), 2016.

A Novel Metal Scheme and Bump Array Design Configuration to Enhance Advanced Si Packages CPI Reliability Performance by Using Finite Element Modeling Technique

Kuo-Chin Chang[*], Mirng-Ji Lii, Steven Hsu, Hao-Chun Liu, Yen-Kun Lai, Sheng-Han Tsai, Chieh-Hao Hsu

Taiwan Semiconductor Manufacturing Company, Ltd.
6, Creation Rd. 2, Hsinchu Science Park, Hsinchu 30077, Taiwan, R. O. C.
[*]Tel: 886-3-5636688 Ext.7076389/7218754;   Fax: 886-3-5773628;   E-mail: kcchange@tsmc.com

Abstract—Flip chip packages with Cu bump have been introduced in recent years to address the needs of advanced packaging for reducing bump pitch and increasing I/O density, which can also enhance performance and offer a cost effective solution with smaller form factor. Moreover, Cu interconnect with extra low-k (ELK) dielectric material was also applied to reduce power consumption and further enhance device performance, especially for advanced silicon nodes. Due to the nature of ELK dielectric material characteristics, it is more sensitive while a flip chip package using Cu bump with ELK dielectric, and subjected to thermal loading conditions. Thermal stresses are induced in Si/package due to the coefficients of thermal expansion (CTE) mismatch among different packaging materials under thermal loading.

ELK dielectric delamination/cracking during chip-package-interaction (CPI) related reliability tests, is a primary concern in the advanced Si packaging development and quick thermal cycling (QTC) test has been widely used to assess the CPI reliability performance for newly developed packages. From the past experience, "white bump" issue, which is related to ELK delamination, is mainly caused by excessive ELK stresses, and appeared within 1~3 bump pitches range from die edge or corner due to CTE mismatch between die (~2.8 ppm/°C) and organic substrate (~17 ppm/°C). The previously proposed solutions include die/substrate co-design, substrate selection, underfill/TIM/adhesive materials selection, and heat spreader design. However, more and more QTC test results indicate that the configuration of top metal layout and bump pattern can also affect the ELK reliability significantly.

This work investigated advanced Si package ELK reliability performance from die attach (using conventional reflow process) to subsequent QTC tests. A three-dimensional (3-D) nonlinear finite element method is applied, and a two-level of specified boundary condition (SBC) of global-local technique are adopted to achieve more accurate resolution. Meanwhile, modeling results were calibrated with experimental package warpage measurements. The modeling predictions were also compared with QTC test data and obtained good agreements. Based on collected modeling and test data, it was revealed that higher density and more uniform pattern in both top metal and bump layouts could relieve ELK thermal stresses significantly.

The proposed methodology in this paper had been validated and design guideline will be proposed upon the findings of this study, which product / package designers can benefit from the superior device / package performance provided from the integration of Cu interconnect and ELK dielectric, meanwhile, alleviate potential CPI related risks while using advanced Si packaging.

Keywords: Cu bump Flip chip package, extra low-k (ELK) dielectric, top metal layout, bump pattern, finite element global-local technique method, chip-package-interaction (CPI), thermal loading, and thermal stress

I. INTRODUCTION

Flip chip packaging technology has been introduced in recent years to address the needs in the electronic packaging industry for increasing density and performance, as well as lightness, thinness, smallness and cost-effectiveness. By employing Cu bump structure (see Fig. 1), it can reduce bump pitch (≤ 90 μm), increase I/O density, enhance performance and offer a cost effective solution with smaller form factor. The use of Cu interconnect with extra low-k (ELK) dielectric material integrated into the advanced Si is an inevitable choice to reduce power consumption and further enhance device performance, which can fulfill continuous demands of computation power. In the surface mount technology (SMT) developed for electronic packaging, the controlled collapse chip connection (C4) using Pb-free instead of eutectic (Sn-Pb) solder bump has been widely adopted in the assembly between the devices and the organic substrate for environmental and health concerns, which is also aligned with the regulation upon European Unions' Restriction of Hazardous Substances (RoHS) [1].

The thermal stress which is induced in packages due to thermal expansion mismatches between the Si die (~2.8 ppm/°C) and the organic substrate (~17 ppm/°C) can be effectively reduced to improve the C4 solder bump reliability [2]. With regard to the packaging reliability for advanced array-type package, such as wafer level packages (WLP) and flip chip packages, the failure of solder joints is not the only reliability concern, the low-k mechanical characteristic also needs to be taken into consideration [2-4].

With these reasons, many literatures had attempted to resolve the complication in reliability that results from the combination of Pb-free solder and Cu/low-k materials through the process of design optimization of electronic package [5-10]. These works gave consideration to the fatigue life prediction of Pb-free solder joint as well as the stresses estimation of low-k multilayer in the flip chip package. Under thermal cyclic loading, the bump will encounter cyclic shear force and transfer stress from substrate to die due to CTE mismatch between die and substrate, which induces the horizontal expansion differences between die and substrate and results in warpage behaviors. Moreover, the Cu bump has higher modulus and more rigid than Pb-free bump, hence Cu bump has smaller deformation which will increase the stress transfer from substrate to die and raise die low-K dielectric stress but can lower the solder strain. Upon above mechanisms, some literatures had investigated the Cu bump impacts on backend-of-line (BEOL) stack reliability and provided solutions through optimized design of Cu bump scheme and package / substrate assembly structure [11-14]. Due to the nature of ELK dielectric material characteristics, it is more sensitive while flip chip package using Cu bump with ELK dielectric, and subjected to thermal loading conditions [15].

Due to the mentioned reliability issues, the previously proposed solutions include die/substrate co-design, substrate selection, underfill/TIM/adhesive materials selection, and heat spreader design [16]. However, more and more quick thermal cycling (QTC) test results indicate that the configuration of top metal layout and bump pattern can also affect the ELK reliability significantly. In this study, a novel metal scheme and bump array design configuration is investigated and proposed to enhance advanced Si package ELK reliability performance from die attach (using conventional reflow process) to subsequent QTC tests. A thermo-mechanical stress analysis is conducted to evaluate the thermally induced stress in the Si/package by performing a three-dimensional (3-D) nonlinear finite element method combining with a two-level of specified boundary condition (SBC) of global-local technique to achieve more accurate resolution. Moreover, the appropriateness of finite element models is demonstrated by comparing the predicted die thermal deformation in flip chip package analysis with experimental results measured by Shadow Moiré and the modeling predictions were also compared with QTC test data.

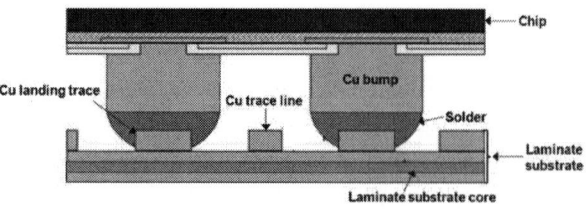

Figure 1. Schematic of Cu bump flip chip package configuration.

II. GLOBAL-LOCAL FINITE ELEMENT MODELING

A three-dimensional (3-D) nonlinear finite element model is constructed using the commercial software, ANSYS®, to perform the package thermo-mechanical analysis. Herein, the global-local finite element analysis was applied to the three-dimensional Cu bump flip chip package model for reducing the number of elements in the mesh model and the total simulation time. The specified boundary condition (SBC) was used for performing the global-local analysis, which involves the following stages [17].

(a) Creating the global and local meshes. In the present study, the global mesh refers to the mesh that models the three-dimensional flip chip package with actual Cu bump pattern structure. The local mesh models focus on the critical Cu bump region of three-dimensional Cu bump flip chip package structure and is more detail than the global mesh. The local mesh was created for analyzing the ELK stress.

(b) Extracting and interpolating nodal displacements at the global-local boundary. Nodal data must be interpolated from the global mesh solution because the global mesh has fewer nodes than the local mesh at the global-local boundary. Extracting and interpolating the nodal displacements of the global model gives the nodal displacements at the boundary of the local mesh.

(c) Analyzing the local problem. The interpolated nodal displacements constitute the boundary condition of the local problem. The effects of various package parameters on ELK stress are studied at this stage.

The package considered in this present work is a Cu bump on trace (CuBOT) flip chip package with 2605 I/Os in an area-arrayed format. The die size of the Cu bump on trace flip chip package is 10×10.8 mm^2 with a thickness of 100 µm (4 mils). The size of the laminate substrate is 14×14 mm^2, and it has a thickness of 230 µm.

Fig. 2(a) displays the three-dimensional global finite element mesh model of the Cu bump on trace flip chip package with actual Cu bump pattern. The eight-node brick element with three degrees of freedom per node was used in the thermo-mechanical stress analysis. Owing to the double symmetry of the three-dimensional Cu bump on trace flip chip package, only one quarter of the package was modeled for computational efficiency. The global finite element model considers the silicon die, laminate substrate, Cu bump and excludes the remainder of the package to simplify the analysis. In the global model, a uniform ELK layer within the die was considered, and no interconnect is included. The laminate substrate was modeled as a sandwich structure consisting of the PP and core layers. The symmetry boundary condition of the global analysis model is defined on the symmetry planes, and the node at the bottom of the intersecting line of these two symmetry planes is constrained to move in the z direction. The global analysis aims to obtain the nodal displacements at the

978-1-7281-1500-9/19 $31.00 © 2019 IEEE

boundary of the detailed local analysis. The local finite element mesh model of the global package model focusing on the critical Cu bump region located on the die corner with much finer meshes is shown in Fig. 2(b). The local mesh model contains the silicon die, Cu bumps, the laminate substrate with PP and core layers, Cu trace on laminate substrate, die Cu/ELK structures and other under-bump layers. The ANSYS built-in cut boundary technique was used during the global-local modeling. The package in the finite element analysis is subjected to a reflow temperature loading: 220°C to 25°C. The stress-free temperature is assumed to be 220°C. All of the materials in the package assembly are assumed to be linearly elastic except for the nonlinear elastoplastic properties of Cu bump and solder in the model are considered to be temperature dependent for both Young's modulus and coefficient of thermal expansion (CTE).

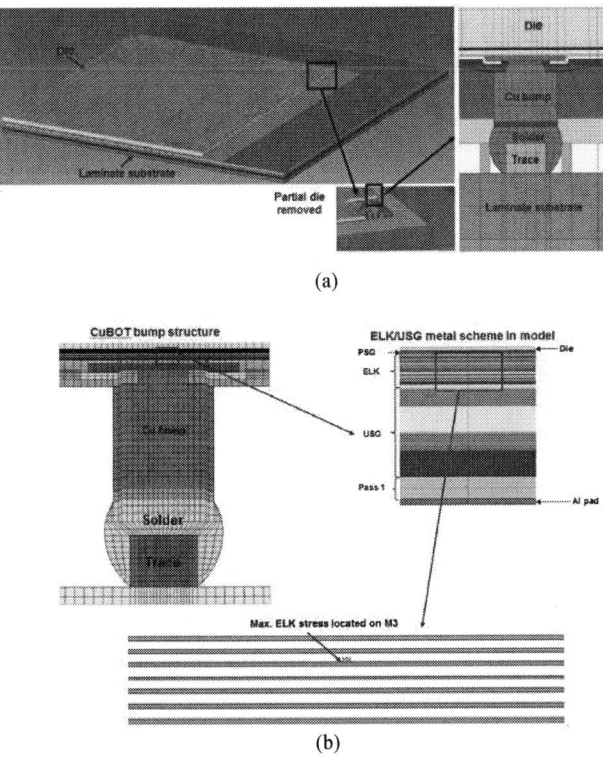

(a)

(b)

Figure 2. Quarter of three-dimensional global-local finite element mesh model of the Cu bump on trace (CuBOT) flip chip package: (a) global model with actual Cu bump pattern and (b) local model.

III. SHADOW MOIRÉ AND FEA MODELING CALIBRATION

In this paper, Finite Element Analysis (FEA) modeling was used to investigate how top metal density and Cu bump density affect the ELK reliability performance in advanced Si package. In general, it is difficult to have precise material properties through experimental measurement, especially in a wide temperature range of data needed, like Young's modulus, CTE and residual stress, which will affect

package stress mostly, those imprecise material properties cause gap between modeling predicts and actual experimental results. In order to verify the modeling methodology correctness and obtain more accurate results for the following investigation topics, a calibration was carried out to adjust the modeling with Shadow Moiré warpage measurement of a bare die test vehicle.

A. Shadow Moiré Experiment

Shadow Moiré is a non-contact technique using moiré fringe patterns resulting from a quartz glass etched with grating and its shadow on a warped test sample, to measure relative vertical displacement. For the shadow setup, a white light illuminates the gratings, and then the gratings generate a shadow on the top surface of the warped test sample. As the surface of the sample warps, moiré patterns are produced by the optical interference between grating and the shadow of gratings on the surface of the warped test sample. The Shadow Moiré measurement system diagram is shown as Fig. 3. In addition, the technique known as phase stepping [18] is applied to in-house experiment tool Akrometrix AXP to increase measurement resolution, so that the vertical resolution of Akrometrix AXP can be enhanced to about 1 μm.

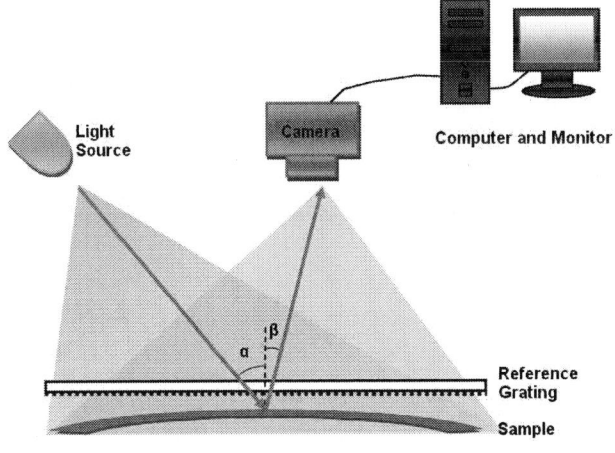

Figure 3. Shadow Moiré measurement system from Akrometrix.com.

The test vehicle die thickness is 100 μm, and the die size is 10.0 x 10.8 mm^2. A total of three samples of test vehicle had been measured in a temperature range from 250°C cool down to 25°C to observe the warpage behavior. The sample with polyimide (PI) face upward, and the warpage is defined as 'crying-face' when the warpage value is positive, and is defined as 'smiling-face' when the warpage value is negative.

The experimental data is shown as Fig. 4, the warpage results showed that when the temperature cools down from 250°C to 25°C, the warpage of the bare die turned over from crying-face warpage to smiling-face warpage. The row measurement data is noisy from temperature to temperature,

978-1-7281-1500-9/19 $31.00 © 2019 IEEE   399

caused by the poor roughness of bare die surface and the horizontal resolution limitation of experiment tool, therefore the measurement data was processed by averaging and regression fitting to eliminate the noise for modeling is easier to calibrate.

B. FEA Modeling Calibration

A FEA model of the bare die test vehicle had been constructed and the dimensions were exactly the same as the experiment sample. The bare die structure cross-section diagram is shown as Fig. 5 which indicates material of each layer. Material properties of Young's modulus and coefficient of thermal expansion (CTE) were in-house measured, and will be calibrated by fitting with experiment results, as well as material residual stress. A three-dimensional nonlinear modeling was performed to compare with Shadow Moiré warpage measurement data from temperature 250°C to 25°C. Fig. 6 shows the modeling warpage results at 210°C, 75°C and 25°C, the warpage was in a crying-face shape at 210°C, and turned into smiling-face shape at 75°C and 25°C, the modeling prediction warpage shows the same trending with experimental results, variations between modeling and experimental results are listed in TABLE I. The FEA modeling shows the good match with experiment results of bare die warpage after calibration, and considered with the measurement resolution of experiment tool, the modeling accuracy was reasonable as well. This modeling methodology will be used for the following investigation.

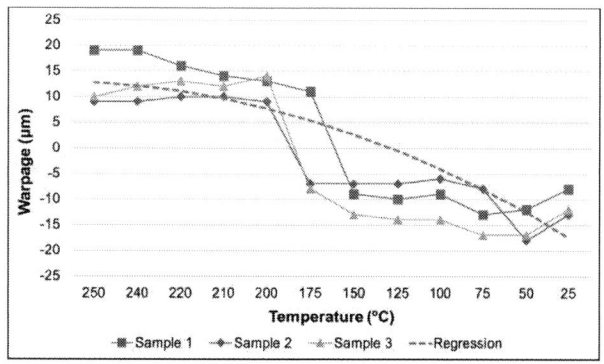

Figure 4. Shadow Moiré warpage measurement for different temperatures.

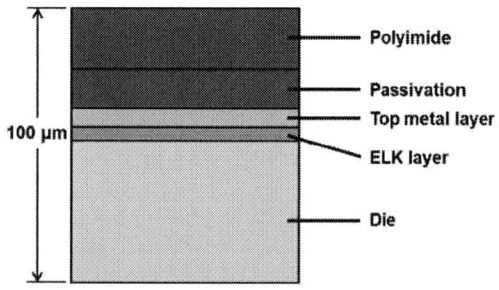

Figure 5. FEA model structure cross-section for warpage calibration.

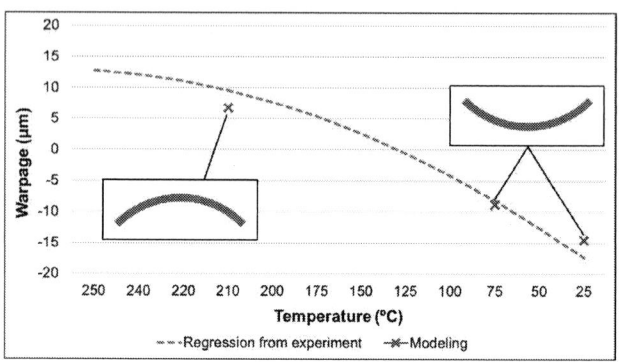

Figure 6. Modeling warpage results compared to experimental data.

TABLE I. Bare die warpage variations between FEA modeling and experimental results.

	Bare die warpage		
Temperature (°C)	210	75	25
FEA modeling (µm)	7.76	-9.57	-15.83
Experiment (Regression) (µm)	9.59	-8.11	-17.38
Variation (µm)	-1.83	-1.46	1.55

IV. RESULTS AND DISCUSSIONS

A. Validation of FEA Modeling Results

To verify FEA modeling results, the results from the thermo-mechanical simulation were compared to failure analysis (FA) results obtained from Cu bump on trace flip chip package QTC test data which are shown in Fig. 7 and TABLE II. The Cu bump on trace flip chip package with 15% of top metal density and low Cu bump density at chip corner suffered QTC ELK delamination issues, while modeling predicted normalized ELK stress has 6% of increment when decreasing top metal density from 25% to 15%.

The stress transmits easily with less protection of top metal and makes higher risk of ELK layer failure. According to ELK stress contour from thermo-mechanical modeling, which is shown in Fig. 7(b), maximum ELK stress is located at first corner Cu bump and the maximum stress is located at under bump metallization (UBM) edge of Cu bump toward to chip corner. In Fig. 7(a), region of ELK delamination was found in top ELK layer and above UBM edge of Cu bump. The maximum stress location obtained from modeling prediction results has good agreement with failure analysis from QTC test data.

B. Metal Density and Layout Effect

Based on the findings from failure analysis and simulation, less top metal density will induce higher ELK stress, which may result in ELK delamination in Cu bump flip chip package. In this study, the top metal density and top metal size were analyzed to identify possible solutions for relieving ELK stress.

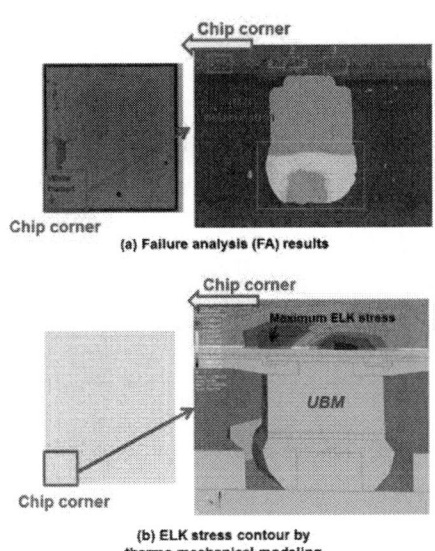

(a) Failure analysis (FA) results

(b) ELK stress contour by thermo-mechanical modeling

Figure 7. Comparison between failure analysis data and ELK stress contour by thermos-mechanical modeling: (a) Failure analysis (FA) results and (b) ELK stress contour by thermo-mechanical modeling.

TABLE II. Comparison between QTC test data and normalized ELK modeling stress.

Bump density (%)	10.7	10.1
Top metal density (%)	25	15
QTC test data	Passed	ELK delamination
Normalized ELK modeling stress	1.00	1.06

(a) Metal density effect

The ELK thermal stress with top metal density changing from 10% to 70% was investigated by thermo-mechanical modeling and is shown in Fig. 8. In general, higher top metal density forms better protection of ELK layer and reduces ELK thermal stress significantly. For instance, increasing top metal density from 40% to 60% will reduce ELK stress by ~11%.

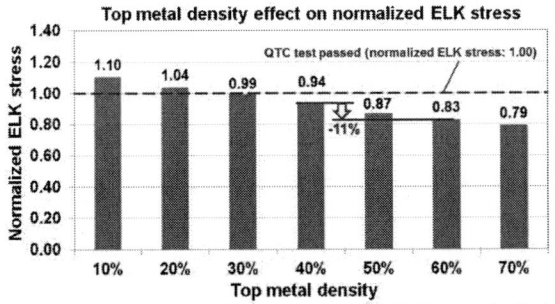

Figure 8. Top metal density effect on normalized ELK stress in Cu bump flip chip package.

(b) Metal size and space effect

With model assumption in this paper, uniform dummy metal size of 3×3 μm^2 were embedded in top metal layer in local model as baseline. Three dummy patterns, 2×2 μm^2, 3×3 μm^2, and 4×4 μm^2, were studied with the same top metal density which equals to 40%. Fig. 9 demonstrates three top metal dummy patterns modeling configuration. Under the same top metal density, larger dummy size has larger dummy space between dummies, and the larger dummy space will induce ELK stress concentration under top metal gap and increase ELK stress. The modeling study of dummy size in Fig. 10 reveals 2×2 μm^2 dummy size has lower ELK stress (down 11%) compared to 3×3 μm^2 dummy size. However, 4×4 μm^2 dummy size will increase ELK (up 4%) due to larger dummy gap.

Top metal dummy

Dummy size = 2 x 2 µm²   Dummy size = 3 x 3 µm²   Dummy size = 4 x 4 µm²

Figure 9. Three top metal dummy patterns in modeling: 2×2 μm^2, 3×3 μm^2, and 4×4 μm^2.

C. Bump Density and Array Pattern Effect

Bumps density and array pattern also have significant effect on ELK thermal stress since bumps provide interconnect between die and substrate which induce thermo-mechanical stress by CTE mismatch between die and substrate.

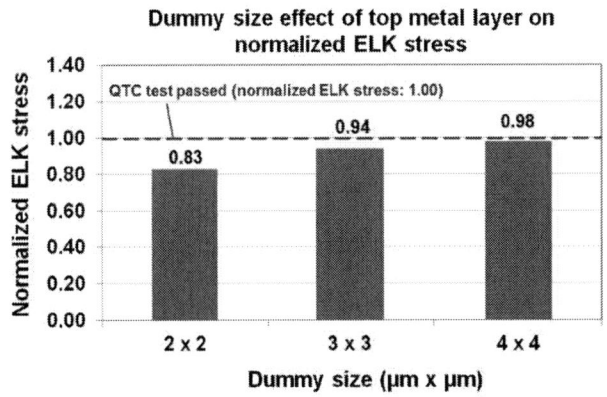

Figure 10. Top metal dummy size effect on normalized ELK stress in Cu bump flip chip package.

(a) Bump density effect

Bump density changing from 5% to 20% in die corner

978-1-7281-1500-9/19 $31.00 © 2019 IEEE

region of Cu bump flip chip package based on 40% of top metal density was simulated and shown in Fig. 11. Since higher Cu bump density indicates more Cu bumps can share thermal stress from die / substrate CTE mismatch, Cu bump flip chip package with higher Cu bump density can reduce ELK stress. Compared to Cu bump flip chip package with 10% of bump density, 15% of bump density decreases ELK stress (down 5%), and 20% of bump density can reduce more ELK stress (down 8%).

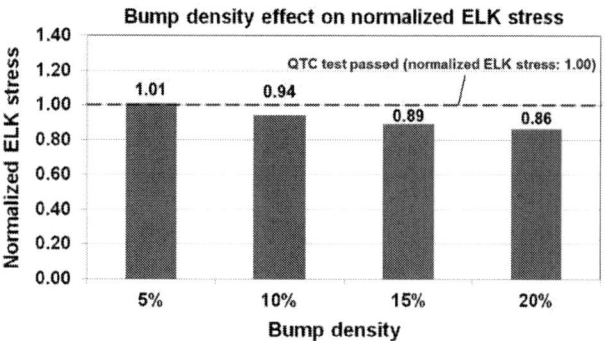

Figure 11. Cu bump density effect on normalized ELK stress in Cu bump flip chip package with 40% top metal density.

(b) Bump array pattern effect

Two different Cu bump array patterns including full array and stagger array were simulated and shown in Fig. 12. The results reveal that stagger bump array pattern has higher ELK stress (up 3%) than full array pattern. Maximum ELK stress locates in the bump with 0 degree orientation rather than the bump with 45 degree orientation. The bump with 0 degree orientation is isolated under stagger array pattern and induces higher ELK stress even both patterns have the same 10% of bump density. The ELK stress contour comparison between full array bump pattern and stagger array bump pattern is shown in Fig. 13.

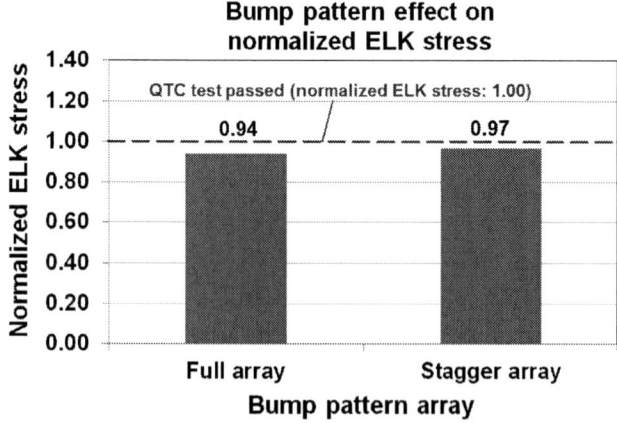

Figure 12. Cu bump array pattern effect on normalized ELK stress in Cu bump flip chip package.

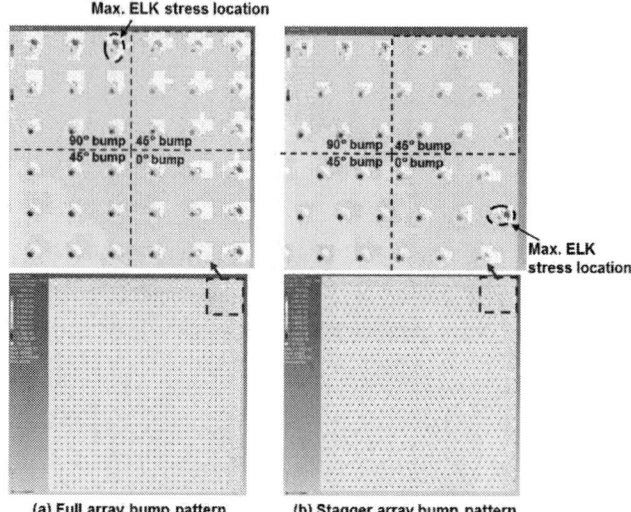

Figure 13. ELK stress contours for (a) Full array bump pattern and (b) Stagger array bump pattern.

D. Proposed Metal Scheme and Bump Array Design Guideline

With this comprehensive assessment, high top metal density and high bump density are two key factors to alleviate ELK stress. For addressing possible bump pattern and top metal layout designs, a stress contour based on percentage of bump density and top metal density was developed to define a safe zone which can mitigate ELK delamination risk. It shows that low top metal density with high bump density may be a design choice. As listed in TABLE III, a 20% of high bump density design along with the top metal density down to 20% can still reduce ELK stress by 5%.

Likewise, it is recommended that low bump density should be designed with high top metal density to increase ELK margin. For instance, a 5% of low bump density design should come with at least 50% of top metal density to have sufficient ELK stress margin (down 8%) as compared to QTC test passed model. Therefore, a design guideline can be summarized based on modeling results as listed below and TABLE IV.

(1) When bump density is larger than 5% and smaller than or equal to 10%, suggested top metal density is larger than or equal to 50%.

(2) When bump density is larger than 10% and smaller than or equal to 15%, suggested top metal density is larger than or equal to 30%.

(3) When bump density is larger than 15% and smaller than or equal to 20%, suggested top metal density is larger than or equal to 20%.

Based on above design guideline, we could define a safe zone of ELK stress with specific bump density and top metal density which is shown in Fig. 14.

TABLE III. Normalized ELK stress for different top metal densities and bump densities.

Bump density	Top metal density						
	10%	20%	30%	40%	50%	60%	70%
5%	1.17	1.08	1.06	1.01	0.92	0.88	0.85
10%	1.10	1.04	0.99	0.94	0.87	0.83	0.79
15%	1.05	0.99	0.95	0.89	0.82	0.77	0.75
20%	1.02	0.95	0.91	0.86	0.79	0.74	0.71
Normalized ELK stress 1.00 based on QTC test passed model							

TABLE IV. Design guideline for bump density and top metal density.

Bump density	5% < density ≤ 10%	10% < density ≤ 15%	15% < density ≤ 20%
Top metal density	≥ 50%	≥ 30%	≥ 20%

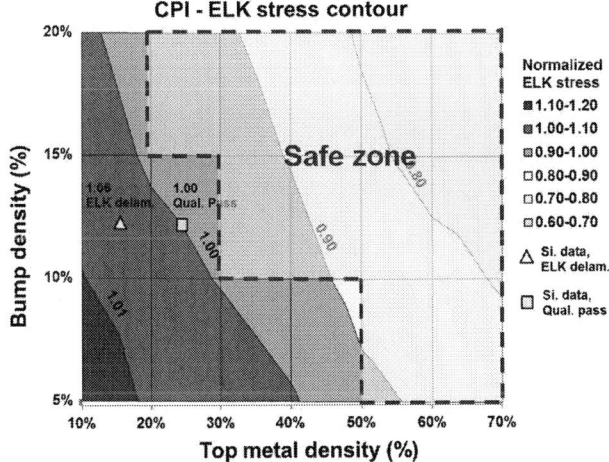

Figure 14. Definition of safe zone for ELK stress with various bump densities and top metal densities.

V.   CONCLUSIONS

This study examines the effects of metal scheme and bump array design configuration of Cu bump flip chip package on the advanced Si package ELK reliability under thermal loading. A three-dimensional (3-D) nonlinear finite element method is used, and a two-level of specified boundary condition (SBC) of global-local technique are adopted to simulate the thermo-mechanical behaviors of the ELK dielectric material. The following conclusions can be made based on the analytical results. Firstly, the proposed methodology in this paper is based on a given test vehicle

design with a known Si / ELK material properties, substrate design and materials properties, and comprehensive QTC test results. Modeling indicated the maximum first principal stress of ELK locates at first corner bump and above UBM edge of bump toward to chip corner, which agrees with QTC test data. In addition, with modeling and test data, higher density and more uniform pattern in both top metal and bump layouts could ease ELK stresses significantly. Finally, the findings of this study can offer designer a design guideline to benefit from the superior device/package performance provided from the Cu/ELK integration and prevent potential CPI risk while using advanced Si packaging.

REFERENCES

[1]   EU Directive 2002/95/EC, Restriction on Hazardous Substances (RoHS) in electrical and electronic equipment, 2003.

[2]   J. H. Lau et al, "Failure analysis of solder bumped flip chip on low-cost substrates," IEEE Transactions on Electronics Packaging Manufacturing, Vol. 23, pp. 19-28, 2000.

[3]   G. B. Alers et al, "Interlevel dielectric failures in copper/low-k structures," IEEE Transactions on Device and Materials Reliability, Vol. 4, pp. 148-152, 2004.

[4]   G. Q. Zhang et al, "Integrated thermo-mechanical design and qualification of wafer backend structures," Microelectronic Reliability, Vol. 44, pp. 1349-1354, 2004.

[5]   D. G. Yang et al, "Parametric study on flip chip package with Pb-free solder joints by using the probabilistic designing approach," Microelectronic Reliability, Vol. 44, pp. 1947-1955, 2004.

[6]   W. Y. Seung et al, "150μm pitch Pb-free flip chip packaging with Cu/low-k interconnects," Proceedings of the IEEE 55th Electronic Components and Technology Conference, pp. 100-106, 31 May-3 June 2005, San Diego, CA, USA.

[7]   Uchida et al, "Low-stress interconnection for flip chip BGA employing Pb-free solder bump," Proceedings of the IEEE 57th Electronic Components and Technology Conference, pp. 885-891, 29 May-1 June 2007, Reno, NV, USA.

[8]   Chang-Chun Lee, Kuo-Chin Chang, and Ya-Wen Yang, "Lead-free solder joint reliability estimation of flip chip package using FEM-based sensitivity analysis," Soldering & Surface Mount Technology, Vol. 21, No. 1, pp. 31-41, 2009.

[9]   S. Raghavana, I. Schmadlakb, and S. K. Sitaramana, "Interlayer Dielectric Cracking in Back End of Line (BEOL) Stack," Proceedings of the IEEE 62nd Electronic Components and Technology Conference,

pp. 1467-1474, 29 May-1 June, 2012, San Diego, CA, USA.

[10] S. Raghavan, I. Schmadlak, G. Leal, and S. K. Sitaraman, "Study of Chip–Package Interaction Parameters on Interlayer Dielectric Crack Propagation," *IEEE Transactions on Device and Materials Reliability*, Vol. 14, pp. 57-65, 2014.

[11] V. Lin, N. Kao, D. S. Jiang, and C. S. Hsiao, "Stress Simulation and Design Optimal Study for Cu Pillar Bump Structure," *Proceedings of IEEE 15th Electronics Packaging Technology Conference*, pp. 598-601, 11-13 December, 2013, Singapore.

[12] F. X. Che, J. K. Lin, K. Y. Au, H. Y. Hsiao, and X. Zhang, "Stress Analysis and Design Optimization for Low-k Chip With Cu Pillar Interconnection," *IEEE Transactions on Components, Packaging and Manufacturing Technology*, Vol. 5, pp. 1273-1283, 2015.

[13] C. Sart, S. G. Garreignot, V. Fiori, O. Kermarrec, C. Moutin, C. Tavernier, and H. Jaouen, "Experimental and Numerical Investigations on Cu/low-k Interconnect Reliability during Copper Pillar Shear Test," *Proceedings of the IEEE 65th Electronic Components and Technology Conference*, pp.1594-1598, 26-29 May, 2015, San Diego, CA, USA.

[14] P. Lianto, H. Y. Li, R. Balamurugan, J. Wei, N. B. Jaafar, L. C. E. Wai, and A. Sundarrajan, "CPI Parametric Investigation of UBM-Al Interface for Cu Pillar Flip-Chip Application," *IEEE Transactions on Components, Packaging and Manufacturing Technology*, Vol. 6, pp. 1120-1126, 2016.

[15] E. Ouyang, M. S. Chae, S. G. Chow, R. Emigh, M. Joshi, R. Martin, and R. Pendse, "Improvement of ELK Reliability in Flip Chip Packages using Bond-on-Lead (BOL) Interconnect Structure," *International Microelectronics and Packaging Society (IMAPS)*, Vol. 2010, No. 1, pp. 197-203, 2010.

[16] K. C. Chang, Y. Li, C. Y. Lin, and M. J. Lii, "Design Guidance for the Mechanical Reliability of Low-K Flip Chip BGA Package," *Proceedings of International Microelectronics and Packaging Society (IMAPS) Topical Workshop and Exhibition on Flip Chip Technology*, June 21-24, 2004, Austin, Texas, USA.

[17] S. R. Voleti, N. Chandra, and J. R. Miller, "Global-Local Analysis of Large-Scale Composite Structure Using Finite Element Methods," *Computers & Structures*, Vol. 58, No. 3, pp. 453-464, 1996.

[18] Y. Wang, P. Hassell, "Measurement of Thermally Induced Warpage of BGA Packages/Substrates Using Phase-Stepping Shadow Moire," *IEEE CPMT Electronic Packaging Technology Conference*, pp 283-289, 1997.

Assessment of CMP Fill Pattern Effect on the Thermal Performance of Interconnects in Integrated Circuits BEOL

Assaad El Helou
Mechanical Engineering Department
Southern Methodist University
Dallas, USA
ahelou@smu.edu

Archana Venugopal
Analog Technology Development
Texas Instruments Inc.
Dallas, USA
avenugopal@ti.com

Peter E. Raad
Mechanical Engineering Department
Southern Methodist University
Dallas, USA
praad@smu.edu

Abstract—This work analyzes the thermal characteristics of CMP square pattern layers and the added cooling effectiveness for activated interconnects in a multi-layered BEOL stack. An equivalent thermal resistive analysis is used to obtain the effective in-plane and through-plane thermal conductivities of the patterned MET layer. An extensive numerical thermal model is developed for modeling the full device using an ultra-fast self-adaptive multi-grid thermal engine TMX T°Solver®. The pattern-filled design is compared to the base design void of CMP fills. The simulation results are validated with experimental temperature measurements of samples using Thermoreflectance imaging system TMX T°Imager®. The results show visible thermal improvement in the stack with potential of using the fill patterns for embedded heat sink designs to improve heat dissipation of thermally critical components in the BEOL.

Keywords-thermal; back-end-of-line; interconnects; characterization; chemical-mechanical planarization; dummy fill;

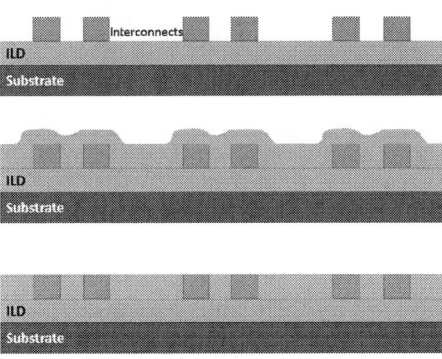

Figure 1. Schematic showing CMP process at a BEOL MET layer (TOP) after oxide deposition (MIDDLE) which gives a level surface for further fabrication (BOTTOM)

I. INTRODUCTION

The back-end-of-line consists of several metalizations (MET) layers forming the multi-level interconnect network layouts that connects and powers all devices at the front end. Chemical-mechanical planarization (CMP) process aids the fabrication of the BEOL stack by leveling the oxide fill around the metalization[1], [2]. This process comes with two major surface defects in the form of thickness loss, both for the metal (dishing), and oxide (erosion) regions. Incorporating a dummy MET fill pattern improves the surface uniformity of the planarization process and increases the fabrication yield[3]. This metallic fill replaces some ILD material and thus adds some parasitic electrical effects, such as capacitance coupling and signal delay[4], [5]. Mechanical stress is also noticeably increased in the stack[6] due to the increased adhesion sites of different materials. However, from the heat transfer perspective, more thermally conductive material replacing the ILD would improve the thermal conductivity of the stack, thus reducing self-heating and improving reliability.

In comparison to the back-end of line interconnects that have been thermally studied extensively, the CMP fill pattern thermal effect studies remain somewhat lacking. Chiang et

al. studied the cooling effect of vias, and using a similar approach studied the impact of dummy thermal vias. However the spacing, location and via differ from that of CMP fill patterns. Datta et al.[7] performed simulations to assess what thermal improvements the CMP fill pattern has on the heat conduction of the metal layers. The authors reported an improvement in the in-plane conductivity and in the temperature planarity. The study considered different fill pattern densities and different interlayer dielectric (ILD) materials (air, polymer, SiO_2) but no validation was performed with actual sample device measurements.

This work studies the effects of added CMP dummy fills on the thermal characteristics of the back-end of line of ICs. A thermal simulation model is developed for a microresistor device to serve as a heat source and located in a representative BEOL structure. The heating profile then used to assess the advantage of added CMP fill pattern in reducing the operating temperature and evaluate the added thermal allowance in terms of electrical activation rates. The model is validated with specially fabricated sample devices where the heating rate is used to extract the effective thermal conductivity of the stack and the allowable activation rates to compare with simulations results.

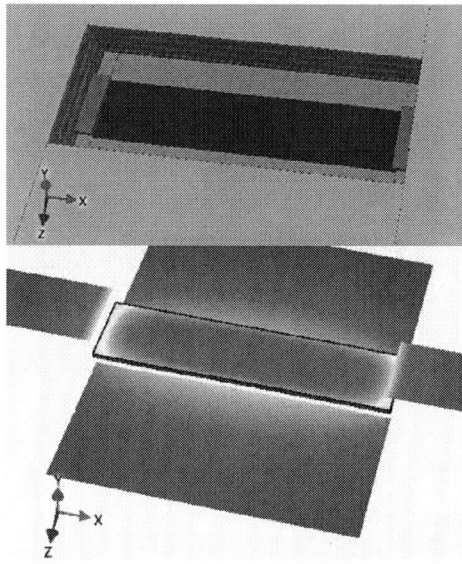

Figure 2. Top: Thermal Simulation Model showing microresistor device with surrounding layers of inter-layer oxide and dummy fills. Bottom: Sample temperature distribution of active microresistor, from which an average temperature rise and effective thermal conductivity of the stack can be obtained.

II. METHODOLOGY

A representative model of the microresistor test device within the BEOL structure shown in Fig. 2. TMX T°Solver® is used as the thermal modeling software for this study. The software provides ultra-fast simulations with a self-adaptive multi-grid meshing for large variations in spatial and temporal scale[8] and is specifically designed for thermal modeling of active IC structures. The effect of the CMP fill on the heat dissipation within the stack is analyzed by comparing the case with fill patterns to the base case of bare ILD passivation.

The modeled device is a microresistor located in second level of a BEOL stack. The first layer is filled with CMP PolySilicon pattern. The layer containing the dummy fill pattern is modeled with uniform material of effective in-plane thermal conductivity ($\kappa_{xz,eff}$) and through-plane conductivity ($\kappa_{y,eff}$) obtained from a representative thermal resistance circuit analysis, as shown in Fig. 3. The effective conductivities are calculated for a square $fill$ pattern configuration of side width w_f and conductivity κ_f in a $dielectric$ unit cell of width w_d and conductivity κ_d. The equations for the resistive circuit equivalents are represented in Eq. 1 and the obtained values are presented in Table I.

$$\frac{\kappa_{xz,eff}}{\kappa_{ILD}} = 1 - \frac{w_f}{w_d} + \left(\left(\frac{w_d}{w_f} - 1 \right) + \frac{\kappa_d}{\kappa_f} \right)^{-1} \quad (1)$$

$$\frac{\kappa_{y,eff}}{\kappa_{ILD}} = 1 - \left(1 + \frac{\kappa_f}{\kappa_d} \right) \left(\frac{w_f}{w_d} \right)^2$$

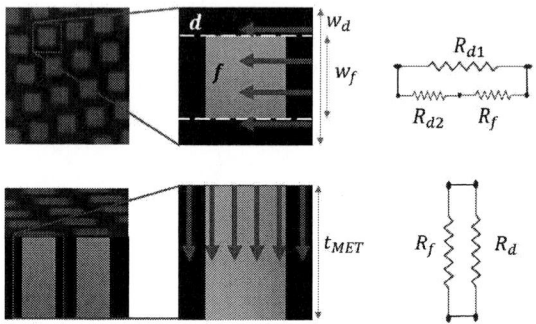

Figure 3. Thermal resistance circuit model for unit cell of CMP fill pattern for square fill used in this study of density w_f^2/w_d^2, showing top view (Top) and side view (Bottom)

The effective material density and specific heat are obtained from a volume average for the square pattern fill as seen in Eq. 2.

$$\frac{c_{p,eff}}{c_{p,d}} = 1 - \frac{w_f^2}{w_d^2} \left(1 - \frac{c_{p,f}}{c_{p,d}} \right) \quad (2)$$

$$\frac{\rho_{eff}}{\rho_d} = 1 - \frac{w_f^2}{w_d^2} \left(1 - \frac{\rho_f}{\rho_d} \right)$$

To validate the thermal model, representative BEOL stack with embedded micro-resistors of different widths 1, 3, and 10 μm are activated and their temperature rise measured using the sub-micron thermoreflectance imaging system TMX T°Imager®. From each of the performed thermal maps, the average temperature rise is extracted from a central region in the resistor away from the edges to avoid the via cooling/heating effects. The temperature variation across the width of the micro-resistor is neglected and the lumped approach is assumed as the thermal conductivity of the resistor material is much larger than the neighboring oxide medium. The temperature is then plotted for several activation levels from which the overall thermal resistance can be obtained as $R_{th,eq} = \Delta T/Q$. The activation limit, in terms of allowable heating rate Q_{all} or electrical current density J_{RMS}, is then obtained for an allowable temperature rise ΔT_{all} as shown in Eq. 3, where ρ_0 is the material sheet resistivity.

$$Q_{all} = \frac{\Delta T_{all}}{R_{th,eq}} \quad (3)$$

$$J_{RMS} = \sqrt{\frac{Q_{all}}{\rho_0 V}}$$

The CMP dummy fill patterns are present at the same level as the MET layer and is of the same material composition as shown in Fig. 5. The patterns are blocked within a "keep-off" distance around device components to minimize parasitic

978-1-7281-1500-9/19 $31.00 © 2019 IEEE

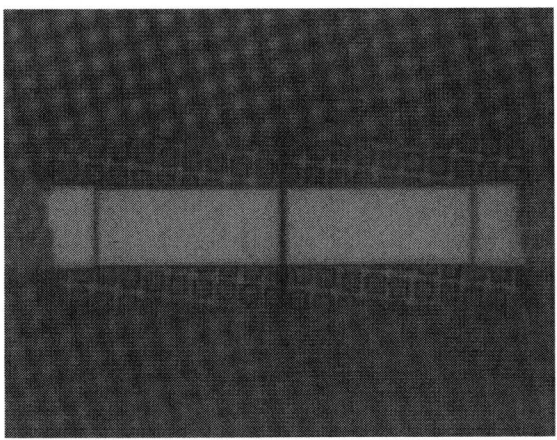

Figure 4. Schematics showing CMP fill pattern at the different levels of the BEOL

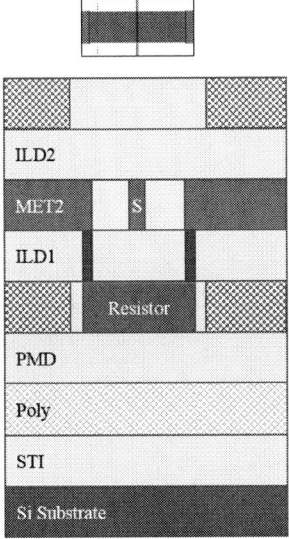

Figure 5. Schematics showing the multi-level stack structure with the CMP fill regions (hatched) at the different MET levels

Table I
EFFECTIVE CONDUCTIVITY OBTAINED FROM EQUIVALENT THERMAL RESISTIVE CIRCUIT ANALYSIS

Fill Material	In-Plane κ $(W/m \cdot K)$	Thru-Plane κ $(W/m \cdot K)$	Density ρ (kg/m^3)	Specific Heat c_p $(J/kg \cdot K)$
AlCu	153	153	2700	894
PolySi	15	15	2330	712
ILD	1.2	1.2	2185	741
AlCu+ILD	2.76	68.67	2414	809
Poly+ILD	2.47	7.33	2249	728

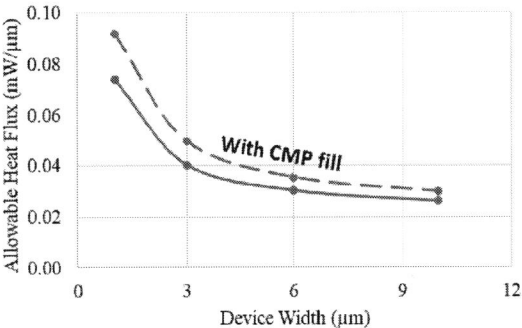

Figure 6. Simulation results for allowable activation limit for a ΔT of 15 °C for devices in the three MET levels of the BEOL as function of width

electrical effects. The fill pattern, as shown in Fig. 4 consists of squares of side dimensions of 2 μm spaced 1 μm apart. The lattice unit cell is a $3 \times 3\mu m$ square that extends in height to the whole thickness of the MET layer, resulting in a fill density of 44% by volume.

A. Results

The results of the equivalent thermal resistive circuit analysis show an increase in the in-plane conductivity of the ILD fill from 1.2 $W/m \cdot K$ to 2.76 and 2.47 $W/m \cdot K$, and to 68.67 and 7.33 $W/m \cdot K$ for the through-plane conductivity, when adding the AlCu or Poly dummy fill respectively. The improvement in in-plane conductivity is less than the through plane due to the geometric configuration of the fill pattern where bumps do not extend to the full width

of the unit cell, thus there is no continuous in-plane path for heat conduction without passing through ILD material. In contrast, the through-plane conductivity sees better improvement since the fill extends to the full thickness of the MET layer, allowing a high conductivity medium for heat dissipation across the whole MET thickness. However, the through-plane effective conductivity is for the MET layer only and does not consider the ILD layers separating the successive MET layers. The MET fill layers are in series with the ILD layer thus adding more resistance within the stack between the device and the substrate. Consequently, the higher the device MET level the higher the ILD thermal resistance and the more critical the self-heating becomes.

The simulation results in Fig. 6 shows the allowable heating generation terms Q' instead of electrical activation J_{RMS} to represent the stack thermal characteristics and to be applicable for materials of different electrical conductivities. A first observation reveals a reduction in the activation limit for wider interconnects. As the width increases, the heat generation region widens and lateral heat dissipation is less effective.

The CMP pattern fill has a visible cooling effect as seen from Fig. 6 and tabulated in Table II. The fill pattern surrounding the microresistor of different widths enhances the heat conduction by increasing the effective conductivity of the insulating medium, with a cooling of 12.8% and 19.4% for the widest and narrowest.

Table II
COOLING EFFECT OF THE ADDED CMP DUMMY FILL AND THE
INCREASE IN ALLOWABLE ELECTRICAL ACTIVATION LEVEL

$Width(\mu m)$	$\Delta T/T(\%)$	$\Delta J_{RMS}(\%)$
1	-19.4	10.2
3	-19.0	10.0
6	-14.6	7.6
10	-12.9	6.6

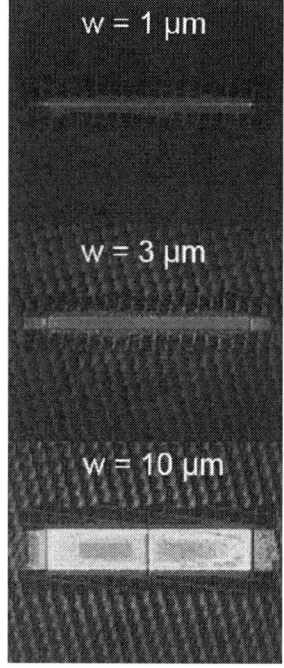

Figure 7. Sample temperature maps obtained using thermoreflectance imaging showing microresistor heating for three widths configurations

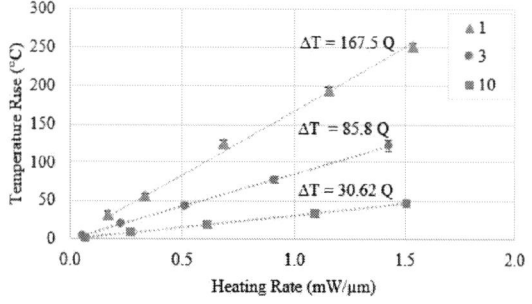

Figure 8. Experimental Heating Profile for three microresistor devices of width (1, 3, and 10 μm)

The simulation results for the three BEOL MET levels and for the different widths cases are compared to the experimental measurements as shown in Fig. 9. A sample

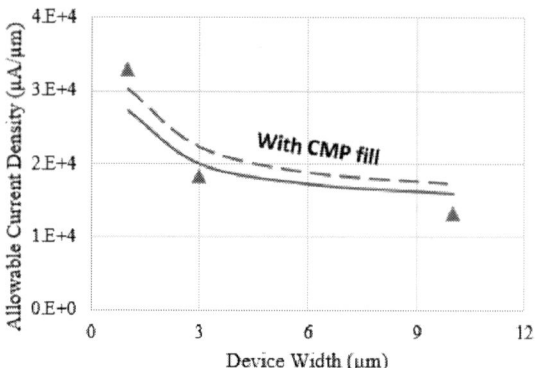

Figure 9. Comparison of simulation results with experimental measurements of the allowable current density J_{RMS}

temperature measurement map is shown in Fig. 7 and the heating profile for the three widths are shown in Fig. 8. The allowable activation results are in good agreement with the experimentally observed values. The experimental devices show a higher thermal resistance in the stack which could be associated with the added interface resistance within the pattern. It is important also to point out that simulation accuracy is always as good as that of the model's input parameters such as the thermal conductivity of the materials and the interface resistance. For thin films, these parameters often differ from the bulk values and require in-situ measurements for representative samples. A small variation from the actual values could lead to some validation error. Given that ballpark values reported in the literature are used, the experimental validation error is within acceptable range.

III. CONCLUSION

In addition to the fabrication yield benefits of CMP dummy fills, this study presents an investigation of their thermal benefit to BEOL regions with critical self-heating. An initial computational study assesses the added cooling effectiveness of the CMP fills and the increase in the allowable activation rates. Sample devices are fabricated and measured to validate the built 3D thermal model. The results show a markedly marginal improvement in heat dissipation achieved by square pattern. More promising potential is attainable if the fill pattern is optimized for cooling and used in developing embedded heat spreaders [9] for thermally critical interconnects in the IC back-end.

REFERENCES

[1] Parshuram B. Zantye, Ashok Kumar, and A. K. Sikder. Chemical mechanical planarization for microelectronics applications, 2004.

[2] Mahadevaiyer Krishnan, Jakub W. Nalaskowski, and Lee M. Cook. Chemical mechanical planarization: Slurry chemistry, materials, and mechanisms. *Chemical Reviews*, 2010.

978-1-7281-1500-9/19 $31.00 © 2019 IEEE

[3] Alex; Chang, Li-Fu; Fan, Zhong; Lu, Daniel; Bao. No Title. In *Improving copper CMP topography by dummy metal fill co-optimizing electroplating and CMP planarization*. Proceedings of SPIE - The International Society for Optical Engineering, Apr 2010, Vol.7641, 2010.

[4] Lei He, Andrew Kahng, King Tam, and Jinjun Xiong. Design of integrated-circuit interconnects with accurate modeling of chemical-mechanical planarization, 2005.

[5] U.a Katakamsetty, C.a Hui, L.-D.b Huang, L.b Weng, and P.b Wu. Timing-aware metal fill for optimized timing impact and uniformity. In *Proceedings of SPIE - The International Society for Optical Engineering*, 2009.

[6] Aditya P. Karmarkar, Xiaopeng Xu, Victor Moroz, Greg Rollins, and Xiao Lin. Analysis of performance and reliability trade-off in dummy pattern design for 32-nm technology. In *Proceedings of the 10th International Symposium on Quality Electronic Design, ISQED 2009*, 2009.

[7] Basab Datta and Wayne Burleson. On temperature planarization effect of copper dummy fills in deep nanometer technology. In *Proceedings of the 10th International Symposium on Quality Electronic Design, ISQED 2009*, 2009.

[8] Peter Raad, James Wilson, and Donald Price. Adaptive modeling of the transients of submicron integrated circuits. *IEEE Transactions on Components, Packaging, and Manufacturing Technology*, 21(3):412–416, 1998.

[9] Assaad El Helou, Peter E. Raad, Dhishan Kande, and Archana Venugopal. Thermal Modeling and Experimental Validation of Heat Sink Design for Passive Cooling of BEOL IC Structures. In *2018 24rd International Workshop on Thermal Investigations of ICs and Systems (THERMINIC)*, pages 1–3. IEEE, sep 2018.

Three-dimensional Simulation of the Thermo-Mechanical Interaction between the Micro-bump Joints and Cu Protrusion in Cu-filled TSVs of the High Bandwidth Memory (HBM) Structure

Jie-Ying Zhou[1], Shui-Bao Liang[1], Cheng Wei[1,2], Wen-Kai Le[1], Chang-Bo Ke[1], Min-Bo Zhou[1], Xiao Ma[1], Xin-Ping Zhang[1,*]

[1] School of Materials Science and Engineering, South China University of Technology, Guangzhou 510640, China
[2] School of Civil Engineering and Transportation, South China University of Technology, Guangzhou 510640, China
*Email: mexzhang@scut.edu.cn; Tel: +86-20-22236396

Abstract—Based on High Bandwidth Memory (HBM) structure and taking into consideration the direct interconnection between the micro-bump and the end of the Cu-filled TSV, there is an interaction between the micro-bump joint and the Cu protrusion. To clarify this issue, in this study two geometric finite element models (i.e., 4-hi stack HBM structure and 8-hi stack HBM structure) are constructed and used to characterize the von Mises stress, volume fraction of plastic deformation region of the Cu-filled TSV, height of Cu protrusion and fatigue life of micro-bump joints with taking into account the interaction between the micro-bump joint and Cu protrusion. The simulation results manifest that micro-bump joints in 4-hi stack HBM model suffer larger average stress than those in 8-hi stack HBM model, and the volume fraction of plastic deformation region of Cu-filled TSV in the former model is also higher than that in the latter one. Interestingly, the height of Cu protrusion becomes equal at upper and lower ends of Cu-filled TSVs in upper TSV layers of 8-hi stack HBM model. From fatigue property point of view, 8-hi stack HBM model is superior to 4-hi stack HBM model.

Keywords- HBM, Cu-filled TSV, Cu protrusion, micro-bump joint, fatigue life, finite element simulation

I. INTRODUCTION

As one of the most important approaches for realizing three-dimensional (3D) integration, the through silicon via (TSV) technology enables massive silicon chips to stack in parallel and to be interconnected vertically. TSV technology has exhibited many advantages, for instance, reducing the length of interconnect, achieving higher density of interconnections and lower consumption [1], and has been applied in advanced products, such as 3D stacking Dynamic Random Access Memory (DRAM) in High Bandwidth Memory (HBM1 and HBM2) [2, 3], and can also meet growing demands for high performance computing and artificial intelligence (AI) market [4]. However, owing to the significant mismatch in coefficient of thermal expansion (CTE) among different materials in Cu-filled TSVs [5], high thermal stress will be generated in TSV structures under thermal cycling, which may cause plastic deformation of Cu fillers when the stress exceeds the yield strength of Cu [6], consequently resulting in protrusion of Cu at the mouth of the via. The plastic deformation regions and residual Cu protrusion will not disappear because of their irreversibility under service circumstance. Nowadays, with the increasing demand for multi-layered chips in 3D integration, micro-bumps are usually used to connect the TSV chips by stacking

perpendicularly, and the bump's feature size may be smaller than 10 μm so as to meet the increasing packaging density with continuously reducing dimension of TSV structures [7]. Considering that the direct interconnection between the micro-bump joint and the end of Cu-filled TSV, there exists an interaction effect between the micro-bump joint and the Cu protrusion, i.e., the irreversible plastic deformation of Cu and subsequent Cu protrusion in the TSV under the alternating and/or cyclic stress will definitely affect the micro-bump joint. Heretofore, most studies [8-10] focused on the Cu protrusion of TSVs, almost nothing is known about the thermo-mechanical interaction between the Cu protrusion and the micro-bump joint, in particular for the HBM which has very complex geometric structure consisting of a large number of TSVs and micro-bump joints.

HBM was adopted as the industry standard (JESD235) in 2013, which was developed and updated to JESD235B in 2018 [11]. In 2015, AMD launched the Radeon™ R9 Fury series graphics, in which HBM was first applied to GPUs. Although HBM can be seen in the market, its reliability is still a concern and remains less well understood, in particular for TSV reliability issues mentioned above. Due to complicated configuration of HBM and very small dimensions of its internal components and structures, it is very difficult to evaluate and predict the reliability of HBM structures by conventional experimental methods. Under such circumstances, the finite element (FE) method is extremely powerful for tackling reliability issues. In this work, two geometric FE models, i.e., 4-hi stack and 8-hi stack HBM structures, are constructed and employed to characterize the von Mises stress, volume fraction of plastic deformation regions of Cu-filled TSVs, height of Cu protrusion and fatigue life of micro-bump joints with taking into account the interaction between the micro-bump joints and Cu protrusion of TSVs in 3D HBM structures.

II. NUMERICAL METHOD

HBM2 products, i.e., the 2-, 4- and 8-GB DRAMs are designed for different applications, such as high performance computing, networking and graphics. Considering the trend of adding more stacked dies, 4-hi and 8-hi stack HBM structure models are constructed, which consist of 4 cores (dies) and 8 cores (dies) stacked on a base die respectively, as shown in Fig. 1. To simplify two FE models, the barrier layer, insulation layer of TSV and the underfill are neglected. The sizes of whole structures are scaled down to save calculation, i.e., the size of cores (dies) and base die is 550 μm×350 μm×50 μm and 600 μm×400 μm×50 μm, respectively, except that die 4 and die 8 are 550 μm×350 μm×100 μm in size. TSVs and micro-bump joints have the same pitch of 40 μm. The micro-bump joint is thought as an integration and taken as homogeneous material. To save computing time, the number of TSVs (or micro-bump joints) is reduced, i.e., TSVs with 3×10 distributed matrix are arranged in the specific area marked by black dotted-line boxes in Fig. 1 (a2) and (b2). Due to symmetry of two models, 1/4th symmetric FE models are used in simulations, as indicated by red dashed-line rectangles in Fig. 1. Geometric parameters and critical materials of 4-hi stack HBM model are shown in Fig. 2, which are the same as those in 8-hi stack HBM model. All degrees of freedom (DOFs) on the bottom surface of base die are constrained.

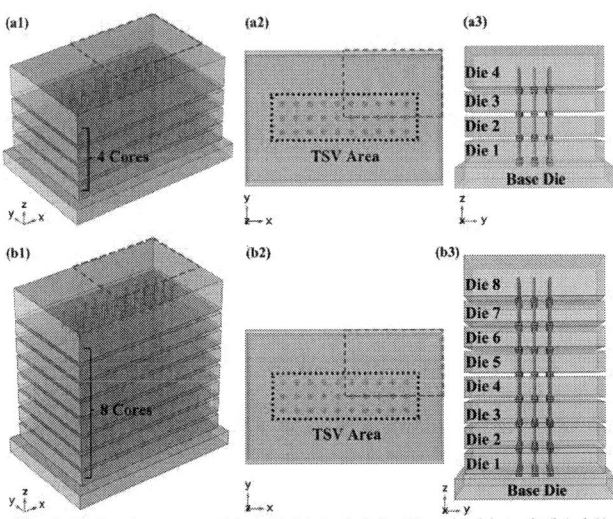

Fig. 1. Finite element models of 4-hi stack (a1–a3) and 8-hi stack (b1–b3) HBM structures, in which 1/4th symmetric models are marked by red dashed-line rectangles.

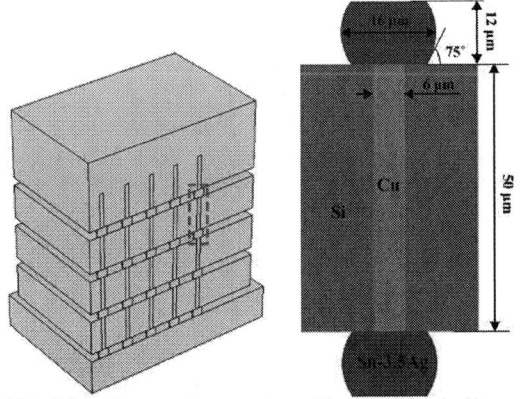

Fig. 2. Materials and geometric parameters of the region marked by a red dashed-line box in the 1/4th symmetric 4-hi stack HBM model (left), which are the same as those in 8-hi stack HBM model.

Accelerated temperature cycling (ATC) condition was used in the FE simulation. Considering the actual working condition and the melting point of Sn–3.5Ag, the temperature profile was adopted based on JESD22-A104E standard [12] but with a modification of the minimum temperature in order to investigate primarily Cu protrusion of TSV at high temperature in this study. ATC was conducted for four cycles with the profile depicted in Fig. 3. Properties of materials mentioned in Fig. 2 are listed in Table I [13, 14]. Cu is considered as elastic-plastic material and analyzed by a bi-linear model with the tangent modulus. Sn–3.5Ag is thought to be viscoplastic material, so the Anand constitutive model [15, 16] is used, and there are nine material parameters: A, Q, ξ, m, S_0, S_{init}, h_0, a and n, which are listed in Table II [17].

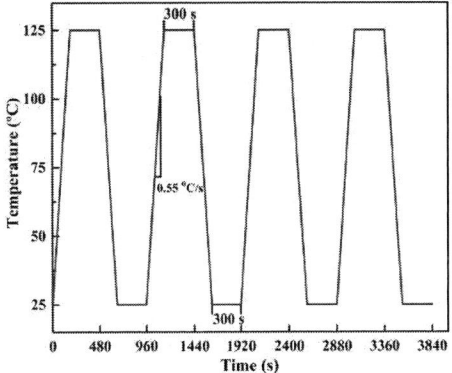

Fig. 3. The profile of accelerated temperature cycling.

TABLE I. Material Properties Adopted in Simulations [13, 14]

Material Properties	Materials		
	Cu	Si	Sn–3.5Ag
Young's modulus (GPa)	117	169	
Passion's ratio	0.30	0.26	
Density (kg/m³)	8930	2329	Data from COMSOL Multiphysics 5.3a
Thermal conductivity (W/(m · K))	385	156	
Heat capacity at constant pressure (J/(kg · K))	385	713	
Coefficient of thermal expansion (×10⁻⁶ K⁻¹)	16.70	2.30	

TABLE II. Nine Parameters of Anand Constitutive Model Used for Sn–3.5Ag [17]

Parameters	Values	Definition
A (s⁻¹)	2.23×10^4	Pre-exponential factor
Q (J/mol)	85459	Activation energy/Boltzmann constant
$\tilde{\xi}$	6	Multiplier of stress
m	0.182	Strain rate sensitivity of stress
S_0 (MPa)	39.090	Coefficient for deformation resistance saturation
S_{init} (MPa)	73.810	Initial value of deformation resistance
h_0 (MPa)	3321.150	Hardening constant
a	1.820	Strain rate sensitivity of hardening
n	0.018	Sensitivity for deformation resistance

Owing to large numbers of TSVs and micro-bump joints in 4-hi stack HBM and 8-hi stack HBM models, we need to number them clearly and concisely, so as to make analyses easier and more efficient. Taking $B_{25\text{-}L1\text{-}M4}$ as an example, as shown in Fig. 4, B stands for a specific group of "micro-bump joints", 25 represents micro-bump joints with distribution of two rows and five columns, L1 means the first layer (from the bottom) of micro-bump joints, M4 denotes 4-hi stack HBM model. Each of micro-bump joints has its own number, e.g., $b_{hi\text{-}Lj\text{-}Mk}$ (b: an individual micro-bump joint in B, h: the row number, i: the column number, j: the layer number, k: the model type). Similar numbering method is used for identification of TSVs, for instance, for $T_{25\text{-}L1\text{-}M4}$, T stands for a specific group of "TSVs", other numbers refer to TSVs with distribution of two rows and five columns in the first layer (from the bottom) in 4-hi stack HBM model, and an individual TSV is expressed as $t_{hi\text{-}Lj\text{-}Mk}$ (t: an individual TSV in T). This numbering method is also applied to 8-hi stack HBM model.

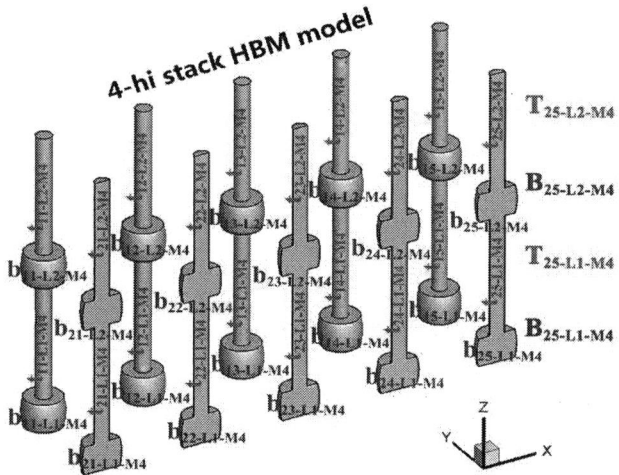

Fig. 4. Schematic diagram for identification of TSVs and micro-bump joints with taking two layers of TSVs and micro-bumps in 4-hi stack HBM model as an example.

To predict fatigue life of micro-bump joints, Darveaux's methodology [18] is used in this study, in which the experimental measurements of low-cycle crack initiation and crack propagation rates were linked to the viscoplastic performance of the solder [19]. It is also an energy based method with work condition consisting of time-dependent plasticity and time-independent viscosity [18, 19]. According to this methodology, the dissipation energy density is calculated in each cycle during temperature cycling, and then is used to calculate number of cycles to initiate a crack as well as number of cycles to spread crack across a solder joint. The expressions are listed below [18-21]:

$$N_0 = K_1 (\Delta W_{ave})^{K_2} \tag{1}$$

$$\frac{da}{dN} = K_3 (\Delta W_{ave})^{K_4} \tag{2}$$

$$N_p = \frac{a}{da/dN} \tag{3}$$

$$N_f = N_0 + N_p \tag{4}$$

$$\Delta W_{ave} = \frac{\sum_{i=1}^{N} v_i \Delta W_i}{\sum_{i=1}^{N} v_i} \tag{5}$$

where N_0 represents the number of cycles initiating crack, da/dN denotes the crack propagation rate per cycle, a is the characteristic area of the fatigue crack and N_f corresponds to the number of cycles to failure when the fatigue crack has propagated across the whole solder. ΔW_{ave} denotes the average density of dissipation energy calculated at each cycle, ΔW_i is the density of dissipation energy calculated in every element per cycle, v_i stands for the volume of element i and N is the total number of elements in a whole solder joint. K_1, K_2, K_3 and K_4 are constants in Darveaux's model, as given in Table III [18].

TABLE III.   THE CRACK GROWTH CORRELATION CONSTANTS [18]

The Crack growth correlation constants			
K_1	K_2	K_3	K_4
56300	-1.62	3.34×10^{-7}	1.04

III.   RESULTS AND DISCUSSION

A. Distributions of Stress and Plastic Deformation Region in HBM Structures during Thermal Cycling

Simulation results indicate that at both 125 °C and 25 °C the stress distribution patterns in 4-hi stack HBM and 8-hi stack HBM models are similar. For convenience, a local area, as shown in Fig. 5, is selected for analysis. At 125 °C, the high stress areas mainly locate at the Cu/Si interface, which is caused by large mismatch of CTEs between Si and Cu. Meanwhile, stress concentration appears at both ends of the Cu-filled TSV, as shown in Fig. 5 (a1), where the maximum stress is 173.0 and 173.1 MPa in 4-hi and 8-hi stack HBM models respectively, both exceed the yield strength of Cu (172.3 MPa), resulting in Cu in the areas yielding and deforming plastically. Likewise, the high stress zones appear on top and bottom surfaces of the micro-bump joint and gradually spread to the inside of the joint, as presented in Fig. 5 (a2). At 25 °C, von Mises stress concentrates in most areas of the micro-bump joint, whereas the stress remains relatively high only at both ends of the Cu-filled TSV. Further, it is important to characterize the change of the average stress with time. Fig. 6 presents the change of the average von Mises stress in micro-bump joints in two models. Obviously, despite slight differences in the average stress between two models, there is a distinct difference in stress level among micro-bump joints. The average stress in all micro-bump joints in 4-hi stack HBM model is higher than that in 8-hi stack HBM model. Notably, the stress in $B_{25\text{-}L1\text{-}M4}$ is the highest while the second highest in $B_{25\text{-}L1\text{-}M8}$, meaning that $B_{25\text{-}L1\text{-}M4}$ and $B_{25\text{-}L1\text{-}M8}$ are more likely to fail first.

As presented above, Cu fillers deform plastically due to higher von Mises stress in some areas over the yield strength of Cu. Fig. 7 exhibits distribution of plastic deformation regions in some Cu-filled TSVs in 4-hi stack HBM model. Clearly, plastic deformation regions mainly appear at the ends of Cu fillers, as marked by red dashed-line circles in Fig. 7, specifically at the edge of Cu filler ends, i.e., the corner between the Cu filler and Si with high stress. To characterize quantitatively the plastic deformation behavior, the volume and volume fraction of plastic deformation regions in Cu-filled TSVs are calculated, as shown in Fig. 8. As time (cycle) proceeds, the volume of plastic deformation regions grows and exhibits ladder-like, in particular, the volume fraction is larger in 4-hi stack HBM model than in 8-hi stack HBM model after four temperature cycles. The irreversible plastic deformation can directly cause Cu protrusion and has an adverse effect on Cu-filled TSVs, so 8-hi stack HBM model seems to be superior to 4-hi stack HBM model.

Fig. 6. Change of the average von Mises stress of micro-bump joints in 4-hi stack HBM and 8-hi stack HBM models with time.

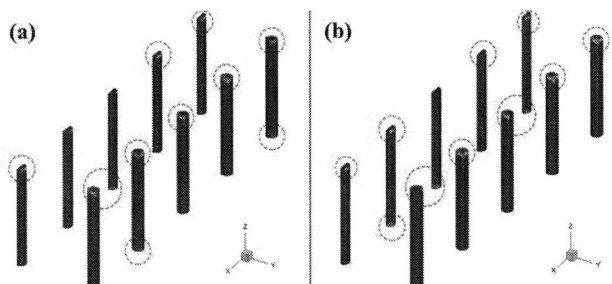

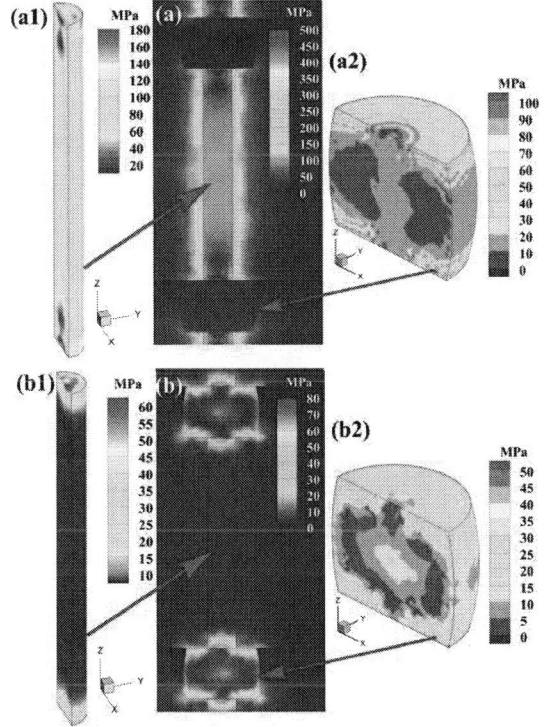

Fig. 5. Distribution of von Mises stress in the Cu-filled TSV (a1, b1) and the micro-bump joint (a2, b2) at 125 °C (a1, a2) and 25 °C (b1, b2).

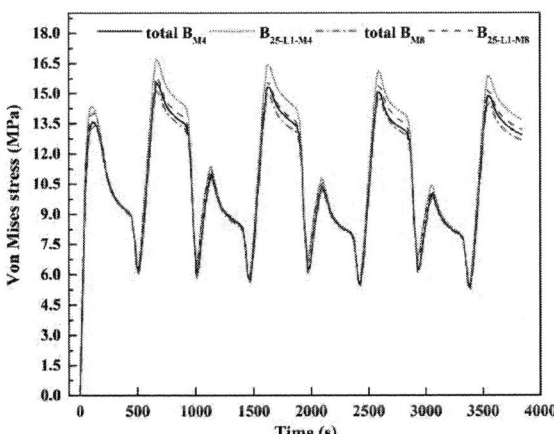

Fig. 7. Distributions of plastic deformation regions in $T_{25-L4-M4}$ (a) and $T_{25-L8-M8}$ (b) marked by red dashed-line circles.

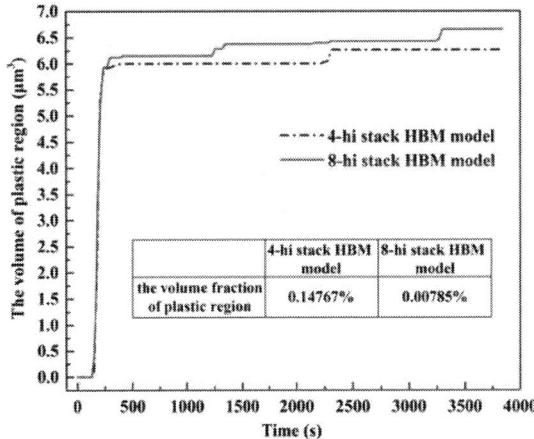

Fig. 8. Change of the volume of plastic deformation regions of Cu-filled TSVs in 4-hi and 8-hi stack HBM models with time, and the corresponding volume faction after four temperature cycles.

B. Protrusion of Cu Fillers in TSV Structures

At elevated temperature (125 °C), the expansion of the Cu filler towards radial direction is constrained by surrounding Si, which leads to the filler expanding along the Z axis easily and causes Cu protrusion at the mouth of the via. Simulation results reveal that Cu protrusion height mainly depends on the displacement in the Z direction at 125 °C and the residual height of Cu protrusion is almost equal to the displacement along the Z axis at 25 °C. To depict this phenomenon more clearly, $T_{25-L1-M4}$ is selected for analysis, as exhibited in Fig. 9 (Note that other TSVs in 4-hi and 8-hi stack HBM models have the similar pattern). The enlarged drawing in Fig. 9 (i.e., the inset) shows that with increasing number of temperature cycle, the height of Cu protrusion increases gradually, meaning that applying more cycles will lead to increase in Cu protrusion height and thus may cause serious reliability concern in TSVs, which cannot be neglected in applications.

Furthermore, variations of the Cu protrusion height of different TSV layers in 4-hi and 8-hi stack HBM models at 125 °C and 25 °C, respectively, are calculated and shown in Fig. 10. As the upper ends of $T_{25-L4-M4}$ and $T_{25-L8-M8}$ are not directly interconnected with micro-bump joints, there is no need to characterize Cu protrusion heights corresponding to these ends in Fig. 10. In fact, Cu protrusion generated at 125 °C consists of two parts, elastic protrusion and plastic protrusion, while it comprises only plastic protrusion at 25 °C as the elastic protrusion disappears. Obviously, the height of Cu protrusion at upper ends of Cu-filled TSVs is larger than that at lower ends and the differences become smaller gradually with the site of TSV layer going up (see example data for $T_{25-L1-M4}$, $T_{25-L2-M4}$ and $T_{25-L3-M4}$, as well as for $T_{25-L1-M8}$ and $T_{25-L2-M8}$). This is because all DOFs on the bottom surface of the base die are constrained, thus, at high temperature (125 °C) the whole structure expands upward more easily than downward. Likewise, Cu-filled TSVs expand upward more freely, which means the heights of Cu protrusion at upper and lower ends are different. Remarkably, however, the opposite trend is observed at 25 °C, i.e., the height of Cu protrusion at upper ends of Cu-filled TSVs is smaller than that at lower ends, and the former increases with the location of TSV layer going up while the latter decreases. Notably, in 8-hi stack HBM model the heights of Cu protrusion at upper and lower ends of $T_{25-L3-M8}$, $T_{25-L4-M8}$, $T_{25-L5-M8}$, $T_{25-L6-M8}$ and $T_{25-L7-M8}$ are almost equal, as depicted in Fig. 10 (b1), which means these TSV layers including their ends suffer almost the same stress conditions. To have a detailed look at Cu protrusion, $T_{25-L5-M8}$ is extracted and contour plots of Cu protrusion at 125 °C and 25 °C are shown in Fig. 11. The red areas denote Cu protrusion and the surrounding green areas are the surfaces of Si. Apparently, Cu fillers protrude out of the Si matrix at mouths of vias, exhibiting arc-shaped surface. It should be indicated that despite a high scale factor of 3000 in Fig. 11 (c) and (d), Cu protrusion is not obvious, while displaying evidently arc-shaped at 600 scale factor in Fig. 11 (a) and (b), which implies the height of Cu protrusion at 25 °C is smaller than that at 125 °C, as shown clearly in Fig. 10.

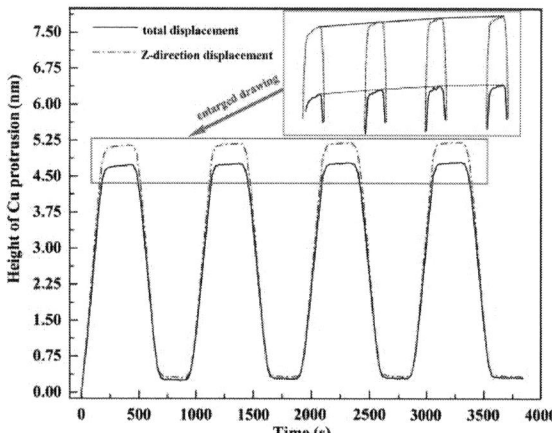

Fig. 9. Height of Cu protrusion of upper ends of $T_{25-L1-M4}$ along Z direction and total displacement direction.

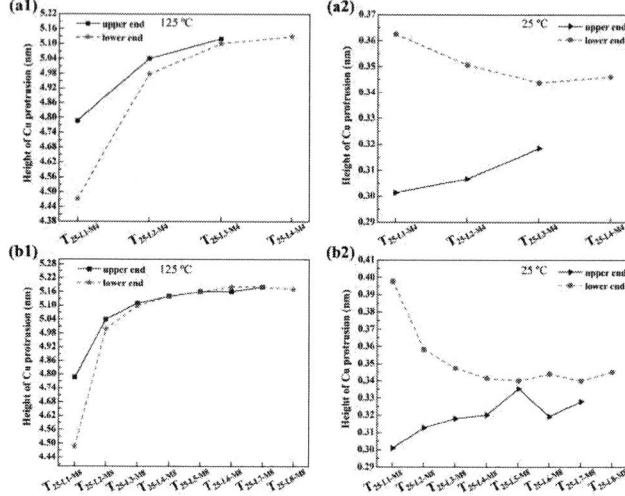

Fig. 10. Height of Cu protrusion of different TSV layers in 4-hi (a1, a2) and 8-hi (b1, b2) stack HBM models at 125 °C (a1, b1) and 25 °C (a2, b2).

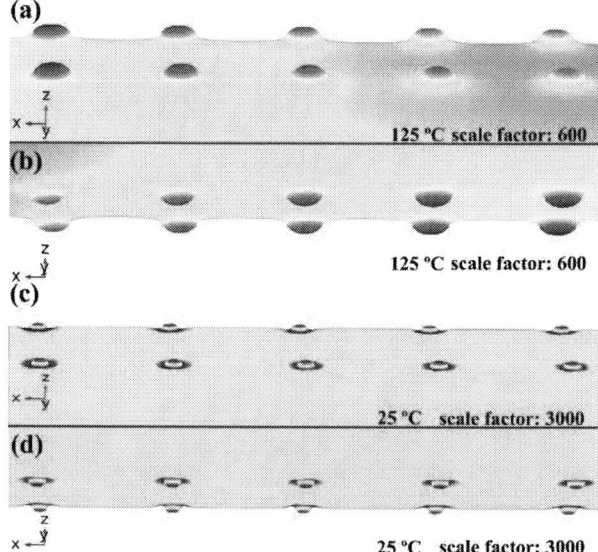

Fig. 11. Contour plots of Cu protrusion of $T_{25-L5-M8}$ at 125 °C (a, b) and 25 °C (c, d) at upper (a, c) and lower (b, d) ends.

C. Influence of Cu Protrusion on fatigue life of Joints

In TSV structure, the ends of Cu fillers are directly interconnected with micro-bumps to form the joints, then Cu protrusion will inevitably affect the interface between the ends of TSVs and micro-bump joints. Undoubtedly, the volume strain of each micro-bump joint is greater than zero owing to its own expansion at elevated temperature (125 °C). But the volume strain at the interface of the micro-bump joint is relatively small. Fig. 12 presents distribution of the volume strain in micro-bump joints ($B_{25-L5-M8}$), where the interfacial strain is very small as marked by a red dashed-line box and a blue dotted-line box in Fig. 12 (a). In contrast, at 25 °C the volume strain in the side of the micro-bump joint is negative, as shown in Fig. 12 (b), which is caused by shrinking freely, while the strain on both top and bottom surfaces is still greater than zero owing to limited shrinkage as a result of constraint from Si and TSV. Due to the effect of residual Cu protrusion, the strain value at the interface is relatively high, as marked by a purple dashed-line box and a black dotted-line box in Fig. 12 (b), in which the pit-like deformation appears in the joint under 1500 scale factor.

Moreover, the thermal fatigue of the micro-bump joints is another important reliability concern. According to above (5), in the simulation the density of fatigue dissipation energy (ΔW_i) is calculated by viscoplatic dissipation energy density in every element per cycle and ΔW_{ave} is the average density of fatigue dissipation energy each cycle. The results of fatigue life of micro-bump joints in different layers of 4-hi and 8-hi stack HBM models are shown in Fig. 13, where red dashed-line boxes indicate the micro-bump joints with the lowest fatigue life in each layer and their fatigue life values, viscoplastic strain and average density of fatigue dissipation energy are given in Table IV. Clearly, $b_{24-L1-M4}$ and $b_{13-L1-M8}$ are the most dangerous joints in two models respectively, and $b_{24-L1-M4}$ is more dangerous than $b_{13-L1-M8}$. The fatigue life of micro-bump joints is related to viscoplastic strain but depends on average density of fatigue dissipation energy, e.g., fatigue life of $b_{24-L1-M4}$ is the lowest whereas its energy density is the largest. From this point of view, seemingly 8-hi stack HBM model is superior to 4-hi stack HBM model. The dangerous micro-bump joints generally locate at the first row of each layer in two models, e.g., $b_{1i-Lj-Mk}$ (1: the first row, i: the column number, j: the layer number, k: the model type), which is different from the situation of BGA joints [20]. It can be understood in a way that BGA joints in [20] are distributed symmetrically over a square area and the joints at four outmost corners are more likely to fail as a result of large displacement and high dissipation energy density. In this work, however, micro-bump joints distribute in a rectangular area with relatively low symmetry, and the joints lying the outmost row in contact with bulk Si have relatively larger displacement compared with inner joints.

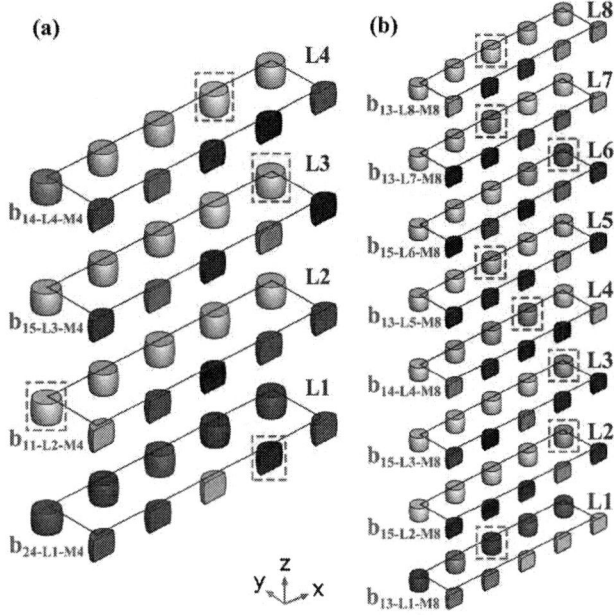

Fig. 13. Fatigue life of micro-bump joints in different layers of 4-hi (a) and 8-hi (b) stack HBM models (the micro-bump joints with the lowest fatigue life in each layer are marked by red dashed-line boxes).

TABLE IV. FATIGUE LIFE, VISCOPLASTIC STRAIN AND AVERAGE FATIGUE DISSIPATION ENERGY DENSITY OF MICRO-BUMP JOINTS MARKED IN FIG. 13

	$b_{24-L1-M4}$	$b_{11-L2-M4}$	$b_{15-L3-M4}$	$b_{14-L4-M4}$
Fatigue life (cycles)	662.72	1183.50	1109.30	1138.20
Viscoplastic strain	0.00507	0.00415	0.00420	0.00420
Average fatigue dissipation energy density (J/m^3)	27998	19565	20364	20043
	$b_{13-L1-M8}$	$b_{15-L2-M8}$	$b_{15-L3-M8}$	$b_{14-L4-M8}$
Fatigue life (cycles)	775.91	1063.90	1066.40	1041.40
Viscoplastic strain	0.00511	0.00422	0.00423	0.00419
Average fatigue dissipation energy density (J/m^3)	25398	20897	20867	21175
	$b_{13-L5-M8}$	$b_{15-L6-M8}$	$b_{13-L7-M8}$	$b_{13-L8-M8}$
Fatigue life (cycles)	1089.70	962.72	1046.80	1144.60
Viscoplastic strain	0.00417	0.00422	0.00419	0.00418
Average fatigue dissipation energy density (J/m^3)	20598	22228	21107	19974

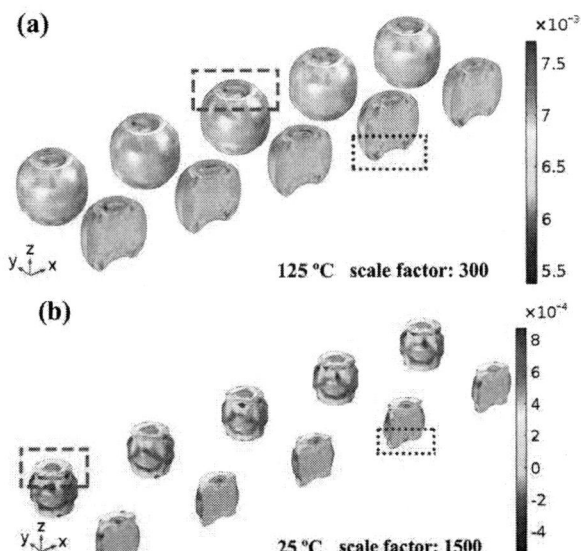

Fig. 12. Distribution of the volume strain in micro-bump joints ($B_{25-L5-M8}$) at 125 °C (a) and 25 °C (b).

IV. CONCLUSIONS

On the basis of a comprehensive FE simulation study, the conclusions are summarized as follows:

1) For 4-hi stack HBM structure the stress in $B_{25-L1-M4}$ micro-bump joints is much higher than in other micro-bump joints. Micro-bump joints in 4-hi stack HBM model bear higher average stress than those in 8-hi stack HBM model.

2) The volume fraction of plastic deformation regions of Cu-filled TSVs in 4-hi stack HBM model is larger than in 8-hi stack HBM model. The volume fraction in both models increases with increasing number of temperature cycle.

3) Cu protrusion height mainly depends on the Z-direction displacement and becomes equal at upper and

lower ends of TSVs in upper TSV layers of 8-hi stack HBM model.

4) Fatigue life of $b_{24-L1-M4}$ micro-bump joint is the lowest due to the largest average density of fatigue dissipation energy. From the fatigue performance point of view, 8-hi stack HBM model seems to be superior to 4-hi stack HBM model.

ACKNOWLEDGMENT

This research is supported by the National Natural Science Foundation of China under grant Nos. 51775195, 51275178 and 51405162.

REFERENCES

[1] J. H. Lau, "Overview and outlook of through-silicon via (TSV) and 3D integrations," *Microelectronics International*, vol. 28, no. 2, pp. 8-22, 2013.

[2] D. U. Lee, K. W. Kim, K. W. Kim, H. Kim, J. Y. Kim, Y. J. Park, J. H. Kim, D. S. Kim, H. B. Park, J. W. Shin, J. H. Cho, K. H. Kwon, M. J. Kim, J. Lee, K. W. Park, B. Chung, S. Hong, "A 1.2V 8Gb 8-Channel 128GB/s High-Bandwidth Memory (HBM) Stacked DRAM with Effective Microbump I/O Test Methods Using 29nm Process and TSV," *Proc. International Solid State Circuits Conference (ISSCC), IEEE Press*, Feb. 2014, San Francisco, pp. 432-434.

[3] J. C. Lee, J. Kim, K. W. Kim, Y. J. Ku, D. S. Kim, C. Jeong, T. S. Yun, H. Kim, H. S. Cho, Y. O. Kim, J. H. Kim, J. H. Kim, S. Oh, H. S. Lee, K. H. Kown, D. B. Lee, Y. J. Choi, J. Lee, H. G. Kim, J. H. Chun, J. Oh, S. H. Lee, "A 1.2V 64Gb 8-channel 256GB/s HBM DRAM with peripheral-base-die architecture and small-swing technique on heavy load interface," *Proc. International Solid State Circuits Conference (ISSCC), IEEE Press*, Jan. 2016, San Francisco, pp. 318-320.

[4] J. C. Lee, J. Kim, K. W. Kim, Y. J. Ku, D. S. Kim, C. Jeong, T. S. Yun, H. Kim, H. S. Cho, S. Oh, H. S. Lee, K. H. Kwon, D. B. Lee, Y. J. Choi, J. Lee, H. G. Kim, J. H. Chun, J. Oh, S. H. Lee, "High Bandwidth Memory(HBM) with TSV Technique," *Proc. International Soc Design Conference (ISOCC), IEEE Press*, Oct. 2016, New York, pp. 181-182.

[5] A. Heryanto, W. N. Putra, A. Trigg, S. Gao, W. S. Kwon, F. X. Che, X. F. Ang, J. Wei, R. I Made, C. L. Gan, K. L. Pey, "Effect of copper TSV annealing on via protrusion for TSV wafer fabrication," *Journal of Electronic Materials*, vol. 41, no. 9, pp. 2533-2542, 2012.

[6] T. Jiang, W. Cheng, S. Laura, I. Jay, T. Nobumichi, K. Martin, S. Ho-Young, G. K. Byoung, H. Rui, H. S. Paul, "Plasticity mechanism for copper extrusion in through-silicon vias for three-dimensional interconnects," *Applied Physics Letter*, vol. 103, no. 21, pp. 602-633, 2013.

[7] C. Chen, D. Yu, K-N. Chen, "Vertical interconnects of microbumps in 3D integration," *MRS Bulletin*, vol. 40, no. 3, pp. 257-263, 2015.

[8] F. X. Che, W. N. Putra, A. Heryanto, A. Trigg, X. W. Zhang, C. L. Gan, "Study on Cu protrusion of Through-Silicon Via," *IEEE Transactions on Components, Packaging and Manufacturing Technology*, vol. 3, no. 5, pp. 732-739, 2013.

[9] D. Smith, S. Singh, Y. Ramnath, M. Rabie, D. Zhang, L. England, "TSV residual Cu step height analysis by white light interferometry for 3D integration," *Proc. 65th Electronic Components and Technology Conference (ECTC), IEEE Press*, May 2015, San Diego, pp. 578-584.

[10] S. Chen, T. An, F. Qin, P. Chen, "Microstructure evolution and protrusion of electroplated Cu-filled Through-Silicon Vias subjected to thermal cyclic loading," *Journal of Electronic Materials*, vol. 46, no. 10, pp. 5916-5932, 2017.

[11] High Bandwidth Memory (HBM) DRAM, JESD235B, JEDEC, 2018.

[12] Temperature Cycling, JESD22-A104E, JEDEC, 2014.

[13] C. Okoro, Y. Yang, B. Vandevelde, B. Swinnen, D. Vandepitte, B. Verlinden, I. De Wolf, "Extraction of the appropriate material property for realistic modeling of through-silicon-vias using μ-Raman Spectroscopy," *Proc. International Interconnect Technology Conference (IITC), IEEE Press*, June 2008, Burlingame, pp. 16-18.

[14] T. Chellaih, G. Kumar, K. N. Prabhu, "Effect of thermal contact heat transfer on solidification of Pb-Sn and Pb-free solders," *Materials & Design*, vol. 28, no. 3, pp. 1006-1011, 2007.

[15] L. Anand, "Constitutive equations for hot-working of metals," *International Journal of Plasticity*, vol. 1, no. 3, pp. 213-231, 1985.

[16] S. B. Brown, K. H. Kim, L. Anand, "An internal variable constitutive model for hot working of metals," *International Journal of Plasticity*, vol. 5, no. 2, pp. 95-130, 1989.

[17] G. Z. Wang, Z. N. Cheng, K. Becker, J. Wilde, "Applying Anand Model to represent the viscoplastic deformation behavior of solder alloys," *Journal of Electronic Packaging*, vol. 123, no. 3, pp. 247-253, 1998.

[18] R. Darveaux, "Effect of simulation methodology on solder joint crack growth correlation," *Proc. 50th Electronic Components and Technology Conference (ECTC), IEEE Press*, May 2000, Las Vegas, pp. 1048-1058.

[19] B. A. Zahn, "Finite element based solder joint fatigue life predictions for a same die size-stacked-chip scale-ball grid array package," *Proc. 27th Annual IEEE/SEMI International Electronics Manufacturing Technology Symposium, IEEE Press*, July 2002, San Jose, pp. 274-284.

[20] H. B. Qin, H. H. Yuwen, M. B. Zhou, X. P. Zhang, "Influence of geometry of microbump interconnects on thermal stress and fatigue life of interconnects in copper filled through silicon via structure," *Proc. 14th International Conference on Electronic Packaging Technology (ICEPT), IEEE Press*, Aug. 2013, Dalian, pp. 1019-1024.

[21] H. B. Qin, W. Y. Li, M. B. Zhou, X. P. Zhang, "Low cycle fatigue performance of ball grid array structure Cu/Sn–3.0Ag–0.5Cu/Cu solder joints," *Microelectronics Reliability*, vol. 54(12), pp. 2911-2921, 2014.

978-1-7281-1500-9/19 $31.00 © 2019 IEEE

Study of design optimization method for ultra-low power micro gas sensor.

Eiji Nakamura, Keiji Matsumoto, Hiroyuki Mori
IBM Research - Tokyo
IBM Japan, Ltd.
Kawasaki, Japan
e-mail: e36055@jp.ibm.com

Andrea Fasoli, Luisa Bozano
IBM Research - Almaden
IBM Corporation
San Jose, United States
e-mail: andrea.fasoli@ibm.com

Abstract—In this paper, we propose a design optimization scheme of metal oxide gas sensor with suspended beams. Our design scheme comprises two parts, design of suspended membrane and beam membrane, and design of heater geometry. These processes are based on mathematical analysis. In the suspended membrane and beam design, power consumption at steady state and transient state as well as response speed of microheater are considered. In the heater geometry design, a mathematical method for realizing an arbitrary temperature profile is proposed. After these steps, the electro-thermal performance and the mechanical properties of the suggested designs are investigated by finite element analysis. Here, we demonstrated this scheme for designing gas sensors with two suspended beams. Calculated design operates at 388°C with 7.88 mW power consumption and have uniform temperature profile. In addition, the relation between the dimensions of suspended beams and their mechanical properties has been investigated. By utilizing our scheme, design process of low-power gas sensors satisfying constraints on mechanical properties and footprint can be rapidly obtained.

Keywords-component; gas sensor; low power; metal oxide semiconductor; microheater; suspended beam; thermal analysis

I. INTRODUCTION

Metal oxide (MOX) gas sensors have been widely used for detection of harmful gases including CO, NOx and volatile organic compounds (VOCs). MOX gas sensors utilize a microheater element to activate surface reactions which, in turn, cause changes in the sensitive material resistivity. Most of the power to operate the device is consumed by the micro heater element. Therefore, MOX gas sensors typically require relatively higher power than ordinary electronic components. Even the lowest-power commercially available gas sensor requires around 10 mW to operate at constant temperature.

On the other hand, there are potential needs for further reduction of the power consumption. For example, mobile applications for detecting harmful gases or health monitoring requires battery-operation for hours, while most commercially available gas sensors need power supply from wired power source. In addition, smart odor recognition system, called electronic nose (EN), has been increasingly investigated as machine learning algorithms progress [1,2]. As EN systems generally have several sensors for expanding odor recognition capability, power consumption of total system increases proportionally to the number of sensors. Ultimately, mobile electronic nose systems can be

envisioned if power consumption can be dramatically reduced.

Toward reduction of the power consumption, there have been many attempts including development of new sensitive materials [3-6], pulsatile heating modes [7] and design optimization of microheater element [8]. Regarding the design optimization of the heater, suspended beam structure is one of the promising ways to reduce the power consumption because suspended-beam structure can realize better thermal isolation between suspended membrane and silicon support. Although suspended-beam microheater gas sensors have been already commercialized, for further power saving, suspended beam structure requires some technical consideration. First, a trade-off between thermal isolation and power loss at circuits on suspended beams should be investigated. Narrower beam results in better thermal isolation, but also causes larger power loss because of higher electrical resistance of the circuit above the suspended beams. Second, suspended beam structure has a concern regarding mechanical reliability. In general, evaluation of mechanical reliability is more complicated because material properties depend on fabrication processes and most membrane materials have residual strain at room temperature.

In this paper, we studied these two aspects by mathematical analysis and finite element analysis (FEA). For the first challenge, a mathematical model to optimize the suspended beam design is established. By this model, dimensions of suspended beam are optimized for minimizing power consumption in a steady-state. Transient response is also calculated, although it is not considered in optimization criteria of this work. In addition, a design scheme of heater geometries which realize uniform temperature profile is proposed. Temperature uniformity is also an important requirement for maximizing sensor sensitivity because sensitivity of sensitive materials is dependent on temperature [9,10]. Although there have been many studies which realized uniform temperature distribution by optimizing heater geometry [11-14], those didn't show mathematical principle for those designs. Hence, in this study, a mathematical scheme for finding optimal heater geometries, discrete design method, is established. Geometries created by this method is finally validated by FEA.

FEA is also utilized for exploring structures with lower mechanical stress as the final step of our optimization scheme. Mechanical reliability is determined by the dimension of suspended beam, heater size, and temperature distribution. Here, we evaluate mechanical properties by relatively comparing several geometries.

II. Mathematical Analysis

A. Model for optimization of suspended beam dimensions

The design flow is shown in Fig. 1. Here, a square-shape membrane structure with two suspended beams is first investigated as a model case. Fig. 2 is a schematic illustration of the square shape membrane structure. The thermal behavior of the suspended membrane and the supporting silicon is described by the equations shown below:

$$C_m \frac{dT_m}{dt} = \left(\frac{V_0}{R_m + R_b}\right)^2 \cdot R_m - \frac{T_m - T_{Si}}{\theta} - A_m \alpha (T_m - T_{amb}) \quad (1)$$

$$C_{Si} \frac{dT_{Si}}{dt} = \frac{T_m - T_{Si}}{\theta} - A_S \alpha (T_{Si} - T_{amb}) \quad (2)$$

$$\theta = \frac{1}{\frac{k_h t_h x}{L} + \frac{k_i t_i y}{L}} \quad (3)$$

$$\frac{V_0^2}{R_h + R_b} = P \quad (4)$$

Equation (1) and (2) are heat transfer equations of the suspended membrane and the silicon support, respectively. Equation (3) represents the thermal resistance of the suspended beam, and (4) represents the total power consumption. C_m and C_{Si} is the heat capacity of suspended membrane and silicon support, respectively, T_m and T_{Si} are the temperatures of the suspended membrane and the silicon support, which are assumed to be uniform in each region, V_0 is the applied voltage, R_m and R_b are the electrical resistances of the circuit above the suspended membrane and the circuit on the suspended beams, respectively, θ is the thermal resistance of the suspended beams, A_m and A_{Si} are the surface areas of the suspended membrane and silicon support, T_{amb} is the temperature of the ambient air, k_{Pt} and k_{SiN} are the thermal conductivities of platinum and silicon nitride, t_{Pt} and t_{SiN} are thicknesses of platinum heater and silicon nitride membrane, L is the length of the suspended beam, P is the power consumption in a steady-state for keeping the target temperature. α is the heat transfer coefficient which is assumed to be constant value (100 W/Km²) here, although the heat transfer coefficient to the ambient is generally a function of temperature. The material properties used here are shown in Table 1. In this model, the radiation from the suspended membrane and convection from the suspended beam to the air are not taken into account because their effect is negligible at around 400°C.

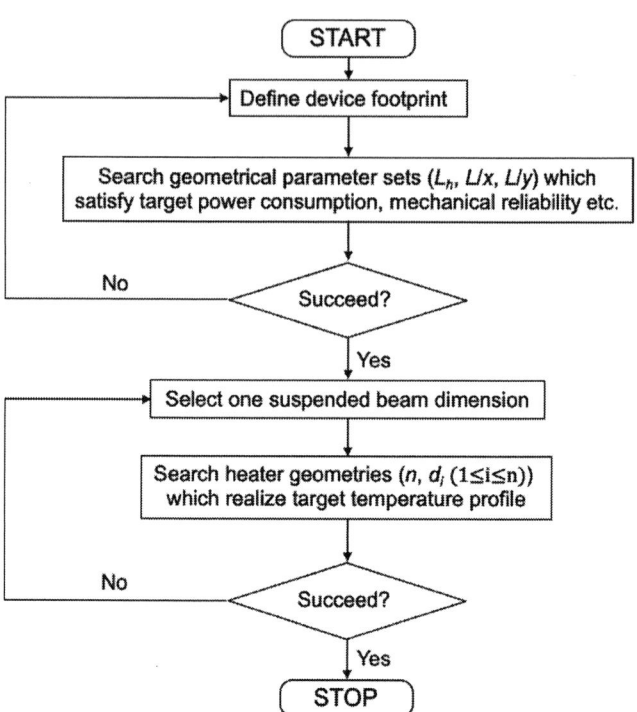

Figure 1. Design optimization flow.

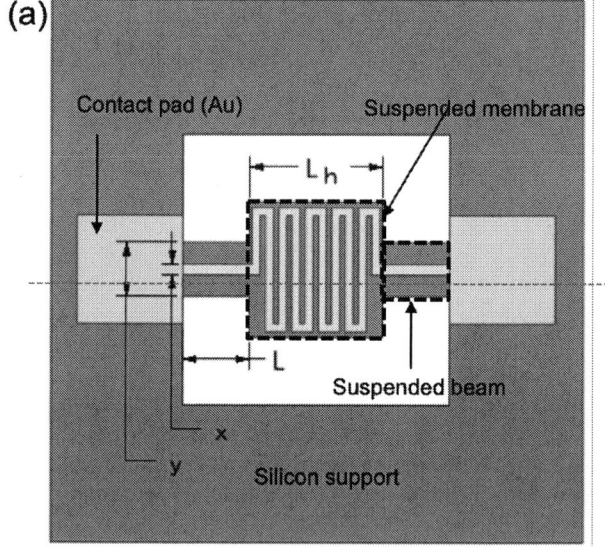

Figure 2. Schematic illustration of gas sensor model. (a)Top view and (b) cross section along with the red dotted line in (a).

TABLE I.   Material Properties Used In This Paper

Property	Material		
	Silicon	*Silicon nitride*	*Platinum*
Density [Kg m⁻³]	2330	2900	21500
Specific heat [J kg⁻¹ K⁻¹]	678	300	134
Thermal conductivity [W m⁻¹ K⁻¹]	130	6	72
Electrical resistivity [Ω m]	-	-	$3.62 \times 10^{-10} \times T$ $+9.99 \times 10^{-8}$ [15]

978-1-7281-1500-9/19 $31.00 © 2019 IEEE

By assuming a steady state (dT_m/dt=0, dT_{Si}/dt=0, T_m=400°C), P and R_h for keeping the target temperature (400°C) in steady state can be represented as a function of L_h, L/x and L/y. Here, aspect ratios (L/x, L/y) are used as variables instead of circuit linewidth x and beam width y. Parameters are varied in the following order; the size of the suspended membrane (L_h), the aspect ratio of the silicon nitride (SiN) suspended beam (L/y), and finally aspect ratio of the circuit on the suspended beam (L/x). It should be noted that the number of suspended beams doesn't have to be determined at this point, because the same thermal resistance can be realized by different numbers of suspended beams. For example, splitting one beam into 2 beams doesn't affect total thermal resistance.

Fig. 3 shows the relation between power consumption, L/x, and L/y under constant L_h value (300μm). Although the shape of this relation varies drastically with L_h, (L/x, L/y) = (22.0, 22.0) is optimal point for low power operation just in this case. The value of L, x, y can be designed to satisfy other requirements including device footprint and mechanical reliability.

Fig. 4 shows the relation between L_h and minimum power consumption. At each data point, aspects of the suspended beam and the circuit are optimized for minimizing power consumption as discussed above. Power consumption monotonically increases as L_h increases, which is because larger heater size leads to increased energy loss due to convection to the ambient air. Therefore, if we need lower-power microheater, heater size should be smaller just from the viewpoint of power consumption.

It should be noted that response time of the temperature of membrane becomes longer with higher thermal resistance of the suspended beam. This is indicated by the fact that T_m and T_{Si} is divided by thermal resistance in right hand side of (1) and (2). Fig. 5 shows a transient response of microheaters with different dimensions. How to handle this response time depends on situation. In most cases, this response time is in the order of milliseconds, as shown in Fig. 5. As this is much shorter than the chemically-driven response time of typical chemiresistors, this parameter may be not require further optimization. On the other hand, in the case of operational modes based on transient response aimed at enhancing recognition capabilities[16-18], or for sensors subject to pulsed heating, the response time should be designed precisely. In this paper, simple resistance measurement in a steady state is assumed so that response time is not to be optimized.

In this section, we select (L_h, L/x, L/y) = (300 μm, 22.0, 22.0) as a demonstration. According to Fig. 3, power consumption is supposed to be 7.3 mW at 400°C. Although L_h is 300 μm is a conservative size of suspended membrane, it is suitable for the first demonstration because it would not be faced with problems regarding the manufacturability. Fig. 7a is a design of suspended membrane and beam in the case of L = 890 μm. The heater geometry above the suspended membrane is designed in the next section.

B. Optimization of heater design

In order to calculate precise temperature distribution generated by microheater, FEA is indispensable. Nevertheless, using FEA for each design is time-consuming.

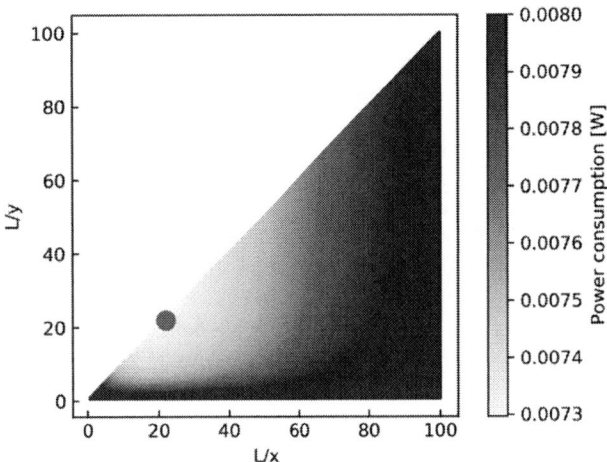

Figure 3. Color map of power consumption with 300 μm suspended membrane as a function of aspect ratios of the suspended beam and the circuit on the suspended beam. The red dot indicates the point which gives minimal power consumption. Upper-left region is not calculated because the circuit width should be smaller than the width of suspended beam.

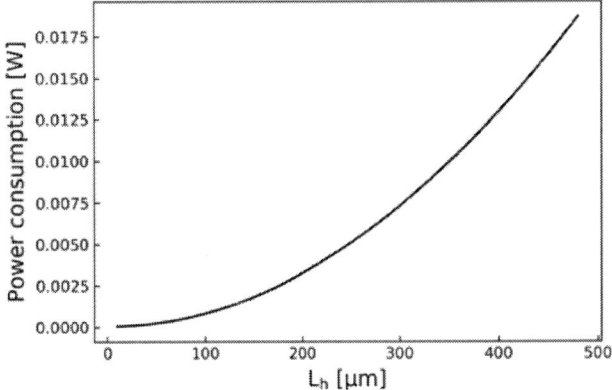

Figure 4. Minimum power consumption versus the size of the suspended membrane (L_h).

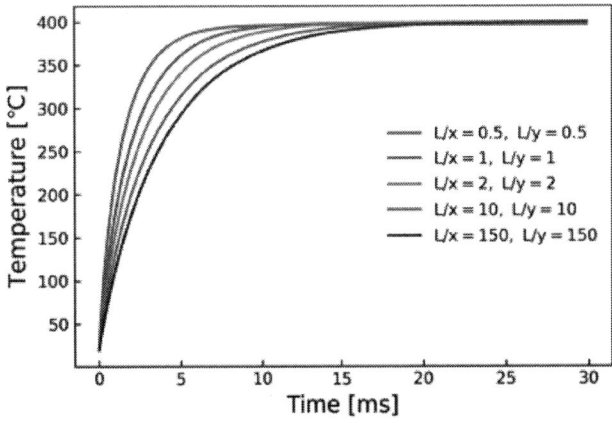

Figure 5. Transient response of the temperature of suspended membranes with different dimensions.

Hence, we established an alternative efficient mathematical scheme, DDM, to find optimal heater geometry. First, target temperature profile is configured as shown in Fig. 6. Arbitrary profile including linear shape and quadratic shape can be selected. In the next step, heater element is divided into n areas (called area i) in which it is assumed that single heater circuit is running vertically. The shape of divided areas should be adjusted for heater shape. In the case of square shape suspended membrane, rectangular shape is reasonable. On the other hand, for spiral-shape heater, ring shape is reasonable. Here, designing of a meander-shape heater is demonstrated. To sustain the target temperature profile in a steady state, the following equations describing the heat balance of each divided area, must be satisfied simultaneously:

$$P_i + F_i = F_{i+1} + A_i \alpha (T_i - T_{amb}) \quad (1 \le i \le n) \tag{5}$$

$$P_i = \frac{\rho \cdot 2x_i}{w_i \cdot t_h} I^2 \tag{6}$$

$$F_i = -ki \frac{dt}{dx} \cdot L_h \cdot t_i \tag{7}$$

Equation (5) is a heat transfer equation of each divided area in a steady state. P_i is the joule heat generated in area i, F_i is the heat transfer between area i-1 and area i (F_0 is defined as 0), A_i is the surface area of area i, T_i is the temperature of area i, ρ is electrical resistivity, x_i is the x coordinate of the right side of area i, w_i is the circuit width of area i (variable), I is the electrical current, R_i is the thermal resistance of area i, which is a function of w_i. As shown in (6), P_i is a function of the circuit width in area i, and F_i is a function of the circuit width of area i. Exceptionally, T_{n+1} is defined as the temperature of support silicon and R_{n+1} is defined as the thermal resistance of a suspended beam, which are both constant. Therefore, (5) for $i = 1, \ldots, n$ has n unknown quantities ($w_1, w_2, \ldots, w_n$) can be determined uniquely.

It is possible that there is no combination of ($w_1, w_2, \ldots, w_n$) satisfying (5) and the constraint condition described above. This is because there is a constraint condition such that w_i is narrower than the width of area i ($x_i - x_{i-1}$) for isolation of the circuits of divided areas. Hence, practically, iterations over division number n and current I are necessary. Qualitatively, if n is too big, the upper limit of w_i ($x_i - x_{i-1}$) shrinks and A_i become narrower and, as a result, it may be impossible to balance the joule heat and the heat dissipation through convection. On the other hand, when n is sufficiently small, it is possible to find a design. However, if n is too small, the temperature gradient inside divided areas can't be ignored. Therefore, maximum number of n which give possible heater design should be selected. In addition, for utilizing the result of suspended beam optimization discussed in previous section, applied current or voltage should be the same as the voltage we used in previous section. If there is no design which satisfies equations above and can be operated by V_0, a different suspended beam dimension has to be selected again.

Fig. 7b is an output from the equations described above and an implementation of the output. The dimensions of the

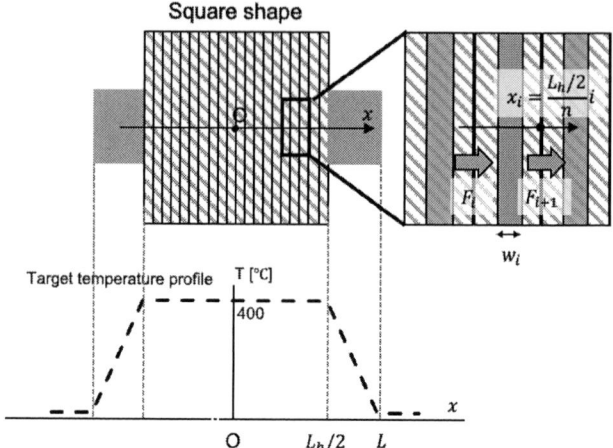

Figure 6. Schematic illustration of discrete design method and target temperature profile.

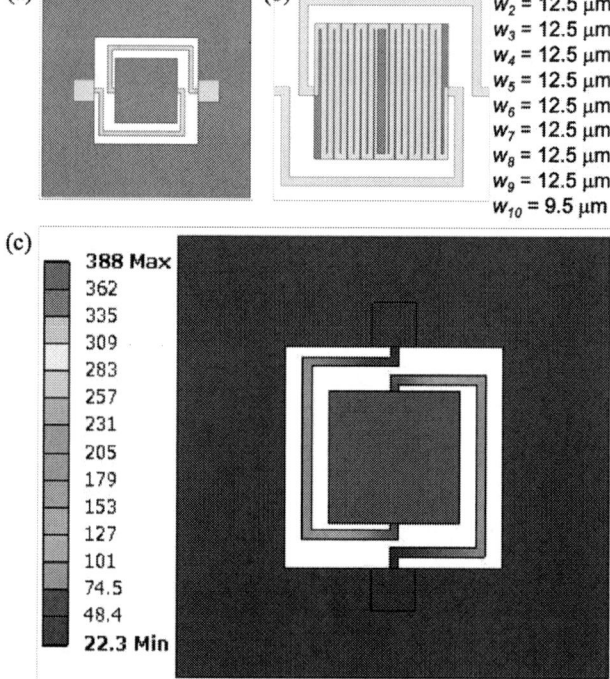

Figure 7. (a)Schematic illustration of an optimized suspended beam structure. (b)An output of the optimization scheme of heater geometry. (c)The temperature profile of the optimized microheater design.

suspended membrane and beam are (L_h, L/x, L/y) = (300 μm, 22.0, 22.0) as calculated in the previous section. The circuit of each divided area has the same width, with the exception of w_{10} that is narrower than the other circuits because it has to generate larger joule heat in order to compensate the larger heat dissipation through the suspended beam. Although w_1 is 12.5 μm, the actual width of the circuit of area 1 in the implementation is 6.25 μm. The purpose of this modification is to keep the heater geometry a point symmetry, which is necessary for avoiding stress concentration while heating. Because the number of the divided area is even in our current scheme, two circuits of w_l have to be connected in parallel. In order to cancel the effect of the parallel connection, the circuit width is designed as half of the calculated width. Besides, the problem about keeping point symmetry can be solved by making the mathematical scheme compatible with odd number of the division.

III. FINITE ELEMENT ANALYSIS

A. Validation of electro-thermal performance

The performance of the design proposed in the previous section is validated by FEA in this section. FEA simulations are performed using ANSYS Mechanical 19.0. Fig. 6c shows the temperature profile generated by the microheater shown in Fig. 6b at 3V. The highest temperature in the suspended membrane is 388°C, which shows good agreement with the target temperature (400°C), and temperature distribution across most of the membrane is within 10°C. The electrical current and the power consumption is 2.10×10^9 A m^{-2} and 7.88 mW, respectively, while the target is 2.43×10^9 A m^{-2} and 7.30 mW. The slight difference between results of the mathematical analysis and the FEA is due to approximation of the mathematical model.

For example, the microheater model in the mathematical analysis is assumed to have circuits in one direction, but the FEA model has connections between adjacent circuits. In addition, the suspended beams of the FEA models have corners which affect the effective length of the suspended beams.

B. Mechanical properties of suspended-beam structures

Here, three structures with different length and width of suspended beams (L) are investigated in terms of mechanical property. Fig. 8a-c show the temperature profile of microheaters with (L, x, y) = (440 μm, 10 μm, 10 μm), (890 μm, 20 μm, 20 μm), (1315 μm, 30 μm, 30 μm), respectively. Fig. 8d-f show the distributions of von-Mises stress of each structure. It should be noted that the residual stress of the membrane structure is not incorporated in this simulation, therefore only the effect of thermal stress is considered.

Because those structures are equivalent from the viewpoint of the aspect ratio of the suspended beam, the temperature profiles on the suspended membrane are comparable. The major difference between the three profiles is the thermal distribution on the suspended beam. In Fig. 8a, the peak temperature is located on the suspended beam. This is because of the higher electrical resistance and thermal resistance of the suspended beam. This local hotspot causes the relatively higher von-Mises stress in the suspended beam as shown in Fig.8d. Therefore, the suspended beam should

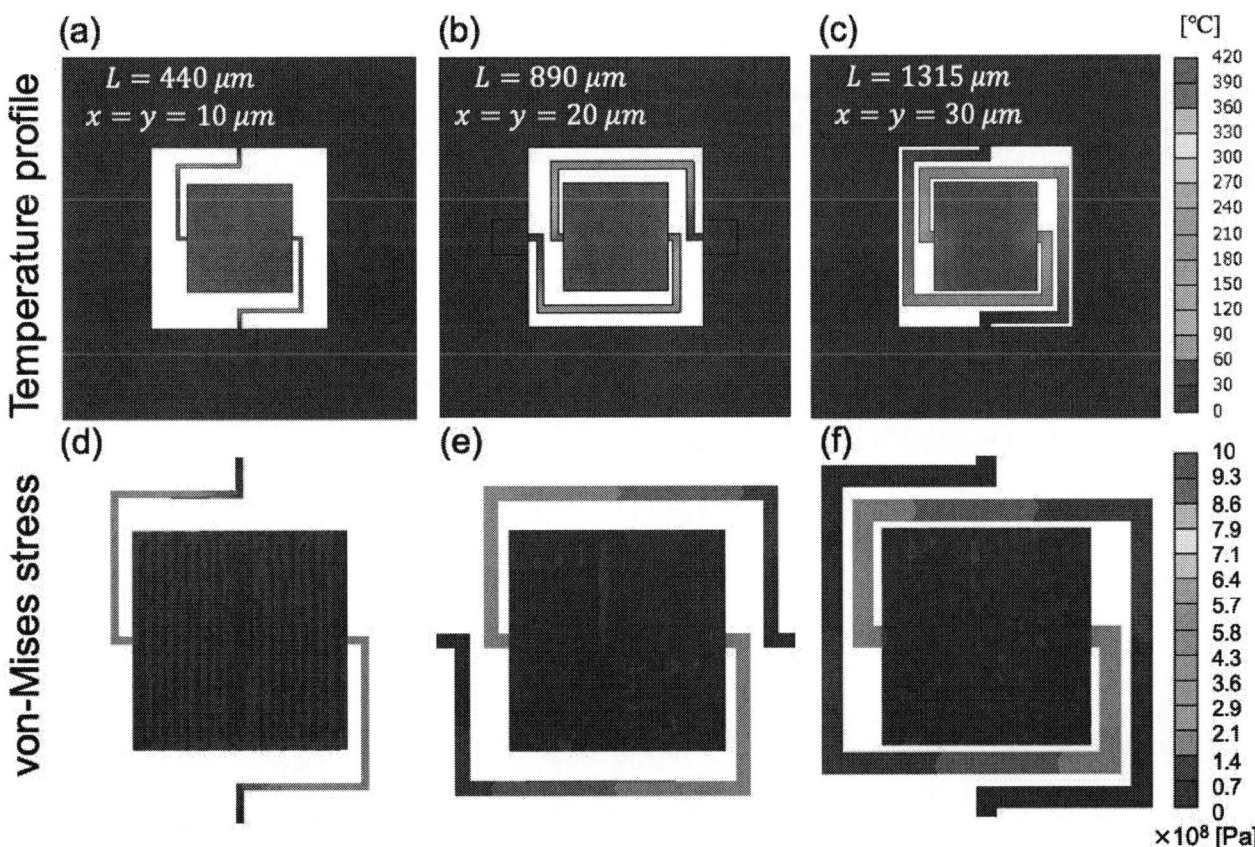

Figure 8. (a)-(c)Temperature profiles and (d)-(f)von-Mises stress distributions of three microheater structures with different dimensions of suspended beams.

have a certain width for preventing the hotspot and stress concentration. However, there is a concern also for the wider suspended beam. Fig. 8c shows a thermal gradient near the connection of the suspended membrane and beam, which is caused by the larger heat dissipation through the convection from the suspended beam to the ambient air. The suspended beam structure shown in Fig. 8c and 8f has the larger surface area of the suspended beam so that the temperature of the beam is lower than those of other microheater structures. As a result, heat dissipation through the suspended beam is larger than other structures. This issue concerning the thermal gradient can be solved by adjusting the circuit width to increase the joule heat near the connection, but comes at the expenses of larger power consumption. Eventually, the structure shown in Fig. 8b and e has a good balance of electro-thermal performance and mechanical property.

IV. CONCLUSION

We have established a novel scheme for finding microheater designs which realize lower power consumption and uniform temperature profile. In the first step of the mathematical analysis, an optimal aspect ratio of suspended beams is provided, and we are able to design the actual width and length freely, satisfying the aspect ratio. As a demonstration of our scheme, we designed a microheater which has 300 μm × 300 μm suspended membrane and two 20 μm × 890 μm suspended beams. In the next step, discrete design method is used for calculation of optimal heater circuit geometries. Finally, the electro-thermal performance of the output design of the scheme is validated by the FEA. This designed microheater operates at 388℃ with 7.88 mW, which is significantly lower than commercially available gas sensors. The power consumption can be further decreased by using the smaller suspended membrane, while a conservative size was used in this paper. In addition, by comparing the three structures, it has been inferred that the mechanical properties of microheaters having the same aspect ratio of suspended beams, are dependent on the length and the width of the suspended beam. In summary, the optimal suspended beam dimensions are obtained by balancing the electrical resistance and the thermal resistance.

ACKNOWLEDGMENT

The authors would like to thank the technical staff of Cybernet Systems Co., Ltd. for giving us technical supports regarding FEA model creation.

REFERENCES

[1] A. D. Wilson, "Advances in Electronic-Nose technologies for the detection of volatile biomarker metabolites in the human breath," Metabolites, vol. 5, Mar. 2015, pp. 140-163, doi:10.3390/metabo5010140.

[2] M. Baietto and A. D. Wilson, "Electronic-nose applications for fruit identification, ripeness and quality grading" Sensors, vol. 15, Jan. 2015, pp. 899-931, doi:10.3390/s150100899.

[3] L. Vigna, A. Fasoli, M. Cocuzza, F. C. Pirri, L. D. Bozano and M. Sangermano, "A flexible, highly sensitive, and selective chemiresistive gas sensor obtained by in situ photopolymerization of an acrylic resin in the presence of MWCNTs," Macromol. Mater. Eng., vol. 304, Oct. 2018, pp. 1800453, doi:10.1002/mame.201800453.

[4] R. Kumar, O. AI-Dossary, G. Kumar and A. Umar, "Zinc oxide nanostructures for NO_2 gas–sensor applications: A review," Nano-Micro Lett. vol. 7, Dec. 2014, pp. 97–120, doi:10.1007/s40820-014-0023-3.

[5] K. Toda, R. Furue and S. Hayami, "Recent progress in applications of graphene oxide for gas sensing: A review," Anal. Chim. Acta, vol. 878, Feb. 2015, pp. 43-53, doi:10.1016/j.aca.2015.02.002.

[6] J. Zhang , X. Liu, G. Neri and N. Pinna, "Nanostructured materials for room-temperature gas sensors", Adv. Mater., vol. 28, Dec. 2015, pp. 795-831, doi: 10.1002/adma.201503825.

[7] D. Spirjakin, A. M. Baranov, A. Somov and V. Sleptsov, "Investigation of heating profiles and optimization of power consumption of gas sensors for wireless sensor networks," Sens. Actuator A-Phys., vol. A247, Jun. 2016, pp. 247-253, doi:10.1016/j.sna.2016.05.049.

[8] Q. Zhou, A. Sussman, J. Changa, J. Donga, A. Zettla and W. Mickelson, "Fast response integrated MEMS microheaters for ultra low power gas detection," Sens. Actuator A-Phys., vol. A223, Jan. 2015, pp. 67-75, doi: 10.1016/j.sna.2014.12.005.

[9] J. G. Partridge, M. R. Field, A. Z. Sadek, K. Kalantar-zadeh, J. Du Plessis, M. B. Taylor, A. Atanacio, K. E. Prince and D. G. McCulloch, "Fabrication, structural characterization and testing of a nanostructured tin oxide gas sensor," IEEE Sens. J., vol. 9, May 2009, pp. 563-568, doi: 10.1109/JSEN.2009.2016613.

[10] N. Yamazoe and N. Miura, "Some basic aspects of semiconductor gas sensors," Chem. Sensor Technol., vol. 4, 1992, pp. 19–42.

[11] O. Sidek, M. Z. Ishak, M. A. Khalid, M. Z. Abu Bakar and M. A. Miskam, "Effect of heater geometry on the high temperature distribution on a MEMS micro-hotplate," 3rd Asia Symposium on Quality Electronic Design, Jul. 2011, pp. 100-104, doi:10.1109/ASQED.2011.6111709.

[12] W. J. Hwang, K. S. Shin, J. H. Roh, D. S. Lee and S. H. Choa, "Development of micro-heaters with optimized temperature compensation design for gas sensors," Sensors, vol. 11, Mar. 2011, pp. 2580-2591, doi:10.3390/s110302580.

[13] A. Lahlalia, L. Filipovic and S. Selberherr, "Modeling and simulation of novel semiconducting metal oxide gas sensors for wearable devices," IEEE Sens. J., vol. 18, Mar. 2018, pp. 1960-1970.

[14] S. Yu, S. Wang, M. Lu and L Zuo, "A novel polyimide based micro heater with high temperature uniformity," Sens. Actuator A-Phys., vol. 257, Apr. 2017, pp. 58-64, doi:10.1016/j.sna.2017.02.006.

[15] G. K. WhiteM and L. Minges, "Thermophysical properties of some key solids: An update", IJT, 1997.

[16] R. E. Cavicchi, J. S. Suehle, K. G. Kreider, M. Gaitan and P. Chaparala, "Fast temperature programmed sensing for micro-hotplate gas sensors," IEEE Electron Device Lett., vol. 16, Jun. 1995, pp. 286-288, doi: 10.1109/55.790737.

[17] R. G. Osuna, H. T. Nagle and S. S. Schiffman, "Transient response analysis of an electronic nose using multi-exponential models," Sens. Actuator B-Chem., vol. 61, Dec. 1999, pp. 170-182, doi: 10.1016/S0925-4005(99)00290-7.

[18] R. Gutierrez-Osuna, A. G. Galvez and N. Powar, "Transient response analysis for temperature-modulated chemoresistors," Sens. Actuator B-Chem., vol. 93, Aug. 2003, pp. 57-66.

A More than Moore Enabling Wafer Dicing Technology

Jeroen van Borkulo, Rogier Evertsen, Richard van der Stam

ASM Laser Separation International B.V.

Beuningen, The Netherlands

E-mail: jvanborkulo@alsi.asmpt.com

Abstract—As the materials that the wafer dicing process need to singulate become more complex, a diverging current Process of Record (PoR) dicing technologies are not able to meet the quality and/or cost requirements. Laser provides the solution to dice all these different materials but has the challenge to achieve the die strength level specified. In this paper, we will elaborate on the advances made to apply a laser full cut process while achieving the required die strength.

Index Terms—Laser, plasma, dicing, die strength, thin wafer, heterogeneous integration

I. INTRODUCTION

Wafer singulation in the semiconductor industry has transformed in recent years [1] from having to dice predominantly Si to a complex stack of various materials in which the Si substrate is a relatively small portion of the total stack thickness [2]. As Moore's law (1.0) is running out of steam, more focus is put on advanced packaging to keep the momentum going. Heterogeneous Integration, 2.5 and 3D packaging, are some of the technologies which have further accelerated this market trend in recent years.

For these types of wafers, a traditional blade dicing process is encountering serious yield issues. These issues can be addressed by applying hybrid dicing technologies such as DBG, SDBG or Plasma dicing. However, as the wafers are becoming thinner and have many other materials in the dicing street than just Si, these process flows are not providing the yield, cost, flexibility and productivity required.

For wafer thicknesses in the range of 100 µm up to 250 µm the laser ablation singulation process has become the process of record in many semiconductor applications areas. For the wafer thickness regime below 100 µm, it is considered to have limited capability due to a reduced die strength (related to the alternative separation technologies). More specifically, as the ratio between die size and wafer thickness is > 10:1, traditional laser ablation is not considered as a separation technology.

Advances in the laser sources used for dicing as well as applying post-processes created a full cut solution which meets all the different requirements for advanced (packaging) singulation.

As presented in ECTC 2018 [3] a study has been done to compare three different laser based dicing technologies for thin Si wafers. Each of the three processes had their own characteristics and benefits, however the multi beam laser full cut solution followed by a plasma etch process demonstrated the highest productivity, process window and die strength performance.

The previous study showed that in order to improve die strength, the laser-processed areas need to be separately treated in order to anneal or remove defects. Currently, high volume manufacturing solutions use post-treatment methods such as wet etching and laser irradiation. This latter method, such as that described in US9312178, is particularly attractive since it potentially increases productivity and reduces costs. However, for Si based wafers this wet etching process can not be applied and therefore alternative methods need to be developed.

In this paper, the parameter regime of the laser + plasma etch process is further investigated. The scope of the investigation is a comparison of different plasma sources, Radio Frequency (RF) vs. Micro Wave (MW) source. We will explore the influence of different etch gasses. In addition, we will investigate various parameter scans. The output parameters analyzed are die strength, productivity, undercut and overall quality.

Unfortunately, at the time of writing this paper not all experimental data was available. During the presentation at ECTC 2019 more data and results will be shared.

II. PROCESS FLOW

The process flow for the laser + plasma process is the following (as shown in Figure 1). A wafer is loaded into an integrated coat and cleaning system which performs a pre-clean process step. The pre-clean process step applies a demineralized water (DI) water jet over the wafer surface to remove dust and particles from the wafer surface. Once this step is completed and the wafer is dried through spin drying the second step is to apply a water-based polymer coating on the wafer surface.

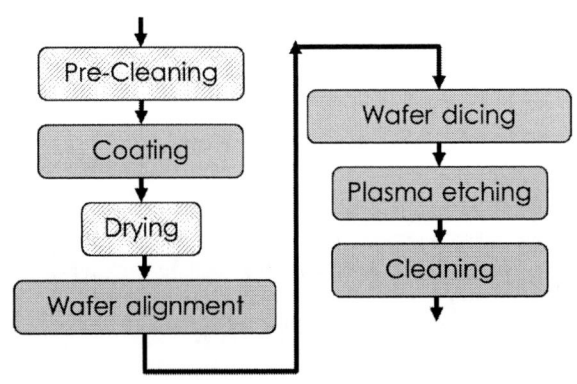

Figure 1. Laser + Plasma etch process flow

The typical polymer used in the coat material is an ASMPT proprietary developed coat material. However, additional trials have been done with other commercially available protective coat materials and these have also shown to work. More details will be provided in chapter V, Protective Coating Materials. Once the wafer is dried, it is loaded onto the dicing chuck and wafer alignment is executed. Once this is completed, the multi beam laser dicing is executed. After the laser dicing is completed, the wafer is (currently) transferred back to the cassette and loaded onto a separated plasma etch system. Within the plasma etch system the wafer is automatically loaded into the plasma etch chamber. After the plasma etch cleaning, the wafer is loaded into the clean station which removes the water-based coating material and remaining residue that may be present on the wafer. Subsequently, the wafer is transferred back to the cassette. This process flow results in a fully singulated wafer with the Heat Affected Zone (HAZ) removed. Below (Figure 2) is a Transmission Electron Microscope (TEM) image, as presented during ECTC 2018 [3], of a laser diced Si wafer prior to plasma etching. In the TEM image, the HAZ is clearly visible and a strong contributor to a reduced die strength of the singulated device. The plasma etch post process needs to remove the HAZ and thus recover the die strength. At the same time, it will remove the burr and recast from the side wall.

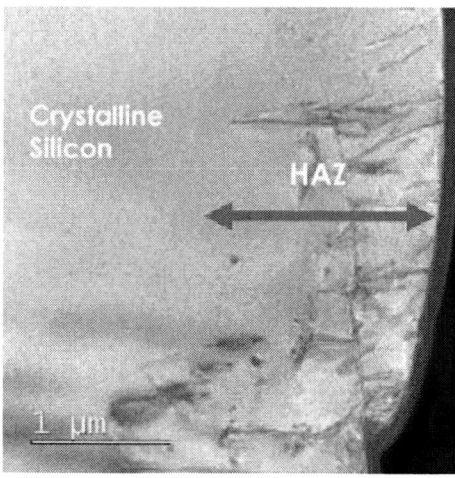

Figure 2. TEM image cross-section of the side wall after standard laser dicing demonstrating a 2um Heat Affected Zone (HAZ).

III. PLASMA SOURCE

Plasma is known as the forth state of matter after solid, liquid and gas. Plasma results from the ionization of gas. Using a plasma source to create radicals that will be able to etch Si is a well-known process in semiconductor industry.

An example of a chemical reaction is provided in the underneath Figure 3 [4].

$$CF_4 + O_2 \rightarrow 4F^* + CO + other \qquad (1)$$
$$Si_{(s)} + 4F^* \rightarrow SiF_{4(g)} \qquad (2)$$

Figure 3. Chemical reaction of plasma etching using CF_4 as the etch gas to etch Si.

The plasma etching process used is utilizing a high-density plasma (10^{14} to 10^{16}) in which a chemical reaction is used to etch the Si side wall. This in contrast to another typical etch process called Deep Reactive Ion Etching (DRIE) [5]. The DRIE process uses a relatively low plasma density flow. A bias between the source and the target (wafer) is applied to accelerate the ions towards the wafer surface at which they bombard the Si and therefore are capable to remove it. The benefit of this process flow is the capability to perform directional etching (anisotropic). The drawback of this process is the requirement of a hard mask (e.g. a photoresist) and the heat generated in the work piece (wafer) during the etching process.
Using ae chemical etch process with a high plasma density has the advantage that it is a "cold" process (40°C), and no bias is needed. It has the disadvantage that it is an isotropic etch (see Figure 4 below).

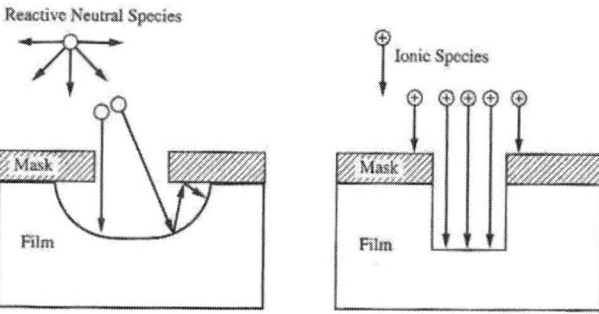

Figure 4. Comparison between isotropic etch (left) and an anisotropic etch on the right side (e.g. DRIE etching) [5].

However since the etch rate required is only 2-3um from the side wall of the die (thickness of the HAZ), there is limited impact on the etch shape.
When moving forward with this plasma etching technology there are basically two choices to make. To use a micro wave (MW) plasma source or a radio frequency (RF) plasma source. In this study we have investigated both.

When first comparing the hardware a MW source is approximately a factor 2 higher in cost compared to a RF source due to the complexity of the electronics.
From a process point of view the MW produces more heat which impacts the productivity and cooling requirements of the system.
The benefit of the MW is a higher Si etch removal rate vs. an RF source (2um/min vs. 1um/min respectively). The higher etch rate of an MW source is beneficial from a productivity

point of view. However, a negative side effect is the associated undercut as shown in picture 5. The study has found that the MW source produces a stronger isotropic etch behavior resulting in a stronger undercut of the top section of the die side wall. With the RF source it is approximately 50% less (6.8um vs 4.5um). The undercut is seen as a significant quality issue and needs to be kept as narrow as possible to prevent delamination and chipping of the active top structures.

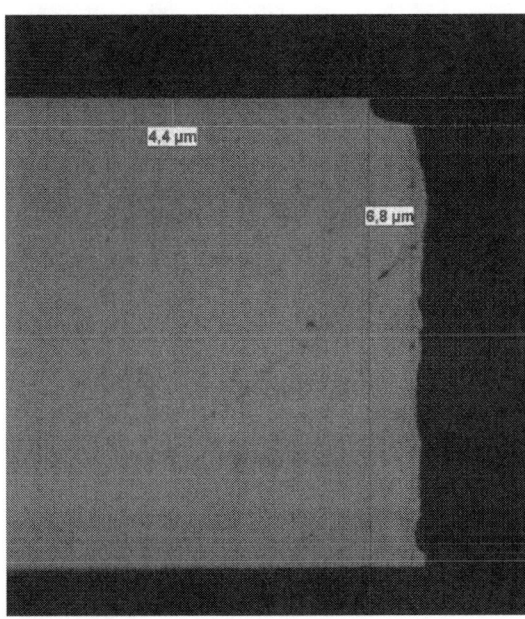

Figure 5. Side wall undercut of 6.8um when using an MW plasma source to etch die side wall

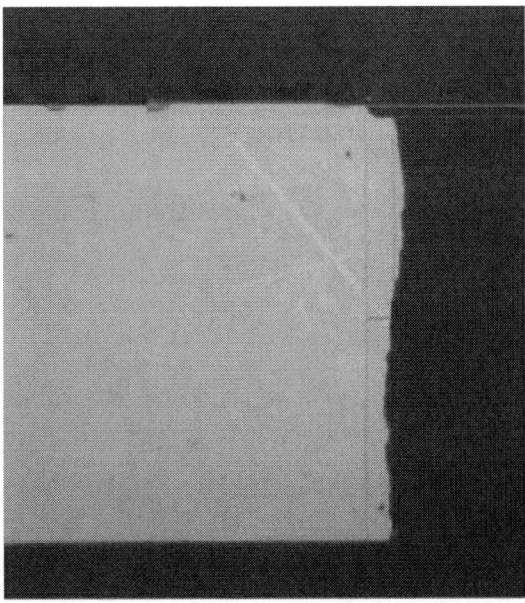

Figure 6. Side wall undercut of 4.5um when using an RF plasma source to etch die side wall

The above pictures are crosscuts of blank Si wafer sample. However, when assuming there are active layers present on top of the wafers (SiO, SiN, Low-K, Metal), it is not desired to have these "free standing" which leads to a potential chipping. ThusXXX

In table 1 the comparison result between the two plasma source types are evaluated and based on these results the conclusion is made to continue the developments with the RF source.

Source	CapEx	Etch rate	Undercut	Complexity
MW	$$	++	--	-
RF	$	+	-	+

Table 1. Comparison between a MW and a RF plasma source.

IV. ETCH GASES

There are several options to use for the plasma etching process as an etch gas.

Most common gases used are CF_4 [6] or SF_6, besides the alternatives NF_3 (toxic) and CHF_3. The main similarity between the gases is that plasma source breaks down the gas and creates the fluorine radicals which etchthe Si.

However, during side wall etching not only Si needs to be etched but also other materials such as SiO or SiN. When there is a strong selectivity it does not provide a homogenous etch process.

At the time of writing of this article we have only tested CF_4.

Etch rate CF4	
Si	0.5um/min
SiO	0.1um/min
SiN	1-1.5um/min

Table 2. Etch rates of CF_4 for various materials.

During the presentation at ECTC 2019, we will share more results for other etch gases. However, based on the results for CF_4 we can already conclude that when etching IC wafers which have a stack on top of the wafer of consisting of SiO_x, Si_yN_z among other materials, this results in an etch rate-dependent pattern resulting from the type of etch gas. It seems like this etch pattern induced structure results in a low die strength even though the HAZ in the Si is removed. Part of this study is to find etch gases which provide a similar etch rate for the various materials.

V. PROTECTIVE COATING MATERIALS

As described in the previous chapter the plasma etching process not only etches the HAZ in the Si side wall but also etches into the active top structures and or passivation. As this is not allowed the top surface and active structures of the devices need to be protected. Regardless of the post process of plasma etching, the standard Process of Record (PoR) for laser ablation dicing is using a water based protective coating. The initial purpose of this coating is to prevent (molten) particles which are ejected from the laser processing to adhere to the wafer surface. With the coat protection these particles can easily be removed after laser dicing with a water cleaning process. The typical thickness of this coating is 1-2um thick. This same coat material also has the characteristic that it will protect the wafer surface from the plasma etch process. As the etch rate in coat material is low compared to Si (>1:10) a 1-2 um coat thickness layer is enough to protect the top surface of the devices when 2-3um of Si is etched away from the side wall.

This is significant difference from a DRIE process. For a hard-baked Photo Resist (PR) the etch rate of PR vs Si is 1:10 [7]. When plasma dicing through a 100um thick Si wafer a 2um PR layer is not enough. For the plasma dicing application a photo resist is required to protect the surface from the ion bombardment.

During the complete process investigated in the paper the wafer is not cleaned after the laser dicing step is completed. The wafer can directly be loaded into the vacuum chamber and the plasma etching step can be applied. After the plasma etching step is completed the wafer is cleaned using a high-pressure water jet which is a standard way of cleaning wafers when using a laser grooving or dicing process flow.

In this study we have investigate several coating materials. One of them (BMK) is developed by ASMPT and made commercially available to its customers. Other commercially coating materials are also going to be investigated to determine compatibility to both the laser dicing process as well as the plasma etching process. For the laser dicing process, it is critical that the coating adheres properly to the wafer surface and does not delaminate from the surface and provide sufficient protection from the particles which are redeposited on the wafer surface. More results will be made available during the presentation at ECTC 2019.

VI. PARAMETER SCAN

In order to optimize the plasma etching process performance (die strength, undercut, side wall smoothness) as well as optimize the productivity several parameters can be varied to achieve this.

Key parameters are temperature, gas flow, pressure, time and gas mix as already discussed in chapter IV.

Temperature is a critical parameter as a higher temperature will help to increase the etch rate and therefore efficiency and productivity. However, when the temperature becomes too high it may create heat damage to the die and or dicing tape or die attach foil (DAF) underneath the die. Therefore, finding the optimum window for the operating temperature is important. Heat and cooling of the chuck on which the wafer is placed is required as well as the heating of the vacuum chamber wall.

In the underneath graph 1 the temperature of the chuck is varied vs. the die strength measured on the sample.

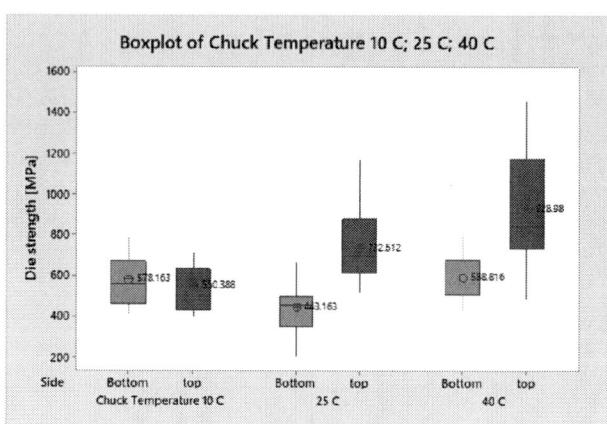

Graph 1. Die strength vs. chuck temperature.

This data demonstrates that a higher chuck temperature (40°C) provides an improvement in the die strength compared to the room temperature condition of 25°C.

In another experiment we have varied the vacuum chamber wall temperature, see graph 2.

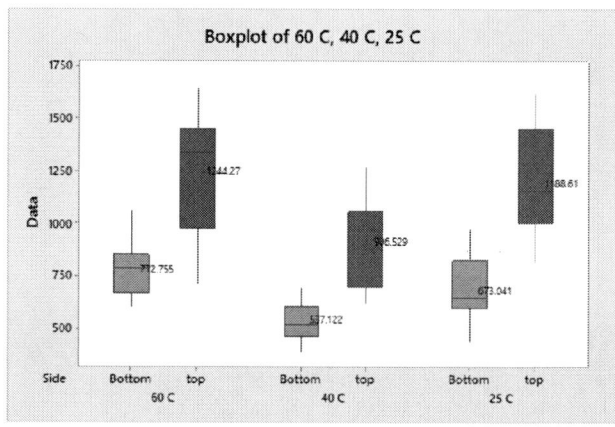

Graph 2. Die strength vs. vacuum chamber temperature.

This data shows no significant influence of the temperature on the die strength measured.

Further analysis of this data is required, and other parameter scans need to be executed. One of them is the influence of the gas flow. A higher flow will create more radicals. The increase in radicals may results in too strong etching on the top side vs. the back side of the device which may result in a non-uniform etch rate between top and bottom side wall and therefore a not optimum die strength recovery. In addition, it can lead to an undercut. More results and analysis will be presented during the ECTC 2019.

VII. CONCLUSIONS

Advanced packaging trends such as heterogeneous integration are increasing the complexity of the wafer stack while at the same time the total thickness and specifically the SI thickness needs to come down. This is trend requires wafer dicing technology which can enable this type of packaging. The main benefit of using laser ablation is the fact that it has the capability to dice through many different types of materials (Si, passivation, metals, polymers, etc). Yet equally important is the requirement of having a high die strength and is this area a typical laser ablation process is lacking (300MPa). Combining a laser singulation process with a plasma etching process addresses the issue of the die strength. In this study the plasma etch process conditions as well as hardware configuration has been investigated. Preliminary data shows that a combination of the plasma source and process parameters can lead to a die strength well over 1000MPa. Further investigation of the process parameter and how they impact both quality (undercut), die strength and productivity will be executed in the coming months and presented at the ECTC in 2019 in Las Vegas, USA.

REFERENCES

[1] W.-S. Lei, A. Kumar, and R. Yalamanchili, "Die singulation technologies for advanced packaging: A critical review," *Journal of Vacuum Science and Technology B*, vol. 30, no. 4, p. 040801, 2012.

[2] "Thin wafer processing and dicing equipment market 2016," Yole Development, Tech. Rep., 2016.

[3] J. van Borkulo and Richard van der Stam "Laser Based Full Cut Dicing Evaluations for Thin Si wafers" in *Electronic Components and Technology Conference, 2018 ECTC 2018. 68th*, May 2018

[4] C. J. Mogab, A. C. Adams, and D. L. Flamm "Plasma etching of Si and SiO2 The effect of oxygen additions to CF4 plasmas" *Journal of Applied Physics* 49, 3796 (1978)

[5] Vincent M. Donnelly and Avinoam Kornblit, "Plasma etching: Yesterday, today, and tomorrow", *J. Vac. Sci. Technol.* A 31(5), Sep/Oct 2013

[6] Young H. Lee and Mao-Min Chen, "Silicon etching mechanism and anisotropy in CF4+O2 plasma", *Journal of Applied Physics* 54, 5966 (1983)

[7] D A Porter and T A Berfield, "Die separation and rupture strength for deep reactive ion etched silicon wafers" *J. Micromech. Microeng.* 23 (2013) 085020 (8pp)

Plasma Dicing Integration Schemes for Scribe Lane Layout and the Impact on Die Strength

D. Parker, E. Gourvest and B. Bouillard

Back-End Manufacturing & Technology R&D
STMicroelectronics, 12 rue Jules Horowitz
38019, Grenoble, France
e-mail: david.parker@st.com

Abstract—For all applications where a semiconductor device is used, there is a demand for improved performance under all environmental conditions. One such application, which is in regular use, is the smartcard chip that is found predominantly in bank or travel cards and passports. These may endure harsh physical treatment in normal usage and the demands for severe robustness criteria for these card-based applications are continually increasing to ensure that the packaged chips are able to continue functioning without breaking. Key to achieving these exacting standards is the singulation of the chips in a way that does not generate any inherent mechanical weakness. The standard blade sawing process has so far met existing requirements but the technology roadmap for such devices has presented new challenges with the deployment of fragile ultra-low k BEOL dielectrics. This has prompted the introduction of laser grooving before dicing to minimize damage to these materials caused by the sawing blade but this, in combination with mechanical dicing, can itself result in weakening of the die and a higher risk of failure. Plasma dicing has the potential to improve die breakage strength significantly, essentially as there is no mechanical element to the dicing action. However, this technique requires that the silicon be exposed in the dicing street, which must be free of metals and dielectrics before etching. Under some circumstances this can be achieved by mask design and dedicated processing but in practice laser grooving is again a prime candidate for meeting these conditions. To be compatible with the plasma etching process the laser recipe needs to be optimized according to both the presence and layout of test structures in the dicing street. An influential factor is that the options for the street layout are themselves dependent upon the front-end technology and its complexity level. This paper will examine how laser grooving process adjustments devised to promote successful plasma dicing integration can affect the die strength whilst adapting to different structures and materials contained in the dicing street. We will compare results between alternative plasma dicing integration schemes to show how design layout for the scribe lane influences their choice, especially where die strength is a priority. We will also investigate the impact of varying some of the etch process output parameters to seek an understanding of whether the etching step itself has a role to play in die robustness improvement for the practical, future application of plasma singulation to real cases on an industrial scale.

Keywords-plasma dicing, laser grooving, die strength, integration

I. INTRODUCTION

As the need for higher levels of security in ATM banking transactions and personal identification have evolved so has the technology used to fabricate credit and debit cards, telephone SIM cards, passports, drivers licenses and identity cards. Many such semiconductor applications involve the use of assembled modules that are designed to be thin and reasonably flexible yet must withstand a significant amount of handling on a daily basis. Regular physical manipulation of these modules is a fundamental characteristic of how the products are intended to be used and stored, with cards inserted into readers, kept in back pockets and generally subjected to harsh treatment over significant lifetimes. Another example is the rising use of RFID wristbands that are now becoming established for identification at schools, hospitals, sports events and other public gatherings.

To ensure the robustness of the packages used for these applications a stringent evaluation method has been developed and is now accepted as an industry standard known as Card Quality Management (CQM). Assembled modules are put through a series of mechanical trials that simulate the effects of regular handling by exerting twisting, bending and compressive forces and are then retested to ensure that they remain fully functional. In order to meet the criteria imposed by these tests it follows that the silicon devices at the heart of these modules must also survive the stresses they experience during the procedure.

Silicon is inherently a brittle material and a number of causes of die weakness have been identified and studied. One of the most significant mechanisms is the propagation of cracks on the edge of the die and it has been noted that the most severe defect rather than the number of defects has a supreme effect [1]. One of the most important contributory factors to the severity of edge defects is the dicing method used to separate the finished wafer into individual devices. Historically, in the semiconductor industry this operation has been dominated by sawing, using a rotating, diamond-tipped blade to remove silicon from the dicing street of a tape-mounted wafer. However, the mechanical action of the sawing blade is a major source of crack defects and edge chipping that result in die weakness. Although there are some techniques for reducing this effect, including careful blade selection and slowing the feed rate to the blade [2, 3], a compromise is usually necessary to meet industrial objectives. Alternative methods such as laser full-ablation and stealth dicing have been introduced and studied for their impact on die strength but also have some drawbacks with regard to this property related to crystal dislocations caused by localized heating effects [4]. A more recent development in wafer dicing is the application of the well-characterized, 3-phase Bosch DRIE process shown in Fig. 1 for the vertical etching of deep silicon holes and trenches. Known as Plasma Dicing, this technique is promising for a number of applications and for a variety of reasons. One of the principal incentives for using it is the inherent improvement in die sidewall quality that derives from employing an etching process to remove the silicon from the scribe lanes rather than a mechanical one. Singulation by plasma etching does not produce cracks or other crystalline dislocations on the die edge that can act as an origin for die breakage under stress. Instead, characteristic horizontal scallops are formed in the sidewall during the etch reaction as shown in Fig. 2. This 'soft' dicing action is able to produce a sidewall profile with a much lower defect density so the die strength is expected to be significantly enhanced.

However, there are two key requirements for plasma dicing to be effective. One is that the silicon surface in the scribe lane be exposed to the plasma reactants and the other is that the active devices on the wafer be suitably protected from them. Thus, the ease of implementation of this approach is strongly dependent on how it is integrated into existing fabrication sequences. From a purely technical point of view the choice of integration scheme is influenced by a number of factors, notably pertaining to the construction and content of the dicing streets, i.e. width, design, materials present and layer thickness, but also the die size and surface topography. These in turn depend on the nature of the front-end technology and the type of product, so in practice plasma dicing may be integrated in several ways ranging from the simple, where the wafer to be singulated requires no specific preparation, to the more complex involving additional operations such as lithography patterning or laser grooving [5].

The objective of this work is to examine how the presence of random metallic and/or dielectric structures in the dicing street affects potential for realizing high die strength values through the constraints they impose on plasma dicing integration schemes, mostly based on laser grooving. In this case, laser power is adjusted according to the layers and materials comprising the dicing street and the effects studied. The initial step of the plasma dicing process, intended to clear away any non-silicon debris in the groove before the main etch, is also varied to understand whether this too can influence the measured die strength. The most promising plasma dicing integration schemes for obtaining robust die performance are thus proposed.

II. PLASMA DICING INTEGRATION: DISCUSSION

The Bosch deep reactive-ion etch (DRIE) process [6] has been firmly established in front-end semiconductor device fabrication processes for many years and is well characterized in this domain [7, 8], for example where deep trenches are used to create thin, vertical features for MEMS applications [9] or where deep holes in silicon are needed to make through-silicon vias (TSVs), widely used in CMOS image sensors. Further innovation and significant progress in etch chamber hardware development to increase etching rates has since led to the deployment of this process in wafer dicing applications [10, 11, 12].

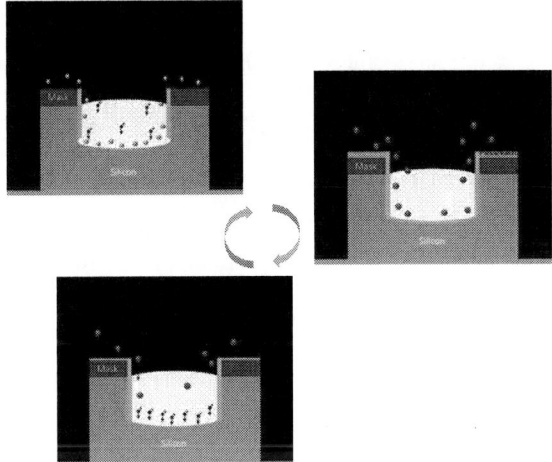

Fig. 1. Bosch DRIE plasma silicon etch process description

Fig. 2. Scallop formation on the die sidewall after plasma dicing

There is a general trend amongst equipment suppliers to support the classic 'dice-after-grind' method that is typical of many back-end assembly plants by facilitating the etching process on thinned wafers mounted on dicing tape and frames. The intention is thus to simply substitute the plasma dicing process for the existing wafer sawing step without significantly altering the overall sequence of operations in the assembly plant. Despite this, some additional steps are likely to be necessary. As the width of the dicing streets has to be compatible with the singulation process they have traditionally been used advantageously to allow a large variety of test, metrology and alignment structures to be included in the mask layout with minimal impact on the die per wafer count of active devices. Many of these elements contain metal and dielectric features, including stacked probing pads with multiple aluminum and copper layers, interconnections and large-area plates.

As stated in the introduction, plasma dicing requires that the silicon in the scribe lane be fully exposed to the plasma reactants at the start of the etching step. If dielectrics are present the process as optimized for silicon etching will attack SiO_2 with an etch rate of about 0.3µm/min, which implies long processing times for removing typical BEoL process layers. More importantly, regardless of conditions, it is in practice incapable of etching metallic layers, yet the complex structures present cannot reasonably be etched prior to dicing even with individually optimized steps. In addition to this, the active devices must be protected in the plasma step using a suitable masking layer with sufficient resistance to erosion during the etching sequence.

These circumstances force us to examine carefully the options for plasma dicing integration. Laser grooving has been adopted as a preparatory step before blade sawing to ensure that fragile ultra-low k dielectric layers are not damaged by contact with the rotating surface of the blade. With contemporary laser tools the depth of the groove is controlled by adjusting the power of the applied beam and the number of laser passes, so it is possible to ablate the BEoL layers, including metals, completely, so as to expose the silicon in the dicing street. The advantage here is that existing products can be singulated using the plasma etching process without modification. In addition, the organic coating that is customarily deposited on the active surface of the wafer to catch the debris ejected from the groove can also function adequately as a masking layer. However, the inclusion of a laser grooving step to ablate the materials in the dicing street prior to plasma dicing has ramifications for die robustness. It has previously been shown that die strength measurements obtained after laser grooving and blade sawing are strongly impacted by the laser fluence applied during this operation, as this in turn determines the structure and width of the heat-affected zone [13]. Fluence is defined as shown in (1), where fluence is in J/cm², pulse energy in J and beam spot area in cm².

$$\text{Fluence} = \text{Pulse energy} / \text{Beam spot area} \qquad (1)$$

The relationship of the laser pulse energy to the laser power is shown in (2), where peak power is in W, laser pulse energy in J and pulse duration in s.

$$\text{Peak Power} = \text{Laser pulse energy} / \text{Pulse duration} \qquad (2)$$

Through laser heating effects, increasing the fluence generates a higher density of defects in the silicon at the edge of the die and weakens the structure, resulting in lower die strength values when the active surface is placed under tension. This phenomenon is independent of the subsequent dicing method but it may be expected that the soft action of plasma dicing would contribute far less to crack propagation into the silicon than the mechanical vibrations experienced during blade sawing.

The minimum laser fluence (threshold) required to obtain total ablation of the BEoL layers depends on the construction of the BEoL stack and the thickness of its components. In practice, for blade sawing, the power and frequency of the laser are adjusted in the recipe so that the resulting groove penetrates a certain depth into the silicon. For plasma dicing this would meet the requirement for the silicon surface to be exposed for etching but the process is less tolerant of metal or dielectric residues in the laser groove than classical sawing. Care is thus needed to ensure that sufficient energy is applied. Additionally, the fast-changing landscape of the dicing street as the pulsed laser beam traces its path through the BEoL layers is a source of discontinuity at the edge of the die after the etching process. As the energy is constant for a given laser pass, alternating metallic and dielectric-only features result in a variations in both the shape of the side of the groove and in the depth profile in the travelling direction of the beam. Local variations in heat conduction through the metallic structures could also alter the heat-affected zone in a random manner providing another defect-generating mechanism whose impact is difficult to assess or control. An alternative approach to this integration scheme is more radical but also closely aligned to the objectives for which plasma dicing is potentially an attractive solution. This involves redesigning the mask layout for the product so that scribe lanes are entirely empty of ancillary structures. In this way the complexity of the construction of the dicing street is significantly reduced, which simplifies the challenges for the laser grooving operation. In this configuration the materials encountered by the laser beam are uniformly distributed and the groove obtained under constant fluence conditions of an even depth throughout the scribe lane. Variation in heating effects should be minimized and lead to fewer discontinuities.

III. EXPERIMENTAL DETAILS

A. Sample Construction

To study the influence of the plasma dicing integration scheme with laser grooving, a range of 200mm wafer samples was generated with different levels of complexity in the BEoL construction. Dicing streets containing different populations of metal features were represented by production-like wafers fabricated using a CMOS E²PROM memory technology with four aluminum metal levels and low k dielectric layers. To simulate metal-free scribe lanes, partially-processed blank silicon wafers were prepared with deposited dielectric layers representative of the BEoL architecture of a related front-end technology. The blank wafers were divided into two groups composed of either the entire BEoL dielectric stack or processed up to but excluding the deposition of the final device passivation layers. The latter was designed to represent an integration scheme with a thinner dielectric layer in the scribe lane obtained by adjusting the final steps of the front-end flow, allowing further reduction in the laser energy needed to form the groove opening. To obtain data for an integration scheme without any laser steps blank wafers with a simple deposition of 0.5µm of TEOS-based SiO_2 were patterned using an i-line photolithography process then subjected to a typical CF_4 based plasma oxide etch process to expose the silicon in simulated dicing streets. The details of the construction of the various samples are shown in Fig. 3.

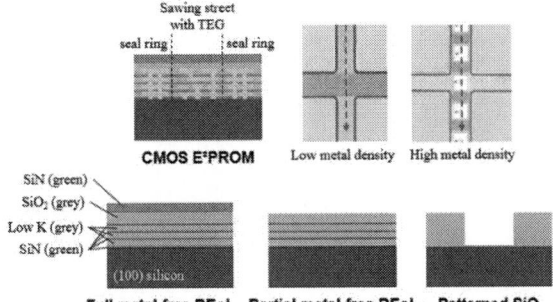

Fig. 3. Sample construction before laser grooving step or etch patterning

All wafers were thinned to a final thickness of $140\pm3\mu m$ using an identical two-step back-grinding process on a standard wafer grinding tool. This was completed with a final dry polishing step but without any additional stress-relief treatment. The wafers were then mounted on 400mm stainless steel dicing frames using a commercially available $80\mu m$ thick polyolefin dicing tape. The samples representing the plasma dicing integration scheme including laser grooving were then spin-coated with a suitable protective material, especially developed for plasma processing after laser treatment. After air drying the final thickness of this film was approximately $9\mu m$, this being estimated from previous spin curve measurements.

TABLE I. SCRIBE LANE METAL FEATURE POPULATION DENSITY FOR DIFFERENT SAMPLES

Metal Population Density (N°/100μm)	Metal 1	Metal 2	Metal 3	Metal 4	All Metals
Low	1.49	0.97	1.41	0.03	1.95
Medium	5.14	4.65	2.14	2.04	13.97
High	19.67	19.67	10.66	0.00	50.00

B. Laser Grooving

Laser grooving was performed using a Disco DFL7161 tool with a frequency-tripled Nd:YAG UV laser source producing a pulsed beam in the nanosecond range at a wavelength of 355nm. The laser trace directly defines the final shape and size of each sample so the grooving pattern on each finished device wafer was programmed to follow a chosen dicing street in order to obtain samples with different populations of metallic structures at the edge of die to be tested. Population density levels were estimated using the mask layout data for the relevant scribe lane and are summarized in Table I. In all cases the sample dimensions were adjusted to be compatible with the 3-point bending procedure used to measure the die strength.

As the laser fluence required to fully ablate the material in any given dicing street depends on the material composition of the structures in the street, laser recipes were developed to optimize the fluence for each sample type. The width of the groove was influenced by the width of the opening of the mask available on the laser tool, this being normally used in an industrial context. In this case the mask opening width was $40\mu m$. Feed speeds were chosen so as to avoid overlapping laser pulses. For each recipe used, the total fluence was calculated using an application specifically developed by the laser supplier for the equipment used. This

is based on the standard calculations but with adjustments for the real shapes of the different Gaussian beams involved. For the scribes containing metal structures a known POR was used, having been qualified industrially for laser grooving with blade dicing and subsequently demonstrated successfully with the plasma process. The total fluence needed to ablate the materials in this case was 118.7 J/cm².

For the blank wafers representing a simpler scribe lane composition without metal features laser grooving recipes were developed so that the total fluence was minimized in each case. To determine suitable conditions for the samples with the complete BEoL dielectric stack with a measured thickness of about $5\mu m$ a Design of Experiments (DoE) matrix was defined and the resulting laser groove profiles analyzed. Fig. 4 shows the profile selected on the basis of full ablation of the BEoL layer with a shallow penetration into the silicon to expose it for etching. All samples were then processed with the optimized recipe with a total fluence of 91.0 J/cm². The visual appearance of the final groove is shown in Fig. 6 after removal of the protective coating with deionized water to produce a clearer image.

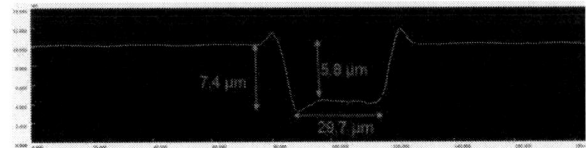

Fig. 4. Laser groove profile selected for full BEoL dielectric stack

The objective for the third category of samples was to investigate the die strength performance using a scheme that would permit further reduction of the laser fluence whilst utilizing the plasma dicing chamber to remove a thin layer of silicon dioxide left at the base of the groove. Using a thinner BEoL stack to minimize the energy required to ablate the dielectric materials, the fluence of the laser step was reduced further so as to avoid penetration of the groove into the silicon. In this case a simpler low power laser recipe was derived from a series of split tests by assessment of the resulting profiles. A laser recipe with a total fluence of 47.2 J/cm² was thus selected. Fig. 5 shows the measured profile of the groove obtained, which penetrates $2.52\mu m$ at its deepest point into the sample layer of original thickness estimated at $3.05\mu m$ from SEM measurements. However, given the 'W' shape of the groove the remaining dielectric layer is about $1.5\mu m$ thicker in the center than at the edge. Fig. 6 compares the final appearance of both grooves after coating removal, indicating the presence of a layer of dielectric at the base of the partial groove.

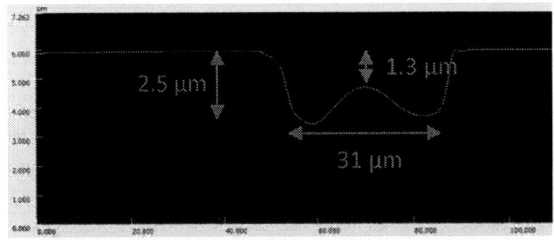

Fig. 5. Profile of groove for reduced thickness BEoL sample

Fig. 6. Image of final partial groove for reduced BEoL sample (left) and groove into silicon for full thickness BEoL sample (right)

C. Plasma Dicing

After completion of laser grooving all wafers were submitted to the plasma singulation process on an Mosaic fxP plasma dicing cluster tool with a Rapier 300S etch chamber (provided by Orbotech's SPTS Technologies) configured for 400mm frame-mounted wafer handling. Etching recipes were adjusted according to the sample construction and a few variations were made to some main parameters to understand whether there was an influence on die strength. The basic structure of the process recipe consists of two main etching steps. The first of these, typically called a descum step, is designed to remove any remaining debris from the bottom of the laser groove using near vertical ionic bombardment through the application of a strong electric field between the plasma and the wafer. The second step contains the cyclic process sequence that defines the Bosch method. The effective end of this step is determined using an appropriate end-point signal by optical emission spectroscopy, then an over-etch equivalent to a fixed percentage of the duration of the main etch is applied to ensure complete singulation.

As the first etching step interacts directly with the silicon at the top edge of the die where defects have the most influence on the die strength some variations were introduced to understand whether the etching process could moderate the effects of the laser. To singulate the device wafers with silicon exposed after laser grooving, the basic recipe was used with a descum step of 30 seconds based on previous tests. For the blank wafers, selected changes were made to the basic recipe structure according to the sample type. For the samples with the full-thickness BEoL stack, again with exposed silicon, recipes were run with descum steps of 0, 30 and 60 seconds respectively. For the samples with the reduced BEoL layers where a thin film of dielectric was left after laser grooving, recipes with different initial steps were applied. In addition to a 60 second standard descum process to remove this layer samples were run with dedicated SiO_2 etch conditions to understand whether exposing the silicon for plasma dicing in this way could help to improve the die strength. Descriptions of the recipes used for each sample are summarized in Table II.

TABLE II. INITIAL PLASMA ETCH PROCESS STEP BY SAMPLE TYPE

Sample Type	Dicing Street Construction		Material to Etch	Initial Etch Step	
	Metals	BEoL Stack		Oxide RIE	Descum
Device	Yes	Full	Silicon	-	30 secs
Blank	No	Full	Silicon	-	-
				-	30 secs
				-	60 secs
Blank	No	Partial	SiO₂	-	60 secs
				-	140 s 2x platen power
				280 secs	-
				280 secs	30 secs
Blank¹	No	0.5µm SiO₂	Silicon	-	30 secs

¹ Etched Groove

After plasma processing and inspection the protective film used as the etch mask was stripped from all laser grooved wafers in a dedicated, automatic module using deionized water at room temperature after which the wafers were spin-dried.

D. Die Break Strength (DBS) Measurement

DBS measurements were performed on samples from each of the processed wafers following the SEMI standard 3-point bending method [14]. The method is already well documented, consisting of a gradually increasing loading force applied through a loading blade to the center of a sample resting on two supporting blades with a span, l. For this evaluation the equipment used was a Mecmesin MultiTest 1-i with a 50N load cell and an adjustable jig to set the span to 6.1mm. The loading speed was set to 15mm/min. These test conditions were chosen having been validated in previous die strength studies [15]. In this configuration the applied force measured at the point of breakage was converted into flexural stress, σ, using the standard equation (3), which represents the die strength expressed in MPa

$$\sigma = 3Fl \, / \, 2wt^2 \qquad (3)$$

where F is the measured loading force at breakage, l is the span and w and t are the width and thickness of the sample respectively. The set-up is shown in Fig. 7. To obtain statistical data on the die strength 15 samples were extracted from five different locations on the wafer (Top, Centre, Bottom, Left and Right) for each dicing process and for both top (active) and bottom side measurements. The load force was applied from above the sample so the surface to be put under tension was placed face down on the two supports.

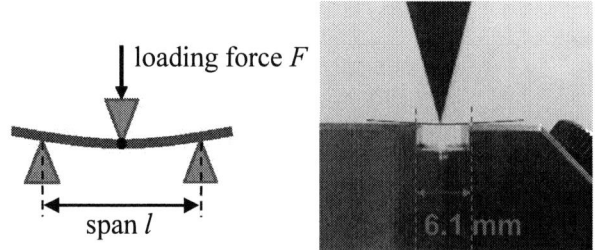

Fig. 7. 3-point bending measurement for die strength evaluation

IV. RESULTS AND DISCUSSION

Unless otherwise stated, the die strength results in this section refer to measurements at the front-side, or active surface. Fig. 8 indicates the die strength results for three levels of population density of metallic structures in the scribe lane where the layers have been ablated using the same laser recipe with a total estimated fluence of 118.7 J/cm². The same initial descum step with a duration of 30 seconds was used for all samples during plasma singulation. The sample with a high density of small, frequently occurring metal features produced a die strength of 224.5 MPa with a standard deviation of 32.2 MPa, significantly lower than that given by a medium density of longer structures (398.6 MPa, S.D. 98.7 MPa) or a low density of simpler structures (498.6 MPa, S.D. 86.3 MPa). This is consistent with the number of discontinuities created at the bottom of the groove caused by different rates of ablation between metals and dielectrics at the same fluence. With metals present these irregularities are replicated by the directional silicon etching process and can be observed as vertical striations in the die sidewall. These are shown in Fig. 9. Where there are many small features, such as the tiling used to facilitate Chemical Mechanical Polishing (CMP) in inter-metal dielectric planarization processing, an extremely high density of these defects is generated. Longer structures, or ones comprising just one or two metal layers, produce fewer of these sites of weakness.

To examine the effect of reducing the fluence when metal features are present, the lowest population density configuration was selected and processed with the laser grooving recipe developed for the tests with the full BEoL dielectric stack on blank wafers. This applied a total fluence of 91.0 J/cm², denoted medium fluence for reference. The comparison with the higher fluence recipe is shown in Fig. 10. No difference is noted between the die strength obtained from the two test conditions. As seen from the previous samples any point where metal features transition into purely dielectric layers creates a discontinuity when plasma etching is applied. The presence of even a small number of metal structures in the dicing street appears to effectively limit any improvement in die strength that might be obtained through a decrease in the laser-induced defect density in the silicon with diminished fluence.

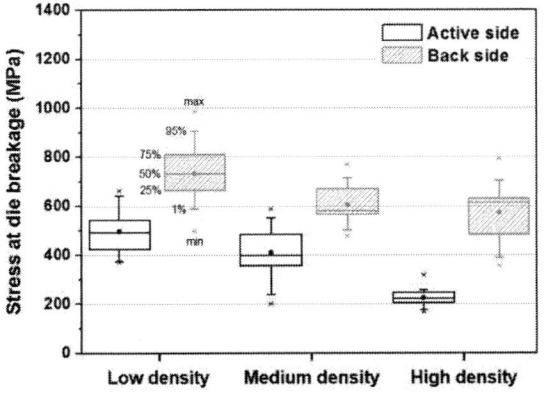

Fig. 8. Die strength comparison for different population density of metallic structures in dicing street

Fig. 9. Vertical striations caused by inefficient ablation of periodic metal features

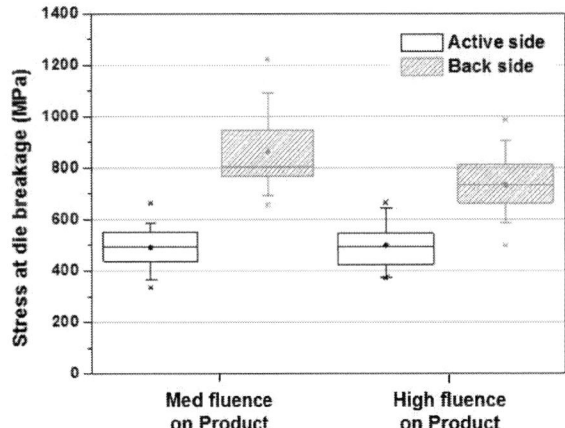

Fig. 10. Die strength comparison for different total fluence with a low density of metallic structures in dicing street

The die strength results for the laser grooved samples with metal-free dicing streets are shown in Fig. 11. Here the silicon was fully exposed but the fluence was adjusted to minimize penetration into the silicon. The data show die strength values in the range of 250 to 500 MPa. The effects of the pulsed laser grooving can be seen as discontinuity in the scallop formation with some vertical striations and a distinct indentation pattern at the top edge of the silicon. The frequency of this pattern, shown for a 30 second descum in Fig. 12, has been correlated precisely with the laser pulse frequency. A comparison of the die sidewall by SEM after etching is shown in Fig. 13. Here there is a general trend as die strength values increase with longer descum duration in the plasma dicing recipe. The sample with 60 seconds of descum has a smoother appearance to a depth of about 10μm into the silicon with fewer indentations than those with shorter descum times, suggesting that a longer treatment with high-energy ionic bombardment acts to remove some of the defects induced by the laser at the top edge prior to the main dicing step.

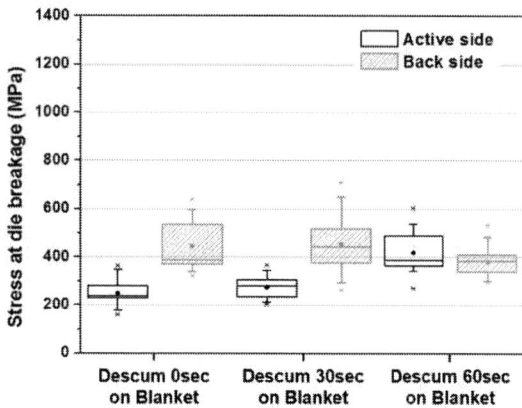

Fig. 11. Die strength data for laser grooving vs. SF_6 descum time in metal-free scribe lanes

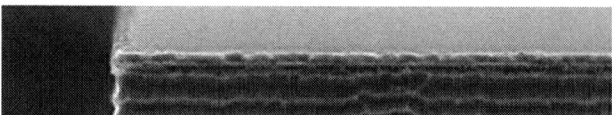

Fig. 12. Indentation at the top of the groove aligned to laser pulse frequency

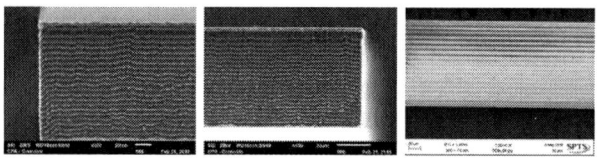

Fig. 13. SEM photographs of die sidewalls after plasma dicing: Laser groove with initial descum for 30 s (left), 60 s (center) and plasma etched groove without laser step (right)

For the samples processed with the laser recipe giving the lowest total fluence of 47.2 J/cm² it was necessary to use the initial etching step, in the place of the descum step in the other experiments, to ensure that the thin oxide layer at the bottom of the groove was completely etched prior to starting the main silicon etch sequence. Potential conditions were defined based on experience with the plasma etching process and the chamber hardware, and the etch rate of silicon dioxide was evaluated for each set using blank wafers with a SiO_2 layer deposited on silicon. After set-up trials four sets of conditions were selected and the resulting die strength data are plotted in Fig. 14. Also shown here are measurements obtained for the samples plasma diced after lithography and RIE oxide etch patterning in SiO2, which allow comparison of the use of related etch processes, with and without laser steps, to expose the silicon before dicing. It can be noted that the mean die strength of 1709.5 MPa (S.D. 394.3 MPa) obtained through the latter integration scheme is considerably higher than all other values observed, so this is a useful reference. The SEM of the sidewall in this case, also presented in Fig. 13, has a smooth appearance with no pronounced defects or striations.

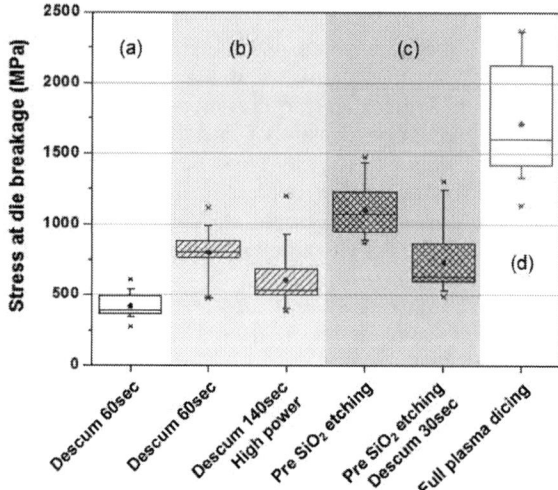

Fig. 14. Die strength obtained with partial laser groove leaving a thin oxide layer on the silicon at the bottom of the groove, (a) Full BEoL removed by medium fluence laser & Descum, (b) Partial BEoL removed by low fluence laser & SF_6 descum, (c) Partial BEoL removed by low fluence laser & SiO_2 etching, (d) Full BEoL removed by dedicated FE patterning & SiO_2 etching

Returning to the other samples in this set, using the typical SF6 descum conditions as the initial step for 60 seconds to remove the oxide layer by ionic bombardment produced a mean DBS value of 799.0 MPa (S.D. 170.2 MPa). This is significantly higher than the result obtained with an identical descum step at the medium fluence of 91.0 J/cm² (419.0 MPa, S.D. 90.7 MPa), with the groove into the silicon. With a longer SF_6 descum time of 140 seconds and a higher platen power at the wafer to further increase the energy of the impinging ions, the mean die strength was measured at 603.8 MPa (S.D. 216.0 MPa). It should be noted that this is higher than any of the mean values produced from samples with the groove into the silicon so the effect of reducing the fluence and avoiding direct damage to the silicon surface remains evident. The physical impact of the longer, high energy descum step is visible in the SEM photograph in Fig. 15 where the silicon has been etched to depth of about 15μm before the start of the main etch, marked by the position of the first scallop. The highest mean die strength in this group of 1102.8 MPa (S.D. 194.4 MPa) was attained by replacing the descum step with conditions dedicated to SiO_2 etching and with a higher selectivity to silicon. A suitable process uses plasma produced from a mixture of C_4F_8, O_2 and Ar, all standard in the plasma dicing equipment configuration. The sidewall is shown in Fig. 16 where, although some striations are visible, the top edge profile is quite smooth and there is no indication of indentations from the laser step. Finally, these conditions were repeated but with a 30 second SF_6 descum before the main etch. The resulting lower die strength of 731.6 MPa (S.D. 254.0 MPa) suggests that despite having a beneficial influence when laser-induced dielectric defects are present, the effect of adding ionic bombardment to etch silicon after having removed the oxide layer by softer means can introduce defects into the silicon at the top edge of the die.

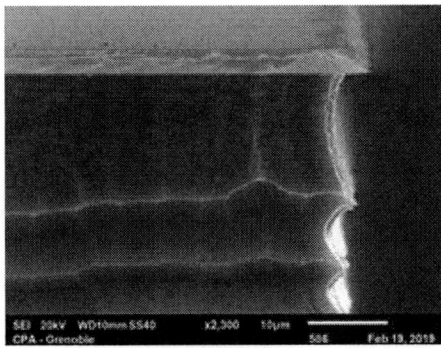

Fig. 15. SEM photograph of top edge of die with 140 seconds descum and high platen power showing depth of initial silicon etch

Fig. 16. SEM photograph of sample with partial laser groove where the thin oxide layer was removed with a dedicated oxide etch step

Die strength measurements were also evaluated with the back-side of the sample under tension for information but this was not the main aim of the work. Usually these values are strongly influenced by backside finishing, such as polishing and stress relief treatment. When plasma dicing is performed there is an additional factor to consider. This is the possible occurrence of so-called 'notching', a phenomenon that may occur in the over-etch step at the end of the process. This is caused by erosion of the polymer passivation on the die sidewall at the bottom of the trench where the silicon meets the dicing tape. This allows the reactive species to etch into the silicon causing defects at the back-side edge of the die. A few such defects were observed during analysis but it must be considered that the die size used for DBS measurement was very large, which reduces the sensitivity of the end-point signal. Here no specific attention was paid to optimize this because the need for the evaluation was to ensure complete singulation. Indeed values are quite dispersed from sample to sample, which is consistent with variability in this aspect of the process. This merits a separate investigation that is outside the scope of this paper.

V. CONCLUSION AND FURTHER WORK

The influence on die strength of adjusting the laser grooving fluence according to the construction of the dicing streets before singulating samples by plasma dicing has been studied. Although the back-side die strength was also measured, the main objective was to examine the die strength with the 'active' surface of the wafer under tension. Data have been obtained for several levels of complexity, with and without metallic structures, and compared with results for an integration scheme using lithography and etching instead of laser patterning. Indeed, the highest die strength of all was attained through this method where both the mask opening and the singulation were entirely realized through etching steps without any laser processing, thus avoiding any risk of damage from local heating effects. This value with a mean of 1.7 GPa was used as a comparative reference for the alternative schemes where laser grooving was used.

Within the laser grooved samples a distinction was made between a groove that penetrates into the silicon surface and one in which the BEoL layers are only partially ablated to leave a thin layer of oxide at the base of the groove. For the former a typical plasma dicing etching recipe can be used as the silicon is already exposed at the outset. In the latter case, where the laser fluence is reduced to obtain a shallower groove, the oxide layer must be removed in the first step of the etching recipe to expose the silicon. This was achieved either through a longer sequence of high energy ion bombardment or by using plasma conditions specifically adapted to oxide etching, albeit at a lower etch rate. It has been found that this partial groove technique delivers significantly higher die strength values than those from samples where the groove reaches the silicon surface and independently of the initial etch step used to remove the oxide layer. The highest mean die strength value obtained with this integration scheme was 1.1 GPa, this result coming from the sample where a dedicated oxide etching step was employed through C_4F_8, O_2 and Ar plasma. This suggests that laser damage at the silicon surface has been reduced by this method and that the etching of the oxide by reactive means gives a lower defect density. This appears to be a promising approach to deploying plasma dicing with existing and typically available laser tools, especially if die strength is an important consideration for the application. This does imply some modification to the front-end masking layout that could be absorbed, or even exploited for small die sizes through kerf width reduction, but may imply a cost adder for larger chips. In the latter case, or where plasma dicing is to be introduced for existing products without mask changes, the presence of test or alignment structures in the scribe lane, many of which contain metallic features, requires a laser grooving step with a higher fluence to ensure sufficient ablation to expose the silicon.

The samples used to investigate this configuration showed decreasing mean die strength results as the population density of the metallic features increased, with the same laser grooving recipe in each case to keep the total fluence invariable. The lowest value of 224.5 MPa derived from the sample with stacked metal tiling composed of smaller, closely spaced structures. Larger and more widely spaced metal structures produced values above 400 MPa. This effect may be attributable to local variations in laser

978-1-7281-1500-9/19 $31.00 © 2019 IEEE

heating effects between alternating metal and dielectric features and so independent of the plasma dicing process. However, the silicon etching conditions are very sensitive to metallic residues or debris left at the edge of the laser groove after ablation as this masks the vertical etching process to leave striations and discontinuity defects in the die sidewall. Regarding the integration of plasma dicing in this case it seems clear that the scribe layout must be considered as a factor influencing die strength but further investigation is needed to understand whether the sidewall defects play a role in weakening the die edge.

Results from the samples representing a uniform scribe lane construction without any metallic elements, where the aim of the laser step was to ablate the dielectric layers and penetrate into the silicon, were considerably lower than expected with most data points below 300 MPa. The highest mean value extracted was 419.0 MPa with a high dispersion (S.D. 90.7 MPa) compared to other samples from the same group. This came from the test to examine a longer SF_6 descum time of 60 seconds leading to the hypothesis that the reduction in fluence in dielectric-only samples resulted in more fused debris being left at the bottom of the groove and shorter descum times were less effective at removing this. However, this contrasts sharply with higher values, of close to 490 MPa, obtained from the device wafer samples with low metal density and processed with the same laser recipe. It is possible that apparently minor differences between BEoL layers on the device and blanket wafers could affect the ablation rates between the two groups but more work is needed to understand fully these data.

Further work is also suggested to investigate the effect on die strength of increasing the number of laser passes but with a reduced fluence in each individual pass. In this case care would be required to ensure that the threshold fluence for the ablation of the materials contained in the scribe lane was attained but this approach could allow silicon damage to be minimized. It should be noted that the data presented concern laser processing with a nanosecond pulse source. For future studies an area of particular interest in this context is the application of ultra-fast pulse lasers. The ablation mechanism with this type of source is significantly different, involving rapid ionization and plasma creation with minimal heating of the materials adjacent to the groove. This translates as reduced damage to the layers in the surrounding area, including the silicon surface. As the potential for die strength above 1 GPa has been demonstrated here with a specific technique to reduce defect-inducing mechanisms and using a nanosecond pulse laser, it may be expected that a short-pulse laser process can improve on this result. As access to such equipment is becoming wider a demonstration of its performance in this context is feasible in the near term.

ACKNOWLEDGMENT

The authors wish to extend thanks to the following:

- SPTS Technologies, Newport, UK for providing equipment, technical support and resources for processing samples and for permission to use the SEM image in Fig. 12.
- Disco Hi-Tec Europe GmbH, Munich, Germany for technical discussion.
- Physical Analysis Laboratory, STMicroelectronics, Grenoble, France.

REFERENCES

[1] D.Y.R Chong, W.E Lee, B.K Lim, J.H.L Pang and T.H Low, "Mechanical characterization in failure strength of silicon dice," Inter Society Conference on Thermal Phenomena, 2004.

[2] M. Vagues, "Analysing backside chipping issues of the die at wafer saw," San Jose State University, May 2003, unpublished.

[3] M.Xue, T.Chen, X.Zhang, L.Gao and M.Li, "Effect of blade dicing parameters on die strength," 19th International Conference on Electronic Packaging Technology, August 2018.

[4] O Haupt, F Siegel, A Shoonderbeek, L Richter, R Kling and A Ostendorf, "Laser dicing of silicon: Comparison of ablation mechanisms with a novel technology of thermally-induced stress," JLMN-Journal of Laser Micro/Nanoengineering vol. 3, No. 3, 2008

[5] F. Wei, T. Tabuchi, T. Lazerand, C. Johnston, K. Mackenzie and M. Notarianni, "Plasma dicing fully integrated process-flows suitable for BEOL advanced packaging fabrications," IEEE 67th Electronic Components and Technology Conference, 2017.

[6] F. Laermer and A. Schilp, Robert Bosch GmbH, "Method of anisotropically etching silicon," U.S. Patent N°5,501,893A, 1994.

[7] R. Abdolvand and F.Ayazi, "An advanced reactive ion etching process for very high aspect-ratio sub-micron wide trenches in silicon," Sensors & Actuators A: Physical, vol.144, Issue 1, 2008

[8] T.Xu, Z.Tao et al, "Effects of deep reactive ion etching parameters on etching rate and surface morphology in extremely deep silicon etch process with high aspect ratio," Advances in Mechanical Engineering vol.9, Issue 12, 2017

[9] F. Laermer, A. Schilp, K. Funk and M. Offenberg, "Bosch deep silicon etching: Improving uniformity and etch rate for advanced MEMS applications," Proceedings of the IEEE International Conference on Micro Electro Mechanical Systems (MEMS), February 1999

[10] R. Barnett, O. Ansell and D. Thomas, "Considerations and benefits of plasma etch based wafer dicing," IEEE 15th Electronics Packaging Technology Conference, 2013.

[11] K.D. Mackenzie, D. Pays-Volard, L. Martinez, C. Johnson, T. Lazerand and R. Westerman, "Plasma-based die singulation processing technology," IEEE 64th Electronic Components and Technology Conference, 2014.

[12] N. Matsubara, R. Windemuth, H. Mitsuru, H. Atsushi, "Plasma dicing technology," 4th Electronic System-Integration Technology Conference, 2012

[13] D.S. Finn, Z. Lin, J. Kleinert, M.J. Darwin, and H. Zhang, "Study of die break strength and heat-affected zone for laser processing of thin silicon wafers," Journal of Laser Applications, vol. 27, 032004, 2015.

[14] SEMI G86-0303, "Test method for measurement of chip (die) strength by means of 3-point bending," 2003

[15] E. Gourvest, I. Raid, O. Robin, S. Gallois-Garregnot and J-E. Luan, "Experimental and numerical study on silicon die strength and its impact on package reliability," 20th Electronics Packaging Technology Conference, 2018.

Advanced Dicing Technologies for Combination of Wafer to Wafer and Collective Die to Wafer Direct Bonding

Fumihiro Inoue, Alain Phommahaxay,
Arnita Podpod, Samuel Suhard, Erik Sleeckx,
Kenneth June Rebibis, Andy Miller, Eric Beyne
imec
Kapeldreef 75, 3001, Leuven, Belgium
Fumihiro.Inoue@imec.be

Hitoshi Hoshino, Berthold Moeller
Disco Hi-Tech Europe GmbH
Liebigstrasse 8, D-85551 Kirchheim b. Munich,
Germany

Abstract— Feasibility study of alternative dicing technologies for collective die to wafer direct bonding combined with wafer to wafer direct bonded dies has been performed. Several dicing technologies such as blade dicing, laser grooving + plasma dicing, laser grooving + stealth dicing and laser grooving from backside were evaluated for this integration scheme. For the case of blade diced dies, the collective die to wafer direct bonding are not succeeded. This was due to particle interruption, caused by remaining particles from dicing. For the case of laser grooving + plasma dicing and laser grooving from backside, successful die to wafer direct bonding were observed. However, the die edge was not bonded for the case of laser grooving + stealth dicing. This was attributed to the occurrence of the laser recast caused during laser grooving. Based on the characterization of dicing techniques for this approach, we have achieved successful integration of collective die to wafer bonding combined with wafer to wafer bonded dies.

Keywords-component; Dicing; collective die-to-wafer bonding; hybrid bonding

I. INTRODUCTION

With the increase demand of high computing in a single chip, multi-stack heterogeneous 3D integration is more and more attractive [1-4]. There are several crucial 3D stacking integration methods for each component such as wafer to wafer (W2W) bonding and die to wafer (D2W) bonding. W2W bonding offers a massive connection in a bonding, which decrease the processing cost in particular for small chips. However, there is a limitation that the die must be equal size for the pair wafer. Furthermore, no Known Good Die (KGD) selection is possible in W2W bonding, which can have a negative impact on stacking yield. On the other hand, D2W is able to handle different die sizes, which allow great flexibility in 3D stacking configurations. In addition, the capability to select the KGD ensures the yield. The drawback is the significantly long process time as per wafer, which raise the total cost of ownership. In order to decrease the processing time for D2W stacking, massive D2W bonding by using temporary carrier wafer with populated dies (so called collective D2W bonding) is introduced as an intermediate solution [4-6]. With respect to the high yield and cost reduction for multi-stack heterogeneous 3D integration, combination of W2W and collective D2W bonding has a curious candidate to explore.

For the interconnection technology between different components by D2W, solder base flip chip bumping is comprehensively developing. However, there is a certain limitation for scaling below 5 µm of the bump pitches due to the structure. In order to overcome the issue, implementation of direct bonding technology, which is an already established technology solution for W2W bonding, can be an alternative [7-9]. In this approach, two surfaces are prepared with a planarized dielectric layer. One of the challenges of D2W direct bonding is the cleanness control on die level. Clean surface is required following many processing steps that introduce particles such as dicing.

In this study, alternative dicing technologies for collective D2W direct bonding combined with W2W direct bonded dies has been investigated. Several dicing technologies such as blade dicing, laser grooving + plasma dicing, laser grooving + stealth dicing and laser grooving from backside were evaluated for this integration scheme.

II. COLLECTIVE DIE TO WAFER DIRECT BONDING

A. Bonded wafer preparation

Figure 1 show the schematic illustration of process flow for the preparation of direct bonded wafer. All the tests were done on 300 mm Si wafers (775 µm thickness as initial). The dielectric layers (SiCN/SiO$_2$) were deposited by plasma enhanced chemical vapor deposition (PECVD). Sequentially, the thermal annealing of 20 min at the 420 °C was performed to densify the films. Chemical mechanical polishing (CMP) of the surface dielectric layer is used to planarize and to smoothen the dielectric layer. After planarization, the thicknesses of dielectric layers are SiCN (100 nm) / SiO$_2$ (300 nm). Then the top wafer is edge-trimmed to avoid edge chipping during thinning process [10,11].

Prior to direct bonding, the pairing wafers are treated by N$_2$ plasma activation. Wafer bonding takes place at room temperature in an atmospheric pressure in a clean room ambient followed by an annealing at 250 °C for 2 h to enhance the interfacial adhesion strength (> 2.3 J/m^2) [12-14]. After the direct bonding, the top wafer was thinned

978-1-7281-1500-9/19 $31.00 © 2019 IEEE

down to 5 µm (extreme thinning), by using grinding, CMP and dry etching, respectively [14-16]. After the extreme thinning, another $SiCN/SiO_2$ layers are deposited, densified and planarized on the backside of extreme thinned wafer.

After the planarization of the SiCN, the process flow is differed for each dicing steps. The process flows for each step are described in the later figures (Fig 5-7). For the thinning of the bottom wafer till 100 µm, CONDOx wafer support system has been utilized [17].

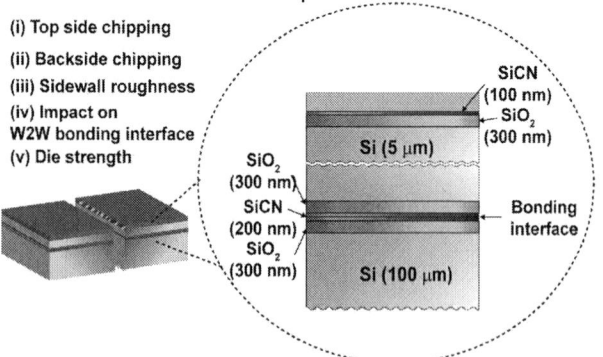

Figure 1 Schematic images of process flow for preparation of wafer to wafer direct bonded wafer

B. Dicing evaluation

The wafer structure for dicing evaluation is described in Figure 2. The top Si thickness is 5 µm and the bottom Si is 100 µm. The extreme thinned Si (5 µm) is facing up at the dicing evaluation. A polyolefin base UV curable tape was used for all the dicing techniques. The die size is 10 x 10 mm^2. For the case of laser dicing process, water soluble protective layer is spin-coated on the top surface. The chippings and the sidewalls were observed by optical microscope (OM) and scanning electron microscope (SEM). The bonding interface were revealed by using focused ion beam (FIB), then observed by scanning ion microscope. The die strength was measured by using 3 point bending method, where the distance of bottom pins are 8 mm.

Figure 2 Schematic images of wafer at dicing evaluation

C. Die population on temporary carrier wafer

After dicing, the dies are picked from dicing tape with protective layer on the die. Then, the dies are placed on another 300 mm Si carrier wafer with temporary glue (Population) as described in Fig 3. All the population were done in a single die to wafer bonding system with double flip function. The pitch of the die was 20.2 mm. In total, 152 dies are placed on a temporary carrier wafer. On the wafer, the extreme thinned wafer (= protective layer) is faced up. Then, the protective layer is stripped by deionized water (DIW) using full-auto 300 mm wafer level cleaning tool. The particle inspection after protective layer strip was done by dark filed image.

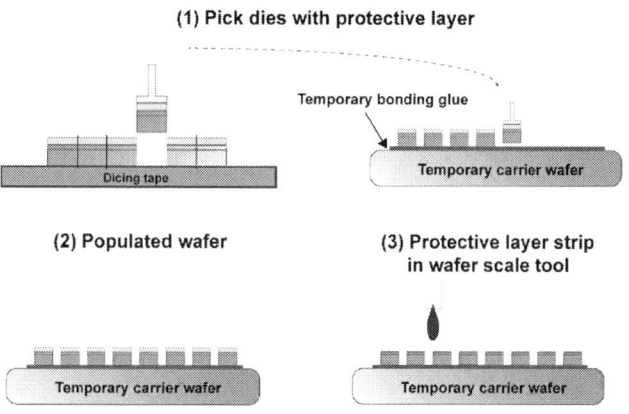

Figure 3 Schematic images of population and protective layer strip steps

The populated wafers are used for collective D2W bonding test. Figure 4 (a) shows the schematic image of collective D2W bonding with W2W bonded dies. As like initial direct wafer bonding, PECVD SiCN was deposited and planarized by CMP on another Si wafer (Target wafer). Both of the wafers are treated by N_2 plasma and then cleaned by megasonic DIW. The collective D2W bonding is done in a wafer level bonding tool.

After the wafer bonding, scanning acoustic microscope (SAM, Tepla AutoWafer 300) was used to detect voids in between dies and target wafers. Figure 4 (b) shows the schematic image of wafer after collective D2W bonding. As described in the figure, there will be 3 different Si layers after collective D2W bonding with 2 different dielectric bonding interfaces.

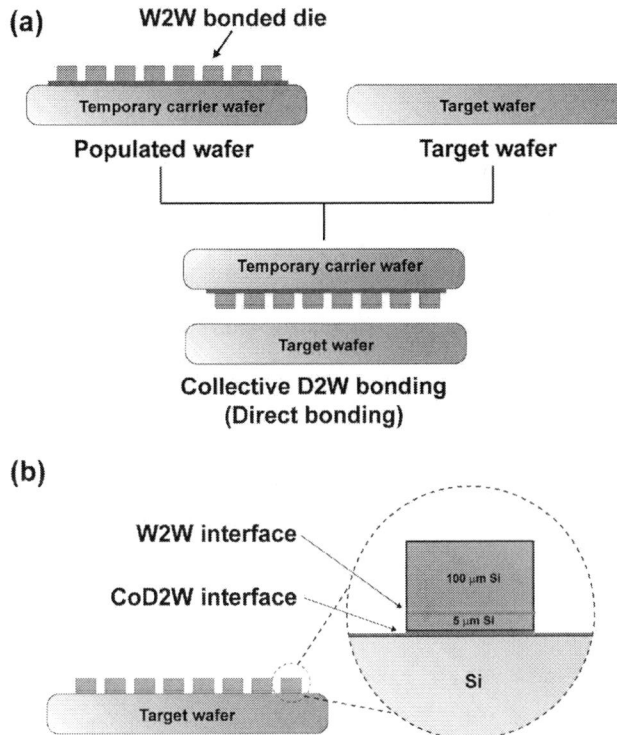

Figure 4 Schematic images of collective die to wafer direct bonding (a) at bonding step (b) after bonding and temporary carrier debonding

III. ADVANCED DICING ON DIRECT BONDED WAFERS

Each dicing process has been effectively adapted to bring out distinct advantages. For the case of blade dicing, it was executed after backside grinding (BG) and CMP on the backside of bottom wafer (not shown in figures). A single cut process with a blade in grid size of #4500 was used (Disco, DFD6361). In this study, no protective layer was usesd for the case of blade dicing.

A. Laser grooving followed by plasma dicing (LG + PD)

Figure 5 (a) shows the process flow of laser grooving followed by plasma dicing. As the first step, water soluble layer was spin-coated on the planarized dielectric surface. Then, CONDOx BG tape stack was laminated under vacuum onto the protective layer. CONDOx is a wafer support system which is embedding front side topography without using a glue layer to avoid any kind of glue residues on the bonding surface (Disco, DPM2190CX and Takatori, DTM-300B). Glue is only used in the 2 mm edge exclusion area. The embedding of the topography is done by an UV curable resin, which is covered by another PO base film. [17]. With the CONDOx tape, backside thinning on bottom wafer was executed. Then the wafer was transferred onto dicing tape to remove CONDOx and followed by dicing. Figure 5 (b) shows the schematic illustrations of the dicing

process. Firstly laser grooving was executed with the target width 20 μm and depth 10 μm from the surface. With the LG process, Si in the bottom wafer is revealed which is necessary for plasma dicing. Then, plasma dicing was performed on the grooved and revealed Si at bottom wafer. Bosch dry etching was used for the plasma dicing.

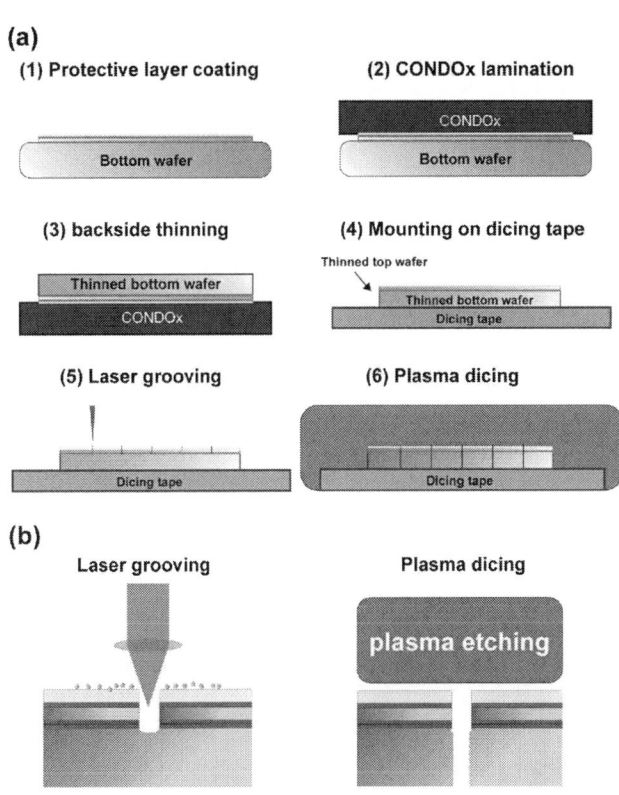

Figure 5 Schematic images of (a) process flow and (b) dicing details for LG + PD.

B. Laser grooving followed by stealth dicing before grinding (LG + SD)

Stealth dicing consists of laser process and tape expand process for the die singulation. The actual die separation is relied on the crystal orientation of the base substrate. Direct bonded wafers as like used in this study, the orientation of 2 pair wafers may not completely match. In addition, there is an uncertainty that the crack may not be able to propagate through the $SiO_2/SiCN/SiO_2$ W2W interface. Here, SD process within the 5 μm Si is rather challenging. Therefore, grooving process on top 5 μm Si is applied for this study. For the case of combination of laser grooving and stealth dicing, dicing before grinding (DBG or SDBG in this case) approach was used.

Figure 6 (a) shows the process flow of Laser grooving followed by stealth dicing before grinding. At first,

978-1-7281-1500-9/19 $31.00 © 2019 IEEE

protective layer coating and laser grooving were processed before thinning of bottom wafer. then, CONDOx tape was laminated on the grooved surface with protective layer. Then, laser damage (stealth dicing) was implemented from the backside of bottom wafer. The laser damage depth was 50-80 µm from the surface of top wafer. then backside thinning step was executed on bottom wafer. After the wafer transferred onto dicing tape, expansion step was applied to separate all the dies. Figure 6 (b) shows the detail of laser grooving process and stealth dicing process.

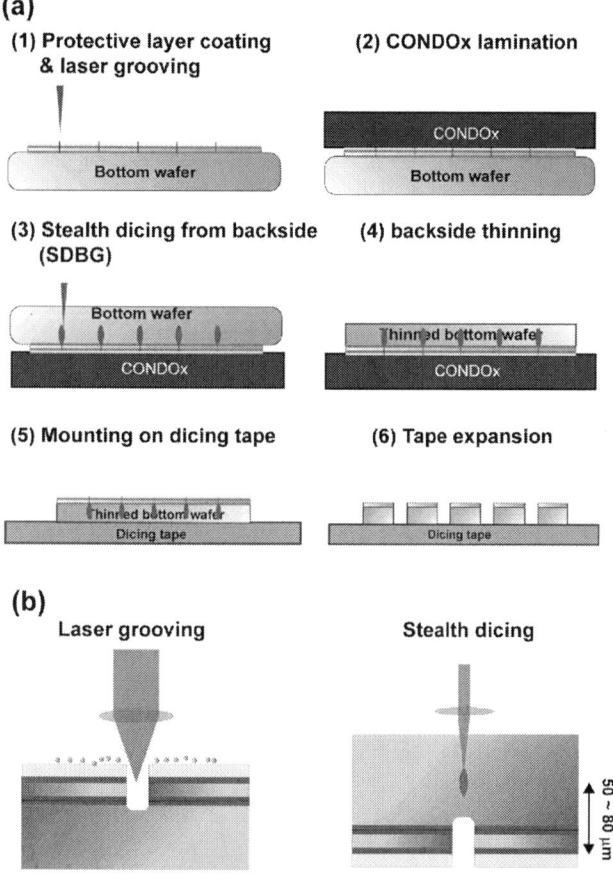

Figure 6 Schematic images of (a) process flow and (b) dicing details for LG + SD.

C. Blade dicing followed by laser dicing from backside of the wafer (LG from backside)

In the dicing of low-κ device wafer, combination of laser grooving followed by blade dicing is standardly used, which reduces delamination at interface of fragile dielectric layers. However, it requires wide kerf width at laser grooving. In addition, the residues caused during laser process (e.g. debris at edge) will stay at the surface. Therefore, in this study, we applied the dicing step from backside of the wafer in a reverse sequence.

Figure 7 (a) shows the process from laser grooving from backside. Until the backside thinning of the bottom wafer, the process is exactly same as laser grooving + plasma dicing. After backside thinning, dicing step was done on CONDOx as support film on dicing tape. As the initial dicing step, blade dicing was applied from the backside of thinned bottom wafer. The target depth is 80 µm, which cut in only bottom wafer. the rest of the stacks were singulated by laser grooving through the blade diced path. After dicing, the wafer was transferred to another dicing tape, then the CONDOx was de-taped. Figure 7 (b) shows the schematic images of the dicing process. the blade dicing is to be used to remove bulk Si on the bottom wafer, afterwards the singulation of bonding interface and 5 µm top Si is done by laser grooving. With this process sequence, the laser residues will stay at the sidewall, but not on the surface. In addition, the width of the laser grooving can be minimized, which will eventually minimize kerf loss in the design.

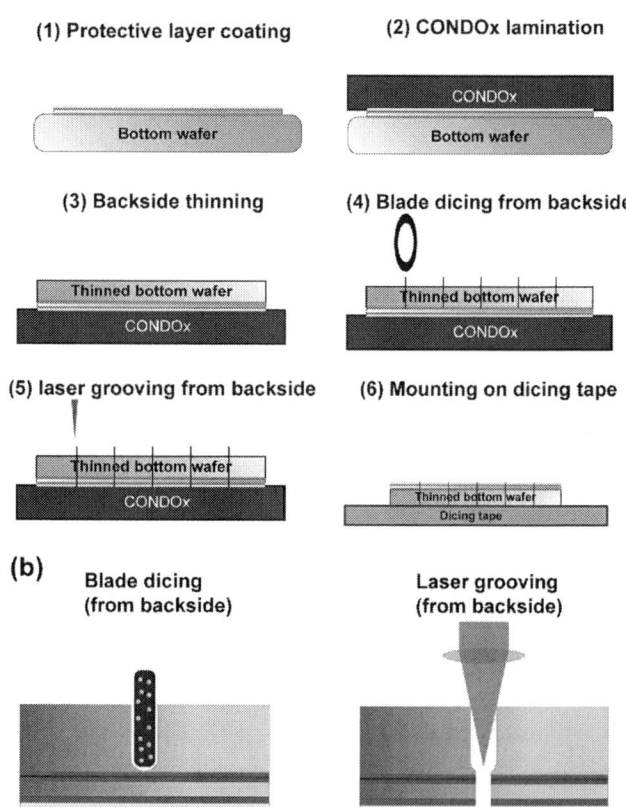

Figure 7 Schematic images of (a) process flow and (b) dicing details for LG from backside.

IV. RESULTS AND DISCUSSION

A. Impact of dicing steps on die surface, sidewall, bonding interface and die strength

Figure 8 shows the surface and sidewall images of die after each dicing processes. For the inspection of the top surface, the protective layer is stripped. For the case of blade dicing, a lot of particles can be seen on the surface of the dies (Fig 8 (a)). In addition, few tens of micron meters of chipping can be seen on the edge of the surface, which is unavoidable for diamond abrasive process. Furthermore, large backside chipping can be recognized for blade diced die bottom (Fig 8 (b)). This is also an well-known drawback of blade dicing process.

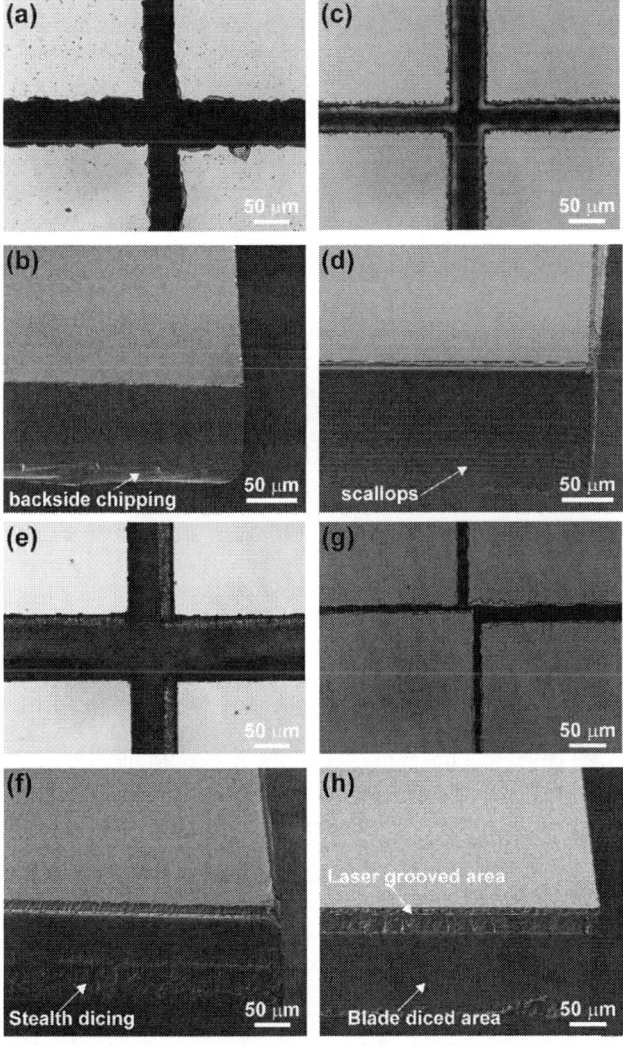

Figure 8 Top view OM images and tilt SEM vies from sidewall (a) and (b) blade dicing, (c) and (d) LG + PD, (e) and (f) LG + SD, (g) and (h) LG from backside

On the other hand, there is no chipping for other case on the surface, since the surface is removed by laser ablation for all other cases. In LG + PD and LG + SD, a step is formed at near the surface by laser grooving. At the sidewall after LG + PD, wavy topography is formed (so called scallop) due to the repeat of etching and passivation during both deep etching process (Figure 8 (d)). For the case of LG + SD, laser damaged region is visible at the sidewall (Fig 8 (f)). The stealth dicing induced damaged area is 50 – 80 μm depth from the surface, which is matching with the target depth irradiated from the backside. For laser grooving from the backside, the dicing was done before transferred to dicing tape, therefore the mis-location of dies due to die movement during mounting is seen (Fig 8 (g)). However, it was no issue for picking process. The blade diced area and laser grooved area is clearly distinguished at the sidewall (Fig 8 (h)). The depth of the laser area was about till 20 μm. Since the blade dicing need to be wider than the laser grooving, the shape of the dies is opposite as usual diced die shape (top slightly wider than bottom).

In order to assess the impact of dicing process on W2W direct bonding interface, the bonding interface near die edge are exposed by FIB. Figure 9 show the FIB images after each dicing process. Even though, blade dicing gives huge mechanical stress onto Si to remove by the diamond abrasive, no delamination of bonding interface is observed (Fig 9 (a)). This indicates that blade dicing is still applicable for integration with W2W bonding when the optimization of chipping can be done. This also indicates that our SiCN-SiCN direct bonding is having sufficient adhesion strength in compatible with packaging process. Figure 9 (b) shows the SIM image after LG + PD. Although there is small undercut (1.5 - 3 μm) below the bonding interface and the top surface, there was no delamination. For the case of LG + SD (Fig 9 (c)), there was also no delamination of bonding interface. In addition, there is no undercut for this case. Figure 9 (d) shows the image after LG from the backside. Although the laser grooving was irradiated from the backside through blade diced path, there was no damage into bonding interface is observed as like other laser process irradiated from the top surface.

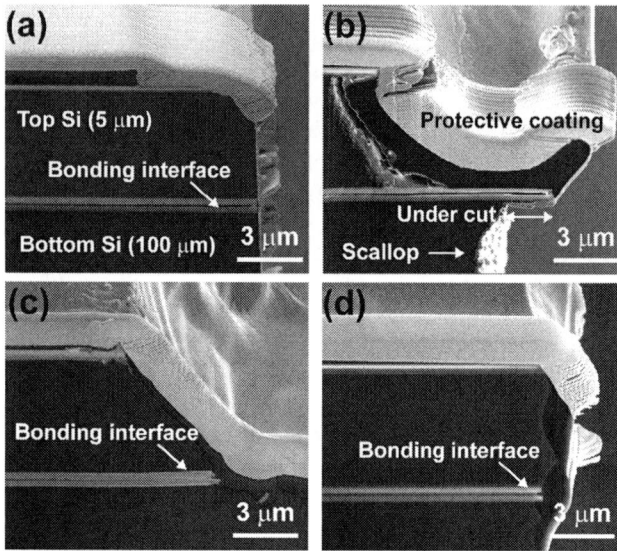

Figure 9 Cross-sectional FIB images at die top edge (a) blade dicing (b) LG + PD (c) LG + SD (d) LG from backside

The die strength was measured by three point bending method. Figure 10 show the results of die strength for each dicing process. For the average in 25 dies are 880, 1294, 525, 696 MPa for blade dicing, LG + PD, LG + SD and LG from backside, respectively.

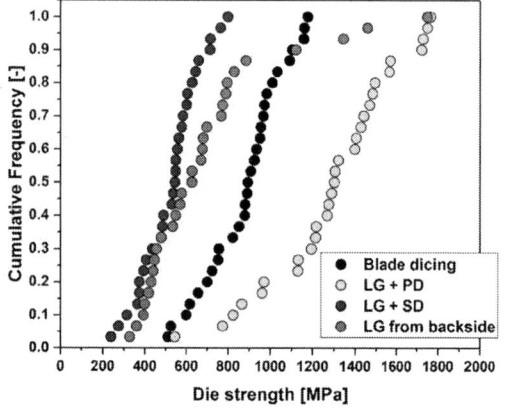

Figure 10 Die strength for different dicing techniques

For the case of LG + SD and LG from backside, the die strength are weaker than blade dicing, which can be attributed the heat affect zone (HAZ) caused by laser process. It is well-known that laser ablation turns crystalline Si into amorphous or polycrystalline by the heat [18-20]. The subsurface defects at the sidewall generates huge strain/stress on entire die which makes the die fracture strength drastically weak. The transformed Si is also created by

mechanical abrasive process like blade dicing, however the impact of laser is much higher. For the case of LG + PD, the HAZ caused by laser grooving can be removed during plasma etching. Therefore, the highest die strength was obtained. For the collective die to wafer bonding, the die strength is very important to avoid any residue coming from broken dies. From this point of view, utilizing plasma process to remove HAZ is a key to improve the yield of collective die to wafer bonding.

B. Die population and collective die to wafer direct bonding

The dies are picked from dicing tape and placed on carrier wafer using die to wafer flip chip bonder. The yield of the population for each dicing process are 96, 100, 100, 96% for blade dicing, LG + PD, LG + SD and LG from backside, respectively. For the case of blade dicing, huge particles might be transferred from die surface to the pick head, which makes die breakage for neighbor dies. Resultant, some dies are partially broken or some case not well placed on the wafer. Therefore some of missing dies are seen after population as shown in Fig 11 (a). For the case of LG + PD and LG + SD, there was no issue for pick and place. For LG from backside, some dies are picked with neighbor dies due to adhere by protective layer (Stripped and re-applied protective layer before population, resulting pick and place failure.). This can be improved when the process flow is optimized for this dicing process. On the other hand, it indicates that the kerf width of the dicing technique is very narrow as it can be filled by submicron thickness of protective layer. Figure 11 (e) – (h) show the DF inspection images after population and strip protective layer. In these images, defects can be identified in white dots. The blade diced dies are fully covered with plenty of particles caused by blade dicing, therefore, the surface is entirely white. For the other cases, most of the residue caused during dicing process were stripped with protective layer from the surface. Therefore, almost no particles are seen on the die surface. Although some residues are seen on temporary glue, it does not affect for collective die to wafer bonding.

Then, the populated wafers and target wafers are bonded. After collective bonding (before the temporary carrier is de-bonded), SAM images are taken. Figure 11 show the SAM images after collective die to wafer bonding. Here, bonding voids and open space/gap are shown in white. As shown in Fig 12 (a), blade diced dies are not bonded to target wafers. This should be due to interruption of direct bonding by particles caused during blade dicing. For the other case, dies are bonded to target wafer with dielectric-dielectric bonding. Some minor voids are still visible in between die and target wafer, which may mainly come from imprint of pick head during population. As shown in Fig 12 (b) – (d), all the dies are bonded to target wafer. In particular for LG + PD and LG from the backside, high SAM yield in die are obtained (Figure 12 (e) and (g)). However, the die edges were not bonded for all the dies from LG +SD. The average of detected bonded area in die for LG + SD is 72.5 % (= about 0.75 mm all around the edge is not bonded).

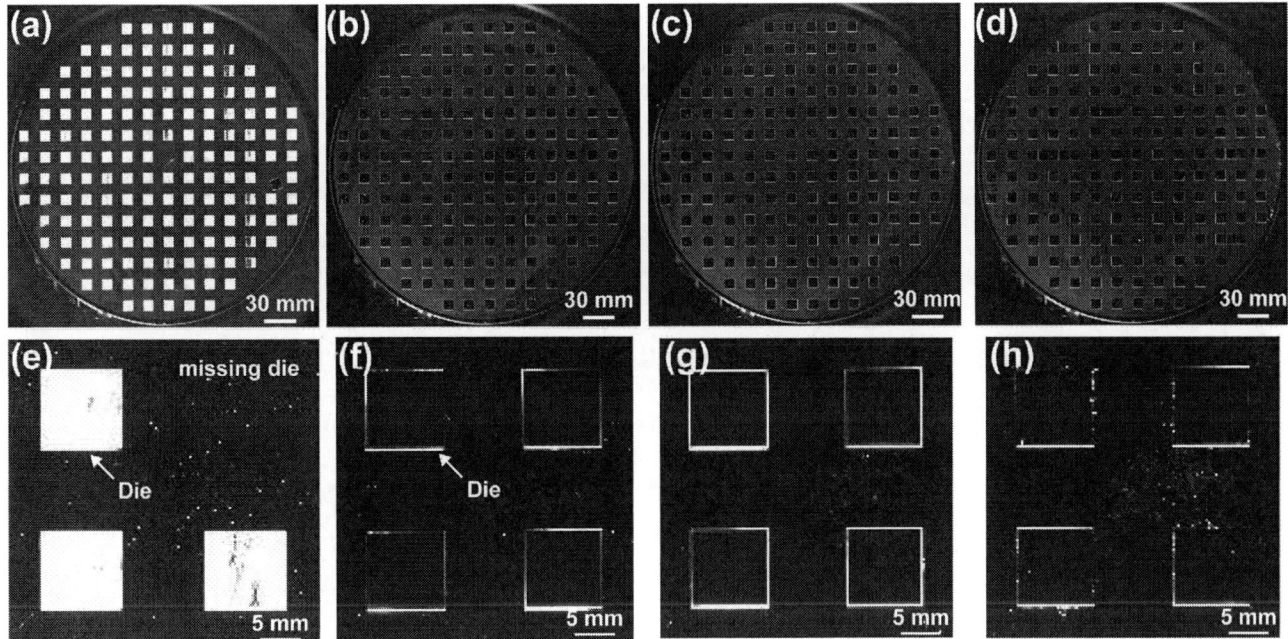

Figure 11 DF images after population (a) and (e) blade dicing, (b) and (f) LG + PD, (c) and (g) LG + SD, (d) and (h) LG from backside

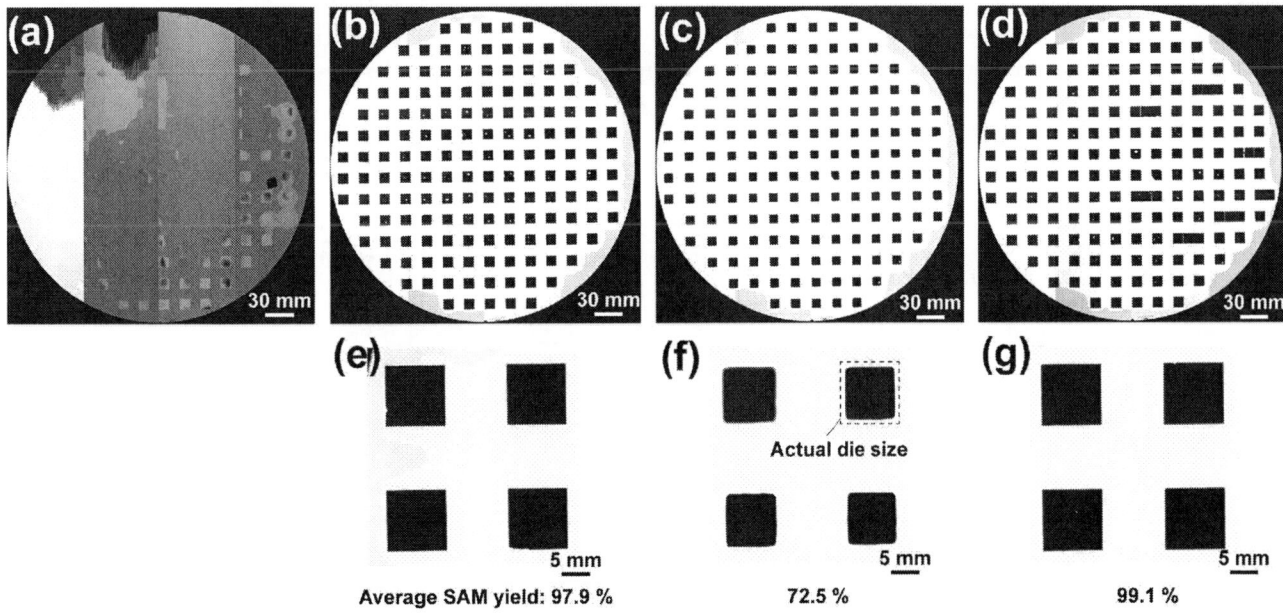

Figure 12 SAM images after collective die to wafer direct bonding (a) blade dicing, (b) and (e) LG + PD, (c) and (f) LG + SD, (d) and (g) LG from backside

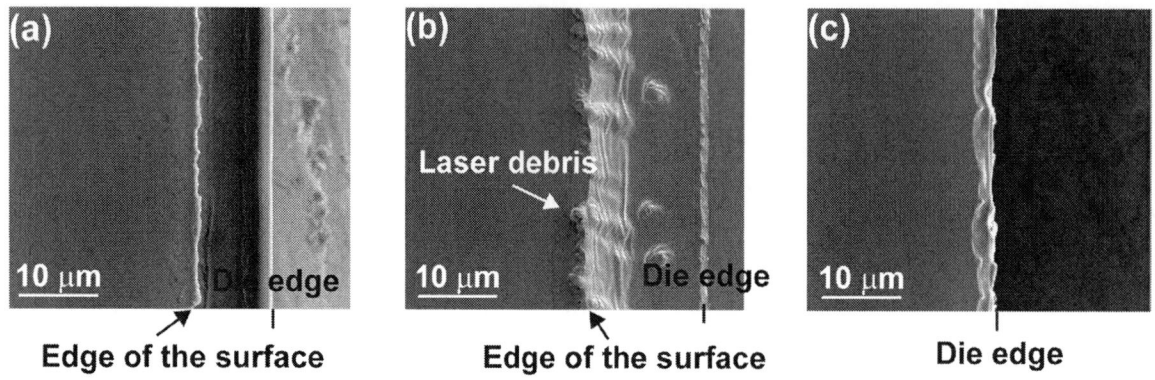

Figure 13 Top view SEM images after dicing (a) LG + PD, (b) LG + SD and (c) LG from backside

In order to find out the root cause of the edge non-bonded area cause on LG + SD, inspection of die edge before bonding were done. Figure 13 show the top view SEM image of the die edge for LG + PD, LG + SD and LG from backside. These images are taken after protective layer removal. For the case of LG + PD, the surface near the edge is slightly rougher than inner surface (Fig 13 (a)). This can be a signature of removed laser debris/recast during plasma etching. For the case of LG + SD, a lot of laser debris/recast still remain at the die edge even after protective layer strip (Fig 13 (b)). This can be the root cause of the edge non-bonded area seen after collective die to wafer bonding. It is well known in direct bonding that the formed void size become much bigger than the residue size on the bonding surface. For this case the laser debris/recast is only ~2 μm at the edge, however it turns to ~ 0.75 mm of voids after bonding. For the case of LG from the backside, the surface is protected by protective layer and dicing tape. Therefore no laser debris/recast were formed on the surface (Fig 13 (c)). Therefore, when the dies are used for D2W direct bonding with laser dicing process, the recast/debris must be removed. The most promising process to remove from the surface is to use plasma process. The other way to avoid this issue is to irradiate the laser from the wafer backside with protection of top surface by protective layer and dicing tape. Then the laser recast/debris will not stay on the bonding surface.

C. Bonding Interfacial analysis of die to wafer direct stacking

Figure 14 (a) shows cross-sectional TEM image of die edge after collective die to wafer bonding by using LG + PD. As shown in Fig 8 (d) and Fig 12 (a), The die edge is bonded towards very edge of the surface. No voids are seen at the interface of collective die to wafer. Furthermore, there is no subsurface strain/stress area and HAZ on 5 μm Si sidewall. It indicates that the defects caused by laser grooving was removed by plasma dicing. In addition, the TEM image also support that there is no delamination onto W2W bonding interface by this dicing technique towards the very edge of the die

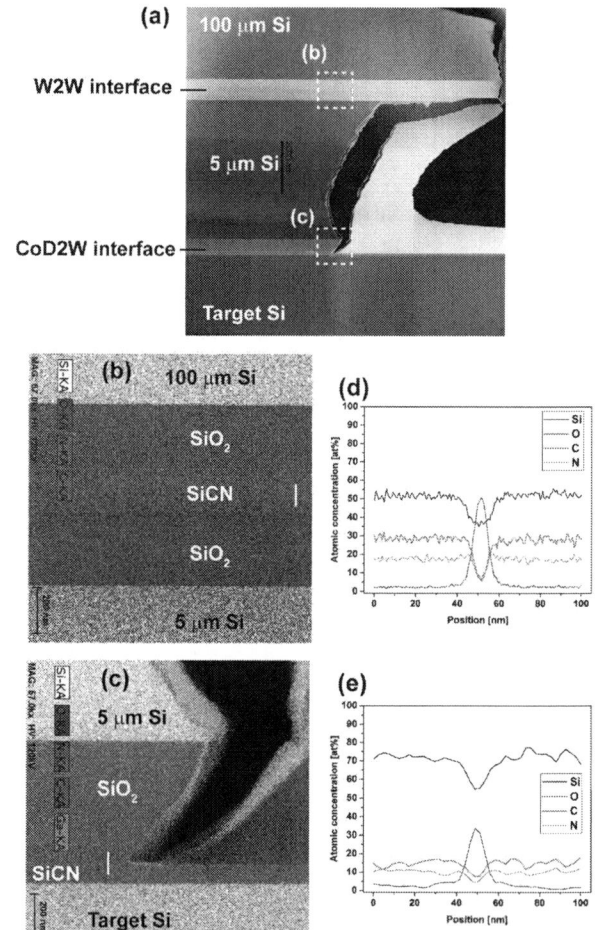

Figure 14 (a) Cross-sectional TEM image after collective die to wafer direct bonding with LG + PD. (b) EDS overlap map image at W2W interface (c) CoD2W interface (d) and (e) EDS line scan results at interfaces.

Figure 14 (b) and (c) show the high magnification scanning TEM images with EDS color maps on W2W bonding interface and CoD2W bonding interface. As we reported in previous work, the majority of W2W bonding interface created by SiCN-SiCN is Si and O [12-13]. There is no difference for the interface even after extreme thinning (5 μm), backside preparation and collective die to wafer bonding. Similarly, the formation of Si and O rich interface is identified at the interface of collective die to wafer bonding. this is more visible when the line scan at the interface is compared (Fig 14 (d) and (e)). It indicates that the dielectric bonding with SiCN-SiCN occurred successfully even for the collective die to wafer bonding towards very edge of the die.

V. CONCLUSIONS

Based on the characterization of dicing techniques, we have identified dicing methods for successful integration of W2W and collective D2W direct bonding combination. Several dicing techniques for combination of wafer to wafer direct bonding and collective die to wafer direct bonding have been evaluated. Although blade dicing is still applicable for singulation of direct bonded wafer having 5 μm Si on top, chipping on front and backside will be an issue for pick and place. On the other hand, LG + PD, LG + SD, LG from backside showed dicing performance without particles on the die surface. Furthermore, these techniques can separate device layers by laser grooving, which is compatible with W2W bonding, as well as low-κ layer. With respect to the die strength, LG + PD shows the highest, this can be explained by stress relief effect during plasma dicing.

Blade dicing clearly showed the limitation to be applied into collective D2W direct bonding due to the particle. Laser grooving also having a laser residue issue when it is applied to collective die to wafer direct bonding. However, the use of protective layer and plasma dicing to remove the laser recast/debris can solve the issue. On top of that, to use laser grooving from the backside is an option to avoid any issues caused by laser residues on the die to wafer bonding interface. These alternative dicing technologies enables multi-stack heterogeneous 3D integration which offers high computing in a single chip, with high yield and cost reduction.

ACKNOWLEDGMENT

We are grateful to Herbert Struyf, Lan Peng, Serena Iacovo, Julien Bertheau, Giovanni Capuz, Pieter Bex, Ferenc Fodor for their contribution.

The authors would like to thank Hideyuki Sandoh, Makoto Saito, Kei Tanaka, Koji Watanabe, Tsubasa Obata, Satoshi Kumazawa, Kentaro Odanaka, Hiroyuki Takahashi, Kazuki Higashiyama, Masatoshi Wakahara from Disco Japan for their valuable input and support during experiments.

REFERENCES

[1] E. Beyne, "The 3-D Interconnect Technology Landscape," *IEEE Des. Test*, vol. 33, no. 3, pp. 8–20, 2016.

[2] M. Koyanagi *et al.*, "Three-Dimensional Integration Technology Based on Wafer Bonding With Vertical Buried Interconnections," *IEEE Trans. Electron Devices*, vol. 53, no. 11, pp. 2799–2808, 2006.

[3] E. Beyne, "3D system integration technologies," *2006 Int. Symp. VLSI Technol.*, pp. 1–9, 2006.

[4] T. Fukushima *et al.*, "New Three-Dimensional Integration Technology Based on Reconfigured Wafer-on-Wafer Bonding Technique," *in Technical Digest - International Electron Devices Meeting, (IEDM 2007)*, vol. 3, no. 6, 2007.

[5] T. Wang *et al.*, "3D IC assembly using thermal compression bonding and wafer-level underfill - Strategies for quality improvement and throughput enhancement," in *Proceedings of the 2016 IEEE 18th Electronics Packaging Technology Conference, (EPTC 2016)*.

[6] L. Sanchez *et al.*, "Collective Die Direct Bonding for Photonic on Silicon" *ECS Transaction*, vol. 86, no. 5, pp. 223–231, 2018.

[7] E. Beyne *et al.*, "Scalable, Sub 2μm Pitch, Cu/SiCN to Cu/SiCN Hybrid Wafer-to-Wafer Bonding Technology," *in Technical Digest - International Electron Devices Meeting, (IEDM 2017)*.

[8] Y. Kagawa *et al.*, "Novel Stacked CMOS Image Sensor with Advanced Cu2Cu Hybrid Bonding," in *Technical Digest - International Electron Devices Meeting, (IEDM 2016)*.

[9] Y. Kagawa *et al.*, "An Advanced CuCu Hybrid Bonding For Novel Stacked CMOS Image Sensor," *Proceedings of the 2017 IEEE International Interconnect Technology Conference (IITC 2017)*.

[10] F. Inoue *et al.*, "Edge trimming for surface activated dielectric bonded wafers," *Microelectron. Eng.*, vol. 167, pp. 10–16, 2017.

[11] F. Inoue *et al.*, "Edge Trimming Induced Defects on Direct Bonded Wafers," *J. Electron. Packag.*, vol. 140, no. September, pp. 1–6, 2018.

[12] S.-W. Kim *et al.*, "Permanent wafer bonding in the low temperature by using various plasma enhanced chemical vapour deposition dielectrics," in *Proceedings of the 3D Systems Integration Conference (3DIC)*, 2015, p. TS7.2.1-TS7.2.4.

[13] L. Peng *et al.*, "Advances in SiCN-SiCN Bonding with High Accuracy Wafer to Wafer (W2W) Stacking Technology," *Proceedings of the 2018 IEEE Int. Interconnect Technol. Conf.*, pp. 179–181, (2018).

[14] F. Inoue *et al.*, "Influence of Si wafer thinning processes on (sub)surface defects," *Appl. Surf. Sci.*, vol. 404, pp. 82–87, 2017.

[15] N. Rassoul *et al.*, "RIE dynamics for extreme wafer thinning applications," *Microelectron. Eng.*, vol. 192, pp. 30–37, 2018.

[16] A. Jourdain *et al.*, "Extreme wafer thinning optimization for via-last applications," in *2016 IEEE International 3D Systems Integration Conference, (3DIC 2016)*.

[17] C. Epple, "Laser-lift-off (LLO) and CONDOx for wafer ultra-thinning process for 3D stacked devices, TSV, eWLB and WLCSP wafers" in *2019 3D & Systems Summit* (2019).

[18] W.-S. Lei *et al.*, "Die singulation technologies for advanced packaging: A critical review" *J. Vac. Sci. Technol. B App.*, vol. 301, no. 819, pp. 40801–231901, (2012).

[19] K. Rahim *et al.*, "A Review on Laser Processing in Electronic and MEMS Packaging," *J. Electron. Packag.*, vol. 139, pp. 03801-1-10 (2017).

[20] A. Podpod *et al.*, "Investigation of Advanced Dicing Technologies for Ultra Low-k and 3D Integration," *in Proceedings Electronic Components and Technology Conference 2016 (ECTC2016)*.

Active control of NCF fillet shape for 3D CoW by multi beam laser bonder

Keiko Ueno, Kazutaka Honda, Tsuyoshi Ogawa, Toshihisa Nonaka
Packaging Solution Center, Hitachi Chemical Co., Ltd.
7-7 Shinkawasaki, Saiwai-ku, Kawasaki-shi, Kanagawa, 212-0032, Japan
e-mail: kei-ueno@hitachi-chem.co.jp

Abstract— Non-conductive film fillet control during the thermal compression bonding using a multi laser beam bonder was investigated comparing with the case of using a conventional ceramic pulse heater bonder. The laser bonder can control each laser beam power individually, which can change the temperature distribution in the heated die. Fillet coverage index was defined as the ratio of the fillet width of the die corner and the maximum point of the die peripheral. The optimization of the temperature distribution of the die by the laser bonder could enlarge FCI from 0.25 with even reducing the fillet width at the maximum point of the die peripheral from 167 µm in comparison to those of 0.33 and 187 µm, which were obtained using a conventional bonder with a ceramics pulse heater, respectively.

Keywords – laser bonder; fillet shape; multi beam; High productivity; NCF

I. INTRODUCTION

3D stacked DRAM packages with through silicon via (TSV) have been developed and in mass production. The dies usually have more than one thousand bumps and pads, which are aligned with several tens micrometer pitch. Such dies are generally interconnected by TCB (Thermal compression bonding) technology, which is good at making solder joint with precise alignment. Non-conductive film (NCF) is also commonly used in the TCB for TSV-3D stacking as a pre-applied underfill. [1-4]

It is usually laminated on the wafer first and then individualized. The dies covered with the same size NCF are three dimensionally stacked by TCB. During the process NCF material is squeezed out from the die edge, which makes the fillet. It should cover the die edge to provide the high reliability of the package. The stress often becomes largest at the corner area of the die. The coverage of the die corner by underfill material is very important to secure the reliability. The amount of the squeezed out NCF material is usually very small at the corner comparing with the middle area of the die peripheral by a conventional TCB process. The amount of the fillet of the middle area usually becomes undesirable rich to attain enough fillet coverage at the corner. Too large fillet hinders the high density die stacking of chip on wafer (CoW) assembly because the fillet must not invade the adjacent die stacking area. Exceeded fillet also happens the creeping up of the NCF material on the die top surface, which brings about the sticking of the material to the bonding tool.

A conventional TCB bonder usually equipped with a ceramic pulse heater. When such a bonder is used, the central part of the die is heated higher than the die corner, so the shape of the NCF fillet tends to be circular. Although a temperature distribution in a die which is heated by such a conventional TCB tool is passive concentric, the multi beam laser bonder can control the distribution actively. It is equipped with 32-fiber array connected to individual laser light source, and can expose the laser light onto the die with independently controlled laser power, resulting in providing the temperature gradient over the die.

In this paper, active control of the fillet shape of NCF was evaluated using multi beam laser bonder.

II. LASER BONDING

The assembly process flow with NCF is as follows. The first step is NCF lamination on an active side of the top wafer. The second step is NCF-laminated wafer dicing. The third step is picking up the diced chip and bonding to bottom die by flip chip (FC) bonder as pre bonding. The fourth step is main bonding, in which the multi beam laser bonder "FDB250" (Shibuya Corporation) was used. Fig. 1 shows the structure of the laser head of FDB250, which is equipped with 32-fiber array connected to individual laser light source. It can expose the independently controlled 32 laser lights onto the die, resulting in providing the temperature gradient over the die.

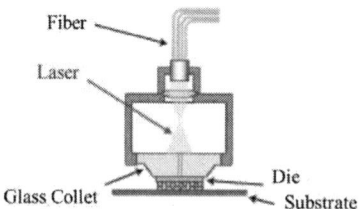

Figure 1. Structure of laser bonder [5]

Compared to conventional TCB equipment with a ceramics pulse heater, the laser bonder has three advantages. One is that it can achieve approximately 4 times higher speed in heating and cooling of the die as shown in Fig. 2 (a) and (b), which can shorten the process time significantly. It can also control the head position in height because of the extremely low heat expansion of the head as shown in Fig. 2 (c). The third advantage is that it can control the temperature

978-1-7281-1500-9/19 $31.00 © 2019 IEEE

gradient over the die, which means it can also achieve uniform heating all over the die. Examples of the laser irradiation pattern are shown in table I. When irradiating the entire surface on the die uniformly, the temperature difference between the die corner and the die center was 2.2°C, whereas the temperature difference was 12.7°C by controlling the intensity of the laser light.

TABLE I. TEMPERATURE DISTRIBUTION IN A DIE

Heating	The entire heating	The intensive heating
Laser irradiation pattern	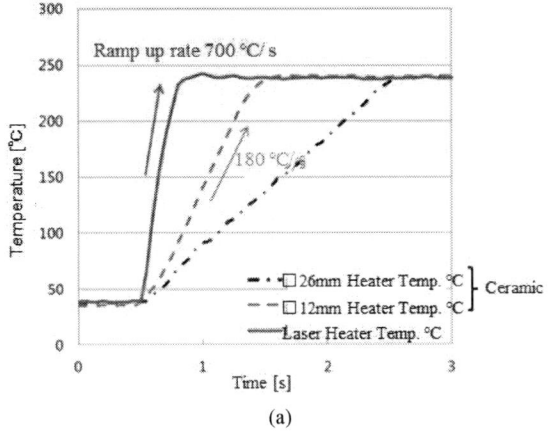	
Thermal image		
Temperature difference	2.2 °C	12.7 °C

In this evaluation, we focused on the structure of the laser head, which can control the intensity of each laser beam and therefore control the temperature gradient in the die. It is supposed that the fluidity of the NCF, which has a dependency on the temperature, can be controlled by applying the temperature gradient. Therefore, the fillet width and shape can be tuned.

III. EXPERIMENTS

A. Test Vehicle

We used the test vehicle (TV) described in Table II and Fig. 3. The size of the top die was 7.3 mm square. The thickness was 50 µm. The die had peripheral bumps with 80 µm pitch and full-array bumps with 300 µm pitch. The bumps were formed with Cu post and Sn/Ag solder. The peripheral bumps were aligned staggered in two rows. The size of the bottom die was 10 mm square. The thickness was 100 µm. The Cu pads were plated with Ni/Au.

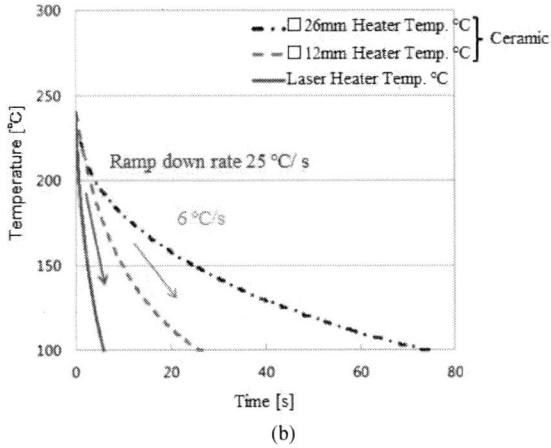

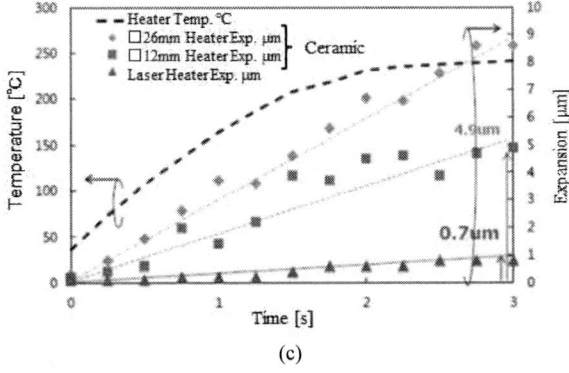

Figure 2. Head tempearture at rapid heating and cooling are shown in (a) and (b), respectively. Thermal expansions of the ceramics and laser heaters at heating up process [6].

TABLE II. TV SPECIFICATIONS

Top die	7.3 mm × 7.3 mm, 50 µm^t Passivation : SiN Peripheral bump : 80 µm pitch, 648 pin Full array bump : 300 µm pitch, 400 pin Bump height : Cu Pillar (30 µm^t) + Sn/Ag Solder (15 µm^t)
Bottom die	10 mm × 10 mm, 100 µm^t Passivation : SiN Pad : Ni/Au plating

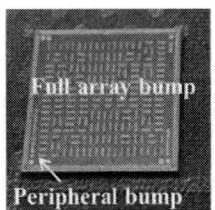

(a) Top die (b) Bottom die

Figure 3. TV example

978-1-7281-1500-9/19 $31.00 © 2019 IEEE

B. Conditions of sample preparation

NCF thickness was 40 μm. The preparation procedure of bonding sample is shown in Fig. 4. The first step was NCF lamination on the bump side of the top wafer by vacuum laminator at 80°C for 120 seconds. The second step was the NCF laminated wafer dicing. The third step was picking up the diced die at low temperature and pre-bonding to plated wafer at 80°C for 3 seconds by a conventional TCB tool. The fourth step was main bonding by using the laser bonder.

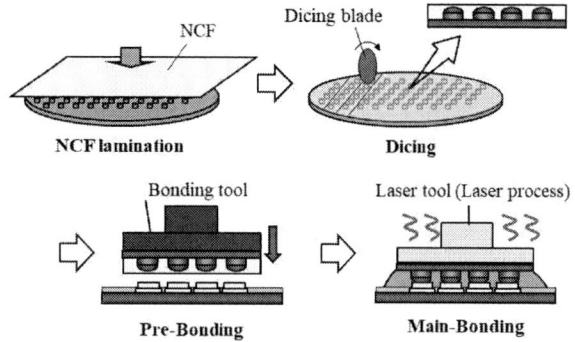

Figure 4. Preparation procedure of bonding sample

The laser output power of the 32 beams were categorized and set at 3 levels of 0, 25 and 100% in this evaluation. Three types of the laser irradiation distribution pattern were evaluated, which are shown in Fig. 5.

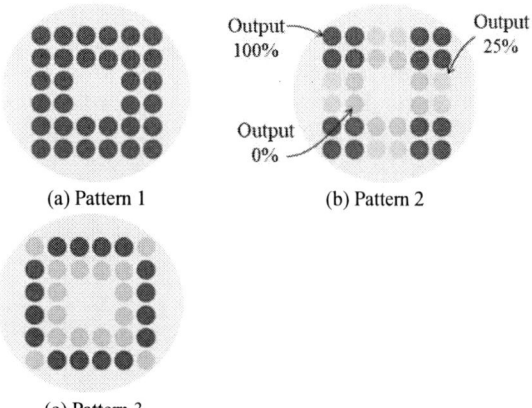

Figure 5. Laser irradiation distibution pattern

C. Fillet width

The maximum length of the NCF fillet that protruded from the die edge was defined as "fillet width at side". An example of the measured point is shown in Fig. 6.

Figure 6. Measurement location of fillet width

D. Fillet Coverage Index (FCI)

The fillet width at the point of 200 μm from die edge was measured and defined as "fillet width at the corner", which is shown in Fig. 7. Fillet coverage index (FCI) is defined as the ratio of "fillet width at side" and "fillet width at the corner" in the following equation (1). The closer the FCI is to 1, the fillet coverage at the corner of the die becomes better.

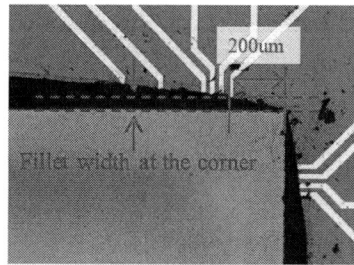

Figure 7. An example of the measured point of "fillet width at the corner"

$$FCI = \frac{Fillet\ width\ at\ the\ corner}{Fillet\ width\ at\ side} \qquad (1)$$

IV. RESULTS AND DISCUSSIONS

A. Filler width and coverage characteristics by laser irradiation pattern

The main bonding was evaluated under the condition of 230°C for 3 seconds by the 3 laser irradiation patterns. The condition of 230°C is the temperature at the die center. The temperature distribution within the die was measured 1 s later when the temperature of the die center reached at 230°C. The results of the temperature measurement by the 3 laser irradiation patterns are shown in Table III. By increasing the laser power at the die corner as in pattern 2, the maximum temperature difference in the die became up to 46.0°C.

978-1-7281-1500-9/19 $31.00 © 2019 IEEE

TABLE III. TEMPERATURE DISTRIBUTION

Pattern	1	2	3
Laser irradiation pattern			
Thermal image			
Temperature difference	10.7 ℃	46.0 ℃	19.5 ℃

Table IV shows the results of "fillet width at side" and FCI by the 3 irradiation patterns described in Table III, where bonding time was 3 seconds. The count of the bonding time started when the die center reached 230℃. Pattern 1 and 3 provide small temperature gradient in the die and small FCI, while pattern 2 provides large temperature gradient in the die and large FCI. It was supposed that the fluidity of the NCF around the corners increased more than that of the center and FCI becomes higher by pattern 2.

TABLE IV. MEASUREMRNT RESULTS OF "FILLET WIDTH AT SIDE" AND FCI BY 3 IRRADIATION PATTERN..

Pattern	1	2	3
Fillet width at side (μm)	146	108	143
FCI	0.28	0.31	0.28

FCI can be generally smaller as the "fillet width at side" is smaller. However, FCI in pattern 2 did not change even the "fillet width at side" was reduced. Therefore, we found pattern 2 is the most effective for fillet minimizing and FCI enlarging.

Fig. 8 (a) shows SAM (Scanning Acoustic Microscope) image and the cross sectional views observed by SEM (Scanning Electron Microscope) are shown in Fig. 8 (b) and (c) after laser bonding at 230℃ for 3 seconds using laser irradiation pattern 2. Although some voids were observed in Fig. 8 (a), the excellent solder joint was formed, which were shown in Fig. 8 (b) and (c).

(a)

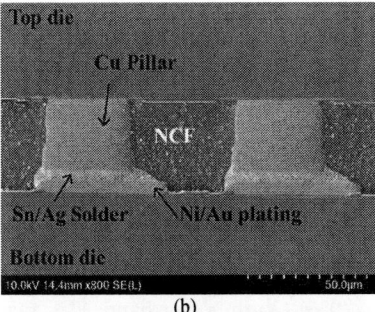

(b)

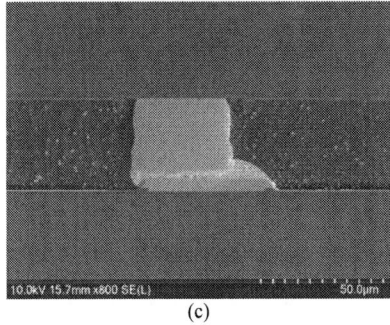

(c)

Figure 8. Bonding results after laser bonding at 230℃ for 3 s using laser irradiation pattern 2. SAM image is shown in (a) and cross sectional SEM images of peripheral and cetral bumps are shown in (b) and (c), respectively.

The temperature distribution in the die with the bonding time of 0, 1 and 3 seconds using laser irradiation pattern 2 are shown in Table V. As the bonding time elapsed, the peak temperature around the corner became lower while the central part temperature did not change much, so the temperature gradient in the die became uniform.

TABLE V. TEMPERATURE DISTRIBUTIONS IN THE DIE WITH THE BONDING TIME OF 0, 1 AND 3 SECONDS USING LASER IRRADIATION PATTERN 2.

Bonding time	0 second (Temprature rise)	1 second	3 seconds
Thermal image			
Center temperature	225.8 ℃	234.3 ℃	234.4 ℃
Peak temperature	319.5 ℃	280.3 ℃	273.7 ℃
Temperature difference	93.7 ℃	46.0 ℃	39.3 ℃

B. Evaluation of optimum conditions

The small amount of void was observed by laser bonding at 230°C for 3 seconds using laser irradiation pattern 2. The optimization of the bonding condition with pattern 2 was performed.

In the evaluation, the bonding temperature was changed from 210 to 280°C with the fixed bonding time of 3 seconds. The results of "fillet width at side" and FCI are shown in Table 6. The fillet width became larger as the temperature increased from 210 to 260°C and became almost constant from 260 to 280°C. On the other hand, FCI increased as the temperature increased from 210 to 230°C and became constant from 230 to 280°C. The coverage does not change at 230°C or higher. It was found that the reduction of "fillet width at side" and the fillet coverage at die corner hardly depend on the bonding temperature above 230°C.

The voids were observed by SAM and IR microscope. The results are shown in Fig. 9 and 10, respectively. Many big voids were observed in the samples of bonded at 210°C. The voids became small over 230°C and still remained around the bumps. We suppose that the voids remained because the bonding time is short, and the pressure to the die was released before NCF was sufficiently cured. From the viewpoint of the void elimination, there is little difference between the results at 260 and 280°C. At this time 260°C was selected for the further optimization.

TABLE VI. MEASUREMENT RESULTS OF "FILLET WIDTH AT SIDE" AND FCI FOR 3 SECONDS.

Bonding temperature	210 °C	230 °C	260 °C	280 °C
Fillet width at side (μm)	79	147	154	158
FCI	0.22	0.33	0.34	0.32

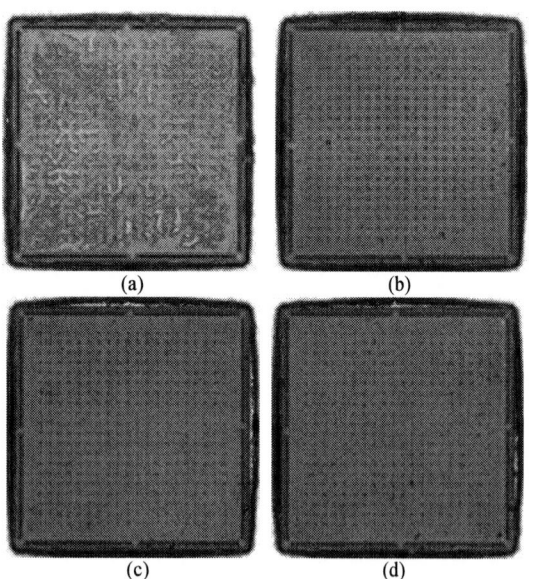

Figure 9. SAM images of the samples bonded for 3 seconds at the tempearture of 210, 230, 260 and 280°C are shown in (a), (b), (c) and (d), respectively.

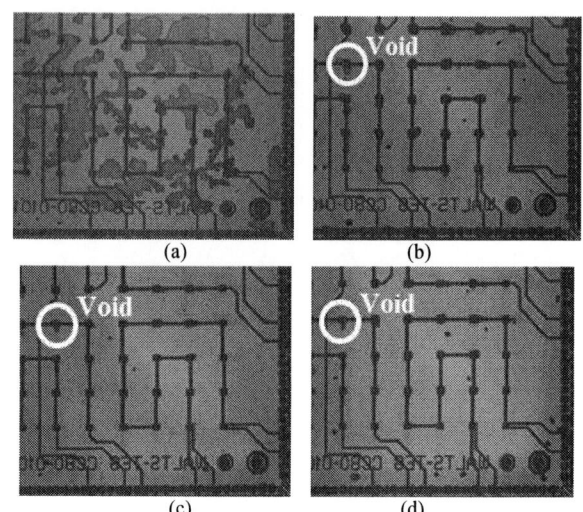

Figure 10. IR microscope images of the samples bonded for 3 seconds at the tempearture of 210, 230, 260 and 280°C are shown in (a), (b), (c) and (d), respectively.

The bonding experiments were performed for 1, 3, 5 and 7 seconds at fixed temperature of 260°C. Table VII shows the results of "fillet width at side" and FCI. As a result, both "fillet width at side" and FCI did not change by the bonding time.

The observation results by SAM and IR microscope are shown in Fig. 11 and 12. The results indicated that the void disappeared for 5 seconds. It was found that the optimum bonding condition using laser irradiation pattern 2 was 260°C for 5 seconds.

TABLE VII. RESULTS OF "FILLET WIDTH AT SIDE" AND FCI BODEND AT 260°C.

Bonding time	1 second	3 seconds	5 seconds	7 seconds
Fillet width at side (μm)	148	154	158	160
FCI	0.32	0.34	0.33	0.31

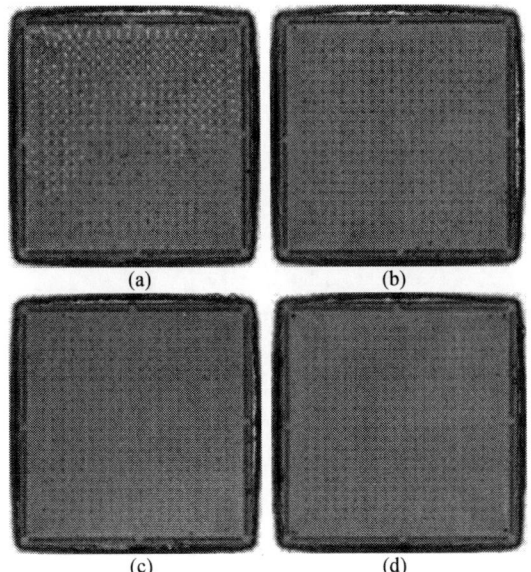

Figure 11. SAM images of the samples at 260°C of the bonding temperature for 1, 3, 5 and 7 seconds are shown in (a), (b), (c) and (d), respectively.

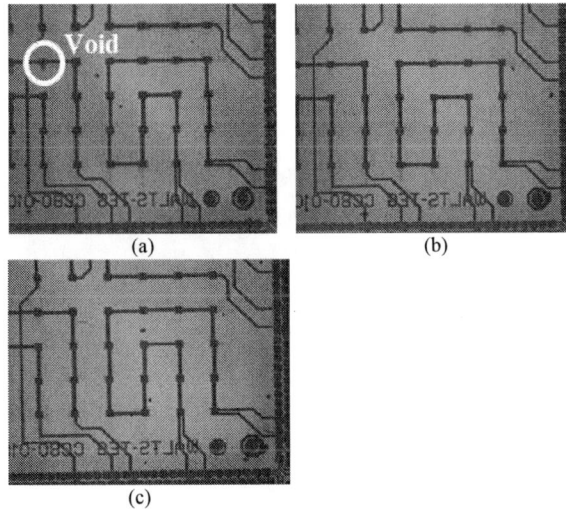

Figure 12. IR microscope imagesof the samples bonded at 260°C for 3, 5 and 7 seconds, are shown in (a), (b) and (c), respectively.

C. Comparison between TCB process and laser bonding process

Comparison of the laser bonding with the conventional TCB process was carried out. A conventional flip chip bonder was used for the latter process. Table VIII shows "fillet width at side" and FCI results of the conventional TCB process and the laser bonding process. Fig. 13, 14 and 15 show the observation results by SAM, IR microscope and the cross sectional views by SEM, respectively. Compared to the conventional TCB process, it was found that the laser bonding process can afford both the smaller "fillet width at

side" and the larger FCI. Since the small solder creeping on the side of the bumps was found from the SEM image, further optimization of the laser bonding condition will be examined.

TABLE VIII.   RESULTS OF "FILLET WIDTH AT SIDE" AND FCI

Process	Conventional TCB process	Laser bonding process
Fillet width at side (μm)	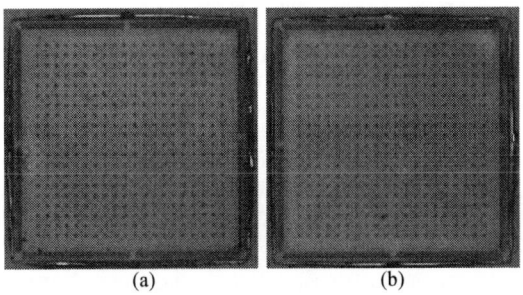 187	158
FCI	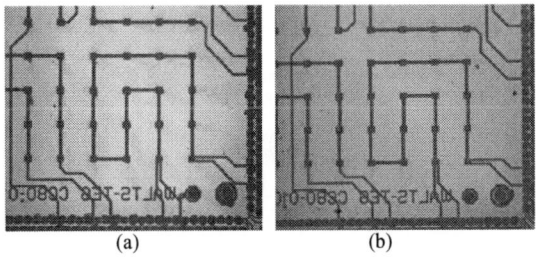 0.25	0.33

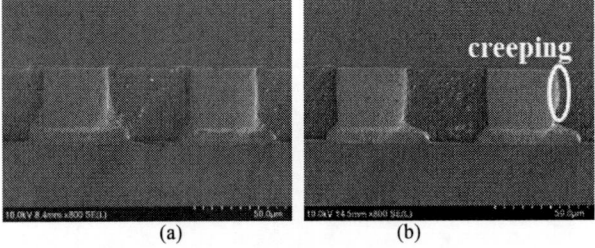

Figure 13. SAM images of the samples made by conventional TCB tool and laser bonder are shown in (a) and (b), respectively.

Figure 14. IR microscope images of the samples made by conventional TCB tool and laser bonder are shown in (a) and (b), respectively.

Figure 15. SEM images of the samples made by conventional TCB tool and laser bonder are shown in (a) and (b), respectively.

978-1-7281-1500-9/19 $31.00 © 2019 IEEE

V. CONCLUSION

The laser bonding for the fillet coverage improvement around the die corners with the fillet reduction at the middle area of the die peripheral was studied. As a result, by controlling the temperature at the die corners to be higher than the center during bonding, the high temperature gradient in the die was generated so that the fillet width could be reduced and the fillet coverage around the die corners could be improved. By using this bonding technology, the increasing of dies per wafer in CoW assembly can be expected because the small fillet at the middle area of the die peripheral with good fillet coverage at the die corner can offer the higher density die assembly.

ACKNOWLEDGMENT

The authors would like to thank Shibuya Corporation for the useful discussion and the data collection efforts of the laser bonding.

REFERENCES

[1] John H. Lau, "TSV manufacturing Yield and Hidden Costs for 3D IC integration", Proceedings of 2010 Electronic Components & Technology Conference, pp. 1031-1042.

[2] N. Asahi, Y. Miyamoto et al., "High Productivity Thermal Compression Bonding for 3D-IC", Proceedings of IEEE 2015 Interational 3D System Integration Conference(3DIC), pp. 129-133.

[3] Kazutaka Honda, Tetsuya Enomoto et al., "NCF for Wafer Lamonation Process in Higher Density Electronic Packages", Proceedings of 2010 Electronic Components & Technology Conference, pp.1853-1860.

[4] Kazutaka Honda, Akira Nagai et al., "NCF for Pre-Applied Process in Higher Density Electronic Package Including 3D-Package", Proceedings of 2012 Electronic Components & Technology Conference, pp.385-392.

[5] Provided by Shibuya Corporation.

[6] Provided by Shibuya Corporation.

Ultrafast laser Scribe: An Improved Metal and ILD Ablation Process

Julia Chiu, Aaron Gore, Tyler Osborn
Assembly and Test Technology Development
Intel Corporation
Hillsboro, USA
Julia.chiu@intel.com

Daragh Finn, Zhibin Lin, David Lord, Jon Mellen
Semi-Conductor Products Group
Electro Scientific Industries
Portland, USA
finnd@esi.com

Abstract— Traditional laser scribe which utilizes a nanosecond laser to ablate metal and interlayer dielectric layers (ILD) has been widely adopted by the microelectronics packaging industry as the gold standard for laser scribe processing in die prep singulation. Shrinking device size and thickness are driving increased demand for high die quality or die break strength. Next generation laser scribe tools are being developed to focus on minimizing thermal laser damage by using ultrafast lasers with short pulse widths in the picosecond to femtosecond range. This paper reports a robust ultrafast laser scribe process utilizing a femtosecond ultrafast laser platform developed through ESI and Intel collaboration. This first-of-a-kind equipment and process deliver significant improvement in die break strength and demonstrate a substantial reduction in bulk silicon cracking or voiding as seen in the nanosecond laser scribe heat affected zones (HAZ). Additionally, this ultrafast laser platform offers precise control over beam placement and scribe depth with onboard monitoring capability, key components to delivering a minimally needed scribe depth without sacrificing silicon integrity and processing time. This ultimately leads to a gentle coat-free ablation process that affords a low cost of ownership compare to its peers. Detailed scribe quality characteristics, process controls and overall system manufacturability will be discussed.

Keywords-ultrafast; femosecond; laser grooving; laser scribe; ablation;

I. INTRODUCTION

Rapid advances in microelectronics packaging industry have trended towards smaller, thinner, and more complex multi-chip packaging architectures which demand significant improvement in die quality to sustain increased packaging stresses at reduced silicon dimensions for individual dies. Additionally due to shrinking in device size and metal layer separation, fabrication of devices in new technology nodes has moved towards very low-k (relative dielectric constant) interlayer dielectric (ILD) materials between the metal stack [1]. Processing these novel low-k materials poses major challenges for die prep assembly and packaging due to decreased thermomechanical strength and low adhesion compared to predecessors. To address high risk of interlayer delamination during wafer singulation, a laser ablation processing step to gently remove the ILD and metal layers around the active die prior to mechanical saw cut was

introduced in the 90nm technology node which first adopted low-k carbon doped oxide in the ILD [2]. Since first implementation in high volume manufacturing by the packaging industry, the evolution of laser ablation or laser grooving technology has been relatively steady and largely focused on nanosecond lasers when first used in early 2000s. It is well known that nanosecond laser grooving process can impart significant heat and thus damage to the device substrates which is manifested in low die break strength and can lead to reliability failures. For reasons mentioned above, laser processing equipment manufacturers as well as device manufactures in microelectronics industry have focused development efforts on employing ultrafast lasers with pulse widths in ~10s of picoseconds to 100s of femtoseconds that can significantly reduce thermal impact during the so-called cold ablation process. Intel Corporation and ESI have worked jointly on developing the industry's leading HVM (high volume manufacturing) ultrafast laser (UFL) grooving tool which exploits favorable laser to material interactions afforded by femtosecond lasers. This paper is focused on process and design challenges, equipment capability, process quality, and process controls on this UFL platform.

II. BACKGROUND AND CHALLENGES

Design rules for the IC (integrated circuit) device play an important role in finding the optimal scribe technology for an IC wafer. The characteristics of the IC street will govern the specification, scribe quality and cost of ownership for scribe technology. The following characteristics are key in optimizing or enabling new scribe technology: street width, dielectric and metal thickness, passivation type and test element group (TEG) content. These characteristics will differ across IC applications and device manufactures IC test strategies. For example, memory devices will typically have very thin dielectric and metal layers, logic devices tend to have thicker dielectrics and metals, the majority of RF devices will have thick nitride passivation layers and deep metal test features. Traditional nanosecond scribe is essentially a brute force thermal process that digs deep into the silicon substrate in order to remove all the dielectric and metal contents, this aggressive process defines current scribing specifications in the industry but also results in quality concerns such as ILD delamination, HAZ and debris. The properties of femtosecond lasers provide a different mechanism for ablation of the IC circuitry enabling higher quality scribe processes with a large process window across IC applications.

Optimal scribe process parameters minimize the damage to the silicon substrate and improve the mechanical yields during downstream package process steps. The ideal scribe process would only remove the dielectrics and metals from the street to allow the saw blade to cut the silicon un-impeded. This is essentially a "Stop-on-Silicon" (SoS) process, ultrafast scribing can enable a practical SoS scribe process with appropriate street design, where a minimal amount of silicon is ablated during the scribing process. However, there are challenges to implement SoS scribe process in the packaging industry due to performance and capability metrics are established based on nanosecond technology as the process of record (POR). Current nanosecond grooving process scribes into the silicon ~5-10 μm below the lowest layer of the IC. Over-processing creates excess heat and damage to the silicon, reducing the mechanical strength of the die. New disruptive technologies are frequently evaluated against the POR specification prior to adoption or evaluation in the factory. For example, applying UFL process to achieve similar grooving depth as nanosecond process may result in sub-optimal run rates, DBS and cost of ownership. Redefining the specification typically does not occur unless there is customer demand or technical need to evaluate and implement a new technology. The improved ablation efficiencies and process control of the ultrafast scribe process avoid unnecessary ablation of the silicon substrate, resulting in higher mechanical strength and faster processing time.

A. Nanosecond and Femtosecond Ablation Theory

Nanosecond laser processing has been well researched for laser machining in many industrial applications such as cutting, scribing, dicing and drilling [3-5]. The ablation of materials by nanosecond laser pulses typically involves heating, melting, vaporization of irradiated target, formation of vapor plume and ejection of liquid droplets, as well as re-deposition of ablation products on the target. As the pulse duration of a nanosecond pulse is much longer than the electron-lattice equilibration time, which is typically on the order of one to tens of picoseconds, the electron and lattice remain in equilibrium during the laser irradiation. The absorbed laser energy first increases the surface temperature of the target to the melting point and then to boiling temperature. The portion of the material above the boiling temperature at the surface will vaporize, creating a vapor plume above the target surface. Material removal in nanosecond laser process relies largely on the molten material expulsion by the recoil vapor pressure. During high-power nanosecond laser process, plasma plume may also be produced through vapor ionization during the pulse and be reheated by the incoming laser beam, thus effectively shields, attenuates and/or steers the laser pulse, reducing the efficiency and precision of the process [6]. On the other hand, the molten material remaining in the target quickly re-solidifies prior to the next pulse for a typical PRF of 50-400kHz in industrial nanosecond lasers. The energy left in the target is lost as heat conduction into the material. For nanosecond laser pulses, the HAZ, which is proportional to the square root of the pulse duration, is significantly larger as compared to the ones created by femto- or pico-second pulses. Furthermore, it has been found that the HAZ generated in multi-pulse nanosecond process is composed of multi-layer amorphous recast containing significant micro-cracks that could be related to the generation of large recast and laser-induced shock wave in the irradiated target during nanosecond laser process.

Femtosecond laser process is fundamentally different from nanosecond laser process. In this case, the pulse duration is shorter than the electron-lattice equilibration time. Thus, the laser pulse ends before the electrons thermally equilibrates with any ions or lattice. During laser irradiation, the laser energy is first absorbed by the free electrons in metals or by promoting valence electrons to the conduction band in semiconductors or dielectrics through nonlinear photoionization and/or avalanche ionization once excited elections obtain enough kinetic energy to excite other bound electrons [7]. In femtosecond laser process, extremely high temperature and pressure can be achieved within a small sub-μm depth, creating a superheating condition where the material in the laser-excited region quickly decomposes into small clusters and vapor mixture without much of recast layer left on the target. In addition, thanks to its short pulse duration, the ultrafast laser pulse does not directly interact with the laser-induced plasma plume which only develops after the pulse. Therefore, the plasma shielding effect in femtosecond laser process is negligible [8]. In principle, a femtosecond process could have "zero" HAZ, as the heat diffusion outside of the irradiated area is minimized in a single pulse process [9, 10]. However, practical applications of industrial machining requires using multi-pulse processes to leverage the high average powers and frequencies of the commercial lasers. For ultrafast laser process, there is a PRF-dependent accumulative pulse-to-pulse average heating effect which raises the temperature of the substrate, as the pulse count increases for a given area, this effect can reduce the ablation thresholds of the substrates and improve the machining efficiencies and material ablation rates [11].

There is also a significant difference in the per pulse ablation rates of nanosecond vs. femtosecond lasers. Nanosecond process leverages the thermal diffusion and molten material expulsion to remove a higher volume of material. Thus, nanosecond laser process typically has an order of magnitude higher material removal rate on a per pulse basis for the same energies and fluences. Industrial nanosecond lasers typically operate in the pulse repetition frequency (PRF) ranges of 50-400kHz, whereas femtosecond lasers can operate over PRF ranges of 50kHz-5MHz allowing for much higher pulse counts per unit area on the substrate to compensate for the low per pulse ablation rates. Practically speaking, both nanosecond and femtosecond processes can produce large HAZ and low mechanical strengths if the process parameters are not optimized. However the ultrashort femtosecond or picosecond processes have the potential for a much lower HAZ if the key parameters are optimized correctly [12].

Another significant differences between nanosecond and femtosecond processes is the formation and distribution of particles and debris during laser ablation. Nanoparticle generation scales as a function of pulse width, generally shorter pulse widths produce smaller sizes of nanoparticles and debris. [13, 14]. Smaller nanoparticles can be advantageous by enabling cleaner ablation processes and allowing optimization of downstream process steps.

B. Laser Wavelength Ultraviolet vs Green

Before Current commercial laser scribe processes leverage ultraviolet (UV) nanosecond lasers. In the early 2000's when wafer scribe was first developed, nanosecond lasers were a mature technology with sufficient ablation rates for high volume manufacturing to achieve acceptable costs of ownerships. The UV wavelength was chosen over green for the high absorption into semi materials and small thermal diffusion lengths relative to Green and IR. Key factors for selecting laser appropriate wavelengths for an application are; ablation rates, ablation thresholds, optical penetration and thermal diffusion. In this paper we will discuss and evaluate UV vs Green femtosecond scribe process through empirical testing.

For femtosecond laser pulses the optical penetration depth at UV is two orders of magnitude smaller than the one at Green wavelength, ~0.01 um vs. ~1 um, based on the absorption coefficients of silicon [15], however the effective penetration depth for ultrafast laser pulses is challenging to model as it also depends on laser intensity and the carrier density through free-carrier absorption. Practically speaking, for the multi-pulse process, HAZ is governed by thermal diffusion and heat accumulation effect when a high-repetition-rate laser is used. Given the similar thermal diffusion length between UV and Green pulses based on its theoretical dependence on pulse duration and the thermal diffusivity of silicon, the actual HAZ would be determined primarily by the laser process parameters. Since high average power UV lasers are not commercially available to achieve cost competitive throughputs, and optic damages is more likely to occur at UV, Green ultrafast lasers becomes a more practical choice for industrial high-power ultrafast laser processing in silicon.

While it has been shown [16] that for ultrashort (3ps) pulse ablation, the ablation threshold for UV (343nm) is less than half of that for Green (515nm) wavelength, the ablation threshold for Green reduces by 35% when the substrate temperature is raised from room temperature to 320 C degrees, approaching the ablation threshold for UV which remains mostly temperature independent. This is due to the fact while the linear absorption is dominating the absorption process at both UV and Green wavelengths, the linear absorption is independent of temperature at UV, whereas it increases exponentially with temperature for Green. This observation has an important implication for high-repetition-rate green ultrafast laser processing in our study as the heat accumulation effect on the silicon substrate will lead to an enhanced absorption for higher repetition rate process which is also illustrated by our thermal modeling experiment discussed below.

C. Ultrafast Surface Temperature modeling

To support the empirical DBS evaluation of the UFL scribe process, a thermal modeling experiment was designed (Central Composite Design) to investigate the key factors in the ablation depths and substrate surface temperatures during scribing. The model is a zero-order approximation, focusing on conductive heating through the lattice for high overlap multi-pulse scribe scenarios for a Gaussian spatial profile. We only considered the Green wavelength for the thermal model, the absorption of a femtosecond pulse in silicon is treated as linear absorption since two-phonon absorption can be neglected at this wavelength. Temperature dependent heat capacity and thermal conductivity of silicon and latent heat of melting and vaporization are considered within the thermal model. Key factors in the design were pulse [400-800fs] width, spot size [20-35μm], fluence [0.2-1 J/cm2] and repetition rate [200-2000kHz]. The responses were the ablation depth [μm] and max and min surface temperatures for center and edge of the Gaussian energy profile. The surface temperatures can be used to identify the melt zones for the scribe process. The max and min temperatures were measured at 20μs for each leg. A least-square regression model was used to analyze create a model and estimate the key parameters.

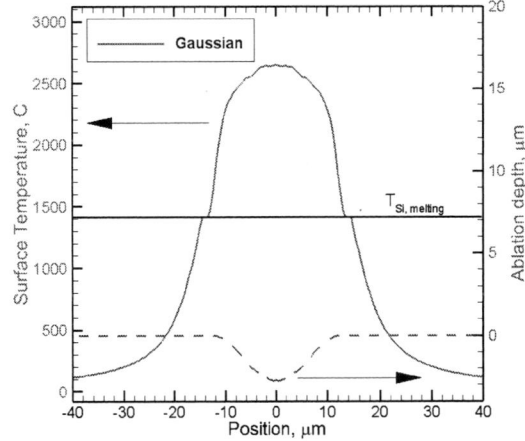

Figure 1. A typical spatial surface temperature and ablation depths across the Gaussian energy profile as obtained from the thermal model.

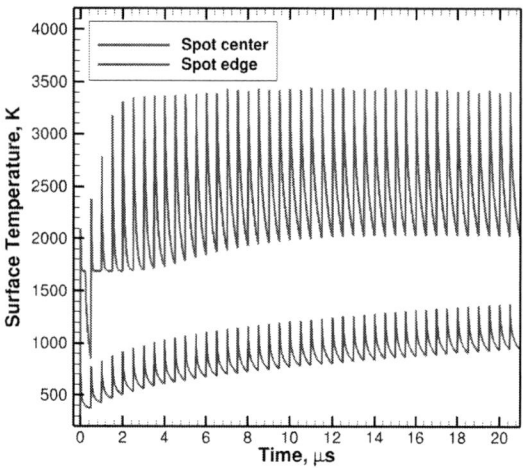

Figure 2. Temporal surface temperature over time for the center and edge of the Gaussian profile (1/e2) for accumulative femtosecond pulses.

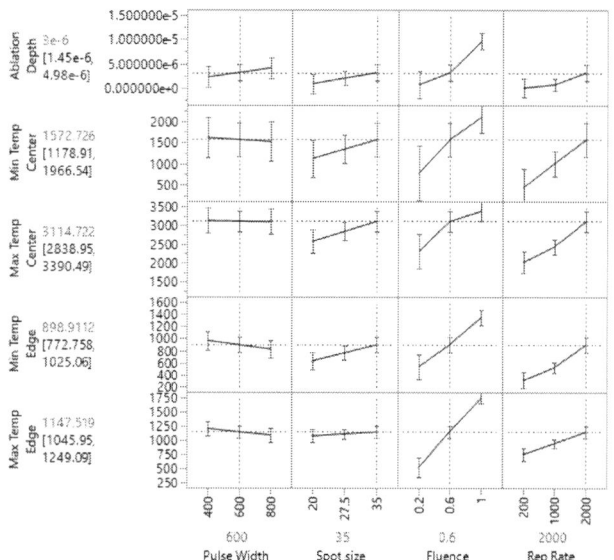

Figure 3. Model predicted response profiles for ablation depth, center temps (max and min) and edge temps (max and min).

Table 1. Key parameter estimate for the thermal ranked in order of impact to the model responses.

Source	LogWorth		PValue
Fluence(0.2,1)	10.375		0.00000
Rep Rate(200,2000)	5.743		0.00000
Fluence*Rep Rate	3.843		0.00014
Spot size*Rep Rate	2.679		0.00209
Spot size(20,35)	2.060		0.00870 ^
Fluence*Fluence	1.822		0.01507
Spot size*Fluence	1.580		0.02628
Pulse Width*Spot size	0.797		0.15975
Pulse Width*Rep Rate	0.660		0.21870
Rep Rate*Rep Rate	0.652		0.22262
Pulse Width(400,800)	0.508		0.31070 ^
Pulse Width*Fluence	0.380		0.41694

Fluence was found to have the most significant impact on the ablation rates and surface temperatures, followed by rep

rate, followed by a cross effect between fluence and rep rate (average power). Interestingly pulse width was not found to be a significant factor in this experiment. This modeling experiment supports the empirical DBS test findings that higher fluences and repetition rates will produce higher ablation rates and therefore better throughput, resulting in lower cost of ownership without degrading the DBS or scribe quality. The model also suggests that the increased local surface temperature above the silicon melting threshold (1414°C) via the inter-pulse thermal accumulation is beneficial to the ablation rates.

III. RESULTS AND DISCUSSION

Technical characterization such as scribe quality and die break strength across key process parameters are presented in this section. It is followed by a brief discussion on process controls and high volume manufacturability. Finally, total cost of ownership (CoO) is estimated and provided for three configurations: UFL UV, UFL green, and HVM UFL green configuration without coat and with inline metrology enabled.

A. UV vs Green

As discussed earlier, a pertinent study regarding utilization of UFL for grooving is the characterization of wavelengths on the scribe quality due to its relevance to throughput. The first set of experiments conducted on ultrafast platform compares grooving quality between UV and green wavelengths.

Figures 4 shows optical images of laser groove surfaces carried out by UFL UV and green wavelengths. Some observations can be drawn based on optical inspection of the scribe surface. First, the overall groove quality looks good in both images, that is, UV and green wavelength both demonstrate good ablation without incurring large HAZ or other visual defects. Scanning across the streets on these device wafers confirm no metal remaining.

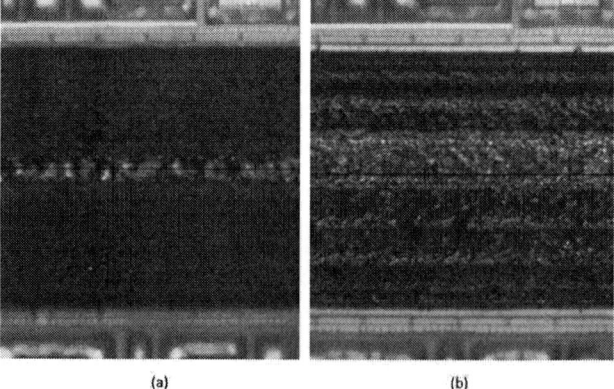

Figure 4. Optical image of groove surface for (a) femtosecond UV and (b) femtosecond green process.

Additionally, the number of "lines" across the groove surface (i.e. street) visible on both images corresponds to the number of passes utilized in each process. As discussed

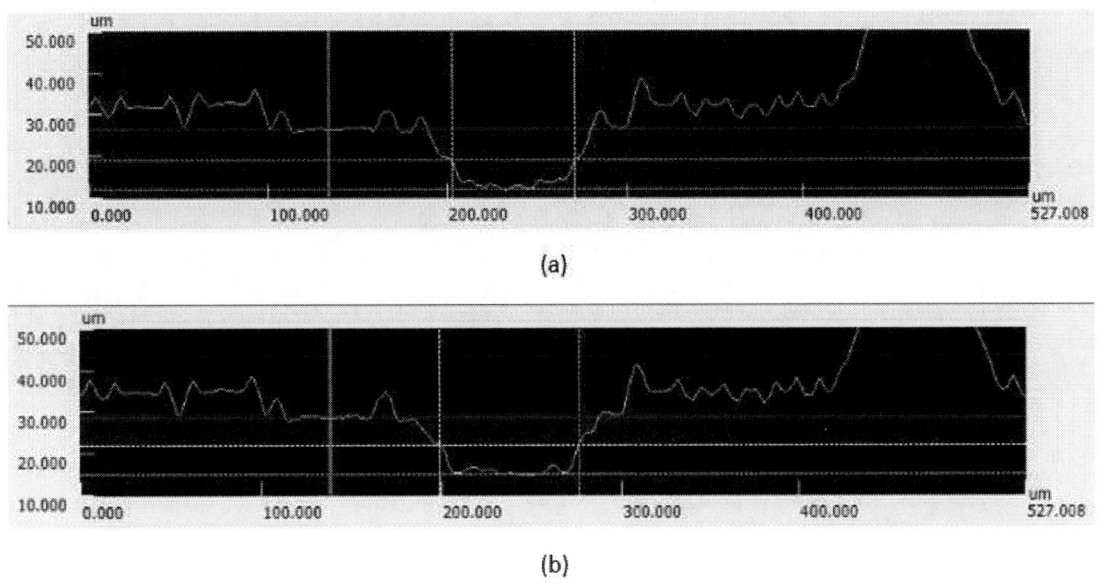

Figure 5. Groove depth profiles for (a) femtosecond UV and (b) femtosecond green process.

earlier, due availability of higher power in green wavelength compared to UV from the same laser head, throughput improvement using green wavelength can be achieved by utilizing a larger spot size at the work surface which results in fewer passes required to complete the same ablation width at the same fluence.

To confirm ablation efficiency is matched between the processes, confocal microscopy was used to collect depth profile across the groove surface. Referring to Fig. 5, grooving depth can be estimated from the distance between two horizontal dashed lines and is approximately ~7um in both cases for this device wafer. Additionally, focusing on the edge of the groove profile as indicated by the vertical dashed lines, no significant burr can be detected. This is later confirmed with SEM shown in following section.

B. Femtosecond vs Nanosecond

To characterize qualitative improvement for femtosecond laser grooving, a UFL green process was compared against a traditional nanosecond process. Following laser grooving, blade dicing was carried out to singulate dies which allow individual dies to be subsequently picked for SEM analysis. Using oblique corner imaging, micrographs of representative dies from each process are taken and are presented in Fig. 6. As captured clearly in the images, the grooving quality differs significantly between the two processes. Referring to Fig. 6(a), ultrafast laser presents an improved grooving profile with a well-defined damage-free die corner. This is sharply contrasted with Fig. 6(b) which shows rough grooving surfaces with sidewalls containing gouges and voids in the bulk silicon. The defects in Fig. 6(b) are widely recognized and reflective of heat affect zones from thermal damage in a nanosecond brute force ablation process.

C. Femtosecond vs Nanosecond

To characterize qualitative improvement for femtosecond laser grooving, a UFL green process was compared against a traditional nanosecond process. Following laser grooving, blade dicing was carried out to singulate dies which allow individual dies to be subsequently picked for SEM analysis. Using oblique corner imaging, micrographs of representative dies from each process are taken and are presented in Fig. 6. As captured clearly in the images, the grooving quality differs significantly between the two processes. Referring to Fig. 6(a), ultrafast laser presents an improved grooving profile with a well-defined damage-free die corner. This is sharply contrasted with Fig. 6(b) which shows rough grooving surfaces with sidewalls containing gouges and voids in the bulk silicon. The defects in Fig. 6(b) are widely recognized and reflective of heat affect zones from thermal damage in a nanosecond brute force ablation process.

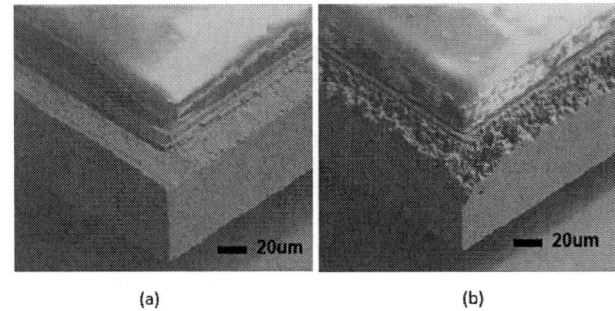

Figure 6. SEM image of die corner from (a) UFL process and (b) nanosecond process.

978-1-7281-1500-9/19 $31.00 © 2019 IEEE

As mentioned earlier, based on historical expectations, one area for ultrafast laser ablation that draws concerns is that the depth which is able to achieve due to low ablation rate at the laser energy afforded by femtosecond lasers in the market for high volume semiconductor manufacturing. Indeed, looking at Fig. 6(a) and 6(b) which are taken at the same magnification for the same substrate type, the groove depth of femtosecond ablation is quite different from nanosecond ablation. A closer look at the groove surface morphology, nanosecond process has rougher surface with some areas showing deeper groove across the surface compared to UFL process. A 3-dimensional depth profile presented in Fig. 7 is consistent with this observation. By contrast, UFL groove profile is reflective of the earlier mentioned SoS process which provides consistent groove depth across the entire ablation area and delivers optimal metal/ILD ablation with minimal silicon ablation or damage.

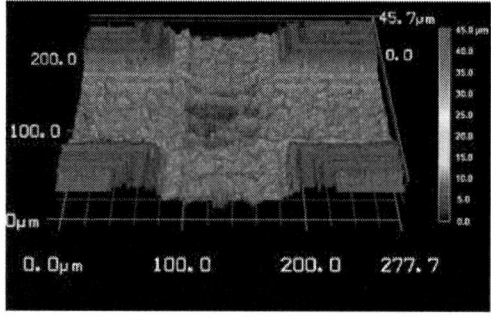

(b)

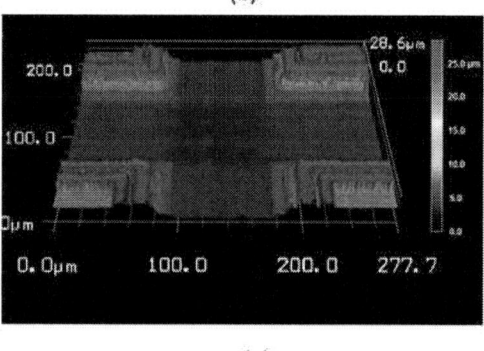

(a)

Figure 7. 3-D depth profile for (a) femtosecond process and (b) nanosecond process

To ensure complete metal removal and no concerns for ILD delamination, focus ion beam (FIB) SEM was performed over the groove edge region on the optimized HVM process and shown in Fig. 8. As shown, no metal layer can be seen under the groove surface indicating all metal layers are removed, additionally, the sub-surface high magnification image shows a flat damage-free bulk silicon with no voids or cracks which is characteristics of a SoS process. Finally, a closer look of the laser groove edge at the metal and ILD layers (not shown) confirms no ILD delamination is taking place.

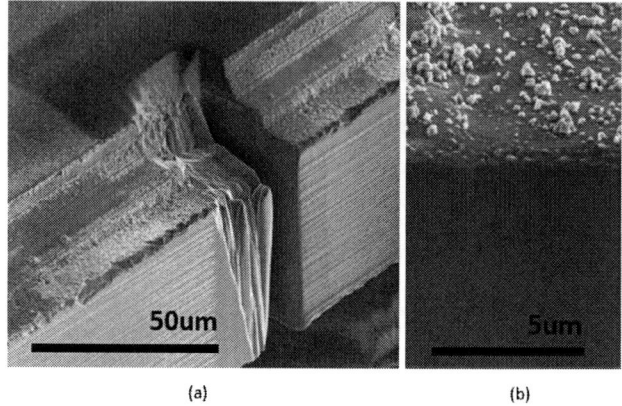

(a) (b)

Figure 8. SEM of ULF groove with (a) FIB and (b) high magnification

To improve throughput, experiments to vary rep rate from ~0.5MHz to 2.0MHz in UV and green wavelengths while keeping same net Fluence were also carried out. Summarized in Fig. 9, results show no degradation to DBS with rep rate increase and similar average values between UV and green were observed across the range studied. Based on this result, it was demonstrated low risk to implement rep rate up to 2.0MHz for the UFL platform.

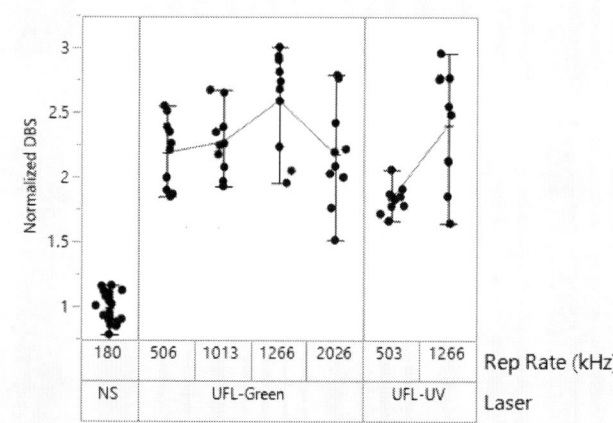

Figure 9. Normalized DBS for various repetition rates and laser types.

D. Coat vs No-coat

For the next set of experiments, debris contamination on wafer surface with and without protective coating material during laser grooving process was evaluated. Fig. 10 shows a comparison of UFL process debris generation to a typical nanosecond laser grooving process on test silicon substrate for the case of no protective coating. As can be seen, some level of debris next to the grooving edge is observed in both cases, however, the amount and size of the debris particles are significantly higher in the nanosecond case, which is contrasted with very fine particles produced by UFL grooving process. A typical laser grooving tool is equipped with debris management system at the process point to attempt to remove debris caused by ablation. Additionally, depending on the size of the particles, it may be acceptable for device performance based on device requirement.

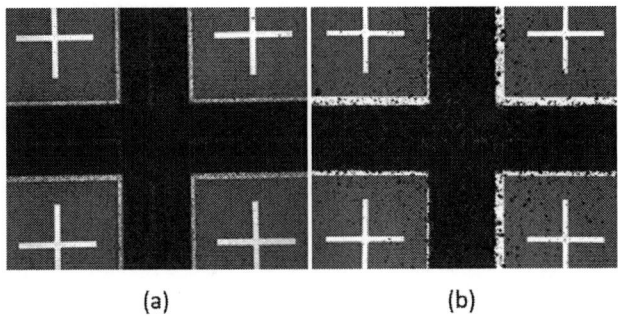

Figure 10. Typical process debris for (a) femtosecond process and (b) nanosecond process.

To investigative the feasibility of eliminating use of protective coating and relying on process exhaust for debris extraction during UFL laser grooving for a particular application, experiments were performed on a 14m node device wafers with and without coat using the HVM optimized green process. Optical images of the as-grooved street surfaces along the edge of active dies are presented in Fig. 11. In contrast to Fig. 10 above and not surprisingly, the difference between coat and no-coat ultrafast processes in Fig. 11 is not as clear which indicates a no-coat process could be feasible from debris contamination point of view. For the device wafer evaluated here in this paper, it is demonstrated and a no-coat process is viable. However, elimination of coat for ULF process cannot be generalize for all device types and applications, therefore additional evaluation will be needed for each device type with respect to specific customer requirements.

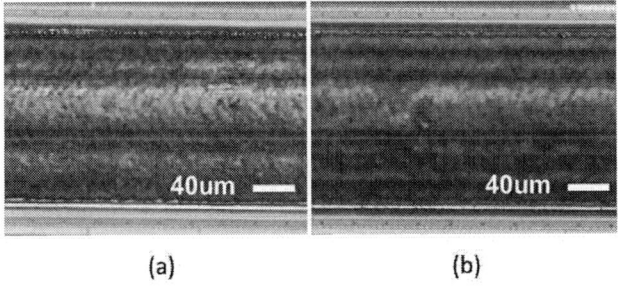

Figure 11. Optical image for (a) UFL coated and (b) un-coated process.

E. Process Control and HVM Manufacturability

Due to utilization of high PRF laser, to ensure good process control, the UFL platform employs a number of novel hardware systems to monitor laser power, beam profile, and ultimately laser groove placement accuracy. Using onboard fine camera and a fine control Z-stage, inline kerf check capability was developed which can specifically handle ultrafast groove process without coat or beam masking, two factors that can pose challenge for a traditional metrology using 2-dimensional image analysis approach.

The inline metrology results on device wafers from the UFL platform consistently show 1) process is capable for placement accuracy of +/-3um or better, that is CpK>1.33 for 3 times standard deviation of placement raw values; and 2) improved Cpk compared to traditional offline metrology using simple 2-D approach. It is important to note that placement accuracy on integrated device wafers often has higher variability and measurement error caused by the groove topology and ILD stack have on the vision algorithm. Therefore, from equipment capability perspective, accuracy monitoring standards can be used to de-confound substrate or process impacts and characterize unbiased system performance. Using monitoring standards, the UFL platform has been shown to achieve placement accuracy of +/-1.5um or better.

In addition to better process control, inline kerf check metrology also provides cost savings by improving metrology time and thus overall process throughput as well as offline metrology equipment cost elimination. These factors are included for total cost of ownership analysis that follows.

F. Cost of ownership (CoO)

Reducing the total CoO while maintaining high DBS was a key metric to successful HVM implementation of the UFL program. To calculate the UFL equipment CoO, only wafer grooving costs were considered. Fig. 12 quantifies the CoO in terms of normalized cost progressing from UV configuration to the final HVM green configuration. As shown, the availability of laser heads having higher repetition rates and power harnessed by UFL green process significantly improved the UFL equipment output while reducing the overall cost from UFL UV progress by approximately 40%. The ability to eliminate the sacrificial coat process material reduced the CoO by additional 10% while reducing environmental and integrated wafer impacts. Included in the reduced CoO was the benefit of inline metrology. Specifically, this inline metrology capability benefits directly improved equipment utilization rates, eliminated external metrology equipment, and reduced operator headcount.

Overall a reduction in total cost of ownership is approximately 50% by utilizing UFL green with high repetition rate. This significant reduction in CoO has allowed the UFL platform to be cost competitive compared to traditional nanosecond process and thus making it extremely attractive for HVM implementation purely from cost perspective. Additional technical improvements in grooving quality and DBS as presented above further increase the value proposition for the ULF platform.

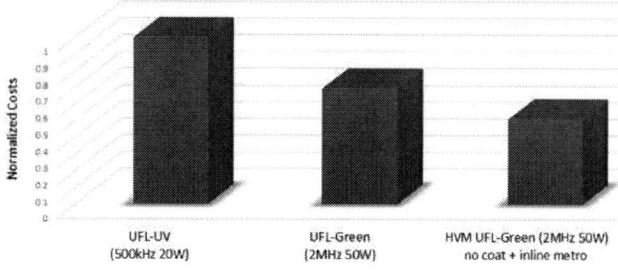

Figure 12. UFL normalized cost of ownership.

IV. CONCLUSION

An improved ultrafast grooving process was demonstrated on a new platform equipped with femtosecond lasers and optics. Surface temperature modeling was used to understand and optimize key process parameters for the ultrafast process. Grooving quality for UV and green wavelengths both showed comparable good results with similar die break strengths which are significantly higher than traditional nanosecond process. To further extend process throughput and decrease total cost of ownership, effect of PRF on DBS, elimination of coat, and inline metrology performance were evaluated. Results showed matching DBS for the range of PRF studied (up to 2MHz) and no reliability failures on any samples tested indicating a robust HVM process that is process capable at +/-3um accuracy for the device wafers characterized.

As femtosecond laser technology continue to advance, higher power or higher PRF laser heads can provide the next tier improvement for throughput. At the same time, device manufacturers are moving towards smaller and newer materials in device stack, thus demands for higher fluence (ie laser power) and varying spot size may arise to handle complex stacks. Further improvements for process monitoring as part of inline metrology including groove depth profiling will be forthcoming. Additional optimization across key processing parameters including PRF, spot size, overlap, ability to eliminate coat will need to be re-examined for upcoming new device stacks and challenges.

REFERENCES

[1] H. S. Lee, A. S. Lee, K. Y Baek, S. S. Hwang, Low Dielectric Materials for Microelectronics (Intech, 2012)

[2] W. S. Lei, A. Kumar, R. Yalamanchili, Die singulation technologies for advanced packaging: A critical review, Journal of Vacuum Science & Technology B 30, No. 4, Jul/Aug 2012

[3] C. R. Phipps, Laser Ablation and Its Applications (Springer, New York, 2007).

[4] N.B. Dahotre, S.P. Harimkar, Laser Fabrication and Machining of Materials, Springer, New York, 2008.

[5] M.R.H. Knowles, G. Rutterford, D. Karnakis, A. Ferguson, Micro-machining of metals, ceramics and polymers using nanosecond lasers, Int. J. Manuf. Technol. 33 (2007) 95-102.

[6] O. A. Ranjbar, Z. Lin, A. N. Volkov, One-dimensional kinetic simulations of plume expansion induced by multi-pulse laser

irradiation in the burst mode at 266 nm wavelength, Vacuum, 157, 361-375, 2018

[7] R. R. Gattass, E. Mazur, Femtosecond laser micromachining in transparent materials, Nature Photonics 2, 219–225 (2008)

[8] LaHaye, N.L., S.S. Harilal, P.K. Diwakar and A. Hassanein, The effect of laser pulse duration on ICP-MS signal intensity, elemental fractionation, and detection limits in fs-LAICP-MS. Journal of Analytical Atomic Spectrometry, 2013. 28(11): pp. 1781–1787.

[9] Chichkov, B.N., C. Momma, S. Nolte, F. von Alvensleben and A. Tünnermann, Femtosecond, picosecond and nanosecond laser ablation of solids. Applied Physics A, 1996 63(2): pp. 109–115.

[10] Momma, C., B.N. Chichkov, S. Nolte, F. von Alvensleben, A. Tünnermann, H. Welling and B. Wellegehausen, Short-pulse laser ablation of solid targets. Optics Communications, 1996. 129(1): pp. 134–142.

[11] H. Matsumoto, Z. Lin, J. Kleinert, Ultrafast laser ablation of copper with ~GHz bursts, Proc. SPIE. (2018) 1051902.

[12] D. S. Finn, Z. Lin, J. Kleinert, M. J. Darwin, and H. Zhang, "Study of die break strength and heat-affected zone for laser processing of thin silicon wafers," Journal of Laser Applications 27(3), p. 032004, 2015.

[13] Jeon, J.-W.; Yoon, S.; Choi, H.W.; Kim, J.; Farson, D.; Cho, S.-H. The Effect of Laser Pulse Widths on Laser—Ag Nanoparticle Interaction: Femto- to Nanosecond Lasers. Appl. Sci. 2018, 8, 112.

[14] Chakravarty, U., P. Naik, C. Mukherjee, S. Kumbhare and P. Gupta, Formation of metal nanoparticles of various sizes in plasma plumes produced by Ti: sapphire laser pulses. Journal of Applied Physics, 2010. 108(5): p. 053107.

[15] D. W. Bäuerle, Laser Processing and Chemistry (Springer-Verlag, Berlin, 2011

[16] J. Thorstensen and S. E. Foss, J. Appl. Phys. 112, 103514 (2012)

978-1-7281-1500-9/19 $31.00 © 2019 IEEE

Reliability and Benchmark of 2.5D Non-molding and Molding Technologies

Yu-Hsiang Hsiao, Che-Ming Hsu, Yi-Sheng Lin, Chien-Lin Chang Chien

Advanced Semiconductor Engineering, Group, Inc., Kaohsiung, Taiwan
Email: Hunter_Hsiao@aseglobal.com

Abstract—**Non-molding grinding technologies that apply in 2.5-dimensional integrated circuits (2.5D ICs) packaging were investigated. Die chipping failures were found after back-side wafer grinding without molding, if the standard 2.5 IC assembly process flow is followed without any optimization. The finite element analysis (FEA) model used to analyze the stress concentration and summarize as: (1) decreasing the outer grinding force and (2) increasing the die fracture strength to avoid the die chipping. The factors can be divided into three parts: (1) Process, (2) Structure and (3) Material. The optimization of the process are: the grinding recipe, the underfill (UF) fillet height, the UF bleed out distances and the different types of the UFs can be achieved and succeed to build the non-molding 2.5D IC package with back-side wafer grinding without die chipping. This non-molding with back-side wafer grinding test vehicle which we call the 2.5D Plus (2.5D+) is inspected by Scanning Acoustic Tomography (SAT), cross-section and Scanning Electron Microscopy (SEM) observation. There are no delaminations for each interface. There are good qualities of the Controlled Collapse Chip Connection (C4) bumps and micro-bump joints post assembly. The 2.5D+ also shows good reliability performance which passes 1200 thermal cycles test (TCT). The reliability results indicated that the changing of process, structure and material would not induce any side effects to damage the 2.5D+. The comparison between the 2.5D_M and 2.5D+ can be summarized: 3D flow compatible is the advantage for 2.5D_M. The lower cost (10~20%), the shorter cycles time and others application of non-molding packages are the advantages for the 2.5D+.**

Keywords-2.5D IC, Non-molding, Back side Grinding

I. INTRODUCTION

The heterogeneous integration of electronic products using 2.5D IC packaging is mass productions currently. [1-6] Lots of advanced technologies are requested. Such as micro-bump join composition, warpage control, passivation quality, back side co-planar and so on. [7-13] The back-side co-planar of different components is very important which not only impact the heat dissipation system attachment but also impact the function test (FT). The solution from Taiwan Semiconductor Manufacturing Company (TSMC) is Chip on Wafer on Substrate (CoWoS) which mold the components on the interposer wafer (we call 2.5D_M) and follow by back-side molding wafer grinding till the back-side of

components are co-planar. [8-22] In this paper, the interposer with Application-Specific IC (ASIC) and High Bandwidth Memory (HBM) without molding (non-molding wafer) is used to investigate the non-molding wafer back-side grinding technology.

II. FINITE ELEMENT ANALYSIS (FEA) AND DIE FRACTURE STRENGTH TEST

Figure 1(a) shows the wafer level warpage behavior of the test vehicle which maximum warpage is about 3158 μm. The finite element analysis (FEA) model was made for the warpage and stress concentration distribution. The warpage data calibrated with FEA model and the results as shown in Figure 1(b), the maximum warpage is 3150 μm. The validated FEA model was established as the grinding stress concentration distribution which shown in Figure 2(a). The highest stress location is at the ASIC side wall toward HBM direction which we call T-zone as shown in Figure 2(b). The T-zone is the highest risk location and the die chippings were occurred during back-side wafer grinding without molding when the process followed the standard 2.5D IC process. The FEA model and OM observations would focus on the T-zone to see if the actions improve the die chipping or not.

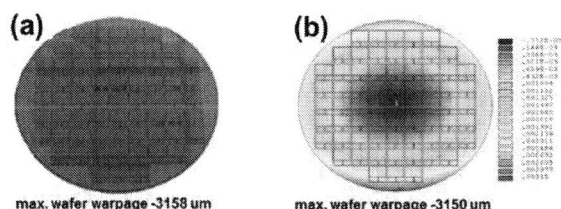

Figure 1. (a) Wafer level warpage and (b) FEA model for wafer level warpage simulation

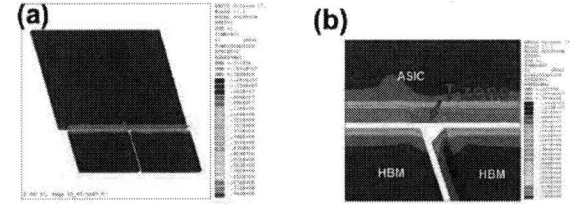

Figure 2. (a) FEA model for grinding stress concentration distribution and (b) the highest stress location at T-zone

The UF process plays an important role for the die stress during back-side wafer grinding without molding. The schematic of the UF fillet height and the UF bleed out distance are shown in Figure 3. The UF bleed out distance is defined from the die edge to the UF bleed out edge. The UF fillet height is defined from the die surface to the UF clam to the height of the die side wall. Figure 4(a) shows the stress concentration of the UF fillet height 100%. The highest stress location is at T-zone and the stress level reaches 58 MPa. If we changed the UF fillet height to be 50%, the stress level reduce to 3.9MPa as shown in Figure 4(b). The lower stress level at T-zone was observed when the UF fillet heights change from 100% to 50%. This stress level is much lower than the UF fillet height 100% and suspect that would improve die chipping issue.

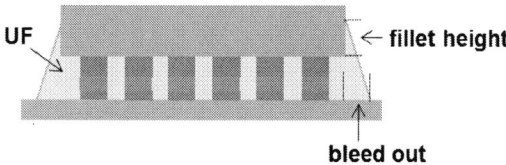

Figure 3. The schematic of the UF fillet height and the UF bleed out distances

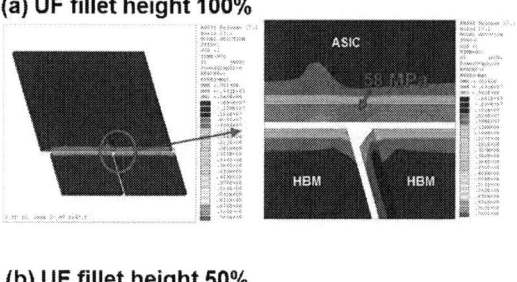

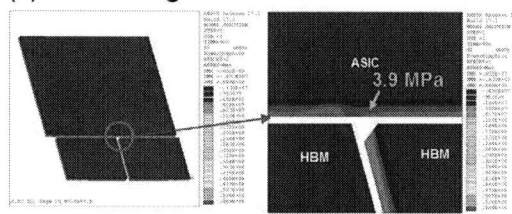

Figure 4. (a) Stress concentration of the UF fillet height 100% and (b) stress concentration of the UF fillet height 50%

Figure 5 shows the stress level versus the UF bleed out distance at T-zone. The stress level could be separated into two parts and the boundary stress level is around 44.1 MPa. The left hand side is the "Pass" area which stresses level was lower than 44.1 MPa with the UF bleed out distances less than 118 μm. The right hand side is die chipping area which stresses level was higher than 44.1 MPa with the UF bleed out distances larger than 118 μm, From the FEA results, the significant reducing stress level would be improved when the UF process control the bleed out distances less than 118μm.

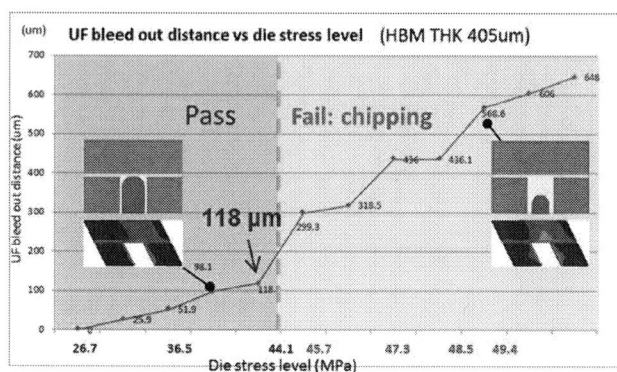

Figure 5. Stress level vs. UF bleed out distance at T-zone

Figure 6 shows the die principle stress based on two kinds of the UF. As we know that the lower die principle stress would improve the die chipping issue. The type A is the standard UF which used for the original UF process and the die principle stress level is about 60 MPa. The type B UF is a new UF which would replace the type A and follow the back-side wafer grinding without molding process. The properties of the type B UF are the lower modulus and the lower coefficient of thermal expansion (CTE) than the type A UF which would result the lower die principle stress and avoid the die chipping.

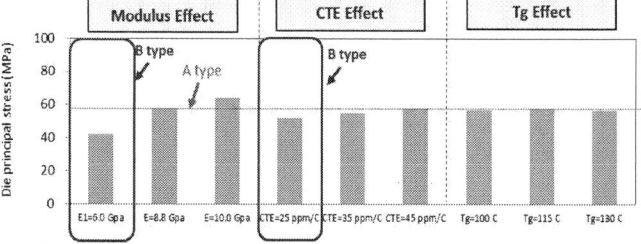

Figure 6. Die principle stress for two kinds of the UFs

Figure 7 shows the die surface roughness versus the die fracture strength for two kinds of die thickness. From the results, the die fracture strength can separate into two parts. One is around 30~110 MPa and the other is larger than 160 MPa. The lower die fracture strength means that the die would easy to be damaged during back-side wafer grinding. In other words, the higher die fracture strength means that the die would be more robust. In Figure 7, the lower die fracture strength are not only for 6 mils die thickness but also for 11 mils die thicknesses when the mean of the die surface roughness (Rz) is larger than 0.19 μm (marked with red frame in Figure 7). These results indicate that the die surface roughness would be the major factor for the die fracture strength, no matter the die thickness is 6 mils or 11 mils. If the back-side wafer grinding recipe could be controlled and make the Rz less than 0.19 μm, the die fracture strength would be increased and the die chipping issue would be improved.

Die Fracture Strength

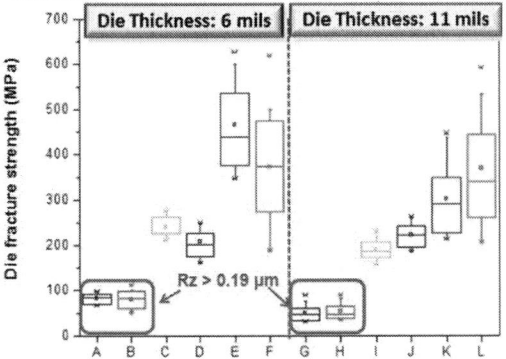

Figure 7. Die surface roughness vs. die fracture strength

III. EXPERIMENT DESCRIPTION

An interposer with ASIC and HBMs is using to do the back-side wafer grinding process without molding. ASIC and HBM are flip-chip on interposer and follow reflow, flux clean, UF process, UF curing and back-side wafer grinding process. Then, the combo die wafer would be sawed to combo die unit and flip chip on substrate then attach the passives and the ring on the 2.5D package as shown in Figure 8. The impact factors: (1) Process, (2) Structure and (3) Material would be optimized based on FEA results. The Design of Experiments (DoEs) are grinding recipes, UF fillet height, UF bleed out distance and UF types. Finally, the test vehicle was subjected to reliability check post assembly followed by Thermal Cycles Test (TCT) and analyzed by Optical Microscope (OM), Scanning Acoustic Tomography (SAT), cross-section and Scanning Electron Microscopy (SEM).

Figure 8. 2.5D IC with back-side wafer grinding without molding

IV. RESULTS & DISCUSSION

A. Grinding Recipe (Process)

Figure 9(a) shows the OM image for the original grinding recipe at T-zone and the surface roughness Rz is larger than 0.19 μm. The serious die chipping was occurred on the ASIC edge but was not occurred on the HBMs after back-side wafer grinding. Figure 9(b) shows the OM image for optimization grinding recipe and surface roughness Rz is

about 0.05~0.11μm. There were no die chipping on the ASIC and the HBMs at T-zone. As mentioned before, the surface roughness is high related to the die fracture strength. When the optimum grinding recipe (Rz was controlled < 0.11μm) is using for back-side wafer grinding, the die chipping issue would be solved.

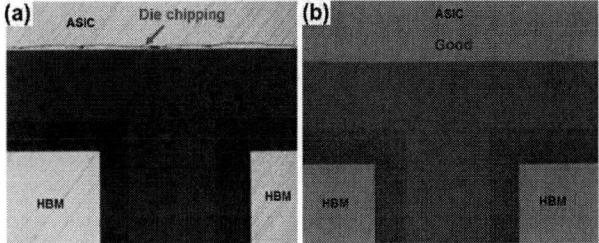

Figure 9. OM images at T-zone (a) original grinding recipe (Rz > 0.19μm) and (b) optimization grinding recipe (Rz < 0.11μm)

B. UF Process (Structure)

Figure 10(a) shows the OM image for the UF fillet height 100% at T-zone. The die chipping was observed on the ASIC after back-side wafer grinding. Figure 10(b) shows the OM image for the UF fillet height 50% at T-zone. The results are good and no die chipping on the ASIC and the HBM at T-zone. These results are consistence with the FEA results. The UF fillet height decreases from 100% to 50%, the stress level decreases from 58MPa to 3.9MPa and this stress level would not induce the die chipping.

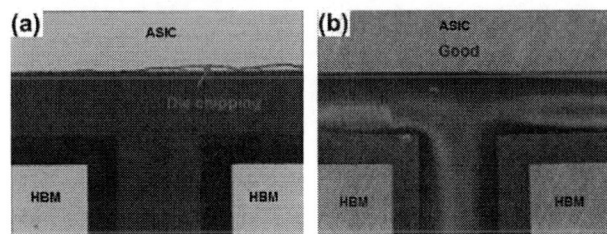

Figure 10. OM images at T-zone (a) UF fillet height 100% and (b) UF fillet height 50%

Figure 11(a) shows the OM image for the UF bleed out distances less than 118 μm at T-zone and Figure 11(b) shows the OM image for the UF bleed out distances larger than 118 μm at T-zone. No matters the UF bleed out distances larger or less than 118 μm, the back-side wafer grinding results are all good and there are no die chipping on the ASIC and HBMs at T-zone. The FEA results point out that to control the UF bleed out distance would reduce the stress level (<44.1 MPa). But the experimental results indicate that the UF bleed out seems not to be the key impact factor of the structure. Therefore, the optimization back-side wafer grinding will control the 50% UF fillet height but didn't control the UF bleed out distance.

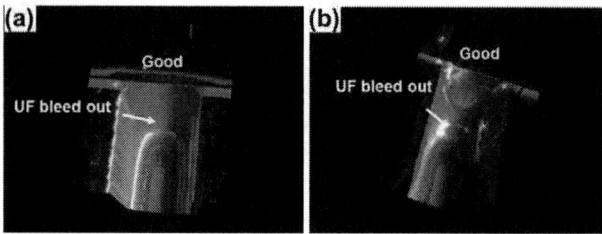

Figure 11. OM images at T-zone (a) UF bleed out distances <118 μm and (b) UF bleed out distances >118 μm

C. UF Types (Material)

Figure 12(a) shows the OM image for type A UF (original) at T-zone. The serious die chipping was occurred on the edge of ASIC but didn't observe any crack or chipping on the HBMs. Figure 12(b) shows the OM image for type B UF at T-zone. The serious die chipping was observed on the edge of ASIC but there are no cracks on HBMs. Comparison the properties of two kinds of UFs, the type B UF are lower modulus and the lower CTE than the type A UF. Suspected that the better principle properties would enhance the stress level and avoids the die chipping. In fact, the experimental results are different from FEA. It maybe because the real reducing rate of the stress level for the type B UF is not enough to prevent the die chipping during back-side wafer grinding without molding.

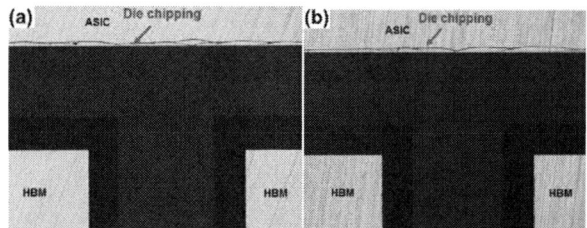

Figure 12. OM images at T-zone (a) Type A UF and (b) Type B UF

D. Post Assembly (T0) and TCT 1200 Cycles

From previously experimental results, the process would be optimized: control the grinding recipe to reach Rz < 0.11μm, control the 50% UF fillet height and use the type A UF to do the 2.5D test vehicle with backside wafer grinding without molding. Figure 13(a) shows the OM image for the test vehicle with optimization process recipe at T-zone. There are no cracks and die chipping on the ASIC and the HBMs. The optimization process, structure and material solved the die chipping issue. Figure 13(b) shows the SAT image (C-scan) of the test vehicle with the optimization process recipe. There are good qualities of each interface and didn't observe any delamination post assembly. The SAT results indicate that different grinding recipe and UF fillet height would not induce additional stress on the 2.5D package. Figure 13(c) shows the SEM image of the C4 bumps with the optimization process recipe. There are good qualities of the C4 bumps post assembly. Figure 13(d) shows the SEM image of the micro-bumps with the optimization

process recipe. There are also good qualities of the micro-bumps post assembly. There are no cold joints and non-joints. The good quality of C4 bumps and micro-bumps are as the evidence that the warpage would not get worse when the structure and material were changed. The preliminary results show that the qualities with the optimization back-side wafer grinding process recipe post assembly are all good.

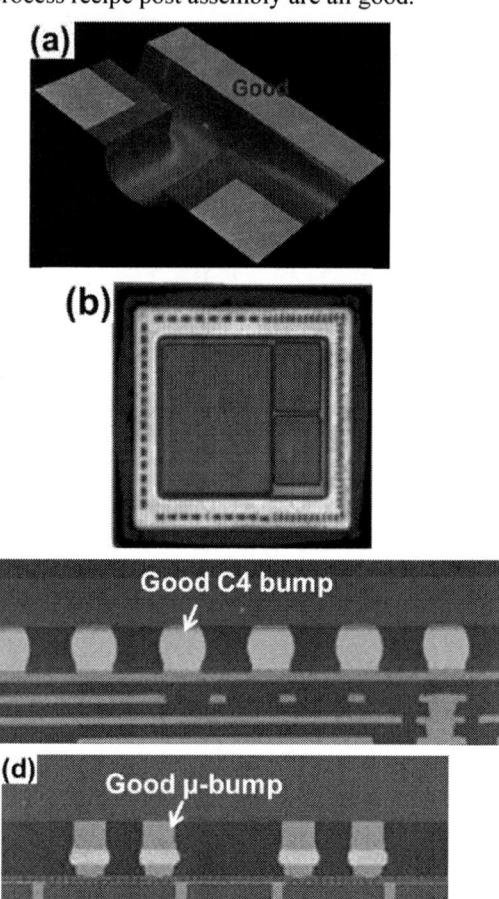

Figure 13. (a) OM image at T-zone, (b) SAT, (c) C4 bump and (d) micro-bump post assembly

Figure 14 shows the construction analysis results for the test vehicle after TCT 1200 cycles. Figure 14(a) shows the OM image at T-zone, Figure 14(b) shows the SAT image (C-scan) of test vehicle, Figure 14(c) shows the SEM image of the C4 bumps and Figure 14(d) shows the SEM image of the micro-bumps, respectively. There are no cracks and die chipping at T-zone and good qualities for each interface after 1200 cycles thermal stressing. The good joints of C4-bump and micro-bumps were also observed after 1200 cycles thermal stressing. The TCT results are as the evidences that the changing of the grinding recipe and structure would not induce additional side effects during thermal stressing. Furthermore, the test vehicle is robust and ready for qualification.

978-1-7281-1500-9/19 $31.00 © 2019 IEEE

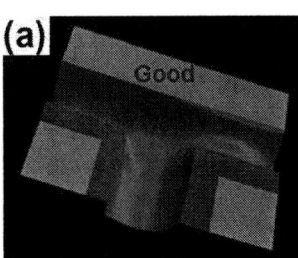

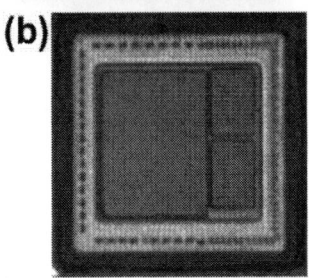

Figure 14. (a) OM image at T-zone, (b) SAT, (c) C4 bump and (d) micro-bump after TCT 1200 cycles

V. CONCLUSION

Good quality of non-molding 2.5D IC package with back-side wafer grinding is built-up which we call 2.5D plus (2.5D+). The surface roughness (Ra) and UF fillet height are the key factors to avoid the die chipping. The surface roughness (Ra) should be controlled smaller than 0.11 μm and the UF fillet height should be controlled 50%. The UF bleed out distances and the UF types do not play the major roles for die chipping. In others words, control the UF bleed out distances would change the structures and reduce the stress level. But the impact of the UF bleed out distances on stress level is slight. No matter the UF bleeds out distances over 118 μm at T-zone or not, there are no die chipping. The material properties of the type B UF are more suitable for non-molding back-side grinding due to the lower modulus and CTE. But the die chippings were observed for two kinds of UFs. The properties of type B UF would reduce the stress level but the effect is too slight to avoid the die chipping. These results indicated that the material properties would not impact the die chipping and Build of Materials (BOM) can have more selections of the UF types if the process has others application.

The process capabilities and BOMs are ready for 2.5D+ and the 2.5D+ meet the 2.5D quality requirement post assembly. The 2.5D+ also shows good reliability which passes the TCT 1200 cycles without any delamination and failures. The total yield of post assembly plus the function test is higher than 99%. The comparison between 2.5D_M and the 2.5D+, we can find that the Chip first process, the same die thickness and 3D flow compatible are the advantages of 2.5D_M. The chip middle process, the lower cost (10~20%), the shorter cycles times and the same die thickness are the advantages of the 2.5D+. Otherwise, if the packages are required the same die thickness but can't be molded in epoxy molding compound, the 2.5D+ would be the only one solution.

ACKNOWLEDGMENT

The authors would like to thank ASE Corporate R&D Product Characterization Laboratory for excellent support and FEA model built. Also many thanks to ASE Corporate R&D assembly team for valuable inputs in developing a comprehensive test vehicle and thanks to William Chen for his valuable suggestions and inputs for this paper.

REFERENCES

[1] Y. Morikawa, T. Murayama, T. Sakuishi, M. Sato, A. Suzuki, K. Suu, "Highly accurate wiring fabrication technologies by plasma dry processes for 3D-IC and fan-out packaging", Pan Pacific Microelectronics Symposium (Pan Pacific), 2018

[2] L. Wang, G. Fountain, B. Lee, G. Gao, C. Uzoh, S. McGrath, P. Enquist, S. Arkalgud, L. Mirkarimi, "Direct Bond Interconnect (DBI®) for fine-pitch bonding in 3D and 2.5D integrated circuits", Pan Pacific Microelectronics Symposium (Pan Pacific), 2017

[3] A. Agrawal, S. Huang, G. Gao, L. Wang, J. DeLaCruz, L. Mirkarimi, "Thermal and Electrical Performance of Direct Bond Interconnect Technology for 2.5D and 3D Integrated Circuits", Electronic Components and Technology Conference (ECTC), 2017

[4] C.C. Lee, C.P. Hung, C. Cheung, P.F. Yang, C.L. Kao, D.L. Chen, M.K. Shih, C.L. Chang Chien, Y.H. Hsiao, L.C. Chen, "An Overview of the Development of a GPU with integrated HBM on Silicon Interposer", Electronic Components and Technology Conference (ECTC), 2016

[5] S. Wang, R. Wang, K. Chakrabarty, M. B. Tahoori, "Multicast Test Architecture and Test Scheduling for Interposer-based 2.5D ICs", Electronic Components and Technology Conference (ECTC), 2016

[6] G. Katti, S. W. Ho, L. H. Yu, Z. Songbai, R. Dutta, R. Weerasekera, "Fabrication and Assembly of Cu-RDL-Based 2.5-D Low-Cost Through Silicon Interposer (LC–TSI)", IEEE Design & Test, 2015 pp. 23-31

[7] L. Li, P. Ton, M. Nagar, P. Chia, "Reliability Challenges in 2.5D and 3D IC Integration", Electronic Components and Technology Conference (ECTC), 2017

[8] Y.H. Hsiao, C.L. Chang Chien, C.C. Lee, H.J. Chang, "2.5D IC Micro-bump Materials Characterization and IMCs Evolution Under Reliability Stress Conditions", Electronic Components and Technology Conference (ECTC), 2016

[9] J.C Liao, A. Liao, S. Peng, G.T Lin, T. Lu, S. Chen, "Die Bonding with Non-Clean Flux in Fine Pitch Copper Pillar Bump Study and Reliability Performance for 2.5D IC Package", Electronic Components and Technology Conference (ECTC), 2016

[10] C.L. Lai, H.Y. Li, A. Chen, T. Lu, "Silicon Interposer Warpage Study for 2.5D IC without TSV Utilizing Glass Carrier CTE and Passivation

978-1-7281-1500-9/19 $31.00 © 2019 IEEE

Thickness Tuning", Electronic Components and Technology Conference (ECTC), 2016

[11] H.C. Cheng, T.H. Cheng, W.H. Chen, T.C. Chang, H.Y. Huang, "Board-Level Drop Impact Reliability of Silicon Interposer-Based 2.5-D IC Integration", IEEE Transactions on Components, Packaging and Manufacturing Technology, 2016, pp. 1493-1504

[12] R. Chaware, G. Hariharan, J. Lin, I. Singh, G. O'Rourke, K. Ng, S.Y. Pai, "Assembly Challenges in Developing 3D IC Package with Ultra High Yield and High Reliability", Electronic Components and Technology Conference (ECTC), 2015

[13] W.C. Chiou, K.F. Yang, J.L. Yeh, S.H. Wang, Y.H. Liou, T.J. Wu, J.C. Lin, C.L. Huang, S.W. Lu, C.C. Hsieh, H.A. Teng, C.C. Chiu, H.B. Chang, T.S. Wei, Y.C. Lin, Y.H. Chen, H.J. Tu, H.D. Ko, T.H. Yu, J.P. Hung, P.H. Tsai, D.C. Yeh, W.C. Wu, A.J. Su, S.L. Chiu, S.Y. Hou, D.Y. Shih, Kim H. Chen, S.P. Jeng, C.H. Yu, " An ultra-thin interposer utilizing 3D TSV technology", Symposium on VLSI Technology Digest of Technical Papers, 2012

[14] G. Hariharan, R. Chaware, I. Singh, J. Lin, L. Yip, K. Ng, S.Y. Pai, "A Comprehensive Reliability Study on a CoWoS 3D IC Package", Electronic Components and Technology Conference (ECTC), 2015

[15] W.S. Liao, C.H. Chang, S.W. Huang, T.H. Liu, H.P. Hu, H.L. Lin, C.Y. Tsai, C.S. Tsai, H.C. Chu, C.Y. Pai, W.C. Chiang, S.Y. Hou, S.P. Jeng, Doug Yu, "A manufacturable interposer MIM decoupling capacitor with robust thin high-K dielectric for heterogeneous 3D IC CoWoS wafer level system integration", IEEE International Electron Devices Meeting, 2014

[16] L. Lin, T.C. Yeh, J.L. Wu, G. Lu, T.F. Tsai, L. Chen, A.T. Xu, "Reliability Characterization of Chip-on-Wafer-on-Substrate (CoWoS) 3D IC Integration Technology", Electronic Components and Technology Conference (ECTC), 2013

[17] B. Banijamali, C. C. Chiu, C.C. Hsieh, T.S. Lin, C. Hu, S.Y. Hou, S. Ramalingam, S.P. Jeng, L. Madden, C.H. Yu, "Reliability Evaluation of a CoWoS-enabled 3D IC Package", Electronic Components and Technology Conference (ECTC), 2013

[18] Y.L. Chuang, C.S. Yuan, J.J. Chen, C.F. Chen, C.S. Yang, W.P. Changchien, Charles C. C. Liu, F. Lee, "Unified Methodology for Heterogeneous Integration with CoWoS Technology", Electronic Components and Technology Conference (ECTC), 2013

[19] J. Sun, P. Chen, F. Qina,T. Ana, H. Yua, B. Hec, "Modelling and experimental study of roughness in silicon wafer self-rotating grinding", Precision Engineering, 2018, pp. 625-637

[20] J. Sun, F. Qin, P. Chen, T. An, "A predictive model of grinding force in silicon wafer self-rotating grinding", International JournalofMachineTools&Manufacture, 2016, pp. 74-86

[21] S. Gaoa, Z. Dong, R. Kanga, B. Zhang, D. Guo, "Warping of silicon wafers subjected to back-grinding process", Precision Engineering, 2015, pp. 87-93

[22] S. Gaoa, Z. Dong, R. Kanga, B. Zhang, D. Guo, "Changes in surface layer of silicon wafers from diamond scratching", CIRP Annals, 2015, pp. 349-352

Laser-induced trench design, optimisation and validation for restricting capillary underfill spread in advanced packaging configurations

Gul Zeb, David Danovitch

Electrical & Computer Engineering,
Université de Sherbrooke
Sherbrooke QC, Canada
gul.zeb@usherbrooke.ca

Eric Turcotte

IBM Canada
Bromont QC, Canada

Abstract— Spreading of flip-chip capillary underfill material into regions of other components can complicate their assembly and/or integrity. We propose a novel, cost-effective means to control this spread through the use of thin, linear trenches of controlled depth on solder mask surfaces. To optimally exploit this method, an in-depth study is conducted to understand the underlying mechanism.

Using high resolution 3D optical profilometry, trench profile is first correlated to key laser process parameters. Trench profiles are then evaluated by means of a custom designed underfill loading test vehicle in order to determine their relative effectiveness. Characterization of both the trench and restricted underfill profiles demonstrate correlation to the Gibbs' inequality relationship for surface tension. A trench profile is identified for chip assembly that balances high underfill restricting capacity with acceptable width and depth to suit the specificity of a substrate solder mask.

The restriction of underfill spreading is then investigated in the critical dispense region of a Package on Package (PoP) application comprising stringent spacing criteria between a flip chip device and proximal BGA connections. Using trenches proposed by the laser parameter study and placed as close as 0.7 mm from chip edge, successful dispense processing and subsequent underfill flow bounding are demonstrated.

Finally, underfill spread and fillet formation in the presence of trenches on the non-dispense (exit) sides of a chip assembly is investigated to determine the limits of trench proximity to chip edge. Control of underfill spread is demonstrated at trench lines as close as 0.2 mm from the chip edge on the exit sides. Using comparable samples with unrestricted underfill flow, a reduction of 0.4 mm in underfill spread on each exit side is observed. Considering the possible contact angle of the underfill at the trench edge, one can model how close the trench lines should be placed to achieve a fillet height satisfying the design specifications.

Keywords- heterogeneous integration; capillary underfill; system in package; flip chip; ultraviolet laser

I. INTRODUCTION

As transistor scaling struggles to cost effectively respect its historical pace, heterogeneously integrated packaging in 2D and 3D configurations is increasingly regarded as the next wave of performance improvement. Responding to such an expectation has placed an unprecedented focus on package density along two principle thrusts- interconnection pitch and

device/component proximity. For the former, flip chip bonding has evolved as the de facto standard for advanced devices owing to its superior electrical performance and area array interconnect capability. At the same time, other components with less stringent I/O requirements leverage alternative connection means, such as wire bonding or peripheral surface mounting, for optimal cost-performance. Bringing these disparate components closer together introduces new issues, one of which concerns the flow of capillary underfill used to reinforce flip chip connections.

Underfill improves the flip-chip reliability by reducing the stress on solder bumps and the device caused by the thermal mismatch between substrate, solder bumps and die, and therefore, it is one of the key factors affecting the performance of flip chip packages. In this context, the underfill flow characteristics become very crucial. Key requirements for underfill include good wetting of the underfill around the solder joints without any voids, adequate flow and fillet shape [1]. Spreading of this material into regions of non-flip chip components can complicate their assembly and/or integrity, for example, any underfill material flowing into ball grid array regions or onto bond pads may cause insulation, non-sticking or even reliability issues.

In manufacturing, increasing the number of underfill dispense passes and the number of die sides along which the underfill is dispensed, using smaller dispense needle diameters and decreasing the process temperature may reduce the wetted substrate area before underfilling occurs [2]. However, combining all factors in the most favorable fashion may result in manufacturing compromise yet may still not bring the boundary of "keep-out zone" (no-go areas for underfill on the substrate) to levels being sought by the application.

Adjusting the surface roughness by physical means or the application of chemical coating by fluorocarbons on the substrate have been cited as means to avoid underfill flow in the restricted regions [3-5]. However, these methods require an additional patterning scheme and may become complicated in practical use. Alternatively, a plasma treatment is claimed to improve the fluidity of underfill in the selected regions by increasing surface roughness and thus reduce backflow of underfill toward restricted regions [6]. Similarly, however, the robustness of the process and the requirement of a mask step would limit the application of such a method.

Another strategy to control the underfill flow is based on the introduction of physical barriers. An array of patterned structures extending from under the chip through the "keep-

978-1-7281-1500-9/19 $31.00 © 2019 IEEE

out zone" provide a reservoir for the underfill material [7]. A combination of channel and dam material is also reported to control outward flow of the underfill material to prevent excess underfill material from covering the "keep-out" zone [8]. The underfill flow is controlled by material filling the channels and being stopped by the dams' walls.

The ability of edges to obstruct liquid spreading is well established [9]. Continuous, closed undercut edges have been fabricated in substrates to confine liquids in well-defined shapes [10]. Trenches in a masking layer on top of solder mask, or alternately, a frame pattern have been proposed to inhibit the encapsulant flow in glob top applications [11]. More recently, we have demonstrated proof of principle for a cost-effective and minimally intrusive means to restrict the flow of capillary underfill using laser-engraved trenches [12]. An ultraviolet laser source (λ=355 nm), which is typically employed for wafer dicing in the semiconductor industry, is adapted to fabricate thin, linear trenches of controlled depth on solder mask.

Trench geometry (depth, width and edge profile) is crucial in developing a robust underfill spread control process. Trench edge sharpness determines its ability to impede the fluid flow based on a geometrical rule, also known as Gibbs' criterion. Suppose that a fluid with an equilibrium contact angle of θ_0 ($0<\theta_0<180°$) meets a perfectly sharp edge, such as a trench edge. Gibbs' criterion describes the limit of the contact angles at the edge (θ), such as $\theta_0 \leq \theta \leq \theta_0+(180°-\phi)$, where ϕ ($0<\phi<180°$) is the angle subtended by the two surfaces forming the solid edge (Figure 1a) [9]. The flow front can stay pinned at sharp edges until the contact angle exceeds the critical value of $\theta_0+(180°-\phi)$, e.g., through the addition of fluid. This pinning leaves an opportunity to control underfill flow by engineering trenches with sharp edges with smaller ϕ. The edge sharpness depends on laser parameters and substrate material, and a perfectly sharp edge may not be practically achievable. Since we obtain round tapered edges with a laser engraving process (see later in this paper), we define an entry radius of curvature (r_e) of a trench edge to simulate macroscopic edge angle (Figure 1b). A smaller r_e is analogous to a smaller ϕ, which is required for increasing the capability of a trench to pin the underfill.

It has been experimentally shown that there also exists a critical step height below which liquid spreading is no longer inhibited [13]. On the other hand, the exposure of the metallic conductive traces buried under the substrate solder mask even to indoor atmospheric conditions such as through a harsh engraving process is undesired from reliability standpoint [14]. Developing a trench-engraving process within the constraints of allowable trench depth for a given substrate is therefore important for overall product robustness (Figure 1c).

Lastly, trench width becomes relevant for two obvious reasons. Too narrow a trench may not offer enough resistance to the fluid flow as the fluid may contact the farthest edge and continue to spread. On the other hand, maximum allowable trench widths may be dictated by the proximity of other components to the chip (Figure 1d). As such, trenches of <100 μm width are desirable for several applications that bring flip-chip device in close proximity to other components or interconnect features.

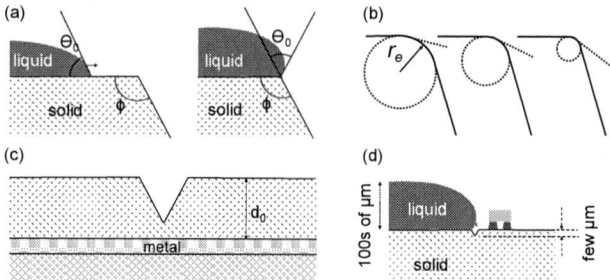

Figure 1. Schematic showing critical trench parameters for underfill spread control.(a): edge angle (b): entry radius of curvature of edge – r_e (c): depth (d): width.

This paper develops the understanding of the underlying mechanism of the laser-assisted trench formation and means to leverage the same for optimal control of underfill on a substrate solder mask. We first discuss the relationship of laser parameters with trench geometry and engineer an optimum trench profile for underfill spread control using one of the industry's standard laser processing tools. Later, we validate the trench-engraving process in two applications in a flip chip assembly configuration where the placement of laser trenches defines the boundary of "keep-out" zone and the width and height of the fillet.

II. RELATIONSHIP BETWEEN LASER PARAMETERS AND TRENCH GEOMETRY

The first part of our study related laser process parameters (power, line velocity and focus depth) to trench profile (depth, width and r_e) using high resolution 3D optical profilometry. The process consisted of a high-speed laser engraving of laminates on an ICA 1204 (Advanced Laser Separation International N.V.) laser dicing system equipped with a 355 nm UV laser source. Laminates were 31×31 mm^2 and comprised a 15–20 μm thick solder resist. Optical profilometry of the trenches was performed on a 3D laser scanning confocal microscope (Keyence, VK-X250) using a 408 nm violet laser.

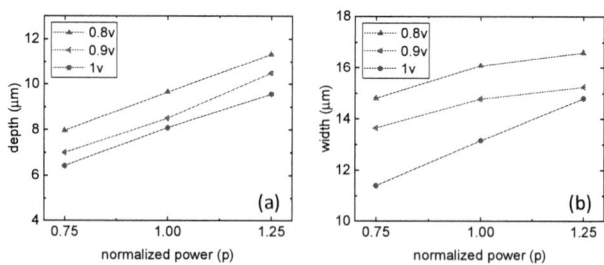

Figure 2. Variation in trench depth (a) and width (b) with laser power at different laser line velocities.

Figure 2 shows the effect of increasing laser power on trench depth and width, respectively, at various normalized velocities (v) of the laser beam. We observed an increase in trench depth and width with increasing power. Increasing the laser power increases the incident energy per unit area, which explains the higher rate of material removal. The same holds true with a decrease in laser beam velocity. For a given power

978-1-7281-1500-9/19 $31.00 © 2019 IEEE

setting, we observed material removal in the order of $1v < 0.9v < 0.8v$, which corresponded to progressively higher incident energy per unit area. With the help from these plots, appropriate laser parameters can be selected to fabricate the trench profiles which satisfy design guide requirements.

The effect of changing focus depth on the trench depth, width and radius of curvature at trench edge (r_e) is graphically shown in Figure 3. Again, p and v denote normalized laser power and velocity respectively. There is a strong correlation between focus depth and trench depth. We observed the highest trench depth for an in-focus laser beam (f=0), as both an upward or downward shift in focus increased beam radius which reduced the energy per unit area for a given parameter set. By the same token, the increased beam radius of the defocus resulted in higher widths (Figure 3b). On the other hand, we did not observe a clear trend in r_e versus focus in the -1f to 1f range. Thus, it may be useful to study focus depth across a higher range.

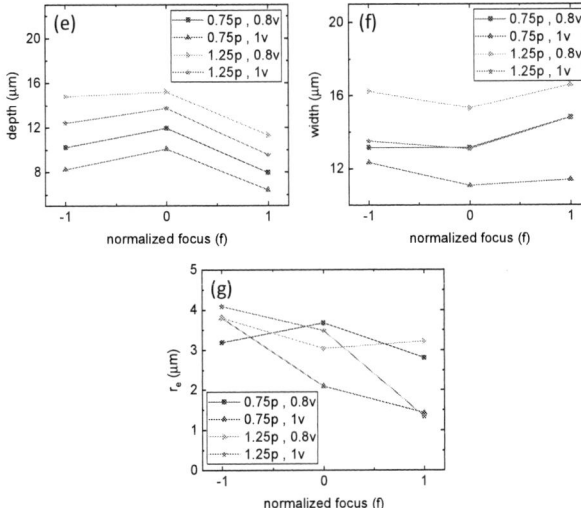

Figure 3.   Effect of laser focus depth on trench depth, width and r_e at different powers and velocities.

Trench profiles obtained by 3D confocal microscopy are shown in Figure 4 for laser focus depths up to 4f for a constant laser power and velocity. Cross sectional profiles averaged over the shown regions are also provided (in red). The in-focus laser beam formed a narrower and deeper trench and the roughness (waviness) of trench edges was very noticeable (a). This inhomogeneity in trench edge is undesirable for underfill spread control as the microscopic regions of the edge with higher radii of curvature will be less efficient toward stopping the "leakage" of underfill into the trench. It is well-understood that edges with ragged geometry and with more rounded edges do not inhibit liquid spreading appreciably [13]. As the laser focus was shifted to 2f, the beam spread out and the trench edges became more homogenous and uniform. At the same time, as previously described, the trench width increased and trench depth decreased. It is also noted that r_e assumed a relatively low value at 2f which, according to Gibbs' criterion, is required to optimally stop the underfill flow. At 4f focus

depth though, r_e assumed a larger value (d), and thus may become unsuitable for practical applications.

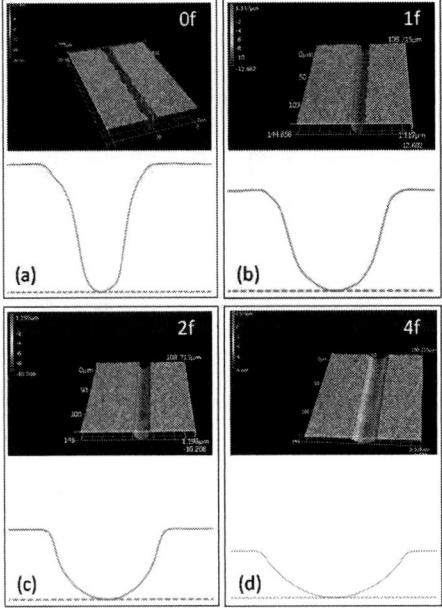

Figure 4.   The effect of changing focus depth (f=in-focus, 1f, 2f and 4f) on the trench geometry. Laser profilometry of trenches showing increase in trench width and decrease in depth at larger defocus depths. The trench edges become more uniform at 2f. (Average cross sectional profiles at the bottom (in red) are not on the same scale.)

III.   ENGINEERING OPTIMUM TRENCH FOR UNDERFILL SPREAD CONTROL

Underfill restriction responses of various trench profiles were evaluated by means of a custom designed underfill loading test vehicle, relating both underfill volume and edge profile to trench effectiveness. In a typical design, trench lines were marked on substrates to form squares of 7×7 mm^2 dimensions through cross hatching (Figure 5). To replicate the underfill process in manufacturing, a standard plasma process was carried out prior to underfill dispense. A computer controlled automatic dispensing system Camalot FX-D (Precision Placement Machines Inc., USA) was used to dispense the capillary underfill dropwise at a drop weight of ~1 mg in the middle of the trench squares at 110 °C. The effectiveness of trenches in stopping the underfill was related to the maximum underfill loading (in mg) within a square without having the underfill pass over the trench lines. Accurate weight measurements were performed using a high precision laboratory balance before curing the samples for further characterization.

Underfill restriction response of trenches with varying widths (11 μm, 13.5 μm and 16.5 μm) at nearly equal r_e values (2.5±0.2 μm) is shown in Figure 6a. Trench depths were between 7 and 11 μm, which were well within the safety margin to avoid exposure of the underlying metallurgy. It can be seen that underfill loading per test square progressively increased from 11 mg for the 11 μm trench to 30 mg for the 16.5 μm trench. These results indicate that a more robust

978-1-7281-1500-9/19 $31.00 © 2019 IEEE

trench design toward stopping the underfill can be fabricated by increasing the trench width. However, obtaining a trench wider than ~17 μm while maintaining an r_e of ~2.5 μm was not experimentally achievable. Within the tested experimental conditions, trenches >17 μm had r_e considerably greater than 2.5 μm and exhibited a decline in underfill loading, presumably because of the increasing r_e. Figure 6b shows underfill restriction response of trenches with varying r_e at a fixed width of 17 μm. We observed an increase in underfill loading, from 30 mg to 40 mg while decreasing r_e to ~1.3 μm. This observation is in line with the geometrical Gibbs' criterion discussed previously. However, for 17 μm wide trench, it was experimentally not possible to obtain $r_e < 1$ μm both within and beyond the tested conditions presented in this paper. The profile with the largest width (17 μm) and smallest r_e (1.3 μm) experimentally achievable provided the highest underfill loading of 40 mg per square.

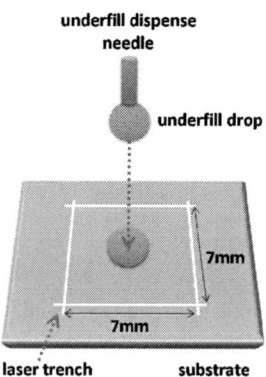

Figure 5. Schematic representation of the test vehicle used to evaluate underfill spread control of various trench geometries, as determined in terms of maximum underfill loading in mg on a custom designed area of 0.49 cm².

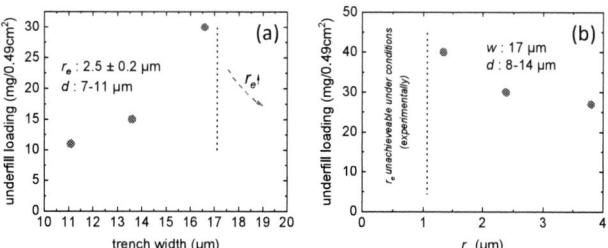

Figure 6. Maximum underfill loading increases with increasing trench width (a) and with decreasing r_e (b).

IV. MULTI-PASS ENGRAVING PROCESS

We used a multi-pass engraving approach to increase trench width. For this study we selected a trench with an r_e of ~1.3 μm and width of ~17 μm (depth = ~9 μm) which had previously exhibited an underfill loading of 40 mg per test square. We prepared samples with 2 to 6 juxtaposed laser passes. Figure 7a shows average profiles of the one-, two- and three-pass trenches only. The trench depth was ~9 μm on the initial pass and ~10 μm on the subsequent passes. Still, the depth was well within the safety margin for a typical solder

resist thickness of 15–20 μm. Figure 7b shows the nearly linear relationship between total trench width and number of laser passes. By controlling other process parameters, it was possible to keep the initial r_e unchanged while increasing the trench width during multiple passes, thus overcoming the limitation depicted in Figure 6a. Figure 7b exhibits nearly constant r_e during the multi-pass approach.

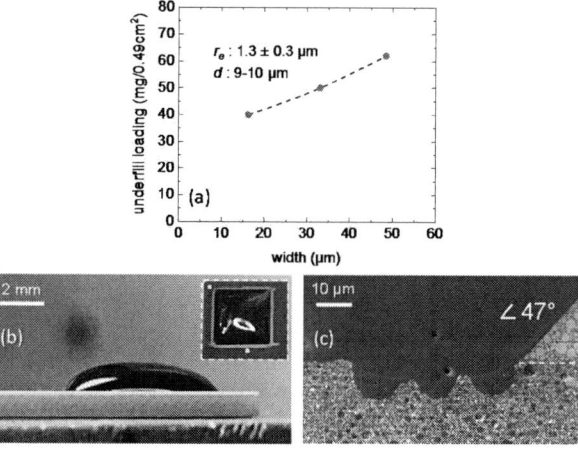

Figure 7. (a): Laser profilometry of one-, two- and three-pass trenches (b): increase in trench width and stability of r_e as a function of number of passes.

Figure 8. (a): Maximum underfill loading for various trench widths obtained through multi-pass engraving approach. The underfill loading increases with increasing trench width. (b): Side-view and top-view (inset) of 62 mg of liquid underfill confined inside a 7×7 mm² square. (c) Cross-sectional micrograph at the trench/underfill interface after curing.

Figure 8a shows underfill loading for the one-, two- and three-pass trenches. We observed a considerable increase in underfill loading for the two- and three-pass trenches in comparison with that for the one-pass trench. The two-pass trench showed an underfill loading of 50 mg, while the three-pass trench showed a loading of 62 mg per test square. The ability of a trench to stop the underfill at a high angle, as implied by very high underfill loading, is of potential interest for practical applications demanding robustness. It is also noteworthy that we did not observe any further increase in underfill loading for trenches with more than 3 passes (data not shown here). Figure 8b shows the side view and top view (inset) of the 62 mg underfill confined inside the three-pass trench test square before curing. The underfill edge shape, that is uncharacteristic for normal spread on the solder resist surface, demonstrates a high robustness of the multi-pass strategy for underfill spread control. Indeed, as shown in the cross-sectional micrograph of the trench/underfill interface in Figure 8c, the underfill angle at which it was stopped by the trench is 45-50°, whereas the angle of the unrestricted flow front with the substrate was merely ~11° (shown in our prior work, [12]). Another important observation is that the underfill stopped right at that first trench edge which came in contact with the flow front, while the resin bleed (the resin part of the underfill without filler particles) filled the trenches and flowed farther away (see inset of 8b). A more comprehensive investigation into determining the exact mechanism of underfill flow in multi-pass approach is presently underway.

V. FLIP-CHIP ASSEMBLY APPLICATIONS: UNDERFILL SPREAD CONTROL IN DISPENSE REGION

In this section and the following section, we discuss two applications that exploit the engraving process in a flip chip assembly configuration. This section investigates a configuration where controlling the underfill flow in the underfill dispense region is critical. The following section discusses the underfill flow on the non-dispense (exit) sides of the chip assembly including the formation of the exit fillet.

The control of underfill spread in the critical dispense region was related to a Package on Package (PoP) application requiring stringent spacing criteria between a flip chip device and proximal ball grid array (BGA) connections. Figure 9a shows the top view of a dispense end of the chip which was in close proximity to a row of the BGA. The underfill had to be dispensed between the chip edge and BGA connections, as the dimensions of the part were such that the other three sides did not offer sufficient space for underfill dispense without inducing unwanted material flow onto the substrate edge.

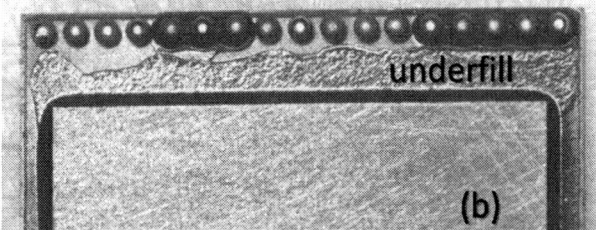

Figure 9. A chip-substrate assembly prior to (a) and after (b) underfill process showing uncontrolled flow of underfill toward the BGA in the absence of any control measures.

The limited distance between the BGA and the chip edge restricted underfill dispense to a very slow rate in order to control the underfill quantity at each pass. The distance between the dispensing needle and chip edge was adjusted to avoid underfill flow onto the top of the chip. It is also undesirable from a manufacturing viewpoint that the underfill flows towards the BGA connections as it would risk rendering them non-wettable. However, under all tested dispense conditions it had not been possible to restrain the underfill from flowing toward the BGA (Figure 9b).

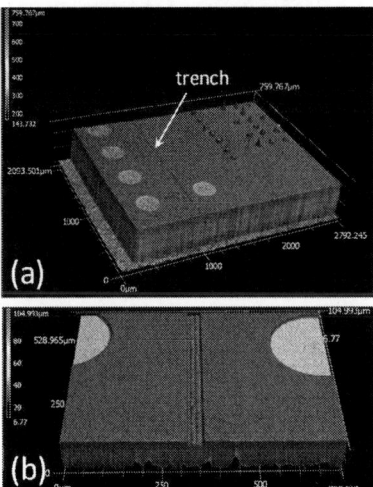

Figure 10. Application of the engraing process at the underfill dispense side in chip-substrate assembly. 3D profiles of the trench placed between BGA pads and the first chip interconnect row at different magnifications.

We addressed this issue by placing a multi-pass trench line between chip edge and BGA pads. The trench was formed prior to BGA attachment, chip assembly and oxygen plasma treatment. 3D laser microscopy profiles of one of the substrate edges at different magnifications (Figure 10) show the placement of the trench line between BGA pads and the first chip interconnect row. After BGA attachment, chip assembly and oxygen plasma treatment, the underfill was dispensed between the trench line and chip edge in a mode of repetitive dispense lines. The optical micrograph of the part in Figure 11 confirms adequate spread control of the underfill at the edge of the trench. Moreover, this process enabled a significantly faster underfill dispense rate and hence an increase in throughput.

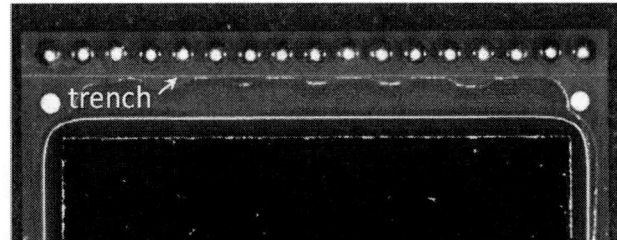

Figure 11. Successful application of the engraving process at the underfill dispense side of a chip-substrate assembly where a trench between chip edge and BGA row serves as underfill stop line.

VI. FLIP-CHIP ASSEMBLY APPLICATIONS: UNDERFILL SPREAD CONTROL AT NON-DISPENSE SIDES

Underfill control on the non-dispense (exit) sides of the chip can enable safe placement of adjacent components closer to the chip, thereby increasing packaging density. Experiments were therefore conducted to investigate underfill spread control below 500 μm on the exit sides and the impact of such control on fillet formation.

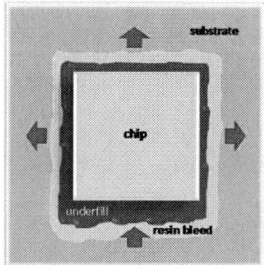

Figure 12. Schematic representation of the flip-chip assembly without trenches after underfill dispense.

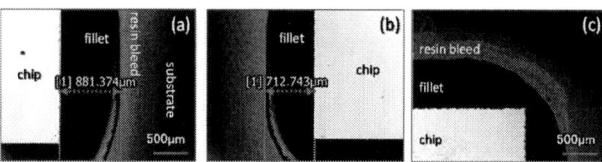

Figure 13. Laser microscopy images of randomly selected fillet regions on the flip-chip assemblies without trenches at high (a), medium (b) and low (c) underfill loading.

We prepared three parts without trenches at different amounts of underfill loading to estimate the underfill spread in uncontrolled conditions. The arrows on the schematic diagram (Figure 12) show the underfill entry and exit sides on the flip-chip assembly. Underfill was dispensed only on one side in the mode of repetitive dispense lines and the samples were cured before performing microscopy. Results of the microscopic observations are shown in Figure 13. The fillet at the exit sides of the chip laterally extended to a distance of ~900 μm from the chip edge on the substrate with highest loading of underfill material (Figure 13a). While the presence of resin bleed is obvious, this bleed can be readily removed using reactive ion etching (RIE) plasma as previously reported [12]. On the samples with medium and low underfill loading, the fillets laterally extended to distances of ~700 μm and ~600 μm, respectively. These results show the extent of underfill spread in the absence of any control measures.

We then prepared three samples with three-pass trenches on the underfill dispense and exit sides (Figure 14). On the first sample, the trench lines on the three exit sides were placed at ~550 μm from the chip edge, and for the dispense side, at ~1000 μm after which the assembly was loaded with a relatively high amount of underfill material. On the second sample, the trench lines on the three exit sides were placed at only ~200 μm and ~800 μm from the chip edge for the exit and dispense sides, respectively. A relatively low underfill

amount was dispensed because the underfill confinement area was smaller than for the first sample as the trenches were closer to the chip edge. On the third sample the trench lines were placed at intermediate distances (~300 μm and ~1000 μm) and a medium amount of underfill was dispensed. These three samples were comparable in underfill amounts to the ones shown in Figure 13 with unrestricted underfill flow. The samples were cured prior to performing laser microscopy.

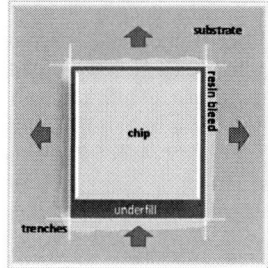

Figure 14. Schematic representation of the flip-chip assembly with trenches after underfill dispense.

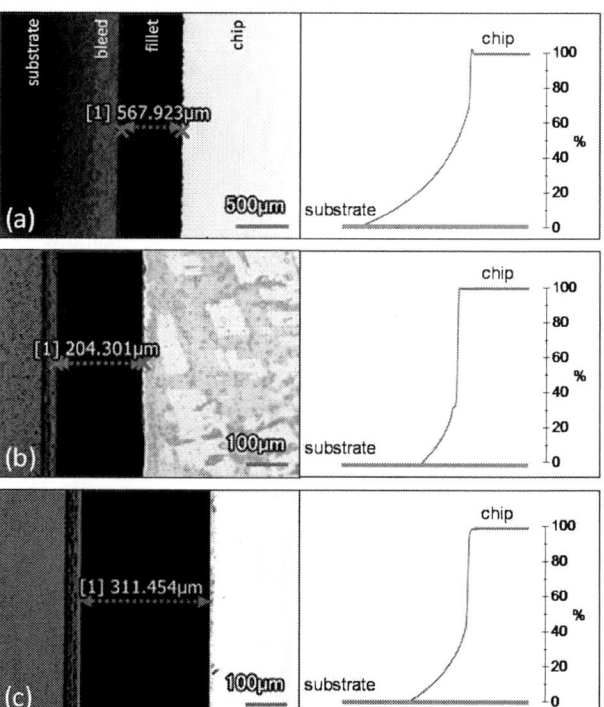

Figure 15. Laser microscopy images (left) and average profiles (right) in the fillet regions of flip-chip assemblies with target trench-chip edge distances of 550μm (a), 200μm (b) and 300μm (c). Underfill loading was adjusted accordingly.

Figure 15a shows a section of the flip-chip assembly with trenches at ~550 μm (theoretical) from the chip edge on the exit sides. Given the available area, a relatively high quantity of underfill could be confined within trench lines as seen on the micrograph (left). The average 3D profile (right) obtained on this micrograph shows the fillet shape. The fillet extended from the chip edge right up to the trench edge and a fillet height of ~70% of the chip was achieved. The fillet width on the sample with comparable underfill loading without spread control was ~900 μm (Figure 13a).

Figure 15b demonstrates that the underfill on the exit sides can be restricted to extremely small distances from the chip edge. The trench lines placed at only 200 μm from the edge on the exit sides and 800 μm on the dispense side successfully restricted the underfill within the designated area. A relatively high underfill contact angle was observed. This angle was comparable to the one observed in Figure 8c suggesting the limit of maximum loading had been achieved. We observed a fillet height of approximately one third of the chip height which is lower than for the samples with higher underfill loading but trench lines farther from the chip edge. These results suggest a tradeoff between underfill weight (or in other words, fillet height) and how close the trench lines can be placed. Balancing the two can provide a chip edge to trench distance to achieve sufficient fillet height.

Figure 15c shows the underfill confinement with the trench placed at an intermediate distance (~300 μm) from the chip edge on the underfill exit side. Here a fillet height closer to the application target was achieved. It can thus be surmised that, considering the underfill contact angle at the trench-underfill interface, one can model how close the trench lines should be placed to achieve a required fillet height. The underfill spread on the comparable sample with unrestricted flow was ~700 μm (Figure 13b), thereby demonstrating a fillet width difference of ~400 μm on each exit side between trench controlled and uncontrolled conditions. Considering the stringency in component density on various application, this difference in fillet width at three exit sides becomes very relevant.

VII. Conclusion

We propose a simple and cost-effective strategy to restrict the lateral flow of capillary underfill using trenches formed by ultraviolet laser in the solder mask of a substrate. Underfill restriction response of trenches depend on its geometry, as the trench width and edge sharpness turn out to be the two most relevant responses for optimal control. By controlling the laser parameters and devising a multi-pass strategy, we increase the trench width to ~50 μm while maintaining a relatively small radius of curvature of ~1 μm at the trench edge. This trench geometry is able to pin the underfill at an angle of 50° in comparison with the 11° angle measured for the free flowing underfill after curing.

The laser-engraving process is then integrated in an industrial process flow at a critical dispense region of a PoP application comprising stringent spacing criteria between a flip chip device and proximal BGA connections. Successful underfill dispense and spread control in chip-trench regions were demonstrated using trenches placed as close as 700 μm from chip edge.

Finally, underfill spread and fillet formation in the presence of trenches on the non-dispense sides of the chip is studied to determine the limits of trench proximity to chip edge. Control of underfill spread is demonstrated at trench lines placed as close as 200 μm from the chip edge on the exit sides. Using comparable samples with unrestricted flow of underfill, it is observed that the trench control affords a reduction of 400 μm in underfill spread on each exit side. Based on fillet height measurements, it can be surmised that trigonometric relationship for fillet shape could be employed to approximate how close the trench lines should be placed to achieve a desired fillet height.

Acknowledgment

The authors would like to thank David Bolduc, Geneviève Beaulieu, Valérie Oberson, Linda Waid and Lise Brault from IBM Canada, and the financial support of the NSERC-IBM Industrial Research Chair in Smarter Microelectronics Packaging for Performance Scaling, PROMPT Quebec and Mitacs.

References

[1] L. Nguyen and H. Nguyen, "Effect of underfill fillet configuration on flip chip package reliability," Proc. 27th Annual IEEE/SEMI International Electronics Manufacturing Technology Symposium, July 2002, pp. 291-303, doi: 10.1109/IEMT.2002.1032769.

[2] A. Lewis, A. Babiarz, C. Q. Ness, and J. Klocke, "Best practices in automated underfill dispensing," Prod. 2nd Electronics Packaging Technology Conference, 1998, pp. 63-68, doi: 10.1109/EPTC.1998.755980.

[3] M. Farooq, T. Lombardi, J. N. Filtreau, S. Bradley, C. Blais, and R. Indyk, "Surface treatments for underfill control," Patent US20070099346A1, 2005.

[4] S. A. Kulinich and M. Farzaneh, "On wetting behavior of fluorocarbon coatings with various chemical and roughness characteristics," Vacuum, vol. 79, no. 3, 2005, pp. 255-264, doi: 10.1016/j.vacuum.2005.04.004.

[5] G. Subramanian, N. Chakrapani, L. D. Decesare, S. Gokhale, J. M. Murphy, and J. Wang, "Controlling flow of underfill using polymer coating and resulting devices," Patent US20080142996A1, 2006.

[6] V. Gupta and C. A. Odegard, "Methodology to control underfill fillet size, flow-out and bleed in flip chips (FC), chip scale packages (CSP) and ball grid arrays (BGA)," Patent US20070269930A1, 2006.

[7] A. Munding, "Semiconductor device including structure to control underfill material flow," Patent US20160365258A1, 2015.

[8] K. Lee, K. Jang, and J. Kim, "Semiconductor device and method of forming adjacent channel and DAM material around die attach area of substrate to control outward flow of underfill material," Patent US8399300B2, 2010.

[9] J. F. Oliver, C. Huh, and S. G. Mason, "Resistance to spreading of liquids by sharp edges," Journal of Colloid and Interface Science, vol. 59, no. 3, 1977, pp. 568-581, doi: 10.1016/0021-9797(77)90052-2.

[10] V. Liimatainen, V. Sariola, and Q. Zhou, "Controlling liquid spreading using microfabricated undercut edges," vol. 25, no. 16, 2013, pp. 2275-2278, doi: 10.1002/adma.201204696.

[11] P. C. Celaya and J. R. Kerr, "Electronic component assembly having an encapsulation material and method of forming the same," Patent US5808873A, 1997.

[12] C. Faucher-Courchesne, D. Danovitch, L. Brault, M. Paquet, and E. Turcotte, "Controlling underfill lateral flow to improve component density in heterogeneously integrated packaging systems," Proc. 2018 IEEE 68th Electronic Components and Technology Conference (ECTC), 2018, pp. 1206-1213, doi: 10.1109/ECTC.2018.00186.

[13] Y. H. Mori, T. G. M. van de Ven, and S. G. Mason, "Resistance to spreading of liquids by sharp edged microsteps," Colloids and Surfaces, vol. 4, no. 1, 1982, pp. 1-15, doi: 10.1016/0166-6622(82)80085-1.

[14] D. Persson and C. Leygraf, "Metal carboxylate formation during indoor atmospheric corrosion of Cu, Zn, and Ni," Journal of The Electrochemical Society, vol. 142, no. 5, 1995, pp. 1468-1477, doi: 10.1149/1.2048598.

Effect of substrate preheating treatment on thermal reliability and micro-structure of Ag paste sintering on Au surface finish

Zheng Zhang, Chuantong Chen and Katsuaki Suganuma*
Institute of Scientific and Industrial Research
Osaka University
Mihogaoka 8-1, Ibaraki, Osaka, Japan
E-mail: zhangzheng@eco.sanken.osaka-u.ac.jp

Seigo Kurosaka
Central Research Laboratory
C. Uyemura & Co., Ltd.
1-5-1, Deguchi, Hirakata, Osaka, Japan
E-mail: seigo-kurosaka@uyemura.co.jp

Abstract—Sintering Ag paste on Au surface finished substrate (Ag-Au joint) is a hot topic due the Au surface finish is one of the mostly used surface methods in industry. However, to realize a reliable Ag-Au joint is not easy due to massive Ag-Au grain boundary diffusion being able to degrade the joint. In this work, we reported a preheating treatment of the substrate for improving the reliability of Ag-Au joints. The Ag-Au joints with the preheated substrate remained a shear strength of 16.5 MPa after 1000 h thermal aging at 250 °C, which is superior to the Ag-Au joints without the preheating treatment and other reported works. The fracture surface of both types of joints was also investigated. The preheating treatment can also drastically alleviate the diffusion of Ni-P to Au surface, which leaves black spots on fracture surface. The improvement of thermal reliability and the change of micro-structure are attributed to the decrease of Au grain boundaries, which is caused by the growth of Au grains during preheating treatment.

Keywords-Ag paste sintering; Ag-Au joint; thermal aging; preheating treatment; micro-structure

I. INTRODUCTION

Sintering Ag paste is an important connection method for lead-free die attachment, which is compatible with high power electronics applications. The Ag paste, usually consisting of Ag particles and organic solvents, can be sintered into a solid porous structure below 250 °C, whereas the sintered structure possesses a high melting point (961 °C) [1, 2]. Meanwhile, Ag is an excellent electrical and thermal conductor [ρ =15.87 nΩ·m (at 20 °C); k = 429 W/(m·K)], making the die attachment possible to work even at heavy currents or high temperatures[1, 3, 4]. These merits perfectly address the requirements of die attachment connection materials used for power electronics, which is difficult to obtain by traditional solder paste. In the last decade, many groups reported a favorable bonding strength (above 20 MPa) by sintering Ag paste under a pressureless, low-temperature, and atmospheric sintering condition[5-7].

To realize a robust die attachment by sintering Ag paste, the surface finish of substrate, not just the Ag paste itself, is also important. Unlike the solder joining that is realized by intermetallic compound (IMC), the bonding between sintered Ag and substrate is accomplished by the interdiffusion between Ag and surface finish[2, 8]. As a consequence, the substrate is needed to coat with a metallization layer to match Ag sintering process. At present, gold (Au) surface finish is one of the most applicable surface treatments for metallizing the substrates because Au surface finish has (i) mature processing technology, (ii) excellent electrical and thermal property, (iii) superior surface durability. Recently, many researchers are focused on sintering Ag paste on Au surface finish. For instance, Yu *et. al* realized Ag sinter joining on electroplated Au substrate with an average bonding strength above 15.3 MPa[9]. In their case, however, assisted pressure was required during the sintering process, which inevitably complicates the manufacturing process. Wang *et. al* compared the bonding performance of Ag sinter joining on immersion and electroplated Au substrates by using a Nano Ag paste[10]. Their results indicate that the electroplated Au substrate is beneficial for achieving a robust Ag-Au joint due to the larger Au grains and less (111) crystal orientation. Not long ago, our group reported on Ag sinter joining on immersion Au substrate with the use of a micron and submicron Ag particles mixed paste. The initial bonding strength achieved about 14.2 MPa, and the bonding strength can be further improved to 26.3 MPa by a one-step preheating treatment[11].

Even though many works have been working on acquiring a robust Ag-Au joint, there a still one problem that perplexes the application of Ag-Au joints — thermal reliability. According to the related works from Paknejad *et. al* and Yu *et. al* unmodified industry standard plated Au surface finish is not appropriate for high temperature applications with sintered Ag paste[12, 13]. During the thermal aging process, there is a void free layer at Au surface followed by a high porosity Ag structure that degrades the joint. The formation of weak Ag structure is mainly ascribed to the fast grain boundary diffusion between Ag and Au, which consumes enormous Ag and forms narrow necks above the void free layer. To solve this problem, changing the structure of Au surface finish a reasonable method. Usually, the thermal treatment can significantly alter the structure of metallization layer due to the growth of grains and elimination of grain boundaries at high temperatures[14, 15]. Therefore, it is believed that the thermal reliability of Ag-Au joint can be improved by the preheating treatment of the Au surface finished substrate. However, there is no related work on this method currently.

In this work, a preheating treatment of Au surface finished substrate was proposed for improving the bonding strength and thermal reliability of Ag-Au joints. The sintered joints were thermally aged at 250 °C up to 1000 h for analyzing thermal reliability. Meanwhile, the micro-structure of the joints with and without the preheated substrate was investigated by SEM and EDS to further understand the effect of preheating treatment. The findings might provide a convenient way for improving the thermal reliability of Ag-Au joints.

II. EXPERIMENTS

A. Materials

Micro-sized Ag flakes (AgC-239, Fukuda Metal Foil and Powder Co. Ltd.) and CELTOL-IA ($C_xH_yO_z$, $x > 10$, the boiling point-approximately 200 °C, Daicel Corporation) were selected as the Ag particles and an organic solvent, respectively. The weight ratio of Ag flakes and solvent was about 8 to 1. The precursor of Ag paste was mixed in a planetary vacuum mixer for 10 min at a rotating speed of 2000 rpm. The substrate surface metallization layers were prepared by a company (C. Uyemura & Co., Ltd, Japan). The Cu substrate was initially coated with an electroless Ni layer with a thickness of 7 μm and then an autocatalytic Au layer with an initial thickness of 0.8 μm. Si chips (3 × 3 × 0.8 mm) were chosen as the dummy chips and were sputtered with Ti (0.1 μm) and Ag (2 μm) successively.

B. Methods and Characterizations

For the preheating treatment, the substrates were initially preheated on a hotplate for 60 min at 250 °C. After that, the screen-printing stencil method was used to print the Ag paste. Each sample was mounted with 6 Si chips and then sintered on a hotplate at 200 °C for 60 min under an atmospheric pressureless condition. The sintered joints were thermally aged in a constant temperature oven at 250 °C for 100, 250, 500 and 1000 h.

Bonding strength of sintered joints was measured by a shear tester (Dage 4000, Japan) with a shear speed at 50 μm/s. Cross section of joints was polished by an ion milling machine and then investigated by a scanning electron microscope (SEM, SU-8020; Hitachi, Japan) and energy dispersive spectroscopy (EDS). Optical pictures of cross-section of joint fracture were token by Leica DMC camera.

III. RESULTS AND DISCUSSION

A. Shear Strength of Sintered Ag-Au Joints

Figure 1 shows the shear strength of thermally aged (250 °C) Ag-Au joint. The initial shear strength of Ag-Au joints with the original substrate was about 30.3 MPa, while the shear strength of joint was improved to 38.3 MPa after the preheating treatment. With the increase of thermal aging time, the shear strength of all Ag-Au joints decreased simultaneously. However, Ag-Au joints with the preheated substrate constantly present a superior shear strength than those with the original substrate. After 1000 h thermal aging, the shear strength of joints degraded to 13.8 (original) and

16.5 MPa (preheated), which is still better than the reported results[12, 13]. This result not only indicates that our Ag paste is suitable for Ag-Au joint—exceeding 30 MPa at initial before thermal aging, but also manifests that the thermal stability of Ag-Au joint can be improved by a substrate preheating treatment.

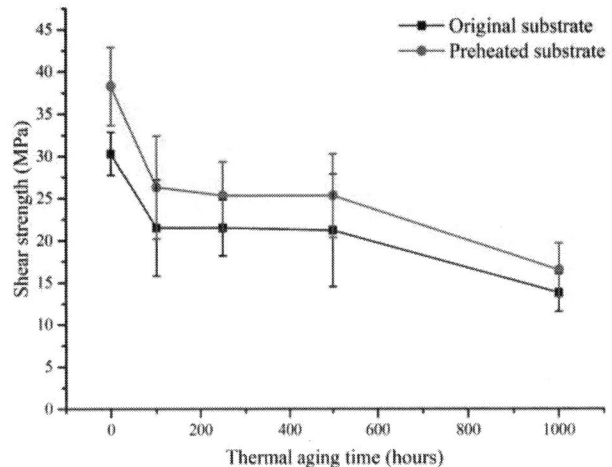

Figure 1. Shear strength of sintered Ag-Au joint with original and preheated substrate aged from initial to 1000 h.

B. Fracture Surface and Cross Section of Ag-Au joint

Figure 2 presents the optical fracture surface images of Ag-Au joints after shear test. As depicted from Figure 2 (a)-(c), the initial fracture surface on the original substrate appears a uniform morphology, suggesting the fracture happened at the interface between the substrate and the sintered Ag. After 100 h thermal aging, massive black spots are observed on the Ag-Au joint fracture surface, and the number of the black spots is increased as thermal aging time reached 1000 h. Interestingly, fracture surface on the preheated substrate shows a different phenomenon in Figure 2(d)-(f). The fracture surface shows an uneven morphology with a few sintered Ag structures. It suggests that some fracture happened at sintered Ag due to the robust bonding between sintered Ag and Au surface at interface. As the thermal aging time is up to 100 h, the fracture surface remains the uniform morphology with no black spots. Even after 1000 h, very few black spots occur on the joints fracture surface. The different fracture surface reveals that the preheating treatment might change the structure of Au layer, which affects the diffusion process during thermal aging.

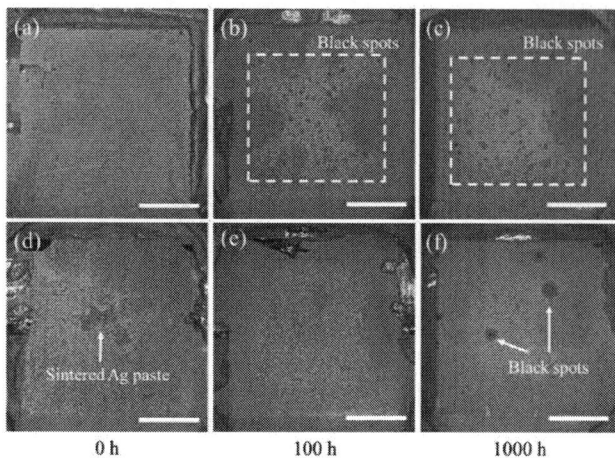

Figure 2. Optical images of surface fracture of Ag-Au joints after 0, 100 and 1000 h thermal aging storage. (a)-(c) Ag-Au joints with the original substrate; (d)-(f) Ag-Au joints with the preheated substrate.

To further verify the structure of fracture surface and composition of black spots, the fracture surfaces of the Ag-Au joint after 100 h thermal aging were investigated by SEM and EDS analysis. In Figure 3(a), most parts of fracture surface present ductile deformations of sintered Ag structure. However, a few areas are covered by fine particles, which congregate into a round shape. These rounds are supposed to correspond to black spots observed in optical images (see Figure 2). The center part of the black spot shows a dark contrast region in Figure 3(b) by the backscattered electron (BSE) mode, which suggests different elements appear in the black spot. As shown in Figure 3(c), massive ductile deformations uniformly dispense on the fracture surface when the substrate was preheated before conducting die attachment. Meanwhile, the BSE mode of SEM picture shows the same contrast in all area, suggesting there are no new elements appeared on the fracture surface. The different micro-structures of the fracture surface may be affected by these new elements appeared on the surface.

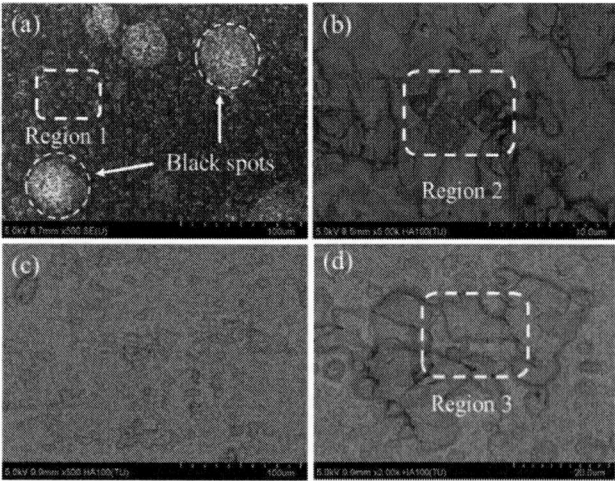

Figure 3. Fracture surface morphology of Ag-Au joint by SEM observation. (a) and (b) the original substrate; (c) and (d) the preheated substrate.

Region 1, 2 and 3 were analyzed by EDS area scanning to identify chemical composition of the fracture surface, and the quantitative results are listed in Table 1. Region 1 and 3 exhibit a similar chemical composition that mainly consists of Au ang Ag, which is due to the interdiffusion between Ag and Au. In region 2, a large amount of Ni, O, and P were detected as well. The appearance of these elements indicates amorphous Ni (Ni-P) diffuses on Au surface and gets oxidized into nickel oxide and phosphorus oxide. These oxides spread around and affect the interdiffusion and the Ag coarsening process, which results in the black spots on the fracture surface.

TABLE I.   CHEMICAL COMPOSTION OF REGION 1, 2 AND 3.

Region	Chemical Composition (At%)				
	Au	Ag	Ni	O	P
1.	24.4	72.8	0.3	1.3	1.2
2.	9.6	51.6	12.3	19.8	6.7
3.	25.8	72.1	0.2	0.9	1.0

Figure 4 presents the cross section of Ag-Au joints before and after 1000h thermal aging. As depicted in Figure 4(a) and (b), the Ag paste in as bonded Ag-Au joints with either the original or preheated substrate is sintered into a porous structure. This sintered Ag structure tightly bonds to the Si chip and the substrate, which provide a favorable shear strength above 30 MPa. A magnified view of the interface between sintered Ag and Au layer are shown in Figure (e) and (f). The Au layer of original substrate shows a fragmentized structure with some cracks inside. This fragmentized structure is attributed to grain boundary diffusion of Ag. The Ag prefers to diffuse along with Au grain boundaries due to the fast diffusion speed of Ag and the lower activation energy, thus damaging the Au layer[16]. Also, these grain boundaries might provide potential pathways for Ni-P diffuse to the Au surface. Nevertheless, the Au layer exhibits an intact structure with no cracks inside after being preheated. The morphology of cross section indicates the preheating treatment can alleviate the grain boundary diffusion during sintering process, which might be ascribed to structure changes of Au layer during preheating. The cross sections of the thermally aged Ag-Au joint with the original substrate are shown in Figure 4(c) and (g). On the surface of Au, there is a very thick layer, which is Ag-Au solid solution caused by Ag-Au interdiffusion. Sintered Ag appears a distinctive morphology according to the low magnification SEM image. The coarsening of Ag can be clearly observed in the most region of the sintered structure due to the growth of Ag grains at high temperature[17]. However, the Ag in some region remains an uncoarsening structure, which is associated with Ni-P diffusion. Figure 4(g) reveals morphology at the center region of the uncoarsening structure. The Au layer shows a broken structure with a big crack in the center. It is also worthwhile to notice that some voids occur beneath the crack. These voids are attributed to the oxidization of Ni and P due to O atoms being able to diffuse through the cracks during thermal aging. Simultaneously, some Ni can P can also diffuse onto Au

978-1-7281-1500-9/19 $31.00 © 2019 IEEE

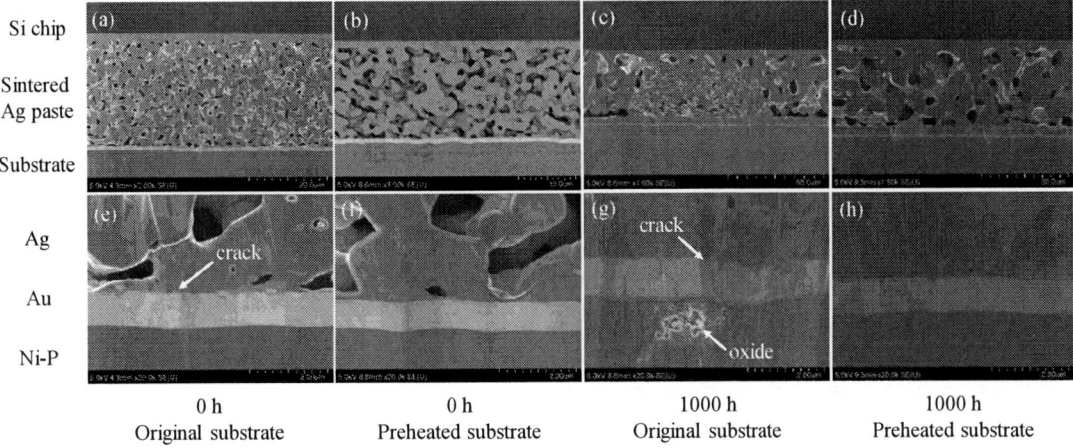

Si chip			
Sintered Ag paste	(a)	(b)	(c)	(d)
Substrate				
Ag	(e)	(f)	(g)	(h)
Au				
Ni-P				

| 0 h | 0 h | 1000 h | 1000 h |
| Original substrate | Preheated substrate | Original substrate | Preheated substrate |

Figure 4. Cross section of Ag-Au joint at initial stage and after 1000 h thermal ageing.

surface through the crack, forming nickel and phosphorus oxide as shown in Figure 3(b). These oxides can prohibit the further growth of Ag, resulting in a uncoasening area inside the sintered Ag structure. The sintered Ag structure of Ag-Au joint with the preheated substrate presents a consistent morphology in Figure 4(d). All the Ag structure have coarsened due to the thermal aging. The magnified view in Figure 4(d) shows that no oxide exists in the joint. The Au layer keeps intact structure even after 1000h thermal aging as well. The cross sections of thermally aged Ag-Au joints further testified that preheating treatment can affect the thermal stability and micro-structure of Ag-Au joint.

C. Surface Morphology of the Substrate

Figure 5(a) and (b) depict the surface morphology of Au surface without the preheating treatment in SE and BES mode, respectively. According to the SE mode SEM image, the substrate exhibits an uneven interfacial morphology, where enormous craters cover the Au surface. The BES mode image shows a more obvious observation of these craters because they have a darker contrast. After being preheated for 1 h at 250 °C, the substrate surface shows an even and uniform morphology in Figure 5(c). It is also of interest that most areas appear a similar contrast while some areas with different contrast can be found in the SEM image in BSE mode. These different contrast areas are supposed to be annealing twins that are quite common in the face-centered crystal after the annealing process[18]. Also, the appearance of annealing twin demonstrates that the growth of Au grains, which can smooth the surface and reduce grain boundaries by eliminating the smaller Au grains. When the Ag-Au joint with the preheated substrate was conducted for thermal aging. The Au layer with larger Au grains and fewer grain boundaries significantly lower the potential of Ni-P diffusing to the Au surface due to the fewer grain boundaries. Therefore, there were very few black spots even though the Ag-Au joint was thermally aged up to 1000 h.

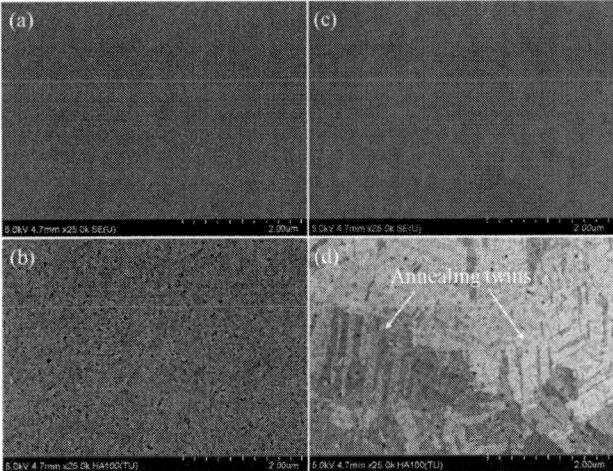

Figure 5 Surface morphology of substrates in SE and BSE mode. (a) and (b) the original substrate; (c) and (d) the preheated substrate

D. Effect of Preheating on the Ag-Au joint

As shown in the picture of fracture surface and cross section, the preheating treatment did affect the thermal reliability and the micro-structure of Ag-Au joint. This is attributed to the structure change of Au layer during the preheating (see Figure 5). It is known to that the Au surface finished is full of grain boundaries due to poly-structure of Au layer. In one aspect, these grain boundaries can lead to fast Ag-Au boundaries diffusions, resulting in weak bonding necks inside the Ag-Au joint[13, 19]. In another aspect, Au grain boundaries provide potential pathways for Ni-P diffusing onto Au surface during the thermal aging process. The Ni-P gets the reaction with O and forms nickel oxide and phosphorous oxide on the Au surface, appearing as black spots in Figure 2. When the substrate is processed with a preheating treatment, the number of Au grain boundaries is reduced due to the growth of Au grains. The preheated substrates not only afford a favorable shear strength for Ag-Au joints but also reduce the possibility of Ni-P diffusing onto Au surface. However, it is still unclear that the effect of

978-1-7281-1500-9/19 $31.00 © 2019 IEEE

the black spot on the shear strength. As shown in the tendency of shear strength with thermal aging time (see Figure 1). The initial increase of the shear strength is attributed to the growth of Au grains. With the increase of the thermal aging time, both types of Ag-Au joints show a decrease in shear strength. But Ag-Au joints did not exhibit a more severe decrease tendency in shear strength due to the appearance of black spots. This might suggest that the appearance of these black spots has no effect on the shear strength of joints. At present, the effect of black spots still needs to be investigated.

IV. CONCLUSIONS

In this work, we studied the thermal aging reliability of Au-Ag joints and proposed a feasible method for improving the sintered joints—preheating substrate before conduct Ag sintering. The as bonded joint of Ag-Au joint reached a shear strength of 30.3 MPa, while the joint with the preheated substrate was improved to 38.3 MPa. After 1000 h thermal aging, the shear strength of joint with preheated substrate remained at 16.8 MPa, which was superior to the joint with original substrate. The micro-structure of cross section, Au surface, and fracture surface suggested that the preheating treatment increased the Au grains, which can alleviate Ag-Au grain boundaries diffusion and prevent the Ni-P diffusing to Au surface. These findings provide a convenient way for improving the reliability of Ag-Au joints.

ACKNOWLEDGMENT

This work was supported by the JST Advanced Low Carbon Technology Research and Development Program (ALCA) project "Development of a high frequency GaN power module package technology" (Grant No. JPMJAL1610). The author acknowledges the financial support from the Ministry of Education, Science and Culture of Japan for his doctoral program in Osaka University, and is also thankful to the Comprehensive Analysis Center of Osaka University for use of TEM, Daicel Company in Japan for providing the solvent, and the Network Joint Research Centre for Materials and Devices, Dynamic Alliance for Open Innovation Bridging Human, Environment and Materials.

REFERENCES

[1] K. Suganuma, S. Sakamoto, N. Kagami, D. Wakuda, K.-S. Kim, and M. Nogi, "Low-temperature low-pressure die attach with hybrid silver particle paste," *Microelectronics Reliability,* vol. 52, pp. 375-380, 2012.

[2] V. R. Manikam and K. Y. Cheong, "Die attach materials for high temperature applications: A review," *IEEE Transactions on Components, Packaging and Manufacturing Technology,* vol. 1, pp. 457-478, 2011.

[3] Y. Yang, S. Ding, T. Araki, J. Jiu, T. Sugahara, J. Wang, *et al.,* "Facile fabrication of stretchable Ag nanowire/polyurethane electrodes using high intensity pulsed light," *Nano Research,* vol. 9, pp. 401-414, 2016.

[4] L. Ye, Z. Lai, J. Liu, and A. Tholen, "Effect of Ag particle size on electrical conductivity of isotropically conductive adhesives," *IEEE Transactions on Electronics Packaging Manufacturing,* vol. 22, pp. 299-302, 1999.

[5] C. Chen, S. Noh, H. Zhang, C. Choe, J. Jiu, S. Nagao, *et al.,* "Bonding technology based on solid porous Ag for large area chips," *Scripta Materialia,* vol. 146, pp. 123-127, 2018.

[6] P. Peng, A. Hu, A. P. Gerlich, G. Zou, L. Liu, and Y. N. Zhou, "Joining of Silver Nanomaterials at Low Temperatures: Processes, Properties, and Applications," *ACS Appl Mater Interfaces,* vol. 7, pp. 12597-618, Jun 17 2015.

[7] V. R. Manikam and E. N. Tolentino, "Sintering of Ag paste for power devices die attach on Cu surfaces," in *Electronics Packaging Technology Conference (EPTC), 2014 IEEE 16th,* 2014, pp. 94-98.

[8] K. Zeng, V. Vuorinen, and J. K. Kivilahti, "Interfacial reactions between lead-free SnAgCu solder and Ni (P) surface finish on printed circuit boards," *IEEE Transactions on Electronics Packaging Manufacturing,* vol. 25, pp. 162-167, 2002.

[9] F. Yu, J. Cui, Z. Zhou, K. Fang, R. W. Johnson, and M. C. Hamilton, "Reliability of ag sintering for power semiconductor die attach in high-temperature applications," *IEEE Transactions on Power Electronics,* vol. 32, pp. 7083-7095, 2017.

[10] X. Wang, Y. Mei, X. Li, M. Wang, Z. Cui, and G.-Q. Lu, "Pressureless sintering of nanosilver paste as die attachment on substrates with ENIG finish for semiconductor applications," *Journal of Alloys and Compounds,* vol. 777, pp. 578-585, 2019.

[11] C. Chen, Z. Zhang, C. Choe, D. Kim, S. Noh, T. Sugahara, *et al.,* "Improvement of the Bond Strength of Ag Sinter-Joining on Electroless Ni/Au Plated Substrate by a One-Step Preheating Treatment," *Journal of Electronic Materials,* vol. 48, pp. 1106-1115, 2019.

[12] F. Yu, R. W. Johnson, and M. C. Hamilton, "Pressureless sintering of microscale silver paste for 300° C applications," *IEEE Transactions on Components, Packaging and Manufacturing Technology,* vol. 5, pp. 1258-1264, 2015.

[13] S. Paknejad, G. Dumas, G. West, G. Lewis, and S. Mannan, "Microstructure evolution during 300 C storage of sintered Ag nanoparticles on Ag and Au substrates," *Journal of Alloys and Compounds,* vol. 617, pp. 994-1001, 2014.

[14] B. N. Kim, K. Morita, J. H. Lim, K. Hiraga, and H. Yoshida, "Effects of preheating of powder before spark plasma sintering of transparent MgAl2O4 spinel," *Journal of the American Ceramic Society,* vol. 93, pp. 2158-2160, 2010.

[15] D. Raoufi and T. Raoufi, "The effect of heat treatment on the physical properties of sol–gel derived ZnO thin films," *Applied surface science,* vol. 255, pp. 5812-5817, 2009.

[16] J. Pan and R. Balluffi, "Diffusion induced grain boundary migration in AuCu and AuAg thin films," *Acta Metallurgica,* vol. 30, pp. 861-870, 1982.

[17] S. Sakamoto, S. Nagao, and K. Suganuma, "Thermal fatigue of Ag flake sintering die-attachment for Si/SiC power devices," *Journal of Materials Science: Materials in Electronics,* vol. 24, pp. 2593-2601, 2013.

[18] S. Mahajan, C. Pande, M. Imam, and B. Rath, "Formation of annealing twins in fcc crystals," *Acta materialia,* vol. 45, pp. 2633-2638, 1997.

[19] T. Fan, H. Zhang, P. Shang, C. Li, C. Chen, J. Wang, *et al.,* "Effect of electroplated Au layer on bonding performance of Ag pastes," *Journal of Alloys and Compounds,* vol. 731, pp. 1280-1287, 2018.

Package Material Selection Criteria for High Temperature Automotive Applications

R.T.H Rongen [a], A. Mavinkurve [a], G.M. O'halloran [a], N. Owens [b], Y. Weber [c], P. Oberndorff [a],

M-L Farrugia [a], E. van Olst [a], M. van Soestbergen [a]

NXP Semiconductors

[a] Gerstweg 2, 6534 AE, Nijmegen, Netherlands
rene.rongen@nxp.com

[b] 134 Avenue du Général Eisenhower, 31100 Toulouse, France

[c] 1300 N Alma School Rd, Chandler, AZ 85224, US

Abstract— **In this paper, a structured approach is shown for the selection of reliable package materials for microelectronic devices to be applied in high-temperature automotive applications. The principles of Physics-of-Failures are followed in which the application use conditions play a leading role. Primary focus is on the 1st bond (contact from ball to bond pad).**

It will be shown that Au-wire has intrinsic capability which is limited by current density and direction. The alternative Cu-wire may have constraints too, because of its enhanced sensitivity for corrosion. Dependent on wire type, surrounding molding compound and temperature under use life application, two different corrosion mechanisms must be considered: galvanic corrosion (only present for Pd-coated Cu-wire) and interfacial Cu-Al intermetallic corrosion (occurring for all types of Cu-wire).

Life time prediction models and material characterization are used to study the dynamics and physics of corrosion mechanisms during High-Temperature Operating Life, which is a reliability test intended to address all kind of degradation mechanisms in the silicon die itself in the first place.

Finally, using the developed fundamental understanding, pros and cons of various material combinations will be discussed.

Keywords: bare Cu-wire, Pd-coated Cu-wire, IMC corrosion, galvanic corrosion, mission profile, operational life, Sulphur

I. INTRODUCTION

Application of microelectronics in harsh environments – such as high-temperature (high-T) automotive – do require built-in reliability by appropriate material selection. Solid understanding of the high-T induced failure mechanisms and their dependency on material properties as a function of temperature is a prerequisite.

The usage conditions of microelectronic devices in an application are typically characterized by the mission profile, which is a collection of relevant environmental and functional loads that devices will be exposed to during their full life cycle. In performing the assessment of the mission profile and identifying the relevant failure mechanisms (by the semiconductor supplier), it is important to define the temperature distribution over life in as much detail as possible (by the semiconductor user).

The vehicle for this study is a power device in a common, leaded medium power plastic package for an assumed, but realistic high-T automotive application. The objective is to ensure a sufficiently large reliability margin. This margin will experimentally be explored by High-Temperature Operating Life (HTOL), a reliability test used in automotive qualifications to validate primarily the functional reliability performance of the silicon (Si) die but will also address the robustness of the bond contacts to the die.

An interesting first step is the wire material selection. It will be shown by life time calculations following the model in [1], that Au-wire has its limitations. The dominant failure mechanism in this case, is void formation at the Gold (Au) wire to Aluminum (Al) bond pad, better known as Kirkendall voiding, which is enhanced by current density and current direction [1][2][3].

One alternative is Copper (Cu) wire for which this mechanism does not occur [3], but interfacial corrosion of the contact area between Cu-wire and Al bond pad determines the life time of the Intermetallic Compound (IMC) between both materials [4]. A second option is Palladium (Pd) coated Cu-wire. Concerning the interfacial IMC corrosion, this wire type in most cases behaves very similar and, in limited cases, even better. However, at elevated temperatures one must also consider galvanic corrosion at Pd to Cu interfaces, resulting in large voids that may start influencing the degradation of the IMC contact [5].

Finally, one may enter a temperature regime in which potential degradation mechanisms in the Epoxy Molding Compound (EMC) must be considered. On the one hand, additives may start to dissociate [6] and released species may enhance ongoing degradation mechanisms like corrosion. On the other hand, EMC cracking should be considered, when devices are operated at these temperatures for a long time [7].

This study will conclude with a comprehensive overview and discussion of advantages and disadvantages of different material combinations. Apart from reliability capability considerations in the process of selecting package materials, other aspects may have to be taken into account, for instance manufacturability. These are outside the scope of this study.

978-1-7281-1500-9/19 $31.00 © 2019 IEEE

II. TEST VEHICLES AND APPLICATION MISSION PROFILE

Test vehicles in this study use silicon that is manufactured in 0.28 and 0.13 μm technology with High Voltage (HV) capability of 80 and 90 V. The package used for this device is a leaded thermally enhanced Quad Flat Package (QFP) with exposed pad. The EMC type will be described in detail in section IV, Table IV. It is specially formulated with Sulphur (S) containing additives needed to enhance adhesion to the lead frame while simultaneously maintain good flowability to keep wire sweep under control.

For this study, the reliability relevant items of the mission profile are shown in Table I: detailed profile of the junction temperature (T_{junc}) and maximum applied bias (V_{max}) next to maximum current (I_{max}) – per wire – during operation of the device. T_{junc} is defined as the highest temperature in the die during operation. It is important to note that the T_{junc} profile is assumed for the sake of this study; yet based on existing mission profiles and more severe than typical for an Automotive AEC Grade 1 application – denoted in this paper as Grade 1+ [8]. Characteristic for AEC Grade 0 or 1+ is that the operating junction temperatures can exceed 150 °C.

TABLE I.   ASSUMED APPLICATION MISSION PROFILE

Temperature-time intervals		Stimuli & Loading
Junction Temperature (T_{junc}) in °C	Duration (t) in h [1]	Operational Status [2]
195	30	"On" I_{max} = 1.0 A (per wire) V_{max} = 40 V
185	30	
175	100	
145	1400	
115	6000	
40	123000	"Standby" V_{max} = 40 V
-20	840	

1. The total amount of hours (use life) is 131400 h, which is equivalent to 15 y.
2. The total amount of hours in "on" status adds is 7560 which is equivalent to about 500 h/y.

III. AU-WIRE VERSUS CU-WIRE

At elevated temperatures and high current densities, contacts to Al bond pads with Au-wire degrade differently compared to those with Cu-wire. In Au-Al contacts, the IMC grows faster than in the Cu-Al case, and the dominant degradation mechanism is Kirkendall voiding for which the growth rate depends on the current direction. For high electric currents flowing out of the device, electromigration supports Kirkendall voiding since the electron flow is in the opposite direction of the electric current, thus from the Au bond ball into to the Al bond pad, which results in faster degradation of the Au-Al interface [9]. However, this enhanced degradation appears to start only above a specific threshold current (I_{th}), which for a 50 μm Au wire is about -1200 mA, where the negative (-) sign indicates that the current is flowing out of the device [1]. Moreover, below I_{th} (i.e., larger but negative current), the degradation is following Black's power current

acceleration model for electromigration, whereas above I_{th} there is only temperature but no current dependency. This is summarized in a simple overview in Table II.**Error! Reference source not found.**

For high-current conducting wires in the vehicles used in the present study, a diameter of 30 μm is considered. So, I_{th} needs to be determined. It is assumed that I_{th} scales with the area of the IMC contact. From IMC contact diameter measurements using X-sections through bond wire to Al pad connection, this ratio turns out to be about 2.7. The resulting I_{th} is also shown in Table II.

TABLE II.   AU-AL IMC CONTACT DEGRADATION MODELS [1]

Current Regime (50 μ wire)	Model	Model Parameters [1]
I ≥ I_{th}	Arrhenius	E_A = 0.89 eV
I < I_{th}	Black	n = 4.79 E_A = 1.14 eV
I_{th} = -1200 mA (50 μm wire) I_{th} = -450 mA (30 μm wire)		

1. E_a = activation energy, n = current acceleration factor

The lifetime for currents above I_{th}, can simply be taken from in-situ HTSL experiments using a method described in [10]. For Au and AuPd wire in a typical halogen-free, Au-wire compatible, compound and at a temperature of 175 °C, these values are 4000 h and 5400 h respectively. These values have been determined on Kelvin test structures comparable to the those in [1]. The failure criterion applied is 0.1 Ω increase and 0.1 % cumulative number of fails.

The reason to consider the life time performance at 175 °C, is closely related to the mission profile. It is common to transform a detailed T_{junc} profile, as shown in Table I, to conditions for and duration of the HTOL test. The junction temperature in HTOL test ($T_{junc,HTOL}$) is usually adjusted to a value which results in a reasonable overall test duration.

To that extent, first the effective junction temperature in the application ($T_{junc\text{-}eff,app}$) has to be calculated from an Arrhenius weighted % of use-life ($\%_{life\text{-}aw,i}$) for each Temperature-time interval "i" (in Table I, 7 in total):

$$\%_{life\text{-}AW,i} = \exp[-E_a/(k_b \cdot T_i)] \cdot t_i/\textstyle\sum_i t_i \ (1)$$

where t_i is the duration of, and T_i (in K) the temperature during temperature-time interval "i", while E_a is the activation energy and k_B Boltzmann constant. Then, the resulting $T_{junc\text{-}aff,appl}$ is the equal to:

$$T_{junc\text{-}eff,app} = -[E_a/\ln(\textstyle\sum_i \%_{life\text{-}AW,i})]/k_B - 273 \ (2)$$

Next the Acceleration Factor between HTOL and Application Use-life ($AF_{HTOL\text{-}appl}$) is calculated using the standard Arrhenius model:

$$AF_{HTOL\text{-}app} = \exp[E_a/k_b \cdot (1/T_{junc\text{-}eff,app} - 1/T_{junc,HTOL}) \ (3)$$

where $T_{junc\text{-}eff,app}$ and $T_{junc,HTOL}$ are in K.

And finally, this allows to calculate the HTOL test duration (t_{HTOL}):

$$t_{HTOL} = \sum_i t_i / AF_{HTOL-app} \quad (4)$$

So far, the value of E_a was not discussed. In case of HTOL testing, many failure mechanisms are accelerated simultaneously with E_a ranging from 0.3 eV to values as high as 1.5 eV. It is common-practice in industry, to use 0.7 eV in cases where the dominant failure mechanism – the one that determines the onset time to wear-out – is not (yet) known. And this explains the selection of 175 °C for the HTOL test, because according to equations (1) to (4), the total duration would "only" take about 1000 h compared to almost 3000 h in case 150 °C would be used (which is most commonly used in HTOL).

Figure 1 can be constructed, with the Au-Al degradation model in Table II, the observed lifetime at 175 °C *without current* (4000 and 5400 h for Au and AuPd respectively), and the decrease in life time *with current* below I_{th} (-450 mA).

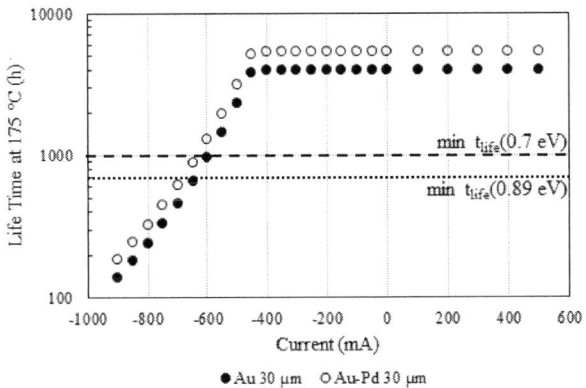

● Au 30 µm   ○ Au-Pd 30 µm

Figure 1.   Life time as a function of current for 30 µm Au- and AuPd-wire at 175 °C. Negative currents flow "out of the device" (meaning electron current flows from the Au bond ball into to the Al bond pad).

The horizontal lines indicate the minimum required life time at 175 °C to cover the mission profile in Table I, in case of unknown dominant failure mechanism (0.7 eV, dashed line) and for Au-Al IMC degradation (0.89 eV, dotted line). With current values going up to 1.0 A in some of the wires, the conclusion is that Au-wire is not an option for these devices at equivalent wire diameter.

In previous studies, it was shown that for Cu-wire, the degradation mechanism of the Cu-Al IMC is different [3]. The dominant failure mechanism is interfacial IMC corrosion. It progresses from the periphery of the IMC contact towards the center and it is *independent* of current direction and current density. In the remaining section, it will be shown that Cu-wire does have the required reliability capability for the harsh mission profile discussed in this paper. But one has to deal with additional challenges: in order to create enough reliability margin, only specific material combinations can be applied.

IV.   RELIABILITY CAPABILITY OF CU-WIRE

In this study, the performance for two types of products from the same family are compared during extended HTOL testing at 175 °C. Type A contains Pd-coated Cu-wire, and type B bare Cu-wire. First fails for Type A, are observed before 1000 h, and therefore the mission profile is seemingly not covered. Type B does not reveal any fail until 1300 h.

A.   Failure Mechansims

Physical analyses of type A failures after 1300 h (i.e. 30% use life margin) were performed, and lifted ball bonds are observed. An example is shown in Figure 2. Interfacial Cu-Al IMC corrosion [4] is clearly visible over the complete contact area on the bond pad (Fig. 2a). Inspection of the contact side of the ball reveals a large void next to severe cracks along the ball periphery. In order to analyze the voids in more detail, X-sections by ion-milling are prepared followed by SEM-EDX analysis. The result is shown in Figure 3. In the large void, clearly traces of S and Oxygen (O) can be observed.

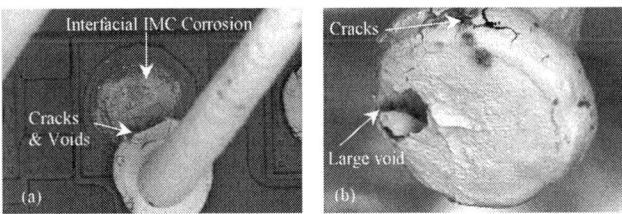

Figure 2.   (BS) SEM analysis of lifted ball showing: (a) Interfacial IMC corrosion in complete contact area (b) associated with voids and cracks along the periphery of the ball.

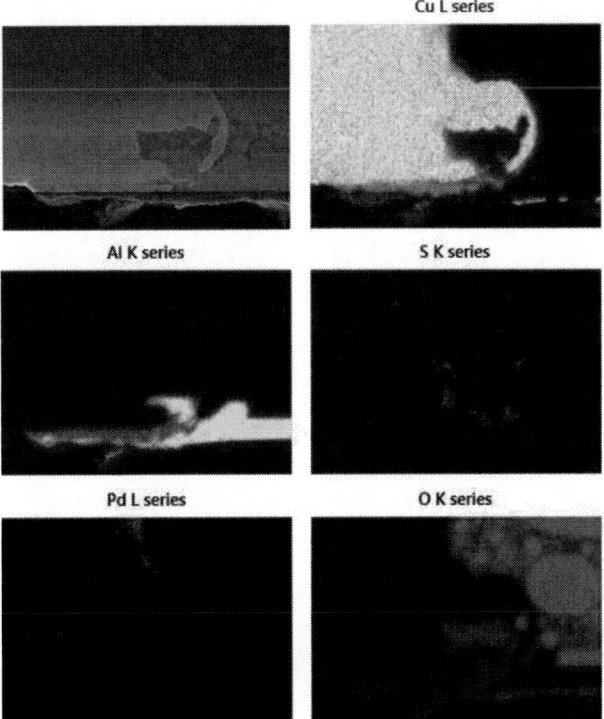

Figure 3.   X-section of failing conact by ion milling followed by SEM-EDX analysis of Cu, Al, S, Pd and O.

978-1-7281-1500-9/19 $31.00 © 2019 IEEE

At elevated temperatures and after long time exposure, Cu in contact with Pd will start to galvanically corrode and migrate, while leaving back voids and cracks at the original interface [10]. In this mechanism, S is known to act as a catalyst, so it will enhance the ongoing galvanic corrosion [5]. The presence of S inside the voids, can be explained by the fact that S is a required constituent of the adhesion promotor in the formulation of the EMC. Now, S is known to be released during aging at very high temperatures. Previous studies have described this paradoxical role of S and the high temperature reaction kinetics in EMCs in detail [6]. It is currently hypothesized that Cu-cations produced during the galvanic corrosion will migrate from inside the ball to the ball surface and further along the wire where it is forming oxide and sulfide. This has been observed for Au-Pd wires exposed to high temperatures, where Cu is migrating from the stitch area to the Au-Al IMC contact [12]. Several attempts to visualize this have been made, but so far without positive results. This is most likely explained by the relatively small amounts of Cu compared with the large surface over which it can easily migrate and distribute, because of the presence of Pd.

Revisiting the interfacial IMC corrosion, both S and residual halides such as chloride will act as a catalyst, and therewith S will contribute to the degradation of IMC contacts in the Type B device too. Therefore, devices from both product types were subjected to BS SEM investigations to determine the amount of degradation [4], [13]. Two bond pads of each type are shown in Figure 4 for comparison.

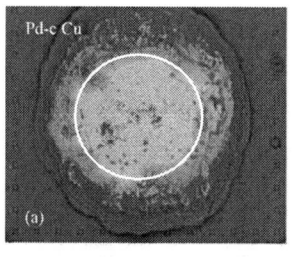 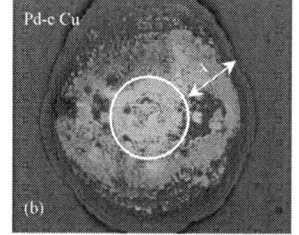

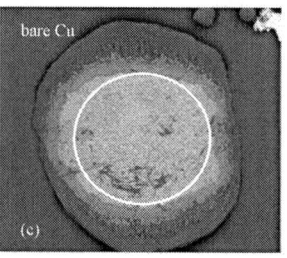 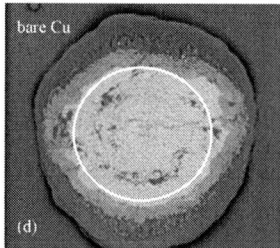

Figure 4. Planar BS SEM analysis of specially prepared IMC contact area (after 1300 h HTOL aging at 175 °C), to determine the progression of the corrosion front from the periphery of the contact towards the center (white circles): (a), (b) Device Type A with Pd-coated Cu and (c), (d) Device Type B with bare Cu [13].

It is obvious that the IMC degradation of the Pd-coated Cu-wire contact is slightly higher and that the spread in degradation is larger. This is explained by enhanced interfacial IMC corrosion via voids and cracks resulting from simultaneously on going galvanic corrosion. To illustrate this, the degraded area has been quantitively characterized, by

measuring the corrosion progression length "x" in µm (Fig. 4). The result is shown in Figure 5. Both mean value and standard deviation for Pd-coated Cu-wire are significantly larger. According to [4], "x" is progressing with the square root over time "t":

$$x(t) = \sqrt{D_0 . \exp[-E_a/(k_b \cdot T)] \cdot t} \qquad (5)$$

where D_0 is a pre-exponential constant (in µm²/h), which is only dependent on EMC properties [4]. The *square root over time* relation can be explained by the fact that the corrosion process is diffusion controlled. Limiting factor is expected to be transport of oxygen through the EMC, which is a required consumable in the corrosion [3][7]. This is irrespective of whether the catalyst is halide- or Sulphur-based.

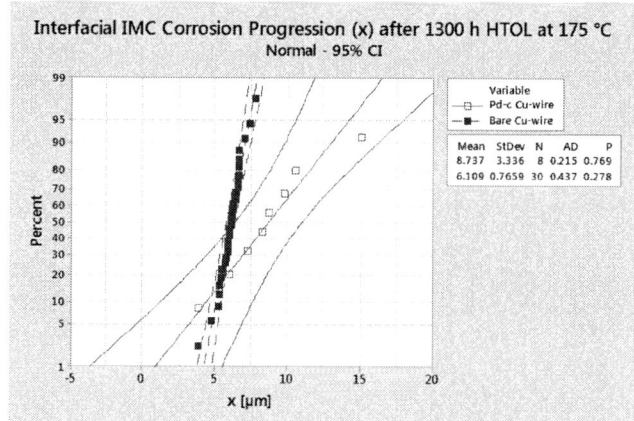

Figure 5. IMC degradation, characterzied by corrosion progression over distance x (Fig. 4) for Pd-coated Cu-wire and bare Cu-wire.

B. Re-assessment of the Mission Profile

Diving deeper into the reaction kinetics of the S decomposition in the EMC, it becomes clear that up to temperatures of about 155 °C, there is no significant reduction of S on going [6]. One should note that the larger part of the mission profile in Table I is below this temperature; i.e. about 99%. So, this could mean that the effect of galvanic corrosion is exaggerated when exposing parts for a long period above 150 °C; i.e., hundreds of hours at 175 °C. To address this observation, the data as reported in [6] are presented differently in Figure 6: the residual amount of S in the EMC – so, the amount which is not reduced yet – is shown as a function of temperature. Between about 155 °C and 200 ° C a significant decrease is observed.

However, the amount of S being released as function of temperature is needed. To that extent, the reduction rate of S at each temperature was fitted to an assumed power law equation. The Arrhenius relation is used to make the empirical power law relation with time, temperature dependent as follows:

$$[S](t) = [S](0) \cdot t^{-(n - \frac{m}{T+273})} \qquad (6)$$

where [S](t) is the amount of S after time "t", [S](0) is the initial amount, and both "m" and "n" are fit parameters. Fit was performed within temperature range of 150 to 200 °C.

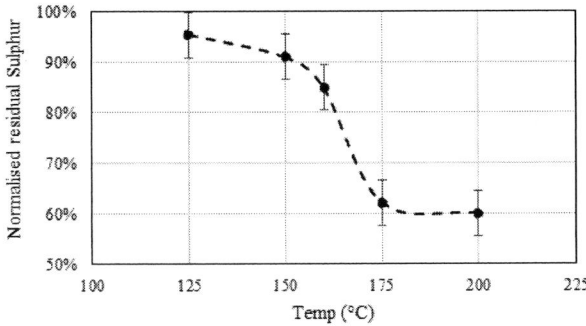

Figure 6.   Normalized concentration of residual S as a function of temperature (measurement error is about 5 %).

TABLE III.   APPLICATION MISSION PROFILE ASSESSMENT TAKING INTO ACCOUNT THE RELEASE OF S IN EACH INTERVAL

Mission profile Temperature-time intervals			Re-defined HTOL Temperature-time intervals		
T_{junc} in (°C)	t in (h)	$[S]_{rel}$ in (a.u.)	T_{junc} in (°C)	t in (h)	$[S]_{rel}$ in (a.u.)
195	30	44.0	185	30	38.2
185	30	3.9			
175	100	1.5	175	272	12.2
145	1400	0	155	1000	0
115	6000	0			
40	123000	0			
-20	840	0			
Total $[S]_{rel}$: 50.4 a.u.			Total $[S]_{rel}$: 50.4 a.u.		

Equation (6) is now used to determine the amount of S released ($[S]_{rel}$) during the temperature-time intervals above 150 °C for the mission profile of Table I. (3 in total) in the sequence high to low temperatures:

$$[S](t)_{rel} = [S](t) - [S](0) \qquad (7)$$

This is shown in the left half of Table III. The right half is showing the proposal for the redefined HTOL test with temperature-time intervals calculated such that:

1. The total amount of released S, $[S]_{rel,tot}$, is equal to that of the application mission profile and originates from the first temperature-time intervals, starting with the highest possible temperature in HTOL. This is about 185 °C, due to hardware limitation. Using these high temperatures at the beginning of the HTOL test sequence, ensures that the released S is able to migrate also for temperatures below 155 °C and is therewith worst case.

2. The mission profile coverage is ensured by applying equations (1) through (4). Note that an activation energy of 0.9 eV is used which is at the lower bound of values found in literature [10].

C. Results of Pd-coated Cu-wire with re-defined HTOL test

Type A products with Pd-coated Cu-wire, have been used to repeat the HTOL test with the redefined temperature-time intervals. No fails are observed after the final interval of 1000 h at 155 °C (1302 h in total). So, with this re-defined HTOL test, the mission profile can be covered. To investigate the reliability margin, the test has been continued and first fails start to occur after an additional 500 h (i.e. 50% use life margin). Detailed analysis via normal X-sections (Figure 7) clearly reveals that galvanic corrosion proceeds also at 155 °C, but voiding is limited to the outer ball periphery. Progression of Interfacial IMC corrosion is also visible in the X-sections shown in Figure 7.

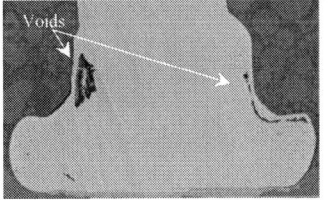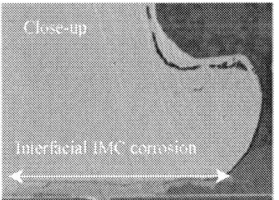

Figure 7.   X-section after 1500 h at 155 °C for the re-defined HTOL test. Galvanic corrosion is clearly visible but voids are limited to the ball preriphery and much smaller than after the original HTOL test at 175 °C for 1300 h (Fig. 2).

Finally, to quantify the reliability margin, equation (5) for the progression of "x" over time is used. For the estimation of D_0 applicable to the EMC in this product family, data of various other EMC formulations as reported in previous work [6] are re-used. An overview of the compounds is given in Table IV.

TABLE IV.   EMC FORMULATION USED IN THIS STUDY COMAPRED TO THOSE IN PREVIOUS WORK

EMC Type	Characteristic description
D [6]	Multi-aromatic epoxy from Supplier 2
F [6]	Multi-aromatic/Biphenyl epoxy from Supplier 2
B(1) [6]	Multi-aromatic epoxy from Supplier 1
I (this study)	Multi-aromatic/Biphenyl epoxy from Supplier 1

Figure 8 graphically shows the relation between total (S + Cl) content and D_0, for EMC B(1), D and F [6]. The value of D_0 for EMC I, is estimated based on scaling the amount of S actually being liberated under long time exposure to high temperatures of EMC B(1). In that sense, it is an improvement compared with the approach used in [6].

Applying equations (1) to (4), the corrosion progression per temperature-time interval can be predicted using equation (5). For E_a, a conservative value of 0.96 eV, which was determined for EMC B(1), is re-used because its formulation is close to EMC I. In addition, previous studies [4][5][6], show that the values of activation energy are typically in the range of 0.9 to 1 eV, which justifies the use of this value.

Figure 9 shows the comparison of observed and predicted corrosion progression. Note that the time axis is converted to use life, i.e., the application mission profile as given in Table I. The model fits the data well. The deviation of average measured and predicted value is larger for stretched life times. To explain this, the same argumentation as used for the wider distribution of Pd-coated Cu-wire in Figure 5 applies, which is the enhanced transportation of reaction species – essential for the interfacial IMC corrosion – via the voids and cracks created by the simultaneously ongoing galvanic corrosion.

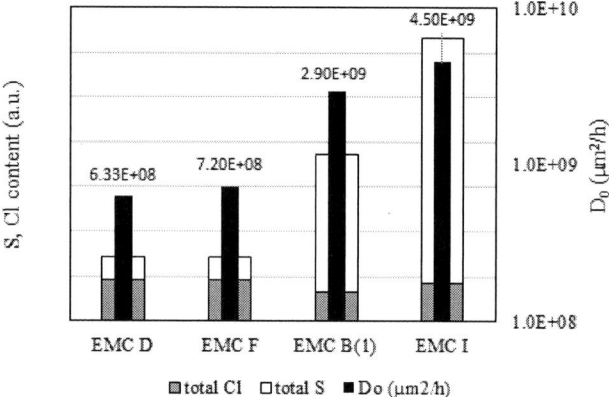

Figure 8. S and Cl content (individual and combined) in relation to pre-exponential constant D_0 – refer to equation (5) – for compounds listed in Table IV, using an improved approach based on previous results [6].

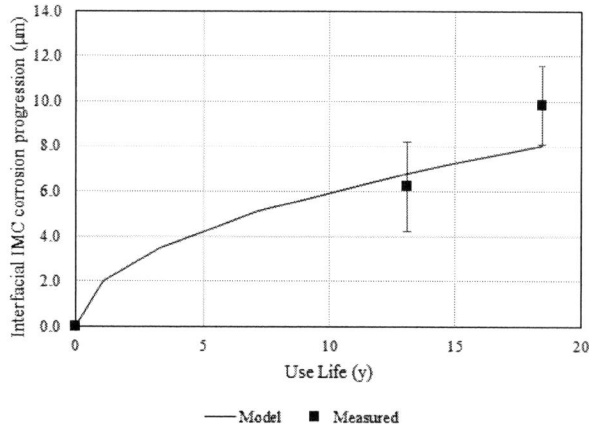

Figure 9. Interfacial IMC corrosion progression: measured vs predicted, using model of equation (6), as a function of use life accoding to mission profile in Table I. Error bars indicate the standard deviation of the normal distribution fit for measured values of "x" (like shown in Figure 5). Note that the time-axis is directly the use life time domain of the application mission profile in Table I

V. DISCUSSION

In Table V an overview is given of possible combinations of EMC type and wire material, to be used for typical AEC Grade 0 or 1+ mission profiles as for the example given in this study.

TABLE V. AEC GRADE 0 OR 1+ APPLICATION MISSION PROFILE SELECTION OF EMC / WIRE COMBINATIONS

EMC Type	Dominant Failure Mechanism	Wire Type selection
D	Interfacial Cu-Al IMC corrosion	both Pd-coated Cu-wire, and bare Cu
F	Interfacial Cu-Al IMC corrosion	both Pd-coated Cu-wire, and bare Cu
B(1)	Interfacial Cu-Al IMC corrosion	Pd-coated Cu-wire preferred
I	Pd-Cu Galvanic corrosion	Bare Cu-wire preferred

In the following assessment, the internal condition of the bond ball (degree of voiding and cracking) and its potential effect on the Cu-Al IMC interfacial corrosion (larger spread in corrosion progression) after long time exposure to elevated temperatures is considered for each combination.

For Pd-coated Cu-wire, the galvanic corrosion reliability margin depends on the interactions of the wire and additives in the mold compound composition. EMC materials with lower S content will enlarge the reliability margin with respect to galvanic corrosion. For EMC type D and F, the S content is very low, galvanic corrosion is hardly observed, and therefore both Pd-coated as well as bare Cu-wire can be used.

An interesting EMC type is B(1), for which the S-content is moderate. In case of Pd-coated wire, Cu-Al interfacial IMC corrosion is retarded due to the consumption of mobile Sulphur (e.g. in H_2S form) by competing reactions at the Pd-Cu interface [6]. Overall, this results in a significantly slower IMC corrosion for Pd-coated wire compared to bare Cu and is therefore preferred.

If the S-content further increases, as is the case for EMC type I, the competing reaction of galvanic corrosion results in severe voiding which starts to negatively influence the corrosion at the Cu-Al IMC as shown in this study. To better control the degradation of the IMC to Cu interface, bare Cu-wire is preferred.

A prerequisite for sufficient reliability margin for any of the combinations, is a large and continuous Cu-Al IMC contact at 0 h (directly after wire bonding). It will ensure that interfacial IMC corrosion progression will closely follow the *square root over time model* of equation (6). The amount of IMC required to meet AEC Grade 0 or 1+ for the material combinations in Table V is increasing the larger the value of D_0 (Figure 8). For EMC type D and F this is about the same, and lower than for B (1). The largest IMC contact is required for EMC type I.

In these material combination considerations for AEC Grade 0 or 1+, the robustness of the second or stitch bond should also be included. Low S content alternatives require attention regarding delamination performance. Delamination preventing measures like surface treatment of the lead frame may be needed. On the other hand, high S content, may also cause galvanic corrosion in the second bond area and thereby weakening the stitch weld [15]. In general, the temperature in the second bond area will be lower compared to the first bond contact and this will considerably slow down the reaction rate.

Finally, the corrosion performance under biased humidity conditions is of concern. The material combinations as discussed in this paper, can all pass twice the requirement for any Automotive AEC Grade as provided in [16].

VI. CONCLUSIONS

This paper shows the strength of applying knowledge on the Physics-of-Failure (and Degradation), in selecting package materials for high temperature automotive applications. Three different failure mechanisms must be considered: current enhanced Kirkendall voiding (Au-wire only), Pd-Cu galvanic (Pd-coated wire only), and Cu-Al IMC corrosion (both Pd-coated and bare Cu). Transforming the application mission profile into HTOL time-temperature profiles by taking into account the degradation kinetics in relation to material characteristics, allows for proper validation of reliability capability and margin. A structured way to do this, is demonstrated for an assumed but realistic mission profile, typical for an Automotive AEC Grade 0 or 1+ application. Finally, it was discussed, what considerations need to be made concerning corrosion mechanisms in the first and second bonds using Cu-wire, when selecting suitable molding compound and wire material combinations for these harsh mission profiles. Other aspects that must be considered, such as manufacturability where Pd-coated wire can in some cases be superior, were outside the scope of this study.

REFERENCES

[1] B. Krabbenborg, "High current bond design rules based on bond pad degradation and fusing of the wire", *Microelectronics Reliability*, vol39, p77-88, 1999.

[2] E. Zin, et al., "Mechanism of Electromigration in Al/Cu/Al films observed by transmission electron microscopy", *Electronic Components and Technology Conference*, ECTC 59th 2009, pp 943-947.

[3] R.T.H. Rongen, Arjan van IJzerloo, Amar Mavinkurve, G.M. O'Halloran, "Degradation of Cu-Al Wire Bonded Contacts under High Current and High Temperature Conditions using In-situ Resistance Monitoring", *Electronic Components and Technology Conference*, ECTC 65th 2014, pp 1396-1402.

[4] R.T.H. Rongen, G.M. O'Halloran, Amar Mavinkurve, Leon Goumans, Mark-Luke Farrugia, "Lifetime prediction of Cu-Al wire bonded contacts for different mold compounds", *Electronic Components and Technology Conference*, ECTC 64th 2014, pp 411-418.

[5] Chu-Chung (Stephen) Lee, Tu Anh Tran, Varughese Mathew Rusli Ibrahim and Poh-Leng Eu, "Copper Ball voids for Pd-Cu Wires: Affecting Factors and Methods of Controlling", *Electronic Components and Technology Conference*, ECTC 66th *2016*, pp 606-613.

[6] A. Mavinkurve, L. Goumans, M. L. Farrugia, E. van Olst, M. van Soestbergen, B. Bumrungkittikul & R.T.H. Rongen, "The paradoxical role of Sulphur in molding compounds: influence on high temperature reliability of Cu-Al wire bond interconnects", *Electronic Components and Technology Conference*, ECTC 67th 2017, pp 1172-1178.

[7] A. Mavinkurve, L. Goumans & J. Martens, "Epoxy molding compounds for high temperature applications", *European Microelectronics Packaging Conference*, EMPC 2013, ThA3.

[8] AEC-Q100 rev-H, "Failure Mechanism Based Stress Test Qualification for Integrated Circuits", *Automotive Electronics Council*, AEC 2014.

[9] H.T. Orchard and A.L. Geer, "Electromigration effects on intermetallic growth at wire bond interfaces", *Journal of Electronic Materials,* vol. 35, pp. 1961-1968, Nov 2006.

[10] F.W. Ragay, J.A. v.d. Pol & J Naderman, "In-situ monitoring of dry corrosion degradation of Au ball bonds to Al bondpads in plastic packages during HTSL", *Microelectronics Reliability* Vol 36 (11-12), 1996 pp 1931-1934.

[11] R. Villa, C.M. Villa, C. Passagrilli, A. Mancaleoni, "Extended Lifetime Study of Cu and Cu-Pd Bonding Wires", *Automotive Electronics Council Reliability Workshop*, AEC Workshop 2015.

[12] KC Hsieh and Theo Martens, "Ag and Cu Migration Phenomena on Wire Bonding", *Journal of Electronic Materials*, Vol 29, Bo 10, 2000, pp 1229-1232.

[13] G.M. O'Halloran, Arjan van IJzerloo, Rene Rongen, Frank Zachariasse, "Planar Analysis of Copper-Aluminium Intermetallics", *Proc. International Symposium for testing and Failure Analysis (ISTFA)*, San Jose, CA, Nov 3-7 2013, pp.297-300.

[14] Chu-Chung (Stephen) Lee, TuAnh Tran, Dan Boyne, Leo Higgins, Andrew Mawer, "Copper versus Palladium Coated Copper Wire Process and Reliability Differences" *Electronic Components and Technology Conference*, ECTC 65th 2014, pp 1539-1548.

[15] J.C. Krinke, D. Dragicevic, S. Leinert, E. Friess, J. Glueck, "High temperature degradation of palladium coated copper bond wires", *Microelectronics Reliability*, Volume 54, Issues 9–10 (2014), pp 1995-1999

[16] AEC-Q100 rev-A, "Qualification Requirements for Components using Copper (Cu) Wire Interconnections", *Automotive Electronics Council*, AEC 2015.

Solder Joint Reliability of Double-side Mounted DDR Modules for Consumer and Automotive Applications

Dongji Xie*, Joe Hai, Zhongming Wu, and Manthos Economou
Nvidia Corp. USA
2701 San Tomas Expressway, CA 95050, USA
*Contact E-mail: Dongjix@nvidia.com; Tel: +408 486 8630; Fax: +408 486 2919

Abstract— This paper describes solder joint reliability studies for DDR memories using single side and double side mount modules in the application of consumer and automotive fields. The types of DDRs include LPDDR4 and GDDR5. The components are from well-known memory manufacturers. Both experimental work and numerical simulation are employed to understand the reliability and failure mechanisms. It is found the reliability of DDRs changes with different DDR types as well as suppliers. LPDDR4 has much lower reliability as compared to that of GDDR5. The reason is the low ball profile which has increased the thermal stress for LPDDR4. However, the most critical factor is the double side mount vs. single side mount configuration. Both experimental and FEA results show corner fill may be a better choice in handling both mechanical and thermal stresses. To enhance the solder joint reliability, one effective way is to employ corner fill, edge bond or underfill. However, in order to get better reliability, corner fill and underfill are normally recommended.

Key Words: memory, double sided, DDR, automotive, glue, underfill, edge bond, corner fill, reliability, thermal cycling, solder joint

I. INTRODUCTION

DDR (Double Data Rate) Memory modules are widely used in the computer, mobile and networking products including GPU cards, embedded devices, etc. DDR technology has evolved from DDR1 to DDR6 to increase capability and speed. Another series of DDR, a low power type DDR or LPDDR, are also widely used in mobile and embedded applications. To further increase the density, DDRs are often mounted in double side mirrored configuration on both sides of the PCB. However, there are some challenges in PCB assembly as well as reliability concerns when using double mounted DDRs. As shown in earlier studies, there is a clear reliability degradation for solder joints in double side mounted DDRs as compared to single side mounted case [1]. Main reason being the stiffness of double mount module is much higher than single side mount which means the component has no compliancy for flexing.

On the other hand, PCB warpage may exist due to mechanical assembly and global expansion as shown in Fig.1. Fig 1a, shows a x-section view of a double side mount DDR where solder balls are mirror imaged or clam shelled across the PCB board. Fig. 1b shows the solder joint details

of double side mount and Fig. 1c shows the board warps towards top package as it happens in most cases in actual product assembly. In this case, the solder joint behavior will be different when comparing top and bottom. The solder joints on the bottom package shall be stretched in the corner regions. Therefore, the solder joints shall be subjected to a tensile stress which may result in faster degradation during thermal cycling. The reliability for this type of double side mount DDRs shall be degraded significantly. Earlier research from one of the authors [1] has also shown the PCB materials, board thickness, and via structure may have less impact.

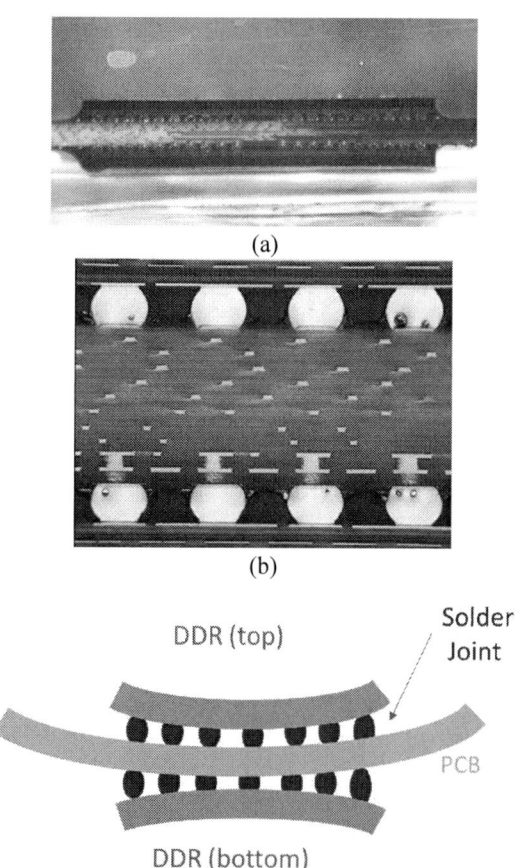

(a)

(b)

(c)

Fig. 1 (a) A cross-section view of DDR (global view), (b) solder joints of the cross-section view and (c) schematic view of double side mount DDR.

To enhance the reliability of double side mount DDRs for high reliability requirements such as automotive applications, one can utilize adhesives such as corner fill, edge bond and underfill. However, the selection of those adhesives is very different compared to typical applications such as mobile phones and other consumer applications. One important factor is that automotive applications must have high reliability not only mechanically but also thermo-mechanically to fulfill warranty operation of 10,000 to 45,000 hours in 15 years. This is especially true for double side mount DDRs, where thermal cycling reliability degradation is the root cause.

To enhance solder joint reliability in electronic automotive modules to meet harsh field environments, a wide range of glues are used in mass production and listed in Table 1. A variety of glues are available and used in the production of PCB assemblies. One main driving force to utilize glue is PCB assembly improvement for drop and shock events similar to a consumer mobile application. In these cases, the solder joints start failing from the corner region due to the high stress concentration. Edge bond and corner fill are two most common types of glue applications selected because of their low cost and effectiveness in reducing mechanical stresses due to handling and field use. Edge bond can be applied on each corner in a L-shape to protect the solder joints in the corner regions. In some special cases, it can be applied as a dot if only the corner-most solder joints are concerned. Corner fill is a new glue category because it is applied on the corner in a L-shape, but the glue flows underneath the BGA and fills the corner balls. This method is usually non reworkable. Underfill is another type of glue which originated from flip chip BGA (ball grid array, fcBGA) packaging. The performance of glues for reliability enhancement of BGA solder joints is determined by the material properties, especially T_g (glass transition temperature), CTE (coefficient of thermal expansion) and E (tensile modulus). For glues used in edge bond, corner fill and underfill, there is a wide selection of T_g, CTE and E. Comparing the types of glues, corner fill and edge bond are easier to apply and offer good processability. Underfill usually takes longer time for large packages and may involve preheating the board before underfilling.

All those glues shall improve reliability performance of BGA in bending, drop, vibration, etc. However, depending on the selection of materials, the thermal fatigue life may easily end up worse than without glue as demonstrated by P. Borgesen [2-7]. All present underfills reduced the thermal cycling performance, while edge bonding improved it by up to 50% [7].

Apart from FEA, experimental work has been conducted using DDR builds with LPDDR4 and GDDR5 respectively. DDR packages from two suppliers have been demonstrated for both single side and double side mount PCBAs. The experimental results shall then be compared with FEA results.

Table 1 Common adhesives used in Electronic Assembly

Glue Type	a) Edge bond	b) Corner Fill	c) Underfill
Where to apply (BGA)	Edge L-shape	Corner and partial underfill	Underfill whole gap
Cure	Oven	Oven	Oven
CTE	Low to medium	Low to medium	Low to high
Modulus	High	Medium to high	Low to high
Reliability	Low to medium	Medium to high	Low to high
Ability to Rework	Depends	Hard	Hard
Processability	Good	Good	Bad
Cost	Medium	Medium	High

II.   EXPERIMENTAL SETUP AND METHODOLOGY

In Fig. 2, the test board is designed to represent a real functional product. The test unit has one GPU and four DDRs, two on the top and two on the bottom in a mirror image mount (double side mount). Fig 2a, shows a partial loaded board (bottom side). Fig. 2b shows full loaded board (bottom side) where a large ceramic connector with LxWxH = 67mm x 7.6mm x 5mm. It also includes several other packages such as PMIC (power management IC), WiFi modules, etc. The board is a typical FR-4 with a thickness of 1.6mm. The DDR package structure is as shown in Table 2 and two types of DDRs are listed from two suppliers. LPDDR4 has a ball count of 200 with body size = 10mm x 15mm. GDDR5 has a ball count of 190 and body size = 10mm x 14mm.

The adhesive materials in this study are listed in Table 3. Three types of adhesives are utilized: edge bond, corner fill and full underfill. It is noted the same materials may be used in different applications. For example, CF2 and CF3 can be used for full underfill. In general, underfill can be used for corner fill because of its low viscosity. However, edge bond and corner fill material may not be used as underfill because the flowability is limited.

FEA was also employed for comparison purposes. The elastic-plastic model of solder (Alloy: SAC305) was chosen. Thermal fatigue modeling starts from modeling the DDR assembly under stress condition such as thermal cycling test. An elastic-plastic-creep study was employed to assess the creep-fatigue of the solder joints. The material properties used in this FEA are listed in Table 3. For solder materials, the Anand creep model for SAC alloy is employed and described by Darveaux [6]. The life prediction using creep energy density, as described by A. Syed et al. [8], is shown in Equation 1.

$$N_f = (0.0019\Delta W)^{-1} \qquad (1)$$

where, N_f is the characteristic life of the solder joint and ΔW is the cyclic creep strain energy density (averaged) along the crack path, which is calculated by FEA. Thermal cycling test was also evaluated for various DDRs and glues in the range of -40 °C to +105 °C. The cycle time was 43 minutes, ramp rate was about 12 °C/minute and dwell time was 10 minutes at both extremes per JESD22-A104E [9] and IPC-9701[10].

Table 2 DDRs used in this study. (all dimension in mm)

DDR Type	Body Size	Ball Size	Pitch	Ball Number	SRO Diameter	Standoff (post SMT)	Mold Thickness	Substrate Thickness	PCB Pad Size	Remark
LPDDR4	10*15*0.9	0.3	0.8/0.65	200	0.28	0.22	0.249	0.106	0.3	Supplier1 & 2
GDDR5	10*14*1	0.4	0.65/0.65	190	0.3	0.28	0.3	0.18	0.3	Supplier3

Table 3 Adhesive materials used in this study

Material Group	Materials Code	T_g(°C)by TMA	CTE1, (ppm/°C)	CTE2, (ppm/°C)	E, Tensile Modulus, (MPa)	Remark
Edge Bond	EB1	90	35	106	4	Reworkable
Edge Bond	EB2	134	30	104	3.1	Reworkable
Corner Fill	CF	135	43	123	6.7	Non-reworkable
Underfill	UF	161	28	104	7.6	Non-reworkable

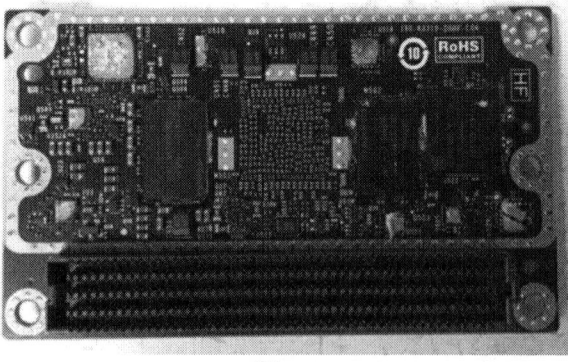

(b)

Fig. 2. A consumer module with DDRs and other ICs. (a) partial loaded board and (b) full loaded board.

III.   RESULTS AND DISCUSSION
3.1 Thermal Fatigue Life of DDR—Single Side Mount vs. Double Side Mount

Thermal fatigue life is studied using both experimental work and FEA. Thermal cycling experiments are performed for LPDDR4 using single side mount and double side mount. The experimental results clearly show the difference between those two types of assembly. A typical solder joint check by Dye and Pry (DnP) has been employed and shown in Fig. 3. Fig. 3a shows there is no crack at 1000 cycles under thermal cycling (TCT) in single side mount DDR and some severe cracks are observed in double side mount DDRs. As a part of statistically meaningful results, a Weibull plot of unreliability vs. thermal cycles is shown in Fig. 4. Reliability results from both LPDDR4 and GDDR5 with single side mount and double side mount are illustrated. Fig. 4, clearly shows the double side mount DDRs show significant degradation in thermal cycling reliability as compared to that of single side mount DDRs. A summary of results is also shown in Table 4. The results show the degradation (Nf ratio) is about 0.71 and 0.55 respectively. GDDR5 has generally much higher reliability than LPDDR4. The main reason could be due to ball size which is much smaller in LPDDR4 as compared to that of GDDR5, resulting in higher stress at the solder joint thermal cycling.

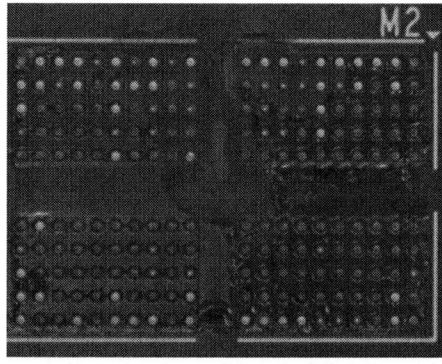

(a)

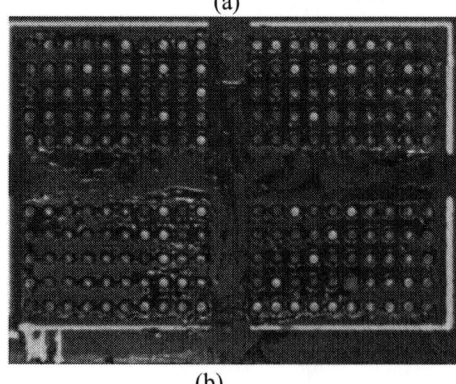

(b)

Fig. 3 Dye & Pry pictures for LPDDR4 solder joints post 1000cycles of TCT showing crack check for single side mount (a) and double side mount (b) long term thermal cycling. Single side mount DDR does not show cracks while double side mount DDR has severe cracks at the corner and edge rows.

978-1-7281-1500-9/19 $31.00 © 2019 IEEE

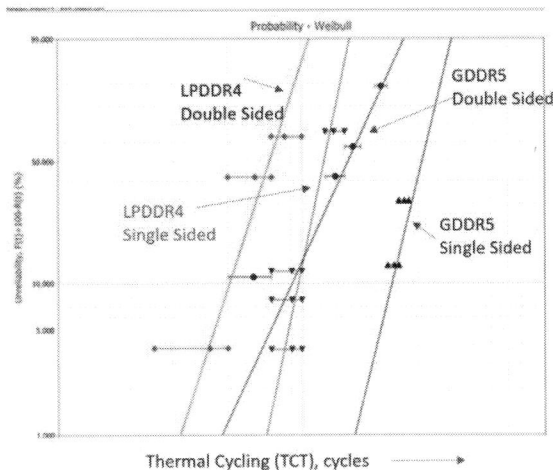

Fig 4 Solder joint failure-Weibull Plot in DDR from thermal cycling Test (TCT, from -40C to 105C).

Table 4 Thermal fatigue life of DDRs—comparison between single mount and double mount.

DDR Type	Nf ratio, Double sided/Single sided, experimental	Nf ratio, Double sided/Single sided, FEA	Remark
LPDDR4-Supplier1	0.71	0.75	DDR Supplier1
LPDDR4-Supplier2	0.15	0.23	DDR Supplier2
GDDR5--supplier3	0.522	0.41	DDR Supplier3

3.2 Thermal Cycling Fatigue of Solder Joint—Impact of Edge Bond, Corner Fill and Underfill

Thermal cycling tests from -40 °C to 105 °C were also performed at the board level to study the impact of DDR glue enhancement. The pass/fail determination for the units before and after thermal stress test was determined by functional test (FCT). FCT was conducted every 250 cycles until 3,000 cycles or failure. The glue selection by FEA is also conducted prior to actual experiments. Strain energy density of each solder ball is calculated using an FEA tool (Abaqus) and the results are shown in Fig. 5. It illustrates creep strain energy density (CENER) for four different cases of double mount DDRs. Similar results have been obtained for single mount DDRs. Contour of CENER of a double side mount DDR is shown in four cases: (a) no glue; (b) edge glue; (c) corner fill and (d) full underfill. Both die and glue areas are shown. As shown in Fig. 5, the ball at the corner has highest CENER and hereby is the weakest ball across the whole package. When solder balls are glue protected, the weakest ball may move inwards to where no glue is observed. Life prediction is calculated by Equation 1 and

the results are listed in Table 5. For the purpose of relative comparison, only a solder joint fatigue life ratio is shown in reference to that of no-glue cases. Interestingly enough, as shown in Table 5, the edge bond has increased the thermal fatigue life for around 10% for double side mount DDRs while corner fill may have increased the reliability by more than double. Among all glues, underfill is still the best for enhancing the thermomechanical reliability which increases more than 10x. In the experimental testing, no failures are observed during limited test time (up to 2000 cycles). Two reasons for the underfill better performance over corner fill: (a) all balls are covered by underfill and (b) underfill used is a hard material with very low CTE. However, corner fill is still a viable selection for most applications where the reliability target is in the range from medium to high. For single side mount DDR, both corner fill and underfill can enhance the reliability significantly by more than 4x.

Typical cross section views of solder joints from corner fill (CF) and underfill (UF) are shown in Fig. 6. Fig. 6a and 6b show a crack observed in the bottom DDR in double side mount DDR from a unit after long term thermal cycling which is equivalent to more than 15 years in the field for automotive. The crack is in the bulk solder close to the package side. However, good solder joints are also found for the same group as shown in Fig. 6c. On the other hand, no crack is found for the double side mount DDR after long term thermal cycles as shown in Fig. 6d.

The correlation between the glue properties can be described in Fig. 7. Fig 7a shows the reliability enhancement of double side mount DDRs using various glues. Fig. 7b shows that edge bond materials provide limited benefit and corner fill is much better while the underfill is best. This is partially explained because the CF and UF selected in this study have lower CTE as compared to edge bond. If the modulus and CTE impact is combined, we may get a better picture as shown in Fig. 7b. Because a glue with a higher modulus and lower CTE will give better support mechanically for reducing the thermal strain in the solder joints, combining E and CTE as in E/CTE may be used as reliability enhancement index for the glue. As shown in Fig. 7b, higher E/CTE shall give a higher reliability (Nf ratio).

Table 5 Glue enhancement of solder joints fatigue life ratio for DDRs—comparison of experimental work for thermal cycling test results. (TCT: -40 °C to 105 °C).

Material Group	Materials Code	CTE1, (ppm/°C)	E/CTE, MPa-C/ppm	Nf ratio, single side DDR	Nf ratio, double side DDR
Edge Bond	EB1	35	114		1.1
Edge Bond	EB2	30	103		1.1
Corner Fill	CF	43	156	4.5	2.1
Underfill	UF	28	271	5.0	12.7

3.2 Solder Joint Reliability of Double Side Mount DDR—Impact of Board Flexing

Flexing and warpage may also have a major impact on reliability. This is also true for double side mount DDRs. As shown in Fig. 2, a full embedded module is used as a plug-in device which has a 400pin ceramic connector. The connector is the largest component on the board mounted on the bottom side of the board. It is known the board is warped significantly due to presence of the connector. The reliability impact from the connector is shown in Fig. 8. The Weibull plot in Fig. 8 shows the reliability degradation is about 3x as compared to the regular mount DDRs. In this case, corner fill and/or underfill may be needed to enhance reliability to meet various field applications. Fig. 8 also shows that corner fill is effective in increasing reliability by 3x as compared to that of no glue. This module has been in production for some time.

IV. CONCLUSIONS

Reliability studies on double side mount DDRs has been executed by extensive experimental work. To understand the failure mechanism, FEA (finite element analysis) is also employed for stress/strain simulation of the solder joints for both single side and double side DDRs. Experimental results also show the double side DDRs are usually degraded by more than 50%~80% as compared to their single side counterpart. FEA simulation results have also demonstrated same trend of solder joint reliability between double side DDRs and single side DDRs. A reasonable correlation of the fatigue lives predicted by FEA and experimental work are obtained. Both FEA and experimental work show that edge bond has limited reliability enhancement for double side mount DDRs and CF or UF shall be used for a high reliability application. However, the glue materials should have a low CTE and high modulus to ensure the effectiveness of the reliability enhancement. E/CTE ratio may be used as selection index for glue materials. Corner fill may be sufficient for general application reliability while underfill may be needed for higher reliability requirements. Reworkable adhesives are also tested but may only be recommended for double side mount DDRs. Furthermore, DDR structure is also a dominant factor in determining solder joint reliability. Compared to LPDDR4, GDDR5 has a much higher reliability because it has a larger ball size and solder joint standoff. The board warpage from other components such as large ceramic connectors may also play a major role impacting the reliability.

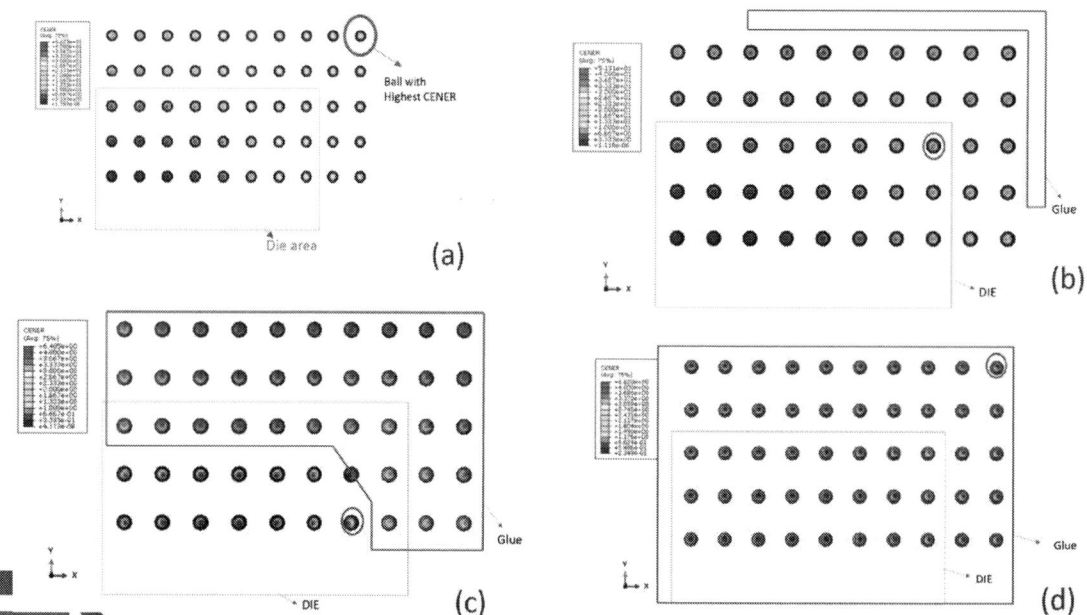

Fig. 5 Contour of creep strain energy density of a double side mount DDR (a) no glue; (b) Edge Glue; (c) Corner Fill and (d) full Underfill. Both die and glue area are shown.

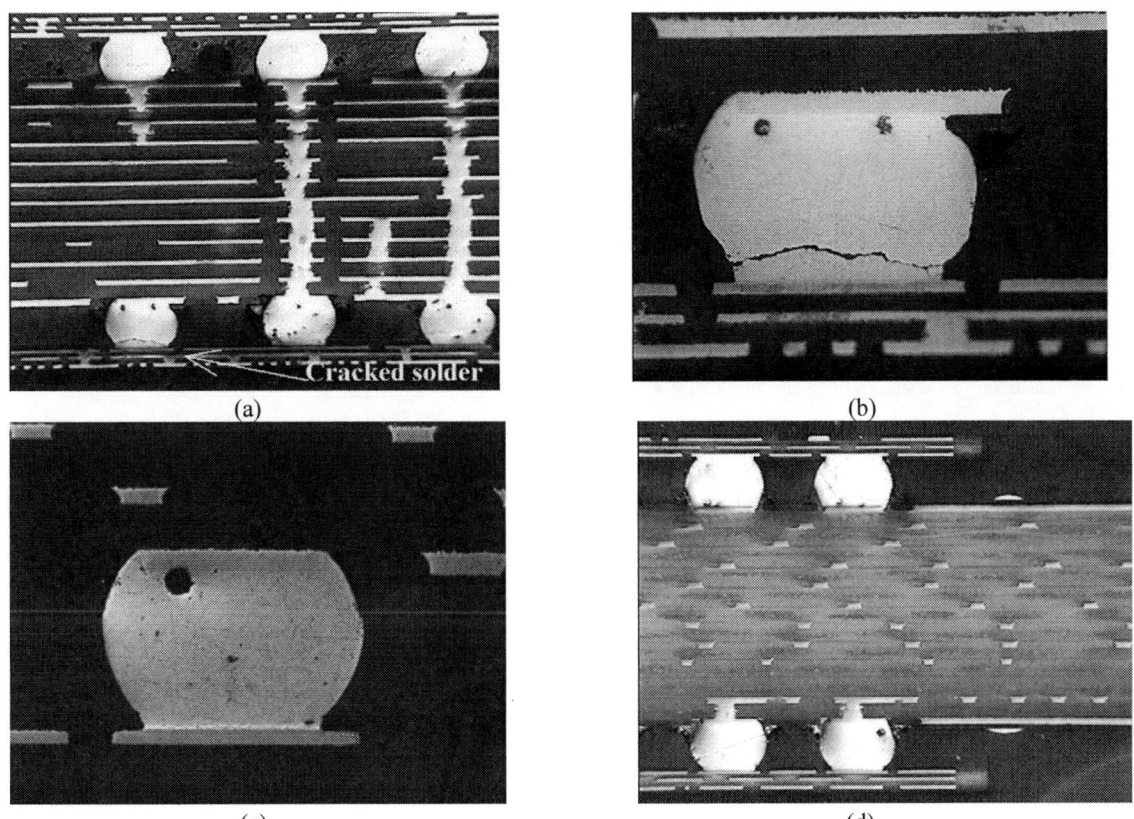

Fig. 6 Crack check for long term thermal cycling. (a) A cracked solder joint found in bottom DDR using CF after long term TCT cycles;   (b) zoom in view of the cracked solder joints of (a); (c) Good solder joints from CF bonded long term of TCT, no crack and (d) UF, after long term TCT, no crack.

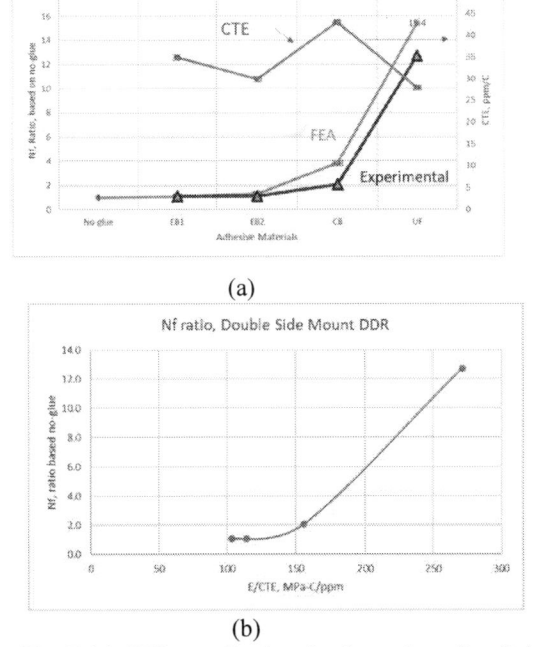

Fig. 7 (a) CTE contribution for thermal cycling fatigue life of solder joints for glue materials, (b) correlation of E/CTE vs. Nf ratio.

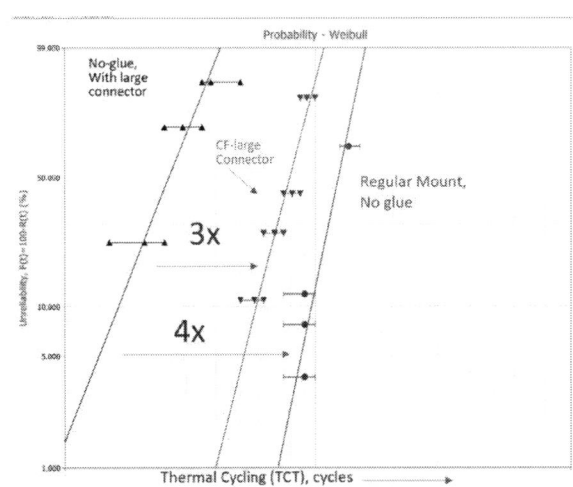

Fig. 8 Double Side Mount DDR Reliability--Impact from a Large Ceramic Connector

978-1-7281-1500-9/19 $31.00 © 2019 IEEE

ACKNOWLEDGEMENTS

The authors wish to thank Sylvia Wong, Long Huynh, Hung Nguyen and Luis Alvarez for their kind assistance and generous contribution in functional testing, sample preparation and destructive physical analysis at Nvidia.

REFERENCES

[1] Dongji Xie and Sammy Yi, Reliability studies and design improvement of mirror image CSP assembly, Microelectronics Reliability 42(12): 1931-1937. December 2002.

[2] Dongji Xie, Zhongming Wu, Joe Hai and Manthos Economou, Reliability Enhancement of Automotive Electronic Modules Using Various Glues, Proceedings of Electronics Component Technology Conference, 2018.

[3] Hannan, P. Viswanadham, K. Kulojarvi & J. Ahlstedt, Proceedings of the SMTAI, Rosemont, Illinois, pp. 858-870, Sept 2000

[4] Ghaffarian, Proceedings of APEX 2000, Long Beach, California, pp. P-AD2/3-1 to P-AD2/3-7, March 2000

[5] Peter Borgesen, J. Electron. Packag *134(1), 011010 (Mar 19, 2012)*

[6] Ahmer Syed, TMS Annual Meeting & and Exhibition, Feb. 11-15, 2001.

[7] R. Darveaux, "Effect of Simulation Methodology on Solder Joint Crack Growth Correlation," Proc 2000 ECTC, pp. 1048-1058.

[8] Dhruv Bhate, Dennis Chan, Ganesh Subbarayan, Chiu Tz-Cheng, Vikas Gupta and Darvin Edwards, Constitutive Behavior of Sn3.8Ag0.7Cu and Sn1.0Ag0.5Cu Alloys at Creep and Low Strain Rate Regimes, IEEE Transaction on CPMT, 2007.

[9] Ahmer Syed, Accumulated Creep Strain and Energy Density Based Thermal Fatigue Life Prediction Models for SnAgCu Solder Joints, Proceedings of Electronics Component Technology Conference, 2004, pp737-746.

[10] JESD22A104E, Temperature Cycling, Oct 2014.

[11] IPC9701, Performance Test Methods and Qualification Requrirements of Surface Mount Solder Attachment, Jan., 2002.

Reliability Investigation of Extremely Large Ratio Fan-Out Wafer-Level Package with Low Ball Density for Ultra-Short-Range Radar

P.S. Huang[1], C.K. Yu[1], W.S. Chiang[1], M.Z. Lin[1], Y.H. Fang[1], M.J. Lin[1], N.W. Liu[2], Benson Lin[2], Ian Hsu[2]

1. Quality Assurance
2. Advanced Package Technology
MediaTek Inc.
HsinChu, Taiwan
Ps.huang@mediatek.com

Abstract—Driven by aggressive development of electronic products with high robustness demand for automotive application to endure severe usage environment, both component-level and board-level reliabilities have to be concerned more for safety assurance. In this paper, a system-on-chip millimeter-wave ultra-short range radar (mmWave USRR) realized in complementary metal-oxide-semiconductor (CMOS) technology and assembled with fan-out wafer level packaging (FOWLP) technology was introduced, and the board-level reliability (BLR) was studied experimentally on the risk of chip-to-board interaction (CBI). The factors of solder ball material, package thickness and underfill material, thought to dominate on CBI performance, were studied experimentally.

First of all, two solder materials were studied to evaluate their capabilities for this FOWLP to against board level thermal cycling and drop tests. It was found that the solder with higher elastic modulus performed much better on board-level thermal cycling (BLTC) reliability. Moreover, no difference was found in board level drop test since no failure occurred in both solder materials. Both package thicknesses of 425 μm and 580 μm were studied on the board level reliabilities, and the results revealed that the design with both thicker Si die and thicker molding material significantly improved the BLTC reliability. Both epoxy-based materials – one is low-CTE underfill material and the other is edge-bond glue, were applied to know the workability of enhancing the BLTC performance on the FOWLP. The experiment results showed that both the epoxy materials miserably decreased the BLTC performance, and severe solder crack and bulk underfill crack were found.

Since vibration test is indispensable and of much concern for automotive electronics, the stringent test condition of sine-wave frequency swept from 20 Hz to 2,000 Hz and peak acceleration of either 50g or 20g, was applied to evaluate anti-vibration property of the FOWLP mTV mounted on daisy-chain PCB. From the results of 50g peak acceleration vibration test, high resistance was found in the specific daisy-chain loop which electrically connects corner solder balls. From the failure analysis it could be found that delamination existed at the interface of redistribution layer (RDL) and under-bump metallization (UBM) of component side and PCB Cu trace crack. It is noteworthy that all the failures only happened on the package located at the 5x3 array corner while subjecting to Z-axis vibration. From experience, poorly fixing the PCB on vibration platform potentially causes more bending stain on PCB during Z-direction vibration and further concentrates much higher

stress singularly nearby the corner. Moreover, the board-level vibration test with 20g peak acceleration was also implemented, and there wasn't any failure found.

Finally, the BLR was thoroughly studied for the extremely large area-ratio FOWLP, and the package was proved its capability of meeting AEC-Q100 compliant stringent reliability tests.

Keywords : FOWLP, board level reliability, USRR

I. INTRODUCTION

Owing to more safety concerns for drivers, the automotive manufacturers are continuously expanding their vehicle developments by including advanced driver-assistance system (ADAS). The automotive mmWave radar is already well adopted as an active safety part of ADAS [1]. Owing to the characteristic of parasitics reduction with short interconnection of thin-film RDL and high-precision assembly process, the FOWLP technology emerging as a high-density packaging technology, has been proven to realize better performance over conventional wire-bond and flip-chip packages [2-7]. The mmWave radar with antenna integrated by FOWLP technology in this study, as shown in figure 1, has the size of 6.5mm*6mm which features an unprecedented large package-to-die area ratio of 4.3 and low solder ball density. Even though with the advantage of good electrical performance, fan-out structure inherently exhibits weaker mechanical properties due to the substrate-less process which is more sensitive to thermal-mechanical stress induced by thermal-expansion coefficient (CTE) mismatching among composed materials and PCB. Moreover, the features of both large package-to-die ratio and low solder ball density were thought to be key factors causing possible failures during BLR stressing, especially board level thermal cycling (BLTC). Hence, study on the high latent risk of chip-to-board interaction (CBI) would become critical for the successful development of this large fan-out ratio product, especially applied to automotive field.

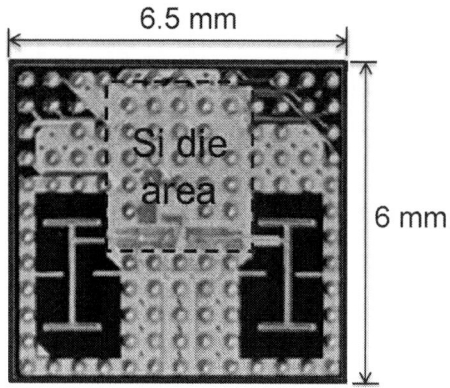

Figure1. Picture of the USRR device in FOWLP in this study

II. EXPERIMENTAL METHOD

A. Specimen Preparation

To have excellent board-level reliability of the FOWLP, which inherently has the features of large fan-out ratio and low-ball density, the structure and solder material were both studied with mechanical test vehicle (mTV) to understand their impacts on BLR performance. As shown in table 1, Si die thickness (350 μm and 400 μm); mold gap (the molding compound thickness above Si die, 45 μm, 150 μm, and exposed die), solder material (SAC_A and SAC_B, the later has higher elastic modulus), and epoxy material (underfill and edge-bond-glue) were all experimentally studied. Since underfill material is always applied for enhancing the strength of the joint between IC component and PCB to resist BLRT stressing, in this study a low-CTE underfill was evaluated the effectiveness on the BLTC performance improvement even though the existence of underfill does largely deteriorate the radar performance. Additionally, the edge-bond-glue (as shown in figure 2) without deteriorating the radar performance were dispensed abreast on both sides of the FOWLP on PCB for studying the BLTC performance as well.

In the experiments, board-level thermal cycling, drop, and vibration were carried out to find out the most appropriate structure and solder material for reliability assurance. In the mTV two daisy-chain loops, including die loop and BGA loop (as shown in figure 3), were designed individually to monitor the resistance changes during BLRT for detecting the failure inner and outer die area respectively. Apart from the both loops, additional one critical loop (named corner loop) connecting only the 4 corner balls was specifically designed since the corner ball crack is most frequently found and always dominates the failure during board-level reliability.

Table1 Specimen preparation in this study

Leg	1	2	3	4	5	6
Underfill or edge-bond glue			NA		Underfill	Edge-bond glue
Si thickness (μm)	350	350	400	400	400	400
Backside lamination (μm)		NA		25		NA
Mold gap (μm)	45	45	150	Exposed Die/BSC	150	150
Package thickness (include 30um RDL, μm)	425	425	580	425	580	580
Solder Material	SAC_A	SAC_B	SAC_B	SAC_B	SAC_B	SAC_B

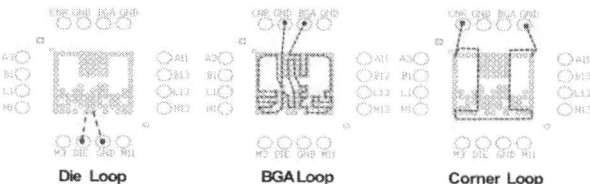

Figure2. Cross-section view of FOWLP on board with edge-bond glue

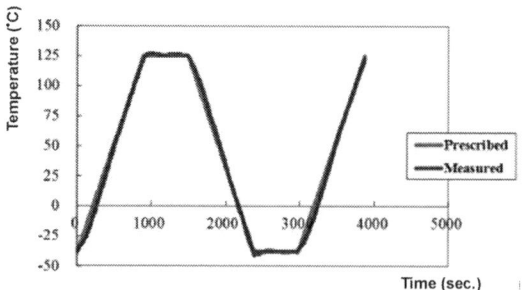

Figure 3. Daisy-chain loops designed in mTV

B. Board-Level Temperature Cycling Test

Since the solder joint reliability is always the most predominant issue in BLTC especially for the packages with large size or with specific structure (like large fan-out ratio and low ball density in this study), the BLTC of condition G (-40 °C to 125 °C, as shown in figure 4) was executed to understand the reliability of the solder joint between the FOWLP and PCB, and further to find out the most appropriate one from the 6 DoE legs. During the test, daisy-chain resistance was in-situ monitored to have the data of cycle to fail with failure criteria of 20% resistance increase.

Figure 4. Temperature profile during BLTC

C. Board-Level Drop Test

In addition to assessing the ability of the FOWLP against the persistent temperature change, the test on evaluating the ability of this FOWLP to resist dropping is always indispensable, especially for the package of handheld or automotive electronic products. In this study the drop test condition of 1500G with 0.5ms duration time was applied to evaluate the robustness of solder joint against drop impact, and the electrical failure was judged when the resistance monitored in-situ by event detector exceeds 1000 ohm.

D. Board-Level Vibration Test

Since vibration test is indispensable and always of much concerned for automotive electronics by tier-1 customers. In this study the stringent test condition of sine-wave frequency swept from 20 Hz to 2,000 Hz and peak acceleration of either 50g or 20g, was applied to evaluate anti-vibration property of the FOWLP. The test was performed four times in each of the three orientations X, Y, and Z, and each sweep shall be completed for 4 minutes. That's to say, the vibration test was conducted first in X orientation for 16 minutes, and then followed by Y and Z orientations in sequence. The failure criteria was defined as the resistance exceeds 1,000 ohms lasting for 1 microsecond or longer. Figure 5 shows the setup of the vibration test.

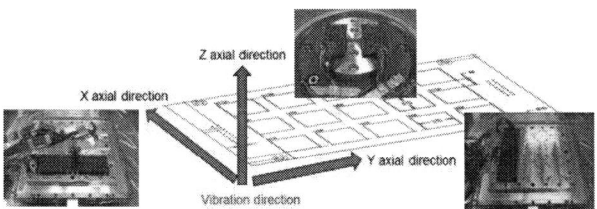

Figure 5. Picture of experiment setup of vibration test

III. RESULTS AND DISCUSSION

A. Board-Level Temperature Cycling Test

Table 2 shows the BLTC results of the 6 DoE legs, and the data reveals that the SAC_B with higher elastic modulus (leg 2, 3 and 4) performs much better reliability on BLTC with the temperature range from -40°C to 125°C for 3000 cycles. In contrast, solder A (in leg 1) with lower elastic modulus suffered 100% failure in corner ball loop and even 77% failure in die loop, which indicated apparent disparity exists in solder reliability as compared with the results in leg 2. Both package thicknesses of 425 μm and 580 μm were studied as well and the results revealed that the design of leg 3 with both thicker Si die and thicker molding significantly improved the BLTC since there's no failures found after 3000-cycle BLTC. Figure 6 shows the FA result of solder joint failure mode in corner-solder loop from DoE leg4, and the IMC crack at PCB side was found. Finally, it could be concluded that the package with the features of thicker Si die and thicker molding, and with the adoption of higher-elastic-modulus solder, performs the best BLTC reliability. That's to say leg 3 is the most appropriate one among the first 4 legs to resist BLTC reliability.

The epoxy-based underfill frequently thought to serve as a key element of increasing board level reliabilities, especially the drop test, was evaluated as well in this study. Both epoxy materials – one is low-CTE underfill material (in leg 5) and the other is edge-bond glue (in leg 6), were applied to the package designed as leg 4 for knowing the effectiveness of enhancing the BLTC performance on this FOWLP with large fan-out area ratio and low ball density. As shown in table 2 and figure 7, the experiment results showed the DoE leg 5 (with low-CTE underfill) suffered 100% failure in corner loop with 1st fail at 289 cycles, 100% failure in BGA loop with 1st failure at 227 cycles, and even 50% die loop failure with 1st failure at 496 cycles, when undergoing only 1000-cycle BLTC reliability. Figure 7 also reveals severe bulk solder crack, IMC fracture, underfill delamination and underfill crack in DUT#3 of leg 5 after 1000-cycle BLTC. Hence, the underfill material, even with low CTE, miserably decreased the BLTC performance of this FOWLP. Furthermore, the edge-bond glue with high viscosity and thixotropic index dispensed abreast at both sides of the FOWLP was also found to worsen the BLTC performance. As shown in table 2 and figure 8, 1000-cycle BLTC resulted in 8% failure in corner loop and even die loop, and IMC crack near PCB side was found. Finally, it could be concluded that underfill and edge-bond glue were both found to be harmful to the BLTC reliability of this FOWLP featuring large fan-out ratio and low ball density.

Table 2 BLTC results

Batch	Leg	Loop	Test cycle	DUT	Failure	Failure Rate
1st (w/o underfill)	1	Corner	3000	48	48	100%
		BGA		48	37	77%
		Die		48	1	2%
	2	Corner		48	2	4%
		BGA		48	0	0%
		Die		48	0	0%
	3	Corner		48	0	0%
		BGA		48	0	0%
		Die		48	0	0%
	4	Corner		48	1	2%
		BGA		48	0	0%
		Die		48	0	0%
2nd	5 (Underfill)	Corner	1000	8	8	100%
		BGA		8	8	100%
		Die		8	4	50%
	6 (Edge-bond epoxy)	Corner		24	2	8%
		BGA		24	1	4%
		Die		24	2	8%

Figure 6. IMC crack near PCB pad side at corner solder ball of the specimen in DoE leg4 after 3000-cycle BLTC

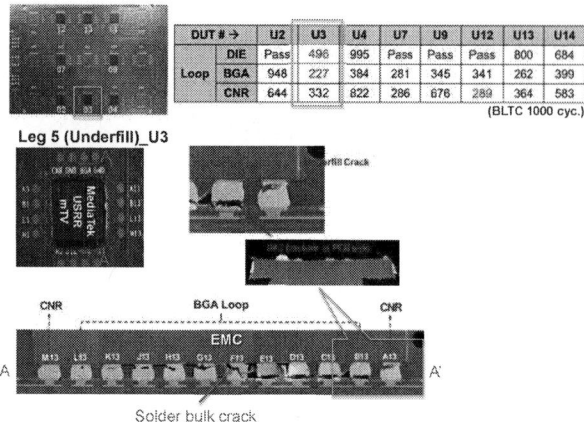

DUT # →		U2	U3	U4	U7	U9	U12	U13	U14
	DIE	Pass	496	995	Pass	Pass	Pass	800	684
Loop	BGA	948	227	384	281	345	341	262	399
	CNR	644	332	822	286	676	289	364	583

(BLTC 1000 cyc.)

Figure 7. BLTC result of leg 5 with low-CTE underfill and failure mode (bulk solder crack, IMC fracture, underfill delamination and crack)

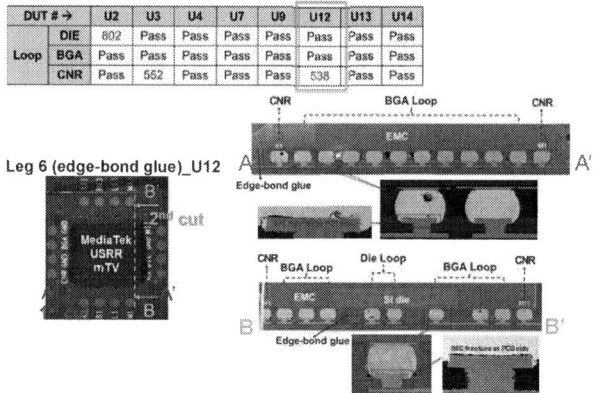

DUT # →		U2	U3	U4	U7	U9	U12	U13	U14
	DIE	802	Pass	Pass	Pass	Pass	Pass	Pass	Pass
Loop	BGA	Pass	Pass	Pass	Pass	Pass	Pass	Pass	Pass
	CNR	Pass	562	Pass	Pass	Pass	538	Pass	Pass

Figure 8. BLTC result of leg 6 with edge-bond glue and failure mode

B. Board-Level Drop Test

Since underfill and edge-bond glue were both verified experimentally to be harmful to the BLTC performance of this FOWLP, only the first 4 legs were tested on the BL-drop in this study. From the result, as shown in table 3, it could be seen that there's no difference found in board level drop test since no failure occurred after 400 drops. It could be concluded that owning the excellent capability to withstand board-level drop could be attributable to the lightweight nature of the FOWLP.

Table 3 Board-level drop test result of DoE leg 1~4

Leg	Loop	Number of failures (15 units/board, 4 boards/leg)				
		Time zero	30 drops	150 drops	400 drops	1st failure
1	Corner	0/60	0/60	0/60	0/60	NA
	BGA	0/60	0/60	0/60	0/60	NA
	Die	0/60	0/60	0/60	0/60	NA
2	Corner	0/60	0/60	0/60	0/60	NA
	BGA	0/60	0/60	0/60	0/60	NA
	Die	0/60	0/60	0/60	0/60	NA
3	Corner	0/60	0/60	0/60	0/60	NA
	BGA	0/60	0/60	0/60	0/60	NA
	Die	0/60	0/60	0/60	0/60	NA
4	Corner	0/60	0/60	0/60	0/60	NA
	BGA	0/60	0/60	0/60	0/60	NA
	Die	0/60	0/60	0/60	0/60	NA

C. Board-Level Vibration Test

Figure 9 shows the results of 50g peak acceleration vibration test accompanied with in-situ resistance monitor, and high resistance even open circuit was found in the daisy-chain corner and BGA loops. It is noteworthy that all the failures only happened on the package located at the corners from the 5x3 array while subjecting to Z-axis vibration. From experience, poorly fixing the PCB on vibration platform potentially causes more bending stain on PCB during Z-direction vibration and further concentrates much higher stress singularly nearby the corner. The failure analysis showed that delamination existed at the interface of RDL and UBM of component side and crack existed at PCB Cu trace, as shown in figure 10. And the Cu crack initiated from the edge of the Cu pad and propagated to the Cu trace at the base of via. To verify the PCB fixing issue, another four PCBs (15 packages on each PCB) were tested again the board-level vibration of 50g peak acceleration while being with more concern on the fixing quality, and the result showed 100% passing the test. That is to say, in this study all the packages mounted on the PCB could endure the stringent vibration test in horizontal axes (X and Y), and the package defect while undergoing Z-axis vibration is strongly relevant to location and fixing issues rather than reliability. Thus, it's believed that this FOWLP is robust enough to against the component-level vibration test with the test condition meeting the requirement described in AEC-Q100. Finally, the board-level vibration test with 20g peak acceleration was implemented as well, and certainly there's no failure found.

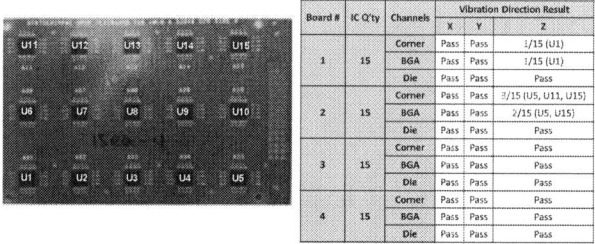

Board #	IC Q'ty	Channels	Vibration Direction Result		
			X	Y	Z
1	15	Corner	Pass	Pass	1/15 (U1)
		BGA	Pass	Pass	1/15 (U1)
		Die	Pass	Pass	Pass
2	15	Corner	Pass	Pass	3/15 (U5, U11, U15)
		BGA	Pass	Pass	2/15 (U5, U15)
		Die	Pass	Pass	Pass
3	15	Corner	Pass	Pass	Pass
		BGA	Pass	Pass	Pass
		Die	Pass	Pass	Pass
4	15	Corner	Pass	Pass	Pass
		BGA	Pass	Pass	Pass
		Die	Pass	Pass	Pass

Figure 9. Board-level vibration test results

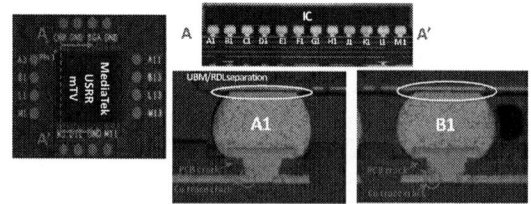

Figure 10. RDL/UBM interfacial delamination and PCB Cu trace crack after BL-vibration test with 50g peak acceleration

IV. CONCLUSIONS

In this paper, the mm-wave USRR realized in CMOS technology and assembled with FOWLP, was studied experimentally on the BLRT. Two solder materials were studied and the results showed that the solder with higher elastic modulus performed much better on BLTC reliability. Moreover, no failure occurred in both solder materials while undergoing board level drop test, and it could be attributable to the lightweight nature of the FOWLP. Regarding the factor of package structure, the results revealed that the design featuring both thicker Si die and thicker molding significantly improved the BLTC reliability. Both epoxy materials, low-CTE underfill material and edge-bond epoxy, were found to miserably decrease the BLTC performance and severe solder crack and bulk underfill crack were found. Board-level vibration test with the frequency sweeping from 20 Hz to 2,000 Hz, accompanied with peak acceleration of either 50g or 20g, was performed to evaluate anti-vibration property of the FOWLP mTV on daisy-chain PCB. The results showed that the FOWLP could endure the severe condition of 50g peak acceleration which is compliant with AEC-Q100.

Finally, the BL-TC, drop, and vibration tests were thoroughly studied for the extremely large area-ratio FOWLP, and the package was proved its capability of meeting stringent reliability tests which is compliant with AEC-Q100.

ACKNOWLEDGMENT

The authors would like to thank STATS ChipPAC for jointing the works of development, manufacturing, and reliability verification for this FOWLP.

REFERENCES

[1] Berthold Hellenthal, "Inventing the Automotive Future," Semicon Korea, Jan 2016.

[2] M. Brunnbauer, E. Fürgut, G. Beer, T. Meyer, H. Hedler, J. Belonio, E. Nomura, K. Kiuchi, K. Kobayashi, "An Embedded Device Technology Based on a Molded Reconfigured Wafer", 56th Electronic Components and Technology Conference (ECTC 2006), June 2006.

[3] M. Brunnbauer, E. Fürgut, G. Beer, T. Meyer,"Embedded Wafer Level Ball Grid Array (eWLB)," 8th Electronics Packaging Technology Conference (EPTC 2006), Dec. 2006.

[4] M. Wojnowski, R. Lachner, J. Böck, C. Wagner, F. Starzer, G. Sommer, K. Pressel, and R. Weigel, "Embedded Wafer Level Ball Grid Array (eWLB) Technology for Millimeterwave Applications," Proc. 13th Electronic Packaging Technology Conference (EPTC 2011), Singapore, Dec. 2011.

[5] G. Haubner, W. Hartner, S. Pahlke, and M. Niessner, "77 GHz Automotive RADAR in eWLB Package: From Consumer to Automotive Packaging," Microelectronics Reliability, Sep. 2016.

[6] M. Wojnowski, M. Engl, M. Brunnbauer, K. Pressel, G. Sommer, and R. Weigel, "High Frequency Characterization of Thin-Film Redistribution Layers for Embedded Wafer Level BGA," in Proc.

9th Electronic Packaging Technology Conference (EPTC 2007), Singapore, Dec. 2007.

[7] M. Wojnowski, M. Engl, B. Dehlink, G. Sommer, M. Brunnbauer, K. Pressel, and R. Weigel, "A 77 GHz SiGe Mixer in an Embedded Wafer Level BGA Package," in Proc. 58th Electronic Components and Technology Conference (ECTC 2008), Lake Buena Vista, USA, May 2008.

Fatigue Behaviour of Lead-Free Solder Joints Under Combined Thermal and Vibration Loads

Karsten Meier, Maria Winkler, Karlheinz Bock
Technische Universität Dresden
Institute of Electronic Packaging Technology
Dresden, Germany
e-mail: karsten.meier@tu-dresden.de

David Leslie, Abhijit Dasgupta
University of Maryland
Center for Advanced Life Cycle Engineering
Maryland, USA
e-mail: dasguptag@umd.edu

Abstract—The increasing demand for highly reliable electronic devices, even though they are exposed to harsh use conditions, is one of the main drivers for the development of electronic systems. System development process relies on the selection of materials, technologies and a proper design to meet the mission profile's demands. Among many others, the lead-free solder alloy SnAg1.0Cu0.5 (SAC105) is widely used for many electronic assemblies deployed for various applications. The fatigue behaviour of SAC105 under thermal loads (namely temperature cycling and shock testing) and drop testing has been covered extensively in the literature. Work on damage accumulation under vibration conditions has been accomplished but primarily at room temperature. Therefore, this work aims to expand knowledge of the fatigue behaviour of SAC105 under combined thermal and vibration loading. In this work, vibration durability experiments were conducted at temperatures from -40°C to +125°C and vibration peak-to-peak amplitudes from 0.6 mm to 1.6 mm. Currently, specimens have been subjected to tests with durations of 75×10^6 or 150×10^6 vibration cycles. Cross sections were analysed to relate damage locations and severity to stress conditions (temperature and vibration amplitude). As expected, damage levels were observed to increase with increasing temperatures and vibration amplitudes.

Keywords-Automotive & harsh environment reliability, drop/dynamic mechanical reliability, failure analysis techniques & materials characterization, reliability/life test methods & models, board and system level reliability

I. INTRODUCTION

The increasing demand for electronic devices to perform with high reliability while also being exposed to harsh use conditions is a well-known challenge and one of the main drivers for the development of electronic packages and systems that have to be deployed in vibration environments [1], [2], [3]. Package and system development processes rely on the selection of materials, technologies and processes as well as a robust design to meet the mission profile's demands and cost requirements simultaneously. However, to be able to accomplish the material selection process, comprehensive modelling and detailed knowledge of material properties are essential. Thermal and mechanical loads are especially critical in mission profiles for automotive, aerospace, industrial and military applications [4]. Humidity is an additional major factor but is not the focus of this study [5]. Thermal environments (isothermal and temperature cycling) and mechanical loads (quasi-static flexure, vibration and shock) occur simultaneously, rather than separately, and pose a significant risk to the reliability of solder joints at the 1st and 2nd packaging levels [6], [7]. Hence, there is a significant demand for knowledge about the deformation and damage behaviour of commonly used solder alloys. Material models used to describe the deformation behaviour under thermal and mechanical stress have been researched in several investigations [2], [8], [9], [10], [11], [12]. However, damage under combined thermal and mechanical loads for SAC105 solder have so far received less attention in the literature [10]. Lifetime data and modelling approaches also require further attention. As presented in Meier's [13] and Leslie's [10] works, experimental setups have been developed for fatigue investigations on solder joints of chip resistors. This paper focuses on Meier's test specimen design, and discusses further improvements to the design regarding the development of closed-loop temperature and vibration amplitude control. The improved specimen configuration has been consequently used to perform fatigue investigations at various temperature and amplitude levels.

II. EXPERIMENTAL SETUP AND PROCEDURE

The vibration experiments are conducted with shaker setups at two different laboratories – one at the Technische Universität Dresden and one at the University of Maryland. Although the setups have a different environmental and acceleration control ([10], [13]), the most important parts of the experimental setup, which are the specimen mounting fixture and the specimen itself, are identical. Both setups use a fixture which was introduced by Meier [13] and are made of the same high strength aluminium alloy. In order to be able run experiments at temperatures below room temperature, the shaker setup at the University of Maryland is equipped with a modified thermal chamber. As the specimen itself is as important as the specimen fixture, the specimen introduced in Meier [13] is used in all experiments (see Fig. 1).

To ensure a stable and constant vibration amplitude during an experiment, a closed loop control has been established. An optical sensing system was applied for experiments at room and high temperature conditions, whereas acceleration sensors were used to enable a closed loop amplitude control at low temperature. Both control concepts have been applied to control one side of the specimen (see Fig. 1). In addition, the vibration amplitude of the opposite side of the specimen was measured and the deformation mode was found to be almost symmetric about the central clamp (see Fig. 2). Measured deflection was found to be within 5% of the desired set-point at the examined loading conditions. This minor variation is caused by a small difference of the resonant frequency between the two specimen sides due to inevitable tolerances of specimen geometry, local material properties and clamping. The eventual accelerometer placement on both sides is subjected to position tolerances and thus can cause a shift of the resonance frequency. An investigation of the interconnect damage at comparable locations on the controlled and measured side, further confirmed that there is no significant stress difference due to near-symmetric deformation modes. 10 out of 13 examined specimens showed damage difference of less than 5% with respect to crack length. The remaining 3 specimens showed damage difference of 12% to 28%.

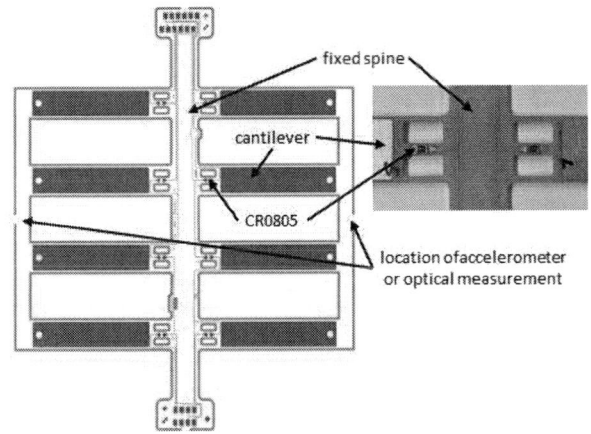

Figure 1. DESIGN SCHEMATIC AND FOTOGRAPH OF AN ASSEMBLED SPECIMEN USED IN THE VIBRATION EXPERIMENTS.

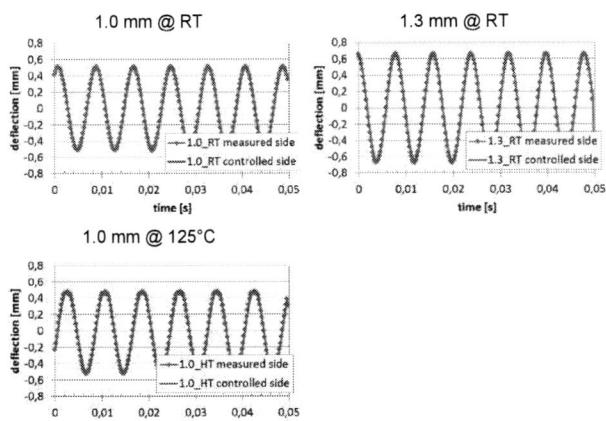

Figure 2. COMPARISON OF SPECIMEN DEFLECTION AT THE CONTROLLED SIDE (RED) AND MEASURED SIDE (BLUE) SHOWS CLOSE MATCH AT DIFFERENT EXPERIMENTAL CONDITIONS.

While the recent experiments at room and high temperature were conducted using an open loop temperature control, the specimen design has been improved meanwhile to enable a closed loop version without needing a temperature controlled chamber. This involves the addition of a Pt100 temperature sensor and PID control unit. A SMD version of the temperature sensor was selected so that it could be integrated into the specimen layout. The CR0603 sized component was placed in very close proximity to one of the tested components. This allows very precise temperature control of the solder joints throughout the entire experiment duration. A temperature deviation of less than 2 K was observed in a trial run covering a temperature interval from 30°C to 100°C.

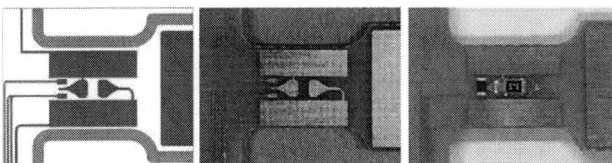

Figure 3. LAYOUT SCHEMATIC AND FOTOGRAPHS OF A SPECIMEN IN THE MANUFACTURED STATE AND THE ASSEMBLED STATE USED IN TRIAL EXPERIMENTS.

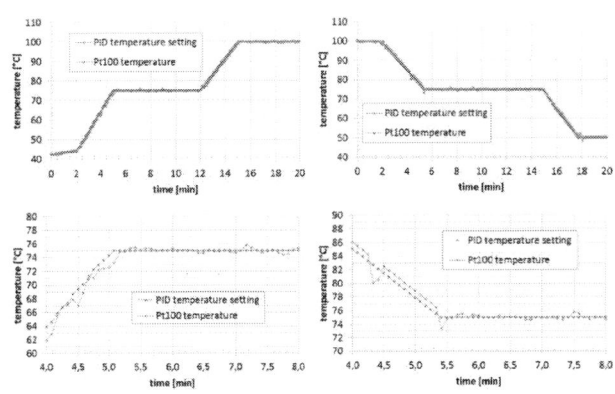

Figure 4. TEST OF THE CLOSED LOOP TEMPERATURE CONTROL FOR HEATING, COOLING AND DWELL PHASES OF AN EXPERIMENT.

The experimental setups were used to conduct vibration experiments covering harmonic vibration with peak-to-peak deflections ranging from 0.6 mm to 1.6 mm at temperatures from -40°C to +125°C lasting for 75×10^6 vibration cycles (see Tab. I). A small set of experiments at specific conditions was also conducted with a cycle count of 150×10^6. Deflections of less than 0.6 mm were examined as well, but no damage or indication of fatigue was noticed. Therefore, these low deflection experiments are not considered for further evaluation. In order to perform all experiments under comparable conditions, an excitation frequency of 125 Hz was used in all experiments reported here.

TABLE I. OVERVIEW OF COMPLETED VIBRATION EXPERIMENTS IN TERMS OF LOAD CONDITIONS

Specimen deflection (peak-to-peak) [mm]	Specimen Temperature [°C]			
	-40	*RT*	*+100*	*+125*
0.6			X	X
1.0	X	X	X	X
1.3	X	X		
1.5		X		
1.6	X			

TABLE II. OVERVIEW OF COMPLETED VIBRATION EXPERIMENTS IN TERMS OF LOAD CONDITIONS AND APPLIED VIBRATION CYCLES DURING ISOTHERMAL VIBRATION TESTING (VALUES GIVEN IN 10⁶)

Specimen deflection (peak-to-peak) [mm]	Specimen Temperature [°C]			
	-40	*RT*	*+100*	*+125*
0.6			75	75, 150
1.0	75	75, 150	75	75
1.3	75	75		
1.5		75, 150		
1.6	75			

All specimens were fabricated using ITEQ IT-180 high T_g material ($T_g = 175°C$) and assembled using SnAg1.0Cu0.5 solder alloy. All specimens were pre-aged for 100 h at 125°C immediately after the SMD assembly. Each test specimen carried 8 CR0805 components which were individually evaluated by means of cross sectioning. Damage locations and relative crack lengths in the solder interconnects of each component were determined and documented.

III. RESULTS

A. Observed Damage Phenomena

The first damage phenomenon observed was a roughened surface at the outer meniscus (see Fig. 5). This is a clear indicator for cyclic plastic deformation taking place within the solder joint volume. Dislocation motion near the solder joint surface causes intrusion and extrusion bands, thus changing the meniscus surface. More interestingly, a roughened meniscus surface was not noted for specimens tested at -40°C. Though this finding indicates a deformation and damage scenario incorporating less plastic deformation mechanisms (which is consistent with the strengthening expected at low temperature), a more conclusive statement on this requires a higher number of experiments accomplished at low temperatures.

Indications for further fatigue damage of the solder joint meniscus were found as well. Micro-cracks were found in the outer meniscus of components, which is indicative of cyclic fatigue damage. The area which contains the micro-cracks will be referred to as the fatigue zone. Fig. 6 depicts an example of such damage phenomenon. Solder joints showing this fatigue zone also had a roughened surface which indicates a sequential order of fatigue damage. While the roughened meniscus surface is the first indication of severe plastic deformation, the fatigue zone denotes further weakening on a microscopic scale. Fatigue cracks

progressing through the fatigue zone support this hypothesis. Cross sectioning artifacts are not believed to be the cause for the observed fatigue zone.

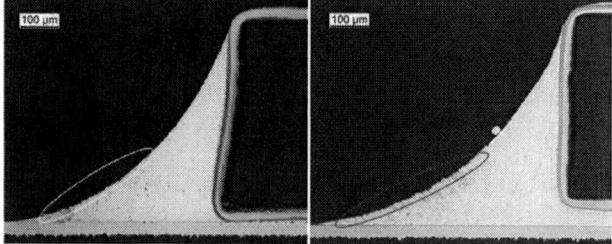

Figure 5. SOLDER JOINTS WITH INITIATED (LEFT) AND FURTHER PROGRESSED (RIGHT) MENISCUS SURFACE ROUGHENING AS AN INDICATOR FOR FATIGUE DAMAGE.

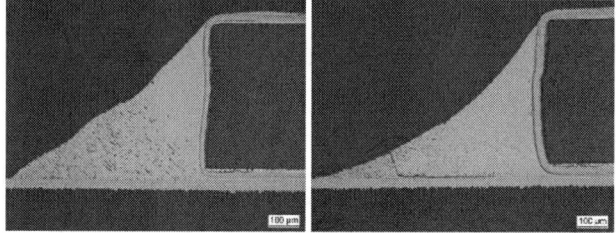

Figure 6. SOLDER JOINTS WITH INDICATORS OF FATIGUE DAMAGE: MICRO-CRACKS IN THE OUTER PART OF THEIR MENISCUS (LEFT AND RIGHT) AND A FATIGUE CRACK THROUGH THE FATIGUE ZONE (RIGHT).

When comparing the fatigue cracks observed within solder joints stressed at different temperatures, two features were noted (see Fig. 7). First, the crack initiates at a higher point of the meniscus at higher temperature tests. For tests conducted at -40°C, cracks started at the outermost edge of the solder joint meniscus in very close proximity to the interface. This indicates a shift in the location of highest stress in the solder joint which is most likely due to the fact that the material properties of the FR4 substrate, the copper pad and the solder alloy are all temperature dependent. While a change in temperature greatly alters the elastic properties of the FR4, it also strongly effects the solder alloy's viscoplastic behaviour. Thus, a more detailed analysis of this scenario should be accomplished by means of FEA.

Figure 7. SOLDER JOINTS WITH CRACKS OBSERVED AFTER VIBRATION TESTING AT -40°C, RT AND 125°C RESPECTIVELY.

Second, the crack path and width suggest that at elevated test temperature the crack propagates along a zig-zag path and shows greater distributed damage (see Fig. 8). This indicates that more severe plastic deformation of the solder is occurring at elevated temperatures. Additionally, at elevated temperature, the crack propagates through the interior of the solder volume rather than approaching the interface with the PCB pad, as seen at low temperatures.

Furthermore, a higher risk of copper cracking in tear drop part of the copper pad (seen in Fig. 3) at -40°C test conditions must be mentioned. This is potentially due to the higher stresses generated by the higher yield strength of copper at low temperature. This finding is limited to experiments with specimen deflections of 1.6 mm and did not lead to pad cratering. Hence, stresses and strains occurring within the tested solder joints were not significantly affected and the data gained from these experiments could be still considered valid. This unfortunately does have the potential to corrupt future solder fatigue failure data from these test specimen if these particular failures are not carefully excluded.

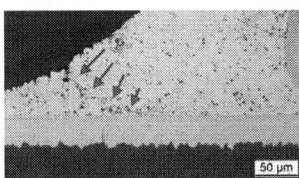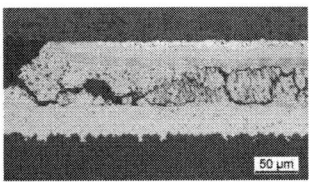

Figure 8. TWO DIFFERENT SOLDER JOINTS TESTED AT 125°C SHOWING CRACKS AT THE MENISCUS AND THE STAND-OFF RESPECTIVELY.

B. Evaluation of the Solder Joint Fatigue

The crack length in each individual solder joint was determined after cross sectioning. This absolute measure was related to the anticipated effective total crack length, as reported earlier [14]. Due to the specimen design, equal loading of all CR0805 solder joints of the specimen during testing can be assumed. Based on this assumption, an average crack length for each specimen was calculated taking all damaged solder joints into account. Averaging the observed crack lengths scales down the effect of inevitable variability from stochastic variability of solder volume, solder microstructure, component position with respect to the PCB pads, stand-off height, pad size etc.

The effect of the deflection was assessed first by focusing on the experiments performed at room temperature. Although experiments at lower deflections, i.e. at 0.6 mm, were performed, no cracks were found. As can be seen in Fig. 9, the observed crack length increases with increased peak-to-peak deflection. The observed crack length at a deflection of 1.5 mm was noted on a specimen with a comparatively high solder volume which is assumed to reduce solder joint stress. It follows that a longer crack length would have been seen if a lower solder volume was present for this deflection level. This result shows that the solder joint damage progress depends strongly on the enforced deflection amplitude.

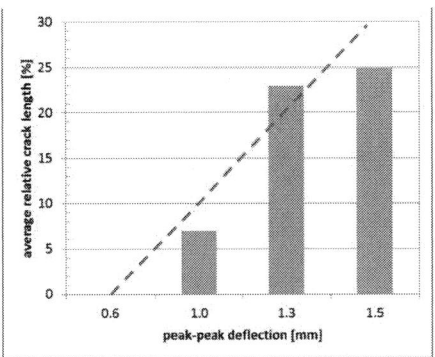

Figure 9. AVERAGE RELATIVE CRACK LENGTH OBSERVED AT SOLDER JOINTS AFTER VIBRATION TESTING WITH DIFFERENT DEFELCTIONS AT ROOM TEMPERATURE.

The effect of temperature on solder joint damage behaviour can be addressed by comparing experiments performed at -40°C, room temperature, +100°C and +125°C (see Fig. 10). Clearly, the solder joint damage significantly increases at elevated temperatures. A tip deflection of 0.6 mm showed moderate temperature sensitivity with no cracks at room temperature, and 4% increase in crack length (from 8% to 12%) between +100°C and +125°C. As the tip deflection is increased to 1.0 mm, the fatigue damage shows significantly stronger and nonlinear temperature sensitivity, with negligible temperature sensitivity at the low temperatures (6% crack length at both -40°C and room temperature), but a stronger temperature sensitivity at high temperature (12% increase in crack length from 50% to 62%, between +100°C and +125°C). Increasing the deflection even further to 1.3 mm, the fatigue damage starts to exhibit temperature sensitivity even at the low temperatures (with 5% increase in crack length from 17% to 23% between -40°C and room temperature). This nonlinear temperature sensitivity of fatigue damage implies that damage acceleration is influenced not only by the temperature difference but also the temperature level. As an example, while a temperature increase of 60K causes a damage increase of only 6% at lower temperatures, a temperature increase of only 25K causes a damage increase of 12% at higher temperatures. This nonlinear temperature sensitivity is believed to be related to the increase in viscoplastic behaviour of solder at high temperature, which has been extensively investigated for slow strain rates (e. g. [11], [12]) but has received limited amount of attention in the literature at high strain rates (e. g. [15]).

In summary, these findings confirm that the damage level is possibly too low at 1.5 mm deflection at room temperature. The damage acceleration has a nonlinear dependence on temperature. Furthermore, there is also clearly a combined synergistic effect between temperature and deflection that further accelerates fatigue damage.

978-1-7281-1500-9/19 $31.00 © 2019 IEEE

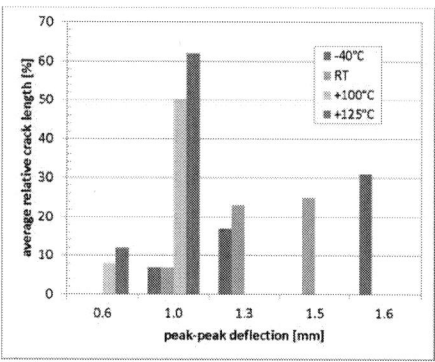

Figure 10. Overview of the crack lengths observed from tests at various deflections and temperatures.

The effect of the vibration exposure (75×10^6 vs. 150×10^6 cycles) on solder joint damage has been examined only at a few selected load conditions so far (see Table III). When tested at 1.0 mm tip deflection, the fatigue crack length increases slightly with the cycle count at room temperature. Similar behaviour is seen at 1.3 mm tip deflection, although the absolute crack length values are significantly higher than at room temperature. In order to understand this behaviour, the solder volume of the component joints has to be considered. As can be seen from Fig. 11, the solder volume can differ significantly. Solder joints with a low solder volume apparently show a classical concave meniscus rather than a straight fillet observed for a high solder volume. Coincidentally, all the experiments terminated at 75×10^6 vibration cycles happened to be on specimens with low solder volume while all experiments terminated at 150×10^6 vibration cycles happened to be on specimens with high solder volume. Considering the decelerating effect of a higher solder volume (as described before), increased damage due to an increased cycle count can still be assumed. Such behaviour was proven in earlier experiments [16]. However, a reduction in the effect of the cycle count might still be possible for higher temperatures. In order to better compare these results, a FEA would need to be performed to estimate the strain in the solder due to the differing geometry which then could be related to the fatigue damage. This needs to be researched in further experiments.

TABLE III. Average Relative Crack Lengths Observed from Experiments Conducted with Low and High Cycle Count at Different Deflections and Temperatures

Specimen deflection (peak-to-peak) [mm]	Specimen Temperature [°C]			
	RT		+125	
	Applied Vibration Cycles [10^6]			
	75^a	150^b	75^a	150^b
0.6	-	-	12%	10%
1.0	7%	9%	-	-
1.3	23%	25%	-	-

^a…conducted using specimens with low solder volume (see Fig. 11)
^b…conducted using specimens with high solder volume (see Fig. 11)

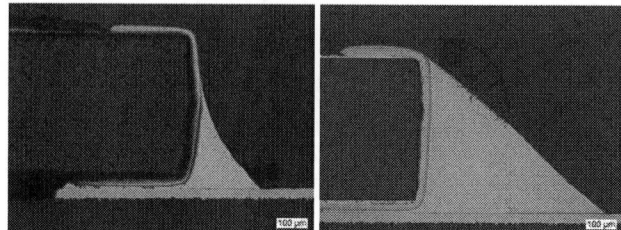

Figure 11. Solder joints with Low (left) and High (right) Solder Volume.

IV. Discussion

The fixture and specimen was designed to minimize the thermal expansion stress caused by temperature changes when the specimen is clamped, even when a temperature cycling profile is superimposed on the vibration excitation. A temperature-independent stress-free clamping enables a well-defined solder joint loading condition throughout such a combined experiment. This is a prerequisite for combined vibration and temperature cycling tests with the aim of fatigue analysis of solder joints. Loading conditions, and therefore the corresponding stresses, need to be well understood to be able to connect them to the observed failure mode and figures. The proposed specimen design complies with this need. The most recent version offers closed loop control of both the deflection and temperature as well as the option to apply strain gauges directly adjacent to the solder joints. The addition of the strain gauges allows the board strains to be monitored and used for finite element model calibration. Solder joint stress results from the calibrated finite element simulations can then be used for fatigue modelling.

The newly developed closed loop temperature control allows precise temperature adjustment and can be used for combined temperature cycling and vibration experiments. The measured temperature approximates the solder joint temperature as the sensor is located in close proximity to it. Using the temperature control with the local heaters enables temperature profiles from room temperature to +125°C with ramp up rates of up to 10 K/min. However, investigations on stable closed loop vibration tests that depend on the rate of an applied temperature change revealed a limit of 2 K/min. Thus, there is still need for improvement to enable temperature cycling profiles according to standard requirements.

Durability results revealed that fatigue damage progression is explicitly dependent on the solder joint shape. A higher solder volume leads to a straight fillet which causes significantly less damage. Therefore, specimens with this higher solder volume showed significantly higher vibration endurance, i.e. less cracking due to vibration fatigue damage, than joints with low solder volume. Accordingly, there is a need to carry out fatigue investigations with uniformly well-controlled specimen geometries. On the other hand, this observation also shows the need to directly examine the solder volume effect on fatigue damage in some future study.

978-1-7281-1500-9/19 $31.00 © 2019 IEEE

Solder joints exposed to closed loop controlled vibration loads show indications of fatigue which clearly depict the progress of fatigue damage throughout an experiment. Surface roughening, followed by meniscus fatigue, and finally fatigue cracking was observed when applying vibration loads with increasing deflection and temperature, respectively, and also when increasing these stresses in combination.

The fatigue damage observed during an isothermal vibration experiment is seen to increase with the test temperature. Location of crack initiation and crack path are temperature dependent. These findings can be traced to the temperature dependent properties of the involved materials. The temperature sensitivity of the solder viscoplastic deformation mechanisms are assumed to have a major influence on the findings.

Although the copper cracks are assumed to not significantly change the solder joint stress, future studies will focus on loading conditions which do not cause such failures. Copper cracks were found at -40°C and deflections of 1.6 mm. Therefore, for low temperature experiments, applied tip deflections should be less than 1.6 mm. Deflection limits for higher temperatures have yet to be determined.

The conducted isothermal vibration experiments show a significant correlation between the applied load on the components by means of deflection and temperature. It was not possible to fully evaluate the effect of the vibration cycle count within this work due to different solder joint volumes between the specimens used in different cycle counts. However, summarising all recent findings, there is strong reason to assume that solder joint damage will be significantly affected by the vibration cycle count. If vibration damage is compared across test specimens with different solder geometries, a FEA would be needed to estimate the strain in the solder, which then can be related directly to the fatigue damage. As stated earlier, a slow fatigue progression and crack growth rate is related to high temperature plastic deformation mechanisms rather than fast crack propagation based on elastic deformations [16]. This has to be carefully considered for future determination of fatigue models.

V. CONCLUSIONS

The developed experimental setup comes with various features enabling well defined isothermal vibration experiments:

- A temperature range from -40°C to +125°C can be covered. Vibration deflections of up to 1.6 mm can be applied depending on test temperature, as there is an increased risk of copper cracks at lower temperature.
- A temperature-independent stress-free clamping of the specimen has been accomplished by the recent specimen design.
- Vibration deflections are well controlled due to a closed loop control. Strain measurement options enable a detailed verification of finite element models.
- Isothermal vibration experiments are possible due to a closed loop temperature control based on the actual

solder joint temperature. Local and global heating options are available. Experiments at temperatures below room temperature are limited to the use of an oven which means global cooling.

Isothermal vibration experiments were performed covering temperatures from -40°C to +125°C and peak-to-peak deflections from 0.6 mm to 1.6 mm. Experiments show considerable influence of loading conditions and specimen geometry on damage of lead-free solder joints of LCR components:

- Solder damage is initiated at the outer meniscus surface and progresses through the bulk of the solder volume at higher temperature or in close proximity to the interface of the copper pad at lower temperatures towards the solder joint stand-off.
- While severe solder joint damage, *i.e.* crack lengths of >50%, was seen for 1.0 mm deflections at +125°C, crack lengths at room temperature, and -40°C did not exceed 10% for the same deflection. Crack lengths of about 30% were observed at room temperature and -40°C for deflections of 1.5 mm or more.
- The sensitivity of solder joint damage on temperature is increased significantly at higher temperatures.
- Copper cracks are only observed with deflections of 1.6mm at -40°C. Solder joint stresses are not affected by the copper cracks.

Though there is still need for further research on solder joint fatigue under isothermal vibration loading, the developed setup can also be used for combined vibration and temperature cycling experiments in the future.

ACKNOWLEDGMENT

The authors very much appreciate the help of Dr. Mike Roellig and his team at the Fraunhofer IKTS-MD by means of fruitful discussions and experimental support. The work of Yifan Liu as a research assistant at Technische Universität Dresden and of Tamara A. H. Storz as a student assistant at University of Maryland is gladly mentioned as well.

REFERENCES

[1] P. Fruehauf, A. Munding, K. Pressel, M. Vogt, P. Schwarz, *Chip-package-board reliability of System-in-Package using laminate chip embedding technology based on Cu leadframe*, Proc. IEEE Electronics System-Integration Technology Conference (ESTC 2018), Dresden, 2018, pp. 1-7.

[2] Y. Maniar, G. Konstantin, A. Kabakchiev, P. Binkele, S. Schmauder, *Experimental Investigation of Temperature and Mean Stress Effects on High Cycle Fatigue Behavior of SnAgCu-Solder Alloy*, Proc. IEEE Electronics Components and Technology Conference (ECTC 2018), San Diego, 2018, pp. 1645-1652.

[3] S. Kumar, *Challenges and requirements for 5G packaging*, Panel Talk at IEEE Electronics Packaging and Technology Conference (EPTC 2017), Singapore, 2017.

[4] R. Schwerz, *Zuverlässigkeit von gekapselten Bauelementen in der 3D-Leiterplattenintegration*, Dissertation Thesis, Technische Universität Dresden, TUDpress, ISBN 978-95908-140-5, 2018, pp. 1-11.

[5] K. Piotrowska, M. Grzelak, R. Ambat, *No-Clean Solder Flux Chemistry and Temperature Effectson Humidity-Related Reliability of*

978-1-7281-1500-9/19 $31.00 © 2019 IEEE

Electronics, Journal of ELECTRONIC MATERIALS, Vol. 48, No. 2, 2019, https://doi.org/10.1007/s11664-018-06862-4.

[6] E. George, M. Osterman, M. Pecht, R. Coyle, R. Parker, E. Benedetto, *Thermal Cycling Reliability of Alternative Low-Silver Tin-Based Solders*, Journal of Microelectronics and Electronic Packaging, October 2014, Vol. 11, No. 4, pp. 137-145.

[7] R. Coyle, M. Reid, C. Ryan, R. Popowich, P. Read, D. Fleming, M. Collins, J. Punch, I. Chatterji, *The Influence of the Pb-free Solder Alloy Composition and Processing Parameters on Thermal Fatigue Performance of a Ceramic Chip Resistor*, Proc. IEEE Electronics Components and Technology Conference (ECTC 2009), San Diego, 2009, pp. 423-430.

[8] K. Meier, *Beiträge zur Charakterisierung des Verformungsverhaltens von bleifreien Lotwerkstoffen unter dynamischen Beanspruchungen*, Dissertation Thesis, Technische Universität Dresden, Verlag Dr. Markus A. Detert, ISBN 978-3-934142-77-0, 2015, pp. 3, 92-165.

[9] P. Lall, V. Yadav, J. Suhling, D. Locker, *Anand Parameters for Modeling Prolonged Storage on High Strain Rate Mechanical Properties of SAC-Q Leadfree Solder at High Operating Temperature*, Proc. IEEE Electronics Components and Technology Conference (ECTC 2018), San Diego, 2018, pp. 448-459.

[10] D. Leslie, T. Heid, A. Dasgupta, *Effect of Temperature on Vibration Fatigue of SAC105 Solder Material after Extended Room Temperature Aging*, Proc. IEEE International Conference on Thermal, Mechanical & Multi-Physics Simulation and Experiments in Microelectronics and Microsystems (EuroSimE 2018), Toulouse, 2018, pp. 1-5.

[11] M. Roellig, *Beiträge zur Bestimmung von mechanischen Kennwerten an produktkonformen Lotkontakten der Elektronik*, Dissertation

Thesis, Technische Universität Dresden, Verlag Dr. Markus A. Detert, ISBN 978-3-934142-28-2, 2008.

[12] S. Wiese, *Verformung und Schädigung von Werkstoffen der Aufbau- und Verbindungstechnik - Verhalten im Mikrobereich*, DOI 10.1007/978-3-642-05463-1, Springer Verlag, 2010.

[13] K. Meier, R. Metasch, M. Roellig, K. Bock, *Fatigue Measurement Setup under Combined Thermal and Vibration Loading on Electronic SMT Assembly*, Proc. IEEE International Conference on Thermal, Mechanical & Multi-Physics Simulation and Experiments in Microelectronics and Microsystems (EuroSimE 2017), Dresden, 2017, pp. 1-7.

[14] K. Meier, M. Roellig, G. Lautenschlaeger, A. Schiessl, K.-J. Wolter, *Lifetime Assessment for Bipolar Components under Vibration and Temperature Loading*, Proc. IEEE International Conference on Thermal, Mechanical & Multi-Physics Simulation and Experiments in Microelectronics and Microsystems (EuroSimE 2013), Wroclaw, 2013, pp. 1-7.

[15] P. Lall, D. Zhang, V. Yadav, D. Locker, *High Strain-Rate Constitutive Behavior of SAC105 and SAC305 Leadfree Solder During Operation at High Temperature*, Proc. IEEE International Conference on Thermal, Mechanical & Multi-Physics Simulation and Experiments in Microelectronics and Microsystems (EuroSimE 2013), Wroclaw, 2013, pp. 1-11.

[16] K. Meier, M. Roellig, A. Schiessl, K.-J. Wolter, *Reliability Study on Chip Capacitor Solder Joints under Thermo-Mechanical and Vibration Loading*, Proc. IEEE International Conference on Thermal, Mechanical & Multi-Physics Simulation and Experiments in Microelectronics and Microsystems (EuroSimE 2014), Gent, 2014, pp. 1-7.

Prognostication of Accrued Damage and Impending Failure Under Temperature-Vibration in Leadfree Electronics

Pradeep Lall[1], Tony Thomas[1], Jeff Suhling[1], Ken Blecker[2]

[1]Auburn University
NSF-CAVE3 Electronics Research Center
Department of Mechanical Engineering
Auburn, AL 36849
[2] US Army Combat Capabilities Development Command - Armament Center
Picatinny Arsenal, NJ
Tele: +1(334)844-3424
E-mail: lall@auburn.edu

Abstract— **Electronics in defense applications may be exposed to sustained period of high temperature vibration. A number of mission critical applications are enabled through electronics control systems including navigation, guidance and control. Prognostication methods for assessment of accrued damage and identification of impending failure are needed under combined temperature vibration environments. In this paper, two-formats of leadfree board assemblies have been used to study the effect of prognostication sensor-location and use-conditions on the detectability of accrued damage and identification of impending failure. Transient strain deformation histories have been acquired using strain gauges fixed on the backside of the PCB, in the footprint of the front-side part. Feature vectors have been identified and their evolution studied in the pre-failure space. Effect of sensor location on the fidelity of early-identification of damage onset has been quantified. The feature-vectors were analyzed using a combination of statistical and frequency-based techniques. The frequency content of the strain signal is analyzed and patterns corresponding to before and after failure of electronic components at different conditions of vibration loads. A comparison of different feature vectors and change in the behavior of feature vectors with damage evolution under temperature and vibration has been quantified.**

Keywords-prognostics health management, feature vectors, temperature-vibration, reliability, leadfree solder joints, remaining use-life.

I. INTRODUCTION

Electronics has been used extensively in military and defense applications –which require sustained operation in harsh environments involving exposure to high-g levels of acceleration. Use of prognostic health management (PHM) methods for electronics can improve the functional safety of systems. Degradation in electronic systems is difficult to detect because of the small length scales [Braden 2014]. Electronics may be used for a number of mission critical functions where knowledge of the system readiness is valuable even after prolonged periods of storage. In defense applications, electronics may be used for ground-based vehicles, aerospace, missile fuzing and weapon systems.

Electronics in these applications may experience sustained exposure to high temperature, thermal cycling, high humidity, and high-g loads during transportation, storage and normal operation. In addition, electronics in defense applications may be subjected to stresses of transport and handling prior to deployment. Assurance of high reliability and low defect density is desired in a presence of new packaging architectures and miniaturization of electronics with new materials. It is not uncommon to have electronics designs in which prior historical data is not available to allow for the computation of field failure rates. Prognostics health monitoring methods can allow for the detection of incidence of defects and the pre-mature onset of expedited damage progression prior to catastrophic failure.

A number of approaches exist for prognostic health monitoring of electronic systems including the use of resistive opens, use of self-diagnostics using built-in self-test, and the use of fuses and canaries which fail prior to the catastrophic failure of the device. Previously, the authors have developed leading indicators of failure using a number of approaches including use of feature vectors based on mahalanobis distance, wavelet packet energy decomposition [Lall 2006a,b], joint time-frequency analysis in the time-frequency window [Lall 2007a,b], autoregressive moving average based on time and spectral domain feature vectors [Lall 2008], Karhunen Loeve Transform (KLT) [Lall 2009]. Electronics in high-g applications may be subjected to a multiplicity of failure mechanisms.

In this paper, feature vectors have been developed based on strain based measurements using principal components of instantaneous Frequency and the principal components of discrete-time Fourier transform. Different feature vectors were modelled based on the characteristics of the strain signal using frequency and statistical based techniques. This analysis also looks at the reliability of the feature vectors in predicting failure based on two different test boards with different package configurations. Selection of feature vector is done by filtering the input-signal to remove the noise present in the measured signal. The two test boards selected for this vibration experiment are intended for two specific purposes. Test board-1 is used to identify the survivability of the two feature vectors based on the distance of the strain signal from the packages. Test board-2 is used to test the two feature vectors at varying conditions of package configurations where the test board-2 has twelve packages and all package fail at different time during vibration.

II. TEST SETUP

Test setup for this experiment consists of a shaker and two test boards with the data acquisition system. First test board consists of a single package as shown in Figure 1 and the second with 12 packages on a PCB board as shown in Figure 3. The vibration of the test board is carried out on the top bed of the shaker which can be seen from Figure 2. The initial step during this experiment is to find the natural frequency of the test board and set the frequency range with a small bandwidth. The natural frequency of test board-1 is found to be at 340 Hz and the band width of vibration was set to be from 330 Hz to 350 Hz. Similarly, the natural frequency of the test board-2 was found to be at 380 Hz and the bandwidth of vibration for test board -2 was set to be from 370 Hz to 390 Hz. This bandwidth of vibration for these two board are set to accommodate the changes in the natural frequency of the two test boards during vibration due to board-to-board variations and changes in the frequency of vibration resulting from accrued damage. The vibration profile of the shaker was set to be a sinusoidal with an acceleration level of 5G for test board-1 and test board-2.

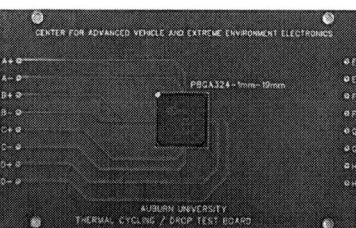

Figure 1: Test board-1

Figure 3: Test Board-2

Figure 2: Shaker System

Figure 4: Strain-Signal Conditioning Amplifier   Figure 5: Digital Storage Oscilloscope

The strain signals from the test boards are acquired using strain gages and strain rosettes as seen in Figure 6 and processed using a strain-signal conditioning amplifier from Figure 4. Data is acquired at regular intervals during vibration from all the strain gages. This signal is then stored using an oscilloscope seen in Figure 5 and analyzed using MATLAB for identification of feature vector. There are a total eight strain gauges fixed on test board-1. Strain rosette is fixed to the corner of the package to capture the prominent failure represented by C1, C2 and C3. A strain gauge is also fixed to the opposite side of the strain rosette at the corner of the package represented by C4. Two other strain gauges are fixed to the distant locations of the package C5 and C6 to study the survivability of the feature vector in predicting failure. Location of the strain signals C1 to C6 on the test board are shown in Figure 6 and location of the strain signals C7 and C8 on the test board are shown in Figure 7.

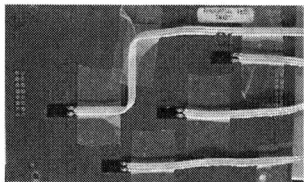
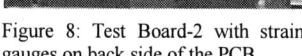

Figure 6: Test board-1 with strain gauges on the back side   Figure 7: Test board-1 with strain gauges on the front side

The strain gage positions of the test board-2 are shown in Figure 8 and Figure 9. There are a total four strain signals from test board-2 on the board at the locations of the package-footprint. The strain gage locations on the back-side of the printed circuit board are shown in Figure 8. An additional 4-strain gage are attached on the top of packages (Figure 9). These four locations of the test board are selected by considering the symmetry of the test board along the x and y directions. From Figure 9 it is seen that the strain gages are fixed to the corner of the package-footprint on top of the package, which is where the propensity for solder joint failure is the highest in out-of-plane deformation encountered in vibration. The location of the package strain gages is in the corner that is opposite to the back-side gages as seen in Figure 9.

Figure 8: Test Board-2 with strain gauges on back side of the PCB   Figure 9: Test board-2 with strain gage locations on different packages

The resistance values of the package for both the test boards are measured at interval of 1-minute till failure to identify the exact time of failure of the package. Further analysis on the strain signals are based on the failure time identified from the change in the resistnace values. A 20-percent increase from the initial resistance value of the package is considered to be the failure criterion for the analysis.

III. DAMAGE PROGRESSION IN THE PRE-FAILURE SPACE (TEST BOARD-1)

The experimental results on the test board-1 are based on the eight strain signals acquired during the different time-instance of vibration. The strain signals are acquired at a sampling frequency of 10,000 Hz and for a sampling time of 100 seconds during the vibration. The sampling time of 100 seconds is taken to capture the complete behavior of the strain signals during the upsweep and down sweep profile of the shaker. A sinusoidal profile is selected with an amplitude of 5G acceleration levels and a frequency range from 330 Hz to 350 Hz is selected as the profile for the shaker. This frequency range is selected to accommodate the natural frequency of the board at 340 Hz with a bandwidth of 20 Hz.

A. Variation of Strain Signal

The variation of different strain signals from C1 to C4 are plotted in Figure 10. C1 to C3 are the strain rosette signals in the vicinity of the package corner footprint. The C4 strain signal is located at the next package-footprint corner opposite to the strain rosette. Figure 11 shows the strain signals from C5 to C8 where C5 and C6 are from the back-side of the PCB and C7 and C8 are from the locations on the top of the package. The purpose of C5 and C6 is to understand the sensitivity of the feature vector to the sensor location and distance from the package and the purpose C7 and C8 is to understand the difference in the strain vectors acquired from back and front side of the package.

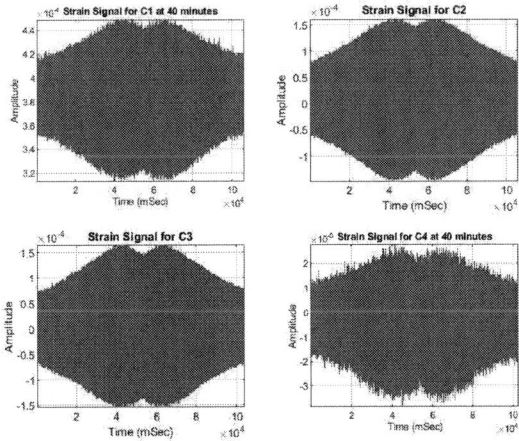

Figure 10: Strain plot of test board-1 for C1 to C4

It can be seen from Figure 10 and Figure 11 that the time-domain plots of the strain signals have different properties as the position of the strain gages on the test boards are different. The strain signals at the center of the test board have higher amplitude of vibration compared to those at locations that are farther from the center. A comparison of the strain signals C1-C6 with the strain signals C7-C8 shows the difference in the strain signature on the printed circuit board from the strain signature on top of the package. In addition, the strain signals C7 and C8 show a higher incidence of noise in the signal (Figure 11).

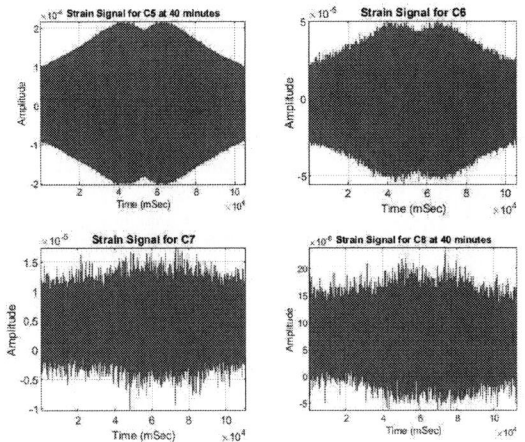

Figure 11: Strain plot of test board-1 for C5 to C8

B. Variations in Frequency content

The frequency content of the strain signal gives more insight in to the damage progression than the time-domain signal. In this analysis the variation in the frequency content is the principal characteristic in identifying the failure of package on the board. The frequency content of the strain signal from 1 Hz to 3000 Hz of the frequency spectrum is taken for consideration and a plot of (1-3000) Hz for strain signals C1 to C4 is plotted in Figure 12 at a vibration time of 40-minutes. Since the natural frequency of the board is previously found to be at 340 Hz, the maximum amplitude in the frequency content is seen at around 350 from Figure 12. The frequency content in Figure 12 seems to be similar for all the strain signals C1 to C4.

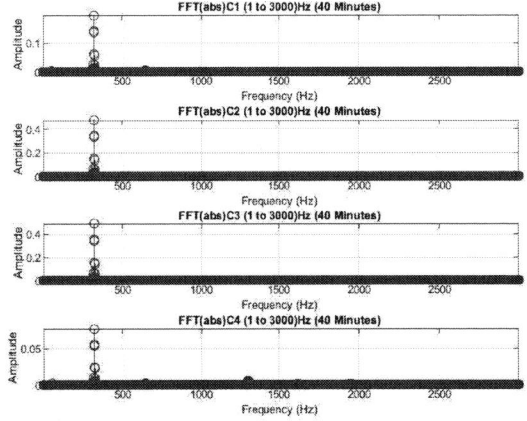

Figure 12: FFT (1-3000) Hz for C1 to C4 at 40 Minutes

Figure 13 is the frequency plot of the strain signals from C1 to C4 from 500 Hz to 3000 Hz at a vibration time of 40-minutes. Given the richness of features at frequencies higher than 500Hz, it is anticipated that feature vectors that are useful in predicting the failure of the package will include frequency content above 500 Hz. An examination of Figure 13 shows distinct differences in the frequency content at frequencies higher than 500 Hz unlike the pattern shown in Figure 12. This is because that the higher amplitude frequency content masks the variation of the lower amplitude frequency content. In order to understand the variation of lower amplitude frequency components, frequency plots from 500 Hz to 3000 Hz are plotted at various vibration times.

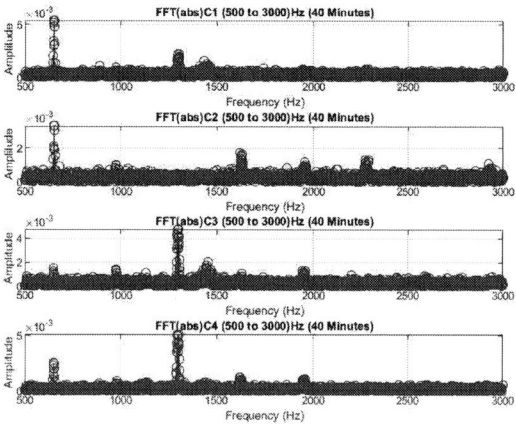

Figure 13: FFT (500-3000) Hz for C1 to C4 at 40 Minutes

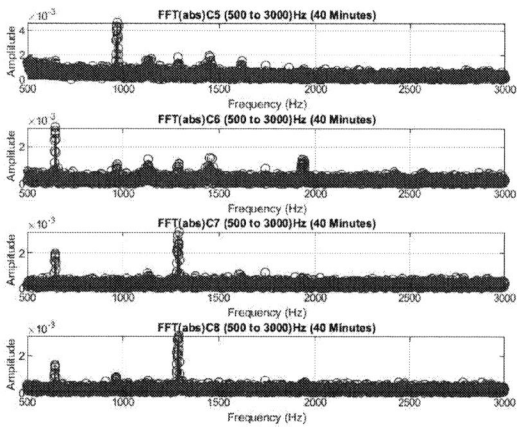

Figure 14: FFT (500-3000) Hz for C5 to C8 at 40 Minutes

Figure 14 shows the frequency-domain strain signal for strain gage locations C5-C8. A comparison of Figure 13 and Figure 14 shows that the distinct differences in the characteristics of the strain signal based on the position of the strain gauge. It is hypothesized differences in the evolution in the measured strain at various locations with vibration time can be used for the analysis of failure through statistical analysis. Figure 15 and Figure 16 are the frequency plots of the strain signals from C1 to C8 for 500 Hz to 3000 Hz of the frequency spectrum at a vibration time of 450 minutes. Comparing Figure 13 and Figure 15 plots are useful in understanding the changes in the frequency content for before after failure of the package during vibration of strain signals C1 to C4. Similarly, Figure 14 and Figure 16 shows a comparison of the strain signals, C5 to

C8, for a test time of 40-minutes and a test time of 450-minutes. Packages on test board-1 were found to fail at times in the vicinity of 420-minutes. Thus, the comparison of Figure 14 and Figure 16 provides a direct comparison of the frequency content of the C5 to C8 strain signals before failure and after failure.

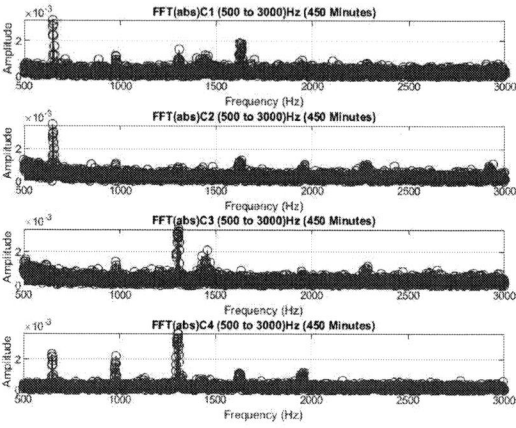

Figure 15: FFT (500-3000) Hz for C1 to C4 at 450 Minutes

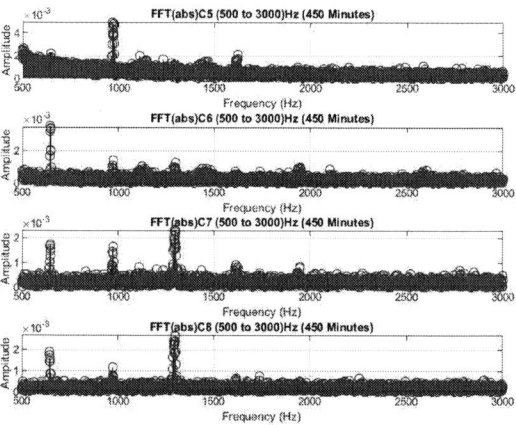

Figure 16: FFT (500-3000) Hz for C5 to C8 at 450 Minutes

C. Identification of Feature Vectors for Test Board -1

Frequency analysis of the strain signals from C1 to C8 was able to identify that there are changes in the frequency content of the strain signal from 500 Hz to 3000 Hz of the frequency spectrum. A combination of statistical technique and frequency content of the strain signal has been used to quantify the variation in the frequency content. Principal component analysis of the frequency content was carried out to understand the variation of the principal components with vibration time. Two features vectors were identified which could indicate the failure of the package as a function of vibration time. The first feature vector is the PCA of the frequency components from (500-2000) Hz of the strain signals and the second is the PCA of the instantaneous frequency. The variation of these two feature vectors with vibration time are discussed below to predict the failure.

1) Test Matrix for statistical analysis

The test matrix for the statistical analysis is derived from the original strain signal $[x]_{1\times10^6}$, which is of size 1×10^6. This is because the sampling frequency of the strain signal is set to be at 10,000 Hz and the sampling time is at 100 seconds which makes the length of the signal to become 10^6. The frequency content of the strain signal is $[f]_{1\times10^4}$ which is of size 10^4 as the sampling frequency is 10,000 Hz. When all the strain signals are considered then the total size of the strain signal is $[X]_{50\times10^6}$ and the total frequency content is $[F]_{50\times10^4}$ as the total number of strain signals considered for analysis are 50.

2) PCA of FFT (500-2000) Hz –(Feature vector-1)

PCA analysis on the frequency components of the strain signal from 500 Hz to 2000 Hz is referred to as the first feature vector. Figure 17 shows the plots of the first a second principal component of the frequency matrix from 500 Hz to 2000 Hz. The variation in Figure 17 is a scattered plot of red and blue dots and each dot has number on it. The red dots indicate the before-failure strain signals and blue dots represent the after-failure strain signals and the serial numbers indicate the order of strain signals corresponding to the vibration time. Figure 17 does not have a pattern for before and after failure strain signals. Figure 18 is the graph that plots vibration time and the variance of the first 10 principal components for the frequency matrix from 500 Hz to 2000 Hz. The variance of the PCA components of the FFT matrix shows a trend that indicates an increase in the variance just before failure of the components. This behavior of the feature vector is identified as the indicator for failure.

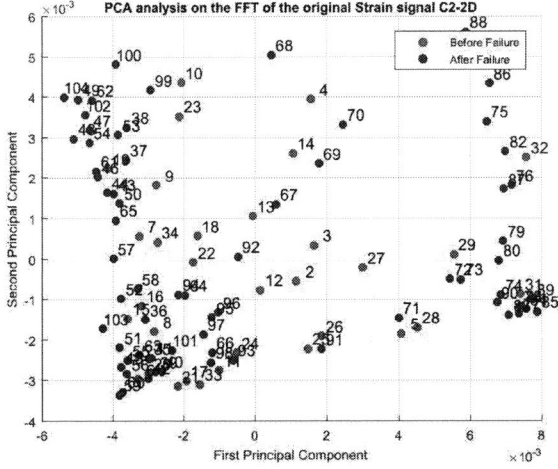

Figure 17: 2D plot of the first two PCA for FFT (500-2000) Hz

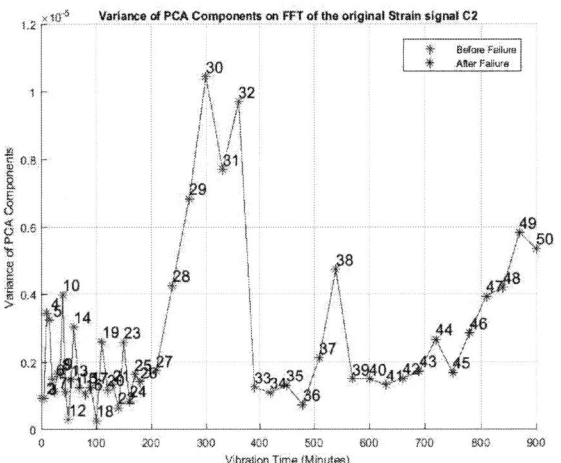

Figure 18: Variance of the first ten PCA for FFT (500-2000) Hz

3) PCA of IF–(Feature vector-2)

The second feature vector is the instantaneous frequency matrix of the strain signal. The PCA analysis on the instantaneous frequency matrix is plotted in Figure 19 where the variation of the first and second principal component doesn't have a useful behavior in identifying the before and after failure strain signals. Figure 20 plots the variance of the first ten principal components with the vibration time. The variance of the principal components of instantaneous frequency show a sudden increase just before failure of the component.

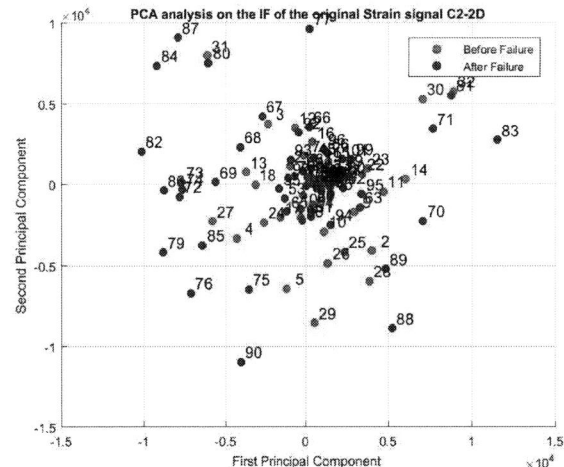

Figure 19: 2D plot of the first two PCA for IF

This behavior is identified as the second feature vector for failure of the package. Comparing these two feature vectors, both shows a transition phase of increase in variance just before failure of the package which helps in identifying the failure of the package. Feature vector-2 has a steep increase in the value, higher in magnitude than that of feature vector-1 – however, there is presence of more noise for feature vector-2. The variation of feature vector-1 and feature vector-2 have been studied for the eight strain signals acquired from different positions of the board.

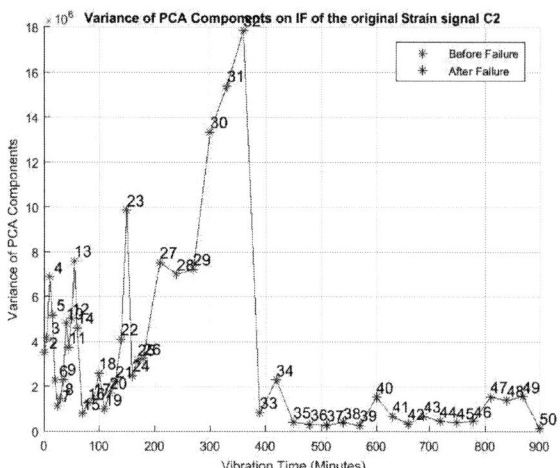

Figure 20: Variance of the first ten PCA for IF

D. Comparison Analysis based on Feature Vectors

Comparison study looks at the variation of the two feature vectors with the eight strain signals. These eight strain signals are used to study the reliability of the two feature vectors in identifying the damage of the package during vibration. The comparison study on the eight strain signals are divided into three analysis. Analysis -1 looks at strain signals C1, C2 and C3 whereas analysis-2 looks at strain signals C4, C5 and C6 and finally analysis-3 studies the strain signals C7 and C8.

1) Analysis-1

Analysis-1 looks at the strain signals from strain rosettes where C1 and C3 are shear strain signals and C2 is the in-plane strain perpendicular to the axis of vibration. The purpose of this analysis is to identify whether the two feature vectors are useful in identifying failure from normal strain and shear strain.

a) Strain rosette C1

The strain signal C1 is the in-plane shear strain acquired from the back-side corner of the package which is a potential location for solder joint failure and has a high propensity for accrual of damage during vibration.

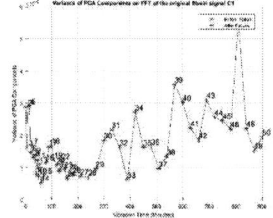

Figure 21: Variation of feature vector-1 on strain signal C1

Figure 22: Variation of feature vector-2 on strain signal C1

The variation of the two feature vectors are not indicative in identifying the failure as there is not feature change in the

behavior around the failure time as seen from Figure 21 and Figure 22. Hence the strain signals C1 is not suitable in predicting the failure based on these two feature vectors.

b) Strain rosette C2

Strain signal C2 is the normal strain at the back-side corner of the package in the in-plane direction perpendicular to the direction of vibration of the board. The variation of the two feature vectors are suitable for predicting the failure of the package. The sudden increase in the variance of both the feature vectors makes it suitable for the identification of the onset of failure as seen from Figure 23 and Figure 24.

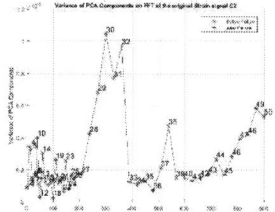 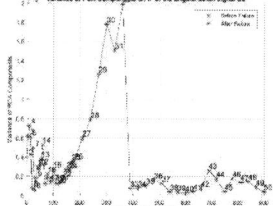

Figure 23: Variation of feature vector-1 on strain signal C2

Figure 24: Variation of feature vector-2 on strain signal C2

c) Strain rosette C3

The strain signal C3 is the shear strain that is at an angle of 45 degrees to the plane perpendicular to that of vibration. There is an increasing trend in the feature vector-1 after the manifestation of failure, which is not suitable for the prediction of failure as seen in Figure 25. Figure 26 has a behavior which is suitable for prediction as the increase in the variance occurs prior to the manifestation of failure and there is also a presence of transition period for failure. Summarizing analysis-1, it is found that normal strains are better in predicting failure and secondly the second feature vector is more reliable in predicting the failure compared to feature vector-1.

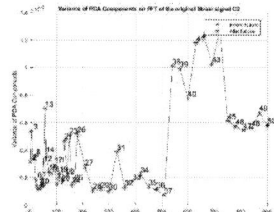

Figure 25: Variation of feature vector-1 on strain signal C3

Figure 26: Variation of feature vector-2 on strain signal C3

2) Analysis -2

The second analysis focuses on the study of variation of the strain signals C4, C5 and C6 in order to gain an understanding of the reliability of the feature vectors versus the location of the strain measurement and distance from the package. C4 is the closest strain gauge to the package and C5 is situated at the middle of the PCB and C6 at the far end.

a) Strain Gauge-C4

The location of the strain gauge C4 is opposite to the strain signal C2 and the only difference in the strain signal is that C2 is from a strain rosette and C4 is from a strain gauge. Figure 27 and Figure 28 shows the variation of the feature vectors one and two and feature vector-2 was the one to provide a behavior which is suitable for the identification of failure before the actual failure occurring. Figure 28 shows that there is a big increase in the variance of the principal component feature vector-2 which can be used in the health monitoring of the package.

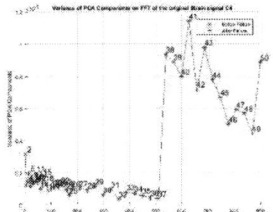 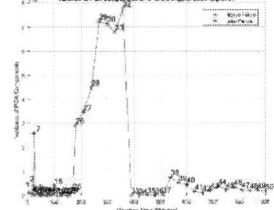

Figure 27: Variation of feature vector-1 on strain signal C4

Figure 28: Variation of feature vector-2 on strain signal C4

b) Strain Gauge-C5

The strain signal C5 is acquired from the middle position on the PCB. Figure 29 and Figure 30 shows the variation of the two feature vectors with vibration time. Measurements and analysis indicates that feature vector-2 is able to identify impending failure through an increase in variation of strain components as seen in Figure 30. Feature vector-1 was not able to give a behavior helpful in predicting failure in the case of C5 – as the feature vector shows an increase in variance after the manifestation of failure.

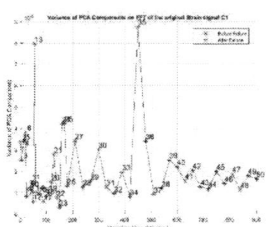 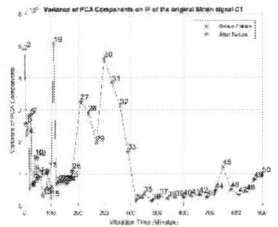

Figure 29: Variation of feature vector-1 on strain signal C5

Figure 30: Variation of feature vector-2 on strain signal C5

c) Strain Gauge-C6

The strain signal C6 is the farthest strain signal which is located at the corner of the PCB as seen Figure 6. Both feature vectors are not able to identify the failure of the package during vibration which is evident from Figure 31 and Figure 32. It is hypothesized that the lack of predictive ability is because as the distance from the package is increased the ability of the frequency components to capture the failure is lost. The second analysis provides insight into the survivability of the two feature vectors with the increase in distance of the strain signals from the package. The second feature is more reliable in predicting the failure which was conclusive in the analysis-1 as well. Finally, the strain signals C6 was not able to capture the failure with both the feature vectors indicating the limit in the ability of these feature vectors with distance from the package.

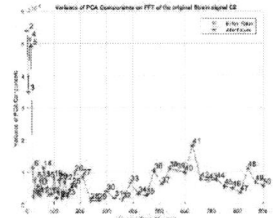 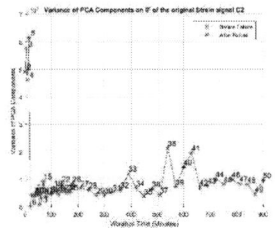

Figure 31: Variation of feature vector-1 on strain signal C6

Figure 32: Variation of feature vector-2 on strain signal C6

3) Analysis-3

Analysis-3 focuses on the strain signals from the front side corner of the package as seen from Figure 7. This analysis investigates the differences in the strain signal captured from the front side of the PCB and that of the back side. Analysis-1 and analysis-2 looks at the strain signals captured from the back side of the PCB while analysis-3 looks at the strain signals captured from the front side of the package.

a) Strain Gauge-C7

The variation of the two feature vectors for the strain signal C7 are shown in Figure 33 and Figure 34. Both the feature vectors are not able to predict the failure of the package from the strain signals. There is no increase in the variance of the feature vector one and two just before failure of the package which could indicate the failure. This inability of the strain signal may be due to the presence of higher amount of noise as seen in Figure 11.

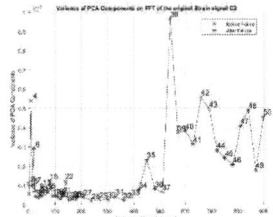 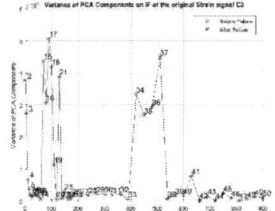

Figure 33: Variation of feature vector-1 on strain signal C7

Figure 34: Variation of feature vector-2 on strain signal C7

b) Strain Gauge-C8

Strain signal C8 is also from the front side corner position of the PCB which is in parallel position to the strain signal C7 which is displayed in Figure 7.

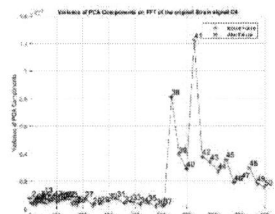 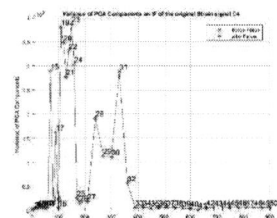

Figure 35: Variation of feature vector-1 on strain signal C8

Figure 36: Variation of feature vector-2 on strain signal C8

Feature vector-1 of the strain signal has no increase in the variance value before the failure of the package which is not useful in predicting failure. Feature vector -2 of the strain signal has multiple increase in variance values which are also not suitable for the identification. Summarizing the analysis of strain signals from the front side of the PCB compared to the back side there are no relevant patterns possible for strain signals from the front side with the use of the existing two feature vectors selected for analysis. The presence of higher amount of noise in the strain signal may be one of the reasons for the lack of predictive patterns.

IV. DAMAGE PROGRESSION IN THE PRE-FAILURE SPACE (TEST BOARD -2)

The experimental results based on the test board-2 are based on the four strain signals captured from the symmetric locations as shown in Figure 8. The natural frequency of the board was found to be at 370 Hz and the bandwidth of vibration of the shaker was set from 360 Hz to 380 Hz vibrating with acceleration level of 5G. The sampling frequency and sampling time for the experiment for test board-2 was set to 10,000 Hz and 100 seconds same as that of test board-1. The primary difference between the two test boards are the increase in the number of components for the test board-2.

A. Variation of feature vector-2 for test board-2

Analysis on both the feature vectors for test board-1 gave insights about the ability of the same at different locations of the strain signal. It was also found that the feature vector-2 had better ability in predicting failure at varying conditions of position of strain gauge on the PCB and distances from the package. Hence the analysis of test board-2 is only based on the variation of feature vector-2. The analysis on test board-2 is presented by considering two cases where on one board only two packages failed during vibration and on the second board eight packages failed during vibration.

1) Case-1

Case-1 of the test board-2 looks at a test board where only two packages fail during vibration. This condition is important as the failed package is also far away in position on the PCB so as the failure of one package does not affect the feature vector of the other. The two strain signals taken for analysis are C2 and C4 where C2 fails at 450 minutes and C4 fails at 2750 minutes as seen from Figure 37.

Failure time of the Packages

C2-
450 minutes

C4-
2750 minutes

Figure 37: Test board-2 with case-1

a) Variation of feature vector-2 for C2 and C4

The variation of feature vector-2 for strain signal C2 with vibration time is displayed in Figure 38. The red-dots in the plots represent the before failure strain signals and blue-dots represent the after failure strain signals. Figure 38 shows that there in an increase in the variance of the feature vector just before failure which will help in the identification of failure of the package. The second strain signal analyzed was C4. The variation of the feature vector-2 for strain signal C4, with increase in the duration of vibration is shown in Figure 39. This package was able to give a better variation in identifying failure as there is a steep increase in variances just before failure (which occurs at 2750-min) which is optimum in identifying failure. The steep increase in the variance values just before failure of the package is the desirable behavior in predicting the failure and in absence of interactions from the nearby package provides a way for identification of the onset of degradation.

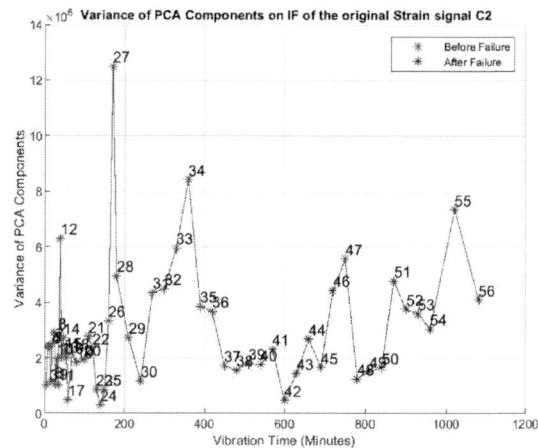

Figure 38: Variation of Feature vector-2 of C2 for test board-2 at case-1

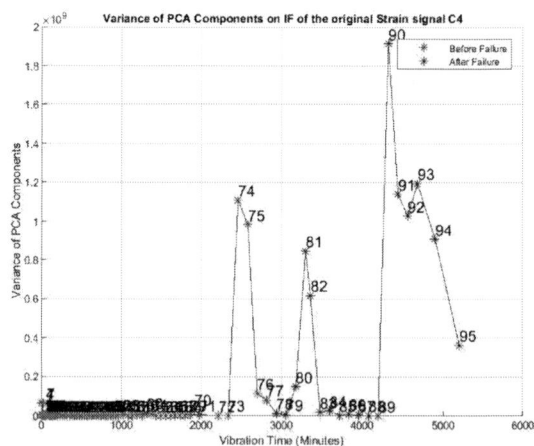

Figure 39: Variation of Feature vector-2 of C4 for test board-2 at case-1

2) Case-2

Case-2 of the test board-2 has eight packages failed out of the total twelve packages and three packages with a strain signal corresponding to the failure times shown in the Figure 40. The strain signal C1 fails at 600 minutes and signal C2 fails at 270 minutes and C3 fails at 27 minutes respectively. Careful examination of the failure time of strain signal C1 shows that there are four other packages closer to this package-location, which fail earlier than C1. This is similar for the strain signals C2 where there are nearby packages which fail earlier than the failure of C2. Failure of C3 is at 27 minutes and there are no other package failures before this vibration time.

Figure 40: Test board-2 with case-2

Failure time of the Packages

C1- 600 minutes

C2- 270 minutes

C3- 27 minutes

a) Variation of feature vector-2 for C1, C2 and C3

The variation of the feature vector-2 for the strain signals C1 is plotted in the Figure 41 and it is evident that there is a gradual increase in the variance to a point where the peaks starts to appear. The different peaks of the graphs may be due to the multiple failure of the surrounding packages. Since the vibration at the corner position of the package are relatively small compared to that at the center the effect of increase in minimal. Figure 42 shows the variation of the feature vector-2 for the strain signal C2. There are packages that fail before 270 minutes which is the failure time of the strain signal C2.

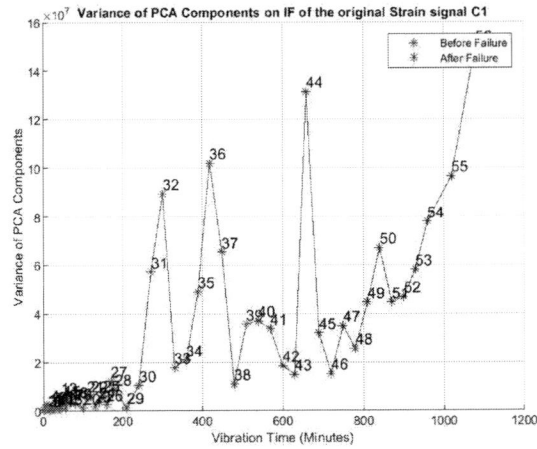

Figure 41: Variation of Feature vector-2 of C1 for test board-2 at case-2

These failures increase the variance of the feature vector-2 with vibration time and could not show a very high change in the variance value just before failure. Figure 43 shows the failure pattern of the strain signal C3 which fails at 27 minutes of vibration. There are no packages present in the vicinity that fails earlier than the failure time of C3 and hence it has a pattern which shown an increase in the variance just before failure. Even if the failure time of the package is considerably less, the feature vector is able to capture the increase in the variance of the feature vector-2 with vibration time.

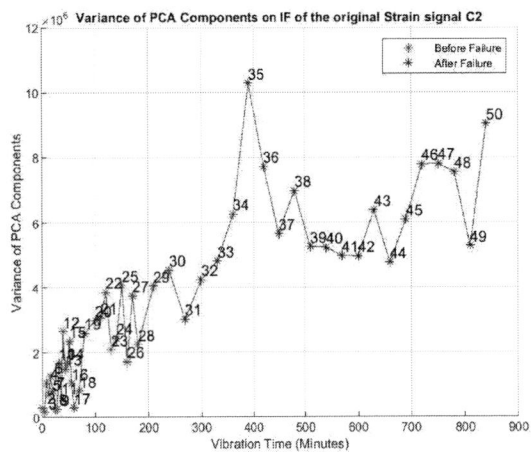

Figure 42: Variation of Feature vector-2 of C2 for test board-2 at case-2

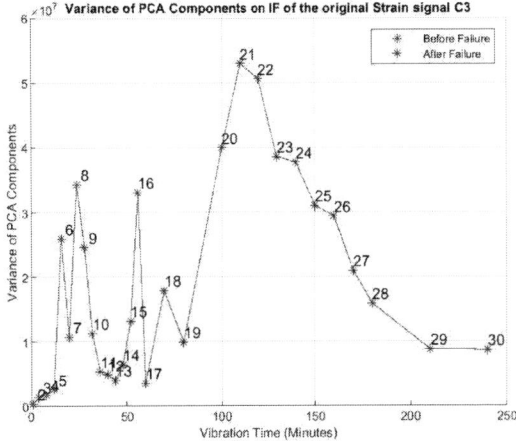

Figure 43: Variation of Feature vector-2 of C3 for test board-2 at case-2

V. CONCLUSIONS AND DISCUSSION

The vibration analysis on two test boards with different configurations of packages were tested with two feature vectors for predicting failure. Test board-1 had one package at the center and eight strain signals from different positions of the PCB. Three type of analysis were carried out on test board-1 to understand the effectiveness of the feature vectors in predicting failure. First analysis looked at the normal and shear strain captured from the back of the PCB to understand whether normal and shear strain could predict based on the two feature vectors and it was found that shear strains are not capable of predicting from the above feature vectors. The second analysis looked at whether the distance of the strain signal from the package had an effect on the ability to predict failure and it was found as the distance increases the ability of the features to predict failure decreases and feature vector-2 outperformed feature vector-1 at some conditions. The third analysis looked at the difference in the strain signals acquired from the front and back of the PCB in predicting and it was found that the back side strain signals were able to predict the failure more precisely. The vibration analysis on the test board-2 was to understand the ability of the feature vectors in predicting failure if there many components on the PCB. Two case of test board-2 was taken into consideration where case-1 had only two packages failed during vibration and case-2 has eight packages failed during vibration. The feature vector-2 was only used in this analysis as it was the most reliable one form test board-1. The failure patterns of case-1 was good enough to predict the failure without any disturbance in the pattern but for case-2 there were different package failures in the surrounding of the strain signals which gave rise to an increasing behavior of the feature vector in predicting failure. Finally, even though there was an increasing behavior of the feature there was a sudden change in the variance values that can be used in the failure prediction of multiple components in failure.

ACKNOWLEDGMENT

SERC ART-001 titled – Characterization of Emerging Technologies in Military Environments from Combat Capabilities Development Command - Armament Center, supported the project. The Research was conducted at the Auburn University's NSF-CAVE3 Electronics Research Center.

REFERENCES

[1] Braden, D. and Harvey, D., "A Prognostic and Data Fusion Based Approach to Validating Automotive Electronics," AE 2014 World Congress & Exhibition, SAE Technical Paper 2014-01-0724, 2014

[2] Lall, P., Choudhary, P., Gupte, S., Health Monitoring for Damage Initiation and Progression during Mechanical Shock in Electronic Assemblies, 56th Electronic Components and Technology Conference, San Diego, California, pp.85-94, May 30-June 2, 2006[a].

[3] Lall, P., Panchagade, D., Liu, Y., Johnson, W., Suhling, J., Models for Reliability Prediction of Fine-Pitch BGAs and CSPs in Shock and Drop-Impact, IEEE Transactions on Components and Packaging Technologies, Volume 29, No. 3, pp. 464-474, September 2006[b].

[4] Lall, P., Choudhary, P., Gupte, S., Suhling, J., Hofmeister , J., Statistical Pattern Recognition and Built-in Reliability Test for Feature Extraction and Health Monitoring of Electronics under Shock Loads, 57th Electronic Components and Technology Conference, Reno, Nevada, USA, pp. 1161-1178, May 29-June 1, 2007[a].

[5] Lall, P., Panchagade, D., Iyengar, D., Shantaram, S., Suhling, J., Schrier, H., High Speed Digital Image Correlation for Transient-Shock Reliability of Electronics, 57th Electronic Components and Technology Conference, Reno, Nevada, pp. 924-939, May 29 – June 1, 2007[b].

[6] Lall, P., Gupta, P., Kulkarni, M., Panchagade, D., Suhling, J., Hofmeister , J., Time-Frequency and Auto-Regressive Techniques for Prognostication of Shock-Impact Reliability of Implantable Biological Electronic Systems, 58th Electronic Components and Technology Conference, Orlando, Florida, pp. 1196-1207, May 27-30, 2008.

[7] Lall, P., Gupta, P., Panchagade, D., Angral, A., Fault-Mode Classification for Health Monitoring of Electronics Subjected to Drop and Shock, 59th Electronic Components and Technology Conference, San Diego, California USA, pp. 668-681, May 25-29, 2009.

Electrochemical Impedance Spectroscopy (EIS) for monitoring the water load on PCBAs under cycling condensing conditions to predict electrochemical migration under DC loads

Simone Lauser, Theresia Richter
Robert-Bosch GmbH, Automotive Electronics
71701 Schwieberdingen, Germany
simone.lauser@de.bosch.com

Vadimas Verdingovas, Rajan Ambat
Center for Electronic Corrosion, Department of
Mechanical Engineering, Section of Materials and
Surface Engineering, Technical University of Denmark,
2800 Lyngby, Denmark

Abstract—Humidity induced failures like metallic dendrite formation are a major problem for automotive electronic components. The harsh environment, where operating conditions in terms of temperature and humidity vary, can repeatedly provoke thin water layers on the surface of Printed Circuit Board Assemblies (PCBAs). The presence of a water film on electronics enables various corrosive processes. The understanding of the film formation and its effects is therefore crucial for assessing the humidity robustness of a specific setup. In this work, we conducted temperature and humidity load experiments with test boards containing interdigitated copper traces of different gap sizes on FR-4 substrate material. We repeatedly provoked condensation and evaporation conditions on the boards' surfaces by temperature cycling between 25 °C and 55 °C at 97 %rH. Electrochemical impedance spectroscopy (EIS) was employed as testing approach to detect the water film formation and respectively its evaporation. An AC excitation of 10 mV over a frequency range between 1 kHz and 100 kHz was used. Simultaneously, the commonly used SIR (Surface Insulation Resistance) test method was conducted at 5 V DC. This method lacks in delivering information on the actual water layer build up, but it detects the growth of dendrites, for which the DC voltage is required. The evaluated results of the EIS testing show, that the magnitude of water present can be depicted by the change in phase shift in the high frequency domain. We could also detect the water film closing for different gap sizes upon condensation. The DC measurements showed a correlation in terms of dendrite formation upon certain water load conditions.

Keywords- electrochemical impedance spectroscopy; surface insulation resistance; water film formation; electronic reliability; humidity, corrosion, electrochemical migration

I. INTRODUCTION

Corrosion processes enabled by a thin layer of water are a common problem when dealing with electronics operating under humid conditions. Once a voltage bias between two electrodes and a connecting electrolyte layer come together, electrochemical migration (ECM) is likely to happen. This failure mode describes the electro-dissolution of metal ions from a positively biased electrode, their transport through an electrolyte layer and their electrodeposition at the negatively biased electrode [1, 2]. The precipitated material grows in the form of dendrites and can ultimately bridge the gap between the oppositely biased metallic materials. The lifetime of such a dendritic structure can be a few seconds, if they burn due to the sudden current pulse running through it, but they can also remain permanently [3]. Their functioning as unwanted conductive bridge between isolated tracks or contacts on PCB (Printed Circuit Board) setups therefore is an issue regarding the lifetime of electronic devices. Especially in times when layout development is driven by miniaturization, the humidity reliability is increasingly difficult to fulfill [4]. The common method in terms of evaluating electronics' humidity robustness is the SIR (Surface Insulation Resistance) method. According to this technique, a DC voltage is applied to a test structure and the leakage current is recorded. This current comprises the bulk's as well as the surface's conduction and can also be expressed as resistance. Monitoring the resistance value over time under different humidity and temperature conditions will give an insight into climatic reliability. A drop in SIR equates corrosion processes. In many cases, it is attributed to dendrites that result from ECM and bridge the supposedly insulating gap over which the voltage bias is applied [5].

There are different aspects to consider when striving for improvement of temperature and humidity robustness of PCBAs (Printed Circuit Board Assemblies). Aside from the mentioned trend towards smaller distances and the respective considerations that should be made in terms of design and layout, the used materials in the PCBA have a significant impact. When choosing materials, care should be taken on the reactivity of the alloys in distinct environmental conditions as well as their electrochemical behavior upon coupling. Also, the choice of base material and potential coating materials have an impact on the reliability in harsh environments. The surface's morphology and its hydrophilic or respectively hydrophobic character will influence the water adsorption [6]. Cleanliness is another factor to optimize. Contamination, for example flux residues remaining on boards after a soldering process can favor dendrite growth under high humidity [7]. This corrosion process preferably happens due to the attraction of humidity in the air to such residues and the fact that their dissociation into ions to a forming water layer improves the conductivity of this electrolyte [8]. The aspects that determine a PCBA's humidity robustness all interlink in terms of their impact on condensate formation. The build-up kinetics of the water layer, its continuity and electrochemical properties are the crucial impact factors for electrochemical migration. A general understanding of how the water film characteristics are influenced by all of the mentioned circumstances is difficult to attain. The water layer formation itself is not detectable by common SIR methods; however, they reflect synergistic effects of water layer build up and DC voltage. The application of a bias voltage polarizes the metal electrodes, which prevents identification of the dynamic humidity adsorption. It is also an intensive experimental approach considering cost and time, as the test underlies a statistically occurring failure mode as a result of several influencing factors. The failure mode can change by creation of leak currents e.g. the burning of dendrites, and therefore lacks in explanation of failure cause. To overcome the limitations of the DC measurement techniques, this work uses the AC technique Electrochemical Impedance Spectroscopy (EIS) as method to assess humidity robustness of PCBs.

When performing EIS, an AC potential with a small amplitude is applied to metal electrodes and a regulated frequency spectrum is scanned. Therefore, in contrast to conventional electrochemical techniques, that measure for example current or electrode potentials over time, the current response measured in EIS is shifted by a phase angle φ with respect to the excitation signal. Determining the current density, the complex impedance of the test structure under specific conditions can be calculated. The complex impedance includes resistance, capacitance, inductance, and hybrid forms of the components. [9, 10]. This generic representation is one of EIS' various advantages. Due to the broad frequency scanning possibilities, the information content of the experimental data is substantial. It is especially useful when determination of system kinetics and when consideration of complex dimensions like capacitors or inductivities is required. [10, 11].

The scope of application is very broad. It can be used to detect interfacial processes like redox reactions at electrodes and to identify limitations of the reactions due to mass transfer or charge transfer. In addition, geometric impacts can be specified by EIS measurements. In battery, fuel cell and semiconductor development, it is a regularly employed technique [10]. In the framework of corrosion studies on PCBAs, AC methods like EIS are employed as well. EIS can be applied to determine the influence of organic acids in flux systems on their hygroscopic properties [12]. The technique also comes to use when investigating the corrosion rate of different metallic materials under a thin electrolyte layers and can be the basis for non-destructive tests for dendrite probability. The non-destructiveness due to the high frequency excitation signal of low amplitude is considered one of EIS' most important benefits [13].

In this work, we employ EIS to observe the formation of a water layer on a PCB upon condensation and its drawback during evaporation. The process of the water film building up, until a bridge between differently biased electrodes forms, is what ultimately enables a dendrite formation. An evaluation of the phase shift for different frequency domains delivers information on the status of the water layer for particular temperature/humidity conditions. As EIS is able to monitor the water film and its electrolytic properties but does not provoke dendrite growth itself, we conducted simultaneous DC measurements with the same test boards. A comparison of the EIS results and the leak current values obtained from DC application enable assertions on ECM probability in cyclic condensation scenarios.

II. MATERIALS AND METHODS

A. Test Boards

The EIS as well as the SIR measurements were performed with interdigital copper comb patterns with OSP (Organic Solderability Preservative)-finish on glass-fibre reinforced epoxy resin (FR-4) of 1.6 mm thickness. One test PCB was of the size of 125 mm x 60 mm and comprised two different comb structures. Both of them had the same copper width of 200 μm, whereas one had a gap size of 100 μm and the other one of 300 μm (Fig. 1).

Figure 1. FR-4 test board with copper comb structures of 200 μm width and gap sizes of 100 μm (left) and 300 μm (right) used for EIS and SIR testing.

Prior to EIS/SIR-testing, the test boards were subjected to a 20 minute cleaning procedure in a bath comprised of 75% Isopropyl alcohol and 25% water, tempered at 45 °C (based on IPC-TM-650, Method 2.3.25). Following the cleaning process, the boards were stored to dry in a desiccator filled with silica gel for a duration of at least twelve hours.

B. Climatic Chamber

For exposure of the samples to cyclic temperature conditions under controlled humidity, a climatic chamber (PL-3 J, ESPEC Corp., Japan), providing a temperature range between −40 °C and 150 °C and relative humidity (rH) between 20 % and 98 %, was used.

To achieve condensing conditions on the samples' surfaces, a variant of a standard Temperature Humidity Cycling test (IPC-TM-650, No. 2.6.6) was performed. The profile the test boards were exposed in the climatic chamber is displayed in Fig. 2.

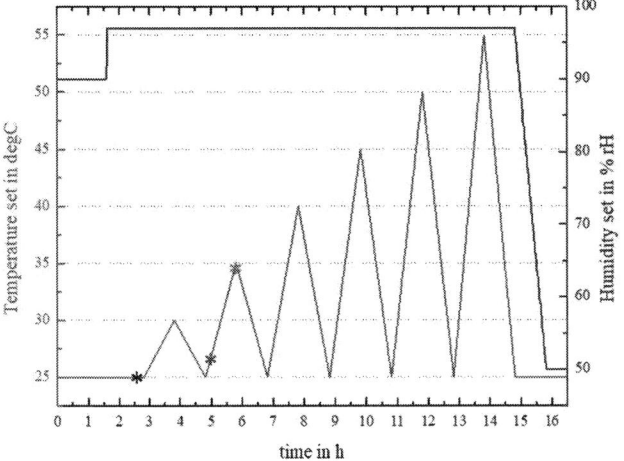

Figure 2. Climatic profile input of temperature and humidity chamber to provoke repeated condensation on the test board surfaces. Marked points refer to measurement data displayed in Fig. 3.

Within the chamber, the samples were mounted on a thermal mass (aluminum heat sink), using a heat conducting paste (THERM-A-GAP GEL25NS, Parker Chomerics, Germany), to enhance the condensation process by a time delay in sample temperature and ambient temperature during cycling. After an equilibration time at 25 °C and 90 %rH, the humidity was further increased up to 97 %rH and kept for 30 minutes. Subsequently a stepwise increasing temperature cycling profile was initiated. The temperature was ramped up over the course of one hour and respectively ramped down again over the course of one hour each time for 30/35/40/45/50/55 °C peak temperatures. This rapid cycling i.e. an increase of up to 30 °C/h caused condensing conditions on the board surfaces.

C. Electrochemical Workstation

For electrochemical measurements performed during temperature cycling to display the formation of a water layer and respective dendrite formation under DC- loads, a potentiostat (VMP3, multi-channel potentiostat, BioLogic Science Instruments SAS, France), enabling EIS and SIR measurements at the same time, was used. The workstation is equipped with four channels, two of which were used for EIS testing and two for simultaneous SIR testing of each a 100 μm and 300 μm gap comb structure. Both measurement types were conducted in the two-electrode setup.

For SIR measurements a DC voltage of 5 V was applied to the boards. The EIS tests were conducted with an AC voltage of 10 mV amplitude, the frequency was scanned from 100 kHz to 1 kHz with 6 measurement points per decade and 5 measures averaged for each of the respective frequencies. After the equilibration period, SIR and EIS measurements were started simultaneously.

III. RESULTS

A. Evaluation of Bodeplots for different condensation conditions

The results of the EIS measurements at run time of the climatic profile for both the 100 μm (upper graph) and the 300 μm gap size (lower graph) are presented in the form of Bode plots in Fig. 3. Exemplarily, the impedance |Z| and phase shift values over the frequency are displayed for three different points in time during temperature cycling.

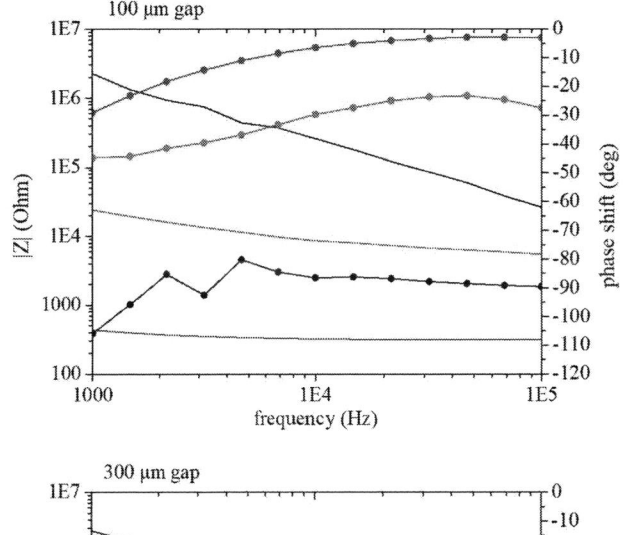

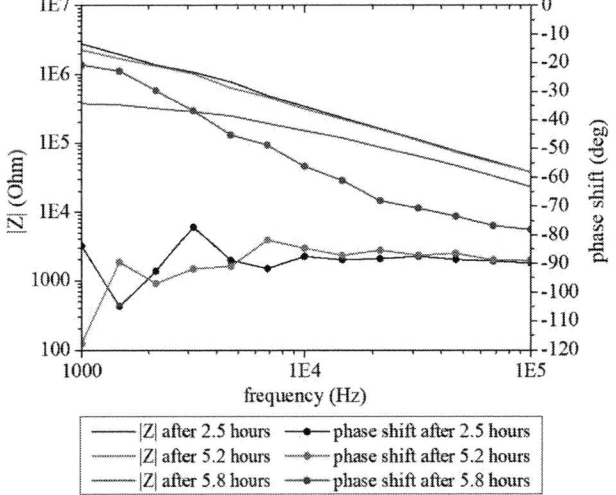

Figure 3. Bode plots obtained from the 100 μm gap (upper graph) and 300 μm gap (lower graph) test boards at distinct points in time during temperature cycling.

Each first curves (black solid for |Z|, black solid with filled circles for phase shift) are obtained after 2.5 hours into the profile (cf. marked points in Fig. 2), corresponding to an ambient condition of 25 °C and 97 %rH. At that time, the samples are still considered dry. For the 100 μm gap structure, the impedance at 1 kHz is at a level of $2.26 \cdot 10^6$ Ohm, steadily attaining smaller values for higher frequencies. The phase shift is in the range of −90°, being particularly stable in the high frequency region between 10 kHz and 100 kHz. For the

300 μm gap, the impedance at 1 kHz is 2.78·10⁶ Ohm, so slightly higher due to the increased distance between the copper tracks. The phase shift is likewise confirming a dry sample surface, with values in the −90° region. The magenta curves were measured during the ramp up to 35 °C after 5.2 hours, at a chamber ambient temperature of 27. °C. In the case of the 100 μm gap structure, for a frequency of 1 kHz, an impedance drop of almost two orders of magnitude down to 2.4·10⁴ Ohm occurs. Also, a change in phase shift is significant. The largest variation takes place at a frequency of 46.4 kHz, where the phase goes from −88.5° to −23.2°. This transition already suggests a humidity absorption on the test boards that cannot yet be recognized for the 300 μm gap sample at the same point in time.

Looking at the impedance curve, the decrease at 1 kHz amounts merely 0.53·10⁶ Ohm. For higher frequencies, a change in impedance compared to the spectrum obtained during equilibration time is not apparent. The same applies to changes in the phase shift, where the variation compared to the dry state spectrum is negligible. The lilac curves (5.8 hours into the temperature cycling profile) display the frequency spectrum obtained right before the peak temperature of 35 °C was reached. The impedance curve of the 100 μm gap shows another major drop, the most distinct at 1 kHz. Here it reaches a value of 310 Ohm, whereas the impedance for higher frequencies now also is found in the same range (mean value of 340 Ohm with a standard deviation of 38.5 Ohm over all measured frequencies). The phase shift has further increased towards 0° especially in the high frequency region, reaching its largest value of −2.9° at a 68 kHz. For the test board with 300 μm gap, now close to the 35 °C peak region of the temperature cycle, also a change in the Bode plot is distinguishable. The condensation process induced by the delay in temperature adjustment of sample to chamber ambient can now be depicted best in a drop of impedance down to 3.73·10⁵ Ohm at 1 kHz. The phase shift shows slight increases in the high frequency domain; at low frequencies, a large shift up to −21° is visible.

B. Evaluation of the phase shift for different frequencies over the course of temperature cycling

To evaluate more clearly the impact of gap size on the formation of a closed water layer and to stress the adequacy of the frequency range that is regarded, the phase shift over the course of the whole temperature cycling profile is plotted in Fig. 4 for both gap sizes. Here, the frequencies 100 kHz (black), 46 kHz (red), 22 kHz (green) and 10 kHz (blue) are chosen. For a better understanding, the temperature is also documented in green. In the upper graph, that displays the measurement results for the 100 μm gap test board, one can see a clear traceability of the temperature peaks in the phase shift development. It can be seen that the first ramp up towards 30 °C, starting after approximately 1 hour results in a change towards −80° best visible for 10 kHz. Only looking at 100 kHz, the shift in phase barely differs from the background noise of the signal at −90°. The next temperature peak (35 °C, starting after 3 hours) shows a change in phase shift of up to −15° for 46 kHz, whereby also the higher and lower depicted frequencies reach this value of up to 4° lower. It can be seen

that the phase shift of the 10 kHz as lowest of the displayed frequencies reacts first to the provoked condensation. Also, upon ramp down, the phase shift of this frequency takes the longest (up to 4.8 hours in the profile) to return to its dry base value of about −88°. For the following temperature peaks, the four frequencies are hardly differentiable in their signal information. Phase shift values of up to −2° are obtained. It is also noticeable that with larger ΔT/Δt in the cycling profile, the period of time in which the phase shifts are close to 0° also gets broader.

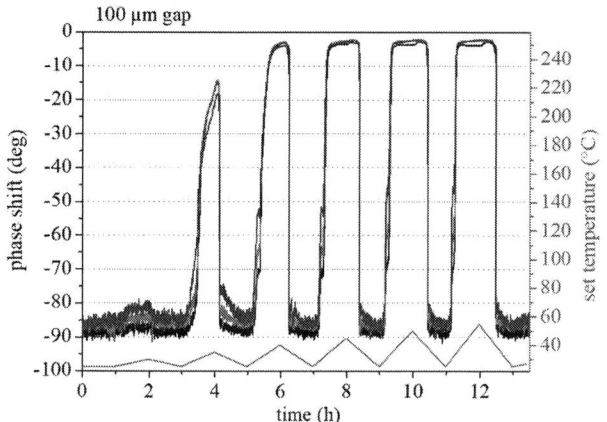

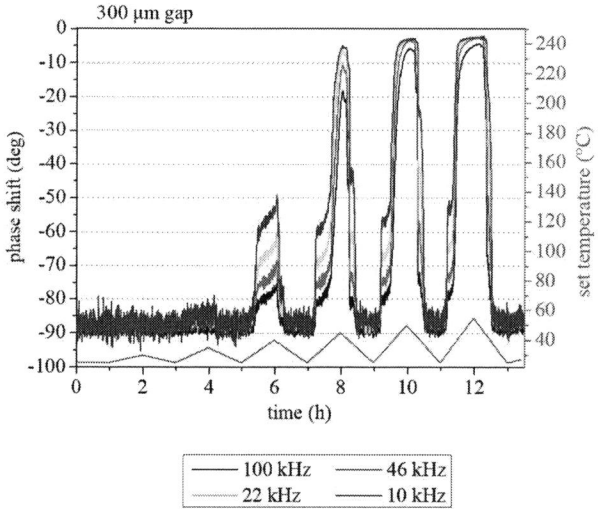

Figure 4. Phase shift development for 100 kHz, 46 kHz, 22 kHz and 10 kHz during temperature cycling for the 100 μm test board (top graph) and the 300 μm gap test board (bottom graph).

Looking at the obtained phase shift information from the 300 μm gap sample, one can see that the first temperature ramp up towards 30 °C does not produce any noticeable changes in phase shift for none of the displayed frequencies. Also the second condensation provoking peak, that already leads to incisive phase shift changes for the 100 μm gap only shows a small deviation from the signal noise for the 10 kHz curve. Regarding the ramp up towards the 40 °C peak (after 5 hours in Fig. 4), it can be seen that ΔT/Δt of 15 °C per hour is needed to form enough condensate for the 300 μm gap test board to show condensate formation. During this ramp, it is particularly well illustrated, that the different frequencies display the available humidity on the surface in different phase shift values. After 6.1 hours, the 100 kHz curve reaches a phase angle of −75°, the 46 kHz −69°, the 22 kHz curve −61° and the 10 kHz reaches the highest value at −49°. Also for the following peaks of 45 °C, 50 °C and 55 °C, the different frequencies the phase shift is displayed for are distinguishable. The 10 kHz frequency reaches the largest values in terms of phase angle change, but it also comprises the highest signal to noise ration. While for the 100 µm gap structure the dry baseline phase shift was already ranging −79° to −88°, for the 300 µm gap it is increased once more, ranging between values from −79° and −95°.

C. Comparison of EIS and leakage current measurements

The DC measurements that were conducted in parallel to EIS measurements on 100 µm gap and 300 µm gap structure are compared in terms of current development with the signal from the EIS measurements. For this, the results from the DC tests are displayed in the form of current in Fig. 5 (upper graph), while the impedance |Z| and the phase shift at 22 kHz from the EIS measurement are plotted in the lower graph.

This representation shows that for the DC measurement, the successive temperature cycles are represented in the signal evolvement. The black curve portraying the current response from the 100 µm gap structure changes significantly during the first ramp up towards 30 °C. It exceeds a baseline value of about $2 \cdot 10^{-5}$ mA more than two orders of magnitude up to $4 \cdot 10^{-3}$ mA on average during the first peak. Comparing with the EIS signal, there is no impedance drop evident, solely the previously mentioned slight shift in the phase. Upon ramp down (at a point of approx. 3 hours in the graph), the current drops again to $10 \cdot 10^{-6}$ mA, indicating that the provoked condensation did not form a water layer that was able to produce dendrite growth across the 100 µm gap. In the following cycle (around 4 hours), the current of the 100 µm gap structure increases up to 3.6 mA. During ramp down, the current does not drop to baseline level again but remains at about 0.1 mA, hinting to the gaps remaining connected by a mixture of water and conductive corrosion product, and possibly dendrites. Looking at the EIS results (|Z| and phase shift displayed in Fig. 5), at the same time the impedance signal exhibits a drop down to 3 kOhm. The phase shift reaches a maximum angle of −18°. During the subsequent cycles, the current profile keeps slightly increasing; the maximum values reached up to 15 mA at the temperature peaks during the last cycle. Also, the minimum current during ramp downs in the temperature profile increases likewise. Comparing with the impedance and phase shift from the EIS, the values for |Z| also reach a seemingly absolute minimum of about 220 Ohm during the last temperature cycles, as well as a maximum phase shift of −2.2°.

Looking at the 300 µm gap test board (brown curve), the current already shows a deviation from the baseline during the first temperature ramp up to $6 \cdot 10^{-4}$ mA. Comparing with the EIS signal in the lower graph, a change neither in impedance, nor in phase shift can be recognized. For the next peak, the current remains below 0.01 mA for most of the ramp up time, aside from a short increase of up to 0.3 mA. Once the

temperature is reduced, the baseline current values are reached again. The EIS signal during this time shows slight changes in form of an impedance drop and an increase in phase shift towards less negative values. During the following cycles, the same behavior as already described for the 100 µm gap current can be recognized in terms of steadily increasing maximum current (at 6, 8, 10 and 12 hours) and slightly increasing current at the lowest temperature point (at 6.5, 8.5, 10.5 and 12.5 hours). Comparing with the EIS information, it is noticeable that while the phase shift for the last three peaks does not change in maximum value anymore (goes up to −1°) but only in extension of the plateau, the impedance still keeps dropping for every cycle, down to 1 kOhm for the 55 °C peak.

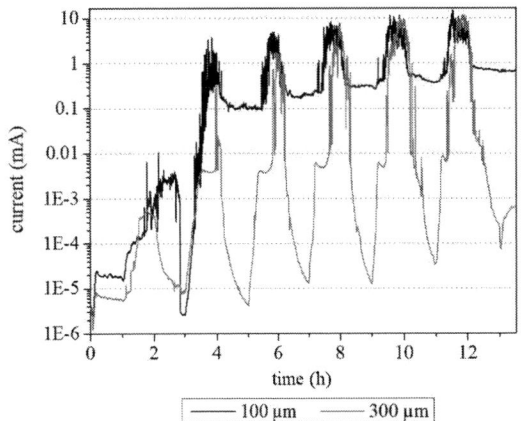

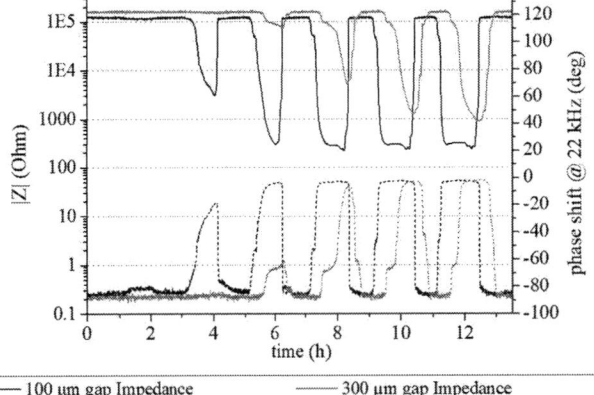

Figure 5. Current signal for the 100 µm gap and 300 µm gap test boards obtained from DC measurements (upper graph) and impedance |Z| and phase shift at 22 kHz for 100 µm gap and 300 µm gap test boards obtained from EIS measurements during temperature cycling (bottom graph).

Also, the occurrence of the peculiar shoulders in the curve shape of the 300 µm gap current as in an initial increase of about three decades, than hold of a value around $3 \cdot 10^{-3}$ mA until the current rapidly increases again up to the mA region also occurs to some extent in the EIS signal.

Based on the graphs in Fig. 5, it is also possible to compare the response of the two different gap sizes more precisely. In the bottom graph that displays the EIS results, it cannot only be seen that the signal changes evoked by condensation are smaller for the 300 µm gap test board over the whole course of the cycling profile. As already described in more extent in the previous section for the phase shift, also for the impedance a signal change deviating from the baseline fails to appear for the first two peaks for the 300 µm gap, while it does already make a change for the second peak when considering the 100 µm gap structure. It is also clear to see that once both structure show a reaction in curve progression to the temperature cycling, the change in signal occurs delayed for the larger gap.

The test boards were examined by optical methods after the measurement. Fig. 6 displays microscope images of all of the samples. The EIS samples (100 µm gap in Fig. 6a), 300 µm gap in Fig. 6c)) do not show changes compared to initial appearance, meaning they remain unharmed by the test method. The SIR samples ((100 µm gap in Fig. 6b), 300 µm gap in Fig. 6d)) show the formation of dendrites across the gaps.

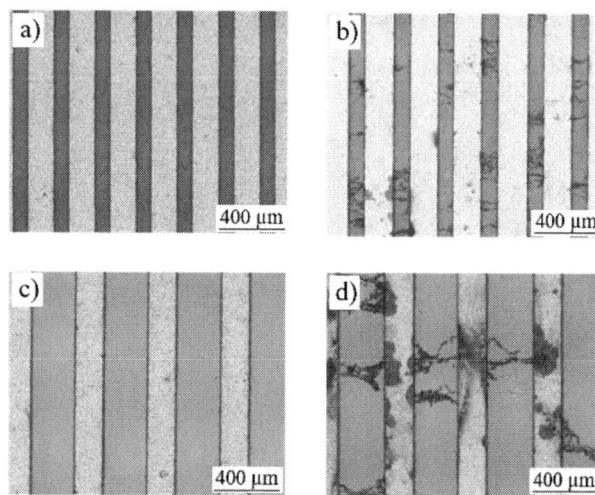

Figure 6. Microscope images of test boards after measurement. a): 100 µm gap test board used for EIS measurements, b): 100 µm gap test board used for SIR measurements, c): 300 µm gap test board used for EIS measurements, d): 300 µm gap test board used for SIR measurements.

IV. DISCUSSION

Overall, the conducted experiments make it possible to relate the water film formation status that is obtained from EIS measurements to a current flow due to dendrite formation in the case of a DC application. As the EIS technique has a high data content, we first evaluated the experiments in terms of applicable analysis possibilities.

A. Evaluation of substantial EIS values

From looking at the Bode plots of both of the gap sizes during the course of cyclic climatic changes (Fig. 2), we are able to display the provoked condensation. The representation over the complete recorded frequency spectrum also points at distinct ranges that demonstrate the condensation. When looking at the impedance, the largest changes are displayed for the values of the 1 kHz frequency. The phase shift

information content is highest for 46.4 kHz. The 22 kHz was recognized as a frequency that provided a sufficient degree of stability for both impedance and phase shift. Impedance response at frequencies lower than 10 kHz was seen to have negligble variation, thereby did not deliver information the for harsh temperature cycling profile the samples were exposed to.

B. Impact of gap size

We see from looking at the EIS signal evolvement during cycling (Fig. 4 and Fig. 5) that the larger gap size sample reacts to the imposed climatic changes with a delay and a less pronounced signal change compared to the smaller one. We showed in a previous study [9] that this was the case for slow condensation processes. With the current experiments, we can see that also for fast and repeated dewing and evaporation, the different gap sizes are distinguishable. A difference can be made in terms of a temporal shift, as the larger gap needs more humidity to absorb to the surface for displaying a partial closing of the film. This can also be expressed in terms of a smaller divergence of impedance or respectively phase shift from the baseline.

C. Hysteresis in build up and evaporation of water layer

The results in Fig. 5 indicate that that there is some hysteresis behavior in terms of dewing and evaporation. The phase shift and impedance profile during the rising temperature ramp is more extended in time than the signal change during ramp down. During formation of the water layer, the condensate first starts forming separate islands that grow larger until they interlink, forming a closed film across the gaps. This we can detect in EIS by a phase shift change towards $0°$, respectively the impedance drop. As we cool down again the water layer starts to evaporate. We believe that this process is detectable by EIS once the film that interconnects the tracks loses its continuity, meaning that the cohesive forces between the molecules are overcome by the evaporation energy. Hence we see a faster signal change during evaporation of the water film. As the formation, as well as evaporation of a continuous water film are processes that imply fluctuating transition states we think that the respective developments need to be resolved temporally more in detail to receive and improved correlation of EIS values and obtained leakage current.

D. Comparison of EIS and DC information

Comparing EIS and leakage current measurements, we can see from the 100 µm gap size signal in Fig. 6, that meeting a value of up to $4·10^{-3}$ mA in a DC test does not signify the formation of a closed water layer. We can also tell from the fact that the current drops to baseline level again, that no dendrites are formed under this condition, which confirms the EIS information that no continuous water film was built up. If however the phase shift exceeds $-20°$ for the 100 µm gap structure, the leakage current of the respective DC sample increases up to a mA level and does not drop below 0.01 mA anymore, indicating dendrite formation. This is congruent with the findings we made in our previous study [9], where we demonstrated that a phase shift of about $-15°$ due to a water layer is equating a single digit kOhm resistance.

For the 300 µm gap test board, the deviation in current from the baseline after the temperature cycles is not as pronounced as for the 100 µm gap size (c.f. Fig. 6). Still the image of the sample after the measurement clearly show dendrite formation. The fact that we do not have any inline information on the dendrite status during the course of cycling does not allow for determination of a point in time at which dendrite formation started. To correlate more precisely the necessary water film conditions obtained from EIS with the start of dendrite growth, experiments with slower condensation scenarios or respectively in-situ observation of the samples are required. The establishment of an allocation of dendrite current for distinct gaps is also conceivable.

V. CONCLUSION

From the EIS results we obtained information on the water layer present at every stage of the cycling process, meaning the closing of the film and its withdrawal upon drying. We carried out a comparison of the required temperature/humidity impact that is needed for different gap sizes to form a water layer of a certain thickness. We compared the EIS values for distinct frequencies with the leakage current development over the time course of the temperature cycles to correlate the respective water amount present on the sample surface with the dendrite formation it enables under DC loads.

This way, the results we obtained provide a general guideline to the assessment of ECM probability under cyclic temperature and humidity loads by EIS measurements.

ACKNOWLEDGMENT

The authors want to thank Robert Bosch GmbH and the Center for Electronic Corrosion (CELCORR) for financial support of the research as well as all the practical help received during experiment runs.

REFERENCES

[1] He, X., Azarian, M., & Pecht, M. (2010). Effects of solder mask on electrochemical migration of tin-lead and lead-free boards. *IPC printed circuit Expo, APEX & Designer summit proceedings.*

[2] Krumbein, S. J. (1995). Electrolytic models for metallic electromigration failure mechanisms. *IEEE transactions on reliability, 44*(4), 539-549.

[3] Minzari, D., Jellesen, M. S., Møller, P., & Ambat, R. (2011). On the electrochemical migration mechanism of tin in electronics. *Corrosion Science, 53*(10), 3366-3379.

[4] Huang, H., Guo, X., Zhang, G., & Dong, Z. (2011). The effects of temperature and electric field on atmospheric corrosion behaviour of PCB-Cu under absorbed thin electrolyte layer. *Corrosion Science, 53*(5), 1700-1707..

[5] Pauls, D. (1997). Test vehicles in surface insulation resistance testing. *Circuit World, 23*(3), 35-39.

[6] Ambat, R. (2008). Perspectives on climatic reliability of electronic devices and components. *Proc. Int. Microelectron. Packag. Soc. IMAPS Nordic*, 1-18.

[7] Schmitt-Thomas, K. G., & Schmidt, C. (1994). The influence of flux residues on the quality of electronic assemblies. *Soldering & Surface Mount Technology, 6*(3), 4-7.

[8] Piotrowska, K., Jellesen, M. S., & Ambat, R. (2017). Thermal decomposition of solder flux activators under simulated wave soldering conditions. *Soldering & Surface Mount Technology, 29*(3), 133-143.:

[9] Lauser, S., Eckold, P., Richter, T., Verdingovas, V., & Ambat, R. (2018). Implementation of electrochemical impedance spectroscopy (EIS) for assessment of humidity induced failure mechanisms on PCBAs. In *EUROCOR 2018-Applied science with contstant awareness.*

[10] Lasia, A. (2002). Electrochemical impedance spectroscopy and its applications. In *Modern aspects of electrochemistry* (pp. 143-248). Springer, Boston, MA.

[11] Zou, L. C., & Hunt, C. (2009). Characterization of the conduction mechanisms in adsorbed electrolyte layers on electronic boards using AC impedance. *Journal of the Electrochemical Society, 156*(1), C8-C15.

[12] Verdingovas, V., Jellesen, M. S., & Ambat, R. (2015). Solder flux residues and humidity-related failures in electronics: relative effects of weak organic acids used in no-clean flux systems. *Journal of Electronic Materials, 44*(4), 1116-1127.

[13] Nishikata, A., Ichihara, Y., & Tsuru, T. (1995). An application of electrochemical impedance spectroscopy to atmospheric corrosion study. *Corrosion science, 37*(6), 897-911.

Micro-fabricated SERF Atomic Magnetometer for Weak Gradient Magnetic Field Detection

Xiang Yue[1,2], Jintang Shang*[1,2], Chen Ye[1,2]

1. Key Lab of MEMS of Education Ministry, Southeast University
2. Quantum Information Research Center, Southeast University

Sipailou 2, Nanjing, China

* Corresponding author, +86 13913869603, Email: jshang@seu.edu.cn

Abstract—**Weak gradient magnetic field detection is significant. This paper presents a novel micro-fabricated spin-exchange-relaxation-free (SERF) atomic magnetometer for weak gradient magnetic field detection, and the atomic vapor cell is fabricated by micro electro mechanical systems (MEMS) technology. Compared to the single-channel magnetometer, the novel magnetometer's signal-to-noise ratio has increased by 3.63 times. The magnetic field gradient noise of the novel magnetometer reaches 45.33 fT·Hz$^{-1/2}$·cm^{-1} in the frequency range of 8 to 15 Hz, and the novel magnetometer has a CMRR (common mode rejection ratio) of 6.1. At the same time, the relationship between the magnitude of the magnetic field gradient and the peak of the dispersive signal has been verified, which provides an idea for the detection of weak gradient magnetic field.**

Keywords-weak gradient magnetic field; SERF atomic magnetometer; micro-fabricated;sensitivity

I. INTRODUCTION

Magnetic phenomena are one of the most common physical phenomena in nature, carrying abundant information. Nowadays, the application range of magnetic detection has been extensive: it has been widely used in high magnetic field (~μT) detection, such as iron ore detection [1], explosive detection [2], metal crack detection [3], etc. In terms of weak magnetic field (~fT or ~pT) detection, people are also exploring in depth, such as magnetoencephalography (MEG) [4], magnetocardiography (MCG) [5], and nuclear magnetic resonance (NMR) [6], etc. In this paper, what we call the application of weak magnetic field detection generally refers to the application of biological magnetic field detection.

In 1972, Cohen [7] first used the superconducting magnetometer (SQUIDs) [8] to measure the magnetic field generated by intracranial current changes, opening up a chapter on the exploration of biological magnetic fields. Although the measurement of brain magnetism can be achieved, a huge volume and costly cooling device makes this method difficult to apply widely. Compared to SQUIDs that require expensive and large cryogenic cooling equipment, the Spin-Exchange-Relaxation-Free (SERF) [9] atomic magnetometer has significant advantages due to its small size and low power consumption. The SERF atomic magnetometer operates at low magnetic fields and high

atomic density, has high sensitivity, which also measure the vector of the magnetic field. In 2006, the team of M. V. Romalis first realized the development of a potassium atom magnetometer based on the SERF system for the detection of human brain magnetic fields [10]. Two laser beams were adopted in the fabrication and with a sensitivity of 10 fT/Hz$^{1/2}$. V.Sha from National Institute of Standards and Technology (NIST) demonstrated the first micro-fabricated SERF atomic magnetometer with a single laser [11], with a sensitivity of 65 fT/Hz$^{1/2}$.

Weak biological magnetic fields provide a lot of useful information, such as whether the heart is healthy and whether the brain is at rest. However, the weak magnetic field that needs to be measured is interspersed with many stray magnetic fields, which are many orders of magnitude larger than the magnetic field to be measured, would affect the accuracy and accuracy of the measurement. There are also a lot of magnetic field information hidden in the gradient magnetic field [12]. By detecting the gradient magnetic field, the effect of stray magnetic fields is reduced, the distribution information of the magnetic field is better obtained, and the measurement accuracy is improved. Despite the high spatial resolution and the high sensitivity of the SERF atomic magnetometer, it is difficult for a single SERF atomic magnetometer to acquire information hidden in the magnetic field gradient. It is necessary to obtain the spatial distribution information of the magnetic field by detecting the weak gradient magnetic field and improve the measurement accuracy.

The focus of this paper is to design a novel Micro-fabricated SERF atomic magnetometer with a high sensitivity for detecting the gradient magnetic field distribution. The novel atomic magnetometer consists of several modules including the laser diode, micro rubidium vapor cell, heating modules, magnetic coils, photo-detector parts and the circuit modules. Rubidium atomic vapor cells are the core of the entire system, whose fabrication method is based on a standard micro electro mechanical systems (MEMS) process. The magnetic field gradient noise and sensitivity of the novel atomic magnetometer were measured. The common mode rejection ratio (CMRR) was tested and compared with a single atomic magnetometer to characterize its performance. The measurement of weak gradient magnetic field signals is discussed.

978-1-7281-1500-9/19 $31.00 © 2019 IEEE

II. EXPERIMENT

The novel SERF atomic magnetometer is a combination of two single-channel SERF atomic magnetometers called an atomic gradient magnetometer. The following describes the integration of the system and the preparation of the experiment.

A. Integration of the System

As the core of the entire system, alkali atomic vapor cells are vital [13]. The chip-scale vapor cell is fabricated by anodic bonding technique [14] as shown in Fig.1. The alkali metal atoms used in the vapor cell is rubidium. Two nearly identical vapor cells will be placed in both channels. The vapor cell is composed of two chambers, including one with through hole and one with blind hole, which connected to each other by micro channels. Chamber with blind hole serves as a rubidium atom storage space, and chambers with through hole as working spaces for the interaction. The fabrication process of the rubidium atomic vapor cell has been described in previous work [15].

(a)　　　　　　　(b)

Figure 1. (a) Negative side of the alkali atomic vapor cell. (b) Positive side of the alkali atomic vapor cell.

The schematic of the novel magnetometer is shown in Fig.2. It is mainly composed of an optical module, two vapor cells, a magnetic compensation module, a photodetector module, and a signal processing module. In this magnetometer, a single beam is used as the pump beam and the detect beam. The laser controller parameters are adjusted, and a 795nm distributed feedback (DFB) laser is generated by a fiber-coupled output DFB laser diode, which is then split into two beams of similar power by a fiber coupler. The beams are connected to the optical module through the fiber. The optical module includes a fiber collimator, a polarizer, and a quarter wave plate. The distance between the centers of the two fiber collimators is maintained at 15 mm to ensure that the distance between the two beams is 15 mm. The fiber collimator concentrates the divergent beam into parallel beam, and the polarizing plate causes the beam to become linearly polarized light, which becomes circularly polarized light after passing through the quarter wave plate. The fiber-coupled-output DFB laser diode separates the laser control circuit from the magnetometer to prevent the magnetic field generated by the laser control circuit from affecting the performance of the magnetometer. The two beams pass through the quarter-wave plate, interact with the rubidium atoms in the vapor cell, and are then collected by two photodiodes. The collected signals passes through the transimpedance amplifier and finally enters the lock-in amplifier for processing. The processed signal waveform is displayed on the oscilloscope.

The entire system is integrated into a cylinder which is manufactured by 3D printing and has the advantage of being detachable, with a volume of approximately 100 cm³. The outer surface of the cylinder is wound with coils for generating a compensating magnetic field.

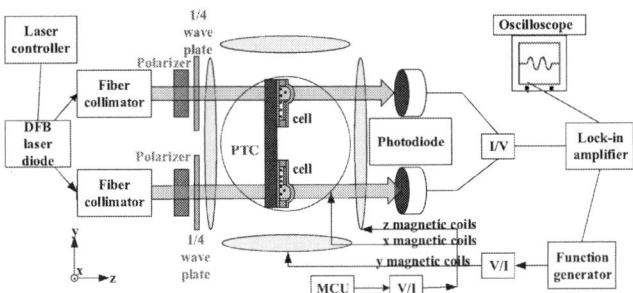

Figure 2. Schematic diagram of the atomic magnetometer. The fiber coupler for splitting which between the DFB laser diode and the laser collimator is not drawn. "V/I" represents a voltage-controlled constant current source for converting a voltage signal into a current signal. "I/V" means a transimpedance amplifier for converting the current signals output from the photodiode into voltage signals.

B. Experimental Condition

When the rubidium atoms achieve the SERF state, the following relationship needs to be satisfied:

$$\omega_0 \ll f \tag{1}$$

Where f is the spin exchange collision frequency of rubidium atoms, proportional to temperature T, ω_0 is the Larmor precession frequency of rubidium atoms, could be expressed as

$$\omega_0 = \gamma B \tag{2}$$

Where γ is the gyromagnetic ratio of the rubidium atoms, which is a constant (4.7 Hz/nT), B is the magnetic flux density.

It can be known from (1) and (2) that in order for the rubidium atoms to rapidly enter the SERF state, two conditions must be met. First, the atomic density must be high enough to ensure that the spin exchange collision frequency is much greater than the Larmor precession frequency, which could be achieved by increasing the temperature. Second, the magnetic field must be sufficiently weak (~nT).

In general, rubidium atoms can reach the SERF state at around 150°C. In the magnetometer, a PTC (positive temperature coefficient) heater which driven by 480 KHz AC current is used to raise the temperature of the rubidium atomic vapor cell. The reason for using high frequency current heating is that the magnetometer is not sensitive to high frequency signals, which can eliminate the influence of heating current on the experiment. During the heating process, the rubidium atoms gradually change from solid to gas, entering the chamber with through hole, interacting with the circularly polarized beam passing through the chambers.

The low magnetic field environment could be achieved by shielding and magnetic compensation. A shielded cylinder with four layers of shielding was used to isolate the stray magnetic field. Several sets of saddle coils are wrapped around the outer surface of the 3D printed cylinder to compensate for the residual magnetic field in both the X and Y directions in the shield cylinder. A five-layer toroidal coil is used to compensate for the residual magnetic field in the z-direction within the shield cylinder. Firstly, the current of the x-direction coil is adjusted by the MCU (Microcontroller Unit), so that the light intensity detected by the photodiode is maximum, and the total magnetic field in the x direction is zero. Then, the coil current in the z direction is adjusted by the MCU, the light intensity detected by the photodiodes continue to increase until the maximum, at this time, the magnetic field in the z direction is also zero. Finally, the function generator generates the y-direction coil current to compensate the magnetic field in the y direction, so that the magnetometer is in a near-zero magnetic environment, ensuring that the rubidium atoms are in the SERF state.

After the rubidium atoms enter the serf state, the existing heating temperature is kept constant and measurement is started.

III. RESULT AND DISCUSSION

A. The Performance of Single Channel

First we need to determine if the two channels of the magnetometer are working properly. Compensating for the magnetic field in the x and z directions, a sinusoidal signal with a frequency of 1.19 K Hz and a peak-to-peak value of 100 mV is added in the y direction, and the DC offset is changed to obtain a dispersive signal to check whether the magnetometer is working normally. The DC bias voltage is varied by a computer controlled function generator to ensure that the bias voltage varies uniformly.

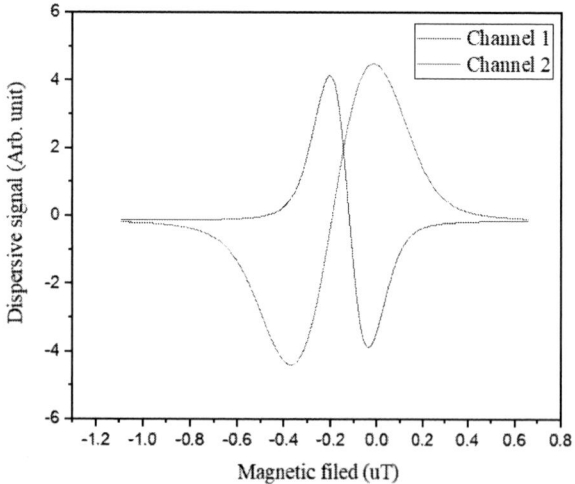

Figure 3. The dispersive signals of channel 1 and channel 2. Both channels are work normally. The solid black line is the channel 1 dispersive signal, and the red solid line is the channel 2 dispersive signal.

Fig.3 has shown the dispersive signals of channel 1 and channel 2. This indicates that both channels are in normal working condition.

After confirming that the magnetometer is working properly, measure the sensitivity of the single-channel magnetometer. First adjust the current value in the x, z direction to maximize the lock-in amplifier's display, then change the DC offset generated by the function generator until the lock-in amplifier's display becomes zero. At this time, the single-channel magnetometer is in a zero-magnetic environment, keeping all the parameters unchanged, collecting 50 seconds of data with an oscilloscope, and performing FFT transformation on the data to obtain a noise spectrum. Then, the slope near the zero point is obtained from the dispersive signal curve, and the sensitivity of the single channel magnetometer is obtained.

As shown in Fig.4, the sensitivity of a single-channel magnetometer is 247 fT/Hz$^{1/2}$ in the frequency range of 8 to 15 Hz.

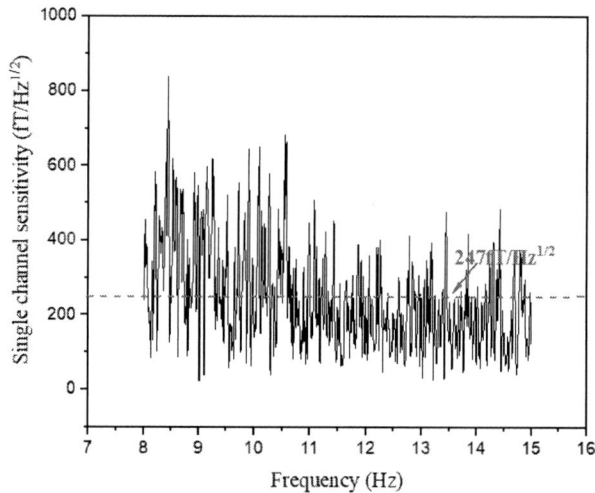

Figure 4. The sensitivity of channel 1 in the frequency range of 8 to 15 Hz. The average is 247 fT/Hz$^{1/2}$.

B. The Performance of Gradient Magnetometer

It is necessary to accurately adjust the parameters of the two channels in order to get the performance of the gradient magnetometer. First, the residual magnetic field in the x and z directions in the shield cylinder is compensated. Then use the single-channel output mode of the lock-in amplifier to observe the dispersive signals of the two channels, and adjust the two channels separately so that the peak-to-peak value and bandwidth of the dispersive signals of the two channels are as equal as possible. Finally, the DC bias voltage of the function generator is changed, and the dispersive mode of the differential mode is obtained by using the A-B mode (differential mode) of the lock-in amplifier.

In a gradient-free magnetic field environment, there should be no difference between the two channels of the gradient magnetometer. The result of the differential output should be a straight line. The actual graphics are shown in Fig.5. This indicates that there is a gradient in the magnetic field in the y-direction of the shielding cylinder, which leads to a phenomenon that is inconsistent with the ideal situation.

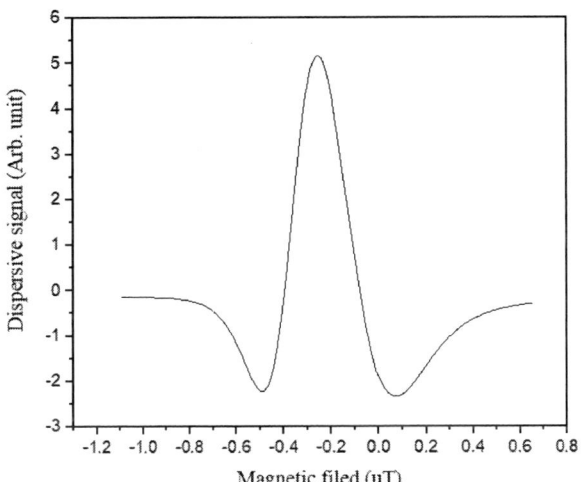

Figure 5. The dispersive signal obtained by the differential output. This shows that there is a gradient in the residual magnetic field in the y direction of the shielding cylinder.

The noise of the gradient magnetometer is shown in Fig.6. It could be seen that the average noise is 68 fT/Hz$^{1/2}$ in the frequency range of 8 to 15 Hz. The distance between the two channels is 1.5cm, so the gradient noise of the gradient magnetometer is 45.33 fT·Hz$^{-1/2}$·cm^{-1}.

Fig.7 shows the magnetic field noise of channel 1 and the magnetic field noise of a gradient magnetometer under the same conditions. It could be known that the gradient magnetometer has a signal-to-noise ratio that is 3.63 times higher than that of a single-channel magnetometer.

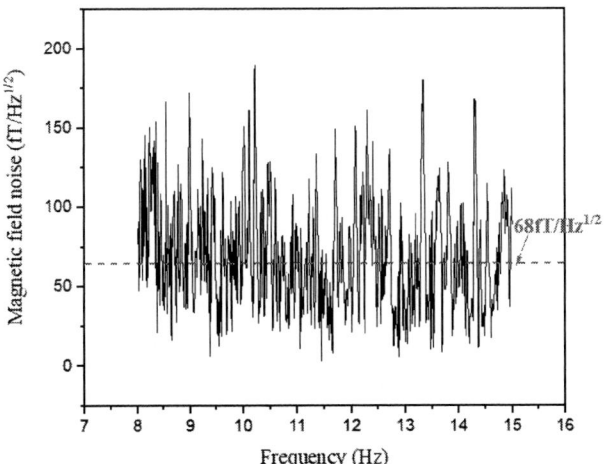

Figure 6. Magnetic field noise of the gradient magnetometer in the frequency of 8 to 15Hz.

In order to verify the common mode rejection of the gradient magnetometer, a coil is wound around the outside of the 3D printed cylinder. The coil produces substantially the same magnetic field at both channels. A square wave signal is manually generated by the MCU and input to the coil, and the response pattern of the channel 1 and the response pattern of the gradient magnetometer are respectively obtained, as shown in Fig.8.

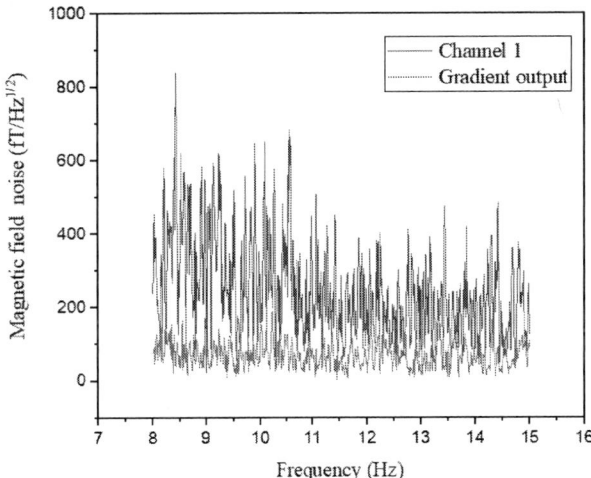

Figure 7. Comparison between magnetic field noise of channel 1 and magnetic field noise of gradient magnetometer. The black solid line is the channel 1 magnetic field noise, and the red solid line is the magnetic field noise of the gradient magnetometer differential output.

As seen from Fig.8, the peak-to-peak value of the response signal of channel 1 is about 1.1, and the peak-to-peak value of the gradient output signal is about 0.18. The calculated CMRR is about 6.1.

For the dispersive signal obtained in Fig. 5, C.Affolderbach[16] thinks that the higher peak indicates the gradient of the environment in which the magnetometer is located and is proportional to the magnitude of the gradient of the magnetic field, the positive and negative peaks indicate the direction of the gradient. To verify this idea, a gradient coil for generating a y-direction gradient magnetic field was designed outside the 3D printed cylinder. The dispersive signal of the gradient magnetometer in three cases is drawn by the computer control function generator. The first case is that the gradient coil does not apply any current, as shown by the solid black line in Fig.9. In the second case, a current of "-2000" is applied through the MCU, and a solid red line as shown in Fig.9 is obtained. The third case is to apply a current of "+2000" through the MCU, and the pattern shown by the solid blue line in Fig. 9 is obtained.

978-1-7281-1500-9/19 $31.00 © 2019 IEEE   525

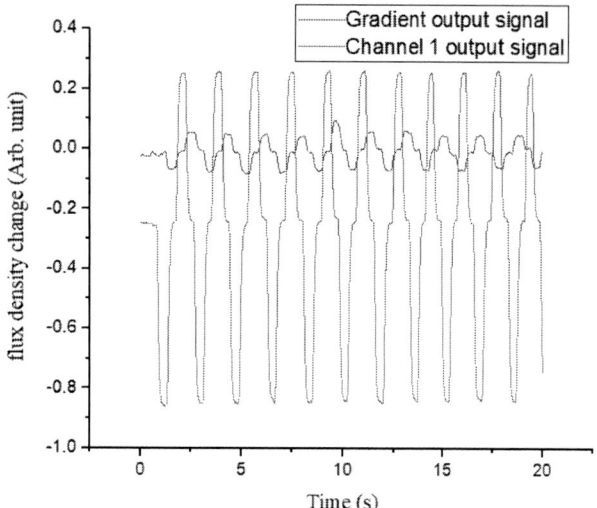

Figure 8. The output of the common mode suppression experiment. The solid black line is the output of the gradient magnetometer, and the solid red line is the output of channel 1.

It can be seen that the peak of the "-2000" red solid line is significantly smaller than the peak of the black solid line. The reason why the blue solid line is almost equal to the peak of the solid black line may be that the gradient in the y direction is already large enough, continue to increase the gradient, the peak change is not obvious.

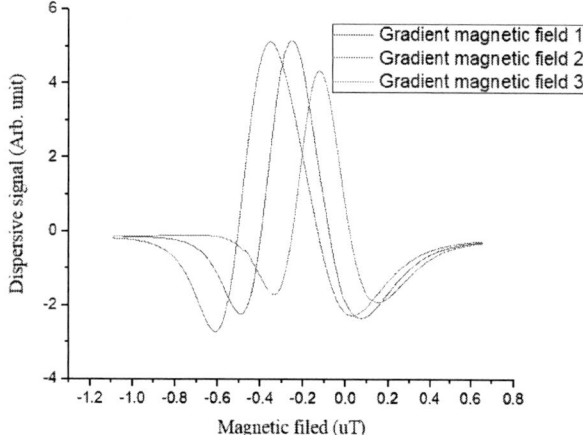

Figure 9. The dispersive signal of a gradient magnetometer under different magnetic field gradients. The solid black line indicates the dispersive signal when the MCU output current is "0". The solid red line is "-2000". The solid blue line is "+2000".

In order to verify the proportional relationship between the magnetic field gradient and the peak value, a current is applied to the gradient coil by the MCU, and the magnetic field gradient is changed to obtain the relationship diagram shown in Fig.10. As the gradient coil current increases, the peak value of the dispersive signal gradually increases from a negative value to a positive value, indicating that the applied magnetic field gradient cancels out the original magnetic field gradient and reverses the magnetic field

gradient. This provides a method and idea for weak gradient magnetic field detection.

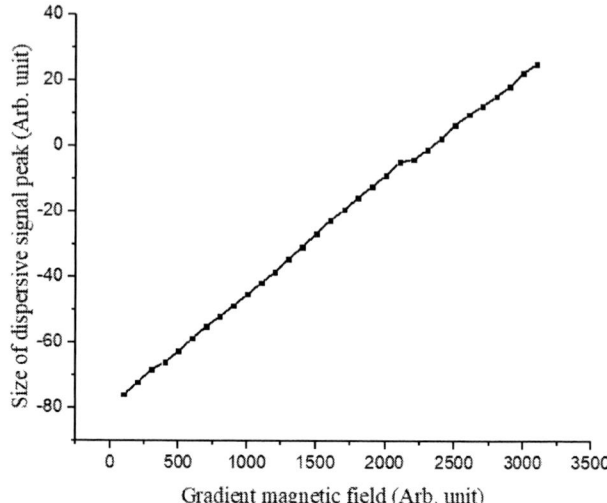

Figure 10. The relationship between the peak value of the dispersive signal and the magnitude of the gradient magnetic field. The horizontal axis of the coordinate axis represents the output current of the MCU.

IV. CONCLUSION

This paper shows a novel magnetometer that can be used for weak gradient magnetic field detection. The atomic vapor cell is based on MEMS process technology, the package housing is completed by 3D printing to achieve miniaturization, low cost, low power consumption. The production and integration process of the novel magnetometer probe has also been introduced. The magnetic field gradient noise of this novel magnetometer reaches 45.33 fT·Hz$^{-1/2}$·cm^{-1} in the frequency range of 8 to 15 Hz. Compared with the single-channel magnetometer, the novel magnetometer has a signal-to-noise ratio that is 3.63 times higher. At the same time, the novel magnetometer has a CMRR of 6.1. Finally, the relationship between the magnetic field gradient and the peak value of the differential output dispersive signal is verified, which provides an idea for the detection of weak gradient magnetic field.

Of course, this is only the first exploration of our group's weak gradient magnetic field. There are still many problems and deficiencies, which will be gradually improved in the future work.

ACKNOWLEDGMENT

This work is supported by National Science Foundation of China (No. 51675102, No. 51275091). The funding for Specially-Appointed Professors by Universities in Jiangsu

Province and "333 Projects" of Jiangsu Province are also acknowledged.

REFERENCES

[1] A. E. Levanto, "A three-component magnetometer for small srill-holes and its use in ore prospecting," Geophysical Prospecting, vol. 7, no. 2, pp. 183–195, Jun. 1959.

[2] G. Schultz, R. Mhaskar, M. Prouty, and J. Miller, "Integration of micro-fabricated atomic magnetometers on military systems," in Detection and Sensing of Mines, Explosive Objects, and Obscured Targets XXI, 2016, vol. 9823, p. 982318.

[3] C. Deans, L. Marmugi, S. Hussain, and F. Renzoni, "Electromagnetic induction imaging with a radio-frequency atomic magnetometer," Appl. Phys. Lett., vol. 108, no. 10, p. 103503, Mar. 2016.

[4] E. Boto et al., "Moving magnetoencephalography towards real-world applications with a wearable system," Nature, vol. 555, no. 7698, pp. 657–661, Mar. 2018.

[5] K. Kamada, Y. Ito, and T. Kobayashi, "Human MCG measurements with a high-sensitivity potassium atomic magnetometer," Physiol. Meas., vol. 33, no. 6, p. 1063, 2012.

[6] M. Jiang, R. P. Frutos, T. Wu, J. W. Blanchard, X. Peng, and D. Budker, "Magnetic Gradiometer for Detection of Zero- and Ultralow-Field Nuclear Magnetic Resonance," arXiv:1808.02743 [physics], Aug. 2018.

[7] D. Cohen, "Magnetoencephalography: Detection of the Brain's Electrical Activity with a Superconducting Magnetometer," Science, vol. 175, no. 4022, pp. 664–666, Feb. 1972.

[8] R. L. Fagaly, "Superconducting quantum interference device instruments and applications," Review of Scientific Instruments, vol. 77, no. 10, p. 101101, Oct. 2006.

[9] S. J. Seltzer and M. V. Romalis, "Unshielded three-axis vector operation of a spin-exchange-relaxation-free atomic magnetometer," Appl. Phys. Lett., vol. 85, no. 20, pp. 4804–4806, Nov. 2004.

[10] H. Xia, A. Ben-Amar Baranga, D. Hoffman, and M. V. Romalis, "Magnetoencephalography with an atomic magnetometer," Appl. Phys. Lett., vol. 89, no. 21, p. 211104, Nov. 2006.

[11] V. Shah, S. Knappe, P. D. D. Schwindt, and J. Kitching, "Subpicotesla atomic magnetometry with a microfabricated vapour cell," Nature Photonics, vol. 1, no. 11, pp. 649–652, Nov. 2007.

[12] F. G. Shellock, E. Kanal, and T. B. Gilk, "Regarding the Value Reported for the Term 'Spatial Gradient Magnetic Field' and How This Information Is Applied to Labeling of Medical Implants and Devices," American Journal of Roentgenology, vol. 196, no. 1, pp. 142–145, Jan. 2011.

[13] L.-A. Liew, S. Knappe, J. Moreland, H. Robinson, L. Hollberg, and J. Kitching, "Microfabricated alkali atom vapor cells," Appl. Phys. Lett., vol. 84, no. 14, pp. 2694–2696, Apr. 2004.

[14] H. Henmi, S. Shoji, Y. Shoji, K. Yoshimi, and M. Esashi, "Vacuum packaging for microsensors by glass-silicon anodic bonding," Sensors and Actuators A: Physical, vol. 43, no. 1, pp. 243–248, May 1994.

[15] Y. Ji, J. Shang, Q. Gan, L. Wu, and C. Wong, "Micro-fabricated spherical rubidium vapor cell and its integration in 3-axis atomic magnetometer," in 2015 IEEE 65th Electronic Components and Technology Conference (ECTC), 2015, pp. 946–949.

[16] C. Affolderbach, M. Stähler, S. Knappe, and R. Wynands, "An all-optical, high-sensitivity magnetic gradiometer," Applied Physics B: Lasers and Optics, vol. 75, no. 6–7, pp. 605–612, Nov. 2002.

Novel solder pads for self-aligned flip-chip assembly

Yves Martin, Swetha Kamlapurkar, Nathan Marchack, Jae-Woong Nah and Tymon Barwicz

IBM T. J. Watson Research Center, Yorktown Heights, NY 10598
ymartin@us.ibm.com

Abstract - **Self-alignment via solder-surface tension in flip-chip bonding opens the door to low-cost, high-throughput assembly of components with sub-micron accuracy. This is especially impactful to integrated photonics as used for high speed optical communication and sensors [1,2]. Assembly yield hinges on the details of solder-induced forces and on the geometry of the melted solder surface. Low curvature of melted solder is best to balance solder forces for optimal re-alignment yield but leads to shallow contact angles and solder de-wetting on traditional solder pads. We introduce and demonstrate the concept of recessed solder pads with shallow angled edges. Such geometry enables arbitrarily-low curvature of the molten solder surface and even flat or slightly concave shapes. The solder stays anchored at the angled edges of recessed pads and can be made to flow in long and narrow conduits. Both aspects are key to widening the fabrication and process window for the solder-induced chip-alignment technology.**

Keywords - flip-chip assembly; heterogeneous assembly; Silicon & III-V photonics; solder reflow; formic acid

I - INTRODUCTION

Assembly of chips with sub-micron alignment accuracy is of growing importance to a variety of applications. In particular, precision assembly of single-mode optoelectronic components constitutes a substantial fraction of the overall cost of integrated photonic sub-systems. Solder-induced self-alignment of chips offers the capability to reduce these costs,

at the same time as providing a scalable path to achieving sub-micron precision of waveguide-to-waveguide interfaces in complex heterogeneous optoelectronic solutions.

Despite early demonstrations of solder-induced alignment in the literature [3-6], substantial challenges remain for it to become a high-yield manufacturable process. This is especially true for three-dimensional solder-alignment, which has proved substantially more challenging than solder-alignment of lower dimensionality. We have previously reported on yield challenges from solder-volume control which we addressed with solder reservoirs for solder volume self-balancing [7,8]. This was then followed by addressing metallurgic challenges to enable thin-solder flow from solder reservoirs to chip-joining pads [9]. Solder wetting challenges and formation of intermetallics hinder solder flow and prevent the needed solder self-balancing from occurring during the melting phase.

While these previous contributions allowed substantial progress towards manufacturability by addressing issues of variability in solder volume and in wettability of solder pads, we encountered situations where dynamics of solder flow yielded effective solder contact angles that were incompatible with reliable operation. These occur in various parts of the solder pad structures, and especially in solder reservoirs and links. This paper describe novel solder pad designs that radically overcome wetting, and de-wetting, defects.

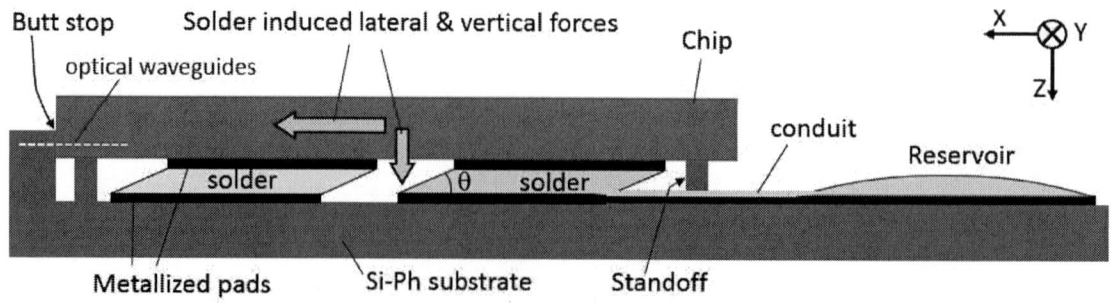

Figure 1. Schematic for chip 3D self-alignment on silicon photonic substrate via melted solder.

II- SOLDER INDUCED CHIP ALIGNMENT & LIMITS OF TRADITIONAL SOLDER PADS

Figure 1 shows a cross-sectional diagram of a chip soldered on a substrate with features adapted for solder-induced alignment. Metallized pads of chip and substrate are purposely offset in the X & Y lateral directions so as to induce X and Y forces on the chip during solder melting, via surface tension of the solder. The chip rests on standoffs that are part of the substrate and result from precise micro-electronic fabrication techniques, which include photo-lithography and reactive-ion etching. The solder forces are designed to move the chip against the standoffs and slide it toward mechanical stops in X and in Y. Precision of alignment rests primarily on the manufacturing of the mechanical stops for both chip and substrate, in the 3 directions X, Y and Z. Owing to the maturity of microelectronic manufacturing, sub-micron precision is readily attainable.

A key factor of success in solder alignment lies in the relative magnitude of the solder-induced forces. By nature, these forces are one to two orders of magnitude larger than the weight of the chip which can be ignored during the alignment process. The lateral surface tension forces along X (and similarly along Y) depend primarily on the cosine of the average solder-surface angle, highlighted as θ in Fig.1. By contrast, the vertical force along Z is a function of two distinct parameters:

- A surface tension pulling force that primarily hinges on the sine of the average solder surface angle.

- A hydrostatic force, controlled by the surface curvature of the solder, following the Young-Laplace formula. This hydrostatic force can be either positive or negative, pending on the type of surface curvature: a concave surface curvature generates a pulling force, whereas a convex surface curvature yields a pushing force.

At melting, the solder surface curvature is critically dependent on the amount of solder between chip and substrate pads. An excess solder readily generates a convex curvature and a large pushing force that can lift the chip off the standoffs. A lack of solder yields a concave curvature that strongly pulls the chip down and prohibits X-Y alignment motion due to friction on the standoffs.

Solder reservoirs alleviate this dependency on solder volume. Figure 1 & 2 depict the solder reservoir as an additional and separate metallization pad on the substrate, connected to a chip-joining pad via a narrower conduit pad. Lateral and vertical forces for typical pad dimensions were determined using a model that takes into account solder between the pads and in the reservoir [8]. Calculations include the hydrostatic pressure resulting from solder surface curvature after solder

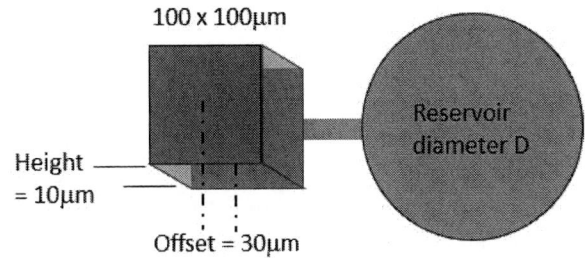

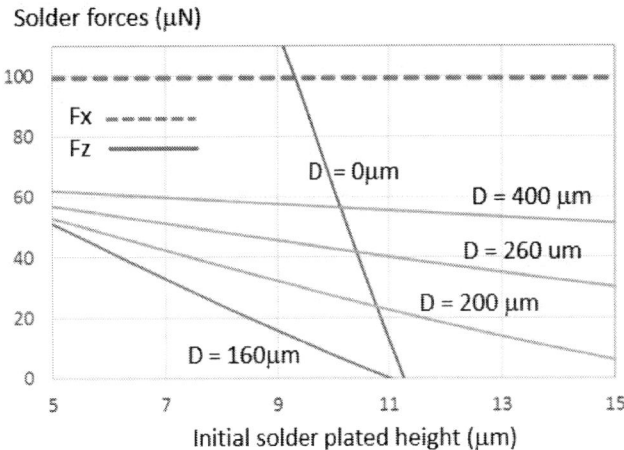

Figure 2. Diagram of solder pads with reservoir, and calculated forces on solder pads. Without reservoir (D=0mm), the vertical force (Fz in solid lines) depends heavily on solder amount. The addition of a reservoir relaxes this dependency, especially when the reservoir diameter is large relative to the solder pads. For comparison, the horizontal force (Fx in dotted line) is approximately constant.

has equilibrated between the reservoir and the space between the pads. The plotted curves highlight the benefit of a reservoir with an area larger than the area of the chip-joining pads: the dependency of the vertical force on the initial solder height is lessened. This effect greatly increases the fabrication tolerances and yield for solder plating by extending the range of solder plating height resulting in the correct balance of solder forces for chip re-alignment.

A large reservoir (ex. D = 300 μm) and commensurate low initial plating height (ex. h = 10 μm) results in a low solder contact angle, around 15 degrees. Sustaining such a low contact angle is difficult for most wetting materials, and we find that sufficient process robustness cannot be reliably ensured. Figure 3 depicts a situation occurring during the melting process when solder becomes liquid. Pictured solder reservoirs are rectangular in shape, but similar effect occur in circular reservoirs also. After melting, solder de-wetting can occur at the edge of a pad where the contact angle is the lowest

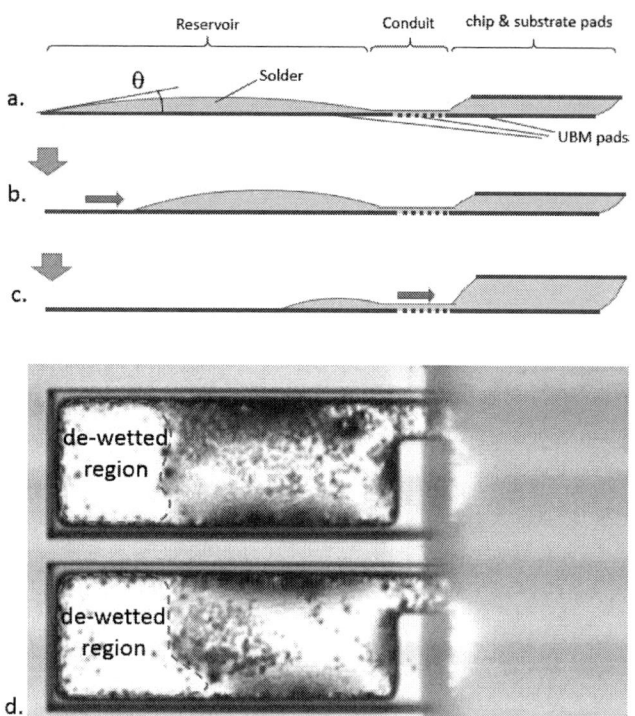

Figure 3. Example of solder de-wetting observed when the solder curvature over the reservoir is small. The cross-section diagrams (a) to (c) illustrate the de-wetting sequence that pushes solder towards the joining pads. Top view of partially de-wetted reservoirs is shown in (d).

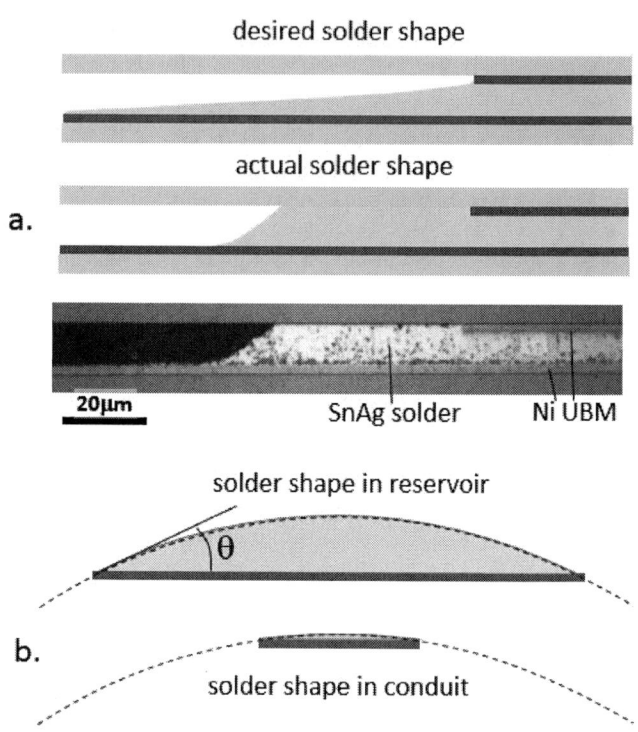

Figure 4. (a) Example of undesired wetting observed at joining pads, viewed in cross section. Such wetting substantially reduces the lateral solder-alignment force resulting in unsatisfactory assembly yield. (b) Diagram of solder shapes, highlighting the vulnerability of narrow conduit pads from the very thin equilibrium solder thickness in that region. Such thin solder shows poor mobility and is often fully consumed by intermetallic formation.

(Fig.3.b). The receding solder line raises the solder curvature and the hydrostatic pressure which in turns pushes solder from the reservoir into the gap between the joining solder pads. The process can be self-perpetuating and end in a runaway situation: as solder leaves the reservoir, its curvature increases which leads to more de-wetting and more pressure that can eventually lift the chip off the standoffs and result in alignment failure in the vertical direction.

Other types of failure induced by low contact angle and low solder curvature are depicted in Figure 4. Fig.4(a) shows cross-sectional diagrams and picture of solder profiles between solder pads, also called Under Bump Metallization (UBM) and made mostly of nickel. The desired solder shape illustrated in the top diagram provides maximum lateral force on the top chip via surface tension, but calls for an unsustainably small contact angle on the substrate UBM. It follows that solder de-wets the substrate and piles up in the region between the two components. As a result, the lateral tension force is substantially reduced from its expected value and the chip fails to move and align in the lateral direction. Figure 4(b) highlights an undesirable effect of the low surface curvature for the solder in narrow conduits. Pressure

equilibrates in the liquid medium, and curvature at any point of the solder surface is maintained through the Young-Laplace equation. Hence only a sub-micron solder layer often remains in the 20 to 40 μm wide conduit. Solder consumption through intermetallic formation can freeze the conduit or expose some of the protruding intermetallic crystals to oxidation, which appears irreversible in our process, and can prevent further solder wetting. As a result, the reservoirs can become effectively disconnected from the chip-joining pads and their benefit can be severely limited.

III- NOVEL PAD DESIGN

In order to sustain the low solder hydrostatic pressure desired for high-yield re-alignment, and its associated low surface curvature, a new UBM pad structure is required to prevent solder de-wetting as well as solder freezing and intermetallics oxidation related to thin solder coverage. We pursued a concept of metallized pads with a raised edge, aimed at better

978-1-7281-1500-9/19 $31.00 © 2019 IEEE   530

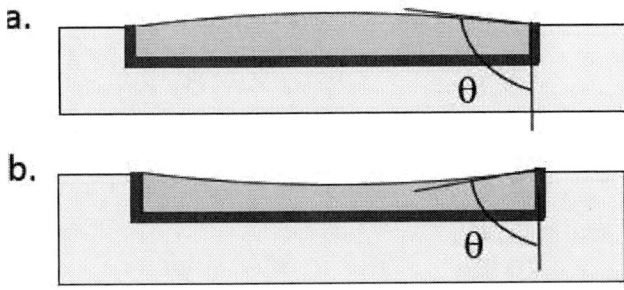

a.

b.

Figure 5. Cross-section diagrams of novel recessed solder pads aimed at maintaining a large effective wetting angle even (a) when the solder curvature is low, or (b) when the solder curvature is slightly negative. This is allowed by the minimum solder wetting contact angle being now in reference to the vertical and not the horizontal surface.

anchoring the liquid to the edge, while also maintaining a minimum depth of solder even for a very shallow curvature. For ease of implementation, we chose to recess the pads from the components' surface as this provides a simple way to create an edge all around. Figure 5 describes the essential feature of the new metallized pads, where the edges include a short vertical portion of the UBM metallization. The vertical portion locally increases the contact angle between the pad and the solder, from an otherwise very small value to nearly

90 degrees. The vertical surface essentially pins the liquid solder boundary to the edge of the pad, and is effective for all types of shallow solder curvature, from convex to flat to slightly concave.

We produced recessed pad structures in both substrate and chip, using similar semiconductor processing techniques as before [7-9], albeit adding extra steps for creating the recesses in the silicon substrates prior to generating the UBM metallization. After patterning via photo-lithography, Reactive Ion Etching (RIE) was applied with the aim of creating slightly slanted sidewalls. This ensures adequate electrical conductivity to the bottom of the recess for electrodeposition of UBM and solder, through sputtered metal seed-layers that are deposited after RIE. The same photo-lithographic mask then served for defining the UBM and solder areas. The photo-resist exposures were modified to produce slightly larger lateral dimensions for the UBMs than for recesses, to ensure coverage of all the recess sidewalls. This also ensured the correct definition of narrow recessed conduits and the coverage of their sidewalls with UBM metallization.

Figure 6 are cross-sectional views of recessed pads filled with varying amounts of solder, after solder was melted and re-solidified. Recess depth was approximately 4 µm in the four examples, and the radii of solder curvatures are indicated in each case. As evidenced from the views, solder stayed anchored at the edges of the pads even for very low curvature of the solder surface. Figure 7 shows examples of chip-to-substrate cross-sectional solder bonds, where both chip and

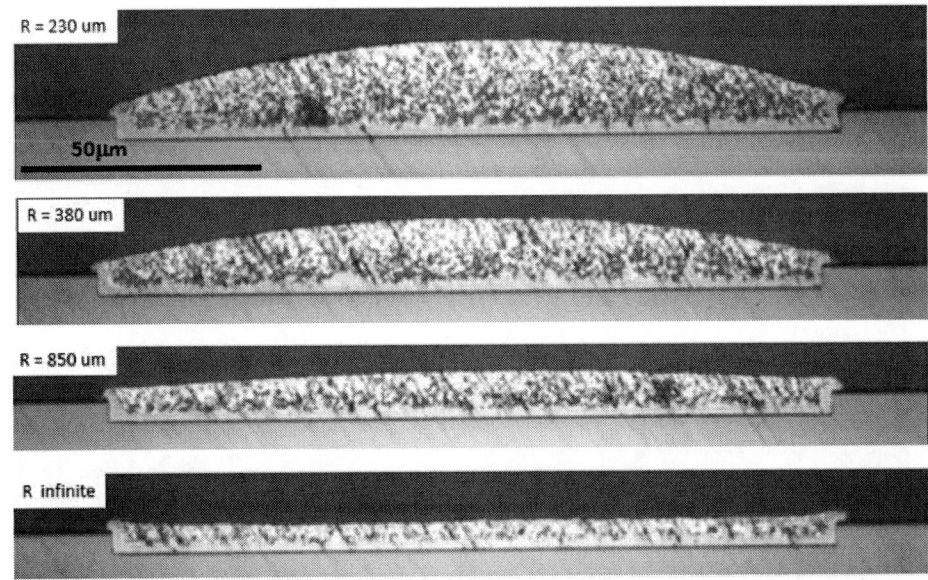

Figure 6. Cross-section views of novel solder pads, serving as reservoirs, for different amounts of plated solder. Solder radii of curvature are indicated in each of the 4 cases.

978-1-7281-1500-9/19 $31.00 © 2019 IEEE

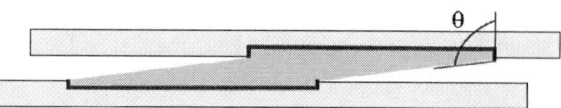

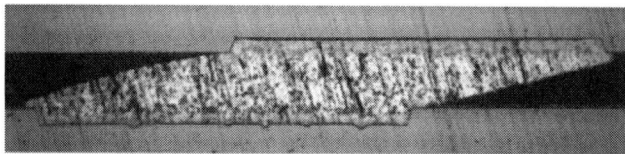

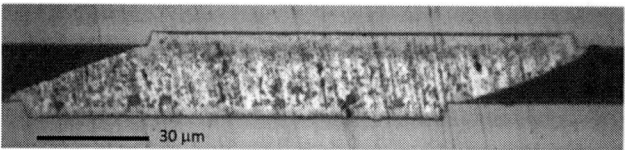

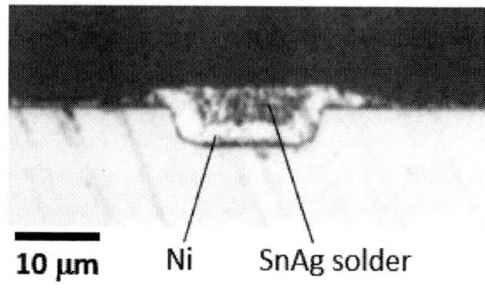

Figure 9. Cross-section of a recessed narrow solder conduit. The recess enables substantial amount of solder to flow in the conduit with close to zero top-curvature, avoiding conduit freeze-out via consumption of ultra-thin solder by intermetallic formation.

Figure 7. Cross-section views of novel solder pads with solder, joining chips and substrates, showing good anchoring of the solder to the edges of the solder pads.

substrate have recessed solder pads. Good solder anchoring is observed in both cases. Solder angle (versus horizontal) is 15 to 20 degrees as a result of the lateral offset between the pads. Since Fig. 7 illustrates the final chip position after alignment, it is understood that even a larger pad offset existed before alignment. We estimate that during the early phase of the melting process, the liquid solder surface angle was as low as 10 degrees and withstood pulling away from the edge of the pad.

Figure 8 exemplifies a most extreme case of solder surface between substrate and chip pads. Here, the left edge of the substrate pad extends significantly beyond the end of the chip pad and serves as an "integrated" reservoir [8,9]: the un-matched portion of the pad can accommodate some excess or lack of solder with lessened impact on the solder surface curvature, i.e. on the hydrostatic pressure in the solder. The effectiveness of this function hinges on the solder not de-wetting the substrate pad and not wetting the surfaces beyond

pads, which is the case depicted here. In addition to recessing the substrate solder pads to prevent de-wetting, we applied changes to the reflow ambient consistent with the study of [10] to help prevent wetting of solder beyond pads. The resulting shape of the solder is optimal for creating a substantial lateral force - an essential requirement for lateral alignment - while providing for a small downward force that ensures the chip remains on the standoffs.

Further evidence of the effectiveness of recessed (or edged) solder pads is observed in the working of solder conduits. Fig. 9 and 10 illustrate the working of conduits in extreme geometry of length relative to width of these narrow channels. In the 2 cases shown, the conduits are approximately 4 μm deep and 17 μm wide, and have lengths of 500 and 1000 μm. A 2 μm thick UBM covers all sides and solder fills the resulting ~ 4 x 13 μm channel. For testing, excess solder was added to the left pads before melting, as shown in Fig. 10(a). In the molten state, liquid solder flows through the narrow channel from the left to the right pads, until identical liquid surface curvature is attained throughout. It follows that these novel recessed conduits allow solder equilibrium to be realized between reservoirs and joining pads even when long paths are required by restricted component geometries.

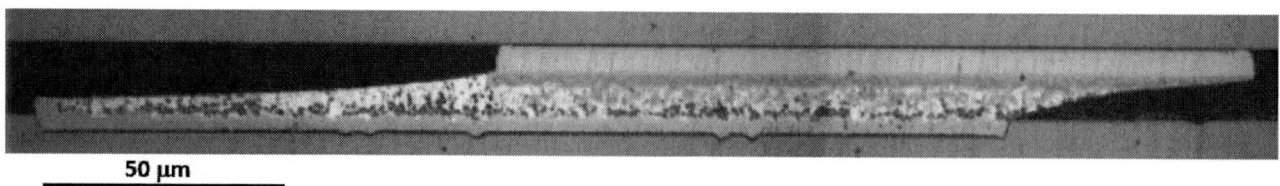

Figure 8. Cross-section views of a novel recessed substrate solder pads with solder that sustains a very thin solder volume with very low angle between chip and substrate. This example successfully contrasts the defect depicted on Fig. 4(a). In addition to recessing the substrate solder pads to address de-wetting, changes in the reflow ambient consistent with the study of [10] contributed to preventing wetting of the solder beyond pads.

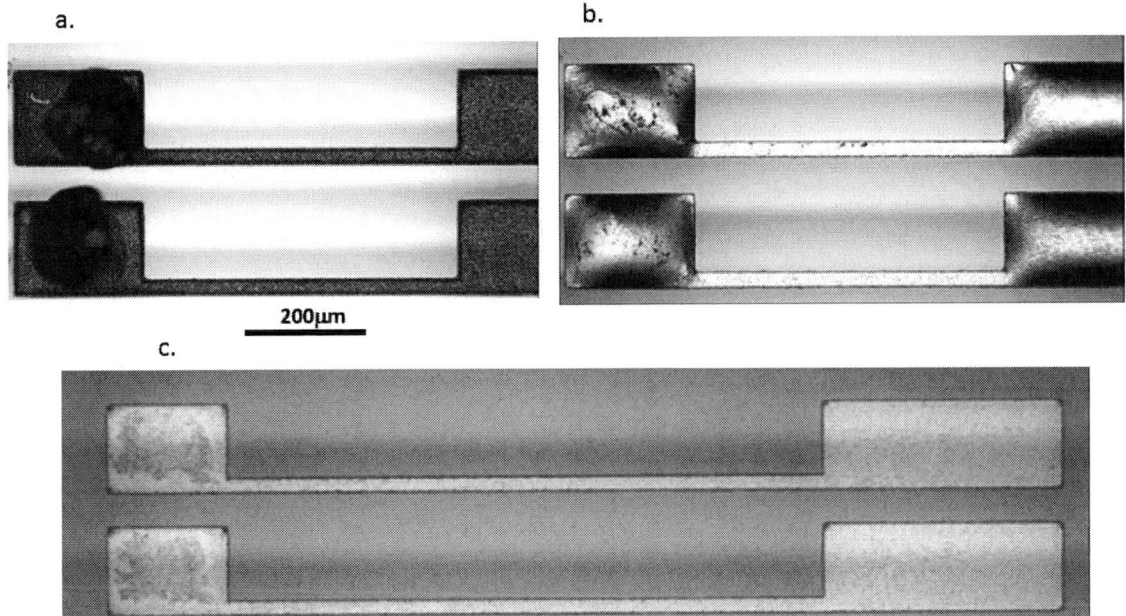

Figure 10. Solder flow through long and narrow recessed conduits. (a) 500 x 13 µm conduits before melting with additional beads of solder placed on the left pads. (b) Pads of (a) after solder reflow. Molten solder flows and equilibrates between left and right pads. (c) Same as in (b) but with 1000 x 13 µm conduits.

IV- CONCLUSIONS

We highlighted limitations of traditional solder pads for solder-induced chip alignment. The requirements for shallow solder curvatures dictate low solder angles which are not sustainable over flat solder pads. Recessed solder pads show effectiveness in anchoring solder in its liquid phase for geometries that are necessary for high lateral forces and low vertical forces, which are key to robust chip alignment. We demonstrated a CMOS-compatible fabrication process for integration of recessed pads with minimal additional manufacturing complexity. The combination of reservoirs and recessed pads dramatically increases the fabrication process window for high-yield three-dimensional solder re-alignment. Consequently, it enables a path toward low-cost, large-scale assembly with sub-micron alignment accuracy.

ACKNOWLEDGEMENTS

We would like to thank the staff of the Microelectronics Research Laboratory (MRL) at the IBM T.J. Watson Research Center where the chips were fabricated. We would like to acknowledge the leadership from the managerial team in our organization, particularly from Wilfried Haensch, William M. J. Green, and Jessie Rosenberg.

REFERENCES

[1] T. Barwicz et al, "Demonstration of Self-Aligned Flip-Chip Photonic Assembly with 1.1dB Loss and >120nm Bandwidth," in OSA Technical Digest (online) of 2016 Frontiers in Optics, paper FF5F.3.

[2] Y. Martin et al, "Flip-Chip III-V-to-Silicon Photonics Interfaces for Optical Sensor", accepted for presentation at the 69th IEEE Electronic Components and Technology Conference (ECTC), Las Vegas, NV, May 28-31, 2019.

[3] K. P. Jackson et al. "A High-Density, Four Channel, OEIC Transceiver Module Utilizing Planar-Processed Optical Waveguides and Fli-Chip, Solder Bump Technology", J. of Lightwave Tech., Vol. 12, No. 7, 1994, pp.1185-1191.

[4] P. J. Nasiatka et al., "Determination of Optimal Solder Volume for Precision Self-Alignment of BGA using Flip-Chip Bonding", in IEEE Proc. 1995 Electron Devices Mtg., pp.6-9

[5] Q. Tan et al., "Soldering Technology for Optoelectronic Packaging", in IEEE Proc. of 1996 ECTC, pp. 26-36.

[6] M. Hutter et al "Precise Flip Chip As sembly Using Electroplated AuSn20 and SnAg3.5 Solder", in IEEE Proc. of 2006 ECTC, pp. 1087-1094.

[7] J.-W. Nah et al, "Flip chip assembly with sub-micron 3D re-alignment via solder surface tension," in IEEE Proc. of 2015 ECTC, San Diego CA, May 26-29, 2015.

[8] Y. Martin et al, "Toward high-yield 3D self-alignment of flip-chip assemblies via solder surface tension", in IEEE Proc. of 2016 ECTC, Las Vegas NV, May 31-Jun 03, 2016.

[9] Y. Martin et al, "Solder mobility for high-yield self-aligned flip-chip assembly", in IEEE Proc. of 2017 ECTC, Orlando FL, May 30-Jun 02, 2017.

[10] M. Samson et al. "Fluxless Chip Join Process Using Formic Acid Atmosphere in a Continuous Mass Reflow Furnace", in IEEE Proc. of 2016 ECTC, pp. 575-579.

Collective curved CMOS sensor process: application for high-resolution optical design and assembly challenges

B.Chambion[1], C.Gaschet[1], M. Lombard[1], M. Fernandez[1], P. Joly[1], S.Caplet[1], F.Zuber[1], A.Vandeneynde[1], P.Peray[1], G.Lasfargues[1], M. Zussy[1], J. Deschamps[1], A. Bedoin[1], D.Henry[1].

[1] Univ. Grenoble Alpes, CEA, LETI, MINATEC campus, F38054 Grenoble, France.
e-mail: bertrand.chambion@cea.fr

Abstract— Curved sensors is a well-adapted technology solution to enhance the vast majority of optical systems. It helps to remove lenses and simplify optical architectures. These advantages open news challenges such as a specific fabrication process applied to curved sensors and new rules for final system integration. In this paper, we first introduce benefits of curved sensors applied on a compact high-resolution camera to define the sensor shape specifications and to reach high performances and compactness (-50% compared to a benchmark system). Mechanical limits and optical modeling are used. Then, based on these curved specifications, a novel collective curving process is described, developed on 1/1.8'' format CMOS image sensors with a radius or curvature target R= 55 mm and R=60 mm. This work includes packaging and assembly steps, optimizations, and morphological characterizations in accordance to optical design requirements. Finally, a dedicated optical test bench is used for Modulation Transfer Function (MTF) characterization of the final camera prototype. All these experiments and optical results introduce new opto-mechanical requirements and demonstrate the feasibility and high performances of systems with curved sensors.

Keywords— Curved CMOS sensors, collective process, Arrays, Imaging systems, Optical design, Petzval Field Curvature, Photography, Mechanical limits, Aberrations, Electro-optical characterization.

I. Introduction

The need for curved focal planes in optical designs have been known by opticians since Joseph Petzval 180 years ago and proved in further researches. Recent studies by Rim et al. [1] and Stamenov et al. [2] demonstrated a potential reduction of lenses, leading to compactness, better resolution and improved illumination with monocentric systems using a curved focal plane. Iwert et al. and Muslimov et al. [3,4] showed new ways to create telescopes and optical instruments with higher resolution for astronomical applications. Reshidko et al. [5] demonstrated the possibility to create miniature lenses for mobile phone applications with greater apertures. All these studies demonstrate the link between the sensor's curvature and the performance of related optical systems.

At the same time, different approaches have been investigated for curved focal planes.

The first one creates an assembly of multiple sensor segments electrically connected to each other [6–11]. This method provides a high curvature because of the removal of stresses due to the segmentation but is limited by dead zones. This approach is incompatible with high pixel density, due to Complementary Metal Oxide Semiconductor (CMOS) designs including periphery areas. The second approach uses a fiber optic bundle to transfer the light from of the curved focal plane to a flat sensor [2]. This solution adds complexity and has a lower resolution than current CMOS sensors. The third approach curves a monolithic standard sensor. Different solutions have been investigated by several teams and companies, as Sony, Microsoft through patents and papers [12, 13]. This approach has also been described to provide functional uncooled and cooled infrared sensors [14, 15]. In [16–18], our developments on the curvature of monolithic sensors show the feasibility to curve CMOS sensors without electro-optical performance loss. Besides, characteristics of a new optical design are shown.

In this work, we propose a collective curved CMOS process applied to a standard 1/1.8'' format image sensor. First, mechanical considerations have to be taken into account to define the boundaries of the reachable curvatures and to be integrated into the optical design step, to specify the sensor surface shape to fabricate. Then, our collective curving process is detailed from a standard CMOS wafer to a curved sensor in its ceramic package. It includes process steps optimizations, packaging steps, and morphological characterization. Finally, electro-optical responses of 6 curved sensors are discussed and a dedicated optical test bench is used for Modulation Transfer Function (MTF) characterization of the camera prototype, compared to a commercial one. New opto-mechanical specifications for the next curved sensor generations and optical assembly are proposed.

II. Optical Design Specification

A. Sensor mechanical limitations

All technologies considered to create the sensor curvature on the entire die are limited by the sensor's mechanical breakage. According to [16, 17], the mechanical behavior of CMOS sensors can be modeled using a Finite Element Analysis software, with three parameters conducting mechanical limits: the size of the die, the thickness of the substrate, and the tensile failure strength criterion $\sigma 1_{max}$. We consider two tensile failure strength criterion: $\sigma 1_{max1} =$

200MPa for a standard chip surface preparation and $\sigma 1_{max2} =$ 500MPa related to advance chip preparation processes [19]. The first step in the methodology is to simulate this behavior for a full range of sensor size: 1/3" format for smart-phone applications, 1/1.8" format, APS-C and full-frame like format (24 mm x 32mm), respectively linked to 5.16 mm x 6.25 mm , 7.74 mm x 8.12 mm, 22.3 mm x 28.1 mm, 28 mm x 40 mm die size, which can be approximated by (100)-oriented silicon chips with a thickness of 100 microns. The reference plot is in Figure 1.

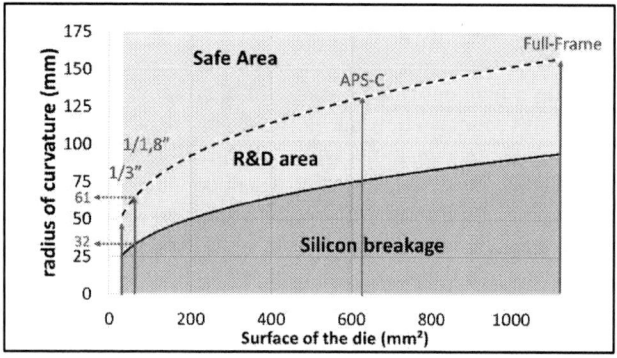

Fig. 1. Results of breakage limits considering two failure criteria. $\sigma 1_{max1}=$ 200MPa (dash line) and $\sigma 1_{max2}=$ 500MPa (solid line), for 1/3" format, 1/1.8" format, APS-C format, Full frame format, at 100 microns of thickness.

As this work is built on a 1/8" sensor format, the minimum allowable radius of curvature R=32 mm for $\sigma 1_{max2} = 500$MPa and R=61 mm for $\sigma 1_{max1} = 200$MPa. These modeling results has been validated by experimental tests, detailed in section IV.A.1.

B. Optical design specifications

The global optical methodology and the prototype demonstration are detailed in [23]. The starting point is the monocentric lens, studied in recent publication [2], and well known for wide field of view capability, compactness, and high-resolution with a curved sensor on the Petzval Surface. The field of view, which is the subtended angle of the sensor, drops to at most 17 degrees at any sensor's size due to mechanical limits $\sigma 1_{max2} = 500$MPa, Fig 1. The goal is to keep high-resolution and compactness and increase the field of view. During the optical optimization process, some lenses are added in order to decrease step by step the subtended angle on the sensor. The final version of this new architecture, showed Fig 2, and equipped with a 1/1.8" format 1.3Mpx global shutter CMOS sensor (Teledyne EV76C560). It has only 6 spherical lenses, is 24 mm long and has an optimized sensor radius of curvature range from 47 to 62 mm, as a function of the object distance D.

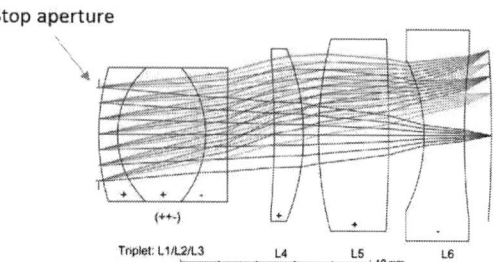

Fig. 2. Optical layout of the compact system.

Regarding the sensor shape error and positioning, curved sensor is now embedded in the optical design optimization process. It brings new considerations for final packaging and assembly, compared to standard design with flat sensors.

- For flat sensor, the sensing area is define by a 2D surface. At infinity object distance, the depth of focus ΔZ is defined as describe in fig 3:

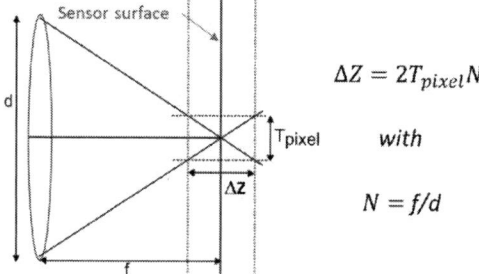

$$\Delta Z = 2T_{pixel}N$$

$$with$$

$$N = f/d$$

Fig. 3: Depth of focus ΔZ explained for a flat sensor design. T_{pixel} is the pixel size, N is the f# (system aperture), f the focal length, and d the exit pupil diameter.

- For curved sensor, the sensing area is define by a 3D surface. Assuming h/R<<1 et $\Delta R/R$<<1 at infinity object distance, the depth of focus ΔZ with a curved sensor is define as describe in fig 4 and converted in (ΔR), (tolerance on R):

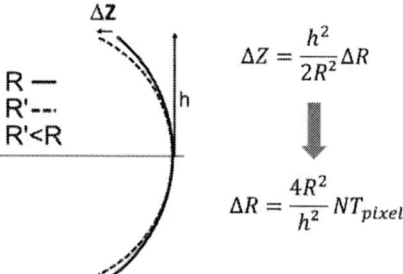

$$\Delta Z = \frac{h^2}{2R^2}\Delta R$$

$$\Delta R = \frac{4R^2}{h^2}NT_{pixel}$$

Fig. 4: Focus depth ΔZ explained for a curved sensor design. T_{pixel} is the pixel size, R and R' the radius of curvature of the sensor, h the sensor height, N is the f# (system aperture), f the focal length, and d the exit pupil diameter.

For assembly tolerances, specifically for our optical design described in figure 2, it gives ΔZ= 27 μm and $\Delta R = 8$ mm (± 4 mm). These optical results give us the tolerances to reach, in curved CMOS sensor fabrication.

978-1-7281-1500-9/19 $31.00 © 2019 IEEE

III. COLLECTIVE CURVED CMOS PROCESS

The curved process detailed below is applied to a commercial 8" CMOS sensor wafer from Teledyne (EV76C560 1.3Mpx global shutter product). The overall process can be divided into 5 phases:

A. Chip preparation from wafer

Initially, the CMOS wafer thickness is 725µm. To give mechanical flexibility to the substrate, the wafer is thinned down to below 100µm using special procedures due to targeted thicknesses with grinding and polishing equipment's. After thinning, the wafer is diced at wafer level and dies are ready for pick and place step. This first phase is summarized in figure 5.

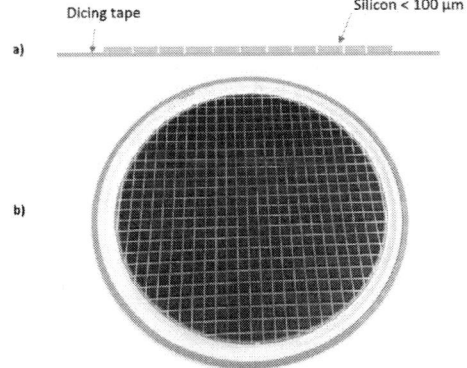

Fig. 5: Chip preparation process from standard wafer to thinned and diced dies, ready for pick and place process. a) Schematic of diced wafer on tape, b) Diced CMOS wafer, ready for pick and place.

B. Curved holders preparation

The final sensor shape is driven by the curved holder. In this work, individual curved holder are used, made up of micro machined Aluminum. Each curved holder is 9*8*1mm and has R=55 mm or R=60 mm. Each substrate radius of curvature and sphere error is characterized using confocal microscopy (see part IV.A.2). Then, curved holders are picked and placed on a collective chuck designed for 12 chips slots in this process version. The collective approach means that all dies are curved at the same time during the process. Collective chuck and curved holder are introduced in Figure 6.

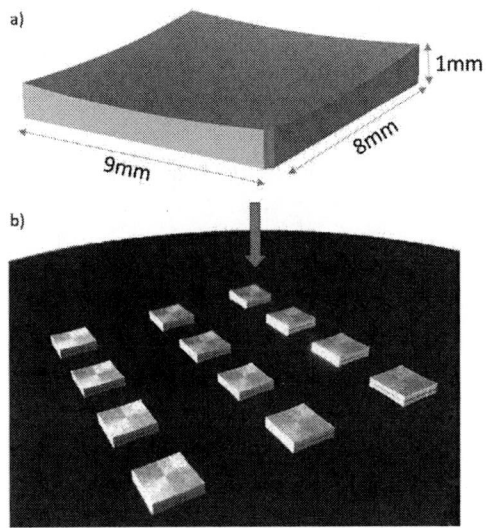

Fig. 6: Schematics of a curved holder (a) and pick and place on the collective chuck (Silicon) (b).

C. Die attach dispense and pick an place

Die attach material is a key point in the curving process. For a standard flat sensor, the die attach has to maintain the die onto the substrate without external mechanical stress. In a curved sensor configuration, the silicon die is bent within its mechanical elastic domain. It means that during operation, the die attach has to keep a good adhesion at the chip/package interface under mechanical stress from the chip. This process flow uses a silver filled epoxy die attach associated to a drop dispense equipment. To target a void free joint, the dispense pattern has been optimized, and the die attach volume is calculated for a final 20µm thick joint. The process is applied on chip side, and the die is flipped and placed on its curved holder (Figure 7).

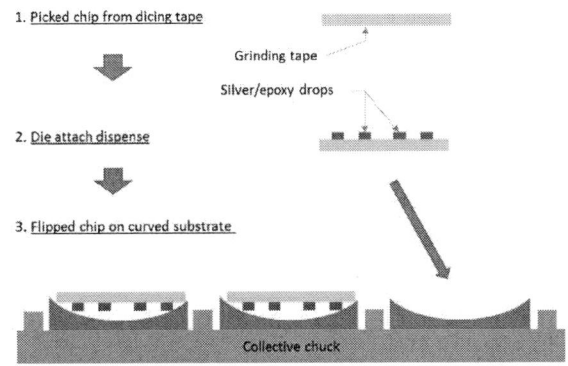

Fig. 7: Screen printing process and chip placement on collective chuck.

D. Curving step process

The collective chuck is now ready for curving process, each curved holder is equipped with thinned sensor (Figure 8). This step consists in applying a mechanical pressure on each die to reach the final curved holder shape. This pressure

P is applied via a pressure tool. During this process, the equivalent hydrostatic pressure used is higher than 2 bar. Once the pressure established, the die attach curing is done via a time-temperature schedule in accordance to the die attach material. Finally, the tool pressure is removed, and curved CMOS sensors are ready for integration in package.

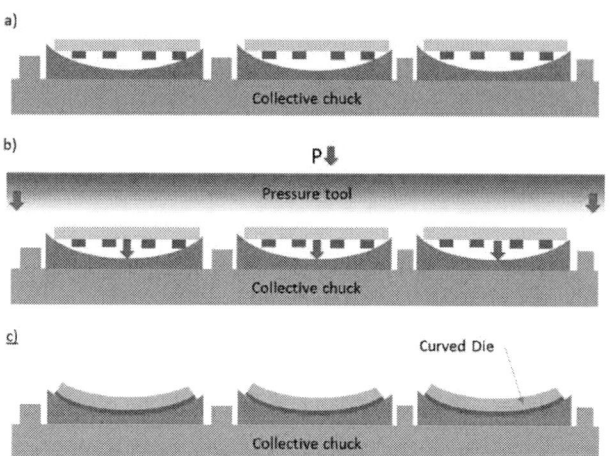

Fig. 8: Collective curving process step details. a) Flipped chip on curved holder. b) Die shaping and die attach curing. c) Pressure tool release.

E. Package integration

Integration in package is done with standard process but using specific tools, because of the curved shape of the chip (figure 9). First, each curved die is glued into the ceramic package. Then, the CMOS circuit is wire bonded to the package and finally, a glass cover is placed onto the package as a device protection. We use a modified ball bonding process due to the curved working surface. Packaged curved sensors are mounted on Printed Circuit Board (PCB) with standard reflow process (Figure 10).

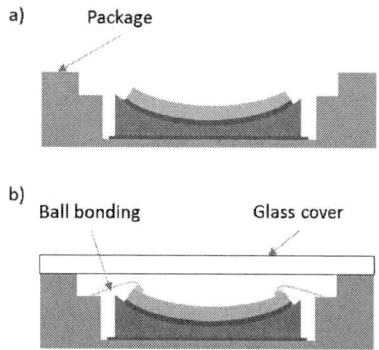

Fig. 9: Final package integration process. a) After pick and place in package. b) After ball bonding and glass cover process.

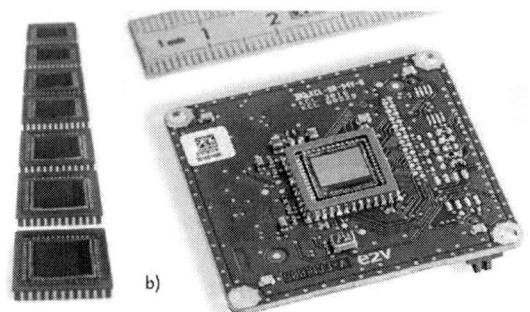

Fig. 10: Curved 1.3Mpx global shutter CMOS prototypes. R= 60 mm. a) several curved prototypes in package. b) Soldered on PCB.

IV. PROCESS CHARACTERIZATIONS

Characterizations are classified in two categories: those related to the packaging process, including mechanical model validation, shape measurement, bondings and die attach, and those related to active measurements, with the electro-optical response and the MTF measurement, associated to the optical design.

A. Collective curved process characterization

1) Mechanical model validation

As describe in section II.A, the model is able to define the minimum radius of curvature we can reach as a function of the die size, die thickness, substrate material, and failure criteria. To validate model predictions, 7.74 mm x 8.12 mm bare silicon dummies chips and real CMOS circuit, 100µm thick, has been curved using our standard process flow. The strength criterion considered is $\sigma1_{max2} = 500$MPa. We use several curved holder to cover a radius of curvature range from 65 mm to 20 mm, with 5 mm pitch. The output data is binary (breakage or not), and the breakage radius of curvature value is recorded. Table 1 summarizes our results between model prediction and breakage data on dummies and CMOS chips. Experimental tests are in good accordance to the model prediction, within 3%.

TABLE I. MECHANICAL MODEL VALIDATION USING CURVED PROCESS WITH DUMMY AND CMOS CHIPS (7.74 MM X 8.12 MM).

	Model prediction	*Dummy chips (Si)*	*CMOS chip*
Population size	-	16	25
Breakage radius of curvature in mm – (std deviation σ)	32	33.1 - (7.4)	31.6 - (4.7)

2) Morphological measurements

Due to optical design requests, sensor shape and shape error to the perfect sphere are key points to measure. Using an Altisurf 530 confocal microscope, we choose two parameters to track: the best sphere fit radius of curvature (using least squares method), and the peak to valley error to that perfect sphere, within the pixel area (Sz, according to ISO25178 standard). The methodology is based on a comparison between the curved holder shape before curving process and the effective sensor shape after curving process. Figure 11 gives an example of sensor surface characterization after curving process.

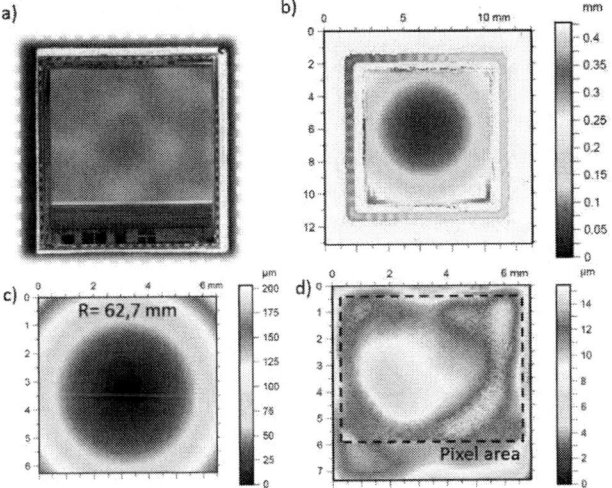

Fig. 11: C001 Curved CMOS sensor shape surface characterization. Target R= 55 mm. a) Visible top view, entire chip. b) Surface altitude characterization (raw data). c) Calculated best sphere fit to raw data (calculated R=62.7mm). d) Best sphere removed to raw surface on pixel area.

Table 2 summarizes our results on curved prototypes, before curving process (curved holder only), and for final assembly (curved chip mounted on package). We note that final shape radius of curvature are in accordance with the optical specifications (section II.B). Sz values are in accordance with Δz spec, except for C003 and C006 sample. Also, the Sz value on final assembly is lower than the curved holder one (except for C002 sample). It means that during the curving process, the substrate mechanical behavior of the thinned sensor has a mechanical smoothing effect on the final sensitive surface shape.

TABLE II.   SHAPE MEASUREMENTS SUMMARY DURING CURVING PROCESS.

Sample	Shape target, radius of curvature (mm)	Curved holder radius of curvature (mm) – (Sz in μm)	Curved sensor radius of curvature (mm) – (Sz in μm)	Variation between substrate and final assembly measurement %(%)
C001	55	55.6 – (37.5)	60.1 – (27)	+8.1 (-28)
C002	55	54.7 – (14)	55.9 – (26)	+2.1 (+85)
C003	60	61.1 – (22.3)	62.0- (28)	+1.4 (+25)
C004	60	62.3 – (14.2)	60.5 – (10.5)	-2.9 (-26)
C005	60	61.8 – (16.6)	60.1 – (9.2)	-2.8 (-45)
C006	60	60.7 – (46)	58.0 – (31.2)	-4.5 (-38)

3) Die attach joint characterization

In order to investigate the joint thickness, repartition and void free characteristics after process, several cross-section has been made along the sensor diagonal. One of these cross section in presented in figure 12. Its shows a thickness variation along the sensor diagonal, 5μm at die corner, compared to 17.9μm for die center. This variation could be explained by the pressure repartition from the pressure tool on the die attach layer during curing process. This is due to the mechanical strength of the die chip during curving process. We note a good density of the die attach material, without voids.

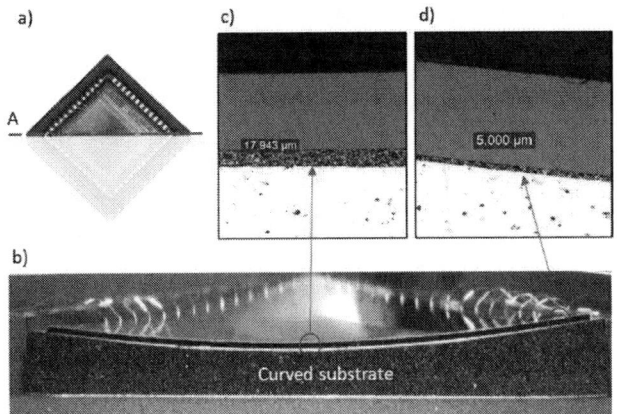

Fig. 12: Packaged curved CMOS sensor cross-section. a) Cross section localization. b) Cross section view along A axis. c) Center die attach thickness measurement e=17.9μm. d) Corner die attach thickness measurement e=5μm.

4) Wire bonding process validation

In order to connect the CMOS circuit to the package, a conventional wire bonding equipment is used (ball/wedge configuration). Due to the non-planar pad surface (curved chip side), the maximum pad incidence is to 6-8°. Consequently, wire bonding parameters need to be adjusted in term of ultrasonic power, bonding force, and bonding time to guaranty a robust welding interface at ball/pad contact. To validate our bonding parameters optimizations, we carried out destructive pull tests to quantify the wire strength resistance. In figure 13 are presented a cumulative failure plot on 56 gold wires (25 μm diameter) as a function of the load force. It shows a minimum failure force > 5gr (5.64gr), which validate our bonding process for prototyping level on curved surface configuration.

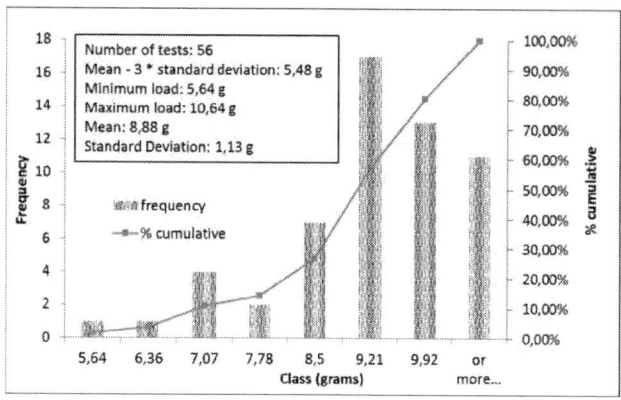

Fig. 13: Curved CMOS wire bonding validation. Cumulative failure plot on 56 ball/wedge gold wires on curved chip. 25µm wire diameter. Minimum load = 5.64gr, mean value = 8.88gr and standard deviation = 1.13 gr.

B. Active characterizations

1) Electro-optical characterization

Electro-optical parameters have been calculated according to the EMVA1288 Standard for Characterization of Image Sensors and Cameras [20]. Sensor output in digital numbers is converted into electrons using conversion factors deduced from measurements of the photon transfer curves [21], PTC, of each sensor. For the PTC measurement, the sensor is illuminated at constant irradiance and exposure time is swept from minimum to saturation. The conversion factor is determined from a fit of the relation between the variance and the mean of the sensor output in the linear range.

The temporal dark noise, $\sigma_{e.dark}$, is calculated as temporal standard deviation of the dark pixel output at minimum exposure time in a series of dark frames. The mean over all pixels in the pixel array is reported.

The photo-response non-uniformity, PRNU, is defined as ratio between the dark corrected spatial standard deviation and the dark corrected mean sensor output at half saturation.

Dark current, DC, is the sensor output in absence of photons. It is mainly caused by thermal charge carrier generation and is consequently exponentially increasing with temperature. DC is measured as slope between exposure time and mean sensor output in dark condition at controlled temperature. Dark current non-uniformity, DCNU, is a measure for the spatial distribution of DC and defined as spatial standard deviation of the DC over all pixels.

In this study, we focus the discussion on DC mean values and the temporal dark noise, $\sigma_{e.dark}$. Electro-optical parameters of the curved sensors are measured after the curving process and compared to those of typical flat reference sensors. We can thus quantify and discuss about deviations of the curved sensor performances from standard ones. Note: no distinction can be made between the impact of the curving process and device-to-device variations.

For dark current measurements, the sensor is covered by blackout fabric in order to assure a dark environment. The sensor is turned on and its temperature is measured (T_{ceram}, for ceramic package temperature). DC, $\sigma_{e.dark}$ and T_{ceram} are measured as a function of the time. In figure 14, is plotted the DC variation as a function of T_{ceram} for 8 samples, 6 curved and 2 flat references. Due to exponential behavior, figure 15 is plotted in log scale. We first note a similar DC response for every devices, either curved or flat configurations. Also, flat configuration have lower DC than the curved prototypes. In this work, the packaging structure of curved prototypes is glued on curved holder whereas reference sensor is flat with bulky silicon substrate. This packaging difference could explained that DC offset due to thermal effect.

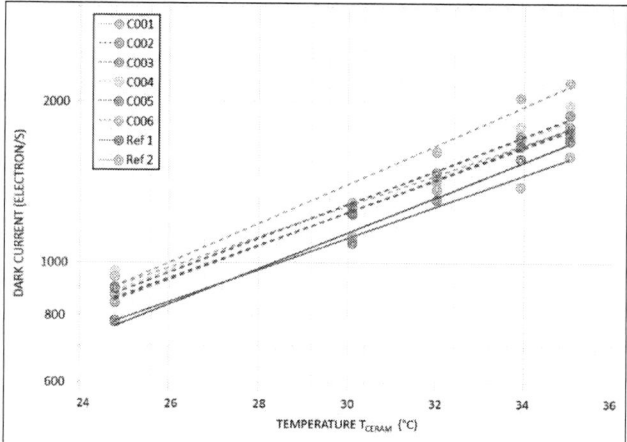

Fig. 14: Dark current of curved and flat CMOS sensors. Behavior comparison between curved and flat references as a function of ceramic package temperature.

In table III are summarized the temporal dark noise $\sigma_{e.dark}$ values for curved and flat reference prototypes. No distinction can be made between curved and flat configurations. It means that curving process has clearly no influence on temporal dark noise.

TABLE III.   MEASURED TEMPORAL DARK NOISE ON CURVED AND FLAT REFERENCE SENSORS.

	Ref1	Ref2	C001	C002	C003	C004	C005	C006
$\sigma_{e.dark}$ (e-)	28.2	30	28.3	29.5	28.7	29	28	28.3

2) MTF characterization and optical bench

MTF is a well-known tool to measure how different spatial frequencies are handled with an optical system. This function is directly link to its resolution and sharpness aspects. MTF can be seen as the contrast reduction evaluation of a periodic pattern through the optical engine. This function is global, and takes into consideration system integrations aspects: sensor, optical parts, and alignment quality. To calculate MTF, we used the slanted edge procedure as described in [22].

To maximize MTF, the theoretical R as a function of object distance D is presented in Figure 15. It shows that for a given R=55mm, it corresponds to a D range according to ΔR = 8mm (green area), defined in optical modeling section. Thus, MTF value should be maximum for 0.4m <D< 1.25m.

Our camera prototype, equipped with C001 curved sensor (R=55mm) is characterized with a dedicated bench for MTF measurement as a function of objet distance D, from 0.4m to 2m. Our custom test chart is designed to capture the same pattern on D range. An automatic routine is used to detect slanted edges and extract MTF value in 9 positions within the image, as described in figure 16.

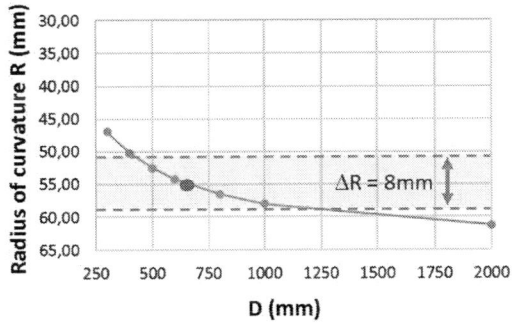

Fig. 15: Theoretical R as a function of D, associated to the optical design in figure 2. Green area: ΔR specification tolerance.

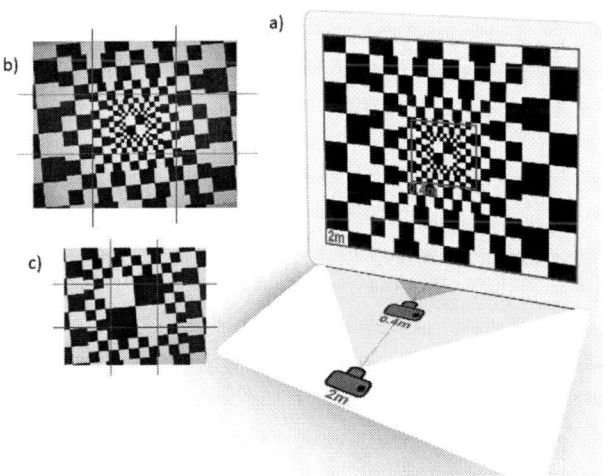

Fig. 16: MTF measurement tool. a) Optical test bench used, for MTF characterization as a function of object distance, range 0.4-2m. b) And c) image segmentation for MTF analyze across the field, respectively for D=2m and D=0.4m.

In order to validate the optical design benefits of curved sensors, we have measured MTF of an equivalent commercial compact design Techspec® lens #63-779, mounted on a flat 1/1.8'' sensor, to compare with our camera prototype.

MTF results for D=1 m are respectively plotted in figure 17 and 18, for compact commercial design, and for our camera prototype MTF measurements show that in both cases, optical performances are similar. However, our prototype has higher curves dispersion than commercial one, which could be explained by misalignment due to manual lenses integration or sensor position compared to the commercial product. These optical results confirm all the advantages of using a curved sensor in optical camera Actually, thanks to curved sensors we obtain the same E/O results as for a flat one but with a dramatically more compact system (60% less length in our case).

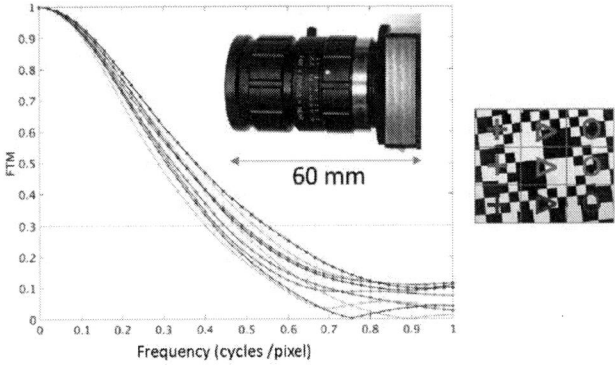

Fig. 17: MTF measurement of a commercial compact design Techspec® lens #63-779 with flat CMOS sensor. D=1m.

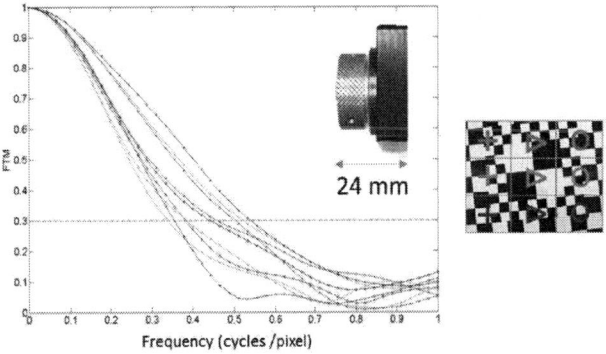

Fig. 18: MTF measurement of a compact camera prototype equipped with curved CMOS sensor (R=55mm). D=1m.

V. CONCLUSIONS

In this paper, we have discussed about a collective curved sensors fabrication process for the design of enhanced performance optical systems. We proposed a global approach from sensor specifications to prototype optical performances validation. First, benefits of curved sensors applied on a compact high-resolution camera are presented, and it gives precise sensor shape specifications to reach high performances and compactness in accordance to our mechanical limits model prediction. Then, a novel collective curving process is described, developed on 1/1.8'' format CMOS image sensors with a radius of curvature target R= 55

978-1-7281-1500-9/19 $31.00 © 2019 IEEE

mm and R=60 mm. Chip preparation, die attach, curved holder preparation, curving step, and final integration in package are detailed. Characterization results are presented and discussed on shape measurement, die attach, wire bonding, and electro-optical response of curved CMOS to validate our process fabrication in accordance to optical specifications. Finally, a dedicated optical bench for MTF measurement is introduced. The compact camera equipped with curved sensor (24 mm long) is compared to a commercial compact design Techspec® lens #63-779 with flat sensor (60 mm long), and it reveals equivalent optical performances. All these experiments and optical results demonstrate the feasibility and high performances of systems with curved sensors. We believe that future curved sensor fabrication processes such as collective or wafer level solutions will offer new possibilities in the design of optical systems for numerous applications from miniaturized devices to large scale systems.

REFERENCES

[1] S. B. Rim, P. B. Catrysse, R. Dinyari, K. Huang, and P. Peumans, "The optical advantages of curved focal plane arrays," Opt. Express 16(7), 4965-4971 (2008).

[2] I. Stamenov, A. Arianpour, S. J. Olivas, I. P. Agurok, A. R. Johnson, R. A. Stack, R. L. Morrison, and J. E. Ford, "Panoramic monocentric imaging using fiber-coupled focal planes," Opt. Express 22(26), 31708-31721 (2014).

[3] O. Iwert and B. Delabre, "The challenge of highly curved monolithic imaging detectors," Proc. SPIE 7742,774227 (2010).

[4] E. Muslimov, E. Hugot,W. Jahn, S. Vives, M. Ferrari, B. Chambion, D. Henry, and C. Gaschet, "Combining Freeform Optics and Curved Detectors for Wide Field Imaging: A Polynomial Approach over Squared Aperture," Opt. Express 25, 14598-14610 (2017).

[5] D. Reshidko, and J. Sasian, "Optical Analysis of Miniature Lenses with Curved Imaging Surfaces," Appl. Opt. 54, E216-E223 (2015).

[6] H. C. Ko, M. P. Stoykovich, J. Song, V. Malyarchuk, W. M. Choi, C. J. Yu, J. B. Geddes 3rd, J. Xiao, S. Wang, Y. Huang, and J. A. Rogers, "A hemispherical electronic eye camera based on compressible silicon optoelectronics, Nature 454(7205), 748-753 (2008).

[7] X. Xu, M. Davanco, X. Qi, and S. R. Forrest, "Direct transfer patterning on three dimensionally deformed surfaces at micrometer resolutions and its application to hemispherical focal plane detector arrays," Org. Electron. 9(6), 1122-1127 (2008).

[8] I. Jung, G. Shin, V. Malyarchuk, J. S. Ha, and J. A. Rogers, "Paraboloid electronic eye cameras using deformable arrays of photodetectors in hexagonal mesh layouts," App. Phys. Lett. 96 (2), 021110 (2010).

[9] Y. M. Song, Y. Xie, V. Malyarchuk, J. Xiao, I. Jung, K. J. Choi, Z. Liu, H. Park, C. Lu, R. H. Kim, R. Li, K. B. Crozier, Y. Huang, and J. A. Rogers, "Digital cameras with designs inspired by the arthropod eye," Nature 497(7447), 95-99 (2013).

[10] 10. T. Wu, S. S. Hamann, A. Ceballos, O. Solgaard, and R. T. Howe, "Design and fabrication of curved silicon image planes for miniature monocentric imagers," in 2015 18th International Conference on Solid-State Sensors, Actuators and Microsystems, TRANSDUCERS 2015 (2015), pp. 2073-2076.

[11] R. Dinyari, S. B. Rim, K. Huang, P. B. Catrysse, and P. Peumans, "Curving monolithic silicon for nonplanar focal plane array applications," Appl. Phys. Lett. 92(9), 091114 (2008).

[12] K. Itonaga, T. Arimura, K. Matsumoto, G. Kondo, K. Terahata, S. Makimoto, M. Baba, Y. Honda, S. Bori, T. Kai, K.Kasahara, M. Nagano, M. Kimura, Y. Kinoshita, E. Kishida, T. Baba, S. Baba, Y. Nomura, N. Tanabe, N. Kimizuka, Y. Matoba, T. Takachi, E. Takagi, T. Haruta, N. Ikebe, K. Matsuda, T. Niimi, T. Ezaki, and T. Hirayama, "A novel curved CMOS image sensor integrated with imaging system," Digest of Technical Papers - Symposium on VLSI Technology, 6894341 (2014).

[13] B. Guenter, N. Joshi, R. Stoakley, A. Keefe, K. Geary, R. Freeman, J. Hundley, P. Patterson, D. Hammon, G. Herrera, E. Sherman, A. Nowak, R. Schubert, P. Brewer, L. Yang, R. Mott and G Mcknight, "Highly Curved Image Sensors: A Practical Approach for Improved Optical Performance." Optics Express 25(12), 13010 (2017).

[14] K. Tekaya, M. Fendler, K. Inal, E. Massoni, and H. Ribot, "Mechanical behavior of flexible silicon devices curved in spherical configurations," in 2013 14th International Conference on Thermal, Mechanical and Multi- Physics Simulation and Experiments in Microelectronics and Microsystems (EuroSimE, 2013), paper 6529978.

[15] D. Dumas, M. Fendler, N Baier, J. Primot, and E. Le Coarer, "Curved focal plane detector array for wide field cameras," Applied Optics 51(22), pp. 5419-5424 (2012).

[16] B. Chambion, C. Gaschet, T. Behaghel, A. Vandeneynde, S. Caplet, S. Gétin, D. Henry, E. Hugot, W. Jahn, S. Lombardo, M. Ferrari, "Curved sensors for compact high-resolution wide-field designs: prototype demonstration and optical characterization", Proc. SPIE 10539, 1053913 (2018).

[17] B. Chambion, L. Nikitushkina, Y. Gaeremynck and W. Jahn, "Tunable curvature of large visible CMOS image sensors: Towards new optical functions and system miniaturization," IEEE ECTC 178-187 (2016). Research Article Applied Optics 9

[18] W. Jahn, M. Ferrari, E. Hugot, B. Chambion, G. Moulin, L. Nikitushkina, C. Gaschet, D. Henry, S. Getin, and Y.Gaereminck, âAœFlexible focal plane arrays for UVOIR wide field instrumentation,âAI Proc. SPIE 10562, 105624Z (2016).

[19] J.-H. Zhao, J. Tellkamp, V. Gupta and D. R. Edwards, "Experimental evaluations of the strength of silicon die by 3-point-bend versus ball-on-ring tests", IEEE Trans. Electron. Packag. Manuf., vol.32(4), pp.248-255 (2009).

[20] EMVA Standard 1288, Release 3.1, December 30, 2016 Available https://www.emva.org/wp-content/uploads/EMVA1288-3.1a.pdf (accessed 2018-06-01)

[21] Janesick, J. R.: Photon Transfer, SPIE Press, doi: 10.1117/3.725073, 2007

[22] Stan Birchfield. Reverse-projection method for measuring camera MTF. Electronic Imaging, 2017(12) :105–112, 2017.

[23] C. Gaschet, W. Jahn, B. Chambion, E. Hugot, T. Behaghel, S. Lombardo, S. Lemared, M. Ferrari, S. Caplet, S. Gétin, A. Vandeneynde, and D. Henry, "Methodology to design optical systems with curved sensors," Appl. Opt. 58, 973-978 (2019).

978-1-7281-1500-9/19 $31.00 © 2019 IEEE

Integration and Characterization of InP Die on Silicon Interconnect Fabric

Eric Sorensen[1], Boris Vaisband[1], SivaChandra Jangam[1], Tim Shirley[2], and Subramanian S. Iyer[1]

[1]Center for Heterogeneous Integration and Performance Scaling (CHIPS)
University of California, Los Angeles, CA 90095
[2]Keysight Technologies, Santa Rosa, CA 95403

Email: esorensen@ucla.edu

Abstract—The silicon interconnect fabric (Si-IF) is a wafer-level packaging platform that enables heterogeneous integration of die at ultra-fine pitch (2 to 10 μm) directly onto a lithographically defined silicon wafer with no intermediate packaging hierarchy. The die are attached with an extremely tight inter-dielet spacing (< 100 μm). The small inter-dielet spacing is especially advantageous in high frequency applications due to reduced loss associated with the transmission line behavior of off-chip interconnects. Since indium phosphide (InP) is a popular technology choice for high frequency applications, the goal of this paper is to investigate the efficacy of direct Au-Au thermo-compression bonding (TCB) of InP die to the Si-IF platform for the first time. To evaluate this process, 84 InP die were successfully bonded to the Si-IF. The sheer strength of the integrated die ranges from 38 MPa to 238 MPa, for die that were attached using pressure ranging, respectively, from 100 MPa to 350 MPa. Daisy chain resistance of the bonded die was measured exhibiting good correlation with calculated theoretical values. After thermal cycling, it was found that 100% of the attached die withstood all thermal stressing despite the thermal mismatch of 2 ppm/K between the die and the Si-IF.

Keywords–Silicon interconnect fabric (Si-IF); fine pitch interconnects; thermo-compression bonding; high frequency integration; wafer-scale packaging; system-on-wafer; heterogeneous integration

I. INTRODUCTION

It is known that power dissipation associated with communication based on serializer/deserializer (SerDes) circuitry, increases exponentially with increasing frequency [1]. Because of this, I/O power in a typical system, which is dominated by SerDes, accounts for a minimum of 30% of the total system power (Figure 1). This problem is exacerbated as clock speeds increase because of the exponential relationship between frequency and SerDes power [2]. There is a need, therefore, for increased parallel connectivity between monolithic microwave integrated circuits (MMICs) and silicon devices as such systems become more complex and require more I/Os. This is one of the key motivations for heterogeneous integration.

Another important barrier for high frequency technologies is the high communication loss and increased latency that is introduced by the off-chip communication links. Due to large discrepancies between chip and package area, and the need for multiple levels of routing driven by the coarse trace pitch on printed circuit boards (PCBs), off-chip links can be several centimeters in length. As a result of this relatively long length,

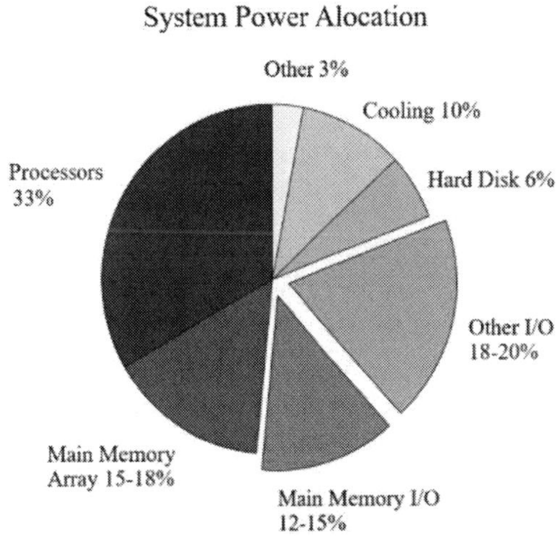

Figure 1 A pie chart depicting the allocation of power in a typical system. Notice that I/O power accounts for at least 30% of the total power consumption.

the inductance of the lines becomes significant, and they exhibit transmission line behavior. These long communication links can result in high losses and increased latency for inter-die communication. Often SerDes are employed to ensure signal integrity by providing channel equalization, boosting signals, and providing error correction; however, the use of SerDes comes at the expense of additional chip area, increased power consumption, and increased latency [3].

The silicon interconnect fabric (Si-IF) is a platform for package-less, wafer-scale heterogeneous integration that enables highly parallel communication between adjacent die at low inter-dielet spacing. The Si-IF approach eliminates the need for SerDes and reduces the power loss and latency associated with long off-chip communication links. The traditional organic packaging, which is at the root of the limited I/O count, is eliminated altogether, and bare die are bonded directly to a silicon substrate at an ultra-fine pitch (2 to 10 μm). This approach enables a greatly increased number of I/Os which eliminate the dependence on SerDes [4].

Due to the extremely high cutoff frequency (~1 THz) of indium phosphide (InP) based heterojunction bipolar transistors (HBTs) and high-electron-mobility transistors (HEMTs), InP substrates are gaining popularity for use in high frequency 5G applications [5]. Because of their high-speed performance and their increasing popularity, InP die are excellent candidates for heterogeneous integration using the

Si-IF platform. Unfortunately, there is a linear thermal expansion mismatch between silicon and III-V materials (2 ppm/K for InP vs. silicon and 3.13 ppm/K for GaAs vs. silicon [6]) giving rise to concerns regarding the reliability of the bonds between the III-V die and the silicon substrate under thermal stressing. Another concern is the lower mechanical strength of such die as InP substrates are often thinned to allow for better heat dissipation. While other investigators have also focused on the efficacy of integrating InP-based devices onto silicon [7]-[9], their approaches do not offer the same simplicity and flexibility as the Si-IF approach to heterogeneous integration. This paper will discuss experiments that were performed to test the integrity of direct Au-Au thermo-compression bonds (TCBs) between individual bare InP die and a Si-IF platform before and after annealing and 500 rounds of thermal cycling. The work described in this paper, *i.e.*, InP die successfully integrated for the first time on the Si-IF, demonstrates the viability of the Si-IF as a platform for true heterogeneous integration of systems that are either newly designed, or assembled using off-the-shelf silicon and III-V components.

The rest of the paper is composed of the following topics: Traditional packaging concepts and a comparison to the Si-IF approach is presented in Section II. The fabrication process of the InP die and the Si-IF wafers employed in the experiments is described in Section III. Testing procedures and experimental results are provided in Section IV. Finally, conclusions and future work are offered in Section V.

II. BACKGROUND AND MOTIVATION

Traditionally, system integration has been achieved by attachment of components to PCBs using ball grid array (BGA) solder balls, which support a minimum pitch of 400 μm. To accommodate this relatively large pitch, individual die are either wire bonded or flip-chip attached via controlled collapse chip connect (C4) bumps to a packaging laminate. The laminate is, effectively, used as a space transformer to redistribute the fine on-chip pad pitch (last metal level) to the coarse pitch BGAs and traces on the PCB. Although C4 bumps support a smaller minimum bonding pitch (~150 μm) than BGAs, their pitch is still more than ten times larger than the pitch of the terminal wiring level on most integrated circuits (2 to 10 μm). These large bonding pitches limit the number of I/Os available in this type of packaging approach. The large package to chip area ratio, resulting from the redistribution laminate, not only increases the real estate required by each component, but also leads to the high inter-dielet spacing (>10 mm) [2]. An image of a typical organic package is shown in Figure 2.

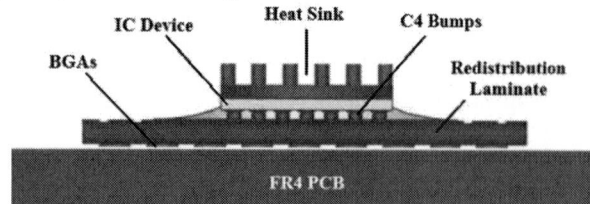

Figure 2 Schematic of a typical organic package.

Another drawback to the traditional organic packaging approach is derived from an empirical trend in system design known as Rent's rule. Rent's rule dictates that the number of necessary I/Os is related to the number of gates on a die as follows

$$P = K \cdot G^r, \tag{1}$$

where P is the number of I/O pins, G is the number of gates on the die, and K and r are constants between 0 and 1 dictated by the class of circuit the die belongs to [10]. Since the 1960s, devices have scaled according to Moore's law by over 1,000x, whereas package dimensions have only scaled by 5x (figure 3). This discrepancy resulted in a lower I/O count on the package than the number that is prescribed by Rent's rule since G has increased significantly faster than P. Moreover, this problem is exacerbated by the fact that up to 90% of available I/O pins are dedicated to power delivery, further reducing the number of available I/Os for signaling [2].

Currently, in most applications, SerDes have been the immediate solution to the I/O problem posed by Rent's rule. To maintain a constant data rate, the frequency of a SerDes that is used to serialize n signals, must be increased n-fold [1]. The increase in the frequency of the SerDes leads to a comparable increase in power consumption, rendering the SerDes a power and area hungry component. An alternative solution to the I/O problem posed by Rent's rule is, therefore, required as operating frequencies and I/O counts increase.

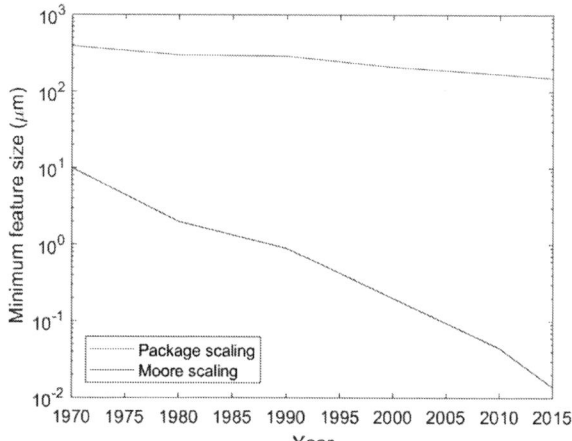

Figure 3 A logarithmic plot showing that CMOS scaling according to Moore's law (shown in blue) has progressed much faster than package scaling (shown in red).

To increase integration and limit the use of SerDes, innovative packaging solutions that offer die attach capabilities at much tighter pitches have been devised. One such approach is an interposer consisting of a thinned silicon substrate that supports attachment of several die at a pitch as low as 40 μm utilizing C2 (chip connect) bumps. Interposers are then bonded to a laminate using C4 bumps and this laminate is bonded to a PCB with BGAs. While this approach goes a long way towards increasing the bandwidth between die on the interposer, it does not address the problem of increasing the number of IOs available for off-interposer communication [2].

The silicon interconnect fabric (Si-IF) eliminates the requirement for a PCB and supports integration of an entire system of unpackaged die and passives onto a full thickness silicon wafer. The die are bonded at a pitch matching that of the terminal wiring layer of most ICs (2 to 10 μm), and can be attached at an extremely tight inter-dielet spacing (<100 μm) [11]. The ultra-fine pitch is made possible by direct bonding using Cu or Au pillars, whereas the tight inter-dielet spacing is realized by abandoning the organic packaging used to fan out connections. The fine bond pitch supports an extremely large communication bandwidth (several Tbps) at low energy per bit (<0.4 pJ/bit). These benefits can be attributed to the large number of short parallel inter-dielet links which increase the number of available physical I/Os [12]. The tight inter-dielet spacing results in low latency (50-100 ps) between neighboring die, and studies have shown that the communication links between die are short enough that they exhibit single pole transfer characteristics consistent with on-chip RC links [13]. These advantages show that the Si-IF approach is ideal for future high frequency technologies, further emphasizing the importance of verifying the compatibility between III-V die and the Si-IF.

III. FABRICATED STRUCTURES AND TESTING PROCEDURE

Five different types of InP die (all thinned to 100 μm thickness) were fabricated for these experiments. Three of these die types are 2x2 mm^2, fully populated with arrays of 2x2, 3x3, and 4x4 pad bundles named, respectively, FC2, FC3, and FC4 (where FC stands for flip-chip). The other two die types are 1.2x1.2 mm^2 and 1.2x2.2 mm^2, with 7x7 pad bundles along the periphery only named, respectively, FCPP2 and FCPP3 (FCPP stands for flip-chip periphery pad). An x by y pad bundle is a region similar to a traditional bond pad where an x by y array of Si-IF pillars are attached to the IC during assembly. Fabricated on the die are 2 μm thick evaporated gold traces, passivated with 1 μm of SiN that has been removed at the locations of the pad bundles. The interconnect fabric (Figure 4) incorporates 3.5 μm thick gold traces and 5x5x5 μm^3 gold pillars deposited using a semi-additive electroplating process. The RMS surface roughness of the pillars is roughly 40 nm.

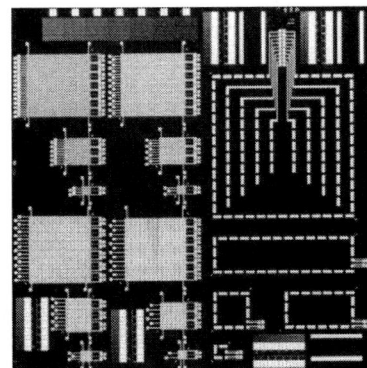

Figure 4 A reticle of the Si-IF used in this experiment.

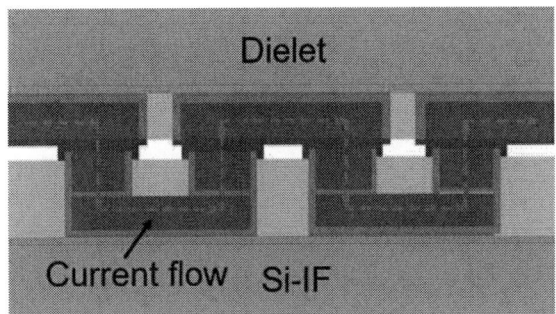

Figure 5 A daisy chain interconnect structure that is typically used for testing the reliability of bonds between two substrates.

The pillars on the Si-IF correspond to the locations of the exposed pads on the die. Both the Si-IF and the die were designed such that after bonding, they formed daisy chain interconnect structures. A daisy chain interconnect structure (Figure 5) typically consists of two sets of dashed traces, one on the substrate and one on the die. They are offset such that when a die is bonded to the substrate, the traces on the die bridge the gaps between interconnects on the substrate to form a continuous path. The resistance of these daisy chains can then be used to determine the integrity of the bond since a correctly bonded substrate would have a continuous conduction path. Abnormally resistive or open daisy chains are classified as a bonding failure, and resistances close to the theoretical values are an indication of successfully bonded die. Four-point Kelvin contacts were fabricated on the Si-IF to ensure accurate resistance measurements (Figure 6).

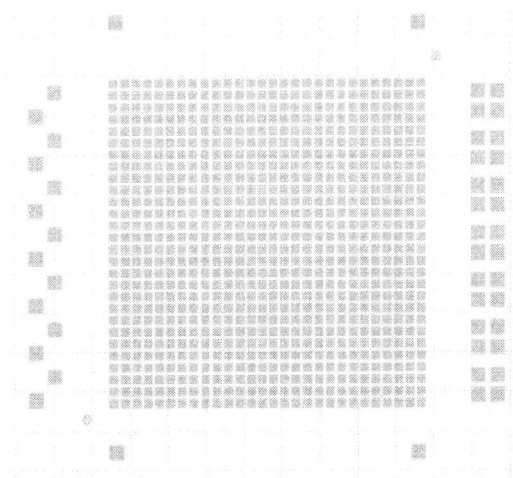

Figure 6 A typical footprint of an FC chip. The large 4 by 4 squares on the right are Kelvin contacts.

Fewer than 24 hours prior to bonding, an Oxford reactive ion etcher was used to do an argon plasma pretreatment of the die and the SI-IF. A recipe with a base pressure of 0.01 Torr, an argon flow rate of 30 sccm, and an RF power of 40 W at 13.56 MHz was used. The purpose of this pretreatment step was to remove any oxides or organic contaminants found at the gold surface which may have inhibited bonding. Studies have shown that this step is crucial for achieving quality bonding using TCB [15].

The die were bonded to the Si-IF individually using a K&S APAMA flip-chip thermal compression bonder. The Si-IF substrate was heated to 150ºC while the IC and bond head were heated to 350ºC, resulting in a temperature of approximately 250ºC at the interface between the Si-IF and the InP die. FC2-FC4 were bonded using pressure ranging from 100 to 400 MPa with 50 MPa increments, and FCPP2-FCPP3 were bonded at, respectively, 350 MPa and 300 MPa. To evaluate the influence of bonding time on the reliability of the bonds, FC2 die were bonded for 10 seconds, FC3 die for 15 seconds, and FC4, FCPP2, and FCPP3 die were all bonded for 20 seconds.

An annealing test was also performed on the bonded structure. The Si-IF was annealed for one hour at 250ºC followed by another hour of annealing at 350ºC. Finally, the integrated system was subjected to 500 iterations of thermal cycling at temperatures ranging from -65ºC to 150ºC. Each thermal cycle had a temperature ramping rate of 20ºC/min and dwells of 6 minutes at each of the thermal extrema. The resistance of the daisy chain structures was measured immediately after bonding and then again after each successive thermal treatment.

A separate Si-IF of identical make up was fabricated on which 17 FC3 die were bonded for 15 seconds with pressure ranging from 100 MPa to 350 MPa at 50 MPa increments. The resistance of these daisy chains was then measured to determine the connectivity of the die. Six of the attached die were sheared using a mechanical shearing tool at a shearing height of 7 µm and shearing speed of 100 µm/s.

IV. TEST RESULTS

84 InP die were successfully bonded to the Si-IF, as shown in Figure 7. Die that were not successfully bonded failed due to various reasons that will be discussed in this section. Die classified as not bonded typically exhibited an open circuit behavior during the first round of electrical measurement. Meanwhile, die that were correctly bonded, had resistances very close to the theoretical resistance (Figure 8) and showed a consistent linear increase in resistance as additional rows of the daisy chain were considered, implying a consistent resistance for each row (Figure 9).

Figure 7 A photograph of the Si-IF with bonded InP die visible.

The theoretical resistance of the daisy chains between the different die and the Si-IF was estimated by using the bulk resistivity of Au and a current crowding approximation [14]. Both the theoretical and measured resistance of the daisy chains are shown in Figure 8. Although the worst case error is 31.4%, this error is obtained for very small absolute resistance difference (up to 1.5 Ω for a daisy chain across an entire die). The measured error is attributed to the current crowding approximation and the discrepancy between the actual resistivity of the plated and evaporated Au and the bulk resistivity.

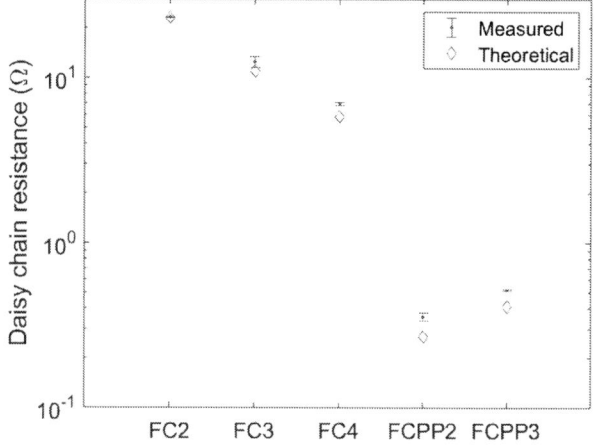

Figure 8 A logarithmic plot of the measured resistance compared to the theoretical calculations.

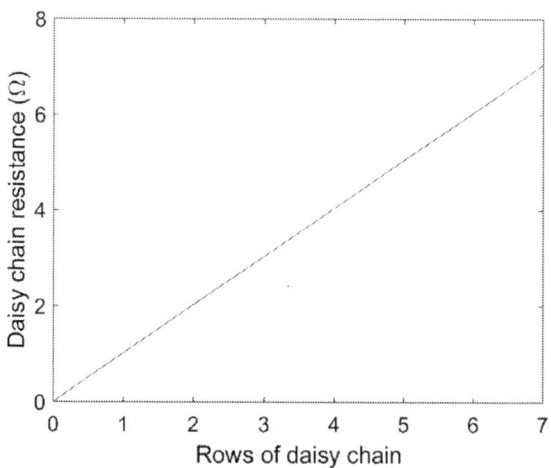

Figure 9 Resistance as a function of the row count for a typical FC4 die exhibiting a linear nature.

The initial quality of bonding was found to be highly sensitive to bonding time. Die bonded for longer periods of time tended to exhibit a much higher success rate as compared to die that were bonded for shorter periods. For instance, only 2 of the 35 FC2 die, which were bonded for 10 seconds, were bonded successfully. As for the FC3 die, 13 out of 35 die were successfully attached after bonding for 15 seconds. Finally, 26 out of the 34 FC4 chips were attached successfully after bonding for 20 seconds. FCPP3 die, which were also bonded for 20 seconds, had a 100% attachment success rate – 29 out of 29 die bonded successfully. Although only 9 out of 30 FCPP2 die were bonded successfully, this low yield can likely be attributed to issues related to planarity during bonding. The dependence of successful bonding on time is expected since successful TCB relies on surface diffusion to reduce the size of voids at the bonding interface that originate from surface roughness [15].

The success rate of bonding was also dramatically impacted by bonding pressure. This trend can be seen most clearly by examining the results of FC3 measurements, though it was true of FC2 and FC4 die as well. Only 1 out of 5 FC3 die were successfully attached at 150 MPa, whereas 3 out of 3 FC3 die attached successfully at 400 MPa. It was also found that the resistance of successfully bonded die was dependent on the bonding pressure (Figure 10). Passing die that were bonded at a higher pressure tended to have a lower resistance than did die bonded at lower pressure.

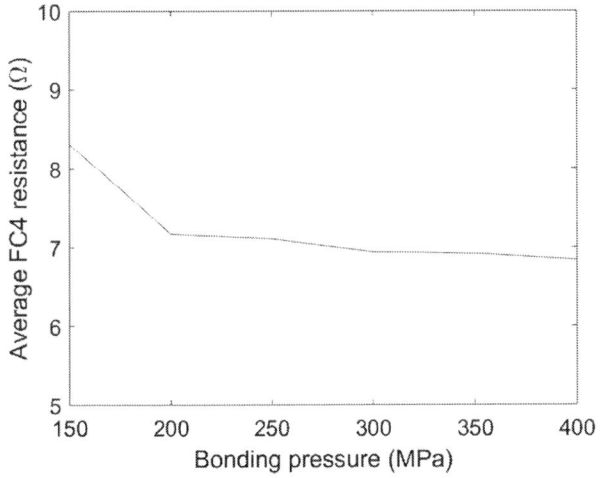

Figure 10 Relationship between bonding pressure and the average resistance of successfully bonded FC4 die.

Of the 84 successfully bonded die, 100% withstood all thermal processing steps despite the thermal mismatch of 2 ppm/K between the die and the Si-IF. This was determined by examining box and whisker plots (Figure 11) of the measured resistance of the die after these steps. While the overall resistance of the die does increase slightly with thermal cycling, this does not appear to be indicative of failures since the variation from baseline testing is very small. This is impressive considering that the maximum expansion mismatch between 2x2 mm^2 die and the wafer due to a bonding temperature of 250ºC and minimum thermal cycle temperature of -65ºC was on the order of 1 µm – corresponding to 20% of the pillar height. Nonetheless, the pillars withstood the effects of thermal cycling indicating that the Si-IF assembly is reliable. The thermal cycles were, however, not completely without effect. Note, in Figure 11, the effect of the annealing steps was not necessarily a

reduction of the median resistance, but instead a tightening of the range of the measured resistance values. The thermal cycling steps, however, seem to have the opposite effect, causing a broadening of the resistance distribution. This result may imply that some of the die suffered from incomplete pillar failures that left connectivity intact. Further statistical level testing is required to verify this.

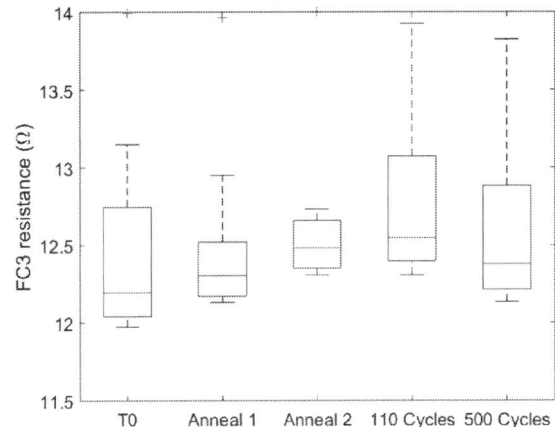

Figure 11 A box and whisker plot depicting the resistance of FC3 die across various thermal treatments.

Out of the 17 FC3 die that were bonded to the second wafer, only one was found to be completely connected. At higher pressure (200-300 MPa), the disconnected die had only 1 out of 6 daisy chain rows fail and this row was consistently near the edge of the die. At lower pressure (100 and 150 MPa) however, it was found that 2 to 5 rows on each die were disconnected. After this measurement, one die from each bonding pressure group was sheared. The shear strength was then calculated from the maximum shearing force by normalizing it to an ideal bond area of 250,000 µm^2. The results show that there is a clear relationship between the shear strength of the die and the bonding force (Figure 12), with greater bond forces resulting in higher shear strength, as expected.

An examination of the die after shearing (Figure 13) reveals the underlying cause of both the poor connectivity and the relatively low shear strength at low bonding pressure. As bonding pressure decreases, the effective bonding area is reduced. This reduction in bond area starts from the corners and works its way into the center of the die as pressure is reduced. Furthermore, even die that were bonded at higher pressure, seem to exhibit some degree of bonding failure at the corners. This corresponds to the finding that many of them had at least one failing row close to the edge of the die, which was the dominant failure mode in the original experiment as well. This is likely the result of a lack of back pressure in these regions during bonding due to the presence of relatively large vacuum holes (~500 µm) on the bond head located in the immediate vicinity of the die corners (Figure 14). To supply the bonding pressure in these regions, the corners of the die must act as cantilevers and the force they generate is evidently insufficient. During bonding of full thickness silicon die, this has not been an issue since the die have had enough mechanical strength to compensate for this problem. It is clear, however, that thinned III-V die will require modified bonding heads with smaller vacuum holes.

In addition, InP die could be initially warped leading to lower success rate of attachment. However, characterization of the die before bonding was not performed.

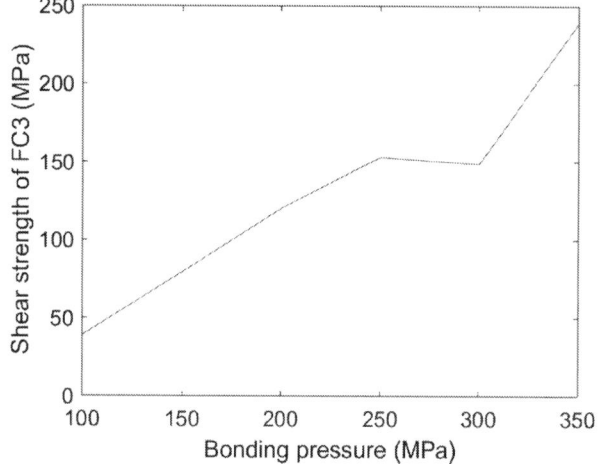

Figure 12 Shear strength vs. bonding pressure for FC3 die. Note, shear strength increases along with bonding pressure, as expected.

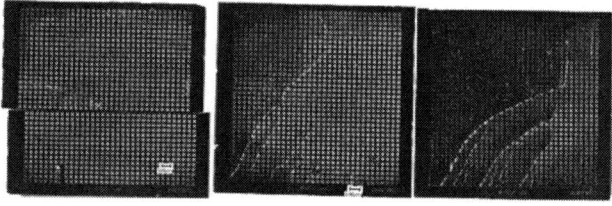

Figure 13 Sheared die bonded at various pressures (350 MPa left, 200 MPa center, and 100 MPa right). Note that the effective bond area becomes smaller as bonding pressure decreases, as can be seen from the darkening of the die from left to right.

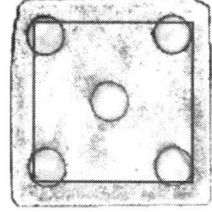

Figure 14 A 2.45x2.45 mm^2 vacuum head with a typical 2x2 mm^2 die footprint overlaid in red. Notice that the corners of the die are not fully covered by the vacuum holes.

V. CONCLUSIONS AND FUTURE WORK

Despite the die attachment failures due to the presence of relatively large diameter vacuum holes on the bonding head, the results of this study are promising. InP die that were successfully bonded withstood all subsequent thermal stressing. This means that despite the 2 ppm/K thermal

expansion mismatch between InP and silicon, it is possible to create stable and reliable bonds at an ultra-fine pitch. This type of bonding in the future will enable truly heterogeneous integration of III-V MMICs and silicon ICs.

The shear strength and resistance of the bonded dies was also evaluated. The shear strength results show very strong attachment, especially for die bonded at higher pressures. The sheer strength of the attached die is 38 MPa, 79 MPa, 120 MPa, 153 MPa, and 238 MPa, for die that were attached using, respectively, 100 MPa, 150 MPa, 200 MPa, 250 MPa, and 350 MPa bonding pressures. The attached die exhibited a low measured resistance (up to 1.5 Ω across an entire die), approaching the theoretical minimum values.

Further statistical level testing is needed to increase confidence in this preliminary evaluation. New samples will be fabricated and the effect of bonding pressure and time on yield will be explored. New bonding heads with smaller diameter vacuum holes are expected to improve these results dramatically. In addition, damascene Si-IF samples are also being fabricated. It is expected that the lower roughness and increased planarity of Si-IFs fabricated using a damascene process (RMS roughness ~1 nm) will result in higher yielding bonds at lower bonding time and pressure. In addition, characterization of the die before bonding will be performed.

It will also be necessary to conduct similar reliability testing with GaAs substrates since they exhibit a greater thermal expansion mismatch with silicon and are also extremely popular for use in high frequency technologies. Finally, once the reliability of the bonding process between III-V die and the Si-IF platform has been properly assessed, it will be critical to conduct RF measurements on III-V structures bonded to Si-IFs to determine the scattering parameters of links on the Si-IF.

ACKNOWLEDGMENT

This work was supported by the industrial sponsors of the UCLA CHIPS Consortium. The authors would like thank Keysight for their support throughout this work.

REFERENCES

[1] J. D. Rockrohr, A. Mohammad, C. R. Ogilvie, K. Dramstad, M. A. Sorna, J. T. Mechler, and D. R. Stauffer, *High Speed Serdes Devices and Applications.* Boston, MA: Springer US, 2009.

[2] S. S. Iyer, "Heterogeneous Integration for Performance and Scaling," *IEEE Transactions on Components, Packaging and Manufacturing Technology*, Vol. 6, No. 7, pp. 973–982, 2016

[3] R. Bansal, *Handbook of engineering electromagnetics: applications.* New York: Marcel Dekker, 2006.

[4] A. A. Bajwa, S. C. Jangam, S. Pal, N. Marathe, T. Bai, T. Fukushima, M. Goorsky, and S. S. Iyer, "Heterogeneous Integration at Fine Pitch (≤ 10 µm) using Thermal Compression Bonding", *Proceedings of the IEEE Electronic Components and Technology Conference*, Orlando, FL, pp. 1276-1284, 2017.

[5] M. Urteaga, M. Seo, J. Hacker, Z. Griffith, A. Young, R. Pierson, P. Rowell, A. Skalare, M. J. W. Rodwell, "InP HBT integrated circuit technology for terahertz frequencies", *Proceedings of the IEEE Compound Semiconductor IC Symposium*, pp. 1-4, 2010-Oct.

[6] A. Dargys and J. Kundrotas, *Handbook on physical properties of Ge, Si, GaAs and InP*. Vilnius, Lithuania: Science and Encyclopedia Publishers, 1994.

[7] Z. Wang, M. Pantouvaki, G. Morthier, C. Merckling, J. van Campenhout, D. van Thourhout, G. Roelkens, "Heterogeneous Integration of InP devices on silicon", *Compound Semiconductor Week (CSW)*, June 2016.

[8] O. Moutanabbir, U. Gösele, "Heterogeneous integration of compound semiconductors", *Annual Review of Materials Research*, pp. 469-500, 2010.

[9] -Aitken, P. Chang-Chien, D. Scott, K. Hennig, E. Kaneshiro, P. Nam, N. Cohen, D. Ching, K. Thai, B. Oyama, J. Zhou, C. Geiger, B. Poust, M. Parlee, R. Sandhu, W. Phan, A. Oki, R. Kagiwada, "Advanced Heterogeneous Integration of InP HBT and CMOS Si Technologies", *Proceedings of the IEEE Compound Semiconductor Integrated Circuit Symposium*, pp. 1-4, Oct 2010.

[10] B. Landman and R. Russo, "On a Pin Versus Block Relationship For Partitions of Logic Graphs," *IEEE Transactions on Computers*, Vol. C-20, No. 12, pp. 1469–1479, 1971.

[11] A. A. Bajwa, S. Jangam, S. Pal, B. Vaisband, R. Irwin, M. Goorsky, and S. S. Iyer, "Demonstration of a Heterogeneously Integrated System-on-Wafer (SoW) Assembly," *P ceedings of the IEEE Electronic Components and Technology Conference*, 2018.

[12] S. Jangam, S. Pal, A. Bajwa, S. Pamarti, P. Gupta, and S. S. Iyer, "Latency, Bandwidth and Power Benefits of the SuperCHIPS Integration Scheme," *Proceedings of the IEEE Electronic Components and Technology Conference*, 2017.

[13] S. Jangam, A. A. Bajwa, K. K. Thankkappan, P. Kittur, and S. S. Iyer,

Interconnects in Silicon-Interconnect Fabric," *Proceedings of the IEEE Electronic Components and Technology Conference*, 2018.

[14] M. Horowitz and R. W. Dutton, "Resistance Extraction from Mask Layout Data," *IEEE Transactions on Computer-Aided Design of Integrated Circuits and Systems*, Vol. 2, No. 3, pp. 145 – 150, July 1983.

[15] Y.-H. Wang and T. Suga, "Metal surface cleanliness and its improvement on bonding," *2010 11th International Conference on Electronic Packaging Technology & High Density Packaging*, 2010.

Y-Branched Multimode/Single-Mode Polymer Optical Waveguides for Low-Loss WDM MUX Device: Fabrication and Characterization

Takaaki Ishigure, Tomoki Nakayama, Fukino Nakazaki, and Hiroki Hama
Faculty of Science and Technology, Keio University
Yokohama, Japan
e-mail: ishigure@appi.keio.ac.jp

Abstract— In this paper, we represent low-loss Y-branched multimode polymer optical waveguides with graded-index (GI) cores applicable to an optical coupler as a MUX device in coarse wavelength division multiplexing links. We apply the Mosquito method as well as the imprint method to form the Y-branched structures in polymers. We experimentally confirm that the insertion loss (including the coupling, propagation, bending, and multiplexing losses) in the multimode GI-core Y-branched waveguide is approximately 1-dB lower than the same Y-branched waveguide with step-index (SI) core. In addition, we also succeeded in fabricating Y-branched single-mode polymer waveguides applying the Mosquito method. In order to satisfy the single-mode condition, the core diameter needs to decrease to 10 μm and less.

Keywords-component; polymer optical waveguide; coarse wavelength division multiplexing (CWDM); the Mosquito method

I. INTRODUCTION

Over the last couple of decades, optical fiber links have replaced the legacy electrical wiring even in short-reach networks. For instance, 50-μm core multimode fibers (MMFs) are widely deployed in short-reach networks in high-performance computers and datacenter at a data rate of 10 Gbps, which is one of the Ethernet standards (10Gbase-SR). To upgrade the link performance, it has been already confirmed by several groups that current MMF links can support a data rate as high as 28 Gbps for 100-m distance by improving the modal dispersion of fibers and refining the launch conditions. However, the Ethernet standard next to 10 Gbps had to be 40 or 100 Gbps. Furthermore, in these days, the network architecture of datacenter is shifting from legacy fat-tree type to leaf-spine type in order to address the bottleneck against the increasing data traffic. So, some large-scale datacenters start employing even single-mode fibers (SMFs) with an approximately10-μm core diameter in order to meet the demand for higher bandwidth distance product of the link.

Meantime, for covering such a high data rate as 100 Gbps and beyond, *coarse wavelength division multiplexing* (CWDM) technology is employed [1], by which 75- to 150-m long MMF links (corresponding to OM3 to OM5, respectively). In the CWDM MMF links, directly modulated vertical cavity surface emitting lasers (VCSELs) are utilized in the optical transmitter, where the emitting wavelengths of these VCSELS are around 850-nm region with slight differences of 30-nm channel spacing. Since wide tolerances in the position alignment have been the key to allow low cost optical devices for the large core MMF links, simple and low loss multiplex/demultiplex (MUX/DEMUX) devices are highly required on the transceivers for the CWDM-MMF links.

As these MMF links are supposed to be installed in the inter-rack connections between servers and switches, optical transceivers are normally placed at the edge of the racks. Meanwhile, the copper links between the chip on board and the optical transceivers at the board edge have been a concern from the bandwidth density point of view. To replace the copper wires on the board for optical circuits, polymer optical waveguides are drawing attentions. In this configuration, the optical transceivers are able to be placed as close to the chips as possible, and then, polymer optical waveguides connect the transceivers and MMFs at the board edge. Therefore, polymer waveguides should be multimode or single mode to connect them to MMF or SMF, respectively, with high coupling efficiencies. In this paper, we focus on multimode polymer waveguides for the MUX device in CWDM-MMF links, first.

In the past ECTC conferences [2,3], we have demonstrated a very simple fabrication technique for polymer waveguides named the Mosquito method, by which an almost the same waveguiding structure as MMF (50 μm circular GI core) can be formed in polymer. Therefore, we have emphasized that the coupling loss between the waveguide and MMF can be remarkably reduced as well as the propagation loss.

In this paper, we represent low-loss Y-branched multimode polymer optical waveguides with graded-index (GI) cores applicable to an optical coupler as a MUX device in CWDM links. We apply the Mosquito method as well as the imprint method to form the Y-branched structures in polymers. We experimentally confirm that the insertion loss (including the coupling, propagation, bending, and multiplexing losses) in the multimode GI-core Y-branched waveguide is approximately 1-dB lower than the same Y-branched waveguide with step-index (SI) core.

In addition, we also succeeded in fabricating Y-branched single-mode polymer waveguides applying the Mosquito method. In order to satisfy the single-mode condition, the core diameter needs to decrease to 10 μm and less.

II. DESIGN FOF LOW-LOSS MULTIMODE COUPLER

The optimum structure of the Y-branched waveguide to work as a low-loss MUX device is designed using theoretical

978-1-7281-1500-9/19 $31.00 © 2019 IEEE

simulation applying a beam-propagation method (BPM) [4]. In the simulation, we suppose the signal light is coupled to one of the two cores in two-port side of a Y-branched waveguide and propagates through the core to the single port end. Then, the output light from the waveguide is coupled to a GI circular core MMF (50GI-MMF).

The light propagations through a variety of Y-branched structures are simulated. The simulation model is shown in Fig. 1. Here, square cores whose refractive index distribution $n (x, y)$ is approximated applying the power-law form well known to express the index profile in a circular core. When the core center is placed on the origin (0, 0) in the x-y coordinate, the profile is described by Eq. (1).

$$n(x,y) = n_1 \left[1 - 2\Delta \left(\left|\frac{x}{a}\right|^g + \left|\frac{y}{a}\right|^g - \left|\frac{xy}{a^2}\right|^g\right)\right]^{\frac{1}{2}} \quad (1)$$

$$\text{Relative index difference} \Delta = \frac{n_1^2 - n_2^2}{2n_1^2} \quad (2)$$

Here, n_1 and n_2 are the refractive indices of the core and cladding, respectively, a is the half width (height) of the core, and g is the index exponent. The values of n_1 and n_2 shown in Table 1 are based on the measured values of the materials at 850 nm, that are used for the waveguide fabrication. The refractive index distribution expressed by equation Eq. (1) is appropriate for approximation for square core so that the refractive index distribution spreads symmetrically in the longitudinal, transverse and oblique directions in rectangular core cross sections [4]. The index exponent is set to $g = 2$, which provides the ideal refractive index distribution in the case of circular core.

Under these conditions, the insertion loss is calculated for the three different multiplexing structures: V, U, and S shape, and already found that the S-shape provided the lowest insertion losses. The detailed structural designs are already expressed in our previous publication[5]. As shown in Fig. 1, the multiplexing length L is largely varied, which is defined as the length required to merge the two cores into one core. The calculated insertion loss of the GI square core Y-branched waveguides with S-shaped structures over the multiplexing length L is shown in Fig. 2, where the loss of square SI core counterparts is compared.

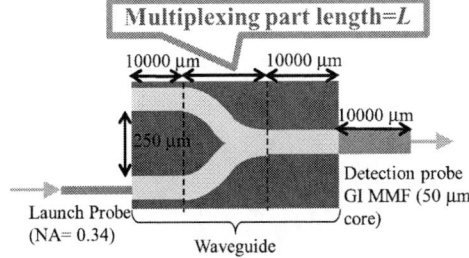

Figure 1. Y-branched waveguide design for simulating the insertion loss.

TABLE I. SIMULATION CONDITIONS

n_1	1.584
n_2	1.557
Core size	45 μm×45 μm
Launch condition	Wavelength 850 nm NA = 0.34 Gaussian beam Beam spot radius 2.2 μm

It is noted that such a huge insertion loss as 10 to 40 dB is observed when L is shorter than 2,000 μm (2 mm) in both SI and GI core waveguides. This is because the multiplexing length L is so short that the small bending radius and large bending angle are required, resulting in high bending loss. Furthermore, the core has an axially tapered shape in the multiplexing region in order to merge two cores in one. Hence, if L is not large enough, the taper angle needs to be large to cause light leakage. Meanwhile, if L exceeds 3,200 μm (3.2 mm), the insertion loss of both waveguides remains low.

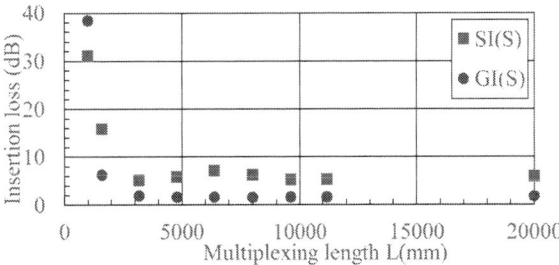

Figure 2. Calculated insertion loss of GI square core Y-branched waveguides with S-shape structures compared to the SI square core counterpats [5].

Fig. 2 indicates that GI core waveguides mostly show lower insertion loss than the SI core waveguides independent of the multiplexing length L. After a detailed analysis on the simulated results, it is found that the loss due to the light leakage from the multiplexing region in GI core waveguide is lower due to the strong light confinement at the core center in GI core [5].

The lowest loss is observed when the S-shape is formed in GI core Y-branch waveguides with an L of 8,000 μm, as shown in Fig. 2. In this design, the bending radius and angle are relaxed enough due to the sufficient multiplexing length L. Since the bending loss is negligible, the light leakage from the taper shaped core in the multiplexing region could dominate the total insertion loss, as well as the coupling losses at both end of the waveguide. These loss factors: coupling loss at waveguide ends and light leakage from taper shaped core in multiplexing region are influenced by the waveguide NA. In the simulation to obtain the results in Fig. 2, the NAs of launch, the waveguide, and detection probe (50 GI MMF) are 0.340, 0.294, and 0.220, respectively. Here, the waveguide NA is

978-1-7281-1500-9/19 $31.00 © 2019 IEEE

varied from 0.100 to 0.34 and the insertion loss is calculated in the same manner. The result is shown in Fig. 3.

It is obvious that with increasing the waveguide NA, the insertion loss of the GI core Y-branched waveguide monotonically decreases. Since the launch NA (supposing VCSEL source) of 0.340 is higher than the waveguide NA supposed in this simulation (from 0.100 to 0.340), the coupling loss at the launch end decreases with increasing the waveguide NA. In contrast, the coupling loss at the detection end increases with increasing the waveguide NA, since the NA of the detection probe (0.220) is not high enough. However, the high NA of the waveguide allows to tightly confine the light even in the tapered core in multiplexing region, which contributes to decrease the light leakage even when the waveguide NA is higher than that of the detection probe. Hence, the NA of the waveguide should be high as far as the appropriate materials combination for the core and cladding exists.

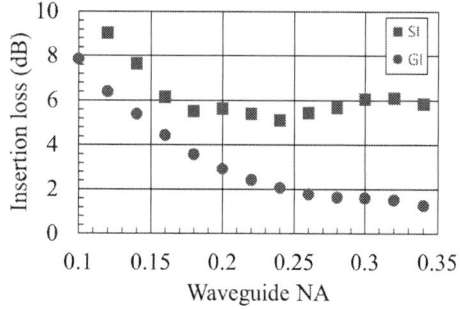

Figure 3. Y-branched waveguide design for simulating the insertion loss.

III. FABRICATION AND CHANRACTERIZATION FOR Y-BRANCHED POLYMER WAVEGUIDES

A. Imprint Method for Square Core Waeguides

In order to fabricate the Y-branched waveguides with S-shaped structures designed in the above section, we employ an imprint method that is possible to accurately form the desired channel (core) patterns using a master mold fabricated based on photolithography. It is noted that the imprint method is possible to form the cores with SI and GI profiles, separately. We already reported that channel shuffling multimode waveguides with both SI and GI cores, which include the core bending and crossing were successfully fabricated by means of the imprint method [6].

Fig. 4 shows the waveguide fabrication procedure using the imprint method. The process is divided into three steps: 1) under-cladding fabrication, 2) filling up the grooves for core on the under-cladding with a liquid core monomer, and 3) over-cladding fabrication. First, an under-cladding is fabricated using a polydimethylsiloxane (PDMS) based flexible and optically tranparent mold on which Y-branched core patterns (convex shape) are formed. The core patterns on

the PDMS mold is pressed over a viscous cladding monomer coated on a glass plate. After, the under-cladding pressed with the PDMS mold is UV cured from both sides: the glass substrate and the mold sides, the PDMS mold is peeled off to obtain the under-cladding.

Next, the grooves with the Y-branched core patterns on the under-cladding are filled with a liquid state core monomer using doctor-blade method. After filling up the grooves, the core monomer is cured under UV exposure.

Finally, the cladding monomer is uniformly applied on the under-cladding with core, and then the monomers are completely cured under UV exposure to obtain a waveguide.

In this paper, GI profiles are formed by applying an interim time of 15-minutes after filling the groove with the core monomer followed by another 20-minute interim time after coating the over-cladding monomer to allow the monomer diffusion between the core and cladding. As the materials for the Y-branched multimode waveguides, organic-inorganic hybrid resins, SUNCONNECT® series supplied by Nissan Chemical Corporation are used: NP-208 ($n = 1.557$) is used for the cladding, while NP-001 ($n = 1.584$) is used for the core.

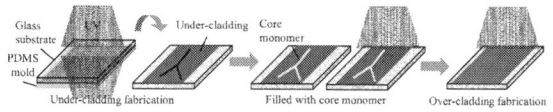

Figure 4. SI and GI-core Y-branched waveguide fabrication applying the imprint method.

B. The Mosquito Method for Circular Core Waveguides

As mentioned in the introduction section, we developed a unique method to fabricate polymer optical waveguides with circular cores named the *Mosquito method*. The procedure of the Mosquito method is shown in Fig. 5. First, a liquid state cladding monomer is casted on a substrate (a). Next, another liquid monomer for core is placed in a syringe, and then the tip of the syringe needle is inserted into the cladding monomer and horizontally scans following the path of core patterns to be formed, while dispensing the core monomer from the tip, as shown in Fig.5(b). Finally, both the core and cladding monomers are simultaneously UV cured to fix the core pattern, as shown in Fig. 5(c). As various monomers of UV curable resins can be applied to the Mosquito method, in this paper, acrylate materials are used to fabricate multimode Y-branched waveguides with circular cores. The monomers of acrylate-based UV curable resins supplied by Kyoritu Chemical & Co., Ltd. are used for both the core and cladding.

Various core patterns can be formed by programming the needle-scan path on the CAD software. However, it has been supposed that the core crossing and merging / splitting structures would be difficult to form, because the cores dispensed earlier could be disarrayed by the needle scan for dispensing the cores to be crossed, merged and split.

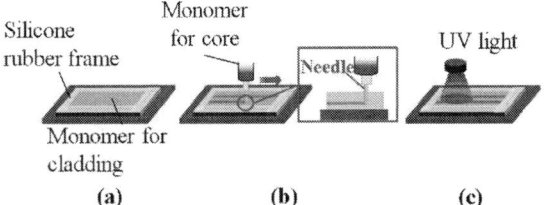

Figure 5. Experimental procedure of the Mosquito method. (a): Monomer for the under-cladding coated on a glass substrate (b): Monomer for the core is dispensed from the needle tip. (c): UV exposure for curing both monomers.

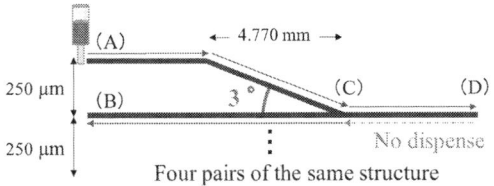

Figure 6. Needle-scan path in the Mosquito method for fabricating Y-branched waveguide

However, we find that a *unicursal needle-scan path* shown in Fig. 6 allows to form Y-branched structures successfully. In this scan path, the needle tip is inserted into the cladding monomer at (A) to scan all the way to (D) via (C), and then, return to (C) and proceed to (B) without pulling out the needle from the cladding. Here, on the scan path between (C) and (D), a round-trip scan is needed, where the core monomer is not dispensed during the returning path from (D) to (C) to keep the core diameter constant in whole pattern.

C. Charcterization of Square Core Y-branched Waveguides

Fig. 7 shows the top-view of the Y-branched waveguides with an S-shaped multiplexing structure, whose core size is 45 μm × 45 μm for multimode.

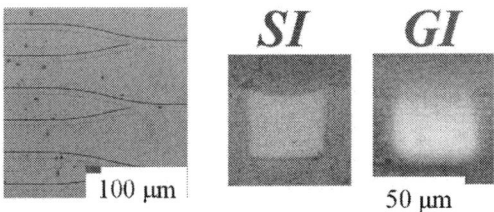

Figure 7. Top-view and cross-sectional photos of square core Y-branched waveguides with S-shaped strucure

The insertion loss of the fabricated Y-branched waveguides is measured at 850-nm wavelength. In the measurement setup, an ultra-high NA single-mode fiber (UHNA 4, NA = 0.340) is used to launch the waveguide core, which models a direct coupling condition with a VCSEL chip having an emitting area of a few micrometer in diameter with a large divergence angle (an NA of 0.34 should be almost the worst case). Here,

in order to evaluate the insertion loss dependence on the multiplexing length L, the total waveguide length of all the fabricated waveguides is fixed to 4 cm in which a 1-cm long input part (including no bending for S-shape structure) is maintained, while the length of the output port after two cores are merged is adjusted with L. Finally, the output light from the Y-branched waveguide is coupled to a 50GI-MMF.

The results are shown in Fig. 8. It is noted that both SI and GI core S-shaped Y-branched waveguides exhibit the lowest insertion loss when a multiplexing length is 9,600 μm (9.6 mm). As shown in Fig. 2, it is theoretically predicted that the insertion loss decreases and remains the lowest level when L exceeds 8,000 μm. The experimental results in Fig. 8 well agree with the simulated results. In Fig.8, the insertion loss of 4-cm long *straight* waveguides (no branched and bending structures) with SI and GI cores are indicated by dotted lines for comparison. These waveguides are fabricated in the same way as the Y-branched waveguides using the same materials.

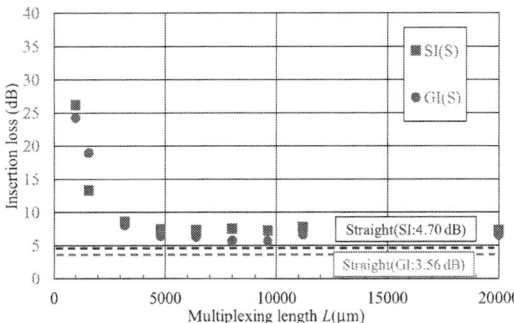

Figure 8. Measured insertion loss of fabricated Y-branched waveguide with different multiplexing length L at 850-nm wavelength [5]

The insertion loss of 4.70 dB is measured in just the straight waveguide with an SI core, while more than 1 dB lower loss (3.56 dB) is observed in GI core waveguide. The lowest loss of the SI core Y-shaped waveguide is 7 to 9 dB while that of the GI core Y-branched waveguide is slightly lower (about 5 to 7 dB), which are observed when the multiplexing length is larger than 4,000 μm. Here, the GI core Y-branched waveguide shows the lowest insertion loss of 5.60 dB when L = 9,600 μm. The loss value is not necessarily low enough as a MUX device for CWDM links. Hence, the loss in the measured Y-branched waveguides are classified into four factors: the coupling loss at input, bending loss at S-shape core, the multiplexing loss at the two-core junction, and the coupling loss at the output end. For this analysis, reference waveguides: straight waveguides with the same length as the Y-branched waveguide and S-shaped curved waveguides (the same structure as one port in the Y-branched waveguide) are separately fabricated. Fig. 9 compares the breakdown of the insertion loss in GI core Y-branched waveguide to that of SI core counterpart. It is found from Fig. 9 that the bending loss is quite low, because the bending angle and radius are sufficiently relaxed by setting the multiplex length L as large as 9,600 μm. Compared to the SI core waveguide, slightly higher bending loss (0.17 dB in GI, while 0.02 dB in SI) is observed, which could attribute to the difference in the numerical aperture (NA) between the two waveguides. The NAs of the fabricated waveguides are discussed later.

Meanwhile, a coupling loss of approximately 4 dB in total is observed at both input and output ends of the GI core waveguide, which is almost the same as that in the SI core waveguide. Here, the coupling loss at the input end is higher than output end in the GI core waveguide, while the magnitude relation between the input and output is opposite in the SI core waveguide. It should be noted that the multiplexing losses in the SI and GI core are almost identical, although it is theoretically confirmed that the GI core exhibits lower multiplexing loss due to its light confinement effect at the core center[5].

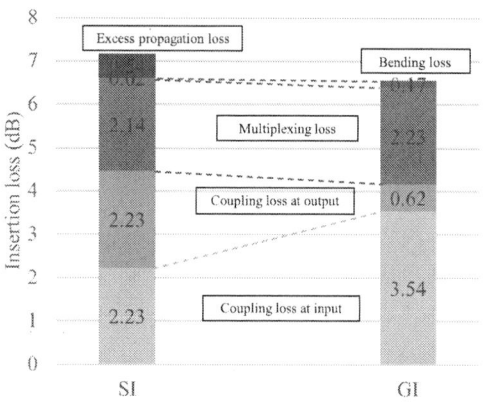

Figure 9.   Breakdown of the loss factor in SI and GI core Y-branched polymer optical waveguides

From the results in Fig. 9, it is confirmed that the lower insertion loss observed in the GI core Y-branched waveguide is attributed to remarkably low coupling loss at the output end as well as no excess propagation loss. However, theoretically estimated insertion loss of GI core Y-branched waveguide is much lower than that of SI core counterpart. The small difference in the insertion loss between the SI and GI core waveguides is explained by the difference of the waveguide NA, as mentioned above.

From the combination of the core-cladding materials used for fabricating these waveguides, the NAs are calculated to be approximately 0.294. Meanwhile, the experimentally measured waveguide NAs of SI and GI core Y-branched waveguides are summarized in Table 2. The GI core shows an NA close to just 0.2 which is approximately 0.1 lower than the theoretically calculated value. In contrast, the NA of the SI core is close to the calculated value. The high NA of the SI core waveguide allows low loss in coupling at input and

multiplexing. On the contrary, the low NA in GI core waveguide lowers the light confinement in the core, resulting in higher coupling loss at the input in addition to high multiplexing loss as shown in Fig. 9.

In order to clearly exhibit the advantage of GI core in the insertion loss, the waveguide should have higher NA, as theoretically confirmed in Fig. 3.

TABLE II.   MEASURED WAVEGUIDE NAS

SI	0.271
GI	0.197

D. Charcterization of Circular Core Y-branched Waveguides

Since a liquid core monomer is dispensed into another liquid cladding monomer in the Mosquito method, the mutual monomer diffusion allows to form index profiles in circular cores. The diffusion rate (diffusion constant) is influenced by the monomer miscibility and viscosity. Since the monomers of UV curable acrylate resins used for fabricating the Y-branched waveguides are highly miscible, after multiple needle scans, the monomer diffusion highly progresses, which makes the waveguide NA too low. Hence, the temperature of the cladding monomer is maintained as low as 13 ºC to reduce the diffusion rate. Then, two Y-branched waveguides are fabricated under the low temperature condition and their insertion losses are measured. Here, the branching angles of them are set to 3° and 1° to explore the structure exhibit the lowest insertion loss. In the fabricated samples, four-set of the same branched waveguides are formed in order to investigate the reproducibility.

A top-view (the branching angle is 3°) and the insertion loss measurement results at 850-nm wavelength are shown in Fig. 10 and 11, respectively. Here, the waveguide length is 2 cm for the insertion loss measurement. In Fig. 10, it is confirmed that the Y-branched waveguides are successfully fabricated without any structural disorders. In Fig. 11, the lowest insertion loss is observed in Ch. 8 of waveguide with 1° branching angle. In this measurement, the launch NA is slightly lower than that for the square core waveguide shown in Fig. 8, but the loss as small as 3 dB is observed, which is lower than the lowest loss observed in Fig. 8. Even if the length of the square core waveguide is 2 cm longer, the propagation loss for the excess 2-cm long waveguide cannot compensate the insertion loss difference.

Figure 10. Top-view photo of the fabricated circular core Y-branched waveguide with a branch angle of 3°

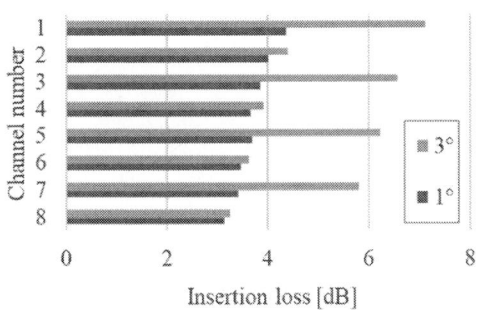

Figure 11. Measured insertion losses of the waveguides circular core Y-branched waveguides with a different branch angles.

It is noted that the waveguide with 1° branching angle is lower than that with 3° because the abrupt change of needle-scan direction at (C) in the scan path shown in Fig. 6 is not needed.

We already reported the capability of fabricating *single-mode polymer waveguides* applying the Mosquito method. Hence, the single-mode waveguides with Y-branch structures are also fabricated[7]. In order to satisfy the single-mode condition, the core diameter should be reduced to 10 µm and less. In our previous investigations [3, 7], we already found that thinner needle, low dispensing pressure for the core monomer, and the high needle-scan velocity allowed to decrease the core diameter. Single-mode Y-branched waveguides are also fabricated applying these conditions. Fig. 12 shows a top-view and cross-sections of a fabricated Y-branched waveguide. In order to maintain the low insertion loss, the branched angle is set to such a small value as 0.1°. Although the channel spacing between (A) and (B) is quite small, the existence of a Y-branched core is visually confirmed in Fig. 12. In addition, from the cross-sections, the core diameter is successfully reduced to 5 µm to satisfy the single-mode condition.

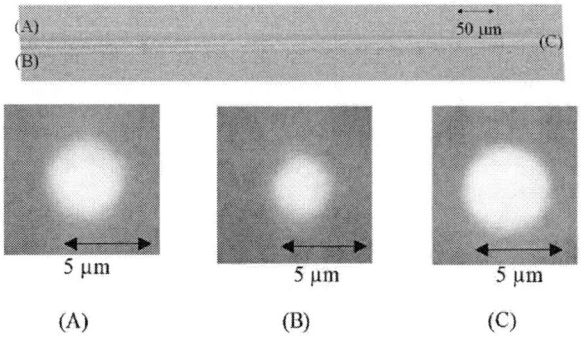

Figure 12. Top-veiw and cross-sectional photos of single-mode Y-branched polymer optical waveguides fabricated using the Mosquit method.

From the output NFP, the single-mode operation of the waveguide is also experimentally confirmed.

The single-mode Y-branched polymer optical waveguides could be applied to a MUX device for mode division multiplexing system using few mode fibers [8].

IV. CONCLUSION

In order to apply to CWDM MMF links, multimode GI core Y-branched polymer optical waveguides are designed to exhibit low insertion loss, and experimentally fabricated applying two methods: imprint method for square core and the Mosquito method for circular core. It is both theoretically and experimentally confirmed that S-shaped Y-branched structure with multiplexing length longer than 3,200 µm allows low insertion loss. In particular, the strong light confinement to the core center in GI core contributes to decrease the light leakage from the multiplexing region (taperd core), resulting in lower insertion loss than SI core counterpart.

Although core branching structure had been supposed to be difficult to fabricate applying the Mosquito method, unicursal needle-scan paths make it possible to fabricate Y-branched waveguide with circular cores, not only multimode but also even single mode. The lowest insertion loss is observed in multimode circular core Y-branched waveguide, which suggests another advantage of circular GI cores fabricated using the Mosquito method.

REFERENCES

[1] J. Lavrencik, S. Varughese, V.A. Thomas, and S. E. Ralph, "Scaling VCSEL-MMF links to 1 Tb/s using short wavelength division multiplexing," J. Lightw. Technol, vol. 36, pp. 4138-4145, 2018.

[2] T. Ishigure, H. Masuda, K. Date, C. Marushima, and T. Enomoto, "Direct fabrication for polymer optical waveguide in PMT ferrule using the Mosquito method," Proc. Electron. Compon. Techonol. Conf. (ECTC2018), pp. 1103 – 1108 (2018)

[3] T. Ishigure, K. Katori, H. Toda, and K. Yasuhara, "Axially tapered circular core polymer optical waveguides enabling highly efficient light coupling," Proc. Electron. Compon. Techonol. Conf. (ECTC2017), pp. 1601 – 1605 (2017)

[4] T. Kudo and T. Ishigure, "Analysis of interchannel crosstalk in multimode parallel optical waveguides using the beampropagation method," Opt. Express, vol. 22, no. 8, pp. 9675-9686, April 2014

[5] F. Nakazaki and T. Ishigure. "Fabrication and Evaluation for Polymer Waveguide coupler devices using the imprint method," Proc. IEEE CPMT Symo. JPN (ICSJ2018), pp. 175 –178 (2018)

[6] K. Abe, Y. Oizumi, and T. Ishigure, "Low-loss graded-index polymer crossed optical waveguide with high thermal resistance," Opt. Express, vol. 26, no. 4, pp. 4512-4521, Februaly 2018

[7] K. Yasuhara, F. Yu, and T. Ishigure, "Circular core single-mode polymer optical waveguide fabricated using the Mosquito method with low loss at 1310/1550 nm," Opt. Express, vol. 25, no. 8, pp. 8524-8533, April 2017

[8] N. Hanzawa, K. Saitoh, T. Sakamoto, T. Matsui, K. Tsujikawa, T. Uematsu, and F. Yamamoto, "PLC-based four-mode multi/demultiplexer with LP11 mode rotator on one chip," J Lightw. Tehcnol., vol. 33, no. 6, pp. 1161-1165, March 2015

Vertically Stacked and Directionally Coupled Cavity-resonator-integrated Grating Couplers for Integrated-optic Beam Steering

Shogo Ura and Junichi Inoue
Faculty of Electrical Engineering and Electronics
Kyoto Institute of Technology
Kyoto, Japan
ura@kit.ac.jp

Kenji Kintaka
Inorganic Functional Materials Research Institute
AIST
Ikeda, Japan
kintaka.kenji@aist.go.jp

Abstract—Combination of an integrated-optic chip launching a light beam from variable position on a waveguide surface and a Fourier transform lens will provide a microoptic beam-steering device. An array of switching grating couplers in a channel waveguide is a possible candidate for varying the beam launching position with miniaturized size. Utilization of a cavity-resonator-integrated grating coupler is discussed theoretically. A resonator waveguide with a grating coupler is stacked on a bus waveguide. Vertical directional coupling between the two waveguides occurs only when a resonance wavelength coincides with that of an incident guided wave. Vertically transferred optical wave in the resonator is coupled out by the grating coupler. The vertical directional coupling can be electrically tuned by utilizing electrooptic or thermooptic effects. A design model was developed on the basis of the coupled mode analysis. Coupling characteristic of design examples using silicon waveguides were discussed. Selective coupling was predicted with the radiation efficiency of 30% and the FWHM of 1.4 x 10^{-3} in the effective refractive index of the cavity waveguide. Difference between neighboring peaks of radiation efficiency was predicted to be 5.2 x 10^{-2} indicating the resolution power of 37 for cavity length of 15 μm. These characteristics show good agreement with simulation results by the finite-difference time-domain method.

Keywords-component; integrated optics; grating couplers; waveguide resonator; directional couplers; silicon photonincs; beam steering

I. INTRODUCTION

An integrated-optic beam steering device will be a key component for constructing future optical systems including a compact light detection and ranging system, a micro projector, a retinal imaging laser eyewear, etc. Various types based on silicon photonics technology have been investigated and their operation principles were demonstrated [1-6]. Phase arrays are the most popular means for controlling beam propagation direction. Two-dimensional (2-D) phase arrays have been demonstrated as direct solutions for 2-D beam steering [1,3], but it will need integration of a large number of waveguide channels and electrodes. Then a phase array is usually employed for beam steering along lateral direction with respect to guided-wave propagation and combined with a different type along longitudinal direction for 2-D beam steering. It is a key issue to realize the longitudinal beam steering. Utilization of wavelength dispersion of grating coupler is simple [2,4-5], but large wavelength variation is required for steering with sufficient angle variation. Extraordinary wavelength dispersion in photonic crystal is attractive to enhance the steering angle [6], but it is difficult to obtain lateral beam steering because large channel spacing is required for forming the photonic crystal and is not suitable for the phase array. It is thus not good idea to control an output angle from the grating coupler in practical applications. Introduction of other mechanism is essential for sufficient steering angle.

Optical Fourier transform is an attractive candidate, where a light emitting position is converted to a light propagation direction. Combination of an integrated-optic chip launching a light beam from variable position on a waveguide surface and an outer lens for the Fourier transform will serve as a microoptic beam steering device. An array of grating couplers with radiation switching function will provide a variable beam launching device. A conventional grating coupler is not suitable for the switching function. There is a strict coupling condition of phase matching for the input coupling from radiation modes in the air/substrate to the guided mode in the waveguide since the radiation mode is continuous while the guided mode is discrete with respect to its propagation constant along the guided mode propagation. This strict condition can be utilized for varying the input-coupling efficiency. It is not so difficult to control the input-coupling efficiency by electrooptic or thermooptic effects of the waveguide. On the other hand, the output coupling occurs from the discrete guided mode to continuous radiation modes to be nonselective.

Cavity-resonator integration is a potential candidate for variable output coupling [7,8]. A grating coupler is integrated inside a waveguide cavity resonator. Coupling efficiency of the cavity-resonator-integrated grating coupler (CRIGC) depends on the relative position of the grating coupler with respect to the waveguide resonator. Push-pull tuning of refractive indices between the grating coupler and both of cavity mirrors can cause an effective displacement of the grating coupler. An electrical switching of the coupling efficiency is expected by the electrooptic or thermooptic effects. Incident guided-wave power is coupled to radiation wave into the air from the grating coupler at an on-resonance state but reflected back in the waveguide at an off-resonance state. The back reflection prevents this switching coupler from being utilized for our application. In order to vary the beam launching position along the guided-wave propagation direction, we have to integrate a number of switching grating couplers in line in a channel waveguide and activate only one coupler for the beam launching. In other words, the incident guided wave should transmit through off-state couplers. Simple CRIGC is not enough for the current application.

Combination of vertical directional coupling is discussed in this paper. A CRIGC is vertically stacked on a bus waveguide. A guided wave in the bus waveguide is coupled to the waveguide of the CRIGC only when the resonance wavelength is just the same as that of the incident guided wave propagating in the bus waveguide. Integration of multiple CRIGCs of different resonance wavelengths along the waveguide and an electrical tuning of the effective refractive index of the CRIGC waveguide will provide an integrated-optic device for light beam launching with variable position. This paper describes a design model based on the coupled mode analysis (CMA) and theoretical prediction of the coupling characteristics of the vertically stacked CRIGC with simulation by the finite-difference time-domain (FDTD) method.

II. Basic Configuration of Integrated-Optic Device for Beam Steering

A basic configuration is illustrated in Fig. 1 for a microoptic 2-D beam steering. An integrated-optic device and an outer cylindrical lens are combined. The integrated-optic device consists of an array of channel waveguides, phase shifters made of thin film heaters, and stacked CRIGCs. An incident guided wave is divided and coupled to multiple channel waveguides. The guided waves are coupled out by CRIGCs to be a radiation beam into the air. An output angle of the radiated beam can be controlled laterally by phase retardation of the guided waves as shown in Fig. 2 (a). Since a wavefront of the radiation wave is formed of many segments launched from waveguide channels, the phase retardation tilts the wavefront and propagation direction. The phase

retardation of the guided waves is generated by the integrated phase shifters. This is a popular means for one dimensional beam steering. An array of CRIGCs of perpendicular out-coupling are integrated on each channel (bus) waveguide. If only one of the CRIGCs can be active, the guided wave would be coupled out only by the active CRIGC. In other words, position of beam launching can be controlled by selective activation of the CRIGCs. Displacement of the wave launching position is converted to propagation angle by Fourier transform function of the outer lens as shown in Fig. 2 (b). As a result, 2-D beam steering can be expected by this configuration.

The key issue is the selective and switchable coupling of CRIGCs. This time, we investigated utilization of directional coupling of guided modes between the resonance and bus waveguides. The CRIGCs have resonance wavelengths different from one another. Resonance wavelengths can be tuned simultaneously by a single heater via thermooptic effect so that only one CRIGC shows the wavelength matching with the bus waveguide and becomes active to couple the guided wave to radiation wave into the air.

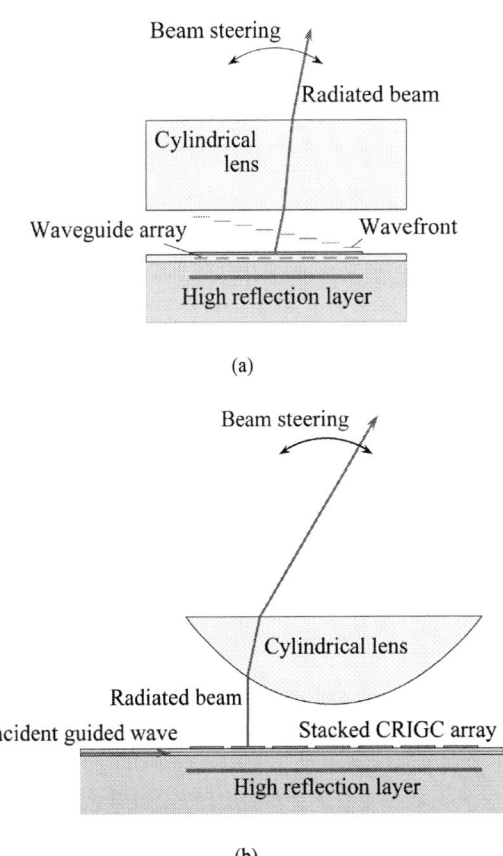

(a)

(b)

Figure 2. Operation principles of two-dimensional beam steering with a phased array (a) and array of switchable grating couplers (b).

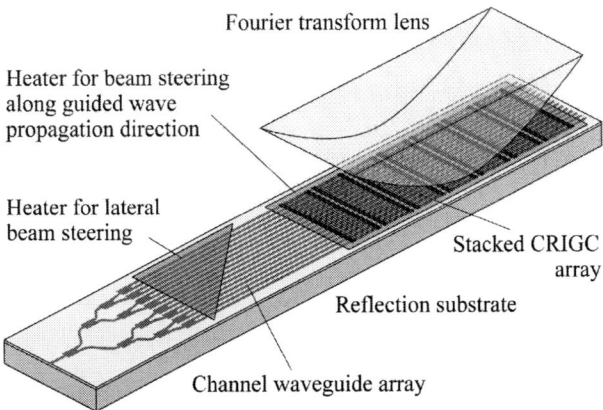

Figure 1. Concept image of a microoptic beam-steering device combined by an integrated optic chip and an outer cylindrical lens. The integrated opitc chip consists of a phased array of channel wavegduides and an array of switchable grating couplers.

978-1-7281-1500-9/19 $31.00 © 2019 IEEE

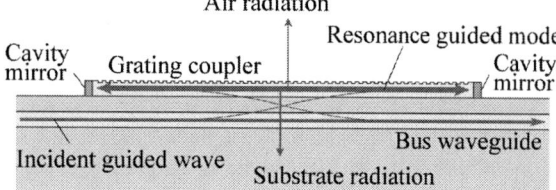

Figure 3. Cross-sectional structure and wave couling of a basic configuration for a swithcable output coupler. A CRIGC is stacked on a bus waveguide.

III. DIRECTIONALLY COUPLED CRIGC

Basic configuration for a switchable output coupler is illustrated in Fig. 3. A leaky resonator waveguide of CRIGC is stacked on a bus waveguide. The CRIGC consists of a cavity resonator of a pair of waveguide mirrors and a grating coupler. Each waveguide would support only a single guided mode when it were alone, but a directional coupler is formed by the two waveguides. Propagation constants of the waveguides are designed to be almost the same. Thickness of a layer separating the waveguides is determined to give a sufficiently small coupling coefficient for the directional coupler.

An incident guided wave propagating in the bus waveguide couples to the resonator waveguide only when the wavelength satisfies a resonance condition of the waveguide resonator. The resonance condition is expressed by

$$N_R L_R = m \frac{\lambda_0}{2}, \qquad (1)$$

where N_R, L_R, m, and λ_0 are the effective refractive index of the guided mode in the resonator waveguide, the cavity length, a mode number of the resonance, and the wavelength in free space, respectively. The coupled and accumulated wave in the resonator waveguide is vertically radiated out by the grating coupler into the air. The incident guided wave in the bus waveguide passes through when (1) is not satisfied. We can switch both states electrically by varying N_R via electrooptic or thermooptic effects.

IV. DESIGN MODEL BASED ON CMA

Coupling behavior in a simple CRIGC has been already discussed and demonstrated [9,10]. However, directional coupling and resonance occur simultaneously in our case. Wave coupling in such a complicated configuration has not been analyzed yet. It is then crucial to develop an analytical model for efficient design. The coupled mode analysis (CMA) was utilized to give a simple and perspective model for predicting coupling characteristics.

Coordinate axis of z is set to the propagation direction of the incident guided mode in the bus waveguide. We consider TE$_0$ modes. Propagation constants in the resonator and bus waveguides are denoted by β_A and β_B, respectively. Electric

fields of forward and backward propagating guided waves in the resonator waveguide without grating coupler are expressed by $A_f(z)\exp(-j\beta_A z)$ and $A_b(z)\exp(j\beta_A z)$, respectively. Electric fields of guided waves propagating to z and $-z$ directions in the bus waveguide are indicated by $B_f(z)\exp(-j\beta_B z)$ and $B_b(z)\exp(j\beta_B z)$, respectively. Coupled wave equations are expressed by

$$\frac{dA_f(z)}{dz} = -\alpha A_f(z) - j\kappa_{AB} B_f(z) e^{-j2\Delta z},$$

$$\frac{dB_f(z)}{dz} = -j\kappa_{BA} A_f(z) e^{+j2\Delta z},$$

$$-\frac{dA_b(z)}{dz} = -\alpha A_b(z) - j\kappa_{AB} B_b(z) e^{-j2\Delta z}, \qquad (2)$$

$$-\frac{dB_b(z)}{dz} = -j\kappa_{BA} A_b(z) e^{+j2\Delta z},$$

where α indicates the radiation decay factor of the grating coupler and κ_{AB} and κ_{BA} represent the coupling coefficient of the directional coupling from the bus to resonator waveguides and from the resonator to bus waveguides, respectively. Phase mismatching factor Δ is given by

$$2\Delta = \beta_B - \beta_A. \qquad (3)$$

General solutions of (2) can be written as

$$B_f(z) = \left(B_f^+ e^{-j\kappa_f z} + B_f^- e^{+j\kappa_f z} \right) e^{-\frac{\alpha}{2} z} e^{j\Delta z},$$

$$A_f(z) = \left(c_f^+ B_f^+ e^{-j\kappa_f z} + c_f^- B_f^- e^{+j\kappa_f z} \right) e^{-\frac{\alpha}{2} z} e^{-j\Delta z},$$

$$B_b(z) = \left(B_b^+ e^{-j\kappa_b z} + B_b^- e^{+j\kappa_b z} \right) e^{\frac{\alpha}{2} z} e^{j\Delta z}, \qquad (4)$$

$$A_b(z) = \left(c_b^+ B_b^+ e^{-j\kappa_b z} + c_b^- B_b^- e^{+j\kappa_b z} \right) e^{\frac{\alpha}{2} z} e^{-j\Delta z},$$

$$c_f^+ = \frac{\kappa_f - \Delta - j\frac{\alpha}{2}}{\kappa_{BA}}, \quad c_f^- = \frac{-\kappa_f - \Delta - j\frac{\alpha}{2}}{\kappa_{BA}},$$

$$c_b^+ = \frac{-\kappa_b + \Delta - j\frac{\alpha}{2}}{\kappa_{BA}}, \quad c_b^- = \frac{\kappa_b + \Delta - j\frac{\alpha}{2}}{\kappa_{BA}}, \qquad (5)$$

$$\kappa_f = \sqrt{\kappa_{BA}^2 + \Delta^2 + j\alpha\Delta - \left(\frac{\alpha}{2}\right)^2},$$

$$\kappa_b = \sqrt{\kappa_{BA}^2 + \Delta^2 - j\alpha\Delta - \left(\frac{\alpha}{2}\right)^2}, \tag{6}$$

where we used a relation of $\kappa_{AB} = \kappa_{BA}$.

When the resonator cavity is located at $0 < z < L$ and an incident wave only comes into the structure from $z = 0$ in the bus waveguide, boundary conditions are given by $B_f(0) = 1$ and $B_b(L) = 0$ and rewritten to

$$B_f^+ + B_f^- = 1,$$

$$B_b^+ e^{-j\kappa_b L} + B_b^- e^{+j\kappa_b L} = 0. \tag{7}$$

Boundary conditions at the waveguide ends of the resonator waveguide are given by $A_f(0) = rA_b(0)$ and $A_b(L)\exp(j\beta_A L) = rA_f(L)\exp(-j\beta_A L)$ with the reflection coefficient r of the cavity mirrors and rewritten to

$$c_f^+ B_f^+ + c_f^- B_f^- = r\left(c_b^+ B_b^+ + c_b^- B_b^-\right),$$

$$\left(c_b^+ B_b^+ e^{-j\kappa_b L} + c_b^- B_b^- e^{+j\kappa_b L}\right)e^{\frac{\alpha}{2}L}e^{j\beta_A L} \tag{8}$$

$$= r\left(c_f^+ B_f^+ e^{-j\kappa_f L} + c_f^- B_f^- e^{+j\kappa_f L}\right)e^{-\frac{\alpha}{2}L}e^{-j\beta_A L}.$$

We can obtain B_f^+, B_f^-, B_b^+, and B_b^- from (7) and (8). Guided mode behavior can be predicted by substituting them to (4).

V. DESIGN EXAMPLES

Devices were designed with silicon waveguides for a wavelength of 1.55 μm. The bus waveguide consists of a Si guiding core layer of 220-nm thickness on a SiO₂ optical buffer layer of 3.00-μm thickness on a Si substrate. A Si guiding core layer of 222-nm thickness for a CRIGC is stacked on a SiO₂ separation layer of 500-nm thickness on the bus waveguide. A grating coupler of the CRIGC is formed by corrugation of a 6-nm Si layer on the core layer. The fill factor of Si against the air clad is 0.5. Refractive indices of Si layer and SiO₂ layer are 3.50 and 1.46, respectively.

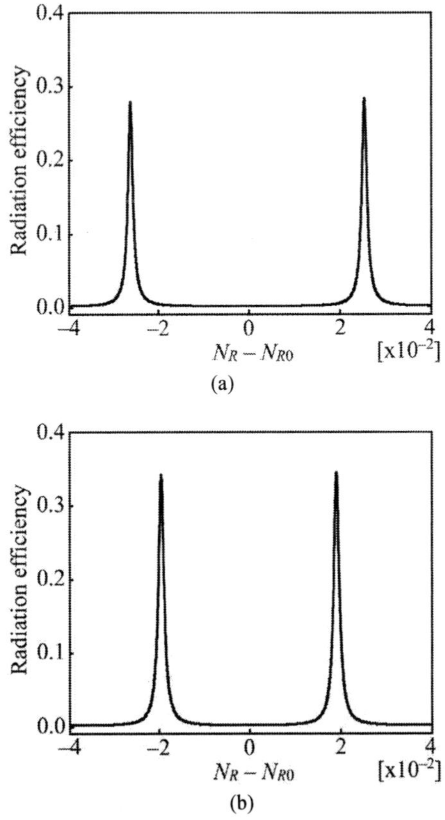

Figure 4. Calculated dependence of radiation effciency on a variation of N_R. with L = 15 μm (a) and 20 μm (b).

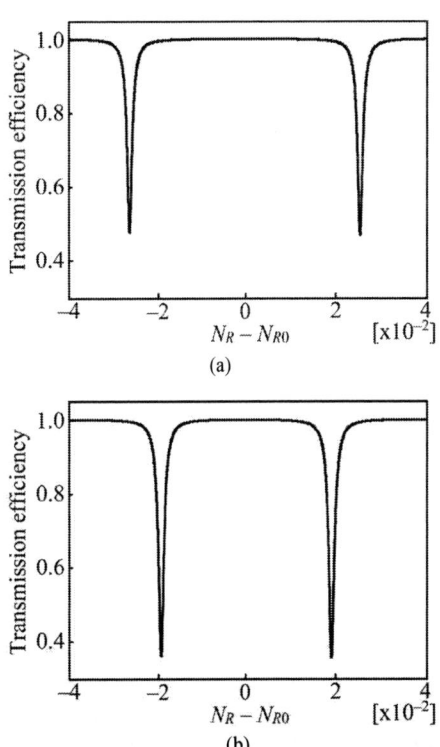

Figure 5. Calculated dependence of transmission effciency on a variation of N_R. with L = 15 μm (a) and 20 μm (b).

The thicknesses of the guiding core layers were determined to give almost the same guided mode indices for TE$_0$ modes in the resonator and bus waveguides. The mode indices were calculated to be 2.873. The radiation decay factor α was calculated to be 1.25 mm^{-1}. The coupling coefficients κ_{AB} and κ_{BA} for directional coupling were the same and calculated to be 10.8 mm^{-1}. The reflection coefficient r of the cavity mirror of the resonator was assumed to be 0.99.

Coupling efficiency from the incident guided wave to the wave radiated by the grating coupler was calculated. Figure 4 shows typical examples of dependence of the coupling efficiency on a variation of the effective refractive index N_R from the design value of N_{R0}. Cavity length L_R was assumed to be the same as L in this section and 15 μm for Fig. 4 (a) and 20 μm for Fig. 4 (b). The efficiency shows sharp dependence on N_R and has peaks at the resonance condition of (1). The full width at half maximum (FWHM) δN_R of the efficiency peak was estimated to be 1.4 x 10^{-3}. The half width at $1/e^2$ was 1.7 x 10^{-3}. We can then control the output coupling efficiency by tuning N_R. The peaks appear periodically against N_R. Index deviation between the neighboring peaks ΔN_R was 5.2 x 10^{-2} for Fig. 4 (a) and 3.8 x 10^{-2} for Fig. 4 (b). It is worth to note that ΔN_R is inversely proportional to L as predicted from (1) while δN_R does not depend on L. This means that the quality factor of the CRIGC is inversely proportional to L. On the other hand, the maximum coupling efficiency is proportional to L. These predictions are consistent with the fact that the most of the cavity loss is the radiation from the grating coupler. It is then predicted that lower r results in larger discrepancy from the proportional or inversely proportional relations.

Transmission efficiency from the incident guided wave represented by $B_f(0)$ to a transmitted wave given by $B_f(L)$ in the bus waveguide are depicted in Fig. 5 (a) for $L = 15$ μm and Fig. 5 (b) for $L = 20$ μm. Transmission efficiency is expected to be almost 100% except for sharp drops at N_R satisfying (1). The incident wave power can transmit through off-state CRIGCs without serious loss. This is required for constructing a beam-launching device of controllable position. Resolution power of the beam steering angle is given by the number of the beam-launching positions. A ratio $\Delta N_R / \delta N_R$ gives the potential number of positions, namely the resolution power, and are calculated to be 37 and 27 for L of 15 and 20 μm, respectively.

VI. SIMULATION RESULTS BY FDTD METHOD

Coupling characteristics of the proposed and designed device were simulated with use of FDTD method. A structure for the simulation is depicted in Fig. 6. Specifications of the waveguide structure are the same as designed in V. The grating period of the grating coupler Λ was determined to be 540 nm for giving vertical radiation. The coupling length of the grating coupler was chosen to be 28 Λ = 15.1 μm. A waveguide cavity mirror is a key component as well as the grating coupler in practical device. Five slots formed in the guiding core are used to be the mirror. Slot and land widths were 150 and 218 nm, respectively. High reflection coefficient can be expected thanks for high refractive index

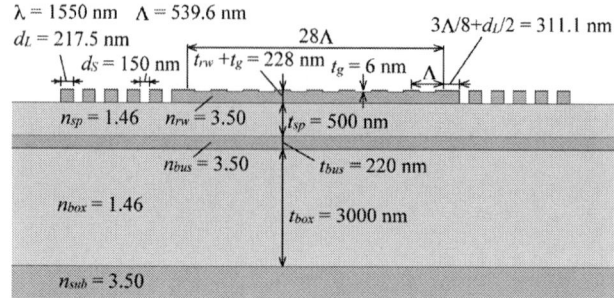

Figure 6. A CRIGC using cavity mirrors of multiple slots stacked on a SOI substrate.

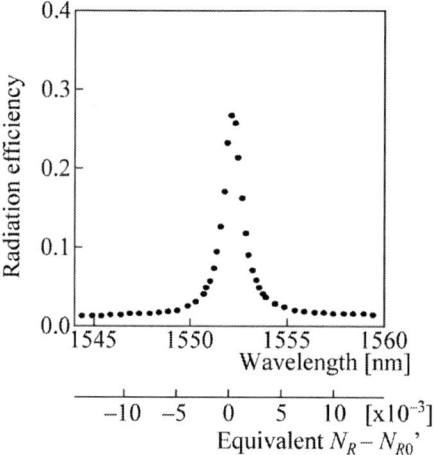

Figure 7. Radiation efficiency dependecy simulated by FDTD method.

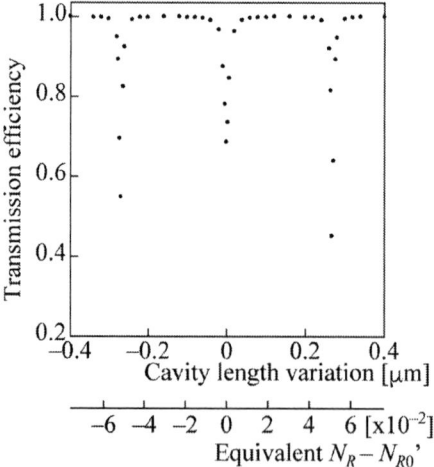

Figure 8. Transmission efficiencies simulated by FDTD method. Cavity length was varied from 15.1 μm.

contrast between the air and Si. Distance between the mirror and grating coupler was designed to be $3/8\,\Lambda$ for matching the refractive index boundaries of the grating grooves to the nodes of the standing wave formed in the cavity resonator.

Calculated dependence of the radiation efficiency on wavelength λ is plotted in Fig. 7. The maximum radiation efficiency is given at $\lambda = 1552.3$ nm to be 27 % which is almost the same as predicted in Fig. 4(a). It suggests that the coefficient r of the slot mirror would be similar value to 0.99. Equivalent N_R values from the value of N_{R0}' giving the maximum radiation for a fixed λ of 1552.3 nm is also shown as another lateral axis. The line width δN_R was estimated to be 2.0×10^{-3}. This is a little bit larger than CMA prediction, indicating lower quality factor of the resonator resulting from some excess loss.

Transmission efficiencies for various cavity lengths were calculated and are summarized in Fig. 8 with an additional lateral axis of equivalent N_R variation. Transmission efficiency of almost 100% was obtained for off-resonance condition. The FDTD simulation results show good agreement with the CMA based design.

VII. DISCUSSION FOR MONOLITHIC INTEGRATION OF FOURIER TRANSFORM LENS

A Si has a considerably large thermooptic coefficient of $1.8 \times 10^{-4}\,K^{-1}$ and is one of attractive materials as the resonator waveguide for thermooptic tuning of N_R. Integration of thin film heater of the minimized size is required for temperature control with low power consumption and fast switching time. However, insertion of the heater film beneath the resonator waveguide is not easy for the device structure illustrated in Fig. 6 from viewpoints of operation principle and fabrication convenience. Another problem must be hybrid integration of

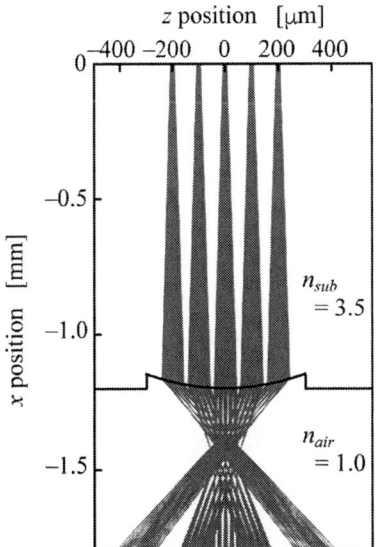

Figure 9. Diagram of ray tracing for a device integrating CRIGCs on the top and Fourier transform lens on the bottom of the Si substrate.

the Fourier transform lens for constructing a beam steering device as depicted in Fig. 1.

One of potential solutions would be a configuration utilizing substrate radiation. Thin film reflection layer and heater are formed with another separation layer on top of the stacked CRIGC. Upwardly radiated beam is reflected downward and vanished by the layer. The downwardly reflected beam will transmit the substrate to be radiated into the air from the bottom side of the substrate. A Fourier transform lens can be formed on the bottom side. An example of ray tracing for a device utilizing such configuration is shown in Fig. 9. Aperture size of a CRIGC was assumed to be 15 μm. A focal length of the lens in the substrate was set to be 1.2 mm. Optical rays are drawn from CRIGCs located at $z = -200, -100, 0, 100$ and 200 μm. Optical beams propagate to $-x$ direction with a diffraction angle determined by the aperture size and the refractive index of the substrate. The beams are refracted at a lens surface and deflected to be collimated beams in the air. Propagation angle ranges from −40 to 40 degrees. Optical beam steering of 80 degrees are expected by thermooptic tuning.

VIII. CONCLUSIONS

A new scheme of integrated-optic device utilizing the optical Fourier transform is discussed and characterized theoretically for 2-D beam steering. The integrated-optic device consists of a series of CRIGCs of different cavity lengths integrated along the guided-wave propagation on the buss waveguides. A guided wave in the bus waveguide can be directionally coupled to guided wave supported in a stacked waveguide of CRIGC in a resonance condition. A free-space wave is vertically launched from a selected CRIGC by tuning the resonance condition.

A design model of directionally coupled CRIGC was developed on the basis of CMA model. Selective coupling was predicted with the radiation efficiency of 30% and the FWHM of 1.4×10^{-3} in the effective refractive index of the cavity waveguide. Difference between neighboring peaks of radiation efficiency was predicted to be 5.2×10^{-2} indicating the resolution power of 37 for the cavity length of 15 μm. These characteristics show good agreement with simulation results by FDTD method. Combination of a basic configuration of a directionally coupled CRIGC and a thermooptic material for a cavity waveguide can serve as an attractive component to construct a microoptic 2-D beam steering device.

REFERENCES

[1] K. V. Acoleyen, H. Rogier, and R. Baets, "Two-dimensional optical phased array antenna on silicon-on-insulator," Opt. Express, vol. 18, pp. 13655-13660, June 2010.

[2] K. V. Acoleyen, W. Bogaerts, and R. Baets, "Two-dimensional dispersive off-chip beam scanner fabricated on silicon-on-insulator," IEEE Photon. Technol. Lett., vol. 23, pp. 1270-1272, Sept. 2011.

[3] J. Sun, E. Timurdogan, A. Yaacobi, E. S. Hosseini, and M. R. Watts, "Large-scale nanophotonic phased array," Nature, vol. 493, pp. 195-199, Jan. 2013.

[4] D. Kwong, A. Hosseini, J. Covey, Y. Zhang, X. Xu, H. Subbaraman, and R. T. Chen, "On-chip silicon optical phased array for two-

dimensional beam steering," Opt. Lett., vol. 39, pp. 941-944, Feb. 2014.

[5] J. K. Doylend, M. J. R. Heck, J. T. Bovington, J. D. Peters, L. A. Coldren, and J. E. Bowers, "Two-dimensional free-space beam steering with an optical phased array on silicon-on-insulator," Opt. Express, vol. 19, pp. 21595-21604, Oct. 2011.

[6] K. Kondo, T. Tanabe, S. Hachuda, H. Abe, F. Koyama, and T. Baba, "Fan-beam steering device using a photonic crystal slow-light waveguide with surface diffraction grating," Opt. Lett., vol. 42, pp. 4990-4993, Dec. 2017.

[7] S. Ura, K. Mori, R. Tsujimoto, J. Inoue, and K. Kintaka, "Position dependence of coupling efficiency of grating coupler in waveguide

cavity," Proc. 67th Electron. Compo. Technol. Conf. (ECTC 2017), Orlando, Florida, pp. 1619-1626, May 2017.

[8] R. Tsujimoto, K. Mori, K. Kintaka, J. Inoue, and S. Ura, "Output position variation in grating coupler integrated in waveguide resonator," Tech. Dig. Microopt. Conf. (MOC 2017), Tokyo, Japan, pp. 318-319, Nov. 2017.

[9] S. Ura, S. Murata, Y. Awatsuji, and K. Kintaka, "Design of resonance grating coupler," Opt. Express, vol. 16, pp. 12207-12213, Aug. 2008.

[10] K. Kintaka, Y. Kita, K. Shimizu, H. Matsuoka, S. Ura, and J. Nishii, "Cavity-resonator-integrated grating input/output coupler for high-efficiency vertical coupling with a small aperture," Opt. Lett., vol. 35, pp. 1989-1991, June 2010.

2019 IEEE 69th Electronic Components and Technology Conference (ECTC)

CiB(Chip in Board) Optical Engine Module Using Advanced Fan-out Package Technology.

Sang Yong PARK, Ju Hyun NAM, Ji Ni SHIM, Jun Kyu LEE, Yong Tae KWON, Chang Woo LEE,
Jong Heon KIM, Nam Chul KIM

nepes Corporation
Gwahaksaneop-2ro, Ochang-eup, Cheongwon-gu, Cheongju-si, South Korea 28116
psy0408@nepes.co.kr

Abstract— **In this paper, the development of a new optical chip in board (CiB) package adapting Fan-out technology that offers thermal, electrical and thin structure benefit was reported. Optical CiB package contain 4 optical die and is made smaller and thinner than market with the redistribution layer technology of FO-WLP. The key advantages such as high production yield, low cost and simple process steps surpass the conventional optical packages that depends on high precision alignment but always difficult to achieve good performance. Through finally demonstrated that communication is possible at the target speed of 10Gbps/Ch through actual measurement. In this paper describes the structural features of Embedded optical CiB package with integration of nepes'Fan-out technology.**

Keywords-component; Fan Out Package, COBO, OSA, Optical Engine, Transceiver

I. INTRODUCTION

The utilization of optical interconnections has become popular, as to meet the increasing demand pertaining to 5G data traffic and to overcome the limitations of electrical interconnects due to its excellent advantages such as a short signal delay, a light weight with lower power consumption, and immunity to electromagnetic interference. In addition, optical interconnects improve the speed and quality of data transmission in short and long range applications widely used in data processing units, optical storage applications and switches in data centers. Especially for high-storage data processing units used in data centers in which high-speed data transmission for rack to rack, board to board and chip to chip interconnections is necessary, design with high-capacity and compact pluggable optical modules for bidirectional optical has to be top priority in consideration, hence, meticulous design of Tx/Rx modules incorporating optical components is desirable to minimize the optical assembly costs whilst maximizing the package performances.

Also, this work is to response to the explosively growing demands for more advanced package in semiconductor industry - the development of a new optical Chip-in-Board (CiB) package concept adapting Fan-Out technology that hold key advantages such as high production yield and low cost was demonstrated. Optical CiB (Chip-in-Board) package begins with the idea of embedding several chips in a board to minimize the distance from the chip to signal output, thereby reducing the signal delay and unwanted parasitic effects is expected. In addition, new optical CiB package is made smaller and thinner than conventional market with the integration of redistribution layer (RDL) technology of Fan-Out WLP. In addition, to tackle the high precision alignment requirement for optical package, Pick and Place process was introduced when placing the optical interposer (MOI) onto the CiB module.

II. CIB GEOMETRY

Optical Engine include Tx/Rx, requiring a low cost and large volume production capability, are particularly suitable to the adoption of optical interconnects. Considering that the optical coupling beetween the optoelectronic devices and the MMF is the most critical for an OSA, our main consideration points on the new package design were based on the alignment tolerance of optical device to MMF (Multi-Mode fiber) coupling and having short electrical interconnection. In the end of this paper, an optical sub assembly (OSA) using VCSELs has been extensively studied [1~6]. To implement those modules, a plastic injection molding and passive assembly had been also actively studied in order to achieve low cost volume production. Several previous research papers showing about multi-channel modules using step index plastic optical fibers with 25Gbps per channel had been reported, yet they did not accomplish a fully passive alignment based on a pick and pick scheme.

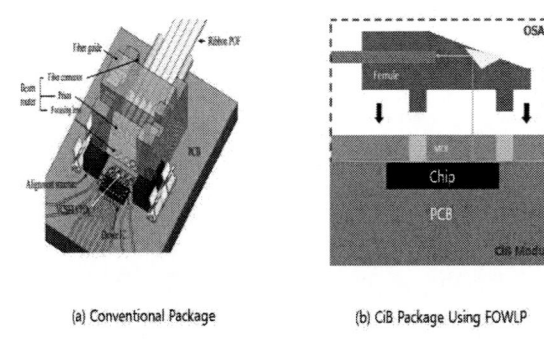

(a) Conventional Package   (b) CiB Package Using FOWLP

Figure 1. Proposed CiB Structure.

978-1-7281-1500-9/19 $31.00 © 2019 IEEE

Figure 1 shows the images Converntional [1] and CiB Module Package. In this figure, conventional package is built with relatively complex structure and manual assembly is done for OSA and electrical interconnection, but CiB structure is simple and is designed for high volume production prospect.

The basic configuration of proposed four-channel optical Tx/Rx CiB module is illustrated in Figure 1 (b), consisting of:

- Optical Module Part includes IC (Driver IC, TIA) and Optical Device (VCSEL, Photo Diode)
- PCB include cavity area
- OSA includes MOI (Mechanical Optical Interconnection) and Ferrule.

Items		Specification
Target Speed		10Gbps/ch
Chip	Electrical IC	Driver, TIA
	Optical IC	VCSEL, PD
PCB Panel	Dimension	299 mm
	Thickness	0.55mm
Cavity	Dimension	4.82× 4.00mm
	Thickness	0.15mm
Module	Dimension	22 × 22mm
	Thickness	0.58mm

Table 1: CiB Prototype Sample Specifications.

Table 1 shows the specifications of the prototype sample and was developed on a 12-inch FOWLP platform. Figure 2 shows the product layout and optical sub assembly components of the prototype. This time, a commercial product ferrule "PRIZM Light Turn (PLT)" form US Conec Ltd was used.

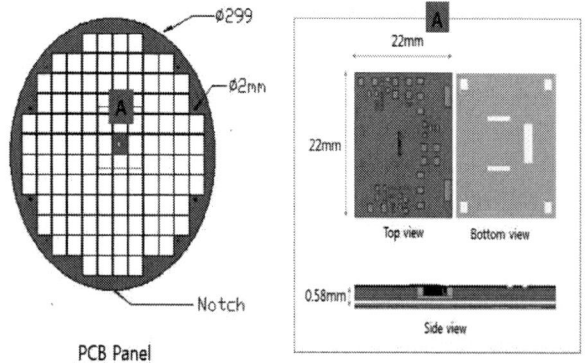

(a) POD for CiB Panel and Module

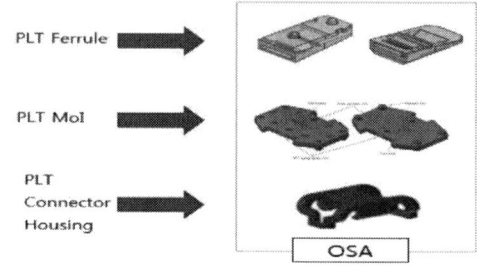

(b) OSA components (Prizm Light Turn)

Figure 2.   Product of design

III.   CiB Platform

A.   CiB Manufacturing Process Flow

In Fig. 3, there are 4 major divisions in fabricating the prototype samples and majority processes were easily performed in Fan-Out WLP production line.

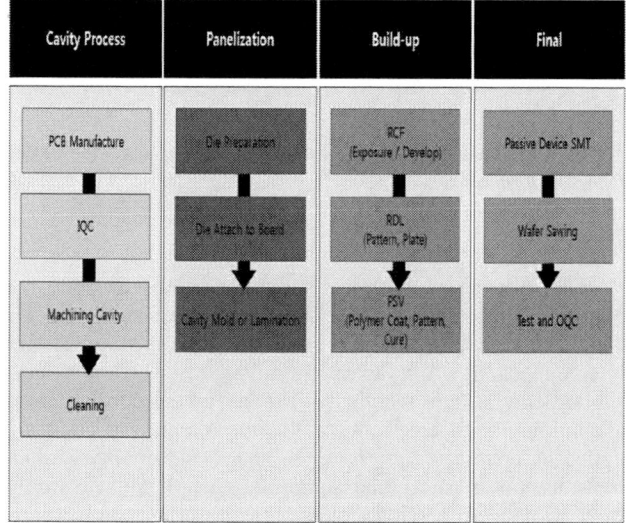

Figure 3.   CiB Process Flow

In the first 'Cavity Process' division, 4.8 × 4.0 × 0.15mm cavity was formed in the PCB by Machining process. To enhance the insertion and attachment of the chips into the cavity, the top exposed Cu surface has to be smoothly done. Figure 4 (a) shows the images before and after cavity formation.

In the second 'Panelization' division, the chips were singulated and then well aligned into the CiB cavity area in Face-Up position by Pick and Place process (Fig. 5). The main advantage to apply PnP process is due its excellent alignment accuracy and tighter process control which provide advantage in terms of overall design and package

978-1-7281-1500-9/19 $31.00 © 2019 IEEE

performance. This also could facilitate the assembly of MOI/OSA in the later process.

In the third 'Build Up' division, the careful selection of dry dielectric film layer was firstly laminated. During laminating process, it filled the space between the chips and the PCB without void formation in the cavity gap.

Next, photo patterning was formed on 1st dielectric layer, followed by 5um metal redistribution layer was built to electrically connect the PCB and chips. Finally, last dielectric layer was deposited to protect the cu metal layer. Figure 4 (c) shows the top view images build-up process.

In the last division, passive components were mounted on the PCB using SMT method and then followed by singulation process and panel cleaning process. Finally, CiB module (without OSA) was accomplished. As shown in Fig. 5, this semi-final CiB product consists of several integrated passive components and a CiB package but only take up the size of 22×22mm only ("small form factor" is possible).

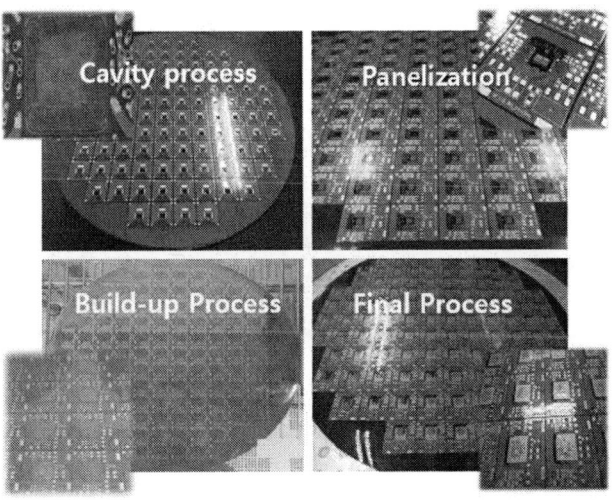

Figure 4.   Process Figure. : (a) Cavity (b) Panel (c) Build-up (d) Final

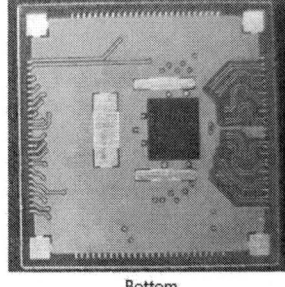

Figure 5.   After SMT and Sawing Process

B.   Assembly optical CiB OSA Concept

The conventional OSA process is extreme-expensive and time consuming process for optical products. Unlike the existing technology available in the market, we do not need wire bonding in our CiB product, hence thinner package is possible compared to CoB package. To assembly the OSA onto the CiB module, 2 steps are required:-

First Step: Place the MOI on the CiB module in 300mm panel using PnP process. As the design of MOI integrates a focusing lens and mechanical connection hole, it simplifies the alignment between CiB Module and Ferrule without major concern considering PnP process has the superb capability of 5um process tolerance. Through process optimization, until now 0.8 ± 0.3um tolerance was possible to achieve. Figure 6 shows the High Accuracy PnP Process.

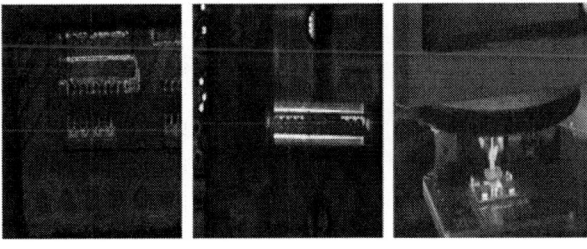

Figure 6.   MOI high Accuracy Pick and Place Process

Second Step: Assemble the Ferrule to MOI after first step completes. Ferrule consists of 45 degree prism, fiber guide, focusing lens and MPO which ease the assembly process of ferrule and MOI. It has an advantage that the design could be changed according to customer requirements and product applications, hence substantially down on the number of individual components so as facilitate the passive alignment desirable for mass production. In our prototype sample development, a commercial product ferrule "PRIZM Light Turn" form US Conec Ltd was used.

The output form each VCSEL and Photo Diode is collimated by the MOI so that it is focused by photonic connector lens for coupling into attached multimode fiber. So, the distance between the MOI and the optical element greatly affects the optical characteristics. Figure 7 shows a view image after MOI assembly process.

In prototype development, we used spacer on CiB module so that MOI can be position on the target distance (Optical element to MOI Lens). Figure 8 shows prototype side view. As a result, it was positioned at the target distance (290 ± 10um).

978-1-7281-1500-9/19 $31.00 © 2019 IEEE

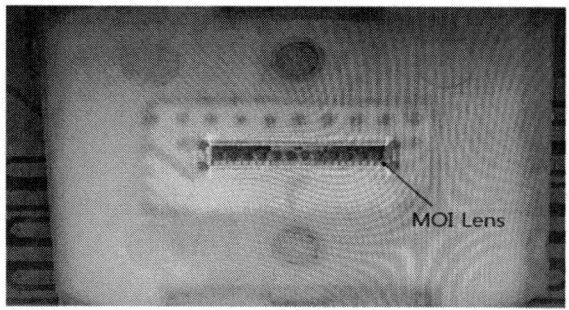

Figure 7. After MOI Pick and Place

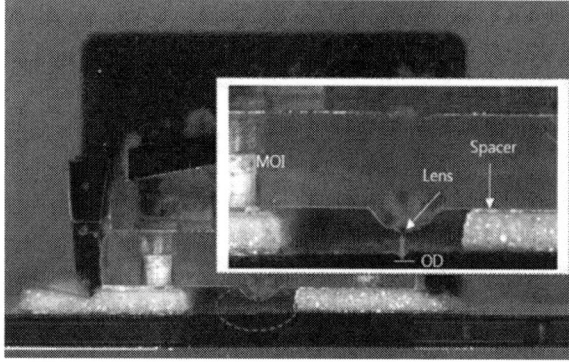

Figure 8. Detailed Side View of CiB Module

IV. CiB PERFORMANCE

A. Electrical Charaterization for CiB Structure

In order to understand electrical behavior and performance of the newly CiB concept, 3 types of packages (COB, FOWLP w/ Via Frame, CiB) were studied and compared using electrical simulation tool, i.e. Ansys HFSS EM Simulator.

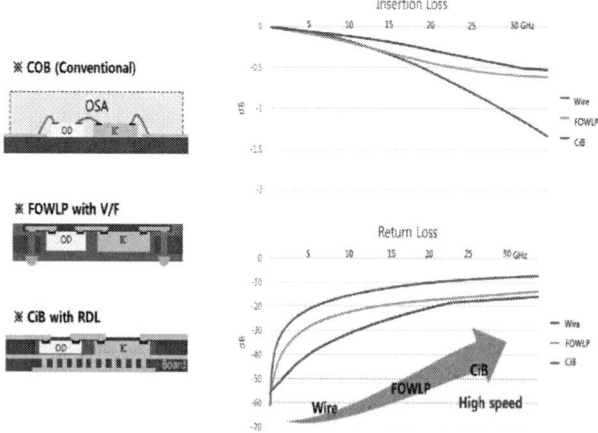

Figure 9. Electrical Chariteristics Comparison According to Different Packages (COB, FOWLP with VF, CiB)

The insertion loss and return loss graphs clearly indicates that CiB possessed the best performance. The design of the CiB package to have the shortest physical and electrical path from die to PKG Signal out pad lead to excellent signal transmitting. Therefore, it is concluded that it serves as the most advantageous technology to realize high speed and low power consumption that are crucial in 5G environment.

B. Optimized electrical charaterization for CiB module

The IC and Optical device used in this development supports bit rates up to 10Gbps/Ch. Hence, by combining all the 4 channels, we expect this product able to achieve maximum speed communication of 40Gbps.

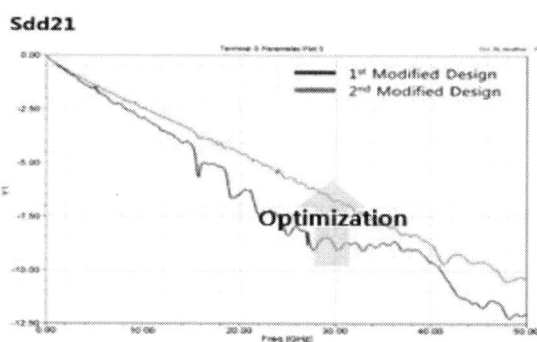

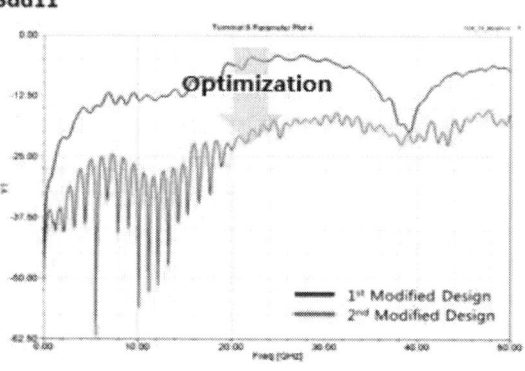

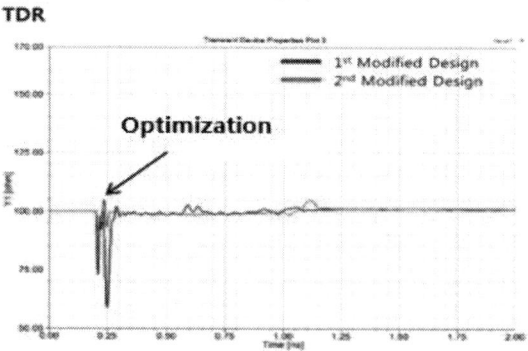

Figure 10. Electrical Chariteristics Optimization

However, through the CiB concept, we implemented impedance matching optimization as shown in Fig. 10 and discovered that a minimum 25 Gbps/ch in optical engine with total of speed 100Gbps or more is viable and seems promising.

C. Performance Test

For the purpose of investigating the modulation characteristic, the CiB Optical modules were then linked together via a MPO connector to attempt for high-speed data transmission using over 10Gbps. When evaluating the characteristics and performance of the CiB Module, it was attached to the evaluation board and Driver and TIA was controlled by I2C communication using Arduino (Fig. 11).

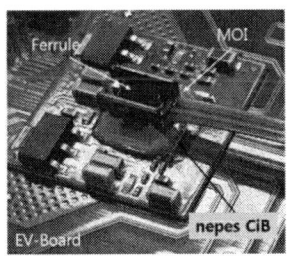

Figure 11. Setup of CiB Module on EV Board

The experimental setup for detecting the signal is illustrated in Figure 12. The performance for the Tx was obtained by monitoring the optical output form MPO Connector with a pulse pattern generator (PPG) Anritsu MP 1800A, an evaluation board, a detector and digital communications analyzer, Tektronix DSA8300. Through the evaluation, we confirmed the feasibility and stable performance characteristics of the CiB.

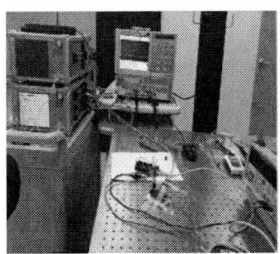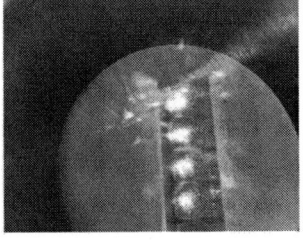

Figure 12. Experimental Setup

D. Reliability Test

Component level environment level test is carried out at given test conditions to validate performance and robustness at electrical test and visual inspections. Table 2 shows the package level reliability result and it passed JEDEC (Joint electron device Engineering Council) standard package reliability tests such as MSL1 (Moisture sensitivity Level).

Items	Condition	S/S	Results
Precon. (MSL1)	Bake: 24hr@125±5°C Soak: 85±2°C,85±3%RH,168Hr Reflow: ≥260°C, 3 cycles	81ea	0/81 Pass
T/C	500 cycles /-55 ~125°C	15ea	0/15 Pass
uHAST	96 hr /130°C/85%RH/230kPa	15ea	0/15 Pass
PCT	168 hr /121°C/100%RH/2atm	15ea	0/15 Pass
HTS	500hr/150°C	15ea	0/15 Pass

Table 2: Component Level Environment Test Results

V. CONCLUSION

In this paper, existing Fan-Out technology had further demonstrated a breakthrough and setting the bar high in more advanced packaging level. The advantages of our proposed optical CiB Structure using Fan-Out technology includes excellent electrical characteristics but also not limited to high productivity and cost-saving benefits. The simple alignment between the VCELs/PDs and the OSA was efficiently conducted with the application of Pick and Place process. In the meantime, the metal connection or routing lines connecting the both optical ICs to the PCB was successfully demonstrating the feasibility of obtaining thinner CiB package than conventional CoB package. Throughout the numerous design changes and experiments, the proposed concept was thoroughly simulated, eventually leading to an excellent performance, reliability and great possibility achieving speed of over 25Gbps/ch. The proposed optical CiB concepts will be applied a transceiver optical engine and mid board optical switch module in datacenter as future works.

Figure 13. Potential Application (Optical Swich module)

ACKNOWLEDGMENT

The authors would like to express their thanks to the Technology Innovation Program (20000868, Development of AI 3D IC Fabrication Process Technology using a FO-package) founded by the Ministry of Trade, Industry & Energy (MOTIE, Korea) for supporting this project.

REFERENCES

[1] Hak-Soon. Lee, Jun-Young Park, Sang-Mo Cha and Sang-Shin Lee, "Ribbon plastic optical fiber linked optical transmitter and receiver modules freaturing a high alignment tolerance," OSA 2011. vol. 19, pp. 4302–4309, Feb 2011.

[2] Jamshid. Sangirov, Gwan-chong Joo, Jae-Shik Choi and Do-Hoon Kim, "40 Gb/s optical subassembly module for a multi-channel optical link" OSA 2014. Vol. 22, pp. 1768-1783, Jan 2014.

[3] Eun kyu. Kim, Yong woo Lee, Sooraj Ravindran and Jun ki Lee, "4 channel x 10Gb/s bidirectional optical subassembly using silicon optical bench with precise passive optical alignment" OSA 2016. vol. 24, pp. 10777–10785, May 2016.

[4] S. Chuang, D. Schoellner, A. Ugolini, Wakjire, and G. Wolf, "Development and Qualification of a Mechanical-optical Interface for Parallel optical Links" Phopo Optical instrumentation Engineers 2015.

[5] Jong Heon Kim, Yong Tae Kwon, Young Ho Kwon and Yong Woon Yeo, "Fan out package: Performance and Scalability Perspective" ECTC 2018. vol. 68, pp. 1194–1199, 2018.

[6] Hak-Soon Lee, Sang Shin Lee and Yung sung Son, "CWDM based HDMI interconect incorporating passibely aligned POF linked optical subassembly module" OSA 2011. vol. 24, pp. 10380–10387, May 2016.

Active interposer technology for chiplet-based advanced 3D system architectures

Perceval Coudrain, J. Charbonnier, A. Garnier, P. Vivet, R. Vélard, A. Vinci, F. Ponthenier, A. Farcy*, R. Segaud,
P. Chausse, L. Arnaud, D. Lattard, E. Guthmuller, G. Romano, A. Gueugnot, F. Berger, J. Beltritti*, T. Mourier,
M. Gottardi, S. Minoret, C. Ribière, G. Romero, P.-E. Philip, Y. Exbrayat, D. Scevola, D. Campos*, M. Argoud,
N. Allouti, R. Eleouet, C. Fuguet Tortolero, C. Aumont, D. Dutoit, C. Legalland, J. Michailos*, S. Chéramy, G. Simon
Univ. Grenoble Alpes, CEA, LETI, 38000 Grenoble, France
* STMicroelectronics, 850 rue Jean Monnet, 38926 Crolles Cedex, france
perceval.coudrain@cea.fr

Abstract—We report the first successful technology integration of chiplets on an active silicon interposer, fully processed, packaged and tested. Benefits of chiplet-based architectures are discussed. Built up technology is presented and focused on 3D interconnects process and characterization. 3D packaging is presented up to the successful structural test and characterization of the demonstrator.

Keywords: 3D integration, active interposer, chiplet partitioning, through silicon via (TSV), Cu pillar, assembly

I. INTRODUCTION

Targeting high performance computing (HPC) applications, large many-core three-dimensional (3D) systems with multiple chiplets integrated on a silicon active interposer have been proposed [1]. The use of fine pitch 3D interconnects increases chip-to-chip bandwidth and limits overall power consumption, while a chiplet approach allows yield optimization with smaller chips and reusable IP blocks. Chiplet partitioning has already been achieved successfully and widely on passive interposers [2], but the association with active circuits on the interposer allowing smart features such as advanced network-on-chip (NoC) interconnects, fast IOs for off-chip communication, embedded power management and system-on-chip (SoC) infrastructure has not yet been reported. In this paper, we detail the first successful technology integration of chiplets on an active silicon interposer fully packaged and tested.

The paper first describes the approach of chiplet-based architectures for computing and explains its benefits in terms of memory wall breaking and cost effectiveness. Furthermore, the first part details the advantages of introducing an active circuit in the interposer, especially the capability of extended communication by using network-on-chip within the interposer. The practical case of INTACT prototype will then be presented. This object integrates a total of 96 cores offering a low power computing fabric with innovative cache-coherent architecture and wide voltage range. The 3D stack is composed of six identical multiprocessor 22 mm² chiplets, fabricated in 28 nm fully depleted silicon-on-insulator (FDSOI) technology and face-to-face stacked on a 200 mm² 65 nm CMOS active interposer by means of ultra-fine pitch Cu pillars. These Cu pillars offer large chiplet-to-chiplet communication bandwidth through the interposer, with a reduced impact on chiplet floorplan.

Section II explores the 3D technology deployed to achieve this demonstration. It focuses on the fabrication and optimization of the 3D interconnects leading to a successful integration. In particular, it details the metallization of the through silicon vias (TSV) with an investigation on diffusion barrier. Process optimizations leading to higher TSV process robustness are discussed based on a dedicated systematic defectivity study after TSV planarization. The realization of μ-pillar with ultra-fine pitch (20 μm) on Al pads is described. Backside process including copper nail reveal, the formation of a redistribution layer (RDL), an organic passivation and the realization of solder bumps, is exposed.

The assembly of the 3D system is presented in section III. Interposer is assembled on a 10-layer ball grid array (BGA) by means of Cu/SnAg solder bumps and a mass reflow process. Chiplets of 600 μm thick are stacked by thermocompression before a metal lid is added on the top and balls are attached on the BGA. The full 3D stack is characterized by cross sections and X-ray tomography.

The fourth and last part of this article describes the characterization of the entire technology. The impact of the TSV process on CMOS devices in the interposer is discussed. 3D interconnects performance is assessed from parametric electrical measurements conducted on TSV Kelvin structures and RDL/TSV/back end of line (BEOL) daisy chains. Finally, the design-for-test architecture and test strategy of the INTACT system is discussed and illustrates the first successful integration of chiplets on an active interposer.

A. Chiplet on interposer concept

Targeting high performance computing applications, large many-core three-dimensional (3D) systems with multiple chips integrated on a silicon interposer have been proposed. Two main concerns are currently driving this innovation: memory wall, and system cost.

Memory wall constraint. Stacking high bandwidth memories (HBM) onto passive silicon interposer and using standardized logical and physical memory interface achieves large DRAM memory capacity (up to 8 GBytes) and large memory bandwidth (up to 256 GBytes/s) with reduced energy cost [3]. This is achieved with 55 μm pitch μ-bumps. HBM memories are widely used in graphics processing unit (GPU) and field-programmable gate array (FPGA) devices; they may be used for other massively parallel computing architectures with large memory demand (Fig. 1).

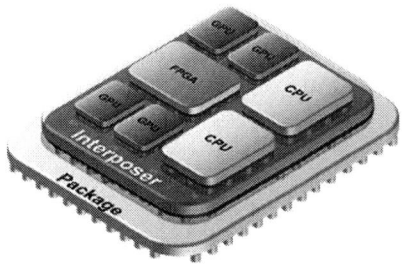

Figure 1. Chiplet on interposer topology.

Cost and form factor. Partitioning a single large die in a multitude of smaller dice leads to reduced costs in advanced nodes [4]. Stacking known-good-die (KGD) after electrical wafer sorting (EWS) reduces final system cost, as it was the case for the first FPGA partitioned in duplicated slices.

Nevertheless, there are also other qualitative reasons leading to chiplet based partitioning, heterogeneity and specialization. Instead of integrating all functions in a single CMOS node, one can use adequate technology for the right design and architecture. The case of HBM memories could be applied further, *e.g.* by integrating high-end computing cores in advanced nodes and complex analog and circuit IO physical layers (PHY) in more mature technologies. This could be achieved with a service IO chiplet offering all required system IOs. As a result, computing chiplets could be smaller and cost effective without integrating any analog and PHY.

Specialization also leads to genericity. This concept offers highly configurable and optimized chiplet, using adequate technologies: generic computing cores chiplet, GPU cores chiplet, FPGA fabrics chiplet, artificial intelligence (AI) accelerators chiplet for energy efficient execution of AI kernels. Finally, more dedicated functions can be integrated on specialized chiplets, designed and fabricated on demand, for system specialization.

The final system is then fully scalable by adding more dice or larger dice onto the interposer, while still keeping genericity, specialization and control of the system cost. Many technical challenges are still ahead, such as standardization of chiplet interfaces, system partitioning to get early system performance estimates, system level design methodologies, etc. In addition, chiplets are still currently integrated onto passive 2.5D interposers, which offer energy efficiency but rather basic chip-to-chip interfaces, and no opportunity for interposer specialization.

B. Benefits of an active interposer

By adding active circuits within the interposer, this latter becomes a complete circuit. Specialization and system optimization is then offered not only by the chiplet but also by the active interposer. Fig. 2 presents the overview of an active interposer for a computing many-core architecture.

Regarding technology partitioning and system cost, since the active interposer is assumed large, it is required to use a mature technology for the interposer to ensure high system yield, while the computing chiplets will be implemented using an advanced node.

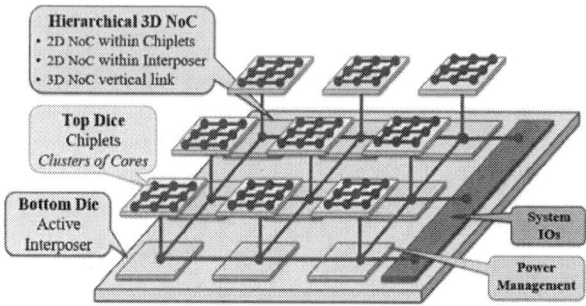

Figure 2. Active Interposer Partitioning for Many-core [1].

In terms of architecture, contrarily to passive interposer with system level communication restricted to chip-to-chip side-by-side communication, active interposers offer the possibility of extended communication capability. By using network-on-chip (NoC) within the interposer, a hierarchical NoC can be obtained, allowing communication from any chiplet to any chiplet, which is mandatory to support scalable cache coherent protocols [1][5]. With finer 3D interconnect pitch, communication bandwidth and density are increased, thus increasing the system level communication capacity.

Active interposers may also integrate system IOs as a specialization of the interposer for system level communication, and primarily with the memory.

Active interposers may also integrate power management features. The objective is to deliver the power as close as possible to computing chiplets. This allows reducing voltage drops and converter reaction time, and increases energy efficiency. Fully 3D integrated power management can be envisioned without any external passive components [6].

Regular circuit services such as clocking, reset, probes, and design-for-test (DFT) logic for testing the chiplet individually and the interposer can be easily and safely implemented within the interposer (see section IV.B).

Finally, a low logic density of 10-20 % has to be considered within the active interposer. It should be enough to provide all required and aforementioned features, while still ensuring a cost effective 3D stack [7].

C. INTACT prototype: computing with 96 cores on a chip

The INTACT prototype is a general purpose computing architecture offering a total of 96 cores (Fig. 3). An overview of its system architecture is shown in Fig. 3. The 3D stack is composed of six identical multiprocessor chiplets [8]. Each 22 mm² chiplet, fabricated in 28 nm FDSOI technology, integrates 16 cores, offering a low power computing fabric, with wide voltage range (0.6 V – 1.2 V). The chiplets are face-to-face stacked onto a 200 mm² 65 nm CMOS active interposer by means of ultra-fine pitch (20 µm) Cu µ-pillars [9]. A total of 150,000 Cu pillars are used on the active interposer to connect the six chiplets, including power-supplies. The 100 µm thick interposer contains 14,000 through silicon vias (TSV) mid-process of 10 µm in diameter. It integrates power units (switched cap DC-DC converters), system interconnects (network-on-chip), but also design-for-test, clocking and monitoring functions.

978-1-7281-1500-9/19 $31.00 © 2019 IEEE

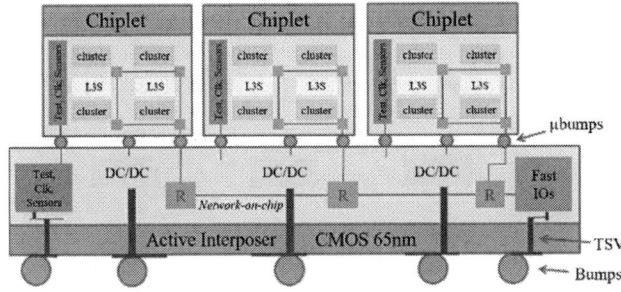

Figure 3. INTACT system architecture overview.

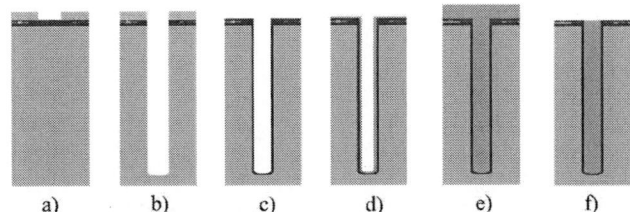

Figure 4. TSV process flow: a) lithography, b) dielectric & Si etch, c) TSV insulation, d) barrier/seed deposition, e) Cu fill, f) TSV planarization.

II. 3D TECHNOLOGY DEVELOPMENT

In this section the process technology built up for the realization of INTACT prototype is presented. Because it constitutes the vertebral column of the 3D architecture, linking chiplets, interposer and substrate, the 3D interconnects will be described in details. As already mentioned in section I, for INTACT prototype six chiplets are stacked above an active interposer by means of Cu pillar with a minimum pitch of 20 µm [10]. The 100 µm thick interposer holds TSV with a diameter of 10 µm that connect frontside BEOL to backside redistribution layer (RDL) and solder bumps. Attention will be paid to the interposer fabrication. Chiplet process will be limited to the realization of µ-bumps while BGA process will not be discussed. The interposer process flow is the following:

- CMOS 65nm front end of line (FEOL)
- TSV middle
- CMOS BEOL with 7 metal layers and Al pads
- Cu pillar above Al pads
- Temporary wafer bonding and silicon thinning
- Cu nail reveal, insulation and opening
- Redistribution layer (RDL) and passivation
- Solder bumps

A. Through silicon via development

1) TSV middle process flow

TSV middle flow is chosen for the realization of the INTACT prototype (Fig. 4). Details on this flow have been published in previous papers for 10x80 µm TSV, as in [11][12]. After front end of line (FEOL) processing up to pre-metal dielectric (PMD), a 40 nm SiN layer was deposited by chemical vapor deposition (CVD) as stop layer for future TSV planarization sequence. Lithography was done with 6 µm thick resist, with a unique diameter of 10 µm. Nitride and oxide layers were opened with dielectric etch, down to the silicon. Deep reactive ion etching (DRIE) was then used to etch the silicon with a depth of 105 µm. Etching processes were tuned to avoid silicon undercut under dielectric stack, to optimize via profile for the further metallization sequence, improving continuity of metal layer. TSV insulation was performed with the deposition of 350 nm sub-atmospheric CVD and 80 nm plasma enhanced CVD oxides. Thicknesses measured on TSV cross sections are given in Table I.

TABLE I. TSV INSULATION THICKNESS MEASUREMENTS

TSV region	Wafer Center	Wafer Edge
Sidewall Top	378 nm	379 nm
Sidewall Middle	330 nm	347 nm
Sidewall Bottom	319 nm	321 nm
TSV Bottom	343 nm	334 nm

After insulation, barrier and seed layers were deposited. A widely used TaN-based barrier was compared to an alternative TiN-based barrier, studied for high aspect ratio TSV integration. Among possible alternatives to TaN/Ta, that could be limited in terms of step coverage, TiN is an interesting candidate, as described in [13]. Low temperature metal organic chemical vapor deposition (MOCVD) with tetrakis(diethylamino)titanium (TDEAT) has proven high conformality with excellent barrier properties against copper diffusion [14].

For the TaN-based barrier TSV, the deposition sequence consisted in pysical vapor depositions (PVD) of 10 nm TaN, 120 nm Ta and 600 nm Cu seed layers. For the TiN-based barrier TSV, the deposition sequence consisted in 20 nm MOCVD TiN, 100 nm PVD Ti and 1500 nm ionized PVD (iPVD) Cu seed layers. Conformality of the iPVD process being limited for aspect ratio higher that 10x80 µm, an additional 200 nm electro-grafted Cu layer was deposited to ensure proper seed continuity across the full TSV depth for the TiN flow. Note that the Ti cap was added on the TiN to improve the adherence of the Cu seed on the barrier.

In both cases, electro chemical deposition (ECD) of Cu was processed with an optimized recipe allowing perfect bottom-up filling of the TSV, as illustrated in Fig. 5 and Fig. 6. Consecutive to ECD Cu fill, an overburden reduction step (OBEB) was done *in situ* with diluted sulfuric peroxide (DSP) to reduce the Cu thickness on the field from ~5 µm to less than 2 µm. This process helped at reducing the stress on TSV interfaces during the subsequent Cu anneal.

TSV was then annealed at 400°C with a heating profile also optimized to minimize stresses in the interfaces. Finally a three-step chemical mechanical polishing (CMP) process was performed to remove Cu, TaN or TiN barriers and SACVD oxide respectively, with selective stop on the SiN layer above PMD. At this stage the wafer surface was perfectly flat on the PMD dielectric and the remaining SiN. No adaptation was nedded for TaN and TiN-based wafers as uniform stop on the SiN was achieved in both cases.

978-1-7281-1500-9/19 $31.00 © 2019 IEEE

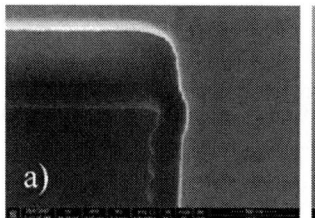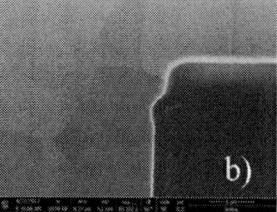

Figure 5. Cross sections of TSV with (a) TiN/Ti, (b) TaN/Ta barrier.

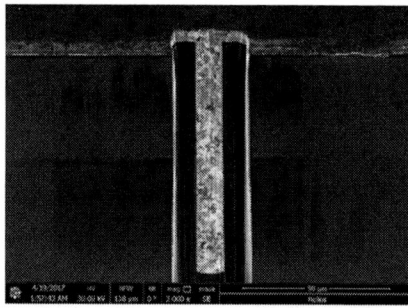

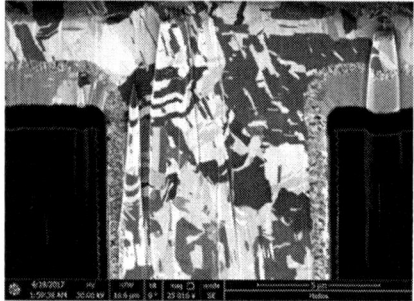

Figure 6. TSV filled with TiN/Ti barrier flow.

2) TSV defectivity management

Because wafers were re-entering the CMOS flow for BEOL processing, particular attention was paid to the control of the defectivity associated to the TSV. A systematic defectivity inspection was implemented after CMP. In addition to FEOL patterns, over 2.8 million TSVs were inspected per wafer and allows quantitative analysis. An example of data is given in Fig. 7 (a) where 938 defects are localized on a wafer. We clearly observe on Fig. 7 (b), showing all defects superposed on one field, that the main defectivity is related to the TSV, as the pattern directly corresponds to TSV layer layout. Inspection revealed that 95 to 99% of these defects, depending on the wafers, are located at the TSV surface. They have a homogeneous distribution across the wafer with a size ranging from 50 nm to 10 µm, allowing a possible classification by size, as illustrated in Fig. 7 (c). *Class 1* includes purely cosmetic defects with diameter <1 µm, *Class 2* contains defects <3 µm with no consequence for the post process and *Class 3* contains bigger defects with possible impact on BEOL process. Finally, *Class 4* (not shown) depicts large defects as cracks, layer delamination or unfilled TSV. In addition to TSV defects, random defects, estimated between 50 and 200 per wafer, are found outside TSV vicinity. They correspond to embedded particles, FEOL defects and micro-scratches.

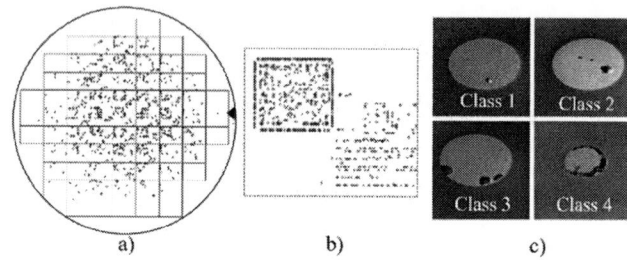

Figure 7. (a) 938 defects localized on wafer, (b) defects superposed on one field showing relation with TSV layout, (c) defect classification by size.

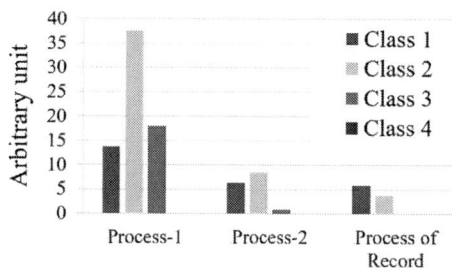

Figure 8. Defectivity analysis allowed improving overall TSV process flow quality (data from TiN-barrier wafers).

Data analysis allowed validating compatibility of wafers with the requirements of subsequent BEOL process. It also allowed improving the overall TSV flow by assessing optimizations of etch, metallization and anneal processes as shown in Fig. 8 with the distribution of defects for three evolutions of the technology. Final process of record (POR) gives lowest defect density with only *Class 1* and *2* defects.

Wafers with TiN barrier exhibit a higher defect count than TaN barrier, with a ratio reaching 5:1 on some wafers, but with lower criticality, essentially *Class 1* and *2* defects. Wafers with TaN barrier however exhibit, a larger amount of *Class 3* defects with several *Class 4* that were not detected on TiN wafers. Localized cracks and layer delaminations in relation with TSV localization were observed at the edge of two wafers with TaN barrier and caused wafers rejection. This difference seems to be related to a higher stress in the TSV region due to the TaN/Ta barrier. Full field X-Ray diffraction conducted at the European Synchrotron Radiation Facility (ESRF) confirmed this assumption. TiN-based flow may suffer from micro-voids at Ti/Cu interface due to a weakness of iPVD Cu process, leading to *Class 1* and *2* defects after anneal. This issue is not met with PVD Cu in TaN-based flow.

B. Interposer/Chiplets Interconnect Technology

1) µ-pillars on interposer

µ-pillars were processed on Al pads on interposer wafers. TiW barrier and Cu seed layers of 130 nm and 200 nm, respectively, were deposited by PVD. Lithography was achieved with 12 µm thick positive resist, with a critical dimension (CD) of 10 µm and a pitch of 20 µm. Considering this minimal pitch, a stepper process was preferred to mask aligner to avoid misalignment on Al pads and further stacking issues with the chiplets.

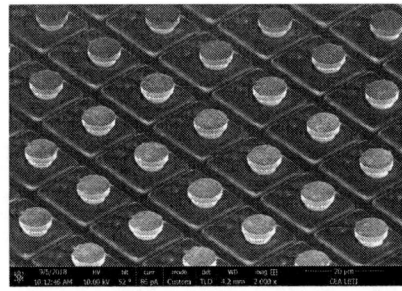

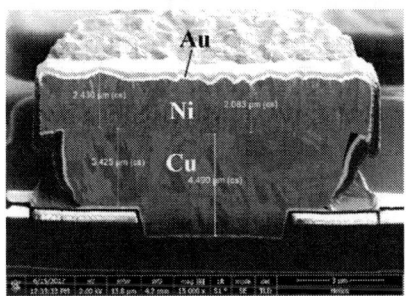

Figure 9. (top) μ-pillars array on interposer, (bottom) cross section of a μ-pillar on Al pad, composed of Cu 3 μm / Ni 2 μm / Au 0.1 μm.

A descum was done before ECD step which consists in Cu 3 μm / Ni 2 μm / Au 0.1 μm growth. After resist stripping, μ-pillars heights was measured by confocal interferometry on 78 points per wafer (see Table II for typical result). Cu seed and TiW barrier were etched by diluted sulfuric peroxide (DSP) and H_2O_2, respectively. μ-pillars after full process are illustrated in Fig. 9 in an array (top), and in cross section (bottom). Reduction of Cu diameter under Ni occurred during Cu seed etching step.

2) μ-bumps on Chiplets

μ-bumps are formed above Al pads on chiplet wafers, with a process similar to μ-Pillars. ECD here consisted in Cu 3 μm / Ni 2 μm / SnAg 5 μm growth. After resist stripping, μ-bumps were measured (see Table II). Cu seed and TiW barrier were etched with diluted phosphoric peroxide (DPP) and H_2O_2, respectively. A reflow process was conducted at 255°C under N_2 atmosphere to obtain the semi-spherical shape illustrated in Fig. 10.

C. Interposer/BGA interconnect technology

1) Si thinning and Cu nail reveal

Temporary wafer bonding was achieved with 20 μm thick adhesive and controlled by acoustic microscopy. Thinning was achieved by a combination of coarse grinding, edge trimming, fine grinding and CMP, with the objective of final Si thickness of 112 μm. Si thickness was monitored by infrared interferometry with 4000 points per wafer. Average variation from the target was inferior to 2 μm and typical standard deviation was inferior to 1 % within a wafer.

After proper cleaning, Cu nails were revealed with a SF_6 etch process with end point detection. Nails height ranged from 2 to 5 μm after etching. A 2 μm TEOS oxide was deposited at 150°C. Finally CMP was done to open the Cu for contact, with an oxide consumption of less than 300 nm.

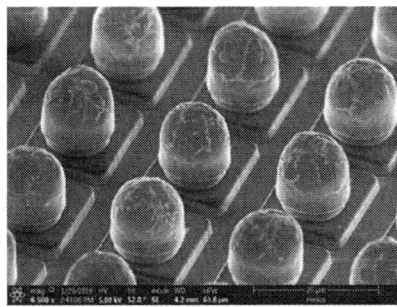

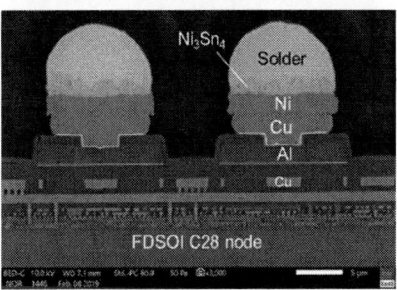

Figure 10. (top) μ-bumps arreay on chiplet, (bottom) cross section of a μ-bump composed of Cu 2 μm / Ni 2μm / SnAg 5μm after reflow at 255°C.

2) Redistribution layer and passivation

Before redistribution layer (RDL) process, Cu was slightly recessed to ease alignment of RDL lithography. 100 nm Ti barrier and 200 nm Cu seed layers were deposited. Lithography was done with a 12 μm thick resist, with minimum line width of 10 μm (Fig. 11). After descum 2.5 μm Cu were grown by ECD. RDL height was measured by confocal interferometry (see Table II for typical result) before Cu/Ti etching. ALX passivation was patterned on RDL with a mask aligner lithography, and annealed at 190°C for 2 h under N_2 atmosphere. Passivation openings with 55 μm diameters defined locations of the solder bumps.

TABLE II. 3D INTERCONNECTS TYPICAL THICKNESS MEASUREMENTS (78 POINTS PER WAFER)

Interconnect	Thickness	Std. dev
μ-pillar (interposer)	5.52 μm	0.22 μm (3.9%)
μ-bump (chiplet)	9.61 μm	0.47 μm (4.9%)
RDL (interposer)	2.37 μm	0.08 μm (3.3%)

3) Backside solder bumps process

Solder bumps lithography was performed with a 100 μm thick negative dry film laminated under vacuum after deposition of 100 nm Ti barrier and 300 nm Cu seed layers. Lithography was done by mask aligner with a CD of 78 μm. Bumps were grown by ECD of 4 μm Cu and 68 μm SnAg. Stripping was carefully optimized for the thick film. Cu/Ti layers were etched and wafers underwent a reflow process to obtain a quasi-spherical shape with ~80 μm in diameter, as shown in Fig. 12.

Wafers were then debonded and cleaned to remove adhesive residues; thin interposers with both sides processed were obtained.

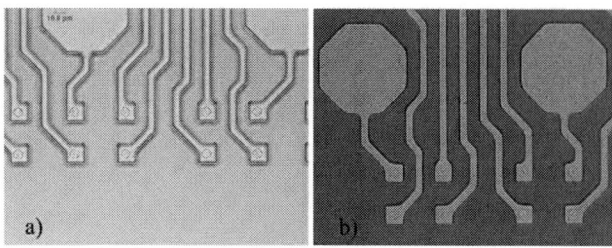

Figure 11. RDL with minimal CD of 10 μm. (a) after lithography, recessed Cu nails are visible. (b) after ECD, stripping and seed/barrier etching.

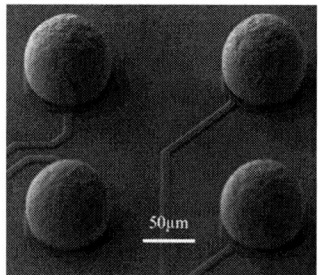

Figure 12. Solder bumps after seed etch.

III. 3D SYSTEM PACKAGING

This section details the packaging sequence used for INTACT demonstrator illustrated in Fig. 13. The interposer was first assembled on a laminate substrate by standard mass reflow (MR) process and capillary underfill (CUF). Then six 600 μm thin chiplets were stacked on the interposer frontside by thermocompression (TC) bonding followed by CUF process. Highlight is done on chiplets stacking with 20 μm pitch μ-bumps. In this process, an accurate alignment with optimization of the parallelism was needed for each chiplet.

A. Substrate description

The laminate substrate is a 40 x 40 mm² 10-layer flip chip thermal enhanced ball grid array (FCTEBGA) with 1521 backside interconnects using 1 mm pitch. Its total thickness is 1.27 mm without balls. Capacitances were mounted on the substrate before interposer assembly.

B. Interposer stacking on substrate

Prior to assembly, substrate and interposer warpages were measured as a function of temperature. Whereas substrate warpage was found constant with temperature, with an amplitude of 5 μm in interposer area and 20 μm on the whole diagonal, interposer warpage was concave with a maximal value of ~70 μm at 30°C (Fig. 14), and 40 μm at 230°C. Successful assembly was achieved by pick and place (P&P) and mass reflow. Hence, 80 μm diameter bumps on interposer and 15 μm solder-on-pad on substrate overcame ~70 μm warpage at 30°C and ~40 μm warpage at 230°C. After interposer assembly, X-ray microscopy (XRM) characterization was carried out, typical observation is given in Fig. 15. Attention was paid to track defects such as open and short circuits in solder bumps connections. Analysis only revealed a few defects while major part of assembled interposers exhibited perfect connections.

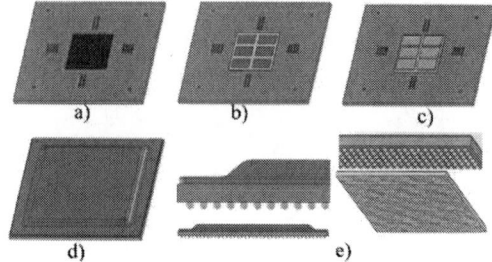

Figure 13. INTACT packaging flow: a) surface mounted devices (SMD), b) interposer assembly, c) chiplet assembly, d) lid bonding, e) balling.

Figure 14. Interposer warpage at room temperature.

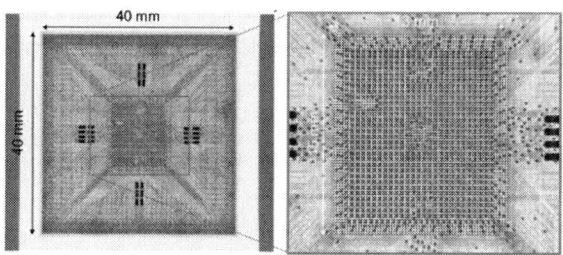

Figure 15. XRM view of the interposer on BGA, zoom on bumps location.

C. Chiplet stacking on interposer

Warpage of interposer assembled on substrate was measured as a function of temperature. In chiplet locations the warpage at 30°C was found negligible and increased to 15 μm at 230°C with a concave shape. Contact and soldering at 230°C with only 5 μm solder available on μ-bumps was then challenging. Thus thermo-compression was recommended. Successful chiplet assembly was achieved by TC on a SET FC300 tool, enabling efficient and repeatable alignment performance. This tool also enabled parallelism tuning on each chiplet. Liquid flux was used to enable solder wetting and improve the soldering during TC. Parallelism was tuned using chiplets corners. Cross sections were performed at the edge and in the center on all chiplet to check alignment performance and observe contact between chiplets and interposer. Proper connection and alignment was demonstrated on all chiplets (Fig. 16).

D. Full assembly characterization

SEM cross-sections were carried out to image all 3D interconnects levels from BGA to the chiplets (Fig. 16). X-Ray tomography also revealed the internal structure of the 3D stack with a clear view of 3D interconnects (Fig. 17). Both techniques confirmed perfect alignment and contact of μ-pillars and μ-bumps during chiplet stacking.

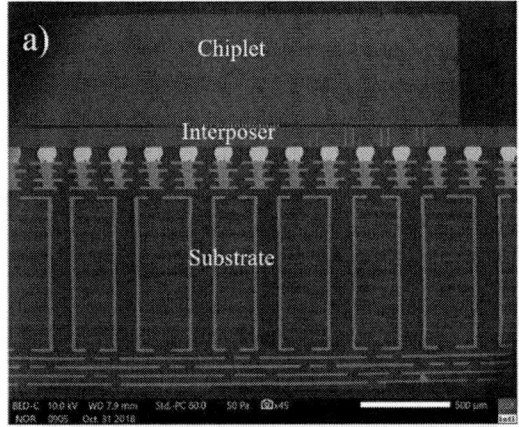

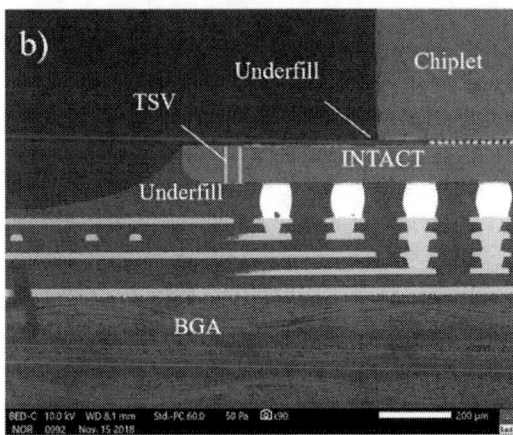

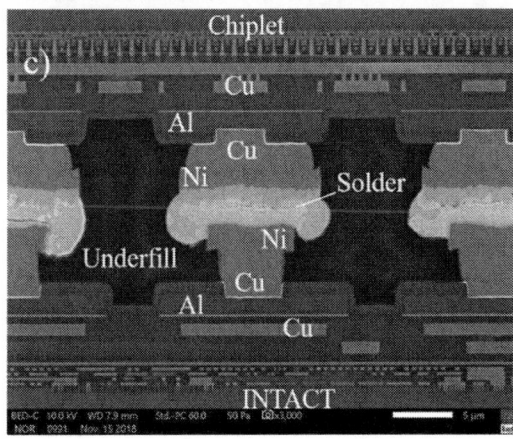

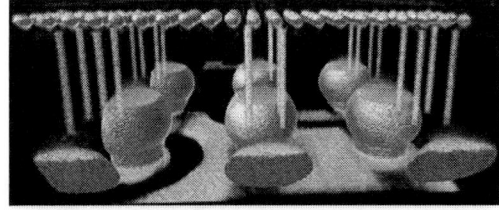

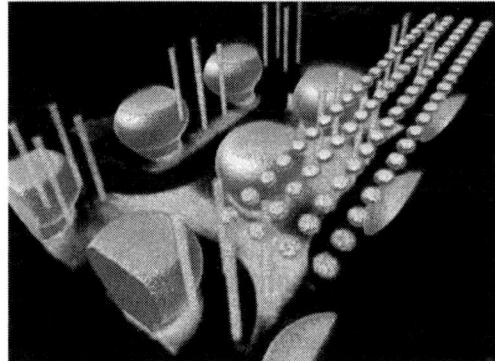

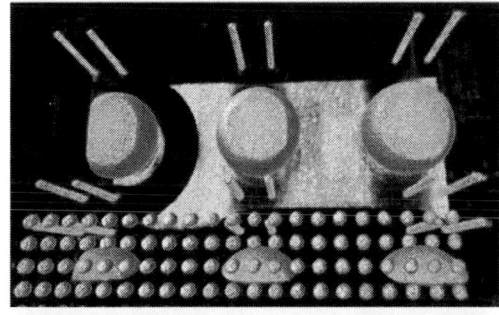

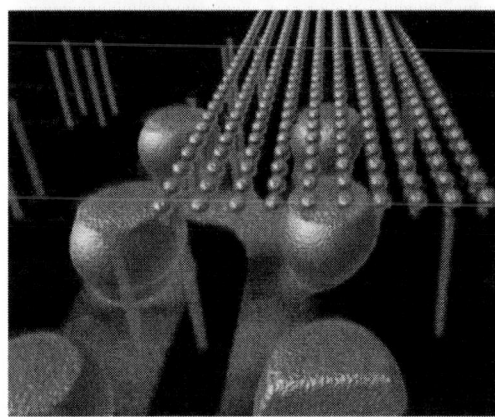

Figure 16. INTACT prototype after 3D assembly. (a) 600 μm thick chiplet on 100 μm thin interposer on BGA, (b) focus on the interposer with TSV, (c) 20 μm pitch / 10 μm diameter μ-pillars/μ-bumps connections.

Figure 17. INTACT full 3D stack after assembly seen through X-ray tomography where all 3D interconnects layers are simultaneously visible. Perfect alignement and contact of μ-pillars and μ-bumps is observed.

978-1-7281-1500-9/19 $31.00 © 2019 IEEE   575

IV. Characterization and Test

A. 3D process technology assessment

1) TSV impact on circuit

Parametric Test (PT) was conducted on interposer wafers during BEOL processing. TSV impact on circuit features such as MOS devices, SRAM cells, contact chains, metal layers, MIM parasitic capacitance etc. was evaluated and compared to reference lots without TSV. Fig. 18 illustrates this comparison with I_{OFF} and I_{ON} currents measured on CMOS transistors. No detrimental effect of TSV was noted during PT. However, slight differences were still observed between TaN and TiN barriers, for which we suspect a mechanical origin, explained by a lower stress level for TiN barriers.

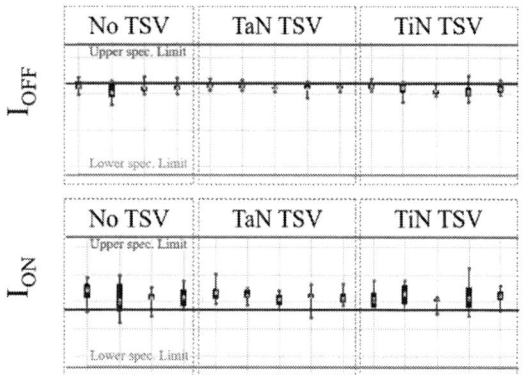

Figure 18. I_{OFF} & I_{ON} currents of MOS transistors tested at PT. TSV wafers with either TaN or TiN show equivalent behaviour than reference wafers.

2) 3D interconnects electrical characterization

This part discusses electrical characterizations of 3D interconnects conducted on RDL pads after passivation. Test structures include TSV Kelvin and RDL/TSV/BEOL daisy chains. The test was run on 75 dice per wafer, for nine wafers within three batches: batch A and batch B were processed with TaN barrier and batch C with TiN barrier.

TSV resistance was measured on Kelvin structures, avoiding resistive contributions from contacts, RDL and BEOL. Five Kelvin structures were tested on each die. Measurement results are given in Table III for the nine wafers. The unitary TSV resistance is found around 26.5 mΩ. Theoretical value for a perfect cylinder, considering 100 μm high TSV, Cu effective diameter of 9.2 μm and copper resistivity of 18 mΩ.μm, leads to 27 mΩ. These results indicate well continuous TSV without filling voids for all wafers. A limited dispersion is observed on the values within wafers and between wafers, with a maximal variation between wafers inferior to 2 %. Dispersion on TSV resistances mostly corresponds to silicon thickness variation, which is coherent with measurements conducted after wafer silicon thinning. Fig. 19 illustrates the results obtained on four wafers from batch A and B in the form of cumulative frequency graph. Measurement yield reaches 100 % for three wafers, while 7 % defective structures on wafer B1 were linked to a process issue on wafer edge.

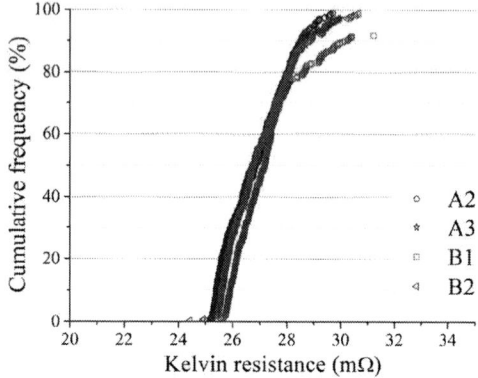

Figure 19. Kelvin resistance measured on 375 structures per wafer for A2, A3, B1 and B2 wafers.

TABLE III. RESISTANCE MEDIAN VALUE FOR THE TEN WAFERS

Wafer	Barrier	Kelvin mΩ	DC 300 mΩ / TSV	DC 870 mΩ / TSV
A1	TaN	26.2	169.1	169.2
A2	TaN	26.4	170.1	170.1
A3	TaN	26.4	169.9	169.8
B1	TaN	26.9	168.9	170.0
B2	TaN	26.4	169.6	169.5
C1	TiN	26.2	174.2	170.8
C2	TiN	25.7	171.9	171.4
C3	TiN	26.0	169.6	171.1
C4	TiN	26.7	173.6	172.7

Daisy chains (DC) of 30, 50, 150 and 870 TSV with 40 μm pitch were tested in 4-probe configuration to characterize the electrical yield of RDL/TSV/BEOL path. Measurement results are given in Table III for the nine wafers, and in Fig. 20 where the results on the two best wafers are illustrated in the form of cumulative frequency graph exhibiting 100 % yield on all TSV chain. Maximal variation of the resistance between wafers is about 2 % with a median value around 169 mΩ per TSV node, including RDL and BEOL resistances. A good linearity is obtained over the four chain lengths. A slight increase of the chain resistance is observed for wafers of batch C, and is explained by a thinner RDL. The higher resistance for 300 TSV DC compared to 870 TSV DC for wafers C1 and C4 indicates that some nodes in the chain are more resistive (their relative contribution in the resistance per node is lower for a long chain as DC 870). We attribute this increase to a non-optimal contact between RDL and Cu nail.

No noticeable difference was observed in TSV or chain resistance between TaN/Ta and TiN/Ti integrations, which indicates that from an electrical point of view the barrier is of second order. The characterization yield, *i.e.* the amount of functional structures, however, is lower for TiN wafers, which indicates that the integration is still not as mature as with TaN. As for the defectivity analysis, we attribute these failures to the iPVD Cu layer which presents continuity weakness. Nevertheless, the overall results exhibit high performance for full technology from FEOL to RDL.

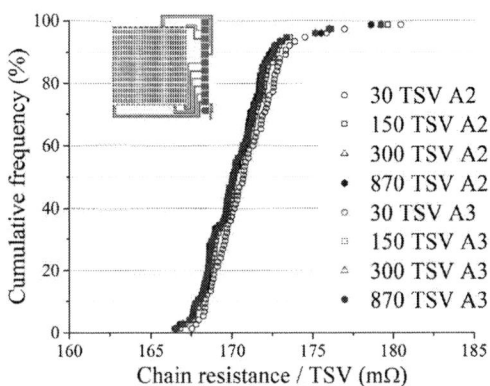

Figure 20. 30/150/300/870 TSV daisy chains measured for the two best wafers A2 & A3 (resistance is given by TSV node).

B. Active interposer demonstrator test

For such chiplet based 3D architecture, testability is a clear challenge. With the standard test challenges but also with new defect sources (TSV cracks, µ-bump misalignment, etc), a known good die (KGD) test strategy has to be defined [15]. The 3D test architecture has to offer pre-bond test of the chiplets at wafer level, before 3D assembly, and the final test after 3D assembly, while testing all features to ensure coverage of logic, memories, and 3D interconnects. Another challenge concerns the test access in the case of ultra-fine pitch 3D interconnects.

1) 3D Test Architecture

For testing such large systems with high level of maturity, a complete 3D design-for-test architecture is proposed within the active interposer [16]. It is based on two main structural test structures. The first is a IJTAG test interface, using IEEE1687 standard, which offers access to boundary scan chains for individually testing all 3D interconnections (passive or active links), and for controlling memories built-in-self-test (BIST). The second is a full scan network using test compression, to test full logic stack, with a reduced pin count at 3D interface. The main design innovation is coming from chiplet footprint proposal, allowing to individually test each chiplet, in standalone for pre-bond test or mounted on the active interposer for final test, as presented in Fig. 21.

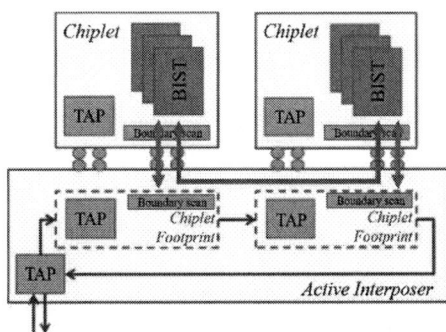

Figure 21. 3D design-for-test architecture principle.
TAP stands for test access ports.

2) Test strategy

Specific care has to be paid to provide required test access structures, namely test pads for EWS probe card. Since 20 µm pitch µ-bumps cannot be directly accessed by any standard available probe card, test is performed on a dedicated standard IO ring, integrating test signals and required power supplies.

For INTACT demonstrator, the FDSOI 28nm chiplet IO ring presents a 36 µm pitch, which is compatible with advanced probe card cantilever technology. Fig. 22 presents chiplet IOring where EWS test is performed, and corresponding chiplet probe card for wafer level testing.

Since the active interposer is using a mature 65 nm technology and a low logic density, yield is assumed to be very large and no test is done in pre-bond phase at wafer level (however test architecture and IO test pads are still available). After 3D assembly, final test of INTACT demonstrator is done in package using dedicated probe card and socket as shown in Fig. 23.

3) Test results

The FDSOI 28nm chiplets were successfully tested at wafer level. Binning strategy was defined according to test results (Fig. 24). For final test, the test vectors have been successfully applied to the demonstrator using an automated test equipment (ATE). It has been possible to test individually: i) the active interposer and its connection to the package, ii) the chiplet connectivity, either for chiplet-to-interposer links or for chiplet-to-chiplet links.

Correct 3D connection was achieved between the package and the interposer, including package, RDL and TSV, and between the interposer and the chiplets through µ-bumps and µ-pillars. More advanced tests are still on-going onto the ATE for full chip qualification, and for INTACT demonstration onto an application board.

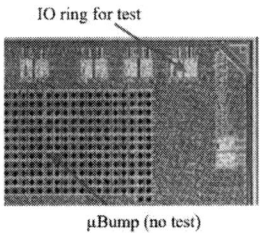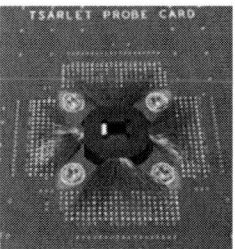

Figure 22. Chiplet test access and associated EWS probe card.

Figure 23. INTACT package and corresponding socket probe card.

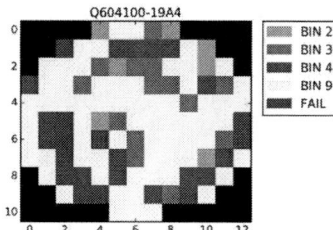

Figure 24. Chiplet EWS binning results.

V. CONCLUSION

Breaking with the System-on-Chip paradigm, future systems will consist in smaller, less expensive and independently designed components called chiplets, connected by high bandwidth interconnections over silicon interposers. We have explained the benefits of stacking chiplets on an active interposer for achieving large and scalable many-core systems for high performance computing. The INTACT prototype, embedding 96 cores within six chiplets, was introduced. We have detailed the full fabrication technology to enable turning this concept into reality: 3D interconnects process including TSV and ultra-fine picth μ-pillars and the 3D packaging flow. The entire technology was assessed morphologically and electrically with specific 3D test structures. Finally, the structural test of the INTACT prototype was reached and is starting to show results well within expectations. The full extraction of the system performance associated to the architecture details is still on-going. These results pave the way to the design of future high efficiency systems for high performance computing.

ACKNOWLEDGMENT

This work was supported by the French National Program "Programme d'Investissements d'Avenir, IRT Nanoelec" under Grant ANR-10-AIRT-05.

The authors thank teams of STMicroelectronics, Grenoble and Crolles sites (France) for their support and help on circuit manufacturing, assembly and physical characterization. The authors thank the Optronic department at CEA-Leti for his support on the assembly process and characterization. The authors finally thank ESRF member Tao Zhou for his contribution to the understanding of TSV metallization impact on the mechanical stress in the interposer.

REFERENCES

[1] P. Vivet, C. Bernard, E. Guthmuller, I. Miro-Panades, Y. Thonnart and F. Clermidy, "Interconnect Challenges for 3D Multi-cores: From 3D Network-on-Chip to Cache Interconnects," 2015 IEEE Computer Society Annual Symposium on VLSI, Montpellier, 2015, pp. 615-620.

[2] C. Lee *et al.*, "An Overview of the Development of a GPU with Integrated HBM on Silicon Interposer," 2016 IEEE 66th Electronic Components and Technology Conference (ECTC), Las Vegas, NV, 2016, pp. 1439-1444.

[3] D. U. Lee *et al.*, "25.2 A 1.2V 8Gb 8-channel 128GB/s high-bandwidth memory (HBM) stacked DRAM with effective microbump I/O test methods using 29nm process and TSV," 2014 IEEE International Solid-State Circuits Conference Digest of Technical Papers (ISSCC), San Francisco, CA, 2014, pp. 432-433.

[4] D. Stow, Y. Xie, T. Siddiqua and G. H. Loh, "Cost-effective design of scalable high-performance systems using active and passive interposers," 2017 IEEE/ACM International Conference on Computer-Aided Design (ICCAD), Irvine, CA, 2017, pp. 728-735.

[5] J. Yin *et al.*, "Modular Routing Design for Chiplet-Based Systems," 2018 ACM/IEEE 45th Annual International Symposium on Computer Architecture (ISCA), Los Angeles, CA, 2018, pp. 726-738.

[6] G. Pillonnet, N. Jeanniot and P. Vivet, "3D ICs: An opportunity for fully-integrated, dense and efficient power supplies," 2015 International 3D Systems Integration Conference (3DIC), Sendai, 2015, pp. TS6.4.1-TS6.4.8.

[7] D. Gitlin *et al.*, "Generalized cost model for 3D systems," 2017 IEEE SOI-3D-Subthreshold Microelectronics Technology Unified Conference (S3S), Burlingame, CA, 2017, pp. 1-3.

[8] E. Guthmuller *et al.*, "A 29 Gops/Watt 3D-Ready 16-Core Computing Fabric with Scalable Cache Coherent Architecture Using Distributed L2 and Adaptive L3 Caches," ESSCIRC 2018 - IEEE 44th European Solid State Circuits Conference (ESSCIRC), Dresden, 2018, pp. 318-321.

[9] D. Lattard *et al.*, "ITAC: A complete 3D integration test platform," 2016 IEEE International 3D Systems Integration Conference (3DIC), San Francisco, CA, 2016, pp. 1-4.

[10] A. Garnier *et al.*, "Electrical Performance of High Density 10 μm Diameter 20 μm Pitch Cu-Pillar with Chip to Wafer Assembly," 2017 IEEE 67th Electronic Components and Technology Conference (ECTC), Orlando, FL, 2017, pp. 999-1007.

[11] J. Colonna *et al.*, "Electrical and morphological assessment of via middle and backside process technology for 3D integration," 2012 IEEE 62nd Electronic Components and Technology Conference, San Diego, CA, 2012, pp. 796-802.

[12] P. Coudrain *et al.*, "Towards efficient and reliable 300mm 3D technology for wide I/O interconnects," 2012 IEEE 14th Electronics Packaging Technology Conference (EPTC), Singapore, 2012, pp. 330-335.

[13] C. Aumont *et al.*, "12:1 aspect ratio mid-process TSV integration and electrical tests using advanced metallization processes", International Conference and Exhibition on Device Packaging, Fountain Hills, AZ, 2018.

[14] L. Djomeni *et al.*, "Study of low temperature MOCVD deposition of TiN barrier layer for copper diffusion in high aspect ratio through silicon vias", Microelectronic Engineering, Volume 120, 2014, pp. 127-132.

[15] C. Papameletis, B. Keller, V. Chickermane, S. Hamdioui and E. J. Marinissen, "A DfT Architecture and Tool Flow for 3-D SICs With Test Data Compression, Embedded Cores, and Multiple Towers," in IEEE Design & Test, vol. 32, no. 4, pp. 40-48, Aug. 2015.

[16] J. Durupt, P. Vivet and J. Schloeffel, "IJTAG supported 3D DFT using chiplet-footprints for testing multi-chips active interposer system," 2016 21th IEEE European Test Symposium (ETS), Amsterdam, 2016, pp. 1-6.

2019 IEEE 69th Electronic Components and Technology Conference (ECTC)

Process Development of Power Delivery Through Wafer Vias for Silicon Interconnect Fabric

Meng-Hsiang Liu, Boris Vaisband, Amir Hanna, Yandong Luo, Zhe Wan, and Subramanian S. Iyer

Center for Heterogeneous Integration and Performance Scaling (CHIPS),
Henry Samueli School of Engineering, University of California, Los Angeles
{mhliu@ucla.edu}

Abstract–At UCLA Center for Heterogeneous Integration and Performance Scaling (CHIPS), we have been developing a fine pitch heterogeneous wafer-scale platform with a single level of hierarchy called the silicon interconnect fabric (Si-IF). The Si-IF is a platform for heterogeneous integration of different bare dies at fine pitch (2 to 10 μm) and close proximity (<100 μm die spacing). The Si-IF platform can accommodate an entire 50 kW data center on a single 300 mm diameter wafer. Power delivery and heat extraction are fundamental challenges. To minimize the overhead of power conversion, current at mission (point-of-load) voltage is planned to be delivered directly to the assembly; this requires a uniform delivery of tens of kilo-amperes. Our approach is to deliver the current from the back of the Si-IF, using cooled Cu fins and through wafer vias (TWVs), to the front side of the wafer, where the dies are assembled facedown. TWVs are a key component of this power delivery system and are required to penetrate through the entire thickness of the Si-IF (500 – 700 μm). A process for fabrication of large-sized (100 μm diameter) TWVs for the Si-IF is described in this paper. The TWVs are etched in 500 μm Si wafer (aspect ratio of 1:5) and are designed to enable back-side power delivery to the integrated system. Each TWV exhibits a resistance of 1.1 mΩ with an extracted resistivity of $1.73 \cdot 10^{-8}$ Ω·m. The scale and performance of these large-sized TWVs supports high current density for power delivery applications.

Keywords–*Through wafer via; silicon interconnect fabric (Si-IF); silicon etch; copper electroplating*

I. INTRODUCTION

Reducing communication power and latency, drives the need for both, high performance devices and improved integration (*i.e.*, high interconnect density and fine spaced die assembly) [1]. Conventional packaging and integration technologies, however, are based on solder attachment of chips to printed circuit boards (PCBs). In this approach, the large interconnect pitch on the PCB, typically 0.4 to 1 mm, significantly limits the data bandwidth [1-3].

We have developed the silicon interconnect fabric (Si-IF), a heterogeneous integration wafer-scale platform with a single level of hierarchy. The Si-IF supports fine pitch and solder-less metal-metal integration. In the Si-IF technology, bare dies are attached to the Si wafer [2] using thermal compression bonding (TCB) [1]. The Si-IF supports heterogeneous integration of bare dies at fine pitch (2 to 10 μm) and high proximity (<100 μm spacing), as compared to the large pitch and spacing on PCBs [1,3], leading to reduces communication energy and latency [4]. The Si-IF platform is expected to accommodate an entire 50 kW data center on a single 300 mm diameter wafer. Heat dissipation and power delivery are,

therefore, key challenges of the Si-IF approach. Our approach is to deliver the current from the back side of the Si-IF using cooled copper (Cu) fins and through-wafer vias (TWVs) to the front side of the wafer, where the dies are assembled facedown. To ensure high quality of power delivered to the front side of the Si-IF, a process for TWVs within a full thickness Si wafer (500 μm) is developed. A schematic of the Si-IF construct including TWVs is shown in Figure 1.

The development of through-silicon vias (TSVs) originated from the evolution of three-dimensional (3D) integration, which shows the potential to ensure scaling of the next generation electronic devices with feasible cost-efficiency [5-7]. In the 3D integration scheme, two or more layers of active electronic components are integrated vertically using TSVs enabling reduced delay and increased bandwidth [6]. Since thermal management of 3D integrated systems can limit the application space, TSVs also double as thermal conduits for heat extraction. Nowadays, TSV-based 3D integration techniques have been widely-accepted to improve system performance and reduced form factor [8].

As TWVs are used for full-thickness wafers, they are much deeper as compared to TSVs fabricated in thinned silicon substrates. TWV holes can be formed using several methods, including laser etch, wet etch, and dry etch. Deep reactive-ion etching (DRIE) is distinguished from other etching or ablation techniques [10], since this anisotropy etch provides the least pattern alignment mismatch and deformation, the highest aspect ratio, and, comparatively, a better sidewall profile of TWVs. In addition, different profiles

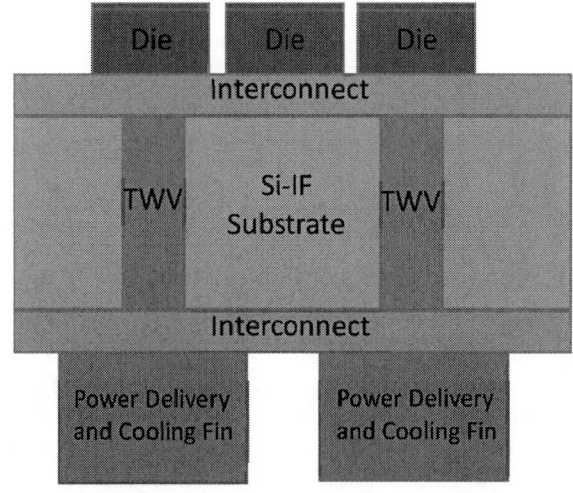

Fig. 1. A schematic of TWVs supporting back-side power delivery within the Si-IF [9]. Power is delivered to the back side of the Si-IF using large Cu fins that double as thermal conduits to remove heat away from the Si wafer.

978-1-7281-1500-9/19 $31.00 © 2019 IEEE

of sidewalls within the substrate can be achieved by adjusting the amount of etchant gases, such as vertical or tapered sidewalls [11]. Cu is chosen as the fill material of the TWV due to higher conductivity in the target dimension (~100 μm), higher melting point, and superior electromigration resilience as compared to aluminum. The preferable characteristics of Cu led to a wide adoption of Cu electroplating by the industry, as a method to fill up large via holes (>50 μm).

Cu electroplating setups require a Cu anode, and the target wafer coated with a Cu seed layer to serve as the cathode. Both anode and cathode are immersed in a solution containing Cu cation (Cu^{2+}), originating from a Cu sulfate ($CuSO_4$) electrolyte solution, to form a closed circuit [12]. We then apply a direct current to the system reducing the Cu^{2+} in the solution to Cu that is plated unto the target wafer, meanwhile, Cu atoms on the anode are oxidized and become Cu^{2+} that dissolve into the solution to maintain electrical neutral.

In the following sections, we present the entire TWV process flow, recipe development, and scanning electron microscope (SEM) images of the fabricated TWVs. We also discuss the results of the electrical characterization, annealing and temperature cycling tests. Moreover, we will show the measured capacitance, inductance, and IR-drop of the TWVs.

II. Fabrication Process

In this section, we will introduce the process flow of TWVs with detailed description for the deep silicon etching and plating processes. After showcasing the process development, we will also present metrology results from an optical microscope and an SEM. The process flow for fabricating TWVs is presented in Figure 2.

A. Through Wafer Via Etch

The TWV process begins with a standard p-type single-side polished 100 mm (4 inches) wafer with a <100> orientation, 500 μm in thickness, and resistivity of 10-20 Ω·cm. To optimize for power delivery, we design 100 μm diameter TWVs to support large currents and as efficient heat extraction tools.

Before silicon etch, we use plasma-enhanced chemical vapor deposition (PECVD) to deposit 4 μm of silicon oxide serve as a hard mask. A thick silicon oxide hard mask is required to achieve a good sidewall profile during the etching process of 500 μm of silicon. The via holes are formed by using dry reactive ion etch (DRIE), similar to the Bosch process [13,14]. The Bosch process includes cyclic isotropic etching and fluorocarbon-based protection film deposition by quick gas switching. The SF_6 plasma (gas etchant) cycle etches silicon, and the C_4F_8 plasma cycle creates a protection layer as passivation [15]. By controlling the number of cycles, i.e., etching time, we can also control the depth of the silicon trench. During DRIE etching, we begin with a low etch rate DRIE recipe (~5 μm/min) to etch through the first 100 μm of the Si substrate. Then, a second recipe is selected with a faster etch-rate (11 μm/min) to etch out the rest of the TWV hole leading to a cumulative taper angle of ~2 degrees. The low

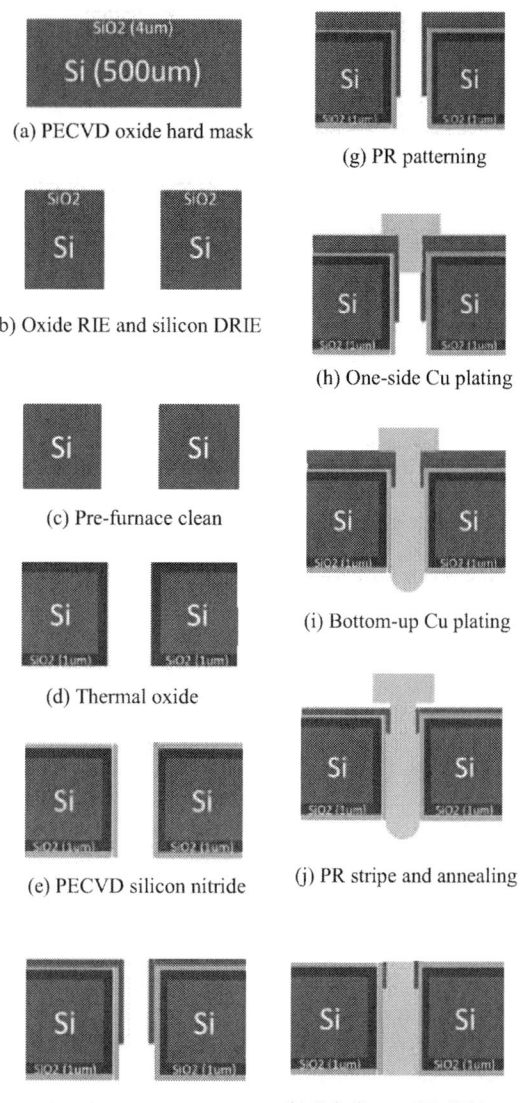

Fig. 2. TWV fabrication process.

(a) PECVD oxide hard mask
(b) Oxide RIE and silicon DRIE
(c) Pre-furnace clean
(d) Thermal oxide
(e) PECVD silicon nitride
(f) TiN/Ti/Cu layer deposit
(g) PR patterning
(h) One-side Cu plating
(i) Bottom-up Cu plating
(j) PR stripe and annealing
(k) Grinding and Polishing

and high etch rate DRIE recipes include a similar flow rate of C_4F_8 and SF_6 etching gases, but different deposition time (15 vs. 80 minutes), plasma power (2,500 W vs. 3,000 W), applied voltage bias (550 V vs. 450 V) and, oxide/silicon etching selectivity (1:180 vs. 1:250). The low etch rate recipe exhibits a smaller tapering effect, but also lower selectivity as compared to the high etch DRIE recipe. The two processes are optimized to significantly reduce tapering of the 500 μm-deep TWV profile. An example of a poor TWV profile, fabricated using a single DRIE step process is shown in Figure 3, which shows significant pattern distortion during the first 100 μm of the TWV etching, and a tapered profile throughout the entire via hole.

B. Barrier Layer Deposition and Copper Electroplating

RIE lag [16], which is a slowdown in the etch rate at

978-1-7281-1500-9/19 $31.00 © 2019 IEEE

Fig. 3. TWVs fabricated using a single DRIE step exhibit poor sidewall profile after Cu electroplating.

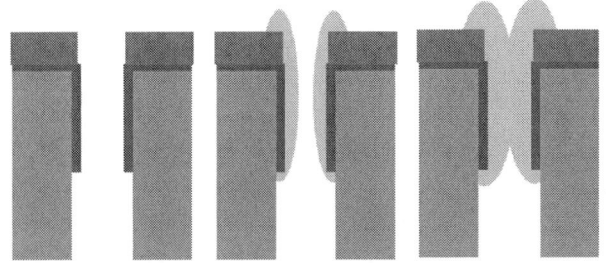

Fig. 4. Schematic of the Cu electroplating process [16]. Electroplating begins (left) with a Cu seed layer (brown). Electroplated Cu (yellow) extends along the width and the length of the TWVs (center). Finally, one side of the TWV hole is sealed by the Cu (right). Note, photoresist is in pink and silicon is in gray.

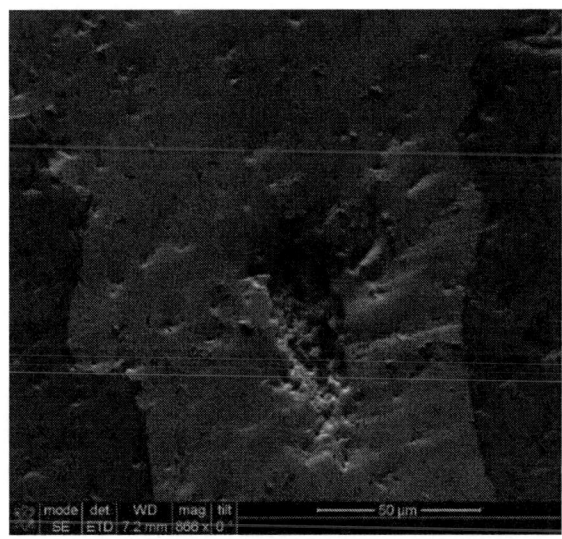

(a)

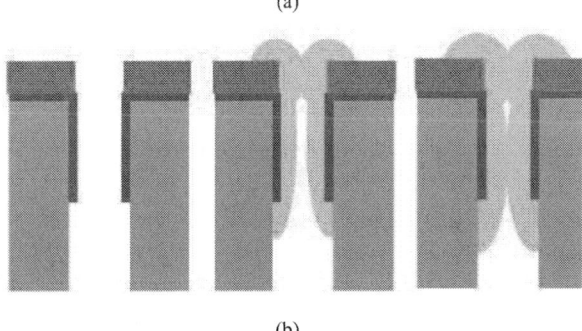

(b)

Fig. 5. (a) A 10 μm x 40 μm void is identified within the TWV after using a 150 mA electroplating current. (b) Schematic of Cu electroplating at high current (≥150 mA). The electroplating rate is faster on the substrate side (with the seed layer), leaving a void within TWV.

greater depth, is also observed during processing. For example, the theoretical (without considering the RIE lag) two-step DRIE process time for 500 μm etching is approximately 50 minutes, whereas in practice, in our experiments, it takes approximately 80 minutes to etch the 500 μm TWV holes. In addition, the etch rate varies at different locations of the wafer. The etch time is roughly 12% longer (70 vs. 80 minutes) at the edge of the wafer than at the center, which does not affect the quality of the center vias due to the protection provided by the thick oxide hard mask.

Prior to seed layer deposition, 1 μm of wet thermal oxide is formed as an electrical isolation layer for the TWVs. Next, a 400 nm of silicon nitride polish stop layer is deposited using PECVD on both sides of the wafer. Then a 30 nm titanium nitride (TiN) layer is deposited using atomic layer deposition, serving as a diffusion barrier [17]. The diffusion barrier is used to prevent Cu and Ti from diffusing into the silicon substrate and the thermal oxide, during high temperature processing and high current propagation. A 50 nm seed layer of Ti is deposited using DC bias sputtering on the top side of the substrate, as illustrated in Figure 2(f), serving as the adhesion layer [18] between the TiN diffusion barrier and the Cu. Another 250 nm of Cu seed layer is sputtered on the same side (top). The Cu seed layer reaches deep along the TWV side walls, enabling the Cu electroplating process to deposit metal on the side walls.

The TWVs are electroplated using a Cu damascene electroplating process that utilizes brightener and carrier as additives. Brighteners, typically organic-sulfates such as 3-mercapto-1-propanesulfonate (MPS) or bis(3-sulfopropyl) disulfide (SPS) [19], enable a uniform current density at the interface of $CuSO_4$ electrolyte and the sidewall of via.

Whereas carriers, such as polyethylene glycol (PEG) or Polyalkylene Glycol (PAG), create and maintain a defined diffusion layer at the Cu anode to control (slow down) the electroplating rate.

The electroplating of the vias, is performed in two steps. The first electroplating step is a damascene electroplating process with a layer of semi-additive AZ 4620 photoresist

(PR) [20]. The AZ 4620 PR has a sufficient thickness of 10 μm to both protect the Cu seed layer from electroplating and prevent erosion of the PR within the acidic Cu electroplating setup. PR covers most of the Cu seed layer that was previously sputtered throughout the wafer. Then, the through holes are lithographically patterned, enabling electroplating of Cu only at the opening and sidewalls of the TWVs, as depicted in Figure 4. During this first electroplating step, a low electroplating current of 70 – 100 mA is applied to ensure deposition of high-quality Cu. Cu is deposited on both the substrate surface (top side where the Cu seed layer was deposited), and the via hole sidewalls, simultaneously, with a faster electroplating rate at the substrate surface. The result of electroplating using a high current (150 mA) is shown in Figure 5. The electroplating rate is faster on the substrate side (with the seed layer), leading to voids inside the TWV.

During the second electroplating step, the wafer is flipped (top side down), and a bottom-up electroplating process is used to fill the remaining volume of the TWV. After electroplating, the wafer is annealed at 250°C for one hour in a nitrogen ambient. Finally, CMP is performed to remove the excess copper overburden, and reduce the surface roughness of the TWV.

C. Physical Structure and Characterization

We have discovered voids inside TWVs when we elevate electroplating current up to 150mA or more. Figure 5 presents schematic Cu plating at high plating current and the SEM

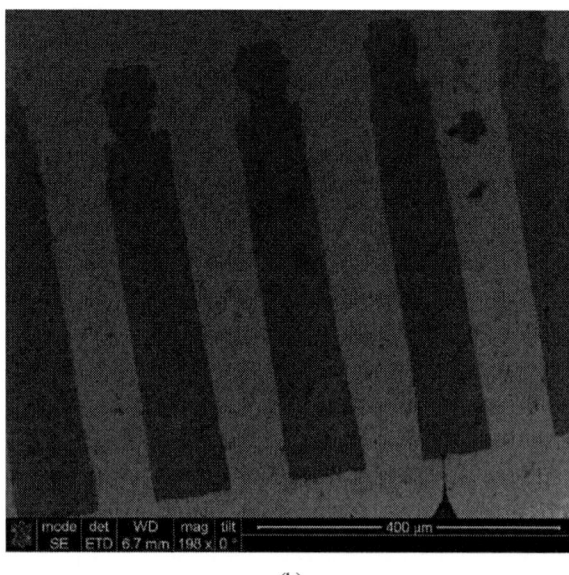

(a)

(b)

Fig. 6.   Cross section SEM images of TWVs with different process condition. (a) 2-step DRIE using low current (50 mA) electroplating, and (b) 1-step DRIE using medium current (100 mA) electroplating.

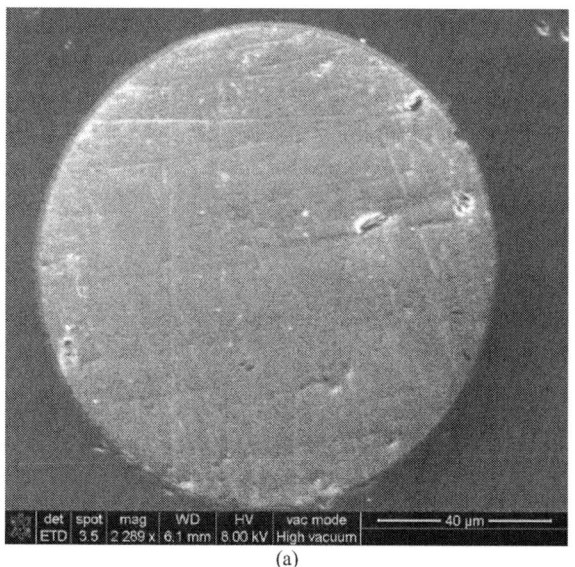

(a)

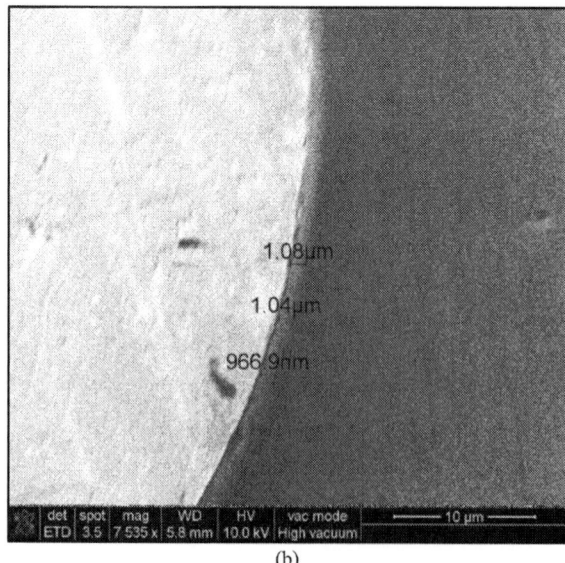

(b)

Fig. 7. A top-view SEM image of a TWV. (a) A circular profile of the TWV, and, (b) zoom-in on the 1 μm thermal oxide isolation layer between the silicon and the copper.

image of void. Large voids lead to TWV breakdown at high power delivery, we will later discuss in the section of TWV reliability analysis.

The profile from our proposed two-step DRIE is shown in Figure 6(a) with a tapering angle of 0.34 degrees, which is an improvement over the profile of the one-step DRIE from Figures 3 and 6(b) (tapering angle of 2.5 degrees). A top-view SEM image of a TWV (Figure 7) shows the distinct layers of Cu, liner oxide, and Si.

III. RESULTS OF ELECTRICAL AND RELIABILITY CHARACTERIZATION

In this section, we show characterization results for the TWVs, including resistance, inductance, and capacitance, by using a Keysight E4980A precision LCR meter with a four-point probe setup. In addition to electrical characterization, we also examine the reliability of the TWVs using temperature cycling testing, and electrical stressing by passing a large current (>2 A) through the via.

Due to the difficulty in simultaneously accessing both sides of the TWV during electrical characterization, pairs of TWVs are shorted from the bottom side using the large overburden created during the electroplating. The TWVs are separated by a 1 μm oxide isolation and 100 μm of Si substrate,

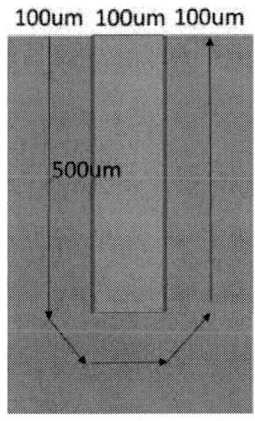

Fig. 8. A schematic of two TWVs shorted on the back side of the test structure (orange). The TWVs are isolated by a 1 μm thermal oxide (blue) along with 100 μm of Si (gray). Note that the Cu thickness at the bottom is 250 μm with a width of 4 mm (perpendicular to the cross section).

as shown in Figure 8. The shorted part at the bottom, is 250 μm in height and 4 mm in width (perpendicular to the cross section in Figure 8). The resistance of this Cu short is significantly lower (~8 μΩ) than the resistance of the TWV and can, therefore, be neglected. A four-point probe measurement of the two adjacent TWVs (connected in series) is performed, and electrical parameters of a single TWV are extracted. After measuring 20 pairs of shorted TWVs (40 TWVs), we extract the average measured resistance of the TWV to be 1.25 mΩ, and the smallest resistance is 1.05 mΩ. The extracted smallest resistivity of TWV Cu is $1.73 \cdot 10^{-8}$ Ω·m, less than 3% higher than the theoretical resistivity of Cu

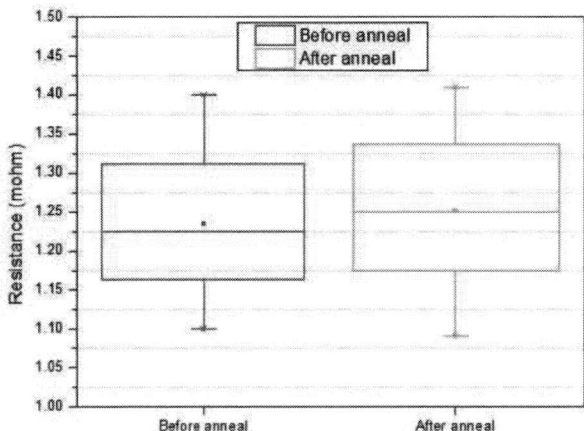

Fig. 9. A statistic bar graph of extracted TWV resistance before and after annealing at 250°C.

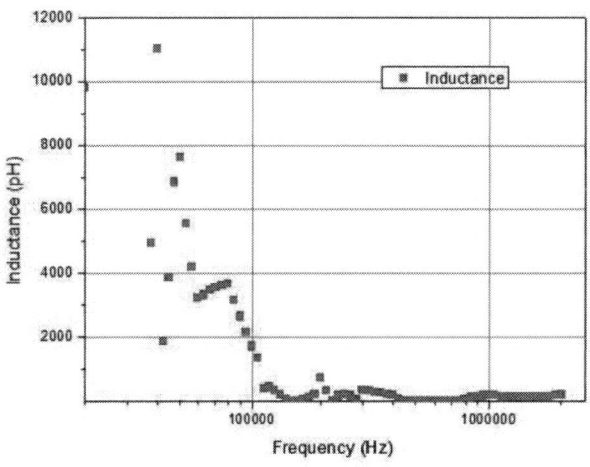

Fig. 10. The inductance of two TWVs connected in series versus frequency.

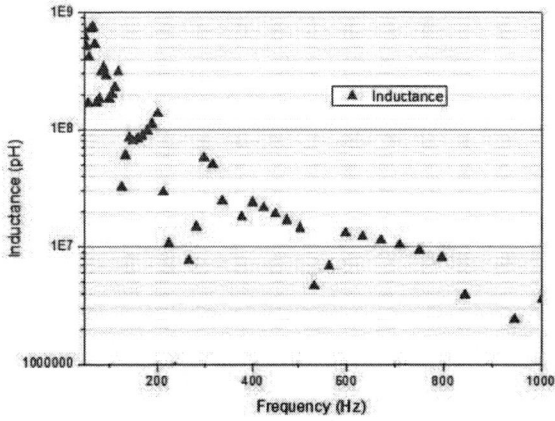

Fig. 11. The inductance of two TWVs connected in series versus frequency. Zoom-in on the very low frequency region (<1 kHz)

($1.68 \cdot 10^{-8}$ Ω·m), indicating a high quality of electroplated metal. Post annealing (at 250°C) resistance measurement shows negligible change as compared to the measurement before annealing (Figure 9), indicating the high-quality of the electroplated metal, and the excellent protection of TiN/Ti diffusion barrier.

The inductance of the fabricated TWVs was also measured on a similar set up (Figure 8), using the same LCR meter. The measured self-inductance of the two TWVs in series is 1.7 nH at 10 kHz, 194 pH at 1 MHz, and 205pH at 2 MHz, with a minimum of 13 pH at 0.47 MHz. The data is shown in Figures 10 and 11.

At the low frequency region (<1 kHz), shown in Figure 11, the inductance of the TWV-pair (connected in series) decreases with increasing frequency. Compatible with the semiconductor physics of a large-scale MOS capacitor. Alternatively, at the higher frequency range (0.1 to 2 MHz), the inductance is approximately constant (from 0.1 to 0.5 MHz).

The measured capacitance of the TWVs is extracted form a pair of TWVs in parallel using a TWV-dielectric-substrate model [21]. The theoretical capacitance of a pair of TWVs is 2.73 pF. To experimentally verify the capacitance of the TWVs, the excess of Cu on the bottom of the test structure was removed, resulting in two isolated TWVs (Figure 12). The measured capacitance of the TWV pairs is 2.8 pF (at 20 Hz) and 1.7 pF (at 1 MHz), with a minimum of 1.4 pF (at 0.54 MHz). The capacitance, plotted in Figure 13, decreases with increasing frequency, exhibiting concord with device physics of MOS capacitors. The measurement results agree well with the simulated values that are based on the model from [21], exhibiting a low parasitic effect.

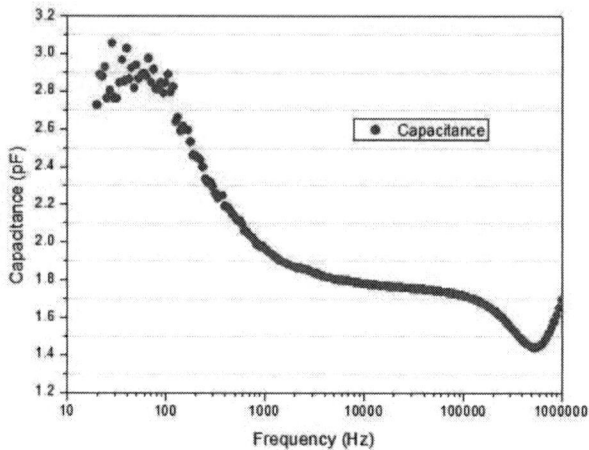

Fig. 13. Capacitance measurement versus frequency between two isolated TWVs.

temperature cycling test (TCT) from -40°C to 125°C (conforming to the JESD22-A104D standard) with 0, 50, and 100 cycles are performed, showing stable resistance under different current. Because the coefficient of thermal expansion (CTE) is very different between air (3690ppm/°C) and copper (16.5ppm/°C), voids within via increase the probability of TWV breakdown under TCT. The TCT results are provided in Figure 14. It can be observed that TWV resistance remains stable after the TCT, indicating that the vias have been filled up with high quality copper. In summary, TCT and large current test show good reliability of the TWVs made from our process.

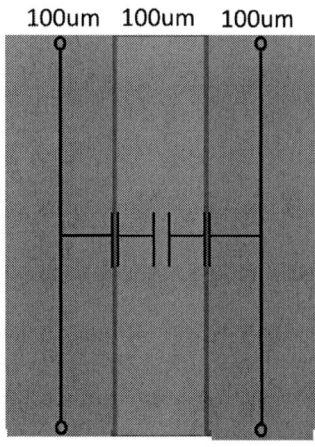

Fig. 12. A simplified schematic figure of two isolated TWVs (orange) for capacitance measurement [20]. Vias are separated by a 1 μm thermal oxide (blue) and a 100 μm Si (gray).

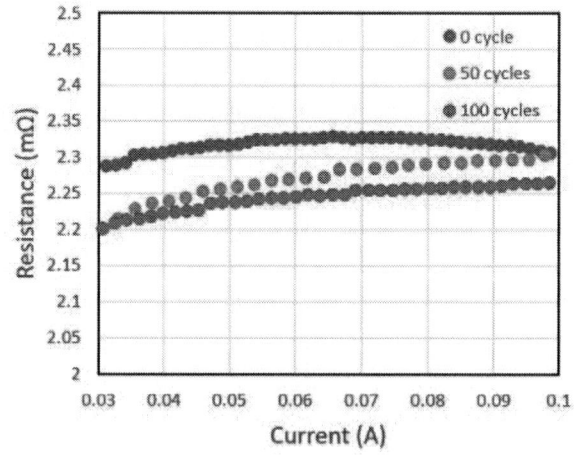

Fig. 14. Resistance measurement of two shorted TWVs after temperature cycling tests.

The fabricated TWVs were subjected to reliability testing. Current carrying capacity of the TWVs was evaluated using the TWV-pair test structure (Figure 7). The result shows that each TWV is capable of carrying at least 2.5 A of current at DC (limited by experiment equipment) without degradation, corresponding to a current density of 79 A/mm². In addition,

In addition to the electrical characterization, a small circuit is designed to showcase the power delivery of the fabricated TWVs. A green LED is connected to two isolated arrays of TWVs using tin solder. Each leg of the LED is connected to a separate TWV array. Due to the large size of the solder, all of TWVs in each of the arrays are connected, but arrays are still isolated from each other, through the LED. An image of the setup is shown in Figure 15. Using this simple circuit, the IR drop and power dissipation within the TWVs are also measured [22]. We have measured an IR-drop of 1.8 mV across the wafer with 20 TWVs. The extracted IR-drop of a single TWV is therefore 90 µV. This experiment demonstrates that the TWV arrays (20 TWVs) dissipate only 36 µW of power (1.8 µW per TWV) in this circuit.

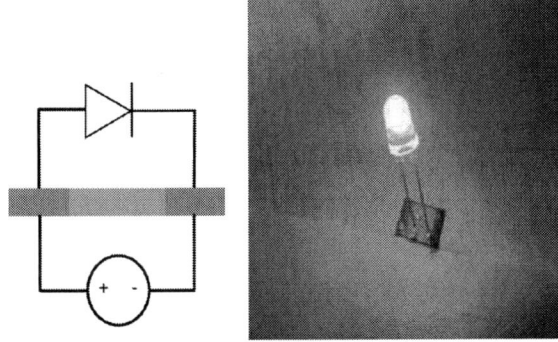

Fig. 15. A schematic (left) and a photograph (right) of a green LED operating at the front side of the silicon substrate. The current is supplied from the back side of the substrate and delivered to the LED using TWVs. Note that the orange color (in schematic) stands for a set of 20 TWVs, and the gray color is the silicon wafer.

IV. CONCLUSIONS

In this paper, we have presented the process flow of through-wafer vias for high quality power delivery across a full thickness Si wafer. The dry etch process is optimized to reduce the tapering angle from 2.5 degrees to 0.34 degrees. Moreover, reliability tests of the fabricated TWVs are performed using high current and temperature cycling testing and exhibit promising results. We measured a resistance of 1.1 mΩ, an inductance of 97 pH at 1 MHz, and a capacitance of 1.7 pF at 1 M Hz, for a single TWV, indicating a high-quality of the electroplated copper and excellent device performance. Finally, we fabricated a simple circuit using an LED and TWVs to show the functionality of the TWVs for power delivery. An IR-drop of 1.8 mV was measured, and the power dissipated in a single TWV is 1.8 µW. In conclusion, the power delivery TWVs within the Si-IF platform, exhibit good performance and reliability.

ACKNOWLEDGMENT

This work was supported by the industrial sponsors of the UCLA CHIPS Consortium. We also thank Prof. Mark Goorsky, Dr. Umesh Mogera, SivaChandra Jangam, Pranav Ambhore, Niloofar Shakoorzadeh, and Jonathan Cox for valuable discussions.

REFERENCES

[1] S. Jangam, A. Bajwa, K. K. Thankappan, P. Kittur, and S. S. Iyer, "Electrical Characterization of High Performance Fine Pitch Interconnects in Silicon-Interconnect Fabric," *Proceedings of the IEEE Electronic Components and Technology Conference*, pp. 1283 – 1288, 2018.

[2] A. Bajwa, S. Jangam, S. Pal, N. Marathe, T. Bai, T. Fukushima, M. Goorsky, and S. S. Iyer, "Heterogeneous Integration at Fine Pitch (≤ 10 µm) Using Thermal Compression Bonding," *Proceedings of the*

IEEE Electronic Components and Technology Conference, pp. 1276 – 1284, 2017.

[3] S. S. Iyer, "Heterogeneous Integration for Performance and Scaling," *IEEE Transactions on Components, Packaging, and Manufacturing Technology*, Vol. 6, No. 7, pp. 973 – 982, July 2016.

[4] S. Jangam, S. Pal, A. Bajwa, S. Pamarti, P. Gupta, and S. S. Iyer, "Latency, Bandwidth and Power Benefits of the SuperCHIPS Integration Scheme," *Proceedings of the IEEE Electronic Components and Technology Conference*, 2017.

[5] J. H. Lau, "Overview and Outlook of Through-Silicon Via (TSV) and 3D Integrations," *Microelectronics International*, Vol. 28, No. 2, pp. 8 – 22, 2011.

[6] J. H. Lau, "Evolution, Challenge, and Outlook of TSV, 3D IC Integration and 3D Silicon Integration," *Proceedings of the IEEE International Symposium on Advanced Packaging Materials*, pp. 462 – 488, October 2011.

[7] R. S. Patti, "Three-Dimensional Integrated Circuits and the Future of System-on-Chip Designs," *Proceedings of the IEEE*, Vol. 94, No. 6, pp. 1214 – 1224, June 2006.

[8] P. Ramm, A. Klumpp, J. Weber, N. Lietaer, M. Taklo, W. De Raedt, T. Fritzsch, and P. Couderc, "3D Integration Technology: Status and Application Development," *Proceedings of the IEEE European Solid-State Circuits Conference*, pp. 9 – 16, September 2010.

[9] Y. Luo, *Developing Through-Wafer Via (TWV) and Plasma Dicing Process for Silicon Interconnect Fabric (Si-IF)*, M.S. Thesis, 2018.

[10] D. Vasilache, S. Colpo, F. Giacomozzi, S. Ronchin, S. Gennaro, A. Q. A. Qureshi, and B. Margesin, "Through Wafer Via Holes Manufacturing by Variable Isotropy Deep RIE Process for RF Applications," *Microsystem Technologies*, Vol. 18, No. 7-8, pp. 1057 – 1063, August, 2012.

[11] D. Chung, J. Korejwa, E. Walton, and P. Locke, "Introduction of Copper Electroplating Into a Manufacturing Fabricator," *Proceedings of the IEEE/SEMI Advanced Semiconductor Manufacturing Conference*, pp. 282-289, Spetember 1999

[12] R. Li, Y. Lamy, W. F. A. Besling, F. Roozeboom, and P. M. Sarro, "Continuous Deep Reactive Ion Etching of Tapered Via Holes for Three-Dimensional Integration," *Journal of Micromechanics and Microengineering*, Vol. 18, No. 12, pp. 125023, November 2008.

[13] F. Laermer, A. Schilp, K. Funk, and M. Offenberg, "Bosch Deep Silicon Etching: Improving Uniformity and Etch Rate for Advanced MEMS Applications," Technical Digest of the *IEEE International Conference on Micro Electro Mechanical Systems*, pp. 211 – 216, January 1999.

[14] F. Marty, L. Rousseau, B. Saadany, B. Mercier, O. Francais, Y. Mita, and T. Bourouina, "Advanced Etching of Silicon Based on Deep Reactive Ion Etching for Silicon High Aspect Ratio Microstructures and Three-Dimensional Micro- and Nanostructures," *Microelectronics Journal*, Vol. 36, No. 7, pp. 673 – 677, July 2005.

[15] F. Laermer and A. Schilp, "Method of Anisotropically Etching Silicon," United States Patent, No. 5501893A, March 26, 1996.

[16] Z. Wang, L. Wang, N. T. Nguyen, W. A. H. Wien, H. Schellevis, P. M. Sarro, and J. N. Burghartz, "Silicon Micromachining of High Aspect Ratio, High-Density Through-Wafer Electrical Interconnects for 3-D Multichip Packaging," *IEEE Transactions on Advanced Packaging*, Vol. 29, No. 3, pp. 615 – 622, August 2006.

[17] H. Kizil, G. Kim, C. Steinbrüchel, and B. Zhao, "TiN and TaN Diffusion Barriers in Copper Interconnect Technology: Towards a Consistent Testing Methodology," *Journal of Electronic Materials*, Vol. 30, No. 4, pp. 345 – 348, April 2001.

[18] M. J. Wolf, T. Dretschkow, B. Wunderle, N. Jürgensen, G. Engelmann, O. Ehrmann, A. Uhlig, B. Michel, and H. Reichl, "High Aspect Ratio TSV Copper Filling with Different Seed Layers" *Proceedings of the IEEE Electronic Components and Technology Conference*, pp. 563 – 570, 2008.

[19] W.-P. Dow, H.-S. Huang, and Z. Lin, "Interactions Between Brightener and Chloride Ions on Copper Electroplating for Laser-Drilled Via-Hole

978-1-7281-1500-9/19 $31.00 © 2019 IEEE

Filling," *Electrochemical and Solid-State Letters*, Vol. 6, Issue 9, pp. C134 – C136, 2003.

[20] P. M. Raj, C. Nair, H. Lu, F. Liu, V. Sundaram, D. W. Hess, R. Tummala, "'Zero-Undercut' Semi-Additive Copper Patterning – A Breakthrough for Ultrafine-Line RDL Lithographic Structures and Precision RF Thinfilm Passives," *Proceedings of the IEEE Electronic Components and Technology Conference*, pp. 402 – 405, 2015.

[21] J. Kim, J. S. Pak, J. Kim, and Manho Lee, *Electrical Design of Through Silicon Via*, Springer, 2014.

[22] M. Jung and S. K. Lim. "A Study of IR-Drop Noise Issues in 3D ICs with Through-Silicon-Vias," *Proceedings of the IEEE International 3D Systems Integration Conference*, pp. 1 – 7, November 2010.

Active Through-Silicon Interposer Based 2.5D IC Design, Fabrication, Assembly and Test

Jayasanker Jayabalan, Vivek Chidambaram, Sharon Lim Pei Siang, Wang Xiangyu, Jong Ming Chinq, Surya Bhattacharya

Institute of Microelectronics, A*STAR (Agency for Science, Technology and Research)
2 Fusionopolis Way, #08-02, Innovis, Singapore 138634
Email: jayasanker_jayabalan@ime.a-star.edu.sg

Abstract— **Active Through-Silicon Interposer (ATSI) based 2.5D/3D IC packaging is a solution to extend Moore's law beyond the limitations inherent in 2D packages. We present the implementation of an ATSI platform for providing Analog to Digital converter (ADC), Digital to Analog converter (DAC) and embedded Power Management Unit (ePMU) functions to support high performance logic, fabrication of 140 micron pitch Via-Last Through-Silicon Via (TSV) of 40 micron height, assembly of Chip-on-Chip-on Substrate, functional test and reliability assessment. The active interposer fabricated in 130nm CMOS easily supports the I/O, Analog, Electro Static Discharge (ESD), De-cap functions with via-last TSV. This approach enables significant die-size reduction of the top die (usually in expensive 16nm CMOS or below tech. node) to achieve system miniaturization and cost reduction.**

Keywords- Active Through-Silicon interposer (ATSI), 2.5D IC, 3D IC, TSV, TSI

I. INTRODUCTION

Speed up of package scaling through heterogeneous integration has more than compensated the slowing down of semiconductor scaling to achieve performance and cost objectives at system level [1-4]. Integrating the passive devices and active circuits on the same silicon interposer have been shown to enhance signal integrity, power integrity and lower power consumption [5] while lowering the cost [6-7]. The main objective of this work is to transfer i/o, analog, Electro Static Discharge (ESD) and decoupling capacitor functions from the leading edge CMOS top-die to an active interposer (as against a passive interposer [8-9]). There are many challenges associated with the development of Active Through-Silicon Interposer (ATSI) based 2.5D IC, shown in Fig. 1, which include: (i) Logic partitioning between the top dies and active TSI; (ii) "Via Last" TSV fabrication process; (III) Material selection and process flow for handling of and assembly on thin interposer wafer, which is 40 micron in this work; (iv) electrical performance and reliability. ATSI design described in Section II includes the active interposer and substrate design. Via Last TSV fabrication process integration with main stream chip fabrication process is discussed in Section III followed by assembly process in Section IV. Parametric and Functional electrical tests including reliability stress are detailed in Section V.

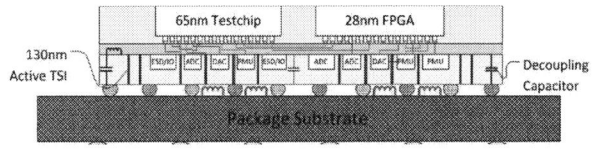

Figure 1. 2.5D IC Schematic.

II. 2.5D IC PACKAGE DESIGN

The Active Interposer Package consists of an FPGA (28nm, 12.65 mm x 12.34 mm size, programmable SoC), two IO chips (65nm, 4mmx 4mm, thin-gate Split IO and ESD) and ATSI (130nm, 22.8 mm x 16.4 mm containing thick-gate Split IO, Split ESD, ePMU, ADC/DAC, High-K MIM Cap, Via-Last TSV) as shown in Fig. 2.

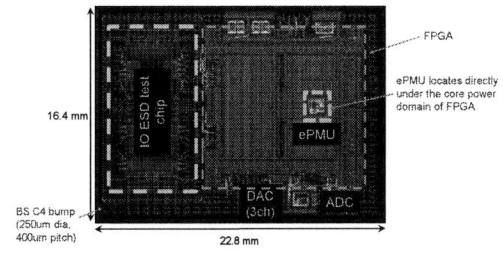

Figure 2. Layout Floorplan of 2.5D IC.

Top level RDL is implemented with 6 metal layer stack as shown in Fig. 3. TSV is arranged in staggered array with pitch of 140um. ATSI noise suppression is done with grounded TSVs. Parallel with TSVs are the C4 bumps to increase current carrying capacity. Substrate C4 bumps are arranged in regular array with 400um pitch. TSV is directly connected to the UBM of C4 bump and there is no need for backside RDL for routing, as shown in Fig. 4. Layout design

978-1-7281-1500-9/19 $31.00 © 2019 IEEE

and verification are done with Cadence Virtuoso together with Calibre and Assura. For circuit simulation, Cadence Spectre is used.

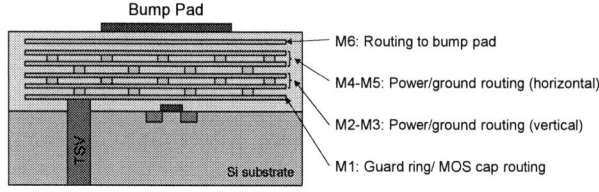

Figure 3.   ATSI Interposer Metal stack for power/ground plane routing.

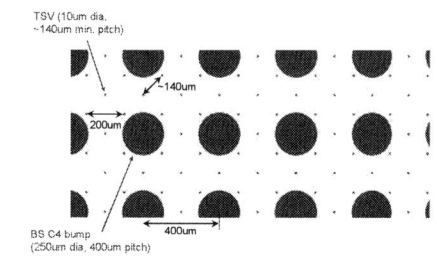

Figure 4.   ATSI TSV and C4 Bump Allocation.

ATSI is flip-chip bonded to an organic Substrate to form the 2.5D IC as shown in Fig.5. Geometric details of the IC are shown in Table I. The substrate size is 31mmx31mm made of 4 layer 1-2-1 stack up. The substrate core thickness is 0.8mm. The design rule used build-up and core line width/space of 50micron/50micron. Substrate top pad size/opening/IO Counts are 250um/ 180um/ 1969 and BGA side pad size/ Opening/ IO Counts are 700um/ 600um / 529 respectively.

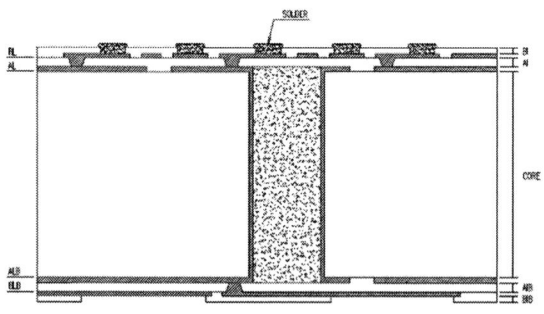

Figure 5.   Substrate cross-section.

TABLE I.   GEOMETRIC DESIGN DETAILS OF 2.5D IC PACKAGE

FTV Design Parameter	ATSI: FPGA	ATSI: Split IO	ATSI: C4 back	Substrate C4 Front	Substrate LGA
IO pitch	150um (min)	165um	400um	400um	1.27mm
IO count	3932	458	1969	1969	529
Metal pad	110um (Oct)	110um (Oct)	220um	250±20 um	700±20um
Opening	80um	80um	180um	180±20u m	600±20um
UBM	110um	110um	200um	N.A.	N.A
UBM/pad	Metal 1	Metal 3	Metal 5	Sn/3Ag/ 0.5Cu	(Ni/Pd/Au) (6/0.05/0.0 5um)
C4 bump	Metal 2	Metal 4	Metal 6	SOP: SAC305	N.A

III.   FABRICATION PROCESS

Fabrication of Via-Last TSV consists of hard mask deposition, TSV patterning and Si etching, Liner oxide deposition, silicon oxide etch back and clean, barrier and seed layer deposition, copper Electro-Chemical Plating (ECP) and Chemical Mechanical Planarization (CMP) as depicted in Fig. 6.

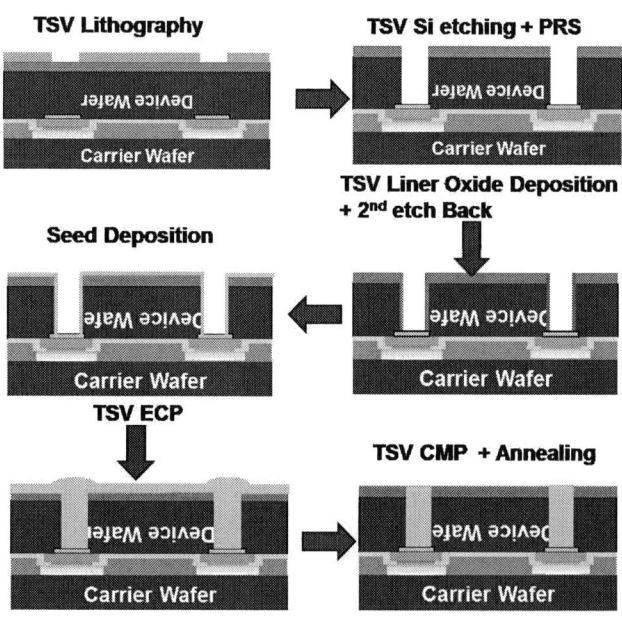

Figure 6.   Via-Last TSV process steps.

IV.   ASSEMBLY PROCESS

Assembly process, as shown in Fig.7, consists of solder ball attachment to the substrate via laser jetting process, flip-chip bonding of interposer to substrate by reflow, second level under filling, Au stud-bumping for IO chip with Al metallization, followed by SnAg solder jetting the IO and

FPGA dies and then attaching the top dies together to the interposer by reflow and first level under filling. 1.3mil Aw99 gold wire was used. Ball diameter and shear strength were being examined in the DOE runs for capillary selection. The criteria for bonded ball size formation is targeted in the range of 80um to 90um. Ball shear is >5.5gf/mil2 as per Mil standard 883. The assembled 2.5D 529-pin (23x23) LGA package is shown in Fig. 8.

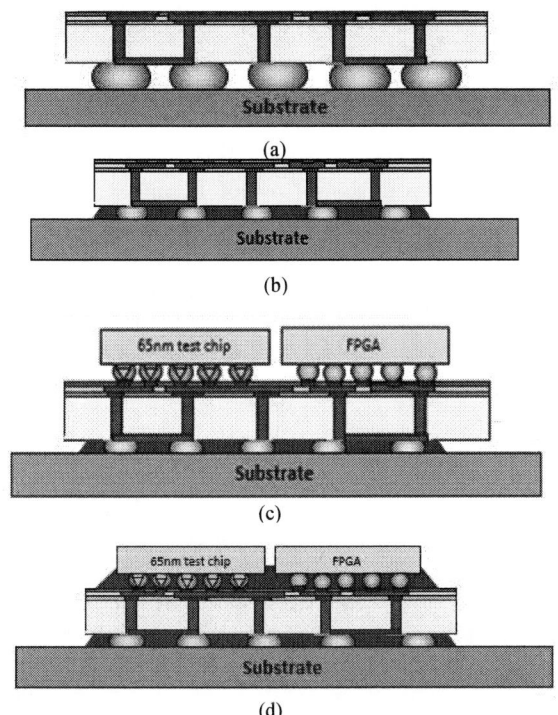

Figure 7. 2.5D IC assembly process steps (a) Interposer bonding (b) second level underfill (c) Top die bonding (d) First level underfill.

Figure 8. Chip-on-Chip on Substrate (with 2 IO dies and one FPGA die on top of Si interposer).

V. ELECTRICAL TEST AND RELIABILITY ASSESSMENT

A. Assembly Test (Substrate + Interposer)

Package level diode continuity through electrical resistance measurements were conducted with two different flip-chip bonded interposer-substrate assembly dies of which one of them had been subjected to reflow process and the other to Thermo-Compression Bonding (TCB). Six digital pins were tested for connections via TSVs through diode continuity measurements using a precision measurement unit (PMU) by forcing ~ 1 milli-amp current and measuring the resulting voltage and hence the resistance. The measurement circuit of an input buffer cell named XMHB is shown in Fig. 9, as an example. Measurement data obtained from the tests are given in Table II. For the package with reflow process, all the test pins passed continuity consistently while for the package with TCB, 4 of the six pins showed intermittent connections. Excessive stresses introduced in TCB process renders the interposer fragile due to its extremely thin size (40 microns) whereas the reflow process is found to be naturally more effective for bonding as seen from the test results.

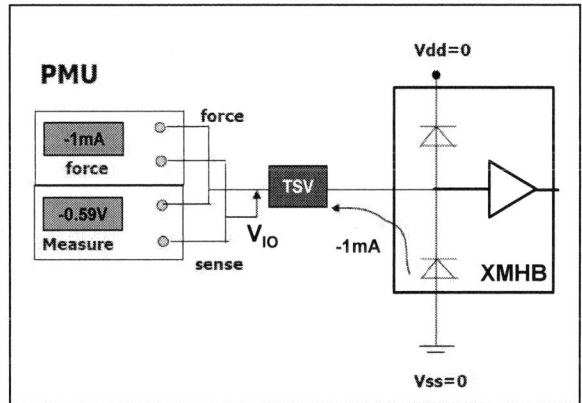

Figure 9. Diode continuity test.

TABLE II; DIODE CONTINUITY MEASUREMENT RESULTS

Test Pin	TCB (Resistance, Ohms)	Reflow (Resistance, Ohms)
Digital In 1	Intermittent/991	989
Digital Out1	intermittent/570	535
Digital In 2	intermittent/992	991
Digital Out2	intermittent/536	532
Digital In 3	562	561
Digital In 4	567	562

B. Assembly Test (Substrate + interposer + top dies)

Package level daisy chain continuity electrical resistance measurements were conducted with two different Dies-on-Interposer-on-Substrate packages subjected to reflow process. Four daisy chains in each package were tested for connections via TSVs using two-probe ohmic continuity measurements on HP 34401A. For the longer daisy chains, the resistance per TSV is between 5.5 and 6.2 ohms, while for shorter chains, the resistance is between 3.9 and 4.2

Ohms. The larger ohmic value for the chains is attributed to the additional length of RDL metal lines, resistance of micro-bump interface and contact resistance. The measured data are shown in Table III.

TABLE III. TABLE TYPE STYLES

No	# of TSVs	Resistance (Ohms)		Resistance Per TSV (Ohms)	
		Package 1	Package 2	Package 1	Package 2
Chain 1	31	128	121	4.129032258	3.903225806
Chain 2	40	240	223	6	5.575
Chain 3	30	125	117	4.166666667	3.9
Chain 4	39	239	230	6.128205128	5.897435897

C. Functional Test

Functional testing of the 2.5D IC has been carried out in two stages: (1) Block level testing of DAC, ePMU and ADC without TSV; (2) System level functional, reliability tests with TSV. Block test of DAC results show that the worst DNL and INL results as DC test among six DAC samples are 0.58 LSB and 0.65 LSB, respectively. The worst SFDR and THD result among six DAC samples are 63.9 dBc and 61.1 dBc, respectively. Block test results of ePMU show that the tested chip has an average end-to-end efficiency of 75 %. Line regulation of Buck alone, LDO alone and the full system are 0.0754 ΔV/V, 0.0064 ΔV/V & 0.0707 ΔV/V respectively. The designed band-gap reference voltage, required for operating the Buck converter and the LDO, provides a stable output voltage of 1.2V. The overall, average end-to-end efficiency of the system is measured to be 76 %. Standalone buck converter efficiency is measured to be 80 %. The e-PMU system functions as expected as a standalone power supply/ power management unit in the ATSI interposer. Block ADC is tested with Fclk ≈ 400MHz, Fin ≈ 100KHz, resulted in SNDR = 33.56dB, SFDR = 40dB and Noise floor -90dBc for 10Bit.

System level IO loop back functional test was carried out by sending a digital bit stream through the input pin of IO chip that passed through a series of buffers and captured at the output pin of the IO chip and analyzed by the tester for error-free reception to indicate pass status. Keysight 33600 Waveform generator was used for generating digital bit stream and Agilent Infinium 54832D MSO, 1 GHz, 4 GSa per second oscilloscope was used to capture the waveforms. System level setup and buffer configuration of loop back functional test are shown in Figs. 10 and 11

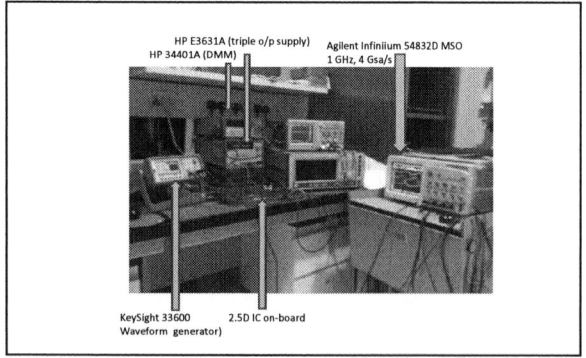

Figure 10. System Setup of IO loop back test.

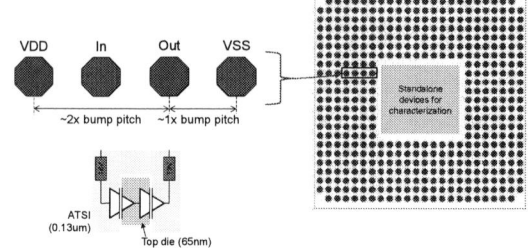

Figure 11. IO buffer loop back configuration.

Functional test eye diagram with 1 second persistence at 50 MHz, 2V p-p is shown in Fig. 12 and at 3V p-p with infinite persistence is shown in Fig. 13. At 3V swing level, signal rise of 200 picoseconds per volt is observed, validating the suitability of the ATSI interposer for Gigabit data transfer applications.

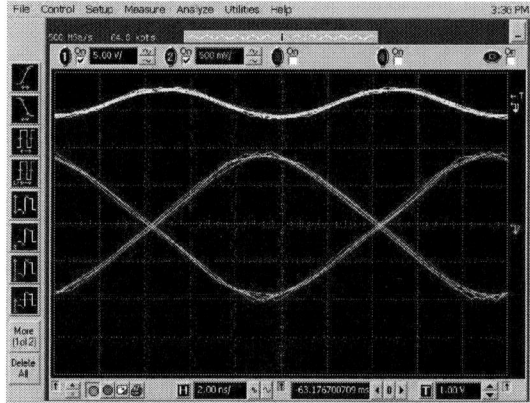

Figure 12. Eye diagram at 2Vp-p.

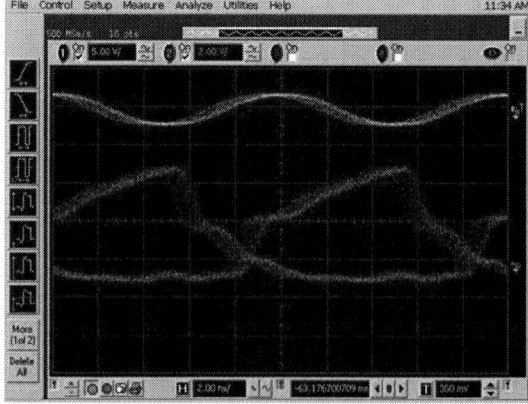

Figure 13. Eye diagram at 3Vp-p.

D. High voltage (HV) Stress Test

High voltage (HV) stress test was performed at 50 Volt, according to IEC 61000-4-2 standard [10]. Two parts were stressed at the input & output pins in each device. Output and input buffer circuits subjected to stress test are shown in Figs. 14 and 15 respectively. The pins were stressed with respect to ground. Test setup is shown in Fig. 16. Both devices passed continuity, functional and leakage tests. The HV test results are shown in Table IV.

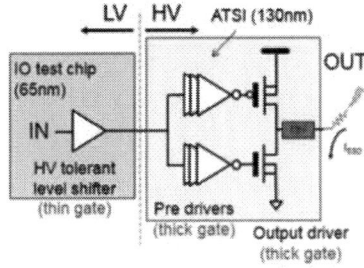

Figure 14. Output buffer subjected to HV stress.

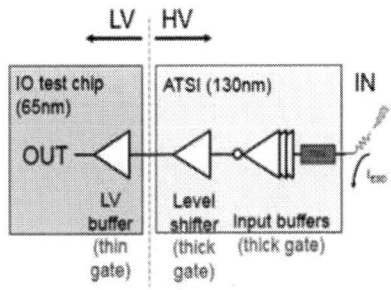

Figure 15. Input buffer subjected to HV stress.

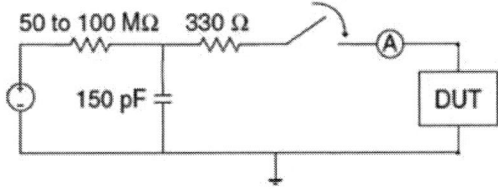

Figure 16. HV Stress Test Setup

TABLE IV. HV STRESS TEST AND MEASUREMENT RESULTS

Parameter	Device #1		Device #2	
	Pre-stress	Post-stress	Pre-stress	Post-stress
Digital Input Diode resistance (Ohms)	986	986	991	990
Digital Input Diode resistance (Ohms)	539	540	536	538
Functional (go/no-go)	pass	pass	pass	pass
IDDL (Vddt=1.1V))	315uA	318uA	270uA	268uA
IDDH (VddH=3V)	137mA	138mA	125mA	125mA

E. Reliability Tests

Reliability tests have been performed on 10 random ATSI samples through ohmic and diode resistance measurements at time zero and then subjected to stress conditions (HAST, TC, MSL3) before repeating the electrical measurements. The test data, before and after reliability stress, are compared to confirm the closeness between post-stress and the pre-stress results.

Four devices were subjected to Thermal Cycling (TC) test for 100 cycles according to JESD22-A104-G standard at -40°C to +125°C with 15 min dwell time. There are 28 test points of which 1 to 4 were subject to daisy chain ohmic continuity and 5 through 28 were subject to diode continuity. All 4 devices passed continuity after TC. Pre and Post TC test results are shown in Fig. 17.

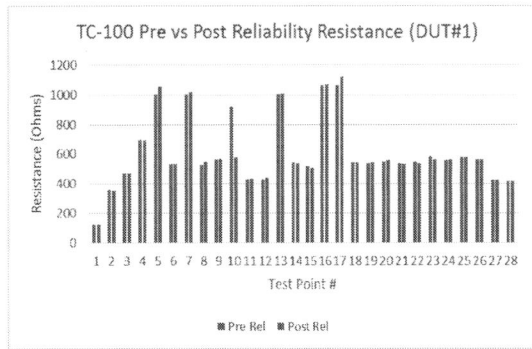

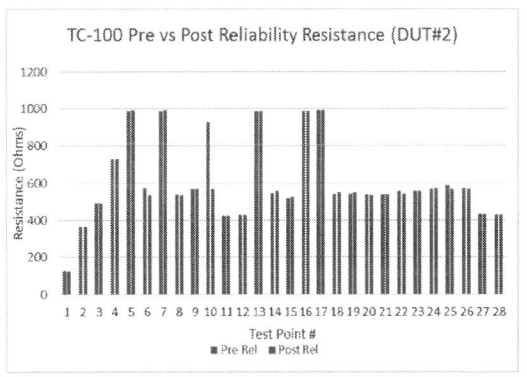

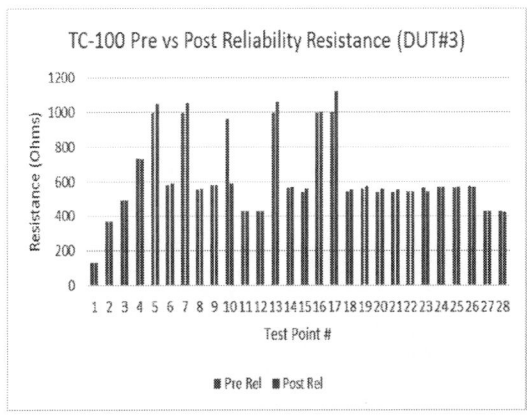

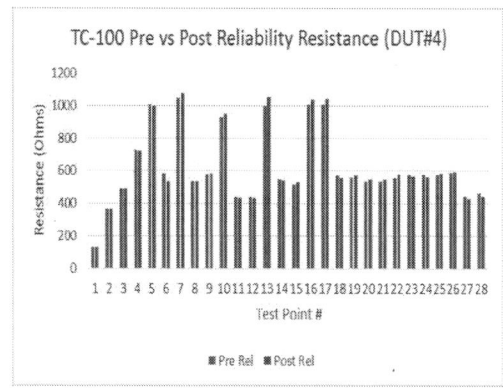

Figure 17. TC results of DUT #1 through #4

Three devices were subjected to Highly Accelerated Stress Test (HAST) as per JESD22-118A standard. There are 28 test points of which 1 to 4 were subject to daisy chain ohmic continuity and 5 through 28 were subject to diode continuity. All 3 devices passed continuity after HAST. Pre and Post HAST test results are shown in Fig. 18.

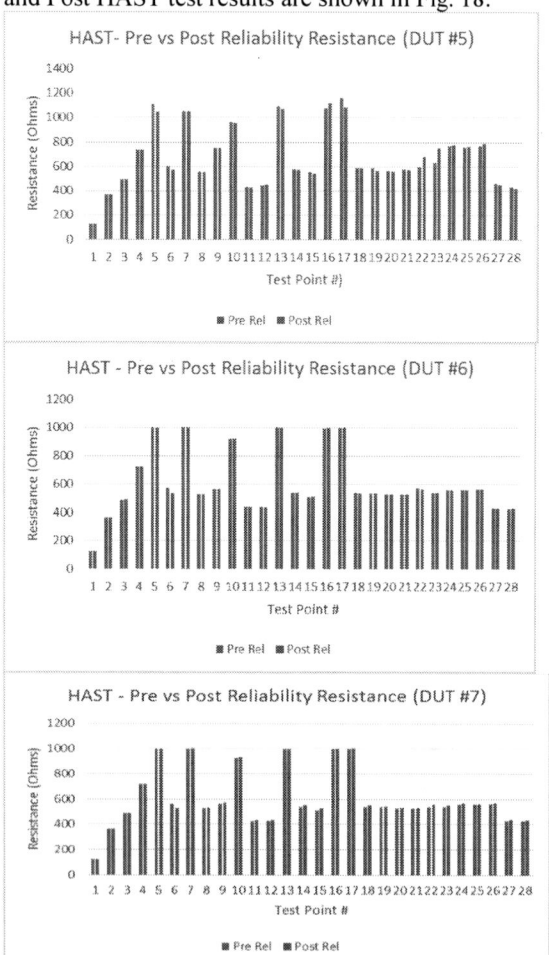

Figure 18. HAST results of DUT #5 through #7

Three devices were subjected to Moisture Sensitivity Level (MSL) test according to JESD 22-A-113-D MSL3 standard, with triple-reflow at 260°C. There are 28 test points of which 1 to 4 were subject to daisy chain ohmic continuity and 5 through 28 were subject to diode continuity. All 3 devices passed continuity after MSL3. Pre and Post MSL test results are shown in Fig. 19.

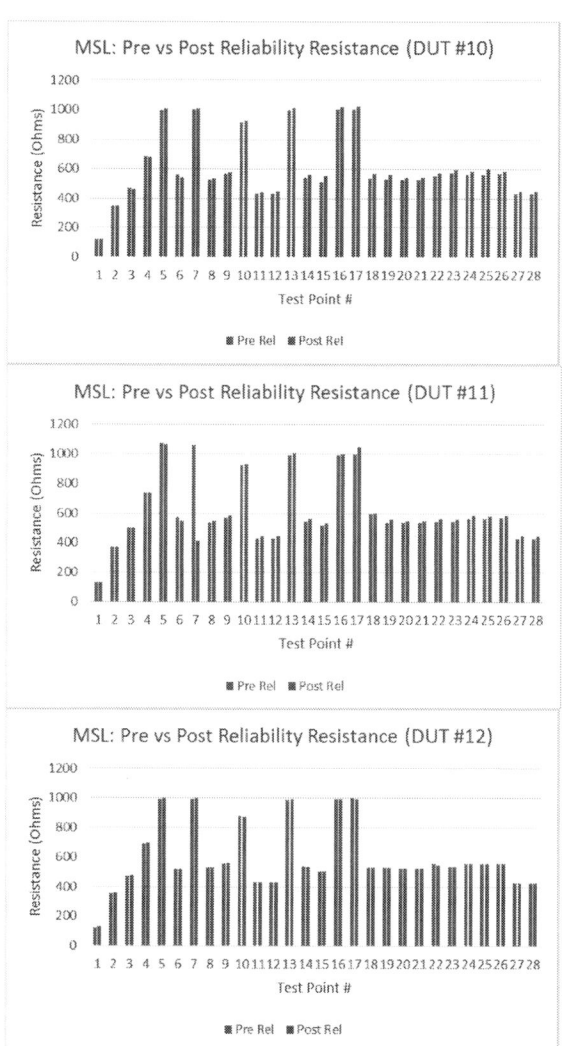

Figure 19. MSL3 results of DUT #10 through #12

VI. CONCLUSION

The development of the ATSI system has made a few significant achievements. Some of the important results are summarized below:

1 System scaling/miniaturization by implementing key system functions into active interposer has been achieved.

2 Partitioning of System-on-Chip (SoC) into smaller dies with lower cost, higher yield has also been achieved.

3   Heterogeneous integration involving 28 nm node logic, two 65 nm node IO chips flip chip bonded to 130 nm active thin interposer has been successfully demonstrated.

4   All 10 random test vehicle samples have passed the reliability tests (HAST, MSL 3, TC).

5   System level functional testing of ATSI has also been demonstrated for Gigabit data transfer application.

ACKNOWLEDGMENT

This work was partly funded by Institute of Microelectronics industry consortium project on Active Through-Silicon Interposer for System Scaling. The authors greatly appreciate the members' support and encouragement throughout the course of the project.

REFERENCES

[1] Xiaowu Zhang, Jong Kai Lin, Sunil Wickramanayaka, Songbai Zhang,Roshan Weerasekera, Rahul Dutta, Ka Fai Chang, King-Jien Chui, Hong Yu Li, David Soon Wee Ho, Liang Ding, Guruprasad Katti, Suryanarayana Bhattacharya, and Dim-Lee Kwong, "Heterogeneous 2.5D integration on through silicon interposer," APPLIED PHYSICS REVIEWS 2, 021308, pp. 1-58, 2015

[2] Dylan Stow, Yuan Xie, Taniya Siddiqua, Gabriel H. Loh, "Cost-effective design of scalable high-performance systems using active and passive interposers," IEEE/ACM International Conference on Computer-Aided Design (ICCAD) 2017

[3] Subramanian S. Iyer, "Heterogeneous Integration for Performance and Scaling," IEEE Transactions on Components, Packaging and Manufacturing Technology, Volume: 6 , Issue: 7 , pp. 973 – 982, July 2016.

[4] N. Khan, H. Y. Li, S. P. Tan, S. W. Ho, V. Kripesh, and D. Pinjala, "3-D packaging with through-silicon via (TSV) for electrical and fluidic interconnections," IEEE Trans. Compon., Packag., Manuf. Technol. 3(2), 221, 2013.

[5] Joungho Kim, "Active Si interposer for 3D IC integrations," International 3D Systems Integration Conference (3DIC), Sendai, pp. TS11.1.1-TS11.1.3, 2015.

[6] G. Hellings et al., "Active-lite interposer for 2.5 & 3D integration," Symposium on VLSI Circuits, Kyoto, pp. T222-T223, 2015.

[7] M. Scholz et al., "ESD protection design in active-lite interposer for 2.5 and 3D systems-in-package," 37th Electrical Overstress/Electrostatic Discharge Symposium (EOS/ESD), Reno, NV, pp. 1-10, 2015.

[8] C. Erdmann et al., "A Heterogeneous 3D-IC Consisting of Two 28 nm FPGA Die and 32 Reconfigurable High-Performance Data Converters," IEEE Journal of Solid-State Circuits, vol. 50, no. 1, pp. 258-269, Jan. 2015.

[9] M. Detalle et al, "Interposer technology for high band width interconnect applications", IEEE Electronic Components and Technology Conference (ECTC), pp. 323-328, 2013.

[10] V. Vashchenko and M. Scholz, System Level ESD Protection, pp. 54-55, Springer International Publishing, Switzerland, 2014.

2019 IEEE 69th Electronic Components and Technology Conference (ECTC)

System on Integrated Chips (SoIC™) for 3D Heterogeneous Integration

F.C. Chen, M.F. Chen, W.C. Chiou, Doug C.H. Yu

Research and Development, Taiwan Semiconductor Manufacturing Company
Ltd. 166, Park Ave. 2, Hsinchu Science Park, Hsinchu, Taiwan 30075, R.O.C.
Phone: 886-3-5636688 Ext. 722-3896, Email: mfchen@tsmc.com

Abstract—**A brand new 3D integrated circuit (3DIC) solution, Sy̲stem o̲n I̲ntegrated C̲hips (SoIC™), has been successfully developed to integrate active and passive chips into a new integrated SoC system to meet ever-increasing market demands on higher computing efficiency, wilder data bandwidth, higher functionality packaging density, lower communication latency, and lower energy consumption per bit data. 3D packaging is challenging and requires overcoming three major challenges – thermal, power delivery, and yield. The SoIC, as industry-first 3D logic-on-logic and memory-on-logic chiplet stacking technology platform, enables the heterogeneous integration (HI) of known good dies (KGDs) with different chip sizes, functionalities and wafer node technologies, all to be integrated in a single, compact new system chip. From external appearance, SoIC looks like a general SoC chip with multiple pre-designed heterogeneous functional chips embedded. As SoIC is fabricated using "front-end" process, it can be holistically integrated into variant "back-end" advanced packaging technology platforms such as flip chip, integrated fan-out (aka InFO), 3DIC, and 2.5D with Si interposer (e.g. CoWoS™) [1-2] to provide a miniaturized and highly integrated HI SiP for the future HPC, AI, 5G, and edge computing applications.**

With the innovative bonding scheme, SoIC enables the strong bonding pitch scalability for chip I/O to realize a high density die-to-die interconnects. The bond pitch starts from sub-10 μm rule. Short die-to-die connection of SoIC has the merits of smaller form-factor, higher bandwidth, better power integrity (PI), signal integrity (SI), and lower power consumption comparing to the current industry state-of-the-art packaging solutions. In this paper, we demonstrated for the first time an integration of SoIC chip into InFO_PoP without increasing its form-factor. The SoIC was made on a logic-on-logic stacking to validate the design rules, process maturity, and reliability.

Keywords—3DIC; SoIC; PI/SI; 5G/AI; InFO_PoP; data bandwidth; sub-10um IO

I. INTRODUCTION

We have previously developed 3DIC TSV technology to directly stack logic on logic, and wide I/O memory on top of an application processor on an organic substrate for low power consumption, high data bandwidth, low latency, and improved PI/SI [3-5]. This was an industry-first effort led by leading foundry involving logic chips and/or memory chip stacking using TSV technology and fine pitch Cu pillar flip chip assembly technology ready prior 2012. Recent years, 3DIC integration technology has evolved from substrate package

level to wafer level system integration (WLSI). This WLSI leads the semiconductor industry into a new era of system scaling beyond Moore's Law. High density integrated fan-out (aka InFO_PoP) in smartphone and high performance silicon interposer 3D chip stacking (aka CoWoS) in HPC and AI have been widely adopted in market segments for years [6-8]. The challenges of Moore's Law of advanced node SoC become more stringent and some critical functions in the SoC become more difficult to scale with the core devices. Those issues can be mitigated by SoIC technology with partition and chiplets approach [9-12].

II. SoIC™ 3D INTEGRATION

SoIC technology enables the system scaling with better merits of cost and performances by implementing SoC partition and re-integrating through the stacking of I/O logic chip and core circuits chip. SoIC also offers the design and integration flexibility for mix and match of heterogeneous chips in different technology nodes, materials, functionalities, and chip sizes to create a true heterogeneous 3DIC. The SoIC envelope to expanding its chip size opens new opportunities for highly integrated system as shown in Fig.1. With the closest proximity between two chips through direct chip-to-chip interconnect, the SoIC creates a superior power integrity, signal integrity and a much lower communication latency with more than 20Tbps memory bandwidth to support the future HPC, AI, 5G, and edge computing applications [13]. The ultra-fine pitch of SoIC bond can be scalable down to sub-10 micron to enable chip functional block level ultra-high density interconnects that cannot be achieved by the typical 3D IC TSV with flip chip bond technology [7-8, 14, 15].

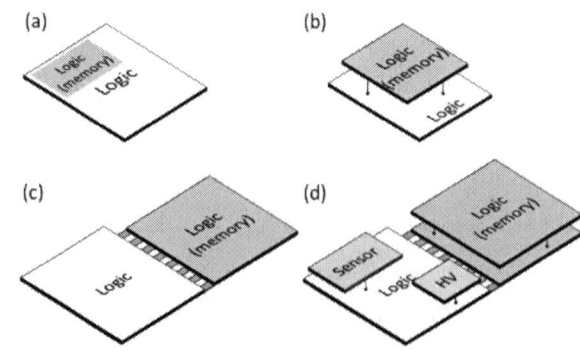

Figure 1. (a) SoC before chip partition; (b), (c), (d) Variant partitioned chiplets and re-integrated schemes enabled by SoIC

Leveraging leading foundry wafer front-end process and WLSI infrastructure, the SoIC manages the challenging thin

978-1-7281-1500-9/19 $31.00 © 2019 IEEE

die handling and bonding in advance and leverages well-established WLSI package integration to achieve high integration yield and reliability beyond the typical 3DIC stacking approach as shown in Fig. 2.

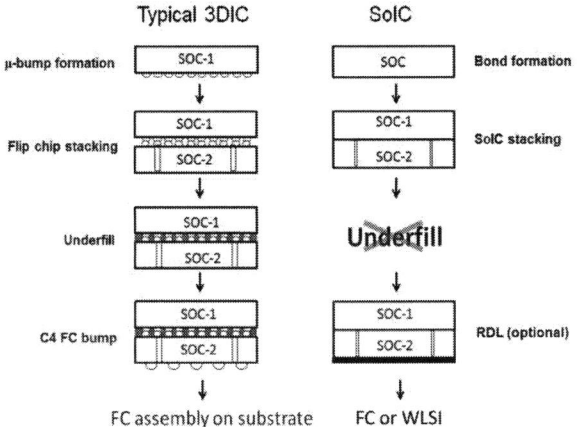

Figure 2. Comparison of integration flow between typical 3DIC package and SoIC package

In 3D stacking, chiplets are interconnected in face-to-face (F2F), face-to-back (F2B), and a combination of both. SoIC employs KGDs for 3D heterogeneous integration. Chiplets with different sizes, different functions/technology nodes can be integrated in 2D side-by-side manner for mix and match design flexibility and in 3D stacking manner for integration extendibility. Fig. 3a-3b demonstrates the design flexibility and integration capability of SoIC with different chiplets sizes. More than that, Fig. 3c demonstrates the extendibility of SoIC in 3-tiers of logic-on-logic stacking, which can be theoretically extended to more tiers depending on design needs. Fig. 3d shows the typical 3DIC flip chip stacking with μbumps, which leads to an obvious higher profile than SoIC does.

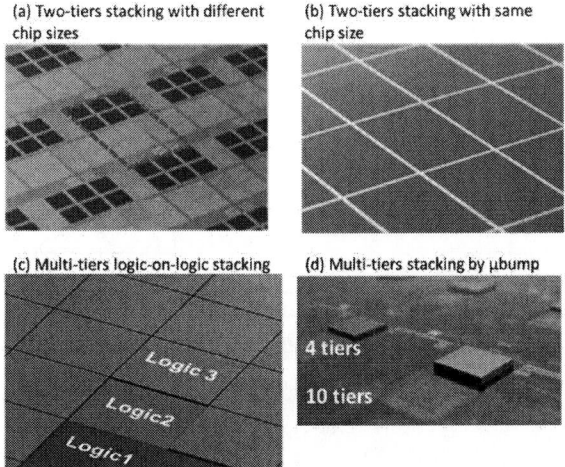

Figure 3. SoIC bond/stacking with (a) different small chiplets (b) the same large chiplets, (c) 3-tiers logic-on-logic integration, (d) typical 3DIC stacking with μbump bond

III. SoIC™ PACKAGE INTEGRATION

SoIC integrates both homogeneous and heterogeneous chiplets into a single SoC-like chip with a smaller footprint and thinner profile, which can be holistically integrated into any advanced package platforms such as flip chip, and WLSI (aka CoWoS [16] and InFO [17-18]). From external appearance, the SoIC is just like a general SoC chip yet embedded with desired and heterogeneously integrated functionalities (see Fig. 4)

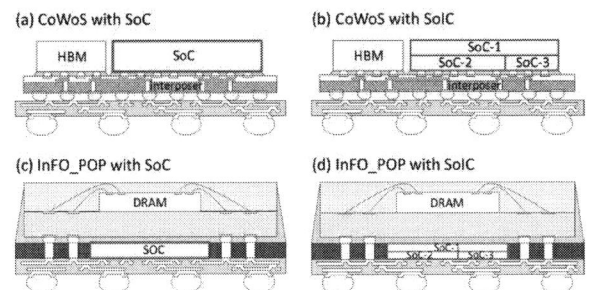

Figure 4. (a) CoWoS with SoC, (b) CoWoS with SoIC integrating partitioned SoCs (c) InFO_PoP with SoC, (d) InFO_PoP with SoIC integrating partitioned SoCs

IV. REALIZATION OF InFO_PoP WITH SoIC™

In this paper, we have successfully demonstrated the industry-first SoIC with logic-on-logic stacking integration in InFO_PoP platform. Compared to the typical 3DIC PoP, the SoIC-embedded InFO_PoP offers higher interconnect I/O bonding density, lower power consumption and thinner package profile. Fig. 5 shows the system scaling of bump/bond density can be significantly boosted by SoIC technology to unleash the limitation of flip chip bonding density beyond $10K/mm^2$. The limitation of bump/bond density on flip chip bond lies in two bottlenecks. One is the fine bump pitch limitation due to the inadequacy of bumping tools and materials. The other is the reliability issue due to the so-called chip-package-interaction (CPI) in fine bump pitch flip chip assembly.

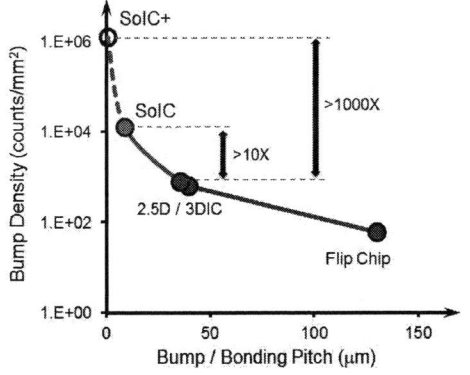

Figure 5. Leapfrog boost of SoIC on system bump/bond density

With foundry front-end process control, the integration of SoIC achieves high bonding yield. Fig. 6 shows the measured high bonding yield upscale to several millions SoIC bonding I/O counts.

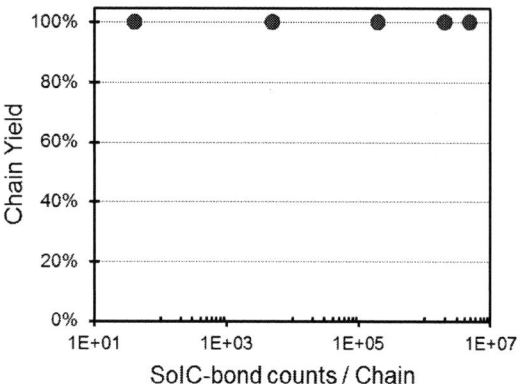

Figure 6. SoIC achieves high bonding yield upscale to several millions I/O counts/ chain

V. SYSTEM PERFORMANCE BENCHMARK

System performances such as interconnect bump/bond density, electrical performances (aka PI/SI), and thermal performance are critical to meet mobile and HPC system designer's demands on high computing efficiency, data bandwidth, low latency, low energy consumption per bit data operation. In the next session, the comparisons between SoIC bond and typical 3DIC stacking with µbump flip chip bond on aforementioned system performances are presented to realize the enablement of SoIC technology beyond the 3DIC stacking.

A. Interconnect Bump/ Bond Density

Fig. 7 shows the comparisons of interconnect bump/bond pitch and density between typical 3DIC and SoIC. According to our analysis, SoIC bond outperforms µbump of 3DIC at least by 4X smaller in bond pitch, and minimum by 16X higher in bond density, respectively in normalized scale. Higher interconnect bonding density allows chip designer and system designer to have the flexibility and extendibility on designing a new chip and new system device adopting the most state of the art integration technology for cost as well as performances.

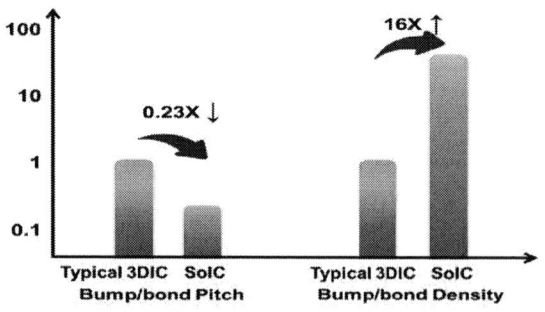

Figure 7. Bump/Bond comparison between SoIC bond and µbump of typical 3DIC

B. Electrical Performance

In adopting a new interconnect technology, the electrical resistance and its stability are critical to package designer. Fig. 8 compares the excellent cumulative percentage variations of electrical resistance of SoIC bond, and TSV/TDV with foundry mature Cu dual damascene via. The result reflects the compatibility and process readiness/robustness of SoIC bond, TSV, and TDV for customer designs and applications.

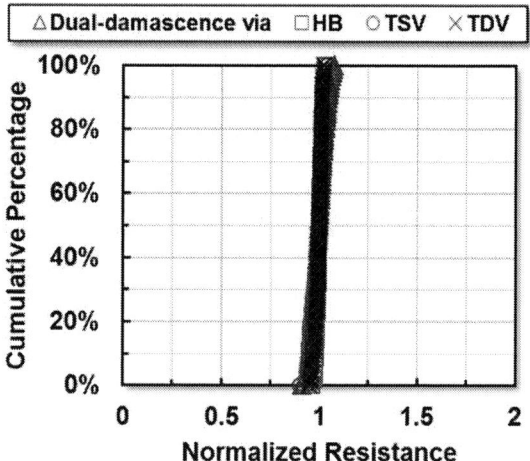

Figure 8. Normalized electrical resistance comparison between Cu dual damascene and SoIC bond/TSV/TDV

Low insertion loss is essential to enable the interconnect performance in high speed, high frequency data transmission between two SoC chips, particularly in 5G communication applications. Fig. 9 compares the insertion loss between SoIC bond and 2D side-by-side flip chip bond at the frequency less than 30 GHz. SoIC bond exhibits almost no insertion loss throughout the 30 GHz spectrum, while flip chip bond extends the insertion loss as frequency increases.

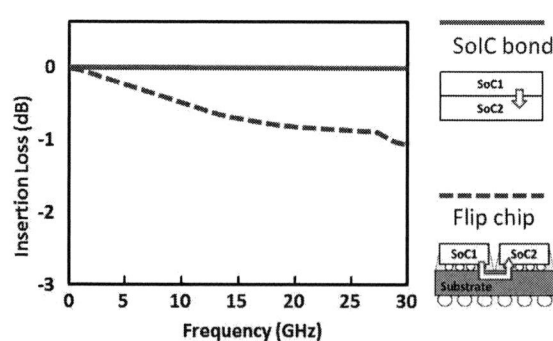

Figure 9. Insertion loss of SoIC and flip chip spans 30GHz frequency spectrum

Aside from the insertion loss, the interconnect wiring parasitics of R, L, and C for fine pitch, high density bonding is very important to ensure the high quality of signal integrity and power integrity with low RC delay and low IR drop. Fig. 10 compares the wiring parasitics between the µbump of typical

978-1-7281-1500-9/19 $31.00 © 2019 IEEE

3DIC stacking and SoIC bond. According to our analysis, SoIC bond outperforms µbump of 3DIC, at 2GHz, by 12.5X in R, by 100X in L, and by 12.5X in C, respectively in normalized scale. As a result, SoIC bond leads µbump bond by 156X in RC delay, and by 12.5X in IR drop.

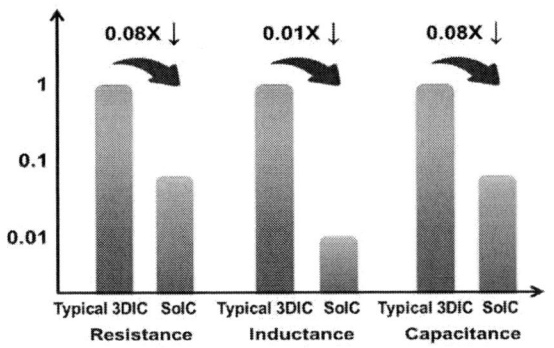

Figure 10. R, L, C wiring parasitics comparison between SoIC bond and µbump of typical 3DIC

C. Thermal/Power Performance

Thermal management becomes increasingly important in heterogeneous system integration, particularly for 3D chip stacking, due to a high thermal power density challenge. Exceeding the allowable working temperature, the chip may fail to function normally and cause severe leakage current issue. Thus engineers from advanced packaging community have exhausted means to resolve the thermal issues from 3D chip stacking. Compared to the typical 3DIC stacking, SoIC by its nature offers higher metal routing density and enables thermal flow in both upward and downward directions, and thus can dissipate the thermal flux more effectively (see Fig. 11). According to our analysis, SoIC bond outperforms µbump of 3DIC by 11X in energy consumption/ bit data, and by 1.25X~2X in thermal resistance, respectively in normalized scale. The significant reduction in the amount of energy required to move data around is very meaningful. At a fixed power envelope, the less energy spent per bit data, the more bits can be transferred. Alternatively, saved energy can be spent on other resource uses.

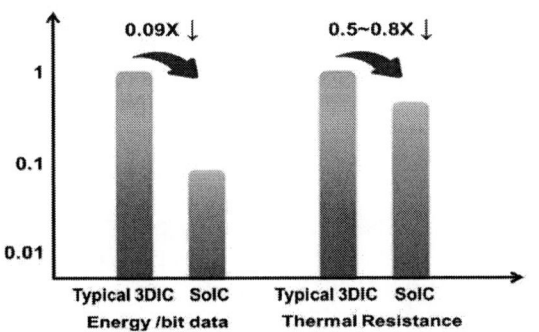

Figure 11. Comparison of thermal performance between SoIC bond and µbump of typical 3DIC

VI. RELIABILITY

To ensure the structural integrity of fine pitch, high density interconnects on SoIC bond, TSV, and TDV (through dielectric via), the chip level reliability of InFo_PoP with SoIC on electro-migration (EM), stress-migration (SM), and break down voltage (Vbd) were conducted. The test results are plotted in Fig. 12, Fig. 13, and Fig. 14, respectively. According to test results, TSV, TDV, and SoIC bond demonstrate robust reliability performances to pass TSMC internal requirements on EM, SM, and Vbd throughout the test conditions. Fig. 15 further shows SoIC to pass the PCT/168hrs, HTS/1000x, TCC/1000x, and uHAST/168hrs.

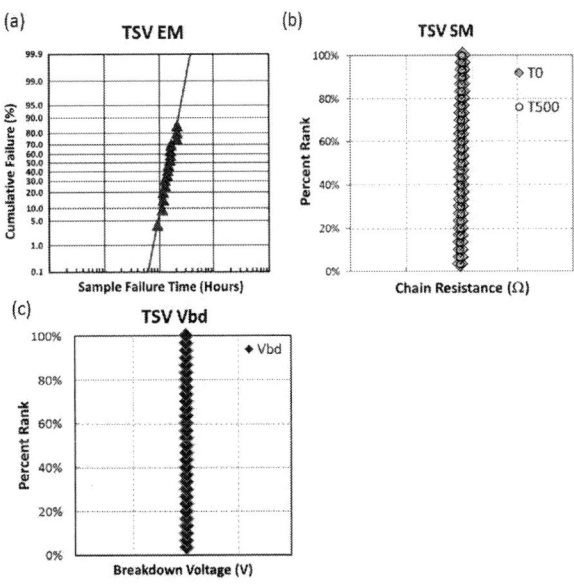

Figure 12. TSV reliability on (a) EM, (b) SM, and (c) Vbd

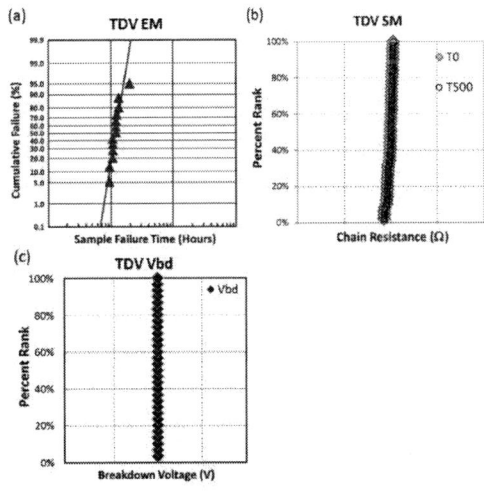

Figure 13. TDV reliability on (a) EM, (b) SM, and (c) Vbd

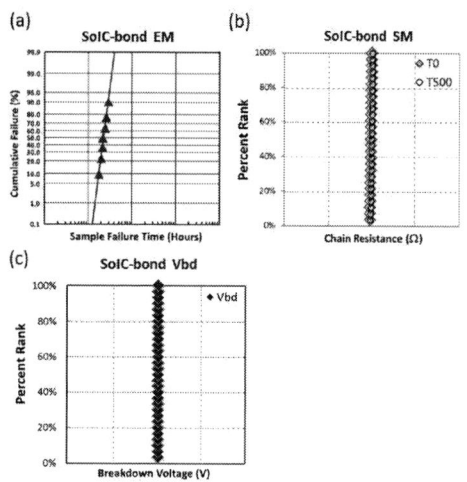

Figure 14. SoIC bond reliability on (a) EM, (b) SM, (c) Vbd

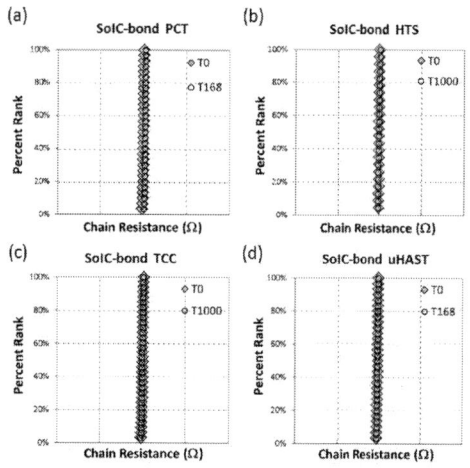

Figure 15. SoIC bond reliability on (a) PCT, (b) HTS, (c) TCC, and (d) uHAST

At package level, Table 1 tabulates various reliability tests being carried out on SoIC-embedded InFO_PoP. All test samples have passed the standard JEDEC test conditions and tests with extended test cycles and time. No failures were found after TC-B (1000x) and uHAST (192 hours).

Table 1. SoIC in InFO PACKAGE LEVEL RELIABILITY

Reliability	Test Conditions	Duration/Cycle	Result
Quick Torture	MR (multi-reflow @ 260°C)	10x	Pass
	MR10x + TCC (-65~150°C)	200x	Pass
	MR3x + uHAST	96 hours	Pass
CLR	MSL	Level 3	Pass
	TC-B (-55~125°C)	700x	Pass
		1000x	Pass
	uHAST (130°C, 85% RH, 33.3-psi VP)	96 hours	Pass
		192 hours	Pass

VII. CHALLENGES

There are challenges on design and process of SoIC integration. These challenges are very critical yet manageable for us by leveraging the well-established WLSI infrastructure in know-hows, materials, tools, and design rules. Many studies were conducted to characterize the impacts of TSV along with the size of keep out zone on IC transistors from thermal and mechanical aspects. In advanced node SoC, the silicon asset is precious for function blocks and IP designs. Large size TSV will consume more silicon area and increase the fabrication cost of advanced node SoC. Thus design rules with finer feature size of TSV are critical challenges to be addressed.

VIII. CONCLUSIONS

We have successfully developed and demonstrated the industry-first logic-on-logic and memory-to-logic 3D stacking using the SoIC technology. In comparison to the typical 3DIC stacking, SoIC offers higher I/O bonding density beyond the limitation of μbump flip chip bond, and enables lower energy consumption/ bit data, lower electrical parasitics in RC delay/IR drop, lower thermal resistance to unleash the boundary of IC designer on future heterogeneous integrations in 5G, AI, mobile, and HPC applications. Furthermore SoIC technology adopts the foundry front-end 3D process and leverages the well-established infrastructure of 3D WLSI in materials, tools, processes, design rules to create a holistic 3D WLSI family via CoWoS, InFO, and flip chip to shorten the development cycle time with high yield and competitive cost. To this end, we have demonstrated in this paper a SoIC-embedded InFO_PoP with verified system performances, robust reliability, and high stacking yield. SoIC becomes one of the most important WLSI technologies in future 5G and AI-driven applications for years to come.

ACKNOWLEDGEMENT

Authors would like to acknowledge R&D C.C. Kuo, K.C. Yu, C.H. Tung, C.T. Wang, Fab, Reliability colleagues for their contributions to this paper.

REFERENCES

1. Doug C.H. Yu, "Advanced System Integration Technology Trend", SEMICON Taiwan SiP Global Summit, Taipei, Taiwan, 2018/9/6
2. Doug C.H. Yu, "WLSI and Wafer Foundry Growth with Moore's Law and More-than-Moore, and Vice Versa", IWLPC Keynote speech, San Jose, CA, 2018/10/23
3. T. Lo et al. "Thinning, Stacking, and TSV Proximity Effects for Poly and High-K/Metal Gate CMOS Devices in an Advanced 3D Integration Process", in 2012 IEDM, pp 793-796
4. K.F. yang et al. "Yield and Reliability of 3DIC Technology for Advanced 28nm Node and Beyond", VLSI

978-1-7281-1500-9/19 $31.00 © 2019 IEEE

5. "A Wide I/O Memory-on-Logic Product Prototype Enabled by Through-Silicon Stacking Technology", iMAPs 2013

6. Doug Yu, "A new integration technology platform: Integrated fan-out wafer-level-packaging for mobile applications," in IEEE VLSI-T, 2015, T46-T47

7. H. Pu, H. J. Kuo, C. S. Liu and D. C. H. Yu, "A Novel Submicron Polymer Re-Distribution Layer Technology for Advanced InFO Packaging," in 2018 IEEE 68th Electronic Components and Technology Conference (ECTC), San Diego, CA, 2018.

8. S. Y. Hou., et al. "Wafer-Level Integration of an Advanced Logic-Memory System Through the Second-Generation CoWoS Technology," *IEEE Transactions on Electron Devices*, vol. 64, no. 10, pp. 4071-4077, Oct 2017.

9. M. T. Bohr and I. A. Young, "CMOS Scaling Trends and Beyond," *IEEE Micro*, vol. 37, pp. 20-29, November 2017.

10. A. Wei., et al. "Challenges of analog and I/O scaling in 10nm SoC technology and beyond," in *2014 IEEE International Electron Devices Meeting*, San Francisco, CA, 2014.

11. J. Hoentschel and A. Wei, "From the present to the future: Scaling of planar VLSI-CMOS devices towards 3D-FinFETs and beyond 10nm CMOS technologies; manufacturing challenges and future technology concepts," in *2015 China Semiconductor Technology International Conference*, Shanghai, 2015.

12. A. L. S. Loke., et al. "Analog/mixed-signal design challenges in 7-nm CMOS and beyond," in *2018 IEEE Custom Integrated Circuits Conference (CICC)*, San Diego, CA, 2018.

13. J. Y. Sun, "System scaling for intelligent ubiquitous computing," in *2017 IEEE International Electron Devices Meeting (IEDM)*, San Francisco, CA, 2017.

14. L. Xie., et al. "High-Throughput Thermal Compression Bonding of 20 um Pitch Cu Pillar with Gas Pressure Bonder for 3D IC Stacking," in *2016 IEEE 66th Electronic Components and Technology Conference (ECTC)*, Las Vegas, NV, 2016.

15. W. Koh, B. Lin and J. Tai, "Copper pillar bump technology progress overview," in *2011 12th International Conference on Electronic Packaging Technology and High Density Packaging*, Shanghai, 2011.

16. 19. W. S. Liao., et al. "A high-performance low-cost chip-on-Wafer package with sub-µm pitch Cu RDL," in 2014 Symposium on VLSI Technology (VLSI-Technology): Digest of Technical Papers, Honolulu, HI, 2014.

17. 20. C. C. Liu., et al. "High-performance integrated fan-out wafer level packaging (InFO-WLP): Technology and system integration," in 2012 International Electron Devices Meeting, San Francisco, CA, 2012.

18. C. Tseng, C. Liu, C. Wu and D. Yu, "InFO (Wafer Level Integrated Fan-Out) Technology," in *2016 IEEE 66th Electronic Components and Technology Conference (ECTC)*, Las Vegas, NV, 2016.

Die-to-Wafer (D2W) Processing and Reliability for 3D Packaging of Advanced Node Logic

Luke England, Daniel Fisher, Katie Rivera,
Bill Guthrie
GLOBALFOUNDRIES
United States
luke.england@globalfoundries.com

Ping-Jui Kuo, Chang-Chi Lee, Che-Ming Hsu,
Fan-Yu Min, Kuo-Chang Kang, Chen-Yuan Weng
Advanced Semiconductor Engineering (ASE)
Taiwan
Calvin_Lee@aseglobal.com

Abstract— In order to support emerging applications such as machine learning, where large amounts of fast access memory are required, the use of 3D packaging is inevitable. Previous work on 3D packaging with advanced node logic has shown that the technology is ready for implementation. In this paper, GF and ASE have demonstrated a Die-to-Wafer (D2W) process using 50um thickness logic wafers as the base. The 3D package also includes integrated thermal structures for heat removal from the base logic die. The process flow will be reviewed in detail, and challenges that were faced and overcome will be discussed. Reliability performance of the 3D package will also be reported. In addition, extensive thermal modeling was completed to understand the impact of two competing solutions for heat removal, which will also be reviewed in detail.

Keywords – 3D Packaging, TSV, Die-to-Wafer, D2W, Chip-to-Wafer, C2W

I. INTRODUCTION

As emerging applications and markets are continuing to push the boundaries and require increasing amounts memory integration, "new" packaging technology must also be adopted to support this trend. 3D/TSV based packaging is an obvious candidate to satisfy high bandwidth and low power requirements of these new applications. Previous publications have discussed system level benefits of stacking both HBM (DRAM) and SRAM on logic devices [1,2], and each type of memory has application spaces that can benefit. Both types of memory can reap power reduction benefits by residing directly on the logic, at a mere 50um distance from the logic circuits rather than several millimeters for other packaging types such as 2.5D or standard MCM.

Although 3D/TSV based packaging is not necessarily considered a new technology, it has not yet gained widespread adoption outside the DRAM industry due to various implementation challenges, which will be addressed in this study. These include factors such as process standardization, TSV and packaging reliability, and thermal dissipation. This paper will first discuss thermal considerations for 3D packaging that may drive application based decisions in terms of package structure and methodology. This includes heat removal options and thermal simulations quantifying the performance of each. In addition, a complete description of the test vehicle structure, process development learning, and reliability performance will be discussed.

II. THERMAL CONSIDERATIONS

Heat removal from 3D packages, especially those with high performance computing devices that consume large amounts of power, poses a significant challenge. The simple act of stacking a memory die on a logic die prevents direct contact of the package lid and thermal interface material (TIM) to the high power logic die. Multiple options are available to address this challenge, depending on the package configuration, each option with their own pros and cons. With proper logic die floorplanning in the design phase to strategically place the lowest power IP blocks of the logic die directly underneath the memory stack, all 3D thermal solutions available to us provide roughly equivalent heat removal performance. Figure 1 contains images and sketches that illustrate three thermal solutions available to satisfy the thermal demands of high performance 3D packages containing memory on logic.

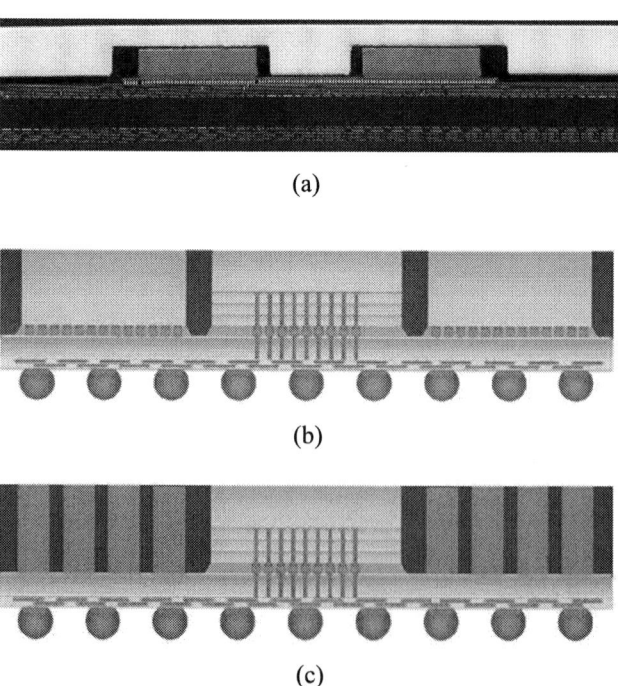

(a)

(b)

(c)

Figure 1: (a) Conformal Lid (b) Dummy Si Dies (c) Through Mold Vias (TMV)

Conformal Lid: Packaging and reliability of a large 3D package utilizing a conformal lid strategy as shown in Fig 1a has been discussed previously [3]. Direct contact of the lid with the high power bottom logic die provides a direct heat removal path. The advantage of this methodology is that lid attach falls generally within existing standard processes. TIM dispense patterns must be optimized for smaller areas of lid contact area, but a high tolerance cavity lid can be machined with a minimal price adder over a standard lid, making this an attractive option. The biggest negative with this method is that it is obviously not possible for molded 3D packaging approaches, which is where the following two options become useful.

Dummy Si Dies: The use of dummy Si dies as a heat removal medium for 3D packaging is shown in Fig 1b. Direct heat flow from the bottom logic die is achieved through the dummy die micropillar structures into the Si, which acts as an intermediate head spreader. Micropillar attach points are easily integrated with the standard UBM pad formation process as part of the MEOL module (Middle End of Line) in the assembly flow. The thermal interface from the backside of the bottom logic die can be considered limited by the amount of Cu present as micropillars through the polymer underfill layer. Since a typical flip chip die attach pattern equates in the 20-25% area fraction, which is Cu in this case, the interface provides a significant amount of thermal flow capability. The disadvantage of this methodology is highlighted in terms of manufacturing logistics, where an inventory of dummy Si die must be created and maintained. In most cases, two Si dummy die sizes will be required to fill gaps between the top die and package edge in the x and y dimensions. As more and more products are manufactured, managing a complete dummy die inventory to satisfy each product becomes a daunting task.

Through Mold Vias (TMV): Integration of TMV structures directly on the backside of the bottom logic die essentially provides the same effect as the dummy Si die approach described previously, in that thermal flow is defined by the Cu area fraction. Using TMV structures, however, provides some useful advantages. Since a solder based flip chip attach process is not used, an extra degree of freedom opens in terms of TMV pattern. Cu density can be pushed higher than a typical flip chip die attach pattern, providing the opportunity for higher levels of heat removal. In addition, the TMV diameter can be adjusted to help minimize spacing between Cu heat pipe elements for more uniform cooling. Additionally, the use of TMV structures for heat removal eliminates the need to manage a large library of dummy Si designs, which is a huge advantage in terms of manufacturing supply chain management. Since the TMV structures are typically electroplated directly on the backside of the logic die as an additional process step in the MEOL flow, integration is straight forward, but may limit the use to top die(s) that can be reasonably thinned. For instance, an HBM memory stack is ~750um tall, which would mean that electroplating of a TMV >750um tall would be required. This is not necessarily feasible, and would drive the selection

of a dummy Si die process. Studies will need to be completed in the future to understand the height limitation of the TMV structures. An additional challenge that needs to be overcome is warpage control during the D2W assembly process. Since we are essentially replacing a low CTE Si dummy die with a high CTE epoxy mold compound / Cu matrix, material selection and wafer level molding process control will be critical to maintain a flat enough D2W sub-assembly to handle downstream. In addition Cu diffusion protection must be integrated onto the backside of the active top die to prevent failures due to potential Cu transfer from the TMV structures to the backside of the active die during the composite backgrinding process to planarize all materials.

A series of heat flow simulations was completed for the three thermal solutions described previously, based on the high performance computing test vehicle to be described in Section III. A total logic die power of 200W was assumed, and two configurations were compared:

- Uniform Power: The full 200W was evenly distributed across the entire logic die.
- Power Map: The 200W was mapped to place high power content (i.e. SerDes) toward the logic die edges, while placing lower power content directly under the HBM locations. The power map used is shown in Figure 2, and the power density for the bottom die by area is labeled by color.

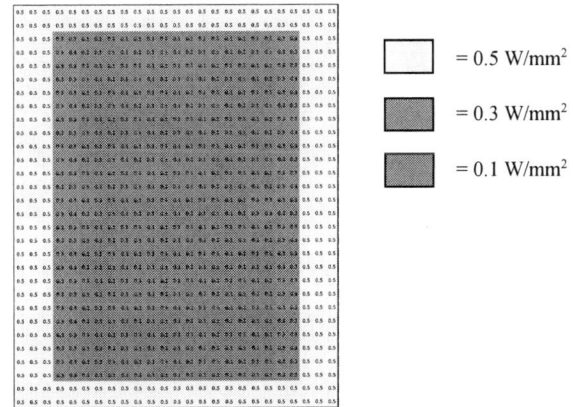

Figure 2: Power map used for the HBM-on-Logic thermal simulations.

Thermal simulation results are shown in Figure 3, estimating Logic and HBM bottom die temperatures for the three thermal solution methods described previously. In addition, a uniformly distributed power is compared to a "smart" power map based on location of IP blocks of varying power. For example, in an ASIC device for a high performance data center or networking application, the die edges typically are comprised of SerDes content. When studying the simulated temperatures of each condition in Figure 3, the following conclusions can be drawn:

- There is only a minor difference in thermal performance between the cooling methods.

Although the TMV solution is predicted to perform the best, there is only a ~5°C difference from the worst case conformal lid. This small difference is likely to be negligible in a real life scenario, where factors such as process controls can cause significant variation (i.e TIM bond line control after lid attach).

- When compared to a uniform power distribution, a smart power mapped layout provides significantly better cooling performance of the Logic and HBM dies, by a margin of roughly 10-15°C. This highlights the importance of architecture and layout planning early on in the product design phase to ensure the best possible cooling can be achieved.

- An ambient temperature of 55°C was used in these simulations. At this temperature, the memory die is on the cusp of meeting the maximum operating temperature of 105°C. In a high performance data center application where liquid cooling is used, better margins can be achieved.

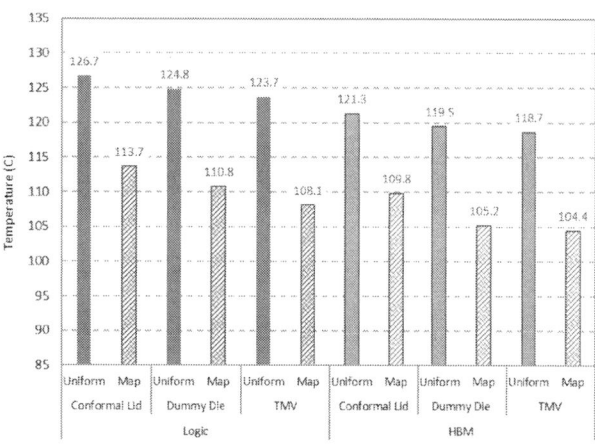

Figure 3: Plot showing simulated thermal performance of an HBM-on-Logic application. Logic and HBM bottom die temperatures are shown for uniform and smart power mapped logic configurations.

III. TEST VEHICLE DESCRIPTION

The 3D packaging test vehicle in the current study was designed to emulate the rough configuration of two HBM memory stacks on a high performance logic die. The test chips were built in GLOBALFOUNDRIES' 14LPP technology node with the bottom die sized at 19.2x21.3mm, and the top dies at 6x10mm. TSV structures of 5x55um size were integrated into the bottom die at the M1 level, between the FEOL and BEOL phases. A full suite of chip package interaction (CPI) test structures are present on the bottom die, as well as one of the two top dies, which has CPI structures wired out through TSV connections in the bottom die. It should be noted that the second top die was not wired out electrically due to BGA limitations of the package, and due to overall symmetry in the design, this was deemed acceptable. MEOL processing resulted in a final logic die thickness of 50um with a single layer of backside

redistribution (BS-RDL), and a top die thickness of 350um, all in a 50x50mm size. Based on the discussion points from the Thermal Considerations section, a dummy die configuration was selected for this application qualification space since the height of the top die is such that TMV plating would be unreasonable. A sketch highlighting the key features of the test vehicle is shown in Figure 4.

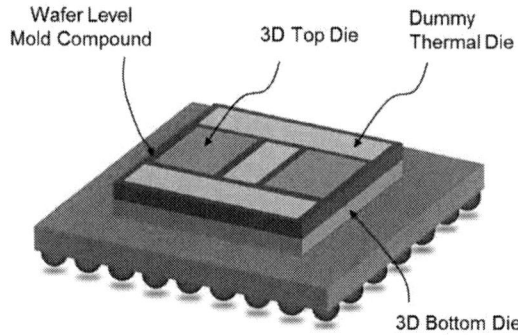

Figure 4: Sketch illustrating the 3D package used in this study. Note the two different dummy thermal die dimensions, which is typical of a single package configuration (requiring supply chain management).

Figure 5: Schematic of the micropillar bumping process flow used for top die wafers.

IV. 3D PACKAGING PROCESS

A. Micropillar Bumping

The 3D top die in the package are connected to the backside of the bottom die through a micropillar interconnect. A minimum pitch of 50um was used in a full array configuration across the top die to promote capillary underfill flow and prevent voids, and the diameter was half the pitch at 25um. The micropillar metallurgy stack consisted

978-1-7281-1500-9/19 $31.00 © 2019 IEEE

of 15um Cu + 3um Ni + 2um Cu + 16um SnAg for a total nominal thickness of 36um. The Cu/Ni/Cu/SnAg metallurgy allows for a robust micropillar solder joint through formation of a (Ni,Cu)Sn intermetallic compound layer that is very stable through high temperatures. A standard electroplated bumping process flow was used, as shown in Figure 5.

B. MEOL

The MEOL module contains several key process steps in the 3D packaging flow, including wafer thinning, TSV reveal, and backside metallization. These processes have been well established for both Si interposer and logic wafers containing TSV, and it must be remembered that there is one significant difference between TSV wafer types that can severely impact reliability. Logic wafers utilizing TSV contain active devices, whereas interposer wafers do not. A typical interposer MEOL process employs a rather simple TSV reveal technique through backgrinding into the TSV structures, exposing the Cu features and planarizing with the Si in one step. This process results in Cu smearing on the wafer backside surface, which is not an issue with Si interposers, but can result in Cu migration through the bulk Si causing device failures in logic wafers. Therefore, a "soft reveal" process must be employed to prevent such Cu diffusion failures.

The soft reveal process used in this study is shown in Figure 6. It is achieved by first mounting the bumped logic wafer to a glass carrier wafer. Since the logic wafer C4 bumps were ~80um tall, a carrier adhesive layer thickness of 90-100um was used to fully absorb the entire bump structures. Thick carrier adhesive processing can pose several risks such as TTV control and wafer breakage during backgrind or CMP, but the carrier attach process was fully optimized to prevent any downstream issues. Following temporary carrier attach, mechanical backgrinding to a final wafer thickness just above the bottom position of the TSV structures was performed. A post-backgrind CMP process acts to smooth the Si surface in preparation for the following dry etch step, preventing etch non-uniformities. A Si dry etch process using a F containing gas is then performed to recess the Si below the level of the TSV features. At this point, the TSV dielectric liner is still in place, preventing Cu exposure. Low temperature CVD processing follows, which deposits an oxide/nitride dielectric layer encapsulating the TSV structures. Finally, a CMP process is performed to planarize the oxide/nitride dielectric and TSV structures. By waiting to expose the TSV Cu material at this phase, we can prevent downstream Cu diffusion into the bulk Si and subsequent reliability failures since the wafer backside CVD dielectric acts as a diffusion barrier.

Finally, a UBM layer to receive top die for stacking is formed using standard electrodeposition processes. A backside RDL layer can also be added if the application requires it. In the current study, one layer of Cu RDL was used. The final logic wafer thickness after all processing was ~50um. Post-MEOL inspection of the 50um thick wafers showed that there were no processing issues across multiple lots that were ran. All layers met thickness targets across the wafers (i.e. silicon, dielectric, UBM, RDL). In addition, no

wafer breakage or edge chipping issues were observed, highlighting a robust process with high levels of repeatability.

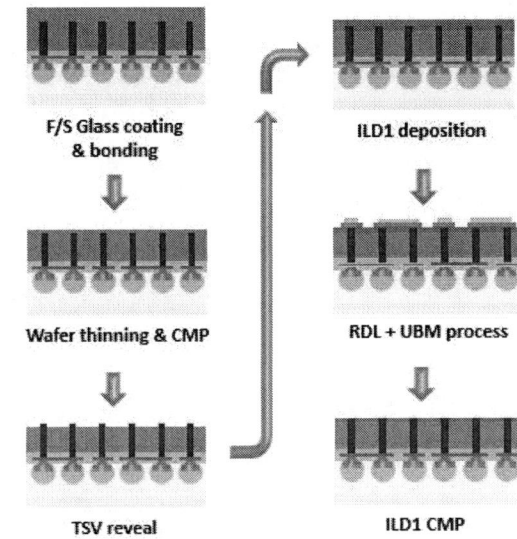

Figure 6: MEOL process utilized for the TSV bottom wafers. A "bump first" process flow was utilized.

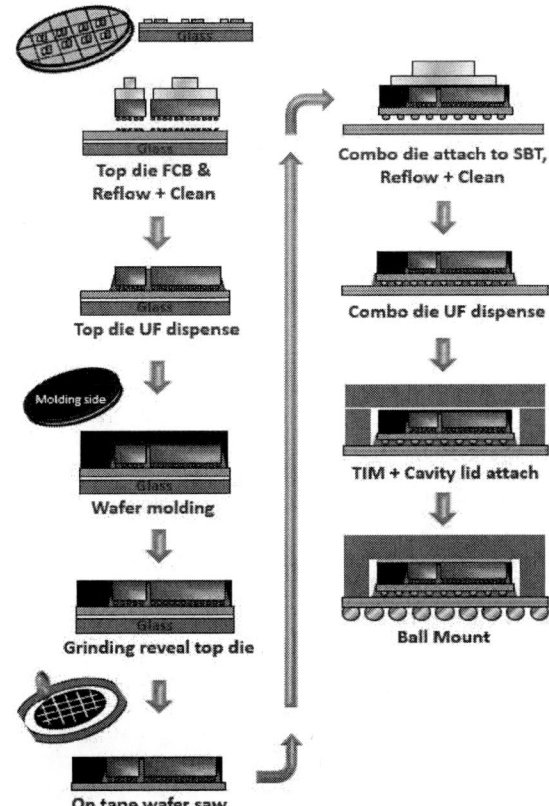

Figure 7: Die-to-Wafer assembly flow utilized for 3D packaging in the current study.

C. Die-to-Wafer Assembly

Following MEOL processing, a Die-to-Wafer (D2W) process was utilized to stack top dies on the logic wafer. The detailed process flow is shown in Figure 7. To ensure successful solderability of the micropillar interconnects of the top die to the bottom wafer UBM pads, a flux dipping process was used prior to both active and dummy die placement. A mass reflow process was utilized, which has a significantly higher throughput and lower cost than a thermocompression bonding process. Although the micropillars were placed on a 50um pitch, no issues with mass reflow attach were observed. Die attach was followed by a standard capillary underfill process. DOEs were performed with multiple underfill materials containing fine filler particles, and the best performing was selected. CSAM inspection was completed after underfill cure, revealing no signs of voids in the underfill layer. An optical inspection showed good control of underfill fillet height and lack of underfill bleed from the edge of the top dies, as shown in Figure 8.

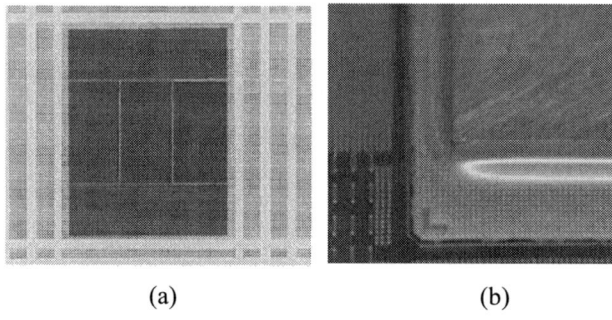

(a)   (b)

Figure 8: Post-die attach inspection images showing a) x-ray highlighting good die alignment and no solder voids, and b) good underfill fillet control.

Wafer level molding was performed after underfill, covering the entire stacked die wafer structure. A backgrind planarization step was done to expose the top dies, resulting in epoxy mold compound presence only in the gaps between top dies. DOEs were performed with multiple molding and post-mold grinding conditions to ensure warpage control during the D2W assembly. Warpage measurements across the D2W assembly are shown in Figure 9. It can be seen that directly after wafer molding with the full epoxy molding compound cap over the top dies, there was roughly 500um of bow in the D2W subassembly. After thinning to planarize and expose the Si top dies, the wafer bow dropped to roughly 350um. Multiple samples were measured, and it was determined that the distribution was sufficient to guarantee downstream processing.

Following D2W processing, the molded wafer assembly was removed from the glass carrier. A standard blade dicing process was used, followed by "combo die" attach to the substrate (combo die meaning the combination of top and bottom die in one sub-assembly). A mass reflow process was used, followed by combo die to substrate underfill application. A highly conductive thermal interface material

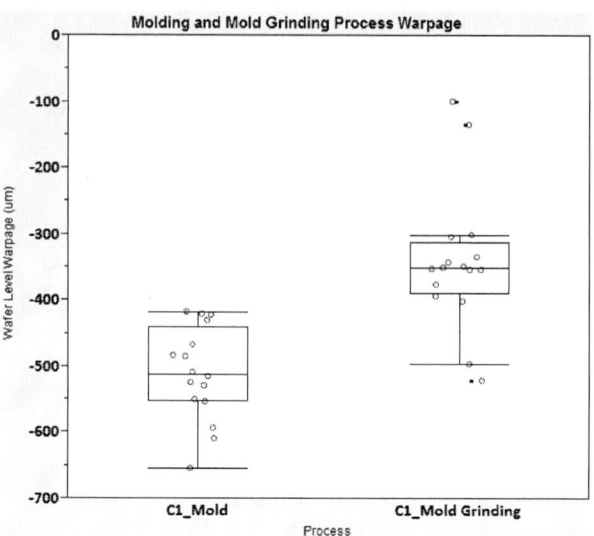

Figure 9: Warpage measurements across the D2W assembly after wafer molding and thinning indicate excellent warpage control.

Figure 10: Completed 3D module assembled using the D2W process flow described previously.

(TIM) was applied to the top of the combo die subassembly, and a Cu lid was applied. After substrate ball mount, the packages were ready for testing and electrical characterization. Of the 254 3D modules assembled for the qualification efforts, 243 passed final electrical screening equating to an assembly yield of 95.7%. An image of a final 3D package (without the Cu lid) is shown in Figure 10.

A package warpage study was completed using shadow moire testing through a temperature range of RT – 260C – RT to understand any potential impacts on downstream processing, including future board mounting of the package. Various stages throughout the D2W assembly flow are investigated, and results are shown in Figure 11. Following combo die removal from the glass carrier wafer, the molded and background sub-assemblies are relatively flat throughout the reflow process. Upon attachment of the combo die to the substrate, the package clearly becomes warped, which is dominated by the substrate. This being said, the warpage is only ~150um. Upon attaching the lid, the final package warpage is lowered further to ~80um at room temperature. The lid acts as a package stiffener and keeps it relatively flat throughout the reflow process, helping to ensure future success in 2nd level attachment of the package to board.

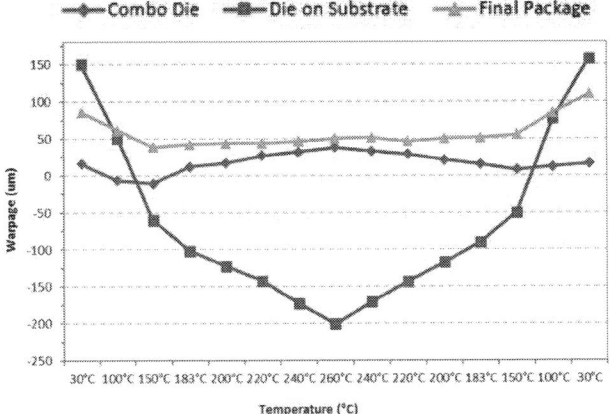

Figure 11: Shadow moire warpage measurements through various stages of the D2W assembly process.

Cross section analysis of the packages was completed, and a representative image is shown in Figure 12. It can be seen that micropillar solder joints are fully joined, with no Cu sidewall wetting or voids. C4 solder joints to the substrate are adequate with minimal voids present at the substrate UBM interface that are well within the allowed specifications. Although not shown, when observing the left, center, and right areas of the package, well controlled bond lines are present for both bottom and top die interfaces.

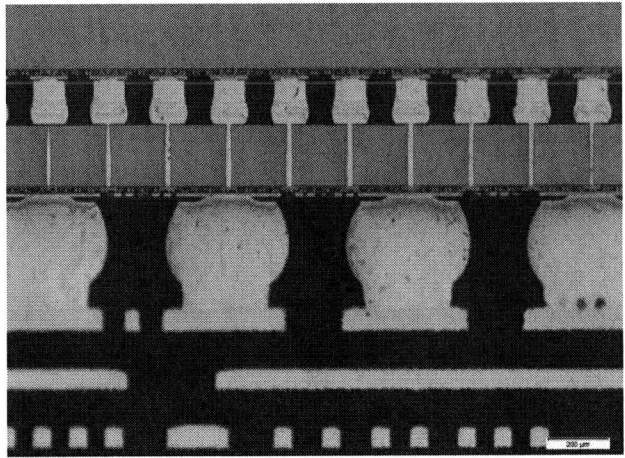

Figure 12: Representative cross section of a 3D D2W package.

V. RELIABILITY

Upon completion of post-assembly testing and achieving 95.7% yield, all good modules were subjected to MSL3 preconditioning with no failures found after post-precon testing. Modules were randomly selected for stress testing and subjected to JEDEC stress conditions as shown in Table 1. As of writing, the modules have successfully completed TC-G through the 500 cycle readpoint and HTS through a 280hr readpoint with no failures, providing an early indication that the 3D packaging structure is very robust. The remaining results will be shared after completion in the upcoming ECTC presentation.

TABLE I. PACKAGE RELIABILITY STRESS PLAN AND RESULTS

Stress Test	Conditions	Sample Size	Results
Precon (MSL3)	30°C/60RH, 336hrs + 3x Reflow, 245°C	214	214 / 214
TC-G	-40 - 125°C 1200cyc	80	200cyc: 80/80 500cyc: 80/80
HTS	150°C 1000hrs	80	280hrs: 80/80
HAST	110C/85%RH/3.6V 264hrs	54	TBD

VI. CONCLUSIONS

A robust and highly manufacturable 3D packaging process has been developed and demonstrated based on a molded die-to-wafer flow, using a test vehicle that simulates an HBM-on-Logic configuration for high performance computing. A "soft reveal" MEOL process was integrated to prevent backside Cu contamination failures after wafer thinning to 50um. Warpage characterization was completed at various points throughout the process flow. The results show a very flat final package in the range of 50-80um warpage, which is extremely small considering the 50x50mm package size. Early package stress readpoints through TC-G and HTS indicate that the 3D packaging solution developed will be very reliable.

Thermal modeling of three different cooling approaches was completed, with the results showing that adequate performance can be achieved when using a smart logic die layout, where high power IP blocks are placed away from the HBM footprint area. The final packaging type and end application will determine which of the three cooling solutions are to be used.

The outcome of this development project demonstrates that 3D integration is ready for customer adoption to help continue technology scaling through increased devices per volume rather than area.

978-1-7281-1500-9/19 $31.00 © 2019 IEEE

ACKNOWLEDGMENTS

Although this project's success was achieved through cooperation of many different individuals, the authors wish to highlight the following teams for their critical support: the GF reliability and test organization, and the ASE Assembly, MEOL, and Wafer Bumping teams. In addition, the following individuals provided key supporting roles: Wei-Hang Tai (ASE Underfill Process Engineer) and Wei-Chih Tseng (ASE Packaging Engineer), Pradip Pichumani (GF Failure Analysis Engineer), and Sukesh Kannan (former GF Packaging Engineer).

REFERENCES

[1] L. England, et al., "High Performance Processing Drivers for 2.5D and 3D Packaging", SEMI European 3D Summit, Jan 2017.

[2] L. England & I. Arsovski, "Advanced Packaging Saves the Day! - How TSV Technology Will Enable Continued Scaling," International Electronic Devices Conference (IEDM), 2017, pp. 3.5.1-3.5.4.

[3] R. Agarwal, et al, "3D Packaging Challenges for High-End Applications," IEEE 67th Electronic Components and Technology Conference (ECTC), 2017, pp. 1249-1256.

Enabling Ultra-Thin Die to Wafer Hybrid Bonding for Future Heterogeneous Integrated Systems

Alain Phommahaxay, Samuel Suhard, Pieter Bex, Serena Iacovo, John Slabbekoorn, Fumihiro Inoue, Lan Peng, Koen Kennes, Erik Sleeckx, Gerald Beyer and Eric Beyne

imec
Leuven, Belgium
phomma@imec.be

Abstract— The recent developments of wafer-to-wafer bonding technology based on direct assembly of inorganic dielectric materials is offering a path for the continuous need for higher integration density and lower interconnect pitches. However, numerous applications could benefit of a higher degree of design flexibility offered by a die-to-wafer approach. The achievement of high yielding die-to-wafer bonding with micron range die overlay is an essential element to unlock the potential of heterogeneous integration.

Keywords: Wafer bonding, Dieletric Bonding, Hybrid Bonding, Die to Wafer bonding, Heterogeneous Integration, Temporary Bond Material

I. INTRODUCTION

Among the technological developments pushed by the emergence of 3D integrated systems, numerous chip interconnection and bonding technologies have been introduced over the past years. Die to die stacking, wafer to wafer bonding and die to wafer bonding have emerged to fulfill a large range of die interconnect density requirements.

Hence, depending on the interconnect pitch and overlay specification, one would prefer the use a flip chip/die bonder or a wafer bonder to realize the stacking scheme. Whilst wafer to wafer bonding can achieve overlay below a micron, it creates a certain degree of constrains for system design as it is limited to equal die size.

The design flexibility offered through die by die based stacking and the possibility to maximize yield by selecting known good dies, explains the wide spread of the technique Today.

Imec has been working on the development of high density die to die/wafer bonding approaches with interconnect pitches reaching sub 20µm. It requires the joint exploration of novel polymeric bonding materials but also optimization of the thermocompression equipment.

On the other hand, for extremely low interconnection pitches, imec has been developing extremely precise wafer to wafer bonding based on SiCN inorganic dielectric bonding, with overlay now reaching well below 500nm and with extremely high bond strength at low temperature.

To answer the need for high flexibility designs and higher needs for interconnect density, we are proposing to explore a die to wafer bonding approach based on inorganic dielectrics that would fit the overlay gap achievable by polymer-based die bonding and wafer to wafer bonding.

Associating the advantages of direct inorganic dielectric bonding with the flexibility of die based bonding seems appealing but it is combining also two critical aspects: the high surface cleanliness needed for dielectric bonding and the need to singulate dies prior to bonding.

Whilst some previous art demonstrated the initial feasibility study of such die to wafer scheme [1-7], very few were dealing with a more representative case with thinned dies. Indeed, thin die handling in itself is highly challenging but when combined with the requirement of cleanliness linked to inorganic dielectric bonding becomes a technology roadblock that need to be answered.

This study will therefore deal with this problematic and the demonstration of 50µm thin dies bonding to a 300mm substrate using SiCN-SiCN inorganic dielectrics.

To answer these challenges, a new D2W collective bonding scheme will be introduced. The problematic and solutions to keep dies clean during and after singulation especially during handling will be also discussed.

II. SEQUENTIAL DIE TO WAFER HYBRID BONDING

Various contributions have described the concept of die-to-wafer direct/hybrid bonding. [1-7]. Such schematic concept flow is depicted in Fig. 1.

First the die source substrate is prepared by depositing a dielectric, typically SiO_2, followed by dielectric annealing and planarization of the surface roughness by CMP.

Upon singulation of the wafer info dies (Fig 1. a)), the dies are picked up from dicing tape as illustrated in step b). Depending on the equipment used, multiple die manipulation can occur in order to get the die on the proper side for the assembly to the wafer. This step is illustrated in c). Finally, the die is bonded to the wafer and the final dielectric annealing occurs after all dies have been assembled.

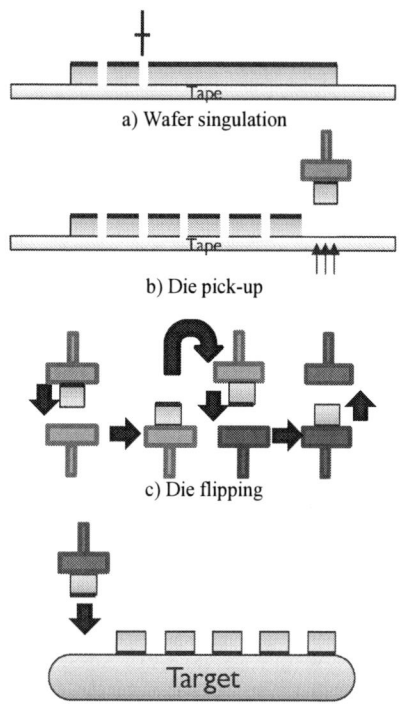

a) Wafer singulation

b) Die pick-up

c) Die flipping

Target

d) Die alignment and bonding to target substrate

Figure 1. Schematic of D2W sequencial hybrid bonding

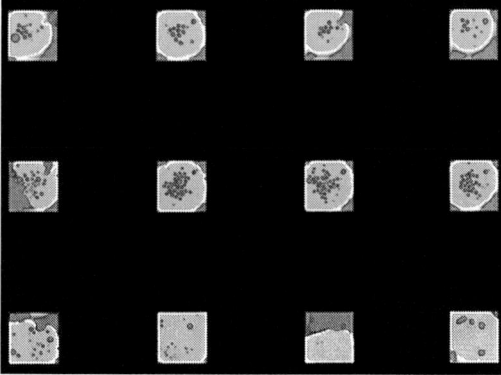

Figure 2. Scanning Acoustic Microscope inspection of dies directly bonded to a target substrate

To illustrate the kind of defects that could occur during such kind of die-to-wafer (D2W) sequential bonding process, we have performed a bond test and inspected samples using Scanning Acoustic Microscope (SAM) inspection as illustrated in Fig. 2.

When dealing with acoustic imaging, one must be careful with the interpretation. Indeed, SAM is based on the detection of acoustic reflection caused by some acoustic impedance change (i.e. change of material) as illustrated in Fig. 3. A SAM image is usually black and white with defects highlighted by the white zones. This is not entirely correct.

We have illustrated the various cases that can occur when inspecting samples through SAM in Fig. 3. The acoustic reflection coefficient is given by equation (1). It is computed for various interfaces.

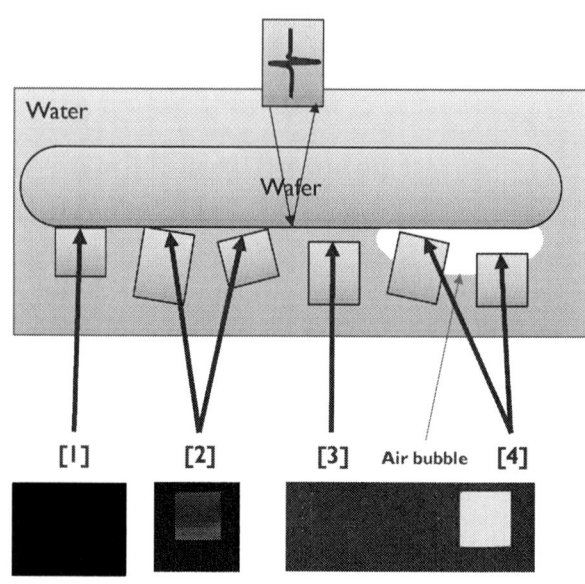

Figure 3. D2W Scanning Acoustic Microscope interpretation

TABLE I. ACOUSTIC REFLECTION FACTOR FOR VARIOUS INTERFACE

Interface	
Silicon-Air	100%
Silicon-Water	85%
Silicon-Silicon	0%

$$\Gamma = \left| \frac{Z2 - Z1}{Z2 + Z1} \right| \tag{1}$$

There are mainly four cases:

- [1] with a perfect bond interface, there is no acoustic reflection, the image is black
- [2] as SAM is performed in water, the liquid can creep into the interface, the SAM image is indicating a greyscale variation
- [3] a slight defocus of the inspection can lead to a very dark looking interface compared to [2]
- [4] a more common interface with air gap, leading to total sound reflection at the interface and a white image

Wafer-to-wafer (W2W) inspection is typically falling into cases [1] and [2] as no water can get into the interface. For our case, D2W inspection needs careful interpretation as the 4 scenarios can co-exist.

In all subsequent SAM inspections, we have been focused through the target wafer to minimize defocus issues linked to die to die thickness variation. We have used several systems from PVA Tepla Analytical Systems, equipped with 175 MHz transducers and 8 mm focal length. The inspections have been carried at 20 μm pixel size.

978-1-7281-1500-9/19 $31.00 © 2019 IEEE 608

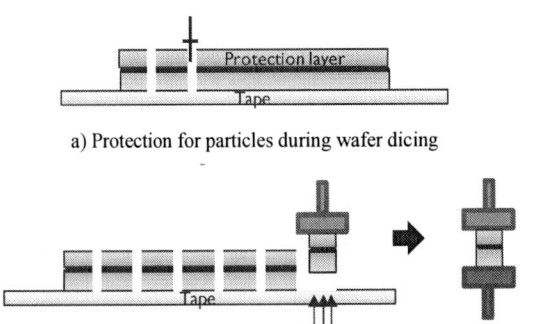

a) Protection for particles during wafer dicing

b) Protection from chucking defects during die manipulation

Figure 4.   Key purposes of proposed protection layer

The SAM inspection illustrated in Fig. 2 indicates a large amount of defect on the bond interface. Some random patterns are suspected to be generated during dicing singulation (particles). Some other ones are more located at systematic locations, hence probably related to chucking pattern and die handling.

Whilst equipment is still being optimized for such kind of D2W sequential hybrid bonding [7-8], we are proposing the introduction of a protection layer as depicted in Fig. 4. The purpose of this layer is to prevent particles and chuck defects to affect the bond interface.

III.   COLLECTIVE DIE TO WAFER HYBRID BONDING CONCEPT

The introduction of a protective layer for die singulation and handling is naturally leading to a collective D2W transfer approach illustrated in Fig. 5.

First an inorganic dielectric material is deposited, densified and planarized by CMP. A protection material is then coated prior to singulation as indicated in step a).

Dies are then picked as shown in b), manipulated, aligned and placed onto a temporary carrier substrate pre-coated with an adhesive material as shown in c).

The protection material is then removed selectively to the adhesive material using wafer level equipment, step d) prior to the dielectric surface activation illustrated in e) by wet and plasma processes.

The die carrier is then aligned and bonded to a target substrate f). Finally, the carrier substrate is debonded and the temporary adhesive is stripped off as illustrated in steps g) and h) respectively prior to annealing of the dielectric.

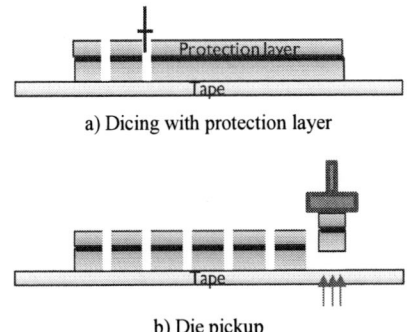

a) Dicing with protection layer

b) Die pickup

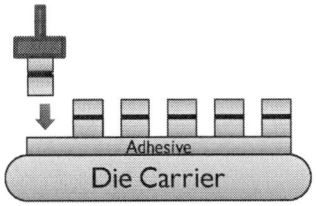

c) Die to carrier alignment and placement on adhesive material

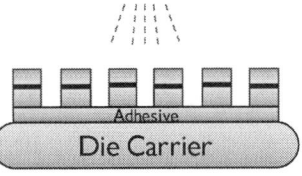

d) Collective protection layer removal selectively to adhesive material

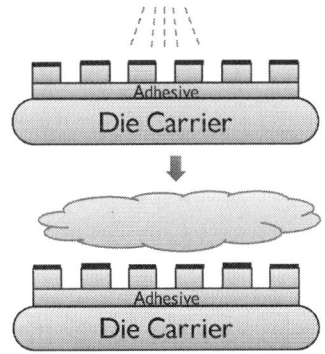

e) Collective surface preparation for hybrid bonding

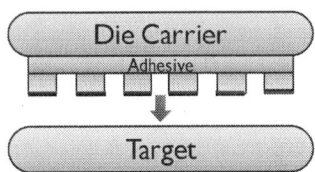

f) Carrier to target substrate alignment and wafer to wafer bonding

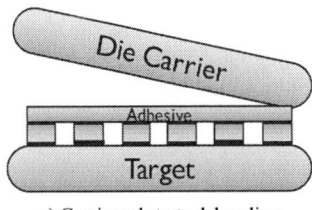

g) Carrier substrate debonding

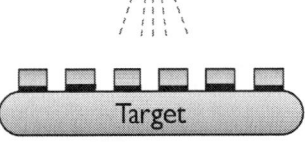

h) Adhesive removal

Figure 5.   Schematic concept flow for collective die-to-wafer hybrid bonding

978-1-7281-1500-9/19 $31.00 © 2019 IEEE   609

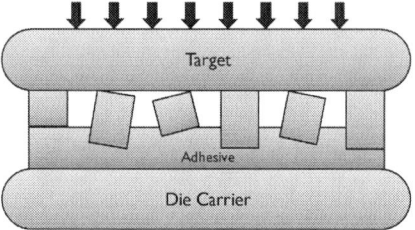

a) Situation before compression bonding

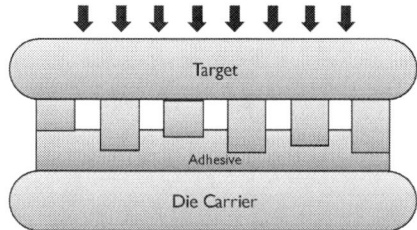

b) Situation after compression bonding

Figure 6. Illustration of coplanarity compensation by the adhesive layer during wafer-to-wafer bonding

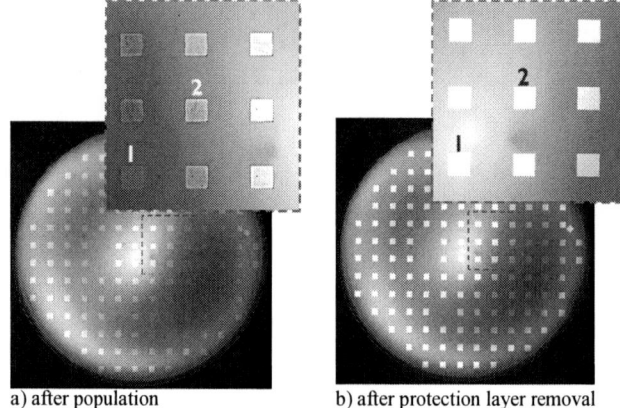

a) after population   b) after protection layer removal

Figure 7. Brightfield inspection after die population and protection layer removal

Defect count before and after die protective layer strip

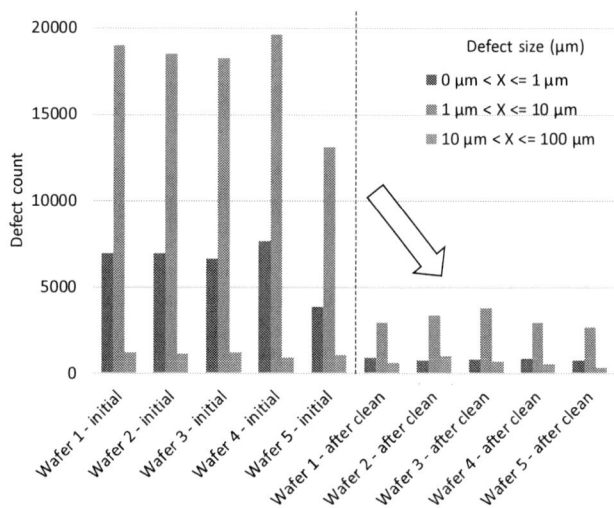

Figure 8. Illustration of defect reduction using a protection layer on top of dies

As dies could be sourced from various substrates and combined on one wafer, die thickness variation can be expected. The collective bonding of the dies as described in Fig. 5 f) will require a compression step as illustrated in Fig. 6 to compensate for any coplanarity difference and tilt.

Hence the role of the adhesive material is not only to maintain the dies in place during die attach and subsequent surface cleaning steps but also to provide compliance during compression bonding.

The adhesive material is chosen depending on the level of coplanarity to compensate. The protection layer is then selected mainly based on its stripper selectivity towards the adhesive material and its ability to survive a singulation step.

Based on these requirements we have used a mechanical debondable adhesive and a photoresist as first protection layer. Indeed, temporary bonding materials are typically designed to withstand the typical chemicals used in semiconductor manufacturing. The key properties of the material are summarized in Table II.

Contrary to the previously reported D2W hybrid bonding contribution [4], we are now using SiCN instead of SiO_2. This alternative dielectric material offers a higher bond strength when compared to typical oxides [9-10] especially with low temperature annealing below 250°C.

TABLE II. MATERIAL USED IN EXPERIMENTS

Material	*Type*	*Remover*
Protection layer	Photoresist (5-10μm)	Solvent A
Adhesive	Temporary bond material (30μm)	Solvent B
Carrier	Si	
Dielectric	SiCN	

978-1-7281-1500-9/19 $31.00 © 2019 IEEE

The concept validation first starts with the die placement onto a temporary carrier substrate followed by the removal of the photoresist protection layer (steps c) and d)). A defect inspection has been carried out after these two steps using a Nanometrics Spark FX3000 system. The brightfield output of the inspection can be found in Fig. 7. The analysis of the darkfield data lead to the generation of a defect map and quantification of the defects. This output is summarized in Fig. 8 and indicates a large reduction of the defects after photoresist strip.

The dies have then been bonded to a target substrate after water rinse and N_2 plasma activation using a SÜSS Microtec XBC300-Gen 2 bond cluster (steps 5 e) and f)). The high-resolution SAM inspection conducted right after W2W bonding is illustrated in Fig. 9. A center to edge variation can be noticed and indicates the need to optimize the compression pressure uniformity. Nevertheless, a further image analysis allows a quantification of the results as shown in Fig. 10. In this case an image segmentation has been performed and enables a relative comparison of each die bond surface compared to a perfect die. In this first experiment at this step 44% of the dies are perfectly bonded.

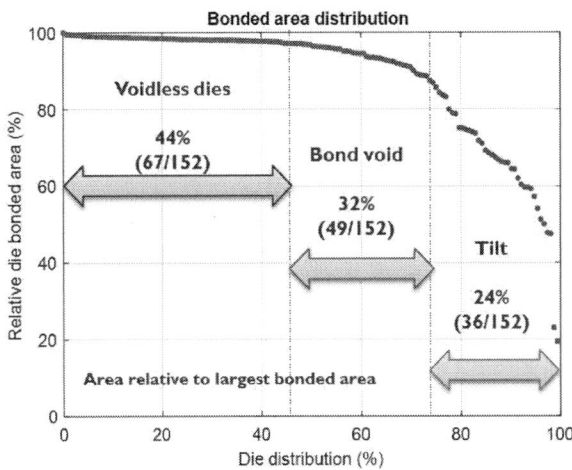

Figure 10.   Die bond yield distribution derived from Scanning Acoustic Microscope image analysis

Figure 11.   Photograph picture of transferred dies after carrier debonding and adhesive cleaning

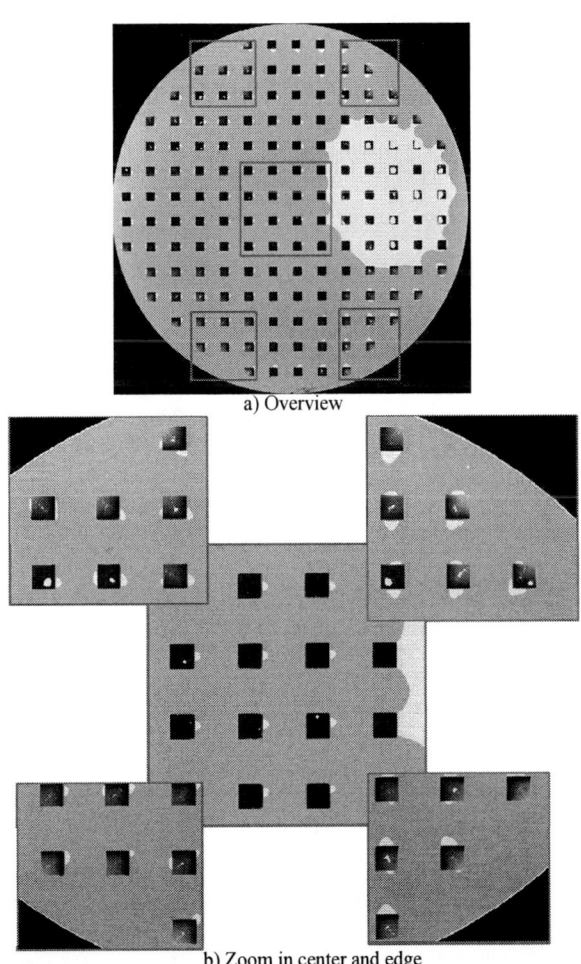

a) Overview

b) Zoom in center and edge

Figure 9.   SAM inspection after D2W collective bonding

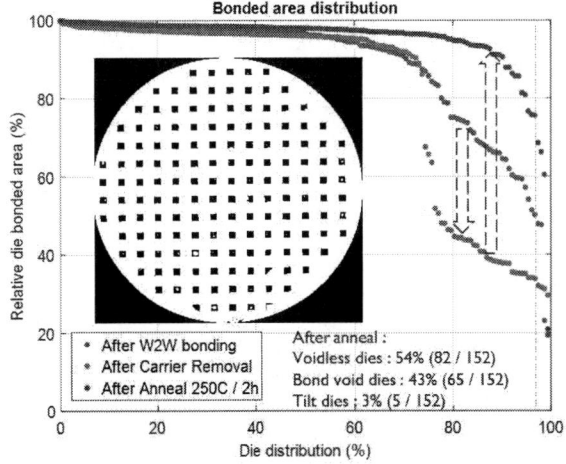

Figure 12.   Bond yield evolution after debonding and final annealing

The carrier substrate has then been mechanically debonded and the adhesive cleaned using the SÜSS Microtec DB12T and AR12 respectively. This corresponds to the process steps described in Fig. 5 g) and h). A photograph picture of the wafer as shown in Fig. 11 indicate a 100% physical die transfer success rate.

Further inspection by SAM of the interface after mechanical debonding reveal however a SAM yield loss. Indeed, at this step, the SiCN dielectric has not been annealed yet after bonding. Mechanical debonding of the carrier substrate induced a peeling force and affected the die as shown in the SAM yield in Fig. 12. However final annealing of the dielectric at 250°C for 2h recovered some of the yield lost during mechanical debonding.

IV. VALIDATION WITH THIN DIES

The collective D2W transfer concept has been further evaluated using thin dies including SiCN as dielectric material.

In order to prepare the dies, wafers with SiCN have been temporary bonded using the BrewerBOND© 305 material, followed by subsequent thinning to 50 μm. The thin wafer has then been transferred to another carrier substrate over allowing to re-access the SiCN side using a similar approach as described in [11].

This need to debond the temporary carrier used for thinning and carrier transfer result in a configuration where the protection layer (photoresist) is facing the dicing tape in Fig. 5 a) and b). Further die flipping inside the die bonder enables the rest of the steps c) to h) to be identical.

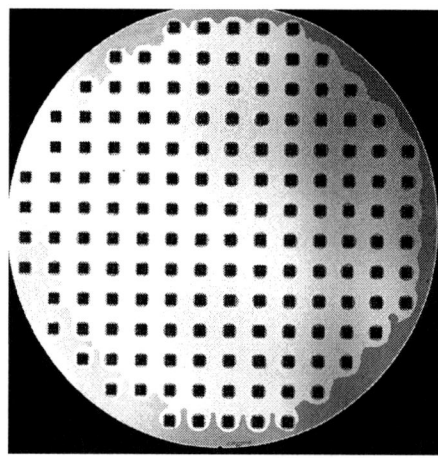

Figure 13. Scanning Acoustic Microscope inspection after collective die-to-wafer bonding of 50 μm thin dies

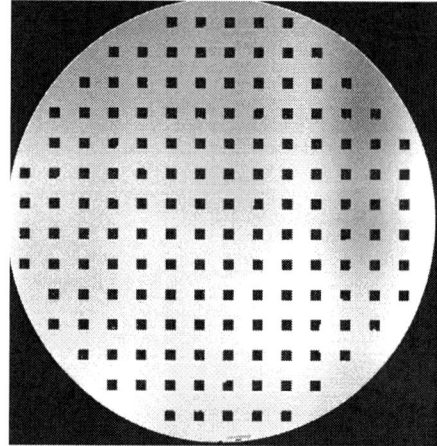

Figure 15. Scanning Acoustic Microscope inspection of thin dies after carrier debonding

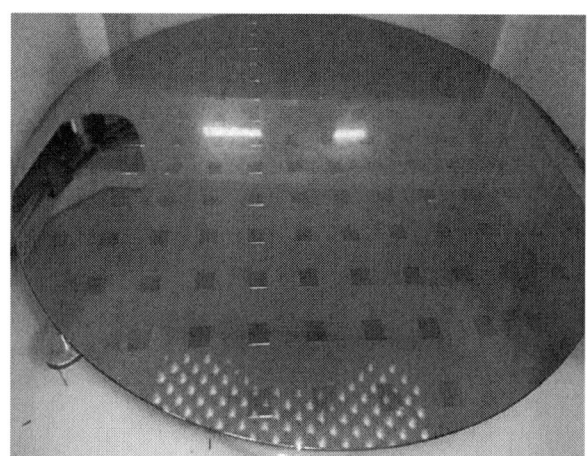

Figure 14. Photograph picture of transferred 50-μm thin dies after carrier debonding and adhesive cleaning

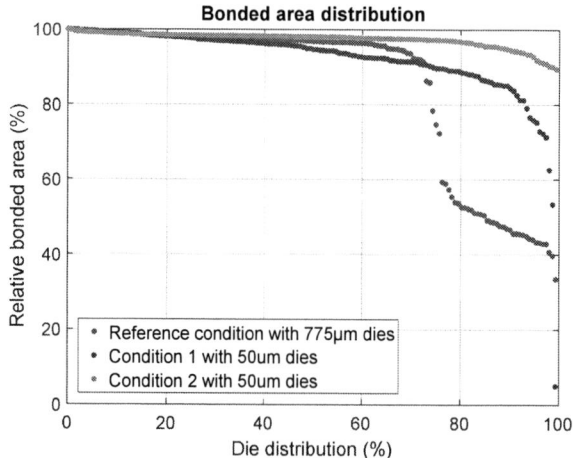

Figure 16. Bond yield distribution derived from Scanning Acoustic Microscope image analysis after carrier removal

Similar inspections steps have been carried out and are illustrated in Fig. 13 to 16. After some process optimization, it has been possible to reach a very uniform bond after W2W bonding as indicated in the SAM image in Fig. 13. The center to edge variation seen in Fig. 9 has been indeed eliminated.

Further on, the mechanical debonding of the carrier wafer has also resulted in a full transfer rate of the thin dies on the target substrate as shown in the photograph in Fig. 14. The final SAM inspection after carrier debonding in Fig 15 and analyzed in Fig 16, indicates a very promising approach. Indeed, very little defects can be detected and the majority can be attributed to the use of force during mechanical carrier separation.

The move to a laser debondable carrier system is promising to offer a higher process robustness level and will be more adapted to other dielectrics that have lower bond energy compared to SiCN.

V. CONCLUSION

We have successfully demonstrated a new collective D2W transfer. The combination of high bonding strength dielectric, SiCN and this transfer approach opens the way to heterogeneous integration and a higher flexibility compared to classical W2W bonding. Further studies will include an evaluation of the die to target overlay and interconnect bonding yield.

ACKNOWLEDGMENT

The authors would like to thank Frank Lauterbach, Mike Soules, Hans Mathee and Stefan Lutter from SÜSS MicroTec, Jakob Visker, Koen Smolders, David Huls from imec for their support during the experiments.

We are also grateful for the support provided by Brewer Science Inc. and Fujifilm.Electronic Materials.

This work is carried within the frame of the imec 3D System Integration Industrial Affiliation Program and within a Joint Development Project between imec and SÜSS MicroTec, between imec and Besi and between imec and PVA Analytical Systems GmbH.

REFERENCES

[1] L. Di Cioccio et al., "An innovative die to wafer 3D integration scheme: Die to wafer oxide or copper direct bonding with planarised oxide inter-die filling," 2009 IEEE International Conference on 3D System Integration, San Francisco, CA, 2009, pp. 1-4.

[2] P. Leduc et al., "First integration of Cu TSV using die-to-wafer direct bonding and planarization," 2009 IEEE International Conference on 3D System Integration, San Francisco, CA, 2009, pp. 1-5.

[3] L. Sanchez et al., "Chip to wafer direct bonding technologies for high density 3D integration," 2012 IEEE 62nd Electronic Components and Technology Conference, San Diego, CA, 2012, pp. 1960-1964.

[4] Teng Wang et al., "On the feasibility of die-to-wafer inorganic dielectric bonding," 2016 6th Electronic System-Integration Technology Conference (ESTC), Grenoble, 2016, pp. 1-5.

[5] G. Gao et al., "Scaling Package Interconnects Below 20μm Pitch with Hybrid Bonding," 2018 IEEE 68th Electronic Components and Technology Conference (ECTC), San Diego, CA, 2018, pp. 314-322.

[6] G. Gao et al., "Development of Low Temperature Direct Bond Interconnect Technology for Die-To-Wafer and Die-To-Die Applications-Stacking, Yield Improvement, Reliability Assessment," 2018 International Wafer Level Packaging Conference (IWLPC), San Jose, CA, 2018, pp. 1-7.

[7] P. Metzger et al., "New Flip-Chip Bonder Dedicated To Direct Bonding For Production Environment," 2018 7th Electronic System-Integration Technology Conference (ESTC), Dresden, 2018, pp. 1-6.

[8] H. Pristauz, "Disruptive Developments for Advanced Die Attach to Tackle the Challenges of Heterogeneous Integration," 2018 IEEE 68th Electronic Components and Technology Conference (ECTC), San Diego, CA, 2018

[9] S. Kim, L. Peng, A. Miller, G. Beyer, E. Beyne and C. Lee, "Permanent wafer bonding in the low temperature by using various plasma enhanced chemical vapour deposition dielectrics," 2015 International 3D Systems Integration Conference (3DIC), Sendai, 2015, pp. TS7.2.1-TS7.2.4.

[10] L. Peng et al., "Advances in SiCN-SiCN Bonding with High Accuracy Wafer-to-Wafer (W2W) Stacking Technology," 2018 IEEE International Interconnect Technology Conference (IITC), Santa Clara, CA, 2018, pp. 179-181.

[11] A. Phommahaxay et al., "Advances in Thin Wafer Debonding and Ultrathin 28-nm FinFET Substrate Transfer," 2017 IEEE 67th Electronic Components and Technology Conference (ECTC), Orlando, FL, 2017, pp. 740-745.

The Thermal Dissipation Characteristics of The Novel System-In-Package Technology (ICE-SiP) for Mobile and 3D High-end Packages

Taejoo Hwang, Dan(Kyung Suk) Oh, Jaechoon Kim, Euseok Song, Taehun Kim, Kilsoo Kim, Joungphil Lee, Taehwan Kim

Package Development, Test & System Package
Samsung Electronics Co., Ltd.
Cheonan-si, Republic of Korea
taejoo.hwang@samsung.com

Abstract— As information technologies evolve with the 4th industry revolution, such as artificial intelligence and 5G mobile communication, much more computing power and data bandwidth are required for both mobile and server systems. However, one-dimensional thermal packaging solutions such as a heat spreader or high conductive materials are not sufficient to solve the heat dissipation problems for the system-in-packages. In this research, a novel thermal dissipation technology based on two-dimensional heat flow was studied for the 5G high thermal power system-in-package modems and high performance computing logics. By applying a high thermal conductive material such as silver paste to conventional epoxy mold compound structures and creating direct high thermal dissipation paths from a bottom logic die to the heat spreader, it can bypass memory die that is more sensitive to the temperature rise than the logic die. The thermal performance of this novel technology was demonstrated using actual 5G modem system-in-packages comprised of a modem and two LPDDR4x dice. In conclusion, two-dimensional heat dissipation technique using thermal chimney is effective to reduce thermal crosstalk between top memory and bottom logic dice. Consequently, the overall system thermal performance was able to be improved by reducing heat flow through top memory dice.

Keywords; SiP, 5G, AI, thermal, 3D TSV, HBM, MCM

I. INTRODUCTION

Recently, emerging technologies such as 5G and artificial intelligence require more computing performance and higher memory bandwidth than conventional computing systems at mobile and server applications because data processing throughput is a critical factor in evaluating system performance [1]. Thus, the high computing power and high memory bandwidth lead to the demand of high thermal performance packaging technologies as shown at Fig.1 [2] For mobile applications, 5G modem devices are expected to consume higher thermal power than long term evolution (LTE) modem. In addition, leading-edge smart phones require thinner and smaller modem package form factors. This results in unavoidable thermal performance drop for heat dissipation. Furthermore, strict form factor requirements force 5G modem devices to adopt a system-in-package (SiP). 3D through silicon via (TSV) SiP with high bandwidth memory is also an alternative package option with high throughput and low latency data processing suitable for

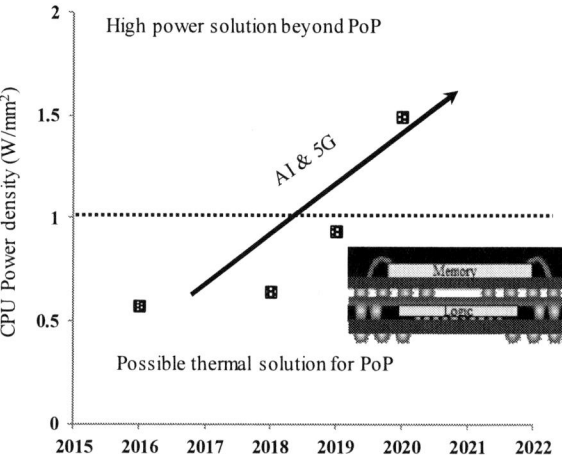

Figure 1.   Thermal challenges for mobile packages by AI and 5G

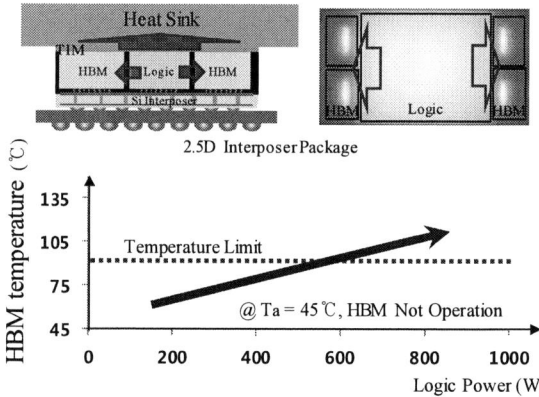

Figure 2.   Thermal crosstalk for server packages

artificial intelligence server systems. However, 3D TSV packaging configuration has a challenge of delivering full system performance due to thermal crosstalk between logic and high bandwidth memory dice, as indicated in Fig. 2 [3,4,5]

Hence, SiP is a more suitable package solution for heterogeneous integration for the near term 5G and artificial intelligence technology era [6]. For that reason, this novel heat-dissipation technology of ICE-SiP was studied in this

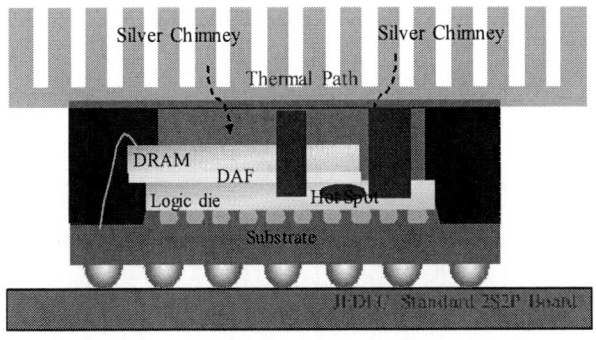

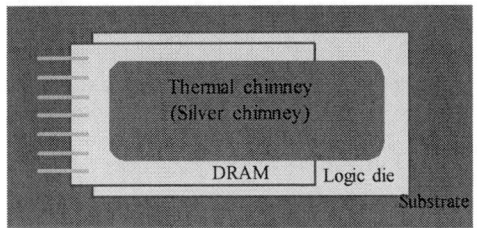

Figure 3. "ICE-SiP" conceptual diagram with a "thermal chimney"

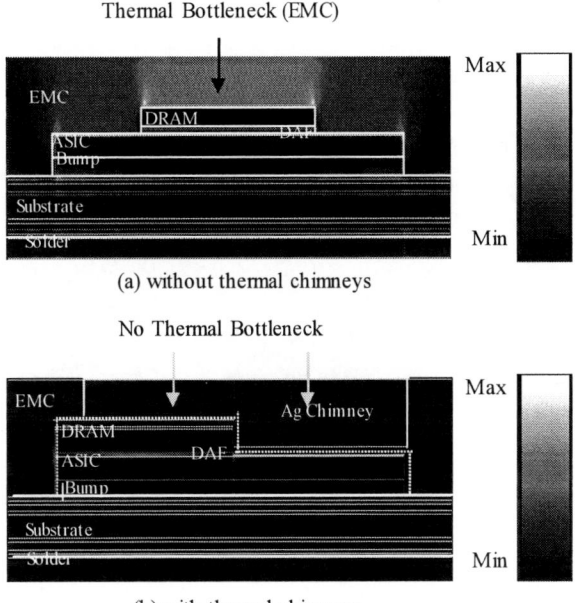

(a) without thermal chimneys

(b) with thermal chimneys

Figure 4. The effects of "thermal chimney" in SiP

paper to provide thermal solutions for the high performance SiP.

II. CONCEPTS AND EFFECTS

The adaption of SiP is steadily increasing in mobile applications because a SiP with vertical die stacks provides a smaller form factor and lower thermal resistance than a package-on-package (PoP). However, a top silicon die gets quickly heated by a bottom die in a vertical-stacked SiP since two vertical stacked dice directly contact each other with low thermal resistance in comparison to a PoP. Also, a hot spot in the bottom die is closer to the top die than that in a PoP. Therefore, the top die is more vulnerable to the hot spot of the bottom die. If the top die is a DRAM device, overall thermal performance of the SiP can degrade in comparison to a logic-logic stacked SiP due to the operating temperature limit.

The concept of ICE-SiP technology illustrated at Fig.3 is to make high thermal conductive paths from a bottom die to the top surface of a SiP by reducing the heat flow through top memory dice. The high thermal conductive path is called a "thermal chimney" indicated in Fig.4. A 5G modem SiP package in this study places multiple DRAM devices on the top of a bottom modem die. Because those DRAM dice are connected into the modem dice by wires, epoxy molding compound (EMC) is necessary for encapsulating the top surfaces of SiP. Since the thermal conductivity of EMC is relatively lower than silicon or metal in packages, the temperature of DRAM dice rapidly rises. The thermal chimney of ICE-SiP divides the overall heat flow into two directions. One is to flow through top DRAM dice and the other is to flow through the thermal chimneys, which are made of higher thermal conductivity material than conventional EMC. The reduced heat flow through DRAM dice helps the temperature drop of DRAM dice in the SiP.

Furthermore, the thermal chimneys can replace EMC on the top surface of DRAM dice with high conductive material. It is easy to remove the heat from the DRAM top dice.

In this study, three concepts of the thermal chimney in ICE-SiP were demonstrated as shown in Fig.5. The first concept is called a trench chimney. This concept makes trenches through EMC using various ablation methods like a laser drilling and filling the trenches with high conductive material. This concept is effective for complex structures that have many wire bonding connections and multiple top dice. The second concept is a silicon chimney. This concept places silicon dice on top and bottom dice with die attach film material, which works as a thermal chimney. Silicon dummy dice has a thermal conductivity of about 120 W/mK comparing to that of conventional EMC and are easy to handle in conventional packaging process. However, die attach film with a silicon chimney increases thermal resistance due to low thermal conductivity. The third concept is a silver chimney. This concept dispenses silver paste on top and bottom dice. The overall thermal performance of the silver chimney is better than the silicon chimney t because of its relatively low interfacial thermal resistance comparing to EMC and die attach film (DAF).

In order to evaluate thermal performance of the silver chimney, temperature of top and bottom dice in logic and memory SiP is simulated. It was assumed that the simulated SiP model consisted of one LPDDR4x DRAM die and one logic die in Fig. 6. The logic die was bonded to the substrate by flip chip process and the DRAM die was attached on the logic die with DAF. Active side of the DRAM die faced up and bonding pads are interconnected to the substrate by wire bonding. Package size was 50-60 ㎟ and the bottom die size was 30-40 ㎟. Conventional EMC, high thermally

978-1-7281-1500-9/19 $31.00 © 2019 IEEE

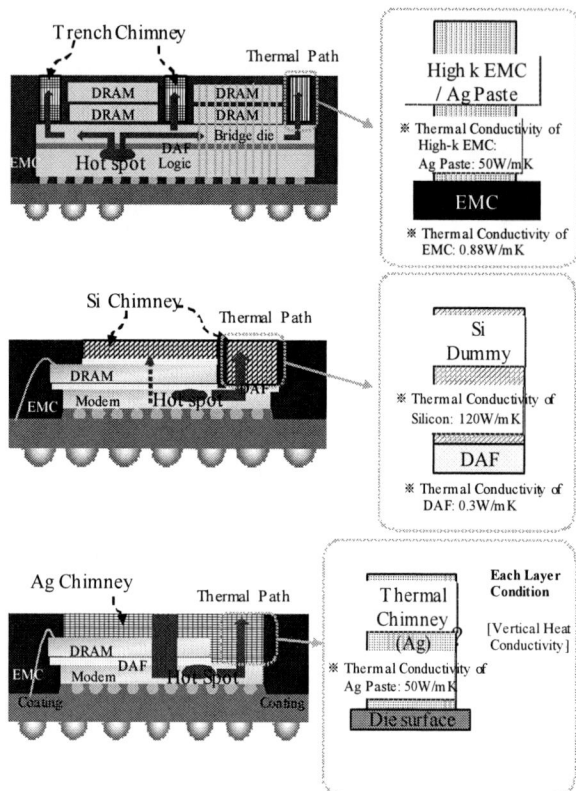

Figure 5. Three thermal chimney concepts of "ICE-SiP"

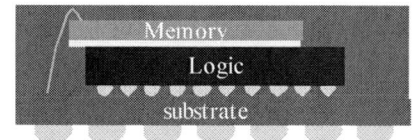

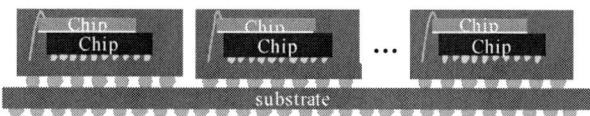

Figure 6. The system model of "ICE-SiP" for simulation

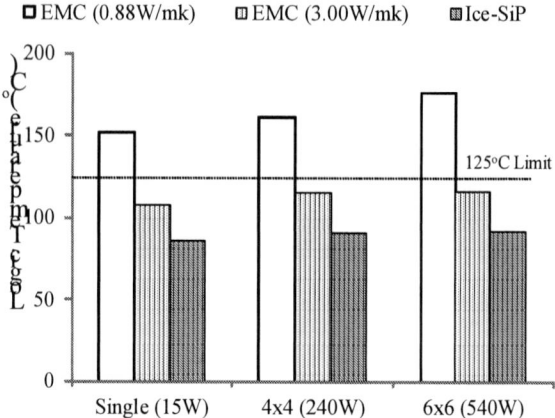

Figure 7. Simulated temperature of logic dies for each system

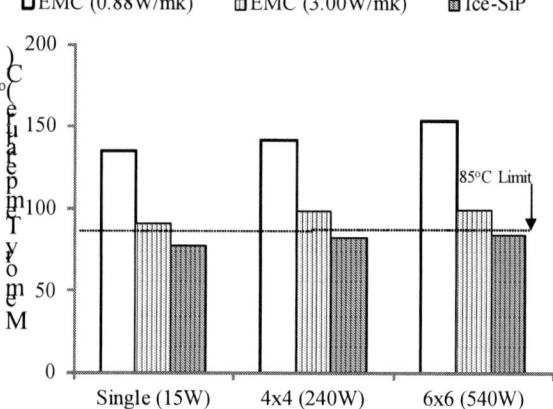

Figure 8. Simulated temperature of memory dies for each system

succeeded in keeping both temperatures below the specified operating temperatures. The same tendency was shown for the multiple package matrix modules comprised of individual SiPs, as shown in Fig.7 and Fig. 8. This study showed that the concept of ICE-SiP works for single SiP and multiple package modules in the server applications.

It is not easy to increase the thermal conductivity of EMC due to its effect on modulus, but high thermal conductivity material can be selectively placed on hot spots or specified positions.

III. PROCESS AND MATERIAL

A. Material

Silver paste is a key material to carry the heat flow from silicon dice to the top package surface with low thermal conductivity and to cover complex die surface morphology by reducing contact thermal resistance. The thermal conductivity of silver paste was enhanced up to 50W/mK by optimizing curing conditions and adjusting filler contents and size. High sintering temperature of silver paste lead to high thermal conductivity, but the sintering temperature

conductive EMC and the silver chimney were applied in the SiP simulation model. Package configuration ranging from single package to multi-matrix package module was to investigate the distributed computing system performance as illustrated in Fig. 6. Considering server environment, case temperature is set to 50℃.The top die of DRAM was set to 85℃ temperature limit and the bottom logic die was set to 125℃ temperature limit for this evaluation as suggested in Fig. 7. Thermal power of single SiP was assumed to be 15W. As shown in Fig. 8, conventional EMC material was not able to satisfy both logic and memory temperature requirements. High thermally conductive EMC of 3W/mK mitigated overheating of the bottom logic die, but it failed to reduce the temperature of the top DRAM die below the 85 ℃. In contrast, the ICE-SiP technique of the silver chimney

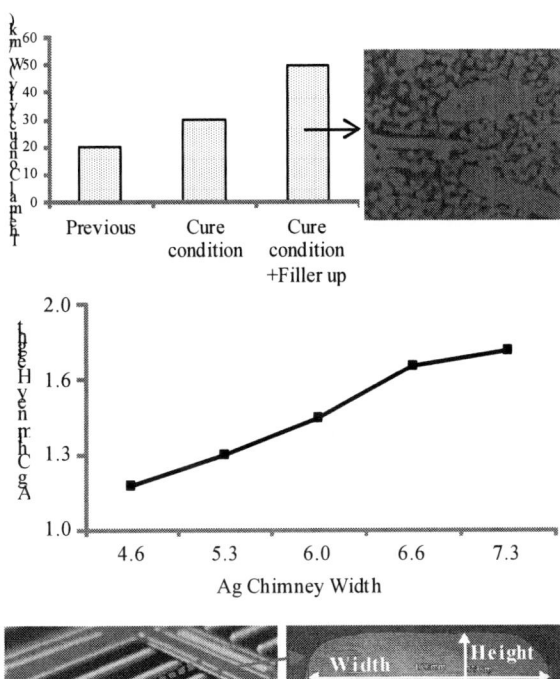

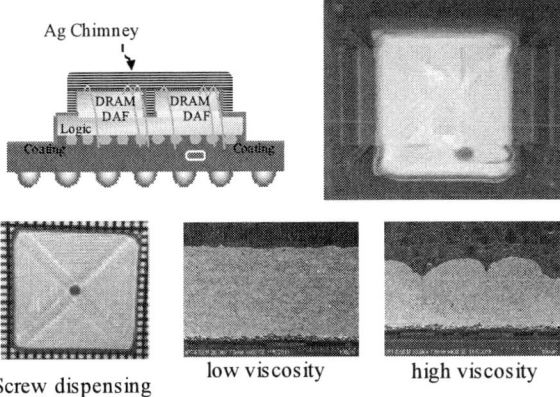

Figure 9. Dispensed silver paste pattern and aspect ratio

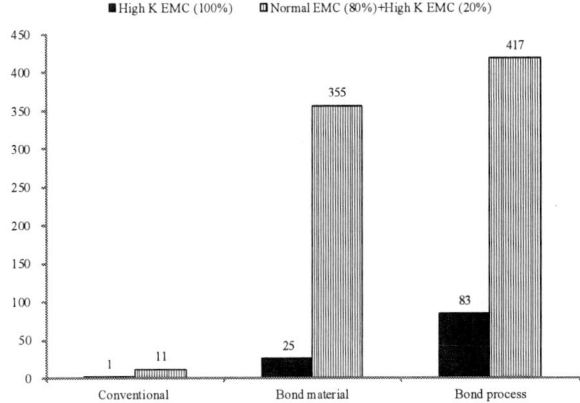

Figure 11. Blade life time comparison for High K EMC

since the top surface of silver chimney was ground and polished to flat.

Moreover, high thermal conductivity EMC (High K EMC) further improves the overall thermal performance in conjunction with the ICE-SiP technology. However, changes of filler material could lead to several problems in reliability and assembly process. Memory devices are susceptible to alpha-ray radiated from the filler material, so the high K filler material has to be purified to less than $0.002CPH/cm^2$.

Also, high K filler material can reduce the life time of package sawing blades significantly. That is one of the reasons why it's difficult to use high K EMC as SiP encapsulation. Further study needs to be done to extend life time of the sawing blades.

B. Process

As presented in Fig. 12, silver chimney formation was

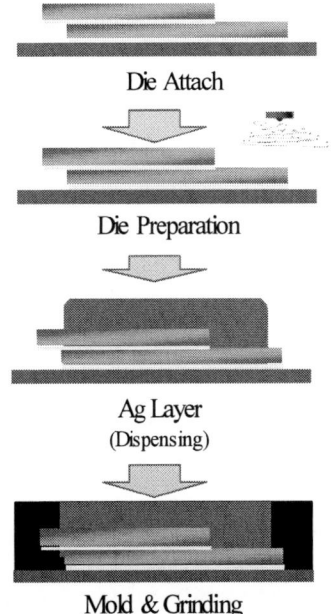

Figure 12. "Silver chimeny" process for SiP assembly

Figure 10. Dispensed silver chimney on dies in SiP

higher than conventional EMC cure temperature could affect device yields. High aspect ratio of the dispensed silver paste makes it easy to control the shape of dispensed silver paste, as shown at relative length ratio in Fig. 9. Controllability of the silver paste is important to prevent paste overflow and contacting wires and to improve throughput of dispensing process. In terms of reliability, interfacial adhesion between silver paste and EMC was tested with the moisture sensitivity level testing of level 3 A dispensing machine was used to dispense the high viscosity silver paste with rough surface morphology as presented in Fig. 10. However, it did not affect the final package shape and thermal performance

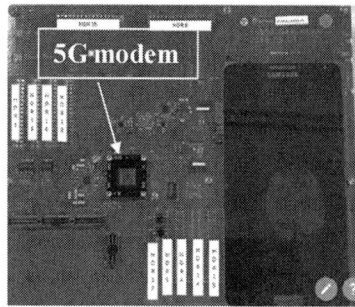

Figure 13. Test system configurations with 5G modem

performed after interconnections of the flip chip and wire bonding. Before dispensing silver paste over the top and bottom dice, die preparation processing was performed to prevent the silver paste from contaminating pads, wires and others. The silver paste was dispensed s by dispensing machine to cover the sidewall between top and bottom dice. Dispensing process is easy to control with proper viscosity and thixotropic index optimization. In case of 5G modem SiP, thin package height is generally required, so silver paste dispensing process is high enough to build silver chimney above top and bottom dice. In this study, silver paste was dispensed over those dice and they were encapsulated using EMC to cover the whole silver chimney. Then, excessive EMC with the silver paste over the specified package height was removed by grinding process. After the grinding process, top surface of SiP was ready for attaching a heat spreader.

IV. EVALUATION RESULTS

A. Test-beds setup

For empirical verifications, 5G modem devices were packaged in four different types of packages and their thermal performance was evaluated in our test beds as shown in Fig. 13. We demonstrated and compared each thermal solution in Fig. 14 in terms of their relative thermal resistance and elapsed time when the junction temperature reached from 25℃ to 85℃. Before packaging, tested chips were selected to have the same dynamic and leakage power dissipation characteristics. Each type of the package samples was assembled and implemented into the Samsung mobile development kit, which is a socket type test board designed for the firmware-level thermal and power testing. Dhrystone, one of the most popular CPU benchmark program, was used for a thermal testing vector. We changed the operation frequency of Dhrystone in order to change thermal test power conditions at 60Hz rate. The package junction temperatures were acquired from thermal sensors which are directly attached on chips and the sampling rate is at 60Hz.

B. Results

Fig. 16 indicates relative comparison of thermal performance among the tested packages. We evaluated four types of SiP: standard SiP, silicon chimney SiP, silicon chimney SiP with thermally enhanced DAF, and silver chimney SiP. From these results, we concluded that the concept of ICE-SiP was effective to improve the thermal performance of the vertically stacked SiP. Silver chimney showed the best thermal performance among the tested packages despite the fact that silicon chimney has higher thermal conductivity than silver paste. Comparing silicon chimney with silver chimney SiP, there is less interfaces between high K thermal chimney and die surface, which brought better thermal performance. Previously discussed in Fig. 5, silicon chimney has DAF between two surfaces and its thickness and thermal conductivity will affect the overall thermal performance of silicon chimney in ICE-SiP. Since the silicon chimney with high K DAF shows better thermal performance, we assume that interfacial thermal resistance from DAF can be a decisive factor for the overall thermal performance. However, the difference between the simulated data and the actual data for the thermal performance must be considered for improvement of interfacial resistance, for DAF was attached to actual surfaces not ideal. Therefore, it is very difficult to deliver perfect adhesion among those surfaces. As a result, the conformity of silver paste in

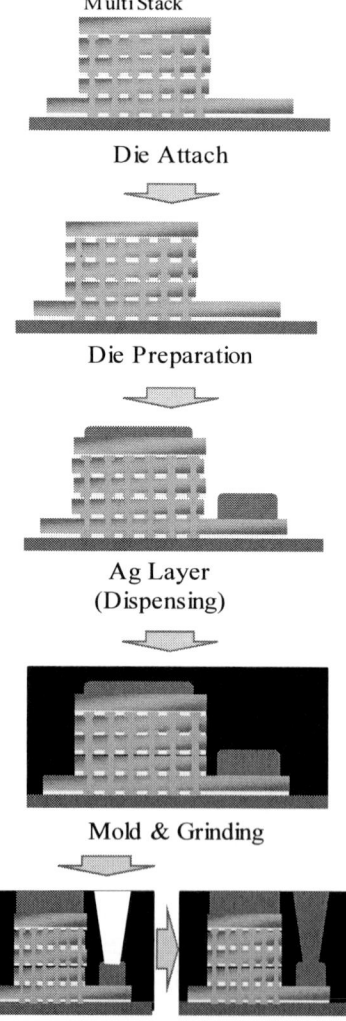

Figure 14. The formation of thermal chimneys for 3D TSV stacks

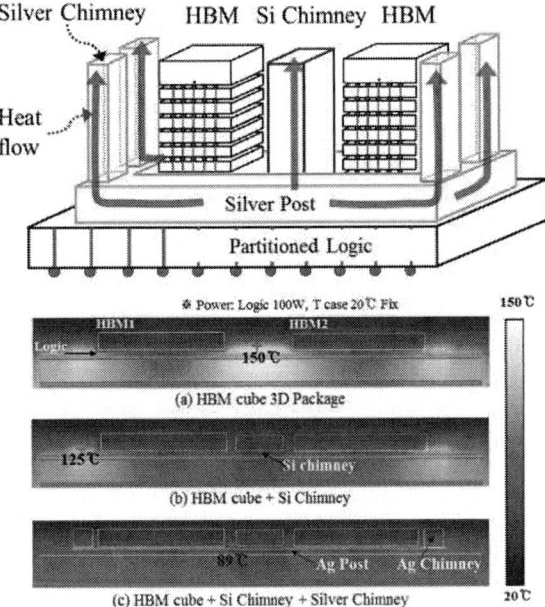

Figure 15. Heat flows of 3D ICE-SiP

chimney can provide relatively low interfacial defects and deliver more heat conduction between silicon die and chimney, as shown in Fig. 16.

V. FUTURE STUDY

In this paper, we studied the novel thermal dissipation solution of ICE-SiP and verified that it was an effective thermal solution for vertically stacked SiP of DRAM and logic dice. This solution is able to release the limit of memory operating temperature on bottom logic dice. In addition, we can apply this ICE-SiP concept onto 3D TSV packages that were presented in Fig. 15. Future applications such as cloud computing, artificial intelligence training, and inference will require more I/O density and shorter interconnection between memory and logic dice, so vertically stacked 3D TSV packaging will be a viable option.

However, thermal limitation of top memory dice over the bottom logic dice will be unavoidable, as demonstrated in previous studied. Fig. 15 is explaining how to resolve 3D TSV packaging thermal problems using ICE-SiP concepts. In Fig. 15, we placed a silicon chimney in the middle of two high bandwidth memories, and built silver chimney on the surface of bottom logic dice. Consequently, our results indicate that ICE-SiP concepts improve the thermal performance of 3D TSV packages as shown in Fig. 15. For future applications, we need to maintain the study of "ICE-SiP" for 3D TSV applications.

ACKNOWLEDGMENT

We would like to appreciate their contribution by Gynkyu Kim, Pyongwan Kim, Hwail Jin, Seungtae Hwang and all other members in package development

REFERENCES

[1] R.K.Singh, D.Bisht, and R.C.Prasad, "Development of 5G Mobile Network Technology and Its Architecture", International Journal of Recent Trends in Engineering & Research (IJRTER), Vol.3(10), pp.196-201, 2017.

[2] S.H.You, S.H.Jeon, D.Oh, K.S.Kim, J.C.Kim, S.Y.Cha, and G.B.Kim, "Advanced Fan-Out Package SI/PI/Thermal Performance Analysis of Novel RDL Packages", in IEEE 68th Electronic Components and Technology Conference (ECTC), pp. 1295-1301, 2018.

[3] R.Agarwal, S.Kannan, L.England, R.Reed, Y.Song, W.Lee, S.H.Lee, and J.K.Yoo, "3D Packaging Challenges for High-End Applications", in IEEE 67th Electronic Components and Technology Conference (ECTC), pp. 1249-1256, 2017.

[4] G.Shan, S.Kannan, D.Smith, R.Agarwal, and R.Alapati, "TSV integration with 20nm CMOS technology for 3D-IC enablement", in IEEE 17th Electronic Packaging and Technology Conference (EPTC), pp. 1-5, 2015.

[5] L.Li, P.Chia, P.Ton, M.Nager, S.Patil, and J.Xue, "3D SiP with Organic Interposer for Asic and Memory Integration", in IEEE 66th Electronic Component Technology Conference (ECTC), pp. 1445-1450, 2016.

[6] A.Martins, M.Pinheiro, A.F.Ferreira, R.Almeida, F.Matos, J.Oliveira, H.M.Santos, M.C.Monteiro, H.Gamboa, and R.P.Silva, "Heterogeneous Integration challenges within Wafer Level Fan-Out SiP for Wearables and IoT", in IEEE 68th Electronic Component Technology Conference (ECTC), pp. 1485-1492, 2018.

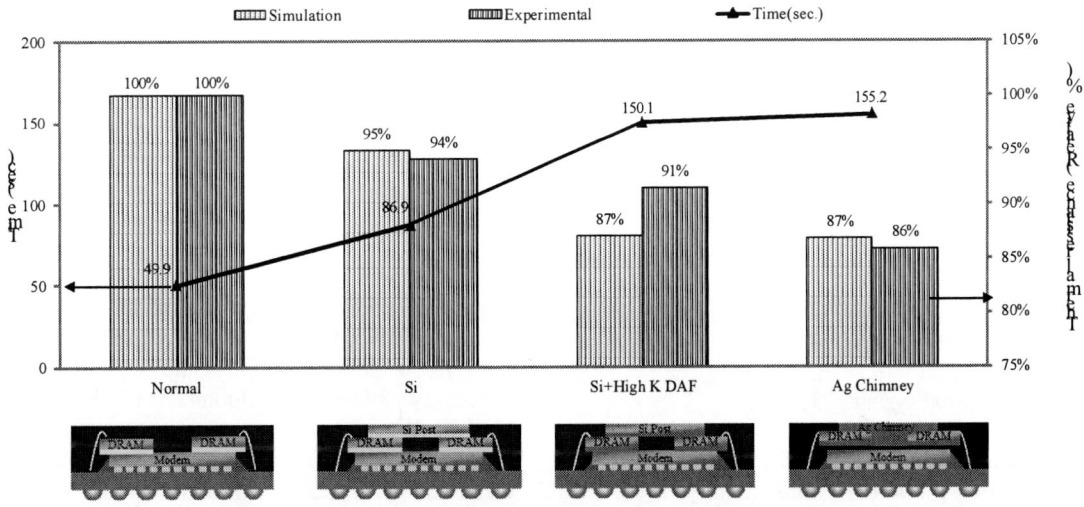

Figure 16. Thermal performance comparison among various package thermal solutions

Fine-Pitch (≤10 μm) Direct Cu-Cu Interconnects using In-situ Formic Acid Vapor Treatment

SivaChandra Jangam[1], Adeel Bajwa[2], Umesh Mogera[1], Pranav Ambhore[1], Tom Colosimo[2], Bob Chylak[2] and Subramanian S. Iyer[1]

[1]Center for Heterogeneous Integration and Performance Scaling (CHIPS),
Samueli School of Engineering, University of California Los Angeles
[2]Kulicke & Soffa Industries Inc, Fort Washington, PA, USA
sivchand@ucla.edu, abajwa@kns.com

Abstract—We demonstrate a solderless direct copper-copper (Cu-Cu) thermal compression bonding (TCB) process for die-to-wafer assembly in ambient environment using a novel in-situ formic acid vapor treatment. We show that this approach produces excellent Cu-Cu bonds with an average shear strength of >150 MPa. Using this TCB process, we demonstrate dielet assemblies on the Silicon-Interconnect Fabric (Si-IF) platform with fine-pitch (≤ 10 μm) Cu-Cu interconnects. Further, we show electrical continuity across multiple dies on the Si-IF with an interconnect specific contact resistance of <0.7 Ω-μm².

Keywords- Silicon-Interconnect Fabric; 10 μm Fine Pitch Interconnects; Cu-Cu Thermal Compression Bonding

I. INTRODUCTION

Conventional systems today use packaged dies assembled on a printed circuit board (PCB) using solder-based (Sn) interconnects. The interconnect pitch in these substrates is typically 0.4-1 mm, which is more than 1000 times the top-level wiring pitch on a die. As a result, the interconnect pitch on PCBs is a major bottleneck in achieving high data-bandwidth and highly efficient systems. Recent developments like the silicon interposers use solder-capped copper pillars to achieve moderate interconnect pitch of ≥50 μm [1]. However, to achieve a System-on-Chip (SoC) like performance, the interconnect pitch should be scaled to 2-10 μm [2], [3]. But, further scaling of the solder-based interconnects is extremely challenging due to limitations like the solder extrusion, and bridging. In addition, the formation of brittle Cu-Sn intermetallics cause several reliability concerns. Therefore, a direct metal-metal bonding process is essential to eliminate the use of solder and scale the interconnect pitch to ≤10 μm.

Earlier, we successfully demonstrated a package-less, fine-pitch, highly scalable heterogeneous integration platform called the Silicon Interconnect Fabric (Si-IF), where unpackaged dielets were assembled at small inter-dielet spacings (≤100 μm) using fine-pitch interconnects (≤10 μm) [3]-[7]. To achieve these fine-pitch interconnects (≤10 μm), we developed a low temperature (250 ºC), solder-less, direct metal-metal (Au-capped Cu) TCB process that overcomes the limitations of solder. In addition, we showed the fine-pitch assemblies on the Si-IF lead to significant improvements (≥100X) in performance and energy efficiencies when compared to conventional systems [3], [6].

Ideally, direct Cu-Cu bonding is desirable because Cu is the de-facto metal in a die metal stack and has excellent electrical and thermal properties. Consequently, using fine pitch Cu-Cu interconnects would seamlessly attach the dies to the Si-IF like vias in a metal stack. However, Cu is prone to oxidation in ambient conditions making direct Cu-Cu bonding extremely challenging. Currently, Cu-Cu bonding is reliable only in wafer-to-wafer TCB processes in a controlled environment such as vacuum or forming gas, with relatively high interface temperatures (300-400 °C), and large bonding times (15-60 min) [8]-[11]. These approaches, however, are not appropriate for die-to-substrate attachment in practice, primarily because creating vacuum in a large machine is difficult and maintaining inert environment requires extremely high flow rates (1000-1500 L/min) of gases. Further, the throughput of these processes is extremely low for dielet assembly, inflating the assembly costs.

In this work, we propose a novel approach for Cu-Cu TCB in ambient conditions using in-situ cleaning of copper oxides by creating a reducing environment locally using formic acid vapor. The rest of the paper is organized as follows: Section II presents some previous work on direct metal-metal TCB. In Section III, the theory of formic acid vapor treatment, tool setup and process are described. Section IV presents the mechanical strength characterization of the Cu-Cu bonds. In Section V, we demonstrate the fine-pitch (≤10 μm) assembly of dies on the Si-IF and discuss the electrical characterization results. Finally, the conclusion is presented in Section VI.

II. PREVIOUS WORK ON OXIDATION PREVENTION

For a successful TCB, it is essential to have flat, and pristine metal surfaces. Flat Cu surfaces with low roughness (<1 nm rms) can be easily achieved using the matured chemical mechanical polishing (CMP) techniques. However, achieving pristine Cu surfaces is challenging since Cu readily oxidizes even in ambient conditions. We observed an oxide layer of >1 nm on bare copper surface within an hour [4]. Therefore, it is necessary to reduce these oxides to achieve successful bonds. There were several approaches in the past to circumvent this issue.

A. Au passivation of Cu

Earlier in ECTC 2017, we have demonstrated that Cu-oxidation can be prevented on mating surfaces i.e. the die-pads and the Si-IF pillars, by capping them with Ti/Au (20/200 nm) layers [4]. This enabled us to implement a TCB process for fine-pitch interconnects under ambient conditions [3]-[7]. We showed that this method of passivation is effective [5], and the thin Au layer does not affect the electrical performance significantly [6]. But this assembly technique requires additional processing on the dies which do not have Au pads (especially Si) and is therefore logistically cumbersome. Moreover, the shear tests of the bonded dies revealed failures at the interface of Ti and Cu instead of the Au-Au bonding interface illustrating poor adhesion of the thin films to Cu.

B. Pre-treatment and bonding in controlled environment

Recently, several different passivation, pre-treatment, and in-situ treatment techniques were investigated for direct Cu-Cu bonding [8]-[11]. Direct Cu-Cu bonding has been demonstrated reliably for wafer-wafer bonding in vacuum for 3D integration applications [8]. The Cu surfaces must be pre-treated to reduce the copper oxides and immediately transferred to a vacuum chamber for bonding. Alternatively, the Cu surfaces can be passivated to prevent oxide formation using self-assembled monolayers (SAM) like hexaethiol [9], [10]. These monolayers can be desorbed at elevated temperatures and bonding was demonstrated in inert environment. Further, Cu-Cu bonding was also demonstrated using Ar plasma to clean the Cu surface, called surface activated bonding (SAB). In addition, Ar/H_2 and Ar/N_2 plasmas were also shown to clean the surface and form copper hydrides and copper nitrides respectively which passivate the surface, preventing further oxidation [10]-[11]. However, these techniques with vacuum or controlled environment work only for wafer-wafer bonding and cannot be extended to die-to-substrate attachment where each die must be sequentially aligned and bonded. Having such a system entirely in a controlled environment is not practical.

C. Ultra-sonic bonding

Ultrasonic bonding relies on the vibration energy to break the copper oxide and clean the Cu surface during the bonding process [12]. However, achieving fine pitch using this process is difficult because it requires tall Cu pillars >20 μm. Moreover, the bonding yield is low, and the bonding interfaces were shown to consist of microscale voids.

D. Formic acid pre-treatment

Authors in [13], have demonstrated Cu-Cu bonding using formic acid pre-treatment method to clean the Cu surface. The samples were placed in a nitrogen inert chamber and the formic acid was purged just prior to bonding. This approach showed good Cu-Cu bond quality. However, the cleaning time was 10 min which is substantial for die-to-substrate attach, and therefore is detrimental to the process throughput. Furthermore, the loading/unloading of the die and substrate from the chamber can be tedious adding to assembly time.

We developed a novel approach for in-situ treatment using formic acid vapor that reduces the copper oxides and cleans the Cu surfaces locally at the target bond sites. This helped to significantly reduce the cleaning times to <10 s, thus, improving the throughput for a TCB process.

III. IN-SITU FORMIC ACID VAPOR TREATMENT

A. Formic acid reduction mechanism

The main chemical reactions through which formic acid vapor can reduce the Cu surface are given in (1), (2) [14]-[16].

$$2HCOOH_{(g)} + CuO_{(s)} \rightarrow (HCOO)_2Cu_{(s)} + H_2O_{(g)} \quad (1)$$

$$(HCOO)_2Cu_{(s)} \rightarrow Cu_{(s)} + 2CO_{2(g)} + H_{2(g)} \quad (2)$$

In gaseous form, the formic acid reacts with the copper oxide layer and forms copper formate and water vapor. These reactions although exothermic require high activation energy. Therefore, these reactions take place between 100-150 °C. The thin copper formate layer covers the bare copper surface which needs to be removed in a subsequent step. When temperature of the surface is raised above 200 °C, the copper formate layer dissociates into carbon dioxide and hydrogen gas, while producing pure Cu metal on the surface.

B. Formic acid vapor treatment setup

The tool setup is shown in Fig. 1 (a). The entire setup is at ambient atmospheric pressure and does not require any vacuum or inert environment. The formic acid vapor is obtained by passing a carrier gas (nitrogen (N_2)) through a bubbler containing formic acid (HCOOH 95%) solution. As a result, saturated formic acid vapor is obtained at the output of the bubbler which is then transferred to the bond head. The percentage of the formic acid in the carrier gas can be altered by diluting with N_2 gas.

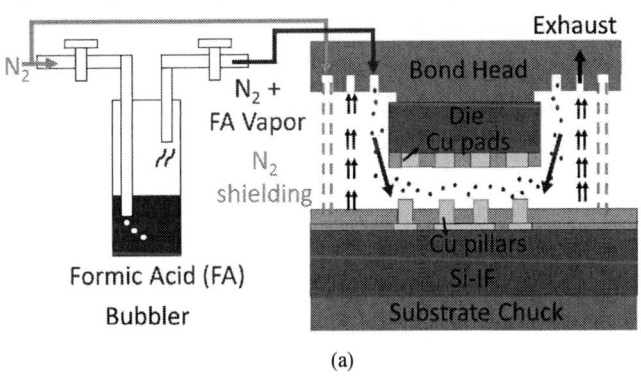

(a)

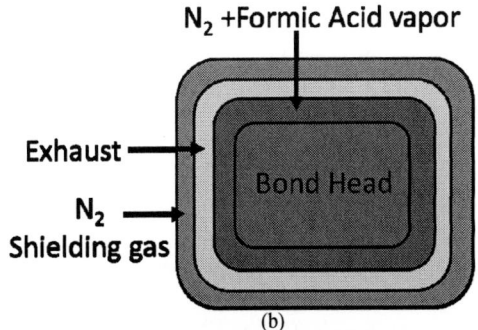

(b)

Figure 1. (a) Schematic of the formic acid vapor treatment setup. The N_2 flows through the bubbler containing formic acid to provide saturated formic acid vapor. (b) Top view of the bond head shroud showing the three channels for shielding N_2 gas, exhaust and formic acid vapor.

The bond head was modified to include a shroud consisting of three channels as shown in Fig. 1 (b). The innermost channel is used to purge the formic acid vapor that cleans the Cu surfaces locally just prior to bonding. The middle channel provides vacuum and acts as an exhaust for the formic acid vapor, and other reaction products during the process. The outermost channel delivers N_2 as a shielding gas around the shroud. This helps contain the formic acid vapor and other products inside the target area, eliminating the need for any controlled environment in the bonding chamber. The flow rates in these channels were optimized to lower the bonding cycle time. The flow rates are adjusted such that the shielding gas has higher flow than the exhaust, which in turn has higher flow than the formic acid vapor. This ensures that the formic acid vapor and reactant products are exhausted without dispersing into the surrounding chamber. The N_2 carrier gas flow-rate into the bubbler was 10 L/min, which corresponds to a formic acid vapor flow of ~100 sccm. The exhaust suction flow was adjusted to be 11 L/min and the shielding gas flow rate was 15 L/min.

C. Process flow for Cu-Cu TCB

As discussed earlier, the formic acid vapor attacks the copper oxide layer and forms copper formate which dissociates at elevated temperatures (>200 °C). However, raising and maintaining the temperature of entire substrate region (diameter >300 mm) is technologically challenging. Further, since the chamber is at atmospheric pressures, the Cu pillars on the substrate (Si-IF) will oxidize considerably if the chuck is held at high temperatures. Therefore, we developed a novel process where the substrate is held at lower temperatures (100 °C) and the die is used to transfer heat to the substrate locally during the bonding process to help dissociate the formates. The overall bonding process consists of three steps, illustrated in Fig. 2, and described below.

Step 1. Alignment and formic acid trigger

First, the die and the substrate are held at 100 °C and aligned using the fiducial marks [4]. Then, the bond head moves down to the touch the Cu pillars on the substrate (Si-IF) which triggers the formic acid vapor upon contact.

Step 2. Oxide reduction

In the second step, the bond head moves vertically above the substrate by an appropriate distance e.g.1-3 mm. This allows the distribution of formic acid vapor over the target region. The temperature of the bond head is now raised to ~350 °C to allow the dissociation of copper formates on the Cu pads of the die. While maintaining 350 °C, a low force pad-to-pillar contact is established for reducing oxides on the Cu pillars. Finally, the residual formic acid vapors and other reaction products are sucked up and pristine Cu surfaces are available for TCB.

Step 3. Thermal Compression Bonding

Once the oxides are reduced locally from the target pads and pillars, a conventional metal-metal thermal compression bonding step is implemented. A bonding pressure of

250 MPa is applied for 10 s with an interface temperature of ~240 °C. The selection of the process parameters is highly dependent on surface morphology, i.e. roughness and planarity, and material related factors such as rigidity, surface oxidation etc. Furthermore, the force and thermal budgets for the TCB process are dictated by underlying applications.

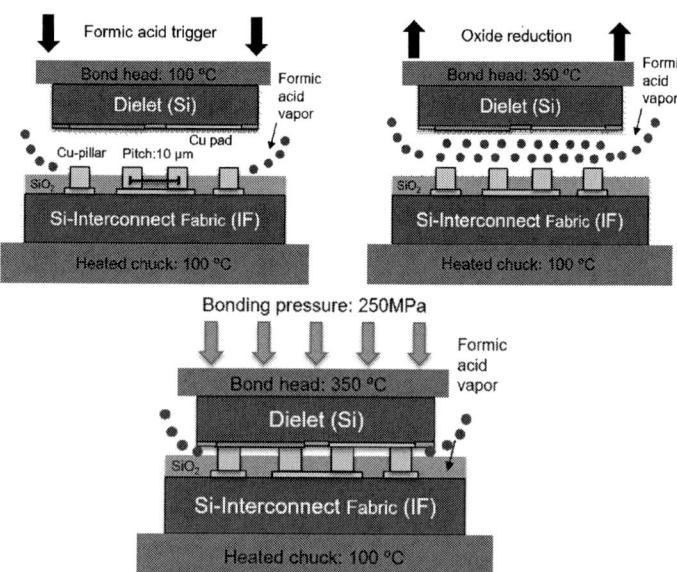

Figure 2. Schematic of the assembly process illustrating the steps of in-situ cleaning and the TCB process. 1) Formic acid trigger (top left), 2) Oxide reduction (top right) and 3) Thermal compression bonding (bottom).

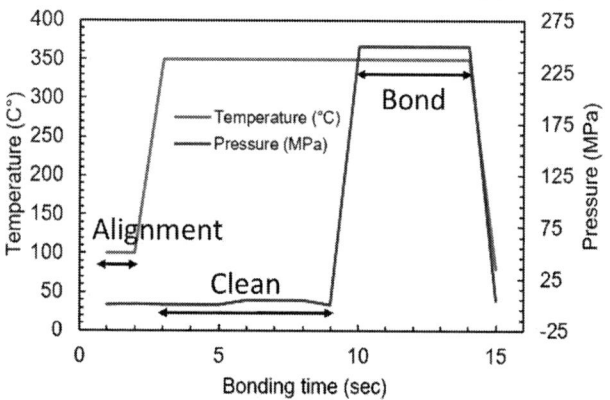

Figure 3. Assembly process flow showing the bonding of a single die. The die temperature and bonding pressure during the assembly are plotted.

IV. MECHANICAL STRENGTH OF CU-CU BONDS

In the case of soldered interconnects, intermetallic compounds are formed at the interface that are brittle and undergo fatigue cracks which fail during thermal cycling. In contrary, direct Cu-Cu bonding eliminates these intermetallics and forms strong bonds. We performed shear tests of Cu-Cu bonded dies to characterize the mechanical strength of the bonds and optimize the process parameters.

A. Shear strength vs TCB process parameters

The test vehicles consisted of Si dies with Cu pillars bonded to blanket Cu layer on a Si substrate. The Cu pillars were distributed across a 7x7 mm² die. The Cu pillars were 30 µm in diameter and 30 µm tall. The pillars were planarized using fly-cut and the resulting surface had a roughness of <9 nm (rms). The effective pillar-to-pad contact area was ~1 mm². The blanket Cu substrate was treated with acetic acid for 20 s prior to assembly process to passivate the Cu surface with Cu acetate which is stable at 100 °C for several minutes. We optimized the process to reduce the bonding and cleaning times to <10 s each. The average shear strength of the bonds was chosen as the criterion for performance comparison. The variation of the average shear strength for different bonding times and bonding pressures are shown in Fig. 4 (a) and Fig. 4 (b) respectively. The best shear strengths were achieved for a bonding pressure of ~250 MPa and interface bonding temperature of ~240 °C. Accordingly, these parameters produced Cu-Cu bonds with an average shear strength of >150 MPa for bonding times as low as 5 s.

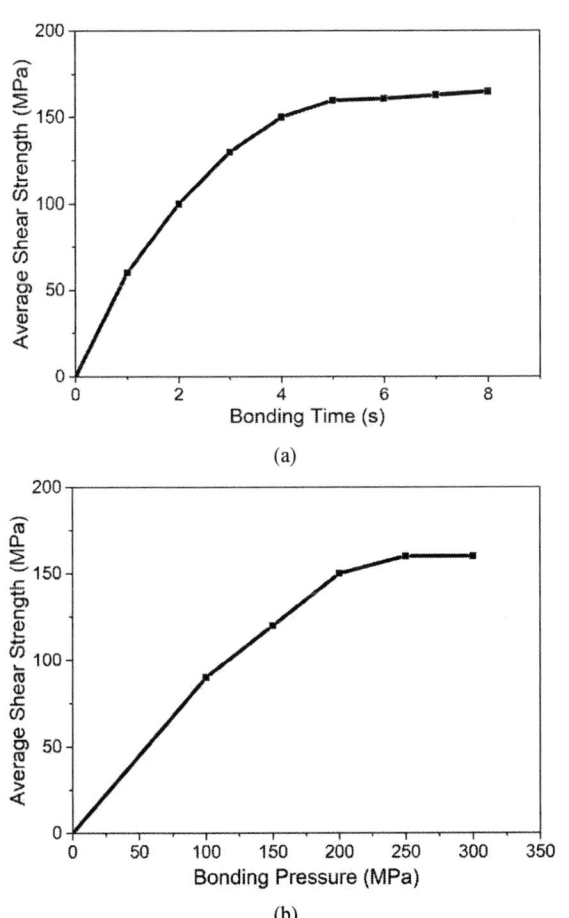

Figure 4. (a) Average shear strength vs bonding time for bonding pressure: 250 MPa and bonding temperature: 240 °C. (b) Average shear strength vs bonding pressure for bonding time: 5 s and bonding temperature: 240 °C.

B. Shear strength of individual Cu pillars

Although, the die shear values give an overall strength of the flip chip connections, they may not indicate the variation of individual pillar shear strength. It is extremely important

to ensure the strength of all bonds for fine-pitch very high density (1×10^6 cm^{-2}) interconnects. In order to quantify the individual pillar strength, test dies were designed where Cu pillars were plated on a sacrificial aluminum layer. These dies were bonded to a blanket Cu surface on a Si substrate like before. After bonding, the assemblies were treated with KOH/DI water solution (10/90 w/w %) to dissolve the sacrificial aluminum layer and release the die as illustrated in Fig. 5. Subsequently, all the pillars remained bonded to the target Cu surface. A shear tester was then used to measure the shear strength of the individual pillars. We observed the bonding interface remained intact after shearing and the tool sheared through the bulk of the pillar as shown in Fig. 6.

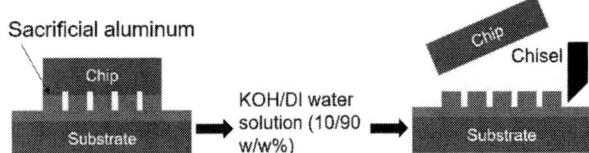

Figure 5. Schematic of the process to singulate Cu pillars from the die.

Figure 6. SEM image of a sheared copper pillar. The interface remains intact and the tool shears through the bulk of pillar.

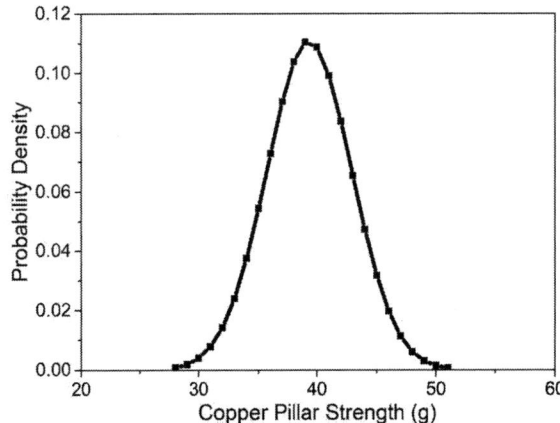

Figure 7. Normal distribution of 300 individual Cu pillar strength. Average shear strength: 39 grams, Standard deviation: 3.6 grams.

The distribution of 300 individual Cu pillar shear strengths across the die is plotted in Fig. 7, showing an average shear strength of 39 g per Cu pillar with a standard deviation of 3.6 g. For performance comparison, the shear strength of individual Cu pillars is compared against the shear strength of the entire die as well as other interconnect technologies, shown in Fig. 8. We observe that Cu-Cu bonding offers 3-4X better shear strength compared to other interconnect technologies.

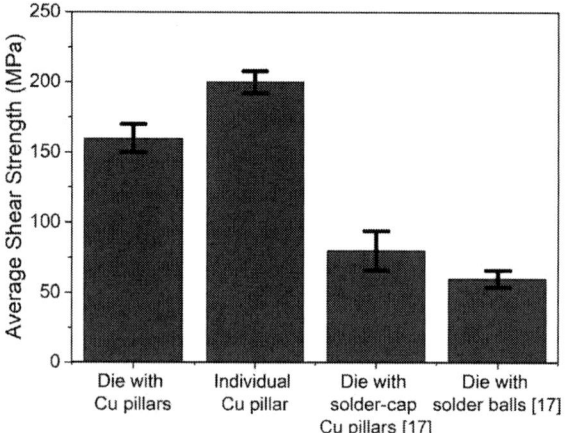

Figure 8. Comparison of various interconnect technologies along with individual Cu pillar strength.

C. Ion-Milled Cross Section

The ion-milled cross section of the Cu-Cu bond showed excellent adhesion at the interface. In our die-to-substrate bonding process, we did not apply any additional annealing step to the bonded samples. Fig. 9 shows a Cu-Cu bonded interface with a total process time of 13 s.

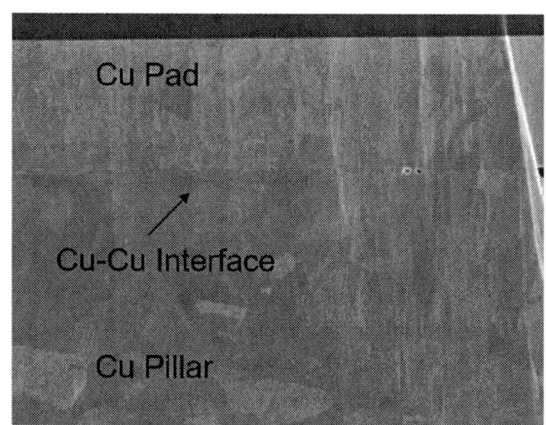

Figure 9. Ion-milled cross section of a Cu-Cu interface.

V. DEMONSTRATION OF (≤10 μM) FINE-PITCH CU-CU INTERCONNECTS

A. Test structures

To characterize the electrical continuity of the fine pitch (10 μm) Cu-Cu bonded interconnects, we designed the Si-IF and the dies to form daisy chain structures when attached, as shown in Fig. 10. The dies and Si-IF were fabricated according to the process in [4] and are shown in Fig. 11, 12

respectively. The Si-IF consists of fine-pitch Cu pillars (Ø: 5 μm and pitch: 10 μm) that are alternately connected using Cu traces below. The total height of the Cu pillars was 5 μm, however, the exposed Cu pillar height above the dielectric layer was ~1.5 μm. The dies consist of Cu pads that form a daisy chain along the horizontal direction when attached to the Si-IF. The daisy chain can be extended to include multiple dies in series as shown in Fig. 10. The inter-dielet spacing between adjacent dies is 100 μm. The dies were 2x2 mm² and correspondingly bond to the Si-IF using 32,400 fine-pitch Cu pillars. This corresponds to an interconnect density of >1x10⁶ cm⁻². Each assembled dielet consists of 180 horizontal daisy chains and each chain consists of 180 Cu pillars. However, due to the limitation of the probe pad size, only 15 of these daisy chains can be tested. The testable chains are distributed evenly across the die. Further, it should be noted that every Cu pillar in a daisy chain must be bonded for electrical continuity.

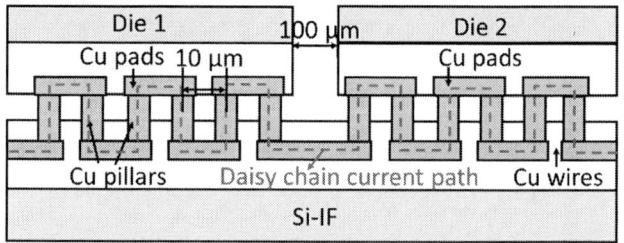

Figure 10. Schematic of the daisy chain test structures consisting of Cu wires on the Si-IF and pads on the dies that are attached using 10 μm pitch Cu pillars. Multiple dies can be assembled at 100 μm spacing to extend the daisy chain in series.

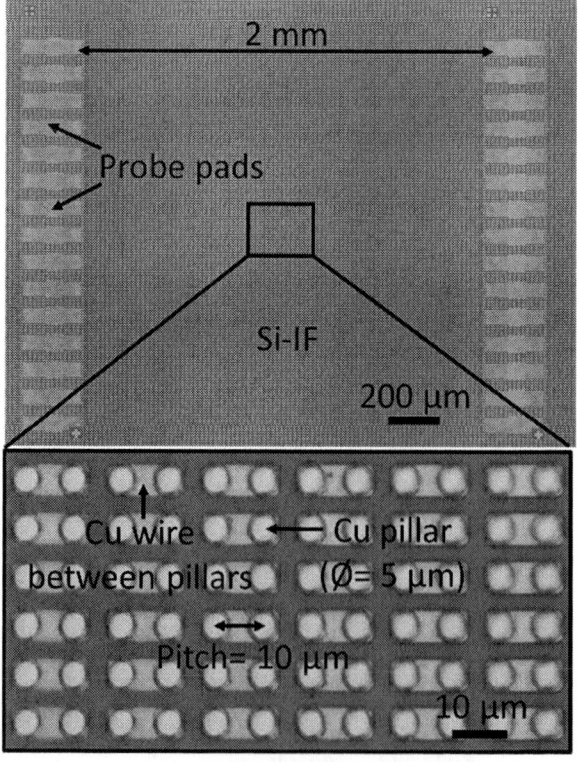

Figure 11. Micrograph of the fabricated Si-IF. Inset shows the 10 μm fine-pitch Cu pillars on the Cu wires.

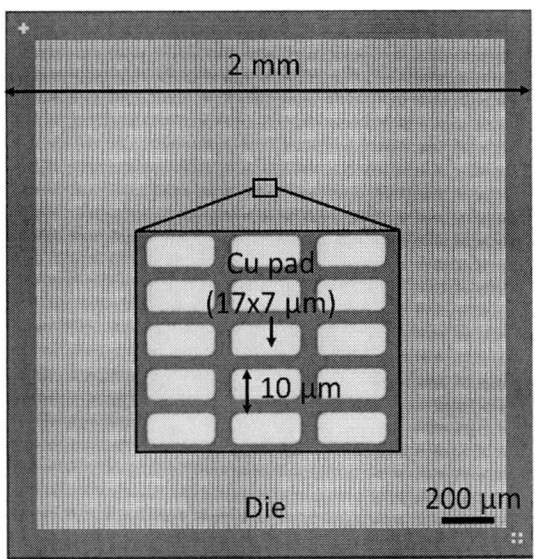

Figure 12. Mircrograph of the fabricated 2x2 mm² die consisting of Cu pads.

B. Alignment accuracy

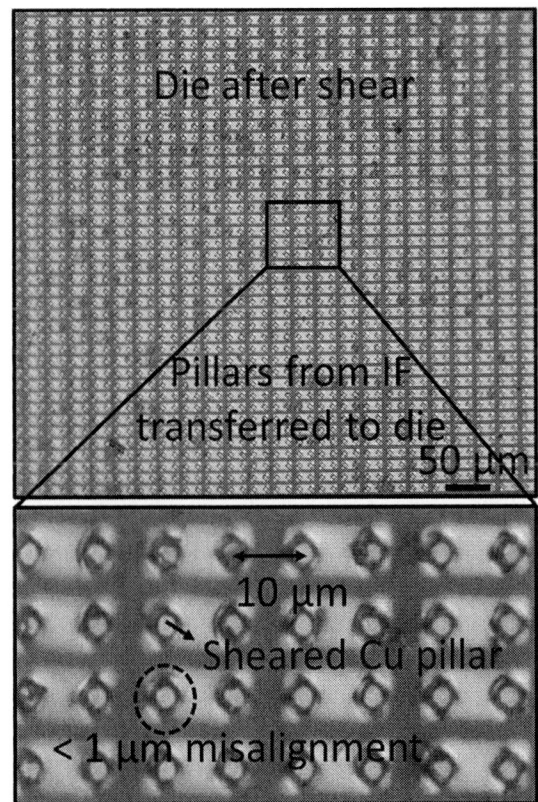

Figure 13. Die sheared after bonding showing all the Cu-pillars from the Si-IF transferred to the die. The failure was not at the Cu-Cu bond interface but at the interface of pillar and Cu traces on the Si-IF. The pillars are aligned to the Si-IF with <1 μm misalignment.

To scale the interconnect pitch to ≤10 μm, it is crucial to achieve the die to substrate alignment of <1 μm. To verify the alignment, dies were bonded using the process established in previous sections and then sheared to observe the interface.

We observed that the failure was not at the bond interface but the Cu pillar broke and transferred from the Si-IF to the pads on the die. Also, in many cases we observed the shearing resulted in the Si die facture into pieces. This demonstrates the excellent bond strength achieved through this process. The average shear strength of the samples was >100 MPa. The sheared die had <1 μm misalignment, shown in Fig. 13.

C. Electrical continuity

We characterized the 4-point electrical resistance of the daisy chains of a single die and two dies in series. The dies and the Si-IF were treated with Ar plasma for 3 min prior to assembly to clean the surface of contaminants. The Si-IF was pre-treated with acetic acid for 20 s to temporarily passivate and reduce the oxidation of the Cu pillar surface as mentioned earlier. The dies were bonded using a 10 s clean and 10 s bond process as shown in Fig. 3. Note that there was no wet cleaning between the bonding of different dies. As a result, the Si-IF was held at 100 °C for >20 min during the assembly process. However, we did not observe noticeable difference in the electrical performance between the dies demonstrating the effectiveness of the formic acid in-situ oxide reduction. The assembled dies on the Si-IF is shown in Fig. 14.

We observed 100% connectivity and all the 15 testable daisy chains were connected for both the single die and two dies case. The current-voltage (I-V) plots of the daisy chains of a single die and two dies in series are shown in Fig. 15 (a) and Fig. 15 (b) respectively. The different colors represent different daisy chains tested. Both the measurements show well behaved resistance of the Cu-Cu interconnects. The variation in the measurements between chains can be due to the misalignment during bonding and measurement errors.

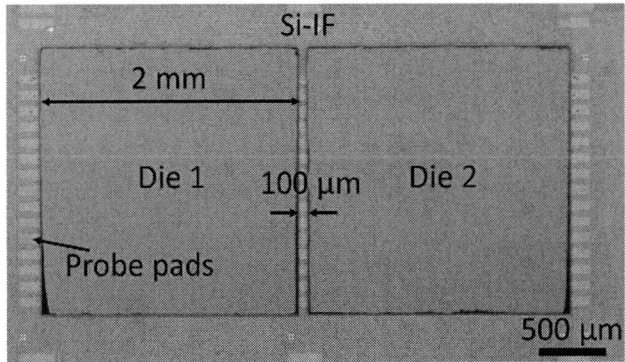

Figure 14. Two dies assembled on the Si-IF at 100 μm inter-dielet spacing. The daisy chains pass through both the dies with a total 360 fine-pitch (10 μm) Cu-Cu bonded interconnects per chain.

D. Contact Resistance

The contact resistance of the individual Cu pillar was extracted from the daisy chain resistance by de-embedding the fan-out wire, Cu trace, and Cu pad resistances. The average resistance per pillar was ~35 mΩ which is the 20% lower than the earlier reported Au-capped Cu pillar resistance [7]. This corresponds to an effective specific contact resistance of ~0.685 Ω-μm². However, we observed the pillar

978-1-7281-1500-9/19 $31.00 © 2019 IEEE

resistance varied from 28 mΩ to 80 mΩ which can be attributed to the misalignment during the bonding process. A comparison of the contact resistance of various interconnects and geometries is presented in table I.

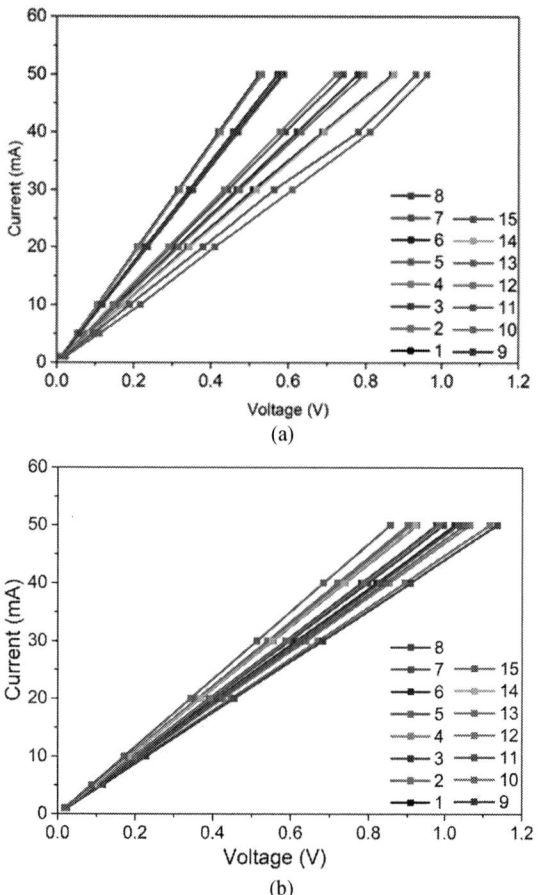

Figure 15. (a) Current vs voltage plot for daisy chains of a single die on the Si-IF. (b) Current vs voltage plot for daisy chains of two dies assembled in series on the Si-IF. The different colors represent different daisy chains tested. The average contant resistance was 35 mΩ.

TABLE I.   GEOMETRIC AND ELECTRICAL PROPERTIES [18], [19], [20]

Interconnect type	Ø [μm]	Contact pad area [μm²]	Contact resistance [mΩ]	Effective specific contact resistance [Ω-μm²]	Material
C4 bump [18]	100	~7800	10	78	PbSn
C4 bump [18]	50	~1950	25	48.7	PbSn,
μ-bump [19]	23	~415	47	19.5	CuSn
μ-bump [19]	16	~201	43	8.64	CuSn
Cu pillar [20]	11.2	~100	12	1.2	Cu
Cu-Pillars [4], [6]	5	~19.6	42	0.82	Au-capped Cu
Cu-Pillars (This work)	*5*	*~19.6*	*35*	*0.685*	*Cu*

VI. CONCLUSION

We successfully demonstrated a direct Cu-Cu TCB process in ambient conditions using a novel in-situ formic acid vapor treatment. By locally spraying the formic acid vapor under the bond head, the copper oxides are reduced in the target area, eliminating the need for any controlled chamber environment. We show that this process results in excellent Cu-Cu bonds with an average shear strength >150 MPa, which is 3-4X higher than solder-capped Cu pillar interconnects. Further, using this process, we show a fine-pitch dielet assembly on the Si-IF using 10 μm pitch Cu-Cu interconnects. We demonstrate electrical continuity across multiple dies on the Si-IF and show low specific contact resistance of <0.7 Ω-μm² for these fine-pitch interconnects. Furthermore, we optimized the assembly process with both the cleaning times and bonding times of < 10 s. Therefore, this approach significantly improves the throughput for a direct Cu-Cu TCB process, making it particularly viable for manufacturing.

ACKNOWLEDGMENT

This work was supported in part by DARPA, Semiconductor Research Corporation (SRC), ONR, UC-MRPI, and the UCLA CHIPS Consortium. We thank Niloofar Shakoorzadeh and Eric Sorensen for their help.

REFERENCES

[1] J. Lee, C. Y. Lee, C. Kim and S. Kalchuri, "Micro Bump System for 2nd Generation Silicon Interposer with GPU and High Bandwidth Memory (HBM) Concurrent Integration," 2018 IEEE 68th Electronic Components and Technology Conference (ECTC), San Diego, CA, 2018, pp. 607-612.

[2] S. S. Iyer, "Heterogeneous Integration for Performance and Scaling," in *IEEE Transactions on CPMT*, vol. 6, no. 7, pp. 973-982.

[3] S. Jangam et. al, "Latency, Bandwidth and Power Benefits of the SuperCHIPS Integration Scheme," 2017 IEEE 67th ECTC, 2017.

[4] A. A. Bajwa, S. Jangam, S. Pal, N. Marathe, M. Goorsky, T. Fukushima, S. S. Iyer, "Heterogeneous Integration at Fine Pitch (≤10 μm) using Thermal Compression Bonding", 2017 IEEE 67th Electronic Components and Technology Conference (ECTC), Orlando, FL, 2017.

[5] N. Shakoorzadeh, S. Jangam, P. Ambhore, H. Chien, A. Hanna, K. Rahim, S. S. Iyer, "Improving Reliability of Si Interconnect Fabric (Si-IF)", IEEE 69th Electronic Components and Technology Conference (ECTC), May 28-31, 2019, Las Vegas, NV.

[6] S. Jangam, A. A. Bajwa, K. K. Thankkappan, P. Kittur and S. S. Iyer, "Electrical Characterization of High Performance Fine Pitch Interconnects in Silicon-Interconnect Fabric," 2018 IEEE 68th Electronic Components and Technology Conference (ECTC), San Diego, CA, 2018, pp. 1283-1288.

[7] A. A. Bajwa et al., "Demonstration of a Heterogeneously Integrated System-on-Wafer (SoW) Assembly," 2018 IEEE 68th Electronic Components and Technology Conference (ECTC), San Diego, CA, 2018, pp. 1926-1930.

[8] Ko, Cheng-Ta, and Kuan-Neng Chen, "Low temperature bonding technology for 3D integration.", Microelectronics reliability 52, no. 2, 2012, pp. 302-311.

[9] Tan, Chuan Seng, Dau Fatt Lim, Xiao Fang Ang, J. Wei, and K. C. Leong, "Low temperature CuCu thermo-compression bonding with temporary passivation of self-assembled monolayer and its bond

strength enhancement", Microelectronics Reliability 52, no. 2, pp. 321-324, 2012.

[10] Tanaka, Koki, Wei-Shan Wang, Mario Baum, Joerg Froemel, Hideki Hirano, Shuji Tanaka, Maik Wiemer, and Thomas Otto, "Investigation of Surface Pre-Treatment Methods for Wafer-Level Cu-Cu Thermo-Compression Bonding", Micromachines 7, no. 12, pp. 234, 2016.

[11] S. L. Chua, G. Y. Chong, Y. H. Lee and C. S. Tan, "Direct copper-copper wafer bonding with Ar/N2plasma activation," 2015 IEEE International Conference on Electron Devices and Solid-State Circuits (EDSSC), Singapore, pp. 134-137, 2015.

[12] Y. Arai, M. Nimura and H. Tomokage, "Cu-Cu direct bonding technology using ultrasonic vibration for flip-chip interconnection," 2015 International Conference on Electronics Packaging and iMAPS All Asia Conference (ICEP-IAAC), Kyoto, pp. 468-472, 2015.

[13] W. Yang, M. Akaike and T. Suga, "Effect of Formic Acid VaporIn SituTreatment Process on Cu Low-Temperature Bonding," in IEEE Transactions on Components, Packaging and Manufacturing Technology, vol. 4, no. 6, pp. 951-956, June 2014.

[14] Schmeißer, Martin, "Reduction of Copper Oxide by Formic Acid an ab-initio study", Fraunhofer, Chemnitz university of technology, 2012.

[15] Youngs, T. G. A., S. Haq, and M. Bowker, "Formic acid adsorption and oxidation on Cu (110)", Surface Science, no. 10, pp. 1775-1782, 2008.

[16] Wei Lin and Y. C. Lee, "Study of fluxless soldering using formic acid vapor," in IEEE Transactions on Advanced Packaging, vol. 22, no. 4, pp. 592-601, Nov. 1999.

[17] T. Huang et al., "Demonstration of Next-Generation Au-Pd Surface Finish with Solder-Capped Cu Pillars for Ultra-Fine Pitch Applications," 2016 IEEE 66th Electronic Components and Technology Conference (ECTC), Las Vegas, NV, 2016, pp. 2553-2560

[18] S. L. Wright, R. Polastre, H. Gan, L. P. Buchwalter, R. Horton, P. S. Andry, E. Sprogis, C. Patel, C. Tsang, J. Knickerbocker, J. R. Lloyd, A. Sharma, and M. S. Sri-Jayantha, "Characterization of microbump C4 interconnects for Si-carrier SOP applications," Electron. Components Technol. Conf., pp. 633–640, 2006.

[19] B. Dang, S. L. Wright, P. S. Andry, C. K. Tsang, C. Patel, R. Polastre, R. Horton, K. Sakuma, B. C. Webb, E. Sprogis, G. Zhang, A. Sharma, and J. U. Knickerbocker, "Assembly, characterization, and reworkability of Pb-free ultra-fine pitch C4s for system-on-package," in Electronic Components and Technology Conference, 2007, pp. 42–48.

[20] L. Di Cioccio, P. Gueguen, R. Taibi, T. Signamarcheix, L. Bally, L. Vandroux, M. Zussy, S. Verrun, J. Dechamp, P. Leduc, M. Assous, D. Bouchu, F. De Crecy, L. L. Chapelon, and L. Clavelier, "An innovative die to wafer 3D integration scheme : Die to wafer oxide or copper direct bonding with planarised oxide inter-die filling," in 2009 IEEE International Conference on 3D System Integration, 3DIC 2009, 2009, pp. 7–10.

Low Temperature Cu Interconnect with Chip to Wafer Hybrid Bonding

Guilian Gao, Laura Mirkarimi, Thomas Workman, Gill Fountain, Jeremy Theil, Gabe Guevara, Ping Liu, Bongsub Lee, Pawel Mrozek, Michael Huynh

Xperi Corporation
3025 Orchard Parkway, San Jose, CA 95134
e-mail: Guilian.Gao@Xperi.com

Catharina Rudolph, Thomas Werner and Anke Hanisch

Fraunhofer Institute for Reliability and Micro-Integration, IZM – ASSID
Ringstr. 12, 01468 Moritzburg, Germany

Abstract— **Current DRAM advanced chip stack packages such as the high bandwidth memory (HBM) use through-silicon-via (TSV) and thermal compression bonding (TCB) of solder capped micro bumps for the inter-layer connection. The bonding process has low throughput and cannot overcome the challenge of scaling below 40 µm pitch. These are compelling reasons to seek an alternative approach such as hybrid bonding. The pursuit of fine pitch die stacking with TSV interconnect using hybrid bonding is pervasive in the packaging industry today due to the promise of improved performance. Specifically, the Cu interconnect provides improved thermal and electrical performance and the all inorganic interface of the complete die stack offers enhanced thermal-mechanical performance and reliability in the final chip stack.**

Direct Bond Interconnect technology, also known as low temperature hybrid bonding, forms a spontaneous dielectric-to-dielectric bond at room temperature and then establishes metal-to-metal connection (usually Cu-to-Cu bond) by a low temperature batch annealing process (150 – 300ºC). The direct bond process eliminates the need for solder and underfill and associated problems.

While the hybrid bonding exists today in wafer-to-wafer (W2W) format in high volume manufacturing, chip to wafer (C2W) bonding developed for future product lines is making significant process in the past three years. A bonding process with high throughput has been demonstrated with electrical test yield above 90% with a daisy chain structure that covers 50mm² of bonding area. The bonded parts showed superior reliability performance in temperature cycling, high temperature storage and autoclave testing.

This paper presents the latest development in C2W hybrid bonding and demonstrates the low temperature annealing capability and integration with TSV.

Keywords- Cu-Cu interconnect, low temperature, TSV integration, direct bond interconnect, hybrid bonding

I. INTRODUCTION

Requirements for higher I/O density and performance at lower cost is projected to drive the 2.5D and 3D interconnect pitch to 20µm and below. Since the solder interconnect technology faces a fundamental challenge to deliver a high-volume manufacturing solution to meet such requirements [1], the search for alternatives continues. Among the options explored, Cu-to-Cu bonding is considered to be a most promising technology.

Cu-to-Cu bonding can be achieved through two means: Cu-to-Cu thermal compression bonding and room temperature direct bond interconnect, commonly referred to as hybrid bonding.

Cu thermal compression bonding bonds Cu pillars protruded from the dielectric surface. The pads on the opposite side of the device pair to be bonded (C2W or chip-to-chop, C2C)) are held at high temperature (normally 350-400ºC) and high pressure to force diffusion of Cu atoms across the interface to form permanent Cu-to-Cu bonding. A common believe was that the high temperature is necessary to destabilize the Cu oxide that forms on the Cu surface at room temperature thereby enabling Cu diffusion across the interface. For applications such as DRAM packaging, the process temperature needs to be under 250ºC to avoid degrading the device performance. The majority of research work in low temperature Cu-to-Cu bonding have focused on modification of the Cu surface to achieve bonding temperature reduction.

Panigrahi and Chen reviewed research effort in this area up to 2017 [2]. A large volume of work has been directed to modify the Cu surface prior to bonding to either prevent or reduce Cu oxidation. Methods reviewed include: acid cleaning immediately prior to bonding in vacuum condition; In-situ formic gas environment to reduce Cu oxide during bonding; Organic self-assembled monolayer (SAM) temporary passivation; Ti or Pd layer over Cu for permanent passivation; engineered metal alloy layer for surface passivation, etc. The work cited in the review paper are early stage studies with a single goal of reducing bonding temperature and each method has some disadvantages. The complications include addition of different materials or surface residue, increase in contact resistance, need of vacuum bonding etc.

However, other fundamental challenges for high volume manufacturing (HVM), such as coplanarity of Cu pillars, reliability, and manufacturing cost, have not been addressed. The process relies on metal-metal contact with temperature and pressure, therefore, height compensation in traditional solder joints is no longer available. Solder joints can compensate for total thickness variation (TTV) of a few microns to tens of microns, depending on the solder volume and interconnect pitch. For direct Cu-to-Cu bonding, the TTV requirement drops by 2 orders of magnitude to the nanometer range. Traditional Cu bump plating technology cannot meet such requirement, making the technology fundamental unattractive. Second, exposed fine Cu pillars are prone to oxidation and corrosion and encapsulation technology of such joints has not been developed. Finally, a bonding process that requires precision alignment and long bonding time under heat and pressure means low throughput from high cost bonders and is inherently expensive.

Direct Bond Interconnect (DBI) technology, also known as low temperature hybrid bonding [3, 4, 5], has distinctive advantages over thermal compression bonding. This technology bonds metal pads slightly recessed from the surrounding inorganic dielectric (oxide for this study) surface. The bonding is a two-step process. The first step forms a dielectric-to-dielectric bond (SiO_2-to-SiO_2 for this study) at room temperature and then the second step establishes metal-to-metal connection (usually Cu-to-Cu) through a batch annealing process (150–300°C). Since oxide bonding takes place at room temperature, Cu oxidation during bonding is not a concern. The bonded dielectric layer surrounding the Cu interconnect encapsulates the joints from environment in the annealing oven; thus minimizing Cu oxidation during the anneal process. The bonded oxide surface also hermetically seals the Cu interconnect during device operation.

The technology was first commercialized in wafer-to-wafer (W2W) bonding applications. Since the process bonds two dielectric surfaces at atomic scale, the key challenges for W2W bonding are sub-nanometer surface roughness, surface with low level of contamination, and precise control of Cu recess on the bonding surfaces. A well-controlled chemical mechanical polishing (CMP) process can solve the above challenges. In fact, wafer-to-wafer (W2W) direct bond interconnect technology has been in high volume manufacturing for several years for pad size up to 3μm diameter [6].

The hybrid bonding technology is also ideally suited for high volume chip-to-wafer (C2W) and chip-to-chip (C2C) assembly. Since the dielectric bond process is spontaneous, it has the same speed as die pick and place and is therefore well suited for high throughput. The batch anneal to achieve Cu-Cu interconnects is also acceptable for high throughput volume manufacturing. Extending the technology to the C2W application involves solving some additional challenges. For C2W applications, a larger pad size is desirable to accommodate the alignment accuracy available in HVM die bonder to maintain low assembly cost. Current technology for flip chip bonder can achieve alignment accuracy in the 1-10 μm range. The lower accuracy machines can meet much higher throughput requirements than the high accuracy models. Wafer singulation is also a dirty process; therefore, maintaining a clean die surface for bonding is challenging.

In our previous publications, we have reported development of a high-volume production-ready C2W hybrid bonding process [7, 8, 9, 10, 11]. We have demonstrated C2W bonding at a 10μm pitch with electrical test yield up to 92% of the die on wafer have 100% of the daisy chain covering a 50mm² area fully connected. We have also demonstrated a process for cleaning and activating die on a dicing tape in a dicing frame. We then picked die directly from the dicing frame for bonding and have demonstrated throughput of 1636 die per hour with a single head bonder.

We have also demonstrated superior reliability in environmental stress tests including temperature cycling, high temperature storage and autoclave testing. All parts showed superior performance with no increase in resistivity, no crack initiation or defect growth. We have also demonstrated of a 20-die stack with hybrid bonding. The parts shown in the previously published work were annealed at 300°C. This paper presents the continuation of our hybrid C2W bonding development effort for low temperature applications.

II. TEST VEHICLE DESIGN AND FABRICATION

Two different designs were used for this study. Design A shown in Fig. 1 is a daisy chain die of 7.96 mm x 11.96 mm. This die size was chosen to mimic an HBM DRAM die foot print. It has Cu bond pads embedded in SiO_2. Its fabrication includes two Cu damascene processes for the two metal layers. The image shown in Fig. 1 includes a bonding layer, which consists of circular 10 μm diameter Cu bond pads embedded in silicon dioxide, and a Cu trace layer in the shape of a grid for daisy chain connection. Since the oxide is transparent, both layers of metal are visible in the optical image. The daisy chain pitch is 40μm. This design can tolerate 5μm misalignment for assembly.

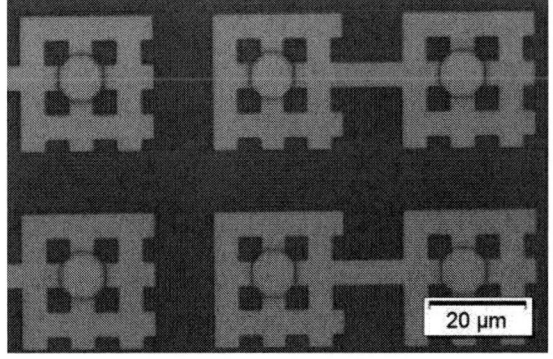

Figure 1.   Design A: Daisy chain test vehicle top-down optical image.

The mating die size on the host wafer is larger in the x-axis to accommodate the probe pads for resistance measurement. The host wafer is 200mm size. Test areas on the bonded D2W structure are shown in Fig. 2. The main daisy chain in test area 1 has 31356 links and covers 50mm² of bonding area. Daisy chain continuity results reported in this paper are from this test area.

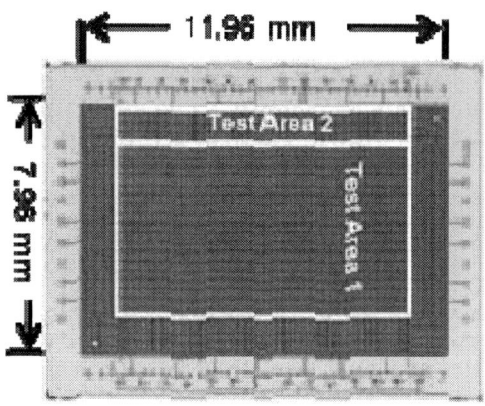

Figure 2. Design A daisy chain test vehicle test areas illustration.

The design B has a single metal layer with 10 μm diameter circular bond pads embedded in oxide with a pitch of 56 μm, as shown in Fig. 3. The mating part is a TSV test vehicle with 10 μm TSV array mirroring the Cu bonding pads shown in Fig 3. Both the TSV wafer and single metal wafer were fabricated by Fraunhofer Institute for Reliability and Micro-Integration, IZM – ASSID. The process for the TSV fabrication has been reported in detail elsewhere [12]. Both the TSV wafer and the single metal wafer for this design are 300mm in diameter.

The hybrid bonding process requires the oxide surface roughness to be less than 1nm. It also requires shallow and uniform Cu recess for low temperature anneal. For bonding pads of 10 μm diameter, most CMP processes used in the BEOL fabrication create Cu recess too deep for low temperature anneal. Xperi has developed a special CMP process to produce shallow and uniform Cu recess [8, 10, 11]. The process was developed on a commercial 200mm CMP tool and transferred to Fraunhofer. The special CMP process scaled well to the 300mm CMP tool and was used for the final surface finish of the 300mm wafers used in this study.

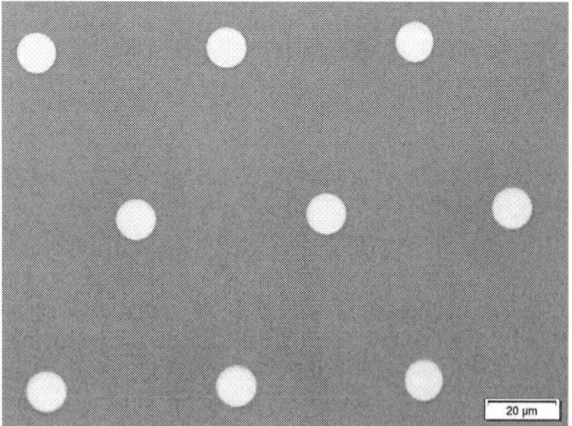

Figure 3. design B test vehicle top down optical image: 10 μm circular bonding pads with 56 μm pitch.

III. WAFER AND CHIP PREPARATION FOR BONDING

The last step of wafer fabrication process is a CMP process to condition the surface to meet the oxide roughness and Cu recess specification for hybrid bonding. The wafer surface after CMP is clean. A rinse in de-ionized (DI) water and a plasma treatment complete the host wafer preparation process.

The wafer used to generate chips for bonding must go through the dicing process. The process generates particles, contaminants, and edge defects. Chip handling during preparation also adds contamination if the process is not well designed and controlled. We applied a protective coating to the wafer surface prior to singulation. After the singulation, the coating is removed with a wet chemical process while the chips are still on the dicing tape. To minimize contamination from the chip preparation steps, the entire clean and preparation process is carried out with the chips on dicing tape. Such process is also well suited for low-cost high-volume manufacturing. Fig. 4 shows a diced wafer at the completion of chip preparation step and is ready for bonding.

Figure 4. Picture of a diced wafer ready for C2W bonding.

IV. CHIP TO WAFER BONDING

We used a Datacon Evo 2200 machine to pick and place chips from the dicing frame shown in Fig. 4. Bonding takes place in a 1K clean room ambient environment. The host wafer stage is also at room temperature. The process is similar to a standard flip chip pick-and-place process without a flux dipping. Spontaneous bonding takes place as soon as the die surface touches the wafer surface. The bonding process can run at the designed speed of the bonder. The alignment accuracy specification of the Datacon Evo 220 bonder is +/-7µm. We slowed the machine down to achieve better alignment accuracy of +/- 5um on the two designs shown in Fig.1 and Fig. 3. We have demonstrated 1636 unit per hour with our single head machine. This is much faster than the thermal compression bonding process for solder capped micro bumps.

For the design A samples, after population of the entire host wafer with desired number of chips, the wafer was cleaved into 4 quarters for an anneal DOE at various temperatures. The anneal process was carried out in a convection oven. Post-anneal samples were characterized by co-focal scanning acoustic microscopy (CSAM) to check for bonding interface voids. Resistance measurement was carried out to determine full continuity joints between the bonded C2W pairs. Representative bonded C2W samples were also cross-sectioned for examination of Cu-to-Cu metallurgical bond interface.

For design B, the C2W samples were examined by cross-section after annealing.

V. RESULTS

Fig. 5a shows a CSAM image of a Design A sample piece after it was annealed at 200ºC for 2 hours. All 29 C2W pairs on the sample piece showed no unexpected voiding. The small white dots at the opposite corner of each chip were voids caused by the fiducial marks used for the C2W bonding. These small voids are stable through reliability testing and do not grow.

Fig. 5b is the resistance measurement results for the same sample piece from the test area 1 shown in Fig. 2. This test area covers 50mm^2 of bonded surface and has 31356 daisy chain links. A "pass" during the electrical test means all 31356 bonding interfaces in the chain are electrical connected. All 29 C2W pairs on the test piece passed the electrical test. The 100% correlation between the CSAM and electrical measurement data is a good confirmation that the 200ºC anneal is sufficient to form Cu-to-Cu bonding.

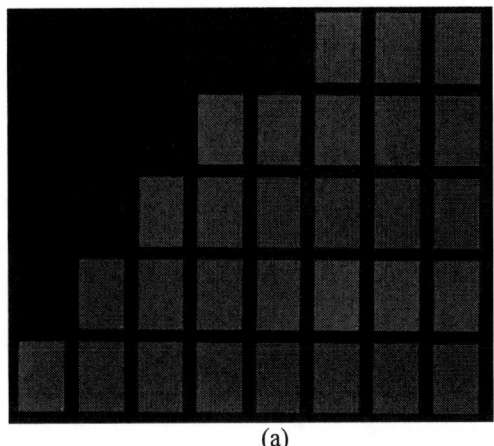

(a)

					Pass	Pass	Pass
			Pass	Pass	Pass	Pass	Pass
		Pass	Pass	Pass	Pass	Pass	Pass
	Pass	Pass	Pass	Pass	Pass	Pass	Pass
Pass	Pass	Pass	Pass	Pass	Pass	Pass	Pass

(b)

Figure 5. (a) CSAM image of representative Design A C2W bonded sample after 200ºC/ 2 hour anneal. (b) Dasiy chain resistance measurement results of the same sample piece for test area 1 shown in Fig. 2 with 31356 links.

After the experiment with 2 hour anneal at 200ºC, we have shortened the anneal time at 200ºC to 1 hour for multiple sample pieces from multipe assembly lots spanning a couple of months. As shown in Table 1, 4 out of the 5 sample pieces showed 100% electrical test yield on void free C2W pairs. One sample set showed electrical test yeild of 92%. This result is very encouraging.

TABLE I. ELECTRICAL TEST RESULTS OF DESIGN A SAMPLES ANNEALED AT 200ºC FOR 1 HOUR

Build	Sample ID	Anneal Temperature (C)	# Die Pass/# Die No Defect	% Yield
1	#1-1	200	24/26	92
2	#2-1	200	20/20	100
3	#3-1	200	19/19	100
3	#3-2	200	13/13	100
3	#3-3	200	15/15	100

We have also cross sectioned representative samples from each build to examine the Cu-to-Cu interface quality. Fig. 6 shows a cross-section image from sample #1-1 in Table 1. The misalignment between the top and bottom cu pads is within 5 µm. The overlapping Cu surfaces showed no visible gaps between the top and the bottom Cu

978-1-7281-1500-9/19 $31.00 © 2019 IEEE

bonding pads indicating good electrical bonding. The electrical resistance was very close to theoretical resistance and was statistically equivalent to resistance measured on test samples annealed at 300°C.

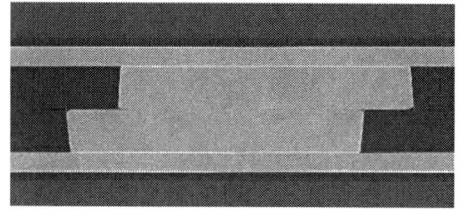

Figure 6.   Cross-section image of a sample from build #1 in Table 1 showing good Cu-to-Cu bonding.

We have collected some preliminary results from the Design B test vehicle. Fig. 7 shows a cross-sectional image of a Cu-to-Cu interface between the single layer metal pads and TSVs after 200°C anneal for 1 hour. The approximately 5 μm misalignment between the Cu pad on the bottom and the TSV on the top was due to the limited alignment accuracy of the Datacon Evo 2200 bonder used for the assembly. Despite some micro voids at the interface, a quality Cu/Cu interconnect is achieved.

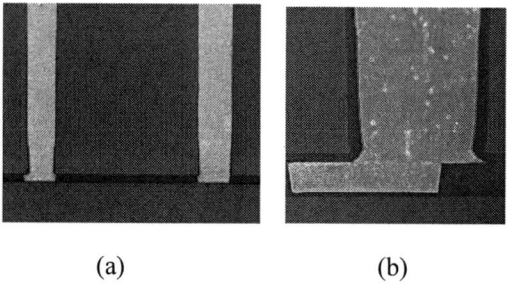

(a)   (b)

Figure 7.   Cross-section images of a hybrid bonded C2W sample from the Design B TSV test vehicle annealed at 200°C for 1 hour. (a): Lower magnification showing bonding interface of two TSVs to pads; (b) at Higher magnification image showing the bonding interface of a single TSV to pad.

VI. DISCUSSIONS

The hybrid bonding is a two-step process that includes a dielectric-dielectric instantaneous bond at room temperature, followed by a heated batch anneal process. During the batch anneal process, the Cu features (either Cu pads or TSVs) expand more than the surrounding silicon oxide due to the large differential of coefficient of thermal expand (CTE) between the two materials. Due to the confinement of the oxide, the Cu expands directionally at the free surface and physically bridge the small gap between the two surfaces that exists at room temperature after the oxide-to-oxide bonding. Once the

two surfaces are in physical contact, Cu atoms diffuse across the interface to form permanent metallurgical bond. Once the Cu-to-Cu bonding is formed, it will not separate when the part is returned to ambient temperature.

The theory behind Cu-to-Cu thermal compression bonding is: Cu oxide formed at room temperature acts as barrier layer for Cu diffusion during Cu to Cu bonding. It is a common believe that by heating the Cu surface to above 200°C, the Cu oxide formed at room temperature becomes thermodynamically unstable, allowing Cu atom to diffuse through for successful grain growth. However, we believe that the impact of Cu oxides formed at room temperature is highly exaggerated since Cu oxides at much lower rate at room temperature than at elevated temperature used for thermal compression bonding. Keil, Lützenkirchen-Hecht and Frahm have shown that Cu oxide thickness formed at room temperature is a two-layer structure with an outer CuO layer of approximately 1.3 nm thickness and an inner Cu_2O layer of about 2.0 nm [13], giving a total oxide thickness of approximately 3nm. However, according to Lee, Hsu and Tuan, oxidation thickness reached 150 nm after only 1 minute at 200°C and 300nm after only 1 minute at 300°C when exposed to an atmospheric oxygen environment [14]. Clearly, the oxide formed at room temperature at the Cu surface is not the main barrier for Cu diffusion during the bonding process, but rather oxidation at elevated temperature is more likely the root of the problem.

For the hybrid bonding technology, the Cu features to be bonded are completely surrounded by bonded oxide surface at room temperature and isolated from oxygen and/or other oxidizing agents completely. During the heated anneal, a Cu-Cu metallurgical bond is formed as the differential thermal expansion between the Cu and surrounding oxide is sufficient to bridge the gap between the two Cu surfaces and form a compression force at the interface. Additionally, the thin oxide formed at room temperature does not impede the Cu inter-diffusion and the formation of a strong bond. This explains why 200°C one hour anneal is sufficient to form solid Cu-to-Cu bonding.

The large electrical chain in Design A fully exercises the hybrid bonding technology. The longest daisy chain on a die covers an area 5.36 mm x 9.36 mm (50mm²) and has 31,356 links. By comparison, the HBM2 DRAM design has only approximately 4,000 interconnects between dies covering an area of 0.8mm x 6.1mm (4.9mm²) in the center strip of a die. As shown in Figs. 5a & 5b, we achieved 100% void free bonding and 100% electrical connection on this test piece. Other lots shown in Table I that were annealed at 200C for 1 hour also showed electrical test yield around 100%.

978-1-7281-1500-9/19 $31.00 © 2019 IEEE

The key to the void free bonding is a well-controlled clean chip and wafer surface prior to bonding. We have developed processes to clean the chips sufficiently to routinely achieve >90% void free bonding in a prototype environment. The key to high electrical test yield is the uniform Cu recess in the final CMP process. As shown in our previous publication, we have developed a CMP process to produce a Cu recess variation of ≤2.5nm across the entire 200mm wafer [8, 10], a distinctive advantage of the hybrid bonding over the thermal compression bonding.

As illustrated in Fig 8a, for hybrid bonding, the oxide surface defines the plane of bonding interface while the total thickness variation (TTV) of Cu pads is the recess depth of Cu pads from the oxide surface. With the special CMP technology that we have developed, we control the Cu recess variation across the entire wafer to be less than 3nm. For any given C2W pair, the sum of TTV on the wafer side and the chip side will be less than 6nm.

In contrast, The TTV of Cu pillars across a wafer is on the order of several μm. Fig. 8b illustrates the TTV of Cu pillars in thermal compression bonding. The Cu pillars protrude above the oxide surface and the height of Cu pillars is defined by the thickness of the plating resist and electroplating uniformity. A process variation of a few microns across a wafer is common. Although TTV can be reduced by a mechanical planarization process, it is still in the micron range. One publication showed TTV of 1.3 μm across a 300mm wafer after fly cutting [15], which is several hundred times larger than what is achieved in hybrid bonding through CMP. Without a compliant layer (such as a solder), or a soft metal that can deform easily, it is challenging to achieve high assembly yield for large I/O count applications with such a large TTV.

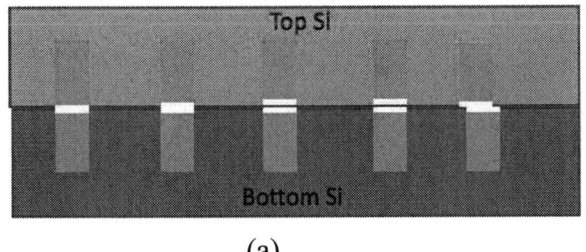

(a)

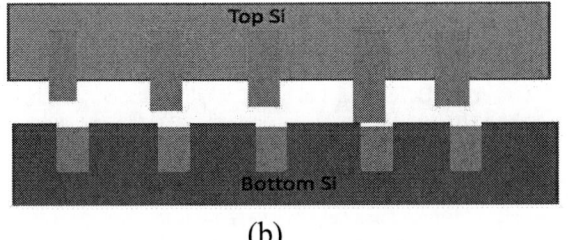

(b)

Figure 8.   (a) illustration of Cu pad TTV in hybrid bonding. The TTV is the sum of Cu recess variation across the die, shown as the white gap between the Cu pads embedded in the oxide of the top and bottom die; (b) Illustratio of Cu pillar TTV in thermal compression bonding.

For die stacking applications such as high-performance DRAM, Cu thermal compression bonding faces more fundamental challenges than just the high TTV from electroplating process.

Currently, a temporary bonding technology using a spin coated polymer layer is commonly used for the backside processing of a thin wafer. TTV of 1-3 μm is inherent in the spin coating process. Consequently, the thinned wafer will have 1-3 μm TTV in the best case. After wafer singulation, the worst case TTV for each die is the total sum of wafer thickness TTV and Cu pillar height TTV. For die stacking, the TTV also accumulates through the stack. The die thickness of current stacked package is around 50 μm, the technology development trend is to reduce die thickness further to increase die count in a stack which will further exacerbate the problem.

In addition, bonding each layer requires substantial time under temperature and pressure. The first die in the stack will see multiple heating cycles that can significantly degrade device performance. There has been research effort to use a layer of patterned organic materials to glue the die together before the thermal compression process to form Cu-to-Cu bonding. However, the organic material can flow into the gap between Cu features to be bonded during the bonding process and interfere with Cu-to-Cu joint formation.

In comparison, hybrid bonding for the die stacking application is straightforward. Since the TTV for Cu-to-Cu bonding is defined by the sum of Cu recess on both sides of the bonding surface, the die thickness TTV resulted from the temporary bonding process has no impact. Further die thickness reduction increases the flexibility of the die and actually improves hybrid bonding yield. In addition, the two-step nature of the hybrid bonding process is ideal for die stacking application. The oxide-to-oxide bonding which occurs spontaneously at ambient temperature is strong enough to hold the die in place with no additional force. As a result, only one anneal process is required to form the Cu interconnect through the entire stack.

Integration of chip with TSV for stacking using hybrid bonding is also straightforward. Fig. 9a shows a typical solder capped micro-bump thermal compression bonding joint taken out of a System Plus Consulting's report. Multiple lithographic and electroplating steps are required to form a solder pad on the TSV side of the bottom Si and Cu micro bump with solder cap on the device side of the top Si. These processes increase manufacturing cost. In addition, assembly throughput is low and requires underfill. Fig. 9b is an illustration of the configuration used in the Fig. 7 of C2W bonding with TSV. Compared to the solder capped micro bump TCB process, the following steps are eliminated: 1) fabrication of solder pads in the bottom Si with TSV; 2) multiple step plating process to form the under bump metallurgy (UBM), Cu pillar and solder over the Cu pillar. Instead, a single damascene process is used to form the Cu pads embedded in a dielectric layer (oxide in this study).

Elimination of the solder and underfill from the joint simplifies the process and reduces cost. Additionally, it eliminates the formation of brittle intermetallic compounds and warpage due to CTE mismatch between the silicon and underfill. These two factors are the largest driving force of solder joint crack initiation and growth. We have shown superior reliability of C2W bonded parts without any TSV [10,11]. Work is continuing to demonstrate reliability of parts with integrated TSVs.

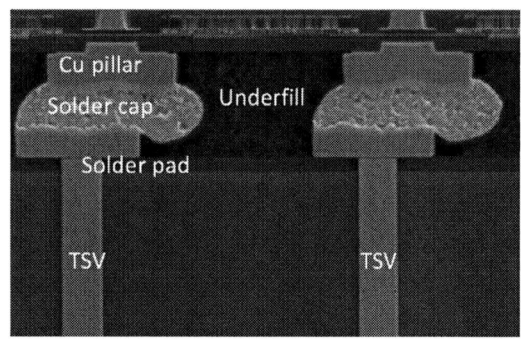

(a)

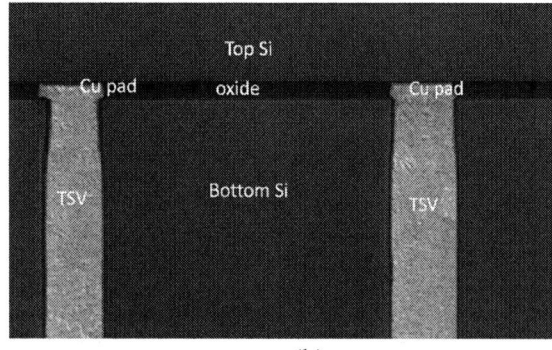

(b)

Figure 9. (a) Typical Cross-section image of a die with TSV joined to another die using solder capped Cu micro bump from a System Consulting's report. (b) Cross-section image of Cu-to-Cu joints in a C2W sample formed by hybrid bonding, demonstrating joint simplificationm.

VII. CONCLUSIONS

The low temperature direct bond interconnect technology promises to enable generations of interconnect pitch scaling below 40 µm. We have demonstrated a C2W technology suitable for applications with 200°C anneal temperature restriction. A daisy chain chip with a test area covering 31356 links and 50mm² surface area has been assembled with 1 hour of anneal at only 200°C. Electrical testing of void-free die has showed 100% of the links were connected. SEM images of the cross section showed a solid Cu-Cu metallurgical connection across the bonding interface.

Integration of TSVs into the C2W bonding has also be demonstrated. The fabrication process to form a Cu bonding layer on one side of the bonding surfaces represents major simplification compared to fabrication of solder pads solder capped micro-bumps required to achieve 40 µm pitch with the micro bump thermal compression bonding technology.

REFERENCES

[1] S. Arkalgud, G. Gao, B. Lee, "Addressing Challenges in 2.5D and 3D IC Assembly: Assembly Process Window Analysis", 3D ASIP, Burlingame, Ca, Dec. 2014

[2] A. K. Panigrahi, K. N. Chen, "Low Temperature Cu-Cu Bonding Technology in 3D Integration: An Extensive Review", Journal of Electronic Packaging 140(1), November 2017 P.

[3] Enquist, "High density direct bond interconnect (DBI) technology for three dimensional integrated circuit applications", MRS Proceedings, 970, 0970-Y01-04. doi:10.1557/PROC-0970-Y01-04, 2006.

[4] P. Enquist, G. Fountain, C. Petteway, A. Hollingsworth & H. (2009), "Low cost of ownership scalable copper Direct Bond Interconnect 3D IC technology for three dimensional integrated circuit applications", 2009 IEEE International Conference on 3D System Integration, 3DIC 2009. 1-6. 10.1109/3DIC.2009.5306533.

[5] P. Enquist, "Advanced direct bond technology" in 3D Integration for VLSI Systems, S. Koester, C. S. Tan and K. N. Chen Eds, CRC Press, 2012, pp175-214

[6] http://www.chipworks.com/about-chipworks/overview/blog/samsung-galaxy-s7-edge-teardown

[7] G. Gao, et al., "Direct bond interconnect (DBI®) technology as an alternative to thermal compression bonding", IWLPC, San Jose, CA, Oct. 2016.

[8] G. Gao et al, "Development of hybrid bond interconnect technology for Die-to-Wafer and Die-to-Die applications", IWLPC, San Jose, CA, Oct. 2017.

[9] A. Agrawal, S. Huang, G. Gao, L. Wang, J. DeLaCruz and L. Mirkarimi, "Thermal and electrical performance of direct bond interconnect technology for 2.5D and 3D integrated circuits", IEEE 67th Electronic Components and Technology Conference, pp989-998, 2017.

978-1-7281-1500-9/19 $31.00 © 2019 IEEE

[10] G Gao, L Mirkarimi, G Fountain, L Wang, C Uzoh, T Workman, G Guevara, C Mandalapu, B Lee and R Katkar, "Scaling Package Interconnects Below 20μm Pitch with Hybrid Bonding", IEEE 68th Electronic Components and Technology Conference, p314, 2018.

[11] G. Gao, et al, "Development of Low Temperature Direct Bond Interconnect Technology for Die-to-Wafer and Die-to-Die Applications—Stacking, Yield Improvement, Reliability Assessment", IWLPC, San Jose, CA, Oct. 2018.

[12] P. Saettler, M. Boettcher, C. Rudolph, K.J. Wolter, " Bath chemistry and copper overburden as influencing factors of the TSV annealing", IEEE 63rd Electronic Components and Technology Conference, p1753, 2013.

[13] P. Keil, D. Lützenkirchen-Hecht, and R. Frahm, "Investigation of Room Temperature Oxidation of Cu in Air by Yoneda-XAFS", http://www.slac.stanford.edu/econf/C060709/papers/144_WEPO79.PDF

[14] H K Lee, H C Hsu and W H Tuan, "Oxidation Behavior of Copper at a Temperature below 300oC and the Methodology for Passivation", Mat. Res. vol.19 no.1 São Carlos Jan./Feb. 2016 Epub Feb 05, 2016.

[15] F. Wei, V. Smet, N. Shahane, H. Lu, V. Sundaram and Rao Tummala' "Ultra-Precise Low-Cost Surface Planarization Process for Advanced Packaging Fabrications and Die Assembly", IEEE 66th Electronic Components and Technology Conference, p1740, 2016.

978-1-7281-1500-9/19 $31.00 © 2019 IEEE

Cu Microstructure of High Density Cu Hybrid Bonding Interconnection

SeokHo Kim, Pilkyu Kang, Taeyeong Kim, Kyuha Lee, Joohee Jang, Kwangjin Moon, Hoonjoo Na, Sangjin Hyun, Kihyun Hwang

Semiconductor Research Center, Samsung Electronics

1, samsungjeonja-ro, Hwasung-si, Gyeonggi-do, 18448, Republic of Korea

e-mail: seokho06.kim@samsung.com

Abstract— The scaling of semiconductor device below 10nm has faced the higher process difficulty and longer development periods. Three-dimensional integrated circuits (3D IC) using chip partitioning and wafer-to-wafer bonding have been acknowledged as the next generation semiconductor stacking technology because of smaller form factor, higher density integration and higher performance compared to same-node devices. Wafer-to-wafer bonding is widely used in stacked CMOS image sensor, that is, the bonding between pixel and logic wafer, and this technology has the potential to apply other semiconductor devices. Cu-Cu hybrid bonding has achieved by simultaneous wafer bonding of metal (Cu-Cu) and dielectric materials. In this study, it is investigated on the microstructure of Cu pad for Cu-Cu bonding after post-electroplating and post-bonding annealing process. The Cu grain size distribution and orientation are analyzed with different anneal temperature, which is applied on electroplated Cu, and with additional heat treatment as post-bonding process. The effect of pad size as well as the position within pattern array on Cu microstructure is also studied as the bonding pad is required smaller and smaller size for high density bonding. After Cu-Cu bonding, the cross-section analysis of bonding interface is carried out to see inter-diffusion of Cu atoms across the opposite Cu pad. Cu-Cu hybrid bonding is applied to test vehicle having the daisy chain of 2.4 million. The electrical resistance is measured before and after thermal stress and the Cu-Cu bonding interface is confirmed as robust structure.

Keywords- Hybrid bonding, Cu-Cu bonding, Cu interface, Cu interdiffusion, Cu grain orientation, reliability, stress migration

I. INTRODUCTION

The interconnection for a semiconductor chip stack requires more number of connections and a smaller pitch interconnection at the reducing chips. Gold wire bonding technology is commonly used for memory multi-stacking. However, it is necessary more spaces as the stacked chips increase. Bump and microbump technology enable multi-chip stacking with the pitch between 10um and 100um. It is combined with TSV (Through Si Via), which makes more interconnection for 2.5D interposer or 3D integration.

Cu-Cu bonding is promising technology under 20um pitch.[1] Cu-Cu bonding can be classified into die-to-wafer bonding and wafer-to-wafer bonding with the chip stacking method.

The application of Cu-Cu hybrid wafer bonding for semiconductor devices may be following: homogeneous stacking for high density memory devices, heterogeneous stacking for advanced performance such as CMOS image sensor [2][3] and stacking memory to logic.

In Cu-Cu hybrid bonding, it exists three different bonding interface - Cu to Cu, Cu to dielectric layer and dielectric to dielectric layer. CMP (chemical mechanical planarization) and bonding are key processes to bond completely without voids at the bonding interface.

The flat surface by CMP process is essential, so the global erosion of dielectric layer is important. The surface roughness of Cu and dielectric layer is also key factor, which is directly related to bonding strength. In the bonding process, plasma treatment condition and bonding alignment are important. Plasma treatment can increase the bonding strength of the dielectric layer [4] and also reduce the oxidation on the Cu surface. Alignment also becomes more important as the Cu pad pitch and size decrease.

Cu properties such as grain orientation, grain size, and Cu oxide, are also dominant factors for robust connection. In this study, Cu grain orientation and grain size are analyzed and the influencing factors are investigated. In addition, Cu-Cu bonding process is verified through electrical resistance and reliability evaluation

II. EXPRIMENTAL

Figure 1 shows the process flow for Cu-Cu hybrid wafer bonding. The wafer preparation was followed damascene Cu BEOL (back-end-of-line) integration process flow. In order to construct the daisy chain by bonding Cu pad of the top and bottom wafer, at least two metal layers are required for each wafer. Dielectric layers were stacked on Si wafer to form a landing metal, and photo patterning and etching were carried out. Then, barrier metal and Cu seed metal were deposited by PVD (physical vapor deposition) process and Cu electroplating was performed. Anneal process and CMP were applied. The Cu pad as the interconnect of the upper and lower wafer was formed in mirror-design by the same process flow. The anneal process after electroplating and CMP was evaluated to investigate the effect of the annealing temperature on the grain size of the Cu bonding pad.

Plasma treatment was performed for the surface activation of dielectric layer to enhance the bonding strength.

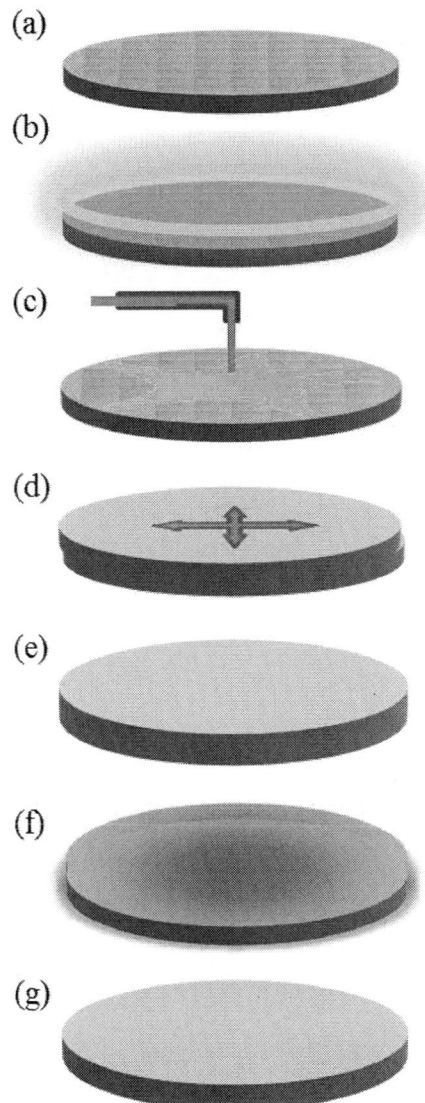

Fig. 1 Cu-Cu bonding process flow: metallization (a), plasma treatment (b), clean & hydration (c), wafer alignment (d), wafer bonding (e), Heat treatment (f), top wafer thinning and backside metallization (g)

Process optimization was achieved by adjusting the plasma power, vacuum pressure, and time to control the oxidation of the surface of the dielectric layer and the Cu pad. Cleaning with de-ionized water after plasma treatment removed the particle and hydrated the wafer surface. The top and bottom wafer were loaded on the bonding module and aligned well and bonding together. The anneal process strengthens the bonding energy and makes Cu diffusion between the contacted Cu pad.

Two wafers were bonded face-to-face, so one wafer of the bonded pair was thinned and back-side metallization was fabricated.

Cu grain size and orientation were analyzed by EBSD (Electron Backscatter Diffraction). Top-view analysis was

used by the as-CMP sample, and cross-section analysis of Cu-Cu bonding interface was applied by ion milling with FIB (Focused Ion Beam).

The electrical characterization of the Cu-Cu hybrid bonding was investigated by using a test vehicle which is dedicated to measure the electrical chain resistance. The pitch of Cu to Cu pad is 4um and the number of connections is 1million to 2.4 million.

The stress migration reliability was evaluated by wafer level test and the chain resistance was compared before and after wafer baking for 500 hours.

III. RESULTS AND DISCUSSION

A. Grain Orientation and grain size

Electroplated Cu has different grain size and orientation depending on electroplating and annealing conditions [5][6]. It is important to understand the change of grain orientation and its size because the diffusivity of Cu atoms varies with grain orientation [7]. In our experiment, anneal process is

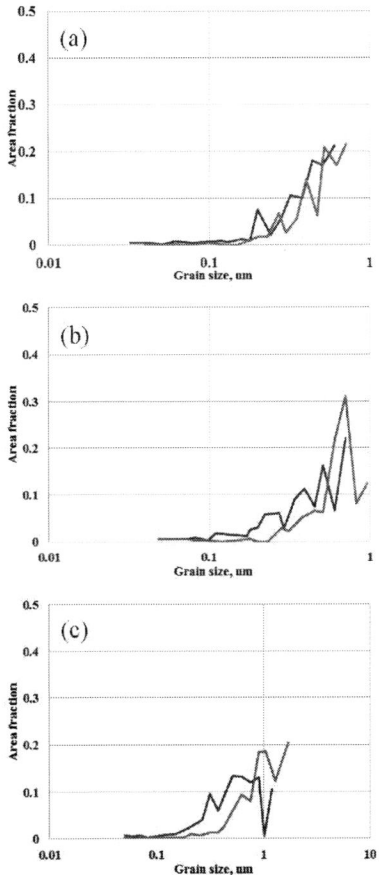

Fig. 2. Grain size distributions with Cu pad size, 0.8um (a), 1.2um (b), 2um (c), at the low (blue line) and high (red line) temperature anneal after electroplating

applied after electroplating and after bonding. So, it is analyzed the effect of each annealing process on grain size and orientation.

Figure 2 shows the grain size distributions with Cu pad

size and the anneal temperature after electroplating. As the Cu pad size increases, the grain size also increases at the higher anneal temperature after electroplating. However, at the small (0.8um) Cu pad size, it has similar grain size distribution regardless of anneal temperature. It is also

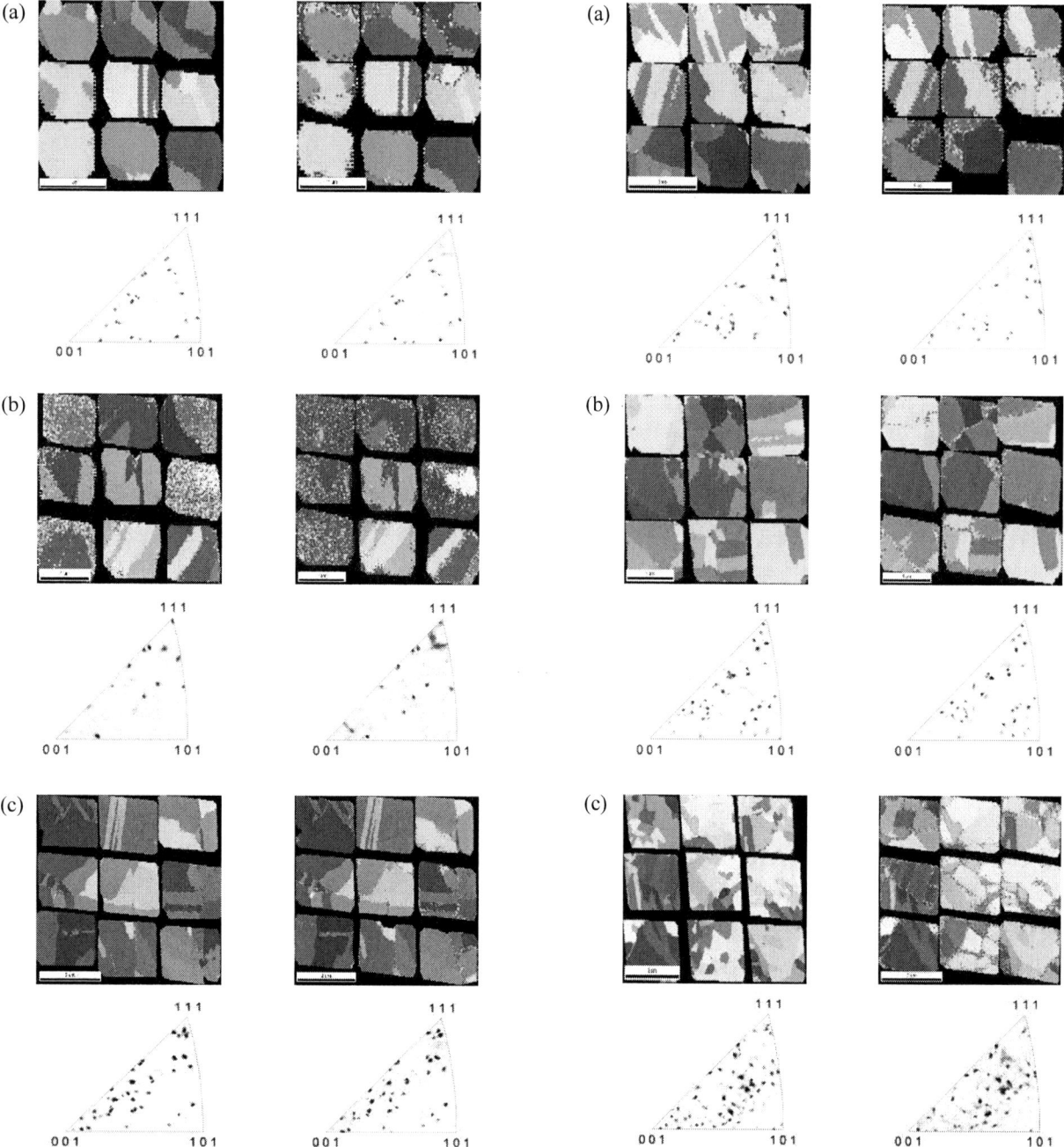

Fig. 3. Grain orientation of low temperature post-electroplating anneal (left) and after CMP and annealed process (right) with Cu pad size, 0.8um (a), 1.2um (b), 2um (c)

Fig. 4. Grain orientation of high temperature post-electroplating anneal (left) and after CMP and annealed process (right) with Cu pad size, 0.8um (a), 1.2um (b), 2um (c)

analyzed the influence of annealing after bonding on grain size and orientation. The anneal process is actually applied after CMP to see the change of grains. The sample in figure 3 is subjected to low temperature annealing after electroplating, followed by CMP. The sample in figure 4 is subjected to high temperature annealing after electroplating. After the first EBSD analysis, both samples are annealed at the temperature after bonding for second analysis.

Figure 3 shows that the grain size and orientation before and after the post-CMP anneal are almost similar and grain orientation is randomly distributed the regardless of the Cu pad size. Figure 4 also has same tendency even if the post-electroplating anneal temperature is higher. It is indicated that the change of grain size is affected by post-electroplating annealing process rather than the annealing after CMP. So, it is more effective to setup the post-electroplating anneal temperature for the control of grain size.

The average Cu grain size constituting one pad is also influenced by the Cu pad pattern array. Figure 6 shows the dependence of grain size on the position of the Cu pad pattern array, regardless of the Cu pad size. The average grain size of the Cu pad located at the edge of the pattern array is the smallest, and its size increases in the order of the side and inside of the pattern array.

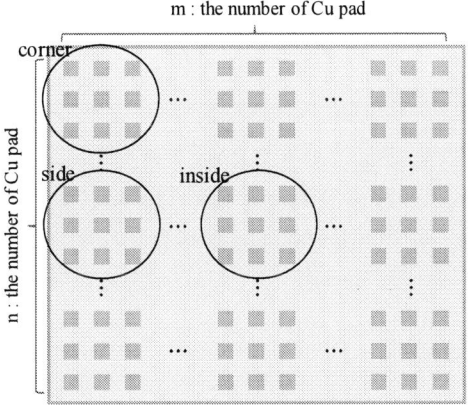

Fig. 5. The analysis position in the pattern array

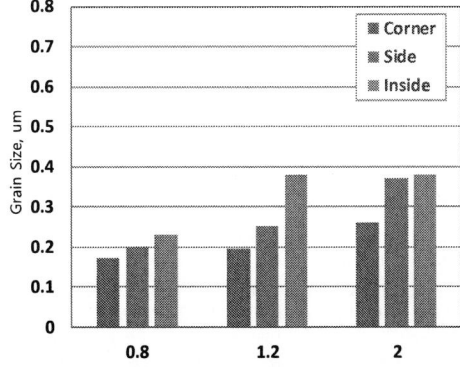

Fig. 6. Average grain size (average) with the position within pattern array

The grain orientation is also affected with the position of the pattern array for each Cu pad size. Figure 7 shows the dependency of grain orientation on the grain size as well as the position in the pattern array. The number of grain orientations decreases when the grain size is less than 0.2μm and it is located on outside the pattern array. However, it has a high indexed grain orientation. On the contrary, the grain orientation appears random as the grain size is more than 0.35μm.

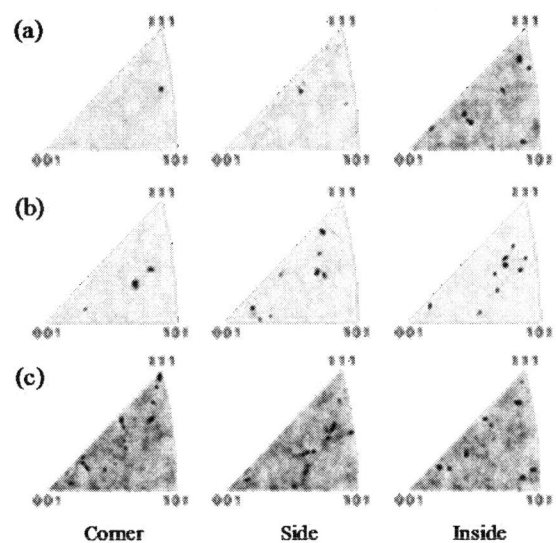

Fig. 7. Grain orientation with the position within pattern array, Cu pad size 0.8um (a), 1.2um (b), 2um (c)

It is assumed that this results are related on the different current density concentration with position of the pattern array during the Cu electroplating process. The current is concentrated more on the outskirts, especially at the corner, rather than the inside of the pattern, so that the difference in the Cu growth rate appears and the average grain size is changed. However, the formation of strong grain orientation is limited due to the additive of the plating solution, which makes the grains having random orientation. The average grain size does not exceed 0.4μm, which means that one Cu pad is composed of four or more grains with different grain orientations.

B. Cu-Cu hybrid bonding interface

Figure 9 shows the grain orientation of cross-sectioned Cu-Cu bonding interface which is indicated by dashed line. The cross-section SEM image was not clear due to the excessive electron charging on the dielectric layer during the sample analysis, but the grain orientation between the interfaces was clear, which has (100) and (111) orientation, mainly. Cu atoms can easily diffuse during anneal after bonding if the grains of the same orientation are contacted each other. However, one Cu pad has several grains with different grain orientation, it is not easy to remove the boundary between two Cu pad completely by Cu atom diffusion, as shown in Figure 10. The grains of the Cu surface before bonding are mixed with (100), (110) and (111) orientation and Cu-Cu interface remains unchanged after bonding. It is necessary to increase annealing temperature or time after bonding or to remove oxide of Cu surface before bonding.

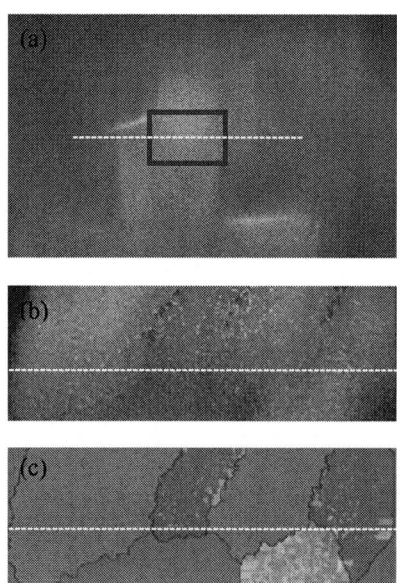

Fig. 8. Cross-section FIB image (a), EBSD (raw image) (b), processed EBSD image (c)

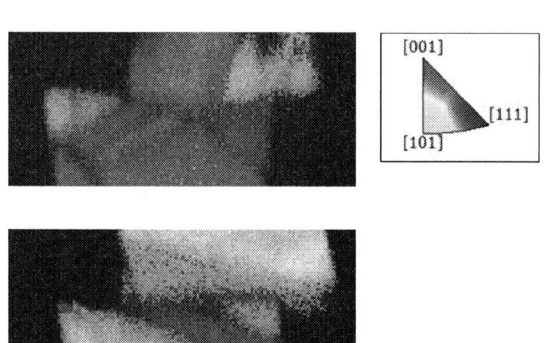

Fig. 9. Grain orientation analysis of Cu-Cu interface

C. Cu-Cu hybrid bonding connectivity

The electrical connectivity of Cu-Cu bonding was investigated by using a test vehicle which can be evaluated the electrical resistance and reliability of Cu-Cu interconnection.

Figure 10 shows the chain resistance distribution at the wafer fabrication process and after the annealing for stress-migration test. The minimum pitch of Cu to Cu pad is 4um and the number of connections is 2.4 million. The Cu-Cu bonding shows good connection and the chain resistance distribution within the wafer was less than 5% at the wafer fabrication process. For the stress migration test, the test wafer was annealed for 500 hours. the chain resistance was found to be within 3% variation compared to initial state.

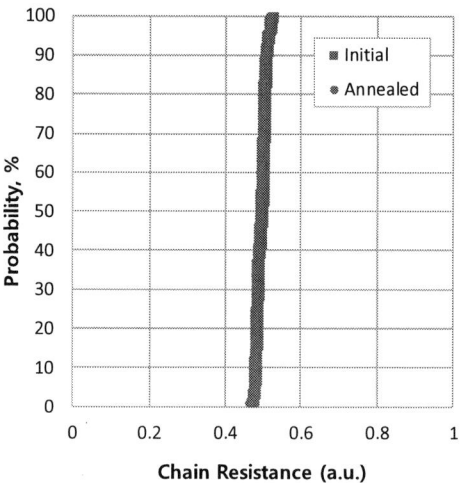

Fig. 10. Contact resistance before and after stress-migration test

IV. CONCLUSION

For Cu-Cu hybrid bonding, it is investigated on Cu grain size and orientation. The grain size changes by post-electroplating annealing temperature but has a little effect on the annealing after CMP. The grain orientation is randomly distributed the regardless of the Cu pad size and annealing temperature. The Cu-Cu bonding interface shows mixed orientation and it is easy to diffuse Cu atoms as the same grain orientation is adjacent. The daisy chain with Cu-Cu bonding shows a good connection and robust reliability results.

REFERENCES

[1] L. Xie, S. Wickramanayaka, S.C. Chong, V.N. Sekhar, D. Ismeal, Y.L. Ye, "6um Pitch High Density Cu-Cu Bonding for 3D IC Stacking," ECTC, 2016, pp.2126-2133.

[2] Y. Kagawa, N. Fujii, K. Aoyagi, Y. Kobayashi, S. Nishi, N. Todaka, S. Takeshita, J. Taura, H. Takahashi, Y. Nishimura, K. Tatani, M. Kwamura, H. Nakayama, T. Nagano, K. Oho, H. Iwmoto, S. Kadomura, T. Hirayama, "Novel Stacked CMOS Image Sensor with Advanced Cu2Cu Hybrid Bonding," IEDM, 2016, pp. 208-211.

[3] S. Lhostis, A. Farcy, E. Deloffre, F. Lorut, S. Mermoz, Y. Henrion, L. Berthier, F. Bailly, D. Scevola, F. Guyader, F. Gigon, C. Besset, S. Pellissier, L. Gay, N. Hotellier, A.-L. Le Berrigo, S. Moreau, V. Balan, F. Fournel, A. Jouve, S. Chéramy, "Reliable 300mm Wafer Level Hybrid Bonding for 3D Stacked CMOS Image Sensors, ECTC, 2016, pp. 869-876.

[4] B. Rebhan, T. Plach, S. Tollabimazraehno, V. Dragoi, M. Kawano, "Cu-Cu wafer bonding: An enabling technology for three-dimensional integration," ICEP, 2014, pp.475-479.

[5] S.-C. Chang, J.-M. Shieh, B.-T. Dai, M.-S. Feng, Y.-H. Li, "The effect of plating current densities on self-annealing behaviors of electroplated copper films," J. Electrochem. Soc, 149 (9), pp. G535-G538 (2002).

[6] A. Hobbs, S. Murakami, T. Hosoda, S. Ohtsuka, M. Miyajima, S. Sugatani, T. Nakamura, "Evolution of grain and micro-void structure in electroplated copper interconnects," Mater. Transact.43 (7), pp.1629-1632 (2002).

[7] Y.-C. Chu, C. Chen, "Anisotropic grain growth to eliminate bonding interfaces in direct copper-to-copper joints using <111>-oriented nanotwinned copper films," Thin Solid Films, 2018, vol. 667, pp. 55-58.

Low-resistance and high-strength copper direct bonding in no-vacuum ambient using highly (111)-oriented nano-twinned copper

Jing Ye Juang[1,*], Kai Cheng Shie[1], Po-Ning Hsu[1], Yu Jin Li[1], K N Tu[1,2], and Chih Chen[1]

[1]Department of Materials Science and Engineering, National Chiao Tung University,
Hsin-Chu, 30010, Taiwan.

[2]Department of Materials Science and Engineering, University of California at Los Angeles,
Los Angeles, California, 90095-1595, USA

*. Tel : +886-963-063507; E-mail: david.mse03g@g2.nctu.edu.tw

Abstract - In this study, we fabricated (111)-oriented nt-Cu microbumps with 30 μm in diameter, and bonded them together using chip-to-chip scheme in N_2 ambient, without vacuum. A well bonded interface in the Cu-to-Cu joint was identified by focused ion beam (FIB). Scanning electron microscope (SEM) images showed a void-less bonding interface within the bonded Cu joint. In addition, a die shear test was conducted. The test results revealed that the Cu joint has a robust bonded Cu joint and the shear strength is 124 MPa, which is nearly two times higher than the traditional SnAg solder joint (64 MPa). In addition, fracture analysis showed that the joint fractured in a ductile manner. Besides, we also performed the resistance measurement by using Kelvin probes on the bonded chip-to-chip test vehicles. The resistance is 4.12 mΩ for a single joint resistance and 4.26 × 10^{-8} Ω·cm² in contact resistivity. More than 30% resistance reduction has been confirmed as compared to the traditional SnAg solder joint (6.32 mΩ). Moreover, we can further reduce the joint resistance by the post-annealing process. It can be brought down to 3.27 mΩ with a resistivity of 3.14 × 10^{-8} Ω·cm². There is a nearly 50% resistance reduction, as compared to SnAg solder joint. The resistance value after the second annealed Cu joint is close an ideal Cu joint. In summary, low-resistance and high-strength copper direct bonding in no-vacuum ambient using highly (111)-oriented nano-twinned copper has been successfully achieved.

Keywords- Cu-to-Cu direct bonding, nanotwinned Cu, surface diffusion creep, grain growth, Cu joint resistance, shear strength, fracture modes.

I. INTRODUCTION

Solder microbumps have been adopted for the vertical interconnects between stacked chips, because of the low melting point and self-alignment process [1], [2]. Currently, the dimension of the microbumps is about 20 μm, and it continues to scale down due to the high I/O demand in the semiconductor industry. However, as the dimension of the solder microbumps continue to shrink, it results in serve yield and reliability issues due to the high volume percentage of brittle Cu-Sn or Ni-Sn intermetallic compounds (IMCs) [3]-

[6], which prevents the solder microbumps from going beyond 10 μm in diameter. Therefore, an alternative solution for ultra-fine pitch packaging is desperately needed. Cu-to-Cu direct bonding emerges to be the best solution for the ultra-fine pitch packaging [7]-[10], because Cu microbumps can be fabricated below 1 μm and it has excellent electrical and thermal conduction. Previous studies reported that the Cu-to-Cu direct bonding can be achieved under a non-vacuum atmosphere by using (111)-oriented nt-Cu [11]-[14]. It is because that the surface diffusion of Cu on (111) planes possesses a high surface diffusivity than other planes [15], [16]. Thus, the rapid surface diffusion characteristic on (111) oriented nt-Cu surface has enabled low-temperature direct bonding in N_2 ambient.

In this study, we fabricated (111)-oriented nt-Cu microbumps with 30 μm in diameter, and bonded them together using chip-to-chip scheme in N_2 ambient, without vacuum. A well bonded interface in the Cu-to-Cu joint was identified by focused ion beam (FIB). Scanning electron microscope (SEM) images showed a void-less bonding interface within the bonded Cu joint. In addition, a die shear test was conducted. The test result revealed that the Cu joint has a robust bonded Cu joint and the shear strength is higher than the traditional SnAg solder joint. In addition, fracture analysis showed that the joint fractured in a ductile manner. Besides, we also performed the resistance measurement by using Kelvin probes on the bonded chip-to-chip test vehicles. A significant resistance reduction has been confirmed as compared to the traditional SnAg solder joint. Moreover, we can further reduce the joint resistance by the post-annealing process. The resistance of the post-annealed Cu joint is close the value of bulk Cu. In summary, the chip-to-chip copper direct bonding has been successfully achieved and low resistance Cu-to-Cu joints has been realized by using (111) oriented nt-Cu in no-vacuum ambient.

II. EXPERIMENTAL PROCEDURES

Fabrication of the test dies and pre-treatment of Cu surface

In this study, (111)-oriented nt-Cu pillar was electroplated on the top and bottom die wafers. In the first step, thermal oxidation process was applied to produce a thin silicon oxide layer on the surface of Si wafers. Then, a 100 nm thick Ti diffusion barrier and 200 nm thick Cu seed-layer were sputtered on oxidized wafers by physical vapor deposition (PVD) processes. Subsequently, the processed wafers were subjected to photolithography and electroplating processes to form the nt-Cu pillar bump arrays. The nt-Cu pillars are 30 μm in diameter and 10 μm in height. Owing to the rough surface condition of nt-Cu pillar, the chemical-mechanical-polishing (CMP) process was performed to flatten the Cu surface. The measured root mean square roughness values (Rq) were ranging from 3 nm to 5 nm. After the nt-Cu fabrication processes, the top and bottom die wafers were diced, respectively. For the top die, the dimension is 6.0 mm × 6.0 mm; for the bottom die, its size is 15.0 mm × 15.0 mm. The die thickness for both top and bottom die is 500 μm. Daisy chain loop wiring with 400 Cu joints and kelvin structure linking with single Cu joint were designed in the test dies. Thus, we can measure the chain loop resistance and single joint resistance after the bonding. Fig.1a to 1b display the as-fabricated top die wafer and bottom die wafer, respectively. The specifications for top and bottom die were listed in Table I. Wet etching was performed to remove organic contaminants and the oxide layer prior to the bonding process. The test dies were rinsed with deionized water, followed by a short immersion in a mixed solution of citric acid and deionized water (in the ratio 133 g/100 ml) at 60 °C for about 30s. Then, they were rinsed again with deionized water and dried by N_2 purging before bonding.

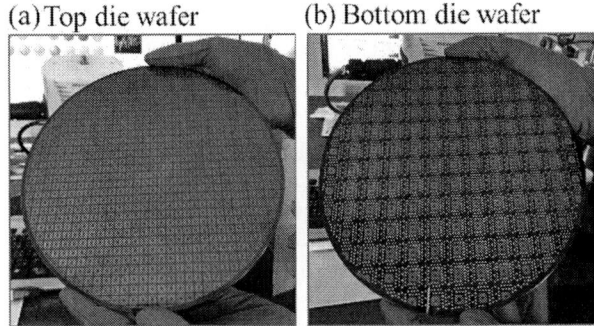

Figure 1. The images show the as-fabricated (a) top die wafer and (b) bottom die wafer, respectively.

TABLE I. SPECIFICATION FOR TOP AND BOTTOM DIE

Cu-to-Cu direct bonding in a two-steps process

Cu-to-Cu direct bonding with two-step processes was applied. In the first step, thermal compression bonding (pressure = 64 MPa) was performed in a N_2 purging atmosphere with a temperature gradient between the top and bottom die. The top die was at 300°C and the bottom die was at 100°C. Thus, there was a temperature gradient between top and bottom die during the bonding. The bonding with a temperature gradient is a typical scheme for the chip-to-chip or chip-to-wafer process. The bonding time is 20 minutes. In the second step, bonded samples were subsequently conducted to an isothermal annealing process under 400°C in a vacuum oven for 60 minutes. The reason for studying the second annealing process will be given later. The flow of the Cu-to-Cu direct bonding processes is shown in Fig. 2.

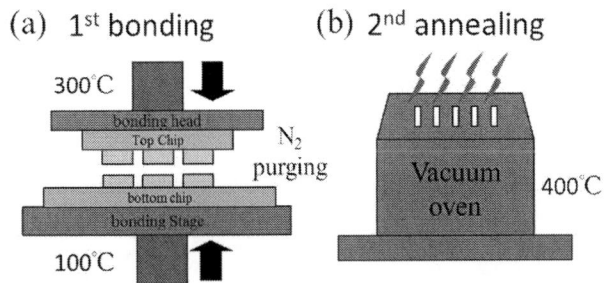

Figure 2. The two-step bonding processes were conducted by (a) 1st bonding at N_2 purging ambient, and (b) 2nd annealing process in a vacuum oven.

TABLE II. PARAMETERS FOR THE DIE SHEAR STRENGTH TEST

Items	Unit	Setting
Shear load	Kg	100 (Max)
Shear speed	μm/s	100
Shear height	μm	200
No. of tested samples	pcs	6

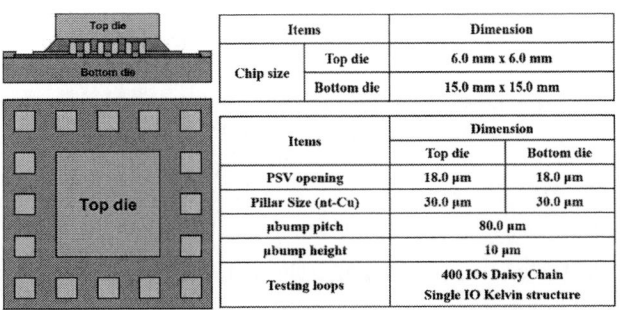

Items		Dimension	
Chip size	Top die	6.0 mm x 6.0 mm	
	Bottom die	15.0 mm x 15.0 mm	

Items	Dimension	
	Top die	Bottom die
PSV opening	18.0 μm	18.0 μm
Pillar Size (nt-Cu)	30.0 μm	30.0 μm
μbump pitch	80.0 μm	
μbump height	10 μm	
Testing loops	400 IOs Daisy Chain Single IO Kelvin structure	

Bonding interface characterization, resistance measurement, and shear strength evaluation.

3D X-ray imaging was carried out to the bonded samples for the Cu interconnects inspection. Scanning electron microscope incorporated with focused ion beam was employed to observe the cross-sectioned Cu-to-Cu bonded interface and grain morphology in the bonded Cu joints. Kelvin sensing by 4-point probe method was applied to measure the daisy chain and single joint resistance. This technique is very suitable for measuring the low resistance interconnect. To evaluate the bonded Cu joint strength, die shear strength test was also carried out. The bonded samples were shear tested by the die shear tester. Shear force was applied to the top silicon die directly. The test conditions are shown in Table II. Shear speed in this test was 100 µm/s and the corresponding shear height was 200 µm above the surface of bottom die. Six bonded die samples were tested. After the test, SEM observations were utilized to analyze the fracture modes of Cu joint in the shear tested samples.

III. RESULTS & DISCUSSIONS

The Cu direct bonding using (111) oriented nt-Cu in N$_2$ purging ambient.

Fig. 3a shows the demonstration of the of Cu direct bonding in N$_2$ purging ambient (first step bonding). The test dies were aligned and bonded by a semi-auto bonder (BONDTECH Inc.). During the bonding, a low level of particulates environment is required. Thus, a 1000 class cleanroom cabin was utilized to avoid the airborne contaminations, such as particles or organisms. After the bonding, resistance measurement was performed to the bonded test samples. Fig. 3b shows the outlook of bonded and encapsulated samples.

(a) Cu-Cu direct bonding in N$_2$ purging ambient

(b) Outlook of the bonded and encapsulated sample

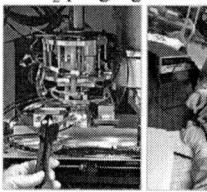  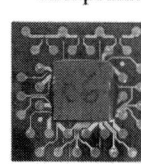 

NCTU Test Vehicle

Figure 3. (a) The demonstration of Cu direct bonding in N$_2$ purging ambient, and (b) the outlook of the Cu-Cu bonded sample, are shown respectively.

For the microstructure analysis, the cross-sectioning process and microstructure observation were carried out to the bonded Cu joint. Fig. 4a displays the cross-sectional view of Cu joints which were wired in series. It reveals that the Cu interconnects can continuously established. However, an alignment shift was found. It was due to the manual operation error during the bonding process. Nevertheless, there is no negative influence for the Cu-Cu joining. However, an optimized alignment operation is needed to diminish this issue in the future study. Fig. 4b shows the enlarged SEM image for a single Cu joint. A well-bonded interfaces can be observed. No voids were found at the bonding interface. Importantly, the Cu joint can be formed without severe oxidation during the non-vacuum bonding. In summary, a successful Cu direct bonding by using (111)-oriented nt-Cu is achieved. The microstructure analysis of the bonded Cu joint at first step bonding process has been conducted.

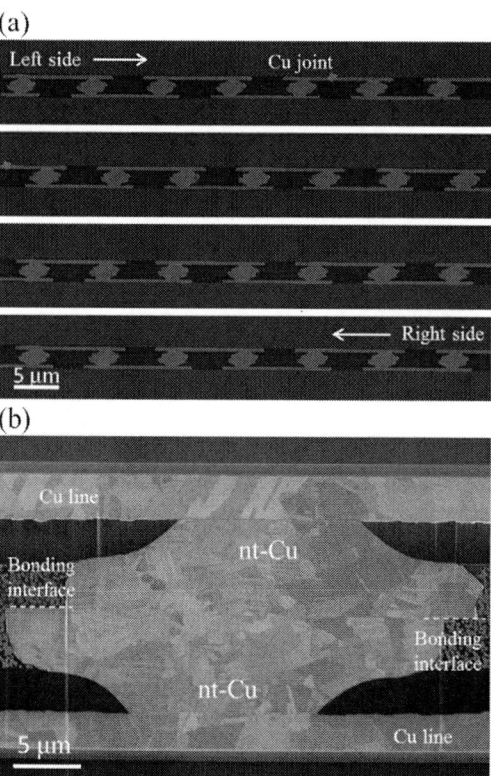

Figure 4. SEM images show the cross-sectional view of (a) Cu joints connected in series, and (b) a single Cu joint.

Simulated bonding interfacial temperature under a temperature gradient

Finite element analysis (FEA) was carried out to simulate the thermomechanical behavior of the Cu to Cu direct bonding structure. Because, it is essential to figure out the bonding interfacial temperature during the bonding under a temperature gradient. The FEA tool, was used. An eight-node high order thermal element named Plane77 was adopted in the finite element model. The thermal conductivity for Copper and Silicon are 401 W/(m·K) and 148 W/(m·K), respectively. The total number of nodes and elements in this model were 1,136,806 and 812,111. The boundary conditions for the thermal analysis were 300°C at the top silicon die and 100°C at the bottom silicon die. Fig. 5 shows the simulated result for a single Cu joint. The simulated temperature gradient was ranging from 136 °C (lower side of Cu joint) to 154 °C (Upper side of Cu joint). The simulated interfacial temperature was at the range of 144 °C to 148 °C. The temperature in a Cu joint is far below 200 °C under a thermal bonding with temperature gradient. This indicated that the low temperature bonding can be achieved.

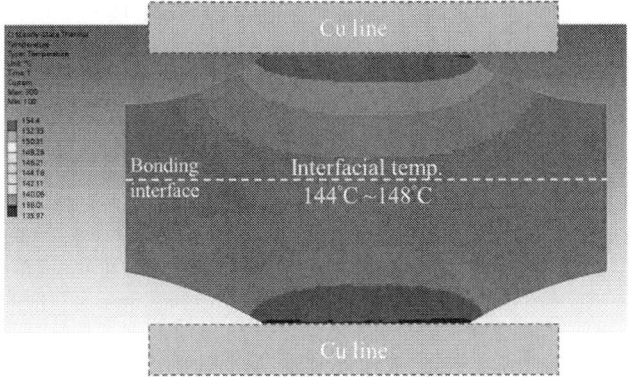

Figure 5. FEA result shows the simulated interfacial temperature and temperature distribution in the Cu joint during the 1st bonding process.

Surface diffusion creep bonding mechanism

Stress-induced surface diffusion (creep) occurred simultaneously at the bonding interface when the thermal compression was applied. It is similar to the model of Nabarro–Herring creep and Coble creep [17], [18]. In these creep models, the atomic flux is driven by the stress potential gradient [19]. Hence, atoms or vacancies can migrate either within the grains or along the grain boundaries. In the present case, two nt-Cu surfaces are forced to contact together under the compression, as depicted by the schematic drawing in Fig. 6a. Thus, there is a stress potential gradient between the contacted regions and the non-contacted regions along the interface. Therefore, the stress potential induces surface diffusion creep to migrate the substances (atoms and voids) from the strained region to the unstrained region, as shown in Fig. 6b. As a result, new atomic bonds across Cu-to-Cu the interfaces would be generated by the stress-induced surface creep.

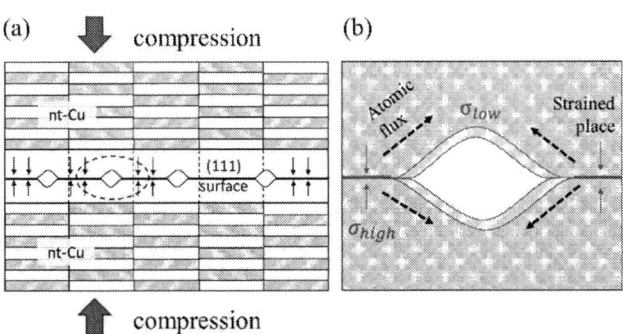

Figure 6. The schematic drawings of nt-Cu to nt-Cu direct bonding mechanism show (a) a less perfect surfaces contact during the compression process, (b) atoms migrate from strained places (high stress) to the non-strained place (low stress) along the interface due to the stress gradient exists between these two sites.

Resistance performance of bonded Cu-Cu joint

The resistance measurement was conducted after first and second step bonding process. For the first step bonding process, the mean values acquired from daisy chain loop (400 joints) and Kelvin structure (single joint) were 5.28 Ω and 4.14 mΩ, respectively. The contact resistivity is 3.98×10^{-8} Ω·cm². For the second step annealing process, we got the results from the testing circuits are 4.94 Ω and 3.27 mΩ. It resistivity is 3.14×10^{-8} Ω·cm². After the measurement, we made a comparative study between Cu joint and solder joint. Test samples with SnAg solder joints were also prepared and tested. By the same measuring process, we got the results from daisy chain loop resistance and single joint resistance are 6.43 Ω and 6.32 mΩ. For a comparison on the single joint resistance, the Cu joint has 34% resistance reduction as compared to SnAg joint. With the aid of the annealing process, the Cu joint can have further resistance reduction to 48%. It is a significant resistance reduction as compare to solder joint. For a comparison on the daisy chain loop resistance, there are only 18% and 23% resistance reduction, as compared to solder joint. The less sensitivity in resistance reduction is due the resistance value was mainly contributed by the long Cu lines, instead by the Cu joints itself in the daisy chain loop.

To better understand the beneficial of two-steps bonding process, a computational simulation work for a Cu joint was conducted. According to the simulation result as shown in Fig. 7, we can calculate the ideal Cu joint resistance. The simulated voltage potential is 2.13×10^{-5} volt and its input current is 0.01 A. Thus, an ideal Cu joint resistance value of 2.13 mΩ was acquired. It was found that the resistance after the second step annealing process approaches to the ideal condition. It reveals that two-steps process can provide significant bonding improvement of resistance reduction. The Comparison of single joint resistance among different bonding materials and bonding processes are listed in Table III.

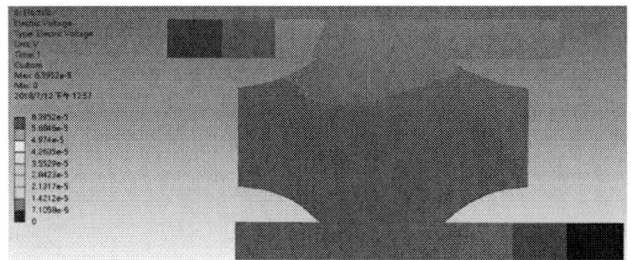

Figure 7. FEA result shows the voltage potential for a single Cu joint.

TABLE III. COMPARISON OF SINGLE JOINT RESISTANCE AMONG DIFFERENT BONDING MATERIALS AND BONDING PROCESSES.

Resistance	SnAg joint	nt-Cu joint		
		1st bonding	2nd annealing	Simulated
Single joint value (mΩ)	6.32	4.12	3.27	2.13

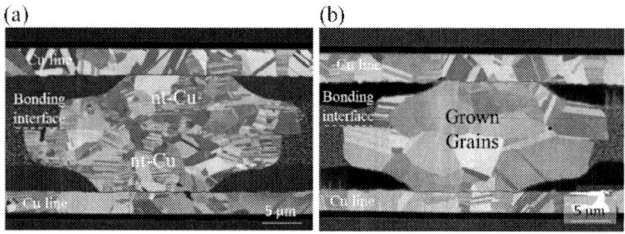

Figure 8. Microstructure changes before and after the second annealing. (a) cross-sectioned SEM image of Cu joint after 1st bonding process, and (b) cross-sectioned SEM image of Cu joint after 2nd thermal annealing process.

Microstructure changes during the two-steps direct bonding process

To investigate the grain growth behavior and the cause of the resistance reduction after the annealing process, microstructure analysis on the Cu-Cu joint before and after annealing process was carried out. Fig. 8a shows a well bonded Cu-Cu joint with a void-less bonding interface after the first step direct bonding in no-vacuum ambient. In addition, no Cu oxide layer was found. It was also found that the nano-twinned columnar grains remained in the Cu joint. The bonding interface can be observed. For the second step annealing process, a significant microstructure change was identified. Recrystallization and grain growth phenomena were observed, as shown in Fig. 8b. Most of the nano-twinned grains disappeared and the nearly all of the grains merged to form a few large grains. In addition, fewer grain boundaries were also observed. Besides, we also found the grain growth went across the bonding interface and its boundary movement eliminated the bonding interface. The resistance of a Cu joint reduced from 4.14 mΩ to 3.27 mΩ after the second step annealing. We speculated that the removal of the original bonding interface is the main cause for the reduction in resistance and specific contact resistivity.

Die shear strength and failure mode of Cu-Cu joints

The die shear test results for Cu-Cu joints and SnAg solder joints are listed in Table IV. The shear strength of the Cu-Cu joints is about 124 MPa, and the SnAg joints is about 64 MPa. By comparing these two different structures, it was found that the Cu-Cu joints had nearly twice the strength of the SnAg solder joints of the same bump diameter. There was a significant improvement when we substituted Cu-Cu joints for SnAg joints. In previous studies, solder joints had four types of failure modes after the shear test [20]–[22]. They were identified as ductile mode (100% area with solder), brittle mode (almost without solder), quasi-ductile mode (or <50% area with exposed pad), and quasi-brittle mode (or >50% area without solder). In this study, only the ductile mode was observed after the shear test. To investigate the fracture surface of the Cu-Cu joints, the SEM observation was carried out on the shear tested samples. The shear direction was from the right to left. Fig. 9a shows the schematic drawing of two fractured positions after the shear strength test. For the first type of joint fracture, it was found that the joint broke at the upper side of the Cu joint. From the enlarged SEM image in Fig. 9b, the phenomena of necking and broken joints with a highly deformed and elongated Cu tail can be observed. For the second type of joint failure, the joint was peeled off from its original place, which left a dimple at the polyimide surface. The area corresponding to the second failure condition is shown in Fig. 9c. Scratches on the PI surface can be found, and part of the Cu pillar remained in the dimple hole.

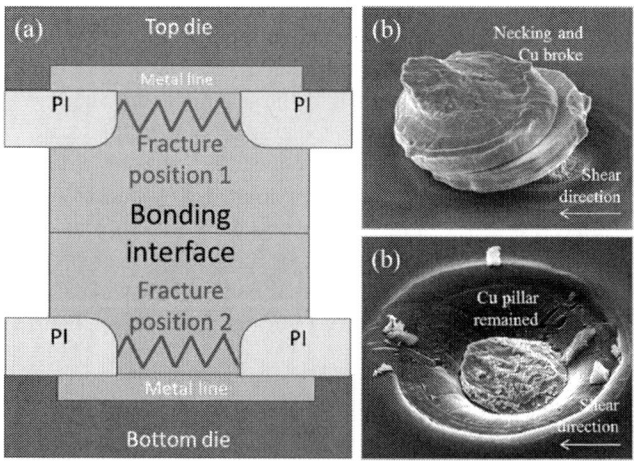

Figure 9. (a) schematic drawing shows the fractured positions after the shear test.; The corresponding the fracture modes show (b) a fractured occurred at the upper side of Cu joint, and (c) a joint peeled off from the bottom die.

TABLE IV. COMPARISON OF DIE SHEAR STRENGTH BETWEEN SNAG AND NT-CU JOINT

Joint Materials	SnAg solder	nt-Cu

Shear strength (MPa)	64	124

In summary, two fracture positions were identified. No joint fracture was found at the bonding interface. It was also found that both of these two joint fractures failed in a ductile manner. Thus, a robust Cu-Cu joint has been confirmed.

IV. CONCLUSIONS

A non-vacuum Cu-to-Cu direct bonding using (111)-oriented nt-Cu has been achieved through thermal compression bonding under a temperature gradient. A surface diffusion creep-assisted bonding mechanism has been proposed to account for the observed direct bonding. The resistance measurement was conducted after first and second step bonding process. For the first step bonding, it shows a low resistance value of 4.12 mΩ for a single Cu joint and 3.98 × 10^{-8} $\Omega \cdot cm^2$ in contact resistivity. With the aid of a second step annealing process, the resistance can be further reduced. It can be brought down to 3.27 mΩ with a resistivity of 3.14 × 10^{-8} $\Omega \cdot cm^2$. There is a nearly 50% resistance reduction, as compared to SnAg solder joint. To investigate the grain growth behavior under the different process steps, microstructure observations in the Cu joint were carried out. It was found that columnar grains in the Cu joints were merged into a few large grains, and its boundary movement went across the bonding interface. Thus, no significant boundary or border was found at the original bonding interface. Both bonding interface removal as well as grain growth evolution, should ease current flowing through the Cu joints and reduce the resistance value. Moreover, the shear strength of the Cu-Cu joints was about 124 MPa, nearly twice the strength of the SnAg solder joints. It is concluded that Cu-Cu joints are a viable option for solving the fabrication and reliability problems faced when soldering decreasing. Cu joints are superior to traditional SnAg solder joints in the areas of advanced packaging technology and heterogeneous chips integration, and thus should be one of most promising candidates to connect the different functions IC chips in a vertical way.

To sum up, a well bonded interface as well as low resistance and high shear strength joint can be achieved by using (111)-oriented nt-Cu microbumps.

ACKNOWLEDGMENT

We acknowledge the financial support from the Ministry of Science and Technology of Taiwan under contracts MOST-105-2221-E-009-008-MY3 and MOST-107-3017-F-009-002, and "Center for Semiconductor Technology Research" from The Featured Areas Research Center Program within the framework of the Higher Education Sprout Project by the Ministry of Education (MOE) in Taiwan.

The authors would like to thank BONDTECH Co., Ltd (JAPAN) for their great assistance on the development of copper-to-copper direct bonding process in N$_2$ ambient (BONDTECH, http://www.bondtech.co.jp).

The authors appreciate Mr. Yen-Tao Tseng's collaboration on development of Cu surface planarization process.

Thanks for the assistance on the analytical equipment from the Center for Micro/Nano Science and Technology (CMNST) at National Cheng Kung University. Thanks for Namics' cooperation to provide underfill materials and information.

REFERENCES

[1] K.Tanida, M.Umemoto, N.Tanaka, Y.Tomita, and K.Takahashi, "Micro Cu Bump Interconnection on 3D Chip Stacking Technology," *Jpn. J. Appl. Phys.*, vol. 43, no. 4B, pp. 2264–2270, Apr. 2004.

[2] B. Dang, S. L. Wright, P. S. Andry, E. J. Sprogis, C. K. Tsang, M. J. Interrante, B. C. Webb, R. J. Polastre, R. R. Horton, C. S. Patel, A. Sharma, J. Zheng, K. Sakuma, and J. U. Knickerbocker, "3D chip stacking with C4 technology," *IBM J. Res. Dev.*, vol. 52, no. 6, pp. 599–609, Nov. 2008.

[3] 11. J. Y. Juang, S. T. Lu, C. J. Zhan, S. C. Chung, C. W. Fan, J. S. Peng, T. H. Chen,, "Development of 30 μm pitch Cu/Ni/SnAg micro-bump-bonded chip-on-chip (COC) interconnects," in *2010 5th International Microsystems Packaging Assembly and Circuits Technology Conference*, pp. 1–4, 2010.

[4] S. T. Lu; J. Y. Juang, H. C. Cheng, Y. M. Tsai, T. H. Chen, W. H. Chen, Y.-M.Tsai, T.-H.Chen, andW.-H.Chen, "Effects of Bonding Parameters on the Reliability of Fine-Pitch Cu/Ni/SnAg Micro-Bump Chip-to-Chip Interconnection for Three-Dimensional Chip Stacking," *IEEE Trans. Device Mater. Reliab.*, vol. 12, no. 2, pp. 296–305, Jun. 2012.

[5] D. R. Frear and P. T. Vianco, "Intermetallic growth and mechanical behavior of low and high melting temperature solder alloys," *Metall. Mater. Trans. A*, vol. 25, no. 7, pp. 1509–1523, Jul. 1994.

[6] H. Huebner, S. Penka, B. Barchmann, M. Eigner, W. Gruber, M. Nobis, S. Janka, G. Kristen, and M. Schneegans, "Microcontacts with sub-30 μm pitch for 3D chip-on-chip integration," *Microelectron. Eng.*, vol. 83, no. 11–12, pp. 2155–2162, Nov. 2006.

[7] J. U. Knickerbocker, P. S. Andry, B. Dang, R. R. Horton, M. J. Interrante, C. S. Patel, R. J. Polastre, K. Sakuma, R. Sirdeshmukh, E. J. Sprogis, S. M. Sri-Jayantha, A. M. Stephens, A. W. Topol, C. K. Tsang, B. C. Webb, and S. L. Wright, "Three-dimensional silicon integration," *IBM J. Res. Dev.*, vol. 52, no. 6, pp. 553–569, Nov. 2008.

[8] Q. Y. Tong, "Room temperature metal direct bonding," *Appl. Phys. Lett.*, vol. 89, no. 18, pp. 98–101, 2006.

[9] T. Suga ; F. Yuuki ; N. Hosoda, "A NewApproach to Cu-Cu Direct Bump Bonding," 1997, pp. 146–151.

[10] C. S. Tan, D. F. Lim, S. G. Singh, S. K. Goulet, and M. Bergkvist, "Cu-Cu diffusion bonding enhancement at low temperature by surface passivation using self-assembled monolayer of alkane-thiol," *Appl. Phys. Lett.*, vol. 95, no. 19, p. 192108, Nov. 2009.

[11] J. Y. Juang, C. L. Lu, K. J. Chen, C. C. A. Chen, P. N. Hsu, C. Chen, and K. N. Tu, "Copper-to-copper direct bonding on highly (111)-oriented nanotwinned copper in no-vacuum ambient," *Sci. Rep.*, vol. 8, no. 1, pp. 13910, Dec. 2018.

[12] J. Y. Juang, C. L. Lu, Y. J. Li, K. N. Tu, and C. Chen, "Correlation between the Microstructures of Bonding Interfaces and the Shear Strength of Cu-to-Cu Joints Using (111)-Oriented and Nanotwinned Cu," *Materials (Basel).*, vol. 11, no. 12, pp. 2368, Nov. 2018.

[13] J. Y. Juang, C. L. Lu, K. J. Chen, T. C. Chang, and C. Chen, "Copper-to-copper direct bonding on highly (111) oriented nano-twinned copper in no-vacuum ambient," in *Proceedings of Technical Papers - International Microsystems, Packaging, Assembly, and Circuits*

Technology Conference, IMPACT, 2018, vol. 2017 Oct., no. 1, pp. 199–201.

[14] J. Y. Juang, C. L. Lu, K. J. Chen, T. C. Chang, and C. Chen, "Copper-to-copper direct bonding on highly (111) oriented nano-twinned copper in no-vacuum ambient," in *2017 12th International Microsystems, Packaging, Assembly and Circuits Technology Conference (IMPACT)*, pp. 199–201, 2017.

[15] C. M. Liu, H. W. Lin, Y. S. Huang, Y. C. Chu, C. Chen, D. R. Lyu, K. N. Chen, and K. N. Tu, "Low-temperature direct copper-to-copper bonding enabled by creep on (111) surfaces of nanotwinned Cu," *Sci. Rep.*, vol. 5, p. 9734, May 2015.

[16] C. Chen, C. M. Liu, H. W. Lin, Y. S. Huang, Y. C. Chu, D. R. Lyu, K. N. Chen, and K. N. Tu, "Low-temperature and low-pressure direct copper-to-copper bonding by highly (111)-oriented nanotwinned Cu," in *2016 Pan Pacific Microelectronics Symposium (Pan Pacific)*, pp. 1–5, 2016.

[17] C. Herring, "Diffusional viscosity of a polycrydtalline solid," *J. Appl. Phys.*, vol. 21, no. 1950, pp. 437–445, 1950.

[18] M. Science and R. L. Coble, "Diffusion Models for Hot Pressing with Surface Energy and Pressure Effects as Driving Forces," *J. Appl. Phys.*, vol. 41, no. 12, pp. 4798–4807, Nov. 1970.

[19] K. N. Tu, *Electronic Thin-Film Reliability*. Cambridge: Cambridge University Press, 2010.

[20] T. T. Mattila and J. K. Kivilahti, "Reliability of lead-free interconnections under consecutive thermal and mechanical loadings," *J. Electron. Mater.*, vol. 35, no. 2, pp. 250–256, Feb. 2006.

[21] Y. J. Chen, C. K. Chung, C. R. Yang, and C. R. Kao, "Single-joint shear strength of micro Cu pillar solder bumps with different amounts of intermetallics," *Microelectron. Reliab.*, vol. 53, no. 1, pp. 47–52, Jan. 2013.

[22] F. Song and S. W. R. Lee, "Investigation of IMC Thickness Effect on the Lead-free Solder Ball Attachment Strength: Comparison between Ball Shear Test and Cold Bump Pull Test Results," in *56th Electronic Components and Technology Conference 2006*, pp. 1196–1203, 2006.

This page intentionally left blank.

Sub-10μm Pitch Hybrid Direct Bond Interconnect Development for Die-to-Die Hybridization

John P. Mudrick, Jonatan A. Sierra-Suarez, Matthew B. Jordan, T.A. Friedmann, Robert Jarecki and M. David Henry

Microsystems Engineering, Sciences, and Applications (MESA)
Sandia National Laboratories
Albuquerque, NM, USA
e-mail: jmudric@sandia.gov

Abstract— Direct bond interconnect (DBI) processes enable chip to chip, low resistivity electrical connections for 2.5-D scaling of electrical circuits and heterogenous integration. This work describes SiO_2/Cu DBI technology with Cu interconnect performance investigated over a range of inter-die Cu gap heights and post-bond annealing temperatures. Chemical mechanical polishing (CMP) generates wafers with a controlled Cu recess relative to the SiO_2 surface, yielding die pairs with inter-die Cu gap heights ranging between 9 and 47 nm. Bonded die with different gap heights show similar per-connection resistance after annealing at 400 degrees Celsius but annealing at lower temperatures between 250 and 350 degrees Celsius results in failing or high-resistance interconnects with intermediate gaps showing lowest resistance. Cross-section scanning electron microscope (SEM) image analysis shows that the microstructure is largely independent of post-bond annealing temperature, suggesting that the temperature behavior is due to nanoscale scale interfacial effects not observable by SEM. The bond strength is affirmed by successful step-wise mechanical and chemical removal of the handle silicon layer to reveal metal from both die. This work demonstrates a 2.5-D integration method using a 3 micron Cu DBI process on a 7.5 micron pitch with electrical contacts ranging between 3.8 and 4.8 Ohms per contact plug.

Keywords-fine-pitch interconnects; hybrid bonding; direct bond interconnect; Cu-Cu bonding; CMP

I. INTRODUCTION

The drive for increased chip performance and multi-functionality has led to increasing development of heterogeneous integration (HI) strategies, where functioning die from separate wafers are assembled on a single chip. Die and wafer stacking integration technologies are now being used in mass microprocessor production [1], and the various technical challenges associated with these integration schemes are tracked by the ITRS Heterogeneous Integration Roadmap [2]. Hybrid direct bonding or direct bond interconnect (DBI) technology is an attractive HI option because it has a demonstrated high interconnect density approaching sub-micron pitches, it is a planar technology that does not require underfill or carrier wafer integration, and interconnects are formed at relatively low temperatures $\leq$ 400 °C [3-5]. Hybrid direct bonding is an extension of direct bonding, whereby two appropriately prepared

homogeneous surfaces form a strong covalent bond after being brought into contact with one another at ambient temperature [6,7]. In the case of SiO_2 direct bonding, plasma activation reduces the thermal requirements for bond formation by generating a high density of surface hydroxyl groups which form a strong covalent bond when brought into contact with a similarly treated partner wafer [7]. In the case of SiO_2/Cu hybrid direct bonding the final pre-bond surface consists of slightly recessed Cu interconnects patterned in an SiO_2 field, and bonding takes place in two steps. The low temperature dielectric bond is formed first at ambient temperature, followed by an annealing step at 200 - 400 °C to maximize dielectric bond strength as well as induce Cu expansion and the formation of electrical interconnects. Interconnect density is defined lithographically and is therefore limited by the alignment accuracy of the bonding equipment, which is in the sub-micron regime for state of the art bonders [8].

To date, there have been several reports of SiO_2/Cu DBI process development demonstrating the importance of final bond surface preparation to bond formation, yield, and reliability. Gao et al. reported that low SiO_2 roughness and high Cu recess depth uniformity at the die- and wafer-scale are critical to achieving interconnect functionality and high yield [3,9]. Cu surface morphology control, low surface defect levels, and high precision alignment have also been shown to be essential to interconnect functionality and reliability [5,10]. Simulation of bonding behavior has also revealed physical details of Cu interconnect formation in the hybrid SiO_2/Cu geometry. Sart et al. used finite element modeling to show that interconnect formation is strongly dependent on initial Cu dishing and post-bond anneal temperature, with less than 5 nm dishing and 400 °C annealing temperature required for full closure of a Cu plug pair [11]. Bonding results follow a cohesive interaction model where adjacent Cu surfaces form a bond with bond energy that increases when annealing temperature is increased from 200 to 400 °C [12]. More rigorous modeling of the Cu surface nanostructure shows that lower Cu surface roughness leads to more efficient contact area formation, but further investigation is required at this level of detail [13].

While previous experimental reports have described process conditions required for SiO_2/Cu hybrid DBI formation and modeling results highlight the importance of

978-1-7281-1500-9/19 $31.00 © 2019 IEEE

Cu/SiO₂ surface morphology, no experimental study has systematically investigated the combined effects of post-bond anneal temperature and quantitative Cu recess depth. In this work, we have employed a chemical mechanical polishing (CMP) process to control the Cu recess depth. By combining various die combinations from separate parent wafers, die pairs with a range of inter-die Cu-Cu gap heights ranging from 9 nm to 47 nm have been bonded and characterized by electrical measurements and cross-section microscope imaging. Bonded die with various inter-die Cu gap heights were then sequentially annealed at temperatures ranging from 250 to 400 °C to determine process requirements, tolerances, and sensitivities for interconnect formation.

II. EXPERIMENTAL DETAILS

A. Fabrication Overview

Electrical daisy chain structures were constructed using subtractive metal back end of line microfabrication methods to define Al:Cu(1 wt%) lines encapsulated in SiO₂ deposited via plasma-enhanced chemical vapor deposition (PECVD) at 400 °C; this PECVD SiO₂ comprises the majority of the bonding surface after further processing. After SiO₂ CMP, 3 μm and 3.5 μm diameter openings were etched into the 1μm-thick top SiO₂ layer via fluorocarbon-based reactive ion etch (RIE) followed by photoresist strip and a brief buffered oxide etch (BOE) chemistry to fully remove etch residues. Wafers were then metallized, as illustrated in Fig. 1(a), with bilayer barrier/seed metal consisting of 20/200 nm Ta/Cu deposited via electron beam evaporation followed by electrochemical deposition of 2 μm Cu to over-fill the holes and coat the top SiO₂ surface. A two-step CMP sequence was utilized to remove Cu and clear Ta barrier metal from the surface, leaving a planar SiO₂ surface with a controlled Cu recess depth; this process is described in Section II B.

Following the CMP sequence, wafers were coated with a protective photoresist layer (AZ P4330 hard baked at 125°C). SiO₂ crack propagation during saw-dicing was mitigated by first lithographically defining dicing streets between adjacent die and etching the exposed SiO₂ in a BOE solution to a depth of 0.5 μm. Wafers were then saw-diced with the blade contained within the street area to form top and bottom die for bonding. Following photoresist removal in acetone/isopropanol, die were plasma-activated in a N₂ RIE process (N₂ flow = 45 standard cubic centimeters per minute, Pressure = 13 Pa, D.C. bias = 200 Volts) for 60 seconds, followed by a final acetone/isopropanol solvent clean immediately prior to bonding. Die were precision-bonded in a flip chip bonder (SETNA) at ambient temperature with a force of 50 kN. Typical alignment accuracy of the bonder was less than one micron, and die were leveled to less than one microradian total curvature prior to bonding. A two hour post-bond anneal was performed under a force of 1 kN to complete the

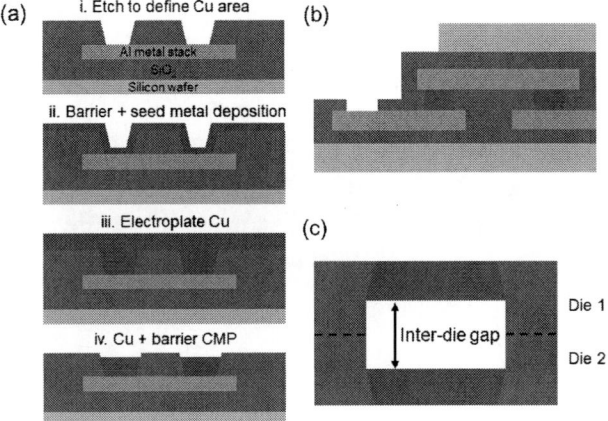

Figure 1. Schematics of process flow and bonded interconnect test structures used in this work. (a) Process flow for generating hybrid SiO2/Cu DBI test die. The "Al metal stack" consists of Ti(20 nm)/TiN(50 nm)/Al:Cu(750 nm)/TiN(100 nm). (b) Partial cross-section of bonded daisy chain die. (c) Illustration of inter-die gap between Cu surfaces prior to annealing.

dielectric bond and induce Cu expansion to form interconnects. A schematic of the hybridized device cross-section is shown in Fig. 1(b) and the detail of the Cu area prior to interconnect formation is illustrated in Fig. 1(c). The larger bottom die has additional metal traces leading to probe pads for electrical characterization.

B. Hybrid DBI Surface Preparation

The dependence of bond interconnect formation on inter-die gap was investigated by developing a two-step CMP process to control the Cu recess depth. This is illustrated schematically in Fig. 2(a). The bulk of the Cu is removed on a single platen utilizing Cabot C8900 series slurry which removes Cu selectively over Ta. The slurry's high selectivity to Cu allows for the platen motor current to be utilized in determining the polish endpoint. Over-polish is empirically determined to synergize with the effective removal during the Ta barrier CMP stage. The Ta barrier layer is removed on a secondary platen utilizing Cabot B7000 series slurry, leaving Cu recessed relative to the SiO₂ field. Alternative slurries may be utilized but will require the selectivity of the materials across both polishes to be optimized.

DBI surface topography was extensively characterized with a Bruker Dimension 5000 series atomic force microscope (AFM) operated in tapping mode and data analysis was performed with Bruker NanoScope Analysis software. Typical scans used for depth determination were 15 x 15 μm and 128 x 128 pixels (corresponding to a 2 x 2 array of Cu plugs) and were collected with real-time and offline planefit corrections turned off in order to eliminate line-by-line flattening. The NanoScope Analysis planefit function was used to correct for scanner drift in the x- and

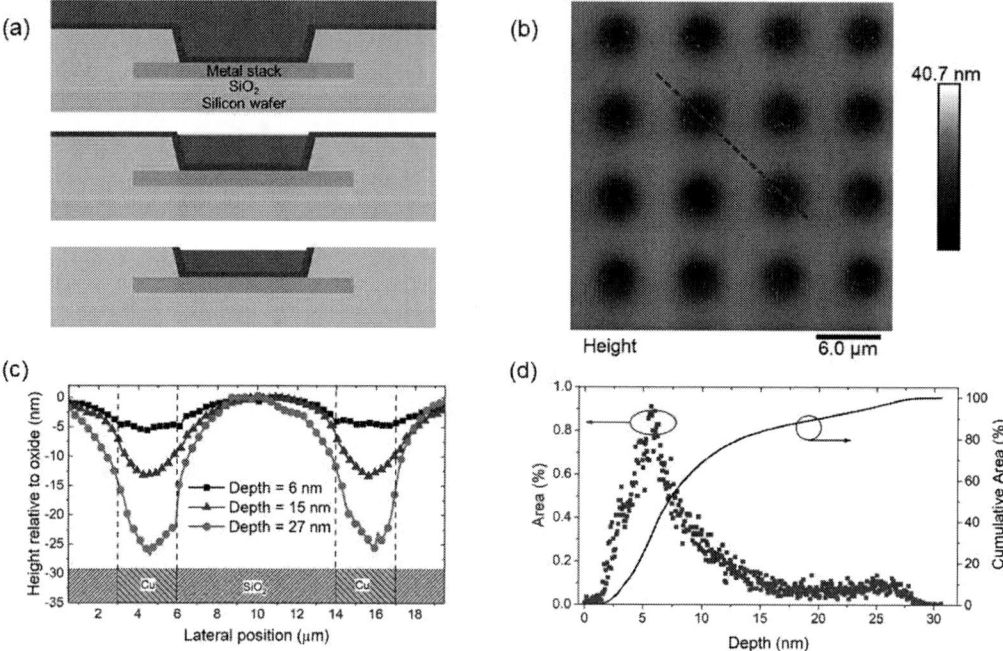

Figure 2. Hybrid bonding surface preparation and characterization. (a) Evolution of surface during CMP process; Ta is shown in gray, and Cu is shown in red. (b) Typical AFM topography scan showing recessed Cu plugs. (c) Line profiles of three wafers with different recess depths corresponding to the blue dashed line in (b). (d) Depth histogram of a 15 x 15 μm scan area consisting of 4 Cu plugs surrounded by SiO₂.

y- directions while preserving the local SiO₂ and Cu recess topographies. A large area AFM scan of a typical post-CMP Cu/SiO₂ surface is shown in Fig. 2(b). Three representative line scans from wafers with different Cu recess depths are shown in Fig. 2(c). The scan data in Figs. 2(b) and (c) show that both the SiO₂ and Cu surface areas have a significant topography range within their respective areas. The curvature is a known systematic which arises from the contact mechanics of the pad and wafer [14]. The CMP polish pad can be represented as an elastic half space, while the wafer is treated as a rigid body. This causes the local pressure to sharply increase whenever there are gradients in the topography. Since the Ta barrier polish is a dominantly Prestonian process, this results in more rapid SiO₂ erosion in the areas nearest Cu features, leaving SiO₂ high points farthest away from each Cu plug. This is manifested in the lightest portions of the AFM scan, Fig. 2(b), at points equidistant from the patterned Cu plugs. A depth histogram of this scan area is presented in Fig. 2(d). The depth range in the SiO₂ field, corresponding to the regions with depth less than 15 nm in Fig. 2(d), increases proportionally with Cu recess depth (data not shown). Detailed local AFM scans of the SiO₂ surface (1 x 1 μm, 256 x 256 pixels) were collected to determine surface roughness of the initial bond area, revealing root mean square (RMS) roughness values < 0.2 nm for all Cu recess depths, within the acceptable range for strong bond formation [3,5,10]. The depth range within the Cu areas also increases in proportion to the recess depth, as

seen in the line scans in Fig. 2(c) and the broad tail spanning from depths of approximately 15 to 30 nm in Fig. 2(d).

Other experimental and simulation studies have concluded that Cu interconnect formation is sensitive to recess depth ranges in the sub-10 nm range [3,5,10]. Since both the SiO₂ and Cu areas have significant topography, it is appropriate to define the recess depth in relation to the SiO₂ and Cu depth ranges. Hereafter the recess depth is defined as the height difference between the highest area within the SiO₂ surface and the bottom of the Cu recess. The Cu bottom is defined as the depth above which 99% of the total scan area lies. Results of CMP depth targeting are presented in Fig. 3. Cu recess depths between 5 and 25 nm are achieved after the first pass or after a single rework step; recess depths vary by less than 3 nm across each wafer and are fully recovered after die singulation, resist strip, and pre-bond preparation steps. Varying the Cu recess depth of top and bottom die allows for the assembly of bonded die with a pre-anneal Cu-Cu gap ranging from 9 to 47 nm.

C. Test Layout Details

Electrical test structures consisted of 3 μm- (bottom die) and 3.5 (top die) μm-diameter Cu plugs patterned on a 7.5 μm pitch. Cu plugs are laterally interconnected in pairs by Al lines. The Al lines are capped by a 100 nm-thick TiN layer, while the Cu plugs have a 20 nm-thick Ta barrier below, making a conductive path consisting of Al/TiN/Ta/Cu. Daisy chains are completed by bonding two die together to form Cu inter-die interconnects, as

978-1-7281-1500-9/19 $31.00 © 2019 IEEE

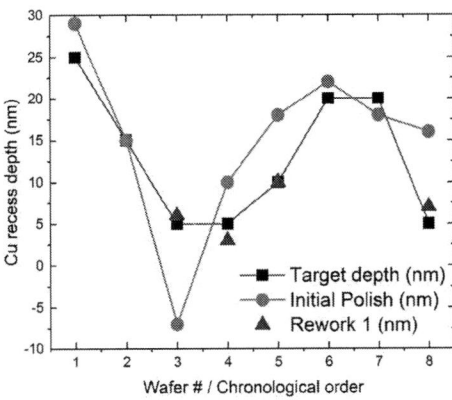

Figure 3. Summary of CMP recess depth targeting.

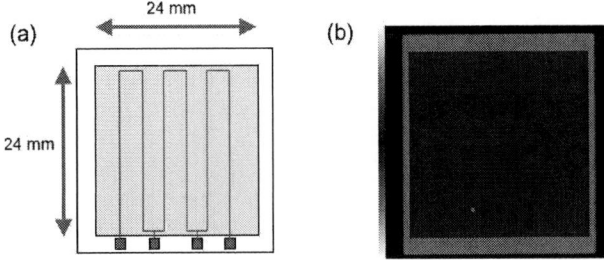

Figure 4. (a) Top-down schematic of serpentine daisy chain pattern. (b) CSAM image of bonded die. The inset black area corresponds to the 24 x 24 mm area of the smaller top die.

illustrated in Fig. 1(b). Daisy chains run in serpentines across the entire 24 mm-wide die as illustrated in Fig. 4(a), with each serpentine line consisting of 3,200 inter-die connections. Electrical resistance was calculated from current-voltage measurements acquired in the 4 probe configuration using an Agilent parameter analyzer with dual voltage sweeps between ± 2 V. Adjacent serpentines are separated by 4.5 µm, so any particle defect results in a large number of failing connections [6]. A series of 10 die pairs were bonded during this study, each having > 98% bonded area as measured by confocal scanning acoustic microscopy (CSAM). Manual die handling after singulation resulted in a small number of bond voids with random spatial distribution within each bonded die pair. Therefore this work primarily reports electrical measurements of the shortest test structure consisting of 6 serpentine lines making 19,200 inter-die DBI connections to investigate the fundamental aspects of the interconnect rather than process robustness and statistical yield. The longest functional chains correlate to failure rates in the parts per million range.

III. RESULTS AND DISCUSSION

A. Bonding and Interconnect Formation

A typical CSAM image of a bonded die pair is shown in Fig. 4(b), where the black region corresponds to bonded areas and white indicates the presence of voids(s). This image confirms that the combined CMP, activation, and clean sequence generates low-defect surfaces that form an efficient bond with low void area. The areas at the outer edges of the top die are void-free indicating the pre-dice SiO₂ etch process mitigates potential particles generated during saw-dicing: residual particles that remain on the final bonding surface would result in bond voids that are 3-4 orders of magnitude larger than the particle size [6]. Such large voids were not observed near the exterior die edges.

Current-voltage data for a die pair measured after annealing for two hours at 350 °C is presented in Fig. 5(a). Daisy chains display Ohmic behavior for all chain lengths

shown, indicating no presence of charge accumulation at the Cu-Cu inter-die interface or within the Al/TiN/Ta/Cu stack. Calculated resistance values are shown as a function of chain length in Fig. 5(b). Chains with up to 1.5 million DBI connections display a linear resistance dependence on chain length, with an average resistance of (4.55 ± 0.04) Ω/connection. This resistance is 1-2 orders of magnitude greater than those observed in other SiO₂/Cu hybrid interconnect studies [5,10,15-17] as well as the expected resistance of approximately 0.1 Ω/connection calculated with standard resistivities and thicknesses of the films used in this work. Possible effects that may result in higher resistance are deposition of a high resistivity Ta phase [18,19], oxidation of the TiN surface prior to Ta deposition, or excessive native oxidation of the Cu surface. Further investigation is required to determine the root cause of this relatively high resistance stack.

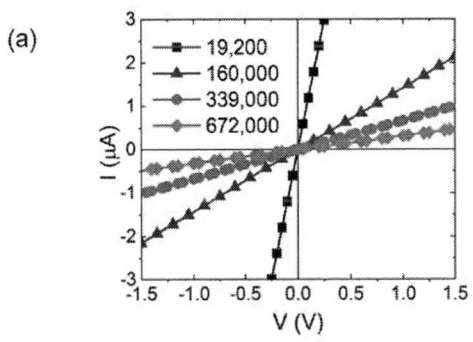

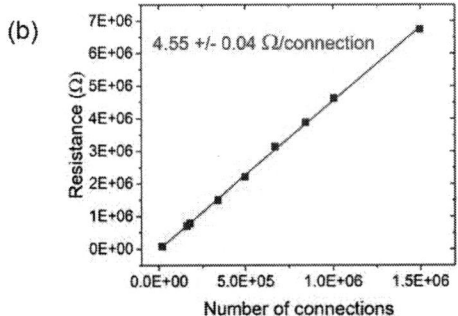

Figure 5. (a) Current-voltage sweeps of daisy chains with number of chains indicated. (b) Resistance as a function of daisy chain length.

Bonded die were further characterized by cross-section scanning electron microscopy (SEM) prepared by focused ion beam (FIB) milling. The FIB-SEM image shown in Fig. 6(a) corresponds to a bonded die pair with 5.1 Ω/connection resistance after a two-hour anneal at 400 °C. The image reveals a mostly closed Cu-Cu interface with small voids present at triple points where Cu grain boundaries intersect with the bond interface; these voids were consistently present at other observed locations within the same sample. Voids at triple points have been assigned to enhanced Cu and Cu oxide diffusivity along grain boundaries [16,20]. Energy dispersive spectroscopy (EDS) images corresponding to the same sample are shown in Fig. 6(b). The Cu emission map shows that the Ta barrier prevents Cu diffusion into the Al lines as well as the surrounding oxide. Furthermore, there is no evidence of Cu diffusion into unbonded area, i.e. the horizontal Cu ledge to the left of the Cu-Cu bonded area. The narrow horizontal Si emission features correspond to the Ta barrier layer, whose emission spectrum overlaps with that of Si.

B. Cu Recess Depth and Annealing Temperature

The evolution of chain resistance with Cu-Cu gap height and post-bond annealing temperature was investigated to determine the process parameter space for hybrid interconnect formation. Die with various Cu recess depths were combined to generate bonded pairs with inter-die gap heights of 9, 18, 32, and 47 nm; each individual depth has an uncertainty range of 5 nm. Resistance values for chains consisting of 19,200 DBI links are plotted in Fig. 7 for these four gap heights and post-bond annealing temperatures ranging from 250 to 400 °C; all pairs were annealed for two hours at each temperature. Open circuits are plotted as 10^4 Ohms. All four heights converge to resistances between 3.8 and 4.8 Ω/connection after annealing at 400 °C; these resistance values are listed in Table I. The assembly with the widest Cu-Cu gap did not have electrical continuity until after the final 400 °C anneal. Die pairs with gaps less than 47 nm were functional after annealing at 350 °C, though resistances were in the range of 12-20 Ω/connection. Samples with 18 and 32 nm gap heights had Ohmic behavior after annealing at 250 and 300 °C, but at elevated per-link resistance. The 9 nm-gap sample diverges from expected behavior: this sample should require the least Cu expansion to form Cu-Cu interconnects. However, the 9 nm gap sample is not electrically connected after annealing at 250 and 300 °C.

Using in-situ TEM annealing, Martinez et al. showed that Cu grain growth across the bond interface is activated at 350 °C and further enhanced after increasing temperature to 400 °C, with grain growth extending more than 100 nm across the interface [20]. Finite element modeling results have also shown that maximum Cu contact area is achieved after higher temperature annealing [11,12]. The independence of ultimate minimum resistance on pre-bond gap height in the present study is in alignment with these previous reports. Furthermore, this result suggests that maximum Cu diffusivity and grain growth is required to minimize electrical resistance for all gap heights considered. However, the precise mechanism for the gap-temperature dependence remains unresolved without further physical and electrical characterization.

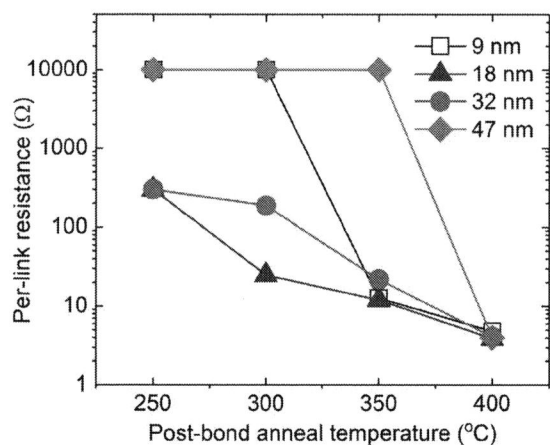

Figure 7. Per-connection resistance as a function of post-bond annealing temperature for 4 different inter-die Cu-Cu gap heights.

TABLE I. MEASURED RESISTANCE VALUES AFTER 400 °C ANNEALING.

Inter-die gap (nm)	Resistance (Ω/connection)
9	4.8 ± 0.2
18	3.9 ± 0.1
32	4.1 ± 0.1
47	4.0 ± 0.1

(a)

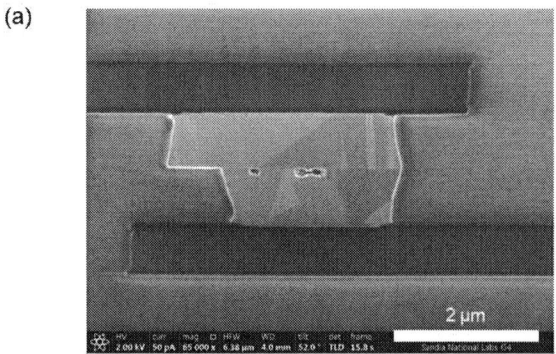

(b)

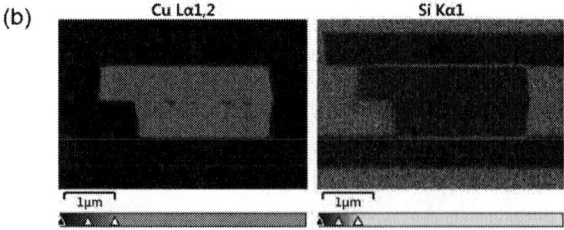

Figure 6. (a) FIB-SEM of bonded die pair. (b) EDS maps of Cu and Si.

C. Annealing Temperature at Fixed Inter-die Gap Height

The 32 nm gap sample was imaged by FIB-SEM after annealing at 250, 350, and 400 °C to determine whether the resistance decrease corresponding to these thermal treatments is caused by microstructure evolution. Fig. 8 shows FIB-SEM images after these three annealing temperatures. While the electrical resistance decreased from 300 to 4.1 Ω/connection, the bulk Cu grain structure was unchanged after increasing the annealing temperature from 250 to 400 °C.

Ion channeling contrast images are shown in Fig. 9 after 350 and 400 °C annealing temperatures to give more insight into the Cu grain structure. Grain growth across the bonding interface is observed in both cases despite the five-fold difference in electrical resistance. Both Figs. 8 and 9 show a nearly complete disappearance of voids at the bond interface after 400 °C annealing. While these voids make up a small percentage of the total bond area, they may correlate with incomplete closure of the interface or insufficient Cu grain growth across the interface for maximum conductivity. Higher resolution imaging such as transmission electron microscopy (TEM) is required to image the Cu-Cu interface at the necessary level of detail and determine whether fine structural and/or chemical differences are present over this annealing temperature range.

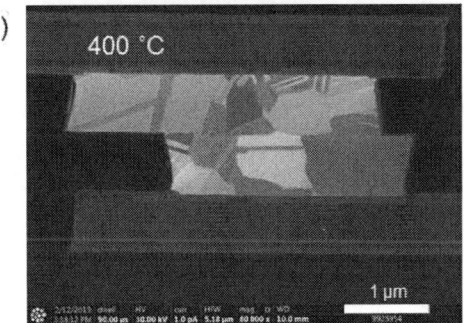

Figure 9. Ion channeling contrast FIB-SEM images of bonded die after annealing at (a) 350 °C and (b) 400 °C.

D. Handle Silicon Removal

To demonstrate the robustness of this hybrid SiO$_2$/Cu bond, a bonded die pair was thinned to reveal the isolation SiO$_2$ layer on the smaller die. Thinning took place in two steps: first a precision mechanical grind and polish process was used to thin the die to 30 - 40 μm remaining silicon, after which a high density SF$_6$ plasma etch was used to remove the remaining silicon and land the etch on SiO$_2$. The resulting bonded die with handle silicon removed is shown in Fig. 10. The microscope image in Fig. 10(b) shows the metal lines of both top and bottom die still intact after this processing. The per-connection resistance was also unchanged after this thinning procedure. This demonstration is particularly appealing because the bond does not require underfill or bonding/separation to/from a surrogate carrier wafer. Furthermore, this represents a possible 3D integration option with greater feature density than through-silicon via processing.

IV. CONCLUSION

Controlled Cu recess depth daisy chain wafers were created in order to bond die with various Cu recess depths, forming bonded pairs with a range of inter-die Cu-Cu gap heights. All heights studied formed functional interconnects with per-connection resistances between 3.8 and 4.8 Ω/connection after annealing at 400 °C. The temperature dependence diverged at lower temperature, with 18 and 32

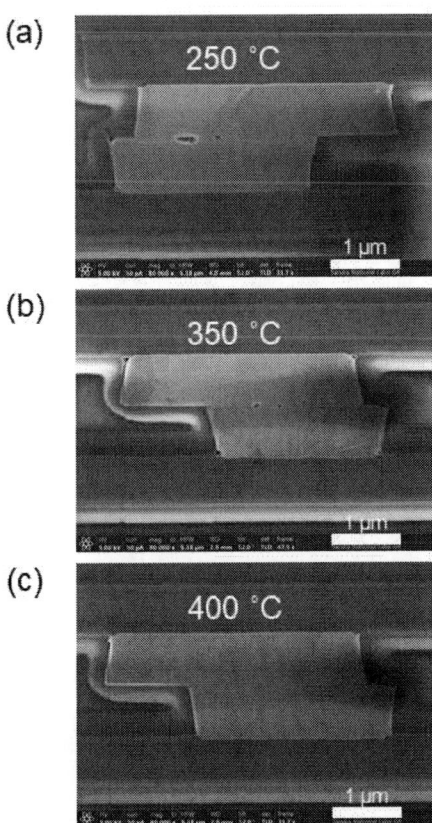

Figure 8. FIB-SEM images showing microstructural evolution after annealing for two hours at the temperatures indicated.

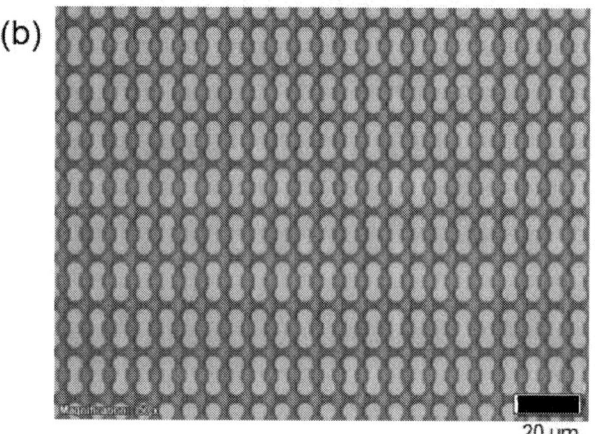

Figure 10. Handle silicon removal from bonded die. (a) Photograph of a still-functional bonded die pair. (b) Top-down optical microscope image showing top and bottom daisy chain metal.

nm gaps having similar functionality at elevated resistance while the 9 nm gap sample did not follow this trend, as it was not functional after annealing at 250 and 300 °C. The temperature dependence was probed on the 18 nm gap sample by FIB-SEM analysis after multiple temperature anneals. While no changes in the Cu microstructure were observed between 250 and 400 °C, the density of Cu-Cu interface voids decreased drastically after annealing at 400 °C. Finally, the robustness of this hybridization scheme was demonstrated by fully removing the handle silicon from one die and verifying both structural integrity and electrical functionality.

ACKNOWLEDGMENTS

The authors are grateful to C. Sennett for extensive CMP development and AFM measurements, L. Caravello for process integration and AFM measurements, the MESA Fab team for process development and device fabrication, as well as to M. Rye for FIB-SEM imaging.

Sandia National Laboratories is a multimission laboratory managed and operated by National Technology & Engineering Solutions of Sandia, LLC, a wholly owned subsidiary of Honeywell International Inc., for the U.S.

Department of Energy's National Nuclear Security Administration under contract DE-NA0003525. This paper describes objective technical results and analysis. Any subjective views or opinions that might be expressed in the paper do not necessarily represent the views of the U.S. Department of Energy or the United States Government.

REFERENCES

[1] "Intel steps toward heterogeneous integration," https://www.eetimes.com/document.asp?doc_id=1334073

[2] Heterogeneous Integration Roadmap, IEEE, https://eps.ieee.org/technology/heterogeneous-integration-roadmap.html

[3] G. Gao, et al. "Direct Bond Interconnect (DBI) Technology as an Alternative to Thermal Compression Bonding," Proc. 13th IWLPC (2016).

[4] A. Jouve, et al. "1µm pitch direct hybrid bonding with <300nm wafer-to-wafer overlay accuracy," Proc. 2017 IEEE S3S (2017).

[5] L. Arnaud et al. "Fine pitch 3D interconnections with hybrid bonding technology: from process robustness to reliability," Proc. 2018 IEEE IRPS (2018).

[6] M. Reiche, "Semiconductor wafer bonding," Phys. Stat. Sol. (a), vol. 203, pp. 747-759 (2006).

[7] T. Plach, K. Hingerl, S. Tollabimazraehno, G. Hesser, V. Dragoi, and M. Wimplinger, "Mechanisms for room temperature direct wafer bonding," J. Appl. Phys., vol 113, p. 094905 (2013).

[8] T. Uhrmann, J. Burggraf, M. Pires, and M. Eibelhuber, "Heterogeneous integration by collective die-to-wafer bonding," Chip Scale Review, vol. 22, p. 10-12 (2018).

[9] G. Gao, et al. "Scaling package interconnects below 20µm pitch with hybrid bonding," Proc. IEEE 68th ECTC (2018).

[10] S. Lhostis, et al. "Reliable 300 mm wafer level hybrid bonding for 3D stacked CMOS image sensors," Proc. IEEE 66th ECTC (2016).

[11] C. Sart, et al. "Cu/SiO2 hybrid bonding: finite element modeling and experimental characterization," Proc. IEEE 6th ESTC (2016).

[12] Y. Beilliard, et al. "Thermomechanical finite element modeling of Cu-SiO$_2$ direct hybrid bonding with a dishing effect on Cu surfaces," Int. J. Sol. Struc., vol. 117, p. 208-220 (2017).

[13] T. Wlanis, et al. "Cu-SiO$_2$ hybrid bonding simulation including surface roughness and viscoelastic material modeling: A critical comparison of 2D and 3D modeling approach," Micro. Rel., vol. 86, p. 1-9 (2018).

[14] J. A. Sierra Suarez and C. F. Higgs III, "A contact mechanics formulation for predicting dishing and erosion CMP defects in integrated circuits," Trib. Lett., vol. 59, p. 36 (2015).

[15] A. Agrawal, et al. "Thermal and electrical performance of direct bond interconnect technology for 2.5D and 3D integrated circuits," Proc. IEEE 67th ECTC (2017).

[16] Y. Beilliard, et al. "Advances toward reliable high density Cu-Cu interconnects by Cu-SiO$_2$ direct hybrid bonding," Proc. 3DIC (2014).

[17] Z. Liu, J. Cai, Q. Wang, H. Jin, and L. Tan, "Room temperature direct Cu-Cu bonding with ultrafine pitch Cu pads," Proc. IEEE 17th EPTC (2015).

[18] K. Stella, D. Bürstel, S. Franzka, O. Posth, and D. Diesing, "Preparation and properties of thin amorphous tantalum films formed by small e-beam evaporators," J. Phys. D: Appl. Phys., vol. 42, p. 135417 (2009).

[19] J. Colin, G. Abadias, A. Michel, and C. Jaouen, "On the origin of the metastable β-Ta phase stabilization in tantalum sputtered thin films," Acta Mater., vol. 126, p. 481-493 (2017).

[20] M. Martinez, et al. "Mechanisms of copper direct bonding observed by in-situ and quantitative transmission electron microscopy," Thin Solid Films, vol. 530, p. 96-99 (2013).

Cu Pillar with Nanocopper Caps: The Next Interconnection Node beyond Traditional Cu Pillars

Ramón A. Sosa, Kashyap Mohan, Rao Tummala,
Vanessa Smet
3D – Systems Packaging Research Center
Georgia Institute of Technology
Atlanta, GA, United States
vanessa.smet@prc.gatech.edu

Luu Nguyen
Texas Instruments Inc.
Santa Clara, CA, United States
Luu.Nguyen@ti.com

Antonia Antoniou
Woodruff School of Mechanical Engineering
Georgia Institute of Technology
Atlanta, GA, United States
antonia.antoniou@me.gatech.edu

Abstract—Off-chip interconnection pitch scaling has been aggressively driven over the last several decades by the continuous need for higher bandwidth and computing power in smaller form factors in emerging high-performance computing systems. It is expected to reach below 10 micron I/O pitch in the near future, beyond the fundamental limits of traditional solder-based interconnection technologies. While the Cu pillar with solder caps technology remains attractive in chip-to-substrate (C2S) applications as it can accommodate substrate and chip non-coplanarities during assembly through melting of the solder, all-Cu interconnections are now pursued as the next interconnection node for their pitch and performance scalability. However, direct Cu-Cu bonding faces several key challenges that have hindered large-scale adoption in C2S, including its relatively high elastic modulus, giving low compliance in assembly. To address this challenge, a novel interconnection technology – Cu pillar with nanoporous copper (np-Cu) caps – is proposed where a solid-state sub-20 GPa modulus np-Cu cap is introduced to replace the solder caps and retain solder-like compliance in assembly, while achieving bulk-like properties through densification in low-temperature sintering. This paper presents the design of this new interconnection system, the developed wafer bumping process, compatible with current industry infrastructures, and a first assembly demonstration where a seamless interface was achieved.

Keywords- All-copper interconnectionss; nanoporous copper; nanoporous metal; direct Cu-Cu bonding

I. INTRODUCTION

The growing need for emerging high-performance computing systems has been continuously driving the semiconductor industry towards increasing bandwidth and computing power of chips in smaller form factors. Faced with limitations to transistor scaling, new packaging solutions are being considered to meet these demands. Emerging high-performance systems are predicted to require off-chip interconnection pitches below 10μm in the next 5 years with average current densities in excess of 10^5 A/cm^2, thereby increasing the expected power densities and operating temperatures of chips well over 100°C. [1] This requirement is pushing beyond the fundamental limits of industry-standard solder-based interconnection technologies, with direct Cu-Cu bonding now aggressively pursued as the next interconnection node. While traditional Sn-based solders can only sustain current densities up to 10^4 A/cm^2 and relatively low operating temperatures to prevent massive creep, they have remained the power-horse of micro-systems assembly due to their ease of processability. Melting of solders during the bonding process gives desirably high compliance and tolerance to non-coplanarities and warpage, self-alignment capability as well as high diffusivity rates for fast metallic joint formation. [2]

The "holy grail" of interconnection technologies, (solderless) all-Cu interconnections, has been implemented in Si-to-Si bonding – die-to-die, die-to-wafer, 3D-ICs – but faces its own set of limitations in terms of scaling *up* the technology for chip-to-substrate applications. Bulk Cu offers excellent electrical and thermal properties but, with a ~120 GPa elastic modulus, is much stiffer than solders and will not melt during assembly, giving little tolerance to non-coplanarities. Thus, it usually relies on expensive chemical-mechanical planarization (CMP) processes to eliminate non-coplanarities and ensure that atomic-level contact can be achieved during assembly. Direct Cu-Cu bonding typically requires bonding temperatures and pressures exceeding 250°C and 200 MPa, respectively [3], and still has poor self-diffusivity in the bulk so enhancing reactivity of the bonding interfaces via surface activation, surface metallization, or a reducing atmosphere is necessary to ensure formation of strong metallurgical joints at reasonable bonding temperatures and pressures. [4]

Efforts have been made toward enabling solder-free interconnections using nanoscale pastes or inks typically made of silver or copper, that can be patterned via screen printing [5] or dipping [6] processes, then sintered to form densified joints with bulk-like properties. The use of highly reactive nanomaterial systems is attractive because they offer similar compliance to solders and allow for a significant

978-1-7281-1500-9/19 $31.00 © 2019 IEEE

reduction in bonding temperatures and pressures due to their high surface area. Though they can achieve excellent thermal and electrical conductivities as well as high-temperature stability, these paste and ink solutions still face challenges in pitch scalability as viscous phases, along with reliability concerns due to residual porosity, filler entrapment, organics corrosion, and shrinkage-induced interconnection geometries that aggravate stress concentration. [7,8]

The 3D Systems Packaging Research Center (PRC) at Georgia Tech is pioneering a novel interconnection technology – the Cu pillar with np-Cu caps – that comprehensively addresses the aforementioned challenges to bridge this technology gap by replacing the traditional solder caps with np-Cu caps [9]. Being a solid-state nanoscale system, these np-Cu structures can be used without the risk of bridging, and have been demonstrated to be sinterable in reducing atmospheres at temperatures well below their bulk melting points [10]. Moreover, their low modulus as-synthesized (<20 GPa) enables significant deformation at minimal bonding pressures to accommodate system-level non-coplanarities during assembly, thus providing solder-like compliance during assembly [11]. Furthermore, because they are highly deformable and highly reactive, no surface modifications like CMP or noble metal passivation are required to enable formation of a strong Cu-Cu metallurgical bond between the bonding interfaces.

Np-Cu can be fabricated using industry-standard semi-additive processes and can be patterned across a wide range of feature sizes. Previous work from PRC has focused on the development and optimization of the synthesis of np-Cu from the selective dealloying of co-deposited Cu-Zn alloys, as well as understanding the role of Cu-Zn plating conditions and bath chemistry on the morphology and composition of the as-plated alloy and resulting nanoporous structures [12].

This paper goes beyond by demonstrating the first successful implementation of this technology on a daisy-chain test vehicle at 100 μm pitch, introducing a wafer-scale bumping process, shown in Fig. 1, and first assembly demonstration. The paper discusses the patterning of Cu-Zn caps in fine features, the dealloying of these alloy caps to fabricate np-Cu caps, as well as highlights some fabrication challenges encountered in process development, particularly

related to seed layer etching. Finally, a generic process for the fabrication of dies with np-Cu capped Cu pillars is presented, along with a first assembly demonstration and characterization of the all-Cu joints formed with this technology.

II.   WAFER BUMPING PROCESS: CU PILLARS WITH NANOPOROUS COPPER CAPS

There are several techniques used to synthesize nanoporous metal films, the most common being chemical dealloying. In this technique, the precursor is an alloy system with two or more components, and the more reactive component/s is/are selectively etched out while the most noble component self-assembles into a 3D network of interconnected, self-supported ligaments [13]. Synthesis of the initial alloy can be accomplished via arc or furnace melting, co-sputtering, or co-electrodeposition. Co-electrodeposition was chosen as the best method to form the patterned features desired for its ability to fabricate films with variable thicknesses and uniformly control the composition and microstructure of alloy across the complete thickness. Also, plating is already a well-established industry practice for semi-additive build-up processes [14], thereby making the industry adoption of the technique simpler. Synthesis of np-Cu from Cu-Zn alloys was previously demonstrated [12] and this system was selected as front-up approach due to the low cost of Zn and well-established plating chemistries and processing for brass, although synthesis of np-Cu requires Zn-rich alloys (>65 atomic weight% Zn) as precursors.

A. Test Vehicle Fabrication Process

Schematics of the daisy-chain test vehicle design used in this study are provided in Fig. 5. While the proposed interconnection technology is inherently scalable to pitches below 10 μm, this first demonstration was carried out at a coarser pitch to better understand fundamentals of the fabrication and assembly processes. The prototype test vehicle is a 5x5 mm die with 760 I/Os distributed in 3 peripheral rows at 100 μm pitch and center area array at 250 μm pitch. The bumps are 30 μm in diameter.

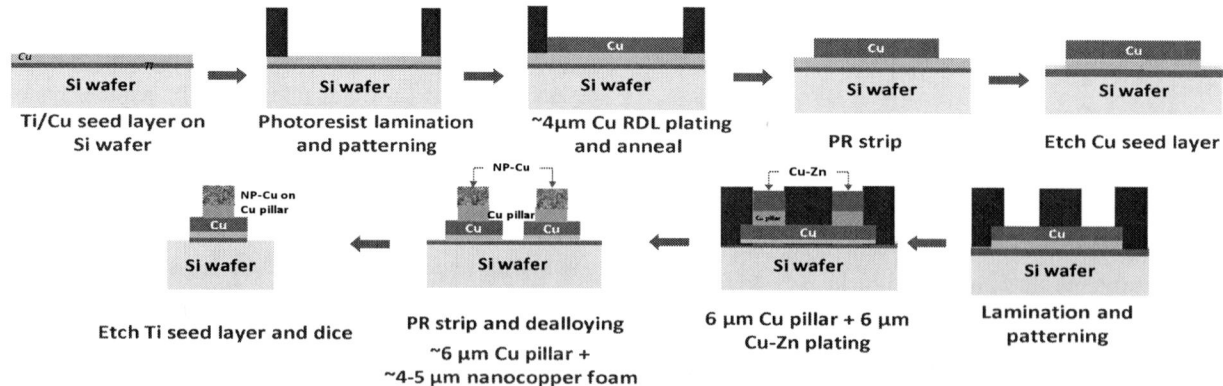

Figure 1. General process flow for the fabrication of NP-Cu caps on Cu pillar from dealloyed Cu-Zn alloy.

The optimized process flow for the fabrication of the dies is shown in Fig. 1. Silicon wafers with SiO_2 were first sputtered with 50/150 nm thick Ti/Cu seed layers. Liquid plating photoresist provided by JSR Corporation was then spin-coated, exposed, developed and hard baked at 120°C to form the dog-bone shaped RDL wiring pattern. A plasma descum process was carried out prior to plating 4-6 μm of Cu. Subsequently, the photoresist was stripped, and the Cu seed layer was etched using citric acid-based Copper etchant 49-1 sourced from Transene Co, leaving only the Ti seed layer as conductive layer for the bumping process. After etching of the Cu seed layer, the wafers were again spin-coated, exposed, developed and baked to form the bumping patterns on top of the RDL layer. After another plasma descum process, Cu was then plated into the patterns to form Cu pillars 6-10 μm in height. The Cu posts were then capped with 8-10 μm of Cu-Zn through a co-deposition process. The bath chemistry and plating parameters were optimized such that the alloy has a starting composition with greater than 65% Zn content. The photoresist was then stripped, and the wafer was dealloyed in 0.75-1.5 wt % HCl solution for 4-10 hours to selectively etch Zn and form the np-Cu caps on top of the Cu pillars. Finally, the Ti seed layer was etched using diluted HF, with minimal effect on the np-Cu structure, after which the coupons were diced for assembly.

The process flow is compared to that of wafer bumping with Cu pillars and solder caps in Fig. 2, to highlight similarities in processing, with the reflow step substituted for a dealloying step as the sole difference, highlighting its compatibility with industry's best practices. Key individual process steps are detailed in the next sub-section.

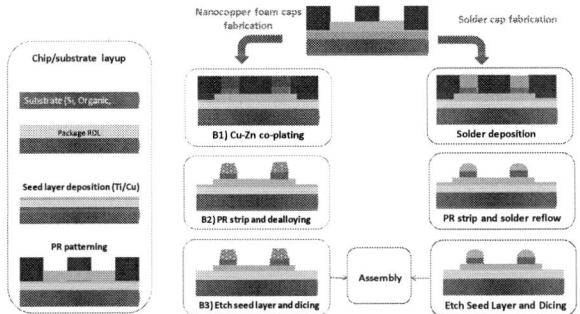

Figure 2. Process flow comparison between fabrication of NP-Cu caps and traditional Cu pillars with solder caps.

B. Patterning of np-Cu by semi-additive processing

1) Co-electrodeposition of Cu-Zn caps

Cu-Zn plating was carried out in a custom-made plating setup with Zn sheets as the counter electrodes. The electrolytic bath was composed of 0.35 M potassium pyrophosphate (PP), 0.15 M zinc sulfate heptahydrate, and 0.0025 M copper sulfate pentahydrate. This chemistry is strongly basic; when made, the solution has a pH of ~9.5. Previous attempts to plate in large area coupons and patterned features resulted in partial delamination of dry film and liquid plating resists which in turn gave poor control over plating as well as plating under the photoresist as illustrated in Fig. 3a. It is known that most photoresists are unstable in highly basic medium and a possible reason for delamination of photoresists during Cu-Zn plating could be the localized increase in pH inside the patterned features. Sulfuric acid was added to the bath chemistry until its pH was within the range of 7-8, virtually eliminating underplating and delamination, allowing for better control of plating parameters. Skip plating and edge effects were additional challenges that were encountered, likely because of wettability issues, that were mitigated by running a plasma descum. Successful patterning on 3x3 mm coupons after optimization of bath conditions can be seen in Fig. 3b.

Figure 3. (a) Sample that experienced severe delamination and uncontrolled underplating. (b) Successful plating into 3x3 mm square coupons.

Previous work has shown that better nanoporous structures are dealloyed from alloys containing greater than 65% Zinc [12]. As such, samples were plated under a constant potential. The plating potential can be varied to control the plated alloy composition.

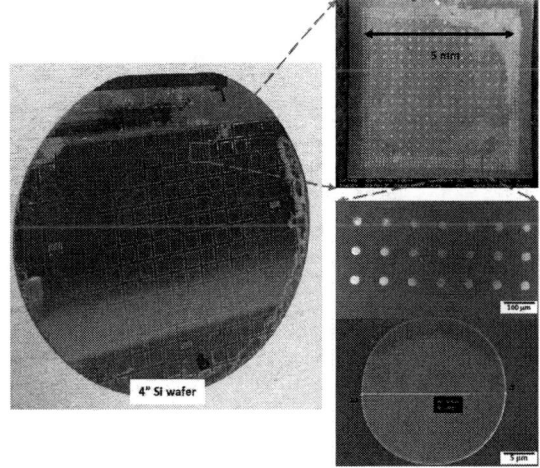

Figure 4. (a) Fully patterned 4" wafer, some underplating observed; (b) SEM images reveal smooth Cu-Zn plating in all features.

Higher negative potentials favor deposition of more Zinc, giving more Zn rich alloys. Higher plating rates are also obtained at more negative potentials. However, it comes at the expense of plating uniformity and possible hydrogen evolution, detrimentally impacting the plating quality.

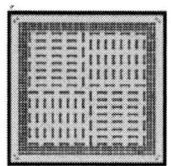 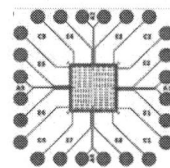

Figure 5. Test vehicle design. Left: Die design;
Right: Substrate design.

After successfully patterning on small pieces, work was done to scale up the plating process to full 4" wafers with the 30 µm-diameter bump patterns. Plating was carried out at a constant potential of -1.65V. Photoresist delamination was encountered during plating along the edges of the wafer but did not negatively affect patterning in the majority of the wafers. Cu-Zn was plated in all the exposed holes, with uniform composition (~65% Zn) across the whole wafer. About 4-6 µm thick Cu-Zn caps were plated in 90 minutes over 3-4 µm of Cu pillar. A fully patterned 4" wafer can be seen in Fig. 4, as well as SEM images of the plated Cu-Zn pillars.

2) Chemical dealloying of plated Cu-Zn structures

After the Cu-Zn plating, the next step is to chemically dealloy the Cu-Zn caps to form the np-Cu caps on Cu pillar. Samples were taken from the 4" wafer and dipped into diluted hydrochloric acid, ranging from 0.75 wt% to 5 wt%. Dealloying times ranged from 20 minutes in the strongest acid concentration, to 10 hours in the most diluted concentration. It was found that there is a very strong relationship between the quality of the as plated Cu-Zn features and the resulting dealloyed structure and its composition. SEM and EDXS were used to characterize the morphology of the nanoporous structures and determine the compositions of structures before and after dealloying.

Surface characterization of the nanoporous features can be seen in the topographical measurements shown in Fig. 6, taken using a 3D confocal microscope. In samples where the Cu content exceeded 30% the plated structures were very smooth (1.5 µm tolerance) and showed little excess edge plating. After dealloying, the resulting np-Cu caps also showed a very smooth surface profile and nanoporous morphology.

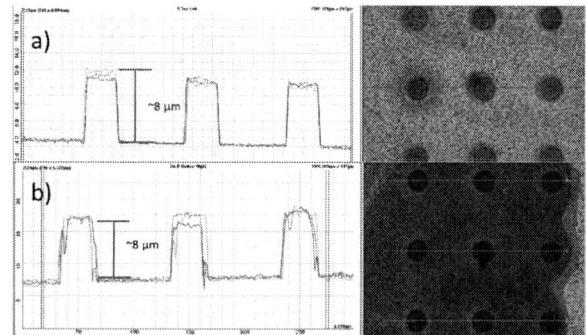

Figure 6. (a) As plated, smooth Cu-Zn caps on Cu pillar; (b) Dealloyed np-Cu caps on Cu pillar. Uniform plating carries into np-Cu cap uniformity.

However, these samples also contained residual Zinc in excess of 6-7% after 120 minutes of dealloying. In samples where the Cu content was below 30% the plated structures were rougher and less well defined, and "islands" of excess Cu-Zn were overplated on the edges of the bump. This roughness was also evident after dealloying, as the np-Cu caps were also rough and less uniform. Within 30 minutes of dealloying the residual Zn was below 2%, suggesting that higher initial Zn concentrations enable better etching of Zn in the np-Cu structures. Allowing for more dealloying time to etch out additional Zn came at the expense of the np-Cu structure coarsening and increased np-Cu ligament size, which is expected to have deleterious effects on sintering and densification. Experiments to determine the effect of ligament morphology and of residual Zn on the interconnection performance of this system requires further exploration. However, it has been shown that minor addition of Zn (<0.6%) improves the electromigration stability of Cu in SnAg solder-based systems [15].

3) Seed layer etching

All previous experiments and trials were completed using plain Cu substrates, or on Si wafers with a Ti/Cu seed layer. It was observed that, due to their higher reactivity, any Cu-Zn or np-Cu already on the wafer would be etched away before the Cu seed layer was completely removed. Initial attempts to etch the Cu seed layer using standard Piranha or sulfuric acid solutions showed that the etchant would preferentially attack the plated alloy or the more reactive NP-Cu caps. Milder etchants such as Transene Cu Etch 49-1, typically chosen for its compatibility with plated solder, also preferentially attacked the Cu-Zn alloy and np-Cu caps. Additionally, the HF that is normally used to etch the Ti seed layer preferentially attacks the Zn in the alloy structures, but was not reactive to Cu. Essentially, the Cu seed layer would have to be etched before Cu-Zn plating, and the Ti seed layer would have to be etched after the Cu-Zn had been dealloyed. The use of an alternative seed layer metallization, such as Au, can also be considered, but it was not feasible to implement this solution in this study due to lab safety concerns with its cyanide based etchant chemistry.

Plating of Cu-Zn after etching the Cu seed layer, suing Tu as the conducting layer was successfully demonstrated and can be seen from results in Fig. 7. The process flow described in Fig. 1 was used to fabricate a daisy-chain test vehicle, with np-Cu foam caps on 6-8 µm-thick RDL. Incidentally, this also solved the problem of underplating and photoresist delamination, as the Cu-Zn did not plate onto the underlying Ti seed layer. After the photoresist was stripped, the Cu-Zn

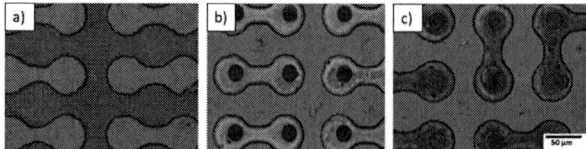

Figure 7. (a) Cu pads after Cu seed layer has been etched; (b) Cu-Zn plated onto all pads using Ti as conductive layer; (c) NP-Cu caps after dealloying.

structures were dealloyed to produce np-Cu caps on the wiring patterns. The Ti seed layer was also etched using HF without any significant damage to the np-Cu caps. Further optimization is required to better control feature uniformity.

III. ASSEMBLY DEMONSTRATION USING PATTERNED NANOPOROUS COPPER CAPS

After dies and substrates fabrication, the Cu pillars with np-Cu caps bumps were assembled onto Si substrates with blanket Cu metallization using a thermocompression bonding process. A limited number of samples were used for this preliminary demonstration. The Cu metallizations on substrate side were purposefully roughened to $R_a \sim 1.07$, an order of magnitude higher than the roughness of production DBC substrates ($R_a \sim 0.1$), to highlight the compliance given by the solid-state np-Cu caps in assembly. Both substrates and dies were dipped in 1wt% HCl for 15 seconds before bonding to remove the native Cu oxides present on the samples. Assembly was carried out using a Finetech Matrix Fineplacer flip-chip bonder at 300°C for 30 min in 2% forming gas atmosphere. A bonding pressure of 10-20 MPa was used for assembly. These bonding parameters are very conservative to ensure successful assembly demonstration but are now being optimized to lower both bonding temperatures and pressures.

The SEM images of Fig. 8 demonstrate that good contact was made across the row of bumps, with metallurgical bonding achieved between the np-Cu on the die side and the roughened substrate metallization to form a seamless interface. Through the bonding process, the np-Cu caps sintered and densified to achieve bulk-like density. EDXs analysis of the assembly cross-section is also reported in Fig. 8. The element map showed less than 0.2% residual Zinc in the bonded interface, indicating very good control over the residual Zn.

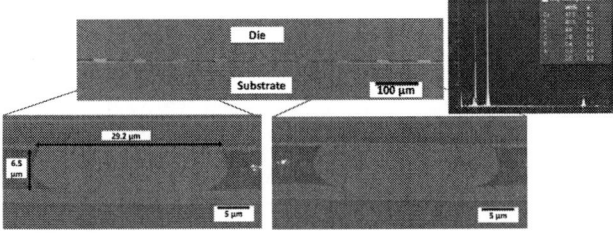

Figure 8. Cross-section of fully densified np-Cu joint after thermocompression bonding; contact was made on all bumps and a seamless interface is observed. EDX mapping indicates less than 0.2% residual Zinc.

Shear testing was carried out with a XYZec Condor Sigma Bond tester at 15 µm/s, and an average die shear value of ~22 MPa over 3 bonded samples was achieved. As seen in Fig. 9, densified np-Cu was observed on both fracture interfaces after shear, indicating that failure occurred within the sintered nanoporous layer and that there was sufficient deformation of the nanoporous structure to make contact between the np-Cu

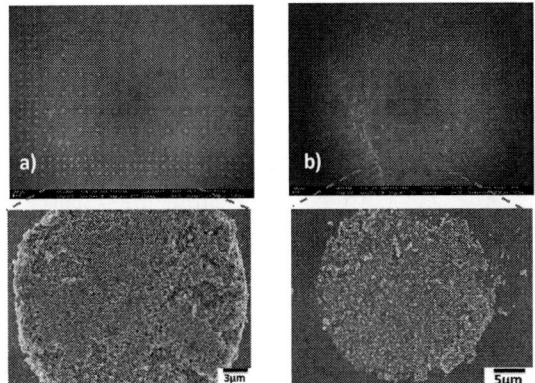

Figure 9. (a) Die side failure interface; (b) Substrate side failure interface. These are 30 µm diameter bumps.

and Cu metallization to establish a metallurgical bond at the bonding interface itself.

IV. CONCLUSIONS AND FUTURE WORK

This paper presented the first demonstration of wafer-level fabrication and assembly using the novel Cu pillars with np-Cu caps technology to enable more manufacturable all-Cu interconnections. Patterning of Cu-Zn alloys using a co-electrodeposition process, with and without the use of a conductive Cu seed layer, was shown in 30 µm diameter features. Successful chemical dealloying was also carried out to yield uniform np-Cu caps with good control of surface uniformity, nanoporous structure morphology and composition, and limited residual Zn. Complete removal of seed layers for electrical isolation was accomplished without damaging the np-Cu caps, laying the ground work for characterization of functional devices.

The bumped test dies were assembled onto Si substrates with Cu metallization using thermocompression bonding under conservative bonding temperatures and pressures in reducing atmosphere. SEM and EDXS analysis of bump cross-sections revealed densified joints, with outstanding accommodation of surface roughness and non-coplanarities. Metallurgical bonding was achieved between all bumps and bulk Cu throughout the die area. Shear testing showed that failure occurred within the densified np-Cu and not at the bonding interface as often seen in sintered joints, suggesting interdiffusion at the interface.

Future work will focus on understanding how the nanoporous structure morphology and dealloying conditions affect assembly performance, as well as enabling batch processing of parts at lower temperatures and pressures. Assembly trials will continue to determine electrical properties of these sintered joints, including resistance measurements and electromigration performance. Thermal ageing, thermal cycling, and thermal shock tests are also being planned to examine thermomechanical performance of this novel interconnection technology.

ACKNOWLEDGMENT

This study was supported by the Interconnections and Assembly (I&A) industry program at Georgia Tech PRC and the Semiconductor Research Corporation (SRC, Task #2661). The authors are grateful to the industry sponsors, especially Texas Instruments and JSR Corporation, for their funding and services. The authors would like to all the SRC liaisons, as well as our colleagues at PRC for their technical guidance.

REFERENCES

[1] P. Totta, "History of Flip Chip and Area Array Technology," in Area Array Interconnection Handbook, K. Puttlitz and P. Totta, Eds., ed: Springer US, 2001, pp. 1-35.

[2] Li Ming, Y. and Q. Wang Chun. Solder joints design attribute to no solder bridge for fine pitch device. in Fifth International Conference on Electronic Packaging Technology Proceedings, 2003. ICEPT2003. 2003.

[3] Tang, Y.-S., Y.-J. Chang, and K.-N. Chen, Wafer-level Cu– Cu bonding technology. Microelectronics Reliability, 2012. 52(2): p. 312-320.

[4] Po-Hao, C., et al., Enhanced Cu-to-Cu direct bonding by controlling surface physical properties. Japanese Journal of Applied Physics, 2017. 56(3): p. 035503.

[5] S. Fu, Y. Mei, G.-Q. Lu, X. Li, G. Chen, and X. Chen, "Pressureless sintering of nanosilver paste at low temperature to join large area (1 100 mm2) power chips for electronic packaging," Materials Letters, vol. 128, pp. 42-45, 8/1/ 2014.

[6] J. Zurcher, K. Yu, G. Schlottig, M. Baum, M. M. Visser Taklo, B. Wunderle, et al., "Nanoparticle assembly and sintering towards allcopper flip chip interconnects," in Electronic Components and Technology Conference (ECTC) , 2015 IEEE 65th, 2015, pp. 1115-1121

[7] Dai, Y., et al., Enhanced copper micro/nano-particle mixed paste sintered at low temperature for 3D interconnects. Applied Physics Letters, 2016. 108(26): p. 263103.

[8] L. D. Carro, J. Zurcher, U. Drechsler, I. Clark, G. Ramos, and T. Brunschwiler, "Low-temperature dip-based all-copper interconnects formed by pressure-assisted sintering of copper nanoparticles," IEEE Transactions on Components, Packaging and Manufacturing Technology, pp. 1–1, Jan. 2019.

[9] N. Shahane et al., "Novel High-Temperature, High-Power Handling All-Cu Interconnections through Low-Temperature Sintering of Nanocopper Foams," 2016 IEEE 66th Electronic Components and Technology Conference (ECTC), Las Vegas, NV, 2016, pp. 829-836. doi: 10.1109/ECTC.2016.352

[10] Shahane, N., et al. Enabling Chip-to-Substrate All-Cu Interconnections: Design of Engineered Bonding Interfaces for Improved Manufacturability and Low-Temperature Bonding. in 2017 IEEE 67th Electronic Components and Technology Conference (ECTC). 2017.

[11] K. Mohan, N. Shahane, P. M. Raj, A. Antoniou, V. Smet and R. Tummala, "Low-temperature, organics-free sintering of nanoporous copper for reliable, high-temperature and high-power die-attach interconnections," 2017 IEEE Applied Power Electronics Conference and Exposition (APEC), Tampa, FL, 2017, pp. 3083-3090. doi: 10.1109/APEC.2017.7931137

[12] K. Mohan et al., "Demonstration of Patternable All-Cu Compliant Interconnections with Enhanced Manufacturability in Chip-to-Substrate Applications," 2018 IEEE 68th Electronic Components and Technology Conference (ECTC), San Diego, CA, 2018, pp. 301-307. doi: 10.1109/ECTC.2018.00053

[13] Ferrando, R., J. Jellinek, and R.L. Johnston, Nanoalloys: from theory to applications of alloy clusters and nanoparticles. Chemical reviews, 2008. 108(3): p. 845-910

[14] K. Mohan, N. Shahane, R. Liu, V. Smet, and A. Antoniou, "A Review of Nanoporous Metals in Interconnects," Jom, vol. 70, no. 10, pp. 2192–2204, 2018.

[15] M. Lu, D.-Y. Shih, S. K. Kang, C. Goldsmith, and P. Flaitz, "Effect of Zn doping on SnAg solder microstructure and electromigration stability," Journal of Applied Physics, vol. 106, no. 5, p. 053509, Sep. 2009.

Cu-Cu bonding by low-temperature sintering of self-healable Cu nanoparticles

Junjie Li, Qi Liang, Chen Chen, Tielin Shi, Guanglan Liao, Zirong Tang

State Key Laboratory of Digital Manufacturing Equipment and Technology
Huazhong University of Science & Technology
Wuhan, China
e-mail: junjieli@hust.edu.cn

Abstract—The Cu-Cu bonding temperature by using Cu nanoparticles is mainly influenced by the size and the purity of Cu nanoparticles. To remove the oxides of Cu, reducing atmosphere is always introduced into the sintering and bonding process. In this paper, a new Cu-Cu bonding method by sintering of self-healable Cu nanoparticles was proposed. With this method, the surface oxidation layer of Cu nanoparticle can be removed without reducing atmosphere at sintering and bonding process. In order to research the self-healing properties of the surface oxidized Cu nanoparticles, the sintering and bonding experiments were carried out under an Ar atmosphere. With self-healable Cu nanoparticles, the electrical resistivity of sintered Cu film can be reduced to lower than 5 $\mu\Omega\cdot$cm after sintering, and a high shear strength Cu-Cu joint over 25 MPa can be achieved after bonding at 250 °C. The oxygen content was also significantly reduced during the sintering and bonding process, which reflected the excellent self-healing property of Cu nanoparticle paste. The high Cu-Cu bonding strength and no requirement for reducing atmosphere indicate that the proposed self-healable Cu nanoparticle paste is promising to be wildly used in advanced electronics packaging.

Keywords-Cu-Cu bonding; Cu nanoparticles; oxidation; self-healable; sintering

I. INTRODUCTION

For the past years, Sn based solders are always used as interconnection materials due to their low costs and low process requirements [1]–[3]. However, the disadvantages of low melting point, whisker growth and strict electromigration have limited the further application of Sn in high-density or high-power devices [4]–[10]. Therefore, high performance interconnection material is becoming an urgent need for packaging industry. Few years ago, Ag was researched as a Cu-Cu bonding media by many groups, because Ag has high electrical and thermal conductivity, and excellent antioxidation property [11]–[16]. However, the high cost and severe electromigration problems of Ag make it impossible to be widely used in packaging industry.

In recent years, all Cu interconnection technology was emerged as a new solution for advanced packaging. Without IMC formation, the Cu-Cu interconnect shows better electrical and thermal conductivity, better anti-electromigration property, higher power density, higher mechanical strength and higher thermal reliability. Nevertheless, the high melting point and oxidizable characteristics of Cu make the all Cu interconnection technology difficult to match the conditions of current packaging process [17]–[20]. In our previous work, a Cu-Cu bonding method by Cu nanoparticles was proposed, which can significantly reduce the bonding temperature to 300 °C [17], [21]. High purity Cu nanoparticles and reducing atmosphere were required in the bonding process to ensure the diffusivity of Cu atoms and the integrity of Cu-Cu joints [22]–[24]. However, the reducing atmosphere cannot fully contact the nanoparticles at the bonding interface, to precisely control the synthesis process for collecting high purity Cu nanoparticles and to introduce the reducing atmosphere for antioxidation will also enhance the costs, which will limit the large-scale use of Cu nanoparticles as high-performance interconnection materials.

Self-reduction of Cu MOD (Metal Organic Deposition) inks was proposed as a new concept in the research field of printed electronics. With this method, the Cu metal film can be formed just by heating the MOD inks at a proper temperature, without any reduction atmosphere [25]–[29]. From this concept, we have proposed a Cu-Cu bonding method by using a new kind of self-healable Cu nanoparticle paste. With this kind of paste, the surface oxidation of Cu nanoparticles can be reduced without reducing atmosphere at bonding temperature, which can lead the formation of pure all Cu joints.

In this paper, the self-healable Cu nanoparticle paste was fabricated. The sintering and Cu-Cu bonding performance by the paste were investigated at 250 °C under Ar atmosphere. With less oxides on the surface of Cu nanoparticles, the diffusivity of Cu atoms was enhanced, leading an excellent interconnection of Cu at low bonding temperature. No strict requirements of reducing atmosphere and high purity Cu nanoparticles make the proposed self-healable Cu nanoparticle paste have a good application prospect.

II. EXPERIMENTAL METHOD

A. Preparation of self-healable Cu nanoparticle paste

In this paper, all chemical reagents and materials were purchased and used without further purification. The self-healable Cu nanoparticle paste was prepared by Cu nanoparticles, formic acid, MIPA (iso-propanol amine), butanol and methanol. Cu nanoparticles were purchased from Nanjing XFNANO Materials Tech Co., China, and all other reagents were from Sinopharm Chemical Reagent Co. LTD, China.

The self-healable Cu nanoparticle paste was fabricated by three steps. Firstly, the Cu nanoparticles were immersed in formic acid and dispersed by ultrasonic for 2 min, then the excessive amounts of formic acid were removed by evaporation at 40 °C in a vacuum oven. Secondly, the organic solution was prepared by mixing the MIPA, butanol and methanol under magnetic stirring for 10 min. Thirdly, the self-healable Cu nanoparticle paste was then prepared by mixing the formic acid treated Cu nanoparticles and the organic solution by ultrasonic dispersion. In this paper, the proportion of Cu nanoparticles in the self-healable paste is 40 wt.%.

For comparison, the Cu nanoparticle paste without self-healing property had also been prepared. This type of paste was fabricated by just mixing the purchased Cu nanoparticles and organic solutions prepared by butanol and methanol.

B. Sintering and bonding experiments

The sintering and bonding experiments were designed to research the performance of self-healable Cu nanoparticle paste. In order to investigate the electrical performance of the sintered Cu film, the self-healable Cu nanoparticle paste and the non-treated Cu nanoparticle paste were coated on non-conductive glass slides separately. Then the coated glass slides were heated at 250 °C for 30 min, under an Ar atmosphere. After sintering, the sintered Cu films on glass slides were used for resistivity measurement and morphology observation.

The Cu-Cu bonding sample was prepared by an interlayer of Cu paste and two sides of Cu substrates, which were fabricated by sputtering Ti and Cu on Si substrates. To simplify the measurement process of Cu-Cu bonding joint, two sides of Cu substrates were cut into two sizes of 5×5 mm^2 and 10×10 mm^2. The bonding temperature, heating time and bonding environment were the same as sintering process, the bonding pressure was chosen to be 0.6 MPa.

C. Characterizations

The morphological features of Cu nanoparticles, the microstructures of sintered Cu films and the Cu-Cu bonding interface were observed by SEM (Scanning Electron Microscope). The oxygen content in Cu nanoparticles and sintered films were characterized by EDX. The resistivity of the sintered Cu film was measured and determined by 4-Point Probes Resistivity Measurement System. The shear strength of Cu-Cu joint was measured by Micro Materials Testing Platform.

III. RESULTS AND DISCUSSION

The sintering and bonding properties by normal and self-healable Cu nanoparticle paste were characterized for comparison, the electrical and mechanical enhancement brought by self-healing of surface oxidized Cu nanoparticles will be discussed in this section.

Figure 1. The SEM of purchased Cu nanoparticles

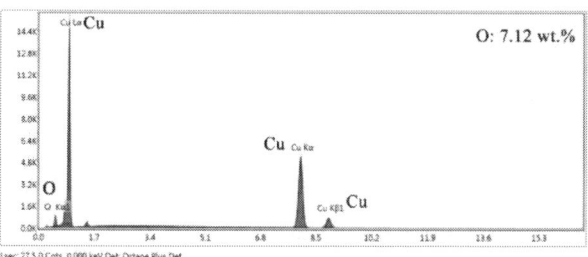

Figure 2. The measured chemical element content of purchased Cu nanoparticles by EDX

A. Original Cu nanoparticles

In our previous work, different sizes of Cu nanoparticles had been synthesized and used for Cu-Cu bonding [17], [23], [30]. From our research results and other relative research papers, we can conclude that decreasing the size of Cu nanoparticles is critical for lower the Cu-Cu bonding temperature, due to the size effect. However, it is difficult to synthesize small size Cu nanoparticles with high productivity, high purity and high dispersion property, which hinders the wide application of Cu nanoparticles in Cu-Cu bonding. To enhance the application value, the easily oxidized commercial Cu nanoparticles need to be used.

Fig. 1 shows the SEM image of commercial purchased Cu nanoparticles. The size of these nanoparticles are around 80 nm, without obvious agglomeration. From the EDX spectrum in Fig. 2, we can observe that the oxygen content is measured to be 7.12 wt.%, which represents the slightly oxidation of the commercial purchased Cu nanoparticles.

The characterization of commercial Cu nanoparticles indicates that the antioxidation property of these particles are not excellent, which may influence the diffusivity of Cu atoms while sintering and bonding.

978-1-7281-1500-9/19 $31.00 © 2019 IEEE

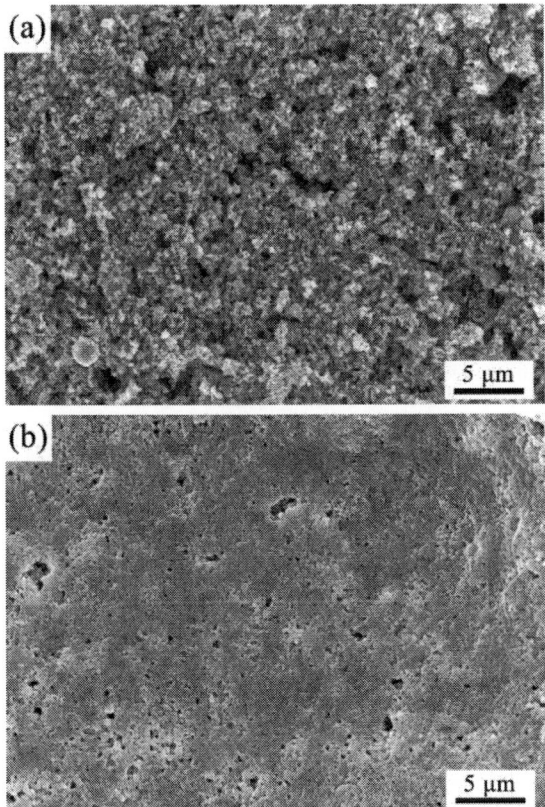

Figure 3. Low magnification SEM images of the 30 min sintered (a) normal Cu nanoparticle paste and (b) self-healable Cu nanoparticle paste.

B. Sintering performance of self-healable Cu nanoparticle paste

To investigate the electrical performance enhancement of the self-healing proposal, the normal Cu nanoparticle paste and the self-healable Cu nanoparticle paste were sintered for 30 min, under an Ar atmosphere.

Fig. 3 shows the low-magnification SEM images of sintered Cu films by two different types of Cu nanoparticle paste. From Fig. 3(a), it can be observed that the morphological features of sintered Cu nanoparticles are almost the same as the original commercial Cu nanoparticles. Adjacent nanoparticles are not connected to each other, and numerous voids can be observed in the sintered film. The features of the sintered self-healable Cu nanoparticle paste are totally different. As observed in Fig. 3(b), the SEM image of the sintered Cu nanoparticles represents a high density morphology. Cu nanoparticles were effectively connected to form an integrated Cu film, and the porosity of the sintered Cu film decreased significantly.

Fig. 4 shows the high-magnification SEM images of sintered Cu films by two different Cu nanoparticle pastes. From Fig. 4(a), we can observe that the size and shape of every single Cu nanoparticle have no obvious changes, compared with those original Cu nanoparticles presented in Fig. 1. Details of the sintered morphology of normal Cu nanoparticle paste indicate that this sintering process did

not lead to effective diffusion at the interface between Cu nanoparticles. From Fig. 4(b), distinct morphological changes of Cu nanoparticles can be observed. No Cu nanoparticle can exist independently after sintering. Sufficient Cu atom diffusion was achieved at the interface of every two adjacent Cu nanoparticles, and a lot of nanoparticles were merged and grown into a larger cluster.

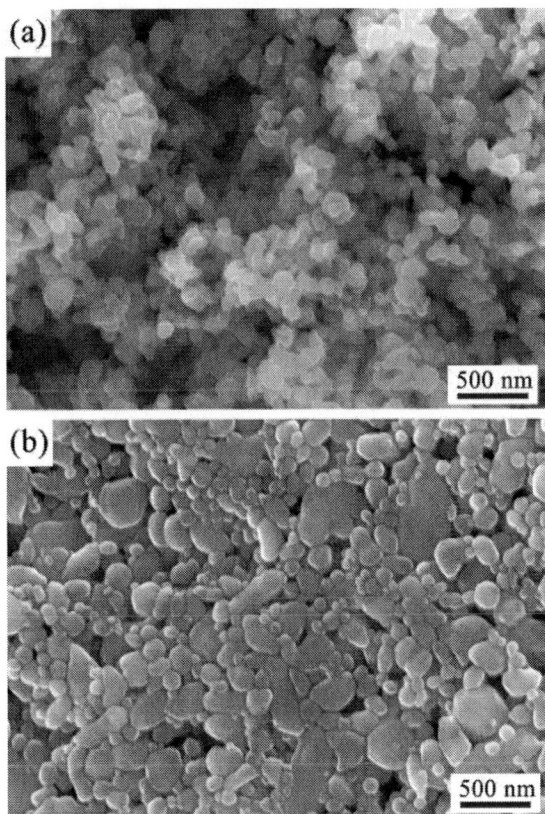

Figure 4. High magnification SEM images of the 30 min sintered (a) normal Cu nanoparticle paste and (b) self-healable Cu nanoparticle paste.

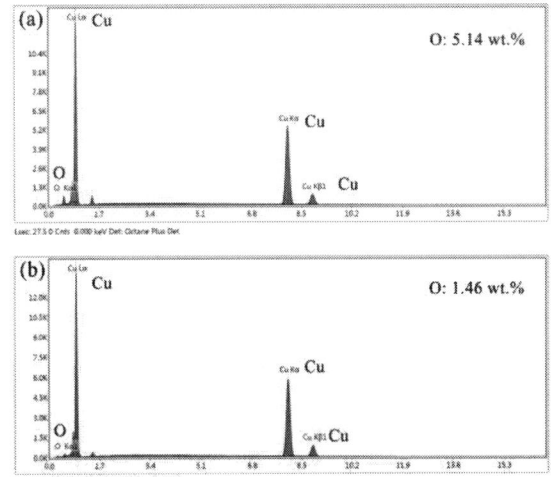

Figure 5. EDX spectrum of the 30 min sintered (a) normal Cu nanoparticle paste and (b) self-healable Cu nanoparticle paste.

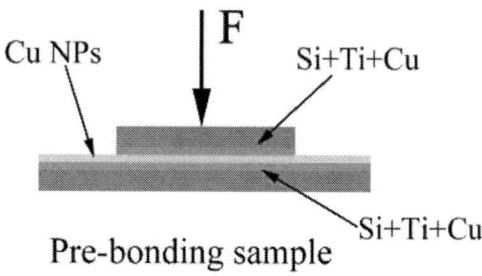

Figure 6. The schematic diagram of Cu-Cu bonding by using Cu nanoparticle paste.

Fig. 5 illustrates the measured content changes of Cu and O after sintering of two different Cu nanoparticle pastes. As shown in Fig. 5(a), the O content in the sintered non-treated Cu nanoparticle paste is measured to be 5.14 wt.%, which is a little lower than the measured value of 7.12 wt.% in original purchased Cu nanoparticles. The slight change of O content after sintering at 250 °C may induced by the movement of Cu atoms during the sintering process. Under the sintering temperature, a few numbers of Cu atoms inside Cu nanoparticles may have diffused and penetrated the oxidation layer to the nanoparticle surface, leaded the slight decrease of oxygen. However, a prominent decreasing of O content achieved by self-healable Cu nanoparticle paste after sintering. As illustrated in Fig. 5(b), the O content was decreased to 1.46 wt.%, which reflected the excellent self-healing property brought by the pre-treatment of formic acid and MIPA.

From the SEM images and the EDX measured O contents of sintered Cu films, we can conclude that the oxides in Cu nanoparticles have seriously influenced the diffusivity of Cu atoms. After sintering, the electrical performance of sintered Cu films had also been tested. Without any treatment, the measured electrical resistivity of sintered normal Cu nanoparticle paste was higher than 300 $\mu\Omega\cdot$cm, which showed a weak electrical conductivity. However, by using the fabricated self-healable Cu nanoparticle paste, the sintering performance had been enhanced significantly. The resistivity of the sintered self-healable Cu nanoparticle paste was measured to be lower than 5 $\mu\Omega\cdot$cm, only 2 or 3 times higher than bulk Cu. Therefore, the oxides in Cu nanoparticles not only influenced the sintering morphological features, but also the sintering electrical performance. To reduce the oxides is meaningful for using Cu as an interconnection material.

C. Cu-Cu bonding performance

The Cu-Cu bonding performance by using Cu nanoparticle paste was also investigated. As illustrated in Fig. 6, the Cu-Cu pre-bonding sample was prepared by three parts, two bonding substrates with different sizes and an interlayer of Cu nanoparticle paste. The pre-bonding samples were heating at 250 °C for 30 min, under the atmosphere of Ar and a bonding pressure of 0.6 MPa to achieve the Cu-Cu bonding. Then the microstructures and the shear strength of bonded samples were characterized.

Figure 7. The Cu-Cu bonded sample by self-healable Cu nanoparticle paste.

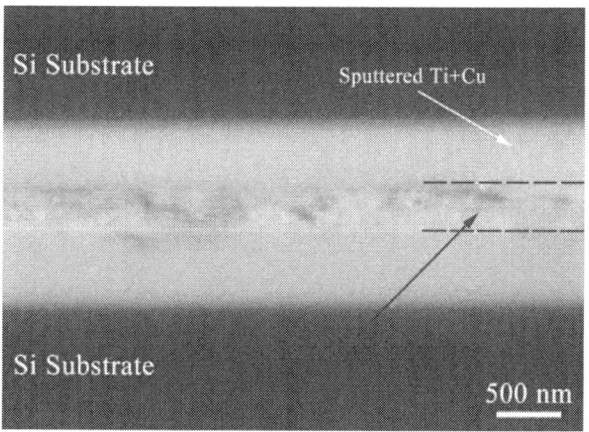

Figure 8. The SEM image of Cu-Cu bonding interface.

Fig. 7 shows the photo of a bonded Cu-Cu sample by using the self-healable Cu nanoparticle paste as an interconnection media. After bonding, the partially oxidized Cu nanoparticle paste were reduced, two Cu surfaces of bonding substrates and the sintered Cu nanoparticle paste fused into an integrated steady joint. However, using the non-treated normal Cu nanoparticle paste cannot achieve such a reliable Cu-Cu joint. With the normal Cu nanoparticle paste, the Cu-Cu bonding substrates can be easily separated under a weak force, and the sintered Cu nanoparticle paste showed a powdery morphology with insufficient diffusion.

Fig. 8 shows the cross-sectional SEM image of Cu-Cu bonding interface by using the self-healable Cu nanoparticle paste. Under a bonding pressure, the sintered Cu nanoparticles were compressed to a compact thin layer. The interface between the sintered Cu nanoparticle paste and the Cu substrate disappeared, which indicates that Cu atoms in the Cu nanoparticle paste layer have sufficiently diffused into the sputtered Cu film on bonding substrate, leading the formation of a steady interconnection structure. Therefore, by using the self-healing Cu nanoparticle paste, pure Cu-Cu joints with almost no Cu oxides can be formed.

Figure 9. The fracture morphology of Cu-Cu joint after shear strength measurement.

After bonding, the shear strength of Cu-Cu bonded joints were investigated by Micro Materials Testing Platform. As described above, the measured bonding strength of the Cu-Cu joint by using the non-treated Cu nanoparticle paste can only achieve a low value less than 2 MPa, which is too weak for practical use.

By self-healing of Cu nanoparticle paste, the Cu-Cu bonding can achieve a high shear strength. As shown in Fig. 9, the Cu-Cu bonded sample was fractured after shear strength test, and the fracture occurred at bonding substrates. However, the Cu-Cu interconnection joint is still sturdy even the fracture happens, which proves the sufficient diffusion between Cu nanoparticle paste layer and Cu substrates. The highest tested value of shear strength was already over 25 MPa before the Cu-Cu sample breaks, indicating that the Cu-Cu bonding method by sintering the self-healable Cu nanoparticle paste is reliable.

From the experiments and characterizations, it can be concluded that sintering and bonding properties of self-healable Cu nanoparticle pastes are much better than that of non-treated normal Cu nanoparticle paste. Therefore, reduction of Cu oxides is a key point to enhance the diffusion of Cu atoms at low sintering temperature, which is valuable for achieving a steady Cu-Cu interconnect.

IV. CONCLUSION

In summary, a new low-temperature Cu-Cu bonding method by sintering of self-healable Cu nanoparticle paste was proposed. The self-healable Cu nanoparticle paste can be fabricated by commercial purchased Cu nanoparticles, formic acid, reduction reagents and some other organic solvents. The sintering and bonding properties of self-healable Cu nanoparticle paste and non-treated Cu nanoparticle paste were investigated at 250 °C, under Ar atmosphere. By using self-healable Cu nanoparticle pastes, the Cu oxides on the surface of commercial purchased Cu nanoparticles can be efficiently reduced at sintering and bonding process. Compared with non-treated Cu nanoparticle pastes, a low Cu film electrical resistivity below 5 μΩ·cm and a high shear strength over 25 MPa can be also achieved after sintering and bonding by self-healable Cu nanoparticle pastes, showing significant advantages. With the proposed self-healable Cu nanoparticle paste, high purity Cu nanoparticles and reduction atmosphere are no longer necessary, which greatly reduces the technological requirements and costs of sintering and bonding, representing a strong application prospect.

ACKNOWLEDGMENT

This work is supported by National Natural Science Foundation of China (Grant No. 51805197), Initiative Postdocs Supporting Program of China (Grant No. BX20180103), National Basic Research Program of China (Grant No. 2015CB057205) and China Postdoctoral Science Foundation (Grant No. 2018M632836).

I also would like to thank Prof. Tielin Shi, Guanglan Liao and Zirong Tang for their supporting, and thank Qi Liang and Chen Chen for their help in all experiments.

REFERENCES

[1] K. L. Lin, E. Y. Chang, and L. C. Shih, "Evaluation of Cu-bumps with lead-free solders for flip-chip package applications," *Microelectron. Eng.*, vol. 86, no. 12, pp. 2392–2395, 2009.

[2] M. R. Lueck, J. D. Reed, C. W. Gregory, A. Huffman, J. M. Lannon, and D. S. Temple, "High-density large-area-array interconnects formed by low-temperature cu/sn-cu bonding for three-dimensional integrated circuits," *IEEE Trans. Electron Devices*, vol. 59, no. 7, pp. 1941–1947, 2012.

[3] J. Wang, Q. Wang, D. Wang, and J. Cai, "Study on Ar(5%H2) Plasma Pretreatment for Cu/Sn/Cu Solid-State-Diffusion Bonding in 3D Interconnection," *Proc. - Electron. Components Technol. Conf.*, vol. 2016–Augus, pp. 1765–1771, 2016.

[4] C. Y. Liu, J. T. Chen, Y. C. Chuang, L. Ke, and S. J. Wang, "Electromigration-induced Kirkendall voids at the Cu/Cu3Sn interface in flip-chip Cu/Sn/Cu joints," *Appl. Phys. Lett.*, vol. 90, no. 11, pp. 1–4, 2007.

[5] K. Zeng and K. N. Tu, "Six cases of reliability study of Pb-free solder joints in electronic packaging technology," *Mater. Sci. Eng. R Reports*, vol. 38, no. 2, pp. 55–105, 2002.

[6] Y. Yamada *et al.*, "Reliability of wire-bonding and solder joint for high temperature operation of power semiconductor device," *Microelectron. Reliab.*, vol. 47, no. 12, pp. 2147–2151, 2007.

[7] K. N. Tu, H. Y. Hsiao, and C. Chen, "Transition from flip chip solder joint to 3D IC microbump: Its effect on microstructure anisotropy," *Microelectron. Reliab.*, vol. 53, no. 1, pp. 2–6, 2013.

[8] K. Suganuma *et al.*, "Sn whisker growth during thermal cycling," *Acta Mater.*, vol. 59, no. 19, pp. 7255–7267, 2011.

[9] X. Liu, S. He, and H. Nishikawa, "Thermally stable Cu3Sn/Cu composite joint for high-temperature power device," *Scr. Mater.*, vol. 110, pp. 101–104, 2016.

[10] W. Zhang, P. Limaye, Y. Civale, R. Labie, and P. Soussan, "Fine pitch Cu/Sn solid state diffusion bonding for making high yield bump interconnections and its application in 3D integration," *Electron. Syst. Integr. Technol. Conf. ESTC 2010 - Proc.*, pp. 1–4, 2010.

[11] J. Yan *et al.*, "Effect of PVP on the low temperature bonding process using polyol prepared Ag nanoparticle paste for electronic packaging application," *J. Phys. Conf. Ser.*, vol. 379, no. 1, 2012.

[12] P. Peng, A. Hu, B. Zhao, A. P. Gerlich, and Y. N. Zhou, "Reinforcement of Ag nanoparticle paste with nanowires for low temperature pressureless bonding," *J. Mater. Sci.*, vol. 47, no. 19, pp. 6801–6811, 2012.

[13] J. Yan *et al.*, "Sintering Bonding Process with Ag Nanoparticle Paste and Joint Properties in High Temperature Environment," *J. Nanomater.*, vol. 2016, pp. 1–8, 2016.

[14] D. Wakuda, K. S. Kim, and K. Suganuma, "Ag nanoparticle paste synthesis for room temperature bonding," *IEEE Trans. Components Packag. Technol.*, vol. 33, no. 2, pp. 437–442, 2010.

[15] J. Yan *et al.*, "Pressureless bonding process using Ag nanoparticle paste for flexible electronics packaging," *Scr. Mater.*, vol. 66, no. 8, pp. 582–585, 2012.

[16] P. Peng, A. Hu, A. P. Gerlich, G. Zou, L. Liu, and Y. N. Zhou, "Joining of Silver Nanomaterials at Low Temperatures: Processes, Properties, and Applications," *ACS Appl. Mater. Interfaces*, vol. 7, no. 23, pp. 12597–12618, 2015.

[17] J. Li *et al.*, "Low-Temperature and Low-Pressure Cu–Cu Bonding by Highly Sinterable Cu Nanoparticle Paste," *Nanoscale Res. Lett.*, vol. 12, no. 1, pp. 0–5, 2017.

[18] Z. Wu, J. Cai, Q. Wang, and J. Wang, "Low temperature Cu-Cu bonding using copper nanoparticles fabricated by high pressure PVD," *AIP Adv.*, vol. 7, no. 3, pp. 1–7, 2017.

[19] Y. T. Yang *et al.*, "Low-Temperature Cu-Cu Direct Bonding Using Pillar-Concave Structure in Advanced 3-D Heterogeneous Integration," *IEEE Trans. Components, Packag. Manuf. Technol.*, vol. 7, no. 9, pp. 1560–1566, 2017.

[20] A. K. Panigrahy and K.-N. Chen, "Low Temperature Cu–Cu Bonding Technology in Three-Dimensional Integration: An Extensive Review," *J. Electron. Packag.*, vol. 140, no. 1, p. 010801, 2018.

[21] J. Li *et al.*, "Low-Temperature and Low-Pressure Cu-Cu Bonding by Pure Cu Nanosolder Paste for Wafer-Level Packaging," *Proc. - Electron. Components Technol. Conf.*, pp. 976–981, 2017.

[22] J. Liu, H. Chen, H. Ji, and M. Li, "Highly Conductive Cu-Cu Joint Formation by Low-Temperature Sintering of Formic Acid-Treated Cu Nanoparticles," *ACS Appl. Mater. Interfaces*, vol. 8, no. 48, pp. 33289–33298, 2016.

[23] J. J. Li *et al.*, "Surface effect induced Cu-Cu bonding by Cu nanosolder paste," *Mater. Lett.*, vol. 184, pp. 193–196, 2016.

[24] J. Li *et al.*, "Depressing of Cu-Cu bonding temperature by composting Cu nanoparticle paste with Ag nanoparticles," *J. Alloys Compd.*, vol. 709, pp. 700–707, 2017.

[25] D. H. Shin *et al.*, "A self-reducible and alcohol-soluble copper-based metal-organic decomposition ink for printed electronics," *ACS Appl. Mater. Interfaces*, vol. 6, no. 5, pp. 3312–3319, 2014.

[26] Y. Farraj, M. Grouchko, and S. Magdassi, "Self-reduction of a copper complex MOD ink for inkjet printing conductive patterns on plastics," *Chem. Commun.*, vol. 51, no. 9, pp. 1587–1590, 2015.

[27] W. Li, S. Cong, J. Jiu, S. Nagao, and K. Suganuma, "Self-reducible copper inks composed of copper-amino complexes and preset submicron copper seeds for thick conductive patterns on a flexible substrate," *J. Mater. Chem. C*, vol. 4, no. 37, pp. 8802–8809, 2016.

[28] S. Cho, Z. Yin, Y. K. Ahn, Y. Piao, J. Yoo, and Y. S. Kim, "Self-reducible copper ion complex ink for air sinterable conductive electrodes," *J. Mater. Chem. C*, vol. 4, no. 45, pp. 10740–10746, 2016.

[29] Y. Kawaguchi, Y. Hotta, and H. Kawasaki, "Cu-based composite inks of a self-reductive Cu complex with Cu flakes for the production of conductive Cu films on cellulose paper," *Mater. Chem. Phys.*, vol. 197, pp. 87–93, 2017.

[30] J. Li *et al.*, "Design of Cu nanoaggregates composed of ultra-small Cu nanoparticles for Cu-Cu thermocompression bonding," *J. Alloys Compd.*, vol. 772, pp. 793–800, 2019.

978-1-7281-1500-9/19 $31.00 © 2019 IEEE

Electrical Performance Limits of Fine Pitch Interconnects for Heterogeneous Integration

Ahmet C. Durgun, Zhiguo Qian, Kemal Aygun,
Ravi Mahajan
Assembly and Test Technology Development
Intel Corporation
Chandler, AZ, USA
ahmet.c.durgun@intel.com

Tim Tri Hoang, Sergey Yuryevich Shumarayev
Programmable Solutions Group
Intel Corporation
San Jose, CA, USA
tim.tri.hoang@intel.com

Abstract—**Heterogeneous integration facilitates faster design cycles with optimal functional IP module and silicon node combinations, but requires ultra-high bandwidth for the die-to-die communications. Fine pitch interconnects can meet such high bandwidth demands with simpler circuits, lower power and less latency. Hence, it is of utmost importance to understand the performance of these interconnects at different speeds and channel lengths. This paper focuses on a parametric study over the basic design parameters of a generic fine pitch interconnect, to explore the electrical performance limits. As a result of this study, practical guidelines are provided for the die-to-die channel design.**

Keywords-heterogenous integration; embedded multi-die intrconnect bridge (EMIB); silicon interposer; very fine pitch interconnect

I. INTRODUCTION

In the era of artificial intelligence and big data, the number of devices connected to the network and the data generated by these devices increase very rapidly, resulting in a significant boost in bandwidth demand between the devices. To cope with this demand and sustain a healthy communication and computing network, the system capacity and speeds need to scale commensurately. It is not cost-effective to achieve this with monolithic system-on-chip (SoC) solutions, because of longer design cycles, coupled with the fact that different transistor characteristics of digital and analog devices lead to higher design, prototyping and process costs. One of the faster and lower cost solutions to meet the scaling demand is to rely on heterogeneous integration [1] utilizing very fine pitch interconnect technologies such as silicon interposers [2], silicon bridges [3,4], or fine pitch organic interposers [5,6]. This approach not only addresses the bandwidth scaling problem but can also facilitate faster delivery of the new silicon nodes to the market via modular design architectures, where smaller chiplets can be stitched together using fine pitch interconnects [7]. However, this is not a straightforward task because of the variety of silicon chiplets and I/O interfaces. It is of utmost importance to standardize the die-to-die interfaces that can work on a wide-range of interconnect solutions and enable faster design cycle for fully functional

modular systems. For this purpose, DARPA has initiated a government-industry-academia "Common Heterogeneous Integration and IP Reuse Strategies (CHIPS)" research program, which focuses on the development of an interface and a framework which is compatible with all commercially available IP blocks with no restrictions [8]. As a part of this project, we have studied and report here the electrical performance limits of fine pitch interconnects for die-to-die communication at speeds up to 10 Gbps.

Similar studies have been reported in the literature showing the impact of line width and line spacing on insertion loss and eye width margins [9], providing design guidelines for glass and silicon interposers [10], and comparing the energy efficiency and bandwidth of different communication protocols [11]. However, the relation between the design parameters and performance limits, in terms of maximum channel length and bandwidth, has not been previously addressed in detail. In most of the previous studies, the parameter sweeps and contrasting was performed for specific cases only, keeping either the channel length or the data rate as constant.

This paper builds on the prior work in terms of the range of the parameters studied and includes a thorough discussion on design trade-offs. The electrical performance of fine pitch interconnects is examined using RLGC parameter extraction and voltage transfer function simulations. The impact of multiple sets of design parameters, including trace width, trace spacing, metal and dielectric thickness, and signal to ground ratio on the eye width margin are reported. In light of how these parameters are related to each other, we provide general design guidelines and compliance curves that will help in the design and standardization of these fine pitch interconnects. Additionally, the impact of scaling on other performance metrics such as channel power and latency are also addressed. Finally, a set of recommendations on driver requirements, such as die capacitance and driver strength, for more efficient bandwidth scaling are provided.

The rest of the paper is organized as follows. Section II introduces the interconnect and channel models together with the performance metric definitions. The results of the parametric study are summarized in Section III, which is followed by a discussion on practical design guidelines.

978-1-7281-1500-9/19 $31.00 © 2019 IEEE

II. PROBLEM DEFINITION

A. Interconnect Model

In this analysis, we utilized a generic 4-layer interconnect model, where first and third layers from the top are assigned as signal (S) layers, and the rest are ground (G) layers. This layer assignment is referred to as SGSG. Fig. 1 shows the cross section of the interconnect for different signal to ground ratios (SGRs). The numbers after each letter shows the number of signal or ground traces for each repeating block. For instance, S2G1 refers to an SGR of 2. The ground layers are shown by the green color, whereas the cyan colored boxes represent the signal traces. Between the signal lines, there are green colored guard traces, which are shorted to the ground. Finally, the red parts show the dielectric regions, with a dielectric constant (Dk) of 4.0. The models have 10 signal traces per layer, which are numbered as shown in Fig. 1. The signals on the top layer suffer higher crosstalk since there is not any ground plane above them. Hence, the 15th trace is selected as the worst case victim line, which is outlined by the blue box. For the sake of generality of this study, these models only include the horizontal routing section of the interconnect.

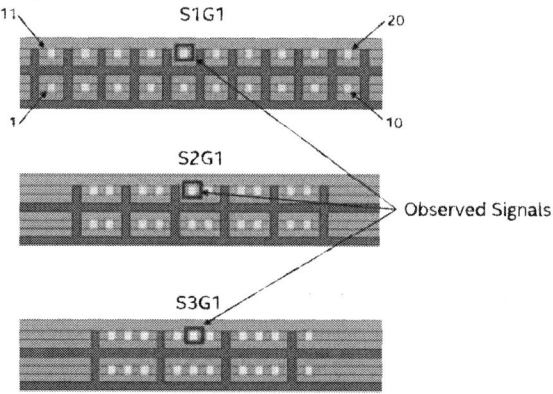

Figure 1.   Interconnect models with different signal to ground ratios.

B. Channel Model

A short range die-to-die channel is typically unterminated to reduce the power consumption. Therefore, our channel model does not include any termination resistance. The channel is composed of a linear driver with an output resistance (R_{on}), the die capacitances on the transmitter ($C_{die\ TX}$) and receiver ($C_{die\ RX}$) sides, and the interconnect, as illustrated in Fig. 2(a). For the initial simulations, R_{on} and C_{die} were assumed to be 60 Ω and 0.5 pF, respectively. They usually need to be lower to support higher data rate and longer channel reach. To investigate the impact of the circuit parameters on the electrical performance, these values were later included in the parametric sweep. The maximum voltage swing (V_{swing}) of the source was set to 1.0 V.

C. Performance Metrics

The conventional performance metric of a channel, in frequency domain, is the S-parameter matrix. By definition, to compute the S-parameters, the channel is assumed to be terminated by matched loads. However, the short die-to-die channel is usually unterminated. Hence, we defined a new set of performance metrics based on the channel configuration. Since we would like to measure the performance of the interconnect only, we defined the new metrics to remove the impact of the resistive and capacitive loading on the interconnect. Fig. 2(b) illustrates the channel configuration when the interconnect is removed, where the transmitter is loaded by C_{die} only. The voltage on the die capacitance is defined as $V_{TX\ loaded}$. Similarly, when the interconnect is included, as shown in Fig. 2(a), the voltage on the die capacitance is denoted as V_{out}. Thus, the voltage transfer functions for these two cases can be defined as V_{out}/V_{in} and $V_{TX\ loaded}/V_{in}$, respectively, in dB scale, and the channel loss (L) is expressed as in (1).

$$L = 20log_{10}(V_{out}/V_{in}) - 20log_{10}(V_{TX\ loaded}/V_{in}) \quad (1)$$

A similar approach was followed for the timing impact of the interconnect, which was defined as the difference between the eye width (EW) of the full channel at the output node and the EW of the die capacitance loaded transmitter at the transmitter load, as given in (2). The EW is measured at ±0.1V_{cc} (200 mV$_{pp}$) vertical eye opening.

$$Timing\ Impact = EW_{out} - EW_{TX\ loaded} \quad (2)$$

In this paper, the timing budget assumed for the interconnect is limited to 0.1 unit interval (UI).

To assess crosstalk impact, far-end crosstalk (FEXT) was defined as the power sum of the ratios of the output signal voltage on the victim line to the input voltage of the aggressor lines in dB scale, for the full channel. We did not follow a difference approach for calculation of FEXT since crosstalk occurs through the interconnect only.

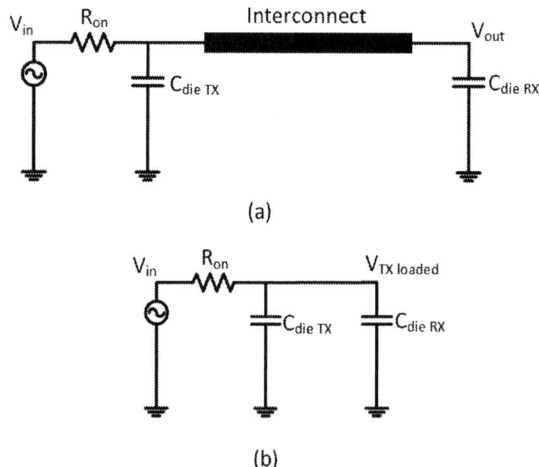

Figure 2.   Schematic drawings of the channel models: (a) Driver loaded with the interconnect and the receiver (b) Driver loaded with the receiver.

978-1-7281-1500-9/19 $31.00 © 2019 IEEE

III. Parametric Study

The electrical performance of a die-to-die interface depends on many design parameters, which can be separated into two groups: (a) Interconnect and (b) circuit parameters. This section provides an extensive discussion on the impact of each parameter on the maximum data rate and channel length. For this study, we first attempted to find the optimum design configuration that satisfies the specifications of the CHIPS program. Later, we expanded the analysis to cover more futuristic design targets.

A. Interconnect Parameters

1) Routing Density and Design Rules: One of the key performance metrics of a die-to-die interface is the bandwidth density, which is determined by the operating data rate of the driver and the routing density of the interconnect. Thus, the routing density is a significant boundary condition for the interconnect design process. In the CHIPS program, the target bandwidth density was selected as 1 Tbps/mm, which can be achieved by many different combinations of data rate and routing density. Some of these combinations, which were also utilized in this parametric study, are summarized in Table I.

Once the routing density is determined, the optimum trace width (tw) and trace spacing (ts) set can be selected, based on SGR. Fig. 3 depicts the timing impact of a 3 mm-long S3G1 channel at different data rates. Note that for each data rate, the routing density, and trace to trace pitch, was kept contant as the trace width varied. The results showed that, for all data rates, the timing impact is smaller when trace width is small and trace spacing is large. This is mainly because of the crosstalk and the trace capacitance. Therefore, in this study, the interconnects were designed with maximum trace spacing, as shown in Table I. At 10 Gbps, the eye is completely closed for traces wider than 3 µm. That is why these points were not included in the plot.

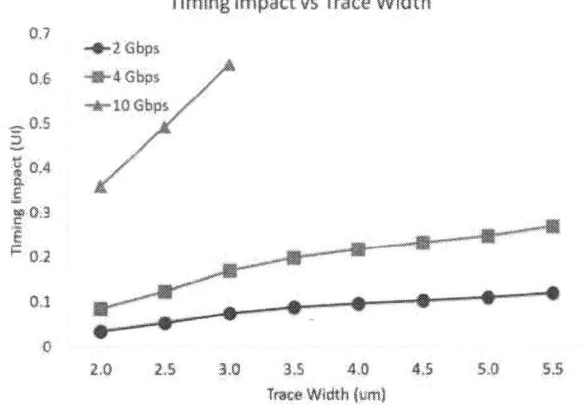

Figure 3. Timing impact of a 3 mm-long S3G1 channel, as a function of trace width, for different data rates. Routing density was kept constant at 100 IO/mm as the trace width varied.

TABLE I. Design Rules for 1 Tbps Bandwidth Density

Data Rate (Gbps)	Routing Density (IO/mm/layer)	Trace Width/ Trace Spacing (µm)		
		S1G1	*S2G1*	*S3G1*
1	1000	0.25/0.25	0.25/0.41	0.25/0.50
2	500	0.50/0.50	0.50/0.82	0.50/1.00
4	250	1.00/1.00	1.00/1.64	1.00/2.00
8	125	2.00/2.00	2.00/3.28	2.00/4.00
10	100	2.00/3.00	2.00/4.60	2.00/5.50

2) Stack-up: Other important design parameters of an interconnect are the metal and dielectric thicknesses, which are mainly determined by manufacturing capabilities and/or height requirements. Therefore, performance optimization within these boundaries is of great interest. In the literature, metal and dielectric thicknesses of typical die-to-die interconnects are on the order of several microns. Our analysis showed that for relatively lower data rates and shorter channel lengths, the optimum metal thickness is between 1.0 µm and 1.5 µm, which also depends on SGR. However, as we go to higher data rates and/or longer channels, the line resistance starts to dominate the performance. Hence, thicker metal layers give better performance. Although it is not easy to generalize for all cases, as a rule of thumb, one can say that for data rates higher than 4 Gbps and/or channels longer than 3 mm, a metal thickness of 2 µm gives better results.

At the same time, thicker dielectric layers always have a better performance because of the lower line capacitance. Therefore, for the rest of this study, the metal and dielectric thicknesses were both selected as 2 µm.

After the physical rules and stack-up optimization, the performance of different design options were examined for the target bandwidth density of 1 Tbps/mm. Fig. 4 illustrates the maximum channel length, that is the length at which the timing impact is equal to 0.1 UI, at different data rates (also routing densities) and for different SGRs. It can be seen that the S1G1 design has the longest reach because of better isolation between the signal traces. Moreover, the largest channel length was obtained for the low data rate/high routing density combination. Having said this, higher density interconnects may not be the easiest to manufacture, especially for the cases where the trace height-to-width aspect ratios are relatively larger.

Since channel loss and/or crosstalk increase with pitch reduction, the maximum channel length and data rate that can be supported by the interconnect decrease with pitch. This results in a maximum channel length vs. bandwidth trade-off, as shown in Fig. 5. In this figure, we compared only the highest two routing density designs to observe the electrical performance limits of the fine pitch interconnects. Similar results can be obtained for other design options. Consistent with previous results, S1G1 case gives the longest reach for all speeds and almost for all cases and the

978-1-7281-1500-9/19 $31.00 © 2019 IEEE

interconnect with the routing density of 1000 IO/mm has a better performance. However, at 1 Tbps/mm bandwidth density, the interconnect with the routing density of 500 IO/mm has a better performance for S2G1 and S3G1 cases. Furthermore, beyond 4 Tbps/mm, the maximum channel length drops below 0.5 mm, when R_{on} and C_{die} are 60 Ω and 0.5 pF, respectively.

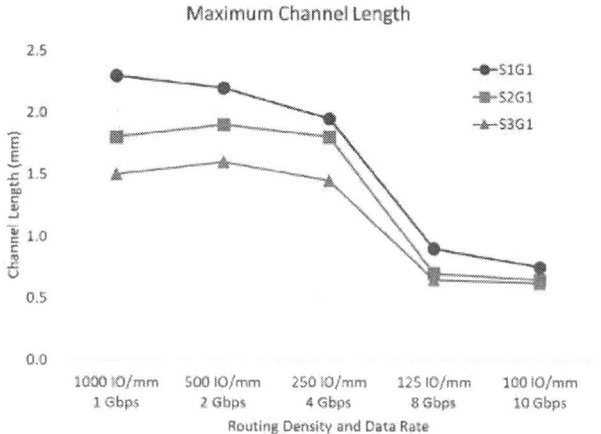

Figure 4. Maximum channel length as a function of data rate and SGR. The bandwidth density is kept constant at 1 Tbps/mm.

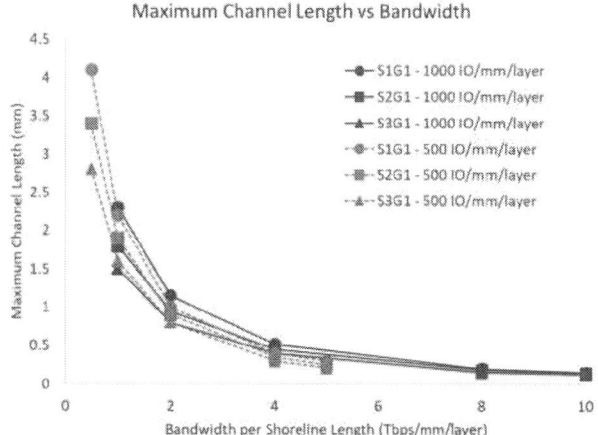

Figure 5. Trade-off between the maximum channel length and bandwidth density.

3) Material: Because of the fine cross-sectional dimensions and short lengths, these die-to-die interconnects generally operate in the RC regime. Therefore, the trace capacitance, and consequently the Dk of the dielectric material, plays a significant role in the overall performance of the interface, including the channel reach, power efficiency, and latency. This makes the dielectric material selection a key design knob to enable larger bandwidth densities and longer channel lengths.

To analyze the impact of Dk on the channel reach, we focused on designs that can support bandwidth densities of 2 Tbps/mm and 4 Tbps/mm. Also, only the results for S1G1

are shown here, as they have the best performance. Fig. 6 shows how the maximum channel length changes with Dk. The solid and dashed lines represent the 2 Tbps/mm and 4 Tbps/mm designs, respectively. As expected, the maximum channel length increases as Dk decreases. For 2 Tbps/mm, there is approximately 0.1 mm increase in channel length per every 0.5 decrease in Dk. For 4 Tbps/mm, the length improvement reduces to 0.05 mm.

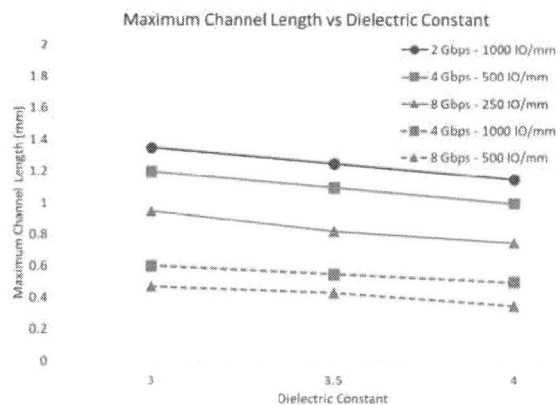

Figure 6. Maximum channel length as a function of dielectric constant, R_{on} = 60 Ω and C_{die} = 0.5 pF.

In addition to the maximum channel length and data rate, two other performance metrics for fine pitch interconnects are power efficiency and latency. It is known that the interconnect power consumption is proportional to the line capacitance and data rate [4]. Likewise, the latency is proportional to the RC time constant. Hence, routing density, design rules, metal and dielectric layer thicknesses, SGR, and material selection have a direct impact on these performance metrics. As expected, lower routing density and dielectric constant results in a lower line capacitance. When scaled with the operating frequency and the number of lines, we observe that, for the same bandwidth density, the lowest power consumption occurs when we have a low routing density interconnect operating at higher data rates. Also, for the same routing density, capacitance decreases as SGR increases because of larger trace spacing. Table II shows the normalized interconnect power consumption of several design options for the same bandwidth density. According to the table, the highest power consumption occurs when the routing density is 1000 IO/mm, SGR is 1 and Dk is 4.0. Interconnect power consumption can be reduced by 80% if an S3G1 250 IO/mm design is utilized with a dielectric constant of 3.0.

On the other hand, line resistance depends on the trace cross section only and increases with the routing density. Therefore, low routing density, low Dk, and large SGR should give the lowest time constant and latency. Table III gives the normalized interconnect latency values of different design options. As expected, the largest time constant is obtained when the routing density is 1000 IO/mm, SGR is 1 and Dk is 4.0. The latency can be improved by 95% by selecting the right design combinations.

978-1-7281-1500-9/19 $31.00 © 2019 IEEE

TABLE II. NORMALIZED INTERCONNECT POWER CONSUMPTION FOR SAME BANDWIDTH DENSITY

Dielectric Constant	SGR	Power Consumption		
		250 IO/mm	500 IO/mm	1000 IO/mm
Dk = 3.0	S1G1	29%	44%	75%
	S2G1	21%	30%	48%
	S3G1	20%	26%	41%
Dk = 3.5	S1G1	34%	51%	87%
	S2G1	25%	34%	56%
	S3G1	23%	30%	48%
Dk = 4.0	S1G1	39%	59%	100%
	S2G1	29%	39%	64%
	S3G1	26%	34%	54%

TABLE III. NORMALIZED INTERCONNECT LATENCY FOR SAME BANDWIDTH DENSITY

Dielectric Constant	SGR	Latency		
		250 IO/mm	500 IO/mm	1000 IO/mm
Dk = 3.0	S1G1	7%	22%	75%
	S2G1	5%	15%	48%
	S3G1	5%	13%	41%
Dk = 3.5	S1G1	8%	26%	87%
	S2G1	6%	17%	56%
	S3G1	6%	15%	48%
Dk = 4.0	S1G1	10%	20%	100%
	S2G1	7%	20%	64%
	S3G1	7%	17%	54%

Overall, we can say that for a typical driver, lower SGR and higher routing density are preferred for bandwidth density, while higher SGR and lower routing density are preferred for power and latency. Moreover, lower Dk is always preferred. All of these results demonstrate that there is not one global optimum point that gives the best performance in terms of the data rate, power efficiency and latency. For each specific application and set of boundary conditions, a separate design process has to be performed.

B. Circuit Parameters

1) Driver Strength: Besides the interconnect related design parameters, the electrical performance of a die-to-die interface can also be improved by utilizing better drivers, which corresponds to smaller R_{on} and C_{die}. Fig. 7 depicts the impact of circuit parameters on the channel reach for different S1G1 interconnect designs that can support bandwidth densities of 2 Tbps/mm and 4 Tbps/mm. As observed earlier, for relatively weaker drivers, low data rate/high routing density combination exhibits a longer channel reach. However, for stronger drivers, the behavior is exactly the opposite. The maximum channel length of the 250 IO/mm design is significantly higher than the rest of the designs when R_{on} is 30 Ω and C_{die} is 0.3 pF. Also, by comparing the slopes of the lines, one can conclude that the

benefit we get from stronger drivers is more significant at higher data rates.

Further improvement is possible when stronger drivers are used with low-Dk materials. For instance, the maximum channel length of a 2 Tbps/mm interconnect at 8 Gbps data rate can be increased up to 3 mm, when Dk is equal to 3.0. Also, it can be observed that the impact of material change is more noticeable with stronger drivers. For instance, for the 2 Tbps/mm interconnect, every 0.5 decrease in Dk increases the channel length by 0.3 mm. Thus, from manufacturing cost point of view, it makes more sense to improve the driver strength before changing the material set.

2) Receiver Sensitivity: Another important circuit parameter is the receiver sensitivity, which is the height of the eye mask. Because of the timing impact definition we adopted in this study, all of the previous results were obtained for a receiver sensitivity of 200 mV_{pp}. When the receiver sensitivity is improved to 100 mV_{pp}, 0.2 mm to 0.4 mm increase of channel length is possible. Combining the improvements of Dk, driver and receiver together, the maximum channel length of the 250 IO/mm interconnect can go up to 3.4 mm at 8 Gbps, as shown in Fig. 7. Similarly, an interconnect with a routing density of 500 IO/mm can reach up to 1 mm at 10 Gbps.

Analogous results can also be obtained for other with different SGR and/or layer assignments, such GSS or GSSS. Yet, because of the inferior signal to signal isolation, the maximum channel length is expected to be shorter for these cases.

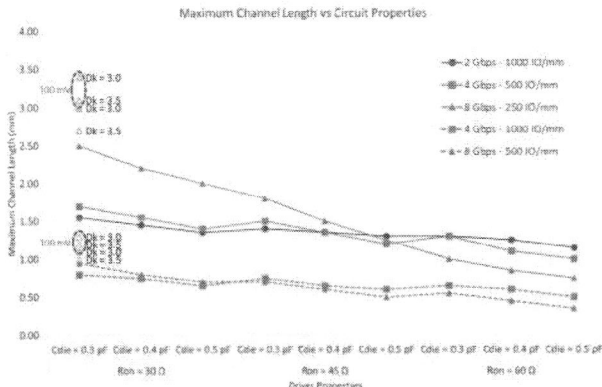

Figure 7. Maximum channel length of 2 Tbps/mm and 4 Tbps/mm designs as a function of circuit parameters. The encircled operating points show the maximum channel length of the interconnects when all of the circuit and material improvements are combined.

IV. DESIGN GUIDELINES

The parametric study of the previous section shows the impact of each design parameter on the timing impact or the channel reach, in detail. However, a quantification of the impact of each parameter on the timing impact is still missing. Obviously this is not an easy task given the number of parameters and the nonlinear nature of the system. To find the operating margins of an interconnect, the designers are required to run computationally expensive time domain simulations for each specific case. Thus, it would be useful to come up with some general design guidelines that can enable a faster design process. In general, in a communication channel, the eye closure occurs as a result of loss and crosstalk. Most often, the direct relation between these frequency domain metrics and the timing impact is unknown. But, if an empirical relation between these metrics can be established, some pass/fail decision can be made without running time domain simulations. This may also help with the standardization of die-to-die fine pitch interconnects.

One can use the contour plots given in Fig. 8 to determine these relations. Fig. 8(a) illustrates the timing impact and channel loss at the Nyquist frequency of the S1G1 design with a routing density of 250 IO/mm/layer, with respect to data rate and channel length, when the crosstalk is excluded. The circuit parameters were selected as R_{on} = 30 Ω and C_{die} = 0.3 pF, and the dielectric constant was 3.0. If these contour plots are compared, we can see that there is a very consistent correlation between the 0.1 UI timing impact and 2.5 dB loss lines. When the crosstalk is included in the analysis, the same correlation occurs around 1.5 dB loss line, as shown in Fig. 8(b). For the S2G1 and

S3G1 designs, the eye closure happens even at lower loss levels, because of the higher crosstalk. Therefore, we can conclude that when there is not any signal to signal coupling, the maximum tolerable channel loss is 2.5 dB. This value gets smaller as the crosstalk increases.

With the help of these contour plots, we can find the maximum channel length for each data rate and obtain the crosstalk/loss operating points at these lengths. Fig. 9 depicts a collection of these operating points and a trend line which was attained by curve fitting. This trend line can be considered as a compliance curve which shows the pass/fail boundary of an interconnect. As expected, when the crosstalk level is small, the channel can tolerate losses up to 2.5 dB. The maximum loss decreases as the crosstalk increases and when the crosstalk reaches the level of -10 dB, it is impossible to have a working channel. According to this data, any interconnect that operates below the trend line has a high likelihood to have a timing impact less than 0.1 UI. In mathematical terms, this can be expressed as in (3). A more accurate expression can be obtained with more data points and a higher order curve fitting.

$$FEXT \leq -16.4 \times L - 9.7 \qquad (3)$$

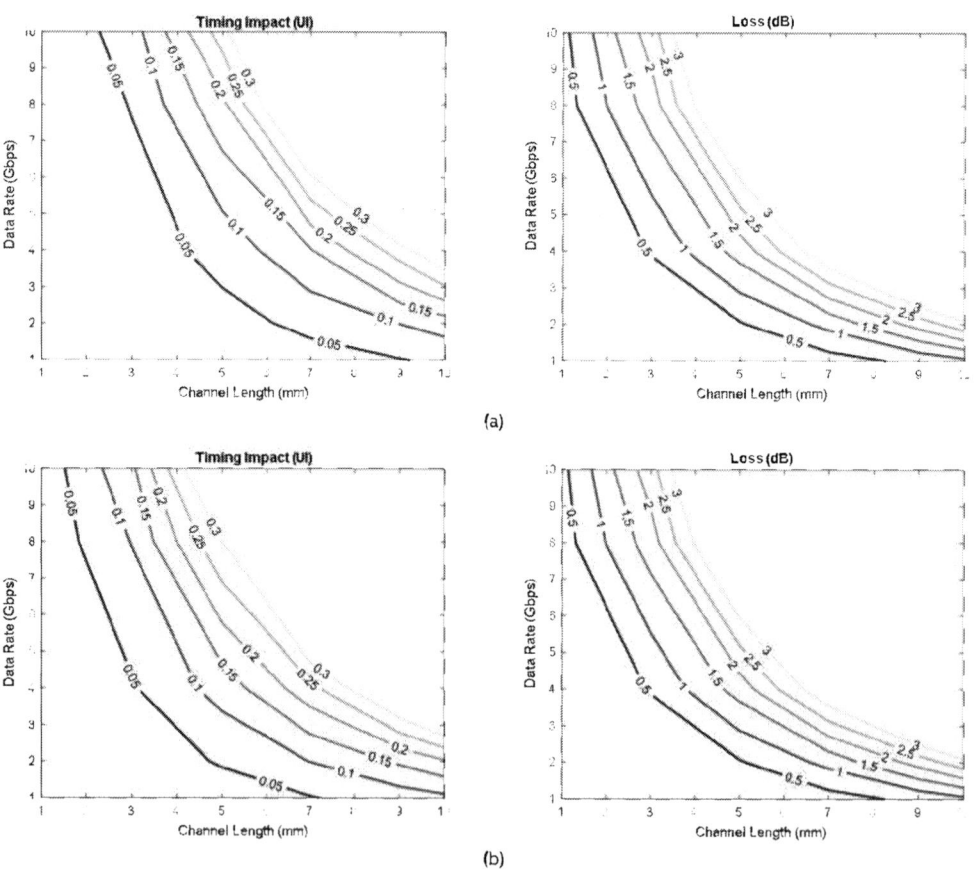

(a)

(b)

Timing margin and channel loss contour plots of the S1G1 design with a routing density of 250 IO/mm/layer, with respect to data rate and channel length. R_{on} = 30 Ω, C_{die} = 0.3 pF and Dk = 3.0. (a) Crosstalk excluded (b) Crosstalk included.

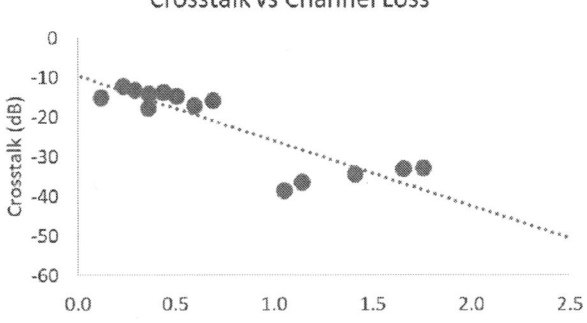

Figure 8. Design curve for fine pitch interconnects. The interconnect should operate below the compliance line to meet the timing budget.

V. CONCLUSIONS

We performed a parametric study to examine the impact of design parameters on the timing margin of very fine pitch interconnects and explored their electrical performance limits. It is verified that the optimum operating point is application dependent, as a result of trade-off between the maximum channel length, bandwidth density, power and latency. We also observed that with weak drivers, higher density interconnects operating at lower speeds exhibit higher bandwidth density. Nevertheless, this trend is reversed with stronger drivers. The path to increase the bandwidth density of these interconnects to larger than 1 Tbps/mm were shown as a function of interconnect design parameters. The performance can further be scaled by utilizing more aggressive drivers and sensitive receivers. Low-Dk materials always increases bandwidth while reduces power consumption and latency. Finally, we provided a practical design guideline, which specifies the limits of loss and crosstalk of the interconnect to support the designated data rate. This guideline can be helpful in terms of enabling faster design cycles and interface standardization for heterogeneous integration applications.

ACKNOWLEDGMENT

This research was developed with funding from the Defense Advanced Research Projects Agency (DARPA) under CHIPS program (Contract Number: HR00111790020). The views, opinions and/or findings expressed are those of the author and should not be interpreted as representing the official views or policies of the Department of Defense or the U.S. Government.

REFERENCES

[1] S. S. Iyer, "Heterogeneous Integration for Performance and Scaling," in IEEE Transactions on Components, Packaging and Manufacturing Technology, vol. 6, no. 7, pp. 973-982, July 2016, doi: 10.1109/TCPMT.2015.2511626.

[2] M. Sunohara, T. Tokunaga, T. Kurihara and M. Higashi, "Silicon interposer with TSVs (Through Silicon Vias) and fine multilayer wiring," 2008 58th Electronic Components and Technology Conference, Lake Buena Vista, FL, 2008, pp. 847-852, doi: 10.1109/ECTC.2008.4550075.

[3] H. Braunisch, A. Aleksov, S. Lotz and J. Swan, "High-speed performance of Silicon Bridge die-to-die interconnects," *2011 IEEE 20th Conference on Electrical Performance of Electronic Packaging and Systems*, San Jose, CA, 2011, pp. 95-98, doi: 10.1109/EPEPS.2011.6100196.

[4] R. Mahajan et al., "Embedded Multi-die Interconnect Bridge (EMIB) -- A High Density, High Bandwidth Packaging Interconnect," 2016 IEEE 66th Electronic Components and Technology Conference (ECTC), Las Vegas, NV, 2016, pp. 557-565, doi: 10.1109/ECTC.2016.201.

[5] Shimizu N., et al., "Development of organic multi chip package for high performance application," International Symposium on Microelectronics, October 2013, pp. 414 – 414, doi: 10.4071/isom-2013-TP65.

[6] K. Oi *et al.*, "Development of new 2.5D package with novel integrated organic interposer substrate with ultra-fine wiring and high density bumps," *2014 IEEE 64th Electronic Components and Technology Conference (ECTC)*, Orlando, FL, 2014, pp. 348-353, doi: 10.1109/ECTC.2014.6897310.

[7] B. Bayraktaroglu, "Heterogeneous Integration Technology," *AFRL/RYDD*, Tech. Rep., May 19, 2017, [Online], Available: https://apps.dtic.mil/dtic/tr/fulltext/u2/1035357.pdf.

[8] DARPA Microsystems Technology Office, "Broad Agency Announcement Common Heterogeneous Integration and IP Reuse Strategies (CHIPS)," *DARPA Microsystems Technology Office*, DARPA-BAA-16-62 September 29, 2016, [Online], Available: https://www.darpa.mil/program/common-heterogeneous-integration-and-ip-reuse-strategies.

[9] K. Cho *et al.*, "Design optimization of high bandwidth memory (HBM) interposer considering signal integrity," *2015 IEEE Electrical Design of Advanced Packaging and Systems Symposium (EDAPS)*, Seoul, 2015, pp. 15-18.

[10] H. Kalargaris and V. F. Pavlidis, "Interconnect design tradeoffs for silicon and glass interposers," *2014 IEEE 12th International New Circuits and Systems Conference (NEWCAS)*, Trois-Rivieres, QC, 2014, pp. 77-80.

[11] S. Jangam, A. A. Bajwa, K. K. Thankkappan, P. Kittur and S. S. Iyer, "Electrical Characterization of High Performance Fine Pitch Interconnects in Silicon-Interconnect Fabric," *2018 IEEE 68th Electronic Components and Technology Conference (ECTC)*, San Diego, CA, 2018, pp. 1283-1288.

978-1-7281-1500-9/19 $31.00 © 2019 IEEE

A high-bandwidth fine-pitch 2.57Tbps/mm in-package communication link achieving 48fJ/bit/mm efficiency

N. Pantano[1,2], G. Van der Plas[1], P. Bex[1], P. Nolmans[1], D. Velenis[1], M. Verhelst[2], E. Beyne[1]

[1]imec - Leuven, Belgium

[2]MICAS-ESAT KU Leuven Leuven, Belgium

Email: nicolas.pantano@imec.be

Abstract—Memory bandwidth is the main bottleneck to improve the performance of today's computing systems, and the demand for bandwidth is expected to grow exponentially in the coming years. The development of advanced packaging solutions making use of a silicon bridge such as Embedded Multi-Die Interconnect Bridge (EMIB) and Fan-Out Wafer Level Package (FO-WLP) are promising solutions to achieve high bandwidth density and to bring more memory closer to the computing units. This work demonstrates a 0.3V-swing 7mm long link over a silicon bridge, running at a bitrate of 9Gbps. It achieves 48fJ/bit/mm power efficiency on 3.5um pitch wires, resulting in a bandwidth density of 2.57Tbps/mm.

Keywords-interconnect, chip-to-chip, links, low-swing, high bandwidth density

I. INTRODUCTION

With the advent of big-data, deep-learning and more generally, the exponential growth of data processing in modern systems, the I/O bandwidth is one of the bottlenecks that limits the performance of current devices. Today's high-end GPUs reach an aggregated memory bandwidth of 1TB/s using HBM2 memories, and the demand for bandwidth is expected to grow exponentially in the coming years. To fulfill this demand, the data rate and the amount of memory available in future systems must increase, and the power consumption kept to a minimum. The development of advanced packaging solutions such as EMIB [1] and FO-WLP [2] that use silicon bridges, allowing very fine line pitches and dense buses to interconnect logic dies together, could help to sustain the high demand for bandwidth.

This paper presents a high-bandwidth interconnect solution combining a driver and a receiver chip interconnected over 7mm through a silicon bridge. Figure 1 shows a picture of the test vehicle. The two CMOS dies are stacked on the silicon bridge which is wire-bonded to a PCB.

This work aims to study the performance of two high-bandwidth density links in terms of power distribution, energy efficiency, data rate and signal integrity. Furthermore, the effect of neighboring lines on an eye diagram measured at the receiver is also examined. Reducing the energy per bit is critical in future systems to sustain the higher bandwidth requirements while keeping the power budget constant. For this purpose, two I/O interfaces have been designed and

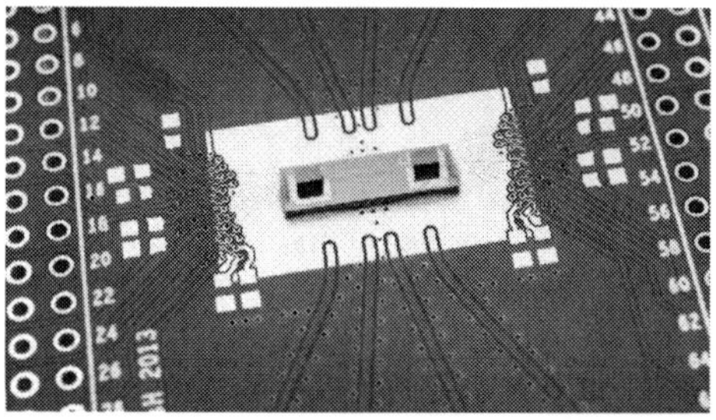

Figure 1: Picture of the demonstrator with the two CMOS dies stacked on the silicon bridge fixed on a PCB.

compared: a high-voltage swing interface based on the high-bandwidth memory standard [3] and a low voltage swing interface.

The first section of this paper describes the design of the silicon bridge and the optimization procedure to size the transmission lines used to send data between two CMOS dies so that the highest possible bandwidth density can be achieved. In the second section, a description of the CMOS dies is given with details about the implementation of the high-swing and low-swing interfaces. Finally, in the third section, the performance of the two interfaces is analyzed and compared to other state-of-the-art publications.

II. SILICON BRIDGE OPTIMIZATION AND DESIGN

Two CMOS dies are stacked on an interposer and interconnected together through a 7mm long line. The same CMOS dies are used, but one is rotated 180° so that the same edges are facing each other as depicted in Figure 2. The layouts of the regions on the interposer where the CMOS dies are located are almost identical. The input data (clock, external PRBS data) is provided by RF probing pads. These pads are in a ground-signal-ground configuration to ensure high-speed operation, and a clean signal at the input of the CMOS dies. The bottom probe pads are used to capture the data received by the interfaces with a single tip high impedance active probe.

The silicon bridge is a passive die with a Back-End of Line (BEOL) composed of 4 metal layers: 3 copper layers and one aluminum layer on top. Figure 4 shows a cross-

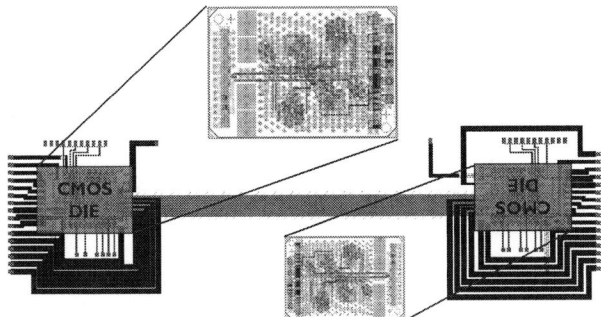

Figure 2: Layout of the interposer with the location of the two CMOS die facing each other.

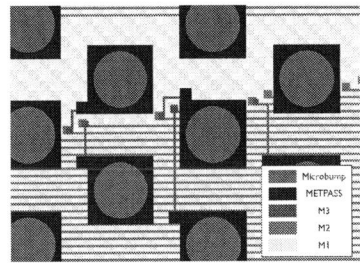

Figure 3: Connection between the fine-pitch interconnect and the micro-bumps.

section of the interposer with the first 3 metal layers. The bus that connects the two chips is in a microstrip configuration. Metal 1 (M1) is used for the ground plane and metal 2 (M2) for the signal line. The two other metal layers are used for power routing and redirecting signals from micro-bumps to transmission lines on the interposer as illustrated in Figure 3.

The dimensions of the signal lines on the interposer are optimized to maximize the bandwidth density. To this end, a calibrated 2D field solver is used to extract the RLCG parameters of the lines for various interposer line dimensions. The field solver also calculates an estimation of the RC time constant, the bandwidth density and the far-end crosstalk coefficient, similarly to what has been done in [4]. The RC constant is obtained by

$$\tau_{RX} = 2.3[R_s(C_{TX}+C_l l+C_{RX})+R_l C_R Xl]+R_l C_l l^2 \quad (1)$$

where $R_s = 50\Omega$ is the driver output resistance, C_{TX} and C_{RX}, both equal to 200fF, are the driver and receiver capacitances, R_l and C_l are the line resistance and capacitance per unit length, and l=7mm is the line length. The bandwidth density, in Tbps/mm, can be derived from this expression and is given by $BWD = \frac{1}{\tau_{RX} \times linepitch}$. To achieve good signal integrity, the amount of cross-talk between the lines should be kept low. The far-end cross-talk coefficient is defined as

$$K_f = 0.5(C_m C_t - L_m L_s) \quad (2)$$

where C_m is the sum of all mutual capacitances, C_t the total capacitance, L_m the mutual inductance and L_s the self-inductance of the line. This model is conservative and gives a lower estimation of the bandwidth density. Figure 5 shows the estimated bandwidth density as a function of the line pitches and thicknesses. A layer thickness and a line width of 1um with a spacing of 2.5um yields the maximum bandwidth while keeping the far-end cross-talk coefficient below 0.15.

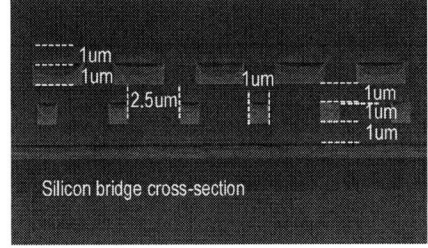

Figure 4: Cross-section of the interconnection between the two CMOS dies on the silicon bridge.

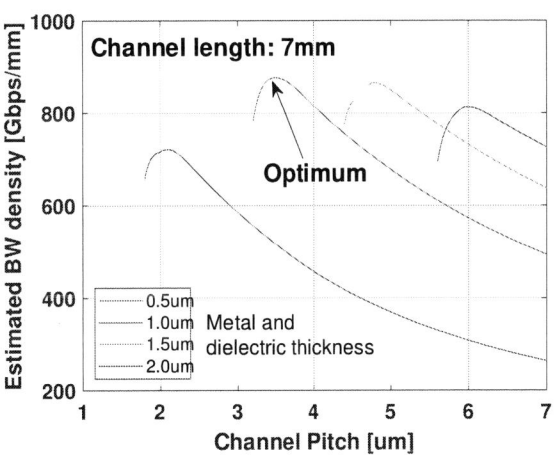

Figure 5: Estimated bandwidth density as a function of the line dimensions.

III. CMOS DESIGN

The CMOS test chip is made using a 14nm FinFET technology and contains two I/O interfaces designed to operate at a bitrate of at least 4Gbps. The high-swing interface has an output swing of 1.2V; it is designed with I/O devices (thick gate oxide FinFETs) and it is compliant with the specifications of the HBM2 standard from Jedec. This interface is already used in commercial products and for that reason, it is considered as the industry standard. The second interface is a low-swing I/O designed with core devices (thin gate oxide FinFETs) where the output swing can be tuned between 0.15V and 0.4V. Each I/O is duplicated five times and connected to five parallel transmission lines, as shown in

Figure 6. The four outer lines are used as aggressors while the central one is used to carry the signal to be transmitted.

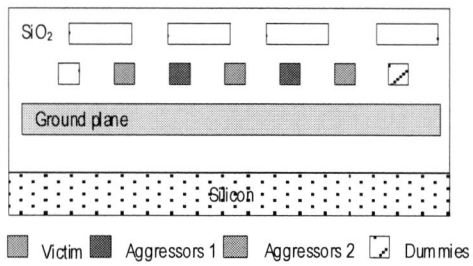

Victim ■ Aggressors 1 ■ Aggressors 2 ■ Dummies ◨

Figure 6: Bus of 5 bits on the silicon bridge.

A. High-swing interface

This interface is designed to evaluate the performance of a typical in-package 1.2V swing logic-to-memory interface based on the HBM2 standard but operating at a faster data rate of 4Gbps. The block diagram of one I/O is given in Figure 7.

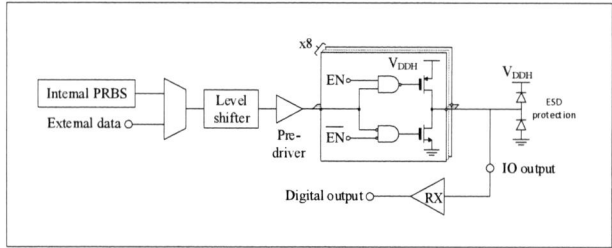

Figure 7: Block diagram of the high-swing I/O interface.

1) Transmitter: The input data can be provided externally or internally with an on-chip 7-bit Pseudo-Random Bit Sequence (PRBS) and a Multiplexer (MUX), which is used to selects which input to forward to the driver. In commercial products, the input data is coming from a digital circuit operating at a different supply voltage of 0.8V. Therefore, the data need to be converted to 1.2V before it is sent to the transmitter. This conversion is performed by a level-shifter; a detailed schematic of this circuit is provided in Figure 8.

The first part of the circuit consists of digital standard cells that are used to convert the single-ended input signal into a differential signal, and the second part of the circuit, made up of IO transistors, is used to level shift the signal from VDD = 800mV to VDDH = 1.2V. Differential signals enable level shifter circuits to operate much faster and more efficiently, which is the reason why the standard cells in the first part of the circuit shown in Figure 8 are used. Because the input goes through the same number of standard cells, there is no delay between the digital signals at the output

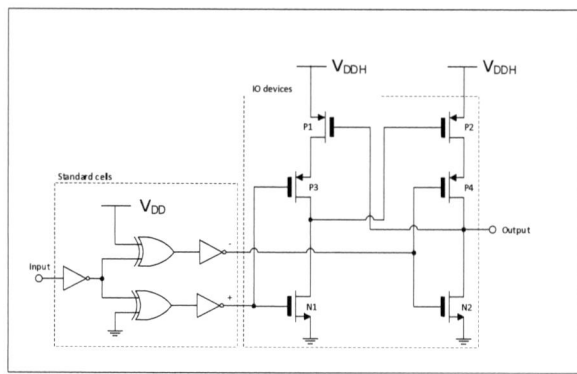

Figure 8: Schematic of the 0.8V to 1.2V level shifter (VDD=0.8V and VDDL=1.2V).

of the positive and negative branches of this first part of the circuit. The second part of the circuit is the one that actually does the conversion between two voltage domains, between 0.8V and 1.2V in this case. A simple level shifter is made of the 4 transistors (N1, N2, P1 and P2). In order to achieve faster switching times, a more elaborate level shifter topology has been used instead. In the simple topology, transistors N1 and N2 can never be fully turned on as the voltage at their gates can not exceed VDD = 800mV, unlike the voltages at the gates of the PMOS pull-up transistors, which vary between 0V and VDDH = 1.2V. This results in lower switching speeds, as the NMOS transistors cannot sink as much current as the PMOS devices. By adding transistors P3 and P4 to the simple level-shifter topology, this issue can be solved. Indeed, when a high voltage VDD = 800mV is applied, a PMOS transistor (either P3 or P4) limits the amount of current that needs to be sinked. As a consequence, the speed at which the voltage at the output of an inverter drops increases, making it possible to use this level-shifter at higher frequencies.

An issue that is common to all level shifters topologies is that they cause signals to be distorted. As a consequence, the duty cycle of the signal that is input to the level-shifter is not preserved. To compensate for this distortion, a Duty-Cycle Compensation Circuit (DCCC) is inserted in the first stage of the pre-driver as shown in Figure 9. As depicted in Figure 7, a pre-driver directly follows a level-shifter in a high swing IO interface. The DCCC changes the output resistance of the pull-up or pull-down transistors of the first stage of the pre-driver. This makes it possible to change the rise and fall times of the signals that propagate through this circuit, and therefore to control their duty cycle. The following stages gradually increase in size so as to drive 8 identical tri-state output drivers. Each of these output drivers can be individually turned on and off, and are designed to deliver a peak output current of 3mA. A detailed schematic of an output driver is shown in Figure 10. By changing the number of output drivers configured in high impedance

978-1-7281-1500-9/19 $31.00 © 2019 IEEE

mode, the drive strength of the transmitter can be controlled. It is therefore possible to adjust the intensity of the signal sent to the receiver.

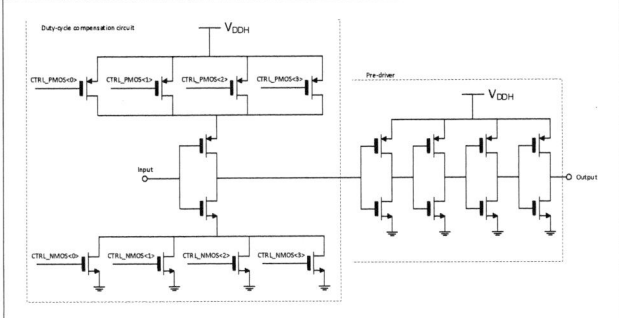

Figure 9: Schematic of the Duty-Cycle Compensation Circuit (DCCC) and the pre-driver.

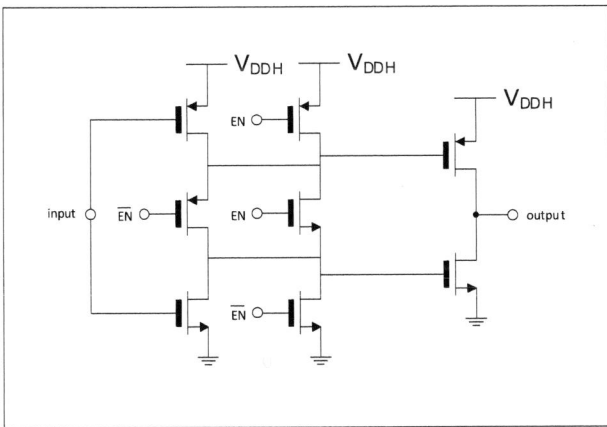

Figure 10: Schematic of the tri-state 1.2V output driver.

2) Receiver: When the I/O is configured as a receiver (all the output drivers are configured in high impedance mode), it needs to convert a 1.2V input signal back into digital data between 0V and 0.8V. This conversion is done by the circuit shown in Figure 11. The first three inverters form what is called a down converter. It converts the input signal, which has a swing of 1.2V, into a digital signal with a swing of 0.8V. The first two stages of the down-converter reshape the input signal such that this signal has very sharp transitions. The third stage converts the signal into one that has a 0.8V voltage swing and sends this signal to a chain of inverters.

The layout of the I/O is shown in Figure 12 and has a footprint of $1118um^2$.

B. Low-swing interface

Compared to the high-swing interface, the low-swing interface is designed to consume less energy-per-bit by lowering the voltage swing of the output signal. In the case of this demonstrator, the low swing interface is designed such that its output swing can be adjusted between 0.15V and 0.4V. This circuit only makes use of core device (thin

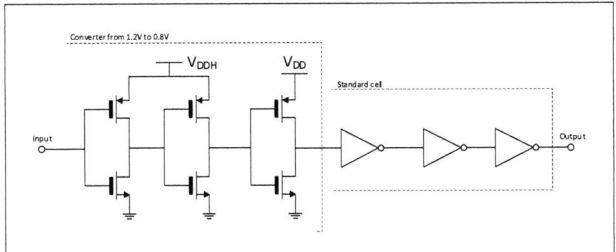

Figure 11: Schematic of the receiver that converts the input signal from 1.2V to 0.8V.

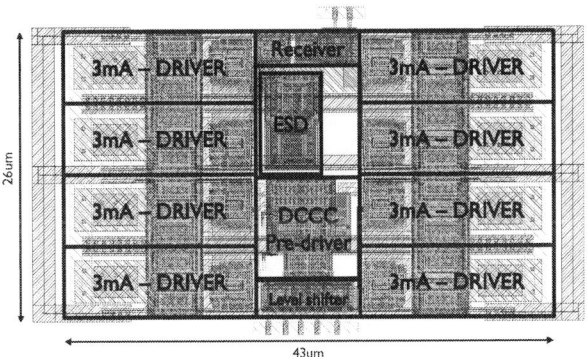

Figure 12: Layout of the high-swing I/O with the level-shifter, the DCCC, the pre-driver, the eight 8mA tri-state drivers, the receiver and the ESD protection.

gate oxide transistors) and is able to operate at a bitrate of at least 4Gbps.

The block diagram of the interface is given in Figure 13. Just like for the high-swing interface described previously, the data can either be provided internally or externally. The receiver is however slightly more complex because as the data is sent with a reduced swing, it needs to be amplified before it can be input to any digital circuit. Moreover, the receiver is separated from the transmitter using a pass gate, which is necessary in order to isolate the analog amplifier in the receiver from all the digital circuitry in the transmitter.

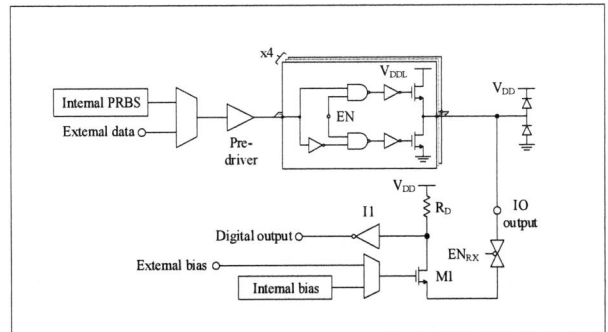

Figure 13: Block diagram of the low-swing I/O interface.

1) Transmitter: The data from the internal or the external source is sent to a pre-driver made of a chain of inverters that gradually increase in size in order to drive the output stage. The output stage consists of 4 identical tri-state drivers that can be individually activated to control the drive strength, each providing a peak output current of 3mA. To operate at low-swing, the driver is composed of two NMOS transistors supplied between 0.15V and 0.4V (VDDL). Since $V_{gs_{N1}} < V_{gs_{N2}}$, the pull up transistor (N1) is less conductive than the pull-down transistor (N2) and $R_{ON_{N1}} > R_{ON_{N2}}$. To overcome this higher resistance, the pull-up transistor is designed with twice the channel width of the pull-down transistor. The digital circuitry that precedes the output stage uses standard cells operating at core voltage (0.8V). A detailed schematic of the pre-driver and the output stage is given in Figure 14.

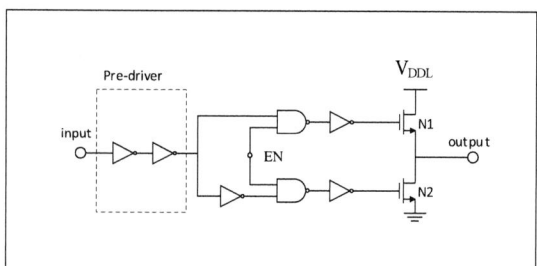

Figure 14: Schematic of the low-swing pre-driver and output stage.

2) Receiver: To recover the data at the receiver side, the signal is amplified by a Common gate amplifier (CGA), that consists of an NMOS transistor and a resistor, followed by a chain of inverters as shown in Figure 15. The CGA amplifies the signal and centers it around the threshold voltage of the inverter at its output, which provides the additional gain needed to raise the signal swing to VDD. The CGA has a DC gain of approximately 11dB and a bandwidth of roughly 9Ghz. To control the bias on the gate of the CGA, a DC voltage can be provided externally or internally. The internal bias control circuit in Figure 15 is a replica of the receiver connected in a negative feedback loop, in which N2 is half the size of N1, and resistor R2 is two times larger than R1. The bias voltage at the drain of N2 is controlled with a bank of resistors connected to the source of N2. To avoid any oscillation in the bias control circuit, a capacitor of 1pF is added in the feedback loop and acts as a low-pass filter. To prevent the receiver from influencing the transmitter by injecting current into the signal line, it is isolated from the output by a transmission gate and only activated when the I/O is in receiver mode (see Figure 13). The width of the NMOS and PMOS transistors that make up the pass gate are made large enough to avoid the channel resistance deteriorates the signal quality.

The layout of the I/O is shown in Figure 16 and has a

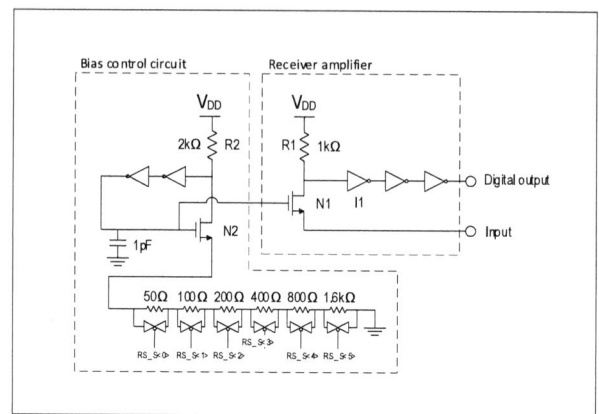

Figure 15: Low-swing receiver with the internal bias control circuit.

footprint of $1480um^2$. It occupies a larger area than the high-swing interface because it includes by-pass pass-gates that have been implemented for testing purposes, but these gates could be removed in future designs.

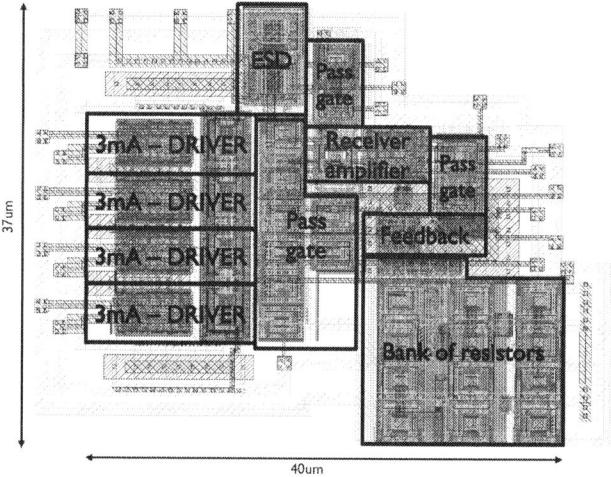

Figure 16: Layout of the low-swing I/O with the four 3mA tri-state drivers, the amplifier, the feedback loop with the bank of resistors, the ESD protection and the pass-gates to control the I/O.

IV. CHARACTERIZATION

To evaluate the performance of our demonstrator, 3 external uncorrelated PRBS generators provide data to a transmitting die with a multi-channel RF probe as shown in Figure 17. One PRBS transmits on the central line whereas the other two send data on the adjacent aggressor lines. The signal at the output of the receiver is captured with a high-impedance active probe (single probe tip). From this measurement, the peak-to-peak jitter and the eye opening can be extracted. The power supplies are also monitored to measure the energy efficiency and the power consumption of the transmitter and the receiver.

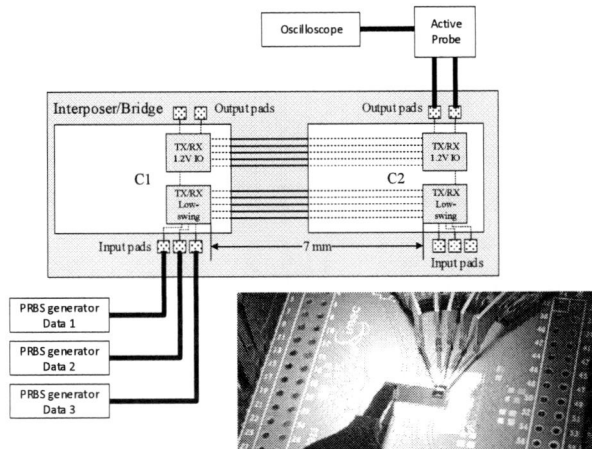

Figure 17: Characterization of the test vehicle.

Measurement results are given in Tables I and II for the high-swing and low-swing interfaces respectively. The jitter and the eye opening are measured at various bitrates and for different aggressor configurations. When the closest aggressors to the victim line are enabled, the jitter on the signal at the output of the receiver increases. This is the case for both interfaces, but the low-swing interface is less sensitive to the impact of the neighboring lines than the high- swing interface. The value of the jitter measured at the output of the low swing interface is half of that measured at the output of the high-swing interface for a bitrate of 4Gbps. In terms of data rate, the high-swing interface can operate at bitrates up to 5Gbps, without degrading signal integrity and maintaining a wide eye opening, as shown in Figure 18. The low-swing interface, on the other hand, is able to operate at a higher data rate of 9Gbps (see Figure 19). This difference is due to the fact the 2 interfaces do not share the same link architecture. The high-swing uses I/O devices that have a large gate length and a thicker gate oxide. These transistors have a slower switching time than the core devices (minimum gate length transistor). Moreover, the input impedance of the high-swing receiver is much higher than for the low-swing interface. Both differences make the low-swing interface faster than its high-swing counterpart.

Besides signal integrity, the energy efficiency, defined as the energy-per-bit per unit of interconnect length, is improved by more than 50% in the case of the low swing interface. At 4Gbps, the high swing interface has an efficiency of 150fJ/bit/mm while for its low-swing equivalent, the efficiency improves to 70fJ/bit/mm. At 9Gbps, the efficiency is further improved and reaches 48fJ/bit/mm. Figure 20 shows the energy efficiency of the interfaces as a function of the bitrate. The output driver of the low-swing I/O was designed such that the signal voltage swing can be

High-swing I/O	4Gbps		5Gbps	
	Jitter	Eye opening	Jitter	Eye opening
All Aggressors disabled	75ps	0.7UI	85ps	0.58UI
Aggressors 2 enabled	85ps	0.66UI	94ps	0.53UI
Aggressors 1 enabled	113ps	0.54UI	105ps	0.48UI
All aggressors enabled	147ps	0.41UI	137ps	0.32UI

Table I: Measured jitter and eye-opening at 4Gbps and 5Gbps for the high-swing interface.

Low-swing I/O (0.3V swing)	4Gbps		9.5Gbps	
	Jitter	Eye opening	Jitter	Eye opening
All Aggressors disabled	58ps	0.77UI	82ps	0.22UI
Aggressors 2 enabled	66ps	0.74UI	84ps	0.20UI
Aggressors 1 enabled	67ps	0.73UI	88ps	0.16UI
All aggressors enabled	72ps	0.71UI	89ps	0.15UI

Table II: Measured jitter and eye opening at 4Gbps and 9.5Gbps for the low-swing interface for 300mV of swing.

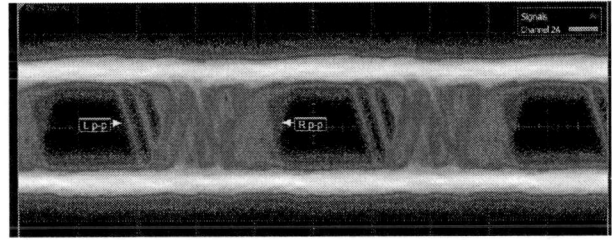

Figure 18: Eye diagram at the output of the receiver of the high-swing interface at 5Gbps.

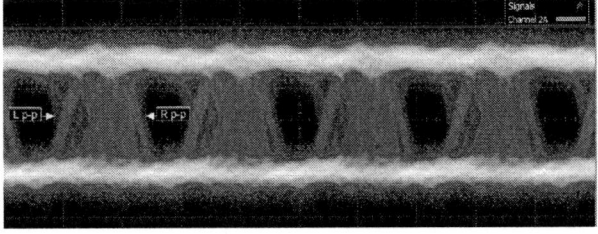

Figure 19: Eye diagram at the output of the receiver of the low-swing interface at 9Gbps.

varied between 0.15V and 0.4V. As shown by the energy distribution plot in Figure 21, the output driver consumes only 4% of the total energy, and the output swing has a negligible impact on the energy efficiency, which varies between 45.7fJ/bit/mm at 0.15V and 49.5fJ/bit/mm at 0.4V at a bitrate of 9Gbps. One of the objectives of this study is to achieve high bandwidth density, defined as the ratio between the data rate and the interconnection pitch expressed in Terabits per second per millimeter. In this work, the pitch of the lines is 3.5um (1um width and 2.5um spacing). At the maximum operation frequency, the high-swing I/O achieves a bandwidth density of 1.43Tbps/mm, while in the case of the low-swing interface, the bandwidth density increases to 2.57Tbps/mm.

In summary, compared to the 1.2V (high swing) interface, the low-swing alternative achieves 80% more bandwidth density and 66% better energy efficiency. Table III presents a summary of our results and a comparison that shows we can achieve 8.6 times more bandwidth density (up to 2.57Tbps/mm) than [5] while maintaining the same energy efficiency.

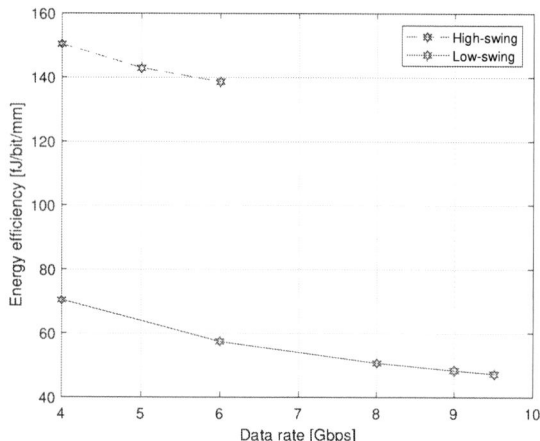

Figure 20: Energy efficiency of the two interfaces as function of the data rate.

V. CONCLUSION

To explore the high bandwidth density capabilities of advanced packaging solutions such as the FO-WLP described in [2], a test vehicle was designed to assess the performance of high-bandwidth density in-package chip-to-chip links. For this purpose two CMOS dies were stacked on the same silicon bridge similar to what is done in FO-WLPs. The bridge is a passive die with 4 metal layers: 3 copper layers and one aluminum layer. The two CMOS chips communicate together over a distance of 7mm with microstrip transmission lines. The lines have a thickness and width of 1um, with adjacent lines spaced apart by 2.5um. The CMOS dies communicate together with two different I/O interfaces: a 1.2V swing interface compliant to the HBM2 standard and a low-swing (0.15V to 0.4V) I/O. The high-swing interface operates at an optimum data rate of 5Gbps with an energy efficiency of 142fJ/bit/mm and a bandwidth density of 1.43Tbps/mm. To improve the energy efficiency, a low-swing interface was designed; it can operate

at an optimum data rate of 9Gbps with an energy efficiency of 48fJ/bit/mm (at 0.3V swing) and a bandwidth density of 2.57Tbps/mm. High-bandwidth density and high energy efficiency are key to sustain the never-ending increase in the demand for bandwidth. This work shows the potential for high bandwidth density links for future systems. Moreover, the development of advanced packaging solutions will help reduce the communication distance between chips and hence help to improve the bandwidth density and the energy efficiency further.

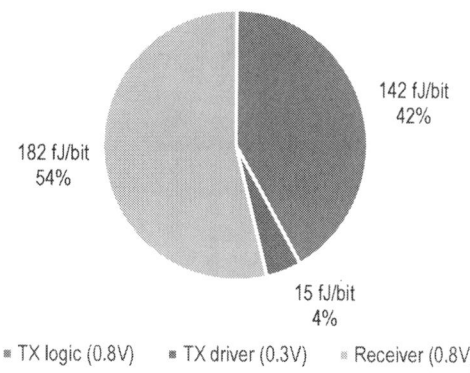

Figure 21: Distribution of the energy at 9Gbps with 0.3V swing for the low-swing interface.

	This work		[6]	[5]	[7]
	1.2V I/O	0.3V I/O			
Technology [nm]	14		40	28	65
Interposer/bridge technology	Si		Si	Orga.	Si
Signal swing [V]	1.2	0.3	0.3	0.15	1.1
Interconnection length [mm]	7		1	6	15
Interconnect pitch [um]	3.5		28.6	66.7	n/a
Bitrate per pin [Gbps]	5	9	1	20	16.8
Bandwidth density [Tbps/mm]	1.43	**2.57**	0.03	0.3	n/a
Energy efficiency [fJ/bit/mm]	142	**48**	105	**49**	393
Energy per bit [fj/bit]	1000	338	105	395	5900

Table III: Comparison of the two interface with some state-of-the-arts publications.

REFERENCES

[1] R. Mahajan *et al.*, "Embedded multi-die interconnect bridge (emib) – a high density, high bandwidth packaging interconnect," in *2016 IEEE 66th Electronic Components and Technology Conference (ECTC)*, May 2016, pp. 557–565.

[2] A. Podpod, J. Slabbekoorn, A. Phommahaxay, F. Duval, A. Salahouelhadj, M. Gonzalez, K. Rebibis, A. Miller, G. Beyer, and E. Beyne, "A novel fan-out concept for ultra-high chip-to-chip interconnect density with 20-m pitch," in *2018 IEEE 68th Electronic Components and Technology Conference (ECTC)*, May 2018, pp. 370–378.

[3] Jedec, "High bandwidth memory (hbm) dram (jesd235b)," JEDEC Solid State Technology Association, Tech. Rep., 2018.

[4] N. Pantano, C. R. Neve, G. V. der Plas, M. Detalle, M. Verhelst, M. Heyns, and E. Beyne, "Technology optimization for high bandwidth density applications on 3d interposer," in *2016 6th Electronic System-Integration Technology Conference (ESTC)*, Sep. 2016, pp. 1–6.

[5] W. J. Turner *et al.*, "Ground-referenced signaling for intra-chip and short-reach chip-to-chip interconnects," in *2018 IEEE Custom Integrated Circuits Conference (CICC)*, April 2018, pp. 1–8.

[6] H. Lee *et al.*, "A 16.8 gbps/channel single-ended transceiver in 65 nm cmos for sip-based dram interface on si-carrier channel," *IEEE Journal of Solid-State Circuits*, vol. 50, no. 11, pp. 2613–2624, Nov 2015.

[7] M. Lin *et al.*, "A 1 tbit/s bandwidth 1024 b pll/dll-less edram phy using 0.3 v 0.105 mw/gbps low-swing io for cowos application," *IEEE Journal of Solid-State Circuits*, vol. 49, no. 4, pp. 1063–1074, April 2014.

A new SI-PI co-simulation approach for efficient consideration of coupling between PDN and SDN

Heesok Lee, Jisoo Hwang, Hoi-Jin Lee, and Youngmin Shin

System LSI Division, Samsung Electronics, Co. Ltd.

1-1 Samsungjeonja-ro, Hwaseong-si, Gyeonggki-do, Korea

hees.lee@samsung.com

Abstract— **As the clock frequency of SOC (system on a chip) including core circuits like CPU and GPU and various IOs like LPDDR4, LPDDR5, and SERDES, increases, the impact of SDN (signal delivery network) and PDN (power delivery network) on the functional stability and low power operation of corresponding circuit blocks becomes more and more important. However, there has been the challenge to the design and verification with co-simulation of SI (signal integrity) and PI (power integrity) considering the coupling between SDN and PDN due to a painful long simulation time, which is usually not allowed in the short design cycles required by modern electronics especially in mobile hand-held devices. This paper will provide a matrix formulation for SI-PI co-simulation in frequency-domain to reduce the simulation time and consider the coupling efficiently.**

Keywords-Signal Integrity, Power Integrity, SI-PI co-simulation, Frequency Domain

I. INTRODUCTION

Increasing the clock frequency of core circuits like CPU and GPU and various IOs like LPDDR4, LPDDR5, and SERDES in SOC (system on a chip) has stressed the importance of SI (signal integrity) and PI (power integrity) on the functional stability and low power operation of the circuit blocks. However, the challenge to design and verification of SDN (signal delivery network) and PDN (power delivery network) with SOC die, package, and system-level board is rising higher and higher with tough limit of cost budget and design cycle-time. In the other hand, it takes pretty much computation time to get S-parameter model of coupled network including both SDN and PDN by EM (electromagnetic) field solver, and to solve the time-domain SPICE simulation for considering SI-PI co-simulation. Although the conventional SI-PI co-simulation considering the coupling between SDN and PDN is possible as presented by several authors through [1-15], the long computation time and resources required for SI-PI co-simulation has given challenges to design and verification engineers developing core circuits, IOs, packages, modules and systems. Modeling the complex PDN with critical signal lines and solving large-size circuit model including SDN and PDN can result in painful long simulation time, which is sometimes not allowed in the short design cycles required by components or systems for mobile hand-held devices.

Although accurate simulation of the interaction between both SDN and PDN is essential for estimating system reliability and stability before fabricating SOC, package, and system-PCB, it is strongly asked to adequately ignore the coupling between SDN and PDN for urgent study and analysis especially in design stage close to tape-out date due to the lack of resource and time. Therefore, the impact of coupling between SDN and PDN should be carefully predicted and studied. When PDN is neglected or separately considered in SI-simulation and PI-simulation for the design, it is strongly required to understand what is missing for adequate risk management in SI-PI perspective. However, the limited design period and computation resource have restricted the SI and PI simulations to a few channels of the SDN without taking into account the influence of noise propagating within the power delivery network (PDN). In fact, the PDN design and analysis is performed with a separate effort.

In this paper, the method given by authors will be useful to review the coupling between SDN and PDN, based on which proper way to consider the coupling can be taken. The matrix formulation will be given by authors with the behavior model of switching circuit network (SCN) constructed by CMOS transistors in SOC and the admittance matrix (Y) model of SDN and PDN given by EM field solution. The time-variant system behavior of SCN and linear time invariant system property of SDN and PDN are combined by convolution matrix to represent voltage and current waves. Based on the matrix formulation given in this paper, it will be addressed that SDN and PDN can give critical impacts on PI, although there is no coupling between SDN and PDN. A simulation work flow for SI-PI co-simulation will be presented.

II. A NEW APPROACH TO CO-SIMULATION

In this section, formulation will be derived to present a new approach for SI-PI co-simulation. As shown in Fig.1 and 2, a black box model with an admittance matrix, Y, will be used to represent the electrical characteristics of power and signal delivery network (PSDN), based on which the coupling will be also considered.

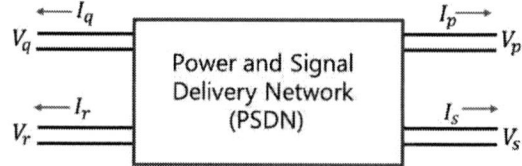

Figure 1. Black-box model to represent a power and signal delivery network (PDSN)

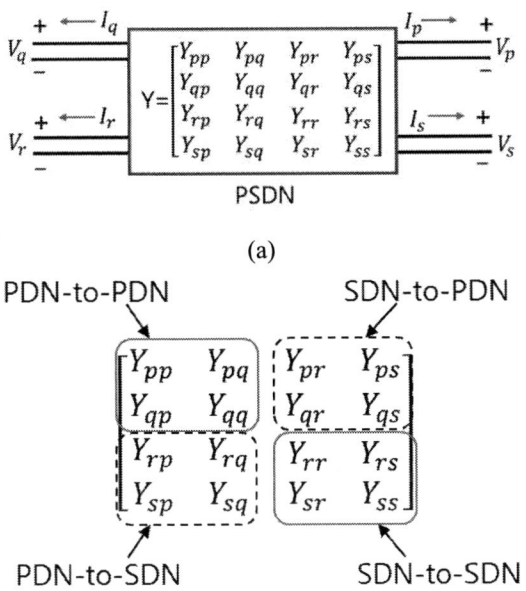

(a)

PDN-to-PDN ——— SDN-to-PDN

PDN-to-SDN ——— SDN-to-SDN

(b)

Figure 2. Black-box model will be represented by Y-matrix, by which coupling between power delivery network (PDN) and signal delivery network (SDN) will be described. While voltage and current vectors with subscripts, p and q are those for power, subscript, s and r denote signal ports. While Y matrix is given by (a) with the definition of nodes, currents and voltages, the coupling is represeted by Y as (b) presents.

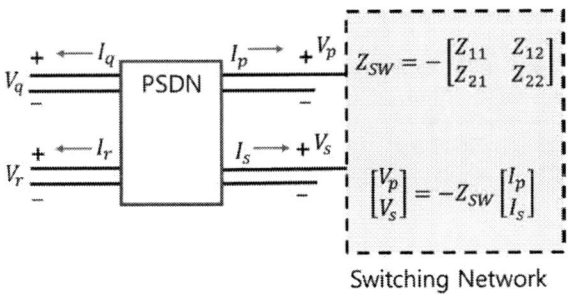

Figure 3. Switching circuit network (SCN) representing output drivers or core circuits should be added to PSDN. Nodes denoted by subscript, p, are connected to power supply node of the switching network. Nodes denoted by subscript, s, are connected to output nodes of switching network. While nodes denoted by subscript, r, represent nodes of receivers, nodes denoted by subscript, q, are connected to be DC power supply.

$$
\begin{bmatrix} I_p \\ I_q \\ I_r \\ I_s \end{bmatrix} = \begin{bmatrix} Y_{pp} & Y_{pq} & Y_{pr} & Y_{ps} \\ Y_{qp} & Y_{qq} & Y_{qr} & Y_{qs} \\ Y_{rp} & Y_{rq} & Y_{rr} & Y_{rs} \\ Y_{sp} & Y_{sq} & Y_{sr} & Y_{ss} \end{bmatrix} \begin{bmatrix} V_p \\ V_q \\ V_r \\ V_s \end{bmatrix} \quad (1)
$$

The equation (1) is the admittance matrix. In the left side of (1) is current column vector. While I_p, and I_q represent

current column vector for power delivery network (PDN), I_r, and I_s represent current column vector for signal delivery network (SDN), as shown in Fig.1 and2, Voltage column vector including V_p, V_q, V_r, and V_s follows the same manner. In (1), Y_{pp}, Y_{pq}, Y_{qp}, and Y_{qq} represent PDN, and Y_{rr}, Y_{rs}, Y_{sr}, and Y_{ss} represent SDN, itself. To describe the coupling between PDN and SDN, Y_{pr}, Y_{ps}, Y_{qr}, Y_{qs}, Y_{rp}, Y_{rq}, Y_{sp}, and Y_{sq} are used as shown in Fig.2 (b). When the number of nodes or ports presented by p, q, r, s is N(p), N(q), N(r), and N(s), respectively, the size of admittance matrix Y given by (1) is [N(p) + N(q) + N(r) + N(s)] x [N(p) + N(q) + N(r) + N(s)]. V_p, V_q, V_r, V_s, I_p, I_q, I_r, and I_s are [N(p) + N(q) + N(r) + N(s)] x 1 column vectors.

As given in Fig. 3, SCN representing output drivers or core circuits should be added to PSDN shown by Fig.1 and 2. SCN is generally implemented by CMOS circuits. Nodes denoted by subscript, p, are connected to power supply node of the switching network. Nodes denoted by subscript, s, are connected to output nodes of switching network. While nodes denoted by subscript, r, represent nodes of receivers, nodes denoted by subscript, q, are connected to be DC power supply.

$$
\begin{bmatrix} I_p \\ I_s \end{bmatrix} = \left\{ \begin{bmatrix} Y_{pp} & Y_{ps} \\ Y_{sp} & Y_{ss} \end{bmatrix} \begin{bmatrix} V_p \\ V_s \end{bmatrix} + \begin{bmatrix} Y_{pq} \\ Y_{sq} \end{bmatrix} V_q + \begin{bmatrix} Y_{pr} \\ Y_{sr} \end{bmatrix} V_r \right\} \quad (2)
$$

$$
I_r = 0 \quad (3)
$$

$$
V_r = - \left\{ \begin{bmatrix} Y_{rr}^{-1} Y_{rp} & Y_{rr}^{-1} Y_{rs} \end{bmatrix} \begin{bmatrix} V_p \\ V_s \end{bmatrix} + Y_{rr}^{-1} Y_{rq} V_q \right\} \quad (4)
$$

From equation (1), the extracted current column vector including I_p and I_s can be rewritten by (2). Since the nodes denoted by subscript, r, are open circuit, equation (4) can be obtained from (1) with (3). As V_r in (2) can be replaced by using (4), we can get equation (5) as following.

$$
\begin{bmatrix} I_p \\ I_s \end{bmatrix}
= \left\{
\begin{bmatrix} Y_{pp} & Y_{ps} \\ Y_{sp} & Y_{ss} \end{bmatrix}
\begin{bmatrix} V_p \\ V_s \end{bmatrix}
+ \begin{bmatrix} Y_{pq} \\ Y_{sq} \end{bmatrix} V_q \\
- \begin{bmatrix} Y_{pr} \\ Y_{sr} \end{bmatrix}
\begin{bmatrix} Y_{rr}^{-1} Y_{rp} & Y_{rr}^{-1} Y_{rs} \end{bmatrix}
\begin{bmatrix} V_p \\ V_s \end{bmatrix} \\
- \begin{bmatrix} Y_{pr} \\ Y_{sr} \end{bmatrix} Y_{rr}^{-1} Y_{rq} V_q
\right\}
\tag{5}
$$

$$
\begin{bmatrix} T_{pp} & T_{ps} \\ T_{sp} & T_{ss} \end{bmatrix}
=
\begin{bmatrix}
Y_{pp} - & Y_{ps} - \\
Y_{pr} Y_{rr}^{-1} Y_{rp} & Y_{pr} Y_{rr}^{-1} Y_{rs} \\[2mm]
Y_{sp} - & Y_{ss} - \\
Y_{sr} Y_{rr}^{-1} Y_{rp} & Y_{sr} Y_{rr}^{-1} Y_{rs}
\end{bmatrix}
\tag{6}
$$

By using equation (6), equation (5) can be simplified to be equation (7), in which there are just only currents on nodes represented by p and s and voltages on nodes denoted by p, s, and q. As shown in Fig.3, the switching network constructed by CMOS circuits will be represented by (8) and (9) to describe the function of voltages and current on power supply nodes with voltages and currents on output driver nodes of drivers. The matrix, Z_{SW}, given by (8) should be a convolution matrix in frequency domain.

$$
\begin{bmatrix} I_p \\ I_s \end{bmatrix}
= \begin{bmatrix} T_{pp} & T_{ps} \\ T_{sp} & T_{ss} \end{bmatrix}
\begin{bmatrix} V_p \\ V_s \end{bmatrix}
+ \begin{bmatrix} Y_{pq} - Y_{pr} Y_{rr}^{-1} Y_{rq} \\ Y_{sq} - Y_{sr} Y_{rr}^{-1} Y_{rq} \end{bmatrix} V_q
\tag{7}
$$

$$
Z_{SW} = - \begin{bmatrix} Z_{11} & Z_{12} \\ Z_{21} & Z_{22} \end{bmatrix}
\tag{8}
$$

$$
\begin{bmatrix} V_p \\ V_s \end{bmatrix} = Z_{SW} \begin{bmatrix} I_p \\ I_s \end{bmatrix}
\tag{9}
$$

By using (9), equation (7) can be rewritten by (10), through which voltage column vector, V_p and V_s, are removed. Consequently, unknown variables in equation (10) are just current column vector, I_p and I_s, because V_q is the voltage column vector for DC power supply, which is known variable and the elements of matrix left are given by (1) and (8). Equation (10) can be also rewritten by (11) by using (8).

Since the equation (10) can be rewritten by (12), equation (13) can be easily obtained. Equation (13) addresses that current variables on power supply node and signal output node of drivers can be obtained by using the admittance matrix, (1), representing PSDN, macro model, (8), representing switching network, and power supply, V_q. Generally, (1) can be obtained by EM field solver. The admittance matrix, Y, can be easily calculated from the scattering parameter obtained by EM field solver. Fig. 4 shows simulation process for SI-PI co-simulation considering the coupling between SDN and PDN with coupled PSDN passive network model represented by (1).

$$
\begin{bmatrix} I_p \\ I_s \end{bmatrix}
= \left\{
- \begin{bmatrix} T_{pp} & T_{ps} \\ T_{sp} & T_{ss} \end{bmatrix}
\begin{bmatrix} Z_{11} & Z_{12} \\ Z_{21} & Z_{22} \end{bmatrix}
\begin{bmatrix} I_p \\ I_s \end{bmatrix} \right. \\
\left. + \begin{bmatrix} Y_{pq} - Y_{pr} Y_{rr}^{-1} Y_{rq} \\ Y_{sq} - Y_{sr} Y_{rr}^{-1} Y_{rq} \end{bmatrix} V_q
\right\}
\tag{10}
$$

$$
\begin{bmatrix} I_p \\ I_s \end{bmatrix}
= \left\{
\begin{bmatrix} T_{pp} & T_{ps} \\ T_{sp} & T_{ss} \end{bmatrix}
Z_{SW}
\begin{bmatrix} I_p \\ I_s \end{bmatrix} \right. \\
\left. + \begin{bmatrix} Y_{pq} - Y_{pr} Y_{rr}^{-1} Y_{rq} \\ Y_{sq} - Y_{sr} Y_{rr}^{-1} Y_{rq} \end{bmatrix} V_q
\right\}
\tag{11}
$$

$$
\left\{ I + \begin{bmatrix} T_{pp} & T_{ps} \\ T_{sp} & T_{ss} \end{bmatrix}
\begin{bmatrix} Z_{11} & Z_{12} \\ Z_{21} & Z_{22} \end{bmatrix} \right\}
\begin{bmatrix} I_p \\ I_s \end{bmatrix} \\
= \begin{bmatrix} Y_{pq} - Y_{pr} Y_{rr}^{-1} Y_{rq} \\ Y_{sq} - Y_{sr} Y_{rr}^{-1} Y_{rq} \end{bmatrix} V_q
\tag{12}
$$

$$\begin{bmatrix} I_p \\ I_s \end{bmatrix}$$

$$= \left\{ I + \begin{bmatrix} T_{pp} & T_{ps} \\ T_{sp} & T_{ss} \end{bmatrix} \begin{bmatrix} Z_{11} & Z_{12} \\ Z_{21} & Z_{22} \end{bmatrix} \right\}^{-1} \quad (13)$$

$$\times \begin{bmatrix} Y_{pq} - Y_{pr}Y_{rr}^{-1}Y_{rq} \\ Y_{sq} - Y_{sr}Y_{rr}^{-1}Y_{rq} \end{bmatrix} V_q$$

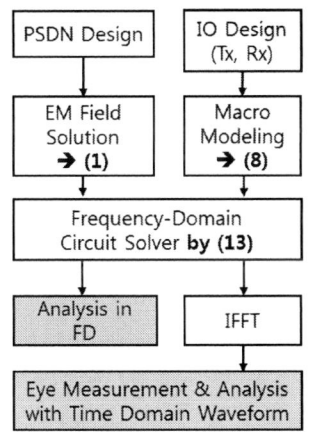

Figure 4. Simulation process based on the new approach summarized by equation (13)

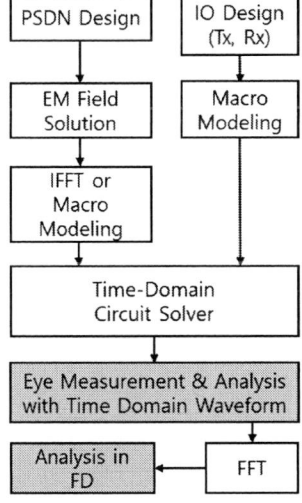

Figure 5. Conventional simulation process of time-domain SI-PI co-simulation. [14]

For reference, the conventional simulation process of time-domain SI-PI co-simulation is presented in Fig. 5, in which inverse fast Fourier transform (IFFT) or macro modeling of passive interconnect model representing SDN and PDN with proper coupling between SDN and PDN should be done before time-domain circuit solving.

Generally, SPICE has been used for time-domain circuit simulation. However, the proposed method in this paper makes the frequency-domain circuit simulation for SI-PI co-simulation possible.

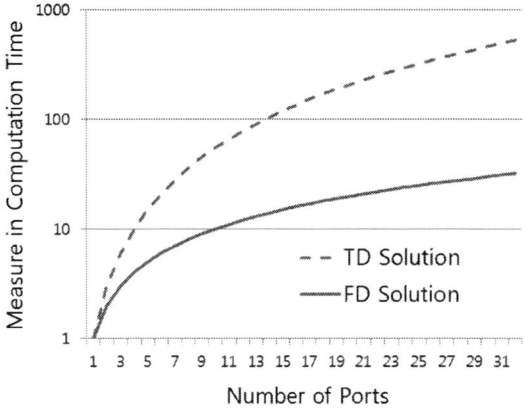

Figure 6. The number of IFFT required for SI-PI co-simulation. The blue dotted line and the red solid line represents that for time-domain solution and frequency domain solution, respectively.

In the conventional time-domain co-simulation, IFFT should be done with entire elements of scattering matrix, impedance matrix or admittance matrix. However, the proposed method with equation (13) needs IFFT only for node variables. When M is the number of ports or nodes of PSDN, IFFT should be done M times for the proposed to get time-domain waveform for SI-PI co-simulation. In other hands, IFFT should be performed M (M+1) / 2 times for the conventional SI-PI co-simulation. The comparison of the new approach based on equation (13) and the conventional time-domain SI-PI co-simulation is presented by Fig.6 in terms of computation time. As the number of nodes to be included in SI-PI co-simulation increases, the new approach proposed by authors becomes more and more efficient.

With the condition that there is no coupling between SDN and PDN, (13) can be rewritten to be (16) with (14) and (15). When the coupling between SDN and PDN is negligible, Y_{pr}, Y_{ps}, Y_{qr}, Y_{qs}, Y_{rp}, Y_{rq}, Y_{sp}, and Y_{sq} are nearly zero, by which equation (6) becomes (14) and (15). With (14), equation (13) can be rewritten by (16).

$$\begin{bmatrix} T_{pp} & T_{ps} \\ T_{sp} & T_{ss} \end{bmatrix} \cong \begin{bmatrix} T_{pp}{}' & 0 \\ 0 & T_{ss}{}' \end{bmatrix} \quad (14)$$

978-1-7281-1500-9/19 $31.00 © 2019 IEEE

$$\begin{bmatrix} T_{pp}' & 0 \\ 0 & T_{ss}' \end{bmatrix}$$
$$= \begin{bmatrix} Y_{pp} & 0 \\ 0 & Y_{ss} - Y_{sr}Y_{rr}^{-1}Y_{rs} \end{bmatrix} \quad (15)$$

$$\begin{bmatrix} I_p \\ I_s \end{bmatrix}$$
$$= \left\{ I + \begin{bmatrix} T_{pp}' & 0 \\ 0 & T_{ss}' \end{bmatrix} \begin{bmatrix} Z_{11} & Z_{12} \\ Z_{21} & Z_{22} \end{bmatrix} \right\}^{-1} \quad (16)$$
$$\times \begin{bmatrix} Y_{pq} - Y_{pr}Y_{rr}^{-1}Y_{rq} \\ Y_{sq} - Y_{sr}Y_{rr}^{-1}Y_{rq} \end{bmatrix} V_q$$

If Y_{pr}, Y_{ps}, Y_{qr}, Y_{qs}, Y_{rp}, Y_{rq}, Y_{sp}, and Y_{sq} are zero, then it is easy to use (16), which saves much computation time and resource. However, Y_{pr}, Y_{ps}, Y_{qr}, Y_{qs}, Y_{rp}, Y_{rq}, Y_{sp}, and Y_{sq} are not zero in most of practical cases in the real design cases. Therefore, we need to find a way to judge the amount of coupling to be able to be ignored, for which authors suggests the conditions with equation (17), (18), and (19). Although Y_{pr}, Y_{ps}, Y_{qr}, Y_{qs}, Y_{rp}, Y_{rq}, Y_{sp}, and Y_{sq} are not zero, equation (17), (18), and (19) can be used as a isolation index for SI-PI co-simulation.

$$Y_{pp} \gg Y_{pr}Y_{rr}^{-1}Y_{rp} \quad (17)$$

$$Y_{sp} - Y_{sr}Y_{rr}^{-1}Y_{rp} = 0 \quad (18)$$

$$Y_{ps} - Y_{pr}Y_{rr}^{-1}Y_{rs} = 0 \quad (19)$$

III. CASE STUDIES

The method proposed in the previous section will be demonstrated with PI simulation and SI-PI co-simulation with coupled PSDN.

A. Power Integrity Simulation

While conventional PI simulation by using CPM (chip power model) has been utilized for a PDN design and analysis of core IPs including CPU and GPU, CPM model representing current for core circuits is not dependent of PDN constructed by package and system-PCB. Although PDN can change power supply current itself, fixed CPM has been usually used for simple analysis, which should be also

carefully studied. In this study, typical PDN model is used as given in Fig 7. In Fig.8, the impact of PDN including equivalent inductance from capacitors in package to bumps in die on the rise time and peak value of current for core power is given by (a) and (b), respectively. In this analysis, equation (16) is used. Generally, PDN and SDN of core circuits are well isolated, which allows (15). Reading (16) shows that the current of power supply on bumps is function of PDN and SDN. As shown by the result in Fig.8, it is needed to consider the variation of the rise time and peak value of current with PDN design. When PI simulation and design with fixed CPM, the analysis could include wrongful results or non-physical data.

B. SI / PI Co-simulation with coupling in SDN and PDN

For SI-PI co-simulation, the proposed method suggests that the isolation index given by (17), (18), and (19) needs to be reviewed for considering the coupling between SDN and PDN. As shown in Fig.9, (18) and (19) incrase with the increase of the coupling.

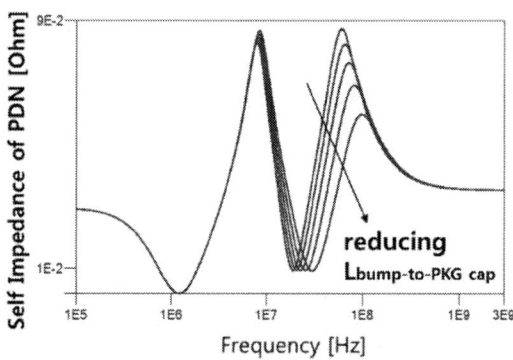

Figure 7. An example of self impedance of PDN with various equivalent inductance value representing path from capacitors in package to bumps on silicon die.

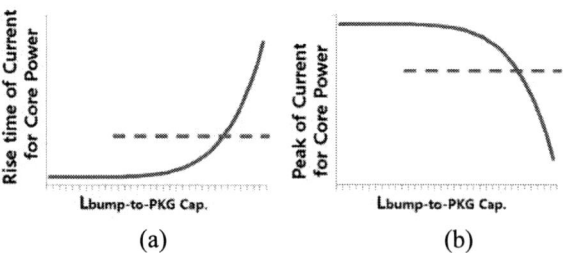

Figure 8. The impact of PDN on the rise time and peak value of current for core power is given by (a) and (b), respectively. The red dotted line is the upper limit for design, based on which the allowable maximum inductance is given.

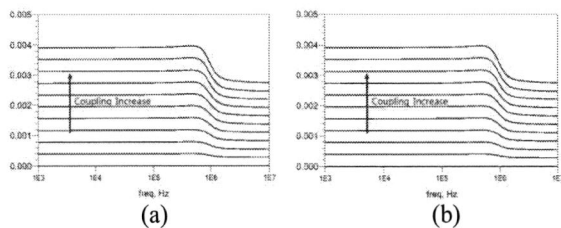

(a)　　　　　　　(b)

Figure 9.   In (a) and (b), (18) and (19) are given with the typical SI-PI co-simulation based on (13). As coupling between SDN and PDN increases, left terms in (18) and (19) increase.

IV.   CONCLUSION

The matrix formulation has been developed by authors with the behavior model of switching circuit network (SCN) constructed by CMOS transistors in SOC and the admittance matrix (Y) model of SDN and PDN given by EM field solution. The time-variant system behavior of SCN and linear time-invariant (LTI) system of SDN and PDN are combined by convolution matrix to represent voltage and current waves, which gives the closed form matrix formulation in the frequency domain. Based on the matrix formulation given in this paper, it has been also addressed that PDN and SDN can give critical impacts on PI, although there is no coupling between SDN and PDN.

A way to monitor the coupling is given, based on which SI-PI co-simulation can ignore the coupling between SDN and PDN. In addition, while SI/PI co-simulation has been conventionally performed by SPICE in time-domain as previously presented by the several papers, authors' matrix formulation proposed the frequency-domain SI-PI co-simulation, which saves the computation time with PSDN having huge number of nodes.

REFERENCES

[1] G. K. Rangaswamy, et al., "Signal Integrity Analysis with Power Delivery Network," Proc. IEEE Workshop on Signal Propagation on Interconnects, 2006, pp205-207

[2] A. Ciccomancini Scogna, et al., "SIPI co-extraction and SPICE co-simulation for package on-die decap optimization," Proc. IEEE 20th Workshop on Signal and Power Integrity (SPI), 2016, pp. 1-4

[3] M. E. Kowalski, and P. Codd, "Co-Simulation of IC, Package and PCB Power Delivery Networks in Ultra-Low Voltage Power Rail Designs," Proc. IEEE 57th Electronic Components and Technology Conference (ECTC), 2017, pp.798-803

[4] A. C. Scogna, et al., "Modeling Methodologies for Multi-Level PCB-Package Co-Simulation & Co-Design," Proc. " Proc. IEEE 22nd Conference on Electrical Performance of Electronic Packaging and Systems (EPEPS), 2013, p.57

[5] A. E. Engin, et al., "Analysis for Signal and Power Integrity Using the Multilayered Finite Difference Method," Proc. IEEE International Symposium on Circuits and Systems (ISCAS) 2007, 1493-1496

[6] G. Signorini, et al., "Power and Signal Integrity co-simulation via compressed macromodels of high-speed transceivers," Proc. 2015 IEEE 19th Workshop on Signal and Power Integrity (SPI), 2015, p.1-4

[7] M. Lai, et al., "SI-PI cosimulation analysis of dual referencing and VSS-Referencing memory bus," Proc. 2014 IEEE International Symposium on Electromagnetic Compatibility (EMC), 2014, pp. 821-826

[8] Z. Chen, "A general co-design approach to multi-level package modeling based on individual single-level package full-wave S-parameter modeling including signal and power/ground ports," Proc. Electronic Components and Technology Conference (ECTC) 2012, pp. 1687-1694

[9] M. Umekawa, "Optimum configuration of SI/PI Co-Simulation using electro-magnetic simulator," Proc. 2015 International Conference on Electronics Packaging and iMAPS All Asia Conference (ICEP-IAAC), 2015, pp.890-893

[10] T.-M. Winkel, et al., "Framework for co-simulation of signal and power integrity in server systems," Proc. IEEE 21st Conference on Electrical Performance of Electronic Packaging and Systems (EPEPS), 2012, pp.21-24

[11] S. Wane, et al., "Dynamic power and signal integrity analysis for chip-package-board co-design and co-simulation,". Proc. 2009 European Microwave Integrated Circuits Conference (EuMIC), 2009, pp. 527-530

[12] H. C. Shu, et al., "Full system power delivery analysis for single ended interface," Proc. 5th Asia Symposium on Quality Electronic Design (ASQED), 2013, pp. 144-147

[13] J. E. Schutt-Aine, et al., "Comparative Study of Convolution and Order Reduction Techniques for Blackbox Macromodeling Using Scattering Parameters," IEEE Transactions on Components Packaging and Manufacturing Technology, vol. 1, pp. 1642-1650, 2011

[14] R. Achar and M. S. Nakhla, ""Simulation of high-speed interconnects", Proc. IEEE, vol. 89, no. 5, pp. 693-728, May 2001.

[15] M. S. Tanaka, et al., "Early stage chip/package/board co-design techniques for System-On-Chip," Proc. IEEE 20th Conference on Electrical Performance of Electronic Packaging and Systems (EPEPS), Oct. 2011, pp.21-24

Signal Integrity of Submicron InFO Heterogeneous Integration for High Performance Computing Applications

Chuei-Tang Wang, Jeng-Shien Hsieh, Victor C. Y. Chang, Shih-Ya Huang,
T. Ko, Han-Ping Pu, and Douglas Yu
Research and Development, Taiwan Semiconductor Manufacturing Company, Ltd.
166, Park Ave. 2, Hsinchu Science Park, Hsinchu 30075, Taiwan, R.O.C.
ctwangm@tsmc.com

Abstract—Heterogeneous integration has attracted much attention for high performance computing (HPC) since artificial intelligence (AI) accelerators surged. The technologies for heterogeneous integration, such as silicon interposer (2.5D), fan-out wafer-level-packaging (FOWLP), and organic substrate, have been proposed to integrate logic-logic or logic-HBM chips in the AI system for performance and cost benefits. However, the tremendous data flow in 5G era requires higher data rate and bandwidth for the extensive die-to-die communication. Therefore, a BEOL-scale re-distributed layer (RDL) technology should be developed to satisfy the requirements. In this paper, a novel ultra-high-density InFO (InFO_UHD) technology with submicron RDL is developed to provide high interconnect density and bandwidth for logic-logic system. The bandwidth density can achieve record high 10 Tbps/mm at line width and spacing (L/S) of 0.8/0.8 μm and length of 500 μm, for a logic-logic system using simplified IO driver. Using the technology in logic-memory system, we found that the scaling of RDL thickness, L/S, and dielectric thickness can mitigate ring-back problems in the eye diagram of organic substrate. Given HBM2 specification, the bandwidth density can achieve more than 0.4 Tbps/mm from dramatically improved signal integrity. Finally, power efficiency, in the metric of energy per bit, of the interconnect technology under simplified IO driver and HBM2 driver condition was calculated and compared with other technology, respectively.

Keywords-InFO; submicron; RDL; bandwidth; data rate; signal integrity; heterogeneous integration; HPC

I. INTRODUCTION

Heterogeneous integration has attracted more attention from semiconductor industry since artificial intelligence (AI) applications emerged. Advanced heterogeneous integration technology for applications from edge to cloud was developed to meet the stringent performance requirements [1]. Innovative packaging technologies such as silicon interposer (2.5D) and fan-out wafer-level-packaging (FOWLP) become new paradigm for the semiconductor industry to realize the system integration. Packaging of silicon interposer, such as CoWoS, was used to integrate logic chips or logic and memory chip/cube with TSVs, fine pitch re-distributed layer (RDL), and micro bumps. InFO, a FOWLP technology, was proposed to remedy with lower RC delay, while without expensive TSV for 3D integration [2]. The paper also reported to use the technology for 2D multi-chips integration (3 chips put side by side) with a fine RDL width/spacing of 2/2 μm to enable homogeneous chip partition as well as heterogeneous integration for mobile computing and high power, high performance applications. The 2D InFO technology was

applied to a high speed SERDES system with two homogeneous chips integration [3]. However, to meet the ever-increasing bandwidth demands, a novel Ultra-High-Density Integrated Fan-Out wafer level packaging (InFO_UHD) was then developed [4]. The advanced interconnect and integration technology with submicron RDL line width and spacing is expected to support new high performance computing (HPC) applications in the future.

At the same time, logic IC with high bandwidth memory (HBM) system has been successfully implemented by CoWoS [5] and EMIB [6] technologies already, which remarkably enhance the HPC and AI accelerator performance [7]. In addition, an organic substrate solution for logic and HBM integration was also proposed as an alternative to seek for performance and cost benefits [8]. However, a ring-back in eye diagram was found in the organic substrate solution which will be a showstopper for HBM memory to scale up data rate from 2.0~2.4 to 4.0 Gbps. The paper mentioned thinning Cu thickness and dielectric thickness are the effective solutions to resolve the ring-back issue. Our proposed InFO_UHD technology with submicron-scale RDL could be a feasible solution to mitigate the ring-back problems in the eye diagram.

In this paper, the signal integrity and bandwidth density of the InFO_UHD technology are studied for the above-mentioned needs in logic-logic or logic-memory system. The rest of the paper is organized as follows. Section II introduces the InFO_UHD structure and technology maturity. Section III shows the intrinsic electrical properties, maximum data rate, bandwidth density and power efficiency of the interconnects. The effect of interconnect resistance on the ring-back, called

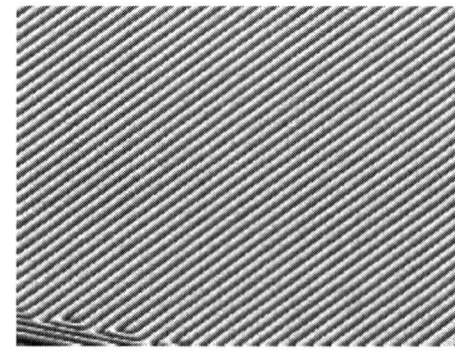

Figure 1. RDL line-width of 0.8 μm with RDL line-spacing of 0.8 μm is demonstrated [4].

ringing noise, of the eye diagram of HBM2 is studied, too. Section IV reports the insertion loss data of a co-planar waveguide (CPW) transmission line. In addition, the measurement is correlated with simulation results for verification. At last, Section V concludes the study of signal integrity and bandwidth on the InFO_UHD.

II. INFO_UHD STRUCTURE

In this section, the InFO_UHD structure is briefly depicted and its technology maturity is introduced. Its process and metrological tool sets are based on what currently used in InFO technology, instead of those for Cu/low-k BEOL used in semiconductor foundry. Hence, this technology enhances die-to-die communication performance for multi-die configuration and exhibits cost advantages for competitiveness in commercial market. The bird eye view of the InFO_UHD structure with L/S=0.8/0.8 μm of RDL is shown in Fig. 1.

The maturity of the InFO_UHD technology has been reported in 2018 ECTC [4]. The devices under test (DUTs) include comb/meander structures for sheet resistance and leakage current, Kelvin via structure for via resistance Rc, and via-chain to via-chain structure for continuity check. Additional assessment on reliability also has demonstrated its robust performances under either interconnect level or package level. For interconnect level, electro-migration (EM), stress-migration (SM), breakdown voltage (Vbd), and time-dependent dielectric breakdown (TDDB) were tested and verified. For package level, component-level multiple reflow (MR) and temperature cycling (TC) were checked. These wafer-level properties with nearly 100% yield warrant the electrical study for bandwidth and signal integrity in next section.

III. ELECTRICAL PROPERTY OF INFO_UHD

In this section, the intrinsic properties of different layout schemes are studied first. And then, the application for the logic-logic and logic-memory interfaces will be further discussed.

A. Intrinsic Property

The resistance (R), inductance (L), and capacitance (C) for a single microstrip structure at different layout dimensions are studied first. Afterward, the important index of RC delay is included.

The R is mainly affected by line width and thickness, as shown in Fig. 2 (a). The effective resistance becomes larger with line width narrower or conductor thickness thinner. The normalized line thickness is changed from 1.0x to 1.5x and the line width is swept from the narrowest line width 1x (0.8 μm), to 2.5x.

For the L and C effects, the line width and dielectric thickness contribute more significantly. In Figs. 2 (b) and (c), the range of line width in question is kept the same as that for

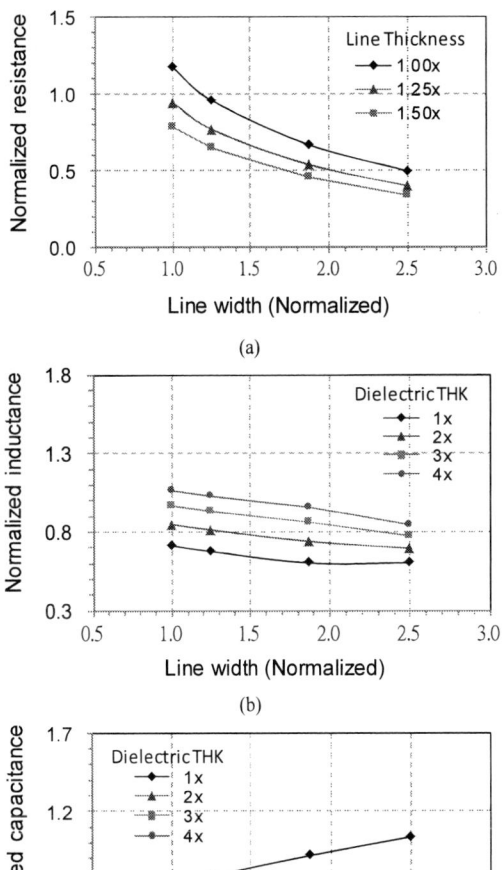

Figure 2. Intrinsic property characterization. (a) resistance effect affected by line width and thickness, (b) inductance effect affected by dielectric thickness and line width, and (c) capacitance effect affected by dielectric thickness and line width.

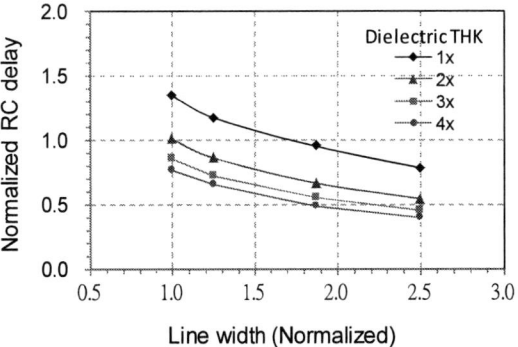

Figure 3. RC delay affected by dielectric thickness and line width.

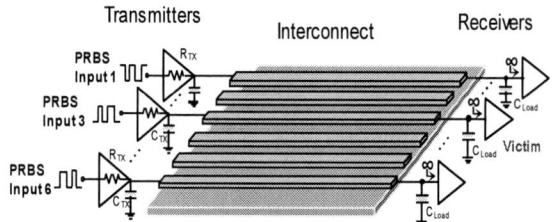

Figure 4. Simulation model for logic-to-logic interface.

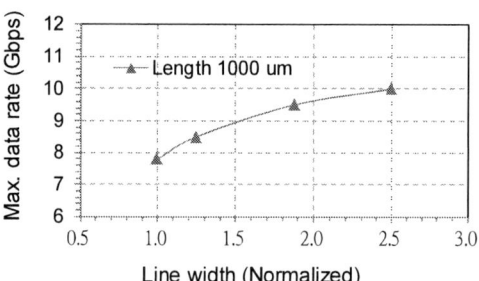

(a)

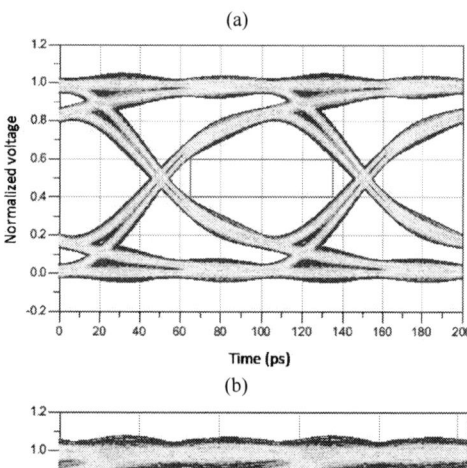

(b)

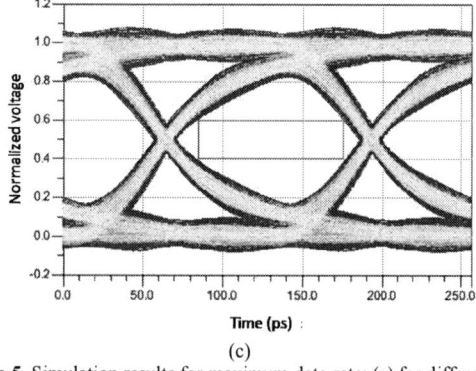

(c)

Figure 5. Simulation results for maximum data rate: (a) for different line widths, (b) the eye diagram for 2.5x of line width, and (c) eye diagram for 1x of line width.

resistance study. The dielectric thickness is changed from the thinnest (1x) of InFO_UHD to 4x. The L result in Fig. 2(b) shows that narrower line width and/or thicker dielectric

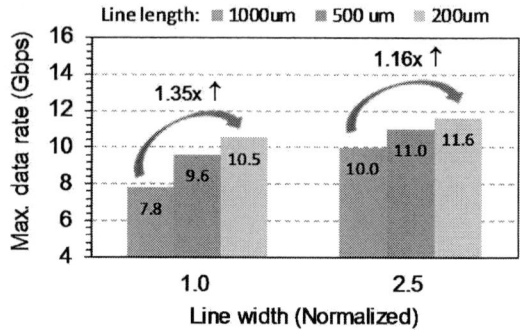

Figure 6. Line length effects on maximum data rate of interconnect, of which the line width and spacing are the same.

thickness cause larger inductance. On the contrary, the capacitance has different trend. The capacitance increases with the increase of line width or the decease of dielectric thickness.

In Fig. 3, the RC delay trend implies that narrower line width or thinner dielectric thickness will give larger product of R and C. For a given dielectric thickness, the RC delay is dominated by resistance. For example, the blue line for the 2x dielectric thickness shows that RC delay increases about 2 times when the line width decreases from 2.5x to 1x. On the other hand, if lower RC delay is desired for a narrower line, thicker dielectric thickness can be applied to help achieve the goal. The reason is that thicker dielectric makes the capacitance smaller.

B. Logic to Logic Interface

For logic to logic interface, the total bandwidth density is one of the criteria to determine the system performance. Two important factors to affect the bandwidth density are maximum data rate and line density. In I/O interface design, the maximum data rate is affected by the line width/spacing and length, while the line density is determined by the feature size of line width/spacing of the interconnect process. For the advanced technology, InFO_UHD, the line density is high from innovative process capability. In practical, narrower line may result in lower maximum data rate, but the bandwidth density could still be higher, since the line density is much higher.

To analyze the logic to logic interface, a multiple line structure with 6 microstrip lines and simplified transmitter/ receiver models are used, as shown in Fig. 4. In the simulation, the line width and spacing of the 6 microstrip lines are varied from 1x to 2.5x in scale for the interconnect model. Simplified transmitter models [6] are on the left hand side of the figure. The internal impedance of the transmitter is set as 50 Ω and the output pad capacitance is assumed to be 0.4 pF. For the interconnect model, six PRBS (Pseudo Random Bit Sequence) patterns are generated at transmitter side. On the right hand side of this model, there are six identical simplified receivers.

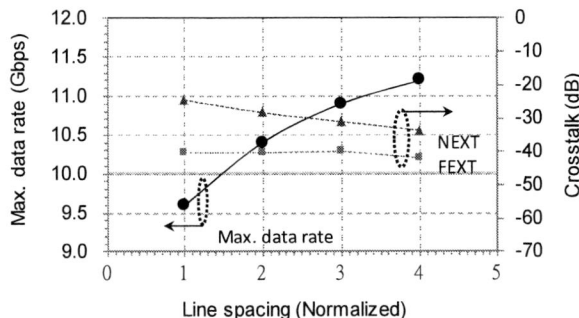

Figure 7. Line spacing effects on maximum data rate and crosstalk with line width of 1x of narrowest line.

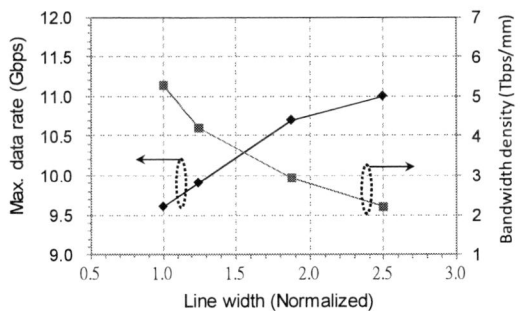

Figure 8. Bandwidth density performance for the scenario with line length of 500 μm.

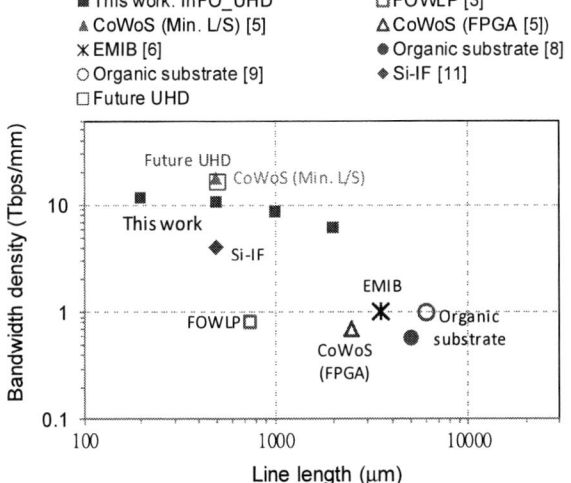

Figure 9. Bandwidth density of this work, InFO_UHD, and various interconnect technologies.

For the receiver models, the load capacitance is set as 0.4 pF and the input impedance of the receiver is infinite.

With aforementioned models and setup, the maximum data rates for lines with 1000 μm length and various line width/spacing are studied and the results are shown in Fig. 5(a). The maximum data rate for the 1x line width/spacing is about 8 Gbps, and for 2.5x line width/spacing, the maximum data rate could reach about 10 Gbps, and the corresponding eye diagrams are shown in Fig. 5(b) and (c), respectively. The maximum data rate is determined by the criterion that the eye diagram of the maximum operation frequency passes the eye mask. The eye mask is defined to be 0.7 times of unit interval of eye width and between 0.4 to 0.6 times of DC voltage in eye height.

One method to increase the maximum data rate is to shorten the line length. In some applications, the length between transmitter and receiver may be shorter than 1000 μm in Fig. 5(a). In Fig. 6, the 1x and 2.5x line width/line spacing cases with different line lengths from 1000 μm to 200 μm are studied. The maximum data rate is improved from 7.8 to 10.5 Gbps when the line length is shortened from 1000 μm to 200 μm for 1x case. One can use more advanced technology to shorten the die-to-die spacing to decrease line length to enhance the maximum data rate, when doing the logic to logic IO design.

Fig. 7 shows the effect of line spacing on the maximum data rate for the 1x line width at 500 μm line length. If the design specification for maximum data rate is 11 Gbps, it could be achieved by increasing the line spacing from 1x to 3x to reduce the crosstalk. In Fig. 6, the maximum data rate could be obtained to 11 Gbps for line width/spacing of 2.5x at the line length of 500 μm condition. The line pitch is 5x from the line width/spacing. But for the 1x line width and 3x line spacing condition in Fig. 7, the line pitch is 4x. Therefore, there is a 1x line pitch saving if we use 1x line width and 3x line spacing for 11 Gbps data rate design. The 1x line pitch saving could provide more design flexibility.

For the bandwidth density performance, besides maximum data rate discussed before, the line density is also critical. The bandwidth density and maximum data rate for various line width/spacing at 500 μm line length are depicted in Fig. 8. It can be found that the bandwidth density for the 1x line is still higher than the 2.5x line, even though the maximum data rate is lower for the 1x line. For example, the bandwidth density for 1x line is about 2.5 times better than that for 2.5x line. Namely, more than 5 Tbps/mm of bandwidth density can be achieved as long as the line length is less than 500 μm.

At last but not least, to compare the bandwidth density for various packaging technologies, a general survey is done and the results are captured in Fig. 9. It is noticed that, in the InFO_UHD technology, the bandwidth density can achieve record high 10 Tbps/mm, assuming two critical routing RDL layers. However, in other technologies and applications, including CoWoS (FPGA), the bandwidth densities range from 1 to 4 Tbps/mm only. Hence, the InFO_UHD is the technology to provide the highest bandwidth density. When we further compare the technology to CoWoS with minimum

TABLE I.   VARIOUS SCHEMES FOR LOGIC TO MEMORY INTERFACE

Scheme	Technology	Line THK (*) (Normalized)	Dielectric THK (Normalized)
1	*Organic substrate* [8]	2.50x	2.5x
2		1.50x	1.0x
3	*This work*	1.25x	1.0x
4		1.00x	1.0x

(*) Both line width and spacing equal 2μm for all schemes.

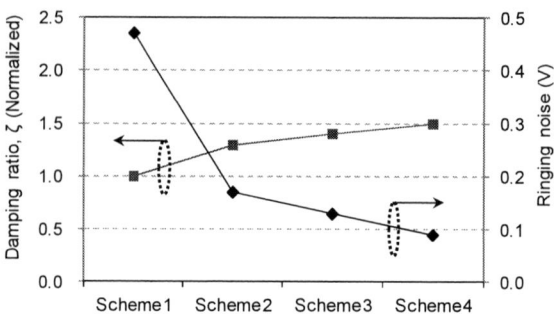

Figure 11. Line thickness effects on damping ratio and on ringing noise mitigation.

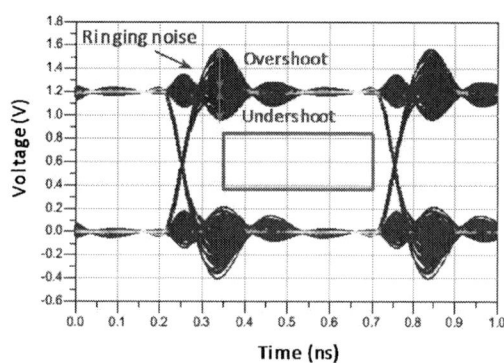

(a)

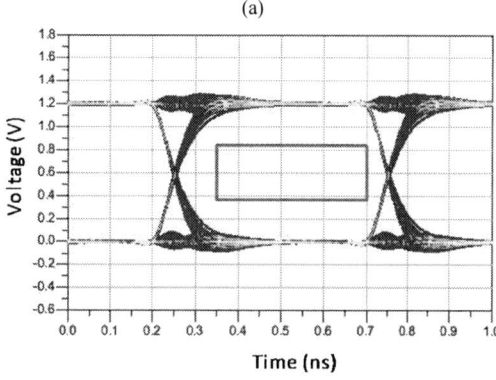

(b)

Figure 10. Eye diagrams of logic to memory interface, where (a) is for scheme 1 and (b) for scheme 4.

L/S using the same simplified IO driver, the bandwidth density of the InFO_UHD is lower than that of CoWoS with min. L/S. However, we can obtain higher bandwidth density when we scale down the InFO_UHD, denoted as Future UHD in Fig. 9.

C.   Logic to Memory Interface

For logic to memory interface, ringing noise of eye diagram of HBM2 on the InFO_UHD technology is studied. There was a study of the ring noise of the eye diagram on an organic substrate [8]. The ringing noise is caused by the resonance effect between the internal impedance of HBM2

driver and the parasitic R, L, and C of package interconnects on the organic substrate. The ringing noise will result in overshoot and undershoot of eye diagram, as shown in Fig. 10(a), which will lead to the failure of the signal transmission. In our study, we find the InFO_UHD technology could resolve this problem.

The ringing noise could be mitigated by adjusting the Cu line thickness and dielectric thickness. For logic to HBM2 interface, the bandwidth density is not as critical as that for logic to logic interface. Hence, the line pitch of 4 μm with equal line width and spacing is used for various schemes of the line thickness and dielectric thickness in our study. Table I shows four schemes in the simulation. The line thickness and dielectric thickness of Scheme 1 are similar to those of the organic substrate [8], while for Scheme 2 to 4, they are using thinner line and dielectric thicknesses in the InFO_UHD technology. The eye diagram result for Scheme 1 could be found in Fig. 10 (a) with obvious ringing noise. In contrast, the ringing noise is mitigated by applying Scheme 4 condition, as shown in Fig. 10 (b).

The success to mitigate ringing noise comes from the increase of damping ratio in the InFO_UHD by thinning of the Cu line and dielectric thickness. The damping ratio is defined as

$$\zeta = \frac{R}{2}\sqrt{\frac{C}{L}}, \qquad (1)$$

where R is the Cu line resistance with driver source resistance, C is the line capacitance with driver load capacitance, and L is the line inductance. The R, C, and L of the interconnect line determine the ζ value. When $\zeta = 1$, it is performed critically damping [10], where there is no ringing noise issue for eye diagram. When the damping ratio is much smaller than 1, the ringing noise is happened to become an issue. For example, when the damping ratio of $\zeta < 0.7$ is performed, a big ringing noise is happened in the eye diagram, like the example as shown in Fig. 10(a).

978-1-7281-1500-9/19 $31.00 © 2019 IEEE   692

To mitigate the ringing noise, the damping ratio should be closer to 1. High R and C will be required to achieve the $\zeta = 1$ condition according to (1). In the InFO_UHD technology, the line resistance and capacitance are higher than those of conventional organic substrate due to the thinner Cu line and dielectric thickness, as described in section III-A. In Fig. 11, the damping ratios and corresponded ringing noises for the four schemes are plotted. It can be seen that if the line thickness is changed from 2.5x to 1x and the dielectric thickness is reduced from 2.5x to 1x, the damping ratio of the interconnect is increased so that the ringing noise could be reduced by 0.4 V from 0.5 V to 0.1 V. The result indicates the InFO technology can provide better interconnect properties to resolve the ringing noise issue. Given HBM2 specifications, bandwidth density can achieve more than 0.4 Tbps/mm from the dramatically improved signal integrity. Finally, power efficiency of InFO_UHD technology, in the metric of energy per bit, will be discussed in next section.

D. Power Efficiency Performance

Power efficiency, in terms of energy per bit, is another important metric for HPC system performance. A 0.06 pJ/bit of power efficiency was reported in a silicon interconnect fabric (Si-IF) technology [11], where 2.5x line width and 500 μm line length are used, for logic to logic interconnect. In the InFO_UHD technology, the power efficiency is simulated using simplified transmitter and receiver models with PRBS input, as mentioned in section III-B. The power efficiency is 0.061 pJ/bit at 1.0x line width and 500 μm line length condition, based on 0.1pF capacitance specified for both of the transmitter and the receiver. Even though, the power efficiency is similar for the two interconnect technologies, as shown in Fig. 12, but the line width of InFO_UHD is two-fifths of that of the Si-IF technology. The InFO_UHD fine line does not consume more interconnect power.

For logic to HBM2 memory interface, InFO_UHD interconnect of 2.5x line width for typical RDL length exhibits 0.62 pJ/bit of power efficiency, based on the memory driver model with 2 Gbps data rate. For comparison, power efficiency of silicon interposer interconnect under typical dimensions is studied. The power efficiency is 0.83 pJ/bit, as shown in Fig. 12. The power efficiency of the InFO_ UHD technology is 25% better than that of the silicon interposer technology in logic to HBM2 memory system, due to lower capacitance in the interconnect.

IV. MEASUREMENT AND CHARACTERIZATION

In this section, the InFO_UHD RLC electrical properties explored previously are characterized by test patterns on a test vehicle. The insertion loss of coplanar waveguide (CPW) transmission lines on a single-layer RDL is measured. Two-port Vector Network Analyzer (VNA) is used to take S-parameters from the DUT up to 50 GHz. For measurement

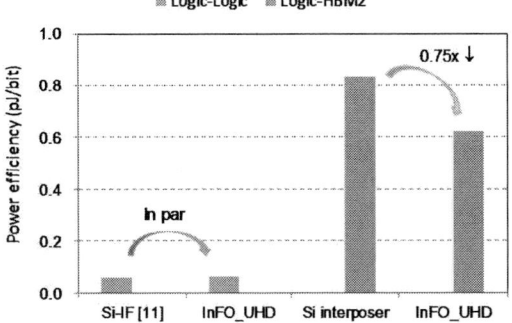

Figure 12. Power efficiency performance of Logic-Logic and Logic-Memory interfaces.

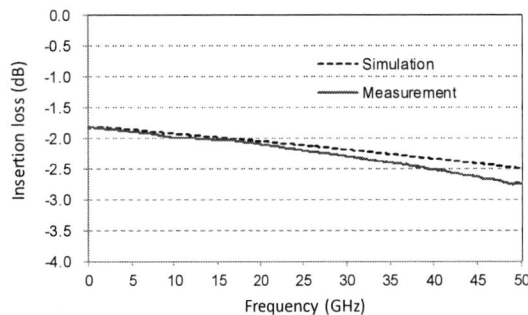

Figure 13. Line loss characterization for CPW line of 500 μm length.

accuracy of the submicron RDL, L-2L de-embedding techniques [12] are deployed.

The DUTs of L-length and 2L-length are two distinct CPW lines with length of 500 and 1000 μm, respectively. Fig. 13 shows the de-embedded measurement results of the L-length CPW line (500 μm). The insertion loss is smaller than -3 dB with measurement frequency up to 50 GHz for the submicron RDL. The measurement and simulation data are in good agreement. The result gives us the confidence on the electrical properties of the submicron RDL in the InFO_UHD technology.

V. CONCLUSIONS

The signal integrity properties of the novel ultra-high-density InFO (InFO_UHD) technology with submicron RDL are studied. The technology can provide more interconnects and higher bandwidth than conventional fan-out technology for logic to logic chip integration and the bandwidth density can achieve record high 10 Tbps/mm when the line length is 500 μm. For logic to memory interface, the signal integrity can be dramatically improved by the mitigation of ringing noise from the scaling of RDL thickness, L/S, and dielectric thickness. In the last, the technology provides superior power efficiency from the lower interconnect parasitic capacitance for HPC system applications.

REFERENCES

[1] Douglas C. H. Yu, "Advanced packaging with greater simplicity," 2017 IEEE International Electron Devices Meeting (IEDM), IEEE Press, Dec. 2017, pp. 3.6.1-3.6.4, doi: 10.1109/IEDM.2017.8268321.

[2] C.-F. Tseng, C. S. Liu, C.-H. Wu, and Douglas Yu, "InFO (wafer level integrated fan-out) technology," 2016 IEEE 66th Electronic Components and Technology Conference (ECTC), IEEE Press, May 2016, pp. 1-6, doi: 10.1109/ECTC.2016.65.

[3] N.-C. Chen, et al., "A novel system in package with fan-out WLP for high speed SERDES application," 2016 IEEE 66th Electronic Components and Technology Conference (ECTC), IEEE Press, May 2016, pp. 1495-1501, doi: 10.1109/ECTC.2016.43.

[4] H.-P. Pu, H. J. Kuo, C. S. Liu, and Douglas C. H. Yu, "A novel submicron polymer re-distribution layer technology for advanced InFO packaging," 2018 IEEE 68th Electronic Components and Technology Conference (ECTC), IEEE Press, May 2018, pp. 45-51, doi: 10.1109/ ECTC.2018.00015.

[5] S. Y. Hou, et al., "Wafer-level integration of an advanced logic-memory system through the second-generation CoWoS technology," IEEE Trans. Electron Devices, vol. 64, no. 10, pp. 4071–4077, Oct. 2017, doi: 10.1109/TED.2017.2737644.

[6] J. Cho, et al., "Electrical characterization of embedded multi-die interconnect bridge (EMIB) and interposer considering system bandwidth and I/O power consumption," 2017 DesignCon, Santa Clara, Jan. 2017.

[7] W. Shi, Y. Zhou, and S. Sudhakaran, "Power delivery network design and modeling for high bandwidth memory (HBM)," 2016 IEEE 25th Electrical Performance of Electronic Packaging and Systems (EPEPS), IEEE Press, Oct. 2016, pp. 1283-1288, doi: 10.1109/EPEPS.2016.7835405.

[8] V. Heyfitch, et al., "High bandwidth memory interface on organic substrate: challenges to electrical design," 2018 IEEE 68th Electronic Components and Technology Conference (ECTC), IEEE Press, May 2018, pp. 1289-1294, doi: 10.1109/ECTC.2018.00198.

[9] W. J. Turner, et al., "Ground-referenced signaling for intra-chip and short-reach chip-to-chip interconnects," 2018 IEEE Custom Integrated Circuits Conference (CICC), IEEE Press, April 2018, pp. 1-8, doi: 10.1109/CICC.2018.8357077.

[10] K.-B. Wu, et al., "Novel RDL design of wafer-level packaging for signal/power integrity in LPDDR4 Application," IEEE Trans. Compon., Packag., Manuf. Technol., vol. 8, no. 8, pp. 1431–1439, Aug. 2018, doi: 10.1109/TCPMT.2018.2850528.

[11] S. Jangam, A. Bajwa, K. K. Thankkappan, P. Kittur, and S. S. Iyer, "Electrical characterization of high performance fine pitch interconnects in silicon-interconnect fabric," 2018 IEEE 68th Electronic Components and Technology Conference (ECTC), IEEE Press, May 2018, pp. 1283-1288, doi: 10.1109/ECTC.2018.00197.

[12] Q.-H. Bu, et al., "Evaluation of a multi-line de-embedding technique for millimeter-wave CMOS circuit design," 2010 IEEE Asia-Pacific Microwave Conference (APMC), IEEE Press, Dec. 2010, pp. 1901-1904.

2019 IEEE 69th Electronic Components and Technology Conference (ECTC)

28GHz Through Glass Via (TGV) Based Band Pass Filter Using Through Fused Silica Via (TFV) Technology

Renuka Bowrothu, Seahee Hwangbo, Todd Schumann and Yong-Kyu Yoon
University of Florida
Gainesville, USA
rbowrothu93@ufl.edu ykyoon@ece.ufl.edu

Anthony Ng'Oma, Cheolbok Kim
Corning Incorporated, Corning
NY, USA

Abstract—We present the design, fabrication, and characterization of a band pass filter with a center frequency of 28 GHz using a Half Mode Substrate Integrated Waveguide (HMSIW) architecture on a smooth and thin fused silica substrate for the reduction of insertion loss. Filter properties are implemented by using a single Complimentary Split Ring Resonator (CSRR) structure. High out-of-band attenuation is achieved by using four spur line structures embedded in the microstrip feed line, which contributes to overall size reduction of the device. Design is simulated using High Frequency Structural Simulator (HFSS). Devices are fabricated using laser machining and microfabrication. The measured insertion loss is 3dB and the total area of the device is 5.719 mm².

Keywords- Band pass filter; Half Mode Substrate Integrated Waveguide (HMSIW); Complimentary Split Ring Resonator (CSRR); High Frequency Structural Simulator (HFSS); Laser Maching; Microfabrication

I. INTRODUCTION

With swift growth in data usage and transmission on wireless communications, bandwidth shortage is one of the main challenging problems in today's global market [1] and hence it leads to the advent of the fifth generation (5G) communication technology. The 5G will be using higher frequency spectra compared to 4G communications such as millimeter wave (mmW) frequency bands, multiple frequency and massive Multiple-Input-Multiple-Output (MIMO) antennas, and higher data rates. Possible drawback is that rain and atmosphere might attenuate the mmW signals. But it is also observed that atmospheric absorption does not create significant attenuation if cell sizes are in the order of 200 m especially for mm waves in 28-38 GHz. Hence frequencies between 28-38 GHz can be one of the realizable bands for future 5G communications [1].

Although the mm wave spectrum has many advantages, the performance of the device is hindered by mainly 3 types of losses i.e. dielectric, conductor and surface roughness losses. Previous literature demonstrated [2-3] low loss metaconductors which exhibit less resistance than

conventional copper in a certain frequency range. The dielectric loss is due to the material properties and this can be minimized by adopting materials with low loss tangent. Surface roughness effects will be significant especially when the roughness of the substrate is in the similar order of magnitude of the skin depth in the operating frequency.

The skin depth δ of a conductor as a function of frequency is given by

$$\delta = \frac{1}{\sqrt{\pi f \mu \sigma}}$$

where f is the frequency, μ is the magnetic permeability, and σ is the electrical conductivity. The skin depth drops in inverse square root as a function of frequency as shown in Figure 1. In the low sub GHz region, the roughness effects are not very dominating due to the thick skin depth. But as the frequency increases and the skin depth decreases, the effective conductivity of the metal is reduced as explained in the Groiss model and the Hammerstad and Jensen model [4-5].

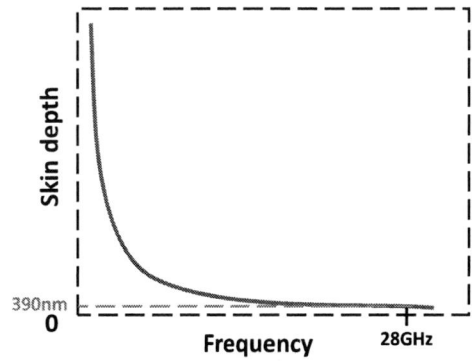

Figure 1. Skin depth of copper as a function of frequency

In this paper, a band pass filter is demonstrated on a 100 μm thick fused silica ($\varepsilon_r = 3.8$) substrate whose surface roughness is typically in the order of 10s of nm and whose dielecrtric loss is small. Design is performed using High Frequency Structural Simulator (HFSS) and prototype devices

978-1-7281-1500-9/19 $31.00 © 2019 IEEE

are microfabricated. Parameters like the thickness of metal, surface roughness, return and insertion loss of the device are characterized.

II. ANALYSIS AND DESIGN

Microstrip lines become very leaky at high frequencies especially when the substrate thickness becomes one tenth of its wavelength [6]. Air filled waveguides demonstrated good performance at high frequency, but it is challenging to integrate with other planar circuitry. Substrate Integrated Waveguide (SIW) [7] is another alternative where it is generally fabricated with two rows of cylindrical vias embedded in the substrate, filled with metal and thereby connecting two parallel plates and acting as magnetic side walls. Advantages of SIW based structures include low loss and integrability with other printed circuitry. A schematic of SIW is shown in Figure 2.

In order to have continuous and smooth electrical boundary, the via diameter (r) and the pitch (b) of the vias are usually taken much smaller than the guided wavelength. It is assumed that there is a magnetic boundary at the center of the waveguide where the symmetrical E field distribution is

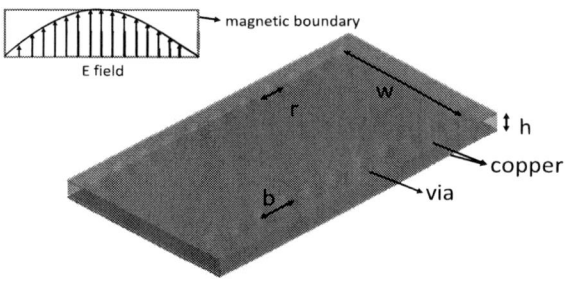

Figure 2. Substrate Integrated Waveguide

around it [8]. This leads to the existence of Half Mode Substrate Integrated Waveguide (HMSIW) whose field distribution is as the same as the conventional SIW if the cutting plane is serving as the magnetic wall, i.e. the center line of the waveguide. Due to the high width (w) to height (h) ratio, the open end acts like a magnetic wall [8]. The HMSIW structure and E field distribution is shown in Figure 3.

HMSIW has $TE_{0.50}$ as the dominating mode which is like half of the TE_{10} dominant mode in SIW. The cut off frequency of the HMSIW $TE_{0.50}$ mode is calculated using the width w/2 as a corresponding quarter wavelength [10-11]. Dimensions of an implemented HMSIW are given in Table 1.

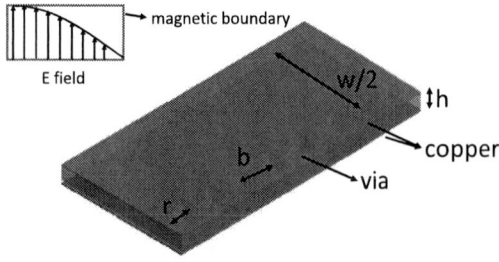

Figure 3. Half Mode Substrate Integrated Waveguide

Table 1. Dimensions of HMSIW

Parameter	Dimension (mm)
h	0.1
w/2	1.65 mm
b	0.15
r	0.1

Metamaterial structures are a type of artificial materials with unique electromagnetic properties that are not readily found in nature. Along with negative ε and μ values these structures shows the inverse Snell law, the inverse doppler effect and backward Cherenkov radiation. Complimentary Split Ring Resonator (CSRR) is also a type of metamaterial, which can be modeled using a parallel resonant circuit. When it is implemented on a HMSIW, it can create a pass band even when the waveguide is in the evanescent mode, i.e. it is operated below its cutoff frequency [10].

Figure 4 and Table 2 present an implemented CSRR structure operating at 28GHz.

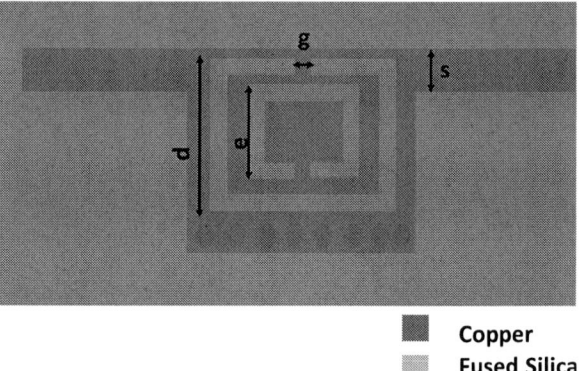

Copper
Fused Silica

Figure 4. CSRR structure on fused silica

Table 2. Dimensions of CSRR

Parameter	Dimensions (mm)
d	0.599
e	0.389
g	0.035
s	0.3

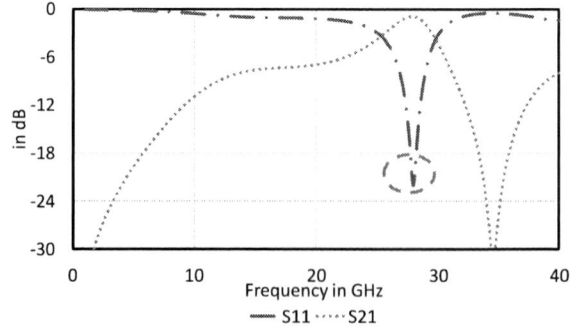

Figure 5. HMSIW CSRR Simulation

978-1-7281-1500-9/19 $31.00 © 2019 IEEE

In order to have large out-of-band attenuation or sharp band pass characteristics, different designs have been used previously [12-13]. One way is to use multiple CSRR structures which are spaced equally at certain distance. To have a compact device size and the effective use of feedlines, spurline structures are implemented on the microstrip line. Spurline can be realized by etching an L shape slot on the microstrip line [14]. Its length is usually chosen as $\lambda_g/4$ and it creates notch filter characteristics. Implementation of spurlines integrated with an HMSIW CSRR structure is shown in Figure 6.

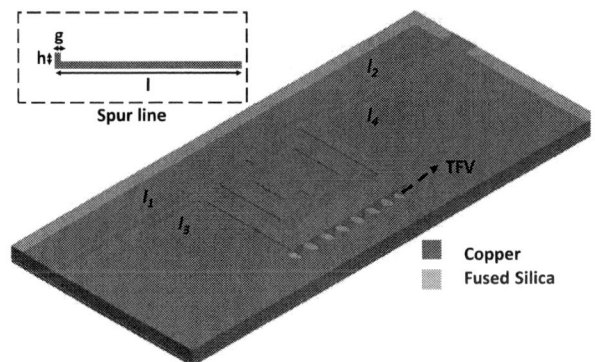

Figure 6. HMSIW CSRR with spur line structures
Inset: spurline schematic

Four of these spur line structures are added to create wide notch band characteristics below and above the CSRR resonant frequencies. $l1$ and $l2$ create lower stop band properties, $l3$ and $l4$ do higher stop band ones. Simulated and designed dimensions are given in Table 3.

Table3: Dimensions of Spur lines

Spurline	l (mm)	g (mm)	h (mm)
$l1$	2.1	0.05	0.07
$l2$	2.38	0.08	0.05
$l3$	1.2	0.06	0.03
$l4$	0.99	0.04	0.03

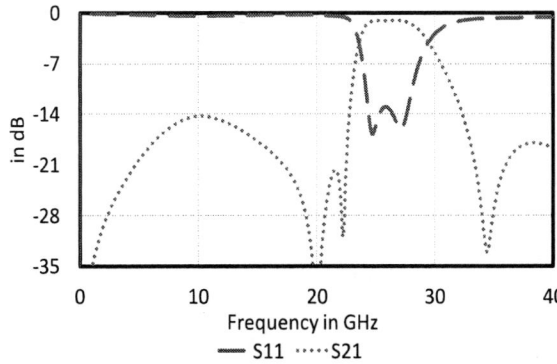

Figure 7 S_{11} and S_{21} of an HMSIW CSRR filter with spur lines

The entire structure is simulated using high frequency structure simulator (HFSS, ANSYS Inc.) and scattering parameters such as S_{11}, S_{21} are shown in Figure 7. In an ideal condition with a smooth surface and a thick copper conductor (e.g. 10 μm thick), the insertion loss is calculated as 1 dB.

III. FABRICATION

The vias are laser drilled using OXFORD LASERS. The diameter on top of the substrate is 100 μm but due to the edge slope, the bottom via diameter is less than 100 μm. Figure 8 shows the laser machined vias.

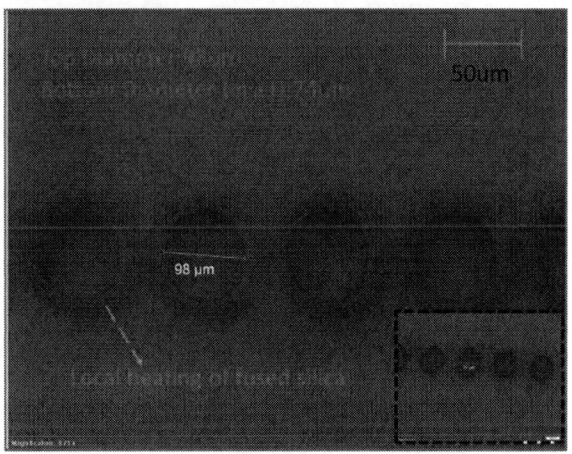

Figure 8. Laser Machined vias on fused silica

40% of total laser power is used to drill the via holes. Due to high thermal power dissipation, local heating of the substrate is observed. A difference of 24 μm in via diameter is observed on the top and bottom of the substrate at the end of process. Since the thickness of glass (fused silica) is very thin and to avoid warpage during electrodeposition, both sides of the substrate (design and the ground plane) are electroplated simultaneously. Electroplating parameters are given in Table 4. Complete fabrication flow is shown in Fig9.

Table 4. Electroplating Parameters

Parameter	Value
Current	10 mA
Temperature	38^0C
Speed	60 RPM
Time	20 mins

. The final fabricated device is shown in Figure 10. For measurement analysis, the coplanar waveguide (CPW) probing technique is used. The signal line width is 130 μm and the gap between the signal and ground lines is 22 μm. The total area of the fabricated device with and without the CPW probe pads is 10.55 mm^2 and 5.719 mm^2, respectively.

978-1-7281-1500-9/19 $31.00 © 2019 IEEE

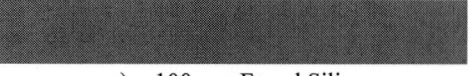

a)   100 μm Fused Silica

b)   Via drilling using Laser Machining
followed by Piranha clean

c)   30 nm Ti followed by 300 nm Cu
Deposition using sputtering technique

d)   20 μm dry film deposition using laminator
followed by lithography

e)   Top and bottom electroplating

f)   PR removal using acetone and IPA

g)   300 nm Cu seed layer removal

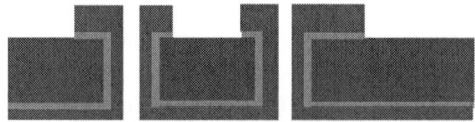

h)   30 nm Ti removal

Figure 9. Fabrication flow

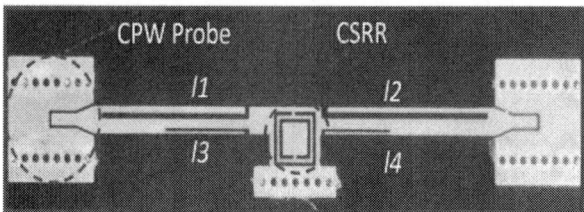

Figure 10. Fabricated Device

IV.   MEASUREMENT RESULTS AND ANALYSIS

Ground-signal-ground (GSG) 3 probe measurements are carried out with a probe station and a vector network analyzer (VNA). The tool is calibrated out using CS-5-150 calibration kit. Measured results are shown in Figure 11.

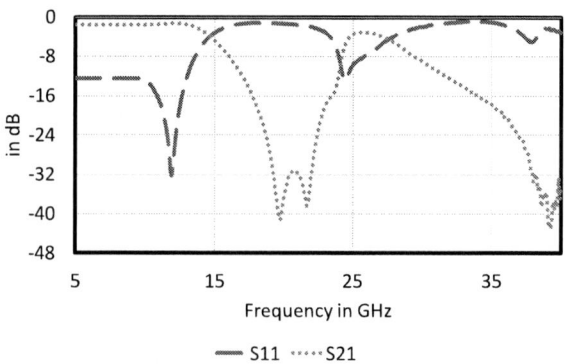

Figure 11. Measured scattering parameters

From measured data, it is observed that the insertion loss is nearly 3dB. The operating center frequency is shifted due to fabrication tolerance. Measured S_{11} is nearly 12 dB at 24.5 GHz. A difference of 2 dB in insertion loss is observed between the ideal simulation conditions and measured data.

Optical profilometer measurements are taken to calculate the thickness of copper. Figure 12 depicts the measured copper thickness.

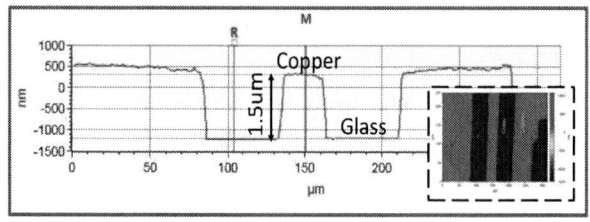

Figure 12. Measured copper thickness (Inset: Scanned area)

Also, the average roughness of fused silica is measured as approximately 600 nm. Figure 13 shows the scanned roughness of fused silica.

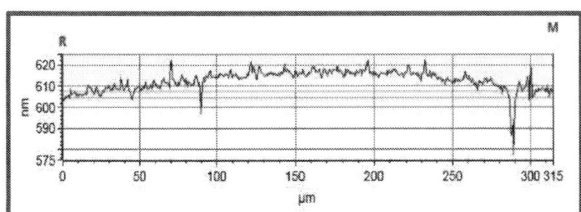

Figure 13. Surface Roughness on Fused Silica

The measured copper thickness and surface roughness of the fused silica substrate are inserted back into HFSS. The newly simulated and measured insertion loss data are plotted in Figure 14.

It can be observed that when both thickness of copper and roughness data are added to the simulation, the insertion loss at 26 GHz is 2.6 dB which is in the similar range of the measured insertion loss.

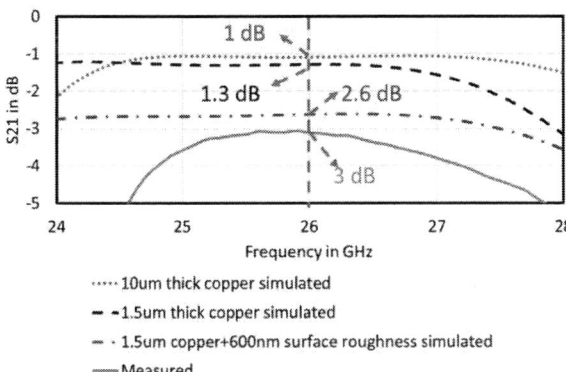

····· 10um thick copper simulated

– –1.5um thick copper simulated

– · 1.5um copper+600nm surface roughness simulated

——Measured

Figure 14. Comparison of various simulated and measured data

V. CONCLUSION

A bandpass filter with HMSIW, single CSRR, and spurlines is designed on fused silica. Via hoes are drilled using laser technique and a complete device is microfabricated. Shift in frequency is observed due to fabrication tolerance. Deviation from simulations in insertion loss is due to the conductor loss of thin copper and the roughness loss of the substrate.

ACKNOWLEDGMENT

The authors would like to thank Corning Inc. for test samples, Dr. Mark Sheplak group from Interdisciplinary Microsystems Group for Laser equipment at University of Florida is acknowledged. Dupont Inc. is acknowledged for the donation of dry film photo resist. Microfabrication is performed in Nanoscale Research Facility (NRF) at University of Florida.

REFERENCES

[1] T. S. Rappaport *et al.*, "Millimeter Wave Mobile Communications for 5G Cellular: It Will Work!," in *IEEE Access*, vol. 1, pp. 335-349, 2013. doi: 10.1109/ACCESS.2013.2260813J.

[2] A. Rahimi and Y. Yoon, "Study on Cu/Ni Nano Superlattice Conductors for Reduced RF Loss," in *IEEE Microwave and Wireless Components Letters*, vol. 26, no. 4, pp. 258-260, April 2016. doi: 10.1109/LMWC.2016.2537780

[3] S. Hwangbo, A. Rahimi and Y. Yoon, "Cu/Co metaconductor based high signal integrity transmission lines for millimeter wave applications," *2017 IEEE MTT-S International Microwave Symposium (IMS)*, Honolulu, HI, 2017, pp. 707-710.

[4] S. Groiss, I. Bardi, O. Biro, K. Preis, and K. Richter, "Parameters of lossy cavity resonators calculated by the finite element method," *IEEE Trans. Magn.*, vol. 32, no. 3, pp. 894–897, May 1996.

[5] E. Hammerstad and O. Jensen, "Accurate models for microstrip computer-aided design," in *IEEE MTT-S Int. Microw. Symp. Dig.*, Washington, DC, May 1980, pp. 407–409.

[6] D. Nghiem, J. T. Williams, D. R. Jackson and A. A. Oliner, "Existence of a leaky dominant mode on microstrip line with an isotropic substrate: theory and measurements," in *IEEE Transactions on Microwave Theory and Techniques*, vol. 44, no. 10, pp. 1710-1715, Oct. 1996.

[7] M. Bozzi, L. Perregrini, K. Wu, and P. Arcioni, "Current and future research trends in substrate integrated waveguide technology," *Radioengineering*, vol. 18, pp. 201–209, Jun. 2009.

[8] L.Ouseph *et al* "Substrate Integrated Waveguide without Metallized Wall Posts", *Progress In Electromagnetics Research Letters, Vol.77*, 2018.

[9] W. Hong *et al.*, "Half Mode Substrate Integrated Waveguide: A New Guided Wave Structure for Microwave and Millimeter Wave Application," *2006 Joint 31st International Conference on Infrared Millimeter Waves and 14th International Conference on Teraherz Electronics*,Shanghai,2006,pp.219-219.

[10] Q. Lai, C. Fumeaux, W. Hong and R. Vahldieck, "Characterization of the Propagation Properties of the Half-Mode Substrate Integrated Waveguide," in *IEEE Transactions on Microwave Theory and Techniques*, vol. 57, no. 8, pp. 1996-2004, Aug. 2009.

[11] D. Deslandes and Ke Wu, "Accurate modeling, wave mechanisms, and design considerations of a substrate integrated waveguide," in *IEEE Transactions on Microwave Theory and Techniques*, vol. 54, no. 6, pp. 2516-2526, June 2006.

[12] L. Qiang, Y.-J. Zhao, Q. Sun, W. Zhao, and B. Liu, "A compact uwb hmsiw bandpass filter based on complementary split-ring resonators," *Progress In Electromagnetics Research C*, Vol. 11, 237-243, 2009.

[13] C. Kim, D. E. Senior, A. Shorey, H. J. Kim, W. Thomas and Y. K. Yoon, "Through-glass interposer integrated high quality RF components," *2014 IEEE 64th Electronic Components and Technology Conference (ECTC)*, Orlando, FL, 2014, pp. 1103-1109.

[14] J. R. Loo-yau, O. I. Gomez-pichardo, and F. Sandoval-ibarra, "Spurline structures and its application on microwave coupled line filter," *Revista Mexicana De Fisica*, vol. 57, no. 3, pp. 184-187, 2011.

Innovative Packaging Solutions of 3D Double Side Molding with System in Package for IoT and 5G Application

Mike Tsai*, Ryan Chiu, Dick Huang, Feng Kao, Eric He, J. Y. Chen, Simon Chen, Jensen Tsai and Yu-Po Wang

Siliconware Precision Industries Co. Ltd.
No. 153, Sec. 3, Chung-Shan Rd. Tantzu, Taichung 427, Taiwan, R.O.C.
* E-mail: miketsai@spil.com.tw; Tel: 886-4-25341525 ext 6718

Abstract—Recently, based on next generation wireless connectivity system evolution, there are more and more components combined into smartphone of Radio Frequency (RF) and Front-End Module (FEM) for up-coming 5G application. Also, the Internet of Things (IoT) continue to grow up due to the electronics industry is moved maturely on the mobile computing market for now. Both of IoT and 5G connectivity devices are required small form factor and high thermal performance. A 3D System in Package (3D SiP) including different approach, such as the double side molding technology and antenna in package (AiP) which is a combination solutions for these requirements.

In this paper, the 3D SiP package platform will use dual side Surface Mount Technology (SMT) technology and 3D structure of double side molding to shrink overall package size of 3D SiP module. The calculation of package size can be shrunk around 60% area, package size can be reduced from 8 x 8mm to 6 x 6mm. From warpage and thermal performance are proceed simulation and measurement. And experiment including the DOE (Design of Experiment) study for molding process with different high thermal epoxy molding compound (EMC) selection to verify warpage performance. By utilizing advanced package structure solutions such as high speed SMT placement, Cu substrate with thermal pad for high thermal, double side molding, a 3D double side SiP module can provide a unique opportunity to address cost, performance, and time-to-market. Considering the limitations of power consumption and form factor, smart phone front end module will become the major requirements for SiP platforms.

The characterization analysis will utilize simulation methodology and measurement correction for warpage and thermal performance comparison. Also, will proceed the typical reliability testing (Temperature Cycle Test, High Temperature Storage Test, un-bias HAST) results as a verification for 3D double side SiP structure. Finally, this paper will find out the suitable 3D SiP structure and feasibility data for future IoT and 5G devices application.

Keywords : IoT, 5G mmWave, 3D System in Package (3D SiP), Heterogeneous Integration, FEM Module, Double Side Molding, GaAs & GaN, Reliability

I. INTRODUCTION

From the electronics industry, the next fast growing opportunity market will be the IoT and 5G application in the near future. This advanced technology had used 3D SiP for small form factor and a low cost system solution for many wireless connectivity applications [1]. From mainly 2D single-die and side-by-side multi-die solutions to 3D stacked, daul side, multi-die solutions, this need drives the development of a new set of capabilities for SiP integration technology [2-3]. The solution of embedded module also offers 3D packaging options to customers for small package and attractive cost requirement [4]. High I/O and better electrical performance included high frequency and high speed product which was drove for fine pitch and fine component to component space attach process [5]. The SMT process for 3D SiP modules with high-density and heterogeneous integration on package level approach [6]. Normally, the frequencies band of 5G applications will be much lower than mmWave, we just call this low band as Sub-6GHz to combine 3G and 4G generation communications. In other words, mmWave bands defined as the new radio in the 5G mobile networks [7-11].

In this paper, the 3D integration of double side SiP package can be shrink package size and enhance the system power performance for Signal Integrity (SI)/ Power Integrity (PI) comparing with traditional single side SiP package. The 3D SiP structure was studied and demonstrated on the 6 x 6mm module size. This structute was designed with Cu substrate with thermal pad for better thermal performance requirement, such as power amplifier (PA) from FEM application in smart phone. The standard of RF was defined to the range of 3 KHz to 300 GHz. For example, frequency was separated to three groups as Low-band, Mid-band and High-band for RF connectivity application. There are 30 to 40 of difference frequency band for 3G/4G; when new 5G coming, more than additional 30 bands for sub-6GHz frequency band and also 3 of new high frequency for 5G mmWave bands, refer to Figure 1. for different application of frequency band introduction.

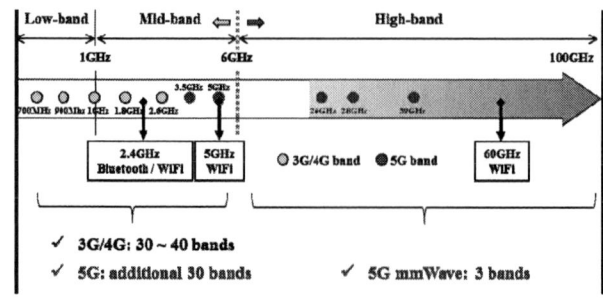

Figure 1. Different application of frequency band

When Low-band group which was started from 30-300 MHz, the major application was used for FM radio, television and amateur radio and so on. During the Mid-band, there are a lot of consumer electrical communication devices such as mobile phones, wireless WiFi and Bluetooth. For example, the mobile communication technology which used a lot of different frequency band for 3G, 4G and Sub-6GHz portion of 5G in different country area . There are 30~40 bands from 3G/4G requirement and also addition 30 bands for 5G. High-band, this is extra high frequency and millimeter wave (mmWave) for 5G high data rate transmission of mobile connectivity. There are definded the 3 bands for 5G mmWave range including 26GHz, 28GHz and 39GHz from 3GPP SPEC [13]. Refer to Table 1 for spectrum frequency aplication summary.

TABLE 1. SUMMARY OF MAJOR SPECTRUM APPLICATION

Groups	Spectrum Frequency		
	(1) Low-band	(2) Mid-band	(3) High-band
Applicati on	✓ 700MHz ~ 2.6GHz for 3G and 4G mobile ✓ 700MHz ~ 6GHz for 5G of Sub-6Ghz etc	✓ 26GHz, 28GHz and 39GHz for 5G mmWave	NA
	✓ 2.4GHz, 5GHz for WiFi (ex. 802.11ac/ax)	NA	✓ 60GHz for WiFi (ex. 802.11 ad/ay)

II. TEST VEHICLE DESCRIPTION

In this paper, this test vehicle (TV) was designed with advanced 3D SiP technology. The major package size was shrunk from 8 x 8mm to 6 x 6mm which was used two major new assembly technology as below items.

✓ 3D SiP of double side assembly technology
✓ 3D SiP of double side molding technology

To meet the high integration requirement , all components (IC, Resistor, Inductor, Capacitor of components) were put into double side assembly. This is a 3D integration & compart approach comparing with 2D side by side assembly methodology to reduce the space during actual product design stage. This 3D double side SiP module had tight component to component spacing distance and one time molding technology, the basic comparison information was shown in Table 2.

TABLE 2. COMPARISON OF DIFFERENT SiP DESIGN STRUCTURE

Item	2D SiP	3D Double Side SiP
Dimension	8 x 8mm	6 x 6mm
Package Structure	2D SiP	3D SiP Thermal Pad
Passive Component	Side by Side	Double Side Assembly
Thermal Solution	NA	Thermal Pad Design

In this table, to compare with traditional SiP structure, there are IC chips and components which was placed by 2D structure arrangement as side by side layout. The major benifites was used mature SMT process, but 2D layout arramgement was limited due to the placement size of 2D layout is one of key challenge concern for product design when small form factor and also thinner thickness requirement. The double side SiP module technology will be developed for different requirements such as higher integation solution, high electrical performance. The basic information comparison between traditional 2D SiP & 3D double side SiP module structure was shown in Table 3.

TABLE 3. BASIC INFORMATION COMPARISON BETWEEN 2D SiP & 3D DOUBLE SIDE SiP MODULE STRUCTURE

Item	2D SiP	3D Double Side SiP
Dimension (mm)	8 x 8	6 x 6
Layout Simulation (Top/Back side)	Top side ... 8mm ... 8mm ... Back side	Top side ... 6mm ... 6mm ... Back side
3D Structure Photo		
Area (mm²)	64	36
Size Ratio	100%	~60%
Component	Total 32pcs (IC chip: 2pcs; R/ L/ C: 30pcs)	

From the 8 x 8mm of traditional 2D SiP module size, the overall placement area is 64mm² which included BOM (Bill of Material) list such as IC (2pcs) and R/L/C components (30pcs), all components are integrated into the SiP module. The new approach idea for 3D double side SiP technology which was based on same compound quantity to re-arrange the component location from single side to double side SMT process. The calculation of package size was shrink to 6 x 6mm and overall got the ~40% package size reduction which also provided low assembly cost and substrate cost benefits.

III. PROCESS FLOW

The major process flow step and sequences were defined in Figure 2. Firstly, this was used the standard SMT process and also passive components were put in the TnR (Tape & Reel) to provide high speed loading during SMT machine. The detail single side of SMT process which included the solder printing, solder paste inspection, SMT component attachment, automated optical inspection, thermal reflow process for solder melting and de-flux cleaning process; the process flow shows in (1) ~ (2) step of Figure 2.

978-1-7281-1500-9/19 $31.00 © 2019 IEEE

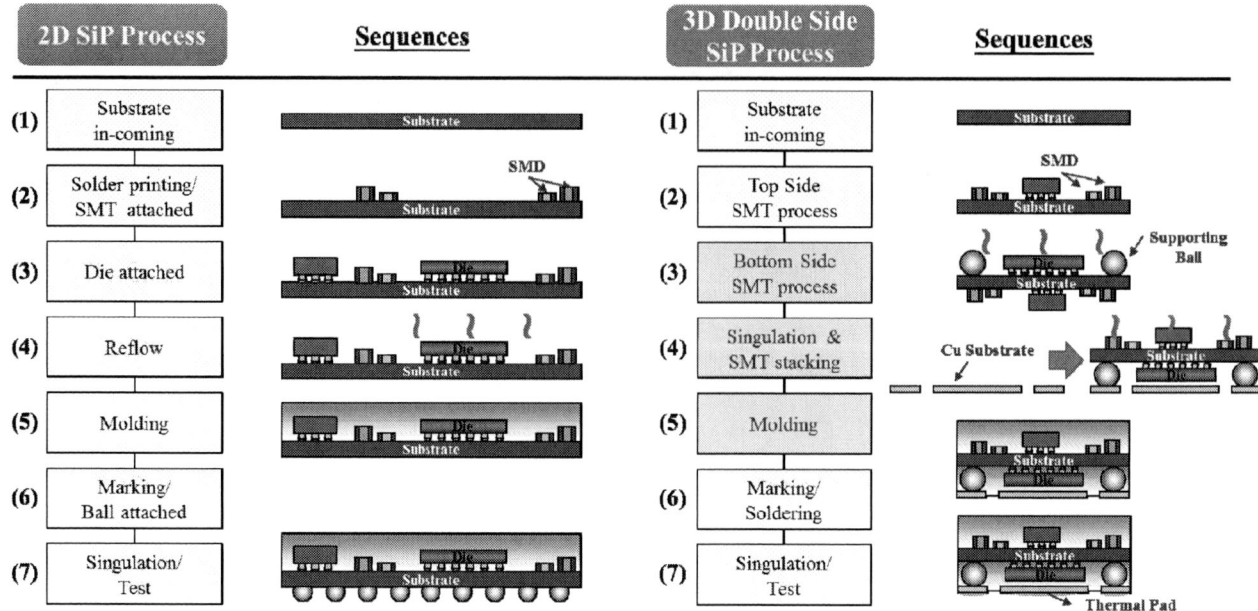

Figure 2. Process flow comparison between 2D SiP and 3D double side SiP structure

One of major challenge is double side SMT process, the process flow shown in (3) step of Figure 2. Comparing with top side of SMT process, the bottom side of handling methodology will use specific tooling to control the SiP module when flip to bottom substrate. The supporting ball will support as spacer for double side molding process. After finished the bottom side of supporting ball placement, will proceed the die bonding process on the bottom side then go through thermal reflow process for supporting ball & Cu pillar bump of solder melting. Before SMT stacking on the Cu substrate, the finished of top module will proceed sawing process for singulation purpose. The top module will do SMT stacking process on Cu substrate, refer to (4) step of Figure 2.

Then, proceed the double side molding process for SiP technology, this double side molding methodology to finish top mold & bottom mold at same time to get low cost benefits and fast assembly cycle time, refer to (5) step of Figure 2. The major material used epoxy molding compound (EMC), when proceed double side molding process, the major key process factors are focus on strip warpage control and mold void verification. All study factor need to be optimized during molding DOE study to define the optimized parameters. Finally, this 3D double side SiP will finish other standard process such as laser marking, pre-solder on Cu substrate, singulation process and testing for all assembly process; final process flow shows in (6) ~ (7) step of Figure 2.

Especially for step (4), the Cu substrate was manufactured by punching process. The standard punching process is a forming process which was proceed force with a steel punch tool with through into raw metal material. The major benefits for Cu substrate were mature manufacturing process to create multiple shaped holes at same time and high

speed for cost effective. Refer to Figure 3. for the example of Cu substrate finished sample and also the Cu substrate was designed the thermal pad for good thermal performance.

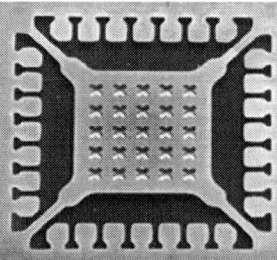

Figure 3. Example of Cu substrate and thermal pad design

The top side of SMT component placement area had shown in Figure 4, the top side solder printing in (a) photo and the passive component attachment after reflow for solder melting & flux clean process in (b) photo.

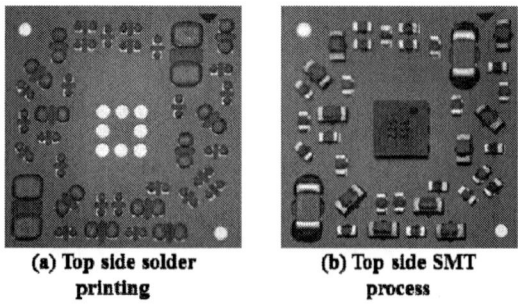

(a) Top side solder printing

(b) Top side SMT process

Figure 4. Top side 3D SiP of SMT process result

978-1-7281-1500-9/19 $31.00 © 2019 IEEE

The major process for bottom side substrate portion was ball placement which call supporting ball, the purpose of supporting ball was ensured the space for 3D SiP structure. To proceed the die attached process and capillary underfill process to protect the bottom side of IC chip, refer to Figure 5. for bottom side ball placement (c), bottom side die attached (d).

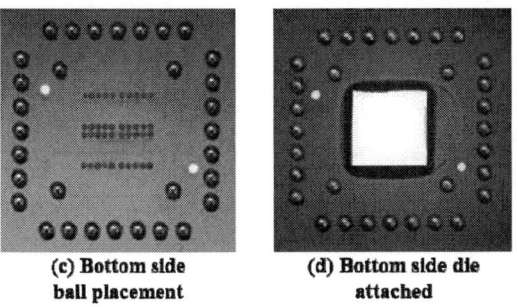

(c) Bottom side ball placement　　**(d) Bottom side die attached**

Figure 5. Bottom side 3D SiP of placement process result

The top finished module was proceed SMT stacking process on Cu substrate, refer to Figure 6 for different die locations. From the demonstration result of 3D double side SiP was shown in Figure 7; refer to (a) for top side molding photo; refer to (b) for bottom side of Cu substrate with large thermal pad design for good thermal performance purpose, refer to (c) for side view of double side molding result which was used one time molding to finish top side and bottom side together.

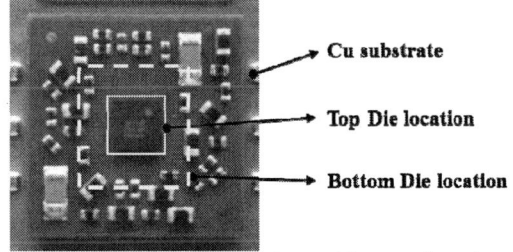

Figure 6. Top finished module stacking on Cu substrate

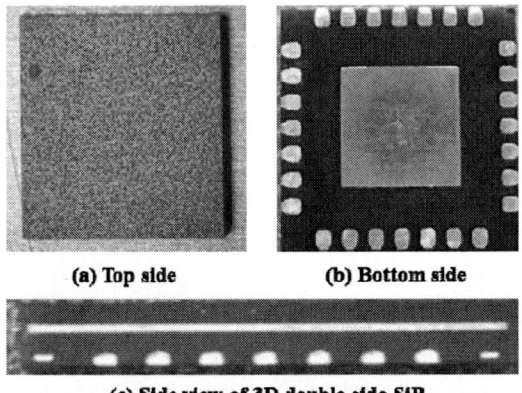

(a) Top side　　**(b) Bottom side**

(c) Side view of 3D double side SiP

Figure 7. Demonstration result of 3D double side SiP

IV. EXPERIMENT PLAN

(1) Electrical Performance for 3D Double Side SiP

Typically, RF power amplifiers will drive the antenna of a transmitter. Major design goals including gain, power efficiency, impedance matching, and heat dissipation. Normally, the PA had different wafer solution (GaAs, GaN) to get high frequency of RF singel data transmission. A standard RF PA is a type of electrical power amplifier that convert a signal data from a low power RF signal into a higher power signal data for good quality connectivity requirement. The RF microstrip signal design will need to specific design control to get the Ground/Signal/Ground (GSG) design requirement and also for shorter trace routing design by via stacking from die to die interconnection to get better RF electrical performance for 3D double side SiP structure. To comparing with traditional 2D SiP which had longer distance with multi-substrate layer design, refer to Table 4 for the microstrip of RF design comparison.

TABLE 4. MICROSTRIP OF RF DESIGN COMPARISON

Item	2D SiP	3D Double Side SiP
Package Structure		
Microstrip Design	Multi-substrate layer	Using Via only
Distance	Longer	Shorter

(2) Thermal Performance Verify for 3D Double Side SiP

The thermal simulation was used Finite Element Method (FEM) model and also assumed the JEDEC standard of PCB condition to simulate the thermal dissipation performance. Refer to Figure 8, the simulation model from ANSYS analysis software and thermal simulation result.

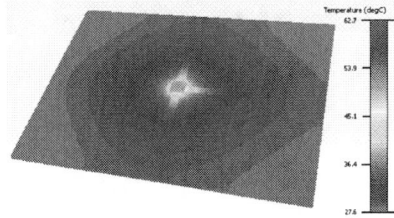

Figure 8. Thermal simulation model & simulation result

Based on the simulation result, the junction temperature for 3D double side SiP structure was provided good thermal performance, the simulation result got the Theta JA=37.7 C/W and the Theta JB=21 C/W. When used high thermal EMC (K=2.5w/mK) that the simulation result got ~15% JA thermal enhancement and ~30% JB thermal enhancement comparing with normal EMC (K=1.0w/mK), the simulation result got the Theta JA=31.0 ℃/W and Theta JB=14.5 ℃/W). In this paper, the double side SiP thermal TV with 6 × 6mm package size that attached with 2 × 2mm thermal test die was applied on the bottom side of this test vehicle. The purpose of thermal test die was simulated the PA working function of thermal dissipation verification. The thermal test die was manufactured by Kokomo, refer to Figure 9. There are consisted of thermal resistors and thermal-couple sensor was designed inside bumped wafer (a) to measured voltage value from the heating of die. The layout of thermal test die was shown in (b) of Figure 9.

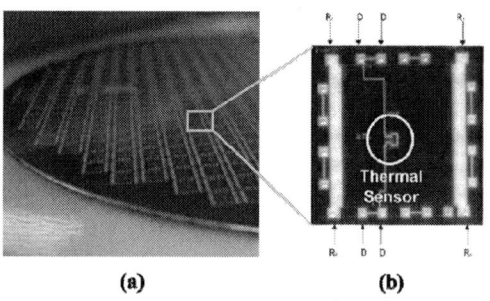

(a)   **(b)**

Figure 9. Thermal test wafer structure; (a) **thermal wafer with Cu pillar bump, (b) Thermal test die illustration**

The junction temperature of thermal test die was measured through EMC and Cu substrate to verify the thermal dissipation performance and result is directly correlated to the simulation. The thermal PCB was designed 4L PCB (4"x4.5") and still air condition with power 1W for measurement. When mounted one unit on thermal PCB board and connected with measured output IO as shown in Figure 10. for the 3D double side SiP on thermal PCB board result.

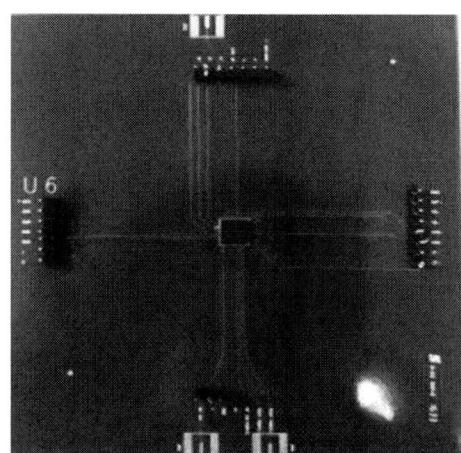

Figure 10. 3D double side SiP mounted on thermal measurement PCB board

According to Table 5 of measurement result, the actual measured result had 95~97% accuracy comparing with simulation result from normal EMC type. When used high thermal EMC, the result had 93~95% accuracy comparing

with simulation result from high thermal EMC type. Overall result on **double side SiP that could be enhanced 1.3X ratio by using high thermal EMC type to get good thermal dissipation.**

TABLE 5. THERMAL DISSIPATION COMPARISON RESULT

EMC type	Normal Thermal		High Thermal	
Result	Measured	Simulation	Measured	Simulation
Theta JA (ºC/W)	38.7	37.7	33.5	31.0
Theta JB (ºC/W)	22.1	21	15.2	14.5
Accuracy (%)	Baseline	95~97%	Baseline	93~95%
Thermal enhanced	1X		~1.3X	

(3) Warpage Performance for 3D Double Side SiP

The key challenge of this paper was warpage performance from 3D double side SiP structure and need to consider the structure balance of warpage result. The DoE design factor was defined as thermal EMC type factor and EMC shrinkage factor, the DoE table shown as below Table 6.

TABLE 6. DIFFERENT EMC FACTOR DoE STUDY TABLE

DoE #	Thermal Factor		Shrinkage Factor		Sample Size	Output Result
	Normal	High	Normal	High		
Leg 1	V		V		100pcs/ leg	A. Strip warpage B. Unit warpage (shadow moiré)
Leg 2		V	V			
Leg 3	V			V		

The purpose of this DoE is to determine the warpage performance after molding process. The output result was strip warpage and unit warpage. The strip warpage result & photo data was shown in Table 7. The high shrinkage of EMC got the worse strip warpage result (7~8mm) comparing with normal shrinkage of EMC legs (<1mm). The high shrinkage EMC with lower filler content which was not provide balance structure for 3D double side SiP.

TABLE 7. WARPAGE COMPARISON OF EMC FACTOR STUDY RESULT

DoE #	Thermal Factor		Shrinkage Factor		Comparison Result (Delta between RT/HT)		
	Normal	High	Normal	High	Strip Warpage	Photo	Judgement
Leg 1	V		V		0.2~0.3 mm		Pass
Leg 2		V	V		0.4~0.6 mm		Pass
Leg 3	V			V	7~8 mm		Fail

The Shadow Moiré methodology as shown in Figure 11. Shadow Moiré was measured by white light through a reference grating on surface of sample with temperature heating, and recorded warpage shadow data by camera [14].

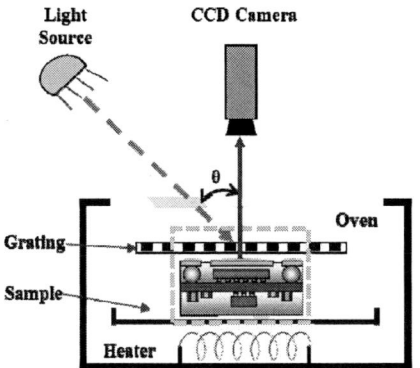

Figure 11. Shadow Moiré measurement methodology

The package level warpage performance by using Shadow Moiré measurement result shown in Table 8, Leg1 & Leg2 were got better warpage result when used normal shrinkage factor of EMC legs (25um & 34um respectively).

TABLE 8. DIFFERENT EMC FACTOR OF STUDY RESULT

DoE #	Thermal Factor		Shrinkage Factor		Comparison Result (Delta between RT/HT)	
	Normal	High	Normal	High	Unit Warpage	Judgement
Leg 1	V		V		25um	Best
Leg 2		V	V		34um	Good
Leg 3	V			V	52um	Normal

According to signed warpage chart result, those three legs are within JEDEC warpage requirement (Max. 80um). The Leg3 had more smiling face warpage (-) during room temperature and crying face warpage (+) during high temperature condition, the warpage chart comparison result shown in Figure 12. Based on EMC selection point of view, both of normal EMC and high thermal EMC could meet warpage requirement for 3D double side SiP structure.

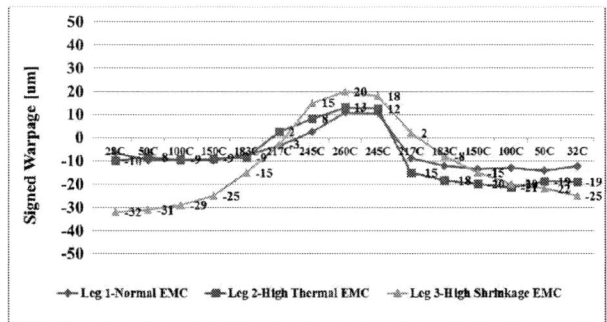

Figure 12. Warpage result of Shadow Moiré data comparison

V. RESULTS

First, this paper had demonstrated the 3D SiP structure with double side molding technology and integrated on the 6 x 6 mm module size. As shown in Figure 13, the x-section of 3D SiP with double side molding demonstration result. The major purpose of supporting ball was provided the space for double side molding process and good molding flow-ability result.

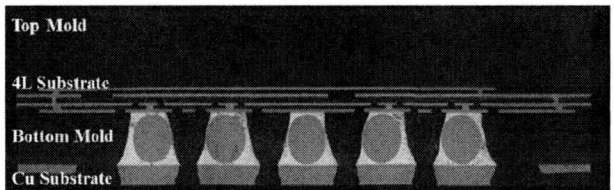

Figure 13. 3D SiP with double side molding of x-section SEM result

The 3D SiP of thermal die and thermal pad design of x-section was observed in Figure 14. The SEM image of top side molding had IC and components and molded by molded underfill (MUF) technology; the bottom side molding had thermal die (PA) and thermal pad design for thermal enhancement. The Cu pillar bump joint result of thermal die x-section SEM result was shown in Figure 15.

Figure 14. 3D SiP of thermal die and thermal pad design of x-section SEM result

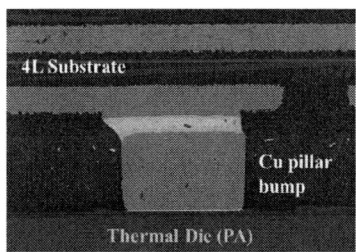

Figure 15. Thermal die of Cu pillar bump joint x-section SEM result

Finally, 3D double side SiP was proceed the standard reliability test items includes (moisture soaking level 3, temperature cycling test -55 ℃ to 125 ℃ 1000 cycle, un-bias high accelerated stress test 96hrs and 150 ℃ high temperature storage 1000hrs). Each read point of reliability test data, O/S testing and SAT inspecation were to confirm the quality result and showed all passed and summary reliability test results as shown in Table 9.

TABLE 9. RESULT OF RELIABILITY TEST

Reliability Test Items	Read Point	Sample size	O/S Test and SAT Result
Time Zero	T0	0 / 231pcs	All Pass
MSL3	Precon	0 / 154pcs	All Pass
TCT (-55℃ ~+125℃)	1000 Cycles	0 / 77pcs	All Pass
u-HAST (130℃/85%RH)	96 Hours	0 / 77pcs	All Pass
HTS (150℃)	1000 Hours	0 / 77pcs	All Pass

VI. CONCLUSION

This paper had demonstrated the 3D SiP structure with double side molding technology included structure and feasibility data for future IoT and 5G application. Comparing with traditional SiP structure, the integration density of components was used tight spacing design rule for small form factor requirement, heterogeneous integration, low cost trend and high electrical performance. The 3D double side SiP structure provided good RF electrical signal design comparing with 2D SiP structure, also the good thermal enhanced performance by using thermal pad design. The key factor of EMC selection result, the high thermal EMC provided ~30% thermal enhancement comparing with normal EMC thermal material by actual thermal die measurement result. Completed package level reliability test for following test conditions, passed MSL 3, TCT1000, u-HAST96 and HTS1000.

ACKNOWLEDGMENT

Authors would like to thank members of SPIL SiP technology development team and reliability team for their help on sample preparation and experimental setup, and substrate suppler partners for test substrate manufacturing.

REFERENCES

[1] Mike Tsai, Albert Lan, Chi Liang Shih, Terence Huang, Ryan Chiu, S. L. Chung, J. Y. Chen, Frank Chu, Cheng Kai Chang, Sheng Ming Yang, Daniel Chen and Nicholas Kao, "Alternative 3D Small Form Factor Methodology of System in Package for IoT and Wearable Devices Application", in *Proc. 67th Electronic Components and Technol. Conf. (ECTC)*, 2017, pp.1541-1546.

[2] A. Martins, M. Pinheiro, A. F. Ferreira, R. Almeida, F. Matos,J. Oliveira, H. M. Santos, M. C. Monteiro, H. Gamboa and R. P. Silva, "Heterogeneous Integration challenges within Wafer Level Fan-Out SiP for Wearables and IoT", in *Proc. 68th Electronic Components and Technol. Conf. (ECTC)*, 2018, pp.1485-1492.

[3] Jin-Yuan Lai, Tang-Yuan Chen, Ming-Han Wang, Meng-Kai Shih, David Tarng and Chih-Pin Hung, "Characterization of Dual Side Molding SiP Module", in *Proc. 67th Electronic Components and Technol. Conf. (ECTC)*, 2017, pp.1039-1044.

[4] Shichun Qu, Jihwan Kim, Glen Marcus and Matt Ring, "3D Power Module with Embedded WLCSP", in *Proc. 63th Electronic Components and Technol. Conf.*

(ECTC), 2013, pp.1230-1234.

[5] Mike Tsai, Albert Lan, Yan Han Yao, Meng Yueh Wu, Cheng Kai Chang, Roger Lo and Eason Chen, "Alternative Fine Pitch Solution of Low Cost and High Throughput Thermal Compression Bonding by using Capillary Underfill", in *Proc. 65th Electronic Components and Technol. Conf. (ECTC)*, 2015, pp.465-469.

[6] Min Miao, Yufeng Jin, Runiu Fang, Fangqing Mu, Shichao Guo, Xiaoqing Zhang, Yang Zhang, Duwei Hu, Zhensong Li and Wei Xiang, "Investigation of Micromachined LTCC Functional Modules for High-density 3D SIP based on LTCC Packaging Platform", in *Proc. 63th Electronic Components and Technol. Conf. (ECTC)*, 2013, pp.1815-1822.

[7] Mike Tsai, Ryan Chiu, Eric He, J. Y. Chen, Royal Chen, Jensen Tsai and Yu-Po Wang, "Innovative Packaging Solutions of 3D System in Package with Antenna Integration for IoT and 5G Application", in *Proc. 20th Electronics Packaging Technology Conf. (EPTC)*, 2018.

[8] John Dzarnoski and Susie Johansson, "Ultra Small Hearing Aid Electronic Packaging Enabled By Chip-In-Flex", in *Proc. 64th Electronic Components and Technol. Conf. (ECTC)*, 2014, pp.157-164.

[9] Hiroki Shibuya, Tatsuaki Tsukuda, Hiroko Suzuki, Tadashi Shimizu, Masahiro Dobashi, Shinji Nishizono, Mikio Baba, Hideki Sasaki and Katsushi Terajima, "A Wireless Charging and Near-field Communication Combination Module for Mobile Applications", in *Proc. 64th Electronic Components and Technol. Conf. (ECTC)*, 2014, pp.763-768.

[10] Jong-In Ryu, Se-Hoon Park, Dongsu Kim, Jun-Chul Kim and Jong-Chul Park, "A Mobile TV/GPS Module by Embedding a GPS IC in Printed-Circuit-Board", in *Proc. 62th Electronic Components and Technol. Conf. (ECTC)*, 2012, pp.1668-1672.

[11] Yi He, Fengman Liu, Anmou Liao, Jun Li, Xiaomeng Wu, Peng Wu, Liqiang Cao and Dongkai Shangguan, "Design and Implementation of a 700-2600MHz RF SiP for Micro Base Station", in *Proc. 64th Electronic Components and Technol. Conf. (ECTC)*, 2014, pp.2131-2136.

[12] JEDEC Solid State Technology Association. JESD22-A104C: Temperature Cycling; 2005.

[13] 3GPP "NR; User Equipment (UE) radio transmission and reception; Part 2: Range 2 Standalone (Release 15)" 3GPP TS 38.101-2 V15.2.0.

[14] JEDEC Solid State Technology Association, JESD22-B112A: Temperature Cycling Package Warpage Measurement of Surface-Mount Integrated Circuits at Elevated Temperature; 2009.

Enhancing Efficiency of Antenna-in-Package (AiP) by Through-Silicon-Interposer (TSI) with Embedded Air Cavity and Polyimide Dielectric Micro-substrate

Yunna Sun, Yunting Sun, Jiangbo Luo, Huiying Wang, Zhuoqing Yang*, Yan Wang*, Guifu Ding*

School of Electronic Information and Electrical Engineering, Shanghai Jiao Tong University, Dong Chuan Road 800, Shanghai 200240, China
yzhuoqing@sjtu.edu.cn; wyyw@sjtu.edu.cn; gfding@sjtu.edu.cn

Kwangwoo Han

Samsung Electronics Co., Ltd., 613-8, Gobul-ro, Baebang-eup, Asan-si, Chungcheongnam-do, Republic of Korea 31581. Korea

Abstract—A compact antenna is designed as a CPW fed monopole with T shape, Y shape and meander shape. Based on the larger range of the adjustable parameters of this antenna, the optimized antenna is relative compact with high radiation performance. In order to enhance the radiation performance of the antenna-in-package, an antenna-in-package based on through-silicon-interposer with embedded air cavity and polyimide dielectric micro-substrate is proposed. The shielding structure is covered among the RF device to reduce the influence of the electromagnetic interference of the AiP. Under this design, the radiant efficiency is enhanced more than 10% compared with the Si substrate antenna within large bandwidth and gain.

Keywords-antenna-in-package; relative permittivity; radiation performance; frequency bands;

I. INTRODUCTION

In recent years, wireless communication systems have been developed widely and rapidly in the modern world, thus, many engineers focus their interests on the design and fabrication of antenna with characteristics of compact size, wide bandwidth, multi-bands, omnidirectional pattern, high communication rate and low-cost, etc. The monopole antenna possessing different configurations, such as meander T-shape, V-shape, Y-shape, and C-shape, is meaningful to realize an enough larger length for mm-wave communication with limited compacted size, [1]. The monopole antennas which has compactness along with multi-bands, low profile lightweight and higher bandwidth is widely designed and applicated, such as, [2]--[4]. Due to the advantages like lower radiation leakage and less dispersion, single metallic structure and easy integration with microwave integrated circuits, coplanar waveguide feed (CWP) is the widely used uniplanar feeding technique, [4]--[6].

The mm-wave communication as a kind of advanced high communication rate technology has been paid wide attentions by more and more researchers. However, there are several challenges in fully exploiting the potential of mm-wave communication for the practical applications, including specific integrated circuits (IC) design, multiband switching and interference management, especially antenna packaging and optimization. In the current state, mainstream technical route about antenna integration includes the antenna-on-chip (AoC) and the antenna-in-package (AiP), [7]--[10]. In order to obtain the high-level integration density, highly matched wireless module and feasible packaging solution, AiP is considered as a better choice for its flexibility in realizing wafer level packaging to construct an antenna array. However, the AiP not only requires a small footprint, but also needs to minimize the additional parasitic effects and have low insertion/return loss, excellent matching and good isolation in packaging for more wider practical applications.

In the completion of AiP, the insertion loss, return loss and bandwidth is mainly determined by through substrate via, bonding wire, interconnections and package resonances. Therefore, we have to do great efforts to minimize the line loss and improve the isolation by optimizing the configuration of the package, including designing more reliable structure and introducing new materials with good dielectric performance. The through-silicon-interposer (TSI) is usually a good choice to satisfy the needs of the AiP for the shortest vertical interconnection, small footprint size and less parasitic effect and loss, [11]--[12]. However, the permittivity of Si substrate is still higher although some engineer has used the high permittivity Si as the substrate, which will reduce the radiation efficiency performance of the antenna. With regards to this, in the present work an improved design is proposed, which will utilize an embedded TSI composite unit consisting of high permittivity Si substrate, polymer micro-substrate, air cavity and protecting structure for shielding the high-performance components and enhancing the radiation efficiency of AiP.

978-1-7281-1500-9/19 $31.00 © 2019 IEEE

II. Optimal Design of Antenna

Based on the forma Eq(1) about bandwidth (B), an extremely small relative permittivity (ε_r) of the substrater benifites the bandwidth, gain and radiation efficiency.

$$B = 3.77 \frac{(\varepsilon_r - 1)Wh}{\varepsilon_r^2 L \lambda_g} \cdot 100\% \qquad (2)$$

where, λ_g, W, and L are the guide wavelength, width and length of the radiator. Thus, it is the goal in this work to obtain a much larger ε_r for the substrate.

With formula of the resonant frequency of monopole radiator shown in Eq(2), it can be obtained that a relative smaller effective relative permittivity (ERP) will enlarge the effective length of the monopole radiator at the fixed resonant frequency. In other words, the relative smaller ERP will make the size of the antenna to be larger without a fine design. In terms of the ERP analyzed in reference [13]--[14], the permittivity variation of the layer next to the ground plane affects the resonant frequency most significantly.

$$f = \frac{C}{\sqrt{\varepsilon_{eff}} \cdot \lambda_g} \approx \frac{C}{\sqrt{\varepsilon_{eff}} \cdot 4L_{eff}} \qquad (1)$$

where, f, C, ε_{eff}, and L_{eff} are the resonant frequency, the velocity of electro-magnetic waves in free space, ERP and effective length of the monopole radiator. Thus, a CPW fed monopole with T shape, Y shape and meander shape is proposed to satisfy the adjustment and integration with TSI.

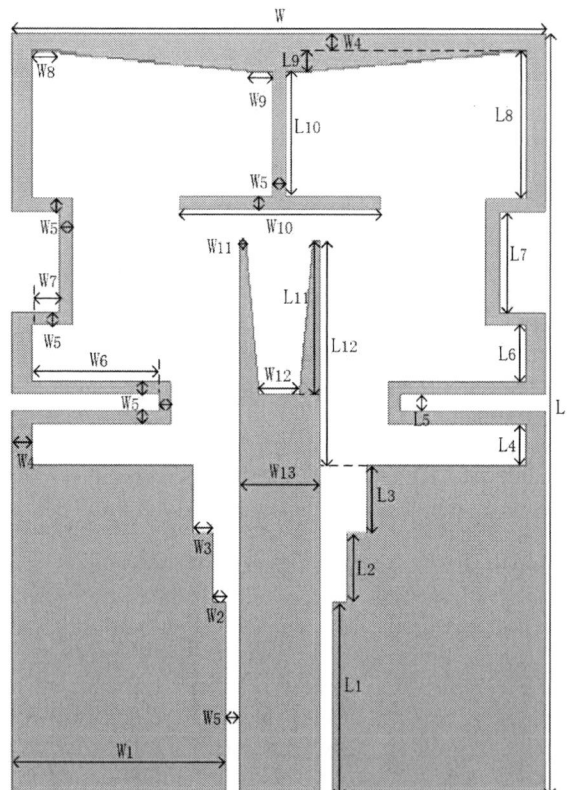

Figure 1   Profile of the proposed CPW fed monopole.

A. Profile of the propsed CPW fed monopole

The basic profile of the proposed CPW fed monopole is shown in Figure 1. The basic parameters consist by about 14 widths (W) and 13 lengths (L) listed in Figure 1. This monopole antenna is symmetric with extended irregular ground. The CPW fed monopole is designed on 0.5 mm thickness FR-4 substrate or high permittivity Si with relative permittivity of 4.4 and 18.6 respectively, and the overall size of the compact monopole is 20×27 mm^2. The CPW feed line consists a signal strip with 3.0 mm (W_{13}) width and gap distance (W_5) of 0.5 mm, corresponding to 50 Ω characteristic impedance.

In order to obtain the optimized parameters of the propose antenna, a series of sweeping simulations among 2--4 GHz are carried out by HFSS. By comparing the comprehensive performances on return loss (S_{11}), bandwidth, gain and radiation efficiency based on the FR-4 substrate and the Cu CPW fed monopole with 5 μm thickness, the optimized parameters of this design are listed in TABLE I. The corresponding bandwidth of the 10 dB return loss is composed of 550 MHz (3.16--3.71 GHz) and larger than 1310 MHz (5.59--7.0++ GHz), as shown in Figure 2. The Radiation patterns at 3.4 GHz and 5.9 GHz are given in Figure 3. The corresponding 3.4 GHz and 5.9 GHz gains are 2.2 dB and 3.2 dB, respectively. And the radiation efficiencies for 3.4 GHz and 5.9 GHz are above 95%.

TABLE I.   DIMENSIONS OF THE CPW FED MONOPOLE

Parameters	W	W_1	W_2	W_3	W_4	W_5	W_6	W_7
Width (mm)	20	8	0.5	0.75	1	0.5	4.5	1
Parameters	W_8	W_9	W_{10}	W_{11}	W_{12}	W_{13}		
Width (mm)	1.0	3.75	7.5	0.25	1.5	3		
Parameters	L	L_1	L_2	L_3	L_4	L_5	L_6	L_7
Length (mm)	27	7.2	2.4	2.4	1.5	0.5	2	5.5
Parameters	L_8	L_9	L_{10}	L_{11}	L_{12}			
Length (mm)	5	1	4.5	5	7.5			

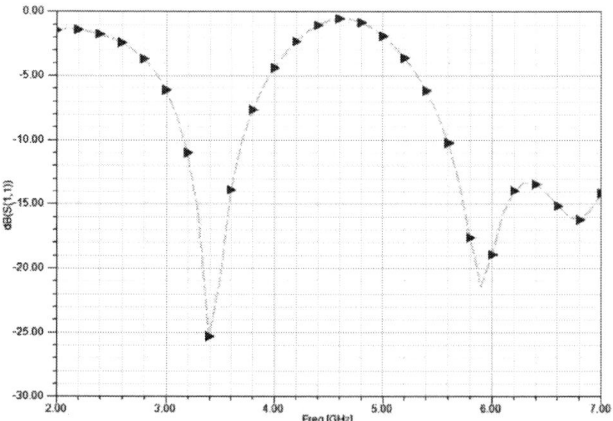

Figure 2   Return loss of the proposed antenna.

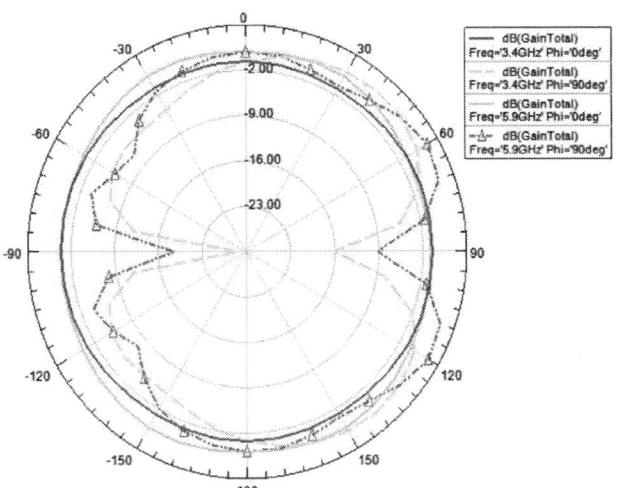

Figure 3   Radiation patterns for the proposed antenna at 3.4, 5.9 GHz.

B. Numerical analysis of the optimazed antenna on Si substrate

As shown in Figure 4, the return loss of the optimized antenna on Si substrate is relatively poor compared with that of the FR-4 substrate. In other words, the proposed antenna possesses more than four frequency bands for adjusting the parameters to satisfy different radiation needs.

Thus, the antenna based on Si substrate, which is taken consideration of the integration and packaging technology, shall be optimized. With the HFSS simulation by adjusting the key parameters of the proposed antenna, the optimized results on the Si substrate is plotted in Figure 5--Figure 6. The corresponding bandwidth of the 10 dB return loss is composed of 540 MHz (2.97--3.61 GHz), 320 MHz (4.74--5.06 GHz), 70 MHz (5.26--5.33 GHz) and 480 MHz (6.09--6.57 GHz), as shown in Figure 5. The Radiation patterns at 3.2 GHz, 4.9 GHz and 6.3 GHz are given in Figure 6. The corresponding 3.2 GHz, 4.9 GHz and 6.3 GHz gains and radiation efficiency are 1.89 dB, 1.89 dB and 5.1 dB, and above 83%, respectively. In terms of the emerging fifth generation wireless systems, the frequency bands are below 6.0 GHz. Thus, the optimized antenna shall be optimized

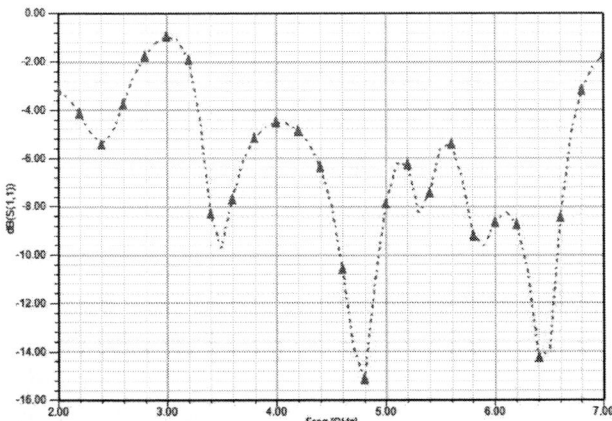

Figure 4   Return loss of the proposed antenna based on Si substrate.

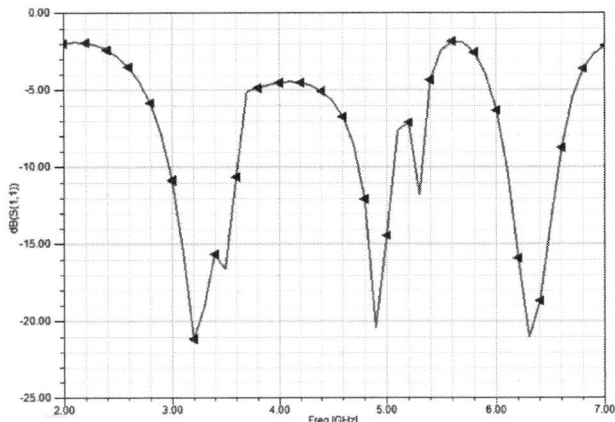

Figure 5   Return loss of the optimazed antenna based on Si substrate.

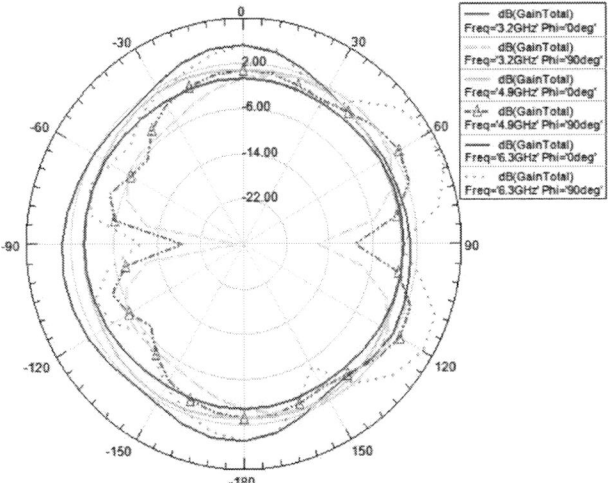

Figure 6   Radiation patterns for the optimazed antenna based on Si substrate.

in further during the AIP stage for the efficiency, bandwidth and gain are all reduced compared with the FR-4 substrate.

III.   ANTENNA-IN-PACKAGE (AIP) WITH THROUGH-SILICON-INTERPOSER

Given to the advantages of through silicon via, the AiP technology is designed with TSI based on the high precision micro-nano processing technology. AiP based on TSI is basically composed, antenna, RF MEMS, RF IC and other components and IC. The influence of electromagnetic interference on the antennas, RF MEMS and IC, and high-performance components is serious, thus, shielding strategies has been taken, [15]--[16]. During the design and simulation stage, the electromagnetic interference has to be considered. A shielding structure is designed on the high-performance components in this AiP.

The proposed AiP is composed of TSI, proposed antenna and RF IC, as exhibited in Figure 7. The high permittivity Si substrate (500 μm depth, serving as the bottom micro-substrate of antenna with 200 μm depth by etching), embedding air cavity (top micro-substrate of antenna, 200 μm depth) and polyimide micro-substrate (middle micro-substrate of antenna, 200 μm depth) together form the substrate of the antenna. Based on this design, the relative permittivity of this composite substrate is reduced to about 2.3 from 18.6 (high permittivity Si). A shield structure is designed among the RF device with size of 1500×750×750 μm³. The radius of through via is 50 μm, which is used to interconnect the shield structure and ground, and the top and bottom IC together. Thus, radiation performances of the proposed AiP, including the return loss, gain and radiation efficiency, are evaluated with HFSS.

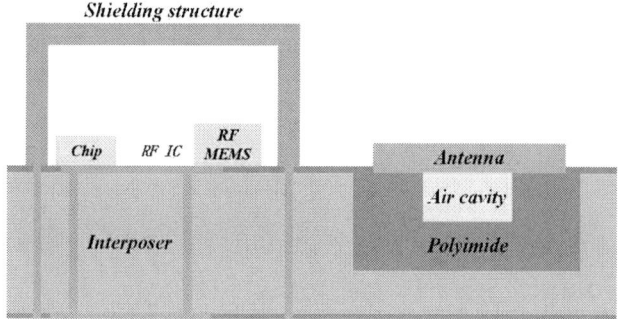

Figure 7   Schemic of AiP with Embedded air cavity and polyimide dielectric micro-substrate.

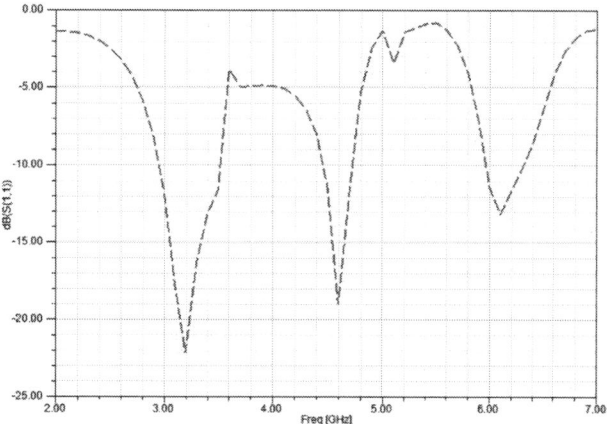

Figure 8   Return loss of the traditional AiP.

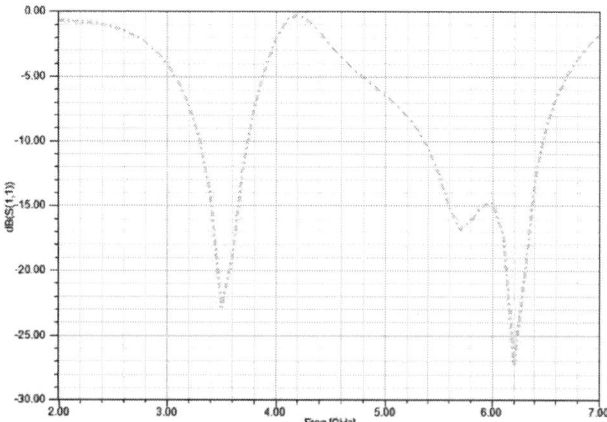

Figure 9   Return loss of the proposed AiP.

Comparing with the traditional AiP with high permittivity TSI and the antenna on high permittivity Si substrate, the proposed AiP with optimized antennas has smaller return loss and larger bandwidth 340 MHz (3.3--3.74 GHz) and 1.24 GHz (5.36--6.5 GHz), illustrated in Figure 8--Figure 9. After packaging and integration, the return loss of the antenna is reduced, plotted in Figure 8. Furthermore, the proposed AiP is also with reliable gain, 1.9 dB @3.3GHz which does not reduce after packaging, and the traditional is about 1.86 dB. The radiant efficiency is 97%, which is enhanced more than 10% compared with the Si substrate antenna.

IV.   CONCLUSIONS

In this design, the footprint of the AiP is ensured for the shortest vertical interconnection firstly. The total thickness of the TSI is 500 um. Secondly, the dielectric performance the substrate of antenna is enhanced by selecting the high permittivity Si, embedding air cavity and polyimide micro-substrate. The embedded substrate for the antenna is made up of Si (bottom micro-substrate), polyimide (middle micro-substrate) and air cavity (top micro-substrate). Based on this design, the relative permittivity is reduced to about 2.3 from 18.6 (high permittivity Si). Furthermore, the distance between the antenna and the ground can be easily adjusted and fabricated by filling polyimide and electroplating for enhancing the matching. Thirdly, the inter-electromagnetic-interference between the integrated antennas and high-performance components is reduced by designing a shield structure on the high-performance IC. Based on the above optimizing on the AiP, the radiant efficiency of antenna is enhanced more than 10% compared with the Si substrate antenna.

ACKNOWLEDGMENTS

The authors would like to thank supports from the Shanghai Professional Technical Service Platform for Non-Silicon Micro-Nano Integrated Manufacturing. This work is supported by Project funded by China Postdoctoral Science Foundation (2018M630440) and the National Defense Science and Technology Innovation Special Zone project (1816321TS00107401).

REFERENCES

[1] Hu, W., Yin, Y., Fei, P., & Yang, X. (2011). Compact triband square-slot antenna with symmetrical l-strip for wlan/wimax applications. IEEE Antennas and Wireless Propagation Letters, 10, 462-465.

[2] Naidu, V., & Kumar, R. (2015). Design of compact dual-band/tri-band cpw-fed monopole antennas for wlan/wimax applications. Wireless Personal Communications, 82(1), 267-282.

[3] Liu, P., Zou, Y., Xie, B., Liu, X., & Sun, B. (2012). Compact cpw-fed tri-band printed antenna with meandering split-ring slot for wimax/wlan applications. IEEE Antennas and Wireless Propagation Letters, 11, 1242-1244.

[4] Wang, Y., Ying, Z., & Yang, G. (2017). A compact CPW-fed wideband antenna design for 5G/WLAN wireless application. In 2017 IEEE International Symposium on Antennas and Propagation & USNC/URSI National Radio Science Meeting, , 1775-1776.

[5] Heydarpanah, B., Ghobadi, C., Nourinia, J., & Beigi, P. (2018). A novel printed small antenna with L-shaped structure for multiband applications. Journal of Instrumentation, 13(08), P08016.

[6] Zhai, H., Ma, Z., Han, Y., & Liang, C. (2013). A compact printed antenna for triple-band wlan/wimax applications. IEEE Antennas & Wireless Propagation Letters, 12(1921), 65-68.

[7] Zhang, Y., & Liu, D. (2009). Antenna-on-chip and antenna-in-package solutions to highly integrated millimeter-wave devices for wireless communications. IEEE Transactions on Antennas and Propagation, 57(10).

[8] Lin, T., Chiu, T., & Chang, D. Design of Dual-Band Millimeter-Wave Antenna-in-Package Using Flip-Chip Assembly. IEEE Transactions on Components Packaging & Manufacturing Technology 4.3(2017):385-391

[9] Tsai, C., Hsieh, J., Lin, W., Yen, L., Hung, J., Peng, T., & Lei, Y. (2015). High performance passive devices for millimeter wave system integration on integrated fan-out (InFO) wafer level packaging technology. In 2015 IEEE International Electron Devices Meeting (IEDM), 25-2.

[10] Dussopt, L., Lamy, Y., Joblot, S., Lanteri, J., Salti, H., & Bar, P., et al. (2012). Silicon interposer with integrated antenna array for millimeter-wave short-range communications. IEEE Microwave Symposium Digest.

[11] Luo, J., Wang, G., Sun, Y., Zhao, X., & Ding, G. (2018). Fabrication and characterization of a low-cost interposer with an intact insulation layer and ultra-low tsv leakage current. Journal of Micromechanics and Microengineering.

[12] Sun, Y., Luo, J., Yang, Z., Wang, Y., Ding, G., & Wang, Z. (2018). Development of a Polyimide/SiC-whisker/nano-particles composite with high thermal conductivity and low coefficient of thermal expansion as dielectric layer for interposer application. In 2018 IEEE 68th Electronic Components and Technology Conference (ECTC), 1537-1542.

[13] Bernhard, J., & Tousignant, C. (1999). Resonant frequencies of rectangular microstrip antennas with flush and spaced dielectric superstrates. IEEE transactions on Antennas and Propagation, 47(2), 302-308.

[14] Zhong, S., Liu, G., & Qasim, G. (1994). Closed form expressions for resonant frequency of rectangular patch antennas with multidielectric layers. IEEE transactions on Antennas and Propagation, 42(9), 1360-1363.

[15] Jin, C., Li, R., Hu, S., Zhang, S., Chang, K., & Zheng, B. (2014). Self-shielded circularly polarized antenna-in-package based on quarter mode substrate integrated waveguide subarray. IEEE Transactions on components, packaging and manufacturing technology, 4(3), 392-399.

[16] Yan, N., Ma, K., & Zhang, H. (2016). A novel self-packaged substrate integrated suspended line quasi-Yagi antenna. IEEE Transactions on Components, Packaging and Manufacturing Technology, 6(8), 1261-1267.

Low-Loss Glass Substrates Formulated with a Variety of Dielectric Characteristics for Millimeter-Wave Applications

Kazuatka Hayashi, Nobutaka Kidera and Yoichiro Sato

AGC Incorporated
Yokohama, Japan
kazutaka-hayashi@agc.com

Abstract—**In this paper, we will introduce the dielectric characteristics of newly developed glasses and discuss the effect of electrical properties of the glasses on the characteristics of glass-employed devices through numerical simulation.**

Keywords-high speed communication; 5G,non-alkali glass; substrate; transmission loss; waveguide; SIW; band pass filter

I. INTRODUCTION

Fifth-generation (5G) wireless communication technology will not be limited to the progress of mobile phones but be driven by a new set of technology services, from autonomous driving to IoT. One of the key features of 5G is the extremely high speed and large data capacity for streaming, while maintaining ultra-low latency. To meet these targets, the usage of millimeter wave (mm-wave) will need to be considered.

Currently, polymeric resins and ceramics are commonly used materials for substrates in devices in mm wave range. However, these materials are considered to have electrical and physical disadvantages. When designing in devices in the mm wave range, very fine features are required and these features must have high positional accuracy in wide array of operating conditions and formats. For example, polymeric substrates have a physic-chemical challenges due to their lack of thermal instability, high moisture absorption. Moreover, polymeric substrates with high thermal expansion coefficient, normally anisotropic, results in high stress and unexpected changes in the electrical properties. Ceramic substrates including low temperature co-fired ceramics (LTCC) suffer from their low-dimensional accuracy during the fabrication process, high processing cost due to the complex manufacturing processes, small formats and the need for complicated processing facilities.

Compared with those materials, glass is the ideal material for the next-generation substrates in mm wave range. Glass has an excellent environmental durability, is ultra-flat and surface roughness in the nanometer range enabling formation of ultra-fine and accurate structures resulting in extremely low loss allowing the engineer a high degree of flexibility in design electrical devices.

The most important feature of the glass is the ability to adjust electrical and physical its properties by means of compositional design. Most glasses are multi-component materials consisting of the combination of several substances, typically oxides such as SiO_2, Al_2O_3, and CaO etc. By combining molar fraction of each oxides, and addition of minor ingredients (colorants, refining agents etc.), many properties can be controlled depending on the specific application or frequency. Electrical properties can be also varied by adjusting the glass composition to obtain ideal dielectric properties for specific application. Table I summarizes the advantages and disadvantages of candidate materials for a substrate of high speed communication application.

Highly efficient signal transmission with low-loss is vital for high-frequency transmission. That requires materials' loss tangent (dissipation factor, D_f) to be low enough in mm wave range to meet performance requirements. To reduce the parasitic capacitance and maintain certain size of thetransmission line width with good transmission characteristics, a low dielectric constant (D_k) is required. In contrast, high dielectric constant is preferred for applications where form factor of each component is critical, such as mobile applications. Both applications required totally difference dielectric constants while maintaining a low D_f. Fig. 1 shows a general relationship between dielectric constant, D_k and loss tangent, D_f for typical for a variety of materials. As clearly seen, in the same materials groups, D_k and D_f have a correlation. The higher the D_k is, the higher the D_f is. Thus, there is a limitation on utilization of materials.

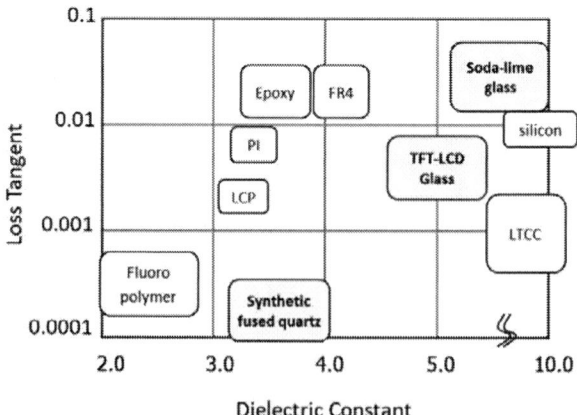

Figure 1. Dielectric constant and loss tangent relationship of existitng materials for high spped communication (LCP: Liquid Crystal Polymer, PI: Poly Imide, LTCC: Low-temperature Co-firing Ceramics, TFT-LCD: thin film transistor liquid crystal display)

TABLE I. COMPARISON OF CANDIDATE MATERIALS FOR SUBSTARATE OF HIGH SPEED COMMUNICATION DEVICES

	Fused Quartz	Borosilicate Glass	Window Glass (soda-lime glass)	LTCC	Fluoro-polymers	Glass Epoxy Board	New Glasses
D_k tunability	Poor	Poor	Poor	Good	Poor	Poor	tunable (4~8)
Loss tangent, D_f	Good	Acceptable	Poor	Good	Good	Poor	Good to acceptable (0.001~0.007)
Heat resistance	Good	Good	Good	Good	Fair	Poor	>500 C
CTE tunability	Poor	Poor	Poor	Poor	Fair	Fair	tunable (3~8 ppm/K)
Chemical resistance	Good	Good	Good	Good	Good	Fair	Good
Stiffness	Good	Good	Good	Good	Poor	Fair	Young's modulus 70~80 GPa
Sheet formability	Poor	Good	Good	Poor	Poor	Fair	Conventional sheet forming process applicable
Surface quality (roughness, flatness)	Good	Good	Good	Poor	Poor	Poor	R_a<2nm can be achieved
Total evaluation	Poor High cost	Fair can be applicable	Poor too large Df	Poor Surface quality	Poor Low stiffness	Poor Surface quality	Good
Note	Alkali free	Alkali containing	High alkali content				Alkali content: TFT-LCD substrate equivalent

In addition to the dielectric properties, availability of the material as the sheet shape is important. From the view point of production cost, and despite its excellent dielectric loss, silica glass (quartz) is not suitable due to its poor process affinity for sheet fabrication, which causes its high price. In addition, its low thermal expansion causes thermal stress when bonding with other materials. Some glass sheets, such as soda-lime silicate glass for windows for automobiles and buildings, aluminosilicate glasses for mobile display covers are commercially available, however, those glasses have high loss tangent due to their relatively high mobile ion content, such as sodium and potassium. Non-alkali glass substrates for liquid crystal displays (LCD) are available with large size, however, their dielectric characteristics are similar and there are almost no variation which is not convenient in designing circuits on glass with specific requirements or targets.

A new generation of glasses have been developed to meet the need of mm devices. These glasses are formulated as series of non-alkali glasses with variety s of dielectric characteristics. These glasses contain less than 0.1 mass % alkali oxide since alkali migration affect reliability of the electrical circuit or device, such as thin film transistor (TFT) [1,2]. The alkali content is as low as in the glass used for TFT-LCD glasses on which electronic devices are formed for commercial products.

In this paper, we reports the possibility of application of newly developed glasses to the high speed communication. First we describe the properties of the glasses and then discuss the electrical characteristics by using numerical simulation. Numerical simulation was performed for two types of configuration. One is a simple micro strip line and the other is a substrate integrated waveguide band pass filter.

II. METHOD

A. Preparation of Glass Samples

Glass samples for the measurement of glass properties were prepared by conventional melting method. Composition of the glasses were designed to have a forming ability as a thin sheet glass using float process. Raw materials for each components, such as oxides, carbonates, hydroxides etc., were weighed and then mixed to form batches. Each batch was then melted in a platinum crucible at 1400~1600 °C for several hours depending on the duration necessary to remove the bubbles. The glass melt was then poured onto a carbon mold. The glass was immediately transferred into a pre-heated electric furnace following with annealing to remove strain. After maintaining glass at the temperature slightly higher than its glass transition temperature, the glass block was cooled down to room temperature at a cooling rate of 1 °C/min. Subsequently, the glass block was shaped into an appropriate form by mechanical processing.

B. Measurement of Glass Properties

Thermal expansion coefficient (CTE) of the glass was measured using a dilatometer (Netzsch DIL-420C) for the temperature range from room temperature up to 300 °C. Heating rate was 4 K/min. Young's modulus was measured by the ultrasonic pulse method using an ultrasonic thickness gauge (Olympus 38DL Plus). Strain point was measured by the fiber elongation method specified in JIS R3103-2: 2001 "Viscosity and viscometric fixed points of glass". Dielectric constant and loss tangent were measured using split-post dielectric resonator (SPDR) method at 10 GHz [3].

978-1-7281-1500-9/19 $31.00 © 2019 IEEE

C. Numerical calculations at mm-wave range

Transmission line is most fundamental but important for radio frequency (RF) circuit at higher frequency of mm-wave band. Microstrip line (MSL), stripline and co-planar waveguide are commonly used RF transmission lines. We chose MSL and calculated its transmission loss using MoM (method of moment).

Fig. 2 shows the MSL model for transmission loss calculation. MSL's characteristic impedance Z was set as 50 Ohm optimized using a finite element method (FEM). Optimized MSL dimensions are listed on Table II. Values of W_m, H and L_m were listed in Table II. We set the conductor condition as lossless. The reason was that we can verify glass material characteristics effect itself on transmission loss. We calculated S21 of MSL.

We also tried verification of the relationship between glass permittivity and mm-wave device characteristics. Substrate integrated waveguide (SIW) band pass filter (BPF) was adopted as a mm-wave device [5]. The operational frequency of the SIW-BPF was designed around 28 GHz. It is applicable for the 5G frequency range.

SIW BPF was designed using finite element method (FEM). Borosilicate glass (BS) and specially formulated glass, GL8 were selected as substrates to compare the effect of D_k on the optimum dimension of SIW. Glass thickness was 0.3 mm. and the via hole diameter of 100 μm was selected. The via hole pitch should be designed to half-length of λ_g (guided-wavelength of the SIW) but this has the drawback of increasing the number of vias. Therefore, we chose 400 μm as the SIW via pitch. In our previous report [5], the irises consist of a pair lines of vias, but we modified the iris [6]. Fig. 3 shows the designed SIW BPF. It has 4 resonators in series. Topology of this BPF is shown in Fig. 4. Coupling efficiency between resonators is regulated by edge vias separation Win. Suffix n denotes the number of irises and its maximum is less than actual number of irises because of filter symmetry.

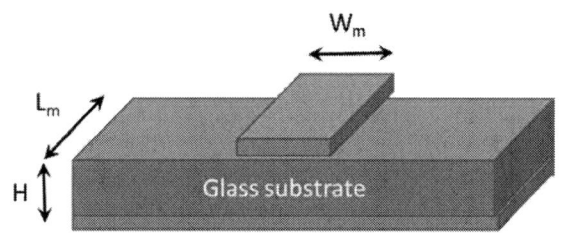

Figure 2. Microstrip line (MSL) model for transmission loss calculation

TABLE II. DIMENSIONS OF MSL MODEL FOR TRANSMISSION LOSS CALCULATION

	W_m [μm]	H [μm]	L_m [μm]
GL6	340		
GL7	300		
GL8	260	300	2000
SLS [4]	280		

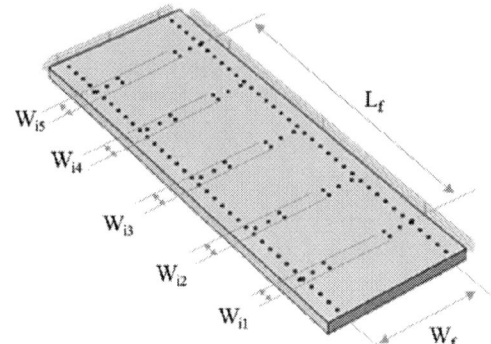

Figure 3. Schematic of designed SIW BPF (4 resonator in series)

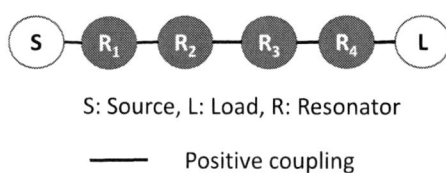

S: Source, L: Load, R: Resonator

————— Positive coupling

Figure 4. Topology of SIW BPF (4 resonators in series)

Lower D_f is preferable for mm-wave passive device because it can directly attain lower insertion loss (higher device efficiency). We examined relation between D_f and insertion loss using our SIW BPF design concept.

In addition, we compared the filter size in terms of Dk. Guided wavelength λ_g is proportional to inverse of square root of Dk. We defined the filter width and length as W_f and L_f respectively.

III. RESULT AND DISCUSSION

A. Physical Propertis of Glasss Substrates

Typical physical properties of the newly developed glasses, GL6, GL7 and GL8 are shown in Table II. Other two types of typical commercially available glasses, soda-lime silicate glass (SLS) and borosilicate glass (BS) are also indicated in the table. In this study borosilicate glass is alkaline containing glass which are generally applied for heat resistant glass ware. One of the feature of GL6 to 8 is their alkali content and it is less than 0.1 mass%. The content is as low as that in the glasses used as thin film transistor liquid crystal display (TFT-LCD) substrates. The alkali content affects the electrical reliability [2]. The risk of ion migration and shortening is reduced when one uses GL6 to 8 as a substrate [1]. In addition, CTE of those glasses are higher than conventionally used TFT-LCD glasses. Due to higher CTE, stress generation caused by CTE mismatch with typical molding compounds containing inorganic filler.

The dielectric, D_fs of GL6, GL7 and GL8 are 50 % lower than that of Soda-lime Glass (SLS). In addition, GL series formulated to have wide range of D_ks from 6.7 to 8.8 providing a high degree of freedom. Fig. 5 shows the relationship between D_k and D_f for both newly developed glass and existing glasses. Newly developed glasses successfully widen the D_k range with keeping relatively low D_f.

B. Transmission loss estimation

Fig. 6 shows the frequency dependence of calculated transmission loss. Vertical axis indicates S21 per 1 mm length of MSL. The newly developed glasses, GL6, GL7 and GL8 had almost the same transmission loss. Typical soda-lime glass, SLS was calculated to have higher transmission loss at the frequency range higher than 20 GHz compared to that of GL6 to 8. The reason is mainly due to higher D_f of SLS glass. According to this calculation result, the impact of D_k on the S21 was relatively small. Fig. 7 shows S21 for those glasses at 28.5 GHz, which might be a candidate frequency for next generation high speed communication. Therefore, using GL6, 7 or 8, low transmission loss circuit can be obtained with changing the characteristics of the substrate, such as D_f, CTE, etc.

TABLE III.   TYPICAL PHYSICAL PROPERTIES OF GLASS SAMPLES

Sample Name	GL6	GL7	GL8	SLS	BS
CTE [ppm/K] (30-220°C)	5.9	6.7	8.2	8.3	3.2
Young's modulus [GPa]	77	78	78	72	64
Strain point [°C]	700	680	680	510	510
Alkali content [mass%]	< 0.1	< 0.1	< 0.1	13	~5
D_k (10GHz)	6.7	7.6	8.8	8.1[4]	4.5
D_f (10GHz)	0.0064	0.0063	0.0070	0.014[4]	0.0084

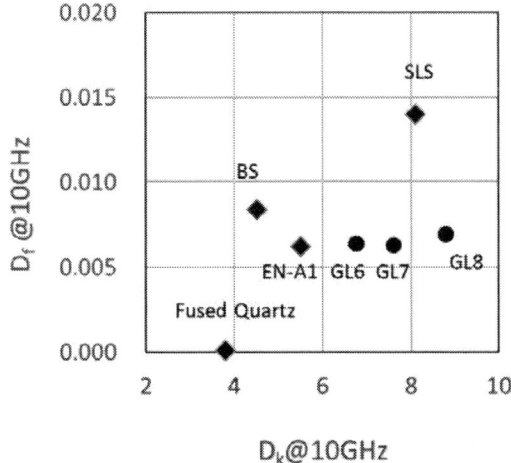

Figure 5. D_k and D_f plot for newly developed glasses

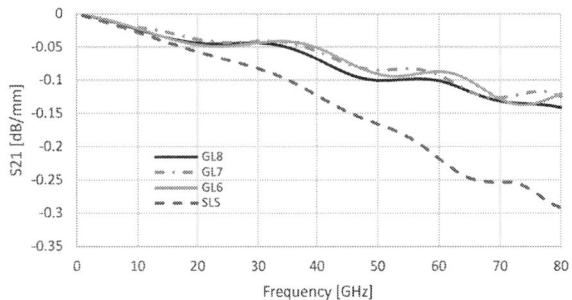

Figure 6.   Microstrip S21 (calculation)

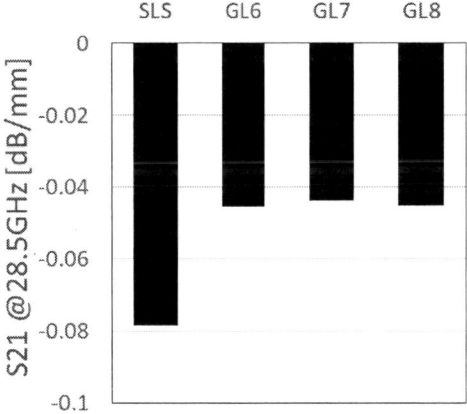

Figure 7.   Cmparison of calculated S21 at 28.5 GHz for examined glasses

C. Effect of D_k and D_f on device dimensions and its performances

D_k is directly proportional to the dimension of the passive device since the wavelength in the medium depends on D_k. Dimensions are proportional to square root of the reciprocal ratio of D_ks of the dielectric materials. Table IV shows the estimated dimensions of SIW BPFs which satisfy resonance condition.

As illustrated in Fig. 3, W_f is the width of filters and corresponds to SIW width. L_f is the filter length, it corresponds to the total length of four series resonators. The square root of the reciprocal ratio of D_ks of the dielectric materials are 71% between borosilicate glass and GL8 glass. Designed SIW BPF has similar size as D_k value estimation. Because of the increase of D_k, the area of SIW device employing GL8 glass can be reduced by half of that of borosilicate glass. Using high D_k glass, passive high frequency device size can be reduced.

Fig. 8 and 9 indicate calculated SIW BPF characteristics for borosilicate glass and that of GL8 respectively. Each SIW BPF has pass bandwidth of about 3GHz. It has return loss below 10 dB within the pass band.

Next, we calculated SIW BPF insertion loss characteristic varying the D_f value for the GL8. Fig. 10 shows the S21 for several D_f values at the center frequency of 28.5GHz of pass band. SIW BPF's S21 was proportional to D_f from Fig. 10. Measured D_f of the GL8 glass is 0.007. In this case the calculated insertion loss of the SIW BPF was 1.81dB. In the case of larger D_f than 0.008, insertion loss will exceed the 2dB. D_f to attain reasonable insertion loss filter. So D_f <0.008 is a milestone of obtaining low insertion loss filter.

TABLE IV. GLASS D_k VALUE AND ITS EFFECT ON THE DEVICE DIMENSION OF 5G 28GHZ BAND SIW BPF

	D_k	$W_f \times L_f$ [mm]	Size ratio	Area ratio
BS	4.5	4.0 × 10.89	100% × 100%	100%
GL8	8.8	2.92 × 7.67	73% × 73.2%	51%

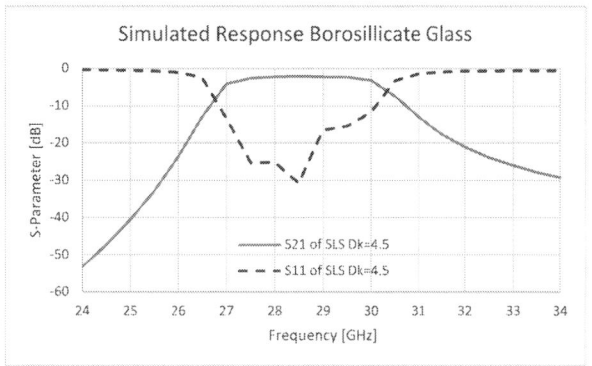

Figure 8. Calculated SIW BPF characteristics for BS glass

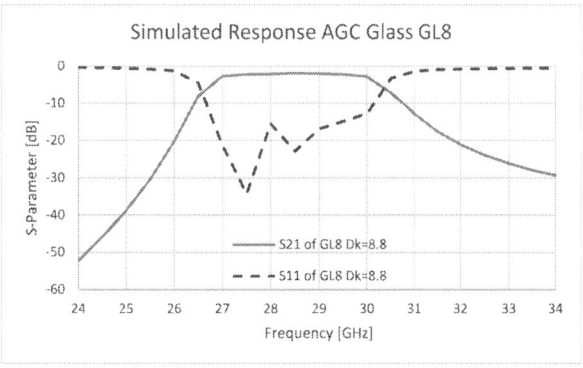

Figure 9. Calculated SIW BPF characteristics for GL8 glass

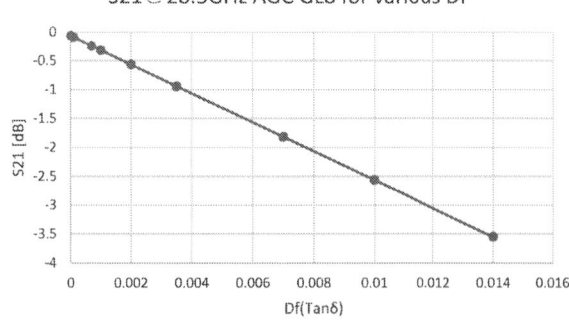

Figure 10. Calculated insersion loss of SIW BPF at 28.5GHz for vairous D_f

D. Advantages of glass substrate at mm-wave application

As we shown in section III A to C, the possibility of newly developed glass substrates for an application to high speed communication are confirmed. In addition to the dielectric characteristics, glass substrate of newly developed glass has some advantages as follows;

- Ultimately smooth surface achieved by mechanical polishing allows suppression of transmission loss caused by skin effect when micro strip line is formed on the surface of the substrate.
- Rigid substrate can be obtained thanks to their high Young's modulus.
- Variation of CTE, which is close to that of molding resins with inorganic filler can suppress the stress generation when used in heterogeneous integration condition.
- Dimension of the SIW filter can be suppressed depending upon the D_k of the glass. Up to 50% in area can be reduced by use of GL8 compared with conventional borosilicate glass substrate. Fig. 11 shows the comparison of the dimension of SIW for BS and GL8.

IV. CONCLUSION

We have demonstrated that new types of glasses with unique D_k and D_f properties can be designed by careful selection of specialized chemical components. These newly developed glass substrates with a wide range of electric properties can be used as a substrate for high speed communication application. This paper has shown the dielectric constant can be formulated to provide a D_k range from 4 to 9 while minimizing the loss tangent. Typical loss tangents of 0.005 to 0.007 at 10 GHz can be achieved. As for thermal characteristics, these glasses have sufficient thermal properties to perform direct sheet fabrication process, such as float process, for example. Mechanical and thermal properties are also superior compared with resins. Young's modulus is about 70 to 80 GPa and strain point is higher than 650 °C.

We also demonstrated the numerical simulation for the application of mm wave devices using relatively simple configuration.

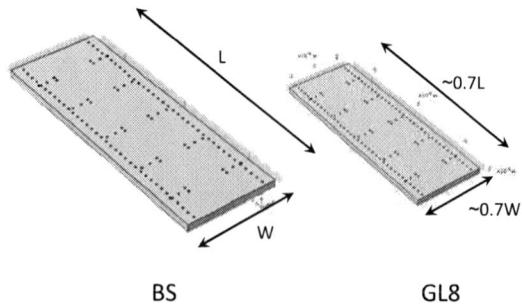

Figure 11. Comparison of the optimized structure of SIW BPF using comventional BS ans newly developed GL8

ACKNOWLEDGMENT

The authors thank to Mr. Tomonori Ogawa for fruitful discussion and advice.

REFERENCES

[1] K. Hayashi, S. Sawamura, S. Nomura, N. Suzuki and N. Naoya, "Development and evaluation of carrier glass substrate for fan-out WLP/PLP process", Proc. IEEE Electronic Components and Technology Conf. (ECTC), 2017, pp896-901

[2] G. Baek, A. Krasnov, W. den Boer, J. Kanicki, "Top Gate Amorphous In−Ga−Zn−O Thin Film Transistors Fabricated on Soda−Lime−Silica Glass Substrates", SID Symposium Digest, vol. 45, 2014, pp 1035-1038

[3] http://www.qwed.com.pl/resonators_spdr.html.

[4] L. Navias and R. L. Green, "Dielectric properties of glasses at ultra-high frequencies and their relation to composition," J. Am. Ceram. Soc., vol. 29, No. 10, 1946, pp. 267-277.

[5] Y. Sato and N. Kidera "Demonstration of 28GHz band pass filter toward 5G using ultra low loss and high accuracy through quartz vias," Proc. Electronic Components and Technol. Conf. (ECTC), San Diego, USA, May 29-1 Jun, 2018, pp.2237-2241.

[6] N. Kidera, "Filter," January 2019. WO Patent App.

Evaluation Of Fine-Pitch Routing Capabilities Of Advanced Dielectric Materials For High Speed Panel-RDL In 2.5D Interposer And Fan-Out Packages

Shreya Dwarakanath, Pulugurtha Markondeya Raj,[#] Amit Agarwal,[***] Daichi Okamoto,[**] Atsushi Kubo,[*] Fuhan Liu, Mohan Kathaperumal, Rao R. Tummala

3D Systems Packaging Research Center, Georgia Institute of Technology, Atlanta, GA, USA

[***] Microchips, United States

[#]Florida International University, United States

[**] TAIYO INK MFG. CO., LTD. Japan

[*]Tokyo Ohka Kogyo Co., Ltd., Japan

Email: shreyad001.sd@gmail.com

Abstract—**High-bandwidth computing with low power requires high-density low-loss interconnects with advanced design rules that cannot be realized with standard epoxies. This paper evaluates the critical material properties required to meet next-generation RDL needs for 2.5D interposer and fan-out packages. It demonstrates that shifting towards lower D_k materials can support higher data rates and maintain high wiring density, thus enabling overall system bandwidth improvement. This paper provides a complete analysis of the material property requirements, critical design metrics, and process options to qualify dielectrics for next-generation high-bandwidth demand.**

Keywords-High-bandwidth, polymer dielectrics, interposer, fan-out, IO density, panel-scale processes, microstrip, crosstalk, data rate, transmission lines, wiring, high-speed electrical interconnections

I. INTRODUCTION

The rapid increase of devices with artificial intelligence (AI) capabilities is driving the demand for growth in compute performance. The need for high-performance compute, while expanding in the traditional cloud computing and data center applications, is extending into new dimensions with specialized ASICs. These are required for four classes of applications: computationally intensive deep-learning and neural network based algorithms, wearable and other devices integrated into the internet-of-things, data centers for block-chain, and autonomous driving. These applications are driving the need for low-cost power-efficient high-bandwidth advanced packaging. As transistor scaling is facing physical and economic limitations for scaling below the 14nm node, system-level integration is gaining importance and ways to bridge the gap between semiconductor package (10^3 I/Os/cm^2) and chip (10^8 I/Os/cm^2) interconnect density are coming to light [1]. System-integration involves putting together dissimilar chips using advanced packaging architectures with superior performance in terms of signal integrity and lower power consumption, low-cost and reduced form factor. Various integration schemes have been developed to meet the need for heterogeneous integration, such as the 2.5D interposers, 3D fan-out type architectures, Intel's EMIB and multi-die "Chiplet" type approaches.

Of special interest, is high bandwidth memory-CPU integration. A typical high-performance logic device will require approximately 2,000 connections to one high bandwidth memory (HBM) stack. High-density interposers with fine-pitch interconnections offer a solution for meeting the interconnection requirement. The current approaches to 2.5D Interposers are typically either organic and silicon based. The organic substrates use sequential build-up processes (SBU) and are limited by wiring density and overall system reliability at large package sizes [2-4]. The Si-based interposers are based on wafer-scale BEOL processes and tools and can achieve high-wiring densities, reducing the CLI (chip level interconnection) pitch and interconnect length but are limited in scaling to large panels given cost considerations [5-8].

Georgia Tech Packaging Research Center and its industry partners pioneered the introduction of the glass interposer that combines the benefits of high-density interconnections of silicon with the low-cost and large panel processing scalability of organic laminates [9]. The smooth surface finish and low TTV (through thickness variation) of glass substrates enables scaling down to 1-2 µm line/spaces across a panel to drive higher interconnect density and thereby, higher bandwidth. The peak bandwidth of an I/O link is the product of number of data lanes and the data rate. Along with bandwidth the performance of an I/O link is also measured by the power consumption. This involves having dense interconnections at a short distance that are power-efficient in transferring large amounts of data. Currently, there is an interconnect gap between organic substrates which are limited to 6 µm line/space and Si wafer BEOL processes which are capable of sub-micron line/spaces. BEOL RDL processes have high resistive losses due to fine CD and low aspect ratio (0.5-2) of the traces. Embedded-trench based process have shown high aspect ratio (2-5) with lower line resistance enabling higher data rates [10]. However, high aspect-ratio structures can increase capacitance between traces and there is a need to use low-D_k dielectrics to reduce overall time-delay. Advanced low D_k panel-scalable dry-film dielectrics that are capable of reducing interconnect signal power and latency on a glass interposer which supports low-loss and high wiring density enables a manifold increase in system-bandwidth.

Si BEOL typically used in Si interposers suffers from high conductor and dielectric losses because of the restrictions on dielectric material (SiO_2). Organic-RDLs that are based on low-D_k polymers offer a compelling alternative as they can meet the I/O density requirement, while also reducing the resistance and capacitance of the interconnects, thereby allowing for higher data rates. This idea is explored in this paper through modeling and fabrication to demonstrate advanced design rules. The second section looks at ideal dielectric material properties along with a comparison of different material classes. This is followed by an upfront material selection based on properties and processing options. The previously-mentioned interconnect gap and bottleneck in scaling is explored using electrical simulations of transmission line structures in the third section. Dielectric thickness control for impedance matching is another important consideration. Ideal height of the dielectric for matching the characteristic impedance of 50 Ω is evaluated. As the speed of data transfer continues to increase, the crosstalk is a limiting factor in reducing the spacing between wires. This section looks at the effect of dielectric constant (D_k) on crosstalk mitigation to provide guidance on material selection. Further, the high signal data rates that are supported by the package drive the need for low-loss, low-k dielectrics. The eye closure at different data rates for 3 material candidates is evaluated to estimate the highest data rates for each candidate. Next-generation RDL routing requirements are moving towards fine-pitch 1-2 µm line/space and 3 µm photo-vias. Two processing techniques for achieving fine-pitch routing are discussed in the fourth section.

This paper discusses key material properties required for next-generation RDL requirements, impact of these material properties on signal integrity and fabrication processes for fine-pitch features with selected material candidates.

II. NEXT GENERATION POLYMER DIELECTRIC MATERIAL PROPERTIES

The choice of polymer dielectric is driven by technology trends towards a)high-density fine-pitch routing b)system level thermo-mechanical reliability c)panel-scalable processing. The key material properties required to drive these material properties are given in Table 1. Low-loss tangent and low dielectric constant are critical for maintain signal integrity at high data-rates. Planar thin-films are required for fine-pitch wiring and for impedance matching. Higher elongation to failure is favorable for multi-layer RDL with a higher copper coverage on each layer. Residual stress of the polymer is dependent on the young's modulus and co-efficient of thermal

expansion (CTE). RDL fabrication processes involve processing steps such as metallization, annealing and curing which induce stresses within the polymer thin-film. Further, CTE mis-match with the copper wiring and overall substrate warpage influence the amount of compressive stress seen by the polymer. There exists a critical limit of stress tolerance for fabrication of multi-layer RDL which depends on the adhesion strength of the polymer. It is essential that the material properties of the polymer are controlled to maintain the residual stress within tolerance. The moisture absorption of the polymer is critical for long-term system-reliability, as often interfacial moisture absorption is responsible for delamination. Good adhesion at the polymer/metal interface is needed to fabricate multi-layer wiring structures. This adhesion is usually achieved by means of either chemical or mechanical interlocking. Recent trends towards 5G and high-frequency applications place restrictions on the copper surface roughness needed to achieve higher adhesion. This imposes restrictions on the material selection and shifts the choice towards polymers with more bonding groups that can interact with copper. The process parameters thereby narrow the selection to polymers that are compatible with existing packaging infrastructure.

Although these properties are stated for the testing conditions, for future applications in automotive and other harsh environments, as reliability standards become more aggressive it is critical that these properties remain stable over a range of temperature, humidity and frequency.

Table 1 Ideal Properties of Next-Generation Polymer Dielectrics

	Parameters	Objective	Prior Art
Electrical	D_k	<3.0 (1 MHz-GHz)	>3.0
	D_f	<0.01	>0.014
Physical	Thickness	2-5 µm	10-75 µm
	DOP	<1-3 µm	-
Mechanical	Elongation to failure	>30%	2.5– 45%
	Residual Stress	<25 MPa	-
Chemical	Moisture absorption	< 0.2 wt. %	0.2 – 1.5 wt.%
	Adhesion	>0.3 kgf/cm	0.2 – 1 kgf/cm
Processes	Resolution	< 2 µm line/space	> 2 µm line/space
	Via	1-5 µm via	> 5 µm via

The present dielectric candidate options include Epoxy (Photo-imageable and non-photo-imageable), Polyimide (PI), BCB, Fluoropolymers (FP), Hydro-carbons (HC), PBO and metal oxides. BCB, FP, PBO and HC have a D_k <3.0 while most Epoxies have permittivities in the 3.0-3.5 range the permittivities of polyimides are higher. Epoxy, BCB and PIs are available in thin dry-film rolls while HCs are usually thick prepreg type materials. PBO and FPs are most commonly available as spin-on liquid polymers. PIs and PBO have a very

Table 2 Dielectric material properties

		Material A	Material B	Material C	Material D
Electrical	Dk	3.2 (5.8 GHz)	3.1 (1 MHz)	3.5 (1 MHz)	2.65 (1 KHz- 1 MHz) 2.55 (1 GHz)
	Df	0.01 (5.8 GHz)	0.017 (1 MHz)	0.022 (1 MHz)	0.0008 (1 KHz- 1 MHz) 0.002 (1 GHz)
Physical	Thickness	> 5 um	> 5 um	> 5 um	> 1 um
Mechanical	Elongation to failure		12-13 %	20%	
	CTE	30 ppm/K	40-45 ppm/K	45 ppm/K	45 ppm/K
	Young's modulus	7.5 GPa	3 – 3.5 Gpa	1.64 Gpa	2.9 GPa
Chemical	Curing Temp	171 C	180-185 C	250 C	210-250 C
	Moisture Absorption	0.6 wt.%	0.8 wt.%	1.5 wt%	< 0.2 wt.%
Processes	Resolution	<5 um line/space	2 um line/space	2 um line/space	<5 um line/space
	Via	4 um via (Excimer)	3 um via (PID)	5 um via(PID)	10 um via

high elongation to failure of more than 30% usually while BCB, epoxy and other polymers are lower. BCB is estimated to have a stress of 30MPa, while PIs and PBO are much higher in the 40MPa range while typical epoxies show stresses in the 20-30 MPa range. PIs have a much higher moisture absorption coefficient than other polymers, with epoxy being in the mid-range and all other polymers feature very low moisture absorption.

Given this understanding of material property requirements and material classes available, four material candidates were selected to evaluate the paper. Our selection consists of a state-of-the-art epoxy dry film candidate (Material A), two photo-imageable epoxy dielectrics (Material B&C) and a low-D_k non-epoxy polymer candidate (Material D) as shown in *Table 2.*

III. ELECTRICAL PERFORMANCE OF DIELECTRIC CANDIDATES

A. Design of Test-Structures

The typical-lengths of connections between logic-HBM dies are 3-6mm [11] and the high-density wiring features are of 2 μm line/space or higher. These wiring lines are usually microstrip or stripline transmission lines, wherein the signal conductors are either above a ground plane or sandwiched between two ground planes. In this paper, we looked at a surface microstrip transmission line model to study the effect of different material properties and the design rule was maintained as shown in Figure 1. Si BEOL processes were taken as the baseline to benchmark with and thus SiO₂ is considered the standard. Simulations were performed for coupled microstrip line structures in Advanced Design System (ADS) environment by Keysight. Line-Calc was used to calculate the height of the dielectric required for impedance matching to 50Ω and the values are given in

Figure 1(b). It can be seen that Material D with the lowest D_k has the least dielectric height or effective polymer thickness, which means the signal-to-ground plane distance is the least.

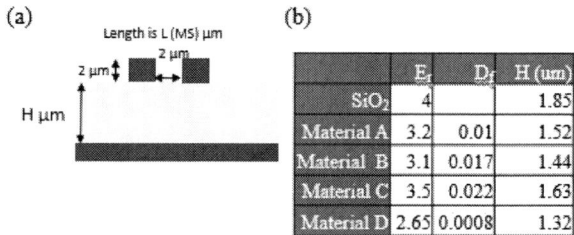

Figure 1 Test Structures of Transmission lines used in simulations

B. Effect of dielectric constant on Crosstalk

Crosstalk becomes important when the spacing between the traces is reduced as in fine-pitch features. Crosstalk can cause coupling between interconnects, induces jitter in the system and can upset circuit logic. This is primarily attributed to the mutual inductance and capacitance between the conductors. In this analysis, a single aggressor and victim line was used. A time-domain source with an input pulse from 0-2.5V with 0.1ns rise time, 0.1ns fall time and 2ns period was simulated using transient analysis for 100ns with a time step of 1ps for all 4 material candidates and the baseline. All ports were terminated with a 50 Ω resistor to minimize crosstalk because of reflections and line lengths were held constant at 5mm. The substrate properties were taken from Figure 1. The effect of the near-end (NEXT) and far-end (FEXT) crosstalk is shown in Figure 2. It can be seen that the FEXT values for Material D is significantly lower than the baseline case. The value of the FEXT is expected to fluctuate with changes in rise time, simulations at 10ps, 50ps and 0.1ns were run and

978-1-7281-1500-9/19 $31.00 © 2019 IEEE

the same trend of lower crosstalk was observed for Material D as compared to the Baseline. This shows the potential for lower crosstalk when moving towards low D_k materials for high-density RDL wiring requirements.

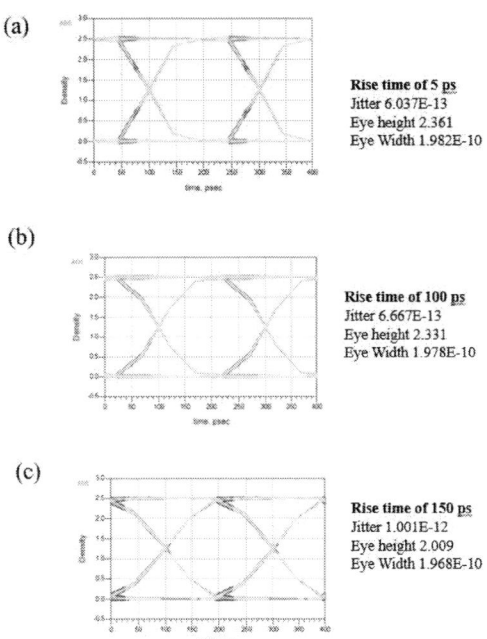

Figure 2 Crosstalk for dielectric candidates

C. Effect of material properties on signaling data rates

The primary motivation for choosing RDL materials with low D_k is their potential to signal at higher data rates while maintaining signal integrity. This along with fine-pitch routing lines will increase the overall bandwidth.

An Eye diagram is a graphical tool that is indicative of signal integrity and electrical performance. It can be used to measure the jitter and distortion of the signal. Of more importance here, is data-dependent jitter which is caused by many factors including inter-symbol interference (ISI), crosstalk, etc. Also, for an ideal eye-opening, the eye-height should be equal to the amplitude. However, noise will cause the eye to close. A threshold value can be set to determine if the eye height is acceptable depending on the designer's budget.

In this analysis, a coupled microstrip transmission line of length 5mm was analyzed using a channel simulation which is designed for rapid signal integrity analysis of linear channels. A PRBS input source was used for a rise time of 10ps, with V_{high} of 2.5V. The connector was terminated at 50 Ω to minimize reflection loss. A crosstalk driver was used that models both synchronous and random crosstalk which inherited properties from the source.

Data rates of 2Gbps, 5Gbps, 9Gbps, 10Gbps, and 12Gbps were studied for a D_k of 4 (SiO$_2$), 3.2 (Material A), and 2.65 (Material D). The properties of Material A are taken to be representative of all epoxy dielectrics although there may be some variation. *Figure 4* shows the eye-height and the complete eye-diagram at three data rates for three material options. A threshold value of 20% of the V_{high} was used to determine eye-closure. At 10Gbps, the eye-height for the

SiO$_2$ and Epoxy are both below 2V while the low-k dielectric is at 2.025V. This suggests that there is potential to signal at higher data rates when we use a low D_k material, as initially hypothesized.

Figure 3 Effect of rise time on the eye height and jitter (a)5ps (b) 100ps and (c) 150ps

However, the eye-opening is dependent on the rise-time of the source. To consider the impact of varying the rise-time, simulations were run at a fixed data rate of 5Gbps and with rise times ranging from 5ps – 150ps for an epoxy material candidate. It can be seen from Figure 3 that a larger rise time decreases the eye-height to a huge extent but does not influence the jitter as much. It is expected that with a higher data rate, the effect of increasing rise time on jitter will be more significant. The eye height changes from 2.361V to 2.009V on increasing rise time. This indicates that the rise time effects are significant at data rates >=5Gbps. The next consideration was the effect of line length on the eye properties. For this, we considered line lengths of 3mm and 6mm, which are within the range of the expected length of connections between HBM-Logic devices. These simulations were run at 5Gbps with the same input source and a 10ps rise time for an epoxy material candidate. From Figure 5, we observe that the eye-properties change with increasing line length. Jitter and noise become worse for longer lengths as is seen from the reduction in eye-height and increase in jitter.

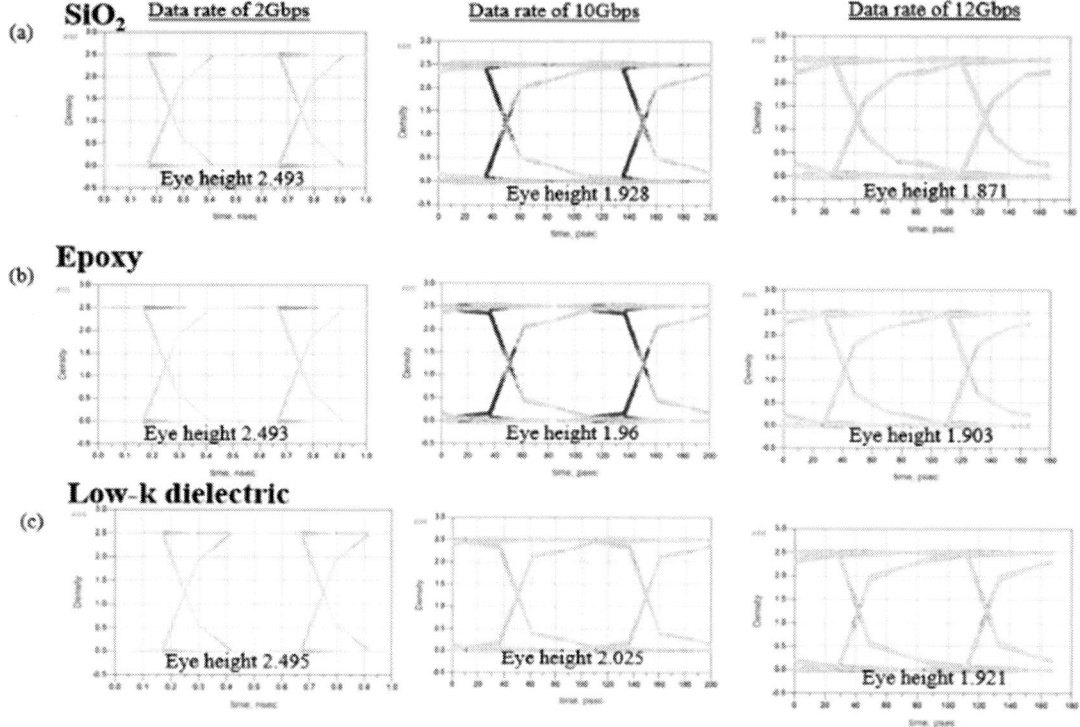

Figure 4 Estimated maximum data-rate for three dielectric material candidates

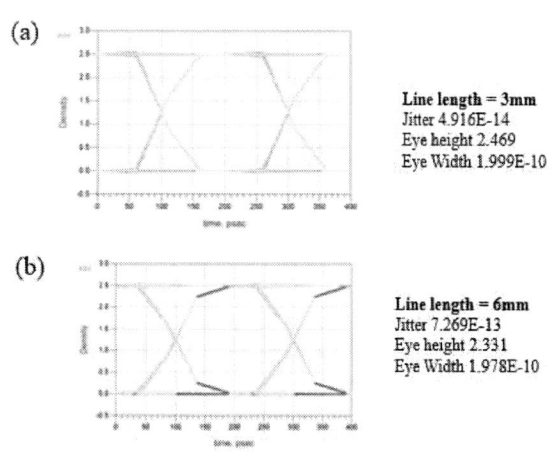

Figure 5 Effect of line length on eye properties (a) 3mm (b) 6mm for epoxy dielectrics.

To summarize, we have simulated the crosstalk between coupled microstrip lines for interposer-like design rules and analyzed the maximum data rate acceptable for given dielectric constant. The results indicate that there is atleast a 10% reduction in crosstalk when you shift to a material with lower dielectric constant as compared to SiO_2. These values are expected to increase when you consider longer routing wires. The maximum data rate for signaling using a low dielectric constant was estimated to be around 10Gbps which overcomes the limitations faced by conventional high-D_k epoxy's and BEOL alternatives. This suggests that it would be beneficial to use low-Dk materials in the high-density routing layers. Current HBM-logic data rates are limited to 1-2 Gb/s as outlined by JESD235A standard as the data rate/pin but it is expected to increase based on future requirements of linear bandwidth density of 10 Pb/s.m at the die edge.

IV. FABRICATION OF FINE-PITCH FEATURES

In this section we look at the processability of our material candidates and the process flows required to achieve fine-pitch routing features using established process flows. To support a 35 µm bump-pitch at the chip-level, an interconnect density of 225 I/Os/mm/layer is needed with 2 µm line/space. This section looks at the process flows and the parameters required to achieve the wiring density.

A. Semi-Additive Process Advances

The process flow for fabricating fine-pitch features is as shown in Figure 6. Material A and D were fabricated using the SAP process flow. The process parameters for Material A will be described, followed by changes required for Material D. The substrates were treated with silane to make the surface more hydrophilic. Silane (3-Aminopropyltriethoxysilane, 95%, ACROS Organics) was vaporized at 90 ^{0}C for 25 minutes in a nitrogen oven. A vacuum hot-press was used to laminate the dry-film polymer dielectric to the substrate.

Lamination is completed by first generating vacuum for 90s and then hot-pressing at 0.6MPa for 30s at 100 °C. The polymer was cured at the recommending curing temperature. A seed layer of Cu was deposited by electroless copper plating with a desmear+electroless line provided by Atotech. The samples were annealed for 30 mins at 150 °C. They were patterned for the high-density features using a stepper panel-scale lithography tool. The samples were electroplated upto the required height using a bath provided by Atotech. The seed layer was etched using a differential seed layer etching tool. An example of the 5 μm line/space structure, fabricating with this process, is shown in Figure 7 (a). The 4 μm via as shown in Figure 7 (c) was drilled using an excimer laser at a partner company.

For Material D, instead of silane coating an adhesion promoter was used. The seed-layer metallization, with a stack of Ti/Cu, was performed using the Denton Discovery RF/DC Sputtering tool. The sputtering parameters were controlled to reduce interfacial stresses from delamination. A special dry-film photoresist capable of 1 μm line/space was used to pattern the dielectric in the Ushio UX 44101 stepper tool. Figure 7 (b) shows the feature fabrication after photoresist striping.

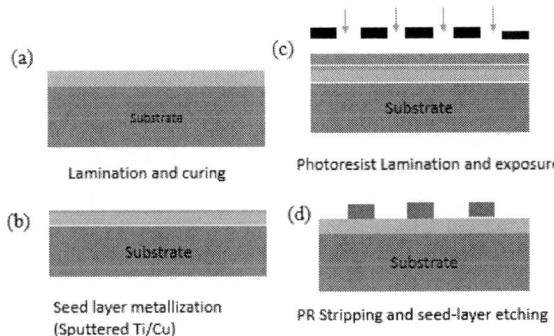

Figure 6 Semi-additive process (SAP) flow for Material A, B and D

B. Advances in embedded RDL

The process flow for fabricating traces using photo-imageable dielectrics is shown in Figure 8. Material B and C can be patterned using this process. As show in Figure 9, the photo-vias, with plated copper of 5 μm and 3 μm, were fabricated using this process flow. A 365nm i-line stepper tool was used to pattern these features. As shown in Figure 10, 2 μm line/spaces, was also patterned via photo-lithography. The key idea here is to use nano-sized fillers with low CTE to achieve high resolution. The advantage with using PIDs is a) potential to reduce the number of fabrication steps and enable high-throughput manufacturing b) fine-pitch features. Currently, the resolution is limited by either material constraints or panel-scale tool restrictions. With advanced lithography techniques such as laser direct-write and improving optics in panel-scale stepper tools current resolution limits can be pushed to further <1 μm feature sizes.

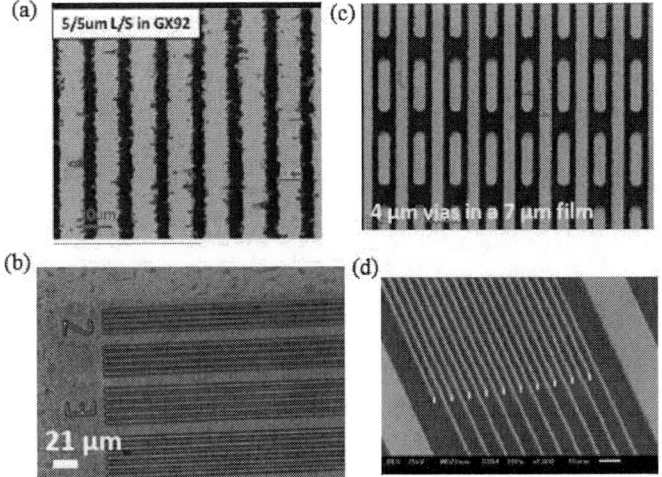

Figure 7 (a) 5 μm line/space in GX92 GX 92 [11] (b) 2 μm line/space in material D (Seed layer present) (c) 4μm vias drilled with excimer laser in Material A [12] (d) 2 μm line/space in Material B

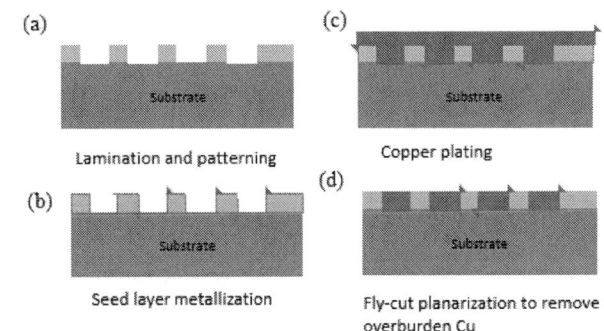

Figure 8 Embedded Conductor Process Flow for PIDs

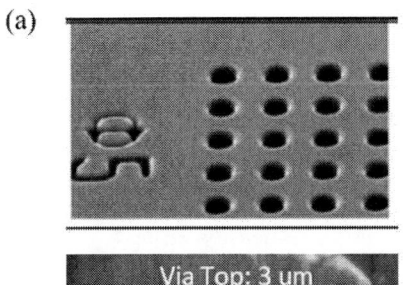

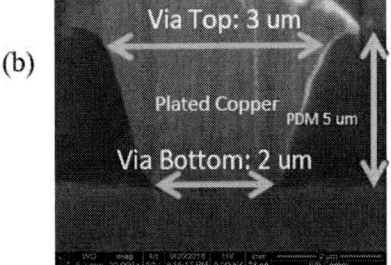

Figure 9 (a) 5 μm photovia in Material C (b) 3 μm photovia in Material B

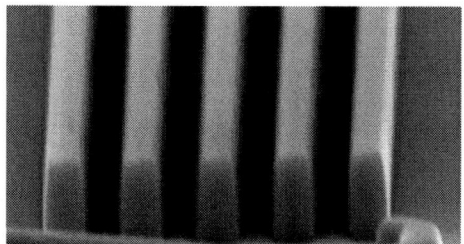

Figure 10 2 μm line/space in Material C

V. CONCLUSIONS

Critical electrical performance metrics such as crosstalk, jitter and eye-opening are investigated for the upfront material candidates In Section 3, it is observed that there is a reduction in cross-talk for low-D_k materials as compared to SiO_2. The eye-diagrams helped quantify the maximum scalable data-rate for current RDL candidates. It clearly illustrates the advantages of a low-D_k dielectric material in terms of supporting higher data rates of nearly 10Gbps which surpasses the capability of current epoxy and BEOL dielectric candidates. In Section 4, advances in fabricating fine-pitch line/space structures using low-D_k materials was reported as well as ultra-small via resolutions reported using PIDs. These advances have been demonstrated with two different processing techniques both of which are panel-scalable and capable of scaling to <1 μm features required for next-generation of RDL. This paper demonstrates that using a low-D_k panel-scalable dielectric increases I/O density and can support higher data rates thus enabling a two-prong increase in bandwidth of the system. This supports the case that the 2.5D/3D Glass Interposer with low-resistance and low-capacitance RDL using low-D_k dielectric materials will be the most suitable option for the next-generation of high bandwidth heterogeneous packaging solution.

In summary, advanced dielectrics capable of meeting next-generation high-performance computing requirements are discussed and design rules for enabling impedance matching for these high-density interconnects are evaluated. This paper introduces (a) next-generation low D_k dielectric materials (b) design consideration for high signal speed and 50 Ω impedance matched high-density wiring (c) fine-pitch fabrication processes for panel-scalable low-D_k dielectrics and advanced PIDs . The paper thereby demonstrates multiple advances in materials and processes to achieve high-speed panel-RDL required for packaging architectures which support computing needs for high-bandwidth.

ACKNOWLEDGMENT

The authors acknowledge PRC supply chain partners in supporting this research effort, particularly Corning, Inc. and Asahi Glass Company for providing glass panels, and Atotech for their electroplating process. The authors would like to thank Siddharth Ravichandran for help with ADS, Chandra Nair for contributing to the process development for 2 μm RDL and Jenefa Kanan for helping with fabrication. The authors would also like to acknowledge Dr. Venky Sundaram for initial discussions and guidance. This work was enriched and inspired by discussions with the Intel SPTD Dielectrics and Litho team. The authors are grateful to Chris White and Lila Dahal for tool training and maintenance.

REFERENCES

[1] R. Tummala, *System on package*. McGraw-Hill Professional, 2008.

[2] K. Oi *et al.*, "Development of new 2.5 D package with novel integrated organic interposer substrate with ultra-fine wiring and high density bumps," in *2014 IEEE 64th Electronic components and technology conference (ECTC)*, 2014, pp. 348-353: IEEE.

[3] M. Ishida, "APX (Advanced Package X)-Advanced Organic Technology for 2.5 D Interposer," in *2014 CPMT Seminar, Latest Advances in Organic Interposers*, 2014, pp. 27-30.

[4] L. Li *et al.*, "3D SiP with organic interposer for ASIC and memory integration," in *2016 IEEE 66th Electronic Components and Technology Conference (ECTC)*, 2016, pp. 1445-1450: IEEE.

[5] T. G. Lenihan, L. Matthew, and E. J. Vardaman, "Developments in 2.5 D: The role of silicon interposers," in *2013 IEEE 15th Electronics Packaging Technology Conference (EPTC 2013)*, 2013, pp. 53-55: IEEE.

[6] Y. Kim *et al.*, "SLIM (TM), high density wafer level fan-out package development with submicron RDL," in *2017 IEEE 67th Electronic Components and Technology Conference (ECTC)*, 2017, pp. 8-13: IEEE.

[7] K. Saban, "Xilinx stacked silicon interconnect technology delivers breakthrough FPGA capacity, bandwidth, and power efficiency," *Xilinx, White Paper,* 2011.

[8] L. Lin *et al.*, "Reliability characterization of chip-on-wafer-on-substrate (CoWoS) 3D IC integration technology," in *2013 IEEE 63rd Electronic Components and Technology Conference*, 2013, pp. 366-371: IEEE.

[9] B. Sawyer, B. C. Chou, S. Gandhi, J. Mateosky, V. Sundaram, and R. Tummala, "Modeling, design, and demonstration of 2.5 D glass interposers for 16-channel 28 Gbps signaling applications," in *2015 IEEE 65th Electronic Components and Technology Conference (ECTC)*, 2015, pp. 2188-2192: IEEE.

[10] F. Liu *et al.*, "Next Generation Panel-Scale RDL with Ultra Small Photo Vias and Ultra-Fine Embedded Trenches for Low Cost 2.5 D Interposers and High Density Fan-Out WLPs," in *2016 IEEE 66th Electronic Components and Technology Conference (ECTC)*, 2016, pp. 1515-1521: IEEE.

978-1-7281-1500-9/19 $31.00 © 2019 IEEE

[11] A. Martwick and J. Drew, "Silicon interposer and TSV signaling," in *2015 IEEE 65th Electronic Components and Technology Conference (ECTC)*, 2015, pp. 266-275: IEEE.

[12] C. Nair *et al.*, "Reliability Studies of Excimer Laser-Ablated Microvias Below 5 Micron Diameter in Dry Film Polymer Dielectrics for Next Generation, Panel-Scale 2.5 D Interposer RDL," in *2018 IEEE 68th Electronic Components and Technology Conference (ECTC)*, 2018, pp. 1005-1009: IEEE.

Attenuation of high frequency Signals in Structured Metallization on Glass: Comparing Different Metallization Techniques with 24 GHz, 77 GHz and 100 GHz Structures

Martin Letz[1], Matthias Jost[8], Brandon T. Gore[2], William J. Kozlovsky[2], Romeo Premerlani[3], Alex Bruderer[3], Manuel Martina[4], Thomas Gottwald[4], Tetsuya Onishi[5,6], Shigeo Onitake[6], Siddharth Ravichandran[7], Holger Maune[8], Mathias Mydlak[1]

[1]SCHOTT AG, Hattenbergstr. 10, 55122 Mainz, Germany, [2]Samtec, Colorado Springs, CO 80919 USA , [3]Varioprint AG, Mittelbissaustrasse 9, CH-9410 Heiden, Switzerland, [4]Schweizer Electronic AG, Einsteinstrasse 10, 78713 Schramberg, Germany, [5]Grand Joint Technology Ltd., Hong Kong, [6]KOTO Electric Co., 2-17-3 Ryusen, Taito-Ku, Tokyo 110-0012, Japan, [7]Packaging Research Center, Georgia Tech, 813 Ferst Drive NW, Atlanta, GA 30332, USA, [8]Institute for microwave engineering and photonics, TU Darmstadt, Merckstr. 25, 64283 Darmstadt, Germany
e-mail: Martin.Letz@schott.com

Abstract— A number of glass substrates made from the SCHOTT glass AF32_eco in a thickness of 0.1mm are metallized with different processes. On the metallized glass substrates, microstrip lines as well as ring resonator structures are realized which are optimized for frequencies of 24 GHz, 77 GHz and 100 GHz. A careful analysis of the S-parameters of these structures allows to extract the influence of the metallization on the attenuation of microstrip lines as well as quality factors of the resonating structures. The particular metallization process plays a crucial role for the GHz performance and the 'best-of' metallization performs comparable with corresponding structures realized on Rogers RO3003 substrates.

Keywords: glass substrate, high-Q, glass package, integrated passives, 5G, mobile communication

I. INTRODUCTION

The demand for applications at higher GHz frequencies is strongly increasing. It is driven by the race for bandwidth in all types of wireless communication applications which are beginning to fill the frequency range above 6 GHz. It is further driven by applications like automotive radar around 77 GHz or gesture recognition around 60 GHz. Even applications above 100 GHz are more and more discussed in literature. With increased crowding of applications and usage of such frequency bands, the geometric accuracy for all designs and high frequency (HF) structures becomes more crucial.

Glass is a class of materials which is produced with extremely high homogeneity. For example, in standard optical glass production, glasses with dimensions of order of 30 cm are made with refractive index variations which are smaller than $\Delta n/n < 10^{-5}$. Since the dielectric constant is the square of the refractive index $\varepsilon = n^2$, this information is also relevant for electronic applications [1]. Glasses are stiff and show (practically) no plastic deformation under mechanic load and temperature cycling. Glasses have low temperature coefficients of all material properties. In particular they have a thermal expansion which can be adjusted in the range between 0 ppm/K and 12 ppm/K. The higher the frequencies requires that more accurate manufacturing be done. This means that very homogeneous substrates materials with small temperature coefficients and intrinsic stiffness become increasingly important at higher frequency ranges.

In the present study, we choose the glass SCHOTT AF32_eco for a comparative study of different metallization techniques. SCHOTT AF32_eco has a dielectric constant ε_r = 5.1 as well as a dielectric loss at several frequencies of $\tan\delta = 3.5\times10^{-3}$ at 1 GHz, $\tan\delta = 4.9\times10^{-3}$ at 5 GHz, $\tan\delta = 9.0\times10^{-3}$ at 24 GHz and $\tan\delta = 1.1\times10^{-2}$ at 77 GHz. The thermal expansion of the glass is α = 3.2 ppm/K, and the temperature coefficient of the dielectric constant at room temperature is $\tau_\varepsilon = 1/\varepsilon \ d\varepsilon/dT$ = 143 ppm/K. In Table 1, we show how it compares to other glasses and materials.

In addition to signal transmission lines, resonating and radiating elements as well as integrated passives [5][4][11][12][13] can be integrated in a single advanced package based on glass [9]. Examples for resonating elements are filter elements or L-C network components whereas antenna elements are radiating structures [7]. Even antenna arrays for massive MIMO (multiple in multiple out) antenna arrays with beam steering capability can be realized on a glass substrate. Fields of application for such approaches are e.g. automotive radar at 77 GHz [8] and gesture recognition around 65 GHz.

The higher the frequencies, the more crucial absorption processes, surface roughness and material interfaces become. The dielectric properties of the glass, especially the dielectric loss tangent, $\tan\delta$, can be directly measured, e.g. with resonant spectroscopic methods. The surface roughness of as-drawn glasses from standard hot-forming processes is extremely small. A typical value for a glass like AF32_eco produced in a so called "down-draw" process has a root mean square (rms) value of the roughness of order 0.8 nm. Material interfaces are a more subtle topic and are investigated in the current work. When metallizing glasses adhesion promotors like Ti, Cr, Pd, … which are polyvalent ions and can exist in different oxidation states play a crucial role on the interfaces. Therefore,

978-1-7281-1500-9/19 $31.00 © 2019 IEEE

the glass-metal interface is one of the largest and unquantified sources of loss in a given structure.

Table 1: <u>Dielectric properties</u> (permittivity ε_r, dielectric loss tanδ and thermal expansion α) of different SCHOTT glasses in comparison with high end RF substrate materials. (*preliminary data, **source data sheet Rogers, *** source data sheet Panasonic, **** source data sheet Isola)

Material	ε_r	tanδ	α / ppm/K
SCHOTT AF32_eco	5.1	4.9×10^{-3} (@5GHz)	3.2
SCHOTT D263T_eco	6.3	10.1×10^{-3} (@5GHz)	7.2
SCHOTT B270D_eco	6.7	7.7×10^{-3} (@5GHz)	9.4
SCHOTT MEMPAX	4.4	7.3×10^{-3} (@5GHz)	3.3
SCHOTT AS87_eco	7.2	17.2×10^{-3} (@5GHz)	8.7
SCHOTT Borofloat33	4.5	7.3×10^{-3} (@5GHz)	3.3
SCHOTT new development	3.9*	1.7×10^{-3}* (@5GHz)	3.3*
Rogers RO3003	3.0**	1.0×10^{-3}** (@10GHz)	17 (xy-dir.)** 25 (z-dir.)**
Panasonic Megtron 7	3.61***	$2.0 \, 10^{-3}$*** (@12GHz)	15 (xy-dir.)*** 42 (z-dir.)***
Isola Astra MT77	3.0****	$1.7 \, 10^{-3}$**** (@10GHz)	12(xy-dir.)**** 50 (z-dir.)****

The glasses were metallized and structured at different commercial and academic partners. The metallization consists of copper with a very small layer of adhesive agents and has an overall thickness of at least 6 μm. The differences of these theoretically identical structures are investigated in terms of the metallization quality, propagation and resonating properties. As a reference, all structures were also realized on well-established Rogers RO3003 substrates with a substrate thickness of 0.125 mm.

As a result, the performance of the best metallized samples of SCHOTT glass AF32_eco comes close to the one of the Rogers RO3003 substrate both for the attenuation of the microstrip line (at 24 GHz: 78 dB/m for RO3003 and 83 dB/m for AF32_eco) as well as for the quality factor of the ring resonator (at 24GHz: $Q = 73$ for RO3003 and $Q = 68$ for AF32_eco). This is a surprising result, since the dielectric loss of the PTFE-based substrate in RO3003 ($\varepsilon_r = 3.0$ and tan$\delta = 0.0013$ at 10 GHz) is almost five times lower than for the glass AF32_eco. This proves that this glass is well suited as substrate material for GHz applications but a carefully optimized metallization solution is essential. The reason for the excellent performance of the glass is probably due to the extremely smooth glass surface with rms values smaller than 1 nm as well as the well-defined glass to metal interface. Since glasses can be produced with much smaller dielectric loss if large enough industry demand is present (see e.g. line 7 in Table 1) glass substrates are possible which clearly

outperform the electric properties of the best available high end PCB materials like RO3003.

II. DESIGN OF TEST STRUCTURES

The structures were designed to allow (i) contacting of any of the given test structures with standard probe heads from the top side of the substrate. Further, the test structures were (ii) designed with a line impedance of 50 Ω to minimize reflection losses. Also (iii) transmitting structures, microstrip lines of different lengths were designed as well as resonating structures, ring resonators, which showed different weak coupling to the attached microstrip lines.

Contact pads in coplanar waveguide topology were included to the design for RF characterization by means of on-wafer probes. Radial stubs ensure a capacitive coupling of the coplanar waveguide's ground planes to the one of the microstrip line at the back side of the substrate. A microstrip line was chosen since a large part of its electric field is confined in the dielectric substrate, the glass. All designs were conducted using CST Studio Suite. The designs were optimized for 24 GHz, 77 GHz and 100 GHz separately on 0.1 mm thick glass in order to match the line impedance of 50 Ω at the targeted frequency of operation.

Also ring resonators were designed with a resonance close to 24 GHz, 77 GHz and 100 GHz. At 24 GHz, it is designed for the fundamental mode, while for 77 GHz and 100 GHz it is designed for the first and second harmonic, respectively. These ring resonators are weakly coupled to microstrip lines on both sides, separated by small gaps. While the radius of the ring influences the resonance frequency, the width of the gaps between the ring and the microstrip feeding lines are influencing the coupling into the resonator. The smaller the gap, the stronger is the coupling, but at the same time, the loaded quality factor of the resonator is decreasing. Besides that, the quality factor of the resonator contains information on the dielectric loss of the substrate and of the quality of the metallization. A qualitative comparison allows to evaluate and to distinguish between different metallization methods on the same glass substrate.

III. EXPERIMENTAL SETUP

Glass Substrate Design

The metallized structures were designed as microstrip lines with different lengths. An example of such a microstrip line with coplanar coupling is shown in Figure 1.

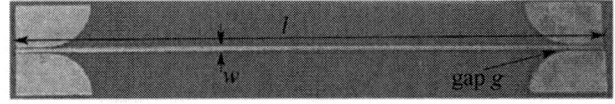

Figure 1: Design of a typical microstrip line with coplanar coupling.

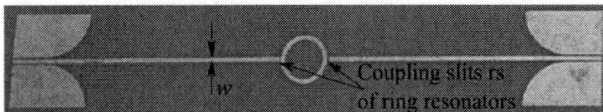

Figure 2: Design of a typical ring resonator with coplanar coupling.

978-1-7281-1500-9/19 $31.00 © 2019 IEEE

Critical points for this design are the widths of the two gaps in the coplanar waveguide contact pad design and the overall width of the microstrip line. For all structures the width of the copper (Cu) metallization was targeted at $w = 163$ μm to achieve a line impedance of 50 Ω on a 0.1 mm thick AF32_eco glass. For the 24 GHz microstrip lines, the gaps in the coplanar waveguide contact pads were targeted as $g = 76$ μm. The lengths of the structures were chosen to be $l = 37$ mm, 32 mm and 27 mm, respectively. For the 24 GHz ring resonators, the outer diameter of the ring was targeted to be 2.163 mm and the inner diameter 1.837 mm. The gap between the ring and the microstrip line was designed to be rs = 12 μm and rs = 48 μm. For the 77 GHz and 100 GHz structures, the lengths of the microstrip lines was chosen to be 17.1 mm, 12.1 mm and 7.1 mm, respectively. The gap in the coplanar coupling was chosen for both types of structures as 20.3 μm. Also for the ring resonators, the width of the slits between the ring and the microstrip line was targeted for both types of structures as 10 μm and 40 μm. The only difference between the two structures for 77 GHz and 100 GHz is the diameter of the ring resonators. For the 77 GHz structures it was chosen to be 1.373 mm for the outer diameter and 1.0519 mm for the inner diameter to obtain a resonance in the ground mode at 77 GHz. For the 100 GHz structures the diameter of the rings was 1.563 mm outer diameter and 1.241 mm inner diameter to resonate at a second harmonic at 100 GHz.

Measurement procedure

The glass substrate and the substrate of the reference material RO3003 were contacted on a wafer prober from Cascade Microtec with a climate chamber to ensure identical environment for all measurements with respect to temperature and humidity. The contacting for the 24 GHz structures was done using a pair of probe heads with 250 μm pitch (ACP40), see Figure 3. A short-open-load-thru calibration was performed prior to the measurements and the S-parameters were measured using a Keysight PNA-X vector network analyzer (VNA) for the frequencies below 65 GHz and a Keysight PNA combined with two Anritsu 3740A-EW transmission-reflection modules for the frequency range from 65 GHz to 110 GHz. As an example we show the S-parameter S_{11} and S_{12} for three 24 GHz microstrip lines with overall lengths of 27 mm, 32 mm and 37 mm in Figure 4 and Figure 5. In Figure 4 it can be seen that all three lines are well matched to 50 Ω, since the S_{11} parameter circles very close around the center of the Smith chart, the point (1,0). In Figure 5, the three measurements should be parallel lines for ideal measurements where no reflection occurs from impedance mismatch at the contacts. The small reflection is seen in the slight oscillations of the S_{21} measurement. From the difference in the transmission S-parameter S_{12} one can derive the attenuation of the microstrip line in the following way. First the operating power loss

$$T_{\text{eff}} = \frac{|S_{21}|^2}{1 - |S_{11}|^2} \tag{1}$$

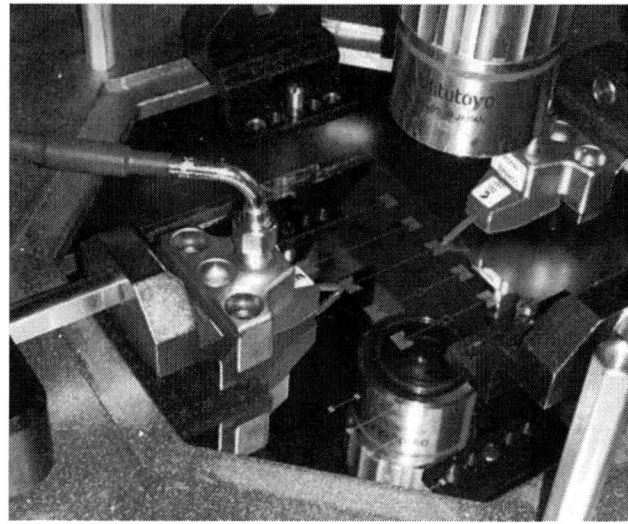

Figure 3: Contacting of the substrate using 250 μm pitch probe heads.

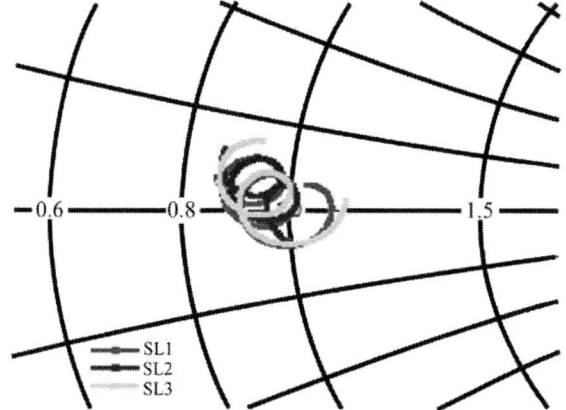

Figure 4: S_{11} parameter plotted in the complex plane (Smith chart) for the three microstrip lines SL1-3 on substrate SM2.

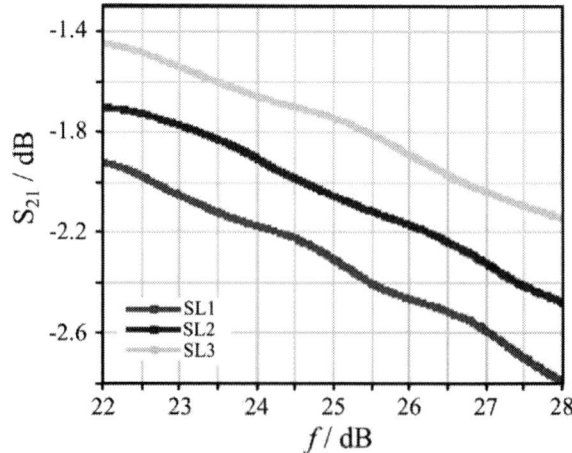

Figure 5: S_{21} parameter plotted as a function of frequency for the three microstrip lines SL1-3 on substrate SM2.

Table 2: Derived attenuation per unit length a of the 24 GHz microstrip lines for the different glass substrates and for the Rogers RO3003 substrate (R1).

Substrate	measured g / μm	targeted g / μm	a / dB/m
R1	79	76	78
AF32_1	78	76	83
AF32_2	75	76	238
AF32_3	86	76	143
AF32_4	76	76	116
AF32_8	73	76	118
AF32_9	70	76	106
AF32_10	70	76	109

is calculated from the S-parameters. It corrects the transmission by the reflection terms. In an ideal world it should give three straight lines, which are equidistant, as a function of frequency for the three different lengths of the microstrip lines. The data are fitted to a linear relationship, and the difference gives the loss of the microstrip line. In Table 2, we summarize the derived attenuation per unit length a for seven different glass substrates which use five different methods for metallization and structuring and compare it to corresponding structures on a 125 μm thick copper laminated RO3003 (R1). The attenuation which we obtain is roughly a factor of three larger than the attenuation reported by Rogers [2]. In [2], a value of 0.7 dB/inch = 27 dB/m is reported for laminate without any climate stress, whereas we measure a loss of 78 dB/m. This discrepancy might be a hint for non-perfect structures but it contradicts to our excellent match of the 50 Ω impedance. The attenuation of the structured glasses ranges between 83 dB/m and 238 dB/m which means that the details of the metallization and structuring processes differ a lot in high frequency performance. This also means that optimization of the metallization is crucial for high frequency performance. The solution of AF32_1 is an excellent compromise since in this case the glass is coated on both sides with a thin (15 μm) polymer layer of low-loss dry film dielectrics (Ajinomoto ABF GY11 with $\varepsilon = 3.2$, $\tan\delta = 0.0042$ at 5.8 GHz). Besides low dielectric loss it also shows good metal adhesion. From the structures where the metal was directly coated on the glass the substrates AF32_4 and AF32_8-10 show the best results. AF32_8-10 are metallized using the same process and demonstrate the reproducibility of the results.

As a next step the S-parameters for the 24 GHz ring resonators are measured. In this case a direct quantitative measure of the substrate quality is more difficult since the loaded Q-factor of the ring resonator depends on the dielectric losses, the metallization losses and the attenuation due to the vicinity of the microstrip line. A basic understanding of the last point is obtained by structuring always a pair of ring resonators with two different gaps (see Figure 2). Typically rs = 12 μm and rs = 48 μm, were targeted for structuring. In Figure 6 and Figure 7 we show representative S-parameters

for the 24 GHz ring resonator structures. The Eigenresonance at 24.5 GHz can be clearly

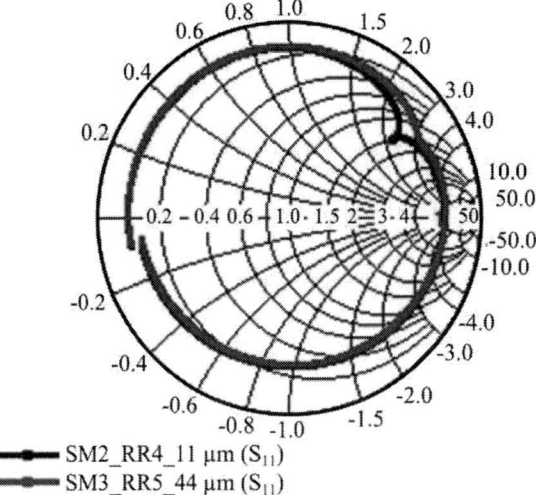

SM2_RR4_11 μm (S_{11})
SM3_RR5_44 μm (S_{11})

Figure 6: S-parameter S_{11} of the ring resonators RR4 and RR5 on substrate SM2=AF32_9 and SM3=AF32_10 with a slit width rs = 11 μm and rs = 44 μm, respectively.

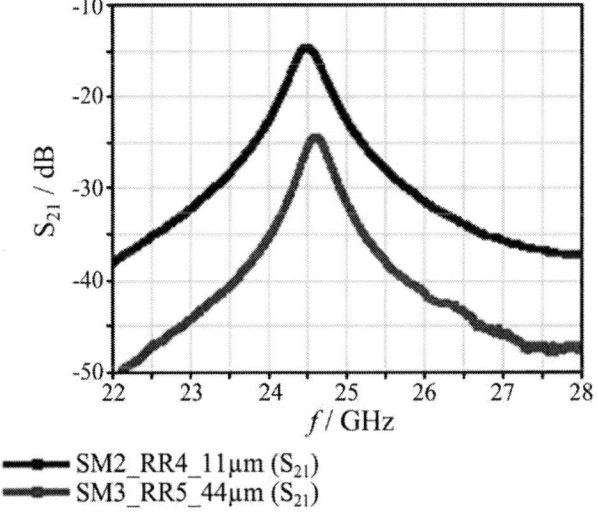

SM2_RR4_11μm (S_{21})
SM3_RR5_44μm (S_{21})

Figure 7: S-parameter S_{21} of the two ring resonators RR4 and RR5. It can be clearly seen that the smaller slit with rs = 11 μm allows to couple more energy into the ring resonator whereas the wider slit with rs = 44 μm allows to couple less energy into the resonator but improves the overall loaded quality factor Q of the resonator.

observed. The damping of the resonator and the energy which is coupled into the resonator differs strongly, depending on the width of the slit rs. The loaded quality factor Q of each ring resonator is obtained by fitting the S_{21} parameter to a Lorentzian line shape

$$S_{21}(\omega) \;=\; \frac{a}{\left((\omega - \omega_0 + \frac{\omega_0}{8Q^2})^2 + (\frac{\omega_0}{2Q})^2\right)} \qquad (2)$$

Table 3: Slit width rs and obtained loaded Q-factor of the 24 GHz ring resonators are shown. The Q-factors are only slightly larger than the ones of the RO3003 substrate (R1).

Substrate – No. of Ring Resonator	measured rs / µm	targeted rs / µm	Q
R1 – RR 4	56	19	73
R1 – RR 5	58	6	73
AF32_1 – RR 4	50	19	58
AF32_1 – RR 5	50	19	59
AF32_1 – RR 5a	5	6	44
AF32_2 – RR 5	46	48	61
AF32_3 – RR 5	95	48	68
AF32_4 – RR 4	35	12	65
AF32_5 – RR 5	69	48	68
AF32_8 – RR4	11	12	56
AF32_8 – RR5	48	48	65
AF32_9 – RR4	11	12	56
AF32_9 – RR5	44	48	66
AF32_10 – RR4	10	12	56
AF32_10 – RR5	44	48	66

with an Eigenresonance ω_0 and a quality factor Q. The results for the obtained ring resonators are shown in Table 3. As an overall result of the 24 GHz structure we conclude that it is possible to make 24 GHz resonating and transmitting structures with good quality on glass. However the metallization process has to be chosen carefully.

IV. MEASUREMENT AT 77 GHz AND 100 GHz

A similar procedure was performed for the frequency dependent measurement in the frequency range between 65 GHz to 110 GHz. In this case the structures were smaller and the contacting was done using a pair of infinity w-band probe heads with 100 µm GSG pitch. Due to the larger frequencies a hollow waveguide was needed to connect the probe heads. To determine the loss of a microstrip line, lines were manufactured with lengths of 7.1 mm, 12.1 mm and 17.1 mm. As an example we show the S-parameter S_{11} and S_{21} of six lines in the frequency range between 70 GHz to 105 GHz in Figure 8 and Figure 9. In the frequency range between 75 GHz to 100 GHz, a straight line was fitted into the T_{eff} value (see Eq. (1)). From that, the loss of the microstrip lines in units of dB/m was determined. Unfortunately, structures on the Rogers substrate were not measurable but for some glass structures reliable and reproducible values for the dielectric loss of the microstrip line were obtained. The results for 77 GHz and 100 GHz are listed in Table 4. A loss of several hundred dB/m can clearly not be neglected but allows to integrate small structures in the frequency range up to 110 GHz on such a glass substrate.

Also ring resonators were processed on the substrates. For 77 GHz a ring resonator with an inner diameter of 1.05 mm and an outer diameter of 1.37 mm was structured.

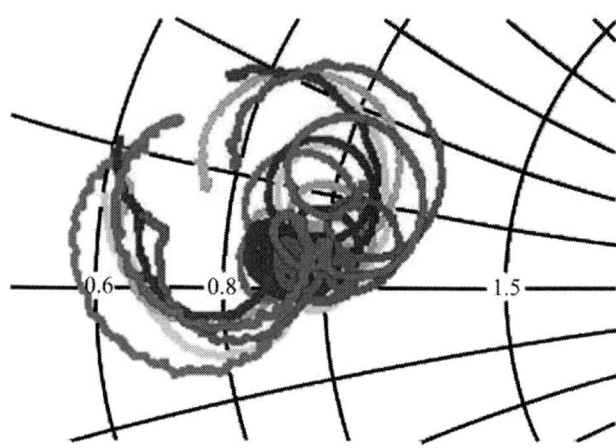

Figure 8: S_{11} parameter plotted in the complex plane (Smith chart) for the six microstrip lines S1-3 and S5-7 on substrate AF32_10.

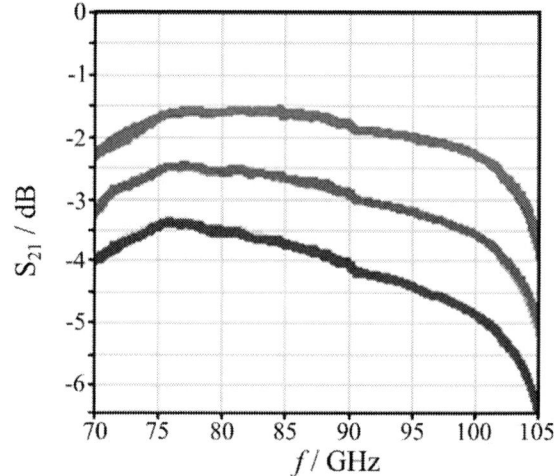

Figure 9: S_{21} parameter plotted as a function of frequency for the six microstrip lines S1-3 and S5-7 on substrate AF32_10.

The first harmonic of such a ring resonates at 77 GHz according to field simulation investigations using CST studio.

The results of the 77 GHz ring resonators are summarized in Table 5. Depending on the details of the metallization and structuring process and depending on the width of the gap on the ring (see Figure 2) loaded Q-values of up to $Q = 54$ were obtained. The substrates AF32_5-7 and AF328-10 are each metallized with identical processes and demonstrate the reproducibility of the results.

The 100 GHz resonating structures were designed in a little different way. The rings were made with an inner diameter of 1.4 mm and an outer diameter of 1.56 mm. In this way they show a resonance in the first harmonic around 69 GHz and for the second harmonic around 100 GHz. The results are shown in Table 6.

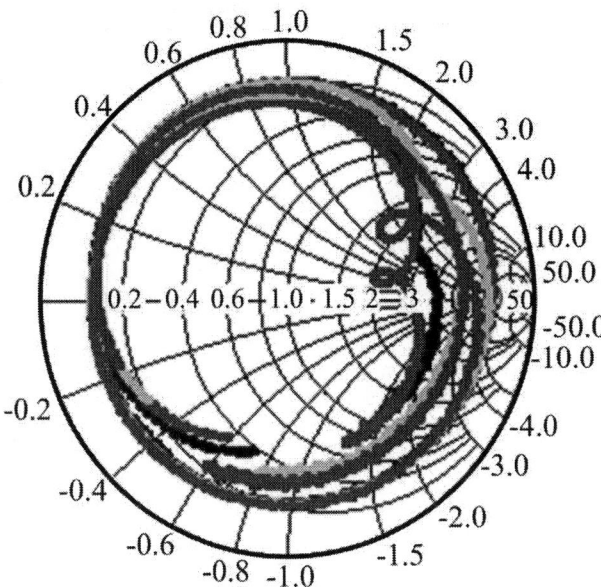

Figure 10: S_{11} parameter plotted in the complex plane (Smith chart) in the range between 65 GHz and 110 GHz for the four ring resonators RR4,5,8 and 9 on substrate AF32_10.

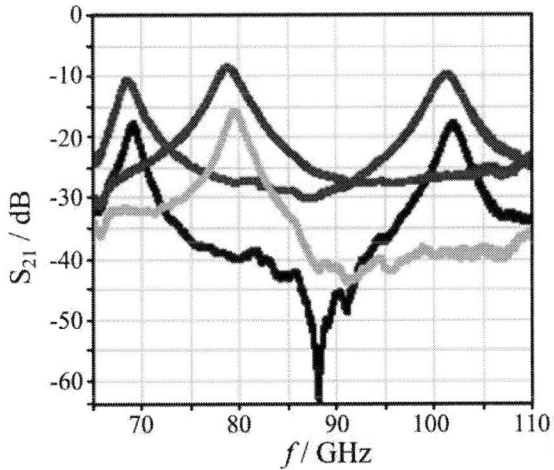

Figure 11: S_{12} parameter plotted as a function of frequency for the four ring resonators RR4,5,8 and 9 on substrate SM3. The 77 GHz resonance with the two different couplings to the ring resonator can be clearly observed as well as the two resonances around 69 GHz and around 100 GHz of the 100 GHz structures.

Table 4: Attenuation a of the microstrip lines of identical glass substrates with different metallization procedures. The realized and targeted slit width g of the GSG coplanar contact pads are given.

Substrate	measured g / μm	targeted g / μm	77 GHz a / dB/m	100 GHz a / dB/m
AF32_1	17	20	309	420
AF32_1	17	20	318	477
AF32_2	12	20	391	549
AF32_5	33	20	359	519
AF32_5	33	20	354	500
AF32_6	53	20	386	504
AF32_6	53	20	335	483
AF32_7	39	20	353	511
AF32_8	24	20	359	502
AF32_9	22	20	356	512
AF32_10	24	20	360	513

Table 5: Properties of the ring resonators around 77 GHz

Substrate – No. of Ring Resonator	measured rs / μm	targeted rs / μm	Q
AF32_1 – RR 9	4	19	26
AF32_2 – RR 10	49	40	53
AF32_5 – RR 10	64	40	43
AF32_5 – RR 4	49	10	48
AF32_5 – RR 5	67	40	54
AF32_5 – RR 9	47	10	47
AF32_6 – RR 10	78	40	35
AF32_6 – RR 4	63	10	49
AF32_6 – RR 5	84	40	50
AF32_6 – RR 9	61	10	52
AF32_7 – RR8	43	40	43
AF32_7 – RR9	39	10	53
AF32_7 – RR4	51	10	48
AF32_7 – RR5	68	40	51
AF32_8 – RR8	40	40	41
AF32_8 – RR9	13	10	29
AF32_9 – RR8	43	40	42
AF32_9 – RR9	15	10	28
AF32_10 – RR8	43	40	43
AF32_10 – RR9	13	10	30

Table 6: Properties of the ring resonators around 100 GHz

Substrate – No. of Ring Resonator	measured rs / μm	targeted rs / μm	Q
AF32_1 – RR 9	4	19	24
AF32_2 – RR 10	49	40	52
AF32_5 – RR 10	64	40	57
AF32_5 – RR 9	47	10	53
AF32_6 – RR 10	78	40	53
AF32_6 – RR 9	61	10	56
AF32_7 – RR 9	39	10	53
AF32_7 – RR 10	65	40	56
AF32_8 – R4	36	40	49
AF32_8 – R5	7	10	32
AF32_9 – R4	45	40	50
AF32_9 – R5	19	10	34
AF32_10 – R4	44	40	49
AF32_10 – R5	14	10	34

Table 1. Properties of the ring resonators around 100 GHz.

V. CONCLUSION

In summary we have shown that efficient structures with reasonable small dielectric loss at frequencies between 24 GHz and 100 GHz can be built using 100 μm thin SCHOTT glass AF32_eco as a substrate material. Also resonating structures which show resonances at 24 GHz, at 77 GHz and at 100 GHz in combination with reasonable Q-factors were demonstrated. A careful optimization and a careful choice of the metallization process and of the structuring method are needed. Larger data rates also contain higher frequency components and become important for e.g. interposers and high power computing. Using glass as a substrate, can make use of the results presented in the current work.

ACKNOWLEDGMENTS

We acknowledge support from the IAB program of the PRC. We thank Rolf Jakoby from the TU Darmstadt for his support of this project.

REFERENCES

[1] M. Letz, M. Jotz, R. Sprengard, „Glass: a universal material class for semi-conductor packaging and further integration", Chip Scale Review, 21(3), 37 (2017)

[2] Rogers product information, MVP "New laminates enable 79 GHz Technology Advancements", Microwave Journal 58(12) 2015

[3] C. S. Lam, "A review of the timing and filtering technologies in smartphones," 2016 IEEE International Frequency Control Symposium (IFCS), New Orleans, LA, 2016, pp. 1-6. doi: 10.1109/FCS.2016.7546724

[4] Y. H. Chen et al., "20" x 20" Panel Size Glass IPD Interposer Manufacturing," 2016 IEEE 66th Electronic Components and Technology Conference (ECTC), Las Vegas, NV, 2016, pp. 2146-2150. doi: 10.1109/ECTC.2016.400

[5] M.Letz et al. "Glass in Electronic Packaging and Integration: High Q Inductances for 2.35 GHz Impedance Matching in 0.05 mm Thin Glass Substrates" 2018 IEEE 68th Electronic Components and Technology Conference DOI: 10.1109/ECTC.2018.00167

[6] S. Onitake and T. Onishi, "Direct copper metallization on glass technology + x-substrate," 2017 IMAPS Nordic Conference on Microelectronics Packaging (NordPac), Gothenburg, 2017, pp. 35-37. doi: 10.1109/NORDPAC.2017.7993159S.

[7] Z. Wu, J. Min, V. Smet, M. R. Pulugurtha, V. Sundaram, R. R. Tummala, "Ultra-miniaturized 3D IPAC Packages with 100 μm Thick Glass Substrates for RF Front-end Modules", *Journal of Electronic Packaging,* 2017, Vol. 13, No. 3, pp. 041001.

[8] J. Hasch, E. Topak, R. Schnabel, T. Zwick, R. Weigel and C. Waldschmidt, "Millimeter-Wave Technology for Automotive Radar Sensors in the 77 GHz Frequency Band," in IEEE Transactions on Microwave Theory and Techniques, vol. 60, no. 3, pp. 845-860, March 2012.

[9] R. Tummala, "Moore's law for packaging to replace Moore's law for IC's", Chip Scale Review, 23(1), 31 (2019)

[10] A. Shorey et al., "Addressing Next Generation Packaging and IoT with Glass Solutions", IMAPS Additional Conferences (Device Packaging, HiTEC, HiTEN, & CICMT), DPC, 1-19 (2017)

[11] A. O. Watanabe et al., "First Demonstration of 28 GHz and 39 GHz Transmission Lines and Antennas on Glass Substrates for 5G Modules," 2017 IEEE 67th Electronic Components and Technology Conference (ECTC), Orlando, FL, 2017, pp. 236-241.

[12] Z. Wu, J. Min, M. S. Kim, M. R. Pulugurtha, V. Sundaram, R. R. Tummala, "Design and Demonstration of Ultra-thin Glass 3D IPD Diplexers", in *2016 Proceedings 66th ECTC*, pp. 2348-2352, Las Vegas, NV, 2016.

[13] Jun-Bo Yoon, Bon-Kee Kim, Chul-Hi Han, Euisik Yoon, Kwyro Lee and Choong-Ki Kim, "High-performance electroplated solenoid-type integrated inductor (SI/sup 2/) for RF applications using simple 3D surface micromachining technology," International Electron Devices Meeting 1998. Technical Digest (Cat. No.98CH36217), San Francisco, CA, USA, 1998, pp. 544-547. doi: 10.1109/IEDM.1998.74641

The Highly Effective EMI Shielding Materials for Electric and Magnetic Fields over the Wide Range of Frequency in Near-Field Region

Yoon-Hyun Kim[1], Kisu Joo[1], Kyu Jae Lee[1], Jung Woo Hwang[1],
Seung Jae Lee, Se Young Jeong[1, *] and Hyun Ho Park[2]

[1] Ntrium Incorporation, 54-42, Dongtanhana 1-gil, Hwaseong-si, Gyeonggi-do, Korea
[2] Department of Electronic Engineering, The University of Suwon, Hwaseong 18323, South Korea

Corresponding author's e-mail: snuven@ntrium.com, hhpark@suwon.ac.kr

Abstract — **The effects of the electrical resistivity and permeability of conductive paste and magnetic paste materials on the near-field shielding effectiveness (SE) were investigated using strip-line and loop-probe methods. The silver paste of $3.4 \times 10^{-5}\,\Omega \cdot cm$ of volume resistivity showed more than 30dB of electric field SE and the SE increased as the thickness of conductive layer increased. For the magnetic field SE showed max 60dB as the thickness increased. However, in low frequency below 10MHz, the magnetic field SE was max 3dB in the case of 10 μm thickness. Magnetic sheets made of soft magnetic materials having permeability of 200 μ showed 11dB of magnetic field SE at 1MHz. The hybrid paste having conductivity of $1.8 \times 10^{-4}\,\Omega \cdot cm$ and permeability of 23.7μ showed SE of 13.5dB for the magnetic field and 63.9dB for electric field at 10MHz, respectively. The paste having conductivity of $3.7 \times 10^{-2}\,\Omega \cdot cm$ and permeability of 53.4μ, showed 7.6dB for magnetic field at 10MHz. In low frequency below 10MHz, the permeability had dominant influence on the magnetic field shielding in the near-field region. As the frequency increases, the SE is mainly affected by the conductivity of hybrid pastes.**

Keywords- *EMI shielding, spray, near field shielding efeectiveness, hybrid paste*

I. INTRODUCTION

Conventional EMI shielding solution used to protect the devices has been shield-can type structures. However, in order to be in accordance with ultra-thin thickness of product, new EMI shielding methods have been intensively on the rise in these days. In order to meet these demands, recently, many researchers have been actively studied on electromagnetic shielding methods such as spray, electroplating, and sputtering methods. Among these, the spray methods are considered to have relative advantages in terms of the byproducts generated during the process, the cost for the initial investment, and the performance and capability of the shielding layer. Also, near field shielding as well as far field shielding is focused by package engineers. It has been known that the near magnetic field shielding is difficult to shield effectively.

In this paper, we have investigated the effects of the electrical resistivity and permeability of EMI shielding conductive paste and magnetic paste materials on the SE level by strip-line and loop-probe methods. To control the permeability of magnetic materials, we have prepared various pastes formulations by different magnetic fillers and mixture ratio. Also, we prepared the hybrid pastes which have high permeability as well as conductivity for shielding near magnetic and electric fields.

II. EXPERIMENTAL DETAILS

A. Conductive paste with epoxy resin

Ag paste was carefully formulated by combining the components: resin, additives, and fillers. In the resin part, based on multifunctional resin is mixed to achieve appropriate property of sprayed film. The leveling additives are the portion that cares leafing speed of fillers sprayed on target substrate, which affects the electrical resistivity and side coverage. In the filler part, flake shaped Ag powder was prepared and blended with other parts such as resin and additives. All the contents of pastes are very similar to those of commercial paste (NSA-F3-R4) which is based on epoxy resin matrix and introduced in the previous paper published in ECTC 2017 [1].

B. Magnetic paste with magnetic powder and epxoy resin

Magnetic paste was prepared by similar methods to that of Ag paste. Soft magnetic materials were employed as magnetic fillers having a specific surface area of 0.7~1.2 m²/g.

As the size of fillers increases, the permeability of magnetic material increases. However, as the particle size becomes larger, the film properties after spray coating deteriorate and the thickness becomes inevitably thicker. If the filler sizes are smaller, the coating properties are better but the permeability is lowered, resulting in disadvantages to magnetic shielding efficiency.

C. Hybrid paste with condutive and magnetic filler

Hybrid paste, including both conductivity and permeability, was manufactured to shield the near field in a wide range of frequency. The hybrid paste was made by mixing a conductive paste and a magnetic paste. The mixing ratio was set based on the new parameter 'x' in consideration of the specific surface area of the conductive powder and the magnetic powder.

Where x is,

A_c = specific area of conductive filler
P_c = mass of the conductive filler / total filler's mass
P_m = mass of the magnetic filler / total filler's mass
A_m = specific area of the magnetic filler

(1) Summation of the conductive filler's specific area =
   $A_c \times$ mass of the conductive filer
(2) Summation of the magnetic filler's specific are =
   $A_m \times$ mass of the magnetic filler
(3) The specific surface ratio of conductive and magnetic fillers
   $\equiv X = (1) \div (2) = (Ac \times Pc) \div (Am \times Pm)$

After the preparation of the paste, the characteristics were analyzed. The new parameter 'x' was changed from 1 to 6. The conductivity, permeability, and SE characteristics were analyzed according to the change of x values.

D. Measurement of the electrical resistivity and permeability.

As shown in Figure 1, a micro-ohm meter (HIOKI, RM3548) was used to measure the electrical properties such as the resistivity and resistance. All the samples were sprayed with the spray equipment and cured at 190°C for 60 minutes under atmosphere. The electrical properties were measured in the shortest direction. The thickness of sprayed conductive film was measured with FE-SEM. Then, the electrical resistivity was calculated by measured resistance from micro-ohm meter and measured dimension from FESEM. By utilizing the result of resistance values measured by micro-ohm meter and thickness information measured by FESEM, we can calculate the electrical resistivity of sprayed shielding layer.

The permeability was measured based on the method of ASTM A342 / A342M-14 using impedance analyzer in frequency range of 1MHz~1GHz. The samples were fabricated by spray process with 50 μm of thickness. As for 200μ sheet, hot pressing process was added after curing for high permeability.

E. Measurement of near-field SE

Figure 2 shows the near field electric and magnetic shielding measurement setups for shielding materials coated on two source patterns on the test boards. A patch source pattern and a 50Ω matched micro strip-line source were used to generate the electric field and magnetic fields, respectively. For the electric shielding measurement as

shown in Figure 2(a), a strip-line made from a thin brass plate was employed to capture a near electric field leakage from the coated materials. Both ends of the strip-line and one end of the patch source were connected to each port of a vector network analyzer (VNA) through subminiature version A (SMA) connectors and RF coaxial cables. The near electric-field shielding SE of planar coating materials can be extracted from 3-port S-parameters measurement [2]. For the magnetic shielding measurement as shown in Figure 2(b), a loop-probe is employed to capture a near magnetic field leakage from the coated materials. One end of the micro-strip line and the loop probe are connected to the port 1 and port 2 of VNA, respectively. The other end of micro-strip line is terminated by a 50 Ω resistor. Using 2-port S-parameters measurement, the near magnetic-field SE of planar magnetic sheets can be simply extracted [3].

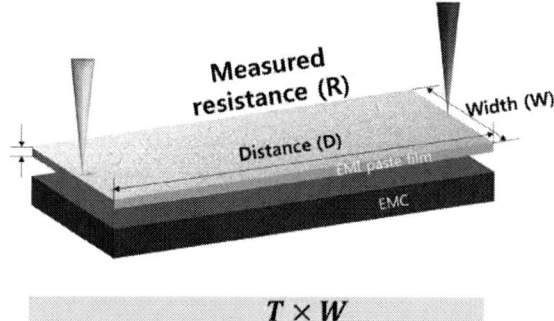

$$\rho = R \cdot \frac{T \times W}{D}$$

Figure 1. Calculation of the electrical resistivity of shielding material

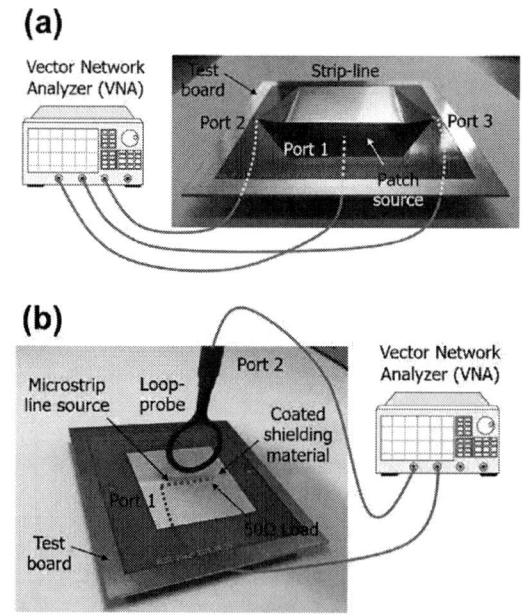

Figure. 2 Electric and magnetic shielding measurement setups of (a) strip-line and (b) loop-probe method, respectively.

978-1-7281-1500-9/19 $31.00 © 2019 IEEE

F. Measurement of far-field SE

The EMI SE in far-field region was analyzed based on ASTM D-4935 method. The measurement specimens were prepared by spray coating a conductive paste, a magnetic paste and a hybrid paste on a 15cm × 15cm sized PI film and then cured at 190°C for 1 hour. The measurements were performed by using impedance analyzer. The frequency was changed from 30 MHz to 1.5 GHz.

III. RESULTS AND DISCUSSION.

A. Effect of conductivity on SE

The SE can be calculated by the following Simon's equation (Equation 1) in the far field region where the impedance of the electric wave and the magnetic wave are 377 Ω

$$SE = SE_{reflection} + SE_{absorption} + SE_{multiple\ reflection} \quad (1)$$

If the SE is more than 10 dB, the multiple reflections terms could be ignored. Then, the SE can be summarized as follows. [4]

SE [dB]= $50 + 10\log(\rho \times f)^{-1} + 1.7t\ (f/\rho)^{1/2}$
ρ: volume resistivity (Ω·cm)
f: frequency (MHz)
t: thickness of sample (cm)

Equation 1. Simon's equation for shielding effectiveness of EMI

Therefore, the SE of the specimen showing the same conductivity depends on the film thickness and frequency, and can be considered as a function of the conductivity. These can be confirmed by our EMI SE results measured in the far-field region.

Samples with thicknesses of 10, 30 and 50μm were prepared and compared by using a conductive paste showing volume resistance of $3.0 \times 10^{-5}\Omega \cdot cm$. Moreover, additional sample was fabricated, which shows the 5 times higher volume resistivity, but the coating thickness was increased 5 times to obtain the same bulk resistance characteristics.

We found that even though the specimens showing the same volume resistivity, as the film thickness became thicker, shielding efficiency was changed. 55dB, 60dB, and 65dB of SE were observed in the range of 30MHz to 1.5GHz.

In addition, the 150μm-thickness sample showed a similar SE with that of the 10μm-thickness sample showing similar sheet resistance. We found that conductivity is the critical factor to determine the SE in the far-field region.

As the film thickness increases, the SE also increases to 55dB, 60dB and 65dB in the range of 30MHz to 1.5GHz. The samples showed a similar SE with that of 10μm

samples with low resistance. It is confirmed that the conductivity is a very important factor for far-field SE.

In the near-field region, the wave impedance is changed depending on the electric and magnetic fields. The distance from the source to the shielding material and the frequency affects the SE. The shielding characteristics of the electric field and the magnetic field according to the sheet resistance of the conductive silver paste are shown in Figures 3 and 4.

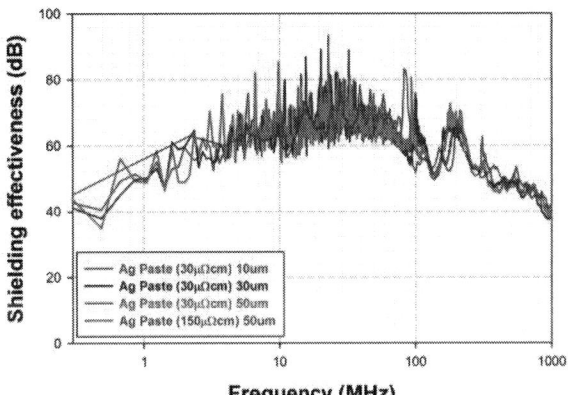

Frequency	300KHz	1MHz	10MHz	100MHz	1GHz
Ag Paste 10μm (30μΩ·cm)	42.7	49.2	62.0	65.9	38.4
Ag Paste 30μm (30μΩ·cm)	41.2	50.1	61.5	75.7	37.8
Ag Paste 50μm (30μΩ·cm)	43.8	48.9	75.6	60.2	39.7
Ag Paste 50μm (150μΩ·cm)	45.2	56.1	63.5	51.9	41.0

Figure 3. Near-field SE results of various samples with the electric field sources. The samples were prepared as a function of sheet resistance.

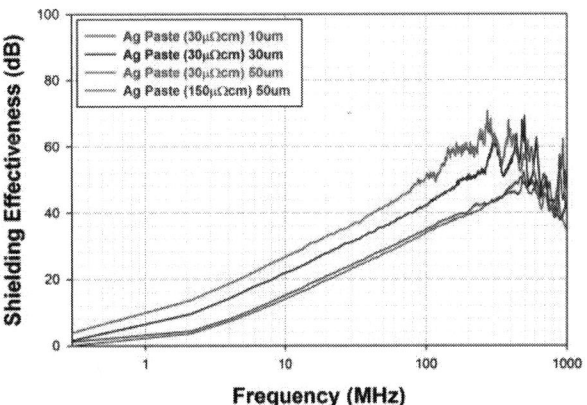

Frequency	300KHz	1MHz	10MHz	100MHz	1GHz
Ag Paste 10μm (30μΩ·cm)	1.19	3.4	14.8	34.9	41.7
Ag Paste 30μm (30μΩ·cm)	1.53	6.6	21.7	42.7	43.0
Ag Paste 50μm (30μΩ·cm)	3.88	10.0	26.3	50.3	36.0
Ag Paste 50μm (150μΩ·cm)	0.04	2.3	13.7	33.9	47.0

Figure 4. Near-field SE results of various samples with the magnetic field sources

As shown in Figure 3, when the thickness of the silver layer was 10μm, the SE was about 42.7 dB at 300 kHz. As the frequency increases, the SE increases gradually to about 65.9 dB at 100 MHz. After that, it gradually decreases to 38.4dB at 1 GHz. At the frequency less than 30 MHz, the SE increases with increasing absorption loss. However, the SE decreases at the frequency more than 30 MHz according to the decrease of reflection loss. The reflection loss in the electric field can be summarized by the following equation [5].

$$R_E = C_3 - 10\log\frac{\mu_r f^3 r^2}{g_r}$$

Where

$C_3 = 322$ if r is in meters
$\quad\ = 354$ if r is in inches
R = distance from electromagnetic source to shield
μ_r = permeability
g_r = conductivity

Equation 2. Reflection loss in electric field [5]

When the volume resistances of each sample were the same, the electric field SE increased slightly as the film thickness increased in the low frequency range. Whereas, the film thickness did not affect SE in the high frequency range. As a result, we found that the sheet resistance determined the electric field shielding performance. In the near-field region, it seems that electric field is mainly shielded due to the reflection loss.

In case of magnetic shielding as shown in Figure 4, it is affected by the absorption loss rather than the reflection loss. Accordingly, as the frequency increases, the absorption loss increases, and then, the SE increases.

$$A = K_1 l \sqrt{f \mu_r g_r}$$

$K_1 = 131.4$ if l is in meters
$\quad\ = 3.34$ if l is in inches
l = shield thickness
f = frequency
μ_r = permeability
g_r = Conductivity

Equation 3. Numerical calculation of absorption loss [5]

The SE of 50μm-thickness specimen is about 10dB at 1MHz. As the frequency increases, the SE also increases gradually. However, after showing 60 ~ 70dB at 300MHz, it gradually decreases and shows about 40dB at 1GHz. As the film thickness decreases, the SE also decreases. In the case of the 10 μm specimen, the SE is about 3 dB at 1 MHz, 15 dB at 10 MHz, and 50 dB at 300 MHz.

B. Effect of magnetic permeability on SE

As shown in Figure 5, the magnetic permeability curves of the magnetic paste containing the magnetic filler were

measured as a function of frequency. The permeability μ' of the magnetic paste shows the highest value at 1 MHz and tends to decrease gradually as the frequency increases. Especially, the permeability of specimens with relatively high permeability decreases rapidly from 200μ' at 6 MHz to 100μ' at 100 MHz.

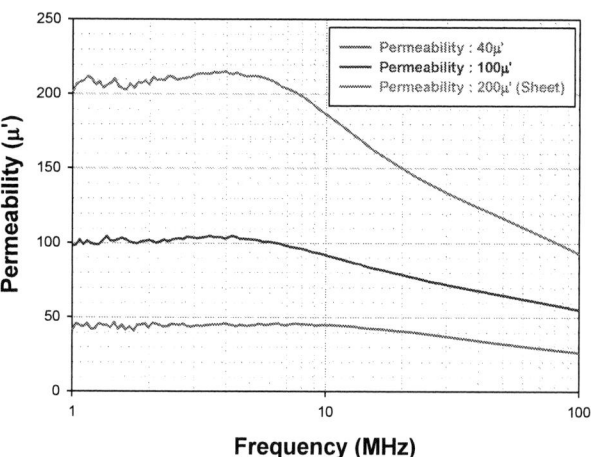

Figure 5. Permeability curves of (a) 40μ' (b) 100μ' and (c) 200μ'

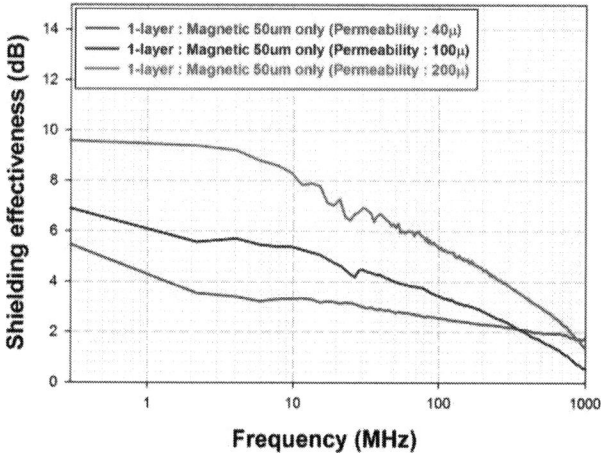

Figure 6. Magnetic field SE of magnetic sheets made of (a) 40μ' (b) 100μ' and (c) 200μ', respectively.

Magnetic paste has very low reflection loss of the electric field, because it does not have a conductivity. Therefore, shielding performance of the electric field is very low. On the other hand, for the magnetic field, the SE is high at the low frequency due to the high permeability of the magnetic sheet. The SE decreases as the frequency gradually increases, and shows a shielding rate as low as 2 dB or less at 1 GHz. This tendency is caused by the reduction of permeability of the magnetic sheet as a function of the frequency. It means that the SE due to the high permeability of magnetic material is very effective at the low frequency below 10 MHz. However, the conductive layer is required for the near-field shielding at the high frequency range.

C. Near-field SE of hybrid pastes

High conductivity and permeability are simultaneously required to shield both electric and magnetic fields in a wide range from low to high frequencies. In this study, a hybrid paste, which was mixed with conductive paste and magnetic paste, can shield the electric field and magnetic field from the low frequency to high frequency. Its shielding characteristics were analyzed.

As the mixing ratio of conductive filler and magnetic filler increases, the resistivity and permeability increase proportionally. Thus, the optimum mixing ratio is important. In order to optimize the permeability while maintaining the conductivity by the contact of the conductive powder, the x value is a very important parameter, which is the content ratio according to the specific surface area of each filler.

In Figure 7, x = 1 indicates that the ratio of the specific surface area to the weight ratio of each filler is 1. At this time, the permeability is 54.6 μ' at 1 MHz. As the frequency increases, the permeability decreases gradually from 50.7μ 'at 10 MHz and to 30.7μ at 100 MHz. As the x increases, the content ratio of the conductive filler increases and the permeability gradually decreases. The permeability at x=6 is 17.8μ' at 1 MHz, and the permeability is very low at 23.7μ' at 10 MHz and 10.8μ' at 100 MHz.

Frequency	1MHz	10MHz	100MHz
X = 6	17.8	23.7	10.8
X = 5	28.5	27.5	11.9
X = 4	32.0	32.6	13.4
X = 3	34.7	37.3	17.4
X = 2	42.8	43.5	22.3
X = 1.5	50.7	50.7	27.1
X = 1	54.6	53.4	30.7

Figure 7. Permeability curves of hybrid paste with x =1.0 to 6.0

X value	-	6	5	4	3	2	1.5	1
Volume resistivity (μΩ·cm)	22.5	180.4	257.6	323.0	720.2	2,646.7	4,439.4	36,700

Table 2. Volume resistivity data of hybrid paste with x =1.0 to 6.0

On the other hand, as shown in Table 2, the conductivity of the hybrid pastes increases as the x value increases. The volume resistivity of the paste having the x value of 1 is $3.67\times10^{-2}\Omega\cdot$cm. In addition, the conductivity gradually decreases as the x value increases. The volume resistivity is $1.80\times10^{-4}\Omega\cdot$cm at the x value of 6.

As shown in Figure 8, in the far-field region, the SE increases as the x increases. The SE is about 25~27dB at x = 1, and it does not change depending on the frequency.

As the x value increases, which means the content of conductive filler increases, the SE also increases. When the x value is 6, the SE is 64dB at 30 MHz and 80dB at 1 GHz, respectively. It seems that the SE depends on the conductivity of the paste, similar to the far-field EMI shield results of conductive pastes.

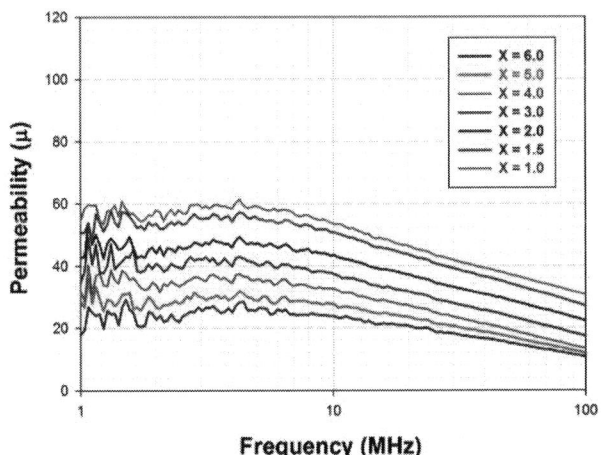

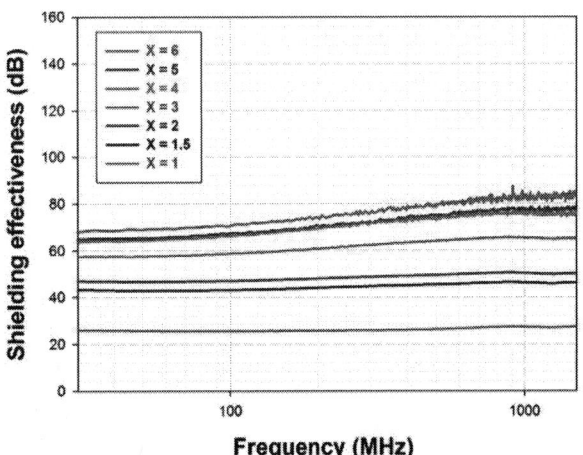

Frequency	30MHz	100MHz	1GHz	1.5GHz
X = 6	68.0	70.8	83.4	84.5
X = 5	64.9	67.5	77.0	79.1
X = 4	63.8	66.3	75.3	76.4
X = 3	57.4	58.8	65.5	65.4
X = 2	46.9	46.9	50.2	50.1
X = 1.5	43.1	43.0	46.2	46.2
X = 1	26.0	25.5	27.3	27.3

Figure 8. Far-field SE results of hybrid paste with x =1.0 to 6.0

Frequency	100KHz	300KHz	1MHz	10MHz	100MHz	1GHz
Ag 30um	12.7	25.1	38.6	64.5	84.2	54.0
X = 6	13.4	25.7	39.2	64.9	80.5	55.3
X = 5	13.1	24.4	36.9	60.8	71.2	45.0
X = 4	14.0	25.4	38.0	61.9	74.0	51.2
X = 3	13.8	23.7	34.5	55.3	61.5	36.5
X = 2	14.0	22.8	32.5	50.8	56.0	32.0
X = 1.5	14.2	21.0	28.6	43.0	47.6	29.5
X = 1	14.3	19.4	24.8	35.2	39.2	27.0

Figure 9. Near-field SE results of hybrid paste with x =1.0 to 6.0 in the electric source

The SEs of electric field and magnetic field in the near-field region were measured as shown in Figures 9 and 10, respectively, and the data at some frequency points were also listed in tables.

The electric field shielding of the hybrid paste improves with the increase of x value. In the high frequency region of 30 MHz or more, the SE is improved with an increase of x value. When the x is 1 and 6, the SEs are about 40dB and 80dB, respectively.

At the frequency less than 1 MHz, especially 100 kHz, the SE becomes lower. When the x is 1 and 6, SEs are 14.3dB and 13.4dB, respectively. These are very similar results, considering the tolerance of measurement. However, the degradation of SE is needed to be reviewed.

The magnetic field shielding results in the near-field region are plotted in Figure 10. In the high frequency range, as the x value increases, the permeability decreases and the conductivity increases and the SE increases. When the x is 1, the SE is 7dB at 50 MHz. The SE increases to 30dB when the x is 6. Because the absorption loss of magnetic field shielding is dependent on the conductivity of shielding materials and the frequency, the SE increases as the conductivity and the frequency increase although the permeability decreases. Whereas, in the low frequency below 1MHz, the SE decreases as the conductivity of shielding materials increases. When the x is 1 and 6 at 100kHz, the SEs are 7.6dB and 0.92dB, respectively. It seems that the SE is reduced due to the eddy current loss of the magnetic filler by the conductive fillers. In the case of the magnetic paste alone, it shows a constant SE in the low frequency range due to the permeability [6].

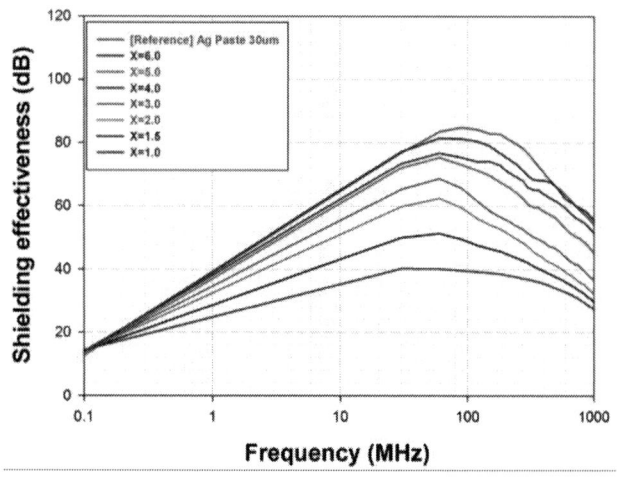

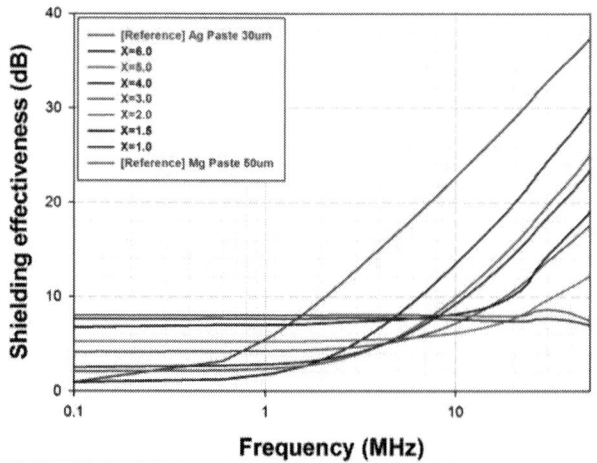

978-1-7281-1500-9/19 $31.00 © 2019 IEEE

Frequency	100KHz	1MHz	10MHz	50MHz
Ag 30um	0.91	5.90	22.73	37.42
X = 6	0.92	1.83	13.53	30.01
X = 5	2.05	2.37	9.91	25.06
X = 4	2.52	2.86	9.28	23.44
X = 3	4.14	4.26	7.21	17.63
X = 2	5.24	5.26	6.18	12.21
X = 1.5	6.76	7.02	8.17	19.05
X = 1	7.62	7.67	7.60	7.02
Mg 50um	8.03	8.04	8.01	7.46

Figure 9. Near-field SE results of hybrid paste with x =1.0 to 6.0 in the magnetic source

IV. CONCLUSION

We have analyzed the effect of conductive layer thickness having the same volume resistivity on near-field electric and magnetic shielding. We also have fabricated test PCBs coated by silver paste having $3.4\times10^{-5}\Omega\cdot$cm of volume resistivity using spray process and measured electric field SE and magnetic field SE of the samples having thickness of 10 μm, 30 μm, and 50 μm. For the electric field, the SE increased as the thickness of conductive layer increased and showed above 60dB at 10MHz. However, the SE decreased to 40dB at 1 GHz as the frequency increased. For the magnetic field, the SE showed better performance of max 60dB as the thickness increased. However, in the low frequency below 10 MHz, the SE was max 14.8dB in the case of 10 μm thickness.

We have prepared magnetic sheets using various pastes having the permeability from 40µ to 200µ and measured the magnetic field SE to investigate the effect of permeability on magnetic field shielding. As the permeability of magnetic layer, the magnetic SE increased to 9.5dB at 1 MHz.

The hybrid pastes with high conductivity and permeability were fabricated and their near-field SE were measured. As the mixing ratio of conductive filler and magnetic filler increased, the conductivity increased and permeability decreased. For the electric field, the hybrid paste with the x=6 showed max 80.5dB at 100 MHz and the SE decreased as the x value decreased. For the magnetic field, the paste with x=1 showed 7.67dB SE at 1MHz and the SE decreased as the x value increased.

REFERENCES

[1] K. Joo, T. R. Kim, J. W. Hwang, J. H. Yoon, S. Y. Jeong and M. J. Yim, "Package-Level EMI Shielding Technology with Silver Paste for Various Applications," *2017 IEEE 67th Electronic Components and Technology Conference (ECTC)*, Orlando, FL, 2017, pp. 1736-1741. doi: 10.1109/ECTC.2017.327

[2] H. H. Park, J.-D. Lim, H.-B. Park, and J. Kim, "Near-field shielding measurement of small shield cans in metallic mobile devices for RF interference analysis," *Electronics Letters*, vol. 52, no. 11, pp. 980-982, 26 May 2016.

[3] H. H. Park, J. H. Kwon, and S. Ahn, "A simple equivalent circuit model for shielding analysis of magnetic sheets based on microstrip line measurement," *IEEE Trans. Magnetics*, vol. 53, no. 6, Art. ID 9401504, June 2017

[4] S. Yang, K. Lozano*, A. Lomeli, H. D. Foltz, and R. Jones, "Electromagnetic interference shielding effectiveness of carbonnanofiber/LCP composites" *Composites Part A: Applied Science and Manufacturing*, 2005, 36, (5), pp. 691-697.

[5] S. Loya and Habibullakhan, *"International Journal of Electromagnetics and Applications*, 2016, 6, (2), pp. 31-41

[6] Cheng-Kuang Liu and Cheng-Yie Chou, "RF interference effects on PIN photodiodes" *IEEE Transactions on Electromagnetic Compatibility*, 1995, 37, (4), pp. 589-592

Low loss NCF material for high frequency device

Kazutaka Honda, Keiko Ueno, Tsuyoshi Ogawa, Toshihisa Nonaka

Packaging Solution Center, Hitachi Chemical Co., Ltd.
7-7 Shin-Kawasaki, Saiwai-ku, Kawasaki-shi, Kanagawa 212-0032, Japan
E-mail:kz-honda@hitachi-chem.co.jp

Abstract— **Low dielectric loss non-conductive film has been developed. The dielectric dissipation factor was 0.006. In the assembly process, non-conductive film afforded both excellent lamination-ability to the bumped wafer and dicing-ability of non-conductive film laminated wafer simultaneously. Furthermore, by changing curing reaction system, it showed high productivity which can form solder joint well without void in non-conductive film for 3 s of the bonding time. In addition, it was confirmed that there was no failure in reliability test.**

Keywords - Low loss material; High productivity; NCF; D_f

I. INTRODUCTION

Pre applied non-conductive film (NCF) combined with a thermal compression bonding (TCB) is good at fine pitch bump interconnection. The technology has been used for the mass productions of the packages of chip-on-board and memory die stacking with through silicon via (TSV), which are 3 dimensional stacking dynamic random access memory (3DS DRAM), high band width memory (HBM) and so on [1-4].

On the other hand 5G era is coming just around the corner, which enables 10 Gbps or higher telecommunication speed. The use of frequencies above 30 GHz is expected to expand. Devices for 5G system, which are radio frequency (RF) antenna in package, base station devices for RF and etc., are eagerly being developed. Advanced driver assistant systems (ADAS) and autonomous system also usually use the radar devices with the frequency of around 77 GHz. Signal transmission loss is in proportion to the frequency. Dielectric materials for such high frequencies electronic packages and modules are demanded to be having low dielectric dissipation factor (D_f), which are substrate, encapsulant, underfill and so on [5-7].

A typical assembly process flow with NCF is as follows. Firstly, NCF is laminated on an active side of the top wafer and then it is singulated to dies by blade saw. The dies are picked up and bonded to the bottom dies by thermal compression bonding (TCB). TCB equipment has a pulse heater head, which can heat up the die very rapidly. The NCF laminated die is picked up and placed at the relatively low temperature to prevent the hardening of the NCF. After the die is touched down, the head heats up the specimen over the solder melting temperature with suitable pressure and makes solder joints between bumps and pads. Simultaneously, the NCF is also cured. Then the head is released from the specimen and cooled down to the temperature for next die picking up step. The temperature

gap is usually greater than 100°C. This is a serial process as mentioned above and the cooling is a kind of idling step for production. The productivity by using only one head is at most several hundreds units per hour [8-9]. Therefore, the bonding step is divided in pre and main bonding in order to decrease the temperature gap as shown in Fig.1. The temperature of the head at the pre bonding is closed to that at the die pick-up step, which is relatively low. Making solder joint and NCF curing are carried out in main bonding process at higher temperature [10-12].

In this paper, the newly developed NCF with low D_f resin system is reported. The TCB process compatibility, detail material properties including dielectric performance in RF region and the reliability of the assembled package were investigated.

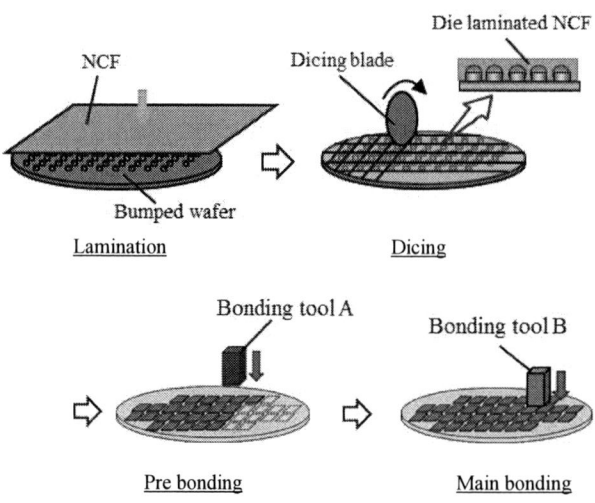

Figure 1. Assembly procedure with NCF

II. EXPERIMENTS

2-1. Test vehicle

The specifications and the pictures of the test vehicle (TV) used in this study are shown in Table 1 and Fig. 2, respectively. Evaluation of the assembly process and the reliability of the fabricated package were performed with the TV.

The size of the top die was 7.3 mm square and 50 μm thick. The die had peripheral and central array bumps with 80 and 300 μm pitches, respectively. The former ones were aligned

978-1-7281-1500-9/19 $31.00 © 2019 IEEE

staggered in two rows. All of the bumps were composed of Cu post and Sn/Ag solder cap. Cu Pads covered by plated Ni/Au, which correspond to the bumps of the top die, were formed on the bottom die surface. It had the size of 10 mm square and 100 μm thick.

TABLE 1. Specifications of the TV

Top die (Upper die)	Size: 7.3 mm × 7.3 mm Thickness 50 μm Passivation : SiN Peripheral bump : 80 μm pitch, 648 pin Full array bump : 300 μm pitch, 400 pin Bump height : Cu Pillar (30 μm) + Sn/Ag Solder (15 μm)
Bottom die (Lower die)	Size : 10 mm × 10 mm Thickness 100 μm Passivation : SiN Pad : Cu + Ni/Au plating

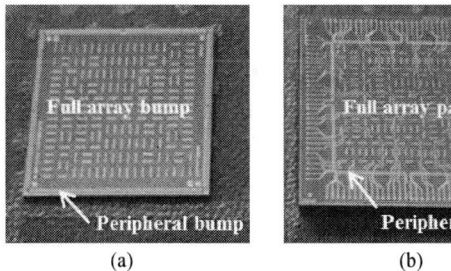

(a)   (b)

Figure 2. Photos of the test vehicles of top (a) and bottom die (b).

2-2. Assembly process and the condition of pre-bonding

The thickness of the NCF for the evaluation was 40 μm. The pre-bonding process is illustrated in Fig. 3, which was a serial one described below. The NCF was laminated onto the bumped surface of the top wafer and then it was singulated by blade saw. The die was picked up at a low temperature and pre-bonded to the bottom die by a flip chip bonder. The lamination was carried out with a diaphragm vacuum laminator heated at 80°C for 60 s. The thickness of the blade used for the singulation was 50 μm. The cutting and rotation speeds of the blade were 20 mm/s and 30,000 rpm, respectively. Temperature of the head and the stage were 130 and 80°C, respectively. The bonding time was 3 s.

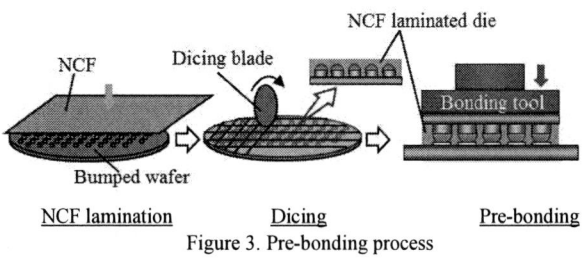

Figure 3. Pre-bonding process

2-3. Main bonding

Main bonding is implemented after the pre-bonding with a TCB bonder. The set condition used in the experiment is shown in Table 2. When the set temperature was 260°C, the actual temperature of the NCF was about 240°C. The actual temperature was measured using a thermocouple inserted between the NCF laminated top die and the bottom die by a reference bonding.

TABLE 2. Set conditions of main bonding.

		Contact condition	Main bonding condition
Bonding temperature	°C	80	260
Bonding time	s	-	3-10
Stage temperature	°C	80	80

2-4. Curing reaction rate measurement of NCF

Curing reaction rate of the NCF was calculated from the measurement results of differential scanning calorimetry. The calculation formula is as follow.

$$(\Delta H 1 - \Delta H 2) / \Delta H 1 \times 100 = \text{Curing reaction rate (\%)}$$
$$\Delta H 1 : \text{Integral value of initial heat flow peak}$$
$$\Delta H 2 : \text{Integral value of heat flow after heating}$$

2-5. Measuremnet of D_k and D_f of NCF

D_k and D_f of the NCF were measured at the frequency of 10 GHz in SPDR (Sprit Post Dielectric Resonators) method. The film thickness was about 200 μm [13].

2-6. Reflow test after moisture absorption treatment

The moisture absorption treatment was performed under JEDEC level 2 condition. The specimen after the treatment was passed through the reflow furnace for three times (Maximum temperature: 260°C).

2-7. Connection reliability test (Thermal cycle test)

TCT (Thermal cycle Test) was done for 1000 cycles between -55 and 125°C. The electrical resistance of the daisy chain was measured at 25 and 125°C in every 200 cycles.

III. RESULTS AND DISCUSSIONS

3-1. TCB compatibility

TCB compatibility of a conventional NCF in each bonding time was evaluated. The NCF-laminated top die was bonded to the bottom die by the pre-bonding process and then the main bonding was carried out with bonding time of 3, 5 and 10 s. (a)-(c) in Fig. 4 show SAM (Scanning Acoustic Microscope) images after main bonding and the cross sectional views are also shown in (d)-(c).

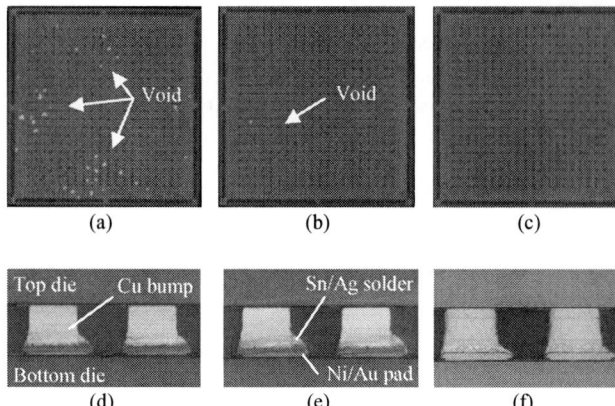

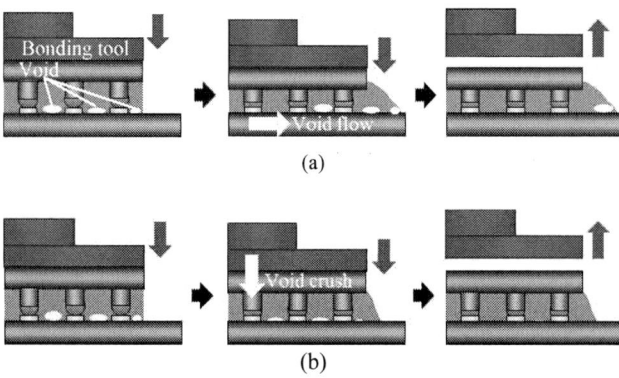

Figure 4. Bonding results of a conventional NCF. SAM images after main bonding with the bonding time of 3, 5, 10 s are shown in (a) to (c), respectively. Cross sectional micrographs after main bonding with the bonding time of 3, 5, 10 s are shown in (d) to (f), respectively.

As shown in Fig. 4, the solder joints were well formed in every bonding time from 3 to 10 s. The SAM images show the void decreased as the bonding time became longer. So, we supposed the removing trapped void in pre-bonding is essential. The countermeasure is illustrated in Fig. 5 [14-16].

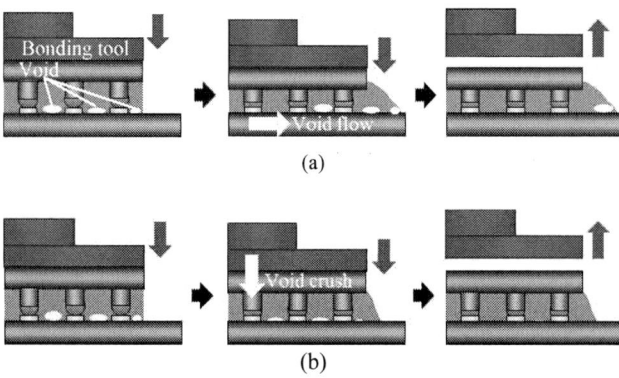

Figure 5. 2 concepts of the void suppressions. (a) Void flow out by lowering of the NCF viscosity. (b) Void crushing by increasing the curing speed of the NCF

It is thought that there are 2 ways to suppress the void in the NCF during TCB. One is to flow out the void to outside form the NCF edge. In that case a low viscosity NCF can be suitable (Fig. 5 (a)). Such a NCF tends to generate large fillet, invade the adjacent package and creep up the top die edge which is a cause of the tool contamination. The other way is to crush the void by the internal pressure of the NCF which can be generated by heat and press to rapidly hardened material. Rapid cure NCF can also prevent the excess fillet and shorten the bonding time (Fig. 5 (b)).

Based on the latter idea the faster curing NCF was designed and prepared. Fig. 6 shows the curing reaction rate of the faster curing and a conventional NCF. The curing system was changed to radical polymerization in the faster curing NCF. The reaction rate of the faster curing NCF with a bonding time of 3 s was 80% or more, which is greater than that of a conventional NCF with a bonding time of 10 s. TCB compatibility of the faster curing NCF was also evaluated. The bonding results were shown in Fig. 7. The faster curing NCF could suppress the void and the solder joint were well formed with a bonding time of 3 s.

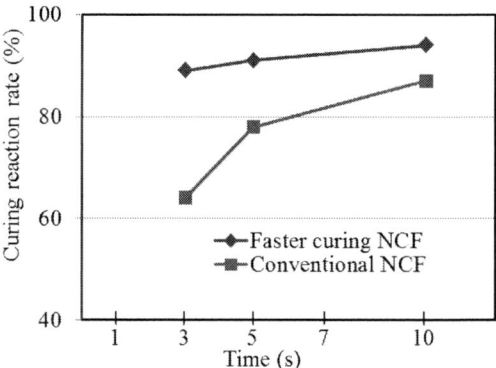

Figure 6. Curing reaction rates of the faster curing and a conventional NCFs at 240°C.

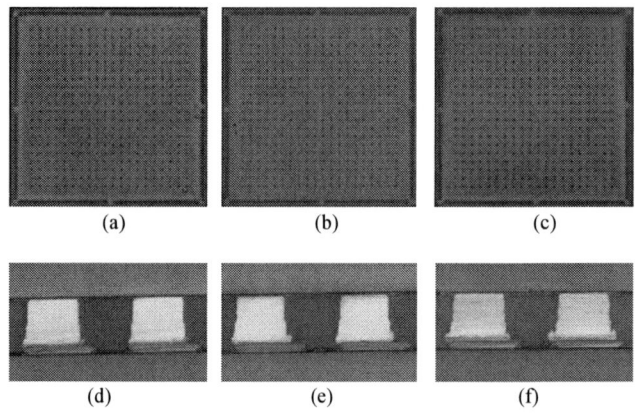

Figure 7. Bonding results of the improved NCF. SAM images after main bonding with the bonding time of 3, 5, 10 s are shown in (a) to (c), respectively. Cross sectional views after main bonding with the bonding time of 3, 5, 10 s are shown in in (d) to (f), respectively.

3-2. Low dielectric dissipation factor

D_f of a conventional NCF was 0.022. The design concept of the low loss NCF is shown in Fig. 8. NCF was composed of main polymer, curing system, filler and so on. The same radical polymerization system as the faster curing NCF mentioned above was applied to the low loss NCF to attain the fast curing reaction rate for good TCB capability in short bonding time. The new low loss material for the main polymer has been developed. Fig. 9 shows measurement results of D_k and D_f of a conventional and the low loss NCFs consisting of the new material and the fast curing system. The values of D_k and D_f of the low loss NCF were smaller than those of a conventional NCF. D_f of the low loss NCF was 0.006.

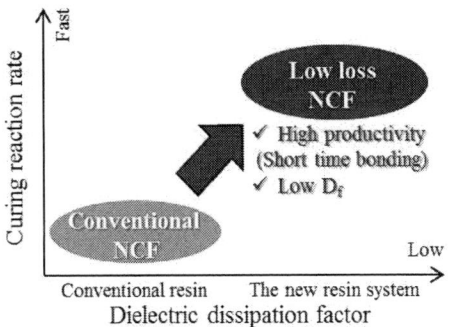

Figure 8. The design concept of the low loss NCF

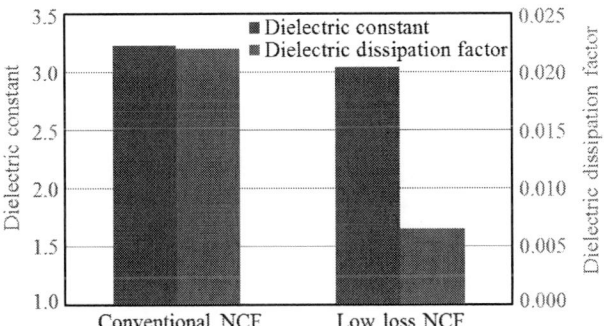

Figure 9. Measurement results of D_k and D_f of a conventional and the new low loss NCF

3-3. Assembly evaluation of the new low loss NCF

The assembly evaluation results of the low loss NCF was summarized as follows. In the lamination and the dicing process, the NCF could fill among the bumps of the top die without void and then it was diced without delamination of the NCF or chipping as depicted in Fig. 10. The die cracking and the burr also didn't occur as shown in Fig. 11.

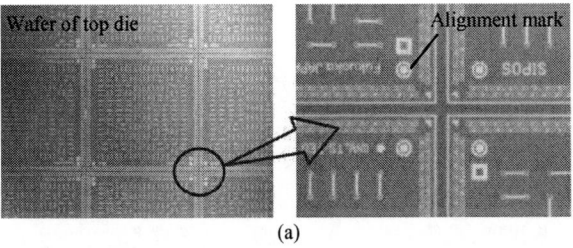

(a)

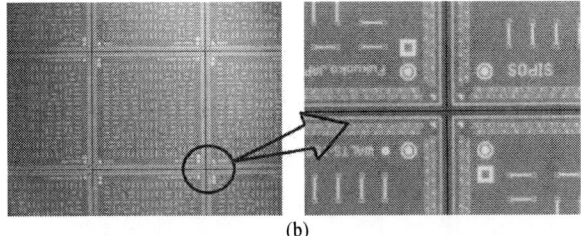

(b)

Figure 10. Observation results of the lamination and the dicing process with the low loss NCF on the TV wafer. Micrographs before and after dicing are shown in (a) and (b), respectively.

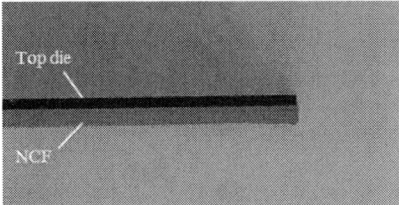

Figure 11. Cross sectional view after dicing

In the pre-bonding step, the NCF is required to have the certain transparency. The alignment mark on the top die must be detected through the NCF by the bonding tool. It had enough transparency and the alignment marks on the TV could be seen clearly as shown in Fig10. The evaluation results after main-bonding were shown in Fig. 12. NCF successfully archived the appropriate void removal and solder joint with a bonding time of 3 s.

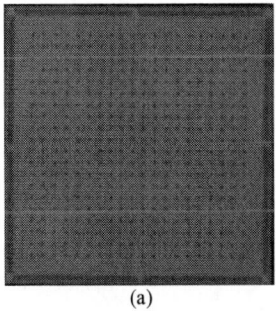

(a)

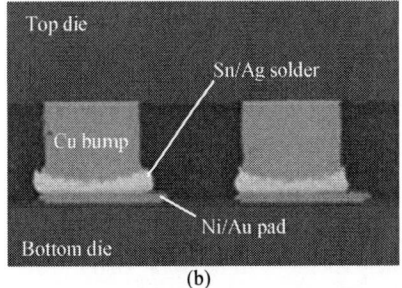

(b)

Figure 12. Observation results after main bonding. SAM and cross sectional SEM images are shown in (a) and (b), respectively.

3-4. Reflow test

The assembled specimen was post-cured at 175°C for 2 h and then the moisture reflow treatment was made. It was passed through the reflow furnace for three times. The obtained SAM images are depicted in Fig. 13.

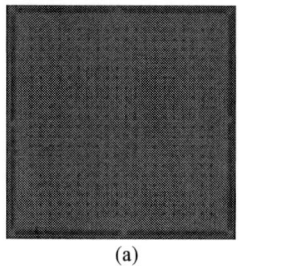 

(a)   (b)

Figure 13. Reflow reliability results with the low loss NCF. SAM images before and after reflow test are shown in (a) and (b), respectively.

3-5. TCT

Interconnection reliability by TCT was evaluated. The electrical connection resistances of the daisy chain at 25 and 125°C were measured in every 200 cycles up to 1,000 cycles as shown in Fig. 14. No electrical failure was detected to 1,000 cycles [17-18].

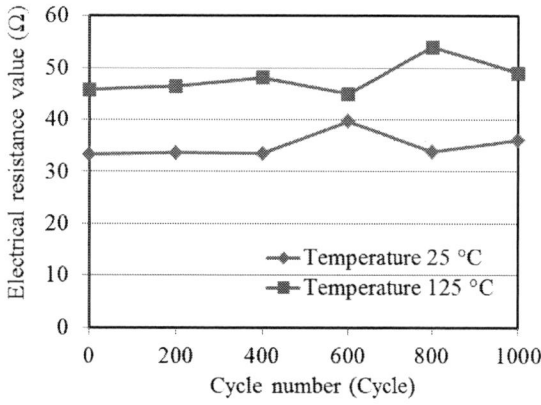

Figure 14. TCT results of the package fabricated with the low loss NCF

IV. CONCLUSIONS

The low-loss NCF which can afford high productivity was developed. It showed the excellent TCB process compatibility to form the solder joint without void in NCF at bonding time of 3 s. Df of the NCF was 0.006. The assembled package with the NCF also showed the no failure at 1, 000 TCT between 25 and 125°C reliability or moisture reflow test.

V. REFERENCES

1. John H. Lau, "TSV Manufacturing Yield and Hidden Costs for 3D IC Integration", Proceedings of 2010 Electronic Components & Technology Conference, pp. 1031-1042.
2. N. Asahi, Y. Miyamoto, M. Nimura, Y. Mizutani and Y. Arai, "High Productivity Thermal Compression Bonding for 3D-IC", Proceedings of IEEE 2015 Interational 3D System Integration Conference(3DIC), pp. 129-133.
3. Kazutaka Honda, Tetsuya Enomoto, Akira Nagai, and Nozomu Takano, "NCF for Wafer Lamonation Process in Higher Density Electronic Packages", Proceedings of 2010 Electronic Components & Technology Conference, pp.1853-1860.
4. Kazutaka Honda, Akira Nagai, and Makoto Satou, "NCF for Pre-Applied Process in Higher Density Electronic Package Including 3D-Package", Proceedings of 2012 Electronic Components & Technology Conference, pp.385-392.
5. Vincens Gjokaj, John Doroshewitz, Jeffrey Nanzer, and Premjeet Chahal "A Design Study of 5G Antennas Optimized Using Genetic Algorithms", Proceedings of 2017 Electronic Components & Technology Conference, pp.2086-2091.
6. Sukhadha Viswanathan, Tomonori Ogawa, Kaya Demir, Timothy B. Huang, P. Markondeya Raj, Fuhan Liu,Venky Sundaram and Rao Tummala, "High Frequency Electrical Performance and Thermo-Mechanical Reliability of Fine- Pitch, Copper - Metallized Through-Package-Vias (TPVs) in Ultra – thin Glass Interposers", Proceedings of 2017 Electronic Components & Technology Conference, pp.1510-1516.
7. Atom O. Watanabe, Muhammad Ali, Bijan Tehrani, Jimmy Hester, P. Markondeya Raj, Venky Sundaram, Manos M. Tentzeris, and Rao R. Tummala, "First Demonstration of 28 GHz and 39 GHz Transmission Lines and Antennas on Glass Substrates for 5G Modules", Proceedings of 2017 Electronic Components & Technology Conference, pp.236-241.
8. D. W. Kim, R. Vidhya, B. Henderson, U. Ray, S. Gu, W. Zhao, R. Radojcic, and M. Nowak, "Development of 3D Through Silicon Stack Assembly for Wide IO Memory to Logic Devices Integration", Proceedings of 2013 Electronic Components & Technology Conference, pp.77-80.
9. H. Clauberg, A. Marte, Y. Yang, J. Eder, T. Colosimo, D.Buergi, A.Rezvani, and B. Chylak, "High Productivity Thermocompression Flip Chip Bonding," Proceedings of 2015 Electronic Components and Technology Conference , pp. 22-29.
10. SeokGeun Ahn, HwanKyu Kim, Dong Wook Kim, David Hiner, KeunSoo Kim, TaeKyeong Hwang, MinJae Lee, DaeByoung Kang, and JuHoon Yoon, "Wafer Level Multi-Chip Gang Bonding Using TCNCF", Proceedings of 2016 Electronic Components & Technology Conference, pp.122-127.
11. Hitoshi Onozeki, Hiroshi Takahashi, Naoya Suzuki, Kumpei Yamada, and Yuta Koseki, "In plane collective CoS assembly by NCF-TCB enabled using the newly developed bonding force leveling film", Proceedings of 2016 Electronic Components & Technology Conference, pp.1691-1697.
12. A. Horibe, F. Yamada, C. Feger, and J.U. Knickerbocker, "Inter Chip Fill for 3D Chip Stack", Transactions of The Japan Institute of Electronics Packaing, Vol. 2, No.1.2009, pp160-162.
13. Chiho Ueta, Kazuya Okada, Toko Shiina, Tadahiko Hanada, Nobuhito Ito, "Development of Solder Resist with Improved Adhesion at HTSL (175 deg C for 3000 Hours) and Crack resistance at TST for Automotive IC Package", Proceedings of 2016 Electronic Components & Technology Conference, pp.122-127.
14. Hyeong-Gi Lee, Se-Yong Lee, Yong-Won Choi, and Kyung-Wook Paik, "Thermo-compression bonding using Non-Conductive Films (NCFs) for 3-D TSV Micro-bump Interconnection", Proceedings of 2016 Electronic Components & Technology Conference, pp1809-1815.
15. Rahul Agarwal, Wenqi Zhang, and Piet Limaye, "Cu/Sn Microbumps Interconnect for 3D TSV Chip Stacking", Proceedings of 2010 Electronic Components & Technology Conference, pp858-863.

16. Ren-Shin Cheng, Kuo-Shu Kao, and Jing-Yao Chang, "Achievement of Low Temperature Chip Stacking by a Wafer-Applied Underfill Material", Proceedings of 2011 Electronic Components & Technology Conference, pp1858-1863.

17. Weidong Xie, Tae-Kyu Lee, and Uuo-Chuan Liu., "Pb-Free Solder Joint Reliability of Fine Pitch Chip-Scale Packages", Proceedings of 2010 Electronic Components & Technology Conference, pp. 1587-1590.

18. Yasumitsu Orii, and Kazushige Toriyama, "Micro Structure Observation and Reliability Behavior of Peripheral Flip Chip Interconnections with Solder-Capped Cu Pillar Bumps", Transactions of The Japan Institute of Electronics Packaing, Vol. 4, No.1.2011, pp73-86.

In-situ Redox Nanowelding of Copper Nanowires with Surficial Oxide Layer as Solder for Flexible Transparent Electromagnetic Interference Shielding

Xianwen Liang[1,2,*], Jianwen Zhou[1], Gang Li[1], Tao Zhao[1], Pengli Zhu[1,*], Rong Sun[1]

[1]Shenzhen Institute of Advanced Electronic Materials
Shenzhen Institutes of Advanced Technology
Chinese Academy of Sciences
Shenzhen, China
[2]Shenzhen College of Advanced Technology
University of Chinese Academy of Sciences
Shenzhen, China
liang.xw@siat.ac.cn, pl.zhu@siat.ac.cn

Ching-ping Wong[3]
[3]School of Materials Science and Engineering
Georgia Institute of Technology
Atlanta, USA
cp.wong@mse.gatech.edu

Abstract—Silver nanowire (AgNW) transparent electrode stands out as a promising candidate to replace indium tin oxide (ITO), whereas the high cost and electromigration of silver ions overshadow the applications of AgNWs in optoelectronics. Copper nanowire (CuNW) is attracting increasing interest and attentions due to its high intrinsic electrical conductivity, earth abundance and lower prince, but the oxidation of CuNW severely prohibits its practical applications, which is an issue to be solved urgently. Herein, nanowelding of CuNWs is achieved via an in-situ redox approach. In this welding process, the copper oxide on the surface of CuNWs as a natural solder is reduced by sodium borohydride ($NaBH_4$) to generate Cu atoms, which selectively aggregate at the intersection of CuNWs and merge the junction owing to the positive site here. The sheet resistance of welded CuNW (W-CuNW) transparent conducting films drop obviously without sacrificing its transmittance, which thereby significantly promotes the optoelectronic performance of the film. Poly(3,4-ethylenedioxythiophene)/poly(styrenesulfonate) (PEDOT:PSS) as a protective layer is coated onto the W-CuNW film to prepare PEDOT:PSS/W-CuNW film. The optoelectronic properties of the PEDOT:PSS/W-CuNW film show excellent stability in ambient atmosphere for 30 days. Beside, no obvious change in the sheet resistance of the PEDOT:PSS/W-CuNW film is observed after 5000 bending cycles under a bending radius of 2 mm, indicating the outstanding mechanical flexibility. Finally, electromagnetic interference (EMI) shielding effectiveness (SE) of the PEDOT:PSS/W-CuNW film is measured within the frequency range from 8.2 GHz to 12.5 GHz. The PEDOT:PSS/W-CuNW film with a EMI SE value above 27 dB and transmittance of 85% underlines the great potential applications in displays, touch panels, airborne optoelectronic pods and aviation camcorders.

Keywords- copper nanowires; copper oxide; nanowelding; transparent electromagnetic interference shielding

I. INTRODUCTION

Transparent electrodes has been widely applied in a variety of optoelectronics such as solar cells, touch screens, displays, light emitting diodes (LED), etc [1]. ITO is considered as today ubiquitous transparent electrode due to its high electrical conductivity (~10 ohm/sq) and optical transmittance (~90%) [2]. However, the drawbacks of ITO, including the trait of brittle ceramic, scarcity of indium, and complicated and high-cost fabricating process severely hamper its future applications [3]. Motivated by these issues, many efforts have been devoted to developing alternatives to ITO, such as carbon based materials (carbon nanotubes and graphene), conducting polymers, metal meshes, and metal nanowires [3]. Among these materials, AgNW is regarded as a promising candidate to replace ITO owing to its comparable optoelectronic performance to ITO, ease of process, mechanical flexibility, and so on. Unfortunately, the high cost of noble metal and electromigration of silver ions overshadow widespread applications of AgNWs.

In the past few years, CuNW has drawn ever-increasing interests thanks to its high intrinsic electrical conductivity (comparable conductivity to silver: $\sigma_{Ag}=6.30\times10^7$ S m^{-1}, $\sigma_{Cu}=5.96\times10^7$ S m^{-1}), earth abundance (1000 times of silver) and a lower price (1/100th of silver) [4]. In order to further improve the optoelectronic performance of CuNW transparent electrodes, it is of great importance to effectively lower the sheet resistance of CuNW films without affecting its original transparency. The contact resistance at the junction of CuNWs governs the sheet resistance of CuNW films, which is derived from the contact interface between CuNWs [5]. Welding CuNWs can eliminate the contact interface between CuNWs and form the intimate contact at the junction, and thereby reinforce electrical conductivity and mechanical flexibility of CuNW films [2]. Recently, some efficacious strategies have been proposed to weld CuNWs, including plasmonic laser, photonic sintering, annealing, photochemical reducing, etc [6-9]. These methods can promote the optoelectronic performance of CuNW films effectively, but some drawbacks limit their applications. Spot-beam laser inhibits its industrial application due to the time and energy-consuming shortcoming [6]. Photonic sintering requires high energy and thereby limits its usage in mass production [7]. Annealing is unsuitable for polymer substrates and may damage the delicate structure of some devices [8]. Besides, severe safety concern exists in the thermal treatment under hydrogen atmosphere [8]. High

978-1-7281-1500-9/19 $31.00 © 2019 IEEE

energy as well as chemical agent are required in photochemical reducing welding, which results in an energy-consuming and complicated process [9]. Consequently, it is imperative to seek a facile, efficacious scenario to weld CuNWs to enhance the optoelectronic performance of CuNW films.

In this paper, we demonstrated nanowelding of CuNWs through an in-situ redox process, during which copper oxide on the surface of CuNWs as a natural solder was reduced by sodium borohydride to generate copper atoms. These newly generated copper atoms selectively aggregated at the junction of CuNWs due to the active site of the intersection, and thereby welded CuNWs. The significant drop in the sheet resistance of the W-CuNW film was observed with a negligible change in the original transmittance (sheet resistance: 6×10^5 ohm/sq to 5×10^3 ohm/sq, transmittance: 94.2% to 94%), as a result of which the optoelectronic performance was enhanced remarkably (figure of merit: 0.01 to 1.20). The CuNW film with a sheet resistance of 17 ohm/sq and transmittance of 85% was obtained via the in-

situ redox nanowelding. PEDOT:PSS was selected as a protective layer to encapsulate W-CuNW network to fabricate PEDOT:PSS/W-CuNW film. The electronic properties of the PEDOT:PSS/W-CuNW film displayed an outstanding stability in the air for 30 days. Besides, the sheet resistance of this film almost kept a flat changing trend during 5000 bending cycles under a bending radius of 2 mm, suggesting the excellent mechanical flexibility. Lastly, the PEDOT:PSS/W-CuNW film showed an EMI SE value above 27 dB at a transmittance of 85% in the measured frequency range from 8.2 GHz to 12.5 GHz, which indicated this film held a great promise as a flexible transparent electromagnetic interference shielding material applied in optoelectronic devices.

II. EXPERIMENTAL SECTION

A. *In-situ redox nanowelding of CuNWs and fabrication of PEDOT:PSS/W-CuNW film*

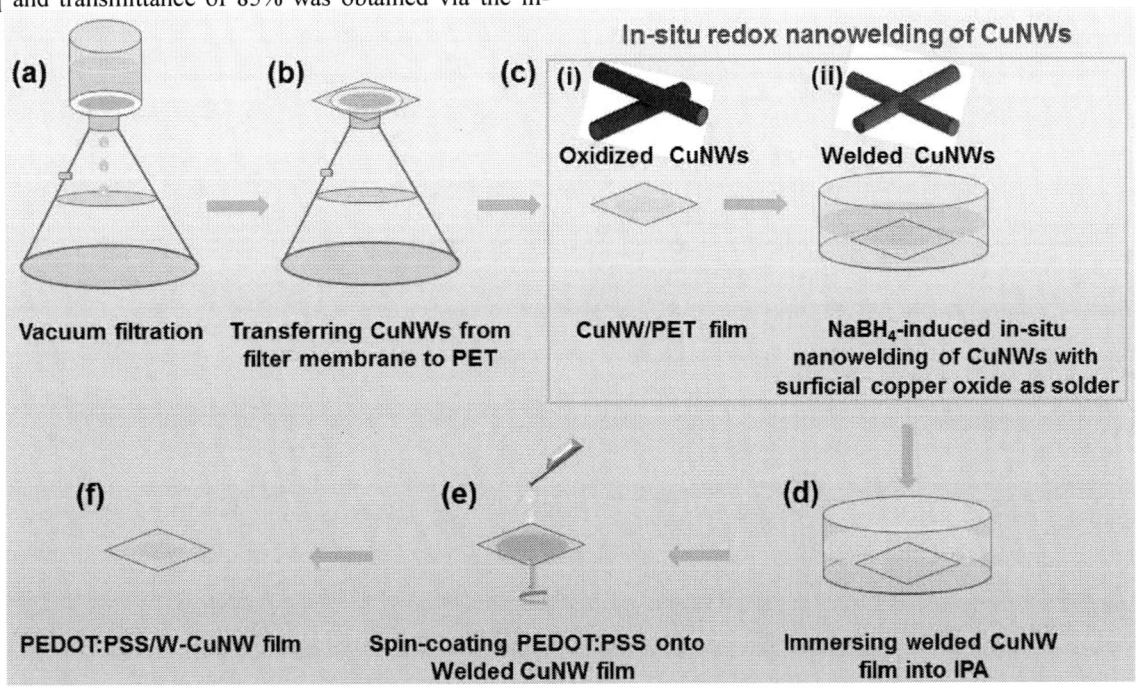

Figure 1. Schematic illustration of in-situ redox nanowelding of CuNWs and fabricating process of PEDOT:PSS/W-CuNW film.

Fig. 1 schematically illustrates in-situ redox nanowelding of CuNWs and fabricating process of PEDOT:PSS/W-CuNW transparent conducting film. CuNWs were synthesized according to previously reported literatures [9]. As-synthesized CuNWs were respectively rinsed three times with deionized water and isopropanol (IPA), and redispersed into IPA for further use. For the in-situ redox nanowelding of CuNWs, CuNW solution was poured into a glass funnel fixed at a suction base, and CuNWs were vacuum filtrated onto a Teflon filter membrane (0.45 μm pore size, 47 mm in diameter) (Fig. 1a). After the filtration, the wet filter membrane supporting CuNWs was absolutely dried in ambient atmosphere (temperature: 25 ℃, relative humidity:

65%). Polyethylene terephthalate (PET) substrate was put on the filter for 60 min under a vacuum suction (Fig. 1b). Then, the filter membrane was torn off carefully, and CuNWs were transferred to PET to obtain CuNW transparent conducting film, as depicted in Fig. 1c-i. Actually, The CuNWs onto PET had been oxidized to form surficial copper oxide layer during drying of the filter membrane and transferring of CuNWs (Fig. 1c-i). Afterwards, the CuNW film was immersed into $NaBH_4$ aqueous solution (20 mL, 0.265 M of $NaBH_4$, 0.013 M of NaOH) at 80 ℃ for 10 min, during which surficial copper oxide served as the natural solder and was reduced into copper atoms to in-situ weld CuNWs (Fig. 1c-ii). For the fabrication of the PEDOT:PSS/W-CuNW film, the W-CuNW film was immersed into IPA at room temperature for 3 min to remove excessive NaBH$_4$ aqueous solution onto the film, as shown in Fig. 1d. After drying, PEDOT:PSS as a protective layer was spin-coated onto the W-CuNW network and cured to obtain the PEDOT:PSS/W-CuNW film (Fig. 1e and f).

B. Measurement and characterization

The morphology of CuNWs was observed using scanning electron microscope SEM and optical microscope OM. The composition of CuNWs was analyzed by X-ray diffractometer and EDS. The sheet resistance and transmittance of the transparent conducting films were measured via multimeter and UV-visible spectrometer, respectively. Bending fatigue test of the films was carried out by a biaxial stretching testing machine at a frequency of 0.1 Hz. A vector network analyzer was used to measure EMI shielding effectiveness SE. The films were sandwiched between two waveguide holders of this instrument, and then S parameters were measured. The total SE, absorption SE

and reflection SE can be obtained in terms of the measured S parameters through the following equations [3]:

$$R = |S_{11}|^2, \quad T = |S_{21}|^2, \quad A = 1 - R - T \tag{1}$$

$$SE_{ref}(dB) = -10\log(1-R) \tag{2}$$

$$SE_{abs}(dB) = -10\log(T/(1-R)) \tag{3}$$

$$SE_{total}(dB) = 10\log\left(\frac{P_I}{P_T}\right) = SE_{ref} + SE_{abs} \tag{4}$$

Where R, T and A are respectively reflection, transmission and absorption coefficient. P$_I$ and P$_T$ are separately the incident and transmitted power. According to abovementioned equations, the EMI shielding effectiveness SE$_{total}$ of the films can be calculated.

III. RESULTS AND DISCUSSION

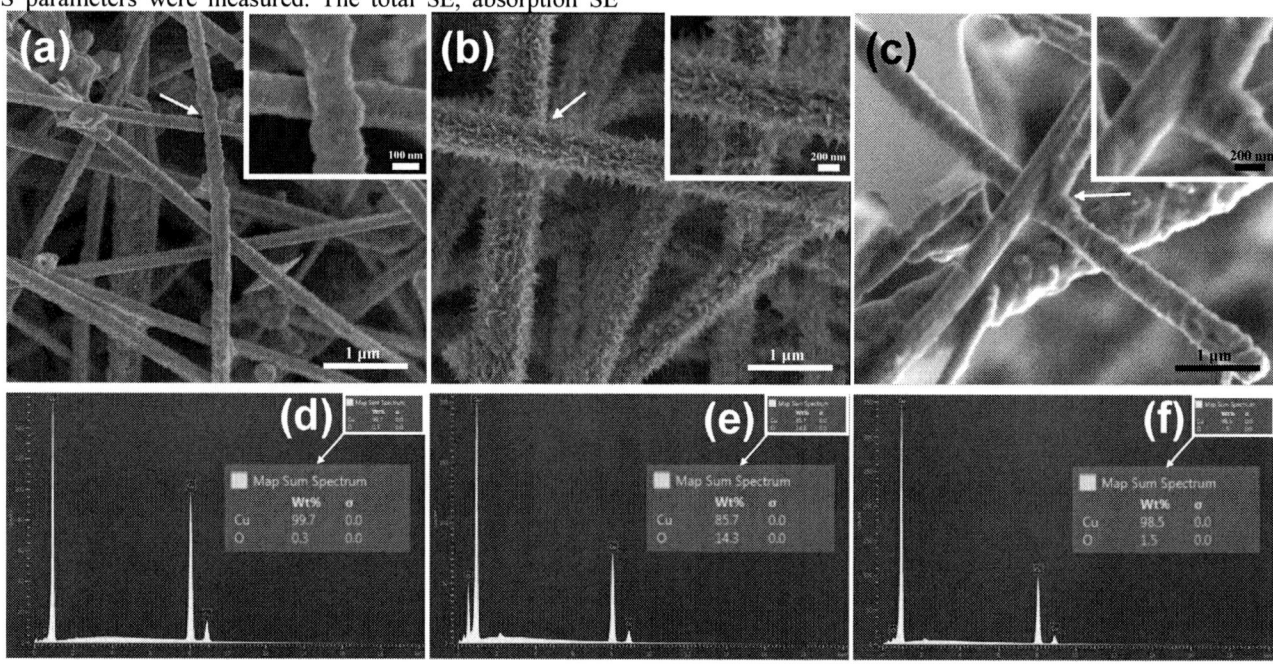

Figure 2. SEM images of (a) pristine, (b) oxidized and (c) reduced CuNWs. The insets in (a-c) show SEM images of the representative CuNW junctions. EDS spectra of (d) pristine, (e) oxidized and (f) reduced CuNWs.

Fig. 2 shows SEM images and EDS spectra of pristine, oxidized and reduced CuNWs. Pristine CuNWs stacks loosely and naturally (Fig. 2a), and simply contact each other at the junctions, which results in a weak connection between CuNWs, as shown in the inset in Fig. 2a. The pristine CuNWs are exposed in the air and oxidized to generate copper oxide on the surface (Fig. 2b), and the junction still remains loose (the inset in Fig. 2b). Aforementioned oxidized CuNWs are subjected to NaBH$_4$, after which surficial copper oxide disappears and fused junction is observed (the inset in Fig. 2c). This implies that the copper oxide on the surface of the CuNWs is reduced, and the nanowires are welded during the treatment of NaBH$_4$.

Besides, the amount of oxygen is very low (0.3 wt%) for the pristine CuNWs, as presented by the EDS spectrum in Fig. 2d. It is up to 14.3 wt% when the pristine CuNWs suffer from air, indicating the CuNWs are oxidized (Fig. 2e). The oxygen content drops to 1.5 wt% again when the oxidized CuNWs are treated by NaBH$_4$ (Fig. 2f), suggesting surficial copper oxide is in-situ reduced and consumed. Taken together, these observations display that nanowelding of CuNWs is realized through an in-situ redox process of CuNWs with surficial copper oxide as a natural solder and NaBH$_4$ as reductant. The mechanism behind the CuNW nanowelding can be demonstrated detailedly as follows: Firstly, the CuNWs are oxidized in the air to generate

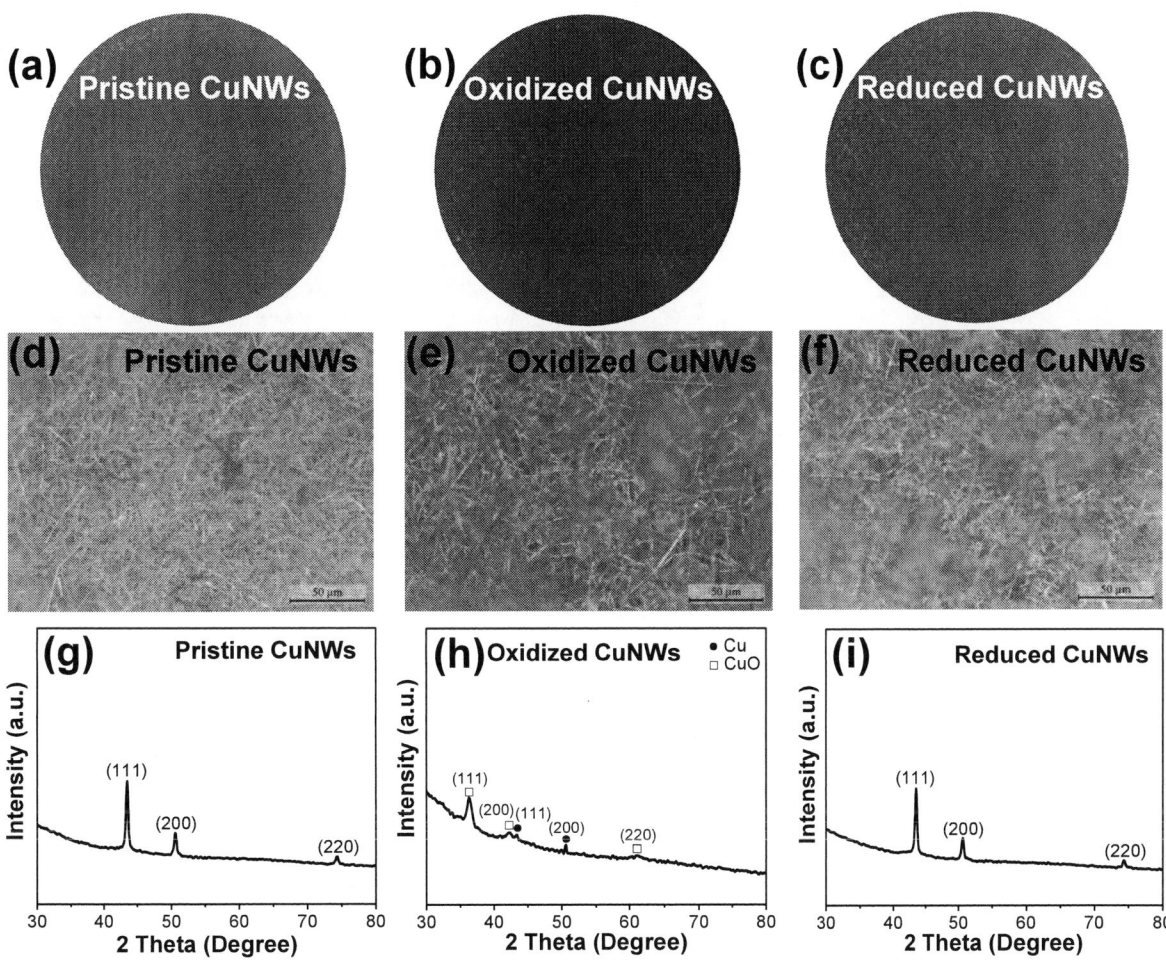

Figure 3. Digital photos of (a) pristine, (b) oxidized and (c) reduced CuNWs. OM images of (d) pristine, (e) oxidized and (f) reduced CuNWs. XRD patterns of (g) pristine, (h) oxidized and (i) reduced CuNWs.

surficial copper oxide layer during fabrication of CuNW/PET film. Afterwards, surficial copper oxide is reduced into copper atoms under the treatment of NaBH$_4$. The newly generated copper atoms selectively aggregate at the junction of CuNWs due to high energy of the active intersection, and thereby fuse the CuNW junction [10]. Lastly, nanowelding of the CuNWs is achieved through the in-situ redox process, during which surficial copper oxide acts as the natural solder to provide copper source for welding.

Fig. 3 shows digital and OM images, and XRD patterns of pristine, oxidized and reduced CuNWs. The pristine CuNWs are reddish (Fig. 3a and d), and the peaks from XRD pattern at 43.5°, 50.6° and 74.3° are respectively originated from the (111), (200) and (220) planes of face-center-cubic (fcc) copper phase, as shown in Fig. 3g [4]. The black oxidized CuNWs are observed following exposure of the pristine CuNWs in the air (Fig. 3b). And the OM image in Fig. 3e shows that the oxidized CuNWs are dark brown. In addition to the diffraction peaks from fcc copper, the peaks at 36.3°, 42.3° and 61.2° separately correspond to the (111), (200) and (220) planes of CuO, as depicted in Fig. 3h [4].

These observations demonstrate that the pristine CuNWs are oxidized and the surficial copper oxide is largely CuO. After the treatment of NaBH$_4$, the oxidized CuNWs convert from black to red (Fig. 3c and f), and the diffraction peaks from copper oxide disappear (Fig. 3i), which suggests surficial copper oxide is reduced and consumed by NaBH$_4$. The aforementioned provides further evidence that CuNW nanowelding is performed by an in-situ redox process with surficial copper oxide as a natural solder and NaBH$_4$ as reductant.

Effect of the CuNW nanowelding on sheet resistance and transmittance of CuNW films is shown in Fig. 4. Significant reduction in the sheet resistance of the W-CuNW films is observed in contrast to the original films with a negligible change in transmittance owing to remarkable drop in contact resistance at the welded CuNW junctions. It is even achieved to two orders of magnitude (sheet resistance: 6×10^5 ohm/sq to 5×10^3 ohm/sq, transmittance: 94.2% to 94%), as presented in Fig. 4. The CuNW film with a low sheet resistance of 17 ohm/sq and a high transmittance of 85% can be obtained via the in-situ redox nanowelding of CuNWs.

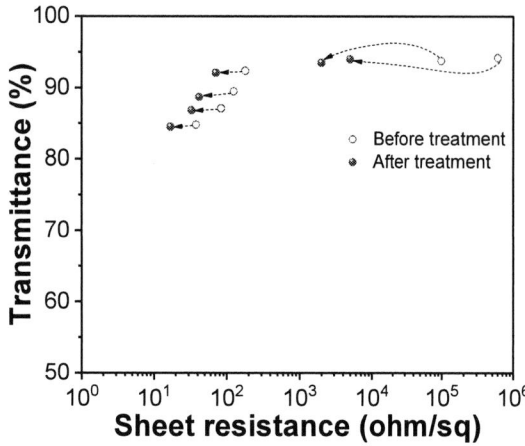

Figure 4. Sheet resistance and transmittance (@550 nm) of the CuNW films before and after welding treatment.

Contribution of the CuNW nanowelding to optoelectronic performance of CuNW films is presented in Fig. 5. Figure of merit is generally used to evaluate optoelectronic performance of transparent conducting film, and defined as σ_{dc}/σ_{opt} (σ_{dc} and σ_{opt} are DC electrical and optical conductivity of the film, respectively). It can be calculated in terms of the equation [3]:

$$\sigma_{dc}/\sigma_{opt} = \frac{188.5\sqrt{T}}{R_s(1-\sqrt{T})} \qquad (5)$$

Where R_s and T are sheet resistance and transmittance (at a wavelength of 550 nm) of the film, respectively. High values of figure of merit are desired for the films, meaning excellent optoelectronic performance. According to the result in Fig. 4, figure of merit of the CuNW films before and after welding treatment can be calculated, as shown in Fig. 5. It rises obviously and even increases by two orders of magnitude (0.01 to 1.20) (Fig. 5), demonstrating the optoelectronic performance of the CuNW films is significantly enhanced by the nanowelding.

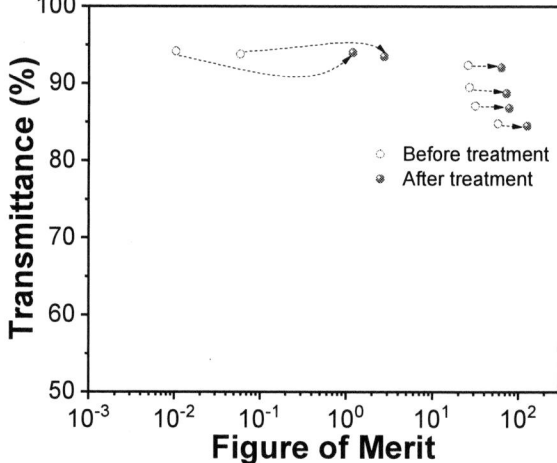

Figure 5. Figure of merit and transmittance (@550 nm) of the CuNW films before and after welding treatment.

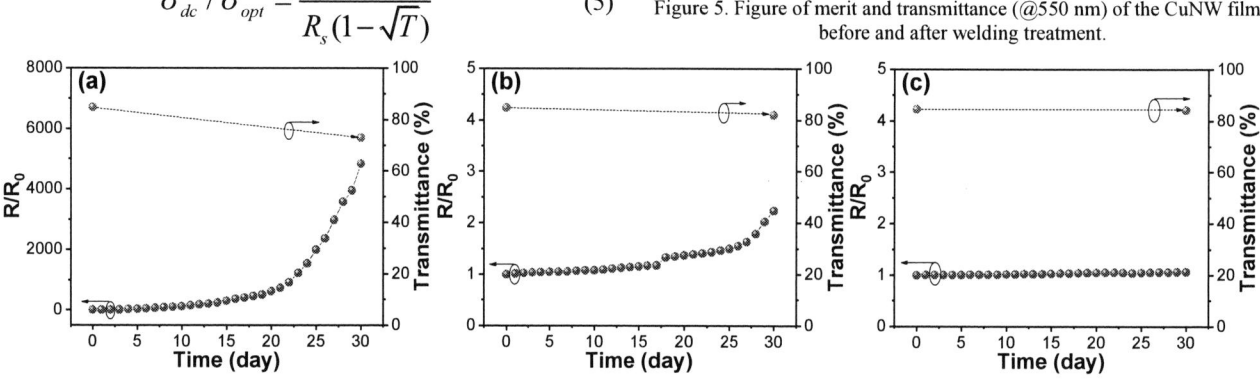

Figure 6. Variation of optoelectronic properties of (a) pristine CuNW, (b) PEDOT:PSS/pristine CuNW and (c) PEDOT:PSS/W-CuNW films in ambient condition for 30 days.

Air-stability test of optoelectronic properties of pristine CuNW, PEDOT:PSS/pristine CuNW and PEDOT:PSS/W-CuNW films in ambient condition is shown in Fig. 6. The sheet resistance of the pristine CuNW film rises dramatically, and is approximately 5000 times larger than the initial value when the film is exposed in the air for 30 days, as depicted in Fig. 6a. And the transmittance of the pristine CuNW film decreases from 84.6% to 72.9% (Fig. 6a). The degradation of optoelectronic properties of the film is caused by the oxidation of CuNWs (Fig. 2b and e, and Fig. 3b, e and h). For the PEDOT:PSS/pristine CuNW film, the sheet resistance only increases 2.3 fold and the transmittance softly lowers from 84.5% to 82.1% thanks to PEDOT:PSS as a protective layer, as shown in Fig. 6b. Almost no change in the optoelectronic properties of the PEDOT:PSS/W-CuNW

film is found when the film is placed in ambient condition for 30 days (Fig. 6c), which is due to the contribution of residual reductive $NaBH_4$ and protective PEDOT:PSS. This demonstrates the excellent air-stability of the PEDOT:PSS/W-CuNW film.

Mechanical flexibility of pristine CuNW, PEDOT:PSS/pristine CuNW and PEDOT:PSS/W-CuNW films is compared in Fig. 7. The sheet resistance of the pristine CuNW film increases about 13 fold after 5000 bending cycles as a result of the poor connection at the loose junctions of CuNWs (Fig. 7a and Fig. 2a). Compared with the pristine CuNW film, the flexibility of the PEDOT:PSS/pristine CuNW film is obviously improved due to the adhesion of PEDOT:PSS to PET substrate, as shown in Fig. 7b. The sheet resistance of the PEDOT:PSS/W-

978-1-7281-1500-9/19 $31.00 © 2019 IEEE

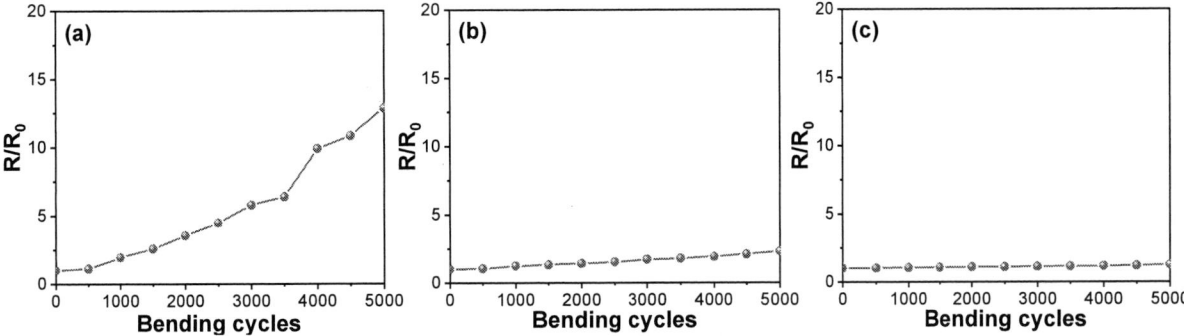

Figure 7. Bending fatigue test of (a) pristine CuNW, (b) PEDOT:PSS/pristine CuNW and (c) PEDOT:PSS/W-CuNW films under a bending radius of 2 mm.

CuNW film almost retains a constant value during 5000 bending cycles (Fig. 7c), which is attributed to the intimate contact of CuNWs at the welded junctions plus the adhesion between PEDOT:PSS and PET (Fig. 2c).

Fig. 8 demonstrates electrical conductivity of the PEDOT:PSS/W-CuNW film under deformation. As shown in Fig. 8, the film is in series connected with a LED light. The LED keeps a constant luminance when the film changes from original, bended to folded state, indicating the electrical conductivity can be maintained well under deformation. This further evidenced the mechanical flexibility of the PEDOT:PSS/W-CuNW film.

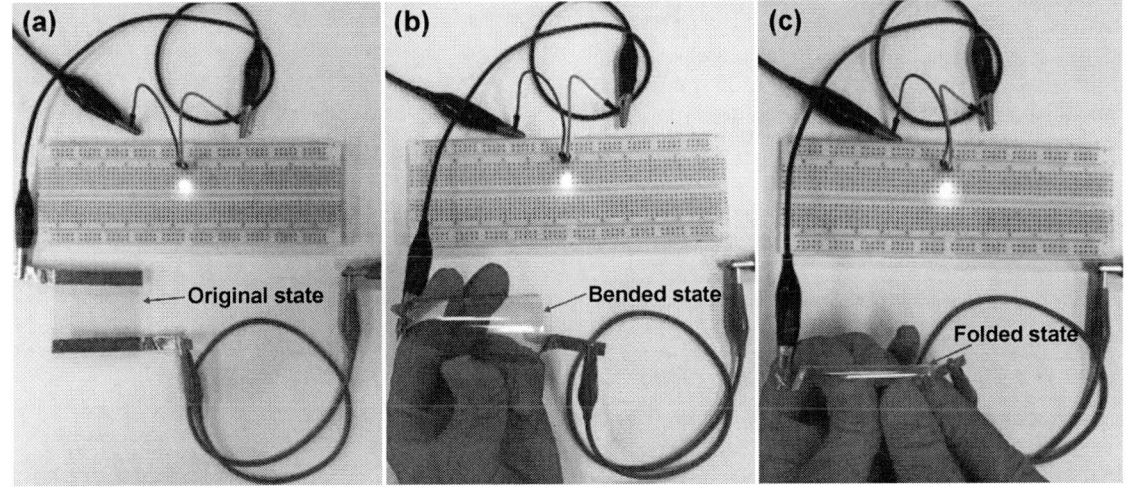

Figure 8. Demonstration of electrical conductivity of the PEDOT:PSS/W-CuNW film under (a) original, (b) bended and (c) folded state, respectively.

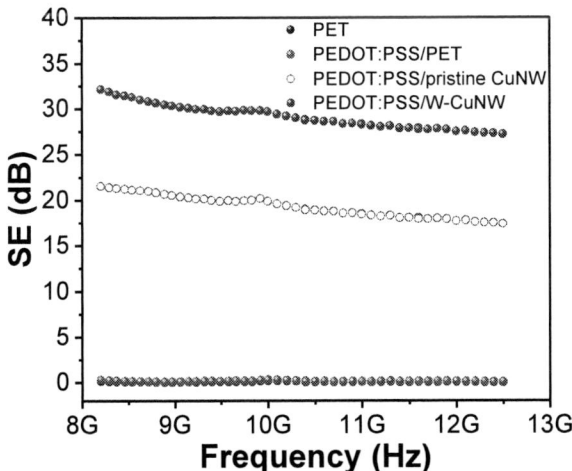

Figure 9. EMI shielding effectiveness of PET, PEDOT:PSS/PET, PEDOT:PSS/pristine CuNW (sheet resistance: 38 ohm/sq, transmittance: 85%, before treatment) and PEDOT:PSS/W-CuNW films (sheet resistance: 17 ohms, transmittance: 85%, after treatment).

EMI shielding effectiveness of the PEDOT:PSS/W-CuNW film is presented in Fig. 9. EMI SE values of PET substrate and PEDOT:PSS/PET film are both close to 0 dB. The SE values of the PEDOT:PSS/pristine CuNW and PEDOT:PSS/W-CuNW film are respectively ranged from 17 dB to 22 dB and 27 dB to 32 dB (Fig. 9). Thus, EMI shielding effectiveness of the PEDOT:PSS/CuNW film increases by ~10 dB through the in-situ redox welding, as presented in Fig. 9. EMI SE more than 20 dB can meet the requirements of commercial applications, indicating the PEDOT:PSS/W-CuNW film with SE above 27 dB at a transmittance of 85% underscores a significant potential as a flexible transparent EMI shielding material applied in touch panels, keyboards, airborne optoelectronic pods and aviation camcorders, etc [3].

IV. CONCLUSION

Nanowelding of CuNWs is realized via an in-situ redox process with surficial copper oxide as a natural solder and $NaBH_4$ as reductant. Significant reduction in the sheet resistance of CuNW films are achieved with a negligible change in the original transmittance, and is even up to two orders of magnitude (6×10^5 ohm/sq to 5×10^3 ohm/sq, transmittance: 94.2% to 94%). The optoelectronic performance of the CuNW films is enhanced remarkably by the CuNW nanowelding, and figure of merit even increases two orders of magnitude (0.01 to 1.20). The CuNW film with a sheet resistance of 17 ohm/sq and a transmittance of 85% is obtained via the in-situ redox nanowelding. As-fabricated PEDOT:PSS/W-CuNW film shows standout air-stability and mechanical flexibility, and EMI SE value above 27 dB at a transmittance of 85% in the frequency range from 8.5 GHz to 12.5 GHz, which may function as a flexible transparent EMI shielding material for optoelectronics.

ACKNOWLEDGMENT

This research was supported by the National Natural Science Foundation of China (21571186), National Key R&D Project from Minister of Science and Technology, China (2016YFA0202702), Chinese Academy of Sciences Key Research Projects of Frontier Science (QYZDY-SSWJSC010), Shenzhen Basic Research Plan (JCYJ20170818162548196), Youth Innovation Promotion Association (2017411). Xianwen Liang and Jianwen Zhou contributed equally to this work.

REFERENCES

[1] Z. Q. Niu, F. Cui, E. Kuttner, C. L. Xie, H. Chen, Y. C. Sun, A. Dehestani, K. Schierle-Arndt and P. D. Yang, "Synthesis of Silver Nanowires with Reduced Diameters Using Benzoin-Derived Radicals to Make Transparent Conductors with High Transparency and Low Haze," Nano. Lett., vol. 18, Jul. 2018, pp. 5329-5334, doi: 10.1021/acs.nanolett.8b02479.

[2] Y. Liu, J. M. Zhang, H. Gao, Y. Wang, Q. X. Liu, S. Y. Huang and C. F. Guo, "Capillary-Force-Induced Cold Welding in Silver-Nanowire-Based Flexible Transparent Electrodes," Nano. Lett., vol. 17, Jan. 2017, pp. 1090-1096, doi: 10.1021/acs.nanolett.6b04613.

[3] X. W. Liang, J. B. Lu, T. Zhao, X. C. Yu, Q. S. Jiang, Y. G. Hu, P. L. Zhu, R. Sun and C. P. Wong, "Facile and Efficient Welding of Silver Nanowires Based on UVA-Induced Nanoscale Photothermal Process for Roll-to-Roll Manufacturing of High‐Performance Transparent Conducting Films," Adv. Mater. Interfaces, vol. 6, Feb. 2019, 1801635, doi: 10.1002/admi.201801635.

[4] S. Han, S. Hong, J. Yeo, D. Kim, B. Kang, M. Y. Yang and S. H. Ko, "Nanorecycling: Monolithic Integration of Copper and Copper Oxide Nanowire Network Electrode through Selective Reversible Photothermochemical Reduction," Adv. Mater., vol. 27, Nov. 2015, pp. 6397-6403, doi: 10.1002/adma.201503244.

[5] S. Han, S. Hong, J. Yeo, D. Kim, B. Kang, M. Y. Yang and S. H. Ko, "Nanorecycling: Monolithic Integration of Copper and Copper Oxide Nanowire Network Electrode through Selective Reversible Photothermochemical Reduction," Adv. Mater., vol. 27, Nov. 2015, pp. 6397-6403, doi: 10.1002/adma.201503244.

[6] S. Han, S. Hong, J. Ham, J. Yeo, J. Lee, B. Kang, P. Lee, J. Kwon, S. S. Lee, M. Y. Yang and S. H. Ko, "Fast Plasmonic Laser Nanowelding for a Cu-Nanowire Percolation Network for Flexible Transparent Conductors and Stretchable Electronics," Adv. Mater., vol. 26, Sep. 2014, pp. 5808-5814, doi: 10.1002/adma.201400474.

[7] S. Ding, J. T. Jiu, Y. Gao, Y. H. Tian, T. Araki, T. Sugahara, S. Nagao, M. Nogi, H. Koga, K. Suganuma and H. Uchida, "One-Step Fabrication of Stretchable Copper Nanowire Conductors by a Fast Photonic Sintering Technique and Its Application in Wearable

Devices," ACS Appl. Mater. Interfaces, vol. 8, Feb. 2016, pp. 6190-6199, doi: 10.1021/acsami.5b10802.

[8] A. R. Rathmell and B. J. Wiley, "The Synthesis and Coating of Long, Thin Copper Nanowires to Make Flexible, Transparent Conducting Films on Plastic Substrates," Adv. Mater., vol. 23, Nov. 2011, pp. 4798-4803, doi: 10.1002/adma.201102284.

[9] J. H. Park, S. Han, D. Kim, B. K. You, D. J. Joe, S. Hong, J. Seo, J. Kwon, C. K. Jeong, H. J. Park, T. S. Kim, S. H. Ko and K. J. Lee, "Plasmonic-Tuned Flash Cu Nanowelding with Ultrafast Photochemical-Reducing and Interlocking on Flexible Plastics," Adv. Func. Mater., vol. 27, Aug. 2017, pp. 1701138, doi: 10.1002/adfm.201701138.

[10] X. W. Liang, T. Zhao, P. L. Zhu, Y. G. Hu, R. Sun and C. P. Wong, "Room-Temperature Nanowelding of a Silver Nanowire Network Triggered by Hydrogen Chloride Vapor for Flexible Transparent Conductive Films," ACS Appl. Mater. Interfaces, vol. 9, Nov. 2017, pp. 40857−40867, doi: 10.1021/acsami.7b13048.

Compartmental EMI Shielding with Jet-Dispensed Material Technology

Xuan Hong, Qizhuo Zhuo, Xinpei Cao, Dan Maslyk, Noah Ekstrom, Juliet Sanchez, Selene Hernandez, Jinu Choi

Henkel Electronic Materials, LLC.
Irvine, California USA
xuan.hong@henkel.com

Abstract— **Device miniaturization continues using System-on-Chip (SoC), System-in-Package (SiP), multichip module (MCM), and heterogeneous integration to deliver a wider range of functionalities without sacrificing valuable space on a substrate. With multiple integrated circuits and MEMS sensors integrated into a thin single module to perform as a full electronic system, the need for more compact and effective electromagnetic interference (EMI) protection between various baseband and wireless, RF, analog, and power management components is greater than ever before. Fortunately, as device miniaturization has accelerated, so has the development of novel shielding technologies to accommodate for the higher density package structures. Jet-dispensed compartment shielding is an integrated package-level solution that allows for much smaller semiconductor form factors and is achieved using a fully automatic assembly process with high performance.**

Keywords: Integrated EMI shielding; package-level EMI shielding; compartment shielding; in-package RF isolation; jet dispensing; conductive paste for trench filling; shielding effectiveness; voiding performance

I. INTRODUCTION

With the proliferation of internet of things (IoT) and internet of everything (IoE), interconnected devices are increasingly becoming more advanced with a wide array of integrated functionality for higher performance, smaller form factor, and lower power consumption. Furthermore, with the commercialization of 5G technology, additional shielding needs are introduced for RF module designs as increasing numbers of integrated antennas and front-end components are needed to support a wider range of frequency bands with minimal signal attenuation.

In order to address these challenges in single module while maintaining a compact form factor, a novel compartment shielding technology has been developed[1]. The shielding partition is created at the back end of the packaging process using laser-ablation into the epoxy mold compound (EMC) followed by jet-dispensing of a highly conductive material into the laser-cut trench, after which conformal shielding encapsulation takes place to complete the Faraday cage as shown in Fig. 1[2-3].

This compartment shielding method enables high-density package designs using customizable isolation structures which allows flexible designs for optimal performance. The trenches also have a high aspect ratio allowing closer component-to-component placement inside a single module.

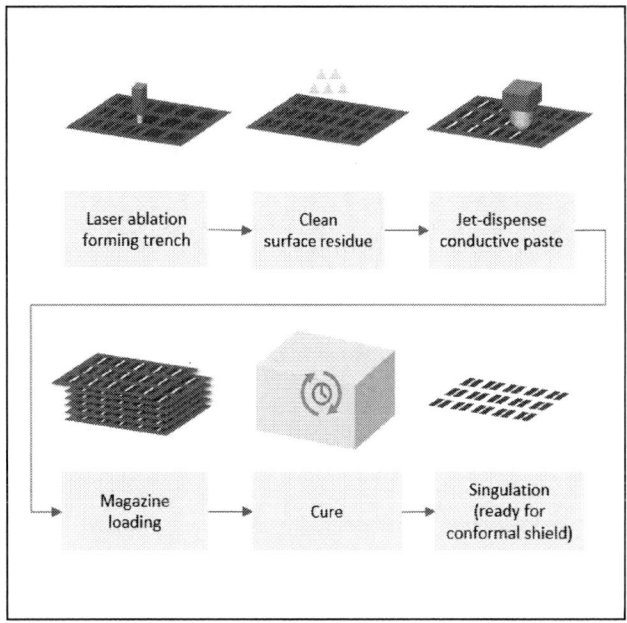

Figure 1. Compartment shielding process

This paper will discuss the development of the trench filling conductive paste designed for jet-dispensing into narrow trenches, covering the characterized material properties, filling performance, shielding effectiveness, and reliability performance.

II. EXPERIMENT DETAILS

A. Material

A Henkel product, LOCTITE® ABLESTIK® EMI 3620FA was used for the material and filling study. The material is a silver-based paste designed to achieve a complete and void free fill while maintaining high electrical and thermal conductivity requirements of package-level EMI shielding. The physical properties of the novel material are shown in Table 1.

B. Jet-Dispensing

The material was dispensed via jetting, using Nordson Asymtek Axiom equipment, into trenches formed in the EMC, which, after curing creates a thin metallic wall.

Figure 2. Trench-filling process

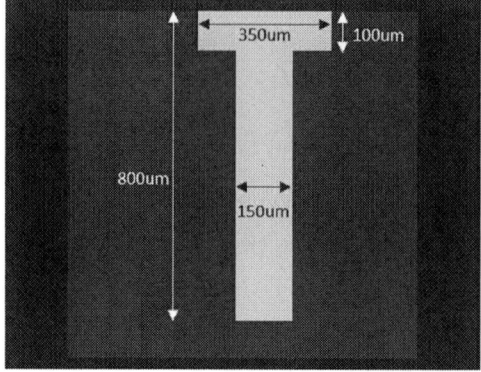

Figure 3. Test vehicle trench dimensions

Processing steps, shown in Fig. 1, begin with molding the system-in-package (SIP) formfactor. Next, through laser ablation of the EMC, open face trenches are created all the way down to the grounding plane of the substrate below. Dry ice (or an equivalent method) cleaning of the trench removes any remaining ablation residue and maintains a contaminant free environment. The material can then readily be jet dispensed into the trench design on the SIP, as shown in Fig. 2 and sent to cure. This, in conjunction with a conformal solution, will shut down communication between adjacent RF or otherwise sensitive components in the SIP.

The use of jet dispensing in the semiconductor industry is a well-known and understood process, used in a variety of applications. With jet dispensing, pulses from a needle in the fluid body are used to eject droplets of material from the nozzle and onto the substrate. The equipment allows for heating on both stage and nozzle to control temperature for optimization of viscosities and flowrate. Pulse duration and head speed are also adjusted parameters that allow control over dispense performance. Finally, dispense height control will assist in the aesthetics by minimizing splash. These parameters, when optimized together, can create maximum jetting performance for the EMI trench fill material. For this experiment Henkel has used an internal test vehicle with dimensions shown in Fig. 3.

C. Viscosity and Thixotropic Index Testing

Brookfield P51 equipment with 1inch cone and plate was used to test viscosity at 0.5rpm and 5rpm with water batch temperature at 25oC. The material was added into the gap and held for 1minute. The thixotropic index is calculated by the dividend of viscosity at 0.5rpm and 5rpm.

D. Die Shear Strength Adhesion Measurement

EMI trench conductive paste material was dispensed on a glass surface using 50um spacer to control the coating thickness. A 3x3mm EMC package was dipped into the paste material and then placed on EMC or copper substrate. Material adhesion was collected by die shear strength using Dage 4000 die shear measurement equipment. Sample sized ranged from 6-8 pieces.

E. Volume Resistivity Measurement

4-point probe method was used for volume resistivity measurements. The probe array was placed on the center of the sample: the two outer probes are used for sourcing current and the two inner probes are used for measuring the resulting voltage drop across the surface of the sample. The volume resistivity is calculated as follows:

$$VR = R/l \,(w.t) \qquad (1)$$

Where:
  VR= volume resistivity (ohm-cm)
  R = measured resistance (ohm)
  W = width
  t = thickness
  l = length

F. Material Shielding Effectiveness Measurement

To analyze comparative bulk material shielding performance, far-field measurements were used following the ASTM D5568-14 for 1.6-2.6GHz. The ASTMD5568 standard is a method for characterizing the complex permittivity and permeability of materials. This method utilizes a network analyzer and standard waveguide sets to evaluate specific frequency bands as shown in Fig. 4. To evaluate shield effectiveness, the sample is inserted into the waveguide spanning the entire orifice of the rectangular waveguide. S21 is evaluated as shield effectiveness. With this technique, since the measurement is in a very controlled and calibrated environment and is also in far field, higher shielding values will be observed compared to other near field techniques.

Figure 4. Far field testing equipment

G. Voiding

The trench fill material was inspected for voids by microscopy after performing cross-sections on both parallel and perpendicular planes to the trench.

III. RESULTS

Traditional high-conductive, solvent-free material uses silver flake in combination with various resin systems[4]. In order to achieve high electrical conductivity, additional flake-shaped fillers are commonly added into these systems. Higher filler loading is needed to achieve higher conductivity. The following are some general challenges for compartment shield material development.

A. Filling Performance

Typical conductive paste used for high-speed dispensing has viscosity over 9000cps at 5rpm, and thixotropic index (TI) around 4-6. This is a good indicator of how the material will flow and level after dispensing. Material with high viscosity and high TI has more difficulty flowing into a narrow trench design. A typical trench formed for compartment shielding has a depth of 0.7-1mm with a width of 120-150μm. This provides a height-to-width aspect ratio of approximately 1:6 to 1:10 as shown in the test vehicle dimension in Fig. 3. To fully fill the high aspect ratio trench with jet dispensing, the material needs to have a lower viscosity and thixotropic index with good flow capability which could successfully flow into the trench bottom and push out the air up. Otherwise, a high viscosity and high thixotropic index material will have difficult to flow into trench bottom and be easy to trap air.

B. Voiding

One of the technical challenges for trench material design is to minimize and eliminate voiding. The root causes for voids are quite complicated in the semiconductor material. From a material design point of view, flow capability, wetting performance, outgassing, and reaction kinetics can all impact voiding performance. Because of the narrow trench width and high aspect ratio, non-optimized material

will not perfectly reach the trench bottom. Wetting is very important; if the surface chemistry between the material and contact interface doesn't match well, material flowability along the incompatible surface may be limited. The non-wetted area could delaminate or show up as a trapped air void after cure. Fig. 5 shows air trapped during jetting process.

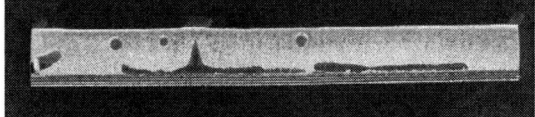

Figure 5. Cross-section of trapped air

During the cure the air increases in pressure and tries to escape upward out of the trench. This increases the voids size and, if unsuccessful at escaping, the void can be encapsulated.

Outgassing is another challenge area. In order to achieve high electrical conductivity, material requires high silver loading. Silver particle volume content must reach the percolation threshold or higher to ensure the homogenous system has excellent electrical conductivity, promoting good shielding effectiveness. However, the high filler loading will come with high bulk viscosity; diluent or solvent can be introduced to reduce viscosity, but they will evaporate during the cure in oven which could form voids in the trench. Fig. 6 shows a material with high outgassing. The outgassing was trapped in the trench and couldn't escape the bulk resin during the cure out because of the vapor pressure was not sufficiently high enough. Optimized cure profile could improve outgassing voiding on some level level, but high solvent content could also bring serious adhesion and reliability issue.

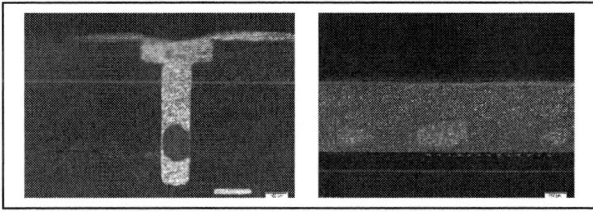

Figure 6. Voids with high outgassing

C. Splashing

Splash is a common issue for the discrete droplet jetted into the narrow trench, especially for lower viscosity material. Material viscosity and surface tension with inner integration optimization can play a role in minimizing splash. Additionally, it is critical to optimize equipment jetting parameters to further control splashing. Fig. 7 shows a material with severe splashing on the molding surface after jetting compared to a material with good splashing performance. Splashing can impact aesthetics of post-jetted conformal coatings.

978-1-7281-1500-9/19 $31.00 © 2019 IEEE

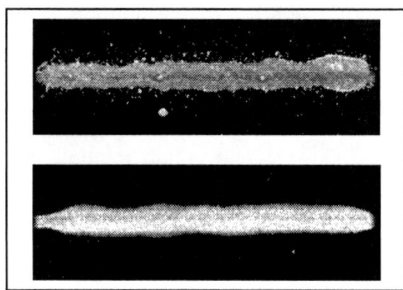

Figure 7.　High splash vs. low splash

D.　Shrinkage

Another issue caused by high outgassing is the curing shrinkage. Fig. 8 shows the indentation or concave surface created by significant curing shrinkage. For trench material, the cure shrinkage needs to be controlled in resin cure and solvent outgassing. The deep dent from shrinkage can cause cosmetic and performance issue for EMI conformal coating in the later package assembly process steps. The high cure shrinkage can also cause high stress, resulting in package warpage, delamination, cracking, and failed reliability.

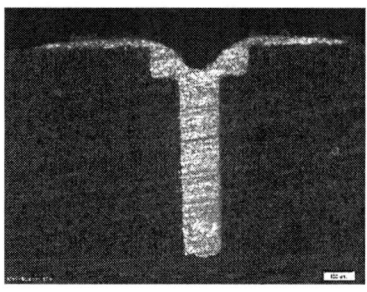

Figure 8.　Indentation caused by high cure shrinkage

LOCTITE ABLESTIK EMI 3620FA is a high silver loading solvent free low viscosity with high electrical conductivity paste material. Table 1 lists all the preliminary material properties and Fig. 9 shows good voiding performance. The material has only 4% outgassing with 175 °C cure for 1 hour yielding a flat top surface for a more uniform conformal coating.

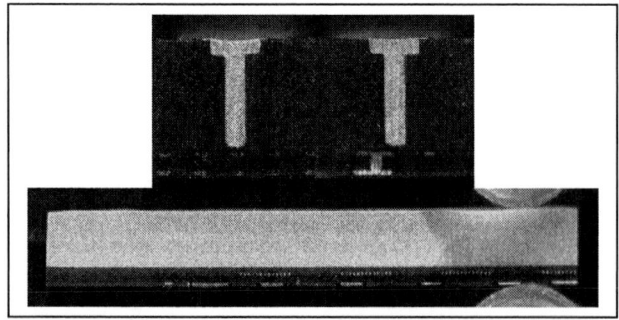

Figure 9.　Perpendicular and parallel cross-section for good trench filling material

TABLE I.　PROPERTIES OF NOVEL MATERIAL

Material Name	LOCTITE ABLESTIK EMI 3620FA
Technology	Electrically Conductive
Application Method	Jetting/Dispensing
Volume Resistivity ($\Omega \cdot$cm)	9×10^{-5}
Thermal Conductivity (w/mk)	7
Viscosity, 5rpm (cps)	5000
Thixotropic Index	1.3
Curing Temperature and Time	175°C, 1 hour, in air
Outgassing	4%
Far-field Shielding Effectiveness (dB)	60-85 dB with 20um thickness @ 1.6-2.6GHz
Adhesion on EMC (3x3mm die kg/die)	12
Linear Shrinkage (%)	5.2
Volume Shrinkage (%)	14.8

E.　Shielding Performance

1)　Effect of Volume Resistivity on Shielding Effectiveness

For EMI trench filling material, logically high EMI shielding effectiveness is essential to fulfill the package performance requirements. Silver is an excellent conductor of electricity and can shield from electromagnetic interference. Current trench-filling material also uses silver as its filler to achieve highest levels of conductivity. The volume resistivity or electrical conductivity of the jetted paste is directly related to electromagnetic shielding performance within the range of frequency being tested.

For the trench material, cure profile affects volume resistivity; the higher temperature cure commonly increases curing shrinkage leading to better silver contact and achieving greater electrical conductivity. Table 2 confirms this by showing the material's resistivity when cured with two different temperature profiles. Fig. 10 shows the test results of shield effectiveness for the material cured with the two temperatures. The shielding effectiveness is well correlated to electrical conductivity. Material cured at higher temperatures shows better electrical conductivity and in turn has 6-8dB higher shielding effectiveness.

TABLE II.　IMPACT OF CURING PROFILE ON VOLUME RESISTIVITY

Cure Profile	VR (ohm-cm)
30min ramp to 150min and hold for 60min	0.0002
30min ramp to 175min and hold for 60min	0.0001

978-1-7281-1500-9/19 $31.00 © 2019 IEEE

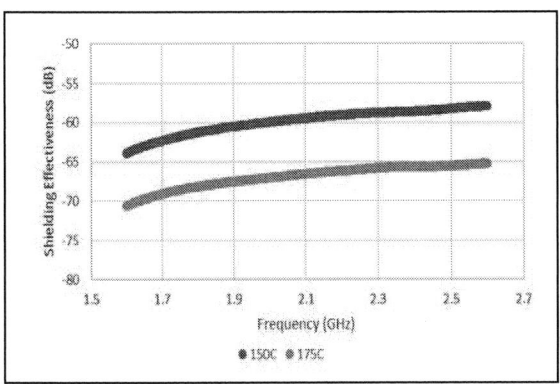

Figure 10. Far-field shielding effectiveness of curing temperature at 1.6-2.6GHz

2) Effect of Thickness on Shielding Effectiveness

The trench material bulk material shielding effectivity is directly related to the material thickness as shown in Fig 11. The thicker material, the more signal is attenuated with high shielding Effectiveness. The equipment test limitation is 120dB, with VR of 0.00008 ohm-cm, material thickness than 40um could achieve over 90dB for the bulk material. Current trench thickness commonly is more than 100um. The material could well shield the electromagnetic signal and provides good protection for the device.

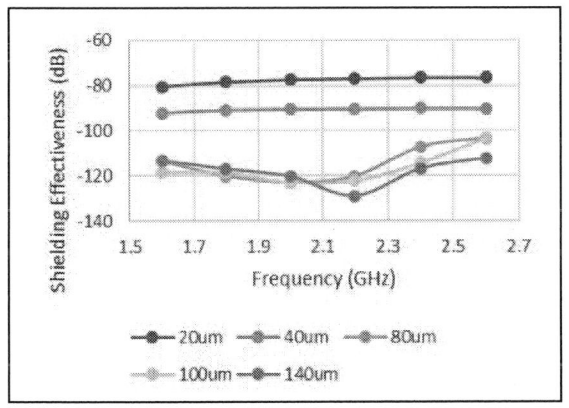

Figure 11. Far-field shielding effectiveness vs. coating thickness at 1.6-2.6GHz

F. Reliability Performance

Critical reliability tests were conducted to assess semiconductor packages after the compartment material was filled into the trench using internal test vehicles. The parts were cured with recommended cure profile 30 minutes ramp to 175C and hold for 60 minutes in conventional oven. The parts were tested under severe conditions during the assembly process. The results showed the material passed Moisture Sensitivity Level 3 (MSL3) followed by temperature cycling (TCT) using the JEDEC JESD22-A104 standard. The results also passed up to 1000 cycles at -40°C to 150°C and 1000 hours at High Temperature Storage (HTS) as shown in Fig. 12. No delamination and cracking were observed at any stage of reliability testing.

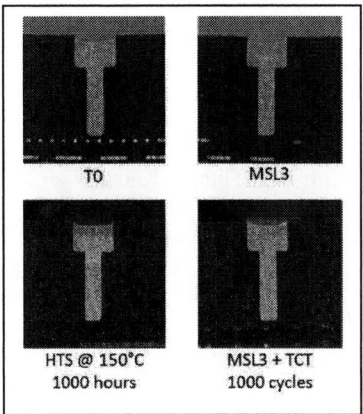

Figure 12. Material adhesion post reliability testing

IV. CONCLUSION

With proven performance, trench-filled compartment shielding using jet-dispensed conductive paste enables highly integrated compact package designs incorporating various sensitive components in a single package. Optimal material properties combined with refined jetting parameters and curing profiles provide good filling results overcoming challenges of voids, splashes, and shrinkage when being dispensed into a narrow trench. The material's volume resistivity and trench width have a direct impact on the isolation structure's shielding performance at standard frequency bands within the electromagnetic spectrum. Such shielding structure also passes stringent reliability requirements at time zero, post-HTS, and post-MSL3-TCT making it suitable for various integrated shielding needs.

Post-laser-ablation, jet-dispensed compartment shielding allows in-module RF isolation using a high width-to-depth aspect ratio of up to 1:10 which enables high-density miniaturization for SiP, SoC, MCM, and other heterogenous integration structures. With the advantages of dense form factor implementation, flexible isolation path designs, and a wide process window, trench-filled compartment shielding is a highly valuable technique to be considered for all integrated packaging structures.

V. REFERENCES

[1] JL Leou, V.Chen, J, Chen, T. Wang, H.Chang, "Miniaturized LTE Modem SiP Using Novel Multiple Compartment Shielding", IEEE CPMT Symposium, pp51-54, Japan 2014

[2] S. Geetha, K.K. Satheesh Kumar, Chepuri R.K. Rao, M. Vijayan, D., "EMI Shielding: Methods and Materials—A Review", Journal of Applied Polymer Science, vol. 112, pp. 2073-2085, Jul. 2007

[3] N. Karim, , "Electromagnetic Shielding for RF and Microwave Packages," RF and Microwave Microelectronics Packaging II pp 43-62, Springer, 2017.

[4] K. Joo, T.R.Kim, J. Hwang, J. Yoon, S. Jeong, M.Yim, "Package-Level EMI Shielding Technology with Silver Paste for Various Applications", 2017 IEEE 67th Electronic Components and Technology Conference (ECTC), pp1736-1741, 2017

2019 IEEE 69th Electronic Components and Technology Conference (ECTC)

Highly (111)-oriented Nanotwinned Cu for High Fatigue Resistance in fan-out wafer-level packaging

Yu-Jin Li[1], Chih-Han Theng[1], I-Hsin Tseng[1], and Chih Chen[1]
[1]Department of Materials Science and Engineering, National
[1]Chiao Tung University
[1]1001 Ta Hsueh Road, Hsin-Chu, Taiwan 30010, ROC
chih@mail.nctu.edu.tw

Benson Lin[2], Chia-Cheng Chang[2]
[2]MediaTek Inc
[2]PT, MediaTek Inc., Hsinchu 300, Taiwan
kris.chang@mediatek.com

Abstract—Fan out wafer level packaging is a potential solution for combining high electrical performance and low cost. However, with the shrinkage of the package, the mechanical reliability of the copper redistribution layer becomes an important issue. However, based on the processing of fan out packaging, the copper lines were annealed several times during the process. The annealing process is harmful to traditional strengthen mechanisms, but not to columnar grains with high density twinning structure. In this study, the technology of electroplating highly <111>-oriented copper lines was introduced to solve the reliability issue in copper RDLs. By the nature of nano-twinned Cu, the performance of thermal cycling test (TCT) lifetime is much better than normal copper, and the failure mechanisms were discussed in this study.

Keywords-nano-twinned, Cu, fan-out, TCT (key words)

I. INTRODUCTION (HEADING 1)

For mobile device, there are some requirements for the chip packaging such as low energy consumption, better electrical performance and smaller in size, and fan-out wafer level packaging(FOWLP) is potential to be the best solution. In 2012[1], TSMC introduced its integrated fan our wafer level packaging (InFOWLP) which takes several advantages over traditional flip chip packaging [2-4]. Nowadays, InFOWLP is widely used in mobile devices and its commercial potential is still growing. However, there are some reliability issues. One is the thermal cycling test lifetime of the copper redistribution layers (RDLs). Due to the structure of the FOWLP, the epoxy molding compound (EMC) is needed for redistributing the I/O. To connect the I/O on the Si die and EMC, the copper lines must cross over the edge of Si die. Generally, this structure is robust when the width and thickness of the copper lines are larger than 10μm. However, with the development of technology, the size of packaging becomes smaller and smaller. It is no doubt that the pitch of RDLs would decrease. In this condition, the mechanical property of the copper lines plays an important role in the reliability performance of the packaging. In 2017 [5], Yu, C. K.'s group reported that the cracks of copper lines were likely to be observed at the edge

of the Si die during the TCT. In their study, the failure mechanism is attributed to the mismatch of the coefficient of thermal expansion (CTE) among Si, EMC, polyimide and copper. Therefore, during the TCT, the thermal stress would be introduced to the copper lines which has the lowest dimension in the packaging, and finally cause the failure.

As mentioned above, the copper lines of the fan out packaging would undergo several times of annealing which are applied for the curing of layers of polyimide. Generally, work hardening, precipitation hardening, solid solution strengthening and grain refining are defined as classical strengthening mechanisms. However, these mechanisms are not thermally stable. By combining the requirements of thermal stability and mechanical strength, the nano-twinned Cu is very likely to solve this problem. In previous study [6-8], the nano-twinned Cu showed high tensile strength, reasonable ductility, excellent electrical conductivity and less Kirkendall voids in the reaction with solder. In 2008, Anderoglu, O., et al. [9] reported that the <111>-oriented nano-twinned Cu is thermally stable. The high twinning structure can stand high temperature annealing (annealing at 800 ° C for 1 h) without serious de-twinning or grain growth. In previous study, there are two main methods to fabricate nano-twinned Cu films: (1) sputtering [10-13] and (2) electroplating [6-8,14-15]. The sputtering nano-twinned Cu has very low impurity, low twin spacing and perpendicular columnar grains. However, the slow deposition rate is not suitable for mass production. In 2012, Liu, Tao-Chi, et al. [14] reported the technology of electroplating <111>-oriented nano-twinned Cu by direct current which is high efficiency in industry level.

In present study, the nano-twinned copper was electroplated on the RDL test vehicles for TCT. After the TCT, the resistance of the copper lines was measured by four probe station. To clarify the mechanism of the failure modes, the measurement of the tensile test was introduced. Finally, it would be the discussion of the role of nt-Cu in fan out packaging.

II. EXPERIMENTAL

978-1-7281-1500-9/19 $31.00 © 2019 IEEE

In this study, the copper lines were electroplated on the well-designed fan-out test vehicles as shown in figure 1. The line width is control to be 5um. Figure 2. shows the process of the sample preparation. Before the electroplating, the test vehicles were cleaned by citric acid to remove the oxide of the copper seed. The electroplating bath is made of 50g/ L Cu^{2+}, 50g/L H_2SO_4, and 50 ppm Cl^-. The stirring rate of the plating bath is 1200 rpm. After electroplating, the focused ion beam system (FIB) was used to check the microstructure of the copper lines.

For the thermal cycling test, the temperature ranged from -55°C to 125°C, and the ramping rate is 15°C/min. After 200, 700 and 1000 cycles, the SEM was employed to observed the cracks of the copper lines.

Top view

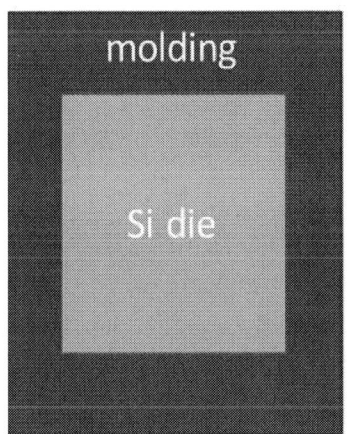

Cross section

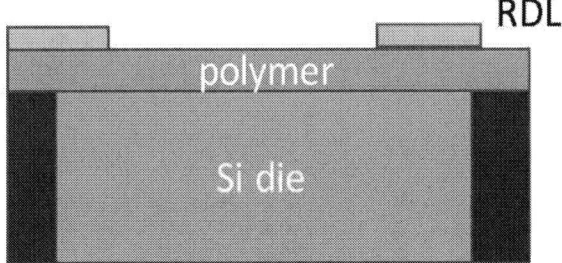

Figure 1. The illustration of the fan-out packaging test vehicle. As mentioned above, the failure was likely to be observed around the edge of Si die. Therefore, as shown in the cross section view, this study will focus on the RDLs cross over the Si die and molding compound.

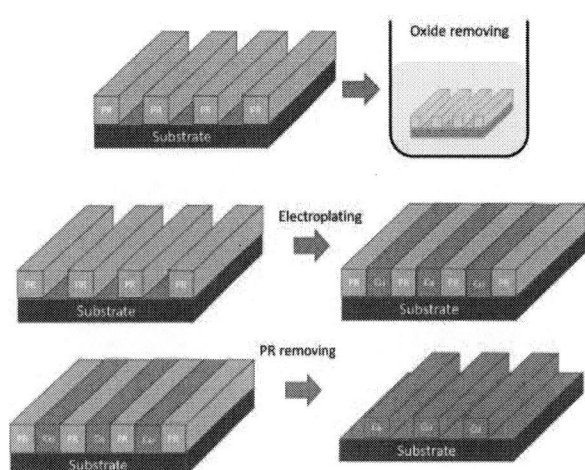

Figure 2. the process flow of electroplating copper lines. The oxide was removed by high concentration citric acid. The copper was deposited in the pattern defined by the photoresistance (PR). After electroplating, the PR was removed by acetone.

For the tensile test, the sample was punched into the shape of dog-bone as shown in Figure 3. The thickness for the tensile sample is 20μm. The tensile test was operated at room temperature with the strain rate of 4.17×10^{-3} (1/s).

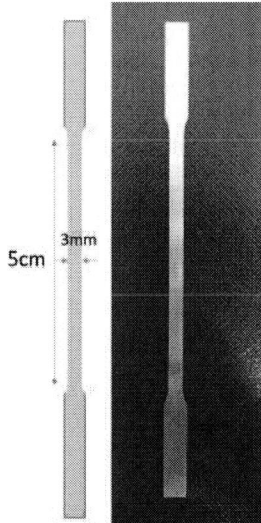

Figure 3. The dimension of the standard sample for the tensile test. The gauge length is 5cm and the width is 3mm.

III. RESULTS AND DISSCUSSION

A. Thermal cycling test of fan-out packaging

After 200 cycles TCT, the nano-twinned copper lines showed no crack. In contrast, there were 27 % of the normal copper lines were observed with cracks. The figure 4 shows the top view of the copper lines. After 1000 cycles, there

978-1-7281-1500-9/19 $31.00 © 2019 IEEE

were only 0.4% of the copper lines were observed with cracks; More than 30% of the regular copper lines were observed with cracks. In the sample with regular copper lines, the copper lines peeled off from the substrate as shown in figure 5.

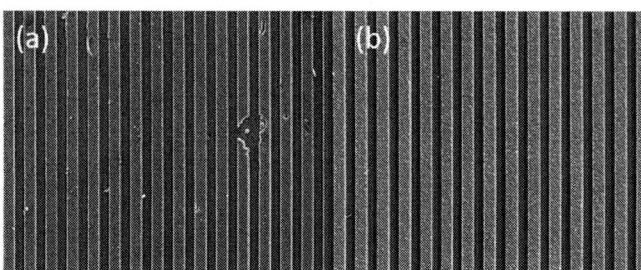

Figure 4. (a) regular copper and (b) nt-Cu lines after 200 cycles TCT.

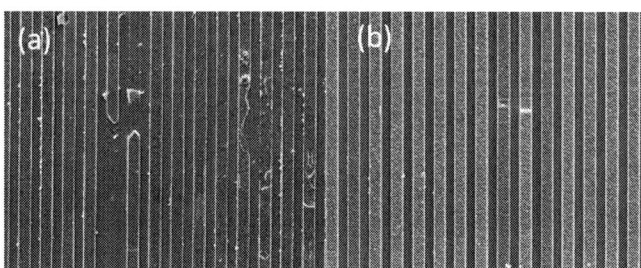

Figure 5. (a) the regular copper and (b) nt-Cu lines after 1000 cycles TCT. The regular copper lines were seriously damaged after the test.

The results proved that the nano-twinned structure could enhance the reliability of the copper lines in fan-out packaging.

B. TCT of RDLs on Si substrate

To understand the reliability of the copper RDLs in the next generation, series of Si substrate test vehicles with 5μ to 2μm line width were employed. The condition of TCT is similar to the previous section. In this study, the resistance of the copper lines was also measured by four probe station to quantify the failure level. To check the microstructure, the ion images which were took by FIB system were shown in figure 6. In this study, the regular copper was electroplated

by commercial additives which provided random grain size and orientation in the copper lines.

Figure 7 shows the resistance change after 250, 500 and 750 cycles TCT.

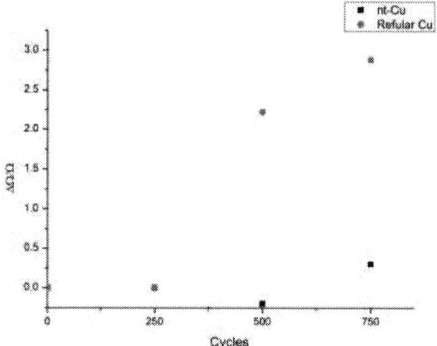

The failure level is much lower than fan-out packaging. No cracks were observed in both nt-Cu and regular copper lines, and the resistance increased very slow. This might be attributed to the structure of the test vehicles. In fan-out packaging, the copper lines must cross the boundary between Si die and the molding compound. Therefore, the complicated CTE mismatch among the materials would cause the failure of the Cu lines. However, in this section, the copper lines were deposited on the Si substrate without crossing other materials, so the thermal stress distribution is predicted to be more uniform.

C. The mechanical properties of the copper films

To clarify the failure mechanisms of the Cu lines, the tensile test was employed. In previous study, scientists had reported the tensile behavior of as-sputtered <111>-oriented nano-twinned copper [10-11]. However, as mentioned above, the annealing process is necessary for PI curing. Therefore, in this study, the copper films were tested after annealing at 250°C for 3h. Figure 8,9 shows the stress strain curve of the regular copper and nt-Cu.

Before annealing, the both of the regular copper and nt-Cu showed high strength. High strength is the well-known nature of nt-Cu. The high strength of regular copper is due to the random twinning and grain refining effect. After annealing, the strength of nt-Cu was still higher than 650MPa. In contrast, the tensile strength of regular copper

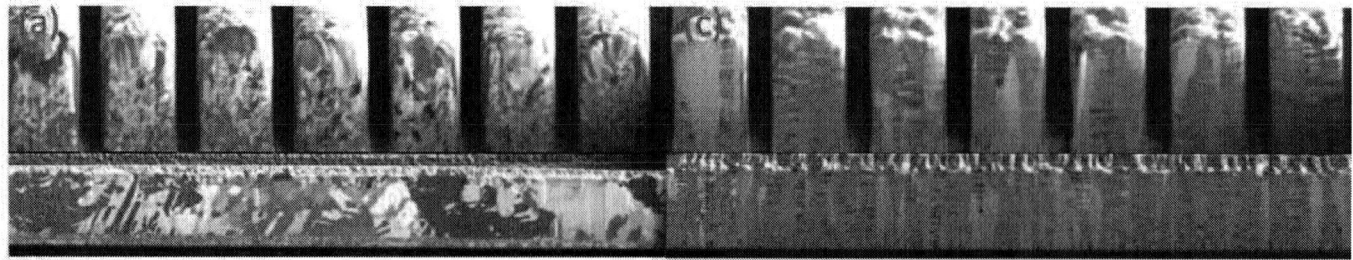

Figure 6. the initial microstructure of the (a), (b) regular copper lines, and the (c),(d) nt-Cu lines.

978-1-7281-1500-9/19 $31.00 © 2019 IEEE

dropped dramatically to about 250MPa.

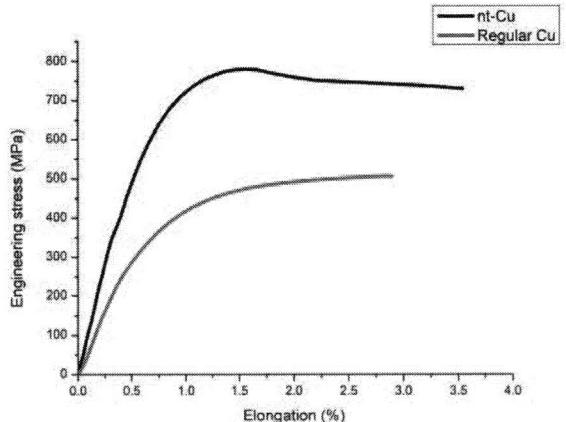

Figure 8. The stress strain curve of as-fabricated nt-Cu and regular copper films.

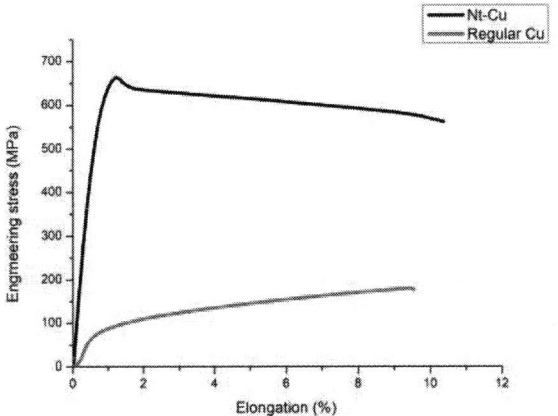

Figure 9. The stress strain curve of nt-Cu and regular copper films after annealing at 250°C for 3h.

From the evolution of the stress strain curve, it could be concluded that the nt-Cu is suitable for fan-out RDLs. The toughness of regular copper kept around 30MJ/m³ after annealing but it of nt-Cu increased from around 30MJ/m³ to 60MJ/m³. The increasing toughness might be attributed to the consumption of small grains at the bottom of the Cu foils. In 2014, Huang, Yi-Sa, et al. observed the grain growth behavior of <111>-oriented nano-twinned copper. In their study, the columnar nt-Cu grain is stable below 250°C, and it could consume the random copper grains by grain growth. However, the grain growth of the normal copper removes the strengthen factors so the yielding point dropped.

The higher toughness would make it harder to fracture the material. During the TCT, the accumulation of thermal stress and the propagation of fatigue cracks would contribute to cause the failure. The increasing of toughness

is proposed to be the best solution to strengthen the copper RDLs in fan-out packaging.

IV. CONCLUSIONS

In this study, the TCTs of copper lines on the fan-out test vehicles and Si substrate were employed. In fan-out packaging, serious fracture of Cu lines was observed in the group of regular Cu. In contrast, the nt-Cu lines were robust in the TCT. There were only 0.4% of the nt-Cu lines were observed with cracks. For the TCT of RDLs on Si substrate, the resistance behavior is still better than regular. Based on these results, the reliability of the nt-Cu was proved. The tensile test was performed to understand the failure of the copper lines during the TCT. In the tensile test, the yielding points of as-fabricated nt-Cu and regular copper were 750MPa and 450MPa respectively. After annealing at 250° C for 3h, the yield point nt-Cu film decreased slightly to 650MPa, but the elongation increased 2 times. In contrast, the elongation of the random copper also increased 2 times, but the yielding point drops to 100MPa. The toughness of the nt-Cu is 2 times larger than the regular copper after annealing. The results of tensile test could prove that the nt-Cu is suitable and helpful for fan-out packaging industry.

ACKNOWLEGEMENT

This work was financially supported by the "Center for the Semiconductor Technology Research" from The Featured Areas Research Center Program within the framework of the Higher Education Sprout Project by the Ministry of Education (MOE) in Taiwan. Also supported in part by the Ministry of Science and Technology, Taiwan, under Grant MOST-108-3017-F-009-003.

REFERENCES

[1] Liu, Christianto C., et al. "High-performance integrated fan-out wafer level packaging (InFO-WLP): Technology and system integration." 2012 International Electron Devices Meeting. IEEE, 2012.

[2] Yu, Douglas. "A new integration technology platform: integrated fan-out wafer-level-packaging for mobile applications." 2015 Symposium on VLSI Technology (VLSI Technology). IEEE, 2015.

[3] Tseng, Chien-Fu, et al. "InFO (wafer level integrated fan-out) technology." 2016 IEEE 66th Electronic Components and Technology Conference (ECTC). IEEE, 2016.

[4] Hsu, Feng-Cheng, et al. "3D Heterogeneous Integration with Multiple Stacking Fan-Out Package." 2018 IEEE 68th Electronic Components and Technology Conference (ECTC). IEEE, 2018.

[5] Yu, C. K., et al. "A unique failure mechanism induced by chip to board interaction on fan-out wafer level package." *Reliability Physics Symposium (IRPS), 2017 IEEE International*. IEEE, 2017.

[6] Lu, Lei, et al. "Ultrahigh strength and high electrical conductivity in copper." Science 304.5669 (2004): 422-426.

[7] Lu, L., et al. "Revealing the maximum strength in nanotwinned copper." Science 323.5914 (2009): 607-610.

[8] Hsiao, Hsiang-Yao, et al. "Unidirectional growth of microbumps on (111)-oriented and nanotwinned copper." Science 336.6084 (2012): 1007-1010.

[9] Anderoglu, O., et al. "Thermal stability of sputtered Cu films with nanoscale growth twins." Journal of Applied Physics 103.9 (2008): 094322.

[10] Hodge, A. M., Y. M. Wang, and T. W. Barbee Jr. "Mechanical deformation of high-purity sputter-deposited nano-twinned copper." Scripta materialia 59.2 (2008): 163-166.

[11] Hodge, A. M., et al. "Shear band formation and ductility in nanotwinned Cu." Scripta Materialia 65.11 (2011): 1006-1009.

[12] Zhang, X., and A. Misra. "Superior thermal stability of coherent twin boundaries in nanotwinned metals." Scripta Materialia 66.11 (2012): 860-865.

[13] Zhao, Yifu, et al. "Thermal stability of highly nanotwinned copper: The role of grain boundaries and texture." Journal of Materials Research 27.24 (2012): 3049-3057.

[14] Liu, Tao-Chi, et al. "Fabrication and characterization of (111)-oriented and nanotwinned Cu by DC electrodeposition." Crystal Growth & Design 12.10 (2012): 5012-5016.

[15] Xu, Luhua, et al. "Through-wafer electroplated copper interconnect with ultrafine grains and high density of nanotwins." Applied physics letters 90.3 (2007): 033111.

WLCSP Package and PCB Design for Board Level Reliability

Jason Chiu, K.C. Chang, Steven Hsu, Pei-Haw Tsao and M. J. Lii
Taiwan Semiconductor Manufacturing Company, Ltd.
6, Creation Rd. 2, Hsinchu Science Park, Hsinchu, Taiwan 300-77, R.O.C.
Email: sschiud@tsmc.com

Abstract—**WLCSP packaging is wildly use in portable electronic products such as phone, watch, and intelligent bracelet. The advantages of WLCSP package are parasitic inductance minimized, reduced package size, and enhanced thermal conduction characteristics. To enable these benefits regardless of the die's functional complexity, we adopted Cu with ELK (extreme Low-K) material as inter-metal-dielectric native to advanced silicon fabrication technology, and WLCSP packing with large die size, thus fulfilling requirements for high speed & low power consumption. Investigating WLCSP package board level reliability is essential and critical for product launch and reducing field return risk. Test vehicles were used with combinations in PBO2 opening, PCB thickness, and PCB metal gradient, to understand stress on ELK behavior and potential impact on board level reliability. Liquid-to-liquid thermal shock (LLTS) 75 cycles of -65^{0}C~150^{0}C will be a quick stress methodology which have ~1.9x acceleration factor compared with TCB stress and used for shortening experiment cycle time. A 6x6 mm^2 test vehicle was used for different WLCSP package PBO2 opening, PCB thickness and PCB metal design to assess board level reliability impact. LLTS 75cycles result showed larger PBO2 opening will get die edge ELK delamination defects. Higher PCB metal gradient board (more than 50%) & more thick (1mm) also got higher fail rate. For better WLCSP board level reliability structure, smaller WLCSP package PBO2 opening, thinner PCB thickness and uniform PCB metal distribution are recommended.**

Keywords-WLCSP; WLCSP PBO openingl; WLCSP PCB metal density; WLCSP package reliability

I. INRODUCTION

As WLCSP packaging is wildly use in portable electronic products such as phone, watch, and intelligent bracelet. The advantages of WLCSP package are parasitic inductance minimized, reduced package size, and enhanced thermal conduction characteristics. Not only the numbers of WLCSP device used were increased, the WLCSP package size was also obviously increased due to higher level of functional requirement. Larger than 5x5 mm^2 WLCSP size becomes unavoidable in fulfilling high-end product performance demands. Knowing the WLCSP structure, shown in Fig. 1, has few thin dielectric layers on chip surface as stress buffer between chip and printed-circuit-board (PCB). The large WLCSP chip's board level reliability will be very challenge due to high coefficient of thermal expansion (CTE) mismatch between the package and PCB. To improve the reliability performance of large chip WLCSP, best WLCSP package design and PCB board design was used as important practices in the industry. To manage the potential side effect and effectiveness of action taken for advanced WLCSP is quite important. Typical on-board defect modes of larger WLCSP package reported recently, are package edge ELK delamination defect, shown in Fig. 2. Therefore, this study was to understand PCB thickness/PCB metal distribution/WLCSP package PBO2 opening effect and potential impact on the board reliability performance of WLCSP using advanced wafer technology.

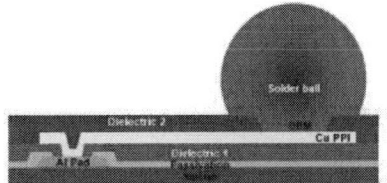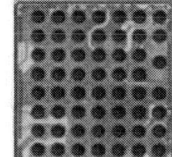

Figure 1. A WLCSP package schematic.

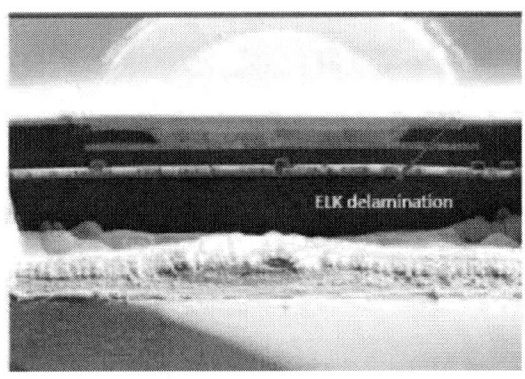

Figure 2a. WLCSP package edge ELK delamination post board level reliability test.

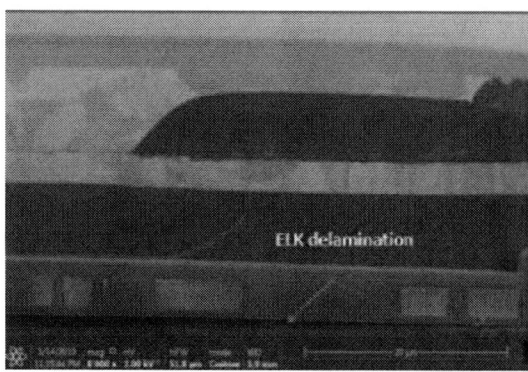

Figure 2b. WLCSP package edge ELK delamination post board level reliability test.

TABLE I. QUICK TEST OF LIQUID-TO-LIQUID THERMAL SHOCK CONDITION VS. TEMPERATURE CYCLE TEST

	Liquid-to-liquid Thermal Shock test	Temperature Cycle Test
Temperature range	$-65^{0}C \sim 150^{0}C$	$-50^{0}C \sim 125^{0}C$
Cycle time (minute)	10	30

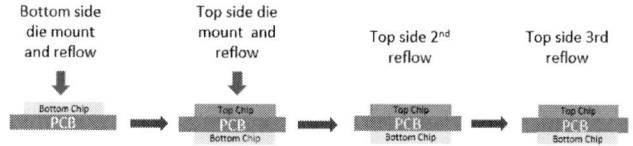

Figure 3. SMT die attachment sequence including top die (test die) bottom die (more stress) then perform reflow

II. EXPERIMENT

Test vehicle, with 6x6 mm² die size and consisting of full backend interconnection Cu metal layers, were processed by advanced wafer fabrication technology using ELK IMD material. The WLCSP packages were built in 270um die thickness using the standard production process flow of assembly house with 2 WLCASP package design PBO2 opening 130um, 190um. The PCB design were designed 3 thickness including 0.65mm, 1.0mm, and 1.2mm. The 0.65mm PCB was designed non-uniform Cu distribution design, as named "High metal gradient" PCB in this study. And the 1.0mm/1.2mm PCB were designed uniform Cu distribution, as named "Low metal gradient" PCB. In order to simulate real portable devices PCB behavior having multi chips on top and bottom side, the test samples were mounted on the top side and bottom side of JEDEC PCB without underfill which can have higher stress, as shown in Fig3. The samples were processed by standard SMT process with 3x260⁰C reflow in representing the worse SMT condition. There were 6 DOE legs for WLCSP package PBO2 opening, PCB thickness and PCB Cu distribution as shown in Table II. Each leg sample size is 45ea.

In order to shorten the experiment cycle time, liquid-to-liquid thermal shock test (LLTS) was used as quick stress method by its wider temperature range and short cycle time compared with in-situ temperature cycle test, Table I. Based on Coffin Manson equation and the Coffin Manson exponent of 3 for ductile material, 1.9 acceleration factor is expected to allow to obtain TCB200 equivalent results in one day. The on boarded samples were subjected to perform LLTS75x stress.

6x6mm² test vehicle mounted on the JEDEC board samples were used for quick stress method validation by benchmark with in-situ monitored board level TCB reliability stress test.

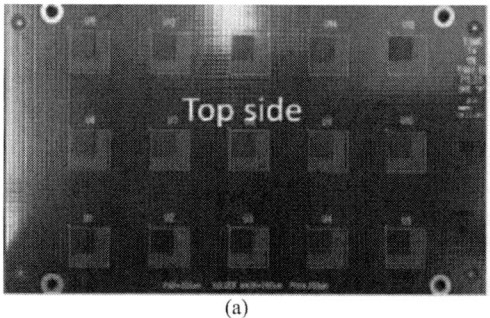

(a)

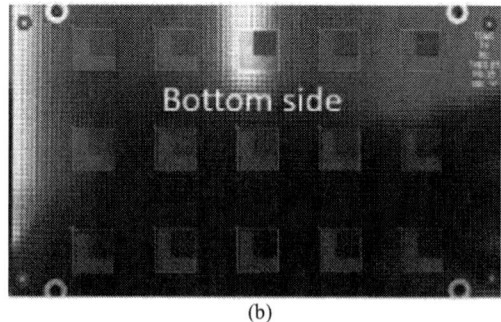

(b)

Figure 4. A picture of board mounted 6x6 mm2 test samples
(a) PCB top side (real test dies)
(b) PCB bottom side (stress enhancement dies)

TABLE II. DOE OF SPLIT LEGES INCLUDING PBO2 OPENING, PCB THICKNESS AND PCB DESIGN METAL GRADIENT (EACH LEG TEST 45EA SAMPLES)

Leg	PM2 CD opening(um)	PCB thickness(mm)	PCB metal gradient
1	130	0.65	High
2	130	1.0	Low
3	130	1.2	Low
4	190	0.65	High
5	190	1.0	Low
6	190	1.2	Low

III. EXPERIMENTAL RESULTS

This research intended to deliver a WLCSP board level reliability assessment methodology with minimum test chip and test board constraint, and also not being limited by using electrical open/short measurement as pass/fail criteria. So the pass/fail criterial will be judged by CSAM.

LLTS75x test results with different PCB thickness (1.0mm vs. 1.2mm) & PBO2 opening (130um vs. 190um) were shown in Fig5. Thicker PCB thickness will have higher fail rate than thinner PCB thickness. Larger PBO2 opening will have higher fail rate also. LLTS75x test results with different PCB thickness (0.65mm, 1.0mm & 1.2mm) & PCB metal gradient (high vs. low) were shown in Fig6 (a) & (b). In thinner PCB thickness condition less than 1.0mm, PCB design with higher metal gradient have higher fail rate. If PCB thickness is more than 1.0mm(ex:1.2mm), the major contribution will be PCB thickness than PCB metal gradient.

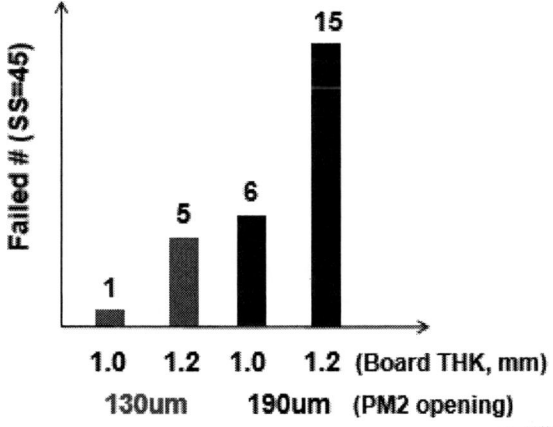

Figure 5. CSAM check result of 6x6mm² die split with different PCB thickness & PBO2 opening.

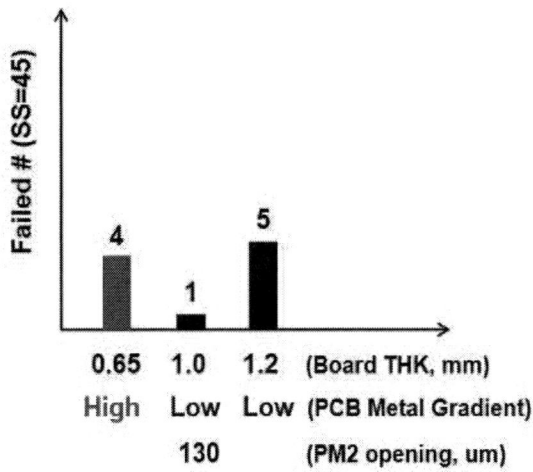

Figure 6a. CSAM check result of 6x6mm² die split with different PCB thickness & PCB metal gradient with 130um PBO2 opening design.

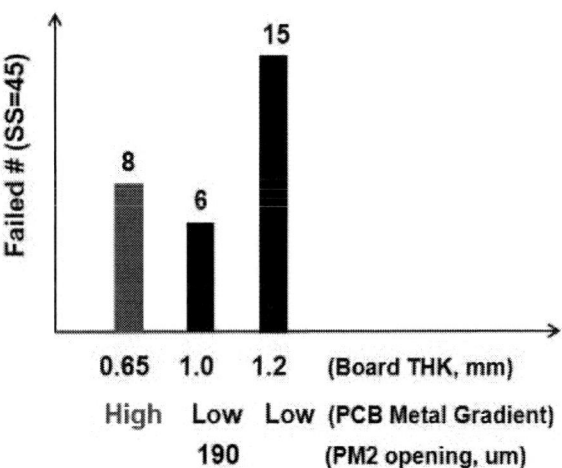

Figure 6b. CSAM check result of 6x6mm² die split with different PCB thickness & PCB metal gradient with 190um PBO2 opening design.

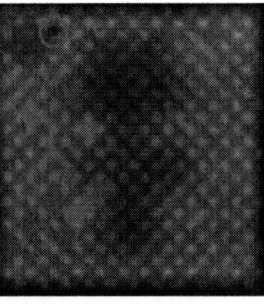

Figure 7a. CSAM fail images which showed ELK delamination defect.

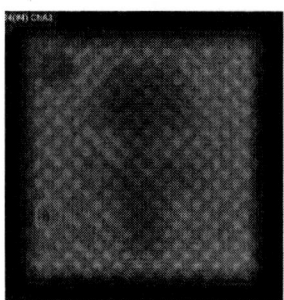

Figure 7b. CSAM fail images which showed ELK delamination defect.

IV. DISCUISSION

Small PBO2 opening showed better board level reliability performance in same PCB thickness condition with 6x6 mm² test vehicle. The major technical reason is PBO is a good stress buffer dielectric material. With more PBO material (small PBO2 opening), the stress from PCB board side (through solder ball) can be reduced by PBO thus less stress transferred into die interior.

Thinner PCB thickness with same PBO2 opening showed better board level reliability. The technical reason is thicker PCB thickness will have larger stiffness between die & PCB which induce more stress.

Higher PCB metal gradient (50%) showed higher ELK fail rate after board level reliability in PCB thickness under 1.0mm. The PCB metal gradient definition means in small area (4x4 mm²), the PCB metal uniformity (max-mini). In general PCB design rule, more uniform is better. Higher PCB metal gradient will generate loading effect when performing board level reliability.

To have better understanding on PBO2 opening stress behavior on WLCSP board level reliability performance, besides performing the experiment. Stress modeling with different PBO2 opening was shown in Table III. The modeling temperature loading is from SMT reflow 220⁰C(stress) to 25⁰C (stress calculated). The Modeling results showed smaller PBO2 opening have less ELK stress. ELK stress decrease 8% from PBO2 opening 190um to 130um. The 3-D modeling focused on ELK layer at die edge, Fig. 9. Stress modeling with different PCB thickness was shown in Table IV. If the PCB thickness is increase from 1.0mm to 1.3mm, the stress will increase 15%. 3-D stress simulation of thicker PCB and chip interaction was shown in Fig. 8 and Fig. 9. Stress modeling with 2 PCB Cu layout design was shown in Table V. PCB with 25% metal gradient got 11% ELK stress reduce than PCB with 50% metal gradient design. 3-D stress simulation of thicker PCB and chip interaction was shown in Fig. 10.

The simulation result showed good agreements with experiment data that small PBO2 opening of package, thinner PCB thickness and more uniform PCB metal distribution (means less metal gradient) can have best CPBI (chip package board interaction) for ELK layer.

TABLE III STRESS MODELING WITH DIFFERENT PBO2 OPENING

Leg	1	2	3	4	5	6	7
UBM (um)	220	220	220	220	220	220	220
PBO2 (um)	190	170	150	130	110	90	70
PBO2/UBM size ratio	0.86	0.72	0.68	0.59	0.50	0.41	0.32
RDL pad (um)	240	240	240	240	240	240	240
PBO Stress (RDL edge)	1.00	0.99	0.98	0.97	0.96	0.96	0.96
PBO Stress (UBM edge)	0.75	0.71	0.71	0.71	0.71	0.71	0.71
ELK Stress	1.00	0.98	0.95	0.92	0.90	0.88	0.87

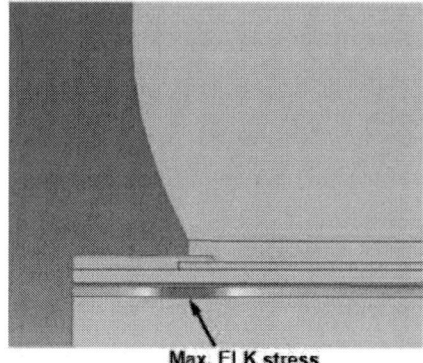

Max. ELK stress

Figure 7. 3-D package model for stress simulation.

TABLE IV STRESS MODELING WITH DIFFERENT PCB THICKNESS

Simulation	Condition-1	Condition-2
Die size (mm²)	7.2x7.2	7.2x7.2
Die thickness (um)	325	325
Ball pitch (um)	400	400
Ball size (um)	250	250
Ball material	SAC405	SAC405
PCB thickness (mm)	1.0	1.3
Normalized die ELK stress	1.00	1.15

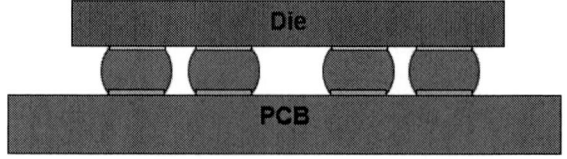

Figure 8. Thicker PCB thickness will have larger stiffness between die & PCB which induce more stress.

978-1-7281-1500-9/19 $31.00 © 2019 IEEE   766

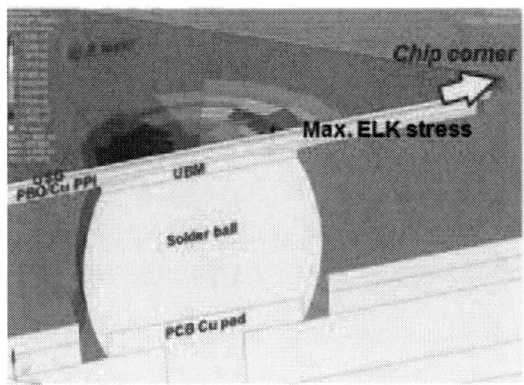

Figure 9. 3-D package model for Thicker PCB

TABLE V Stress Modeling With Different Pcb Metal Gradient Design

Simulation	Condition-1	Condition-2
Die size (mm²)	7.2x7.2	7.2x7.2
Die thickness (um)	325	325
Ball pitch (um)	400	400
Ball size (um)	250	250
Ball material	SAC405	SAC405
PCB thickness (mm)	1.0	1.0
PCB metal gradient (%)	25%	50%
Normalized die ELK stress	1.00	1.11

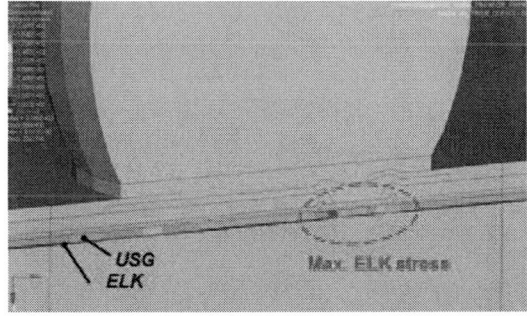

Figure 10. 3-D package model for High PCB Metal Gradient

V. Conclusion

Since the trend of WLCSP package towards large chip size with advanced wafer process using ELK as inter metal dielectric material, the assessment of WLCSP board level reliability is very critical and need to be carefully investigated. Two major key factors, package design and PCB design, impacting WLCSP board level reliability performance were studied.

For WLCSP with ELK die, the package design of PBO2 opening was shown having significant impact on board level reliability due to inducing potential ELK delamination defect.

Larger PBO2 opening design will have higher ELK fail rate since the less PBO material under UBM which will have less stress buffer. The minimum PBO2 opening need to check the process capability and current requirement. For portable devices, thinner PCB thickness will be a toward trend. High PCB thickness also have higher ELK fail since the mismatch between die and PCB. For PCB design, more uniform metal design is recommended. Since the stress loading effect will occur when performing board level reliability.

A quick WLCSP board level reliability test methodology, SMT 3x 260°C plus liquid-to-liquid thermal shock test (LLTS) 75 cycles with CSAM check, was demonstrated as an effective way for reliability assessment with less constrain on test vehicle and test board and much shorter time. Then we can base on the result to enhance the WLCSP package or PCB design. The results of this study provide a reference for WLCSP package board level reliability enhancement.

Acknowledgment

The authors would like to thank the support of TSMC CPI (chip package interaction) team and Tsmc QR team for test vehicle preparation, reliability testing, defect analysis and PCB design.

References

[1] Jeffrey C. B. Lee et al., "The solder joint characterization in green WLCSP," Proc. IEEE Electronic Components and Technology Conference, pp. 1914-1920, May 2004.

[2] Kejun Zeng and Amit Nangia, "Thermal cycling reliability of SnAgCu solder joints in WLCSP," Proc. 2014 IEEE 16th Electronics Packaging Technology Conference, pp. 503-511, 2014.

[3] Peng Sun et al. "Package & board level reliability study of 0.35mm fine pitch wafer level package," Proc. 2017 18th International Conference on Electronic Packaging Technology, pp. 322-326, 2017.

[4] Tak-Sang Yeung et al., "Material characterization of a novel lead-free solder material – SACQ," Proc. IEEE Electronic Components and Technology Conference, pp. 518-522, May 2014.

[5] Wei Lin et al., "SACQ solder board level reliability evaluation and life prediction model for wafer level packages," Proc. IEEE Electronic Components and Technology Conference, pp. 1058-1064, May 2017.

[6] Tung Ching Lui, "Lifetime prediction of viscoplastic lead-free solder – a new solder material, SACQ," 2017 International Workshop on Integrated Power Packaging (IWIPP), 2017

[7] P.H. Tsao, "Board Level Reliability Enhancement of WLCSP with Large Chip Size," Proc. 2018 IEEE 68th Electric Components and Technology Conference

Assessing the Reliability of Highly Stretchable Interconnects for Flexible Hybrid Electronics

R.S. Sivasubramony, A.V. Zachaiah, M. Alhendi,
M. Yadav, P. Borgesen, M.D. Poliks
Dept. of Systems Science & Industrial Engineering
Binghamton University, Binghamton, NY - 13905
Email: rsivasu1@binghamton.edu

N.C. Stoffel, D.M. Shaddock, L. Yin
General Electric Global Research
1 Research Circle,
Niskayuna, NY - 12309

Abstract—**Highly stretchable, bio-compatible interconnects are of particular interest for medical and military applications as Wearable Performance Monitors (WPMs) and sensors. Screen printed trace interconnects on highly compliant Thermoplastic Polyurethanes (TPU) provides a low cost, viable option. But these stretchability has high ramification on the reliability aspects of WPM construction.**

In this paper, we perform the reliability testing of two screen printed inks under repeated mechanical loads 'as-printed' and after exposure to temperature and humidity.

Keywords-**Reliability; Stretchable Interconnects; Silver Trace; Screen Printed; Flexible Hybrid Electronics**

I. INTRODUCTION

A major concern in many flexible electronics applications is the ability of interconnects to both maintain electrical connection during major stretching and survive repeated deformation without degradation of the electrical conductivity. A low-cost technique aimed at addressing this challenge relies on functionalizing the surface of a Thermoplastic polyurethane (TPU) substrate by screen-printing (SP) a stretchable conductive silver ink designed to offer strong stretch recovery [1]. The resulting screen-printed traces are silver particles (flakes) in a binding polymer matrix that provides adhesion to the TPU substrate once sintered. TPU substrates offer particular comfort for wearable applications with have proven bio-compatibility [2], [3]. They are easily stretched, but that also means that they offer limited protection against strain localization in traces on them [4].

This paper reports on the behavior of two conductive silver inks on TPU substrates under various loading and environmental conditions. The commercially available conductive inks are both stretchable-washable silver inks from the same supplier – with one claiming the best stretch recovery and the other the best washability. We find that the TPU can be stretched to high strains without loss of electrical connectivity, although the trace electrical resistance increases extremely quickly with increasing strain. The two inks are found to respond very differently to stretching at different strain rates, but neither was negatively affected

by exposure to elevated temperature and high levels of humidity.

II. EXPERIMENTS AND METHODOLOGY

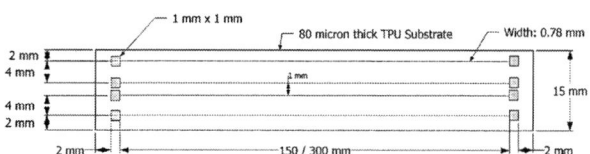

Figure 1: Screen printed Ag traces on Thermoplastic Polyurethane (TPU) substrate.

For the experiments, commercially available screen printable inks (two inks from the same supplier) and TPU substrates were used. The test vehicle had a TPU substrate that is 3mil (75μm). Two screen printable inks were stretchable and washable - one with better stretchablity and the other with better washability. The ink with better washability had a higher trace thickness and initial trace resistance than the other. Traces of both the inks were about 750μm wide. Traces were made at two length - 150mm and 300mm. Figure 1 shows the test vehicle that is 300mm long with 4 traces in each test coupon, but only either one of the middle traces were used while testing.

Figure 2 shows the mechanical testing configuration with *in situ* electrical measurements using 4-wire measurements done using Keithley 2100 Digital Multimeter. Instron 3344 Single Column Load Frame mechanical tester was used for mechanical testing and accompanying Bluehill software was used for recording load and extension responses during testing.

Strain-based fatigue cycling was performed on the test vehicles. While cycling the loads on the traces are not directly measured, whereas the strain on both the trace and the substrate is considered same. Figure 3a shows the from the top optical micrograph image of trace and substrate revealing that the TPU substrate is slightly porous. The screen printed trace is mixture of silver flakes in a

Figure 2: Experimental set-up: test vehicle mounted between two pressure grips on Instron mechanical tester with *in situ* 4-wire electrical measurements through pogo-pin probes.

(a)

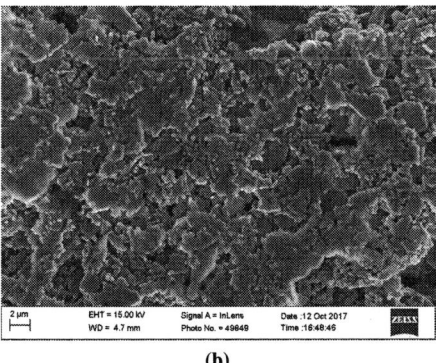

(b)

Figure 3: (a) Optical micrograph of the trace from the top shows that the TPU substrate is micro-porous; and (b) SEM micrograph of the trace at higher magnification shows boundary-fused Ag flakes.

polymer matrix. Figure 3b shows Ag flakes on the top surface of the screen printed trace. All the annealing/drying (both air and N$_2$) was done in conventional convection

ovens. Humidity/Temperature exposures were carried out in a Tenney Benchmaster Test Chamber.

III. RESULTS AND DISCUSSION

A. Mechanical Behavior

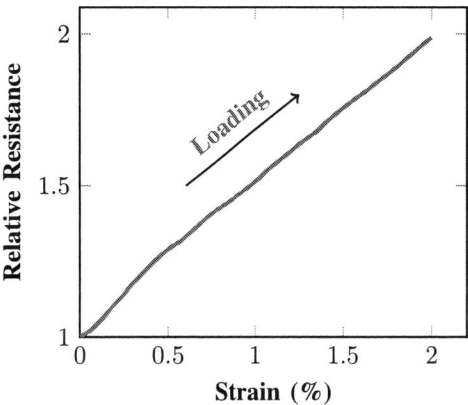

Figure 4: Relative Resistance *vs.* engineering strain up to 2% of ink 1 SP trace stretched at a rate of 0.0006s^{-1}.

1) Behavior of SP Traces on TPU: TPU substrates are usually chosen in wearable applications for reasons of comfort, but the lack of significant resistance to stretching means that conductive traces on them must be highly stretchable as well. The two inks considered in the present study were both designed to survive stretching to multiple times the original trace length and return to a reasonably low resistance upon unloading. That does, however, not mean that the resistance remains low **during** stretching. Stretching an ink 1 trace to 20%, for example, relatively quickly did lead to a 36× increase in its resistance. We are, however, focused on the behavior at lower levels of deformation.

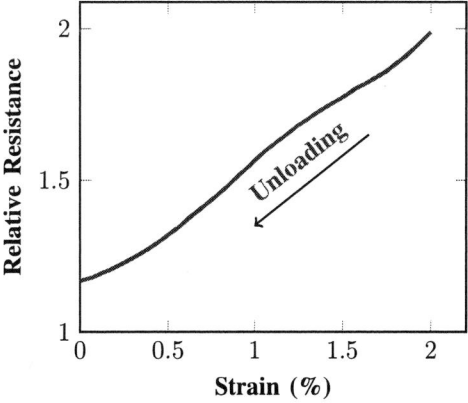

Figure 5: Relative resistance of ink 1 trace during unloading from 2% to nominal 0% at a rate of 0.0006s^{-1}.

Figure 4 shows the relative increase in resistance of a trace of ink 1, the one claiming the best stretch recovery,

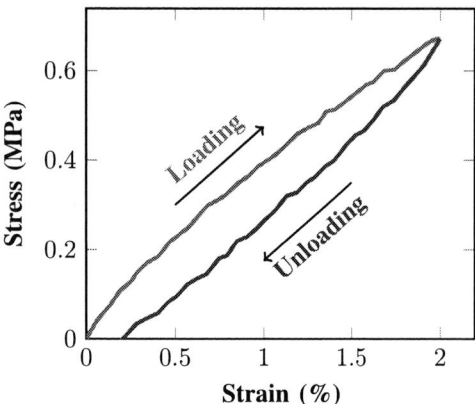

Figure 6: TPU substrate stress *vs.* strain during loading to 2% and return to nominal 0% strain.

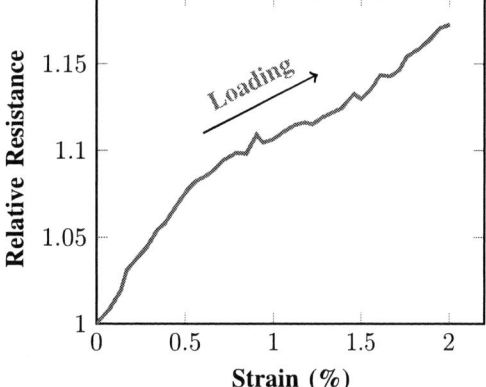

Figure 7: Relative resistance *vs.* engineering strain up to 2% of ink 2 SP trace stretched at a rate of $0.0006s^{-1}$.

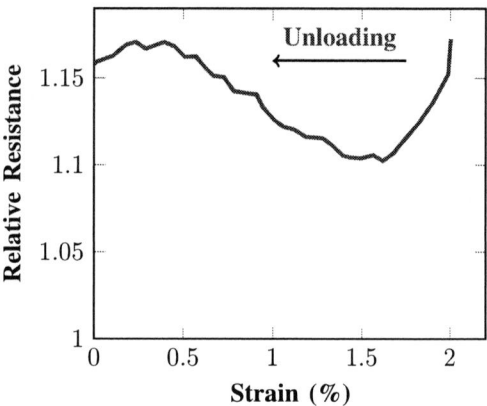

Figure 8: Relative resistance of ink 2 trace during unloading from peak 2% strain to nominal 0% at a rate of $0.0006s^{-1}$.

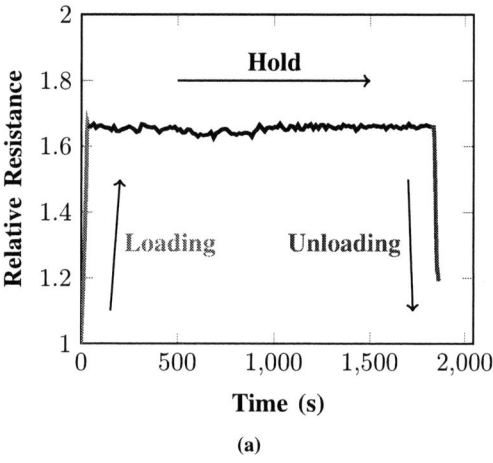

(a)

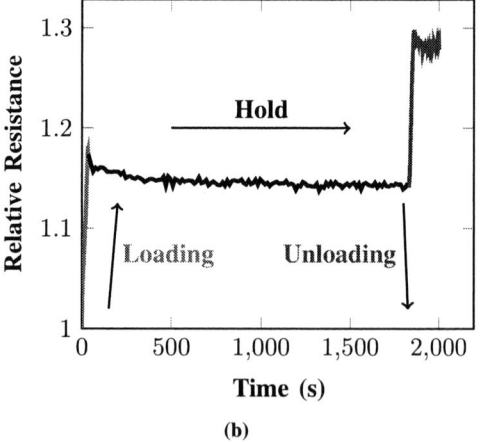

(b)

Figure 9: Trace resistance *vs.* time while the substrate is strained to 2%, held at 2% for 30mins and unloaded to 0% nominal strain: (a) Ink 1 trace; (b) Ink 2 trace.

when stretched by only 2% at a strain rate of 0.0006 s^{-1}. The resistance of other printable inks commonly used on substrates such as Kapton or PET typically increases by 20% or less, and the resistance of an electroplated Cu trace increased by only 4% [5]. The resistance of the ink 1 trace, however, is seen to double!

This does not mean that the trace is damaged very much. Figure 5 shows the resistance to recover to 'only' 17% above the initial resistance upon unloading. Even this is not all damage. Figure 6 shows the stress vs strain during the cycle. The stress is seen to return to zero at a nominal strain of 0.17%, reflecting a remaining viscoelastic strain in the TPU. This means that at a nominal strain of zero the **trace** is still (temporarily) strained to 0.17%. Left unloaded for a few hours the viscoelastic strain relaxes and the initial resistance is largely recovered.

Ink 2 is also referred to as stretchable, although with less recovery. In fact, stretching a trace of it to 2% strain, the resistance only increases by 17% (Figure 7) The resistance vs strain curve is more irregular than for ink 1 (Figure 4), reflecting changes in the material as it is being stretched.

These changes are much more obvious during subsequent unloading to nominal 0% strain (Figure 8). Here the resistance first drops, but then it increases again during further unloading, and it ends up close to the resistance at the peak 2% strain.

This difference is also reflected in the behaviors during and after a 30 minute hold at a fixed strain of 2%. Figure 9a shows the resistance increase of an ink 1 trace to be less than the factor of two in Figure 4 because the strain rate was lower (see below), but the main point is that the resistance changes very little during the hold and upon unloading the resistance drops to the same level as in Figure 5. When stretched to a much higher strain (20%) the resistance of ink 1 traces did drop significantly during a 30 minute hold, undoubtedly reflecting rearrangement of the Ag particles and the organic matrix, but there was no sign of this at a strain of 2%. In contrast, an ink 2 trace held at 2% clearly changed. Figure 9b shows how the resistance drops by about 3% during the hold, but much more significantly the resistance now **increases** by about 15% upon unloading. This behavior was reproduced for different strains (including at 20%) and hold times.

2) Effect of Fatigue Cycling: In-situ monitoring of the resistance of an ink 1 trace during repeated loading and unloading with a strain amplitude of 2% revealed a systematic increase from cycle to cycle (Figure 10a). Some of the initial increase in the resistance at zero load is a result of a cycling induced increase in the viscoelastic strain on the TPU (Figure 11), but this is seen to level off after about 30 cycles. Anyway the resistance at the peak of each cycle, which is not affected by the viscoelastic strain, shows the progression of fatigue damage in the trace (Figure 10b).

Cycling of ink 2 traces also led to a systematic increase (Figure 12a), but in this case the variation in resistance with strain within a given cycle varies much less, as also reflected in Figure 8. Anyway, the resistance at the peak here also increases systematically with cycles (Figure 12b).

After 200 cycles the resistance at the peak strain is higher for ink 1 than for ink 2, but this is because of the much larger temporary effect of stretching. A better comparison of cycling induced damage is probably the evolution of the peak resistance divided by the peak resistance in the first cycle. Figure 9. shows the relative peak resistance to increase at similar rates for the first 150 cycles but after that it increases faster for ink 2 than for ink 1, *i.e.* overall ink 2 traces appear to **damage** faster in cycling to a 2% strain amplitude at a strain rate of $0.0006s^{-1}$. In fact, the difference is much stronger than suggested by Figure 9. Figure 14 shows the recovery of the resistances as the samples are allowed to recover at zero load over a period of 3 hours. After relaxation of both the TPU and the trace the resistance of the ink 1 trace ends up almost $3\times$ lower than the resistance of the ink 2 trace.

3) Effect of Strain Rate: Not surprisingly, considering the very different behavior of the two inks so far, they also respond very differently to changes in strain rate. Figure 15a and 15b shows effect of lowering strain rate by a factor

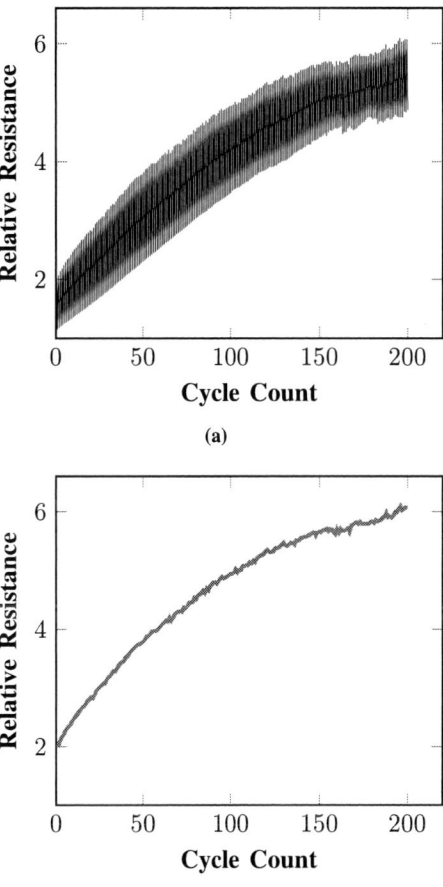

Figure 10: (a) Relative resistance *vs.* time of ink 1 trace during cycling to strain amplitude of 2%; (b) Relative resistance at peak (2%) strain *vs.* number of cycles for the same sample.

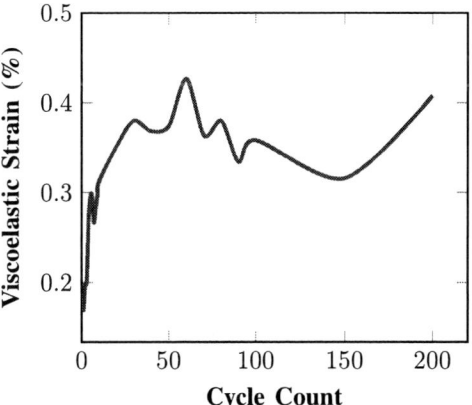

Figure 11: Residual viscoelastic strains in TPU substrate after each fatigue cycle with 2% strain amplitude.

of 6, from $0.0006s^{-1}$ to $0.0001s^{-1}$, on the first cycle.

In the case of ink 1 slower straining leads to less of an increase in resistance, and in spite of the viscoelastic

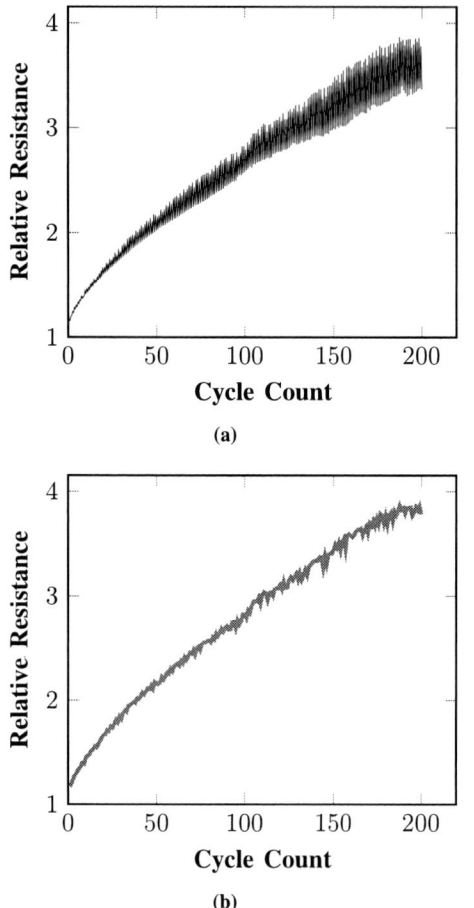

(a)

(b)

Figure 12: (a) Relative resistance *vs.* time of ink 2 trace with during cycling to strain amplitude of 2%; (b) Relative resistance at the peak strain (2%) *vs.* number of cycles for the same sample.

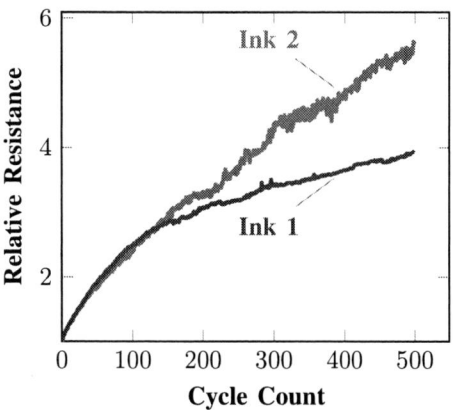

Figure 13: Resistance at the peak 2% strain relative to the first cycle peak for ink 1 and ink 2 *vs.* number of cycles to strain amplitude of 2%.

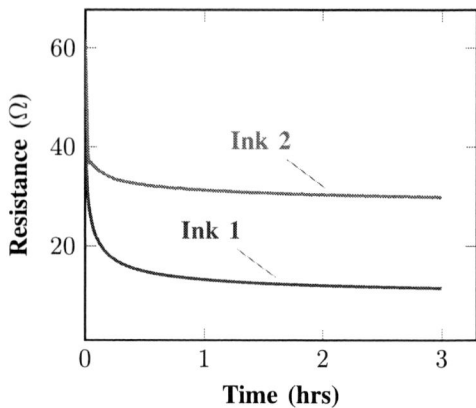

Figure 14: Trace resistance *vs.* time during 3hrs dwells at nominal 0% strain after 500 fatigue cycles at 2% strain amplitude for ink 1 and ink 2 trace.

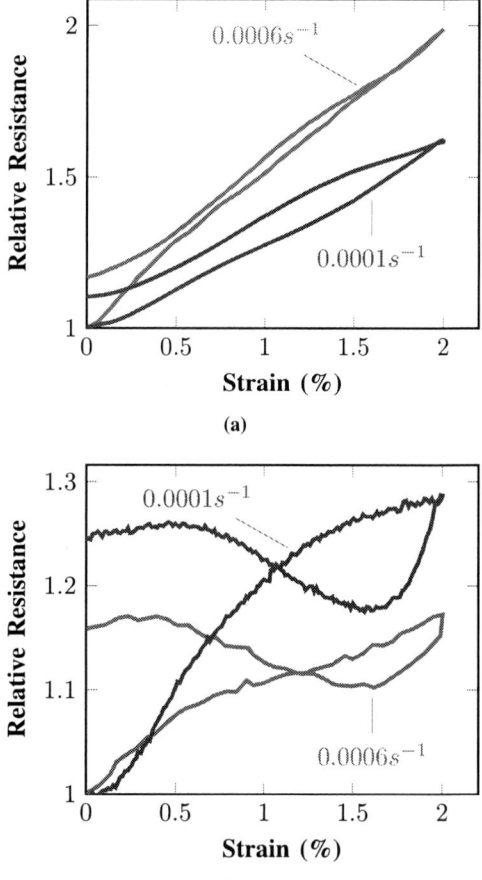

(a)

(b)

Figure 15: Trace resistance *vs.* strain during first cycle to 2% and back to nominal 0% at strain rates of $0.0006s^{-1}$ and $0.0001s^{-1}$: (a) Ink 1 trace, and (b) Ink 2 trace.

strain on the TPU ending up slightly higher the resistance after unloading is lower as well (Figure 15a). In contrast, consistent with the indications of changes in the material as it is stretched (Figure 7), slower straining of ink 2 allows for more time for this, leading to a higher resistance at 2%

978-1-7281-1500-9/19 $31.00 © 2019 IEEE

strain (Figure 15b). Also, like in Figure 8 unloading does not lead to a drop in resistance. In fact, the non-trivial shape of the unloading curve is remarkably similar for both strain rates.

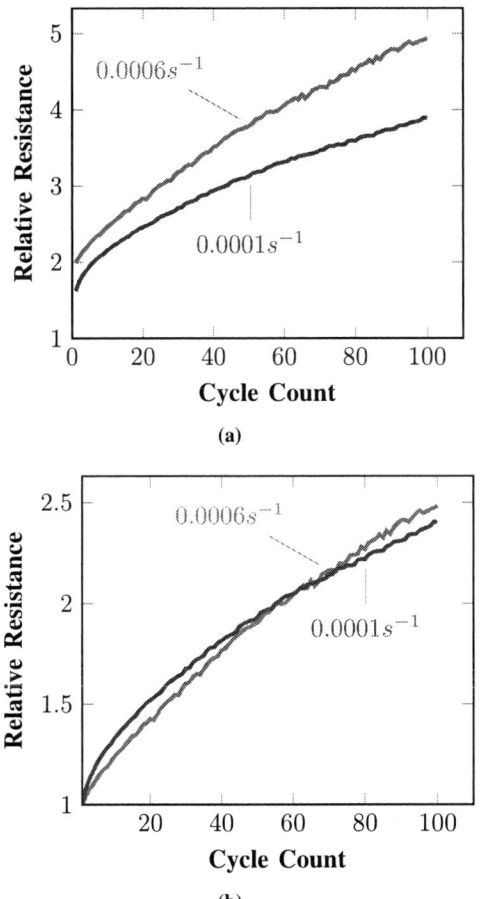

Figure 16: Resistance of ink 1 trace at peak strain of 2% for two strain rates of $0.0006\mathrm{s}^{-1}$ and $0.0001\mathrm{s}^{-1}$: (a) relative to the initial resistance before stretching; and (b) relative to the first peak resistance.

Accordingly, the effects of the strain rate on the fatigue performances of the two inks are also significantly different. Figure 16a shows the lower peak resistance of ink 1 for the lower strain rate to be followed by a slower increase in cycling as well. Normalizing to the resistance at 2% in the first cycle the relative increase in subsequent cycling is seen to be quite insensitive to the strain rate (Figure 16b). The *higher* peak resistance of ink 2 for the lower strain rate is also followed by a slower increase in cycling (Figure 17a). Normalizing to the resistance in the first cycle this difference becomes even clearer (Figure 17b).

4) Effect of Strain Amplitude: The resistance of ink 1 continues to increase close to linear when stretched to 10%, but that of ink 2 increases faster at higher strains (Figure 18), reaching 3.5× as opposed to a 17% increase at a strain

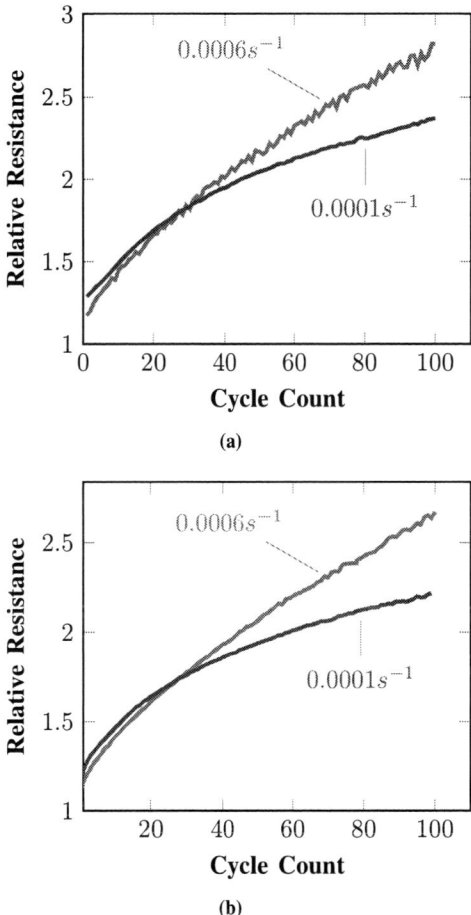

Figure 17: Resistance of ink 2 trace at peak strain of 2% at two strain rates of $0.0006\mathrm{s}^{-1}$ and $0.0001\mathrm{s}^{-1}$: (a) relative to the initial resistance before stretching; and (b) relative to the first peak resistance.

of 2%. Notably, unlike after stretching to 2% (Figure 8) unloading from a 10% strain leads to a recovery to 74% above the initial resistance.

The remaining viscoelastic strain does not only vary with number of cycles (Figure 11), it also increases strongly with amplitude so the evolution of the resistance at a nominal strain of zero (really at the viscoelastic strain) leads to an overestimate of the effect of amplitude on damage [5]. Noting that the viscoelastic strain does not exceed 2% even after 100 cycles to 10% we therefore monitor the resistance at 2% strain in each cycle instead. Comparing ink 1 traces in 100 cycles with an amplitude of 10% to traces in 200 cycles with an amplitude of 2% we find the higher amplitude to damage the traces 3-4× faster. The sensitivity of ink 2 traces to amplitude is much stronger.

Cycling an ink 2 trace at a strain amplitude of 10%, the resistance was seen to have increased by 40-50 times after 50 cycles and more than 100 times its initial resistance after 100 cycles (Figure 19). Comparing the evolution of the resistance

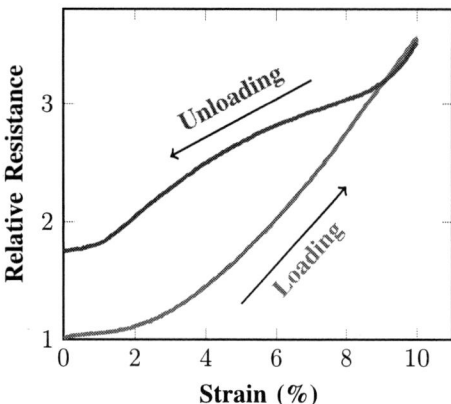

Figure 18: Relative resistance of ink 2 trace *vs.* strain in first cycle to 10% strain at a rate of $0.0006s^{-1}$.

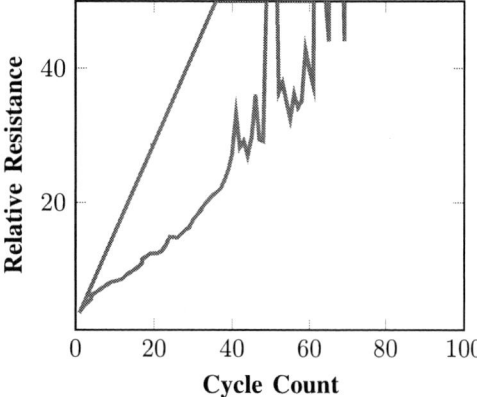

Figure 19: Relative resistance of ink 2 trace at peak *vs.* number of cycles to 10% strain at a rate of $0.0006s^{-1}$.

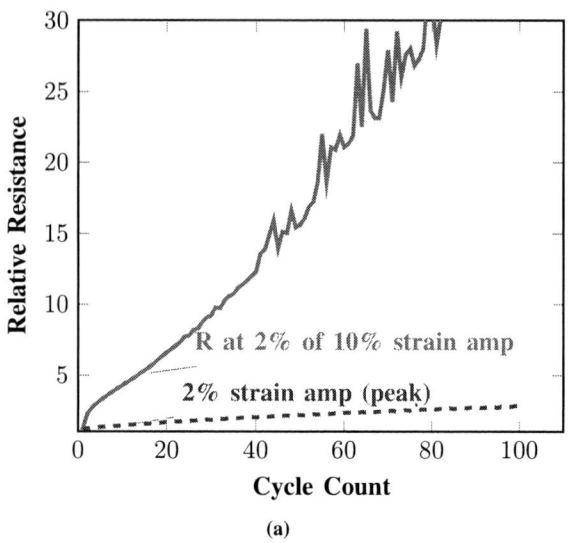

(a)

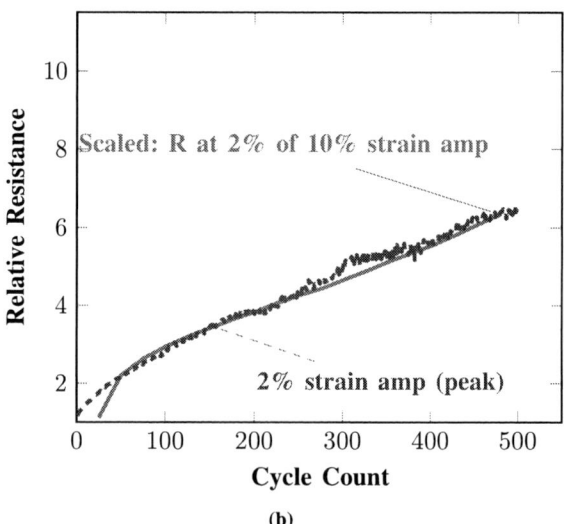

(b)

Figure 20: (a) Relative resistance of ink 2 measured at 2% *vs.* number of cycles to strains of 10% and 2% strain amplitude at a rate of $0.0006s^{-1}$; and (b) 10% strain amplitude curve scaled onto 2% strain amplitude curve.

at 2% strain to that for cycling with an amplitude of 2% (Figure 20a) we notice a large difference indeed. In fact, multiplying the number of 10% cycles by a factor of 25 we see that cycling to 10% leads to 25× faster damage than cycling to 2% (Figure 20b).

B. Ambient Exposure

Although ink 2 claims best 'washability', exposure of **unencapsulated** traces to humidity, as well as to thermal excursions, had very similar effects on the resistance and relative performance in subsequent tensile cycling for the two inks.

Exposure of either ink to 85% humidity at 75°C for 86 hours first of all led to a 65-75% reduction in resistance. This was not simply an annealing effect. Figure 21 shows annealing of ink 2 for 80 hours in air at 75°C to only reduce the resistance by 35%. Similar reductions were found after exposure to 85% humidity for 3.5 days at 75°C and for 7 days at 85°C, but it is not clear that the effect would be active at lower temperatures and humidity levels. We suggest that the humidity causes plasticization and swelling

of the organic matrix in the trace, forcing the Ag particles into better contact. This also had major effects on the complex behavior of ink 2 under subsequent deformation. Compared to Figure 15b, Figure 22a shows major changes in the relationship between resistance and strain in a first cycle to 2% strain. These changes could not be reversed by drying out in nitrogen for 2 days at 75°C after the humidity exposure (Figure 22b), further supporting our suggestion.

Whatever the reason, we finally note that the exposure to humidity did not have a negative effect on the propagation of damage in cycling. Figure 23a shows resistances, measured at 0.5% strain, to remain low in 100 cycles with an amplitude of 2%, and Figure 23b shows the relative resistance

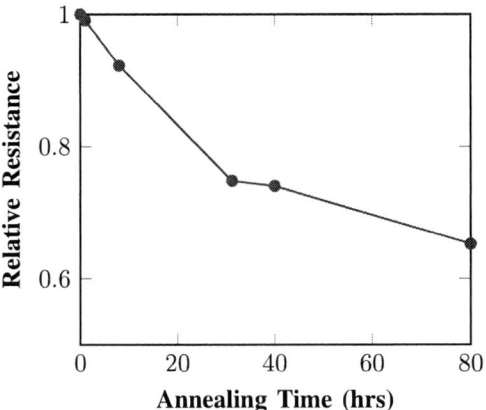

Figure 21: Resistance of ink 2 trace *vs.* time when annealed in air at 75°C.

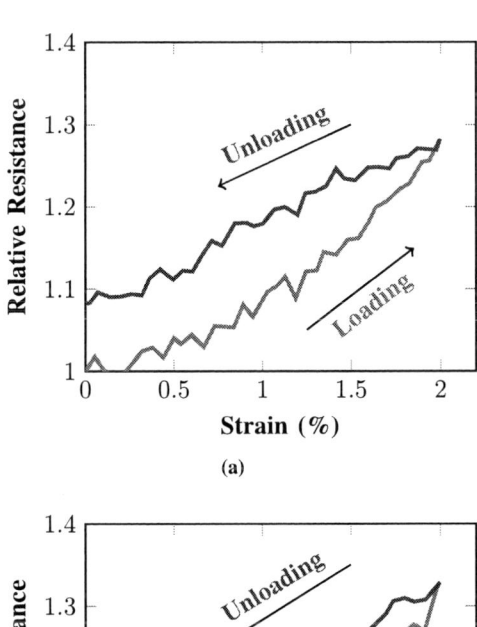

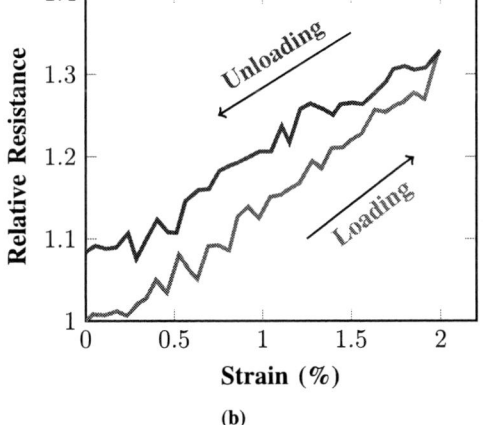

Figure 22: Relative resistance of ink 2 *vs.* strain in first cycle to 2% after exposure to (a) humidity at 75°C/85%RH for 8.5days; and (b) humidity at 75°C/85%RH for 8.5days followed by drying in N₂ at 75°C for 2 days.

(normalized to first cycle peak) to increase at very similar

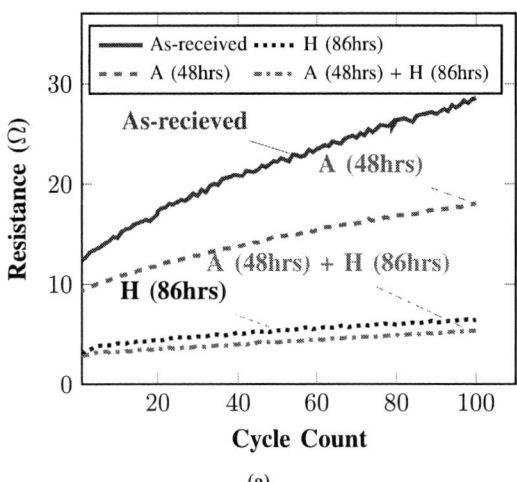

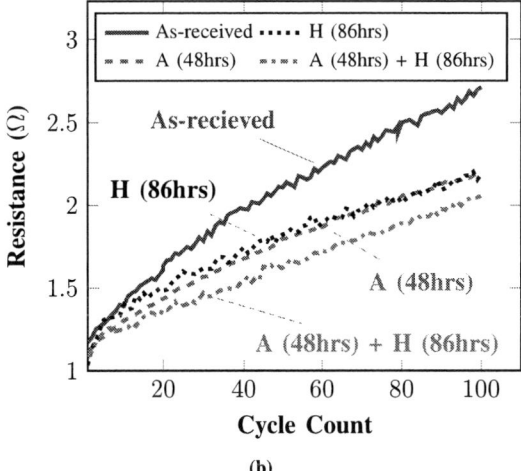

Figure 23: Resistance of ink 2 trace measured at 0.5% *vs.* number of cycles to a strain of 2% 'as-received' and after exposure to humidity (H), annealing (A) or both (A + H) (a) relative to the initial resistance before stretching; and (b) relative to resistance in the first cycle peak.

rates as for initial unexposed traces.

IV. SUMMARY AND CONCLUSION

TPU offers particular comfort for wearable applications, allowing for extreme stretching if necessary, but this has consequences for the robustness and reliability of conductive traces on it. TPU substrates offer very limited protection against strain localization in these traces when deformed. Screen printed traces of two inks from the same supplier formulated for use with such substrates, both offering very strong stretch recovery, were considered in terms of response to repeated stretching and effects of temperature and humidity.

The behavior of the two inks was very different. Stretching to only 2%, at a moderate rate, led to doubling of the resistance of one but most of that recovered upon unloading.

978-1-7281-1500-9/19 $31.00 © 2019 IEEE

The resistance of the other ink increased by less than 20% but did not recover, so that they both ended up at the same elevated resistance right after unloading. Dependencies on strain rate were in opposite directions, as were effects of time at a given strain. Repeated stretching led to rapid degradation, but fatigue resistances and dependencies on strain amplitude were quite different. Exposure to elevated temperature, thermal excursions and humidity had very similar effects, leading to reductions in electrical resistance both before and after subsequent fatigue cycling.

In conclusion, comparisons between alternative inks in simple tests may be greatly misleading. The potential for very different effects of loading parameters means that realistic use conditions may need to be carefully accounted for.

ACKNOWLEDGMENT

This material is based on research sponsored by Air Force Research Laboratory under agreement number FA8650-15-2-5401. The U.S. Government is authorized to reproduce and distribute reprints for Governmental purposes notwithstanding any copyright notation thereon. The views and conclusions contained herein are those of the authors and should not be interpreted as necessarily representing the official policies or endorsements, either expressed or implied, of Air Force Research Laboratory or the U.S. Government.

REFERENCES

[1] A. Mohammed and M. Pecht, "A stretchable and screen-printable conductive ink for stretchable electronics," *Applied Physics Letters*, vol. 109, no. 18, p. 184101, 2016.

[2] B. Karaguzel, C. Merritt, T. Kang, J. Wilson, H. Nagle, E. Grant, and B. Pourdeyhimi, "Flexible, durable printed electrical circuits," *The Journal of The Textile Institute*, vol. 100, no. 1, pp. 1–9, 2009.

[3] J. Suikkola, T. Björninen, M. Mosallaei, T. Kankkunen, P. Iso-Ketola, L. Ukkonen, J. Vanhala, and M. Mäntysalo, "Screen-printing fabrication and characterization of stretchable electronics," *Scientific reports*, vol. 6, p. 25784, 2016.

[4] S. Chen, X. Shan, W. Tang, B. Mohaime, M. Goh, Z. Zhong, and J. Wei, "Mechanical and electrical characteristics of screen printed stretchable circuits on thermoplastic polyurethane," in *2017 IEEE 19th Electronics Packaging Technology Conference (EPTC)*. IEEE, 2017, pp. 1–4.

[5] R. S. Sivasubramony, N. Adams, M. Alhendi, G. S. Khinda, M. Z. Kokash, J. P. Lombardi, A. Raj, S. Thekkut, D. L. Weerawarne, M. Yadav *et al.*, "Isothermal fatigue of interconnections in flexible hybrid electronics based human performance monitors," in *2018 IEEE 68th Electronic Components and Technology Conference (ECTC)*. IEEE, 2018, pp. 896–903.

The how and why of biased humidity tests with copper wire

A. Mavinkurve, R.T.H. Rongen, L. Goumans, M-L Farrugia, E. van Olst, Orla O'Halloran and M. van Soestbergen

NXP Semiconductors

Gerstweg 2, 6534 AE, Nijmegen, Netherlands
amar.mavinkurve@nxp.com

Abstract- **HAST at 110 or 130 °C is often the biased humidity test of choice when qualifying copper wire, due to its short execution time. However, failure mechanisms, like corrosion, can be over-accelerated during HAST compared to a milder test like THB at 85 °C. This is especially the case under certain conditions (e.g. high bias or electrical fields) due to changes in material behavior during the HAST test. In this paper, an approach to determine an acceleration model is proposed based on an observed incubation time before the onset of corrosion. Although HAST at 130°C combined with high electrical field is an extremely stressful test condition, delaying incubation until well beyond the JESD47 requirement of 96h (or AEC Q006 requirement of 192 h for copper wire) can be achieved. Parts that pass this test condition have a large robustness margin in electronic packages.**

Keywords- copper wire, biased humidity test, HAST, THB, acceleration model

I. INTRODUCTION

The introduction of copper wirebonding on aluminium bond pads as a mainstream inter-connect technology in microelectronics packaging has faced several challenges. One of the major reliability challenges has been overcoming the corrosion sensitivity of the Cu-Al system at elevated temperature [1] in the presence of humidity and under bias. Several papers describe the factors relevant to avoid corrosion-related fails during the High Accelerated Stress Test (HAST) [2-4]. However, the HAST test condition may be too harsh when it comes to accelerating mechanisms under useful life conditions. HAST test conditions may lead to non-representative behavior in commonly used packaging materials due to chemical decomposition or drastic changes in physical properties relevant for chemical reactive species [5-7].

One of the advantages of the HAST test is its shorter duration (in equivalent terms 96 h or 264 h requirement at 130°C or 110°C respectively) compared to 1000h for Temperature Humidity Bias (THB) test at 85°C [8]. The basis of the test duration is an acceleration factor derived from the Peck model, that was applicable for Al bond pad corrosion induced by chlorides in older generation packages and package materials [9]. This paper will clarify how to deal with two questions: (1) Does passing or failing the HAST test have any relation with intrinsic product reliability? and (2) Is it possible to tune packaging materials to pass HAST requirements?

To answer the first question, it is relevant to determine what "survival" of such a severe test means in relation to the useful life of the product and its corresponding application mission profile. Based on empirical data and theoretical considerations, the validity of the commonly used Peck acceleration model to predict actual lifetime based on the accelerated test conditions is investigated. The novelty of this work will be an approach to define a new acceleration model which account for changes in material properties across the glass transition temperature (Tg) of the Epoxy Mold Compound (EMC). This model will be used to make a proposal for suitable test conditions and durations for biased humidity tests with Cu wire.

Regarding the second question, this paper shows how packaging materials can effectively be optimized so that parts may pass HAST even at the conditions of 130°C and 85 %RH under high Electrical fields (E-fields) with Cu wire. The EMC material characteristics to be tuned are background levels of inherently present ions and their mobility, halides (e.g. chloride) (EMC) [11].

In addition, it will be shown that passing JEDEC requirements for the HAST test imparts a much larger robustness margin than previously assumed. Packages that pass this requirement could be considered for even harsher environments such as maritime or military applications.

II. EXPERIMENTAL

The materials investigated in this work are shown in Table 1.

Type	Characteristic description	Tg (°C)	Silica filler content (wt %)
D	Multiaromatic/ Biphenyl epoxy from Supplier 2	130	88
D (1)	Multiaromatic/ Biphenyl epoxy from Supplier 2 (upgraded version)	130	88
G	Multiaromatic/Multifunction al epoxy from Supplier 2	175	82

Table 1: EMC's used in this study, to maintain consistency with previous publications [1, 11] the same naming convention of EMC's and supplier is retained.

A. Ionic extraction

Details of the procedure used are given in [11]. Ionic levels of the EMC's were determined by Ion Chromatography after extraction with demineralized water in a pressurized vessel for 24 hours at 121°C. The compound was milled to fine powder to maximize the extracted ionic quantities. In some cases, the extraction time and temperature were varied to assess the impact of material stability during extraction.

B. Volume resistivity

Details of the procedure used have also been described previously [11]. The resistivity measurement was performed with a Keithley 6487 picoammeter at a supply voltage of 100V. The schematic set up is shown in Fig 1.

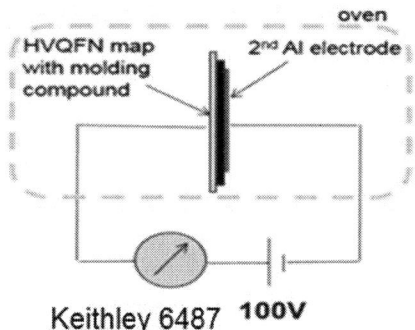

Fig 1: Volume Resistivity (V.R.) set-up

Samples were preconditioned in a climate chamber (e.g. under 85°C/85%RH for 168 h) prior to the resistivity measurement at different temperatures, e.g. 85°C, 110°C and 130°C. Volume resistivity (V.R.), ρ is determined from the sample geometry as follows:

$$\rho = (R \times L \times W)/t \qquad [1]$$

where R is the resistance of the sample, W (width) x L (length) is the area of the electrode (24.5 cm^2), and t is the sample thickness (0.07 cm).

C. Moisture absorption

The procedure used for moisture characterization has been described in detail in [5]. In addition to the moisture absorption data generated under atmospheric conditions, data was also generated under extreme conditions (i.e., under pressure), to simulate what happens during HAST testing which takes place above the boiling point of water (> 100°C).

D. Reliability tests

Standard reliability testing equipment was used to perform the biased humidity tests, HAST and THB, using boards and biasing schematic in line with internal NXP and JEDEC requirements [8]. The progression of Cu-Al IMC degradation was monitored using the method developed to quantify degraded IMC based on Back-Scatter SEM [12].

III. BACKGROUND OF BIASED HUMIDITY TESTS

A. History

The first accelerated test to assess the resistance of microelectronic packages to corrosion induced by ionic background species in EMCs like halides under the influence of bias, was the THB test [13] performed at 85°C/85% RH. The standard test duration, simulating e.g. 10 years of useful life in the actual application, is 1000 h. The aim of this test is to accelerate the exposure of the product to a combination of moisture and bias in useful life, which is typically when the product is in the "standby" mode subjected to bias under ambient conditions where high moisture levels are encountered. The reference condition is taken as typical "tropical ambient" condition of 30°C/95%RH, which is worst case when referring to humid conditions. To drastically shorten the 1000 h test duration, the HAST test was proposed [14] under an extreme condition of 110°C/85% RH or even 130°C/85% RH.

However, one must keep in mind that when this test was devised, the microelectronic packages under consideration were using older generation molding compounds and diffusion (wafer fab) processes. The failure mechanisms studied, e.g. corrosion of metallization due to diffusion of moisture through defects in the passivation, were quite different from the mechanisms observed currently, especially with Cu wirebonds. Details of failure mechanisms with Cu wire have been described elsewhere [1]: the first mechanism in case of Cu wire is Cu-Al interfacial corrosion, starting within the Al solid solution in the Cu, followed by "bulk" corrosion of Cu_xAl_y Intermetallic Compound (IMC).

In any case, the sequence of events is similar: the process starts with diffusion of moisture into the package. Moisture acts as a medium to form an electrolyte due to ionic background levels inherently present in the molding compound (e.g. halides like Cl⁻, which is often associated with corrosive degradation of aluminum-based metal and intermetallic systems [1-4, 9, 15]. In the presence of E-fields due to alternating (positively) biased and grounded pads, ions like Cl⁻ are transported towards the bond pad and initiate corrosion of the Cu-Al interface after a threshold concentration is exceeded which is dependent on the local pH [2, 16].

B. Acceleration models

The most widely accepted acceleration model to relate the acceleration test to useful life conditions, was proposed by Peck [9]. Without any comment on the exact cause of failure

978-1-7281-1500-9/19 $31.00 © 2019 IEEE

and testing principle he reported a survey of data on electrolytic corrosion failures of aluminum metallization in products. The data was fitted to the following empirical model:

$$MTTF \propto (RH)^n exp\left(\frac{E_a}{k_B T}\right) \qquad [2]$$

Where RH is the relative humidity, n a power exponent, E_a the activation energy (eV), K_B the Boltzmann constant (8.617×10^{-5} eV/K) and T the absolute temperature (K). Peck found values of -2.66 and 0.79eV for the n and E_a respectively. He concluded that the observed median life during testing correlated well with the calculated life according to equation [2]. However, he noticed large deviations of the equation when testing close to Tg of the epoxy encapsulation material. Hence, Peck cautioned about assuming validity of the model when extrapolating across the Tg. In addition to this, neither the Peck Model, nor successive models [10] cover the acceleration by bias voltage or related E-fields during biased humidity tests.

A method developed to monitor progression of interfacial corrosion in the Cu-Al contacts [12] led to an acceleration model for Cu-Al degradation driven by temperature and moisture [1, 11] in the absence of bias, which was found to be effective in lifetime predictions of Cu-Al interconnects based on the degraded length along the interface as follows:

$$x = \sqrt{A \cdot \exp(a \cdot RH) \cdot \exp(-Ea/k_B T) \cdot t} \qquad [3]$$

where is x the corrosion progression distance, also called "crack length", A is the pre-exponential constant (μm^2/h) dependent on the chemistry of the system/molding compound [17], a is the unbiased humidity coefficient (/%RH), E_a the activation Energy (eV), k_B is the Boltzmann constant, and T the absolute temperature (K). However, even when disregarding the impact of moisture, the existence of an incubation time was reported during High Temperature operating life when testing was carried out at lower temperatures (below the Tg of the EMC) [18]. An incubation time for based humidity testing was also reported in [11] when comparing HAST conditions at 110°C and 130°C. The existence of incubation, is reflected by a modification of eq. [3] as follows:

$$x = \sqrt{A . \exp(a . RH) . \exp(-\frac{E_a}{k_B T}) . (t - t_0)} \qquad [4]$$

For the temperature dependence of the incubation time , we assume an Arrhenius relation, according to:

$$t_0 = t_{00} \cdot \exp\left(-E_{a'}/k_B T\right) \qquad [5]$$

Where t_{00} is a pre-exponential constant (h), and $E_{a'}$ is the activation energy (eV) for incubation. This was based on

comparing the incubation times between different HAST conditions (e.g. 110° and 130°C).

IV. MATERIAL BEHAVIOUR DURING HAST TESTING

A. Glass transition temperature

Most mainstream EMC's in the industry today, have Tg's between 120°C and 140°C, based on the midpoint of the transition, usually reported by the position of the peak of tan delta in a DMA (Dynamic Mechanical Analyzer). Note that the glass transition region usually takes place across a temperature range, sometimes starting as low as 100°C. Moisture absorption can further reduce the Tg by about 10 – 15°C [19]. This implies that most mainstream EMC's are into, or even above, their glass transition during HAST.

Conventional EMC's usually had a much higher Tg, between 150°C and 175°C. High Tg materials suffer from a higher moisture absorption related to the chemistry (i.e. the contain more polar groups) and the limitations with silica filler loading to maintain a balance with processability considerations. Fig 2 shows a comparison of a mainstream material (EMC D) vs a more recent high Tg variant (EMC G) in a DMA scan. It is seen that EMC D is well into the transition region during HAST at even 110°C, whereas EMC G is well above it.

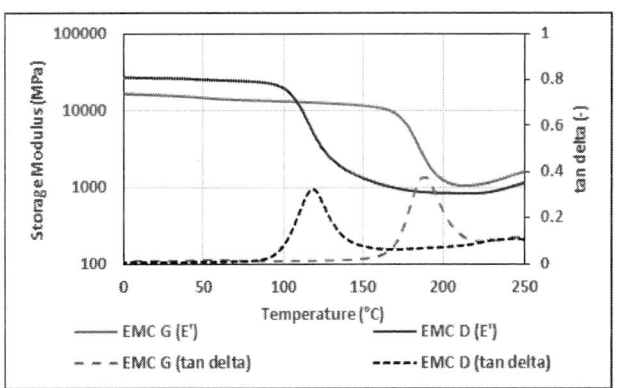

Fig 2: DMA curves if EMC D and G showing the position of the glass transition regions, E': Storage Modulus.

B. Moisture absorption

Moisture absorption data generated under extreme conditions and compared to that collected under more standard conditions for EMC D and G [5] is shown in Table 2. Due to its lower filler content, EMC G absorbs a significantly higher level of moisture compared to EMC D.

The data in Table 2 can be re-calculated to show the moisture absorption (as %) corrected for resin content alone since the silica fillers do not absorb any moisture. The results in Fig 3 show that below the Tg of EMC D, the moisture absorption of both compounds is essentially a function of filler content. Above the Tg of EMC D (still below the Tg of EMC G), the moisture absorption shows an inflection point

	C_{sat} (kg/mm³)	
Condition	EMC D	EMC G
30°C/95%RH	0.0020	0.0026
85°C/85%RH	0.0040	0.0054
110°C/85%RH	0.0060	0.0068
130°C/85%RH	0.0070	0.0086

Table 2: Moisture absorption data under HAST and THB conditions vs standby under tropical ambient conditions for EMC D and EMC G [5].

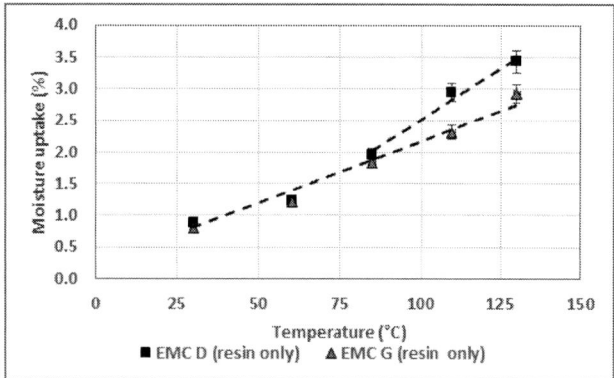

Fig 3: Moisture absorption of Table 2 recalculated to show the moisture absorption (in %) as function of temperature by the resin alone under saturated conditions (100% RH).

towards higher values than what would be predicted based on extrapolation.

C. Material degradation

Literature indicates that irreversible degradation of packaging materials can occur during reliability testing, specifically under the severe HAST conditions due to combined high temperature and moisture loading [20]. The impact of temperature and moisture on the behavior of EMC D was assessed by performing ionic extraction of organic ions, sodium and chloride as well as hydrogen ions as function of time and temperature. The results are shown in Figs. 4 - 6.

Whereas the behavior at 85°C is relatively stable, at higher temps like 110°C and especially 130°C the material starts to degrade due to hydrolysis, leading to progressively higher levels of organic anions like acetate and formate (Fig 4). This also correlates with the evolution of pH towards the lower side, especially at 130°C (Fig 6). Fig 5 shows a similar trend on chloride content, increasing to levels that are close to the total bound chloride content of this material (~ 50 ppm) [17], which implies that additional mobile chloride is released into the system during HAST at 110°C and especially 130°C.

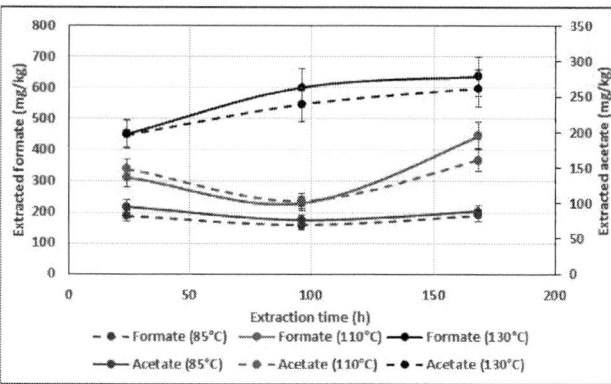

Fig 4: Formation of organic anions like formate and acetate as function of extraction time at 85°C, 110°C and 130°C.

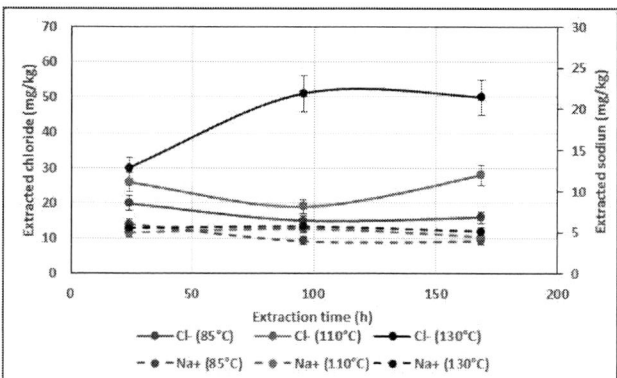

Fig 5: Release of chloride and sodium ions s function of extraction time at 85°C, 110°C and 130°C.

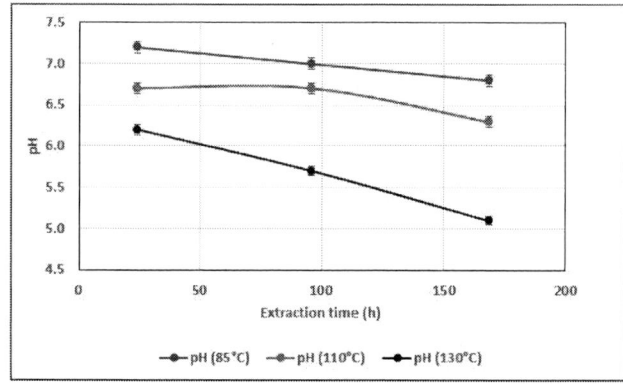

Fig 6: Evolution of pH as function of extraction time at 85°C, 110°C and 130°C.

As is known from the literature [2, 16], a lower pH in combination with increasing levels of chloride accelerates corrosion of the Cu-Al system. In our terms this could cause a shortening if the incubation time or time before the threshold concentration of chloride is reached at which corrosion starts during HAST.

D. Volume resistivity

Volume resistivity (V.R.) of EMC's as a measure of ionic mobility has been published previously [11]. To verify ohmic behavior, the resistance was recorded at bias voltages ranging between 50 – 500V, plotted as function of E-field across the sample in Fig.7. It cannot be ruled out that a part of the conduction is electronic, but it is expected that the dominating mechanism is the transport of ions. This is because moisture has a significant impact on V.R and it is feasible to make significant changes in the values of V.R. by changing the contaminant level and by using ion trappers.

Fig 8 shows the V.R. trend vs temperature for the 3 EMC's under discussion. A comparison of the green and red curves shows the significant impact of moisture on V.R., related to an increase in availability of mobile ions in the system. As shown in Table 2, under HAST conditions the actual moisture uptake will be higher than the 85°C/85%RH condition by about 75%, implying that under the HAST condition the impact of moisture on lowering the V.R. can be expected to be even higher. However, the dominant parameter is the temperature.

For the lower Tg materials, the V.R. undergoes a significant drop in value when the material passes the Tg (orders of magnitude), see the grey, green and red curves corresponding to EMC D in Fig 8. The higher Tg material EMC G on the other hand (blue curve) shows a lower value due to higher moisture absorption (see Table 2), however the value changes gradually across the same temperature range where the material is still below its Tg. When the material is in its glass transition region, the free volume within the resin matrix increases, thereby allowing ions to move more easily through the system. It is expected that the large change in V.R. across the Tg is not only due to an increase in mobile chloride but also other species that can be transported more quickly through the network.

The same holds for EMC D(1), which is an upgraded version of EMC(D) formulated by the supplier with a lower chloride content and better ion trappers to restrict the mobility of chloride ions and other anions. Despite the change in V.R. across the Tg, the value above Tg is significantly higher than those for EMC D. A restriction of mobility will result in a significant increase in incubation time even at 130°C.

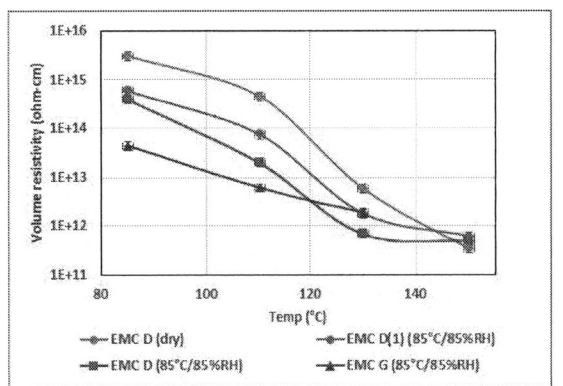

Fig 8: V.R. of EMC D (dry and moist), EMC G (moist) and EMC D(1) moist) at 130°C.

The value at 130°C in relation to HAST capability at 130°C, even under high E-fields will be described in more detail in the next section.

V. HAST 130°C CAPABILITY

A. Motivation

In most cases, no failures are observed during HAST testing at 130°C provided the package is robust (no delamination or open penetration paths), the choice of materials (EMCs) is carefully considered (low halide levels, neutral pH) and the Cu-Al IMC contact is well formed (large, dense and continuous area). However, despite taking these measures, there is no guarantee of passing the HAST test, especially at 130°C, in certain products, e.g. due to high bias and related E-fields. In such cases, an alternative test like THB is worth considering, based on the robustness margin it imparts (section VI). Based on an assessment of the mission profile it can also be shown that THB covers the requirements, not only based on the readpoints specified in the standards [8]. However, in some cases, due to customer requirements based on legacy reasons, it is still needed to pass JEDEC requirements of the HAST test at 130°C, even for the most stringent products (high bias/E-field).

B. Material capability

Based on insights published previously [11], EMC's can be modified to meet this requirement. Key properties that need to be tuned are ionic content and especially ion mobility. As shown previously, EMC D (1) was modified by the supplier in this way. Verification prior to HAST testing was done using V.R. at 130C after moisture soak at 85°C/85%RH for 168 h.

HAST validation under the 130°C/85%RH condition performed on a test product with a very high E-field exceeding 1000 V/mm showed a clear difference between IMC degradation rate between EMC D and EMC D(1). The results (highest positively biased pad) are shown in Fig 9.

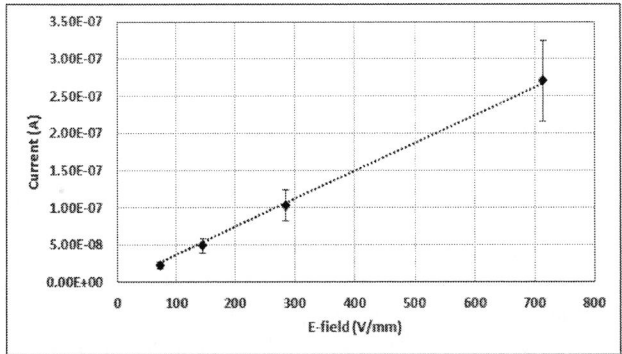

Fig 7: I-V curve (vs E-field) to show ohmic behavior

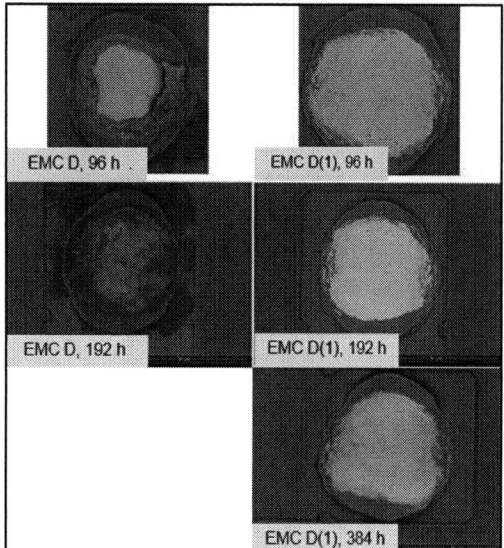

Fig 9: Cu-Al IMC contacts after HAST at 130°C in a high voltage (E-field) product comparing EMC D and EMC D(1). Only the positively biased pads are shown, as expected there was no degradation on the unbiased pads.

EMC D shows significant degradation after 96 h, and complete degradation after 192 h. This indicates an incubation time of < 96 h. EMC D(1) on the other hand, shows no degradation until even 384 h indicating an incubation time of > 384 h.

VI. DISCUSSION

As discussed in section III B and in previous publications [7, 11], finding a relation between the incubation time before the onset of corrosion and parameters like the V.R. of the EMC, the E-field, moisture level (or RH) and temperature will provide a basis for an acceleration model. It will also provide an approach to define and predict the most optimum testing condition needed for a given combination of EMC, temperature, RH and E-field.

E-field	THB 85°C		HAST 110°C		HAST 130°C	
	EMC D	EMC D(1)	EMC D	EMC D(1)	EMC D	EMC D(1)
Low	>2500	>2500	~ 500	>500	~ 192	>384
Medium	>2500	>2500	170	>500	102	>384
High	>2500	>2500	<<264	>500	<<96	>384

Table 3: Incubation times during HAST vs THB EMC D and EMC D(1), trend for EMC D based on results reported previously [7].

Table 3 shows the results previously reported [7] in schematic form for EMC D. Results for EMC D(1) are based on this work; results under milder conditions are assumed to be clean based on the results of the most stringent condition. One striking observation is that THB always passes independent of EMC type and E-field, whereas passing of the HAST test depends on EMC type or E-field.

In some high voltage (E-field) applications, the appearance of a degraded IMC, see Fig 10, shows that there is a preferential direction along which the IMC degradation is initiated. Physically, what happens is that the E-field between the biased pad and conductive trace connected to ground drives migration of chloride ions towards the IMC region preferentially from one direction. In other cases, as shown in Fig 9, the IMC degradation is more axisymmetric since the supply of chloride ions is from all sides (e.g. due to the distribution of E-field around the wire), as was shown in [9]. In either case, when the chloride concentration at the Cu-Al interface exceeds a critical concentration (dependent on local pH) corrosion is initiated.

Fig 10: Preferential Cu-Al IMC degradation from one side due to supply of chloride ions driven by the E-field between the ground pad and positively biased pad (left to right).

The following approach can be taken to find a preliminary relation between incubation time and the other physical parameters like V.R. Based on the values of diffusion constant for moisture (D') as determined in [5], it can be shown that the timescale for moisture to reach the die surface across a distance "h" $\sim h^2/D'$ and is well below the incubation time (e.g. at 110°C the corresponding time is about 8 h). The magnitude of the charge flux is proportional to the product of the inverse of the volume resistivity and the electrical field strength as follows:

$$J \propto E/\rho \qquad [6]$$

Where J is the ionic flux density, ρ is the V.R. of the molding compound, and E is the electrical field strength [21]. To compare the ion mobilities of compounds in the presence of moisture, one cannot quantitively compare the values of V.R. at different temperatures even for a given compound [11]. The conductivity at each temperature is a function of mobility at that temperature (related to free volume) and overall concentration of ions (not only chloride) that may change as a function of the temperature. However, as a first order estimate the incubation time is assumed to be proportional to the total ionic flux density: meaning that either a higher E-field and/or lower V.R. can enhance transport of chloride ions to the bond pad, thereby reducing the incubation time. The E-field is calculated based on the maximum applied positive

978-1-7281-1500-9/19 $31.00 © 2019 IEEE

bias voltage and minimum spacing to closest adjacent ball bond. Finite element simulations planned to estimate the E-fields around the affected region in more detail are not within the scope of this paper. To account for differences in moisture absorbed at each temperature, the flux density is "corrected" by dividing by the C_{sat} values given in Table 2. Values of flux density obtained in this way were normalized to the maximum values obtained under the most stringent condition. The data points for HAST at 110°C and 130°C are included in the same curve since it is the same material which is above Tg under both conditions.

The resulting experimental points shown in Table 3 are included in the graph of incubation time vs normalized flux based on this relation, see Fig 11. The result shown in Fig 10, HAST performance at 130°C/85%RH for EMC D(1) was also included in the graph. Note that the blue point at 2500 h (85°C/85%RH) and black point at 384 h (130°C/85%RH) for EMC D(1) are even higher since no signs of degradation were observed under those conditions.

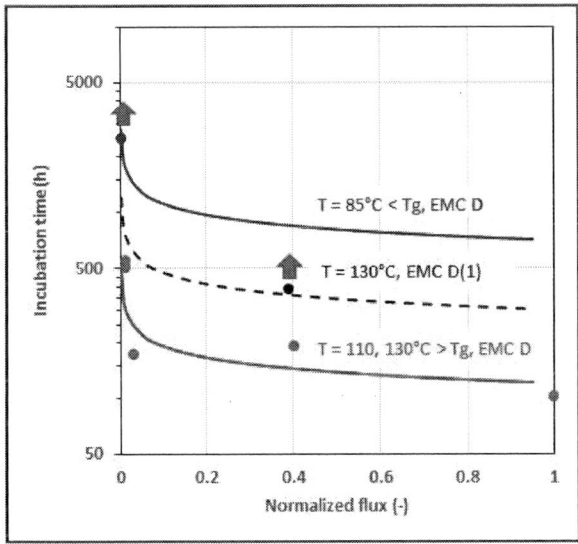

Fig 11: Incubation time vs normalized flux density, experimental data from Table 3 are shown as data points. The arrows indicate a margin on the upper side for two cases.

Although more data needs to be consolidated into the overview, the main conclusions from the graph are as follows. There is a strong dependence of incubation time on the ratio between E-field under HAST conditions and the V.R. of the EMC. The use of an EMC with higher V.R. during HAST or performing THB (85°C/55 RH) will increase the incubation time significantly indicating a large robustness margin under the relevant test condition. It is also seen that at a comparable value of normalized flux (~ 0.4) the incubation time for EMC D(1) is significantly higher. The fraction of flux related to chloride for EMC D(1) is expected to be significantly lower due to a lower level of mobile chloride and more effective ion trappers.

A tentative way to determine the acceleration factor (A.F.) between life conditions is to compare the values of flux density under the stress conditions (HAST or THB) vs actual useful life conditions, e.g. standby conditions. Based on eq. 6, this implies comparing the values of V.R. under different conditions. This is done for EMC D and EMC G and compared to the A.F. as determined by the Peck model (eq. 2). EMC D was specifically selected to represent mainstream low Tg materials because incubation time data is available. EMC D(1) does not show any signs of starting corrosion up to 4 x JEDEC requirements, so may need to be subjected to unrealistically long test times to observe starting corrosion. The summary is shown in Table 4.

		V.R. ratio	
Condition	A.F. (Peck)	EMC D	EMC G
30°C/95%RH	1	1	1
85°C/85%RH	~ 72	~ 12	~ 12
110°C/85%RH	~ 320	250	~ 84
130°C/85%RH	~ 915	~7140	~ 263

Table 4: A.F. based on the Peck model compared to ratios of the V.R. between the stress test and the standby condition.

It is seen that the values for a high Tg EMC G across the entire range are of the same order of magnitude compared to the A.F. as determined by Peck, whereas the value for EMC D which undergoes a large transition is much larger. This implies that for conventional higher Tg materials that do not undergo any transitions, the Peck model may still hold when determining the A.F. between actual application and the stress test, even for Cu-Al corrosion. For a mainstream low Tg material, undergoing the transition implies that enhanced ion generation (moisture absorption) and transport can over-accelerate failure mechanisms like corrosion.

As per the current requirements, in a worst-case mission profile in an Automotive application assuming 22.5 h standby for 15 years would equate to 1680 of THB, 380 h of HAST at 110°C or 133 h of HAST at 130°C based on the Peck model [10]. Assuming a factor of ~ 10 over-acceleration during HAST at 130°C compared to THB based on Table 4, the actual requirement would equate to only 13 h, say about 24 h if you consider time needed to saturate the package with moisture. A Jedec requirement of 96 h would imply the product is being stressed to 4 times of its equivalent useful life. If the Automotive AEC Q006 requirement is to be met (2 x JEDEC requirement), the product would be stressed up to 8 times as long.

VII. CONCLUDING REMARKS

HAST is often the biased humidity test of choice because of its relatively short duration. THB is an accepted alternative but takes significantly longer to perform; the Peck model is used to translate one condition to another. This paper clearly shows that this model can be applied for traditional compounds with Tg's significantly above the HAST temperature. However, for the mainstream EMC's with low Tg developed specifically for Cu wire, the Peck model collapses for HAST because under this condition the acceleration is much higher than predicted. For THB, which is carried out below the Tg of these compounds, this model is still valid.

This paper has also shown how significant property changes take place under HAST conditions for the low Tg EMC's, e.g. step changes in V.R., higher levels of moisture absorption compared to what would be predicted based on extrapolation and chemical decomposition. A combination of these factors leads to enhanced transport of (chloride) ions under HAST conditions, modulated at higher E-fields.

Consequently, using HAST conditions based on current requirements may result in fails which are not relevant for the reliability performance in the useful life. When these conditions are applied, considerations should be made for the test duration requirements in relation to material properties like V.R. and critical E-field corresponding to the maximum applied bias and critical distance (in this paper assumed to be between two ball bonds). While the Peck model was based on the occurrence of electrical fails due to corrosion, the basis for the model explained in this paper is time between the start of the test and the onset of Cu-Al degradation, viz. the incubation time.

This paper does not claim to conclude on an acceleration model, which will be part of future work, but it clearly shows that the industry-wide accepted HAST duration requirements are too severe. We have shown that is the case for a worst-case automotive application could be stressed up to 8 times the actual requirement if it is to meet the AEC Q006 requirement.

It is also shown that material properties of mainstream EMC's can be adapted to meet the current HAST duration requirements. Such changes may result in tradeoffs like manufacturability, cost or development time. An alternative could to be agree on reduced test durations for which a proposal has been discussed in this study, considering moisture saturation time during the first hours of the test.

On a final note, test conditions should be carefully compared with the actual application, especially in harsh environments where a combination of high moisture, high temperature and high bias voltage (internal E-fields) exist. Whereas HAST testing above Tg should be executed with caution [8], it can still be used provided one understands the dynamics of the mechanisms in detail.

ACKNOWLEDGMENTS

The authors would like to gratefully acknowledge support and contributions from Bongkoj Bumrungkittikul, G.W. Peng, Jeroen Zaal, Varughese Mathew & Niels Schutte.

REFERENCES

[1] Rongen R. et al., "Lifetime prediction of Cu-Al wire bonded contacts for different mould compounds", Electronic Components and Technology Conference, ECTC 64th 2014, pp 411-418.

[2] Peng Su et al "An Evaluation of Effects of Molding Compound Properties on Reliability of Cu Wire Components". In Proc. 61st IEEE CPMT ECTC; 2011

[3] H. Abe, et al., "Cu wire and Pd-Cu wire package reliability and molding compounds", in Proc. IEEE Electronic Components and Technology Conference (ECTC), San Diego, May 29-June 1 2012, 2012, pp. 1117-1123

[4] C.L. Gan, et al., "Extended reliability of gold and copper ball bonds in microelectronic packaging", Gold Bull vol 46, pp. 103-115, 2013.

[5] A. Mavinkurve et al, "Moisture absorption by molding compounds under extreme conditions: impact on accelerated reliability tests", ESREF2016.

[6] Yi he & Zina Alam, "Moisture absorption and diffusion in an underfill encapsulant at T> Tg and T < Tg" J. Therm. Anal. Calorim., 2013.

[7] Zaal J. et al., "Over-acceleration of corrosion mechanisms during Reliability testing: a method to relate Biased HAST tests and application conditions for Cu wire products", Proc 16th Electronics Packaging Technology Conf, Singapore, Dec. 2014. pp. 226-231

[8] JESD47

[9] D. Peck. IEEE int. reliability physics symposium proceedings 1986 44-50

[10] JEP122

[11] A. Mavinkurve et al "Biased Humidity Degradation Model for Cu-Al Wire Bonded Contacts Based on Molding Compound Properties", European Microelectronics Packaging Conference (IMAPS) 2015.

[12] O'Halloran G.M., et al., "Planar Analysis of Copper-Aluminium Intermetallics", in Proc. International Symposium for testing and Failure Analysis (ISTFA), San Jose, CA, Nov 3-7 2013, pp.297-300

[13] JESD22-A101

[14] Jeffrey E. Gunn et al, "Highly accelerated temperature and humidity stress test technique (HAST)." Reliability Physics Symposium, 1981. 19th Annual. IEEE, 1981.

[15] R. van Gestel, "Reliability related research on plastic IC-packages: A test chip approach" (PhD Dissertation) ISBN: 90-6275-960-2 Delft university press 1994.

[16] V. Mathew et al. IMAPS 2013 - 46th International Symposium on Microelectronics.

[17] A. Mavinkurve et al, "The paradoxical role of sulphur in molding compounds: influence on high temperature reliability of Cu-Al wirebond interconnects", Electronic Components and Technology Conference, ECTC 67th 2017.

[18] Rongen R. et al., "Degradation of Cu-Al Wire Bonded Contacts under High Current and High Temperature Conditions using In-situ Resistance Monitoring", Electronic Components and Technology Conference, ECTC 65th 2015, pp1396-1402

[19] J. de Vreugd, "The effect of aging on molding compound properties", (PhD Dissertation, TU Delft) ISBN: 978-94-91211-60-7.

[20] M. van Soestbergen et al, "AnomalousWater Absorption by Microelectronic Encapsulants due to Hygrothermal-Induced Degradation," Applied Polymer Science, 2014.

[21] van Soestbergen M. et al., "Electrical characterization of plastic encapsulations using an alternative gate leakage test method" Int. Reliability Physics Symposium (IRPS), Phoenix, AZ, April 2008.

Author Index

Aasmundtveit, Knut E 141
Abe, Takatoshi 1272
Abhijit, Dasgupta 498
Abrami, Avner 270
Abrol, Amrit 370, 1347
Agarwal, Amit 718
Ahari, Arman 1099
Ahasan, Kawkab 1860
Ahmed, Omar 1106
Ahmed, Sudan 1347
Ahn, Geun-Sik 197
Akahoshi, Tomoyuki 1294
Akashi, Takahiro 1022
Akazawa, Miyuki 94
Akejima, Shuzo 1140
Akiyama, Kentaro 1641
Alam, Arsalan 277
Alam, Mohammad S. 1815, 1958
Albertinetti, Andrea 2091
Albrecht, John 948
Alexandre, Giry 1279
Alhendi, M. 1946
Alhendi, Mohammed 768, 1581
Ali, Muhammad 960
Alizadeh, Azar 1581
Allain, Fabienne 1622
Allouti, Nacima 569
Alvarez, Claudio 1672
Amandine, Jouve 225
Amano, Takeru 2042
Ambat, Rajan 515
Ambhore, Pranav 620, 800, 1605
Amiran, Johnny 995
An, Sang-Ho 204
Anai, Kei 76
Andersson, Rickard 1870
Andreini, Antonio 2194
Andriani, Yosephine 1543

Antoniou, Antonia 655
Antretter, Thomas 1509, 2029
Apriyana, Anak Agung Alit 1735
Araki, Hitoshi 346
Araki, Naoko 1002
Araki, Noritoshi 175
Arayama, Chika 1022
Argoud, Maxime 569
Arnaud, Lucile 569, 1926
Arnold, Kim 340
Aschenbrenner, R. 861
Aslani Amoli, Nahid 249, 1939
Assous, Myriam 1622
Audet, Jean 1179
Aue, Maximilian 1883
Auer, Benedikt 1789
Aumont, Christophe 569
Aygun, Kemal 667
Ayssar, Serhan 1279
Ayukawa, Michael 2117
Azizi, Arad 1970
Azzopardi, Stephane 2173
Baelmans, Martine 126
Bajwa, Adeel Ahmed 620
Bakir, Muhannad S. 1803
Balaraman, Devarajan 163
Banerjee, Deepayan 1770
Barth, Maximilian 868
Bartl, Ulf 855
Barut, Atila 825
Barwicz, Tymon 528, 1060
Baty, Greg 1099
Bauwelinck, Johan 1052, 1757
Bea, J.C 1047
Becker, Karl-Friedrich 861
Bécu, Stéphane 168
Bedjaoui, Messaoud 995
Bedoin, Alexis 535

Behr, Andy	1272
Beica, Rozalia	14, 903
Beigné, Edith	1926
Bejugam, Vinith	47
Bellaredj, Mohamed	1672, 1939
Bellaredj, Mohamed L. F.	249
Beltritti, Jérôme	569
Bengsch, Sebastian	1883
Berger, Frédéric	569
Bernstein, Gary H.	2072
Bertheau, Julien	340
Bex, Pieter	340, 607, 674
Beyer, Gerald	340, 607, 1035, 2206
Beyne, Eric	126, 340, 437, 607, 674, 1035, 1215, 2206
Bharath, Krishna	2180
Bhattacharya, Surya	587, 917
Bhuvanendran Nair Gourikutty, Sajay	1227
Bicer, Mehmet	1622
Bilgen, Halim	1926
Billard, Christophe	1680
Biscarrat, Jérôme	168
Bito, Jo	896
Blachier, Denis	168
Blackshear, Edmund	1179
Blattau, Nathan	2103
Blecker, Ken	505
Boddu, Vijaya	2180
Bolkhovsky, Vladimir	1611
Bonam, Satish	2156
Boo, Hyunpil	277
Borgesen, P.	1946
Borgesen, Peter	768, 1581
Bouillard, Boris	428
Bowrothu, Renuka	695, 983, 2085, 2337
Bozano, Luisa	417
Brandl, Elisabeth	1789
Brandstätter, Birgit	1789
Braun, Tanja	363, 861, 861
Bravin, Julian	1789
B. Reddy, Vishnu V.	1333
Bruckner, Gudrun	878
Bruderer, Alex	726
Brun, Jean	995
Bu, Lin	1152, 1419, 1735
Butylkov, Sergey	1312
Bylund, Maria	1870
Byong Jin, Kim	1457
Cahu, M.	1339
Cai, Biao	1200
Cai, Jian	69, 1306, 2234
Cain, Stephen R.	2150, 2331
Campos, Didier	569
Cao, Liqiang	2318
Cao, Xinpei	753
Capecchi, Simone	910
Caplet, Stéphane	535
Carlsson, Mats	968
Cartier, Mathilde	1535
Castagné, Laetitia	1575
Castan, Clément	225
Cecchetto, Luca	2194
Chacon, Oswaldo	1396
Chahal, Premjeet	113, 948, 1240, 1687
Chai, Tai Chong	21, 1419
Chambion, Bertrand	535
Chan, Alex	1272, 1902
Chan, Chin-Wei	1751
Chandrasekhar, Arun	2180
Chang, Chia-Cheng	758, 1328
Chang, Chieh-Lin	14
Chang, Grace	1595
Chang, K.C.	763
Chang, Keng Tuan	41
Chang, Kuo-Chin	397
Chang, Megan	2079
Chang, Tao-Chih	1463

Chang, Victor C. Y.	688
Chang, Yu-Chen	1359
Chang Chien, Chien-Lin	461
Chao, Tz-Yuan	1170
Charbonnier, Jean	569, 1622
Charles, Matthew	168
Chausse, Pascal	569
Che, Fa Xing	842
Che, F. X.	1152, 2126
Chen, Allenyl	1426
Chen, Chen	661
Chen, Cheng-Chih	1826
Chen, Chih	642, 758, 1328
Chen, Chi-Jen	7
Chen, Chi-Yuan	289
Chen, Chuantong	474
Chen, C.S.	931, 1595
Chen, Dao-Long	1710, 1902
Chen, Fang-Cheng	594
Chen, J.H.	1693
Chen, Jie	1221, 1653
Chen, Jing	2061
Chen, J. Y.	700
Chen, Kang	1165
Chen, KarenYU	1710
Chen, Kuan-Neng	1463
Chen, Kuan-Ta	1704
Chen, Liangbiao	1521
Chen, Louis	1426
Chen, Ming-Fa	594
Chen, Qianwen	1246
Chen, Qiaoli	1200
Chen, Rui	249, 1939
Chen, S.	1629
Chen, Shujing	1564
Chen, Si	1324
Chen, Sihai	2186
Chen, Simon	700

Chen, S.M.	931
Chen, Tangsheng	1842
Chen, Tianfang	1983
Chen, Tony	14, 325, 903
Chen, Wei-Han	1877
Chen, Xi-Hong	1359
Chen, Xu	1889
Chen, Yan-Hao	1729
Chen, Y. F.	1175
Chen, YH	903
Chen, Y. H.	14
Chen, Yih-Sin	1170
Chen, Y.N.	1595
Chen, Yu-Hua	1413, 1463
chen, zhaohui	1543
Chen, Zhaoqing	1989
Chen, Zhiwen	63, 850, 1377
Chen, Zihao	917
Cheng, Po Wen	1246
Cheng, Shau-Fei	1877, 1877
Cheng, Ta-Chien	910
Cheng, Wei-Yuan	1877, 2009
Chenhsiu, Sung	1933
Chéramy, Séverine	569
Cheramy, Severine	225, 1926
Cherman, Vladimir	126
Cheung, Y. M.	14, 903
Chew, Ly May	87
Chew, Nam Piau	1735
Chi, Yenyao	1165
Chiang, Jack	2079
Chiang, K.N.	1515
Chiang, W.S.	493
Chien, Feng-Lung	1170
Chien, Feng Lung	1635
Chien, Han	800
Chien, Jason	1902, 2079
Chim, Weng Tuck	1457

Ching, Eva Wai Leong 917	Chylak, Bob 620
Chiou, Wen-Chih 594	Cibié, Anthony 168
Chiu, Chi-Tsung 1426	Colonna, Jean-Philippe 168, 1535
Chiu, Jason 763	Colosimo, Tom 620
Chiu, J. M. 1175	Con, Celal 2219
Chiu, Julia 453	Connolly, Brian 1200
Chiu, Ryan 700	Coquand, Rémi 1622
Chiu, Steve 2009	Coudrain, Perceval 168, 569
Chiu, Tz-Cheng 1359	Crawford, Lara S. 1312
Chiu, Yihsiang 1503	Cromwell, Kevin 1883
Chiu, Yung-Da 1426	Crump, Cameron 948
Chiu, Yu-Shan 235	Cui, Xiaole 1983
Cho, Cheng-Lin 258	Cyr, Élaine 1074
Cho, Jae Kyu 910	Daeumer, Matthias A. 1970
Cho, NamJu 289	Dahl, David 2240
Cho, Sung-Il 204	Dahlbäck, Robin 968
Cho, Youngsang 300	Dalmia, Sidharth 294, 954, 1666
Choi, Heejung 300	Dang, Bing 1246
Choi, Hyun-Seok 204	Danovitch, David 467, 2117, 2252
Choi, Jinu 753	Das, Rabindra 1611
Choi, Kwang-Seong 197	David, Leslie 498
Choi, Kwang Won 1179	Day, Doug 1933
Choi, TaeJin 2278	Decker, Michael 1509
Choi, Won Kyung 968, 1165	Dede, Ercan M. 1437
Chong, Ser Choong 21, 191	Deep, John 1366
Choong, Chong Ser 2126	Dehe, Alfons 855
Chou, P. H. 1515	De Heyn, Peter 1757
Chow, Eugene M. 1312	Del Nero, Daniel 1258, 1848
Chow, Justin H. 785	Deloffre, Emilie 1926
Chow, Seng Guan 1165	DeProspo, Bartlet 334, 1588
Chowdhury, Md Mahmudur 792	DeProspo, Bartlet H. 924
Christie, Leroy 2349	Deschamps, Jerôme 535
Chu, Fu-Cheng 977	Desmaris, Vincent 1870
Chuang, Chun-Hsiang 258	Desmet, Andres 1757
Chuang, Oscar 2009	De Vos, Joeri 1035
Chuang, Wallace 2067	De Wolf, Ingrid 126
Chuh, Erich 2180	Dhandapani, Karthik 819
Chung, C. Key 7, 1287	Dias, Rajen 163

Di Cioccio, Léa	168
Dincau, Brian	1860
Ding, Guifu	707
Ding, Kunpeng	1306
Ding, Qian	1897
Ding, Xuanyi	2186
Docanto, Manuel	1611
Dohi, Kazuhiro	2162
Dolores-Calzadilla, Victor	1060
Dorduncu, Mehmet	825
Dornala, Kalyan	1366
Dreissigacker, Marc	861
Dreps, Daniel	1200
Dressler, Marc	1113
Drouin, Dominique	1396, 2252
Du, Ke	69
Dubis, Monique	87
Dunkel, Christian	1498
Duran-Martinez, Adriana Carolina	1258
Durgun, Ahmet C.	667
Dutoit, Denis	569
Dwarakanath, Shreya	718
Ecker, Melanie	1258
Economou, Manthos	486
Edouard, Deschaseaux	1279
Eichhammer, Yann	1052
Eid, Aline	896
Ekstrom, Noah	753
EL Amrani, Abderrahim	2252
Eleouet, Raphaël	569
Elger, Gordon	2324
El-Mekki, Zaid	1035
Elmogi, Ahmed	1757
En, Yunfei	1324
England, Luke	600
Enomoto, Tetsuya	352
Eom, Yong-Sung	197
Erhart, Andreas	1833

Escoffier, René	168
Eto, Motoki	175
Evans, John	2309
Evans, John L.	792
Evertsen, Rogier	423
Exbrayat, Yorrick	569
Ezawa, Hirokazu	1140
Ezhilarasu, Goutham	277, 1470
Fager, Christian	1405
Fan, Nelson	14, 903
Fan, Xuejun	806
Fan, Zhineng	1200
Fana, Jilei	81
Fang, Bo-Siang	1432, 1704
Fang, Runiu	2168
Fang, Sheng-Po	1647, 1809
Fang, T.J.	931
Fang, Y.H.	493
Farcy, Alexis	569, 1926
Farrugia, M-L	479, 777
Fasoli, Andrea	417
Feng, Qingming	878
Fernandez, Hector	1106
Fernandez, Maïlys	535
Fernandez-Zelaia, Patxi	2349
Fettke, Matthias	47, 210
Feuchter, Michael	2029
Filipp Fuchs, Peter	2029
Finn, Daragh	453
Fischer, Thomas	855
Fischer, Thorsten	1475
Fisher, Daniel	600
Fisher, Timothy	1605
Fisher, Timothy S	277
Fitzgerald, Padraig	1660
Flaim, Tony	1722
Fleischman, Martin	811
Fortier, Paul	1074

Fortin, Clément	306
Fountain, Gill	628, 1041
Fournel, Frank	225
Fowler, Michelle	363
Franiatte, Rémi	225, 1622
Franieck, Erick	811
Fraschke, Mirko	218
Friedmann, T. A.	648
Friedrich, Georg	47, 210
Fritsche, Carola	1475
Fu, Haley	318
Fu, Xing	1324
Fuchs, Peter Filipp	1509
Fuguet Tortolero, César	569
Fujimagari, Junichiro	1641
Fujinaga, Tetsushi	358, 1865
Fujisaki, Hidehiko	1294, 1952
Fujiwara, Atsushi	1641
Fukuda, Takafumi	1140
Fukui, Kei	1294
Fukuomori, Minoru	1451
Fukushima, Takafumi	264, 1047
Furuya, Akira	1067
Gagnon, Pascale	306, 1744
Galbraith, Christopher	1611
Gao, Guilian	628, 1041
Garnier, Arnaud	569
Gaschet, Christophe	535
Geissler, Christian	855
George, Jinto	2117
Gerber, Mark	1902
Gernhardt, Robert	363
Ghannam, Ayad	1789
Ghosh, Tamal	2156
Giesen, Kyle	1200
Gillot, Charlotte	168
Gjokaj, Vincens	948
Glodde, Martin	1060

Goemare, Charlotte	1870
Goggin, Ray	1660
Goller, Bernd	855
Gong, Dan	1503
Goodelle, Jason	28
Goorsky, Mark	1605
Gordon, Seth	2309
Gore, Aaron	453
Gore, Brandon T.	726
Gorrell, Robin	330
Goto, Yoshio	101
Gottardi, Mathilde	569
Gottwald, Thomas	726
Goumans, L.	777
Gourvest, Emmanuel	428
Graap, Pascal	861
Graham, Samuel	1977
Green, Ryan B.	1782
Green, William M.J.	1060
Gromala, Przemyslaw	811, 1529
Gschwandl, Mario	1509, 2029
Gu, Han	243, 1916
Guarino, Lucrezia	2194
Guerrero, Alice	340
Gueugnot, Alain	569
Guevara, Gabe	628
Guidoni, Luca	1735
Guo, Huaixin	1842
Gupta, Sunil	1194, 2097
Gupte, Omkar	1028
Guthmuller, Eric	569
Guthrie, Bill	600
Hagn, Josef	954
Hah, Jinho	1977, 2349, 2359
Hai, Joe	486
Hajjar, Jean-Jacques	1660
Hama, Hiroki	550
Hamasha, Sa'd	792

Hamasha, Sa'd	2309	Hillman, Craig	2103
Han, Bongtae	811, 1382, 1529	Hirabayashi, Keiichi	1179
Han, Jeong Sam	2246	Hirano, Mitsuharu	1067
Han, Kwangwoo	707	Hirose, Masakazu	2162
Han, Yong	21, 1543	Hirt, Etienne	868
Hanada, Tadahiko	2112	Ho, Bin-En	1693
Hanisch, Anke	628	Ho, Cheng-Yu	977
Hanna, Amir	277, 579, 800, 1470	Ho, David	1432
Haque, Mohammad Aminul	2073	Ho, David Soon Wee	917
Harrison, Todd	1653	Ho, Soon Wee	21
Hasegawa, Yasuo	101	Hoang, Tim Tri	667
Hashimoto, Keika	346	Hoelck, Ole	861
Hashmi, Mohammad	1770	Hokari, Ryohei	1764
Hassan, KM Rafidh	1815, 1958	Honda, Kazutaka	446, 740
Haumesser, Paul-Henri	168	Hong, Xuan	753
Hayashi, Kazutaka	712	Hook, Michael David	2219
Hayashi, Toshihiko	1641	Hooshmand, Nasrin	1588
Hazellah, Muhammad Hadhari	1457	Hopsch, Fabian	314
He, Eric	700	Hoque, Mohd Aminul	792
He, Jiangling	135	Horibe, Akihiro	1921
He, Peng	2022	Hoshino, Hitoshi	437
He, Quanfeng	1716	Hosseini, Seyedmahmoud	1258
He, Xuanke	119	Hou, Xinnan	69
Heinig, Andy	314	Hou, Zhuangzhuang	1716
Heisig, Stephen J	270	Houston, Paul	2349
Hejase, Jose	1200	Hsiao, Andy	1099
Helbig, Stephan	855	Hsiao, Hsiang-Yao	21
Helou, Assaad	405	Hsiao, H. Y.	1515
Henrion, Yann	1926	Hsiao, Yu-Hsiang	461
Henry, David	535	Hsieh, Chia-Ping	2009
Henry, M. David	648	Hsieh, Jeng-Shien	688
Hensley, Dale	2073	Hsieh, Ming-Che	289
Herbert, Robert	1233	Hsieh, Ricky	977
Hernandez, Natalie	2103	Hsieh, Tsun-Lung	41
Hernandez, Selene	753	Hsieh, Yi-Chen	1751
Herrmann, Matthias	855	Hsu, C.C.	1595
Hester, Jimmy	896	Hsu, Che-Ming	461, 600
Hikita, Masayuki	1451	Hsu, Chieh-Hao	397

Hsu, Chih Chung	318
Hsu, Chih-Hsun	7
Hsu, Chung-Yi	2200
Hsu, C.K.	931
Hsu, C. K.	1550
Hsu, F.C.	931
Hsu, Fussen	1413
Hsu, Hsiang-Han	1074
Hsu, Ian	289, 493
Hsu, Po-Ning	642
Hsu, Steven	397, 763, 1175
Hsu, Y.N.	1693
Hu, Ian	1710
Hu, Yang	2234
Hu, Yougen	243, 1916
Hu, Yuan	277
Huang, Baron	1722
Huang, Chen-Yu	1287
Huang, Chih-Yi	41
Huang, Dick	700
Huang, Dinos	1710
Huang, G.C.	1595
Huang, H.L.	1595
Huang, Mian	1306
Huang, Mingliang	2022
Huang, Mingliang L.	1774, 2036
Huang, Pei-Chen	1413
Huang, P.S.	493
Huang, Rocky	1200
Huang, Shih-Ya	688
Huang, Shin-Yi	1463
Huang, Yifan	1200
Huesgen, Till	1443
Hung, Han-Tang	1729
Hung, Mi-Chun	41
Huo, Jia Ren	1485
Huo, Yongjun	150
Huynh, Michael	628
Hwang, Jisoo	300, 682
Hwang, Jung Woo	733
Hwang, Kihyun	636
Hwang, Kyo-sung	330
Hwang, Lih-Tyng	1751, 2200
Hwang, Taejoo	614
Hwangbo, Seahee	695, 983, 2085, 2337
Hyun, Sangjin	636
Iacovo, Serena	607, 2206
Ihori, Atsuhito	1865
Iida, Kenji	1952
Iizuka, Tomonori	1451
Im, Yunhyeok	300
Inaba, Takayuki	1952
Inamdar, Adwait	811
Ingelhag, Per	1405
Inoue, Fumihiro	437, 607, 2206
Inoue, Junishi	556
Irwin, Randall	277
Ishigure, Takaaki	550
Islam, Nokibul	325
Itawi, Ahmad	1575
Iwai, Toshiki	1952
Iwamoto, Hayato	1641
Iyer, S. S.	277
Iyer, Subramanian	620, 1470
Iyer, Subramanian S.	543, 579, 800, 1605, 2225
Jacquemond, Achille	264
Jacques, Patrick	1074
Jain, Ritesh	2180
Jalilvand, Golareh	1106, 1909
Jalink, J.	1339
Jamieson, Geraldine	1035
Janek, Florian	868
Jang, Joohee	636
Jangam, SivaChandra	543, 620
Jangam, Siva Chandra	800
Jani, Imed	1926

Janta-Polczynski, Alexander	1074
Jarecki, Robert	648
Jayabalan, Jayasanker	587
Jean-Philippe, Michel	1279
Jemaa, Salwa Ben	1744
Jeng, Shin-Puu	931, 1550
Jeon, HyeongIl	1457
Jeong, James	300
Jeong, leeseul	197
Jeong, Minsu	1146
Jeong, Se Young	733
Jhong, Ming-Fong	41
Ji, Hongjun	183
Jiang, Don-Son	1704
Jiang, Han	1569
Jiang, Jing	1485
Jiang, Tengfei	1106, 1909
Jiang, Yih-Jenn	7
Jin, Yufeng	1503, 1983, 2016, 2061, 2168
Jo, Chanmin	1188
Jo, Jung-Lae	76
Jo, Paul K.	1803
John Akkara, Francy	2309
Johnson, Leonard	1611
Joly, Pierre	535
Jong, Ming Chinq	1543
Joo, Jiho	197
Joo, Kisu	733
Jordan, Matthew B.	648
Joshi, Rahul	806
Joshi, Shailesh N.	1437
Joshi-Imre, Alexandra	1258, 1848
Jourdon, Joris	1926
Juang, Jing Ye	642
June Rebibis, Kenneth	437
Jung, Jin-San	204
Jung, Kwang-Ho	2290
Jung, Seung-Boo	2290

Jung, Seung-Yoon	283
Kabir, Mohammed	1870
Kahle, Ruben	861
Kalappurakal Thankappan, Kannan	2225
Kalb, Jamie	1312
Kalnitsky, Alex	1595
Kalyanam, Huthasana	2180
Kam, Nicholas	2219
Kamimura, Rikiya	1451
Kamlapurkar, Swetha	528, 1060
Kanagawa, Naoki	1022
Kandanur, Sashi	1588
Kaneko, Junichi	1933
Kang, Kuo-Chang	600
Kang, Minsoo	1977
Kang, Pilkyu	636
Kang, Qiushi	1266
Kannan, Jenefa	334
Kao, C. Robert	235, 1729, 2258
Kao, Feng	700
Kao, Hsuan-Ling	258
Karim, Karim S.	2219
Karlheinz, Bock	498
Karsten, Meier	498
Karuppuswami, Saranraj	113, 1240
Katagiri, Shunsuke	1009
Kathaperumal, Mohan	718
Kathaperumal, Mohanalingam	1796, 2112
Kathaperumal, Mohananlingam	334
Katkar, Rajesh	1041
Kavle, Pravin	931
Kawanabe, Naoki	1451
Kawano, Masaya	1996
Kaynak, Mehmet	218, 942
Ke, Chang-Bo	410
Ke, C.N.	1595
Keith Newman, Keith	806
Kelly, Mike	163

Kencana, Sagung Dewi	2067
Kennes, Koen	607
Kenney, Christopher	2072
Kerepesi, Peter	218, 942
Kersjes, Sebastiaan	1789
Keser, Beth	1159
Khazaka, Rabih	2173
Khim, Jin Young	1457
Khinda, G.S.	1946
Khurana, Gaurav	924
Kida, Tsuyoshi	1009
Kidera, Nobutaka	712
Kilger, Thomas	855
Kim, Changsu	1382
Kim, Choong-Un	1316
Kim, Dogeun	937
Kim, Dongsu	1647
Kim, Dong wook	204
Kim, Gahui	937, 2246
Kim, Hae-In	983, 2085
Kim, Haein	2337
Kim, Jaechoon	614
Kim, Ji-Hye	2272
Kim, Ji-Min	204
Kim, Jong Heon	35, 563
Kim, Jong-Hoon	1860
Kim, Ju hyeon	197
Kim, Jung Hak	197
Kim, Junghwa	300
Kim, JunMo	1146
Kim, Kilsoo	614
Kim, Kyoung-Tae	1809
Kim, KyuHyoun	1200
Kim, Nam Chul	35, 563
Kim, Nam-Seog	2246
Kim, Seokho	636
Kim, Soon-Wook	2206
Kim, Taehun	614
Kim, Taehwan	614
Kim, Taek-Soo	1146
Kim, Taeyeong	636
Kim, Yi-Ram	1316
Kim, Yoon-Hyun	733
Kim, Young-Cheon	2246
Kim, Young Ho	35
Kim, Youngja	2349
Kim, Young-Ja	2359
Kino, Hisashi	264
Kintaka, Kenji	556
Kirchner, Lisa	1722
Klengel, Robert	175
Klengel, Sandy	175
Klingler, Hannes	1789
Knickerbocker, John	270, 1246
Ko, Cheng-Ta	14, 903, 1413, 1463
Ko, T.	688
Kobayashi, Naoki	1599
Kodama, Shoichi	1002
Kohl, Paul A.	249
Koide, Masateru	1294
Kojima, Ryoji	1933
Kokash, M.Z.	1946
Kolbasow, Andrej	47, 210
Kong, Yuechan	1842
Kothari, Nakul	1347
Koyama, Koichi	1067
Koyama, Toshinori	1599
Koyama, Yutaro	346
Koyanagi, Mitsumasa	1047
Kozlovsky, William J.	726
Kraetschmer, Daniel	1113
Kraft, Jochen	1052
Krishna, Bhogaraju Sri	2324
Krivec, Thomas	1509
Kröhnert, Kevin	1475
Krumbein, Ulrich	855

Ku, Harry 1175, 1595, 1693
Ku, Terry 1
Kuah, Eric 14, 903
Kuang, Jiameng M. 2036
Kubo, Atsushi 334, 718, 924
Kubsch, Timo 210
Kudo, Hiroshi 94
Kudo, Tomoya 1015
Kuechenmeister, Frank 910
Kulick, Jason 2072
Kulterman, Ron W. 318
Kumar, Deepak 1687
Kumazawa, Yune 1009
Kuo, C.C. 1595
Kuo, C.H. 1595
Kuo, Hung-Chun 41
Kuo, Kuei Hsiao (Frank) 1635
Kuo, Ping-Jui 600
Kuo, Yu-Lin 2067
Kurihara, Kazuma 1764
Kurosaka, Seigo 474
Kurz, Helmut 218, 942
Kwon, Odal 55
Kwon, Yong Tae 35, 563
Kyung, Youjin 1146
Labarbera, Christine 2186
Lai, Chia-Chu 1704
Lai, Chieh-Lung 1170
Lai, Hsin-Cheng 1877
Lai, P. C. 1550
Lai, T.M. 931
Lai, Yen-Kun 397
Lall, Pradeep 370, 505, 792, 1087, 1347, 1366, 1815, 1958
Lambert, Renée 1611
Lambrecht, Joris 1757
Lan, Jia-Shen 2144
Lang, Klaus-Dieter 861, 1475, 1853
Langlois, Richard 1074

Larsson, Andreas 141
Lasfargues, Gilles 535
Lattard, Didier 569, 1926
Lau, Boon Long 1419, 1543
Lau, Chun Sean 1387
Lau, John 903
Lau, John H. 14
Laugier, Maxence 225
Lauser, Simone 515
Lavrik, Nickolay V. 2073
Le, Thanh Long 2173
Le, Wen-Kai 410
Leblanc, Alexandre 2117
Lee, Bob 2079
Lee, Bongsub 628, 1041
Lee, Chang-Chi 600
Lee, Chang-Chun 1413, 2009
Lee, Chang Woo 35, 563
Lee, Chia-Hsin 1463
Lee, Chin C. 150, 2302
Lee, Choong-Jae 2290
Lee, Chul-Hee 197
Lee, Chul Hyo 35
Lee, HanMin 1146, 2278
Lee, Heeseok 300
Lee, Heesok 682
Lee, Hohyung 2343
Lee, Hoi-jin 682
Lee, Hung-Ho 1287
Lee, Hungping 1503
Lee, Hyeong Gi 204
Lee, Hyun-Seop 1382
Lee, Hyun Seop 1529
Lee, Jae Cheon 35
Lee, Jeffrey ChangBing 1826
Lee, Joungphil 614
Lee, Jun Kyu 35, 563
Lee, Kang Hai 1165

Lee, K.C.	931
Lee, K. C.	1550
Lee, Kwang-Hee	197
Lee, Kwangjoo	197, 1146
Lee, Kyuha	636
Lee, Kyu Jae	733
Lee, NC	903
Lee, N. C.	14
Lee, Ning-Cheng	2186
Lee, Rick	1635
Lee, Ricky	14, 903
Lee, Sangil	2349
Lee, Seok-hyun	937
Lee, Seung Jae	733
Lee, SeYong	2278
Lee, Sung Hyuk	35
Lee, Tae-Ik	1146
Lee, Tae-Kyu	1099, 1106
Lee, Yisang	1047
Lee, Yuh-Zheng	1877
Leever, Ben	370, 1347
Legalland, Corinne	569
Li, Gang	81, 746
Li, Guanglin	243
Li, Hong Yu	1735
Li, Ji	135
Li, Jiahui	2003
Li, Jiaxiong	2296
Li, Junjie	661
Li, Kunkun	1983
Li, Ming	14, 903
Li, Mingyu	183
Li, Na	1983
Li, Ping	28
Li, Wen-Yang	7
Li, Yu-Jin	758, 1328
Li, Yu Jin	642
Li, Zhang	14, 903

Liang, Qi	661
Liang, Shui-Bao	410
Liang, Xianwen	746
Liao, Guanglan	661
Liao, Kuo-Hsien	1902
Liao, Marvin	1175, 1595, 1693
Liao, Siyuan	81
Lii, Mirng-Ji	397
Lii, M.J.	763
Lim, Francis Chee Peng	968
Lim, Jun Su	204
Lim, Ruiqi	1227
Lim, Sharon Pei Siang	1543
Lim, Simon Siak Boon	21, 1543
Lim, Sze Pei	14, 903
Lim, Teck Guan	917, 1419
Lim, Yeow Kheng	1165
Lim, Yew Kheng	968
Lim, Yu Dian	1735
Lim Sharon, Pei Siang	191
Lin, Ang-Ying	1463
Lin, Benson	493, 758, 1165, 1328
Lin, C.H.	931
Lin, Chang-Fu	7, 1287
Lin, Cheng Ping	924
Lin, Curry	14
Lin, Gu-Yan	1170
Lin, Marc	14
Lin, M.J.	493
Lin, M.Z.	493
Lin, Puru Bruce	1413, 1463
Lin, P. Y.	1550
Lin, P.Y.	931
Lin, Stanley	289
Lin, Tiesong	2022
Lin, Tong-Hong	896, 960
Lin, Vito	1287
Lin, W. Y.	1550

Lin, Yi-Hang	931
Lin, Yi-Sheng	461
Lin, Yu-Min	1463
Lin, Zhibin	453
Liou, Yan-Yu	1413
Litzenberger, Lorenz	1443
Liu, Canyu	63, 850
Liu, Changqing	63, 850, 1569
Liu, Chan-Yuan	1902
Liu, Chun-Chen	1693
Liu, C. S.	1175
Liu, Fuhan	334, 718, 924, 1796, 2112
Liu, Handa	135
Liu, Hao-Chun	397
Liu, Huan	1983, 2168
Liu, Hui	1897
Liu, Johan	1564
Liu, Kai	1246
Liu, K.C.	1595, 1693
Liu, K. C.	1175
Liu, Li	63, 850, 1377
Liu, Liyuan	2054
Liu, Meng-Hsiang	579
Liu, M.S.	931
Liu, N.W.	493
Liu, NW	1165
Liu, Penglin	1897
Liu, Ping	628
Liu, Sheng	850, 1377
Liu, Shengfa	63
Liu, Songlin	1543
Liu, Weidong	28
Liu, Weifeng	1272, 1826
Liu, Xiao	1463, 1722
Liu, Ya	1564
Liu, Yanghe	1437
Liu, Yingia	1716
Liu, Yong	1521
Lo, ChangHo	1826
Lo, I-Fang	1751
Lo, Jeffery	14, 903
Lo, Penny	14, 903
Lodermeyer, Johannes	855
Loerke, Friederike	1113
Loh, Wei Keat	318
Lombard, Marc	535
Lombardi, Jack	1581
Lombardi, J.P.	1946
Lord, David	453
Lowe, Ryan	1366
Lu, Calvin	1175, 1693
Lu, Chun-Lin	1697
Lu, JengPing	1312
Lu, Tan	1916
Lu, Tao	2054
Lu, Tian	2072
Lu, Ying-Wei	1704
Luo, Bin	890
Luo, Daojun	2054
Luo, Jiangbo	707
Luo, Yandong	579
Luo, Yu	1246
Luu, Thi-Thuy	141
Luu Trung Duong, Pham	834
Ma, B.H.	1432
Ma, H.T.	1629
Ma, Kun	1377
Ma, Li	28
Ma, Lulu	806
Ma, Shenglin	1503
Ma, Shuying	28
Ma, Xiao	410
Macaisa, Dexter	2117
Machida, Hideki	1067
Mackowiak, Piotr	1475
Madanipour, Hossein	1316

Madenci, Erdogan	825	Meth, Jeffrey	785
Maeda, Toru	1933	Meunier, Philippe	1789
Maehara, Masataka	1641	Miao, Min	1983, 2168
Maetani, Shinji	1002	Michailos, Jean	569
Mahajan, Ravi	667	Michel, Jean-Philippe	1680
Maier, Dominic	855	Michihiro, Toshiaki	2042
Makita, Toshiyuki	924	Miki, Shota	1599
Mandal, Rathin	834	Miller, Andy	340, 437, 1035, 2206
Manepalli, Rahul	1588	Miller, Scott	370, 1347
Mansoor, Bilal	1081	Milton, Basil	55
Mantysalo, Matti	1252	Min, Fan-Yu	600
Marchack, Nathan	528	Min, Kyung Deuk	2290
Maria, Winkler	498	Ming Chinq, Jong	587
Marnat, Loic	1535	Minoret, Stéphane	569, 1622
Martin, Letz	726	Mirkarimi, Laura	628, 1041
Martin, Yves	528, 1060	Mishra, Dibyajat	1316
Martina, Manuel	726	Missinne, Jeroen	1757
Martineau, Donatien	2173	Mitchell, Nicholas C	2134
Maslyk, Dan	753	Mitev, Ivaylo	1509
Massey, John P.	363	Miura, Seiya	101
Masuda, Koji	1074	Miyazawa, Risa	1921
Masuda, Yuki	346	Miyazawa, Yoshinori	1952
Matsukawa, Daisaku	352	Mizutani, Daisuke	1294, 1952
Matsumoto, Keiji	417	Moehrle, Martin	1060
Matthias, Jost	726	Moeller, Berthold	437
Maune, Holger	726	Mogera, Umesh	620
Mavinkurve, A.	479	Mogera, Umesha	1605
Mavinkurve, Amar	777	Mohan, Kashyap	655
Mayer, Michael	2219	Mohd. Ghazali, Mohd. Ifwat	1687
Mayr, Andreas	1492	Mohd Ghazali, Mohd Ifwat	113
McCann, Scott	2150, 2331	Momozawa, Aya	334
McFarlane, Nicole	2073	Mondal, Saikat	113, 1240, 1687
Mei, Ping	1312	Montmayeul, Brigitte	225
Meiler, Josef	942	Moon, Kwangjin	636
Melin, Peter	1405	Moon, Kyoung-Sik	157, 1977, 2134, 2140, 2296, 2349
Melkote, Shreyes	2349	Moon, Kyoung-Sik (Jack)	2359
Mellen, Jon	453	Moon, Seok Hwan	197
Mercier, Denis	1680	Moon, Sungwook	1188

Mori, Daichi	1933	Nakayama, Tomoki	550
Mori, Hiroyuki	417, 1921	Nakazaki, Fukino	550
Mori, Ken-Ichiro	101	Nam, Ju Hyun	563
Mori, Kentaro	1140	Nam, Seungki	1188
Mori, Kiyoharu	1047	Narayanan, Rajeev	270
Mori, Takahiro	1921	Naseem, Sadia	2079
Morikawa, Yasuhiro	1865	Nauroze, Syed Abdullah	119
Morisako, Isamu	1451	Ndip, Ivan	1475
Motobe, Takeharu	352	Nedumthakady, Nithin	1588
Motoyoshi, Makoto	1047	Nemeth, Csaba	811
Mourier, Therry	569	Neumeyr, Christian	1052
Mourier, Thierry	1622	Ng, Daniel	1287
Moussodji Moussodji, Jeff	1396	Ng, Eric	14
Mrozek, Pawel	628, 1041	Ngo, Ha-Duong	1475
Mu, Fengwen	989	Nguyen, Hoang-Vu	141
Mudrick, John P.	648	Nguyen, Luu	655, 1316, 1333
Muehlbauer, Franz-Xaver	855	Nguyen, Thong	1889
Muga, Karthik	1035	Nieh, Simon	1246
Müller, Ernst	868	Nilsson, Torbjörn M. J.	1405
Murakami, Yasunori	1067	Nishikawa, Hiroshi	1081, 2003
Murayama, Takahide	1865	Niu, Mengnian	1246
Murray, Bruce T.	1970	Niu, Yuling	819
Murtagian, Gregorio	1028	Noguet, Dominique	1535
Murugan, Rajen	1221, 1653	Nolmans, Philip	674
Murugesan, Murugesan	1047	Nonaka, Toshihisa	446, 740
Mydlak, Mathias	726	Noriki, Akihiro	2042
Na, Hoonjoo	636	Oates, Daniel	1611
Na, Nanju	1208	Oberndorff, P.	479
Nachiappan, Vivek Chidambaram	587	Ogawa, Tsuyoshi	446, 740
Nagai, Koji	1599	Ogura, Nobuo	972
Nagamatsu, Tatsuo	1933	Oh, Dan(Kyung Suk)	614
Nah, Jae-woong	528, 1246	Oh, KwangSeok	163
Nahalingam, Kirthika	1666	O'Halloran, G.M.	479
Nair, Chandrasekharan	334, 924	O'Halloran, Orla	777
Nakamura, Ai	1047	Ohba, Takayuki	1002
Nakamura, Eiji	417	Ohde, Christian	106
Nakamura, Takuya	1641	Ohkubo, Tomohiro	1641
Nakamura, Tomonori	1933	Oi, Kiyoshi	1599

Öjefors, Erik 968
Okada, Kazuya 1015
Okamoto, Daichi 718, 2112
On, JY 1710
Onishi, Tetsuya 726
Onitake, Shigeo 726
Oo, Aung Kyaw 968
Oppermann, Hermann 1052
Oprins, Herman 126
Orcutt, Jason S. 1060
Ortega, Carlos 2072
Osborn, Tyler 453
Ötzlinger, Herbert 1498
Owens, N. 479
Pacot, Guilhem 1870
Paeck, Marcus 1853
Paik, Kyung-Wook 283, 1146, 2022, 2213, 2266, 2272, 2278
Palmer, Jordan 2186
Palys, Anna 47
Pan, Jhih-Yuan 14
Pan, Ke 2343
Pan, Ponder 1693
Pancoast, Leanna 1246
Pang, Ponder 1175
Panigrahy, Asisa Kumar 2156
Panikkanvalappil, Sajanlal 1588
Pantano, Nicolas 674, 1215
Pantouvaki, Marianna 1757
Papapolymerou, John 948
Paradis, Etienne 2252
Paranjpe, Ajit 1470
Pares, G. 1279
Pares, Gabriel 1535, 1622
Parikh, Bakul 1121
Park, Gyu-Tae 2246
Park, Hyun Ho 733
Park, Jae-Hyeong 2213
Park, Jongcheol 2213

Park, JoonYoung 163
Park, Sang Yong 35, 563
Park, S.B. 2150, 2331
Park, Seungbae 1130, 2343
Park, SooIn 2278
Park, Yong-Jin 204
Park, Yong Sung 204
Park, Young-Bae 937, 2246
Parker, David 428
Parthasarathy, Srivatsan 1660
Paul, Jens 910
Pei, Yu 2234
Pei Siang, Sharon Lim 587
Peng, lan 607, 2206
Peray, Patrick 535
Petzold, Matthias 175
Pfost, Martin 1509
Pham, Van-Lai 1130
Pham, Vanlai 2150, 2331, 2343
Phansalkar, Sukrut 1382
Philip, Pierre-Emile 569
Phommahaxay, Alain 340, 437, 607, 2206
Pierre, Ferris 1279
Pietryga, Christoph 1159
Plant, Jason 1611
Plochowietz, Anne 1312
Podpod, Arnita 340, 437
Polezhaev, Vladimir 1443
Poliks, Mark D. 768, 1581, 1946
Ponthenier, Fabienne 569
Posthill, John 1041
Pozzobon, Fiorella 2194
Premerlani, Romeo 726
Prenger, Luke 1463
Prisacaru, Alexandru 811
Pristauz, Hugo 1492
Proschwitz, Jan 1159
Pu, Han-Ping 688

Pu, Li 1716
Puligadda, Rama 1722
Pulugurtha, Markondeya Raj 960
Pulugurtha, P. Raj 1300
Qi, Tao 1509
Qian, Zhiguo 667
Qiang, Song Guan 1485
Qiao, Y.Y. 1629
Qin, Ivy 55
Raad, Peter 405
Raghavan, Nagarajan 834
Rahim, Kaysar 800
Raj, Anto 2309
Raj, P. Markondeya 718, 972
Rajagoapal, Varun 334
Ramon, Hannes 1757
Rao, Vempati Srinivasa 842, 1152, 2126
Rasilainen, Kimmo 1405
Rastogi, Ravi 1611
Ravichandran, Siddharth 726, 1796
Ravinder, Pal Singh 1419
Raychaudhuri, Sourobh 1312
Raynaud, Christine 1680
Rebhan, Bernhard 218, 942
Refai-Ahmed, Gamal 2150, 2331, 2343
Ren, Chao 1977, 2296
Ren, Linlin 1556
Ren, Qin 1996
Ribière, Céline 569, 1622
Richter, Theresia 515
Rivera, Katie 600
Robertson, Stuart 63, 1569
Rogers, Jeff 270
Rolland, Emmanuel 225
Romano, Giovanni 569
Romero, Gilles 569
Rongen, Rene T.H. 479
Rongen, R.T.H. 777, 1339

Ross, Joseph 1121
Roucou, R. 1339
Rovitto, Marco 2091
R. Tummala, Rao 718
Rudolph, Catharina 628
Rupp, Bradley B. 1312
Ruzicka, Klaus 868
Ryu, Dong Su 1457
Ryu, Hyodong 2246
Sakai, Taiji 1952
Sakakibara, Shiori 352
Sakaue, Takahiko 76
Sakuma, Katsuyuki 270
Sakuyama, Seiki 1952
Salahoueldhadj, Abdellah 340
Salcedo, Javier 1660
Saleem, Amin 1870
Saleh, Rafat 868
Sammakia, Bahgat G. 1970
Sanchez, Juliet 753
Sanchez, Loïc 225
Santerre, Francis 1396
Sarangapani, Murali 2048
Sato, Muneyuki 1865
Sato, Nobuaki 1451
Sato, Yoichiro 712
Savage, Eric 2117
Saxena, Antra 1770
Scevola, Daniel 569
Schares, Laurent 1060
Scheller, Britta 106
Schellkes, Eckart 2067
Schempp, Fabian 1113
Schiffmacher, Alexander 1443
Schiffres, Scott N. 1970
Schingale, Angelika 1509
Schischka, Jan 175
Schmid, Maximilian 2324

Schmitt, Wolfgang	87
Schneider-Ramelow, Martin	861
Schroeder, Raoul	1498
Schulze, Gary	55
Schulze, Sebastian	218, 942
Schumann, Todd	695, 1647, 1809
Schuster, Christian	2240
Schutt-Aine, Jose	1889
Schwarz, Mark	392, 819
Schwenk, Erika	87
Seefisch, Henning	2284
Segal, Julie	2072
Segaud, Roselyne	569
Sekhar, Vasarla Nagendra	842
Sekiguchi, Masahiro	1140
Selhofer, Hubert	1492
Selvanayagam, Cheryl	834
Serebreni, Maxim	2103
Shaddock, David M.	768
Shah, Aashish	55
Shah, Ujash	1605
Shahane, Ninad	1316
Shakoorzadeh, Niloofar	800
Shambach, William	2186
Shang, Jintang	522, 890
Shangguan, Dongkai	1272
Shapiro, Dmitri	1611
Sharma, Himani	1300, 1588
Sharon, Gil	2103
Sharon Lim, Pei Siang	21
Sheikhi, Roozbeh	150
Sheikhnejad, Ommeaymen	243
Shelton, Douglas	101
Shen, Yu-An	1081, 2003
Shi, Aihua	884
Shi, Tielin	661
Shibahara, Hiromi	1933
Shibasaki, Yoko	2112

Shibata, Daisuke	2112
Shie, Kai Cheng	642
Shigetou, Akitsu	235
Shih, Andy	1246
Shih, Meng-Kai	1710, 1902
Shika, Seiji	1009
Shim, Ji Ni	563
Shim, Moo-Sup	197
Shimada, Sawako	1015
Shimatsu, Takehito	989
Shimazu, Takayuki	1764
Shimizu, Kan	1641
Shimizu, Koji	1451
Shin, SangMyung	1146, 2278
Shin, Youngmin	300, 682
Shiraiwa, Tomio	163
Shirley, Tim	543
Shoji, Hideaki	163
Shoji, Yu	346
Shreve, Matthew	1312
Shumarayev, Sergey Yuryevich	667
Shunmugasamy, Vasanth	1081
Sidorov, Victor	1052
Sierra-Suarez, Jonatan A.	648
Sigl, Alfred	855
Sigmund, Ariane	1060
Sikka, Kamal	1121
Silberer, Gerald	942
Simmons, Jacob C.	1970
Simon, Gilles	569
Singh, Chrandeep	1130
Singh, Shiv Govind	2156
Sirbu, Bogdan	1052
Sitaraman, Suresh	1521
Sitaraman, Suresh K.	249, 785, 1939
Sitaraman, Suresh K	382
Sivapurapu, Sridhar	249, 1939
Sivasubramony, Rajesh S.	1581

Sivasubramony, Rajesh Sharma	768
Slabbekoorn, John	340, 607
Sleeckx, Erik	340, 437, 607, 2206
Smet, Vanessa	655, 972, 1028, 1796
Smith, Stephen	1200
So, R.	14
So, Raymond	903
Soares, Francisco	1052
Son, Ho-Young	2246
Son, Kirak	937, 2246
Song, Bo	157, 2134, 2140
Song, Changming	2234
Song, Euseok	614
Soon-Wook, Kim	1215
Sorensen, Eric	543
Sosa, Ramón A.	655
Souriau, Jean-Charles	1575
Southard, Arthur	1722
Sover, Raanan	294
Spurney, Robert G.	1300
Sridhar, Sharath	2309
Srinivasan, Sriram	2180
Stegmann, Tamira	87
Steinhorst, Rachel	1240
Stephan, Tino	175
Stewart, Benjamin G	382
Stoffel, Nancy C.	768, 1946
Stone, Bill	392
Stone, David	1179
Stucchi, Michele	1035
Su, An-Jhih	1
Su, C.H.	1175
Su, Jay	1463
Su, Ming-Sin	1175
Su, Peng	1099, 1106
Su, Sinan	792, 2309
Su, Zhaoxi	890
Subbiah, Nilavazhagan	878
Sueoka, Kuniaki	1921
Suga, Tadatomo	989
Suganuma, Katsuaki	474
Sugiura, Kazuhiko	1451
Suhard, Samuel	437, 607
Suhling, Jeff	505, 1347
Suhling, Jeffery	2309
Suhling, Jeffrey C.	792, 1087, 1815, 1958
Sulkis, Michael	1977
Sun, Hongyu Y.	2036
Sun, Rong	81, 243, 746, 1556, 1916
Sun, Teng	1300
Sun, Xiao	1215
Sun, Yangyang	392
Sun, Yunna	707
Sun, Yunting	707
Sung, Yun Hyun	35
Surillo, Emanuel	334
Suryoatmojo, Heri	1751
Susumago, Yuki	264
Suzuki, Akiyoshi	1865
Suzuki, Takuya	1009
Suzuki, Yui	2162
Suzuki, Yuya	1015
Swaminathan, Madhavan	249, 1672, 1939
Syed, Ahmer	392, 819
Sylvestre, Julien	1744
Symonds, Ken	2103
Tai, Jui-Feng	7
Tak, Coen	2029
Takahashi, Noriyuki	264
Takano, Takamasa	94
Takiar, Hem	1387
Tal, Sharon	294
Talebbeydokhti, Pouya	294, 954, 1666
Tamura, Akira	1952
Tan, Chuan Seng	1735
Tan, KH	325

Tan, Kim Hwee	14, 903
Tanaka, Kazunori	1067
Tanaka, Masaya	94
Tanaka, Tetsu	264
Taneda, Hiroshi	1599
Tang, Gongyue	1419
Tang, Junyan	1200
Tang, Tom	1432
Tang, Zirong	661
Tani, Daisuke	1933
Tan Swee Seng, Eric	2048
Tao, Jing	1735
Tao, Mian	14, 903
Tao, Qi	2029
Tao, Wang Jun	1485
Tarng, David	1426, 1710
Tasi, Mike	1704
Tatsumi, Kohei	1451
Tehrani, Bijan	896
Tekin, Tolga	1052
Tentzeris, Manos M.	119, 896, 960, 972
Teoh, Kristie	1028
Tetsuya, Onishi	1498
Teutsch, Thorsten	47, 210
Thai, Trang	294, 954
Theil, Jeremy	628, 1041
Theng, Chih-Han	758
Theuss, Horst	855
Thirugnanasambandam, Sivasubramanian	2309
Thomas, Tony	505
Thorsell, Mattias	1405
Thukral, Varun	1339
Tian, Yanhong	1266
Tissier, Pierre	1622
To, Hing "Thomas"	1208
Toda, Keiji	1451
Toepper, Michael	1853
Tokunari, Masao	1074
Tollefsen, Torleif A	141
Tolunay Wipf, Selin	942
Tomikawa, Masao	346
Tomita, Yasunari	1022
Tomohiro, Fukao	1272
Topsakal, Erdem	1782
Tremble, Eric	1179
Trombley, Django	1221
Tsai, Chung-Hao	1
Tsai, Clair	1175
Tsai, Jensen	700, 1432
Tsai, Mike	700
Tsai, Sheng-Han	397
Tsao, Pei-Haw	763
Tschoban, Christian	1475
Tseng, I-Hsin	758, 1328
Tseng, Yi-Hsiu	1359
Tseng, Yu-Chou	1902
Tsfati, Yossi	954
Tsunoda, Masatoshi	2042
Tsuruta, Kazuhiro	1451
Tu, K N	642
Tu, K. N.	1716, 2003
Tummala, Rao	334, 655, 1028, 1300, 1588, 1796
Tummala, Rao R.	924, 960, 972, 2112
Tunga, Krishna	1121
Tuominen, Samuli	1252
Turcotte, Eric	467
Tutunjyan, Nina	1035
Twiefel, Jens	2284
Ueda, Kazutoshi	1451
Ueno, Keiko	446, 740
Ueno, Kenichi	2162
Ueta, Chiho	1015
Ume, I. Charles	1333
Uomoto, Miyuki	989
Ura, Shogo	556

Uresti, Tiffani	1081
Utano, Tetsuya	1933
Uy, William	1272
Vadimas, Verdingovas	515
Vaisband, Boris	543, 579, 1605, 2225
van Borkulo, Jeroen	423
Van Campenhout, Joris	1757
Vandendaele, William	168
Vandeneynde, Aurélie	535
Van der Plas, Geert	674, 1215
van der Stam, Richard	423
van Haare, Niek	1789
Van Huylenbroeck, Stefaan	1035
Vanjari, Siva Rama Krishna	2156
van Olst, E.	479, 777
van Soestbergen, M.	479, 777
Van Steenberge, Geert	1757
Varga, Edit	2072
Vasarla, Nagendra Sekhar	2126
Vélard, Rémi	569
Velenis, Dimitrios	674
Venkataraman, Srikrishnan	2180
Venugopal, Archana	405
Verdonck, Patrick	2206
Veres, Agnes	811
Verhelst, Marian	674
Verrun, Sophie	1622
Viehweger, Kay	1833
Vijayakumar, Swathi	1666
Vinci, Andrea	569
Viswanath, Ram	2180
Vitello, Dario	2091
Vivet, Pascal	569, 1926
Vladimirova, Kremena	168
Vobl, Matthias	855
Voges, Steve	363, 861
Voit, Walter	1848
Voit, Walter E.	1258
von Waechter, Claus	855
Wada, Keiko	1451
Wagner, Juergen	855
Wagner, Thomas	1159
Wai, Leong Ching	21, 1419
Waidhas, Bernd	1159
Wan, Weikang	2318
Wan, Zhe	579
Wang, Chang-Ning	1175
Wang, Chen-Chao	41, 977
Wang, Chengqian	28
Wang, Chenxi	1266
Wang, Chuei-Tang	688
Wang, Chun-Min	1463
Wang, Daixing	2016
Wang, Huayan	1130, 2150, 2331, 2343
Wang, Hui	243
Wang, Huiying	707
Wang, J. H.	1550
Wang, Jin	69
Wang, Jing	1130, 2150, 2343
Wang, Jiunn Jie	1635
Wang, Kirin	1175, 1595, 1693
Wang, Lejun	392
Wang, Liyuan	1983
Wang, Nan	1564
Wang, Peiren	135
Wang, Qian	69, 1312, 2234
Wang, Qidong	2318
Wang, Rung-De	1693
Wang, Shiu-Chih	1426
Wang, Tai-Jui	1877
Wang, Wei	392, 819, 2016
Wang, Xiangy-Yu	1996
Wang, Xiaobai	1543
Wang, Xin	2234
Wang, Xinying	1889
Wang, Xueqiao	2140

Wang, Yan 707
Wang, Yang 135
Wang, Yen Neng 1635
Wang, Yiteng 972
Wang, Y.P. 1629
Wang, Yu 1312
Wang, Yu-Cheng 1693
Wang, Yunda 1312
Wang, Yu-Po 700, 1432
Wang, Zhijie 819
Watanabe, Atom 924, 960, 1300
Watanabe, Atom O. 972
Watanabe, Manabu 1294
Watanabe, Naoki 924
Watanabe, Osamu 1933
Watanabe, Takuro 1764
Watariguchi, Shigeru 1047
Webb, Bucknell 270
Weber, Y. 479
Weerawarne, Darshana L. 1581
Weerawarne, D.L. 1946
Wei, Cheng 410
Wei, Tiwei 126
Weichart, Johannes 1833
Weichart, Jüergen 1833
Weir, Terence 1611
Weiss, Thomas 306
Weng, Chen-Yuan 600
Weng, I-An 1729
Wen Kong, Ling 1485
Werner, Thomas 628
Widiez, Julie 168
Wietstruck, Matthias 218, 942
Wigger, Benedikt 868
Wijewardena, Kanishka 1687
Wilde, Juergen 878, 1113, 1443
Williamson, Jaimal 1333
Wipf, Christian 942

Wohrmann, Markus 363
Wöhrmann, Markus 861, 1853
Wolfberger, Archim 2029
Wong, Chee Wei 277
Wong, Ching-Ping 81, 243, 746, 1556, 1916, 2296, 2349
Wong, CP 157, 2134
Wong, C. P. 1977, 2140, 2359
Wong, Nelson 55
Wong Chin Yeung, Jason 2048
Workman, Thomas 628
Wu, An-Tai 1426
Wu, C.M.L. 1629
Wu, Dapeng 968
Wu, Fan 157, 2134
Wu, Hsing-Hui 14
Wu, Jiaqi 2302
Wu, Jing 1087
Wu, Jui-Yang 2258
Wu, Mei-Ling 2144
Wu, Sheng-Tsai 1463
Wu, W. C. 1175
Wu, Yang 2022
Wu, Y.H. 1693
Wu, Zhichao 1306
Wu, Zhongming 486
Wurz, Marc Christopher 1883
Xi, Cao 14, 903
Xiang, Gengzhao 135
Xiang, Hui 63
Xiangyu, Wang 587
Xiao, Hui 2054
Xiao, Zhiyi 884
Xie, Dongji 486
Xie, Hong 28
Xie, Yong 1221
Xiong, Chi 1060
Xiong, Yaoxu 243
Xu, Hui 55

Xu, Iris	14, 903
Xu, Jianbin	1556
Xu, Jiefeng	1130, 2150, 2331, 2343
Xu, Jikai	1266
Xu, Xirui	884
Xue, Mei	2318
Yadav, M.	1946
Yadav, Manu	768
Yagi, Hidekazu	1933
Yahyaei-Moayyed, Farzaneh	2180
Yakabe, Sho	1764
Yalagach, Mahesh	2029
Yamada, Shuhei	1796
Yamada, Takashi	175
Yamamoto, Kazunori	842, 1419, 2126
Yamashita, Soichi	1140
Yamauchi, Sinichi	76
Yamawaki, Seigo	1294
Yamazaki, Noriyuki	352
Yang, C. C.	1550
Yang, Chen-Tsai	1877
Yang, Cheol-Woong	2246
Yang, Chi-Hau	2200
Yang, Fan	850
Yang, Henry	14, 903
Yang, Jenn-Ming	2258
Yang, Ming	2022
Yang, Rolance	1175
Yang, S.B.	1595
Yang, Sean	1729
Yang, Tilo H.	235
Yang, T. L.	1175
Yang, Xiaobing	28, 884
Yang, Yang	1983, 2168
Yang, Yong	1716
Yang, Yong-suk	330
Yang, Yu-Hsiang	811
Yang, Yu-Ting	1359
Yang, Zhuoqing	707
Yao, Ruohe	1324
Ye, Chen	522
Ye, Lilei	1564
Ye, Ning	1387
Ye, Yong Liang	1419
Yee, Kuo-Chung	1
Yeh, Meng-Kao	1697
Yeh, Shu-Shen	1550
Yen, Yee-Wen	2067
Yeo, Woon-Hong	1233
Yess, Kim	340
Yew, M.C.	931
Yew, M. C.	1550
Yi, Luyun	1200
Yildiz, Ömer Faruk	2240
Yin, Liang	768
Yin, Xin	1052
Yook, Jongmin	1647
Yoon, Dal-Jin	2266, 2272
Yoon, Gil-Sang	197
Yoon, Seung Wook	325, 968, 1165
Yoon, Yong-Kyu	695, 983, 1647, 1809, 2085, 2337
Yoshida, Shu	1009
Yoshihiro, Furukawa	1300
Yotsuyanagi, Hiroko	352
Young Suk, Kim	1002
Youssef, Toni	2173
Yu, C.K.	493
Yu, C.T.	931
Yu, Daquan	28, 884
Yu, Doug C.H.	594
Yu, Douglas	1, 688
Yu, Hai-Yang	235
Yu, Jambo	28
Yu, Ji-In	204
Yu, Shiang-Hwua	1751
Yu, Ta-Jen	289

Yuan, K.S.	1595
Yue, Xiang	522
Zaal, J.J.M.	1339
Zachariah, Ashwin Varkey	768
Zarr, Scott	1611
Zeb, Gul	467
Zeng, Qinghua	2061
Zeng, Xiaoliang	1556
Zhang, Baotan	81
Zhang, Bowei	1306
Zhang, Dongxiao	63
Zhang, Eric J.	1060
Zhang, Hong	1722
Zhang, Hongqing	1246
Zhang, Jianfeng	890
Zhang, Jincan	1983
Zhang, Shuye	2022
Zhang, Wenwu	183
Zhang, Xiaowu	1152, 1419, 1543
Zhang, Xin-Ping	410
Zhang, Xuefeng	392
Zhang, Yao	1377
Zhang, Zheng	474
Zhao, Cong	2309
Zhao, Liguo	1569
Zhao, Lily	392
Zhao, Lixin	69
Zhao, N.	1629
Zhao, Peng	1735
Zhao, Tao	81, 746, 1916
Zhao, Weiwei	183
Zhao, Xiuchen	1716
Zhao, Yang	2234
Zheng, Fengxia	28
Zheng, Hao	1377
Zheng, Ting	1803
Zhengyang, Qian	264
Zhou, Bin	1324

Zhou, Han	1569
Zhou, Jianwen	746
Zhou, Jie-Ying	410
Zhou, Min-Bo	410
Zhou, Shicheng	1266
Zhou, Shiqi	1081, 2003
Zhou, Yi	249, 1521, 1939
Zhou, Zhaoxia	63, 850, 1569
Zhu, Pengli	243, 746, 1916
Zhua, Pengli	81
Zhuo, Qizhuo Zhuo	753
Zoberbier, Ralph	106
Zoschke, Kai	1475
Zou, Lin	1774
Zou, Yichao	884
Zuber, Fabien	535
Zullino, Lucia	2194
Zussy, Marc	535

IEEE
445 Hoes Lane
Piscataway, NJ 08854-4141

ISBN 978-1-7281-1500-9